ISBN 978-0-266-82284-4
PIBN 10894668

A NEW
GAZETTEER
OF
THE UNITED STATES
OF
AMERICA;

CONTAINING

A COPIOUS DESCRIPTION

OF THE

STATES, TERRITORIES, COUNTIES, PARISHES, DISTRICTS, CITIES AND TOWNS—MOUNTAINS, LAKES, RIVERS AND CANALS—COMMERCE, MANUFACTURES, AGRICULTURE, AND THE ARTS GENERALLY, OF THE UNITED STATES;

EMBRACING ALSO

THE EXTENT, BOUNDARIES, AND NATURAL PRODUCTIONS OF THE PRINCIPAL SUBDIVISIONS, THE LATITUDE AND LONGITUDE OF CITIES AND TOWNS, AND THEIR BEARING AND DISTANCE FROM IMPORTANT PLACES;

INCLUDING

OTHER INTERESTING AND VALUABLE

GEOGRAPHICAL, HISTORICAL, POLITICAL, AND STATISTICAL INFORMATION;

WITH THE POPULATION OF 1830.

BY WILLIAM DARBY

AND

THEODORE DWIGHT, JR.

HARTFORD,
PUBLISHED BY EDWARD HOPKINS.

1833.

LIBRARY OF CONGRESS
DUPLICATE
EXCHANGED

PREFACE.

The collection, and arrangement of the materials for this Gazetteer, were commenced in April, 1830, and have been steadily continued to the present time, Nov. 1832. It must therefore be evident that great labour has been expended upon the work—how satisfactory to the public the result of the undertaking will prove, remains to be decided. It is not with a view to enhance the value or importance of these labors, but to explain one of the principal causes of delay attendant upon the publication of this volume, that some remarks are introduced on the manner, and protracted time of publishing the census, which was not available until June of the current year, or about two years after it was taken; this fact, together with the confused, and utter want of arrangement in that document, renders unnecessary any farther apology for the delay to which we advert. Indeed as this document is published, it is to a convenient analysis of the population of the U. S., what stones in a quarry are to a building; and deserves to be estimated much as the material, on the ground where an edifice is to be erected. In every state, and territory, with the exception of the northern district of New-York, the eastern district of Pennsylvania, Delaware, Maryland, the western district of Virginia, North Carolina, Georgia, Kentucky, the western district of Tennessee, and Ohio, an alphabetical arrangement of the census table was generally omitted, and of course to find any given county or town, required an examination of the whole mass. The inconvenience of this arrangement may be estimated, when it is known that five hundred and thirty counties, with their towns, are thus defective in arrangement. Of some states, nothing is given but the aggregate of the counties; and in Georgia, no city, town, or village is named;—the returns of some other states are equally imperfect. These facts will account for numerous apparent deficiencies in population, and explain their true cause. The post offices, particularly in the middle, southern, southwestern, and western states, have been inserted from the official list of 1831; and great pains have been taken, with the aid of the best maps, to locate the most obscure of them. The qualifying term, "post road" has been adopted, as by that the postage of letters, &c. is regulated, altho' in innumerable cases these much exceed more direct routes. In regard to rivers, it may be doubted whether their extent should be estimated by the meanders of the streams, or by the length of their basins, or vallies. Except in New England, N. York, and N. Jersey, we have chosen to estimate them by the latter method, the length of the surface which they drain. Tanner's new and excellent map of the U. S., a proof sheet of which was early and obligingly forwarded to Mr. Darby, has been the guide generally followed in regard to geographical position; with the aid of this, extensive additions have been made to the geography of the western country, particularly that of the states of Ohio, Illinois, and Indiana, and the territories of Michigan and Huron.

The extended plan of this Gazetteer, seemed at first to promise the compiler of that part of it embracing New York, New Jersey, and the New England states, an opportunity to introduce large details of the intellectual and moral institutions of the country; and the hope of accomplishing this object was one great motive for his engaging in the work. Experiment however soon proved, that the prescribed limits, although large, would not allow the introduction of many such details, without excluding others more practically indispensable, in a work of this kind. He acknowledges his obligations to the authors of the following works, from some of which he has derived much statistical information; Greenleaf's Survey and Map of Maine; Tanner and Moore's Gazetteer of N. Hampshire; Thompson's Gazetteer of Vermont; Spofford's Gazetteer of Massachusetts; Pease and Niles's Gazetteer of Connecticut and Rhode Island; Spafford's Gazetteer of New York, and Gordon's Map of New Jersey; with particular gratitude he also acknowledges his obligations to James Parker, Esq. of Amboy, N. J., for valuable statistics of different parts of that state, which have been embodied in the work.

In the compilation of the Gazetteer of the U. S., numerous authorities beside those already mentioned, have been consulted; and accuracy has ever been a prominent object of its compilers. But some recent sources of information have been deficient;—the census was formerly an invaluable source of various statistical matter; that of 1830, however, has proved to be entirely useless on every subject other than population. Inconsistencies and contradictions in orthography, and in statistics of different kinds, found even in works of the highest reputation, may perhaps have induced some errors and omissions here. Indeed to say that such will not be found in the work, would be presumption; from the very nature of it, perfect accuracy cannot reasonably be insisted on in every detail, by the reader. Such errors and omissions as have been discovered, on a rapid review of the work, have been noticed in the Appendix. We may say, however, what is undeniable, that much has been added to the geography of the country, in the present work; that numerous new counties, and towns, have been embodied in it, and that it contains many and important corrections of some similar and respectable works. On the whole, as a convenient and safe book of reference, extended in its detail far beyond any work of the kind heretofore published, and to a great extent original, we hope, and confidently believe, that it will prove both useful, and valuable, to those who have occasion to consult its pages.

ABBREVIATIONS.

Ark.—Arkansas.
Ala.—Alabama.
Conn.—Connecticut.
Del.—Delaware.
Dist. Col.—District of Columbia.
Flor.—Florida.
Geo.—Georgia.
Ind.—Indiana.
Il.—Illinois.
Ky.—Kentucky.
La.—Louisiana.
Mass.—Massachusetts.
Md.—Maryland.
Me.—Maine.
Mich.—Michigan.
Miss.—Mississippi.
Mo.—Missouri.
N. C.—North Carolina.
N. H.—New Hampshire.
N. J.—New Jersey.
N. Y.—New York.
N. W. Ter.—North West Territory.
O.—Ohio.
Phil.—Philadelphia.
Pa.—Pennsylvania.
R. I.—Rhode Island.
S. C.—South Carolina.
Ten.—Tennessee.
U. S.—United States.
Vt.—Vermont.
Va.—Virginia.
W. C.—Washington City.

cap.—capital.
co.—county.
dist.—district.
isl.—island.
lat.—latitude.
long.—longitude.
ms.—miles.
mtn.—mountain.
pop.—population.
p-o.—post office
p-t.—post town.
p-v.—post village.
p-r.—post road.
r.—river.
s-p.—sea port.
sq. ms.—square mile
st. jus.—seat of justic
t.—town.
ter.—territory.
tsp.—township.

A few other abbreviations used in the work will at once be intelligible to the reader.

A

GAZETTEER

OF THE

UNITED STATES OF AMERICA.

AARONSBURG, p-v. Centre co. Penn.; situated on one of the highest branches of Penn's creek, 18 miles E. of Bellefonte, and by post road 79 miles N. W. from Sunbury.

ABERDEEN, p-t. Brown co. O.

ABBEVILLE, district of, South Carolina; extending along and from Savannah river: bounded N. W. by Anderson dist.; N. E. by Saluda river separating it from Laurens dist.; S. E. by Edgefield dist.; and S. W. by Savannah river separating it from Lincoln and Elbert counties in Georgia. It lies in form very nearly a square of 31 miles each side; area 960 square miles. Extending in lat. from 33° 50′ to 34° 28′ N.; and in long. from 4° 56′ to 5° 42′ W. from W. C. The southern and central parts of Abbeville slope toward and are drained into Savannah river, with a southern declination. A narrow zone along the northeastern border slopes to the southeastward towards the main stream of Saluda river. By the census of 1820, this district contained a population of 23,167; but in the ensuing ten years it had increased to 28,149. Of the latter 7,680 were white males; 7,181 white females, and the residue, 13,288, were people of color. Distributive population by the census of 1830, was 30 to the square mile. Chief town Abbeville.

ABBEVILLE, p-v. and seat of justice, Abbeville dist., S. C.; situated near the centre of the district, on one of the branches of Little river, at N. lat. 34° 11′; long. W. from W. C. 5° 20′; 63 miles N. N. W. from Augusta in Georgia, and by post road, 100 m. a very little N. of W. from Columbia.

ABBEVILLE, or Abbeyville, p-v. Mecklenburg co. Va.; situated on the left bank of Staunton or Roanoke river, about 10 miles above the influx of Dan river, and by post road, 126 miles S. W. from Richmond, and 227 m. S. S. W. from W. C.

ABBOTTSTOWN, p-v. Adams co. Penn.; situated on almost the eastern line of the county, and on a branch of Conewago creek, 15 miles N. E. by E. from Gettysburg, very nearly an equal distance S. W. by W. from the borough of York; and by post road 86 m. N. from W. C.

ABBOTT'S MILLS, and post office, Rutherford co. Tenn.; by post road, 40 miles southeastward from Nashville.

ABINGDON, p-v. Harford co. Md.; 22 miles northeastward from Baltimore.

ABINGDON, p-v. and seat of justice, Washington co. Va.; situated at the southeastern side of a mountain ridge, about mid-distance between the two main forks of Holston river and about 7 miles distant from each, also on the Great Valley road. According to Tanner's map of the U. S. this place stands about 8 miles northwardly from the northern boundary of Tennessee, N. lat. 36° 42′, long. 4° 58′ W. from W. C., by post road 385 miles S. W. by W. from W. C., and 309, a little S. of W. from Richmond.

ABINGTON, p-t. Plymouth co. Mass.; 22 miles S. E. of Boston, contains 2 ponds, one emptying into North river and the Atlantic ocean, the other into Taunton river and Narraganset bay. Spirited resolutions against the right claimed by the British Parliament to tax the colonies were passed here, 1770. Pop. 2,428.

ABINGTON, p-v. Windham co. Conn.

ABINGTON, p-v. Luzerne co. Penn.; 15 m.

N. E. from Wilkes Barre, and by post road 137 miles N. E. from Harrisburg, and 245 m. N. N. E. from W. C.

Abington, p-v. Wayne co. Ind.; by post road, 76 miles N. E. by E. from Indianapolis.

Abram's Creek, Columbia co. N. Y.; is formed by the junction of Kinderhook and Claverack creeks, and after a course of half a mile falls into Hudson river 4 miles above Hudson.

Absecombe, p-v. Gloucester co. N. J.

Accomac, the northernmost of the two counties which constitute together, that section called, "The Eastern Shore of Virginia." This county extends from the Atlantic ocean to Chesapeake bay,—is bounded by the Atlantic ocean E.; Northampton co. Va. S.; Chesapeake bay W.; Pocomoke bay N. W.; and Worcester co. Md. N.; extending in latitude from 37° 28′ to 38° 02′ N. and in long. from 1° 24′ to 1° 46′ E. from W. C. Greatest length from S. S. W. to N. N. E. 48 miles; mean width about 10 miles, area 480 square miles. Much of the surface is sand banks, or islands along the Atlantic coast; the real arable superficies is about 400 square miles. The surface is level. Chief town, Drummondstown. Pop. 1820, 15,966, and 1830, 19,656. Of the latter, were white males 4,495; white females 4,969, total 9,458; and the residue people of color.

Accomac, court-house. See Drummondstown.

Accord, p-v. Ulster co. N. J.

Achor, p-v. in the northern part of Columbiana co., O.

Acra, p-v. Greene co. N. J.

Acton, t. Windham co. Vt.; 32 miles N. E. of Bennington, 18 N. W. of Brattleboro', has an uneven surface, well watered by brooks, but without good mill streams. Pop. 176.

Acton, p-t. Middlesex co. Mass.; 24 miles N. W. of Boston, has a good soil. Assabet river, a chief tributary of Concord river, passes through a part of it, as well as its 2 branches, and the post road from Boston and Concord to Groton and Keene. Pop. 1,128.

Acworth, p-t. Cheshire co. N. H.; 50 miles W. of Concord, 87 W. of Portsmouth, 93 N. W. of Boston. Cold river rising at Cold Pond, affords mill seats. The town is agricultural, has a good soil, and raises flax, and cattle. It is on the post road and turnpike from Charlestown to Concord and Boston. Pop. 1,401.

Adair, one of the southern counties of Ken.; bounded by Russell S. and S. E.; Monroe S. W.; Barren W.; Green N. W.; and Casey E. The greatest length is about 40 miles from N. E. to S. W.; mean breadth 17, and area 680 square miles. Extending in lat. from 36° 51′ to 37° 28′ N. and in long. from 7° 50′ to 8° 30′ W. from W. C. The northern, central, and indeed the far greater part of this county slopes to the N. W. and is drained in that direction by confluents of Green river. The southern part declines towards and is drained by creeks falling into Cumberland river. In 1820, this county was much more extensive than it is at present, being lessened by the intermediate creation of Russell county. Pop. of Adair, 1830, 8,220. Chief town Columbia.

Adairsville, p.v. Logan co. Ken.; by post road 181 miles S. W. from Frankfort; and 10 in a similar direction from Russellville, the county seat.

Adams, Cape of the United States, on the Pacific ocean. It is the Point Ronde of Lapeyrouse, and the southern point at the mouth of Columbia river. It is thus described in a geographical sketch of Oregon Territory, published at Boston, 1830.—"Point Adams forms the south side of the river Columbia. It is a low projection of land, bearing S. E. about seven miles from the Cape (Disappointment), and thinly wooded. From it sand banks extend within one mile of the Cape (Disappointment), and inside of the bank which runs out from the Cape." As laid down by Tanner, in his map of the United States, the lat. is 46° 17′ N. and long. 46° 50′ W. from W. C. Variation of the magnetic needle at, 22° 40′ E. See Columbia river.

Adams, t. Coos co. N. H.; in a romantic situation at the E. base of the White Mountains, is uneven and partly rocky, but has a rich soil. It contains Black, Baldface and Thorn Mountains, and 2 branches of Ellis' river, which falls into Saco river.

Adams, p-t. Berkshire co. Mass.; 125 miles N. W. of Boston, 20 Pittsfield, 40 E. of Albany, was named after Samuel Adams, is divided from Williamstown on the W. by Saddle Mountain. Hudson's branch, a mill stream, comes from Vermont, and falls into Hoosick river through a channel in one place cut 60 feet into a White marble quarry, leaving a natural bridge, 12 or 15 feet long, 10 wide and 62 high. Some of the marble is clouded. Fort Massachusetts was on the N. E. end of Saddle Mountain. The Marquis de Vaudreuil attacked it, August 26, 1746, with 900 French and Indians, but was resisted, with 45 killed, for 24 hours, by 33 men, women and children, under Colonel Hawkes, who obtained an honorable capitulation. August 2, 1748, it was attacked by 300 French and Indians who were repulsed by Colonel Williams. The Adams and Hoosick cotton and woollen manufactories were incorporated 1809, and 2 more in 1814. There are about 25 cotton and woollen in all. There is a turnpike to Claremont. There are many fine dairies. Graylock, a peak of Saddle mountain and highest land in Massachusetts, is 3,580 feet above Hudson river at Albany. It has two villages, N. and S. Pop. 2,648.

Adams, p-t. Jefferson co. N. Y.; 166 miles W. of Albany, is 6 miles square, has very rich arable land, good for grain. N. branch of Big Sandy creek and Stony creek furnish mill seats. Many remains of ancient mounds are found here, with coarse earthen pipes, stone hearths many feet under ground, &c. 7 tumuli have been observed, with ditches round them, enclosing from a half to two acres. Pop. 2,995.

ADAMS, one of the southern counties of Penn.; bounded by Franklin co. W.; Cumberland N.; York N. E. and E.; and Frederick co. Md. S. Length 25 miles, mean breadth 18, and area 450 square miles. Extending in lat. from 39° 42′ to 40° 2′, and in long. from the meridian of W. C. to 0° 30′ W. This county is table land, and nearly equally divided between the basins of Potomac and Susquehanna. The southern part slopes towards the former stream, and is drained by the sources of Monocacy, whilst the northern section gives source to Conewago, and declines towards the Susquehanna. The surface of the whole county is hilly, but soil generally fertile. Chief town Gettysburg. Pop. 1820, 19,370, and in 1830, 21,379.

ADAMS, one of the southwestern counties of the state of Miss.; bounded N. E. by Jefferson; E. by Franklin; on the S. by Homochitto river which separates it from Wilkinson; and on the W. it is separated from the parish of Concordia in Louisiana, by the Mississippi river. From S. S. W. to N. N. E. along the general course of the Mississippi the length is about 40 miles, mean breadth 15, and area 600 square miles. Extending in lat. from 31° 15′ to 31° 46′, and in long. from 14° 16′ to 14° 43′ W. of W. C. The general slope is to the southwestward. Surface broken by hills, which though very numerous are of inconsiderable elevation; except some bottoms along the Mississippi and Homochitto, there is but little level land in the county. The soil is, though of various qualities in different parts, generally productive. Staple, cotton. Chief towns Natchez and Washington. Pop. 1820, 12,073, and in 1830, 14,919.

ADAMS, one of the southern counties of Ohio; bounded by Brown co. W.; Highland N. W.; Pike N. E.; Scott E.; and by the Ohio river separating it from Lewis and Mason counties of Kentucky, S. Length 28, breadth 22, and area 616 square miles. Extending in lat. from 38° 37′ to 40° N., and in long. from 6° 12′ to 6° 36′ W. from W. C. The general slope of this county is to the southward and towards the Ohio river; it is hilly but fertile. Chief town West Union. Pop. 1820, 10,406, and in 1830, 12,278.

ADAMS, one of the western counties of Illinois; bounded as laid down on Tanner's map on the N. by Hancock; E. by Schuyler; S. E. and S. by Pike; and on the W. is separated from Marion county and a section of the unappropriated part of the state of Missouri, by the Missisippi river.—Length from S. to N. 32, mean width 24, and area 768 square miles. Extending in latitude from 39° 42′ to 40° 11′, and in longitude from 13° 52′ to 14° 26′ W. from W. C. The western and central parts of this county slope westward towards the Mississippi river, whilst the eastern border gives source to creeks, the water of which is finally discharged into Illinois river. Chief town Quincy. Pop. 1830, 2,186.

ADAMS, p-v. Seneca co. O.; by post road, 98 miles northward from Columbus, and 412 N. W. by W. from W. C.

ADAMS, p-v. Decatur co. Ind.; by post road 48 miles S. E. by E. from Indianopolis.

ADAMS BASIN, p-v. Monroe co. N. J.

ADAMSBURG, p-v. Westmoreland co. Penn.; on the main road from Greensburg to Pittsburg, 6 miles W. from the former, and by post road 176, westward from Harrisburg, and 198 N. W. from W. C.

ADAMS, OLD, court house and post office, Adams co. Miss.; 9 miles from Natchez.

ADAMS' MILLS, and post office, Pulaski co. Ky.; by post road 82 miles a little E. of S. from Frankfort.

ADAMSTOWN, p-v. near the northeastern border of Lancaster co. Penn.; 23 miles N. N. E. from the city of Lancaster, and 12 S. S. W. from Reading.

ADAMSVILLE, p-v. Washington co. N. Y.

ADAMSVILLE, post office, Berks co. Penn.; 9 miles from Reading and 61 eastward from Harrisburg.

ADAMSVILLE, p-v. Frederick co. Md.; by post road 48 miles N. W. from W. C.

ADAMSVILLE, p-v. Marlborough dist., S. C.; by post road 110 miles eastward from Columbia, and 398 S. S. W. from W. C.

ADDISON, t. Washington co. Me.; 15 miles W. Machias. Pop. 741.

ADDISON Co. Vt.; bounded by Chittenden co. N.; Chittenden, Washington and Orange counties E.; Windsor co. S. E.; Rutland co. S.; Lake Champlain W. Pop. 1820, 20,469; 1830, 24,940. It is crossed by Otter creek S. to N., and by the Green mountains E. The county town is Middlebury. It has a good port on the lake at Basin Harbor.

ADDISON, p-t. Addison co. Vt.; E. of Lake Champlain, opposite Crown Point, New York, 83 miles N. Bennington, 40 S. W. Montpelier. This was probably the first settlement by Europeans in this state W. of the Green mountains. In 1731 the French built a fort at Crown Point, and occupied this shore. The English first came in 1770. It is low and generally level, with few streams. Snake mountain is in S. E. corner. Otter creek and one of its branches, with Mill and Pike rivers, (falling into Lake Champlain,) are within the town. Sulphuret, and magnetic oxide of iron are found. Pop. 1,306.

ADDISON, p-t. Steuben co. N. Y.; 25 miles S. of Bath, N. of Pennsylvania Line, has Canisteo (navigable for boats) and Tuscarora creeks, but the land is broken, and has little value, except for timber. Grindstones are here made of sandstone. Pop. 944.

ADDISON POINT, p-v. Washington co. N. Y.

ADELPHIA, p-v. in the northeastern part of Ross co. O.; by post road 46 miles southwardly from Columbus, and by the common road 20 N. E. from Chilicothe.

ADGATES' FALLS, New York. (See Chesterfield, N. Y.)

ADRIAN, p-v. Lenawee co. Michigan Ter.; by post road 10 miles from Tecumseh the county seat, 73 s. w. by w. from Detroit, and 502 n. w. by w. from W. C.

Agamenticus Mountain, York, York co. Maine.

Adriance, p-v. Dutchess co. N. Y.

Agawam, p-v. Hampden co. Mass.; 2 miles s. w. Springfield.

Agawam r. Mass. See Westfield river.

Agnew's Mills, and post office, Venango co. Pa.; by post road 248 m. n. w. from W. C.

Ahosky Ridge, post office, northern part of Hertford co. N. C.; by post road 121 miles n. e. by e. from Raleigh, and 240 very nearly due south from W. C.

Ahpmoojeenee-Gamook, lake, Me.; n. of Moosehead lake, empties St. John's river.

Akron, p-v. Portage co. O.

Alabama river, the great northeastern constituent of Mobile river, is formed by the confluent streams of Coosa and Tallapoosa rivers, and receives as a tributary, the Cahaba from the north. Of the three constituents of Alabama, the Cahaba is the only one entirely in the state of Alabama. Rising between the vallies of Black Warrior and Coosa rivers, in the counties of Jefferson and St. Clair, the Cahaba assumes a course a little w. of s. and which it maintains over Shelby, Bibb, Perry and Dallas cos. falling into Mobile at the town of Cahaba in the latter county. The valley of Cahaba is about 120 miles in length with a mean breadth of 20, and with an area of 2400 square miles.

Coosa or the main constituent of Alabama, has it highest and most remote source in Tennessee, interlocking sources with those of Hiwassee and Chattahooche. The most northern sources of Coosa are at n. lat. 35° 05', and are the most northern fountains, the water of which is finally discharged into the Gulf of Mexico e. from the Mississippi basin. There known by the name of Connessauga, it flows first west, but curving to s. s. w. about 70 miles receives from the east the Etowah. The latter rises in Georgia, between the sources of Hiwassee and Chattahooche, and only separated about 15 miles from the Turoree branch of Savannah river, by spurs of the Blue Ridge. Issuing from this elevated region the Etowah, with a sweeping curve to the southward, pursues a general course of s. w. by w. 100 miles to its union with the Connessauga or Oostenalah. Having their fountains and channels in the comparatively high Appalachian vallies, the superior branches of Coosa are rapid mountain streams. Below the junction of Etowah and Connessauga, the united waters henceforth known as the Coosa flow 8 or 10 miles westward, when, leaving Georgia and entering Alabama, the now considerable volume by an elliptic curve inflects first to s. w. thence s. and finally s. s. e. to its junction with the Tallapoosa to form the Alabama, at Coosauda, n. lat. 32° 28', long. 9° 22' w. from W. C. The entire comparative course of the Coosa is about 240 miles, but by the bends may not fall much short of 400 miles. The valley it drains is about 200 miles in length with a mean breadth of 45; area 9000 square miles.

Talapoosa, or eastern branch of Mobile, rises between the vallies of Etowah and Chattahooche, and principally in Carroll county of Georgia, at n. lat. 34 nearly.—Flowing s. s. w. it enters Alabama, and continuing that course, 130 miles, turns abruptly to the west 25 miles, unites with the superior volume of the Coosa as already noted. The valley of Talapoosa lies entirely between those of the Coosa and Chattahooche; it is about 150 miles in length with a mean width of 25, and area 3750 square miles.

In one striking feature the Coosa and Talapoosa have strong resemblance to each other. In the lower part of their respective courses in Alabama, neither receive tributaries above the size of a large creek.

Alabama, formed thus by the union of the Coosa and Talapoosa, assumes a general western course to the influx of Cahaba, and thence curving to the s. s. w. to its junction with Tombigbee to form the Mobile. By a comparative course from the mouth of Coosa to that of Tombigbee, the length of the Alabama is 130 miles, but so tortuous is its channel that the navigating length falls little if any under 250 miles. The valley of the Alabama proper is about 120 miles by 30, with an area of 3600 square miles. Combining the area of all the sections of the Alabama valley we find it comprises 18,750 square miles.

From the great difference of height between the sources and mouth of the assemblage of confluents, the streams of this valley are rapid. At times of flood they are all, however, navigable for down stream vessels from near their sources. Schooners of 5 feet draught are navigated into Alabama and as far as the lower falls at Claiborne, 50 miles above the mouth.

Alabama, one of the United States, bounded W. by the state of Mississippi; N. by the state of Tennessee; E. by Georgia; S. by Florida; and S. W. by the Gulf of Mexico. This state has an outline in common:

	Miles.
With the state of Mississippi . . .	330
" " Tennessee . . .	153
" " Georgia	306
Along N. lat. 31°, and in common with Florida	150
Down Perdido river, from N. lat. 31° to its mouth	60
Along the Gulf of Mexico to place of beginning	60
Having an entire outline of . .	1059

The area of Alabama, is 51,770 square miles, equal to 33,132,800 statute acres. Greatest length of the Gulf of Mexico to the Tennessee line 336 miles; the mean breadth from E. to W. 154. Geographically, this state lies between N. lat. 30° 10', and 35°, and in long. between 8° 05', and 11° 30' W. from W. C.

This state lies, with the exceptions of its southeastern and southwestern angles, in the valley of Tennessee and basin of Mobile. If taken under a general view, it is subdivided into two unequal physical sections. The northern and smaller section is comprised in the valley of Tennessee. That river winding by a general western course, but with a sweeping curve to the south, enters at the northeastern angle of the state, and issues from it at the southeastern.

The southern, and by far the most extensive section, has a slope very nearly due south, and is drained by the main streams, and numerous confluents of Tombigbee, Black Warrior, Alabama, Mobile, Conecuh, Choctawhatchie and Chattahooche rivers.

Northern, or the Tennessee river section of Alabama, contains the counties of:

	Square miles.	Population, 1820.	Population, 1830.
Franklin . .	684	4,988	11,078
Jackson . .	1040	8,751	12,700
Lauderdale	672	4,963	11,781
Lawrence .	816		14,984
Limestone .	600	9,871	14,807
Madison, and	648	17,481	27,990
Morgan . .	600		9,062
Amount . .	5060	46,054	102,402

The surface upon which resided the respective populations of 1820, and 1830, was nearly the same, we therefore find that northern Alabama gained in the 10 intermediate years 222 per cent.

Passing the ridge which separates the sources of the southern creeks of Tennessee river, from those of Coosa, Black Warrior and Tombigbee, we are on the northern and higher border of the great southern slope of Alabama, down which spread the counties of:

	Square miles in 1830.	Population, 1820.	Population, 1830.
Autauga . .	1080	3,853	11,874
Baldwin . .	2000	1,713	2,324
Bibb . . .	800	3,676	6,306
Blount . .	1650	2,415	4,233
Butler . . .	1000	1,405	5,650
Cataco . .		5,963	
Clarke . .	1200	5,839	7,595
Conecuh . .	1531	5,713	7,444
Covington .	1664		1,522
Dale . . .	1600		2,031
Dallas . .	1064	6,003	14,017
Fayette . .	1250		3,547
Greene . .	836	4,554	15,026
Henry . .	1344	2,638	4,020
Jefferson . .	1040		6,855
Lowndes			9,410
Marengo . .	960	2,933	7,700
Marion . .	1140		4,058
Mobile . .	2250	2,672	6,267
Monroe . .	960	8,838	8,782
Montgomery .	1500	6,604	12,695
Perry . . .	966		11,490
Pickens . .	648		6,622
Pike . . .	1750		7,108
St. Clair . .	720	4,166	5,975
Shelby . .	1100	2,416	5,704
Tuscaloosa .	858	8,220	13,646
Walker . .	1500		2,202
Washington .	840		3,474
Wilcox . .	1200	2,917	9,548
	33,451	81,847	207,125
Add N. Alabama	5060	46,054	102,402
Amount .	38,511	127,901	309,527

The preceding area of 38,511 square miles comprises that part of Alabama, yet purchased from the Indians, and organized into counties. But there is on the northeastern border a section of about 600 square miles still in the possession of the Chickasaws. On the western border, and nearly opposite the middle of the state, there is a second tract, possessed by the Choctaws, which comprises about 1800 square miles. Again, there is a region of about 11,000 square miles along the eastern and northeastern side of the state still possessed by the Creeks and Cherokees. The President of the United States in his message to Congress, at the opening of the present session, informs that body that measures have been taken under the laws of the United States, "By which the whole of the state of Mississippi, and the western part of Alabama, will be freed from Indian occupancy, and opened to a civilized population. The treaties with these tribes are in a course of execution, and their removal, it is hoped, will be completed in the course of 1832."

By the preceding elements we are shown that in the decennial period from 1820 to 1830, the population of Alabama had gained 234 per cent.

We may premise, that in the preceding physical division of Alabama, into northern and southern sections, we have not pursued the same limits of division, adopted in taking the recent census, but the difference is not material to any general result. The following tabular statements are from the census of 1830.

Recapitulation, exhibiting the general aggregate amount of each description of persons in the northern district of Alabama.

Free White Persons.

Males under 5 years of age	9,459
" of 5 and under 10 years of age	6,727
" " 10 " 15 "	5,281
" " 15 " 20 "	4,369
" " 20 " 30 "	7,036
" " 30 " 40 "	4,458
" " 40 " 50 "	2,516
" " 50 " 60 "	1,382
" " 60 " 70 "	773
" " 70 " 80 "	346
" " 80 " 90 "	63
" " 90 " 100 "	11
100 and upwards	60
	42,311

Free White Persons.

Females under 5 years of age	8,984
" of 5 and under 10 years of age	6,428
" " 10 " 15 "	4,987
" " 15 " 20 "	4,404
" " 20 " 30 "	6,308
" " 30 " 40 "	2,885
" " 40 " 50 "	2,125
" " 50 " 60 "	1,186
" " 60 " 70 "	575
" " 70 " 80 "	189
" " 80 " 90 "	69
" " 90 " 100 "	16
100 and upwards	7
	38,862
	42,311
Total number of Free White Persons	81,173

White persons included in the foregoing who are deaf and dumb.

Under 14 years of age	11
14 " and under 25 .	12
25 " and upwards .	8
Blind	30
Foreigners not naturalized . . .	20

Slaves.

Males under 10 years of age	8,252
" of 10 and under 24 years of age	7,318
" " 24 " 36 "	4,247
" " 36 " 55 "	1,957
" " 55 " 100 "	604
" " 100 and upwards	5
Total males	22,383
Females under 10 years of age	7,974
" of 10 and under 24 years of age	7,152
" " 24 " 36 "	4,209
" " 36 " 55 "	1,897
" " 55 " 100 "	507
" " 100 and upwards	8
Total Females	21,747
Amount of Slaves	44,130

Free Colored Persons.

Males under 10 years of age	75
" of 10 and under 24 years of age	54
" " 24 " 36 "	66
" " 36 " 55 "	39
" " 55 " 100 "	19
" " 100 and upwards	00
Total Males	253
Females under 10 years of age	54
" of 10 and under 24 years of age	48
" " 24 " 36 "	31
" " 36 " 55 "	19
" " 55 " 100 "	16
" " 100 and upwards	1
Total Females	169
Amount Free Colored	422
Total aggregate population of Northern Alabama	125,725

General aggregate amount of each description of persons in Southern Alabama.

Free White Persons.

Males under 5 years of age	13,305
" of 5 and under 10 years of age	8,755
" " 10 " 15 "	6,908
" " 15 " 20 "	5,209
" " 20 " 30 "	10,404
" " 30 " 40 "	6,941
" " 40 " 50 "	3,513
" " 50 " 60 "	2,092
" " 60 " 70 "	968
" " 70 " 80 "	345
" " 80 " 90 "	84
" " 90 " 100 "	8
" " 100 and upwards	3
Total White males	58,535
Females under 5 years of age	12,376
" of 5 and under 10 years of age	8,375
" " 10 " 15 "	6,165
" " 15 " 20 "	5,547
" 20 " 30 "	8,151
" 30 " 40 "	4,894
" 40 " 50 "	2,560
" 50 " 60 "	1,545
" " 60 " 70 "	744
" 70 " 80 "	250
" 80 " 90 "	75
" 90 " 100 "	13
" " 100 and upwards	3
Total White Females	50,698
Total number of Free White Persons	109,233

Persons included in the foregoing, who are deaf and dumb, under 14	34
Persons of 14 and under 25	13
Persons of 25 and upwards	11
Persons Blind	38
Foreigners not naturalized	45

Slaves.

Males under 10 years of age	13,585
" of 10 and under 24 years of age	12,936
" " 24 " 36 "	6,853
" " 36 " 55 "	3,901
" " 55 " 100 "	8,091
" " 100 and upwards	22
Total of Male Slaves	36,787
Females under 10 years of age	13,418
" of 10 and under 24 years of age	12,517
" " 24 " 36 "	6,879
" " 36 " 55 "	3,001
" " 55 " 100 "	805
" " 100 and upwards	18
Total Female Slaves	36,632
Total of Slaves	73,419

Free Colored Persons.

Males under 10 years of age	209
" of 10 and under 24 years of age	148
" " 24 " 36 "	112
" " 36 " 55 "	85
" " 55 " 100 "	37
" " 100 and upwards	00
Total Free colored Males	591
Females under 10 years of age	191
" of 10 and under 24 years of age	161
" " 24 " 36 "	100
" " 36 " 55 "	65
" " 55 " 100 "	40
" " 100 and upwards	2
Total of Free colored Females	559
Total number of Free colored Persons	1,150

Slaves and colored persons included in the foregoing who are deaf and dumb, under 14 years of age	5
" " " " of 14 and under 25	3
" " " " of 25 and upwards	5
" " " " blind . . .	26

SUMMARY.

Northern Alabama, whites . . .		81,173
Southern " " . . .		109,233
Total Whites		190,406
Slaves and Free colored . . .	44,552	
" " "	74,569	
Total Slaves and Free colored		119,121
Total Population of Alabama, by the Census of 1830		309,527

Physical Features.—The surface of Alabama is divided into two very unequal sections, as we have shown under the head of population. The northern and smaller plain is a part of the valley of Tennessee River, the main volume of that stream entering at the northeastern, and leaving the state at the northwestern angle, flowing in the intermediate distance by a general course to the westward, but with a sweeping curve to the south. The Tennessee enters Alabama in a mountain valley immediately below the influx of Sequatche River, flows thence by comparative courses 60 miles in a direction to the s. w.; thence abruptly inflecting to north-west by west 130 miles, passes a chain of the Appalachian system by the noted pass called the Muscle Shoals; below which, curving gradually more to the northward, leaves the state at the influx of Bear Creek. In this distance of 190 miles, Tennessee receives but one tributary stream deserving the name of a river; that is Elk River which enters from the right; but the confluent creeks are numerous on both sides. Northern Alabama is finely, indeed beautifully variegated by hill, dale, and in some places by plain. It was the first part of the state inhabited by whites, if we except a few spots along Mobile and Tombigbee Rivers, above and below the town of Mobile; and is yet far most densely settled; the mean to the square mile being by the census of 1830, a small fraction above 20, whilst that of Southern Alabama, did not amount but to a little above the one third, or not quite 7 to the square mile.

The southern and much most extensive zone, or inclined plain, falls by a very gentle declivity from N. lat. 34° 20′ to the Gulf of Mexico, over something more than four degrees of latitude. This slope is drained and finely variegated by the numerous branches of the Coosa, Tallapoosa, Cahaba, Black Warrior, Tombigbee, and Alabama, all contributing to form the Mobile. The southeastern angle of the state declines to the southwestward and is drained in that direction by Choctaw, Yellow Water, Conecuh rivers. A very narrow strip of the southeastern border, declines eastward, and is drained into and bounded by the Chattahooche river.

The state is over both physical sections, very advantageously supplied with navigable rivers, though possessing within its own limits but one outlet to the Gulf of Mexico by Mobile Bay. The tides being moderate, not exceeding two and a half feet at a mean. The entrance of sea vessels of any draught, is arrested by rapids at Claiborne, in Alabama, and St. Stephens, in Tombigbee river; both places being about one hundred miles direct distance above the outlet of Mobile Bay, into the Gulf of Mexico. In common winters, the rivers of even northern Alabama, are but little impeded by ice, but there do occur seasons, and that of 1831—32, is an example, when Tennessee river and its confluents are completely frozen. The streams flowing southwardly, towards the Gulf of Mexico, are still more rarely, and below N. lat. 33°, perhaps never rendered unnavigable by frosts. The excessive droughts of summer are, indeed, far more frequently the cause of impeding navigation in, not only the rivers of Alabama, but all the streams between the Mississippi and the Atlantic ocean below N. lat. 35°.

The seasons at the extremes of Alabama, in regard to mean and extreme temperature differ greatly. The border along the Gulf of Mexico may be called tropical when compared with the valley of Tennessee. Between lat. 30° 10′, and 31° 30′, or below tide water, excessive frost or lying snows are rare; but the temperature changes sensibly advancing towards and into the valley of Tennessee. No part, however, of the state reaches sufficiently to the southward to admit the profitable cultivation of sugar cane. Cotton is the general staple of the state. Indian corn or maize is the usual crop; but in the middle and northern part small grain succeeds well.

The arable land of southern Alabama, lies mostly along or near the water courses, and is composed of two species. Alluvion, properly so called, and Interval land. The latter a kind of intermediate soil between the alluvial river bottoms and the open pine woods. The latter description of land, is sterile, and comprises the much greater part of the surface of the state, more comparatively extensive, however, on the southern than on the northern section.

Constitution of Government, Judiciary.—The territory now constituting the state of Alabama was formerly included in the western territory of Georgia, afterwards in the Mississippi territory. (See Georgia and Mississippi.) In March, 1817, the Mississippi territory was divided by an Act of Congress, by which authority was given to form the western section into a state, and subsequently the eastern part was formed into a territory under the name of Alabama. The increase of population was so rapid as to entitle it to admission as a state government in 1818, and application was made to Congress accordingly. A bill was brought in and a law passed, March, 1819, empowering the people of Alabama to form a Constitution; under the authority of which, a Convention met at Huntsville in Northern Alabama, and on August 2, 1819, adopted a Constitution of State Government, which was ratified by Congress December 1819, and the state admitted into the Union.

The Constitution of Alabama provides:—that "The powers of the government of the state of Alabama shall be divided into three distinct departments; and each of them confided to a separate body of magistracy, to wit, Legislative, Executive, and Judicial. No person, or collection of persons, being of one of those departments, shall exercise any power properly belonging to either of the others, except in the instances hereinafter expressly directed or permitted."

"The legislative power of this state shall be vested in two distinct branches: the one to be styled the Senate, the other the House of Representatives, and both together the General Assembly of the state of Alabama.

"The members of the House of Representatives shall be chosen by the qualified electors, and shall serve for the term of one year. No person shall be a representative unless he be a white man, a citizen of the United States, and shall have been an inhabitant of this state two years next preceding his election; and the last year thereof, a resident of the county, city, or town, for which he shall be chosen, and shall have attained the age of twenty-one years.

"Senators shall be chosen by the qualified electors, for the term of three years, at the same time, in the same manner, and at the same places, where they may vote for members of the House of Representatives; and no person shall be a Senator unless he be a white man, a citizen of the United States, and shall have been an inhabitant of this state two years next preceding his election, and the last year thereof a resident of the district for which he shall be chosen, and shall have attained to the age of twenty-seven years."

"Every white male person of the age of twenty-one years, or upwards, who shall be a citizen of the United States, and shall have resided in this state one year next preceding an election, and the last three months within the county, city or town, in which he offers to vote, shall be deemed a qualified elector: Provided, that no soldier, seaman, or marine, in the regular army or navy of the United States, shall be entitled to vote at any election in this state." Electors protected from arrest in civil cases, going to, attendance at, or return from the places of election.

Art. 4.—"The supreme Executive power of this state shall be vested in a chief magistrate, who shall be styled the governor of the state of Alabama. The governor shall be elected by the qualified electors, at the time and places when they shall respectively vote for representatives. He shall hold his office for the term of two years, from the time of his installation, and until his successor shall be duly qualified; but shall not be eligible for more than four years in any term of six years. He shall be at least thirty years of age, shall be a native citizen of the United States, and shall have resided in this state, at least four years next preceding the day of his election."

"The Judicial power of this state shall be vested in one Supreme Court, Circuit courts to be held in each county of the state, and such inferior courts of law and equity, to consist of not more than five members, as the general assembly may, from time to time, direct, ordain, or establish. No person who shall have arrived at the age of seventy years, shall be appointed to, or continue in the office of Judge in this state."

Education.—The following section is made part of the Constitution of Alabama. "Schools and the means of education, shall forever be encouraged in this state; and the general assembly shall take measures to preserve, from unnecessary waste or damage, such lands as are or hereafter may be granted by the United States, for the use of schools, within each township in this state, and apply the funds which may be raised from such lands, in strict conformity to the object of such grant. The general assembly shall take like measures, for the improvement of such lands as have been or may be hereafter granted by the United States to this state, for the support of a seminary of learning, and the monies which may be raised from such lands, by rent, lease or sale, or from any other quarter, for the purpose aforesaid, shall be and remain a fund for the exclusive support of a state university, for the promotion of the arts, literature, and the sciences; and it shall be the duty of the general assembly, as early as may be, to provide effectual means for the improvement and permanent security of the funds and endowments of such institution."

The following provision is proof decisive of the progress of liberality and humanity:—"In the prosecution of slaves for crimes, of a higher grade than petty larceny, the general assembly shall have no power to deprive them of an impartial trial by a petit jury."

The Constitution of Alabama may be amended or revised, whenever two thirds of each house of the general assembly propose such amendments or revision. The proposed changes duly published three months before the next general election, when and where the voice of the people is taken, and if "it shall appear that a majority of all the citizens of this state, voting for representatives, have voted in favor of such proposed amendments: and two thirds of each house of the next general assembly, shall after such an election and before another ratify the same, they shall be valid, to all intents and purposes, as parts of this constitution."

History.—This state derives its name from that of one of the noble rivers which channel its surface. Early in the eighteenth century, the French, in founding the colony of Louisiana, formed small settlements on Mobile river, and built a fort where the city of Mobile now stands, but a large share of what is now Alabama remained in possession of the native Indians for about a century after the founding of Louisiana. The original charter of Georgia covered the whole zone from 31° to 35° N.; of course four degrees wide of Alabama was included in Georgia. In 1802, a cession was made by Georgia to the United States, of all her western territory between Chattahooche and Mississippi rivers, as far up the former as near lat. 33°, and from thence to lat. 35°, by the existing line of demarcation between Georgia and Alabama. Alabama continued a part of the Mississippi territory until separated in the manner stated under the head of Constitution; where also the material facts in the history of the state are also given.

For seat of government see Tuscaloosa.

ALABAMA, p-t. Genesee co. N. Y. Pop. 783.

ALACHUA co. Flor.; bounded N. by Duvall co. N. E. by St John's co.; E. S. E. and S. by the country of the Seminole Indians, and W. by the Gulf of Mexico. Extending from south to north along the Gulf from N. lat. 27° 28′ at Sarazota Entrance, to N. lat. 29° 24′, at the mouth of Suwanne river; and in long. from 5° 10′ to 6° 20′ W. from W. C. In length it is about equal to the difference of its extremes of lat. or about 140 miles; the breadth varies greatly, but may be about a mean of 30 miles; area, 4,200 square miles. The surface slopes westward, but the declivity is slight, and discharges with no great rapidity of descent, Hillsboro', Anaclote, Amasura, and Suwanne rivers. It is generally flat, part marshy, some prairie. Soil mostly sterile. Court House at Dells.

ALACHUA SAVANNA, grassy plain in Alachua co. Florida; "lies about 4 miles above Orange lake; its length is 7 miles, and its breadth 3 miles. The great body of water in this Savanna is represented as losing itself in a large sink, supposed to be at the northern side, and to discharge itself through a subterranean passage into Orange lake. Whatever may be the case, this Savanna exhibits but the appearance of a level, watery meadow, covered with a thick growth of aquatic grass, a circumstance which causes it to be called, in the idiom of the country, *a grassy lake*. Its outlet, the Chechale, which flows into Orange lake, is of rather small size. The Alachua Savanna is lined with hammocks, in which the live-oak and water-oak are predominating."

ALAPAPAHA, a river of Geo. and Flor. the eastern branch of Suwanne river. See Suwanne.

ALAQUA, a small but remarkable river of Flor. in Walton co. rises in a ridge of hills near the centre of the county, and in two branches which unite to form Alaqua, which, assuming a southern course, falls into Choctaw bay. This stream admits the entrance of vessels drawing 5 feet water, for a distance of 15 miles to the margin of a fine body of fertile land, already extensively settled and cultivated. "Alaqua," says Williamson in his Florida, "is the largest stream that enters from the Ridge. The springs of the two eastern branches of this river rise gradually in cane patches, and flow through a beautiful undulating country of good land."

ALAQUA, p-v. and seat of justice for Walton co. Flor. is situated on Alaqua river, 70 miles a little N. of E. from Pensacola, and by post road 161 miles a little N. of W. from Tallahassee. N. lat. 30° 38': long. 9° 20' W. from W. C.

ALATAMAHA, a fine river of Georgia, draining the central parts of that state, and the space between the branches of the Flint, Chattahooche, and Savannah rivers. The Oconee and Oakmulgee rivers, are the great constituent streams of the Alatamaha. The two former rising southeastward from the valley of the Chattahooche; the Oconee in Hall, near Gainsville, and the Oakmulgee in Gwinnett and De Kalb counties. Both streams rise so near as from 10 to 15 miles from the main volume of Chattahooche, but both assuming a nearly parallel course of S. S. E., about forty miles asunder. The Oconee, or eastern branch, pursues the original course with but slight general inflections, 170 miles to its junction with Oakmulgee, between Montgomery and Appling counties. The western, or Oakmulgee branch, pursues the original course, 160 miles, to near Jacksonville in Telfair co., where it curves round to N. E., 30 miles, to its union with Oconee, almost exactly on N. lat. 32°, and directly W. from the mouth of Savannah river. Below the junction of Oconee and Oakmulgee rivers, the united waters form the Alatamaha, which, maintaining nearly the course of Oconee S. S. E. by comparative courses 90 miles, falls into the Atlantic ocean by one main and several smaller outlets, between St. Simon's and Sapeloo islands, and between McIntosh and Glynn counties. The entire length of the basin of Alatamaha is 250 miles, with a mean width of 50, and area 12,500 square miles; and lying between lat. 31° 15' and 34° 28', and between long. 4° 22' and 7° 20' W. from W. C. The sources of Alatamaha interlock with those of St. Ills, Cambahee, Ogeeche, and Savannah rivers, flowing into the Atlantic ocean, and with those of Chattahooche, Flint, and Suwanne, flowing into the Gulf of Mexico.

As a navigable channel, Alatamaha has 14 feet water on its bar. Boats of 30 tons are navigated to Milledgeville on the Oconee, and to an equal or greater distance up the Oakmulgee. Down stream navigation is practicable on both rivers from near their sources.

The climate of this basin differs very much between the extremes, from difference of latitude and of level. The lat. differs near 3¼°, and the level not less than 1000 feet, giving an entire difference of temperature, of upwards of 5° of Fahrenheit's Thermometer. The vegetable productions, both natural and exotic, present a corresponding variety, with the extent of climate. On this basin, near the mouth of Alatamaha, the orange tree and sugar-cane are cultivated, and on its higher branches, the apple, peach, and pear; wheat, rye, oats and meadow grapes flourish. The staple vegetable, however, over the whole basin, which is most cultivated and valued, is Cotton. Tobacco, Indigo and rice, are also occasionally produced.

ALBA, p-v. Bradford co. Pa.

ALBANY, p-t. Oxford co. Me. 18 miles N. W. of Paris. Pop. 387.

ALBANY, p-t. Orleans co. Vt. 6 miles square, 34 miles N. Montpelier, contains several ponds, and part of Black river. The market road from Boston to Montreal passes through it. Pop. 683.

ALBANY co. N. Y. bounded by Schenectady and Saratoga counties N.; Rensselaer E.; Greene S.; Schoharie W.; 22 by 21 miles; 462 square miles; N. lat. 42° 21', long. 20' E. and 15' W. New York city, is W. of Hudson river, about 144 miles N. of New York; contains 9 townships. Albany is the chief town. Formations, transition and secondary, on slate rock, over which lies graywacke, especially in the west and middle. There are also shell lime stone and sand stone. Loose primitive rocks lie on the surface; and the minerals are numerous. The soil is various, low and rich on Hudson river, sand plains in the interior, hilly W., rocky N., much land is uncultivated. Norman's, Coeyman's, Bethlehem and Ten Miles creeks are small mill streams. Catskill river rises in the S.

The Erie and Champlain canals unite and terminate in this county, and the Mohawk and Hudson rail road (for which see *Albany*,) is entirely within it. One of the first settlements in the state was here. There is an agricultural society for the county; at Watervliet on the canal, is the United States arsenal, the principal depot for arms in the northern States. Population, 1820, 38,116; 1830, 53,560.

ALBANY, city, Albany co. N. Y. CAPITAL of the state; on the W. bank of Hudson river, 144 miles N. of New-York, 165 W. of Boston, 230 S. of Montreal, 30 N. of Hudson, 15 S. E. of Schenectady, is the second city in the state in population, trade, wealth and resources. It has been greatly increased and enriched by the operation of the Erie and Champlain canals, which unite 8 miles north of the centre of the city, and terminate at the Basin, which is formed in Hudson river by a pier 4,300 feet in length, along the upper part of the city, by which sloops, tow boats and canal boats are brought side by side, or have their cargoes exchanged over the pier. The amount of canal tolls received at Albany in 1830, was $212,056; 1831, $269,443. The whole amount received since the opening of the canals, $1,273,219 13.

The Capitol, which stands on a fine square at the head of State street, facing E. is a large and spacious stone building, and has two spacious and richly furnished chambers for the Senate and Assembly. In it are also rooms for the Superior Court, the Chancellor's Court, an office for the Governor, Committee, Jurors, and other rooms. The City Hall, situated on the same square, and facing west, is a noble building of white marble, where are held the Courts of the U. S. Circuit, of Common Pleas, the Mayor's, &c. The building is surmounted with a large dome, richly gilded, which marks its site to the traveller when many miles distant. This building, together with the Academy which stands on the same square, and which is a beautiful structure of free-stone, are a just source of pride to the citizens, and are evidence of the taste, wealth and enterprise of the inhabitants. Here are a Female Seminary and an Academy, to which are apportioned, from the school fund, to the former, $115 50 and to the latter, $360 68. There are 5 Banks, 3 Insurance Offices, an Institute, Library and Athenæum. The number of travellers passing through this city is very great, at all seasons. While the river is navigable, four elegant steamboats arrive here from N. York daily, and as many depart, transporting hundreds of passengers; while the travelling is great in all other directions, by tow-boats, canals, stages, &c. A steam boat line is also to begin this year between New York and Troy. Steam tow-boats and sloops transport a vast amount of merchandize for and from the canals. The *Mohawk* and *Hudson Rail Road*, designed to avoid the locks and circuit of Erie canal, was partly in use in 1831, and is now completed. It extends in a straight line from Albany to Schenectady, 14 miles, over an elevated sandy plain, with an inclined plane at each end. On that near Schenectady, a stationary engine is to be placed 130 ft. above the canal: 2 of the 3 sections are level; the others slope very gently towards Albany. It is thought that 600 passengers will pass on this road daily, and many more during the travelling season. The annual expenses are estimated at $14,600. Rail-road routes have been surveyed from Albany to Boston, and it is proposed to construct one either by the 8th Mass. turnpike, through Blanford, or by the Pontoosuc turnpike. Another is proposed, to West Stockbridge, Mass. at an estimated expense of $500,000, to connect part of the valley of Housatonic r., with Hudson r. A charter of a N. York city and Albany rail-road has been granted, to run E. of Hudson r. Pop. 26,000.

ALBEMARLE Sound, a deep bay of N. C. is the estuary of the Roanoke and Chowan rivers, extending 60 miles in length from east to west, along N. lat. 36° with a mean breadth of 8 miles, but protruding several deep minor bays. The Roanoke enters from the west and the Chowan from the northwest at the extreme interior of Albemarle, which spreads below the entrance of those rivers in a shallow expanse of water, with a level, or rather flat country along each shore. Every small inlet has its own comparative broad bay, by one of which the Pasquotank, a navigable inland communication by the Dismal Swamp canal, has been formed between Albemarle sound and Chesapeake bay. Albemarle sound is separated from the Atlantic ocean, by long, low, and narrow reefs of sand; but having two channels of connexion southward with Pamlico Sound, one on each side of the Roanoke Island, and on the northward an opening to the ocean by Currituck Sound and inlet; both rivers are navigable to near their sources. The climate of this basin differs very much between its extremes both from difference of latitude and of level. The latitude differs near 3½ degrees, and the level not less than 1000 feet, giving an entire difference of temperature of upwards of 5 degrees of latitude. The vegetable productions, both natural and exotic, have a corresponding variety with the extent of climate. On this basin, near the mouth, the orange and sugar cane are cultivated; and on its higher branches, the apple, and wheat, rye, oats, and other cerealia. The staple vegetable, however, both on the interior and islands contiguous to this basin, is cotton, though admitting a very wide range of staple, such as tobacco, indigo, &c. Rice is extensively cultivated.

ALBEMARLE, one of the central counties of Va. bounded N. W. by the Blue Ridge which separates it from Augusta and Rockingham, on the N. E. by the western part of Orange, on the E. by Louisa and Fluvanna, on the S. by James River, which separates it from Buckingham, and on the S. W. by Nelson. Length from S. W. to N. E. 35 miles, mean width 20, and area 700 square miles, N. lat. 38°, passes over very nearly the middle of the county, which is again nearly divided into two equal parts by long. 1° 30′ W. from W. C. The body of this county is drained by the constituent creeks of Rivanna River, which uniting below Charlotteville, pass through the south-west mountain, and a few miles lower enter Fluvanna. The face of this county is elegantly diversified by hill and dale, whilst the Blue Ridge affords a fine north-western border, and the more humble ridges of the south-west mountain decorate the eastern part. The soil, of course, partakes of the variety of feature from mountain, rocky and sterile, to productive river alluvion. Chief towns, Charlotteville, Scottsville, and Warren. Pop. 1820, 19,750, and in 1830, 22,618.

ALBERTSONS, post office, Duplin Co. N. C. 70 miles S. E. from Raleigh.

ALBION, p-t. Oswego co. N. Y. Pop. 669.

ALBION, p-v. and seat of justice, Edwards co. Illinois, situated between Bon Pas and Little Wabash rivers, 44 miles S. W. from Vincennes, and by post road 733 miles westward from W. C. and 92 S. E. by E. from Vandalia; N. lat. 38° 23', and long. 11° 07' W. from W. C.

ALBRIGHTS, post-office, Orange co. N. C. 70 miles N. W. from Raleigh.

ALBURG, p-t. and port of entry, Grand Isle co. Vt. 33 m. N. Burlington, is of a triangular form, 10 ms. long, and on an average 3½ wide, with Lower Canada line N. Missisque Bay E. and L. Champlain W., forming a point S. It was settled in 1782, by refugees who returned from Canada. The surface is very level, and there are no mountains or considerable streams. The soil is rich: timber, cedar, elm, beech and maple. There is a mineral spring, used in scrofulous cases. Pop. 1,239.

ALDEN, p-t. Erie co. N. Y. 22 ms. E. Buffalo. Pop. 1,257.

ALDIE post-office, Loudon co. Va. by post road 39 miles westward from W. C.

ALEXANDER, p-t. Genesee co. N. Y.; 8 m. S. S. W. Batavia, is crossed by Tonnewanta cr. with several branches, and has a gently varied surface, with a soil yielding grain, grass, &c. beech, maple, elm, ash, &c. The village is on Tonnewanta cr. Pop. 2,331.

ALEXANDER t. Washington co. Me. 30 ms. N. Machias. Pop. 334.

ALEXANDER, p-v. Athens co. O. by p-r. 79 miles S-E. by E. from Columbus, and 350 N-W. by W. from W. C.

ALEXANDER p-v. southern part of Montgomery co. O. by p-r. 73, S-W. by W. half west from Columbus, and 469 ms. N-W. by W. from W. C.

ALEXANDER'S mills, p-o. Fleming co. Ky.; by p-r. 86 miles estrd. from Frankfort.

ALEXANDRIA t. Grafton co. N. H. 30 m. N. by W. Concord, 72 N. W. Portsmouth. On Smith's and Fowler's rivers are 2000 acres of interval land, good for flax, potatoes and grass. Wheat and maize grow well in other parts. The mountainous tracts are rocky. Pop. 1,083.

ALEXANDRIA, p-t. Jefferson co. N. Y. opposite the 1000 Isls. in St. Lawrence r. which belong to it. The shore, which extends 9 ms. is high, rocky and varied, with many bays. The St. L. is 2 ms. wide. It contains Indian r. Mullet, Otter, Hyde, Plesses and Crooked creeks, with other mill streams; and 12 ponds, 3-4 to 3 ms. long. There are many falls, and numerous fish. At the falls on Ind. r. (6) ft.) is Theresa r. That r. is navigated by boats to Rossie. The timber is maple, beech, bass, elm, red and white oak, black and white ash, hemlock, pine, &c. It is crossed by the military road from Plattsburg to Sacket's Harbor. Pop. 1,523.

ALEXANDRIA, p-t. Hunterdon co. N. J. N. W. Musconetcong r.; S. W. Delaware r.; Mus. con mt. W.

ALEXANDRIA city, s-p. p-t. and st jus. for the co. of the same name, D. C. situated on the right bank of the Potomac, 7 miles below the capitol in Washington. The public buildings in Alexandria are several churches, Court-house, Academy, &c. The streets run at right angles to each other, and nearly so to the Potomac. This city is the centre of a considerable commerce, particularly in the article of flour. The harbor admits vessels of any draught, from ships of the line downwards. The meridian of Washington passing through the capitol leaves the central part of Alexandria, near 3 minutes to the E. lat. Alexandria 38° 48" N.

Population 1810—	
Free white males	2525
Do. Females	2378
Total white pop. 1810	4903
All persons except Indians not taxed	836
Slaves	1488
	7227

In 1820—	
Free white males	2667
Do. Females	2948
Foreigners not naturalized	153
Total white pop. 1820	5768
Free colored males	461
Do. Females	707
Slaves, male,	606
Do. Female	829
Total population in 1820	8371
Engaged in Agriculture	22
Do. Commerce	331
Do. Manufactures	699

Pop. 1830—See W. C.—Table 2d.

ALEXANDRIA co. D. C. comprising that part of the District ceded by Virginia and lying S. W. from the Potomac. From S. E. to N. W. it is 10 miles in length widening from the lower part of the city of Alexandria where it is a mere point, to a width of 4 miles from the little Falls of Potomac, area about 36 sq. ms. The surface with but little exception hilly, and soil thin. It is connected with the city of Washington by a wooden bridge over the Potomac. See D. C. pop. 1830, exclusive of the city, 1345, and including the city, 9608.

ALEXANDRIA p-v. and st. of jus. of Passide Parish, situated on the right bank of Red r. about half a mile below the lower Rapids in the bed of that stream, and by water 350 ms. above New-Orleans; and by a similar means of conveyance 65 ms. below Natchitoches. The t. is mostly comprised in a single line of houses along the river. Pop. about 800. Lat. 31° 18′ N. long. from W. C. 15° 39′ W.

ALEXANDRIA p-v. Huntingdon co. Penn. on Frankstown branch of the Juniatta, 10 miles N. N. W. from Huntingdon and by p-r. 96 ms. N. W. by W. from Harrisburg.

ALEXANDRIA p-v. Smith co. Ten. by p-r. 86 ms. N. E. from Nashville.

ALEXANDRIA p-v. Campbell co. Ky. by p-r. 75 ms. N. E. from Frankfort.

ALEXANDRIANA p-v. Mecklenburg co. N. C. by p-r. 151 ms. S. W. by W. from Raleigh.

ALFORD, t. Berkshire co. Mass. 125 ms. W. Boston, E. N. Y. line, is watered by Green r. and another branch of Housatonic r. and crossed by a turnpike. It is on the E. declivity of Tanghkannuck mts. Pop. 512.

ALFORDSVILLE, p-v. Robeson co. N. C. by p-r. 109 ms. S. S. wstrd. from Raleigh.

ALFRED, p-t. and half-shire, York co. Me. 24 ms. N. York. Pop. 1,453.

ALFRED, p-t. Alleghany co. N. Y. 10 ms. E. Angelica, W. Steuben co. is well watered by streams of Canacadea cr. Pop. 1,416.

ALLEGHANY, important river of western Pennsylvania and New-York, and one of the constituents of the Ohio. In strictness of Physical Geography, the Alleghany is the Ohio, of which the Monongahela is only a branch. By the Indians of different tribes, the words Ohio and Alleghany, mean the same thing, clear or fine river, of which native names, the French term *Belle Riviere* was a literal translation. By the Shawnese, the Alleghany was called Palawa Thepika.

The remote sources of this stream are in McKean and Potter co. Pa., from whence, flowing northward, they unite and enter the state of New York, over which it curves 40 miles in Cataraugus co., and re-enters Pennsylvania, within which, in Warren co., it receives a large branch, the Conewango, from Chatauque co. N. Y., and assuming a S. W. direction to Venango, where the main stream is again augmented by a considerable branch, French Creek, from the N. W. Now a considerable stream, Alleghany by a large eastern curve flows 120 miles by comparative courses, to Pittsburg, where it receives the Monongahela, and forms the Ohio. Below the influx of French Creek, the Alleghany receives no further large branch from the right, but from the left it receives Clarion river, Red Bank and Mahoning creeks, and a still more important tributary, the Kiskiminitas. The entire length of the basin of the Alleghany, extends 173 miles, nearly along long. 2° W. from Washington. The higher sources of Stoney Creek, branch of the Kiskiminitas, rise at N. lat. 42° 26′. The mean breadth of the basin, about 70 miles, area 12,110 square miles. The position of this river is admirably calculated to render it a great navigable chain of intercommunication. The main volume is a boatable stream at Hamilton, in the state of New York, within 25 ms. from the navigable water of Genesee r., at Angelica. French Creek becomes navigable at Waterford, with a portage of only 14 ms. between that point to the navigable waters of lake Erie. These lines have already become of commercial importance, but on the eastern side of the basin, the channel of the Kiskiminitas is becoming that of a part of the Pa. canal. The latter great work commences on the Conemaugh, at Johnstown, and follows that stream and its continuation, the Kiskiminitas to the Alleghany, which it crosses and following the right bank to the mouth, re-crosses, and passing through Pittsburg, enters the Monongahela by Lukes run. A rail road has been designed to continue from the latter canal line at Johnstown, and crossing the main ridge of the Alleghany mountains, terminates at Frankstown on the Juniatta, from whence the chain of canal recommences, and continues by the basins of the Susquehanna and Schuylkill to the city of Philadelphia.—*See Ohio basin.*

ALLEGHANY, mountain of the Appalachian system. It is an unanswerable objection to giving the name of Alleghany to the whole system, that it has been appropriated to a particular chain in Pennsylvania, Maryland, and Virginia. From the basin of the Kenhawa, to that of the western branch of the Susquehannah, through four degrees of latitude, the Alleghany is a dividing ridge between the waters flowing into the Atlantic, from those of the Ohio, giving source, estrd. to the branches of James r. and the Potomac, and on the other side, or wstrd. to those of the Kenhawa, Monongahela, Youghioghany, and Kiskiminitas. The ridges which form the particular chain of the Alleghany, are not very distinctly defined, though the entire chain constitutes so remarkable a feature in the geography of the United States. The length of the Alleghany is, from Monroe co. in Va. in the valley of the Kenhawa, to Centre co. in Pa. in the valley of Susquehannah, 300 miles. The height varies, but may be stated at 2500 feet, as a mean. Similar to all other chains of the Appalachian system, that of Alleghany does not rise into peaks, but stretches in parallel ridges, which, to the view from either side, presents gentle rounded, and swelling knolls, or elegantly defined lines, which bound the distant horizon. The component material of the Alleghany, is mostly graywake, though limestones and other rocks occasionally occur. This chain is rich in iron and bituminous coal. Some ridges have naked summits, but this feature is rare, the ridges generally being clothed with timber in all their height. A few mountain plains with tolerable soil for agriculture occur, but mostly the soil is rocky and barren, and often marshy. Pine and oak the most abundant timber.—*See Appalachian System.*

ALLEGHANY co. N. Y. in the w. part of the state, N. Pennsylvania line, 260 ms. w. Albany, bounded by Genesee and Livingston cos. N. Steuben E., Pa. S., Cataraugus and Genesee w., 40 ms. by 28, has 1120 sq. ms. It contains 13 towns, and is divided nearly equally by Genesee r. running N. with two falls 90 and 60 feet. Much of the soil is good alluvion, and the rest a gently varied surface, generally better for grass than for grain. Iron is mined in the N. There is a good number of mills and manufactories. Pop. 1820, 9,330; 1830, 26,218.

ALLEGHANY co. of Pennsylvania, having Washington co. S. and S. W.; Beaver N. W.; Butler N.; and Westmoreland E. Length from E. to W. 32 miles, mean breadth, 18, and area 575 sq. ms. The face of this county is in a peculiar manner diversified. Though not traversed by any mountain chain, the hills are high and swelling. The soil is fertile to the highest summits, and in its natural state covered with a very dense forest. But it is the rivers and their variegated banks which form the true decoration of this fine county; the Alleghany enters from the N. E., and the Monongahela from the S. E., and uniting at Pittsburg, form the Ohio; the latter winding to the N. W. traverses 14 miles of the western part of the county. To the rivers may be added Chartiers and many other creeks which contribute to drain and fertilize the fine vales which spread over this beautiful country. For lat. and long. see Pittsburg. In 1820 the population was, including Pittsburg, 34,-921, in 1830, pop. 50.552 including the city of Pittsburg, and exclusive of Pittsburg, 37,-984.

ALLEGHANY, extreme western county of Maryland, bounded by Washington in Md. E. west branch of the Potomac, or Hampshire and Hanay counties in Va. S., Randolph and Preston counties of Va. W., and Fayette, Somerset, and Bedford counties in Pa. N. Extreme length along the Pennsylvania line is 65 ms. but the breadth is very irregular, on Randolph and Preston counties, it is about 35 ms.; at Cumberland about 7; the mean breadth 12½, superfices 800 sq. ms. This county is traversed by the main chain of the Alleghany which discharges the higher branches of the Youghioghany to the N. W. and those of the west branch of the Potomac to the S. E. The entire face of the country is excessively broken and rocky, yet there is much excellent arable land, some tracts of good soil even on the mountain plains. The U. S. road commences at Cumberland, and the elevation of the country may be seen, by the fact that Cumberland stands elevated above tide water 537 feet, and following the U. S. road 9 miles to Frost town, the rise is 1255 feet and is 26 ms. to the highest summit, 2289 feet. Bituminous coal abounds in the mountain bowels of this county, which, should one or both the Baltimore and Ohio rail road, and Chesapeake and Ohio canal be completed even to Cumberland, must become of great importance. For lat. and long. see Cumberland. In 1820, pop. 8,654, in 1830 it had risen to 10,609.

ALLEGHANY co. of Va., bounded by Monroe S. W.; by the Alleghany mtn. or Green Briar N. W.; Bath N. E; and Botetourt S. E. Length from N. E. to S. W. 28 ms.; mean breadth, 18 ms. area 500 sq. ms. This county occupies a high mountain valley, drained by some of the higher branches of James river. Dunlap's and Polt's creeks, rising in Monroe, flow N. E. into Alleghany, unite with Jackson's river near Covington, and form the main western branch of James river, which continuing N. estrd, receives Cowpasture on the N. E. border of the county. Lying between lat. 37° 35′ and 38° N. and 3° W. from W. C. and rising to a mean level exceeding 1000 feet above the ocean tides, the seasons of this county are essentially influenced by mountain exposure and by native height. Staples, grain, flour, &c. Chief town, Covington. Pop. 1830, 2,816.

ALLEGHANY BRIDGE p-o. McKean co. Pa. by p-r. 210 ms. N. W. from Harrisburg.

ALLEMANCE, p-v. Guilford co. N. C. 77 ms. N. W. from Raleigh.

ALLEN, p-t. Alleghany co. N. Y.; by p-r. 276 ms. W. from Albany. Pop. 898.

ALLEN, p-v. Cumberland co, Pa.; 16 ms. from Harrisburg.

ALLEN, southern and border co. of Ky; bounded by Simpson W., Warren N., Barren N. E., Monroe E., and Jackson, in Ten. S.—Length from N. to S. 23 ms., mean width 17, and area about 400 sq. ms. Surface generally level; soil middling. It is drained by the higher creeks of Big-Barren branch of Green river. Chief town, Scottsville, N. lat. 36° 45′ and long. 8° 50′ W. intersect in this co. Pop. 1820, 5,327.

ALLEN, p-o. Miami co. O., 66 ms. wstrd from Columbus.

ALLEN'S CREEK, Genesee co. N. Y. 140 ms. long, enters Genesee river in Wheatland. It waters a fertile country, which it supplies with many mill seats. It was named after an Indian robber.

ALLEN'S FERRY, and p-o. Warren co. Ten.; by p-r. 76 ms. S. E. by E. from Nashville.

ALLEN'S FRESH, p-o. Charles co. Md, by p-r. 43 ms. sthrd. from W. C.

ALLEN'S SETTLEMENT, p-o. Natchitoches parish, La.; by p-r. 252 ms. N. W. from New Orleans.

ALLEN'S STORE, p-o. Randolph co. N. C. by p-r. 78 ms. wstrd. from Raleigh.

ALLENSTOWN, Merrimack co. N. H. E. Suncook r., 11 ms. S. E. Concord, 38 W. Portsmouth, 58 N. N. W., Boston; contains 12,225 acres of ordinary land. The timber is oak and pine, great quantities of which are sent to Boston. Bear brook furnishes mill seats. Catamount hill supplies building granite.—Pop. 484.

ALLENSTOWN, v. w. corner of Upper Freehold, Monmouth co. N. J. 11 ms. E. Trenton, on Doctor's creek, a branch of Croswick's creek.

ALLENSVILLE, p-o. Mifflin co. Pa. by p-r. 76 ms. N. watwrd. from Harrisburg.

ALLENTON, p-o. Montgomery co. N. C. by p-r. 123 ms. s. w. by w. from Raleigh.

ALLENTON, p-v. in the s. part of Wilcox co. Ala. 19 ms. southwardly from Canton, the seat of justice, and by p-r. 132 ms. a little E. from Tuscaloosa, and 931 s. w. from W. C.

ALLENTON, p-v. in the northern part of Greene co. Ky. 15 ms. northwardly from Greensburg, the st. jus. and by p-r. 75 ms. s. w. from Frankfort.

ALLENTOWN, p-v. borough and st. jus. for Lehigh co. Pa. The proper legal name of this borough is Northampton, which see.

ALLIGATOR, p-o. in the northern part of Alachua co. Flor. by p-r. 110 ms. s. E. by E. from Tallahasse, and 851 s. s. w. from W. C.

ALLOWAY'S CREEK, Salem co. N. J. empties into Delaware r. about 6 ms. s. of Salem. It rises in the same co.

ALLOWAY'S CREEK, *Lower* t. Salem co. N. J. between Alloway and Slow creeks, N. E. Delaware r. is almost connected with low lands and swamps.

ALLOWAY'S CREEK, *Upper* p-t. Salem co. N. J. on the upper parts of Alloway's creek.

ALMIRANTE p-o. Walton co. Flor. by p-r. 132 miles from Pensacola.

ALMOND, p-t. Allegany co. N. Y. 12 m. N. E. Angelica, has pretty good land, with maple, beech, bass, elm, oak, ash and other timber. Canadea creek and branches furnish mill seats. Pop. 1804.

ALNA, p-t. Lincoln co. Me. 10 ms. N. Wiscasset. Pop. 1175.

ALSTAIN'S p-o. St. Mary's co. Md. by p-r. 52 ms. s. s. E. from W. C.

ALSTEAD, p-t. Cheshire co. N. H. 12 ms. s. E. Charlestown, 14 N. Keene, 50 w. Concord; 24,756 acres; has mill seats on Cold r. &c. Warren's and other ponds afford many perch and pickerel. The soil is strong. There are 2 public libraries. Pop. 1559.

ALTON, p-t. Strafford co. N. H. 22 ms. N. E. Concord, 25 N. w. Dover, s. Winnipissogee lake and bay; contains 23,843 acres. The soil is rough and rocky, but strong; the timber is oak, beech, maple, pine, &c. It contains Mount Major and Prospect Hill, and part of Merrymeeting Bay, which receives Merrym r. Pop. 1993.

ALUM BANK, p-o. Bedford co. Pa. by p-r. 120 ms. wstrd. from Harrisburg.

AMASURA, AMAZURA, or AMAXURA r. of Flor. rising in the Seminole country, interlocking sources with those of St. John's and Hillsborough rivers, N. lat. 28° 25′, and flowing thence 45 ms. to the N. w. curves to the w. enters Alachua co. in which, after a course of 30 ms. it falls into the Gulf of Mexico.

AMBOY, t. Oswego co. N. Y. Pop. 669.

AMBOY, (or Perth Amboy,) city and p-t. Middlesex co. N. J. has a good harbor, which is sometimes resorted to to avoid the port regulations of N. Y. It stands on a point at the junction of Raritan r. and Arthur Kill Sound, 35 ms. s. w. N. Y. 74 N. E. Phil. A rail road hence to Camden will soon be completed, and afford an important route for travelling between N. Y. and Phil. (See *Camden.*)

AMELIA co. of Va. bounded N. w. by Cumberland, or the Appomattox r. N. by the Appomattox or Powhatan, N. E. by Appomattox or Chesterfield; s. E. by Namazine creek or Dinwiddie; s. by Nottoway, and w. by Prince Edward; length from s. E. to N. w. 30 miles, with a mean width of 10, area 300 sq. ms. It is drained by various creeks flowing to the north estrd. into Appomattox. The surface is pleasantly diversified; soil of middling quality. For lat. and long. see next article. Pop. in 1820, 11,106; in 1830, free whites, 3293; slaves, 7518; free colored, 220; total, 11,031.

AMELIA Court House and p-o. Amelia co. Va. 40 ms. s. w. by w. from Richmond. N. lat. 37° 13′, long. 10° 11′ w. from W. C.

AMELIA ISLAND, of Flor. forming the eastern part of the co. of Nassau, and the north eastern of the Territory. It is 13 or 14 ms. long, with a mean breadth of about 1 m; surface low and sandy. The position of this narrow strip is a little inclining from N. and s. towards N. w. and s. E. extending from the mouth of St. Mary's r. to that of Nassau. Chief town Fernandino.

AMENIA, p-t. Duchess co. N. Y. w. Conn. 24 ms. E. N. E. Poughkeepsie, 12 ms. by 4½; contains Great Oblong, Kent and West mts. which are of a poor soil. It has a number of mills, &c. Pop. 2,389.

AMERISCOGGIN r. (See Androscoggin.)

AMESBURY, p-t. Essex co. Mass. 40 ms. N. E. Boston, 4 w. Newburyport, N. side of Merrimac r. s. N. H. It is celebrated for the manufacture of flannel. Powow r. runs through the township, and affords excellent seats for manufactories. It is navigable to the falls, where ships of 300 tons have been built, and sent into Merrimac r. through a draw bridge at the mouth of the Powow r. Pop. 2,445.

AMHERST, p-t. and half shire, Hillsborough co. N. H. 28 ms. s. Concord, 47 N. w. Boston, 60 w. Portsmouth. Souhegan river passes through it, and furnishes mill seats. There are 3 ponds, called Babboosuck, Little do. and Jo English. The soil on Souhegan r. is excellent; on the hills it is generally good. 50 years ago iron ore was wrought here. The alluvion is sand, with loose masses of primitive rock on the surface; the hills primitive. Pop. 1657.

AMHERST, p-v. Hampshire co. Mass. 8 ms. N. E. Northampton, 85 w. Boston, 7 E. Conn. r. Is hilly, with a very good soil. *Amherst College* was founded in 1821, and incorporated in 1825. Rev. Heman Humphrey is president. The expences are from $93 to $118 a year to each student. There are two parallel courses of study, one embracing the ancient languages and literature, the other modern. There are 6 professors and 2 tutors. The apparatus was obtained in Europe. Board costs $1 or $1 50 a week. Students in 1831-2, 195. Pop. 2631.

The Amherst Academy and Mount Pleasant Institutions are also in this town. The former, where youth are fitted for college, has a building 40 by 60 feet, 3 stories. The former, on the plan of a German gymnasium, has a building 200 feet long.

AMHERST, Erie co. N. Y. 7 ms. E. Buffalo, 5 or 6 ms. by 17. Tonawanta creek is on the N. and its tributaries, Ellicott's, Cayuga, Conjocketa, Skyajocketa, Seneca and Cazenove creeks run through the town. The soil is a gravelly loam, nearly level, bearing small oaks, and chesnuts, &c. with stony tracts. Limestone is found, with nodules of gun-flint. Williamsville v. 11 m. N. E. Buffalo v. has a toll bridge, 210 feet long, and many mills. On some reserved land S. W. are some Seneca Indians. Pop. 2489.

AMHERST co. of Virginia; bounded by Bedford co. or James river S. W.; on the S. by James river, or Campbell co.; S. E. by James river or Campbell and Buckingham counties; E. and N. E. by Tye river, or Nelson co.; and N. E. by Blue Ridge or Rockbridge county. The form of this county is that of a parallelogram, 22 by 19 ms.; area 418 sq. ms. It is enclosed on 2 sides S. W. and S. E. by James river, and entirely drained by the confluents of that stream; general slope of the county to the southward; oceanic elevation, from 500 to 800 feet; with a tolerable fertile soil, the face of Amherst is beautifully variegated by a mountain, hill, and river scenery. For lat. and long. see next article. Pop. in 1820, 10,483, in 1830, 12,071.

AMHERST Court house and p-o. Amherst co. Va. 102 ms. W. from Richmond, and 15 N. N. E. from Lynchburg, at N. lat. 37° 29′ long. 2° 12′ W. from W. C.

AMISSVILLE, p-o. Culpepper co. Va. 86 ms. S. W. from W. C. and 100 N. N. W. from Richmond.

AMITE river, of the states of Miss. and La.; rises in the former, 40 ms. S. E by E. from Natchez, and in Franklin co. over which and Amite co. it flows sthrd. 35 ms. and enters La.; continuing a southwardly course 50 ms. farther it receives the Iberville from the W. and turning abruptly to the East, falls into lake Maurepas, after an entire comparative course of 100 ms. Schooners drawing 5 feet water are navigated up this stream to Galvezton, at the mouth of the Iberville.

AMITE, one of the southwestern counties of the state of Mississippi; bounded W. by Wilkinson; N. by Franklin; E. by Pike; and S. by East Feliciana and St. Helena parishes in La. It is in form of a parallellogram, 30 ms. from E. to W. and 24 from N. to S. area 720 sq. ms. Surface moderately hilly, and though some good land is found along and near the streams, the great body of the county is sterile and covered with fine timber. The Honochitto river forms a few miles of its boundary on the N. W. and the Tangipa rises in its S. E. angles, but the central and far greater part of the surface is drained by the Amite r. Lat. 31° 15′ N. long. 14° W. from W. C. Pop. 1820, 6,859, and in 1830, 7,934.

AMITY, Alleghany co. N. Y. Pop. 872.

AMITY, p-v. of Washington co. Pa. on Bane's Fork of Ten Mile creek, 10 ms. S. from Washington, the county town, and 248 ms. wstrd. from Harrisburg.

AMITY tsp. of Berks co. Pa. Pop. 1810, 1,090, in 1820, 1,279 and in 1830, 1,384.

AMITY p-v. in the S. W. part of Trumbull co. Ohio; 10 ms. S. W. from Warren, the st. jus. and 147 by p-r. N. E. by E. from Columbus.

AMMONOOSUCK, Lower or great r. chiefly in Grafton co. N. H. rises on the W. side of the White mts. with one of its sources coming from near the summit of Mount Washington, within a few rods of the source of the Saco r. which runs into the Atlantic. This stream has a S. W. course of about 50 ms. and after receiving the wild Ammonoosuck, 2 ms. from its mouth, enters Conn. r. between Bath and Haverhill. It has a clear current except in floods, and a valley half a m. wide. It has a fall of 40 feet 6½ ms. from the notch in the mts.

AMMONOOSUCK, Upper, or Little, Coos co. N. H. rises N. of the White mts. runs near to Androscoggin r. and after an irregular course of about 50 ms. through a romantic valley 7 or 8 ms. wide, enters Conn. r. in Lancaster.

AMOSKEAG, p-v. Hillsboro co. N. H. 16 ms. Concord.

AMOSKEAG FALLS, Merrimack r. N. H. 48 ft. 3 in. descent in ½ m. are dammed in several places, and have a canal of 1 m. with 9 locks. Cost $35,000.

AMSTERDAM, p-t. Montgomery co. N. Y. N. Mohawk r. 6 ms. by 12. Soil various, generally fertile, with alluvion on the r. Here was fort Johnson. Chuctenanda cr. falls 120 ft. in 100 rods from the Mohawk, at Amsterdam village. Pop. 3,354.

AMSTERDAM, p-v. Botetourt co. Va. 5 ms. S. from Fincastle and by p-r. 204 W. from Richmond.

AMWELL, p-t. Hunterdon co. N. J. 16 ms. N. Trenton. Pop. 7,385.

AMWELL, tsp. Washington co. Pa. Pop. in 1810, 1,673; in 1820, 1,825; and in 1830, 1,733.

ANASTATIA, isld. of Flor. on the Atlantic ocean, extending 18 miles from the harbor of St. Augustine, to Mantanzas Inlet, with a breadth not exceeding a mean of half a mile. It is a low, sandy, ocean beat reef, separated from the main land by one of the Rigolets so very common on the Atlantic and gulf coasts of the U. S. According to Tanners' U. S. the signal tower on the northern end of the island, is at N. lat. 29° 50′, long. 4° 29′ W. from W. C.

Ancocus (or Rancocus) cr. Burlington co. N. J. runs N. W. enters Delaware r. 6 ms. S. Burlington, and is navigable 16 ms.

Ancram, p-t. Columbia co. N. Y. 21 ms. S. E. Hudson, 51 S. Albany, formerly was included in Livingston's Manor. Ancram cr. traverses it, and supplies mills, &c. particularly Ancram Iron Works, the ore for which, is brought from Salisbury, Conn. The soil is various but generally good, and is held chiefly on life leases. Charlotte Pond empties into Dove cr. a branch of Ancrum cr. Pop. 1,533.

Andalusia, p-v. Bucks co. Pa. by p-r. 96 ms. E. from Harrisburg.

Anderson, one of the Western Districts of S. C. lying between the Savannah and Saluda r. having Picken's District, N. W.; Grenville, N. E.; Abberville, S. E.; and Savannah r. or Elbert co. in Geo. S. W.; in form of a rhomb of 28 ms. each perpendicular, and about 800 sq. ms. The two new dists. of Anderson and Pickens, were formed from Pendleton which was abrogated. Though bounded by Saluda on the N. E. the far greater part of Anderson is drained by the confluents of Savannah r.; the slope sthrd. The centre of the dist. is at N. lat. 34½° very nearly, and W. long. 5° 40′ from W. C. Chief t. Pendleton. Pop. in 1830, 17,169.

Anderson co. of Ky. bounded by Mercer S. Spencer W. Shelly N. W. Franklin N. and Kentucky r. which separates it from Woodford E. Length 18 ms. mean breadth 10, and area 180 sq. ms. lying between lat. 37° 58′ and 38° 8′ and long. 7° 46′ and 8° 8′ W. from W. C. Though bordering on Kentucky river, Anderson is drained in most parts by the branches of Salt river, which flows wstwrd. Chief town, Lawrence.

Anderson co. of Ten. bounded S. and S. W. by Roan co. W. by Morgan, N. E. by Campbell, and S. E. by Knox. The outline is very irregular, but from S. E. to N. W. the length is 30 ms. and the mean breadth about 25; the area is 750 sq. ms. The surface is very rugged, though much of the soil is highly productive. It forms part of the valley between Cumberland mountain and Copper Ridge, with the northern angle extending into the valley of Cumberland river. The main volume of Clinch r. passes Clinton, the st. jus. traversing the S. E. angle of the co. Pop. 1820, 4,674, and in 1830, 5,310.

Andersonborough, p-v. Perry co. Pa. 37 ms. N. W. from Harrisburg.

Anderson Court House, p-o. Anderson District, S. C. by p-r. 148 ms. N. W. by W. from Columbia.

Anderson's p-o. Wilson co. Ten. 34 miles eastward from Nashville.

Andersontown, (or *Andersonton*,) p-v. and st. jus. Madison co. Ind. situated on White r. by p-r. 41 ms. N. E. from Indianapolis, N. lat. 40° 7″, long. 8° 42′ W. from W. C.

Andersonville, p-v. in the N. western angle of Franklin co. Ind. 20 ms. N. W. from Brooksville, the st. jus. and by p-r. 50 ms. S. E. by E. from Indianapolis.

Anderson's Store, and p-o. Morgan co. O. by p-r. 84 ms. eastward from Columbus.

Anderson's Store, p-o. Caswell co. N. C. 56 ms. N. W. from Raleigh.

Andersonville, p-v. of Pickers District, S. C. situated on the point of junction of Trigaloo and Seneca rivers, on which the two latter streams unite to form the Savannah r. by p-r. 152 ms. S. W. by W. from Columbia.

Andes, p-t. Delaware co. N. Y. 15 ms. S. Delhi, 55 W. Kingston, 10 ms. by 15, is watered in the E. by Delaware r. which is navigable in rafts to Philadelphia. Tremperskill r. and branches spread over much of the town, affording mill seats. There are also streams of the Little Delaware, and the Plattekill. The surface is hilly, and the soil rich, bearing maple, beech and hemlock. Pop. 1,860.

Andes, p-t. Delaware co. N. Y. 90 ms. S. W. Albany.

Andover, p-t. Oxford co. Me. 61 ms. W. of Augusta, one of the most N. townships in the co. named. It is mountainous, a little N. of Androscoggin r. and is crossed by one of its branches. There is but one tsp. between it and N. H. Pop. 399.

Andover, *West Surplus*, t. Oxford co. Me. Pop. 119.

Andover, *North Surplus*, Oxford co. Me. Pop. 76.

Andover, p-t. Merrimack co. N. H. 18 ms. N. W. Concord, has Pemigewasset r. E. and Blackwater r. S. W. which, with their branches, supply many mill seats. There are also six ponds of pure water, with pleasant shores. The surface is very uneven, and in some places, rocky and barren, but generally of good soil. Ragged mountains are N. and Kearsearge has its base in the western part. There is an academy, endowed with a legacy of $10,000 by the late Mr. Joseph Noyes. Pop. 1,324.

Andover, t. Windsor co. Vt. 20 ms. S. W. Windsor, 68 S. Montpelier, 37 N. E. Bennington; first settled 1768. Markham's mnt. and mt. Terrible W. It has only a few head streams of Williams r. Pop. 975.

Andover, p-t. Essex co. Mass. 20 ms. N. Boston, 16 W. N. W. Salem, 20 S. W. Newburyport, on the main p-r. from Boston to Haverhill and Concord, S. E. Merrimack r. incorporated 1646, is a beautiful town, hilly, generally with good soil and farms. There is a pond 7 ms. round, whose outlet into Merrimack r. furnishes mill seats, as does Shawsheen r. There is a bank, with woollen and cotton manufactories. *Phillips Academy*, founded in 1778, by Samuel and John Phillips, contains about 130 students, in a building adjoining the Theological Academy, 40 feet by 80, and has funds to the amount of more than $50,000. Franklin Academy is in the N. parish.

The Theological Seminary here is the oldest in the U. S. established in 1807, chiefly by the donations of Wm. Bartlett, Moses Brown, Phebe Phillips, John Phillips, Samuel Abbot, John Norris, and his lady. It has a president, 4 professors, and 3 brick buildings on a commanding elevation. The private donations which support the Institution amount to about $400,000, from which the students are partly or wholly instructed. A class of 40 left this Institution in Sept. 1831. There is also a seminary for teachers here, established about 1830, for common and other English Schools and practical men. The usual English branches may be pursued in it, with mathematical and natural sciences, moral and intellectual philosophy, and even modern languages, apparatus, lectures on school keeping, &c. It has four terms in a year, at $5 to $8 for each branch per term, including vocal music. Boarders in commons work on the farm 2 hours a day, which reduces the price to 77 cents a week. Pop. 1830, 4,540.

ANDOVER, p-v. Tolland co. Conn. 17 ms. E. Hartford.

ANDOVER, p-v. Alleghany co. N. Y. 252 ms. W. Albany.

ANDOVER, v. Byram, Sussex co. N. J. 30 ms. N. Trenton, 40 ms. W. N. W. N. Y. N. Musconetcong r.

ANDOVER, t. Alleghany co. N. Y. Pop. 598.

ANDROSCOGGIN, r. (or Ameriscoggin,) Me. The outlet of lake Umbagog, rises in N. H. (one branch, Peabody's r. flowing from the White mts.) then passing into Me. runs E. then S. and joins Kennebec r. at Merrymeeting bay, 18 ms. from the sea. It falls 30 feet, near Lewistown. It flows between two mountainous ranges.

ANGELICA, p-t. and cap. Alleghany co. N. Y. 260 ms. W. Albany, 40 W. Bath, 9 ms. by 30, is divided N. and S. by Genesee r. whose branches supply mill sites, and is slightly undulated with a large share of good soils, of various descriptions, and many kinds of forest trees. Belvidere v. has a bridge over Genesee r. and Angelica v. contains the county buildings, 2 ms. E. of that r. Some iron ore is found here. Pop. 1830, 998.

ANGLINTON, p-v. Gwinnet co. in Georgia, by p-r. 98 ms. N. W. from Milledgeville.

ANNAPOLIS city, seat of government fo Md. and of justice in and for Ann Arundc co. is situated on the right bank of the Severn r. 3 ms. above its mouth into Chesapeake Bay, 37 ms. N. 76 E. from W. C. and 30 a little E. of S. from Baltimore. The seat of government was fixed in this city in 1699 and has remained there ever since. In 1820, the pop. amounted to 2260; in 1830, the number of inhabitants was 2623. N. lat. 38° 58′, long. from W. C. 0° 31′ E.

St. John's College in Annapolis was founded and endowed in 1784, and is now in full operation, as appears from an advertisement of the President, A. C. Magruder, inserted in the National Intelligencer, March 24th, 1831.

The charges in the regular college bills, as stated in the President's expose, are as follows:—payable quarterly in advance. English department, $24 per annum. Preparatory classes 40 do. Freshmen and Sophomore classes, 40 do. Senior and Junior classes, 50 do. There is no commons in the college; but good boarding, it is stated, can be procured in the city for $120 per annum. "An abatement of the college bills may be made to necessitous students; and provision is secured for the gratuitous instruction of 10 students who may bring the proper testimonials for that purpose."

ANNAPOLIS, p-v. Jefferson co. Ohio, 16 ms. N. W. by W. from Steubenville.

ANN ARBOUR, p-t. and st. jus. Washington co. Mich. situated on Huron r. by p-r. 42 ms. W. from Detroit, N. lat. 42° 18′, long. 6° 45′ W. from W. C.

ANN ARUNDEL co. of Md. bounded by Calvert co. S. Patuxent r. or Prince George's co. W. Patuxent r. or Montgomery co. N. W. Patapsco r. or Baltimore co. N. and Chesapeake bay E. This county lies in form of a Crescent 60 ms. in length from Fish creek at the N. E. angle of Calvert, to where it touches Frederick co. at Poplar Spring; mean breadth about 12, and area 720 sq. ms. Face of Ann Arundel is either hilly or rolling; the soil is varied, but in general rather of a middling quality. The eastern part on the Chesapeake is very much indented by small rivers and bays. Pop. in 1820, 27,165; in 1830, 28,295.

ANNSBURG, p-t. Washington co. Me. 30 ms. N. W. Machias.

ANNSVILLE, Oneida co. N. Y. Pop. 1830, 1481.

ANNSVILLE, p-v. Dinwiddie co. Va. by p-r. 57 ms. southwardly from Richmond.

ANNVILLE, p-v. Lebanon co. Pa. 17 ms. E. from Harrisburg.

ANSON, p-t. Somerset co. Me. 12 ms. N. W. Norridgewock, 40 Augusta. Pop. 1533. It lies W. Kennebeck r. and is crossed by 7 mile brook, in the S. part of the co. E. Saratoga co. N. Y.

ANSON, border co. of N. C. bounded by the co. of Mecklenburg W. Rocky r. or Montgomery N. Yadkin r. or Richmond E. and by Chesterfield district in S. C. S. Length from E. to W. 33 ms. mean breadth 23, and area 760 sq. ms. The slope of this county is to the N. E. and estrd; face of the country broken. Chief town Sneadsborough. N. lat. 35° and long 3° W. from W. C. intersect in the eastern part of the co. Pop. 1820, 12,534, and in 1830, 14,095.

ANSON gold mines and p-o. Anson co. N. C. about 140 ms. S. W. by W. from Raleigh.

ANTHONY'S KILL, a mill stream, running from Long and Round lakes into Hudson r. between Half moon and Stillwater.

ANTHONY'S NOSE, mt. N. Y. There are 3 or 4 hills or mts. by this name, in different parts of the state, which bear some resemblance to a human nose; 2 in the Highlands, E. Hudson r. 1 in Johnstown, N. Mohawk r.

ANTHONY'S CREEK and p-o. north eastern part of Green Briar co. Va. by p-r. 195 ms. N. W. by W. from Richmond.

ANTIETAM CREEK of Pa. and Md. rises in Franklin co. of the former, and entering the latter, flows sthrd. past Hagerstown in Washington co. falls into the Potomac.

ANTIOCH, p-o. Marengo co. Ala. about 80 ms. a little w. of s. from Tuscaloosa.

ANTISTOWN, p-v. Huntingdon co. Pa. by p-r. 124 ms. wstrd. from Harrisburg.

ANTRIM, p-t. Hillsborough co. N. H. 30 ms. w. by s. Concord, 20 N. W. Amherst, 67 N. W. Boston; contains 21,743 acres, w. Contoocook r. is hilly with much good land, especially the river alluvion: N. branch supplies mill seats. The w. part is mountainous, but good pasturage. Greggs' Pond covers 200 acres, and is 80 feet deep, with many perch and pike; there are 5 others. The soil is generally gravelly loam, good for grass, corn, oats, flax, &c. apples, maple, beech, birch, ash, elm, hemlock, &c. with very little white pine. First settled 1768. Pop. 1309.

ANTRIM, township of Franklin co. Pa. bordering on Washington co. Md. Pop. 1810, 2864, in 1820, 4120, in 1830, 3831.

ANTWERP, p-t. Jefferson co. N. Y. 170 ms. N. W. Albany, 8 ms. by 13, has an uneven surface, with much good soil for grass and grain, well watered by Oswegatchie and Indian rivers. It contains 2 p-vs. Antwerp and Oxbow. Pop. 2411.

APLINGTON, p-v. and st. jus. Columbia co. Georgia, 21 ms. N. N. W. from Augusta, and 70 ms. N. E. by E. from Milledgeville in Geo. N. lat. 33° 32′, and long. w. from W. C. 5° 20′.

APOLLO, p-v. Armstrong co. Pa. by p-r. 222 ms. wstrd. from Harrisburg.

APPALACHEE BAY. This term has been given to a rather undefined expanse of water spreading from Appalachiola bay and St. George's island, estrd. to the coast of Madison co. in Flor. bounded inward by the Ocklockonne and St. Mark's bays, and merging outwards by no assignable limit into the Gulf of Mexico. As a navigable inlet, Appalachee bay is of little consequence; but it gains importance as being the channel of entrance to Tallahasse, and affording 7 feet water to St. Mark's.

The wide and shoaly bank, which obstructs the coast from Espiritu Santo to Vassassaousa, continues uninterrupted to the bay of Appalache, where its breadth is reduced to about 3 ms. and a channel found to enter the river St. Mark. This channel is accessible to vessels drawing 10 feet, and affords to them a good anchorage 8 miles from St. Mark. Vessels drawing 8 feet, can, at high tide, reach St. Mark. The distance along the coast, from Vassassousa bay, mouth of Suwannee, to Appalache bay, mouth of St. Mark's r. is about 95 ms.; and the latter bay offers the only anchorage to be found from the bay of Espiritu Santo, that is to say, on an extent of coast of about 230 ms.—(*Vide Engineer's Report.*)

APPALACHEE, small r. of Geo. and w. branch of Oconee, rises about N. lat. 34° in Jackson and Gwinnette counties, flows S. E. 50 ms. and falls into the Oconee between Magan and Green counties.

APPALACHICOLA BAY, or the estuary of Appalachicola r. This bay, properly speaking, is composed of two connected parts. St. George's Sound, and Appalachicola Bay proper.

St George's Sound. The shoal which extends all along and parallel with the coast, makes out abruptly at Ocklockony bay to the s. for a distance of about 12 ms. and projects from South Cape. West of this Cape, it reassumes its parallel course to the coast, and may be considered as terminating at Cape San Blas: on this distance it forms the basin of the islands which cover St. George's Sound. These islands are three in number. To the N. E. towards Ocklockony, the first is Dog Island, 6 ms. long. Thence, stretching first s. w. by. w. and thence with a projection to N. W. extends to St. George's Island, 30 ms. in length. Continuing the course of the N. W. curve of St. George's Island, extends to St. Vincent's Island, 9 ms. long. The water distance from the E. to the w. end of the Sound is about 50 ms. The width at its eastern extremity, is about 4 ms. and N. from Cape St. George, about 9 ms. The Sound can be entered by any of the passes; one round the eastern point of Dog Island, the second between this island and St. George's Island, the third between St. George's Island and St. Vincent's Island. The first, called the eastern pass, offers a depth of 14 feet at low tide, and this depth is found on a width of 600 yards, on a bar which lies at about 2½ ms. from the Sound. The second or middle pass, has a least depth, at low tide, of 15 feet on the bar; its channel is contracted between the shoals which make out of the two islands; the width on the bar, for 15 feet water, is about 650 yards; the entrance is on Dog Island side. The distance between Dog and St. George's Islands is three miles. The third pass, called main pass, has a width, on the bar, of 300 yards, for a depth of 14 or 15 feet at low tide. The bar lies s. eastward of Flag Island, and about 2 ms. s. w. of the entrance into the Sound.

Between St. Vincent's Island and the main shore, called the Western or Indian pass, the depth, it is said, admits only canoes. From the eastern to the middle pass, St George's Sound affords from 9 to 18 feet water; but hence wstrd. to the main pass, it is so obstructed by banks as to admit vessels of 6 feet draught only. Mean rise of tide, 2¼ feet, as 15 days observation; s. w. wind prevailing.

Appalachicola Bay is the mere opening of the r. of that name into St. George's Sound. Its entrance into the sound is upwards of 5 ms. wide. The distance in a s. w. direction, from the main pass between St. George's and St. Vincent's islands to the w. point on the bay, is about 12 ms., but intervening obstructions compel vessels bound up or down, to curve estrd. which increases the distance 6 ms. Vessels drawing 10 feet can anchor in the bay, but only 7½ can be carried up and into the river.

Vessels drawing from 5 to 5½ feet, can be navigated between Appalachicola bay and Ocklockony bay, by keeping not far from the shore, and entering St. George's Channel at its eastern end. Except the Capes of St. George and St. Blas, the out shore of the islands covering the sound may be considered as bold, within a short distance from the land. (*Vide Engineer's Report.*)

APPALACHICOLA, r. and basin of the U. S. in Ala. Geo. and Flor. The r. is formed by the united streams of the Chattahoochee and Flint rivers. The Chattahoochee rises in Habersham co. of Geo. N. lat. 34° 50′, and between long. 6° 7′ W. from W. C. interlocking sources with Savannah, Hiwassee, branch of the Tennessee, and Etowah branch of Coosa r. Flowing thence S. W. by comparative courses 200 ms. to N. lat. 33°, where it assumes a southern course of 200 ms. to its junction with Flint r. at N. lat. 30° 42′. It is a singular feature in the character of Chattahoochee valley that in a length of 320 ms. it is in no one place 50 miles wide, and does not average a width of more than 30 ms. giving to this lengthened valley only a superficies of 9,600 sq. ms. Flint r. rises in Cowota, Campbell, Fayette, and Henry counties of Georgia, between the main volume of Chattahoochee, and the sources of the Oakmulgee, and flowing thence S. S. E. by comparative courses, 120 ms. nearly parallel to the latter, thence inflects to S. S. W. 100 ms. to its junction with Chattahoochee. Though comparatively wider, the valley of Flint, like that of Chattahoochee, is narrow. In Derby and Lee counties of Geo. where, in N. lat. 32° the valley of Flint is widest, the breadth is only 65 ms. and the entire valley does not exceed a mean of 35 ms. or about 7000 sq. ms.

The Appalachicola is formed by the united waters of Chattahoochee and Flint, which, assuming a course of a little E. of S. receiving from the N. W. the small river Chissola, and not even a large creek from the opposite slope, after flowing 70 ms. falls into Appalachicola bay by several channels at N. lat. 29° 46′ and exactly on long. 8° from W. C. The Mississippi and Appalachicola, are the only confluents of the Gulf of Mexico, in the U. S. which protrude deltas at their mouths, and the latter, compared with its length of course, of all rivers of the U. S. presents the greatest variety of climate. This basin stretches through 5½° of lat. and rising on a mountainous region, elevated at least 2000 feet above the surface of the Mexican gulf, or an equivalent in relative height to 5° of lat. the difference of temperature, at the extremes, must equal near 10°. The entire superficies in the basin of Appalachicola, is 19,700 sq. ms. of a dry, and, except near the coast of the G. of Mexico, a broken, and near the superior sources, a mountainous country. Though less in volume than the Mobile, it is comparatively a more navigable stream. Sea vessels ascend to the junction of the Flint and Chattahoochee. The Appalachicola r. has three outlets into the bay of the same name. The first, a straight channel, close to the right shore, and through which 5 feet only can be carried at low tide. Second, the main channel, which is very crooked, bends in a semicircle towards the E. and affords, up to the river, a depth of 7½ feet. Thirdly, the swash, or N. E. channel, which comes into the main channel, 3 ms. from the bar, and presents a depth of 5 feet through. It must be remarked, that the tide makes earlier in the first channel than in the others. The mean rise of tide has been 1¾ feet for ten days observation.

The bay of Appalachicola cannot admit, at low tide, vessels drawing more than 10 feet. (*Vide Engineer's Report.*)

APPLE RIVER, small stream rising in the N. W. Ter. flows southwardly into Il. traverses Joe Daviess co. and falls into the Mississippi, about 20 ms. below Galena.

APPLE RIVER, p-v. on the river of the same name, Joe Daviess co. Il. about 20 ms. S. E. by E. from Galena, and by p-r. 310 ms. N. N. W. from Vandalia, and 974 ms. from W. C.

APPLETON, p-t. Waldo co. Me. 35 ms. N. E. Wiscasset, 35 W. S. W. Augusta, between St. George's and Muscongus rivers, which pass Warren and Waldoboro' to the sea. Pop. 735.

APPLING co. of Geo. bounded S. by Ware, S. W. by Irwin, W. by Telfair, N. W. by Montgomery, on the Oakmulgee r. N. and N. E. by Tatnall, or the Altamaha r. and E. by Wayne. Length along the sthrn. boundary 60, mean width 25, and area 1,500 sq. ms. The northern part is drained by small creeks flowing into the Oakmulgee or Altamaha rivers, but the central, southern, and more than half the whole area is watered by various branches of Santilla r. (For lat. and lon. see Applingville.) Pop. in 1830, 1,468.

APPLINGTON, (or *Columbia Court House*,) p-v. and st. jus. Columbia co. Geo. 21 ms. N. W. by W. from Augusta, and by p-r. 81 ms. N. E. by E. from Milledgeville; N. lat. 32° 32′, long. 5° 20′ W. from W. C.

APPLINGVILLE, (or *Appling Court House*,) now Holmesville p-o. and st. jus. for Appling co. Geo. N. lat. 31° 16′, long. 5° 28′ W. from W. C. in a direct line about 100 ms. S. W. by W. from Savannah, and 120 S. S. E. from Milledgeville. Letters intended for this place, ought to be directed to Holmesburg, or Appling Court House.

APPOMATTOX, r. of Va. rising in Buckingham and Prince Edward counties, and flowing thence by a very crooked channel, but by a general eastern course, with the counties of Prince Edward, Amelia, Dinwiddie, and Prince George on the right, and Buckingham, Cumberland, Powhatan, and Chesterfield on the left, falls into James r. after a comparative course of 90 ms. The tide ascends the Appomattox, to the falls of Petersburg, about 20 ms. above the mouth, and thus far contains depth of water for large merchant vessels. This stream drains a very fine section of Virginia, between lat. 37° 33′ and 37° 26′ N.

APPOQUINIMINK, small r. or creek of New Castle co. Del. rising in Kent and Cœcil cos. Md. interlocking sources with Sassafras r. of Chesapeake bay, and flowing thence eastwardly, falls into Delaware bay, below Reedy island.

AQUACKANOCK, p.v. Essex co. N. J. w. Passaic r., 10 m. N. Newark. Boat navigation extends to this place.

AQUASCO, p.v. Prince George's co. Md. situated on the right bank of Patuxent r. 34 m. S. E. from Washington city.

AQUIA p.o. Stafford co. Va. on a small creek of the same name, by p-r. 42 ms. S. S. W. from W. C.

ARARAT, p.o. Patrick co. Va. by p-r. 311 ms. S. W. by W. from W. C.

ARBELA, p.o. Lancaster co. Pa. by p-r. 47 ms. estrd. from Harrisburg.

ARCADIA, p-v. Wayne co. N. Y. Pop. 3,774.

ARCADIA, p.v. Morgan co. Ill. by p-r. 122 ms. N. W. from Vandalia.

ARCOLA, p.v. Lawrence co. Ind. by p-r. 82 ms. S. S. W. from Indianopolis.

ARD'S BLUFF p.o. on Pearl r. Marion co. Miss. about 120 ms. S. E. by E. from Natchez.

ARGYLE, t. Penobscot co. Me. Pop. 326.

ARGYLE, p-t. Washington co. N. Y. 6 ms. S. Sandy hill, 44 N. Albany, E. Hudson r. has an undulated surface, pretty good soil, with several ponds and mill streams, and 2 villages, Argyle and Fort Miller, where is a bridge over Hudson r. Pop. 3,459.

ARKANSAS, r. of the U. S. and of the Mexican province of Texas, rises in the eastern vallies of the Rocky or Chippewayan mts. about N. lat. 42°, and long. 31° w. from W. C. interlocking sources with those of Lewis r. branch of Columbia, the main volume of Platte r. of Missouri, opposite to those of St. Buenaventura, of the Pacific ocean, and the Rio Grande del Norte. Flowing 200 ms. S. E. nearly parallel to the Rio Grande, the Arkansas assumes an easterly course of 400 ms. to its junction with the Negracke, and returns to the course of S. E. 250 ms. to the entrance of the great western confluent, the Canadian r. This r. is the great southwestern confluent of the Arkansas, and as laid down by Maj. Long, rises in the mountains of New Mexico, 27° of long. w. from W. C. and between N. lat. 33° and 37°. Rising in the eastern vallies of a rather elevated chain, the two main, and numerous minor branches of the Canadian pursue a general eastern course over 9° of long. the various branches first uniting into one stream, which joins the Arkansas at N. lat. 35°, and long. 18° w. from W. C. The valley of the Canadian r. is in length about 500 ms. with a mean breadth of 100 or area 50,000 sq. ms.

The far greater part of this physical section is an open, unwooded series of plains, or rolling prairie. The soil is sterile, and partakes much of the character of the steppes of northern Asia. Maj. Long, in 1817, found the bed of this r. almost dry, and himself and party were exposed to great suffering from want of water on the banks of a river, following its windings, perhaps a thousand miles from its source. The valley of the Canadian is in fact a part of a real North American desert. The Canadian drains the space eastward from the sources of Rio Grande del Norte, and between that of Arkansas and Red r. About one half of this valley is E. of the 100th deg. w. from Greenwich, of course in the U. S. Below their junction the united waters of the Arkansas and Canadian rivers pierce the Arkansas mts. and flow S. E. 250 ms. to their final union with the Mississippi, at N. lat. 33° 56′, and long. 14° 10′ w. from W. C.

Of the confluents of the Arkansas, the first rank is due to the Canadian r. The latter, though long known by name, has been but recently sufficiently explored to enable geographers to assign its due rank. As laid down from Maj. Long's discoveries, this river, by two great branches, rises in the mts. of New Mexico, as far wstrd. as 28° from W. C. and within 40 ms. from the main stream of the Rio del Norte, between the towns of Santa Fe and Taos, and between the sources of Arkansas proper and those of Red r. and between N. lat. 34° 30′ and 37° 30′. From this mountainous origin, the Canadian by two branches, called relatively North and South Canadian rivers, flows over the wide central plains of North America by a course of E. a little S. the two streams generally at from 40 to 60 or 70 ms. asunder, gradually approach each other, and finally unite, after a comparative course of upwards of 700 ms. Besides the two middle or principal branches of Canadian r. there are beside many smaller, one the S. E. Fork, a stream 300 ms. long, rising between the southern Canadian and the branches of Red r. and falling into the main Canadian below the junction of its two great constituents. Below the union of all these confluents, the Canadian, after an easterly course of about 50 ms. joins the Arkansas at N. lat. 35°, and 18° w. from W. C. and directly on the western side of the Arkansas mts. Between the Kansas and Osage branches of Missouri, the Arkansas mts. and Arkansas r. a space of about 10,000 sq. ms. is drained by Verdigris, Neoscheo and Illinois rivers, which flow southward into Arkansas, which they join wstrd. from the Arkansas mts. The entire surface contained in the Arkansas valley, above the mountains, is at least 170,000 sq. ms. Below the mountain pass, the Arkansas valley, in a length of 250 ms. does not average a width above 40 ms. or 10,000 sq. ms. Combining the two sections, the entire surface drained by the Arkansas and all its confluents, amounts to 178,000 sq. ms. The geographic position of this valley is between lat. 34° and 42° N. In comparative geography, the river of Europe, which in course and extent most nearly approaches the Arkansas, is the Danube. By actual calculation, the Arkansas flows S. 65° E. a fraction above 1,300 statute ms. in a direct line from source to mouth. The Danube flows S. 82°, E. within an inconsiderable fraction of 1,470 ms. The subvallies of the Danube are something more extensive than those of Arkansas, but in a general view, if even the inflections of those two great rivers are compared, the resemblance is very strong, and if White r. was included in Arkansas valley, as it might be without much violence to correct analogy, the two valleys would present features peculiarly similar. But with the outline and general course of the main and minor volumes of the Danube and Arkansas, all resemblance between these rivers ceases. The latter, indeed, in many essential physical features, bears a striking likeness to the Nile, though taken as a whole, the Arkansas stands alone among the rivers of the earth. Rising on a mountainous table land of probably 5,000 feet elevation, and entering the Mississippi where the river level does not exceed that of the ocean 150 feet, the actual depression of slope is equal to 12 degrees of lat. giving a difference of temperature of 20 degrees. The far greater part of the inclined plane falling from the Chippewayan towards the Arkansas mts. is naked of timber and sterile, scarce of fountain water and presenting a perfect model of the central steppes of Asia. In the winter and spring, or in the rainy season, the streams bear down immense masses of water, whilst in Autumn and in the early part of winter, they are dwindled to mere rills where any stream continues; but in the Canadian, even that feature ceases, and the bed of the river becomes in September, October, and November, a lengthened line of dry sand, and in many places, miles intervene without even a remaining stagnant pond.

Those harsh features are softened rather than changed, below the mountains marked in Tanner's maps as Ozark mts; prairies still continue, though less extensive and less sterile than those more wstrd. Springs of fresh water also increase, approaching the inundated plains near the Mississippi, but in a general view, the entire valley of Arkansas may be regarded as naked and rather sterile. To this character, the most striking exception is afforded by the inundated border near the Mississippi. Without including minute bends, the Arkansas flows 300 ms. from the mouth of Canadian to its own influx into the Mississippi. This part of the valley does not exceed a mean breadth of 50 ms. the tributary streams are of course small; prairies gradually cease, and approaching the great recipient, an annually overflowed and deeply wooded plain, terminates the lengthened valley of Arkansas. The inundations of this great, though secondary river, are as regular as those of the Mississippi, but from the position of the valley, the discharge of the Arkansas is less in quantity in a given time, than might be expected. The water of the Missouri, a milky blue, and that of the Ohio, a blackish green, are very distinct from the ochreous red water of the Arkansas, which latter again differs from the brighter red volume of Red r. The great Arkansas flood reached Delta in May, and early in June preceding the tide from Missouri and the higher Mississippi, and nearly simultaneous with that of the Ohio.

Arkansas Ter. of the U. S. bounded w. by the yet unorganized territory formerly a part of La. n. by the state of Mo. e. by the Miss. r. and s. by the state of La. On Tanner's map of the U. S. the ter. of Ark. is made commensurate on the w. with the state of La. or 17° 30′ w. long. from W. C. having outlines in common with La. on n. lat. 33° of 190 ms. on the Mexican province of Texas about 40 ms. along the wstrn. ter. of the U. S. 210, in common with the state of Mo. along n. lat. 36° 30′, 266 ms. down St. Francis r. to n. lat. 36°, 50 ms. along n. lat. 36° to the right bank of the Miss. r. 34 ms. down the right bank of the Miss. to the n. e. angle of La. 360 ms. entire outline 1140, lying between lat. 33° and 36° 30′ n. and between lon. 12° 44′ and 17° 30′ w. from W. C. The ter. within these outlines embraces an area of 54,860 sq. ms. lying so very compact as to approach a sq. of 234 ms. each side. Ark. is naturally divided into three distinct sections. The e. section along the Miss. White, St. Francis, and Ark. rs. is mostly level, and in winter and spring, except the margin of the streams, liable to inundation. With some prairie, this e. section is covered with a dense forest. The soil, where arable, is very productive. The middle section watered by the higher branches of White r. by the main volume of Ark. the higher branches of Ouachitta, and by Red r. is broken, in part, rather mountainous, and presenting extensive prairies. The w. and n. w. section is mountainous, with extensive prairies. Though an interior region, Ark. is in a high degree supplied with navigable streams. Beside the Miss. which forms the eastern boundary, St. Francis rises in Mo. and flowing s. enters Ark. over which it flows 120 ms. falls into the Miss. about 60 ms. above the mouth of White r. The latter, a much more considerable stream, rises in the s. part of Mo. by two great branches, White r. proper and Black r. which, uniting in Ark. flows below the junction 120 ms. and falls into the Miss. 15 ms. above the mouth of Arkansas r. Though impeded by shoals, the Ark. is a navigable r. far above the limits of the ter. to which it gives name. The Ouachitta rises by numerous branches in the Masserne mountains, between Ark. and Red rs. and is navigable at high water, with boats of considerable size, by the bends 200 ms. within Arkansas. Red r. traverses the extreme s. w. angle of the ter. It may be doubted whether in this large ter. there is one spot, 100 ms. from a navigable water course. The St. Francis, White, and Ark. rs. form navigable channels within the ter. into the Miss.; whilst the Red r. and its confluent, the Ouachitta, leave the ter. traverse La. and finally join the same great recipient. The mineral wealth of Ark. has been too imperfectly developed to admit definite notice. The immense salt prairies or steppes of the interior, give a brackish taste to the water of Ark. Numerous salt springs are scattered over the country. Indications of lead and iron are numerous. Some lead has been smelted but not in large quantity. As a whole, Ark. may be regarded as a barren country, though scattered over the entire extent, tracts of exuberantly fertile soil occur. If we compare it with La. and from what I have seen of both these two regions they appear remarkably similar in the relative extent of arable to unproductive soil in each, respectively, the cultivateable surface would be about as one to four. The difference of lat. and elevation give to the respective sections of Ark. a much greater range of temperature, than from its extent could be expected. Cotton can be cultivated to advantage on the s. and s. e. sections; but advancing up the streams and rising to a much higher elevation, the temperature falls, and small grain succeeds well. On the whole, the climate of Ark. may be considered as excessively variable; that of the s. resembles La. whilst towards the n. and w. the seasons approach those of Mo. Amongst the curiosities of Ark. may be particularly mentioned the Hot Springs. These fountains, with a temperature near that of boiling water, are 47 ms. s. w. by w. from Little Rock, on the Ark. and near the main stream of Washitau. They became a place of resort for invalids about 1800, and have increased in reputation. The adjacent country is rocky, sterile and mountainous, the water pure, and air elastic, affording a most delicious retreat from the summer and autumn temperature of La. and S. Ark. The Washitau is navigable with steam boats to within 30 ms. of the springs. Ark. being yet a ter. the government is therefore not permanently fixed, which precludes the propriety of inserting the existing form in this place.

For counties, their population and chief towns, see Appendix.

Arkansas co. of the territory of the same name, lying on both sides of Ark. r. having Chicot co. s. Clark w. Pulaski n. w. White r. or Philips co. n. e. and the Miss. r. e. Length from n. to s. 60 ms. mean width 40, and area 2,400 sq. ms. This co. is traversed obliquely from n. w. to s. e. by the Ark. r. and the White r. bounding it to the n. e. renders the lower part subject to annual overflows. The Bayou, Barthelany, and Boeuf rs. branches of the Washitau, rise in the s. part, flow s. over Chicot co. enter La. The w. side rises into eminences, but the body of the county is level. (For lat. and long. see next article.)

Arkansas, p-v. and st. jus. for Ark. co. Ark. Ter. This is the same village formerly called "The Post of Ark." and was founded by the French in 1685. It stands on the n. bank of Ark. r. at n. lat. 34°, long. w. from W. C. 14° 25′. The site is a high narrow bluff of poor soil, with the inundated flats of White r. in the rear. Distant about 100 ms. s. e. from Little Rock, and by the bends of the r. 50 ms. from the Miss.

Arkiopolis, p-v. and seat of government Ark. ter. (See Little Rock.)

Arkport, p-v. Hornellsville, Steuben co. N. Y. 20 ms. s. Bath, 260 ms. w. by s. Albany, on Canister r.

Arkwright, Chautauque co. N.Y. Pop. 926.

Arlington, p-t. Bennington co. Vt. 12 ms. n. Bennington, 106 Montpelier, and 40 from Troy, Saratoga Spa, Whitehall and Rutland, was first settled 1763, and is watered by Roaring Branch, Mill Creek, Warm Creek and Green r. as well as Battenkill r. of which they are branches. There is plenty of mill seats and much fine interval land. North and Red Mountains in the w. are divided by Battenkill r. and bear white, red and black oak, white and black birch, chesnut, walnut, &c. The soil is rich, and bears grain. Much white marble is quarried here for tombstones, &c. and lime-stone is burnt. There is a ferrugineous spring and a curious cavern in this t. Pop. 1,207.

Armagh, p-v. s. part of Indiana co. Pa. about 60 ms. e. from Pittsburgh, and 35 n. e. from Greensburg.

Armstrong's p-o. Wabash co. Illi. by p-r. 111 ms. s. e. by e. from Vandalia.

Armstrong co. of Pa. bounded e. by Jefferson co. s. e. by Indiana, s. by Kiskiminitas r. or Westmoreland co. w. by Butler, and n. by Clarion r. or Venango co. Length 45, mean width 20, and area 900 sq. ms. The face of this county is delightfully diversified, with hill, dale, and river scenery. Alleghany r. enters at the n. w. angle, and sweeping over the country by an elliptical curve, leaves it at the s. w. angle receiving within the co. from the e. Clarion r. Red bank, Mahoming, and Crooked creeks, whilst the Kiskiminitas enters as a s. border. n. lat. 41° and long. 2° 30′ w. from W. C. intersect a little n. of the centre of the co. Chief t. Kittanning. Pop. 1820, 10,524, in 1830, 17,695. The traverse section of the Penn. Canal, crosses Kiskiminitas at Warren t. and enters Armstrong from Westmoreland co. and thence follows the right bank of the Alleghany r. The canal is then carried over the latter stream, and down its right bank to Freeport, where it leaves Armstrong and enters Alleghany co.

Arnold's, old place p-o. Fauquier's co. Va. by p-r. 56 ms. w. from W. C.

Arnoldstown, p-v. Campbell co. Va. by p-r. 131 ms. s. w. from W. C.

Arnold's p-o. Rush co. Ind. by p-r. 50 ms. a little s. e. from Indianopolis.

Aroostic, (or Aroostook) r. Me. rises near the w. bound of Penobscot co. which it crosses n. e. in a devious course, receiving several branches, and falls into St. John's, in New-Brunswick. It passes through lands almost unsettled, but which will probably soon become valuable. Its head waters are 20 ms. from those of Penobscot r. and part of the intervening country is very hilly, with numerous ponds, the rest marshy. A road is making from this r. to the Mattawankeag, a branch of Penobscot r. which is to be extended 60 ms. n. to Madawasca. This will facilitate emigration.

Arthursburg, p-v. Dutchess co. N. Y. 77 ms. s. e. Albany.

Asbury, p-v. Warren co. N. J. 35 ms. n. Trenton n. w. Musconetcong r.

Ascension, parish of La. bounded by Assumption s. e. St. James e. Amite r. n. e. Ibberville n. w. and Atchafalaya r. w. length from n. e. to s. w. 40 ms. mean width about 10 and area 400 square ms. The face of this parish is one great plain and most part liable to annual submersion, similar to every other part of the delta. The bank of the Miss. is the highest part of the parish, and that winding across it near the middle, contains the far greater part of the population. This parish is rendered remarkable from containing the outlet of the La Fourche. This stream is in fact a mouth of the Miss. from which it flows at Donaldsonville, (*see La Fourche*). What soil of Ascension is sufficiently elevated for cultivation, is highly fertile. Staple, sugar and cotton. For lat. and lon. see Donaldsonville. Pop. 1820, 3,728, in 1830. 5,426, chief town Donaldsonville.

Ascutney, mt. Windsor co. Vt. between Windsor and Weathersfield, is 3,320 feet higher than tide water, and 3,116 above Conn. r. at Windsor. It is nearly bare, except on the n. e. side, and consists of granite. The view from the summit is very fine. The ascent is performed in 2 and a ½ hours. From Windsor the route is 4 ms. to the base, half a mile up is a moderate ascent, after which it is steep.

Ashburnham, p-t. Worcester co. Mass. 55 ms. n. w. Boston, was granted to certain soldiers from Dorchester, in a Canada expedition in 1690. It is watered from ponds, by mill streams of Miller's r. which falls into Conn. r. and Nashua and Souhegan rs. which fall into Merrimack r. It is hilly, with a strong soil producing grass, fruit and forest trees. Two incorporated companies manufacture leather and soap stone in large quantities. $500 worth of chairs are made here weekly, and many nails, clapboards, &c. Pop. 1,402.

Ashby, p-t. Middlesex co. Mass. 50 ms. n. w. Boston, is handsomely varied with hill and vale, with good farms, producing grass and fruit, and has a fine mill stream, a branch of Nashua r. rising in it. Pop. 1,240.

Ashe, northwestern co. of N. C. bounded by Surrey co. of the same state n. e. the main spine of Blue Ridge or Wilkes co. s. e. Buncombe s. w. and Carter co. of Ten. w. Ashe co. occupies part of the elevated valley between Bald mtn. and Blue Ridge, and from the courses of the rivers, must be the highest section of the valley, and at least 2500 feet above the Atlantic tides.

The n. e. part of of the co. slopes towards Virginia, and is drained by the extreme sources of Great Kenhawa whilst the discharge of the southwestern part is formed by the sources of the Watauga branch of Holston. The extreme length of Ashe, from n. e. to s. w. is 70 ms. with a mean breadth of 12 ms. area 840 sq. ms. The surface is broken and soil of middling quality. The climate is peculiarly fine, air pure and bracing. In long. this co. lies very nearly between 4° and 5° w. from W. C. while the central part is about n. lat. 36° 20′. Pop. 1820, 4,335, and in 1830, 6987. Chief town, Jeffersonton.

Ashfield, p-t. Franklin co. Mass. 15 ms. n. w. Northampton, 105 w. Boston, is on high land between Deerfield and Westfield rs. to each of which it sends a tributary. It was settled and garrisoned about 1754 and called Huntstown, but abandoned in 1755, and re-settled in 1763. Pop. 1,732.

Ashford, new v. Berkshire co. Mass.

Ashford, p-t. Windham, co. Conn. 31 ms. e. Hartford, about 9 ms. long and 7 broad, 59 sq. ms. is hilly, with a gravelly soil, but favorable for grazing, and raises many cattle. Still, Bigelow and Mount Hope rs. are small. Crystal pond, 1 m. by ½ m. is between this town and Pomfret. There are two small villages. First settled in 1700. Pop. 2,660.

Ashford, p-t. Cattaraugus co. N. Y. 282 ms. w. by s. Albany. Pop. 631.

Ashland, p-v. in the northern part of Richland co. O. by p-r. 88 ms. northeastward from Columbus.

Ashley's Mills, p-o. Telfair, co. Geo. 102 ms. by p-r. sthrd. from Milledgeville.

Ashpalaga, p-t. village of Jackson co. Florida, on the Appalachicola r. by p-r. 35 ms. westrd. from Tallahassee.

Ashtabula, northeastern co. of O. bounded n. e. by Erie co. Penn. e. by Crawford co. Penn. s. by Trumbull. O. w. by Geauga and n. by Lake Erie. Greatest length along Penn. 32 ms. on the western border the length is about 26 ms. mean length 29 and breadth 28, area 812 sq. ms. extending in lat. 41° 32′ to 41° 58′ and in long. from 3° 30′ to 3° 58′ w. from W. C. The southeastern angle of this co. slopes to the s. e. and gives source to the Shenango and other branches of Big Beaver. The western and central sections have a northwestern declivity and are drained by the sources of Grand river of Lake Erie, while the northeastern part slopes northwardly and is watered by Conneaut and Ashtabula rivers. From the preceding elements we find this co. occupying a part of the table land between the vallies of Erie and O. Chief town Jefferson. Population 1830, 14,584.

Ashtabula, p-v. on the r. of the same name at its eflux into Lake Erie, northern part of Ashtabula co. O. about 40 ms. s. w. by w. from Erie in Penn. and by p-r. 191 ms. n. e. from Columbus.

Ashville, p-v. and st. jus. for Buncombe co. N. C. stands on French Broad r. 40 ms. n. n. w. from Rutherfordton and by p-r. 273 ms. wstrd. from Raleigh n. lat. 35° 32′, long. 5° 28′ w. from W. C.

Ashville, p-v. and st. jus. St. Clair co. Ala. on Canoe creek (branch of the Coosa) by p-r. 90 ms. n. e. from Tuscaloosa and about 70 s. wstrd. s. e. from Huntsville, n. lat. 33° 48′ long. 9° 21′ w. from W. C.

Ashuelot r. Cheshire co. N. H. rises from a pond in Washington; runs s. to Keene, then s. w. to Conn. r. 3 ms. from the line of Mass. To make it navigable for boats, from

Keene to the mouth, considerable exertions were made several years ago; several places were locked.

Assiniboin, large r. of North America partly in the U.S. and partly in the British territories. It is formed by two great branches, Assiniboin proper and Red r. The Assiniboin rises by numerous branches between lake Winnipec and the northeastern curve of Missouri. Flowing s. e. upwards of 300 ms. it receives from the sthrd. a remarkable confluent, Morse or Sourie r. The latter rises in the United States and, what is remarkable, within 1 m. of the main volume of the Mo. Below the influx of Morse r. the Assiniboin curves to the estrd. about 100 ms. to its junction with Red r. Red r. or the southern branch of Assiniboin, rises in the United States, on the marshy and extended plain which gives source to the higher confluents of Miss. but the extreme southern fountain of the former, as laid down by Tanner, is at n. lat. 45° 37', interlocking sources with St. Peter's r. and extending 1° 39' of lat. south of the northern source of the Miss. Issuing from this elevated table land, Red r. assumes a general northern course, but with an elliptic curve to the westrd. receiving from the eastern side descending, Otter Tail, Buffalo, Wild Rice, Plum, Sand Hill, Red Fork, Swamp, Salt and several smaller streams. From the westrd. this r. is augmented also in descending, by Ipse, Thienne, Elve, Goose, Turtle, Saline, Park Tongue and Pembina rs. Immediately below the influx of the latter the main volume crosses n. lat. 49° enters the British territories, inflects to the northeastward and finally joins the Assiniboin very nearly on n. lat. 50°' and about 40 ms. above the influx of the latter, into the sthrn. extreme of Lake Winnipec. The direct length of Red r. is about 300 ms. but its comparative course no doubt exceeds 350. The valley of Red r. extends between those of Mo. and Miss. and it is remarkable, that the sources of streams flowing into Red r. from the wstrd. rise close upon the margin of Mo. near the Mandan villages, (see Pembina.) Red r. is rendered remarkable for another physical feature; its source is the extreme southern extension of the great northern inclined plain of the continent of N. Am. The following note inserted in Tanner's United States will serve to illustrate the singular features of the region from which Red, Miss. and Rainy Lake rs. have their sources. The note alluded to, and quoted below, is inserted on the map between the extreme sources of Miss. Grand Fork of Rainy lake r. and Red lake, from which flows the Red Fork of Red r. "An uninterrupted water communication between the Grand Fork of Rainy Lake r. and Winnipec Lake, is said to exist at this place."

Rainy Lake r. is another very singular water course, the drain of innumerable lakes between lakes Superior and Winnipec, which though entering the latter lake by a separate mouth from the Assiniboin may be correctly noticed under that head. The Plateau, from which the sources of the Miss. and Assiniboin flow, is continued to the northeastward to within 200 miles of Hudson's Bay. Amongst the countless lakes scattered over this elevated but comparatively flat space, there are two chains to the northwest of Lake Superior, which, after discharging their waters from one reservoir into another, converge and unite about n. lat. 48° 20' and long. 15° w. from W. C. Thus far the general course is westrd. but inflecting to the n. w. dilating into Sturgeon lake, again into a r. and thence once more opening into the large Lake Rain. From the southwestern margin of Lake Rain, issues a large navigable stream, the proper Rain Lake r. This stream, flowing a little north of wstrd. about 80 ms. by comparative course, falls into another considerable sheet of water called Lake of the Woods. From the source of the southern branch of Rainy Lake r. to the Lake of the Woods, the lakes and intermediate straits or rs. form the boundary between the United States and the British territories. The southern extremity of the Lake of the Woods is traversed by the 49th degree of n. lat. and the lake and its discharge into Lake Winnipec, assuming a northwestern course, the residue of the valley of Rainy Lake r. is in the British territories.

Assonet, p-v. Berkley t. Bristol co. Mass. 42 s. Boston.

Assumption, or Assomption parish of La. bounded by the Miss. r. n. e. St. Johns Baptist e. La Fourche interior s. e. and s. Atchafalya r. w. and n. w. by the parish of Ascension. Similar to all other parts of the delta, Assumption is a plain slightly raised above the common level of the rs. The Miss. forms one of the borders, whilst La Fourche entering from Ascension, winds in a direction from n. n. w. to s. s. e. dividing Assumption into two nearly equal parts. Length from s. w. to n. e. 40 ms. mean width 15 and area 600 sq. ms. Most of the arable land is on the two above mentioned rs. as the surface falls gradually, from the La Fourche towards the Atchafalya, and the banks of the latter rise in very few places above annual overflow. Much of Assumption, is, indeed, open, untimbered, and overflow marsh. The inhabitants reside in great part on the margin of the Miss. and La Fourche. The arable margin on the right bank of the Miss. at the efflux of La Fourche, is continued, down the latter though narrower than along the former stream. The soil is of similar quality on both rs. and crops are specifically similar. Though cotton is the common staple, some sugar farms are scattered along the banks of La Fourche in Assumption. The orange tree also begins to flourish in this parish. n. lat. 30°, and long. 14' w. from W. C. intersect in the northern part of Assumption. No town of note. Pop. 1820, 3,576, in 1830, 5,669.

Assumption, Court House, and p-o. on La Fourche, by p-r. 91 ms. from N. O.

Asylum p-o. Bradford co. Penn. by p-r. 184 ms. northward from Harrisburg.

ATACAPAS, large settlement of La. lying along both banks of Teche r. and extending southeastward from Opelousas to the Gulf of Mexio. Under the French and Spanish governments, Atacapas was under the jurisdiction of a commandant; it is now subdivided into the parishes of St. Martin's and St. Mary's, which see.

ATCHAFALAYA, r. of La. and though a mouth of Miss. may be correctly regarded as the continuation of Red River. It is only about a m. and five tenths from the inlet of the Red River to the outlet of Atchafalaya, the latter leaving the main stream at N. lat. 31° almost exactly. Many erroneous opinions exist respecting the Atchafalaya, and the quantity of its drain from the Miss. has been greatly overrated. It is only indeed at high water in the latter, that any current passes into the former. The writer of this article has been at the efflux of the Atchafalaya at almost every stage of weather, and has seen the current actually passing into the Miss. At its outlet, the Atchafalaya is only 111 yds. wide, but dilates below to a general breadth of about 200 yds. Leaving the Miss. the course is S. W. 2 ms. from whence with a very tortuous channel, but general course to the S. it flows 50 ms. then turns S. E. 10 ms. and thence E. 25 ms. to where it receives from the estrd. the Plaquemine, another, though much smaller outlet of the Miss. Receiving the Plaquemine, the Atchafalaya curves to the southward and continues in that direction 60 ms. into the Gulf of Mexico, having an entire comparative course of 147 ms. In its general appearance, the Atchafalaya is a miniature picture of the Miss. though very little soil on the banks of the former, is exempt from annual overflow. Except merely on the margin of the r. the adjacent country for many ms. is a congeries of bayous and lakes; but to apply to this annually inundated tract the term of swamp, is a great misnomer. So far from being swamp the general surface when not flooded is excessively dry and hard. In reality, the basin of the Atchafalaya, if denuded of timber would appear, in season of high water, an immense irregular lake, with lines of land merely rising above the common surface. The whole surface is, however, except the lakes and streams, and near the sea coast, covered with a very dense forest. Timber along the Atchafalaya, cotton wood, willow, ash, red flowing maple, different species of oak, and hickory, &c. whilst in the remote recesses of the overflow, cyprus and tupeloo, intermingle with the oaks and willows. When this region was surveyed by the writer of this article, in 1809 and 1810, the great raft in Atchafalaya commenced 26 ms. from the outlet, and continued down that river at broken intervals to within 5 ms. above the entrance of the Teche. A small fragment of the lower part of the raft broke loose from the main body about 1774, and again lodged immediately below the mouth of Courtableau. The raft is not a stationary mass; I have myself seen large fragments break loose, and again lodge. This much misunderstood obstruction was formed in the first place by floating timber from the Mississippi, and is perpetuated by accessions from the same source. The current of the Atchafalaya, at its issue from the Mississippi, is excessively rapid, but deadens, falling towards the interior low lands, which circumstance, taken in connexion with the very tortuous channel of the former, accounts for the original formation and perpetuity of the raft. The vulgar tales respecting the raft, deserve some notice. So lonely and so seldom visited is that region, that even at Opelousas, within 15 miles direct from the spot, I have heard it asserted that so compact was the raft, that men and horses had passed it without knowing a river was beneath. Myself and assistants did frequently, and in numerous places pass the river by the raft, but never without danger and difficulty. In fact, the timber, thrown together by accident, lies in all directions, interlaced by roots and branches, but still with so much interval and so liable to partial removal, that accumulations of mud and young trees are prevented. The body of the raft rises and falls with the rise and fall of the river, and from the sediment fixed on the logs, is in autumn covered with a rank growth of weeds, which at a distance appears a flowery plain, but approached is found to be a very dangerous and uneven surface to tread. In both the Atchafalaya and Courtableau, the raft impedes navigation. In the latter the people of Opelousas have effected a partial removal; but still the direct line of intercommunication between New Orleans and the upper part of Opelousas, continues in some measure broken. The very near approach of the general plain of Louisiana to a level is shown by the tides in the Atchafalaya and its confluents. In season of high water in the Mississippi, the tide from the higher parts of the basin overcomes the more feeble tide from the Gulf of Mexico; but in Autumn, when the Mississippi falls far within its own channel, the gulf tide not more than 2 feet at an average, penetrates the Atchafalaya above the lower raft, and into the Plaquemine to within 4 miles by the channel from the Mississippi, and up the Courtableau to the prairies of Opelousas. The water of the Atchafalaya, usually turbid, becomes very highly pellucid, when the outlet from the Mississippi is interrupted for some weeks. (See articles Courtableau, Teche, Opelousas, and Atacapas.) A long expected change in the bed of the Mississippi above the outlet of Atchafalaya and influx of the Red river has taken place recently; and the Atchafalaya can no longer be regarded as an outlet of the Mississippi. (See article Mississippi and Red River.)

ATCHAFALAYA, bay of La. the estuary of the river of the same name, an elliptical sheet of water 25 ms. from N. W. to S. E. with a mean width of about 8 ms. Two long, narrow and low points of land, Point Au Fer to the S. E. and Point Chevreuil to the N. W. distant from each other 7 or 8 ms. terminate this bay. A bar with 9 feet water extends from point to point, but within, and far in both rivers above the junction of Atchafalaya and Teche, a sufficient depth of water exists at all seasons for vessels of any tonnage. Bar of Atchafalaya bay is at N. lat. 29° 25′, long. 14° 40′ w. from W.C. The river enters the N. E. side of the bay. The whole adjacent country is a dead level, and grassy plain, interrupted with numerous lakes and interlocking water courses, and nearly destitute of timber. The surface so near that of the gulf, as to be flooded at every tide. The first arable land rises above diurnal tides about 5 ms. below the mouth of Teche.

ATHENS, p-t. Somerset co. Me. 22 ms. N. E. Norridgewock, 45 Augusta, 50 N. Hallowel, is crossed by the branches of a small stream of Kennebeck r. Pop. 1200.

ATHENS, p-t. Windham co. Vt. 10 ms. S. Bellows' Falls, 25 N. E. Brattleborough; contains 7628 acres, and was first settled 1779. It is uneven, with a good soil for grazing, and apple trees; with a native growth of beech, birch, bass, maple, ash, hemlock, and spruce. A small stream affords mill sites. Pop. 415.

ATHENS, p-t. Greene co. N. Y. w. side Hudson r. opposite Hudson city, 5 ms. N. Catskill, 28 S. Albany, is watered by Potock, Corlear's and Catskill creeks, has Hoogeberg, or High Hill w. but pretty good alluvial soil E. with some pine sands.

ATHENS, v. extends 1½ ms. on the Hudson r. and about the same distance w. The channel is near the shore; there is a flat in the r. through which a channel is kept open for the ferry boats. Pop. 2425.

ATHENS, or Tioga Point, p-v. of Bradford co. Pa. is most beautifully situated on the point above the junction of the Susquehannah and Tioga rs. The village extends in a single street up the peninsula; the site an undulating plain, but environed by mountain and river scenery. Athens is but little improved, but from its proximity to two navigable rivers, the fertility of the river soil, and from the picturesque vicinity, nature has done her part in forming a most inviting residence to man. Pop. of the township in 1820, 1108, in 1830, 1249. Lat. 41° 56′ N.

ATHENS, p-v. st. jus. and of Franklin college, Clarke co. Geo. is situated on the right bank of Oconee r. at N. lat. 33° 58′, long. W. C. 6° 40′ w. 85 ms. N. W. from Augusta. Franklin college was located here in 1803, with a faculty of a president, 4 professors, and two tutors, supported by a fund of $100,000 bank stock, and 50,000 acres of land; $12,000 were appropriated to purchase a library and philosophical apparatus. The situation has been represented as in an eminent degree agreeable and healthy.

ATHENS, p-v. Fayette co. Ken. by p-r. 33 ms. from Frankfort.

ATHENS, p-v. and st. jus. Limestone co. Ala. is situated near the centre of the co. about 25 ms. a little N. of w. from Huntsville, and 120 ms. N. N. E. from Tuscaloosa, N. lat. 34° 58′, long. 10° 2′ w. from W. C.

ATHENS, p-o. Giles co. Vir. by p-r. 265 ms. a little S. of w. from Richmond.

ATHENS, co. of O. bounded S. by Meigs, S. W. by Galia and Jackson, W. by Hocking, N. W. by Perry, N. E. by Morgan, E. by Washington, and S. E. by O. r. separating it from Wood co. of Vir. The longest line that can be drawn over this irregular formed county is diagonal from S. W. to N. E. 40 ms. Mean breadth 20 ms. and area 800 sq. ms. extending in lat. from 39° 5′ to 39° 33′, and in long. from 4° 44′ to 5° 30′ w. from W. C. The southern margin of this co. declines to the sthrd. and gives source to some creeks flowing in that direction into O. r. The central and rather most extensive section is watered by the Hocking r. and its confluents. The Hocking enters at the extreme northwestern angle, and flowing to the S. E. diagonally over the county falls into the O. r. at the extreme southwestern angle. The surface of Athens co. of O. is excessively hilly, but fertile. Chief town, Athens. (See Appendix, article Ohio.)

ATHENS, p-t. and st. jus. Athens co. O. situated on Hocking r. near the centre of the co. 41 ms. a little S. of w. from Marietta, 50 ms. E. from Chilicothe; and by p-r. 73 ms. S. E. by E. from Columbus. N. lat. 39° 20′, and long. 5° 6′ w. from W. C.

ATHOL, p-t. Worcester co. Mass. 30 ms. N. Worcester, 70 w. Boston; is uneven, with a strong soil, and is supplied with water power by Miller's r. Pop. 1325.

ATHOL, p-t. Warren co. N. Y. 18 ms. W. N. W. Caldwell, is mountainous, with a poor soil. Pop. 909.

ATKINS, p-v. Bucks co. Pa. 20 ms. N. E. from Phil. and 10 ms. S. W. from Trenton.

ATKINSON, p-t. Penobscot co. Me. 79 ms. from Augusta. Pop. 418.

ATKINSON, p-t. Rockingham co. N. H. 4½ ms. by 5, with 6839 acres, 30 ms. S. W. Portsmouth, 32 S. E. Concord; has an uneven surface, with very good soil. First settled 1727. Dr. Belknap says that a piece of ground, 7 or 8 acres, in a meadow, floats when the meadow is flooded. There is an old and respectable academy here. Pop. 554.

ATKINSON, p-o. Monroe co. Mich. by p-r. 43 ms. S. W. from Detroit.

ATLANTIC Ocean, that great expanse of water which separates Africa and Europe from America. Under this general term very different extremes north and south have been understood by writers on Geograhy. In the present article we include all the ocean surface comprised between a line drawn from the extreme southern point of America to the extreme southern point of Africa, and another line drawn from the western point of Nova Zembla, through the islands of Spitzbergen to the coast of Greenland. Thus defined, the Atlantic mingles on the S. with the great Austral ocean, and to the north with the Frozen or Arctic Ocean. In extent, including its seas, the Atlantic spreads over above thirty millions of sq. ms. The narrowest part between the coasts of Norway and Greenland exceeds one thousand ms. and the widest part on a line nearly at right angles to its general length, something less than five thousand seven hundred ms. from the mouth of the Senegal to that of Rio Grand del Norte. It is far the widest at the northern extreme of the two Oceanic connections between the Polar regions of the Earth, a circumstance most powerfully influential on the respective climates of North America and Europe.

The general phenomena, on the two opposing sides of the Atlantic have great resemblance. To the deep indenting of the Baltic is opposed the much more extensive Mediterranean composed of Hudson's and Baffin's Bays. To the south, relative extent is reversed. To the West Indian Mediterranean is opposed that most interesting Gulf between Europe and Africa, to which that descriptive name has been given, and which has been incorporated with the history and science of mankind in all ages. But departing from a general to a special view, in this article, we shall confine ourselves to the features and sections of the Atlantic connected with the United States.

By actual calculation, a line drawn from the Point of Florida to the Straits of Belle Isle deflects from the Meridians of the Earth 35° fifty-one minutes, say 56° extending 2654 statute ms. Coast of Africa and Europe from Cape Verd of the former, to Cape North of the latter, deflects from the meridians 24° forty five minutes in a distance of 4232 statute ms. These two lines, opening upon each other within a small fraction of 11°, enclose the great body of the northern Atlantic, and if extended to strike the continent of America, will include, with the exception of the Gulf of Mexico, the sphere of action of that immense whirlpool called the Gulf Stream.

The Earth moving around its axis with a maximum of rotation at the equator, and lessening, advancing along the meridians. It is very remarkable that the equator nearly coincides with the deep indenting of Africa on the east and the eastern protrusion of America on the west, and here, particles on the Earth's surface are moved with uniform motion estrd. at the hourly rate of 1042 ms. nearly. The distance is about 7700 ms. from where the equator intersects western Africa to the mouth of Rio Grande del Norte, and if we assume 1000 ms. as the mean intermediate motion, objects are borne through the intermediate space in about 5 ¾ hours. The *vis inertiæ* of matter produces a perpetual retardation, which effects a current of rotation setting from Africa to America in opposition to the horary motion. This current striking the extreme eastern cape of South America is divided into two bodies. That to the south is wafted and dispersed over the Southern Ocean, but that to the north meeting the continent of America is deflected northwardly and augmented by accessions from the northern side of the Torrid Zone, is gradually carried into the Gulf of Mexico, and from thence by its own weight, rushes towards the Atlantic between Cuba and Florida, but meeting part of the orig'nal stream flowing westrd. along the northern shores of St Domingo and Cuba, the whole body is turned northeastwardly along the coast of the United States; flows thus towards the polar section of the Atlantic until again deflected by the northern currents, is swept down the coasts of Europe and Africa, into its original fountain.

An opinion prevails, which was once shared by the author of this article, that the Gulf Stream exerts a great influence on the climate of the United States. More extensive and accurate observation has modified that opinion in his individual case. Under the general head of United States, it may be seen that the wind over the United States, and that part of the Atlantic ocean between North America and Europe, blows with a uniformity from the western points, which vies in steadiness with those from the eastward within the tropics. If, therefore, we regard the atmosphere over the Atlantic ocean nthrd. of the equator, to N. lat. 50°, we discover two powerful currents setting in opposite directions over the extremes; and directing our attention to the incumbent ocean, we find the aquatic current corresponding to the aerial. The effects of these phenomena on the climate of both continents will be shown under the head of climate, under the general article United States.

Commercially, or nautically, to speak in more generic terms, the combined effect of the Gulf Stream and prevalent western winds over the northern Atlantic has had, and as long as the present order of things endures, must have a most extensive agency in the affairs of mankind. The share exerted by the Gulf stream in the mean period of voyages over the Atlantic, has been overrated, whilst that of the wind has been in a corresponding degree overlooked. The Gulf of Mexico, we have shown, is the great reservoir from which the Gulf Stream is supplied, and the difference of level between that reservoir and the Atlantic ocean E. of the peninsula of Florida, has been determined at 3 or 4 feet, by the U. S. engineers; who observe in page 39 of their report, "should the surveys be perfectly accurate, the level of low tide in the Gulf would be 2 65-100 feet above that of low tide in the Atlantic. On another hand it has been shown, in describing the surveys relating to the St. Mary's route, that the result of the levelling has given 3 55-100 for the difference of level between the Gulf and the Atlantic. We must therefore consider it as probable, that at low tide, the elevation of the Gulf at the mouth of the Suwannee, is not more than 3 or 4 feet above low tide at the entrance of the St John's into the Atlantic, and that such might be the limit of the rise caused at this point of the Gulf, by the tropical trade wind."

Such a rise, or double such height, could

not sustain the actual current of the Gulf Stream in the Atlantic ocean, unless aided by other causes. Such a cause, however, does exist in the northern *Trade* winds. The latter current operates so greatly on navigation, as to make the mean of voyages from America to Europe, 23, whilst the mean of those in an opposite direction demand 40 days. Such is the steadiness and intensity of these western winds, that the orchards and forests of the U. S. are bent towards the Atlantic in the U. S. and from it along northwestern Europe. (*See Art. U. States, head of Climate.*)

The Atlantic coast of the U. S. presents an elliptic curve, if taken in its entire extent, with three intermediate and similar curves. Advancing from south to north, the first partial curve has its axis from Cape Florida to Cape Hatteras, about 700 ms. with an ordinate or depth of near 200 ms. the chord deflecting from the meridians by an angle of 25°. Issuing from the Bahama channel, the Gulf stream flows towards cape Hatteras very nearly along the chord of the intermediate bay.

The second, or middle bay, sweeps from cape Hatteras to the outer capes of Massachusetts, 500 ms. with considerably less deflection from the meridians, and less comparative depth from its chord. The Gulf stream, turned from the coast by Cape Hatteras, has its axis beyond the capes of Mass. and, indeed, more estrd. than the general line of the coast.

The third, or northern bay of the U. S. is, in fact, formed by the coasts of Massachusetts, New Hampshire, Maine, New Brunswick and Nova Scotia; therefore, only the southwestern part really appertains to the U. S.

The respective coasts of these three physical sections of that part of the U. S. bordering on the Atlantic ocean, have each its appropriate and very distinct character. The rivers of the southern bay flow generally from N. W. to S. E. with narrow and shallow mouths, made in a low, flat, and inundated coast. The harbors are shallow, and tides moderate and not exceeding a mean of 5 feet.

The rivers of the second bay mingle with the tides in wide sounds, and one, the Hudson, carries its tides inland beyond the Appalachian system of mts. The tides are along this coast of increased elevation, having an average of perhaps 7 feet. The solid land ceases, gradually, advancing northwardly, to be inundated by the Ocean tides or river floods, to any great extent.

The third bay presents phenomena, in a very striking manner distinct. The coast becomes more bold, and the general slope more abrupt. Though the tides so enormously increase in height as to exceed a mean of 25 feet, yet this great swell is arrested within from 15 to 30 ms. from the coast.

Opposite to the United States, the Atlantic ocean no where deepens very rapidly, but on the contrary, admits soundings in every place near the shores. From the course and origin of the Gulf Stream, the ocean water every where, and at all seasons of the year, exceeds the mean temperature of that in the land bays. It is evident that the increased height of the tide advancing along the coast from south to north, arises from the decreased velocity, and wider dispersion of the Gulf Stream, as the magnitude of the two sets of phenomena is reverse to the extremes where they occur.—(*See Articles Gulf Stream, United States, and Gulf of Mexico.*)

ATLAS, p-v. in the S. W. part of Pike co. Il. about 100 ms. following the land route, about N. N. W. from St. Louis, and by p-r. 148 ms. N. W. by W. from Vandalia.

ATSION, v. Gloucester co. N. J. 60 ms. S. E. Philadelphia.

ATTICA, p-t. Genesee co. N. Y. 12 ms. S. Batavia, 250 W. Albany, is watered by Tonewanta cr. and branches. It has a pleasant v. on the cr. and much timber. Pop. 1830, 2,492.

ATTICA, p-v. in the N. part of Seneca co. O. by p-r. 92 ms. northwardly from Columbus.

ATTLEBOROUGH, p-t. Bristol co. Mass. 29 ms. S. W. Boston, 9 N. Providence, has 3 p-vs. is furnished with water power by a branch of Pawtucket r. and has many manufactories. The Falls cotton factory was incorporated in 1813, cap. $100,000. This was an early settlement on the frontier of King Philip's country, and was exposed to much difficutly with the Indians, in his war. Pop. 1830, 3,219.

ATTLEBOROUGH, p-v. Bucks co. Pa. 20 ms. N. E. from Phil. and 10 ms. S. W. from Trenton.

ATWATER, p-v. Portage co. O. by p-r. 137 ms. N. E. from Columbus.

AUBURN, p-v. and cap. in the t. of Aurelius, Cayuga co. N. Y. on the outlet of Owasco lake, is very pleasant and flourishing; 175 ms. W. Albany, 7 S. Erie Canal. It contains the county buildings, and one of the state prisons, the wall of which is of hammered limestone and encloses 5 acres, on the bank of Owasco cr. whose water moves the machinery of the prison. Earnings, the year ending Sept. 30, 1831, $36,209 14 cts. Expenditures, $34,405 61. Balance in favor of the prison, $1,803 83. Receipts, $43,762 81. Number of convicts, Dec. 1, 1831, 646, 60 of whom were received from the Sing-Sing prison. Of 133 discharged that year, 95 had been intemperate. Of the number remaining, 209 were grossly intemperate, 258 regular drinkers, occasionally intoxicated, 132 temperate drinkers. Of these, 346 were under the influence of ardent spirits at the time of the commission of their crimes. The labor performed is on contracts, chiefly for the city of New-York. It has been occupied about 12 years. Only 2 suicides have occurred. Service is performed in the chapel on the sabbath, and a Sunday school is kept by the students of the theological seminary of the Presbyterian church, which is established at Auburn. Pop. 1830, 4,486.

AUBURN, p-v. in the S. W. part of Geauga co. O. by p-r. 143 ms. N. E. from Columbus.

Auburn, p-v. Oakland co. Mich. by p-r. 30 ms. n. w. from Detroit.

Augusta, p-t. and cap. Kennebec co. Me. and cap. of the state; 2 ms. n. Hallowell, 56 n. e. Portland, the third in the state in Pop. (increase in 10 yrs. 61 pr. ct.) is divided by Kennebec r. 47 ms. from its mouth, across which is a bridge. The v. is on an elevated plain and a declivity near the r. and contains the state house, county buildings, 60 stores, 3 churches, an academy, &c. Vessels of 100 tons come up to this place, the head of sloop navigation. The state house, somewhat resembles that at Boston, but is smaller. It stands on Capitol Hill, e. front 150 feet, with 2 wings of 33 feet, and 54 deep; central part, 84 feet and 56 deep. It is of granite—and 8 Doric columns, also of granite, 21 feet high, weighing 10 tons each, form a portico of 89 feet, one story. The dome is 54 feet square, with a cupola. Extreme height, 114 feet 2 in. The back country is very extensive and fertile. Pop. 1830, 3,980. The Kennebec road, hence to Quebec, was travelled in 1831, in carriages, in 3 days.

Augusta, p-t. Oneida co. N. Y. 17 ms. s. w. Utica, 6 ms. by 7. Oriskany, Oneida, and Shanando creeks, furnish mill-seats. The surface is gently swelling, with a pretty good soil for grain and grass. First settled, 1794. Pop. 1830, 3,058.

Augusta, p-v. Frankford, Sussex co. N. J. 79 ms. n. from Trenton.

Augusta, p-v. Northumberland co. Pa. by p-r. 57 ms. n. from Harrisburg.

Augusta, city and seat of justice, Richmond county, Georgia, situated on the right bank of Savannah r. 120 ms. n. w. from the city of Savannah, n. lat. 33°, 28′, long. w. from W. C. 4°, 58′. Pop. 1820, 4,000; 1830, 6,690.

Augusta co. Va. lying w. from the Blue Ridge, is bounded s. w. by Rockbridge, w. by Bath, n. w. by Pendleton, n. e. by Rockingham, and s. e. by the Blue Ridge, or Albemarle and Nelson. It lies in form of a square of about 30 ms. each side, of course, contains about 900 sq. ms. The surface is broken, and in part mountainous, and elevated at a mean of about 1200 ft. above the Atlantic ocean. n. lat. 38°, and long 2° w. from W. C. intersect in the s. e. part of this county. Though some of the higher creeks of James r. rise in and flow s. from Augusta, the body of the co. is drained by the confluents of the Shenandoah. The soil is productive in small grain, and the principal staple is flour. Chief town, Staunton. Pop. 1820, 16,742; in 1830, 19,926.

Augusta, p-v. and st. jus. Perry co. Miss. situated on Leaf r. or the w. branch of Pascagoula, 160 ms. a little s. of e. from Natchez, and by p-r. 137 ms. s. e. from Jackson. n. lat. 31°, 12′, long. 12°, 12′ w. from W. C.

Augusta, p-v. in the s. w. part of Columbiana co. O. by p-r. 141 ms. n. e. from Columbus.

Augusta, p-v. and st. jus. Bracken co. Ky. situated on Ohio r. on an elevated bank, affording a very pleasing site. n. lat. 38°, 45′, long. from W. C. 7° w. Distant about 70 ms. n. e. from Frankfort.

Augusta Springs, p-o. Augusta co. Va. 17 ms. n. n. w. from Staunton, and by p-r. 137 ms. n. w. by w. from Richmond.

Augustine, St., city of Flor. situated on the Atlantic shore of that ter. opposite the n. extremity of the isl. of Anastasia, n. lat. 29° 48′, long. w. from W. C. 4°, 31′. This is the oldest city of either the U. S. or Canada, having been founded by the Spaniards in 1565, upwards of forty years before the establishment of the English at Jamestown. It is situated on a plain, resting on a limestone base, along a safe and commodious harbor, with a depth of water from 28 to 30 feet, at low tide. The following notice appeared in the Nat. Intelligencer, Oct. 17, 1825, and from the importance of this city and harbor, is inserted entire in this Gazetteer. "The entrance is covered outside by sand banks and rocks, extending out more than one third of a league, between which, there are openings and passes which form the channel leading to the interior of the harbor. The south channel is the one vessels generall enter. There is not less than from 20, 25, to 30 ft. of water at low tide; but before you enter this channel you have to pass a bar, over which there is not more than 9 ft. water, at low tide. You can moor your ship outside the bar, and wait till the tide comes in. This pass can easily be distinguished by the breaking of the sea on the rocks s. and n. and thereby plainly marking out the passage, which is at least 300 fathoms wide. After passing the bar you find from 18 to 20 ft. water. You will range along the island of *Anastasia*, within 2 cables lengths; you may anchor near the n. e. point, opposite a battery, in 26 ft. water, within one cable's length and a half of the land. You double the n. point of the island, and then come to anchor before the town. These passes and entrances into the harbor are very advantageously situated; for all the vessels coming from Mexico, Louisiana, and Havanna, are obliged to pass through the Bahama channel, at a short distance from St. Augustine, the port of which can be made with ease." Pop. 1830, 1,377.

Aurelius, p-t. and cap. Cayuga co. N. Y. between Cayuga and Owasco lakes, 159 ms. w. Albany, contains no streams except the outlets of these lakes. Agriculture is prosperous. There are 3 villages, Auburn, Cayuga, and Union Springs. Pop. 1830, 2,767.

Aurora, p-t. Cayuga co. N. Y. Pop. 1830, 2,423; 171 ms. w. from Albany.

Aurora, p-t. Erie co. N. Y. Pop. 1830, 2,421.

Aurora, p-v. Dearbon co. Ind. about 25 ms. wstrdly. from Cincinnati, and by p-r. 102 ms. a little e. of s. e. from Indianopolis.

Aurora, p-v. in the northern part of Portage co. O. 15 m. n. n. w. from Ravenna, the st. jus. and by p-r. 131 ms. n. e. from Columbus.

AU-SABLE, p-v. Essex co. N. Y. 167 ms. N. Albany.

AUSTERLITZ, p-t. Columbia co. N. Y. 17 ms. E. Hudson, 130 E. S. E. Albany; has a broken surface, with a moist, loamy soil, and is crossed by Klinekill and Green r. a branch of Housatonic r. It has two villages, Green River, and Spencertown. Pop. 1830, 2,245.

AUSTINSBURG, p-v. in the northwestern part of Ashtabula co. O. by p-r. 185 ms. N. E. from Columbus.

AUTAUGA, one of the central counties of Ala. bounded E. by Coosa r. S. by Alabama r. W. by Dallas, Perry, and Bibb cos. and N. by Shelby. It approaches a square of 32 ms. each side, with an area exceeding 1000 sq. ms. Surface very varied and broken, with a soil as varied; that along the rivers of first rate quality, but deteriorating from thence to sterile ridges. Staple, cotton. Beside the two fine rivers which form the eastern and southern limits, this county is drained by the Autauga and other large creeks, mostly flowing sthrd. into the Alabama r. Chief t. Washington. Pop. 1820, 3,853, in 1830, 11,784. Central lat. 32° 35′ N. long. W. from W. C. 9° 35′.

AVENTSVILLE, p-v. Nash co. N. C. by p-r. 44 ms. estrd. from Raleigh.

AVERILL, t. Essex co. Vt. has a broken surface and a very sterile soil. It is 6 ms. square, and is watered by a branch of Nolhegan r. with several streams which fall into Connecticut r. and others flowing into Canada.

AVERYSBOROUGH, p-v. Cumberland co. N. C. situated on the left bank of Cape Fear r. 25 ms. N. N. E. from Fayetteville, and about 40 ms. S. from Raleigh.

AVERY'S GORE, Addison co. Vt. Pop. 1830, 33.

AVERY'S GORE, Franklin co. Vt. Pop. 1830, 22.

AVON, p-t. Somerset co, Me. 35 ms. N. from Norridgewock, 50 ms. N. N. W. from Augusta, on Sandy r. Pop. in 1830, 745.

AVON, p-t. Hartford co. Conn. 10 ms. W. Hartford, W. Talcott mtn. has much good level land on the banks of Farmington r. Pop. 1830, 1,025.

AVON, p-t. Erie co. N. Y. 10 ms. N. E. Geneseo, 21 W. Canandaigua, E. Genesee r. S. Honeyco creek, is well watered by these streams and their tributaries, and contains some of the best land in the state, as it includes a tract of the alluvion on Genesee r. Boats go down to Rochester and the Erie canal, 20 ms. Pop. 1830, 2362.

AVON, p-v. in the N. E. part of Lorraine co. O. by p-r. 149 ms. N. N. E. from Columbus.

AVOYELLES, prairie and parish, La. bounded S. by Opelousas, E. by Atchafalaya, Mississippi, and Red r. N. by Red r. W. parish of Rapide, S. W. by part of Opelousas. The extreme length from the mouth of Bayou Rouge, to that of the Ouchitta r. is about 40 ms. The breadth from E. to W. a mean of 20 ms. having an area of 800 sq. ms. Avoyelles designates a district of country composed of two unequal and very different bodies of land. The western part is a plain, elevated 15 or 20 feet above the highest flood of the adjacent rivers. This comparative table land rises at once and on all sides, from the annually inundated low lands, but is nevertheless an almost undeviating plain, of tolerable good soil, covered in part by forest, but in great part prairie. Below the prairie plain, expands the uninhabitable overflow, forming more than two thirds of the whole parish. This overflowed tract is so uniform in its general features, as to admit little variety of description. We may therefore refer to article "Atchafalaya," for a view of the Avoyelles overflow. Bayou de Glaize, one of the outlets of Red r. flows to the estrd. passing the southern extremity of the Avoyelles table land, and by a very winding course, falls into the Atchafalaya. By the Bayou de Glaize, a belt of soil is carried across the deeper inundation, which, except at very high floods, admits a road from Avoyelles to Red r. This tortuous zone contributes in part to produce a phenomenon very remarkable, and yet very little known even in Louisiana. The far greater share of the annual flood of Red r. is prevented by the banks of the de Glaize, from flowing sthrd. towards Opelousas, but is thus directed southeastward, into the outlet of the Atchafalaya, and again on the opposite side of the Mississippi, the incumbent water is in a similar manner turned wstrd. so that the entire body of flood brought down by both rivers is forced into a passage of at most 5 miles wide. This accumulation of water above the Bayou de Glaize, however, contributes to render that part of Avoyelles in a particular manner exposed to *submersion*. On the habitable table land, the staple is cotton—the port of the settlement, Red r. Pop. in 1820, 2245, in 1830, 3484. N. lat. 31°, and W. long. from W. C. 13°, intersect near the centre of this parish.

AYLETTS, p-o. King William co. Va. by p-r. about 36 ms. N. E. from Richmond. It is situated near the right bank of Mattapony r. opposite Dunkirk.

B.

BACHELDOR, t. Oxford co. Me. 20 ms. W. Paris, E. New Hampshire, and just south Androscoggin r. has several mtns. but no streams.

BACHELOR'S RETREAT, p-v. Picken's dist. S. C. 20 ms. a little S. of W. from Pendleton, and by p-r. 154 ms. N. W. by W. from Columbia.

BACK CREEK VALLEY, p-o. Frederick co. Va. by p-r. 88 ms. wstrd. from W. C.

BAGGS, p-v. McIntosh co. Geo. by p-r. 130 ms. S. E. from Milledgeville, and 35 from Darien.

BAHAMA CHANNEL, a narrow sea or sound between Florida and the Bahama Islands. This sound is usually considered as commencing to the sthrd. about N. lat. 24°, where the Florida and Santerim channels unite, and thence extending due N. to Matinilla Reef, about N. lat. 28°. The entire length 280 ms. with a general width of 60 ms. The Bahama channel is a great ocean river, formed by the Gulf Stream flowing estrd. between Cuba and Florida, and which meeting the current from the Santerim channel, the whole turns northward, and flows with a velocity from 2 to 5 ms. per hour, between the Florida coast, and the numerous keys, islands and shoals, known under the general name of Bahama. When the winds are from any southern point, the navigation of Bahama channel is delightful, but on the contrary, when northern winds meet and contend with the powerful current of this sound, the waves of the sea are in a peculiar manner dreadful.

BAILEYSBURG, p-v. Surrey co. Va. by p-r. 72 ms. S. E. from Richmond.

BAINBRIDGE, p-t. Chenango co. N. Y. 20 ms. S. Norwich, 110 W. Albany; contains 48,000 acres. It is crossed diagonally by Susquehannah r. and has 3 post-offices. It has a good soil, well watered, and little waste land. Bainbridge v. w. Susquehannah. r. is pleasant. Pop. in 1830, 3040.

BAINBRIDGE, p-v. Lancaster co. Pa. situated on the left bank of Susquehannah r. opposite the mouth of the southern Conewago creek, 18 ms. below Harrisburg, and about the same distance a little N. of W. from Lancaster.

BAINBRIDGE, p-v. Jackson co. Geo. by p-r. 88 ms. northward from Milledgeville.

BAINBRIDGE, p-v. situated on the left bank of Tennessee r. 5 ms. above, but on the contrary side from Florence, and in the N. E angle of Franklin co. Ala. 100 ms. N. from Tuscaloosa.

BAIRD'S FORGE, p-o. Burke co. N. C. by p-r. 187 ms. W. from Raleigh.

BAIRDS or BAIRDSTOWN, p-v. and st. jus. Nelson co. Ky. situated near the Buck Fork of Rolling r. 41 ms. S. S. E. from Louisville, and 47 ms. S. W. from Frankfort, N. lat. 37° 48', long. W. from W. C. 8° 25'. Pop. in 1820, about 600, in 1830, 1629.

BAKER'S RIVER, Grafton, co. N. H. unites with Pemigewasset r. near Plymouth v.

BAKER'S ISLAND, Essex co. Mass. of Salem harbor.

BAKER'S FALLS, on Hudson r. Sandy Hill t. Washington co. N. Y. about 100 ms. from the source of the r. and 44 above tide. The descent, including the upper and lower Falls, and the rapids is here 70 feet in 100 rods. The dam at Fort Edward has raised the water 10 feet at the foot of the falls.

BAKER COUNTY, Geo. bounded W. by Early, N. W. by Randolph, N. by Lee, N. E. by Dooley, E. by Irwin and Thomas, and S. by Decatur. Length 43 ms. breadth 35. Area 1500 sq. ms. Flint r. traverses Baker diagonally, from N. E. to S. W. It is a new, and in great part, unsettled tract, extending in lat. from 31° 7' to 31° 43', and in long. from 7° 10' to 7° 45' W. from W. C. Chief town, Byron. (For pop. see appendix, Georgia.)

BAKERSFIELD, p-t. Franklin co. Vt. 30 ms. N. E. Burlington, 38 N. N. W. Montpelier; contains 26,000 acres, and was first settled about 1789. It is hilly, with a productive soil, bearing much hard wood, and watered by Black creek and other streams of Missiscoui r. with few mill sites. Pop. in 1830, 1087.

BAKERSTOWN, p-v. Alleghany co. Pa. 14 ms. from Pittsburg on the Butler road.

BAKERSTOWN, p-v. Burke co. N. C. by p-r. 220 ms. wstrd. from Raleigh.

BALCONY FALLS and p-o. in the western part of Rockbridge co. Va. by p-r. 153 ms. W. from Richmond, and 210 S. W. by W. from W. C.

BALD EAGLE, chain of Appalachian system of mtns. in Pa. extends in a direction nearly N. E. and S. W. between the main and west branch of Susquehannah r. separating Northumberland and Columbia cos. from Lycoming, and stretching into Luzerne.

BALD EAGLE, large creek or river, of Centre and Lycoming cos. Pa. rises by numerous branches in the former, which by a general course of N. E. unite, and entering Lycoming, fall into the W. branch of the Susquehannah, at Dunnstown, after a comparative course of 50 ms.

BALD EAGLE, p-v. Lycoming, co. Pa. by p-r. 126 ms. from Harrisburg.

BALD HEAD, cape, Me. in long. 70° 35', lat. 43° N.

BALD HILL, p-v. Cumberland co. Me. 39 ms. from Augusta.

BALDRIDGE'S p-o. Buncombe co. N. C. by p-r. 283 ms. wstrd. from Raleigh.

BALDWIN, p-t. Cumberland co. Me. 26 ms. N. W. Portland, 80 W. S. W. Augusta, W. Sebago pond, N. Saco r. Pop. in 1830, 947.

BALDWIN, p-v. Tioga co. N. Y.

BALDWIN co. Geo. on both sides Oconee r. having Putnam and Hancock N. Hancock N. E. Washington S. E. Wilkinson S. and Jones W. Length from W. to E. 18 miles, mean breadth 12, and area 216 sq. ms. The face of the country rolling, and soil second rate generally, though some of first rate skirts the streams. Pop. in 1820, 5665, (for 1830, see appendix, Georgia.) N. lat. 33° traverses the southern extremity of this county, and the centre is about 6° 20' W. from W. C. Chief town, Milledgeville, the seat of government of the state of Georgia.

BALDWIN co. Ala. bounded by Perdido r. or Escambia co. in Flor. E. and S. E., by Conecuhco, in Ala. N. E., by Munroe N., Alabama r. N. W., Mobile r. and bay W., and S by the Gulf of Mexico. In a direction from N. to S. in Tanner's map of the U. S. this co. is 72 ms. in length, with a mean width of 28, and an area of 2,000 sq. ms. Along the rivers there is some good land, but taken as a whole, this county is sterile. The surface rises very gradually, from the sandy plain near the Gulf of Mexico, to the interior pine ridges. Staple, Cotton. It extends from N. lat. 30° 13' to 31° 16', and is traversed by long. 11° w. from W. C. Chief town, Blakeleys. Pop. 1820, 1,713, and in 1830, 3,324.

BALDWIN'S CROSS ROADS, p-o. Ann Arundel co. Md. on the road from W. C. to Annapolis, 48 ms. from the former, and 12 from the latter place.

BALDWINSVILLE, p-v. Worcester co. Mass. 59 ms. from Boston.

BALLSTON, p-t. Saratoga co. N. Y. 23 ms. N. Albany, about 5 ms. square, is principally on an elevated, open, champaign country, with gentle swells—the soil a strong gravelly loam, and some sand and clay. It produces grass and grain, orchards, and numerous forest trees. It has Ballston lake or Long pond, S. E. 4 ms. long, 90 rods wide. The outlet and Mournkill are the principal streams, which supply a few mills. There is a small post-village, with an academy and a high school. This place was an early settlement, and was attacked and burnt by a party of Indians from Canada, during the Revolution. The Ballston Springs are not in this township, but in the neighboring one of Milton. Pop. 1830, 2,113.

BALLSTON SPA, p-v. t. of Milton, and cap. Saratoga co. N. Y. 18 ms. N. W. Waterford, 7 S. W. Saratoga Springs, 15 N. Schenectady, 26 N. by. w. from Albany.

BALLARD'S BRIDGE, and p-o. in the northern part of Chowan co. N. C. 16 ms. N. from Edenton; and by p-r. 268 ms. sthrd. from W. C. and 155 ms. N. E. by E. from Raleigh.

BALLARDSVILLE, p-o. Logan co. Va. by p-r. 390 ms. S. W. by W. from W. C.

BALLARDSVILLE, p-v. w. part of Oldham co. Ky. 31 ms. N. by w. from Frankfort.

BALTIMORE, t. Windsor co. Vt. Pop. 1830, 179.

BALTIMORE co. Md. bounded by Chesapeake Bay, S. E., Patapsco r. separating it from Ann Arundel co. S. W., Frederick co. Md. W., York co. Pa. N., and Harford co. Md. N. E. Greatest length from S. E. to N. W. 36 ms. mean width, 25 ms. and area, 900 sq. ms. It extends in lat. from 39° 12' to 39° 42½' N.; and in long. from 0° 7' w. to 0° 39' E. from W. C.

This county contains two natural sections in itself, being very nearly commensurate with the vallies of Gunpowder and Patapsco rivers. The surface is finely varied; no part is mountainous, but the greater part is hilly. The great primitive ledge, which, except in the St. Lawrence and Hudson, arrests the Atlantic tides, traverses Baltimore county, leaving a minor section of sea-sand alluvion between the bays of Gunpowder and Patapsco. Though the section below the head of the tides is not very broken, it is far from level. Above the head of tide water the country rises, and the hills in many places are elevated, and near the streams abrupt. The general surface at Reisterstown, 17 ms. N. W. from Baltimore, has gained a relative elevation of 500 feet, and at the sources of the Patapsco and Patuxent, along the southeastern foot of the dividing ridge between Baltimore and Frederick counties, the farms are about 800 feet above tide water. This difference of height produces a very sensible difference of temperature. At and near Baltimore, spring time and harvest are from a week to 10 days in advance of similar seasons at the western and northwestern parts of the county. This and the two adjacent counties, Frederick and Ann Arundel, afford a very central example of the middle climate of the United States, on and near the Atlantic ocean. The great variety in the soil and sensible extremes of temperature arising from relative level, are productive of a corresponding diversity of vegetable genera and species. To enumerate the cultivated plants of Baltimore co. alone, would be to give a list of almost every vegetable reared in the middle climate of the United States.

In commercial rank and importance, Baltimore county is a very interesting section of the U. S., and one that is rapidly increasing in wealth and population. Beside the rising city of Baltimore, the great Western rail-way now in progress, will pass along the whole southwestern margin of this county. This interesting work has been already extended to the city of Frederick, and Point of Rocks on Potomac river, 71 ms. But a recurrence to the census of the city and county of Baltimore, gives the most decisive evidence of the augmentation of both within the last ten years. Pop. 1820, exclusive of the city, 33,463, and in 1830, 40,250. In 1820, aggregate of city and county, 96,201, but in 1830 it had risen to 120,870, and yielded an increase of 126 per cent. The county now, 1832, no doubt contains 136 to the sq. m.

BALTIMORE city, port of entry, p-t. and st. jus. Baltimore co. Maryland, is situated on the primitive ledge, and on a small creek or bay of Patapsco r. 14 ms. above its mouth into Chesapeake bay, 38 ms. N. E. from W. C. within a small fraction of 100 ms. sthrd. from Phil.. and by p-r. 30 ms. a little w. of N. from Annapolis. N. lat. 39° 17', long. 0° 26' E. from W. C.

Baltimore, named from the title of the original founder of Maryland, stands on an admirably well chosen site, at the head of the tide. The city, similar in that respect to Philadelphia, occupies in part the margin of primitive rock. The northern and most elevated part of the ground plan, is composed of rounded hills, rising to 80 or 100 feet above tide level. The slope from the alluvial section of the city, now the basin, is not regular, but is in no place very abrupt. The lower part rests on a real recent alluvion, around a basin into which only vessels of 200 tons can enter. Southward from the main body of the city, and over the basin, rises a conical hill, on which stands Fort McHenry, the citadel of Baltimore, and below which the harbor widens and deepens, so as to admit ships of

600 tons burthen to Fell's Point, or the lower and southeastern part of the city.

Connected with the adjacent country by only the ordinary roads, Baltimore is well situated for internal commerce. It is more contiguous to the valley of Ohio, to all western Maryland, and also to a large section of Pa. than is Phil. and having the advantage of a more southern climate, the harbor of the former, is not so liable as that of the latter to annual obstructions from ice. In shipping tonnage Baltimore is the third city in rank in the the United States. The buildings, public and private, vie in elegance with those of Phil., New York, or Boston. There are within the city between thirty and forty places of public worship; an exchange, several splendid hotels, and two monuments. The battle monument, in N. Calvert street, is a chaste obelisk. Washington's monument rising in the intersection between N. Charles and Monument streets, is by far the most magnificent edifice in the U. S. of that class. It is surmounted by a colossal statue of the Father of his country. This imposing figure is visible from the surrounding country, at an immense number of different points; and is an honor to the city.

There are in Baltimore 10 banks, 4 market houses, a prison, the state penitentiary, a museum, library, three theatres, a great number of private schools, and two colleges.

Of the edifices and institutions of Baltimore those of the most decided utility, are those dedicated to education, and of these, the most remarkable are Baltimore, and St. Mary's colleges. In 1807, a medical college was founded, but subsequently connected with the university of Maryland. All those institutions are in activity. St. Mary's college belongs to the Roman Catholics, and is in a flourishing state. The following notices of this school may be regarded as official, as they are extracts from an address to the public, dated the 3d March, 1831, and undersigned by Samuel Eccleston, the president.

"In the month of January, 1805, this institution was raised to the rank of 'University of Maryland,' and vested with power to hold public commencements and grant degrees."

"St. Mary's college enjoys the advantage of a most healthful and pleasant situation, in the most northwestern part of the city of Baltimore. The buildings are sufficient for the accommodation of one hundred and fifty boarders, and afford the facility of appropriating a separate room to each class of the various literary departments."

"The system of instruction embraces the various arts and sciences usually taught in the most extensive colleges. Latin, Greek, and the Mathematics, are considered as the ground work of general scholarship."

From the residue of the address, too lengthy for insertion in this Gazetteer, it appears, that the means of a complete classical and liberal education, are all embraced in this College. Boarding is at $140, for full and $70 for half boarders, annually. Tuition per annum, $60, paid half yearly in advance. Day scholars $15 per quarter; and entrance fee $5.

The manners of the people of Baltimore are those of business and industry. Habits of mere pleasure or amusement, have gained but little force. In literary acquirement the people of this city are perhaps in the rear of some others of the large commercial capitals of the U. S., but certainly in advance of their modest claims. In solid prosperity, Baltimore has probably no second in the United States. The advance of this emporium is best seen by a single glance on the following table of progressive population.

In 1790	13,503
1800	26,514
1810	35,583
1820	62,738
1830	80,625

This table shows an increase of nearly 600 per cent in 40 years, and if the ratio of increase in the decennial period from 1820, to 1830, has been preserved, this city now (April 1832) contains about 85,000.

BALTIMORE, Hundred, of Sussex co. Del. containing a population of 2176, by the census of 1830.

BALTIMORE, a small village in the middle and eastern part of Fauquier co. Va. about 50 m. a little s. of w. from W. C.

BANKSBRIDGE, and p-o. in the northern part of Richmond co. N. C. by p-r. 5 m. nthrd. from Rockingham, the co. seat, and 108 ms. s. w. from Raleigh.

BANGOR, p-t. and cap. Penobscot co. Me. w. side Penobscot r. 35 m. N. Castine, 66 N. E. Augusta, 52 from Owl's Head point, a cape of Penobscot bay, is the most flourishing town in the state. The place was a wilderness 30 years since. It stands at the head of navigation, and is easily accessible except in the winter. The Maine charity school, a theological seminary, was established here in 1815, and prepares young men for the desk. The course of study embraces 4 years. There is also a mechanic association. Bangor is destined to be the centre of business within the state, for more than half of Me., and it will command the trade of 9000 sq. ms. or more than ½ without competition. The water power in the vicinity is said to be superior to that of any other town in the U. S. Great expectations are entertained of the growth of this place. The increase of Pop. between 1820 and 1830, was about 130 per cent. and the surrounding regions have been rapidly peopled from different parts of the country, though the greater part of the Penobscot valley is still unoccupied. When the lumber shall have been removed, the soil will be arable and valuable. The following recapitulation, the exports of one year, will show the importance of this branch of business to the place.

Boards, plank & joists, surveyed 23,473,180 ft.
Do. shipped without survey, 3,354,000
26,827,180

This item, at $8,14 the thousand, amounts to $218,471; besides which 4338 tons of timber, at $2,75, 99,671 feet ranging timber at $2,50, shingles, clapboards and laths to the value of $96,000, and staves, oars, and other small lumber to the value of $7,000 more, make a total sum of $335,891.

The transportation employs many vessels, their own, and from other districts also. Building goes on fast, but rents are high. Above 3,000,000 bricks were made in 1831. A bridge crosses the Kenduskeag. 4 churches have been built in 3 years, and a jail and court h. are intended to be built this season. Navigation in the summer, and sleighing in the winter, give great activity to the place. Pop. 1830, 2,867.

Bangor, p-t. Franklin co. N. Y. 210 ms. from Albany, 6 ms. by 48, is but little settled, has a sandy loam, bearing beech, maples, bass, elm, cherry and cedar. Little Salmon r. is the principal stream. Pop. 1830, 1,076.

Banister, r. of Va. rising by numerous branches in Pittsylvania co. flows 25 ms. in a northeasterly direction, enters Halifax co. and inflects to the s. e. about 30 ms. and falls into Dan r. about 10 ms. above the junction of the latter, with Roanoke. Banister drains most part of the peninsula between Dan and Roanoke rs. below the e. boundary of Henry and Franklin cos.

Banister, named in the post office list, Halifax C. H. p-v. and st. jus. Halifax co. Va. situated on Banister r. by p-r. 130 ms. s. w. by w. from Richmond, and 220 s. w. from W. C. n. lat. 36° 44', long. 1° 58', w. from W. C.

Barbersville, p-o. Jefferson co. Ind. by p-r. 95 ms. s. e. from Indianopolis.

Barboursville, p-v. Orange co. Va. at the northwestern foot of south mtn. by p-r. 113 ms. s. w. from W. C. and 88 n. w. from Richmond.

Barboursville, p-v. and st. jus. Cabell, co. Va. situated on the right bank of Great Guyandot, 5 ms. above its entrance into the Ohio r. by p-r. 355 ms. s. w. by w. from W. C. and 344 ms. a little n. of w. from Richmond, n. lat. 38° 24', long. 5° 12', w. from Washington City.

Barboursville, p-v and st. jus. Knox co. Ky. situated on the right bank of Cumberland r. by p-r. 122 ms. s e. from Frankfort, and 533 ms. s. w. by w. from W C n. lat. 36° 55', long. 6° 47' w. from W. C. Pop. 1830, 138.

Barges, p-o. Wilcox co. Ala. by p-r. 81 ms. sthrd. from Tuscaloosa.

Baring, p-t. Washington co. Me. 209 ms. e. from Augusta.

Bark Camp, p-o. Burke co. Geo. by p-r. 67ms. estrd. from Milledgeville.

Barkhamsted, p-t. Litchfield co. Conn. 23 ms. n. w. Hartford, contains about 32 sq. ms. 5 ms. by 6 and a half, and is rough and stony, divided n. and s. by two high granite ridges which run far n. A little iron ore and free stone are found. The soil is hard and dry and generally not good for tillage, except along the streams. The grazing is good, and considerable butter and cheese are sent to market, as well as cattle and sheep. Oak, chestnut, sugar, maple, beech, pine, &c. were formerly abundant, but have been destroyed partly by fire and wind. Still r. and another main branch of Farmington r. supply mill seats. First settled, 1744. Pop. 1830, 1715.

Barksdale, p-v. in the western part of Halifax co. Va. by p-r. 246 ms. s. w. from W. C. and 156 s. w. by w. from Richmond.

Barnard, p-t. Windsor co. Vt. 21 ms. n. w. Windsor, 37 s. Montpelier, lies between Queechy and White rs. and has a pond of 100 acres, near the centre. Locust cr. and other streams afford mill seats. There is a small v. in the centre; and in the e. part, a bog of marl. First settled, 1744. Pop. 1830, 1881.

Barnard's Station, p-o. Buncombe co. N. C. about 200 ms. wstrd. from Raleigh.

Barnegat, v. Poughkeepsie, Duchess co. N. Y. is remarkable for the large quantities of limestone which is burnt and transported to different parts of the U. S.

Barnegat, bay and inlet, Monmouth co. N. J. n. lat. 39° 47'; long. 74° 13' w. The bay is about 20 ms. long, separated from the ocean by a beach, and united with it by the Inlet.

Barnes's Store, and p-o. Pike co. Geo. by p-r. 33 ms. w. from Milledgeville.

Barnestown or Barnesville, p-v. in the northwestern part of Montgomery co. Md. on the road from W. C. to New Market, 15 ms. s. e. by e. from Frederick, and by p-r. 40 ms. n. n. w. from W. C.

Barnesville, p-v. in the western part of Belmont co. O. 30 ms. s. w. by w. from Wheeling in Ohio co. Va. and by p-r. 297 ms. n. w. by w. a little w. from W. C. and 128 ms. e. from Columbus. Pop. 1830, 408.

Barnet, p-t. Caledonia co. Vt. w. Connecticut r. opposite Lyman, N. H. 35 ms. e. Montpelier, 65 n. Windsor, contains 40 sq. ms. has Passumpsic and Stevens' rs. which supply mills, particularly Stevens' mills, where the fall is 100 feet in 10 rods. The 15 m. falls of Connecticut r. are in this town, and below them are 21 islands, one of which contains 90 acres. The soil is generally rich, and good for grazing and tillage. There is much slate, and some iron ore. Boats come up Connecticut r. to this place. There are 3 ponds, 1 of 100 acres. Pop. 1830, 1,764.

Barnett's Mills and p-o. Fauquier co. Va. by p-r. 56 ms. wstrd. from W. C.

Barneysville, p-v. Bristol co. Mass. 43 ms. from Boston.

Barns' Mills and p-o. western part of Monongalia co. Va. by p-r. 233 ms. n. w. by w. a little w. from W. C.

Barnstable co. Mass. is the easternmost land in the state, and comprehends the peninsula of Cape Cod. It is bounded by Massachusetts bay n. Buzzard's bay s. w. the ocean e. and s. e. and Plymouth co. w. connected by a narrow neck, and is almost entirely sandy and barren. The inhabitants are almost all fishermen. Barnstable is the chief t. Great quantities of salt are made here from sea water, by solar evaporation; the water being pumped by wind into large vats exposed to the sun's heat, and covered with moveable roofs from dew and rain. Some parts of this county are like an Arabian desert. Pop. 1830, 28,-514.

Barnstable, p-t. s-p. and cap. Barnstable co. Ms. 66 ms. s. e. Boston. The township extends across Cape Cod, and is 9 ms. by 5, with a better soil than most of the Cape, chiefly producing oak and yellow pine. The tide rises from 8 to 14 ft. and supplies many salt vats. The town is on a declivity, s. Barnstable bay, with considerable shipping, and at the mouth of the port is a bar with 6 or 7 ft. water at low tide. Incorporated 1639. Pop. 1830, 3,974.

Barnstead, p-t. Strafford co. N. H. 36 ms. n. w. Portsmouth, 26 w. by n. Dover, 20 e. Concord; has a swelling surface, good soil, bearing pine, oak, beech, maple, &c. Suncook, Brindle, and Halfmoon ponds are stocked with fish, and enter into Suncook r. First settled, 1767. It has a social library. Pop. 1830, 2,047.

Barnwell, one of the southwestern districts of S. C. bounded by Edgefield dist. n. w. South Edisto r. separating it from Orangeburg n. and n. e., Colleton e., Beaufort s. e., and Savannah r. separating it from Scriven co. of Geo. s. w. and from Burke co. w. Greatest length by a line along the general course of South Edisto, 60 ms. mean breadth 28, and area 1,680 sq. ms. Extending in lat. from 32° 50′, to 33° 37′ n. and in long. from 3° 48′ to 4° 50′ w. Both the two rivers which bound this district on the southwest and northeastern sides, as well as the two Salkehatchies which rise within it, flow to the s. e. the general slope of the district is therefore in that direction. Soil in general of middling quality. Chief staple, cotton. Chief t. Barnwell. Pop. of the dist. 1820, 14,750, and in 1830, 19,236.

Barnwell, court house, and p-v. and st. jus. Barnwell dist. S. C. situated on Great Salkehatchie r. near the centre of the dist. 90 ms. n. w. by w. from Charleston, and by p-r. 62 ms. s. s. w. from Columbia, and 562 ms. s. s. w. from W. C. n. lat. 33° 13′, and long. 4° 20′ w. from W. C.

Barre, p-t. Washington co. Vt. 50 ms. n. w. Windsor, contains 31 sq. ms. First settled 1788. The soil is a dry, warm loam, without stones, with an uneven surface, and the people are good farmers. Cobble and Millstone hills in the s. e. yield much granite for building and mill stones. The latter are sent to different parts of the U. S. Spanish brown and alum are found. Stevens' and Jail branch, uniting and joining with Onion r. give excellent mill seats. There are 2 villages, and 14 school districts. Pop. 1830, 2,012.

Barre, p-t. Worcester co. Mass. 66 ms. w. from Boston, named after Col. Barre, a distinguished friend of America in the British parliament, is one of the first agricultural towns in the co. Ware r. passes through it, and there are here several of its branches. The ground is high, dividing the head waters of Ware, Blackstone, and Nashua rs.

Barre, p-t. Orleans co. N. Y. 13 ms. n. Batavia, 5 Montpellier, is on the mountain Ridge, and is watered by small branches of Oak, Orchard and Sandy creeks. Pop. 1830, 2,503.

Barren, Big and Little, the two southern confluents of Green r. Ky. Big Barren r. has its numerous sources in Jackson, Smith and Sumner cos. Ten. Flowing thence in a n. n. w. direction, it enters Ky. receiving confluents from Barren, Monroe, Allen, Simpson and Warren cos. they unite in one channel, in the latter, near Bowling Green. Assuming thence a n. w. course, Big Barren, now a navigable stream, joins Green r. on the border between Warren and Butler cos. The valley of Big Barren is about 70 ms. in length, by a mean width of 30; area 2,100 sq. ms. Little Barren, a much inferior stream to the preceding, rises in the eastern part of Barren, and western of Adair counties, and flowing in a n. n. w. direction, first separates Barren from from Green, and thence Green from Hart cos. finally joining Green r. near Sinking Spring, after a comparative course of about 30 ms.

Barren, one of the southern counties of Ky. bounded s. by Monroe, w. by Warren, n. by Hart, n. e. by Green, and e. by Adair. Length from e. to w. 30 ms. mean breadth 18, and area 540 sq. ms. extending in lat. from 36° 52′ to 37° 11′ n. and in long. from 8° 30′ to 9° 02′ w. from W. C. The eastern part of this co. is drained by Little Barren, but the greater part of the surface is drained by the sources of Big Barren. Surface rather level and soil far more fertile than the ill-chosen name would induce the reader to suppose. Chief town, Glasgow. Pop. in 1820, 10,328, and in 1830, 15,079.

Barren Creek Springs, p-o. in the extreme northern part of Somerset co. Md. about 80 ms. s. e. from Baltimore, and 80 s. e. by e. from W. C.

Barren Hill, p-o. Montgomery co. Pa. by p-r. 93 ms. n. e. from W. C.

Barrington, p-t. Strafford co. N. H. 20 ms. n. w. Portsmouth, 30 e. Concord, 65 n. Boston, has a surface somewhat broken and rocky, with much gravelly loam. The oak ridges have a sandy loam, good for tillage. There are 13 ponds, whose streams furnish mill seats. The rocks are granite, &c. and there are bog iron ore, crystals, &c. The Devil's Den is a curious cavern. First settled, 1732, Pop. 1,895.

Barrington, p-t. Bristol co. R. I. 7 ms. s. e. Providence, s. Massachusetts, n. e. Narraganset r. is about 3 ms. by 3, with a surface nearly level, and a light sandy soil, good for grain. Sea weed is used as manure. A bridge crosses to Warren. There is plenty of fish, and some salt is made. Pop. 1830, 612.

Barrington, Great, p-t. Berkshire co. Mass. (See Great Barrington.)

Barrington, p-t. Steuben co. N. Y. e.

Crooked lake, 21 ms. N. E. Bath, 219 w. Albany, 6 ms. by 5, has a good soil, nearly level, with chestnut, oak, walnut, and some pine. Pop. 1830, 1,854.

BARRON'S, p-o. Perry co. Ala. 52 ms. S. E. from Tuscaloosa.

BARRYVILLE, p-v. Sullivan co. N. Y.

BARRYVILLE, p-v. in the northern part of Stark co. O. by p-r. 127 ms. N. E. from Columbus, and 322 ms. N. W. by W. from W. C.

BART, tsp. and p-o. eastern part of Lancaster co. Pa. by p-r. 54 ms. southeastward from Harrisburg. Pop. 1820, 1,423, and in 1830, 1,470.

BARTHOLOMEW co. of Ind. bounded by Johnson N. W., Shelby N. E., Decatur E., Jennings S. E., Jackson S., and on the W. boundary not known. Length 24 ms. mean width 20, and area 480 sq. ms. Extending in lat. from 39° 03′ to 39° 21′ N. and in long. from 8° 38′ to 9° 08′ w. from W. C. The slope of this co. is nearly southward. In it unite Clifty, Flat Rock, and Blue rs., confluents of Driftwood, fork of White r.

BARTLETT, p-t. Coos co. N. H. 45 ms. S. W. Lancaster, 75 N. by E. Concord, 85 N. by W. Portsmouth, lies at the S. base of the White mtns. and contains about 13,000 acres of rough land, but little cultivated, with some good, on Saco r. Named after Gov. B. Pop. 644.

BARTON r. Orleans co. Vt. rises near the source of Lamoille r. and runs N. to lake Memphremagog, watering about 160 sq. ms.

BARTON, p-t. Orleans co. Vt. E. Montpelier, is crossed by Barton r. and has a part of Willoughby's r. a branch of it, with several ponds. Bellwater pond, a source of Barton r. has an outlet with mills, and a village. First settled, about 1796, from N. H. and R. I. Pop. 1830, 729.

BARTON, p-t. Tioga co. N. Y.

BARTON'S p-o. Lauderdale co. Ala. by p-r. 164 ms. northwardly from Tuscaloosa.

BASCOBEL, p-v. Jackson co. Geo. by p-r. 85 ms. a little W. of N. from Milledgeville.

BASON HARBOR, p-v. Ferrisburg t. Addison co. Vt. is one of the best harbors on Lake Champlain.

BASKING RIDGE, p-v. Bernard, Somerset co. N. J. near the head waters of Passaic r. 7 ms. S. W. Morristown, is situated on elevated land, and is memorable for the capture of Gen. Lee by the British in the revolutionary war. Part of the house is still standing near the stage road. Stage coaches go daily for New York by 2 routes, except in winter.

BATAVIA, p-t. and cap. Genesee co. N. Y. 256 ms. W. Albany, 20 S. Erie canal, 36 E. Buffalo, is crossed by Tonawanta creek, and has small head streams of Oak, Orchard and Black creeks. It has level land, good for farms, and is on the road from Albany to Buffalo. The village is very neat. Here are the county buildings, and the house of the agent of the Holland land company. Pop. 4264.

BATAVIA, p-v. and st. jus. Clermont co. O. situated on the E. branch of Little Miami r. 24 ms. a little S. of E. from Cincinnati, and by p-r. 109 ms. S. W. from Columbus, and 496 W. from W. C. N. lat. 39° 2′, long. 7° W. from W. C.

BATESVILLE, p-v. in the eastern part of Guernsey co. O. by p-r. 119 ms. E. from Columbus.

BATESVILLE, p-v. and st. jus. Independence co. Ark. situated on the left bank of White r. 87 ms. N. N. E. from Little Rock, and about 250 ms. S. S. W. from St. Louis. By p-r. 1044 ms. S. W. by W. from W. C. N. lat. 35° 44′, long. 14° 18′ W. from W. C.

BATH, p-t. s-p. Lincoln co. Me. W. Kennebec r. 37 ms. Augusta 15 ms., ocean 14 ms. S. W. from Wiscasset, 34 N. E. Portland, 153 N. E. Boston, is the highest place to which the r. is navigable in winter. It stands on an acclivity and is almost isolated by some of the numerous arms of the sea which penetrate that part of the coast. Pop. in 1830, 3773.

BATH, p-t. Grafton co. N. H. E. Conn. r. 42 ms. N. E. Dartmouth college, 81 N. N. W. Concord, 148 ms. N. N. W. Boston, has some rich land, and is sheltered E. by the White mtns. and W. by the Green mtns. It has many good mill seats on Ammonoosuc r. and a bridge across it 350 feet. There is a majestic fall in Conn. r. near this place. Perch pond, 100 acres, is also here. Gardner's mtn. about 500 feet high, which divides the t. consists of granite, slate, &c. and contains alum, pyrites, some iron and silver, and a stone which dissolves in warm water. The hills have a red loam, or marl, with maple, beech, birch, oak, &c. The vallies alluvial, with white pine, hemlock, spruce, elm, &c. Pop. in 1830, 1623.

BATH, p-t. and cap. Steuben co. N. Y. 240 ms. W. Albany, 41 S. S. W. Geneva, 41 E. Angelica; is crossed by Conhocton r. which winds through it for 30 ms. and its streams furnish mill seats. The soil is various and irregular. The village is on the E. side of Conhocton r. which is 75 feet wide. Pop. 1830, 3387.

BATH, p-v. Northampton co. Pa. 10 ms. N. N. W. from Easton, and 8 a little W. of N. from Bethlehem.

BATH, one of the central cos. of Va. bounded by Alleghany co. S. and S. W. by Alleghany mtn. separating it from Greenbrier on the W. and Pocahontas on the N. W. by Pendleton N. E. by the Great North mtn. separating it from Augusta on the E. and Rockbridge on the S. E. Length from N. E. to S. W. 36 ms. Mean breadth 21, and area 756 sq. ms. extending in lat. from 37° 50′ to 38° 25′ N. and in long. from 2° 18′ to 3° W. from W. C. Bath occupies part of a high mountain valley, with a slope to the southwestward. The extreme northwestern constituents of James r. Cowpasture and Jackson's rs. have their sources along the southwestern border of Pendleton, but flow into and traverse Bath in a southwestern direction. North r. another branch of James r. issues from the eastern part of this co. and traversing Rockbridge, joins its recipient at the western foot of the Blue Ridge. The mean elevation of the cultivated surface of Bath, exceeds 1200 feet, or an equivalent to 3° of lat. The face of the country is very broken and in part mountainous; and with the latitude and elevation produces winters severe and lengthened. The cultivated vegetables also evince a climate of much lower mean temperature than on like latitudes near the Atlantic coast. Soil in general sterile. Chief town, Warm Springs. Pop. 1820, 5237, and in 1830, 4002.

Bath, court house, or Warm Springs, p-v. Bath co. Va. by p-r. 170 ms. N. N. W. from Richmond, and 226 s. w. from W. C.

Bath, p-v. in the eastern part of Beaufort co. N. C. situated on a small bay near the north shore of Pamlico r. 16 ms. S. E. by E. from the port of Washington, and by p-r. 138 ms. a little s. of E. from Raleigh.

Bath, p-v. in the northeastern part of Medina co. O. by p-r. N. E. from Columbus, and 344 ms. N. W. by W. from W. C.

Bath, co. of Ky. bounded by Morgan S. E. Montgomery S. W. and W. Nicholas N. W. and Licking r. separating it from Fleming N. and N. E. Length from S. E. to N. W. 25 ms. mean breadth 10, and area 256 sq. ms. Extending in lat. from 38° to 38° 17′ N. and in long. from 6° 30′ to 6° 50′ W. from W. C. The slope of this co. is to the northwestward towards Licking r. Chief towns, Owingsburg, and Sharpsburg. Pop. 1820, 7961, and in 1830, 8799.

Bath Iron Works, and p-o. in the southern part of Rockbridge co. Va. by p-r. 167 ms. W. from Richmond, and 209 s. w. by w. from W. C.

Baton Rouge, p-v. in the northern part of Chester dist. S. C. by p-r. 66 ms. N. from Columbia.

Baton Rouge, East, parish of La. bounded by the parish of Iberville S. E. the Mississippi r. separating it from the parish of W. Baton Rouge on the S. and W. and from that of Point Coupee on the N. W., E. Feleciana N. and the Amite r. separating it from St. Helena E. Length from S. to N. 23 ms. mean width 15, and area 345 sq. ms. extending in lat. from 30° 18′ to 30° 37′ N. and in long. 14° to 14° 22′ W. from W. C. This parish contains the first land which rises above the level of the delta, ascending the Mississippi. Immediately above the efflux of the Iberville outlet, the surface begins to swell into eminences or hills of humble elevation. The soil of the parish, in a state of nature, was covered with a very dense forest. The slope either way is but slight, that eastward towards Amite r. is widest. Chief town, Baton Rouge. Pop. 1820, 5220, and in 1830, 6698.

Baton Rouge, (Red Stick,) p-v. and st. jus. for East Baton Rouge, parish of La. is situated on the left bank of the Mississippi r. at N. lat. 30° 31′, and long. 11° 18′ W. from W. C. 30 ms. above Donaldsonville, and 117 above New Orleans, and by p-r. 1237 ms. S. W. by W. from W. C. This town was founded on the lower bluff or high land which reaches the bank of the Mississippi, and extends mostly in one street along the margin of the hill. The site of Baton Rouge is, however, only a hill comparatively, as it does not rise above high water mark more than 25 feet; but contrasted with the uniform plain along the margin of the river, the apparent elevation of Baton Rouge is real and pleasing deception in vision. Pop. about 1000.

Baton Rouge, West, parish of La. bounded by the parish of Iberville S. E. the Atchafalaya r. separating it from the parish St. Martin's S. S. W. and W. and St. Landre' or Opelousas N. W. parish of Point Coupee N. and the Mississippi r. separating it from East Baton Rouge on the E. Length from E. to W. 35 ms. mean breadth 20, and area 700 sq. ms. Extending in lat. from 30° 12′ to 30° 32′ N. and in long. from 14° 15′ to 14° 55′ W. from W. C. The surface being a part of the delta of the Mississippi, is an almost undeviating plain, with a very slight southern declivity, and the far greater part liable to annual submersion. The arable part lies along the streams, and the soil being invariably fertile, every spot which can be brought under the plough is highly productive. Cotton is the common staple; sugar has been attempted, but the situation is a little too far N. for the sugar cane. In the post-office list of 1831, no post-office is named in this parish; nor does it contain a village worthy of notice. The settlements border the streams, and are principally upon the Mississippi. Pop. 1820, 2335, and in 1830, 3084.

Battahatcha, r. of Ala. and Miss. rising in Marion co. of the former, having interlocking sources with Bear creek, branch of Tennessee r. and with those of the northwestern sources of Black Warrior r. and flowing thence southwestwardly enters Monroe co. Miss. falling into the Tombigbee after a comparative course of 70 ms.

Battenkill, r. rising in Vt. and crossing Washington co. N. Y. joining Hudson r. half a mile above Fish creek, is a good mill stream, 50 ms. long.

Battletown, p-v. Frederick co. Va. situated on Opequhar creek, 6 ms. S. E. from Winchester, and by p-r. 68 ms. a little N. of W. from W. C.

Battle Creek, and p-o. southwestern angle of Marion co. Ten. by p-r. 107 ms. southeastward from Nashville.

Bayou Chicot, p-o. in the northwestern part of the settlement of Opelousas or parish of St. Landre', La. about 30 ms. N. W. from the village of St. Landre', and by p-r. 1298 ms. S. W. by W. from W. C.

Bay Settlement, and p-o. along the Erie shore or eastern part of Monroe co. Mich. The settlement extends northwestward from Maumee Bay, and by p-r. the p-o. is 46 ms. S. W. from Detroit.

Bazetta, p-v. in the western part of Trumbull co. O. 7 ms. N. W. from Warren, and by p-r. 162 ms. N. E. by E. from Columbus, and 302 N. W. by W. from W. C.

BEAL'S ISLAND, Washington co. Maine. Pop. 55.

BEALLSVILLE, p-v. on the U. S. road, south-eastern part of Washington co. Pa. 14 ms. from Washington, the co. seat, by p-r. 218 ms. N. W. by W. from W. C. and 300 ms. wstrd. from Phil.

BEALLSVILLE, p-v. in the western part of Monroe co. O. 10 ms. wstrd. from Woodsfield, and by p-r. 140 ms. a little S. of E. from Columbus, and 294 ms. wstrd. a little N. from W. C.

BEAN'S STATION, p-o. Granger co. Ten. by p-r. 204 ms. estrd. from Nashville.

BEAR CREEK, a small confluent of Tennessee r. rising in Marion and Franklin cos. Ala. flows thence westward into the state of Miss. in which it inflects to the nthrd. and falls into Tennesse r. after a comparative course of 70 ms. This small r. gains importance from forming the boundary on and near Tennessee r. between the states of Alabama and Mississippi.

BEARD'S FERRY, and p-o. western part of Morgan co. Il. 7 ms. wstrd. from Jacksonville the co. seat, and by p-r. 140 ms. N. W. by W. Vandalia.

BEARD'S Store and p-o. Anson co. N. C. by p-r. 102 ms. S. W. from Raleigh.

BEATIE'S BLUFF and p-o. southern part of Madison co. Mo. by p-r. 50 ms. S. from St. Louis, and 1085 S. W. by W. from W. C.

BEATIE'S FORD, and p-o. Lincoln co. N. C. by p-r. 151 ms. estrd. from Raleigh.

BEATTY'S BRIDGE, and p-o. New Hanover co. N. C. by p-r. 114 ms. southeastward from Raleigh.

BEAUCOUP, p-v. Washington co. Il. No location given in p-o. list.

BEAUFORT, co. N. C. on both sides of Pamlico r. bounded by Pamlico Sound E., Craven co. S. and S. W., Pitt W. and N. W., Martin N. and Hyde N. E. Length 40 ms. mean breadth 17, and area 670 sq. ms. N. lat. 35° 30′ and the meridian of W. C. intersects in the western part of this co. The whole surface is a plain, and liable to periodical submersion. Staples, cotton, rice, &c. Chief towns, Washington, and Baths. Pop. 1820, 9900, and in 1830, 10969.

BEAUFORT, p-t., pt. ent. and st. jus. Beaufort dist. S. C. situated on the western bank of Port Royal, r. 14 ms. N. from Port Royal Entrance, by the land road 50 ms. N. E. from Savannah, 75 S. W. from Charleston, and by p-r. 148 ms. a little E. of S. from Columbia. N. lat. 32° 25, long. 3° 42′ W. from W. C. The harbor of Beaufort is spacious, and of more than sufficient depth to admit the entrance of any vessel which can pass the bar of Port Royal Entrance. Steam and other vessels of small draught have an inland passage by Beaufort into Coosan r. The whole of St. Helena parish in which this seaport is situated, contained in 1830, a population of 8788.

BEAUFORT, extreme southern dist. of S. C. bounded by the Atlantic ocean S. E., Savannah r. which separates it from Scriven, Effingham, and Chatham co. Geo. S. W., Barnwell dist. N. W., and Cambahee r. separating it from Colleton dist. N. E. Greatest length in a direction from south to north, and from the outerside of Turtle Island to the S. E. boundary of Barnwell dist. 61 ms.; mean breadth 30, and area 1830 sq. ms. extending in lat. from 32° 03′, to 34° and in long. 3° 30′ to 4° 30′ W. from W. C. The slope of this district is to the southeastward, but the declivity is very slight, being from the utmost extent inland a level plain, terminating towards the Atlantic ocean in numerous interlocking streams enclosing, with many others of lesser note, Hilton Head, Turtle, Hunting, Reynold's, St. Helena, and Port Royal Islands.

Though no entrance into Beaufort is of great depth, it is remarkably well supplied with commercial inlets for vessels of moderate draught, of which the principal are Savannah r. Callibogue Sound, Port Royal Entrance, and St. Helena Sound or the outlet of Cambahee river. Commercially, Savannah in Chatham co. Geo. is a port of Beaufort, admitting vessels of 16 feet draught. Vessels drawing 14 feet are navigated to the port of Beaufort. Beside the two rs. which bound this dist. on the S. W. and N. E. sides, the central parts are drained by the Coosawhatchie river, which, rising in Barnwell and entering Beaufort, flows southeastward 35 ms. to where it divides into two channels, which a few ms. lower, again separate. The northern channel under the name of Coosaw river flows estrd. into St. Helena Sound, whilst that of the south maintains the original direction to the southeastward, gradually widens into Broad r. and finally contributes to form Port Royal Entrance. The position of this dist. and its moderate elevation give it a climate which admits the culture of sugar, rice, cotton, and in some places the orange tree. Chief town Beaufort, though the seat of justice is at Coosawhatchie. Pop. 1820, 32,199, and in 1830, 37,032.

BEAUFORT, s-p. and p-t. Carteret co. N. C. situated on the mainland at the head of Onslow Bay, and opposite Old Topsail Inlet, about 11 ms. N. W. from Cape Lookout, and by p-r. 164 ms. S. E. from Raleigh, N. lat. 34° 47′, long. 0° 18′ E. from W. C. This is one of the best ports of N. C. admittting vessels of 14 feet draught, and affording complete shelter. It is well situated to become a mart of considerable consequence. A canal has been proposed across the intermediate peninsula to unite Onslow Bay with Neuse r. and if such a work was executed Beaufort must become the southern of its depots. In the Census lists of 1830, the population of this place not given separate from Carteret co.

BEAVER, Big r. of Pa. and O. is composed of two branches, the Mahoning and Shenango. The Mahoning rises by numerous branches in Columbiana, Portage and Trumbull cos. O. which, uniting in the latter co. bends from N. E. to S. E. and after an entire comparative course of 60 ms. enters Beaver co. Pa. in which it receives the Shenango from the N. The latter rising in Crawford and Mercer cos. Pa. flows by a general course of sixty ms. southwardly, to its junction with the Mahoning. Below the union of its main constituents, Beaver flows a little E. of S. 20 ms. receiving from the estrd. in the intermediate distance, the Conequenessing, and falls into the Ohio at the t. of Beaver, after an entire comparative course of 80 ms. This stream rises on a table land elevated from 800 to 1200 feet, and is precipitated over a fall of about 44 feet entire descent, about 1½ ms. above its entrance into Ohio. The elevation of the farms near the mouth, is about 700 feet above the Atlantic tides. The valley of this river lies between lat 40° 44′ and 41° 45′, and comprises an area of 3850 sq. ms. The channel from the mouth to Warren, in Trumbull co. O. forms a part of the route along which a canal has been projected to unite the Ohio r. to lake Erie, or to unite with the Ohio canal.

The valley of Beaver has that of the Alleghany r. to the N. E.; that of Cayahoga to the N. W. and that of the Tuscarawas branch of Muskingum to the S. W. and W. To the N. it has interlocking sources with those of Conneaut, Ashtabula and Grand rs. flowing into lake Erie.

Beaver, co. of Pa. bounded by Mercer co. N., Butler E., Alleghany S. E., Washington S., Ohio co. Va. S. W., and Columbiana co. O. N. W. Greatest length along the state of Ohio 36 ms. mean breadth 18, and area 648 sq. ms. Extending in lat. from 40° 30′ to 41° 02′ N. and in long. from 3° 06′ to 3° 30′ W. from W. C. The O. r. enters the southeastern border and flowing thence N. N. W. 10 ms. to the influx of Big Beaver, inflects thence to S. W. by W. leaving the co. below the mouth of Little Beaver, and dividing it into two unequal sections. The smaller section comprising about one third of the entire surface lies south of Ohio r. and slopes northwardly towards that stream. The northern section comprised in the vallies of Big and Little Beaver, has a southern declivity towards Ohio r. The surface of the whole co. on both sides of Ohio r. is very hilly, but the soil almost uniformly productive. Chief town, borough of Beaver. Pop. 1820, 13,340, and in 1830, 24,183.

Beaver, borough, p-t. and st. jus. Beaver co. Pa. situated on the right bank of Ohio r. and on the point below the mouth of Big Beaver 30 ms. by water below Pittsburg; 43 a little W. of N. from Washington Pa. 35 ms. N. E. from Steubenville O. and by p-r. 251 ms. N. W. by W. from W. C.; and 229 a little N. of W. from Harrisburg. Lat. 40° 44′ N. long. 3° 23′ W. from W. C. Pop. of tsp. 1820, 351; of the borough alone in 1830, 914.

Beaver, cr. N. Y. a mill stream, rises in Ulster and Sullivan cos. and runs through Delaware co. into a branch of Delaware r.

Beaver Creek, p-o. Anderson co. Ky. about 15 ms. S. W. from Frankfort.

Beaver Dam, p-v. in the central part of Erie co. Pa. 7 or 8 ms. southwardly from the borough of Erie, and by p-r. 325 ms. N. W. from W. C. Pop. of the tsp. 1820, 142, in 1830, 443.

Beaver Dam, p-o. Goochland co. Va. near the northern bank of James r. by p-r. 25 ms. above Richmond, and 139 S. S. W. from W. C.

Beaver Dam, p-v. in the eastern part of Rutherford co. N. C. 19 ms. estrd. from Rutherfordton, and by p-r. 204 ms. a little S. of W. from Raleigh.

Beaver Meadows, post office, in the northern part of Northampton co. Pa. 21 ms. N. from Easton, and by p-r. 211 ms. N. E. from W. C.

Beavertown, p-v. in the southern part of Union co. Pa. by p-r. 59 ms. a little W. of N. from Harrisburg. Pop. of the tsp. in 1830, 2280.

Becket, p-t. Berkshire co. Mass. 17 ms. S. E. Lenox, 110 W. Boston, is on high land of the Green mtn. range, and contains 4 ponds which give rise to branches of Westfield, Farmington and Housatonic rs. The inhabitants are chiefly farmers. Pop. 1830, 1063.

Beckhamsville, p-v. Chester dist. S. C. situated on Catawba river, by p-r. 56 ms. N. from Columbia, and 479 S. S. W. from W. C.

Beddington, t. Washington co. Me. 35 ms. N. W. Machias, is the fourth town from the coast, and is crossed by the upper parts of Pleasant, and Narraguasus rs. It adjoins Hancock co.

Bedford, p-t. Hillsboro' co. N. H. W. side Merrimack r. 8 ms. N. W. Amherst, 18 S. Concord, 52 N. W. Boston, contains 20,660 acres. Merrimack and Piscataquoy rs. water the town. Piscataquoy v. is in the N. E. part. The W. part is uneven and stony, but warm. The E. is a pine plain, and some intervals. White, red, and black oak, walnut, chesnut, birch, pine and hemlock are the trees. Much white oak has been sent by the r. and Middlesex canal, for ship timber. Hops have been raised here to a large amount. Cotton and woollen are manufactured. There is a brook which falls 200 feet. Iron ore, black lead, copper, pyrites, gneiss, &c. are found. The t. was granted by Mass. in 1733, to officers, soldiers, &c. of Philip's War. First settled, 1737. Pop. in 1830, 1563.

Bedford, p-t. Middlesex co. Mass. 16 ms. N. W. Boston, S. W. Concord r. Shawsheen r. a good mill stream, rises here. It is a small town, and half shire. Pop. 1830, 685.

Bedford, p-t. Westchester co. N. Y. 44 ms. N. New York, 130 S. Albany, S. E. Croton r. 6 ms. square. Several small streams supply many mill sites. It is elevated land, with various soils for farms, well watered, and producing much excellent rye. It belonged to Conn. till 1700. The village was burnt in the revolutionary war. Hon. John Jay lived here many years, and here died. Pop. 1830, 2,750.

BEDFORD co. Pa. bounded w. by Somerset, N. w. by Alleghany mtn. separating it from Cambria co., N. E. by Huntingdon co., E. by Franklin, S. E. by Washington co. Md., and S. and S. W. by Alleghany co. Md. Greatest length from the Maryland line to the northern angle, 52 ms., containing 1612 sq. ms. extending in lat. from 39° 42′ to 40° 30′, and in long. from 0° 53′ to 1° 44′ w. from W. C. The surface of this large co. is lined with chains of mtns. extending from S. S. W. to N. N. E. with very productive intervening valleys. In regard to the rivers, the central part is a table land, from which creeks flow southwardly into the Potomac r. The northern and larger section declines to N. N. E. and gives source to both the Raystown and Frankstown branches of Juniata. To travel over Bedford without giving other than a cursory glance, the country would appear a congeries of rocks and mountain ridges, yet the valley soil is generally good, and in many places very productive. Many of the mountain ridges have broad table lands of excellent quality on their summits.

There is no part of Bedford co. which does not exceed an elevation of 500 feet above the Atlantic level, and the farms along the eastern margin of the Alleghany mtn. to at least 1700 feet. The mean height of the arable land is from 1200 to 1300 feet, or an equivalent to at least three degrees of latitude. The winters are long and severe, but the highly diversified features of this region render it a delightful summer residence. The mineral springs near the borough of Bedford are much frequented. The northern extreme of Bedford is traversed or rather touched by one of the rail-roads of Pa. a link in the great chain of internal improvement, advancing to completion in that state. Staples of the county, grain, salted provision, live stock, &c. to which iron, and mineral coal of the bituminous species may be added. Chief towns, Bedford and Mc'Connellsburg. Pop. 1820, 20,248, and in 1830, 24,502.

BEDFORD, p-v., borough, and st. jus. Bedford co. Pa., situated on one of the higher branches of Juniata r. 105 ms. S. W. by W. from Harrisburg, 200 ms. almost due W. from Phil., and by p-r. 126 northwestward from W. C. According to Tanner's U. S. the borough stands exactly on N. lat. 40°, and very near on 1° 30′ w. from W. C. This village had its origin in a fort, and was laid out as a town in 1771. The situation is truly romantic. The main body of the village stands on a globular swell in a beautiful mountain valley. The scenery has great variety of feature; softness is blended with grandeur; there is nothing of sublime, but a pleasing boldness and richness strikes the eye on all sides. The village is, in summer, a place of fashionable resort by persons visiting the mineral waters in its vicinity. It does not, however, appear to gain rapidly in permanent Pop. In 1820 the inhabitants were 789, and only 869 by the census of 1830.

BEDFORD co. of Va. bounded E. and N. E. by Campbell co., by Roanoke r. separating it from Pittsylvania, S., Franklin, S. W., by Blue Ridge separating it from Botetourt W. and N. W., and by James r. separating it from Amherst, N. E. Length between James and Roanoke rs. 30 ms. mean width 22, and area 660 sq. ms. extending in lat. from 37° to 37° 32′ N., and in long. from 2° 10′ to 2° 50′ W. from W. C. The declivity of this co. is to the southeastward; the northern part containing, in the peaks of Otter, the highest elevation of the Appalachian system, 4,250 feet, unless the White mtns. of N. H. are included. From this elevated part of Blue Ridge, Bedford slopes with the course of the great bounding rs. and with that of the two small intermediate streams, Goose and Otter. Chief town, Liberty. Pop. 1820, 19,305, and in 1830, 20,246.

BEDFORD co. of Ten., bounded W. by Maury, N. W. by Williamson, N. by Rutherford, N. E. by Warren, S. E. by Franklin, and S. by Lincoln. Length from E. to W. 40 ms. mean width 25, and area 1,000 sq. ms. Extending in lat. from 35° 20′ to 35° 44′ N., and in long. from 9° 02′ to 9° 47′ W. It is nearly commensurate with the higher part of the valley of Duck r. and with the general course of that stream, slopes wstrd. Chief town Shelbyville. Pop. 1830, 30,396.

BEDFORD, p-v. in the northern part of Oldham co. Ky. 10 ms. N. E. from Westport, the st. jus. and by p-r. 53 ms. S. W. by W. Frankfort.

BEDFORD, p-v. and st. jus. Lawrence co. Ind., situated on the right bank of the east fork of White r. at and above the mouth of Salt cr., 60 ms. N. W. by W. from Louisville Ky., and by p-r. 73 ms. S. W. from Indianopolis, and 633 ms. wstrd. from W. C. N. lat. 38° 52′.

BEDFORDVILLE, p-v. Burke co. N. C. by p-r. 205 ms. W. from Raleigh.

BEDMINSTER, t. Somerset co. N. J. 20 ms. N. W. New Brunswick. Pop. 1830, 1,453.

BEECH GROVE, p-v. Luzerne co. Pa. by p-r. 86 ms. N. N. E. from Harrisburg.

BEECH GROVE, p-v. Bedford co. Ten. 48 ms. sthrd. from Nashville.

BEECH HILL, p-v. Jackson co. Ten. 47 ms. N. E. from Nashville.

BEECH PARK, p-o. Gallatin co. Ky. by p-r. 34 ms. northwardly from Frankfort.

BEEKMAN, p-t. Dutchess co. N. Y. 12 ms. E. Poughkeepsie. The Clove is a fine valley between W. mtn. and Oswego ridge 1 or 2 ms. by 6, with a warm gravelly loam, much improved by gypsum, and watered by Clove-kill, flowing through it and turning mills. Bushkill joins it, and they form Fishkill r. on which is Beekman v. Great Pond covers about 300 acres, and empties into the r. as does Sprout cr. Many human bones found here, probably were left in some Indian fight.

BEEKMANTOWN, p-t. Chiston co. N. Y. 6 ms. N. Plattsburg, 165 N. Albany, W. Lake Champlain, has fertile land in the E. part, and most of its population farmers. Population 2,391.

Beelan's Ferry and p-o. Perry co. Pa. by p-r. 31 ms. from Harrisburg.

Beeler's Station and p-o. in the southern part of Ohio co. Va. by p-r. 250 ms. n. w. by w. something w. of W. C. and 350 ms. n. w. by w. from Richmond.

Belair, p-v. Lancaster dist. S. C. by p-r. 81 ms. nthrd. from Columbus.

Belbrook, or Bellbrook, p-v. in the southwestern part of Greene co. O. 9 ms. s. w. from Xenia, the st. jus. for the co. and 40 n. n. e. from Cincinnati.

Belchertown, p-t. Hampshire co. Mass. 80 ms. w. from Boston, 12 e. Northampton, has hills of gentle declivity, with loam and gravel. The village is large, handsome and on high ground. Pop. in 1830, 2,491.

Belew's Creek and p-o. in the northeastern part of Stokes co. N. C. by p-r. 127 ms. n. w. by w. from Raleigh.

Belfast, p-t. s.p. capital of Waldo co. Me. 40 ms. e. Augusta, 9 w. Castine, from which it is separated by Penobscot r. which, although about 30 ms. from the sea, is very broad. A wide but short stream crosses the tsp. with several of its branches. It is 40 ms. n. e. Hallowell. The harbor has considerable advantages, and the coasting trade is great. Pop. 1830, 1,743.

Belfast, p-v. Alleghany co. N. Y. 262 ms. w. Albany. Pop. 743.

Belford, p-v. Nash co. N. C. by p-r. 50 ms. e. from Raleigh.

Belgrade, p-t. Kennebec co. Me. 10 ms. n. Augusta, is situated between 3 lakes, or large ponds. Pop. 1830, 1,375.

Belgrade Mills, p-v. Kennebec co. Me. 16 ms. n. from Augusta.

Bellborough, p-v. Williamson co. Ten. by p-r. 32 ms. southwardly from Nashville.

Bellefontaine, p-v. and st. jus. Logan co. O. 125 ms. n. n. e. from Cincinnati, and by p-r. 458 ms. n. w. by w. from W. C. and 62 ms. n. w. from Columbus. n. lat. 40° 21′ long. 6° 43′ w. from W. C. Pop. 1830, 266. This place stands on the table land between the valleys of Sciota and Great Miami rs. the water sources discharging from its vicinity like radii from a common centre.

Bellefonte, borough, t. p-v. and st. jus. Centre co. Pa. situated on Spring creek, a branch of Bald Eagle r. 45 ms. s. w. from Williamsport, 35 ms. n. n. e. from Huntingdon, and by p-r. 85 ms. n. w. from Harrisburg, and 192 n. n. w. from W. C. n. lat. 40° 55′. Pop. 1820, 433, and in 1830, 698.

Bellefonte, p-v. and st. jus. Jackson co. Ala. situated on Paint Rock r. 25 ms. a little n. of e. from Huntsville, and by p-r. 171 ms. n. e. from Tuscaloosa, n. lat. 34° 43′, long. 9° 20′ w. from W. C.

Belle Haven, p-o. in the southeastern part of Accomac co. Va. 20 ms. s. of Drummondstown, and by p-r. 194 ms. s. s. e. from Annapolis, and 226 s. e. from W. C.

Belle Isle, p-o. in the northern part of Matthews co. Va. by p-r. 98 ms. a little s. of e. from Richmond, and 182 ms. s. s. e. W. C.

Belle Valley, p-o. in the northeastern part of Rockbridge, co. Va. by p-r. 147 ms. w. from Richmond, and 188 ms. s. w. by w. from W. C.

Belle View, p-v. in the northern part of Calhoun, co. Il. by direct distance, about 100 ms. but by p-r. 146 ms. n. w. by w. a little w. of Vandalia, and 926 ms. w. from W. C.

Belleville, p-v. Essex co. N. J. 3 ms. n. Newark, w. Passaic r. is in a pleasant situation, and has a large printing, bleaching and dying factory, for cotton, woollen and silk. The principal building is of hewn stone, 263 ft. long and of 3 stories.

The bleaching and print works are now upon the largest scale, producing upwards of four thousand pieces weekly, of a variety of styles of goods.

Belle Ville, p-v. on the western border of Mifflin co. Pa. 20 ms. n. e. from Huntingdon, and by p-r. 77 ms. n. w. by w. from Harrisburg, and 169 ms. n. n. w. from W. C.

Belle Ville, p-v. on the left bank of Ohio r. at the mouth of Lee's creek 30 ms. below Marietta, and by p-r. 314 ms. westwardly from W. C.

Belle Ville, p-v. in the northern part of Conecuh co. Ala. 12 ms. northwardly from Sparta, the st. jus. and by p-r. 193 ms. a little e. of s. from Tuscaloosa, and 959 ms. s. w. by w. of W. C.

Belleville, p-v. in the eastern part of Roane co. Ten. 9 ms. estrd. from Kingston, and by p-r. 148 ms. a little s. of e. from Nashville, and 567 s. w. by w. from W. C.

Belleville, p-v. in the southern part of Richland co. O. 10 ms. s. from Mansfield the st. jus. and by p-r. 62 ms. n. n. e. from Columbus, and 368 ms. n. w. by w. from W. C.

Belleville, p-v. and st. jus. St. Clair co. Il. 15 ms. s. e. from St. Louis, Mo. and by p-r. 71 ms. s. w. from Vandalia, and 843 ms. westwardly from W. C.

Bellingham, p-t. Norfolk co. Mass. 26 ms. s. w. Boston, n. R. I., is a manufacturing place. The Bellingham cotton and woollen factory, incorp. 1814, cap. $15,000, and Boston do. 1813, $400,000, are moved by the water of Charles r. The soil is sandy. Pop. 1830, 1102.

Bellona, Arsenabard p-o. on the right bank of James r. 14 ms. above and wstrd. from Richmond, but in Chesterfield co. Va.

Bellows Falls, p-v. Westminster, Windham co. Vt. about 40 m. n. e. Bennington, 80 s. Montpelier, is in a pleasant and picturesque situation at the place where Connecticut r. after flowing tranquilly through Charlestown meadows, with a current 350 ft. wide and 25 deep, rushes down a fall of 44 feet in ½ a m. amongst rocks of hard gray granite, into which the water has worn deep holes. In summer, the r. is here only 16 feet wide. Above the falls is a bridge; and a canal ¾ ms. long, was dug through the rocks, some years ago, at much expense, for the passage of flat bottomed boats and rafts. A steam boat has

6

once or twice gone through, and navigated the r. above. There is an ancient bed of the river. w. of the village. Bones and weapons of Indians are found here, and a rude specimen of their art is seen on a rock. It was once a great fishing place for salmon.

Bell's Landing, and p-o. on Alabama r. in the northern part of Monroe co. Ala. by p-r. 136 ms. s. from Tuscaloosa, and 935 s. w. by w. from W. C.

Belmont, p-t. Waldo co. Me. 20 ms. w. Castine, 34 Augusta, is near the centre of the co. Pop. 1830, 3077.

Belmont, one of the eastern counties of O. bounded s. by Monroe, w. by Guernsey, n. w. by Harrison, n. e. by Jefferson, and e. by Ohio r. separating it from Ohio co. Va. Greatest length along the northern border 30 ms. the breadth 20, and mean length 26, the area 520 sq. ms. Extending in lat. from 39 52′ to 40° 10′ and in long. from 3° 40′ to 4° 14′ w. from W. C. This co. is divided into two very nearly equal sections, by the dividing table land between the vallies of Ohio and Muskingum, from which flow estrd. small creeks into Ohio r. and issue wstrd. the sources of Will's and Stillwater branches of Muskingum. The surface is a series of hills, with deep and narrow intervening vallies. Soil almost uniformly fertile. Chief town, St. Clairville. Pop. 1820, 20,329; and in 1830, 28,412.

Belmont, p-v. in the central part of Belmont co. O. 17 ms. w. from Wheeling in Va. and by p-r. 119 ms. e. from Columbus, and 288 ms. n. w. by w. of W. C. Pop. 1830, 142.

Belmont, p-v. in the eastern part of Wayne co. Miss. 10 ms. eastwardly from Winchester, the st. jus. and about 180 ms. e. from Natchez.

Belpre, p-v. in the southern part of Washington co. O. situated on the right bank of Ohio r. opposite the mouth of Little Kenhawa r. and the village of Parkersburg, Wood co. Va. 12 ms. below Marietta, and by p-r. 111 s. e. from Columbus, and 316 miles west from W. C.

Belvernon, p-v. on the right bank of Monongahela r. and in the extreme northwestern angle of Fayette co. Pa. 20 ms. a little w. of n. from Uniontown, the co. seat, and by p-r. 194 ms. westrd. from Harrisburg, and 217 ms. n. w. by w. from W. C.

Belvidere, t. Franklin co. Vt. 32 ms. n. e. Burlington, 32 n. Montpelier, on the west range of the Green mtns. contains 30,100 acres, is watered by 2 branches of Lamoille r. but has much mountainous land. Pop. 1830, 185.

Belvidere, p-t. and cap. Warren co. N. J. 11 ms. n. e. Easton, 70 w. N. York, is on the e. side of Delaware r. at the mouth of Pequest r. on a broad plain.

Benedict, p-v. on the right bank of the Patuxent, in the extreme eastern angle of Charles co. Md. by p-r. 49 ms. s. s. e. from W. C. and 54 s. from Annapolis.

Benevola, p-v. in the northwestern part of Henry co. Ky. by p-r. 566 ms. westrd. from W. C. and 47 ms. n. w. Frankfort.

Bennett's Branch, small cr. and p-o. in the northwestern part of Clearfield co. Pa. 100 ms. n. e. from Pittsburg, and 120 n. w. from Harrisburg.

Bennett's Store, and p-o. Halifax co. Va. by p-r. 131 ms. s. w. from Richmond.

Bennettsville, p-v. and st. jus. Marlborough dist. S. C. situated between the Little and Great Pedee, 25 ms. s. e. from Sneadsboro' in N. C. and by p-r. 406 ms. s. s. w. from W. C. and 102 n. e. by e. from Columbus n., lat. 34° 36, long. 2° 40′ w. from W. C.

Bennettsville, p-v. in the northern part of St. Clair co. Ala. 13 ms. nthrdly. from Ashville, the st. jus. and by p-r. 735 ms. s. w. by w. from W. C. and 142 ms. n. e. from Tuscaloosa.

Bennington, co. Vt. s.w. corner of the state is bounded by Rutland co. n., Windham co. e., Massachusetts s., and New York, w. It lies between 42° 44′ and 43° 18′ n. lat; and 3° 46′ and 4° 10′ e. long.; it is 20 ms. by 39, and contains 610 sq. ms. a large part of it mountainous. Streams flow out on all sides, into Deerfield, Hoosac, Battenkill, and West rs. and Otter and Wood creeks, and give good mill sites. The low lands are good for tillage. A range of limestone crosses the co. n. and s. and good marble, chiefly white & fine grained, is taken from several quarries. Some iron and lead ore, is also found. Bennington and Manchester are the shire and chief ts. This was the first co. settled in Vt. w. of the Green mtns. Most of the committees of safety in the state, in the revolutionary war, were held here. Pop. 1830, 17,468.

Bennington, p-t. and ½ shire, Bennington co. Vermont, 100 miles southwest Montpelier, 110 west by north Boston, 33 n. e. Albany, 160 n. New York city, 375 e. by n. W. C. was chartered by Gov. Benning Wentworth, 1749, and was first settled by separatists under Samuel Robinson, 1761, who first resisted the authority of New York. The battle of Bennington, 1777, occurred near the line of this town in Washington co. N. Y. Iron ore, marble, &c. are found here in large quantities. There are mills and factories, large iron works, &c. The village is on a hill, with a court house, academy, &c. Pop. 1830, 3419.

Bennington, p-t. Genessee co. N. Y. 18 ms. s. w. Batavia, 250. w. Albany, has Ellicot's cr. on the n. Cayuga branch of Buffalo cr. s. w., Tonawanta, s. e. &c. The land is nearly level, and pretty good. Pop. 1830, 2224.

Bennington, p-v. in the northeastern part of Delaware co. O. situated between Big Walnut and Alum crs. 41 ms. a little e. of n. from Columbus, and 60 ms. n. w. from Zanesville.

Bensboro', p-v. Pitt co. N. C. 115 ms. estrd. from Raleigh.

Benson, p-t. Rutland co. Vt. 57 ms. n. Bennington, 84 Montpelier, e. Lake Champlain,

½ to 1½ ms. wide, contains 25,214 acres, is poorly watered, but has a pond, Hubbardton r. &c. A stream which flows from a swamp runs under a hill ½ a mile. First settled 1783. Timber, chiefly pine, with hemlock, beech, maple, walnut, &c. Pop. 1830, 1493.

BENT CREEK, and p-o. in the southwestern part of Buckingham co. Va. by p-r. 196 ms. s. w. by w. from W. C. and 112 wstrd. from Richmond.

BENTLEYVILLE, p-v. on Pigeon creek, eastern part of Washington co. Pa. 13 ms. s. E. by E. from Washington, the co. seat, 9 ms. N. w. by w. from Brownsville, and by p-r. 222 ms. N. w. by w. from W. C.

BENTON, p-t. Yates co. N. Y. 18 ms. s. E. Canandaigua, w. Seneca lake, 6 ms. by 8, has the outlet of Crooked lake and Cushong creek, with mills, &c. slightly varied surface, a warm, gravelly soil, producing fruit trees, and forests of oak, chesnut, maple, and beech. The soil rests on argilaceous lime and slate. It has Hopeton and Dresden villages, and part of Penn-Yan. Pop. 1830, 3,957.

BENTON, p-v. and st. jus. Yazoo co. Miss. by p-r. 45 ms. nthrd. from Jackson, and about 120 N. N. E. from Natchez.

BENTON, p-v. and st. jus. Scott co. Mo. about 25 ms. N. w. by w. from the mouth of Ohio r. and 40 ms. N. from New Madrid, on Mississippi r. N. lat. 37° 05', long. 12° 38' w. from W. C.

BENT's p-o. in the extreme southwestern angle of Washington co. O. by p-r. 320 ms. wstrd. from W. C. and 107 s. E. from Columbus.

BERGEN, p-t. Genesee co. N. Y. 16 ms. E. N. E. Batavia, 240 Albany, 8 s. Erie canal, has a varied surface, and soil very favorable for fruit. Black creek supplies some mill seats. Pop. 1830, 1,508.

BERGEN, co. N. J. bounded by Orange co. N. Y. N., Hudson r. E., Pequannock and Passaic rs. s. w., with Newark bay and Kill-van-Kuhl, s. E., and Essex co. N. w., is crossed by Hackensack, Saddle, and Ramapo rs. running southwardly, and many small streams, with falls. The Short hills enter the co. near the middle, and the Pompton or Ramapo mtns. cross it above. There are large marshy meadows in the s. E. Along the Hudson extends a remarkable trap precipice, called the Pallisadoes. The N. part is mountainous, with many iron mines and forges, some of which are not in operation. The lower part of the co. is pleasantly situated opposite New York city, from which it derives much advantage, having communication by steamboats. 9 townships. Pop. 1830, 22,412.

BERGEN, p-t. Bergen co. N. J. w. Hudson r. opposite and 3 ms. from N. York city, has a gently varied surface, with extensive meadows along the head of Newark bay; has a village. The soil is pretty well cultivated, and supplying vegetables, &c. to the city. It was settled from Holland, and the inhabitants still speak Dutch. Pop. 1830, 4,651.

BERGER's STORE, and p-o. Pittsylvania co. Va. by p-r. 164 ms. s. w. by w. from Richmond.

BERKLEY, p-t. Bristol co. Mass. 35 ms. s. Boston, E. Taunton r. Pop. 1830, 9?7.

BERKLEY, co. Va. bounded by Jefferson s. E., Frederick, s. w., Morgan, w. and N. w., and Potomac r. separating it from Washington co. Md. N. E. Greatest length from N. E. to s. w. 28 ms., mean breadth, 14 ms. and area 392 sq. ms. Extending in lat. from 39° 14' to 39° 35' and long. from 1° 14' to 2° 11' w. from W. C. The slope of this co. is to the N. E. and it is drained in that direction by Back and Opequhan creeks. The mean elevation of the arable surface of the farms, from 500 to 700 feet above tide water. Surface, broken and mountainous. Chief t. Martinsburg. Pop. 1830, 10,528.

BERKLEY SPRINGS, otherwise called Bath, p-v. and st. jus. Morgan co. Va. situated 8 ms. s. s. w. from Hancockstown, Md. 45 ms. N. w. by w. from Harper's Ferry, and by p-r. 93 ms. N. w. by w. from W. C. and 186 ms. N. N. w. from Richmond.

BERKS, co. Pa. bounded N. E. by Lehigh co. E., by Montgomery, s. E by Chester, s. by Lancaster, s. w. by Lebanon, and w. and N. w. by Kittatinny mtn. separating it from Schuylkill. Breadth 30 ms. the northeastern and southwestern sides being parallel, mean length 34, and area 1,020 sq. ms. Extending in lat. from 40° 09' to 40° 42' N., and in long. from 0° 30' to 1° 24' E.

The face of this fine county is greatly diversified; bounded on the northwest by the Kittatinny, and s. E. by the s. E. chain of mtns. it contains a part of two mountain vallies. The general slope is to the southeastward, and it is traversed in that direction by the Schuylkill r. which breaks through the Blue Ridge at Reading. The latter chain traverses Berks in a southwestern direction, dividing the co. into two unequal valley sections. The lower and lesser, lies between the Blue Ridge and Southeast mtn. and widens from N. E. to s. w. from 3 to 15 ms. The section above Blue Ridge has a nearly equal breadth of 18 ms. something more than a third of which is on the great limestone strata which flanks the Blue Ridge on the northwestern side. The limestone tract is the most fertile, but the general character of the soil of the co. is that of productiveness, and the staples are numerous and valuable. The mean level of the arable soil of Berks is about 300 feet above the level of tide water in Delaware r. The seasons of inflorescence, foliage, and of harvest, are sensibly different above and below Blue Ridge.

Under the head of canals and roads, the great improvements which traverse Berks are noticed. Chief t. Reading. Pop. 1820, 37,327, in 1830, 53,152.

BERKSHIRE, p-t. Franklin co. Vt. 50 ms. N. w. Montpelier, 31 N. E. Burlington, has a variety of good soils, and is watered by Missisque and Pike rs. and many brooks which supply water power. The timber is beech and maple. First settled, 1792. Pop. 1830, 1,308.

BERKSHIRE, co. Mass. the w. co., is bounded by Vt. N., Hampshire, Hampden and Franklin E., Connecticut S., New York W. The people are engaged in agriculture, and a variety of manufactures; Housatonic and Hoosic rs. are the principal streams, whose waters are divided from those of Connecticut r. by the Green mtn. range. Much white marble is obtained from the quarries. Lenox is the co. town. Pop. 1830, 37,835.

BERKSHIRE, p-t. Tioga co. N. Y. 14 ms. N. Oswego, 160 W. Albany, E. of W. branch of Oswego cr., 8 ms. by 14, has a gently varied surface, with soil favorable for grain and grass, especially the meadows on the E. branch of Oswego cr. First settled, about 1793, from Berkshire co. Mass. It is well watered, and the land is held in fee. Pop. 1830, 1,883.

BERKSHIRE, p-v. towards the eastern side of Delaware co. O. 10 ms. E. from Delaware, the st. jus. and by p-r. 23 ms. N. from Columbus. Pop. of the twp. of Berkshire, 1830, 1,057.

Berlin, t. Oxford co. Me. Pop. 1830, 478. 45 ms. N. W. Augusta.

BERLIN, t. Washington co. Vt. 4 ms. Montpelier, near the centre of the state, contains 21,855 acres, much broken but good land, and is watered by Onion r., on N. line Dog r., and has a pond 2 ms. long. First settled, about 1786. Pop. 1830, 1,664.

BERLIN, p-t. Worcester co. Mass. 33 ms. W. from Boston. North brook furnishes mill seats. There is a quarry of building stone. Pop. 1830, 692.

BERLIN, p-t. Hartford co. Ct. 11 ms. S. Hartford, 23 N. New Haven, about 5 ms. by 8, with 40 sq. ms., is uneven, with mtns. S. W. and W. The rocks are clay, slate, and greenstone, of the range extending from Middletown to Northfield Ms., and some coal, carb. lime, iron pyrites, silver, &c. have been found. The soil is generally a gravelly loam, and is very good for grass, grain, and orchards. *Tin ware.*—This manufacture was carried on for many years here, to a great extent, and the products sent to all parts of the U. S. and other countries. It was introduced here, and into the U. S. by Edward Patterson, an Irishman, about the time of the revolutionary war. The neighboring towns are now engaged in it, and manufacturers have gone to the southern states; and the business is now carried on in almost all parts of the country. Other manufactures are also carried on here. Pop. 1830, 3,047.

BERLIN, p-t. Rensselaer co. N. Y. 20 ms. E. Albany, 20 E. S. E. Troy, 7 ms. by 8, is hilly and partly mountainous, with much poor soil, many marshes and evergreen forests, and some fine vallies. It is watered by Little Hoosac cr. and some small streams. The soil is generally permanently leased by S. Van Rensselaer, Esq. for about 10 bushels of wheat for 100 acres. A few Germans settled here in 1764. Pop. 1830, 2,019.

BERLIN, p-v. and borough of Somerset co. Pa. on the main road from Cumberland in Md. to the borough of Somerset, 25 ms. N. W. of the former, and 10 S. E. from the latter place, and by p-r. 157 ms. N. W. by W. from W. C.

BERLIN, p-v. Worcester co. Md. by p-r. 150 ms. S. E. by E. from W. C.

BERLIN, p-v. in the eastern part of Maury co. Ten. by p-r. 47 ms. a little W. of S. from Nashville.

BERLIN, p-v. in the northeastern part of Holmes co. O. 8 ms. a little N. of E. from Millersburg, the co. seat, and by p-r. 333 ms N.W. by W. from W. C. and 88 ms. N. E. from Columbus.

BERMUDIAN, p-v. York co. Pa. 18 ms. sthrd. from Harrisburg and 96 nthrd. W. C.

BERNARDSTON, p-t. Franklin co. Mass. 96 ms. N. W. Boston on high land between Green and Connecticut rs. is uneven, with pretty good soil. First settled 1746, and was attacked by Indians the same year. Pop. 1830, 918.

BERNARD, p-t. Somerset co. N. J. watered by Raritan and Passaic rs. Pop. 1830, 2062.

BERNE, p-t. Albany co. N. Y. 20 ms. W. Albany, on the high lands between Albany and Schoharie cos. has some of the Helderberg mtns., small streams, 2 ponds of 100 acres, fine vallies of calc. loam, marshes and beds of clay, turf and marl. There are many marine petrifactions in the rocks. Pop. 1830, 3,607.

BERRIEN, p-v. and st. jus. Dooley co. Geo. situated on the table land between Oakmulgee, and Flint rs. by p-r. 739 ms. S. W. from W. C. and 97 ms. S. S. W. from Milledgeville, N. lat. 32° 03′, long 6° 48 W. from W. C.

BERRIEN, co. Mich. the position and boundaries of which are uncertain. Chief town, Niles.

BERRY'S p-o. Wayne co. Ky. 11 ms. sthrd. from Monticello, the co. seat, and by p-r. 121 ms. a little E. of S. from Frankfort.

BERRYSVILLE, p-v. in the northern part of Knox co. Ind. 13 ms. nthrd. from Vincennes, the co. seat, and by p-r. 113 ms. S. W. from Indianopolis.

BERTIE, co. N. C. bounded N. W. by Northampton, N. by Herford, E. by Chowan r. separating it from Chowan co., and by Roanoke r. separating it from Washington S. E., Martin S. and S. W., and Halifax W. Greatest length 40 ms. from Chowan point at the head of Albermarle sound and junction of Roanoke and Chowan rs. to the extreme northwestern angle on Roanoke r., mean breadth 25, and area 1000 sq. ms., extending in lat. from 35° 40′, to 36° 15′ N., and in long. from 0° 20′ E., to 0° 21′ W. from W. C. The slope is to the southeastward. Surface generally level, and in part marshy; staples, cotton, tobacco, &c. Chief town Windsor. Population 1820, 10,805, and 12,262, in 1830.

BERWICK, p-t. York co. Me. 16 ms. N. W. Portsmouth N. H. 103 S.W. Augusta, on Salmon Fall r., contains an academy, and has considerable trade in lumber, &c. Pop. 1830, 3168.

Berwick, South, p-t. York co. Me. 17 ms. n. Portsmouth, 103 s. s. w. Augusta, is situated at the falls of Salmon Falls r. e. New Hampshire.

Berwick, p-v. and borough of Columbia co. Pa. situated on the right bank of Susquehanna r. and on the extreme eastern border of the co., 28 ms. below and s. w. from Wilkes, Barre, and by p-r. 86 ms. n. e. from Harrisburg, and 196 n. n. e. from W.C. Population, 500.

Bethania, p-v. Stokes co. N. C. 5 ms. sthrd. from Germantown, the co. seat, and 122 ms. n. w. by w. from Raleigh.

Bethany, p-t. New Haven co. Conn. 45 ms. s. w. Hartford.

Bethany, p-t. Genesee co. N. Y. 8 ms. s. e. Batavia, 240 w. Albany, is on the highest ground between Black and Tonawanta crs. and has a surface favorable for farming. Pop. 1830, 2374.

Bethany, p-v. borough and st. jus. Wayne co. Pa. situated on Dyberry cr. a branch of Lackawaxen r. about 80 ms. a little w. of n. from Easton, 3 ms. n. from Honesdale on the Lackawaxen canal, and by p-r. 265 ms. n. n. e. from W. C. and 162 n. e. from Harrisburg. n. lat. 41° 37′, long. 1° 42′ e. from W. C. Pop. 1830, 327.

Bethany, p-v. Brooke co. Va. 26 ms. n. w. from the borough of Washington, Pa. and 282 ms. n. w. by w. of W. C. and 375 n. w. Richmond.

Bethany Church, and p-o. western part of Iredell co. N. C. 6 ms. westrd. from Statesville, the co. seat, and by p-r. 402 ms. s. w. from W. C. and 152 w. from Raleigh.

Bethel, p-t. Oxford co. Me. 18 ms. n. w. Paris, 63 from Augusta, s. and e. Androscoggin r.

Bethel, p-t. Windsor co. Vt. 30 ms. s. Montpelier, 30 n. w. Windsor; contains 23,060 acres, has a mountainous surface, but generally a warm, and productive soil, watered by White r. and its second and third branches; with 2 rs. e. and w. First settled about 1780. Pop. 1830, 1240.

Bethel, p-t. Fairfield co. Conn.

Bethel, p-t. Sullivan co. N. Y. 16 ms. w. Monticello, 10 ms. by 15, e. Delaware r. and Pa. has Collakoon and Mongaup crs. White lake, 1 m. long, with a p-o. mills, &c. at the outlet. Cochecton v. is on Delaware r. Pop. 1830, 1203.

Bethel, tsp. and p-o. Berks co. Pa. by p-r. 80 ms. estrd. from Harrisburg. Pop. 1830, 1491.

Bethel, p-v. in the southwestern part of Hartford co. N. C. 14 ms. southwestward Winton, the co. seat, and by p-r. 254 ms. s. from W. C. and 143 n. e. by e. Raleigh.

Bethlehem, p-t. Grafton co. N. H. 100 ms. n. from Concord, is crossed by Great Ammonoosuc r., contains 28,608 acres, Round and Peaked mtns.; produces good crops of grain and grass, pine and sugar maple. Some mineral springs, and mtn. and bog iron ore, exist. First settled 1790. Pop. 1830, 673.

Bethlehem, p-t. Albany co. N. Y. ½ m. s. w. Albany, w. Hudson r., contains 96 sq. ms. part of Vlamanskill, Norman's cr., Vlykill, and a part of Helderberg hills. There is much rich alluvial land near Hudson r. inhabited by descendants of early Dutch settlers. There are several caverns. The seat of Gen. Schuyler, distinguished in the revolutionary war, is near the r. New Scotland, and Rensselaer's mills, (mouth of Norman's creek) are villages. Population in 1830, 6082.

Bethlehem, t. Hunterdon co. N. J. Pop. 1830, 2,032.

Bethlehem, p-v. and borough, Northampton co. Pa. situated on a fine acclivity rising from the Lehigh r. below the mouth of Manocasy cr. 48 ms. a little w. of n. Phil. 12 ms. s. w. by w. Easton, and by p-r. 184 ms. n. e. W. C. n. lat. 40° 37′, long. 1° 46′ e. from W. C. Bethlehem was founded on land purchased by the Unitas Fratrum, United Brethren, under Count Zinzendorf, 1741. The Moravians, as the United Brethren are usually called, have retained the ownership, and have produced a very neat and flourishing borough and seat of female education. The body of the village extends up the acclivity from the Lehigh bridge. The houses are neat, substantial, and though not splendid, have a fine appearance when seen from the vicinity. This village, from the date of its foundation, has been the seat of a female school, and in which many of the most accomplished women of the middle states of the U. S. have received their education. There is but one house of public worship, in which divine service is performed in English and German. Though from having to contend with so many other respectable seminaries, that of Bethlehem has not maintained its relative rank, still it may be safely asserted, that the solid and useful elements of female education can be obtained at this school with a cheapness, and moral and bodily health, exceeded by but very few, if any others in the U. S. The manners of its inhabitants and the richly varied scenery of its neighborhood render Bethlehem a very pleasant place of visit to the traveller.

Progressive population.—In 1800 the v. contained 543, and the tsp. 1343. In 1810, the tsp. contained 1436; in 1820, they had risen to 1860, and in 1830, to 2430. The v. alone contains about 1000 or 1200 inhabitants.

The Lehigh canal passes along the river bottom at the lower extreme of Bethlehem.

Bethlehem, p-v. on Tuscarawas r. and Great O. canal, Stark co. O. about 60 ms. n. w. by w. Steubenville, 60 ms. n. n. e. Zanesville, and by p-r. 329 ms. n. w. by w. W. C. and 114 n. e. by e. Columbus.

Bethlehem, p-v. on the right bank of O. r. and eastern part of Clarke co. Ind. 26 ms. above Louisville, Ky. and by p-r. 592 ms. w.

W. C. and 101 s. s. e. from Indianapolis

Bethlem, p-t. Litchfield co. Conn. 38 ms. w. s. w. Hartford, 33 n. w. N. Haven, 4 ms. by 4½, about 18 sq. ms., is hilly, with granite rocks, and a gravelly loam, good for grazing and grain, bearing oak, maple, nut-wood, &c. Branches of Pomperaug r. supply mill seats. Pop. 1830, 906.

Bettsburg, p-v. Chenango co. N. Y. 120 ms. w. Albany.

Bevansville, p-o. in the eastern part of Alleghany co. Md. by p-r. 109 ms. n. w. W. C. and 142 n. w. by w. Annapolis.

Beverly, p-t. Essex co. Mass. 17 ms. n. e. Boston, n. Salem harbor, connected with Salem by a bridge 1500 feet by 32, has excellent soil well cultivated, and is much engaged in fisheries. Pop. 1830, 4073.

Beverly, p-v. and st. jus. Randolph co. Va. situated on Tygart's Valley r. or the eastern fork of Monongahela r. about 60 ms. very nearly due s. from Morgantown, 45 s. e. Clarksburg, by p-r. 221 a little s. of w. W. C. and 210 n. w. by w. Richmond. n. lat. 38° 50′, long. 2° 55′ w. from W. C.

Bibb, co. of Geo. bounded by Tehocunn or. separating it from Houston s. and part of Crawford s.w., by the northern part of Crawford w., Monroe n. w., Jones n. e., and the Ockmulgee r. separating it from Twiggs e. Length parallel to the general course of the Ockmulgee 30 ms., mean breadth 15, and area 450 sq. ms. extending in lat. from 32° 35′ to 33° n. and in long. from 6° 36′ to 7° w. from W. C. The general slope is southeastward. The Ockmulgee r. enters at the extreme northern angle, and flowing within the co. about 20 ms. becomes thence a boundary between it and Twiggs co. Chief town, Macon. Pop. 1830, 7154.

Bibb, co. of Ala. bounded by Perry s. Tuscaloosa w. and n. w Jefferson n. Shelby n. e. and e. and Autauga s. e. Length 40 ms. mean breadth 20, and area 800 sq. ms. Extending in lat. from 32° 46′ to 33° 20′, and in long. from 10° 3′ to 10° 30′ w. from W. C. The general slope is s. southwestward, and is drained in that direction by Cahaba r. Chief town, Centreville. Pop. 1820, 3676, and in 1830, 6306.

Biddeford, p-t. York co. Me. 38 ms. n. e. York, s. Saco r. It extends to the sea, and has Fletcher's neck, off the mouth of Saco r. and near Wood isl. with a revolving light. Pop. 1830, 1995.

Big Black, or Chitteloosa r. of Miss. rises in the country of the Choctaw Indians, interlocking sources with those of the Oaknoxabee, branch of Tombigbee, and with those of Pearl river, and flowing thence about 60 miles westward along north lat. 33° 10′; enters Yazoo co. inflects to s. w. and with a very tortuous channel continues that direction 100 ms. to its junction with the Mississippi between Warren and Claiborne cos. at n. lat. 32° 2′, and long. 14° 7′ w. from W. C. after a comparative course of 160 ms. The valley of Big Black lies between that of Yazoo, and the higher part of that of Pearl.

Big Bone Lick, p-v. Boone co. Ky. situated on a creek of the same name, in the southern part of the co. about 30 ms. s. s. w. from Cincinnati, and by p-r. 66 ms. nearly due n. from Frankfort.

Bigbyville, p-v. southern part of Maury co. Ten. 7 ms. southwardly from Columbia, the co. seat, and by p-r. 42 ms. s. s. w. Nashville.

Big Creek, p-o. southeastern part of Logan co. Va. about 70 ms. s. s. e. Charleston, on Great Kenhawa, by p-r. 396 ms. s. w. by w. W. C. and 338 a little s. of w. Richmond.

Big Creek, a small branch of White r. Ark. rises in St. Francis co. and flowing sthrd. over Philip's falls into white r. about 15 ms. above the mouth of the latter. The valley of Big Creek lies between those of White and Mississippi rs.

Big Creek, p-o. on the last noted stream, by p-r. 106 ms. s. e. by e Little Rock, and 1056 ms. s. w. by w. W. C.

Big Creek, p-o. on a small creek of the same name, northern part of Shelby co. Ten. by p-r. 221 ms. s. w. by w. Nashville, and 918 ms. in a nearly similar direction from W. C.

Big Darly, p-v. in the n. w. angle of Pickaway co. O. 13 ms. s. w. Columbus.

Big Eagle, p-v. in the northeastern part of Scott co. Ky. by p-r. 34 ms. n. e. Frankfort.

Big Flats, p-t. Tioga co. N. Y. 10 ms. n. w. Elmira, 218 w. s. w. Albany, has rich alluvial lands on Cheming r. but the hills are sterile, bearing only pitch and white pine, and shrub oak. Pop. 1830, 1149.

Big Hatchy, r. of Miss. and Ten. rises in the Chickasaw country, and northern part of the former, and flows thence n. into Hardiman co. Ten. and inflecting to northwestward, pursues that course over Hardiman and Wood into Tipton co. In the latter the channel curves round to s. w. to its final discharge into the Miss. above the second Chickasaw Bluff; after an entire comparative course of something above 100 ms. The valley of Big Hatchy lies between those of Loosahatchie and Forked Deer rs.

Big Horn, r. great southern branch of Yellow stone r. has its remote sources in the Chippewayan or Rocky mtns. and as laid down by Tanner, has interlocking sources with those of Arkansas and Platte rs. on the sthrd. and with those of Yellow Stone northward. Its extreme southwestern fountains are the springs which feed Biddle lake, n. lat. 42° 20′, long. 32° w. from W. C. From this elevated region Big Horn flows 150 ms. to the n. e. receives by the influx of Stinkingwater r. from the wstrd. a large accession, and inflecting to a course of n. n. e. 200 ms. joins the Yellow Stone at Manuel's Fort, n. lat. 46°, after a comparative course of 450 ms. The valley of Big Horn lies between those of Yellow Stone proper and Tongue rs. The country it drains is generally composed of open arid plains. (See Yellow Stone.)

Big Island, tsp. and p-v. Marion co. O. By p-r. the p-o. is 51 ms. a little w. of N. Columbus. Pop. 1830, 470.

Big Lick, p-v. in the southeastern part of Botetourt co. Va. 53 ms. a little s. of w. Lynchburg, 173 wstrd. from Richmond, and by p-r. 250 ms. s. w. by w. from W. C.

Big Mills, and p-o. Dorchester co. Md. 14 ms. from Cambridge, the co. seat, and by p-r. 113 ms. s. E. by E. W. C. and 76 s. E. Annapolis.

Big Prairie, and p-v. in the southwestern part of Wayne co. O. by p-r. 77 ms. N. E. from Columbus, and 357 N. w. by w. from W. C.

Big River Mills and p-o. in the eastern part of St. Francis co. Mo. about 70 ms. s. of St. Louis, 30 s. w. from St. Genevieve, and 9 estrd. from Farmington, the co. seat

Big Sandy, r. of Va. and Ky. having its most remote sources in the northwestern slopes of Clinch mtn. but receiving tributaries from a distance of 70 ms. along the upper parts of Russel, Tazewell and Logan cos. Va. The eastern or main branch rises in Logan and Tazewell, but the higher streams uniting, the main channel becomes for a distance of 30 ms. a line of demarcation between those two cos. to where it passes Cumberland mtn. From the latter point to its influx into Ohio, the channel of Big Sandy separates Ky. from Va. flowing between Logan and Cabell of the latter, and Floyd, Lawrence, and Greenup of the former state. The main or eastern branch of Big Sandy has interlocking sources with those of Guyandot, Bluestone, branch of Great Kenhawa, Clinch, branch of Tennessee, and its own West Fork.

The West Fork of Sandy rises in Russel co. Va. flows thence westward, traverses Cumberland mtn. and enters Pike co. Ky. Passing over Pike into Floyd in the original direction, the channel curves to nthrd. and unites with the eastern branch between Lawrence, of Ky. and Cabell of Va. The valley of Big Sandy is in its greatest length from s. s. E. to N. N. w. about 100 ms. with a mean breadth of about 30; area 3000 sq. ms.; bounded to the wstrd. by the vallies of Kentucky and Licking rs. to the northeastward by that of Guyandot, and estrd. by that of New r. or the upper waters of Great Kenhawa. The main stream enters the Ohio at Catlettsburgh in Greenup co. Ky. and opposite to the extreme southern angle of the state of Ohio. N. lat. 38° 24′ and long. 5° 33′ w. W. C.

Big Sioux, r. (*See Sioux r.*)

Big South Fork, of Cumberland r. rises in Morgan co. of Ten. and flowing thence by a general course a little w. of N. enters Wayne co. Ky. which it traverses to its final influx into Cumberland r. having a comparative course of about 40 ms.

Big South Fork, p-v. in the southern part of Wayne co. Ky. by p-r. 28 ms. s. E. Monticello, the co. seat, and 138 ms. s. s. E. Frankfort.

Big Spring, and p-o. western part of Giles co. Va. 26 ms. wstrd. from the court house or Parisburg, by p-r. 324 ms. s. w. by w. W. C. and 266 a little s. of w. Richmond.

Big Spring, p-v. in the northwestern part of Hardin co. Ky. 17 ms. wstrd. Elizabethtown, the co. seat, 45 ms. s. s. w. Louisville, and by p-r. 98 ms. s. w. by w. Frankfort.

Big Spring and p-o. in the western part of Montgomery co. Mo. about 80 ms. a little N. of w. St. Louis, and by p-r. 64 ms. N. E. by E. Jefferson, and 952 wstrd. W. C.

Big Swamp, and p-o. northwestern part of Montgomery co. Ala. by p-r. 112 ms. s. s. E. Tuscaloosa.

Big Walnut r. one of the easterly branches of Sciota r. rises in Delaware co. O. and flowing thence southwardly, traverses the eastern parts of Delaware and Franklin, and joins the Sciota in the northern side of Pickaway co. after a comparative course of about 55 ms. Big Walnut has interlocking sources with those of White Womans r. a branch of Muskingum.

Billerica, p-t. Middlesex co. Mass. 20 ms. N. w. Boston, is supplied with mill sites by Concord and Shawsheen rs. and has a handsome v. in the centre, on a fine eminence, with an academy. Pop. 1830, 1,374.

Billsburg, p-v. in the northwestern part of Randolph co. Va. by p-r. 224 ms. w. from W. C. and 240 N. w. by w. from Richmond.

Bingham, t. Somerset co. Me. 26 ms. N. Norridgewock, 55 N. Augusta, E. Kennebec r. opposite Concord. It has a few brooks. Pop. 1830, 535.

Bingham, formerly Rose's, p-v. in the northwestern part of Potter co. Pa. 15 ms. from Cowdersport, and by p-r. 192 ms. N. w. Harrisburg, and 298 ms. N. N. w. from W. C.

Binghamton, p-v. and cap. Broome co. N. Y. 40 ms. S. W. Norwich, at the junction of Chenango and Susquehannah rs.

Birchardsville, p-o. in the eastern part of Susquehannah co. Pa. about 9 ms. from Montrose the co. seat, and by p-r. 172 ms. N. N. E. from Harrisburg.

Birch Pond, p-o. in the western part of Fayette co. Ten. 12 ms. from Somerville, the co. seat, and by p-r. 196 ms. s. w. by w., Nashville and 885 ms. in the same general direction from W. C.

Birch River, and p-o. in the northern part of Nicholas co. Va. 17 ms. N. w. from the st. jus. of the co. and by p-r. 327 ms. a little s. of w. W. C.

Birdsall, p-v. Alleghany co. N. Y. 240 ms. w. Albany. Pop. 1830, 543.

Birdsong's Bluff, and p-o. on the southern branch of Forked Deer r. and western part of Madison co. Ten. 9 ms. wstrd. from Jackson, the co. seat, and by p-r. 156 ms. s. w. by w. from Nashville.

Birdsville, p-v. western part of Burke co. Geo. 16 ms. wstrd. from Waynesboro', the st. jus. and E. Milledgeville.

Birmingham, p-v. on the Little Juniata r. northwestern part of Huntingdon co. Pa. 15 ms. N. w. from the borough of Huntingdon, and by p-r. 105 ms. N. W. by W. Harrisburg, and 163 ms. N. N. W. W. C.

BISCANE, bay of Florida on the eastern or Atlantic side of that peninsula. This elliptical sheet of water opens from the Bahama channel, having at the extremes of its entrance, cape Florida N. and Ellis island S. The small Paradise islands or Keys lie scattered between the two capes. The centre of the bay is about N. lat. 25° 35′ and long. 3° 20′ W. from W. C.

BISHOPSVILLE, p-v. near the extreme northern angle of Sumpter dist. S. C. about 20 ms. E. Camden, and by p-r. 67 ms. N. E. by E. Columbia, and 477 S. S. W. Washington City.

BISSELL'S, p-o. in the southwestern part of Geauga co. O. about 20 ms. S. S. W. Chardon, the co. seat, and by p-r. 136 ms. N. E. Columbus, and 333 N. W. by W. W. C.

BISTINEAU, lake of La. in the valley of Red r. This very remarkable sheet of water is, however, only the most extensive of a series of similar reservoirs along both sides of the main channel of Red r. for a distance of 120 ms. below its entrance into La. The author of this article surveyed the country on both sides, in the region of these lakes, and from actual observation came to the conclusion that the Cado, Coshatta, Spanish and Cassin lakes on the right, and Bodeau, Bistineau, Black, Saline, and Noix, on the left, are all of comparative recent formation, and formed by the operation of one general cause. Taking Bistineau as an example, since their features are common, it extends along a valley, between hills of considerable elevation, for a distance of about 35 ms. varying from half a mile to 3 miles wide. Into the head of this lengthened body of water Dacheet r. enters, and at the opposite or lower extreme it narrows into a river channel connecting it with Red r. It is completely evident that what is now a permanent lake, lies over what was formerly the valley and low lands, or bottom ground of Dacheet, but the alluvial deposit brought down by Red river, gradually formed bars between the base of the hills and effected natural dams.

These lakes contribute most efficiently to mitigate the floods of Red r. over La. In the latter, summer, autumn, and early winter, as the streams become low, a considerable part of the water of Bistineau and similar lakes drains out, and extensive natural meadows skirt the margin of the contracted lakes. On the contrary, at the season of winter snows, thaws, rains, and spring flood, the lakes open immense depositories into which the surplus water of the main stream is poured. (See Ocatahoola, and delta.)

BLACK r. Windsor co. Vt. 35 ms. long, joins Connecticut r. at Springfield, after passing through many ponds, and watering 160 sq. ms.

BLACK r. Orleans co. Vt. runs 30 ms. N. into the S. bay of lake Memphremagog, watering 150 sq. ms.

BLACK LAKE, (see Oswegatchie lake).

BLACK r. N. Y. is the third r. in size which is wholly in the state. Rising near the sources of Hudson r. E. Canada cr. after a crooked course and receiving many branches, it enters Black r. bay, an arm of Chaumont bay, near the outlet of lake Ontario. Long Falls extend 14 ms. below Wilna, and High Falls, at Turin, are 45 ms. below these. The water looks dark, and is deep and slow, and the land on the low part is good. It is 18 rods wide at Louville, 40 ms. from its mouth.

BLACK r. of N. C. one of the eastern branches of Cape Fear r. rises in the northeastern part of Cumberland co. between the vallies of Neuse and Cape Fear rivers. Flowing thence about S. S. eastwardly over Cumberland, Sampson, Bladen and New Hanover counties, it falls into the main stream of Cape Fear r. between New Hanover and Brunswick counties, after a comparative course of 90 ms.

BLACK r. of S. C. the S. western branch of Great Pedee r. has its most remote source in Kershaw district, and flowing thence over Sumpter and Williamsburg, falls into Great Pedee 3 ms. above the harbor of Georgetown, in Georgetown district, after a course of 110 ms. The higher confluents of this stream drain the greater part of Sumpter and Williamsburg districts.

BLACK LAKE, and r. of La. The river rises out of the pine forests of the parish of Claiborne, interlocking sources with those of the Terre Bonne branch of Washitau, and flowing thence sthrd. expands into a lake and again contracts to a river, which joins the Saline to form the Rigolet de Bondieu. The valley of Black lake and r. lies between those of Saline and Bistineau.

BLACK r. of La. separating the parishes of Concordia and Rapides, though thus locally designated, is in reality lower Washitau. Where the higher Washitau receives the Tensau from the N. E. and the Ocatahoola from the W. the united waters take the local name of Black r. which by a very tortuous channel unites with Red r. about 30 ms. above the junction of the latter with the Mississippi. The soil along both banks of Black r. of La. is highly fertile, and yet rendered generally uncultivatable from annual overflow. The channel of Black r. is navigable except at very low water.

BLACK r. of Mo. and Ark. rises by numerous branches in the former, interlocking sources with those of St. Francis, Maramee and Gasconnade rs. and flowing thence by two main branches, Current r. to the W. and Black r. proper to the E. The latter winds, first S. S. E. but curving gradually to the S. W. enters Arkansas, and unites with Current r. in Lawrence co. after an entire comparative course of 110 ms. After their junction, the name of Current r. is lost in the general term Black r. which, turning to S. S. W. and continuing in that direction by comparative courses 60 ms. in turn looses its name in that of White r. at their union in Independence co. It may be remarked, that in the much greater part of its course, Black r. receives no tributary of consequence from the left. The channel seems to be a common recipient for numerous streams from the N. W. Of these tributaries, Eleven Points, Spring and Strawberry rs. enter below the influx of Current r. The valley of Black r. including all its confluents, extends from lat. 35° 32′ to 37° 40′ N. and is about equal to a parallelogram of 150 by 60 or with an area of 9000 sq. ms. very nearly traversed centrally by long. 14° W. from W. C.

BLACKBYS, p-v. in the western part of Indiana co. Pa. by p-r. 202 ms. N. W. W. C. and 170 a little N. of W. Harrisburg.

BLACKBYVILLE, p-o. in the northern part of Wayne co. O. 8 ms. northwardly from Wooster, the co. seat, and by p-r. 94 ms. N. E. Columbus, and 355 N. W. by W. from W. C.

BLACK HORSE, tavern and p-o. Chester co. Pa.

BLACK ROCK, Conn. (*See Fairfield, Conn.*)

BLACK ROCK, p-v. Buffalo, Erie co. N. Y. 2 ms. N. Buffalo, has a large artificial harbor, made by a pier, intended as the canal harbor in lake Erie. A large amount of money was expended in constructing and repairing it, as it was repeatedly injured by storms, &c. For a few years the village was very flourishing while it was expected to become an important place; but the capital and business have since been chiefly transferred to Buffalo. Black Rock was burnt by the British during the last war, and but one house left standing. There is here a ferry to Waterloo, on the Canada side about ¾ m.

BLACKS and WHITES, p-o. Nottaway co. Va. by p-r. 60 ms. S. W. Richmond.

BLACK'S BLUFF and p-o. on the right bank of Alabama r. about 105 ms. N. N. E. Mobile, and by p-r. 126 ms. S. from Tuscaloosa.

BLACKSBURGH, p-v. in the northern part of Montgomery co. Va. 9 ms. northwardly from Christiansburg, the co. seat, and by p-r. 290 ms. S. W. by W. from W. C. and 215 ms. westwardly from Richmond.

BLACK STOCKS, p-o. in the S. western part of Chester dist. S. C. by p-r. 46 ms. N. N. W. Columbia.

BLACKSTONE r. or Pawtucket, rises in Worcester co. Mass. and after a S. W. course across the N. E. corner of R. I. enters Providence r. on the line of Mass. It supplies a great number of factories and mills, and adds vastly to the wealth of the region through which it flows. The Blackstone canal lies very nearly along its course.

BLACKSTONE, p-v. Worcester co. Mass. 41 ms. S W. Boston.

BLACKSTONE CANAL, Mass. and R. I. extends from Worcester, Mass. to Providence R. I. It was commenced about 1826, and was first navigated in 1829. It is 45 ms. long, and is supplied principally with water from Blackstone r. It is of great service in the transportation of raw cotton, foreign merchandize, &c. into the country, and of manufactured articles and produce to Providence. The cost has been $700,000.

BLACKSTONE, p-t. Worcester co. Mass. 41 ms. S. W. Boston.

BLACKSVILLE, p-o. Monongalia co. Va. by p-r. 243 ms. a little N. of W. W. C.

BLACK WALNUT, p-o. Halifax co. Va. by p-r. 112 ms. S. W. Richmond.

BLACK WARRIOR, (Tuscaloosa) r. of Ala. is formed by two main branches, Locust Fork to the eastrd. and Mulberry r. to the wstrd. Both of these constituent streams have their sources in the ridge of hills which separate the Mobile and Tennessee vallies, and between the confluent streams of Coosa and Tombigbee. The Locust Fork, or eastern branch, rises in Blount co. within 20 ms. from Coosa r. at the mouth of Will's cr. and about the same distance from the extreme southern bend Tennesse r. Flowing thence southwestwardly over Blount co. it unites with Mulberry r. in the western part of Jefferson co. after a comparative course of 80 ms.

Mulberry r. though with a more contracted length of course than the Locust branch, is however the main stream. Deriving its numerous confluent crs. from Lawrence, Walker, and the western and central part of Blount cos. the general course of the Mulberry is from N. to S.; comparative length 60 ms. Below the union of its two great constituents the Black Warrior receives no tributary worthy notice in a comparative course of 80 ms. to its final junction with Tombigbee. The valley of this r. approaches the form of a triangle; base 150 ms. from S. W. to N. E. and from the source of Locust r. to the mouth of the main stream; greatest breadth 65, and area 4,875 sq. ms. extending in lat. from 32° 32′ to 34° 25′, and in long. from 9° 15′ to 11° W. W. C. In this valley are contained all the cos. of Blount and Walker, great part of Jefferson, Tuscaloosa and Greene, with about one third of Fayette. It has the valley of Tombigbee W., Tennessee N., Coosa N. E. and Cahaba, S. E.

BLACKWATER, r. Merrimack co. N. H. joins Contoocook r. in Hopkinton.

BLACKWATER, r. of Va. has its extreme source in Prince George's co. and within 8 or 10 ms. sthrd. from the influx of Appomattox into James r. Flowing thence southeastward over Surry and Sussex, inflects to the southward and separating Southampton on the right from the isle of Wight and Nansemond on the left, falls into the Nottaway r. very nearly on the border between Va. and N. C. after a comparative course of 70 ms.

BLACKWELL'S isl. in the East r. opposite N. York city, near Hurl Gate, is occupied by the city prison or penitentiary, where about 200 convicts are employed in hammering stone &c. and the institution has become a source of profit to the corporation.

BLACKWELL'S MILL and p-o. Fauquier co. Va. by p-r. 60 ms. S. W. by W. W. C. and 116 N. N. W. Richmond.

BLADEN, one of the southern cos. of N. C.

bounded N. W. by Cumberland, by Black r. separating it from Sampson N. E., and from New Hanover E., Brunswick S. E., by White Marsh cr. separating it from Columbus S. W., and by the eastern branch of Lumber r. separating it from Robison W. Length from S. E. to N. W. 40 ms. mean breadth 30, and area 1200 sq. ms. Extending in lat. from 34° 14′ to 34° 42′ and in long. from 1° 15′ to 2° 04′ W. W. C. The main volume of Cape Fear r. enters the northwestern border and winding southeastward divides Bladen into two not very unequal sections. The general slope is sthrd. Much, indeed most of its surface is flat and marshy. Chief town, Elizabethtown. Pop. in 1820, 7,276, and in 1830, 7,814.

BLADENSBURG, p-v. in the northwestern part of Prince George's co. Md. and on the bank of the E. branch of Potomac, 6 ms. N. E. from the general p-o. W. C. and 31 ms. W. Annapolis. It is a village extending chiefly in one street, along the main road from Baltimore to W. C.

BLAIR, p-v. and st. jus. Harford co. Md. 23 ms. N. E. Baltimore, N. lat. 39° 33′, long. 0° 40′ E. W. C.

BLAIR'S CROSS ROADS, and p-o. Grainger co. Ten. by p-r. 191 ms. estrd. from Nashville.

BLAIR'S FERRY, and p-o. in the eastern part of Roane co. Ten. by p-r. 20 ms. from Kingston the co. seat, 179 a little S. of E. Nashville, and 546 ms. S. W. by. W. W. C.

BLAIR'S GAP, and p-o. in the extreme western part of Huntingdon co. Pa. 35 ms. N. from the borough of Bedford, and by p-r. 120 ms. wstrd. from Harrisburg, and 158 N. N. W. W. C.

BLAIRSVILLE, borough and p-v. on the right bank of Conemaugh r. southern part of Ind. co. Pa. by p-r. 161 ms. a little N. of W. from Harrisburg, and 189 N. W. W. C. This borough stands on the Pa. canal, and is a flourishing village. Pop. 1830, 957.

BLAIRSVILLE, otherwise Bellville, p-o. in the southwestern part of York dist. S. C. 9 ms. S. W. from Yorkville, the st. jus. of the co. and by p-r. 86 ms. a little W. of N. Columbia, and 441 ms. S. W. W. C.

BLAKELY, p-v. in the northeastern part of Stokes co. N. C. by p-r. 316 ms. S. W. W. C. and 132 N. W. by W. Raleigh.

BLAKELY, p-v. and st. jus. Early co. Geo. situated between Flint and Chattahooche rs. by p-r. 869 ms. S. W. W. C. and 227 ms. in a nearly similar direction from Milledgeville. N. lat. 31° 22′, and long. 8° W. W. C.

BLAKELY, p-v. port of entry, and st. jus. Baldwin co. Ala. situated on the left bank of Tensaw r. or eastern arm of Mobile r. 8 ms. N. E. and on the opposite side from the city of Mobile, and by p-r. 228 ms. a little W. of S. Tuscaloosa, and 1,020 S. W. by W. W. C. N. lat. 30° 44′, W. long. 11° 04′. By the returns of the census of 1830, the population of this place is not given separate from that of Baldwin co. but is supposed about 500

BLAKESBURG, PLANTATION, Penobscot co. Me. 20 ms. N. Bangor, 90 N. N. E. Augusta, one tsp. S. Piscataqua r. crossed by Dead creek, both branches of Penobscot r. Pop. 1830, 403.

BLAKESBURG, p-v. Putnam co. Ind. 12 ms. from Green Castle, the st. jus. and by p-r. 614 ms. wstrd. W. C. and 54 wstrd. Indianopolis.

BLANCHARD'S FORK, the eastern constituent of the Au Glaize branch of Maumee r. O. The extreme source is in the central part of Hardin co. interlocking sources with those of Sciota and Sandusky, and flowing thence northwardly 30 ms. into the central part of Hancock co. where abruptly inflecting to the wstrd. it crosses Hancock into Putnam, and finally unites with Au Glaize, near the western border of the latter co.

BLANFORD, p-t. Hampden co. Mass. 15 ms. W. Springfield, 116 S. W. Boston, S. W. Westfield r. two branches of which rise here among hilly country, and are subject to sudden floods. It is a good farming town, and was principally settled from N. of Ireland. Pop. 1830, 1,590.

BLANDING, p-v. Orangeburg dist. S. C. by p-r. 61 ms. southwardly from Columbia.

BLEDSOE, co. of Ten. bounded S. W. by Marion, by Cumberland mtn. separating it from Warren W., and White N. W., by Fentress N., Roane N. E., and by Walden's Ridge, separating it from Rhea co. E. Length from S. W. to N. E. 40 ms. mean breadth 15, and area 600 sq. ms. Extending in lat. from 35° 25′ to 36° N. and in long. from 7° 48′ to 8° 36′ W. W. C. This co. occupies part of the comparatively elevated valley between Cumberland mtn. and Walden's Ridge; the central part being a table land from which flows southwestwardly the sources of Sequatchie, and northeastwardly those of Emery's r. The surface hilly and in part mountainous. Chief town, Pikesville. Pop. 1830, 4,648.

BLENDON, p-v. and tsp. of Franklin co. O. The tsp. in 1830, contained a population of 666. The p-o. is within a few ms. from Columbus, the seat of government, but the exact distance is not given in the post office list.

BLENHEIM, p-t. Schoharie co. N. Y. 44 ms. W. Albany, 38 N. N. W. Catskill; is supplied by Schoharie cr. and branches, with many fine mill seats, includes the W. part of the Catsberg hills, bears oak, walnut, beech, maple, birch, &c. Pop. 1830, 2,280.

BLISSFIELD, p-o. Lenawee co. Mich. by p-r. 67 S. W. by W. Detroit.

BLOCKERSVILLE, p-v. Edgefield dist. S. C. by p-r. 64 ms. a little S. of W. Columbia, and 553 ms. S. W. W. C.

BLOCK ISLAND, or New Shoreham, isl. and t. Newport co. R. I. lies in the Atlantic, 15 ms. S. S. W. Point Judith, 12 S. of the nearest part of the continent, 3 ms. by 8, has an uneven surface, generally high, with a chain of ponds from N. to centre, has no forests and is devoted to tillage, though formerly to pasture. It is a heap of loose earth, with separate masses of granite, and is constantly washing away by the sea. It has no harbor, and boats are secured only by being drawn upon shore.

Long. 71° 30′ w., lat. 41° 8′ N. It is proposed by the U. S. gov't. to form a harbor here.

BLOCKLEY, tsp. along the right bank of Schuylkill r. Philadelphia co. Pa. 3 ms. Philadelphia. Pop. 1810, 1,618, 1820, 2,655, and in 1830, 3,401.

BLOODY BROOK, p-v. Franklin co. Mass. 90 ms. w. Boston, the scene of a battle in 1675.

BLOODY RUN, p-v. Bedford co. Pa. on the left bank Juniata r. 8 ms. E. borough of Bedford, and by p-r. 118 ms. N. W. W. C.

BLOOM, p-v. in the S. E. part of Seneca co. O. by p-r. 423 ms. N. W. by W. W. C. and 83 northward of Columbus.

BLOOMFIELD, p-t. Somerset co. Me. 7 ms. N. E. Norridgewock, 33 N. Augusta, s. Kennebec r. at the bend. Has an academy. Pop. 1830, 1,072.

BLOOMFIELD, p-t. Ontario co. N. Y. 13 ms. w. Canandaigua, 12 ms. by 6, has an agreeable variety of surface, with a good and well cultivated soil, and is supplied with many mill seats by Honeoye outlet and Mud cr. It produces grain, grass, and plenty of apples, &c. East and West Bloomfield are p-vs. In the latter is an academy. Pop. 1830, 3,861.

BLOOMFIELD, p-t. Essex co. N. J. 5 ms. N. w. Newark. Pop. 1830, 4,309.

BLOOMFIELD, p-v. and tsp. of Crawford co. Pa. 16 ms. watrd. Meadville.

BLOOMFIELD, p-v. Loudon co. Va. by p-r. 51 ms. N. W. by W. W. C.

BLOOMFIELD, p-v. in the northeastern part of Nelson co. Ky. by p-r. 44 ms. s. w. Frankfort, and 595 wstrd. W. C. Pop. 1830, 301.

BLOOMFIELD, p-v. in the southeastern part of Oakland co. Mich. 19 ms. N. W. Detroit.

BLOOMFIELD, p-v. and st. jus. Greene co. Ind. situated on the west branch of White r. by p-r. 618 ms. w. W. C. and 76 below and s. w. Indianopolis. N. lat. 39° 06′ and long. 10° w. from W. C.

BLOOMFIELD, p-v. Edgar co. Il by p-r. 120 ms. N. E. Vandalia, and 689 ms. w. W. C.

BLOOMINGBURG, p-v. Sullivan co. N. Y. 23 ms. w. Newburgh on Shawangunk cr.

BLOOMINGBURG, p-v. in the N. part of Fayette co. O. by p-r. 44 ms. s. w. Columbus. Pop. 1830, 100.

BLOOMINGDALE, p-v. in the w. part of Jefferson co. O. 14 ms. from Steubenville, and by p-r. 274 N. W. by W. W. C. and 135 N. E. by E. Columbus.

BLOOMING GROVE, p-t. Orange co. N. Y. 12 ms. w. West Point, is broken by mountainous ranges of the Highlands, and watered by Murdner's or Murderer's creek. Skunnemunk mtn. has several commanding eminences. S. E. Salisbury mills is a village and 2 ms. w. of it Washingtonville. Craigsville and Oxford, are small villages. Pop. 1830, 2,099.

BLOOMING GROVE, p-v. in the N. E. part of Tazewell co. Il. by p-r. 772 N. W. by W. half w. W. C. and 169 northwardly Vandalia.

BLOOMINGTON, p-v. and st. jus. Monroe co. Ind. by p-r. 627 ms. w. W. C. and 51 s. w. Indianopolis; N. lat. 39° 12′, long. 9° 34′ w. from W. C.

BLOOMINGTON, p-v. and st. jus. McLean co. Il. Neither the co. or village is located by either Tanner's United States, or the post office list of 1831.

BLOOMINGVILLE, p-v. in the N. part of Huron co. O. 15 ms. N. Norwalk, the co. seat, and by p-r. 415 ms. N. W. by W. W. C. and 108 a little E. of N. Columbus.

BLOOMSBURG, p-v. situated on the right bank of the main or E. branch of Susquehannah r. and in the S. E. part of Columbia co. Pa. by p-r. 75 ms. above Harrisburg, and 196 a little E. of N. W. C.

BLOOMSBURG, p-v. in the southern part of Halifax co. Va. 13 ms. s. Banister, the co. seat, and by p-r. 233 s. s. w. W. C. and 143 s. w. by w. Richmond.

BLOOMVILLE, p-v. Delaware co. N. Y. 70 ms. s. w. Albany.

BLOUNT, co. of Ala. bounded by Jefferson s., Walker s. w. and w., Morgan N., the Cherokee country N. E., and St. Clair co. E. and s. E. Greatest length from E. to w. 55 ms. mean breadth 30, and area 1,650 sq. ms. Extending in lat. from 33° 45′ to 34° 16′, and in long. from 9° 14′ to 10° 12′ w. from W. C. The N. and N. E. boundaries of this co. are along the dividing ridge between the vallies of Tennessee, and Tombigbee branch of Mobile r.

From this rather elevated tract, the higher sources of both branches of Black Warrior r. are poured southwardly, giving a general slope to the county in that direction. The surface is hilly, and soil with some exceptions rather sterile. Chief t. Blountsville. Pop. 1820, 2,415, 1830, 4,233.

BLOUNT, co. of Ten. bounded by Tennessee r. separating it from the Cherokee country s. w., and Monroe co. of Tenn. w., by Holston r. separating Roane co. N. W., and Knox N., Sevier co. N. E. and E., and the Iron mtn. separating it from Haywood co. N. C. s. E. Length 48 ms. mean breadth 14, and area 672 sq. ms. Extending in lat. from 35° 28′ to 36° 53′, and in long. from 6° 24′ to 7° 12′. The slope of this co. is N. wstrd. towards Holston r. The surface is broken, but soil good. It is among the most ancient settlements of Ten. by the whites. Chief t. Marysville. Pop. 1820, 11,258, and in 1830, 11,028.

BLOUNT SPRING, p-o. in the s. part of Blount co. Ala. by p-r. 90 ms N. E. from Tuscaloosa, and 789 s. w. by w. W. C.

BLOUNTSVILLE, p-v. Jones co. Geo. 17 ms. w. Milledgeville.

BLOUNTSVILLE, p-v. and st. jus. Blount co. Ala. situated on Locust branch of Black Warrior r. 56 ms. southward from Huntsville, and by p-r. 748 ms. s. w. by w. W. C. and 110 N. E. Tuscaloosa. N. lat. 34° 05′, w. long. 9° 35′ from W. C.

BLOUNTSVILLE, p-v. and st. jus. Sullivan co. Ten. 107 ms. N. E. by E. Knoxville, and by p-r. 409 s. w. by w. W. C. and 306 a little N. of E. Nashville. N. lat. 36° 32′, long. 5° 18′ w. W. C.

BLUE HILL, p-t. Hancock co. Me. 12 ms. N. E.

Castine, 78 Augusta, on a large Bay—it has an academy. Pop. 1830, 1486.

Blue House, p-v. in the southern part of Collcton dist. S. C. by p-r. 93 ms. s. Columbia, and 588 s. w. W. C.

Blue Mountain. This undistinguishing term has been applied to several chains of the Appalachian system in the U. S. but more particularly to that one called by some tribes of Indians "Kaatatin Chunk," or Endless mountain. If we turn our attention to the Appalachian chain we find them often only interrupted, where a cursory survey would lead us to place a termination. Whether the Kittatinny Chain or "Blue Mountain" could be detected eastward from the Hudson we are unprepared to determine, but westward of that river, this chain is found distinct in the Shawangunk, near Kingston, in Ulster co. N. Y. It thence ranges s. w. meets and turns Delaware r. at the extreme northern angle of N. J. and continues its original direction to the Del. Water Gap, where the mountain chain is traversed by the river, and the former curves more westward, enters Pennsylvania, over which it ranges about 150 ms. to the northern angle of Franklin co. after having been pierced by the Lehigh, Schuylkill, and Susquehannah, rivers. Between Franklin and Bedford cos. the Kittatinny reassumes nearly its original direction in the state of N. Y. and though in some places confounded with the Alleghany, really continues a distinct chain over Md. Va. N. C. and Ten. into Ala. s. w. of Susquehanna, the Kittatinny rises, and extending first nearly w. between the confluents of Coredogwinet and Shoreman's rivers, is thence broken into ridges bounding on the w. the valley of Conecocheague, gradually curves to the southward, and reaches Potomac, extending very little w. of s. Rising again beyond the Potomac, between the Opequan and Black creeks, it runs nearly parallel with the Blue ridge, is passed by the North Fork of Shenandoah, and extends thence between the two main branches of that river. Though scarcely appearing distinctly on our best maps, the chain of Kittatinny is completely distinct and continues over Rockingham, Augusta, and Rockbridge cos. Va. into Botetourt, to where it is traversed by James river, below the mouth of Craig's creek. Rising again beyond James r. the chain stretches along the higher sources of James and Roanoke rs. to the centre of Montgomery co. near Christiansburg. Here it leaves the Atlantic slope, and merges into the valley of O. by entering the subvalley of New river or Upper Kenhawa.

Thus far, in all its range from the Hudson, the Kittatinny chain is broken into links by the higher sources of the Atlantic rivers, and similar to the Southeast mountain and Blue ridge, the base gradually rises, ascending the vast inclined plain obliquely, until it reaches the highest apex between the sources of Roanoke and those of Little river branch of New river. In this region the lowest gap through which measurements have been made for a projected canal, is 2049 feet above the level of the Atlantic ocean. The base of the chain now commences to depress and inflecting to a course considerably west of southwest, is traversed by New river or Upper Kenhaway. Beyond the latter stream, under the local name of Iron mountain, and discharging to the eastward the confluents of New river, and from the opposite flank those of the south branch of Holston and Watauga, reaches the extreme northeastern angle of Ten. At the latter point, the chain assumes a direction very nearly s. w. and under the various local names of Iron mountain, Bald mountain, Smoky mountain, and Unika mountain, is pierced in succession by Watauga, Doe, Nolechucky, French Broad, Big, Pigeon, Tennessee, Proper, and Hiwassee rs. & merges according to Tanner's map of the United States, into Blue ridge, in the northern part of Georgia, between the sources of Coosa and Hiwassee rivers.

If the whole body of the Kittatinny and its mean elevation is compared with the body and elevation of Blue ridge, the former exceeds in both respects, from the Hudson to their termination in Georgia, though at the High lands on the Hudson and in the Peaks of Otter, the Blue ridge rises to a superior elevation from their respective bases.

As a distinct and defined chain the Kittatinny is upwards of eight hundred miles in length. The height above the ocean varies from 800 to 2,500 feet. All the ridges in their natural state were wooded to their summits, though the trees are generally stunted in growth at any considerable height. In the vallies along both flanks the timber is often very large and lofty; particularly the pines, oaks, hemlocks, and liriodendron. On some of the ridges good arable soil is found on the summits, but sterility is the general character of the soil. Amongst the peculiar features of this chain, one may be remarked, which gives it a very distinct character. In all its length, it is no where strictly a dividing limit between river sources. Without assuming any connexion with the mountains eastward of the Hudson, the Kittatinny is pierced by the Delaware, Lehigh, Schuylkill, Susquehannah, Potomac and James rivers, flowing into the Atlantic ocean, and by the Great Kenhawa, and various branches of Tennessee flowing into the valley of Ohio, or basin of the Mississippi.

Blue Ridge; of the distinctive chains of the Appalachian system, and indeed of all the sections of this system, the Blue ridge stands most apart and prominent, though of much narrower base, and of less mean elevation than either the Kittatinny or Alleghany. On a colored map of Virginia the Blue ridge has a very striking appearance, arising from the fact of being a county limit in all its range over that state. Without tracing a probable but hypothetical identity, between the mountains of Mass. Conn. and Vt. with the Blue ridge, we first meet this chain distinct at West Point on the Hudson river. Thence it rises into broken but continuous ridges over N. Y. and N. J. to the Delaware, in a southwesterly direction. Traversed by the Delaware immediately below the influx of Lehigh, and inflecting similar to the Kittatinny, to s. w. by w., it is pierced by the Schuylkill at Reading, by the Susquehannah below the mouth of Swatara, by the Potomac at Harper's Ferry, by James river, between Bedford and Amherst cos. Virginia, and by the Roanoke between Bedford and Franklin cos. in the same state. In its further progress s. w. from Roanoke, the Blue ridge becomes the limit of river source to its final extinction in Ala. The length of this chain from the Hudson to Roanoke, 450 ms. and from Roanoke to where it ceases to be a distinct chain in Ala. 350 ms. having an entire length of 800 ms. s. w. from the Hudson. The Hudson does not, however, terminate the Blue ridge to the N. eastward. Many river passages through mountains have been noticed and celebrated, and, amongst others, the passage of this chain by the Potomac at Harper's Ferry; but it may be doubted whether from all the attendent circumstances, any similar phenomenon on earth combines so many very remarkable features as the tide stream of the Hudson through the two chains, the southeast mountain and Blue ridge.

Profoundly deep, far below the utmost draught of the largest vessels of war, the flux and reflux of the tides rush along a narrow and tortuous channel, on both sides bounded by enormous craggy and almost perpendicular walls of rock, rising from one thousand to twelve or fifteen hundred feet above the water surface. Sailing along this astonishing gorge the mind involuntarily demands by what operation of nature has this complication of wonders been produced? Again, what in an eminent degree enhances the surprise and admiration, is the fact, that this great river pass is made directly through a mountain nucleus. In all the chains of the Appalachian system, masses rise at different places, far above the ordinary height, and spreading much wider than the mean base of the chain in which they occur. The peaks of Otter—the peaks in the Catsbergs, in Windham, Green co. N. Y. several peaks of the Green mts. in Vt. and above all, the White mts. of N. H. are examples. The Highlands, pierced by the Hudson, and passed by the tide from the ocean, are however, every thing considered, by much the most remarkable of these mountain peaks or groups to be found, not only in the U. States, but probably on this planet. Receding from the highlands, either to the s. w. or N. E. the chain depresses so much, that on our maps, the continuity in either direction, is generally not represented. There is, nevertheless, in the vicinity of the Hudson, no real interruption of either the S. E. mtn. or Blue Ridge, along their lines of direction. The highest peaks being in the Blue ridge on both sides of the river. Of these peaks, the highest is Butter Hill, which rises 1,535 feet above the ocean tides, and rising abrubtly from the water, affords a very fine and extended landscape to the N. W. and N.

After leaving the Hudson, Blue Ridge continues to N. E. about 20 ms. and then, similar to other chains of the same system on both sides of that river, rapidly inflects to a course a very little N. of E., a direction which it maintains above 250 ms. in the states of N. Y. Mass. and Vt. For the first 70 ms. of its northerly course, the Blue Ridge discharges from its eastern flank numerous branches of Housatonic, and from the opposing slope, Fishkill, Wappingers, Jansen's or Ancram, and Kinderhook creeks, flowing wstrd. into the Hudson. With the sources of Housatonick and Hoosack rs. the features of Blue Ridge change; hitherto from the Hudson, a line of river source, it now loses that character, and is broken into innumerable ridges by the higher sources of Hoosack and Batten Kill, flowing into the Hudson, and thence by those of Paulet, Otter, Onion, La Moille, and Missisque rs. falling into lake Champlain. All these latter streams rise in the S. E. mountain, and flowing down a western slope pass the Blue Ridge.

A hypothesis may be hazarded that what is designated Green mountains in the southern part of Vt. and the ridge or series of ridges known by the same term in the northern part of the same state, are fragments of two separate chains, though generally represented as the continuation of one & the same chain. Regarding the great western chain E. of the Hudson, in the State of N. Y., Mass. and Vt. as the continuation of Blue Ridge, the whole length of the chain in the U. S. exceeds 1,000 ms. In relative elevation, the Blue Ridge is humble, though in one part, Bedford co. Va. the peaks of Otter rise to 4,200 feet above tide water. Generally, the ridges are from 700 to 1,000 feet above their bases, and the base rising with the mountain, when the ridges are seen from the elevated table land, from which flow Roanoke, Kenhawa, Yadkin, and Tennessee, they are, in fact, less imposing than when seen from the Hudson, Delaware, Susquehannah, Potomac or James rs. though at the former region, the real oceanical elevation is more than double to what it is near the more northern rivers.

From its prominence, and southwestwardly from the Hudson, its isolution, Blue Ridge has been, though very erroneously, regarded and delineated as the extreme southeastern chain of the system; in reality, however, it is the third distinct chain advancing from the Atlantic ocean. (See art. Southeast mtn.)

BLUE ROCK, p.v. in the western part of Muskingum co. O. 12 ms. from Zanesville, and by p-r. 71 E. Columbus, and 318 N. W. by W. W. C.

BLUE SPRING GROVE, p.o. Barren co. Ky. 12 ms. N. Glasgow, the st. jus. and by p-r. 126 S. W. Frankfort.

BLUE STONE, small r. of Va. in Tazewell and Giles cos. rises in the latter, interlocking sources with those of Clinch and Big Sandy, flows thence N. E. down a mountain valley into New r. which it enters about five miles above the influx of Green Brier r.

BLUE STONE, p-o. on the Blue Stone r. southeastern part of Tazewell co. Va. by p-r. 279 ms. a little S. of W. Richmond, and 337 S. W. by W. W. C.

BLUE SULPHUR SPRING, and p-o. Green Brier co. Va. by p-r. 203 ms. W. Richmond, and 264 S. W. by W. W. C.

BLUFFDALE, p-o. in the W. part of Greene co. Il. 10 ms. W. Carrollton, the st. jus. for the co. and by p-r. 116 ms. a little N. of W. Vandalia, and 897 W. W. C.

BOALSBURG, p-v. in the southern part of Centre co. Pa. 15 ms. S. Bellefonte, by p-r. 81 N. W. Harrisburg, and 183 N. N. W. W. C.

BOARDMAN, p-v. on Mahoning r. and in the S. E. part of Trumbull co. O. 11 ms. S. E. and below Warren, the st. jus. and 161 ms. N. E. by E. Columbus.

BODCAU, r. of Ark. and La. The Bodcau rises in Hempstead co. of the former, and flowing thence S. over La Fayette, enters Claiborne parish, La. expands into a lake and again into a river which falls into Red r. opposite the great raft. The valley of Bodcau lies between those of Bistineau and Red r. The lake of Bodcau is similar in its features to that of Bistineau, which see.

BOEUF, large bayou or creek of La. has its source in the pine forests, of the parish of Rapide, 8 or 10 ms. S. W. from the village of Alexandria on Red r. This remarkable water course flows first to the N. E. towards Red r. and entering the alluvial tract near that stream, approaches within a few hundred yards of that arm of Red r. called the Rapide bayou, but the Boeuf retires to the S. E. about 15 ms. where it divides into two streams, one of which, the left or eastern, winds eastwardly and communicates with Red r. by an outlet from that stream ; again divides the right branch, forming the bayou de Glaize which unites with the Atchafalaya, 5 ms. below the outlet of the latter from the Mississippi.

The main or right branch of the Boeuf, after the division of its waters as noticed above, assumes a course of S. S. E. by direct distance thirty miles, but perhaps a third more following the windings to its junction with the Crocodile, to form the Courtableau r. The lands along the Boeuf are exuberantly fertile. By its channel, at seasons of high flood, a navigable connexion exists for small vessels, between Red r. and the streams of Opelousas and Attacapas. The writer of this article made this navigation in a very large piroguo, built on lake Bistineau, brought down Red r. and through the intermediate bayou into the Boeuf, and down the Boeuf into the Courtableau to Lemelle's landing, 4 ms. from the village of St. Lardie in Opelousas.

BOEUF, r. of Ark. and La. rises in the former and in Arkansas co. and within 7 or 8 ms. from Arkansas r. Interlocking sources with those of the Barthelemy, and flowing at a mean distance of about 20 ms. from the Miss. r. over Ark. and Chicot cos. the Boeuf enters La. Bending to S. S. W. about 70 ms. and again curving to a little E. of S. joins the Washitau, twenty direct miles north of the junction of the latter with Ocatahoola and Tensaw. The entire comparative course of the Boeuf is 170 ms. The valley of Boeuf generally separates the pine forest land from the annually overflown tracts along the Mississippi and Tensaw rs.

A boatable channel could be easily formed to unite the Mississippi with the Washitau by means of the Boeuf, and by a canal from Grand lake immediately above the northern boundary of Louisiana.

BOGLE'S, p-o. in the S. W. part of Iredell co. N. C. by p-r. 167 ms. W. from Raleigh.

BOGLE'S, p-o. Perry co. Ala. by p-r. 50 ms. S. E. Tuscaloosa.

BOGUE CHITTO, r. of the states of Miss. and La. is the S. W. branch of Pearl r. rising in Lawrence co. of the former, flows thence by a course a little E. of S. over Pike, and enters Washington parish, La. Inflecting to southeast 35 ms. over Washington, it thence bends still more E. and for a distance of 25 ms. separates the latter from St. Tammany parish to its final influx into Pearl r. after an entire comparative course of about 90 ms. The Bogue Chitto has interlocking sources with those of Bayou Pierre, and Homochitto, but higher part of the valley of Bogue Chitto lies principally between that of Pearl and Tangipaha, and the lower part between that of Pearl and Chifunate.

BOHEMIA, large cr. in the southern part of Coecil co. Md. This cr. rises in Newcastle co. Del. and flowing thence W. opens into a comparatively large bay, which communicates with the Elk r. between 4 and 5 ms. above the opening of the latter into Chesapeake bay. The Bohemia heads with the Appoquinimink.

BOLIVAR, p-v. Alleghany co. N. Y. 265 ms. W. Albany.

BOLIVAR, p-o. Westmoreland co. Pa. by p-r. 189 ms. N. E. by W. W. C. and 166 W. Harrisburg.

BOLIVAR, p-v. in the S. part of Robeson co. N. C. by p-r. 101 ms. a little W. of S. from Raleigh.

BOLIVAR, p-v. Washington co. Miss. about 100 ms. N. N. E. Natchez.

BOLIVAR, p-o. and st. jus. Hardiman co. Ten. situated on Big Hatche r. 70 ms. a little N. of E. Memphis, on Mississippi r. and by p-r. 849 ms. S. W. by W. W. C. and 158, in nearly a similar direction from Nashville. N. lat. 35° 16′ and long. 12° W. from W. C.

BOLIVAR, p-v. on Tuscarawas r. in the northern angle of Tuscarawas co. O. 10 ms. above and north from New Phil. the st. jus. and by p-r. 111 ms. N. E. by E. Columbus, and 324 N. W. by W. W. C.

BOLIVIA, p-v. in the N. W. part of St.

Genevieve co. Mo. by p-r. 894 ms. a little s. of w. W. C. and about 65 a little w. of s. St. Louis.

Bolsters' Mills, p-v. Cumberland co. Me. 81 ms. from Augusta.

Bolton, p-t. Chittenden co Vt. 17 ms. n. w. Montpelier, 17 s. e. Burlington, is very mountainous, and has but a small part habitable, on the w. range of the Green mtns. crossed by Onion r. and several branches. Pop. 1830, 452.

Bolton, p-t. Worcester co. Mass. 33 ms. w. Boston, has a varied surface, with good soil, between Concord and Nashua rs. Pop. 1830, 1258.

Bolton, p-t. Tolland co. Conn. 14 ms. e. Hartford, 3 ms. by 5, is on the granite range which divides the waters of Conn. & Thames rs., has a coarse gravelly loam, with very good grazing, with oak, walnut, chestnut, &c. It has a branch of Hop r. and one of Salmon r. Pop. 1830, 744.

Bolton, p-t. Warren co. N. Y. 14 ms. n. Caldwell, 76 n. Albany, w. Lake George, e. Scaroon r. has a mountainous or hilly surface, with broad vallies. It yields wheat, rye, grass, &c. with fruit trees near the lake. Timber is carried down the lake on rafts. The people are chiefly from N. England. The t. includes Tongue mtn. n. w. Bay, and the Narrows of Lake George, which is spotted with innumerable islands, and offers the most romantic scenery on the lake. The steamboat Mountaineer plies from Caldwell to the bottom of the lake 14 ms. Pop. 1466.

Bond, co. of Il. bounded by Clinton s., Madison w., Montgomery n., and Lafayette e. Length 20, breadth 18, and area 360 sq. ms. Extending in lat. from 38° 44' to 39° 02', and in long. from 12° 16', to 12° 36' w. W. C. The slope is nearly due south, and in that direction is drained by Shoal creek and other smaller streams flowing into Kaskaskia r. Chief town, Grenville. Pop. 1830, 3124.

Bone, p-o. Hopkins co. Ky. by p-r. 210 ms. s. w. by w. Frankfort.

Bonnet Carre', (Square Bonnet,) remarkable bend of the Mississippi r.

Bonnet Carre', p-o. on Bonnet Carre' Bend parish of St. John Baptist, La., 36 ms. above New Orleans, and by p-r. 1241 ms. s. w. by w. W. C.

Bono, p-v. in the southern part of Lawrence co. Ind. by p-r. 84 ms. a little w. of s. Indianopolis, and 631 westward W. C.

Bon Pas, p-v. on the right bank of Wabash r. in the extreme n. e. angle of White co. Il. 45 ms. s. w. Vincennes, and by p-r. 106 ms. s. e. by e. Vandalia, and 747 Westward W. C.

Bon Secours. (*See Mobile bay*).

Boone, p-v. in the w. part of Pickens co. Ala. 11 ms. w. Pickensville, the st. jus. and by p-r. 48 ms. w. Tuscaloosa, and 906 s. w. by w. W. C.

Boone, one of the two extreme northern cos. of Ky. bounded by Grant s., the Ohio river below the mouth of Great Miami, separating from Switzerland co. Ind. w., and Dearborne co. Ind. n. w., by Ohio river above the mouth of Great Miami, separating it from Hamilton co. O. n., and by Campbell co. Ky. e. Length from south to north along the Ohio river, 25 ms., mean breadth 12, and area 300 sq. ms. Extending in lat. from 38° 47', to 39° 08' n., and in long. from 7° 35', to 7° 51' w. from W. C. The slope is westward towards the Ohio r. surface very hilly, but soil productive. Chief towns, Burlington and Florence. Pop. 1820, 6582, 1830, 9075.

Boone, co. of Ind. bounded s. by Hendricks, w. by Montgomery, n. by Wabash, e. by Hamilton, and s. e. by Marion. Length 26, breadth 20, and area 520 sq. ms. Extending in lat. from 39° 57', to 40° 13', and in long. from 9° 13', to 9° 42' w. from W. C. The slope is westward and drained in that direction by Sugar and Raccoon crs. and other streams flowing into Wabash r. Thorntown, the st. jus. is situated in direct distance 35 ms. southwestwardly from Indianopolis, but by p-r. 62, and 598 ms. w. W. C. Pop. 1830, 622.

Boone, co. Miss. bounded on the n. w. by Howard, n. by Randolph, n. e. by Ralls, e. & s. e. by Callaway, and s. and s. w. by Missouri r. separating it from Jefferson. Length from south to north 40 ms.; mean breadth 20, and area 800 sq. ms. Extending in lat. from 38° 38', to 39° 12' and in long. from 15° 03', to 15° 32' w. W. C. It slopes a little w. of s. and is drained in that direction by Rock cr. Cedar cr. and other small streams falling into Missouri r. Chief town, Columbia, situated 24 ms. eastwardly from Franklin, and 130 west St. Louis. Population 1830, 8839.

Boonsboro', or Morganville, p-v. in the eastern part of Washington co. Md. 11 ms. s. s. e. Hagerstown, 16 n. w. by w. Frederick, and by p-r. 59 n. w. W. C.

Boonesboro', p-v. on the left bank of Ky. r. Madison co. Ky. 41 ms. s. e. Frankfort, 13 s. s. e. Lexington, and by p-r. 526 a little s. of w. W. C.

Boone's Mill, and p-o. Franklin co. Va. by p-r. 167 ms. s w. Richmond, and 263 s. w. by w. W. C.

Booneton, p-v. Boone co. Mo. by p-r. 57 ms. northwardly from Jefferson, and 992 ms. westwardly from W. C.

Boonevile, p-v. and st. jus. Warrick co. Ind. situated between Pigeon and Cypress crs. 19 ms. n. e. by e. Evansville, on Ohio r. 55 ms. s. s. e. Vincennes, and by p-r. 187 ms. s. s. w. from Indianopolis.

Booneville, p-v. st. jus. Cooper co. Mo. situated on the right bank of Missouri r. opposite to Franklin in Howard co. by p-r. 185 ms. a little n. of w. from St. Louis, and 51 n. w. by w. Jefferson.

Boonville, p-t. Oneida co. N. Y. 27 ms. n. Utica, is hilly in some parts, is crossed by Black r. of L. Ontario, 10 ms. above High Falls. The v. is in the s. on the Black r. road. Pop. 1830, 2,746.

Boothbay, p-t. Lincoln co. Me. 10 ms. s. e.

Wiscasset, 39 from Augusta, w. Damariscotta r. which, with another arm of the sea w. almost insulate it. Pop. 1830, 2,286.

BORDENTOWN, p-t. Burlington co. N. J. 6 ms. s. Trenton, 24 N. Phil., E. Delaware r. has a level surface, considerably elevated. By means of the river it has a ready communication with the Phil. market. The seat of the Count de Survilliers, Joseph Bonaparte, is here, on the bank of the Del.

BORDENVILLE, p-o. Carteret co. N. C. by p-r. 177 ms. s. E. Raleigh, and near the Atlantic coast.

BORDLAY, p-v. in the western part of Union co. Ky. by p-r. 753 ms. w. W. C., 215 a little s. of w. from Frankfort, and 10 w. from Morganfield, the co. seat.

BORGNE, lake of, as usually denominated, is a bay, and the western extreme of Pascagoula sound. The sheet of water particularly called lake Borgne, lies between the parish of Plaquemines in La. and Hancock co. Miss. It communicates on the N. E. with Pascagoula sound by the pass of Christian, between Cat island and the main shore of Mississippi, on the E. with the gulf of Mexico by the pass of Marian, and to the N. W. with lake Pontchartrain, by the two passes of Rigoletts and Chef Menteur. From its position lake Borgne is important, as through it exists the inland or Pontchartrain entrance to New Orleans. (*See Pascagola sound.*)

BOREDINO, p-v. in the N. W. part of Wayne co. Mich. by p-r. 547 ms. N. W. by W. W. C. and 30 westwardly from Detroit.

BOSCAWEN, p-t. Merrimack co. N. H. 8 ms. N. N. W. Concord, 52 N. W. Portsmouth, E. Merrimack r. 7 ms. by 7, contains 32,230 acres, and is partly watered by Blackwater r. on which are fine meadows and mill sites, and it has Great and Long ponds. It has extensive meadows on Merrimack r. The hilly country (five sevenths of the t.) is fertile and bears oak, &c. It is remarkably healthy. First settled 1734, and the inhabitants lived in a garrison 22 years. Pop. 1830, 2,093.

BOSTIC'S MILLS, and p-o. in the s. part of Richmond co. N. C. by p-r. 411 ms. s. w. W. C. and 127 s. w. by w. Raleigh.

BOSTON CORNER, t. Berkshire co. Mass. Pop. 1830, 64.

BOSTON, s-p. capital of Mass. and st. jus. Suffolk co. is the largest city in New England, and the fourth in the U. S. in population. It stands on an irregular peninsula, at the bottom of Mass. bay, and is united s. w. to the main land by a narrow neck, which formerly was overflowed by high tides. Length nearly 2¼ ms., original breadth 1 m. but by the addition of streets built out upon the flats is now much wider; so that including South Boston (which is not within the peninsula) the whole extent is about 3 sq. ms. It is distant from W. C. N. E. 436 ms. 300 N. E. Phil., 210 N. E. New York, 100 E. N. E. Hartford, 40 N. E. Providence; long. 70° 58′ 53″ w. lat. 42° 22′ N. Pop. 1765, 15,520, in 1790, 18,038, in 1800, 24,937, in 1810, 33,250, in 1820, 43,298, and in 1830, 61,392. Its harbor is commodious; it might contain 500 ships, and is of depth sufficient for those of the largest size. It is protected from storms by numerous islands. On Governor's and Castle islands, are forts Warren and Independence, which defend the harbor, and in a good degree shelter it from the sea. The entrance is very narrow.

Commerce.—The commerce of Boston has always been great. Many ships are owned and employed in their coasting and foreign trade; imports about $14,000,000 and exports about $10,000,000; amount of revenue secured at the custom house in 1831, $5,299,608. Aggregate tonnage of 3 lines regular coasting vessels to New York 1920, and their cargoes estimated at $6,000,000 per annum. There are also regular packets to Phil. Baltimore, Alexandria, Georgetown, Washington, Charleston, Hartford, Albany, Troy, &c. Large investments have been made by the capitalists here, in the joint stock manufacturing establishments of Mass. in Lowell, Waltham and many other towns; and in N. H., R. I., Conn. and other states.

The interior communication has been much improved. The Middlesex canal, which is 29¼ ms. long, breadth 30 feet, and depth 3 feet, with an elevation of 104 feet, extends to Lowell from Boston harbor, and forms with Merrimack river a navigable channel to Concord, N. H. Large quantities of manufactured goods and the raw material, granite, timber, &c. are transported upon it. A rail road is now constructing to Lowell, and several others are projected; from Boston to Albany, and Ogdensburg, N. Y. to Providence R. I. and Taunton; and surveys were begun in 1831 for one or more to the waters of Narraganset bay.

Numbers of fine stage coaches run regularly on all the principal roads from Boston, and the travelling between that place and Providence is very great, while the steam boats ply on Long Island sound. By steam boats to Augusta, Me. and thence to Quebec by the Kennebec road, travellers have gone through in five days.

Banks, Insurance Companies, &c.—There are 22 Banks, the aggregate capital of which is $15,600,000, including a branch of the U. S. bank, capital $1,500,000, and the State bank, capital $1,800,000. In 1831 their dividends were $770,000. There are 13 Marine Insurance companies, capital $3,375,000, and 10 Fire Insurance companies, capital $2,8 0,000: the Mass. hospital and Life Insurance company, capital, $500,000, and the Mass. Assurance, and Fishing Insurance companies. Seventeen of these companies, in 1831, averaged more than 5 per cent. dividend.

Bridges, &c.—There are seven bridges connecting Boston with the neighboring country. The oldest, Charles r. bridge, 1,503 feet long, 42 wide, built on 75 piers and extending across Charles r. to Charlestown; Warren, nearly parallel with the former, and also running to Charlestown, 1,420 feet long, on piers, and macadamized, at the end of which the Lowell rail road terminates; Craigie's bridge, running in a N. W. direction to Lechmere point, standing on piers also, and macadamized; Cambridge, or West Boston bridge, running nearly E. and W. 3,483 feet in length, and supported by 180 piers, extending to Cambridge port; and the Western Avenue, or Mill dam, so called, 8,000 feet long and 50 wide, running a little S. of W. across to Roxbury. This latter is formed of solid earth, supported by stone walls on the sides; with the addition of a cross dam, two large basins are formed which are alternately filled at ebb and flood tide, by means of which a perpetual water power is created for mills, and other machinery connected with it. These dams were built in 1823—4 and cost over $600,000. There is a branch from Craigie's bridge also, which runs to Charlestown point, near the Mass. state prison. The preceding are all toll bridges. South Boston bridge, running about S. S. E. from the Neck to South Boston, formerly a toll bridge, rendered unprofitable by the erection of the free bridge from Wheeler's point, the S. extremity of the city, has been surrendered to the corporation and is now free.

The wharves are numerous, generally spacious, and offer ample accommodation for shipping, and store houses for merchandize. Long wharf is 1,650 feet in length, and on central wharf, which is 1,240 feet long and 150 wide, is built a uniform range of brick stores, 4 stories high, extending the whole length of the wharf. In the centre of these is a hall and observatory, where the telegraph office is kept, which is conducted on the most approved semaphoric system. Here by means of intermediate stations on Rainsford island, and port Alderton at the mouth of the harbor, intelligence can be conveyed in 3 minutes to and from vessels 50 miles from the city. Commercial and India wharves, also, are very extensive.

Public Buildings, &c.—Boston in the number and extent of its public buildings, stores, &c. is not exceeded by any other city in the U. S. The state house stands on an eminence, the highest in the city, is built of brick, and from the summit of its dome presents a most extensive and beautiful view of the city and surrounding country. In a niche erected for the purpose, on the lower floor, stands Chantry's statue of Washington, a fine specimen of sculpture, erected at an expense of $16,000. Faneuil hall market, said to be the most beautiful building of the kind in the world, was built in 1826. It stands near the principal wharves, in the centre of business; it is built of granite, the centre about 60 feet square, surmounted by a dome; there are 2 wings, having at each extremity 4 massy granite columns, weighing about 25 tons each. Its extreme length is 540 feet, width 50, and is 2 stories high; the upper one has a spacious hall, occupied by the New England society for the encouragement of domestic manufactures. Semi-annual sales of these articles are held here. At the last, cotton and woollen goods, amounting to $452,945, boots and shoes, $61,133, and furniture, $4,876 were sold. The Tremont house is the most elegant and commodious hotel in the U. S. Its front is gray sienite, ornamented with a splendid portico of the Doric order, with fluted pillars. It cost upwards of $100,000. These, with the Tremont theatre, the Mass. general hospital, the masonic temple, Trinity church, built of stone in the Gothic style, at a cost of upwards of $160,000, and many others, are splendid specimens of architecture. Other public buildings are the county court house, which is built of stone, and cost $92,000; Faneuil hall, where town and political meetings are held; the custom house; 41 churches, some of which are very elegant; a house of industry; a house of correction; a county jail, and 10 public school houses. Of the churches 12 are occupied by Unitarians, 10 Congregationalists, 4 Episcopalians, 4 Baptists, 3 Methodists, 3 Universalists, 2 Roman Catholics, 1 Freewill Baptists, 2 African, one of which is Baptist, and the other Methodist. The Swedenborgians also have a society.

Literary, scientific, and charitable institutions.—The medical branch of Harvard university is established in Boston, where the professors reside. The Boston Atheneum has two buildings, one containing a library of 24,000 volumes; the other has two halls, one for the exhibition of paintings, &c. the other for public lectures; there are also rooms for various scientific purposes. Of scientific institutions there are the American academy of arts and sciences; the Mass. historical society; the Mass. medical society, and the mechanic institution, where an annual course of lectures is delivered on the mechanic arts. This institution has a library of about 5000 volumes. Beside these, there are the humane society, the Boston dispensary, by which attendance and medicine are furnished to the poor free of expense; the asylum for indigent boys; the female orphan asylum, and several others.

Schools.—Boston takes an honorable station, at the head of American cities, in public schools. Among these are the Elliot, Mayhew, Adams and Boylston for boys; and Hancock, Bowdoin, and Franklin for girls. There are a latin and grammar school, free to boys from 9 to 15 years of age; 8 grammar and writing schools for boys and girls, in each of which is a master for each branch; a high school, where are taught various mathematical branches, and a course of English education; an African school, and about 60 primary schools for children from 4 to 7 years of age, which are taught by females. These schools are supported at an annual expense of about $55,000. The African school is supported by the interest of a fund of $5000, given by Abiel Smith. The public schools are visited annually by the aldermen and school committee, when medals are distributed; 3,913 children attended the primary schools in 1831.

There are numerous private schools also, highly creditable to their founders, and to the city.

The streets in the older parts of Boston are narrow and crooked, but the more recent ones are generally strait, many of them wide, and well built. Granite brought from the banks of the Merrimack and from Quincy, has been much used for many years past, to the great improvement of public and private edifices. Among the latter are to be found many spacious and truly magnificent structures, unsurpassed, perhaps not to be equalled in our country. The wharves and several streets, are furnished with fine blocks of stores, among which those on each side of Faneuil hall market, and built of the same material, are in the first rank, These ranges of stores are four stories high, about 600 feet long, and are uniformly built.

The principal public square is the common, containing about 50 acres, sloping gradually, yet undulating, from the state house. In the centre is a fine pond, and on two sides it has some of the most elegant buildings in the city. The Mall, extending around it, is a gravelled walk, shaded with many fine elms, and with the common is much admired by strangers.

A cemetery has been formed at Mount Auburn, which is to be planted with shrubs, &c. on the plan of that of Pere la Chaise, near Paris. It is in a secluded valley, near a pond, with serpentine walks, and is named after different trees and shrubs.

There are seven daily newspapers published here, and numerous semi-weekly and weekly prints. There are also many other periodicals, conducted, many of them, with great ability, among which is the North American Review.

Boston was founded 1630, and the first church erected in 1632. This was the birth place of Franklin.

Boston, p-t. Erie co. N. Y. 23 ms. s. s. e. Buffalo, 289 w. Albany, has an uneven, moist loam, elevated, with streams of Canquaga cr. bearing beech, maple, hemlock, linden, &c. best for grass. Pop. 1830, 1,521.

Boswellsville, p-v. Louisa co. Va. about 40 ms. n. w. Richmond.

Botetourt, co. of Va. bounded s. w. by Montgomery and Giles, w. by Potts' mtns. separating it from Monroe, n. w. and n. by Potts mtns. separating it from Alleghany co., n. e. by Rockbridge co., e. by Blue Ridge separating it from Bedford, and s. e. by Blue Ridge separating it from Franklin. Length s. w. to n. e. 40; mean breadth 25, and area 1000 sq. ms. Extending in lat. from 37° 08′ to 37° 46′ n. and in long. from 2° 28′ to 3° 19′ w. W. C. The mountainous country has two slopes, one eastward with the general courses of James and Roanoke rivers, and the other northeastward, down the mountain valley between the Blue Ridge and Kittatinny mtns. Salem, on the Roanoke, and in the southern part of the county, is 1006 feet above tide water, and Pattonsburg on James river in the northeastern part, 806 feet. The acclivity is rapid towards the western border on Potts' mtn. and the mean elevation of the arable soil may be assumed at 1250 feet. The great body of the county is drained by the confluents of James r.; the southern angle is traversed by Roanoke. Both rivers are navigable to tide water. Chief towns, Fincastle and Salem. Pop. 1820, 13,590, and in 1830, 16,354.

Botetourt Springs, and p-v. Botetourt co. Va. by p-r. 11 ms. westward from Fincastle, the co. seat, and 189 w. Richmond.

Bottle Hill, p-v. Chatham, Morris co. N. J. 16 ms. n. w. Elizabethtown, 5 ms. s. w. Morristown, is agreeably variegated with pretty good soil and fine landscapes. It is the residence of several respectable French families. Stage coaches run daily on 2 lines to N. Y.

Bouquet, r. Essex co. N. Y. 35 ms. long, runs e. into lake Champlain, 23 ms. n. Crown point, and is boatable 2 ms. With its branches it affords many mill seats. An entrenchment was thrown up here by General Burgoyne.

Bourbon, co. of Ky. bounded s. by Clark, s. w. and w. by Lafayette, n. w. by Harrison, n. and n. e. by Nicholas, and e. by Montgomery. Length 17 ms. mean breadth 15, and area 225 sq. ms., extending in lat. from 38° 03′ to 38° 22′, and in long. from 6° 56′ to 7° 25′. This highly productive county slopes to a little w. of n. and is drained by various confluents of the south Fork of Licking r. Chief towns, Paris and Millersburg. Pop. 1820, 17,664, in 1830, 18,436, or at the latter epoch, a fraction above 72 to the sq. m. Paris, the st. jus. is by p-r. 43 ms. e. Frankfort.

Bovina, p-t. Delaware co. N. Y. 89 ms. s. w. Albany, 10 s. e. Delhi, is hilly, with good grazing, and contains Fish lake. Pop. 1830, 1348.

Boundbrook, p-v. Warren, Somerset co. N. J. 7 ms. n. w. New Brunswick, has a good level soil, n. Rarritan r. The farms are small, but pretty well cultivated.

Bow, t. Rockingham co. N. H. 6 ms. s. e. Concord, contains about 16,000 acres, s. w. Merrimack r. has an uneven and hard soil, contains Turee pond, and Turkey r. In Merrimack r. are Turkey and Garven's Falls: Bow canal, 3 ms. from Concord, ½ m: long, surmounts a fall of 25 feet, cut through granite; cost $13,000. $2,000 of the 1st income were paid for clearing Turkey falls, &c. It is very healthy. Pop. 1830, 1,065.

Bowdoin, p-t. Lincoln co. Me. 20 ms. s. s. w. Augusta, has no considerable streams. Pop. 1830, 2095.

Bowdoinham, p-t. Lincoln co. Me. 20 ms. s. by w. Augusta, w. Kennebec r. Pop. 1830, 2061.

Bowers, p-v. in the southern part of Southampton co. Va. by p-r. 91 ms. s. s. e. Richmond.

Bowers' Store and p-o. in the northern

part of Ashe co. N. C. by p-r. 374 ms. s. by w. W. C.

Bowersville, p-o. in the southern part of Geo. 10 ms. s. from Carnesville, the st. jus. and by p-r. 124 a little E. of N. from Milledgeville.

Bowler's, p-o. in the southern part of Essex co. Va. by p-r. 62 ms. N. E. by E. Richmond.

Bowling Green, p-v. and st. jus. Caroline co. Va. by p-r. 43 ms. N. N. E. Richmond.

Bowling Green, p-v. in the southern part of Oglethorpe co. Geo. by p-r. 62 ms. N. N. E. Milledgeville.

Bowling Green, p-v. and st. jus. Warren co. Ky. on Big Barren, a branch of Green r. by p-r. 142 ms. s. w. Frankfort, and 77 a little E. of N. Nashville.

Bowling Green, p-v. and st. jus. Clay co. Ind. on Eel r. a branch of the West Fork of White r. 69 ms. s. w. by w. from Indianopolis.

Bowling Green, p-v. and st. jus. Pike co. Mo. by p-r. 84 ms. N. w. from St. Louis, and 132 ms. N. E. by E. from Jefferson.

Bowman's Mills and p-o. in the N. w. part of Rockingham co. Va. by p-r. 23 ms. N. N. w. from Harrissonburg, the st. jus. for the co. and 145 N. w. by w. Richmond.

Bowman's Mountain, or Bald Mountain, local name of that part of the Alleghany chain extending over Lycoming and Luzerne cos. Penn. between the two main branches of Susquehannah river. The general range of the Alleghany chain, including Bowman's mountain, until it merges into the great nucleus of the Catsbergs, is from the southwest by west, to northeast by east. Northeastward of the main branch of Susquehannah it is known as the Tunkhannoc mountain. The ordinary height above its base is about 1000 feet, and resting on a plain, with a mean elevation of 500 feet. The actual oceanic elevation of Bowman's mountain is about 1500 feet. The naked and barren aspect of this ridge, has given it the term of Bald mountain.

Bowman's Valley, drained by Bowman's and Bourn's crs. between Bowman's and Mahoopeny mountains, Luzerne co. Pa. The soil is general sterile.

Bowyer's Bluff, precipice of limestone rock, forming the west point of Washington harbor, Green Bay lake, Michigan, and about 100 ms. s. w. from Fort Mackinaw.

Bowyer Fort was a small stockade water battery placed on the salient angle of Mobile point, Baldwin co. Ala. and erected to defend the entrance into Mobile bay. Here on the 5th of September, 1814, Major W. Lawrence, with a small garrison of 158 men repulsed an attack made by a British squadron, of which the Hermes of 28 guns was destroyed. On the 8th of Feb. 1815, this feeble post was regularly invested by a land and naval force and surrendered to the British by Major Lawrence, but at the subsequent peace was restored to the U. S.

Boxborough, p-t. Middlesex co. Mass. 30 ms. N. w. Boston, between Concord and Nashua rs. is a small t. Pop. 474.

Boxford, p-t. Essex co. Mass. 24 ms. N. E. Boston, is uneven and gravelly, with poor soil, but well cultivated and fruitful, and has a legacy of $2061 left for Latin & grammar schools by Hon. Aaron Wood. Rye straw bonnets are made here by females, in great numbers. In 1830, many were sold in the cities at 10 and $14 as imported, which cost about 2 or $3. Pop. 1830, 935.

Boyd's, p-o. in the eastern part of Henry co. Ind. by p-r. 53 ms. a little N. of E. Indianopolis.

Boyd's Creek, and p-o. Sevier co. Ten. by p-r. 216 ms. a little s. of E. Nashville.

Boydstown, p-t. Penobscot co. Me. Pop. 1830, 123.

Boydton, p-v. and st. jus. Mecklenburg co, Va. near the centre of the co, 88 ms. s. w. Richmond.

Boyerstown, p-o. Berks co. Pa. by p-r. 66 ms. eastward from Harrisburg.

Boyle's Store, and p-o. in the northern part of Stoke co. N. C. by p-r. 156 ms. N. w. by w. from Raleigh.

Boylston, t. Worcester co, Mass. 7 ms. N. E. Worcester. Pop. 1830, 820.

Boylston, t. Oswego co. N. Y. Pop. 1830, 388.

Bozrah, p-t. New London co. Conn. 33 ms. w. s. w. Hartford, 14 N. by w. N. London, 5 w. Norwich, 4 ms. by 4½; 18 sq. ms. is uneven, with granite rocks, and rich gravelly soil, bearing oak, walnut, chestnut, &c. grass, grain and flax. It is watered by Yantic r. Pop. 1830, 1073.

Brackville, p-o. and tsp. Trumbull co. O. by p-r. 155 N. E. by E. from Columbus. Pop. 1830, 584.

Bracken, co. Ky. bounded E. by Mason, S. E. by Nicholas, s. w. by Harrison, w. by Pendleton, and N. by Ohio r. separating it from Clermont co. O. Extending in lat. from 38° 30′ to 38° 47′ and in long. from 6° 50′ to 7° 11′ w. from W. C. Length 20, mean breadth 11 ms. area 220 sq. ms. Though bordering on Ohio r. and of moderate extent, Bracken co. lies mostly on a table land, the southern part declining wstrd. and drained in that direction by the North Fork of Licking. The northern section slopes nthrd. towards Ohio r. Surface very broken, but soil productive. Chief t. Augusta. Pop. 1820, 5,280, and in 1830, 6,518.

Bracken Cross Roads, p-o. in the southern part of Bracken co. Ky. by p-r. 62 ms. N. E. from Frankfort.

Bracken's p-o. in the eastern part of Sumner co. Ten. by p-r. 41 ms. northeastward from Nashville.

Braddock's bay, on lake Ontario, N. Y. (*See Greece.*)

Bradford, p-t. Merrimack co. N. H. 28 ms. w. Concord, midway between Connecticut and Merrimack rs. 31 ms. Amherst, 80 N. w. Boston, contains 19,000 acres, of which 500 are in ponds, furnishing small mill streams.

Todd's pond, the N. branch of Warner r. has floating islands. The t. is partly hilly, partly level, with various soils, and has quarries in the E. part. Pop. 1830, 1,285.

Bradford, p-t. Orange co. Vt. 7 ms. s. Newbury, 25 Montpelier, w. Connecticut r. Pop. 1830. 1,507.

Bradford, p-t. Essex co. Mass. 30 ms. N. Boston, 20 N. Salem, 10 w. Newburyport, s. and E. Merrimack r. is uneven, with much good soil. Johnson's cr. supplies mill sites. A bridge of 800 ft. crosses to Haverhill, on stone piers. Ship building has been carried on. There are several villages and an academy, and shoes are made in great quantities. Pop. 1830, 1856.

Bradford, one of the northern cos. of Pa. bounded E. by Susquehannah, s. E. by Luzerne, s. by the eastern part of Lycoming, w. by Tioga co. of Pa., and N. by Tioga co. of N. Y. Length from E. to w. 40, mean breadth 30 ms. area 1,200 sq. ms. Extending in lat. from 41° 32′ to 42° N. and in long. from 0° 2′ w. to 0° 44′ E. from W. C. This co. is traversed from N. w. to s. E. by the main volume of the Susquehannah, which receives its large tributary, the Chemung or Tioga river on its northern border. The co. is formed by two slopes, the western declining rather E. of N. towards the Susquehannah, whilst the eastern slope falls also towards the Susquehannah. The face of this large co. is mountainous, though much excellent soil skirts the stream. It has a very diversified surface, as regards both land and water scenery. Near the northern border, the main Susquehannah receives the Chemung or Tioga from the northwest, and the united waters after a rugged and tortuous course, s. eastward, through mountain chains, leaves the southeastern angle of Bradford and enters Luzerne. Derived also from elevated sources, the Wyalusing pours down from Susquehannah co. and on the opposite sides, Towanda and Sugar creeks fall rapidly out of the vallies of Bowman's mountain. Bituminous coal is found on the Towanda creek. The mountain vallies abound in excellent timber, consisting of hemlock, oak, pine, cedar, sugar maple, beech, elm, &c.; other staples, grain, flour, live stock, &c. Chief towns, Towanda and Athens. Pop. 1820, 11,554, and in 1830, 19,746.

Bradford Springs, p-o. Sumpter dist. S. C. by p-r. 48 ms. E. Columbia.

Bradleysburg, p-v. in the northern part of Louisa co. Va. by p-r. 64 ms. N. w. by w. Richmond.

Bradley's Store and p-o. in the northern part of Northampton, co. N. C. by p-r. 87 ms. N. E. by E. Raleigh.

Bradleyvale, t. Caledonia co. Vt. crossed by Moose r. is unsettled. Pop. 1830, 21.

Bradleyville, p-v. Litchfield co. Conn. 30 ms. w. Hartford.

Bradshaw, p-v. N. part Giles co. Ten. by p-r, 66 ms. s. w. Nashville.

Brailsoin's Mills, p-v. Washington co. Ten. by p-r, 245 E. Nashville and 80 eastward from Knoxville.

Braintree, t. Orange co. Vt. 21 ms. s. Montpelier. Pop. 1830, 1209.

Braintree, p-t. Norfolk co. Mass. s. of Boston, is the birth place of John Adams, 2d President of the U. S.; is on Montiquot r. Pop. 1830, 1758.

Braintrem, p-v. in N. w. part of Luzerne co. Pa. by p-r. 157 ms. from Harrisburg. Pop. 1830, 722.

Brakabeen, p-v. Schoharie co. N. Y.

Branch, co. of Mich. boundaries and position uncertain. Chief town, Bronson's prairie.

Branchtown, p-v. Phil. co. Pa. 7 ms. from Phil.

Brandenburg, p-v. and st. jus. Mead co. Ky. on Ohio r. by p-r. 108 ms. s. of w. Louisville. Pop. 1830, 331.

Brandon, p-t. Rutland co. Vt. 40 ms. N. w. Windsor, 40 s. w. Montpelier, 65 N. Bennington, is level, except near the Green mtns. in the E. with much light, fertile loam. On Otter cr. is fine alluvion; but there is a large pine plain. Mill r. has good mill sites. First settled, 1775. White and yellow pine, white and red oak, cherry, hard and soft maple, ash and cedar, prevail. Bog iron ore, from a bed, yields 33 per cent of soft metal. Copperas is also found; there are several marble quarries, and 2 curious caverns. There is a circulating library and a Lyceum. Pop. 1830, 1940.

Brandon, t. Franklin co. N. Y. Pop. 1830, 316.

Brandon, p-o. and st. jus. Rankin co. Miss. 16 ms. N E. Jackson.

Brandonville, p-v. Preston co. Va. 268 ms. N. w. Richmond.

Brandywine, a considerable creek of Pa. and Del. It rises in the Welch mountains, between Lancaster and Chester co. of the former, flows thence s. E. by comparative courses 30 ms. and enters New Castle, co. of Delaware. Continuing the original course 10 ms. to Wilmington, it there receives the Christiana creek from the w. The latter an inferior but navigable stream is formed by Red clay, White clay, and Christiana proper. The united water forming the harbor of Wilmington, admits vessels of considerable draft. Brandywine again deserves particular notice for the number of manufactories it serves to move. Beside those of flour, numerous powder and paper mills, and cloth factories are in operation along this stream, which falls from a comparatively high co. to the tide level of the Del.

Brandywine Manor, p-v. Chester co. Pa. by p-r. about 40 ms. westward from Phil.

Brandywine Mills, and p-o. in the northwestern part of Portage co. O. by p-r. 130 ms. N. E. Columbus.

Branford, p-t. New Haven co. Conn. 10 ms. E. N. Haven, 40 ms. s. Hartford, N. Long Island Sound, 5 ms. by 9, 45 sq. ms. is uneven, with gravelly loam, with oak, elm, walnut, butternut, &c. and yields grain. The farms are good; Branford river is small and navigable a short distance in vessels of 50 or 60 tons. A beautiful pond, called Saltonstall's lake, is on the w. boundary; and 2 clusters of islands, Thimble and Indian, in the Sound, belong to the t. Many of the people engage in fishing a part of the year. Pop. 1830, 2332.

Brantingham, t. Lewis co. N. Y. Pop. 1830, 662.

Brasher, t. St. Lawrence co. N. Y. Pop. 1830, 828.

Brashersville, p-o. Perry co. Ky. by p-r. 163 ms. s. e. by e. Frankfort.

Brattleborough, p-t. Windham co. Vt. 30 ms. e. Bennington, 60 n. e. Albany, 75 w. Boston, 80 n. Hartford, on w. bank Conn. r. chief town of the co.; it was the first settlement in Vt. 1724, and called Fort Drummer. It has two villages, e. and w. parishes, an academy, 40 by 56 feet, Great and Little Round mountains, with a varied surface and soil, and two streams, West r. and Whitestone branch. The e. village is large, active, and pleasant, with good mill seats on Whitestone br. near its mouth. There a bridge crosses to Hinsdale, N. H. over Conn. r. which runs rapidly here at the "Swift water." A company has been formed to make a Rail road from the e. side of Brattleboro' to the w. side of Bennington or Pownal, across the state, to connect the Troy and Bennington rail-road with the Boston and Lowell rail-road reaching to Brattleborough. Pop. 1830, 2,141.

Bratton's p-o. Smith co. Ten. by p-r. 58 ms. northeastward from Nashville.

Brattonsville, p-o. York district, S. C. by p-r. 70 ms. n. Columbus.

Breckenridge co. of Ky. bounded n. w. by Ohio r. which separates it from Perry co. in Ind., Meade n. w., Hardin s. e., Rough creek branch of Green river which separates it from Grayson s. and by Hancock w. Length 35, mean breadth 20 ms., area 700 sq. ms. surface not very broken; soil generally fertile. (*For lat. and long. see Hardinsburg, the co. st.*) In 1820, pop. 7485, 1830, 7345.

Bremen, p-v. Lincoln co. Me. 45 ms. from Augusta.

Brentonsville, p-o. Owen co. Ind., about 50 ms. s. w. Indianopolis.

Brents, p-o. Henry co. Ky. by p-r. 36 ms. n. w. from Frankfort.

Brentsville, p-o. and c-h. Prince William co. Va. 31 ms. s. w. from W. C.

Brentwood, p-t. Rockingham co. N.H. 37 ms. s. e. Concord, with 10,465 acres, watered by Exeter r. and other streams, yields grass well. At Pick Pocket Falls, on Exeter r. are several mills and factories. Iron ore and vitriol have been found. Pop. 1830, 770.

Breton, two small islands of La. in the southwestern part of Chandeleur bay, about 28 ms. nearly due n. from the main pass of Miss. and 1½ ms. s. w. from the Grand Gozier. There is a channel with 12 feet water between Cape Breton and Grand Gozier, lat. 29° 26′ n. long. on Tanner's U. S. map, 12° 02′ w. from W. C.

Bretton Woods, Coos co. N. H. at the n. w. base of the White mtns. and adjoining ungranted lands; it is almost uninhabited, is uneven, dreary, and contains 24,640 acres, with part of Pondicherry mtn. n., and Amonoosuc and streams of John's and Israel's rs. Pop. 1830, 108.

Brevard, p-o. Smith co. Ten. 54 ms. n. e. by e. Nashville.

Brevardsville, p-o. Buncombe co. N C. 267 ms. w. from Raleigh.

Brewer, p-t. Penobscot co. Me. 5 ms. s. e. Bangor, 67 n. e. Augusta, e. Penobscot r. n. Hancock co. Pop. 1,078.

Brewster, p-t. Barnstable co. Mass. 88 ms. s. e. Boston, 18 n. w. Plymouth, s. cape Cod bay, and beyond the elbow of the cape; was named after elder Brewster, one of the first settlers of Plymouth, who died 1644. Pop. 1830, 418.

Brickersville, p-o. Lancaster co. Pa. by p-r. 45 ms. estrd. from Harrisburg.

Brickland's Cross Road, p-v. in the northern part of Washington co. Pa. by p-r. 233 ms. wstrd. from Harrisburg, 25 a little s. of w. Pittsburgh, and 11 e. Steubenville.

Brick Meeting House, and p-o. in the northern part of Cecil co. Md. 12 ms. n. w. from Elkton.

Bricksville, p-v. in the s. w. part of Cayahoga co. O. by p-r. 122 ms. n. e. Columbus.

Brickville, p-o. formerly Town Creek Mills, in the n. part of Lawrence co. Ala. by p-r. 128 ms. a little e. of n. Tuscaloosa.

Bridgehampton, p-v. Southampton, Suffolk co. N. Y. 100 ms. e. New York.

Bridgeport, p-t. and borough, Fairfield co. Conn. 3 ms. w. Stratford, 17 w. New Haven, 51 s. w. Hartford, 62 n. e. New York, on the w. side Bridgeport harbor, which is 3 ms. long, from L. I. Sound, ½ to 2 ms. wide. Pequanock r. furnishes mill seats. The harbor is narrow, shoal except in the channel, with a bar, having 13 feet at high water, is easy of access, and has a beacon of 40 ft. There is a lighthouse on Fairweather isl. and a toll and draw bridge at the head of the harbor. It is a thriving and pleasant town, with good soil. Banking capital, $305,500. Pop. 1830, 2,800.

Bridgeport, v. Junius, Seneca co. N. Y. called also W. Cayuga v. at w. end Cayuga bridge, 185 ms. w. Albany. The lake boats touch here.

Bridgeport, p-v. Harrison co. Va. 10 ms. s. e. from Clarksburg.

Bridgeport, borough of Fayette co. Pa. situated on the right bank of Monongahela r. separated from Brownsville by Dunlap's cr. The site of Bridgeport is a high bottom of the river. Pop. 1820, 624, 1830, 727.

Bridgeport, p-v. Pease tsp. n. eastern part of Belmont co. O. 10 ms. from St. Clairsville, and by p-r. 134 ms. e. from Columbus. Pop. 1830, 165.

Bridgetown, Cumberland co. Me. 39 ms. n. w. Portland. It has an academy. Here begins the Cumberland and Oxford canal, at

Long pond, and extends (including Brandy and Sebago ponds and outlets, 27 ms.) to Portland, 50 ms. There are 24 locks. Tolls per mile, for planks, 6 cents per M. feet; shingles, 2 cts. an M.; wood 6 cts. a cord; timber 6 cts. a ton; goods in boats, 6 cts. a ton; boats, rafts, &c. 6 cts. additional for each lock. Pop. 1830, 1,541.

BRIDGETOWN, p-t. and cap. Cumberland co. N. J. 50 ms. S. E. Philadelphia, and 69 from Trenton, on Cohanzey cr. 20 from Delaware bay; contains the co. buildings, and is accessible to vessels of 100 tons.

BRIDGEVILLE, (or *Bridgetown*,) p-v. on, or near the right bank of Nanticoke r. Sussex co. Del. 35 ms. a little W. of S. from Dover, and about 30 ms. nearly due E. from Easton Md.

BRIDGEVILLE, p-v. in the N. E. part Muskingum co. O. by p-r. 68 ms. estrd. Columbus.

BRIDGEWATER, p-t. Grafton co. N. H. 20 ms. N. N. W. Concord, W. Pemigewasset r. E. Newfound pond; yields grass, with no large streams. First settlement 1766. Pop. 1830, 784.

BRIDGEWATER, p-t. Windsor co. Vt. 45 ms. S. Montpelier, 17 N. W. Windsor, 60 ms. N. E. Bennington, 7½ ms. by 8, with 46½ sq. ms. first settlement 1779; is watered by Queechy r. and branches, which supply mills, and is uneven, partly rough and stony, with primitive rocks, and a quarry of soap stone, which is manufactured. Iron ore is also found. Pop. 1830, 2,320.

BRIDGEWATER, p-t. Plymouth co. Ms. 28 ms. S. Boston, 18 ms. N. W. Plymouth, has tolerable soil, but a sandy tract begins here which goes through the S. part of the co. Arms were made here in the revolutionary war—first settled, 1651—burnt in 1676, by Indians. It contains 88 acres of tillage, 1,547 mowing, and 4,904 pasturage. Pop. 1830, 1,855.

BRIDGEWATER, p-t. Oneida co. N. Y. 12 ms. S. Utica, 6 ms. by 4, well watered by head streams of Unadilla r. It has a rich valley E. and is hilly W. Pop. 1830, 1,608.

BRIDGEWATER, Somerset co. N. J. 3 ms. N. Boundbrook. It has a copper mine. Pop. 1830, 3,549.

BRIDPORT, p-t. Addison co Vt. E. lake Champlain, opposite Crown Point, 8 ms. W. Middlebury, 35 S. Burlington, 41 S. W. Montpelier, with 42 sq. ms. It is nearly level, with loam and slaty sandstone, bearing oak, white and Norway pine, on the lake—maple and beech E. It has few streams. The soil contains Epsom salts, which were formerly made from the water. There are wharves and landing places on the lake. First permanent settlement, 1768. Pop. 1830, 1,774.

BRIER CREEK, p-o. Wilkes co. N. C. by p-r. 194 ms. N. W. by W. Raleigh.

BRIGHTON, t. Somerset co. Me. Pop. 1830, 722.

BRIGHTON, p-t. Middlesex co. Mass. 5 ms. W. Boston, S. Charles r. has pleasant hills and vallies, good soil, well cultivated, with 200 acres of marsh. The annual cattle Fair is the most important in N. England. It was commenced in the revolutionary war, and is now under the direction of the Mass. agricultural society, who have a neat building for the exhibition of various articles for which premiums are offered. In 1830, were sold,

Beef cattle,	37,767	Sales	$977,989 75
Stores,	13,685	"	154,564 00
Sheep,	132,697	"	215,618 17
Swine,	19,639	"	70,970 50
Whole number,	203,789		$1,419,142 42.

Pop. 1830, 972.

BRIGHTON, p-t. Monroe co. N. Y. E. Genesee r. opposite Rochester, S. Lake Ontario, 66 sq. ms. has a good soil, and a great amount of water power on Genessee r. with many flour mills, factories &c. Carthage v. or Clyde, 2½ ms. N. Rochester, has considerable business as the landing place of lake vessels below the lower falls of Genessee r. The banks are 200 feet perpendicular, of rock in strata. A wooden bridge was thrown across some years ago, which soon after fell. Erie canal passes through the t. and receives a feeder from Genessee r. above the rapids. Pop. 1830, 6519.

BRIGHTON, p-v. Beaver co. Pa. about 20 ms. from Pittsburg.

BRIGHTON, tsp. and p-o. in the N. part Loraine co. O. by p-r. 116 ms. N. N. E. Columbus.

BRIGHTSVILLE, p-v. in the N. part Marlborough dist. S. C. by p-r. 102 ms. N. E. by E. Columbia.

BRIMFIELD, p-t. Hampden co. Mass. 19 ms. E. Springfield, 75 S. W. Boston, on E. side of the lime range of mtns. with hills and valleys, 6 ms. by 5, and good farms. First settled 1701, with Chickopee and Quinebaug rs. Gen. Eaton, formerly U. S. Consul at Tunis, was born here. Pop. 1830, 1,599.

BRINDLETOWN, p-o. Burke co. N. C. 199 ms. W. Raleigh.

BRINKLEYSVILLE, p-v. in the W. part Halifax co. N. C. by p-r. 83 ms. N. E. Raleigh.

BRISTOL, p-t. Lincoln, co. Me. 13 ms. E. Wiscasset, 45 Augusta. Pop. 1830, 2450.

BRISTOL, p-t. Grafton co. N. H. 90 ms. N. W. Boston, 16 S. Plymouth, 30 N. Concord, is hilly, with good soil. It lies W. Penigewasset r. contains 9000 acres land, besides several ponds of water, one, 2 or 3 ms. by 6, called Newfound pond, which empties by an outlet into Penig. r. a pleasant village stands at the junction. A toll bridge crosses to New Hampton. Black-lead is found here. Pop. 1830, 779.

BRISTOL MILLS, p-v. Lincoln co. Me. 42 ms. from Augusta.

BRISTOL, p-t. Addison co. Vt. 54 ms. S. W. Montpelier, 25 S. E. Burlington, 26,000 acres. First settled, at the close of the revolutionary war; ⅓ of the t. W. Green mtns. is quite level and rich, the rest broken and useless. A range of mtns. crosses N. and S. above New Haven r. it is called the Hog Back; below, South mtn. Pop. 1247.

BRISTOL co. Mass. bounded by Norfolk co. N., Plymouth co. the sea and R. I. S. and R. I.

w. contains 19 towns, of which Taunton is the st. jus. and co. t. New Bedford is large and flourishing. It is divided by Taunton river; whose streams supply mills and manufactories. Pop. 1830, 49,592.

BRISTOL, co. R. I. bounded by Mass. N. E. and N. W., Mount Hope bay S. E., Narraganset bay S. W. 3 ms. by 8, with 25 sq. ms. between two fine sheets of water, a pleasant situation, diversified surface, rich loam, with granite rocks, except N., where it is more level. The harbors are good, and much commercial enterprize has existed here, in proportion to the size of the towns. This was the country of the Indian King, Philip. There are few manufactories. A sealing company was formed here in 1831. Population 1830, 5446.

BRISTOL, p-t. s-p. and cf. t. Bristol, co. R. I. 13 ms. N. Newport, 15 S. Providence, 2 ms. by 5, 12 sq. ms. on a good harbor, E. side of Narragansett bay, W. Mount Hope. It has an agreeable variety of surface, with Mount Hope, a tall eminence, S. E. the chief seat of Metacom, or King Philip, who made a destructive war on the New England colonies, 1675, and was killed here 1676. The land is well cultivated. Many onions are exported. There are 4 banks, capital $361,250. Part of the town was burnt in the revolutionary war. The trade is less than formerly, employing 30 in foreign trade, and 12 coasters. A factory is building for patent wrought nails. There are 5 churches, Baptist, Episcopal, Methodist, Reformed Methodist, and Presbyterian. Pop. 1830, 3054.

BRISTOL, p-t. Hartford co. Conn. 16 ms. W. by S. Hartford, 28 N. New Haven, 5 ms. by 5 ½, about 27 sq. ms., hilly, gravelly loam, good for grain and grass, bearing oak, chestnut, &c. The rocks are granite, with some iron, and copper ore, and the streams small branches of Farmington r. The manufactories are various, including wooden clocks, 30,000 of which were made in 1831. 800 persons are employed in making brass clocks. Pop. 1830, 1707.

BRISTOL, Ontario co. N. Y. 10 ms. S. W. Canandaigua, has an inferior soil, high, broken land between Canandaigua and Honeoye lakes, which discharge E. and W. The source of Mud cr. gives a few mill seats. It has a burning spring. Pop. 1830, 2,952.

BRISTOL, p-t. borough and port, Bucks co. Pa. situated on the right bank of Delaware r. 20 ms. above Philadelphia, and 12 below Trenton. It is a neat and elegant village on a swelling bank, running chiefly in one street, along the river. Pop. 1830, 1,262.

BRISTOL, tsp. S. E. part of Bucks co. Pa. lying around the borough of Bristol. Pop. 1830, 1,534, exclusive of the borough.

BRISTOLVILLE, tsp. and p-o. in the eastern part of Trumbull co. O. the p-o. is by p-r. 167 ms. N. estrd. from Columbus. Pop. 1830, 526.

BRITTON'S STORE, and p-o. Bertie co. N. C. by p-r. 110 N. of E. Raleigh.

BROADALBIN, p-t. Montgomery co. N. Y. 38 ms. W. Albany, 6 E. Mohawk r., 5 ms. by 10, has a strong loam, yields grass, grain, &c. sugar maple, beech, birch, &c. First settled 1776; deserted in revolutionary war. Chuctanunda, Fonda's, Hans, and Frenchman's crs. which give mill sites. The v. on W. line, is 10 ms. from Johnstown. Pop. 1830, 2657.

BROAD, r. of N. and S. C. having its extreme source in the Blue Ridge, and in Burke co. N. C. but draws most of its remote constituents from the valley of Rutherford co. Pursuing thence a southeastern course, the various branches unite and enter S. C. between Yorke and Spartanburg dists. Inclining to a general course of S. S. E. receiving only large creeks from the left, but on the right, augmented by the comparatively considerable streams of Pacolet, Tyger, and Enoree rivers, it finally unites with the Saluda at Columbia, to form the Congaree. The valley of Broad r. including all its confluents, reaches from N. lat. 34° to 35° 30′ interlocking sources with the Catawba, French, Broad and Saluda rivers, and draining a valley embracing 130 ms. by a mean breadth of 35 ms. or an area of 4,550 sq. ms.

BROAD, r. of Geo. one of the western branches of Savannah r. rises in Habersham, Hall, and Franklin cos. pursues thence a S. eastern course through Elbert, Madison, Oglethorpe and Wikes cos. and falls into Savannah r. at the centre of the triangle, between Petersburg, Vienna, and Lisbon, after a comparative course of 70 ms.

BROAD RIVER, S. C. is formed by the tide water part of Coosawatchie r. and is the local name of the inner part of Port Royal entrance, Beaufort dist. The bay, for it is in fact such, called Broad river, inside of Hilton Head, extends in a N. western direction, 20 ms. with a mean breadth of 2 ms. and opens to the ocean 22 ms. N. E. from the mouth of Savannah r.

BROAD CREEK, p-v. on the western shore of Kent Island, Queen Anne co. Md. nearly opposite and 10 ms. distant from Annapolis, and 47 ms. a little N. of E. from W. C.

BROAD MOUNTAIN, one of the Appalachian chains in Northampton and Schuylkill co. Pa. It is the next chain, or rather ridge wstrd. from the Mauch Chunk mtns. and like the latter, contains much anthracite coal. It receives its name from its width on the summit, which differs from 2 to 5 ms. (*See Appalachian system.*)

BROAD MOUNTAIN, p-o. on the Broad mtn. and N. eastern part of Schuylkill co. Pa. 71 ms. N. E. from Harrisburg.

BROCK'S GAP, and p-o. Rockingham co. Va. by p-r. 113 N. W. Richmond.

BROCKPORT, p-v. Sweden, Monroe co. N. Y. 18 ms. W. Rochester on Erie canal, has grown to some importance in a few years. A rail road to Alleghany r. has been proposed, about 85 ms.

BROCKVILLE, p-v. Clearfield co. Pa. by p-r. 139 ms. N. W. from Harrisburg.

BROCKWAYVILLE, p-o. in the S. eastern part

of Jefferson co. Pa. by p-r. 154 ms. N. W. by W. from Harrisburg.

BRONSON'S PRAIRIE, and p-v. Branch co. Mich. by p-r. 133 ms. from Detroit.

BRONX cr. Westchester co. N. Y. runs from Rye pond to East r. about 28 ms. and supplies mills. It has been proposed to lead the water to N. York, to supply the city, to which it is supposed to be adequate.

BROOKE, N. W. co. of Va. bounded W. by Ohio r. which separates it from Jefferson co. Ohio, N. W. by Ohio r. which separates it from Columbiana co. O., E. by Beaver and Washington cos. Pa. It is a mere slip, 30 ms. by 5 ms. and area 150 sq. ms. Surface very hilly, but highly fertile soil. Buffalo, Cross, Harman's, and other creeks rising in Pa. traverse Brooke in their way to Ohio r. The staples are grain; and its products some iron, and bituminous coal. Chief t. Wellsburg. Pop. 1820, 6,611, in 1830, 7,041.

BROOKEVILLE p-v. on a small branch of the Patuxent, Montgomery co. Md. 22 ms. almost due N. from W. C. and 28 ms. S. W. from Baltimore. In this village is an academy in active operation, in which are taught the Latin and Greek languages, with the various branches of an English education.

BROOKFIELD, p-t. Orange co. Vt. 17 ms. S. Montpelier, 40 N. W. Windsor, nearly on the height of land between White and Onion rs. is generally good grass land, with the second branch of White r. and several large ponds. Lime is made here from marl. First settled, 1779. Pop. 1830, 1677.

BROOKFIELD, Strafford co. N. H. Pop. 1830, 671.

BROOKFIELD, p-t. Worcester co. Mass. 18 ms. W. Worcester, 64 S. W. Boston, is beautifully varied, has good soil, with two large and beautiful fish ponds, and three handsome villages. Quaboag r. flows into Chickapee r. Iron ore is found. This was one of the earliest white settlements of Mass. and most suffering: began 1660, burnt by Indians 1675, and deserted for several years. Pop. 1830, 2342.

BROOKFIELD, p-t. Fairfield co. Conn. 33 W. N. Haven, 50 S. W. Hartford, S. W. Ousatonick r. contains 17 sq. ms. It is crossed by Still r. Ousatonic r. has a bridge, and affords fish, particularly shad. It yields wheat and rye, oak, hickory, maple, chestnut, &c. The rocks are limestone, and afford marble. Pop. 1830, 1261.

BROOKFIELD, p-t. Madison co. N. Y. 22 ms. S. by W. Utica, 90 W. Albany; is hilly and fertile, and well watered, with lime rocks of petrified shells. Pop. 1830, 4367.

BROOKFIELD, one of the northern tsps. of Tioga co. Penn. the p-o. by p-r. 185 ms. N. Harrisburg. Pop. 1830, 328.

BROOKFIELD, tsp. and p-v. Trumbull co. O. by p-r. 170 ms. N. E. Columbia. Pop. of the tsp. 1830, 874.

BROOKHAVEN, t. Suffolk co. N. Y. crosses Long Island; 20 ms. long, with 300 sq. ms. has various soils, few inhabitants, and much forest. On the harbors and Long Island Sound, the soil is good. It comprehends several important headlands in the Sound; Crane Neck, Old Field Point, Strong's Point, and Mount Misery. A light house stands on Old Field Point. There is much salt meadow and sand on the Sound, and plenty of fish are taken. On the S. are many trout brooks and mill streams. Interior are pine plains, with plenty of good deer. N. are good small harbors—Stony Brook, Setauket and Drowned Meadow; S. is South Bay. Population, 6095. Setauket, v. 58 ms. E. N. Y. is the oldest, and was once inhabited by a tribe of Indians. There are also the villages of Stony Brook, Drowned Meadow, Old Man's, Miller's Place, Wading r. Coram, Patchogue, Blue Point (celebrated for oysters,) Fireplace, Mastic, the Forge and Morriches. The last 4 are on S. Bay, which extends from Hempstead, Queen's co. to Southampton, Suffolk co. nearly 100 ms. and from 2 to 5 ms. wide. It affords fine fish, and clams; and much pine is carried to N. Y. for fuel. It has been proposed to cut a canal through the W. shore to open it more directly to navigation. Stage coaches run regularly from Patchogue, &c. to Brooklyn. Ronconcoma pond, 3 ms. round, on the W. line, is the centre of Long Island. Pop. 1830, 6098.

BROOKLINE, t. Hillsboro' co. N. H. 45 ms. from Concord, 7 Amherst, 43 N. W. Boston, contains 12,664 acres, 240 of them water; Nisitissit r. runs into Potanipo pond, in the centre, and thence to Nashua r. in Pepperell. Pop. 1830, 627.

BROOKLINE, p-t. Norfolk co. Mass. 5 ms. S. W. Boston, has an agreeably diversified surface, adorned with many well tilled farms, and country seats belonging chiefly to city gentlemen. Pop 1830, 1043.

BROOKLYN, t. Windham co. Vt. 40 ms. S. Windsor, 2 ms. by 8. First settled 1777. Grassy cr. runs through a valley in the centre. Here is a bed of porcelain clay. Pop. 1830, 376.

BROOKLYN, p-t. and st. jus. Windham co. Conn. 14 ms. N. Norwich harbor, 30 E. Hartford, 44 W. Providence, 6 ms. by 8, with 46 sq. ms., has a good soil and is hilly, with primitive rocks, and a quarry of building stone near the c-h. It yields a variety of produce, and walnut, oak, chestnut &c. Willimantic and Nachaug rs. unite and form Shetucket r. and with branches give mill sites, and fish. First settled, 1686. It has 2 societies. Pop. 1830, 1451.

BROOKLYN, p-t. Kings co. N. Y. on the W. end of Long Island, opposite N. Y. city, of which it in fact forms a suburb, is the 3d town in the state in population. The village, which is incorporated, and the largest in the state stands on an acclivity rising from the East r. and an extent of high land above, so that some of the houses overlook the metropolis. The lower streets are narrow and crooked, but the higher strait and agreeable. There are 7 churches, 2 Presbyterians, 2

Baptist, 1 Dutch Reformed, 1 Episcopal, 1 Catholic.

Banking cap. $300,000. It has two markets, court-house, several good private schools, manufactories of different kinds, extensive store houses, &c. A navy yard of the U. S. is a little N. of the village, S. of Wallabout bay, where the largest ships are built, launched, and repaired in security. There are two large buildings for constructing frigates and larger ships, under shelter; the house of the commandant, barracks for marines, and a small village adjacent; 3 steam ferries connect Brooklyn with N. Y. and many merchants reside in the village. Aug. 26, 1776, the British gained a battle near Brooklyn, by which they obtained possession of N. Y. Pop. 1820, 7175, 1830, 15,396.

Brooklyn, p-v. Halifax co. Va. by p-r. 101 ms. S. W. Richmond.

Brooklyn, p-v. Conecuh co. Ala. by p-r. 176 ms. E. of S. Tuscaloosa.

Brooklyn, t-s. and p-v. in the northern part of Cuyahoga co. O. by p-r. 149 ms. N. E. Columbus. Pop, 1830, 646.

Brookneal, p-v. in the S. W. part Campbell co. Va. by p-r. 162 ms. S. W. by W. Richmond.

Brooks, p-t. Waldo co. Me. 51 ms. from Augusta.

Brooksville, p-v. Montgomery co. Ten. by p-r. 54 ms. N. W. Nashville

Brookville, p-t. Hancock co. Me. 8 ms. from Augusta. Pop. 1830, 1089.

Brookville, p-v. in the S. W. part of Albemarle co. Va. 20 ms. S. W. Charlotteville, and by p-r. 101 ms. N. W. Richmond.

Brookville, p-v. and st. jus. Jefferson co. Pa. by p-r. 165 ms. N. of W. Harrisburg.

Brookville, p-v. and st. jus. Franklin co. Ind. on White Water r. 30 ms. N. W. from Cincinnati, and 70 S. E. by E. from Indianopolis.

Broome co. N. Y. bounded by Cortlandt and Chenango cos. N., Delaware co. E., lat 42°, and Pennsylvania S., Tioga co. W. contains 8 townships and about 700 sq. ms. It is watered by Susquehannah r. and some of its branches; has many hills, with a hard pan soil, but large and rich vallies, with gravelly loam. Susquehannah r. is boatable, and falls are numerous. First settled, from W. Massachusetts, about 1790. It is healthy and yields fruit, as well as other productions. Pop. 1830, 17,759.

Broome, p-t. Schoharie co. N. Y. 35 ms. S. W. Albany, E. Schorie cr. includes part of Catskill mts. with good alluvial vallies W. Different streams supply fine mill seats. Most of the land is leased. Livingstonville, p-o. is in S. E. part. Pop. 1831, 3161.

Brothertown, Indian v. Paris, N. Y. 8 ms. S. W. Utica, was granted by the Oneida Indians to the remnant of the Stockbridge and other tribes of N. England. They resided here in considerable numbers, with a church, a missionary, &c. but many of them have recently gone to Green Bay.

Brower, p-o. Berks co. Ten. by p-r. 74 ms. E. Harrisburg.

Brower's Mills, and p-o. in the S. W. part Randolph co. N. C. by p-r. 76 ms. westward Raleigh.

Brown *University*. (*See Providence, R. I.*)

Brown, one of the southern cos. of Ohio, bounded by Ohio r. separating it from Mason and Bracken cos. of Ky. S., by Clermont co. O. W., Clinton N. W., Highland N. and N. E., & Adams E. Length from S. to N. 30, mean breadth 17, and area 512 sq. ms., extending in lat. from 38° 44′ to 39° 17′, and in long. from 6° 40′ to 6° 58′ W. from W. C. The northwestern part between Clermont and Highland cos. slopes to S. E. and is drained by the east Fork of Little Miami. The southern and larger section declines towards Ohio r. and is drained by White Oak and several smaller crs. The surface of the whole co. is broken, but soil excellent. Chief town, Georgetown. Population 1820, 13,356, 1830, 17,867.

Brown, co. of Mich. around and contiguous to Green Bay, embracing the few settlements westward of lake Michigan and Green Bay. Boundaries uncertain. On the p-o. list 1831, Menomonie is named as st. jus.

Brown, p-o. Lycoming co. Pa.

Brown, p-o. in the E. part of Stark co. O. by p-r. 130 ms. N. E. Columbus.

Brownfield, p-t. Oxford co. Me. on Saco r 28 ms. S. W. Paris, 81 Augusta. Pop. 936.

Brownfield, p-o. in the northern part of Belmont, co. O. by p-r. 152 ms. E. Columbus.

Brownhelm, p-o. in the northern part of Loraine co. O. by p-r. 139 ms. N. N. E. from Columbus.

Browningtown, p-t. Orleans co. Vt. 95 ms. N. Windsor, 45 N. E. Montpelier, 57 E. Burlington, has 16,750 acres, with mill sites on Willoughby's r. and branches, which empty N. into lake Memphremagog. Pop. 1830, 412.

Brown's p-o. Fairfield dist. S. C. by p-r. 23 ms. northwardly from Columbia.

Brownsboro', p-o. in the S. part of Montgomery co. Md. 9 ms. from W. C.

Brownsboro' p-v. in the N. part of Madison co. Ala. by p-r. 10 ms. from Huntsville, the st. of jus.

Brownsboro', p-v. in the western part of Oldham co. Ky. by p-r. 41 ms. N. W. Frankfort.

Brownsburg, p-o. Bucks co. Pa. about 27 ms. N. Phil.

Brownsburg, p-v. on Hays' creek in the N. part of Rockbridge, co. Va. by p-r. 143 ms. a little N. of W. Richmond.

Brown's Cove, and p-o. in the N. part of Albermarle co. Virg. by p-r. 109 ms. N. W. Richmond.

Brown's Cove, and p-o. in Jackson co. Ala by p-r. 181 ms. N. E. Tuscaloosa.

Brown's Creek, and p-o. in the E. part of Union dist. S. C. 10 ms. E. Unionville and by p-r. 66 ms. N. N. W. Columbia.

Brown's Ferry and p-o. Limestone co.

Ala. by p-r. 129 ms. a little E. of N. Tuscaloosa.

BROWN'S MILLS, and p-o. Mifflin co. Pa. by p-r. 60 ms. northwestward Harrisburg.

BROWN'S MILLS and p-o. in the N. W. part of Washington co. O. 18 ms. N. W. Marietta.

BROWN'S STORE, and p-o. Caswell co. N. C. by p-r. 99 ms. N. W. Raleigh.

BROWN'S TAVERN, and p-o. Ann Arundel co. Md. 46 ms. N. W. Annapolis.

BROWNSTOWN p-v. and st. jus. Jackson co. Ind. situated on Driftwood Fork of the E. branch of White r. by p-r. 69 ms. a little E. of S. Indianopolis, and 50 N. W. from Louisville, in Ky.

BROWNSTOWN, p-v. in the southeastern part of Wayne co. Mich. 18 ms. S. S. W. from Detroit, and by p-r. 508 ms. N. W. by W. from W. C.

BROWNSVILLE, p-t. Penobscot co. Me. 40 ms. N. Bangor, 97. N. N. W. Augusta, N. Piscataway r. and on Pleasant r. one of its branches, with a large pond E. Pop. 1830, 402.

BROWNSVILLE, p-v. and borough of Fayette co. Pa. founded on a rapid acclivity rising from the Monongahela r. where stood formerly Red Stone Fort. The U. S. road passes along the main street, upon which the bulk of the houses are situated. It is separated from Bridgeport by Dunlap's cr. and stands 12 ms. N. W. from Union, the co. st. 35 ms. a little E. of S. Pittsburg. Pop. borough, 1830, 1222.

BROWNSVILLE, p-o. Frederick co. Md. by p-r. 65 ms. N. W. W. C.

BROWNSVILLE, p-v. in the N. part of Granville county, North Carolina, by p-r. 58 miles N. Raleigh.

BROWNSVILLE, p-v. in the S. part of Marlborough dist. S. C. by p-r. 116 ms. N. E. by E. Columbia.

BROWNSVILLE, p-v. and st. jus. Haywood co. Ten. situated near the centre of the co. by p-r. 175 ms. a little S. of W. Nashville, and 891 ms. S. W. by W. W. C. N. lat. 35° 35′, and 12° 20′ W. from W. C.

BROWNSVILLE, p-v. and st. jus. Edmonson co. Ky. by. p-r. 138 ms. S. W. by W. from Frankfort, and 678 wstrd. W. C. Pop. 1830, 229.

BROWNSVILLE, p-v. in the N. E. part Licking co. O. by p-r. 49 ms. N. E. by E. Columbus. Pop. 1830, 155.

BROWNSVILLE, p-o. Union co. Ind. by p-r. 76 ms. E. Indianopolis.

BROWNSVILLE, p-v. and st. jus. Jackson co. Il. situated on Muddy Creek by p-r. 833 ms. wstrd. W. C., 127 S. Vandalia.

BROWNVILLE, p-t. Jefferson co. N. Y. at the mouth of Black r. N. side, S. E. L. Ontario and Griffin's bay, 6 ms. by 15, has a marly loam, with much limestone, bearing beech, maple, bass, elm, &c. The v. is 3 ms. from the mouth of Black r. on its shore, at the head of navigation, and the lower rapids, with some manufactories. Pop. 1830, 2938.

BRUCETOWN, p-v. Frederick co. Va. 7 ms. wstrd. Winchester.

BRUCKVILLE, or Hendricks' Mills, p-o. on Pipe cr. Frederick co. Md. 18 ms. N. E. the city of Frederick.

BRUINGTON, p-v. in the N. part of King and Queen co. Va. by p-r. 36 N. E. Richmond.

BRUNEL'S p-o. Davidson co. N. C. by p-r. 100 ms. W. Raleigh.

BRUMFIELDVILLE, p-o. Berks co. Pa. by p-r. 9 ms. W. Reading.

BRUNSON'S p-o. Stewart co. Ten. 14 ms. wstrd. Dover, the st. jus.

BRUNSWICK, p-t. Essex co. Vt. 55 ms. N. E Montpelier, 23 sq. ms. W. Conn. r. First settlement 1780, watered by W. branch of Nulhegan r. Wheeler's and Paul's streams cross the town and afford mill sites. A mineral spring flows from the bank of Conn. r. near a pond. Pop. 1830, 160.

BRUNSWICK, p-t. Cumberland co. Me. 30 ms. E. Portland, S. W. Androscoggin r. at the falls, which supply excellent mill sites. Bowdoin college, incorporated 1794, was endowed by the Mass. legislature with 5 townships of land, and $3000 per annum. This sum was continued by the legislature of Me. James Bowdoin, its chief benefactor, gave the college $10,000. It has a president, 6 professors in languages, natural philosophy, chemistry, mineralogy, rhetoric and oratory, intellectual and moral philosophy, with lectureships in sacred literature and political economy.

The course of instruction resembles that of most other colleges in the United States. Students in 1831, 226. A medical academy is attached to it, with 92 students. Pop. 1831, 3587.

BRUNSWICK, t. Rensellaer co. N.Y. 5 ms. E. Troy, N. Sand lake, is high and broken N. W. The land W. is handsome. Poestenkill cr. furnishes good mill sites; there are also Tamhanoc and Wynant's crs. The soil is leased at low rates. Pop. 1830, 2570.

BRUNSWICK, one of the southern cos. of Va. bounded by Mecklenburg W., Lunenburg N. W. and N., Nottaway river, separating it from Dinwiddie N. E., Greensville E., Northampton co. in N. C. S., and Warren county North Carolina, southwest. It is nearly a square, 26 ms. each side; area 676 square ms. Extending in lat. from 36° 32′, to 36° 56′ N. and in long. from 0° 39′, to 1° 04′ W. from W. C. The southwestern angle touches the Roanoke, and a small section is drained southwardly into that stream; but the body of the county is comprised in the vallies of Meherin and Nottaway rivers, and declines eastward. Chief town, Lawrenceville. Pop. 1820, 16,687, and in 1830, 15,767.

BRUNSWICK, extreme southern co. of N. C. bounded by White Marsh creek separating it from Columbus co. of the same state W., by Bladen N. W., by Cape Fear river separating it from New Hanover on the N. E., and E. by the Atlantic ocean S., and by Hony dist. S. C. S. W. Length from the border of S. C. to the Forks of Cape Fear river, 48 ms., mean breadth 28, and area, 1344 sq. ms. Extending in lat. from 33° 53′, to 34° 32′, and in long.

from 1° to 1° 46′ w. from W. C. The slope of this county is nearly southward; the surface generally flat, marshy and sterile. Chief town, Smithville. Pop. 1820, 5480, and in 1830, 6516.

Brunswick, p-v. and st. jus. Glynn co. Geo. by p-r. 733 ms. s. s. w. from W. C. and 200 s. e. from Milledgeville. n. lat. 31° 12′, long. 4° 40′ w. from W. C. It is a seaport, and situated on Turtle river about 10 ms. nearly due w. from the opening between St. Simon's and Jekyll islands.

Brunswick, p-v. and tsp. in the northern part of Medina co. Ohio. The p-o. is about 25 ms. s. w. from Cleaveland, on lake Erie, and by p-r. 356 ms. n. w. by w. from W. C. and 118 n. e. from Columbus. In 1830, the tsp. contained a pop. of 449.

Brushy hill, p-o. in the w. part St. Clair co. Il., by p-r. 84 ms. n. of New Vandalia.

Brutus, p-t. Cayuga co. N. Y. 5 ms. n. Auburn, 153 ms. w. Albany, on Erie canal, s. Seneca r. 5 to 6½ ms. by 10, is uneven, with many gravel hills, but very fertile and well watered, and affords gypsum and good limestone for building. *Weeds Port*, on the canal, 7 ms. n. Auburn, 4 e. Bucksville, 9 e. Montezuma, has a large basin. Pop. 1831, 1,827.

Bryan, co. of Geo. bounded s. e. by the Atlantic, Liberty co. s. w., Bullock n. w. and n., and the Great Ogeechee r. separating it from Effingham and Chatham, n. e. Length from s. e. to n. w. 40 ms.; mean breadth 12, and area, 480 sq. ms. Extending in lat. from 31° 43′ to 32° 12′, and in long. from 4° 08′ to 4° 46′ w. from W. C. Chief town, Hardwick. Pop. 1820, 3,021, 1830, 3,139. This co. includes the sea coast of Ossabaw isl. from St. Catharine's sound, to that of Ossabaw or mouth of Great Ogeechee r.

Bryan's, p-o. in the w. part of Hardiman co. Ten. 12 ms. westward from Bolivar, the st. jus. for the co.

Bryantown, p-v. Charles co. Md. 32 ms. s. W. C. and 10 n. e. Port Tobacco.

Bryant's, p-v. in the s. e. part of Fayette co. Pa. by p-r. 178 ms. s. w. by w. Harrisburg.

Brydie's Store, and p-o. in the n. part of Lunenburg co. Va. by p-r. 91 ms. s. w. Richmond.

Buchannon, p-v. in the n. part of Lewis co. Va. by p-r. 266 ms. n. w. Richmond.

Buck, p-o. in the e. part of Lancaster co. Pa. by p-r. 54 ms. e. Harrisburg.

Buckseytown, p-v. Frederick co. Md. by p-r. 49 ms. n. w. W. C.

Buckfield, p-t. Oxford co. Me. 6 ms. s. Paris, 34 Augusta. Pop. 1830, 1,514; has a mtn. s. w. and is crossed by a stream of water.

Buckhead, p-o. Fairfield dist. S. C. 35 ms. n. Columbia.

Buckhead, p-o. Morgan co. Geo. by p-r. 50 ms. n. n. w. Milledgeville.

Buckhorn, p-o. Columbia co. Pa. by p-r. 79 ms. n. Harrisburg.

Buckhorn Falls, and p-o. Chatham co. N. C. by p-r. 28 ms. w. Raleigh.

Buckingham, p-o. Bucks co. Pa. about 27 ms. n. Philadelphia. Pop. of tsp. of B. 1830, 2,132.

Buckingham, co. Va. bounded by Appomattox r. separating it from Prince Edward, s. w. by Campbell, w. by James r. separating it from Amherst, n. w. by James r. separating it from Nelson, n. by James r. separating it from Albemarle, n. e. by James r. separating it from Fluvanna, and e. by Cumberland. Length 34, mean breadth 24, ms.; area 816 sq. ms. Extending in lat. from 37° 13′ to 37° 45′ n. and in long. from 1° 12′ to 1° 55′ w. from W. C. Though from the southern part of this co. the Appomattox rises and flows eastward, the body of the co. declines northwardly towards James r. which latter stream forms about one half the entire outline. Chief town, Maysville. Pop. 1820, 17,582, 1830, 18,351.

Buckingham, C. H. and p-o. (See Maysville.)

Buckland, p-t. Franklin co. Mass. 105 ms. w. n. w. Boston, 12 w. Greenfield, s. Deerfield r. Pop. 1830, 1,039.

Buckland, p-v. in the n. w. part Prince William co. Va. 5 ms. s. w. from Hay Market.

Bucklin, p-v. Wayne co. Mich. 16 ms. from Detroit.

Bucks, co. of Pa. bounded by Phil. co. s., Montgomery s. w., Lehigh and Northampton cos. n. w., Delaware r. separating it from Hunterdon co. N. J. n. e. and e., and Burlington co. N. J. s. e. Greatest length (from opposite Bordentown to the borders of Northampton and Lehigh), 42 ms. mean breadth 13, area 546 sq. ms. Extending in lat. from 40° 04′ to 40° 36′, and in long. from 1° 35′ to 2° 22′ e. from W. C. The general declivity is eastward and obliquely towards the Delaware r. To this the n. w. angle is an exception; it slopes southwardly, and is drained by the sources of Perkiomen cr. The soil of Bucks is diversified, and moderately fertile, some parts highly productive, and is amongst the best cultivated cos. of Pa. The surface is pleasantly broken into hill and dale, and the northwestern border formed by one of the minor chains of the Appalachian system. Of the large creeks which rise and terminate in this co. the principal are the Neshamany in the southern, and Tohicken in the northern part. A canal is completed, or nearly so, along the Bucks co. bank of Delaware river, from Bristol, to be extended to form a chain with the Lehigh navigation at Easton. The staples of Bucks co. are composed of nearly every species of produce brought to the Phil. market, and which the climate will admit. Chief towns, Doylestown st. jus. and Bristol. Pop. 1820, 37,842, 1830, 45,745.

Bucksport, p-t. Hancock co. Me. 25 ms. e. Castine, 61 Augusta, e. Penobscot r. and just above Orphan island. Pop. 1830, 2,237.

Buck's Store, and p-o. Tuscaloosa co. Ala. by p-r. 32 ms. from Tuscaloosa.

BUCKSVILLE, v. Mentz, Cayuga co. N. Y. 8 ms. N. Auburn, on Erie canal.

BUCYRUS, tsp. p-v. and st. jus. Crawford co. O. on the table land, between the sources of Sandusky and Sciota rs. by p-r. 69 ms. N. Columbus. Pop. 1830, v. 308, tsp. exclusive of the v. 362.

BUFFALO, small r. of Miss. rises in Amite co. between the N. sources of Amite r. and the southern of Homochitto, and flowing thence westwardly, over Wilkinson co. falls into the Miss. above Loftus heights, after a course of 80 ms. over a very broken but highly productive country.

BUFFALO, p-t. port of entry, st. jus. Erie co. N. Y. E. end of lake Erie, at the head of Niagara r. and of the Erie canal, on the N. E. side, and at the mouth of Buffalo cr. It is very advantageously situated, and has rapidly increased since the completion of the canal. It was burnt by British troops, 1814, except one house. The creek affords mill sites of great importance, and a canal has been lately formed from the falls to the town, on which are important hydraulic works. Near its mouth the creek forms a good harbor, with 12 or 14 feet water for a mile. On account of a sand bar, a pier has been built into the lake 1000 feet; there is a light house. Steam boats depart often for the principal ports on the lake, and Detroit; and an excursion or more is made annually to Green Bay. A vast and increasing amount of produce is brought hither from the lake shores, and other articles by canal. There were shipped east in 1829, 3,640, 1830, 149,219, and in 1831, 186,148 bushels of wheat; in 1829, 4,335, 1830, 31,810, and in 1831, 62,968 barrels of flour; received during the same period, 1829, 65,435, 1830, 75,370, 1831, 74,064 barrels salt. Canal arrivals and clearances, in 1829, 1,068, 1830, 2,083, 1831, 2,425. Canal tolls, 1829, $25,873 48, 1830, $48,953 02, 1831, $65,980 71. The other waters are Tonawanta cr. Ellicot's, with branches, and several streams of the lake. Soil and surface various. The v. is large, flourishing, and very pleasantly situated, on an elevation overlooking the lake, with regular streets, a square, fine public houses, and stores. The v. of Black Rock is in this t. (see Black Rock), and a tract of the reserve lands of the Seneca Indians. There Red Jacket lately died. The great road from Albany ends here. Pop. 1820, 2,095, 1830, 8,668.

BUFFALO, p-v. and tsp. of Washington co. Pa. 13 ms. W. from the borough of Washington. Pop. of the tsp. 1830, 1,519.

BUFFALO, p-v. on the right bank of Great Kenhawa, S. E. part of Mason co. Va. about 50 ms. S. E. Mount Pleasant.

BUFFALO, p-v. Lincoln co. N. C. by p-r. 187 ms. a little S. of W. Raleigh.

BUFFALO FORGE, and p-o. in the S. part of Rockbridge co. Va. 8 ms. W. Lexington, the co. seat.

BUFFALO SPRINGS, and p-o. W. part of Amherst co. Va. by p-r. 147 W. Richmond.

BUFORD'S BRIDGE, and p-o. on Salkehatchie r. S. part of Barnwell dist. S. C. 14 ms. S. E. Barnwell.

BULL CREEK, p-o. Wood co. Va. by p-r. 299 ms. W. W. C.

BULLIT, co. Ky. bounded by Jefferson N., Spencer E., Nelson S. E., Salt r. separating it from Hardin and Meade cos. S. W., and by a very narrow point on the Ohio r. above the mouth of Salt r. and opposite Harrison co. Ind. Length from E. to W. 25, mean breadth 10 ms. and area 250 sq. ms. Extending in lat. from 37° 47′ to 38° 03′, and long. from 8° 30′ to 8° 55′ W. from W. C. One of the main confluent streams of Salt r. enters the E. border, and traverses Bullit co. in a S. W. by W. direction, and uniting with Rolling Fork, assumes a course N. W. and falls into Ohio r. at the extreme western angle of the county. Similar to most cos. which border on Ohio, the features are hilly, and soil productive. Chief t. Shepherdsville. Pop. 1820, 5,381, and in 1830, 5,632.

BULLOCK, co. Geo. bounded by Bryan S. E. the Cannouchee r. separating it from Tatnall S. W. Emanuel N. W. and Great Ogeechee r. separating it from Scriven N. E. and from Effingham E. Greatest length from S. E. to N. W. 40 ms. mean breadth 20, and area 800 sq. ms. Extending in lat. from 32° 6′, to 32° 43′, and in long. from 4° 28′ to 5° 10′ W. W. C. Enclosed between two rivers, which both flow to the S. estrd., the general slope of the co. is in that direction. The soil is generally sterile; staple, cotton. Chief town, Statesboro. Pop. 1820, 2,578, in 1830, 2,587.

BULLPASTURE, r. and p-o. in the N. eastern part of Bath co. Va. by p-r. 164 ms. N. W. by W. from Richmond. Bullpasture is the local name of the higher part of Cowpasture r. or the middle constituent of James r.

BULLTOWN, p-v. on Little Kenhawa r. Lewis co. Va. 30 ms. W. of S. Clarksburg.

BUNCOMBE co. of N. C. bounded by the Blue Ridge, which separates it from Greenville and Pickens dist. S. C. on the S. by a mountain chain, which separates it from Haywood co. N. C. on the W. by the main chain of Kittatinny, which separates it from Greene, Washington and Carter cos. of Ten. N. W. by Ashe co. of N. C. on the N. E. and by the Blue Ridge, which separates it from Burke and Rutherford cos. N. C. on the E. Greatest length from S. W. to N. E. along the Blue Ridge, 100 ms.; mean breadth 20, and area 2,000 sq. ms. extending in lat. from 35° 3′, to 36° 8′ N. and in long. from 4° 41′, to 5° 51′ W. W. C. Buncombe occupies a part of the great valley between the Blue Ridge and the Kittatinny. The latter chain, where it separates N. C. from Tennessee, is called by the local names of Bald mountain, or Iron mountain. Within this valley, and as far south as N. lat. 35° 5′ rises the French Broad r. which, receiving tributary creeks from both chains flows in a northwardly direction, with a curve to the E. 55 ms. to its passage through the Bald mountain, after having drained the southern part of Buncombe. Similar to the French-Broad, the Nolachucky rises also in Buncombe, and draining the northern part of the co. by confluent streams from both chains, bends to N. W. and enters the state of Ten. between the Bald and Iron mountains. Buncombe co. comprises the S. E. section of the basin of the Mississippi; the French Broad interlocking sources with those of Santee and Savannah rs. as the Nolachucky does with those of the Catawba.

Buncombe is an elevated region, the lowest point perhaps exceeding 1,000 feet above tide water, and the farms varying from that height to 1,400, or 1,500 feet, which, with a north-western exposure gives to Buncombe a winter as intense, if not more so, as that of southern Maryland. The surface is excessively broken, and soil as greatly varied. The air and water are, however, as fine as that of any other section of the earth. Chief town, Ashville. Pop. 1820, 10,542, and in 1830, 16,281.

BUNDYSBURG, p-v. in the northern part of Geauga co. O. by p-r. 13 ms. northwardly from Chardon.

BURGESS' STORE, and p-o. in the S. part of Northumberland co. Va. by p-r. 101 ms. a little N. of E. Richmond.

BURGETTSTOWN, small p-v. in the N. W. part of Washington co. Pa.

BURKE, p-t. Caledonia co. Vt. 40 ms. N. E. Montpelier, 37 N. Newbury First settled, 1790,—has Passumpsick r. with many mill sites, and Burke mtn. 3,500 ft. S. E. It is uneven, with good soil, and hard wood, and evergreen trees. Magog oil-stones are brought from an island in Memphramagog lake, and manufactured here. Pop. 1830, 866.

BURKE, co. N. C. bounded by the Blue Ridge, which separates it from Buncombe N. W., by Wilkes N. E., Iredell E., Catawba r. which separates it from Lincoln S. E., and by Rutherford and the western part of Lincoln S. This county is commensurate with the upper valley of Catawba, enclosed on three sides by Montague hills, Blue Ridge, and Brushy mtn. The various creeks rising in these mtns. and flowing towards the interior of the county to form Catawba, which, flowing estrd. to the western border of Iredell, then turns abruptly to the S. and continues that course to its final issue from N. C. In lat. Burke co. reaches from 35° 32′ to 36° 08′ N. and in long. from 4° 12′ to 5° 15′ W. from W. C. Greatest length N. E. to S. W. in the general direction of the Catawba valley, 65 ms. mean breadth 25 ms. and area 1625 sq. ms. Similar to most mountain valleys, the soil is of every quality. The elevation renders it a grain district. Pop. 1820, 13,411, in 1830, 17,888. Chief ts. Morgantown and Mackeysville.

BURKE, co. of Geo. bounded by Scrive S. E., Great Ogeechee r. or Emanuel co. S. Jefferson W., Richmond N., and Savannah r. which separates it from Barnwell dist. S. C. on the N. E. Length 40 ms. mean breadth 30 ms. and area 1,200 sq. ms. Beside Savannah and Great Ogeechee, which bound this county, it is watered by Brier cr. which, entering its western border, traverses the co. in a S. E. direction. N. lat. 33° and long. 5° W. from W. C. intersect near the centre of Burke. Chief t. Waynesboro. Pop. 1820, 11,574.

BURKES GARDEN, and p-o. Tazewell co. Va. 20 ms. N. N. W. Evansham.

BURKESVILLE, p-v. in the northern part of Prince Edward co. Va. by p-r. 66 ms. S. W. by W. Richmond.

BURKESVILLE, p-v. and st. jus. Cumberland co. Ky. on the right bank of Cumberland r. by p-r. 152 ms. a little W. of S. from Frankfort.

BURKETSVILLE, p-v. Frederick co. Md.

BURLINGTON, p-t. and port of entry, and st. jus. Chittenden co. Vt. 38 ms. Montpelier, 22 S. E. Plattsburgh, 97 S. Montreal, 75 N. Whitehall, has a fine and advantageous situation, E. lake Champlain, and is large and flourishing. The v. is on a slope of 1 m. to a fine harbor. On the top of the hill is the college, and many of the private houses are in beautiful taste, with large gardens, &c. It has the county buildings, an academy, bank, ($150,000 capital) &c. Many of the vessels on the lake belong to this village. A manufacturing village is at the falls of Onion r. 5 ms. from its mouth, 1½ m. N. E. Burlington. First settled just before the revolution, which interrupted it. It is agreeably uneven, with a soil not very good: hard timber S. W., pine plains N. E. Below the falls is a fine alluvial tract. Limestone abounds, and some iron ore is found. The Champlain glass company here make excellent glass. Very good stage coaches travel by day light to Boston, through Middlebury, Rutland, Keene, and in other directions. A branch of the U. S. bank is located here. Burlington college is ¾ m. E. of the village, overlooking it and much of the lake. Two steam boats which ply between Whitehall, and St. John's L. Canada, touch here, and another runs to Plattsburgh, 25 ms. The road to Rutland is quite level, with fine scenery. Pop. 1830, 3,525.

BURLINGTON, p-t. Middlesex co. Mass. 12 ms. N. W. Boston. Pop. 1830, 446.

BURLINGTON, p-t. Hartford co. Ct. 16 ms. W. Hartford, 5 ms. by 6, 30 sq. ms. is irregular, with a gravelly loam on granite rocks, yielding grain, &c. and is watered by Farmington r. and branches of Poquaback r. Pop. 1830, 1,301.

BURLINGTON, city, port of entry and st. jus. Burlington co. N. J. on E. bank Delaware r. 11 ms. S. Trenton, 17 N. E. Phil., is a very pleasant place, with green banks on the river and some fine country seats, principally of Phil. gentlemen. The co. buildings, a bank, and an academy, are here. The large steam boats touch here several times every day. The soil is good, level, well cultivated, and inhabited by industrious people. It is opposite Bristol, Pa. Pop. 1830, 2670.

BURLINGTON, co. N. J. nearly triangular, bounded by Hunterdon, Middlesex and Monmouth cos. N., Little Egg Harbor S. E., Gloucester co. S., Delaware r. W. It has Rancocus and Croswicks crs. with Assompink cr. N. and Little Egg Harbor cr. S. Burlington is the co. town, and it contains also Bordentown, Mount Holley &c. The land near the Delaware, is level, fertile, well cultivated and sends supplies to Phil. but much of the remainder is poor. The steam boat navigation of Delaware river is important to the co. and one of the great routes of travelling between Philadelphia and New York is by Bordentown, where the Rail-road is to strike the Delaware from Amboy, and whence it is to extend to Camden. Pop. 1830, 31,107.

BURLINGTON, p-t. Otsego co. N. Y. 12 ms. W. Cooperstown, 78 W. Albany, is hilly, arable, productive, and well supplied with mill sites, by Butternuts and Otsego creeks, which have rapid descents. The trees are maple, beech, birch, elm, &c. Pop. 1830, 2459.

BURLINGTON, p-v. in the southeastern part of Bradford co. Pa. by p-r. 162 ms. E. of N. Harrisburg.

BURLINGTON, p-v. Hampshire co. Va. about 160 ms. N. N. W. W. C.

BURLINGTON, p-v. Meigs co. Ohio, 8 ms. westward from Chester, the co. seat.

BURLINGTON, p-v. and st. jus. Lawrence co. O. situated on Ohio r. in the extreme southern point of the state, nearly opposite the mouth of Great Sandy r. by p-r. 135 ms. S. S. E. Columbus. Pop. 1830, 149.

BURLINGTON, p-v. and st. jus. Boone co. Ky. 12 ms. S. W. Cincinnati. Pop. 1830, 276.

BURNHAM'S, p-t. Waldo co. Me. 37 ms. S. Augusta. Pop. 1830, 803.

BURNING SPRING, p-v. in the N. W. corner Floyd co. Ky. by p-r. 126 ms. S. E. Frankfort.

BURNT COAT, Island, Hancock co. Me. off Blue-hill bay and Union r. Pop. 1830, 254.

BURNS, town Alleghany co. N. Y. Pop. 1830, 702.

BURNT CORN, p-v. in the S. part of Monroe co. Ala. by p-r. 183 ms. S. Tuscaloosa.

BURNT CABINS, p-v. Bedford co. Penn. by p-r. 70 ms. S. W. by W. Harrisburg.

BURNT PRAIRIE, p-v. in the S. part White co. Il. 15 ms. from Carmi, the st. jus. for the co.

BURNT TAVERN, p-o. in the S. part of Garrard co. Ky. by p-r. 49 ms. S. S. E. from Frankfort.

BURRILLVILLE, p-t. Providence co. R. I. 24 ms. N. W. Providence, is a new t. E. Conn. line, 5 ms. by 12, with 60 sq. ms. rough, with pretty good timber and grazing land, on primitive soil, watered by the outlet of Allum pond. It is a manufacturing town. Pop. 1830, 2196.

BURROW'S OLD STORE and p-o. in the N. part of Madison co. Ala. by p-r. 15 ms. from Huntsville, the st. jus.

BURRSVILLE, p-v. Caroline co. Md. by p-r. 87 ms. E. W. C. and 50 from Annapolis.

BURTON, t. Strafford co. N. H. 75 ms. N. E. Concord, 45 Guilford, 75 Portsmouth, 5 ms. by 12, with 36,700 acres, supplied by Swift r. a branch of Saco r. and other streams, with mill sites. They once were stocked with otter and beaver. There are high, granite mtns. Chocorna, &c. The soil is generally good, with maple, birch, ash, pine, &c. Pop. 1830, 325.

BURTON, p-v. and tsp. in the E. part of Geauga co. O. 9 ms. S. E. Chardon. Pop. of tsp. 1830, 646.

BURTONSVILLE, p-o. E. part of Orange co. Va. by p-r. 81 ms. N. W. Richmond.

BUSHKILL, p-v. on Del. r. at the mouth of Bushkill cr. Pike co. Penn. 90 ms. E. of N. Phil.

BUSHKILL, the name of two creeks of Penn. one rising near the centre of Pike co. in three branches which rise and fall into the Del. at the village of the same name; the second rises in the Blue or Kittatinny mtns. Northampton co. and falls into Del. r. at Easton.

BUSHVILLE, p-v. Franklin co. Geo. by p-r. 116 ms. nthrd. Milledgeville.

BUSHWICK, t. King's co. N. Y. on Long Island, E. East r. opposite N. Y. is hilly, with a light, fertile loam, with Bushwick and Williamsburgh v. At the latter is a steam ferry boat to N. Y. Pop. 1830, 1020.

BUSKIRK'S BRIDGE, p-v. Cambridge, Washington co. N. Y.

BUSTI, p-v. Chatauque co. N. Y. 334 ms. W. Albany. Pop. 1830, 1680.

BUTLER, t. Wayne co. N. Y. Pop. 1830, 1764.

BUSTLETOWN, p-v. Phil. co. Penn. 11 ms. N. E. Phil.

BUTLER, co. of Penn. bounded by Alleghany S., Beaver W., Mercer N. W., Venango N., and Armstrong E. Length 35, mean breadth 23; and area 800 sq. ms. extending from N. lat. 40° 42' to 41° 11', and in long. from 2° 48' to 3° 14' W. from W. C. Alleghany r. merely touches the N. E. angle of Butler, from which it then recedes, and again touches the S. E. angle at Freeport. The body of the county is drained by the Slippery Rock, and other branches of Connequenessing, flowing wtrd. into this r. Surface very hilly, but soil excellent for grain, fruit, and pasturage. Chief town, Butler. Pop. 1820, 10,251, 1830, 14,683.

BUTLER, p-v. borough and st. jus. Butler co. Penn. on the Connequenessing, 32 ms. E. of N. Pittsburg, and by p-r. 205 ms. wstrd. Harrisburg. Pop. 1830, 567.

BUTLER, co. Ky. bounded by O. co. N. W., Grayson N. E., Warren S. E., Logan S. W. and Muhlenburg W. Length from S. W. to N. E. 38 ms. mean breadth 15, and area, 570 sq. ms. extending from N. lat. 37° to 37° 23', and in long. from 9° 25' to 10° 2' W. Green r. receives its great tributary, Big Barren, on the S. E. margin of this county, and the united waters in a N. W. direction, flow across the county, and then turning S. W. form a common boundary between it and O. co. to the eastern angle of Muhlenburg co. Thus though Butler may be considered an inland co. a fine navi gable channel untites it to O. r. Chief town, Morgantown. Pop. 1820, 3083, 1830, 3058.

Butler, co. Ala. bounded w. by Monroe and Wilcox, n. by Montgomery, e. by Pike, s. by Covington and Conecuh. Length 35, mean breadth 30, and area 1000 sq. ms. extending from n. lat. 31° 30′ to 31° 57′, and in long. from 9° 38′ to 10° 2′ w. from W. C. Surface undulating, and soil generally thin and sterile. It is a table land from which flow many creeks towards Ala. r. but the body of country slopes southwardly, giving source to Patsligala, Pigeon and Supulga, branches of Conecuh r. Staple, cotton. Chief town, Greenville. Pop. 1820, 1405, in 1830, 5650.

Butler's, p-o. Putnam co. Geo. by p-r. 28 ms. nthrd. Milledgeville.

Butler's Mills, p-o. Montgomery co. N. C. by p-r. 100 ms. s. w. by w. Raleigh.

Butler's Ferry and p-o. Jackson co. Ten. by p-r. 94 ms. n. e. by e. from Nashville.

Buttahatche, r. rises in the wstrn. part of Marion co. Ala. and flowing s. s. w. enters and traverses Monroe co. Miss. falls into Tombigbee 12 or 13 ms. above Columbus, after a comparative course of 70 ms. This stream has interlocking sources with Bear creek, flowing into Ten. with the Black Warrior.

Buttermilk Channel, the channel between Governor's Island and Long Island in New-York harbor.

Butternuts, p-t. Otsego co. N. Y. 21 ms. s. w. Cooperstown, 87 w. Albany, bears grain and grass; maple, beech, birch, elm, &c. watered by Unadilla creek, and other streams; contains Louisville v. Gilbertsville v. and Gilbert's v. p-o. Pop. 1830, 3991.

Butts, co. Geo. bounded by Monroe s., Pike w., Henry n. w. and Oakmulgee r. separating it from Newton n. e., Jasper e. and Jones s. e. Length 28 ms. breadth 15, and area 420 sq. ms. extending in lat. from 33° 5′ to 33° 28′, and in long. from 6° 50′ to 7° 13′ w. from W. C. Chief town, Jackson. Pop. 1830, 4944.

Butztown, p-v. Northampton co. Penn. by p-r. 103 ms. n. e. by e. Harrisburg.

Buxton, p-t. York co. Me. e. Saco r. 8 ms. n. w. Saco, 40 n. York, 71 s. s. w. Augusta, bordering n. e. on Cumberland co. Here is a large manufactory on Saco r. at a fall of 79 feet; 7 ms. by 40. Pop. 1830, 2856.

Buygonsville, p-v. eastern part of De Kalb co. Geo. by p-r. 92 ms. n. w. Milledgeville.

Buzzard's Bay, on s. coast Mass. 7 ms. by 40, with Plymouth co. n., Barnstable co. e., Bristol co. w. and Atlantic Ocean s. It extends within 3½ ms. of Cape Cod Bay, 2½ Barnstable, to which a canal has been proposed for coasters through Sandwich. It receives a number of small streams; Elizabeth islands are off the mouth. Seakonet is the w. Point.

Byberry, tsp. and p-o. Phil. co. Penn. by p-r. 11 ms. n. e. Phil.

Byfield, p-v. and parish, Rowley and Newbury, Essex co. Mass. 6 ms. s. w. Newburyport, has a good soil, well cultivated. Parker r. falls 40 feet in 1 m. of tide, with many mill sites, on one of which was the first woollen factory in New England, perhaps in America. Boats go to the ocean. Here is Dummer's academy, founded 1756, by Lt. Governor D.——— with a fine farm, for a Latin and grammar school; opened 1763; the first academy established in Mass.

Byon, p-v. and st. jus. Baker co. Geo. on Flint r. at the Falls, by p-r. 150 ms. s. w. Milledgeville.

Byram, r. on the s. w. bound. of Conn. between it and N. Y. is small, and runs into Long Island Sound.

Byram, t. Sussex co. N. J. has a part of the range of Schooley's mtn. verges upon Morris canal s. with Hop Pond on the e. Pop. 1830, 958.

Byran's p-o. Pike co. Geo. by p-r. 68 ms. w. Milledgeville.

Byron, p-t. Genesee co. N. Y. 10 ms. e. n. e. Batavia, 5 ms. by 6, has level and good land, watered by Black, Bigelow and Spring creeks, with some mill seats. Pop. 1830, 1936.

C.

Cabarras, co. of N.C. bounded s.w. by Mecklenburg, n. w. by Iredell, n. by Rowan, and s. e. by Montgomery. It is in form of a triangle, base from e. to w. along Montgomery and Iredell 30 ms. and salient point sthrd. between Montgomery and Mecklenburg with a perpendicular of 20 ms. which yields 300 sq. ms. Cabarras occupies a mountain valley drained by the higher sources of Rocky river. These streams with a general southern course unite in the southern part of the co. and leaving it at the extreme s. point turn thence eastward, separating Montgomery and Anson co. fall into the Yadkin. The face of Cabarras is broken and in part mountainous, though much of the soil is productive. Lat. from 35° 13′, to 35° 30′ n. and long. from 3° 21′, to 3° 52′ w. W. C Chief town, Concord. Pop. 1820, 7228, 1830, 8810.

Cabell, co. of Va. bounded by Mason co. n. e., Kenhawa e., Logan s. e., Sandy r. which separates it from Lloyd, Lawrence, and Greenwich co. of Ky. w., and by O. r. which separates it from Galia and Lawrence co. O. n. Greatest length from s. w. to n. e. 50 ms. mean width about 20, and area of 1000 sq. ms. Extending from 37° 55′, to 38° 40′ n. and in long. from 4° 45′, to 5° 34′ w. from W. C. Beside this and Sandy rivers which form part of the boundaries of Cabell, it is subdivided into two not very unequal parts, by thé Great Guyandot, which rises in Logan, enters Cabell, over which in a northwestern direction, it reaches the Ohio, r. below Barboursville.

The face of Cabell is very broken and in part mountainous. The soil, except a minor part, rocky and sterile. Chief town, Barboursville. By the census of 1820, Cabell, then including about one third of what is now comprised in Logan, contained a pop. of 4789, in 1830, Cabell, as then restricted, contained 5834.

Cabell, court house, p-o. and st. jus. Cabell co. Va. (*See Barboursville, Cabell co. Va.*)

Cabin Point, p-v. almost on the meridian of W. C. and in the w. part of Surry co. Va. 49 ms. s. e. Richmond.

Cabin Creek, p-o. in the southern part of Lewis co. Ky. by p-r. 83 ms. n. e. by e. Frankfort.

Cabot, p-t. Caledonia co. Vt. 18 ms. n. e. Montpelier, 65 n. Windsor, 6 ms. sq. on the head waters of Onion r. First settlement, 1785. The plain is on the height between Conn. and Onion rivers. The soil is hard and uneven. Zerah Colburn was born here. Pop. 1830, 1304.

Cackley's, p-o. Pocahontas co. Va. by p-r. 202 ms. n. w. by w. Richmond.

Cadiz, p-v. and st. jus. Harrison co. Ohio, 27 ms. s. w. by w. Steubenville, and by p-r. 124 n. e. by e. Columbus. It is a very neat, thriving village. Pop. 1820, 537, 1830, 818.

Cadiz, p-v. and st. jus. Trigg co. Ky. on Little r. a small branch of Cumberland r. about 100 ms. n. w. Nashville, Ten. and by p-r. 218 ms. s. w. from Frankfort.

Caddo, p-v. Clarke co. Ark. by p-r. 75 southward from Little Rock.

Cadwallader, p-v. in the e. part of Tuscarawas co. O. by p-r. 112 ms. n. e. by e. Columbus.

Cahaba, r. Ala. rises in Jefferson and St. Clair cos. flowing s. w. by w. over Bibb, Shelby, Perry, and Dallas cos. falls into Ala. r. at the town of Cahaba, after a course of 120 ms.

Cahaba, p-v. and st. jus. Dallas, co. Ala. on the right bank of Alabama river, immediately below the mouth of Cahaba r. 77 ms. s. s. e. Tuscaloosa, and about 140 ms. n. n. e. Mobile.

Cahokia, p-v. on the left bank of the Miss. r. St. Clair co. Il. 5 ms. s. e. St. Louis.

Cahoos Falls, N. Y. 3 ms. from the mouth of Mohawk river, is one of the greatest cataracts in the U. S. The Mohawk falls very abruptly about 70 feet over a broken precipice of slaty rock, in one sheet of foam, at high water. The banks below are nearly 100 feet high, rocky and perpendicular. Fish abound in the basin. Erie canal, on the s. bank, surmounts the fall by locks, and crosses the river on an aqueduct, a little above. A bridge crosses about 1 mile below, and Champlain canal by a ferry.

Cain's p-o. Lancaster co. Pa. 64 ms. eastward Harrisburg.

Cainsville, p-v. Wilson co. Ten. by p-r. 48 ms. eastward from Nashville, and 700 s. w. by w. from W. C.

Ca Ira, (pronounced Sa Era) small p-v. on Willis river, in the w. part Cumberland co. Va. 62 ms. a little s. of w. Richmond and 45 ms. n. e. by e. Lynchburg.

Calahan's, p-o. Alleghany co. Va. by p-r. 189 ms. w. Richmond.

Cairo, p-t. Green co. N. Y. 10 ms. n. w. Catskill, 11 w. Athens v. 40 s. Albany, has the summits of Catskill mts. on s. bound. nearly 7½ ms. sq., is hilly, with alluvial levels on Catskill creek, and 2 branches, which also supply iron works, mills, &c. Pop. 1830, 2912.

Calais, p-t. Washington co. Me. 30 ms. n.w. Eastport, 204 e. Augusta; below the falls of St. Croix r. it is accessible to navigation through Passamaquoddy bay. It stands a little above and nearly opposite St. Andrews in New Brunswick. Pop. 1830, 1686.

Calais, t. Washington co. Vt. 37 ms. e. Burlington, 12 Montpelier. Population 1830, 1539.

Calcasiu r. of Louisiana, rises in the parish of Natchitoches, between Red and Sabine rivers. Flowing thence in a nearly general southern course, but curving to the east ward, with a remarkable compliance to the course of lower Sabine. From the source of the former to the head of Calcasiu lake, the two rivers maintain a nearly regular distance of about 35 ms. asunder. The Calcasiu rises in a forest of pines, which continues to be the prevailing tree on all its tributaries to where all forest ceases; and is followed by the praries of Opelousas. Issuing from this great body of woods, the Calcasiu similar to the Sabine, expands into a lake, of from 1 to 10 ms. wide, and 30 long, and again contracting into a river falls into the Gulf of Mexico at n. lat. 29° 28,′ long. 16° 20′ w. from W. C. Though the tide flows up this river above the head of its lake, it can scarce be called navigable, as there is not more than 3 feet water at its mouth and not much more in the lake. The soil it waters both in woods & prarie is mostly thin and sterile, though in the former section supplied with abundance of pure and limpid spring water.

Caldwell, p-t. and cap. Warren co. N. Y. 62 ms. n. Albany, at the head of lake George, 7 ms. in extent, is very hilly and picturesque, the scenery on this lake being more admired, than almost any other in the U. S. The head of the lake is in the centre of the township, and towards it slopes a high ridge 2 or 3 ms. s. French mtn. nearly e. and Rattlesnake mtn. 1½ ms. w. leaving little arable land. Part of Scaroon creek, the e. branch of Hudson r. touches the w. bound.

Caldwell village, near the head of lake George w. containing the co. buildings, is pleasant, with a fine view down, and over the sites of Fort George and Wm. Henry. There is a hotel for about 200 persons looking upon a basin and the lake, for hundreds of visiters who resort here every summer. Near it is the line of approaches of Gen. Montcalm, who captured Fort Wm. Henry

1757, and allowed the prisoners to be massacred by Indians. A steamboat plies hence to the bottom of the lake in the warm season, and the excursion surpasses all others of the kind in the country for beauty of scenery. Gen. Johnson was attacked, 1765, where Fort George was built, by Gen. Dieskau, who was defeated. Pop. 1830, 797.

CALDWELL, p-t. Essex co. N. J. a little s. Morris canal, and s. Passaic r. has Short Hills in the E. part. Pop. 1830, 2,004.

CALDWELL, co. of Ky. bounded s. w. by Tennessee r. which separates it from Calla-way and McCracken cos., N. w. by Livingston, N. E. by Tradewater r. which separates it from Hopkins co., and N. E. by Trigg. Length 32 ms. breadth 22, and area 700 sq. ms. Cumberland r. enters the s. E. border, and by a very winding channel, passes over the s. w. part, leaving a strip of 1 to 8 ms. wide between the latter r. and Ten. r. Surface mostly level, and soil productive. Chief towns, Eddyville and Princeton. N. lat. 37°, and long. 10° w. intersect in this co. Pop. 1820, 9,022, 1830, 8,324.

CALDWELL'S p-o. Washitau parish, La. by p-r. 201 ms. N. w. N. Orleans.

CALEDONIA, co. Vt. w. Connecticut r. 700 sq. ms. is crossed in w. part by the height of lands on E. range of Green mtns. between which and Conn. r. is fine country, with Passumpsic r. &c. w. of it, forms Onion r. Limestone, granite, &c. abound, and sulphur springs. Chief and county t. Danville. Incorporated 1792. Pop. 1830, 20,967.

CALEDONIA, p-t. Livingston co. N. Y. 31 ms. w. Canandaigua, 12 N. Genesee, 17 above Rochester, w. Genesee r. which is very crooked, but affords navigation to Erie canal; it has excellent wheat land. 2½ ms. by 8. Pop. 1830, 1,618.

CALEDONIA, p-v. in the s. w. part Moore co. N. C. 20 ms. sthrd. from Carthage, the co. seat, and by p-r. 89 ms. s. w. Raleigh.

CALEDONIA, p-v. in the w. part Henry co. Ten. by p-r. 123 ms. a little N. of w. Nashville.

CALEDONIA, p-v. in the southern part Jefferson co. Ind. by p-r. 97 ms. s. s. E. Indianopolis.

CALEDONIA, p-v. in the western part Washington co. Mo., 15 ms. wstrd. from Potosi, the st. jus.

CALHOUN'S, p-v. and st. jus. McMinn co. Ten. on Hiwassee river, 78 ms. s. w. Knoxville, and by p-r. 159 ms. s. E. by E. Nashville.

CALHOUN'S MILLS, and p-o. Abbeville dist. S. C. by p-r. 136 ms. s. w. Richmond.

CALIBOGUE SOUND, an inlet of S. C. between Dawfuskee and Hilton Head islands. It opens a little E. of N. 7 ms. from Savannah r. entrance, and extending inland is lost in a maze of interlocking inlets, enclosing the numerous islands which chequer the ocean border of Beaufort dist. between Broad and Savannah rivers.

CALLAND'S, p-o. Pittsylvania co. Va. by p-r. 136 ms. s. w. Richmond.

CALLAWAY'S MILL, and p-o. Franklin co. Va. by p-r. 190 ms. s. w. by w. Richmond.

CALLINSBURG, p-v. Armstrong co. Pa. by p-r. 248 ms. wstrd. Harrisburg.

CALLOWAY, co. of Ky. bounded by Graves w., McCracken N., Tennesse r. separating it from Caldwell N. E., Trigg E., Stewart co. Ten. s. E., and Henry co. Ten. s. Length 30, mean width 20, and area 600 sq. ms. Extending in lat. from 36° 30′ to 36° 56′ N., and in long. from 11° 11′ to 11° 35′. The eastern margin of this county declines N. E. towards Tennessee r. but the body of it is drained by Clarke's r. flowing N. into Ohio. Chief t. Wadesborough. Pop. 1830, 5,164.

CALVERT, co. Md. bounded N. by Ann Arundel co., E. by Chesapeake bay, and s. s. w. and w. by Patuxent r. which separates it from St. Mary's, Charles, and Prince George's cos. Length 33 ms., mean width 8, and area 264 sq. ms. Rolling surface, rather than level or hilly. For lat. and long. see article Prince Fredericktown. Pop. 1820, 8,073, 1830, 8,900.

CALVERT, p-o. Franklin co. Ky.

CAMBRIA, p-t. Niagara co. N. Y. 7 ms. N. w. Lockport, 13 E. Lewiston v. is crossed E, and w. by the mtn. ridge, and Erie canal. with small streams of Howell's, Cayuga, and 18 m. creeks. The soil is pretty good, ill watered, and greatly diversified. Lockport, a very important v. is in this t. (See Lockport.) Pop. 1830, 1,712.

CAMBRIA, co. Pa. bounded E. by the Alleghany chain which separates it from Bedford and Huntingdon, N. by Clearfield, w. by Laurel Hill, separating it from Ind. and Westmoreland, and s. by Somerset. Length from s. to N. 36 ms., mean breadth between the two chains of mountains, 20 ms., and area 720 sq ms. Cambria occupies part of the elevated mountain valley, from which the streams flow, like radii from a common centre. The southern part is drained by numerous creeks of Conemaugh river, whilst the northern section gives source to the extreme heads of the w. branch of Susquehannah. By the surveys made on the route of the Pennsylvania canal, it appears that Johnstown, at the forks of Conemaugh in this county, is elevated 1154 feet above the tide water, in Delaware r. This point is at least 150 feet below the common level of the county, which may be assumed at 1300 feet. The surface of the county is hilly, rocky, and in part mountainous, with a soil of middling quality. In lat. it extends from 40° 15′ to 40° 40′ N. and in long. from 1° 22′ to 2° w. from W. C. Pop. 1820, 3,287, in 1830, 7,076. The canal and rail way route, designated the Pennsylvania canal, passes over the southern part of Cambria.

CAMBRIDGE, p-t. Franklin co. Vt. 30 ms. N. w. Montpelier, 22 N. E. Burlington, 28,533 acres, first settled 1783. Lamoille r. runs 12 ms. in it, with branches and mill sites. It is uneven, but fertile, with 3 villages. Pop. 1830, 1,613.

CAMBRIDGE, p-t. Middlesex co. Mass. 3 ms.

N. W. Boston, was first settled 1631. It has 3 principal divisions; Cambridge, containing the university, an arsenal, and several churches, is pleasantly situated on a beautiful plain, extending from Charles r. It is a handsome village, and contains the residences of several officers of the university. Cambridge port, p-v. is a village of considerable business, containing several churches, connected by w. Boston bridge with the city; and East Cambridge, p-v. on Lechmere point, is also a flourishing village, where are various manufactories, among them the largest of glass in the U. S., a court house, jail, and 4 churches. This point is connected with Boston by Craigie's bridge.

Harvard University, the oldest institution of the kind in the U. S. was founded in 1638, and derives its name from Rev. John Harvard, who made the first large donation to it. It was designed to be a nursery for the churches. The buildings are University Hall; an elegant granite edifice, 140 feet by 50, and 42 high; Hollis, Massachusetts, Stoughton and Holworthy Halls; Holden Chapel, of brick, containing a chemical laboratory, anatomical museum, and other lecture rooms, and Divinity Hall, a commodious building, appropriated to theological students. The library is the largest in the U. S. containing 35,000 volumes, exclusive of that for the students, of nearly 5,000. There is also a mineralogical cabinet, and a botanic garden of 8 acres. Connected with the University are also a law and medical schools, and a theological seminary. By large donations from the state, and individuals also, this institution is more richly endowed than any other in the U. S. Several professorships have thus been established. The president's house, and the medical college in Boston, containing a library of 4,000 vols. belong to the institution. Presidents, as inducted,—Dunster, 1640, Chauncey, '54, Hoar, '72, Oaks, '75, Rogers, '82, Mather, '85, Willard, 1701, Leveret, '08, Wadsworth, '25, Holyoke, '37, Locke, '70, Langdon, '74, Willard, '81, Webber, 1806, Kirtland, '10, and Quincy, 1828. Undergraduates, 1830—31,248—medical students, 91—law 31. Commencement is on the last Wednesday in August.

The first printing press in America was located here, and was used by Stephen Day, who printed "The Freeman's Oath." The American army encamped here in 1776, during the siege of Boston, and some of their entrenchments remain. On Copp's hill is a monument bearing date 1625. Pop. 1830, 6072.

CAMBRIDGE, Washington co. N. Y. 12 ms. s. Salem, 35 N. E. Albany. It is in part, hilly, has good farms, on a warm deep gravel. Streams—White creek, with few mill seats. Pop. 1830, 2,319.

CAMBRIDGE, p-v. and st. jus. Dorchester co. Md. on s. side of Choptank bay, about 12 ms. above its mouth, 36 ms. s. E. Annapolis in a direct line, but by p-r. 53.

CAMBRIDGE, p-v. in the E. part of Abbeville dist. S. C. by p-r. 81 ms. a little N. of W. Columbia. Population 1820, about 350.

CAMBRIDGE, p-v. and st. jus. Guernsey co. O. on Wilts creek, 53 ms. w. from Wheeling, and by p-r. 83 ms. a little N. of E. Columbus. Pop. 1830, 518.

CAMDEN, p-t. Waldo co. Me. 12 ms. N. E. Thomaston, 59 s. E. Augusta, on Penobscot bay. Pop. 1830, 674.

CAMDEN, p-t. Oneida co. N. Y. 20 ms. N. W. Rome, 6 ms. by 12, has many mill sites on Fish creek—is uneven, with a fertile, sandy loam, good for grain, bearing beech, maple, bass, and hemlock. Camden and Taberg iron works, are villages. Pop. 1830, 1,945.

CAMDEN, p-v. Newton, Gloucester co. N. J. E. Delaware r. opposite Philadelphia, with a ferry. Here commences the Camden and Amboy rail road, designed to transport travellers and merchandize between New York and Philadelphia. (*See Rail Roads and Canals.*)

CAMDEN, co. of N. C. bounded by Nansemond and Norfolk counties, Va. N., by Curituck co. N. C. N. E., Albemarle sound s. and Pasquotank r. and co. w. Greatest length from s. E. to N. W. 38 ms.; mean breadth 6, and area 228 sq. ms.; N. lat. 36° 15', long. 38' E. from W. C. Surface level, and in part marshy. Pop. 1820, 6,305, 1830, 6,733.

CAMDEN, C. H. and p-o. Camden co. N. C. by p-r. 199 ms. N. E. by E. Raleigh.

CAMDEN, p-v. in the eastern part of Kent co. Del. by p-r. 3 ms. from Dover, and 117 a little N. of E. from W. C.

CAMDEN, p-t. and st. jus. Kershaw dist. S. C. near the left bank of Wateree r. 31 ms. N. E. Columbia, and 123 ms. N. N. W. Charleston. Wateree r. is thus far navigable for boats of 70 tons, which gives to Camden considerable trade. Pop. 1820, about 1,000. It contains an academy, and several places of public worship.

CAMDEN, s. eastern co. of Geo. bounded by St. Mary's r. s. and s. w., Warren co. w., Wayne co. N. W. Scilla r. or Glynn N. E. and the Atlantic Ocean E. without including a long narrow strip in the s. western part of this co. and in the great bend of St. Mary's r.; the body is a parallelogram of 35 by 25, and the whole area about 1,000 sq. ms. Lat. from 30° 21', to 31° 10', and long. from 4° 36', to 5° 24. The surface is in great part a plain, with Cumberland isl. stretching along nearly its whole front. The Santilla r. enters it from Wayne's co. and flowing s. 20 ms. turns abruptly E. pursues the latter course 30 ms. into St. Andrew's sound. The river St. Mary's affords the deepest entrance on the Atlantic coast of U. S. s. of Chesapeake bay, a depth sufficient for ships of war of the first class. Chief towns, St. Mary's and Jefferson. Pop. 1820, 3,402, in 1830, 4,578.

CAMEL'S BACK, or HUMP, mtn. Huntington, Chittenden co. Vt. one of the highest of the Green mtns. 4,188 ft. above tide, 3,960 above Montpelier state house. 17 ms. w. Montp., 25 N. E. Middlebury, 20 s. E. Burlington. It affords a fine view, and is seen from lake Champlain.

CAMERON, t. Steuben co. N. Y. 8 ms. s. of

Bath, watered by Canisteo and Conhocton creeks, has broken land, with some alluvion, and pine, hemlock, maple, beech, &c. Pop. 1830, 924.

Camillus, p-t. Onondago co. N. Y. 10 ms. N. W. Onondago, 160 w. Albany, is supplied by Seneca r. with navigation and mills seats, and has 3 villages, Camillus on Otisco cr., Elbridge on Skeneateles cr., and Jodan on Erie canal, which crosses the town. There are remains of two large works, supposed to be ancient fortifications, 4 ms. from Seneca r. one is of 3 acres on a hill, with a ditch and earth wall, with gate ways. There is also a well. Gypsum is found on Otisco cr. Pop. 1830, 2,518.

Campbell, co. of Va. bounded by Stanton and Roanoke rs. s., separating it from Halifax and Pittsylvania cos., by Bedford w., James r. N., separating it from Amherst, by Buckingham N. E., and by Prince Edward and Charlotte E. Campbell is a rude advance to a sq. of 24 ms. each side, with an area of 576 sq. ms., extending in lat. from 37° to 37° 26′, and in long. from 1° 46′ to 2° 22′ w. from W. C. Surface much broken, but soil productive in grain, fruits, tobacco, pasturage, &c. Chief town, Lynchburg. Pop. 1820, 16,570; 1830, 20,350, including the t. of Lynchburg. Both the bounding rivers of this county are navigable for boats far above its limits, affording an opening by water to Chesapeake bay and Albemarle sound.

Campbell, co. Geo. bounded E. and N. E. by De Kalb, s. by Lafayette and Coweta, s. w. by Carroll, and N. W. by Chattahooche river. Length 30, mean breadth 10; area 300 sq. ms. Extending in lat. from 33° 37′ to 33° 56′, and in long. from 7° 30′ to 7° 53′. The southern, a parallelogram from E. to W. and the northern stretching a triangle up the Chattahooche. General slope s. wstrd. towards Coweta and Carroll cos. Pop. 1830, 3,323.

Campbell, C. H. and p-o., Campbell co. Va. 11 ms. s. s. E. Lynchburg.

Campbell, co. of Ten. bounded s. by a chain of mtns. called Chesnut Ridge, which separates it from Knox, s. w. by Clinch r. which separates it from Anderson, w. by Anderson and Morgan, N. by Wayne, Whiteley, and Knox cos. Ky., and E. by Claiborne and Grainger counties, Ten. Extending in lat. from 36° 07′ to 36° 35′, and in long. from 6° 36′ to 7° 17′ w. from W. C. Length from s. to N. 32, mean width 21, and 672 sq. ms. in area. Powell's r. enters the eastern border and traversing an angle of this co. falls into Clinch r. at Grant's corners. The N. E. part is traversed by Cumberland mtn., from the N. W. side of which the creeks are discharged into the state of Ky. and thence into Cumberland r. Campbell co. therefore is a table land between the vallies of Cumberland and Ten., and has a mean elevation above the Atlantic of at least 800 feet. Chief town, Jacksonboro'. Population 1820, 4,244; 1830, 5,120.

Campbell co. of Ky. bounded by Ohio r. which separates it from Hamilton co. O. on the N., and Clermont co. O. on the E., s. by Pendleton, and w. by Boone cos. Ky. Length 20, mean width 12, and area 240 sq. ms. Extending in lat. from 38° 49′ to 39° 07′, and in long. from 7° 12′ to 7° 32′ w. from W. C. Similar to other cos. of Ky. near O. r. the features of Campbell are hilly, but soil fertile; placed directly opposite Cincinnati, and traversed in its greatest length by Licking r., it is well situated for trade and commerce. Though bordering on Ohio along two sides, the body of the co. is in the valley of Licking, and slopes with the course of that stream to s. s. w. Chief towns, Newport and Covington. Pop. 1820, 9,022; 1830, 9,883.

Campbell's Mills, and p-o. in the wstrn. part of Abbeville dist. S. C. 8 ms. from Abbeville Court House and by p-r. 108 ms. wstrd. Columbia.

Campbell's Station, and p-o. in the s. w. part of Knox co. Ten. on the road from Knoxville to Nashville, 14 ms. wstrd. from the former, and 184 ms. a little s. of E. from the latter place.

Campbellsville, p-v. Giles county, Tennessee, by p-r. 66 ms. a little w. of s. from Nashville.

Campbellsville, p-v. in the N. E. part of Greene co. Ky. 12 ms. s. E. Greenburg, the st. jus. for the co. and by p-r. 78 ms. s. s. w. from Frankfort. Pop. 1830, 122.

Campbellton, p-v. and st. jus. Campbell co. Geo. situated on Chattahooche r. by p-r. 134 ms. N. W. by w. from Milledgeville.

Campbellton, p-v. in the wstrn. part of Jackson co. Flor. by p-r. 96 ms. wstrd. from Tallahasse.

Camp Creek, and p-o. Livingston county, Kentucky.

Campbelltown, p-v. in the s. w. part of Lebanon co. Pa. 15 ms. E. Harrisburg.

Campti, p-o. in the northern part of Natchitoches parish, La. by p-r. 7 ms. nrthd. from the village of Natchitoches.

Campton, p-t. Grafton co. N. H. 27,892 acres, 50 ms. N. N. W. Concord, 75 N. W. Portsmouth, is uneven, with mtns. and rocks. It has Pemigewasset and its branches, Mad and Beebee rs., also W. Branch r. and Bog Branch. There is good soil in the vallies, white oak, pitch pine, iron ore, and many orchards. First settled 1765. Pop. 1830, 1,314.

Camptown, p-v. Orange, Essex co. N. J.

Canaan, p-t. Somerset co. Me. 10 ms. E. Norridgewock, 34 N. by E. Augusta, E. Kennebeck r., bordering on Kennebeck co. Pop. 1830, 1,076.

Canaan, p-t. Grafton co. N. H. 40 ms. N. W. Concord, 16 E. Dartmouth college, on the high land between Conn. and Merrimack rs. It has several ponds and small streams. Heart pond, on high ground, has formed a low bank of earth nearly round its circumference, by the motion of the ice in breaking up in the spring. The soil is pretty good, yielding grain, flax, &c. First settled 1766 or 7. Pop. 1830, 1,428.

CANAAN, p-t. Essex co. Vt. Pop. 1830, 373.

CANAAN, p-t. Litchfield co. Conn. 16 ms. N. N. W. Litchfield, 41 N. W. Hartford, S. Mass., E. Ousatonick r., 6 ms. by 9, with 50 sq. ms. is on granite mtns. with fine vallies. Lime stone is quarried, iron ore is mined, and there are several forges, &c. The soil and timber are various. Branches of Ousatonick r. give many mill seats. Pop. 1830, 2,301.

CANAAN, p-t. Columbia co. N. Y. 24 ms. S. E. Albany, 22 N. E. Hudson, has Williamstown mtns. E. with hills and vales S. W. Gypsum has done much for the soil, much of which is very good. It has generally pure water, some bog iron; Whitney's pond and outlet, Klein kill, &c. and a bed of marl.

CANAAN, p-v. and tsp. in the northern part of Wayne co. O. The p-o. by p-r. 97 ms. N. N. E. from Columbus, and 358 ms. N. W. by W. from W. C. Pop. of the tsp. 1830, 1,030.

CANADA CREEK, EAST, runs 30 ms. into the Mohawk, 9 ms. below Little Falls.

CANADA CREEK, WEST, the largest branch of Mohawk r. 60 ms. long, rises near the head waters of Black r. and enters at the German Flats, 6 ms. above Little Falls. The numerous rapids and cascades make this a beautiful stream, with its lofty banks of dark limestone rock, full of marine petrifactions. It is an important point in the tour of travellers through the state. Two unfortunate visiters have been drowned here within 4 or 5 years. There is a public house near, and pains have been taken to make the difficult passes accessible. Utica is the proper place to proceed from, to pay a visit to this interesting vicinity.

CANADA CREEK, Oneida co. N. Y. 10 or 12 ms. long. N. branch, Wood cr.

CANADAWAY CREEK, N. Y. 15 ms. long, with many falls. Formerly there was a portage of 6 ms. between this and the Cordaga waters, to Alleghany r.

CANADIAN RIVER, (*See Arkansas r.*)

CANAJOHARIE, p-t. Montgomery co. N. Y. S. Mohawk r. at Bowman's cr. The N. Y. Central asylum for the Deaf and Dumb is on the cr. 6 ms. S. of the canal, and 7 N. Cherry Valley. The building is of brick, and there are two boarding houses for the male and female pupils, at $80 a year. 15 ms. S. W. Johnstown, 69 W. Albany. The ground is uneven, the crop chiefly wheat, and there are mill seats on Canajoharie and Plattekill crs. &c. The people are German. The Nose (a hill,) has a large cave. A rail road is projected to Catskill, 75 ms. Pop. 1830, 4,348.

CANAL, DOVER, p-v. in the northern part of Tuscarawas co. O. by p-r. 110 ms. N. E. by E. Columbus.

CANAL, FULTON, p-v. in the northwest part of Stark co. O. by p-r. 117 ms. N. E. Columbus.

CANANDAIGUA, p-t. and capital Ontario co. N. Y. 108 ms. E. Niagara falls, 208 ms. W. Albany, 88 E. Buffalo, on the great road to Buffalo, 6 ms. by 12, contains 8 ms. of the N. part of Canandaigua lake, and part of the outlet, and has fine hills and vallies, with good soil and much wealth. First settled 1790. The village or borough is large, and has a number of fine houses, stores, churches, county buildings, a bank, &c. on a strait, broad street, 1 mile long, on the ascent and summit of a high, gentle hill, gradually rising from the N. end of the lake. There is a flourishing female seminary, where some of the higher branches are taught, in which are about 100 pupils. There is also an academy with a department for the instruction of school teachers, from Aug. 8th, 6 weeks. Pop. 1830, 1,830.

CANANDAIGUA LAKE, Ontario co. N. Y. 14 ms. N. and S. and about 1 m. wide, empties by an outlet N. into Seneca r. The land is handsomely varied on the shores, in some parts high, and near the head well cultivated.

CANANDAIGUA CREEK, or outlet, Ontario co. N. Y. flows from the bottom of Canandaigua lake, to Seneca r. which it enters in Wayne co. 50 ms. long, after receiving Mud and Flint crs. &c. It is navigable from Seneca r. to the block house in Clyde, 12 ms.

CANASAUGA, p-v. near Hiwassee r. in Amoi dist. of that part of the Cherokee territory adjacent to McMinn co. Ten. by p-r. 186 ms. S. E. by E. from Nashville.

CANAVERAL. (*See Cannaveral.*)

CANASERAGA, cr. N. Y. a branch of Chitteningo cr. Another is a branch of Genesee r. which it enters 3 ms. N. Geneseo.

CANDIA, p-t. Rockingham co. N. H. 16 ms. S. E. Concord, 4 ms. by 6, with 15,360 acres, has a hard but well cultivated soil, and a high situation, in view of White Hills, and the lights on Plum Island on the coast. It is very healthy. First settled 1748. Pop. 1830, 1,360.

CANDOR, p-t. Tioga co. N. Y. 8 ms. N. Owego; has streams of Owego, Pipe, and Mud crs. with mill sites. First settled 1796; has pretty good land. Pop. 1830, 2,653.

CANEADEA, p-t. Alleghany co. N. Y. 6 ms. S. W. Angelica, 6 ms. by 12, is supplied with a few mill seats. Genesee r. is in N. E. The soil is pretty good, and bog iron ore is found. Pop. 1830, 780.

CANDICE, t. Ontario co. N. Y. Pop. 1830, 1386.

CANE CREEK, p-o. Chatham co. N. C. by p-r. 52 ms. W. Raleigh.

CANE CREEK, p-o. in the northwestern part of Lincoln co. Ten. 62 ms. S. from Nashville, and 721 ms. S. W. by W. W. C.

CANE HILL, p-o. Washington co. Ark. by p-r. 203 ms. N. W. Little Rock.

CANESTOLA, p-v. Lenox, Madison co. N. Y. on Erie canal, 25 ms. W. Utica, was a wilderness, 1819.

CANESUS, t. Livingston co. N. Y. Pop. 1830, 1690.

CANESUS, lake, Livingston co. N. Y. 9 ms. by 1 and 1½, 6 ms. E. Genesee r. into which it empties, in Avon, by an outlet of nine miles.

CANEY SPRING, p o. Bedford co. Ten. s. s. E. from Nashville.

CANFIELD, p-v. and tsp. in the southern part of Trumbull co. O. 18 ms. sthrd. from Warren, the co. seat, and by p-r. 156 N. E. by E. Columbus. Pop. tsp. 1830, 1249.

CANISTEO, t. Steuben co. N. Y. 18 ms. s. w. Bath, 260 w. Albany; has Canisteo r. which is boatable, with rich flats for grass and grain. Pop. 1830, 620.

CANNAVERAL, Cape of Flor. on the Atlantic ocean, being the salient point of a long, narrow, and low sandy island between Indian r. and the ocean. On Tanner's U. S. it is placed at N. lat. 28° 18', and at long. 3° 23' w. from W. C.

CANNONSBURG, borough and p-v. Washington co. Pa. on the road from the borough of Washington, the co. seat to Pittsburg, 7 ms. a little E. of N. from the former, and 18 s. w. from the latter, by p-r. 219 ms. wstrd. from Harrisburg, and 236 N. w. from W. C. It is situated on Chartiers creek, and on a rather bold acclivity from the valley. Here is located Jefferson college, formerly an academy. The faculty is composed of a president and two professors. It contains a respectable library and philosophical apparatus. Pop. of the borough, 1830, 673. N. lat. 40° 17', and long. 3° 18' w. from W. C.

CANNON'S FERRY and p-o. in the s. w. part of Sussex co. Del. 23 ms. s. w. by w. Georgetown, the st. jus. for the co.

CANNONSVILLE, p-v. Del. co. N. Y. 94 ms. s. w. Albany.

CANNOUCHE, r. of Geo. the western and largest confluent of Great Ogeechee; rises in Emanuel co. and flowing s. E. falls into Great Ogeechee in Bryan co. about 12 ms. s.w. from the city of Savannah. The valley of Cannouchee lies between those of Ogeechee and Altamaha. Length 90, mean breadth 10, and area 900 sq. ms.

CANOE CREEK, and p-o. in the N. w. part of Huntingdon co. Pa. by p-r. 20 ms. wstrd. from the borough of Huntingdon.

CANONICUT, isl. Narraganset bay, R. I.

CANTERBURY, p-t. Merrimack co. N. H. 8 ms. N. Concord, E. Merrimack r. uneven, with grass, small mill streams, and 2 bridges over Merrimac r. Contains 26,345 acres. Shaker's v. s. E. has good gardens, and some manufactures. Pop. 1830, 1663.

CANTERBURY, p-t. Windham co. Conn. 40 ms. E. Hartford, 12 N. Norwich, 4½ ms. by 8, 36 sq. ms. is uneven, with rich gravelly loam, yielding rye, maize, oats, &c. Quinebaug r. enriches its banks by spring floods; yields shad, and affords valuable mill sites. Bates's pond is stocked with fish. Pop. 1830, 1881.

CANTERBURY, p-v. on the head of Mother Kill creek, Kent co. Del. by p-r. 8 ms. a little w. of s. Dover.

CANTON, p-t. Oxford co. Me. 32 ms. from Augusta. Pop. 1830, 746.

CANTON, p-t. Norfolk co. Mass. 14 ms. s. Boston, flat, 200 feet above tide, with little arable land; 2 ponds give rise to two branches of Neponset r. Steep Brook cotton factory here, was incorporated 1815; cap. $50,000. Pop. 1830, 1515.

CANTON, p-t. Hartford co. Conn. 15 ms. N. w. Hartford, 4 ms. by 8; 19,000 acres. The soil is gravelly, chiefly yielding oak, grass, rye, corn, oats, and fine orchards, and is crossed by Farmington r. Pop. 1830, 1437, including Collinsville, which see.

CANTON, p-t. St. Lawrence co. N. Y. Pop. 1830, 2440.

CANTON, p-v. Bradford co. Pa. by p-r. 137 ms. nrthd. from Harrisburg.

CANTON, p-v. and st. jus. Wilcox co. Ala. situated on the left bank of Ala. r. by p-r. 113 ms. a little E. of s. Tuscaloosa, and by the common road, 120 ms. N. N. E. Mobile.

CANTON, p-v. in the sthrn. part of Trigg co. Ky. by p-r. 9 ms. sthrdly. from Cadiz, the st. jus. for the co. and 235 s. w. by w. from Frankfort.

CANTON, p-v. and st. jus. Stark co. Ohio, situated on Nemishillen creek, a branch of Tuscarawas r. by p-r. 116 ms. N. E. Columbus, and about 60 ms. a little E. of s. Cleaveland. Pop. 1830, 1257. This is one of the finest towns of interior O. There are three or four fine bridges over the Nemishillen in the vicinity; and the adjacent country is well cultivated and populous.

CANTONMENT GIBSON, p-o. as laid down on Tanner's map of the U. S. is situated on the left bank of Ark. r. on the point below the mouth of Grand r. N. lat. 35° 47'. long. 18° 9' w. from W. C. In the p-o. list it is stated to be 208 ms. from Little Rock, and is 1359 ms. s. w. by w. from W. C.

CANTONMENT, JESSUP, military station and p-o. in the N. w. part of Louisiana, and on Sabine r. N. lat. 31° 30', long. 16° 42' w. from W. C. and by p-r. 379 ms. N. w. from New Orleans and 1353 ms. s. w. W. C.

CANTONMENT, LEAVENSWORTH, on the Missouri r. Clay co. Mo. by p-r. 354 ms. above, and a little N. of w. St. Louis, 220 from Jefferson City, and 1172 ms. wstrd. W. C.

CANTWELL'S BRIDGE, and p-o. on the Appoquinimink creek in the southwestern part of New Castle co. Del. by p-r. 24 ms. N. N. E. from Dover.

CAPE COD, a peninsula forming part of Barnstable co. Mass. s. side Massachusetts bay, is in shape like a man's arm bent inwards at the wrist and elbow. Length 60 ms. varying from 1 to 20 ms. in width. A large proportion is sandy and barren, without vegetation, yet partly populated. The men are employed at sea, and the boys are put on board the fishing boats. Violent E. winds are gradually wearing it away. Lon. 70° 14', w. lat. 42° 4' N.

CAPE ELIZABETH, t. Cumberland co. Me. 6 ms. s. w. Portland. Pop. 1830, 1696.

CAPE FEAR, a remarkable point of N. C. between Long bay and Onslow bay. The term is extended to the whole cape near the mouth of Cape Fear r. but correctly cape Fear is the extreme southern point of Smith's Isle, and on Tanner's U. S. map, is laid down at N. lat. 33° 55′ and 1° 02′ w. from W. C.

CAPE FEAR, river of N. C. rising between the Yadkin and Dan rivers, in Stoke, Rockingham, and Guilford cos. flows thence 200 ms. in a S. E. direction, receiving numerous smaller tributaries and is lost in the Atlantic ocean, by two mouths, one on each side of Smith's island. The basin of cape Fear r. is 200 by 40 ms. mean width, 800 sq. ms. between lat. 34° and 36° 2′ N. and in long. between 1° 30′, and 2° 18′ w. from W. C.

CAPE GIRARDEAU, co. of Mo. bounded by Scott co. S. E., Stoddard S., Wayne S. W., Madison W., Perry N. and the Mississippi r. separating it from Union and Alexander cos. of Il. E. Length E. to W. 38 ms. mean breadth 30, and area 1140 sq. ms. Extending in lat. from 37° 11′, to 37° 36′ N., and in long. from 12° 30′, to 13° 10′ west from W. C. Though bordered on the E. by the Mississippi r. the greatest part of the surface of this co. is drained to the southward by the sources of White water or eastern branch of St. Francis r. Chief town, Jackson.

CAPE HENRY, opposite and bearing a little W. of S. from cape Charles, is the southeast point of the mouth of Chesapeake bay; on Tanner's U. S. cape Henry is laid down at N. lat. 36° 56′, and in long. 1° 02′ E. from W. C.

CAPE MAY, co. N. J. bounded by Gloucester co. N., Atlantic ocean E. and S., Delaware bay and Cumberland co. W. It forms the south point of the state, terminating in cape May, the N. cape of Delaware bay, on which is a light house. The east coast is lined by sand beaches, dangerous to navigation.—Within it is a stretch of marshy lands, with ponds, inlets and creeks. It contains 4 townships. Pop. 1830, 4936.

CAPE NEDDOCK, York co. Me. 95 ms. S. S. W. Augusta, a rocky, barren, head land, stretching into the Atlantic from a hard and almost uninhabited shore. A few huts shelter a small number of fishermen. It is called in derision the city of Cape Neddock.

CAPEVILLE, p-v. Eastern shore, Va. Northampton co. near cape Charles, 176 ms. from Richmond.

CAPE VINCENT, p-v. Lyme, Jefferson co. N. Y. at the foot of lake Ontario, 21 ms. from Brownville, 8 from Kingston. It is on a broad gravelly point, between St. Lawrence r. and Chaumont bay, with Grenadier and Fox isles off the extremity. The St. Lawrence is here ferried by steam.

CAPTAIN'S ISLANDS, Conn. In Long Island Sound off Horse Neck. On one of them is a light house.

CAPTINA, p-v. on Captina cr. S. W. part Belmont co. Ohio, 20 ms. S. W. Wheeling.

CARBONDALE, a very flourishing village on Lackawana cr. at the western base of Moosic mtn. on the N. E. margin of Luzerne co. Pa. 35 ms. N. E. Wilkesbarre, and 130 due N. from Phil. The site of the village is 874 ft. above tide water. It has 150 houses and log huts, stores, inns, &c. and owes its existence to the Lackawana coal strata, which here is 26 feet in depth. The coal bed is in an area surrounded by forests, and is opened in about 20 places. The coal is conveyed by stationary steam engines—first a distance of 4 ms. with an ascent of 855 feet, and thence over a level of 8,300 feet to the head of 3 inclined planes, down which in cars it proceeds to Honesdale, at the head of the canal. The whole length of the rail road and planes is 91,000 ft. or about 17¼ ms. Total ascent from Carbondale to Rix's Gap, the height of land, 855 ft. and total descent thence to Honesdale 912½ ft.

CAROLINE, p-t. Tompkins co. N. Y. 13 ms. S. E. Utica, 170 W. by S. Albany; Owego, Six Mile cr., and a branch of Mud cr. supply mill seats. Pop. 1830, 2,633.

CAROLINE, co. Md. bounded by Queen Ann N. and N. W., by Dorchester co. S., Rutland and Sussex cos. Del. E., and W. by Talbot co. and Tuckahoe r. Length from S. to N. 30, mean breadth 8 and area 240 sq. ms. Surface undulating. The main branch of Choptank r. rises in Kent co. Del., but flows S. S. W. into Caroline, over which it meanders to its junction with Tuckahoe. In lat. from 38° 40′ to 39° 10′ and long. from 1° 03′ to 1° 18′ E. Chief town, Denton, pop. 1820, 10,108; 1830, 9,070.

CAROLINE, co. Va. bounded by Rappahannock r. which separates it N. from Strafford, and N. E. from King George, E. by Essex, S. E. by King and Queen, and King William, S. W. by N. Anna r. which separates it from Hanover, and N. W. by Spottsylvania. It lies very nearly in form of a parallelogram, 30 ms. from S. W. to N. E. with a breadth of 20 ms. area 600 sq. ms. Extending from lat. 37° 47′ to 38° 16′ N. and in lon. from 0° 02′ to 0° 43′ W. from W. C. Surface very much broken by hills, with a soil of great variety; staples, grain, flour, tobacco, &c. Chief town, Bowling Green. Pop. 1820, 18,008; 1830, 17,760.

CAROLUS, p-v. Vermillion co. Il., by p-r. 697 ms. wstrd. from W. C. and 136 ms. N. E. from Vandalia.

CARONDELET, canal of, extends from Bayou St. John about 2 ms. By this channel, vessels drawing 5 ft. water are navigated from lake Pontchartrain into the city of New Orleans.

CARONDELET, p-v. on the right bank of Mo. r. 6 ms. below St. Louis.

CARPENTER'S MILLS and p-o. Lycoming co. Pa. by p-r 97 ms. northwardly from Harrisburg.

CARRITUNK, p-v. Somerset co. Me.

CARROLL, p-t. Chautauque co. N. Y. 336 ms. W Albany. Pop. 1830, 1,015.

CARROLL, co. of Geo. bounded N. E. by Campbell co.; E. and S. E. by Chattahoochee r. which separates it from Coweta co., S. by Troup co.; W. by the state of Alabama, and N. by the Cherokee nation. Length from S. to N. 40 ms. mean breadth about 20, and area 800 sq. ms. Extending from lat. 33° 15′ to 33°

52′ N. and in long. from 7° 52′ to 8° 30′ from W. C. The extreme sources of both branches of the Talapoosa r. rise in the Cherokee country, but little distance above Carroll co. which they enter and traverse in a s. w. direction. It is a high, dry and broken county. For down-stream vessels, the Chattahoochee is navigable above Carroll. Chief town, Carrolton. Pop. 1830, 3,419.

CARROLL, co. W. Tennessee, bounded N. W. by Weakly; N. by Henry; E. by Humphries and Perry; S. by Henderson, S. W. by Madison, and W. by Gibson. Length from E. to W. 30, breadth 24, and area 960 sq. ms. Extending from lat. 35° 49′ to 36° 08′ and in long. from 11° 15′ to 11° 50′ W. from W. C. This county occupies part of the table land between Tennessee and Miss. rivers; Sandy creek, a small branch of the latter, rises in the eastern part of the county, and flows N. N. E. into Henry, whilst the central and western parts are drained by the head branches of Ohio r. flowing watrd. towards the Miss. Chief town, Huntingdon. Pop. 1830, 9,397.

CARROLTON, p-v. and st. jus. Carroll co. Geo. situated near the centre of the co. on the S. Fork of Tallapoosa r., by p-r. 151 ms. N. W. by W. from Milledgeville, N. lat. 33° 35′, long. 8° 10′ W. from W. C.

CARROLL, p-v. in the southwestern part of Washington co. O., by p-r. 96 ms. S. E. by E. Columbus.

CARROLTON, p-v. in the N. W. part of Fairfield co. O., 20 ms. S. E. from Columbus.

CARROLLTON, p-v. and st. jus. Greene co. Il., by p-r. 106 ms. N. W. by W. Vandalia, and 60 ms. a little W. of N. St. Louis.

CARROLLVILLE, p-v. Wayne co. Ten., by p-r. 97 ms. S. W. Nashville.

CARSONVILLE, p-v. Ashe co. N. C., by p-r. 238 ms. N. W. by W. Raleigh.

CARTER, extreme eastern co. of Ten. bounded N. E. by Washington co. Va., E. by Ashe co. N. C., W. by Washington co. Ten., and N. W. by Sullivan. Length, along the Iron mtn., which separates it from Ashe co. 45 ms., mean breadth 12 ms., and area 540 sq. ms. The whole co. is a mountain valley, drained by and commensurate with the main branches of Watauga r. which flow from this co. northwestwardly into the middle fork of Holston. The surface is mountainous and rocky, extending from N. lat. 36° 05′ to 36° 35′ and in long. from 4° 40′ to 5° 15′ W. from W. C. The elevation of this part of Ten. above the surface of the Atlantic, must be at least 2000 feet. Pop. 1820, 4,835; 1830, 6,414.

CARTER'S STORE and p-o. in the southwestern part of Prince Edward co. Va. 81 ms. S. W. by W. Richmond.

CARTER'S STORE and p-o. in the S. E. part of Nicholas co. Kentucky, by p-r. 65 ms. E. Frankfort.

CARTERSVILLE, p-v. on the right bank of James r. Cumberland co. Va., by p-r. 44 ms. watrd. Richmond.

CARTERET co. of N. C., bounded by Onslow co. or Whittock r. W., by Jones and Craven N., by Pamlico sound N. E., and by the Atlantic S. E., S. and S. W. Length from S. S. W. to N. N. E. 60 ms., mean breadth 10 ms. and area 600 sq. ms. Extending in lat. from Cape Lookout 36° 56′ N. and in long. from 0° 15′ W. to 1° E. from W. C. It is a long and sandy, and in part marshy strip, with sandy isles or reefs in front. Chief town, Beaufort. Pop. 1820, 5,609; 1830, 6,597.

CARTHAGE, p-t. Oxford co. Me. 46 N. W. Augusta. Pop. 1830, 333.

CARTHAGE, p-v. Wilna, Jefferson co. N. Y. 16 ms. E. Watertown, 160 from Albany; contains extensive iron works, E. side Long falls, on Black r.

CARTHAGE, v. Brighton, Monroe co. N. Y. 2½ ms. N. Rochester, at Lower falls of Gennesee r. and 5 ms. S. lake Ontario. (*See Brighton.*) Pop. 1830, 333.

CARTHAGE, p-v. Tuscaloosa co. Ala. 17 ms. from Tuscaloosa.

CARTHAGE, p-v. in the southern part of Campbell co. Ky. by p-r. 79 ms. N. E. Frankfort.

CARTHAGE, p-v. in Mill Creek tsp. Hamilton co. O. 7 ms. from Cincinnati.

CARTHAGE, p-v. and st. jus. Moore co. N. C. 55 ms. S. W. by W. Raleigh, and 42 N. W. Fayetteville.

CARTHAGE, p-v. and st. jus. Smith Co. Ten. situated on the right bank of Cumberland r. directly opposite the mouth of Carey Fork, 47 ms. a little N. of E. Nashville.

CARVER, p-t. Plymouth co. Mass. 28 ms. S. E. Boston, 8 E. Plymouth, is thinly populated, with 642 acres tillage, 361 mowing and 1938 of pasturage; soil not very good. Iron ore is found and wrought. Pop. 1830, 970.

CASCO BAY, Cumberland co. Me. between Capes Elizabeth S. W. and Small Point, 40 miles apart. It has fine anchorage, and islands, popularly reported as many as there are days in the year. Portland harbor is on the N. W. corner.

CASDAGA LAKE, Chatauque co. N. Y. connected with Conewango lake by Casdaga r. which is 40 ms. long.

CASDAGA, p-v. Chatauque co. N. Y. 340 ms. W. Albany.

CASEY, co. Ky. bounded by Estille S., by Adair S. W. and W., Mercer N., Lincoln E., and Pulaski S. E; length from S. to N. 32 ms. mean breath 14 and area 448 sq. ms. Extending in lat. from 37° 08′, to 37° 35′, and in long. from 7° 34′ to 7° 58′. The slope of this co. is to the westrd. From the northern section rise the extreme sources of Salt r. and from the southern those of Green r. The surface is high and broken. Chief town, Liberty. Pop. 1830, 4342.

CASHVILLE, p-v. Spartansburg dist. S. C. by p-r. 110 ms. N. W. Columbia.

CASS, co. Mich. boundaries uncertain. This county embraces a region on both sides St. Joseph's r. of lake Mich. Besides at Edwardsburg, the st. jus. it had in 1831, a p-o. at La Grange and Pocagon. The body of the co. lies a little S. of W. of Detroit about 170 ms. and Edwardsburg 169.

CASTANA, p-v. in the southern part of Seneca co. Ohio, by p-r. 97 ms. northwardly from Columbus.

CASTILE, p-t. Genesee co. N. Y. 30 ms. S. E. Batavia, has pretty good land, is crossed by Genesee r. and contains Gardeau Reservation. Pop. 1830, 2269.

CASTINE, sea port, p-t. and cap. Hancock co. Me. 122 ms. E. N. E. Portland, 78 Augusta, on a promontory, near the head of Penobscot bay, with a good harbor for large vessels, open at all seasons. A narrow isthmus might easily be cut through, and made a powerful fortress, to command the country to St. Croix. Long. 68° 46′ W., lat. 44° 24′ N. Pop. 1830, 1148.

CASTLEMANS, r. a N.E. branch of Youghaghany river rises in Alleghany co. Md. and Somerset co. Pa. the higher branches uniting in the latter, flows N. W. 12 ms., and thence S. W. 25 ms. to its junction with Youghaghany, the eastern side of Laurel Hill. It is a real mtn. torrent, having a fall of upwards of 1000 feet in a comparative course of 60 ms. The valley of this stream is intended as part of the route of the Chesapeake and Ohio canal.

CASTLEMANS, p-o. Gallatin co. Ky. by p-r. 48 ms. N. Frankfort.

CASTLETON, p-t. Rutland co. Vt. 10 ms. W. Rutland, 36 sq. ms., first settled 1769. It is crossed by Castleton r. which here receives the waters of lake Bombazine, 8 ms. long, chiefly in this t. containing an island. The land is good, with oak on the hills, and pine in the vallies, and supplied with mill sites. *The Vermont academy of medicine* is in Castleton, incorporated 1818, degrees being received at Middlebury college. There are two buildings, one of which is 50 feet by 30, 2 stories high, with a dissecting room, and rooms for lectures, the library, chemical laboratory, and anatomical museum. Five courses of lectures are delivered annually, commencing on the first Tuesday in September. The buildings are large, and pleasantly situated. The *Rutland co. Grammar school*, was incorporated 1805. Pop. 1830, 1,783.

CASTLETON, r. Rutland co. Vt. rises in Pittsford, runs S. and then W. and joins Poultney r. It is 20 ms. long.

CASTLETOWN, p-t. Richmond co. N. Y. N. E. corner of Staten isl., S. N. Y. bay, is hilly, with arable land, high and agreeably varied, but lately subject to fever and ague. The v. is near the water, looking E. upon the quarantine ground, and Long isl. and contains the Lazaretto, or quarantine hospital, a fever hospital, the Sailor's Snug harbor, and Marine hospital of New York city. A steamboat runs to New York 5 or 6 times daily; distance 5½ ms. The t. contains Clove hills, and others fortified by the British in the revolution. Pop. 1830, 2,204.

CASWELL, co. of N. C. bounded by Person E., Orange S., Rockingham W., and Pittsylvania co. of Va. N. It is a square of 20 ms. each side, extending from lat. 36° 15′ to 36° 02′ and in long. from 2° 11′ to 2° 33′ W. from W. C. The slope of this co. is to the N. E. and its waters flow in that direction into Dan r. which stream, already navigable, winds estrd. along the northern border of the county. The soil is productive and climate agreeable. Chief t. Leesburg. Pop. 1820, 13,253, 1830, 15,185.

CASWELL, C. H. p-o. and st. jus. Caswell co. N. C. on Lime cr. a branch of Dan r.

CASVILLE, p-v. Iowa co. Mich.

CATAHOOLA, or Ocatahoola, parish of La. bounded by the parish of Washitau N. W. and N. Tensas r. or the parish of Concordia E., Catahooche r. and lake S., and Litttle r. W. Length from S. W. to N. E. 75 ms., mean width 28, and area 2100 sq. ms. Extending from lat. 31° 29′ to 32° 20′ and in long. from 14° 24′ to 15° 24′ from W. C. The face of this large parish differs materially in different parts. The Washitau re-enters it from the N. winding over it in a southern direction, and receiving near its centre the Boeuf, from the N. E. Northward from the Ocatahoola r. and between the Washitau and Little rs. the country rises into hills, covered generally with pine timber, and watered by clear, perennial creeks, but soil sterile, except in confined spots near streams. The entire eastern part of the parish lies within the overflow of the Miss. and except some few strips along the rivers, or on Sicily island, is liable to annual inundation. When the soil of the alluvial part of Ocatahoola is sufficiently elevated for cultivation, it is very productive. Staples, cotton, live stock, and lumber. Chief town, Harrisonburg. Pop. 1820, 2,287, in 1830, 2,581.

CATAHOOLA, r. and lake of La. The r. rises in Clairborne, Natchitoches, and Washitau parishes, flows in a general course S. S. E. about 80 ms. to the extreme S. W. angle of the parish of Ocatahoola, where, at seasons of high water of the Washitau and Mississippi rivers, it expands into a lake of 18 ms. long, and from 2 to 5 wide. At the head of the lake the river turns abruptly to N. E. by E. continues in that direction through the lake, and again contracting to a river of about 80 yards wide, flows 15 ms. to where it joins the Washitau to form Black r. Ocatahoola lake is one of those depressions in the great plain of Louisiana which operate to form reservoirs, filled and emptied annually. The bottom of the lake is below that of even the common inundated lands, and when the Washitau and Miss. are rising, receives a surcharge of water by the channel of Ocatahoola. The reverse takes place when the great streams are falling; then the current flows rapidly from the lake, which is finally drained, and in autumn and early winter, becomes a vast meadow covered with herbage, with the river meandering over its surface. Similar features are presented by Black lake, Natchitoches lake, Spanish lake, Bristineau, Bodeau, &c.

CARDINGTON, p-v. in s. part Marion co. O. by p-r. 42 ms. N. Columbus.

CARLETON, Isl. and p-v. Jefferson co. N. Y. in the St. Lawrence, has a good harbor, and much trade. 10 ms. s. E. Kingston, 30 N. Sacket's harbor.

CARLINVILLE, p-v. and st. jus. Macaupin co. Il. by p-r. 95 ms. from Vandalia.

CARLISLE, t. Middlesex co. Massachusetts 20 miles N. W. Boston. Population 1830, 566.

CARLISLE, p-t. Schoharie co. N. Y. 40 ms. W. Albany, 8 W. Schoharie, 7 ms. by 8, produces grass and grain. Limestone, sulp. barytes, white pine, maple, beech, &c. Pop. 1830, 1,748.

CARLISLE. p-v. borough and st. jus. Cumberland co. Pa. about a mile from the right bank of Conedogwinet r., 18 ms. W. Harrisburg and by p-r. 103 ms. a little W. of N. from W. C. N. lat. 40° 12′ and long. 0° 13′ W. from W. C. Cumberland was made a co. separate from Lancaster, Jan. 1749–50, and Carlisle made the st. jus. It is situated on an undulating plain, amid a very fertile and well cultivated country. The houses are generally of brick or lime-stone. The latter material is easily procured, as the town rests on a soil incumbent over a mass of blue limestone. The streets are at right angles, and the buildings generally commodious. In 1783 a college was established at Carlisle and named in honor of John Dickinson. This seminary, after a long period of languishment, was revived in 1820 by private and legislative donation and is now, 1830, in active operation. Pop. of the borough in 1820 about 3,000, in 1830, 3,707.

CARLISLE, p-v. and st. jus. Nicholas co. Ky. 56 ms. a little N. of E. Frankfort and 38 ms. N. E. Lexington.

CARLISLE, p-o. in the s. part of Sullivan co. Ind. 12 ms. s. E. from Merom, the co. seat and by p-r. 115 ms. s. W. from Indianoplis.

CARLTON, t. Orleans co. N. Y. Pop. 1830, 168.

CARLTON'S STORE, and p-o. King and Queen co. Va. by p-r. 44 ms. E. Richmond.

CARLYLE, p-v. and st. jus. Clinton co. Il. on Kaskaskias r. by p-r. 30 ms. below, and s. s. W. Vandalia and 49 E. St. Louis.

CARLO, p-v. Hopkins co. Ky. by p-r. 172 ms. s. W. by W. Frankfort.

CARMEL, p-t. Penobscot co. Me. 15 ms. W. Bangor, 71 N. E. Augusta, has ponds and streams runing E. to Penobscot. Pop. 1830, 257.

CARMEL, p-t. and st. jus. Putnam co. N. Y. 11 ms. E. West Point, contains the county buildings, is hilly, yields grass and contains Mahopack pond, Croton cr. and other ponds emptying into Peekskill cr. with many mill seats.

CARMEL HILL, and p-o. Chester dist. S. C. by p-r. 74 ms. N. Columbus.

CARMI, p-v. and st. jus. White co. Il. by p-r. 94 ms. s. E. from Vandalia and 75 ms. s. s. W. from Vincennes, Ind.

CARMAN'S, p-o. Harford co. Md. 32 ms. N. E. Baltimore.

CARMEL, p-o. in the Cherokee nation, Geo. by p-r. 224 ms. from Milledgeville.

CARMICHAELS, p-o. Greene co. Pa. by p-r. 190 ms. W. from Harrisburg.

CARNESVILLE, p-v. and st. jus. Franklin co. Geo. by p-r. 110 ms. almost due N. from Milledgeville.

CATARAUGUS, co. N. Y. bounded by Cataraugus cr. N. or Erie and Genesee cos., Alleghany co. E., Pennsylvania s., Chatauque co. W., about 34 by 38 ms. 1292 sq. ms., has Alleghany r. winding through s. part, and streams of Genesee and lake Erie, above which it is 500 to 1,200 feet with high hills. There are white pine tracts and marshes s. but the land is generally firm, with maple, beech, bass, nut, and oak. Grass and grain grow best N. This co. was purchased by the Holland company, 19 townships. Pop. 1820, 4,090, 1830, 16,726.

CATARAUGUS RESERVATION, N. Y. 6 ms. by 12, on Cataraugus cr. was reserved by the Seneca Indians, who here enjoy christian worship, schools, &c. with good habits.

CATHERINE, p-t. Tioga co. N. Y. 18 ms. N. Auburn, 200 W. Albany, 12 ms. sq. gives rise to the inlet of Seneca lake, Newtown cr. &c. It has good land N. with oak and pine, and in other parts beech, maple, bass, elm, &c. Limestone and iron ore are found, and a pigment like Spanish brown. Pop. 1830, 2,064.

CATAWBA, r. of N. and S. C. called Wateree in the lower part of its course, rises in the Blue Ridge by numerous branches, which flow generally eastward over Burke co. unite at Morgantown, and continuing eastward 25 ms. still over Burke, turns abruptly s. s. E. between Iredell and Lincoln. This higher valley of Catawba is about 65 ms. long, with a mean breadth of 20, and nearly commensurate with Burke co. Leaving the latter the Catawba, in a general course of s. s. E. flows 50 ms. in N. C. and 100 in S. C., finally unites with the Congaree to form the Santee. The Catawba is remarkable for the narrowness of its valley, which in a distance of 215 ms. is in no place 60 ms. wide, and at a mean under 20. In length of course it exceeds the Congaree, but in volume the latter is greatly the superior stream.

CATAWISSA, p-v. Columbia co. Pa. situated on the left bank of the East Branch of Susquehannah r. and at the mouth of a creek of the same name.

CATFISH, p-o. on a creek of the same name, Marion dist. S. C. by p-r. 138 ms. eastward Columbus.

CATHEY'S CREEK, and p-o. Buncombe co. N. C. by p-r. 267 ms. wstrd. from Raleigh.

CATLETTSBURG, p-v. on the left bank of the Ohio r. at and below the mouth of Great Sandy r. Greenup co. Ky. It is the extreme N. E. village of the state, by p-r. 159 ms. a little N. of E. from Frankfort.

CATLIN, p-t. Tioga co. N. Y. 18 ms. N. W. Elmira. Pop. 1830, 2,015.

Cato, p-t. Cayuga co. N. Y. 18 ms. n. Auburn, 155 w. Albany, n. Erie canal and Seneca r. has a variety of soil, and swamps and ponds, Cross and Otter lakes, and Parker's pond. Pop. 1830, 1,782.

Catonsville, p-v. Baltimore co. Md. by p-r. 44 ms. n. e. from W. C. and 36 n. from Annapolis.

Catskill, p-t. and st. jus. Greene co. N. Y. 36 ms. s. Albany, 5 s. w. Hudson, w. Hudson r. has gentle hills e. with pretty good soil, Catskill mtns. w. and a high plain and sand and clay n. It is watered by Catskill creek and Keaterskill creek, its branch; with rich meadows, and mill sites; it has 3 villages and 2 banks, capital $250,000. The post v. is the st. jus. 1 mile w. Hudson river, with a pier, where the large steam boats touch a ferry; and the co. buildings. A company has been incorporated to make a rail road from here to Schoharie. Pop. 1830, 4861.

The Pine Orchard, on Catskill mountain, is a favorite resort of travellers in the warm months. A fine hotel has been erected there several years, on the brow of a rock, at a great elevation above Hudson river, with a view embracing about 70 ms. from n. to s. on the valley of Hudson river, and the hilly country e. including a number of peaks of the Green mountain range in Mass. and Vt. Thunder storms are often seen below the spectator, and the air is generally cool. Two ponds in the rear of the house, unite their streams, and the water falls 175 feet, and soon after 85 feet, into an immense ravine between 2 ridges of mtns. A limestone range begins a little w. of Catskill v. reaching 4 ms., w. of which is sand-stone, then graywacke slate, the peaks being pudding stone, conglomerate, &c. Stage coaches take visitors to Pine Orchard; the last part of the road is steep and rough.

Caughnawaga, p-v. Johnstown, Montgomery co. N. Y. 39 ms. w. Albany, n. Mohawk r. once the residence of the Mohawk Indians, (*See Johnstown.*)

Cavendish, p-t. Windsor co. Vt. 10 ms. s. w. Windsor, 60 s. Montpelier. First settlement 1769. It has a fertile soil, with Black r. and 20 mile stream. Black r. at the falls, has its channel worn down 100 ft. Dutton's village has an academy, and Proctorsville has another. Serpentine iron ore, and primitive limestone are found near it. Pop. 1830. 1,498.

Cave Mills, p-o. Warren co. Ten. by p-r. 74 ms. s. e. Nashville.

Cavesville, p-o. Orange co. Va. by p-r. 94 ms. n. w. Richmond.

Cavetown, p-v. Washington co. Md.

Cayuga, lake, N. Y. between Cayuga, Tompkins & Seneca cos. from 1 to 4 ms. wide, 38 long, n. and s., 35 ms. s. lake Ontario, receives Seneca r. near the outlet, which runs n. The shores rise gradually, 100 or 150 ft. but in some places are precipitous. It has Fall, 6 miles, and Main Inlet crs. s. and other fine mill streams, and has several villages on its shores.

Cayuga, co. N. Y. 170 ms. w. Albany, bounded by lake Ontario, Oswego, Onondaga and Cortlandt cos. e., Tompkins co. s., Seneca co. w. It is e. Seneca lake. 23½ by 55 ms. are its greatest dimensions, with about 545 sq. ms. It has a spur of Alleghany hills parallel with Cayuga lake, good soil and very good farms. There is much lime rock, with petrifactions, &c. It is watered by Seneca r. Fall, Salmon, Owasco crs. &c. Owasco lake, and parts of Ontario, Cayuga, Skeneateles and Cross lakes. Erie canal crosses the co. Clay slate, limestone, gypsum, and hydraulic lime are found, and argilaceous oxide of iron is abundant. 19 tsps. Pop. 1820, 38,897, 1830, 47,947.

Cayuga, or E. Cayuga, p-v. Aurelius, Cayuga co. N. Y. 165 ms. w. Albany, at the bridge and w. side of Cayuga lake.

Cayuga, p-v. Claiborne co. Miss. by p-r. 60 ms. n. n. e. Natchez.

Cayuta, v. Newfield, Tioga co. N. Y. 20 ms. n. w. Owego, on Cayuta creek. Pop. 1830, 642.

Cazenovia, p-t. Madison co. N. Y. 113 ms. w. Albany, 11 w. Morrisville, 5 ms. by 12, has Canaseraga lake, of 4½ ms., and Chitteningo and Limestone creeks. It is level, high land, rich loam for grass and grain. First settled, 1793. The village is at the s. end of the lake, and is flourishing. Pop. 1830, 4344.

Cecelius, p-v. Cataraugus co. N. Y. 290 w. Albany.

Cecil, or Cozcil, n. e. co. of Md. bounded s. by Kent co., s. w. by Chesapeake bay, w. by the Susquehannah r., n. w. by Lancaster, n. e. by Chester co. Pa., and e. by New Castle co. Del. Length from s. to n. 22 ms. mean breadth 12, and area 264 sq. ms. Extending from n. lat. 39° 22′ to 39° 42′, and in long. from 0° 50′ to 1° 18′ e. from W. C. The surface of Cecil is undulating, and soil of middling quality. It is in a peculiar manner favorably placed, commercially. To the Susquehannah r. and Chesapeake bay may be added Elk r. and the Chesapeake and Del. canal. Chief town, Elkton. Pop. 1820, 16,048, 1830, 15,432.

Cecilton, p-v. Cecil co. Md. This place was formerly called Savingston.

Cedar Creek, one of the w. branches of the n. Fork of Shenandoah r. and separating Shenandoah and Frederick cos.

Cedar Creek and p-o. in the nthrn. part of Shelby co. Ala. by p-r. 97 ms. n. e. by e. Tuscaloosa.

Cedar Spring and p-o. in the estrn. part of Centre co. Pa. by p-r. 101 ms. n. n. w. Harrisburg.

Cedarsville, p-v. in Perry co. O. by p-r. 101 ms. s. w. Columbus.

Cedar Spring, Spartanburg dist. S. C. 5 ms. s. e. Spartanburg, the st. jus. 90 ms. n. w. Columbia.

Celina, p-v. Overton co. Ten. by p-r. 85 ms. n. n. e. Nashville.

Centre co. Penn. bounded n. and n. e. by Lycoming, w. branch of Susquehannah, which separates it from Clearfield and Lycoming W. and N. W., by Huntingdon and Mifflin S. and by Union E. Length 8 ms. mean breadth 26, and area 1560 sq. ms. Extending from N. lat. 40° 43' to 41° 16', and in long. from 0° 12' to 1° 23' W. from W. C. Bald Eagle and Penn's creek rise in this co. and with the Susquehannah afford some good soil, though the body of the co. is mountainous and rocky. Staples, grain, flour, live stock, lumber, iron, &c. Chief town, Bellefonte. Pop. 1820, 13,786, in 1830, 18,295.

CENTRE, p-v. Guilford co. N. C. by p-r. 77 ms. N. N. W. Raleigh.

CENTRE HARBOR, p-t. Strafford co. N. H. 48 ms. N. Concord, 70 N. W. Portsmouth, 110 N. W. Boston, N. E. corner Winnipiseogee lake. 7550 acres; contains part of Squam and Measley lakes, with part of Winnipiseogee lake; has a varied surface, and some good soil. Pop. 1830, 577.

CENTRES MINOT, p-v. Cumberland co. Me. 42 ms. from Augusta.

CENTREVILLE, p-o. Kent co. R. I. 11 ms. from Providence.

CENTREVILLE, p-t. Allgehany co. N. Y. 16 ms. N. W. Angelica, 6 ms. square, has small streams of Genesee r. Bog iron ore abounds. The soil, a light loam, with few stones, bearing maple, bass, beech, &c. Pop. 1830, 1,195.

CENTRE MORELAND, p-v. Luzerne co. Pa. by p-r. 133 ms. Harrrisburg.

CENTRE POINT, Montgomery co. Pa. by p-r. 96 ms. Harrisburg.

CENTRE VILLE, p-v. Crawford co. Pa. about 100 ms. a little E. of N. Pittsburg.

CENTRE, p-v. Delaware co. O. by p-r. 30 ms. N. Columbus.

CENTRE, p-v. Farmington tsp. Trumbull co. O. by p-r. 167 ms. N. E. Columbus.

CENTRETON, p-v. Halifax co. Va. by p-r. 139 ms. S. W. by W. from Richmond.

CENTREVILLE, p-v. in the S. part of Montgomery co. O. by p-r. 41 ms. N. N. E. Cincinnati.

CENTREVILLE, p-v. and st. jus. Wayne co. Ind. on a branch of White Water r. about 70 ms. N. W. from Cincinnati, O. by p-r. 63 E. from Indianopolis.

CENTREVILLE, p-v. in the S. part of Wabash co. Il. 115 ms. S. E. by E. from Vandalia.

CENTREVILLE, p-v. in the N. part of Newcastle co. Del. 10 ms. N. N. E. from Wilmington.

CENTREVILLE, Queen Ann co. Md. situated on Casica creek, 36 ms. S. E. by E. from Baltimore, and 31 ms. a little N. of E. from Annapolis. It is the seat of an academy.

CENTREVILLE, p-v. in the W. part of Fairfax co. Va. 27 ms. a little S. of W. W. C.

CENTREVILLE, p-v. in the N. E. part of Laurens Dist. S. C. by p-r. 81 ms. N. W. Columbia.

CENTREVILLE, p-v. Wilkes co. Geo. by p-r. 81 ms. N. E. from Milledgeville.

CENTREVILLE, p-v. and st. jus. Bibb co. Ala. situated on the right bank of Cahaba r. 32 ms. S. E. from Tuscaloosa.

CENTREVILLE, p-v. Livingston co. Ky. 20 ms. N. E. by E. from Smithland, at the mouth of Cumberland r. by p-r. 275 ms. S. W. by W. from Frankfort.

CENTREVILLE, p-v. Hickman co. Ten. by p-r. 81 ms. S. W. from Nashville.

CENTREVILLE, p-v. Amite co. Miss. about 45 ms. S. E. from Natchez.

CERESTOWN, p-v. on Oswago cr. a branch of Alleghany r. in the N. E. part of Mc Kean co. Pa. 20 ms. S. E. from Hamilton, 165 N. W. Harrisburg.

CERULEAN SPRINGS, and p-o. Trigg co. Ky. 5 ms. N. E. from Cadiz, the st. jus. and by p-r. 221 ms. S. W. by W. from Frankfort.

CHACTAWS, nation of Indians. (See Choctaws.)

CHAGRIN, r. p-v. and tsp. in the N. E. angle of Cayahoga co. O. The p-v. is near the shore of lake Erie, 16 ms. E. from Cleaveland. In 1830, the tsp. contained 1,275 inhabitants.

CHALK LEVEL, p-v. in the W. part of Humphrey co. Ten. by p-r. 70 ms. W. from Nashville.

CHALK LEVEL, p-o. Pittsylvania co. Va. by p-r. 133 ms. S. W. from Richmond.

CHAMBERSBURG, p-t. and st. jus. Franklin co. Pa. situated on both sides of Conecocheague creek, 82 ms. S. W. from Harrisburg. It is a very thriving borough, situated in a fertile limestone region. Pop. 1830, 2,783.

CHAMBERSBURG, p-v. in the E. part of Fountain co. Ind. by p-r. 66 ms. N. W. by W. Indianopolis.

CHAMPAIGN, co. O. bounded by Clarke S., Miami S. W., Shelby N. W., Logan N., Union N. E., and Madison S. E. Length 29, breadth 16, and area 464 sq. ms. extending in lat. from 39° 58', to 40° 15', and in long. from 6° 52', to 7° W. W. C. Though the extreme sources of Darby's creek, a branch of Sciota r. flows from the eastern border, and some fountains of creeks flowing into the Great Miami, issue from the wstrd. the great body of this co. slopes sthrd. and is included in the valley of Mad r. Chief town, Urbana. Pop. 1820, 8,479, 1830, 12,131.

CHAMPION, p-t. Jefferson co. N. Y. 12 ms. E. Watertown, at the Long Falls of Black r. contains 26,000 acres, with rich loam and sand; first settled from Conn. Pop. 1830, 2,342.

CHAMPLAIN LAKE, between Vt. and N. Y. extends from Whitehall, N. Y. a little beyond the Canada line, 140 ms. nearly N. and S. generally narrow and deep, 12 ms. in the widest part. That part of it from Whitehall to Mount Independence, opposite Fort Ticonderoga, was formerly considered a part of Wood creek. The principal islands are N. and S. Hero, Lamotte, Valcour and Schuyler's. It is navigated by many vessels of 80 and 90 tons, which are generally built to pass the canal; an active trade is carried on from the numerous towns and villages on the shores. Large and elegant steamboats ply daily between Whitehall and St. John's, Lower Canada, which touch at the principal places; and multitudes of travellers for pleasure every season pass this route. The shores are varied and pleasant, generally cultivated in farms near the water, and rising towards the mountains which appear in various directions. The principal eminences of the Green mountains are fine features in the landscape. The outlet of Lake George enters at Ticonderoga, and Chazy, Saranac, Sable and Bouquet rs. w. Wood creek s. and Otter, Onion, Lamoille and Missisque rs. E. The largest bay is South bay, and Cumberland the principal head land. Ticonderoga and Crown p-t. N. Y. at two important bends of the lake, were formerly great fortresses, both used in the French wars, and abandoned at the close of the revolution. Large remains of the works are seen. The lake was discovered 1608, abounds in salmon, trout, sturgeon pickerel, &c.; freezes deep for several months, and is usually travelled with land vehicles from Dec. 10th, to March 15th or 20th. Several new villages have recently grown up on the banks, particularly near iron mines, &c.

CHAMPLAIN, p-t. and port of entry, Clinton co. N. Y. 21 ms. N. Plattsburgh, 188 N. Albany, on Lake Champlain. Great Chazy r. affords mill sites at the v. It contains Pointe-au-fer, and has level land on the lake, generally strong loam or clay, bearing apples, pears, plums. Rouse's Point has lately been taken from this town, and added to Canada. Pop. 1830, 2456.

CHAMPLAIN CANAL. *(See Rail Roads and Canals.)*

CHANCEFORD, p-v. York co. Pa. 35 ms. s. E. the borough of York, and 30 a little w. of s. from Lancaster; the two tsps. of Chanceford, Upper and Lower, contained a Pop. in 1830, of 2213.

CHANCELLORSVILLE, p-o. Spottsylvania co. Va. by p-r. 75 ms. from Richmond.

CHANDLERSVILLE, t. Somerset co. Me. 39 ms. N. Augusta. Pop. 1830, 172.

CHAPINVILLE, p-v. Litchfield co. Conn., 50 ms. w. Hartford.

CHAPLIN, p-t. Windham co. Conn., 32 ms. E. Hartford, recently formed of a part of Mansfield, Tolland co. It is divided by Natchaug r., a branch of Shetucket r., and is a hilly, grazing country. Pop. 1830, 807.

CHAPMAN'S MILLS, and p-o. Giles co. Va. by p-r. 225 ms. a little s. of w. Richmond.

CHAPMANS, p-v. Union co. Pa., by p-r. 53 ms. N. N. W. Harrisburg.

CHAPOLA, r. of Flor. and Ala., rises in Henry co., of the latter, enters Jackson, of the former by several creeks, which uniting, passes under a natural bridge, about 15 ms. within Florida. Issuing thence, it flows about 30 ms. nearly parallel, and from 10 to 20 ms. distant from the Appalachicola r. and is finally merged in Horts lake, after an entire course of 45 ms., in a direction s. s. E. by s. "On its margin," says Williams, "is some of the best land in the country." The most extensive settlements are on its western border, extending from 1 to 5 ms. in width, and 30 in length. The soil is a chocolate colored sandy loam or red clay, supported by limestone. The timber, a mixture of oak, pine, hickory, and dogwood, filled up with cane. Corn, cotton, and sugar are the most important staples.

CHARDON, p-v. and st. jus. Geauga co. O., by p-r. 28 ms. N. E. by E. Cleaveland, on Lake Erie, and 157 ms. N. E. Columbus. Pop. 1830, 881.

CHARITON, r. of Mo. rises about N. lat. 40° between the vallies of the Ravine des Moines and Grand r., and flowing thence by a general sthrn. course 130 ms. falls into Mo. r. between Howard and Chariton cos.

CHARITON, co. of Mo. bounded E. by Randolph, s. E. by Howard, s. by Mo. r. separating it from Sabine co., w. by Grand r., N. boundaries uncertain. Length 32 ms. mean breadth 26, and area 832 sq. ms. Extending in lat. from 39° 11′ to 39° 40′ N., and in long. from 15° 39′ to 16° 16′ w. from W. C. The slope of this co. is to the sthrd. with the general courses of Grand and Chariton rs. Chief town, Chariton.

CHARITON, p-v. and st. jus. Chariton co. Mo. by p-r. 79 ms. N. W. from Jefferson co., and 213 m. N. W. by W. St. Louis. It is on the left bank of Mo. r., at the mouth of Chariton r. Long. 15° 48′ w. from W. C.

CHARLEMONT, p-t. Franklin co. Mass., 14 ms. w. Greenfield, 107 N. N. W. Boston, is watered by Deerfield r., and contained 3 garrisons, erected 1754, against the French and Indians. Pop. 1830, 1,065.

CHARLES r. Mass., rises near R. I. and flows through Norfolk and Middlesex cos., between which it forms part of the boundary, and joins Mystic r. in Boston harbor.

CHARLES, co. of Md. bounded by Potomac r. s. s. w. and w., Prince George's N., Swanson cr., Patuxent r., St. Mary's co., and Wernico r. E. Length 30 ms., mean breadth 15, and area 450 sq. ms. Extending from N. lat. 38° 15′ to 38° 40′, and in long. from 14′ w. to 19′ E. from W. C. Surface broken, and soil of middling quality. Chief town, Port Tobacco. Pop. 1820, 16,500, 1830, 17,769.

CHARLES CITY, co. of Va., bounded by James r., which separates it from Prince George's s., Henrico, N. W., Chickahoming r., which separates it from New Kent, N., and by the latter r. which separates it from James City, E. Length 26 ms. mean breadth 8 and area 208 sq. ms. Extending from N. lat 37° 09′ to 37° 28′ and in long. from 5′ E. to 22′ w. from W. C. Surface rolling. Pop. 1820, 5,255; 1830, 5,500.

CHARLES city, C. H. and p-o. near the centre of the co. 31 ms. s. E. by E. Richmond.

CHARLESTON, dist. of S. C. bounded s. w. by Colleton dist., N. W. by Orangeburgh, N. and N. E. by Santee river, which separates it from Sumpter, Williamsburg and Georgetown, and s. E. by the Atlantic ocean. The greatest length along the Atlantic coast 68 miles, and inland at nearly right angles to the coast, 55 miles. Mean breadth about 33, and area 2,244 sq. ms. The surface of this district is in great part an inundated plain nearly commensurate with the basin of Ashley and Cooper rivers. The part towards the Atlantic presents a net work of interlocking streams and islands. The soil where of sufficient elevation for cultivation is highly productive. Staples—cotton and rice. As a commercial section, Charleston dist. is favorably situated, since beside the harbor of Charleston city, there are many inferior inlets. A canal has been constructed to unite Cooper r. with the Santee opposite Black Oak island. Length 21 ms. embracing in lat. from 32° 32′ to 33° 28′, and in long. from 2° 20′ to 3° 32′ w. from W. C. Charleston is in itself a considerable physical section. From observations made from 1750 to 1789 inclusive, and from 1791 to 1824 inclusive, the mean annual temperature of the city of Charleston N. lat. 32° 44′ is within an inconsiderable fraction of 60° Fahrenheit. This is a temperature higher considerably than that on similar lat. in the valley of the Mississippi. Chief t. Charleston. Pop. 1820, 80,212; 1830, 106,706.

Charleston, city, and s-p. Charleston dis. S. C. situated on the point between Ashley and Cooper rs. 6 ms. from the open Atlantic ocean, 113 ms. s. s. e. Columbia, and by p-r. 539 ms. s. s. w. from W. C., N. lat. 32° 44′, long. as marked on Tanner's U. S. 3° w. from W. C. The bay formed by Ashley and Cooper rs. is about 2 ms. wide, and extending from city point a little s. of e. There are two entrances, the deepest of which admits vessels of 16 ft. draught, but the channel coming close upon the s. w. end of Sullivan's Island, gives a safe means of defence, which was reduced to certainty in the revolutionary war, when on June 28th, 1776, a British fleet under Sir Peter Parker was repulsed and shattered by the cannon of Fort Moultrie, a mere stockade battery. On the w. the harbor of Charleston is united to Stono r. by Wappoo creek, and by the channel of Cooper r. and a canal of 20 ms. it is connected with Santee r. 50 ms. a little w. of N. from the city. Ashley, Cooper and Wando rs. are all navigable for small vessels above the harbor. The whole adjacent country, being a plain, but little elevated above tide water, the city is liable to occasional inundation from ocean swells. It is nevertheless a fine commercial mart, well built and prosperous. Every spot in the vicinity capable of improvement is decorated with plantations in a high state of cultivation. Within the city exist all those institutions which mark a wealthy community. The most noted public edifices are the Exchange, City Hall, 6 Banking Houses, a a Guard House, an Arsenal, 2 College buildings, academical and medical, a large fire proof building, erected for the greater security of public documents, at an expense of $60,000, Court House, numerous places of public worship, among which are some of the most ancient in the U. S., 2 markets, one of which is very extensive, St. Andrew's Hall, an Alms House, an Orphan Asylum, and many other charitable institutions, richly endowed, among which are the St. Andrew's, South Carolina, and Fellowship societies. The Orphan Asylum, in which 150 children are protected, supported and educated, is an honor to the state. The public Library contains 15 to 20,000 volumes.

Charleston offers a delightful residence to the planters, who are widely scattered through the surrounding country, many of whom have fine residences in the city. It is one of the gayest cities of the U. S., and its society is excellent. The progressive population of Charleston is as follows: in 1790, 16,359; 1800, 18,711; 1810, white persons, 11,568, slaves and free blacks, 13,143, total 24,711; 1820, whites, 10,653, slaves and free blacks, 14,127, total 24,780; 1830, whites, 12,928, slaves and free blacks, 17,361, total 30,289.

Charleston, p-v. and st. jus. Jefferson co. Va. 10 ms. s. w. by w. Harper's Ferry and 63 N. w. from W. C.

Charleston, p-v. Cecil co. Md. 10 ms. s. w. by w. Elkton, and about 60 ms. N. E. Baltimore.

Charleston, p-v. and st. jus. Kenhawa co. Va. situated on the right bank of the Great Kenhawa r. and on the point above the mouth of Elk r. about 50 ms. by land above the mouth of Great Kenhawa, and by p-r. 304 N. w. by w. from Richmond.

Charleston, p-v. and st. jus. Clarke co. Ind. by p-r. 105 ms. s. s. e. from Indianopolis, 14 N. N. w. from Louisville in Ky. It is situated on the bank of Ohio river.

Charleston, p-v. and tsp. in the sthrn. part of Portage co. O. by p-r. 132 ms. N. E. from Columbus and 10 s. from Ravenna, the co. st. Pop. 1830, 475.

Charleston, p-t. Penobscot co. Me. 73 ms. N. E. Augusta. Pop. 1830, 859.

Charlestown, p-t. Sullivan co. N. H. 51 ms. from Concord, 100 from Boston, 18 from Windsor Vt., E. Conn. r., contains 21,400 acres. Little Sugar r. and 3 isls. are opposite this town in Conn. r. It has various soils, few mill sites, 1500 acres of rich meadow in one place, and in another a ridge of waste land. There are two villages, the s. very pleasant. Here was a fort built, 1743, above 30 ms. in advance of other settlements, and stood a siege and repeated attacks, till about 1760. Pop. 1830, 1,773.

Charlestown, p-t. and port of entry, Middlesex co. Mass. 1 m. N. of Boston, with which it is connected by a bridge 1,503 feet long. and also by a branch of Craigie's bridge. Chelsea bridge crosses Mystic r. E. nearly 1 m. on the Salem road, and Malden bridge, 2,420 feet, leads to Malden. A bay of Charles r. is w., Mystic r. E., and a narrow neck connects it with the main land N. The surface is irregular, with two fine eminences, Breed's and Bunker's hills. The v. is large and flourishing, one of the suburbs of Boston, with Bunker hill bank, and many other public buildings. It was burnt 1775, by British troops. Soon after the battle of Lexington, while a body of American militia were at Copp's hill, in Cambridge, detachments of them were sent to fortify Breed's hill, to prevent the British troops in Boston from occupying it and Charlestown. June 17, 1775, the latter landed and attacked the American redoubt three times, being repulsed twice with great loss. They finally succeded, and the Americans retreated; but the resistance to regular troops was considered as encouraging as a victory, and greatly animated the people. General Warren lost his life, with many others. A granite obelisk, in commemoration of this memorable event, magnificent in design, has been commenced on the battle ground.

The state prison of Mass. is in Charlestown, near the r., and has been recently rebuilt, on the Auburn plan, with 300 cells, and reorganized, at an expense of $86,000. It had in 1831, 290 convicts, in solitary cells at night and meal times.

The navy yard of the U. S. in the s. E. part of the t. opposite Boston, is surrounded by a wall enclosing about 60 acres, a marine hospital, warehouse, arsenal, powder magazine, and superintendent's house, all brick, with 2 large wooden houses, to shelter frigates and sloops of war on the stocks. The dry dock is the finest in the U. S. $382,104 were paid for materials and labor before Nov. 1831, and it was supposed that $118,000 more would be required to complete it. Pop. 1830, 8,783.

CHARLESTOWN, t. Washington co. R. I. 40 ms. s. w. Providence, s. Charles r., N. Atlantic ocean, nearly 7 miles square, 43 sq. ms. including 3 fresh, and 2 salt ponds, which open to the sea, part of the year. It has plenty of fish, good mill sites, with a rich level tract in the s. and rough land N. It bears nut, maple, ash, birch, white and yellow pine; corn, rye, barley, oats, &c. There is a remnant of Narraganset Indians in this t. Pop. 1830, 1284.

CHARLESTOWN, p-t. Montgomery co. N. Y. 40 ms. w. N. w. Albany, 10 s. Johnstown on Mohawk r. containing 100 sq. ms. is somewhat hilly, with rocks which afford quarries. The soil is generally clay or loam. Arieskill and Schoharie creek give mill sites. There was once an Indian town at the mouth of Schoharie creek, partly settled before the revolutionary war by Dutch; and since by New England emigrants. Charlestown, Voorhies, and Currie, are villages. Pop. 1830, 2148.

CHARLESTON, t. Orleans co. Vt. Pop. 1830, 664.

CHARLESTOWN, p-v. Chester co. Pa. by p-r. 82 ms. E. Harrisburg.

CHARLOTTE, t. Washington co. Me. Pop. 1830, 557.

CHARLOTTE, p-t. Chittenden co. Vt. 10 ms. s. Burlington, 10 N. Vergennes, 48 Montpelier, E. Lake Champlain. First settled 1776, is pleasantly situated, and is watered by Platt r. and Lewis creek. Pine and hemlock grow E.; hard wood on a good soil w. favorable to fruit. There are some high hills. A ferry to Essex, N. Y. Pop. 1830, 1702.

CHARLOTTE, or PORT GENESEE, p-v. and port of entry, Genesee, Monroe co. N. Y. at the mouth of Genesee r.

CHARLOTTE, t. Chatauque co. N. Y. Pop. 1830, 886.

CHARLOTTE, co. of Va. bounded N. w. by Campbell, N. by Prince Edward, E. by Lunenburg, s. E. by Mecklenburg, and s. and s. w. by Stanton or Roanoke r. which separates it from Halifax. Length 33, mean breadth 18, and area 600 sq. ms. Extending from N. lat. 36° 41′, to 37° 16′ and in long. from 1° 33′ to 2° 05′ w. from W. C. The slope of Charlotte co. is to the sthrd. towards the Roanoke. It is in great part drained by Little Roanoke and Cub creeks. Soil generally good and productive in grain, fruits, tobacco, &c. Chief town, Marysville. Pop. 1820, 13,290, in 1830, 15,252.

CHARLOTTE, p-v. and st. jus. Mecklenburg co. N. C. by p-r. 157 ms. s. w. by w. Raleigh.

CHARLOTTE, p-v. and st. jus. Dickson co. Tenn. 36 ms. w. from Nashville.

CHARLOTTE, r. bay, and harbor, west coast of Florida. The r. rises in the interior plains or swamps and flowing westward enters the eastern part of a deep and safe bay, sheltered on the side next the Gulf of Mexico, by a chain of islands or reefs. The adjacent country is low, sandy, marshy, and also sterile. The centre of the bay is about N. lat. 26° 45′ and 5° 20′ w. from W. C.

CHARLOTTE HALL, p-v. near the northern extremity of St. Mary's co. Md. 17 or 18 ms. a little s. of E. from Port Tobacco.

CHARLOTTEVILLE, p-v. and st. jus. for Albermarle co. Va. and also the seat of the central college or university of Virginia. Situated on the right bank of Rivanna river and near the northwestern foot of South West mountain, by p-r. 123 ms. s. w. from W. C. and 81 ms. N. w. by w. from Richmond. N. lat. 38° 03′, long. 1° 35′ w. from W. C. The university of Virginia, was organized in 1825, and in 1828, had 120 students, a library, containing 7000 vols. and a small observatory for the use of the students. Pop. 1830, not given in the tabular returns of the census; supposed about 1000. The arable surface on which this town is located, is elevated from 500 to 700 feet above the Atlantic tides, and the vicinity is regarded as salubrious.

CHARLTON, p-t. Worcester co. Mass. 15 ms. s. w. Worcester, 60 s. w. Boston. A farming town, with hard, rough, but strong soil, destitute of large streams, but well watered. Pop. 1830, 2,173.

CHARLTON, p-t. Saratoga co. N. Y. 25 N. w. Albany, 8 s. w. Ballston Spa, agreeably varied, with a gentle descent s. to Mohawk r. Eel Place creek furnishes mill sites. Farming prevails. Pop. 1830, 2,023.

CHARTIERS CREEK, or small river of Washington and Alleghany cos. Pa. rises by numerous branches in the vicinity of the borough of Washington, and flowing thence a course a little E. of N. about 30 ms. falls into Ohio r. 4 ms. below Pittsburg. At high water it is navigable for down stream vessels from its main Fork 2 ms. below Cannonsburg; and the country it drains is remarkable for fertility of soil, and for immense strata of bituminous coal.

CHATAUQUE, lake, Chatauque co. N. Y. 16 ms. long, 1 to 4 wide, is on high land, with good grazing banks, producing grain in some parts. Maysville, the co. t. on its margin is N. W. 8 ms. from Portland, on Lake Erie.

CHATAUQUE r. Chatauque co. N. Y. flows from Chatauque lake into Conewango creek, a branch of Alleghany r.

CHATAUQUE creek, Chatauque co. N. Y. runs 15 ms. from the ridge, through a deep ravine, into Lake Erie.

CHATAUQUE co. N. Y. the S. W. co. of the state, 360 ms. W. Albany, 60 S. W. Buffalo, bounded by Lake Erie N., Cattaraugus co. E., Pa. S. and W.; contains 659,280 acres, and 22 tsps. It is high land, with streams running into the lake.

Chatauque lake, in this co. flows through Casdaga creek into Conewango, thence into the Alleghany r. at Warren, Pa. which empties into the Mississippi, and thus a boat navigation is opened from within 6 or 8 ms. of Lake Erie to the Gulf of Mexico, and rafts go down every year. From 3 to 10 ms. from Lake Erie is a ridge 800 or 1200 feet high, being a loam on clay, or mica slate, bearing nut, oak, maple, birch, &c. and making good farms. On Lake Erie is a rich alluvion, from 1 to 4 ms. wide. The co. is exposed to cold and damp winds, late springs, and cold winters; but is healthful. Bog iron ore is found, and fruit trees flourish. Maysville, the co. t. is 164 ms. W. Albany. Pop. 1820, 12,568, 1830, 34,057.

CHATAUQUE, t. Chatauque co. N. Y. Pop. 1830, 2,442.

CHATEAUGAY, r. rises in Franklin co. N. Y. flows into Canada and joins St. Lawrence r. a little W. Montreal.

CHATEAUGAY, p-t. Franklin co. N. Y. 12 ms. E. N. E. Malone, 10 ms. wide, by 40 long, has a sandy loam, bearing beech, maple, bass, hemlock, pine, &c. Chatauque r. on which are the High falls and many ponds S. abound in trout. Part of the iron district is in the S. part. Pop. 1830, 2,432.

CHATHAM, t. Strafford co. N. H. Pop. 1830, 419.

CHATHAM, t. Coos co. N. H. on E. side of White mtns., W. Maine boundary, contains 26,000 acres. It is mountainous and rocky, with ponds and streams. Carter's mtn. W. cuts off direct communication with Adams.

CHATHAM, p-t. Barnstable co. Mass. 20 ms. E. Barnstable, at S. E. point Cape Cod, is surrounded by water, except N. W. where it touches Harwick; has a good harbor S., outside of which is a long beach, a moveable sandy soil, without trees, with some salt marshes. Long. 69° 50′ W., lat. 41° 42′ N. Pop. 1830, 2,130.

CHATHAM, t. Middlesex co. Conn. opposite Middletown, 16 ms. S. Hartford, E. Conn. r. about 6 ms. by 9, 56 sq. ms., is hilly, being crossed by the granite range, but has some very good farms. Extensive quarries of freestone are wrought on the shore of Conn. r. and sloops are loaded there, which supply the city of N. York with the best stone of the kind, and transport it also to many other places. A cobalt mine has been wrought on Rattlesnake hill, at different times, but it is not rich enough to bear the expense. At Middle Haddam v. is a good landing for river vessels. Opposite Chatham, Conn. r. turns E. through the narrows, where the ice often stops in the spring, and causes considerable freshets in the r. Pop. 1830, 3,646.

CHATHAM, p-t. Columbia co. N. Y. 18 ms. N. E. Hudson, 18 S. E. Albany, has different soils, with good farms, slaty hills, with tracts of alluvion; it is supplied with mill sites by Lebanon cr. Klein kill, &c. It has several villages, New Britain, New Concord, &c. Pop. 1830, 3,538.

CHATHAM, p-t. Morris co. N. J. 13 ms. N. W. Elizabethtown, 6 S. W. Morristown, W. Passaic r. Pop. 1830, 1,865.

CHATHAM, p-v. Chester co. Pa. 16 ms. S. W. from West Chester, and 40 ms. S. W. by W. from Philadelphia.

CHATHAM, co. of N. C. bounded S. by Moore, W. by Randolph, N. by Orange, E. by Wayne, and S. E. by Cumberland. It is an oblong of 33 ms. E. and W. and 26 N. and S., area 858 sq. ms. Extending in lat. from 35° 30′ to 35° 53′ and in long. from 1° 55′ to 2° 40′ W. W. C. Haw and Deep rs. unite near its S. E. angle, to form Cape Fear r. General slope, S. E. Chief t. Pittsboro'. Pop. 1820, 12,661, in 1830, 15,405.

CHATHAM, co. of Geo. bounded by Ogeechee r. which separates it from Bryan S. W., by Effingham co. N. W., by Savannah r. separating it from Beaufort dist. in S. C. N. E., and by the Atlantic ocean S. E. Length from S. E. to N. W. 27 ms. breadth 15 ms, and area 405 sq. ms. Extending from lat. 31° 50′ to 32° 13′ and in long. 3° 56′ to 4° 26′ W. W. C. The very slight declivity of this co. is S. E. towards the Atlantic ocean. The surface is level, and but slightly elevated above the Atlantic tides. Staples, rice, cotton and sugar. Chief town, Savannah. Pop. 1830, 14,230.

CHATTAHOOCHEE, r. of Geo. Ala. and Flor. the western and main constituent of Appalachicola r. rises on the high table land of the Appalachian system, with sources issuing from Blue Ridge, and either interlocking with, or nearly approaching those of Savannah, Tennessee, Pieper, Hiwassee and Coosa. The higher Chattahoochee is formed by two branches, Chestatee and Chattahoochee proper. Both branches rise in Habersham, and unite on the western border of Hall co. Geo. having flowed in a sthrn. direction about an equal distance, 45 ms. Thence known as Chattahoochee, the r. assumes a southwestern course of 140 ms. in Geo. to Miller's Bend, where it becomes a boundary between Geo.

and Ala. Below Miller's Bend, with partial windings, the general course is 150 ms. very nearly due s. to its union with Flint, to form Appalachicola r. About 20 ms. of the lower part of its course, Chattahoochee separates Flor. from Geo. It is very remarkable that in a comparative distance of 280 ms. from the junction of Chestatee and Chattahoochee, to the mouth of Flint, no tributary stream enters the main recipient above the size of a large creek, and the valley at its widest part does not exceed 50 ms. and the whole, fully estimated at a mean breadth of 35 ms. Entire length of this vale 325 ms. The higher part of the valley of Chattahoochee lies between those of Coosa to the N. W. and Oconee and Oakmulgee to the S. E. It thence, for about 130 ms. intervenes between the vallies of Flint and Tallapoosa, and the lower section between those of Flint and Choctaw rivers. (*See Appalachicola r.*)

CHATICO, p-v. on a small creek of Wicomico r., St. Mary's co. Md. by p-r. 53 m. S. S. E. W. C. and 64 a little W. of S. from Annapolis.

CHATUGA, r. the extreme highest constituent of Savannah r. rises in the southern vallies of Blue Ridge, and in Macon co. N. C. Issuing thence by a course of a little W. of S. and traversing N. lat. 35°, in a distance of 25 ms. separates Pickens dist. S. C. from Rabun co. Geo. to its union with Turoree r. and forms Tugaloo r. This small stream has its sources opposite to those of Tennessee proper, and Hiwassee.

CHAUMONT, p-v. Lyme, Jefferson co. N. Y. 10 ms. from Brownville, at the head of Chaumont bay, carries on a valuable fishery of white fish and siscoes.

CHAZY, p-t. Clinton co. N. Y. 175 N. W. Albany, 12 N. Plattsburgh, S. and W. lake Champlain, is watered by Little Chazy r. It has good land in some parts. The village is 15 ms. N. Plattsburg. The landing on lake Champlain is 1 m. S. Little Chazy r., 3 ms. E. of the village. Pop. 1830, 3,097.

CHAZY, r. Franklin and Clinton cos. N. Y. about 50 ms. long, and a good mill stream, enters lake Champlain, near Port au Fer.

CHAZY, (LITTLE), r. Clinton co. N. Y. 1½ ms. S. Chazy r. 18 ms. long, is a good mill stream.

CHEAT, r. of Va. rising on the border between Randolph and Pocahontas cos. interlocks with Elk and Green Brier branches of Great Kenhawa, and after uniting with the south branch of Potomac, flows thence by a general northern course 70 ms. over Randolph into Preston co., inflecting in the latter co. to N. N. W. 40 ms. to its junction with the Monongahela, at the southwestern angle of Fayette co. Pa. The valley of Cheat lies between those of the Monongahela on the W., Potomac E., and Youghioghany N. E. Length about 100, mean breadth not exceeding 18, area 1800 sq. ms.

CHEEK'S CROSS ROADS, and p-o. Hawkins co. Ten. by p-r. 212 ms. estrd. Nashville.

CHEEKSVILLE, p-o. E. part Marion co. Ten. by p-r. 124 ms. S. E. by E. from Nashville.

CHELMSFORD, p-t. Middlesex co. Mass. 27 ms. N. Boston, S. Merrimack r. formerly contained the present town of Lowell. Pop. 1830, 1387.

CHELSEA, p-t. and st. jus. Orange co. Vt. 20 ms. S. E. Montpelier, 20 S. W. Newbury, 36 sq. ms., first settled 1783; it is watered by the 1st branch of White r., &c., has an uneven surface, but a warm and fertile soil. Pop. 1830, 1958.

CHELSEA, t. Suffolk co. Mass. 3 ms. N. E. Boston, to which Winnesemit ferry crosses, was incorporated 1638. It is N. of Boston harbor, W. of Lynn bay; a strip of land 100 rods wide, stretches 3½ ms. to Reading. A granite hospital was built here in 1827, looking on Boston harbor. With Boston it forms Suffolk co. but has no vote nor expense in co. business. Pop. 1830, 770.

CHELSEA LANDING, p-v. Norwich, Conn. (*See Norwich.*)

CHELSEA, p-v. Cataraugus co. N. Y. 260 ms. W. Albany.

CHEMUNG, r. or Tioga, in Steuben and Tioga cos. N. Y. a large W. branch of Alleghany r.

CHEMUNG, p-t. Tioga co. N. Y. 198 ms. S. S. W. Albany, ; 9 E. Elmira, N. Pennsylvania, is crossed by Tioga or Chemung r. with Cayuta creek on E. side, and has other mill streams. It is hilly, with fine alluvion on Tioga creek, and is gravelly on the hills, bearing yellow pine and some hemlock, oak, beech and maple. The narrows of Chemung r. are wild and singular. Pop. 1830, 1462.

CHENANGO, r. N. Y. rises in Madison and Oneida cos. near the head waters of Oneida, Oriskany and Sadaquada creeks, crosses Chemung co. S. S. W., and Broome co., and joins Susquehannah r. at Binghampton; it is 90 ms. long, with many useful branches.

CHENANGO, co. N. Y. bounded by Madison co. N., Otsego and Delaware E., Broome co. S. and Broome and Cortlandt W.; 35 and 28 ms. are its greatest dimensions; 780 sq. ms. and it contains 19 townships. It is watered by streams of Susquehannah r. as, Chenango, Unadilla, Otselic crs. &c. and is high, hilly, with various soils, generally good for farms. Settled principally from New England. Pop. 1830, 37,238.

CHENANGO FORKS, p-v. Lisle, Broom co. N. Y. at the union of Chenango and Tioughnioga rivers.

CHENANGO POINT, or BINGHAMPTON, p-v. and co. seat, Chenango, Broome co. N. Y. 148 ms. W. S. W. Albany, 40 ms. from Norwich.

CHENANGO, p-v. Beaver co. Pa. 80 ms. N. W. Pittsburg.

CHENANGO, r. of Crawford and Mercer cos. Pa. (See Shenango.)

CHENEYVILLE, p-v. on Bayou Boeuf, S. E. part of Rapide parish, La. by p-r. 32 ms. S. S. E. from Alexandria; st. jus. for the parish.

CHENOWETH'S p-o. wstrn. part of Darke co. O. by p-r. 113 ms. N. of W. Columbus.

CHEPACKET, p-v. Providence co. R. I. 16 ms. from Providence. A flourishing village on the river of its name.

CHERAW, p-v. in the N. E. part of Chesterfield dist. S. C. on the right bank of Great Pedee r. by p-r. 88 ms. N. E. by E. Columbia.

CHEROKEES, or CHELOKEES, nation of Indians, inhabiting a part of northwestern Geo. northeastern Ala. southeastern Ten. and the extreme wstrn. angle of N. C.

"In 1809, by an enumeration made by the agent, the pop. of this people amounted to:

Cherokees, one half of whom were mixed,	12,395
Negro slaves	583
Whites resident,	341
Total	13,319

The following document was published in the National Intelligencer of Aug. 14, 1830; and is literally copied, if we except a different arrangement of the items in the enumeration of the population.

"A statistical table exhibiting the population of the Cherokee Nation, as enumerated in 1824, agreeably to a resolution of the legislative council; also of property, &c.

Males		6,883
" under 18 years of age	3054	
" from 18 to 59	3027	
" over 59	352	
Females		6,900
" under 15 years of age	3010	
" from 15 to 40	3103	
" over 40	782	
Add for those who have since removed into the nation from North Carolina, who were living in that state on reservations,		500
Negroes, { Males,	610	1,277
{ Females,	667	
Total population		15,560

There are 147 white men married to Cherokee women, and 68 Cherokee men married to white women.

Schools	18	Black cattle	22,531
Scholars of both sexes	314	Swine	46,732
Grist mills	36	Sheep	2,566
Saw mills	13	Goats	432
Looms	762	Blacksmith shops	62
Spinning wheels	2486	Stores	9
Wagons	192	Tan yards	2
Ploughs	2923	Powder mill	1
Horses	7683		

Besides many other items not enumerated; and there are several public roads and ferries, and turnpikes in the nation."

Speech of Mr. Everett, on the bill for removing the Indians from the E. to the W. side of the Mississippi, H. of R. 19th of May, 1830.

A discrepancy appears above, on footing the sums attached to the ages of the male and female population, and comparing the amount with the sums total of each, as stated collectively. The latter are presumed to be correct.

By some still more recent document, it appears that the population of the Cherokee nation is on the increase. That part of this people who reside in Georgia, have been made subject to the laws of Georgia by a statute of that state.

CHEROKEE CORNER, p-v. Oglethorpe co. Geo. by p-r. 77 ms. a little E. of N. from Milledgeville. and 611 ms. S. W. from W. C.

CHERRY, p-o. Lycoming co. Pa. 113 ms. nthrd. Harrisburg.

CHERRYFIELD, t. Washington co. Me. 30 ms. W. Machias, E. Hancock co. is crossed by Narraguagus r. Pop. 1830, 583.

CHERRRY RIDGE, p-o. Wayne co. Pa. by p-r. 165 ms. N. E. Harrisburg.

CHERRY TREE, p-o. Venango co. Pa. by p-r. 244 ms. N. W. by W. Harrisburg.

CHERRY VALLEY, p-v. and tsp. in the E. part Ashtabula co. O. The p-o. is by p-r. 5 ms. from Jefferson, the st. jus. for the co. and 192 N. E. Columbus. Pop. of the tsp. in 1830, 219.

CHERRY VALLEY, p-t. Otsego co. N. Y. 53 ms. W. Albany, 14 ms. N. E. Cooperstown, is high and hilly; gives rise to Canajoharrie creek, of Mohawk r. Cherry Valley creek, and of Unadilla r. which runs into Susquehannah r. Mill sites are numerous; rich alluvion abounds in the vallies, and the great number of wild cherry trees gave name to the town. Pop. 1830, 4,098.

CHERRY VALLEY, v. in the above town, situated in a valley with an academy, has an important position on 3 turnpikes 13 ms. S. W. Schoharie; marble is quarried here. The village was destroyed by the French and Indians, Nov 1758. Pop. 1830, 641.

CHESAPEAKE BAY, a deep gulf opening from the Atlantic ocean, between Capes Henry and Charles; lat. 37° and long. 1° E. from W. C. intersecting in the mouth of the bay, near midway between the capes, which are about 15 ms. asunder. The mouth of this fine sheet of water extends wstrd. 20 ms. to the mouth of James river. Curving rapidly, above the influx of James river, the Chesapeake extends almost directly north over one degree of lat. with a mean breadth of 20 ms. having received from the westrd. James, York, Rappahannoc, and Potomac rivers, and from the opposite side, Pocomoke and Nantikoke rivers. Widened by the union of so many confluents, the Chesapeake is upwards of 40 ms. wide from the mouth of the Potomac to that of Pocomoke, and about 35 from the most southern capes of the Potomac to the influx of Nantikoke river. Above the entrance of the two latter streams, the main bay narrows to a mean width of about 10 ms. and at some places under 5 ms., but with an elliptic curve to the wstrd. 115 ms. to its termination at the mouth of Susquehannah river, having received from the westrd. above the Potomac, the Patuxent, Patapsco, Gunpowder and Bush rivers, and from the estrd. Nantikoke, Choptank, St. Michaels, Chester, Sassafras, and Elk rivers. The entire length of Chesapeake Bay is 185 ms.; and it may be doubted whether any other bay of the earth, is, in proportion to extent, so much diversified by confluent streams as is the Chesapeake.

In strictness of geographical language, it is, however, only a continuation of Susquehannah river, of which primary stream all the other confluents of Chesapeake are branches. In the main bay the depth of water continues sufficient for the navigation of the largest ships of war to near the mouth of Susquehannah; and in Potomac that depth is preserved to Alexandria. In the other tributary rivers large vessels are arrested before reaching the head of tide water. If taken in its utmost extent, including the Susquehannah valley, the Chesapeake basin forms a great physical limit; to the s. w. with few exceptions, the rivers, bays and sounds, are shallow, and comparatively unnavigable; but with the Chesapeake commences deep harbors, which follow at no great distance from each other, to the utmost limits of the Atlantic coast of the United States. The entire surface drained into this immense reservoir amounts to near 70,000 sq. ms.

Chesapeake Peninsula. This article is introduced in order to give a general description of a natural section of the United States, the peculiar features of which are lost or confused, in most of our geographical works, amongst the political subdivisions which have been drawn upon its surface. This physical section is bounded by the Atlantic ocean s. e., by Chesapeake bay w., by Delaware bay n. e., and united to the main continent by an isthmus, now traversed by the Chesapeake and Delaware canal n. The latter work has in fact insulated the peninsula, and given it water boundaries on all sides. Thus restricted, the Delaware peninsula extends from Cape Charles n. lat. 37° 08', to the Chesapeake and Delaware canal at n. lat. 39° 32'. Greatest length very nearly in a direction n. and s. 182 ms. The general form is that of an elongated ellipse, which, in component material, features, and elevation, differs in nothing essential from other Atlantic islands scattered along the coast of the United States. Chesapeake Bay is itself divided between Virginia, and Maryland; the shores on both sides s. of the Potomac and Pocomoke rs. belonging to the former, and to the northward to the latter state. The southern part of the peninsula is entirely in Virginia, and is a long narrow promontory 70 ms. by 8 to 10 ms. wide. Above Pocomoke Bay the peninsula widens, and after an intermediate distance of 33 ms. is equally divided between the states of Maryland and Delaware. In the widest part, between Cape Henlopen, Sussex co. Delaware, and the western part of Talbot, Md. the width is 70 ms.; but narrowing towards both extremes the mean breadth is about 27; area 4900 sq. ms. The surface is generally level or very gently undulating. The ocean and Chesapeake shores are strongly contrasted. Along the former, are narrow and low islands, with shallow sounds, and with no stream issuing from the land of any consequence. The opposite shore or Chesapeake is in an especial manner indented by innumerable bays, and compared with the confined width of the peninsula, rivers of great magnitude of volume. The character of the Atlantic shore is extended along the Delaware bay, and entirely round the peninsula; much of the soil is liable to diurnal or occasional submersion from the tides.

The general slope is southwestward as demonstrated by the course of the rs. Pocomoke, Nantikoke, Choptank, Chester, Sassafras & Elk. Politically it contains all Sussex, Kent, and more than one half of New Castle cos. Del.; all Worcester, Somerset, Dorchester, Talbot, Caroline, Queen Ann, and Kent, and one third of Cecil cos. Md. with all Accomac, and Northampton cos. Va.

Chesapeake and Delaware Canal. *(See art. Rail Roads and Canals.)*

Chesapeake, p-v. in the s. part of Cecil co. Md. on Chesapeake Bay, immediately below the mouth of Elk river, by p-r. 35 ms. n. e. by e. Baltimore.

Cheshire, co. N. H. the western co. of the state, bounded by Grafton co. n., Hillsborough, e., Mass. s., and Vt. w., 26 by 54; 1,254 sq. ms., has Conn. r. w. It contains 37 towns, Sunapee and Spafford and Ashauelot and Sugar rivers; Grand Monadnock, above 3,000 feet, Craydon and Grantham mtns. Bellows Falls, on Conn. r. are in this co. There is much good meadow land. Chief towns, Keene and Charlestown. Pop. 1820, 26,753, 1830, 27,016.

Cheshire, p-t. Berkshire co. Mass. 130 ms. w. Boston, is crossed by a branch of the Hoosick r. There are glass manufactories here. Pop. 1830, 1,050.

Cheshire p-t. New Haven co. Conn. 13 ms. n. New Haven, 6 ms. by 7, 40 sq. ms., is watered by Quinipiack r. and a branch. Here is the Episcopal academy of Conn. It has a fund of $25,000, a brick edifice and small library. It is under the direction of a principal and professor of languages. The soil is uneven, with gravelly loam, bearing chestnut, oak, walnut, &c. The Farmington canal passes through. Pop. 1830, 1,780.

Cheshire, p-v. and tsp. Galia co. O. by p-r. 106 ms. s. s. e. Columbus. Pop. of the tsp. 1830, 664.

Chesnut Grove, p-o. in the s. part of Pittsylvania co. Va. by p-r. 13 ms. southardly from Competition, the st. jus. for the co. 180 s. w. by w. Richmond.

Chesnut Hill, (now Shafer's) p-o. Northampton co. Pa. by p-r. 20 ms. n. from Easton.

Chesnut Hill, p-o. Phila. co. Pa. 8 ms. n. Phila.

Chesnut Hill, p-o. Orange co. Va. by p-r. 88 ms. n. w. Richmond.

Chesnut Hill, p-v. Hall co. Geo. by p-r. 113 ms. northward Milledgeville.

Chesnut Level, p-o. in the e. part of Lancaster co. Pa. by p-r. 51 ms. s. e. by e. Harrisburg.

CHESNUT RIDGE, p-o. Stokes co. N. C. by p-r. 152 ms. N. W. by W. Raleigh.

CHESNUT RIDGE, local name given to that part of the western prominent chain of the Appalachian mountains, between Kiskiminitas and Youghioghany rivers, and in Westmoreland and Fayette cos. Pa.

CHESTER, p-t. Rockingham co. N. H. on the Merrimac r. 23 ms. S. E. Concord, 17 Exeter, is crossed by a branch of Exeter r. and contains Massabesick ponds, fine meadows, and several caverns. The rocks are granited gneiss. Pop. 1830, 2,028.

CHESTER, p-t. Windsor co. Vt. 16 ms. S. W. Windsor, has a good soil with hills and vallies, and three streams here form Williams r. An academy. Pop. 1830, 2,320.

CHESTER, p-t. Hampden co. Mass. 20 ms. N. W. Springfield, 120 ms. W. Boston, N. W. Westfield r. is crossed by two of its streams. Pop. 1830, 1,407.

CHESTER, p-v. Saybrook, Middlesex co. Conn. has great water power, and some manufactories near Conn. r.

CHESTER, p-v. Goshen, Warren co. N. Y. 21 ms. N. W. Caldwell. Pop. 1830, 1,284.

CHESTER, p-t. Morris co. N. J. crossed by Black r. Pop. 1830, 1,338.

CHESTER, t. Burlington co. N. J. lies S. E. Delaware r. and has Rankokus creek N. and Pensaukin creek S. Pop. 1830, 2,333.

CHESTER, co. Pa. bounded by Lancaster co. W., Berks N. W., Schuylkill r. which separates it from Montgomery N. E., Delaware co. E., Newcastle co. in the state of Delaware S. E., and Cecil co. Md. S. Length from S. W. to N. E. 44 ms.; mean breadth 18 and area 792 sq. ms. Extending in lat. from 39° 42', to 40° 15', and in long. from 0° 55' to 1° 40' E. W. C. Surface very diversified; the eastern part rolling, rather than hilly, and in some places level; but the central and western part hilly, and in many places even mountainous. The soil is greatly varied from rocky and sterile to highly fertile. The general slope is to the S. E. The largest stream originating in this co. is the Brandywine, which rising on the border between Lancaster and Chester cos. crosses the latter from N. W. to S. E. The extreme northern part is drained into Schuylkill r. by French creek, and from the opposite extreme issue the fountains of Elk river flowing southwardly into Chesapeake Bay. Immense strata of fine marble exist in Chester. It is one of the best and most skillfully cultivated cos. in the U. S. and its staples are numerous and valuable. The difference of level between the extreme N. and S. about equivalent to a degree of lat. This co. is now traversed by a fine rail road, uniting the Schuylkill and Susquehannah rivers. Chief town, West Chester. Pop. 1820, 44,455, 1830, 50,910.

CHESTER, borough, p-t. and st. jus. Delaware co. Pa. situated on the right bank of Delaware river, 15 ms. S. W. from Phila. and by p-r. 121 ms. N. E. from W. C., lat. 39° 50', long. 1° 42' E. W. C. This borough has in great part recovered from the ravages of a destructive fire, which a few years since destroyed a number of the best buildings. Pop. 1820, 657, 1830, 847, showing an increase of 44 per cent in ten years.

CHESTER, r. of Del. and Md. rises in Kent co. of the former, from which it flows westrd. into the latter state, within which it separates Queen Ann from Kent co. The upper part of the course of this stream is very circuitous, and lower down spreading into a large bay, is navigated by small vessels to Chestertown 30 ms. above its mouth, into Chesapeake bay. The valley of Chester river lies between those of Sassafras and Choptank.

CHESTER, dist. S. C. bounded by Broad r. which separates it from Union on the W., York dist. N., Catawba, r. which separates it from Lancaster E., and by Fairfield dist. S. Length from E. to W. 30, breadth N. and S. 20, and area 600 sq. ms. Extending in lat. from 34° 31' to 34° 50', and in long. from 3° 52' to 4° 32' W. W. C. Surface pleasantly broken by hill and dale. Chief town, Chester. Pop. 1820, 14,389, and in 1830, 17,182.

CHESTER, p-v. and st. jus. Chester dist. S. C. near the centre of the dist. 448 ms. S. W. W. C. 56 a little W. of N. Columbia, and about 80 ms. wstrd. from Sneadsboro, in N. C. lat. 34° 42', and long. 4° 12' W. W. C.

CHESTER, p-v. and st. jus. Meigs co. O. by p-r. 343 ms. W. W. C. and 94 ms. S. E. Columbus. It is on Shade creek in the N. E. part of the co. Pop. 1830, 164.

CHESTER CROSS ROADS, and p-o. in the E. part of Geauga co. O. by p-r. 11 ms. E. Chardon, the st. jus. and 157 ms. N. E. Columbus.

CHESTERFIELD, p-t. Cheshire co. N. H. on Connecticut r. opposite Brattleborough. Lat. 42° 53'. From Keene 11 ms., Concord 65, Boston 90. First settled 1761, generally hilly and uneven, has much good upland, well adapted for grazing and production of Indian corn. Chief articles for the market are beef, pork, butter and cheese. Contains Cat's Banebrook, furnishing many mill seats, and Spafford's lake, a beautiful sheet of water, covering 526 acres; the lake enclosing an island of about 6 acres, forming a delightful retreat for the students of the academy in the summer. From its E. side, issues Partridge's brook, sufficient to carry saw mills, &c., a factory with 800 spindles and 40 water looms. A Congregational society founded in Chesterfield, 1771, Baptist 1819, Universalist 1818. It has a flourishing academy. Pop. 1830, 2,045.

CHESTERFIELD, p-t. Hampshire co. Mass. on the E. Green mtn. ridge, 12 ms. W. Northampton, high and finely watered by a branch of Westfield river, produces good crops of grass & corn. The beryl is found here, and the emerald, weighing fm. an ounce to 6 lbs. a hexangular prism, sometimes 12 inches in diameter. The town contains a Congregational and Baptist society. Pop. 1830, 1,416.

CHESTERFIELD, p-t. Essex co. N. Y. on

Lake Champlain, opposite Burlington, has 9 ms. of lake shore; extends about 10 ms. E. to W. mountainous in some part; level along the lake; fertile; soil, a sandy loam, mingled with clay. Its lumber trade is considerable. Contains several small ponds discharging mill streams. Principal r. Sandy or Sable, here exhibiting Adgate's Falls, a curiosity worthy the attention of travellers. Fall, 80 feet into a narrow channel walled on each side by perdicular rock 100 feet high. This channel, a mile in length, evidently worn by the water. A cavern, in the town, furnishes a natural self-storing perennial ice house. Chesterfield abounds in iron ore. Contains a Congregational and Methodist society. Pop. 1830, 1,671.

CHESTERFIELD, t. Burlington co. N. J. Pop. 1830, 2,386.

CHESTERFIELD, p-o. (*See Massena.*)

CHESTERFIELD, co. of Va. bounded by Powhatan N. W., James r. separating it from Henrico N., by a bend of James, separating it from Charles City co. E., Appomattox r. separating it from Prince George S. E., Dinwiddie S., and Amelia S. W. Extreme length from the junction of James and Appomattox rivers to the western angle 38 ms., mean breadth 12, and area 456 sq. ms. Lat. 37° 10', to 37° 31' N., and long. 0° 22' to 1° 05' W. W. C. Surface rather broken. Chief town, Manchester. Population 1820, 18,003, 1830, 18,637.

CHESTERFIELD, court house, and p-o. Chesterfield co. Va. by p-r. 14 ms. S. S. W. Richmond, lat. 37° 19', long. 0° 43' W. of W. C.

CHESTERFIELD, dist. of S. C. bounded N. E. and E. by Great Pedee r. separating it from Marlborough dist., S. E. and S. by Darlington dist., S. W. by Lynche's creek separating it from Kershaw, N. W. by Lynche's creek separating it from Lancaster, and N. by Anson co. N. C. Length 30, mean breadth 25, and area 750 sq. ms. Extending in lat. from 34° 22' to 34° 48', and long. from 2° 53' to 3° 40', W. W. C. The slope of this district is S. S. E. central parts drained by Black creek, flowing into Great Pedee. Chief town, Chesterfield. Pop. 1820, 6,645, 1830, 8,472.

CHESTERFIELD, p-v. and st. jus. Chesterfield dist. S. C. by p-r. 426 ms. S. S. W. W. C., 102 N. E. Columbia, lat. 34° 51', long. 3° 07' W. W. C.

CHESTER SPRINGS, and p-o. Chester co. Pa. by p-r. 76 ms. E Harrisburg.

CHESTERTOWN, s-p. p-v. and st. jus. Kent co. Md. situated on the right bank of Chester r. by p-r. 82 ms. northeastward from W. C. and about 30, a little S. of E. from Baltimore. Lat. 39° 13', long. 0° 58' E. W. C. Population about 800.

CHESTERVILLE, p-o. Kennebeck co. Me. 28 ms. from Augusta.

CHESTERVILLE, (now Millington,) p-v. in the N. E. part of Kent co. Md. 18 ms. N. E. Chestertown, the st. jus. and by p-r. about 40 ms. E. Baltimore.

CHENEY'S SHORE, p-o. Delaware co. Pa. 17 ms. from Phila.

CHETIMACHES, lake of La. between the Teche, and Atchafalaya rivers, is from 1 to 6 ms. in breadth. It is shallow, and on all sides environed by a low, annually inundated, and uninhabitable country. It is supplied by numerous intermediate outlets from Atchafalaya, and discharges its water into that sream near and above its junction with the Teche.

CHEVIOT, p-o. in the S. part of Hamilton co. O. by p-r. 5 ms. W. Cincinnati.

CHICAGO, small, but from its relative position with lake Michigan, and the northern sources of Illinois river, a very important stream of Cook co. state of Illinois. The Chicago heads with the Plain, one of the northern branches of Illinois. Both rs. originate in a flat prairie country, flow nearly parallel to each other, and to the course of the adjacent shore of Lake Michigan, for a comparative distance of 30 ms. Thence diverging, the Plain r. to the S. W., Chicago bending at a nearly right angle, falls into its recipient at the village of Chicago. The mouth is obstructed by a bar, on which there is only 3 feet water, though inside adequate depth is found for ships of almost any tonnage. The portage between Chicago and Plain river is only about 9 ms. and at seasons of high water small vessels are navigated over the intermediate flats. This is one of those positions on which the hand of nature has traced a canal, and left to man the more humble duty of completing the work.

CHICAGO, p-v. and port, on Lake Michigan, at the mouth of Chicago river, Cook co. Il. 300 ms. N. N. E. Vandalia. N. lat. 42° 09', and long. 10° 42' W. W. C. The position of this place is bleak. Behind are extensive prairies; before, the lake, without a harbor for three hundred ms. The land, one m. wide, on the margin of the lake, is a barren sand, thence a rich loam on limestone strata. *Dr. Morse.*

CHICHESTER, p-t. Rockingham co. N. H. Lat. 42° 15' E. Concord, 8 ms. generally level, soil good, richly repaying the tiller. Suncook r. (furnishing mill seats,) and its branches water it. Congregational church organized 1791. Pop. 1830, 4,084.

CHICKAHOMINO, r. of Va. rises between the vallies of Pamunkey and James rivers, about 20 ms. N. W. from Richmond; flowing thence S. E. by E., the cos. of Henrico, and Charles City on the right, and New Hanover, New Kent, and James City cos. on the left, falls into James river after a compartive course of 60 ms.

CHICKAPEE, r. Mass. formed by Ware, Swift and Quaboag, empties into the Conn. 4 ms. N. of Springfield.

CHICKAPEE, p-v. Mass. part of Springfield, contains a factory, having a capital of $400,000, produces daily 11,000 yards of cloth; employing 700 females, who earn from 12 to $21 per month.

CHICKASAW, a nation of Indians inhabiting the country comprising the northwestern angle of Ala. and the northern part of Miss. states. They reside to the northward of the Choctaws, and between Ten. and Miss. rs. The number of this tribe as given by Rev. Jedediah Morse, in his Report on Indian Affairs, 1822, was then 3,625. Mr. M. states in his appendix, page 201, "There are 4 males to 1 female. This inequality is attributed to the practice of polygamy, which is general in this tribe. If the curious fact is truly stated, the reference is at variance with either the causes or effects usually connected with the history of polygamy.

"The nation resides in 8 towns, and like their neighbours, are considerably advanced in civilization."

CHICOT, S. E. co. of Ark. bounded by Clark W., Arkansas, co. N., by the r. Mississippi E., and by the parish of Washitau, La. S. Extending in lat. 33° to 33° 40′ N., and in long. from 14° 5′ to 14° 57′ W. W. C. It approaches the form of a sq. of 50 ms. each side, area 2500 sq. ms. The boundaries indeed to the N. and W. are vague. The slope is southwardly, and in that direction it is bounded by Mississippi, and traversed by Barthelemy and Boeuf rivers. The surface is generally a plain, and most part an alluvial flat. The soil differs extremely in character, much of it liable to annual submersion; parts rising nearly above inundation and exuberantly fertile, whilst in other places the land is sterile. Chief town, Villemont. Pop. 1830, 1,165.

CHIFUNCTE, pronounced Chifunty, r. of the states of Miss. and La. rises in Pike co. of the former, from which it issues a mere creek; enters La. within which it traverses Washington and St. Tammany parishes, falling into the northern side of Lake Pontchartrain, after a comparative course of 55 ms. S. S. E. There is 7 or 8 feet depth of water on the bar of this river, and a safe and deep harbor within its mouth, at Madisonville.

CHILDSBURG, p-v. La Fayette co. Ky. 32 ms. S. E. Frankfort.

CHILHOWEE, p-v. in the E. part of Monroe co. Ten. about 180 ms. S. E. by E. Nashville.

CHILI, p-t. Monroe co. N. Y. on Genesee or Henrietta r. 10 ms. S. W. Rochester, watered by Black creek, a fine mill stream. Pop. 1830, 2,010.

CHILISQUAKE, p-v. on Chilisquake cr. Northumberland co. Pa. 10 ms. above Sunberry, the st. jus. and by p-r. 62 ms. N. Harrisburg.

CHILLICOTHE, flourishing p-t. and st. jus. Ross co. O. on the right bank of Scioto r. at the salient point of the bend above the mouth of Paint creek, by p-r. 404 ms. a little N. of W. W. C. 45 ms. nearly due S. from Columbus, 96 ms. a little N. of E. from Cincinnati, and 56 ms. S. W. by W. from Zanesville. N. lat. 39° 20′, long. 6° W. from W. C. The Grand canal is cut through it.

The site of this town is a plain, but the adjacent country rising into hills of from 200 to 300 feet elevation above the river bottoms, the place seems to occupy the centre of a series of varied and delightful landscapes. This village was laid out on the site of an old Indian town, in 1796, and the rapid advance of population will be shown below. It contained 5 years since, 2 printing offices, 3 banks, and between 30 and 40 mercantile stores, with numerous and flourishing manufactories, oil, fulling, flour and saw mills. Pop. 1810, 1,369, 1820, 2,426 1830, 2,847.

CHILMARK, p-t. Dukes co. Mass. is the S. W. end of Martha's Vineyard. The N. W. point is called Gays head. The S. point Squibnócket. Congregational, Methodist and Baptist society. Pop. 1830, 2,010.

CHILO, p-v. in the extreme southern part of Clermont co. O. on Ohio r. by p-r. 127 ms. S. W. Columbus. Pop. 1830, 128.

CHINA, p-t. Kennebec co. Me. 20 ms. N. Augusta, W. Palermo, Waldo co. Population 1830, 2,233.

CHINA, p-t. Genesee co. N. Y. 32 ms. S. W. Batavia, watered by head streams of the Tonnewanta, Cataraugus and Seneca creeks. Moderately uneven. Heavily timbered with beach, maple, elm, ash, linden, hemlock, &c. sure indications, on all table lands, that the soil and climate are better adapted to grass than grain. Has 4 saw mills, 3 grist mills, 9 schools 5 months in 12. Pop. 1830, 2,387.

CHINA GROVE, p-o. Rowan co. by p-r. 130 ms. W. Raleigh.

CHINA GROVE, p-o. Williamsburg dist. S. C. by p-r. 112 ms. S. Columbus.

CHINA GROVE, and p-o. Pike co. Ala. by p-r. 159 ms. S. E. Tuscaloosa.

CHINA GROVE, p-o. in the N. part of Pike co. Miss. by p-r. 75 ms. S. of E. Natchez.

CHINA HILL, p-o. W. part of Mecklenburg co. N. C. 12 ms. W. Charlotte, the st. jus.

CHINQUIPIN GROVE. (*See Locust creek, Louisa co. Va.*)

CHINQUIPIN RIDGE, p-o. Lancaster dist. S. C. by p-r. 81 ms. N. N. E. Columbia.

CHIPOLA RIVER. (*See Chapola river.*)

CHIPPEWA, p-v. and tsp. in the N. W. part of Wayne co. O. 18 ms. N. E. Wooster, the st. jus. and by p-r. 104 ms. N. E. Columbus. Pop. of the tsp. 1830, 1,498.

CHIPPEWAN. (*See Chippewayan mountains.*)

CHIPPEWAY, r. of the N. W. territory, one of the left branches of the Miss. r. rises interlocking sources with some small streams which fall into the southwestern part of lake Superior, and with the St. Croix to the N. W. and Ouisconsin to the S. E. The sources of this stream are drawn from a flat table land chequered with lakes, one of which, Flambleau lake, is as laid down by Tanner, upwards of 40 ms. in length with a breadth of from one to ten ms. It flows into lake Pepin, lat. 44½°, long. 15° 10′ W. W. C. after a S. W. course of 135 ms.

CHIPPEWAYAN, or Chippewan, mountains of North America. This immense system extends continuous chains from the Isthmus of Darien to the Arctic ocean, through 60 degrees of lat. with a considerable difference of long. between the extremes. Whether this system is connected or separate from the Andes of South America, is a yet unsolved problem in physical geography. In North America it forms the principle spine, from which rivers flow in opposite directions towards the two great oceans which bound the opposite sides of the continent. These corelebra, for they deserve the title, range upwards of five thousand miles. In southern Mexico the system is known by the general term Anahuac, and further north as the mountains of New Mexico. In the U. S. it is designated Rocky Mountains, whilst in British America, it is called by its native name, Chippewan or Chippewayan. In Guatemala, or Central America, and in Mexico, rise from it enormous volcanic summits, elevated far above the region of perpetual snow. Popocatapetl, Citlaltepetl or peak D'Orizaba, Pico Frailes, and Coffre de Perote, all rise above 13,500 feet, and the former to 17,700 feet above the ocean tides. In the U. S. and northwardly, the elevation remains undetermined, but must be considerable, as is shown by the rapid current and great length of course of the rivers which flow from its flanks.

CHITTENDEN, co. Vt. bounded N. by Franklin co., E. by Washington, S. by Addison, W. by lake Champlain. Length 30 ms. breadth 22, area about 500 sq. ms. Onion r. traverses the middle part, falling into lake Champlain at Burlington, Lamoille, the N. W. corner, Laplott, S. part. Lake shore generally level, other parts uneven. Soil varies from light and sandy to rich loam and deep alluvion. Pop. 1830, 21,765.

CHITTENDEN, t. Rutland co. Vt. 30 ms. N. W. Windsor, lat. 43° 44', watered by Philadelphia r., Tweed r. and East creek. In great part mountainous and incapable of cultivation. Contains a mineral spring. Pop. 1830, 610.

CHITTENINGO, creek, N. Y. a fine mill stream, falls into Oneida lake.

CHITTENINGO, p-v. Madison co. N. Y. on the above creek, at the head of a canal 1½ ms. long, completing the navigation from its quarries of gypsum and water lime to the Erie canal. Contains an oil mill, and one for grinding gypsum, and the water cement or water lime.

CHOCONUT, p-o. and on the Choconut creek in the N. W. part Susquehannah co. Pa. by p-r. 175 ms. from Harrisburg. Pop. of the tsp. 1830, 780.

CHOCTAW, r. of Ala. and Florida, rises in Pike co. of the former, flows thence over Henry and Dale cos. into Florida, over which it passes, leaving Walton co. to the right, and Jackson and Washington to the left. It expands into a bay of the same name, after a comparative course of 130 ms. in a direction of nearly S. S. W. The valley of Choctaw river lies between that of the Conecuh and Chattahoochee rivers. Much of the soil of the valley of this stream is described by Williams, in his Florida, as of excellent quality, though much of it is sterile, covered with a pine forest.

CHOCTAW BAY, or the estuary of Choctaw river, extending between Jackson and Walton cos. Florida. In the Report of the board of internal improvement, this bay is called St. Rosa, from which, however, it is distinct. (*See St. Rosa Island and Sound.*) Under the name of St. Rosa the engineers state, "the entrance of this sound and bay lies about 85 ms. W. of Cape St. Blass, and 68 from the mouth of St. Joseph's bay. On the whole (intermediate) distance the sea shore is very bold, and the depth generally 4 fathoms close to the land."

"The pass enters between the eastern point of St. Rosa island and the main; it is called Eastern Pass, it comes in from the S. and affords a depth of 8 feet on the bar. The channel is narrow, and the width on the bar, for 8 feet depth, is about 150 yards. On account of breakers, this pass is not considered safe when southerly winds blow fresh, but the winds being from the land, the channel is easy of entrance."

The bay, according to Williams, is difficult to navigate, from shoals, but admits a depth of 7 feet water, which is continued to Big Spring 60 miles above the bar, and 30 above the mouth of Choctaw r.

CHOCTAW BLUFF, and p-o. Green co. Ala. by p-r. 69 ms. S. S. W. Tuscaloosa.

CHOCTAW, or FLAT HEAD, a nation of Indians, formerly more numerous than at present, and also spread over a much wider surface than they now occupy. They amount to about 25,000 persons, and reside between the white settlements of the state of Miss. and the Chickasaws, and between the Mississippi and Tombigbee rivers. Their country, as restricted by cessions to the U. S. extends from lat. 31° 50' to 35'. Length from S. E. to N. W. 230 ms. with a mean breadth of 80, area 18,400 sq. ms. It is drained by the Big Black and Yazoo rivers flowing S. W. into the Miss. by the sources of the Pearl, and by those of Tombigbee. Many scattered settlements of Choctaws, have been formed, within the last 35 years, to the westward of the Mississippi river. This nation has made some advances in civilization, though not so much improved as the Chickasaws and Cherokees.

CHOCTAW ACADEMY, and p-o. Scott co. Ky. by p-r. 31 ms. E. Frankfort.

CHOCTAW AGENCY, and p-o. Yazoo co. Miss. by p-r. 56 ms. northward Jackson, and 154 N. N. E. Natchez.

CHOICE'S STORE, and p-o. Gwinnett co. Geo. by p-r. 99 ms. N. W. Milledgeville.

CHOTA, p-v. in the W. part of Blount co. Tennessee by p-r. 197 miles south of east Nashville.

CHOPTANK, r. of Del. and Md. is formed by two branches, Choptank proper and Tuckahoe. The former rises in Kent co. Del. from which it flows S. S. W. into Caroline co. Md. and continuing the same course traverses Caroline to its junction with Tuckahoe on the E. border of Talbot; Tuckahoe rises in the northeast part of Queen Ann co. Md. and flowing to the sthrd. separates Queen Ann and Talbot from Caroline, and joins the Choptank after each has flowed about 30 ms. Then assuming a S. S. W. course, gradually swells into a bay, and above Cambridge bends to the N. W. by W. opens into the main Choptank bay between Cook's Point and Tilghman's island. It is navigable for sloops to the Forks, 40 ms. above the mouth.

CHOWAN, r. of N. C. formed by the united streams of Meherin, Nottaway, and Black Water rs. The Meherin rises in Charlotte co. Va. 1° 30′ w. from W. C. lat. 37°, between the vallies of Roanoke and Appomattox, and flowing thence S. E. by E. by comparative courses 80 ms. passes into N. C. between Northampton and Gates cos. and 20 ms. farther unites with the Nottaway, above Winton, between Gates and Hertford cos.

The Nottaway derives its remote sources from Prince Edward co. Va. between those of Meherin and Appomattox. In a general eastern course of 70 ms. the Nottaway separates Lunenburg, Brunswick and Greenville cos. from Nottaway, Dinwiddie and Sussex, and flows into the central parts of the latter. Thence inclining S. E. 40 ms. it receives Black Water r. almost on the bounding line between Va. and N. C. (*See Black Water river, Va.*) Below the junction of the Nottaway and Black Water, the name of the former and course of the latter are preserved, and about 10 ms. within N. C. and in Gates co. meet the Meherin to form Chowan river.

A tide water river, or more correctly a bay, the Chowan, gradually widens, but still retaining a moderate breadth, 25 ms. to the influx of Bemer's creek, there bends to near a sthrn. course and more rapidly widens for 25 ms. to its junction with Roanoke, at the head of Albemarle sound. Lat. 36°, passes up Albemarle sound, and intersects the eastern point of Bertie co. N. C. between the mouths of Chowan and Roanoke, 0° 20′ long. E. of W. C.

Incluing all its confluents or constituents the Chowan drains an area of 3,500 sq. ms. which, as a physical section, comprises the northeastern part of the basin of Roanoke. As a commercial channel the Chowan, Nottaway, and Black Water extend almost directl from the mouth of Roanoke to that of James river. There is at all seasons sufficient depth of water to admit sloops of war to Murfreesboro' on Meherin, about 10 ms. above the entrance of Nottaway river. There is now a struggle between contending interests, whether to extend a rail road from the basin of Roanoke to Chesapeake bay, or a canal and lock navigation by the channels of Chowan, Nottaway and Black Water rivers.

CHOWAN, co. N. C. bounded N. by Gates co., E. by Perquimans, S. E. and S. by Albemarle sound, and S. W. and W. by Chowan river, which separates it from Bertie and Hertford cos., mean width 8, and area, 200 sq. ms. Extending in lat. from 36° to 36° 20′ and in long. from 0° 18′ to 0° 36′ E. from W. C. The slope is slight, but what little declivity there exists in the co. is to the S. W. towards Chowan river. Soil productive. Chief town, Edenton. Pop. 1820, 6,464, and in 1830, 6,697.

CHRISTIAN, co. Ky. bounded by Trigg W., Hopkins, N. W., Muhlenburg N. E., Todd E., and Montgomery co. of Tenn. S. Length from N. to S. 34 ms., mean breadth 18, and area 612 square miles. Extending in latitude from 36° 37′, to 37° 07′, and in longitude from 10° 04′ to 10° 23′ W. from W. C. Though not very elevated, this county is a table land, from which Little r. flows W. into Cumberland r., Pond r. N. into Green r. and the western fork of Red river branch of Cumberland S. into Red river. Chief town, Hopkinsville. Pop. 1830, 12,864.

CHRISTIANA, tide water creek, principally of New Castle co. Delaware, but deriving its remote sources from Cecil co. Md. and Chester co. Pa. It is formed by the junction of Christiana proper with the United streams of White Clay, and Red Clay creeks. The two latter rise in Chester co. Pa., the latter in Cecil co. Md. The general course of Christiana proper and the united stream below the influx of Red and White Clay creeks is from S. W. to to N. E., comparative length 30 ms. to its junction with the Brandywine r. at Wilmington. The tide ascends Christiana, and enables vessels of 6 feet draught to be navigated to Christiana bridge, 10 ms. above Wilmington.

CHRISTIANA, usually called Christiana Bridge, p-v. New Castle co. Del. situated on Christiana creek, 10 ms. above and S. W. Wilmington, by p-r. 47 ms. a little N. of W. Dover.

CHRISTIANA, p-v. in the N. E. part of Butler co. O. by p-r. 88 ms. S. W. by W. Columbus.

CHRISTIANSBURG, p-v. and st. jus. Montgomery co. Va. by p-r. 282 ms. S. W. W. C. and 206 S. of W. Richmond. Lat. 37° 08′, long. 3° 24′ W. from W. C.

CHRISTIANSBURG, p-v. Shelby co. Ky. 14 ms. W. Frankfort.

CHRISTIANSVILLE, p-v. in the northern part of Mecklenburg co. Va. by p-r. 128 ms. S. W. Richmond.

CHRISTMASVILLE, p-v. Carroll co. Ten. by p-r. 105 ms. W. Nashville.

CHUCKATUCK, p-v. Nansemond co. Va. about 30 ms. S. W. Norfolk.

CHUCKY BEND, of Nolechucky r. and p-o. is about 45 ms. N. of E. Knoxville, and in the eastern part of Jefferson co. Ten. by p-r. 215 ms. E. Nashville. The Nolechucky r. and French Broad unite about 5 ms. below the bend.

CHURCH HILL, p-v. Queen Ann co. Md. on a small S. E. branch of Chester river, 10 ms. N. N. E. Centreville, and about 50 ms. S. E. by E. from Baltimore.

CHURCH HILL, p-v. Abbeville dist. S. C. by p-r. 96 ms. W. Columbia.

CHURCH HILL, p-v. Montgomery co. Ala. by p-r. 121 ms. S. E. Tuscaloosa.

CHURCHTOWN, p-v. on a small branch of Conestoga creek, in the N. E. angle of Lancaster co. Pa. 25 ms. N. E. Lancaster, and 55 northwest by west Philadelphia.

CHURCHVILLE, p-v. in the N. W. part of Middlesex co. Va. 7 ms. from Urbanna, the st. jus. by p-r. 76 ms. N. E. by E. Richmond.

CICERO, t. Onandago co. N. Y. Population 1830, 1,808.

CINCINNATI, city of Ohio, and st. jus. for. Hamilton co. situated on the right bank of Ohio river, by p-r. 497 ms. (differing only 13′ of lat. from) due W. from W. C. 112 ms. S. W. by W. Columbus, and 79 a little E. of N. Frankfort, Ky. Lat. 39° 06′, long. 7° 32′ W. W. C. according to Tanner's map of the U. S. but 7° 24′ 45″ according to Flint.

The position of Cincinnati is admirable. It stands on two plains or bottoms of Ohio, the higher elevated about 60 feet above the lower, with a rather steep intermediate bank. To an eye in the vicinity, placed on elevated ground, the city seems to occupy the centre and base of an immense basin, the view being in every direction terminated by swelling hills. The streets, laid out at right angles to each other, present an endless, though rather monotonous variety of landscape. Fourteen of the streets are 66 feet wide, and 396 apart; seven extending each way and crossing the other seven. Thus the intermediate squares comprise 156,816 sq. feet. The public buildings already erected occupy one square and a fraction of another; and that part of the city built upon, approaches the form of a parallelogram. The public buildings are, the Cincinnati college, Catholic athenæum, medical college, the mechanics institute, a theatre, two museums, hospital, and lunatic asylum, United States branch bank, court house, prison, 4 market houses, a bazar, and the Woodward high school in the progress of erection.

Of churches there are 24, of which several are fine buildings, banks 3, the United States branch bank, capital $1,200,000; Commercial bank, capital $500,000, and savings bank, insurance companies, 3 belonging to the city, with two branches of companies at Hartford, Conn. A water company supplies the city with water from Ohio river. It is elevated by steam power to the height of 158 feet above low water mark in the river, and flowing into reservoirs, is thence distributed over the city, at an annual expense of $8 per family at an average.

The public prints are 16, comprising, one Quarterly Medical Journal, one Monthly Magazine, one Monthly Agricultural Journal, two semi monthly, two semi weekly, six weekly, and three daily gazettes. Thirty-two mails arrive weekly. There are two fire companies, and 34 charitable societies, and 25 religious societies.

The progressive pop. of this city is perhaps unequalled on a region where rapid advance is every where remarkable. It was laid out in January, 1789, but until after the treaty of Greenville, 1795, progressed but slowly. In 1810, the total population was 2,540, in 1820, 9,642, in 1826, 16,230, and in 1829, 24,408. "By a very accurate enumeration in 1831, 28,014, with a floating population, not included, of 1,500, making the total at this time (1832) more than 30,000.

By the census tables for 1830, printed at Washington, the population of Cincinnati was composed of white males, 12,485; white females 11,256; free colored males, 528, and females 562; total 24,831.

This city, second only in population to New Orleans, amongst the western cities of the United States, has already become the seat of immense and increasing manufactures, of almost every species known in our country. Of steamboats 111 have been built here. The iron manufactures include nearly every article of that metal demanded by a civilized and active population. Cabinet, hatting, shoe and boot making, saddlery, &c. Imports exceed $5,000,000, of which dry goods are the principle part; and the exports exceed the imports. The latter composed of country produce, and the products of the iron, cabinet, and other manufactures of the city and vicinity, are mostly sent down the Ohio. About 40 manufacturing establishments are propelled by steam. Revenue of the city 1831, was $35,231, and expenditure was $33,858.

Business is the chief object of this young city, but education has not been neglected. There are 27 public teachers of free schools, who give instruction to 2,700 children annually. The private schools are numerous, and many of them very respectable.

Mr. Flint states that 450 substantial buildings have been added yearly, for the three last years.

CINCINNATUS, p-t. Cortlandt co. N. Y. 139 W. Albany; 12 S. E. Homer; soil productive, moderately uneven, indifferently supplied with mill streams. Has 1 distillery, 3 asheries, &c. Pop. 1830, 1,308.

CIRCLEVILLE, p-v. and st. jus. Pickaway co. Ohio, by p-r. 26 ms. S. from Columbus, and 394 ms. a little N. of W. W. C. 19 N. and above Chillicothe, and 20 ms. S. W. by W. from Lancaster. N. lat. 39° 36′, and long. 5° 58′ W. from W. C. This place is situated on the left bank of Sciota r. where that stream is crossed by the Ohio and Erie canal, and on the largest aqueduct on the line of this work. It contains the ordinary co. buildings, a printing office, 10 or 12 stores, numerous mechanics shops, and in 1830, a population of 1,136, which now, 1832, it is probable exceeds 1,200. This town derives its name from several remarkable remains of ancient works, in the ordinary circular form of such antiquities, scattered over the valley of Ohio. They were here very extensive, and before the white settlements were made, were well preserved.

CITY ISLAND, N. Y. (*See Pelham.*)

CITY POINT, port and p-v. on the right shore

of James river, on the point formed at the junction of James and Appomattox rs. in the N. W. part of Prince George's co. Va. 12 ms. below Petersburg, by p-r. 34 ms. S. E. from Richmond.

CIVIL ORDER, p-v. in the N. W. part of Bedford co. Ten. by p-r. 48 ms. S. Nashville.

CLAIBORNE, co. Miss. bounded W. by Miss. r. separating it from Concordia parish in La., N. W. and N. by Big Black r. separating it from Warren co. Miss., N. E. by Hinds, S. E. by Copiah, and S. by Jefferson. It approaches the form of a right angled triangle, hypotenuse along the Miss. and Big Black rs. 38 ms., base on Jefferson 30, area 380 sq. ms. Extending in lat. from 31° 53' to 32° 11', and in long. from 13° 50' to 14° 20' from W. C. Along the Mississippi and Big Black rivers the bottoms are level, extremely fertile, but subject to annual submersion. Rising from this alluvial border, the country is elevated into hills, which towards the rivers are fertile, but receding eastward the pine forest and sterile soil commence. Bayou Pierre (*Stony Creek*), a fine stream bordered with excellent land, flows to the S. S. W. and drains the central part of the co.; staple, cotton. Chief town, Gibsonport. Pop. 1820, 5,963, 1830, 9,787.

CLAIBORNE, parish of Lo. as laid down by Tanner, is bounded E. by the parish of Washitau, S. by Natchitoches, S. W. & W. by Red r. and N. by Lafayette co. of Ark. Length from S. to N. 65 ms.; mean breadth 55, and area 3,575 sq. ms. Extending in lat. from 32° 05' to 33°, and in long. from 15° 51' to 16° 57' W. from W. C. The northeast part is drained eastward by the sources of Bayou Terrebonne flowing into Washitau r. but the great body of the parish declines southward, and is drained in that direction by the sources of Dugdemini, Saline, Black Lake, Dacheet and Bodcau rs.; the western part also contains the lakes Bistineau, and Bodcau. Some of the soil along the streams is of good second rate quality, and wooded with oak, hickory, and elm, but the body of the parish, or at least nine tenths of its surface, is composed of barren hills clothed with pine timber. The border on Red river is partially liable to annual submersion. The writer of this article was the first person who surveyed or indeed explored this section of Louisiana. It was then, (1812) an uninhabited, in great part, and pathless wilderness. It contained then 3 white families, and a small Indian village, on Red river. By the post office list, 1831, there were offices at Allen's settlement and Russellville. Pop. 1830, 1,764.

CLAIBORNE, co. of Ten. bounded by Clinch r. separating it from Hawkins E., Granger S. and Campbell S. W.; on the W. it has again Campbell, on the N. W. Knox co. of Ky. and N. E. Lee, the extreme S. W. co. of Va. It lies nearly in the form of a right angled triangle; base 50 ms. along Va. and Ky. perpendicular on Campbell co. 28 ms. and hypotenuse along Clinch r.; area 700 sq. ms. Extending in lat. 36° 13' to 13° 35', and in long. 5° 52' to 6° 48' W. W. C. Surface mountainous. The northwestern angle is occupied by Cumberland mtn. whilst Powell's mtn. traverses it in its greatest length from N. E. to S. W. Between these two chains flows Powell's r. S. W. whilst the co. has again a river border of 70 ms. along Clinch. Chief town, Tazewell. Pop. 1820, 5,508, 1830, 8,470.

CLAIBORNE, p-v. and st. jus. Monroe co. Ala. on the left bank of Ala. r. by p-r. 949 ms. S. W. from W. C. 157 S. from Tuscaloosa, 80 N. N. E. from Mobile. Lat. 31° 33', long. 10° 40' W. from W. C. Claiborne stands at the lower falls and head of schooner navigation in Ala.

CLAPPS, p-o. Guilford co. N. C. by p-r. 73 ms. N. of W. Raleigh.

CLAREMONT, p-t. Sullivan co. N. H. on Conn. r. opposite Windsor Vt. Area 25,800 square acres. Its surface, a rich gravelly loam, finely undulating, and furnishing the best meadows. Produce in 1820: butter 30,000 lbs.; cheese 55,000; flax 7,500; pearlashes 3 tons. Watered by Conn. and Sugar rs. Religious societies; Congregationalist, Episcopalian, Baptist, Methodist, 1 each. Lat. 43° 23'. Pop. 1830, 2,526.

CLAREMONT, p-v. Picken's dist. S. C. by p-r. 163 ms. N. W. by W. from Columbia.

CLARENCE, p-t. Erie co. N. Y. 18 ms. E. Buffalo. N. boundary, Tonnewanta creek. Soil, a loam, which good husbandry may make very productive. Its rocks, horizontal limestone. Schools 21, 6 months in 12. Distilleries 6. Asheries 13. Pop. 1830, 3,360.

CLARENDON, p-t. Rutland co. Vt. 55 ms. S. Montpelier. Otter creek, Mill and Cold r. furnish numerous mill seats. Alluvial flats, from ½ to 1 m. wide, on Otter creek, very productive, extend through the town. Here is one of the fanciful stalactite caves. Marble or limestone, plenty, and wrought. Religious societies, 2 Baptist, 1 Congregationalist. Mill for sawing marble, 3 distilleries, &c. Lat. 43° 31'. Pop. 1830, 1,585.

CLARENDON, p-t. Genesee co. N. Y. 18 ms. N. E. Batavia, about 6 ms. square, watered by Sandy creek. Soil good. It has 8 schools, 7 months in 12. Distillery 1. Pop. 1830, 2,025.

CLARENDON, t. Orleans co. N. Y. Pop. 1830, 2,025.

CLARIDON, p-v. and tsp. in the sthrn. part of Geauga co. O. by p-r. 327 ms. N. W. W. C. and 155 N. E. Columbus. Pop. 1820, 588, 1830, 637.

CLARION, r. of Pa. usually called Toby's creek, rises by numerous branches in Mac Kean and Jefferson cos. interlocking sources with creeks flowing nthrd. into Alleghany r. and opposite to those of the Sinnamahoning branch of Susquehannah. The different branches unite near the centre of Jefferson, and the main stream, assuming a S. W. course over that co. and thence separating Armstrong from Venango, falls into Alleghany r. at Foxburg, after an entire course of 60 ms.

Clarion, p-v. in the n. part Armstrong co. Pa. by p-r. about 70 ms. n. e. Pittsburg.

Clark, co. of Geo. bounded by Walton w. Jackson n. w. Madison n. e. Oglethorpe e. Greene s. and is separated from Morgan s. w. by Appalache, branch of Oconee r. Length 23, mean breadth 18, and area 414 sq. ms. Extending in lat. from 33° 32′ to 34° 2′, and in long. from 6° 17′ to 6° 40′ w. from W. C. The constituents of Oconee r. unite in Clark, and flowing generally to the s s. e. give that declivity to the co. The Appalache, which bounds it on the s. w. flows also to the s. s. e. Chief towns, Watkinsville and Athens. Pop. 1830, 10,176.

Clarke, co. of Ala. occupying the lower part of the peninsula between the Tombigbee and Ala. rs. bounded n. by Marengo, n. e. by Wilcox, by Ala. r. separating it on the s. e. from Monroe, and s. from Baldwin; by Tombigbee r. separating it from Mobile s. w. and from Washington w. and n. w. Length from the junction of Ala. and Tombigbee rs. and nearly along long. 11° w. from W.C. to the s. boundary of Marengo, 60 ms. Mean breadth 20, and area 1200 sq. ms. Extending in lat. from 31° 10′ to 32°, and in long. from 10° 30′ to 11° 18′ w. W. C. Surface hilly, and soil, except near the streams, sterile, and wooded with pine. Much of the river bottoms liable to occasional inundation. Chief town, Clarkesville. Pop. 1820, 5,839, 1830, 7,595.

Clarke, co. of Ky. bounded s. and s. w. by Ky. r. separating it from Madison co., w. by Lafayette, n. by Bourbon, n. e. and e. by Montgomery, and s. e. by Red r. separating it from Estill. Length 20, mean breadth 15, and area 300 sq. ms. Extending in lat. from from 37° 52′ to 38° 10′, and in long. from 6° 50′ to 7° 18′ w. W. C. This small co. is nevertheless a table land. From the nthrn. side issue the extreme sources of the w. fork of Licking, whilst short creeks flow southwardly into Ky. r. The soil is highly fertile. Chief town, Winchester. Pop. 1820, 11,449, 1830, 13,051.

Clarke, co. O. bounded s. by Green, s. w. by Montgomery, n. w. by Miami, n. by Champaign, and e. by Madison. Length 30 ms. mean breadth 18, and area 540 sq. ms. Extending in lat. from 39° 45′ to 40° 03′, and in long. from 6° 31′ to 7° 5′. From the southern part flows little Miami, whilst the central sections are traversed by Mad r. giving a s. w. slope to the body of the co. Chief town, Springfield. Pop. 1830, 13,074.

Clarke co. Ind. bounded by Floyd s. w. Washington w. Scott n. Jefferson n. e. O. r. separating it from Oldham co. Ky. e. and s. e. and from Jefferson co. Ky. s. It approaches the form of a triangle 28 ms. each side; area 336 sq. ms. Extending in lat. from 38° 18′ to 38° 37′, and in long. from 8° 25′ to 8° 54′ w. W. C. The slope of this co. is almost directly s. towards O. r. Silver creek rises in Clark, and flowing s. falls into O. r. at the lower end of the rapids at Louisville. The surface is broken and hilly, soil fertile. Chief town, Charleston. Pop. 1820, 8,079, 1830, 10,686.

Clarke, co. of Il. bounded s. by Crawford, w. by Shelby, n. by Edgar, by Wabash r. separating it from Vigo co. Ind. e. and from Sullivan co. Ind. s. e. Breadth 24, mean length 45, and area 1080 sq. ms. Extending in lat. from 39° 10′ to 29° 30′, and in long. from 10° 34′ to 11° 30′ w. W. C. Little Wabash rises in the w. part of Clarke, which is traversed also by the w. and e. branches of Embarras r. all those streams flowing to the sthrd. The eastern part slopes south estrd. obliquely towards the Washitau. Chief t. Clark Court House. Pop. 1830, 3,940.

Clarke, co. of Ark. extending along both sides of Washitau r. above the influx of Little Missouri. The boundaries or extent not very well defined, but combining Tanner's map with Flint's description, it has Hempstead co. s. Pope w. Hotsprings n. Pulaski n. e. and Union e. Lat. 34°, and long. 16° w. W. C. intersect near its centre. The slope is to the s. e. down which pour the confluents of Washitau and Little Missouri. The surface is hilly, and in part mountainous. Considerable bodies of good land skirt the streams, though the soil is generally sterile. The road from St. Louis, by Little Rock to Lower Texas passes through it, and on which two villages, Biscoeville and Crittenden are laid down by Tanner. Pop. 1830, 1,369.

Clark, court house, and p-o. Clark co. Ark. by p-r. 87 ms. s. w. Little Rock.

Clark, C. H. and p-o. Clark co. Il. by p-r. 134 ms. n. e. Vandalia.

Clarkesburg, t. Berkshire co. Mass. Has Williamstown on the w. Pop. 1830, 315.

Clarkson, p-t. Monroe co. N. Y. on Lake Ontario, 18 ms. w. s. w. Rochester. Area about 80 sq. ms. Soil excellent. Contains many salt springs. Watered by 3 creeks, one a fine mill stream. The village is 1½ ms. n. of Erie canal. Pop. 1830, 3,251.

Clarkstown, p-t. cap. Rockland co. N. Y. on the w. bank of the Hudson 132 ms. s. Albany, 28 n. N. Y. Here are the Nyak hills, furnishing the red sand stone, of which the capitol at Albany is principally built. Church, 1 Dutch Reformed. In 1808, distilleries 3. Schools kept 11 months in 12. Pop. 1830, 2,298.

Clarksville, N. Y. (See Middlefield.)

Clark's Ferry and p-o. Perry co. Pa. by p-r. 44 ms. n. w. Harrisburg.

Clarksburg, small p-v. Montgomery co. Md. on the road from W. C. to Frederick, 28 ms. n. w. from the former, and 15 ms. s. e. from the latter city. It is a small village of one street along the main road. Pop. about 50.

Clarksburg, p-v. and st. jus. Harrison co. Va. by p-r. 260 n. w. by w. Richmond, and 45 ms. above, and s. s. w. from Morgantown.

It is situated on the right bank of Monongahela r.

CLARKSBURG, p-v. and st. jus. Lewis co. Ky. by p-r. 96 ms. N. E. by E. Frankfort. Pop. 1830, 62.

CLARKSVILLE, p-v. in the N. E. part of Greene co. Pa. situated on the point between and above the junction of the two main branches of Ten Mile creek, 10 ms. S. W. from Brownsville, about an equal distance N. E. from Waynesburg.

CLARKSVILLE, p-o. on the road from Rockville to Baltimore, Ann Arundel co. Md. 20 ms. S. W. by W. Baltimore.

CLARKSVILLE, p-v. Mecklenburg co. Va. by p-r. 99 ms. S. W. Richmond.

CLARKSVILLE, p-v. Spartanburg dist. S. C. by p-r. 111 ms. N. N. W. Columbia.

CLARKSVILLE, p-v. and st. jus. Habersham co. Geo. by p-r. 144 ms. a little W. of N. from Milledgeville, on one of the highest branches of Chattahoochee r. Lat. 34° 35′, and long. 6° 40′ W. from W. C.

CLARKSVILLE, p-v. and st. jus. Clark co. Ala. by p-r. 146 ms. a little W. of S. Tuscaloosa, and 84 ms. a little E. of N. from the city of Mobile.

CLARKSVILLE, p-v. and st. jus. Montgomery co. Ten. by p-r. 46 ms. N. W. by W. from Nashville, situated on the point above the junction, and between Cumberland and Red rs.

CLARKSVILLE, p-v. in Clarke tsp. western part of Clinton co. O. The p-v. is by p-r. 76 ms. S. W. Columbus. Pop. of the tsp. 1830, 1,886.

CLARKSVILLE, p-v. on the right bank of Miss. r. in the E. part of Pike co. Mo. about 75 ms. by the land route above St. Louis, and by p-r. 126 N. E. Jefferson.

CLARK'S p-o. and tsp. Coshocton co. O. by p-r.; the p-o. is 88 ms. N. E. by E. Columbus. In 1830 the tsp. contained 246 inhabitants.

CLARKSBURG, p-v. in the nthrn. part of Ross co. O. by p-r. 44 ms. S. S. W. Columbus. Pop. 1830, 56.

CLARKSFIELD, p-v. and tsp. Huron co. O. The p-v. is by p-r. 121 ms. a little E. of N. Columbus, and 385 ms. N. W. by W. W. C. Pop. of the tsp. 1830, 368.

CLARK'S MILLS, and p-o. in the S. part of Moore co. N. C. by p-r. 108 ms. S. W. from Raleigh.

CLARK'S RIVER. (See Oregon.)

CLARK'S STORE and p-o. in the S. part of Martin co. N. C. by p-r. 106 ms. E. Raleigh.

CLARK'S STORE and p-o. in the S. W. part of Hamilton co. O. by p-r. 13 ms. from Cincinnati.

CLARKSTON, p-o. King and Queen co. Va. by p-r. 50 ms. N. E. Richmond.

CLARKSTOWN, p-v. Wayne co. Pa. by p-r. 158 ms. N. E. Harrisburg.

CLAVERACK, t. Columbia co. N. Y. 5 ms. E. Hudson. Claverack creek, its W. boundary, is a fine millstream. Has along the creeks, rich alluvial flats. Contains good limestone, some slate, some lead, and a mineral spring. Schools 13, 11 months in 12. Distillery 1. Pop. 1830, 3,000.

CLAY, t. Onondaga co. N. Y. Pop. 1830, 2,095.

CLAY, co. of Ky. bounded by Knox S. Laurel W. Estill N. and Perry E. Length from N. to S. 40, mean breadth 22, and area 880 sq. ms. Extending in lat. from 36° 57′ to 37° 33′, and in long. from 6° 18′ to 6° 52′ W. from W. C. Though some of the sources of Rockcastle creek, a branch of Cumberland r. rise along the western border of Clay, the body of the co. is drained by, and nearly commensurate with the valley of the southeast Fork of Ky. r. and slopes northwardly. The soil is generally thin. Chief t. Manchester. Pop. 1830, 3,548.

CLAY, co. of Ind. bounded S. W. by Sullivan, W. and N. W. by Vigo, N. by Parke, N. E. by Putnam, E. and S. E. by Owen, and S. by Greene. Length from S. to N. 30 ms. mean breadth 12, and area 360 sq. ms. Extending in lat. from 39° 12′ to 39° 38′, and in long. from 9° 58′ to 10° 18′ W. from W. C. Slopes to the sthrd. and is drained by Eel r. a branch of the W. fork of White r. Creeks, flowing wstrd. into the Wabash r. rise along the watrn. border of Clay, but the body of the co. is in the valley of Eel r. Chief t. Bowling Green. Pop. 1830, 1,616.

CLAY, co. of Il. bounded S. E. by Edwards, S. by Wayne, S. W. by Marion, N. W. by Fayette, N. and N. E. by Crawford, and E. by Lawrence. Length 32 ms. breadth 21, and area 672 sq. ms. Extending in lat. from 38° 37′ to 38° 54′, and in long. from 11° 9′ to 11° 44′ W. from W. C. The main stream of Little Wabash enters the nthrn. border from Fayette, and inflecting to S. E. receives numerous creeks from, and traverses Clay, issuing from it in the S. E. angle. Chief t. Maysville. Pop. 1830, 755.

CLAY, co. of Mo. bounded on the W. by the W. boundary of the state, and N. and N. E. by country not yet laid out into cos.; E. it has Ray co., and S. the Missouri r. separating it from Jackson co. Breadth from E. to W. 22, mean length from S. to N. 30, and area 660 sq. ms. Extending in lat. from 39° 04′ to 39° 34′, and in long. from 17° 06′ to 17° 28′ W. from W. C. Chief t. Liberty. Pop. 1830, 5,338. The Kansas r. enters the Missouri directly opposite the S. W. angle of this co. at a distance by the p-r. of 1170 ms. W. from W. C.

CLAYSVILLE, p-v. on the U. S. turnpike road, Washington co. Pa. by p-r. 222 ms. W. Harrisburg, and 10 ms. S. W. by W. from the borough of Washington.

CLAYSVILLE, p-v. Guernsey co. O. by p-r. 92 ms. E. Columbus.

CLAYSVILLE, p-v. Washington co. Ind. by p-r. 92 ms. S. Indianopolis.

CLAYSVILLE, p-v. in the E. part Harrison co. Ky. by p-r. 50 ms. N. of E. Frankfort.

CLAYTON or CLAYTONSVILLE, p-v. and st. jus. Rabun co. Geo. by p-r. 611 ms. S. W. a little W. of W. C. and 174 ms. N. Milledgeville. It is situated at the southern base of Blue Ridge, between the Chatuga and Turoree branches of Tugaloo r. and is the most northern co. t. of Geo.

CLAYTONVILLE, p-o. Buncombe co. N. C. 286 ms. wstrd. Raleigh.

CLAY VILLAGE, p-v. Shelby co. Ky. 16 ms. w. Frankfort.

CLEAR CREEK, p-o. Hardiman co. Ten. by p-r. 168 ms. s. w. by w. Nashville.

CLEAR CREEK and p-o. Richland co. O. The p-o. by p-r. 96 ms. E. of N. Columbus.

CLEAR CREEK, p-o. in the N. part of Sangamon co. Il. by p-r. 96 ms. w. of N. Vandalia.

CLEARFIELD, co. of Pa. bounded s. by Cambria, s. w. by Ind. w. and N. w. by Jefferson, N. by Mac Kean, N. E. by Lycoming, E. by the w. branch of Susquehannah r. separating it from Centre, and s. E. by Mushannon creek, separating it from the southwestern part of Centre. Length from s. to N. 45, mean breadth 32, and area 1425 sq. ms. Extending in lat. from 40° 45′ to 41° 24′, and in long. from 1° 3′ to 1° 53′ w. W. C. Clearfield is an elevated, and in great part a mountainous region. Lying wstrd. from the main chain of the Appalachian system, it is on the floetz or level formation. From the western border issue the extreme fountains of the Mahoning and Red Bank creeks, flowing to the w. into Alleghany r., but the far greater part of the area is drained by the main streams and numerous branches of Sinnamahoning, and w. branch of Susquehannah. The dividing ridge of the waters, traversing the N. w. part of Clearfield, is elevated about 1200 feet above the Atlantic tides. From this ridge the extreme western sources of the Susquehannah flow s. E. down the mountain vallies, giving an uncommonly diversified surface to Clearfield. The soil is generally rocky and sterile. Chief t. Clearfield. Pop. 1820, 2,342, and in 1830, 4,803.

CLEARFIELD, p-v. and st. jus. Clearfield co. Pa. by p-r. 201 ms. N. N. W. W. C. 129 N. W. by w. Harrisburg, and about 100 ms. N. E. by E. Pittsburg. It is situated between Clearfield creek and the w. branch of the Susquehannah r.

CLEARFIELD RIDGE and p-o. Clearfield co. Pa. by p-r. 4 ms. s. E. Clearfield v. 125 N. W. by w. Harrisburg.

CLEAR SPRING, p-v. in the w. part of Washington co. Md. by p-r. 82 ms. N. W. W. C.

CLEAVELAND, p-v. and st. jus. Cuyahoga co. O. about 130 ms. N. W. Pittsburg, and by p-r. 366 ms. a little w. of N. W. W. C. 140 N. N. E. Columbus, and 104 by the land route, s. w. by w. from the borough of Erie in Erie co. Pa. N. lat. 41° 32′, long. 4° 42′ w. W. C. The site of Cleaveland is an elevated point below the entrance of Cuyahoga r. into lake Erie. The river here admits vessels of 7 feet draught, and with the outlet of the great canal of Ohio has given advantages and rapid advance to the place. By the census of 1830 it contained a pop. of 1,076, and now, (1832) no doubt the inhabitants exceed 1,200. It has the usual co. buildings, upwards of 40 stores, 9 or 10 groceries, a number of taverns, 200 dwelling houses, and 4 or 5 churches. The future and securely permanent prosperity of this place is evident from its position.

CLEMONSVILLE, p-v. Davidson co. N. C. by p-r. 125 ms. w. Raleigh.

CLEMONTVILLE, p-o. Mac Kean co. Pa. by p-r. 201 ms. N. W. Harrisburg.

CLERMONT, p-t. Columbia co. N. Y. on E. bank of the Hudson, 45 ms. s. Albany. Area 14,000 acres, divided into about 120 farms, and leased to practical farmers. The country seat of the late Chancellor Livingston, is one of the most extensive and elegant in the state. There are 7 schools 8 months in the year. Pop. 1830, 1,203.

CLEVES, p-v. in the s. w. part of Hamilton co. O. 16 ms. wstrd. from Cincinnati.

CLIFTON PARK, t. Saratoga co. Pop. 1830, 2,294. (See Half Moon.)

CLIFTON, p-v. Russell co. Va. by p-r. 339 ms. a little s. of w. Richmond.

CLIFTY, p-v. in the s. part of White co. Ten. by p-r. 102 ms. s. E. by E. Nashville.

CLINCH, r. of Va. and Ten. the great northestrn. constituent of Ten. r. rises in Tazewell co. Va. and flows thence by a general course of s. w. over Russell and Scott cos. 90 ms. Entering Ten. Clinch separates Claiborne co. from Hawkins, Granger and Anderson; Campbell from Anderson, and thence traversing the latter, enters Rean, and unites with the Ten. at Kingston, after an entire comparative course of 180 ms. In the s. part of Campbell co. Clinch receives from the N. E. Powell's r. The latter rising in Russell co. Va. issues thence in a direction almost parallel to the Clinch; traverses Lee co. of Va., enters Ten. crossing Claiborne and Campbell cos., joins the Clinch at Grantsboro after a comparative course of 90 ms. A short distance above its junction with Ten. r. the Clinch receives from the N. W. Emery's r. It may be remarked that the course of the higher branches of Emery's r. is directly the reverse of that of Clinch and Powell's r. Uniting the vallies of Emery's and Clinch r. the whole valley is about 220 ms. long; but the width is contracted comparatively, and fully estimated at 20 ms. Area 4400 sq. ms.

In all their respective courses, Clinch and Holston pursue a parallel direction, in few places 20 ms. asunder, each receiving short creeks, from an intervening mountain chain. On the opposite or right side, Clinch in succession interlocks sources with those of Great Sandy, Ky. and Cumberland rs. The relative elevation of the vallies of Clinch and Holston differ but little from each other, and each stream above their junction, must have, from their remote fountains, a fall of 1000 or 1200 feet.

CLINCH DALE, p-o. Hawkins co. Ten. by p-r. 280 ms. a little N. of E. Nashville.

CLINGAN'S p-o. Chester co. [illegible]2 ms. from Phil.

Clinton, p-t. Kennebeck co. Me. 24 ms. n. Augusta. Pop. 1830, 2,130.

Clinton, co. N. Y. on lake Champlain, bounded n. by Lower Canada, lat. 45°, e. by lake Champlain, s. by Essex co. and w. by Franklin co. Greatest length n. and s. 40½, breadth 31. West part mountainous, well timbered, supplied with mill streams, iron ore, exceeded in richness by none in the world. The lake shore 8 ms. in width, moderately uneven, or quite level, very amply repays the labors of the husbandman. Rivers Saranac, Sable, &c. Capital, Plattsburg. Distilleries 4. Pop. 1830, 19,344.

Clinton, p-t. Duchess co. N. Y. Abounds in slate equal to any in the U. S. The quarries employ 300 hands. Watered by Wappingers creek. Pop. 1820, 12,070, 1830, 19,344.

Clinton, p-v. Oneida co. N. Y. on the Oriskany creek, 9 ms. w.s.w. Utica. The proposed Chenango canal runs through this place. An Universalist seminary is building here, 90 feet long. On a high hill 1 m. w. of it, is Hamilton college, incorporated 1812. In 1825 one 4 story, and one 3 story building was erected. It has 4 professors, 2 tutors, and a college and student's library of 3000 volumes each. Undergraduates in 1831-2, 77. Commencement 4th Wednesday in August.

Clinton, p-v. Hunterdon co. N. J. on the s. branch of Raritan r. 30 ms. w.n.w. New Brunswick, formerly Hunt's Mills.

Clinton, p-v. Alleghany co. Pa. 23 ms. from Pittsburg, and by p-r. 224 ms. w. Harrisburg.

Clinton, p-v. and st. jus. Sampson co. N. C. situated on a branch of Black r. 72 ms. s. s. e. Raleigh, and 18 nearly due e. Fayetteville. n. lat. 35°, w. long. 1° 18′.

Clinton, p-v. and st. jus. Jones co. Geo. by p-r. 665 ms. s. w. W. C. and 23 w. Milledgeville. n. Lat. 33° 01′, and long. 6° 40′ w. W. C.

Clinton, p-v. Greene co. Ala. by p-r. 25 ms. s. Tuscaloosa.

Clinton, p-v. Hinds co. Miss. about 80 ms. n. e. Natchez.

Clinton, p-v. parish of East Feliciana, La. about 50 ms. n. e. St. Francisville, and by p-r. 158 ms. n. w. New Orleans.

Clinton, p-v. and st. jus. Hickman co. Ky. by p-r. 847 ms. s. w. by w. ½ w. W. C. and 308 ms. s. w. by w. Frankfort. Pop. 1830, 82.

Clinton, p-v. and st. jus. Anderson co. Ten. by p-r. 534 ms. s. w. by w. W. C. and 195 almost due e. Nashville. It is situated on the right side of Clinch r. Lat. 36° 06′, long. 7° 8′ w. W. C.

Clinton, co. of O. bounded s. e. by Highland, s. w. by Browne, w. by Warren, n. by Green, and n. e. by Fayette. Length 22, mean breadth 18, and area 396 sq. ms. Extending in lat. from 39° 13′ to 39° 33′, and in long. from. 6° 31′ to 6° 57′ w. W. C. This co. is a real table land, from which creeks flow literally in every direction. On its surface are the sources of Paint creek branch of Sciota, and of East Fork, Todd's Fork, and other branches of Little Miami. The soil is generally productive. Chief t. Wilmington. Pop. 1820, 8,085, 1830, 11,436.

Clinton, p-v. in the n. w. part of Stark co. O. by p-r. 121 ms. n. e. by e. Columbus.

Clinton, co. of Ind. bounded by Boone s., Tippecanoe w., Carroll n. w., the Miamis n. e., and Hamilton co. s. e. Length from e. to w. 24 ms. breadth 15, and area 360 sq. ms. Extending in lat. from 40° 14′ to 40° 28′, and in long. from 9° 12′ to 9° 40′ w. W. C. The slope of this co. is nearly due w. and drained by the eastern branches of Wild Cat r. towards the more considerable stream of the Wabash. Chief t. Frankfort. Pop. 1830, 1,423. The st. jus. of this co. is about 45 ms. n. n. w. Indianopolis.

Clinton, p-v. Vermillion co. Ind. by p-r. 87 ms. w. Indianopolis.

Clinton, co. of Il. bounded by Washington s., St. Clair s. w., Madison n. w., Bond n., Fayette n. e., and Marion e. Length from e. to w. 30 ms., mean breadth 15, and area 450 sq. ms. Extending in lat. from 38° 25′ to 38° 45′ n., and in long. from 12° 10′ to 12° 42′ w. W. C. This co. is traversed from its nthrn. border in a direction of s. s. w. by Kaskaskias r. and by Shoal creek and other of its branches. The chief t. Carlyle, stands on the Kaskaskias, and on the road from Vincennes to St. Louis. Pop. 1830, 2,330.

Clintonville, p-v. Green Brier co. Va. by p-r. 231 ms. a little n. of w. Richmond.

Clintonville, p-v. Bourbon co. Ky. by p-r. 52 ms. estrd. Frankfort.

Clio, p-v. in the s. part of Adams co. Il. by p-r. 178 ms. n. w. by w. Vandalia.

Clockville, p-v. Madison co. N. Y.

Cloutiersville, p-v. in the s. e. part of the parish of Natchitoches, La. about 25 ms. s. e. from the village of Natchitoches.

Clover Bottom, p-o. Iredell co. N. C. by p-r. 156 ms. w. Raleigh.

Clover Creek, p-o. Madison co. Ten. by p-r. 159 s. w. by w. Nashville.

Clover Dale, p-o. Botetourt co. Va. by p-r. 160 ms. w. Richmond.

Clover Garden, p-o. Orange co. N. C. by p-r. 48 ms. n. w. Raleigh.

Clover Hill, p-o. Blount co. Ten. by p-r. 162 ms. s. e. by e. Nashville.

Cloverport, p-v. on O. r. n. w. angle of Breckenridge co. Ky. 11 ms. n. w. by w. Hardensburg, the st. jus.

Clyde, r. Vt. empties into Memphremagog lake in Derby.

Clyde, p-v. Wayne co. N. Y. on Erie canal, 4 ms. n. Waterloo.

Clymer, t. Chatauque co. N. Y. s. w. corner, having Pa. boundary on the w. and s. 2 schools, 3 months in 12. Pop. 1830, 567.

Coal River, a r. of western Va. rises in Logan co. by two branches, called relatively Great and Little Coal rivers. The former rises in the western spurs of the Appalachian ridges, flows N. W. out of Logan into Kanawhay co., receives Little Coal r. from the S. W. and finally falls into the right side of Great Kanawhay, after a comparative course of 70 ms. The valley of Coal r. lies between those of Great Kanawhay and Guyandot r.

COAL RIVER MARSHES, p-o. Logan co. Va. by p-r. 277 ms. N. of W. Richmond.

COALSMOUTH, p-v. Kanawhay co. Va. situated on Kanawhay r. at the mouth of Coal r. by p-r. 12 ms. below and wstrd. from Charleston, the st. jus. for the co.

COAT'S TAVERN and p-o. York dist. S. C. by p-r. 97 ms. nthrd. Columbia.

COATESVILLE, small p-v. on the W. bank of Brandywine creek, Chester co. Pa. 39 ms. W. Phil.

COBBS, p-o. McMinn co. Ten. by p-r. 181 ms. S. E. by E. Nashville.

COBLESKILL or COBELSKILL, p-t. Schoharie co. N. Y. 38 ms. W. Albany; is watered by the Cobuskill, a tolerable mill stream, having a fine alluvion margin. Population of German origin. Pop. 1830, 2,988.

COBURN'S STORE and p-o. Mecklenburg co. N. C. by p-r. 167 ms. S. W. by W. Raleigh.

COCHECO, or Dover r. N. H. a branch of the Piscataqua.

COCHRANSVILLE, p-v. in the W. part of Chester co. Pa. 45 ms. W. Phil.

COCHRANSVILLE, p-v. in the W. part of Abbeville dist. S. C. by p-r. 102 ms. W. Columbus.

COCHRANTON, p-v. in the N. part of Marion co. O. by p-r. 56 ms. N. Columbus.

COCKE, co. of E. Ten. bounded S. E. by the main chain of the Alleghany mtns., here called the Smoky mtns. separating it from Haywood and Buncombe cos. of N. C., S. W. by Sevier, W. and N. W. by Jefferson, and N. E. by Greene. Length from S. to N. 22, mean breadth 17, and area 374 sq. ms. Extending in lat. 35° 40′ to 36° 05′, and in long. from 5° 45′ to 6° 13′ W. W. C. The surface is broken and hilly, being part of a mountain valley, sloping to the N. W. and drained in that direction by French Broad and Big Pigeon r. Chief t. Newport. Pop. 1820, 4,892, 1830, 6,017.

CODORUS, large creek, or rather small r. having its remote source in the N. E. part of Frederick co. Md. flows nthrdly. over York co. Pa. and falls into Susquehannah r. at the village of New Holland after a comparative course of 30 ms.

CODORUS, tsp. and p-o. S. W. part of York co. Pa. by p-r. 32 ms. S. Harrisburg. Pop. of the tsp. 1830, 2,429.

COEYMANS, p-t. Albany co. N. Y. on the Hudson 11 ms. S. Albany, has plenty of limestone, some shell marle, 2 Dutch churches, 1 Methodist. There is a sloop-landing at the mouth of Coeyman's creek. Pop. 1830, 2,723.

COHASSET, p-t. Norfolk co. Mass. 20 ms. S. E. Boston. Cohasset rocks, 3 ms. from its shore, have been fatal to many vessels. Pop. 1830, 1,233.

COFFEE CREEK, p-o. Warren co. Pa. by p-r. 270 ms. N. W. Harrisburg.

COFFEE RUN, p-o. Huntingdon, Pa. 10 ms. S. E. from the borough of Huntingdon, and by p-r. 82 ms. wstrd. Harrisburg.

COFFEEVILLE, p-v. on the left bank of the Tombigbee r. Clarke co. Ala., 16 ms. N. W. Clarksville, the co. town, and by p-r. 120 ms. a little W. of S. Tuscaloosa.

COFFYVILLE, p-v. Clark co. Ky. 35 ms. S. E. by E. Frankfort.

COKALAHISKIT, r. a branch of Clark's r. rises in the Chippewan mtns. opposite to the sources of Dearborne branch of Missouri, and flowing thence to the N. W. falls into Clark's r. after a comparative course of 150 ms. The mouth of this r. is according to Tanner at N. lat. 46° 44′, and long. 36° W. W. C.

COLCHESTER, t. Chittenden co. Vt.; W. boundary is lake Champlain, S. Onion r. separating it from Burlington; timbered by beech, maple, ash, oak, chestnut, walnut, white and pitch pine. Has much pine plain, good mill streams, 1 distillery, 3 churches, and 4 school houses. Pop. 1830, 1,489.

COLCHESTER, p-t. New London co. 23 ms. S. E. Hartford, 15 W. Norwich, borders on 4 cos. 6 ms. by 9, 50 sq. ms. is uneven; primitive good grazing land, watered by Salmon r. &c. and has factories. Bacon academy, founded 1801, has a fund of $30,000. Pop. 1830, 2,068.

COLCHESTER, p-t. Del. co. N. Y. 21 ms. S. Delhi, 91 S. W. Albany, is crossed by E. branch of Del. r. and Beaver creek. Much lumber is rafted for Phil. Pop. 1830, 1,424.

COLDEN, p-t. Erie co. N. Y. Pop. 1830, 464.

COLDENHAM, p-v. Montgomery, Orange co. N. Y. 13 ms. from Goshen.

COLD SPRING, v. Cattaraugus co. N. Y. 14 ms. S. W. Ellicottville.

COLD SPRING, landing, Putnam co. N. Y. opposite West Point. Has the great iron foundry of the U. S.

COLD STREAM MILLS, p-o. Hampshire co. Va. by p-r. 104 ms. N. W. by W. from W. C.

COLD SPRING, p-v. Wilkinson co. Miss. by p-r. about 30 ms. S. from Natchez.

COLD SPRING, p-v. Hardiman co. Ten. about 150 ms. S. W. by W. Nashville.

COLD WATER, p-o. St. Joseph's co. Mich. about 150 ms. a little S. of W. Detroit.

COLE, co. of Il. bounded by Jasper S. E. Effingham, S. W., Shelby W., Macon N. west Vermillion northeast, and Edgar and Clark E. Length from S. to N. 50 ms.; mean breadth 24, and area 1200 sq. ms. Extending in lat. 39° 10′ to 39° 53′, and in long. 11° 02′ to 11° 30′ W. from W. C. This co. contains a table land from which flow the Kaskaskias to S. S. W. and the Embarras to the S. Both these rivers have their sources in the country westward from, and yet attached to, Vermillion co. The Kaskaskias enters and traverses the N. W. angle of Cole, retiring from it to the s. w. The Embarras traverses the co. in its greatest length by a general southern course. The extreme source of Little Wabash is also in the s. w. angle of this co. Chief town, Charleston. Not included in the census of 1830.

COLE, co. of Mo. bounded w. and n. w. by Cooper, n. by the Missouri river, separating it from Boone, n. e. by the river Missouri separating it from Callaway co., e. by the Osage, separating it from Gasconnade co., and s. e. and s. by Osage r. separating it from a country not yet appropriated to co. division. As laid down by Tanner, Cole co. is in form of a triangle, longest side 50 ms. along Cooper, and from the Osage to Missouri river; mean breadth 17, and area 850 sq. ms. Extending in lat. from 38° 09′ to 38° 51′, and in long. from 15° to 15° 34′ w. from W. C. The general slope of this co. is to the eastward, though the two bounding rivers converge the Missouri to the s. e., and the Osage to the n. e. Moreau creek flowing from the westward and entering Missouri one or two ms. above the influx of Osage, divides Cole into two not very unequal sections. Chief town, Jefferson, the capital also of the state. Pop. 1830, 3,023.

COLEBROOK, p-t. Coos co. N. H. 40 ms. n. Lancaster, 25,000 acres, has rich meadows on Conn. r. Mohawk r. and Beaver brook. Incorporated 1790. Pop. 1830, 532.

COLEBROOK, p-t. Litchfield co. Conn. 31 ms. n. w. Hartford, 18 n. e. Litchfield, on high ground, 5 ms. by 6, 30 sq. ms. with granite hills, has a hard soil, pretty good for grazing, with many mill seats on the main branch of Farmington and Sandy rivers. Pop. 1830, 1,332.

COLEBROOK, tsp. and p-o. Ashtabula co. O. by p-r. 191 ms. n. e. Columbus. Pop. of the tsp. 1830, 92.

COLEBROOKDALE, p-o. Berks co. Pa. 11 ms. e. from Reading and 63 ms. in a similar direction from Harrisburg. Colebrookdale tsp. in 1820. contained a pop. of 1,046, in 1830, 1,229.

COLEMAN'S CROSS ROADS, and p-o. Edgefield dist. S. C. by p-r. 50 ms. wstward from Columbia.

COLERAIN, p-t. Franklin co. Mass. 105 ms. n. w. Boston, s. N. H., has two forks of Deerfield river, and was settled about 1736, by a colony from Ireland. Pop. 1830, 1,877.

COLERAIN, tsp. and p-o. Lancaster co. Pa. The p-o. is by p-r. 52 ms. from Harrisburg. In 1820, the tsp. contained a pop. of 1,066, in 1830, 1,194.

COLERAIN, p-v. Bertie co. N. C. situated on the western side of Chowan r. 60 ms. s. s. w. from Norfolk, in Va. by p-r. 174 ms. n. e. by e. from Raleigh.

COLERAIN FORGE, and p-o. on Spruce creek, in the northern part of Huntingdon co. of Pa. 15 ms. n. from the borough of Huntingdon, and by p-r. 106 ms. n. w. by w. from Harrisburg.

COLESVILLE, p-t. Broom co. N. Y. 15 ms. e. Chenango Point, 125 s. w. Albany, has good grazing, though hilly lands; crossed by Susquehannah river. Pop. 1830, 2,387.

COLESVILLE, p-o. Montgomery co. Md. 15 ms. n. from W. C.

COLESVILLE, p-v. in the southwestern angle of Chesterfield co. Virginia, 31 miles s. w. from Richmond.

COLLEGE CORNERS, and p-o. Prebble co. O. situated in the n. w. part of the co. about 60 ms. a little w. of n. Cincinnati.

COLLEGE HILL, p-o. at Columbia college, dist. of Columbia, 2 ms. n. from the general p-o. W. C.

COLLETON, dist. S. C. bounded s. w. by Cambahee r. which separates it from Beaufort, n. w. by Barnwell and Orangeburgh, n. and e. by Charleston, and s. e. by Atlantic ocean. Length from s. e. to n. w. 37 ms. mean breadth 37, and area 2,100 sq. miles. Extending in lat. from 32° 28′ n. to 33° 18′ n. and in longitude from 3° 10′ to 4° 8′ west from W. C.

Colleton is situated almost entirely within the Atlantic tide plain, and is in great part a dead level. Besides the Cambahee which bounds it on the s. w., this district is traversed by the Edisto and gives source and course to the Ashepoo river. The latter uniting with Cambahee, contributes to form St. Helena sound. The Edisto, before reaching the ocean, divides into two branches, encompassing an island which bears the name of Edisto island. This island, chequered by points and traversed by numerous creeks, is mostly subject to daily submersions by ocean tides, where the land of either the island or parts more inland, have been made arable; the soil is productive. Chief staples, rice and cotton. Chief town, Watersboro. Pop. 1820, 26,373, in 1830, 27,256.

COLLETON, s. e. parish of Charleston dist. S. C. This parish is composed of a congeries of islands, of which the principal are Wadmelaw, Jones, Seabrooks, and Kiawaw. These low islands are enclosed on the s. w. by N. Edisto, s. w. by Stono, n. and e. by Stono, and s. e. by the Atlantic ocean. (*See Charleston dist.*)

COLLIE'S MILL, and p-o. in the w. part of Caldwell co. Ky. 12 ms. w. Eddyville.

COLLINS, t. Erie co. N. Y. 32 ms. s. Buffalo, is uneven, with a moist loam, favorable to the dairy, bearing much maple, beech, linden, &c. and watered by Cattaraugus creek and two creeks of lake Erie. Pop. 1830, 2,120.

COLLINSVILLE, p-v. a manufacturing village lying on both sides of Farmington river, at the s. part of Canton, Conn. containing about 800 inhabitants, of whom about 300 men are employed in the edge tool manufactory of Collins' & Co.; established here in 1826. The principle article of manufacture at present is axes, of which about 200,000, of superior quality, are manufactured per annum. The village consists of about 20 buildings, of stone and wood, devoted to the business of the manufactory; very neat and comfortable dwellings, (separate tenements,) for about 150 families; a place of worship, lyceum, and library, for the workmen, and schools, for the children; of the latter, none are employed in the manufactory. This village is entitled to particular notice, from the fact that it has been built up entirely by the enterprise of the firm we have mentioned, to whom it exclusively belongs.

Collins' cross roads, and p-o. in the s. part of Colleton dist. S. C. by p-r. 137 ms. a little e. of s. Columbia.

Collins' Settlement, and p-o. in the w. part of Lewis co. Va. by p-r. 286 ms. n. w. Richmond.

Collinsville, p-o. in the w. part of Huntingdon co. Pa. by p-r. 126 ms. n. of w. Harrisburg.

Collinsville, p-o. in the w. part of Madison co. Il. by p-r. 67 ms. from Vandalia w.

Colon, p-v. Callaway co. Ky. about 260 ms. by p-r. s. w. by w. Frankfort.

Colosse, p-v. Mexico, Oswego co. N. Y.

Colts Neck, p-v. Monmouth co. N. J. on a branch of Shrewsbury r. 5 ms. n. e. Freehold.

Columbia, p-v. Washington co. Me. 18 ms. w. Machias, 128 from Augusta, crossed by Pleasant r. Pop. 1830, 663.

Columbia, p-t. Coos co. N. H. e. Conn. r. 30 ms. n. Lancaster, with Stratford mtns. s. from which flow mill streams, and several ponds, near one of which, great quantities of shells are found, which make lime. Few evergreens grow here. Pop. 1830, 442.

Columbia, p-t. Tolland co. Conn. 22 ms. e. Hartford, 4 ms. by 5, 20 sq. ms. is hilly, with a hard, prime soil, favorable to grazing, and bearing oak, chestnut, &c. Pop. 1830, 962.

Columbia, co. N. Y. e. Hudson r. 30 ms. s. Albany, 130 n. N. Y. city, bounded by Renselaer co. n., Mass. e., Duchess co. s., Green and Ulster cos. w. 18 ms. by 30, 594 sq. ms. is one of the richest towns in the state. It is irregular but not mountainous, with Shistic hills e.; slate abounding, and some limestone. South the soil is warm gravel. Abram's and Lebanon or Claverack creeks flow into Hudson r. The Warm spring at Lebanon is one of the principle watering places of the United States. The manufactures are important.

Livingston's Manor, or Lordship, is in this co. It consisted of several grants made in 1684, '85 and and '86 to Robert Livingston, by the British government and extended 10½ ms. on Hudson r. and e. about 20½. It is owned by his heirs, (except a part forming Germantown,) and includes Clermont, Livingston, Taghkanick and Ancram. At Ancram are celebrated iron works. Population 1830, 38,325.

Columbia, p-t. Herkimer co. N. Y. 10 ms. s. Herkimer. Pop. 1830, 2,181.

Columbia, v. Warren co. N. J. on the Delaware below the Water gap, has glass manufactories, &c. n. w. Belvidere.

Columbia, District of, a territory of 100 sq. ms. ceded in 1790 by Va. and Md. to the U. S. and became in 1800, the seat of government of the U. S. It is laid out in a square of 10 ms. each way, the sides lying in a direction of s. e. and n. w. or s. w. and n. e. Extending in lat. from 38° 46½ to 38° 58′ nearly. The capitol stands, as determined by astronomical observation under an act of Congress, 76° 55′ 30″ w. from the royal observatory at Greenwich.

Of the 100 sq. ms. included in the dist. 36 were taken from Va. and included in the co. of Alexandria, and lies s. of the Potomac. A strip 8 ms. long by about 1½ wide, lying e. from the east branch, and n. from the main bed of the Potomac, is included in Washington co. which contains the cities of Washington and Georgetown. The surface of the dist. is gently undulating, affording fine seats for the cities, within its limits, but the soil in its natural state is sterile, with but little exception. In a commercial view, the situation of the dist. is favorable. Ships of any draft are navigated to Alexandria, and those of large size to the navy yard on the east branch. The Chesapeake and Delaware canal, when completed, will give incalculable advantages to Washington. The existing roads from it in every direction are far from being in a state suited to their importance.

The civil government of the District of Columbia is under the immediate authority of the general government, and the municipal power is exercised by a Mayor and Corporation.

In 1820, the population of the dist. was 33,039, viz. whites, in W. C. 9,607; Georgetown, 4,940; Alexandria, 5,615, and in the two cos. independent of the cities, there were in Washington co. 1,512; Alexandria, 941. Total, whites, 22,615. Colored pop. free, 4,048; slaves, 6,376. In 1830, the population was as follows:—

Washington city,

	Males,	Females		
Whites,	6,581	6,798	13,379	
Colo'd, free,	1,342	1,787	3,129	
Slaves,	1,010	1,309	2,319	18,827

Washington co. without the city,

	Males	Females		
Whites,	1,015	712	1,727	
Colo'd, free,	163	104	267	
Slaves,	606	394	1,000	2,994

Alexandria city,

	Males	Females		
Whites,	2,712	2,969	5,681	
Colo'd, free,	565	816	1,381	
Slaves,	462	739	1,201	8,263

Alexandria co. without the city,

	Males,	Females,		
Whites,	401	401	802	
Colo'd, free	76	101	177	
Slaves,	179	185	364	1,345

Georgetown,

	Males,	Females,		
White,	3,052	3,006	6,058	
Colo'd. free,	500	709	1,209	
Slaves,	521	653	1,174	8,441

Total population of the Dist. 1830, 39,868

For more particular statistical and other details, (*see articles Washington city, Alexandria and Georgetown.*)

COLUMBIA, co. of Pa. bounded by Northumberland W., Lycoming N. W., Luzerne N. E., Susquehannah S. E., and Northumberland S. and S. W. Length from S. to N. 35, and mean breadth 20, and area 700 sq. ms. Extending in lat. from 40° 56' to 41° 16', and in long. from 0° 14' to 0° 50' W. from W. C. The east branch of Susquehannah river enters the eastern border of Columbia, and flows over it in a southwestern direction, leaving about one third of the co. to the S. E. The northern section is nearly commensurate with the valley of Fishing creek, which, rising in Bald mountain, flows southwardly, and falls into Susquehannah at Bloomsburg. The face of the co. is broken by numerous lateral ridges of mtns. extending in a direction of N. E. and S. W. The river soil is highly productive in grain and pasturage. Chief town, Danville. Pop. 17,621, and in 1830, 20,049.

COLUMBIA, p-v. and tsp. Lancaster co. Pa. situated on the left bank of Susquehannah r. 10 ms. W. from Lancaster, and 30 S. E. from Harrisburgh. At this place a fine wooden bridge, resting on stone piers, crosses the Susquehannah and connects the village of Columbia with Wrightville. Population 1830, 2,047.

COLUMBIA, p-v. and st. jus. Fluvanna co. Va. situated on the right bank of Fluvanna r. and near the centre of the co. lat. 37° 46', and in long. 1° 28' W. from W. C. and 52 ms. N. W. by W. from Richmond.

COLUMBIA, p-v. and st. jus. Tyrell co. N. C. situated on a small creek which enters the S. side of Albemarle sound. N. lat. 35° 53', and long. 0° 45' E. from W. C. by p-r. 187 ms. E. Raleigh, and 332 ms. E. of S. W. C.

COLUMBIA, p-t. and st. jus. Richland dist. and of the government of S. C. by p-r. 500 ms. a little E. of S. W. from W. C. The real bearing between the two places, calculated on Mercator's principles, is 33° 20' deviation from the meridians, and the distance 406 ms. Columbia, is 110 ms. N. W. from Charleston, and almost exactly on the intersection of lat. 34° and long. 4° W. from W. C. and directly opposite the union of Saluda and Broad rivers. This town is laid out on a regular plan, with streets at right angles to each other, and 100 feet wide. It contains South Carolina college, a state house 170 by 60 feet, 5 or 6 churches, with other public buildings. The college edifices are spacious and splendid, 3 stories high, but unusually narrow for the length, being 210 by 25 feet. Upwards of $200,000 has been expended by the state on this institution, which also receives an annual grant of $15,000. The college possesses a respectable library and philosophical apparatus. Pop. 1832, 3,500.

COLUMBIA, co. of Geo. bounded by Richmond S. E., Warren S. W., Wilkes N. W., Lincoln N., and Savannah r. separating it from Edgefield dist. S. C. N. E. Length 30, mean breadth about 20, and area 600 sq. ms. Surface waving, and soil productive. It extends in lat. from 33° 20' to 33° 42', and in long. from 5° 1' to 5° 40' W. W. C. Chief t. Applington. Pop. 1820, 12,695, 1830, 12,606.

COLUMBIA, p-v. and st. jus. Henry co. Ala. by p-r. 872 ms. S. W. W. C. and 260 S. E. Tuscaloosa. It is situated on one of the higher branches of Choctawhatchie r. Lat. 31° 22', long. 8° 32' W. W. C.

COLUMBIA, p-v. and st. jus. Marion co. Miss. by p-r. 1097 ms. S. W. W. C. 110 ms. S. E. by E. Natchez, and 100 ms. N. New Orleans. It is situated on the left or eastern bank of Pearl r. at lat. 31° 17', and long. 12° 50' W. W. C.

COLUMBIA, p-v. and st. jus. Maury co. Ten. by p-r. 733 ms. S. W. by W. W. C. and 42 ms. S. S. W. Nashville. It is situated on the left bank of Duck r. at lat. 35° 36', long. from W. C. 10° 01' W.

COLUMBIA, C. H. Columbia co. Geo. (See Applington.)

COLUMBIA, p-v. Monroe co. Il. by p-r. 90 ms. S. W. Vandalia.

COLUMBIA, p-v. and st. jus. Boone co. Mo. by p-r. 992 ms. wstrd. W. C. 57 N. Jefferson, and by the common road 130 ms. N. W. by W. St. Louis.

COLUMBIA RIVER. (See Oregon.)

COLUMBIA CROSS ROADS and p-o. in the N. part Bradford co. Pa. by p-r. 148 ms. E. of N. Harrisburg.

COLUMBIANA, p-v. Shelby co. Ala. by p-r. 60 ms. estrd. Tuscaloosa.

COLUMBIANA, co. O. bounded S. by Jefferson, S. W. by Harrison, W. by Stark, N. W. by Portage, N. by Trumbull, E. by Beaver co. Pa. and S. E. by the O. r. separating it from Brooke co. Va. The length from S. to N. a little exceeds the breadth, but the whole co. approaches to near a square of 30 ms. each side, or 900 sq. ms. Extending in lat. from 40° 32' to 41° N., and in long. from 3° 30' to 4° 5' W. W. C. The central part of Columbiana is a table land, from which issue wstrd. Sandy creek, branch of Tuscarawas r.; from the nthrn. the sources of Mahoning, branch of Big Beaver r.; and from the E. and S. E. sections the sources of Little Beaver. Chief t. New Lisbon. Pop. 1820, 22,033, and in 1830, 35,592.

COLUMBIANA, p-v. N. W. part of Columbia co. O. 160 ms. N. E. by E. Columbus. Pop. of the v. 1830, 172.

COLUMBIAN GROVE, and p-o. Lunenburg co. Va. by p-r. 102 ms. S. W. Richmond.

COLUMBIAVILLE, v. of Hudson and Kinderhook, Columbia co. N. Y. on Kinderhook creek, is a large manufacturing village, near Hudson r. accessible in boats, and on the Albany and N. Y. roads. It has 11 cotton factories, of above 2000 spindles each, and employs above 2000 persons. 350 calico printers are employed at Messrs. Marshalls' factory, where 4000 pieces of 30 yards are made weekly. This is connected with cotton spinning, and weaving; and the capital invested amounts to $450,000.

COLUMBUS, p-v. Luzerne co. Pa. by p-r. 92 ms. N. E. Harrisburg.

COLUMBUS, one of the two most southern cos. of N. C. bounded N. W. by Lumber r. separating it from Robeson, N. and N. E. by Bladen, E. and S. E. by Alacamaw r. separating it from Brunswick, and S. W. by Horry dist. S. C. Length 35 ms., mean breadth 15, and area 525 sq. ms. Extending in lat. from 33° 58' to 34° 30, and in long. from 1° 40' to 2° 11' w. W. C. Surface flat, and in part marshy. Chief t. Whitesville. Pop. 1820, 3,912, 1830, 4,141.

COLUMBUS, p-v. and st. jus. Muscogee co. Geo. on the left bank of the Chattahooche r. 123 ms. S. W. by W. Milledgeville. Lat. 32° 36', long. 8° 10' w. W. C.

COLUMBUS, p-v. Lowndes co. Miss. on the left bank Tombigbee r. at the point where the road to New Orleans separates from that to Natchez, 236 ms. N. E. by E. from the latter, and 276 N. N. E. from the former.

COLUMBUS, p-v. McMinn co. Ten. by p-r. 153 ms. S. E. by E. Nashville.

COLUMBUS, p-v. and st. jus. Hickman co. Ky. situated on the left bank of the Miss. r. above the upper end of Wolf Island, about 25 ms. below the mouth of O. and by p-r. 277 ms. S. W. by W. Frankfort. Lat. 36° 48', long. 12° 12' w. W. C.

COLUMBUS, p-t. and st. jus. for Franklin co. and st. of the state government of O. Lat. 39° 57', long. 6° w. and distant 330 (by p-r. 396) ms. from W. C. Flint gives its relative position 551 ms. from N. Y. 477 from Phil. 755 from Boston, 429 from Baltimore, 991 from New Orleans, 377 from Nashville, and 112 from Cincinnati. It is 216 ms. almost exactly due S. from Detroit. It is seated on the eastern or left bank of Sciota r. immediately below the influx of Whetstone r. the site being a gentle acclivity from the stream. In the spring of 1812, the ground on which this now flourishing town stands was a wilderness. By the census of 1830, the pop. was then 2,435. It contains a state house, 75 by 50 feet, with a cupola 106 feet high, a building for public offices 100 by 25 feet; the necessary county buildings, penitentiary, numerous and respectable private schools, and a classical academy, four printing offices, market-house, and an asylum for the deaf and dumb. A canal of 11 ms. connects this place with the Ohio and Erie canal.

There are three or four places of public worship, and from 340 to 350 dwelling houses. The relative position of this town, being very near the physical centre of the state, almost ensures its permanence as the seat of state government, and having a navigable canal to unite it with the O. r. and lake Erie, gives stability to commercial prosperity.

COLUMBUS, p-v. and st. jus. Bartholemew co. Ind. by p-r. 598 ms. a little N. of W. W. C. and 41 ms. S. S. E. Indianopolis. It is situated on Driftwood Fork of White r. 84 ms. a little N. of W. Cincinnati, O. at N. lat. 39° 14', long. 8° 55' w. W. C.

COLVIN'S TAVERN and p-o. Culpepper co. Va. by p-r. 87 ms. S. W. W. C.

COMAN'S WELL and p-o. Sussex co. Va. by p-r. 68 ms. S. S. E. Richmond.

COMBAHEE, r. of S. C. rising between S. Edisto, and Savannah rs. and flowing thence S. E. 50 ms. receiving from the N. a stream of almost equal length, the Salkehatchie. The united waters continue to flow S. E. 30 ms. and fall into the head of St. Helena Sound. The Combahee in the 50 lower miles of its course separates Colleton and Beaufort districts.

COMFORT, p-v. Jones co. N. C. by p-r. 152 ms. S. E. Raleigh.

COMITE, small r. rising near the line between La. and Miss. enters the former state, and traversing the parish of East Feliciana, falls into Amite r. 12 ms. estrd. Baton Rouge.

COMMERCE, p-v. E. part of Wilson co. Ten. 43 ms. E. Nashville.

COMMUNIPA, v. Bergen co. N. J. W. side N. Y. bay, opposite S. end Manhattan Island, 2 ms. S. W. Jersey City, on low lands; sends oysters, &c. to N. Y. market.

CONCORD, p-t. Somerset co. Me. 55 ms. from Augusta, W. side Kennebec r. Pop. 1830, 391.

CONCORD, p-t. Merrimack co. N.H. cap. of the state, 45 ms. W. N. W. Portsmouth, 62 ms. W. N. W. Boston, 505 Washington. Long. 71° 30' W., lat. 43° 12' N. on both sides of Merrimack r. on which are rich meadows; 40,918 acres, of which 1800 are water. It has 5 ponds, is crossed by Contoocook r. and has Sewalls, Turkey and Garvins falls on Merrimack r. with locks for navigation on the last. The river boating company have stores on the bank, and boat navigation extends through Middlesex canal to Boston. Pine grows on the low grounds. The upland is very good. First settled 1724, and suffered from the Indians 1744. The village of Concord is handsomely built, on 2 principal streets W. Merrimack r. and has the state house and state prison, of granite. A banking capital of $200,000. A saving's bank, large hotels, churches, newspaper offices, &c. Pop. 1830, 3,727.

CONCORD, p-t. Grafton co. N. H. 20 ms. N. E. Haverhill, 28 ms. N. N. E. Lancaster. Contains 29,130 acres, is crossed by Ammonoosuc r. and other streams, with ponds, rich meadows, good uplands, and poor plains. Maple sugar is made, and iron ore, used in the Franconia furnaces, is dug in the E. part of this town. Pop. not in the census.

CONCORD, p-t. Essex co. Vt. N. Conn. r. 38

ms. E. by N. Montpelier. First settled 1788; has an academy, incorporated 1823, is partly watered by Moose r., uneven, with good grazing, and some tillage, 9 school districts. Pop. 1830, 1,031.

CONCORD, r. Middlesex co. Mass. runs N. and joins Merrimack r. at Chelmsford, after serving as the only feeder to the Middlesex canal.

CONCORD, p-t. Middlesex co. Mass. 18 ms. N. Boston, crossed by Concord r.; incorporated 1635; has some good meadows, light soil on the plains, and gravelly loam on the hills. In the battle of Concord, 19th April, 1775, the militia drove back the British light infantry, under colonel Smith and major Pitcairne, who had come from Boston to destroy military stores deposited here. The action was at the bridge, and with that at Lexington, on the same day, caused the first bloodshed in the revolutionary war. The provincial congress met here 1774. Pop. 1830, 2,017.

CONCORD, t. Erie co. N. Y. 32 ms. S. S. E. Buffalo, N. Cattaraugus cr. whose branches water it, with some of Cazenovia creek, &c. has a moist loam, good for grazing and bearing beech, maple, bass, &c. Pop. 1830, 1,924.

CONCORD, p-t. Saratoga co. N. Y. 30 ms. N. W. Ballston Spa, is crossed by Sacandaga creek and Kayderosseras mtn. Pop. 1830, 758.

CONCORD, meeting house, and p-o. Del. co. Pa. 10 ms. N. Wilmington.

CONCORD, small p-v. Franlin co. Pa. situated on the head of Tüscarora creek, near the extreme nthrn. angle of the co. about 45 ms. nearly due W. Harrisburg.

CONCORD, small p-v. on the head of Broad creek, branch of Nantikoke r. Sussex co. Del. 40 ms. S. Dover.

CONCORD, p-v. Campbell co. Va. 118 ms. W. Richmond.

CONCORD, p-v. and st. jus. Cabarras co. N. C. situated on a branch of Rocky r. by p-r. 140 ms. S. of W. Raleigh. Lat 35° 26′, long. 3° 32′ W. W. C.

CONCORD, p-v. Decatur co. Geo. by p-r. 186 ms. S. S. W. Milledgeville.

CONCORD, p-v. and tsp. in the N. E. part of Geauga co. O. by p-r. 163 ms. N. E. Columbus. Pop. 1830, 979.

CONCORD, p-v. in the S. part White co. Il. 10 ms. from Carmi, the st. jus. for the co.

CONCORDIA, parish of La. bounded by Miss. r. E. and S., by Red r. S. W., and by Owachitta and Tensaw rs. W. Length 120, breadth unequal, but average about 10; area about 1200 sq. ms. Extending in lat. from 31° to about 32°, and long. from 14° to 14° 50′ W. W. C. It is a long level peninsula, falling by a very gentle slope from the Miss. towards Owachitta and Tensaw rs. It is much traversed by interlocking lakes and water courses, with an exuberantly fertile soil, but at least nine tenths liable to annual submersion. In its natural state a very dense forest covered the whole land surface. The arable part, as every where else in Louisiana, where annual floods prevail, is composed of narrow strips along the streams. Staple, cotton. Chief t. Concordia. Pop. 1820, 2,626, 1830, 4,662.

CONCORDIA, lake of La. in the parish of Concordia, evidently once a bend of the Miss. r. It is about 5 ms. long, curving to the wstrd. with a breadth of between ¼ and ½ a m. connected with the Miss. by an outlet which leaves that stream directly opposite Natchez.

CONCORDIA, p-v. and st. jus. parish of Concordia, La. situated on the right bank of the Miss. opposite Natchez.

CONCORDIA, p-v. in the W. part of Dark co. O. 109 ms. N. of W. Columbus.

CONECOCHEAGUE, r. of Pa. and Md. rises in the former by two branches, the western in the northern part of Franklin co. interlocking sources with those of Tuscarora creek; the eastern rises in Adam's co. but flowing wstrd. enters Cumberland, interlocking sources with those of the Monocacy, Conewago, and Conedogwinet, passing Chambersburg, the E. branch turns to S. S. W. and uniting below Greencastle with the wstrn. enters Washington in Md. and falls into the Potomac at Williamsport. The valley of Conecocheague is about 40 ms. in length, by a mean breadth of 15; area 600 sq. ms.; but it is important from the almost uniform fertility of soil. From Chambersburg to its mouth, this river, serving nearly as a line of separation, leaves the limestone E. and slate W. (See Kittatinny valley.)

CONECUH, r. of Ala. and Flor. rising by numerous branches in Pike, Butler, Conecuh, and Covington cos. of the former, flow generally to the S. W. unite in Conecuh co. where, turning to the sthrd. enters Florida, about 2 ms. within which it receives an inferior branch, the Escambia, but loses its name in that of an unimportant confluent; the extreme remote sources of the Conecuh rise above lat. 32°, and if we include Escambia, the valley reaches to 30° 25′, with a length of 140 ms. and mean breadth of 25; area 3500 sq. ms. The Conecuh is navigable at high water as high as Montezuma in Covington co. but in general the soil of the valley is sterile and wooded by pine timber.

CONECUH, co. of Ala. bounded by Baldwin co. W., Monroe N. W., Butler N., Covington E., and Escambia co. in Flor. S. Length 53 from S. to N., mean breadth 27; area 1531 sq. ms., in lat. from 31° to 31° 46′ N. and long. from 9° 51′ to 10° 30′ W. W. C. This co. is drained by various branches of Conecuh r. which join the main body of that stream near Fort Crawford. The soil is of middling quality. Chief t. Sparta. Staple, principally cotton. Pop. 1820, 5,713, 1830, 7,444.

CONEDOGWINET, r. of Pa. rising in the N. E. part of Franklin, and S. W. of Cumberland co. leaving the former and entering the latter, gradually curves from N. to N. E. and finally nearly E. passes within little more than a mile from Carlisle, finally falls into Susquehannah, about 2 ms. above Harrisburg, after a comparative course of 80 ms. The vallies of Conedogwinet and Conecocheague united, occupy the greater part of the important mountain valley between the Kittatinny and Blue Ridge, and between the Susquehannah and Potomac rs. The Conedogwinet, like the Conecocheague, very nearly separates the limestone and slate formations. The two streams seem to offer a tempting means of constructing a canal to unite the two fine rivers into which they are respectively discharged.

CONEMAUGH, r. of Pa. rises by numerous branches in the valley between the Alleghany mtn. and Laurel Hill, and in Somerset and Cambria cos. opposite the sources of the W. branch of Susquehannah, Juniata, and a branch of Potomac, and in the same valley interlocking sources with those of the Youghioghany to the S. and those of the W. branch of Susquehannah to the N. After a general course to the N. W. the different branches unite at the lower slope of the valley, and the united waters pierce the Laurel Hill, turn to a N. W. by W. course, traverse the valley between Laurel Hill and Chesnut Ridge, and piercing the latter chain, leave the mountains and enter on the great wstrn. hilly region. Continuing to N. W. by W. and receiving from the N. Cherry r. from Ind. co. and from the S. Loyalhanna, from Westmoreland co. fall into the Alleghany r. at Freeport, after a comparative course of 150 ms. very nearly of similar length with the Youghioghany; the Conemaugh valley is more extensive. That of Youghioghany embracing about 4000, and that of Conemaugh 6000 sq. ms. Independent of the mountain ridges, the elevation of the higher part of the Conemaugh valley is about 1,300 feet, but the fall of its plain so rapid, that from the summit of the Alleghany to Johnstown, where the two main lakes unite in Cambria co. in a direct distance of 50 ms. the descent is 1,137 feet. The Conemaugh r. has gained an importance much beyond its comparative size, as its immediate valley from Johnstown to the mouth, has become the route of the traverse section of the Pa. canal.

CONEMAUGH, late Johnstown, p-v. at the forks of Conemaugh r. Cambria co. Pa. by p-r. 138 ms. W. Harrisburg.

CONEQUENESSING, r. of Pa. composed of the Conequenessing and Slippery Rock crs. The inclined plain extending from the Alleghany r. above Pittsburg to the summit level between the vallies of Ohio and Lake Erie, has its slope of declination to the S., giving source to the numerous branches of Shenango and Conequenessing rs. or the E. confluents of Big Beaver r. These streams rise generally within about 10 or 12 ms. from the Alleghany r. and flow directly from it to the S. W. The valley of the Conequenessing is nearly commensurate with the quadrangular space between Alleghany, Ohio, Big Beaver rs. comprising two thirds of Butler, with part of Alleghany and Mercer cos. embracing a square of about 30 ms. each way, or 900 sq. ms.

CONESTOGOE, r. of Pa. in Lancaster, Berks, and Dauphin cos. This fine stream has its remote sources only in Lebanon and Berks; the greatest part of its valley is in Lancaster. The comparative length of Conestogoe, is about 30 ms. and the breadth of its sources about an equal distance, stretching from the Welsh mtn. to the Conewago Hills. The area of the valley is 450 sq. ms. This small natural section includes the city of Lancaster, the northern and central parts of Lancaster co. and is one of the best cultivated and most productive tracts of the U. S. A canal extends along the Conestogoe valley, from the city of Lancaster to its discharge into Susquehannah r. 10 ms. S. S. W. from that city.

CONESTOGOE, p-o. and tsp. of Lancaster co. Pa. Pop. 1830. 2,152.

CONESUS, t. Livingston co. N. Y. Pop. 1830, 1,690.

CONEWAGO, r. rises by its W. and main branch in Adams co. Pa. and by its eastern confluent in Frederick co. Md. The two branches unite in Adams co. near Abbottstown, and assuming a N. E. course, fall into the Susquehannah, opposite Bainbridge in Lancaster co. after a comparative course of 40 ms. The valley of the Conewago and that of Manocacy united, fill the space between the Blue Ridge and the S. E. range of Appalachian system, and between the Susquehannah and Potomac rs.

CONEWAGO, small creek of Pa. rising in Lebanon co. and flowing thence S. S. W. separating Lancaster from Lebanon and Dauphin cos. and falling into the Susquehannah, opposite York Haven, after a course of 15 ms.

CONEWANGO, r. N. Y. rises between Chatauque and Cattaraugus cos. runs W. to the outlet of Chautauque lake, then S. to Alleghany r. at Warren, Pa. Length 40 ms. and is navigated in boats and rafts parts of the year, which may go within 7 ms. of Lake Erie.

CONEWANGO, t. Cattaraugus co. N. Y. Pop. 1830, 1,712.

CONEWANGO, p-v. N. part of Warren co. Pa. on Conewango creek, by p-r. 222 ms. N. W. Harrisburg.

CONEWINGO, creek and p-o. N. W. angle of Cecil co. Md. 40 ms. N. E. Baltimore. The lower falls in Susquehannah, sometimes, though erroneously, called Conewingo falls. The true Conewingo falls are 6 ms. above the lower falls or head of tide water.

CONGAREE, r. of S. C. formed by the united streams of Broad and Saluda rs. which commingle at Columbia, almost at the point where lat. 34° and 4° W. W. C. intersect. The general and comparative course of the Congaree is S. E. with a sweep to the S., and thence

s. 35 ms.; but by the meanders the length would probably exceed 50 ms.; in a swampy tract, between Orangeburgh, Richland, and Sumpter districts, the Congaree unites with the Wateree from the N. to form the Santee. (See Santee.)

CONHOCTON, creek Steuben co. N. Y. enters Chemung r. at Painted Post.

CONHOCTON, p-t. Steuben co. N. Y. 16 ms. N. W. Bath, gives rise to Conhocton r., bears beech, maple, elm, bass, ash, hemlock and grass. Pop. 1830, 2,711.

CONKLIN, t. Broome co. N. Y. Population 1830, 908.

CONNEAUT, lake and creek of Crawford co. Pa. The lake is about 4 ms. long, and 1 to 2 wide, discharging the creek southeastward in French creek, which it enters about 8 ms. s. from Meadville.

CONNEAUT, small r. of Pa. and O. rises in Crawford co. of the former, near a lake of the same name, and flowing thence 20 ms. N. N. E. enters Erie co. in which it inflects to the W. 15 ms. entering Ashtabula co. Ohio, and again turning abruptly to N. E. 10 ms. falls into lake Erie in the N. E. angle of the state of O. at the p-v. of Conneaut.

CONNEAUT, p-v. in the extreme N. E. angle of Ashtabula co. O. at the mouth of Conneaut creek, by p-r. 203 ms. N. E. Columbus, and 30 ms. s. w. by w. Erie, in Erie co. Pa.

CONNEAUTVILLE, p-v. on Conneaut creek in the N. W. part of Crawford co. Pa. 20 ms. N. W. Meadville.

CONNECTICUT river, the principal and most important stream of New England, rises in the highlands, dividing the United States from Lower Canada, the head waters of which, forming Lake Connecticut, are 1600 feet above the level of L. I. Sound. Within the first 25 ms. of its course, which is s. w., it falls about 600 feet, and afterwards, pursuing a more southerly course to the head of Fifteen Mile falls, it has a farther descent of 350 feet in 20 miles. Between the latter and the foot of Enfield falls, where it meets tide water, are several other descents and rapids, among which the principal are White r. falls at Hanover, and Bellows falls near Walpole, in N. H.; Miller's and Montague's and Hadley falls in Mass.; and Enfield falls in Conn. The descent in these, exclusive of smaller rapids which intervene, is 236 feet. The general course of the river is southerly, dividing the states of Vt. and N. H.; afterwards crossing the western part of Mass., and dividing Conn. almost equally from N. to s. as far as Middletown, whence it curves to the s. E. to Saybrook, between which place and Lyme it empties into Long Island Sound. The length of the Conn. including its windings, is 400 miles, and the valley, not following the course of the stream, is over 300 ms. long.

The tributaries of the Connecticut are numerous; among them are the Pasumsic, a large stream emptying into the Con. at the foot of Fifteen Mile falls; White river at Hanover; Deerfield and Agawam, at the two places from which they derive their names, and Farmington, or Windsor river, at Windsor, Conn. These are the principal tributaries on the w. side. On the E. the most important are Miller's river, which flows into the Conn. at Montague; and at Springfield it receives the Chickapee, its largest tributary.

The valley of the Conn. presents to the eye every variety of scenery; magnificent mountains, and hills, valleys and meadows, unsurpassed in beauty or fertility; upon its banks are some of the most beautiful towns and villages in New England. Nearly two hundred small lakes, from one to three miles in length, are scattered over the higher surfaces, and are generally found at the sources of tributaries of the river. The Mascony in Lebanon, N. H. and the Sunapee, are the largest in the valley; the former being 7, and the latter 12 ms. in length. Among the high lands which bound the valley, are the Green mountains in Vt. with peaks and ridges 4,000 feet high; and on the E. are the White mountains, and Monadnok, in N. H. Mount Washington, of the former, is the highest land between the Atlantic and the Rocky mountains, and is 6,250 feet above the level of the ocean. Ascutney mountain in Vt. lies wholly within the valley, and is 3,000 feet high.

The banks of the Connecticut are annually overflown in the spring, and not unfrequently at other seasons; the extensive meadows lying upon its banks receive at such times a rich, valuable and abundant addition to their soil. Numerous bridges are thrown across the river, the lowest of which is at Hartford. At the N. boundary of Vt. the Conn. is 150 feet wide; 60 miles below, 390 feet; and in Mass. and Conn. it varies from 450 to 1,050 feet in width. Salmon, which formerly were abundant in the Conn. have entirely disappeared; the principal fishery is shad, which is very valuable. Large quantities of other fine fish also abound in it. The Connecticut is navigable to Hartford, 50 ms. from its mouth, for vessels of 8 feet draft, and to Middletown, for those drawing 10 feet of water. Large steam boats ply daily between the former place and the city of N. York, touching at the intermediate places on the river. Above Hartford numerous flat bottom boats of 15 to 30 tons burthen ascend 220 miles above Hartford, to Wells river, by aid of locks and canals around the falls. These are principally towed by small steam boats, six in number, placed on the different sections between Springfield, Mass., and Wells river. Two steam boats, for passengers, also ply daily between Hartford and Springfield.

The improvements recently made, and others contemplated in the navigation of the river, have already given a fresh impulse to business; as is evident from the great increase of merchandise and produce transported upon its waters, and the increasing intercourse between the towns and villages in its vicinity.

CONNECTICUT, one of the United States; bounded N. by Massachusetts, E. by Rhode Island, S. by Long Island Sound, and W. by New York. It lies between 41 and 42° N. lat. and between 71° 50′ and 73° 43′ W. long. It is 90 miles long, 70 broad and contains 4,764 square miles.

Connecticut was first settled in 1635, by emigrants from Massachusetts, who located themselves in Windsor, Hartford and Wethersfield. A charter was granted to them by Charles the 2d, in 1662. New Haven, which was settled by emigrants from England in 1638, and for many years formed a separate colony, was united with Connecticut under this charter in 1665. The people were greatly harrassed by the arbitrary and oppressive conduct of James 2d. In 1687, Sir Edmund Andross, having been appointed governor of New England, came to Hartford, and by royal authority demanded a surrender of the charter. The assembly being then in session, were reluctant to make this surrender, and while the subject was under consideration, the charter was secretly conveyed away, and concealed in the cavity of an old oak tree on the estate of Mr. Wyllys, one of the magistrates of the colony. This charter formed the basis of the government until 1818, when the present constitution was adopted. The powers of the government are now divided into three distinct departments, viz. the legislative, executive, and judicial. The legislative power is vested in a senate and house of representatives. The senate must consist of not less than 18, nor more than 24 members, who are chosen annually in as many districts, by a plurality of votes. The present number is 21. The house of representatives consists of 209 members, who are chosen annually in each town by a majority of votes, 178 towns, (the more ancient ones,) sending two members, 53 towns only one. The executive power is vested in a governor, who must be 30 years of age, and is chosen annually by a majority of the votes of the people. The lieutenant governor is also chosen annually by the people. He is president of the senate, and also performs the duties of governor, in case of his death, resignation, refusal to serve, impeachment, or absence. The legislature has one stated session annually, on the first Wednesday in May, alternately at Hartford and New Haven. The judicial power is vested in a supreme court of errors, a superior court, and such inferior courts as the legislature may from time to time establish. All the judges are appointed by the legislature; those of the supreme and superior courts, hold their offices during good behavior until 70 years of age, subject to impeachment, or removal by the governor, on the address of two thirds of each branch of the legislature. The supreme court of errors is composed of five judges, and is held in each county annually. The superior court is held twice every year in each county, by one of the judges of the supreme court. In each county also, there is a county court, composed of a chief judge and two associate judges, who with justices of the peace, are appointed annually. Every white male citizen of the United States, 21 years of age, who has gained a settlement in the state, resided in the town six months, and having a freehold estate of the yearly value of seven dollars; or having performed military duty; or paid state tax, may be an elector.

The surface of the state is uneven and greatly diversified by hills and valleys. There are three ranges of mountains in the state; one running within 8 or 10 miles of Connecticut river, on the east side, as far south as Chatham, where it crosses the river and terminates at East Haven; the Mount Tom range, which comes from Massachusetts, runs through the whole state on the west side of the Connecticut, and terminates at New Haven in a perpendicular bluff called East Rock; and the Green mountain range, which is still further west, comes from Vermont, passes through the whole state, and terminates in a similar bluff, at New Haven, called West Rock. The land is generally good, and the meadows on Connecticut river are uncommonly fine; but a large part of the state is better adapted to grazing than tillage. The principal productions are, Indian corn, rye, wheat in some parts, oats, barley, flax, grass, potatoes. Butter and cheese are made in large quantities. Sheep are extensively raised, and beef and pork are abundant. The farms are generally small, varying from 50 to 300 or 400 acres. The winters are severe, but the country is healthy. The principal rivers are the Connecticut, the Housatonic, and the Thames. The principal harbors, New London, New Haven, and Bridgeport. Iron ore of excellent quality is found in great abundance in Salisbury, and other places in the north western part of the state. A copper mine was opened and wrought at Simsbury previous to the revolutionary war, but was subsequently abandoned, and for many years occupied as a state prison; after the removal of the prison, a company commenced working it again, who have succeeded in obtaining copper ore of great purity. Superior white marble is found at Washington and New Milford, and beautiful variegated marble of the verd antique species, at New Haven and Milford. There are extensive quarries of excellent free stone, at Chatham and other adjacent towns on the river.

The state is divided into eight counties, Hartford, New Haven, New London, Fairfield, Windham, Litchfield, Middlesex, and Tolland. There are five incorporated cities, Hartford, New Haven, New London, Norwich, and Middletown; and eight boroughs, Danbury, Guilford, Bridgeport, Newtown, Stonington, Stamford, Waterbury, and Killinworth.

The population of Connecticut in 1810

was 261,942, and in 1820, 275,248. In 1830 it was as follows.

Counties.		Counties.	
Hartford,	51,141	Windham,	27,077
New Haven,	43,848	Litchfield,	42,855
New London,	42,295	Middlesex,	24,845
Fairfield,	46,950	Tolland,	18,700

Of which were whites,

	Males.	Females.
Under 5 years,	19,033	18,270
5 to 15	35,679	33,518
15 to 30	42,675	42,518
30 to 50	28,203	31,151
50 to 70	13,346	15,952
70 to 90	4,025	4,988
90 and above,	86	159
Total,	143,047	146,556

Of free colored persons there were as follows:—under 10, 1,019 males, 1,051 females—between 10 and 24, 1,121 males, 1,233 females—between 24 and 36, 771 males, 819 females—between 36 and 55, 624 males, 667 females—between 55 and 100, 313 males, 417 females—100 years and upwards, 2 males, 10 females. Total, 8,047. Blacks not emancipated on account of advanced age or infirmities, 8 males and 17 females. Total 25.

Recapitulation,

Whites.	Free color'd.	Slaves.	Total.
289,603	8,047	25	297,675.

Of the foregoing were whites, deaf and dumb, under 14, 43; between 14 and 25, 152; 25 and upwards, 99; total, 294. Blind, 188; aliens 1481. Of the blacks there are deaf and dumb, 6; blind, 7.

The foreign trade of Connecticut is principally with the West Indies, but it is less extensive than the coasting trade. The exports are beef, pork, horses, mules, cattle, butter, cheese, fish, and various articles of manufactures. New London, Stonington, and some other towns, have recently engaged with much success in the whaling business. Connecticut is extensively engaged in manufactures, consisting principally of cotton and woollen goods, iron, glass, paper, tinware, buttons, clocks, leather, shoes, fire arms, and various other articles. The following is an abstract of the rateable estate and polls in Conn. as returned in 1831.

42,852	Houses,	$21,948,740
2,622,676	Acres of land,	50,782,455
1,572	Mills,	843,511
1,826	Stores,	1,467,748
283	Distilleries,	64,052
1,521	Manufactories,	1,637,149
183	Fisheries,	498,625
34,250	Horses, asses, mules, &c.	1,290,694
237,989	Neat cattle,	3,347,667
271,625	Sheep,	333,657
	Silver plate and plated ware,	10,614
5,196	Riding carriages and wagons,	238,798
22,893	Clocks and watches,	174,843
	Insurance stock,	53,642
	Turnpike stock,	157,362
	Money on interest,	2,087,976
	State bank stock,	3,143,736
	U. S. bank stock,	17,880
25	Quarries, and shares of,	38,350
1	Ferry,	200
		87,737,699

Assessments.

	On professions,	147,683	
34,466	polls, $20 each,	689,320	
			837,003

There are 19 state banks in Connecticut, with a capital, as officially returned, March 1832, of $4,944,100; in addition to which is a branch of the U. S. bank, capital $300,000. There are also 5 banks for savings, and 11 insurance companies.

The principal literary and benevolent institutions are Yale College in New Haven, the Wesleyan University in Middletown, and Washington College, the Deaf and Dumb Asylum, and Retreat for the Insane, in Hartford. A general state hospital has also been recently founded in New Haven. Numerous academies and high schools for both sexes, are established in various parts of the state.

The state prison at Wethersfield deserves to be mentioned as an institution creditable to the state. In its construction and general arrangements, it is similar to the New York state prison at Auburn. The number of convicts in March, 1832, was 192, of whom 18 were females. They are kept at hard labor in workshops by day, and confined in solitary cells by night. A prominent feature in the system of discipline, is the prevention of all intercourse or communication between the prisoners. The prison produces a handsome revenue to the state; the avails of it for the year ending on the 31st March, 1832, after deducting all expenses, amounted to $8,713 53. There is a chaplain connected with the institution; a Sunday school has been organized, and all proper means are faithfully used for the reformation of the convicts.

In no part of the world has more ample provision been made for the instruction of all classes of the people in the elements of useful knowledge than in Connecticut. Her institutions of learning, and provision for the general instruction of the people, have placed Connecticut on a proud eminence among her sister states. By the last estimate of the commissioners, April 1, 1831, the aggregate amount of the school fund of the state amounted to $1,902,957 87; and the whole proceeds for the year ending 31st March, 1832, was $84,173 83. This fund is derived from the sale of western lands, and the proceeds are appropriated to the support of common schools. Her citizens have always been distinguished for their intelligence, industry, economy, and correct moral habits. A spirit of enterprise has led thousands of them to emigrate to distant parts of the country, where they have assisted in the settlement of other states and territories. Perfect religious toleration is enjoyed in Connecticut. No person is compelled to support or be connected with any church or religious association; and although while thus connected, he may be compelled to pay his proportion of the expenses, he may at any time dissolve his connection by leaving a written notice of the same with the clerk of such society. There are various religious sects in the state; Congregationalists, Baptists, Episcopalians, Methodists, Unitarians, Friends, Universalists, Shakers, Catholics, some Free Will Baptists, and a few Christ-ians. The Congregationalists are much the most numerous.

CONNECTICUT FARMS, v. Essex co. N. J. 4 ms. N. W. Elizabethtown.

CONNELLSVILLE, p-v. and tsp. Fayette co. Pa. The village is situated on the right bank of Youghioghany river, 12 ms. N. N. E. Union Town.

CONNERSVILLE, p-v. Boone co. Ky. by p-r. 86 ms. N. Frankfort.

CONNERSVILLE, p-v. and st. jus. Fayette co. Ind. by p-r. 527 ms. W. from W. C. 68 ms. a little S. of E. Indianopolis, and 60 ms. N. W. Cincinnati, O. It is situated on White Water r. at lat. 39° 38′, & long. 8° 10′ W. W.C.

CONOTTON, p-v. in the N. part of Harrison co. O. by p-r. 127 ms. N. E. by E. Columbus.

CONQUEST, p-t. Cayuga co. N. Y. 19 ms. N. W. Auburn. Pop. 1830, 1,507.

CONRAD'S FERRY, over the Potomac, just above the mouth of Goose creek, and p-o. in the W. part of Montgomery co. Md. 4 ms. S. E. by E. Leesburg, Va. and 37 ms. from W. C.

CONRAD'S store, and p-o. Rockingham co. Va. by p-r. 141 ms. N. W. Richmond.

CONSTABLE, t. Franklin co. N. Y. 7 ms. N. Malone, 6 ms. by 9, has a sandy loam, with beech, maple, bass, elm, hemlock, and groves of pine. Bog iron ore is dug. Salmon and Trout rivers supply mill seats. Pop. 1830, 693.

CONSTANTIA, p-t. Oswego co. N. Y. 28 ms. W. from Rome. N. Oneida lake, 7 ms. by 17, is low and level, with good land, and some bog iron ore. It includes the site of Fort Brewerton, at the outlet of Oneida lake. The village on the N. side of the lake has iron works. Pop. 1830, 1,193.

CONTOOCOOK, r. Hillsborough co. N. H. enters Merrimack r. at Concord.

CONWAY, p-t. Stafford co. N. H. 76 ms. N. N. E. Concord, crossed by Saco river W. Me. 6 ms. square, is watered also by Swift, and Pequacokett rivers. A sulphur spring here, is visited by invalids; magnesia and fuller's earth are found. The banks of Saco r. are level and rich; the uplands rocky. The timber is oak, maple, beech, and white pine. Saco r. is subject to sudden floods. Pop. 1830, 1,601.

CONWAY, p-t. Franklin co. Mass. 6 ms. W. Greenfield, 100 W. Boston, S. W. Deerfield r. 7 ms. W. Conn. r. formerly part of Deerfield. Pop. 1830, 1,563.

CONWAYS, co. Ark. ter. bounded S. W. by Arkansas r. which separates it from Crawford, W. by a part of Crawford, N. by Izard, N. E. by Red. r. branch of White r. & S. E. by Pulaski, length 55, mean breadth 30, and area 1650 sq. ms. extending in lat. from 34° 52′ to 35° 40′ and in long. from 14° 55′ to 15° 56′ W. from W. C. Chief t. Lafayette.

CONWAY, p-o. Ark. by p-r. 190 ms. S. W. from Little Rock.

CONWAY'S borough p-v. and st. jus. Horry, dist. S. C. on the right bank of Waccamau r. by p-r. 153, but by direct road, about 100 ms. N. E. from Charleston, and about an equal distance a little W. of S. from Fayetteville in N. C. lat. 33° 49′ and long. 2° 05′ W. from W. C.

CONYNGHAM, p-v. situated at the foot of Buck mtn. Nescopeck valley, and in the southern part of Luzerne co. Pa. 12 ms. a little S. of E. from Burwick and 20 ms. N. W. by W. from Mauch Chunk, on the Lehigh. It is a most romantic situation, surrounded by mts. and stretching in one street across the valley, presents to the traveller a well built village, containing a pop. 1830, of about 300.

COOCHE'S BRIDGE, and p-o. in the N. W. part of New Castle co. Del. 62 ms. W. of N. from Dover.

COOK'S LAW office and p-o. Elbert co. Geo. by p-r. 65 ms. N. N. E. from Milledgeville.

COOK'S, late Broom's p-o. in the S. part of Fairfield co. S. C. 20 ms. N. from Columbia.

COOK'S settlement and p-o. in the W. part of St. Genevieve co. Mo. about 60 ms. S. from St. Louis.

COOKSTOWN, p-v. on the right bank of the Monongahela r. N. W. part of Fayette co. Pa. 28 ms. a little E. of S. from Pittsburg.

COOKSVILLE, p-v. N. part of Ann Arundel co. Md. by p-r. 51, but by actual distance 32 ms. N. from W. C.

COOKVILLE, p-o. Jackson co. Ten. by p-r. 92 ms. N. E. by E. from Nashville.

COOLBAUGH'S p-o. Pike co. Pa. about 21 ms. S. from Milford, the st. jus.

COOL SPRING, p-o. Washington co. N. C. by p-r. 182 ms. E. from Raleigh.

COOL SPRING, p-o. in the E. part of Chesterfield dist. S. C. by p-r. 89 ms. N. E. from Columbus.

COOL SPRING, p-o. Wilkinson co. Geo. by p-r. 44 ms. S. from Milledgeville.

COOL SPRING, p-o. Gibson co. Ten. by p-r. 130 ms. W. from Nashville.

COOLVILLE, p-v. on Hocking r. S. E. part of Athens co. O. by p-r. 24 ms. below Athens, the co. seat.

COOPER, t. Washington co. Me. 164 ms. from Augusta, has a stream on E. border emptying into Cooleacook bay. Pop. 1830, 396.

COOPER, r. of S. C. in reality the drain of a swampy tract semicircle, by Santee r. the various drains uniting about 29 ms. N. from Charleston, form Cooper r. which, flowing S. joins Wards r. form the N. and at Charleston Ashley from the S. all contributing to form the fine harbor of that city. Santee canal unites Cooper and Santee r. extending from N. N. W. to S. S. E. 21 ms. from the Santee at Black-oak isl. to the W. branch of Cooper.

Cooper's p-o. Franklin co. Va. 159 ms. s. w. by w. Richmond.

Cooperstown, p-v. and st. jus. Otsego co. N. Y. 12 ms. w. Cherry Valley, 66 w. Albany, 21 s. Erie Canal, s. end Otsego Lake. There is a deep valley at the outlet of Otsego lake, between high hills. Timber, chiefly pine and hemlock. The village has 3 churches; 1 Episcopal, 1 Presbyterian and 1 Methodist, a court-house, county bank with $100,000 capital; a card factory here, is chiefly worked by dogs. Pop. 1830, 1,115.

Cooperstown, p-v. in the n. w. part of Venango co. Pa. by p-r. 70 ms. n. Pittsburg.

Cooperstown, p-o. in Nancoochy valley, w. part of Habersham co. Geo. by p-r. 12 ms. w. Clarksville, the co. st.

Coos, co. N. H. the largest in the state, bounded by Lower Canada n., Me. e., Stafford co. s., Grafton co. and Vt. w. with 1,600 sq. ms. includes the White mtns. the highest in the U. S. and gives rise to the 3 Ammonoosucks, branches of Conn. r., and Saco, which enters the Atlantic. A great part of the co. cannot be improved by cultivation, and is unoccupied. On Conn. r. are some fine meadows. It contains 25 towns and 47 school districts. Pop. 1820, 5,151, 1830, 8,390.

Coosa, r. of Ten. Geo. and Ala. the n. w. and main branch of Ala. r. The extreme higher sources of Coosa is in Ten. at lat. 35° 05', there known by the name of Connesauga. It flows first w. but curving s. s. w. 70 ms. receives from the n. e. the Etowah r. The two branches have interlocking sources with the Hiwassa branch of Ten. with those of Ten. Proper, and the Chattahooche. Having their fountains amid the elevated Appalachian vallies, the higher confluents of Coosa are rapid perennial streams. Below the junction of Connessauga and Etowah, the united waters flow 8 ms. wstrd. entering Ala. near Fort Armstrong, inflect to s. s. w. receiving but few accessions above the size of a large creek, join the Talapoosa at lat. 32° 28', long. 9° 22' w. W. C. to form Ala. having an entire comparative course of about 240 ms. The valley of the Coosa is about 200 ms. long, and mean breadth 45; area 9,000 sq. ms.

Coosauda, p-v. on the right bank of Ala. r. Autauga co. Ala. 6 ms. below the junction of Coosa and Talapoosa rs. and by p-r. 96 ms. s. e. by e. Tuscaloosa.

Coosaw, r. S. C. is a broad and deep inlet, uniting Coosawhatchie or Broad r. to Combahee r. on St. Helena Sound. In fact Coosa is the northern mouth of Coosawhatchie. In the languages of many southern tribes of Indians, Hatchie or Hatchy signifies river, and has become a suffix to several rivers of the southern states.

Coosawhatchie, r. of S. C. rises in Barnwell dist. but entering Beaufort, flows s. e. 30 ms. to where a branch flows from the main stream to the w. This outlet, a mouth by the name of Cyprus creek, inflects to the s. e. falls into the Atlantic between the mouth of Savannah r. and Galibogue Sound, after a course of 80 ms. The main Coosawhatchie again divides into two channels below the efflux of Cyprus creek, but after a separation of 12 ms. reunites, forming Tullyfinny isl. Below the latter island a third separation of the waters of Coosawhatchie takes place. The principal stream widens into Broad r. and finally opens to the Atlantic by Port Royal Entrance. The northern branch flows s. e. by e. and is known as Coosaw r. (*See Coosaw r. and Beaufort dist.*)

Coosawhatchie, p-v. on the right bank of Coosawhatchie r. Beaufort dist. S. C. 75 ms. s. w. by w. Charleston.

Cootstown, or more accurately from the German geography, Kutztown, a fine well built p-v. Berks co. Pa. 17 ms. n. n. e. Reading, and about an equal distance s. w. by w. Allentown.

Copake, t. Columbia co. N. Y. w. Mass. line; has 2 ponds or lakes emptying into Claverack creek, and has Penobscot co. on 3 sides, Rocleff and Janson's Kills. Pop. 1830, 1,676.

Copenhagen, v. Lewis co. N. Y. on Deer creek, 6 ms. n. Denmark.

Copeland, p-v. Telfair co. Geo. by p-r. 77 ms. s. Milledgeville.

Copiah, co. of Miss. bounded by Franklin s. w., Jefferson w., Clairborne n. w., Hinds n., Simpson e., and Lawrence s. e. It is nearly in form of a square of 28 ms. each side, or area of 784 sq. ms. lying between lat. 31° 36' and 32° 4', and long. 13° 21' and 13° 50'. The water courses flow from this co. estrd. into Pearl r.; n. wstrd. they form the Bayou Pierre, whilst the s. w. section gives source to the Homochitto. The central part is therefore a table land, and the whole surface, with partial exceptions, is composed of sterile soil, covered in a natural state, with pine forests, slightly intermingled with other timber. Chief t. Gallatin. Pop. 1830, 7,001.

Copopa, p-v. in the central part of Lorrain co. O. by p-r. 128 ms. n. n. e. Columbus.

Copperhonk, p-o. Sussex co. Va. 59 s. s. e. Richmond.

Coquille, usually called *Petites Coquilles*, Fort and p-o. on the s. point, where the Regolets flow from Lake Pontchartrain, 25 ms. n. e. by e. New Orleans.

Core a Fabre, p-v. Union co. Ark. position uncertain.

Core Creek, p-o. w. part of Craven co. N. C. by p-r. 104 ms. s. e. by e. Raleigh.

Core Sound and Core Island, Cartaret co. N. C. The island is a long, narrow, and low reef, extending 20 ms. from Cape Look Out, its salient point s. w. to Cedar Inlet. The Sound stretches between the island and mainland, from the n. e. part of Onslow bay to Pamlico Sound, and is 40 ms. long, with a mean breadth of 1 or 2 ms. It is shallow, admitting only small coasting vessels.

Corinth, p-t. Penobscot co. Me. 18 ms. n. w. Bangor, 81 Augusta, is situated near the head waters of many streams flowing into Penobscot r. Pop. 1830, 712.

CORINTH, p-t. Orange co. Vt. 21 ms. s. E. Montpelier, 12 w. Haverhill, N. H., 41 w. Windsor, 6 ms. square. First settled, 1777, is very rough, with good dark loam, and hard wood trees, except the hemlock, spruce and furs, on the streams. Waits brook and others supply mills. Pop. 1830, 1,953.

CORINTH, p-t. Saratoga co. N. Y. 18 ms. N. Ballston Spa, s. w. Hudson river, at the Great Falls 30 feet cataract, and one and a ½ ms. above, has a smooth and sandy land above the falls, with white pine and beech; broken, stony and loamy below. Palmer's town mtn. is s. and Kayadarossoras mtn. w. At Hadley, or Jessups landing, is a village, and rafts go from the sands bank 1½ ms. below. About 100 yards above Great Falls is a chasm, 12 feet wide, 20 long, and very deep, through which the entire river passes at low water. Limestone abounds; and oxides of iron used for paints. Population 1830, 1,412.

CORK, p-v. in the N. part Ashtabula co. O. by p-r. 187 ms. N. E. Columbus.

CORLEARS HOOK, city of New York, the N. E. point of the city, at the turn in the Sound.

CORN CREEK, p-o. Gallatin co. Ky. by p-r. 59 ms. N. Frankfort.

CORNELIUSVILLE, p-o. Boone co. Ky. by p-r. 88 ms. N. Frankfort.

CORNERSBURG, p-v. Trumbull co. O. about 150 ms. N. E. Columbus.

CORNISH, p-t. York co. Me. 50 ms. N. York, 83 Augusta, south Ossipee river, where it joins the Saco. Population 1830, 1,235.

CORNISH, p-t. Cheshire co. N. H. 17 ms. N. Charlestown, 50 Concord, 108 Boston, E. Conn. r. 23,160 acres, is fertile, except near the river. A few mill seats are on Blow-me-down and Briant brooks. Settled 1765 from Sutton, Mass., seceded from N. H. 1778, with 15 other towns. Pop. 1830, 1,235.

CORNISHES' p-o. Lauderdale co. Ala. by p-r. 119 ms. N. Tuscaloosa.

CORNVILLE, p-t. Somerset co. Me. 11 ms. E. N. E. Norridgewock, 38 Augusta, is crossed by a small tributary of Kennebec r. Pop. 1830, 1,104.

CORNWALL, t. Addison co. Vt. on Otter creek, 3 ms. s. w. Middlebury, 75 ms. N. Bennington, 36 ms. s. Burlington. Settled 1774, deserted '77, and resettled from Conn. 1784, is generally level, and crossed by Lemonfair r. has no good mill seats, but a large swamp, 7 school districts. Pop. 1830, 1,264.

CORNWALL, p-t. Litchfield co. Conn. 10 ms. N. Litchfield, 38 w. Hartford, and 48 N. w. New Haven, E. Housatonic river, 9 by 5, 46 square miles, has mtns. and mountainous hills, of granite and limestone, with black lead, porcelain clay. Some of the largest vallies have rich calc loam. It yields oak, chestnut, maple &c. grain, grass, beef, &c. There are two ponds 1 mile long, with pickerel and trout, with many mill sites. The American board of Foreign Missions, formed their school here 1816, and educated many young men from heathen countries. Pop. 1830, 1,714.

CORNWALL, t. Orange co. N. York, 52 ms. N. N. York, 108 s. Albany, w. Hudson river, is mountainous, but has good pasturage, and some level lands north, where Murderers creek supplies mills.

The village landing sends wood and stone to N. York. West Point in this town is a tract of land owned by the United States, bordering on the North river, where is the military academy, and the professors quarters; the barracks and parade ground are on a level 182 feet above Hudson river, above which on a mtn. are the remains of Fort Putnam, and in front, those of Fort Clinton, built in the revolution, when this was an important military post. Sir Henry Clinton forced his passage here in 1777, to cooperate with Gen. Burgoyne, but after burning Kingston, &c. returned to N. Y. Gen. Arnold's treasonable design was to betray West Point to the British. There is a large hotel, and a monument to Kosciusko, erected by the cadets. Pop. 1830, 3,485.

COROWAUGH, creek, swamp, and p-o. s. w. part of the Isle of Wight co. Va. 35 ms. s. w. Norfolk.

CORRINNA, p-t. Somerset co. Me. 53 ms. Augusta. Pop. 1830, 1,079.

CORTLAND, co. N. Y. bounded by Onondaga co. N., Madison and Chenango cos. E., Broome and Tioga cos. s., Tompkins and Cayuga cos. w., an oblong 19 ms. by 25, 475 square ms., has 9 towns, many brooks. Tioughnioga creek nearly through it. Osselie creek in the s. E.; it gives rise to branches of Owego creek and Cayuga lake; mill sites abounds. The soils chiefly yellowish loam, on warm gravel, uneven but excellent for grain and grass, bears maple, elm, bass, butternut, pine, &c. There are some salt and sulph. hyd. and chalybeate springs, and iron ore. The N. w. corner touches the s. end of Skeneateles lake. Cortland village is the capital. Pop. 1820, 16,507, 1830, 23,753.

CORTLAND, t. Westchester co. N. Y. 40 ms. N. N. York, 104 s. Albany, E. Hudson river, has 2 post vs. Cortlandt t. Peekskill, Peekskill creek and Croton river afford many mill sites, and it includes the s. peaks of the Highlands, Verplanks point where was Fort Lafayette, and Tellers point. Pop. 1830, 3,840.

CORTLANDTVILLE, p-t. and cap. Cortlandt co. N. Y. 140 ms. w. Albany, on Tioughnioga at the bend, and contains Cortlandt village and Port Watson. Pop. 1830, 3,673.

CORYDON, p-v. and st. jus. Harrison co. Ind. by p-r. 614 ms. a little s. of w. W. C. 124 ms. s. Indianopolis, and 20 ms. a little s. of w. Louisville, in Ky. N. lat. 38° 15′, long. 9° 08′ w. W. C. Pop. 1830, 459.

COSHOCTON, or COCHECTON, p-v. Bethel Sullivan co. N. Y. 16 ms. w. Monticello, 60 ms. w. Newburgh, on Delaware river.

COSHOCTON, co. of O. bounded southeast

by Guernsey, s. by Muskingum, s. w. by Licking, w. and n. w. by Knox, n. by Holmes, and n. e. and e. by Tuscarawas. Greatest length 30, mean width 20, and area 600 sq ms. Extending in lat. from 40° 10′ to 40′ 27 n. and in long. from 4° 40′ to 5° 12′ w. W. C. The union of Tuscarawas r. with White Woman's creek to form the Muskingum river, is made a little s. e. from the centre of this co. Killbuck creek rising in Medina and Lorrain cos. flows to the southward over Wayne and Holmes into Coshocton, and uniting with White Woman's from the westward, the combined waters inflect to the s. e. to their junction with Tuscarawas river at the village of Coshocton. From the course of the three preceding streams, Coshocton co. is formed out of as many deep river vallies. That of White Woman's inclines to the east; Tuscarawas in an opposite direction, and that of Killbuck southwardly. Below the village of Coshocton the channel of Muskingum river is nearly south to the influx of Will's creek, on the southern border of the co. The Ohio and Erie canal reaches the bank of Muskingum a little below the mouth of Will's creek, and following Muskingum and Tuscarawas rivers, traverses Coshocton between 25 and 30 miles. The northern, and about one third part of what surface was included in Coshocton in 1820, has been since united to a part of Wayne, to form Holmes co. Chief town, Coshocton. Pop. 1830, 11,161.

Coshocton, p-v. and st. jus. Coshocton co. O. by p-r. 336 ms. n. w. by w. W. C. 84 ms. n. e. by e. Columbus, and 26 ms. n. Zanesville. It is situated on the left or east bank of Muskingum river, just below the junction of Tuscarawas river and White Woman's creek. Lat. 40° 15′, long. 4° 54′ w. W. C. Pop. 1830, 333.

Cossitat, p-v. Hempstead co. Ark. by p-r. 1,234 s. w. by w. ½ w. W. C. and 166 miles s. w. by w. Little Rock.

Cotaco, formerly a county of Al. now Morgan co.

Cote Isle, post village, Rapide parish, Louisiana.

Cotoctin, a ridge of the Appalachian mtns. This ridge branches from the south mountain on the southern border of Pa. and between Adams and Franklin counties; stretching thence nearly due s. through Frederick co. in Md. reaches the Potomac river between the mouths of Monocacy river and Cotoctin creek. The same ridge or rather chain rises southward of the Potomac and traverses Loudon co. Va. passing about 2½ ms. westward of Leesburg. In Md. the Cotoctin has gained importance and celebrity from the controversy between the Baltimore and Ohio rail road company, and that of the Chesapeake and Ohio canal. Where the ridge terminates on the Potomac, it is known as the Upper Point of rocks and Lower Point of rocks.

Cotoctin, in the p-o. list Cotocton, p-v. in the w. part of Frederick co. Md. by p-r. 46 ms. n. w. W. C.

Coquille, usually called Petite Coquilles (Little Shells), fort and p-o. at the outlet of the Rigolets from lake Pontchartrain, in the n. w. part of Orleans parish, La. by ship channel 31 ms. n. e. by e. New Orleans.

Cotton Gin Port, p-v. at the union of Tombigbee and Notachucky rivers, and on the left bank of the former in Lowndes co. Miss. by p-r. 188 ms. n. e. Jackson. That part of Monroe containing Cotton Gin Port, has been recently erected into Lowndes co.

Cotton Grove, p-v. Madison co. Ten. 162 miles s. w. by w. Nashville.

Cotton Port, p-v. on the right bank of Tenn. river, in the s. e. part of Limestone co. Ala. 15 ms. s. s. w. Huntsville.

Cottonville, p-v. in the s. part of Lawrence co. Miss. about 80 ms. e. Natchez, and by p-r. 1,119 s. w. W. C.

Cottrellville, p-v. on the right bank of St. Clair river, s. e. part of St. Clair co. Mich. according to Tanner. By the land route round the w. side of lake St. Clair 52 ms. n. e. Detroit, and by p-r. 578 ms. n. w. W. C. Pop. 1830, 230.

Councill's Store, and p-o. by p-r. 231 ms. a little n. o w. Raleigh.

Countsville, p-v. on Preston creek, n. angle of Lexington dist. S. C. 31 ms. n. n. w. Columbia.

County Line, p-o. Rowan co. N. C. by p-r. 138 ms. w. Raleigh.

County Line, p-o. in the w. part Campbell co. Geo. by p-r. 725 ms. s. w. W. C. and 139 n. w. Milledgeville.

Courtableau, river of La. formed by two confluents, the Crocodile, from the pine wood between Opelousas and Rapides, and the Boeuf, from the intermediate space between the Crocodile and the overflowed region of Red and Atchafalaya rivers. The two branches unite about 10 miles n. from St. Landre, and assuming a s. e. course flow 35 miles, falling into Atchafalaya at the lower fragment of the Great Raft. This fine though small stream forms a link in the chain of water intercommunication between Opelousas and the Miss. river.

Courtland, p-v. northern part of Laurence co. Ala. about 50 ms. a little s. of w. Huntsville, and by p-r. 104 ms. n. Tuscaloosa.

Courtwright, p-v. in the w. part Fairfield co. O. by p-r. 18 ms. s. e. Columbus, and 382 n. w. by w. W. C.

Cove Creek, p-o. in the w. part of Ashe co. N. C. by p-r. 432 ms. s. w. W. C. and 238 a little n. of w. Raleigh.

Coventry, town, Grafton co. N. H. 9 ms. e. Haverhill, 70 n. by w. Concord, 100 n. w. Portsmouth, mountainous, with some useless soil, watered by streams of Oliverian brook and Wild Amonoosuc river, has Owl's Head mtn. w. Pop. 1830, 440.

Coventry, p-t. Orleans co. Vt. 49 ms. n. Montpelier. First settled 1800, has s. bay of Memphremagog lake, with good soil, and the lower parts of Barton and Black rivers, which are deep with good mill seats. Pop. 1830, 728.

COVENTRY, t. Kent co. R. I. 15 ms. s. w. Providence, E. Conn., 6 ms. by 6, 72 square miles, rugged, primitive, good for grass, with s. branch of Pawtucket, Flat river and other excellent mill streams. It is much devoted to manufacturing. Pop. 1830, 3,851.

COVENTRY, p-t. Tolland co. Conn. 18 ms. E. Hartford, w. Willimantic river, 6½ ms. by 7, 45 square ms., uneven, with gravelly loam, primitive, bearing oak, walnut, chestnut, &c. grass, grain, &c., crossed by Skunamug r. which forms Hop r. and unites with Willimantic river at s. E. corner; Wangumbog lake is 1 mile by 2. First settled, 1711. Population 1830, 2,119.

COVENTRY, p-t. Chenango co. N. Y. 20 ms. s. w. Norwich, midway between Susquehannah and Chenango rivers. The land is broken, but much that is good, with small streams. Pop. 1830, 1,576.

COVERT, p-t. Seneca co. N. Y. 6 ms. s. Ovid, E. Seneca lake, 5 ms. by 12, has mill seats on Halsey's creek, &c. The land on the lake is excellent for wheat.

COVESVILLE, p-v. in the w. part of Albermarle co. Va. 22 ms. w. Charlottsville, and by p-r. 145 ms. s. w. W. C. and 103 N. w. by w. Richmond.

COVINGTON, p-t. Genesee co. N. Y. 12 ms. s. E. Batavia, has soil of ordinary quality, pretty well watered by Allan's creek and branches. Pop. 1830, 2,716.

COVINGTON, p-v. on Tioga cr. Tioga co. Pa. 65 ms. s. w. by w. Tioga Point.

COVINGTON, p-v. and st. jus. Alleghany co. Va. situated on Jackson r. 260 ms. s. s. w. W. C. and 173 ms. w. Richmond. Lat. 37° 48′, long. 3° 3′ w. W. C.

COVINGTON, p-v. in the s. part Richmond co. N. C. 14 ms. sthrd. Rockingham, the co. st. and by p-r. 413 ms. s. s. w. W. C. and 127 s. w. Raleigh.

COVINGTON, p-v. and st. jus. Newton co. Geo. on Yellow r. a branch of Oakmulgee, by p-r. 67 ms. N. w. Milledgeville. Lat. 33° 32′, long. 6° 58′ w. W. C.

COVINGTON, co. Ala. bounded w. by Conecuh, N. w. and N. by Butler, E. by Dale, and s. by Walton co. in Flor. Length s. to N. 52, mean breadth 32, area 1,664 sq. ms. Extending in lat. 31° to 31° 42′, long. 9° 15′ to 9° 52′ w. W. C. Surface generally sterile. The N. w. angle of this co. is traversed by the two main branches of Conecuh and Pigeon rs.; the central section gives source to Yellow Water r. which flows sthrd. towards Pensacola bay; Pea r. the w. branch of Choctaw r. rises in the Creek country, traverses Pike and Dale, enters and again curves out of the eastern border of Covington. Chief t. Montezuma. Pop. 1830, 1,522.

COVINGTON, co. Miss. bounded s. by Marion, w. by Lawrence, N. by the Choctaw ter. and E. by Jones. Length from E. to w. 30, mean breadth 24, and area 960 sq. ms. Extending in lat. from 31° 26′ to 31° 48′, long. from 12° 28′ to 12° 58′ w. W. C. It is traversed by various branches of Leaf r. which flowing s. E. towards their confluents Pascagoula, afford some good land, but in general the face of the co. is open, sterile, piney woods. Chief t. Williamsburg. Pop. 1820, 2,230, 1830, 2,551.

COVINGTON, p-v. and st. jus. parish of St. Tammany, La. situated on Chifuncte r. 36 ms. a little w. of N. New Orleans.

COVINGTON, p-v. on the bank of Ohio r. on the point below the mouth of Licking r. which separates it from Newport, and opposite Cincinnati, Campbell co. Ky. The great road up the Ohio r. passes through Covington over a bridge into Newport. Pop. 1830, 715.

COVINGTON, p-v. and st. jus. Tipton co. Ten. situated on a small branch of Big Hatchie r. 40 ms. N. N. E. Memphis, and by p-r. 225 ms. s. w. by w. Nashville. Lat. 35° 34′, long. 12° 41′ w. W. C.

COVINGTON, p-v. and st. jus. Fountain co. Ind. by p-r. 654 ms. N. w. by w. W. C. and 81 ms. N. w. by w. Indianopolis. It is situated on the left side of the Wabash r. N. lat. 40° 10′, long. 10° 24′ w. W. C.

COVINGTON, p-v. near the N. border of Washington co. Il. by p-r. 812 ms. w. W. C. and 40 s. w. Vandalia. It is situated on the Kaskaskias r. on the great road from Shawneetown on O. r. to St. Louis, 47 ms. s. of E. from the latter. Lat. 38° 28′, long. 12° 28′ w. W. C.

COWAN'S STORE and p-o. Cabarras co. N. C. 151 ms. w. Raleigh.

COWANSVILLE, p-o. 136 ms. w. Raleigh.

COWANSVILLE, p-o. Rhea co. Ten. by p-r. 170 ms. s. E. by E. Nashville.

COWDERSPORT, p-v. and st. jus. Potter co. Pa. situated on Alleghany r. by p-r. 186 ms. N. w. Harrisburg. Lat. 41° 56′, long. 1° 4′ w. W. C.

COWETA, co. Geo. bounded w. and N. w. by the Chattahooche, which separates it from Carroll, N. by Campbell, E. by Fayette, and s. by Merriwether and Troup. Length from s. w. to N. E. 38 ms. mean breadth 14, and area 532 sq. ms. Extending in lat. from 33° 15′ to 33° 37′, in long. from 7° 40′ to 8° 18′ w. W. C. It lies in form of a triangle, extending its hypothenuse along Chattahooche, and its base E. and w. Chattahooche to Flint r. Chief t. Newman. Pop. 1830, 5,003.

COWETA, p-v. Coweta co. Geo. by p-r. 135 ms. N. w. Milledgeville.

COWPASTURE, r. Va. rising in the mountain valley between the Kittatinny and Warm Spring mtn. interlocking sources with the s. branch of Potomac, but flowing in an opposite direction s. s. w. falling into, or joining Jackson's r. to form James r. after a comparative course of 50 ms.

COWPEN'S FURNACE, p-o. between Pacolet and Broad rs. Spartanburg dist. S. C. by p-r. 124 ms. N. N. w. Columbia.

COWPER HILL, p-o. Robeson co. N. C. by p-r. 92 ms. s. s. w. Raleigh.

Coxsackie, p-t. Greene co. N. Y. 26 ms. s. Albany, 10 n. Catskill, w. Hudson r., e. Catskill creek. Cox's creek is n. Mill sites abound; there are hills, pine plains, some sand and clay. The inhabitants are of Dutch extraction, and hold the land in fee. It has a valley 1 m. w. Hudson r. and 3 landings. Pop. 1830, 3,373.

Cox's Cross Roads, in the s. w. part of Coshocton co. O. by p-r. 357 ms. n. w. by w. W. C. and 70 n. e. by e. Columbus.

Cox's Store and p-o. Sampson co. N. C. by p-r. 95 ms. s. s. e. Raleigh.

Coylesville, p-o. in the w. part of Butler co. Pa. by p-r. 10 ms. w. the borough of Butler, and 226 n. w. W. C.

Crab Orchard, p-v. Lincoln co. Ky. by p-r. 62 ms. e. of s. Hartford.

Crab Run, p-v. in the s. w. part Pendleton co. Va. by p-r. 196 ms. s. w. by w. W. C. and 154 ms. n. w. by w. Richmond.

Crafton, p-v. Pittsylvania co. Va. by p-r. 236 ms. s. s. w. W. C. and 156 ms. s. w. by w. Richmond.

Craftsbury, p-t. Orleans co. Vt. 25 ms. n. Montpelier, 25 s. Canada, half way between Lake Champlain and Conn. r. First settled 1789; gives rise to Black r. which has many mill sites; has Wild Branch and 5 Trout ponds. The village is near the centre, on high ground. The trade is with Montreal. There are 5 school districts. Pop. 1830, 982.

Craig's Creek, p-o. Botetourt co. Va. 8 ms. w. Fincastle, the co. st.

Craig's Creek, or more correctly, Craig's r. is the extreme s. w. confluent of James r. rises in Giles and Montgomery co. Va. interlocking sources with a branch of Great Kenhawa, and with the extreme higher sources of Roanoke; and flowing thence to the n. e. over Botetourt co. falls into James r. after a comparative course of 40 ms.

Craig's Meadow, p-o. Northampton co. Pa. by p-r. 127 ms. n. e. Harrisburg.

Crampton's Gap and p-o. Washington co. Md. by p-r. 60 ms. n. w. W. C.

Cranbury Isles, Hancock co. Me. between Frenchman's Bay and Mount Desert Sound, in the ocean. Pop. 1830, 258.

Cranberry, p-t. Middlesex co. N. J. 9 ms. e. Princeton, n. Millstone r.

Cranberry, p-v. and tsp. w. part of Butler co. Pa. about 30 ms. n. Pittsburg, and by p-r. 244 ms. n. w. W. C. and 213 ms. n. of w. Harrisburg. Pop. of the tsp. 1820, 765, 1830, 1,032.

Cranberry Plain, p-o. Grayson co. Va. by p-r. 251 ms. s. w. by w. Richmond. Cranberry creek is a small branch of Great Kenhawa, rising in the Iron mtn. and flowing sthrd. over the w. angle of Grayson co. into Ashe co. N. C.

Cranesville, p-o. in Williams co. O. by p-r. 524 ms. n. w. by w. W. C. and 188 ms. n. w. Columbus.

Craney Island, a small island in Elizabeth r. Va. only of adequate size for a fort, which commands the entrance to the harbor of Norfolk.

Cranston, t. Providence co. R. I. 5 ms. s. Providence, w. Providence r., n. Pawtucket r., 7 ms. by 4½, 19,448 acres; level e. but poor soil, and uneven w. A mine here has furnished ore for many cannon of the navy. Vegetables are furnished for the Providence market. Pop. 1830, 2,653.

Craven, co. N. C. bounded by Cartaret s. e., Jones s. w., Lenoir n. w. Pitt n., Beaufort n. e., and Pamlico Sound e. Length from s. e. to n. w. 65 ms., mean breadth 17, and area 1,100 sq. ms. Extending in lat. from 34° 48′ to 35° 23′, long. about 35′ on each side of the meridian of W. C. Neuse r. enters this co. on its n. w. border, and after flowing to the s. e. 35 ms. opens a wide bay, which, curving to e. and n. e. expands into Pamlico Sound. The surface is level, and in great part marshy, but with much good soil. Chief t. Newbern. Population 1820, 13,394, 1830, 13,734.

Crawford, t. Washington co. Me. has a large pond emptying into Machias r. Pop. 1830, 182.

Crawford, t. Orange co. N. Y. Pop. 1830, 2,019.

Crawford, co. Pa. bounded n. by Erie, e. by Warren, s. e. by Venango, s. by Mercer, s. w. by Trumbull, and w. by Ashtabula cos. of Ohio. Length 48, mean width 22; area 1,016 sq. ms. Extending in lat. from 41° 29′ to 41° 51′, long. from 2° 42′ to 3° 36′ w. W. C. From the s. w. angle rises Shenango branch of Big Beaver; from the n. w. angle rises the sources of Conneaut, flowing into Lake Erie; the e. part gives source to, and is drained by Oil creek, whilst the central section is traversed by the main volume, and several minor branches of French creek. The declivity of the whole co. is sthrd. Chief t. Meadville. Pop. 1820, 9,397, in 1830, 16,067.

Crawford, co. Geo. bounded by Upson n. w., Monroe n., Bibb e., Houston s. e., and Flint r. which separates it from Marion and Talbot s. w. Length 30, mean breadth 12, and area 360 sq. ms. Extending in lat. from 32° 30′ to 32° 50′, long. from 6° 53′ to 7° 24′ w. W. C. Though limited on one side by Flint r., the central part of this co. is a middle ground, from which the waters flow sth. estrd. by the Chocunno and Chupee crs. into Oakmulgee r., and by various branches s. w. into Flint r. Chief t. Knoxville. Pop. 1830, 5,313.

Crawford, co. Ark. bounded w. by the Indian or Mexican ter., n. by Washington co., n. e. by Conway, s. e. by Pulaski, and s. by Clark and Miller cos. The existing boundaries must, however, be temporary, since as laid down on Tanner's U. S. it stretches 120 ms. from e. to w. with a mean breadth of 65, and area of 7,800 sq. ms. Extending in lat. from 34° 43′ to 35° 36′, long. from 15° 28′ to 17° 30′ w. W. C. The Ark. r. enters this co. on its n. w. border, and winds eastwardly over it about 100 ms.; and thence turning to S E. forms the boundary 40 ms. between it and Conway. The surface is diversified by mountains, prairies, and wood lands near the water courses. Chief town, Marion.

CRAWFORD, co. Mich. on both sides of the Ouisconsin r. and bounded W. by the Miss. r. The outlines of this co. except on the Miss. are uncertain. Prairie du Chien, the st. jus. stands at the point above the entrance of the Ouisconsin r. into the Miss., and derives its name from a Prairie or natural meadow, so called. As laid down by Tanner, the junction of the two rs. is at lat. 43° and 14° 12' W. W. C. The village of Prairie du Chien is stated in the p-o. list of 1831, as being 1,060 ms. distant from W. C.

CRAWFORD, co. of O. bounded by Marion S., Hardin S. W., Hancock N. W., Seneca N., Huron N. E., and Richland E. Length from E. to W. 32, mean breadth 20, and area 640 sq. ms. Extending in lat. from 40° 43' to 41° 02', and in long. from 5° 48' to 6° 24' W. W. C. This co. is nearly commensurate with the higher part of the valley of Sandusky r. and is drained N. by the various constituents of that stream. The surface is level, and is a rather elevated table land. Chief t. Bucyrus. Pop. 1830, 4,791.

CRAWFORD, co. of Ind. bounded by Perry S. W., Dubois N. W., Orange N., Washington N. E., Harrison E., and O. r. separating it from Meade co. Ky. S. Length 24, mean breadth 14, and area 336 sq. ms. Extending in lat. from 38° 07' to 38° 25', and in long. from 9° 18' to 9° 43' W. W. C. The slope is to the S. and towards O. r. The surface very broken. Chief t. Fredonia. Pop. 1830, 3,238.

CRAWFORD, co. Il. bounded S. E. by Lawrence, S. W. by Clay, W. by Lafayette, N. by Clark, N. E. by the Wabash r. separating it from the northern part of Sullivan co. Ind., E. by Wabash r. separating it from the sthrn. part of Sullivan co. Ind., and the Wabash r. separating it from the N. W. angle of Knox co. Ind. Greatest length from E. to W. 50 ms., mean breadth 20, and area 1,000 sq. ms. Extending in lat. from 38° 50' to 39° 10', and in long. from 10° 34' to 11° 30' W. W. C. This co. is traversed in a S. E. direction, and subdivided into two not very unequal sections by Embarras r. Some of the higher sources of the Little Wabash rise in its S. W. angle. The general slope is a little E. of S. Chief t. Palestine Pop. 1830, 3,117.

CRAWFORD, co. of Mo., position uncertain, but supposed to be on the head branches of the Maramec and Gasconade rs., S. from Gasconade and Franklin cos.; about 100 ms. S. W. St. Louis.

CRAWFORD, C. H. and p-o. by p-r. 136 ms. N. W. Little Rock, and 1,204 ms. S. W. by W. W. C. Exact situation uncertain.

CRAWFORD'S p-o. in the E. part of Estill co. Ky. 34 ms. E. Irvine, the st. jus. for the co., and by p-r. 531 ms. S. W. by W. W. C. and 71 ms. S. E. by E. Frankfort.

CRAWFORD'S MILLS and p-o. Del. co. O. by p-r. 29 ms. N. Columbus, and 425 ms. N. W. by W. W. C.

CRAWFORDSVILLE, p-v. and st. jus. Montgomery co. Ind. by p-r. 617 ms. N. of W. W. C. and 44 ms. N. W. by W. Indianopolis. It is on Sugar creek, and on the road from Indianopolis to Covington, in Fountain co. Lat. 40° 03', long. 9° 53' W. W. C.

CRAWFORDSVILLE, p-v. and st. jus. Taliaferro co. Geo. situated between Little r. and Great Ogeechee r. 44 ms. N. N. E. Milledgeville, and 65 a little N. of W. Augusta. Lat. 33° 34', long. 5° 58' W. W. C.

CRAYTONVILLE, p-v. Anderson dist. S. C. 81 ms. N. W. Columbia.

CREAGERS or CREAGERSTOWN, p-v. near the left bank of Monocacy r. Frederick co. Md. 12 ms. a little E. of N. Frederick city.

CREEK AGENCY and p-o. Creek ter. Ala. 181 ms. from Tuscaloosa.

CREEK INDIANS, or Muscogees, stated by Dr. Jedediah Morse in 1820, at 20,000, overrated perhaps, reside principally in Geo. and Ala. but with some scattering bands in Flor. and La. This once comparatively considerable Indian nation has been known under the name of tribes, as Appalaches, Alabamas, Abacas, Cowittas, Coosa, Oakmulgees, Oconees, &c.

CREEK PATH and p-o. in the Cherokee ter. Ala. by p-r. 135 ms. N. E. Tuscaloosa, and 723 ms. S. W. by W. W. C.

CREELSBURGH, p-v. Russell co. Ky. by p-r. 162 ms. S. Hartford.

CRESAPTOWN or CRESAPSBURG, p-v. Alleghany co. Md. near the left bank of Potomac, 6 ms. S. W. by W. Cumberland.

CRICHTON'S STORE and p-o. in the S. part Brunswick co. Va. 18 ms. S. Lawrenceville, st. jus. for the co. and by p-r. 209 ms. S. S. W. W. C. and 87 from Richmond, in a nearly similar direction.

CRIPPLE CREEK, p-o. Greenville dist. S. C. by p-r. 118 ms. N. W. Columbia.

CRITTENDEN, v. on the road from Little Rock to Hempstead co. on Red r. Clark co. Ark. 82 ms. S. W. Little Rock.

CRITTENDEN, co. Ark. bounded E. by the Miss. r. W. and S. W. by St. Francis r., N. by New Madrid co. Mo. Length from S. to N. 105 ms. mean breadth about 20 ms.; area 21,000 sq. ms. Extending in lat. from 34° 35' to 36°, long. from 12° 40' to 13° 45' W. W. C. It is composed of an immense plain, in most part liable to annual submersion; but where the soil is of sufficient elevation to admit cultivation, it is highly productive. Staple, cotton. Chief t. Greenock.

CROOKED CREEK and p-o. N. part of Livingston co. Ky. by p-r. 235 ms. S. W. by W. Frankfort.

CROOKED CREEK and p-o. in the N. E. part of Tioga co. Pa. by p-r. 9 ms. N. Wellsborough, the co. st. 262 ms. N. W. C. and 156 W. of N. Harrisburg.

CROOKED LAKE, N. Y. Steuben and Ontario cos. 18 ms. by 1½, has two branches divided by Bluff Head. An outlet gives mill sites, and runs 6 ms. into Seneca lake.

CROOKED RIVER, Me. flows into Sebago pond.

CROCKETT, p-v. Gibson co. Ten. by p-r. 149 ms. s. of w. Nashville, and 854 ms. s. w. by w W C.

CROMMELIN, p-v. Montgomery co. Md. by p-r. 18 ms. from W. C. and 55 from Annapolis.

CROSS ANCHOR, p-o. in the extreme s. part of Spartanburg dist. S. C. by p-r. 82 ms. N. W. Columbia.

CROSS CANAL, p-o. Cambden co. N. C. by p-r. 151 ms. N. E. by E. Raleigh.

CROSS CREEK, v. and p-o. w. part of Washington co. Pa. 17 ms. N. W. Washington, the co. st.

CROSS KEYS, p-o. Rockingham co. Va. by p-r. 123 ms. s. w. by w. W. C.

CROSS KEYS, p-o. in the w. part of Union dist. S. C. by p-r. 63 ms. N. W. Columbus.

CROSS LAKE, N. Y. Cato, Cayuga co.

CROSS PLAINS, p-r. Robertson co. Ten. 31 ms. N. W. Nashville.

CROSS PLANS, p-v. Ripley co. Ind. by p-r. 88 ms. s. E. Indianapolis, and 560 w. W. C.

CROSS ROADS, p-o. Jones co. N. C. by p-r. 163 ms. s. E. Raleigh.

CROSS ROADS, p-o. Hardiman co. Ten. by p-r. 222 ms. s. w. by w. Nashville.

CROSS ROADS, p-o. Bibb co. Ala. 42 ms. s. E. Tuscaloosa.

CROSS ROADS, p-o. Chester co. Pa. (See New London, Cross Roads.)

CROSS ROADS, p-o in the w. part of Newton co. Geo. by p-r. 10 ms. w. Covington, the co. st. 70 N. W. Milledgeville, and 672 s. w. W. C.

CROSSWICK'S CREEK, N. J. rises in Monmouth co., and running through Burlington, falls into the Del. at Bordentown; is navigable several miles for sloops.

CROSSWICKS, p-v. Chesterfield, Burlington co. N. J. 8 ms. s. E. Trenton, 2 E. Bordentown, on Croswick creek.

CROTON CREEK, Dutchess and West Chester cos. N. Y. runs s. and s. w. 40 ms. into Hudson r. at Tappan bay, with good mill sites. It has been proposed to take the water to N. Y. city by aqueduct.

CROTON, v. Cortlandt, West Chester co. N. Y. at the mouth of Croton creek, where is a fall of 60 or 70 feet.

CROWNPOINT, p-t. Essex co. N. Y. 15 ms. N. Ticonderoga, 18 s. Elizabethtown, 184 s. Montreal, w. Lake Champlain. Level E. mountainous w. Contains the site of the old fortress of Crown Point, which was first occupied as a military position by the French, 1731. Surrendered to the British, 1759, and to the Americans, 1755; evacuated and taken by Gen. Burgoyne, 1777; retaken by Americans the same year. It has been long abandoned, but the earth shows the form of the fortress, which was a spar work, with 5 bastions, the walls of the barracks, &c. on a low level cape, running N. opposite Chimney Point, where the lake is 1 m. wide. Oct. 13th, 1776, the American flotilla, under Gen. Arnold, was destroyed off Crown Point by the British. Pop. 1830, 2,441.

CROWDER'S CREEK and p-o. in the E. part of York dist. S. C. The creek falls into Catawba r. where the road from Charlotte in N. C. passes to Yorkville in S. C. by p-r. 101 ms. N. Columbia.

CROWELL'S CROSS ROADS and p-o. Halifax co. Geo. by p-r. 229 ms. s. W. C. and 99 N. E. Raleigh.

CROW'S FERRY and p-o. parish of Natchitoches, La. This ferry is over the Sabine r. by the common road 33 ms. s. w. by w. Natchitoches or Red r., and by p-r. 405 N. W. by w. New Orleans, and 1,379 s. w. by w. W. C. By the p-o. list of 1831, Crow's ferry was the extreme s, w. p-o. in the U. S.

CROW'S NEST, mtn. Cornwall, Orange co. N. Y. 1,330 feet elevation.

CROWSVILLE, p-v. s. part of Spartanburgh dist. S. C. 72 ms. N. W. Columbia.

CROYDON, t. Cheshire co. N. H. 44 ms. N. w. Concord, 100 ms. Boston, 26,000 acres, is crossed by N. branch Sugar river and Croydon mtn.; though moist and rocky it yields grass and some grain. Pop. 1830, 1,056.

CRYSTAL SPRING, p-o. Lawrence co. Ark. by p-r. 176 ms. N. N. E. Little Rock.

CUBA, p-t. Alleghany co. N. Y. 18 ms. s. w. Angelica, N. Pa. 6 ms. by 18. It has Oil creek and some branches of Genesee river, is cold and wet, bears red oak, ash, maple, beech, some evergreen. Pop. 1830, 1,059.

CUBA, p-v. in the w. part of Clinton co. O. by p-r. 6 ms. w. Wilmington, the co. st. 450 w. W. C. and 73 s. w. Columbus.

CUCKOVILLE, p-v. Louisa co. Va. by p-r. 95 ms. s. w. W. C.

CULBERTSONS, p-o. Mercer co. Pa. by p-r. 293 ms. N. W. by w. Harrisburg.

CULBREATH'S, p-o. Columbia co. Geo. by p-r. 86 ms. N. E. by E. Milledgeville.

CULLEN, p-v. Weakley co. Ten. by p-r. 108 ms. w. Nashville.

CULLODEN'S, p-o. Monroe co. Geo. by p-r. 68 ms. w. Milledgeville.

CULPEPPER, co. Va. bounded by Rappahannoc r. which separates it from Fauquier N. E., by Rapid Ann river, which separates it from Spottsylvania, and Orange s., by Madison s. w., and by the Blue Ridge which separates it from Shenandoah N. W. Length from the junction of Rapid Ann and Rappahannoc rs. to its northern angle on the Blue Ridge, 42 ms.; mean breadth 16 and area 672 sq. ms. Extending in lat. from 38° 15′ to 38° 51′, long. from 0° 35′ to 1° 20′ w. W. C. Besides the boundary streams, Culpepper is watered by Thornton river which, rising in the spurs of Blue Ridge, winds s. E. over the central parts of the co. and falls into Rappahannoc. The surface is finely diversified with hill and dale, with large bodies of excellent land; staples, grain, tobacco, &c. Chief town, Fairfax. Pop. in 1820, 20,942, 1830, 24,027.

Culpepper, court house, (*see Fairfax, Culpepper co.*)

Cumberland, co. Me. bounded by Oxford co. n., Lincoln co. e., the Atlantic s., York and Oxford cos. w. It is one of the smallest cos., but contains Portland, the cap. and in 1820, 52,000 acres under tillage, 17,000 pasturage, 17,000 upland mowing, 950 meadow mowing, 1,000 working horses, and 2,600 working oxen. It contains Sebago Pond and several others, some of which it has been proposed to connect by navigable channels. Population 1820, 49,445, 1830, 60,113.

Cumberland, p-t Cumberland co. Me. on the sea coast, 54 ms. from Augusta. Pop. 1830, 1,558.

Cumberland, t. Providence co. R. I. 8 ms. n. e. Providence, n. e. Pawtucket r. 28 square ms., contains much hilly and rocky grass land, but the rest is generally good. Has Abbot's mill and Peter's rivers. Cotton has been manufactured here for some time; and 700 boats have been made here yearly, generally of oak. Pop. 1830, 3,675.

Cumberland, co. N. J. bounded by Salem and Gloucester cos. n., Cape May co. e., Delaware Bay s., Delaware Bay and Salem co. w. Chief town, Bridgetown, is crossed by Maurice river and Cohansey creek branches. Pop. 1830, 14,093.

Cumberland, mtn. chain of the Appalachian system, and continuation over Va. Ky. Ten. and part of Alabama, of the Laurel chain of Pa. The Cumberland chain, though not so delineated in our defective maps, is continuous from Steuben co., N. Y. into Jackson, Morgan, and Blount cos., Ala. along an inflected line of 800 ms. About the extreme e. angle of Ky. and s.w. Great Sandy, this chain is distinctly known as Cumberland mtn. and ranging s. w. separates Va. from Ky. as far as Cumberland gap, on the northern boundary of Tenn. Continuing s. w. but with an inflection to the n. w., this chain stretches over Tenn. as dividing ridge between the confluents of Cumberland and Ten. rivers. Entering Ala. and crossing Ten. river at its great bend, gradually disappears amongst the sources of Black Warrior river. The Cumberland chain is in no part very elevated, varying from 800 to 1000 feet above the tide level; but though humble as to relative height it maintains otherwise all the distinctive characteristics of other Appalachian chains. Extending in long, regular, and often lateral ridges, passable only at long intervals where gaps occur, or where traversed by rivers. The ridges are wooded to their summits.

Cumberland, r. Ken. & Ten. rises in the former fm. the n. w. slope of Cumberland mtn. interlocking sources with Ky. r. to the n. and Powell r. s.; flowing thence westward by comparative course 120 ms. in Russell co. Here it inflects to s. w. leaving Ky. and entering Ten. and preserving the latter course to Carthage in Smith co. having flowed in a s. w. direction 65 ms. Below Carthage, Cumberland inflects to a western course, which it pursues 100 miles to its great bend in Stewart co. and thence turning to n. w. flows 75 miles to its final junction with the Ohio, after an entire comparative course of 360 ms. The above measurements are made by extending from extreme to extreme of the respective courses, but as the stream is in its particular bends very tortuous, we may without excess allow for its comparative length 200 ms. in Upper Ky., 190 in Ten. and 50 in Lower Ky. or an entire comparative course of 440 ms. At high water it is navigable for boats, to near its source, and for at least one half its length, at all seasons. Without reference to the inflections of the river itself, the valley it drains is 350 ms. long with a mean breadth not exceeding 50 ms. area about 17,500 square ms. holding the third rank in regard to superficies of the confluents of Ohio r. The relative difference of level between the source and mouth of Cumberland, has never been determined but must exceed 1000 feet. The far greater part of this valley lies between lat. 36° and 37° and between long. 6° and 12° w. W. C.

Cumberland, co. Pa. bounded n. w. and n. by the Kittatinny, or as there locally named N. mtn. which separates it from Perry's, Susquehannah river separating it from Dauphin e., York co. s. e.; Adams s., Franklin s. w. Length 34, mean breadth 16, and area 544 square ms., lying between lat. 39° 58′, and 40° 18′, and long. 0° 08′ e., and 0° 40′ w. W. C. This co. is in great part commensurate with the valley of the Conedogwinet, which rising in Franklin enters the s. w. border of Cumberland and by a very winding channel flows n. e. by e. into Susquehannah r. The southern part including the ridges and valleys of s. mountain is watered by Yellow Breeches creek, which also flows n. e. by e. into Susquehannah river. The s. side of Cumberland rests mostly on limestone, whilst the substratum along the Kittatinny is clay slate. The surface moderately hilly, and soil generally very productive in grain, pasturage and fruits. Iron ore abounds in S. mtn. Chief town, Carlisle. Population 1820, 23,606, 1830, 29,228.

Cumberland, p-v. and st. jus. Alleghany co. Md. situated on the left bank of the Potomac river, and on both sides of Wills' creek, 136 ms. n. w. by w. W. C. and 140 a little n. of w. Baltimore. Lat. 39° 38′, long. 1° 46′ w. W. C.

The United States western road has its eastern termination at this village, which is elevated 537 feet above the level of the Atlantic. It is neat and well built, mostly in one street along the main road.

Cumberland, co Va. bounded by Appomattox river which separates it from Amelia s. and Prince Edward s. w., by Buckingham w. and n. w., by James river which separates it from Goochland n. e., and by Powhatan e. Length 32, mean breadth 10, area 320 square miles. Between lat. 37° 12′ and 37° 39′, long. 1° 13′, and 1° 40′ w. W. C. The slope of this co. is N. E. and the central parts drained of Wills' river, a branch of James river. The surface is moderately hilly, and soil productive. Chief town, Carterville. Pop. 1820, 11,023, 1830, 11,690.

CUMBERLAND, court house, (*see Springfield, Cumberland co. Va.*)

CUMBERLAND FORD, and p-o. Knox co. Ky. where the road from Frankfort through Lancaster, Mount Vernon, and Barboursville, into Ten. by Cumberland Gap, passes Cumberland river 16 miles up that stream above Barboursville, and 138 S. S. E. Frankfort.

CUMBERLAND GAP, and p-o. Claiborne co. Ten. The gap is in Cumberland mtn. 15 ms. S. Cumberland Ford.

CUMBERLAND, p-v. Guernsey co. O. by p-r. 91 ms. E. Columbus, and 330 north of west W. C.

CUMMINGTON, p-t. Hampshire co. Mass. 20 ms. N. W. Northampton, 110 W. Boston; has an academy, and several factories on the N. branch of Westfield river. Pop. 1830, 1,261.

CUNNINGHAM'S STORE, and p-o. in the N. W. part of Person co. N. C. by p-r. 75 ms. N. W. Raleigh, and 257 ms. S. W. W. C.

CURRAN, p-v. in the N. part of Gallatin co. Il. by p-r. 118 ms. S. E. Vandalia, and 792 ms. W. W. C.

CURRENT, r. of Mo. and Ark. the western and indeed the main branch of Black r. rises in Miss. interlocking sources with the Black, Merrimack, and Gasconade rivers, forms by its course an eliptic curve, first S. E. then S., and finally S. W. to its junction with Black r. in Lawrence co. Ark. (*See Black river of Mo. and Ark.*)

CURRITUCK, co. N. C. bounded by the Atlantic E., Princess Ann, and Norfolk cos. Va. N., Camden co. N. C. W., and Albermarle Sound S. Length 40, mean breadth 15, area 600 square miles. Lying between lat 36° and 36° 30′, long. 0° 45′, and 1° 25′ E. W. C. It is composed of a plain country, in part marshy, and divided into two sections by Currituck sound. Chief town, Currituck. Pop. 1820, 8,098, 1830, 7,655.

CURRITUCK SOUND, Isle and Inlet, Currituck co. N. C. The sound is a narrow sheet of water extending from Albemarle sound N. about 50 miles, and terminating by North r. bay and Rocky r. bay, in Princess Ann co. Va. It is shallow, with a breadth varying from one to ten miles. Currituck sound is open to the Atlantic by two inlets, S. inlet, and Currituck inlet, enclosing between them a long, low, and sandy reef of 20 miles in length, called Currituck island. According to Tanner's United States, Currituck inlet, is at lat. 36° 26′, 30 miles N. Albermarle sound.

CURRITUCK, p-v. and st. jus. Currituck co. N. C. situated on the W. side of Currituck sound, 35 miles S. S. E. Norfolk in Va. and by p-r. 231 miles N. E. by E. Raleigh. Lat. 36° 24′, long. 1° 02′ E. W. C.

CURWINSVILLE, p-v. Clearfield co. Pa. on the left bank of Susquehannah, 7 miles above and S. W. the borough of Clearfield.

CUSHING, t. Lincoln co. Me. 33 ms. E. Wiscasset, N. W. St. George's river, and indented with coves from the sea. Population 1830, 1,681.

CUTLER, t. Washington co. Me. E. Machias bay on the sea coast. Pop. 1830, 454.

CUYAHOGA, r. of O. This stream though comparatively small is very remarkable in itself, & has gained great importance from having become in part the route of the Ohio and Erie canal. The extreme source of Cuyahoga is near the eastern border of Geauga co. interlocking sources with those of Grand r. and within 20 miles from lake Erie at the mouth of the latter. Flowing thence S. S. W. nearly parallel, though rather inclining from the opposite shore of lake Erie, by comparative courses 45 ms., traversing Geauga and Portage cos. to near the E. border of Medina. Inflecting at more than a right angle upon its former course, the Cuyahoga bends to a little W. of N., is intersected by the Ohio canal near Northampton in Portage co., continues over the latter and Cuyahoga co. to its influx into lake Erie at Cleaveland, after an entire comparative course of 85 miles. The higher part of the course of Cuyahoga river is on a real table land. The summit level of the Ohio and Erie canal, between the vallies of Tuscarawas and Cuyahoga is 973 feet above tide water in the Atlantic, and 408 feet above lake Erie. The elevation of the canal in the aqueduct 18 ms. above the mouth of Cuyahoga, is 704 above the ocean tides, and 139 above the level of lake Erie.

CUYAHOGA, co. of Ohio, bounded by Geauga co. N. E., Portage S. E., Medina S. W., Lorain W., and lake Erie N. Length from east to west 32 ms., mean breadth 17, and area 544 square ms. Extending in lat. from 41° 18′ to 41° 45′, and in long. from 4° 26′ to 5° W. W. C. The surface of this co. is a rather rapidly inclining plain, the farms having a fall of at least 400 feet from the southern and higher, to the lower border on lake Erie. Chief town, Cleaveland. Pop. 1820, 6,328, 1830, 15,813.

CUYAHOGA FALLS, and p-o. Portage co. O. by p-r. 122 ms. N. E. Columbus, and 334 ms. N. W. W. C.

CYNTHIANA, p-v. and st. jus. Harrison co. Ky. by p-r. about 70 ms. a little E. of S. Cincinnati, 513 a little S. of W. W. C. and 38 N. E. by E. from Frankfort. It is situated on the eastern or right bank of the south fork of Licking river, at lat. 37° 23′, long. 7° 17′ W. W. C. Pop. 1830, 975. The adjacent country is fertile and well cultivated.

CYNTHIANA, p-v. in Shelby co. Ohio by p-r. 92 miles N. W. by W. Columbus, and 489 ms. from Washington City in a similar direction nearly.

CYNTHIANA, p-v. in the N. E. part of Posey county, Indiana, by p-r. 157 ms. S. W. Indianopolis, and 718 S. of W. W. C.

D.

DABNEY'S MILLS and p-o. in the E. part of Louisa co. Va. by p-r. 84 ms. w. of s. W. C. and 52 w. of N. Richmond.

DACHEET, r. of Ark. and La. rises in Hempstead, and flowing sthrd. over La Fayette co. of the former, enters the parish of Claiborne in the latter, is rather continued than lost in Lake Bisteneau. The writer of this article made a survey of Lake Bisteneau and Dacheet r. as far as the northern boundary of La. and found ample memorial to demonstrate, that the existence of the lake is recent. The cypress timber, once growing in the valley, is dead; but that timber resisting decay, the stumps remain standing in the water. The lake was, no doubt, formed by gradual accretion of soil, brought down by Red r. and deposited at the ancient mouth of Dacheet, 30 ms. below the existing head of the lake.

DACRESVILLE, p-v. Pickens dist. S. C. by p-r. 133 ms. N. W. Columbia.

DAGGETT'S MILLS, p-o. Tioga co. Pa. by p-r. 155 ms. N. N. W. Harrisburg.

DAGSBORO, p-v. Sussex co. Del. on Pepper creek, a confluent of Rehoboth bay, 18 ms. a little w. of s. Lewistown.

DALE, co. of Ala. bounded by Covington w., Pike N., Henry E., and by Jackson co. of Flor. s. E., and Walton co. Flor. s. w. Length from s. to N. 50 ms., mean breadth 32, and area 1,600 sq. ms. Lying between lat. 31° and 31° 43', long. 8° 46' and 9° 16' w. W. C. Dale co. occupies great part of the valley of Choctawhatchie r. Chief t. Richmond. Pop. 1820, 2,031.

DALE, p-v. Berks co. Pa. 20 ms. N. W. Reading.

DALETOWN, p-v. Wilcox co. Ala. by p-r. 92 ms. s. s. E. Tuscaloosa.

DALLAS, co. Ala. bounded by Marengo w., Perry N. W. and N., Autauga N. E., Montgomery E., and Wilcox s. and s. w. Length 38, mean breadth 28, and area 1,064 sq. ms. Extending in lat. from 32° 03' to 32° 33', long. from 9° 57' to 10° 40' w. W. C. Ala. r. enters the eastern border of Dallas, and winding over it by a very circuitous channel, and receiving the Cahaba near the centre, leaves the co. on the s. w. flowing thence into Wilcox. The river lands are highly fertile. Chief staple, cotton. Chief t. Cahawba. Pop. 1820, 6,003, 1830, 14,017.

DALLAS, p-v. Luzerne co. Pa. by p-r. 116 ms. N. E. Harrisburg.

DALMATIA, p-v. Northumberland co. Pa. 42 ms. N. Harrisburg.

DALTON, p-t. Coos co. N. H. next s. of Lancaster, E. Conn. r. at the head of 15 m. falls. It has John's r. &c., hills w. and s., and good soil on the uplands; 16,455 acres. Pop. 1830, 532.

DALTON, p-t. Berkshire co. Mass. 12 ms. N. N. E. Lenox, 120 w. Boston, near the head of E. branches of Housatonic r., is nearly level, with good soil, and manufactures cotton, paper, &c. Pop. 1830, 827.

DALTON, p-v. in the E. part of Wayne co. O. 15 ms. E. from Wooster, the co. st., and by p-r. 336 ms. N. W. by W. W. C. and 99 N. E. Columbus.

DALEY'S p-o. in the w. part of Montgomery co. Ten. 6 ms. w. Clarksville, and by p-r. 52 N. W. by W. Nashville.

DAMARISCOTTA, r. Me. passes through Lincoln co. and empties between Brothbay and Bristol.

DAMARISCOTTA MILLS, p-v. Lincoln co. Me. 34 ms. from Augusta.

DAMASCUS, p-v. on the right bank of Del. r. Wayne co. Pa. 15 ms. N. E. Bethany. The twp. in 1820 contained a pop. of 366, in 1830, 613.

DAMASCUS, p-v. in the extreme N. part of Montgomery co. Md., and on one of the roads from the city of Baltimore to Frederick t.

DAMASCUS, p-v. in the N. part of Henry co. O. situated on the left bank of Maumee r. by p-r. 485 ms. N. W. by W. W. C. and 161 N. N. W. Columbus.

DAMASCOVILLE, p-v. in the N. W. part of Columbiana co. O. 15 ms. N. W. New Lisbon, the co. st., and by p-r. 158 N. E. by E. Columbus, and 297 N. W. W. C.

DAN, r. Va. and N. C. drains the far greater part of Granville, Person, Caswell, Rockingham and Stokes cos. of the latter state, and of Patrick, Henry, Pittsylvania, and Halifax cos. of the former state. The extreme western sources of Dan r. are in Patrick co. Va. and in the s. E. spurs of the Blue Ridge. The general course almost due E. along the intermediate borders of N. C. and Va. to where the 4 cos. of Pittsylvania, Halifax, Person and Caswell meet. Here entering and flowing in Va. N. E. by E., falls into the Roanoke at Clarksville, having a mean breadth of about 33 ms. This river drains 3,960 sq. ms.

DANA, p-t. Worcester co. Mass. 75 ms. w. Boston, is crossed by a branch of Swift r. Pop. 1830, 623.

DANBORO, p-v. Bucks co. Pa. 30 ms. N. Phil.

DANBURGH, p-v. Wilkes co. Geo. 68 ms. N. E. Milledgeville.

DANBURY, t. Grafton co. N. H. 93 ms. from Boston, 30 from Concord; 19,000 acres, diamond-shaped; is generally hilly, and watered by Smith's r. First settled, 1771. Pop. 1830, 786.

DANBURY, p-t. Fairfield co. Conn. 55 ms. s. w. Hartford, 35 N. W. New Haven, 65 N. E. New York; 6 ms. by 8½; 53 sq. ms.; is fertile, with granite rocks, gravelly loam, undulated, with some marble quarries. There are manufactories on Still r. a branch of

Housatonic. It bears oak, walnut, &c. Hats are made to a great amount. The British burnt the town, 1777, with some military stores, and Gen. Wooster fell in opposing them. Pop. 1830, 4,331.

DANBY, p-t. Rutland co. Vt. 18 ms. s. Rutland, 34 N. Bennington; 39 sq. ms. First settled, 1768; has Otter creek E. and several small branches, and is uneven, with some mountains. It has large dairies. There are several caverns, and lead ore is found. Pop. 1830, 1,362.

DANBY, p-t. Tompkins co. N. Y. 7 ms. s. Ithaca, 11 from Candor, 22 Owego, has pretty good land, watered by Mud creek, and a branch of Cayuga inlet.

DANCEY'S STORE and p-o. Northampton co. N. C. by p-r. 100 ms. N. E. by E. Raleigh, and 200 s. W. C.

DANDRIDGE, p-v. and st. jus. Jefferson co. Ten. situated on the right bank of French Broad r. on the road from Knoxville to Greenville, 32 ms. E. from the former, and 39 s. w. by w. from the latter place. Lat. 38° 58′, and long. 6° 14′ w. W. C.

DANDRIDGE, p-v. Morgan co. Ala. by p-r. 110 ms. N. N. E. Tuscaloosa.

DANIELSVILLE, p-v. Spottsylvania co. Va. by p-r. 78 ms. s. w. W. C.

DANIELSVILLE, p-v. and st. jus. Madison co. Geo. 81 ms. N. Milledgeville. Lat. 34° 10′, and long. 6° 15′ w. W. C.

DANSBY, p-o. Oglethorpe co. Geo. 81 ms. N. Milledgeville.

DANUBE, p-t. Herkimer co. N. Y. 10 ms. s. E. Herkimer, 68 N. w. Albany, s. Mohawk r., E. German Flats, 5½ ms. by nearly 10, has a stiff loam, on hard grit, with many springs, few streams, and waving surface. Nowadaga creek enters Hudson r. at the site of Hendrick's castle; a friendly Mohawk chief was killed at Lake George, 1755. The Indians had a church here. Pop. 1830, 1,723.

DANVERS, p-t. Essex co. Mass. 16 ms. N. E. Boston. The village streets form a continuation of those of Salem, 7 ms. by 8. The first victim of the persecution of witchcraft was a daughter of a clergyman of Salem, living in this town, then a part of it. The soil is good, well cultivated; granite mill stones are made, and different manufactures. There are several creeks from Bass r., two navigable to the two villages. Pop. 1830, 4,228.

DANVILLE, p-t. Cumberland co. Me. 32 ms. from Augusta, s. w. Androscoggin r. Pop. 1830, 1,128.

DANVILLE, p-t. and st. jus. Caledonia co. Vt. 25 ms. N. E. Montpelier, 25 N. w. Newbury, 160 N. by w. Boston. First settled, 1784, from Essex co. Mass.; broken w., hills and vallies E., with fine farms; Merritt's r. &c. supplies mills. The village is pleasant in the centre. Pop. 1830, 2,631.

DANVILLE, (now Wilmington,) p-t. Steuben co. N. Y. 21 ms. N. w. Bath. Pop. 1830, 1,728.

DANVILLE, p-v. on the right bank of Susquehannah r. Columbia co. Pa. 25 ms. above Northumberland.

DANVILLE, p-v. Pittsylvania co. Va. on the right bank of Dan r., and near the s. border of the co. In the natural state, the falls of Dan r. at Danville, were the head of boat navigation, but by a not very expensive canal improvement, the higher part of Dan valley might be opened to an intercommunication by water with the Roanoke.

DANVILLE, p-v. Warren co. Ten. 55 ms. s. w. Nashville.

DANVILLE, flourishing p-v. on the s. E. border of Mercer co. Ky. 41 ms. s. s. E. Frankfort, and 35 a little w. of s. Lexington. Pop. 1820, 1,000, in 1830, 849. Centre college is situated in the immediate vicinity of Danville, a quiet and retired village, where there are few temptations to seduce the young men from their studies, or allure them into vice. The town and surrounding country are remarkable healthy. Danville is 10 ms. distant from the Harrodsburg springs, the favorite watering place of persons from the s. The proximity of these springs, affords to parents from that quarter, a favorable opportunity of occasionally seeing their sons who may be placed there for education. The price of tuition is in the college classes $30, and in the preparatory department $24, per college year, payable half yearly in advance. Boarding of a superior kind in the refectory, including washing and lodging, $1 50 per week. Fuel and light, furnished by the steward, at cost. Whole estimated expense $103 per annum, exclusive of books.

DANVILLE, p-v. in the N. E. part of Knox co. O. by p-r. 362 ms. N. w. by w. W. C. and 59 ms. N. E. Columbus. Pop. 1830, 234.

DANVILLE, p-v. and st. jus. Hendricks co. Ind. by p-r. 593 ms. N. w. by w. W. C. and 20 ms. w. Indianopolis. It is situated on the head of White Lick creek, at lat. 39° 47′, long. 9° 30′ w. W. C.

DANVILLE, p-v. and st. jus. Vermillion co. Il. by p-r. 683 ms. N. of w. W. C. and 150 N. N. E. Vandalia. It is situated on Vermillion r. about 4 ms. from the E. boundary of Il. Lat. 40° 8′, long 9° 42′ w. W. C.

DARBY CREEK, or small r. of O. rises in Champaign and Union cos. flows s. s. E. over Franklin, and falls into the w. side of Sciota r. in Pickaway co. nearly opposite Circleville, having a comparative course of 60 ms.

DARBY, p-v. 7 ms. s. w. Phil., Del. co. Pa. The old town of Darby is divided into Lower Darby and Upper Darby. In 1830, the former contained 1,085, and the latter 1,325 inhabitants. The village of Darby is a fine and pleasant town.

DARBY, p-v. and tsp. on Darby creek, s. part of Union co. O. by p-r. 22 ms. N. w. Columbus, and 418 N. w. by w. W. C. Pop. of the tsp. 1830, 417.

DARBY'S p-o. Columbia co. Geo. 12 ms. from Applington, the co. st. and by p-r. 590 s. w. W. C. and 81 N. E. by E. Milledgeville.

DARBYVILLE, p-v. in Darby tsp. Pickaway co. O. by p-r. 39 ms. S. Columbus, and 407 N. of W. W. C. Pop. of the tsp. 1830, 827.

DARDANELLES, two mountain peaks, so called, Crawford co. Ark. They are situated on the right side, and near Ark. r. 82 ms. above Little Rock.

DARDANELLES, p-o. Ark. near the Dardanelles mtns. Crawford co. (See Tekatoka.)

DARDENNE, small r. of St. Charles co. Mo. rising along the boundary between the latter and Montgomery co. and flowing N. E. falls into the Miss. r. about 5 ms. N. N. W. from the village of St. Charles.

DARDENNE, p-v. in the W. part of St. Charles co. Mo. by p-r. 34 ms. N. W. St. Louis, 100 N. of E. Jefferson, and 897 W. W. C.

DARDENNE BRIDGE and p-o. in the N. E. part of St. Charles co. Mo. about 30 ms. N. W. by W. St. Louis, and by p-r. 886 ms. W. W. C.

DARIEN, p-t. Fairfield co. Conn. 42 S. W. New Haven, N. Long Island Sound, formerly S. E. part of Stamford, and has a good undulating soil. Pop. 1830, 1,201.

DARIEN, important p-t. and sea-port McIntosh co. Geo. on the N. side and principal channel of the Altamaha, 12 ms. above the bar, and 190 by water below Milledgeville. The pop. of Darien has rapidly increased; in 1810, the inhabitants were about 200, in 1820, 2,000. It has a bank, custom house, and many splendid private buildings. A steam boat navigation extends above to Milledgeville, whilst the bar admits vessels of 12 feet draught from the ocean. It is 56 ms. S. S. W. Savannah, and by p-r. 185 S. E. Milledgeville. Lat. 31° 23′, long. 4° 37′ W. W. C.

DARKE, co. O. bounded N. by Mercer, N. E. by Shelby, S. E. by Miami and Montgomery, S. W. by Wayne co. Ind. and W. by Randolph co. Ind. Extending in lat. from 39° 52′ to 40° 27′, and in long. 7° 26′ to 7° 48′ W. W. C. Breadth 21 ms. from E. to W., mean length 36, and area 756 sq. ms. Darke co. is nearly commensurate with the region drained by and giving source to the higher branches of the W. Fork of Great Miami, with a general S. E. slope. Surface pleasantly diversified by hill and dale, soil in part productive. Chief t. Greenville. Pop. 1820, 3,717, 1830, 6,204.

DARKESVILLE, p-v. on Sulphur Springs cr., a branch of Opequau, Berkley co. Va. 25 ms. a little N. of W. Harper's Ferry.

DARLING'S p-o. in the N. E. part of Knox co. O. by p-r. 65 ms. N. E. Columbus, and 362 N. W. by W. W. C.

DARLINGSVILLE, p-o. Pike co. Pa. by p-r. 12 ms. N. Milford, the co. st. and 169 N. E. Harrisburg.

DARLINGTON, formerly Griersburg, p-v. in the N. W. part of Beaver co. Pa. about 10 ms. N. W. from the borough of Beaver, 37 ms. in a similar direction from Beaver, the co. st. and 263 in a similar direction from W. C.

DARLINGTON, p-v. Harford co. Md.

DARLINGTON, dist. S. C. bounded by Kershaw W., Chesterfield N. W. and N., Great Pedee river, which separates it from Marlborough N. E., Marion S. E., and Lynch's creek which separates it from Sumpter S. W. Length 35 ms., mean breadth 30, and area 1,050 square ms. Extending from lat. 33° 58′ to 34° 32′, long. from 2° 40′ to 3° 20′ W. W. C. The slope of this co. is to the S. E.; the surface rather waving than hilly. Chief town, Darlington. Pop. 1820, 10,949, 1830, 13,728.

DARLINGTON, p-v. and st. jus. Darlington dist. S. C. on Black creek, 40 ms. a little N. of E. Camden, and by p-r. 93 N. E. by E. Columbia. Lat. 34° 19′, long. 2° 58′ W. W. C.

DARNESTOWN, p-v. in the W. part of Montgomery co. Md. by p-r. 25 ms. N. W. W. C.

DARRTOWN, p-v. in Milford tsp. western part of Butler co. O. 8 ms. from Hamilton, the co. st. and by p-r. 109 ms. S. W. by W. from Columbus, and 496 ms. W. from W. C.

DARTMOUTH COLLEGE. (*See Hanover, N. H.*)

DARTMOUTH, p-t. and sea port, Bristol co. Mass. 62 ms. S. Boston, 27 S. Taunton, N. Buzzard's bay, E. R. I., has several creeks running S. one of which, Aponiganset river, is navigable nearly to its centre. It was destroyed by Indians 1675. Pop. 1830, 3,866.

DARTMOUTH, p-o. Tioga co. Pa. by p-r. 153 ms. N. Harrisburg.

DARVILLS, p-o. Dinwiddie co. Va. 32 miles S. Richmond.

DAUPHIN, p-v. Dauphin co. Pa. by p-r. 8 ms. from Harrisburg.

DAUPHIN, co. Pa. bounded by Mahantango creek, which divides it from Northumberland N., by Schuylkill N. E., Lebanon E., Conewago creek, which separates it from Lancaster south, and by the Susquehannah river, which separates it from York south, and from Cumberland and Perry west. Length 38 miles, mean breadth 16, area 608 square miles. Extending from lat. 40° 08′ to 40° 40′, long. from the meridian of W. C. to 30′ E. The surface of Dauphin is peculiarly diversified. The lower, southern, and smaller section enclosed by the Susquehannah r., Conewago creek, Kittatinny mtn. and the western boundary of Lebanon, and comprising about 170 square miles, is hilly, but highly fertile, and the southern part resting on a substratum of limestone. This lower section of Dauphin is traversed by the Swatara creek, and through its valley by the Union canal. The higher northern and mountainous part of Dauphin, beyond and including Kittatinny mountain, is formed by a congeries of mountain ridges, lying parallel to each other from N. E. to S. W., with narrow intervening vallies, discharging their streams S. W. into Susquehannah. The soil of Dauphin is productive in fruit, grain, and pasturage, and its mountain vallies abound in excellent timber. Besides the Union canal already mentioned, the transversed section of the Pa. canal follows the left bank of Susquehannah, in Dauphin, from the Conewago, to opposite the mouth of Juniata river. (*See Pa. canal.*) Chief town,

Harrisburg. Pop. 1820, 21,663, in 1830, 26,241.

DAVENPORT, p-t. Delaware co. N. Y. 11 ms. N. Delhi, is hilly; yields grass and lumber, and has Charlotte river of the Susquehannah. The lands are leased. Pop. 1830, 1,778.

DAVIDSON, co. N. C. bounded by Stokes N., N. E. by Guilford, E. by Randolph, S. by Montgomery, and W. by Yadkin r. which separates it from Rowan. Length from S. to N. 40 ms., mean breadth 20, and area 800 square miles. Lying between lat. 35° 30′ and 36° 04′, long. 3° 05′, and 3° 34′ w. W. C. The slope of this co. is southardly, and is drained by different small creeks flowing into the Yadkin. Chief town, Lexington. Pop. 1830, 13,389.

DAVIDSON, co. Ten. bounded S. by Williamson, W. by Dickson, N. by Robertson, N. E. by Sumner, E. by Wilson, and S. E. by Rutherford. Length 30, mean breadth 22, area 660 square ms. Extending from lat. 35° 56′ to 36° 22′, long. from 9° 40′ to 10° 10′ w. W. C. Cumberland r. winds by a very tortuous channel over this co. from E. to W. dividing it into two nearly equal parts; surface moderately hilly, with a very fertile soil, abounding in limestone. Staples, grain, and cotton. Chief town, Nashville. Pop. 1820, 20,154, 1830, 28,122.

DAVIDSONSVILLE, p-v. and st. jus. Lawrence co. Ark. situated on the point above the junction of White and Eleven Points river, by p-r. 169 ms., but by direct road 124 miles N. N. E. Little Rock. Lat. 36° 10′, long. 14° 03′ w. W. C.

DAVIDSONVILLE, p-o. Ann Arundel co. Md. by p-r. 11 ms. from Annapolis and 30 ms. eastward W. C.

DAVIES, co. Ky. bounded by Hancock co. N. E., Ohio S. E., Green r. dividing it from Muhlenburg S., Hopkins S. W., and Henderson W., by a small part of Henderson N. W., and by Ohio river separating it from Spencer co. Ind. N. Extending in lat. from 37° 29′ to 37° 53′, long. from 9° 55′ to 10° 36′ w. W. C. It is nearly a square of 22 ms. each way; 484 sq. miles, surface rather flat, and soil productive. Chief town, Owensburg. Pop. 1820, 3,876, 1830, 5,209.

DAVIS' MILLS, and p-o. Bedford co. Va. by p-r. 138 ms. S. W. by W. Richmond.

DAVIS' MILLS, and p-o. Barnwell district S. C.

DAVIS' MILLS, and p-o. Bedford co. Ten. by p-r. 45 ms. S. E. Nashville.

DAVIS' CROSS ROADS, p-o. Franklin co. N. C. by p-r. 31 ms. N. E. Raleigh.

DAVIS' TAVERN, and p-o. Sussex co. Va. about 50 ms. S. E. Richmond.

DAVIS' STORE, and p-o. Bedford co. Va. by p-r. 152 ms. W. Richmond.

DAVIS' STORE, and p-o. Martin co. N. C. by p-r. 81 ms. E. Raleigh.

DAVISBORO', p-v. in the southern part of Washington co. Geo. by p-r. 39 ms. S. E. Milledgeville, and 657 S. W. W. C.

DAVISVILLE, p-o. Bucks co. Pa.

DAWSON'S, p-o. Alleghany co. Md. by p-r. 16 ms. W. Cumberland, and 148 miles N. W. W. C.

DAWSON'S, p-o. Nelson co. Va. by p-r. 107 miles N. of W. Richmond, and 149 miles S. W. W. C.

DAWSONVILLE, p-o. in the western part of Montgomery co. Md. by p-r. 27 miles N. W. W. C.

DAY, t. Saratoga co. N. Y. Population 1830, 758.

DAYTON, p-v. and st. jus. Montgomery co. O. by p-r. 462 ms. N. W. by W. ½ W. from W. C. 66 ms. a little S. of W. from Columbus, and 52 ms. N. of E. from Cincinnati. Population 1830, 2,950. It is situated on a fine site along the left bank of Great Miami river, directly below the influx of Mad river, and near where the Miami canal connects with Miami river. The water of Mad river is conveyed across the point to the Miami, affording numerous and excellent mill seats. N. lat. 39° 43′, long. 7° 11′ w. from W. C.

The population of Dayton township, was in 1830, 6,828. The town contained 370 houses and mercantile stores, 4 churches, the county buildings, and market house. This place is in a remarkable manner prosperous, arising from an active and intelligent population, rendering available great natural advantages of water power. This power is in one way employed to give motion to saw mills, grist mills, cloth factories, and many other applications of machinery; and on the other, Miami canal has opened a water intercommunication with the Ohio river at Cincinnati; distance between the two places by the canal 67 miles.

DEAD, river, Me. W. branch of Kennebec river, rises on the borders of Lower Canada, and Oxford co.

DEADFALL, p-o. Abbeville dist. S. C. 112 ms. W. Columbia.

DEAL, a part of the sea shore in Monmouth co. N. J. south of Long branch.

DEAL, v. N. J. 7 ms. S. Shrewsbury, W. Atlantic; has a gently varied surface, with much thin sandy soil, and a white sand beach, on which vessels are frequently wrecked in E. storms. The marl, dug here near swamps and creeks, is excellent manure. It often contains sharks' teeth, bits of bones, &c. 30 loads to an acre, after being exposed one winter, converts a sand waste into a garden. The only table land in the United States tillable to the beach of the ocean (with one exception S. in this state,) is in Deal.

DEARBORN, p-t. Kennebeck co. Me. 22 ms. N. Augusta, contains parts of two large ponds, one with several islands. Population 1830, 616.

DEARBORNE, co. Ind. bounded by Switzerland co. S., Ripley W., and Franklin N., again on the N. E. it is bounded by Hamilton co. O., and on the S. E. by the Ohio river, separating it from Boone co. Ky. Length 27, mean breadth 15, and area 405 sq. ms. Extending in lat. from 38° 54′ to 39° 18′, and in long. from 7° 48′ to 8° 08′ w. W. C. The slope of this co. is rather E. of S. E. towards the Miami and Ohio rivers. Surface generally hilly. Chief town, Lawrenceburg. Population 1820, 11,468, 1830, 13,974.

DEAVERTOWN, p-v. York tsp. in the N. part of Morgan co. O. by p-r. 352 ms. N. W. by W. W. C. and 75 ms. S. of E. Columbus. Pop. 1830, 116.

DECATUR, p-t. Otsego co. N. Y. 12 ms. S. E. Cooperstown, 5 ms. by 6, on high land, is good for grazing, watered by the heads of Oaks' and Parker's creeks, rapid streams. It is very healthy. Pop. 1830, 1,110.

DECATUR, S. W. co. of Geo. bounded by Early and Baker N., Ocklockonne r. which separates it from Thomas co. E., by Gadsden co. Flor. S., and by Chatahooche r. which separates it from Jackson, in Flor. and Henry in Ala. W. Length from E. to W. 60 ms. mean breadth 28, area 1,680 sq. ms. Extending in lat. from 30° 42′ to 31° 06′, long. from 7° 11′ to 8° 12′ W. W. C. Flint river enters from Baker and flowing S. W. joins Chatahooche at the southwest angle of Decatur. It has advantages of three navigable rivers, the two bounding streams, and the Flint. The surface is moderately hilly, with a soil generally of second rate. Chief town, Bainbridge. Pop. 1830, 3,854.

DECATUR, p-v. and st. jus. De Kalb co. Geo. by p-r. 680 ms. S. W. W. C. and 117 ms. N. W. Milledgeville. It is situated on the high ground between the waters of Ockmulgee and Chatahooche rivers. Lat. 33° 40′, long. 7° 24′ W. W. C.

DECATUR, court house. *(See Bainbridge, Decatur co. Geo.)*

DECATUR, p-v. on the left bank of Ten. r. and in the N. E. part of Morgan co. Al., about 20 ms. S. W. Huntsville.

DECATUR, p-v. on the western border of Adams co. Ohio, by p-r. 469 miles W. W. C. and 110 W. of S. Columbus.

DECATUR, co. of Ind. bounded S. E. by Ripley, S. by Jennings, S. W. by Bartholomew, N. W. by Shelby, N. by Rush, and N. E. by Franklin. Length diagonally from S. W. to N. E. 30 ms., mean breadth 10, and area 300 sq. ms. Extending in lat. from 39° 07′ to 39° 27′, and in long. from 8° 18′ to 8° 49′ W. W. C. A slip along the E. border gives source to Loughery's creek, flowing S. E. over Ripley and Switzerland cos. into the O., and to Salt creek, entering White Water branch of Great Miami; but the body of the co. declines to the S. W. and is drained in that direction by Sand, Clifty, and Flat Rock creeks, flowing into the E. Fork of White r. Chief t. Greensburg. Pop. 1830, 5,887.

DECATUR, p-v. and st. jus. Macon co. Il. by p-r. 771 ms. N. W. by W. W. C., 70 E. of N. Vandalia, and about 150 ms. N. of W. Indianopolis in Ind. It is situated on Sangamon r. at lat. 39° 55′, long. 11° 50′ W. W. C.

DECKERSTOWN, p-v. Wantage, Sussex co. N. J. on Deep Clove creek.

DEDHAM, p-t. and cap. Norfolk co. Mass. 10 ms. S. W Boston, S. Charles r., W. Neponset r., has different soils, some high and arable; pine, and swamps capable of draining. It has a large village on Charles r. on the turnpike road from Boston to Providence. Mother Brook, a mill stream, runs from Charles r. into Neponset r. First settled from England. Silk is reeled and throwsted here on a limited scale, one of the first experiments of throwsting in the U. S. Pop. 1830, 3,117.

DEEP CREEK and p-o. Norfolk co. Va. The creek is a branch of Elizabeth r., and village is situated at the N. extremity of the Dismal Swamp canal. The village has been indeed created by the canal, and is now a flourishing depot, about 10 ms. S. S. W. Norfolk.

DEEP RIVER, one of the main northwestern branches of Cape Fear r. N. C. rising by several branches in Guildford and Randolph cos., which flowing S. E. unite in the latter. Leaving the N. E. angle of Randolph, and curving along the northern side of Moore, turns to N. E. by E., enters Chatham, and joins Haw r. to form Cape Fear r. The valley of Deep r. lies between those of Yadkin and Haw rs.

DEEP RIVER, p-o. S. W. part of Guilford co. N. C. about 90 ms. N. of W. Raleigh.

DEEP SPRING and p-o. Monroe co. Ten. by p-r. S. E. by E. Nashville.

DEERFIELD, p-t. Rockingham co. N. H. 17 ms. S. E. Concord, 35 N. W. Portsmouth; 28,254 acres; has Shingle, Moulton's and part of Pleasant Ponds, emptying into Suncook and Lamprey rs. It is uneven and hard, bearing maple, beech, birch, red oak, pine, &c.; has Tuckaway, Saddleback, and other mtns. First settled, 1756. Pop. 1830, 2,090.

DEERFIELD, p-t. Franklin co. Mass. 4 ms. S. Greenfield, 17 N. Northampton, 92 W. Boston, W. Conn. r., has rich soil, with fine meadows on Deerfield r. The village is a little elevated above them on a level. First settled, 1670. About 80 men were killed at Bloody Brook, 3 ms. S. returning with loads of wheat from this place, and long after an exposed frontier settlement. It was burnt by French and Indians, except one house, 1704. Many of the inhabitants were at different periods killed or carried captive to Canada. The bell taken from the church, 1704, still hangs in that of St. Regis, above Montreal. There is an academy. Pop. 1830, 2,003.

DEERFIELD, t. Oneida co. N. Y. 96 ms. W. N. W. Albany, N. Mohawk r. opposite Utica, S. W. Canada creek, E. Nine Mile creek. The uplands are good for grain and grass. The vallies have rich sand, loam, and pebbles. Pop. 1830, 4,182.

DEERFIELD, p-t. Cumberland co. N. J. near the source of Cohansey creek. Contains several villages, of which the largest is Bridgeton. Pop. 1830, 2,417.

DEERFIELD, p-v. Warren co. Pa. 100 ms. N. N. E. Pittsburg.

DEERFIELD, p-v. Augusta co. Va. by p-r. 181 ms. S. W. W. C.

DEERFIELD, p-v. and tsp. in the S. E. angle of Portage co. O. The p-o. is situated by p-r. 307 ms. N. W. W. C. and 15 S. E. Ravenna, the co st., and 142 N. E. Columbus. Pop. 1830, 694.

DEERFIELDVILLE, p-v. Union tsp. Warren co. O. by p-r. 4 ms. N. Lebanon, the co. st. 472 W. W. C. and 87 S. W. by W. Columbus. This place is not the same as Deerfield tsp. in the same co. Pop. 1830, 66.

DEERING, p-t. Hillsboro co. N. H. 23 ms. W. by S. Concord, 23 from Hopkinton, 66 from Boston; 20,057 acres, is uneven and favorable for agriculture, with 3 ponds, sources of N. branch of Piscataquog r. First settled, 1765. Pop. 1830, 1,228.

DEER ISLE, p-t. and island, Hancock co. Me. 95 Augusta. Is protected from the sea by numerous small islands, 9 ms. S. E. Castine, in Penobscot bay.

DEER PARK, p-t. Orange co. N. Y. 30 ms. W. Newburgh, 14 W. N. W. Goshen, 110 W. by S. Albany, E. Del. r. and Pa. 6 ms. by 12, W. Shawangunk creek; Navisink creek follows Shawangunk mtns. through E. part, half is uncultivated mountains, quarter stony pasture, quarter pretty good for grain. Shawangunk mtns. are rich and arable W. Pop. 1830, 1,167.

DEERSVILLE, p-v. Stock tsp. Harrison co. O. by p-r. 290 ms. N. W. by W. W. C. and 131 N. E. by E. Columbus.

DEFIANCE, usually called Fort Defiance, on the point between and above the junction of Maumee and Au Glaize rs. and in the S. E. angle of Williams co. O. The p-v. is situated in a tsp. of the same, and is also the st. jus. for the co. distant 511 ms. N. W. by W. W. C. and 175 N. W. Columbus. This was a very important military station during the Indian wars, and is situated in a very fertile, but, as yet, a thinly settled country. Pop. 1830, 52. N. lat. 41° 18′, long. 7° 22′ W. W. C.

DE KALB, co. of Geo. bounded S. by Henry and Fayette, W. by Campbell, N. W. by Chatahooche r. N. E. by Gwinnet co. and S. E. by Newton. Length from S. to N. 30 ms., mean breadth 12, and area 360 sq. ms. Extending in lat. from 33° 41′ to 34° 06′, and in long. from 7° 6′ to 7° 32′ W. W. C. De Kalb is a table land, from which issues one of the extreme branches of Ockmulgee r. flowing S. E. whilst from the N. part of the co. rise brief streams, falling into Chatahooche r. Chief t. Decatur. Pop. 1830, 10,047.

DEKALB, p-t. St. Lawrence co. N. Y. 15 ms. S. Ogdensburgh, 10 ms. square, is crossed by Oswegatchie, 20 ms. above its mouth in St. Lawrence, to which there is a boat navigation from the village and falls, between rich meadows. It affords oak, maple, beech, &c. marble, iron ore, ashes, &c. Settled from Conn. Pop. 1830, 1,061.

DELAWARE, r. N. Y., N. J. and Pa. rises in Schoharie co. N. Y. and in the western spurs of Catskill mtns. by two large branches, the Oquago and Popachton. The Oquago is the most remote and real source of Del. r. flows S. W. 50 ms. reaching within 10 ms. of the Susquehannah, turns to S. E. and flowing in that direction 5 ms. to the N. E. angle of Pa. and 5 ms. still lower, receiving the Popachton from N. E., continues the latter course 70 ms. to the western base of Kittatinny mtns., having for 60 ms. formed the boundary between Pa. and N. Y. Inflected to S. W. by the Kittatinny, and almost washing the base of that chain 35 ms. to the entrance of Broad Head's creek from the W., and from Pike and Northampton cos. it takes a southern course, and pierces the Kittatinny, by the known Del. Water gap. Continuing southwardly 21 ms. it receives its first great confluents from the right, the Lehigh, at Easton; 2 ms. lower it pierces the Blue Ridge, and 5 ms. still lower the S. mtn. having traversed a great part of the Appalachian system obliquely. Below the S. mtn. this now fine navigable r. assumes a course S. S. E. from which, 35 ms. it falls over the primitive ledge, and meets the tide at Trenton; 5 ms. below, opposite Bordentown, it again turns to S. W. Following nearly the range of the primitive rock, the Del. now widening, passes Phil., and 5 ms. below that city, receives its greatest tributary, the Schuylkill, from N. W., Cartney S. W.; 35 ms. still farther it passes Old Chester, Wilmington, and New Castle, to an imaginary line from Cape May to Cape Henlopen. The comparative length of Del. r. from its source to tide water is 185 ms., and 132 ms. from the rapids and head of tide at Trenton, to the Atlantic, having an entire comparative course of 317 ms. It has been already remarked, that from the N. E. angle of Pa. to the bend at the W. base of Kittatinny mtn., the Del. forms the line of separation between N. Y. and Pa., from N. J. and thence to the mouth of the bay, it divides N. J. from Del. The navigation of Del. bay is tortuous and something difficult, but admits the entrance of vessels of the first class to near Phil. Above that city the depth gradually decreases, but small sea vessels are navigated to Trenton. Though above tide water, this river is much impeded by shoals, and at low water by rapids; no falls, properly so called, exist in its bed, it is therefore navigable for down stream rafts and boats from near its source. As a commercial basin, that of Del. is in a rapid and extensive state of improvement. Within Cape Henlopen a breakwater or artificial harbor is in progress, which, when completed, will secure safety to vessels entering in all weather. The bay of Del. is connected with that of Chesapeake by the Chesapeake and Del. canal, extending 14 ms. with a depth of 8 feet, 60 wide at the surface, and 36 at bottom, and the river with Hudson r., by the Del. and Hudson, and the Morris canals. An active business on this stream has originated in the extensive mining districts in its vicinity; and these have chiefly encouraged the construction of the canals to Hudson r. (*See article 'Rail Roads and Canals.'*) Aside from the immense business arising from the coal mines, the Del. is one of the principal channels of internal trade in the U. S.

The basin of Del. r. lies between lat. 38° 45′ and 42° 30′, and long. 0° 42′ to 2° 35′ E. W. C. It is about 250 ms. in length from S. to N. with a mean breadth of 45, area 11,250 sq. ms. The surface greatly diversified, and with considerable difference of relative height. The higher fountains of Oquago and Papachton rs. must be at an elevation of at least 2500 feet, but the fall is rapid, and the general and comparative height of the cultivated sections is as follows:—Port Carbon, on the head of Schuylkill, 620 feet; Mauch Chunk village on Lehigh 534; Easton on Del. at the mouth of Lehigh 170; Del. at the mouth of Lackawana 455.

These points are greatly depressed when compared with the adjacent arable country, we may therefore regard the basin of the Del. as an inclined plain, rising from the alluvial deposites almost on a level with the tides, to 1,200 or 1,800 feet. The difference of climate above and below the Kittatinny chain is very marked, arising from this great change of aerial pressure. In their general inflections the conformity of the Del. and Susquehannah rs. is too great not to have arisen from some common causes. So greatly striking is this unity of course, that where the Del. flows S. between the Water gap and S. mtn. a corresponding southern stretch of the Susquehannah reaches from the mouth of the W. Branch, to that of Juniata. Both rs. receive their great tributaries from the N. W. and receive only comparative creeks from the opposite direction; and both rivers open to their great recipient by wide and deep bays.

DELAWARE, state of the U. S. bounded by Worcester and Somerset cos. Md. S., by Dorchester, Caroline, Queen Ann, Kent, and Cecil cos. of Md. W., by Chester and Del. cos. of Pa. N., by Del. bay N. E., and by the Atlantic Ocean S. E. Outlines: along the Atlantic from Cape Henlopen to Fenwick's isl. 20 ms.; W. along Md. 36 ms.; N. along Md. to the W. cusp of the semicircle round New Castle 87 ms.; along the semicircle to Del. r. 26 ms.; and thence down Del. bay to Cape Henlopen 90 ms. having an entire outline of 259 ms. Length 100, mean breadth 21, area 2,100 sq. ms. Extending in lat. from 38° 27′ to 39° 50′, long. from 1° 17′ to 2° E. W. C.

Natural features.—Del. comprises a comparatively long and narrow inclined plain, with its declivity E. towards Del. bay. Down this slope flow Indian r. Broad Kill, Cedar, Mispohan, Mother Kill, Jones Duck, Apoquinimink and Brandywine, with some lesser streams. This plain includes the whole N. E. and S. E. sections of the state. The S. E. angle slopes to the N. W. and is drained by the sources of Nanticoke and Choptank rs. The N. part of Del. is waving rather than hilly, but these humble elevations gradually depress, and the S. part spreads into an almost general level. The soil, in some places very productive, is, however, generally thin, and in many places marshy. The climate at the two extremes differs much more in temperature than might be expected from so little extent of lat. and small difference in relative height. The staples of the N. part, grain, flour, &c.; near the Atlantic cotton can be profitably cultivated.

Political geography.—Delaware is divided into three cos. New Castle N., Sussex S., and Kent, lying between them. The population of the state in 1790 was 59,094, in 1800, 64,273. The progressive pop. since, has been as follows:

	1810.	1820.	1830.
New Castle,	24,429	27,899	29,710
Kent,	20,795	20,793	19,911
Sussex,	27,750	24,057	27,118
	72,974	72,749	76,739

Of which were white persons,

	Males.	Females.
Under 5 years of age,	4,744	4,647
From 5 to 10,	4,099	4,011
" 10 to 15,	3,919	3,654
" 15 to 20,	3,184	3,381
" 20 to 30,	5,508	5,484
" 30 to 40,	3,206	3,179
" 40 to 50,	2,036	2,047
" 50 to 60,	1,286	1,397
" 60 to 70,	609	630
" 70 to 80,	202	263
" 80 to 90,	43	56
" 90 to 100,	9	6
" 100 and upwards,	0	1
Total,	28,845	28,756

Of the above are deaf and dumb, under 14 years, 6; 14 to 25, 15; 25 and over, 14; blind, 18; foreigners not naturalized, 313.

Colored population as follows:

	Slaves.		Free colored.	
	Males.	Fems.	Males.	Fems.
Under 10 years of age,	580	508	2,627	2,524
From 10 to 24,	853	617	2,259	2,359
" 24 to 36,	245	230	1,303	1,446
" 36 to 55,	83	80	1,180	1,102
" 55 to 100,	42	49	503	526
" 100 and upwards,	3	2	10	16
Total,	1,806	1,486	7,882	7,973

Slaves and colored persons included in the foregoing who are deaf and dumb, under 14 years, 5; 14 to 25, 4; 25 and over, 11.

Recapitulation,

Whites.	Free colored.	Slaves.	Total.
57,601	15,855	3,292	76,748

Constitution of government, judiciary.—The existing constitution of this state was so greatly changed by amendments in convention Dec. 1831, as to render it, in fact, a new instrument, and as it is brief and not to be found in but few editions of the State Constitutions, we have concluded to insert it entire.

1. The representatives are to be chosen for two years; the property qualification abolished.

2. The senators are to be chosen for four years.

3. The legislature is to meet biennially; the first Tuesday of January, 1833, is to be the commencement of biennial sessions.

4. The state treasurer is to be elected by the legislature biennially. In case of his death, resignation, &c. the governor is to fill the office until the next session of the legislature. He is to settle annually with the legislature, or a committee thereof, which is to be appointed every biennial session.

5. No acts of incorporation are hereafter to be passed without the concurrence of two-thirds of each branch of the legislature, except for the renewal of existing corporations—all acts are to contain a power of revocation by the legislature. No act hereafter passed shall be for a longer period than 20 years, without a re-enactment by the legislature, except incorporations for public improvement.

6. The governor is to be chosen for four years, and to be ever after ineligible. New provisions are made for contested elections of governor; and to fill vacancies. He is to set forth in writing, fully, the ground of all reprieves, pardons, and remissions, to be entered in the register of his official acts, and laid before the legislature at its next session:

7. All elections are to be on the second Tuesday of November. Every free white male citizen, who has resided one year in the state, the last month in the county, and, if he be of the age of 22 years, is entitled to vote. All free white male citizens, between the ages of 21 and 22 years, having resided as aforesaid, may vote without payment of tax. No person in the military, naval, or marine service of the United States, can gain such residence as will entitle him to vote in consequence of being stationed in any military or naval station in the state: no idiot, insane person, pauper, or person convicted of a felony can vote; and the legislature is authorized to impose the forfeiture of the right of suffrage as a punishment for crime.

8. The judicial power of the state is to be exercised by four common law judges, and a chancellor. Of the four law judges, one is chief justice, and three associates. The chief justice and chancellor may be appointed in any part of the states—of the associates, one must reside in each county. [The court of civil jurisdiction is styled the supreme court; and is composed of the chief justice and two associates—no associate judge sits in his own county—the chief justice presides in every county. Two judges constitute a quorum.]

The court of general sessions of the peace and gaol delivery, is composed of the same judges and in the same manner as the superior court.

The court of oyer and terminer is composed of the four law judges. Three to constitute a quorum.

The chancellor exercises the powers of the court of chancery. The orphans' court is composed of the chancellor and the associate judge residing in the county. Either may hold the court, in the absence of the other. When they concur in opinion there shall be no appeal, except in the matter of real estate. When their opinions are opposed, or when a decision is made by one sitting alone, and in all matters involving a right to real estate, there is an appeal to the supreme court of the county, whose decision shall be final.

The court of errors and appeals, upon a writ of error to the superior court, is composed of the chancellor, who presides, and two of the associate judges, to wit, the one who, on account of his residence, did not sit in the case below; and one who did sit. Upon appeal from the court of chancery, the chief justice and three associates compose the court of errors and appeals; three of them constitute a quorum. If the superior court deem that a question of law ought to be heard before all the judges, they may, upon the application of either party, direct it to be heard in the court of errors and appeals, which shall then be composed of the chancellor (who presides) and all the judges.

When the chancellor is interested in a chancery case, the chief justice, sitting alone in the superior court, shall have jurisdiction, with an appeal to the three associate judges sitting as a court of errors and appeals.

When there is an exception to the chancellor or any judge, so that a quorum cannot be constituted in court, in consequence of said exception, the governor shall have power to appoint a judge for that special cause, whose commission shall expire with the determination of the cause.

The judges are to receive salaries, which shall not be less than the following sums, to wit—the chief justice $1,200, chancellor $1,100,—the associates, each $1,000. They are to receive no other fees or perquisites for business done by them.

The general assembly may establish inferior courts, or give to one or more justices of the peace, jurisdiction in cases of assaults and batteries, unlicensed public houses, retailing liquors contrary to law, disturbing camp meetings or other meetings of public worship, nuisances, horse-racing, cock-fighting, and shooting matches, larcenies committed by negroes or mulattoes, knowingly receiving, buying, or concealing stolen goods by negroes or mulattoes, &c. This jurisdiction may be granted either with or without the intervention of a grand or petit jury, and either with or without appeal, as the legislature shall deem proper.

The clerk of the supreme court is to be styled the prothonotary. The office of clerk of the supreme court is abolished.

9. But one person is to be voted for as sheriff and one person as coroner, in each county. The term of office in each case is two years. In New Castle and Kent coun-

ties, at the expiration of the term of office of the present sheriffs and coroner, respectively, in 1833, the governor is authorized to fill up the offices for any year, in consequence of there being no election in that year, under the biennial system.

10. Elections for conventions to revise the constitution, are hereafter to be held on the third Tuesday of May in any year. The majority of all the citizens of the state having right to vote, is to be ascertained by reference to the highest number of votes given at any one of the three general elections next preceding, unless the number of votes given on the occasion, shall exceed the number given in any of the three preceding elections, in which case the majority shall be ascertained by reference to the election of itself.

11. No offices are vacated except the chancellor and judges of the existing courts, and the clerks, whose offices will be abolished on the third Tuesday of January next; on which day the new judicial system goes into effect. The offices of registers for wills and justices of the peace are not affected.

The above sketch of the amendments adopted by the convention, does not enter into details, but merely presents a general view of the changes which have been made.

History.—Delaware was first colonized by the Swedes and Fins, under the auspices of Gustavus Adolphus, and was called New Sweden. The Swedes were then too poor, and not sufficiently commercial, to form colonies; therefore New Sweden fell under the power of the Dutch, in 1655. In 1664 the Delaware colony was conquered with all New Netherlands, by the English, and granted by Charles II. to James, duke of York, who in 1682 conveyed it to William Penn. Delaware thus under the same proprietary remained nominally a part of Pennsylvania, until 1775, though really a distinct colony from 1704, when a colonial assembly for the three lower counties met at New Castle. Delaware was amongst the first states in which a constitution of government was formed, in 1776. In 1792, a convention met, and on June 12th of that year, the existing government was adopted.

DELAWARE, co. N. Y. bounded by Otsego co. N., Schoharie and Greene cos. E., Ulster and Sullivan S., Pa. S. W. and Broome and Chenango cos. W., 60 ms. W. Hudson r., 70 W. S. W. Albany. Greatest extent 35 ahd 54 ms., 1,425 sq. ms.; 24 towns, is hilly and mountainous, with rich valleys, well watered, chiefly by N. E. sources of Del. r., to which timber is sent in rafts. E. branch of Susquehannah is N. E.; deserted in revolutionary war. Pop. 1820, 26,587, 1830, 32,933.

DELAWARE, co. Pa. bounded by Chester co. W. and N. W., by Montgomery N. E., by Phil. co. E., by Del. r. separating it from Gloucester co. N. J. S. E., and by New Castle co. S. Length 20, mean breadth 11, area 220 sq. ms. Extending in lat. from 39° 47′ to 46° 05′, long. from 1° 28′ to 1° 48′ E. W. C. The slope of this co. is to the S. E., down which flow Darby, Ridley, and Chester creeks, with sufficient descent to give innumerable sites for water propelled machinery; and so greatly have the facilities of nature been improved, that as early as 1822, there were, from good authority, 144 machines of various descriptions in actual operation. At present, 1830, it is probable that the saw and grist mills, with other manufactories, do not fall much short of one to each sq. m. The surface is gently rolling, and soil productive. To enumerate its staples, would be to give a list of most articles found in Phil. vegetable market, of which the climate admits the culture, and of an indefinite invoice of the product of mills and looms. Chief t. Chester. Pop. 1820, 14,810, 1830, 17,361.

DELAWARE, p-v. Pike co. Pa.

DELAWARE CITY, p-t. New Castle co. Del. situated on Del. r. at the termination on that stream of the Chesapeake and Del. canal, 32 ms. nearly due N. from Dover. Pop. 1830, about 100; it contains several handsome brick houses.

DELAWARE, co. of O. bounded S. by Franklin, W. by Union, N. by Marion, N. E. by Knox, and S. E. by Licking. Greatest length 28, mean breadth 25, and area 760 sq. ms. Extending in lat. from 40° 08′ to 40° 32′, and in long. from 6° 45′ to 7° 18′ W. W. C. The E. border of this co. extends from S. to N. along the summit between the vallies of Muskingum and Hocking on the E., and the Sciota on the W. The extreme sources of White Woman's Fork of Muskingum, and of the Hocking rise along the E. border of Del. and flow estrd. The body of the co. has a slope almost due S., and is traversed in that direction by the two main constituent streams of Sciota, the Whetstone and Sciota Proper; Allum and Walnut creeks, also tributary waters of Sciota, rise in its eastern section. Chief t. Delaware. Pop. 1820, 7,639, 1830, 11,504.

DELAWARE, p-v. st. jus. and tsp. Del. co. O. The st. jus. is by p-r. 419 ms. S. W. by W. W. C. and 23 above, and N. W. Columbus. It stands on the right bank of Whetstone r. Lat. 40° 13′, long. 6° 7′ W. W. C. Pop. of the village, 1830, 527, and of the tsp. including the village, 936.

DELAWARE, co. of Ind. bounded by Randolph E., Henry S., Madison W., and Grant N. W. Length 22, breadth 20, and area 440 sq. ms. Extending in lat. from 40° 05′ to 40° 24′, and in long. from 8° 12′ to 8° 24′ W. W. C. The Mississinawa and White rs. rising in Randolph, traverse Del. in which they diverge, the former to the N. W. and the latter to the W. General slope to the wstrd. Chief t. Munsey. Pop. 1830, 2,374.

DELHI, v. Del. co. N. Y. Pop. 1830, 435.

DELHI, p-t. and st. jus. Delaware co. N. Y. 70 ms. S. W. Albany, 54 W. Catskill, 63 Kingston, 156 square ms., is crossed by Del. r. and Little Del. which joins it near the village. There are hills, mtns., vallies, and on the river fine meadows. Pop. 1830, 2,114.

DELIGHTFUL GROVE, p-o. Spartanburg dist. S. C. by p-r. 111 ms. N. W. Columbia, and 484 S. W. W. C.

DELLS, p-o. Allachua co. Flor.

DELPHI, p-o. Marion co. Tenn. by p-r. 129 ms. S. E. Nashville.

DELPHI, p-v. and st. jus. Carroll co. Ind. by p-r. 661 ms. N. W. by W. W. C. and 88 N. W. Indianopolis. It is situated on Wabash river, just below the mouth of Deer creek. Lat. 40° 37′, long. 9° 40′ W. W. C.

DEMOPOLIS, p-v. Marengo co. Ala. at the left bank of Tombigbee, r. immediately below the mouth of Black Warrior or Tuscaloosa r., 65 ms. S. S. W. Tuscaloosa.

DENMARK, p-t. Oxford co. Me. on Saco r. 30 ms. S. W. Paris, 85 Augusta. Population 1830, 954.

DENMARK, p-t. Lewis co. N. Y. 150 miles N. W. Albany, W. Black river, 22,000 acres; has a rich soil in the vallies. Deer creek has a fall of 175 feet, almost perpendicular, with high limestone banks, and other falls from 10 to 60 feet. Copenhagen is the only village. Pop. 1830, 2,270.

DENMARK, p-v. Madison co. Ten. by p-r. 16 ms. S. W. by W. Nashville.

DENMARK, p-v. and tsp. Ashtabula co. O. by p-r. about 190 ms. N. E. Columbus, and 340 N. W. W. C. Pop. 1830, 169.

DENNINGS, p-o. Frederick co. Md. by p-r. 60 ms. N. W. W. C.

DENNIS, p-t. Barnstable co. Mass. 97 ms. S. E. Boston, 8 from Barnstable, N. Atlantic, E. Bass river, has several ponds, a poor soil, except N., and many works for making salt from sea water, by evaporation. Scargo hills is the highest land in the co. Population 1830, 2,317.

DENNIS, p-t. Cape May co. N. J., has a small harbor. Pop. 1830, 1,508.

DENNIS, p-o. Amelia co. Va. 54 ms. S. W. Richmond.

DENNISVILLE, p-t. Washington co. Me. 17 ms. N. W. Eastport, 172 N. N. E. Augusta, on Denny's river. It is crossed by several streams. Colescook bay lies S. of it. Pop. 1830, 856.

DENNYVILLE, p-v. Wilkes co. N. C. by p-r. 172 ms. N. W. by W. Raleigh.

DENTON, p-v. and st. jus. Caroline co. Md. on the left bank of Choptank river, 18 ms. N. E. Easton, and by p-r. 65 ms. a a little S. by E. Annapolis. Lat 38° 53′, long. 1° 14′ due E. W. C.

DENTONSVILLE, p-o. Hanover co. Va. 42 ms. N. Richmond.

DENVILLE, village, Morris co. N. J. 8 ms. N. of Morristown, on Rockaway creek.

DEPOSIT, p-v. Tompkins, Delaware co. N. Y. 40 ms. S. W. Delhi, 105 W. Catskill, on Delaware river.

DEPTFORD, t. Gloucester co. N. J. 20 ms. S. Burlington, between Bigtimber and Mantua creeks, on Delaware river. Population 1830, 3,599.

DEPTFORD, t. Gloucester co. N. J. on the Delaware, between Mantua creek, which parts it from Greenwich on the S. W. and Gloucester on the N. E. Population 3,599. Woodberry, the st. jus. is in this township.

DERBANE, corrupted from Terre Bonne, the name of several small streams of La.; one, a r. of about 80 miles comparative course, rises in Lafayette co. Ark. and parish of Claiborne La. flows S. E. and falls into Ouachitau river a short distance above the village of Monroe, in the parish of Ouachitau. Another of the same name, giving name to a parish, rises W. from the river La Fourche, and flowing a few ms. S. falls into the Gulph of Mexico, between Timballier and Petite Caillon bays. There are 3 or 4 more, but of too little consequence to deserve particular notice.

DERBY, p-t. Orleans co. Vt. 52 ms. N. E. Montpelier, E. Memphremagog lake, 7½ ms. on Canada line, 23,040 acres. First settled 1745, from Conn. &c. White and Norway pine grows near the lake, with red oak, and rock maple, &c., elsewhere, the soil being generally rich. It has Salem pond, and Clyde river, with mill seats. Pop. 1830, 1,469.

DERBY, p-t. New Haven county, Conn. 8 ms. W. New Haven, at the confluence of Naugautuck and Housatonic rivers, 12 miles from Long Island Sound, navigable for vessels of 80 tons, 4½ miles by 5½, has a varied surface, with some meadows, mill sites, advantages for trade, and shad fisheries. Humphreysville is one of the oldest woollen manufactories in the country; incorporated 1810, with $500,000 capital. There are other manufactures. An agricultural seminary was opened here, 1824, for practical education, with philosophical aparatus, &c. Pop. 1830, 2,253.

DERRY, p-t. Rockingham co. N. H. 28 ms. from Concord. Pop. 1830, 2,176.

DERRY, p-v. and tsp. Columbia co. Pa. The p-o. is 7 ms. E. Danville, the co. st. and by p-r. 77 ms. E. of N. Harrisburg, and 187 ms. from W. C. in a nearly similar direction. Pop. of the tsp. 1820, 1662, in 1830, 1689.

DE RUYTER, p-t. Madison co. N. Y. 21 ms. W. S. W. Morrisville, 123 W. Albany, is hilly, well watered by sources of Tioughnioga creek and yields grass and some grain, 17 miles S. Erie canal. Pop 1830, 1,447.

DETROIT, or "the Strait," river, uniting lakes St. Clair and Erie, and forming part of the limit between Upper Canada and Michigan. At its outlet from lake St. Clair, Detroit river is upwards of a mile wide and divided into two channels by Peach Islands; the course a little S. of W. 8 ms. to the lower extreme of the city of Detroit, where it makes a regular curve to the S. S. W. and continues the latter course 4 ms. to the influx of the river Rouge, from the N. W. One mile below the mouth of the Rouge the river is again divided into two channels, by Grand Turkey Island. The Detroit now rapidly widens to from 3 to 4 miles, and assuming a southern course of 17 miles, finally opens into the extreme N. W. angle of lake Erie. This is indeed a most beautiful, gentle and navigable stream, of 29 miles in length. Though encumbered with islands, and the channel rather intricate, vessels of considerable burthen can be navigated through into lake St. Clair. The shores though not elevated are bold, and being cultivated give a charming appearance in summer to the landscape along both shores. On the right are the city of Detroit and Brownstown, and on the Canada shore Sandwich and Amherstburg. From Michigan enter at the influx of both into Erie, the Huron, and one mile above Grand Turkey Island the Rivierie Rouge; and from the same side the lesser streams of Bauche, Curriere, and Clora. From Canada the only stream which enters the Detroit worthy of notice is the Canard, falling into the main stream 3½ miles above Amherstburg.

The Detroit islands are elevated and are composed of excellent arable soil; two of them, Gros Isle and Grand Turkey Island, exceed 6 miles each, in length, but are comparatively narrow. The whole river is frequently and completely frozen over in winter.

DETROIT, p-t. city, port of entry and st. jus. for Wayne co. and of the government of Michigan, as laid down on Tanner's United States, is at lat. 42° 20′, and exactly 6° west W. C. These relative positions give by calculation a bearing of 52° 50′, and a distance in statute miles, of 416, from W. C. to Detroit. By the post office list of 1831, the distance from W. C. to Detroit is stated at 526 miles.

Detroit is situated on a rising plain along the western or right shore of Detroit r. The streets are laid out at right angles to each other, though something oblique to the course of the stream. The plain has a gentle acclivity from the water to the main street, but spreads thence to the westward nearly level. A remark may be made of Detriot which applies with equal force to many other places in the U.S., of comparatively small population; that is, that the real and commercial wealth of the smaller are far above the proportion which relative numbers would produce between them and places of greater population. Few places can be more admirably situated for a commercial city than Detroit, and few have a more solid promise of permanent prosperity. Pop. 1830, 2,222.

DEVEREAUX, store and p-o. in the S. W. part of Hancock co. Geo. by p-r. 16 ms. N. E. Milledgeville, and 626 S. W. W. C.

DE WITT, p-v. in the western part of Clinton co. Illinois, 18 ms. from Carlyle, the st. jus. and by p-r. 48 S. W. Vandalia, and 820 ms. W. W. C.

DEXTER, p-t. Penobscot co. Me. 30 ms. N. W. Bangor, 67 Augusta, has waters flowing into Penobscot and Kennebec. Population 1830, 805.

DEXTER, p-v. Washtenaw co. Mich. 10 ms. N. W. Ann Arbor, the co. st., and by p-r. 52 W. Detroit, and 545 N. W. by W. W. C.

DIAMOND GROVE, p-v. Brunswick co. Va. by p-r. 73 ms. S. S. W. Richmond.

DIAMOND GROVE, p-v. Northampton co. N. C. by p-r. 25 ms. S. E. Raleigh.

DIANA, t. Lewis co. N. Y. Pop. 1830, 309.

DIANA MILLS and p-o. Buckingham co. Va. by p-r. 125 ms. W. Richmond.

DICKENSON, p-v. Franklin co. Va. by p-r. 199 ms. S. W. by W. Richmond.

DICKENSON'S STORE and p-o. Bedford co. Va. by p-r. 151 ms. W. Richmond.

DICKINSON, t. Franklin co. N. Y. 12 ms. W. Malone, 233 W. Albany, 6 ms. by 48, watered by Little Salmon r., has much sandy loam, with beech, maple, bass, elm, &c. Population 1830, 446.

DICKINSON, p-v. and tsp. in the W. part of Cumberland co. Pa. The p-o. is 36 ms. W. Harrisburg, and 108 W. of N. W. C. Pop. of the tsp. 1830, 2,523.

DICK'S r. Ky. rises in Rockcastle co. interlocking sources with Rockcastle and Green rs., and flowing thence N. W. passes through Lincoln, and thence separating Garrard from Mercer co., falls into the left side of Ky. r. 10 ms. from Harrodsburg.

DICK'S MILLS and p-o. in the W. part of Butler co. O. 8 ms. from Hamilton, the st. jus. for the co. and by p-r. 496 ms. N. of W. W. C. and 109 S. W. by W. Columbus.

DICKSON, co. Ten. bounded by Humphrey's W., Stewart N. W., Montgomery N., and Hickman S. Length 36, mean width 28, area 100 sq. ms. Extending in lat. from 35° 55′ to 36° 20′, long. 10° 09′ to 10° 45′ W. W. C. This co. is a table land, from which the water flows S. into Duck r., W. into Ten., N. into Cumberland, and E. into Harpeth rs. The N. W. and N. E. angles touch Cumberland r., but no stream of consequence flows into the body of the co. Soil of middling quality. Chief town, Charlotte. Pop. 1820, 5,190, 1830, 7,265.

DICKSON'S MILLS and p-o. in the E. part of Parke co. Ind. by p-r. 10 ms. E. Rockville, the co. st., 58 W. Indianopolis, and 630 ms. N. W. by W. W. C.

DIGHTON, p-t. port of entry, Bristol co. Mass. 38 ms. S. Boston, W. Taunton r., has an irregular surface, conglomerate rocks, (boulders,) in diluvial soil, over granite. A rock inscribed by Indians has excited attention. Several coasting vessels are owned here, Taunton r. being navigable for small vessels. Pop. 1830, 1,723.

DILL'S BOTTOM and p-o. in the W. part Belmont co. O. by p-r. 268 ms. N. W. by W. W. C. and 149 E. Columbus.

DILLON'S p-o. in the W. part of Tazewell co. Il. by p-r. 821 ms. N. W. by W. W. C. and 159 W. of N. Vandalia.

DILLON'S RUN, p-o. Hampshire co. Va. by p-r. 16 ms. E. Romney, the co. st. 100 ms. N. of W. W. C. and 179 N. W. Richmond.

DILLONSVILLE, p-v. in the s. part of Mecklenburg co. N. C. by p-r. 196 ms. s. w. by w. Raleigh.

DILLSBERG, or more correctly DILLSTON, p-v. w. part York co. Pa. 20 ms. s. w. Harrisburg.

DILLWORTH'S TOWN, p-v. E. border Chester co. Pa. 7 ms. s. West Chester.

DIMOCKSVILLE, p-v. in the E. part of Susquehannah co. Pa. by p-r. 274 ms. N. N. E. W. C. and 175 ms. E. of N. Harrisburg.

DINGMAN'S CREEK, p-o. and Ferry, over Del. r. 25 ms. above Del. Water Gap.

DINWIDDIE, co. Va. bounded by Nottaway r. which separates it from Brunswick s. w., by Nottaway co. w., Namazine creek, separating it from Andie, N. W., by Appomattox r. separating it from Chesterfield, N. E., by Prince George's co. E., and Sussex and Granville S. E. It lies in nearly the form of a hexagon, equal to a circle of 28 ms. diameter. Area about 616 sq. ms., and is divided into very nearly equal portions by lat. 37°. In long. it lies between 0° 33′ and 1° 3′ w. W. C. About one fourth part on the N. border slopes towards, and is drained into the Appomattox. The other three quarters incline to the S. E. and are drained by Monk's Neck, Stony, Sapony, and other confluents of Nottaway r. Surface waving. Chief t. Petersburg. Pop. 1820, 13,792, 1830, 21,901.

DINWIDDIE, C. H. and p-o. on Stony creek, 15 ms. s. w. Petersburg.

DISMAL SWAMP, a rather undefined, marshy tract, between the s. part of Chesapeake bay and Albermarle sound, occupying a part of Nansemond and Norfolk cos. Va., and of Camden and Pasquotank cos. N. C. The sources of Nansemond and Elizabeth rs. flowing N. in the estuary of James r., those of Pasquotank and Perquiman's entering Albermale sound, and some small creeks flowing S. E. into Currituck Sound, have their heads in Dismal Swamp.

DIVIDING CREEK, a small stream of Va. forming for a few ms. the boundary between Lancaster and Northumberland cos. and then falling into the Chesapeake.

DIXBORO', p-o. in the E. part of Washtenaw co. Mich. by p-r. 540 ms. N. W. by w. W. C. and 37 w. Detroit.

DIXFIELD, p-t. Oxford co. Me. 18 ms. N. E. Paris, N. Androscoggin r., 40 from Augusta. Pop. 1830, 889.

DIXMONT, p-t. Penobscot co. Me. 20 ms. w. of Bangor, 44 of Augusta. Pop. 1830, 945.

DIXON'S SPRINGS and p-o. Smith co. Ten. by p-r. 48 ms. E. Nashville.

DIXVILLE, p-v. Henry co. Va. by p-r. 158 ms. s. w. by w. Richmond.

DIXVILLE, t. Coos co. N. H.; settled 1805, 31,023 acres, with small streams and uneven lands. Pop. 1830, 2.

DOAKS' STAND, and p-o. Yazoo co. Miss. about 120 ms. N. N. E. Natchez.

DOBSONS CROSS ROAD, and p-o. Stokes co. N. C. by p-r. 110 ms. N. W. by w. Raleigh.

DOCKLEY'S STORE, and p-o. Richmond co. N. C. by p-r. 121 ms. s. w. Raleigh.

DODDSVILLE, p-v. Fauquier co. Va. by p-r. 53 ms. from W. C.

DODGEVILLE, p-v. Iowa co. Mich. 75 ms. E. Prairie du Chien, 60 N. N. E. from Galena in Il. and by p-r. 1042 N. W. by w. W. C.

DODSONSVILLE, p-v. Jackson co. Ala. by p-r. 186 ms. N. E. Tuscaloosa.

DOEBAUN, p-o. Chester co. Pa. 14 ms. s. w. by w. West Chester.

DOG, river, E. branch of Pascagoula river, rises in the pine forests between Pascagoula and Tombigbee, and flowing a little w. of s. 90 miles, nearly along the line between Ala. and Miss. falls into Pascagoula, 10 ms. above its mouth.

DOG, river, a much smaller stream than the preceding, rises between it and Mobile bay, and flowing S. E. falls into the latter 10 ms. s. Mobile.

DOGWOOD SPRINGS, and p-v. Pulaski co. Ark. by p-r. 15 ms. westward Little Rock, and 1083 ms. s. w. by w. W. C.

DOHRMANS, p-v. and tsp. in the E. part of Tuscarawas co. O. The p-o. by p-r. is 298 ms. N. W. by w. W. C. and 123 N. E. by E. Columbus. Pop. of the tsp. 1830, 1,161.

DOHERTYVILLE, p-v. Jefferson co. Ten. by p-r. 196 ms. E. Nashville.

DOLBEE'S, p-v. N. W. part Potter co. Pa. 16 ms. from Coudersport, and by p-r. 299 ms. N. N. W. W. C. and 190 N. W. Harrisburg.

DOLINGTON, p-v. Bucks co. Pa. near Delaware river, 9 miles above Trenton.

DONALDSONVILLE, p-v. and st. jus. for the parish of Ascension, and seat of government of La. It stands on the right bank of Miss. r. below the efflux of Lafourche, extending along both rivers. Pop. 1820, 200, 1830, 500. Lat. 30° 05′, long. 14° 03′ w. W. C.

DONEGAL, p-o. Westmoreland co. Pa.

DONORAILE, p-v. Fayette co. Ky. by pr. 31 ms. s. E. Frankfort.

DOOLEY, co. Geo. bounded by Trewin S. E. and s., Flint river separating it from Lee w., Houston N., and by Oakmulgee river separating it from Pulaski N. E., and Telfair E. Length along lat. 32° from Oakmulgee to Flint river 48 ms., mean width 34, and area 1,632 square miles. Extending in lat. from 31° 42′ to 32° 18′, in long. from 6° 21′ to 7° 14′ w. W. C. It must be obvious from the position of Dooley that it is composed of two inclined planes falling towards Flint and Oakmulgee respectively. The extreme source of Savannah river is also in the s. part of this co. Chief town, Berrien. Population 1830, 2,135.

DORCHESTER, p-t. Grafton co. N. H. 50 ms. N. by w. Concord, 23 s. Haverhill, 90 N. W. Portsmouth, 12 E. Connecticut r., 8 w Merrimac river, has rocky highlands, and fertile vallies on several brooks, 8 school districts. Pop. 1830, 693.

DORCHESTER, t. Norfolk co. Mass. 3 ms. s. s. E. Boston, w. Mass. bay, N. W. Neponset river, has a rich soil and many inhabitants w.

E. with few hills; favorable to fruit &c. has Thompson's and Moon's islands, with 600 acres of salt marsh, several factories and dams on Neponset river; was settled, 1630, soon after Plymouth and Salem. There was a fort on Rock Hill. 1636, about 100 persons travelled across the wilderness in 14 days, and settled Hartford, Conn. March 4th, 1776, 1,200 men, sent by Gen. Washington, threw up works on the Dorchester Heights in the night, which commanded Boston harbor, and drove the British army away. Part of Dorchester neck belongs to Boston, to which a bridge extends. Pop. 1830, 4,074.

Dorchester, v. on Maurice river, Cumberland co. N. J. E. Maurice river, 5 miles from its mouth in Delaware bay.

Dorset, p-t. Bennington co. Vt. 27 miles N. Bennington, 41 square ms. First settled 1768, has part of Otter creek and sources of Battenkill and Powlet river, with mill sites, Dorset and Equinox mtns., several caves and some manufactories. Population 1830, 1,507.

Dorchester, co. Md. bounded by Nantikoke bay S., Chesapeake bay S. W., W. and N. W., Choptank river N., Caroline co. N. E., Sussex co. Del. E., and Choptank river which separates it from Worcester co. Md. S. E. Length from S. W. to N. E. 32 miles, mean breadth 20, and area 640 square miles. Extending in lat. from 38° 14′ to 38° 40′, in long. from 0° 36, to 1° 20′ E. W. C. Chief town, Cambridge. Population 1820, 17,700, 1830, 18,686.

Dorchester, p-t. Colleton dist. S. C. on Ashley river, 20 miles above Charleston.

Dorsettsville, p-o. Chatham co. N. C. 20 ms. from Raleigh.

Dorsey's, p-o. southwestern part of St. Mary's co. Md. 4 miles from Leonardstown, and by p-r. 78 miles S. from Annapolis, and 59 S. S. E. from W. C.

Double Branches, p-o. Anderson district, South Carolina, by p-r. 132 miles northwest Columbia.

Double Branches, p-o. Lincoln co. Geo. by p-r. 95 ms. N. E. Milledgeville.

Double Bridge, p-o. Lunenburg co. Va. by post road 118 miles southwest Richmond.

Double Cabins, p-o. in the western part of Henry co. Geo. by p-r. 107 miles N. W. by W. from Milledgeville, and 699 miles S. W. from W. C.

Double Pipe, creek, p-o. N. E. part Frederick co. Md. about 50 ms. a little W. of N. W. C.

Double Wells, p-o. Warren co. Geo. by p-r. 37 ms. N. E. Milledgeville.

Dougherty's, Carroll co. Ten. (*See Lamoresville.*)

Douglass, p-t. Worcester co. Mass. 47 ms. S. Worcester, N. Conn. has Mumford r. a branch of Blackstone r. between which and Shetucket it lies, artificial irrigation is resorted to, with wisdom, and deserves to be practised elsewhere. It has good meadows. Pop. 1830, 1,742.

Douglass, p-v. in the W. part of Logan co. O. by p-r. 10 ms. from Bellefontaine, 468 N. W. by W. W. C. and 72 in a nearly similar direction from Columbus.

Douglass' Mills and p-o. Perry co. Pa. by p-r. 30 ms. a little N. of W. Harrisburg, and 117 N. N. W. W. C.

Douglassville, p-v. in the N. E. part of Berks co. Pa. by p-r. 147 ms. N. E. W. C. and 64. E. Harrisburg. Pop. of Douglass tsp. 1830, 839.

Douthet, p-v. in the N. part of Anderson dist. S. C. by p-r. 531 ms. S. W. W. C. and 139 ms. N. W. by W. Columbus.

Dover, p-t. Penobscot co. Me. 77 ms. Augusta, S. Piscataquis r. Pop. 1830, 1,042.

Dover, p-t. and st. jus. Strafford co. N. H. 10 ms. N. W. Portsmouth, on the E. great road and W. of Piscataqua r. and Me. The town contains a court house, gaol, four public houses, and seven meeting houses.

The Cocheco manufacturing company have a capital of one million five hundred thousand dollars, 4 large brick mills. Three are situated in the centre of the town. They run 24,320 spindles and 780 looms—employ 900 operatives—750 of whom are females. They consume 2,600 bales of cotton, or 1,000,000 lbs., and produce about 100,000 a week, or 5,200,000 yards yearly.

The calico printing, is equal to the best imported. They bleach and print 3,000 pieces, of 28 yards each, a week, equal to 4,368,000 yards per annum.

They consume 4,000 gallons of oil, 500 barrels of flour, 26,000 lbs. of potato starch, 3,000 cords of wood, 2,000 tons of anthracite coal, &c.

The Cocheco is navigable for vessels of 80 tons, up to the landing, in the town. There are 50 shops, some large. Piscataqua r. is formed here of Cocheco and Belamy, or Black rs. which afford fine water power, and supplies many factories. The land swells gently, and is picturesque. First settled 1623, on the neck S. between the rs. by the company of Laconia, from Eng. who entrenched the place, and established a fishery. The population have since collected at Cocheco falls; 4 ms. N. W. the Cocheco descends 32½ft. at the head of navigation 12 ms. from the sea. Here in the village, 1689, Major Waldron was killed by Indians, to revenge the death of 7 or 8 whom he had executed 13 years before. The place often suffered from Indians. Here was the first preaching in N. H. Pop. 1830, 5,449.

Dover, t. Norfolk co. Mass. 7 ms. W. Dedham, 16 S. W. Boston, E. and S. Charles r. is uneven, woody, with some manufactories. Pop. 1830, 497.

Dover, p-t. Duchess co. N. Y. 21 ms. E. Poughkeepsie, 100 S. Albany, W. Conn. 6 ms. by 7, level in centre, where is 10 m. creek of Housatonic r. hilly E. and W. and grain and grass flourish. In this town, near the v. of the Plain, E. of a mtn. is a wild passage cut by a stream among rocks, which in one place meet over head, and also form a hollow, called the stone church, which is 50 ft. long and 30 wide in the broadest place. Pop. 1830, 2,198.

DOVER, t. Monmouth co. N. J. 45 ms. S. N. Y. 20 S. E. Bordentown, with the ocean E. is crossed by Tom's cr. falling into Tom's bay, and Cedar creek. A narrow beach, called Long and Cran beach, forms Barnegat bay, most of which is in this t and receives its waters. Cranberry inlet is now closed, so that the entrance of Bar brook is S. in Stafford; Egg and other islands are in Bar brook. There are 15 or 20 furnaces here, chiefly on Tom's r. Pop. 1830, 2,898.

DOVER, p-v. Morris co. N. J. on the Rockaway, 8 ms. N. of Morristown, containing extensive manufactories of Iron. The Morris canal passes the village.

DOVER, p-v. and tsp. W. part York co. Pa. 24 ms. S. Harrisburg and 94 a very little E. of N. W. C. Pop. of the tsp. 1820, 1,816, 1830, 1,874.

DOVER, p-t. st. jus. for Kent co. and of the government of Del. by p-r. 114 ms. N. E. by E. W. C. It is by the road about 50 ms. S. Wilmington, lat. 39° 09', long. 1° 28' E. W. C. Pop. of the hundred of Dover, 1830, 4,316.

DOVER, p-v. and st. jus. Stewart co. Ten. by p-r. 787 ms. S. W. by W. W. C. and 81 N. W. by W. Nashville. It is situated on the left bank of Cumberland r. lat. 36° 28' long. 10° 52' W. W. C.

DOVER, p-v. and tsp, in the N. W. angle of Cuyahoga co. O. The p-v. is situated on Lake Erie 12 ms. W. Cleaveland, by p-r. 366 N. W. W. C. and 140 N. N. E. Columbus. Pop. of the tsp. 1830, 462.

DOVER FURNACE, and p-o. Stewart co. Ten. by p-r. 7 ms. S. E. Dover, the co. seat 780 S. W. by W. W. C. and 74 N. W. by W. Nashville.

DOVER MILLS, and p-o. Goochland co. Va. 21 ms. N. W. Richmond.

DOWNE, t. Cumberland co. N. J. 60 ms. S. by W. Bordentown, is nearly an island, with Maurice r. E. Nantuxet creek W. and Deleware Bay S. about ½ appears to be swamps, near the water, and Bear Swamp is near the middle. Pop. 1830, 1,923.

DOWN EAST, p-v. Penobscot co. Me. 96 ms. from Augusta.

DOWNINGTOWN, p-v. on the left bank of the N. branch of Brandywine creek, and near the centre of Chester co. Pa. 30 ms. W. Phil. and by p-r. 122 N. E. W. C. The village is small but contains in its vicinity extensive grist mills, and is situated in a very well cultivated and pleasantly diversified country. It stands on the great road from Phil. to Lancaster.

DOWNINGTON, p-v. in the N. W. angle of Meigs co. O. by p-r. 85 ms. S. S. E. Columbus and 356 W. W. C.

DOYAL'S MILLS, and p-o. Jackson co. Ala. by p-r. 670 ms. S. W by W. W. C. and 188 N. E. Tuscaloosa.

DOYLESTOWN, p-v. and st. jus. Bucks co. Pa. by p-r. 171 ms. N. E. W. C. and 107 nearly due E. Harrisburg. By the relative p-o. distances it appears to be 35 ms. from Phil. to Doylestown, whilst the real distance is only about 26. It is situated on a branch of Neshamony creek, lat. 40° 18', long. 1° 56' E. W. C. Pop. of the borough and tsp. 1820, 1,430, 1830, 1,777.

DRACUT, p-t. Middlesex co. Mass. 28 ms. N. N. W. Boston, S. of N. H. line, N. Merrimack r. is pleasant, with pretty good soil, well watered by Beaver brook, &c. A fine bridge crosses Pawtucket falls to Chelmsford, and the growth of Lowel, to which is another bridge 500ft. and roofed, has been useful to Dracut. Pop. 1830, 1,615.

DRAKE'S, p-o. in the N. W. part of Holmes co. O. by p-r. 359 ms. S. W. by W. W. C. and 71 N. E. Columbus.

DRAKEVILLE, vil. Morris co. N. J. on the Morris canal, 12 ms. N. W. Morristown.

DRANESVILLE, p-o. Fairfax co. Va. 17 ms. from W. C.

DRAPER'S VALLEY, and p-o. in the W. part of Wythe co. Va. 18 ms. from Evansham, the co. seat, and by p-r. 310 ms. S. W. W. C. and 225 S. of W. Richmond.

DRESDEN, p-t. Lincoln co. Me. 8 ms. N. W. Wiscasset, 14 from Augusta, on both sides of Kennebec r. Pop. 1830, 1,151.

DRESDEN, p-t. Washington co. N. Y. 20 ms. N. Sandyhill, 72 N. Albany, W. Lake Champlain, E. Lake George, and ends N. at Pulpit point. It is mountainous, with several natural ice-houses. Pop. 1830, 475.

DRESDEN, p-v. and st. jus. Weakly co. Ten. by p-r. 834 ms. S. W. by W. W. C. and 132 ms. a very little N. of W. from Nashville. It is situated on a branch of Obion river, lat. 36° 19' and long. 11° 50' W. W. C.

DRESDEN, p-v. in Jefferson tsp. N. part of Muskingum co. O. by p-r. 14 ms. N. Zanesville, the co. seat, 73 N. of E. Columbus, and 350 N. W. by W. W. C. Pop. 1830, 391.

DRIPPING SPRING, p-v. Edmonson co. Ky. by p-r. 138 ms. S. W. Frankfort.

DROWNED LANDS, Orange co. N. Y. on Wallkill creek, 10 miles long, 3 to 5 broad, have a rich mould, good for hemp when drained.

DROWNED MEADOW, p-v. Brookhaven, Suffolk co. N. Y. 3 ms. E. Setauket.

DROWNING CREEK, and p-o. Burke co. N. C. about 200 ms. W. Raleigh.

DRY CREEK, and p-o. Campbell co. by p-r. 82 ms. N. N. E. Frankfort.

DRYDEN, p-t. Tompkins co. N. Y. 35 ms. S. Auburn, 9 E. Ithaca, 150 W. Albany, 10 ms. sq. is level with much good soil. Good pine abounds, Fish and 6 m. creeks give many mill seats. Pop. 1830, 5,206.

DRY RIDGE, p-v. Grant co. Ky. by p-r. 48 ms. N. N. E. Frankfort.

DRY RUN, p-o. in the N. part of Franklin co. Pa. 23 ms. from Chambersburg, and by p-r. 63 ms. W. Harrisburg, and 113 N. W. W. C.

DUANE, t. Franklin co. N.Y. Pop. 1830,247.

DUANESBURG, p-t. Schenectady co. N. Y. 8 ms. square, s. end of the co. is 400 or 500ft. above Hudson r. at Albany, a little uneven, with good soil, and sources of Norman's and Bowza Kills, which falls 70 ft. Lake Maria is drained by Chuctenunda creek which turns about 20 mills.

DUBLIN, p-t. Cheshire co. N. H. 10 ms. E. s. E. Keene, 50 from Concord, 70 from Boston, 26,560 acres,on high land between Conn. and Merrimac rs. contains most of Grand Monadnock mtn. Centre and North ponds, and is pretty good for grass, 10 school districts. Rev. Ed. Sprague left a fund of $8,000 to public schools, and $5,000 to the congregational church pastor. There are two libraries. First settled 1762. Pop. 1830, 1,218.

DUBLIN, p-v. Bucks co. Pa. 6 ms. N. N. W. Doylestown, the co. seat, and by p-r. 166 ms. N. E. W. C. and 97 E. Harrisburg.

DUBLIN, p-v. in the N. part of Harford co. Md. 32 ms. N. E. Baltimore, and 3 ms. from Conowingo Ferry.

DUBLIN, p-v. and st. jus. Laurens co. Geo. situated on the right bank of Oconee r. near the centre of the co. 55 ms. below, and E. of s. from Milledgeville, lat. 32° 34′ and long. 6° 05′ w. W. C.

DUBLIN, OR DUBLINTON, p-v. in Washington tsp. N. W. part of Franklin co. O. 12 ms. N. N. W. Columbus, and by p-r. 408 N. W. by W. W. C. Pop. 1830, 96.

DUBOIS, co. of Ind. bounded N. E. by Martin and the s. w. part of Orange, E. by Crawford; s. E. by Perry; s. by Spencer; w. by Pike; and N. W. and N. by the East Fork of white river, separating it from Daviess. Length 24 ms. mean breadth 20, and area 480 sq. ms.—Extending in lat. from 38° 14′ to 38° 34′ and in long. from 9° 43′ to 10° 08′ w. W. C. Though this co. bounds on the East Fork of white river, the far greater part of the surface is drained by the Patoka and confluent creeks, and slopes westward. Chief t. Portersville. Pop. 1830, 1,778.

DUBOURG'S, p-v. and st. jus. parish of St. Baptiste, La. 49 ms. above New Orleans.

DUCK r. Ten. having its main sources in Warren and Franklin co. between those of Elk r. a branch of Ten. and a branch of Cumberland, flowing thence through Bedford, Maury, Hickman, Perry, and Humphries cos. falls into Ten. r. in the latter, after an entire comparative course of 130 ms. in a direction N. W. by W. In seasons of high water it is navigable about 100 ms.; the valley of Duck r. is comparatively narrow, not averaging above 25 ms. and in no place above 60 wide. It lies between lat. 35° 10′ and 36° 10′.

DUCK BRANCH, and p-o. Barnwell dist. S. C. The Duck Branch is a small stream near the s. E. border of the dist. forming one of the sources of the Coosawhatchie r. The p-o. is situated on the cr. by p-r. 81 ms. a little w. of s. Columbia.

DUCK CREEK CROSSINGS, and p-o. in the N. W. part of Franklin co. Ind. 10 ms. N. W. Brookville, the st. jus. for the co. and by p-r. 538 ms. W. W. C.

DUDLEY, p-t. Worcester co. Mass. 20 ms. s. Worcester, 55 s. Boston, N. Conn. is well supplied with mill seats by Quneboag and French or Stony rivers, the heads of Thames r. and has wool and other factories. There are several ponds, one nearly 5 ms. long. It was one of the Christian Indian colonies formed in early times. Pop. 1830, 2,115.

DUFF'S FORKS, and p-o. E. part of Fayette co. O. by p-r. 32 ms. s. W. Columbus, and 425 N. of W. W. C.

DUGGER'S FERRY, and p-o. Carter co. Ten. by p-r. 420 ms. s. W. by W. W. C. and 316 N. of E. Nashville.

DUKE'S co. Mass. consists of the islands of Martha's Vineyard, Chippaquiddick, Norman's Land and Elizabeth islands, forming 3 towns —Chief town, Elizabethtown.—The soil is poor. Martha's Vineyard is favorable to commerce and fishing. Pop. 1820, 1,702, 1830, 1,768.

DUKE'S p-o. Dickson co. Ten. by p-r. 44 ms. W. Nashville.

DUMAS'S STORE, and p-o. in the s. part of Richmond co. N. C. by p-r. 18 ms. s. Rockingham, the co. st. 417 ms. s. s. W. W. C. and 131 s. W. Raleigh.

DUMFRIES, p-v. on Quantico creek, Prince William co. Va. 33 ms. s. s. W. W. C.

DUMMER, t. Coos co. N. H. is of little value, watered by Amonoosuck and Amariscoggin rs. Pop. 1830, 65.

DUMMERSTON, p-t. Windham co. Vt. 5 ms. N. Brattleborough, 31 E. Bennington, W. Conn. r.; was one of the first settled in N. H. is watered by West r. &c. with many mill sites. Black mtn. is granite; the roof slate is quarried here, and primitive limestone is found. Pop. 1830, 1,592.

DUNBARTON, p-t. Merrimack co. N. H. 10 ms. N. Amherst, 9 s. W. Concord, has 21,000 acres, few hills, with clear air, good water, chestnut, pine and oak timber, and good soil. Settled from Londonderry, N. H. 1749; and partly by Scotch and Irish. Pop. 1830, 1,067.

DUNBARTON, p-o. in the s. W. part of Adams co. O. by p-r. 450 ms. W. W. C. and 91 s. s. W. Columbus.

DUNCAN'S CREEK, and p-o. in the E. part of Rutherford co. N. C. 18 ms. E. Rutherfordton, the co. st. and by p-r. 467 s. W. W. C. and 206 W. Raleigh.

DUNCAN'S p-v. Thomas co. Geo. by p-r. 120 ms. s. s. W. Milledgeville.

DUNCAN'S p-o. Hardiman co. Ten. by p-r. 190 ms. s. W. by W. Nashville.

DUNCANSVILLE, p-v. Barnwell dist. S. C. is by p-r. 24 ms. from Barnwell, the st. jus. for this dist. 86 W. of s. Columbia, and 584 s. s. W. W. C.

DUNCANTON, p-v. White co. Il. by p-r. 780 ms. s. of W. W. C. and 109 s. E. Vandalia.

DUNDAFF, p-v. in the s. E. angle of Susquehannah co. Pa. 22 ms. s. E. Montrose, the co. seat, by p-r. 256 ms. N. N. E. W. C. and 148 N. E. Harrisburg.

Dunkard Creek, and p-o. in the N. W. part of Monongalia co. Va: about 22 ms. N. W. by W. Morgantown, and by p-r. 247 N. W. by W. W. C.

Dunkirk, p-v. Pomfret, Chatauque co. N. Y. 45 ms. S. W. Buffalo, 45 N. E. Erie, has a good harbor, with 7 ft. water on the bar.

Dunkirk, called in the p-o. list King and Queen C. H., p-v. on the left bank of Mattapony r. at or near the head of tide water, 60 ms. above Yorktown, and by p-r. 140 ms. a little W. of S. W. C. and 54 N. E. Richmond; lat. 37° 50′, long. 0° 11′ W. W. C. Vessels of considerable tonnage are navigated up to Dunkirk.

Dunlapsville, p-v. in the S. W. angle of Union co. Ind. by p-r. 82 ms. S. of E. Indianopolis, and 521 ms. N. of W. W. C.

Dunningstreet, p-v. Malta, Saratoga co. N. Y.

Dunsburg, or Dunstown, p-v. Lycoming co. Pa. on the left bank of the W. branch of Susquehannah, opposite the mouth of Bald Eagle cr. 25 ms. above Williamsport.

Dunnsville, p-o. S. part of Essex co. Va. 56 ms. S. E. Richmond.

Dunstable, p-t. Hillsboro' co. N. H. 12 ms. S. E. Amherst, 40 N. W. Boston, W. Merrimack r. contains 18,878 acres, has a variety of good soils, level E. hilly W. with mill seats on Salmon brook, and rich land on Nashua r. on which is the chief village; first settled in the co. 1672, and was attacked by Indians. Loverell's company went from this t. performed exploits, and were cut off 1724 at Fryeburg, Me. Pop. 1830, 2,414.

Dunstable, t. Middlesex co. Mass. 37 ms. N. W. Boston, S. Merrimack r. has pretty good level land, with pine, oak, and nutwood. Nashua r. on N. W. Pop. 1830, 593.

Duntonville, p-v. W. part of Edgefield dist. S. C. by p-r. 67 ms. W. Columbia.

Duplessis, Landing and p-o. Opelousas, La. by water route 180 ms. N. W. by W. New Orleans.

Duplin co. N. C. bounded W. by Sampson, N. by Wayne, N. E. by Lenoir, E. by Onslow, and S. by New Hanover. Length 30, mean breadth 20, and area 640 sq. ms. extending from lat. 34° 48′ to 35° 12′, and divided into nearly equal portions by long. 1° W. W. C. It is drained by, and nearly commensurate with, the higher part of the valley of the E. branch of Cape Fear r. Soil of middling quality. Pop. 1820, 9,744; 1830, 11,291.

Duplin, C. H. and p-o. by p-r. 86 ms. S. E. Raleigh.

Duplin, old C. H. and p-o. by p-r. 81 ms. S. E. Raleigh.

Durand, t. Coos co. N. H. 77 ms. N. Concord, N. White mtns., contains 26,680 acres, crossed by Israel's and Moose rs., has a pretty good soil.

Durant's Neck, and p-o. Perquimans co. N. C. by p-r. 218 ms. N. of E. Raleigh.

Durham, p-t. Cumberland co, Me. 26 ms. N. E. Portland, 31 Augusta, S. W. Ameriscoggin r. Pop. 1830, 1,731

Durham, p-t. Strafford co. N. H. N. W. Little and Great bays, contains 14,970 acres, has Piscataqua r. and branches; the village is on Oyster r. at the falls, to which the tide flows. The soil is hard, but good, especially on Onion r. Granite is quarried. The place has suffered from the Indians. Pop. 1830, 1,606.

Durham, p-t. Middlesex co. Conn. 7 ms. S. Middletown, 18 N. E. New Haven, 4 ms. by 6, 23 sq. ms., handsomely varied, with hills E. is at the beginning of the argillaceous tract running N., has sand stone quarries, and good soil, especially on Middletown and West rs. Gen. James Wadsworth, of the revolution, was born here. Pop. 1830, 1,116.

Durham, p-t. Greene co. N. Y. 22 ms. N. W. Catskill, 30 S. W. Albany, on the top of Catskill mtns.; greatest dimensions 8 ms. by 17, has various soils, generally good for grass. Pop. 1830, 3,039.

Durham, tsp. and p-o. Bucks co. Pa. It is the extreme northern tsp. of the co. on Del. r. 12 ms. S. from, and below Easton.

Durhamville, p-o. Tipton co. Ten. by p-r. 190 ms. a little S. of W. Nashville.

Dutchess co. N. Y. bounded by Columbia co. N., Conn. E., Putnam co. S., Hudson r. and Ulster co. W., contains 725 sq. ms. has 18 towns, and is one of the richest in the state. The soil is generally warm loam, N. W. clayey and uneven. Mattawan mtns. E. some ridges are bare, and some slate, both are quarried —gypsum has been very useful. It is watered by Wappingers, Fishkill, Fall, Croton and Ancram creeks; Cram, Elbow, &c. It has manufactories. Chief t. Poughkeepsie. Pop. 1820, 46,615, 1830, 50,926.

Dutch Settlement, C. H. and p-o. St. Mary's parish, La. on Teche r. about 120 ms. W. New Orleans.

Dutotsburg, p-v. Northampton co. Pa. situated on Del. r. N. side of the water gap, and at and below the cr. 25 ms. N. Easton.

Dutton, p-t. Penobscot co. Me. 76 ms. from Augusta.

Duval, co. Flor. as laid down on Tanner's U. S. includes all the country from the Atlantic, between St. Johns and Nassau r. to Suwanne r. on the W. St. Mary's N. and on the S. and S. E. by a line from Jacksonville on the St. Johns, to the mouth of the Suwanne into the Gulf of Mexico. This would include a triangle of 125 ms. base, and 40 ms. perpendicular, or 2500 sq. ms. lying between lat. 29° 22′ and 30° 30′, long. from 4° 38′ to 6° 28′ W. W. C. It is probable that only the N. E. part, between Nassau co. and r., the Atlantic, St. Johns, and St. Mary's rs. or about 750 sq. ms. will remain included in Duval co.

Duxbury, p-t. Plymouth co. Mass. 10 ms. N. Plymouth, 38 S. E. Boston, W. Plymouth harbor. Capt. Standish was buried here, 1656. The soil is warm and sandy, good E. and the people live chiefly by trade and fishing. Pop. 1830, 2,716.

Duxbury, t. Washington co. Vt. 13 ms. W. Montpelier, 22 S. E. Burlington, 100 N. Bennington. It is mountainous and unsettled s. Chief population E. on Onion r. over which is a natural bridge, with caves. There are 4 school districts. Pop. 1830, 651.

DWIGHT, p-v. Pope co. Ark. by p-r. 1,146 ms. s. w. by w. W. C. and 71 ms. above, and N. w. by w. from Little Rock. As laid down by Tanner, it is situated on the left bank of the Ark. r. 7 or 8 ms. above, and on the opposite side from the influx of Petite Jean r.

DYER co. Ten. bounded N. by Obion co, E. by Gibson, s. by Haywood and Tipton, and w. by the Miss. r. which separates it from Crittendon and New Madrid cos. Ark. Length from w. to E. 36, mean width 28, and area 840 sq. ms. extending in lat. from 35° 48′ to 36° 10′, long. from 12° 15′ to 12° 46′. Obion r. enters this co. on the N. border, and flowing s. w. falls into the Miss. about 12 ms. s. w. Dyersburg. The main branch of Forked Deer r. enters Dyer from the s. and flowing N. w. receives a large confluent from the E. at Dyersburg, and then abruptly turning to s. w. leaves Dyer, and falls into Miss. r. at Tipton co. at the upper end of the first Chickasaw Bluff. The surface of this co. is rolling, except some alluvial flats along Miss. r. Chief t. Dyersburg. Pop. 1830, 1,904.

DYER, C. H. or more correctly Dyersburg, p-v. and st. jus. Dyer co. Ten. situated on the N. branch of Forked Deer r. about 30 ms. from Miss. r. at the first Chickasaw Bluff, and by p-r. 164 ms. a very little s. of w. Nashville.

DYER'S, p-o. Franklin co. Va. by p-r. 191 ms. s. w. by w. Richmond.

DYER'S, old store and p-o. Albemarle co. Va. by p-r. 101 ms. s. w. W. C.

E.

EAGLE, t. Alleghany co. N. Y. Pop. 1830, 892.

EAGLE, p-o. Franklin co. Geo. by p-r. 101 ms. N. Milledgeville.

EAGLE GROVE, p-o. Elbert co. Geo. by p-r. 93 ms. N. N. E. Milledgeville.

EAGLE ROCK, p-v. Wake co. N. C. 12 ms. from Raleigh.

EAGLEVILLE, p-v. in the N. E. part of Ashtabula co. O. by p-r. 189 ms. N. E. Columbus.

EAKER'S MILLS, and p-o. in the w. part of Graves co. Ky. 15 ms. w. Mayfield, the co. st. and by p-r. 299 ms. s. w. by w. Frankfort.

EARL, tsp. and p-o. Lancaster co. Pa. on Conestoga creek, 12 ms. above Lancaster.

EARLESVILLE, p-v. Anderson dist. S. C. by p-r. 135 ms. N. w. by w. Columbia.

EARLY, co. Geo. bounded N. by Randolph co. Geo., E. by Baker, s. by Decatur, and w. by Henry co. Ala. or by Chattahooche r. Length from s. to N. 40, mean breadth 32, area 1280 sq. ms. extending in lat. from 31° 06′ to 31° 43′, long. from 7° 46′ to 8° 20′ w. W. C. Chief t. Blakely. Pop. 1830, 2,081.

EARLY, C. H. (*see Blakeley,*) Early co. Geo.

EAST BERLIN, tsp. and p-o. Adams co. Pa. on a branch of Conewago, 17 ms. N. E. Gettysburg.

EAST BETHLEHEMS, p-o. Washington co. Pa. 16 ms. w. Washington, the co. st.

EAST BLOOMFIELD, p-o. Crawford co. Pa. 10 ms. N. w. Meadville.

EAST BRIDGEWATER, town, Plymouth co. Mass. Pop. 1830, 1,653.

EAST CENTERVILLE, p-v, in the south part of Columbiana co. O., about 12 ms. s. s. w. New Lisbon, the co. st. 138 N. E. by E. from Columbus.

EASTCHESTER, p-t. Westchester co. N. Y. 8 miles s. White Plains, 20 N. N. York, 2½ ms. by 7; E. Bronx creek, w. East Chester creek and bay, where is a landing, for trade with New York; level, stony, but pretty good soil. Pop. 1830, 1,300.

EAST CLARIDON, p-v. Geauga co. Ohio, by p-r. 174 ms. N. E. Columbus.

EAST FAIRFIELD, p-v. near the eastern border of Columbiana co. Ohio, 8 ms. E. New Lisbon, the co. st. and by p-r. 152 N. E. by E. Columbus.

EAST FARMINGTON, p-v. in the N. part of Oakland co. Mich. by p-r. 40 ms. N. Detroit.

EAST FELICIANA, parish of La. bounded by Amite co. which separates it from St Helena parish E., by East Baton Rouge s., Thompson's creek which separates it from West Feliciana w., and by Wilkinson and Amite co. of Miss. N. Length from s. to N. 28 ms., mean breadth 20, area 560 square ms. Extending in lat. from 30° 37′ to 31°, in long. from 14° to 14° 24′ w. W. C. The slope of this parish is almost directly s. Much of the soil on Amite river, on Comite and Thompson creeks, is excellent; staple, cotton. Chief town, Jackson. Pop. 1830, 8,247.

EAST GREEN, p-v. Kennebec co. Me. 20 ms. Augusta.

EAST GREENWICH, p-t. and st. jus. Kent co. R. I. 13 ms. s. Providence, w. Narraganset bay, 4 ms. by 6, 24 square miles, is rough with primitive rocks, pretty good gravelly loam, making good cider; oak, chestnut, &c. It has a safe harbor, with 15 feet water at high tide. Codfish, &c. are taken, and whaling was once carried on. In the village is a bank, court house, academy, and the legislature has sometimes set here. Major Gen. Green, was from this town. Population 1830, 1,591.

EAST HADDAM, p-t. Middlesex co. Conn. 14 miles s. w. Middletown, 27 s. by w. Hartford, E. Conn. river, 6½ miles by 8, 50 square miles, is rough, with granite rocks, containing garnets, beryl, &c. It is good for grass, with some flats, and good timber. Salmon and Modus rivers N. w., and other streams supply mill seats. There were formerly shakings of the earth, attended with sounds here. The

Indians were considered conjurers. Population 1830, 2,664.

EASTHAM, p-t. Barnstable co. Mass. 24 ms. N. E. Barnstable, 75 S. E. Boston, on Cape Cod, is a narrow strip of sand, E. Cape Cod bay and W. ocean, both of which are seen at once from the road. On the E. is some pretty good land; the rest moveable sand. Salt is made here from sea water. The Nanset Indians had a christian church many years here. First settled from Plymouth, 1644. Pop. 1830, 970.

EAST HAMPTON, p-t. Hampshire co. Mass. 5 miles S. Northampton, 90 W. Boston, W. Conn. river. On the E. side is the proposed route of Farmington canal; the town has a variety of soil, and much pine plain. Pop. 1830, 745.

EASTHAMPTON, p-t. Suffolk co. N. Y. 112 miles E. N. York, 35 E. Riverhead, at E. end of Long Island, includes Governor's Island and Montauk Point, S. Gardner's bay and Long Island sound, N. and W. ocean. Greatest breadth 8 miles, greatest length on main land 24. First settled 1649, from Lynn, Mass. The people are farmers, mechanics, and shoemakers. Clinton academy, founded 1784, with $24,000 given by them. At Montauk 9000 acres of good land are owned in common. The light house was built 1796, for $25,000. Gardner's island contains 2,500 acres; Gardner's bay, a good harbor for a fleet of ships, was used by the enemy during the last war.

EAST HANOVER, tsp. and p-o. on Swatara r. W. part of Lebanon co. Pa. about 17 ms. N. E. by E. Harrisburg.

EAST HARTFORD, p-t. Hartford co. Conn. E. Connecticut river, is connected with Hartford with a bridge; has fine meadows, with level, light soil; fine elms in the village. Pop. 1830, 3,537.

EAST HAVEN, town, Essex co. Vt. 45 N. Montpelier, gives rise to Moose river, and is rough and almost uninhabited. Population 1830, 33.

EAST HAVEN, town, New Haven co. Conn. 4 miles E. New Haven, N. Long Island sound, has fine swells, with light soil; a light house at E. point of New Haven harbor; the town is connected with New Haven by a bridge. Pop. 1820, 1,229.

EAST HEMPFIELD, tsp. and p-o. Lancaster county, Pennsylvania, 34 miles S. E. by E. Harrisburg.

EAST KINGSTON, town, Rockingham co. N. H. 21 ms. S. W. Portsmouth, 39 Concord, 3 square miles, has a good soil for grass and grain, and is crossed by Powow river. Pop. 1830, 442.

EAST LIBERTY, tsp. and p-v. Fayette co. Pa. 34 ms. S. E. Uniontown.

EAST LIBERTY, p-v. Marion county, Tennessee, by post road 138 miles southeast Nashville.

EAST MACHIAS, town, Washington co. Me., crossed N. and S. by a broad stream, and emptying into Machias bay. It has a large pond on its eastern border. Population 1830, 1,065.

EAST NANTMILL, tsp. and p-o. N. part of Chester co. Pa. on the waters of French creek, about 33 miles N. W. Phil.

EAST NEW MARKET, p-v. on the waters of Nantikoke river, E. part of Dorchester co. Md. 16 miles a little N. of E. Cambridge.

EASTON, p-t. Bristol co. Mass. 22 miles S. Boston, has large manufactories of iron, woollen and cotton. A lead and silver mining company was incorporated here, 1825, with a capital of $80,000. Population 1830, 1,756.

EASTON, p-t. Washington co. N. Y. 27 ms. N. Albany, 16 S. W. Salem, E. Hudson river, 6 miles by 12, 70 square miles, has good farms, uneven surface, and various soils. Battenkill N. has a fall of 60 feet, and other mill sites.

EASTON, borough, p-t. and st. jus. Northampton co. Pa. situated on the right bank of Delaware river, between the mouths of Lehigh river and Bushkill creek. The site of this borough is a limestone valley environed on all sides by masses of that rock. Beyond the Lehigh rises the Blue Ridge, which about 2 miles below the town, is traversed by Delaware river. The vicinity along the Delaware, Lehigh, and Bushkill, is finally broken and varied, with a very productive soil, and a soil well cultivated, which adds to the attractive scenery, the charm of abundance. According to information procured on the spot, there were in Easton, 1821, about 2,500 inhabitants. A library containing 1,200 volumes; an academy called the Union academy, three places of public worship, 1 for Presbyterians, 1 for German Lutherans, and 1 for Episcopalians. There were within the borough 6 grist mills, 2 saw mills, 2 distilleries, 3 tan yards, 1 brewery, and 31 dry good stores. Four fine bridges, 1 over the Delaware, 1 over Lehigh river, and 2 over Bushkill creek. The town is laid out at right angles, streets along the Cardinal points, issuing from a central square, in which stands the court house, built in 1758. Since 1821, the advance of Easton has been rapid. The Lehigh and Delaware canals have made it an emporium in reality, from which lines of intercommunication radiate as from a common centre. Pop. 1810, 1,857, 1820, 2,370, 1830, it had risen to 3,529. Lat. 42° 42′, long. 1° 50′ E. W. C.

EASTON, p-v. seaport and st. jus. Talbot co. Md. situated near the centre of the co. at the head of Tread Haven river or bay. Lat. 38° 46′, long. 1° E. W. C. by p-r. 81 miles a little S. of E. W. C. and 41 S. E. by E. Annapolis. Pop. 1820, 2,000.

EAST PENN, p-o. and township, Northampton co. Pa. The office is by p-r. 191 miles N. N. E. W. C. and 91 ms. N. E. by E. Harrisburg. Pop. of the tsp. 1830, 1,007.

EASTPORT, p-t. and port of entry, Washington co. Me. 176 miles E. Augusta, 279 E. N. E. Portland, 41 ms. E. N. E. Machias, in Passamaquoddy bay, on Moose Island, 4 miles long, with bold shores, is an important place for trade, and the easternmost military post of the United States. Lumber trade and fishing are principal branches of business. A ferry of 3 miles crosses to Lubec, and a bridge to Perry. The village is s. Pop. 1830, 2,450.

EASTPORT, p-v. Lauderdale co. Alabama, by p-r. 111 ms. N. Tuscaloosa.

EAST RIVER, King's, Queen's, New York and Westchester cos. N. Y. is a strait, connecting New York bay with Long Island sound, is an important channel for coasting vessels, about 25 miles by 1, navigable for the largest ships, with several isls. and a swift and rocky pass at Hell Gate, or Horl Gatt.

EAST SMITHFIELD, p-o. Bradford co. Pa. by p-r. 188 ms. N. Harrisburg.

EAST SUDBURY, p-t. Middlesex co. Mass. 18 ms. w. Boston, has good soil, several ponds, and is crossed by Sudbury river. Pop. 1830, 944.

EASTVILLE, p-v. and st. jus. Northampton co. Va. situated on the Peninsula between Chesapeake bay and the Atlantic, 18 miles N. Cape Charles, by p-r. as stated in the p-o. list, 254 miles S. S. E. W. C., though in a direct line the distance is only 125 miles. Lat. 37° 30′, long. 1° 15′ E. W. C.

EAST WATERFORD, p-v. in the southwestern part of Juniata co. Pa. about 40 miles in direct road N. of w. Harrisburg, but by p-r. 62 miles.

EAST WHITELAND, township and p-o. Chester co. Pa. on the main road from Philadelphia to Lancaster, 20 miles from the former.

EAST WILLIAMSBURG, p-v. Northampton co. Pennsylvania, by p-r. 128 miles N. E. by E. Harrisburg.

EAST WINDSOR, p-t. Hartford co. Conn. 8 ms. N. Hartford, has rich meadows, a pleasant village on a wide street, lined with fine elms, and many fine farms. It was one of the 4 earliest settlements in the state. First settled 1636. Population 1830, 2,129.

EAST WINDSOR, town, Middlesex co. N. J. Pop. 1830, 1,905.

EATON, p-t. Strafford co. N. H. 71 miles N. N. E. Concord, 41 N. E. Guilford, 7 N. Portsmouth, w. Maine, contains 33,637 acres, has pretty good uplands, and pine on plains, with some iron ore, small mill streams and several ponds. Pop. 1830, 1,432.

EATON, p-v. in the N. W. part of Luzerne co. Pa. by p-r. 29 ms. N. Wilkes-Barre. Population 1830, 599.

EATON, p-v. and st. jus. Preble co. Ohio, 26 miles w. Dayton, 51 w. of N. Cincinnati, 488 miles a little N. of w. W. C. N. lat. 39° 46′, long. 7° 38′ w. W. C. Pop. 1830, 510.

EATON'S NECK, Huntington, New York, on Long Island sound, has a light house.

EATONVILLE, or Eatonton, p-v. and st. jus. Putnam co. Geo. near the centre of the co. 20 ms. N. N. W. Milledgeville. Lat. 33° 19′, long. 6° 28′ w. W. C.

EBENEZER, academy and p-o. a. w. part of York district, South Carolina, 66 miles N. N. W. Columbia.

EBENEZER, village, Effingham co. Geo. on the right bank of Savannah river, 25 miles above Savannah.

EBENSBURG, borough, p-v. and st. jus. Cambria co. Pa. situated on the head waters of Little Conemaugh, 75 ms. a very little N. of E. Pittsburg, and by p-r. 144 miles N. W. by W. Harrisburg. Lat. 40° 31′, long. 1° 40′ W. W. C.

ECHOCUNO, or Tchocunno river, Georgia, rising in Monroe co. between Flint and Chupee rivers, and flowing thence into Crawford, over the N. E. angle of Bibb and Crawford and Bibb and Houston cos. falls into the Oakmulgee, after an entire comparative course of 40 miles, in a southeast direction.

ECHOCONNO, p-o. on Echoconno r., Crawford co. Geo. by p-r. 42 ms. s. w. Milledgeville.

ECONOMY, p-v. Erie co. Pa. about 100 ms. N. Pittsburg.

ECONOMY, p-v. in the eastern part of Wayne co. Indiana, by p-r. 77 ms. E. Indianopolis.

EDDYVILLE, p-v. on the right bank of Cumberland river, Caldwell co. Ken. 12 miles from Princeton, the co. st. Pop. 1830, 167.

EDDINGTON, p-t. Penobscot co. Me. 70 ms. N. E. Augusta, E. Penobscot river, opposite Bangor. Pop. 1830, 405.

EDEN, p-t. Hancock co. Me. 36 miles E. Castine, 92 Augusta, is almost insulated by Frenchman's bay and Mount Desert sound. Pop. 1830, 957.

EDEN, t. Orleans co. Vt. 30 ms. N. Montpelier, 37 N. E. Burlington, 36 sq. ms. was granted to Col. S. Warner, and his regiment, 1781, has many small streams, with the sources of Wild Branch and Green r. Mount Norris, Belvidere and Hadley mtns., 5 school districts. Pop. 1830, 461.

EDEN, p-t. Erie co. New York, 23 miles s. Buffalo, 6 miles square, 7 miles E. lake Erie, has a varied surface, and watered by Canquada creek, with loamy sand and gravel, best for grass; beech, maple, hemlock, &c. Population 1830, 1,066.

EDEN'S RIDGE, and p-o. w. part Sullivan county, Tennessee, by p-r. 297 miles N. of E. Nashville.

EDGARTOWN, p-t. port of entry and st. jus. Duke's county, Mass. 100 miles S. S. E. Boston, 14 miles south main land, has a good and convenient harbor, protected by Chippaquiddick island, a shelter in storm, and has considerable shipping.

EDGECOMB, p-t. Lincoln co. Me. on Sheepscott river, 20 miles from Augusta, is almost insulated by Damariscotta and Sheepscott rivers. Pop. 1830, 1,258.

EDGECOMBE, co. N. C. bounded by Neuse river, which separates it from Wayne s. w., by by Nash w., and N. W. by Halifax, N. and N. E. by Martin, E. and S. E. by Pitt and Greene. Length from S. W. to N. E. 35 miles, mean breadth 18 miles, area 648 square miles. Extending in lat. from 35° 34′ to 36° 06′, long. 0° 27′ to 1° 02′ w. W. C. The two main branches of Tar river enter this co. separate, but unite within it, a short distance above Tarborough, and flow from the s. E. border into Pitt co. The s. part is drained by various branches of Neuse river. The surface level, and soil middling quality. Chief town, Tarborough. Population 1820, 13,276, 1830, 14,935.

EDGEFIELD, p-v. Fauquier co. Va. by p-r. 47 ms. w. W. C.

EDGEFIELD, dist. S. C. bounded by Abbeville N. w., by Saluda river separating it from Newburg N., by Lexington N. E., Orangeburg E., Barnwell S. E., and Savannah river separating it from Richmond, Columbia and Lincoln co. of Georgia, S. w. Length from S. to N. 60 miles, mean breadth 28, area 1,680 square miles. Extending in lat. from 33° 17′ to 34° 11′, long. from 4° 50′ to 5° 20′ w. W. C. The southern part of Edgefield slopes south towards the Savannah river, and is drained by Stephens creek and some minor streams. The southern section has its slope towards Saluda, and is drained in great part by Little Saluda. Surface gently hilly, and soil mostly of second rate quality. Chief town, Edgefield. Pop. 1820, 25,179, 1830, 30,509.

EDGEFIELD, court house and p-v. Edgefield dist. S. C. by p-r. 57 miles S. w. by w. Columbia.

EDGEMONT, p-v. Delaware co. Pa. 123 ms. N. E. W. C. Pop. 1830, 757.

EDINBORO, p-v. Montgomery co. N. C. by post road 97 miles southwest by west Raleigh.

EDINBURGH, p-t. Saratoga co. N. Y. 30 ms. N. w. Ballston Spa, 7 miles by 8, is crossed by Sacandaga river, is hilly east, and level west, and has generally a stiff loam, with good land in the middle, and S. w.

EDINBURGH, p-o. and tsp. in the E. part of Portage co. Ohio, 7 ms. E. Ravenna, the co. st.

EDINBURGH, p-v. in the S. E. angle Johnson co. Indiana, by p-r. 30 ms. S. S. E. Indianopolis. It is situated at the junction of Blue river and Sugar creek, branches of Driftwood fork of White river.

EDISTO, river of S. C. rises by two branches in Edgefield dist. S. Edisto flowing S. E. leaves Edgefield and forming the boundary between Barnwell and Orangeburg districts, receives N. Edisto, and continuing S. E. enters Colleton, and inflecting to the S. reaches the alluvial plain near the Atlantic, where it divides into two channels, again called relatively N. Edisto, and S. Edisto, enclosing Edisto isl. on both sides. The entire comparative length of Edisto by either branch is about 130 miles. Its basin is 130 ms. by a mean breadth of 30, area 3,900 square miles. Lying between the Savannah and Santee rs.

EDISTO ISL., S. C. enclosed by the two outlets of Edisto river and the Atlantic. Length from Clark's inlet on the ocean to the separation of the two Edistos, 12½ miles, mean breadth 7 miles, area 87½ square miles, forming a part of Colleton district. Surface flat and in great part marshy, with numerous interlocking water courses. Soil where fit for culture, highly productive. Central lat. 32° 33′. Staple culture, cotton and rice, though the climate would perhaps admit sugar cane.

EDMONDS, town, Washington co. Me. w. Colescook bay. Pop. 267.

E. EDMONDSON, co. Ky. bounded S. and S. w. by Warren, w. and N. w. by Grayson, and N. E. and E. by Hart. It lies nearly in form of a circle of 18 miles diameter, area about 250 square ms. Extending in lat. from 37° 05′ to 37° 20′, long. from 9° 02′ to 9° 23′ w. W. C. The main volume of Green river winds through this co. from E. to w. receiving a large northern branch, Adin's creek, near the centre. It lies in the limestone range and within what has been called the Barrens of Ky., though in reality the soil is productive. Chief town, Brownsville. Population 1830, 2,642.

EDMONTON, p-v. Barren co. Ky. by p-r. 114 ms. S. w. Frankfort.

EDMESTON, p-t. Otsego co. N. Y. 18 miles w. Cooper's town, 84 w. by S. Albany, E. Unadilla river, has 26,628 acres, is varied in surface and soil, and has mill seats on Unadilla and Wharton's creeks, and has limestone S. E. Pop. 1830, 2,087.

EDMUND'S, p-o. Brunswick co. Virginia, S. S. w. Richmond; position in the county, uncertain.

EDNYVILLE, p-o. Buncombe co. N. C. by p-r. 234 ms. w. Raleigh.

EDONTON, p-v. and st. jus. Chowan co. N. C. situated on a small bay opening S. w. into Chowan bay, and S. E. into Albermarle sound, about 65 miles S. S. w. Norfolk, Va. and by p-r. 183 ms. a little N. of E. Raleigh.

EDSALVILLE, p-o. Bradford co. Pennsylvania, by p-r. 182 ms. N. Harrisburg.

EDWARDSBURG, p-v. and st. jus. Cass co. Mich. by p-r. 643 ms. N. w. by w. W. C. and 169 ms. a little S. of w. Detroit. It is situated near the S. border of the co. and of Mich., and on a branch of St. Joseph's r. Lat. 42° 48′, long. 9° 9′ w. W. C.

EDWARD'S FERRY and p-o. The ferry is over the Potomac where the road crosses that river, between Rockville in Montgomery co. Md. and Leesburg in Va. at and above the mouth of Goose creek, 21 ms. a little N. of w. from the former, 4 ms. N. E. from Leesburg, and 31 ms. N. w. W. C. The p-o. is in Montgomery co. Md.

EDWARDSVILLE, p-v. Salem tsp. in the S. E. part of Warren co. O. by p-r. 460 ms. w. W. C. and 83 S. w. by w. Columbus. Pop. 1830, 48.

EDWARDSVILLE, p-v. and st. jus. Madison co. Il. by p-r. 836 ms. w. W. C. 55 a little S. of w. Vandalia, and by the intermediate road 20 ms. N. E. from St. Louis in Mo.

EDYVILLE, p-v. Caldwell co. Ky. situated on the right bank of Cumberland r. about 35 ms. following the stream above its mouth, and as laid down on Tanner's U. S. exactly on lat. 37°, by p-r. 207 ms. S. w. by w. Frankfort.

EFFINGHAM, t. Strafford co. N. H. 43 ms. N. E. Concord; contains 34,000 acres, has several high mountains, and is crossed by Ossipee r. Pop. 1830, 1,911.

EFFINGHAM, co. Geo. bounded by Great Ogeeche r. which separates it from Bryan s. w., and Bullock w., by Scriven N. W., by Savannah r., which separates it from Beaufort dist. S. C. N. E. and E., and by Chatham S. Length 30, mean breadth 11, area 330 sq. ms. Extending in lat. from 32° 08′ to 30° 33′, long. from 4° 12′ to 4° 31′ w. W. C. Surface level. Chief t. Springfield. Pop. 1820, 3,018, 1830, 2,924.

EFFINGHAM, p-v. Bedford co. Ten. by p-r. 56 ms. s. Nashville.

EFFINGHAM, co. Il. bounded s. by Clay, w. by Fayette, N. by Shelby, N. E. by Coles, and E. by Jasper. Length 22, breadth 18, and area 396 sq. ms. Extending in lat. from 38° 54′ to 39° 12′ w. W. C. The slope is southwardly, and in that direction is traversed by Little Wabash. It was formed from what was formerly the E. part of Fayette, and its central part is about 35 ms. E. Vandalia.

EGG HARBOR, t. Gloucester, co. N. J. on the Atlantic, bounded s. w. and w. by Great Egg Harbor r. Pop. 1830, 2,510.

EGG HARBOR, GREAT, r. and inlet, Gloucester co. N. J. The river is navigable for vessels of large size for some distance from its mouth, which is in lat. 39° 18′, 20 ms. N. of Cape May, and 60 from Phil.

EGG HARBOR, GREAT, port of entry Gloucester co., and the name of a collection district, the tonnage of which, in 1829, was 9,511 tons, 60 ms. S. E. Phil.

EGG HARBOR, LITTLE, bay and inlet, Burlington co. N. J. on the Atlantic ocean, at the mouth of Mullicus r. and about 40 ms. N. of Cape May.

EGG HARBOR, LITTLE, t. Burlington co. N. J. on the sea coast, bounded s. w. by Mullicus r. which separates it from Gloucester co. Pop. 1830, 1,491. It gives name to a collection district, the collector of which resides at Tuckerton. Tonnage in 1829, 2,783 tons.

EGREMONT, p-t. Berkshire co. Mass. 15 ms. s. s. w. Lenox, 130 w. Boston, E. N. Y. on E. declivity of Taughkannuck mtn. tributary to Housatonic r. Pop. 1830, 890.

ELBA, p-t. Genesee co. N. Y. 6 ms. N. Batavia, 10 s Erie canal, is nearly level, good for grazing, and gives rise to Oak Orchard creek, and sends streams s. to Black creek. Pop. 1830, 2,678.

ELBERT, co. Geo. bounded by Broad r. which separates it from Lincoln s. E., Wilkes s., Oglethorpe s. w., and Madison w., by Franklin N. W., and by Savannah r. which separates it from Anderson dist. S. C. N. E., and from Abbeville dist. E. Length along Savannah r. 40 ms., mean breadth 14, and area 560 sq. ms. Extending in lat. from 33° 56′ to 34° 30′, long. from 5° 33′ to 6° 10′ w. W. C. Surface hilly, and soil productive. Slope s. estrd. Chief t. Elberton. Pop. 1820, 11,788, 1830, 12,354.

ELBRIDGE, p-v. Onondaga co. N. Y. 2 ms. s. Erie canal. Pop. 1830, 3,357.

ELBRIDGE, p-v. in the s. E. part Edgar co. Il. by p-r. 116 ms. N. E. by E. Vandalia, and 665 ms. w. W. C.

ELDERTON, p-v. in the s. E. part of Armstrong co. Pa. 13 ms. from Kittaning, the co. st., and by p-r. 202 ms. N. W. W. C. and 170 w. from Harrisburg.

ELDERSVILLE, p-v. Washington co. Pa. on the road from Washington, the st. jus. for the co., to Steubenville, 20 ms. N. W. the former, and 16 s. E. the latter place.

ELDENTON, p-v. Armstrong co. Pa.

ELDREDVILLE, p-o. Lycoming co. Pa. by p-r. 105 ms. N. W. Harrisburg.

ELDRIDGE, p-o. Buckingham co. Va. by p-r. 82 ms. w. Richmond.

ELDRIDGE, p-v. and tsp. in the N. E. part of Huron co. O. The p-o. 397 ms. N. W. W. C. and 124 N. N. E. Columbus. Pop. of the tsp. in 1830, 742.

ELIZABETH RIVER, Va. rises by numerous small branches in Princess Ann and Norfolk cos., flows to the N. W. opening into a wide estuary, terminating in the mouth of James r. The entire length of Elizabeth r. is only about 25 ms., but it gains importance as forming the fine harbor of Norfolk, admitting to that port vessels of 18 feet draught, and again as constituting with the Dismal Swamp canal and Pasquotank r., a chain of inland navigation from Chesapeake bay to Albemarle sound.

ELIZABETH ISLANDS, Duke's co. Mass. are 16 in number, not all inhabited, extend s. w. from Barnstable, forming the s. E. side of Buzzard's bay, s. E. Bristol co., and s. w. Martha's Vineyard. The largest are Nashawn, Nashawenna, and Presque Isle. Gosnold spent the winter of 1602 here with a party of English.

ELIZABETH, p-o. Alleghany co. Pa. by p-r. 234 ms. N. W. W. C.

ELIZABETH, p-v. in the s. part of Harrison co. Ind. 11 ms. s. Corydon, the co. st. and by p-r. 613 s. of w. W. C. and 135 ms. s. Indianopolis.

ELIZABETH CITY, co. Va. bounded w. by Warwick, N. by Black r. separating it from York co., E. by Chesapeake bay, and s. by Hampton roads, or mouth of James r. It lies in the form of a square of 18 ms. each side, area 64 sq. ms. Extending in lat. from 37° 02′ to 37° 08′, long. from 0° 37′ to 0° 47′. Chief t. Hampton. Pop. 1820, 3,789, 1830, 5,053.

ELIZABETH CITY, p-v. and st. jus. Pasquotank co. N. C. situated on the right bank of Pasquotank r. at the point where that stream widens into a bay, 45 ms. s. Norfolk, Va. by p-r. 182 ms. N. E. by E. Raleigh. Lat. 36° 14′, long. 09 52′ E. W. C.

ELIZABETHTOWN, p-t. and st. jus. Essex co. N. Y. 126 ms. N. Albany, 16 w. Essex, w. of N. West bay of lake Champlain, has mtns. with some large and fertile valleys. Pleasant valley is crossed by Bouquet r. It has a village, with co. buildings, state arsenal, &c. The Giant of the valley mountains is

1,200 feet high. There are ores and forges. Pop. 1830, 1,015.

ELIZABETHTOWN, p-t. and borough, Essex co. N. J. 15 ms. w. by s. N. Y. by water 6, s. Newark, 17 N. E. New Brunswick, w. Newark bay, level, with pretty good soil, well cultivated for gardens, &c., supplying many articles for N. Y. market. Was settled from Connecticut, and has a large and handsome village, with a court house, &c., an academy and apprentices' library, 1 m. from the point whence is frequent daily steamboat navigation to N. Y. and Phil. Vessels of 300 tons go to the point, and those of 30 to the village. Pop. 1830, 3,445.

ELIZABETHTOWN, p-v. near the w. border of Lancaster co. Pa. on the road from the city of Lancaster to Harrisburg, about 18 ms. from each.

ELIZABETHTOWN, v. Alleghany co. Pa. on the right bank of Monongahela r. 15 ms. a little E. of s. Pittsburg.

ELIZABETHTOWN, Washington co. Md. (*See Hagerstown.*)

ELIZABETHTOWN, p-v. and st. jus. Bladen co. N. C. situated on the right bank of Cape Fear r. 37 ms. by the road below Fayetteville, and by p-r. 98 ms. s. Raleigh. Lat. 34° 40′, long. 0° 38′ w. W. C.

ELIZABETHTOWN, p-v. and st. jus. Hardin co. Ky. situated on a small creek, N. branch of Nolins creek, 43 ms. s. Louisville, and by p-r. 72 ms. s. w. Frankfort. Lat. 37° 42′, long. 8° 50′ w. W. C.

ELIZABETHTOWN, p-v. and st. jus. Carter co. E. Ten. situated on the waters of Watauga r. about 120 ms. N. E. by E. Knoxville, and by p-r. 270 ms. a little N. of E. Nashville. Lat. 36° 22′, long. 5° 5′ w. W. C.

ELIZABETHTOWN, p-v. White Water tsp. in the s. w. part Hamilton co. O. 17 ms. w. Cincinnati, and by p-r. 514 ms. w. W. C. Pop. 1830, 134.

ELIZAVILLE, p-v. in the w. part of Flemming co. Ky.

ELK, r. stream of Pa. Del. and Md. The extreme source in Chester co. of the former state between Octora and White Clay creeks, and flowing thence s. enters Coecil co. Md. receiving from Del. Back and Bohemia creeks, falls into the head of Chesapeake bay 8 ms. s. s. E. the mouth of Susquehannah r. This small river is important from its position. The lower part below Back creek forms a part of the line of inland navigation by the Chesapeake and Del. canal.

ELK r. Western Va. rises amid the Appalachian Ridges in Randolph and Pocahontas cos., interlocking sources with those of Monongahela, Little Kenhawa, Wheat, Green Brier and Gourly rs. Leaving Randolph and Pocahontas, and traversing Nicholas and Kenhawa cos., it finally is lost in Great Kenhaway at Charleston, after a comparative western course of 100 ms.

ELK RIVER of Ten. and Ala., drawing its remote sources from the N. w. slope of Cumberland mtn. Franklin co. Ten., and flowing thence by a general course s. w. by w. over Franklin, Lincoln, and Giles cos. Ten. enters Ala., traversing limestone, and falling into Ten. r. in the s. E. angle of Lauderdale co., after a comparative course of 110 ms. The valley of Elk r. lies between those of Ten. and Duck rs.

ELK CREEK, p-o. in Elk Creek tsp. N. w. part of Erie co. Pa. by p-r. 306 ms. N. w. Harrisburg.

ELK CREEK, tsp. Erie co. Pa. on the heads of Cussewago, Conneaut and Elk creeks, 17 ms. s. w. the borough of Erie. Pop. 1820, 288, 1830, 562.

ELK FORK, p-v. in the N. part of Jefferson co. O. by p-r. 23 ms. northerly from Steubenville, the co. st., 283 ms. N. w. by w. W. C. and 145 ms. N. E. by E. Columbus.

ELK GROVE, p-v. Iowa co. Mich. by p-r. 1,110 ms. N. w. by w. W. C.

ELKHART, co. of Ind. bounded by La Grange co. E., the Putawatomie territory s. E. and s., St. Joseph's co. w., Berrien co. of Mich. N. w., and Cass co. of Mich. N. E. Length from s. to N. 26 ms., breadth 20, and area 520 sq. ms. Extending in lat. from 41° 25′ to 41° 46′, and in long. from 8° 45′ to 9° 8′ w. W. C. The southwestern angle gives source to the Kankakee branch of Illinois, and delines wstrd. The northern part also declines wstrd., but is traversed in that direction by the main volume of St. Joseph r. Elkhart r., from which the co. derives its name, enters the southeastern angle, and flowing N. N. w. falling into St. Joseph r. and receiving confluents from both sides, gives a slope in that direction to the body of the co. Pop. 1830, 935.

ELK HEART PLAIN, p-v. Wabash co. Ind. by p-r. 616 ms. N. w. by w. W. C. and 196 N. N. E. Indianopolis.

ELK HILL, p-o. Amelia co. Va. by p-r. 59 ms. s. w. Richmond.

ELKHORN, small r. of Ky. rising in Lafayette co. near Lexington, and traversing Scott and Woodford, falls into the right side of Ky. r. in Franklin co. 10 ms. below Frankfort, after a comparative course of 30 ms.

ELKHORN, p-o. Franklin co. Ky. 4 ms. from Frankfort.

ELKHORN, p-v. on a small river of the same name, in the s. w. part of Washington co. Il. The p-o. is by p-r. 824 ms. w. W. C. and 52 ms. s. s. w. Vandalia. The r. is a small stream rising near the northern border of Perry co. and flowing northwstrd. over Washington, falls into Kaskaskias r. near the boundary between Washington and St. Clair cos.

ELKLAND, p-o. Tioga co. Pa. by p-r. 161 ms. N. Harrisburg.

ELK MARSH, p-o. s. part Fauquier co. Va. 22 ms. N. w. Petersburg.

ELK RIDGE, p-o. Giles co. Ten. by p-r. 91 ms. s. s. w. Nashville.

ELK RIDGE LANDING, p-o. Ann Arundel co. Md. on the right bank of Patapsco r. 9 ms. s. w. Baltimore.

ELK RUN, church and p-o. s. E. part of

Fauquier co. Va. 20 ms. N. N. W. Fredericksburg.

Elkton, p-t. and st. jus. Cœcil co. Md. situated on the point between and above the junction of the two main branches of Elk r., very nearly on the direct line and mid distance between Philadelphia and Baltimore, or about 50 ms. following the road from each. Lat. 39° 36′, long. 1° 13′ E. W. C. The importance of Elkton as a travelling station has been lessened by the change of routes, and particularly by the opening of the Chesapeake and Del. canal. It is still, however, a neat village, and the depot of considerable trade.

Elkton, p-v. S. part of Giles co. Ten. situated on the point and above the junction of Elk r. and Richland creek, 10 ms. S. S. E. Pulaski.

Elkton, p-v. and st. jus. Todd co. Ky. situated on Elk creek, a branch of Red r. by p-r. 190 ms. S. W. by W. Frankfort. Lat. 36° 51′, long. 10° 13′ W. W. C.

Ellejoy, p-v. in the S. part of Blount co. Ten. by p-r. 534 ms. S. W. by W. W. C. and 208 S. of E. Nashville.

Ellenburgh, t. Clinton co. N. Y. Pop. 1830, 1,222.

Ellenton, p-v. and st. jus. Elbert co. Geo. about 70 ms. N. W. Augusta, and by p-r. 73 ms. N. N. E. Milledgeville. Lat. 34° 05′, long. 5° 52′ W. W. C.

Ellerslie, p-v. in the N. part of Susquehannah co. Pa. by p-r. 16 ms. N. from Montrose, the co. st., and 287 ms. a little E. of N. W. C. and 179 N. N. E. Harrisburg.

Ellerslie, p-v. in the W. part of Harris co. Geo. by p-r. 776 ms. S. W. W. C. and 134 W. Milledgeville.

Ellery, p-t. Chatauque co. N. Y. 11 ms. S. E. Mayville, 54 sq. ms. with most of Chatauque lake, pretty good land, with oak, ash, bass, &c.; recently settled, has many small streams. Pop. 1830, 2,002.

Ellicott, t. Chatauque co. N. Y. 30 ms. S. E. Mayville, 144 sq. ms., the N. and W. branches of Connewongo creek meet here, and are navigable for rafts. Mill seats abound; the soil is various. Pop. 1830, 2,101.

Ellicotts, or Eleven Mile Creek, Genesee and Erie cos. N. Y., joins Tonawanda creek near Lake Erie.

Ellicotts Mills, p-v. Baltimore co. Md. on the main stream of Patapsco, 10 ms. S. W. by W. Baltimore. The village straggling along the valley, and intermingled with mills and other manufactories, is in both Baltimore and Ann Arundel cos., and on the main road from Baltimore to the city of Frederick. The Baltimore and Ohio rail road, which leaves the city and follows the valley of the Patapsco, generally intersects the turnpike in Ann Arundel, part of Ellicott Mills. The vicinity is broken and romantic, and scenery formerly not suspected to exist 10 ms. from Baltimore, will now command attention, and become a fashionable place of resort, from the facility and pleasure of moving on the rail way.

Elliottsburg, p-v. Perry co. Pa. by p-r. 48 ms. N. W. Harrisburg.

Elliotts' Cross Roads and p-o. Cumberland co. Ky. by p-r. 151 ms. a little W. of S. Frankfort.

Ellicottville, p-t. and st. jus. Cattaraugus co. N. Y. 325 ms. W. Albany, 6 ms. by 15, is supplied with excellent mill sites by Great Valley creek. Pop. 1830, 626.

Ellington, p-t. Tolland co. Conn. 13 ms. N. E. Hartford. Greatest extent 6 by 9 ms., 34 sq. ms., level W., broken E., good for grain, has a pleasant village, near which is Mr. Hall's academy. Pop. 1830, 1,455.

Ellington, t. Chatauque co. N. Y. Pop. 1830, 1,279.

Elliot, p-t. York co. Me. E. Piscataquay r., which divides it from Newington and N. H. 107 ms. from Augusta. Pop. 1830, 1,845.

Ellis, r. Coos co. N. H. joins Saco r. in Bartlett.

Ellisburg, p-t. Jefferson co. N. Y. S. lake Ontario. First settled, 1797, 9 ms. square, level except S. E., watered by Great Sandy creek, and has many mill sites, with a tolerable harbor on the lake Ontario, at the mouth of Great Sandy creek, and a navigation of 2 ms. up each of its branches. Here is a salt spring. Pop. 1830, 5,292.

Ellis Island, low, sandy reef of Flor. on the Bahama channel, and forming the S. E. boundary of Biscane bay. Lat. 25° 24′, long. 3° 20′ W. W. C.

Ellisville, p-v. Warren co. N. C. by p-r. 67 ms. N. N. E. Raleigh.

Ellisville, p-v. and st. jus. Jones co. Miss. situated on a confluent of Leaf r. branch of Pascagoula, about 100 ms. due E. Natchez, and by p-r. 81 ms. S. E. Jackson. Lat. 31° 37′, long. 12° 17′ W. W. C.

Ellsworth, p-t. Hancock co. Me. 24 ms. N. E. Castine, 81 Augusta, crossed by Union r. Pop. 1830, 1,385.

Ellsworth, p-t. Grafton co. N. H. 11 ms. N. N. W. Plymouth, 52 N. N. W. Concord, 84 N. W. Portsmouth; contains 16,606 acres, has Carr's mtn. in N. and centre, much bad soil, but yields grain, maple sugar, clover seed, &c. Pop. 1830, 1,492.

Ellsworth, p-v. Sharon, Litchfield co. Conn. 47 ms. W. by N. New Hartford.

Elmira or Newtown, p-t. and half co. t. Tioga co. N. Y. 32 ms. W. Owego, 16 E. Painted Post, 19 S. head of Seneca lake, 210 W. by S. Albany, crossed by Chemung r., and there are mill seats on this and Elmira creek, with hills, and some good meadows. The land is held in fee. Pop. 1830, 2,962.

Elmore, t. Orleans co. Vt. 17 ms. N. Montpelier, 33 E. Burlington, 6 ms. square. First settled, 1790, from Conn., is uneven, with Fordway mtn. N. W., hard wood and iron ore, sends streams to Lamoille and Onion rs. There are 3 school districts. Pop. 1830, 442.

ELSENBOROUGH, t. Salem co. N. J. on Del. r. s. w. Salem. Pop. 1830, 503.

ELSWORTH, p-v. and tsp. Trumbull co. O. The p-o. is by p-r. 296 ms. N. W. W. C. and 151 N. E. Columbus. Pop. of the tsp. 1830, 803.

ELY, p-v. Jennings co. Ind. by p-r. 574 ms. W. W. C. and 69 S. E. Indianopolis.

ELYRIA, p-v. tsp. and st. jus. Lorain co. O. The village is by p-r. 377 ms. N. W. by W. W. C. and 130 a little E. of N. Columbus. It is situated on Black r. 10 ms. from lake Erie, at lat. 41° 24', long. 5° 6' W. W. C. Pop. of the tsp. 1830, 663.

ELYTON, p-v. and st. jus. Jefferson co. Ala. situated on the road from Tuscaloosa to Huntsville, 48 ms. N. E. the latter, and 88 S. S. W. the former. Lat. 33° 35', long. 10° W. W. C.

EMAUS, p-v. S. part Lehigh co. Pa. situated near Little Lehigh creek, at the N. W. foot of the Blue Ridge, 10 ms. S. W. Bethlehem, and by p-r. 88 ms. N. E. by E. Harrisburg. This village is one of the settlements of the United Brethren or Moravians, and is included in the tsp. of Salisbury. Pop. 1820, about 100.

EMANUEL, co. Geo. bounded by Great Ohoope r. which separates it from Montgomery S. W., by Washington N. W., and Jefferson N. Great Ogeechee r. which separates it from Burke N. E., Scriven E., Bullock S. E., Tatnell S. Length from E. to W. 56 ms., mean breadth 20, and area 1,120 sq. ms. Extending in lat. from 32° 21' to 32° 52', long. from 4° 51' to 5° 48' W. W. C. Surface generally level, soil sandy and barren. Chief t. Swainsboro'. Pop. 1820, 2,928, 1830, 2,681.

EMBREEVILLE, p-o. Chester co. Pa. by p-r. 106 ms. N. E. W. C.

EMERY, r. of Ten. having its source by several streams issuing from the S. E. slope of Cumberland mtn. in Bledsoe and Fentress cos., and flowing thence N. E. enter and unite in the S. angle of Morgan co., and abruptly inflecting to the S., separating Anderson from Roane, falls into Clinch r. opposite Kingston, after an entire comparative course of 60 ms.

EMERY IRON WORKS and p-o. on Emery r. Roane co. Ten. by p-r. 141 ms. E. Nashville.

EMMETTSBURG, p-v. N. part of Frederick co. Md. on the road from the city of Frederick to Gettysburg, Adams co. Pa. 22 ms. a little E. of N. Frederick.

EMINENCE, p-v. in the E. part Greene co. Il. by p-r. 860 ms. from W. C. and 79 from Vandalia, in a nearly similar direction a little N. of W.

EMISON'S MILLS and p-o. Knox co. Ind. 10 ms. from Vincennes, the co. st., and by p-r. 693 ms. W. W. C. and 136 ms. S. W. Indianopolis.

EMPORIUM, p-v. Lycoming co. Pa. not located in the p-o. list.

ELSINGBOROUGH, t. Salem co. N. J. 60 ms. S. W. Trenton, S. Salem r., N. Alloway's cr., E. Del. r., opposite Del. city. Pop. 1830, 503.

EMBDEN, p-t. Somerset co. Me. 46 ms. from Augusta, 16 N. Norridgewock, W. Kennebec r. just above Seven Mile brook. Pop. 1830, 894.

ENFIELD, p-t. Grafton co. N. H. 12 ms. S. E. Hanover, 42 N. W. Concord, 105 N. N. W. Boston, with 24,060 acres, is hilly, with fish ponds and streams. Mascomy pond, 4 ms. long, has many islands, and receives Mascomy r. Here is a Shaker settlement. Pop. 1830, 1,492.

ENFIELD, p-t. Hampshire co. Mass. 81 ms. W. Boston, has several factories on Swift r. Pop. 1830, 1,056.

ENFIELD, p-t. Hartford co. Conn. 16 ms. N. Hartford, S. Mass., E. Conn. r., 5½ ms. by 6, 33 sq. ms., is generally level, but high near the river, has a light, rich soil, which bears oak and walnut, grain and grass. Scantic r. has mill sites and meadows. First settled, 1681, from Salem, as a part of Springfield, Mass. The village is pleasant, with fine elms. There is a settlement of Shakers in this town. At Thompsonville, on Conn. r. is an extensive manufactory of carpets, where Scoth weavers were first employed. Pop. 1830, 2,129.

ENFIELD, p-t. Tompkins co. N. Y. 5 ms. W. Ithaca, is hilly, but has pretty good soil. Pop. 1830, 2,690.

ENFIELD, p-v. King William co. Va. by p-r. 31 ms. N. E. Richmond.

ENFIELD, p-v. on Beach Swamp creek, Halifax co. Va. by p-r. 110 ms. direct line, by the road 88 ms., N. E. Raleigh.

ENFIELD, p-v. Halifax co. N. C. about 15 ms. W. of S. Halifax, the co. st., and by p-r. 228 W. of S. W. C. and 74 N. E. by E. Raleigh.

ENGLISH NEIGHBORHOOD, v. Bergen co. N. J. 12 ms. from N. Y. on the E. branch of Hackensack r., is pleasantly situated on W. bank of Hudson r., with good land, and settled by Dutch.

ENGLISH TOWN, p-v. Monmouth co. N. J. 18 ms. E. Princeton, 21 W. Shrewsbury, on Matchaponix creek, the S. branch of Raritan r.

ENNISVILLE, p-v. Huntingdon co. Pa. by p-r. 79 ms. W. Harrisburg.

ENNOREE, r. of S. C. rising in Greenville dist. interlocking sources with those of Saluda and Tyger rs., and generally about from 5 to 10 ms. distant from the latter, falls into Broad r. after a comparative course of 75 ms.

ENSE, p-v. Orange co. N. C. 6 ms. W. Hillsboro', the co. st. and by p-r. 302 S. S. W. W. C. and 47 N. W. by W. Raleigh.

ENOSBURG, p-t. Franklin co. Vt. 35 ms. N. E. Burlington, 43 N. W. Montpelier. First settled, 1797; is very healthy, with hills and vallies, good for grass, crossed by Missisque and Trout rs. &c., with good mill sites, 12 school districts. Pop. 1830, 1,560.

EPHRATAH, t. Montgomery co. N. Y. Pop. 1830, 1,818.

EPHRATA or TUNKERTOWN, p-v. on a branch of Conestogoe r. Lancaster co. Pa. 15 ms. N. N. E. Lancaster.

EPPING, p-t. Rockingham co. N. H. 20 ms. W. Portsmouth, 30 S. E. Concord, 8 N. Exeter, nearly 20 sq. ms., has good soil, and is crossed by Lamprey and North rs. Pop. 1830, 1,262.

EPSOM, p-t. Rockingham co. N. H. 12 ms. E. Concord, 45 N. W. Portsmouth; contains 19,200 acres, is uneven, with McKoy's fort, Nat's and Nottingham mtns., generally bears grain and grass; has Great and Little Simcook rs. which unite here. Pop. 1830, 1,413.

EQUALITY, p-v. and st. jus. Gallatin co. Il. by p-r. 773 ms. W. W. C. and 137 ms. S. S. E. Vandalia. It is situated at the Forks of Saline r. 12 ms. N. W. by W. Shawneetown, on Ohio r. at lat. 38° 45′, long. 11° 25′ W. W. C.

ERIE, large lake of the U. S. and Upper Canada, forming a link in the great central chain of fresh water seas in the interior of North America. The greatest length of Erie is from the mouth of Maumee to the outlet of Niagara strait, within an inconsiderable fraction of 270 ms. The width varies from 15 to 50 ms. The widest part from Ashtabula co. Ohio, to Middlesex in Upper Canada, narrowing towards both extremes. The depth of Erie is much less than that of either of the other Canadian lakes, not exceeding a mean of 120 feet, or 20 fathoms, and generally very shallow towards its shores. The harbors are mostly obstructed by bars, and none having a depth of more than 6 or 7 feet. From the W. this lake receives the Maumee, Raison, Huron, and Detroit rs.; from the N. only the Ouse or Grand river, but from the S. the Portage, Sandusky, Huron, Cayahoga, Grand Conneaut, Cattaraugus and Buffaloe. Erie is united to Ontario by Niagara, with the Hudson by the Erie canal, with the Ohio, by the Ohio canal, and with the higher lakes, by Detroit and St. Clair straits. With all the impediments to navigation arising from defective harbors, the commerce on lake Erie is already immense, and very rapidly augmenting. The position of Erie lake is in a singular manner favorable to its becoming the centre of an unequalled inland navigation. To the natural, and already completed artificial channels of connexion, may be, amongst some others of less obvious facility of execution, noticed the route through Maumee and Wabash rs. That by the channels of Huron and St. Joseph's into the S. part of lake Michigan, &c.

Commerce on Lake Erie.—The following extract will serve to show the immense and increasing value of navigation of this lake. They are part of the remarks of Mr. Sill, of the house of representatives, on the bill making additional improvements of certain harbors, &c., delivered Feb. 18, 1831. "I have not ascertained the exact amount of the export trade of lake Erie during the past year. I have seen a partial statement of its amount, which proves it to be of great extent, and should it be estimated at 15,000 tons, which is probably below the actual amount, it would swell the aggregate amount of that trade to 40,000 tons.

ERIE COUNTY, N. Y. bounded by Niagara co. N., Genesee co. E., Cattaraugus and Chatauque cos. S., lake Erie and Niagara r. W. 33 ms. by 40, 950 sq. ms., has 16 towns. It has Tonnewanta creek N. with Ellicotts'; Buffalo creek in the middle, with its branches, Cayuga, Seneca and Cazenove creeks, and on the W. Canquaga, Conjocketa, Two Sisters, Smoke's, Delaware creeks, and others. Oaks grow N. on a swelling gravelly loam, with limestone. A wet loam S. with beech, maple, &c., grass, grain, bog iron, limestone, water lime, gun flint, &c. Erie canal passes along W. and N. This co. suffered in the late war. Pop. 1820, 15,668, 1830, 35,710.

ERIE, p-t. Erie co. N. Y. 23 ms. E. N. E. Buffalo, 260 W. Albany. Pop. 1830, 1,926.

ERIE, co. Pa. bounded W. by Ashtabula co. Ohio, N. W. and N. by lake Erie, N. E. by Chatauque co. N. Y., E. by Warren, and S. by Crawford cos. Pa. Greatest length along Crawford co., 45 miles, mean breadth 17, area 765 square ms. Lat. 42°, and long. 3° W. W. C. intersect near the centre of this co. Surface finely diversified by hill and dale, with a very productive soil. Chief town, Erie. Population 1820, 8,553, 1830, 17,027.

ERIE, p-t. borough, port of entry, and st. of jus. Erie co. Pa. It is the same place formerly called Presque Isle by the French, from the peninsula which forms the harbor. The borough extends along the main shore, is well built and increasing. Pop. 1820, 635. The harbor is formed by the main shore peninsula, and a sandy shallow or reef. The opening is to the N. E. having in common only 8 feet water on the reef. The depth within is more than adequate to the draught of any vessel navigated on lake Erie. The lake and inland trade of this place is already extensive and increasing. A turnpike road extends hence to Pittsburg, 136 miles, the two towns lying almost exactly N. and S. from each other. Lat. 42° 08′, long. 3° 10′ W. distant by p-r. 357 ms. N. W.. W. C. and 302 N. W. by W. Harrisburg.

ERIE, p-v. and st. jus. Green co. Alabama, by p-r. 896 ms. S. W. W. C. and 47 S. S. W. Tuscaloosa. It is situated on the left bank of Black Warrior or Tuscaloosa river. Lat. 32° 43′, long. 10° 54′ W. W. C.

ERIE, town, Tioga co. N. Y. 12 miles N. E. Elmira, W. Cayuta creek, has no other mill stream, is hilly. Pop. 1830, 976.

ERNEST'S STORE, and p-o. Butler co. Ala. by p-r. 920 ms. S. W. W. C. and 125 ms. S. S. E. Tuscaloosa.

ERROL, town, Coos co. N. H. on W. side of Umbagog lake, W. Me. contains 35,000 acres,

2,500 of which is water, is crossed by Amerascoggin river, which is here joined by several streams. Population 1830, 82.

ERVINNA, p-v. N. part of Bucks co. Pa. 16 ms. N. Doylestown.

ERWINSVILLE, p-v. Rutherford co. N. C. by p-r. a little s. of w. Raleigh.

ESCAMBIA, river of Florida and Alabama. The small stream called Escambia rises in Monroe co. Ala. and flowing s. over Baldwin, enters Florida, and falls into the much more considerable volume of Conecuh, though below their junction the united waters take the name of the lesser confluent. Now known as Escambia, this stream continues s. 40 ms. with an elliptic curve to the w. and gradually spreads into a bay, which is again lost in the more extensive sheet of Pensacola bay. (*See Conecuch river.*)

ESCAMBIA, extreme western co. of Florida, bounded by Perdido river, or Baldwin co. Ala. w., by Monroe and Conecuch cos. Ala. N., by Walton co. Florida E., and the Gulf of Mexico s. It is nearly a square of 50 miles each side, or with an area of 2,500 square ms. Extending in lat. from 30° 16′ to 31°, long. from 9° 38′ to 10° 48′ w. W. C. The surface rises gradually from the Gulf shore, from sandy plains to ridges of some elevation. The soil with but partial exception is barren, and its natural state wooded with pine. The asperity of soil is in some measure compensated by the fine harbor of Pensacola, and its confluent rivers. These rs. are the Escambia and Yellow Water,(*see these articles, and also Conecuh and Pensaco'a.*) Chief town, Pensacola. Pop. 1830, 3,386.

ESCAMBIA, p-v. Escambia co. Florida, on Escambia river, 78 ms. N. Pensacola.

ESOPUS, town, Ulster co. N. Y. 4 miles s. Kingston, 69 s. Albany, s. Walkill creek, w. side Hudson river, about 2½ miles by 7, 12 square ms. has good land, long cultivated by Dutch descendants. Pop. 1830, 1,770.

ESOPUS, creek, Ulster co. N. Y. runs 58 ms. into Hudson river, at Saugerties, 11 ms. below Catskill.

ESPERANCE, or Schoharie bridge, p-v. Schoharie co. N. Y. 26 ms. w. Albany.

ESPY, p-v. Columbia co. Pa. by p-r. 84 ms. N. Harrisburg.

ESSEX, co. Vt. forms the N. E. corner of the state, and is bounded by Lower Canada N., Connecticut river, (the line of N. H.) E. and S., Caledonia co. S. W., and Orleans co. W., 23 ms. by 45. It is rocky and poor, and has but few inhabitants, and those chiefly on Connecticut river. Nulhegan river and others enter Connecticut river. Passumpsic and Moose rivers S. W., Clyde, &c. run into Canada. Chief town, Guildhall. Pop. 1820, 3,284, 1830, 3,981.

ESSEX, p-t. Chittenden co. Vt. 8 miles N. E. Burlington, 32 w. Montpelier, N. Onion river. First settled 1783, from Salisbury Conn. has few hills, is sandy; bears pine, rye and corn S. and W., elsewhere, grass and hard wood. Onion river has 2 falls; there are also Brown's and Indian rivers, 10 school districts. Pop. 1830, 1,664.

ESSEX, co. Mass. bounded by N. H. state N., the Atlantic ocean E. and S. E., Suffolk co. S. W., Middlesex co. W., has Merrimac river N., Ipswich river in the centre. Parker r. enters Plumb Island sound. Saugus river Lynn bay. The land is highly cultivated. It has an antiquarian and an agricultural society, and contains 27 towns. Pop. 1820, 74,656, 1830, 82,887.

ESSEX, p-t. Essex co. Mass. 12 ms. N. E. Salem, 25 N. E. Boston, is pleasant, has navigation on a creek, fishing and ship building, with a canal thro' the marsh from Ipswich bay, for rafts from Merrimack river. The small and useful coasting craft, called Chebacco boats, derived their name from this place, which the Indians called Chebacco. Pop. 1830, 1,333.

ESSEX, co. N. Y. bounded by Clinton and Franklin cos. N., lake Champlain and Vt. E., Warren co. S, Hamilton and Franklin cos. W., about 41 miles by 43; contains 1,763 square ms., has 16 towns, granite hills and mtns. of 1,200 feet and comprises much of the iron region. It is about half way between N. Y. and Quebec, on navigable waters. It has white and black oak, white and yellow pine, maple, beech, &c., much game and fish, Au Sable, Bouquet, Hudson and Scaroon rivers, water power, particularly the outlet of lake George. Limestone, marble, black lead, asbestos, &c. are found. Population 1820, 12,811, 1830, 19,387.

ESSEX, p-t. Essex co. N. Y. 6 ms. E. N. E. Elizabethtown, 133 N Albany, w. lake Champlain, has pretty good land, landings and trade on the lake. From the village is a ferry to Charlotte village, has good farms and iron ore; there is the curious split rock, and is crossed by Bouquet river. Population 1830, 1,543.

ESSEX, co. N. J. bounded east by Staten Island sound, Newark bay, and Passaick r. which separate it from Staten Island and Bergen co., N. by the Passaick and Bergen, W. by Morris and Somerset, S. by Middlesex. Principal towns, Newark, Patterson, Elizabeth. Pop. 1820, 30,793, 1830, 41,928. Altho' the smallest co. (save one) in N. J. it is the most populous. It is an excellent agricultural district, containing many prosperous manufactories, fine streams, and good facilities for transportation; among which is the Morris canal, which passes through it.

ESSEX, co. Va. bounded S. E. by Middlesex, S. W. and W. by King and Queen, N. W. by Caroline, and by Rappahannoc river which separates it from Westmoreland N., and Richmond E. Length 28, mean breadth 10, area 280 square ms. Lat. 37, is intersected by the meridian of W. C. in the N. W. part of this co. Surface moderately hilly. Chief town, Tappahannoc. Population 1820, 9,909, 1830, 10,531.

ESSEX HALL, and p-o. in the N. part Harford co. Md. by p-r. 22 ms. N. Belair, the co.

st. 83 ms. N. E. W. C. and 45 miles E. of N. Baltimore.

ESTILL, co. Ky. bounded by Madison W., Clarke N W., Montgomery N., Morgan N. E., Perry E. and S. E., and Clay S. Length from S. E. to N. W. 48 miles, mean breadth 18, area 864 square ms. Extending in lat. from 37° 30′ to 37° 34′, long. from 6° 15′ to 7° 04′ W. W. C. Kentucky river, by a very winding channel, traverses Estill, in its utmost length, receiving several confluents, particularly from the south. Chief town, Irvine. Pop. 1820, 3,507, 1830, 4,618.

ESTILLVILLE, p-v. and st. jus. Scott co. Va. on Moccasin creek, between N. fork of Holston and Clinch rivers, by p-r. 445 ms. S. W. by W. W. C. and 348 a little S. of W. Richmond.

ETNA, p-t. Penobscot co. Me. 63 ms. Augusta. Pop. 1830, 362.

ETNA FURNACE, and p-o. Hart co. Ky. by p-r. 96 ms. S. W. Hartford.

ETOWAH, river, Geo. in the Cherokee territory, rises in the western border of Habersham co. flowing by a general course of S. W. but with extensive inflections, 120 miles comparative course to its junction with the Oostenahah to form the Coosa. The valley of the Etowah, lies between those of Chattahoochee and Oostenahah, and between latitude 34° and 35°.

ETOWAH, Indian village, and st. of a p-o. is situated on Etowah river in the Cherokee nation, N. W. part of Geo. about 130 miles N. W. Milledgeville. This place and the river from which it has either derived or communicated its name, is with some absurdity in our books and on the p-o. list changed to High Tower.

EUBANKS, p-o. Columbia co. Geo. by p-r. 88 ms. N. E. Milledgeville.

EUCLID, tsp. and p-v. Cuyahoga co. Ohio. The p-v. is situated in the N. E. part of the co. 10 ms. N. E. Cleaveland, the co. st. and by p-r. 363 ms. N. W. W. C. and 147 N. E. Columbus. By the census of 1830, the tsp. contained a pop. of 1,099.

EUGENE, p-v. in the W. part of Vermillion co. Indiana, by p-r. 658 miles N. W. by W. W. C. and 86 N. W. by W. Indianopolis.

EUTAW SPRINGS, small stream of S. C. falling into Santee river at the point where the line between Charleston and Orangeburg districts intersects that river, about 60 miles N. N. W. Charleston.

EVANS, p-t. Erie co. N. Y. 25 miles south Buffalo, W. lake Erie, N. Cattaraugus creek, has Delaware creek, Two Sisters, &c.; has wet loam, is uneven, with beech, maple, hemlock, bass, &c. Bad for corn, cold and changeable weather. Sturgeon point puts into the lake. Pop. 1830, 1,185.

EVANS CROSS ROADS, p-o. Williamson co. Ten. 32 ms. S. Nashville.

EVANSBURG, p-o. Crawford co. Pa.

EVANSHAM, or Wythe, court house, p-v. and st. jus. Wythe co. Va. by p-r. 329 miles S. W. by W. W. C. and 253 a little S. of W. Richmond. Lat. 36° 56′, long. 4° 05′ W. W. C.

EVANSVILLE, p-v. and st. jus. Vanderburgh co. Indiana, by p-r. 728 miles S. W. by W. W. C., 170 S. W. Indianopolis, and 55 ms. W. of S. Vincennes. It is situated on the right bank of Ohio river, lat. 38°, long. 10° 38′ W. W. C.

EVERETTS, house and p-o. Lewis co. Ky. by p-r. 90 ms. N. E. by E. Frankfort.

EVERETTSVILLE, p-o. Albermarle co. Va. by p-r. 128 ms. S. W. W. C.

EVERTON, p-v. Fayette co. Indiana, by p-r. 534 ms. W. W. C. and 75 ms. S. of E. Indianopolis.

EVESHAM, town, Burlington co. N. J. adjoining Gloucester county, and on the south branch of Rankolm creek. Population 1830, 4,239.

EWING'S MILLS, and p-o. Indiana co. Pa. by p-r. 151 ms. W. Harrisburg.

EWINGSVILLE, p-o. Coecil co. Md.

EWINGSVILLE, p-o. in the western part of Cooper co. Mo. by p-r. 20 ms. W. Booneville, the co. st. 71 W. Jefferson, and 1,043 miles W. W. C.

EXETER, p-t. Penobscot co. Me. 20 miles N. W. Bangor, 75 Augusta. Population 1830, 1,439.

EXETER, p-t. Rockingham co. N. H. situated at the falls of Squamscot or Exeter river. A branch of Pascataqua river, which here meets tide, is navigable for vessels of 500 tons, and affords valuable mill sites. There are several manufactories, and the soil is various. This town was first settled 1638, by Jonathan Wheelright &c. who left Mass. on account of his peculiar religious opinions. It suffered in early times from the Indians. Phillips' academy, founded here 1781, has furnished many valuable men. The building is 2 stories high, 76 feet by 36. The funds amount to $80,000. Pop. 1830, 2,753.

EXETER, town, Washington co. R. I. 24 ms. S. W. Providence, E. Conn. state, about 5 ms. by 12, contains 66 square ms., has primitive rocks, gravelly loam, uneven, good for dairies, and furnished with some mill seats by Wood river and its branches. Population 1830, 2,383.

EXETER, p-t. Otsego co. N. Y. 10 miles N. W. Cooperstown, 73 W. Albany, about 5¼ miles square, gives rise to Butternut and Wharton's creeks, is high and hilly, with good vallies. Pop. 1830, 1,690.

EXETER, tsp. and p-v. Luzerne co. Pa. 10 ms. above Wilkes-Barre.

EXETER, p-v. in the W. part of Morgan co. Il. by p-r. 852 ms. N. of W. W. C. and 130 N. W. Vandalia.

EXPERIMENT MILLS, and p-o. in the N. part Northampton co. Pa. by p-r. 26 ms. N. Easton, the co. seat. 216 N. N E. W. C. and 128 N. E. by E. Harrisburg.

F.

FABER'S MILLS and p-o. in the w. part of Nelson co. Va. by p-r. 170 ms. s. w. W. C. and 103 w. Richmond.

FABIUS, p-t. Onondaga co. N. Y. 20 ms. s. E. Onondaga, 50 s. w. Utica, 125 w. Albany, 5 ms. by 10, crossed by Chitteningo cr. and other streams, which give mill sites. It is high land, good and level N., hilly s., 14 school districts. Marle is found in the N. E. with petrified branches and leaves. Many military enclosures are found N. E. with stumps of palisadoes and bones. Pop. 1830, 3,071.

FACTORYVILLE, p-v. Lincoln co. Me. 29 ms. from Augusta.

FACTORYVILLE, p-v. N. E. part of Luzerne co. Pa. about 20 ms. above Wilkes-Barre, and by p-r. 152 ms. N. E. Harrisburg.

FAIR BLUFF, p-o. Columbus co. N. C. by p-r. 124 ms. s. Raleigh.

FAIRDALE, p-v. Susquehannah co. Pa. by p-r. 271 N. N. E. W. C. and 163 ms. N. E. Harrisburg.

FAIRFAX, t. Kennebec co. Me. 25 ms. N. Augusta.

FAIRFAX, t. Franklin co. Vt. 18 ms. N. E. Burlington, 37 N. W. Montpelier. First settled in 1763, is level, with high soil, good for corn and rye, and watered by Lamoille r. and Brown r. Parmelee's and Stones brooks, the branches of Lamoille, has good mill sites. The great falls of Lamoille are curious, 11 school districts. Pop. 1830, 1,729.

FAIRFIELD, p-t. Somerset co. Me. 9 ms. s. Norridgewock, 26 from Augusta, on Kennebec r. the most southern town in the co. Pop. 1830 2,002.

FAIRFAX co. Va. bounded by the district of Columbia E., by Potomac r. which separates it from Prince George's co. Md. s. E., by Occoquon cr. which separates it from Prince William co. Va. s. and s. w., by Loudon co. N. W. and by Potomac r. which separates it from Montgomery co. Md. N. E. Length from s. E. to N. W. 25 ms. mean breadth 18, and area 450 sq. ms. extending in lat. from 38° 36′ to 39° 03′, long. from 0° 03′ to 0° 33′ w. W. C. The surface of Fairfax is hilly and broken, with some good, but much sterile soil. Chief town, though not the st. jus. Matildaville. Mount Vernon, the resting place of Washington, is on the Potomac r. in the s. E. part of the co. Pop. 1820, 11,404, 1830, 9,204.

FAIRFAX, C. H. and p-o. Fairfax co. Va. 21 ms. s. w. by w. W. C. and 129 N. Richmond.

FAIRFAX, p-v. and st. jus. Culpepper co. Va. 38 ms. a little N. of w. Fredericksburg, and by p-r. 81 ms. s. w. W. C. lat. 38° 26′, long. 1° 01′ w. W. C.

FAIRFIELD, p-t. Franklin co. Vt. 27 ms. N. E. Burlington, contains 60 sq. ms. First settled 1788, has an academy. Black cr. affords good mill sites, and joining Fairfield r. enters Missisque r. in Sheldon. Smithfield pond, 3 miles long, has an outlet, on which are mill sites. It is uneven with good soil.

FAIRFIELD co. Conn. bounded by Litchfield co. N., New Haven co. E., Long Island sound E. and s., New York, s. w. and w. It is w. of Housatonic r. triangular. Mean extent 21 ms. by 30; contains 630 sq. ms. and has 17 towns. It extends 40 ms. along the coast on Long Island sound, which is level, and abounds with bays, points, and harbors. The middle and N. parts are higher, and have some hills. The soil is a primitive gravelly loam, arable, and in Fairfield rich. Still r. falls into Housatonic; Pequonuc, Saugatuck, Ash, Naraton, Mill, Stamford, and Byram rs. into the sound. The best harbors are Bridgeport, and Black Rock, Mill r. Saugatuck, Norwalk, Stamford and Greenwich. The coasting trade, chiefly with New York, is important; and fishing is carried on, on the coast. Hats are made in great quantities at Danbury, and other manufactures exist to some extent. There is some foreign trade. Fairfield and Danbury are co. towns. Pop. 1820, 42,739, 1830, 46,950.

FAIRFIELD, p-t. and port of entry, Fairfield co. Conn. 21 ms. w. New Haven, 58 N. E. N. Y., N. E. Long Island sound, mean extent 6 ms. by 9, contains 54 sq. ms. nearly level, with good land, and a large and pleasant village which was burnt by the British in the revolutionary war. In a swamp 2 ms. w. of the village, the remains of the Poquod tribe, after fleeing from their country, in New London co. were killed or taken prisoners by the Mass. and Conn. troops. The villages of Greenfield, Black Rock, Saugatuck and Mill r. are considerable, and the three latter have harbors and trade. At Greenfield and Saugatuck are academies; Greenfield is on a fine hill, with excellent farms, and an extensive and delightful view. Pop. 1830, 4,246.

FAIRFIELD, p-t. Herkimer co. N. Y. 10 ms. N. E. Herkimer, 76 w. N. W. Albany, E. W. Canada cr. 4 ms. by 8, is high, hilly, well watered, with a productive soil; but few mill sites. It was settled principally from the eastern states. The college of Physicians and surgeons here, has 5 professors. The village, which is situated on an eminence, is well built. Pop. 1830, 2,265.

FAIRFIELD, t. Cumberland co. N. J. 25 ms. E. Salem, E. Cohansey bay on the Del. lies between Cohansey and Nantuxet creeks, and has several smaller streams entering the Del. r. with swamps along the shore. Pop. 1830, 1,312.

FAIRFIELD, village Essex co. N. J. 3 ms. N. Caldwell.

FAIRFIELD, p-v. Adams co. Pa. at the foot of Jacks mtn. 7 ms. s. w. by w. Gettysburg.

FAIRFIELD, p-v. Rockbridge co. Va. on one of the roads from Lexington to Stanton, 13 ms. N. N. E. the former, and 23 s. s. w. the latter.

Fairfield, p-v. Lenoir co. N. C. 87 ms. s. e. by. e. Raleigh.

Fairfield, dist. S. C. bounded n. by Chester, n. e. by Catawba r. separating it from Lancaster and Kershaw, s. e. by that part of Kershaw w. of Catawba r., s. by Richland, and by Broad r. separating it from Lexington s. w., Newberry w. and Union n. w. Length from e. to w. 38 ms. mean breadth 22, and area 796 sq. ms. Extending in lat. from 34° 12′ to 34° 32′, long. from 3° 44′ to 4° 26′ w. W. C. This district, filling the space from the Catawba to the Broad r. is divided into two inclined plains, falling s. w. towards the latter, and n. e. towards the former stream. Chief t. Kinnsboro. Pop. 1820, 17,174, 1830, 21,546.

Fairfield, p-v. Putnam co. Geo. 32 ms. n. w. Milledgeville.

Fairfield, p-v. s. w. part of Spencer co. Ky. 35 ms. s. e. Louisville and by p-r. 40 ms. s. w. Frankfort.

Fairfield, p-v. Amite co. Miss. about 60 ms. s. e. Natchez.

Fairfield, p-v. Bath tsp. Greene co. O. by p-r. 452 ms. a little n. of w. W. C. and 56 s. w. by w. Columbus. Pop. 1830, 137.

Fairfield, p-v. in the n. part of Franklin co. Ind. 3 ms. n. e. Brookville, the co. st. and by p-r. 524 ms. w. W. C., and 70 ms. s. e. by e. Indianopolis.

Fairfield, p-v. and st. jus. Wayne co. Il. by p-r. 756 ms. w. W. C. and 69 s. e. Vandalia. It is situated on a branch of Little Wabash r., 48 ms. s. w. by w. Vincennes in Ind. lat. 38° 28′, long. 11° 30′ w. W. C.

Fairhaven, p-t. Rutland co. Vt. 9 ms. n. e. Whitehall, 60 s. Burlington, 52 n. Bennington, e. N. Y. First settled 1779, from Conn. and Mass., has a variety of soil, with pine, hemlock, birch, maple, nut, &c. Poultney and Castleton rs., and 4 school dists. Castleton r. some years since changed its channel here, left several mills dry, exposed old buried trees, and ruined the harbor. Pop. 1830, 675.

Fairhaven, p-t. Bristol co. Mass. 48 ms. s. Boston, n. Buzzard's Bay, e. Acushnett r. has some commerce, an academy, and a bridge of 3,960 ft. to N. Bedford, several islands between, extend it 2,000 ft. more. First settled 1764, and in 1778, the village was defended against the British, by Major Fearing. Pop. 1830, 3,034.

Fairlee, p-t. Orange co. Vt. 35 ms. n. Windsor, 17 n. Dartmouth college, w. Conn. r. First settled 1768, mountainous, with little arable land, and has high precipices on the river. The trees are pine and hemlock. Pickerel have been introduced into the pond, and greatly multiplied. A bridge crosses the Conn. r. to Orford, N. H. Pop. 1830, 656.

Fairton, village, Cumberland co. N. J. on Cohansey creek, 3 ms. s. of Bridgeton.

Fairmount, p-v. Lancaster co. Pa. by p-r. 117 ms. n. e. W. C. 13 n. e. Lancaster city, and 43 e. Harrisburg.

Fairport, p-v. and port at the mouth of Grand r., and on the southern shore of lake Erie; 32 ms. n. e. Cleveland, and by p-r. 349 n. w. W. C., and 164 n. e. Columbus. The mouth of Grand r. affords a good harbor for vessels drawing about 5 feet water.

Fair View, p-v. Hunterdon co. N. J. 6 ms. n. w. Flemingtown.

Fairview, p-v. and tsp. on lake Erie in the n. w. part Erie co. Pa. The village stands near the lake shore at the mouth of Walnut cr. 9 ms. s. w. from the borough of Erie, and by p-r. 349 ms. n. w. W. C. Pop. of the tsp. 1830, 1,526.

Fairview, p-v. Brooke co. Va. by p-r. 302 ms. n. w. by w. W. C.

Fairview, p-v. in the s. part of Greenville dist. S. C. by p-r. 509 ms. s. w. W. C. and 117 ms. n. w. by w. Columbus.

Fairview, p-v. Oxford tsp. Guernsey co. O. by p-r. 105 ms. e. of Columbus, 22 ms. eastward Cambridge, the co. st. and 294 n. w. by w. W. C. Pop. 1830, 162.

Fairview, p-v. in the e. part of Rush co. Ind. 14 ms. e. Rushville, the co. st. and by p-r. 541 a little n. of w. W. C. and 54 s. e. by e. Indianopolis.

Fall Branch, p-o. Washington co. Ten. about 280 ms. e. Nashville, and 430 s. w. by w. W. C.

Falling Bridge, and p-o. s. e. part of Campbell co. Va. about 20 ms. s. s. e. Lynchburg, and by p-r. 106 ms. s. w. by w. Richmond.

Fall cr. Cayugá co. N. Y. runs 30 ms. into Cayuga lake, at Ithaca. It falls about 100 ft., 1 m. from its mouth.

Fall River, p-v. Bristol co. Mass. at the mouth of Fall r. on Mt. Hope bay. The descent of the stream is 140 ft. in 600 yds. and turns machinery for several factories.

Fallsington, p-v. Bucks co. Pa. 4 ms. s. e. Trenton, 23 n. e. Phil.

Falling Spring Creek, a small branch of Jackson's r. Bath co. Va. in the channel of which there is a fine fall of water, estimated to be near 200 feet perpendicular fall.

Falling Waters, p-o. in the w. part of Berkeley co. Va. by p-r. 79 ms. n. w. W. C.

Falltown, p-o. in the s. part of Iredell co. N. C. 13 ms. from Statesville, and by p-r. 151 w. Raleigh.

Falls, p-o. Lincoln co. N. C. about 170 ms. s. w. by w. Raleigh, and 420 s. w. W. C.

Falls, p-v. Pickens dist. S. C. by p-r. 550 ms. s. w. W. C. and 157 n. w. by w. Columbia.

Falls of Schuylkill, p-v. Phil. co. Pa. 5 ms. n. w. Phil. and 101 e. Harrisburg.

Fallston, p-v. Beaver co. Pa. about 260 ms. n. w. W. C.

Fallstown, Iredell co. N. C. (*see Falltown, Iredell co.*)

Falmouth, s-p. and p-t. Barnstable co. Mass. 19 ms. s. w. Barnstable, 72 s. by e. Boston, n. Atlantic ocean, e. Buzzard's bay, is level, except some hills e. with thin soil, but the best on Cape Cod. It has about 40 ponds, fresh and salt. Waquoit bay is a good harbor, with a narrow and crooked entrance.

Wood's Hole is another, with from 3 to 6 fathoms. The inhabitants are generally in the s. part, which is 6 or 8 ms. from Martha's Vineyard. It has considerable coasting trade. Pop. 1830, 2,548.

FALMOUTH, p.v. on the left bank of Susquehannah r. on the point below the mouth of eastern Conewago creek, and at the extreme w. angle of Lancaster co. Pa. 20 ms. N. w. by w. Lancaster, and 16 s. E. Harrisburg.

FALMOUTH, p.v. on the left bank of Rappahannoc r. in the s. part of Stafford co. Va. directly opposite Fredericksburg, with an intervening bridge, by p-r. 58 ms. s. s. w. W. C. and 68 N. Richmond.

FALMOUTH, p-v. and st. jus. Pendleton co. Ky. by p-r. 502 ms. w. W. C., and 60 N. E. Frankfort. It is situated on the point above the junction of the two main branches of Licking r. about 40 ms. s. s. E. Cincinnati, lat. 38° 40′, long. 7° 18′ w. W. C. Pop. 1830, 207.

FANCY BLUFF, and p-o. in the s. part of Glynn co. Geo. 5 ms. s. Brunswick, the co st. and 738 s. s. w. W. C.

FANCY HILL, and p-o. Rockbridge co. Va. by p-r. 210 ms. s. w. W. C.

FANNETTSBURG, p-v. and tsp. on the w. border of Franklin co. Pa. The village is situated on the main road from Shippensburg to Bedford, 17 ms. w. from the former, by p-r. 105 ms. N. N. w. W. C. and 55 a littte s. of w. Harrisburg. The tsp. of Fannet extends along the fine valley of the w. branch of Conecocheague, and between the cove or Tuscarora and Jordens mountains. Pop. of the tsp. 1820, 1,747, and in 1830, 2,110.

FARM, p-v. in Franklin co. Geo. by p-r. 9 ms. from Carnesville, the co. st. 585 ms. s. w. W. C. and 122 N. Milledgeville.

FARMER, p-v. Ovid, Seneca co. N. Y.

FARMERSVILLE, p-t. Cataraugus co. N. Y. 15 ms. N. E. Ellicottsville, 10 w. Genesee r. 6 ms. by 8, has plenty of mill sites, though the streams are small and few. Pop. 1830, 1,005.

FARMINGTON, p-t. Kennebec co. Me. 30 ms. N. Augusta. Has an academy. Pop. 1830, 2,340.

FARMINGTON, p-t. Stafford co. N. H. 25 ms. E. N. E. Concord, 26 N. w. by. w. Portsmouth, is rough, but productive, with some meadow on Cocheco r. It is crossed by Blue Hills or Frost mtn., and from mt. Washington, the highest point, ships may be seen without a glass, off Portsmouth, and on the other side the White Hills, &c. A rock of about 60 tons is balanced by nature, and can be moved with one hand. Pop. 1830, 1,465.

FARMINGTON, t. Ontario co. N. Y. Pop. 1,773.

FARMINGTON, p-t. Hartford co. Conn. 10 ms. w. Hartford, 30 N. New Haven, has much very rich meadow land on Farmington or Tunxis r. and is one of the richest agricultural towns in the state. The village is pleasant, contains some fine houses, an academy, &c. on a plain a little above the meadows. The Farmington canal affords boat navigation to New Haven, and partly by the Hampshire and Hampden canal to Westfield, Mass. It was intended to extend it to Northampton. The town contains about 70 sq. ms. nearly 7 ms. by 11. Talcott mtn. of the range beginning at Neck Rock, New Haven, and running far N. crosses it. There is much light sandy soil s. Montevideo, the seat of Mr. Daniel Wadsworth of Hartford, in the N. E. on the ridge of Talcott mtn. is a delightful place, with a pond, a country house, &c. and an extensive view N. E. and w. over the vallies Tunxis and Conn. up to mt. Tom, Mass.

FARMINGTON, p-t. Ontario co. N. Y. 9 ms. N. Canandaigua, 6 ms. sq. is gravelly and undulated N., clayey, good for grass, and level s., large tracts of water limestone are found in the centre. It was principally settled by Friends, who were dairy farmers from Cheshire, Mass. The people own the land in fee, and have good schools. Mud creek supplies mill seats. The villages are Salem v. and Brownsville. Pop. 1830, 1,773.

FARMINGTON, p-v. in the s.w. part of Bedford co. Ten. by p-r. 48 ms. s. Nashville, and 707 ms. s. w. W. C.

FARMINGTON, p-v. and tsp. in the N. w. part Trumbull co. O. by p-r. 311 ms. N. w. W. C. and 164 N. E. Columbus. Pop. of the tsp. 1830, 696.

FARMINGTON, p-v. in the N. part of Oakland co. Mich. 26 ms. N. N. w. Detroit, and 564 N. w. W. C.

FARMINGTON, p-v. and st. jus. St. Francois co. Mo. by p-r. 912 ms. a little s. of w. W. C. 152 s. E. Jefferson, and 60 s. s. w. St. Louis. It is situated on the table land, from which flow the Big River, branch of Marramec, northward, the extreme sources of St. Francis to the southward, and some comparatively small creeks N. E. into the Miss.; lat. 37° 47′, long. 13° 25′ w. W. C.

FARM TAVERN, and p-o. Southampton co. Va. about 210 ms. s. W. C. and 90 s. s. E. Richmond.

FARMVILLE, p-v. on the right bank of Appomattox r. N. border of Prince Edward co. Va. by p-r. 81 ms. s. w. by w. Richmond, and about an equal distance w. Petersburg.

FARNHAM, p-v. Richmond co. Va. about 85 ms. N. E. Richmond.

FARROWVILLE, p-v. in the N. w. part of Fauquier co. Va. 64 ms. w. W. C.

FAUQUIER, co. of Va. bounded by Frederick co. N. w., Loudon N. E., Prince William E., Stafford s. E., and by Rappahannoc r. separating it from Culpepper s. w. and w. Greatest length 45 ms., mean breadth 16, and area 720 sq. ms. Extending in lat. from 38° 24′ to 39° 02′, and in long. from 0° 32′ to 1° 5′ w. W. C. The general slope is s. E., and down which flow the higher branches of Rappahannoc and Occoquhan rs. Surface pleasantly broken, and soil tolerably productive. Chief t. Warrenton. Pop. 1820, 23,103, 1830, 26,086.

FAUSSE RIVIERE, (*False River*,) once a

bend of the Miss., but now a lake of Louisiana, in the parish of Point Coupee. In or about 1714, the change was affected, from which both names were taken, that is Fausse Riviere, and Point Coupee, (*Point Cut Off.*) Previous to that era, the Miss. r., below Bayou Sara, made an immense bend to the watrd., curving until it returned so nearly upon itself as to leave only a narrow neck of land. This isthmus, gradually diminished on both sides, was at length worn through, and the vast river shortened its channel upwards of 30 ms. The old bed rapidly filled with alluvion near the new channel, but in all other parts, retained its forms and features, and is now a fine lake, lined with farms and farm houses, with a soil possessing the usual fertility of the Miss. banks.

FAWN GROVE, tsp. and p-v. s. E. part of York co. Pa. The tsp. extends from Muddy creek to the Md. line. The p-o. is by p-r. 49 ms. s. E. Harrisburg, and 22 in a similar direction from York.

FAYETTE, p-t. Kennebec co. Me. 20 ms. w. Augusta, has several ponds on its E. border. Pop. 1830, 1,049.

FAYETTE, t. Seneca co. N. Y. 6 ms. E. Geneva, 3 s. Waterloo, 18 N. Ovid. 188 w. by N. Albany, w. Cayuga lake, E. Seneca lake, s. Seneca r. about 7 ms. by 9. Seneca r. and Canoga creek supply mills. It is nearly level, and has gypsum and limestone.

FAYETTE, p-v. Montgomery co. Va. by p-r. 208 ms. s. w. by w. Richmond.

FAYETTE COUNTY, Pa. bounded by Allegany co. Md. s. E., Preston and Monongalia cos. Va. s., by Monongahela r. which separates it from Green co. of Pa. w., and Washington N. w., by Westmoreland N., and Somerset E. It approaches a square of 28 ms. each side, area 784 sq. ms. Extending in lat. from 39° 42′ to 40° 10′, and in long. from 2° 23′ to 3° 3′. The surface of this co. is every where broken by hills, and the eastern part is traversed from s. w. to N. E. by two chains of mountains. The soil is almost uniformly fertile. The Monongahela, with a rather crooked channel, winds a navigable stream along the western border. The Youghioghany, after piercing a chain of mountains, enters Fayette from the s. E., and breaking through hills and mountains, traverses it flowing to the N. w. The channel of the Youghioghany is part of the projected route of the Chesapeake and Ohio canal. Chief t. Union. Pop. 1820, 27,285, 1830, 29,237.

FAYETTE, p-v. in the N. w. part of Montgomery co. Va. 16 ms. E. Christiansburg, the co. st., by p-r. s. w. W. C., 190 a little s. of w. Richmond.

FAYETTE, co. of Geo. bounded s. by Pike, E. by Henry, N. E. by De Kalb, N. by Campbell, and by Flint r. separating it from Coweta w., and Merriwether s. w. Length 30, mean breadth 18, and area 540 sq. ms. Extending in lat. from 33° 11′ to 33° 40′, and in long. from 7° 28′ to 7° 46′ w. W. C. Chief t. Fayetteville. Pop. 1830, 963.

FAYETTE, co. of Ala. bounded by Marion N., Walker N. E., Tuscaloosa and Pickens s., and Monroe co. Miss. w. Length E. to w. 50 ms., breadth 25, and area 1,250 sq. ms. Extending in lat. from 33° 26′ to 33° 47′, and in long. from 10° 28′ to 11° 25′ w. W. C. This co. slopes to the sthrd., and is drained by the Sipsey, Luxapatilla, and Battahatchy, branches of Tombigbee r. Chief t. Fayette Court house. Pop. 1830, 3,547.

FAYETTE, p-v. and st. jus. Fayette co. Ala. by p-r. 874 ms. s. w. by w. W. C. and 50 ms. N. N. w. Tuscaloosa.

FAYETTE, p-v. and st. jus. Jefferson co. Miss. by p-r. 1,127 ms. s. w. by w. W. C., 93 ms. s. w. Jackson, and 19 N. E. Natchez. It is situated on Coles creek, and on the great road from Natchez towards Tennessee, &c. Lat. 31° 42′, long 14° 18′ w. W. C.

FAYETTE, co. of Ten. bounded w. by Shelby, N. w. by Tipton, N. by Haywood, E. by Hardiman, and s. by the Indian country in the state of Miss. Length from N. to s. 24, and same from E. to w., area 576 sq. ms. Extending in lat. from 35° to 35° 23′, and in long. from 12° 12′ to 12° 39′ w. W. C. Chief t. Sumnerville. Pop. 1830, 8,658.

FAYETTE, co. Ky. bounded s. w. by Jessamine, w. by Woodford, N. w. by Scott, N. by Harrison, N. E. by Bourbon, E. by Clark, and s. E. by Kentucky r. separating it from Madison. Length from N. to s. 25 ms., mean breadth 11, and area 275 sq. ms. Extending in lat. from 37° 51′ to 38° 13′, and in long. from 7° 14′ to 7° 38′ w. W. C. This fine county is a true table land, from the centre of which flow streams like the radii of a circle. These streams all finally discharge into Kentucky, which touching on the s. E., semicircles the co. The chief t. Lexington. Pop. 1830, 25,098, or upwards of 91 to the sq. m.

FAYETTE, co. of O. bounded by Highland s., Clinton s. w., Greene N. w., Madison N., Pickaway N. E., and Ross s. E. Length from s. to N. 26, mean width 16, and area 416 sq. ms. Extending in lat. from 39° 21′ to 39° 43′, and in long. from 6° 16′ to 6° 38′ w. W. C. The slope is a little E. of s. The s., central, and indeed greater part of the surface is in the valley of Paint creek, but some of the higher branches of Little Miami rise and issue from the N. w. angle, as do from the s. E. some creeks entering Deer creek, a tributary of Sciota r. The surface of this co. is rather too level, though the soil is productive. Chief t. Washington. Pop. 1830, 8,182.

FAYETTE, co. of Ind. bounded by Franklin s., Rush w., Henry N. w., Wayne N. E., and Union E. Length from s. to N. 18 ms., mean breadth 10, and area 180 sq. ms. Extending in lat. from 39° 33′ to 39° 47′, and in long. from 8° 3′ to 39° 17′ w. W. C. This co. is traversed by the main stream, and is chiefly comprised in the valley of White Water r. Chief t. Connersville. Pop. 1830, 9,112.

FAYETTE, co. of Il. bounded s. E. by Clay,

s. by Marion, s. w. by Clinton, w. by Bond, n. w. by Montgomery, n. by Shelby, and e. by Effingham. Length from e. to w. 30 ms., mean breadth 24, and area 720 sq. ms. Extending in lat. from 38° 50′ to 39° 12′, and in long. from 11° 44′ to 12° 17′ w. W. C. The slope is to the s. s. e., and traversed in that direction by the main volume of the Kaskakia r., which leaves the co. at its extreme southwestern angle. Chief t. Vandalia, which is also the capital of the state. Pop. 1830, 2,704. The latter aggregate, however, includes also the inhabitants of two recently formed cos. Effingham and Jasper.

Fayette Corner and p-o. in the eastern part of Fayette co. Ten. 9 ms. estrd. from Somerville, the co. st., and by p-r. 865 ms. s. w. by w. W. C. and 173 ms. in a nearly similar direction from Nashville.

Fayetteville, p-v. in the northwestern part of Franklin co. Pa. by p-r. 96 ms. n. w. W. C. and 52 wstrd. Harrisburg.

Fayetteville Village, in the southwstrn. part of Fauquier co. Va. about 50 ms. s. w. W. C.

Fayetteville, p-v. and st. jus. Cumberland co. N. C. by p-r. 347 ms. s. s. w. W. C. 61 in a very nearly similar direction from Raleigh, and by the land road up Cape Fear r. 107 ms. above Wilmington. Lat. 35° 02′, and long. 1° 50′ w. W. C. Pop. 1830, 2,868. This once flourishing depot is situated on the right bank of Cape Fear r. at the head of uninterrupted boat navigation. On May 29, 1831, it was desolated by a most destructive fire, but is again rising from the ruin occasioned by the calamity. We insert the following for the display of a fine moral picture. The humane feelings of the people of other parts of the U. S. were not vainly appealed to on the distress at Fayetteville, and contributions were raised in

Maine,	$125	N. J.	$805	S. C.	$9,100
N. H.	290	Pa.	12,731	Geo.	4,102
Mass.	14,518	Md.	6,820	Ten.	45
R. I.	2,067	Dist. Col.	870	Ohio,	1,158
Conn.	3,002	Va.	8,040	Miss.	1,119
N. Y.	10,648	N. C.	11,406	La.	5,050

an aggregate amount, including fractions, of $91,902 38.

Fayetteville, p-v. and st. jus. Fayetto co. Goo. by p-r. 700 ms. n. w. W. C. and 107 ms. n. w. by w. Milledgeville. It is on a small branch of Flint r. Lat. 33° 27′, and long. 7° 36′ w. W. C.

Fayetteville, p-v. and st. jus. Lincoln co. Ten. by p-r. 722 ms. s. w. by w. W. C. and 73 ms. a little e. of s. Nashville. It is on the right bank of Elk r., and on the direct road from Nashville in Ten. to Huntsville in Ala. Lat. 35° 10′, long. 9° 37′ w. W. C.

Fayetteville, p.v. and st. jus. Washington co. Ark. by p-r. 1,285 ms. s. w. by w. ½ w. W. C. and 217 ms. n. w. by w. Little Rock.

Fayston, t. Washington co. Vt. 25 ms. s. e. Burlington, 16 s. w. Montpelier. First settled about 1798, is on the w. range of Green mtns. broken, with little useful land, almost uninhabited, and has only a few streams of Mad r. Pop. 1830, 447.

Fearing, p-v. in the southern part of Washington co. O. by p-r. 312 ms. a little n. of w. W. C. 114 s. e. Columbus, and 8 Marietta.

Federal Hill, p-o. in the eastern part of Hardy co. Va. by p-r. 125 ms. w. W. C.

Federalsburg, p-v. in the extreme n. e. angle of Dorchester co. Md., and on Marshy Hope creek, 20 ms. a little s. of e. Easton, 25 n. e. by e. Cambridge, and by p-r. 99 a little s. of e. W. C.

Federalton, p-v. in the eastern part of Athens co. O. by p-r. 344 ms. a little n. of w. W. C. and 73 s. e. Columbus.

Feliciana. (*See East Feliciana, West Feliciana.*)

Feliciana, p-v. in the western part of Graves co. Ky. by p-r. 16 ms. wstrd. Mayfield, the co. st., and 823 from W. C. and 284 from Frankfort, in a similar direction s. w. by w. ½ w.

Felicity, p-v. Franklin tsp. Clermont co. O. by p-r. 9 ms. wstrd. Batavia, the co. st. 485 wstrd. W. C. and 116 s. w. Columbus. Pop. 1830, 199.

Femme Osage, p-v. in the wstrn. part of St. Charles co. Mo. 20 ms. wstrd. from St. Charles, the co. st. and by p-r. 896 wstrd. W. C. and about 20 n. w. by w. St. Louis.

Fenner, p-t. Madison co. 12 ms. n. w. Morrisville.

Fenn's Bridge and p-o. Jefferson co. Geo. by p-r. 91 ms. estrd. Milledgeville.

Fenwick's Tavern and p-o. St. Mary's co. Md. by p-r. 82 ms. s. e. W. C.

Ferdinand, t. Essex co. Vt. n. Granby, is a poor tract of mountains and swamps, uninhabited; 23 sq. ms., watered by the great branch of Paul's stream.

Fernandina, p-v., city, s-p., and st. jus. Nassau co. Flor. by p-r. 776 ms. s. s. w. W. C., 8 ms. s. e. from the town of St. Mary's, 80 a little n. of w. St. Augustine, and 181 a little n. of e. Tallahassee. Lat. 30° 40′, long. 4° 41′ w. W. C. It is situated on the northern end of Amelia isl., and nearly opposite to the mouth of St. Mary's river. Pop. 1830, 198.

Ferrisburg, t. Addison co. Vt. 19 ms. s. Burlington, 34 w. Montpelier, e. lake Champlain, settled 1784, from Bennington and Conn. It is supplied with excellent mill sites by Otter, Little Otter, and Lewis creeks. It has Basin harbor, and Otter and Little Otter creeks, whose mouths are 80 rods apart, are navigable 8 and 3 ms. A ferry of 2 ms. crosses the lake from below Little Otter cr. Hilly n. level w. Has afforded excellent timber for Quebec, maple, beech, bass, &c. on uplands; pine, oak, &c. on low lands. It sends out many fat cattle,—11 school districts. Pop. 1830, 1,822.

Fife's, p-o. western part of Goochland co. Va. by p-r. 116 ms. s. s. w. W. C. and 39 a little n. of w. Richmond.

Fincastle, p-v. and st. jus. Bottetourt co. Va. situated near the right bank of Catawba cr. by p-r. 235 ms. s. w. W. C. 176 westward Richmond, and 45 n. e. Christiansburg. Lat.

37° 28', and long. 2° 57' w. W. C. This place contains the ordinary co. buildings, and a pop. of about 1,000.

FINDLAY, p-v. and st. jus. Hancock co. Mo. by p-r 502 ms. s. w. by w. W. C. and 114 N. N. w. Columbus. It is situated on Blanchard's Fork of Auglaize r. on the road from Cincinnati to Detroit, lat. 41° 04', long. 6° 40' w. W. C.

FINDLAYSVILLE, p-v. Mecklenburg co. N. C. by p-r. 363 ms. s. w. W. C. and 111 s. w. by w. Raleigh.

FINEYWOOD, p-v. Charlotte co. Va. by p-r. 162 ms. s. w. by w. Richmond, and 17 from Marysville.

FINLAYVILLE, p-v. in the N. w. part of Washington co. Pa. about 12 ms. N. w. Washington, the co. st. and 30 s. w. Pittsburg.

FINNEY MILLS, and p-o. Amelia co. Va. by p-r. 61 ms. s. w. Richmond.

FISCHLIE'S MILLS, and p-o. in the northern part of Jackson co. Ind. by p-r. 613 ms. westward W. C. and 59 southward Indianopolis.

FISH DAM, p-o. south western part of Wake co. N. C. 12 ms. Raleigh.

FISH DAM, p-o. western part of Union dist. S. C. by p-r. 63 ms. N. w. Columbia.

FISHERSFIELD, p-t. Merrimac co. N. H. 23 ms. N. N. w. Hopkinton, 30 from Concord. Pop. 1830, 797.

FISHER'S ISLAND, N. Y. Long Island sound, 5 ms. s. w. Stonington, (*see Southold.*)

FISH'S STORE, and p-o. in the northern part of Washington co. Geo. by p-r. 37 ms. eastward Milledgeville, and 659 ms. s. w. W. C.

FISHING CREEK, tsp. and p-v. in the northeastern part of Columbia co. Pa. by p-r. 199 from W. C. and 89 from Harrisburg, and in a similar direction a little N. of E. Pop. of the tsp. 1830, 568.

FISHING CREEK, and p-o. Tyler co. Va. The creek rises in Tyler, flows N. w. into Ohio co. and falls into Ohio r. in the southwestern angle of the latter. The p-o. is situated on the head waters of the cr. about 28 ms. w. Morgantown, and by p-r. 238 N. w. by w. W. C.

FISHING CREEK, and p-o. in the northeastern part of Chester dist. S. C. The creek rises in York dist. near Yorkville, interlocking sources with Allison's creek of Catawba, and Bullock's of Broad r. and flowing s. s. E. enters Chester, within which it falls into Catawba r. 20 ms. s. E. Chesterville. The p-o. is by the common road 60 ms. N. Columbia, and by p-r. 442 s. w. W. C.

FISHKILL, N. Y. a branch of Wood cr.

FISHKILL, N. Y. the outlet of Saratoga lake, joins the Hudson at Schuylerville. On the banks of this creek Burgoyne's army surrendered to Gen. Gates, Oct. 17, 1777.

FISHKILL, p-t. Duchess co. N. Y., E. Hudson r. 14 ms. s. Poughkeepsie, 89 s. Albany, and 65. N. of New York. It derives its name from the Fishkill, which runs nearly centrally through it to the Hudson. There are 4 landings on the Hudson. It is a place of considerable business. The village of Fishkill is 5 ms. E. of the Hudson. This town was the first that was settled in the co. Here is the Matteawan cotton factory, the largest in the state, situated on the Fishkill, about half a mile from the Hudson. It produces annually about half a million yards of cloth. Near this factory is Schenck's extensive grain mill, which manufactures 50,000 bushels of wheat per annum. There is likewise an extensive wollen manufactory at this place. The Glenham wollen factory is about two ms. from the Matteawan factory. At this establishment are manufactured superfine blue and black cloths. The other settlements in this town, are the Upper Landing, Low Point or Carthage Landing, Hopewell, New Hackensack, and Middlebush. Pop. 1830, 8,292.

FISHKILL MTNS. (*see Matteawan.*)

FITCHBURG, p-t. Worcester co. Mass. 42 ms. N. w. Boston, 25 N. E. Worcester. Finely watered by Nockege r. a branch of the Nashua, and by two other streams, which render it an excellent town for manufactories; and accordingly it has become an extensive manufacturing place. Pop. 1830, 2,169.

FITCHVILLE, p v. and tsp in the northern part of Huron co. O. p-r. 388 ms. N. w. by w. W. C. and 109 ms. a little E. of N. Columbus. Pop. of the tsp. 1830, 347.

FITZWILLIAM, p-t. Cheshire co. N. H. 13 ms. from Keene, 60 from Concord, and 65 from Boston. It was named in honor of the Earl of Fitzwilliam. This town was the residence of Brigadier Gen. James Reed, a revolutionary patriot. Pop. 1830, 1,229.

FLANDERS, village, Morris co. N. J. on the s. branch Raritan, 12 ms. w. N. w. Morristown.

FLATBUSH, p-t. st. jus. Kings co. Long Island, 4½ ms. s. Brooklyn, 5 s. N. Y. Here is a flourishing academy, denominated Erasmus Hall. A battle was fought near this place, Aug. 27, 1776, in which the Americans were defeated by the British, and suffered a heavy loss. Pop. 1830, 1,143.

FLAT CREEK, p-o. Campbell co. Va. 11 ms. s. w. Linchburg, and by p-r. 119 s. w. by w. Richmond.

FLAT CREEK, p-o. in the western part of Bath co. Ky. by p-r. 494 ms. s. w by w. ½w. W. C. and 67 E. Frankfort.

FLATLANDS, t. King's co. s. side and near w. end Long Island, 7½ ms. a little E. of s. N. Y. and 2 ms. s. Flatbush. Pop. 1830, 596.

FLAT LICK, p-o. Trigg co. Ky. by p-r. 224 ms. s. s. w. Frankfort.

FLAT ROCK, p-o. eastern part of Powhatan co. Va. 24 ms. w. Richmond.

FLAT ROCK, p-o. in the western part of Buncombe co. N. C. by p-r. 285 ms. westward Raleigh.

FLAT ROCK, p-o. in the northern part of Kershaw district, S. C., about 20 miles N. Camden, and by p-r. 53 N. E. Columbia.

FLAT ROCK, p-o. in the eastern part of Bourbon co. Ky. by p-r. 504 ms. westward W. C. and 55 eastward Frankfort.

Flat Rock, p-o. on a large cr. of the same name, southern part of Shelby co. Ind. by p-r. 53 ms. s. e. Indianopolis.

Flat Woods, p-o. in the western part of Lewis co. Ky. by p-r. 304 ms. w. W. C.

Fleetwood, p-v. Hinds co. Miss. by p-r. 1,053 ms. s. w. by w. W. C. and about 100 n. e. Natchez.

Fleming, p-t. Cayuga co. N. Y. 4 ms. s. Auburn. Pop. 1830, 1,461.

Fleming, co. Ky. bounded w. by Nicholas; n. w. by Mason; n. e. by Lewis; e. and s. e. by Lawrence, and s. w by Licking r. which separates it from Bath. Length s. e. to n. w. 36 ms., mean breadth 16, and area 576, sq. ms., extending in lat. from 38° 06′ to 38° 33′, and in long. from 6° 22′ to 6° 55′ w. W. C. The slope of Fleming is to the s., soil productive, and surface moderately hilly. Chief t. Flemingsburg. Pop. 1820, 12,186, and in 1830, 13,499.

Flemingsburg, or Flemingsburg, p-v. and st. jus. Fleming co. Ky. by p-r. 498 ms. a little s. of w. W. C. and 79 eastward Frankfort. It is situated on the table land near the sources of creeks flowing northwards into O., and southwards into Licking r. lat. 38° 25′, and long. 6° 40′ w. W. O. Pop. 1830, 648.

Flemington, p-v. Amwell, Hunterdon co. N. J. 23 ms. n. n. w. Trenton.

Fleming's, p-o. Weakly co. Ten. by p-r. 834 ms. s. w. by w. W. C. and 132 westerly from Nashville.

Fleming's, p-v. in the north-western part of Shelby co. Ind. 5 ms. n. w. Shelbyville, by p-r. 580 a little n. of w. W. C. and 35 s. e. Indianopolis.

Fletcher, t. Franklin co. Vt. 22 ms. n. e. Burlington, and 35 n. w. Montpelier. Pop. 1830, 793.

Flint, r. Ontario co. N. Y., waters Italy, Middlesex, Gorham, and Phelps, where it joins the Canandaigua outlet, at the village of Vienna, its whole course of 32 miles being a good mill stream.

Flint, r. of Geo. (*see Appalachicola, first paragraph.*)

Flint Hill, p-o. Culpepper co. Va. by p-r. 104 ms. s. w. W. C.

Flint Mills, and p-o. in the southern part of Madison co. Ala. 10 ms. s. from Huntsville, by p-r. 716 s. w. by w. W. C. and 165 n. n. e. Tuscaloosa.

Flint Stone, p-o. Alleghany co. Md. by p-r. 119 ms. n. w. W. C.

Flood's, p-o. Buckingham co. Va. by p-r. 180 ms. s. w. W. C., and 96 westward Richmond.

Florence, p-t. Oneida co. N. Y., 20 ms. n. w. Rome. Pop. 1830, 964.

Florence, p-v. and st. jus. Lauderdale co. Ala. by p-r. 796 ms. s. w. by w. W. C., 146 a little w. of n. Tuscaloosa, and 70 westward Huntsville. It is situated on the right bank of Tennessee r. at the mouth of Cypress creek, and below the Muscle Shoals. Lat. 34° 47′, long. 10° 46′ w. W. C. At seasons of high or even moderate height of water, steam boats ascend to Florence. The pop. is estimated at 1,500, and the place so advantageously situated, possesses a flourishing trade.

Florence, p-v. Boone co. Ky. by p-r. 507 ms. westward W. C. and 70 n. Frankfort. Pop. 1830, 63.

Florence, p-v. and tsp. in the western part of Huron co. O. The p-v. is situated on Vermillion r. by p-r. 13 ms. n. e. by e. Norwalk, the co. seat, 395 n. w. W. C. and 127 n. n. e. Columbus. Pop. of the tsp. 1830, 760.

Florida, a canal, town, and p-t. Montgomery co. N. Y., s. shore Mohawk river, 35 ms. n. w. Albany. Pop. 1830, 2,851.

Florida, p-v. Orange co. N. Y. 6 miles s. Goshen.

Florida, p-t. Berkshire co. Mass. 25 miles n. e. Lenox, 120 w. Boston. Hoosic mountain lies between this town and Adams; and Deerfield river rises on the eastern declivity of the mountain, on the Florida side. Pop. 1830, 454.

Florida, cape, promontory of the southeastern coast of Florida, projecting southward, & enclosing on the n. e. the bay of Biscane. On Tanner's U. S. map cape Florida is laid down at lat. 25° 38′, long. 3° 10′ w. W. C.

Florida, extreme southern territory of the United States, bounded n. w. by Alabama, n. by Geo., e. by the Atlantic ocean and Bahama channel, s. by Cuba channel, and w. and s. w. by the Gulf of Mexico. Length, if we follow the curve from Perdido river to cape Sable about 660 miles. The breadth from Perdido river along the northern border, to the Atlantic ocean, is 375 miles, but the mean breadth being only about 84 miles, the area is 55,400 square miles, or 35,456,000 statute acres. Extending in lat. from 25° to 31°, if we include only the continent, but embracing Thompson's island, the southern extreme is at lat. 24° 30′. In long. this territory lies between 3° and 10° 44′ w. W. C.

Florida has a boundary along the Gulf of Mexico, from the mouth of Perdido, to cape Sable, 600 miles; along the Cuba and Bahama channels, and Atlantic ocean from cape Sable to the mouth of St. Mary's river, 450 miles; in common with Geo. from the mouth of St. Mary's to that of Flint river, 240 miles; up Chattahooche river to lat. 31°, 40 miles; in common with Alabama from Chattahooche to Perdido river, 140 miles; down Perdido to its mouth, 40 miles; entire outline, 1,510; with a perimeter exceeding 1500 miles, and extending through 6 degrees of lat. Florida presents some diversity of climate, but the difference of relative level being but slight, the seasons at the extremes more nearly approach an equality of temperature than does any other similar extent of lat. in the United States. The northern and part of central Florida is covered with a dense forest, except the partial clearing of land for agricultural and other purposes of human society, but the southern section presents large spaces of open, grassy, and in part marshy plains. Pine is the prevailing timber, but great variety of other forest trees are intermixed.

Sterility is the true general character of the soil, with, it is true, some favorable exceptions. The value of the soil is, however, in some measure equalized with that of more fertile but more nthrn. land, by the high temperature of the climate of Florida. Of cultivated vegetables the principal species are, of grains, rice and Indian corn; the sweet potatoe is produced in great abundance. Garden vegetables admit of immense variety. Staples are cotton, indigo, and sugar. Of fruits, the orange, lime, several varieties of figs, the peach, pomegranate, and some others flourish. The olive and some species of the *vitis vinefera* (wine producing grape vine) might be produced perhaps to more advantage than in any other section of the U. S. The profitable culture of the coffee plant and date palm, is more doubtful.

That part of Florida extending along the northern shore of the Gulf of Mexico slopes towards that recipient of its rivers, with a general southern course; Florida is traversed, advancing from east to west, by the rivers Suwannee, Oscilla, Ocklockonne, Appalachicola, Choctaw, Yellow Water, and Escambia. The two latter are discharged into the fine sheet of water, which forms the harbor of Pensacola. Choctaw opens into a wide bay of the same name. The Appalachicola forms a delta. Ocklockonne is discharged into the deep bay of Appalachie. The Suwanne has a more southern influx than any of the preceding, and looses its volume in Vacasausa bay, lat. 29° 25′.

With the Vacasausa bay the slope of Florida bends with the peninsula, and Amasura, Anclota, Hillsboro', Charlotte, Gallivan's, and Young's rivers have a general western course. The St. John's of Florida is an anomaly amongst the rs. of the Atlantic coast of the United States. The source of this stream is rather indefinable, being derived from the flat grassy plains, about lat. 28, and flowing thence to the west of north, nearly parallel to the opposite Atlantic coast, has more the appearance of a sound than a river. The long eastern shore of Florida can hardly be regarded as having a slope. The level is general and very slightly broken by elevations of any kind. On this monotonous expanse, a shell bank appears an object of magnitude.

Descending from the more general to the more specific features, we find the long nthrn. parallelogram of Florida offering considerable diversity of surface. The rise from the mouth of St. Mary's to the dividing ground between its basin and that of Suwannee exceeds 200 feet. The ridges or table land, protruded southwardly between the other more western rivers, are perhaps still more elevated, but all imperceptibly decline, approaching the sea coast until merged in the sandy shores. The depth of the harbors may be seen under the respective heads of the rivers and bays. The table land between the basins of St. Mary's and Suwannee rivers stretches southward, and may be regarded as the spine of peninsula Florida, until gradually lost in the plains between the sources of St. John's and Amasura rivers.

Florida as a political subdivision comprises two natural sections. The northern slope, already noticed, extending from the Atlantic ocean to Perdido river, deeply furrowed by the river channels, is followed by the peninsula, properly so called. The two physical sections can have no actually defined line of separation, but a line drawn from the mouth of St. John's river to that of Suwannee, would afford a demarcation, having two natural points of termination. A base of calcareous rock commences in northern, and extends under perhaps all peninsular Florida. This friable stone breaks forth at St. Augustine and many other points, but is overlaid generally by deep superstrata of clay, shells, and sand. We insert the following description as being official. It is extracted from the files of congressional document, and from a Report of the engineer department, on the practicability of a canal across the peninsula of Florida. Read in congress March 26th, 1832.

"The part of the peninsula of Florida, comprehended between the southern boundaries of Georgia, and a line drawn from Tampa bay to cape Cannaveral, is an extensive pine forest, interspersed with numberless lakes, ponds, low savannahs, and cypress swamps, of various sizes. The country, though generally flat, is, however, much undulated in some districts, and even hilly in many places. The ridge which divides the waters emptying into the Atlantic ocean from those running into the gulf, is sloping gradually from N. to S., and seems to become totally depressed south of a line drawn from the bay of Tampa to cape Cannaveral. Indeed all that great tract of country south of this line, is represented, by those best informed, as an extensive marsh, forbidding, during the rainy seasons (between June and October,) any land passage from the gulf to the Atlantic. The elevation of the ridge above the level of the sea, has been found to be 152 feet at the head of St. Mary's river, near the Georgia line, 158 between Kinsley's pond, and Little Santa Fe pond, head of Santa Fe river, and 87 feet between the head branches of the Amaxura and Ocklawaha rivers.

"The soil is generally sand, except at places called hammocks, the soil of which is either a red-yellow, or black clay, mixed with sand. These hammocks are numerous, and much scattered throughout the country; they vary in extent, from a few acres to thousands of acres, and form together but an inconsiderable portion of the peninsula. On them, the growth of tree is red oak, live oak, water oak, dog wood, magnolia and pine; the red oak predominating. Whilst these hammocks, under the auspicious climate of Florida, present a very productive arable land, the pine forests afford every facility to the raising of cattle; and under this point of view, this part of the peninsula may be considered as a most valuable grazing country.

"It is to be observed that, in Florida, the fern grass is exclusively peculiar to low grounds and heads of water courses.

"If the upper stratum of the peninsula, is generally sand on both sides of the ridge, and that to a depth of at least 5 or 6 feet, the substratum is not the same on both sides. On the eastern, it is clay mixed with a great deal of sand; but on the western, it is, throughout, a kind of stratified rotten limestone, which frequently appears at the surface, and which at many places, is undermined by streams sinking abruptly to take their passage through the cavernous parts of the mass, and to resume, at some distance down, their natural course. It is owing to the numerous cavities of this rotten substratum, that the surface of the ground is seen interspersed with numberless inverted comic hollows, called sinks, the size of which varies from a few square yards to many acres.

"The streams which run through the peninsula, present, generally, no flat bottom, or arable fluviatic depqsite along their banks; they force their course through the sandy upper stratum, and are fed more by lateral filtration than by tributaries. However, their margins are often trimmed with trees, such as live oak, water oak, magnolia, and laurel thicket, which receive chiefly their nourishment from moisture.

"The sea along the western coast of the peninsula is shallow, from Tampa bay to Appalachie bay, and on a width varying from 5 to 15 miles. From the latter to cape San Blas, this width diminishes, except at the intervening capes, where extensive shoals project out; but from cape San Blas to lake Pontchartrain, the shore is generally bold, and the coast affords several good harbors.* As to the coast on the Atlantic, the sea is all along shallow, and offers no harbors except at the mouth of St. John's r. and St. Augustine." St. Mary's river ought to be added.

"The shortest distance across the peninsula is about from St. Augustine, to a point on the gulf between the mouths of the Suwannee and Amaxura rivers; this distance is 105 miles. The distance on a straight line from the mouth of the St. John to that of the Suwannee is 130 miles, and from the mouth of St. John to St. Mark, 170 miles."

The long problem in theory seems to be solved, that is, the difference of level between the Atlantic ocean outside, and the Gulf of Mexico inside of the peninsula of Florida. On this subject the topographical engineers observe, "should the surveys be perfectly accurate, the level of low tide in the gulf would be 2 263-100 feet above that of low tide in the Atlantic. On another hand it has been shown, in describing the surveys relating to the St. Mary's route, that the result of the leveling has given 3 55-100 for the difference of the level between the gulf and the Atlantic. We must therefore consider it as probable, that, at low tide, the elevation of the gulf at the mouth of the Suwannee, is not more than 3 or 4 feet above low tide at the entrance of St. John's into the Atlantic.

Political Geography. In the census of 1830, Florida is subdivided into Eastern, Western, Middle and Southern; of which the population was as follows:—

	Whites,	Free col'd,	Slaves,	Total,
Eastern,	4,515	346	4,095	8,956
Western,	5,319	396	3,753	9,468
Middle,	8,173	19	7,587	15,779
Southern,	368	83	66	517
Total	18,375	844	15,501	34,720

The counties and their population, as exhibited in the census, is as follows:—

Eastern Florida,		Western Florida,		Middle Florida,	
Alachua,	2,204	Escambia,	9,468	Gadsden,	4,895
Duval,	1,970	Jackson,		Hamilton,	553
Nassau,	1,511	Walton,		Jefferson,	3,312
Moscheto,	733	Washington,		Leon,	6,494
St. Johns,	2,538			Madison,	595

All southern Florida is included in the county of Monroe. In the census, the population of the counties, composing Western Florida, is not individually given.

Of the foregoing population, there were white persons:—

	Males.	Females.
Under 5 years of age,	1,932	1,807
From 5 to 10,	1,333	1,251
" 10 to 15,	1,015	981
" 15 to 20,	789	923
" 20 to 30,	2,161	1,447
" 30 to 40,	1,536	848
" 40 to 50,	760	484
" 50 to 60,	436	247
" 60 to 70,	194	101
" 70 to 80,	57	45
" 80 to 90,	10	10
" 90 to 100,	2	5
" 100 and upwards,	1	0
Total,	10,296	8,149

Of the foregoing are deaf and dumb, under 14 years, 2; 14 to 25, 0; 25 and upwards, 3; Blind 2.

Colored population, as follows:—

	Free.		Slaves.	
	Males.	Fems.	Males.	Fems.
Under 10 years of age,	138	144	2,501	2,560
From 10 to 24,	109	136	2,482	2,449
" 24 to 36,	47	70	1,830	1,561
" 36 to 55,	56	62	948	766
" 55 to 100,	33	48	224	177
" 100 and upwards,	0	1	0	1
Total,	383	461	7,985	7,516

Deaf and dumb, colored, under 14 years, 1; 14 to 25, 1; over 25, 1.

From the preceding analysis of its distributive population of 1830, Florida contained the largest number of inhabitants of the three United States territories, Arkansas, Michigan, and Florida; and of this aggregate 53 per-cent are whites. The weight of the population of Florida lies along the northern

* This character of coast between Mobile bay and lake Pontchartrain, is only applicable outside of the islands. (*See Pascagoula sound.*)

parallelogram and above lat 28°. The relative distribution must at all future times remain not greatly different from the present, as the productive soil and navigable rivers are in great part confined to the three northern sections. Extensive surveys have been made to determine the practicability of forming a canal across Florida, from the Atlantic ocean to the Gulf of Mexico. These surveys have been productive of much very valuable geographical information, of which we have availed ourselves in this treatise, but the practical construction of the proposed canal remains doubtful, and the doubts arise chiefly from the shallowness of the water on the gulf side of the peninsula.

History.—The fine, sonorous name of Florida, was imposed by the discoverer, John Ponce de Leon, from having made the coast on or about "Pasqua Florida," Palm Sunday, in 1512. The first attempt to form a civilized colony in Florida, was made by the French in 1562, under Francis Ribault, but the colonists were, in 1565, surprised and murdered by the Spaniards. This massacre was severely revenged by a French expedition; but the Spaniards remained masters of the country, and founded in 1565, the city of St. Augustine, in East Florida. West Florida was not colonized until 1699, when Pensacola was founded by Don Andre de la Riola. Though often invaded by French and English armaments, this province remained a part of Spanish America, until 1763, when it was ceded to Great Britain. By the definitive treaty of 1783, it was receded by Great Britain to Spain. When Florida was a colony of Spain, and Louisiana of France, or from 1699 to 1763, the Perdido river was a common boundary, but when in 1769, Louisiana was taken into possession by Spain, under the treaty of cession of 1763, they, for their own convenience, incorporated that part of Louisiana between the Mississippi and Perdido river with Florida. This incorporation of part of Louisiana into Florida, involved Spain in a controversy with the United States, when the latter government gained possession of Louisiana. On virtue of claiming the latter, as held by France previous to 1763, the United States, in 1811, seized Baton Rouge, and all other parts of Florida west of Perdido, except Mobile, which also surrendered in 1812. After a lengthened and interrupted negotiation, Florida was ceded to the United States, February 22d, 1819, by a treaty formed at Washington. This treaty was finally ratified by the King and Cortes of Spain, October 24th, 1820, and February 22d, 1821, was ratified by the congress of the United States. Since the latter period, there has occured no event in the history of Florida, worthy of particular notice.

FLORIDA KEYS, is a chain of islets, rocks, reefs, and sand banks, extending westward from the southwardly part of Florida, stretching in long. from 3½° to 6° 15′ w. W. C. between lat 24° 30′ and 24° 45′. This very dangerous chain, is composed of Ball islands, Matacumbe islands, Cayasbacos, the Pine islands, Thompsons islands, Mule islands, Cayos Marques, and on the extreme w. the Tortugas.

FLORISANT, p-v. in the northern part of St. Louis co. Mo. situated between the Miss. and Mo. rivers, below their junction, 18 miles a little w. of N. St. Louis, and 872 westward W. C.

FLOURNOY'S MILLS, and p-o. Telfair co. Geo. 772 ms. southwestward W. C. and 120 sthrd. Milledgeville.

FLOWING SPRING, p-o. Bath co. Va. 206 ms. s. w. W. C. and 164 a little N. of w. Richmond.

FLOYD, p-t. Oneida co. N. Y. 6 miles E. Rome, and 10 N. Utica. Pop. 1830, 1,699.

FLOYD, eastern co. Ky. bounded s. by Pike, s. w. by Perry, w. by Morgan, N. by Lawrence, and E. by the Tug or eastern branch of Big Sandy, which separates it from Logan co. of Va. Length, from east to west 50 miles, mean breadth 30, and area 1,500 sq. miles. Extending in lat. from 39° 24′ to 37° 55′, and in long. from 5° 02′ to 6° 10′ west W.C. This county occupies an elevated table land, the eastern part sloping northwardly, and down which flow the two main branches of Sandy river. From the southwestern angle flow the higher branches of Kentucky, and from the northwestern, the extreme sources of Licking river. The face of the country is hilly and broken. Chief town, Preston berg. Pop. 1820, 3,207, in 1830, 4,347.

FLOYD, co. of Indiana, bounded by Harrison s. s. w. and w., Washington N. w., Clark N. E. and E., and the Ohio river, separating it from Jefferson county, Kentucky, s. E. Length 20 miles, mean breadth 15, and area 300 square miles. Extending in lat. from 38° 16′ to 38° 31′, and in long. from 8° 44′ to 9° 03′ w. W. C. This co. lies directly opposite Louisville, in Ky. The slope is southward; surface hilly, and soil productive. Chief town, New Albany. Pop. 1831, 6,361.

FLOYDSBURG, p-v. on Floyd Fork, southern part of Oldham co. Ky. 20 ms. N. E. by E. Louisville, and 34 a little N. of w. Frankfort.

FLOYD'S FORK, r. of Ky. rises in Oldham co., flows thence s. s. w. over the eastern part of Jefferson, and northern of Bullitt, falls into salt r. at Shepperdsville, after a comparative course of 35 ms.

FLUKES, p-o. Bottetourt co. Va.

FLUSHING, p-t. Queen's co. N. side Nassau, on Long isl. 15 ms. E. N. Y. Flushing v. which stands at the head of Flushing bay, is a fashionable place of resort. In this town are still remaining two of the white oaks, under whose shade George Fox, the founder of Quakerism, held a religious meeting in 1672. Pop. 1830, 2,820.

FLUSHING, p-v. in Flushing tsp. Belmont co. O. by p-r. 124 ms. E. Columbus, and 275 ms. N. w. by w. W. C. Pop. 1830, 114.

FLUVANNA, co. of Va. bounded N. w. by Albemarle, N. E. by Louisa, S. E. by Goochland,

and s. and s. w. by James r., which separates it from Buckingham. Length along Albemarle 26, mean breadth 16, and area 416 sq. ms. Extending in lat. from 37° 36′ to 40°, and in long. from 1° 12′ to 1° 43′ w. W. C. Fluvanna r. enters it from Albemarle, and flowing s. E. divides it into two nearly equal sections. Surface agreeably broken. Chief t. Columbia. Population 1820, 6,704, 1830, 8,221.

FOGELSVILLE, p-o. Lehigh co. Pa.

FOLLY, p-o. Gates co. N. C. by p-r. 186 ms. N. E. by E. Raleigh.

FORD'S FERRY and p-o. Livingston co. Ky. by p-r. 227 ms. s. w. by w. Frankfort.

FORKED DEER RIVER, stream of Ten. rising in Carroll, Henderson, and McNair cos., and flowing over Madison, Gibson, Haywood and Dyer cos., falls into the Miss. r. in the nthrn. part of Tipton, above the first of the Chickasaw bluffs, after a comparative course of 80 ms. N. w. by w. The valley of Forked Deer lies between those of Obion and Big Hatchie, and embraces an area of about 2,000 sq. ms.

FORK SHOALS, p-o. on Saluda r. Greenville dist. S. C. by p-r. 107 ms. N. w. Columbus.

FORSYTHE, p-v. and st. jus. Monroe co. Geo. situated on Chussee creek, about 60 ms. a little s. of w. Milledgeville. Lat. 33°, and in long. 7° 5′ w.

FORT ADAMS, p-v. on the left bank of Miss. r. at Loftus Heights, Wilkinson co. Miss. 41 ms. by the road s. Natchez.

FORT ANNE, p-t. Washington co. N. Y. derives its name from the fort here erected during the French wars, which stood at the head of batteaux navigation on Wood creek. The Champlain canal runs through this town, 62 ms. from Albany, 10 N. Sandy Hill, and 11 s. Whitehall. The village of Fort Anne stands near the site of the fort, on the Champlain canal. Pop. 1830, 3,200.

FORT BALL, p-v. in the western part of Seneca co. O. by p-r. 446 ms. s. w. by w. W. C. and 85 a little w. of N. Columbus.

FORT BLOUNT, p-v. Jackson co. Ten. by p-r. 660 ms. s. w. by w. W. C. and 70 N. E. by E. Nashville.

FORT CLAIBORNE. (*See Claiborne, st. jus. Monroe, Ala.*)

FORT COVINGTON, p-t. Franklin co. N. Y. on the St. Lawrence, at the mouth of Salmon creek, 15 ms. N. w. Malone, 235 from Albany, and 53 E. N. E. of Ogdensburg. This town and the fort within it derive their name from Gen. Covington, who was mortally wounded in the battle of Williamsburg, during our last war with Great Britain. St. Regis v. (Indian) is situated in this town. The chief of these Indians is a descendant of a daughter of Rev. Mr. Williams, minister of Deerfield, Mass. She was carried into captivity when Deerfield was destroyed by the Indians, in the time of the old French war, and marrying an Indian, refused to return; and so lived and died among the Indians. Pop. 1830, 2,901.

FORT CRAWFORD, Crawford co. Mich. (*See Prairie du Chien.*)

FORT DALE, p-v. Butler co. Ala. 132 ms. N. E. from Blakely, on Mobile r. and by p-r. 152 ms. s. s. E. Tuscaloosa.

FORT DEFIANCE, p-v. in the westert part of Wilkes co. N. C. situated on the Yadkin near its source, 25 ms. above, and s. w. by w. Wilkesboro, the co. st., and by p-r. 428 ms. s. w. W. C. and 200 w. Raleigh.

FORT EDWARD, p-t. Washington co. N. Y. E. Hudson r. near the Great Bend. It is 2 ms. s. Sandy Hill, 16 from Caldwell, and 22 from Whitehall. The old fort, which gives the name to the town, was built by the Americans in 1755. The village is built on the Champlain canal, which here forms a junction with the Hudson. A dam is built across the river at this place, for the purpose of supplying the canal with water, by means of a feeder, which is half a mile long. The dam is 900 feet long, and 27 high. It cost $30,000. Pop. 1830, 1,816.

FORT GAINES, p-v. Early co. Geo. by p-r. 175 ms. s. w. Milledgeville.

FORT JACKSON, p-v. Montgomery co. Ala. situated on the point between Coosa and Tallapoosa rs. immediately above the junction, 96 ms. s. E. Tallapoosa. Lat. 32° 29′, and long. 9° 23′ w. W. C.

FORT GRATIOT, p-v. in the eastern part of St. Clair co. Mich., and on the point w. side of St. Clair r. where that stream issues from lake Huron, and just above the mouth of the river Dulude, by p-r. 597 ms. N. w. W. C. and 71 N. N. E. Detroit.

FORT JACKSON, p-v. Plaquemines parish, La. 75 ms. below New Orleans.

FORT JEFFERSON, p-v. Dark co. O. by p-r. 103 ms. wstrd. Columbus, and 501 a little N. of w. W. C.

FORT LAFAYETTE, N. Y. on a reef of rocks in the Narrows, commanding the entrance of N. Y. bay.

FORT LEE, Bergen co. N. J. a ferry and landing place on the Hudson, 10 ms. above N. Y. near the site of the fort of same name. A turnpike road from Paterson and Hackensack terminates at this place, and a steamboat plies to N. Y. city.

FORT LITTLETON, in the southeastern part Bedford co. Pa. by p-r. 103 ms. N. w. W. C. and 64 wstrd. Harrisburg.

FORT MILLER, p-v. Washington co. N. Y. on the Hudson, E. side, which here has falls, around which there is a canal, with a dam across the river for supplying the same with water, s. Sandy Hill 11 ms.

FORT RICHMOND, N. Y. on the heights, at the s. E. point of Staten isl., on the w. of the Narrows, commanding the entrance of N. Y. bay.

FORT ST. PHILIP, p-o. an important military establishment of the U. S. on the left bank of the Miss. parish of Plaquemines, La. 70 ms. below New Orleans.

FORT SENECA, p-v. in Seneca tsp. Seneca co. O. situated on Sandusky r. by p-r. 437 ms.

N. W. by W. W. C. and 94 a little W. of N. Columbus. Pop. tsp. 1830, 369.

FORT SMITH, military station and p-v. as laid down by Tanner, is situated on the right bank of Arkansas r. on the extreme western border of Crawford co. Ark. by p-r. 235 ms. above, and N. W. by W. Little Rock, and 1,303 S. W. by W. ½ W. W. C.

FORTSMOUTH, p-v. Shenandoah co. Va. by p-r. 92 ms. wstrd. W. C.

FORTSMOUTH, p-v. Page co. Va. by p-r. 83 ms. N. W. Richmond, and 147 wstrd. W. C.

FORT SNELLING, military station and p-o. at the mouth of St. Peters r. and Falls of St. Anthony in Miss. r. The distance by p-r. from Washington City is not given in the p-o. list, but as it is something above 200 ms. above Prairie du Chien, which latter is stated at 1,060 ms. from W. C., fort Snelling must amount to near 1,300 ms. from the seat of the general government. Lat. 44° 53′, long. 16° 13′ W. W. C. These relative positions yield a bearing from W. C. to fort Snelling of 63° 33′ W.; distance 968 statute ms.

FORT TOWSON, or CANTONMENT TOWSON, as laid down in Tanner's U. S. is situated almost on lat. 34° and 18° 07′ W. W. C., about 10 ms. N. from the efflux of Kiameche into Red r. In the p-o. list of 1828, the p-o. at fort Towson is named in Miller co., distant 253 ms. from Little Rock, though on the map, even the road distance falls short of 180 ms. S. W. by W.

FORTUNE'S FORK, p-o. (*See Gretna Green, Halifax co. Va.*)

FORT VALLEY, p-o. Crawford co. Geo. 48 ms. by p-r. S. W. by W. Milledgeville.

FORTVILLE, p-v. Jones co. Geo. 31 ms. wstrd. Milledgeville.

FORTIMES FORK and p-o. Halifax co. N. C. by p-r. 85 ms. N. E. Raleigh.

FORT WASHINGTON, p-v. and military station of the U. S. on the left bank of Potomac r. at the mouth of Piscataway creek, in Prince George's co. Md. 15 ms. below, and very nearly due S. W. C.

FORT WAYNE, p-v. and st. jus. Allen co. Ind. by p-r. 561 ms. N. W. by W. W. C. and 141 N. W. Columbus, and also by the intermediate road 160 ms. S. W. Detroit. This village occupies the very remarkable point above the junction of St. Mary's and St. Joseph's rs. and on the right bank of the former. The united streams here take the name of Maumee. (*See Maumee r.*) Fort Wayne received its name in honor of the old veteran hero, who so much contributed to give peace and security to the new settlements in Ohio valley, and who found a grave at Presqu' Isle, now Erie in Pa. the 15th Dec. 1796. Fort Wayne stands at lat. 41° 04′, long. 8° 7′ W. C.

FORT WINNEBAGO, as laid down by Tanner, is situated on the portage ground between the Ouisconsin and Fox r. of Green bay. In the p-o. list it is placed in Jowaco. In the latter work the relative distances are not given, but measured on Tanner's U. S., it is distant, by the circuitous route of Pektano r., 258 ms. N. W. Chicago, though only about 150 in direct course, 142 above and S. W. fort Howard, at the mouth of Fox r., 150 ms. by the land route above Prairie due Chien at the mouth of Ouisconsin, and by actual calculation N. 57° W., 586 statute miles from W. C.

FOSTER, p-t. Providence co. R. I. 15 ms. W. Providence; well watered, and contains numerous sites for water works. Pop. 1830, 2,672.

FOSTERTOWN, v. Evesham, Burlington co. N. J. 15 ms. E. Phil.

FOTHERINGAY, p-v. Montgomery co. Va. by p-r. 201 ms. S. W. by W. W. C.

FOULKSTOWN, p-v. in the eastern part of Columbiana co. O. by p-r. 166 ms. N. E. by E. Columbus, and 269 N. W. W. C.

FOUNTAIN, co. of Ind. bounded by Tippecanoe on the N. E., Montgomery E., Parke S., Wabash r. separating it from Vermillion, S. W., and Wabash r. again separating it from Warren W. and N. W. Greatest length along the eastern border 30 ms., mean breadth 16, and area 480 sq. ms. Extending in lat. from 39° 58′ to 40° 22′, and in long. from 10° 06′ to 10° 26′ W. W. C. Slope south wstrd. towards the Wabash. Chief t. Covington. Pop. 1830, 7,619.

FOUNTAINDALE, p-v. in the southern part of Adams co. Pa. by p-r. 46 ms. S. W. Harrisburg, and 71 N. W. C.

FOUNTAIN HEAD, p-o. in the eastern part of Sumner co. Ten. 34 ms. N. E. Nashville.

FOUNTAIN OF HEALTH, p-o. Davidson co. Ten. 6 ms. from Nashville.

FOUNTAIN INN and p-o. Chester co. Pa.

FOUNTAIN POWDER MILLS and p-o. Hart co. Ky. by p-r. 101 ms. S. W. Frankfort.

FOUNTAIN SPRING, p-v. Warren co. Ten. by p-r. 72 ms. S. E. by E. Nashville, and 659 S. W. by W. W. C.

FOURCHE A RENAULT, p-v. Washington co. Mo. by p-r. 128 ms. above and N. W. by W. Little Rock, and 921 ms. N. W. by W. ½ W. W. C.

FOUR CORNERS, p-v. Huron co. O. by p-r. 404 ms. N. W. by W. W. C. and 125 N. Columbus.

FOUR MILE BRANCH and p-o. Barnwell dist. S. C. by p-r. 90 ms. S. W. Columbia.

FOUR MILE PRAIRIE, p-o. Howard co. Mo. by p-r. 1,032 ms. W. W. C. and 214 ms. wstrd. St. Louis.

FOWLER, p-t. St. Lawrence co. N. Y. on the Oswegatchie, 36 ms. S. Ogdensburgh. Well watered and timbered, and contains many sites for water works. It contains iron ore, and other valuable minerals. Pop. 1830, 1,437.

FOWLER, p-v. and tsp. in the northern part of Trumbull co. O. The p-o. is by p-r. 12 ms. nthrd. from Warren, the co. st., 309 N. W. W. C. and 169 N. E. Columbus.

FOX RIVER, of Il. and Mich. the main nthrn. branch of Il., rises at lat. 43° 30′, between Rock r. and Manawakee r., and flowing thence by comparative courses s. s. w. 160 ms., falls into Il. at Otawa, 86 ms. s. w. Chicago. Fox r. of Il. receives no tributary streams of consequence; its valley lies between those of Rock r., the upper Il., and wstrd. from lake Michigan. That part of Fox r. which is comprised in Michigan or the higher part of its course, is nearly parallel, and from 20 to 25 ms. distant from the wstrn. shore of lake Michigan.

Fox River, confluent of Green bay, is composed of two main and numerous minor branches. The two principal branches are Fox r. proper, and Wolf r. Fox r. proper is that remarkable stream which derives its source from the level table land estrd. from the Ouisconsin. The two higher branches approach each other in nearly opposite directions, and uniting at fort Winnebago, leave a portage of less than 2 ms. from the navigable channel of the Ouisconsin. These higher constituents of Fox r. from their respective courses, seem to be natural tributaries of Ouisconsin, but reflowing, if we may use the expression, back upon their own courses, deflect to the northward, assume the name of Fox r., and continuing to the nthrd. 20 ms. dilate into Buffalo lake, and bend to the e. Buffalo lake, an intervening strait, and Puckawa lake, occupy 25 ms. of this curious r., which, leaving the latter, inflects abruptly to the wstrd., again estrd., and finally north estrd. The last course is maintained 55 ms. to the influx of Wolf r. from the nthrd. Wolf r. has its sources interlocking with those of Ontonagon of lake Superior. Flowing thence s. s. e. 120 ms. unites with and loses its name in Fox r. Below the mouth of Wolf r., Fox r. inflecting to s. e. dilates into Menomonie lake, and thence with a short intervening strait, opens into the comparatively large Winnebago lake. The latter sheet of water stretches from s. to n. 30 ms. with a breadth varying from 2 to 10 ms. Fox r. enters its western side near the middle, and issues from the northwestern angle. Below Winnebago lake Fox r. inflects again to the n. e., but with an elliptic curve to the estrd. 45 ms. comparative course to the head of Green bay at fort Howard.

The general course of Fox r. from the portage at fort Winnebago to its final efflux into Green bay, is a little e. of n e., and distance, by comparative course, 130 ms. Such is, however, the great and numerous inflections of the stream and lakes that the navigable distance perhaps exceeds 200 ms.

The valley of Fox r. lies between that of Green bay and the upper Ouisconsin. In fact Green bay is the continuation and lower depression of the Fox r. valley, and both contribute to form links of the navigable route from lake Mich. to the Miss. r. by the Ouisconsin. Including Wolf r. the Fox r. valley extends from lat. 43° 30′ to 46°, and in long. from 10° to 12° 12′ w. W. C.

Foxborough, p-t. Norfolk co. Mass. 24 ms. s. w. Boston. Well watered, and contains manufactories. Pop. 1830, 1,165.

Foxcroft, p-t. Penobscot co. Me. 35 ms. n. w. Bangor. Pop. 1830, 677.

Fox, p-v. Clearfield co. Pa. by p-r. 119 ms. n. w. Harrisburg.

Foxburg, p-v. at the extreme southern angle of Venango co. Pa. situated on the point between and above the junction of Alleghany and Clarion rs. about 100 ms. n. n. e. Pittsburg.

Fox's Creek and p-o. Lawrence co. Ala. by p-r. 129 ms. n. Tuscaloosa, and 758 s. w. by w. W. C.

Foxtown, p-o. Madison co. Ky. 6 ms. n. w. Richmond, the co. st., and by p-r. 543 s. w. by w. ½ w. W. C.

Foxville, p-v. Fauquier co. Va. by p-r. 56 ms. wstrd. W. C.

Foy's Store and p-o. Onslow co. N. C. by p-r. 410 ms. s. W. C. and 193 s. e. Raleigh.

Framingham, p-t. Middlesex co. Mass. 20 ms. w. Boston. Finely watered by Sudbury r. a branch of Concord r. The manufacturing business is here carried on on a large scale. Pop. 1830, 2,313.

Francestown, p-t. Hillsboro' co. N. H. 12 ms. from Amherst, 55 from Hanover, 27 from Concord, and 60 from Boston. It contains a quarry of free-stone. Very eligibly situated for business, being on the great thoroughfare from Windsor to Boston. Population 1830, 1,541.

Francisburg, p-v. Union co. Ky. by p-r. 204 ms. s. w. by w. Frankfort.

Franconia, p-t. Grafton co. N. H. 28 ms. from Haverhill, 74 n. Concord, and 140 from Boston. A large proportion of the town is mountainous. There is a singular natural curiosity in this town, called the Profile, situated on a peak about 1,000 feet high, presenting a front of solid rock, a side view of which exhibits a striking profile of the human face, every feature being conspicuous. The town contains an iron mine, said to be inexhaustible, yielding the richest supply in the U. S. There are two iron manufactories in the town; the ore of them is very extensive, manufacturing 12 to 15 tons per week. There is a highly impregnated mineral spring in the town. Pop. 1830, 447.

Franconia, p-v. and tsp. Montgomery co. Pa. The p-o. is 30 ms. n. w. Phil. The tsp. is on the s. side of the n. e. branch of Perkioming creek. Pop. 1820, 848, 1830, 998.

Frankford, p-v., tsp. and borough, Phil. co. Pa. 5 ms. n. e. from the central part of the city of Phil. Pop. of the tsp. 1820, 1,405, 1830, 1,633.

Frankford, p-v. near the right bank of Greenbriar r. Greenbriar co. Va. 12 ms. n. n. e. Lewisburg, by p-r. 257 ms. s. w. W. C.

Frankford, p-v. in the n. w. part of Pike co. Mo. 94 ms. n. w. St. Louis.

Frankfort, t. Sussex co. N. J. Pop. 1830, 1,996.

Frankfort, p-t. Waldo co. Me. w. Penob.

scot r., head of navigation, 26 ms. N. Castine, 12 S. Bangor. Pop. 1830, 2,487.

FRANKFORT, p-t. Herkimer co. N. Y. on the Erie canal, in which town there are 3 locks on said canal, terminating the level from Salina, a distance of 69½ miles without a lock. This town is situated S. of the Mohawk, 8 ms. W. Herkimer, and 86 W. N. W. Albany. Pop. 1830, 2,620.

FRANKFORT, p-v. Beaver co. Pa. 25 ms. a little N. of W. Pittsburg, and 30 a little W. of N. Washington, Pa.

FRANKFORT, p-v. near the right bank of Patterson's creek, northern part of Hampshire co. Va. 15 ms. S. Cumberland, Md. and by p-r. 119 ms. N. W. by. W. W. C.

FRANKFORT, p-t. st. jus. for Franklin co. and of government for the state of Ky. It is situated on the right bank of Ky. river, 24 ms. N. N. W. Lexington, 53 a little S. of E. Louisville, by the p-r. 86 S. S. W. Cincinnati, and 538 S. 83½ W. W. C. At seasons of high water steam boats are navigated to Frankfort, and the Ky. r. is navigable for down steamboats, to near 200 ms. following the stream above that town. In 1810, the population was 1,092, of whom 407 were slaves; in 1820, the aggregate was 1,679, of whom 643 were slaves; in 1830, the population amounted to 1,682. Lat. 38° 12′, long. 7° 52′ W. W. C.

FRANKFORT, p-v. and st. jus. Clinton co. Indiana, by p-r. 620 miles northwest by west ½ W. W. C. and 50 ms. N. N. W. Indianopolis. It is situated on the South Fork of Wild Cat river. Lat. 40° 20′, long. 9° 30′ W. W. C.

FRANKLIN, co. Vt. N. W. part of the state. Bounded N. by Lower Canada, E. Orleans co., S. Chittenden co.; W. Grand Isle co., from which it is separated by a part of lake Champlain, 34 ms. from E. to to W. and about 33 from N. to S., containing 730 square ms. Shire town, St. Albans, a place of considerable business. The Missisque river waters the N. part of this co. and the Lamoille the S. The E. part extends on to the western range of the Green mtns. and is high and broken, the W. part is generally level, and is a very fine farming country. It began to be settled immediately after the close of the revolution. Very fine marble is found in abundance in Swanton, and large quantities of iron ore in Highgate. Population 1820, 20,469, 1830, 24,525.

FRANKLIN, p-t. N. part Franklin co. Vt. 36 ms. N. E. Burlington, and 51 N. W. Montpelier. Much injured by a large pond near the centre. Pop. 1830, 1,129.

FRANKLIN, co. Mass. bounded N. by N. H., E. by Worcester co., S. by Hampshire co., and W. by Berkshire. Watered by Connecticut, Deerfield, and Miller rivers. Few tracts of country exceed this for the extent and value of its water powers. Shire town, Greenfield. Population 1820, 29, 268, 1830, 29,501.

FRANKLIN, p-t. Norfolk co. Mass. 26 miles S. W. Boston. Finely watered by Charles river and its branches, and is a flourishing manufacturing town. The Franklin cotton manufacturing company, was incorporated in 1813. Capital $200,000. In this town, is the private hospital of Dr. Nathaniel Miller, a distinguished physician and surgeon of that place. Here also resides that eminent divine, Rev. Nathaniel Emmons, D. D. Pop. 1830, 1,662.

FRANKLIN, p-t. New London co., Connecticut, 34 miles from Hartford. Diversified with hills and dales, best adapted to grazing. Watered by the Shetucket and a branch of the Yantic. 9 school districts. Population 1830, 1,196.

FRANKLIN, co. N. Y. on the N. line of the state, bounded N. by Lower Canada, E. by Clinton and Essex counties, S. by Essex and Hamilton, and W. by St. Lawrence co. The W. line is 60 miles long. Greatest breadth 30 ms. containing 1,506 square ms. In the S. W. part are some lofty ridges of the Peru mtns. the rest is rather level than hilly. Small streams numerous. A number of small lakes or ponds. Capable of being rendered a pretty good farming country. Here are mines of iron ore, and some indications of other metals. Chief town, Malone. Pop. 1820, 4,439, 1830, 11,312.

FRANKLIN, p-t. Delaware co. N. Y. on the Susquehannah, 13 miles N. W. Delhi. Surface broken, hilly, and mostly mountainous. Well watered, and reputed healthy. Has various kinds of manufactories. Population 1830, 2,786.

FRANKLIN, town, Somerset co. N. J. bounded by the Raritan and Millstone river, and south by the county of Middlesex. This t. includes the N. J. part of New Brunswick. Pop. 1830, 3,352.

FRANKLIN, town, Bergen co. N. J. on the N. Y. line, bounded N. W. by Pompton, S. E. by Harrington. Pop. 1830, 3,449.

FRANKLIN, town, Gloucester co. N. J. adjoins Salem co. E. of Woolwich and Greenwich. Pop. 1830, 1,574.

FRANKLIN, co. of Pa. bounded by Bedford W., Huntingdon N. W., Mifflin N., Perry and Cumberland N. E., Adams E., and Washington, Md. S. Length S. to N. 40 ms., mean breadth 18, and area 720 square ms. Extending in lat. from 39° 43′ to 40° 18′, and in long. from 0° 28′ to 1° 09′ W. W. C. This co. is bounded on the E. by the continuation in Pa. of the Blue Ridge, and is traversed in a N. N. E. direction by several chains of the Appalachian system. The slope of the southern and central parts are towards the Potomac, and down which flow the various branches of the Conecocheague and Antictam creeks, in nearly a southern direction. The Conedogwinnet rises in the northern section, and flows N. E. towards the Susquehannah. The great valley of the Conedogwinnet and Conecocheaque, for in reality these two streams flow in opposite directions along the same valley, has a limestone base towards the Blue Ridge, and one of clay slate towards the Kittatinny, or as there locally called, the

North or Blue mtn. Franklin co. may, with the exception of the N. E. and S. E. sections, be considered as nearly co-extensive with the valley of the Conecocheague. The soil, especially where resting on limestone strata, highly productive in grain, grasses, and fruit. Chief town, Chambersburg. Population 1820, 31,892, 1830, 35,103.

FRANKLIN, p-t. borough and st. jus. Venango co. Pa. situated on the right bank of Alleghany r. and French creek at their junction, 70 ms. N. Pittsburg, and about 65 a little E. of S. Erie. Lat. 41° 24', long. 2° 55' W. W. C. Pop. 1820, 252, 1830, 410.

FRANKLIN, tsp. Adams co. Pa. on Marsh creek, another of York co., on the head of Bermudian cr., another in the northern part of Huntingdon co., another in Westmoreland co. on the head waters of Poketon's and Turtle cr., and another in Green co. on the S. fork of Ten Mile cr.

FRANKLIN, co. of Va. bounded E. by Pittsylvania; S. E. and S. by Henry; S. W. by Patrick; by the the Blue Ridge which separates it from Montgomery W., and Botetourt N., and by Roanoke, r. which separates it from Bedford N. E., length 30, mean breadth 25, and area 750 sq. ms. Extending in lat. from 36° 46' to 37° 13', in long. from 2° 41' to 3° 18' W. W. C. The slope of this co. is to the E. and S. E., and down which flow, beside Roanoke r. Black Water, Pig, and Irwine rs. The elevation of surface, about equal to that of the adjoining co. of Bedford, or about 650 feet above the ocean tides. It comprises a part of the valley between Blue Ridge, and Turkey Cock mtn. and is moderately hilly, soil productive. Chief t. Rocky Mount. Pop. 1820, 12,017, 1830, 14,911.

FRANKLIN, p-t. and st. jus. Pendleton co. Va. on the middle branch of the south fork of Potomac, by p-r. 171 ms. S. W. by W. W. C. and 171 N. W. by W. Richmond. Lat. 38° 42', and long. 2° 26' W. W. C.

FRANKLIN, co. of N. C. bounded S. W. by Wake; N. W. by Granville; N. E. by Warren; and S. E. by Nash. Length 30, mean breadth 18, and area 540 sq. ms., lying between lat. 35° 49' and 36° 16', and long. 1° 02' to 1° 32' W. W. C. Tar r. enters it from the N. W. and flowing southeasterly, divides it into nearly equal sections. The entire slope of the co. is to the S. E.; chief t. Louisburg. Pop. 1820, 9,741, 1830, 10,665.

FRANKLIN, p-v. and st. jus. Haywood co. N. C. by p-r. 311 ms. westward Raleigh and 18 W. Waynesville.

FRANKLIN, co. of Geo. bounded S. E. by Elbert; S. by Madison; S. W. by Jackson; W. by Hall; N. W. by Habersham, and N. E. by Tugaloo r. separating it from Pickens dist. S. C. Length from S. W. to N. E. 34 ms. mean breadth 20 ms., and area 680 sq. ms. Extending in long. from 5° 50' to 6° 33' W. W. C. The slope of this co. is southeastward with the general course of Tugaloo r. The central and western sections are, however, drained by and are nearly commensurate with the higher branches of the north fork of Broad r. Chief t. Carnesville. Pop. 1830, 10,107.

FRANKLIN, p-v. Troup co. Geo. by p-r. 762 ms. S. W. W. C. and 143 W. Milledgeville.

FRANKLIN, one of the northwestern cos. of Ala. bounded N. by Ten. r. separating it from Lauderdale, E. by Lawrence co., S. E. by Walker, S. W. by Marion, and W. by the Chickasaw territory of Ala. Greatest length along the eastern boundary 38 ms. mean width 18, and area 684 sq. ms. extending in lat. from 34° 18', and in long. from 10° 36' to 11° 04' W. W. C. The southern part slopes to the northwestward, and is drained by the sources of Bear cr., the northern part falls to the N. towards Ten. r. Chief ts. Russellville, Tuscambia, and Bainbridge. Pop. 1830, 11,078.

FRANKLIN, p-v. in the northern part of Henry co. Ala. by p-r. 850 ms. S. W. W. C. and 238 S. E. Tuscaloosa.

FRANKLIN, co. of Miss. bounded W. by Adams', N. W. by Jefferson, N. E. by Copiah, E. by Lawrence, S. E. by Pike, S. by Amite, and S. W. by Wilkinson. Length from E. to W. 36 ms. mean breadth 20, area 720 sq. ms. lying between lat. 31° 22' and 31° 40', and long. 13° 40' and 14° 18' W. W. C. It is chiefly drained by the various branches of Homochitto r. though the extreme sources of Amite rise in the southeastern angle; soil near the water courses productive, but in the intervals barren pine woods, staple cotton, surface moderately hilly. Chief t. Meadville. Pop. 1820, 3,881, 1830, 4,622.

FRANKLIN, p-v. Yazoo co. Miss. by p-r. 1037 ms. S. W. by W. W. C. and 120 N. E. Natchez.

FRANKLIN, p-v. and st. jus. St Mary's parish, La. by p-r. 1344 ms. S. W. by W. W. C. and 141 a little S. of W. New Orleans. It is situated on the Teche r. 22 ms. above its mouth, lat. 29° 52', long. 14° 37' W. W. C.

FRANKLIN, co. of Ten. bounded W. by Lincoln, N. W. by Bedford, N. E. by Warren, E. and S. E. by Cumberland mtn. which separates it from Marion, and by Jackson co. in Ala., length from the Ala. line N. N. E. 42 ms., mean breadth 20, and area 840 sq. ms. lying between lat. 35° and 35° 34', and long. 8° 36' and 9° 21' W. W. C. Falling by a not very rapid declivity from Cumberland mtn., this co. discharges from its northern extremity the extreme sources of Duck r. flowing westward, whilst the central and southern parts are drained by the higher branches of Elk, Paint, Rock, and other streams, flowing to the S. W. and S. into Ten. r. Though broken, the soil is productive. Chief t. Winchester, the st. jus. near the centre, Metcalfboro', in the N. E. and Salem S. W. Pop. 1820, 16,571, 1830, 15,626.

FRANKLIN, p-v. and st. jus. Williamson co. Ten. by p-r. 732 ms. S. W. by W. W. C. and 18 ms. a little W. of S. Nashville. It is situated on Harpeth r. Lat. 35° 53', long. 9° 50' W. W. C.

FRANKLIN, co. of Ky. bounded W by Shelby,

N. W. by Henry, N. by Owen, E. by Scott, S. E. by Woodford, and S. W. by Anderson. Length 20, mean breadth 10, and area 200 sq. ms. lying between lat. 38° 06', and 38° 24', and long. 7° 42' and 7° 59' W. W. C. Ky. r. enters from the S. and traversing in a direction nearly N. divides this co. into two not very unequal sections. Elkhorn r. also traverses the E. part, entering from Scott and falling into Ky. r. about 8 ms. below Frankfort. In 1820, the population amounted to 11,024, but it then included what is now comprised in Anderson. In 1830, the pop. was 9,254.

FRANKLIN, p-v. and st. jus. Simpson co. Ky. situated on Drakes cr. branch of Big Banner r. and on the road from Nashville in Ten. to Bowling Green in Ky. 55 ms. a little E. of N. from the former and 22 a little W. of S. from the latter place, and by p-r. 162 ms. S. W. Frankfort. Lat. 36° 44', and long. 9° 29' W. from W. C. Pop. 1830, 280.

FRANKLIN, co. O. bounded S. by Pickaway, S. W. and W. by Madison, N. W. by Union, N. by Delaware, N. E. by Licking, and S. E. by Fairfield. Length 25, mean breadth 22, and area 550 sq. ms. Extending in lat. from 39° 37' to 40° 08' and in long. from 5° 44' to 6° 16' W. W. C. The slope is directly southward, and in that direction the co. is traversed by Whetstone and Sciota rs. which, entering on the northern border, and uniting between Franklin and Columbus, the combined waters, assuming the name of Sciota, continue south over the residue of the co. The eastern side is also traversed in a southern course by Big Walnut, as is the western by Darby cr. The soil is productive. Chief t. Columbus, st. jus. for the co. and capital of the state. Pop. 1820, 10,291, 1830, 14,741.

FRANKLIN, p-v. Warren co. O. (*See Franklinton, same co. and state.*)

FRANKLIN, co. of Ind. bounded by Dearborne S., Ripley S. W., Decatur W., Rush N. W., Fayette, and Union N., and Butler co. of O. E. Length 24, breadth 21, and area 504 sq. ms. Extending in lat. from 39° 17' to 39° 33', and in long. from 7° 52' to 8° 17' W. W. C. This co. is almost entirely comprised in the valley of White Water r. the main volume of which traverses it from N. W. to S. E. The surface rather hilly. Chief t. Brookville. Pop. 1820, 10,763, 1830, 10,190. The apparent decline in pop. of this co. ought to be explained. In 1820, it contained great part of what is now comprised in Fayette and Union counties.

FRANKLIN, p-v. and st. jus. Johnson co. Ind. by p-r. 20 ms. a little E. of S. Indianopolis, and 593 westward W. C. It is situated on a small tributary of the Driftwood fork of White r., lat. 39° 30', long. 9° 05' W. W. C.

FRANKLIN, co. of Il. bounded by Johnson S., Union S. W., Jackson W., Perry N. W., Jefferson N., Hamilton N. E., and Gallatin E., length from S. to N. 36, breath 24, and area 864 sq. ms. Extending in lat. from 37° 37' to 38° 08', and in long. from 11° 47' to 12° 14' W. W. C. The body of this co. is about equi-distant from the Miss. r. on the W., and the Wabash and Ohio on the east. It is a table land from which Muddy creek and its branches flow southwestward into the Miss., and the western confluents of Saline r. southeastward into Ohio r. Though rather level it has therefore two slopes, that on the western side towards the Miss. and that of the east towards the Ohio. Chief t. Frankfort. Pop. 1820, 1,763, 1830, 4,083.

FRANKLIN, p-v. and st. jus. Franklin co. Il. by p-r. 802 ms. a little S. of W. W. C. and 102 a little E. of S. Vandalia.

FRANKLIN, co. of Mo. bounded N. E. by St. Louis co., E. by Jefferson, S. E. by Washington, S. W. co. unknown, W. by Gasconade, N. W. by Missouri r. separating it from Montgomery, and N. by Missouri r. separating it from St. Charles co. Length 36 ms. mean breadth 30, and area 1080 sq. ms. Extending in lat. from 38° 10' to 38° 44', and in long. from 13° 44' to 14° 20' W. W. C. The slope of this co. is to the northeastward, and the surface about equally divided between the vallies of Maramac and Missouri. The southern and southeastern parts are drained by the former, whilst the northern section is drained by short creeks into Missouri. The surface is hilly. Chief town, Union. Pop. 1820, 2,379, 1830, 3,484.

FRANKLIN, p-v. Howard co. Mo. about 188 ms. a little N. of W. St. Louis. Neither distance nor relative position given in the P. O. list. This Franklin is a distinct p-o. from Old Franklin in the same co. (*See Old Franklin.*)

FRANKLIN, p-v. in the northwestern part of Oakland co. Mich. by p-r. 7 ms. N. W. Pontiac, the co. st. 33 N. W. Detroit, and 559 northwesterly from W. C.

FRANKLIN ACADEMY, and p-o. Upson co. Geo. by p-r. 82 ms. westward Milledgeville.

FRANKLINDALE, p-o. Bradford co. Pa. by p-r. 169 ms. N. Harrisburg.

FRANKLIN FURNACE, and p-o. Sciota co. O. by p-r. 106 ms. southward Columbus, and 434 westward W. C.

FRANKLIN MILLS, and p-o. Portage co. O. by p-r. 326 ms. northwesterly from W. C., and 133 N. E. by E. Columbus.

FRANKLIN SETTLEMENT, Chicot co. Ark. by p-r. 111 ms. S. E. Little Rock.

FRANKLIN SQUARE, and p-o. in the northern part of Columbiana co. Ohio, by p-r. 288 miles northwesterly from W. C. and 158 N. E. by E. Columbus.

FRANKLINTON, p-v. on Scota river, Franklin co. Ohio, situated directly opposite Columbus. Pop. 1830, 331.

FRANKLINTON, p-v. in Franklin tsp. N. W. angle of Warren co. Ohio, 11 miles N. N. W Lebanon, the co. st. and by p-r. 481 westrd. W. C. and 84 S. W. by W. Columbus. Pop. 1830, 584.

FRANKLINTOWN, p-v. and st. jus. parish of Washington, La. situated on the Bogue Chitto river, 60 miles a little W. of N. New Orleans, and 84 nearly due E. St. Francisville, lat. 30°

50′, in long. 13° 08′ west Washington City.

FRANKLINTOWN, p-v. in the northwestern part of York co. Pa. by p-r. 17 miles s. s. w. Harrisburg, and by direct road about 22 n. w. the borough of York.

FRANKLINVILLE, p-v. and st. jus. Lowndes co. Georgia by p-r. 829 miles s. s. w. ½ s. w. W. C. and 187 s. Milledgeville.

FRANKSTOWN BRANCH, northwestern constituent of the Juniatta, rises in the eastern slopes of the Alleghany chain, between the sources of Raystown branch and those of Bald Eagle creek. The sources of Frankstown are extended from the n. e. angle of Bedford, over the northwestern part of Huntingdon into Centre county. Flowing like vadii from the circumference to the center of a circle, the numerous branches of this river unite in Huntingdon co. near the village of Petersburg, where assuming a course of s. southeast passes the borough of Huntingdon and two miles below unites with Raystown branch and forms the Juniata. The two constituents of Juniata above their junction drain a space to 5 miles by 25, or 1,625 square miles, composed of lateral mtn. chains and narrow, but in many places highly productive vales. The immediate valley of Frankstown branch has become of great statistical importance from being a part of the route of the Pa. canal.

FRANKSTOWN, tsp. of Huntingdon co. Pa. on Frankstown branch. Pop. 1820, exclusive of the two villages of Frankstown and Holladaysburg, 1,297.

FRANKSTOWN, p-v. Huntingdon county, Pa. in Frankstown tsp. situated on the left bank of Frankstown river, 20 miles a little s. of w. from the borough of Huntingdon. It is at the village of Frankstown that the eastern section of transversed division of the Pennsylvania canal connects with the rail-road over the Alleghany mtn. The village or point of connection is 910 feet above the level of Atlantic tides, and the summit level of the rail road, has a similar relative elevation of 2,291 feet.

FRAZER, p-v. in the western part of Chester co. Pa. by p-r. 128 ms. n. e. W. C. and 74 eastward Harrisburg.

FREDERICA, p-v. Kent co. Delaware, 13 ms. s. Dover. Pop. 1820, 250.

FREDERICA, p-v. and sea port, on St. Simons islands, Glynn co. Georgia, 12 ms. s. Darien, and by p-r. 198 ms. s. e. Milledgeville.

FREDERICK, tsp. of Montgomery co. Pa. on the right side of Perkiomen creek, 9 ms. n. e. Pottstown. Pop. 1820, 927.

FREDERICK, co. of Maryland, bounded n. by Adams, and n. e. by York co. Pa., the s. e. mtn. forms its boundary from the mouth of Monocacy to the Pa. line, separating it on the e. from Baltimore, and on the s. e. from Ann Arundel and Montgomery cos. On the s. w. the Potomac river, between the s. e. mtn. and Blue Ridge, separates it from Loudon co. Va., and on the w. the Blue Ridge constitutes its line of separation from Washington co. Maryland. Length from s. w. to n. e. 42 ms., mean breadth 18, and area 776 square miles. Lying between lat. 39° 14′ and 39° 43′, and between 8′ e. and 39′ w. W. C. Except its higher sources in Adams co. in Pa. the valley of Monocacy is entirely in, and together with the small valley of Cotoctin creek, on the s. w. is commensurate with, Frederick co. in Maryland. A minor ridge, the Cotoctin mtn. detaches from the Blue Ridge, in the n. w. part of Frederick, stretches in a southwardly direction between the Cotoctin and Monocacy vallies, terminates near the Potomac river at the mouth of Monocacy creek. With the exception of the Cotoctin ridge, though Frederick is bounded on two sides by mtns., the surface is not even very hilly, and in places is level. The soil is generally fertile in grain, fruit and pasturage. It is one of the best cultivated sections of Md. Chief town, Frederick. Pop. 1820, 40,459, 1830, 45,793.

FREDERICK, city, p-t. and st. jus. Frederick co. Maryland, situated on the great western road from Baltimore, 47 miles westward from the latter, 2 ms. w. from Monocacy bridge, and by p-r. 44 ms. n. n. w. W. C. Lat. 39° 24′, long. 0° 24′ w. W. C. Pop. 1830, 7,255. Frederick is in size, wealth, and the elegance of its buildings, the second town of Maryland, and is increasing in all respects. The adjacent country is pleasant and well cultivated.

FREDERICK, co. Va. bounded on the n. w. by the Kittatinny chain, in part which separates it from Hampshire, on the n. w. it reaches Morgan, n. Berkley, n. e. Jefferson. The Blue Ridge separates it on the e. from Loudon, and on the s. e. from Fauquier. A direct line from the Blue Ridge to Shenandoah river, and thence up that stream to the mouth of Cedar creek, and along the n. e. border, the mean width 20, and 660 square miles. Lying between lat. 38° 50′, and 39° 25′, long. 0° 48′, and 1° 28′ w. W. C.

The surface of this county is very much diversified by hill, and mtn. scenery, and by diversity of soil. It occupies s. from the Potomac part of the continuation of the great valley, in which are situated Lebanon, the lower part of Dauphin, the greatest part of Cumberland and Franklin counties, Pennsylvania, and Jefferson and Berkley counties, Vaginia. The Shenandoah river traverses the southeastern border meandering along the northwestern base of the Blue Ridge. Opequan, Back and Sleepy creeks, flowing n. n. e. into the Potomac, also rise in Frederick. The slope of the county is of course northestrd. in a similar direction with the streams. The ground near Harpers Ferry and along the Potomac is about 200 feet above tide water, and allowing a similar rise from the Potomac, the mean height of Frederick would be about 400 feet. The soil of this county is highly productive, though the face of the county is considerably broken by mtn. ridges. Chief town, Winchester. Population 1820, 24,706, 1830, 26,046.

FREDERICSBURG, port of entry, p-t. and st. jus.

Spottsylvania county, Virginia, situated on the right bank of the Rappahannoc river, by p-r. 57 miles s. s. w. W. C. and 66 miles a little E. of N. Richmond. Lat. 38° 19′ long. 0° 28′ w. W. C. Placed at the head of tide water this is a very prosperous port; vessels of 140 tons can be navigated to the foot of the falls. The staples of domestic produce, grain, with its products, tobacco, &c. Pop. 1830, 3,308.

FREDERICKSBURG, p-v. on the left bank of Ohio river, in the N. E. angle of Gallatin co. Ky. about 40 miles directly N. Frankfort.

FREDERICKTON, (*see city of Frederick, Frederick co. Maryland.*)

FREDERICKSBURG, p-v. in the northern part of Holmes co. O. 8 ms. N. Millersberg, the co. seat, by p-r. 342 N. w. by w. W. C. and 95 northeastward from Columbus.

FREDERICKTOWN, p-v. on Monongahela r. below the mouth of Ten Mile creek, in the southeastern angle of Washington county, Pennsylvania, 22 miles s. E. Washington the co. seat, by p-r. 213 s. w. by w. W. C. and 206 a little s. of w. Harrisburg.

FREDERICKTOWN, p-v. on Ky. river, northwestern angle of Washington county, Ky. 8 ms. N. w. Springfield, the co. seat, by p-r. 610 miles s. w. by w. ½ w. W. C. and 59 s. w. Frankfort. Pop. 1830, 58.

FREDERICKTOWN, p-v. in Wayne tsp. northern part of Knox co. Ohio, 7 ms. a little w. of N. Mount Vernon, the co. seat, by p-r. 382 ms. N. w. by w. W. C. and 52 N. N. E. Columbus. Pop. 1830, 161.

FREDERICKTOWN, p-v. and st. jus. Madison co. Mo. by p-r. 894 ms. a little s. of w. W. C. 90 ms. s. St. Louis, and 40 ms. s. w. St. Genevieve. It is situated on one of the northwestern branches of St. Francis river, lat. 37° 32′, long. 13° 21′ w. W. C.

FREDONIA, p-v. Chatauque county, N. Y. (*See Pomfret.*)

FREDONIA, p-v. and st. jus. Crawford co. Indiana, by p-r. 632 ms. westward W. C. and 122 ms. s. s. w. Indianopolis.

FREDONIA, p-v. Montgomery co. Tenn. by p-r. 55 ms. N. w. Nashville.

FREEBURG, p-v. Union co. Pa. 10 ms. s. s. E. New Berlin, and by p-r. 48 ms. a little w. of N. Harrisburg.

FREEDINSBURG, p-v. Schuylkill co. Pa. 10 ms. s. w. Orwicsburg, and by p-r. 53 miles N. E. Harrisburg.

FREEDOM, p-t. Waldo co. Me. 28 miles N. w. Augusta. Pop. 1830, 867.

FREEDOM, p-v. Dutchess co. N. Y. 8 miles E. Poughkeepsie, well watered, and has a good supply of mill seats. A good tsp. of land, and highly cultivated. Has various kinds of manufactories.

FREEDOM, p-t. Cataraugus co. N. Y. 18 ms. N. E. Ellicottville. First rate as to soil, timber, and face of the country very level. Well watered. Pop. 1830, 1,505.

FREEDOM, p-v. in the northern part of Baltimore co. Md. about 30 ms. N. N. w. Baltimore, and by p-r. 63 ms. N. N. E. W. C.

FREEDOM, p-v. in the northwestern part of Portage co. Ohio, and in the tsp. of the same name. The p-v. 9 ms. N. w. Ravenna, the co. seat, by p-r. 328 N. w. W. C. and 141 N. E. Columbus. Pop. of the tsp. 1830, 341.

FREEHOLD, or MONMOUTH, p-t. and st. jus. Monmouth co. N. J. 20 ms. s. E. New Brunswick, 30 E. Trenton. Pop. 1830, 5,481. A battle was fought in this town June 28, 1778.

FREEHOLD, UPPER, town, Monmouth co. N. J. bounded N. E. by Freehold, N. w. by Middlesex co., w. by Burlington, s. E. by Dover. Pop. 1830, 4,826.

FREEMAN, p-t. Somerset co. Me. 38 ms. N. w. Norridgewock. Pop. 1830, 724.

FREEMANSBURG, p-v. in the southern part of Northampton co. Pa. by p-r. 187 ms. N. E. W. C. and 97 ms. a little N. of E. Harrisburg.

FREEMAN'S CREEK, and p-o. Lewis county, Virginia, by p-r. 249 ms. westward W. C.

FREEMAN'S STORE, and p-o. Jones co. Geo. 26 ms. westward Milledgeville.

FREEMAN'S STORE. (*See Green Hill, Jones co. Geo.*)

FREEPORT, p-t. Cumberland co. Me. head of Casco bay, 20 miles N. Portland. Pop. 1830, 2,623.

FREEPORT, town, Livingston co. N. Y. 10 ms. s. E. of Geneseo.

FREEPORT, p-v. Armstrong co. Pa. on the Alleghany river above the mouth of Buffalo creek, and about 2 miles below the mouth of Kiskiminitas river, 15 ms. by land below Kittatinny, and 25 above Pittsburg.

FREEPORT, p-v. in Freeport tsp. and in the southwestern part of Harrison co. Ohio, 12 ms. westward from Cadiz, the co. st. and by p-r. 297 s. w. by w. W. C. and 107 a little N. of E. Columbus. Pop. of the village, 1830, 211, and of the tsp. exclusive of the village, 980.

FREEMASON'S PATENT, N. Y. 5000 acres, granted June 12, 1771, then in Albany co. now in Oneida and Herkimer cos.

FREETOWN, p-t. Bristol co. Mass. 40 miles s. Boston, and 9 s. E. Taunton. Pop. 1830, 1,909.

FREETOWN, town, Cortlandt co. N. Y. about 9 ms. s. E. Homer, and 142 w. Albany. Pop. 1830, 1,054.

FRENCH BROAD, river of N. C. and Tenn. formed by two branches, French Broad and Nolachucky. The French Broad rises in the Blue Ridge at the extreme southern part of Buncombe co. N. C. Flowing thence in a northwardly direction 50 ms. comparative course, receiving tributary creeks from the Blue Ridge on one side, and a ridge of hills on the other, turns to N. w. and at the Warm Springs traverses the Bald mtn. and enters Tenn. Continuing the latter course 20 ms. receives Pigeon river from the left, and 5 ms. below joins the Nolachucky. Pigeon river branch rises in Haywood, N. C. and flowing by a course of N. N. w. 50 miles, also traversing the Bald mtn. joins the French Broad as already noticed. The Nolachucky, similar to the French Broad, derives its higher sources from the Blue Ridge, in the northern part of Buncombe county. The various branches traversing the mtn. valley, unite, and breaking through Bald mtn. enters Tenn. where assuming a western course over the southern angle of Washington, and separating Greene and Jefferson from Cocke joins the French Broad almost exactly on lat. 36°. The united stream, by a general western course, but with a sthrn. curve, after traversing Jefferson and Sevier cos. enters Knox, and falls into Holston river, 4 or 5 ms. above Knoxville, after an entire comparative course from the sources of French Broad of 120 ms. The whole valley drained by the various confluents of this stream forms a triangle of 90 ms. base and 70 perpendicular, area 3,150 square ms.

French Creek, a large branch of the Alleghany r. of the O. heads in Clymer, Chatauque co. N. Y. 10 ms. s. lake Erie.

French Creek, absurdly so called, rises in Chatauque co. N. Y. Flowing thence s. w. 20 ms. enters Erie co. Pa. where gradually curving to s. s. w. and s. receives the Cussawago at Meadville, and still curving, turns to s. e. and finally joins the Alleghany r. at Franklin, after a comparative course of 80 ms. having drained part of Chatauquo co. N. Y. the central parts of Erie and Crawford, with the nrthestrn. part of Mercer and the nrthwstrn. of Venango co. Pa. It is navigable into Erie co. within a few ms. from Waterford.

French Creek, and p-o. Lewis co. Va. by p-r. 267 ms. wstrd. from Richmond.

French Grant, p-v. Sciota co. O. by p-r. 420 ms. wstrd. W. C. and 111 sthrd. Columbus.

Frenchman's Bay, Me. between Mt. Desert isl. and the peninsula of Goldsboro, long. 68° w. lat. 44° 20′.

French Mills, v. (*See Fort Covington.*)

French's Mills, and p-o. Bradford co. Pa. by p-r. 268 ms. n. W. C. and 162 n. n. e. Harrisburg.

French's Mills, and p-o. Onslow co. N. C. by p-r. 405 ms. s. W. C. and 188 s. e. by e. Raleigh.

Friend's Grove, p-v. Charlotte co. Va. by p-r. 195 ms. s. s. w. W. C. and 104 s. w. by w. Richmond.

Frenchtown, p-v. Hunterdon co. N. J. on Del. r. 30 ms. above Trenton.

Friendship, t. Lincoln co. Me. 30 ms. e. Wiscasset. Pop. 1830, 634.

Friendship, p-t. Alleghany co. N. Y. 13 ms. s. w. Angelica. Pop. 1830, 1,502.

Friendship, p-v. Ann Arundel co. Md. by p-r. 40 ms. Annapolis.

Friendship, p-v. in the sthrn. part of Sumpter dist. S. C. by p-r. 501 ms. sthwstrd. W. C. and 64 s. e. Columbia.

Friendsville, p-v. in n. w. part of Susquehanna co. Pa. on the road from Montrose to Owego; 12 ms. n. w. from the former, and 25 s. e. from the latter, and by p-r. 166 n. n. e. Harrisburg.

Frog's Point. (*See Throg's Point.*)

Front Royal, p-v. in the extreme sthestrn. part of Frederick co. Va. 20 ms. a little e. of s. Winchester, and by p-r. 74 w. W. C.

Frostburg, p-v. Alleghany co. Md. situated on the united road, and on the Back-bone, or Alleghany mtn. at an elevation of 1792 feet above the Atlantic tides; 9 ms. w. and 1,155 feet above Cumberland on the Potomac, and by p-r. 145 ms. n. w. by w. W. C.

Frost Run, p-o. Lycoming co. Pa. 6 ms. above Williamsport, and 101 above Harrisburg by p-r.

Frost's Iron Works, and p-o. in the wstrn. part of Stokes co. N. C. by p-r. 6 ms. wstrd. from Germantown, the co.-seat, 361 s. w. W. C. and 133 n. w. by w. Raleigh.

Frostville, p-v. in the nrthwstrn. part of Cuyahoga co. O. by p-r. 368 ms. n. w. W. C. and 136 n. e. Columbus.

Frozen Run, p-v. formerly called Lycoming, in Lycoming tsp. Lycoming co. Pa. See Lycoming tsp.

Fruit Hill, p-v. Clearfield co. Pa. by p-r. 179 ms. n. w. Harrisburg.

Fruit's, p-v. in the nthrn. part of Callaway co. Mo. by p-r. 951 ms. w. W. C., 48 n. Jefferson, and 150 n. w. by w. St. Louis.

Fryeburg, p-t. Oxford co. Me. on the Saco, which here has a remarkable bend, winding for 36 ms. through the town. The village of Fryeburg stands on a plain, surrounded on all sides, except towards the south, by lofty mountains. It contains a flourishing academy, whose funds consist of 15,000 acres of land. It is 60 ms. n. w. of Portland, and 120 n. by e. of Boston. Pop. in 1830, 1,353.

Fryeburg, p-v. in the estrn. angle of Lehigh co. Pa. 12 ms. s. e. Allentown; 12 ms. s. Bethlehem; and by p-r. 102 ms. n. e. by e. Harrisburg.

Fulghampton, p-v. in the sthrn. part of Copiah co. Miss. about 50 ms. s. w. Jackson, and an equal distance n. e. by e. Natchez.

Fullwood's Store, and p-o. Mecklenburg co. by p-r. 152 ms. sthwstrd. from Raleigh.

Fulton, p-v. Rowan co. N. C. by p-r. 137 ms. wstrd. from Raleigh.

Fulton, p-o. Sumpter dist. S. C. on the road from Eutaw Springs to Statesburg, by p-r. 51 ms. s. e. Columbia.

Fulton, p-v. in the nrthestrn. part of Tipton co. Ten. by p-r. 218 ms. s. w. by w. Nashville.

Fulton, p-v. Hamilton co. O. 3 ms. estrd. Cincinnati, by p-r. 494 w. W. C. and 109 s. w. by w. Columbus.

Fulton, p-v. and st. jus. Callaway co. Mo. by p-r. 967 ms. w. W. C. 32 n. n. e. Jefferson, and 150 a little n. of w. St. Louis. Lat. 39° 11′ and long. 14° 52′ w. W. C.

Fulton, Fulton co. Il. (*See Lewistown, Fulton co. Il.*)

Fulton, co. of Il. bounded s. w. by Schuyler; w. by Macdonough; n. w. by Warren; n. by Knox; n. e. by Peoria; and s. e. by Illinois r. separating it from Tazewell. Length from s. to n. 33 ms. mean breadth 24, and area 792 sq. ms. Extending in lat. from 40° 12′ to 40° 42′ and in long. from 12° 50′ to 13° 25′ w. W. C. This co. is chiefly drained by the Kickapoo creek. The slope sthestrds. towards Illinois r. The latter stream, opposite to the upper part of the co. dilates into Peoria lake, a sheet of water upwards of 20 miles in length, with a breadth from ½ a mile to 2 miles. The face of the co. is diversified by hill and dale; soil highly fertile. Chief t. Peoria, called in the p-o. list, Fulton. Pop. 1830, 1,841.

FULTONHAM, p-v. in the wstrn. part of Muskingum co. O. by p-r. 345 N. W. by W. ½ W. W. C. and 55 E. Columbus.

FUNDY, BAY OF, sets up between Cape Sable in Nova Scotia, and Mt. Desert isl. Me. The tides at Cumberland, N. B. at the head of the bay, and at some other places, often rise to the height of 70 ft. in the spring; and from 30 to 60 ft. at other points along the bay.

FUNKSTOWN, p-v. on the left bank of Antietam creek, Washington co. Md. 3 ms. S. S. E. Hagerstown, and 22 N. W. from the city of Frederick.

G.

GADSDEN, co. of Flor. bounded S. by the Gulf of Mexico, by the Appalachicola r. which separates it from Washington W., Jackson N. W., N. by Decatur co. Geo. and on the E. the Ocklockonne r. which separates it from Leon co. Flor. Length from Cape St. George, including Appalachicola bay, 75 ms. mean breadth 28, and area 2,100 sq. ms. Chief town Quincy. Pop. 1830, 4,895. Extending in lat. from 29° 39′ to 30° 40′, and in long. from 7° 26′ to 8° 08′ w. W. C.

GAINES, p-t. Orleans co. N. Y. on the canal, 22 ms. N. Batavia. Soil, mostly a rich loam, watered by Otter, Marsh and Sandy creeks. Pop. 1830, 1,833.

GAINESBORO, p-v. Frederick co. Va. 13 ms. S. E. Winchester, and by p-r. 144 N. W. by W. W. C.

GAINESBORO, p-v. and st. jus. Jackson co. Ten. on the right bank of Cumberland r. 68 ms. N. E. by E. Nashville. Lat. 36° 24′, long. 8° 42′ w. W. C.

GAINESBURG, p-v. Dauphin co. Pa. by p-r. 22 ms. from Harrisburg.

GAINE'S CROSS ROADS, p-o. Culpepper co. Va. by p-r. 97 ms. southwesterly from W. C. and 128 N. N. W. Richmond.

GAINE'S CROSS ROADS, p-o. Boone co. Ky. by p-r. 68 ms. a little E. of N. Frankfort, and 25 S. S. W. Cincinnati.

GAINE'S STORE, and p-o. Pike co. Ala. by p-r. 934 ms. S. W. W. C. and 204 S. E. Tuscaloosa.

GAINESVILLE, p-t. Genesee co. N. Y. 28 ms. s. Batavia. Soil a loamy gravel, heavily timbered; bog iron ores. Pop. 1830, 1,934.

GAINESVILLE, p-v. and st. jus. Hall co. of Geo. 167 ms. a little W. of N. Milledgeville. Lat. 34° 22′, long. 6° 42′ w. W. C.

GALEN, t. Seneca co. N. Y. 12 ms. N. Waterloo. Contains about 70,000 acres. Soil principally good, excepting 4000 acres of marsh, being part of the great Cayuga marsh. The Erie canal passes through the whole extent of the town. There is a salt spring on the eastern margin, capable of supplying any quantity of good water.

GALENA, p-v. and st. of jus. Joe-Daviess co. Il. by p-r. 990 ms. from W. C., and 326 N. N. W. ½ N. Vandalia. According to Tanner it is situated on Fever r. 5 ms. above its influx into the Miss. r. lat. 42° 19′, long. 13° 22′ w. W. C. These relative geographical positions, give the bearing from W. C. to Galena N. 71¼° W. very nearly, and the direct distance 771 statute ms. within an inconsiderable fraction.

Galena derives its name from the abundant mines in its vicinity, of the Galena ore of lead, (*see Joe-Daviess co.*)

GALION, p-v. in the northwestern part of Richland co. O. by p-r. 398 ms. N. W. by. W. W. C., and 81 a little E. of N. Columbus.

GALLATIN, p-v. and st. jus. Sumner co. Ten. situated on the road from Nashville to Glasgow, in Ky. 31 ms. N. E. from the former, and 62 S. W. from the latter place. Lat. 36° 20′, and long. 9° 24′ w. W. C.

GALLATIN, co. of Ky. bounded by Grant co. E., Owen S. E., Henry S., Oldham S. W., and by Ohio r. which separates it from Jefferson in Ind. N. W., and from Switzerland Ind. N. The Ohio r. where it bounds Gallatin co. Ky. flows by a course of nearly W. and the greatest length of the co. lies parallel to that stream 36 ms.; mean breadth about 10, and area 360 sq. ms., lying between lat. 38° 33′ and 38° 50′, and long. 7° 45′ and 8° 25′ w. W. C. Ky. r. enters the southern border, and flowing thence N. W. falls into Ohio r. at the co. st. Port William. The surface is hilly but soil fertile. Pop. 1820, 7,075, in 1830, 6,674.

GALLATIN, p-v. and st. jus. Copiah co. Miss. situated on the southern branch of Bayou Pierre 65 ms. N. E. by E. Natchez, and 40 ms. S. W. Jackson. Lat. 31° 51′, long. 13° 35′.

GALLATIN, p-v. in the western part of Parke co. Ind. by p-r. 639 ms. N. W. by W. ½ W. W. C. and 77 W. Indianopolis.

GALLATIN, co. of Il. bounded S. by Pope, W. by Franklin, N. W. by Hamilton, N. E. by White, N. E. by E. by the lower part of Wabash r. separating it from Posey co. of Ind., E. by Ohio r. separating it from Union co. Ky., and S. E. by the Ohio r. separating it from Livingston co. Ky. Greatest length from south to north 40 ms., mean breadth 22, and area 792 sq. ms. Extending in lat. from 37° 27′ to 38°, and in long. from 11° 08′ to 11° 48′ w. W. C.

This large co. is a natural section in itself being nearly commensurate with the valley of Saline river. The slope is to the s. e. in the general direction of Saline r. though the base of its plain, the channels of Wabash and Ohio rivers, extend from north to south very nearly; as the Ohio opposite Gallatin co. of Il. continues the course of lower Wabash. Saline r. derives its name from extensive springs of water, impregnated with common culinary salt (muriate of soda,) which is manufactured in considerable quantities in this co. Chief town, Shawneetown. Pop. 1830, 7,405.

Gallatin's r. one of the extreme southwestern sources of Missouri proper. This stream rises in one of the Chippewayan vallies, about lat. 44°, long. 32° w. W. C., and interlocking sources with those of Yellow Stone r. on the east, Madison's river on the west, and with a mountain chain intervening, opposite to those of Lewis r. Flowing northwards unites with Madison's and Jefferson's rs. to form Missouri. The very elevated valley from which issue these remote fountains of Missouri, is one of those interesting mountain basins, of which Bohemia in Europe, and Mexico in North America, are striking examples. (*See Missouri r.*)

Gallia, co. of O. bounded s. w. by Lawrence, w. by Jackson, n. w. by Athens, n. by Meigs, and n. e., e. and s. e. by Mason co. Va., from which it is separated by Ohio r. Length from south to north 30, mean width 16, and area 480 sq. ms. Extending in lat. from 38° 34′ to 39°, and in long. from 5° 07′ to 5° 30′ w. W. C. This co. lies opposite to the mouth of Great Kenhawa, is traversed by Raccoon creek, and though bounded by the Ohio r. on the east, the slope is parallel to and not towards that stream. The surface of the whole co. with but little exception is broken, the soil various, but in part very productive. Chief town, Gallipolis. Pop. 1820, 7,098, 1830, 9,733.

Gallipolis, p-v. and st. jus. Gallia co. O. by p-r. 362 ms. w W. C. and 108 s. e. Columbus. It is situated on a rather elevated second bottom, on the right bank of Ohio r. nearly opposite Point Pleasant in Mason co. Va., at the mouth of Great Kenhaway. This place and Gallia co. were named by the original French settlers under M. D'Hebecourt. This small colony were fixed on Ohio, about the beginning of the French revolution. The place now contains, according to Flint, a court house, jail, two places of public worship, an academy, three steam mills, one printing office, 80 houses, and 12 mercantile stores. Pop. 1830, 755. Lat. 38° 51′, long. 5° 11′ w. W. C.

Gallivemts Ferry, over Little Pedee, and p-o. in the northwestern part of Hony district S. C. 22 ms. n. w. Conwaybro, and by p-r. 138 ms a little n. of e. Columbia.

Galloway, t. Gloucester co. N. J. Pop. 1830, 2,960.

Galveston, small village of La. in the parish of Iberville, situated on the right bank of Amite r. immediately below the mouth of bayou Iberville, about 20 ms. n. n. e. Donaldsonville, and 25 southeastward Baton Rouge.

Galway, p-t. Saratoga co. N. Y. 10 ms. n. w. Ballston Spa, a good tsp. for agriculture. Gypsum is said to have been discovered in a state of solution in a spring in this town. Pop. 1830, 2,710.

Gamage's, p-o. Bibb co. Ala. by p-r. 38 ms. eastward from Tuscaloosa.

Gambier, p-v. in Pleasant tsp. Knox co. O. by p-r. 5 ms. e. Mount Vernon, the co. st. 370 ms. n. w. by w. W. C. and 50 n. e. Columbus. Pop. 1830, 220.

Gamble's Mills, and p-o. in the n. w. angle of Richland co. O. by p-r. 12 ms. northwards from Mansfield, the co. st. 398 n. w. by w. W. C. and 81 n. n. e. Columbus.

Gamble's p-o. Alleghany co. Pa. by p-r. 223 ms. n. w. W. C.

Gandy's p-o. Morgan co. Ala. by p-r. 104 ms. n. n. e. Tuscaloosa.

Gap, p-v. on the eastern border of Lancaster co. Pa. and on the Phil. road 16 ms. s. e. by e. Lancaster and 48 ms. w. Philadelphia.

Gardner, p-t. Kennebec co. Me. w. Kennebec r. Contains a flourishing Lyceum. Well situated for manufactures, 6 ms. s. Augusta. Pop. 1830, 3,709.

Gardner's Bay and Island, end of Long Island, N. Y. celebrated for its dairies. From 6,000 to 7,000 weight of cheese are made annually. The neat profits of the farm average about $5,000.

Gardner, p-t. Worcester co. Mass. 25 ms. n. Worcester, and 58 n. w. Boston. Face of the town uneven. Soil good for grass, and most other products. Pop. 1830, 1,023.

Gardner's Bridge, and p-o. Martin co. N. C. by p-r. 140 ms. eastward from Raleigh.

Gardner's Cross Roads, and p-o. Louisa co. Va. by p-r. 72 ms n. w. Richmond, and 101 s. w. W. C.

Gardner's Store, and p-o. Randolph co. N. C. by p-r. 51 ms. westward from Raleigh.

Gardner's Tavern, and p-o. Hanover co. Va. 21 ms. from Richmond.

Garland, p-t. Penobscot co. Me. 28 ms. n. w. Bangor. Pop. 1830, 621.

Garner's Ford, and p-o. Rutherford co. N. C. by p-r. 222 ms. s. w. by w. Raleigh.

Garnet, p-v. in the southern part of Henry co. Ind. by p-r. 530 ms. n. w. by w. ½ w. W. C. and 44 a little n. of e. Indianopolis.

Garoga cr. rises in Johnstown, Montgomery co. N. Y. and runs s. w. about 20 ms. to the Mohawk, and is a fine mill stream.

Garrard, co. of Ky. bounded by Rockcastle n. e., Lincoln s. w., Mercer n. w., Ky. r. which divides it from Jessamine n., and by Madison n. e. Length 30, mean breadth 8, and area, 240 sq. ms. Lying between 37° 28′ and 37° 52′, and long. 7° 16′ and 7° 42′ w. W. C. It is composed of the space between Dicks r. and Paint Lick creek, and is a highly productive tract. Chief town, Lancaster. Pop. 1820, 10,851, 1830, 11,871.

Garrettsville, p-v. Portage co. O.

Garwood's Mill, and p-o. southeastern part of Logan co. Ky. by p-r. 10 ms. southeastward Bellefontaine, the co. st. 448 n. w. by w. W. C., and 52 n. w. by w. Columbus.

Gasconade r. confluent of Mo. in the state of Mo. rises interlocking sources with those of the southeastern branches of Osage r. and the numerous branches of White, Black, and Maramec rs. The extreme sources of Gasconade are in Wayne co., but flowing thence towards the n. e. enters Gasconade co. which it traverses to its entrance into Mo. r. at the village of Gasconade, after a comparative course of 140 ms. The valley of this r. lies between lat. 37° and 38° 40′, and between long. 14° 25′ and 16° 0′ w. W. C.

Gasconade, co. of Mo. bounded e. by Franklin, s. and s. w. by counties unknown, n. w. by Osage r. separating it from Cole co., n. w. by Mo. r. separating it from Calaway co. and Mo. r. on the n. e. separating it from Motgomery co. Length from e. to w. 45, mean breadth 28, area 1,260 sq. ms. Extending in lat. from 38° 67′ to 38° 32′, and in long. from 14° 20′ to 15° 08′ w. W. C. This co. is traversed and bisected into two very nearly equal sections by Gasconade r. the general slope being to the n. eastward, toward Mo. r. The southeastern angle giving source to the Bourbeun branch of the Maramec. Chief town, Gasconade. Pop. 1830, 1, 545.

Gasconade, p-v. and st. jus. Gasconade co. Mo., is situated on the point above the junction of Gasconade with Mo. r. 80 ms. a little n. of w. from St. Louis, and by p-r. 47 ms. a little n. of e. Jefferson. Lat. 38° 40′, long. 14° 32′ w. from W. C.

Gassaway's Mills, and p-o. in the eastern part of Monroe co. O. by p-r. 304 ms. westward W. C., and 150 eastward Columbus.

Gates, t. st. jus. Monroe co. N. Y. 236 ms. w. n. w. Albany. The post borough of Rochester, the seat of the co. buildings, is in this town. The Erie canal extends e. and w. through this twp., crossing the Genesee r. at the Falls by a stone aqueduct. Pop. 1830, 1,631.

Gates, county of N. C. bounded east by Pasquotank, southeast by Perquimans, by Chowan or Mohcrin r. which separates it from Hertford s., and Northampton s. w., by Southampton co. Va. n. w., and Nansemond n., lying between lat. 36° 18′ and 36° 30′, and between long. 10 minutes w. and 20 e. from W. C. Pop. 1820, 6,837, in 1830, 7,866. Gates co. is well situated for navigation, as besides being bordered by Chowan r. that stream receives within the co. the united waters of Nottaway and Black water.

Gates, C. H. and p-o. in the forks of Bennett's cr. Gates co. N. C. about 45 ms. s. w. Norfolk in Va. and by p-r. 214 ms. n. e. by e. Raleigh. Lat. 36° 25′, long. 0° 12′ e. W. C.

Gatesville, p-v. and st. jus. Gates co. N. C. by p-r. 254 ms. s. W. C. and 141 n. e. by e. Raleigh. It is situated on the main road from Raleigh to Norfolk in Virginia, and at the forks of Bennett's cr. a small branch of Chowan r. lat. 36° 23′, long. 0° 14′ e. W. C.

Gauley, river of Virginia, rises in Randolph, Pocahontas, and Green Briar counties, by numerous creeks which unite in Nicholas, and flow by a course of a little s. of w. falling into the right side of the Great Kenhawa river, at the head of the Great Falls. The valley of Gauley river is about 60 miles long, and lies between those of Elk and Green Briar rivers.

Gauley Bridge, and p-o. on the Great Kenhawa r. Kenhawa co. Va. by p-r. 278 ms. a little n. of w. Richmond and 344 s. w. by w. W. C.

Gebharts, p-o. in the western part of Somerset co. Pa. by p-r. 175 ms. n. w. from W. C.

Geiger's Mills, and p-o. in the southeastern part of Berks co. Pa. by p-r. 138 ms. n. n. e. W. C. and 63 eastward Harrisburg.

Gelostee, p-v. Kalamazoo co. Michigan, about 140 ms. w. Detroit.

General Pike, p-o. late Phoenixville, Chester co. Pa. by p-r. 132 ms. n. e. W. C. and 77 s. e. by e. Harrisburg.

Genesee, co. N. Y. bounded by lake Ontario n., by Monroe and Livingston counties e., s. by Alleghany and Cattaraugus, and w. by Erie and Niagara cos. Extreme length n. and s. 54 ms., extreme width 29½, containing 1,280 square ms. or 819,200 acres. Lands heavily timbered, on the lake Erie table land. Soil loamy or gravelly, good for wheat and grass. Its surface undulating, extensive champaigns, small swells, and broad vallies. Limestone, iron ore, water lime, salt springs, and various clays are among its mineral productions. It produces vast quantities of maple sugar.

A tract about 40 miles wide, along Genesee river, is the best land in all the state, and equal to any wheat country in the world, and yet till within about 20 years, was unknown as such. Multitudes of New Englanders passed it, and went to settle on cold poor land, where 30 bushels of corn and potatoes only, satisfied them. At length a man named Rogers made an experiment on the alluvial Genesee land with wheat, and raised immense crops. Now it yields 60 bushels of corn to the acre, 25 of wheat; the latter at an expense (rent included) of only 33 and sometime 25 per cent.

Ancient mounds are numerous, mere burying grounds, where bones are found in heaps, much decayed, laid horizontally. Indian skeletons are often washed out of the banks in sitting postures, with implements, &c. Broaches and crosses are sometimes found, which were brought from Canada. Consumption is not known in this region; but cutaneous disorders abound. There are no manufactories except flour mills, &c. Threshing is done by machines, which cost from $80 to $150 each. Population 1820, 39,835, 1830, 51,992.

Genesee, river, rises on the great table land, or *Grand Plateau* of Western Pennsyl.

vania, runs N. across the western part of N. Y. and empties into lake Ontario. Near its mouth, at Carthage, there are falls of 75 feet, and at Rochester, just above, of 96 feet, and some rapids for 2 miles further, from the head of which, the feeder leads into the Erie canal. In the town of Nunda, at the N. end of Alleghany co. are two other falls near each other, of 60 and 90 feet. At the falls at Rochester, the notorious Sam Patch lost his life. It was here, that he made his "last leap," which proved fatal to him.

GENESEO, p-t. st. jus. Livingston co. N. Y. 27 miles S. S. W. Rochester, and 238 from Albany. Surface undulating. In Fall brook is a cascade nearly 100 feet, almost perpendicular. Pop. 1830, 2,675.

GENEVA, village and p-o. (*See Seneca.*)

GENEVA, p-v. and tsp. in the western part of Ashtabula co. Ohio, by p-r. 348 ms. N. W. W. C. and 180 ms. N. E. Columbus. Pop. of the tsp. 1830, 771.

GENEVA, p-v. in the northwestern part of the co. by p-r. 585 ms. W. W. C. and 53 S. E. Indianopolis.

GENITO, p-v. on the left bank of Appamattox river, in the southeastern part of Powhatan, co. Virginia, by p-r. 34 ms. S. W. by W. Richmond.

GENOA, p-v. and tsp. Delaware co. O. The p-v. is in the southern part of the county, 17 ms. northward from Columbus, and 392 ms. N. W. by W. W. C. Pop. of the township, 1830, 659.

GENOA, p-t. Cayuga co. N. Y. 20 miles S. Auburn, and 185 from Albany. Surface gently uneven. Soil remarkably fertile. Well supplied with mill seats. Pop. 1830, 2,768.

GENTRY'S STORE, and p-o. Spencer co. Indiana, about 160 ms. a little W. of S. Indianopolis.

GENTSVILLE, p-o. Abbeville district, S. C. by p-r. 102 ms. a little N. of W. Columbia, and 536 S. W. W. C.

GEORGES STORE, and p-o. Pike co. Ala. by p-r. 168 ms. S. E. Tuscaloosa.

GEORGESVILLE, p-v. Yazoo co. Miss. by p-r. 81 ms. northward from Jackson.

GEORGESVILLE, p-v. in Pleasant tsp. and in the southwestern angle of Franklin co. Ohio, 13 ms. S. W. Columbus, and by p-r. 409 S. W. by W. ½ W. W. C. Pop. 1830, 39.

GEORGETOWN, town, Lincoln co. Me. at the mouth of the Kennebec, 15 ms. S. W. Wiscasset. Pop. 1830, 1,258.

GEORGETOWN, p-t. Madison co. N. Y. 12 ms. S. W. Morrisville, and 106 W. Albany, good for grazing, &c. Pop. 1830, 1,094.

GEORGETOWN, p-v. on the left bank of the Ohio river, immediately above the mouth of Mill creek, and of the Virginia line, Beaver co. Pa. 35 ms. by land W. N. W. Pittsburg.

GEORGETOWN, p-v. and st. jus. Sussex co. Delaware, on the height of land between the sources of Nanticoke and Indian rivers, 37 ms. a little E. of S. Dover. Lat. 38° 43', long. 1° 37' E. W. C.

GEORGETOWN, port of entry, and p-t. at the head of the tide, and on the left bank of Potomac river, Washington co. Dist. Columbia. It extends in length along the Potomac, and in breadth up Rock creek, rising by a bold acclivity from both streams. The Chesapeake and Ohio canal passes through this town. It is the seat of a Roman Catholic college, and of considerable commerce. The progressive population is shown by the subjoined table. (*See article District Columbia.*)

	Whites,	Free Col'd,	Slaves,	Total
1810,	3,235	551	1,162	4,948
1820,	4,940	894	1,526	7,360
1830,	6,057	1,209	1,175	8,441

GEORGETOWN, district of S. C. bounded by Santee river, which separates it from Charlestown dist. S. and S. W., by Williamsburgh dist. W. and N. W., by Horry dist. N. and N. E., and by the Atlantic E. and S. E. Length nearly parallel to the ocean 40 miles, mean breadth 26, and area 1,040 square ms. Lying between lat. 33° 05' and 33° 46', and long. 2° 13' and 2° 50' W. W. C. The surface of this district is a plain, in many places marshy, but much of the river soil is very productive, and as it is in a remarkable manner traversed by rivers, the commercial advantages are extensive. Beside the outlets of Santee river, Winyau bay is the estuary of Waccamaw, Great Pedee, and Black rs. All those confluents of Winyau, unite at, or near Georgetown, the mart and st. jus. of the district. Rice and cotton are the principal staples. Population 1820, 17,603, 1830, 19,943.

GEORGETOWN, p-t. port of entry and st. jus. Georgetown dist. S. C. situated on the point above the junction of Sampit creek and Pedee r.; the body of the town is, however, on the former. About 3 ms. above, the Port Pedee receives Black r. from the W., and directly opposite, and E. from the harbor. Waccamaw comes in from the N. N. E. Vessels of 11 feet draught are admitted over the bar of Pedee and up to Georgetown. Having an extensive and well cultivated interior, Gorgetown carries on an extensive commerce. Lat. 33° 21', and long. 2° 22' W. W. C. Georgetown is distant by the road 70 ms. N. E. Charleston, by p-r. 151 ms. S. E. by E. Columbia, and 480 S. S. W. W. C.

GEORGETOWN, p-v. and st. jus. Scott co. Ky. situated on the N. branch of Licking r. by p-r. 20 ms. E. Frankfort. It contains, besides the ordinary co. buildings, a bank, printing office, and several places of public worship. Lat. 38° 14', and long. 7° 31' W. W. C.

GEORGETOWN, p-v. Copiah co. Miss. about 45 ms. a little N. of E. Natchez, and by p-r. 57 S. S. W. Jackson.

GEORGETOWN CROSS ROADS and p-o. in the N. E. part of Kent co. Md. 15 ms. N. E. Chestertown, and 40 N. N. E. Baltimore.

GEORGETOWN, p-v. and st. jus. Brown co.

O. by p-r. 480 ms. w. W. C., 104 s. s. w. Columbus, and 45 s. e. by e. Cincinnati. It is situated on White Oak creek, and in the southwestern part of the co. Lat. 38° 53′, long. 6° 51′ w. W. C. Population 1830, 325.

Georgia, p-t. Franklin co. Vt. 18 ms. n. Burlington, and 41 n. w. Montpelier, being situated on lake Champlain. Mill privileges numerous. The soil is in general rich and productive. There is a natural bridge over one of the streams. Population 1830, 1,897.

Georgia, one of the U. S., bounded s. and s. w. by Flor., w. by Ala., n. w. by Ten., n. by N. C., n. e. and e. by S. C., and s. e. by the Atlantic. The greatest line that can be drawn in Geo. is from the mouth of St. Mary's r. to the n. w. angle of the state, in a direction n., 40° 41′ w.; 394 statute ms., and carefully measured by the rhomb, the area is found 62,083 sq. ms., therefore the mean breadth is 157½ ms. very nearly. Extending in lat. from 30° 20′ at the extreme southern bend of St. Mary's r. to 35 n. on the Ten. and N. C. line. In long. it extends from 3° 57′, at the mouth of Savannah r., to 8° 42′ w. W. C. at the northwestern angle of Ten. Georgia occupies the great inclined plain, from which the peninsula of Florida is protruded, and from which, on the s. e. the rivers run into the Atlantic, and s. w. into the gulf of Mexico. From the southern border of Geo. this great plain rises by a gradual acclivity from the inundated Atlantic border, to at least 1,200 feet elevation above the tides, without estimating the ridges of mountains. The difference of height being equivalent to 3 degrees of temperature, and the lat. difference amounting to 4° 40′; the whole extreme of temperature included in Geo. is 7°⅔ Fahrenheit. It is found both from vegetable life and from experiments made with the thermometer, that the seasons on the Atlantic coast have at least two degrees higher temperature than those of places on equal height, and the same latitude in the basin of Miss. From all the preceding causes the state of Georgia presents a very marked variety of seasons and of vegetable production. In both these respects, the latter, however, arising from the former, gives to the state a range of vegetable existence wider than that of any other state of the U. S. It is physically divided, like the two Carolinas, into three zones. First, the flat sea border, including numerous small islands; second, the sand hill zone, spreading by an indefinite outline between the sea border, and the third, a hilly and part mountainous tract, beyond the lower falls of the rivers. The sea sand alluvial border, in part diurnally inundated by the ocean tides, with some fertile, but much sterile soil, may be called the tropical climate of Georgia. Here, along the streams, the season of summer is sufficiently long and warm to mature the sugar cane, orange, olive, date, palm, and many other tender plants. The second, or sand hill region, with equal diversity of soil, produces maize and cotton, as the most valuable staples. But the third, the hilly and mountainous section, abounding in excellent soil, pure fountain water, and a more salubrious air, is for human residence, much the finest part of the state. Here the bread gracus, the apple, peach and plum, the green pasture, and rich meadows in summer and autumn, and in winter the denuded forest, announce a climate of northern texture. The description of these zones must, nevertheless, be taken as general; where they separate, the features are so blended as to defy exact demarcation, but on the other hand, if we assume the two extremes, the contrast is indeed strongly marked. No two regions could, in every physical feature, differ much more essentially than does the low, flooded, bilious Atlantic border, cut by the St. Mary's, Santilla, Altamaha, and Ogeechee rs., and the elevated, broken, rocky tract, from which are poured the clear and pure confluents of the Coosa and Ten. One very remarkable circumstance in the climates of the southern sections of the U. S. may be here appropriately noticed. Sweet oranges are reared on the Atlantic coast as high as Beaufort dist. in S. C. or to 32° 30′ n., and the fan palm, and live oak grow as indigenous vegetables as far as the mouth of Cape Fear r. in N. C., lat. 34°. On the Miss. the live oak ceases below 30° 30′ n., and the fan palm (palmetto) at about 31° n. Sugar cane cannot be cultivated to advantage in La. above lat. 30° 30′, whilst that plant flourishes along the entire sea border of Georgia into S. C. In brief, it may be stated that at one extremity Georgia produces wheat, and at the other sugar, and taking the whole state, amongst many more staples of less value, we may enumerate sugar, rice, indigo, tobacco, cotton, wheat, rye, oats, and maize. The range of garden vegetables is also immense. Such are the natural advantages possessed by this extensive state, that its advance in wealth and population since the American revolution has been so rapid, that of the Atlantic states, it has been second only to N. Y. in relative progression. Though from many causes, seminaries of education have languished in the southern states, this primary object of human policy has met with considerable attention in Georgia. Franklin college, at Athens, Clark co., is the incipient step towards a projected university. A branch of their plan of instruction was to have an academy in every co. This has in part been affected; but necessarily remains imperfect in the recently settled cos., many of which were not designated at the taking of the census for 1820.

Table of the free, slave, and aggregate population of the counties and state of Georgia, from the abstract of the returns of the census for 1830, to which is annexed the population of the counties and state in 1820.

Counties.	Free.	Slaves.	total 1830.	1820.
Appling,	1,289	179	1,468	1,264
Baker,	978	275	1,253	
Baldwin,	2,753	4,542	7,295	7,734
Bibb,	4,166	2,988	7,154	
Bryan,	737	2,402	3,139	3,021
Bullock,	1,937	650	2,587	2,578
Burke,	5,191	6,642	11,833	11,577
Butts,	3,261	1,683	4,944	
Camden,	1,492	3,086	4,578	4,342
Campbell,	2,705	618	3,323	-
Carroll,	2,932	487	3,419	
Chatham,	4,649	9,478	14,127	14,737
Clarke,	5,467	4,709	10,176	8,767
Columbia,	4,574	8,032	12,606	12,695
Coweta,	3,631	1,372	5 003	
Crawford,	3,595	1,718	5,313	
Decatur,	2,546	1,308	3,854	
De Kalb,	8,394	1,648	10,042	
Dooly,	1,799	336	2,135	
Early,	1,511	540	2,051	768
Effingham,	1,712	1,212	2,924	3,018
Elbert,	6,589	5,765	12,354	11,788
Emmanuel,	2,208	465	2,673	2,928
Fayette,	4,317	1,187	5,504	
Franklin,	7,737	2,370	10,107	9,040
Glynn,	799	3,968	4,567	3,418
Greene,	5,079	7,470	12,549	13,589
Gwinnett,	10,957	2,332	13,289	4,589
Habersham,	9,762	909	10,671	3,145
Hall,	10,567	1,181	11,748	5,086
Hancock,	4,640	7,180	11,820	12,734
Harris,	2,836	2,969	5,005	
Henry,	7,995	2,571	10,506	
Houston,	5,175	2,194	7,369	
Irwin,	1,071	109	1,180	411
Jackson,	6,221	2,783	9,004	8,355
Jasper,	6,809	6,322	13,131	14,614
Jefferson,	3,662	3,647	7,309	7,056
Jones,	6,516	6,829	13,345	16,560
Laurens,	3,214	2,375	5,589	5,436
Lee,	1,369	311	1,680	
Liberty,	1,609	5,624	7,233	6,695
Lincoln,	2,869	3,276	6,145	6,458
Lowndes,	2,118	335	2,453	
Madison,	3,387	1,259	4,646	3,735
Mac Intosh,	1,204	3,794	4,998	5,129
Marion,	1,327	109	1,436	
Merriwether,	3,028	1,394	4,422	
Monroe,	8,849	7,353	16,202	
Montgomery,	934	335	1,269	1,862
Morgan,	5,226	6,820	12,046	13,520
Muscogee,	2,263	1,240	3,508	
Newton,	8,152	3,003	11,155	
Oglethorpe,	5,670	7,940	13,618	14,046
Pike,	4,376	1,773	6,149	
Pulaski,	3,141	1,765	4,906	5,283
Putnam,	5,554	7,707	13,261	15,475
Rabun,	2,117	59	2,176	524
Randolph,	1,509	682	2,191	
Richmond,	5,398	6,246	11,644	8,608
Scriven,	2,410	2,366	4,776	3,941
Talbot,	3,841	2,099	5,940	
Taliaferro,	2,199	2,735	4,934	
Tatnall,	1,534	506	2,040	2,644
Telfair,	1,571	565	2,136	2,104
Thomas,	2,131	1,168	3,299	
Troup,	3,611	2,188	5,799	
Twiggs,	4,524	3,507	8,031	10,640
Upson,	4,456	2,557	7,013	
Walton,	7,766	3,163	10,929	4,192
Ware,	1,144	61	1,205	
Warren,	6,253	4,693	10,946	10,630
Washington,	5,911	3,909	9,820	10,627
Wayne,	687	276	963	
Wilkes,	5,277	8,960	14,237	
Wilkinson,	5,591	1,922	6,513	
Total,	299,292	217,531	516,823	340,947

Of the foregoing population of 1830, were white persons :—

	Males.	Females.
Under 5 years of age,	33,027	30,958
From 5 to 10,	23,709	22,500
" 10 to 15,	18,584	17,988
" 15 to 20,	15,186	16,452
" 20 to 30,	26,844	24,036
" 30 to 40,	16,156	13,974
" 40 to 50,	9,542	8,427
" 50 to 60,	5,674	5,089
" 60 to 70,	3,083	2,664
" 70 to 80,	1,120	987
" 80 to 90,	290	268
" 90 to 100,	63	65
" 100 and upwards,	10	20
Total,	153,288	143,518

Of the above are deaf and dumb, under 14 years, 50 ; 14 to 25, 51 ; 25 and upwards, 44 ; Blind 150.

Colored population.

	Free.		Slaves.	
	Males.	Fems.	Males.	Fems.
Under 10 years of age,	368	347	38,367	38,108
From 10 to 24,	353	330	34,253	33,917
" 24 to 36,	224	231	19,440	20,527
" 36 to 55,	186	185	12,818	12,225
" 55 to 100,	118	126	3,847	3,765
" 100 and upwards,	12	6	92	78
Total,	1,261	1,225	108,817	108,714

Free colored and slaves who are deaf and dumb, under 14 years, 26 ; 14 to 25, 21 ; 25 and upwards, 12 ; blind, 123.

Recapitulation.

Whites.	Free colored.	Slaves.	Total.
296,806	2,486	217,531	516,823

On comparing the aggregate population of Georgia for 1820, with that of 1830, it will be seen in that decennial period, the ratio of increase has been over 51 per cent.

Government.—The first constitution of Georgia was adopted February, 1777. The second in 1785, which was amended in 1789, and the third, last, and existing constitution in May, 1798. The legislature consists of a senate and house of representatives, elected each annually. To be elegible to the senate, demands one year's residence in the district from which elected, 3 years an inhabitant of

property to the amount of $1,000. The senate is composed of one member from each co. The house of representatives is composed of members from all the cos., which were formed at the date of the constitution, or which might be formed subsequently, according to their respective numbers of free white persons, and including three fifths of all the people of color. Enumerations are made septennially ; each co. to have at least one, and not more than four members. Members of the lower house must be 2 years of age must have been seven years a citizen of the U S., 3 years an inhabitant of Georgia, residing at least ye in the co. immediately preceding his election, from which he may be chosen ; and be possessed, in his own right, of a settled freehold estate of the value of $250, or of taxable property to the amount of $500, within the co., for at least 1 year preceding his election. Absence on the public business of the state, or of the U. S., excuses from the otherwise requisite residence ;

and the required property must be clear of all incumbrance. The executive power is vested in a governor, who holds his office 2 years, and is elected by the general assembly; he must, when elected, have been a citizen of the U. S. 12 years, of the state of Georgia 6 years, have attained to the age of 36 years, and possess 500 acres of land, of his own right, within the state, and other property to the amount of $4,000, and whose estate shall on a reasonable estimation be competent to the discharge of his debts over and above that sum. To exercise the right of suffrage, demands citizenship, 21 years of age, and the actual payment of taxes. The judiciary is composed of a supreme and inferior courts. Judges of the supreme court are elected by the people, for the term of 3 years, and are removeable by the governor on the address of two thirds of both houses of the legislature, or by impeachment. Inferior judges are elected annually. No religious obligation, test, or disqualification is admitted, nor is any person to be denied the enjoyment of any civil right merely on account of his religious principles. Amendments to the constitution are made by vote of two thirds of both branches of the legislature, at two succeeding sessions.

History.—Of the thirteen original states of the U. S. Georgia was settled most recently. The patent under which this colony was established was granted by George II, 1732, to 21 persons, under the title of "the trustees for settling the colony of Georgia." The name was given in honor of the royal grantor; and the first settlers arrived at Charleston in January, 1733, under the command of general James Oglethorpe. In the spring of that year, the foundation of Savannah was laid, but from the blind feudal principles of granting land, and the defective characters of most of the colonists, the advance was very slow in the first years of settlement. Time and experience meliorated these municipal evils; but as a feeble colony, the ravages of war could not be averted. Spain, even in times of peace, claimed the country; and in war, her colonies in Florida, and the West Indies, facilitated an invasion of Georgia, and what was perhaps fully as injurious, retaliation was equally facile. In fact, the first serious attempt at conquest was made in 1740, when general George Oglethorpe made an attempt to seize St. Augustine, and was repulsed with loss. In 1742 the Spaniards in their turn invaded Georgia, and were also defeated in their design. Laboring under so many combined burthens, the exports of Georgia in 1750 fell short of $50,000. In 1752 the charter was changed, and the province became a royal colony, when more liberal principles of trade and tenure were adopted. A general representative assembly was established in 1755, and was in 1763 followed by a cession of all the country between the Altamaha and St. Mary's rivers. The latter grant was one of the meliorating consequences to Georgia, of the cession of Florida to Great Britain. From this epoch Georgia prospered, though vexed and retarded by Indian warfare, and by the war of the revolution. Indeed no other state of the U. S. has suffered more, if so much, from the proximity of the Indian tribes, nor has any other of the original colonies, Virginia excepted, ceded to the U. S. so much of chartered territory. By different conventions, all of the new states of Ala. and Miss. N. of lat. 31°, or about 100,000 sq. ms. have been yielded to the general government. At present, 1830, Georgia holds a respectable rank amongst her sister states. The value of her exports in 1817 amounted to between 8 and 9 millions of dollars, and which has since been gradually augmenting. (*See article U. S.*)

GEREN'S STORE and p-o. Guilford co. N. C. by p-r. 101 ms. N. W. by W. Raleigh.

GERMANNA, p-v. on the right bank of Rapid Ann r. in the N. E. angle of Orange co. Va. 20 ms. by land above Fredericksburg, and by p-r. 72 ms. S. W. W. C.

GERMAN, t. Chenango co. N. Y. 15 ms. W. Norwich, and 115 W. Albany. Contains abundance of fine mill seats. Lands very rich along the streams. Pop. 1830, 884.

GERMAN FLATS, p-t. S. Mohawk, Herkimer co. N. Y., 5 ms. S. E. Herkimer, and 75 from Albany. A remarkably rich soil. It lies on the grand canal. Here stood fort Herkimer. In 1757, the settlements in this town were desolated by fire and sword. Pop. 1830, 2,466.

GERMANS, p-v. Harrison co. O. by p-r. 281 ms. N. W. by W. W. C. and 140 N. E. by E. Columbus.

GERMAN SETTLEMENT, and p-o. Preston co. Va. by p-r. 170 ms. N. W. by W. W. C.

GERMANTOWN, t. Columbia co. N. Y. E. of the Hudson, 12 ms. S. of the city of Hudson. Surface gently undulating, soil good for grass, &c. Poorly watered—remarkably well timbered—noted for fruit. Pop. 1830, 967.

GERMANTOWN, p-v. and st. jus. Hyde co. N. C. situated on a small bay of Pamlico sound, or rather of Pamlico r. about 40 ms. a little S. of E. Washington, at the mouth of Tar r. and by p-r. 149 ms. in nearly a similar direction from Raleigh. Lat 35° 24' and long. 0° 35' E. W. C.

GERMANTOWN, p-v. Phil. co. Pa. It is a double line of houses, with the Reading road as a street, extending upwards of 4 ms. from its commencement, 6 ms. from Phil. Mt. Airy college is located in Germantown. Pop. 1830, 4,628.

GERMANTOWN, p-v. about the centre of Fauquier co. Va. by p-r. 133 ms. N. W. by W. Raleigh.

GERMANTOWN, p-v. in the S. W. part of Mason co. Ky. by p-r. 81 ms. N. E. Frankfort.

GERMANTOWN, p-v. in German tsp. and in the nrthwstrn part of Montgomery co. O. 15 ms. S. W. Dayton, the co. st. by p-r. 487 N. W. by W. ½ W. W. C. and 90 ms. a little S. of W. Columbus. Pop. of the tsp. 1830, 4,700.

GERMAN VALLEY, Morris co. N. J. a beautiful and rich valley, through which runs the s. branch of Raritan r. 16 ms. w. Morristown.

GERRARDSTOWN, p-v. in the sthrn. part of Berkley co. Va. 18 ms. N. Winchester.

GERRY, t. Chautauque co. N. Y. 18 ms. E. Mayville. Well watered. The timber consists of beech, maple, birch, basswood, ash, elm, oak, walnut, hemlock, &c. Pop. 1830, 1,110.

GETTYSBURG, p-v. borough and st. jus. Adams co. Pa. situated on a fine elevated site between Marsh and Rock creeks of Monococy r. 115 ms. a little s. of w. Phil. and by p-r. 44 ms. s. s. w. Harrisburg. Lat. 39° 50′ and long. 0° 14′ w. W. C. It is a very pleasant town, in a well cultivated and delightful vicinnage, extending mostly in a single street along the main and direct road from Phil. to Pittsburg.

GHENT, t. Columbia co. N. Y. 11 ms. E. Hudson. Excellent land, well supplied with mill seats and mills. Pop. 1830, 2,783.

GHENT, p-v. Gallatin co. O. on the left bank of O. r. opposite Vevay, Switzerland co. Ind. by p-r. 52 ms. a little w. of N. Hartford.

GHOLSONS, p-o. Graves co. Ky. by p-r. 259 ms. s. w. by w. Frankfort.

GHOLSONVILLE, p-v. on the left bank of Meherin r. Brunswick co. Va. by p-r. 78 ms. s. s. w. Richmond.

GIBBONS' TAVERN, and p-o. Delaware co. Pa. by p-r. 94 ms. s. E. by E. Harrisburg, and 126 N. E. W. C.

GIBBONSVILLE, (*See Watervliet.*)

GIBRALTAR, p-v. in Iowa co. Mich. by p-r. 1,012 ms. s. w. by w. W. C. This place is not located on Tanner's map, but relatively with Cassville, and Galena in Il. it must be between the two latter, and about 22 ms. above Galena.

GIBSON, p-v. Susquehannah co. Pa. by p-r. 177 ms. N. N. E. Harrisburg.

GIBSON co. of Ten. bounded by Dyer w.; Obion N. w.; Weakly N.; Carroll E.; Madison s.; and Haywood s. w. Length 30 ms. mean breadth 22; area 660 sq. ms. Lat. 36°, long. 12° w. W. C. intersect near the centre of this co. The slope of this co. is wstrd. towards the Miss. r. and down which flow various branches of Forked Deer, and Obion rs. Chief t. Gibbonsville. Population 1830, 5,801.

GIBSON PORT, p-v. of Gibson co. Ten. about 150 ms. a little s. of w. Nashville.

GIBSON PORT, Claiborne co. Miss. (*See Port Gibson, same co.*)

GIBSON, co. of Ind. bounded N. by White r. separating it from Knox co.; N. E. and E. by Pike; s. E. by Warrick; s. by Vanderberg; s. w. by Posey, and w. and N. w. by Wabash r. separating it from Wabash co. Il. Greatest length from E. to w. 38 ms.; mean breadth 16, and area 600 sq. ms. Extending in lat. from 38° 12′ to 38′ 34′ and in long. from 10° 22′ to 11° 04′ w. from W. C. Slope of the nrthrn. and wstrn. section very nearly due w. and in that direction the nrthrn. part is traversed by Patoka r. The sthestrn. angle gives source to Great Pigeon creek, a confluent of O. r., and declines sthrdly. towards that comparatively large recipient. Face of the co. broken. Chief t. Princeton. Pop. 1830, 5,418.

GILEAD, p-t. Oxford co. Me. on the Androscoggin, 30 ms. w. Paris. Pop. 1830, 377.

GILEAD, p-v. and st. jus. Calhoun co. Il. by p-r. 907 ms. w. from W. C.; 126 ms. a little N. of w. from Vandalia, and 50 ms. N. N. w. from St. Louis in Mo. It is situated between the Miss. and Il. rs. near the right bank of the latter: lat. 39° 03′, long. 13° 37′ w. W. C.

GILFORD, t. Strafford co. N. H. s. side lake Winnipiseogee, 23 ms. N. E. Concord. Contains an academy, 11 schools, a valuable paper manufactory, and other useful mills and machinery. Pop. 1830, 1,870.

GILES, co. of Va. bounded N. by Monroe; N. E. by Botetourt; s. E. by Montgomery; s. by Wythe; s. w. by Tazewell; and w. by the Great Flat Top mtn. which separates it from Logan. The form of this co. is a rude approach to a half moon, and the length between the points about 70 ms. Lying between lat. 37° 06′ and 37° 43′ and long. 3° 15′ to 4° 15′ w. W. C. Surface a congeries of mtn. ridges and intervening vallies, extending in a N. E. and s. w. direction. The mtn. vallies are cut and traversed almost at right angles by the Great Kenhawa, which pouring from the elevated vales between the Blue Ridge and Alleghany, in a N. N. E. course, suddenly inflects to N. w. and passing the latter, enters Giles: breaking through several more minor chains, and receiving Greenbriar from Monroe, Kenhawa r. pursues its nrthwstrn. course towards O. At the mouth of Sinking creek into Kenhawa, in the estrn. and upper part of the co. opposite the mouth of Greenbriar, the water surface is 1,333 feet above the Atlantic tides. We may therefore safely assume 1,600 feet, as the mean level of the cultivated land of Giles. This height is fully equivalent to 4 deg. of Fahrenheit, and would give to Giles a winter climate equal to that on N. lat. 41° along the Atlantic margin. Chief t. Parisburg. In 1820, the pop. was 4,522 only, and at that epoch it included a large tract now comprised in Logan co. Pop. 1830, 5,274.

GILES, C. H. Giles co. Va. (*See Parisburg.*)

GILES, one of the sthrn. cos. of Ten. bounded w. by Lawrence; N. w. by Hickman; N. by Maury; E. by Hickman; and s. by Madison and Limestone cos. of Ten. It is very nearly a square of 30 ms. and area 900 sq. ms. Extending in lat. from 35° to 35° 25′ and traversed by long. 10 w. from W. C. Elk r. winding sthrdly. traverses the s. E. angle of this co. but the much greater part is drained by Richland creek, and other branches of Elk r. Surface moderately hilly. Chief t. Pulaski. Pop. 1820, 12,558; 1830, 18,703.

GILL, p-t. Franklin co. Mass. 90 ms. N. w. Boston. Fine land, beautifully situated N. & w. Conn. r. which here makes a bend. Miller's Falls are in the Conn. adjoining this t. Pop. 1830, 1,407.

GILL LAND'S CREEK, N. Y. (*See Willsborough.*)

GILLMANTOWN, p-t. Strafford co. N. H. 17 ms. from Concord, 44 from Portsmouth, 78 from Boston, and 522 from W. C. Very hilly and rocky—well watered—contains iron ore and mineral springs—has a flourishing academy. Pop. 1830, 3,816.

GILSUM, p-t. Cheshire co. N. H. 37 ms. s. Concord. Has good mill privileges. Pop. 1830, 642.

GLADWIN, co. of Mich. bounded by Arena co. E., Midland s. and ter. not yet divided into cos. w. and N. It is a sq. of 24 ms. each side, area 576 sq. ms. Extending in lat. from 43° 50′ to 44° 10′ and in long. from 7° 08′ to 7° 35′ w. W. C. Tittibawassee r. or the nrthrn. branch of Saginaw r. drains the body of this co. flowing by a general sthrn. course. The central part of Gladwin is about 150 ms. N. N. W. Detroit.

GINSENG, p-v. Logan co. Va. about 380 ms. s. w. by w. W. C.

GLADE RUN, p-o. Armstrong co. Pa. by p-r. 214 ms. N. W. W. C.

GLADDEN'S GROVE, and p-o. Fairfield co. S. C. 31 ms. N. Columbia.

GLADY CREEK, Cross Roads and p-o. Randolph co. Va. 60 ms. s. E. Clarksburg, and by p-r. 223 westrd. W. C.

GLASGOW, p-v. on the head of Christiana creek, New Castle co. Del. 15 ms. s. E. Wilmington, and by p-r. 98 ms. N. E. W. C.

GLASGOW, p-v. and st. jus. Barren co. Ky. 116 ms. s. s. w. Frankfort, and 89 ms. N. N. E. Nashville, in Ten. lat. 37° 01′ long. 8° 46 w. W. C.

GLASSBOROUGH, p-v. Gloucester co. N. J. 20 ms. s. E. Phil. Here is a glass factory.

GLASTENBURY, p-t. Hartford co. Conn. E. Conn. r. Timber, oak, chestnut, &c.—well watered—contains cotton, woollen, and iron manufactories, &c. and fine shad fisheries in the Conn.; also a mineral spring. It has 13 school districts. Pop. 1830, 2,980.

GLASTENBURY, t. Bennington co. Vt. 9 ms. N. E. Bennington, 25 N. W. Brattleborough. Land a great part high, broken, and incapable of being settled. Pop. 1830, 59.

GLEN, p-t. Montgomery co. N. Y. on the canal, 8 ms. s. Johnstown. Pop. 1830, 2,451.

GLENCOE, p-v. Hampshire co. Va. by p-r. 124 ms. nrthwstrdly. from W. C.

GLENN'S, p-o. Gloucester co. Va. by p-r. 92 ms. E. Richmond.

GLEN'S FALLS, v. & p-o. (*See Queensbury.*)

GLENVILLE, p-t. Schenectady co. N. Y. N. Mohawk r., 5 ms. N. W. Schenectady. Has 9 school houses. Pop. 2,497.

GLOUCESTER, p-t. and port of entry, Essex, co. Mass. 30 ms. N. E. Boston, and 16 N. E. Salem, situated on cape Ann, N. extremity of Mass. bay. One of the most considerable fishing towns in the state, with a harbor open and accessible to large ships at all seasons. About 10,000 tons of shipping are usually owned in this town. Rocky and uneven. The principal part of Gloucester is a peninsula, connected with the main by a very narrow isthmus, across which is a canal for the passage of small vessels. On the south-east side of the town is Thatcher's island, on which are two light houses. This town is a charming place in the warm season. Pop. 1830, 7,510.

GLOUCESTER, p-t. Providence co. R. I. 16 ms. from Providence. Surface generally uneven. Extensive and valuable forests. Chepachet river runs through the centre of the town, upon which river, near the centre of said town, is a considerable village called Chepachet, where are a number of cotton factories, and some additional water works. Contains twelve schools. Population 1830, 2,522.

GLOUCESTER, co. N. J. extends from the Delaware river to the Atlantic, bounded N. E. by Burlington, s. w. by Salem, Cumberland and cape May. Pop. 1830, 28,431. Principal towns, Woodbury, and Camden. The lands along the Delaware, extending inland, are highly cultivated for fruit, vegetables, &c. for the Philadelphia market. In the interior, are pine lands, and several forges, and manufactories of glass, &c.

GLOUCESTER, town, Gloucester co. N. J. bounded N. E. by Waterford, s. w. by Deptford, s. w. by Gloucester t. Pop. 1830, 2,332.

GLOUCESTER TOWN, t. Gloucester co. N. J. on the Delaware, between Deptford and Newton, bounded N. E. by Gloucester. Pop. 1830, 686.

GLOUCESTER, co. of Va. bounded N. W. by King and Queen, N. by Piankatank river, which separates it from Middlesex, N. E. by North river, which separates it from Matthews, E. by Chesapeake or Mobjack bay, and s. and s. w. by York r. which separates it from York and James city cos. Length 28 miles, mean width 10, and area 280 square ms. Extending in lat. from 37° 15′ to 37° 35′ and in long. from 0° 14′ to 0° 42′ E. W. C. Chief town, Gloucester. Pop. 1820, 9,678, 1830, 10,608.

GLOUCESTER, court house, Gloucester co. Va. p-o. near the centre of the co. by p-r. 88 ms. E. Richmond.

GLOVER, p-t. Orleans co. Vt. 33 miles E. Montpelier. Contains 8 school houses. This town is noted for the following remarkable occurrence. A pond, a mile and a half long, and half a mile wide, situated partly in this town, and partly in Greensborough, on the 6th of June, 1810, on having a small outlet opened, broke loose through the quicksand, of which its bank was in that place principally composed, and in 15 minutes was entirely emptied, its waters rushing forth in a mighty mass, 60 or 70 feet in height, and 20 rods in width, levelling forests and hills, filling up the valleys, sweeping away houses, barns, cattle, &c. and giving the inhabitants time barely to escape with their lives into the mountains. In this manner did it deluge the country for the space of 10 ms. So rapidly flowed the torrent, that it reached lake Memphremagog, 27 ms. distant, in about 6 hours from the time of its getting vent. Nothing now remains of the pond but its bed, a part of which is cultivated, and a part overgrown with bushes and wild grass, with a small brook running through it, which is now at the head of Barton river. Pop. 1830, 902.

GLYNN, co. of Geo. bounded by Camden co. S. W., Wayne N. W., Altamaha river which separates it from McIntosh N. E., and the Atlantic ocean S. E. It lies very nearly in form of a square of 25 ms.; area 625 square ms. Extending in lat. from 31° to 31° 29', and in long. from 4° 22' to 4° 58' W. W. C. St. Simons, and Jekyl islands, constitute the Atlantic border of Glynn. The whole surface is level and cut by interlocking water courses. Where the soil admits of culture, the climate is suitable to rice, indigo, tobacco, sugar cane, &c. The orange tree and fig tree flourish. Chief towns, Brunswick and Frederica. Pop. 1820, 3,418, in 1830, 4,567.

GNADENHUTTEN, p-v. Clay twp. Tuscarawas co. Ohio. The twp. is in the southern part of the co. The post village is situated on Tuscarawas river, 11 ms. S. New Philadelphia, the co. st. Pop. of the p-v. 1830, 49.

GODFREY, Savannah post office, Colleton dist. S. C. by p-r. 114 miles sthrd. Columbia.

GODFREY'S FERRY, and p-o. by p-r. 151 ms. a little S. of E. Columbia. The ferry is over the Great Pedee, about 10 miles above the mouth of Lynches creek or river.

GOFFSBORO', p-o. Washington parish, La. by p-r. 1147 ms. southwestward W. C. and 83 N. N. W. New Orleans.

GOFFSTOWN, p-t. Hillsborough co. N. H. 12 ms. from Amherst, 16 from Concord, and 55 from Boston, W. Merrimack river, at Amoskeag falls. Piscataquog river runs through its centre, and falls into the Merrimac. Good land. The timber, oak, several sorts of pine, hemlock, beech, and maple. A great number of masts, for the English navy, have been furnished from this place. It is the present residence of Hon. David L. Morrill, late governor of the state, and member of congress. Pop. 1830, 2,208.

GOLANSVILLE, p-v. Caroline co. Va. by p-r. 29 ms. S. S. W. W. C. and 56 nearly due N. Richmond.

GOLDEN, p-v. Baltimore co. Md.

GOLDEN GROVE, p-o. Greenville dist. S. C. by p-r. 110 ms. N. W. Columbia.

GOLCONDA, p-v. and st. jus. Pope co. Il. by p-r. 791 ms. S. S. W. ½ W. W. C. and 160 ms. S. S. E. Vandalia.

GOLDSBOROUGH, town, Hancock co. Me. 40 ms. E. Castine. Pop. 1830, 880.

GOLD MINE, p-v. Chesterfield dist. S. C. by p-r. 449 ms. S. S. W. W. C. and 101 ms. N. E. Columbia.

GOOCHLAND, co. of Va. bounded by Fluvanna N. W., Louisa N., Hanover N. E., Henrico S. E., and James river which separates it from Powhatan S., and Cumberland S. W. Length 28, mean breadth 12, and area 336 square ms. Extending in lat. from 37° 31' to 37° 51', and in long. from 0° 47' to 1° 20' W. W. C. Goochland slopes to the S. and is drained by several small creeks falling into James river. Chief town, Hardensville. Pop. 1820, 10,007, 1830, 10,369.

GOOCHLAND, court house, and p-o. Goochland co. Virginia, by p-r. 32 ms. S. W. by W. Richmond.

GOODE'S BRIDGE, and p-o. in the sthrn. part Chesterfield co. Va. 38 ms. S. W. Richmond.

GOODFIELD, p-v. Rhea co. Tenn. by p-r. 151 ms. S. E. by E. Nashville.

GOOD LUCK, formerly Magruder's p-o. southwestern part of Prince George's co. Md. by p-r. 18 ms. S. E. W. C. and 40 ms. S. W. Annapolis.

GOODSON'S, p-o. Montgomery co. Va. by p-r. 299 miles S. W. W. C. and 221 westward Richmond.

GOODSON'S, p-o. Cumberland co. Ky. by p-r. 634 miles S. W. by W. W. C. and 122 S. Frankfort.

GOOD SPRING, p-o. Williamson co. Tenn. 12 ms. S. W. Nashville.

GOODWYNSVILLE, p-o. Dinwiddie co. Va. 7 ms. S. Dinwiddie court house, and 47 S. S. W. Richmond.

GOOSEBERRY ISLAND AND ROCKS, off cape Ann, Mass.

GOOSE CREEK, or river, branch of Roanoke river rising in the southeastern vallies of Blue Ridge, 4 or 5 ms. S. W. from the peaks of Otter, and flowing thence S. E. over Bedford co. falls into Roanoke river, in the western angle of Campbell co. Va. after a comparative course of 30 ms.

GOOSE CREEK, post office, on the preceding creek, in the western part of Bedford co. Va. 10 ms. westward Liberty, the co. st.

GOOSEPOND, p-o. Oglethorpe co. Geo. by p-r. 86 ms. N. N. E. Milledgeville.

GORDONSVILLE, p-v. at the eastern foot of South West mtn. and on the source of North Anna river, Orange co. Va. about 50 ms. S. W. by W. Fredericsburg, by p-r. 92 ms. N. W. Richmond, and 115 S. W. W. C.

GORDONSVILLE, p-v. Smyth co. of Ten. 6 ms. from Carthage, and by p-r. 81 ms. though direct only about 50, eastward Nashville.

GORDONTON, p-o. Person co. N. C. 60 ms. by p-r. N. N. W. Raleigh.

GORE, a tract of land lying W. of Williamstown, Mass. 140 ms. N. W. of Boston, 2 ms. wide at S. end, and tapering to a point at the Vt. line.

GORHAM, p-t. Cumberland co. Me. 9 ms. N. W. Portland. It has a considerable village, in which is an academy. Pop. 1830, 2,988.

GORHAM, p-t. Ontario co. N. Y. 8 ms. S. E. Canandaigua. Flint creek runs across the E. part, and supplies mill seats. It contains 23 school districts. Pop. 1830, 2,081.

GORHAM, p-v. in the eastern part of Daviess co. Ky. 8 ms. S. E. Owensburg, or Owensboro', the co. st. and 150 S. W. by W. ½ W. Frankfort.

Gosham, p-v. Daviess co. Ky. by p-r. 152 ms. s. w. by w. from Frankfort.

Goshen, p-t. Sullivan co. N. H. 42 ms. w. Concord. Soil particularly good for grass. Timber, maple, birch, beech, hemlock, spruce, and some oak. Maple sugar is here manufactured to a considerable extent. Pop. 1830, 772.

Goshen, town, Addison co. Vt. 31 ms. s. w. Montpelier, and 43 n. w. Windsor. Considerably mountainous. Watered by Leicester river. Contains iron ore, and the oxide of manganese; 6 school districts. Population 1830, 555.

Goshen Gore, Caledonia co. Vt. There are two gores of this name, both in this co. the largest contains 7,339 acres, the smaller, 2,828.

Goshen, p-t. Hampshire co. Mass. 115 ms. w. Boston, and 12 n. w. Northampton. Several minerals, among which is the emerald, are found here. Pop. 1830, 617.

Goshen, p. t. Litchfield co. Conn. 32 ms. w. Hartford, and 42 from New Haven. It is the highest land in the state. The sugar maple is the predominant forest tree. It contains various kinds of manufactories—8 school dists. It is remarkably healthy. Pop. 1830, 1,734.

Goshen, p-t. and half-shire town, Orange co. N. Y. 20 ms. w. Hudson r. 110 s. Albany, and 60 n. N. York. Contains 12 schools. Pop. 1830, 3,361.

Goshen, p-v. Monmouth co. N. J. 12 ms. s. Allentown.

Goshen, small village in the sthrn. part of Loudon co. Va. about 35 ms. w. W. C.

Goshen, p-v. Lincoln co. Geo. about 45 ms. above Augusta, and by p-r. 99 ms. n. e. Milledgeville.

Goshen, p-v. in the estrn part of Iredell co. N. C. 11 ms. estrd. Statesville, the co. st. and 157 ms. westrd. Raleigh.

Goshen, p-v. in Goshen tsp. nrthrn. part of Clermont co. O. by p-r. 93 ms. s. w. Columbus. Pop. 1830, 139.

Goshen Hill, p-v. between Ennoree and Tyger rs. sthrn. part of Union dist. S. C. by p-r. 107 ms. n. w. Columbia. The real road distance between those two places about 50 ms.

Goshen Mills, and p-o. Montgomery co. Md.

Goshensville, p-o. Chester co. Pa.

Governeur, p-t. St. Lawrence co. N. Y. 23 ms. s. Ogdensburg. Received its name in honor of Governeur Morris. It has a great diversity of soil, and is situated on the Oswegatchie r. Pop. 1830, 1,430.

Govanstown, p-o. in the sthrn. part of Baltimore co. Md. by p-r. 42 ms. n. e. W. C.

Governor's Island, N. Y. directly s. of the city of N. York, forming the harbor in the East r Belongs to government and is strongly fortified.

Gowansville, p-v. Greenville dist. S. C. by p-r. 121 ms. n. w. Columbia.

Gowdysville, p-v. Union dist. S. C. by p-r. 454 ms. s. w. W. C. and 89 n. w. Columbia.

Graceham, p-v. Frederick co. Md. on the road from Hagerstown to Westminster, 15 ms. n. Frederic, and 81 ms. n. n. w. W. C.

Grafton co. N. H. 58 ms. long, and 30 at its greatest breadth, containing 828,623 acres, besides a large tract of ungranted land.—Bounded n. by Coos co., e. by Strafford, s. by Hillsborough, and w. by Vt. It is watered by Conn. r., Pemigewasset, Lower Amonoosuck, and many smaller streams—somewhat mountainous—contains fine tracts for pasturage, a large proportion of arable land, and on the rivers extensive and fertile intervals—there are in the co. 36 towns—the sessions of the superior court and of the court of sessions are holden alternately at Haverhill and Plymouth. Pop. 1820, 32,989; 1830, 38,632.

Grafton, t. Grafton co. N. H. 36 ms. n. w. Concord, and 13 s. e. Dartmouth college—well watered—rocky, hilly, and mountainous. Pop. 1830, 1,207.

Grafton, p-t. Windham co. Vt. 36 ms. n. e. Bennington, and 22 s. w. Windsor—contains good mill seats—surface uneven—abounds in a great variety of minerals—contains an immense quantity of soap stone—11 school dists. Pop. 1830, 1,439.

Grafton, p-t. Worcester co. Mass. 40 ms. s. w. Boston, and 8 s. e. Worcester, on the Blackstone canal. Watered by Blackstone r.; contains flourishing manufactories, one of which has a capital of $500,000, at which is manufactured twine and duck, from flax and hemp. Pop. 1830, 1,889.

Grafton, p-t. Rensselaer co. N. Y. 11 ms. e. Troy. Soil principally an argillaceous loam; timber, hemlock, white pine, fir, spruce, maple, &c.; contains 10 school houses. Pop. 1830, 1,681.

Grafton, p-v. in the nrthrn. part of Medina co. O. by p-r. 367 ms. nrthwstrdly. W. C. and 129 n. n. e. Columbus.

Graham's Station, and p-v. Meigs co. O. by p-r. 352 ms. w. W. C. and 103 s. e. Columbus.

Graham's Bridge, and p-o. Richmond co. N. C. by p-r. 93 ms. s. w. Raleigh.

Grahamsville, or Grahamstown, p-v. near the centre of Beaufort dist. S. C. 74 ms. s. w. by w. and by a rather circuitous road from Charleston, and 10 s. Coosawhatchie the st. jus. for the dist.

Grainger, co. of Ten. bounded by Clinch r. which separates it from Claiborne co. n. w., by Hawkins co. n. e., by Jefferson s. e. and Knox, s. w. Length 32 ms. breadth 10, and area 320 sqare ms. Extending in lat. from 36° 08′ to 36° 30′ and in long. from 6° 03′ to 6° 40′ w. W. C. This co. being bounded on the n. w. by Clinch, and traversed on the s. e. by Holston r. has some very fine r. soil, but the body of the co. is hilly, and in part rocky and mountainous. A minor ridge called Chesnut Ridge, stretches to the s. w. and n. e. between the two rivers, and divides Grainger into two not very unequal parts. Chief t. Oresville. Pop. 1820, 7,650; 1830, 10,066.

Granberry's, p-o. in the sthrn. part of Twigg's co. Geo. 8 ms. sthrd. Marion, the co. st. and 45 s. w. Milledgeville.

Granby, t. Essex co. Vt. 47 ms. n. e. Montpelier, Vt. Pop. 1830, 97.

Granby, p-t. Hampshire co. Mass. 90 ms. w. of Boston. In a cavern recently discovered here, were found two decayed statues originally formed of wood or earth. The date of 1760 was on the walls. Pop. 1830, 1,064.

Granby, p-t. Hartford co. Conn. bordering on Mass. 17 ms. n. n. w. Hartford. The surface is diversified with mountain, hill, and dale; it has within its limits a great variety of timber, and contains several kinds of minerals. It is pretty well watered, and contains various kinds of manufactories, 16 school dists. and a small village in the centre of each of the located religious societies. Here is the Newgate, once the state prison, which was formerly a cavern, and originally opened and wrought as a copper mine.—Since the removal of the prison, this mine has been purchased by a mining company, and wrought with some success. Pop. 1830, 2,722.

Granby, p-t. Oswego co. N. Y. s. w. Oswego village, 12 ms. above Oswego, 25 n. of Salina, and 155 w. of Albany. Surface moderately uneven; soil generally fertile; timber, pine, oak, maple, beach, &c. The Oswego falls are between this town and Volney; there are 8 schools in the town. Pop. 1830, 1,423.

Granby, p-v. and st. jus. Lexington dist. S. C. situated on the right bank of Congaree r. nearly opposite Columbia. Lat. 33° 58' long. 4° 03' w. W. C.

Grand r. or Neosho r. a branch of Ark. r. rises in the angle between Ark. proper, Kansas, and Osage rs., flows by a general course of s. s. e. upwards of 200 ms. by comparative courses, and falls into Ark. at Cantonment Gibson. The valley of Grand r. stretches from lat. 35° 47' to 38° 40', and though some of its branches rise in the state of Mo. the great body of the valley lies westrd. of that state and of the Ter. of Ark. between 17° and 20° w. W. C.

Grand r. a branch of Mo. r. about lat 42°, between the sources of Raccoon fork of Des Moines r. and Naudaway branch of Mo. and flowing thence sthrdly. about 100 ms. enters the state of Mo.; inflecting thence a little e. of s. 1 0 ms. falls into the left side of Mo. by direct course 200 ms. above St. Louis. This stream and its confluents water upwards of 5000 sq. ms. in the nrthwstrn. angle of Mo. Its valley lies between long. 16° and 17° 30' w. W. C.

Grand r. stream of Mich. Ter. and confluent of Lake Mich. rises, interlocking sources with those of Huron, Lake Erie, Resin, and also with those of St. Joseph's and Kalamazoo rs. flowing into Lake Mich. The extreme sources of Grand r. are on the flat table land of the Mich. peninsula, about 80 ms. a little s. of w. from Detroit, about n. lat. 42°. Flowing thence by comparative courses n. n. w. about 100 ms. the channel abruptly inflects to the wstrd. which latter course is continued 70 ms. to its final influx into lake Mich. at lat. 43° 08'. Some of the nrthestrn. confluents of Grand r. interlock with those of Saginaw r. and the main body of the valley of the former, lies between those of Saginaw and Kalamazoo.

Grand, p-v. and tsp. in the nrthwstrn. part of Marion co. O.; the p-v. is 74 ms. n. n. w. Columbus. Pop. tsp. 1830, 317.

Grand Blanc, p-v. Oakland co. Mich. n. n. w. Detroit.

Grand Cakalin, p-v. Brown co. Mich.—The names or position of these two latter not on Tanner's map; nor is the relative position of either given in the p-o. list.

Grande, p-v. Crittenden co. Ark. 141 ms. by p-r. n. e. by e. Little Rock, and 932 ms. s. w. by w. ½ w. W. C.

Grand Gulf, abrupt and remarkable bend of the Mississippi r. at the influx of Black r.

Grand Gulf, p-v. on the Grand Gulf and Miss. r., and in the nrthwstrn. part of Claiborne co. state of Miss. by the land road 50 ms. above and n. n. e. Natchez.

Grand Isle co. Vt. bounded n. by L. Canada, on the n. line of Alburgh, the rest of the co. consisting of isl's. in lake Champlain. It is 28 ms. long from n. to s. and about 5 ms. wide, containing 82 sq. ms.; streams small, having scarcely a good mill privilege in the co.; surface generally level, and very rich and productive; chief t. North Hero. Pop. 1820, 3,527; 1830, 3,696.

Grand Isle, p-t. Grand Isle co. Vt. 18 ms. n. Burlington. Pop. 1830, 643.

Grand Island, N. Y. in Niagara river 12 ms. long, and 2 to 7 wide, commencing about 3 ms. below Black Rock, and terminating a mile and a half above Niagara Falls; containing 17,800 acres. Soil strong and rich. A large marsh in the centre. It is well wooded. Here was acted the farce of laying the foundation of the Jewish city of Ararat.

Grand or Chilnucook lake, Me. the source of St. Croix r. 30 ms. long, and about 5 broad.

Grand Traverse, Strait and Islands, between lake Michigan and Green Bay. This strait opens at lat. 45° 30', and is filled with small islands, which render the entrance from lake Michigan into Green Bay rather intricate.

Grand Traverse, bay, or the outlet of Ottawa river, Mich. Amongst the confusion of naming rs. by the same name, or by one appropriated to another object, this affords an instance. Directly opposite Grand Traverse Strait and Islands, but on the contrary shore of lake Michigan, opens Grand Traverse bay. The Ottawa r. rising on the peninsula of Michigan, and flowing to the n. w. about lat. 45, widens to a bay, which with a length of 30 ms. terminates in lake Michigan, and is known as Grand Traverse Bay.

Granger, p-v. Caldwell co. Ky. by p-r. 235 ms. s. w. by w. Frankfort.

Granger, p-v. and tsp. Medina co. O. by p-r. 348 ms. n. w. by w. of W. C. and 121 n. n. e. Columbia. Pop. of the tsp. 1830, 676.

Grant, co. of Ky. bounded s. and s. w. by Owen, Gallatin n. w., Boone n., and Pendleton e. Length 23, mean width about 8, and area 184 sq. ms. Extending in lat. from 38° 30′ to 38° 48′, and in long. from 7° 35′ to 7° 47′ w. W. C. Eagle cr. a branch of Ky. r. winds over a part of Grant, and is the only stream of consequence in the co. Chief town, Williamsville, or Williamstown. Pop. 1820, 1,805, 1830, 2,987.

Grantham, t. Sullivan co. N. H. 12 ms. s. e. Dartmouth college, and 45 n. w. Concord. Croydon mtn. runs through the west part of the town. Soil productive. Well watered by numerous brooks and rivulets. Contains a medicinal spring, and a bed of paint. Pop. 1830, 1,079.

Grantley's, p-o. Culpepper co. Va. by p-r. 59 ms. s. w. W. C.

Grant's Lick, and p-o. Campbell co. Ky. by p-r. 76 ms. n. n. e. Frankfort.

Grantsville, p-v. Green co. Geo. by p-r. 52 ms. northward Milledgeville.

Granville, t. Hampden co. Mass. 120 ms. s. w. Boston, and 18 s. w. Springfield—a handsome and flourishing town. Pop. 1830, 1,649.

Granville, p-t. Washington co. N. Y. about 60 ms. n. e. Albany. Soil excellent; surface handsomely diversified, and well watered with springs, rivulets, &c. It contains an academy, and 19 schools. It has a marble quarry; and common slatestone, and limestone are also found. It contains likewise various kinds of manufactories. Pop. 1830, 3,882.

Granville, p-v. on Duncard cr. near the southern border of Green co. Pa., but in Monongalia co. Va., about 12 ms. n.w. Morgantown, and by p-r. 211 ms. n. w. by w. W. C.

Granville, p-v. Monongalia co. Va. 217 ms. n. w. by w. ½ w. W. C.

Granville, co. of N. C. bounded by Warren e., Franklin s. e., Wake s., Orange s. w., Person w., and by Halifax co. of Va. n. w., and Mecklenburg co. of Va. n. Length 36, mean width 23, and area 828 sq. ms. Extending in lat. from 36° 03′ to 36° 30′, and in long. from 1° 20′ to 1° 50′ w. W. C. The southern part of this co. is drained by creeks flowing s. into Neuse r., the centre is traversed by Tar r. whilst the northern section slopes towards, and is drained by creeks flowing into Roanoke. Chief town, Oxford. Pop. 1820, 18,216, 1830, 19,343.

Granville, p-v. and tsp. in the southwestern part of Licking co. O. The village is by p-r. 28 ms. n. e. by e. Columbus, and 6 ms. westward Newark, the co. st. Pop. of the tsp. 1830, 1,784, and of the village 362.

Grape Island, and p-o. Tyler co. Va. by p-r. 273 ms. westward W. C.

Grass r. N. Y. enters the St. Lawrence, opposite St. Regis Island. It is 125 ms. long. It is naturally connected with the Oswegatchie in Canton.

Grass Lake, p-v. Jackson co. Mich. by p-r. 88 ms. w. Detroit.

Grassy Creek, and p-o. Pendleton co. Ky. by p-r. 68 ms. n. n. e. from Frankfort.

Grassy Creek, and p-o. Burke co. N. C. about 200 ms. w. Raleigh.

Grassy Point, p-v. Madison co. O. by p-r. 61 ms. westward Columbus.

Gratiot, co. of Mich. bounded by Saginaw co. e., Clinton s., Montcalm w., Isabella n. w., and Midland n. e. It is a sq. of 24 ms. each side, area 576 sq. ms. Extending in lat. from 43° 08′ to 43° 28′, and in long. from 7° 22′ to 7° 48′ w. W. C. The slope to the southwest, and drained by some of the higher northern sources of Grand r. of lake Michigan. The central part is about 110 ms. n. w. from Detroit.

Gratiots Grove, p-o. Joe-Daviess co. Il. 17 ms. n. e. by e. Galena, and by p-r. 972 n. w. by w. W. C.

Gratis, p-v. 2 ms. from Eaton, the co. st. Prebble co. O., and 94 a little s. of w. Columbus.

Gratz, p-v. in Wiconisco valley, in the northern angle of Dauphin co. Pa. by p-r. 46 ms. a little e. of n. Harrisburg.

Grave Creek, or Elizabethtown, p-v. of Ohio co. Va. 12 ms. below Wheeling, and by p-r. 352 n. w. by w. W. C. This creek and village take their name from very extensive tumuli, scattered over an elevated bottom or plain. The author of this article visited this plain twice in 1794, previous to the plough or other operations of farming having much disturbed the remains. At that epoch, one very large conical mound surrounded by a ditch, was itself environed by numerous and similar, though smaller tumuli. The remains of the roads, sloping down the banks from the plain, were also perfectly distinguishable; as was the trench of a work, in form of a parallelogram.

Graves, co. Ky. between Ten. and Miss. rs., bounded by McCracken n., Calloway e., Weakly co. of Ten. s., and Hickman in Ky. w. Length 33, mean breadth 20, and area 660 sq. ms. Extending in lat. from 36° 30′ to 36° 58′, and in long. from 11° 35′ to 11° 56′ w. W. C. Chief town, Mayfield. Pop. 1830, 2,503.

Gravelly Hill, and p-o. Bladen co. N. C. by p-r. 101 ms. s. Raleigh.

Graves, p-o. in the western part of Madison co. Va. by p-r. 110 ms. s. w. W. C.

Gravesend, t. Kings co. N. Y. 9 ms. s. N. Y. on the coast; a bathing resort. Pop. 565.

Gray, p-t. Cumberland co. Me. 20 ms. n. of Portland. Pop. 1830, 1,575.

Grayson, co. of Va. lying between the Blue Ridge and Iron mtn., bounded w. by the Iron mtn. which separates it from Washington; n. w. and n. by the Iron mtn. separating it from Wythe; n. e. by Montgomery co.; e. and s. e. by Blue Ridge, separating it from Patrick co.; and s. e. by Surrey, and s. w. by Ashe counties N. C. The greatest length of Grayson is about 70 ms. from the extreme western angle on Iron mtn. to the extreme eastern on Blue Ridge; mean width 12, and area 840 sq. ms. Extending in lat. from 36° 33′ to 36° 53′, and in long. from 3° 28′ to 4° 46′ w. W. C. Grayson is the most eastern of the southern cos. of Va. which are comprised in the valley of Ohio r. Great Kenhawa r. rising in Ashe co. of N. C. flows northeastwardly into Grayson, and thence turning eastward about 20 ms. along the line between Va. and N. C. and turning to N. N. E. traverses Grayson, which it leaves by piercing the Iron mtn. This co. is a part of the Great Valley west of the Blue Ridge, and slopes northward, drained by innumerable creeks flowing from the two bounding channels into the Great Kenhawa, here called New River. Comparing the mean elevation of Grayson, with that of Wythe, Montgomery and Giles, we cannot assume for the former less than 1,600 feet above the ocean level. (*see Giles &c.*) Chief town of Grayson, Greensville. Pop. 1820, 5,598, 1830, 7,675.

Grayson C. H. and p-o. (*see Greensville, Grayson co. Va.*)

Grayson, C. H. and p-v. Grayson co. Va. by p-r. 354 ms. s. w. W. C., and 276 s. w. by w ½ w. Richmond.

Grayson, co. of Ky. bounded w. by Ohio co., N. W. by Hancock or by Rough creek, branch of Green r., N. by Rough creek, separating it from Breckenridge, N. E. and E. by Hardin, S. E. by Nolin's creek, separating it from Hart, s. by Edmonson, and s. w. by Butler. Length from E. to w. 40, mean breadth 20, and area 800 sq. ms. Extending in lat. from 37° 10′ to 37° 38′, and in long. from 8° 58′ to 9° 40′ w. W. C. This co. occupies part of the peninsula between Green r. and Rough creek, the central part being a table land, from which small creeks flow into the two bordering streams. The general slope is to the wstrd., as both the bordering rivers flow in that direction. Chief t. Litchfield. Pop. 1820, 4,055, 1830, 2,504. This county must have been divided in the intermediate time.

Gray's Settlement, and p-o. Erie co. Pa. by p-r. 327 ms. N. W. W. C.

Graysville, p-v. Huntingdon co. Pa. by p-r. 96 ms. westward Harrisburg.

Great Bay, Rockingham co. N. H. The western branch of the Piscataqua, 4 miles wide, empties north east through Little Bay.

Great Bay, Strafford co. N. H. connected with Winnipiseogee lake, and heads Winnipisseogee r.

Great Bend, p-v. on the left bank of the East Branch of Susquehannah r. at the mouth of Salt Lick creek, Susquehannah co. Pa. 15 ms. N. N. E. Montrose, and by p-r. 170 N. N. E. Harrisburg.

Great Bridge, p-v. Norfolk co. Va. situated on Southern r. 12 ms. s. s. E. Norfolk, and by p-r. 124 s. E. by E. Richmond.

Great Crossings, p-v. Scott co. Ky. 15 ms. N. E. Frankfort.

Great Kenhawa. (*See Kanhawa.*)

Great Mills and p-o. at the head of St. Mary's r., St. Mary's co. Md. by p-r. 81 ms. s. s. E. W. C. and almost due s. Annapolis.

Great Salt Works, or Saltsburg, at the forks, and on the right bank of Conemaugh r., a p-v. Ind. co. Pa. about 30 ms. a little N. of E. Pittsburg, and 211 wstrd. Harrisburg.

Great Ogeechee. (*See Ogeechee r.*)

Great Valley, p-t. Cattaraugus co. N. Y. 14 ms. s. E. from Ellicottville. Pop. 1830, 647.

Great Works River, Me. enters the Penobscot r. 2 ms. below the Great Falls.

Greece, p-t. Monroe co. N. Y. at the mouth of Genesee r. Pop. 1830, 2,574.

Green River, considerable navigable stream of Ky. having its most remote source in Lincoln co. heading with Dick's r., and with the extreme northern branches of Cumberland r. Flowing thence westwardly, inclining a little to the N., receiving, beside numerous creeks, the comparatively large tributaries of Big Barren from the s. and Nolin and Rough creeks from the N. The main stream enters Ohio r. after a comparative course of about 200 ms. The valley of Green r. extends from N. lat. 36½° to 37° 55′, and is in length 170 ms., with a mean breadth of 40 ms., and area 6,800 sq. ms. It has the vallies of Salt and Ohio rs. N. and that of Cumberland s.

Green, t. Sussex co. N. J., joins Warren co. 6 ms. s. Newton. Pop. 1830, 801.

Green Bank, p-o. Pocahontas co. Va. by p-r. 242 ms. wstrd. W. C.

Green Bay, p-o. Hanover co. Va. by p-r. 94 ms. s. s. w. W. C.

Green Bay, Brown and Chippewa cos. Mich. The sheet of water to which this title has been given is usually regarded as a part of lake Michigan, though in geographical strictness, Green bay is itself a lake connected with Michigan by a strait called Grand Traverse. (*See Grand Traverse Islands.*)

Green Bay, considered as a separate sheet of water from lake Michigan, though connected by a common strait, extends from s. w. to N. E. 120 ms.; the width varies, but is generally about 25, and the mean breadth would be rather underrated at 20 ms. The southwestern extreme branches into two large arms, Sturgeon bay to the estrd. and Fox r. bay to the wstrd. (*See Fox r. of Green bay.*) Besides Grand Traverse Islands, there are other islands in Green Bay, the largest of which, Menomonie island, w. of the Grand Traverse, is about 25 ms. in length, but comparatively narrow. Vessels of 200 tons burthen are navigated into and through Green Bay to the mouth, and some distance up Fox r. Menomonie r. enters Green Bay from the northwstrd., and 50 ms. N. E. from the influx of Fox r.

Green Bay, p-v. Brown co. Mich. by p-r.

1,037 ms. N. W. by W. W. C. and 511 ms. in nearly a similar direction Detroit.

GREEN BRIER, r. of Va. rising in the nthrn. part of Pocahontas co. over which it flows, and entering and traversing Green Brier co., falls into Great Kenhawa, after a comparative southwestern course of 90 ms. Green Brier has its remote sources in the same ridges with those of Cheat r. branch of Monongahela, and those of the South branch of the Potomac. The valley of Green Brier lies between those of James and Ganley rs. It is an elevated region. The water level is from actual admeasurement, 1,333 feet at the efflux of Green Brier into Great Kenhawa. The mean height of the farms above the ocean level cannot fall much, if any, short of 1,500 feet.

GREENBRIER, co. of Va. bounded by Nicholas N. W., Pocahontas N. E., Alleghany E., Monroe S., and Great Kenhawa r. separating it from Logan N. W. and W. Length from S. W. to N. E. 60 ms., mean breadth 22, and area 1,320 square miles. Extending in lat. from 37° 40′ to 38° 18′, and in long. from 3° to 4° 3′ W. W. C. It is principally drained by Green Brier r. and confluents; but from the western margin numerous creeks flow N. westwardly into Gauley r. Surface broken, and in part mountainous. The mean elevation of the farms above the ocean level, at least 1,500 feet. Chief t. Lewisburg. Pop. 1820, 7,040, 1830, 9,006.

GREEN CASTLE, p-v. in the southern part of Franklin co. Pa. situated at mid-distance between Chambersburg and Hagerstown, 11 ms. from each, and 77 N. W. W. C.

GREEN CASTLE, p-v. and st. jus. Putnam co. Ind. by p r. 614 ms. a little N of W. W. C. and 42 W. Indianapolis. Lat. 39° 42′.

GREEN CREEK, p-v. and tsp. northern part of Sandusky co. O. The p-o. is by p-r. 111 ms. N. Columbus, and 434 ms. N. W. by W. W. C. Pop. of the tsp. 1830, 444.

GREENBUSH, p-t. Rensselaer co. N. Y. on Hudson r. opposite Albany. The high ground above the village was an important cantonment during the late war. There is an academy, board and instruction $20 or $25 per quarter. Pop. 1830, 3,216.

GREENE, p-t. Kennebec co. Me. 39 ms. N. Portland, on the Androscoggin. Pop. 1830, 1,324.

GREENE COUNTY, N. Y. bounded by Schoharie and Albany cos. N., the Hudson r. E., Ulster co. S., and Del. co. W. Area about 508 sq. ms. It is crossed by the Catsberg mtns. Pop. 1820, 22,996, 1830, 29,525.

GREENE, southwestern co. of Pa. bounded by Washington co. N., by Monongahela r. separating it from Fayette E., by Monongalia co. of Va. S., Tyler co. Va. S. W., and Ohio co. Va. W. Length E. to W. 32 ms., mean breadth 18, and area 576 sq. ms. Extending in lat. from 39° 42′ to 40° 01′, and in long. from 2° 57′ to 3° 35′ W. W. C. About two thirds of the surface slopes estrd. and is drained by the numerous branches of Ten Mile and Dunkard creeks into Monongahela r. The western side slopes to the wstrd. and is drained by Fish and Wheeling creeks. Surface very broken, and along the line of separation between the confluents of Ohio and Monongahela rs. the aspect is mountainous. The soil is, however, almost invariably productive. Chief t. Waynesburg. Pop. 1820, 15,554, 1830, 18,026.

GREENE, co. of N. C. bounded by Lenoir S., Wayne W., Edgecombe N., and by Sandy creek, separating it from Pitt E. Length 20 ms., mean breadth 12, and area 240 sq. ms. Extending in lat. from 35° 32′ to 35° 40′, and in long. from 0° 35′ to 0° 50′ W. W. C. Chief t. Snow Hill. Pop. 1820, 4,533, 1830, 6,413. Contentney creek or river, a branch of Neuse r. traverses this co. from the N. W. to S. E. giving the surface a general slope in that direction.

GREENE, co. of Geo. bounded N. W. by Clark, N. E. by Oglethorpe, E. by Talliaferro, S. E. by Hancock, and by Oconee r. which separates it from Putnam S. W., and Morgan W. Length 28, mean width 18, and area 504 sq. ms. Extending in lat. from 33° 22′ to 33° 43′, and in long. from 6° 5′ to 6° 31′ W. W. C. The Oconee r. enters the northern border, and receiving Appalache from the N. W., becomes thence a boundary to the extreme southwestern angle of the co. Chief t. Greensboro′. Pop. 1820, 13,589, 1830, 12,549.

GREENE, co. of Ala. bounded by Pickens N. W., Tuscaloosa N. E., Perry E., Marengo S., and by Tombigbee r. which separates it from the Choctaw country S. W. and W. Length 38, mean width 22, and area 836 sq. ms. Extending in lat. from 32° 32′ to 32° 57′, and in long. from 10° 40′ to 11° 20′ W. W. C. This county being bounded by the Tombigbee, and traversed from N. to S. by Tuscaloosa r. its down stream navigable facilities are very great. The surface is hilly, but having considerable river bottom, much of its soil is excellent. Chief t. Erie. Pop. 1820, 4,554, 1830, 15,026.

GREENE, co. of Miss. bounded by Jackson S., by Perry W., Wayne N., and by Mobile co. Ala. E. Length 36, mean width 24, and area 864 sq. ms. Extending in lat. from 30° 55′ to 31° 27′ and in long. from 11° 37′ to 11° 58′ W. W. C. Chickasawhay r. enters the nthrn. border of this co., and winding to the sthrd. receives Leaf r. from the N. W., and the union of the two near the southern border of the co. forms the Pascagoula r. (*See article Chickasaw bay.*) The general feature of the surface of Green co. is that of pine forest, of course most of the soil is sterile. Staple, cotton. Chief t. Greensboro′. Pop. 1820, 1,445, 1830, 1,854.

GREENE, co. of East Ten. bounded by Cocke co. S. W., Jefferson W., Bays mtn., separating it from Hawkins N. W., Washington N. E., and by the Iron mtn., separating it from Buncombe co. N. C. S. E. Length 32, mean width 22, and area 704 sq. ms. Extending in lat. from 35° 52′ to 36° 20′, and in long.

from 5° 35′ to 6° 10′ w. W. C. This co. occupies part of an elevated valley between two Appalachian chains, and is in a peculiar manner diversified by hill, dale, mountain, and r. scenery. The Nolachucky r. rising in Buncombe co. N. C., and in Washington Ten. enters and traverses Greene co. in a western direction, receiving from the nthrd. Lick creek and numerous other streams of lesser size. Chief t. Greenville. Pop. 1820, 11,328, 1830, 14,410.

GREENE, co. of Ky. bounded by Barren s. w., Hart w., Hardin N. w. and N., Casey E., and Adair s. E. Length from s. w. to N. E. 38 ms., mean breadth 12, and area 456 sq. ms. Extending in lat. from 37° 07′ to 37° 30′, and in long. from 8° to 8° 35′ w. W. C. The slope wstrd., and in that direction traversed by the main volume of Green r. which receives within its limits, numerous tributary crs. from the N. E. and s. E. Chief t. Greensburgh. Pop. 1820, 11,943, 1830, 13,138.

GREENE, co. of O. bounded s. E. by Clinton, s. w. by Warren, w. by Montgomery, N. by Clark, N. E. by Madison, and E. by Fayette. Length 28, mean breadth 18, and area 500 sq. ms. Extending in lat. from 39° 30′ to 39° 51′, and in long. from 6° 38′ to 7° 8′ w. W.C. This co. is drained by some of the higher branches of Little Miami. Slope south wstrd. Chief t. Xenia. Pop. 1830, 14,801.

GREENE, co. of Ind. bounded by Daviess s., Knox s. w., Sullivan w., Clay N. w., Owen N. E., Monroe E., and Lawrence s. E. It is a parallelogram. Length 30 ms. from E. to w., breadth 18, and area 540 sq. ms. Extending in lat. from 38° 56′ to 39° 12′, and in long. from 9° 42′ to 10° 17′ w. W. C. Slope a little w. of s., and in that direction traversed, and nearly equally divided by the main stream of the western Fork of White r. Chief t. Bloomfield. Pop. 1830, 4,242.

GREENE, co. of Il. bounded by Morgan N. Macoupin E., Madison s. E., Miss. r. separating it from St. Charles co. Mo. s., Il. r., separating it from Calhoun co. Ind. w., and still by Il. r. separating it from Pike co. Ind. N. w. Extending in lat. from 38° 54′ to 39° 30′, and in long. from 13° 08′ to 13° 35′ w. W. C. The general slope is southwestward towards Il. r., and in that direction is drained by Otter, Macoupin and Apple creeks. Chief t. Carrollton. Pop. 1830, 7,674.

GREENE, tsp. and p-o. Harrison co. O. The p-o. is by p-r. 271 ms. N. w. by w. W. C. and 131 a little N. of E. Columbus.

GREENE MOUNTAINS, Vt. The range begins near New Haven, Conn. and runs nearly parallel to Conn. r. till it passes into Lower Canada. It gave the name to Vermont, through the middle of which it passes. Mansfield North Peak is the highest elevation, 4,279 feet above lake Champlain. Those nearest this in height are Camel's Back, Shrewsbury mtn., Mansfield, South Peak and Killington Peak, the last 3,924 feet. The range is crossed by several turnpike roads. In the s. part of Washington co. the range divides; and a spur called the Height of Land runs N. E. into the w. part of Caledonia co.

GREENE RIVER, p-v. Columbia co. N. Y.

GREENFIELD, tsp. and p-v. Erie co. Pa. Pop. of the tsp. 1830, 654.

GREENFIELD, p-t. Hillsboro' co. N. H. 38 ms. s. w. Concord. Pop. 1830, 946.

GREENFIELD, p-t. Franklin co. Mass. w. side of Conn. r., 21 ms. N. Northampton. Pop. 1830, 1,540.

GREENFIELD, p-t. Saratoga co. N. Y. 36 ms. N. Albany. Pop. 1830, 3,151.

GREENFIELD, p-v. Madison tsp. Highland co. O. It is situated in the northeastern angle of the co., and on Paint creek, 67 ms. s. s. w. Columbus, and 20 N. E. Hillsboro', the co. st. Pop. of the tsp. 1830, 399.

GREENFIELD, p-v. Nelson co. Va. by p-r. 114 ms. wstrd. Richmond.

GREENFIELD, p-v. in the northeastern part of Johnson co. Ind. 10 ms. s. E. Indianopolis.

GREENFIELD, p-v. and st. jus. Hancock co. Ind. This village, called in the p-o. list, Hancock court-house, is situated near the head of Sugar creek, 21 ms. by p-r. N. E. by E. Indianopolis. Pop. 1830, 133.

GREENFORD, p-v. Columbiana co. O.

GREEN GARDEN, p-v. Sumner co. Ten. 37 ms. N. E. Nashville.

GREEN HILL, formerly Freeman's Store, p-o. Jones co. Ga.

GREEN HILL, p-o. Columbiana co. O. 295 ms. N. w. W. C.

GREENLAND, p-t. Rockingham co. N. H. 4 ms. s. w. Portsmouth, on Great bay. Pop. 1830, 681.

GREENMONT, p-o. King William co. Va. 53 ms. from Richmond.

GREENOCK, p-v. and st. jus. Crittenden co. Ark. by p-r. 938 ms. s. w. by w. W. C.

GREEN POND, a beautiful lake in Morris co. N. J. giving name to a ridge of mountains 16 ms. N. Morristown.

GREEN RIVER, p-v. Rutherford co. N. C.

GREEN'S p-o. Jefferson co. Al. by p-r. 66 ms. N. E. Tuscaloosa.

GREEN'S p-o. Grayson co. by p-r. 130 ms. s. w. by w. W. C.

GREENSBOROUGH, p-t. Orleans co. Vt. 27 ms. N. E. Montpelier. Pop. 1830, 784.

GREENSBORO', p-v. Greene co. Pa.

GREENSBORO', p-v. Caroline co. Md. 8 ms. a little N. of E. Denton.

GREENSBORO', p-v. and st. jus. Guilford co. N. C. by p-r. 89 ms. N. w. by w. Raleigh. Lat. 36° 07′, long. 2° 52′ w. W. C.

GREENSBORO', p-v. and st. jus. Greene co. Geo. by p-r. 40 ms. a very little E. of N. Milledgeville. Lat. 33° 33′, long. 6° 12′ w. W. C.

GREENSBORO', p-v. in the southeastern part of Greene co. Al. 40 ms. almost directly s. Tuscaloosa.

GREENBURGH, t. West Chester co. N. Y. 28 ms. N. York, on the Hudson. Pop. 1830, 2,195.

GREENSBURG, p-v. in the northern part of Mecklenburg co. Va. 10 ms. N. N. E. Boydton, the co. st. and by p-r. 93 s. s. w. Richmond.

GREENSBURG, p-v., borough and st. jus. Westmoreland co. Pa. by p-r. 192 ms. N. W. W. C. It is situated on one of the head branches of Sewickly creek, 32 ms. S. E. by E. Pittsburg. It is a neat village, composed in great part of a single street along the great western r. Lat. 40° 18′, long. 2° 34′ w. W. C. Pop. 1830, 810.

GREENSBURG, small village on the left bank of Monongahela river, and in the southeastern angle of Green co. Pennsylvania, 20 ms. by land above Brownsville.

GREENSBURG, p-v. and st. jus. Greene co. Ky. situated on Greene river, 120 ms. N. E. Nashville, in Tenn. and by p-r. 82 ms. s. w. Frankfort. Pop. 1830, 669.

GREENSBURG, p-v. in the northeastern part of Trumbull co. Ohio, by p-r. 304 miles N. W. W. C.

GREENSBURG, p-v. and st. jus. Decatur co. Ind. by p-r. 559 ms. w. W. C. and 55 s. E. Indianopolis. Lat. 39° 16′, long. 8° 30′ w. W. C.

GREEN'S FORK, and p-o. Wayne co. Ind. by p-r. 75 ms. a little N. of E. Indianopolis.

GREENSVILLE, p-v. and st. jus. Grayson co. Va. This place, called in the p-o. list Grayson court house, is situated on the right bank of New river, or the higher part of Great Kenhawa, 25 ms. S. S. E. Evansham, and by p-r. 354 ms. s. w. by w. W. C. Lat. 36° 38′, long. 3° 55′ w. W. C.

GREENTOWN, p-v. in Lake tsp. Stark co. Ohio, 11 ms. N. W. Canton, the st. jus. for the co. Pop. 1830, 85.

GREENTREE-GROVE, and p-o. Stewart co. Tenn. by p-r. 94 miles N. W. by W. Nashville.

GREENUP, northeasterly co. of Ky. bounded S. by Lawrence, W. by Lewis, N. W. by Ohio river, separating it from Sciota co. state of Ohio, N. E., again by Ohio river, separating it from Lawrence co. state of Ohio, and E. by Big Sandy river, separating it from Cabell co. Va. Length from west to east, 48 ms. mean breadth 16, and area 768 square ms. Extending in lat. from 38° 13′ to 38° 44′, and in long. from 5° 30′ to 6° 23′ w. W. C. Slope as of the adjoining co. of Va. Cabell is to the northward. The greatest part of Greenup is drained by Little Sandy and Tyger's creeks. Chief town, Greenupsburg. Pop. 1820, 4,311, 1830, 5,852.

GREENUP, or GREENUPSBURG, p-v. and st. jus. Greenup co. Ky. situated on Ohio river, at the mouth of Little Sandy river, by p-r. 138 ms. N. E. by E. Frankfort. Lat. 38° 32′, long. 5° 46′ w. W. C. Pop. 1830, 204.

GREENVILLE, dist. S. C. bounded by Spartanburg E., Lawrence co. S. E., Anderson S. W., Pickens W., and Buncombe co. N. C. N. Length from S. to N. 47 ms. mean width 15, and area 705 square ms. Extending in lat. from 34° 28′ to 35° 10′, and in long. from 5° 10′ to 5° 40′ w. W. C. The slope nearly to the southward, falling from the Blue Ridge, which bounds it on the north. On the west it is limited in all its length by Saluda river, which separates it from Anderson and Pickens districts. It is drained by the branches of Saluda, Reedy, Ennoreo, and Tyger rivers. The surface is finely diversified by mtn. hill and valley scenery, with much excellent soil. Chief town, Greenville. Population 1820, 14,530, 1830, 16,476.

GREEN VALLEY, p-v. Warren co. Pa. by p-r. 233 ms. N. W. Harrisburg.

GREEN VALLEY, p-v. on Cowpasture river, Bath co. Va. by p-r. 230 ms. s. w. by w. Washington City, and 181 N. W. by W. Richmond.

GREEN VILLAGE, p-v. and tsp. Franklin co. Pa. The p-o. is about 5 ms. N. E. Chambersburg.

GREENVILLE, p-t. Greene co. N. Y. 17 ms. N. W. Catskill, on Catskill river. Population 1830, 2,565.

GREENVILLE, p-v. in the northern part of Luzerne co. Pa. by p-r. 155 ms. N. E. Harrisburg.

GREENVILLE, one of the southern counties of Va. bounded by Brunswick W., Notaway river separating it from Dinwiddie N. W., by Notaway river again separating it from Sussex N., a part of Sussex and a part of Southampton E., and by Northampton co. N. C. on the S. Length 22, mean width 14, and area 308 square ms. Extending in lat. from 36° 30′ to 36° 48′, and in long. from 0° 20′ to 0° 46′ w. W. C. Meherin river entering the western border, traverses it southeasterly and divides it into two not very unequal sections, and being bounded on the north by Notaway river. It is well situated, commercially. Slope eastward with a slight inclination to the south. Chief town, Hicksford. Pop. 1820, 6,858, and in 1830, 7,117.

GREENVILLE, p-v. in the southern part of Augusta co. Va. 11 ms. s. s. w. Stanton, and 136 N. W. by W. Richmond.

GREENVILLE, p-v. and st. jus. Pitt co. N. C. situated on the left bank of Tar river, 23 ms. by land above Washington, and by p-r. 105 ms. a little S. of E. Raleigh. Lat. 35° 35′, long. 0° 24′ w. W. C.

GREENVILLE, p-v. and st. jus. Greenville dist. S. C. situated on, and near the head of, Reedy river, by p-r. 110 ms. N. W. Columbia. Lat. 34° 50′, long. 5° 27′ w. W. C.

GREENVILLE, p-v. and st. jus. Merriwether co. Geo. by p-r. 753 ms. s. w. W. C. and 111 ms. w. Milledgeville.

GREENVILLE, p-v. and st. jus. Butler co. Ala. situated on a creek of the Sapulga branch of Conecuch r. about 120 ms. N. E. Mobile, and by p-r. 151 ms. s. s. E. Tuscaloosa, lat. 31° 42′, long. 9° 46′ w. W. C.

GREENVILLE, p-v. and st. jus. Jefferson co. Miss. situated on a branch of Coles creek, 24 ms. N. N. E. Natchez. Lat. 31° 47′, long. 14° 9′ w. W. C.

GREENVILLE COLLEGE, and p-o. Green co.

Ten. is laid down on Tanner's map of the Uunited States, about 4 ms. a little E. of S. Greenville, the co. st. This is the most ancient collegiate establishment made in the United States, westward of the Appalachian mountains, being founded in 1794, four years previous to Transylvania University, Ky. According to the statement of Mr. W. R. Johnson, head of education, art. U. S. in the Philadelphia edition of Brewster's Encyclopedia, published this year, 1832, Greenville college, had students 32; volumes in the college library, 3,500, and the annual term of instruction included 42 weeks.

GREENVILLE, p-v. and st. jus. Green co. Tenn. situated 71 ms. a little N. of E. Knoxville, and by p-r. 232 ms. E. Nashville, lat. 36° 07.'

GREENVILLE, p-v. and st. jus. Muhlenberg co. Ky. situated on a small branch of Green river, by p-r. 171 ms. S. W. by W. Frankfort. Pop. 1830, 217.

GREENVILLE, p-v. and st. jus. Dark co. O. by p-r. 501 ms. N. W. by W. ½ W. W. C. and 103 westward Columbus. It is situated on Greenville creek, a branch of Great Miami river, lat 40° 06', long. 7° 36' W. W. C. Pop. 1830, 160.

GREENVILLE, p-v. in the western part of Floyd co. Ind. 9 ms. N. W. New Albany, the co. st.

GREENVILLE, p-v. and st. jus. Bond co. Il. 20 ms. S. W. by W. Vandalia, and 801 W. W. C. lat. 38° 53'.

GREENVILLE, p-v. and st. jus. Wayne co. Mo. by p-r. 908 ms. a little S. of W. W. C. and about 120 a very little W. of S. St. Louis, lat. 37° 06'.

GREENWICH, p-t. Hampshire co. Mass. 20 ms. E. Northampton. Pop. 1830, 813.

GREENWICH, p-t. Fairfield co. Conn. 48 ms. W. N. Haven, on Long Isl. Sound. Pop. 1830, 3,805.

GREENWICH, p-t. Washington co. N. Y. on Hudson r., contains a number of manufactories on the Battonkill. Union v. is 5 ms. from the r., 37 ms. N. E. Albany. Pop. 1830, 3,850.

GREENWICH, t. Cumberland co. N. J. on the Del. bounded E. by Cohansey cr., W. by Salem co. Pop. 1830, 912.

GREENWICH, p-v. in the town of the same name, Cumberland co. N. J. on Cohansey cr. 6 ms. S. W. of Bridgeton.

GREENWICH, t. Gloucester co. N. J. on Del. r. bounded N. E. by Deptford, S. W. by Woolwich. Pop. 1830, 2,657.

GREENWICH t. Warren co. N. J., in the S. W. end of the co., bounded on Del. r. & Musconetcunk, 31 ms. S. Newton. Pop. 1830, 4,486.

GREENWICH, p-v. and tsp. in the estrn. part of Huron co. O. by p-r. the p-o. is 384 ms. N. W. by W. W. C., and 105 N. N. E. Columbus.

GREENWOOD, t. Oxford co. Me. 5 ms. N. W. Paris. Pop. 1830, 694.

GREENWOOD, p-v. nrthestrn. part of Columbia co. Pa., by p-r. 92 ms. nrthrd. Harrisburg.

GREENWOOD, p-v. Laurens dist. S. C. 81 ms. N. W. Columbia.

GREGGVILLE, p-v. Loudon co. Va. 54 ms. wstrd. W. C.

GREGSTOWN, v. Somerset co. N. J., 6 ms. N. E. Princeton.

GRETNA GREEN, formerly Fortune's Fork, p-v. Halifax, N. C., by p-r. 216 ms. S. W. C., and 86 N. E. by E. Raleigh.

GRIERSBURG, Beaver co. Pa. (*See Darlington, same co.*)

GRIFFINSBURG, p-o. Culpepper co. Va. 90 ms. S. W. W. C.

GRIGGSBY's Store and p-o., Fauquier co. Va. 61 ms. from W. C.

GRIGGSTOWN, v. Somerset co. N. J., on Milstone r., and the Del. and Raritan canal, 12 ms. W. New Brunswick.

GRIMVILLE, p-o, Berks co. Pa.

GRISWOLD, t. N. London co. Conn. 6 ms. N. E. Norwich, on Quinebaug r. Pop. 1830, 2,212.

GROTON, t. Grafton co. N. H., 45 ms. N. W. Concord. Pop. 1830, 689.

GROTON, t. Caledonia co. Vt. 16 ms. E. Montpelier. Pop. 1830, 836.

GROTON, p-t. Middlesex co. Mass. 34 ms. N. W. Boston. Pop. 1830, 1,925.

GROTON, p-t. N. London co. Conn. at the mouth of Thames r. opposite N. London. Fort Griswold is on the summit of a hill commanding N. London harbor; a monument has been erected there, by subscription, in memory of the capture of the fort, and a cruel massacre, made by British troops under Benedict Arnold, September 6th, 1781. Pop. 1830, 4,750.

GROTON, p-t. Tompkins co. N. Y. 14 ms. N. E. Ithaca. There are mills, &c. on Fall cr. and two villages, Moscow and Peru. Pop. 1830, 3,597.

GROVE, p-v. Tazewell co. Il., by p-r. 196 ms. N. Vandalia, and 748 ms. N. W. by W. ½ W. W. C.

GROVE HILL, p-o. Clark co. Al., by p-r. 127 ms. a little W. of S. Tuscaloosa.

GROVELAND, p-t. Livingston co. N. Y. 6 ms. S. Genesco. On Genesee r. is Williamsburgh v. Pop. 1830, 1,703.

GROVE LEVEL, p-o. Franklin co., Geo. by p-r. 97 ms. N. Milledgeville.

GROVEVILLE, v. Burlington co. N. J. on Crosswick cr. 3 ms. N. E. Bordentown.

GUANOS, (*See Brooklyn, N. Y.*)

GUERNSEY, co. O. bounded S. E. by Monroe; by Morgan S. W.; Muskingum W.; Coshocton N. W.; Tuscarawas N.; Harrison N. E., and Belmont E. Length from sth. to nrth. 28 ms.; mean breadth 25, and area 700 sq. ms. Extending in lat. from 39° 51' to 40° 12'; and in long. from 4° 13' to 4° 43' W. W. C. It is almost commensurate with the higher valley of Wills' cr.; slope nrthwstrdly. in the general course of that stream; surface hilly, with good soil. Chief t. Cambridge. Pop. 1820, 9,292; 1830, 18,036.

GUILDERLANDT, p-t. Albany co. N. Y., 12 ms. W. Albany. Norman's Kill and its branches furnish mill seats; Hamilton v. 8 ms. from Albany. Pop. 1830, 2,742.

GUILDHALL, p-t. and st. jus. Essex co. Vt., opposite Lancaster, with two bridges over Conn. r., 50 ms. N. E. Montpelier; it has mill seats, a court house and jail. Pop. 1830, 481.

GUILFORD, t. Penobscot co. Me., 49 ms. N. E. Norridgewock. Pop. 1830, 655.

GUILFORD, p-t. Windham co. Vt., 31 ms. E. Bennington. 350 acres of land were appropriated to schools, and 500 acres to the governor; the last include Governor's mtn. a barren tract; it has several mills. Pop. 1830, 1,760.

GUILFORD, p-t. and borough, N. Haven co. Conn., 15 ms. E. New Haven, on Long Isl. sound; it has two harbors. Pop. 1830, 2,344.

GUILFORD, p-t. Chenango co. N. Y., 108 ms. w. Albany. Pop. 1830, 2,634.

GUILFORD, p-v. York co. Pa., by p-r. 78 ms. N. W. C.

GUILFORD, p-v. and tsp. Medina co. O., p-o. by p-r. 103 ms. N. N. E. Columbus. Pop. tsp. 1830, 625.

GULF MILLS, p-o. Montgomery co. Pa.

GULF (The), p-o. Chatham co. N. C., 16 ms. wstrd. Pittsboro', the co. st., and 49 ms. wstrd. Raleigh.

GULF STREAM, (*See art. Atlantic ocean.*)

GULL ISLANDS, at the mouth of Long Isl. sound. They are two, Great and Little; on the latter is a light house.

GUNPOWDER, r. of Maryland. This stream has its remote sources near the line of demarcation between Pa. and Md., but enters the latter as mere rills; pursuing a general sthestrn. course over Baltimore co., by comparative distance 30 ms. it receives a large cr. from the nrthestrd., called the Falls of Gunpowder, and having met the tides, inflects to the sthrd. 10 ms. widening into a bay, which is finally merged in the larger sheet of the Chesapeake. Gunpowder is navigable for small vessels to Joppa, at the confluence of the two branches, and near the head of tide water.

GUSTAVUS, p-v. and tsp. nrthrn. part of Trumbull co. O., 22 ms. N. Warren, the co. st. and 319 ms. N. W. W. C.

GUTHRIESVILLE, p-o. wstrn. part Chester co. Pa., by p-r. 72 ms. E. Harrisburg.

GUYANDOTTE, r. of Va., rising in Logan co. from the nrthwstrn. foot of the Great Flat Top mtn. and flowing thence N. N. W. draining a valley between those of Great Kenhawa and Sandy rs., enters Cabell co. and falls into O. r. below Barboursville, after a comparative course of about 100 ms.

GUYANDOTTE, LITTE, cr. of Va., falling into O. r. between the mouths of Guyandotte and Great Kenhawa rs., and for some ms. above its mouth constitutes the boundary between Mason and Cabell cos.

GUYANDOTTE, p-v. Cabell co. Va., by p-r. 401 ms. s. w. by w. ½ w. W. C.

GUY'S MILLS, and p-o., Crawford co. Pa.

GWINN'S MILL, and p-o. Monroe co. Va., by p-r. 277 ms. s. w. by w. W. C.

GWINNET, co. Geo., bounded N. E. by Hall; E. by Jackson; S. E. by Walton; S. by Newton; S. W. and W. by De Kalb; and N. by Chattahooche r. Length 36 ms.; mean width 18, and area 648 sq. ms. Extending in lat. from 33° 50′ to 34° 12′, and in long. from 6° 47′ to 7° 28′ w. W. C. Lat 34° extends across Gwinnet, along very nearly its greatest length, and divides it into two nearly equal portions. This natural geographical limit also passes upon a dividing ridge, from which flow nrthwstrdly. some confluents of Chattahooche, and sthrdly. the extreme sources of Ockmulgee r. Gwinnet, therefore, occupies a part of the summit ridge between the rs. of the Atlantic slope, and those of the Gulf of Mexico. Chief t. Lawrenceville. Pop. 1820, 4,589; 1830, 13,289.

GWYNNED, or Gynned, p-o. and tsp. Montgomery co. Pa. on the heads of Tonamensing and Wissahiccon crs. 18 ms. N. N. W. Phil.

H.

HABERSHAM, co. of Geo., bounded by Turoree r. separating it from Rabun N. E.; the Tugaloo r. separating it from Pickens dist. S. C. E.; Franklin co. of Geo. S. E.; Hall S.; Chestatee r. W.; and Macon co. in N. C. N. Length 38; mean breadth 20, and area 760 sq. ms. Extending in lat. from 34° 27′ to 35°, and in long. from 6° 20′ to 6° 55′ w. W. C. Habersham embraces an elevated and remarkable natural section; from it flows to the N. W. the sources of Hiwassee branch of Ten. r.; from the wstrn. part flows the sources of Etowah, or the most nrthestrly. fountains of Mobile; in the central part rises the extreme nrthrn. sources of Chattahooche; whilst from the estrn. side are discharged the most nrthwstrly. constituents of Savannah r. It is in this co. and in the wstrly. border of Rabun, that the Blue Ridge declines to the w., forming the nucleus from which the streams are discharged like radii from a common centre. The mean height of Habersham, independent of the mtn. ridges, must be at least 1,500 feet, or an equivalent to near 4 degrees of temperature; assimilating the winter climate to that on the Atlantic border, on lat. 39°. Chief t. Clarksville. Pop. 1820, 3,171; 1830, 10,671.

HACKERSVILLE, p-v. Lewis co. Va., by p-r. 246 ms. w. W. C.

HACKETTSTOWN, v. Warren co. N. J., on the w. side of Muskonetcunk r., 22 ms. w. Morristown, 4 ms. N. Schoolley's mount. The Morris canal passes 1 m. N. W. of this place.

HACKINSACK, r. N. J., rises in Rockland co. N. Y. runs 14 ms. and enters N. J., emptying into Newark bay; navigable 15 ms.

HACKINSACK, p-v. and st. jus. Bergen co.

N. J., on the w. side of Hackinsack r., 14 ms. N. N. Y., 14 N. N. E. of Newark, and 7 ms. E. of Paterson. There is a handsome court house, 2 churches, a bank and academy in the v.

HACKNEY'S CROSS ROADS, and p-o. Chatham co. N. C., by p-r. 38 ms. w. Raleigh.

HADDAM, p-t. Middlesex co. Conn., 23 ms. s. Hartford, on the w. side Conn. r. Granite is quarried here. Pop. 1830, 2,830.

HADDONFIELD, v. Gloucester co. N. J., on Cooper's cr. 9 ms. E. Camden.

HADENSVILLE, p-v. Todd co. Ky., by p-r. 188 ms. s. w. by w. Frankfort.

HADLEY, p-t. Hampshire co. Mass. on the E. side Conn. r. opposite Northampton, (to which it is connected by a bridge across the Conn.) and 97 ms. w. Boston; the Hopkins academy here is very respectable and flourishing; this is a fine farming town, and the meadows are the finest in New England; the manufacture of brooms in this town is very extensive; the crop of broom corn in 1831, was estimated at 150 tons, and the value of the brush and seed alone, at $21,750; the whole crop of 1831, was manufactured into brooms within the town, and great quantities of this article are annually scattered hence, through the U. S. Pop. 1830, 1,886.

HADLEY, t. Saratoga co. N. Y. 27 ms. N. Ballstown Springs, 51 N. Albany. Much timber is sawn at the falls on the Mohawk. Pop. 1830, 829.

HADLEY'S MILLS, and p o. by p-r. 43 ms. westward Raleigh.

HAERLEM, p-v. N. York co. N. Y. 8 ms. N. New York. The heights were fortified in the revolutionary war, and in the late war. A canal is projected to cross Manhattan island here.

HAERLEM r. N. Y. co. N. Y. is a strait on the N. side of Manhattan island, 6 ms. long, and from ¼ to ½ m. wide.

HAERLEM, p-v. and tsp. called on the census tables Harlem, in the southern part of Delaware co. O. The p-v. 20 ms. northward from Columbus. Pop. tsp. 1830, 532.

HAGERSTOWN, p-v. and st. jus. Washington co. Md. situated on a fine limestone valley 2 ms. westward from Antietam creek, 72 ms. N. W. by W. from Baltimore, and by p-r. 69 ms. N. W. W. C. Lat. 39° 39', and long. 0° 42' w. W. C. It is a well built and thriving town, with a well cultivated, fertile and wealthy neighborhood. It contains the usual co. buildings, a female academy, numerous private schools, with several places of public worship. By the census of 1830, the pop. of this place stood,

Whites, Males.	Fems.	Free col'd.	Slaves.	Total.
1,307	2,675	396	300	3,371

HAGUE, p-t. Warren co. N. Y. on lake George, 22 ms. N. E. Caldwell. Brant lake and Rogers Rock, are in this town. Pop. 1830, 721.

HAGUE, p-v. eastern part of Westmoreland co. Va. by p-r. 116 ms. s. E. Richmond.

HAILSTONE, p-v. Mecklenburg co. Va. by p-r. 91 ms. s. s. w. Richmond.

HALBERTS' p-o. Tuscaloosa co. Ala. 25 ms. southwestward Tuscaloosa.

HALFMOON, p-t. Saratoga co. N. Y. on Hudson r. 14 ms. N. Albany. Clifton Park, the borough, and Newtown, are small villages. The Erie and Champlain canals, run through the town. Pop. 1830, 2,042.

HALF MOON, tsp. and p-o. southern part of Centre co. Pa. by p-r. 178 ms. N. w. Harrisburg. Pop. tsp. 1830, 1,092.

HALFWAY HOUSE, and p-o. Ann Arundel co. Md. by p-r. 15 ms. from Annapolis.

HALFWAY HOUSE, and p-o. eastern part of York, Va. 84 ms. s. E. by E. Richmond.

HALIFAX, p-t. Windham co. Vt. 9 ms. s. w. Brattleborough, has a male and female school for the higher branches. Pop. 1830, 1,562.

HALIFAX, p-t. Plymouth co. Mass. 13 ms. N. w. Plymouth, 35 s. E. Boston. Pop. 1830, 709.

HALIFAX, co. of Va. bounded by Pittsylvania w., Roanoke r. which separates it from Campbell N., and Charlotte N. E. and E., by Mecklenburg co. Va. and Granville of N. C. s. E., and Person co. of N. C. s. Length 33, mean breadth 23, and area 759 sq. ms. Extending in lat. from 36° 30' to 37° 02', and in long. from 1° 38' to 2° 12' w. W. C. Though the Roanoke curves semicircularly round the northern and eastern border of this co. the slope is almost directly eastward; Dan r. enters at the s. w. angle and flowing N. E. by E. over the co. receives within it Banister r. from the northwest, and Hycootee from the southwest, and thus augmented, joins the Roanoke at the extreme eastern angle of the co. It is a well watered co. with much excellent soil. Chief town, Banister. Pop. 1820, 19,060, 1830, 28,034.

HALIFAX, p-v. and tsp. on the left bank of Susquehannah r. at the mouth of Armstrong cr. Dauphin co. Pa. 18 ms. above, and northwards Harrisburg. Pop. tsp. 1830, 1,772.

HALIFAX, C H. Halifax co. Va. (*See Banister.*)

HALIFAX, co. N. C. bounded by Roanoke r. which separates it from Northampton N. E. and E., and from Bertie s. E., by Martin co. s., by Fishing cr. separating it from Edgecombe and Nash s. w., and by Warren w. Length 45 ms., mean width 16, and area 720 sq. ms. Extending in lat. from 35° 57' to 36° 28', and in long. from 0° 18' to 1° 03' w. from W. C. This co. is crossed by the great primitive ledge which separates the sea sand alluvion, from the hilly, or intermediate region between the tide waters, and the mountain system in the interior of the continent. By the joint exertions of N. C. and Va., a canal and sluice navigation has been completed along the Roanoke, from Weldon in Halifax co. N. C. to Salem in Botetourt co. Va. (*see art. Roanoke, Halifax co. of N. C.*) possesses the advantages of tide and r. navigation, with extensive bodies of fertile soil, and a climate admitting the profitable cultivation of cotton.

Chieftown, Halifax. Pop. 1820, 17,237, 1830, 17,739.

HALIFAX, p-t. port of entry, and st. jus. Halifax co. N. C. situated on the right bank of Roanoke r. 90 ms. s. w. by w. from Norfolk in Va. and by p-r. 103 N. E. from Raleigh in N. C. Lat. 36° 18', long. 0° 38' w. W. C. Vessels of 45 tons ascend to this port and there come in contact with the extensive navigation of Roanoke r. above tide water.

HALL, co. of Geo. bounded by Habersham N., Franklin E., Jackson S. E., Gwinnet s. w., and Chestatee or Chattahooche r. w. Length 35, mean breadth 15, and area 525 sq. ms. Extending in lat. from 34° 03' to 34° 28', and long. 6° 28' to 6° 53' w. W. C. This co. similar to Gwinnet and Habersham, occupies a part of the table land between the waters of the Atlantic and those of the Gulf of Mexico. The whole southeastern side slopes to the S. E. discharging into Franklin, the extreme sources of Broad r. branch of Savannah r. and into Jackson, the higher fountains of Oconee. The western and northern sections are drained by Chestatee and Soquire, uniting near the middle of the co. to form Chattahooche. Chief town, Gainesville. Pop. 1820, 5,086, 1830, 11,748.

HALLOCA, p-v. Muscogee co. Geo. by p-r. 134 ms. s. w. by w. Milledgeville.

HALLOCKSBURG, p-v. Bourbon co. Ky. by p-r. 37 ms. eastward Hartford.

HALLOWELL, p-t. Kennebec co. Me. on Kennebec r. at the head of the tide, 2 ms. below Augusta, 54 N. E. Portland. It is one of the most wealthy, populous, and flourishing towns in the state. The principal village is on the w. bank of the r. It is navigable to this place for vessels of 150 tons. The Hallowell granite is very celebrated, and is extensively quarried and wrought. Pop. 1830, 3,961.

HALLS r. N. H. forms the boundary between L. Canada and N. H. from its source in the highlands, to its junction with the Connecticut r. at Stewartstown.

HALLSBORO, p-v. Chesterfield co. Va. by p-r. 17 ms. from Richmond.

HALLS CROSS ROADS, and p-o. at the head of the N. E. branch of Bush r. Harford co. Md. 30 ms. N. E. from Baltimore.

HALLSVILLE, p-v. Amelia co. Va. by p-r. 33 ms. sthwestrd. Richmond.

HALLSVILLE, p-v. Duplin co. N. C. by p-r. 106 ms. s. E. from Raleigh.

HALLSVILLE, p-v. in the western part of Fairfield district, S. C. by p-r. 60 ms. N. N. W. Columbia.

HALLSVILLE, p-o. Ross co. O. by p-r. 50 ms. southward Columbus.

HALSELLVILLE, p-o. Chester district, S. C. by p-r. 50 ms. N. Columbia.

HALSEYVILLE, p-v. Chester dist. S. C. by p-r. 47 ms. N. from Columbia.

HAMBAUGH'S, p-o. Shenandoah co. Va. by p-r. 82 ms. w. W. C.

HAMBURGH, p-t. Erie co. N. Y. 9 ms. s. Buffalo. Surface variable. Climate always dripping with an overload of moisture. Contains 17 schools. Population 1830, 3,351.

HAMBURGH, village Sussex co. N. J. on the Wallkill 12 ms. N. E. Newton.

HAMBURG, flourishing p-v. on the left bank of the Schuylkill r. immediately below the gap where that stream passes the Kittatinny mtn. Berks co. Pa. 16 ms. above Reading. It is composed in great part of a single street extending along the great western road, and nearly parallel to the r. Pop. 1830, about 500.

HAMBURG, p-v. on Savannah r. directly opposite to Augusta, in Geo. and in the southern part of Edgefield dist. S. C. by p-r. 81 ms. s. w. Columbia. A rail road is in progress from this town to Charleston. (*See rail roads and canals.*)

HAMBURGH, p-v. Calhoun co. Il. by p-r. 136 ms. w. Vandalia.

HAMDEN, t. N. Haven co. Conn. 5½ ms. from N. Haven, and 32 ms. from Hartford. It contains several kinds of minerals. A mass of copper weighing 90 lbs. was once discovered on one of the Greenstone hills of the town. Soil generally fertile. Timber, walnut, oak of the various kinds, and other deciduous trees. It contains numerous mill privileges, an extensive gun manufactory, and other manufactories of various kinds. Pop. 1830, 1,669.

HAMILTON, p-t. Essex co. Mass. 26 ms. N. E. Boston. A neat and pleasant town. Soil good. Surface sufficiently level for beauty or utility. Pop. 1830, 748.

HAMILTON, co. N. Y. bounded N. by St. Lawrence and Franklin cos., E. by Essex, Warren, and a small part of Saratoga cos., S. by Montgomery co., and w. by Herkimer co. It is 60 ms. long N. and S., 30 ms. wide E and w. containing 1800 sq. ms. or 1,152,000 acres. Its surface is elevated. It is traversed by mtns. abounding with swamps. The principal part of the land is of little value. Pop. 1820, 1,251, 1830, 1,325.

HAMILTON, p-t. Madison co. N. Y. situated on the Chenango r., being the seat of the N. Y. Baptist theolog. seminary, founded in 1819 by the Baptist education society of the state of N. Y. It has an edifice of stone, 64 feet by 36, erected at the expense of the inhabitants of the village; several scholarships, each endowed with $1000; and a library, 8 ms. S. E. Morrisville, and 25 s. w. Utica. Pop. 1830, 3,220.

HAMILTON, village, Albany co. N. Y. (*See Guilderlandt.*)

HAMILTON COLLEGE, N. Y. (*See Paris.*)

HAMILTON, river, N. Y. (*See Olean.*)

HAMILTON, town, Gloucester co. N. J. Pop. 1830, 1,424.

HAMILTON'S STORE, and p-o. Loudon co. Va. by p-r 37 ms. w. W. C.

HAMILTON, p-v. Martin co. N. C. by p-r. 120 ms. E. Raleigh.

HAMILTON, p-v. and st. jus. Harris co. Geo. named on the post office list, Harris court house, and stated by p-r. 112 ms. from Milledgeville, course a little s. of w.; on Tanner's U. S. it is laid down at 32° 44', long. 8° 03' w. W. C.

Hamilton, co. of Ten. bounded by Marion w., Bledsoe n. w., Rhea n. e., and Ten. river e. s. e. and s. Extending in lat. from 35° 04' to 35° 41', and long. from 8° to 8° 22' w. W. C. The eastern & southestrn. sections of this co. occupy part of the slope descending easterly from Walden's Ridge, to Ten. r. The northwestern section is a parallelogram of about 8 by 10 ms. sloping westerly from Walden's Ridge to the Sequatche river. The Ten. inclined plain is about 32 ms. in length along the river, with a width of 12, area 384 square ms. The whole co. having a superficies of 464 square ms. Chief town, Hamilton court house. Pop. 1820, 821, in 1830, 2,274.

Hamilton, court house, and p-o. Hamilton co. Ten. about 120 ms. s. e. by e. from Nashville.

Hamilton, co. Ohio, bounded w. by Dearborn co. Ind., n. by Butler co. O., n. e. by Warren, e. by Clermont, s. e. by Ohio river, separating it from Campbell co. Ky., and s. w. by Ohio river, separating it from Boone co. Ky. Length from east to west 30, mean breadth 16 miles, and area 480 square ms. Lat. 39° 02' to 39° 20', long. 7° 18' to 7° 48' w. W. C. This co. occupies a part of that great buttress of hills, which skirt the right bank of Ohio river from its head near Pittsburg, to below the influx of Wabash. Great and Little Miami traverse Hamilton in deep vallies. The surface of the co. is hilly, but the soil in an especial manner productive. Bituminous mineral coal abounds. The Miami canal, connecting the stream of Great Miami with the Ohio, reaches the latter in Cincinnati. In 1830, this co. exclusive of the city of Cincinnati, contained a population of 22,317, and including the inhabitants of that city 52,317, or with the city a distributive population of 109 to the sq. m.

Hamilton, p-v. and st. jus. Butler co. O. 25 ms. n. Cincinnati. It is situated on the left bank of Great Miami, lat. 39° 22'. Pop. 1830, 1,079.

Hamilton, co. Ind. bounded s. e. by Hancock, s. by Marion, w. by Boone, n. by ——, and e. by Madison. It is a square of 21 ms. each way, 441 square ms. Lat. 39° 57' to 40° 13', long. 8° 53' to 9° 15' w. W. C. Slope s. s. w. and in that direction is traversed by the main stream of White river, and also by some of its tributaries. Chief town, Noblesville. Pop. 1830, 1,757.

Hamilton, p-v. and st. jus. Monroe co. Miss. situated on Battahatchee river, 15 ms. n. n. e. Columbus, 237 n. e. Natchez, and 70 n. w. Tuscaloosa, in Ala.

Hamilton, co. of Il. bounded s. by Gallatin, s. w. by Franklin, n. w. by Jefferson, n. by Wayne, and e. by White. It is a square of 24 ms. each way, 576 square ms. Lat. 37° 59' to 58° 16'. Long. 11° 25' to 11° 16' w. W. C. This co. is a table land; from the southwestern angle rises the Raccoon branch of Muddy river, the northern part is drained by Wayne's fork of Little Wabash, whilst the central and rather most extensive section gives source to the north branch of Saline r. The slopes are consequently s. w. towards the Mississippi in the general direction of Muddy river, s. e. towards Ohio river by the course of the Saline, or eastward towards the Wabash, with the branches of Little Wabash. Chief town, MacLeansboro'. Pop. 1830, 2,616.

Hamilton, co. of Florida. This co. is, as laid down on Tanner's U. S. map, bounded n. by Lowndes and Ware counties, Geo., e. and s. by Little Sawannah, and w. by Withlacuchee river. Length 34 ms. mean breadth 17, and area 578 square miles. Extending in latitude from 30° 20' to 30° 29', and in long. 5° 52' to 6° 28' w. W. C. This co. slopes southward, and is drained by the various branches of Suwannee river. Chief town, Micco. Pop. 1830, 553.

Hamilton, village, on the west bank of Schuylkill, opposite to and adjoining Phila. of which city it is really a suburb, extending principally along West Chester, Darby and Lancaster roads. The site rises by a fine acclivity from Schuylkill, and affords elegant seats for houses, many of which are the summer retreats of the citizens of Phila. The village and the city are connected by Permanent Bridge.

Hamlet's, p-o. Stewart co. Tenn. by p-r. 67 ms. s. w. by w. Nashville.

Hamlinton's, p-o. Wayne co. Pa. by p-r. 150 ms. n. e. Harrisburg.

Hamorton, p-o. Chester co. Pa. by p-r. 107 ms. n. e. W. C.

Hamor's Store, and p-o. Delaware co. Pa. 129 ms. n. e. W. C.

Hampden, p-t. Penobscot co. Me. w. Penobscot river, 10 ms. s. s. w. Bangor, 29 ms. n. w. Castine. Pop. 1830, 2,020.

Hampden, co. Mass. contains 19 towns, bounded n. by Hampshire, e. by Worcester, s. by Conn. line, w. by Berkshire. Chief town, Springfield, e. Conn. river. Excellently watered, by the passing of Conn. river through its centre from n. to s., by Chickapee river from the e., and Westfield river from the west. The Farmington canal passes through the width of the co. and opens a direct communication with New Haven.—Steam-boats now pass up the Conn. river through this co. Pop. 1830, 31,640.

Hampden, p-v. Walton co. Geo. by p-r. 82 ms. n. n. w. Milledgeville.

Hampden, p-v. and tsp. northeastern part of Geauga co. Ohio. The p-o. is by p-r. 127 ms. n. e. from Columbus. Pop. of the tsp. 1830, 530.

Hampshire, co. Mass. contains 23 towns. Northampton is the chief, by which runs the Conn. river, and through the centre of the co. from n. to s. A branch of Swift river waters the e. and a branch of Westfield river the w. parts of the co. It is bounded n. by Franklin, e. by Worcester, s. by Hampden,

lying wholly in the valley of the Conn. The soil is of the best quality. Population 1830, 20,210.

HAMPSHIRE, co. of Va. bounded by Morgan N. E., Frederick E., Hardy S. and S. W., and the Potomac river separating it from Alleghany co. in Md. N. W. and N. Length 40, mean breadth 24, and area 960 square ms. Extending in lat. from 1° 28' to 2° 12' w. W. C. The slope of this mountainous co. is to the northeast, traversed in that direction by the south branch of Potomac and several lesser streams, with lateral mtn. ridges intervening. Though so much broken by mtn. much of the soil on the streams is excellent. The lowest part along the two branches of Potomac, exceeds an elevation of 500 feet above tide water. Chief town, Romney. Pop. 1820, 10,889, in 1830, 11,279.

HAMPSTEAD, p-t. Rockingham co. N. H. 24 ms. from Portsmouth, an ill shaped town, having about 30 angles. The soil hard, strong land, favorable to the growth of oak, walnut, and elm, with some chestnut, maple, &c. Pop. 1830, 913.

HAMPSTEAD, town, W. angle of Rockland co. N. Y. 130 ms. S. Albany. Surface broken. Crossed on the W. by Ramapo river, which here receives a stream, that also supplies mill seats, abounding with falls. The iron works in this town employ a great number of hands. Ramapo works, on Ramapo river, employ 300 hands and give support to about 700 persons. Dater's works, 2 ms. above these on the same river, support about 140. The town has various other manufactories, among which is a cotton factory, containing 5000 spindles, and employing 200 women and children.

HAMPSTEAD, p-v. Baltimore co. Md. 25 ms. from Baltimore.

HAMPSTEAD, p-v. in the southeast angle of King George's co. Va. by p-r. 90 ms. S. W. C. and 82 N. N. E. Richmond.

HAMPTON, p-t. Rockingham co. N. H. on the sea coast, 7 ms. from Exeter, 13 miles S. W. Portsmouth, and 50 ms. from Concord. Pleasantly situated. Its beaches are little inferior to the far famed Nahant beach, and have long been the resort of invalids and parties of pleasure. It has a singular bluff called Boars Head. Pop. 1830, 1,102.

HAMPTON, p-t. Windham co. Conn. 8 ms. N. E. Windham, and 37 E. Hartford. Surface uneven. Soil, a gravelly loam, strong and fertile, and well adapted to grazing. Timber, oak, walnut, chestnut, and other deciduous trees. Well watered, and contains some mill privileges, and a variety of manufacturing establishments. It has 10 school districts. Pop. 1830, 1,101.

HAMPTON, town, Washington co. N. Y. 6 ms. S. E. Whitehall, 70 N. N. E. Albany. Timber, maple, beech, &c. interspersed with beautiful groves of white pine. Well watered and healthy, and remarkable for the growth of its apple trees, which produce excellent fruit. It has 7 schools. Pop. 1830, 1,069.

HAMPTON, p-v. Adams co. Pa. by p-r. 90 ms. N. W. C.

HAMPTON, p-v. and st. jus. Elizabeth City co. Va. by p-r. 199 ms. a little E. of S. W. C. and 16 N. N. W. Norfolk. It is a seaport on a small bay of Hampton Roads, on the N. side 3 ms. N. W. Old Point Comfort.

HAMPTON, p-v. Adams co. Pa. by p-r. 31 ms. S. W. Harrisburg.

HAMPTON, p-t. and st. jus. Elizabeth City co. Va. situated on a small bay of Chesapeake bay, or rather of the estuary of James r. 16 ms. N. N. W. Norfolk, and by p-r. 93 S. E. by E. Richmond. Lat. 37° 02', long. 0° 44' E. W. C.

HAMPTON FALLS, p-t. Rockingham co. N. H. on the sea coast, 45 ms. from Concord, 41 from Boston, and 16 S. W. Portsmouth. Soil moderately good. Pleasantly situated. Pop. 1830, 583.

HAMPTON ROADS, local name of the mouth of James r. opposite the mouths of Nansemond and Elizabeth rs. Towards the Chesapeake bay, Hampton Roads is defined on the N. by Old Point Comfort, and on the S. by Point Willoughby; within James r. the termination is indefinite. This sheet of water is sufficiently deep for the largest ships of war. The U. S. commissioners, appointed to examine the lower part of Chesapeake bay in 1818, reported, that, although extensive, Hampton Roads admitted the erection of adequate defences against an enemy's fleet.

HAMPTONVILLE, p-v. Surry co. N. C. by p-r. 151 ms. N. W. by W. Raleigh.

HAMTRAMCK, p-v. Wayne co. Mich. 13 ms. from Detroit.

HANAN'S BLUFF, p-o. Yazoo co. Miss. by p-r. 69 ms. nthrd. Jackson, and about 120 N. N. E. Natchez.

HANCOCK co. bounded by Penobscot co. N., Washington co. E., by the Atlantic S., and by Penobscot bay and r. W. This co. is very irregular, and includes numerous islands off the coast, and several peninsulas. Lat. from about 44° 10' to 45° 10', and long. from 8° 15' to 9° 10' W. W. C. This co. in 1820 contained a pop. of 31,290. Since which the co. of Waldo, W. Penobscot bay and r. has been divided from it. Chief t. Castine. Pop. 1830, 24,347.

HANCOCK, p-t. Hillsborough co. N. H. 19 ms. E. Keene, 35 S. W. Concord, and 22 from Amherst. The soil generally productive; W. part of the town mountainous; the rest agreeably diversified with plain, hill, and dale. Named in honor of governor Hancock of Boston, one of the original proprietors. There are 9 school houses. Here is a manufactory of excellent and elegant fowling pieces and rifles. Pop. 1830, 1,217.

HANCOCK, p-t. Berkshire co. Mass. 130 ms. W. Boston. This is a strip of land about 3 ms. wide, extending along the western boundary of the state for more than 20 ms. A branch of the New Lebanon (N. Y.) Shakers reside within its limits. Pop. 1830, 1,052.

Hancock, p-t. s. angle, Del. co. N. Y. 27 ms. s. w. Delhi, and 65 w. Kingston. A rough, hilly tract of land. The e. branch of the Del. r. runs through the centre of this town, and the principal business of the inhabitants is getting lumber that descends the Del. to Phil. It contains 5 schools. Pop. 1830, 766.

Hancock, flourishing p-v. situated on the left bank of Potomac r. Washington co. Md. 39 ms. a little n. of e. Cumberland, 27 a little n. of w. Hagerstown, and by p-r. 93 ms. n. w. W. C.

Hancock, p-v. Union dist. S. C. (*See Hancockville, same district and state.*)

Hancock, co. of Geo. bounded by Green n. w., Tallaferro n., Great Ogeechee r. separating it from Warren n. e. and e., Washington s., Baldwin s. w., and Oconee r. separating it from Putnam w. Length 30 ms., mean width 20, and area 600 sq. ms. Extending in lat. from 33° 04′ to 33° 32′, in long. from 5° 50′ to 6° 22′ w. W. C. The slope of this co. is to the s. drained by various branches of Oconee and Great Ogeechee rs. Chief town, Sparta. Pop. 1820, 12,734, 1830, 11,820.

Hancock, co. Il. bounded by Warren n., McDonough n. e. and e., Schuyler s. e., Adams s., and by the Miss. r., separating it from the state of Mo. below, and from the unappropriated n. w. territory above the mouth of Des Moines r. Length from s. to n. 33, mean breadth 22, and area 726 sq. ms. Extending in lat. from 40° 11′ to 40° 38′, long. from 13° 52′ to 14° 26′ w. W. C. It may be observed under this head that the valley of Miss. between the mouths of Rock r. and Il. r. is very restricted on the e., the streams rising near, but flowing from that great stream to the southestrd. towards the Il. r. Amongst these tributaries of Illinois, Crooked creek, or more correctly Crooked r. rises in and drains the eastern part of Hancock co. The western part slopes wstrd. towards Miss. r., the central part being a table land between the two vallies. Chief t. Montebello. Pop. 1830, 483.

Hancock, one of two southeastern cos. of Miss. bounded by Pearl r. which separates it from St. Tammany's parish of La. s. w., and Washington parish of the same state w., by the co. of Marion, Miss. n. w., Perry and Jackson cos. n. e. and e., and the Gulf of Mexico, or rather lake Borgne s. Length 60, and breadth 28, area 1,680 sq. ms. Extending in lat. from 30° 12′ to 31°, and in long. from 12′ to 0° 54′ w. W. C. The declivity of this co. is in the direction of the streams, about s. s. e. The surface towards the nthrn. border waving or hilly, but gradually becoming more level towards lake Borgne. The whole co., with very little exception, was in its natural state, covered with pine; the soil thin and sterile. Pearl r., from its length, and the surface it drains, promises more navigable facility than from nature it affords; the mouth is shallow and obstructed. In front of Hancock co. Cat isl. and the two groupes of Marianne and Malheureux, are part of a line of sand banks, which extend along the coast of Flor., Ala., Miss., and merge into the Delta of the Miss. r. in La. (*See lake Borgne.*) It is on the lake shore of Hancock co. that the marshy coast of the Gulf of Mexico, so remarkable along the front of La. is followed by a solid, dry, pine covered, though still low shore. Chief t. Shieldsboro'. Pop. 1820, 1,594, 1830, 1,962.

Hancock, co. Ind. bounded by Rush s. e., Shelby s., Marion w., Hamilton n. w., Madison n., and Henry n. e. Length 20, mean width 18, and area 360 sq. ms. Lat. 39° 42′ to 39° 58′, long. 8° 35′ to 8° 58′ w. Slope sthrd., and in that direction drained by the northwestern sources of the Driftwood fork of White r. Chief t. Greenfield. Pop. 1830, 1,436.

Hancock, C. H. Hancock co. Ind. (*See Greenfield, same co.*)

Hancocksville, marked on the p-o. list Hancock, p-v. in the northern part of Union dist. S. C. by p-r. 86 ms. n. n. w. Columbus.

Hanging Fork, p-o. Lincoln co. Ky. 53 ms. s. Frankford.

Hanging Rock, p-o. Hampshire co. Va. by p-r. 99 ms. n. w. by w. W. C.

Hannibal, p-t. s. w. corner of Oswego co. N. Y. 11 ms. s. Oswego, 160 from Albany. Watered by several mill streams. Surface gently uneven, soil fertile. It has 9 schools. Pop. 1830, 1,794.

Hanover, p-t. Grafton co. N. H. 53 ms. n. w. Concord, 102 from Portsmouth, 114 from Boston, and 495 from W. C., situated on Conn. r. Timbered with maple, beech, birch, ash, &c. Surface agreeably diversified with hill and dale. It contains less waste land than any other town in the co. Crossed by Moose mountain from n. to s. Dartmouth college is located in this town. It received its name from William, earl of Dartmouth, one of its principal benefactors, and was founded 1770. It is situated in a beautiful village, half a mile from the Conn r. The college buildings are, a handsome edifice of wood, 150 feet by 50, three stories high, for undergraduates, and for other purposes; an edifice of brick, called medical house, 75 feet by 32, three stories high, a convenient chapel, and a green house for botanical purposes. Students, 1831-2, 153. Total Alumni 2,250. The college library contains 6,000, and the students 8,000 volumes. Commencement, last Wednesday but one in August. Here is also the N. H. medical school, which is connected with the college. There are three professors, and students, 1831-2, 98. Lectures commence 2 weeks after the college commencement. Pop. Hanover, 1830, 2,361.

Hanover, p-t. Plymouth co. Mass. 22 ms. s. Boston, divided from Pembroke by North r., a stream of some magnitude. Pop. 1830, 1,303.

Hanover, p-t. n. angle Chatauque co. N. Y., on the shore of lake Erie, 30 ms. n. e.

Mayville, 37 s. w. Buffalo. It is washed on the N. by Cataraugus creek, at the mouth of which there is a harbor for small vessels, with about 4 feet water over the bar at the entrance. It is a good tsp. of land, well watered, moderately uneven. It has 20 schools, kept 4 months in 12. Pop. 1830, 2,614.

HANOVER, t. Burlington co. N. J. on the Monmouth line, bounded s. by Northampton, w. by Springfield. Pop. 1830, 2,859.

HANOVER, p-t. Morris co. N. J. on the Passaic, 16 ms. N. W. Elizabethtown. Pop. 1830, 3,718.

HANOVER, p-v. and borough, in the southwestern part of York co. Pa. 20 ms. s. w. the borough of York, and 33 s. s. w. Harrisburg.

HANOVER, co. of Va. bounded by the Chickahomina r. or Henrico s., Goochland s. w., Louisa N. W., North Anna r. or Spottsylvania N., North Anna r. or Caroline N. E., Pamunky r. or King William E., and New Kent s. E. Length 45, mean width 14, and area 630 sq. ms. Extending in lat. from 37° 29' to 38° 05', and in long. 0° 15' w. to 0° 57' w. W. C. North Anna r. is the recipient of the creeks, which drain the northern part of Hanover, whilst the Pamunky enters from the w. traversing the co. in an easterly direction, and after draining the central section, unites with North Anna at the extreme southwestern angle of Caroline, sthrd. from the junction of North Anna and Pamunky; the united water is known by the latter name, and the body of Hanover co. lies between the Chickahomina, and that r. with its general slope N. estrd. The surface is hilly, and soil of every extreme, from best river alluvion to barren sand. Chief t. Hanover. Pop. 1820, 15,267, 1830, 16,253.

HANOVER, C. H. Hanover co. Va. (*See Woodville, same co. and state.*)

HANOVER, p-v. and tsp. in the northeastern part of Licking co. O. by p-r. the p-o. is 41 ms. northestrd. from Columbus. Pop. of the tsp. 1830, 709.

HANOVER, p-v. in the northwestern part of Shelby co. Ind. 23 ms. s. E. Indianopolis.

HANOVERTON, p-v. Hanover co. Va. situated on the right bank of Pamunky r. 31 ms. N. E. Richmond, and by p-r. 94 s. s. w. W. C. Lat. 37° 42', long. 0° 23' w. W. C.

HANOVERTON, p-v. in Hanover tsp. in the central part of Columbiana co. O. The p-v. is 7 ms. s. w. by w. New Lisbon, the co. st., and 145 N. E. by E. Columbus. Pop. of the tsp. 1830, 2,043.

HANSFORD, p-v. Kenhawa co. Va. by p-r. 356 ms. s. w. by w. ½ w. W. C.

HANSON, p-t. Plymouth co. Mass. 24 ms. s. Boston. Pop. 1830, 1,030.

HARBOUR CAPE, the N. extremity of Wells bay, Me. Long. 70° 24' w., lat. 43° 18' N.

HARBOUR CREEK, tsp. and p-o. in the nthrn part of Erie co. Pa. The p-o. is 6 ms. N. E. the borough of Erie. Pop. of the tsp. 1830, 1,104.

HARDIMAN, co. of Ten. bounded by Lafayette w., Haywood N. W., Madison N., McNair E., and the state of Miss. s. It is a regular parallelogram 30 ms. from s. to N., and 24 from E. to w., area 720 sq. ms. Extending in lat. from 35° to 35° 27', and long. from 11° 50' to 12° 14' w. W. C. This co. is entirely drained by the constituent creeks of Big Hatchee r., and slopes to the N. N. W. Chief t. Bolivar. Pop. 1830, 11,655.

HARDIMANS CROSS ROADS, and p-o. Williamson co. Ten. 34 ms. from Nashville.

HARDIN, co. of Ten. bounded by McNair w., Henderson N. W., Perry N. E., Wayne E., Lauderdale in Ala. s. E., and Chickasaw co. in the state of Miss. s. w. Length from s. to N. 32 ms., breadth 24, area 768 sq. ms. Extending in lat. from 35° to 35° 28', and long. from 11° 03' to 11° 28' w. W. C. The Ten. r. enters near the middle of the southern border, flows northwardly with a western curve, to near the northern side of this co., and then turns to N. E. leaving it at the northeast angle. On the western side the valley of Ten. r. is very narrow; the sources of Big Hatchee rising on the border between Hardin and McNair cos. On the eastern side also the slope of Ten. is narrow, and almost confined to Hardin co. The co. is therefore composed of two narrow inclined plains falling towards the Ten. Chief town, Hardensville. Pop. 1830, 4,868.

HARDIN, co. Ky. bounded by Grayson s. w., Breckenridge w., Meade N. W., Rolling fork of Salt r. or Nelson N. E., Washington E., Greene s. E. and Hart s. Length 60, mean width 20, and area 1,200 sq. ms. Extending in lat. from 37° 22' to 37° 52', and in long. from 8° 20' to 9° 18' w. W. C. The general slope of this co. is to the s. w. being in that direction drained by various creeks flowing into Green r., the northern part, however, slopes to the N. and is drained by creeks flowing into O. or Salt r. Chief town, Elizabethtown. Pop. 1820, 10,498, 1830, 12,849.

HARDIN, co. O. bounded by Union s. E., Logan s., Allen w., Hancock N., Crawford N. E., and Marion E. It is very nearly a sq. of 24 ms. each way, 576 sq. ms. in area. Lat. 40° 32' to 40° 51', long. 6° 24' to 6° 52' w. W. C. This co. occupies a table land from which the branches of Sandusky and those of Blanchard's r. flow to the north; the extreme sources of Sciota s. E., those of Sandy creek, branch of Great Miami s. w., and those of Au Glaize r. west. Chief town, Hardy. Pop. 1830, 210.

HARDIN, p-v. in the western part of Shelby co. O. by p-r. 12 ms. N. W. from Sidney, the county seat, and 88 miles N. W. by w. Columbus.

HARDINSBURGH, p-v. and st. jus. Breckenridge co. Ky. 35 ms. w. from Elizabethtown, 29 s. s. w. from Brandenburg, on O. r. and by p-r. 110 ms. s. w. by w. from Frankfort. Lat. 37° 47', long. 9° 28', w. W. C.

HARDINSBURG, p-v. Dearbon co. Ind. by p-r. 98 ms. s. E. Indianopolis.

HARDINS TAVERN, and p-o. Albemarle co. Va. by p-r. 130 ms. s. w. W. C.

HARDINSVILLE, p-v. st. jus. Hardin co. Ten. situated on the right bank of Ten. r. 40 ms. N. W. Florence in Ala. and about 120 ms. s. w. by s. Nashville. Lat. 35° 12', long. 11° 18' w. W. C.

HARDINSVILLE, p-v. southeastern part of Shelby co. Ky. 10 ms. s. w. Frankfort.

HARDISTON, t. Sussex co. N. J. Pop. 18 0, 2,588.

HARDWICK, p-t. w. part of Caledonia co. Vt. 21 ms. N. E. Montpelier, 73 N. Windsor. Surface pleasantly diversified with large swells and valleys. The r. Lamoille runs circuitously through it, furnishing a number of excellent mill privileges. Timbered with maple, beech, birch, &c. Contains sulphur springs, and 9 school districts. Pop. 1830, 1,216.

HARDWICK, p-t. Worcester co. Mass. 70 ms. w. Boston, 20 N. W. Worcester. Surface uneven. Soil fertile, adapted to grass and fruit trees. Pop. 1830, 1,885.

HARDWICK, t. Warren co. N. J. 10 ms. s. w. Newton. Pop. 1830, 1,962.

HARDY, co. of Va. bounded by Hampshire N. E., Shenandoah S. E., Rockingham S., Pendleton S. W., Randolph W., and Alleghany co. of Md. N. W. Length 42, mean width 17, and area 714 sq. ms. Extending in lat. from 38° 43' to 39° 18', and in long. from 1° 43' to 2° 30' w. W. C. The surface of Hardy inclines to N. E. and is traversed in that direction by the south branch, and several other confluents of Potomac, with lateral chains of mtns. intervening, which also extend in a similar direction with the rivers. The surface is indeed excessively broken, rocky, and sterile, though tracts of excellent river lands lie detached between the mtn. ridges. The mean elevation of the arable land perhaps exceeds 1,000 feet above the ocean level. Chief town, Moorfield. Population 1820, 5,700, 1830, 6,798.

HARDY, p-v. Hardin co. O. by p-r. 66 ms. N. W. Columbus.

HAREWOOD, p-v. Susquehannah co. Pa. by p-r. 174 ms. N. E. Harrisburg.

HARFORD, p-v. and tsp. Susquehannah co. Pa. situated between Vanwinkles and Martins branches of Tunkhannock r. 12 ms. S. E. Montrose, and 40 a little E. of N. Wilkesbarre.

HARFORD co. of Md. bounded by Susquehannah r. separating it from Cecil N. E., by the head of Chesapeake bay, separating it from Kent S. E., by Baltimore co. S. W. and W., and by York co. in Pa. N. Length 30, mean width 16, and area 480 sq. ms. Extending in lat. from 39° 19' to 39° 43' nearly. The southern part of this co. is drained principally by Bush r. and slopes to the southward. The northern is drained by Deer cr. or r. and declines eastward, towards the Susquehannah. The soil of this co. is very diversified, from best to worst. Chief town, Belair. Pop. 1820, 15,924, 1830, 16,319.

HARFORD, p-v. and named in the p-o list Harford C. H., is situated at the head of Bush r. bay 26 ms. N. E. Baltimore.

HARLAN, co. of Ky. bounded W. and N. W. by Knox, N. by Perry, N. E. by Pike, and S. E. E. and S. by Cumberland mtn. which separates it from Lee, the extreme southwestern co. of Va. Length 48, mean width 10, area 480 sq. ms. This co. contains the extreme higher sources of Cumberland r. and is in great part confined to a narrow valley between Cumberland mtn. and the Laurel ridge, with an inclination to the S. W. It extends in lat. from 36° 36' to 37°, and long. from 5° 49' to 6° 24' w. W. C. The mean elevation of the arable surface of this co. is propably above 1000 feet above the ocean level, which would yield a mean temperature equal to that of between 38 and 39 on the Atlantic coast. Chief town, Mount Pleasant. Pop. 1830, 2,929.

HARLAN, C. H. and p-o. (*See Mount Pleasant, Harlan co. Ky.*)

HARLANSBURG, (*See Harlensburg, Mercer co. Pa.*)

HARLEESVILLE, p-v. on Little Pedee r. in the northern part of Marion dist. S. C. about 54 ms. S. S. W. Fayetteville in N. C. and by p-r. 121 N. E. by E. Raleigh.

HARLEM, t. Kennebeck co. Me. 16 ms. E. Augusta.

HARLEM, (*See Haerlem, N. Y.*)

HARLENSBURG, p-v. southeast angle of Mercer co. Pa. 50 ms. N. N. W. from Pittsburg.

HARMONSBURG, p-v. Crawford co. Pa.

HARMONY, p-t. Somerset co. Me. 25 ms. E. Norridgewock. Pop. 1830, 925.

HARMONY, p-t. Chatauque co. N. Y. 15 ms. s. Mayville, bounded s. by Pa. Land heavily timbered with beech, maple, ash, butternut, &c.—moderately uneven, better adapted to grass than grain. 12 schools kept 6 months in 12. Pop. 1830, 1,989.

HARMONY, p-v. Warren co. N. J. 12 ms. s. Belvidere.

HARMONY, p-v. on the Conequenessing cr. Butler co. Pa. 14 ms. s. w. by w. Butler, and 28 N. N. W. Pittsburg. This village was founded by the Harmonists.

HARMONY, p-v. York dist. S. C. by p-r. 85 ms. N. Columbia.

HARMONY, p-v. Washington co. Mo. about 60 ms. S. S. W. St. Louis.

HARMONY GROVE, p-v. Jackson co. Geo. by p-r. 56 ms. northward Milledgeville.

HARPERS FERRY, Jefferson co. Va. 22 ms. s. w. by w. Frederic, 25 almost due s. Hagerstown, and by p-r. 65 s. w. by w. W. C. Lat. 39° 29', long. 0° 42' w. W. C. The village is situated on the right bank of Potomac, and on the point above the mouth of Shenandoah r. This place, the seat of one of the U. S. armories, has risen at the justly celebrated pass of the Potomac through the Blue Ridge. The level of low water at the junction of the two rs. is 182 ft. above tide water at Georgetown. The place and vicinity has the romantic aspect of an immense amphitheatre, and is amongst the situations of the U. S. most worthy of a visit, whether the object be science, or the gratification of taste.

HARPERSFIELD, p-t. Delaware co. N. Y., 20 ms. N. E. Delhi, 56 s. w. Albany, and 51 from Catskill. Contains good mill seats; soil well adapted for grass; surface broken, with hills and vallies; land well watered by springs and brooks; timbered with maple, beech, bass-wood, ash, &c.; has 11 schools. Pop. 1830, 1,976.

HARPERSFIELD, p-v. and tsp. in the north wstrn. part of Ashtabula co. O., by p-r. the p-o. is 10 ms. N. w. Jefferson, the co. st. Pop. tsp. 1830, 1,145.

HARPERSVILLE, p-o. (*See Colesville.*)

HARPERSVILLE, p-v. Shelby co. Ala., by p-r. 77 ms. estrd. Tuscaloosa.

HARPETH, small r. of Ten., rising in and draining the greatest part of Williamson co. Flowing thence N. W., enters and traverses the western part of Davidson, and draining the estrn. part of Dickson, falls into Cumberland r. on the border between Dickson and Davidson co. after a comparative course of 55 ms.

HARPSWELL, t. Cumberland co. Me., 40 ms. E. Portland. Pop. 1830, 1,352.

HARRINGTON, t. Bergen co. N. J. Pop. 1830, 2,581.

HARRING'S STORE, and p-o. Hinds co. Miss., about 150 ms. N. E. Natchez.

HARRIS, co. of Geo., bounded by Troup N. W.; Merriwether N. E.; Talbot E.; Muscogee S.; and the Chattahooche r. separating it from the state of Ala. W. It is very nearly a parallelogram, 20 ms. by 22, area 440 sq. ms. Extending in lat. from 32° 35′ to 32° 50′, and long. it is bisected by 8° W. from W. C. The slope of this co. is westrd., and is drained by small water courses, flowing in that direction in Chattahooche r. Chief t. Hamilton. Pop. 1830, 5,105.

HARRISBURGH, p-t. near N. W. corner of Lewis co. N. Y., 20 ms. N. Brownville, and 65 N. Rome. Soil a dark, loose, moist loam, good for grain, but better for grass; timbered with large and heavy maple, beech, elm, &c.; surface pretty level; well watered; contains good mill privileges, and 6 school dists. Pop. 1830, 712.

HARRISBURG, p-v., borough and st. jus. for Dauphin co. and of the government of Pa., 96 ms. N. W. by W. from Phil.; 35 ms. from Lancaster, and 110 a very little E. of N. W. C. Lat. 40° 16′, long. 0° 07′ E. W. C. Harrisburg is built at nearly parallel lines or right angles to the Susquehannah r., on a peninsula between that r. and Paxton cr. The nrthrn. part of the site is a swelling hill, which gradually sinks to a plain towards the mouth of Paxton. Opposite the borough the Susquehannah is divided into two channels by an isl., the widest being that next the town. Over these channels and isl., and extending from near the central street, is a substantial bridge resting on stone piers, but a frame and flooring of wood above, and roofed with the latter material. The bridge, including the isl. is nearly a mile from shore to shore. On the highest part of the same swell on which the town is built, and to the N. of the latter, stands the capitol, a substantial, and as a whole, an imposing building, from the cupola of which is one of the finest panorama views in the U. S. This view cannot properly be called a landscape; it is a circle of landscapes, embracing the swelling and cultivated co. around, relieved by r. and mtn. scenery. A C. H. and number of places of public worship are contained in the body of the borough. Pop. 1820, 2,990; in 1830, 4,312.

HARRISBURG, p-v. Lancaster dist. S. C., by p-r. 82 ms. N. N. E. Columbia.

HARRISBURG, p-v. Haywood co., Ten., situated in the N. E. part of the co. on the s. branch of Forked Deer r., about 150 ms. s. w. by w. Nashville.

HARRISBURG, p-v. Fayette co. Ind., by p-r. 64 ms. estrd. Indianopolis.

HARRIS' GORE, a tract of land of 6,020 acres, s. w. corner of Caledonia co. Vt. Mountainous and uninhabited.

HARRISON, t. Cumberland co. Me., 41 ms. N. W. Portland.

HARRISON, t. Cortland co. N. Y., 15 ms. s. s. E. Homer, and 143 w. Albany. A pretty good township of land.

HARRISON, t. Westchester co. N. Y., 30 ms. from N. Y., and 3 E. White Plains. Land under good cultivation; contains an abundant supply of mill seats; 6 schools 11 months in 12. Pop. 1830, 1,085.

HARRISON, co. Va., bounded s. by Lewis; w. by Wood; N. w. by Tyler; N. by Monongalia, and by Tiggart's Valley r., which separates it from Preston N. E.; and Randolph s. E. Length 50, mean breadth 22, area 1100 sq. ms. Extending in lat. from 39° 03′ to 39° 35′, long. 2° 53′ to 3° 55′ w. W. C. The wstrn. branch of Monongahela r. enters the sthrn. border of Harrison, and winding N. N. E., receives from both sides numerous creeks, which drain the central and much most considerable part of this large co. The wstrn. part, however, declines wstrd. and is drained by the sources of Middle Isl. cr. The surface of the whole co. is very broken, but generally fertile. Chief t. Clarksburg. Pop. 1820, 10,932; 1830, 14,722.

HARRISON, co. of Ky., bounded by Scott s. w.; Owen w.; Pendleton N. W. and N.; Bracken N. E.; Nicholas E; and Bourton s. Length 30, mean breadth 12, area 360 sq. ms. Extending in lat. from 38° 13′ to 38° 34′, long. 7° 04′ to 7° 30′ w. W. C. This fine small co. is traversed in a nrthrly. direction, and divided into two very nearly equal sections, by the w. branch of Licking r. Soil generally fertile. Chief t. Cynthiana. Pop. 1820, 12,271; 1830, 13,234.

HARRISON co. O., bounded s. E. by Belmont; s. w. by Guernsey; w. by Tuscarawas; N. w. by Stark; N. by Columbiana; and E. by Jefferson. Length from sth. to nrth. 27 ms., mean breadth 18, and area 486 sq. ms. Lat. 40° 10′ to 40° 33′, long. 3° 50′ to 4° 20′

w. W. C. The dividing ridge between the vallies of O. and Tuscarawas rs. extends from sth. to nrth. the entire length of this co., dividing it into two unequal sections. The estrn. side slopes to the est. and gives source to creeks flowing over Jefferson into O. r. The wstrn. side declines wstrd. towards the Tuscarawas; surface excessively broken and hilly, but soil highly fertile. Chief t. Cadiz. Pop. 1830, 20,916.

HARRISON, p-v. in Baltimore tsp. and north west border of Hamilton co. O., 25 ms. N. W. Cincinnati. Pop. 1830, 173.

HARRISON, co. Ind. bounded by Crawford N. W.; Washington N.; Floyd N. E.; O. r. separating it from Jefferson co. Ky. E.; and O. r. separating it from Meade co. Ky. S. and S. W. Length from sth. to nrth. 36 ms.; mean breadth 20, and area 720 sq. ms. Lat 38° to 38° 26′ and long. 8° 48′ to 9° 20′ W. W. C. Slope S. W. towards, and at right angles nearly, to that part of Ohio r. between Otter cr. of Ky., and Blue r. of Ind. Surface excessively hilly and broken, but soil excellent. Chief t. Corydon. Pop. 1830, 10,273.

HARRISONBURG, p-v. and st. jus. Rockingham co. Va., 24 ms. N. N. E. Stanton, 40 N. N. W. Charlotteville, and by p-r. 128 ms. S. W. by W. W. C. Lat. 38° 25′, long. 1° 48′ W. W. C.

HARRISONBURG, p-v. and st. jus. Catahoola parish, La. about 40 ms. N. W. by W. Natchez. Lat. 31° 47′, long. 14° 54′ W. W. C.

HARRISONBURG, p-v. and st. jus. Conway co. Ark. Ter., by p-r. 1,104 ms. S. W. by W. W. C., and 40 ms. N. W. Little Rock. Lat. 35° 5′, long. 15° 30′ W. W. C.

HARRISON'S MILLS, and p-o., Charles City co. Va., by p-r. 32 ms. S. E. by E. Richmond.

HARRIS'S LOT, and p-o. Charles co. Md., by p-r. 38 ms. sthrd. W. C.

HARRISON VALLEY, and p-o. Potter co. Pa., 188 ms. N. N. W. Harrisburg.

HARRISONVILLE, p-o. Monroe co. Il., situated on the Miss. r. opposite Herculaneum in Mo., 12 ms. sthwstrd. Waterloo, the st. jus. of the co., and 113 ms. S. W. Vandalia.

HARRISVILLE, p-v. in the extreme N. W. angle of Butler co. Pa., 55 ms. almost due N. Pittsburgh.

HARRISVILLE, p-v. on Nottaway r., in the N. E. angle of Brunswick co. Va., by p-r. 57 ms. a little W. of S. Richmond.

HARRISVILLE, p-v. in the nrthestrn. part of Harrison co. O., by p-r. 8 ms. N. Cadiz, and 132 N. E. by E. Columbus. Pop. 1830, 314.

HARRISVILLE RESERVE, p-v. Medina co. O., 111 ms. N. E. Columbia.

HARRODSBURG, p-v. and st. jus. Mercer co. Ky., situated near the main source of Salt r., 31 ms. a very little E. of S. from Frankfort. Lat. 37° 44′, long. 7° 48′ W. W. C. This place was amongst the most early towns founded in Ky.; in 1830 contained a pop. of 1,051.

HART, co. of Ky. bounded by Edmonson W.; Nolin Fork of Green r. which separates it from Grayson N. W.; Raccoon cr. which separates it from Hardin N.; Greene E.; and Barren S. Length 24, mean breadth 18, area 432 sq. ms. Extending in lat. from 37° 06′ to 37° 25′, and long. from 8° 28′ to 9° 03′ W. W. C. The main body of Green r. traverses Hart in a S. W. by W. direction, and receiving creeks from each side drains the co. The surface is generally level, and in its natural state, in a great part composed of a species of soil deceptively called barrens, as much of it in this and adjacent cos. is highly fertile. Chief t. Mumfordsville. Pop. 1820, 4,184; 1830, 5,191.

HARTFIELD, p-v. wstrn. part of Tipton co. Ten., 10 ms. Covington, the co. st., and by p-r. 207 ms. a little S. of W. Nashville.

HARTFORD, p-t. Oxford co. Me., 12 ms. N. E. Paris. It is very hilly, being at the extremity of a spur from the White mtns. Pop. 1830, 1,294.

HARTFORD, co. Conn., situated in the N. central section of the state, principally within the valley, and on both sides of Conn. r.; is bounded by Hampden co. Mass. N.; by Tolland co. E.; New London co. S. E.; Middlesex and New Haven cos. S.; and the cos. of New Haven and Litchfield W. Lat. 42° traverses its northern, and 4° E. W. C. its wstrn. border. It forms nearly a square, is about 30 ms. in length N. and S., and 25 in width; and comprises an area of about 727 sq. ms., or 465,280 acres. This co., as a whole, will rank before any other in the state; and in many respects before any in N. England. The soil is rich, various and fertile; well adapted to grain, fruit, and almost every thing of which the climate admits, and is for the most part highly cultivated. The co. is intersected nearly in the centre by Conn. r.; W. by the Greenstone (locally Talcott) mtns.; and S. E. by a high range of hills. The surface is undulating, abounding in the fertile and varied scenery common to much of the valley of the Conn., and is timbered with various kinds of the oak, walnut or hickory, elm, maple, ash, &c. &c. It is watered by several streams, among which is the Tunxis or Farmington r. on the W.; and Freshwater, Scantic, Podunk, and Hockanum, on the E. A great variety of manufactories are carried on in this co.; among them, that of cotton, (according to a recent return to the Secretary of the Treasury,) employs a capital of $260,000, consuming annually 509,000 lbs. of cotton; and the woollen manufacture employs a capital of $311,500, consuming about 600,000 lbs. of wool per ann. The towns and villages, which are numerous, are generally pleasant; many of them are populous and wealthy. Pop. 1820, 47,261; 1830, 51,141.

HARTFORD, city and p-t. Hartford co. Conn., one of the seats of government of the state, and st. jus. for the co., is situated at the head of sloop navigation on the W. side of Conn r., 50 ms. from its mouth, at lat. 41° 45′, and long. 4° 15′ E. W. C. It is 123 ms. N. E. N. York, 34 N. N. E. New Haven, 15 N. Middletown, 44 N. W. New London, 71 W. Providence, 100 W. S W. Boston, and 97 S. E Albany. The legislature of the state assembles alternately at this place and New Haven —the odd years at the former. The city is over a mile in length, and ¾ths of a mile wide; surface undulating, sloping gradually from the principal street to the Conn.; it is irregularly laid out, and is divided E. and W. by Mill, or Little r. Across this stream a fine bridge of free stone has been thrown, which connects the two parts of the city. This structure is 100 feet wide, supported by a single arch, 7 feet in thickness at the base, and 3 feet 3 inches at the centre; the chord or span of which is 104 feet; elevation from the bed of the river to the top of the arch, 30 feet 9 inches. Another bridge across the Conn., covered, 1,000 feet long, and which cost over $100,000, unites the city with East Hartford. Hartford is very advantageously situated for business, is surrounded by an extensive and wealthy district, and communicates with the towns and villages on the Conn. above, by small steam boats, (now 8 in number) two of which, for passengers, ply daily between Hartford and Springfield. The remainder are employed in towing flat bottomed boats of 15 to 30 tons burthen, as far as Wells r., 220 ms. above the city. The coasting trade is very considerable, and there is some foreign trade, not extensive, carried on. Three steamboats form a daily line between here and New-York. The manufactures of this city, by a late return made to the Secretary of the Treasury, exceed $900,000 per ann.; among these are various manufactures of tin, copper, and sheet iron; block tin and pewter ware; printing presses, and ink; a manufactory of iron machinery; an iron foundry; saddlery, carriages, joiners tools, paper hanging, looking-glasses, umbrellas, stone ware, a brewery, a web manufactory, cabinet furniture, boots and shoes, hats, clothing for exportation, soap and candles, 2 manufactories of machine and other wire cards, operated by dogs; &c. &c. More than twice as many books are published here, annually, as are manufactured in any other place of equal pop. in the U. S. There are 15 periodicals; 12 weekly newspapers (5 sectarian), 2 semi-monthly and 1 monthly. The city is well built, and contains many elegant public and private edifices. The state house, in which are the public offices of the state, is surmounted by a cupola, and is a very handsome and spacious building. The city hall, built for city purposes, is also spacious, and elegant; it has two fronts, with porticos, supported each by 6 massy columns. In the city are 11 places of public worship—5 for Congregationalists, 1 Episcopal, 1 Baptist, 1 Methodist, 1 Universalist, 1 Rom. Catholic, and 1 African; several of these are very handsome, and the Episcopal, a gothic edifice, is much admired for its elegance. There are 4 banks, including a branch of the U. S. B., with an aggregate capital of $2,856,400; a bank for savings; 3 fire and marine insurance offices, an arsenal, museum, two markets, &c. The American asylum for the deaf and dumb, the Retreat for the insane, and Washington college, are all beautifully located, in the immediate vicinity of the city. The Asylum, the first institution of the kind in America, incorporated in 1816, was founded under the auspices of Rev. T. H. Gallaudet, who visited Europe with that object. The system of deaf and dumb instruction in the U. S., which is uniform, proceeded from this institution, and in some respects differs from any other. By the aid of a considerable fund, pupils are instructed and supported, at a yearly expense of $115; a sum much below the actual cost. Beside the Principal, there are 9 teachers in the institution, which contains 138 pupils, many of whom are taught cabinet making, shoe making, and tailoring; females are taught the latter, as well as boys. The principal building is 130 feet long, 50 wide, and 3 stories high, beside a basement and dormitory. Total number of pupils, 412. The Retreat for the insane was instituted principally by the munificence of the citizens of Hartford, and ranks high among the first institutions of the kind. The edifice is both spacious and elegant; it has an entire front of 254 feet: viz. a centre 50 feet, two wings 70 feet each, and ends each 32 feet; centre and ends 3, and wings 2 stories high; and basement. The grounds belonging to the Retreat are spacious, and highly improved. Washington college, founded in 1826, has two edifices of free stone; one 148 feet long by 43 wide, and 4 stories high, containing 48 rooms; the other 87 feet by 55, and 3 stories high, containing the chapel, library, mineralogical cabinet, philosophical chamber, laboratory and recitation rooms. There are 5,000 vols. in the college library, and 2,500 in the libraries of the different societies. A complete philosophical apparatus, cabinet of minerals, and botanical garden and green house, belong to the institution. The faculty consists of a president, 6 professors, and 2 tutors. Students, about 60. Commencement 1st Thursday in Aug.

The Alms house, with a farm on which the able inmates are employed, is conducted on a plan of remarkable economy, and nearly supports itself. The public schools are numerous, and there are several excellent private schools. Mill r. has several water privileges, which are improved; and about 2 ms. from the city is an extensive quarry of wall stone, suitable for building and other purposes. The location of the city is in every respect delightful; it is surrounded with a fertile and indeed exuberant soil, and is not exceeded by any other inland town in the variety and beauty of its scenery. Hartford was settled 1635, the city incorporated 1784, and is memorable as the seat of the Hartford convention. Pop. city, including the t. 1830, 9,789; city, 7,076.

HARTFORD, p-t. Washington co. N. Y. 54 ms. a little E. of N. from Albany, 8 E. Sandy Hill. Wood cr. runs along the N. W. corner; 15 schools, 7 months in 12. Pop. 1830, 2,420.

HARTFORD, p-v. and st. jus. Pulaski co. Geo. situated on the left bank of Ockmulgee river, about 60 ms. s. s. w. Milledgeville, latitude 32° 20′, long. 6° 30′ w. Washington City.

HARTFORD, p-v. and st. jus. Ohio co. Ky. situated on the left bank of Rough creek, near the centre of the co. by p-r. 147 ms. s. w. by w. Frankfort, and 45 N. Russellville, lat. 37° 25′, long. 9° 56′ w. W. C.

HARTFORD, p-v. and tsp. northeastern part of Trumbull co. Ohio. The p-o. is by p-r. 175 ms. N. E. Columbus. Pop. of the tsp. 1830, 859.

HARTFORD, p-v. Dearborn co. Ind. 100 ms. s. E. Indianopolis.

HARTLAND, p-t. Somerset co. Me. Pop. 1830, 718.

HARTLAND, p-t. Windsor co. Vt. w. Conn. river, 50 ms. s. E. Montpelier, 62 N. E. Bennington, 100 from Boston. A rich farming town, pleasantly diversified with hills and vallies. Watered N. E. by Queechy river, s. by Lull's brook, which afford some of the best mill privileges in the state. The town contains a valuable bed of paint. There are 18 shool disiricts. It has a variety of manufactories. Pop. 1830, 2,503.

HARTLAND, p-t. Hartford co. Conn. 22 ms. N. W. Hartford, bounded N. by. Mass. line. Hilly and mountainous. Tolerable for grazing, but poor for grain. Timbered with beech, maple, chestnut, and evergreen. Watered by the E. branch of Farmington river. Pop. 1830, 1,221.

HARTLAND, p-t. Niagara co. N. Y. on lake Ontario, 12 ms. N. E. Lockport, and 30 E. N. E. Lewiston. Land good, 11 school districts. Croosed by the Ridge road, on which stands Hartland village, 2 ms. N. Erie canal, 10 ms. N. E. Lockport. This town contains salt springs. Pop. 1830, 1,584.

HARTLETON, p-v. and tsp. of Union co. Pa. 12 ms. w. New Berlin, and by p-r. 65 ms. N. N. W. Harrisburg. Pop. of the tsp. 1830, 1,737.

HART'S CROSS ROADS, and p-o. Crawford co. Pa.

HART'S GROVE, p-o. Ashtabula co. Ohio.

HARTSVILLE, p-v. Bucks co. Pa. about 25 ms. Phila.

HARTSVILLE, p-v. on the right bank of Cumberland river, in the extreme southeastern angle of Sumner co. Ten. by p-r. 43 ms. N. E. by E. Nashville.

HARTWELL'S BASIN. (*See Perrinton.*)

HARTWICK, p-t. Otsego co. N. Y. 6½ ms. s. w. Cooperstown, 70 w. Albany. Watered by the Susquehannah, Oak's creek, the outlet of Caniaderaga or Schuyler's lake, and the Otsego creek, which supply a great abundance of mill seats. Surface considerably broken and hilly; well watered by springs and brooks; 14 schools kept 6 months in 12. A literary and theological seminary was established here in 1816, by members of the Lutheran church. Pop. 1830, 2,772.

HARVARD, p-t. Worcester co. Mass. 20 ms. N. E. Worcester, 30 N. W. Boston. Divided by Nashua river from Lancaster and Shirley. Soil good, especially for fruit. Here is a settlement of Shakers. Pop. 1830, 1,600.

HARVARD UNIVERSITY. (*See Cambridge.*)

HARVEY'S p-o. Greene co. Pa. by p-r. 241 ms. N. W. by W. W. C.

HARVEY'S STORE, and p-o. Charlotte co. Va. 108 ms. s. w. Richmond.

HARVEYSVILLE, p-o. Luzerne co. Pa. 94 ms. N. E. Harrisburg.

HARWICK, p-t. Barnstable co. Mass. 79 ms. s. E. Boston, bounded s. by the Atlantic. Pop. 1830, 3,974.

HARWINTON, p-t. Litchfield co. Conn. 23 ms. w. Hartford. Elevated and hilly. The timber consists principally of deciduous trees. Lands best adapted to grazing. Watered by the Naugatuck, and the Lead Mine rivers. Contains various kinds of manufactories, 11 school districts, and an academy. Pop. 1830, 1,516.

HASKINSVILLE, p-v. Gibson co. Ten. by p-r. 151 ms. westward Nashville.

HAT (THE), tavern and p-o. Lancaster co. Pa. by p-r. 54 ms. from Harrisburg.

HATBORO', p-v. situated near the northeastern border of Montgomery co. Pa. and on a branch of Penepack creek, 17 ms. N. Phila.

HATCHERSVILLE, p-o. Chesterfield co. Va. by p-r. 12 ms. from Richmond.

HATCHY, BIG. (*See Big Hatchy river.*)

HATFIELD, p-t. Hampshire co. Mass. w. Conn. river, 5 ms. N. Northampton, 95 w. Boston. Soil various, but valuable. This town unanimously protested against the revolutionary war, and was the head quarters of the Shay's insurrection, but its patriotism has since been undoubted. It is supposed that the farmers of this town now still feed three times as many oxen as were fattened in all the towns of Old Hampshire, 100 years ago. Pop. 1830, 893.

HATFIELD, tsp. of Montgomery co. Pa. 24 ms. N. N. W. Phila. Pop. 1830, 835.

HATTERAS, a very remarkable cape of the Atlantic coast of the United States, in N. C. Pamlico sound is inclosed on the ocean side by a long low reef of sand and rock stretching 65 ms. from Ocracock inlet on the s. w. to New inlet on the N. E. Cape Hatteras, properly so called, is the salient point of this reef, jutting in the Atlantic ocean, at lat. 35° 12′, and long. 1° 35′ E. W. C. This low but stormy promontory, is a true cape of winds and a point of interest and dread in navigating the Atlantic coast of the United States.

HAVANA, p-v. Greene co. Al. by p-r. 26 ms. s. w. Tuscaloosa.

HAVANA, p-v. in the northwestern part of Sangamon co. Il. by p-r. 123 ms. N. N. W. Vandalia.

HAVERFORD, p-v. Delaware co. Pa. 6 ms. from Phila. Pop. of the tsp. 1830, 980.

HAVERHILL, p-t. and half shire town, Grafton co. N. H. 31 ms. N. W. Plymouth, 27 ms. above Dartmouth college, 70 N. Concord, 132 from Boston. It is situated on Conn. river. Watered by Oliverian and Hazen brooks. A pleasant town. Has a handsome village, denominated Haverhill corner, lying at the s. w. angle of the town. Pop. 1830, 2,151.

HAVERHILL, p-t. Essex co. Mass. 30 ms. N. Boston. A handsome and flourishing town, at the head of sloop navigation on the Merrimac, being situated on the N. side of that river. It has an academy. An elegant and costly bridge, 800 feet long, across the river, connects the principal village with Bradford. Four miles below is Rocks village, where is another bridge nearly 1000 feet long. Pop. 1830, 3,896.

HAVERSTRAW, p-t. Rockland co. N. Y. 36 ms. N. N. York, w. Hudson river, which here spreads into Haverstraw bay. It includes Stony point with the old forts, Clinton and Montgomery, together with Dunderberg, or Thunder mtn. It has many good mill seats and several landings, with wharves, sloops, and some trade. There is an academy in the village of Warren. Pop. 1830, 2,306.

HAVRE DE GRACE, p-v. and sea port, situated on the right bank of Susquehannah river, near its mouth, 36 ms. N. E. Baltimore. Lat. 39° 33', long. 0° 58' E. W. C.

HAW, r. of N. C. rises in Rockingham and Guilford cos. and flowing thence eastward, unite in Orange, when turning to S. E. the united stream traverses Orange, and Chatham and in the S. E. angle of the latter at Haywoodboro, joins Deep river to form the main Cape Fear river. The union of Haw and Deep river is in a direct line about 30 ms. s. w. by w. Raleigh.

HAW, river, p-o. in the northwestern part of Orange co. N. C. by p-r. 65 miles N. W. Raleigh.

HAWFIELD, p-v. westrn. part of Orange co. N. C. by p-r. 14 ms. westward Hillsboro', and 55 ms. N. W. by W. Raleigh.

HAWKE, p-t. Rockingham co. N. H. 19 ms. s. w. Portsmouth. Had 500 inhabitants more in 1775, than it has had since. Soil uneven. It has 3 schools. Pop. 1830, 520.

HAWKINSVILLE, p-v. Pulaski co. Geo. about 70 ms. s. Milledgeville.

HAWLEY, p-t. Franklin co. Mass. 120 miles N. W. Boston, 14 s. w. Greenfield. Situated on the Green mtn. range, and well watered by several branches of Deerfield river. Pop. 1830, 1,037.

HAWSVILLE, p-v. and st. jus. Hancock co. Ky. by p-r. 130 ms. a little s. of w. Hartford.

HAYE'S CROSS ROADS, and p-o. Richland co. Ky. 14 ms. from Marsfield, the st. jus. and 71 N. N. E. Columbus.

HAY MARKET, p-v. in the northern part of Prince William co. Va. situated on the head of Occoquon creek, by p-r. 38 ms. a little s. of w. W. C.

HAYNES' p-o. Grainger co. Tenn. by p-r. 231 ms. estrd. Nashville.

HAYNESVILLE, p-v. Lowndes co. Al. by p-r. 120 ms. but by direct road 60 ms. westward Tuscaloosa.

HAYSBORO', village of Davidson co. Ten. on Cumberland river, 7 ms. above Nashville.

HAY'S MILLS, and p-o. Shenandoah co. Va. by p-r. 111 miles westward Washington City.

HAYESVILLE, p-v. Franklin co. N. C. by p-r. 31 ms. N. E. Raleigh.

HAYWOOD, co. N. C. bounded s. by Picken's district S. C., Rabun co. in Geo. s. w., Cowee branch of Ten. which separates it from Macon co. N. C., w. by the Iron or Bald mtn. separating it from Blount, Sevier, and Cocke cos. of Ten. on the N. W., and by a nameless chain of mtns. separating it from Buncombe E. The greatest length of Haywood is from the s. w. to the N. E. angle, 63 ms., the mean width 30, area 1,890 square ms. Extending in lat. from 35° to 35° 46', and long. from 5° 38' to 6° 43' w. W. C. The extreme southern section of Haywood is occupied by the Blue Ridge, from the southern side of which rise and flow southwardly the extreme higher sources of Savannah river. The Blue Ridge, however, forms merely the border of the co., the far greatest part of the surface sloping to the N. W., and is drained on the western side by Cowee and Tukaseegee branches of Ten. and on the eastern by Pigeon river, branch of French Broad. These streams rising in Blue Ridge flow northwestward, and pierce the Iron mtn. about 50 ms. asunder at the extremes of the co. If a correct estimate can be made from the courses of the streams, Haywood co. is amongst the most elevated sections of the United States. The arable vallies must be from 1,500, to 1,800 feet above the ocean level, giving a temperature as low in winter as that on the Atlantic ocean at the mouth of the Del. Chief town, Franklin. In 1820, Haywood comprised what is now Macon, and the census, 4,073, included both cos.; in 1830, Haywood contained 4,578 inhabitants.

HAYWOOD, C. H., N. C. (*See Waynesville, Haywood co. N. C.*)

HAYWOOD, p-v. Chatham co. N. C. (*See Haywoodboro'.*)

HAYWOOD, co. of Ten. bounded s. E. by Hardiman, s. by Lafayette, w. by Tipton, N. W. by Dyer, N. E. by Gibson, and E. by Madison. Length 30, mean width 20, area 600 sq. ms. Extending in lat. from 35° 22' to 35° 48', and long. 12° 07' to 12 32' w. W. C. The southern part of this co. is drained by the Hatchee and its branches, the main stream entering the S. E. angle and traversing the co. in the direction of N. W. by. W. The S. branch of Forked Deer r. enters the eastern border and leaves the co. near the N. W. angle, flowing nearly parallel to the Big Hatchee. The slope of the whole co. is therefore N. W. by w. Chief town, Brownsville. Pop. 1830, 5,334.

HAYWOODSBORO, p-v. Chatham co. N. C. by by p-r. 38 ms. s. w. by w. Raleigh. It is situated in the S. E. part of the co. and the point above the confluence of Haw and Deep rs. and at the head of Cape Fear r.

HAZARD FORGE, and p-o. Hardy co. Va. by p-r. 135 ms. westward W. C.

HAZLEGREEN, p-v. in the northern part of

Madison co. Ala. 12 ms. almost due N. Huntsville.

HAZLE PATCH, p-v. Laurel co. Ky. by p-r. 101 ms. S. S. E. Frankfort.

HAZLEWOOD, p-o. Chester dist. S. C. about 60 ms. N. Columbus. This place was formerly midway.

HEAD OF COOSA, p-o. in the Cherokee ter. state of Geo. by p-r. 196 ms. N. W. Milledgeville, and 643 southwestward W. C.

HEAD OF NAVIGATION, p-v. Spartanburg dist. S. C. by p-r. 107 ms. N. N. W. Columbia.

HEAD OF SASSAFRAS, p-v. in the N. E. part of Kent co. Md. about 50 ms. N. E. by E. Baltimore.

HEAD'S, p-o. southern part of Fayette co. Geo. by p-r. 100 ms. N. W. by W. Milledgeville.

HEALTH SEAT, p-o. Granville co. N. C. by p-r. 58 ms. northward Raleigh.

HEARD, co. Geo. relative position uncertain. The C. H. is given in the p-o. list at 153 ms. from Milledgeville.

HEARD, C. H. and p-o. Heard co. Geo. by p-r. 153 ms. from Milledgeville.

HEATH, p-t. Franklin co. Mass. 125 ms. N. W. Boston, 12 N. W. Greenfield, bounded N. by Vt. line. Pop. 1830, 1,199.

HEBRON, p-t. Oxford co. Me. 35 ms. N. W. Portland. Pop. 1830, 915.

HEBRON, p-t. Grafton co. N. H. 9 ms. from Plymouth, and 40 from Concord. Pop. 1830, 540.

HEBRON, p-t. Tolland co. Conn. 20 ms. S. E. Hartford. Surface uneven—Soil considerably fertile—watered by Hop r. a branch of the Willimantic, and several small streams. It contains a considerable number of manufactories, and 12 school districts. Pop. 1830, 1,939.

HEBRON, p-t. Washington co. N. Y. 52 ms. N. N. E. Albany, 6 N. Salem. Surface broken by hills, some of large size. It has no rivers and is therefore destitute of mill seats. Soil productive, and well watered. 19 schools kept 7 months in 12. Pop. 1830, 2,686.

HEBRON, p-v. in the northern part of Greene co. Ala. by p-r. 38 ms. S. S. W. Tuscaloosa.

HEBRON, p-v. Washington co. Geo. 17 ms. southeastward Milledgeville.

HECKTOWN, p-v. Northampton co. Pa. 191 ms. N. E. W. C.

HECTOR, p-t. Tompkins co. N. Y. lying between Cayuga and Seneca lakes, 17 ms. W. Ithaca and 187 W. Albany. Surface broken by elevated ridges. 24 school districts. Pop. 1830, 5,212.

HEDRICKS, p-v. southwestern part of York co. Pa. 18 ms S. S. W. from the borough of York and 40 N. N. W. Baltimore.

HEIDLESBURG, p-v. Adams co. Pa. 11 ms. N. N. E. Gettysburg and 27 S. S. W. Harrisburg.

HELDERBERG, OR HELLEBERG, ranges of hills of a mountain character, extending from the Catskill mtns. to the Mohawk near Schenectady.

HELENA, p-v. Pickens dist. S. C. by p-r. 149 ms. N. W. Columbia.

HELENA, p-o. and st. jus. Iowa co. Mich. position uncertain.

HELENA, p-v. and st. jus. Phillips co. Ark. situated on the right bank of Miss. r. in a direct course about 100 ms. a little S. of E. but by p-r. 151 from Little Rock. Lat. 34° 28', long. 13° 39' W. W. C.

HELLEN, p-v. Clearfield co. Pa. by p-r. 122 ms. N. W. Harrisburg.

HELLERSTOWN, p-v. in the sthrn. angle of Northampton co. Pa. 4 ms. S. E. Bethlehem.

HELL GATE, OR HURL GATE. (*See Horll Gatt.*)

HELM'S, p-o. Franklin co. Va. by p-r. 281 ms. S. W. W. C.

HEMLOCK LAKE, 6 ms. long. East corner of Livingston co. N. Y.

HEMPHILL'S STORE and p-o. Mecklenburg co. Va. 140 ms. S. S. W. Richmond.

HEMPHILL, p-o. Butler co. Ala. by p-r. 167 ms. S. S. E. Tuscaloosa.

HEMPSTEAD, p-t. Queens co. N. Y. S. side Long Island, 22 ms. a little S. of E. of New York. Hempstead plains lie principally in this town, and are 15 ms. long by 4 broad. 14 school districts, schools kept 10 months in 12. Pop. 1830, 6,215.

HEMPSTEAD, co. Ark. as laid down on Tanner's map of the U. S. is bounded by Red r. separating it from Texas S. W., by Sevier co. of Ark. W., by Clark N. N. E. and E., and by Lafayette S. Length from S. E. to N. W. 75, mean width 15, area 1,125 sq. ms. Extending in lat. 33° 32' to 31°, and long. 15° 42' to 16° 50' W. W. C. The sthrn. and wstrn. part of this co., is drained into Red r., the nthrn. and estrn. into little Mo. branch of Washitau. The surface is generally thin and sterile. Chief t. Washington. Pop. 1830, 2,512.

HEMPSTEAD, C. H. and p-o. (*See Washington, same co. Ark. ter.*)

HENDERSON, p-t. Jefferson co. N. Y. on lake Ontario, 8 ms. S. of Black r. mouth. Soil productive. Timber—oak, walnut, elm, beech, &c. 12 school districts. Pop. 1830, 2,428.

HENDERSON, p-o. Mercer co. Pa. 280 ms. by p-r. N. W. W. C.

HENDERSON'S, p-o. Botetourt co. Va. by p-r. 227 ms. S. W. W. C.

HENDERSON, co. of Ky. bounded by Green r. which separates it from Daviess E., by Hopkins S., Union W., and by O. r. separating it from Posey, Vanderburg, Warwick, and Spencer cos., Indiana, N. Length parallel to the general course of O. r. 40 ms., mean width 18, area 720 sq. ms. Extending in lat. from 37° 30' to 37° 58', and in long. from 10° 20' to 11° 12' W. W. C. The course of Green r. near its discharge into the O., and the general slope of Henderson co. is to the nrthrd. Chief t. Henderson. Pop. 1820, 5,714, 1830, 6,659.

HENDERSON, p-v. and st. jus. Henderson co. Ky. situated on the left bank of O. r. about 44 ms. by water above the mouth of Wabash r. and by p-r. 183 a little S. of W. Frankfort. Lat. 37° 48', long. 10° 42' W. W. C.

HENDERSON, co. of Ten. bounded by Perry

E., Hardin S. E., McNair S. W., Madison W., and Carroll N. It is very nearly a sq. of 28 ms. each side; area 784 sq. ms. Extending in lat. from 35° 24′ to 35° 48′, and in long. from 11° 14′ to 11° 44′ W. W. C. Though the eastern border of Henderson approaches with a mean distance of 10 ms. from Ten. r. the central part is a table land from which the water courses flow, like radii from a common centre. From the southwestern angle issue the sources of the S. branch of Forked Deer r.; from the wstrn. side flows the middle branch of the same stream, whilst, from the N. W. angle, flow the extreme sources of Obion r. The wstrn. slope of the co. is therefore to the wstrd. towards Miss. The estrn. part declines towards Ten. but even there, crs. flow like diverging radii, Sugar cr. northwards, Beech r. estrds., and Doe cr. to the S. E. Chief t. Lexington. Pop. 1830, 8,748.

HENDERSONVILLE, p-v. st. jus. Nottaway co. Va. situated on little Nottaway r. 65 ms. S. W. Richmond. Lat. 37° 04′, long. 1° 18′ W. W. C.

HENDERSONVILLE, p-v. Sumner co. Ten. 44 ms. N. E. Nashville.

HENDRENSVILLE, p-v. Henry co. Ky. by p-r. 40 ms. wstrd. Frankford.

HENDRICKS co. Ind. bounded S. by Morgan; W. by Putnam; N. W. by Montgomery; N. by Boone; and E. by Marion. It is a square of 21 ms. each way; area 441 sq. ms. lat. 39° 40′ to 39° 56′, long. 9° 16′ to 9° 40′ W. W. C. Slope sthrd. and drained by different branches of the main volume of White r. Chief town, Danville. Pop. 1830, 3,975.

HENDRICK'S STORE, and p-o. Bedford co Va. 239 ms. S. W. W. C.

HENDRYSBURG, p-o. Belmont co. O.

HENLOPEN CAPE, the sthrn. salient point at the mouthof Del. r. opposite and (by Tanner's U. S. Pa. and N. J.) between 12 and 13 ms. asunder. By the same authorities, the lat. is by the Pa. and N. J. 38° 45′ and by the U. S. 38° 47′. The long. is by the former map 1° 58′ and by the latter 1° 53′ E. W. C.

HENLOPEN COAL MINES, and p-o. Fentress co, Ten. by p-r. 124 ms. estrd. Nashville.

HENNIKER, p-t. Merrimack co. N. H. 15 ms. W. Concord, 27 from Amherst, 75 from Boston. Watered by Contoocook r. Contains excellent water privileges. Soil as various and fertile as any in the co. Pop. 1830, 1,725.

HENRICO, co. of Va.,bounded S. and S. W. by James r. which separates it from Chickihominy r. which separates it on the N. from Hanover, and N. E. from New Kent; and on the S. E. it has Charles City co. The greatest length from S. E. to N. W. 30, mean width 10, area 300 sq. ms. Extending in lat. from 37° 17′ to 37° 40′, and long. 0° 20′ to 0° 49′ W. W. C. This co. is composed of a central ridge with two narrow inclined plains, falling towards James r. to the S. W., and Chickahominy to the N. E. The surface is broken by waving hills; soil very much diversified. Chief town, Richmond. The entire pop 1820, 33,667, of whom, 12,067 were contained in the city of Richmond in 1830.

HENRIETTA, p-t. Monroe co. N. Y. 11 ms. S. Rochester. Peculiarly good for grazing 12 school districts. Pop. 1830, 2,322.

HENRIETTA, p-v. northwestern part of Lorain co. O. by p-r. 133 ms. N. N. E. Columbus.

HENRY CAPE, of Va. the sthrn. salient point at the mouth of Chesapeake bay. On Tanner's U. S. cape Henry is in lat. 36° 55′, long. 1° 02′ E. W. C. It is the extreme northeastern angle of Princess Ann co. Va.

HENRY, co of Va. bounded by Patrick W., Franklin N., Pittsylvania E. and Rockingham co. N. C. S. It is in form or nearly so of a rhomb,and about equal to a sq. of 20 ms. each side, area 400 sq. ms. Extending in lat. from 36½° to 36° 50′, long. 2° 44′ to 3° 08′ W. W. C. The slope of this co. is rapid, and to the S. E. The extreme sthwstrn. angle is crossed by the two branches of Mays r., but the much greater part of the area of the co. is included in the valley of Irvine r. which enters at the N. W. and leaves the co. at the S. E. angle. Chief t. Martinsville. Pop. 1820, 5,624, 1830, 7,100.

HENRY, co. Geo. bounded by Butler S. E., Pike S., Fayette W., De Kalb, Ockmulgee r. separating it from Newton, N. E. Length 33, mean width 18, and area 594 sq. ms. Extending in lat. from 33° 12° to 33° 42′, and long. 7° 02′ to 7° 28′ W. W. C. This is one of the cos. of Geo. which occupies a part of the dividing plain between the waters of the Atlantic and those of the Gulf of Mexico. The wstrn. part is drained by the sources of Flint r. whilst from the estrn. section the waters flow S. E. into the Ockmulgee. Chief town, McDonough. Pop. 1830, 10,567.

HENRY, sthestrn. co. of Ala. bounded W. by Dale, N. by Pike, E. by Chattahooche r. which separates it from Early co. Geo., S. by Jackson, and S. W. by Walton co. Flor. Length from S. to N. 48, mean width 28, area 1,344 sq. ms. Extending in lat. from 31° to 31° 42′. The northwestern part of this co. is drained by different branches of Choctawhatche; the estrn. and sthrn. by crs. flowing into Chattahooche r.; surface generally covered with pine, and soil sterile. Chief t. Columbia. Pop. 1830, 3,955.

HENRY, co. O. bounded E. by Wood, S. by Putnam, S. W. by Paulding, W. by Williams, and N. by Lenawe co. Mich. Length from S. to N. 32 ms., breadth 27, and area 864 sq. ms. Lat. 41° 06′ to 41° 39′, long. 6° 50′ to 7° 20′ W. W. C. This co. lies entirely in the valley of Great Miami, and is traversed by the main volume of that r. flowing northeastward, and dividing the co. into two not very unequal sections. Chief town, Damascus. Pop. 1830, 262.

HENRY, p-v. Muskingum co. O. by p-r. 79 ms. E. Columbus.

HENRY'S CROSS ROADS, and p-o. Sevier co Ten by p-r. 182 ms. E Nashville.

HERBERT'S CROSS ROADS, and p-o. Harford county, Md. 29 miles N. E. Baltimore.

HERCULANAEUM, p-v. and st. jus. Jefferson co. Mo. by p-r. 886 ms. a little s. of w. W. C. and 30 below, and a little w. of s. St. Louis. It is situated on the right bank of the Miss. r. nearly opposite Harrison in Il. It is the usual landing place from the lead mines in Washington co. Pop. about 300, lat. 38° 15', long. 13° 24' w.

HEREFORD, p-v. Berks co. Pa.

HEREFORD, p-v. Baltimore co. Md. by p-r. 29 ms. from Baltimore.

HEREFORDS, p-o. Mason co. Va. by p-r. 329 ms. N. W. by w. Richmond and 316 westward W. C.

HERKIMER co. N. Y. lies between Oneida and Montgomery cos. embracing the Mohawk r., and is bounded N. by St. Lawrence and Montgomery cos., s. by Otsego, w. by Oneida and Lewis. Greatest length N. and s. 85 ms. greatest width 22, containing 1,290 sq. ms. or 725,600 acres. The Mohawk r. and the grand canal run through the heart of the pop. of this co. its whole width. It has a pretty large proportion of hilly land, and as great a diversity of soil as any in the state. Watered by branches of the Oswegatchie and Black r., and by W. Canada creek. The East Canada creek forms the eastern boundary. Pop. 1830, 35,870.

HERKIMER, p-t. and st. jus. of Herkimer co. N. Y. N. Mohawk, 14 ms. s. E. Utica, 79 w. N. W. Albany, extending along the Mohawk r. nearly 15 ms. Land of a superior quality. The village of Herkimer stands on the w. side of W. Canada creek. The village of Little Falls is situated at the Little Falls of the Mohawk r. The scenery here abouts is grand and interesting; 13 school districts; schools kept 11 months in 12. Pop. 1830, 2,486.

HERMITAGE, p-v. in the western part of Prince Edward co. Va. 87 ms. s. w. by w. Richmond.

HERMON, t. Penobscot co. Me. 7 ms. w. Bangor. Pop. 1830, 535.

HERNDON'S p-o. Orange co. N. C. by p-r. 19 ms. N. W. by w. Raleigh.

HERNDONSVILLE, p-v. Scott co. Ky. 33 ms. from Frankfort.

HERON, PASS OF, the strait uniting Mobile bay to Pascagoula sound. It is enclosed to the N. by the main shore of Ala., and to the s. by Dauphin Isl. At mid tides it admits the passage of vessels drawing 6 feet water.

HERRIN'S p-o. Humphreys co. Ten. by p-r. 81 ms. wstrd. Nashville.

HERRIOTVILLE, p-o. Alleghany co. Pa. by p-r. 211 ms. N. W. W. C.

HERTFORD, p-v. and st. jus. Perquimans co. N. C. about 50 ms. a little w. of s. Norfolk in Va. and by p-r. 200 N. E. by E. Raleigh. Lat. 36° 13', long. 0° 36' E. W. C.

HERTFORD COUNTY, N. C. bounded by Bertie s., by Northampton w., Roanoke r. which separates it from Gates N. and N. E., and Chowan bay, which separates it from Chowan co. s. E. Length 28, mean width 12, and area 356 sq. ms. Extending in lat. from 36° 11' to 36° 27', and in long. from 0° 20' E. to 0° 12' w. W. C. The slope of Hertford is to the N. E. Pollacasty, Loosing, and Pine creeks, all flow in that direction into Roanoke or Chowan r. Chief t. Wynton. Pop. 1820, 7,712, 1830, 8,541.

HETRICK'S p-o. York co. Pa. by p-r. 83 ms. nthrd. W. C.

HIBERNIA, p-v. on Missouri r. opposite Jefferson, Callaway co. Mo. 981 ms. by p-r. w. W. C.

HICKLENS, p-v. Washington co. Geo. 31 ms. from Milledgeville.

HICKMAN COUNTY, Ten. bounded by Wayne s. w., Perry w., Dickson N., Williamson N. E., Maury E., Giles s. E., Lawrence s. Length 38, mean width 28, and area 1,064 sq. ms. Extending in lat. from 35° 23' to 35° 51', long. from 10° 12' to 10° 45' w. W. C. The southern part of Hickman is drained by the Buffalo branch of Duck r., flowing over it westerly, but full three fourths of the whole surface is included in the valley, which also flows westerly, receiving creeks from each side. Chief t. Vernon. Pop. 1820, 6,080, 1830, 8,132.

HICKMAN, southwestern co. Ky. bounded by Mayfield's r. separating it from McCracken N., by Graves E., Weakly co. Ten. s. E., Obion co. Ten. s. w., and w. by the Miss. r. separating it from New Madrid and Scott cos. of Mo. Length 31, mean width 18, and area 540 sq ms. Extending in lat. from 36° 30' to 36° 57'. The slope of this co. is wstrd. towards Miss. r. and is drained in that direction by Mayfield's r. and little Obion. Chief t. Columbus. Pop. 1830, 5,198.

HICKMAN'S p-o. Monongalia co. Va. by p-r. 208 ms. N. W. by w. W. C.

HICKORY, small p-v. Washington county Pa.

HICKORY CREEK and p-o. southern part of Warren co. Ten. The p-o. is 10 ms. sthrd. McMinville, the co. st., and 74 ms. s. E. by E. Nashville.

HICKORY FLAT, p-o. Gwinnet co. Geo. by p-r. 130 ms. N. N. W. Milledgeville.

HICKORY FORK and p-o. Gloucester co. Va. by p-r. 85 ms. E. Richmond.

HICKORY GROVE, p-o. Mecklenburg co. N. C. by p-r. 181 ms. s. w. by w. Raleigh.

HICKORY GROVE, p-o. York dist. S. C. by p-r. 89 ms. N. Columbia.

HICKORY GROVE, p-o. Henry co. Geo. by p-r. 81 ms. N. W. by w. Milledgeville.

HICKORY GROVE and p-o. Montgomery co. Ala. by p-r. 142 ms. s. E. Tuscaloosa.

HICKORY GROVE, p-o. Bond co. Il. 30 ms. wstrd. Vandalia.

HICKORY GROVE, p-o. Montgomery co. Mo. 45 ms. w. St. Louis.

HICKORY HILL, p-v. on Coosaw r. northern part of Beaufort dist. S. C. 70 ms w. Charleston, and by p-r. 93 a little w of s. Columbia.

HICKORY MOUNTAIN, p-v. Chatham co. N. C. by p-r. 46 ms. wstrdly. Raleigh.

HICKSFORD, p-v. and st. jus. Greenville co. Va. situated on the right bank of Meherin r. by p-r. 69 ms. almost exactly due s. Richmond. Lat. 36° 37′, long. 0° 35′ w. W. C.

HICKSTOWN, p-v. and st. jus. Madison co. Flor.

HIGGINSPORT, p-v. in the southern part of Brown co. O. by p-r. 111 ms. s. s. w. Columbus.

HIGHGATE, p-t. Franklin co. Vt. on Missisque bay, 33 ms. N. Burlington. Pop. 1830, 2,038.

HIGH GROVE, p-v. Nelson co. Ky. near Bardstown, and 54 ms. s. w. by w. Frankfort.

HIGHTSTOWN, p-v. Middlesex co. N. J. 3 ms. s. Cranbury, 12 N. E. Bordentown.

HIGHLAND, co. O. bounded s. E. by Adams, s. w. and w. by Brown, N. w. by Clinton, N. by Fayette, N. E. by Ross, and E. by Pike. Lat. 39° 0′ to 39° 22′, long. 6° 16′ to 6° 47′ w. W. C. This co. derives its name from occupying a table land between the Ohio, Sciota, and Little Miami vallies. The southeastern sources of Paint creek flow to the N. E. towards the Sciota; those of Brush and Eagle creeks, sthrd. towards Ohio r., and the estrn. branches of Little Miami wstrd. The surface is hilly and broken. Soil various, but generally productive. Chief t. Hillsboro'. Pop. 1820, 12,308, 1830, 16,345.

HIGHLANDS, N. Y. (*See Matteawan mtns.*)

HIGH PLAINS, p-v. Bledsoe co. Ten. by p-r. 89 ms. estrd. Nashville.

HIGH ROCK, p-o. Rockingham co. N. C. by p-r. 74 ms. N. w. Raleigh.

HIGH SHOALS, p-o. Rutherford co. N. C. by p-r. 234 ms. s. w. by w. Raleigh.

HIGH SPIRE, p-v. Dauphin co. Pa. 6 ms. from Harrisburg.

HIGH TOWER, p-v. Cherokee ter. Geo. on Etowah r. by p-r. 151 ms. N. w. Milledgeville. This is one amongst the many Indian names mutilated to suit the English idiom; it is a corruption of Etowah.

HILHAM, p-v. in the western part of Overton co. Ten. 14 ms. wstrd. Monroe, co. st. and 109 ms. N. E. by E. Nashville.

HILLEGAS, p-o. Montgomery co. Pa. by p-r. 170 ms. N. E. W. C.

HILL GROVE, p-v. Pittsylvania co. Va. by p-r. s. w. by w. Richmond.

HILLHOUSE, p-v. in the northern part of Geauga co. O. by p-r. 185 ms. N. E. Columbus, and 336 N. w. W. C.

HILLIARDSTON, p-v. Nash co. N. C. 10 ms. northestrd. Nash court house, and 54 N. N. E. Raleigh.

HILLSBOROUGH COUNTY, N. H. bounded N. by Grafton co., E. by Rockingham, s. by Mass., w. by Cheshire co. Greatest length 52 ms., greatest width from E. to w. 32 ms.; containing 1,245 sq. ms., or 796,800 acres. Surface generally uneven; mountains, Kearsarge, Ragged, Lyndenborough, Sunapee, Unconoonock, Crotched, and Society Land. Well watered. The Merrimack, the Contocook, the Nashua, the Souhegan, and the Piscataquog, are the principal rivers. It has several mineral springs. It possesses many advantages for manufacturing establishments. Chief towns, Amherst and Hopkinton. Pop. 1820, 35,781, 1830, 37,762.

HILLSBOROUGH, p-t. Hillsborough co. N. H. 23 ms. from Amherst, 24 w. Concord, 70 from Boston. Well watered by Contocook and Hillsborough rs. Land uneven. Pop. 1830, 1,792.

HILLSBOROUGH, t. Somerset co. N. J., lies w. of the Milstone, and s. of the Raritan, 15 ms. w. New Brunswick. Pop. 1830, 2,878.

HILLSBORO', p-v. on the U. S. road, Washington co. Pa. very nearly mid-distance between Washington, Pa. and Brownsville, and 11 ms. from each; by p-r. 221 ms. N. w. W. C. This village stands on ground elevated 1,750 feet above the Atlantic level. It extends in a single street along the road.

HILLSBORO', p-v. in the western part of Caroline co. Md. situated on Tuckahoe creek, about 13 ms. N. N. E. Easton, and 46 s. E. Baltimore.

HILLSBORO', p-v. at the eastern foot of the Blue Ridge, northern part Loudon co. Va. by p-r. 51 ms. N. w. W. C.

HILLSBORO', p-v. and st. jus. Orange co. N. C. by p-r. 41 ms. N. w. Raleigh. Lat. 36° 04′, long. 2° 7′ w. W. C. It is situated on Eno r. one of the higher branches of Neuse r.

HILLSBORO', p-v. Jasper co. Geo. by p-r. 61 ms. N. w. Milledgeville.

HILLSBORO', p-v. Madison co. Ala.

HILLSBORO', p-v. Franklin co. Ten. by p-r. 85 ms. s. s. E. Nashville.

HILLSBORO', v. of Davidson co. Ten. 11 ms. w. Nashville.

HILLSBORO', p-v. and st. jus. Highland co. O. by p-r. 74 ms. s. s. w. Columbus, and 441 ms. w. W. C. It is situated on the head of the s. w. branch of Paint creek, in a fine healthy country. Lat. 39° 12′, long. 6° 35′ w. Pop. 1830, 566.

HILLSBORO', p-v. in the eastern part of Fountain co. Ind. 61 ms. N. w. by w. Indianopolis, and 20 ms. eastward Covington, the co. st.

HILLSBORO', p-v. and st. jus. Montgomery co. Il. by p-r. 28 ms. N. w. by w. Vandalia, and 809 ms. w. W. C. Situated on Shoal creek branch of Kaskaskias r. Lat. 39° 08′, long. 12° 32′ w.

HILL'S BRIDGE and p-o. Halifax co. N. C. by p-r. 83 ms. N. E. Raleigh.

HILLTOWN, post tsp. Bucks co. Pa. about 22 ms. N. Philadelphia. Pop. of the tsp. 1830, 1,670.

HILLSDALE, p-t. Columbia co. N. Y. 16 ms. E. Hudson. Surface broken. 15 schools, kept 9 months in 12. Pop. 1830, 2,446.

HILLSDALE, one of the southern cos. of Mich. bounded by Williams co. O. s., Branch co. Mich. w., Calhoun N. w., Jackson N., and Lenawee E. Length from s. to N. 32 ms., breadth 26, and area 832 sq. ms. Lat. 41° 38′ to 42° 06′, long. 7° 21′ to 7° 50′ w. W. C. Hillsdale co. is in an especial manner a table land. The sources of St. Joseph's branch of Great Maumee rise in and drain the sthrn. and central part of the co. leaving it by a sthrn. course. Along the eastern margin rises Boan or Tiffin's r. another branch of Maumee, flowing also to the sthrd. The northwestern angle gives rise to the extreme sources of St. Joseph's r. of lake Michigan, flowing to the N. W., and finally from the northeastern angle issue to the estrd. the extreme sources of the river Raisin. To the above it may again be added, that the extreme sources of Grand r. of lake Mich. rise on the southern border of Jackson co., and almost on the northern margin of Hillsdale. It must therefore be obvious that the latter occupies the central plateau of Michigan, from which literally, the streams flow like radii from a common centre. Chief t. Sylvanus.

HILL'S GROVE, and p-o. Lycoming co. Pa. 100 ms. northward Harrisburg.

HILL'S STORE, and p-o. Randolph co. N. C. 84 ms. wstrd. Raleigh.

HILLVILLE, p-v. in the northwestern part of Mercer co. Pa. 12 ms. N. W. the borough of Mercer.

HINDSVILLE, p-v. in the southwestern part of Jefferson co. Ind. 17 ms. wstrd. Madison, the co. st. and 82 ms. S. S. E. Indianopolis.

HINDS, co. Miss. bounded S. by Copiah, S. W. by Claiborne, N. W. by Big Black river, separating it from Warren, N. by Madison, and E. by Pearl river, separating it from Rankin. Extending in lat. from 32° 02' to 32° 28', long. 13° 06' to 13° 50' W. W. C. It is in length from S. to N. 30, with nearly the same mean width, area 900 square miles. Hinds is composed of two inclined plains, the westrn. and most extensive, slopeing towards the Miss. and drained by the Big Black river, and Bayou Pierre; the eastern plain declines towards, and is drained into the Pearl river. Chief town, Jackson. Pop. 1830, 8,645.

HINESBURG, p-t. Chittenden co. Vt. 12 ms. S. E. Burlington, 26 W. Montpelier. Fine for farming. Principal streams, Platt river and Lewis creek. Pop. 1830, 1,665.

HINGHAM, p-t. Plymouth co. Mass. 14 ms. S. Boston. A handsome and compact village, at the head of an arm of Mass. bay. Surface broken and unpleasant. Has a respectable academy. Is a place of considerable trade and manufactures, and has some navigation. In 1830, there were 44,878½ bbls. of mackerel packed in this place. Population 1830, 3,387.

HINKLETON, p-v. situated at the forks of Conestogo creek, Lancaster co. Pa. 15 ms. N. E. Lancaster.

HINKLEY, p-v. and tsp. in the northern part of Medina co. Ohio. The p-o. is by p-r. 125 ms. N. E. Columbus. Pop. of the township 1830, 399.

HINSDALE, p-t. S. W. corner of Cheshire co. N. H., E. Conn. river, 75 ms. from Concord, 96 from Boston, 86 from Hartford, Conn. and 86 from Albany. Well watered. Crossed by the Ashuelot, besides which, it has several other streams. It contains iron ore, and some other minerals and fossils. These are found in West River mtn. which, some years since, suffered a slight volcanic eruption. Timber, pitch and white pine, white and yellow oak, chestnut, and walnut. Here is a bridge across the Conn. r. Population 1830, 937.

HINSDALE, p-t. Berkshire co. Mass. 10 ms. E. Pittsfield, 125 W. Boston. Situated on the highlands, near the heads of the Housatonic and Westfield rivers. A farming town. Pop. 1830, 780.

HINSDALE, p-t. Cataraugus co. N. Y. 16 ms. S. E. Ellicottville. Timber, pine, beech, and maple. Pop. 1830, 919.

HIRAM, p-t. Oxford co. Me. on the Saco, 34 ms. S. W. Paris. Pop. 1830, 1,026.

HIRAM, p-v. and tsp. in the northern part of Portage co. Ohio. The p-o. is by p-r. 141 ms. N. E. Columbus. Pop. of the township 1830, 517.

HIX'S FERRY, and p-o. Lawrence co. Ark. by p-r. 1,014 ms. W. C. and 151 northeastward Little Rock.

HOBOKEN, village, Bergen co. N. J. beautifully situated on the Hudson river, opposite N. Y. city, with which there is a constant communication by a steam boat ferry, every 20 minutes; a place of great resort for the citizens of N. Y. in warm weather.

HOCKMAN, p-o. Green Briar co. Va. by p-r. 275 ms. S. W. by W. W. C.

HODGENSVILLE, p-v. Hardins co. Ky. by p-r. 83 ms. southwestward Frankfort.

HOFFSVILLE, p-o. Harrison co. Va. by p-r. 236 ms. W. W. C.

HOGESTOWN, p-v. Cumberland co. Pa. 9 ms. from Harrisburg.

HOGG'S STORE, and p-o. in the northern part of Newberry district, S. C. 63 ms. N. W. Columbia.

HOG ISLAND, on the coast of Northampton co. Va.

HOG MOUNTAIN, p-o. northern part of Clark co. Geo. by p-r. 84 ms. N. N. W. Milledgeville.

HOKESVILLE, p-o. Lincoln co. N. C. by p-r. 178 ms. S. W. Raleigh.

HOLDEN, p-t. Worcester co. Mass. 51 ms. W. Boston. Situated on elevated ground. The main branch of the Blackstone river heads in this town. Well watered and supplied with mill seats. Pop. 1830, 1,719.

HOLDENS, p-o. Lycoming co. Pa.

HOLDERNESS, p-t. Grafton co. N. H. 40 ms. N. Concord, 65 from Portsmouth. Soil hard. Timber, oak, pine, beech, and maple. Well watered, and supplied with mill seats by the Pemigewasset and various other streams. Pop. 1830, 1,430.

HOLLAND, town, Orleans co. Vt. 56 ms. N. E. Montpelier, 61 N. Newburg. Settlement commenced since 1800. Watered by several branches of Clyde river. Land handsome and excellent. Pop. 1830, 422.

HOLLAND, p-t. Hampden co. Mass. 20 ms.

E. Springfield, 75 s. w. Boston. Crossed by the Quinnabaug. Pop. 1830, 453.

HOLLAND, p-t. Erie co. N. Y. 24 ms. s. E. Buffalo. Watered by Cazenovia and Seneca creeks. Land moderately uneven, an easy and rather a moist loam, timbered with maple, beech, linden, hemlock, &c. Population 1830, 1,071.

HOLLAND, p-v. Venango co. Pa. by p-r. 302 ms. N. W. W. C.

HOLLIDAYSBURG, p-v. in the western part of Huntingdon co. Pa. 3 ms. s. w. Frankstown, 40 ms. N. of Bedford and by p-r. 111 a little N. of w. Harrisburg.

HOLLIDAY'S COVE, p-o. Brooke co. Va. on the p-r. about 35 ms. w. Pittsburg, and by p-r. 269 N. W. W. C.

HOLLINGSWORTH FARM, and p-o. Habersham co. Geo by p-r. 137 ms. N. Milledgeville.

HOLLIS, p-t. Hillsborough co. N. H. 8 ms. s. Amherst, 36 s. Concord, 42 N. W. Boston. Watered s. E. by Nashua river, s. w. by Misitissit river. Soil various. A pleasant village near the centre. Pop. 1830, 1,792.

HOLLIS, p-t. York co. Me. on Saco river, 42 ms. N. York. Pop. 1830, 2,272.

HOLLISTON, p-t. Middlesex co. Mass. 25 ms. s. w. Boston. Soil good, and well cultivated. Water privileges valuable. It has several extensive factories, and is a very flourishing town. The shoe manufacturing business is recently carried on extensively in the place. Pop. 1830, 1,304.

HOLLOWAY'S p-o. Edgefield district S. C. by p-r. 89 ms. wstrd. Columbia.

HOLLY GROVE, p-o. Monroe co. Geo. by p-r. 32 ms. westrd. Milledgeville.

HOLLY IRON WORKS, and p-o. Cumberland co. Pa. by p-r. 31 ms. w. Harrisburg.

HOLMES, co. of Ohio, bounded E. by Tuscarawas, s. by Coshocton, s. w. by Knox, N. W. by Richland, and N. by Wayne. Length from E. to w. 30, breadth 18, and area 540 square ms. Lat. 40° 27′ to 40° 40′, long. 4° 42′ to 5° 13′ w. W. C. This co. is traversed from north to south by Kilbuck, branch of White Woman's river, and the western border by Mohiccon river, slope southward. Chief town, Millersburgh. Pop. 1830, 9,133. Holmes co. was formed since the census of 1820, from the southern part of Wayne, and northern of Coshocton.

HOLMESBURGH, village, Philadelphia co. Pa. on the Pennipack and main stage road, 9½ ms. N. E. Philadelphia. There are several manufactories here on the Pennipack.

HOLMESBURGH, p-o. Phila. co. Pa. 10 ms. N. E. from the city of Phila.

HOLMES' HOLE, p-v. Duke's co. Mass. on Martha's Vineyard, 9 ms. from Falmouth, 91 s. E. Boston. Has a safe and spacious harbor, where wind bound vessels often wait for a propitious gale, to waft them safely by the Cape Cod shoals.

HOLME'S MILL, and p-o. Loudon co. Va. by p-r. 46 ms. westerly W. C.

HOLME'S VALLEY, p-v. on Holme's creek, in the northwestern part of Washington co. Florida, 108 ms. a little N. of E. Pensacola, and 71 w. Tallahasse. The tract of country called Holme's valley is described in William's Florida, under the head of Jackson, but in his map it is included in Washington co. It is thus delineated. "Holme's Valley commences near the Choctawhatche river and extends eastwardly 10 or 12 ms. parallel with Holme's creek, from which it is separated by a sand ridge, one or two ms. wide. It contains from 8 to 10 sections of good land, sunk nearly 100 feet below the surface of the surrounding country. The soil is a dark sandy loam, covered with white, black, and yellow oak, white ash, black gum, wild chery, red bay, magnolia, &c." It is already extensively settled along a stream supplied by springs from the adjacent hills.

HOLMESVILLE, formerly called Appling court house, p-v. and st. jus. Appling co. Geo. by p-r. 145 ms. s. E. Milledgeville, and 787 ms. southwestward W. C. N. lat. 31° 43′, long. 5° 32′ w.

HOLMESVILLE, p-v. and st. jus. Pike co. Miss. 56 ms. s. E. by E. Natchez, and 1,128 s. westward W. C. It is situated on the main stream of Bogue Chitto river. N. lat. 31° 12′.

HOLT'S STORE, and p-o. Orange co. N. C. 68 ms. N. W. Raleigh.

HOMER, p-t. Cortlandt co. N. Y. the st. jus. of the co. 26 ms. s. Onondaga, 15 N. E. Ithaca, and 138 w. Albany. It is situated on the Tioughnioga, which, with its numerous branches, supply an abundance of mill seats. The land is good. It contains two considerable and flourishing villages, Homer and Cortland, in each of which is an academy. There are 32 school districts, in which schools are kept 7 months in 12. Pop. 1830. 3,307.

HOMOCHITTO, river of the state of Miss. has its most remote source in Copiah co. from which it flows southwestward into and over Franklin, draining by its confluents nearly the whole surface of the latter co. Leaving Franklin it assumes a w. s. w. course separating Adams from Wilkinson co. and falls into the Miss. after a comparative course of 75 ms. The Homochitto has interlocking sources with those of Amite, Bogue Chitto, and Pearl river, and with those of Bayou Pierre.

HONE FACTORY, and p-o. Randolph co. N. C. 64 ms. westward Raleigh.

HONEOYE LAKE, Richmond, Ontario co. N. Y. about 5 ms. long N. and s., and 1 mile wide. It discharges, at the N. end Honeoye creek, which receives also the outlets of Canadea and Hemlock lakes, and falls into the Genesee at Avon.

HONESDALE, flourishing p-v. Wayne co. Pa. at the junction of Dyberry creek with Lackawaxen river, 24 ms. above the junction of the latter with the Delaware river, at an elevation above the Atlantic tides, of 816 feet. The situation is delightful, and here, as at many other places, canals and roads have produced almost instantly a town with all the attributes of business; wealth, and population possessing intelligence and independence. Honesdale stands at the point of connexion between the Lackawaxen canal and rail road, by the latter 17½ ms. estrd. Carbondale, and 130 N. Phila. lat. 41° 35′, long. 1° 44′ E. W. C. Pop. 1830, 433, now perhaps, 1000.

HONEY BROOK, p-t. tsp. Chester co. Pa. on the heads of Brandywine creek, about 40 ms. wstrd. Phila.

HONEY CREEK, and p-o. Vigo co. Ind. 7 ms. s. Terre Haute, the co. st.

HONEY HILL, and p-o. Monroe co. Ala. by p-r. 152 ms. s. Tuscaloosa.

HONEYVILLE, p-v. Shenandoah co. Va. by p-r. 115 ms. w. W. C.

HOOKERSTOWN, p-v. Greene co. N. C. by p-r. 85 ms. N. E. Raleigh.

HOOKESTOWN, p-v. Beaver co. Pa. by p-r. 258 ms. from W. C.

HOOKSETT, p-t. Merrimack co. N. H., on the Merrimack, 9 ms. s. Concord, 12 from Hopkinton, and 54 from Boston. Here are those beautiful falls, known by the name of Isle of Hooksett falls; the r. descends 16 feet in the course of 30 rods; here too is a bridge across the Merrimack. Pop. 1830, 880.

HOOKSTOWN, v. of Baltimore co. Md., on the Reisterstown road, 6 ms. N. W. Baltimore.

HOOSAC CREEK, or Little Hoosac, waters the fine farming valley in the E. of Rensselaer co. and unites with Hoosac r. at Petersburg, being about 12 ms. in length.

HOOSAC R. rises in the N. W. corner of Mass., and after coursing 45 ms., falls into the Hudson at Schaghticoke point.

HOOSACK MTN., Williamstown, Mass., one of the loftiest summits of the Green mts.

HOOSACK, Hosick, or Hoosick, p-t. Rensselaer co. N. Y., 20 ms. N. Troy, 26 from Albany, and 8 w. Bennington; crossed nearly centrally by Hoosac r.; land broken; it has water privileges in abundance, supplied by Hoosac falls; Bennington battle was fought here, and in Bennington and White Creek, Aug. 16, 1777; contains limestone, brick clay, and slate; here are several nitrogen springs. Pop. 1830, 3,584.

HOP BOTTOM, p-v. Susquehannah co. Pa.

HOPE, p-t. Waldo co. Me. 35 ms. N. E. Wiscasset. Pop. 1830, 1,541.

HOPE, t. S. E. corner Hamilton co. N. Y., 25 ms. N. N. E. Johnstown; land very broken. Pop. 1830, 719.

HOPE, v. Warren co. N. J. 16 ms. s. Newton, 10 N. E. of Belvidere.

HOPE, p-v. Pickens co. Ala., by p-r. 43 ms. wstrd. Tuscaloosa.

HOPEWELL, p-t. Ontario co. N. Y., 5 ms. E. Canandaigua village; excellent land. Pop. 1830, 2,198.

HOPEWELL, t. Cumberland co. N. J., bounded N. and E. by Cohansey cr., and S. W. by Greenwich. Pop. 1830, 1,953.

HOPEWELL, p-t. Hunterdon co. N. J., on Del. r., 11 ms. N. Trenton, 14 w. Princeton. Pop. 1830, 3,151.

HOPEWELL, p-t. tsp. Bedford co. Pa., in the N. E. part of the co., on Rayton branch of Juniata, by p-r. 110 ms. w. Harrisburg.

HOPEWELL, p-v. Mecklenburg co. N. C., by p-r. 173 ms. s. w. Raleigh.

HOPEWELL, p-v. York dist. S. C., by p-r. 64 ms. N. Columbia.

HOPEWELL, p-v. Rock Castle co. Ky., by p-r. 83 ms. s. s. E. Frankfort.

HOPEWELL, p-v. Muskingum co. O., by p-r. 54 ms. estrd. Columbus.

HOPEWELL, Cotton Works, p-o. Chester co. Pa.

HOPKINS, co. of Ky., bounded E. by Pond r. separating it from Mecklenburg; S. E. by Christian; S. W. by Trade water, separating it from Caldwell; W. by Livingston; N. W. by Union and Henderson; and N. by Green r. separating it from Daviess. Length 35, mean width 20, area 750 sq. ms. Extending in lat. from 37° 04′ to 37° 34′, and long. 10° 18′ to 10° 52′ w. W. C. The nrthestrn. and larger section of this co. slopes towards, and is drained into Green r., whilst the sthwstrn. declines towards, and gives source to several creeks, flowing into Trade water. Chief t. Madisonville. Pop. 1820, 5,322; in 1830, 6,763.

HOPKINSVILLE, p-v. st. jus. Christian co. Ky., 81 ms. N. W. Nashville in Ten., 33 W. Russellville, and by p-r. 212 ms. s. w. by w. Frankfort. Lat. 36° 52′, long. 10° 35′ w. W. C.

HOPKINSVILLE, p-o. Warren co. O., by p-r. 88 ms. s. w. by w. W. C.

HOPKINTON, p-t. and one of the sts. jus. in Merrimack co. N. H., 28 ms. N. Amherst, 7 w. Concord, 46 N. E. Keene, 30 S. E. Newport, 50 w. Portsmouth, and 65 N. N. W. Boston. Crossed in S. W. part by Contocook r. Pop. 1830, 2,474.

HOPKINTON, p-t. Middlesex co. Mass., 32 ms. s. w. Boston. The main branch of Concord r. rises in this town, also branches of Charles and Blackstone rs.; land, large swells, well watered, good for grazing and orcharding; it contains two large manufacturing establishments, with a capital of 100,000 dollars each. Pop. 1830, 1,809.

HOPKINTON, p-t. Washington co. R. I., 30 ms. s. w. Providence. There is a seventh day Baptist society in this place. Pop. 1830, 1,777.

HOPKINTON, p-t. St. Lawrence co. N. Y., 40 ms. E. Ogdensburgh, 23 w. s. w. Malone; heavily timbered with maple, beech, elm, bass, butternut, &c. Pop. 1830, 827.

HOPPER'S TAN YARD, and p-o. in the N. E. part Christian co. Ky., by p-r. 202 ms. s. w. by w. Frankfort.

HORLL GATT, Hurl Gate, Hell Gate, a strait in East r. N. Y., 8 ms. from N. Y. city, between the islands of Manhattan and Parsell on the N. W., and L. I. on the S. E. Here are numerous little whirlpools; but vessels may nevertheless pass with the greatest safety, if well piloted. The proper name of this strait is *Horll Gatt*, a Dutch term signifying a whirlpool; but it is sometimes corruptly written and pronounced Hell Gate.

HORNBECK'S, p-o. Pike co. Pa., by p-r. 242 ms. N. N. E. W. C.

HORNELLSVILLE, p-t. Steuben co. N. Y., 20 ms. W. Bath, 260 from Albany; watered by the Canisteo, a boatable stream; good land. Pop. 1830, 1,365.

HORN ISLAND, a long, low, and almost naked bank of sand in the gulf of Mexico, opposite the mouth of Pascagoula r. It is one of that chain of islands, which merely merge above high water, and range from the mouth of Mobile bay in a westerly direction, to the mouth of Pearl r. and are in few places half a mile wide.

HORNTOWN, p-v. Accomac co. Va., situated on the road from Drummondtown to Snow Hill, 26 ms. N. N. E. from the former, and 16 a little W. of N. from the latter place, and by p-r. 188 ms. S. E. W. C.

HORRY, extreme estrn. dist. of S. C., bounded S. by Georgetown; W. and N. W. by Little Pedee r., separating it from Marion; N. E. by Columbus co. N. C.; and S. E. by the Atlantic. Length 50, mean width 20, area 1,000 sq. ms. Extending in lat. from 33° 34' to 34° 17', and long. 0° 44' to 1° 25' W. W. C. The declivity of this dist. is almost due S., and though having 30 ms. of ocean border, none of its streams flow in that direction; Waccamaw r. on the contrary entering from N. C., flows almost parallel to the opposing ocean coast, at from 8 to 15 ms. distance. The whole dist. is, with little exception, a plain; in a great part sandy and marshy. Staples, cotton and rice. Chief t. Conwaysboro. Pop. 1820, 5,025, 1830, 5,248.

HORSEHAM, p-v. and tsp. Montgomery co. Pa., 20 ms. N. Phil. Pop. tsp. 1820, 1,081; 1830, 1,086.

HORSE HEAD, tavern and p-o. Prince George's co., 35 ms. by p-r. sthestrdly. W. C.

HORSE RACE, in the Highlands, 15 ms. below West Point, a zig zag course in the Hudson, between Anthony's Nose and Dunderberg.

HORSE SHOE BOTTOM, p-v. Russell co. Ky., by p-r. 153 ms. S. Frankfort.

HORSE SHOE BRIDGE, and p-o. Colleton dist. S. C., by p-r. 105 ms. S. Columbia.

HORSE WELL, tavern, cross roads and p-o. Barren co. Ky., 9 ms. N. Glasgow, and by p-r. 106 ms. S. S. W. Frankfort.

HOSKINSVILLE, p-o. Morgan co. O., by p-r. 94 ms. a little S. of E. Columbus.

HOTEL CREEK, in Riga, runs into Black creek.

HOT SPRINGS, p-v. Bath co. Va., situated between the Cow Pasture, and the eastern branch of Jackson's r., 40 ms. S. W. by W. Stanton, by p-r. 231 S. W. by W. W. C., and 183 N. W. by W. Richmond.

HOUNSFIELD, t. Jefferson co. N. Y., lying on the E. side of Chaumont bay, of lake Ontario; surface gently uneven; soil principally clay or loam; contains fine mill seats. In this town is Sackett's Harbor, the settlement of which was not commenced till 1801, but which came into general notice, and made a conspicuous figure during the last war between Great Britain and the U. States. It has a most excellent harbor, containing a depth of water sufficient for the largest ships of war; on Navy Point, there is now the "largest ship of war on the stocks that ever was built." Here is laid up the U. S. squadron employed on lake Ontario during the last war, under Commodore Chauncey; and here Gen. Brown gathered his first laurels, by a brilliant and successful defence of the place against the British forces from Kingston. Pop. 1830, 3,415.

HOUSTON, co. of Geo., bounded by Dooley S.; by Flint r., separating it from Marion W.; by Crawford N. W.; Tchocunno r., separating it from Bibb N. E.; and the Ockmulgee r., separating it from Twiggs and Pulaski E. Extending in lat. from 6° 32' to 7° 13' W. W. C. This co. is in length along Dooley from Flint to Ockmulgee r. Pop. 1830, 7,369.

HOUSTON'S, store and p-o. Rowan co. N. C., by p-r. 136 ms. westrd. Raleigh.

HOUSTON'S store and p-o. Morgan co. Ala., by p-r. 114 ms. N. Tuscaloosa.

HOUSTONVILLE, p-v. Iredell co. N. C. by p-r. 172 ms. W. Raleigh.

HOWARD, p-t. Steuben co. N. Y. 10 ms. W. Bath, 254 W. S. W. Albany. Rough and broken; 9 schools, kept 6 months in 12. Pop. 1830, 2,464.

HOWARD, p-t. tsp. on Beach creek, northern angle of Centre co. Pa. about 15 ms. N. Bellefonte, and by p-r. 94 N. W. Harrisburg.

HOWARD, co. Mo. bounded N. W. by Chariton, N. E. by Randolph, E. by Boone, S. by Missouri river, separating it from Cooper, and by the Mo. river separating it from Saline. Length 24 ms., mean breadth 20, and area 480 square ms. Lat. 38° 55' to 39° 17', long. 15° 21' to 15° 50' W. W. C. Slope sthrd. towards that part of Missouri river, which separates it from Cooper. Chief town, Fayette. Pop. 1830, 10,854.

HOWARD'S RACE, p-o. St. Mary's co. Md. by p-r. 60 ms. S. E. W. C.

HOWELL, island, Monmouth county, N. J. bounded E. by the sea, S. by Dover, W. by Freehold, N. by Shrewsbury. Pop. 1830, 4,141.

HOWELL, p-o. Logan co. O. by p-r. 73 ms. N. W. Columbus.

HOWELLVILLE, p-v. Frederick co. Va. by p-r. 74 ms. N. W. by W. W. C.

HOYLESVILLE, p-o. Lincoln co. N. C. by p-r. 200 ms. S. W. by W. Raleigh.

HOYSVILLE, p-v. Loudon co. Va. by p-r. 53 ms. W. W. C.

HOYSVILLE, p-v. Loudon co. Va. 43 ms. N. W. W. C.

HUBBARD, p-v. and tsp. Trumbull co. O. By p-r. the p-o. is 285 ms. N. W. W. C., and 174 N. E. Columbus. Pop. of the tsp. 1830, 1,085.

HUBBARDSTOWN, p-t. N. W. part of Rutland co. Vt. 50 ms. S. W. Montpelier, and 50 N. Bennington. Surface uneven; well watered; well timbered with hard wood. Contains good mill seats. 9 school districts. A part of general St. Clair's army was here defeated on their retreat from Ticonderoga, July, 1777. Hubbardton r. which rises in Sudbury, passes through this town, and falls into E. bay in W. Haven. Its length is about 20 ms. Pop. 1830, 865.

HUBBARDSTON, p-t. Worcester co. Mass. 60 ms. w. Boston, 20 N. E. Worcester. Agreeably diversified with hills and valleys. Pop. 1830, 1,674.

HUDSON RIVER, one of the best for navigation in America, rises in the high mountainous region w. lake Champlain, in numerous branches, and pursuing a straight southerly course for more than 300 ms., unites with the Atlantic below the city of N. Y. It has three large expansions, Tappan bay, Haverstraw bay, and another bay between Fishkill and New Windsor. The Mohawk is its principal tributary. Notwithstanding it flows through a hilly and mountainous country, it is navigable for small sloops to Troy, 166 ms. from its mouth. The combined action of the tides, arriving in the Hudson by the East r. and the Narrows, carries the swell of the river upwards at the rate of 15 to 25 ms. an hour. Swift sailing vessels, leaving N. Y. at new tide, frequently run through to Albany with the same flood-tide. The passage of this river, through the Highlands, is charming and sublime. The Erie and Champlain canals connect this river with lakes Erie and Champlain.

HUDSON, city, p-t., port of entry, and st. jus. Columbia co. N. Y. E. Hudson r., which is navigable to this place for ships of the largest size, 117 ms. N. N. Y., 28 S. Albany. Claverack creek, which forms the E. boundary, affords the best of sites for water works. Factory creek, which forms the boundary towards Kinderhook, has likewise mill privileges. It contains limestone, brick clay, lead, nitre, alum, &c. Its manufactures and commerce are considerable. The city is supplied with water brought in an aqueduct from a spring 2 ms. distant. It is pretty well laid out, the streets generally crossing each other at right angles. Pop. 1830, 5,392.

HUDSON, p-v. in the N. W. part of Caswell co. N. C. by p-r. 86 ms. N. W. Raleigh.

HUDSON, p-v. and tsp. Portage co. O. The p-o. is by p-r. 124 ms. N. E. Columbus, and 336 N. W. W. C. Pop. of the tsp. 1830, 775.

HUDSON'S p-o. Culpepper co. Va.

HUDSONVILLE, p-v. Grayson co. Ky. by p-r. 113 ms. S. W. Frankfort.

HUFFERSVILLE, p-o. Greene co. O. by p-r. 59 ms. S. W. by W. Columbus.

HUGHE'S p-o. Allen co. Ky. by p-r. 165 ms. S. S. W. Frankfort.

HUGHESVILLE, p-v. Chester dist. S. C. by p-r. 53 ms. N. Columbia.

HUGHESVILLE, p-o. Loudon co. Va. 4 ms. from Leesburg, the co. st. and by p-r. 36 ms. N. W. W. C.

HULINGSBURG, p-v. on Piney creek, northeastern part of Armstrong co. Pa. 28 ms. a little E. of N. Kittanning, and 70 N. N. E. Pittsburg.

HULING'S FERRY and p-o. Perry co. Pa. by p-r. 18 ms. northrd. Harrisburg.

HULL, t. Plymouth co. Mass. S. side Boston harbor, on a peninsula 8 ms. long, and from 40 rods to half a mile in width, connected with Hingham by a mill dam, 9 miles E. Boston, 36 N. Plymouth. Population 1830, 198.

HULL'S STORE and p-o. Pendleton co. Va. by p-r. 206 ms. wstrd. W. C.

HULMESVILLE, p-v. on the left bank of Neshaminy creek, 4 ms. N. W. Bristol, and 20 N. N. E. Phil.

HULMESVILLE, p-o. Bucks co. Pa.

HUME, t. Alleghany co. N. Y. 13 ms. N. W. Angelica, crossed S. E. by Genesee r. Has bog iron ore. Pop. 1830, 951.

HUMMELSTOWN, p-v. on the left bank of Swatara creek, Dauphin co. Pa. 10 ms. E. Harrisburg, and 94 N. W. by W. Phil. It is for its size a wealthy village, extending, in good substantial houses, principally in one street along the main road, from Reading to Harrisburg.

HUMPHREY'S CREEK and p-o. in the northwestern angle of McCracken co. Ky. The creek falls into the Ohio r. about 12 ms. above the junction of that stream with the Miss. The p-o. is by p-r. 30 ms. N. W. by W. Wilmington, the co. st. and 309 ms. S. W. by W. ½ W. W. C.

HUMPHREY'S MILLS and p-o. Monroe co. Ten. by p-r. 145 ms. S. E. by E. Nashville.

HUMPHREY'S VILLA, p-o. Holmes co. O. by p-r. 66 ms. N. E. Columbus.

HUMPHREYSVILLE, p-v. Derby, Conn. 10 ms. N. W. New Haven, 15 N. E. Bridgeport. It is a manufacturing village of some extent, and promises much more than it already is, being finely situated on the Naugatuck r. At this place merino sheep were first introduced into the U. S. by general Humphreys, in 1801. It is surrounded by lofty hills, covered with wood, and is considered by visiting strangers as one of the most beautiful and romantic places in the country.

HUMPHREYSVILLE, p-v. Chester co. Pa.

HUMPHRIES, co. of Ten. bounded S. by Perry, W. by Carroll, N. W. by Henry, N. by Stewart, and E. by Dickson. Length 30, mean width 24, and area 720 sq. ms. Lat. 36°, long. 11° W. W. C., intersect about 5 ms. S. of the centre of this co. Ten. r. enters the southern border, and about 3 ms. within it receives Duck r.; thence traversing the co. in a northern direction, divides it into two unequal inclined plains, the most extensive being on the E. towards Dickson co. Chief town, Reynoldsburgh. Pop. 1820, 4,067, 1830, 6,189.

HUNTER, p-t. Greene co. N. Y., W. Kaatsberg, or Catskill mtns. 22 ms. W. Catskill, 58 from Albany. It embraces the highest points of the Catskill mtns. The Kaaterskill falls, of about 300 feet, are in this town. Here is one of the most extensive tanneries in the U. S. Pop. 1830, 1,960.

HUNTERDON COUNTY, N. J. on Delaware r. bounded N. W. by Warren, N. E. and E. by Morris, Somerset and Middlesex, S. E. by Burlington, st. jus. Flemington. Trenton in this co. is the st. of government of the state, where there are several manufactories on the Assanpink. The remainder of the co. is principally agricultural. Pop. 1820, 28,604, 1830, 31,066.

HUNTER'S HALL and p-o. Franklin co. Va. by p-r. 194 ms. S. W. by W. Richmond.

HUNTERSVILLE, p-v. and st. jus. Pocahontas co. Va. by p-r. 219 ms. S. W. by W. W. C. and 186 N. W. by W. Richmond. Lat. 38° 12', long. 3° 1' W. W. C. It is situated on one of the higher branches of Green Briar r. between Green Briar and Alleghany mtns. at an elevation above the Atlantic of upwards of 1,800 feet.

HUNTERSVILLE, p-v. Lincoln co. N. C. by p-r. 185 ms. S. W. by W. Raleigh.

HUNTERSVILLE, p-v. Tippecanoe co. Ind. 6 ms. estrd. Fayette, and by p-r. 64 ms. N. W. Indianopolis.

HUNTINGDON, co. of Pa. bounded S. W. by Bedford, N. W. by Alleghany mtns., separating it from Cambria, N. by Centre, N. E. by Mifflin, and S. E. by Tuscarora mtn., separating it from Franklin. Length, diagonally from S. E. to N. W. 58 ms. mean breadth 22, and area 1,276 sq. ms. Extending in lat. from 40° 03' to 40° 46', and in long from 0° 44' to 1° 35' W. W. C. This co. is composed of lateral chains of mtns., ranging from S. W. to N. E. with very fertile intervening vallies; and it may be remarked that the declivity of its surface is estrd., and the central part south estrd., as evinced by the course of the rivers. The whole co. is embraced in the valley of Juniata, and traversed along the channel of that stream, by the Pa. canal. The mean elevation of the arable surface of Huntingdon must exceed 800 feet. At Frankstown in the western part of the co. the surface of the canal is 910 feet above tide water in Del. r. This co. abounds in iron ore of very superior quality. Chief t. Huntingdon. Pop. 1820, 20,142, in 1830, 27,145.

HUNTINGDON, p-v., borough, and st. jus. Huntingdon co. Pa. 50 ms. N. N. E. Bedford, 60 in a direct line, though by p-r. 92 ms. N. W. by W. Harrisburg. Lat. 40° 31', and long 1° 2' W. W. C. It is situated on the left bank of Frankstown branch of the Juniata, about 2 ms. above the junction of the latter, with the Raystown or Main stream of Juniata. In 1820, this borough contained 841 inhabitants, but being situated on the Pa. canal, its pop. must rapidly increase. Not being given in the census tables (1830) separate from the tsp. in which it is situated, the exact population cannot be given; but supposed 1,200.

HUNTINGDON, tsp. and p-v. on Huntingdon creek, in the extreme western angle of Luzerne co. Pa. about 20 ms. N. W. by W. Wilkesbarre, and by p-r. 105 ms. N. N. E. Harrisburg.

HUNTINGDON, p-v. and st. jus. Carroll co. Ten. by p-r. 109 ms. W. Nashville. It is situated on the S. Fork of Obion r. Lat. 36° 02', long. 11° 28' W. W. C.

HUNTINGTON, t. S. E. part of Chittenden co. Vt. 20 ms. W. Montpelier, 15 S. E. Burlington. Principal stream, Huntington r., which affords some good mill seats. Surface very uneven, consisting of high mountains and deep gullies. Here is that celebrated peak of the Green mtns., called Camel's Rump. Soil in general gravelly and poor. 8 school districts. Huntingdon r., which rises in Lincoln, runs through this town, and joins Onion r. at Richmond, after a course of about 20 ms. Pop. 1830, 923.

HUNTINGTON, p-t. Fairfield co. Conn., W. Ousatonic r. Surface uneven. Soil fertile. 18 school districts, 17 ms. W. New Haven. Pop. 1830, 1,371.

HUNTINGTON, p-t. Suffolk co. N. Y. on Long Isl. 40 ms. E. N. Y. bounded N. by Long Isl. sound. Has an academy and 24 school districts, together with a variety of manufactories. This is a place of resort for strangers in summer, for the purpose of fishing and fowling. Pop. 1830, 5,582.

HUNTINGTON, p-v. Lawrens dist. S. C. by p-r. 64 ms. N. W. Columbia.

HUNTINGTON, p-v. Calvert co. Md. on the road from Prince Frederick to Annapolis, 3 ms. N. the latter, and by p-r. 57 ms. S. the former, and 94 S. E. W. C.

HUNTINGTON, p-v. and tsp. Lorain co. O. By p-r. the p-o. is 105 ms. N. N. E. Columbus. Pop. of the tsp. 1830, 169.

HUNTSBURGH, p-o. Geauga co. O. by p-r. 173 ms. N. E. Columbus.

HUNT'S MILLS, (now Clinton,) v. Hunterdon co. N. J.

HUNTSVILLE, p-t. near S. W. angle of Otsego co. N. Y. 25 ms. S. W. Cooperstown, embracing both sides of the Susquehannah r., whence rafts of timber descend to Baltimore. Pop. 1830, 1,149.

HUNTSVILLE, p-v. Luzerne co. Pa. by p-r. 125 ms. N. E. Harrisburg.

HUNTSVILLE, p-v. and st. jus. Surry co. N. C. situated on the right side of Yadkin r. in the S. E. part of the co. by p-r. 151 ms. N. W. by W. Raleigh. Lat. 36° 09', long. 2° 32' W. W. C.

HUNTSVILLE, p-v. Lawrens co. dist. S. C. 9 ms. S. E. Lawrenceville, and by p-r. 81 ms. N. W. Columbia.

HUNTSVILLE, p-v. and st. jus. Madison co. Ala. 101 ms. almost due S. Nashville in Ten. 146 ms. by the common road, but on the p-o. list stated at 165 N. N. E. Tuscaloosa. N. lat. 34° 44', long. 9° 35' W. W. C. This place has been called the capital of northern Ala. and is a very flourishing village; by the census of 1820, the population stood, whites, 833; colored, 483; total, 1,316. The pop. of 1830 not given in the census.

HUNTSVILLE, p-v. Butler co. O. by p-r. 93 ms. southwstrd. Columbus.

HUNTSVILLE, p-v. and st. jus. Randolph co. Mo. by p-r. 1,042 ms. wstrd. W. C. and 230 N. w. by w. St. Louis.

HURLEY, t. Ulster co. N. Y. 3 ms. w. Kingston, 68 s. Albany, 100 N. N. Y. There is an inexhaustible quarry of variegated marble in this town, composed of petrified shells of a bluish and reddish cast. Population 1830, 1,408.

HURON, one of the five great lakes, which, with many smaller ones, form the inland fresh water sea of North America, usually called the sea of Canada. Huron in its utmost extent has been, until the publication in 1829, of Tanner's map of the U. S., very inaccurately delineated. This extensive sheet of water is there laid down, according to its real natural divisions, into three parts, lake Huron Proper, lake Iroquois, and Manitou bay.

HURON PROPER lies in form of a crescent, the middle curve of which stretches 260 ms. from the Michilimakinak straits to the head of St. Clair r. With the exception of the Saginau, and some other lesser bays, the outline approaches very nearly to a real crescent. Greatest breadth, independent of the bays, about 70 ms. Superficial extent about 20,000 sq. ms. On the s. w. it is limited by the peninsula between lake Huron and Michigan; on the N. w. it has the peninsula between Huron and Superior; and on the s. E. the peninsula of Upper Canada. From the latter protrudes northwardly a smaller peninsula, called Cabot's Head, which is followed as part of the same chain by a series of isls. inflecting to the N. w. towards St. Mary's strait. These islands retaining their Indian name, "Manitou (Great Spirit) islands," form with Cabot's Head so nearly a continuous land barrier as to divide lake Huron into two, and by the northern protrusion of Great Manitou isl. into three bodies of water.

To the northeastward of Huron Proper, and E. of Cabot's Head, spreads a sheet of water, called by Tanner, lake Iroquois. This latter lake is in form of an ellipse: 140 ms. the longer, by 70 the shorter axis. Allowing for the angles, the area is about 7,000 sq. ms.

Separated from Huron Proper by Drummond's, and the Lesser and Greater Manitou isls., and from lake Iroquis by the Great Manitou, stretches another sheet of water, called by Tanner, Manitou bay. This is in length from E. to w. 80 ms. with a mean breadth of 20, and area 1,600 sq. ms.

Taken in all its extent with the islands between the sections, lake Huron fills a physical area of 28,600 sq. ms. having a rude approach to a triangle of 240 ms. base. The main lake is excessively deep, but similar to most part of all the other Canadian lakes; the shores are generally shallow, though some fine harbors exist. Lake Huron is the common recipient of lake Superior, lake Michigan, lake Nipissing, lake Simcoe, and numerous small rs. It is in fact the lower depression of a basin, in form of an equilateral triangle of 300 miles each side, or about 37,500 sq. miles.

That part of the water of the Huron basin not abstracted by evaporation is poured to the sthrd. by the river or strait of St. Clair. The surface of the water at its mean height is about 600 feet above the Atlantic level.

The boundary between the U. S. and Canada passes along the main Huron about 225 ms., and thence between Drummond's and Little Manitou islands, and over the western end of Manitou lake 25 ms., or along 250 ms. from the influx of the northern branch of St. Mary's r. into lake Manitou to outlet of St. Clair r. from lake Huron. As a commercial link in the chain of inland navigation, Huron is of immense importance. In its natural state, the main lake opens a spacious channel into both Superior and Michigan lakes. Being united to lake Simcoe by Matchadash r., a natural channel is thus extended from the extreme sthestrn. angle of lake Iroquois, by the Matchadash r., Simcoe lake, and the Trent r. and lakes, into the extreme nrthestrn. angle of lake Ontario, with only a few intervening portages. This latter route has long attracted attention as one admitting, with moderate comparative expense in improvement, an abridgement of one half in distance from the head of St. Lawrence r. to the head of lake Huron. By the route of Ontario, Erie, Huron, and connecting rs. the distance is 800 ms.; whilst by the Simcoe, and lake Iroquois route it is only 400 ms. The actual execution of the Rideau canal is a practical illustration of what may be expected, in regard to the future artificial and direct union of Ontario and Huron lakes.

HURON, a name given to several rs. one in the northern part of the state of Ohio, rising in Richland co. but flowing nrthds. into lake Erie, drains the greatest part of and gives name to Huron co. Entire comparative course 40 ms.

Another Huron r. of much greater length of course, and draining a greatly larger valley, falls into the extreme nrthwstrn. angle of lake Erie at the outlet of Detroit r. This second Huron has interlocking sources with those of St. Joseph's, Kalamazoo, and Grand rs. of lake Michigan, with those Saginau r. and with a third Huron flowing into lake St. Clair; and again to the sthrd. with those of the r. Raisin. With a comparative course of 65 ms. Huron, or as it might be called middle Huron, rises in Oakland, Ingham, and Jackson cos., but drains nearly all Washtenaw and part of Wayne cos.

The third or nrthrn. Huron, has interlocking sources with those of Raisin, middle Huron, and Saginau rs. It is one of those rivers the breadth of which exceeds the length of course; as across the stream it is 40 ms. but from head to mouth only about 33 ms. It drains great part of Oakland and Macomb cos.

Such a repetition of the same name applied to the same species of object is a serious inconvenience, which is in the present instance enhanced by two rivers of the same name falling into lake Erie, depriving us of the remedy of distinguishing them by their recipients.

HURON, co. of O. bounded E. by Lorain co.; S. by Highland; S. W. by Crawford; W. by Seneca and Sandusky, and N. by lake Erie. Greatest length from S. to N. along the wetrn. border 48 ms., the mean length is about 40 ms., breadth 28 ms; and area 1,120 sq. ms. Lat. 40° to 40° 38', long. 5° 18' to 5° 48' w. W. C. Vermillion r. of Erie flows nrthwrdly. along its eastern border. Sandusky r. after a nrthrn. course over Marion, Crawford, Seneca and Sandusky cos. inflects to the E. widens into a bay, the lower part of which traverses Huron co. and separates Sandusky point from the other parts of the county. Huron river, however, from which the county takes its name, drains the much greater part of its surface. The whole area is an inclined plain falling by gentle slopes nrthwards. towards lake Erie. The soil is generally fertile. Chief town, Sandusky, though Norwalk is the st. jus. Pop. 1830, 13,341.

HURON, p-v. and tsp. at the mouth of Huron r. Huron co. O. The p-v. is by p-r. 125 ms. a little E. of N. Columbus. Pop. tsp. 1830, 480.

HURON, territory of the U. S. This article is introduced to admit a general view of that region of the U. S. extending wstrd. of lake Mich. to the Miss. r. A bill has been several times before Congress, to obtain a law for the formation of a territory of the U. S. W. of that of Mich., and in choosing and imposing a general name, Huron and Ouisconsin have been alternately introduced; the latter will most probably be preferred.

According to information communicated to the author of this article by Austin E. Wing, Esq. delegated from Michigan, that territory when erected into a state, will, it is probable, follow the middle of lake Mich., from the nrthwstrn. part of Ind. to some distance W. of Michilimakinak strait, or about the 8th degree of long. W. W. C., and thence due N. to lake Superior. If this demarcation is adopted, an immense territory will be left between it and the Miss. r. amounting to something above 100,000 sq. ms. In regard to the recipients of its rivers, it is composed of three natural sections; which may be designated the Miss , Michigan and lake Superior slopes.

The declination of the Miss. slope of Huron, is to the sthwst. very nearly at right angles to the general course of that part of the Miss. r. from the influx of the Riviere au Corbeau, (Crow river) to the great bend, opposite the nrthwstrn. angle of the state of Illinois. Down this plain, of 400 ms. length and 140 mean breath in descent, fall, beside many rivers of lesser size, the Owisconsin, La Croix, Black, Chippeway, St. Croix, Rum, Savannah and Meadow rs. Of those tributaries of the Miss. the largest in volume and most important as a commercial channel is the Ouisconsin. The breadth of the plain, down which these streams flow, restricts that of the western shore of Mich. It is rather remarkable that along the W. shore of the Michigan from Green bay sthrds. no r. enters the lake having a comparative length of 60 ms. (*See Rock r. and Fox r. of Illinois.*)

To the nrthrds. of the two preceding slopes extends a third, that of lake Superior, stretching about 500 ms. along the sthrn. side of that lake. This latter plain is comparatively narrow, not having a mean breadth above 60 ms. The rivers are numerous, but brief in their length of comparative course. The whole surface is amongst the least inviting of the sections of the Huron region. The eastern section is a long narrow peninsula enclosed between lake Superior, lake Huron, Green bay, and the nrthrn. part of lake Michigan. In the interior it is wet and marshy, a character of country which applies in good measure to much of Huron. Along the line of separation of the streams flowing into the basin of the Mississippi, from those discharged into the Canadian sea, the smaller lakes and swamps are numerous. The whole region is, it is true, not very minutely known; but as far as explored, if taken as a whole, does not appear so well adapted to agricultural settlement as the valley of O. or the lower part of that of the Miss. proper. Flat, and of course in winter and spring, wet, prairies or savannahs are common.

Over Huron there are two routes, along which nature has afforded facility of water intercommunication. The lower and yet most frequented is that by Fox and Ouisconsin rs. from lake Mich. to the Miss. r. The direction of this route is S. W. by W., and the reverse, about 400 ms. without calculating minute bends from the straits of Michilimakinak to the mouth of Ouisconsin. There is only a short portage between the Ouisconsin and Fox rs. to interrupt the passage of boats along the entire distance; and so nearly is the actual water line complete, and so level the portage, that small craft are navigated at seasons of high water from one r. to the other.

The second route passes through lake Superior in its greatest length, to reach St. Louis r. By the channel of the latter and Savannah branch of the Miss. that great stream is reached above N. lat. 47°, and at no very great distance below its source. Independent of partial bends the distance of the two points of contact with the Miss. exceeds 500 ms. That region from which originates the Miss., Rain Lake r., Red r., branch of Assiniboin, and the St. Louis r. of lake Superior, is an immense elevated plain, from which the water flows slowly or stands stagnant, owing to the very near approach of the whole surface to the curve of the sphere. To the same physical construction arises the many interlocking water courses, which in fact connect the conflu-

the water bends to N. N. W. 15 ms. to where it mingles with that of Plano river. Both the Kankakee and Pickimink, have channels incurving in a very remarkable manner with the outline of the southern part of lake Mich.

Below the union of Plane and Kankakee, the Illinois flows about 60 ms. very nearly W. receiving the Fox river, from the N., and Vermillion from the S. and falling over rapids, inflects abruptly to the S. S. W. The upper part of the Illinois valley encircles the southern part of lake Michigan, about 200 ms. A canal has been proposed to follow the Illinois and Plane, and thence over the intermediate space to Chicago on lake Mich. Nature seems to have done a great share of the necessary labor, to effect this improvement. The canal distance from the rapids to lake Michigan will be 100 ms. The rapids of Illinois are a mere shelf, uniting two plains of no considerable difference of elevation. Lake Michigan is elevated about 600 feet above tide water in the Atlantic ocean, and the higher part of the Illinois valley has but little more relative height. The surface of the country presents no considerable difference of elevation; it is in great part a plain, and much of it naked of timber.

Below the great bend, the Illinois with a considerable western curve pursues a general S. S. W. course 200 ms. to its junction with the Mississippi. If measured by the channel of Fox river, the entire comparative length of Illinois, is very near 400 ms. Below the great bend it is augmented by Spoon river from the west, and by the much larger volume of Sangamon, and also from both sides by numerous creeks, or small rivers. The greatest breadth of the Illinois valley is 120 ms. from the eastern sources of Sangamon to the western of Spoon river, but the northern or rather northeastern part, branching into two long narrow arms, the mean breadth does not exceed 60 ms. The area about 24,000 square ms. Physically this fine valley has lake Michigan N. E., Rock river N. W., Mississippi W. and S. W., Kaskaskias river S. E., and the higher branches of Wabash river, N. E. Below the rapids, the character of the river itself approaches to that of a tortuous canal, in many parts widening into swells that appear similar to lakes.

Politically a small section drained by the higher part of Rock river, is in Huron, area about 700 square ms. Both branches of Kankakee rise in Indiana, draining about two thousand three hundred square miles leaving twenty one thousand square miles in Illinois. No circumstance could exhibit the immense extent of the vast basin of the Mississippi more forcibly than to compare it with the valley of Illinois. The latter exceeding the fourth part of an area equal to the Rhine, does not amount to the fortieth part of the basin of which it forms a section. (*See the tributary rivers, Fox, Kankakee, Sangamon, Spoon, &c. under their respective heads.*)

ILLINOIS, state of the United States, bounded by the Mississippi river W. and S. W., by Huron territory N., lake Michigan N. E., Indiana E., and Ohio river separating it from Kentucky S. E. and S.

For outlines, commencing at the junction of Ohio and Mississippi rivers, and thence up the latter opposite the state of Missouri, to the mouth of Lemoine river, by comparative courses 340 miles; continuing up the Mississippi to lat. 42° 30′ 200 ms.; thence due E. to lake Michigan along the S. boundary of Huron, 167 ms.; along the S. W. part of lake Michigan 60 ms.; thence due S. along the western boundary of Indiana to Wabash river 163 ms.; down the Wabash, opposite Indiana to the Ohio river 120 ms.; down the Ohio river opposite Kentucky, to the Mississippi river, and place of beginning 130 ms.; having entire outline of 1,170 ms.

This state, next to Virginia, and Missouri, is the third in area amongst the states of the United States, extends in lat. from 37° to 42° 30′, and in long. from 10° 36′ to 14° 30′ W. W. C.

The greatest length is exactly on a line with the extreme of its lat. or 382 ms. A similar feature is presented by the extremes of its long. where the breadth is greatest, or 206 ms. Narrowing, however, towards both extremes, and the actual area being 53,480 square miles, the mean width is about 140 miles.

Embracing a zone of 5½ degrees of lat. and with an area of 34,227,200 acres, this state presents, as far as lat. is concerned, the most extended arable surface of any state of the United States. As a physical section it occupies the lower part of that inclined plain of which lake Michigan and both its shores are the higher sections, and which is extended into and embraces the much greater part of Indiana. Down this plain in a very nearly southwestern direction, flow the Wabash and confluents; the Kaskaskias, the Illinois and confluents, and the Rock, and Ouisconsin rivers. (*See article Huron Territory.*) The lowest section of the plain is also the extreme southern angle of Illinois, at the mouth of Ohio river, about 340 feet above tide water in the gulf of Mexico. Though the state of Illinois does contain some hilly sections, as a whole, it may be regarded as a gently inclining plain, in the direction of its rivers as already indicated. Without including minute parts, the extreme arable elevation may be safely stated at 800 feet above tide water, and the mean height at 550. With all the uniformity of its surface and the moderate difference of its relative level, there still exists a great difference in the extremes of its climate. Compared with the temperature on Rock river plains, that near the confluence of the Ohio and Mississippi rivers may be called warm. Cotton can be cultivated to the southward, and the summers are often intense, but a very severe winter climate prevails over the whole state. From actual observation the thermometer of Fahrenheit has fallen frequently below zero, at New Harmony, opposite the southern part of the state. We may here observe, that in making observations with the thermometer, they are made too often almost exclusively whilst the sun is above the horizon, and therefore give, not the mean of all the astronomical day, but that of day light, and consequently the far great number of places are represented as having a mean temperature altogether too high. If compared with other parts of the United States, (*see that article*,) it will be found very doubtful whether any part of Illinois has a mean temperature as high as 53° of the scale of Fahrenheit, and that the mean of the state falls as low, if not lower than 50°.

Soil and Productions. Of the surface of Illinois, it is safe to state as much as 50,000 square miles arable. In respect to soil it bears some resemblance to Ohio and Indiana, but has less broken, sterile, and rocky, or of flat and wet land than either of the latter, even when the respective superficies of the three states are compared. The worst feature of Illinois, is the vast extent of its naked and level plains (prairies) and the consequent scarcity of timber and fountain water. It must not, however, be understood, that the prairies are uniformly level plains; some are rolling and even hilly, and abound in good fountains, but as a general character they are plains in the true meaning of the term. In the article Louisiana the reader will observe that the prairies present all the extremes from fertility to extreme barrenness. The word is French and signifies *meadows*, and not *plains*. They are, and with all their variety of surface and soil, the same as the steppes of northern Asia.

In Illinois as in Louisiana, many of the prairies present alluvial deposites, which prove them to have once been morasses, perhaps lakes. Whatever may have been their origin, the prairies constitute the most striking feature of Illinois, and extend in the general direction of its rivers from the Mississippi to lake Michigan, and indeed stretch south of lake Michigan over Indiana into the state of Ohio, lessening nevertheless advancing eastward. The wooded soil is generally productive, and from what has been already observed, it must be obvious that the state in its vegetable productions assimilates with the northern and middle states, abounding in pasturage, and where cultivated with advantage, with small grain. Fruits common in the middle states grow and flourish, but it has been observed, that from the very great fertility of soil are comparatively vapid in taste and flavor.

Commercial Facilities. In the articles lake Michigan, and the rivers Illinois, Sangamon, Kankakee, Fox, Kaskaskias, Rock, Mississippi, and Wabash, the prodigious natural channels which bound, or traverse Illinois, will be seen. Rich in a productive soil, and every where open to navigable streams, it is not too much to say that ten million of inhabitants will be far from its ultimate population. This fine natural section has been subdivided into the following counties; to which is added the population according to the census of 1830. Those left blank are not named in the census tables.

Counties.		Counties.	
Adams,	2,186	Macdonough & Schuyler,	1,309
Alexander,	1,390	Madison,	6,229
Bond,	3,124	Marion,	2,021
Calhoun,	1,090	Mercer,	26
Clarke,	3,940	Monroe,	2,119
Clay,	755	Montgomery,	2,950
Clinton,	2,330	Morgan,	12,709
Cook,		Schuyler,	included with Macdonough
Crawford,	3,113	Peoria, Putnam,	1,309
Edgar,	4,071	Perry,	1,215
Edwards,	1,649	Pike,	2,393
Fayette,	2,704	Pope,	3,223
Franklin,	4,081	Randolph,	4,436
Fulton,		Rock Island,	
Henry, Knox,	2,156	Saint Clair,	7,092
Gallatin,	7,407	Sangamon,	12,960
Green,	7,664	Shelby,	2,973
Hamilton,	2,620	Tazewell,	4,716
Hancock,	484	Union,	3,239
Jackson,	1,827	Vermillion,	5,836
Jasper,		Wabash,	2,709
Jefferson,	2,555	Warren,	307
Joe Daviess,	2,111	Washington,	1,674
Johnson,	1,596	Wayne,	2,562
Lasalle,		White,	6,091
Lawrence,	3,661		
Macaupin,	1,989		
Mc Lean,			
Macon,	1,122		
		Total,	137,445

Of the foregoing were white persons,

	Males.	Females.
Under 5 years of age,	18,834	17,429
From 5 to 10,	12,753	12,000
" 10 to 15,	10,024	9,246
" 15 to 20,	7,770	8,053
" 20 to 30,	14,706	12,461
" 30 to 40,	8,825	6,850
" 40 to 50,	4,827	3,750
" 50 to 60,	2,833	2,047
" 60 to 70,	1,172	812
" 70 to 80,	384	273
" 80 to 90,	90	77
" 90 to 100,	6	14
" 100 and upwards,	4	1
Total,	82,048	73,013

Of the above, are deaf and dumb, under 14 years, 23; 14 to 25, 27; 25 and upwards 16; blind 35.

Colored population—free,

	Males.	Fems.
Under 10 years of age,	277	305
From 10 to 24,	251	225
" 24 to 36,	136	125
" 36 to 55,	119	106
" 55 to 100,	40	50
" 100 and upwards,	1	2
Total,	824	813

Slaves—males, 347; females, 400; Colored persons, deaf and dumb, 0; blind 4.

Recapitulation.

Whites.	Free colored.	Slaves.	Total.
155,061	1,637	747	157,445

Progressive population has been truly rapid in Illinois. During the decennial period between 1820, and 1830, this advanced at the rate of 185 per cent. *(See closing part of article Indiana.)*

Constitution, government, judiciary. The constitution of Illinois was adopted at Kaskaskias, 26th August, 1818.

Art. 1. *Sec.* 1. The powers of the government of the state of Illinois, shall be divided into three distinct departments, and each of them confided to a separate body of magistracy, to wit: those which are legislative, to one; those which are executive to another; and those which are judiciary to another.

Sec. 2. No person or collection of persons, being one of those departments, shall exercise any power properly belonging to either of the others, except as hereinafter expressly directed or permitted.

Art. 2. *Sec.* 1. The legislative authority of this state shall be vested in a general assembly, which shall consist of a senate and house of representatives, both to be elected by the people.

Sec. 3. No person shall be a representative who shall not have attained to the age of 21 years, who shall not be a citizen of the United States, and an inhabitant of this state, who shall not have resided within the limits of the county or district in which he shall be chosen, twelve months next preceding his election, unless absent on public business, &c.

Sec. 6. No person shall be a senator who has not arrived at the age of 25 years, who shall not be a citizen of the United States and who shall not have resided one year in the county or district in which he shall be chosen immediately preceding his election, &c.

Art. 3. *Sec.* 1. The executive power of this state shall be vested in a governor.

Sec. 2. The governor shall be chosen by the electors of the members of the general assembly, at the same places, and in the same manner that they shall respectively vote for members thereof.

Sec. 3. The governor shall hold his office for four years, and until another governor shall be elected and qualified; but he shall not be eligible for more than four years in any term of 8 years. Must be 30 years of age when elected, a citizen of the United States, and 2 years next preceding his election a resident of Illinois.

Sec. 13. A lieutenant governor shall be chosen at every election for governor, in the same manner, continue in office for the same time, and possess the same qualifications.

Sec. 14. The lieutenant governor is speaker of the senate, &c.

The governor has power to grant reprieves and pardon after conviction, except in cases of impeachment, and has farther the usual powers and duties to perform of governors of the other respective states. From any disability, death, &c. of the governor, the powers and duties of the office devolve on the lieutenant governor.

Art. 4. *Sec.* 1. The judicial power of this state shall be vested in one supreme court, and such inferior courts as the general assembly shall, from time to time ordain, and establish.

Sec. 4. The justices of the supreme court, and the judges of the inferior courts, shall be appointed by joint ballot of both branches of the general assembly, and commissioned by the governor; their offices during good behavior. Removable by impeachment or by address of two thirds of each branch of the general assembly.

By the general provisions of the constitution of Illinois, the right of suffrage is secured to the white male citizens above 21 years of age. No person can be imprisoned for debt unless on refusal to deliver up his property, or on strong suspicions of fraud. The right of trial by jury is to remain inviolate. By article 6th, section 1, neither slavery nor involuntary servitude is to be introduced into the state. All children born in the state, white or colored, become free, the males at 21, and the females at 18. Liberty of the press, of public worship, limited only by the public peace.

History. The early settlements of the French along the Illinois and Mississippi rivers, date back to 1673. The distant and feeble establishments of that nation at any place within the chartered limits of Illinois, never arose to the dignity of colonies. At the close of the revolutionary war, and by the treaty of 1783, the country was claimed under the charter of Virginia, and held by that state until ceded to the United States in 1787. It was then made a part of the territory N. W. of the Ohio river. When the now state of Ohio was made a separate territory in 1800, Illinois and Indiana remained united, and continued one territory, until 1809, when they were separated into two. Indiana lying eastward, and in the direction of the stream of emigration, preceded Illinois, as a state; the former reached that dignity in 1815, and the latter in 1818, as may be seen by the date of her constitution. Since that epoch the history of the state merges in that of the United States.

INDEPENDENCE CREEK, N. Y. about 25 ms. in length, runs from Herkimer co. across Lewis co. to the Black r. in Watson, midway between Beaver and Moose rs.

INDEPENDENCE, p-t. Alleghany co. N. Y. 18 ms. S. E. Angelica. Watered by Crider's, Dike's, and Baker's creeks, good sized mill streams, which fall into the Genesee river. Timbered with deciduous trees of the various kinds. Land better for grass than grain. Pop. 1830, 877.

INDEPENDENCE, t. Warren co. N. J. on Musconetcunk r., and the Sussex line, bounded N. W. by Hardwick, S. W. by Mansfield and Oxford. Pop. 1830, 2,126.

INDEPENDENCE, p-v. in the northeastern

part of Washington co. Pa. about 17 ms. N. W. Washington, the co. st.

INDEPENDENCE, p-v. Autauga co. Ala. by p-r. 85 ms. S. E. Tuscaloosa.

INDEPENDENCE, p-v. and st. jus. Jackson co. Mo. It is not located on either Tanner's map or p-o. list, but from the position of the co. is near the western boundary of the state, and on or near the Mo. r. The post distance 177 ms. wstrd. of Jefferson city.

INDIA KEN, p-v. Ripley co. Ind. by p-r. 87 ms. S. E. Indianopolis.

INDIAN RIVER, Coos co. N. H. one of the principal and most northerly sources of Conn. r., rises in the Highlands near the N. limits of the state, and pursues a S. W. course to its junction, with the E. branch, flowing from lake Conn. 30 ms. long.

INDIAN RIVER, N. Y. rises in Lewis co. winds across Jefferson co., runs through St. Lawrence co., and joins the Oswegatchie, about 4 ms. above its mouth in the St. Lawrence. A very crooked stream, and runs in its whole course probably 100 ms.

INDIAN RIVER, small stream of Sussex co. Del., rises near, and to the sthrd. of Georgetown, and flowing estrd. falls into Rehoboth bay, 10 ms. a little W. of S. Cape Henlopen.

INDIAN RIVER of Flor., is properly a sound, commencing at N. lat. 28° 40′, and stretching within cape Canaveral nearly parallel to the Atlantic coast, with a long, narrow, intervening reef of sand to N. lat. 27° 35′, where it opens to the ocean by Indian r. inlet.

INDIAN RIVER, HUNDRED, of Sussex co. Del. on Indian r. Pop. 1820, 1,887.

INDIANA, co. of Pa. bounded by Armstrong W. and N. W., by Jefferson N., Clearfield N. E., Cambria E. and S. E., and by Conemaugh r., separating it on the S. from Westmoreland. Length 35, mean width 23, and area 800 sq. ms. Extending in lat. from 40° 24′ to 40° 56′, and in long. from 1° 52′ to 2° 30′ W. W. C. This co. lies W. of the Laurel ridge, and its plane of descent is also wstrd. drained by the branches of the Conemaugh r. and of Crooked and Mahoning creeks. The descent of the declivity from the eastern to the wstrn. border of this co. is very rapid. By admeasurements made on the Pa. canal, the level of Conemaugh r. at the S. W. angle of this co. is 1,154 feet, and this point is the lowest part of the co., and of course the whole arable surface rises above an equivalent to a temperature of 3° of lat. Chief t. Indiana. Pop. 1820, 8,882, 1830, 14,252.

INDIANA, p-v. and st. jus. Ind. co. Pa. 35 ms. N. E. Greenburg, 48 N. E. by E. Pittsburg, and by p-r. 180 ms. N. W. by W. Harrisburg. Lat. 40° 40′, long. 2° 12′ W. W. C.

INDIANA, state of the U. S. bounded E. by O., S. by the O. r., separating it from Ky., W. by the state of Il., N. W. by lake Mich., and N. by the ter. of Mich.

Indiana extends along O. r. opposite Ky. from the mouth of Great Miami to that of the Wabash 340 ms. Up the Wabash, opposite the state of Il. to a meridian line, extending from lat. 39° 23′, 150 ms. Along the above meridian line to the southern part of lake Mich. 160 ms. Along lake Mich. according to Tanner, to lat. 41° 47′, 40 ms. Due E. along lat. 41° 47′, to the N. E. angle of the state, 110 ms. Thence due S. to the mouth of Great Miami, and place of beginning, 190 ms. Having an entire outline of 990 ms.

A diagonal line drawn from the S. W. to the N. E. angle of Ind. measures 325 ms., but its greatest length from S. to N. along its western border, from the Ohio r. opposite the mouth of Green r. to lake Mich. is 272 ms. The mean length is very near 260, and mean breath 140, with an area of 36,400 sq. ms. Measured carefully by the rhomb, the area comes out 36,670 sq. ms., and the mean between the two methods is so near 36,000 sq. ms. as to justify the adoption of that superficial area. This state extends in lat. from 37° 50′ to 41° 47′, and in long. W. W. C. from 7° 48′ to 11° 08′.

Much of what has been said respecting the physical features of Il. applies also to Ind. the two states being included in the same physical section. The reader will find great share of the general features of Ind. under the head of Wabash r., that stream and its confluents draining fully the five sixths of the whole state. In features, soil and climate, Indiana forms a connecting link between O. and Il. having the physiognomy of both the contiguous states. Less monotonous in surface than Il., Ind. presents fewer bold and prominent marks than does O.

Commencing on the Ohio r. we find a range of rough and abruptly rising hills, stretching along that great stream from the influx of Great Miami to near that of the Wabash. These hills, so imposing near the Ohio r. are in themselves a true geographical deception. Passing along this river's verge, no creek is found flowing from them of any considerable magnitude. Ascending these heights they are discovered to be the mere relative elevations formed by the deep channel of Ohio, and discharge their waters to the northwstrd. into the sub-valley of White r. or into the valley of Wabash. At the great bend of Ohio r. opposite the mouth of Ky. r., the fountains of White r. rise within 1 m. of the channel of Ohio. Traversing this range of hills the observer finds himself in the beautiful valley of the Wabash, variegated by hill and dale, and presenting one of the finest natural sections of the earth. The surface of the country softens advancing northwardly over the numerous tributaries of White r., and over the main volume of Wabash. A real table land is now reached, flat, level, and wet, giving source to the Tippecanoe and Eel r. branches of Wabash, to the Kankakee and Pickmink branches of Il. r., to the Elkhart, Pigeon and other southern branches of the St. Joseph's r. of lake Mich., and finally to the St. Joseph's branch of Maumee

From the preceding we find that Northern Ind. is a table land, discharging rivers in four, and nearly opposite directions. The settlements cease with the Wabash part, and a zone extending over the elevated plateau from Lenawee and Hillsdale cos. in Mich. into Il., and indeed almost to Il. r. remains in savage hands. The extreme northern section of Ind. drained into lake Mich. has been reclaimed and laid out into the cos. of La Grange, Elkhart, St. Joseph's and La Porte.

Properly speaking, the great western plain of Indiana, commences on lake Erie, between the mouths of Maumee and Raisin rivers, and extends to the junction of the Illinois with the Mississippi river, discharging to the N. W. the various confluents of St. Joseph's river of lake Michigan, and the Kankakee, Pickimink, Vermillion, Mackinaw, Sangamon, and other tributaries of Illinois r.; and on the opposite side giving source to the innumerable branches of Wabash and Kaskaskias rivers. The length of this plateau is from the mouth of the Raisin, to that of Illinois river 400 miles in a direction of S. W. by W. and N. E. by E. General character prairie, as noticed in the description of the state of Illinois. The surface and still more so the sub-soil abound with marine and river shells, with embedded trees, and other memoria of having been once inundated.

Prairies are not however confined to the northern section; they abound over the "White River country," as it is called, and present all the varieties of dry, wet, level, rolling, and of great fertility and barreness. They are generally however productive and are frequently most luxuriantly fertile.

Soil and Productions. It would be mere repetition to give a detail under this head, after what has been said respecting Illinois, the two states having such strong resemblance in both characters. A like remark applies to climate, with the exception that Indiana has less extension north or south than Illinois, the former having nevertheless more variety of features, has also a perceptibly severer winter over its northern plains. Under the article Ohio, the reader will find tables to illustrate the climate of the Ohio valley generally.

Commercial Facilities. Indiana already enjoys a share of the benefits arising from the canal connecting the Ohio and Miami river at Cincinnati. A rail road has been projected from Indianopolis in a nearly northern direction to lake Michigan in La Porte co. The courses of Maumee and Wabash rivers, and the nature of the intermediate country between their sources, invite a canal of connection. The Ohio river borders the southern part of the state, and with the Wabash and confluents offers immense natural commercial channels. (*See the various rivers under their respective heads.*)

Abstract, from the census of 1830, of the population of the counties and state of Indiana.

Counties.	Pop.	Counties.	Pop.
Allen,	996	Lawrence,	9,234
Bartholomew,	5,476	Madison,	2,238
Boone,	621	Marion,	7,192
Carroll,	1,611	Martin,	2,010
Cass,	1,162	Miami,	
Clark,	10,686	Monroe,	6,577
Clay,	1,616	Montgomery,	7,317
Clinton,	1,423	Morgan,	5,593
Crawford,	3,238	Orange,	7,901
Daviess,	4,543	Owen,	4,017
Dearborn,	13,974	Parke,	7,535
Delaware,	2,374	Perry,	3,369
Decatur,	5,887	Pike,	2,475
Dubois,	3,778	Posey,	6,549
Elkhart and ter. attached,	935	Putnam,	8,262
		Randolph,	3,912
Fayette,	9,112	Ripley,	3,989
Floyd,	6,361	Rush,	9,707
Fountain,	7,619	Scott,	3,092
Franklin,	10,190	Shelby,	6,295
Gibson,	5,418	Spencer,	3,196
Greene,	4,242	St. Joseph'h and ter. attached,	267
Grant,			
Hamilton,	1,757	Sullivan,	4,630
Hancock,	1,436	Switzerland,	7,028
Harrison,	10,273	Tippecanoe,	7,187
Henry,	6,197	Union,	7,944
Hendricks,	3,975	Vanderburgh,	2,611
Huntington,		Vermillion,	5,692
Jackson,	4,870	Vigo,	5,766
Jefferson,	11,465	Wabash,	
Jennings,	3,974	Warren, with ter. attached,	2,861
Johnson,	4,019		
Knox,	6,525	Warrick,	2,877
La Grange,		Washington,	13,064
La Porte,		Wayne,	18,571

Of the foregoing were white persons,

	Males.	Females.
Under 5 years of age,	39,789	37,505
From 5 to 10	28,692	27,313
" 10 to 15	22,872	21,072
" 15 to 20	17,653	18,087
" 20 to 30	28,153	26,702
" 30 to 40	17,904	15,703
" 40 to 50	10,306	9,028
" 50 to 60	6,004	4,808
" 60 to 70	3,160	2,275
" 70 to 80	1,059	780
" 80 to 90	240	212
" 90 to 100	49	25
" 100 and upwards,	13	4
Total,	175,885	163,514

Of the above are deaf and dumb, under 14 years, 49; 14 to 25, 59; 25 and upwards 33; blind 150.

Colored population—free,

	Males.	Females.
Under 10 years of age,	617	594
From 10 to 24	544	573
" 21 to 36	307	279
" 36 to 55	240	215
" 55 to 100	138	107
" 100 and upwards,	11	4
Total,	1,857	1,772

There are three slaves only in Indiana, and these are females. Colored persons who are deaf and dumb, under 14 years, 1; 14 to 25, 2; blind 2.

Recapitulation—

Whites,	Free Col'd,	Slaves,	Total.
339,399	3,629	3	343,031

Progressive population in Indiana, was 132 per cent during the 10 years preceding the last census. We have seen that that of Il. was still more in excess, being 185 per cent during the same term. The two states taken together, contain 89,880 square miles, equal to 57,523,200 statute acres. Their joint population amounts to 500,476. Thus on a physical section of 89,880 square ms., in 1830, the population amounted to a small fraction above 5½ to the square mile. Forty times such a distributive population would only a little exceed 20-millions, and fall even then far short of what has already comparatively accumulated on regions greatly less productive in every necessary requisite to sustain a dense population; such are the immense voids to be filled in the central United States.

Constitution, government, judiciary. The constitution of Ind. was adopted on the 10th of June, 1816, and contains the following essential provisions:

Art. 1. Is a Bill of Rights containing 24 sections.

Art. 2. The powers of the government of Indiana shall be divided into three distinct departments, and each of them be confided to a separate body of magistracy, to wit: those which are legislative to one; those which are executive to another; and those which are judiciary to another; and no person, or collection of persons, being of one of those departments, shall exercise any power properly attached to either of the others, except in the instances herein expressly permitted.

Art. 3. *Sec.* 1. The legislative authority of this state shall be vested in a general assembly, which shall consist of a senate and house of representatives, both to be elected by the people.

Sec. 3. The representatives shall be chosen annually, by the qualified electors of each county respectively, on the first Monday in August.

Sec. 4. No person shall be a representative, unless he shall have attained the age of 21 years, and shall be a citizen of the United States, and an inhabitant of this state; and shall also have resided within the limits of the county in which he shall be chosen, one year next preceding his election, &c.

Sec. 5. The senators shall be chosen for three years, on the first Monday in August, by the qualified voters for representatives.

Sec. 7. No person shall be a senator, unless he shall have attained the age of 25 yrs., and shall be a citizen of the U. S., resided two years in the state, and the last year in the county from which elected.

Art. 4. *Sec.* 1. The supreme executive power of this state shall be vested in a governor, who shall be styled, the governor of the state of Indiana.

Sec. 3. The governor shall hold his office during 3 years, or until a successor shall be chosen and qualified.

Sec. 5. He is required to be 30 years of age, a citizen of the United States 10 years, and have resided in the state 5 years next preceding his election. He has the usual power of governors of states. As in Illinois a lieutenant is chosen with the governor, and as in Illinois, the two officers have the same legal relation to each other.

Art. 5. *Sec.* 1. The judiciary power of this state both as to law and equity shall be vested in one supreme court, in circuit courts, and such other inferior courts as the general assembly may, from time to time, direct and establish.

Art. 6. *Sec.* 1.—Every white male citizen of the United States, of the age of 21 years and upwards, who has resided in the state one year immediately preceding such election, shall be entitled to vote in the co. where he resides. All elections by ballot.

Art. 11. *Sec.* 7. There shall be neither slavery nor involuntary servitude in this state.

The other provisions of the constitution of Indiana, have the ordinary features of those charters in other states.

History. The town of Vincennes is the cradle of Indiana, and was founded by the French about 1690. This remote village remained of little consequence, but was the scene of some interesting events in the revolutionary war. It was reached and taken by a British force, and again reached and retaken by a small army under the authority of Virginia and commanded by Col. Rogers Clarke. After the treaty of Grenville, 1795, settlements along the Ohio, Wabash and White rivers, began to extend. What is now Indiana, was severed from Ohio, in 1801, and Illinois constituted a territory. These two latter were separated in 1809, when each became a separate territory. In 1815, having attained the requisite population, Indiana became a state, as may be seen in the sketch of its constitution. Since becoming an independent member of the Union, its history is merged in that of the United States.

INDIANOPOLIS, p-v. and st. jus. for Marion co. and also st. of government for the state of Indiana, is situated on the right or w. bank of White river, by p-r. 573 ms. N. W. by W. ½ W. W. C., 108 N. W. from Cincinnati, and by the common road about 200 ms. N. E. by E. Vandalia, lat. 39° 47′, long. 9° 10′ W. W. C. According to Flint it contains 200 houses and 1,200 inhabitants, with the usual co. and state buildings. These new capitals increase so rapidly as to annually antiquate the description of the year before. At high water White river is navigable from Indianopolis. This town is remarkably near the actual centre of the state, and stands in a country presenting every advantage of soil, and surface.

INDIAN SPRINGS, p-v. in the sthrn. part Butts co. Geo. by p-r. 55 ms. N. W. Milledgeville.

INDIANTOWN, p-v. at the head of North r.

Currituck co. N. C. about 45 ms. a little E. of S. Norfolk Va. and by p-r. 231 ms. N. E. by E. Raleigh.

INDIANTOWN, p-v. on Cedar cr. Williamsburg dist. S.C. about 80 ms. direct, but by p-r. 127 ms. S. E. by E. Columbia.

INDIANTOWN, p-v. Graves co. Ky. by p-r. 262 ms. S. W. by W. Frankfort.

INDUSTRY, p-t. Somerset co. Me. 13 miles W. Norridgewock. Pop. 1830, 902.

INDUSTRY, p-v. Montgomery co. O. wstrd. Columbus.

INGHAM, p-v. Tioga co. Pa. by p-r. 152 ms. nrthrd. Harrisburg.

INGRAHAM'S MILLS, and p-o. Darlington district, S. C. by p-r. 83 ms. estrd. Columbia.

INGRAM'S STORE, and p-o. Randolph co. N. C. by p-r. 84 ms. wstrd. Raleigh.

INTERCOURSE, p-v. Lancaster co. Pa. 12 ms. E. Lancaster, by the common road 48 ms. but by p-r. 54 S. E. by E. Harrisburg.

IOWA, co. of Mich. or more correctly of Huron, bounded S. by Joe Daviess co. of Il. the Miss. r. W., Ouisconsin N., and with indefinite limits E. Lying between N. Lat. 42° 36′ and 43° 10′, long. W. C. 12° to 14° 10′. These limits are given from Tanner's improved map. From the same authority it appears, that a range of high ground separates the lower valley of Ouisconsin from the sources of numerous streams, which, flowing sthrd. into the Miss. or Rock r., traversing Iowa county, fall into their recipients in Joe Daviess co. Il. Limiting this co. by a meridian line running S. from Fort Winnebago, it would have been a length of about 100 ms. with a breadth of 40, or 4000 sq. ms. Chief town, Cassville. Pop. 1830, 1,576. The principal seat of the Indian war, in 1832, was in the estrn. part of this co. on Peektans r., Sugar cr., and Goosewehawn r. (*See Ouisconsin and Rock rs.*)

IPSWICH, the Agawam of the Indians, p-t. port of entry, and one of the shire towns of Essex co. Mass. 27 ms. N. E. Boston. There is a large and compact village on both sides of Ipswich r. about 2 ms. from its mouth, which are united by an excellent stone bridge. Site uneven. Land in most parts of the town excellent. Ships of considerable burthen come up to the lower part of the town, and the falls in the r. above, furnish convenient and extensive water power. It contains a male and female academy; the latter has a department for female teachers. It has long been noted for the manufacture of lace, which was formerly done by hand; but there is now a lace manufactory, with a capital of $150,000. Pop. 1830, 2,949.

IRA, p-t. Rutland co. Vt. 47 ms. N. Bennington, 32 W. Windsor. Somewhat mountainous. Watered by Ira brook and Castleton r. 5 school districts. Pop. 1830, 442.

IRA, t. Cayuga co. N. Y. 24 ms. N. Auburn, 11 ms. N. Erie canal. Poorly watered. Soil light. No marshes, swamps, or ponds. Pop. 1830, 2,193.

IRASBURGH, a post and shire town in the centre of Orleans co. Vt. 40 ms. N. E. Montpelier. Gently diversified with hill and dale. Soil good, and easily cultivated. Watered by Black r. Near the centre of the town is a small village containing a court house, jail, &c. Pop. 1830, 860.

IREDEL, co. N. C. bounded W. by Burke, N. W. by Wilkes, N. E. by Surry, E. by Rowan, S. by Mecklenburg, and S. W. by the Great Catawba r., separating it from Lincoln. Length 40, mean breadth 20, and area 800 sq. ms. Extending in lat. 35° 32′ to 36° 04′, and in long. 3° 45′ to 4° 14′ W. W. C. Iredell, though bounded by the Catawba, slopes in great part towards the estrd. is drained by the S. Yadkin. Chief t. Slateville. Pop. 1820, 13,071, and in 1830, 14,318.

IRONDEQUOT, cr. N. Y. waters W. Bloomfield, Mendon, Victor, Pittsford, Perrinton, and Brighton, where it enters the head of Irondequot or Teoronto bay, of lake Ontario, being about 20 ms. in length. It is a good mill stream. It crosses the Erie canal, on which there is a stupendous work in Pittsford and Perrinton, the great embankment.

IRVILLE, p-v. Muskingum co. O. 46 ms. estrd. Columbus.

IRVINE, p-o. Warren co. Pa. by p-r. 247 ms. N. W. Harrisburg.

IRVINE, p-v. and st. jus. Estill co. Ky. It is situated on Ky. r. 71 ms. N. E. by E. Frankfort. Lat. 37° 43′, long. W. C. 6° 53′ W. Pop. 1830, 91.

IRWIN, co. Geo. bounded W. by Baker, N. W. by Dooly, N. E. by Ockmulgee r., separating it from Telfair, E. by the sthrn. part of Telfair and the wstrn. of Appling, S. E. by Ware, S. by Lowndes, and S. W. by Thomas. Length along the sthrn. border from E. to W. 63 ms., mean width 33, and area 2,079 sq. ms. Extending in lat. 31° 22′ to 32°, and in long. 6° 17′ to 7° 10′ W. C. A very small section of the nrthestrn. part of Irwin, is drained into the Ockmulgee, and another small triangle on the estrn. side, by the extreme higher sources of the Santilla. The sthestrn. and central part is drained by the Suwanne and its confluents, whilst the southwestern section gives source to the Ocklockonne r. The general declivity is S. S. E. Chief t. Irwin. Pop. 1830, 1,180.

IRWIN, C. H. and p-o. Irwin co. Geo. by p-r. 143 ms. a little W. of S. Milledgeville.

IRWINE, r. of Va. and N. C. (*See Smith's r.*)

IRWINTON, p-v. and st. jus. Wilkinson co. Geo. 24 ms. S. Milledgeville. Lat. 32° 50′, long. 6° 18′ W. W. C.

ISABELLVILLE, p-v. Todd co. Ky. by p-r. 186 ms. S. W. by W. Frankfort.

ISCHUA, t. Cataraugus co. N. Y. 11 ms. E. Ellicottville. Crossed by Ischua cr. Soil and surface diverse. Timber principally maple, beech, elm, ash, butternut, &c.

ISINGLASS r. N. H. takes its rise from Long Pond in Barrington, and Bow Pond in Strafford, and after receiving the waters of several other ponds, unites with the Cocheco, near the S. part of Rochester.

tending in lat. 33° 53′ to 34° 17′, and in long. 6° 22′ to 6° 50′ w. W. C. The slope of this co. is to the southeast, and drained by different branches of Oconee and Appalachee rs. Chief town, Jefferson. Pop. 1820, 8,355, 1830, 9,004.

Jackson, p-v, and st. jus. Butts co. Geo. by p-r. 60 ms. though in a direct line only about 45 n. w. by w. Milledgeville; n. lat. 32° 12′, long. 7° 02′ w. W. C. It is situated on Towaubigan cr. a branch of Oconee r.

Jackson, co. Ten. bounded by Overton e., by White s., Smith w., and Monroe co. in Ky. n. Length 30, mean breadth 20, and area 600 sq. ms. Extending in lat. 36° 10′ to 36° 35′, and in long. 8° 27′ to 8° 49′ w. W. C. Cumberland r. enters the nthestrn. angle, and traverses this co. diagonally in a sthwstrn. direction. Chief town, Williamsburg. Pop. 1820, 7,593, 1830, 9,698.

Jackson, p-v. and st. jus. Madison co. Ten. situated on Forked Deer r. by p-r. 147 ms. s. w. by w. Nashville. Lat. 35° 36′, long. W. C. 11° 54′ w.

Jackson, co. of O. bounded s. e. by Meigs, s. by Lawrence, s. w. Sciota, w. by Pike, n. w. by Ross, n. by Hocking, and n. e. by Athens. Length 30, mean breadth 15 and area 450 sq. ms. Lat. 38° 50′ to 39° 17′, long. W. C. 5° 16′ to 5° 45′ w. It is a table land, discharging creeks nrthwstrd. into Sciota r., sthrd. and sthestrd. into O. r. Surface extremely broken. Chief town, Jackson. Pop. 1830, 5,941.

Jackson, p-v. Wayne co. O. by p-r. 98 ms. n. e. Columbia.

Jackson, p-v. on Thompson's cr., E. Feliciana parish of La., 6 ms. n.e. St. Francesville, and 26 a little w. of n. Baton Rouge.

Jackson, co. Ind. bounded s. e. by Scott, s. by the S. branch of White r. separating it from Washington, w. by Lawrence, n. by Bartholomew, and e. by Jennings. Length 30, mean breadth 20, and area 600 sq. ms. Lat 38° 47′ to 39° 03′, long. W. C. 8° 48′ to 9° 18′ w., slope s. w. and traversed by Driftwood, and other northern confluents of White r. Chief town, Brownstown. Population 1830, 4,870.

Jackson, sthest. co. of Miss. bounded s. by the Gulf of Mexico, s. w. and w. by Hancock co. Miss., n. w. by Perry, n. by Greene, and e. by Mobile co. in Alabama. Length 42, mean width 35, and area 1,470 sq. ms. Extending in lat. 30° 13′ to 30° 55′ n., and in long. 11° 32′ to 12° 28′ w. W. C. This co. embraces the lower part of the basin of Pascagoula, that river opening into Pascagoula sound about the middle of the co. Here the pine hills reach the coast of the sound, and with some but partial exceptions along the streams, a pine forest on sterile soil stretches over the whole surface. Staple, cotton. St. jus. Jackson C. H. Pop. 1820, 1,682, 1830, 1,792.

Jackson, p-v. st. jus. Hinds co. and of the government of the state of Miss. It is situated on the w. bank of Pearl r. about 100 ms. n. e. Natchez, and by p-r. 1,035 ms. s..w. by w. W. C. lat. 32° 17′, long. W. C. 13 16′ w. It is an inconsiderable place, and from its position will most likely remain so.

Jackson, C. H. and p-v. Jackson co. Miss. by p-r. 188 ms. s. e. Jackson the seat government for the same state, and by the common road about 180 ms. s. e. by e. Natchez.

Jackson, co. of Il., bounded by Randolph n. w., Perry n., Franklin e., Union s. e., and the Miss. r. separating it from Perry co. Mo. on the s. and s. w. Length 28, mean breadth 25, and area 700 sq. ms. Lat 37° 37′ to 37° 58′, long. 12° 13′ to 12° 40′ w. W. C. Slope sthwstrd. and drained in that direction by Muddy cr. and branches. Chief t. Brownsville. Pop. 1830, 1,828.

Jackson, co. of Ala. bounded by Madison co. in the same state w., by the sthrn. boundary of Ten. separating it from Lincoln co. in the latter state n. w., Franklin n., and Marion e., and by Tumesco r. separating it from the Cherokee country s. e., s., and s. w. Length s. w. to n. e. 52, mean breadth 20, and area 1,040 sq. ms. Extending in lat. 34° 24′ to 35°, and in long. 8° 50′ to 9° 30′ w. W. C. Ten. r. as it passes the boundary between Ala. and Ten., assumes a sthwstrn. course, which it pursues about 52 ms. along Jackson co. and abruptly turns to n. w. by w. traverses Cumberland mtn. and again bounds Jackson 22 ms. to the mouth of Flint cr. Cumberland mtn. leaving Ten. in the sthrn. part of Franklin co. ranges over Jackson co. in a s. s. w. direction, giving source along its wstrn. slope to Paint Rock r. which also traverses Jackson parallel to the mtn. chain. The general slope of the co. is to the s. s. w. It is a hilly and broken region, tho' with a considerable proportion of excellent land. Chief town, Bellponte. Pop. 1820, 8,751, 1830, 12,700.

Jackson, p-v. Clark co. Al. by p-r. 159 ms. s. Tuscaloosa.

Jackson, p-v. situated on the left bank of Tombigbee r. 65 ms. above and a little e. of n. Mobile, and by p-r. 132 ms. a little w. of n. Tuscaloosa.

Jackson, co. Mo. bounded n. by Mo. r. separating it from Clay, e. by Lafayette, on the s.——and on the w. by the w. boundary of the state. The breadth from e. to w. as laid down by Tanner, is 28 ms., but the sthrn. boundary being uncertain, the area cannot be even estimated. Chief town, Independence. Pop. 1830, 2,823.

Jackson, p-v. and st. jus. Cape Girardeau co. Mo. about 120 ms. a little e. of s. St. Louis, and 10 ms. w. Bainbridge on the Miss., n. lat. 37° 26′, long. W. C. 12° 42′ w.

Jackson, co. of Mich. bounded s. e. by Lenawee, s. by Hillsdale, w. by Calhoun, n. w. by Eaton, n. by Ingham, and e. by Washtenau. Length from w. to e. 32, mean breadth 24, and area 768 sq. ms. Lat. 42° 05′ to 42° 26′, long W. C. 7° 08′ to 7° 45′ w. Slope of the sthwstrn. section to the w. and drained by the confluents of Kalamazoo r., and of the residue of the co. to the N. W. giving extreme source to the tributaries of Grand r. of lake Michigan. Chief t. Jacksonopolis.

JACKSON, co. Ark. position uncertain, but supposed to be between the St. Francis and White rs. about 150 ms. to the N. E. of Little Rock.

JACKSON, p-v. given as the st. jus. Lawrence co. Ark. but is most probably the st. jus. of Jackson co. of the same territory. By p-r. 152 ms. N. E. Little Rock.

JACKSONBORO', p-v. and st. jus. Colleton dist. S. C. 34 ms. W. Charleston, and by p-r. S. S. E. Columbia. Lat. 32° 44', long. 3° 31' W. W. C. It is situated on the right bank of Edisto r. about 25 ms. above the mouth.

JACKSONBORO', p-v. and st. jus. Scriven co. Geo. situated on the forks of Brier cr. 62 ms. N. N. W. Savannah, and by p-r. 135 ms. S. E. by E. Milledgeville. Lat. 32° 43', long 4° 33' W. W. C.

JACKSONBORO', p-v. and st. jus. Campbell co. Ten. situated at the sthestrn. foot of Cumberland mtn. 36 ms. N. N. W. Knoxville, and by p-r. 152 ms. a little N. of E. Nashville. Lat. 36° 22', long. 7° W. W. C.

JACKSONBORO', p-v. Butler co. O. by p-r. 96 ms. S. W. by W. Columbus.

JACKSON, C. H. and st. jus. Jackson co. O. situated near the centre of the co. 74 ms. a little E. of S. Columbus, and 387 by p-r. W. W. C. Lat. 39° 02'. Pop. of the tsp. 1830, 329.

JACKSON HALL, p-o. Franklin co. Pa. by p-r. 90 ms. N. W. W. C.

JACKSONHAM, p-o. Lancaster dist. S. C.

JACKSON HILL, p-o. Davidson co. N. C. by p-r. 96 ms. W. Raleigh.

JACKSONOPOLIS, p-v. and st. jus. Jackson co. Mich. by p-r. 77 ms. W. Detroit.

JACKSON RIVER, p-o. Alleghany co. Va. by p-r. 272 ms. S. W. by W. W. C., and 202 a little N. of W. Richmond.

JACKSON'S r., the main constituent stream of James' r., rises by two branches, the N. and S. forks in the sthrn. part of Pendleton co. Va. Flowing thence sthwstrd. and nearly parallel, and between lateral chains of mtns., the two branches traverse Bath co. and entering Alleghany, incline towards each other and unite, but the united stream still pursues a sthwstrn. course, receiving Dunlops creek from the W. and Potts creek from the S. after a comparative course from the source in Pendleton of about 50 ms. With the junction of Potts cr. the whole stream inflects very abruptly to N. E. and flowing in that direction 15 ms. through rugged mtn. passes, unites with Cow Pasture river to form James' r. The valley of Jackson's r. is an elevated region. At Covington, the co. st. of Alleghany co. where Dunlops cr. falls into Jackson's r., the water surface is 1,238 ft. above the Atlantic level; it is therefore probable that the far greatest part of the arable surface of the adjacent country exceeds a comparative height of 1,500 ft. Lat. 38°, and long 3° W. W. C. intersect in the wstrn. part of Bath co. about 6 ms. N. the junction of the two main branches of Jackson's r.

JACKSON'S CREEK, p-o. Fairfield dist. about 5 ms. W. Winnsboro', and by p-r. 31 ms. N. N. W. Columbia.

JACKSON'S GROVE, p-o. in the sthwstrn. part of Abbeville dist. S. C. by p-r. 132 ms. wstrd. Columbia.

JACKSONVILLE, p-v. in the wstrn. part of Lehigh co. Pa. by p-r. 81 ms. N. E. Harrisburg, and by common road 20 ms. from Allentown, and 25 a little E. of N. Reading.

JACKSONVILLE, p-v. Wood co. Va. by p-r. 311 ms. W. W. C.

JACKSONVILLE, p-v. Mecklenburg co. N. C. by p-r. 119 ms. S. W. Raleigh.

JACKSONVILLE, p-v. Sumpter co. S. C. by p-r. 70 ms. Columbia.

JACKSONVILLE, p-v. and st. jus. Telfair co. Geo. by p-r. 111 ms. S. Milledgeville. Lat. 31° 55', long. W. C. 6° 05' W.

JACKSONVILLE, p-v. and st. jus. Duval co. Flor. situated on the left bank of St. John's r. 45 ms. N. W. St. Augustine, and by p-r. 165 ms. a little S. of E. Tallahasse. Lat. 30° 15', long. 5° W. W. C.

JACKSONVILLE, p-v. Bourbon co. Ky. 45 ms. estrd. Frankfort.

JACKSONVILLE, p-v. Dark co. O. by p-r. 99 ms. W. Columbus.

JACKSONVILLE, p-v. and st. jus. Morgan co. Il. by p-r. 115 ms. from Vandalia, and 837 from W. C. Lat. 30° 44', long. W. C. 13° 13' W.

JACKSONVILLE, springs and p-o. in the northeastern part of Washington parish, La. about 70 ms. N. New Orleans.

JACOBSBURG, p-v. Belmont co. Ohio, by p-r. 134 ms. E. Columbus.

JACOB'S STAFF, p-v. Monroe co. Ark. 84 ms. Little Rock.

JACQUES, or James river, a confluent of the Mo. rising between the latter and the Miss. about lat. 47° and flowing thence by a general course to the southward, nearly parallel to and about 60 ms. distant the Mo. into which it falls at lat. 42° 50' after a comparative course of something above 300 ms.

JAFFREY, p-t. Cheshire co. N. H. 62 ms. N. W. Boston, 46 S. W. Concord. The Grand Monadnoc mtn. is situated in the N. W. part of this town and in Dublin. Well watered by streams issuing from the mountain. Contains red and yellow ochre, alum, vitriol, and black lead. Pop. 1830, 1,354.

JAKES PRAIRIE, p-o. Gasconade co. Mo. 80 ms. W. St. Louis.

JAMAICA, p-t. Windham co Vt. 26 ms. N. E. Bennington, 32 S. W. Windsor. Watered by West r. and its numerous branches, which supply numerous and excellent mill privileges. Surface broken and mountainous. Soil in general warm and productive. Contains limestone and the micaceous oxide of iron; 10 school districts. Pop. 1830, 1,523.

JAMAICA, p-t. Queens co. S. side Long Island, 12 ms. E. N. York. Jamaica village has an academy. It is a most charming place; 8 schools kept 11 months in 12. Here is the place selected by the jockeys for horse racing. Pop. 1830, 2,376.

JAMAICA PLAINS, in Roxbury, Mass. remarkable for its beautiful scenery and elegant country seats.

JAMES, river of Virginia and sthrn. stream of the Chesapeake basin. For the two higher constituents of this fine r., see the respecpective articles, Cow Pasture and Jackson rs. Below the junction of its two constituents, the united water is first known as James r. which forcing a passage thro' between Potts and Mill mtns. enters Botetourt, and assumes a sthrn. course 10 ms. to where it receives Craig's creek from the south, and inflecting to s. s. E. flows in that direction 15 ms., thence abruptly turns to N. E. by E. 20 ms. to the western foot of Blue Ridge, and the reception of North river from Augusta and Rockbridge counties. Assuming a s. E. course of 28 ms. James river, now a fine navigable stream, traverses a gap of Blue Ridge, about 15 ms. N. E. the Peaks of Ottor, and in a distance of 30 ms. separating Amherst from Bedford and Campbell counties, and traversing another lateral chain of mtns. near Lynchburg, again turns to N. E. Continuing the latter course 40 ms. and separating Amherst and Nelson from Campbell and Buckingham cos. James river assumes a course of a little s. of E. 70 ms. by comparative course, having on the left the counties of Albemarle, Fluvanna, Goochland, and Henrico, and on the right the cos. of Buckingham, Cumberland, Powhatan and Chesterfield, to the head of tide water and the lower falls at Richmond.

Meeting the tide, James river, similar to most of the Atlantic rs. of the United States, generally widens, and presenting rather the features of a bay than those of a r. turns to a little E. of s. E. 90 ms. by comparative courses, finally merges into Chesapeake bay, between Point Willoughby, and Old Point Comfort. The entire length of Jame's r. from its source in Pendleton to its efflux into Chesapeake, is 368 ms. but following the actual meanders it is probable that this stream flows not much if any less than 500 ms.

The valley of James river, including all its confluents, lies between lat. 36° 40′ and 38° 20′, and in long. extends near 1° E. to 3° 40′ w. W. C. Drawing a line in a s. s. w. direction from Old Point Comfort to the Alleghany mtn. will pass along very near the middle of this valley 225 ms. The broadest part is along the extreme sources, from the fountains of Jackson r. to those of Craig's creek 90 ms., but the mean width amounts to about 45 ms. and the area to 10,125 square ms.

In the natural state James river affords at and for a few miles above its mouth depth of water for ships of any required draught, but the depth gradually shallows so that only vessels of 130 tons can reach Rockets, or the port of Richmond. Though much has been designed above tide water in meliorating the navigation, little has been actually accomplished. A short canal connects the tide below, and the boatable water above the falls at Richmond. The following relative heights will show the gradual rise of the James river. Columbia at the mouth of Rivanna 178 feet; Scottsville, at the southeastern angle of Albemarle co. and below the southeast chain of the Appalachian system 255 feet; Lynchburg, also below the southeast mtn. 500 feet; Pattonsburg, at the great bend above Blue Ridge 806 feet; Covington, at the junction of Dunlap's creek and Jackson's river 1,222 feet; highest spring tributary to Craig's cr. 2,498 feet. Those heights are only the elevation of the water, and at every point must fall short of that of the arable soil. Without any great risk of error, an allowance of winter temperature equal to 6 degrees of Fahrenheit may be made between the extremes of this valley on the same lat.

JAMES CITY, co. Va. bounded by James r. which separates it from Surry s., by Chickahomina r. separating it from Charles City co. w., by N. Kent N. w., by York r. separating it from Gloucester N., by York N. E., and Warwick s. E. Length 23, mean breadth 8, and area 184 square ms. Extending in lat. 37° 09′ to 37° 25′ N., and in long. 0° 03′ to 0° 24′ E. This county is waving, rather hilly. Chief town, Williamsburg. Pop. 1820, 3,161, 1830, 3,838.

JAMESTOWN, an insulated township on Canonicut island, in Narraganset bay, Newport co. R. I. about 3 ms. w. Newport, 30 s. Providence, including the whole of Canonicut island, being about 8 ms. in length from N. to s. and having an average width of nearly a mile, containing about 8 square ms. Soil rich and productive. It has two ferries, the one to Newport, the other to South Kingston. Pop. 1830, 415.

JAMESTOWN, p-v. on Appomattox r. in the northeastern angle of Prince Edward co. Va. by p-r. 86 ms. s. w. by w. Richmond.

JAMESTOWN. It may be noticed as a curious fact that Jamestown, the first Anglo-American settlement, made on Powhatan's, now James r. has no name on the post office list. It stood on a point of land in the sthrn. part of James City co. lat. 37° 12′, long. 0° 14′ E. W. C.

JAMESTOWN, p-v. Guilford co. N. C. by p-r. 147 ms. N. w. by w. Raleigh; the real common road distance must fall short 100 ms.

JAMESTOWN, p-v. Fentress co. Ten. by p-r. 135 ms. E. Nashville.

JAMESTOWN, p-v. and st. jus. Russell co. Ky. by p-r. 123 ms. s. Frankfort.

JAMESTOWN, p-v. Greene co. Ohio, 68 ms. N. w. by w. Columbus.

JAMESVILLE, p-o. in the sthrn. part of Sumpter dist., S. C. by p-r. 6 ms. s. E. Columbia.

JASPER, co. Geo. bounded by Oakmulgee r. separating it from Butts w.; it has Newton N. w., Morgan N. E., Putnam E., and Jones s. Length 30, mean breadth 16, and area 480 square ms. Extending in lat. 33° 09′ to 33° 37′ and in long. 6° 36′ to 6° 56′ w. W. C.

Chief town, Monticello. Pop. 1820, 13,614, 1830, 13,131.

The name of this co. was a just tribute to real and humble merit; it was to perpetuate the name of Sergeant Jasper, who replaced the United States colors on the parapet of fort Moultree at Sullivan's island, near Charleston, S. C. when they where shot away by a British cannon ball, in the attack made on that feeble fortress July 28th, 1776.

JASPER, p-v. and st. jus. Marion co. Ten. situated on the right bank of Sequache r., by p-r. 120 ms. S. E. Nashville, lat. 35° 18', long. 8° 31' W. W. C.

JASPER, co. of Il. bounded by Lawrence and Clay S., Effingham W., Coles N., Clarke N. E., and Crawford E. It is nearly a square of 22 ms. each way, area 484 square ms. Lat. 38° 50' to 39° 05', long W. C. 11° 00' to 11° 21' W. It is traversed by the Embarras r. from N. to S. The centre of this county lies about 50 ms. a little N. of E. from Vandalia.

JAY, p-t. Oxford co. Me. on the Androscoggin, 20 ms. N. E. Paris. Pop. 1830, 1,276.

JAY, town, Orleans co. Vt. 50 ms. N. Montpelier, 50 N. E. Burlington. Has some good mill seats. Being a town on the Canada frontier, its inhabitants, consisting of but five or six families, nearly all left it during the late war with Great Britain. It is now settling slowly. Pop. 1830, 196.

JAY, p-t. Essex co. N. Y. 18 ms. N. W. Elizabethtown, 145 N. Albany; E. and W. borders hilly and mountainous; central part a vale, pleasant and fertile, through which runs Little Au Sable r. Water privileges in abundance, with timber and iron ore; 7 schools kept 7 months in 12. Pop. 1830, 1,629.

JAYNESVILLE, p-o. Covington county, Mich. about 100 ms. E. Natchez.

JEANERETT'S p-o. St. Mary's parish, La. 161 ms. wstrd. New Orleans.

JEFFERSON, p-t. Lincoln co. Me. 28 ms. N. E. Wiscasset. Pop. 1830, 2,074.

JEFFERSON, p-t. Coos co. N. H. 77 ms. N. Concord. Pop. 1830, 495.

JEFFERSON, co. N. Y. situated at the east end of lake Ontario, and on the St. Lawrence r. Bounded N. W. by the St. Lawrence, N. E. by St. Lawrence co., E. by Lewis co., S. by Oswego co., W. by lake Ontario, extending about 65 ms. along the lake and river, containing an area of 600,000 acres. Watered by Black r. running across the centre in a westerly direction; by the Indian r. winding over the E. and N. E. parts, and by Big Sandy cr. and some other mill streams, &c. in the S. W. These waters furnish some navigation, and numerous mill seats. A large proportion of of the soil is of a rich and superior quality. Surface in general waving and undulating. Timbered with maple, beech, birch, oak, walnut, bass, ash, elm, hemlock, groves of pine, &c. It contains iron ore in the E. and S. E. parts. Climate mild and agreeable. Chief town, Watertown, on S. Black r. 4 ms. from navigable waters, 12 E. Sacket's Harbor. Pop. 1820, 32,952, 1830, 48,493.

JEFFERSON, p-t. Schoharie co. N. Y. 20 ms S. W. Schoharie, 48 W. Albany. Soil good for grass. Considerable grain is raised; 11 schools, kept 8 months in 12. Population 1830, 1,743.

JEFFERSON, town, Morris co. N. J. bounded N. W. by Sussex co., N. E. by Bergen co., S. W. by Roxbury, and S. E. by Pequanack. Pop. 1830, 1,551.

JEFFERSON co. Pa. bounded by Indiana S. Armstrong and Venango W., Warren N. W., McKean N. E., and Clearfield E. and S. E. Length 46 ms., mean breadth 26, and area 1,196 square ms. Extending in lat. 40° 55' to 41° 36' N., and in long. 1° 41' to 2° 17' W. W. C. The declivity of this co. is to the S. W. and drained in succession S. to N. by the branches of Mahoning, Redbank, Clarion, and Teomista rs. all flowing towards and finally entering Alleghany r. Surface rocky and hilly, and in part mountainous. Chief t. Port Barnet. Pop. 1820, 561, 1830, 2,025.

JEFFERSON, p-v. on the southern branch of Ten Mile creek, Greene co. Pa. 15 ms. S. W. Brownsville and 9 N. E. by E. Waynesburg.

JEFFERSON, co. Va. bounded by the Blue Ridge, separating it from Loudon S. E., by Frederick S. W., by Berkshire W. and N. W., and by Potomac r. separating it from Washington co. Md. N. E. Length 22, mean breadth 10, and area 220 square ms. Extending in lat. 39° 10' to 39° 28', and in long. 0° 43' to 1° 02' W. W. C. The Shenandoah r. enters the southern angle and traversing this co. in a N. N. E. course along its southeastern border, and parallel to the Blue Ridge, falls into Potomac at Harper's Ferry. The declivity of the co. is to the N. N. E. The water elevation at Harper's Ferry being 182 feet above tide water, that of the arable soil of Jefferson must be greatly higher and cannot fall short of a mean of 400 ft., or an equivalent to a degree of lat. Though the face of this county is broken and even mountainous, it is a very productive tract in grain, pasturage and fruit. Chief towns, Harper's Ferry, and Charleston. Pop. 1820, 13,087, 1830, 12,927.

JEFFERSON, p-v. on the right bank of James r. northern part of Powhatan co. Va. 84 ms. above, and N. N. W. Richmond.

JEFFERSON, co. Geo. bounded by Washington W., Warren N. W., Richmond N. E., Burke E., and Emanuel S. Length 33, mean width 20, and area 660 square ms. Extending in lat. 32° 51' to 33° 20', and in long. 5° 14' to 5° 46' W. W. C. It is traversed on the southwest side by Great Ogeechee, and bounded on the northeast by Brier cr., both flowing to the S. E. in the direction of the general declivity. Chief town, Louisville. Pop. 1820, 7,058, 1830, 7,309.

JEFFERSON, p-v. and st. jus. Jackson co. Geo. situated on one of the higher branches of Oconee, by p-r. 85 ms. a little W. of N. Milledgeville, lat. 34° 07', long. 6° 37' W. W. C.

JEFFERSON, co. Alabama, bounded S. by Bibb, S. W. by Tuscaloosa, W. by Lafayette, N. W. by Walker, N. by Blount, N. E. St. Clair, and S. E. by Shelby. Length S. W. to N. E. 52 ms., mean breadth 20, and area 1,040 sq. ms. Extending in lat. 33° 17′ to 33° 52′, and in long. 9° 37′ to 10° 32′ W. W. C. This co. lies entirely in the valley of Tuscaloosa, the main volume of which entering the northeast border from Blount, flows over the co. in a S. W. direction, receiving near its exit from the southwestern angle Mulberry r. from the southwestward. The main road from Tuscaloosa to the northeastern part of the state passes nearly centrically over Jefferson. Chief town, Elyton. Pop. 1830, 6,855.

JEFFERSON, co. Miss. bounded by Claiborne N., Copiah E., Franklin S. E., Adams S. W., and the Miss. r., separating it from Concordia in Louisiana N. W. Length E. to W. 35, mean width 18, and area 630 sq. ms. Extending in lat. from 31° 37′ to 31° 53′, and in long. from 13° 50′ to 14° 28′ W. W. C. Though bounding on the Miss. r. this co. is a real table land. The bluffs extending parallel to the general course of the Miss. range along the western part of Jefferson, leaning towards the Great r. some annually overflowed and level bottom. With the bluffs commences a very *rolling* country, to adopt an expressive figurative term. From this broken region issue wstrd. Fairchild's and Cole's creeks; to the N. W. branches of the Bayou Pierre, and to the S. W. those of Homochitto r. The Miss. bottoms where capable of being protected from flood, are extremely productive. The soil of the bluff land is also excellent; but advancing estrd. the pine woods gradually expand, so that the two extremes of the co. are also extremes of fertility and the reverse. Staple, cotton. Chief t. Greenville. Pop. 1820, 6,822, 1830, 9,755.

JEFFERSON, parish, La. bounded S. by the Gulf of Mexico, S. W. by the parish of La Fourche Interior, W. and N. W. by St. John Baptiste, N. by lake Pontchartrain, E. by the parish of New Orleans, and E. by that of Plaquemines. Length S. to N. between the Gulf of Mexico and lake Pontchartrain 60 ms., mean width 12, and area 720 sq. ms. Extending in lat. from 29° 17′ to 30° 05′, and in long. W. W. C. from 12° 54′ to 13° 10′. The Miss. r. traverses the northern part of this parish, and with some strips on the sthrn. water courses towards the Gulf of Mexico, affords the only land sufficiently elevated above the tide-level to admit cultivation. Staples, cotton, sugar and rice. Pop. 1830, 6,846.

JEFFERSON, co. Ten. bounded S. W. by Sevier, W. by Knox, N. by Granger, N. E. by Hawkins, E. by Greene, and S. E. by Cocke. Length 28, mean width 28, and area 356 sq. ms. Extending in lat. from 35° 48′ to 36° 11′, and in long. from 5° 54′ to 6° 24′ W. W. C. This co. is bounded on the N. W. by Holston, and on the S. E. by the French Broad, whilst a mountain ridge stretches over it from Knox into Hawkins. The course of the rivers and general slope is to the W. S. W., with a very rugged surface. The soil where arable is productive. Chief t. Dandridge. Pop. 1820, 8,953, 1830, 11,801.

JEFFERSON, p-v. Rutherford co. Ten. 21 ms. S. E. Nashville.

JEFFERSON, co. Ky. bounded by Oldham N. E., Shelby E., Spencer S. E., Bullitt S., Ohio r. which separates it from Harrison in Ind. W., and from Floyd and Clark, Ind. N. Length 28, mean width 18, and area 504 sq. ms. Extending in lat. from 38° 02′ to 38° 22′, and in long. from 8° 25′ to 8° 55′ W. W. C. The slope of this co. is sthwrd. giving source to some of the northeastern branches of Salt r. It is rendered remarkable, as lying opposite the rapids of O., and as containing the canal of Louisville. Chief town, Louisville. Pop. 1820, 20,768, 1830, 23,979.

JEFFERSON, co. Ohio, bounded N. by Columbiana, by the Ohio r. E. separating it from Brooke co. Va., by Belmont, Ohio, S., and Harrison W. Length 27, breadth 20, and area 540 sq. ms. Lat. 40° 10′ to 40° 33′, long. 3° 50′ W. W. C. Slopes estrd. towards Ohio r., and in that direction is drained by Yellow Cross and Short creeks. Surface hilly, but soil fertile and abounding in bituminous mineral coal. Chief t. Steubenville. Pop. 1820, 18,531, 1830, 22,489.

JEFFERSON, p-v. and st. jus. Ashtabula co. O. by p-r. 191 ms. N. E. Columbus, and 325 ms. N. W. W. C. Pop. 1830, 370. It is the most northeastern co. town in the state.

JEFFERSON, co. Ind. bounded by O. r. separating it from Gallatin co. Ky. S. E., Clarke S. W., Scott W., Jennings N. W., Ripley N., and Switzerland N. E. Length 25 by 15, mean breadth 375 sq. ms. Lat. 38° 43′, long. 8° 28′ W. W. C. The features of this co. are remarkable; though bounding on Ohio r., it is drained almost from the margin of that stream, by creeks which flow directly from it into the valley of White r. The surface very broken; soil fertile. Chief t. Madison. Pop. 1820, 8,038, 1830, 11,465.

JEFFERSON, p-v. Clinton co. Ind. about 50 ms. N. W. Indianopolis.

JEFFERSON, co. of Il. bounded by Franklin S., Perry S. W., Washington W., Marion N., Wayne N. E., and Hamilton S. E. It is a square of 26 ms. each way; area 676 sq. ms. Lat. 38° 09′ to 38° 30′, long. 11° 48′ to 12° 09′ W. W. C. The eastern part slopes to the S. E., and gives source to Waynes fork of Little Wabash, the residue slopes sthrd. and gives source to Muddy creek. Chief town, Mount Vernon. Pop. 1830, 2,555.

JEFFERSON, co. Mo. bounded by the Miss. r., separating it from Monroe co. Il. E., St. Genevieve co. S. E., St. Francis S., Washington S. W., Franklin W., and St. Louis N. Lat. 38° to 38° 30′, long. 13° 13′ W. W. C. It may be observed as a curious feature in the geography of this co. that the Big r. branch of Merrimack, traverses the western side to the nthrd. in direct opposition to the course of the Miss. along its eastern boundary. The surface is hilly and broken. Chief town, Mount Vernon. Pop. 1830, 2,592.

Jefferson, co. Flor. bounded e. by Madison co. of the same ter., s. by Appalachee bay of the Gulf of Flor., w. by Leon co. and n. by Thomas co. of Geo. Extending in lat. 30° to 30° 42′, and in long. 7° to 7° 16′ w. W. C. Length 48 ms., mean width 16, and area 768 sq. ms. Chief t. Monticello. Pop. 1830, 3,312.

Jefferson Barracks and p-o. St. Louis co. Mo.

Jefferson City, st. jus. for Cole co., and of the government of Mo. situated on the right bank of Mo. r. about 9 miles above the mouth of Osage river, by p-r. 134 ms. w. St. Louis, and 980 ms. w. W. C. Lat. 39° 32′, long. 15° 06′ w. W. C. It is a new town containing 200 houses and 1,200 inhabitants, and after Little Rock in Ark. the most western state capital of the U. S. The two towns differ but little in long.

Jeffersonton, p-v. near the right bank of the Rappahannoc river, and n. e. angle of Culpepper co. Va. about 33 ms. n. w. Fredericburg, and by p-r. 62 ms. s. w. by w. W. C.

Jeffersonton, p-v. on Santilla r. Geo. 25 ms. n. w. St. Mary's in the same co., and by p-r. 219 ms. s. s. e. Milledgeville.

Jeffersontown, p-v. Jefferson co. Ky. 15 ms. s. e. Louisville, and by p-r. 44 ms. w. Frankfort.

Jeffersonville, p-v. Montgomery co. Pa.

Jeffersonville, p-v. on the North Fork of Clinch r. Tazewell co. Va. 30 ms. n. w. by w. Evansham, and by p-r. 372 ms. s. w. by w. W. C., and 275 a little s. of w. Richmond. Lat. 37° 05′, and long 4° 32′ w. W. C.

Jeffersonville, p-v. Clarke co. Ind. by p-r. 119 ms. a little e. of s. Indianopolis. It is situated on Ohio r. opposite Louisville in Ky. Pop. about 1,000.

Jekyl, small island on the Atlantic coast of Geo. between Cumberland and St. Simon's isl. It is the s. e. part of Glynn co.

Jemappe, p-v. Caroline co. Va. 69 ms. Richmond.

Jena, p-v. Jefferson co. Flor. (*See Lipona.*)

Jenkinton, p-v. in the s. e. part of Montgomery co. Pa. 10 ms. n. Phil.

Jennersville, p-v. Chester co. Pa. 43 ms. s w. by w. Phil.

Jennings, co. Ind. bounded by Jefferson s., Scott w., Jackson n. w., Bartholomew n. w., Decatur n., and Ripley n. e. Length 26, mean breadth 20, and area 520 sq. ms. Lat. 39° n., long. 8° 30′ w. W. C. Slope s. w. Drained by numerous branches of White r. Chief t. Mount Vernon. Pop. 1830, 3,974.

Jenning's Gap, over North mtn. and p-o. in the northern part of Augusta co. Va. by p-r. 162 ms. s. w. by w. W. C.

Jericho, p-t. Chittenden co. Vt. on Onion r. 12 ms. e. Burlington, 26 n. w. Montpelier. Watered by Brown's r. and a great number of smaller streams, which furnish numerous mill privileges. 13 school dists. Pop. 1830, 1,655.

Jeromesville, p-v. n. e. Wayne co. O. 90 ms. n. e. Columbus. Pop. 1830, 133.

Jersey, p-t. Steuben co. N. Y. 12 ms. s. Bath, 228 w. Albany. A broken township, with some good land. Timbered with oak, chestnut, hemlock, beech, maple, &c. 7 schools, kept 6 months in 12. Pop. 1830, 2,391.

Jersey City, or Paulus Hook, p-v. Bergen co. N. J. on the Hudson r. opposite New York.

Jersey Settlement and p-o. in the southwestern part of Rowan co. N. C. by p-r. 133 ms. wstrd. Raleigh.

Jersey Shore and p-o. on the left bank of the W. branch of Susquehannah r. below the mouth of Pine creek, 14 ms. above Williamsport, and by p-r. 108 ms. n. n. w. Harrisburg.

Jersey Town, p-v. near the centre of Columbia co. Pa. 8 ms. a little e. of n. Danville, and by p-r. 86 ms. n. n. e. Harrisburg.

Jerusalem, p-t. s. line of Ontario co. N. Y. 20 ms. s. e. Canandaigua, 18 s. Geneva. Scenery, wild and romantic. Jemima Wilkinson, the founder of a sect denominated, by herself, the Universal Friends, died here in 1819.

Jerusalem, p-v. and st. jus. Southampton co. Va. situated on Nottaway r. 70 ms. s. s. e. Richmond. Lat. 36° 42′, long. 0° 3′ w. W. C.

Jessamine, co. Ky. bounded n. w. by Woodford, n. and n. e. by Lafayette, and on all other sides by Ky. r., which separates it on the s. e. from Madison, s. from Garrard, and s. w. from Mercer. It lies nearly in the form of a square, and would average about 16 ms. each side. Area 256 sq. ms. Extending in lat. 37° 43′ to 38° 01′, and in long. w. W. C. 7° 24′ to 7° 43′. It is almost an undeviating expanse of fertile soil, moderately level. Chief t. Nicholasville. Pop. 1820, 9,297, 1830, 9,960.

Jettersville, p-o. Amelia co. Va. 35 ms. s. w. Richmond.

Joe Daviess, extreme n w. co. of Il. as laid down by Tanner on his recently improved map of the U. S., is bounded w. by Miss. r., n. by Iowa co. of Huron, e. by La Salle co. Il., s. e. by Rock r., and s. by Plum creek, separating it from Rock Island co. Il. It extends about 40 ms. from s. to n., but the outlines towards Rock r. are too undefined to admit an estimate of its superficial area. In lat. it extends from 41° 55′ to 42½° n., and is traversed by long. 13° w. W. C. Peektans r., a branch of Rock r. rises in Iowa co. Huron, and flowing s. e. by e., enters Il., and joins the main stream in the northern part of La Salle co. Joe Daviess co. is composed of an inclined plain between the Peektans and Miss., and is traversed in the direction of s. w. Fever r., and by Apple and Rush crs., with other smaller streams. Joe Daviess co. comprises the lead mines around the chief town, Galena, a name imposed from the abundance of the galena ore of lead found in its vicinity. The same country has recently become painfully interesting as the seat of a desolating Indian war. Pop. 1830, 2,111.

JOE's BROOK, or Merritt's r. Vt. rises near the N. line of Walden, and falls into the Passumsic in Barnet. A rapid stream, furnishing many good mill privileges.

JOHN's r. N. H., has its principal source in Pondicherry pond, Jefferson co., and falls into the Conn. r. about 60 ms. above the head of Fifteen Mile falls, where its mouth is about 30 yards wide.

JOHN's r. or creek, rising in the Blue Ridge, and in the northern part of Burke co. N. C. flows s. into Great Catawba r.

JOHN's r. p-o. or John's r. Burke co. N. C., by p-r. 151 ms. w. Raleigh.

JOHNSBURG, p-t. Warren co. N. Y. 30 ms. N. w. Caldwell. Surface hilly. Soil good for grass and grain. Well watered. Timber mostly maple and beech. 9 schools, kept 5 months in 12. Pop. 1830, 985.

JOHNSON, p-t. Franklin co. Vt. 28 ms. N. w. Montpelier, 28 N. E. Burlington. Crossed by the river Lamoille, which in this town has a fall of 15 feet, called McConnel's falls, and a singular kind of natural bridge. Surface uneven. Soil productive. 6 school districts. Pop. 1830, 1,079.

JOHNSON, co. N. C. bounded N. w. by Wake, N. E. by Nash, Wayne E. and S. E., Sampson S., and Cumberland S. w. Length 30, mean width 22, and area 660 sq. ms. Extending in lat. 35° 15′ to 35° 48′, and in long 1° 4′ to 1° 40′ W. C. Neuse r. winds over this co. in a S. S. E. direction, dividing it into two not very unequal sections. The N. E. part is also traversed by Little r. a branch of Neuse, and flowing on a similar course. Chief town, Smithfield. Pop. 1820, 9,607, and in 1830, 10,938.

JOHNSON, p-v. Pendleton co. Ky. 66 ms. Frankfort.

JOHNSON, co. of Ind. bounded by Bartholomew S., Morgan W., Marion N., and Shelby E. Length 22, breadth 18, and area 396 sq. ms. Lat. 39° 30′, long. 9° w. W. C., lying between the Driftwood Fork of White r., and the main Wabash. Chief t. Franklin. Pop. 1830, 4,019.

JOHNSON, co. of Il. bounded S. w. by Alexander, N. w. by Union, N. by Franklin, E. by Pope, and S. by the Ohio r. separating it from McCracken co. Ky. Breadth 18, mean length 30, and area 540 sq. ms. Lat. 37° 20′, long. 12° w. W. C. Slope sthrd. towards Ohio r. Chief town, Vienna. Pop. 1830, 1,596.

JOHNSONSBURG, v. Warren co. N. J. 9 ms. S. w. Newton, 16 N. E. Belvidere.

JOHNSON's creek of lake Ontario, rises in Niagara co., and falls into the lake at Oak Orchard, after a course of about 20 ms. A good mill stream.

JOHNSON's LANDING, and p-o. Barnwell co. S. C. 127 ms. w. Columbia.

JOHNSON's MILLS, and p-o. Dallas co. Ala. by p-r. 69 ms. S. E. Tuscaloosa.

JOHNSON's SPRINGS, and p-o. Goochland co. Va. by p-r. 82 ms. N. w. Richmond.

JOHNSONVILLE, p-v. Trumbull co. O. 180 ms. N. E. Columbus.

JOHNSTON, t. Providence co. R. I. 5 ms. w. Providence. Surface interspersed with hill and dale. Contains quarries of free stone suitable for building, &c. It likewise contains limestone, and stone suitable for furnace hearths. Soil generally good. Watered by the Wanasquatucket, the Powchassett and Cedar brook; which streams afford numerous water privileges. 7 schools. Pop. 1830, 2,113.

JOHNSTOWN, p-t. st. jus. Montgomery co. N.Y. 40 ms. N. w. Albany. Rich land, agreeably undulated. 33 schools. The village of Johnstown is situated about 4 ms. N. of the Mohawk. It contains an academy. In this town was fought the battle of Johnstown, Oct. 25, 1781, in which the British and Indians, consisting of 600, were defeated by the Americans under Col. Marinus Willett, consisting of 400 levies and militia, and 60 Oneida Indians. Pop. 1830, 7,700.

JOHNSTOWN, p-v. on the point above the junction of Stony cr. and little Conemaugh, in the sthwstrn. part of Cambria co. Pa. 18 ms. S. w. Ebensburg, 60 ms. a little S. of E. Pittsburg, and by p-r. 171 ms. N. w. W. C.

This village stands on ground where the water level of the two contiguous streams is 1,154 ft. above that of the Atlantic tides, and is the point where the wstrn. extremity of the Pa. rail-road joins the Conemaugh section of the Pa. canal. Lat. 40° 20′, long. 1° 55′ w. W. C.

JOHNSTOWN, p-v. Dicking co. O. by p-r. 33 ms. N. E. Columbus.

JOHNSVILLE, p-v. Obion co. Ten. by p-r. 179 ms. wstrd. Nashville.

JONES, co. N. C. bounded by Onslow S., Duplin S. w., Lenoir N. w., Craven N. and N. E., and Carteret E. and S. E. Length 38, mean breadth 10, and area 380 sq. ms. Extending in lat. 34° 48′ to 35° 12′, and in long. 0° 08′ to 0° 44′ w. W. C. It is a part of a level and in great part marshy plain, traversed w. to E. by the small but navigable r. Trent. Chief t. Trenton. Pop. 1820, 5,216, 1830, 5,608.

JONES, co. Geo. bounded S. by Twiggs, S. w. by Bibb, w. by Ockmulgee r. separating it from Monroe and Butts, N. by Jasper, N. E. by Putnam, E. by Baldwin, and S. E. by Wilkinson. Length diagonally S. E. to N. w. 30 ms., mean width 12, and area 360 sq. ms. Extending in lat. 32° 52′ to 33° 10′, and in long. 6° 28′ to 6° 53′ w. W. C. This co. is composed of two inclined plains, the wstrn. inclining sthwrd. is drained into Ockmulgee, and the estrn. sloping estrd. is drained into Oconee. Clinton, the st. jus. is situated near the centre of the co. 22 ms. S. w. by w. Milledgeville. Pop. 1820, 17,410, 1830, 13,345.

JONES co. Miss. bounded by Wayne E., Perry S. Covington w. and by the Choctaw country N. Length 28, mean width 24, and area 672 sq.

ms. Extending in lat. 31° 27′ to 31° 50′, and in long. 12° 05′ to 12° 28′ w. W. C. The slope of this co. is nearly due s. down which flow different branches of Leaf r. Chief t. Ellisville. Pop. 1830, 1,471.

JONESBOROUGH, p-t. Washington co. Me. 12 ms. w. Machias. Pop. 1830, 810.

JONESBORO', p-v. Brunswick co. Va. by p-r. 83 ms. sthwrd. Richmond.

JONESBORO', p-v. and st jus. Washington co. Ten. situated on a branch of Nolachucky r. 26 ms. s. w. by w. Elizabethtown, and about an equal distance N. E. by E. Greenville, and by p-r. 260 ms. E. Nashville. Lat. 36° 17′, and long. 5° 20′ w. W. C.

JONESBORO', p-v. and st. jus. Union co. Il. by p-r. 154 ms. a little w. of s. Vandalia, and about 40 ms. nrthrd. of the mouth of O. Lat. 37° 28′.

JONESBORO', p-v. Saline co. Mo. about 200 ms. wstrd. St. Louis.

JONESBORO', p-v. in the sthrn. part of Jefferson co. Ala. on the road from Tuscaloosa to Elyton, 42 ms. N. E. the former, and 10 s. w. the latter town.

JONESTOWN, p-v. situated on the point at the confluence of the two main branches of Swatara, and in the nrthrn. part of Lebanon co. Pa. by p-r. 31 ms. N. E. by E. Harrisburg.

JONESVILLE, p-v. and st. jus. Lee co. Va. situated on a creek of, and N. from Powell's r., 65 ms. N. E. Knoxville in Ten., 60 s. E. by E. Barbourville in Ky., and by p-r. 491 ms. w. s. w. W. C., and 394 s. w. by w. Richmond. Lat. 36° 40′, long. 6° 02′ w. W. C. It is the most wstrn co. st. of Va.

JONESVILLE, p-v. Surry co. N. C. by p-r. 178 ms. N. w. by w. Raleigh.

JONESVILLE, p-v. Union dist. S. C. by p-r. 112 ms. N. N. w. Columbia.

JONESVILLE, p-v. Monroe co. O. by p-r. 154 ms. estrd. Columbus.

JONESVILLE, p-o. Lenawee co. Mich. by p-r. 103 ms. sthwstrd. Detroit.

JOPPA CROSS ROADS, and p-o. sthrn. part of Harford co. Md. on Gunpowder bay, 16 ms. N. E. Baltimore.

JORDANSVILLE, p-v. Mecklenburg co. N. C. by p-r. 172 ms. s. w. by w. Raleigh.

JOY, t. Kennebec co. Me. 30 ms. N. Augusta.

JUDDSVILLE, p-v. Surry co. N. C. by p-r. 175 ms. N. w. by w. Raleigh.

JUNCTA, (*See Watervliet, and the junction of Erie and Champlain canals.*)

JUNCTION, p-v. Perry co. Pa. 17 ms. Harrisburg.

JUNIATA, r. of Pa. and the sthwstrn. branch of Susquehannah, is formed by 2 confluents, Rayston branch from Bedford, and Frankstown branch from Huntingdon co. These two confluent streams unite in Huntingdon co. (*See the two artic'es Frankstown and Raystown branches.*) After the junction of its forming branches, Juniata assumes a sthestrn. course 12 ms., breaking through several chains of mtns. to where it receives Aughwick cr. from the s., thence inflecting to N. E. flows 28 ms. parallel to the adjoining mtns.; passes Lewistown and turning to s. E. by E. 30 ms. general distance, but much more following the meanders to its junction with Susquehannah. Including the whole valley of Juniata, it drains one half of Bedford, all Huntingdon, Mifflin, and about one third of Perry, and comprises an area of about 2,750 sq. ms. In all its parts it is a true mtn. r., having the remote sources of both the main branches in Alleghany mtn. at an elevation of upwards of two thousand ft. above the ocean tides, and winding its numerous streams along deep mtn. vales or breaking directly thro' the chains. In lat. this valley extends 39° 50′ to 40° 50′, and has now gained permanent interest amongst the streams of the U. S. affording a passage for the Pa. canal, through five considerable chains of mtns.

JUNIATA, p-v. and tsp. in the nrthrn. part of Perry co. Pa. 31 ms. N. w. Harrisburg.

JUNIATA CROSSINGS, and p-o. Bedford co. Pa. 14 ms. E. Bedford, and 91 s. w. by w. Harrisburg.

JUNIATA FALLS, and p-o. nthestrn. part of Perry co. Pa. 21 ms. N. w. Harrisburg.

JUNIUS, p-t. half shire of Seneca co. N. Y. N. end of Seneca and Cayuga lakes, on the Seneca r. 185 ms. w. Albany. Surface level, soil good, and tolerably well watered. Here are limestone, soft slate stone, and gypsum. The Seneca outlet or river, which runs along the s. border of this town, is a very important stream for navigation and for hydraulic works. Its course from the N. end of Seneca lake to the N. end of Cayuga lake, is about 15 ms. In this town are the villages of Bridgeport, Seneca Falls, and Waterloo. 24 schools, kept 7 months in 12. Pop. 1830, 1,581.

K.

KAATSBERGS, or Katsberg, or Catskill mtns. rise boldly, w. side of Hudson r. in Greene co. N. Y. to an elevation little short of 4,000 feet. (*See Matteawan mtns.*)

KAATSKILL, Katskill, or Catskill r. a large and good mill stream, rises in the s. E. of Schoharie co. and runs s. w. through Greene co. to the Hudson, near the village of Catskill. Its whole course may be 35 ms.

KALAMAZOO, r. of the Ter. of Mich. and confluent of lake Mich. rises on the table land of the Mich. peninsula, about 80 ms. s. w. by w. Detroit. It has interlocking sources with those of Raisin, St. Joseph's branch of Maumee, and with those of St. Joseph's and Grand rs. of lake Mich. The general comparative course of the Kalamazoo is about 100 ms. to the N. w. by w. falling into lake Mich. a little N. of w. Detroit, and about midway between the mouths of Grand and St. Joseph's rs.

KALAMAZOO, co. Mich. on Kalamazoo r.

bounded E. by Calhoun, S. by St. Joseph, W. by Van Beuren, N. W. by Allegan, and N. E. by Ionia. Length N. to S. 26 ms., and E. to W. 26, area 676 sq. ms. Extending in lat. 42° 6′ to 42° 27′, and in long. 8° 18′ to 8° 46′ W. W. C. The Kalamazoo r. enters its nthestrn border, and sweeping a large southern curve, leaves the county on its northwestern border. Some of the branches of St. Joseph's r. rise along its sthrn. and sthwstrn. sections and flow thence to S. W. The co. is therefore divided into two inclined plains; one drained by the Kalamazoo, sloping to the N. W., and the other drained by the tributaries of St. Joseph, and sloping to the S. W.

KANE, p-v. Greene co. Il. by p-r. 98 ms. N. W. by W. Vandalia.

KANSAS, or Konsas, large r. of the U. S. rising on the great desert plains between the vallies of Platte and Arkansas r. as far wstrd. as the 27th degree of long. W. W. C. The general course of the Kansas is from W. to E. and in that direction the two main branches flow by comparative courses upwards of 400 ms. then unite, and thence flowing about 150 ms. falls into the Mo. r. at the wstrn. border of the state of Mo. The valley of the Kansas is about 500 ms. from E. to W., but if any thing near correctly deliniated on our maps, the mean width does not exceed 70 ms., area 35,000 sq. ms.; similar to the higher confluents of Arkansas, Red r. of the Miss. and Platte r., the Kansas flows down the inclined and desert plains E. of the Chippewayan or Rocky mtns.

KARTHAUS, p-v. on the left bank of W. branch of Susquehannah r. at the mouth of little or nthrn. Moshannon creek, in the estrn. part of Clearfield co. Pa. 20 ms. N. E. by E. Clearfield, and by p-r. 87 ms. N. W. Harrisburg.

KASEY'S p-o. Bedford co. Va. by p-r. 142 ms. W. Richmond.

KASKASKIA, r. of Il. rises at lat. 41° interlocking sources with those of the Sangamon, branch of Il. r., and with those of the Vermillion, branch of Wabash, and flowing S. W. over the cos. of Vermillion, Edgar, Shelby, Fayette, Bond, Clinton, Washington, St. Clair, and Randolph, falls into the Miss. after a comparative course of 180 ms. The Kaskaskias valley is narrow, about 30 ms. mean width; the higher part lying between those of the Wabash and Sangamon, and the lower between those of Wabash and Ohio estrd. and Il. and Miss. wstrd.

KASKASKIA, ancient village of Il. and st. jus. Randolph co. is situated on Kaskaskias r. on the narrow neck between that stream and the Miss. by p-r. 95 ms. S. W. from Vandalia. The site is very fine, and contains a pop. of about 1000, a bank, printing office, land office, and numerous stores. Lat. 37° 58′, long. W. W. C. 13°.

KATAHDIN, or Ktadne, mtn. Me. the highest mtn. in the state, supposed by some to be as high as the White mtns. in N. H. It lies between the E. and W. branches of Penobscot r. 80 ms. N. Bangor. The Indians considered it the abode of supernatural beings. It is steep and rugged. It is almost isolated. By those who have visited it, this region is spoken of as scarcely rivalled in sublimity of scenery.

KAYADEROSSERAS MTS. an extensive range of primitive mtns. stretching N. N. E. across the N. W. part of Saratoga co. the E. part of Warren, and into Essex and Clinton cos. In the co. of Saratoga, the general elevation of this range from the adjoining plains may be estimated at 300 to nearly 700 ft. Further N. near L. George, some of its summits may be 1200 ft. above the surface of that lake. Their sides are very steep, masses of granite and gneiss, piled almost perpendicularly.

KEARSARGE MTN. Hillsborough county, N. H. between Sutton and Salisbury, extending into both towns. It rises 2,461 ft. above the level of the sea, being the highest mtn. in the county.

KEATING, p-v. McKean co. Pa. by p-r. 186 ms. N. W. Harrisburg.

KEENE, p-t. and half shire of Cheshire co. N. H. on a tongue of land between the two principal branches of the Ashuelot, 14 ms. S. Walpole, 43 from Windsor, 55 W. S. W. of Concord, 95 W. Portsmouth, 79 W. N. W. Boston. A very pleasant village, and a place of considerable business. Pop. 1830, 2,374.

KEENE, p-t. Essex co. N. Y. 12 ms. west Elizabethtown, 138 N. Albany. Surface diversified with mtns., hills, valleys and plains. In the south part, the La Sable or Sandy mtn. rises to a great height, rugged and uncommonly bold. In this town are the extreme sources of the Hudson river. The Saranac lake, 15 ms. in circumference, is on the west line of the town. Well supplied with water privileges. There are extensive iron and steel works. It has some iron ore. Lake Saranac is remarkable for the size and abundance of its trout, many having been caught weighing 40 lbs. A barrel has been filled with them in one hour, taken by the hook and line. Pop. 1830, 787.

KEENE, p-v. Coshocton co. Ohio, by p-r. 89 ms. N. E. by E. Columbus.

KEENER'S MILLS, p-o. Adams co. Pa. 81 ms. N. W. C.

KEESVILLE, p-v. situated on both sides of the Great Au Sable, S. part of Peru, Clinton co. N. Y. 3½ ms. W. Port Kent, 16 ms. from Plattsburgh, and about 4 from the W. shore of lake Champlain. It abounds in iron ore of various qualities, with forests, mountains and fine streams of water.

KELLEY'S creek, village and p-o. in the N. E. angle of Shelby co. Ala. by p-r. 87 ms. N. E. by E. Tuscaloosa.

KELLEY'S-VILLE, p-v. Ohio co. Ky. by p-r. 153 ms. S. W. by W. Frankfort.

KELLOGSVILLE, p-v. Ashtabula co. Ohio, by p-r. 207 ms. N. E. Columbus.

KELLY'S FERRY, and p-o. Rhea co. Ten. by p-r. 137 ms. estrd. Nashville.

KELLYVALE, p-t. Orleans co. Vt. 36 ms. N. Montpelier, 42 N. E. Burlington. Land productive, timbered mostly with hard wood. At the grist mill near the centre of the town the r. passes through a hole in the solid rock. Contains serpentine, chlorite and chlorite slate, bitter spar, talc and magnetic iron, pudding stone, &c. Pop. 1830, 314.

KELLYSVILLE, p-v. Marion co. Ten. by p-r. 120 ms. S. E. Nashville.

KELSO, p-v. Dearborn co. Ind. by p-r. 85 ms. S. E. Indianapolis.

KEMPSVILLE, p-v. Princess Ann co. Va. on the E. branch of Elizabeth r. 10 ms. S. E. by E. Norfolk, N. lat 36° 48', long. 0° 56' E. W. C.

KENANSVILLE, p-v. Duplin co. N. C. by p-r. 83 ms. S. E. Raleigh.

KENDALL'S STORE, Montgomery co. N. C. by p-r. 130 ms. southwestward Raleigh.

KENHAWA, GREAT, r. N. C. and Va. has the most remote source in Ashe co. of the former, between the Blue Ridge and main Appalachian chain, there known by the name of Iron mtn. the two higher branches, after draining the northern part of Ashe, unite near the boundary between North Carolina and Virginia, and continuing their original course to the northeast by north, enters Grayson county of the latter state, breaks through the Iron mtn. between Grayson and Wythe; winds over the latter and Montgomery; thence inflecting to the N. N. W. traverses Walker's and Peter's mtns. Below the latter chain, the course of N. N. W. is continued to the mouth of Gauley r. having received also from the northeast Green Briar.

Above Gauley r. the main volume of Kenhawa is called New river; but receiving the Gauley and turning to N. W. this now large stream, known as the Great Kenhawa, is still farther augmented from the N. by Elk r. and from the S. by Coal r., falls into Ohio river at Point Pleasant, after a comparative course of 280 ms., 100 above Walker's mtn., 100 from the pass thro' Walker's mtn. to the mouth of Gauley r. and 80 from the mouth of Gauley to the Ohio.

The higher branches of New r. have interlocking sources with those of Catawba and Yadkin on the S. E., and with those of Watauga and Holston to the northwest. Below the Iron mtn. the interlocking sources are with those of Clinch and Sandy to the W., those of Roanoke to the E. and those of James r. N. E. as far down as the gorge of Peter's mtn. wstrd. of the latter pass. Green Briar, coming in from the N. has its sources in the same region with those of the Potomac on the northeast, and with those of the Monongahela to the northward. The valley of Kenhawa proper, below Gauley r. lies generally between the valley of Guyandot on the S. W. and that of Little Kenhawa N. E., tho' the sources of Elk r. also reach the vicinity of those of Monongahela.

The entire valley of Great Kenhawa, including that of New r. extends lat 36° 15' in Ashe co. N. C. to 38° 52' at the junction of Kenhawa and Ohio. and in long. 2° 43' at the higher source of Green Briar, to 5° 08' W. W. C. The length of this valley from the Blue Ridge between Patrick and Montgomery cos. Va. in a N. W. direction is 180 ms., the utmost breadth from the sources of New r. to those of Green Briar is 180, but the mean width is about 60, and the area may be stated at 10,800 square ms.

The most remarkable feature in the valley of the Great Kenhawa, as a physical section, is relative height. At the mouth of Sinking creek, between Walker's and Peter's mtns. 120 ms. by comparative courses below the sources, the water level is 1,585 feet above the Atlantic tides, at the mouth of Green Briar 1,333, and at the mouth into Ohio 525 feet. Comparing the fall from Sinking creek to the mouth of Green Briar 252 feet in 30 ms. direct, that above Sinking creek must be 900 feet at least, consequently, the higher branches of New r. in Ashe co. must rise at a comparative height of upwards of 2,500 feet.

KENHAWA, LITTLE, r. Va. rising in Lewis co. and flowing N. W. by W. enters Wood and falls into the Ohio at Parkersburg, after a comparative course of 90 ms. The valley of this r. is nearly commensurate with Wood and Lewis cos. and has that of Great Kenhawa S., Middle Island creek to the N. and that of Monongahela N. E.

KENHAWA, co. Va. bounded by Logan S., Cabell S. W., Mason N. W., Wood N., Lewis N. E., and Nicholas E. Length 60, mean width 40, and area 2,400 square ms. Extending in lat. 37° 53' to 38° 53', and in long. 3° 55' to 5° W. W. C. Great Kenhawa river receives Gauley r. on the eastern boundary, and thence traversing this co. in a northwestern direction, receives within it Elk and Pocatalico r. from the N. E. and Coal r. from the S. E. The general slope is to the N. W. with the Great Kenhawa. The surface very broken, and in part mountainous. Some excellent soil is contrasted with much more of an opposite character. Chief town, Charleston. Pop. 1820, 7,000, 1830, 9,326.

KENHAWA, court house. (*See Charleston, same co.*)

KENHAWA SALINE, p-o. Kenhawa co. Va. by p-r. 320 ms. N. W. by W. W. C. and 300 a little N. of W. Richmond.

KENNEBEC, co. Me. lies on both sides of Kennebec r., and is bounded N. by Somerset, E. by Waldo, S. and S. E. by Lincoln, and W. by the Androscoggin r. and Oxford co. It lies with lat. 44° and 45°, and long. 7° 17' W. W. C. passes thro' the centre of the co. Chief t. Augusta, which is also the seat of government. Pop. 1820, 40,150, 1830, 52,484.

KENNEBEC, r., Me. next to Penobscot the largest in the state. It has two principal branches, the E. rising in Moosehead lake, at the base of the height of land, the W. called Dead r. rising in the highlands which separate Me. from Canada, and uniting with the E. branch about 20 ms. below Moosehead lake. Whole course about 300 ms., navigable for ships 12 ms to Bath; for sloops 45 ms. to Augusta, at the head of the tide; and for boats 60 ms. to Waterville, where the navigation is interrupted by Toconic falls. This r. during its whole course descends about 1,000 feet. The lands are fertile and well adapted to pasturage. On the w. side of the upper part of its course are high mtns. It flows in a great valley, with Penobscot and St. John's rs. 120 ms. long, and about 20 wide. The valley of the Kennebec proper is varied with moderate hills s., mtns. n. Below Somerset co. the hills rise from the banks; above, there are flats; near Dead r. the valley is broken; at Moosehead lake it expands. Here is nearly the level of the sources of the Penobscot and John's rs. Salmon remain in deep holes in the Kennebec most of the year. In the town of Strong, they have been taken in winter from Pierpoles holes in Sandy river. They abound until the spring freshet.

Kennebunk, p-t. and port of entry, York co. Me. mouth of Kennebunk r. which affords a good harbor, 10 ms. s. Saco, 25 s. w. Portland. A place of considerable commerce. Pop. 1830, 2,233.

Kennedy's p-o. Brunswick, co. Va. by p-r. 75 ms. s. s. w. Richmond.

Kennedy's p-o. Garrard co. Ky. by p-r. 57 ms. s. e. Frankfort.

Kennet's Square, and p-v. Chester co. Pa. 35 ms. s. w. by w. Philadelphia, and 18 n. w. Wilmington in Delaware.

Kensington, town, Rockingham co. N. H. 13 ms. s. w. Portsmouth, 40 from Concord, 45 from Boston. Surface pretty even. Pop. 1830, 717.

Kensington, p-v. Philadelphia co. Pen. lies on the Delaware n. e. of the Northern Liberties, and is incorporated; it has numerous ship yards and manufactories. In this town is the spot where Wm. Penn made his treaty with the Indians, and the Elm tree under which the conference was held was not long since standing.

Kent, co. R. I. is an agricultural and manufacturing co. centrally situated, on the w. shore of the Narraganset. Bounded n. by Providence co., e. by the Narraganset r., s. by Washington co., w. by Connecticut. Average length nearly 20 ms., breadth more than 9, comprising an area of 186 square miles. Surface generally uneven; soil in general strong and productive. Forests, deciduous trees. A large portion of n. w. section of the county is watered by the Pawtuxet river and its branches. This r. is a beautiful mill stream, unrivalled for its advantageous sites for manufacturing establishments, and other hydraulic works. The cotton manufactures of this co. claim the first rank. Chief town, Warwick. Pop. 1830, 12,789.

Kent, p-t. Litchfield co. Conn. on the Ousatonic, 45 ms. w. Hartford. Mountainous. Contains iron ore and iron manufactories. Soil various. Timber, oak, chestnut, walnut, ash, &c. Watered by the Ousatonic and its numerous branches, which afford many valuable sites for water works; 10 school districts. Pop. 1830, 2,001.

Kent, p-t. Putnam co. N. Y. 20 ms. s. e. Poughkeepsie. Much broken by high hills and mtns. Timber, oak, chestnut, &c. Well watered and healthy. Pop. 1830, 1,931.

Kent, co. Del. bounded by Duck cr. separating it from New Castle co. n., by the Del. bay e., by Sussex co. of Del. s., and by Caroline, Queen Ann, and Kent cos. of Md. w. Length 32, mean breadth 20, and area 640 sq. ms. Extending in Lat 38° 50′ to 39° 20′, and in long. 1° 18′ to 1° 50′ e. W. C., with a very slight exception along the wstrn. border, on which rise the sources of Choptank and Nantikoke rs., the slope of Kent co. of Del. is estrd. towards Del. bay, and drained by Mispillion, Mother Kill, Jones, and the two Duck creeks. The surface is level or moderately waving. Soil of midling quality. Chief t. Dover. Population 1820, 20,793, in 1830, 19,913.

Kent, co. Md. bounded s. w. and w. by Chesapeake bay, n. by Sassafras r. separating it from Cecil, e. by New Castle, and Kent cos. Del., and s. e. and s. by Chester r. separating it from Queen Ann. Length 30, mean width 8, and area 240 sq. ms. Extending in lat. 39° 01′ to 39° 23′, and in long. 0° 45′ to 1° 18′ e. W. C. This co. is composed of a peninsula curving from the wstrn. boundary of Del. between Sassafras and Chester rs. with the convexity nrthwstrd. towards Chesapeake bay. General slope wstrd. Surface moderately hilly, and soil of varied quality. Chief town, Chester. Pop. 1820, 12,453, in 1830, 10,501.

Kentontown, p-v. Harrison co. Ky. 47 ms. n. e. Frankfort.

Kentucky, Indian name Cutawa, r. Ky. from which the name of the state has been derived, rises in numerous branches from the nrthwstrn. slope of Cumberland mtn. interlocking sources with those of sandy, Powell's and Cumberland rs. Assuming a nrthwstrn. course, the various confluents from Pike and Perry cos. unite in Estill, where inflecting to wstrd., and separating Madison from Clarke, wind to s. w. between Madison and Lafayette, and between Jessamine and Garrard; receives Dick's r. from the s.e. and finally bends to its ultimate n. n. w. course, which is continued to its junction with O. r. at Port William. The general course is very nearly s. e. to n. w. The valley drained by this r. lies in lat. between 37 and 38° 40′, and in long. between 5° 40′ and 8° 10′ w. W. C. Length 175 ms., mean width about 40, and area 7,000 sq. ms. or a small fraction above the one sixth part of the whole state of Ky., and comprising all or part of Gallatin, Henry, Owen, Scott, Franklin, Anderson, Woodford, Jessamine, Mercer, Lincoln, Garrard, Madison, Lafayette, Clarke, Montgomery, Estill, Clay, Perry and Pike cos. The channel of Ky. is a deep chasm, yet steamboats of 300 tons burthen ascend this r. to Frankfort, at times of high water, and at similar seasons, it is navigable for down steam boats from Estill co.; similar to other steams of the same physical section, it is without direct falls, though the current is rapid, and bed rocky.

KENTUCKY, state of the U. S., bounded S. by the state of Ten., S. W. by the Miss. r. separating it from Miss., W. by Ohio r. separating it from Il., N. W. by Ohio r. separating it from Ind., N. by Ohio r. again separating it from the state of Ohio, and E. Sandy r. and Cumberland mtns. separating it from Va.

The longest line that can be drawn in Ky. is 431 statute ms. declining from the meridians 80° 33′ and extending from the S. W. angle on Miss. to the passage of Sandy r. through Cumberland mtn., or the extreme estrn. angle of the state. The broadest part is along the meridian 7° 45′ W. W. C., extending from the N. W. angle of the state, between Cincinnati and the mouth of Great Miami, thro' 148 minutes of lat. or 171½ statute ms. nearly; in lat. Ky. extends 36° 30′ to 39° 06′, and in long. 5° 03′ to 12° 38′ W. W. C. The area of Ky. has been generally underrated. On Tanner's U. S., the extent in sq. ms. is given at 40,500, and carefully measured by the rhombs on the same map, the superficies comes out 40,590, so that we may safely assume 40,500 sq. ms. equal to 25,920,000 statute acres as the area of Ky.

As a physical section Ky. lies entirely in the valley of Ohio, and is part of an immense inclined plain falling from Cumberland mtns. towards and terminating in the O. r. In its extent from Sandy r. to the Ten., inclusive, the direction of descent is to the N. W. The physical section indeed of which Ky. is a part, extends to and includes Ten. r. If we glance over a general map of this part of the U. S. we perceive the rs. at their sources inclining to W. or S. W., and following their courses we find them curving to the nrthrd. and finally joining their common recipient, the Ohio, in a N. N. W. direction. This uniformity of course is perceptible in Ten. Cumberland, Green r. Salt r., Ky. r. and Licking, and even the Ohio itself, from the mouth of Sandy to that of Great Miami, conforms to this remarkable inflection. The rs. flow in channels, more or less deeply scooped from the rocky base of the plain, but with a regularity of course demonstrative of a common cause. The relative elevation of the lower and higher margins of Ky. has never been, it is probable, very accurately determined, but compared with the determined elevations on Great Kenhawa, the arable soil of the higher part of Ky., Pike, Perry, and Harlan cos. must be at least 1,200 ft. above the ocean tides. The extreme southwestern co. Hickerman on Miss. r. is not generally elevated above the 350 ft. above the Gulf of Mexico; therefore, without regarding mtn. ridges, the cultivatable surface of Ky. has a descent of between 800, and 900 ft.

Continuing the difference of level, with that of lat. it is obvious, that the extremes of the state must have a very sensible difference of climate and mean temperature. These extremes of season are still farther widened by the peculiar features of the country.

The rivers in their descent, have abraded the plain, and flow in enormously deep vales, a feature which the Ohio partakes with its confluents. These chasms receiving the rays of the sun in various inclinations following local exposure, produce also local climate. The state is divisible into 3 sections, which, however, so imperceptibly pass into each other as to preclude any very definite lines of separation. Descending from the foot of Cumberland mtn. nthwstrd. down the streams, to a distance of about 100 ms. the country is hilly or rather mountainous. This broken section includes at least one third part of the state and stretches from the state of Ten. to the O. r. Drawing a line from the O. r. opposite the mouth of Sciota to the heads of Big Barren, branch of Green r., it will extend almost exactly parallel to the general course of O. r. between the mouths of Great Miami and Salt r.; and again if the latter course is continued, it will leave Ky. very nearly where the sthrn. boundary is crossed by Cumberland r. These two lines, with the course of Ohio r. from the mouth of Sciota to that of Great Miami, and the boundary between Ky. and Ten. will enclose a rhomb of 90 ms. in width, and 200 mean length, or comprising an area of 18,000 sq. ms. This rhomb is nearly commensurate with the central hilly section of Ky. It is very remarkable nevertheless, that the general surface of this great section is much more broken into hills at its opposite sides towards the O. r. or Cumberland mtn. than in the middle line between the extremes. The whole of this great middle region, may be comparatively regarded as a table land, with a substratum of limestone. The soil in general in a high degree productive, but similar to all other places where carbonate of lime prevails, an unequal distribution of fountain water is amongst the asperities opposed to comfortable human residence.

The sthwstrn. section of Ky. the least extensive, presents a physiognomy very distinct from either of the preceding. The strong bold scenery, so prominent in the two higher regions, is now succeeded by a monotony of feature which advancing wstrd. sinks into a country, which, though not absolutely level, presents relative elevation faintly.

Reversing our survey; if we leave the banks of the Miss. we set out from a plain over which the eye in vain seeks relief from hill and dale. Proceeding obliquely over the vallies of Ten. Cumberland, and Green rs., the face of nature very gradually breaks into indentations which terminate in all the rich variety of hill and river scenery. The hills indeed are not abrupt, but rounded into swells, or terminating in plains or furrowed by the excessively deep chasms along which the rivers wind their devious way. On the right towards Ten. spreads the tract so very improperly called "The Barrens". Here the hills are isolated knobs, wooded with oak, chestnut, and elm. The hills are rounded, and present a striking contrast to the common ridge character of a hilly country. The soil is far from barren, though much of the timber has a stunted appearance. Advancing ntheatrd. the same substratum of limestone continues, but in its natural state the central section of Ky. was remarkable for the excessive growth of forest timber, and undergrowth of reed cane. The surface comparatively level, except the channels of the streams, which were, as has been observed, deep and with abrupt banks.

Turning the r. eatrd., however, towards the sources of Licking, Ky., and Cumberland rs. the ground rises into hills, sharp, steep and rocky. The soil particularly in the vales deteriorates. Fountain water becomes more equally distributed.

Taken as a whole, Ky. may be regarded as not only a political, but physical section, presenting distinct structure and features. The physiognomy of this tract in connection with other parts of the valley will be more particularly noticed under the head of Ohio r.

Politically Ky. is subdivided into the following counties.

	sq. ms.	Pop. 1820,	Pop. 1830,
Adair	800	8,765	8,217
Allen	600	5,372	6,485
Anderson			4,520
Barren	900	10,328	15,079
Bath	340	7,960	8,799
Boone	300	6,542	9,075
Bourbon	176	17,664	18,436
Bracken	264	5,280	6,518
Breckenridge		7,185	7,345
Bullitt	300	5,831	5,652
Butler	825	3,083	3,058
Caldwell	800	9,022	8,324
Callaway			5,164
Campbell	320	7,022	9,883
Casey	360	4,349	4,312
Christian	1,050	10,459	12,684
Clarke	200	11,449	13,565
Clay	1,400	4,393	3,184
Cumberland	1,034	8,058	8,624
Daviess	600	3,876	5,209
Edmondson			2,642
Estill	700	3,507	4,618
Fayette	264	23,254	25,098
Fleming	560	12,186	13,499
Floyd	2,000	8,207	4,347
Franklin	270	11,021	9,254
Gallatin	350	7,075	6,674
Garrard	220	10,851	11,871
Grant	260	1,805	2,986
Graves			2,504
Grayson		4,055	3,880
Greene	400	11,943	13,138
Greenup	537	4,311	5,852
Hardin	1,100	10,498	12,819
Harlan	560	1,961	2,929
Harrison	330	12,278	13,231
Hart	320	4,184	5,191
Henderson	600	5,714	6,656
Henry	400	10,816	11,387
Hickman	675		5,198
Hopkins	750	5,322	6,763
Jefferson	520	20,768	22,979
Jessamine	170	9,297	9,960
Knox	840	3,661	4,315
Lawrence			3,900
Lewis	530	3,973	5,229
Lincoln	450	9,979	11,002
Livingston	720	5,824	5,971
Logan	630	14,423	13,012
Madison	570	15,954	18,751
McCracken			1,297
Mason	250	13,588	16,203
Meade			4,131
Mercer	350	15,587	17,694
Monroe	700	4,956	5,340
Montgomery	420	9,587	10,240
Morgan			2,857
Muhlenburg	580	4,979	5,340
Nelson	510	16,273	14,932
Nicholas	360	7,973	8,834
Ohio	640	3,879	4,715
Oldham			9,588
Owen	240	2,031	5,786
Pendleton	340	3,086	3,863
Perry,	1,000		3,330
Pike	750		2,677
Pulaski	800	7,597	9,500
Rockcastle	380	2,249	2,865
Russell			3,879
Scott	170	14,219	14,677
Shelby	520	21,047	19,030
Simpson	400	4,852	5,815
Spencer			6,812
Todd	450	5,089	8,680
Trigg	450	3,874	5,916
Union	540	3,470	4,764
Warren	700	11,776	10,949
Washington	550	15,987	19,017
Wayne	970	7,951	8,685
Whitley		2,340	3,806
Woodford	160	12,207	12,273
Total,		564,317	687,917

Of the above 165,350 are slaves.

Note.—The area annexed to the respective counties in this table, will not, in all cases, be found to correspond with the text under the co. heads. The frequent subdivisons of cos. in a few years derange any admeasurment of area; the numbers were left, however, as they give a general view.

In the census tables and in the abstract of the census, both afforded by the government of the U. S. there is a discrepancy under the head of Ky.

Census tables, total pop.		688,844
Abstract,	do.	687,917
Difference,		927

History. In 1755, Lewis Evans of Phil. published a map of the middle British colonies in N. America. An edition of this map, with a statistical account of the regions it represented, was published by J. Almon, London March 25, 1776. Both the map, and attending volume, is now lying before the writer of this article. The map reaches as far s. as N. lat. 36° 30′, and as far wstrd. as the meridian of 10° 30′ w. W. C., and from the delineations it would appear that at the period, 1752 to 1776, settlements had reached the sources of Great Kenhawa, Roanoke, Clinch, and Holston, as this region is tolerably well represented, and it is noted on the map, that this was the boundary of white settlement. Receding to the watrs. are laid down, relatively correct, Big Sandy, Licking, Catawa, or Ky. rs. and Bear Grass cr., but the s.w. angle of the map is blank, demonstrating, that in 1776, Ky. might be regarded, as in great part unknown.

In 1767, this country was visited by John Finley, from N. C., and was followed in 1769, by Daniel Boone, and some others. Boone remained there until 1771. In '75, the same brave spirit conducted a small band and effected the first actual civilized settlement. Ky. was truly planted with sweat, and watered with blood and tears. So distressed were the settlers in 1780, as to excite a plan of abandonment, but other adventurers arriving, and aided as they were by the great military talents of Rogers Clark, they laid aside their purpose. In 1777, the legislature of Va. had made it a co. and in 1782, a supreme court was established. With the American war the worst difficulties of the inhabitants terminated. Settlements were rapidly formed, and as early as 1785, projects of separation from Va. were formed, but from various causes not effected until December 1790, when Ky. became independent of Va., and June 1st, 1792, was admitted into the Union.

The existing constitution of Ky. was ratified at Frankfort, Aug. 17, 1799; since which epoch, the history of the state has been merged in that of the U. S.

Government. Legislature composed of a general assembly, and governor. The assembly divided into a senate and house of representatives. To be eligible for governor, the person must be a citizen of the U. States; 35 years of age. and six years next preceding his election, an inhabitant of the state. "The governor," says the constitution, "shall be elected for the term of four years, by the citizens entitled to suffrage, at the time and place where they shall respectively vote for representatives." "The governor shall be ineligible for the succeeding 7 years after the the expiration of the time for which he shall have been elected."

The powers of the governor of Ky. are ample; he is commander of the army and navy of the state, and of the militia, except when called into the actual service of the U. States. He has the power of nomination, and by and with consent of the senate the appointment of most officers of the state; he has power in the recess of the legislature to fill all vacancies, by granting commissions which shall expire at the end of the next session. He has power to remit fines and forfeitures, grant reprieves and pardons, except in cases of impeachment. In cases of treason, he shall have power to grant reprieves until the end of the next session of the general assembly, in which the power of pardoning shall be vested.

The lieutenant governor, bears almost exactly the same relation to the legislature, and governor of Ky., as does the vice president of the U. S. to the senate and president.

Senators are chosen for four years, and divided into four classes, whose seats are filled annually, so that one fourth shall be chosen every year. "No person shall be senator, who, at the time of his election, is not a citizen of the U. S. and who hath not attained to the age of 35 years, and resided in this state six years next preceding his election, and the last year thereof in the district from which he may be chosen."

Members of the house of representatives are elected for one year; and, "no person shall be a representative, who, at the time of his election, is not a citizen of the U. S. and hath not attained to the age of 24 years, and resided in this state two years next preceding his election, and the last year thereof in the co. or town for which he may be chosen."

The judiciary power, both as to matter of law and equity, is vested in one supreme court, styled the court of appeals, and in inferior courts created and established by the general assembly. Judges both of the supreme and inferior, are appointed by the governor and senate, and hold their offices during good behavior; but for any reasonable cause, which shall not be sufficient ground of impeachment, the governor shall remove any of them on the address of two thirds of each house of the general assembly.

Right of suffrage, vested in free white males, who at the time being hath attained the age of 21 years, and resided in the state two years, or in the co. or town, in which he offers to vote, one year next preceding the election.

Staple productions.—Kentucky is essentially a grain country, though hemp and flax of excellent quality are produced, and in the extreme southwestern part some cotton is cultivated. An immense quantity of flour, spirits, salted provisions, and live stock are exported, down the Ohio, and inland to the eastrd. Manufactures of cloth, cordage, &c. have been carried to considerable extent, but the state commercially remains dependent in a great measure on foreign supply, for most articles of domestic use.

Education.—In promoting the requisite institutions to advance learning and science, Kentucky has more than preserved her priority over the other central states which she gained by anterior settlement. Transylvania university was founded, and most correctly named, at an early stage of settlement; organized in 1798 and by a report of the professors dated Feb. 11th, 1822, then contained "all the means requisite for a complete course of medical education, conducted in the usual academical form."

This institution, in 1820, contained undergraduates 143, 200 medical, and 19 law students. Part of the buildings were destroyed by fire a few years since, but the damage has been repaired.

Lectures commence on the 1st Monday of November annually, and terminate in the ensuing March. The professors and students have also the advantage of an extensive library and anatomical museum.

To be eligible as a candidate for a degree of Dr. of Medicine, the applicant must have attained twenty-one years of age, and have attended two full courses of lectures, one of which, at least, in this institution. But any physician, who has practiced reputably his profession 4 years, and attended one course in the Transylvania medical school, may receive a degree of M. D.

KENZUA. (*See Kinzua.*)

KEOWEA, p-v. in the eastern part of Pickens dist., S. C. by p-r. 128 ms. N. W. Columbia.

KERNESVILLE, p-v. on a small creek of Lehigh river, Northampton co. Pa. 12 ms. N. E. W. Bethlehem.

KERSEY'S p-o. Clearfield co. Pa. by p-r. 184 ms. N. W. Harrisburg.

KERSHAW, district, S. C. bounded by Richmond S. W., Fairfield W., Lancaster N., Chesterfield N. E., Darlington E., and Sumpter S. E. and S. Length 33, mean width 24, and area 792 square miles. Extending in lat. 34° 05′ to 34° 35′, and in long. 3° 16′ to 3° 50′ W. W. C. The Catawba, or as there called, the Wateree river, traverses the western part of Kershaw, and the eastern is bounded by Lynch's creek; both streams S. of S. S. E., of course the slope of the district is in that direction. Chief town, Camden. Pop. 1820, 12,442, 1830, 13,515.

KEYSVILLE, p-v. on the head of Meherin r. Charlotte co. Va. by p-r. 96 ms. S. W. by W. Richmond.

KEY WEST, small island of Florida, in the Gulf of Mexico, one of the Florida Keys.

KIDZIES GROVE, and p-o. Lenawee co. Mich. 70 ms. S. W. Detroit.

KILKENNY, town, Coos co. N. H. 8 ms. N. E. Lancaster. A poor tract of country, unfit to be inhabited. Pop. 1830, 27.

KILLINGLY, p-t. Windham co. Conn. on the Quinnebaug, 25 ms. W. Providence, 45 E. Hartford. Surface uneven, but no portion mountainous. Contains several quarries of freestone, and extensive forests, the trees being of the deciduous species. Supplied with numerous water privileges. Shad and salmon are taken in the Quinnebaug. This is a manufacturing town; 21 school districts. Pop. 1830, 3,257.

KILLINGTON PEAK, Vt. a summit of the Green mountain, S. part of Sherburn, 3,924 feet above tide water, 10 ms. E. Rutland.

KILLINGWORTH, p-t. Middlesex co. Conn. on Long Isl. sound, 26 ms. E. New Haven, 38 S. E. Hartford, 26 W. New London. Surface and soil various. Its most considerable streams are the Hammonassett and the Menunketesuck. There is a harbor in the S. part of the town; 15 school districts, and an academy. Pop. 1830, 2,484.

KILLS (THE.) *See Newark bay.*

KILMARNOCK, p-v. on a small creek of Chesapeake bay, Lancaster co. Va. by p-r. 115 ms. N. E. by E. Richmond.

KIMBERTON, p-v. on French creek, in the N. E. part of Chester co. Pa. about 27 ms. N. W. Philadelphia.

KIMBLES, p-v. Lawrence co. Ohio, 139 ms. S. S. E. Columbus.

KINCANNON, iron works, and p-o. Surry co. N. C. by p-r. 139 ms. N. W. by W. Raleigh.

KINDERHOOK, creek, one of the best mill streams in the U. S. is formed by numerous branches, that spread over N. E. of Columbia co. N. Y. and the S. corner of Rensselaer co. which united, run to the S. W. through the town of Kinderhook, where the stream takes its name, which it continues to its junction with Claverack creek, near the Hudson river, when the united streams lose their name for Factory, or Major Abram's creek.

KINDERHOOK, p-t. Columbia co. N. Y. 10 ms. N. Hudson, 20 S. Albany, W. of the Hudson, enjoying the navigation of said r. and having several landings, with stores, sloops, &c. Soil in general good. Surface pretty level. Timber scarce. Contains iron ore, limestone, slate and various kinds of clays. Some red oxides of iron are found. There are a great number of mills of various kinds, watered by Kinderhook creek; 10 school houses and 2 academies. Kinderhook village stands on an extensive and beautiful plain near the centre of the town west of the creek, where is Kinderhook landing. Columbiaville, another village of the town, is on the line between this and Hudson. Pop. 1830, 2,706.

KING AND QUEEN, co. Va. bounded by Caroline N. W., Essex N. E., Piankatank river separating it from Middlesex E., Gloucester S. E., James r. S., and Matapony r. separating it from King William S. W. and W. Length 40, mean width 11, and area 440 sq. ms. Extending in lat. 37° 27′ to 37° 56′ and in long. 0° 18′ E. to 0° 13′ W. W. C. The surface sloping southward towards Matapony r. or southeastward towards Piankatank. Chief town, Dunkirk. Population 1820, 11,798, 1830, 11,644.

KING AND QUEEN, court house, and p-o. (*See Dunkirks, same co.*)

KING, creek and p-o. in the extreme southern angle of Barnwell district, S. C. by p-r. 90 ms. S. S. W. Columbia.

KING GEORGE, co. Va, bounded W. by Stafford, N. and E. by Potomac r. separating it from Charles co. in Md., S. E. by Westmoreland, and S. by Rappahannoc r. separating it from Caroline. Length 18, mean breadth 10, and area 180 square ms. Extending in lat. 38° 11′ to 38° 23′, and in long. 0° 03′ E. to 0° 19′ W. W. C. This co. occupies a hilly region between the two bounding rs. with a varied soil. Chief town, Hampstead. Pop. 1820, 6,116, 1830, 6,397.

KING GEORGE, court house, and p-o. near the centre of King George co. Va. by p-r. 81 ms. a little W. of S. W. C. and 67 ms. N. N. E. Richmond.

KINGS, co. N. Y. comprises a very small area of the w. end of Long Island, immediately opposite N. Y. Bounded N. by East river, E. by Jamaica bay, and Queens co., S. by the Atlantic, W. by N. York bay, and the communication of the Hudson r. with the Atlantic. It contains about 81½ square ms. or 52,160 acres, the whole area not equalling that of a tsp. 10 ms. square. Soil in general very good. Chief town, Flatbush. Pop. 1820, 11,187, 1830, 20,535.

KINGSBRIDGE, village, N. Y. on Haerlem r. which separates the county of Westchester from N. Y. island, 16 ms. N. N. Y. city.

KINGSBURY, p-t. and half shire town of Washington co. N. Y. E. Hudson r. 55 ms. N. Albany. General surface very level. Pretty good for farming. Contains fine groves of pine. Kingsbury v. is situated near the centre of the town, about 2 ms. from which is the spot where Putnam was defeated by the Indians. The village of Sandy Hill is in the S. W. corner of the town, close on the margin of the Hudson, immediately above Baker's falls. The whole descent of these falls, is 76 feet within 60 rods. There is no perpendicular cataract. Here are a number of mills, with a chance of many more. At this village and Salem, are alternately holden the courts of Washington co. The Champlain canal traverses this town from N. E. to S. W. It has 11 schools, kept 8 months in 12. Contains an academy. Pop. 1830, 2,606.

KINGSESSING, p-o. Kingsessing tsp. Phila. co. Pa. 6 ms. S. S. W. Phila. This tsp. is the extreme southern part of the co. lying between Darby creek, and the river Schuylkill. Pop 1820, 1,188, 1830, 1,068.

KING'S FERRY, (over Monongahela r.) and p-o. in the southern part of Monongalia co. Va. 15 ms. by land above Morgantown, and by p-r. 204 ms. N. W. by W. W. C.

KINGSFIELD, p-t. Somerset co. Me. 40 ms. N. W. Norridgewock. Pop. 1830, 554.

KINGS, gap and p-o. Harris co. Geo. 126 ms. westward Milledgeville.

KINGSLEY'S p-o. Crawford co. Penn. 313 ms. N. W. W. C.

KINGS mountain, a ridge or hill, Lincoln co. N. C. and York district, S. C. It was on this mtn. and within York district, that, Oct. 7th 1780, a body of British and tories under Col. Ferguson were defeated, their commander slain, and nearly the whole body killed or captured by three regiments of U. S. militia.

KINGSPORT, p-v. on the road from Knoxville in Tenn. to Abington in Va. situated on the point above the junction of the two main branches of Holston r. and in the N. W. part of Sullivan co. Tenn. 90 ms. by the road N. E. Knoxville, 42 S. W. by W. Abingdon, and by p-r. 246 ms. a little N. of E. Nashville.

KINGSTON, p-t. Rockingham co. N. H. 20 ms. S. W. Portsmouth, 37 ms. from Concord, 6 from Exeter. Contains an academy, some bog iron ore, and red and yellow ochre. Pop. 1830, 929.

KINGSTON, town, Addison co. Vt. 22 ms. S. W. Montpelier, 42 N. W. Windsor. White r. is formed here by the union of several branches, on one of which is a fall of 100 feet, 50 of the lower part of which are perpendicular. A considerable portion of the town is mountainous; 3 school districts. Pop. 1830, 403.

KINGSTON, p-t. Plymouth co. Mass. 32 ms. S. E. Boston. Watered by Jones' r. Has some manufactories of cotton and woollen. Here also are iron works. Soil fertile. Surface agreeably diversified. Pop. 1830, 1,321.

KINGSTON, formerly Esopus, p-t. and st. jus. of Ulster co. N. Y., W. Hudson r., 100 ms. N. New York, 65 S. Albany. Soil good. Almost all the houses are built of lime stone, which is plentiful here. Well supplied with mill privileges by Esopus creek, which waters this town. Has several landings on the Hudson, and is a place of very considerable business; 9 school districts. Kingston village lies on the S. side of Esopus creek, 10 ms. S. of its mouth in the Hudson, and 3 W. of the Hudson, at Kingston Landing. It has an academy. It was burnt by the British under Vaughan, in 1777. It has an elegant court house, which cost $40,000. Population 1830, 4,170.

KINGSTON, village, Middlesex and Somerset co. N. J. on the Millstone r. and main p-r. 3 ms. N. E. of Princeton, 13 ms. S. W. New Brunswick. The Delaware and Raritan canal passes through this village.

KINGSTON, p-v. Luzerne county, Pa. (See *Wyoming*.)

KINGSTON, p-v. southern part of Somerset co. Md. by p-r. 152 ms. S. E. W. C.

KINGSTON, p-v. Morgan co. Geo. 33 ms. N. W. Milledgeville.

KINGSTON, p-v. Adams co. Miss.

KINGSTON, p-v. and st. jus. Roane co. Ten. situated on the point above the junction of Clinch and Holston rs. 43 ms. S. W. by W. Knoxville, and by p-r. 130 ms. a little S. of E. Nashville, lat. 35° 53', long. 7° 26' W. W. C.

KINGSTON, p-v. Hopkins co. Kentucky, by p-r. 216 ms. S. W. by W. Frankfort.

KINGSTON, p-v. Ross co. Ohio, by p-r. 36 ms. S. Columbus.

KINGSTREE, p-v. and st. jus. Williamsburg district, S. C., situated on Black river, 43 ms. N. W. Georgetown, and 71 a little E. of N. Charleston, lat. 33° 37', and long. 2° 55' W. W. C.

KING WILLIAM, co. Va. bounded by Caroline N., by Mattapony r. separating it from King and Queen, N. E. and E., and by Pamunkey river separating it from New Kent S., and Hanover W. Length 38, mean width 14, and area 532 square ms. Extending in lat. 37° 30' to 37° 57' and in long. 0° 09' E. to 0° 19' W. W. C. Chief p-o. King William court house. Pop. 1820, 9,697, 1830, 9,812.

KING WILLIAM, court house, and p-o. King William co. Va. by p-r. 40 ms. N. E. Richmond, and 136 ms. a little W. of S. W. C.

KINGWOOD, town, Hunterdon co. N. J. on

the Del. extends N. E. to the south branch of Raritan. Pop. 1830, 2,898.

KINGWOOD, p-v. and st. jus. Preston co. Va. situated w. Cheat r., 23 ms. S. E. Morgantown, and by p-r. 172 ms. N. W. by W. W. C. Lat. 39° 27', long. 2° 45' w. W. C.

KINNICONICK, creek, and p-o. eastern part of Lewis co. Ky. 100 ms. N. E. by E. Frankfort.

KINSMAN'S p-o. Trumbull co. Ohio, by p-r. 184 ms. N. E. Columbus.

KINZUA, or Kenjua, p-v. on the left bank of Alleghany r. in the N. E. part of Warren co. Pa. 12 ms. by land above Warren and by p-r. 226 ms. N. W. Harrisburg.

KIRBY, town, Caledonia co. Vt. 30 ms. N. Newbury, 36 N. E. Montpelier. Surface uneven, and in many places, ledgy or swampy. Well watered with springs and brooks. Pop. 1830, 401.

KIRKSEY'S Cross roads, and p-o. Edgefield dist., S. C. 15 ms. N. N. W. the v. of Edgefield, and by p-r. 65 ms. a little s. of w. Columbia.

KIRKS, Mills, and p-o. Lancaster co. Pa. 46 ms. E. Harrisburg.

KIRTLAND, Mills, and p-o. Geauga co. O. by p-r. 151 ms. N. E. Columbus.

KISKIMINITAS, r. of Pa. the southeastern and largest confluent of Alleghany r. This stream is more commonly known under the name of Conemaugh. (*See the latter article.*)

KISKIMINITAS, post tsp. in the northern part of Westmoreland county, Pa. about 10 ms. N. Greensburg, and 25 ms. E. Pittsburg.

KITE'S, Mills, and p-o. Rockingham co. Va. by p-r. 141 ms. wstrd. W. C.

KITTANNING, p-v. and st. jus. Armstrong co. Pa. situated on the left bank of Alleghany r., 40 ms. N. E. Pittsburg, and by p-r. 214 ms. N. W. by W. Harrisburg, lat. 40° 51', long. 2° 33' w. Pop. 1820, 318, 1830, 520.

KITTATINNY, mtns. an extensive and important chain of the Appalachian system. In Pa. the Kittatinny is very definite and with an intervening valley between their ranges parallel to the Blue Ridge. It is the same chain, however, which first becomes definite in the state of New York, w. of the Hudson, and there known as the Shawangunk, and extending S. W. over the upper part of New Jersey, enters Pa. at the Delaware Water gap. Thence inflecting to W. S. W. is traversed by the Lehigh at the Lehigh Water gap, by the Schuylkill above Hamburg, and by the Susquehannah, 5 ms. above Harrisburg. From the latter point the chain again inflects still more to the westward, between Cumberland and Perry cos. At the western extremity of those two counties, the chain abruptly bends to a nearly southern course, between Franklin and Bedford counties, enters Md. by the name of Cove mtn. being traversed by the Potomac r. between Williamsport and Hancockstown, and stretches into Virginia, as the Great N. mtn. over Virginia from the Potomac to James r. between Rockbridge and Alleghany cos. This chain tho' broken remains distinct; a similar character prevails from James r, to New r. between Wythe and Grayson cos. After being traversed by New r. the chain again assumes complete distinctness, leaves Virginia, and under the local name of Iron mtns. Bald mtns. Smoky mtns. and Unika mtns. separates N. C. and Ten. to the Unika turnpike on the western border of Macon co. of the former state. Thence continuing a little w. of s. w. crosses the N. W. angle of Geo. enters Ala. and separating the sources of the creeks of Middle Ten. r. from those of Coosa, merges into the hills from which rise the numerous branches of Tuscaloosa.

Thus, defectively as the Kittatinny, called expressly by the Indians *Kataatin Chunk*, or the Endless mtns. are delineated on our maps, it is in nature a prominent and individual chain, N. lat. 34° 31' to 41° 30', and 2° 45' E. to 10° W. long. W. C. Ranging thro' 7° of lat. and almost 13 degrees of long. stretching along a space exceeding 900 statute ms. and varying in distance from the Blue Ridge, between 15 to 25 ms. generally about 20, though in some places the two chains approach, as at Harrisburg, to within less than 10 ms. from each other. In relative height the Kittatinny exceeds the Blue Ridge, but as regards the plain or table land on which they both stand, it rises gradually from tide water in Hudson r. to an elevation of 2,500 feet in Ashe co. of N. C. From James r. to the Hudson, the chain ranges along the Atlantic slope, and is broken by streams flowing through it on their course towards the Atlantic ocean, but passing the higher valley of James r. the Kittatinny winds over the real dividing line of the waters, and is thence traversed by New r., Watauga, Nolechucky, French Broad, and Ten. rs.

KITTATINNY VALLEY, in the most extended sense of the term, is in length commensurate with the mtn. chain from which the name is derived, therefore extends from Hudson r. to the northern part of Ala. varying in width 8 to 25 ms. with generally a substratum of limestone towards Blue Ridge and of clay slate on the side of the Kittatinny. Some of the most flourishing agricultural districts of the U. S. are included in this physical section. The co. of Orange in N. York, Sussex and Warren in New Jersey are nearly all comprised within its limits. In Pa. it embraces the greater part of the lower section of Northampton; nearly all Lehigh, Berks, and Lebanon, the lower part of Dauphin with the greater share of Cumberland and Franklin. In Maryland the eastern and left part of Washington. In Va. a large part of Berkley, Jefferson, Frederick, Shenandoah, Rockingham, Augusta, Rockbridge, Botetourt, Montgomery and Grayson, and in N. C. the cos. of Ashe, Buncombe, Haywood, and Macon.

The lat. and relative elevation of this great zone has already been shown in the preceding article, and the peculiar features of its parts may be seen under the respective heads of the cos. it embraces, in whole or in part.

KITTERY, p-t. York co. Me. at the mouth of the Piscataqua, opposite Portsmouth, N. H. 5 ms. s. York. Pop. 1830, 2,202.

KLINESVILLE, p-v. in the northeastern part of Berks co. Pa. 74 ms. N. E. by E. Harrisburg.

KLINGERSTOWN, p-v in the eastern part of Schuylkill co. Pa. 81 ms. N. E. Harrisburg.

KNOWLTON, town, Warren co. N. J. on Del. r. s. E. of the Blue mtn. Pop. 1830, 2,827.

KNOX, p-t. Waldo co. Me. 25 ms. N. W. Castine. Pop. 1830, 666.

KNOX, p-t. Albany co. N. Y. 20 ms. W. Albany, on the height of land between Albany and Schoharie. Pop. 1830, 2,189.

KNOX, co. of Ten. bounded by Blount S., Roane S. W., Anderson N. W., Campbell N., Granger N. E., Jefferson E., and Sevier S. E. Length 48, mean width 18, and area 864 sq. ms. Extending in lat. 35° 48′ to 36° 15′, and in long. 6° 11′ to 7° 12′ W. W. C. Holston r. enters the eastern border, and winding S. W. by W. receives the French Broad r. from the S. E. and leaves the western part of the county between Roane and Blount. The northern, northwestern and southeastern parts are mountainous, but the central vallies of Holston and French Broad afford extensive tracts of highly productive soil. Chief town, Knoxville. Pop. 1820, 13,034, 1830, 14,498.

KNOX, co. Ky. bounded by Whiteby W., Laurel N. W., Clay N. and N. E., Harlan E., and Claiborne and Campbell cos. of Ten. S. Length 33, mean width 15, and area 495 sq. ms. Extending in lat. 36° 34′ to 37° 02′, and in long. 6° 30′ to 7° W. W. C. This co. is traversed and drained by Cumberland r., slope to the westward. Chief town, Barbourville. Pop. 1820, 3,661, including what is now Laurel co. In 1830, Knox contained 4,315 inhabitants.

KNOX, co. Ohio, bounded S. by Licking, Delaware W., Marion N. W., Richland N., Holmes N. E. and Coshocton E. Length 30, mean width 21, and area 630 sq. ms. Lat. 40° 14′ to 40° 32′, long. W. C. 5½ W., slope southestrd., and drained by the sources of Mohiccon creek. Chief town, Mount Vernon. Pop. 1830, 17,085.

KNOX, p-v. Knox co. Ohio, by p-r. 56 ms. N. N. E. Columbus.

KNOX, co. Ind. occupying the lower part of the peninsula between the Wabash and White rivers, opposite Wabash and Lawrence cos. Illinois, and having Sullivan and Green cos. of Ind. N. Length from the junction of White and Wabash rivers to the N. E. angle on the latter 50 ms., mean breadth, 10, and area 500 square ms. Lat. 38° 40′, long. W. C. 10° 30′ W.

KNOX, co. Il. bounded by Fulton S., Warren W., Henry N., and Peoria E. Length 26, breadth 24, and area 572 sq. ms. Lat. 41° N., long. 13° 10′ W. W. C. Slopes sthrd. and is traversed and drained by Spoon r. This co. is comprised in the military bounty land.

KNOX, C. H. p-v. Knox co. Il. by p-r. 188 ms. N. N. W. Vandalia.

KNOXVILLE, p-v. Tioga co. Pa. by p-r. 165 ms. nthwrds. Harrisburg.

KNOXVILLE, p-v. and st. jus. Crawford co. Geo. situated on a creek of Flint r. 65 ms. S. W. by W. Milledgeville. Lat. 32° 41′, long. 1° 10′ W. W. C.

KNOXVILLE, p-v. and st. jus. Knox co. Ten. situated on the right bank of Holston r., 26 ms. by land above its junction with Ten., 61 ms. a little S. of W. Greenville, and 178 ms. a little S. of W. Nashville. Lat. 35° 56′, long. 6° 43′ W. W. C. This town has been regarded as the capital of E. Ten. An academy has been long in operation, and the general government has lent its aid towards the formation of a college in Knoxville. Pop. 1820, about 2,000, 1830, 3,000.

KNOXVILLE, p-v. Jefferson co. O. by p-r. 160 ms. N. E. by E. Columbus.

KNOXVILLE, p-v. Frederick co. Md. by p-r. 53 ms. N. N. W. W. C.

KORTRIGHT, p-t. Del. co. N. Y. 6 to 12 ms. E. and N. Delhi. Hilly or mountainous. Soil strong, capable of producing good crops. 17 schools, kept 8 months in 12. Pop. 1830, 2,870.

KREIDERSVILLE, p-v. Northampton co. Pa. 12 ms. N. N. W. Bethlehem, and 14 ms. E. Easton.

KUTZTOWN. (*See Cootstown.*)

KYADEROSSERAS creek, a good mill stream of Saratoga co. N. Y. rising in Corinth and Greenfield, and falling into Saratoga lake in the town of Saratoga Springs.

KYKENDALL'S p-o. Henry co. Ten. by p-r. 121 ms. W. Nashville.

KYLERSVILLE, p-v. Clearfield co. Pa. by p-r. 168 ms. N. W. Harrisburg.

L.

LACKAWANNOC, r. rises in Wayne and Schuylkill cos. between the Lackawannoc and Tunkhannoc chains, and flowing S. 20 ms. turns to S. W. 25 ms., falls into Susquehannah r. 9 miles above Wilkesbarre. The valley of Lackawannoc is, in reality, the continuation to the N. E. of the Wyoming valley, and equally remarkable for the great abundance of mineral coal.

LACKAWANNOC, ridge of mtns. in Luzerne and Wayne cos. Pa. the continuation of Wyoming mtn. E. of Wilkesbarre. It ranges in a northestrn. direction, between the sources of Lackawaxen and Lackawannoc rs. The now remarkable Moosuck mtn. traversed by a rail-road between Carbondale and Honeydale, is the N. E. part of the Lackawannoc.

LACKAWAXEN, r. of Pa. rising principally in Wayne co., but after the union of its main branches forming the boundary between Wayne and Pike cos. This comparatively small stream has gained importance from a canal constructed along its valley. This canal commences on the Del. r. at the mouth of Lackawaxen creek, and following the valley of the latter 24 ms. to Honeydale, where it joins a rail-road over Moosuc mountain. (*See articles Honeydale and Carbondale.*) In 1830, there was sold in the city of New-York 23,605 tons of Schuylkill, Lehigh, and Lackawannoc coal.

Laconia, p-v. Harrison co. Ind. 21 ms. s. Corydon, and 145 ms. s. Indianopolis.

Lacy's Spring and p-o. Morgan co. Al. by p-r. 149 ms. n. n. e. Tuscaloosa.

Lady Washington, sign of, and p-o. Montgomery co. Pa. 22 ms. Phil.

Lafayette, p-v. McKean co. Pa. by p-r. 178 ms. n. w. Harrisburg.

Lafayette, p-v. Montgomery co. Va. by p-r. 208 ms. a little s. of w. Richmond.

Lafayette, parish of La. bounded by Mermentau r. w., bayou Queue Fortue, separating it from St. Laudre n., by St. Martin's parish n. e., by Vermillion r. separating it from St. Mary's e., and by the Gulf of Mexico s.; greatest length along the Gulf 55 miles, mean width 30, and area 1,650 sq. ms. Extending in lat. 29° 30′ to 30° 06′. The whole surface of this large parish is a plain, the far greater part a marsh. The very small lines of soil along the Vermillion, and Queue Fortue, with still less on Mermentau, are the only parts admitting cultivation. It is also, with very slight exceptions, an unwooded prairie. The narrow lines of wood along the streams composed of black oak, white oak, live oak, &c. cease before reaching the Gulf. Clumps of live oak are seen rising on shell and sand banks from the marsh. Chief town, Mountenville. Pop. 1830, 5,653.

Lafayette, co. Ten. (*See Fayette co. Ten.*)

Lafayette, co. of Ky. bounded s. w. by Jessamine, w. by Woodford, n. w. by Scott, n. e. by Bourbon, e. by Clark, and s. e. by Ky. r. separating it from Madison. Length 23, mean width about 11, and area 253 sq. ms. Extending in lat. 37° 52′ to 38° 12′, and in long. 7° 15′ to 7° 55′ w. W. C. This co. is chiefly drained by the Elkhorn river, and slopes to the n. w. Surface comparatively level, and soil highly productive. Chief t. Lexington. Pop. 1820, 23,250, and in 1830, 25,174.

Lafayette, co. Ind. (*See Fayette co. same state.*)

Lafayette, co. of Mo. having the Mo. r. n., and extending s. indefinitely to Osage r.; it is mostly uninhabited. Chief t. Lexington. Pop. 1830, 2,912.

Lafayette, C. H. and p-o. Lafayette co. Ark. 182 ms. s. w. Little Rock.

Lafayette, p-v. and st. jus. Tippecanoe co. Ind. 70 ms. n. w. Indianopolis. It is situated on Wabash r. about 10 ms. below the mouth of Tippecanoe r.

Lafayette, southwestern co. of Ark., the limits of which are yet but vaguely defined; on Tanner's map of the U. S., it has Hempstead in Ark. on the n., and the parish of Claiborne in La. s., extending e. and w. Washitau river to the western boundary of the Ter. Length along La. 130 ms. mean width about 35, and area 4,550 sq. ms. It is traversed by Red r. Pop. 1830, 748.

Lafayetteville, p-v. Oldham co. Ky. 43 ms. nthwrd. Frankfort.

Lafourche, (The Fork), r. of La. a mouth of the Miss. This outlet, about 80 yards wide at its efflux from the main stream, issues at and above Donaldsonville, and though remarkable as receiving no tributary water in all its length of, by comparative courses, 90 ms., the stream widens and deepens as it approaches its discharge into the Gulf of Mexico. The general course is very near s. e., and though presenting on a smaller scale, similar features with the Miss. the bends of Lafourche are comparatively less numerous and abrupt in the Lafourche. It enters the Gulf over a bar of 9 feet water, at n. lat. 29° 12′, and long. 13° 09′ w. W. C.

Lafourche Interior, parish of La. bounded by Assumption n. w., St. John Baptiste n., St. Charles and the Gulf of Mexico s. e., and the parish of Terre Bonne s. and w. Length 70, mean width 15, and area 1,050 square ms. Extending in lat. 29° 12′ to 29° 57′, and in long. 13° 07′ to 14° 15′ w. W. C. The Lafourche r. winds through this parish in the direction of its greatest length, and containing on its banks the far greatest part of the arable soil of its surface. The whole, indeed, as part of the delta, is an almost undeviating plain; the banks of the streams rising but little above the interior marshes. The arable margins of the streams contain also most of the timbered land. Where the soil, however, admits cultivation, it is highly productive. Staples, sugar and cotton. Chief t. Thibadeauxville. Pop. 1820, 3,755, 1830, 5,503.

La Grange, p-v. Chester dist. S. C. by p-r. 77 ms. n. Columbia.

La Grange, p-v. and st. jus. Troup co. Géo. situated on a small creek of Chattahooche r. by p-r. 133 ms. very nearly due w. Milledgeville. n. lat. 33° 05′, long. 8° 10′ w. W. C.

La Grange, p-v. Franklin co. Ala. by p-r. 110 ms. n. n. w. Tuscaloosa.

La Grange, p-v. Fayette co. Ten. by p-r. 242 ms. s. w. by w. Nashville.

La Grange, p-v. Oldham co. Ky. marked in the p-o. list as the st. jus. in the list of offices, though in the list of cos. which precedes the offices, Westport is annexed to Oldham, as the co. st. Lafayetteville is not inserted on Tanner's map, but in the p-o. list stated at 43 ms. Frankfort.

La Grange, t. Loraine co. O. by p-r. 119 ms. n. Columbus.

La Grange, p-v. Cass co. Mich. 178 ms. a little s. of w. Detroit.

Lairdsville, p-v. Lycoming co. Pa. by p-r. 92 ms. n. Harrisburg.

Lake George, a beautiful body of water, about 33 ms. long, and nearly 2 wide, principally in the cos. of Warren and Washington,

N. Y. It discharges itself into lake Champlain at Ticonderoga. The outlet is little more than 3 ms. long, and is said to descend 157 ft. This lake is surrounded by high mountains, and is surpassed in the romantic by no lake scenery in the world. Water deep and clear, abounding with the finest of fish. The lake abounds with small isls. It is a fashionable place of resort in summer. In consequence of the extraordinary purity of the waters of this lake, the French formerly procured it for sacramental purposes; on which account they denominated it Lac Sacrament. Roger's rock is on the w. side of the lake, 2 ms. from its outlet. It rises out of the water at an angle of more than 45° to the height of 300 or 400 feet. It received its name from Major Rogers, who, to evade his Indian pursuers, ascended the rock on the land side with snow shoes; and throwing his pack down the precipice on the water side, turned his feet about on his snow shoes, and travelled back with them, they being heel foremost; thus leading the Indians to suppose that two persons had ascended the rock, and precipitated themselves into the lake. This lake was conspicuous during the French and revolutionary wars, forming the most convenient connexion between Canada and the Hudson; hence the establishment of the forts at the head of the lake, and also in part of fort Ticonderoga.

LAKE PLEASANT, p-t. Hamilton co. N. Y. 70 ms. N. W. Albany It is a wild waste of mtn. and swamp lands, abounding with small lakes; so poor in general that nobody inclines to settle in it. The lakes are very numerous and produce immense quantities of very fine large trout. Lake Pleasant is said to be 4 ms. long, with a fine sandy beach. Pop. 1830, 266.

LAKE PORT, p-v. Chicot co. Ark. by p-r. 200 ms., but by direct distance only 130 S. E. from Little Rock.

LAKE PROVIDENCE, and p-o. nrthest. part of the parish of Washitau, La. The p-o. is about 100 ms. N. Natchez. The lake in every respect similar to Fausse Riviere, lakes Concordia, St. Joseph, and Grand lake, is evidently the remains of an ancient bend of Miss. Lake Providence is entirely omitted on Tanner's U. S.

LAMBERTON, village, Burlington co. N. J. on the Del. 2 ms. below Trenton.

LAMBERTSVILLE, village, Hunterdon co. N. J. on the Del. 16 ms. above Trenton, connected by a bridge with New Hope.

LAMINGTON, (Indian, Alamatunk,) v. Sommerset co. N. J.

LAMOILLE, r. Vt. formed by the union of several streams, in Greensborough, and falls into lake Champlain at Colchester.

LAMORESVILLE, p-v. Carroll co. Ten. 118 ms. w. Nashville.

LAMPETER, tsp. and p-o. Lancaster co. Pa. The tsp. lies between Pequea and Mill crs., and the p-o. is about 6 ms. S. E. the city of Lancaster. Pop. of the township in 1820, 3,278.

LAMPREY, r. N. H. rises on the w. of Saddleback mtn. in Northwood, and meets the tide about 2 ms. above the Great bay at Durham.

LAMPTON'S, Clark co. Ky. by p-r. 53 ms. S. E. Frankfort.

LANCASTER, p-t. and st. jus. Coos co. N. H. on the S. E. bank of Conn. r. 110 ms. w. Portland, 130 N. Portsmouth, 95 almost due N. from Concord, and 75 above Dartmouth college. Watered by Conn. r., Israel's r., and several smaller streams; situated near lofty mountains. Pop. 1830, 1,187.

LANCASTER, p-t. Worcester co. Mass. the oldest town in the co., 35 ms. N. W. Boston, 15 N. E Worcester. Finely situated on both sides of Nashua r. Here are found slates of a good quality, andalusite, earthy marl, phosphorate of lime, and several other minerals. Map printing, and comb making, are here carried on, upon an extensive scale. Here is an academy. Pop. 1830, 2,014.

LANCASTER, co. Pa. bounded by the estrn. Conewago cr. separating it from Dauphin w., by Lebanon co. N. W., Berks N. E., Chester E., Coecil co. of Md. S., and Susquehannah r. separating it from York co. S. W. Lines drawn over this co. from its S. E. angle on Octarara cr. a little w. of N. to its extreme nthrn. angle on Lebanon and Berks or N. W. parallel to Susquehannah r. to its extreme wstrn. angle at the mouth of Conewago, are very nearly equal and 43 ms. in length, and the area being within a trifle of 1,000 sq. ms., the mean width will be about 23 ms. Extending in lat. 39° 42′ to 40° 19′, and in long. 0° 19′ to 1° 10′ E. W. C. The slope of this fine co. is towards the Susquehannah, and in a direction of S. S. W. The central and best parts are drained by the Conestoga, but the Conewago and Chiques in the wstrn. angle, and Pequea and Octarara crs. in the sthestrn., are creeks of some size, watering excellent land. Lancaster is one of the best cultivated cos. of Pa. and produces large quantities of live stock, salted meat, hides, leather, grain, flour, fruit, particularly apples, cider, whiskey, &c. Chief t. Lancaster. Pop. 1820, 68,336, and in 1830, 76,631.

LANCASTER, city of, p-t. and st. jus. Lancaster co. Pa. situated in the fine fertile and well cultivated valley of Conestoga, about one mile wstrd. of that stream, 62 ms. Phil. and 36 Harrisburg; N. lat. 40° 03′, and long. 0° 41′ E. W. C. This city was laid out in streets at right angles. The central part is well and closely built. The C. H. is the central edifice, standing at the intersection of the two main streets. The commerce and manufactures of the place are flourishing. Pop. 1810, 5,405, in 1820, 6,633, and in 1830, 7,704.

LANCASTER, co. Va. bounded N. W. by Richmond, N. E. and E. by Northumberland, S. E. by Chesapeake bay, and S. and S. W. by Rappahannoc r., separating it from Middlesex. Length 24, mean width 8, and area about 200 sq. ms Extending in lat. 37° 35′ to 37° 55′, and in long. 0° 22′ to 0° 40′ E. W. C. On the Potomac border it is deeply indented by small but convenient bays. Pop. 1820, 5,517, in 1830, 4,801.

LANCASTER, C. H. and p-o. Lancaster co. Va. situated near the middle of the co. by p-r. 85 ms. N. E. by E. Richmond, and 152 s. s. E. W. C. Lat. 37° 46', long. 0° 30' E. W. C.

LANCASTER, dist. of S. C. bounded by Lynches cr. separating it from Chesterfield s. E., by Kershaw s., by Catawba r. separating it from Fairfield, Chester and York w., and by Mecklenburg co. N. C. N. Length s. to N. parallel to Catawba r. 44 ms.; the sthrn. part is about 23 ms. wide, but to the nthrd. the breadth is reduced to less than 5 ms.; the mean width may be assumed at 12, and area 524 sq. ms. The general slope is sthrd. Chief t. Lancaster. Pop. 1820, 8,716, 1830, 10,361.

LANCASTER, p-t. and st. jus. Lancaster dist. S. C. situated near the centre of the dist. 38 ms. a little w. of N, Camden, and 63 ms. E. of N. Columbia. Lat. 34° 42' long. 3° 47' w. W. C.

LANCASTER, p-v. Smith co. Ten. by p-r. 58 ms. N. E. Nashville.

LANCASTER, p-v. and st. jus. Garrard co. Ky. 37 ms. s. Lexington, 10 a little s. of E. Danville, and 52, s. s. E. Frankfort. Lat. 37° 37', long. 7° 30'. w. W. C.

LANCASTER, p-v. and st. jus. Fairfield co. O. by p-r. 28 ms. s. E. Columbus, and 372 N. w. by w. ½ w. W. C. This is one of the finest interior villages of O., containing from 250 to 300 houses, by the census of 1820, 1,037, but by that of 1830, 1,530, and at present at least 1,600 inhabitants. According to Flint it contains the common co. buildings, an academy, several private schools, 4 churches, 12 stores, 2 printing offices, issuing a weekly English, and weekly German paper; a bank, and is united to the great central canal of O. by a side cut. Lat. 39° 45'.

LANCASTER, p-v. Jefferson co. Ind. by p-r. 76 ms. s. E. Indianopolis.

LANDAFF, t. Grafton co. N. H. 12 ms. E. Haverhill corner, 9 from Concord. Watered by Wild Amonoosuck and Great Amonoosuck rivers. Landaff mtn., Cobble hill, and Bald hill, are the principal elevations. Soil in some parts very fertile. Pop. 1830, 949.

LANDGROVE, p-t Bennington co. Vt. 33 ms. N. E. Bennington, 70 s. Montpelier. Watered by head branches of West r. 3 school districts. Pop. 1830, 385.

LANDISBURG, p-v. on the waters of Shermans cr. Perry co. Pa. 12 ms. N. N. w. Carlisle, and by p-r. 32 ms. N. w. by w. Harrisburg.

LANDSFORD, p-v. in the nthest. part of Chester district, S. C. 10 ms. N. E. Chesterville, and by p-r. 92 ms. N. Columbia.

LANE'S p-o. Mason co. Va. by p-r. 366 ms. wstrd. W. C.

LANESBOROUGH, t. Berkshire co. Mass. 5 ms. from Pittsfield, 14 N. Lenox, 135 from Boston. Part of the waters of this town descend to Long Island sound, and part to the Hudson. Soil, a fine loam. Contains great quantities of white marble. Pop. 1830, 1,192.

LANESBORO', p-v. Anson co. N. C. by p-r. 154 ms. s. w. Raleigh.

LANESBORO', p-v. Susquehannah co. Pa. by p-r. 187 ms. N. E. Harrisburg.

LANESVILLE, p-v. Susquehannah co. Pa. by p-r. 178 ms. N. N. E. Harrisburg.

LANESVILLE, p-o. Floyd co. Ky. by p-r. 154 ms. s. E. by E. Frankfort.

LANGDON, p-t. Sullivan co. N. H. 17 ms. from Keene, 50 w. Concord. Watered by a branch of Cold r. It was named in honor of Gov. Langdon. Pop. 1830, 666.

LANGHORN'S TAVERN, and p-o. Cumberland co. Va. by p-r. 60 ms. wstrd. Richmond.

LANGSBURY, p-v. Camden co. Geo. by p-r. 199 ms. s. s. E. Milledgeville.

LANSING, t. Tompkins co. N. Y., having Ludlowville p-o. on the E. side, near s. end of Cayuga lake, 7 ms. N. Ithaca, 160. w. Albany. Soil of the best quality. Watered by Salmon creek, which has falls. 20 school districts. Pop. 1830, 4,020.

LANSINGBURGH, p-t. Rensselaer co. N. Y. E. Hudson r., 4 ms. N. Troy, 10 N. Albany. There is an elegant bridge across the Hudson, between this place and Waterford, the first as we ascend from the ocean. Here is a very extensive nursery. Epsom salts, and stone for building are found here. The village of Lansingburgh is 2 ms. long, and half a mile wide, being regularly laid out in blocks, or oblong squares 400 by 260 ft. It has a flourishing academy. A dam 11 ft. long, and 9 ft. high, is built across the r. below the village, by which the water has been made sufficiently deep for sloops throughout the season. Vessels ascend through a sloop lock 30 feet wide, and 114 long. Cost of the dam and lock, $92,270. Lansingburgh employs in trade about 12 sloops. Population 1830, 2,663.

LAPEER, co. Mich. bounded s. by Oakland, s. w. Shiawassee, N. w. Saginaw, N. Sanilac, E. St. Clair co., and s. E. by Macomb. Lat. 43°, long. 7° 15' w., slope N. w., and drained by Flint r. branch of Saginaw. It lies N. N. w. about 60 ms. from Detroit.

LARKIN'S FORK, and p-o. Jackson co. Al. about 170 ms. N. E. Tuscaloosa.

LA SALLE, county of Illinois, along both sides of Illinois r. from the junction of the Kankakee and Plane rivers down to below the mouth of Vermillion r. It would be useless to offer a delineation of this county, as it contains, as laid down by Tanner, 110 ms. from s. to N. with a breadth of 50, of course must be rapidly subdivided. That part along the Illinois near the rapids, will probably retain the the title. *(See Ottawa, Fox river of Illinois, &c.)*

LAUDERDALE, northwestern co. of Alabama, bounded on the N. by the cos. of Hardin, Wayne and Lawrence in Tenn., E. by Limestone, Ala., s. by Ten. r. separating it from Lawrence and Franklin, Ala., s.w. by Ten. r. separating it from the Chickasaw territory, in Ala., and w. again by Ten. r. separating it from the Chickasaw territory, in the state of

Miss. Length E. to W. 56 ms., mean width 12, and area 672 square ms. Extending in lat. 34° 43′ to 35°, and in long. 10° 16′ to 11° 15′ W. W. C. Elk river enters Ten. about the middle of the Muscle shoals, and in the extreme southeastern angle of Lauderdale co. The general declivity is southward towards Ten. down which flow Blackwater, Shoal, Cypress, Second and other creeks. The surface is broken and soil excellent. Chief town, Florence. Pop. 1820, 4,963, and in 1830, 11,781.

LAUGHERY, p-v. Ripley co. Ind. situated on a creek of the same name in the S. E. part of the co. by p-r. 81 ms. S. E. Indianopolis.

LAUGHLINTOWN, p-v. Westmoreland co. Pa. 26 ms. E. Greensburg, and 43 N. W. by W. Bedford.

LAUGHRIDGE, p-v. Gwinnett co. Geo. by p-r. 99 ms. N. W. Milledgeville.

LAUREL, p-v. southern part of Sussex co. Del. 58 ms. southward Dover.

LAUREL FURNACE, and p-o. Dickson co. Ten. by p-r. 44 ms. westward Nashville.

LAUREL HILL, or Laurel mountains, a local name given to several of the western chains of the Appalachian system, an absurdity productive of no small share of confusion. The chain in Pa. extending from the Conemaugh to Youghioghany r., and which separates Cambria co. from Westmoreland, and Somerset from Westmoreland and Fayette, is there called "*The Laurel Hill*," whilst another chain westerly and with an intervening valley of 10 ms. wide is called "*The Chestnut Ridge*." Both chains are continued out of Pa. into Va. southwestward of the Youghioghany, but the names are reversed, and the Chestnut ridge of Pa. is the Laurel ridge of Va. Such is the wretched delineation of the Appalachian system on all our maps, that no adequate idea of the respective chains can in many instances be obtained by their assistance. The two chains mentioned in this article, though not so represented, preserve their idenity, similar to the Blue Ridge from the state of N. Y. into Ala.

LAUREL HILL, p-o. Somerset co. Pa. by p-r. 162 ms. N. W. W. C.

LAUREL HILL, p-o. Lunenburg co. Va. by p-r. 112 ms. S. W. Richmond.

LAUREL HILL, p-v. Richmond co. N. C. by p-r. 97 ms. S. W. Raleigh.

LAUREL HILL, p-v. W. Feliciana parish, La. 20 ms. St. Francisville.

LAUREL SPRING, p-v. Fluvanna co. Va. by p-r. 61 ms. N. W. by W. Richmond.

LAURENCE, co. of Ala. bounded E. by Morgan, S. by Walker, W. by Franklin, N. W. by Ten. r. separating it from Lauderdale, and N. E. from Ten. r. separating it from Limestone. Length from N. to S. along its water boundary 38 ms., the breadth 24, mean length 34, and area 816 square ms. Extending in lat. from 34° 18′ to 34° 48′, and in long. 10′ 13′ to 10° 36′ W. W. C. The southern border extends into the higher rim of the valley of Mulberry river, branch of Black Warrior, and slopes southwardly, but the central and northern sections, comprising full two thirds of the whole surface, declines northwardly towards Ten. r. That part of the latter stream which forms the northern boundary of Lawrence, is known as the Muscle shoals. Chief town, Moulton. Pop. 1830, 14,984.

LAURENS, p-t. Otsego co. N. Y. 12 ms. S. W. Cooperstown, 78 W. Albany. Surface broken by hills of a moderate height. Soil a rich loam. Timber, pine, oak, chestnut, walnut, &c. Contains a mineral spring; 14 schools, kept 8 months in 12. Pop. 1830, 2,231.

LAURENS, district of S. C. bounded by Newbury S. E., the Saluda river separating it from Abbeville S. W., Greenville N. W., and the Ennoree river separating it from Spartanburg N., and Union N. E. Length 33, mean width 28, and area 924 square miles. Extending in lat. 34° 12′ to 34° 45′, and in long. 4° 37′ to 5° 18′ W. W. C. The slope of this co. is very nearly southeastward with the course of Ennoree and Saluda rivers. Chief town, Laurensville. Pop. 1820, 17,682, 1830, 20,263.

LAURENS, co. of Geo. bounded by Montgomery E. and S. E., Pulaski S. W. and W., Wilkinson N. W., Washington N., and Emanuel N. E. The greatest length from the southern to northern angles 40 ms., and as the area is about 800 square ms. the mean width will be 20 ms. In lat. it extends 32° 12′ to 32° 45′ and in long. 5° 40′ to 6° 18′ W. W. C. Oconee river traverses this co. in a S. S. E. direction, dividing it into two unequal sections, two thirds to the right and one third to the left of the river. Chief town, Dublin. Pop. 1820, 5,436, 1830, 5,589.

LAURENSVILLE, p-v. and st. jus. Laurens district, S. C. situated near the centre of the district, about 75 ms. almost due N. Augusta in Geo. and by p-r. 81 ms. N. W. by W. Columbia, lat. 34° 31′, long. 5° W. W. C.

LAUSANNE, tsp. and p-o. Northampton co. Pa. by p-r. 13 ms. Mauch Chunck and 132 N. E. Harrisburg. In this township are situated the vast strata of anthracite coal near Mauch Chunk.

LAWRENCE, town, Hunterdon co. N. J. lies N. E. of Taunton, and extends to Somerset and Middlesex. Pop. 1830, 1,433.

LAWRENCE, co. of Ten. bounded by Wayne W., Wickman N., Giles E., and Lauderdale co. in Ala. S. It is a square of 28 ms., area 784 square ms. Extending in lat. 35° to 35° 24′, and in long. 10° 17′ to 10° 45′. This county occupies a table land; from the southern and larger section, the waters flow southward, over Lauderdale co. into Ten. river, whilst the northern discharges to the N. W., the sources of the Buffalo branch of Duck river. Chief town, Lawrenceburg. Pop. 1820, 3,271. and in 1830, 5,411.

LAWRENCE, co. of Ky. bounded by Floyd S., Licking r. separating it from Morgan S. W., and Bath W., Fleming N. W., Greenup N., and Sandy r. separating it from Cabell co. Va. E. Length W. to E. 60, mean width 26, and area 1,560 square ms. Extending in lat. 37° 53′

to 38° 22'. This co. occupies a table land, from which the creeks flow S. W. into Licking, N. towards Ohio, and N. E. into Sandy r. Chief town, Louisa. Pop. 1830, 3,900.

LAWRENCE, extreme southern county of Ohio, bounded by Sciota co. N. W., Jackson N., Gallia N. E., S. E. by O. r. separating it from Cabell co. Va., and S. W. by O. r. separating it from Greenup co. Ky. Length 30, mean breadth 13, and area 390 square ms. It lies directly opposite the mouth of Big Sandy r. slopes southward, and in that direction is drained by Symme's creek, and some smaller streams. Chief town, Burlington. Pop. 1820, 3,499, and in 1830, 5,367.

LAWRENCE, co. Ind. bounded S. by Orange, Martin S. W., Greene N. W., Monroe N., Jackson E., and Washington S. E. It is about 22 ms. square, area 464 square ms. Lat. 39° N., long. 9° 40' w. W. C. It is traversed from east to west by the main volume of the South fork of White r. Chief town, Bedford. Pop. 1830, 9,234.

LAWRENCE, co. of Miss. bounded W. by Franklin, N. W. by Copiah, N. by Simpson, E. by Covington, S. E. by Marion, and S. W. by Pike. Length E. to W. 42, mean width 20, and area 840 square ms. Extending in lat. 31° 24' to 31° 47' N., and in long. 12° 58' to 13° 40' w. W. C. This co. is traversed in a S. S. E. direction by Pearl river, the western part is, however, a table land, from which flow, northwards, the head waters of Bayou Pierre, westward those of the Homochitto, and southward, those of the Bogue Chitto. The surface generally a barren soil, covered with pine timbers. Staple, cotton. Chief town, Monticello. Pop. 1820, 4,916, and in 1830, 5,293.

LAWRENCE, county of Illinois, bounded by Wabash co. S., Edwards S. W., Clay W., Jasper N. W., Crawford N., and Wabash r. separating it from Knox co. Ind. E. It is 20 by 25 ms., area 500 square ms. Lat. 38° 45,' long. W. C. 11° W., slope S. and traversed by Embarras r. Chief town, Lawrenceville. Pop. 1830, 3,668.

LAWRENCE, county of Ark. as laid down on Tanner's United States, is bounded S. by St. Francis co., S. W. by Independence, W. by Izard, N. by Wayne, co. of Miss., E. by St. Francis r. separating it from New Madrid co. of Miss., and S. E. by Crittenden co. in Ark. Length of St. Francis river to the eastern boundary of Izard co. 86 ms., the greatest breadth is near 70, but the mean breadth about 50, area 4,300 square ms. Extending in lat. 35° 30' to 36° 30', and in long. 13° 10' to 14° 40' w. W. C. The large tract included under the name of this co. comprises a very diversified surface. The estrn. section near the St. Francis is flat, and in great part liable to annual submersion. Approaching the centre the surface rises into hill and dale, presenting a fine country and congeries of confluent rs. The Black and Current rs. flowing down in fine copious navigable streams from southern Miss., here unite and at Davidsonville, the seat of justice, receives from the northwestward Eleven Points and Spring rs. Chief town, Davidsonville. Pop. 1820, 5,602, and in 1830, 2,806.

LAWRENCEBURG, p-v. on Alleghany river and in the N. W. angle of Armstrong co. Pa. about 50 ms. N. N. E. Pittsburg and by p-r. 195 ms. N. W. by W. Harrisburg.

LAWRENCEBURG, p-v. and st. jus. Lawrence co. Ten. situated in the forks of Shoal creek, 28 ms. N. N. E. Florence, in Ala. and by p-r. 88 ms. S. S. W. Nashville, lat. 35° 08', long. 10° 35' w. W. C.

LAWRENCEBURG HOTEL, and p-o. in the village of Lawrenceburg, Anderson co. Ky. 10 ms. S. Frankfort.

LAWRENCEBURGH, p-v. and st. jus. Dearborn co. Ind. situated on Ohio r. immediately below the mouth of Great Miami, and by p-r. 98 ms. S. E. Indianopolis, and 23 below Cincinnati. Lat. 39° 04'.

LAWRENCEVILLE, village, Hunterdon co. N. J. 6 ms. N. E. Trenton.

LAWRENCEVILLE, p-v. Tioga co. Pa. by p-r. 151 ms. northrd. Harrisburg.

LAWRENCEVILLE, village, on the left bank of Alleghany r. Alleghany co. Pa. This place is only two ms. above the nthrn. Liberties of the city of Pittsburg. It is the seat of an arsenal and U. S. military depot.

LAWRENCEVILLE, p-v. Madison co. Ohio by p-r. 23 ms. wstrd. Columbus.

LAWRENCEVILLE, p-v. and st. jus. Lawrence co. Il. situated on Embarras r. by p-r. 84 ms. a little S. of E. Vandalia, and 10 ms. W. Vincennes in Ind. N. lat. 38° 45', long. W. C. 10° 45' W.

LAWRENCEVILLE, p-v. and st. jus. Brunswick co. Va. situated on a branch of Meherin r. by p-r. 72 ms. a little W. of S. Richmond. Lat. 36° 48', long. 0° 50' w. W C.

LAWRENCEVILLE, p-v. and st. jus. Montgomery co. N. C., situated on the right bank of Yadkin r. by p-r. 109 ms. S. W. by W. Raleigh. Lat. 35° 25', long. 3° 11' w. W. C.

LAWRENCEVILLE, p-v. and st. jus. Gwinnett co. Geo. situated near the extreme source of Ockmulgee r. by p-r. 87 ms. N. W. Milledgeville. Lat. 33° 58', long. 7° 05' w. W. C.

LAWSON'S, p-o. Logan co. Va. by p-r. 239 ms. wstrd. Richmond.

LAWSVILLE, p-v. in the nthrn. part of Susquehannah co. Pa. about 20 ms. from Montrose, and by p-r. 179 N. N. E. Harrisburg.

LEACOCK, p-v. and tsp. of Lancaster co. Pa. The p-o. is 7 ms. estrd. Lancaster. Pop. of the tsp. 1820, 2,882, 1830, 3,315.

LEADING CR., and p-o. nthrn. part of Lewis co. Va. by p-r. 233 ms. almost due W. W. C.

LEADSVILLE, p-v. Randolph co. Va. by p-r. 218 ms. W. W. C.

LEAF r. wstrn. branch of Pascagoula r. This stream, frequently called from its principal constituent branch, Chickisawhay r. rises in the Choctaw country, state of Miss., and flowing thence in a sthrn. direction over Covington and Jones cos. gradually bends to S. E., unite in Perry, from which the united waters enter Greene and fall into Pascagoula, a short distance below N. lat. 31°. The general feature of the valley of Leaf r. is that of sterile pine woodland.

Leaf r. p-o. (*See Greesboro', Greene co. Miss.*)

Leakesville, p-o. on the right bank of Dan r. northern part of Rockingham county, N. C. 5 ms. n. Wentworth, and by p-r. 105 n. n. w. Raleigh.

Leakesville, p-o. Laurens dist. S. C. by p-r. 92 ms. nthwstrd. Columbia.

Leakesville, p-v. Newton co. Geo. by p-r. 50 ms. n. w. Milledgeville.

Leakesville, p-v. Green co. Miss. by p-r. 152 ms. s. e. Jackson.

Leasburg, p-v. Caswell co. N. C. by p-r. 85 ms. n. w. Raleigh.

Leavenworth, p-o. Crawford co. 126 ms. sthrd. Indianopolis.

Lebanon, p-t. York co. Me. on the Piscataqua, 28 ms. n. w. York. Pop. 1830, 2,391.

Lebanon, p-t. Grafton co. N. H. e. Conn. r. 4 ms. below Dartmouth college. Watered by Conn. and Mascomy rivers. Contains many valuable mill seats. Timbered with white pine, oak, sugar maple, birch, beech, &c. There are falls in the Conn. in this town, which are locked and canalled. Lyman's bridge, across the Conn. connects this town with Hartford, Vt. The principal village is situated on a plain near the central part, at the head of the falls of Mascomy r. In this town is a medicinal spring. Here are also a lead mine, and a vein of iron ore. Pop. 1830, 1,868.

Lebanon, p-t. New London co. Conn. 30 ms. s. e. Hartford. Moderately hilly. Soil a rich, deep, unctuous mould, very fertile, and peculiarly adapted to grass. Timbered principally with chestnut, walnut, and oak. Well watered with brooks and rivulets, some of which afford mill sites. 17 school dists. Pop. 1830, 2,554.

Lebanon, p-t. Madison co. N. Y. 35 ms. s. w. Utica. Surface hilly. The Chenango r. runs through the e. part. Soil light. Timbered with maple, beech, birch, ash, &c. 13 schools, kept 8 months in 12. Pop. 1830, 2,249.

Lebanon, village, Columbia co. N. Y. a beautiful village, famous for its springs, the water of which issues in great abundance from the side of a high hill, and being remarkably clear, soft and tepid, is much used for bathing. The houses of accommodation are excellent, and it is a place of great resort in the summer months; 27 ms. e. Albany; 31 n. e. Hudson.

Lebanon, t. Hunterdon co. N. J. bounded s. w. by Bethlehem and Kingwood, n. w. by the Musconetcunk, n. e. by Morris co. and Jewksbury, s. e. by Readingtown. Pop. 1830, 3,436.

Lebanon, co. Pa. bounded by Dauphin w. and n. w., Schuylkill n., Berks n. e. and Lancaster s. e. The greatest length of Dauphin, is a diagonal 29 ms., the estrn. to the wstrn. angle; mean width 12, and area 348 sq. ms. Extending in lat. 40° 11′ to 40° 32′ and in long. 0° 20′ to 0° 51′ e. W. C. The estrn. angle of this co. gives source to the Tulpehocken, and to the nthwstrn. branches of Conestoga; from the extreme sthrn. part flows the estrn. Conewago, but more than three fifths are included in the valley of the Swatara, and slopes s. s. wstrd. The whole co. is included in the fine valley of Kittatinny, and similar to other parts of this physical region, the side next the Blue Ridge is based on limestone, and that towards the Kittatinny on clay slate. Soil generally excellent. The surface tho' bounded by the Kittatinny n. w., and Blue Ridge s. e. is not even very hilly, no part is however level. To the many natural advantages of Lebanon, may be added the artificial r., the Union canal. This work, pursuing the valley of the Tulpehocken, that of the Quitapahilla, into Swatara, and down the latter into Susquehannah r., passes over the central part and divides Lebanon into two not very unequal sections. Staples are every agricultural product of that part of the middle states of the U. S. included in the same zone of lat., with immense quantities of cast and hammered iron. Chief town, Lebanon. Pop. 1820, 16,988, 1830, 20,557.

Lebanon, p-t. boro', and st. jus. Lebanon co. Pa., situated near the centre of the co. on the Union canal, and on one of the head branches of the Quitapahilla cr., 24 ms. a little n. of e. Harrisburg, 133 n. n. e. W. C., and 77 ms. n. w. by w. Phil. Lat. 40° 20′, long. 0° 35′ e. W. C. This is a very neat, well built, and flourishing town; situated on the limestone part of the Kittatinny valley, with a well cultivated and fertile vicinity. Pop. 1820, 1,437, 1830, 3,555.

Lebanon, p-v. and st. jus. Russell co. Va. situated on a branch of Clinch r. about 130 ms. n. e. by e. Knoxville in Ten., and by p-r. 430 ms. s. w. by w. W. C. Lat. 36° 53′ and long. 5° 03′ w. W. C.

Lebanon, p-v. Washington co. Geo. 21 ms. from Milledgeville.

Lebanon, p-v. and st. jus. Wilson co. Ten. situated on a creek of Cumberland r. 23 ms. a little n. of e. Nashville, and 24 a little s. of w. Carthage; n. lat. 36° 12′, and long. 9° 21 w. W. C.

Lebanon, p-v. and st. jus. Warren co. O. by p-r. 83 ms. s. w. Columbus, 28 s. Dayton, and 31 n. e. Cincinnati. It contains the common co. buildings, a printing office and bank. Pop. 1830, 1,165. Lat. 39° 25′, long. W. C. 7° 12′ w.

Lebanon, p-v. in the nrthestrn. part of St. Clair co. Il. It is situated on Silver cr. 8 ms. n. e. Belleville, and by p-r. 59 miles s. w. Vandalia.

Lebanon, p-v. on Chaplin's fork of Salt r. Washington co. Ky. by p-r. 56 ms. s. s. w. Frankfort.

Ledyard, t. Cayuga co. N. Y. on Cayuga lake, 19 miles s. w. Auburn. Pop. 1830, 2,427.

Lee, t. Hancock co. Me. 25 ms. n. w. Castine.

Lee, p-t. Strafford co. N. H. 13 ms. N. W. Portsmouth. Watered by Lamprey, Little, North, and Oyster rivers. Population 1830, 1,009.

Lee, p-t. Berkshire co. Mass. 5 ms. S. E. Lenox, 120 W. Boston. Finely situated on both sides of the Housatonic r., which is here a large and powerful stream, and affords great facilities for manufacturing purposes. Here are 6 paper mills which annually consume 500 tons of rags. Here too are marble, limestone, and iron ore in abundance. Pop. 1830, 1,825.

Lee, p-t. Oneida co. N. Y. 8 ms. N. Rome. Well watered and supplied with mill seats. Land rich. Most excellent for flax. 10 school houses. Pop. 1830, 2,514.

Lee, extreme sthwstrn. co. of Va. bounded N. E. by Russell, E. by Scott, S. and S. W. by Claiborne co. Ten., and by Cumberland mtns. which separates it from Harlan co. Ky. N. W., and Pike co. Ky. N., length along Cumberland mtn. 60 ms., mean width about 10 ms., and area 600 sq. ms. Extending in lat. 36° 30′ to 37° 06′, and in long. 5° 35′ to 6° 30′ W. W. C. This co. occupies the higher part of Powell's valley, extending from Cumberland to Powell's mtn. The extreme sources of Powell's r., are in Russell, but they unite and form a river in Lee co., which, flowing sthwstrd. divide it into two narrow but steep inclined plains. Chief town, Jonesville. Pop. 1820, 4,256, 1830, 6,461.

Lee, co. Geo. bounded by Baker S., Randolph W., Marion N., and Flint r. separating it from Dooley E. Length 43 ms., mean width 30, area 1,290 sq. ms. Extending in lat. 31° 42′ to 32° 18′, and in long. 7° 8′ to 7° 42′ W. W. C. The slope of this co. is to the S. E. towards Flint r. Chief town, Pindertown. Pop. 1830, 1,680.

Lee, p-v. Athens co. O. by p-r. 82 ms. S. E. Columbus.

Leech's Stream, rises in Averill, Vt. and falls into Conn. r., where it is about 2 rods wide.

Leechburg, p-v. Armstrong co. Pa. 227 ms. N. W. W. C.

Leeds, p-t. Kennebec co. Me. on the Androscoggin r. 20 ms. S. W. Augusta. Pop. 1830, 1,685.

Leeds, village, Gloucester co. N. J. on the Atlantic S. of Great bay, at the mouth of Mulleins r.

Leeds, p-v. Westmoreland co. Va. by p-r. 89 ms. S. S. E. W. C.

Leedsville, p-v. Randolph co. Va. situated on Tygarts valley r. at the passage of that stream through Laurel mtns., 10 ms. N. N. E. Beverly, by p-r. 200 ms. W. W. C.

Leesboro', p-o. Montgomery co. Md. by p-r. 31 ms. W. C.

Leesburg, p-v. and st. jus. Loudon co. Va. by p-r. 35 ms. N. W. W. C., and 158 N. Richmond. Lat. 39° 07′ long. 0° 33′ W. W. C. It is a well built and neat village, situated near a minor ridge of mtns. The environs are waving, well cultivated and delightfully variegated by hill and dale. Pop. 1830, about 1,500.

Leesburg, p-v. Washington co. Ten. on the r. between Greenville and Jonesboro', 18 ms. N. E. by E. the former, and 9 S. W. the latter, and by p-r. 250 ms. E. Nashville.

Leesburg, p-v. in the sthrn. part of Harrison co. Ky. 10 ms. S. W. Cynthiana, and 30 N. E. by E. Frankfort.

Leesburg, p-v. Lancaster co. Pa. by p-r. 44 ms. estrd. Harrisburg.

Leesburg, p-v. Highland co. O. by p-r. 62 ms. S. W. Columbus.

Leesville, p-v. and manufacturing village Mid. Haddam, Middlesex co. Conn. 15 ms. S. E. Middletown.

Leesville, p-v. Campbell co. Va. by p-r. 119 ms. S. W. by W. Richmond.

Leesville, p-v. Robeson co. N. C., by p-r. 101 ms. S. S. W. Raleigh.

Leesville, p-v. in the wstrn. part of Lexington dist. S. C., 31 ms. a little S. of W. Columbia.

Leesville, p-v. Tuscarawas co. O. 123 ms. N. E. by E. Columbus.

Leesville, p-v. Lawrence co. Ind. by p-r. 76 ms. S. S. W. Indianopolis.

Leetown, p-v. in the wstrn. part of Jefferson co. Va. 30 ms. W. Harper's ferry, and by p-r. 84 ms. N. W. by W. W. C.

Lee Valley, p-v. Hawkins co. Ten. by p-r. 277 ms. estrd. Nashville.

Legrand's store and p-o. Anson co. N. C. 140 ms. S. W. by W. Raleigh.

Legro, p-v. Randolph co. Ind. by p-r. 87 ms. N. E. by E. Indianopolis.

Lehigh, r. of Pa. a branch of Del. having its most remote sources in the sthrn. part of Wayne, and the sthestrn. of Luzerne near Wilkesbarre. The general course of its higher constituents, is sthwstrd. to their junction below Stoddartsville, and between Luzerne and Northampton cos. It thence flows 10 ms. by a general S. W. course, but curving to the wstrd. enters Northampton, and turning to nearly a sthrn. course 15 ms. to Lehighton, having in the latter part of its course received numerous mtn. creeks from both sides, and passed the now noted coal depot, Mauch Chunk. Below Lehighton the stream inflects to S. E. 10 ms. to its passage thro' the Kittatinny chains by "The Lehigh Water Gap." Inflecting below "The Gap", to S. S. E. 20 ms. to the reception of Little Lehigh, and N. W. side of the Blue Ridge. Turned by the latter mtns. to N. W. the now beautiful Lehigh flows down its base 15 ms. to Easton, where it is lost in the Del. The Lehigh drains a small sthrn. section of Wayne; the sthest. part of Luzerne; the wstrn. angle of Pike, more than two thirds of Northampton; small sections of Schuylkill and Berks, and seven eighths at least of Lehigh. From the intricacy of its course through numerous ridges of mtns. the real length of this r. is difficult to determine. The valley from S. E. to N. W., and from the Blue Ridge to the mtns. E. of Wilkesbarre, is about 50 ms., the mean breadth is at least 25, and area 1,250 square miles.

To the truly romantic and ever varying landscapes on this stream, it has now gained great celebrity from having become part of the channel of intercommunication from the great coal strata near Mauch Chunk; and the Atlantic tide water. In a distance following the stream 47 ms. from Easton to Mauch Chunk, the rise is 364 ft. This relative elevation is obviated by 57 locks, and 8 dams, as the chain is formed by alternate canals; and slack water ponds. The canals are 60 ft. at top, and 45 at bottom, with 5 ft. depth of water. Locks 22 ft. by 10½ ft. From the termination of the canal chains at Mauch Chunk, a rail road of 9 ms. reaches the great mass of anthracite coal, lying upwards of 1,000 feet above the Lehigh at the village.

The following relative heights will exhibit the rise of the Lehigh valley above the tide water in Del. r. Easton—level of the water at the confluence of Del. and Lehigh rs. above tide water, 170 ft. Ascent from Easton to Mauch Chunk, 364 ft. Ascent from Mauch Chunk to Stoddartsville, 850 ft. Total 1,384.

The Lehigh has interlocking sources to the N. with the Lackawannoc, and sthrn. confluents of Lackawaxen, to the sthrd. it embosoms the sources of Broadheads cr. To the wstrd. the sources of Bear creek, branch of Lehigh, rise within 10 ms. from the Susquehannah at Wilkesbarre, and finally curving from the sources of Nesquehoning, to those of Saucon cr., the confluents of Lehigh are embosomed by those of the Schuylkill.

Lehigh, co. Pa. bounded by Schuylkill co. W., by Northampton N. W., N. and N. E., and by Bucks S. E. Length 28, mean width 13, and area 364 sq. ms. Extending in lat. 40° 25′ to 40° 46′, and in long. 1° 11′ to 1° 43′ E. W. C. The sthestrn. part is a narrow valley between the Blue Ridge, and South mtn., containing the two tsps. of Upper Milford, and Upper Saucon. This truly beautiful vale, or that part contained in Lehigh, is about 10 ms. by 4, or 40 sq. ms. The residue, or the 8-9th of the whole co., lies in the Kittatinny valley, and very nearly subdivided into equal portions by the limestone and slate formations. The soil of the valley is more productive, and the surface less broken on the former rock, but the co. taken as a whole, is amongst the most productive in Pa., in grain, fruit and pasturage. The general elevation above tide water in Del. from about 350 to 500 ft. The highest water level of Lehigh r. at the Water gap, 375 ft. Chief t. Allenton or Northampton. Pop. 1820, 18,895, 1830, 22,256.

Lehighton, or Lehightown, p-v. of Northampton co. Pa. on the road from Bethlehem to Mauch Chunck, 36 ms. N. W. the former, and 3 lower down the Lehigh than the latter. It is a small village situated on a fine acclivity rising from the Lehigh, and about one fourth of a mile on the right of that stream. A little distance below the present village, stood the old Moravian town of Gnadenhutten, on the Lehigh above the mouth of Mahoning cr. This establishment was made about 1742, and here in July, 1752, a treaty of amity was held between the Moravian brethren and Shawnese Indians. On the 24th of Nov. 1755, the settlement was surprised, and the whites mostly massacred by a party of French Indians. A large gravestone, with a very pathetic inscription recording the fact, and the names of the sufferers, was lying on the ground of the old burial place, to the S. E. of Lehighton, when the author visited the place in 1821, and 1823.

Lehigh Water Gap, tavern and p-o. on left bank of the Lehigh, at the mouth of Aquanshicola cr., and immediately above the passage of the Lehigh through the Kittatinny mtn. 20 ms. N. W. Bethlehem.

Leicester, p-t. Addison co. Vt. 9 ms. S. Middlebury, 36 S. W. Montpelier. Principal streams, Otter creek, and Leicester r. Soil a rich sandy loam, interspersed with some flats of clay. 5 school districts. Pop. 1830, 638.

Leicester, p-t. Worcester co. Mass. 6 ms. S. W. Worcester, 46 W. Boston. An uneven town. It occupies an elevated position, its waters running both to the Conn. and Blackstone rivers. Soil deep and strong; clay predominates. Here are manufactures of various kinds, especially that of cards, of which $200,000 worth are manufactured annually. Here is a large and flourishing academy. A society of Jews once resided in this town, who came from Newport, R. I. to avoid the dangers of the war. None are now remaining. Pop. 1830, 1,782.

Leicester, t. Livingston co. N. Y. W. Genesee r., 5 ms. W. Geneseo. Good land, supplied with mill seats. 11 schools, kept 8 months in 12. Pop. 1830, 2,042.

Leighton, p-v. Lawrence co. Ala. by p-r. 104 ms. nthrd. Tuscaloosa.

Leipersville, p-o. Del. co. Pa. by p-r. 97 ms. N. E. W. C.

Leipersville, p-o. Crawford co. O. by p-r. 75 ms. nthrd. Columbus.

Leitersburg, p-v. in the extreme wstrn. part of Washington co. Md. by p-r. 98 ms. N. W. W. C.

Lemay's Cross Roads, and p-o. Granville co. N. C. 26 ms. N. Raleigh.

Lemington, p-t. Essex co. Vt. 64 ms. N. E. Montpelier, W. Conn. r. On a brook in this t. is a cascade of 50 ft. The Monadnock mtn. of Vt. lies in the N. E. corner of this t. 2 school districts. Pop. 1830, 183.

Lempster, p-t. Sullivan co. N. H. 40 ms. W. Concord, 90 from Boston. Surface in general uneven, W. part mountainous. Soil moist, better for grass than grain. Well watered with small streams. Has some water privileges. Pop. 1830, 999.

Lenoir, co. N. C. bounded S. W. and W. by Duplin, N. W. by Wayne, N. by Greene, N. E. by Pitt, E. by Craven, and S. E. by Jones. Length 26, mean width 15, and area 390 sq. ms. Extending in lat. 35° to 35° 23′, and in long. 0° 33′ to 0° 50′ w. W. C. The slight declination of this co. is estrd., the nthrn. part traversed in that direction by Neuse r., whilst the sthrn. gives source to the small r. Trent, flowing also to the estrd. into the Neuse. Chief town, Kingston. Pop. 1820, 6,800, 1830, 7,723.

LENOIRS p-o. Roane co. Ten. by p-r. 143 ms. eastward Nashville.

LENOX, p-t. and shire town of Berkshire co. Mass. 6 ms. s. Pittsfield, 125 w. of Boston. Contains an academy. Surrounded by romantic mountain scenery. Soil excellent. Contains iron ore in great abundance, and has a furnace for casting hollow iron ware. This vicinity abounds with primitive white limestone, and white marble is so plentiful as to be used for door steps and foundations. Pop. 1830, 1,359.

LENOX, p-t. Madison co. N. Y. about 25 ms. w. Utica on Oneida lake. Soil productive. The Erie canal runs through this town. Near the centre, 10 rods from the canal, is a salt spring. Limestone, iron ore, water lime, or water cement, and gypsum are found in abundance; 15 schools, kept 10 months in 12. Pop. 1830, 5,039.

LENOX, p-v. Susquehannah co. Pa.

LENOX, p-v. Ashtabula co. Ohio, by p-r. 190 ms. N. E. Columbus.

LENOX CASTLE, and p-o. Rockingham co. N. C. by p-r. 105 ms. N. E. Raleigh.

LEOMINSTER, p-t. Worcester co. Mass. 46 ms. w. Boston, 19 N. Worcester. Watered and supplied with mill privileges by a principal branch of Nashua river. Pretty level, soil excellent. Contains good stone for building, and good clay for bricks. Combs to the value of $100,000 per annum, are manufactured here. Contains numerous mills and manufactories. Pop. 1830, 1,861.

LEONARDSTOWN, p-v. and st. jus. St. Mary's co. Md. situated on a small tide water creek of Potomac, called Britton's river, 25 ms. S. E. Port Tobacco, and by p-r. 62 ms. S. S. E. W. C. and 72 a very little w. of s. Anapolis, lat. 38° 18′, and long. 0° 24′ E. W. C.

LE RAYSVILLE, p-v. Susquehannah county, Pennsylvania.

LEROY, p-t. Genessee co. N. Y. 10 ms. E. Batavia, 38 w. Canandaigua, 17 s. Erie canal. Good land, watered by Allan's creek; 13 schools, kept 8 months in 12. Population 1830, 3,902.

LEROY, p-v. Medina co. Ohio, by p-r. 109 ms. N. E. Columbus.

LETART FALLS, and p-o. Meigs co. Ohio. The p-o. is by p-r. 109 ms. S. E. Columbus. The falls of Letart are merely rapids, entirely covered at a moderate rise of the Ohio r. the navigation of which except at very low water they but little obstruct.

LEVANT, p-t. Penobscot co. Me. 10 ms. N. w. Bangor. Pop. 1830, 717.

LEVERETT, p-t. Franklin co. Mass. 10 ms. S. E. Greenfield, 85 w. Boston. Pop. 1830, 939.

LEWIS CREEK, Vt. rises near the N. line of Bristol, and falls into lake Champlain in Ferrisburgh, a short distance N. of the mouth of Little Otter creek. The mill privileges on this stream are numerous, and many of them excellent.

LEWIS, co. N. Y. bounded N. E. by St. Lawrence co., E. by Herkimer co., southerly by Oneida co., westerly by Oswego and Jefferson cos. Greatest length N. and S. 54 miles, greatest width 33, containing about 1,008 sq. ms. or 645,120 acres. Watered centrally by Black river, E. by Beaver and Moose creeks, and several other small streams, w. by Deer creek and some other small streams, N. by some branches of Indian and Oswegatchie rs. and Fish creek and Salmon r., w. part a good tract of country. Chief town, Martinsburgh. Pop. 1830, 15,239.

LEWIS, p-t. Essex county, New York, 4 miles north Elizabethtown, 130 north of Albany. Broken by high mountains. Timbered with maple, beech, some oak and walnut, ash, elm, &c. Apples grow abundantly. Well watered, and tolerably supplied with mill sites. Mount Discovery is in this town, from the summit of which, the view is sublimely grand. It is supposed to be 2,000 feet in height. Iron ore abounds; seven schools, kept 7 months in 12. Population 1830, 1,305.

LEWIS, p-v. Sussex co. Delaware, by p-r. 127 ms. N. E. by E. W. C.

LEWIS, co. of Virginia, bounded S. by Nicholas, S. W. by Kenhawa, w. by Wood, N. by Harrison, and E. and S. by Randolph. Length diagonally 70 ms., mean width 21, and area a small fraction above 1,600 square ms. Extending in lat. 38° 38′ to 39° 12′, and in long. 3° to 4° 17′ w. W. C. This co. is composed of two inclined plains, the dividing ground between which is very nearly a diagonal from the southeastern to the northwestern angles. Southwestardly the slope inclines westward and is drained by Little Kenhawa, whilst the northeastern plain gives source to the two main branches of Monongahela and slopes to the northward. The whole surface is rocky, hilly, and even in part rather mountainous. Chief town, Weston. Pop. 1820, 4,247, 1830, 6,241.

LEWIS, co. Ky. bounded by a ridge of hills, separating it from Greenup E. and S. E., by Fleming S. W., Mason w., and by Ohio river which separates it from Adams and Scott cos. Ohio, N. Length diagonally 35 miles, mean width 11, and area 375 square miles. Extending in lat. 38° 22′ to 38° 42′, and in long. 6° to 6° 35′ w. W. C. The general slope of this co. is northeastward towards that part of Ohio river by which it is bounded. Chief town, Clarksburg. Pop. 1820, 3,973, 1830, 5,229.

LEWIS, p-v. Brown co. Ohio, by p-r. 122 ms. S. S. W. Columbus.

LEWIS BAY, Mass. puts up from Hyannis harbor, between Barnstable and Yarmouth, on Cape Cod.

LEWISBERRY, p-v. in the northern part of

York co. Pa. by p-r. 13 ms. southward Harrisburg.

Lewisburg, p-v. on the right bank of Susquehannah river below the mouth of Buffalo creek, Union co. Pa. 8 ms. above, and on the contrary side of the river from Northumberland, and 65 above Harrisburg.

Lewisburg, p-v. and st. jus. Greenbriar co. Va. by p-r. 263 ms. s. w. by w. W. C. and 263 ms. w. Richmond. It is situated near the southern border of the co. on a branch of Greenbriar r. lat 37° 48', long. W. C. 3° 26' w.

Lewisburg, p-v. on the left bank of Green r. and in the northern part of Muhlenburg co. Ky. 10 ms. n. Greenville, and by p-r. 167 ms. s. w. by w. Frankfort.

Lewisburg, p-v. Preble co. Ohio, by p-r. 90 ms. westward Columbus.

Lewisport, p-v. in the northwestern part of Harrison co. Va. about 20 ms. northward Clarksburg, and 247 a little n. of w. W. C.

Lewiston, town, Lincoln co. Me. on the Androscoggin, at the Falls, 30 ms. w. of Wiscasset. Pop. 1830, 1,549.

Lewiston, p-t. Niagara co. N. Y. 27½ ms. n. n. w. Buffalo, 7 s. fort Niagara, 16 w. Lockport. Traversed by the mountain ridge. Land tolerably good. Contains gypsum. Here is a village of the Tuscarora Indians. This tribe came from North Carolina about 1712, and joined the confederacy of the Five Nations, themselves making the sixth. The village of Lewiston was laid waste during the last war between Great Britain and the United States, and likewise the Indian village before mentioned. Lewiston was deserted of its inhabitants from Dec. 1813, to April 1815. It lies on Niagara river, opposite Queenston in Upper Canada. It is situated at the head of navigation, and steamboats ply between this place and Ogdensburgh. Pop. 1830, 1,528.

Lewiston, p-v. and st. jus. Fulton co. Il. about 130 ms. n. w. Vandalia. It is situated on the table land between the Illinois and Spoon rivers, on the military bounty lands.

Lewistown, p-v. usually called Lunenburg court house, Lunenburg co. Va. by p-r. 103 ms. s. w. Richmond, lat. 36° 58', long. 1° 16' w. W. C.

Lewistown, p-v. and st. jus. Mifflin co. Pa. situated on the left bank of the Juniata river, on the point above the mouth of Kishicoquillas creek, 56 ms. by the land road above and n. w. Harrisburg, lat. 40° 36', long. 0° 37' w. W. C. Pop. 1820, 600 and in 1830, 1,480.

Lewistown, p-v. and v. of Sussex co. Del. situated on Del. bay, 3 ms. westward cape Henlopen, and opposite the Del. break water. It is laid down by Tanner at 38° 46' n. lat., 1° 54' e. W. C.

Lewistown, p-v. and st. jus. Montgomery co. Mo. 74 ms. n. w. by w. ½ w. St. Louis, and by p-r. 67 ms. n. e. by e. Jefferson city, lat. 38° 51', long. W. C. 14° 21' w.

Lewisville, p-v. Brunswick co. Va. by p-r. 82 ms. s. s. w. Richmond.

Lewisville, p-v. in the northeastern part of Chester district, S. C. 10 ms. n. e. Chesterville, and by p-r. 72 n. Columbia.

Lexington, p-t. Middlesex co. Mass. 10 ms. n. w. Boston. Surface uneven. Here was shed the first blood in the American revolution. There is a monument on the spot where fell the first victims. Pop. 1830, 1,543.

Lexington, p-t. Greene co. N. Y. 30 ms. w. Catskill, 43 from Albany. Rough and broken. Watered by the Schoharie and Albion creeks, and by several other mill streams. A great amount of leather is made at two very extensive tanneries in this town. Pop. 1830, 2,548.

Lexington Heights, p-v. in the foregoing town.

Lexington, p-v. western part of Erie co. Pa. 22 ms. s. w. the borough of Erie, and 25 n. n. w. the borough of Meadville.

Lexington, post town and st. jus. Rockbridge co. Va. situated on the right bank of North river, branch of James river, about 35 ms. n. w. Lynchburg, and by p-r. 129 ms. a little n. of w. Richmond, lat. 37° 44', long. 2° 21' w. W. C.

The following account of this village was remitted to the author of this article in 1821. It has no doubt both increased in population and improved in other respects in the intervening 9 years. It is distant about half a mile from North river, contains 120 dwelling houses, and 766 inhabitants. Many of the houses are constructed of brick. Beside the ordinary county buildings, and houses of public worship for Presbyterians and Methodists, it contains a state arsenal, in which are deposited about 20,000 stand of arms; this town has become noted for its literary establishments. Washington college doubly deserves its title, as it was endowed by that incomparable man with 100 shares of the stock of James river company, now (1821,) producing an annual income of $2,400. The two college halls, built of brick, are capable of containing and accommodating from 50 to 60 students, and additional buildings are about to be erected. The faculty are a president, two professors, and a tutor. The library, and philosophical apparatus, are tolerably ample. Andrew Smith's academy, for the education of young ladies, occupies a large and handsome edifice in which are teachers of all the requisite branches of such an institution.

Lexington, p-v. and st. jus. Davidson co. N. C. situated on Abbot's creek on eastern branch of Yadkin r. by the common road 109 but by p-r. 136 ms. w. Raleigh, lat. 35° 49,' long. 3° 18' w. W. C.

Lexington, district, S. C. bounded by Edgefield w., Newberry n. w., by Broad r. separating it from Fairfield n. and Richland n. e., by Congaree r. separating it from Richland e., and by Orangeburg s. e. and s. w. Length diagonally from s. to n. 45, mean width 20, and area 900 square ms. Extending in lat. 33° 40' to 34° 15', and in long. 3° 50' to 4° 34' w. W. C. The Saluda r. traverses in an eastern direction the northern part of this district, falling into or joining Broad r. at Columbia to form the Congaree. From the southern part flow the higher branches of North Edisto. The general slope of the whole surface is southeastward towards the Broad and Congaree rs. Chief town, Granby. Pop. 1820, 8,083, 1830, 9,065.

LEXINGTON, court house, and p-o. Lexington district, S. C. by p-r. 15 ms. w. Columbia.

LEXINGTON, p-v. and st. jus. Oglethorpe co. Geo. 76 ms. N. w. by w. Augusta, and 65 a little E. of N. Milledgeville, lat. 33° 53', long. 6° 10' w. W. C.

LEXINGTON, p-v. and st. jus. Henderson co. Ten. situated on Beech creek, a small western branch of Ten. r. 44 ms. s. s. w. Reynoldsburg, and by the p-r. 114 ms. s. w. by w. Nashville, lat. 35° 38', long. 11° 25' w. W. C.

LEXINGTON, post town, and st. jus. Fayette co. Ky. situated on the head waters of Town creek, a branch of Elkhorn r., 24 ms. s. E. by E. Frankfort, and about 80 ms. very nearly due s. Cincinnati, and by p-r. 517 ms. a little s. of w. W. C. On Tanner's United States, it is laid down at 38° 03' lat., long. 7° 28' w. W. C. This now flourishing seat of the arts, law and polished life, the cradle of Kentucky, first began to assume the aspect of a village in 1785, but so slow was its progress during the existence of Indian wars, that in 1795, it contained only about 50 ordinary houses, and at most 350 inhabitants, whilst by the recent census, 1830, it contained 3,757 whites, 230 free colored persons, and 2,100 slaves; total, 6,087 inhabitants. This population is actively engaged on manufactories of cotton, woollen and linen, copper, tin and iron ware, grist mills, paper mills, rope walks, tanneries, breweries, distilleries, printing, bookselling, commerce, agriculture, &c.

Besides numerous private schools, Lexington contains Transylvania university. The incipient steps towards the foundation of this institution were taken before the sepation of Kentucky from Virginia. It was reorganized in 1798, and in 1818, placed under its existing regulations. In 1820, it was under a president, 7 professors, 4 of whom were medical, 5 tutors, and the principal of a preparatory department. The library then contained about 3000 volumes. (*See article Transylvania university.*)

LEXINGTON, p-v. in the southern part of Richland co. Ohio, by p-r. 71 ms. Columbus.

LEXINGTON, p-v. and st. jus. Scott co. Ind. situated in the eastern angle of the co. 30 ms. N. Louisville in Ky., and 89 ms. s. s. E. Indianopolis, lat. 38° 40', long. 8° 40' w. W. C.

LEXINGTON, p-v. and st. jus. Lafayette co. Mo. situated on Mo. r. by p-r. 138 ms. above Jefferson city, and 272 above St. Louis, lat. 39° 05', long. W. C. 16° 44' w.

LEYDEN, p-t. Franklin co. Mass. 117 ms. N. w. Boston, 6 N. w. Greenfield. Pop. 1830, 796.

LEYDEN, p-t. Lewis co. N. Y. 33 ms. N. of Utica, w. of Black r. Surface somewhat uneven. Well watered by small springs. Soil better adapted to grass than to grain. Limestone abounds; 6 school districts. Pop. 1830, 1,502.

LIBERIA, p-v. Prince William co. Virginia, by p-r. 33 ms. s. w. W. C.

LIBERTY, p-t. Sullivan co. N. Y. 22 ms. N. w. Monticello. Watered by the Mongaup and the Collakoon, with their branches. Timbered with beech, maple, ash, &c. 6 schools kept 6 months in 12. Population 1830, 1,277.

LIBERTY CORNER, p-v. Somerset co. N. J. 2 ms. s. w. Baskenridge, 7 ms. s. Morristown.

LIBERTY, post township, Tioga co. Pa. by p-r. 123 ms. northward Harrisburg.

LIBERTY, p-v. and st. jus. Bedford co. Va. on a branch of Otter r. 26 ms. a little s. of w. Linchburg, and by p-r. 140 ms. s. w. by w. Richmond, lat. 37° 17', long. 2° 29' w. W. C.

LIBERTY, co. Geo. bounded by McIntosh s., Alatamaha r. separating it from Appling s. w., Tatnall w. and N. w., Bryan N. and N. E., and the Atlantic ocean s. E. This co. lies in the singular form of a curve or half moon, from St. Catharine's island inclusive to Alatamaha r. embosoming McIntosh co. 66 ms., mean width 10, and area 660 square ms. Extending in lat. 31° 26' to 32° 04', and in long. 4° 16' to 5° 08' w. W. C. That part of this co. bordering on the Atlantic is low and intersected by interlocking tide water courses. The northern part traversed by the most southern branches of Cannouchee r. rises something higher than the ocean border, but the whole co. may be regarded as flat. Chief town, Riceboro.' Pop. 1820, 6,695, 1830, 7,233.

LIBERTY, p-v. eastern part of Talbot co. Geo., by p-r. 105ms. westward Milledgeville.

LIBERTY, p-v. Smith co. Ten. by p-r. 59 ms. N. E. by E. Nashville.

LIBERTY, p-v. and st. jus. Casey co. Ky. situated on Green r. 68 ms. very nearly due s. Frankfort, N. lat. 37° 20', long. 7° 50' w. W. C.

LIBERTY, p-v. Montgomery co. Ohio, by p-r. 74 ms. s. w. by w. Columbus.

LIBERTY, p-v. and st. jus. Union co. Ind. by p-r. 77 ms. E. Indianopolis, and 54 N. N. w. Cincinnati, lat. 39° 40'.

LIBERTY, p-v. and st. jus. Amite co. Miss. situated on Amite r. 50 ms. s. E. Natchez, and by p r. 112 ms. s. s. w. Jackson, lat. 31° 10', long. 13° 58'.

LIBERTY, p-v. Clark co. Al. about 140 ms. southward Tuscaloosa.

LIBERTY, p-v. and st. jus. Clay co. Mo. by p-r. 190 ms. N. w. by w. Jefferson city, and 324 ms. above, and by the land road westward St. Louis, lat. 39° 10', long. W. C. 17° 17' w.

LIBERTY HALL, p-v. Pittsylvania co. Va. by p-r. 121 ms. s. w. by w. Richmond.

LIBERTY HALL, p-v. Morgan co. Geo. by p-r. 45 ms. N. N. w. Milledgeville.

Liberty Hill, p-v. Iredell co. N. C. by p-r. 154 ms. westward Raleigh.

Liberty Hill, p-v. Kershaw district, S. C. by p-r. 40 ms. N. E. Columbia.

Liberty Hill, p-v. Dallas co. Al. by p-r. 114 ms. southward Tuscaloosa.

Liberty Pole, p-v. Northumberland co. Pa. by p-r. 81 ms. N. Harrisburg.

Liberty Town, p-v. Frederick co. Md. 10 ms. N. E. from the city of Frederick, and by p-r. 55 ms. a little N. of W. W. C.

Lick, creek, p-o. Greenbriar co. Va. by p-r. 293 ms. N. W. by W. W. C.

Licking, co. Ohio, bounded S. E. by Perry, S. Fairfield, S. W. Franklin, N. W. Delaware, N. Knox, N. E. Coshocton, and E. by Muskingum. It is 30 ms. from E. to W. and 24 broad, and area 720 square ms. lat. 40° 10′, long. 5° 30′. The slope is eastward, and the whole surface very nearly commensurate with the higher part of the valley of Licking creek, or more correctly river. The great central Ohio canal enters this co. on its southern border near Hebron, sweeping a northern curve past Newark, the st. jus.; this work passes down the Licking valley and leaves the co. near the middle of its eastern side. Though a level country, it is a rather elevated table land; the level of the canal at Newark is 834 feet above mean level of the Atlantic ocean, and 219 feet above that of the Ohio river, at the mouth of Sciota r. The arable land of the co. is from 900 to 1,100 feet above the ocean tides. The excellence of the soil is shewn by progressive population; 1820, 11,861, 1830, 20,714.

Licking, river, a stream of Ky. rising in Floyd co. interlocking sources with those of the W. branch of Sandy, and with those of the northeastern branches of Ky. r., and flowing thence by a general course very nearly N. W. between the vallies of O. and Ky. rs. passing through or touching the counties of Floyd, Morgan, Fleming, Lawrence, Bath, Nicholas, Harrison, Bracken, Pendleton and Campbell, falling into Ohio river, between Covington and Newport, and directly opposite the city of Cincinnati, after a comparative course of about 175 ms. The valley of Licking is narrow, compared with its length, the greatest width falling short of 50 ms. and the mean breadth fully estimated at twenty, and area at 3,500 square miles. (*See Kentucky river.*)

Licking, small but important river of Ohio. This stream has interlocking sources with those of the various eastern branches of Sciota on the S. W. and W., and with those of Owl creek, branch of White Woman's river, on the N. The creeks which form Licking, drain Licking co. uniting at Newark, and flowing thence E. into Muskingum co. inflect to S. E. to the main Muskingum r. at Zanesville. Comparative length 75 miles. (*See Licking co. Ohio.*)

Licking, p-v. Floyd co. Ky. by p-r. 120 ms. S. E. from Frankfort.

Licking Creek, and p-o. southeastern part of Bedford co. Pa. about 25 ms. S. E. from Bedford, and 10 N. Hancockstown, Washington co. Md.

Licking Forge, and p-o. Bath co. Ky. by p-r. 78 ms. eastward Frankfort.

Licking Forge, and p-o. eastern part of Bath co. Ky. 13 ms. E. Owingsville, and by p-r. 85 ms. E. Frankfort.

Lickville, p-v. in the northeastern part of Greenville district, S. C. by p-r. 116 ms. N. W. from Columbia.

Ligonier, p-v. on the r. from Philadelphia to Pittsburg, at the western foot of Laurel hill, and in the eastern part of Westmoreland co. Pa. 19 ms. a little S. of E. Greensburg, and by p-r. 151 ms. W. Harrisburg.

Lilesville, p-v. Anson co. N. C. by p-r. 112 ms. S. W. by W. Raleigh.

Lilly, p-o. Brown co. Ohio, by p-r. 90 ms. S. S. W. Columbus.

Lilly Point, p-o. King William co. Va. by p-r. 36 ms. N. E. Richmond.

Lima, p-t. Livingston, co. N. Y. 13 ms. N. E. Geneseo, 18 W. Canandaigua. Soil good; 9 schools, kept 8 months in 12. Pop. 1830, 1,764.

Lime, p-t. Grafton co. N. H. 6 ms. S. Orford, 54 from Concord. Smart's mtn. lies in the N. E. part of the town. Pop. 1830, 1,804.

Limerick, p-t. York co. Me. 35 ms. N. York, 30 N. W. Portland. Contains a flourishing academy. Pop. 1830, 1,419.

Limerick, p-o. and tsp. Montgomery co. Pa. lying E. Pottstown, and 24 ms. N. W. Phil. Pop. 1820, 1,577, 1830, 1,744.

Limestone, p-v. Armstrong co. Pa. by p-r. 241 ms. N. W. W. C.

Limestone, p-v. Buncombe co. N. C. by p-r. 245 ms. a little S. of W. Raleigh.

Limestone, co. of Ala. bounded by Madison E., Ten. r. separating it from Morgan S., and Lawrence S. W., by Lauderdale W., and by Giles co. of Ten. N. Length 30, mean breadth 20, and area 600 square ms. Extending in lat. 34° 33′ to 35°, and in long. 9° 52′ to 10° 18′ W. W. C. Elk r. entering the northern border traverses the N. western angle of this co. flowing to the S. W. The general slope is a little W. of S. down which flow into Ten. r. several bold fine creeks. Chief town, Athens. Pop. 1820, 9,871, 1830, 14,807.

Limington, p-t. York co. Me. on Saco r. 40 ms. N. Saco. Pop. 1830, 2,317.

Linbank, p-v. Granville co. N. C. by p-r. 48 ms. N. Raleigh.

Lincoln, co. Me. bounded N. by Kennebec, N. E. by Waldo, E. by Penobscot bay, S. by the Atlantic, and W. by the Androscoggin river, which separates it from Cumberland. It is divided by Kennebec river, and the whole southern and S. E. part is composed of numerous islands, and long peninsulas, extending into the ocean. Within the co. are numerous bays and rivers. The surface of the co. in the interior is finely diversified, and soil productive in grain and pasturage. Chief towns, Wiscasset, Warren, and Topham.

Population 1820, 53,189, 1830, 57,181.

LINCOLN, town, Hancock co. Me. 27 ms. N. W. Castine.

LINCOLN, town, Grafton co. N. H. 70 ms. N. Concord. Watered by the middle branch of the Pemigewasset. In the N. part of the town are two large gulfs, made by an extraordinary discharge of water from the clouds in 1774. Pop. 1830, 50.

LINCOLN, town, Addison co. Vt. 21 ms. S. W. Montpelier, 28 S. E. Burlington. Considerably uneven. West part watered by New Haven river, which is formed here; east part by several small branches of Mad river. Timber principally hard wood; 4 school districts. Pop. 1830, 639.

LINCOLN, p-t. Middlesex co. Mass. 16 ms. N. W. Boston. Rather uneven and encumbered with rocks. Pop. 1830, 709.

LINCOLN, co. N. C. bounded by York dist. S. C. S., Rutherford co. N. C. W., Burke N. W. and N., and by Catauba river which separates it from Iredell N. E., and Mecklenberg E. Length south to north 48, mean width 25, and area 1,200 square ms. Extending in lat. 35° to 35° 49′, and in long. 4° to 4° 33′ w. W. C. This co. is very nearly commensurate with the valley of Little Catauba, for though bounded in all its length, by the Great Catauba, the creeks generally enter the former. Chief town, Lincolnton. Pop. 1820, 18,147, 1830, 22,455.

LINCOLN, co. of Geo. bounded by Little r. separating it from Columbia S., by Wilkes W., Broad r. separating it from Abbeville district S. C. N. E., and from Edgefield S. C. E. Length 22, mean width 10, and area 220 sq. ms. Extending in lat. 33° 40′ to 33° 56′ and in long. 5° 16′ to 5° 38′ w. W. C. The slope of this co. is to a little N. of E. Chief town, Lincolnton. Pop. 1820, 6,458, 1830, 6,145.

LINCOLN, co. Ten. bounded by Giles W., Bedford N., Franklin E., Jackson co. Ala. S. E., and Madison co. Ala. S. Length 26, mean breadth 25, and area 650 square ms. Extending in lat. 35° to 35° 24′, and in long. 9° 16′ to 9° 40′ w. W. C. This is composed of two inclined planes, being the opposing slopes of Elk river valley, which stream traverses it flowing S. W. by W. Chief town, Fayetteville. Population 1820, 14,761, 1830, 22,075.

LINCOLN, co. Ken. bounded by Casey S. W. and W., Mercer N. W., Garrard N. E., Rock Castle S. E. and Pulaski S. Length 27, mean width 16, and area 432 square ms. Extending in lat. 37° 17′ to 37° 38′, and in long. 7° 23′ to 7° 44′ w. W. C. This is amongst the central counties of the state, and occupies a table land, from which flow creeks towards Cumberland river S., the extreme sources of Salt and Green rivers W., and Dicks river northward into Kentucky river. Chief town, Stanford. Pop. 1820, 9,979, 1830, 11,002.

LINCOLN, co. Mo. bounded S. by St. Charles, S. W. and W. Montgomery, N. W. and N. Pike, and on the E. by Miss. r. separating it from Calhoun co. Il., very nearly a sq. of 24 ms., area 576 sq. ms. Lat. 39°, long. 14° w. Slope southestrd. and traversed by Cuivre r. by which it is principally drained. Chief t. Troy. Pop. 1830, 4,059.

LINCOLNTON, p-v. and st. jus. Lincoln co. N. C., situated on Little Catauba, 45 ms. N. E. by E. Rutherfordton, and by p-r. 166 ms. a little S. of W. Raleigh. Lat. 35° 28′, long. 4° 16′ w. W. C.

LINCOLNTON, p-v. and st. jus. Lincoln co. Geo. situated near the centre of the co. 40 ms. N. W. Augusta, and by p-r. 91 ms. N. E. Milledgeville. Lat. 33° 44′, long. 5° 28′ w. W. C.

LINCOLNVILLE, p-t. Hancock co. Me. W. side Penobscot bay, 16 ms. W. Castine.

LINDEN, p-v. and st. jus. Marengo co. Ala. by p-r. 78 ms. S. Tuscaloosa. Lat. 32° 20′, long. 10° 56′ w. W. C.

LINDSAY'S Cross Roads and p-o. Fluvanna co. Va. 80 ms. wstrd. Richmond.

LINDSEY'S store and p-o. Albemarle co. Va. 76 ms. W. Richmond.

LINE creek, p-o. wstrn. part of Greenville dist. S. C. N. W. from Columbia.

LINE creek and p-o. Montgomery co. Ala. by p-r. 145 ms. S. E. Tuscaloosa.

LINE LEXINGTON, p-v. Bucks co. Pa. about 23 ms. from Phil.

LINE Mills and p-o. Crawford co. Pa. by p-r. 278 ms. N. W. by W. Harrisburg.

LINGLESTOWN, p-v. Dauphin co. Pa. 8 ms. N. E. Harrisburg.

LINVILLE creek and p-o. Rockingham co. Va. by p-r. 142 ms. wstrd. W. C.

LINVILLE creek and p-o. western part of Burke co. N. C. The p-o. is about 10 miles wstrd. Morgantown, and by p-r. 215 ms. W. Raleigh.

LIONVILLE, p-v. Chester co. Pa. about 20 ms. N. W. Phil.

LIPONA, formerly Jena, p-v. Jefferson co. Flor. 20 ms. E. Tallahasse.

LISBON, t. Lincoln co. Me. on the Androscoggin, 23 ms. W. Wiscasset. Pop. 1830, 2,423.

LISBON, t. New London co. Conn. at the junction of Quinebaug and Shetuck rs. 7 ms. N. Norwich, 45 S. E. Hartford. Uneven and somewhat hilly. Timbered with oak, walnut, chestnut, &c. Soil fertile. Here are several fisheries of shad and salmon. Has several manufactories. Pop. 1830, 1,161.

LISBON, p-t. St. Lawrence co. N. Y. on St. Lawrence r., 3 ms. below Ogdensburgh. Soil very excellent. In this town is a small Indian village. Pop. 1830, 1,891.

LISBON, p-v. Ann Arundel co. Md. situated on the turnpike road from Baltimore to Frederic, 34 ms. N. W. C. It is a small village of a single street along the road.

LISBURN, p-v. on Yellow Breeches creek, southeastern part of Cumberland co. Pa. 12 ms. S. W. Harrisburg.

LISLE, p-t. Broome co. N. Y. 18 ms. N. Chenango Point, 130 from Albany. Watered and abundantly supplied with mill seats by Tioughnioga, Otselic, and Nanticoke creeks.

Soil in general good. Surface uneven. 29 schools, kept 11 months in 12. Pop. 1830, 4,378.

LITCHFIELD, p-t. Lincoln co. Me. 25 ms. N. W. Wiscasset, 10 from Hallowell. Pop. 1830, 2,308.

LITCHFIELD, t. Hillsborough co. N. H. a small fertile tsp. on the E. bank of Merrimack r. 8 ms. from Amherst, 30 S. Concord. 3 school dists. Pop. 1830, 494.

LITCHFIELD co. Conn., an extensive agricultural and manufacturing co. bounded N. by Berkshire co. Mass., E. by Hartford and New Haven cos., S. by New Haven and Fairfield cos., W. by N. Y. Average length 33 miles from N. to S., average width, nearly 27 ms., containing about 885 sq. ms., being the largest co. in the state. Seat of justice, Litchfield. Principal part of the co. elevated and mountainous. Prevailing soil a gravelly loam, strong and fertile. Watered abundantly by the waters of the Ousatonic and Tunxis rs. The iron manufacture is carried on more extensively in this co. than in any other section of the state. The ore is obtained within the co. Pop. 1820, 41,267, 1830, 42,858.

LITCHFIELD, p-t. and st. jus. of Litchfield co. 30 ms. W. Hartford, 36 N. W. New Haven, 100 from N. Y. An elevated tsp. diversified with hill and dale. Mount Tom is in the W. part of this town; height 700 feet above the margin of Naugatuck r. Contains a quarry of inferior slate stone, and a good quarry of free stone. Prevailing soil, a dark colored, gravelly loam, deep, strong, and fertile. Well supplied with forests, consisting of sugar maple, beech, button wood, oak, birch, &c. Well watered and supplied with excellent hydraulic privileges, by the Naugatuck and Shepaug rs., and the Bantam waters. Litchfield great pond, the largest in the state, is a beautiful sheet of water, comprising about 900 acres. At its outlet are numerous and valuable mill seats. The manufacture of iron is here carried on, on an extensive scale. 26 school districts, and a most respectable academy. Contains a medicinal spring. Litchfield v. is delightfully situated on an elevated plain, surrounded with interesting scenery and charming landscapes. Here is a very celebrated law school. A manual labor high school has recently been incorporated here. Pop. 1830, 4,458.

LITCHFIELD, p-t. Herkimer co. N. Y. 11 ms. S. W. Herkimer, 11 S. Utica. Situation elevated. 11 school dists., schools kept 8 months in 12. Pop. 1830, 1,750.

LITCHFIELD, p-v. Bedford co. Pa. by p-r. 153 ms. W. Harrisburg.

LITCHFIELD, p-v. and st. jus. Grayson co. Ky. 69 ms. S. S. W. Louisville, 26 S. W. Elizabethtown, and by p-r. 105 S. W. by W. Frankfort. Lat. 37° 28′, long. 9° 15′ W. W. C.

LITCHFIELD, p-v. Jackson co. Ark.

LITHOPOLIS, p-v. Fairfield co. O. 10 ms. N. W. Lancaster, the co. st. and 18 S. E. Columbus. Pop. 1830, 161.

LITIZ, small, but neat p-v. 7 ms. N. the city of Lancaster, Pa. This village was founded by the United Brethren or Moravians in 1757.

LITTLE BEAVER bridge and p-o. eastern part of Columbiana co. O. by p-r. 169 ms. N. E. by E. Columbus.

LITTLE BRITAIN, extreme southern p-tsp. of Lancaster co. Pa. The p-o. is situated 22 ms. S. S. E. Lancaster, and by p-r. 58 ms. S. E. Harrisburg.

LITTLE CAPE CAPON creek and p-o. Hampshire co. Va. 188 ms. N. W. W. C.

LITTLE COMPTON, p-t. Newport co. R. I. situated in the S. E. extremity of the co. and state, 30 ms. S. E. Providence. Soil, a deep, rich loam. Surface pleasantly diversified. 7 schools. Pop. 1830, 1,378.

LITTLE FALLS, p-v. Herkimer co. N. Y. on the Mohawk, derives its name from the falls in the river at this place, which descend in the course of about a m., 42 feet. For about half a mile, it passes through a fissure in the rocks, which rise on each side 500 feet, and seem formerly to have been united, and to have constituted the barrier of a lake extending far to the W. Here is a canal on the N. side of the r. round the falls, three quarters of a mile long, through an uncommonly hard rock. This canal is now connected with the Erie canal, on the opposite side of the r. by an aqueduct 170 feet long, and 30 above the stream. The Erie canal here descends 40 feet in 1 mile. The village of Little Falls stands on this canal, and is 72 ms. W. Albany, and 22 E. Utica. It is the centre of one of the best grain and grazing districts in the state. It is large and well built. Materials for the erection of factories, &c. are on the premises in large quantities. No other place in the Union combines greater advantages for the economical and profitable operation of all kinds of machinery. Pop. of the vil. 1832, 1,500.

LITTLE FLAT ROCK, p-o. Rush co. Ind. by p-r. 57 ms. S. E. by E. Indianopolis.

LITTLE GUNPOWDER creek and p-o. eastern part of Baltimore co. Md. by p-r. 16 ms. N. E. by E. Baltimore.

LITTLE HOCKHOCKING, small stream and p-o. in the southwestern part of Washington co. O. 7 ms. below Belpre, and by p-r. 104 ms. S. E. Columbus.

LITTLE MIAMI, r. of O., has its extreme sources in Clark co. interlocking with those of Mad r., and with those of Deer and Paint creek branches of Sciota, and flowing thence by comparative courses 120 ms. to the S. S. E. to its entrance into Ohio r. about 10 ms. by water above Cincinnati. The course of the Little Miami is very nearly parallel to that of Great Miami, the former deriving its principal tributaries from the cntrd. and draining great part of Green, Clinton, Warren, Clermont, with parts of Brown, Clark, and Hamilton cos. Rising on a comparatively elevated tract, the fall is rapid, rendering this r. one of the best in O. for mills.

LITTLE MISSOURI is the name of two small,

and from each other, distant rs. One is a branch of Mo., and the next of any consequence from the sthrd. below the Yellow Stone r. As laid down by Tanner, Little Mo. rises at lat. 45°, about 200 ms. s. w. of the Mandan villages, and has thence a course of N. N. E. 200 ms. nearly parallel to and about 60 ms. distant from Yellow Stone r.

Little Missouri is the name also of the principal wstrn. confluent of Ouachita r. The valley of this stream lies between that of Ouachita proper, and Red r. It drains part of Hempstead and Clark cos. Ark.

LITTLE PEDEE, r. of N. and S. Carolina. (*See Pedee and Lumber rs.*)

LITTLE PINEY, p-v. and st. jus. Crawford co. Mo. by p-r. 97 ms. S. S. E. Jefferson City, and about an equal distance s. w. St. Louis. Little Piney cr. is an estrn. branch of Gasconade r.

LITTLE PLYMOUTH, p-v. in the sthrn. part of King and Queen co. Va. by p-r. 56 ms. N. E. by E. Richmond.

LITTLE RED RIVER, p-v. Pulaski co. Ark. 11 ms. wstrd. Little Rock.

LITTLE RIVER, is a name given to numerous streams in the United States. Little r. one of the branches of Pedee. Little r. branch of Savannah r. which falls into its recipient, 30 ms. above Augusta, after having drained a part of Wilkes, Warren, Columbia and Lincoln cos. Geo. Little r. also in Geo., falls into Oconee from the wstrd., 12 ms. above Milledgeville. Little r. of Trigg and Christian cos. Ky., falling into Cumberland r. below Cadiz. Little r. of the south, as it is there called, a small stream falling into Red r. from the wstrd. in the sthwstrn. angle of La. Red r. of the north another, and much more considerable branch of Red r., joining that stream between Sevier and Hempstead cos. Ark. Red r. of the north, is a stream of some size, having a comparative length of upwards of 100 ms. There are some other rivers bearing the same title, but of too little consequence to merit particular notice. The Ocatahoola r. La. is frequently in that country called Little r., above its lake and below the mouth of Dugdomony r.

LITTLE r. small r. of Montgomery co. Va. rises in the wstrn. vallies of the Blue Ridge, and flowing to the N. W. about 25 ms. comparative course, falls into New r., 12 ms. s. w. by w. Christiansburg.

LITTLE r. S. C. rises on the border between Anderson and Abbeville dists., between the Saluda and Savannah rs., and flowing sthrd. drains by its confluents, the central and larger part of Abbeville, and falling into Savannah r. opposite Lincoln co. Geo.

LITTLE r. p-v. or p-o. on Little r. Henry dist. S. C. 120 ms. N. E. Charleston, and by p-r. 179 ms. E. Columbia.

LITTLE r. Geo. rising by numerous branches between Ockmulgee and Oconee rs., and draining part of Morgan, Jasper, Jones, Putnam and Baldwin, falls into Oconee r. between the two latter cos.

LITTLE r. La. rises in the parishes of Claiborne, and Ouachita, flows S. S. E. into Rapides parish and falls into Ocatahoola lake. The valley of Little River lies between those of Red and Ouachita.

LITTLE r. of the north, rises in Texas, and flowing S. E. enters the Ter. of Ark. and falls into Red r. between Hempstead and Sevier cos., draining by its confluents the greatest part of Miller and Sevier cos.

LITTLE r. Ken. rises in Christian, flows into Trigg and falls into Cumberland r. below Cadiz.

LITTLE r. inlet, a small opening at the mouth of an inconsiderable creek of the Atlantic Ocean, but gaining importance from forming the limit on the Atlantic Ocean between N. and S. Carolina.

LITTLE r. p-v. wstrn. part of Burke co. N. C. by p-r. 220 ms. wstrd. Raleigh.

LITTLE r. p-o. Marion co. Miss. about 110 ms. S. E. by E. Natchez.

LITTLE ROCK, p-v. and st. jus. Pulaski co. and of the government Ark. Ter. situated on the right bank of Arkansas river, and about 120 ms. by land above the mouth of that stream. Lat. 34° 42′, long. 15° 15′ w. W. C. The course and distance between W. C. and Little Rock by a mercator's calculation, is s. 71° 10′, w. 980 statute ms., the p-r. as stated on the p-o. list gives a distance of 1,111 ms.

It was intended to give the name of Acropolis to Little Rock, but the people of the country playfully called it by its present name from the enormous rocks in the vicinity. The site is a high rocky bluff on the right bank. Steamboats are safely navigated thus high, about 300 ms. from the Miss. by the bends of Ark. r.

LITTLE SANDUSKY, p-v. nthrn. part of Crawford co. O. by p-r. 71 ms. nthrds. Columbus.

LITTLE SANDY, p-v. on Little Sandy r. in the sthrn. part Greenup co. Ky. by p-r. 132 ms. a little N. of E. Frankfort. The p-o. is at the salt works, 20 ms. above the mouth of Little Sandy at Greenupsburg.

LITTLETON, p-t. Grafton co. N. H. on Conn. r. at the Fifteen Mile falls, 18 ms. below Lancaster, 30 from Haverhill corner, 100 N. Concord. Timbered with sugar maple, beech, birch, bass, &c. Amonoosuck r. waters the s. part, on which, in this town, is the pleasant village of Glynville, where there are falls. Pop. 1830, 1,433.

LITTLETON, p-t. Middlesex co. Mass. 28 ms. N. W. Boston, 10 N. W. Concord. Pop. 1830, 947.

LITTLETON, p-v. Sussex co. Va. by p-r. 36 ms. sthrd. Richmond.

LITTLETON, p-v. Warren co. N. C. by p-r. 67 ms. N. N. E. Raleigh.

LITTLE VALLEY, p-t. Cataraugus co. N. Y. on the Alleghany r., 12 ms. s. w. Ellicottville. Land in general of a superior quality, moderately uneven, timbered with hickory, oak, chesnut, &c. Pop. 1830, 336.

LITTLE YADKIN, p-v. in the sthwstrn. part

of Stokes co. N. C. by p-r. 167 ms. N. w. by w. Raleigh.

Little Yadkin, r. N. C. and one of the wstrn. confluents of Great Yadkin, rises in Iredell co. most of which it drains, having its remote sources within 3 ms. from Great Catauba, though flowing from it estrdly. towards the Yadkin over Iredell and Rowan cos.

Little York, p-v. Hardin co. Ky. sthwstrd. Frankfort.

Little York, p-v. nthestrn. part Montgomery co. O. by p-r. 74 ms. wstrd. Columbus.

Livermore, p-t. Oxford co. Me. on the Androscoggin, 18 ms. N. E. Paris, 78 from Portland. Pop. 1830, 2,453.

Livermore, p-v. Westmoreland co. Pa. 22 ms. S. E. Pittsburg, and by p-r. 172 ms. w. Harrisburg.

Liverpool, p-v. on the right bank of Susquehannah r. in the nthestrn. part of Perry co. Pa. 29 ms. N. Harrisburg.

Liverpool, p-v. Medina co. O. by p-r. 124 ms. N. E. Columbus.

Liverpool, p-v. Yazoo co. Miss. about 20 ms. N. N. E. Natchez.

Livingston, co. N. Y., situated on Genesee r. bounded N. by Genesee and Munroe cos., E. by Munroe and Ontario cos., S. by Steuben and Alleghany cos., W. by Alleghany and Genesee cos.; containing 460 sq. ms. or 294,400 acres. Watered by Genesee r. on the W., Canasaraga and Cashque creeks S., Honeoye creek &c. E. and N. Surface a pleasing variety. Limestone and clay slate abound. Soil a good variety. Iron ore is found almost every where, not in beds, but in lumps, in the soil or subsoil. Chief town, Geneseo. Pop. 1820, 19,196, 1830, 27,719.

Livingston, p-t. Columbia co. N. Y. on the Hudson, 12 ms. below Hudson. Pop. 1830, 2,087.

Livingston, t. Essex co. N. J. adjoins Passaik river, 54 ms. N. E. Trenton. Pop. 1830, 1,150.

Livingston, co. Ky. bounded N. E. by Tradewater r. separating it from Union, E. by Hopkins, S. E. by Caldwell, S. W. by Ten. r. separating it from McCracken, by O. r. separating it from Posey co. Il. W., and again by O. r. seperating it from Gallatin Il. N.; length from N. E. to S. W. 40 ms.; mean width 20, and area 800 sq. ms. Extending in lat. from 36° 04′ to 36° 30′, and in long. 10° 52′ to 11° 35′ W. from W. C. In a navigable point of view this county is in a peculiar manner advantageously placed, beside Ten., Ohio, and Trade water rs. by which it is bounded. Cumberland r. traverses the sthrn. part falling into O. at Smithland. The surface of the co. is mostly level or moderately hilly, with fertile soil. Chief t. Salem. Pop. 1820, 5,824, 1830, 5,971.

Livingston, p-v. and st. jus. Madison co. Miss., N. N. E. from Natchez.

Livonia, p-t. Livingston co. N. Y. 8 ms. E. Geneseo. Land pretty good. Contains some small streams. 12 schools, kept 9 months in 12. Pop. 1830, 2,665.

Livonia, p-v. Washington co. Ind. by p-r. 103 ms. S. Indianopolis.

Lloyd's, p-o. Essex co. Va., by p-r. 84 ms. N. E. Richmond.

Loch Rauza, p-v. Montgomery co. Ala. by p-r. 82 ms. S. E. Tuscaloosa.

Locke, p-t. Cayuga co. N. Y. 21 ms. S. S. E. Auburn, 152 W. Albany. Excellent land, handsomely diversified with easy swells, hill and dale, and extensive alluvial flats. Well watered with springs and brooks. 13 schools kept 6 months in 12. Pop. 1830, 3,310.

Lockport, p-t. and st. jus. of Niagara co. one of the results of the Erie canal, 31 ms. by that canal N. E. Buffalo, at the E. extremity of the Buffalo level, 20 ms. E. Lewiston, 63 W. Rochester. When the route of the canal was established in 1821, this place was a wilderness. It is now a large and flourishing town! The canal here descends the terrace called the Mountain ridge, or Ontario Heights, by 5 double locks, each of 12 feet descent, to the Genesee level. These locks being double, one line of boats can ascend while another descends. Above the locks, the canal is cut through rock to the depth of 20 ft. for the distance of 3 ms. The Genesee level extends eastward from this place to the distance of 65 ms. The locks at Lockport are the only ones from lake Erie to Genesee r. which by the canal route is a distance of 96 ms. The descent of the canal down the Mountain ridge at this place, is truly a fine spectacle. This is the same ridge over which roll the thundering torrents of Niagara, constituting the Niagara falls. Pop. 1830, 1,801.

Lock's, village, and p-o. Franklin co. Miss. about 25 ms. E. from Natchez, and by p-r. 86 from Jackson.

Locust, r. Ten. the nthestrn. branch of Black Warrior, rising from the table land between the basins of Mobile and Ten., and within 15 ms. from the great bend of the latter, where it traverses Cumberland mtns., draining the estrn. part of Blount and five sixths of Jefferson cos., and flowing to the S. W. by comparative courses 75 ms. it joins the Mulberry to form the Black Warrior. (*See Mulberry r.*)

Locust, cr. p-o. formerly Chinquipin Grove, Louisa co. Va. 101 ms. S. W. W. C.

Locust Dale, p-v. Culpepper co. Va. by p-r. 86 ms. S. W. W. C.

Locust Grove, p-o. Orange co. Va. by p-r. 81 ms. S. W. W. C.

Locust Grove, p-o. Perry co. Ten. by p-r. 99 ms. S. W. by W. Nashville.

Locust Hill, p-o. Butler co. Ala. by p-r. 160 ms. S. E. Tuscaloosa.

Locust Shade, p-o. Overton co. Ten. by p-r. 81 ms. N. E. by E. Nashville.

Lodi, t. Bergen co. N. J. between the Hackensack and Passaic rs., S. of New Barbadoes and Saddle r. Pop. 1830, 1,356.

Lodi, p-v. Abbeville dist. S. C. by p-r. 86 ms. N. W. by W. Columbia.

Lodi, p-v. Washtenau co. Mich. by p-r. 47 ms. W. Detroit.

Lodimont, p-v. wstrn. part of Abbeville dist. S. C.

Logan, p-v. Centre co. Pa. by p-r. 92 ms. wstrd. Harrisburg.

Logan, co. Va. bounded n. w. by Cabell, n. by Kenhawa, n. e. by New r. or Great Kenhawa, separating it from Nicholas and Greenbriar, e. by the Great Flat Top mountain, separating it from Giles, w. by the estrn. branch of Sandy r. separating it from Floyds co. Ky., and s. by Tazewell. Length 70, mean breadth 55, and area 3,850 sq. ms. Extending in lat. 36° 13′ to 37° 10′, and in long. from 3° 50′ to 5° 22′ w. W. C. The surface of this very broken and extensive co. is from the Great Flat Top mtn. to n. n. w., beside Great Kenhawa and Sandy rs. Logan is drained by Guyandot and Coal rs. Chief t. Logan. Pop. 1830, 3,680.

Logan, C. H. and p-o. Logan co. Va. by p-r. 338 ms. w. Richmond.

Logan, co. Ken. bounded by Todd w., Muhlenburg n. w., Butler n., Wayne e., Simpson s. e., and Robertson co. Ten. s. Length 30, mean breadth 20, and area 600 sq. ms. Extending in lat. from 36° 36′ to 37° 02′, and in long. from 9° 33′ to 10° 03′ w. W. C. Logan occupies a part of the table land between the vallies of Cumberland and Green rs. From the sthrn. section flow the nthrn. branches of Red river of Cumberland, and from the nthrn. part flow creeks towards the n. into Green r. Chief town, Russellville. Pop. 1820, 14,423, 1830, 13,012.

Logan, co. Ohio, bounded s. by Champaign, w. Shelby, n. w. Allen, n. Hardin, and e. Union, lat. 40° 25′, long. 6° 45′ w., slope southward and principally drained by Sandy creek branch of Great Miami and the sources of Mad river. Chief town, Bellefontaine. Pop. 1830, 6,440.

Logan, p-v. and st. jus. Hocking co. Ohio, by p-r. 47 ms. s. e. Columbus. It is situated on Hockhocking river, near the northeastern angle of the co., lat. 39° 33′, long. W. C. 5° 24′ w. Pop. 1830, 97.

Logansport, p-v. and st. jus. Cass co. Ind. by p-r. 113 ms. a little w. of n. Indianopolis. It is situated at the junction of the main Wabash with Eel r., lat. 40° 45′, long. W. C. 9° 20′ w.

Loganville, p-v. York co. Pa. by p-r. 89 ms. northward W. C.

Log House Landing, p-o. southern part of Beaufort co. N. C. by p-r. 170 ms. a little s. of e. Raleigh.

Log Lick, p-o. eastern part of Clark co. Ky. by p-r. 51 ms. s. e. by e. Frankfort.

Lombardy, p-v. Amelia co. Virginia, by p-r. 50 ms. s. w. Richmond.

Lombardy, p-v. Columbia co. Geo. by p-r. 64 ms. n. e. by e. Milledgeville.

Lombardy Grove, p-o. Mecklenburg co. Virginia, by p-r. 81 ms. s. w. Richmond.

London, p-v. formerly Hazel Patch, st. jus. Laurel co. Ky. by p-r. 102 ms. s. e. Frankfort. It is situated on a tributary of Rock Castle, branch of Cumberland r., lat. 37° 13′, long. 6° 56′ w. Pop. 1830, 15.

London, p-v. and st. jus. Madison co. Ohio, by p-r. 27 ms. s. w. by w. Columbus, lat. 39° 50′, long. 6° 28′ w. Pop. 1830, 249.

London Bridge, p-v. in the northeastern part of Princess Ann co. Va. 15 ms. a little n. of e. Norfolk, and 8 ms. s. w. Cape Henry.

London Grove, post tsp. Chester co. Pa. between New Garden and Oxford. The p-o. is about 40 ms. s. w. by w. from Phila.

Londonderry, p-t. Rockingham co. N. H. 15 ms. n. Haverhill, Mass., 35 s. w. Portsmouth, 25 s. Concord. It is a valuable agricultural township, and contains an academy, with a fund of $14,000, the donation of Maj. John Pinkerton, after whom the academy is named. This town is celebrated for the longevity of its inhabitants. Population 1830, 1,467.

Londonderry, p-t. Windham co. Vt. 30 ms. n. e. Bennington, 27 s. w. Windsor. Watered by West and Winhall rivers, Utley brook and another considerable mill stream. Mill privileges are numerous. Contains a bed of very fine clay, two villages, and 9 school districts. Pop. 1830, 1,302.

Londonderry, p-v. Guernsey co. O. by p-r. 102 ms. eastward Columbus. Pop. 1830, 54.

Long Bottom, p-v. Meigs co. Ohio, by p-r. 102 ms. s. e. Columbus.

Longbranch, Monmouth co. N. J. The sea shore about 6 ms. s. of Shrewsberry river, a place of great resort for sea bathing and fishing, having several large and well kept boarding houses, 30 ms. s. New York.

Long Creek Bridge, and p-o. New Hanover co. N. C. by p-r. 128 ms. s. e. Raleigh.

Long Falls Creek, and p-o. Daviess co. Ky. by p-r. 165 ms. s. w. by w. Frankfort.

Long Hollow, p-o. Sumner co. Ten. by p-r. 14 ms. n. e. Nashville.

Long Island, N. Y. extends from the narrows, below New York city, in an easterly direction, 140 ms. to Montauk Point. Average width 10 ms. Contains 1,400 square ms. Divided into 3 counties, Kings, Queens, and Suffolk. It belongs wholly to the state of N. Y. Bounded s. by the Atlantic, separated from the continent on the n. by Long Island Sound, and East river. Much indented with bays. There is a rocky ridge denominated the spine of Long Island, extending from the w. end to River Head, the highest point of which is 319 feet above the level of the tide, situated in N. Hempstead. Land on the n. side of this ridge, rough and hilly, on the s. side level and sandy. Waters stored with a vast abundance and variety of fish, and the island has long been celebrated for its wild fowl, and various forest game. A beach of sand and stones runs along the s. side of the island 100 ms. with various inlets, admitting vessels of 60 or 70 tons. The long narrow bay formed by the beach is in the widest places 3 ms. broad.

Long Island Sound, an inland sea, from 3 to 25 ms. broad, and about 140 long, dividing Long Island from Conn. It communicates with the ocean at the n. end, and with

N. York harbor at the s. and affords a very safe and convenient passage.

Long Lick, p-o. Scott co. Ky., 23 ms. estrd. Frankfort.

Long Meadow, p-t. Hampden co. Mass., e. Conn. r., 6 ms. s. Springfield, 97 s. w. Boston; beautifully situated; soil fine. Pop. 1830, 1,257.

Longmires, Store and p-o. Edgefield dist. S. C., 76 ms. s. w. by w. Columbia.

Long Old Fields, p-o. Prince George's co. Md., 14 ms. estrd. W. C., and 26 wstrd. Annapolis.

Long Pond, Me. chiefly in Bridgetown, 10 ms. long and 1 broad, connected by Sungo r. with Sebago lake.

Long Pond, lake on the line between N. Y. and N. J., principally in the former; discharges through Long pond and Pompten rs. into the Passaic.

Long Prairie, p-o. Hempstead co. Ark., by p-r. 175 ms. s. w. Little Rock.

Long Run, p-o. Jefferson co. Ken., by p-r. 25 ms. w. Frankfort, and about 17 e. Louisville.

Long's Bridge, and p-o. Hancock co. Geo., 10 ms. n. e. Milledgeville.

Long's Mills, and p-o. Orange co. N. C., by p-r. 81 ms. n. w. Raleigh.

Long Street, p-v. Moore co. N. C.

Long Street, p-v. Lancaster dist. S. C., by p-r. 64 ms. n. n. e. Columbia.

Long Swamp, p-tsp. Berks co. Pa., situated on the head waters of Little Lehigh. The p-o. is 18 ms. n. e. Reading.

Longtown, p-v. Davidson co. N. C., by p-r. 152 ms. wstrd. Raleigh.

Longwood, p-v. Albemarle co. Va., by p-r. 86 ms. n. w. by w. Richmond, and 151 ms. s. w. W. C.

Loop, p-v. Logan co. Va., by p-r. 390 ms. s. w. by w. ½ w. W. C.

Lorain, co. O., bounded n. e. by Cuyahoga co.; e. Medina; s. e. Wayne; s. w. Richland; w. Huron, and n. lake Erie. From s. to n. 40 ms., mean breadth 15, and area 600 sq. ms. Lat. 41° 15', long. 5° 10' w. Slope almost due n. and drained by Black r. and some smaller streams. Chief t. Elyria. Pop. 1830, 5,686.

Lorenz, Store and p-o. Lewis co. Va., by p-r. 261 ms. w. W. C.

Loretto, p-v. Cambria co. Pa., 7 ms. n. e. Ebensburg, 75 ms. estrd. Pittsburg, and by p-r. 116 ms. n. w. by w. Harrisburg.

Lorraine, p-t. Jefferson co. N. Y., 16 ms. s. w. Watertown, 150 w. n. w. Albany, 9 e. lake Ontario. Healthy, and well watered by a number of small creeks, of a tolerable size for mill streams, and a great variety of small springs and rivulets. Pop. 1830, 1,727.

Lorretto, p-v. Essex co. Va., by p-r. 81 ms. n. e. Richmond.

Lost Prairie, p-o. Lafayette co. Ark , by p-r. 152 ms. s. w. Little Rock.

Lost r., local name of the higher part of Great Cacopon r. of Va.

Lost r., p-o. on Lost r., estrn. part of Hardy co. Va., by p-r. 130 ms. w. W. C.

Lott's mills and p-o., Copiah co. Miss. about 75 ms. n. e. Natchez, and 39 s. Jackson.

Lottsville, p-v. Warren co. Pa., by p-r. 235 ms. n. w. Harrisburg.

Loudon, p-t. Merrimack co. N. H., 7 ms n. e. Concord; furnished with valuable mill privileges by Soucook r.; timbered with sugar maple, beech, pine, oak, and chestnut. Pop. 1830, 1,642.

Loudon, p-v. in the wstrn. part of Franklin co. Pa., 15 ms. a little s. of w. Chambersburg, and 63 s. w. by w. Harrisburg.

Loudon, co. Va., bounded s. e. by Fairfax; s. by Prince William; s. w. by Fauquier; by the Blue Ridge separating it from Frederick w.; and Jefferson n. w; and by Potomac r. separating it from Frederick co. in Md. n.; and Montgomery co. Md. n. e. Length from s. e. to n. w. 22, mean breadth 21, and area 462 sq. ms. Extending in lat. 38° 49' to 39° 18' n., and in long. 0°, 20' to 0° 54' w. W. C. The declivity of this co. is to the n. e. towards the Potomac; surface broken and even in part mountainous; much of the soil excellent. Chief t. Leesburg. Pop. 1820, 22,702; 1830, 21,939.

Loudonville, p-v. Richland co. O., by p-r. 67 ms. n. e. Columbus.

Louisa, co. Va., bounded by Hanover s. e.; Goochland s.; Fluvanna s. w.; Albemarle w.; Orange n.; and Spottsylvania n. e. Length 36, mean breadth 16, and area 576 sq. ms. Extending in lat. 37° 45' to 38° 6', and in long. 0° 48' to 1° 28' w. W. C. The declivity of this co. is towards the s. e., down which flow numerous branches of N. and S. Annanvers. Chief t. Louisa C. H. Pop. 1820, 13,746; 1830, 16,151.

Louisa, p-v. and st. jus., Lawrence co. Ky., by p-r. 127 ms. e. Frankfort; lat. 38° 12', long. 6° w. Pop. 1830, 87.

Louisa, usually called Louisa C. H., p-v. and st. jus., Louisa co. Va., by p-r. 110 ms. s. w. W. C., and 54 ms. n. w. Richmond; and on Tanner's map U.S. exactly on the intersection of lat. 38° and 1° w. W. C.

Louisburgh, p-v. and st. jus., Franklin co. N. C., 30 ms. n. n. e. Raleigh; lat. 36° 06', long. 1° 18'.

Louisiana, state of the U. S., bounded s. by the Gulf of Mexico; e. and n. e. by the state of Miss.; n. w. by the ter. of Ark.; and w. by the Mexican province of Texas. Louisiana, without including the partial indentations of the coast, extends along the Gulf of Mexico 400 ms.; up Sabine r. from the mouth of that stream, to where intersected by n. lat. 32°, 190 ms.; thence along one degree of lat. 32° to 33°, 69½ ms.; thence due e. along lat. 33°, to the right bank of Miss. r. 168 ms.; thence down the latter r. to where it is crossed by lat. 31°, 220 ms.; thence along lat. 31° from the Miss. to Pearl r. 105 ms.; thence down Pearl r. to the mouth, 60 ms.; having an entire outline of 812½ ms. The longest line that can be drawn over La., is a diagonal from the s. pass of Miss. to the N. W. angle, 380 ms., and the area being 48,320 sq. ms., the mean breadth is about 127 ms. In lat. this state extends 28° 56′ to 33°, and in long. 11° 55′ to 17° 25′ w. W. C. In regard to natural features, and to the intrinsic qualities of soil, La. is divisible into four distinct sections. The Delta, is the first which is indefinitely connected with the great prairies of Attacapas and Opelousas. The latter is followed by the immense pine and oak forests of the nrthwstrn. part of the state. The fourth, though the least extensive, is in many respects the most interesting section of La.; that is the fine slope formerly part of west Florida, between the Pearl and Miss. rs., and s. of lat. 31°.

In a general view, the Miss. r. is the most conspicuous and important feature in the topography of La. By a very winding channel, that great river forms a boundary between the states of Miss. and La. between lat. 31° and 33°, but below the former lat. enters entirely into La. Assuming a course of a little E. of S. E., but still with a very sinuous channel, the Miss. winds over La., embracing by its numerous inlets or mouths, an alluvial region, to which in nature, and even in outline, the name of Delta is not unaptly applied.

Between lat. 31° and 33°, the general course of the Miss. is along the bluffs, or wstrn. margin of a comparative table land. Here it receives numerous small streams from the left, whilst on the right stretchés a narrow, annually inundated tract. When swelled by spring floods, the superabundant water of the Miss. flows out by innumerable channels, which are discharged into the Tensaw, Black, and Red rs., and by the latter borne back into the main stream. But, as if disdaining to receive into its bosom the rejected water, less than 1½ mile below the mouth of Red r., the Atchafalaya is discharged to the left, forming the upper mouth of the Miss.; below which on that side, the adjacent surface of the land being lower than that of the surcharged r., all water which escapes from the main stream returns to it no more, but slowly seeks a recipient in the Gulf of Mexico. On the estrn or left, the bluffs are continued on or near the Miss. to a few ms. below Baton Rouge, where the outlet of Iberville terminates high land, and commences the Delta on that side also.

If we regard the efflux of Atchafalaya, as its head, and the Gulf of Mexico as its base, the Delta stretches over two degrees of lat. and three degrees of long. The utmost length from the outlet of the Atchafalaya, to the mouths of the Miss., 220 ms. Its widest part from the Point Timballier to the Pass of Manchar, between lakes Pontchartrain and Maurapas, 100 ms. The breadth, however, varies from ten miles to the utmost width. From the generally well defined outline, the limits of the Delta are distinct; but from the great indentations of that outline, the area is difficult to estimate accurately, but amounts to at least the one fourth of the state, or 12,000 sq. ms.

The lower, or sthrn. and sthestrn. part of the Delta, is with very trifling exceptions sea marsh, naked of timber, and flooded with every flow of the tide, and with very few spots or strips of arable soil. Advancing nrthwstrdly. up the streams, the surface very slowly rises, and the arable borders along the rivers increase in width, and become more continuous. The unwooded sea marsh is followed by a dense forest, but which stands on a plain in a great part annually inundated by the spring floods of the Miss., Ouachitta, and Red rs. The very gradual and trifling acclivity of the Delta, is demonstrated by the fact, that in autumn, when the rivers are reduced to their lowest level, the tides of only about two feet mean height, are sensible in Atchafalaya and Iberville. I have myself seen the current of the former flowing into the Miss.

West from the Delta, the sea marsh is continued, and the prairie or grassy plains rise from the great forest overflow of Atchafalaya. The acclivity from the sea marsh of Opelousas and Attacapas is so much more abrupt as to raise the surface of the prairies above annual overflow, but even here the rise is very gradual and so small as to admit the tides in autumn, as high as Lemell's landing, on Courtableau river. From actual observation I doubt whether a single spot of southwestern La. below lat. 31° is elevated 50 feet above high tide; the far greater part is, I am confident, under ten feet comparative elevation.

The surface of the prairies of La. has been very greatly overrated, and these plains have also been confounded with the sea marsh. Though contiguous, and similar in the single feature of being void of timber, in all other respects these two sections differ from each. The prairies though approaching a dead level, are composed of solid, and arable soil. If an eye sufficiently elevated could scan the whole surface from the Gulf of Mexico to the forests of Red and Sabine rivers, the streams would be seen issuing from those forests and carrying lines of woods along their banks, and which wooded borders gradually narrowing would terminate at different distances from the sea coast, in most cases about the line of separation between the Prairie and sea marsh, small wooded spots isolated from the great forest would be seen dotted along the sea marsh, the timber of which composed in part of live oak retaining leaf throughout the winter. But on the sea margin of the Sabine and Calcasire, even the live oak ceases, and the great grassy marsh expands.

The western sea marsh of La. may be regarded as extending from the Atchafalaya to the Sabine 160 ms., the mean breadth about 25, and area 4,000 square ms. The prairie section reaches from the junction of Techo and

Atchafalaya, to the Sabine, 160 ms. along the sea marsh, but lies in form of a triangle, the apex at the head waters of Mumentau, perpendicular 60 ms. and area 4,800 square ms. In the latter superficies are, however, included large bodies of woods; the real prairie does not exceed the sea marsh in extent, and the aggregate of both may be safely assumed at 8,000 square ms.

Lying northward from the prairies of Opelousas, and westward of the inundated margin near the Miss. spreads what may, from its prevailing timber, be called the pine section of La. This extensive region, embracing about 24,000 square ms. is watered by Ouachitta, Red, Calcasin and Sabine rivers. The surface considerably broken into hills, though of moderate elevation. In this tract some rocks and even water falls appear. The low grounds near streams are clothed with various species of oaks, elms, hickory, sweet gum, honey, locust, and cypress, but leaving the water courses, pine prevails to such an extent, that from five to twenty ms. may be travelled over in one unbroken pine forest. The river soil on Red and Ouachitta, generally productive; fertile water margins occur in other places, but the general character of the soil is sterility.

The fourth and least extensive natural section of La. is that of the former W. Florida. In general character, the latter bears a very exact resemblance to that of the northwest. Pine becomes again so much the prevailing timber, that at least nine parts in ten of the whole surface is covered with this tree. The arable soil is on or near the streams and confined in extent.

Taken as a whole, La. is composed of inundated and noninundated land. The tract of soil liable to annual submersion, is narrow above Red river, but widening below that stream, expands like a fan, and finally embraces the whole gulf border. What soil is of adequate elevation for cultivation within the inundated region, is of the very best quality, and towards the gulf, the climate admits the very profitable growth of sugar cane.

The Gulf's grassy border is followed inland by a forest which, from the peculiar nature of the soil, must remain many ages but partially disturbed. Of forest, in point of relative quantity, pine, oak, sweet gum, and hickory predominates, but admixed with an indefinite number of other trees, such as maple, liriodendron, cypress, black gum, ash, persimon, black walnut, honey, locust, elm, dog wood, &c. On the margin of overflow, immense brakes of reed cane rise amid the forest, but this gigantic grass, contrary to common opinion, never flourishes where the surface is liable to periodical submersion. Where the cane abounds, so do various species of grape vine and smilax, rendering those forests most difficult to penetrate. Below lat. 31° and on land partially liable to overflow, are extensive brakes of palmetto, or dwarf palm. The latter vegetable, though capable of supporting the inundations longer than the reed cane, cannot, however, exist where the ground is liable to deep and annual overflow. In the latter case indeed the ground produces few weeds and the lofty trees are the only vegetables of any consequence which rise from the saturated earth.

Climate and seasons.—In a country where the extremes of latitude are only 4 degrees, and those of height perhaps less than two hundred feet, it might be supposed that very little difference of seasons, would be perceptible, but with both these causes of equality, the mean and extreme temperature of the Delta, and that of the northwestern section differs far beyond what could be expected. The relative temperature is more decisively shown by indigenous vegetables than by observations made with a thermometer. Amongst those vegetable indicia the live oak affords the most conclusive data. This tree is found to abound most in the lower part of the Delta, and to decrease ascending to the N. W. It would appear from places where this tree flourishes, that its existence must depend more on the relative temperature than on soil. From the bay of Mobile westward to the Teche, the live oak is limited northward by N. latitude 30° 25′ very nearly. Passing the Teche, where the northwestern winds have free access over the prairies, live oak ceases in great part above lat. 30°, and on the Calcasin and Sabine does not exist. On the Atlantic coast of the U. S. the live oak is found as far N. as 34°.

The cultivation of sugar cane ceases in Louisiana at about N. lat. 30° 10′, but on the Atlantic coast can be made a profitable crop two degrees higher. Similar remarks apply to the orange tree and some other exotic trees, which are restricted on the Delta of the Miss. and contiguous places between two and three degrees lower than on the Atlantic coast.

At Natchez, lat. 31° 33′ the thermometer has fallen to 12° above the zero of Fah't. I have myself seen the creeks and ponds of La. at New Orleans, frozen, and once, January 1812, saw snow at Opelousas 11 inches deep. These phenomena are rare, but their occurrence exhibits a severity of climate much greater than is experienced on similar latitudes along the Atlantic ocean.

In regard to staple productions, sugar and rice in La. will, it is probable, be always restricted to the lower sections, whilst cotton can be cultivated over the whole surface, as may be maize, tobacco, and indigo.

Of fruit trees, the peach and fig are those which seem most congenial to the climate. The apple can be cultivated, but not to advantage; the cherry is utterly unproductive. The latter circumstance is the more curious as the wild cherry tree grows to the size and elevation of a forest tree of large magnitude, not unfrequently of 50 or 60 feet shaft, and from 2 to 3 feet diameter.

If we assume New Orleans as a stationary

point, and allow a mean temperature of 60° Fah't. probably rather too high, we have a temperature very nearly similar to that of Charleston, S. C. (*See the latter article.*) By recurring to Dr. Lovell's tables of relative mean temperature,& comparing the mean range of thermometer at Cantonment Jessup, on the Sabine, Baton Rouge, Pensacola, Tawpa Bay, St. Augustine and Charleston, we find the curious result, that the lowest depression at Cantonment Jessup was plus 7, Baton Rouge plus 18, Pensacola plus 11, and at Charleston plus 19°. In brief, combining vegetable physiology with the thermometrical results, the seasons of Charleston, lat 32° 42', are milder than at New Orleans, lat. 30°.

Political Geography.—For civil or municipal purposes Louisiana is subdivided as follows :—

Parishes.	Chief Towns.	Pop. 1830.
Ascension,	Donaldsonville,	5,426
Assumption,	Assumption,	5,669
Avoyelles,	Marksville,	3,484
Catahoola,		2,581
Claiborne,		1,764
Concordia,	Concordia,	4,662
East Baton Rouge,	Baton Rouge,	6,698
East Filiciana,	Jackson,	8,247
Iberville,	Iberville,	7,049
Jefferson,		6,846
Lafayette,		5,653
Lafourche Interior,	Thibadeauxville,	5,503
Natchitoches,	Natchitoches,	7,905
Orleans,	New Orleans,	49,838
Plaquemines,	Plaquemines,	4,489
Point Coupee,	Point Coupee,	5,936
Rapides,	Alexandria,	7,575
St. Bernard,		3,356
St. Charles,		5,147
St. Helena,	St. Helena,	4,028
St James,	Bringier's,	7,646
St. John Baptist,	Duboуу's,	5,677
St. Landry,	St. Landry,	12,591
St. Martins,	St Martinville,	7,205
St. Mary's,	Dutch Settlement,	6,442
St. Tammany,	Covington,	2,864
Terre Bonne,		2,121
Washitau,	Monroe,	5,140
West Baton Rouge,	Mt. Pleasant,	3,084
W. Feliciana,	St. Francisville,	8,629
Washington,	Franklinton,	2,286
Total.		215,541

Of the above, 109,600 are slaves.

Principal towns.—The only city of consequence, is New Orleans, which see. Donaldsonville, Baton Rouge, St. Martins, St. Landry, Alexandria, and Natchitoches, are small villages, which will be found noticed under their respective heads. Donaldsonville is at present the seat of legislation.

Constitution of government.—The legislative power is vested in a senate and house of representatives. To be eligible to the senate demands a landed estate, in full right of $1,000. The members of the senate shall be chosen for the term of 4 years. Senators divided by lot into two classes ; the seats of the senators of the first class, shall be vacated at the expiration of the second year, & of the second class at the expiration of the fourth year ; so that a rotation shall be chosen every year, and one half thereby be kept up perpetually. No person shall be a senator, who, at the time of his election, is not a citizen of the United States, and who hath not attained the age of 27 years, resided in this state four years, next preceding his election, and one year in the district in which he may be chosen.

No person shall be a representative, who, at the time of his election, is not a free white male citizen of the United States, and hath not attained the age of 21 years, and resided in the state two years next preceding his election, and the last year thereof in the county of which he may be chosen, and who must hold landed property to the value of 500 dollars, according to the tax list in the county or district for which he is chosen.

The supreme executive power is lodged in the hands of a governor, chosen for 4 years, and ineligible for the succeeding 4 years after the expiration of the time for which he shall have been elected. He shall be at least 35 years of age, and a citizen of the United States, and have been an inhabitant of this state, at least six years preceding his election, and shall hold in his own right a landed estate of $5,000 value, agreeably to the tax list. No member of congress, or person holding any office under the United States, or minister of any religious society, shall be eligible to the office of governor.

The general powers of the governor of La. in extent and limitation, are very similar to those of the president of the United States.

The judiciary power is vested in a supreme and inferior courts The supreme court having appellate jurisdiction only, extending to all civil cases where the matter in dispute shall exceed the sum of $300. The supreme court shall consist of not less than three judges, nor more than five ; the majority of whom shall form a quorum. The legislature is authorised to establish such inferior courts as may be convenient to the administration of justice.

The judges, both of the supreme and inferior courts, shall hold their offices during good behavior. Removeable by address of both houses of the legislature, or by impeachment by the lower house before the senate, and in both cases a concurrence of two thirds requisite for removal or conviction.

To enjoy the right of suffrage, it is requisite, to be a free white male citizen of the United States, to have attained the age of 21 years, resided in the county where he offers to vote, one year next preceding the election, and within the last six months prior to the said election, have paid a state tax.

No person, while he continues to exercise the functions of a clergyman, priest, or teacher of any religious persuasion, society, or sect, shall be eligible to the general assembly, or to any office of profit or trust under this state.

History.—The term Louisiana, once so comprehensive, including all Arkansas, Missouri, the undefined regions on the waters of Miss. and the region now included, under the name of La. was imposed by the French in honor of Louis XIV. Confining therefore this brief notice to the state to which the name is now exclusively appropriated, we may observe that M. de la Salle, a French officer, made its first known civilized dicovery, in 1683. In 1699 M. d'Iberville laid the foundation of the first French colony. The local knowledge of the country was so defective, that the first settlements were very injudiciously made along the barren coast east and west from Mobile, and so slow was the advance, that in 1712, the inhabitants amounted to only 400 whites, and 20 negroes.

Hitherto a royal colony, in 1712 La. was ceded to Crozet, who after abortively expending large sums, in 1717 surrendered the government to the Miss. company. In the latter year the permanent base of the colony was laid by the foundation of New Orleans. Under the Miss. company La. flourished, though ultimately ruinous to the company itself, who in 1731, ceded their powers to the crown. Again a royal colony, La. slowly augmented in population and wealth, but until its cession to Spain by France, in 1762, and its being taken into actual possession by the former in 1769, the colony afforded no important matter for history.

France after her cession of La. regretted the step, and by a secret treaty with Spain, signed Oct. 1st, 1800, La. was receded to her former parents. The reacquisition of the colony availed nothing to France in the accomplishment of its original intention, but it enabled her to negociate a sale to the United States in consideration of 60,000,000 of francs. These negociations were consummated April, 1803. In the following December, the Spanish commissioners transferred the country to France; the authorities of the latter duly transferred it to the United States.

By an act of Congress, passed March, 1804, La. was definitively subdivided; the northern part above lat. 31° was named "*The Territory of Miss.*" the lower section, "*The Territory of Orleans.*" The latter in 1811 was authorised to form a constitution of government, and that part of West Florida, west of Pearl river, subsequently annexed. Thus bounded as noticed in the first part of this article, La. in 1812, was formally received into the Union as a sovereign state.

A powerful British fleet and army invaded La. December, 1814, but after some partial actions, the army was utterly defeated, January 8th, 1815, and the seige of New Orleans raised. This event gave to La. a classic interest in the history of the United States, and left her to the peaceable pursuit of the arts of social life, and the cultivation of her soil.

The progressive population of this state, has been regular though not comparatively rapid.

In 1810, the inhabitants amounted to		86,000
1820,	. . . .	153,000
1830,	. . . .	215,541

Louisiana, p.v. on the right bank of the Miss. at the mouth of Salt r. Pike co. Mo. 12 ms. n. e. Bowling Green, and 90 ms. n. w. St. Louis.

Louisville, p-t. St. Lawrence co. N. Y. on St. Lawrence r., 30 ms. below Ogdensburgh. Soil a rich loam, gently uneven, well watered with Racket and Grass rs. Has vast forests of pine, cedar, &c. Williamsburgh in Canada, where was fought the battle of Williamsburgh, Nov. 11, 1813, lies opposite this town. 10 schools, kept 7 months in 12. Pop. 1830, 1,076.

Louisville, p-v. and st. jus. Jefferson co. Geo. situated on or near the left bank of Great Ogeechee r. 58 ms. a little s. of e. Milledgeville, and 43 ms. s. w. Augusta. Lat. 33° 02', long. 5° 22' w. W. C.

Louisville, p-v. or city, Jefferson co. Ky., is situated on the left bank of O. r. between the head of the Rapids and the mouth of Bear Grass cr., 52 ms. a little n. of w. Frankfort, and 112 ms. s. w. and by the land route from Cincinnati, n. lat. 38° 17', and long. 8° 45' w. W. C.

The site of Louisville is a swelling bank, rising by a gentle acclivity from the r. and from Bear Grass cr. The streets are laid out at right angles to each other. The advance of this port, for such it is in fact, has been very rapid. In 1800, the pop. amounted to 1,357, and by the census of 1830, 10,196. The manufacturing establishments are numerous and valuable, as are the buildings for judical, commercial, and religious purposes. The most important works, however, ever attempted near Louisville, is a canal on the Ky. side to pass the Rapids. This enterprise, so interesting not alone to the people of the wstrn. states, but those of the whole Union, either direct or indirectly, advances towards completion. In brief, this flourishing town exhibits all the attributes of a prosperous commercial depot.

Louisville and Portland Canal. (*See article rail roads and canals.*)

Louisville, p-v. Blount co. Ten. by p-r. 168 ms. a little s. of e. Nashville.

Louisville, p-v. Pike co. Ala. by p-r. 144 ms. s. e. by e. Tuscaloosa.

Loutre (*Otter*) Island, and p-o. sthrn. part of Montgomery co. Mo., 75 ms. wstrd. St. Louis, and by p-r. 59 ms. estrd. Jefferson City.

Loutre Lick, and p-o. Montgomery co. Mo. 78 ms. wstrd. St. Louis.

Lovell, p-t. Oxford co. Me. 20 ms. n. Paris. In this town are Lovell falls, which have been discovered within a few years, and are an object of great natural curiosity. Where the water makes over into the tremendous basin below, it falls perpendicularly 40 feet. Above the falls, there is a chain of 8 ponds, partly in Lovell, and partly in Waterford, connected by small natural dams one or two rods in width, through which there are sluice-ways, which will admit the passage of a common sail boat. The scenery of the mountains and ascending lands in the vicinity, is rural and beautiful. Pop. 1830, 697.

LOVELL'S POND, N. H., the head of the E. branch of the Piscataqua.

LOVELY, co. of Ark. This co. is named in the p-o. list, but if it exists, it has been omitted on Tanner's U. S.

LOVETTSVILLE, p-v. Loudon co. Va. in the N. W. part of the co. by p-r. 55 miles N. W. W. C.

LOVEVILLE, p-v. New Castle co. Del. 103 ms. N. E. W. C.

LOVINGTON, p-v. and st. jus. Nelson co. Va. situated on a branch of Tye r. by p-r. 94 ms. a little N. of W. Richmond, and 171 ms. S. W. W. C. Lat. 37° 44′, and long. 1° 52′ W. W. C.

LOWELL, the American Manchester, situated at the confluence of Merrimack and Concord rs. Middlesex co. Mass. This place is undoubtedly destined to be a manufacturing city. Its growth for a few years past has been almost unparalleled. The foundation of the second factory was laid here in 1822, at which time, the territory now included in the town, exclusive of one factory establishment, contained less than 100 inhabitants. There are now 9 manufacturing cos. viz. the Merrimack, Appleton, Hamilton, Lowell, Hurd's (formerly), Jackson, Tremont, Suffolk, and Lawrence. In 1831, when only the first five were in operation, from 12 to 14 million yards of cloth were manufactured in a year, equal to 1 yard per second. The Lowell company make carpets, which are equal to the imported. There are 15 houses of worship, and 3 newspapers. 200 houses were built between April and November 1831. $500,000 worth of land was sold that year, and $270,000 in 4 years. Land rose 100 per cent. in 1831. Rents afford a higher profit than in any other New England town. The Merrimack manufacturing company have a capital of $1,500,000, with 5 large brick factories, containing 26,000 spindles, and about 1000 looms. They employ from 3 to 400 males, and from 8 to 900 females, and use 5,000 bales of cotton, or about 1,500,000 lbs. annually. They manufacture, bleach, and print, 6,500,000 yards yearly. The Hamilton and Lawrence manufacturing companies have each a capital of $1,200,000. Lowell manufacturing company $600,000; Appleton, Middlesex, and Tremont, $500,000 each; Suffolk $450,000. Then there are the locks and canal company with a capital of $600,000, who own the water privileges, and dispose of them as they are wanted. This company own a machine shop 150 ft. by 40, and 4 stories high, in which are employed about 200 hands. The stock of this co. is 160 per cent. advance. The great water power is produced by a canal a mile and a half long, 60 ft. wide, and 8 ft. deep, from its commencement above the head of Pawtucket falls on the Merimack, to its termination in Concord r. The entire fall is 32 ft. The water is taken from this canal by smaller canals, and conveyed to the factories, and thence into the Merrimack. There are room and water power sufficient for 50 huge additional factories! In the suburbs of Lowell, near the canal, is a settlement called New Dublin, which occupies upwards of an acre of ground. It contains not far from 500 Irish people, and about 100 cabins, from 7 to 10 ft. high, built of slabs and rough boards, a fire place made of stones in one end, topped out with several flour barrels or lime casks. In a central situation is the school house, built in a similar style, turfed up to the eaves, with a window in one end, and small holes in two sides for the admission of air and light,—all this under the eye of capitalists having their seven millions invested in establishments along side of them! There is a canal round the falls of the Merrimack, 90 feet wide and 4 deep; which however is no longer used for boat navigation. On the Concord r. about one m. from the town, are powder works, at which powder of a very superior quality is made. 30,000 kegs, 25 lbs. each, are made annually. Lowell communicates with Boston by means of the Middlesex canal, and a rail road between the two places is in progress. It lies 25 ms. N. W. Boston. The village of Belvidere, on the opposite side of Concord r. has grown up along with Lowell, and, from its contiguity, seems but a part of the latter place. Lowell continues rapidly to increase, and is becoming a rival of the manufacturing towns of England. Pop. in 1830, 6,474. At the next census, it will probably contain 20 or 30,000.

LOWER BEAVER, p-v. wstrn. part of Beaver co. Pa. about 26 ms. N. W. Pittsburg.

LOWER BLUE LICK, and p-o. Nicholas co. Ky. by p-r. 65 ms. N. E. by E. Frankfort.

LOWER CHANCEFORD, p-v. York co. Pa. 20 ms. nthestrd. the boro' of York, 16 S. S. W. Lancaster, and 36 S. E. Harrisburg.

LOWER CR. p-o. Burke co. N. C. 14 ms. N. E. Morgantown, and by p-r. 219 ms. wstrd. Raleigh.

LOWER MARLBOROUGH, p-v. on the left bank of Patuxent r. in the nthwstrn. part of Calvert co. Md. by p-r. 49 ms. S. E. W. C., and about a similar distance S. W. Annapolis.

LOWER MERION, p-v. Montgomery co. 14 ms. nthrds. Phil.

LOWER PEACH TREE, p-o. Wilcox co. Al. by p-r. 129 ms. sthrd. Tuscaloosa.

LOWER SALEM, p-v. Washington co. O. by p-r. 118 ms. S. E. Columbus.

LOWER SANDUSKY, p-v. and st. jus. Sandusky co. O. by p-r. 103 ms. N. Columbus. Lat. 41° 21′, long. 6° 10′ W. Pop. 1830, 351.

LOWER SAUCON, p-v. and tsp. sthrn. part of Northampton co. Pa. by p-r. 51 ms. nthrd. Phil. 97 ms. estrd. Harrisburg, and 187 ms. N. E. W. C. Pop. tsp. 1830, 2,308.

Lower Smithfield, p-v. Northampton co. Pa. by p-r. 15 ms. n. n. e. Easton.

Lower Three Runs, p-o. southern part of Barnwell dist. S. C. by p-r. 81 ms. s. w. Columbia.

Lowe's p-o. Robertson co. Ten. 23 miles nthrd. Nashville.

Lowhill Port, p-v. eastern part of Lehigh co. Pa. by p-r. 76 ms. n. e. by e. Harrisburg, and 179 n. n. e. W. C.

Lowman, p-v. in the northern part of Lewis co. Va. by p-r. 256 ms. w. W. C.

Lowndes, co. of Geo. bounded w. by Thomas, n. by Irwin, e. by Ware, s. by Hamilton co. in Flor., and s. w. by Madison co. Flor. Length from s. to n. 52, breadth 40, and area 2,080 sq. ms. Extending in lat. 30° 38′ to 31° 22′, and in long. 6° 6′ to 6° 46′. The slope of this co. is almost directly s., and is entirely drained by various confluents of Suwanee r. Pop. 1830, 2,453.

Lowndes, C. H. and p-o. Lowndes co. Geo. by p-r. 165 ms. s. Milledgeville.

Lowndes, co. Ala. on Ala. r. bounded n. e. and e. by Montgomery, s. e. by Pike, s. by Butler, s. w. by Wilcox, w. and n. w. by Dallas, and n. by Ala. r. separating it from Autauga. Greatest length diagonally from s. e. to n. w. 50 ms., mean breadth 32, and area 1,600 sq. ms. Extending in lat. 31° 51′ to 32° 23′, and in long. 9° 21′ to 10° 04′ w. W. C. The general slope of this co. is to the n. n. w. towards the Ala. r. Chief t. Lowndes C. H. Pop. 1830, 9,410.

Lowndes, co. of Miss. bounded by Battalatche r. separating it from Monroe on the n. w., by Lafayette, Ala. n. e., by Pickens Ala. s. e., and Tombigbee r. separating it from the ter. of the Chickasaws s. w. and w. Length along Ala. line 36 ms., mean breadth 9, and area 324 sq. ms. Extending in lat. 32° 18′ to 50° 54′ n., and in long. 11° 21′ to 11° 36′ w. W. C. The slope of this co. is westrd. towards Tombigbee r. Chief town, Columbus. Pop. 1830, 3,173.

Lowndes, p-v. Rankin co. Miss. by p-r. 118 ms. n. e. Natchez, and 6 from Jackson.

Lowrey's Mills and p-o. Chesterfield dist. S. C. by p-r. 143 ms. n. e. Columbia.

Lowville, p-t. Lewis co. N. Y. 35 ms. n. Utica, 150 from Albany, w. Black r. Well watered. Eligibly situated. 10 school dists. Contains an academy and a handsome village. Pop. 1830, 2,334.

Loyalsock, small r. of Pa. rising in the sthrn. part of Bradford county, interlocking sources with Mahoopenny and Towanda crs. Entering Lycoming co. and flowing s. w. between the vallies of Muncy and Lycoming creeks, it falls into the w. branch of Susquehannah r. 4 ms. below the boro' of Williamsport.

Lubec, p-t. and port of entry, Washington co. Me. in Passamaquoddy bay, lying however on the main land, and possessing a spacious harbor, sheltered from every wind, and never closed by ice. The first settlement was made here no longer ago than 1815. A valuable lead mine has recently been discovered at this place. Pop. 1830, 1,535.

Lucastown, p-v. Limestone co. Ala. by p-r. 132 ms. a little e. of n. Tuscaloosa.

Lucasville, p-v. Sciota co. O. by p-r. 79 ms. s. Columbus. Pop. 1830, 45.

Ludlow, p-t. Windsor co. Vt. 16 ms. w. Windsor, 61 s. Montpelier. Watered by Black and Williams rs. Mountainous. Well situated for trade with the surrounding country. Land in general well timbered. Amethyst in crystals has been found here, three fourths of an inch long, and an inch in diameter. 10 school dists. Pop. 1830, 1,227.

Ludlow, p-t. Hampden co. Mass. 90 ms. s. w. Boston, 10 n. e. Springfield. Watered by the Chickapee and several smaller streams. Here is a glass manufactory, with a capital of $40,000. Pop. 1830, 1,327.

Ludlow, Morgan co. O. (*See Olive Green.*) The office is now called permanently Ludlow, and is situated by p-r. 85 ms. s. e. by e. Columbus.

Ludlowville, p-v. (*See Lansing.*)

Lumber, r. of N. and S. C., rises in Montgomery and Moore cos. of the former, the boundary between which it for some distance forms, flowing in a s. s. e. course. Thence assuming a southern course between Richmond and Cumberland, and between Richmond and Robeson. Turning to s. e. and entering and traversing Robeson, it once more inflects to s. s. w., and separating Robeson and Columbia cos. finally enters S. C. between Marion and Horry districts, joins little Pedee after a comparative course of about 100 ms. Lumber r. is the northeastern and main branch of Little Pedee.

Lumberland, t. Sullivan co. N. Y. 14 ms. s. w. Monticello, on Del. r. It is well named, being in reality lumber land. Pop. 1830, 953.

Lumberton, p-v. and st. jus. Robeson co. N. C., situated on the left bank of Lumber r. 32 ms. s. s. w. Fayetteville, 33 w. Elizabethtown, and by p-r. 92 ms. s. s. w. Raleigh. Lat. 33° 41′, long. 2° 10′ w. W. C.

Lumberville, p-v. on the right bank of Del. r. Bucks co. Pa. 10 ms. n. e. Doylestown, and 35 n. Phil.

Lumpkin, formerly called Randolph C. H., p-v. and st. jus. Randolph co. Geo. by p-r. 170 ms. s. w. Milledgeville.

Lunenburgh, p-t. Essex co. Vt. 45 ms. e. n. e. Montpelier. Some parts very stony. Timber generally hard wood. Conn. r. waters the s. e. part, besides which the town is watered by Neal's and Catbow branch, which are considerable mill streams. 9 school districts. Pop. 1830, 1,054.

Lunenburg, p-t. Worcester co. Mass. 45 ms. n. w. Boston, 26 n. Worcester. Watered by several branches of Nashua r. Pop. 1830, 1,317.

Lunenburg, co. Va. bounded by Meherin r. separating it from Mecklenburg s., by Charlotte w., Prince Edward n., Nottaway r. separating it from Nottaway co. n. e., and by

Brunswick E. Length 26, mean width 16, and area 416 sq. ms. Extending in lat. 36° 46′ to 37° 04′, and in long. 1° 8′ to 1° 32′ w. W. C. The slope of this co. is to the S. E. by E. Chief town, Lewistown. Pop. 1820, 10,662, 1830, 11,957.

LUNENBURG, C. H. and p-o. (*See Lewistown, Lunenburg co. Va.*)

LUNEY'S creek and p-o. Hardy co. Va. 10 ms. wstrd. Moorfields, and by p-r. 133 wstrd. W. C.

LURAY, p-v. Shenandoah co. Va. by p-r. 132 ms. wstrd. W. C.

LUSK'S FERRY and p-o. Livingston co. Ky. by p-r. 254 ms. s. w. by w. Frankfort.

LUTHERSBURG, p-v. Clearfield co. Pa. by p-r. 212 ms. N. W. W. C.

LUZERNE, p-t. Warren co. N. Y. E. Hudson r. at Hadley falls, 12 ms. s. w. Caldwell, 12 w. Sandy Hill. Pop. 1830, 1,362.

LUZERNE, co. Pa. bounded by Columbia s. w., Lycoming w., Bradford N. W., Susquehannah N., Wayne N. E., Pike E., Northampton S. E., and Schuylkill co. s. Length s. to N. 50 ms., mean breadth 36, and area 1,800 sq. ms. Extending in lat. 40° 56′ to 41° 43′, and in long. 0° 40′ to 1° 36′ E. W. C. This is perhaps the most diversified co. in the U. States. The N. E. branch of Susquehannah enters it from Bradford at the N. W. angle, and pursuing a S. E. course forces its passage through numerous mountain chains, by comparative courses 35 ms. to the entrance of Lackawannoc r. from the N. E. It here enters Wyoming valley, and turning at right angles, flows down the mountain vallies again 35 ms. to the mouth of Nescopeck creek, where it leaves Luzerne and enters Columbia. The peculiar features of Susquehannah r. will be seen under its own head. It is evident from the course of Susquehannah r. through Luzerne, that the western part of the co. is included in a concavity of that stream, and vice versa, on the opposite side. In the concave section the creeks are small, but on the convex side two streams of considerable comparative magnitude enter from the N. E., the Tunkhannoc and Lackawannoc. Below the latter, and estwrd. from Wilkesbarre, the higher sources of Lehigh r. are within 5 ms. from the Susquehannah bank. Luzerne is composed of narrow vallies and intervening mountain chains, both extending from N. E. to S. W. That part of the Susquehannah valley above the entrance of Lackawannoc being the only exception to the foregoing arrangement.

The vallies of Luzerne are narrow, but contain the great body of the population. The central and principal valley is that of Wyoming, between Bullock and Shawaney mtns. The distance from ridge to ridge about 5 ms., but the real arable part of the valley less than 2 on an average. This fine vale commences about 15 ms. below Wilkesbarre, and extending to the N. E. is continued by the Lackawannoc into Wayne co., and besides the great fertility of soil in most of its length, this great valley abounds in interminable strata of mineral coal, both along the Susquehannah and Lackawannoc.

Tunkhannoc valley in the northern part of the co. is narrow, not exceeding 1 mile; also very irregular, but extends into Susquehannah co. Nescopec valley is in the extreme southern part of the co.; this valley abounding in excellent soil, it is remarkable, is detached from the Susquehannah r.; the Nescopeck cr., after winding about 20 ms. to s. w. by w. turns abruptly N., leaves the arable valley by passing through a mountain chain.

On the concave or western part of Luzerne, the two principal vallies are those of Mohoopenny and Bowman's.

Of the whole surface of this large co. about one fortieth is perhaps already cultivated, and at most one third admits of being so; in pasture more might be made useful, but more than one half is irreclaimable.

In mineral coal this co. is peculiarly affluent. The writer has himself visited many of the mines already opened. The quantity seems to admit of indefinite supply, and more recent and deeper examinations have given enlarged expectations of the value and abundance of this mineral treasure.

The mountain timber of Luzerne, similar to that of most other parts of the Appalachian system, is gigantic in the vallies, and diminishing in size ascending the mountain heights. In the northwestern section, in the vallies, beech and sugar maple abound, admixed with hemlock of very large growth. The hemlock is indeed the production of all varieties of soil, and ascending the Susquehannah, commences to be found in large quantities in Luzerne. In the southern or lower part of the co. the prevailing trees are, yellow and white pine, oaks of several species, beech, hickory, and more rare, black walnut and sycamore. The staples of this co. are coal, lumber, grain and flour. Chief ts. Wilkesbarre, Kingston and Stoddartsville. Pop. 1820, 20,027, 1830, 27,380.

LYCOMING, co. of Pa. bounded by Clearfield s. w., McKean w., Potter N. W., Columbia S. E., and Union and Centre s. Length from E. to w. 106 ms., and the area being 2,332, the mean breadth must be 22 ms. Extending in lat. 41° 04′ to 41° 36′, and in long. 1° 18′ w. to 0° 45′ E. W. C. The junction of the w. branch of the Susquehannah r., with the large confluent from the N. W., the Sinemahoning is formed in the western part of Lycoming. The united waters, now a fine navigable river, winds eastwardly over the co. about 80 ms. by comparative courses to Pennsboro' where it winds to the sthwrd., leaving Lycoming between Union and Northumberland co. In its passage over Lycoming, the w. branch receives from the N. Pine creek, Lycoming, Loyalsock and Muncey creeks, and from the S. Bald Eagle creek. The face of this co. is very broken, and similar to Luzerne, traversed by several mountain chains stretching from S. W. to N. E. Though the general slope is estwrd. the western part of Luzerne actually declines towards the Appalachian system. In general features the resemblance is very strong between Luzerne and Lycoming; the arable part of both being narrow river or mountain vallies. The proportion of good soil in Lycoming is perhaps rather more, comparatively, than in Luzerne. In 1820, Lycoming contained 13,517 inhabitants, and in 1830, 17,636. Chief t. Williamsport.

LYCOMING, tsp. and p-v. Lycoming co. Pa., extending from the W. Branch up Lycoming cr. The p-o. by p-r. 6 ms. nrthwstrdly. Williamsport, and 101 N. N. W. Harrisburg. (*See Frozen run.*)

LYELL'S, store and p-o. Richmond co. Va., by p-r. 60 ms. S. S. E. W. C.

LYMAN, p-t. York co. Me., 25 ms. N. York. Pop. 1830, 1,503.

LYMAN, p-t. Grafton co. N. H., E. Conn. r., 13 ms. above Haverhill, 90 from Concord, 155 from Boston; prevailing forest trees are pine and hemlock; contains a considerable elevation, called Gardner's or Lyman's mtn. Pop. 1830, 1,320.

LYME, N. H. (*See Lime.*)

LYME, p-t. New London co. Conn., E. Conn. r. at its mouth, opposite Saybrook, 40 ms. S. E. Hartford, and about the same distance E. New Haven; it is a maritime town; surface strikingly diversified; prevailing soil, a gravelly loam; timber, deciduous trees; among the vegetable productions, are ginseng and Virginia snake root; waters abundant; the town is accommodated with several good harbors; the fishing business is carried on extensively; 24 school dists. Population 1830, 4,084.

LYME, t. Jefferson co. N. Y., S. E. St. Lawrence r., S. Chaumont bay, W. lake Ontario; predominant soil, clay or marl; timber, a lofty growth, consisting of white pine, white oak, beech, sugar maple, hickory, &c.; 6 school dists., schools kept 6 months in 12. The fisheries of Chaumont bay are important; excellent white fish are here taken in abundance. Pop. 1830, 2,873.

LYME, p-v. nrthrn. part of Huron co. O., by p-r. 103 ms. N. Columbus.

LYME RANGE, a branch of the White mtns., commencing a little below Northampton, Mass., and running S. along the E. bank of Conn. r., at the distance of 8 or 10 ms., till it terminates at Lyme on Long Island sound.

LYNCHBURG, large and flourishing p-t. Campbell co. Va., situated on the right bank of James r., at the great bend below the southestrn. chain of the Appalachian system. By p-r. 108 ms. a little S. of W. Richmond, and 206 ms. S. W. W. C.. lat. 37° 19', long. 2° 05' W. W. C. This town was incorporated in 1805, and has risen to the rank and importance of a flourishing commercial mart. James r. being navigable for batteaux, for a considerable distance above, and below to tide water, gives to Lynchburg the advantages of a sea port. The buildings, public and private, are substantial and elegant. In addition to numerous stores and groceries, there are in this town 4 book stores, and a marble manufactory, marking the advance of wealth, intelligence and taste. Pop. 1830, 4,630.

LYNCHBURG, p-v. sthrn. part of Lincoln co. Ten., by p-r. 70 ms. sthrdly. Nashville.

LYNCH'S cr., one of the numerous misnomers so disgraceful to our nomenclature. It is a river of N. and S. Carolina, rising in Mecklenburg and Anson cos. of the former, between Yadkin and Catawba rs., and flowing thence enters S. Carolina between Lancaster and Chesterfield dists.; continuing a S. E. course, separates Kershaw from Chesterfield and Darlington, Sumpter from Darlington, and Williamsburg from Marion, and falls into Great Pedee, after a comparative course of about 120 ms.; more than one half of which course it is navigable.

LYNCHWOOD, p-v. in the nrthrn. part of Chesterfield dist. S. C., by p-r. 55 ms. N. N. E. Columbia.

LYNDEBOROUGH, p-t. Hillsborough co. N. H., 10 ms. from Amherst, 35 S. Concord; divided by a mtn. from E. to W.; soil deep and strong, excellent for grazing. Pop. 1830, 1,147.

LYNDEN, or Marengo, p-v. and st. jus., Marengo co. Ala., situated on Chickasaw cr., by p-r. 72 ms. a little W. of S. from Tuscaloosa; lat. 32° 22', and long. 10° 51' W. W. C.

LYNDON, p-t. Caledonia co. Vt., 34 ms. N. E. Montpelier; watered by Passumpsic r. At the Great Falls in this r. near the S. part of the town, the water descends 65 ft. in the distance of 30 rods; at the Little Falls, one mile above, the water descends 18 ft., affording excellent water privileges; agaric mineral is found in this town; it is a tolerable substitute for chalk, and a good one for Spanish white; this is a valuable township; soil a rich loam, easy to cultivate; 14 school dists. Pop. 1830, 1,822.

LYNESVILLE, p-v. nrthrn. part of Granville co. N. C., by p-r. 60 ms. N. Raleigh.

LYNN, p-t. Essex co. Mass., on the coast, 10 ms. N. E. Boston, 6 S. W. Salem; it has long been noted for the manufacture of ladies shoes; the number of shoes manufactured at this place annually, is from 1,500,000 to 2,000,000; they are sent in large quantities to the southern states, and to the W. Indies. Lynn beach connects Nahant with the main land, and is a favourite place of resort in the summer. Lynn has a small and convenient harbor; soil of the first quality; it has a town house and academy. Pop. 1830, 6,138.

LYNN CAMP, p-v. nrthwstrn. part of Knox co. Ky., by p-r. 112 ms. S. E. Frankfort.

LYNN, cr. p-o. sthwstrn. part of Giles co. Ten., by p-r. 112 ms. S. S. W. Nashville.

LYNNFIELD, t. Essex co. Mass., 10 ms. W. Salem, 12 N. E. Boston. Pop. 1830, 617.

LYNNVILLE, on the p-o. list, (but Linville on Tanner's maps), p-v. nrthwstrn. part of Lehigh co. Pa., 20 ms. N. W. Allentown, and 30 a little E. of N. Reading. Lynn tsp. contained in 1820, 1,664 inhabitants, in 1830, 1,747.

LYONS, p-t. and st. jus. Wayne co. N. Y., 16

ms. N. Geneva, 205 from Albany; soil good; has an abundance of mill seats; situated on the Erie canal, which here crosses the r. Clyde by an aqueduct of 90 feet; contains 20 school dists. Pop. 1830, 3,603.

LYSANDER, p-t. Onondaga co. N. Y., 15 ms. N. N. W. Onondaga, 24 S. S. E. Oswego; watered by Seneca r.; 10 school dists. Pop. 1830, 3,228.

M.

Note.—Under the letter M. those names which usually begin with the abbreviation Mc, are in this work given in their proper place, and spelled at full length.

MACALLISTER'S, cross roads and p-o. Montgomery co. Ten., by p-r. 67 ms. N. W. Nashville.

MACALLISTERSVILLE, p-v. Mifflin co. Pa., by p-r. 42 ms. N. W. Harrisburg.

MACARTHURSTOWN, p-v. Athens co. O., by p-r. 71 ms. S. E. Columbus.

MACAUPIN, cr. of Il., rising in and giving name to Macaupin co., flows a little S. of W. over Montgomery co., and falls into Il. r. nearly opposite Gilead, in Calhoun co.

MACAUPIN, co. Il., bounded S. by Madison; W. Greene; N. W. Morgan; N. E. Sangamon; and E. Montgomery. Length 38, mean width 25, and area 950 sq. ms. Lat. 39° 20′, long. 13° W. W. C. The sthrn. part slopes S., and is drained by the sources of Cahokia cr.; but the body of the co. slopes sthwstrd., and is drained by the numerous fountain streams of Macaupin cr. Chief t. Carlinville. Pop. 1830, 1,990.

MACAUPIN POINT, p-o. wstrn. part of Montgomery co. Il., 51 ms. wstrd. Vandalia.

MACCALL'S, cr. and p-o. Franklin co. Miss., about 40 ms. E. Natchez.

MACCLELLANDSTOWN, p-v. in the wstrn. part of Fayette co. Pa., 8 ms. W. Uniontown, and 10 S. Brownsville.

MACCLELLANSVILLE, p-v. Camden co. Geo., by p-r. 219 ms. S. S. E. Milledgeville.

MACCONNELSBURG, p-v. in the estrn. part of Bedford co. Pa., on the r. from Chambersburg to Bedford, 18 ms. W. the former, 31 a little S. of E. the latter place, and by p-r. 70 S. W. by W. Harrisburg.

MACCONNELSVILLE, p-v. and st. jus. Morgan co. O., situated on a branch of Muskingum r., 30 ms. N. N. W. Marietta, 70 ms. S. E. by E. Columbus, and 30 ms. S. S. E. Zanesville; lat. 39° 40′, long. W. C. 4° 46′ W. Pop. 1830, 267.

MACCRACKEN, co. Ky., bounded by Calloway S. E.; Graves S.; Hickman S. W.; the Miss. r. separating it from Scott co. Miss. W.; Ohio r. separating it from Alexander co. of Il. N. W.; Johnson co. of Il. N.; and Posey co. of Il. N. E.; and by Ten. r. separating it from Livingston and Caldwell cos. of Ky. E. Length along 37th deg. of N. lat. from the junction of Ohio and Miss. rs., to the eastern border of Ten. r. 54 ms.; mean width 14, and area 756 sq ms.; lat. 37°, and long. 12° W. W. C. intersect in the sthwstrn. part of this co. The surface is generally level, part annually submerged, but the soil, where suitable for culture, highly productive. Chief t. Wilmington. Pop. 1830, 1,297.

MACCULLOUGH'S p-o. Jefferson co. O. by p-r. 140 ms. N. E. by E. Columbus.

MACDONOUGH, t. Chenango co. N. Y. 11 ms. N. Norwich. Named in honor of Com. Macdonough. Soil good. Heavily timbered with maple, beech, basswood, elm, &c. Well watered by springs and brooks, and a branch of the Chenango r. Pop. 1830, 1,232.

MACDONOUGH, p-v. and st. jus. Henry co. Geo. situated on Towanligan cr., a branch of Ockmulgee, 67 ms. by p-r. N. W. by W. Milledgeville. Lat. 33° 26′, and long. 7° 17′ W. W. C.

MACEDON, p-t. on the canal, Wayne co. N. Y. 20 ms. W. Lyons. Pop. 1830, 1,989.

MACEDONIA, p-v. Carroll co. Ten. by p-r. 121 ms. W. Nashville.

MACEWENSVILLE, p-o. Northumberland co. Pa. 70 ms. N. Harrisburg.

MACHIAS r. Me. formed of two branches, which unite at a place in Machias called the Rim; when the r. widens into a bay, called Machias bay, which communicates with the ocean 6 ms. below. There are falls on each of the branches, about 3 ms. above their confluence, which afford numerous mill seats.

MACHIAS, p-t. port of entry, and st. jus. of Washington co. Me. on Machias bay, 221 ms. N. E. Portland. The principal settlement is at the falls of the E. branch of Machias r. At the falls of the W. branch is another considerable village. A bridge is erected across Middle r. between the two villages, which, with the causeway, is 1,900 feet long. Machias has an academy, which is situated in the eastern village. There is a p-o. at each of the settlements. It is a thriving town, has considerable trade, principally lumber, and has 26 saw mills, which cut 10,000,000 feet of boards annually. Pop. 1830, 2,774.

MACINTOSH, co. Geo. bounded by the Altamahah r. separating it from Glynn and Wayne S. W., by Liberty N. W., N. and N. E., and by the Atlantic ocean S. E. Length 50, mean breadth 12, and area 600 sq. ms. Extending in lat. 31° 08′ to 31° 55′, and in long. 4° 58′ W. W. C. This co. besides some of lesser note, includes the two islands on the Atlantic coast of St. Simon's and Sapelo. The whole surface is an almost unbroken plain, inclining sthwstrdly. towards the Altamahah r. Staples, cotton, rice, and sugar. Chief town, Darien. Pop. 1820, 5,129, 1830, 4,998.

MACKEAN, co. Pa. bounded by Potter E., Lycoming S. E., Clearfield S., Jefferson S.W.,

Warren w., Cattaraugus co. of N. Y. N., and Alleghany co. of N. Y. N. E. Length 42, mean width 32, and area 1,344 sq. ms. Extending in lat. 41° 24′ to 42°, and in long. 1° 16′ to 2° 03′ w. W. C. This co. occupies part of an elevated table land, from which flow S. E. the higher sources of Sinnamahoning branch of Susquehannah, S. W. the sources of Clarion r. branch of Alleghany. The Alleghany r. rising in Potter co., flows wstrd. into MacKean, and winding N. N. W. enters Cattaraugus co. of N. Y. Within the latter co. this stream forms an elliptic curve and again re-enters McKean by a southern course. In the semicircle or segment of an ellipsis thus formed, the water courses radiate from the central parts of McKean, flowing in different directions, into Alleghany as a common recipient. The surface is hilly, in part it is mountainous and soil generally of middling quality. Pop. 1820, 728, 1830, 1,439.

MacKeans, old stand, and p-o. Westmoreland co. Pa. about 20 ms. N. W. Greensburg, and by p-r. 199 ms. a little N. of W. Harrisburg.

MacKeansburt, p-v. Schuylkill co. Pa. 5 ms. N. E. Orwicksbury, and by p-r. 64 N. E. Harrisburg.

MacKees Port, p-v. situated on the right side of Youghioghany and Monongahela rs. at their junction, in Alleghany co. Pa., 11 ms. by land S. E. Pittsburg, and by p-r. 189 ms. W. Harrisburg. It is a village composed in great part of a single street along both rs., and on a high bottom. The situation is a most delightful one for a town, as far as natural scenery is concerned.

MacKee's Half Falls, and p-o. Union co. Pa. by p-r. 56 ms. nthwrd. Harrisburg.

Mackinac, on the p-o. list, Mackinaw, on Tanner's U. S. map, the old Michilimakinak, port, and st. of jus. co. of Michilimakinaw. As given on the p-o. list it is 321 ms. N. N. W. Detroit. (*See Michilimakinac island and co.*)

Mackinaw, r. of Il. rising on the plains, near the centre of the state, interlocking sources with those of Vermillion branch of Il. and with those of Sangamon. Flowing thence S. W. falls into Il. r. about 5 ms. above the influx of Spoon r. The valley of Mackinaw lies between those of Vermillion and Sangamon, and comprises most part of McLean, and Tazewell cos.

Mackinaw, p-v. and st. jus. Tazewell co. Il. is situated on Mackinaw r. by p-r. 149 ms. N. Vandalia. Lat. 39° 33′, long. 12° 18′ w. W. C.

MacKinstry's, mills and p-o. nthwstrn. part of Frederick co. Md. by p-r. 68 ms. N. N. W. W. C.

Mackville, p-v. in the nthrn. part of Washington co. Ky. 13 ms. N. W. by W. Harrodsburg and by p-r. 34 ms. S. S. W. Frankfort.

Mackville, p-v Franklin co. Geo. by p-r. 100 ms. N. N. E. Milledgeville.

MacLean, co. Il. bounded by Vermillion co. E., Macon S., Sangamon S. W., Tazewell W., Putnam N. W., and La Salle N. Extent as laid down on Tanner's improved map, 50 from N. to S. and 40 E. to W., area 2,000 sq. ms. Lat. 40° 40′, and long. 12° w. W. C. Slope wstrd. and drained by the sources of Salt cr. branch of Sangamon, and those of the Vermillion branch of of Illinois. The surface in great part open grassy plains. Chief t. Bloomington.

MacLeansville, p-v. Jackson co. Ten. by p-r. 77 ms. nthestrd. Nashville.

MacMinn, co. of Ten. bounded by Rhea co. N. W.; it merely touches Roane on the N., is again bounded by Monroe N. E. and E., and by Hiwassee r. S. Length 38, mean width 16, and area 608 sq. ms. Extending in lat. 35° 15′ to 35° 44′, and in long. 7° 23′ to 7° 52′ w. W. C. It lies in a form approaching a triangle, but the sthrn. side curving outwards along Hiwassee r. The whole surface an inclined plain, sloping towards the Hiwassee sthwstrdly., and drained by numerous crs. which flow into that r. Chief town, Athens. Pop. 1820, 1,623, 1830, 14,460.

MacMinville, p-v. and st. jus. Warren co. Ten. situated on a creek of Caney fork, branch of Cumberland r., 65 ms. S. E. by. E. Nashville. Lat. 35° 44′, long. 8° 48′ w. W. C.

MacNairy, co. of Ten. bounded by Hardiman W., Madison N. W., Henderson N., Hardin E., and the Chickasaw country in the state Miss. S. Length 30, mean width 24, and area 960 sq. ms. Extending in lat. 35° 26′, and in long. 11° 26′ to 11° 50′ w. W. C. This co. occupies part of the table land between the Ten. and Big Hatchee r., the wstrn. part drained by the sources of the latter, the northwestrn. by the sources of the Forked Deer r., and the estrn. by small creeks flowing into Ten. r. Chief town, Purdy. Population 1830, 5,697.

Macomb, co. of Mich bounded by Wayne S. W., Oakland W., Lapeer N. W., St. Clair co. N. E., and St. Clair lake E. and S. E. Length 32, mean breadth 12, and area 384 sq. ms. Lat. 42° 35′, long. 5° 50′ w. W. C. This co. is drained by the main stream and numerous branches of the Huron of lake St. Clair, and slopes to the E. Chief t. Mount Clemens. Pop. 1830, 2,413.

Macomb's, p-v. Abbeville dist. S. C. about 100 ms. W. Columbia.

Macon, extreme wstrn. co. of N. C., bounded S. by Rabun and Habersham cos., and S. W. by the Cherokee country in Geo.; W. by Amoi district, in Tenn.; N. W. by Unika mtn. separating it from Amoi district in Ten., N. E. and E. by Tenn. r. separating it from Haywood co. N. C. Length from E. to W. 45 ms.; mean width 20, and area 900 sq. ms. Extending in lat. 35°, to 35° 28′, and in long. 6° 20′ to 7° 09′ w. W. C. This co. is very elevated; the surface above 1,500 ft. mean height from the ocean. The general slope is to the nthwstrd. Hiwassee r. rises in Geo. but enters and traverses in a N. W. by W. direction Macon co.; the extreme sthestrn. branch of Tenn. also rises in Geo. but flowing N. enters N. C. forming its boundary between Macon and Haywood cos. (*See Tenn. r.*) If allowance is made for relative height, Macon must have a winter climate similar to that of N. lat. 39°, on the Atlantic, or like that near the mouth of Del. r. Chief t. Franklin. Pop. 1830, 5,333.

Macon, p-v. Franklin co. N. C. 35 ms. N. E. Raleigh.

Macon, p-v. and st. jus. Bibb co. Geo. situated on the right bank of Ockmulgee r. 3 ms. S. W. by W. Milledgeville, lat. 32° 52' and long. 6° 42' W. W. C.

Macon, p-v. Bedford co. Tenn. about 50 ms. S. E. Nashville.

Macon, co. Il. bounded by Cole S. E., Shelby S., Sangamon W., MacLean N., and Vermillion N. E. Extent 40 by 35, or area 1400 sq. ms. Lat. 39° and long. 12° W. W. C. intersect not far from the centre. From the sthestrn. angle flow some of the higher sources of Kaskaskias r. The main volume of Sangamon, rising in MacLean and Vermillion enters the nthestrn. angle and winding over it diagonally, divides it into two not very unequal sections. The sthwstrn. part is drained by the sources of Salt cr. branch of Sangamon. General slope of the co. S. W. Chief t. Decatur. Pop. 1830, 1,122.

Mac Williamstown, p-o. sthwstrn. part of Chester co. Pa. by p-r. 63 ms. sthestrd. Harrisburg.

Macungy, p. t-ship. on Little Lehigh r. Lehigh co. Pa. 5 ms. S. W. Allentown, and by p. r. 95 ms. N. E. by E. Harrisburg.

MacVeytown, p-v., Mifflin co. Pa. by p-r. 68 ms. from Harrisburg.

Madbury, t. Strafford co. N. H. 11 ms. N. W. Portsmouth. Contains bog iron ore, and red and yellow ochre. Pop. 1830, 510.

Madison, p-t. Somerset co. Me. on the Kennebec r. 9 ms. N. Norridgewock. Population 1830, 1,272.

Madison, co. N. Y., bounded N. and N. E. by Oneida lake and co., E. by Otsego co., S. by Chenango co., W. by Cortland and Onondaga cos. containing 616 sq. ms. or 394,240 acres. Morrisville is the st. jus. of the co. Its mineralogical productions are no where exceeded in the western cos. Pop. 1820, 32,-208—1830, 39,038.

Madison, p-t. Madison co. N. Y. 7 ms. E. Morrisville, 95 W. Albany. Pop. 1830, 2,544.

Madison, tsp. of Columbia co. Pa. 5 ms. N. Danville.

Madison, co. Va. bounded N. W. by the Blue Ridge, which separates it from Shenandoah co., N. E. and E. by Culpepper, and S. E., S. and S. W. by Rapid Ann r. separating it on all these sides fm. Orange co. Length S. to N. 28; mean brdth. 12, and area 336 sq. ms. Extending in lat. 38° 14', to 38° 38', and in long. 1° 09' to 1° 30' W. W. C. This co. slopes to the S. S. E., drained by various branches of Rapid Ann. The surface is moderately hilly, and soil of middling quality. Chief. t. Madison. Pop. 1820, 8,490—1830, 9,236.

Madison, p-v. and st. jus. Madison co. Va. situated near the centre of the co., by p-r. 99 ms. S. W. by W. W. C., and 95 ms. N. N. W. Richmond. N. lat. 38° 22', and long. 1° 15' W. W. C.

Madison, p-v. Rockingham co. N. C., by p-r. 32 ms. N. W. Raleigh.

Madison, co. Geo. bounded S. by the S. Fork of Broad r., which separates it from Oglethorpe, S. W. by Clark, W. by Jackson, N. W. and N. by Franklin, and E. by Broad R. separating it from Elbert. Length 28 ms. mean wdth. 9 and area 252 sq. ms. Extending in lat 34° to 34° 15', and in long. 6° to 6° 28' W. W. C. The slope of this co. is generally S. E. towards Savannah r., but the extreme estrn. part is drained by Sandy creek, a branch of Oconee flowing into Altamahah. Chief t. Danielsville. Pop. 1820, 3,735,—1830, 4646.

Madison, p-v. and st. jus. Morgan co. Geo. situated near the centre of the co. 40 ms. N. N. W. Milledgeville, lat. 34° 09' long. 6°. 14' W. W. C.

Madison, co. Ala. bounded E. by Jackson, S. by Ten. r., separating it from Morgan, W. by Limestone, N. W. by Giles co. in Ten., and N. by Lincoln co. in Ten. Length 36, mean width 18, and area 648 sq. ms. Extending in lat. 34° 30', to 35° N. and in long. 9° 24' to 9° 54' W. W. C. This co. slopes sthrd. towards Ten. r. The surface moderately hilly and soil highly productive. Principal staple cotton. Chief t. Huntsville. Pop. 1820, 17,481—1830, 27,990.

Madison, co. Miss., bounded by Big Black r. separating it from Yazoo co. N. W., by the Choctaw co. N. E. and E. by Rankin co. S. E., and Hinds S. W. Length S. W. to N. E. 55, mean wdth. 12, and area 660 sq ms. Extending in lat. 32° 28' to 33°, and in long. 12° 58' to 13° 38' W. W. C. This co. is bounded on the N. W. as has been shown, by Big Black r., and the sthestn. part is traversed by Pearl r., the two streams flowing to the S. W. at a distance of 18 to 20 ms. asunder, at this particular part of their respective courses. The great road from Natchez to Florence, Nashville, &c. also traverses this co. between Pearl and Big Black rs. Chief t. Madisonville. Pop. 1830, 4,973.

Madison, co. of Ten. bounded S. by Hardiman, W. by Haywood, N. Gibson, N. E. Carroll, E. Henderson, and S. E. by MacNairy. Length 28, width 24, and area 672 sq. ms. Extending in lat. 35° 24' to 35° 47'. The declivity of this co. is to the N. N. W., drained towards Miss. r. by different branches of Forked Deer r. Chief t. Jackson. Pop. 1830, 11,-549.

Madison, p-v. and st. jus. Monroe co. Ten. situated near the centre of the co. by p-r. 168 ms. S. E. by E. Nashville, lat. 35° 27', long. 7° 18' W. W. C.

Madison, co. of Ky. bounded by Estill E., Laurel S. E., Rock Castle S. W., Garrard W., and Ky. r. which separates it from Jessamine N. W., Lafayette N., and Clark N. E. Length 40, mean width 13, and area 520 sq. ms. Extending in lat. 37° 23′ to 37° 54′, in long. 6° 48′ to 7° 30′ w. W. C. The extreme sthrn. part of this co. declines sthrd., and is drained in that direction by the sources of Rock Castle, branch of Cumberland r. The residue of its surface slopes nthrd. towards Ky. r. Chief t. Richmond. Pop. 1820, 15,954, 1830, 18,751.

MADISON, co. O., bounded by Lafayette s., Green s. w., Clarke w., Champaign N. w., Union N., Franklin E., and Pickaway s. E. Lat. 40°, long. 6° 24′ w. W. C. Length from s. to N. 30, mean breadth 14, and area 420 sq. ms. Slope sthestrd., and principally drained by Darby's cr. Chief town London. Pop. 1820, 4,799, 1830, 6,190.

MADISON, p-v. Geauga co. O. by p-r. 173 ms. N. E. Frankfort.

MADISON, co. Ind. bounded by Hancock s., Hamilton w., Grant N., Delaware N. E. and Henry s. E. Length 30, breadth 18, and area 540 sq. ms. Lat. 40° 10′, long. 8° 42′ w. W. C. Slope s. w. and drained by the main stream and various branches of White r. Chief t. Andersontown. Pop. 1830, 2,238.

MADISON, p-v. and st. jus. Jefferson co. Ind. by p-r. 76 ms. s. E. Indianapolis. It is situated on the Ohio r. 46 ms. above Louisville, and 18 below Vevay, lat. 38° 43′, long. W. C. 8° 24′ w.

This place was commenced in 1811, and has been very prosperous; according to Flint, it contained in 1829, from 40 to 50 brick buildings, an insurance company, and did extensive mercantile business. It contains two printing offices, and in brief has all the appearance of a wealthy mart.

MADISON, co. Il. bounded on the N. by Greene, Macaupin, and Montgomery; Bond E., Clinton s. E., St. Clair s., Mississippi r. separating it from St. Louis co. Mo. s. w., and the Mississippi r. again separating it from St. Charles co. Mo. N. w. Lat. 38° 45′, long. 13° w. W. C. Slope s. s. w., and drained by Cahokia, and the sources of Silver cr. branch of Kaskaskias r. Chief t. Edwardsville. Pop. 1830, 6,221.

MADISON, co. Mo. bounded by Washington N. w., St. Francis N., Perry N. E., Cape Girardeau E., Wayne s., and w. uncertain. It is a square of 30 ms., area 900 sq. ms. Lat. 37° 25′, long. 13½° w. W. C. Traversed from N. to s. by the main stream and drained by the branches of St. Francis r. Chief t. Fredericktown. Pop. 1830, 2,371.

MADISON, Cross Roads and p-o. Madison co. Ala. by p-r. 21 ms. from Huntsville, and 171 N. N. E. Tuscaloosa.

MADISON Springs, p-v. Madison co. Geo. by p-r. 75 ms. N. Milledgeville.

MADISONVILLE, p-v. seaport and st. jus. parish of St. Tammany, La. situated on or near the mouth of Chifuncte r. about 28 ms. N. fm. the city of New Orleans, and on the opposite side of lake Pontchartrain, lat. 30° 24′.

MADISONVILLE, p-v. and st. jus. Madison co. Miss. about 137 ms. N. E. Natchez.

MADISONVILLE, p-v. and st. jus. Hopkins co. Ky., situated on the table land between the vallies of Green and Tradewater rs. 65 ms. nearly due N. Nashville in Ten., 53 sthestrd. Shawneetown on Ohio r. and by p-r. 191 ms. s. w. by w. Frankfort. Lat. 37° 20′, and long. 10° 30′ w. W. C.

MADISONVILLE, p-v. Hamilton co. O. by p-r. 106 ms. s. w. by w. Columbus.

MADRID, p-t. St. Lawrence co. N. Y., s. St. Lawrence r., 110 ms. above Montreal, 60 below Kingston, 250 N. w. Albany. Level, fertile, and well watered. Timbered with beech, maple, &c. 13 schools, kept 7 months in 12. Pop. 1830, 3,459.

MAD RIVER, rises in Grafton co. N. H., and falls into the Pemigewasset, near the centre of Campton.

MAGNOLIA, p-v. on St. Marks r. Leon co. Flor. 16 ms. s. s. E. Tallahassee.

MAGRUDER'S, p-o. Prince George's co. Md., by p-r. 15 ms. estrd. W. C. and 31 wstrd. Annapolis.

MAHANOY, three townships in the sthrn. part of Northumberland co. Pa., called relatively Little Mahanoy, Lower Mahanoy, and Upper Mahanoy.

MAHANOY, p-o. Northumberland co. Pa. by by p-r. 45 ms. N. Harrisburg.

MAHANOY, r. of Pa. rising in the nthrn. part of Schuylkill co. interlocking sources with the Cattawissa creek and Schuylkill r. it assumes very nearly a wstrn. course, and entering Northumberland, falls into Susquehannah r. about 11 ms. below Sunbury. The valley of the Mahanoy is between those of Mahantango and Shamokin.

MAHANTANGO, mtn. a ridge of the Appalachian system in Pa., extending from the left bank of the Susquehannah r. along the nthwstrn. part of Dauphin into Schuylkill co. It is a continuation of the chain known to the s. w. of Susquehannah, as the Tuscarora mtn. or Cove mtn.

MAHANTANGO, r. Pa. rises by two main and several smaller branches in the nthwstrn. angle of Schuylkill co. and in the vales of the Mahantango mtn. Assuming a course of a little s. of w. the branches unite on the border of Schuylkill, and flowing along the western side of the mtn. of the same name, forms a boundary between Northumberland and Dauphin cos. to its influx with the Susquehanna. The valley of Mahantango lies between those of Mahanoy and Wiconisco.

MAHONING, r. Pa. formed by two branches. The main stream, or Mahoning proper, rises in the wstrn. part of Clearfield co. and flowing to the wstrd., traverses the sthestrn. angle of Jefferson, and nthwstrn. of Indiana, receiving the sthrn. branch in the latter co. at Nicholsburg, where entering Armstrong, the united waters continue to flow wstrd. to their confluence with Alleghany r. The Mahoning has interlocking sources with the extreme wstrn. branches of the Susquehannah river.

MAHONING, r. of Pa. and Ohio, rises by numerous branches in Columbiana, Stark, Portage, and Geauga cos. of the latter state, which unite in Trumbull, and assuming a s. E. course enters Beaver co. Pa., and there joins the Shenango, forming the Big Beaver river.

MAHONING, p-v. in the extreme nthwstrn. angle of Ind. co. Pa. on Mahoning creek, 18 ms. a little w. of N. the boro' of Indiana.

MAHONING, p-v. Stark co. O. by p-r. 135 ms. N. E. by E. Columbus.

MAIDEN, cr. one of the nthestrn. branches of Schuylkill r. rising in the extreme wstrn. angle of Lehigh co. and flowing sthwrd. into Berks, falls into the Schuylkill r. 8 ms. above Reading.

MAIDEN, cr. p-o. near the mouth of Maiden creek, Berks co. Pa. 8 ms. N. Reading, and by p-r. 60 ms. E. Harrisburg.

MAIDSTONE, t. Essex co. Vt. w. side of Connecticut r. 53 ms. N. E. Montpelier, contains Maidstone lake, which is small. It was chartered 1761, 1st settled 1770, contains 17,472 acres, and is watered by Paul's stream. Pop. 1830, 236.

MAINE, the easternmost and northernmost of the United States; bounded N. W. and N. by Lower Canada, E. by New Brunswick, S. E. and E. by the Atlantic Ocean, and W. by New Hampshire. It lies between 43° 5′ and 48° N. lat. and between 66° 49′ and 70° 55′ W. long.

Piscataqua river forms the S. W. boundary for about 35 miles, and the N. line runs by treaty along the highlands which divide the St. Lawrence from the ocean. The S. line reaches from Kittery point, to Quoddy head, about 221 miles. The whole area is about 33,223 square miles, including a large tract in the N. E. which has been claimed by Great Britain. In 1621, the W. boundary of Nova Scotia, as was definitively ascertained by treaty, was the St. Croix river, and a line running from its source N. to the St. Lawrence river. In 1691, the E. boundary of Me. was fixed at the W. boundary of Nova Scotia. In 1763, N. Brunswick and Maine, which had before extended to the St. Lawrence r., were reduced on the N. and fixed at the highlands which separate the waters of the St. Lawrence river from the ocean. These bounds were repeatedly acknowledged by parliament down to 1774, and were never doubted until 1814, when the British plenipotentiaries proposed to discuss and revise the boundary so as to prevent future uncertainty and dispute. They stated that they desired a direct communication from Quebec to Halifax, and left it to the Americans to demand an equivalent. This was refused, on the ground that the territory sought, was undoubtedly American. The tract alluded to, includes most of the country watered by the St. John's river, Mars Hill, S. of that stream, being considered by the British as a part of the "height of land," though in fact it is far distant, disconnected from it, and of very inferior elevation. The king of Holland as umpire in the case, has decided in favor of Great Britain; but it is believed that his decision will not be submitted to, as at the time of making the award, he was not an independent sovereign. In 1831, Madawasca, and a tract S. of St. John's river were incorporated by the State, as well as the disputed territory N. of it, though without any design of taking forcible possession. In October, 1831, in consequence of the election of municipal officers at Madawasca, the lieutenant governor of N. Brunswick and other officers, with a military force, arrested a number of persons and took them prisoners to Frederickton, but soon after released them. The subject is now in the hands of commissioners for arrangement. It was originally granted in 1606 by James 1st. to the Council at Plymouth, by whom in 1624, a grant was made to Gorges & Mason, of all the country from Merrimac to Sagadahok. This claim was purchased by Mass. for £1,250. The first permanent settlement was made in 1630. From 1674, to 1763, Mass. had to defend it from the Indians, with little profit. In 1691, Mass. obtained a confirmation of the charter, which added Maine, Nova Scotia. &c., to her territory, and through long disputes with the French and Indians, those additions were still maintained. From its first settlement Maine was a district of Mass. In 1820, when its present constitution was adopted, it was separated from Mass. and admitted into the Union as an independent state. Its government now consists of three distinct departments. The legislative power is vested in a senate and house of representatives, the former at present containing 20, the latter 153 members. The members are chosen annually, and are proportioned to the population. Their regular yearly meeting commences on the 1st Wednesday of January. The executive power is vested in a governor, who is annually chosen by the people, and a council of seven elected by the legislature. The governor, who must be at least 30 yrs. of age, has a qualified negative on the laws proposed by the legislature, he has also the control of the official patronage, and together with the council exercises the pardoning power. In case of a vacancy, the president of the senate acts as governor. The judiciary consists of a supreme judicial court, and a court of common pleas, each of three judges. The judges are appointed by the governor and his council; they hold their offices during good behavior until 70 yrs. of age, and are removable only by impeachment. Justices of the peace are appointed for 7 yrs. The time of annual elections is the 2d Monday in September. Any changes may be made in the constitution by a vote of two thirds of both houses of the legislature, if such vote be ratified by the people.

The surface of the state is generally diversified, and moderately hilly. A tract on the west side, east of the White mountains in New Hampshire, and also a small district in the north extremity, are mountainous; some few elevations are above ordinary vegetation. The range of high land which crosses Vermont, and New Hampshire, enters the N. W. corner of Maine, passes round Chaudiere r. and running nearly parallel with the St. Lawrence river, at the distance of 15 or 20 miles, terminates on the gulf of St. Lawrence, near cape Rozier. This is the "height of land," or the "N. E. Ridge," spoken of in the treaties as the N. boundary of Maine, and though of gradual elevation, is in some places 4,000 feet above the ocean. The mountains of Maine lie in irregular groups, with a line drawn from S. part of Oxford co. E. of Androscoggin lakes, then N. on W. side of Kennebec river, and Moosehead lake, to the mountains among the W. sources of Penobscot river. These mountains belong to the Alleghany range, and the White mountain spur. There are several subordinate spurs, the S. one extending 40 miles from the White mountains. North of Androscoggin river, and to Dead r. is a rough range, including some of the highest peaks in the state; the principal are Speckled mountain, White Cap, Saddleback, &c. about 4,000 feet above the sea. Bald mountain ridge, with peaks of the same elevation, lies between Moose river and the S. W. branch of Penobscot river. Kennebec, Penobscot, and St. John's rs. run through a broad irregular valley 20 ms. by 120, which is bordered by ranges of mountains. Beside those mentioned, are several other groups, as the Spencer, &c. Katahdin mountain has been found by barometrical observation, to be 5,335 feet above the ocean, and 4,685 above W. branch of the Penobscot river. It is the highest peak E. of the Miss. except a few of the White mountains of New Hampshire. The view from the summit is fine and varied, and extends over 80 or 100 miles; from it may be seen 63 lakes which are tributary to the Penobscot river, and others, the heads of the St. John's, and Kennebec. It has high table land on three sides, 4 miles in width and covered with forests, is inaccessible on the E., S. and on part of the W. sides, and is covered with broken rocks, and overgrown with spruce trees, which gradually diminish towards the top, leaving its summit bare.

The principal rivers of Maine, are the Penobscot, Kennebec, Saco, Androscoggin, St. John's, and St. Croix, which with their branches water most of the state. The Saco waters 650 square miles, the Androscoggin 3,300, the Kennebec 5,280, and the Penobscot, which is navigable to Bangor, by the largest merchant vessels, 8,200. Those portions of the country near the sea, are watered by the Piscataqua, Kennebec, Sheepscot, Damariscotta, Muscongus, St. Georges, Union, Narragaugus, Machias, and other rs. Between the Penobscot and Kennebec rs. a distance of 50 ms. on the sea shore, there are 4 considerable rs. beside innumerable inlets, so that almost every town has its particular channel of communication with the sea.

The soil of Maine is generally equal, and in some places superior to that of the other northern states. The tract of country along the sea coast from 10 to 20 ms. wide, though it embraces all the varieties of sandy, gravelly, clayey, and loamy soils, is for the most part poor. The principal productions of this section, are maize, rye, barley, grass &c. In the tract lying north of this, and extending from 50 to 100 ms. into the interior, the soil is more fertile, and produces maize, wheat, barley, rye, oats, millet, flax, hemp, grass, and most northern plants. The land between the Penobscot and Kennebec rivers, is well adapted to the purposes of agriculture, and as a grazing country, is one of the finest in New England. Land of average quality, yields with good cultivation, 40 bushels of maize to the acre, 20 to 40 bushels wheat, rye, oats, &c., and 1 to 3 tons hay. Agriculture, until recently, has been much neglected; the forests and fisheries being very productive, now 5-6 of the people are supported by it, most of the inhabitants being farmers, but many, merchants and manufacturers. Apple, pear, plum, cherry trees, melons, &c. succeed; peach trees do not. The extreme season of vegetation, is between April 21st, and October 16th; vigorous vegetation from June 3d, to September 12th. The climate of the state is subject to great extremes of heat and cold. In all parts, the air is pure and salubrious, but most so, as well as most mild, where the forests have been cleared away. The winters are very severe. Snow lies in some parts 5 months, near the sea but 3 or 4. Many sheep are raised. The west and old counties raise food, &c. for their consumption, and send out some, the east counties not enough. Cattle and swine are sent into, and through New Hampshire; and to New Brunswick, Hamilton, and St. Johns. The trees are various. White pines are the most abundant, and are found chiefly on the sources of the Penobscot, Kennebec, and Aroostook rivers. Iron is abundant, and of excellent quality. Lime is made in great quantities, at Thomaston, and Cambden. Fine marble is found on the west branch of the Penobscot river. Granite and slate are abundant. Salt and fishing are profitable on the coast.

Maine enjoys great facilities for navigation and commerce. The sea shore abounds in excellent harbors, and the settled parts of the country are mostly near to markets, where produce is readily exchanged for money. The centres of interior trade, are Portland, Hallowell, Bangor, Calais, Brunswick, Belfast, &c., which being always open to navigation, enjoy some advantages over other ports. Saco, Machias, and Eastport, are important harbors. The tonnage of Maine, is ⅟ of that of the whole United States, though its population is but one thirtieth. The principal exports are timber, lumber of various kinds, dried fish, salt meat, lime, beef pork. butter, pot and pearl ashes; & some grain. The tonnage of Maine, entered in the year ending September, 1830, was 74,741; departed, 97,791; value of imports, $572,666; ex ports, domestic produce, $643,435 ; foreign, $27,087 ; total exports, $670,522.

Manufactures are very few. The direct revenue is chiefly derived from assesments on polls and estates, laid equally, 1 per cent on bank stock, and the indirect, from duties on litigation.

Maine is rapidly increasing in population. In 1810, there were 228,705 inhabitants ; in 1820, 298,335 ; and in 1830, 399,437. The state is divided into 10 counties, the population of each of which, for the years 1820 and 1830, are given below. Waldo county was formed from Hancock, since the census of 1820.

Counties.	Pop. 1820	Pop. 1830
York,	46,283	51,722
Cumberland,	49,445	60,102
Lincoln,	53,189	57,183
Kennebec,	42,623	52,484
Oxford,	27,104	35,211
Waldo,		29,788
Somerset	21,787	35,787
Penobscot,	13,870	31,530
Hancock,	31,290	24,336
Washington,	12,744	21,294
Total,	298,335	399,437

Of the foregoing population of 1830, were whites,

	Males.	Females.
Under 5 years of age,	34,052	32,471
Between 5 and 15,	54,265	51,743
" 15 and 30,	57,385	57,949
" 30 and 50,	36,248	36,443
" 50 and 70,	15,184	15,234
" 70 and 90,	3,458	3,600
" 90 and over,	95	140
Total,	200,687	197,573

Of which 153 are deaf and dumb, 154 are blind, and foreigners not naturalized 2,489. Of the deaf and dumb, 8 are supported by the state, at the American Asylum, at Hartford.

Of colored persons, there are, free,

	Males.	Fems.
Under 10 years of age,	159	140
Between 10 and 24,	169	171
" 24 and 36,	111	117
" 36 and 55,	105	91
" 55 and 100,	52	52
" 100 and and over,	2	
Total,	600	571

Slaves—males, none ; females, 6 ; colored deaf and dumb, 16 ; blind 1.

Recapitulation.

Whites.	Free colored.	Slaves.	Total
398,260	1,171	6	399,437

Since Maine became a separate state, an improvement has been made upon the old school system of Mass. The school fund of the state, consists of the proceeds of 20 townships of land, on interest ; the balance of money to be received from Mass. over the debts of the state, and the proceeds of land required to be reserved for the ministry, which last is applied for schools where the land is. Besides this, every town is obliged to raise 40 cents a year, for each inhabitant, which is paid for free schools in proportion to the number of persons between 4 and 21 years. The amount of money raised by the new method, is much greater than formerly. The districts build school houses, &c., and the parents furnish books ; all have equal right to the schools. In 1825, $137,878 were expended for schools, in 2,499 districts ; 101,325 children attended, 4½ months in the year ; average wages of teachers, $12 per month. Expense of each scholar per year $1,35, and annual increase of scholars, 6,000. In 1831, the number of school districts, was 2,500 ; amount expended $200,000, and the number of students at public schools, not supported by voluntary contributions 100,000. The whole number of academies was 35, and the students 900 ; students at colleges, including medical school, 260. Grammar schools have been superseded by academies, founded by private persons and supported by land granted by the state. In 1819, there were 28, 24 of which were incorporated by Mass. with capital invested, $220,000, annual income $9,500 ; receipts for tuition, $8,000, number of pupils 950, for 10 months in the year ; average expense $50. The Maine Wesleyan seminary, Waterville college, and Bowdoin college are superior institutions. Bangor theological seminary, and Gardiner lyceum have both ceased operations for the present. Bowdoin college, at Brunswick, was founded in 1794, by Hon. J. Bowdoin, and has lands from Mass. and an annuity from Maine. A medical school is attached to it. In Maine, there is one child at school, for every 4 inhabitants.

The state prison, at Thomaston, is on an eminence, a few yards from navigable water. The convicts are employed in quarrying limestone on the grounds, and in hammering granite which is brought by water. In its construction and general arrangements the late improvements have been introduced. The convicts have separate cells, in which they are confined at night ; they are all instructed, and taught to read the scriptures. The proceeds of their labor in 1831, exceeded the expenses, (exclusive of the officers' pay,) more than $400. There is but one county prison in Maine, and by a recent law, the state has abolished imprisonment for debt, which saves annually about 1000 imprisonments.

An act was passed in March, 1832, to encourage agriculture, horticulture, and manufactures, authorising the payment to each incorporated agricultural society, or horticultural society, as much money as it raises by subscription or otherwise, not exceeding $300, in each co. These societies are authorized to offer annual premiums for improving animals, tools, implements of husbandry, or manufacture, trees, plants, &c. The legislature has also, at different times, appropriated money to improve the road through the White mountains of N. Hampshire, as it affords an important channel of transportation, for the produce of the interior.

The state expenses, in 1830, were about $297,000, $50,000 of which was raised by direct taxation. $5,000 is annually appropriated for the education of indigent deaf and dumb persons, at the American Asylum, in Hartford, Conn.

There is in Maine, a state temperance society, and a historical society, with a depository, &c. A marine hospital is to be erected by the government of the United States, at Portland, and $15,000 have been appropriated for it.

The following are the names of the principal places, and their distances, on the new road through Maine to Canada, beginning at the capital. From Augusta to Waterville, 183 miles, Fairfield 187½, Bloomfield 198, Madison, Wherf's, 205, Solon, Boies', 215, Bingham, Goodridge's, 223, Moscow, Spaulding's, 235, forks of Kennebec river, Temple's, 245, Parlin pond, Baker's, 260, Moose river, Holden's, 275, Hilton's camp in township No. 5, 3d range, 286, St. Charles, Owen's, 317, St. Francis, Boldue's, 326, St. Joseph's, Suponsey's, 340, St. Mary's, Slaven's, 352, St. Henry's, 370, Point Levi, McKensey's, 382, over the St. Lawrence river to Quebec 383. From the Canada line, to Point Levi, 117 miles, the road is perfectly smooth, and there are no very steep hills.

There are various religious sects in the state. Baptists, 210 churches, 136 ministers, 22 licentiates, and 12,936 communicants; Congregationalists, 156 churches, 107 ministers, 9,626 communicants; Methodists, 56 ministers, 12,182 communicants; Free Will Baptists, 50 congregations; Friends, 30 societies; Unitarians, 12 societies, 8 ministers; Episcopalians, 4 ministers; Roman Catholics, 4 churches; New Jerusalem church, 3 societies; beside some Universalists.

Maine Paint, creek, and p-o. Fayette co. Ohio, by p-r. 53 ms. s. s. w. Columbus.

Mainsburg, p-v. Tioga co. Pa. by p-r. 144 ms. n. Harrisburg.

Malaga, p-v. Monroe co. Ohio, by p-r. 142 ms. e. Columbus.

Malcolm, p-o. Jefferson co. Miss. 15 ms. n. Natchez.

Malden, p-t. Middlesex co. Mass. 4 ms. n. e. Boston, 4 miles by 2½. The bridge to Charlestown is nearly 2,500 feet long. In the south, are about 1,000 acres of salt marsh. North part uneven. First settled 1648. Pop. 1830, 2,010.

Malloryville, p-v. in the eastern part of Wilkes co. Geo. by p-r. 71 ms. n. n. e. Milledgeville.

Malone, p-t. and st. jus. Franklin county, N. Y. on Salmon river, with a court house, and state arsenal. It is 50 miles w. n. w. of Plattsburg. Several ponds afford trout, as well as the streams. Pop. 1830, 2,207.

Malta, p-t. Saratoga co. N. Y. 4 ms. s. e. Ballston Spa, 25 n. Albany; contains Round lake and part of Saratoga lake, with few mill seats, and is a good farming town, with a village called Dunning Street. Pop. 1830, 1,517.

Malta, p-v. Morgan co. O. by p-r. 70 ms. s. e. by e. Columbus.

Mamakating, t. Sullivan co. N. Y. Villages, Bloomingsburg, (on Shawangunk creek) Burlingham and Mamakating. It is 7 or 8 ms. by 15, and is crossed by Shawangunk mtn. or Blue Ridge. The streams afford trout and pike. Pop. 1830, 3,070.

Mamaronec, p-t. West Chester co. N. Y., has a harbor on Long Island sound for vessels of 100 tons. Two creeks afford mill seats, 23 ms. n. e. N. Y., 2½ ms. by 3. Pop. 1830, 838.

Mamgunk, p-v. Phil. co. Pa.

Manahawken, p-v. Monmouth co. N. J. near the sea, on a creek of the same name, 6 ms. n. e. Tuckerton, 50 s. Freehold.

Manasquan r. Monmouth co. N. J., falls into the ocean 30 ms. s. Sandy Hook, 4 n. Barnegat bay, is navigable for small vessels.

Marichar, p-o. parish of E. Baton Rouge, La. situated on the left bank of Miss. r. at the efflux of Ibberville cr., 11 ms. s. Baton Rouge.

Manchester, t. Hillsborough co. N. H. e. side of Merrimac r.; contains part of Massabesick pond, and several small streams. A canal of 1 m. passes the Amoskeag falls in Merrimack r., which descends 45 ft. Made in 1816, and cost $60,000. General Stark died here. 16 ms. s. Concord. Pop. 1830, 877.

Manchester, p-t. and half shire, Bennington co. Vt. Battenkill r., and its branches give many mill seats. Equinox mtn. is 3,706 feet above tide. It has 2 villages, white marble quarries, a jail, court house, academy, &c., 22 ms. w. Bennington. A turnpike road crosses the Green mtns. Pop. 1830, 1,525.

Manchester, p-t. Essex co. Mass. 8 ms. e. n. e. Salem, 27 n. e. Boston. Incorporated in 1645. The inhabitants are much employed in the fisheries, which are valuable. It has a good harbor, and lies sloping to the water s. Pop. 1830, 1,236.

Manchester, p-t. Hartford co. Conn. 10 ms. e. Hartford. The village is pleasantly situated on the great route from Hartford to Boston, and the town, which contains much fine soil, has several manufactories. Pop. 1830, 1,576.

Manchester, p-t. Ontario co. N. Y. Canandaigua outlet affords mill seats, 199 ms. w. Albany. The Clifton springs are sulphureous, rising through lime rocks, 10 ms. n. Canandaigua. Pop. 1830, 2,811.

Manchester, p-o. York co. Pa. between the boro' of York and Harrisburg.

Manchester, p-v. in the northwestern angle of Baltimore co. Md. 33 ms. n. n. w. Baltimore.

Manchester, p-v. very pleasantly situated on James r. opposite Richmond, and in Chesterfield co. Va. Mayo's bridge over the Rapids of James r. unites Richmond to Manchester.

Manchester, p-v. in the western part of Sumpter dist. S. C. It is situated about 8 ms. n. n. e. the junction of Wateree and Congaree rs., and by p-r. 40 ms. s. e. by e. Columbia.

MANCHESTER, p-v. and st. jus. Clay co. Ky. situated on a branch of the South fork of Ky. r. by p-r. 126 ms. s. E. Frankfort. Lat. 37° 10', and long. 6° 38' w. W. C.

MANCHESTER, p-v. Adams co. O. by p-r. 110 ms. s. s. w. Columbus.

MANCHESTER, p-v. Dearborn co. Ind. by p-r. 89 ms. s. E. Indianopolis.

MANCHESTER, p-v. St. Louis co. Mo. by p-r. 20 ms. wstrd. St. Louis.

MANDARIN, p-v. Duval co. Flor. by p-r. 267 ms. estrd. Tallahassee.

MANGOHICK, p-o. King William co. Va. 40 ms. sthwrd. W. C.

MANHATTAN, isl. (*See N. Y. city.*)

MANHATTAN, p-v. Putnam co. Ind. by p-r. 52 ms. wstrd. Indianopolis.

MANHATTANVILLE, v. E. side of Hudson r. 9 ms. N. N. Y., included within the bounds of the city and co. of N. Y.

MANHEIM, p-t. Herkimer co. N. Y. 69 ms. w. N. w. Albany, 14 E. Herkimer; contains a number of mills, and very good land. Pop. 1830, 1,937.

MANHEIM, p-v. Lancaster co. Pa. 10 ms. N. w. the city of Lancaster, and about 30 a little s. of E. Harrisburg.

MANLIUS, p-t. Onondaga co. N. Y. 10 ms. E. Onondaga, 137 w. Albany; contains many mill seats on Limestone, Chitteningo and Butternut creeks. A branch of the first falls 100 feet. There are sulphur springs. The Erie canal passes through the town. 5 villages, Manlius, Fayetteville, Orville, Eagleville and Jamesville. Pop. 1830, 7,375.

MANNBORO', p-v. Amelia co. Va. by p-r. 48 ms. N. w. Richmond.

MANNINGHAM, p-v. Butler co. Ala. by p-r. 152 ms. s. s. E. Ala.

MANNINGTON, t. Salem co. N. J. 50 ms. s. w. Trenton; has Salem creek N. and w., and is crossed by Mannington creek. Pop. 1830, 1,172.

MANOR, p-o. Lancaster co. Pa. 6 ms. s. s. w. Lancaster.

MANOR HILL and p-o. Huntingdon co. Pa. by p-r. 163 ms. N. W. C.

MANSFIELD, t. Chittenden co. Vt. 20 ms. N. w. Montpelier; contains much uninhabitable mountain land. Pop. 1830, 1,726.

MANSFIELD, t. Bristol co. Mass. 12 ms. N. Taunton. Pop. 1830, 1,172.

MANSFIELD, p-t. Tolland co Conn. 28 ms. E. Hartford. A larger quantity of silk is manufactured here than in any other place in the U. S. This branch of industry was introduced into the country by Dr. Aspinwall of this place, above 70 years ago, who established the raising of silk worms in New Haven, Long Island and Phil. Assisted by Dr. Stiles, half an ounce of mulberry seed was sent to every parish in Conn., and the legislature for a time offered a bounty on mulberry trees and raw silk; 265 lbs. were raised here in 1793, and the quantity has been increasing ever since. In 1830, 3,200 lbs. were raised. Here is a small silk factory, under an English manufacturer, with swifts, for winding hard silk; 32 spindles for doubling; 7 dozen of spindles for throwing; 7 do. of spindles for spinning; 32 spindles for soft silk winding, and 2 broad and 1 fringe silk looms. There is machinery enough to keep 30 broad silk looms, and 50 hands in operation. Pop. 1830, 2,661.

MANSFIELD, t. Alleghany co. N. Y. 245 ms. w. by s. of Albany. Pop. 1830, 378.

MANSFIELD, p-t. Warren co. N. J., is hilly, crossed lengthwise by Morris canal and Pohatcong creek. It is bounded s. E. in its whole length by Musconetcong r., and is 7 ms. s. E. Oxford and 35 N. Trenton.

MANSFIELD, t. Burlington co. N. J. 8 ms. s. Trenton; has Del. r. N. w., Blacks creek N. E., and is crossed by Crafts creek, on which are several mills. It is opposite Newbold's isl. in Del. r.

MANSFIELD, p-v. and st. jus. Richland co. O. by p-r. 71 ms. N. N. E. Columbus. Lat. 40° 47', long. 5° 53' w. W. C. Pop. 1830, 840.

MANSFIELD, p-v. Tioga co. Pa. by p-r. 140 ms. N. N. w. Harrisburg.

MANSKER'S creek and p-o. western part of Davidson co. Ten. 25 ms. wstrd. Nashville.

MANTUA, p-v. Portage co. O. by p-r. 137 ms. N. E. Columbus.

MAPLE GROVE, p-o. Armstrong co. Pa. 231 ms. N. w. W. C.

MAPLESVILLE, p-v. Bibb co. Ala. by p-r. 35 ms. estrd. Tuscaloosa.

MARAMEC, r. of Mo. interlocking sources on the sthrd. with those of St. Francis, and on the w. with those of Gasconade r. It is composed of two branches, Maramec Proper, and Big r. Maramec rises in Crawford and Washington cos., and flowing thence N. E. traverses Franklin, receiving the Bourbeuse, a large tributary from the wstrd. Having reached to within 8 ms. from Mo. r., the Maramec curves to the E. and receives Big r. between St. Louis and Jefferson cos. Still inflecting, this stream finally assumes a s. E. course to its influx into Miss. r. 20 ms. below St. Louis.

Big r. rises in the Iron mtns., and in Washington and St. Francis cos., and flowing thence N. over Jefferson falls into the main stream of Maramec at Lawrenceton.

It may be observed as a curious fact in physical geography, that the general course of the Maramec is directly contrary to, and very nearly parallel to that of the Miss. from St. Louis to the influx of Kaskaskias. The valley of Maramec is 100 ms. in length, with a mean breadth of 35, or area 3,500 sq. ms. comprising the space between the lower Mo. and St. Francis, and between the Gasconade and Miss. rs. Lat. 38°, and long. 14° w. intersect between the main Maramec and Big r.

MARAMEC, p-v. Gasconade co. Mo., about 70 ms. s. w. by w. St. Louis.

MARATHON, p-t. Cortland co. N. Y., 145 ms. s. by E. of Albany. Pop. 1830, 895.

MARBLEHEAD, p-t. and port, Essex co. Mass., 16 ms. E. Boston, 4 s. E. Salem, 1 m.

by 3½ on a neck of land. The harbor is 1½ ms. long, ½ broad, safe and defended by fort Sewell. The town is large and handsome, with a fine square, custom house, bank, and other public buildings, on a rocky neck. The chief business is the cod fishery on the banks; fifty-seven vessels and 412 men were employed in the cod and mackerel fishery, from Marblehead, in 1831; the number of fish taken was 1,132,650, weighing 55,000 quintals, and the whole proceeds valued at $160,490. The coast is rocky and barren, and there are but few spots of good soil in the town; it has water on three sides. Pop. 1830, 5,149.

MARBLE HILL, and p-o. Prince Edward co. Va., 83 ms. s. w. Richmond.

MARBLETOWN, p-t. Ulster co. N. Y., 10 ms. s. w. Kingston. Esopus and Rondout crs. pass through it, and with their branches afford mill seats. Clouded marble is quarried here; the Delaware and Hudson canal passes through it. Pop. 1830, 3,223.

MARBURYVILLE, p-v. parish of W. Feliciana, La., 8 ms. estrd. St. Francisville, 83 N. w. by w. New Orleans.

MARCELLUS, p-t. Onondaga co. N. Y., on Skeneateles lake, 10 ms. w. Onondaga, includes half of Skeneateles and Otisco lakes, with many mill seats. Pop. 1830, 2,626.

MARCUS HOOK, p-v. on the right bank of Del. r., and in the extreme sthrn. angle of Del. co. Pa., 20 ms. below Phil.

MARENGO, co. of Ala., bounded N. by Greene; N. E. by Perry; E. Dallas; S. E. Wilcox; S. Clark; and W. Tombigbee r., separating it from the Choctaw country. Length S. to N. 40 ms., width 24, and area 960 sq. ms. Extending in lat. 32° to 32° 35′, and in long. 10° 40′ to 11° 41′ w. W. C. This co. extending down the Tombigbee from the influx of Black Warrior r., is composed of an inclined plane, and declining wstrd. towards the latter stream; the estrn. and sthestrn. limits, though straight lines, follow nearly the dividing ridge between the vallies of Tombigbee and Ala. rs. Similar to the contiguous cos., the greatest share of the surface of Marengo is covered with pine, and with a sterile soil; in 1820, what is now Greene co. was included in Marengo, and the whole had a population of 3,933; in 1830, the latter contained 7,700. Chief t. Marengo.

MARENGO, p-v. and st. jus. Marengo co. Ala. (*See Lynden.*)

MARGALLAWAY r. N. H., rises on the line of Maine and L. Canada, and is the head stream of Androscoggin r.

MARGARETTA, Furnace and p-o. York co. Pa.

MARGARETTA, p-v. Huron co. O., by p-r. 119 ms. N. Columbus.

MARIANA, p-v. on Cupola r., in the nrthrn. part of Jackson co. Flor., about 140 ms. N. E. by E. Pensacola, and 70 ms. N. W. by w. Tallahasse.

MARIETTA, p-v. on the left bank of Susquehannah r., above the mouth of Chiques cr. Lancaster co. Pa., 13 ms. w. the city of Lancaster. In 1820, the twp. contained 1,545 inhabitants.

MARIETTA, p-v. and st. jus. Washington co. O., situated on the point above the junction of Ohio and Muskingum rs., about 60 ms. S. S. E. Zanesville, and by p-r. 304 ms. a little N. of w. W. C., and 106 S. E. by E. Columbus. The site is pleasant, but the lower part near the point liable to occasional inundation. This town was the cradle of the state of O., and was founded in 1787 by a colony from Mass., whose descendants have maintained the industrious and frugal habits of their parents. The town now contains an academy, several private schools, the common co. buildings, two printing offices, a bank, and two or three churches. Pop. 1830, 1,207, distributed over three wards. The scenery of the vicinity is peculiarly fine, even on the Ohio.

MARION, dist. S. C., bounded E. and S. E. by Lumber r. or Little Pedee, separating it from Horry; on the S. by Great Pedee, separating it from Georgetown; S. W. by Lynches cr., separating it from Williamsburg; W. by Darlington; N. W. by Marlboro'; and N. by Robeson co. in N. C. Length from the junction of Great and Little Pedee, to the extreme nrthrn. angle on N. C., 67 ms.: mean width 18, and area 1,200 sq. ms. nearly. Extending in lat. 33° 41′ to 34° 36′, and in long. 2° 10′ to 2° 50′ w. W. C. The general slope of this large dist. is sthrd., down which flow the Great and Little Pedee, and numerous smaller streams; the surface is mostly level, much of it flat and marshy. In a navigable point of view, Marion has great advantages; it lies open to the ocean by 4 boatable streams, which are finally united at its extreme sthrn. angle. Chief t. Marion. Pop. 1820, 10,201; 1830, 11,008.

MARION, p-v. and st. jus. Marion dist. S. C., situated near the centre of the dist., about 65 ms. N. Georgetown, and by p-r. 116 ms. a little N. of E. Columbia; lat. 34° 11′, long. 2° 28′ w. W. C.

MARION, co. Geo., bounded S. by Lee; S. W. by Randolph; W. Muscogee; N. Talbot; and E. Flint r. separating it from Crawford N. E.; and Houston E. Length 35 ms., mean width 20, and area 700 sq. ms. Extending in lat. 32° 18′ to 32° 35′, and in long. 7° 12′ to 7° 46′ w. W. C. The slope of this co. is E. towards the Flint. Chief t. Marion C. H. Pop. 1830, 1,436.

MARION, p-v. and st. jus. Twiggs co. Geo., by p-r. 37 ms. S. W. Milledgeville; lat. 32° 42′, and long. 6° 30′ w. W. C.

MARION, C. H. and p-o. Marion co. Geo., by p-r. 174 ms. S. W. by W. Milledgeville.

MARION, co. Ala. bounded N. by Franklin; E. Walker; S. Lafayette; W. Monroe co. in the state of Miss.; and N. W. by the Chickasaw country in Ala. Length 38, mean width 30, and area 1,140 sq. ms.; lat. 34°, and long. 11° w. W. C. intersect very near the centre of this co. The nrthrn. part is drained by the sources of Bear cr. flowing into the Ten. r.; the greater part however slopes sthrd., and is drained into Tombigbee by the different branches of Buttahatche and Sipey rs. Chief t. Pikeville. Pop. 1830, 4,058.

Marion co. Miss., bounded w. by Pike; n. w. by Lawrence; n. by Covington; e. by Perry; s. e. by Hancock; and s. w. by the parish of Washington, La. Length 42, mean width 30, and area 1,260 sq. ms.; extending in lat. 31° to 31° 27′, and in long. 12° 28′ to 13° 17′ w. W. C. The estrn. part of this co. slopes to the s. e., and is drained by Leaf r. and Black cr., branches of the Pascagoula r., whilst the watrn. section is traversed by Pearl r. in a s. s. e. direction. Most of the surface is covered with pine, and soil sterile; the margin of the streams, however, affords good soil; staple, cotton. Chief t. Columbia. Pop. 1820, 3,116; 1830, 3,691.

Marion co. Ten., bounded by Cumberland mtn. which separates it from Franklin w. and n. w.; by Bledsoe n.; Hamilton e.; and Ten. r. s. Length 30, mean width 20, and area 600 sq. ms. Extending in lat. 35° 10′ to 35° 26′, and in long. 8° 15′ to 9° 09′ w. W. C. The Sequatchie r. enters this co. on the nrthestrn. border, and traverses it in the greatest length, flowing to the s. w. into Ten. The co. lies in most part in the valley of the Sequatchie; surface hilly. Chief t. Jasper. Pop. 1820, 3,888; 1830, 5,508.

Marion, co. O., bounded by Del. s.; Union s. w.; Hardin w.; Crawford n.; and Richland e. Length from e. to w. 32, mean breadth 15, and area 480 sq. ms.; lat. 40° 40′, long W. C. 6° w. Slope sthrd. and drained by the higher branches of Sciota r. Chief t. Marion. Pop. 1830, 6,190.

Marion, p-v. and st. jus. Marion co. O., by p-r. 47 ms. a little w. of n. Columbus. Pop. 1830, 287.

Marion, co. Ind., bounded s. by Johnson; Morgan s. w.; Hendricks w.; n. w. Boone; n. Hamilton; and e. Hancock. It is a square of 20 ms. each side; area 400 sq. ms. The central lat. 40° 45′, long. 9° 09′ w. W. C. It is traversed in a direction of s. s. w. by the main stream of White r., which divides it into two not very unequal sections. It is a surface which, every thing considered, is perhaps unsurpassed. Chief t. Indianopolis, the capital of the state. Pop. of the co. 1830, 7,192.

Marion, p-v. Shelby co. Ind., by p-r. 25 ms. s. e. Indianopolis, and 5 ms. from Shelbyville.

Marion, co. Il., bounded by Jefferson s.; Clinton w.; Fayette n.; Clay n. e.; and Wayne s. e. It is a square 24 ms. each side, 576 sq. ms.; lat. 38° 40′, long. W. C. 12° w. It is a table land, from which flow watrd. some branches of the Kaskaskias, and south estrd. the extreme sources of Waynes fork of Little Wabash. Chief t. Salem. Pop. 1830, 2,125.

Marion, co. Mo., bounded s. by Ralls; on the w. and n. uncertain; by the Miss. r. e., separating it from Adams and Pike cos. Il. Breadth 20, mean length 24, and area 480 sq. ms. Lat. 39° 45′, long. 14½° w. W. C.; the slope estrd. and drained by several creeks. Chief t. Palmyra. Pop. 1830, 4,837.

Marion, p-v. in the nrthwstrn. part of Cole co. Mo. It is situated on the right bank of Missouri r., 15 ms. above Jefferson city.

Marksborough, v. Warren co. N. J., on Paulingskill, 15 ms. n. n. e. Belvidere.

Marksville, p-v. and st. jus. parish of Avoyelles, La., situated 55 ms. by the road n. St. Landre, in Opelousas; 35 ms. s. e. Alexandria in Rapid co., and as marked in the p-o. list, 1,308 ms. W. C.; lat. 31° 05′, and long. 15° 08′ w. W. C.

Marksville, p-v. Shenandoah co. Va., by p-r. 125 ms. wstrd. W. C.

Marlborough, p-t Cheshire co. N. H. 5 ms. s. e. Keene, 55 from Concord, contains several ponds, emptying into Ashuelot r. with rocky soil, good for grain, flax and grass. First settled 1760. Pop. 1830, 822.

Marlborough, p-t. Windham co. Vt. 24 ms. e. Bennington, 44 s. w. Windsor. First settled 1763; has the w. branch of West r., Whetstone branch, and Green r., which give good mill seats. 2 ponds supply trout. Centre mtn. is in the middle. The soil is rich, and products and minerals numerous. 12 school districts. Pop. 1830, 1,218.

Marlborough, p-t. Middlesex co. Mass. 16 ms. e. Worcester, 28 ms. w. Boston. Incorporated 1660, on a branch of Concord r., has very good land and a varied surface, where many cattle are fattened. First settled 1654. Here was the Christian Indian t. of Okamakamesit; when a part of Sudbury, in 1676, suffered from the Indians. Pop. 1830, 2,077.

Marlborough, p-t. Hartford co. Conn. 4 ms. by 5½; 22 sq. ms., is hilly and stony, best for grass, with good mill seats on small streams. Black lead is found here. Pop. 1830, 704.

Marlborough, p-t. Ulster co. N. Y., w. Hudson r., 23 ms. s. e. Hudson, 3 ms. by 6, 18 sq. ms. well cultivated, and has many inhabitants of English extraction. Pop. 1830, 2,273.

Marlboro', Lower. (*See Lower Marlboro'.*)

Marlboro', Upper. (*See Upper Marlboro'.*)

Marlow, p-t. Cheshire co. N. H. 15 ms. from Keene, 45 Concord, 15,937 acres, is crossed by Ashuelot r., has a wet soil, but fertile meadows, and produces much grain. Pop. 1830, 645.

Marquis, p-v. Tippecanoe co. Ind. by p-r. 77 ms. n. w. Indianopolis.

Marrowbone, p-v. Cumberland co. Ky. by p-r. 128 ms. sthrd. Frankfort.

Mars, p-v. Guilford co. N. C. by p-r. 95 ms. n. w. by w. Raleigh.

Mars, p-v. Bibb co. Ala. by p-r. 26 ms. estrd. Tuscaloosa.

Mars Bluff, and p-o. on the left bank of Great Pedee r. Marion dist. S. C., where the road passes from Darlington to Marion C. H. by p-r. 118 ms. a little n. of e. Columbia.

Marsh Island, Penobscot co. Me. in Penobscot r. 4 ms. above Bangor.

Marsh, p-o. Chester co. Pa. by p-r. 136 ms. n. e. W. C.

Marshall, p-t. Oneida co. N. Y. 110 ms. w. Albany. Pop. 1830. 1,908.

Marshalls' Ferry, and p-o. Grainger co. Ten. by p-r. 248 ms. e. Nashville.

Marshallville, p-o. Wayne co. O. about 90 ms. n. e. Columbus.

Marshalton, p-v. Chester co. Pa. 4 ms. w. West Chester, and 28 w. Phil.

Marshfield, p-t. Washington co. Vt. 12 ms. n. e. Montpelier, 16 s. w. Danville. It is crossed by Onion r., and is uneven, with slate and granite rocks. It contains 6 school districts, and was granted to the Stockbridge Indians in 1782, and sold by them to Isaac Marsh 1789. Pop. 1830, 1,271.

Marshfield, p-t. Plymouth co. Mass. 30 ms. s. e. Boston; incorporated in 1640; is pleasantly situated on the ocean, with North and South rs., and a small harbor. It first belonged to Plymouth. Pop. 1830, 1,565.

Mars Hill, Me. 1 m. 16 chains w. from the e. bound of U. S.; has been recently assumed by the British as the n. w. angle of Nova Scotia. It is isolated, with 2 peaks, 1,506, and 1,363 ft. above St. Johns r.

Marshpee, Indian t., Barnstable co. Mass. 170 ms. s. e. Boston, has a harbor and some shipping. It was an Indian town, and has some remains of the original inhabitants. It has Pomponesset bay e., a light soil, with much wood. Here was an Indian christian congregation.

Marthasville, p-v. Montgomery co. Mo. about 55 ms. wstrd. St. Louis.

Martha's Vineyard, island, Dukes co. Mass. contains 3 towns, Edgartown, Tisbury, and Chilmark. The court of common pleas is held at Edgartown for Barnstable and Dukes cos. Soil poor, but many cattle and sheep are raised; and the fisheries are valuable. The people are much engaged as pilots, seamen, and as fishermen, and are hardy and enterprising. An Indian church was formed here in 1666, by Cotton Mather. It is a little w. Nantucket, 21 ms. by 6.

Marticville, p-v. in the tsp. of Martic, Lancaster co. Pa. 8 ms. sthrd. the city of Lancaster. In 1820, the tsp. contained 1,701 inhabitants.

Martin, co. N. C. bounded by Washington e., Beaufort s. e., Pitt s. w., Edgecombe w., Halifax n. w., and Roanoke r. separating it from Bartie n. and n. e. Length 40, mean width 12, and area 481 sq. ms. Extending in lat. 33° 40' to 36° 02', and in long. 0° 16' e. to 0° 28' w. W. C. The surface level, and in part marshy, with a slight declination to the n. e. towards Roanoke r. Chief town, Williamstown. Pop. 1820, 6,320, 1830, 8,539.

Martin, co. Ind. bounded s. by Dubois, Daviess w., Greene n., Lawrence n. e., and Orange s. e.; breadth 15, mean length 20, and area 300 sq. ms. Lat. 38° 40', long. 9° 50' w. W. C. It is traversed in a s. s. w. direction by the east branch of White r. Chief town, Hindostan. Pop. 1830, 2,010.

Martinsburg, p-t. and st. jus. Lewis co. N. Y. Roaring branch, has good mill seats. It contains the county buildings, and is 48 ms. n. Utica. Pop. 1830, 2,382.

Martinsburg, p-v. in the nthrn. part of Bedford co. Pa. 27 ms. a little e. of n. the boro' of Bedford, and by p-r. 112 ms. s. w. by w. Harrisburg.

Martinsburg, p-v. and st. jus. Berkley co. Va. 24 ms. n. n. e. Winchester, 21 n. w. Harpers Ferry, and by p-r. 84 n. w. W. C. Lat. 39° 27', and long. 0° 58' w. W. C.

Martinsburg, p-v. in the sthrn. part of Monroe co. Ky. 14 ms. from Tompkinsville, and by p-r. 151 s. s. w. Frankford.

Martinsburg, p-v. Knox co. O. by p-r. 55 ms. n. e. Columbus.

Martinsburg, p-v. Washington co. Ind. by p-r. 103 ms. sthrd. Indianopolis.

Martins, cr. and p-o. The creek falls into the Del. r. in Northampton co. Pa. 10 ms. above Easton, and the p-o. is near it.

Martin's Mills, and p-o. Richland co. O. by p-r. 86 ms. n. n. e. Columbus.

Martin's Store, and p-o. Montgomery co. N. C. by p-r. 133 ms. s. w. by w. Raleigh.

Martinville, p-v. and st. jus. Morgan co. Ind. situated on white r. 30 ms. below, and s. s. w. Indianopolis. Lat. 39° 26' long. W. C. 9° 24' w.

Martinsville, p-v. and st. jus. Henry co. Va. situated near the left bank of Irvine or Smith's r. about 70 ms. s. w. Lynchburg, and by p-r. 151 s. w. by w. Richmond.

Martinsville, p-v. Guilford co. N. C. situated on the Reedy fork of Haw r., by p-r. 94 ms. n. w. by w. Raleigh.

Martinsville, p-v. in the nthestrn. part of Warren co. Ky. 29 ms. from Bowling Green, and by p-r. 113 ms. s. w. by w. Frankfort.

Maryland, p-t. Otsego co. N. Y. 16 ms. s. Cooperstown, 66 w. Albany, has much good grazing; but Cromhorn mtns. are barren. 10,000 acres of the tsp. belong to the state. It has mills, manufactories, &c. Pop. 1830, 1,834.

Maryland, one of the states of the U. S., bounded by the state of Del. e., the Atlantic ocean and the estrn. shore of Va. s. e., Chesapeake bay s., Potomac r. separating it from Va. s. w., a part of Va. w., and Pa. n. w. and n. Maryland extends along the Atlantic ocean from the sthestrn. angle of the state of Del. to the sthestrn. angle of Maryland 35 ms.; between Md. and Va., on the estrn. shore 15; from the mouth of Pokomoke r. to that of Potomac 40; up Potomac r. to the source of its n. branch 320; thence due n. to the sthrn. boundary of Pa. 36; along the limit between Pa. and Md. 200; and along the limit between Del. and Md. to the place of beginning on the Atlantic ocean 124; having an outline of 770 ms.; and extends from lat. 38° to 39° 43' very nearly, and in long. 1° 56' e. to 2° 24' w. W. C.

From the great irregularity of its outline, and from including in its superficies that of Chesapeake bay, the area of the land surface of Md. has been, by most geographers, overrated. The subjoined table gives the area of the counties, and the aggregate of the whole.

	Lth.	mn. wth.	area.	Pop. 1820,	Pop. 1830,
Alleghany,	60	12¼	812¼	8,654	10,602
A. Arundel,	60	12	720	27,165	28,295
Baltimore,	36	25	900	96,201	120,876
Calvert,	33	8	264	8,073	8,899
Caroline,	30	8	240	10,041	9,070
Cecil,	22	12	264	16,048	15,432
Charles,	30	15	450	16,500	17,666
Dorchester,	32	20	640	17,755	18,685
Frederick,	42	18	776	40,459	45,793
Harford,	30	16	480	15,924	16,315
Kent,	30	8	240	11,453	10,502
Montgomery,	28	18	500	16,400	19,816
Prince Geo's.	30	17	510	20,216	20,473
Queen Ann,	40	10	400	14,952	14,396
St. Mary's	38	10	380	12,974	13,455
Somerset,	35	15	500	19,579	20,155
Talbot,	25	8	200	14,389	12,947
Washington,	40	12	480	23,075	25,263
Worcester,	30	20	600	17,421	18,271
Total,			9,356	407,279	446,913

Of the area, the estrn. shore contains 3,084 sq. ms. with a pop. in 1820, of 121,638, or something above 39 to the sq. m., whilst the wstrn. part, comprising 6,272 sq. ms., and including the two cities of Baltimore and Frederick, contained in 1820, 285,641 inhabitants. In the last decennial period, the aggregate pop. of the state, has increased a small fraction above 9½ per cent., having gained an increment of 39,634, but this augmentation has been entirely w. of Chesapeake bay. Of the 8 estrn. cos. 4 have decreased, and taken together, the aggregate is 2,171 less than that of 1820. On the contrary, every one of the wstrn. cos. has gained more or less, and the whole has augmented from 285,641 to 327,446. Of the increase, 17,887 was in the city of Baltimore.

The progressive population, of the state since 1790, has been as follows :—

Date,	Whites,	Free col'd.	Slaves,	Total col'd.	Total,
1790	208,647	8,043	103,036	111,079	319,728
1800	221,998	19,987	107,707	127,094	349,654
1810	235,117	33,927	111,502	145,429	380,546
1820	260,222	39,730	107,398	147,128	407,350
1830	291,093	52,912	102,873	155,820	446,913

Of the pop. in 1830, were :—whites, blind, 156; deaf and dumb, 132; colored, blind, 117; deaf and dumb, 82.

The free white pop. has within the last 10 years augmented from 260,222 to 291,093, having gained 30,871 or about 12 per cent. The free colored increase is 13,182 or at the rate of 33⅓ per cent. The slaves have decreased 4,520, or at the rate of 4½ per cent. The total increase of colored pop. since 1820, is 8,652, or at the rate of 5¾ per cent. There were in 1790, 183 whites to one colored,—1800, 175,—1810, 162,—1820, 177, and in 1830, 187.

Natural Geography.—Md. is naturally subdivided into three sections; eastern, middle and western.

The estrn. called locally, "the eastern shore," separated from the middle by Chesapeake bay, comprises a part of that remarkable peninsula between the Del. and Chesapeake bays. Except in size and in being united to the continent on the N. by a neck of land of about 20 ms. the Chesapeake peninsula differs in nothing essential from the other insular strips along the Atlantic coast of the U. S. Both Long Island and Staten Island, are indeed much more relatively elevated than any part of the peninsula of Chesapeake. (*See Chesapeake peninsula.*)

Of this peninsula, Md. comprises the wstrn. slope from Pokomoke bay, to the junction of Susquehannah r. with Chesapeake bay. The estrn. shore of Md. is peculiarly indented by bays and chequered with small islands. Pokomoke bay is an expanse of water spreading from the mouth of a small r. of the same name, and is followed nthwstrdly. by Tangier island and sound, leading into Fishing bay, below the mouth of Nanticoke r.; with the Tangier islands and the mouth of Potomac, Chesapeake bay abruptly contracts from a width of 25 to about 10 or 12 ms. Above the mouth of Nanticoke r., with the peninsular forming Dorchester co., intervening, opens Choptank bay, separating Dorchester from Talbot co. The latter co. is again subdivided into several fragments by Tread Haven, Broad and St. Michael's bays, and is followed by Chester bay and r., separating Queen Ann from Kent co. All those bays and numerous creeks intersect the coast, in the space of one degree of lat. between lat. 38° and 39°.

From lat. 39°, Kent co. sweeps a semicircular peninsular, between Chester and Sassafras rs. with its convex on Chesapeake bay. Sassafras bay is followed in quick succession by Elk and North rs., and finally by the great discharge of Susquehannah r.

We may regard the Chesapeake and Del. canal, as a natural limit, since, though artificial and of recent construction, it must remain permanent, and insulates the natural section under review. The eastern shore of Md. is alluvial. The surface of the country is either waving or level, and in no place sufficiently elevated to be correctly designated hilly. The soil varies, but in its general character may be set down as above middling quality. The climate from the nthrn. part of Cecil, to the sthrn. of Somerset and Dorchester cos., differs in temperature much more than might be expected in 103 minutes of lat., over a region so little diversified in relative elevation. On the lower cos. cotton can be cultivated to advantage. The very numerous inlets, and the proximity of Baltimore, give this section of Md. great commercial advantages, and yet, as we have seen, the distributive pop. is on the decrease. If, however, we abstract the combined pop. of Baltimore and Frederick cities, the mean density of the two shores of

Md., stand by the census of 1830, very nearly equal at 38 to the sq. m., but with the cities of Frederick and Baltimore, the wstrn. side has a distributive pop. of 52 to the sq. m. The primitive ledge, and Susquehannah r. enter Md. together, but extend at almost exactly right angles to each other; the r. flowing sthestrd. to the head of tide water, and the primitive ledge inclining sthwstd. to the head of tide water in Potomac r., in the dist. of Columbia. Along the shores of Chesapeake bay, from the mouth of Potomac, to that of Susquehannah, the components of soil, formation and aspect of the surface, do not materially differ from the opposite or estrn. shore, but advancing to the nthwstrd. from Chesapeake, the country gradually rises, becomes more and more broken and rocky, until the primitive ledge is attained. This great physical boundary has a mean elevation of at least 400 ft., and divides the state into two sections, and also into two very distinct zones of soil. The primitive is not very definite in its termination to the s. e., and is still less distinctly traceable on the opposite side. Its lower visible boundary is generally determined by the head of tide water, and all the streams which traverse it, pass through gorges with cataracts of more or less descent.

The primitive ledge is in fact a part of an Appalachian chain. Mere elevation excepted, the ridge has every trait of other ridges of the system to which it belongs. The Sugar Loaf mtn. which rises in Md., near and below the junction of Potomac and Monocacy rs., is another portion of a nameless chain, which extends nthestrd. separating Frederick from Montgomery, Anne Arundel, and Baltimore cos. The immediate valley, about 20 ms. wide, between the lower primitive and Sugar Loaf chain, is a real mtn. valley, comprising in Md. great part of Montgomery, the upper part of Anne Arundel, Baltimore, and Harford cos., and which is followed nthwstrd. from the Sugar Loaf chain, by the fine valley of Monocacy. That part of the latter valley contained in Md., is commensurate with Frederick co. Westrd. from Frederick the two cos. of Washington and Alleghany, are composed of narrow but generally very fertile vales, between lateral ridges of the Appalachian system.

Independent of the mtn. ridges, the surface of Maryland gradually rises from the Chesapeake ba to the sources of Potomac, or from the level of tide water to near 2,000 feet. The relative height from tide water in the basin at Baltimore to the dividing ground in Alleghany co. Md., between the sources of Potomac and Youghioghany rs. is as follows: from the forks of Patapsco r. about midway from tide water at Balt., to the second or Sugar Loaf ridge, about mid tide, 385 ft. Sources of Patuxent, Patapsco, flowing sthestrd. and Liganore and Pipe cr. branches of Monocacy, flowing sthwstrd. and near where the great road passes from Baltimore to Frederick, from 600 to 850 ft. Country adjacent to Frederick, from 300 to 500 ft. Harman's gap, over Blue Ridge or Catoctin mtn., about 10 ms. E. Hagerstown, 1,550 ft. Lower part of the vallies of Antietam and Conecocheague around Hagerstown, 460 to 800 ft. Arable ground along the Potomac near Cumberland, 550 ft. Arable ground in the valley between Will's and Savage mtns., and between Cumberland and Frostburg, drained into Potomac by Will's and George's crs., from 800 to 1,000 ft. Arable vallies, between the numerous ridges of mtns. from which flow on one side the sources of Potomac, and on the other those of Youghioghany rs., from 1,600 to 2,000 feet.

From the above elements, given in round numbers, it is shewn that wstrn. Md. forms part of an inclined plane rising from tide water in a distance of about 150 statute ms. air measure, to 2,000 ft. If 400 ft. is assumed as equivalent to a degree of temp. on Fahrenheit's scale, the relative height will equal 5 degrees, or give to the extreme wstrn. part of the state a climate in winter similar to that on the Atlantic coast in lat. 44° 43′.

There is much good soil existihg in every section of this state, but the most productive is grain and fruit in some of the limestone tracts in the three wstrn. cos.; vegetation, however, either indigenous or exotic, is greatiy influenced by the extremes, if lat. and relative height are combined, of upwards of 6¼ degrees of temp. On the low sandy plains of Worcester, Somerset and Dorchester cos. between lat. 38° and 38° 40′, cotton can be cultivated, whilst the elevated vales of Alleghany co., though of a highly productive soil, are almost too cold for wheat.

As Md. occupies a nearly middle latitude amongst the states of the U. States, the following meteorological tables may serve to elucidate not alone the climate of that, but of the middle Atlantic states generally. The observations were made and recorded by the author of this article, at his residence near Sandy Spring, Montgy. co., about 20 ms. N. W. C. at lat 39° 09′ and at an elevation above tide water of 400 ft.

No. 1. Table of the mean and extreme monthly temp. at the White Cottage near Sandy Spring, from observations made during two years, 1829 and 1830.

	1829	1830	Mean	Highest	Lowest
Jan.	30. 30	30. 03	30. 3	49	8
Feb.	25.	34. 66	29. 83	46	2
March	37. 23	43. 88	45. 55	67	20
April	51 37	54. 49	52. 93	89	30
May	64. 16	65. 18	64. 67	78	37
June	69. 73	68. 94	69 33	90	50
July	70. 72	79. 01	74. 91	91	55
August	73.	72. 72	72. 81	90	52
Sept.	61. 42	63. 58	62. 05	82	36
Oct.	52. 85	55. 48	54. 16	77	26
Nov.	39. 68	51. 77	45. 22	70	21
Dec.	42. 5	35. 53	39. 01	67	6
Mean	51. 496	54. 47	53. 435		

No. 2. Table of the monthly prevalent winds from observations made on 786 consecutive days, from January 1st. 1829 to February 28th, 1831, inclusive.

Months	N.	N. W.	W.	S. W.	S.	S. E.	E.	N. E.
Jan.	5	40	4	17	2	9	2	10
Feb.	3	43	3	13	4	8	1	9
March	2	30	0	12	4	12	0	2
April	5	24	1	12	3	8	4	3
May	1	17	1	18	2	13	2	8
June	4	23	5	13	4	10	0	1
July	5	14	7	23	2	7	0	4
August	5	17	3	15	3	11	0	8
Sept.	5	23	1	13	4	9	1	5
Oct.	3	17	3	13	3	10	0	13
Nov.	3	23	1	14	1	8	4	6
Dec.	2	18	5	17	3	9	0	8
	43	289	34	180	35	104	14	77
Reduced to proportions of 1000.	50	367	43	241	44	145	17	90

The results of table No. 2, correspond in a very striking manner with those of similar observations made in the nthrn. temp. zone on the continent of N. America, from the Pacific to the Atlantic ocean, on the Atlantic ocean, and on wstrn. Europe.

Statement of mean temperature of the seasons at Sandy Spring. Winter of 1828—29,—mean temp. from winter solstice 1828, to vernal equinox 1829, 28. 39. Spring of 1829, —mean temp. from vernal equinox, 1829, to summer solstice, 1829, 58. 22. Summer of 1829,—mean temp. from the summer solstice, to the autumnal equinox, 1829, 69. 31. Autumn of 1829,—mean temp. from autumnal equinox, to winter solstice, 1829, 46. 96. Winter of 1829–30,—mean temp. from winter solstice 1829, to vernal equinox 1830, 35. 63. Spring of 1830,—mean temp. from vernal equinox, to summer solstice 1830, 58. 14. Summer of 1830,—mean temp. from summer solstice, to autumnal equinox 1830, 71. 46. Autumn of 1830,—mean temp. from autumnal equinox, to winter solstice 1830, 49. 23. Winter of 1830–31,—mean temp. from winter solstice 1830, to vernal equinox 1831, 29. 88. Spring of 1831,—mean temp. from vernal equinox, to summer solstice 1831, 59. 64. Summer of 1831,—mean temp. from summer solstice, to autumnal equinox 1831, 69. 95. Autumn of 1831,—mean temp. from autumnal equinox, to winter solstice 1831, 41.81. Mean of the 12 seasons, 51. 63.

The winter of 1831–1832, though remarkable for occasional low temperature gave a mean of plus 33. 00.

At Sandy Spring on the morning of Dec. 16th, 1831, three thermometers at from 2 to 3 miles distance, yielded a mean of 13° below zero. By the same instruments, the mean of January 26th, 1832, was at zero, and on the morning of the 27th, the whole three were again down to minus 13°, and the mean of the whole day was minus 1 16-100°.

Internal Improvements.—The political subdivisions and relative extent and pop. of Md. has been already given. The chief city of this state, Baltimore, has assumed a very respectable rank among the emporia of the U. S. Besides the great wstrn. turnpike road, extending from Baltimore through Fred. to join the U. S. road at Cumberland, several other leading roads connect Baltimore with W. C. on one side and with several of the most productive cos. of central Pa.

A rail-road on a plan of unusual magnitude has been projected to extend from Baltimore to the Ohio river. But though a part of this road has been so far completed as to admit road cars to travel over it, too little comparatively has been done to admit a general, much less a specific description. It is, however, a work commanding so much of public interest as to justify some desultory extracts from the different Reports made by the board of directors of this road. (*See Chesapeake and Ohio Canal.*) The charter for the latter work is of prior date to that for the Baltimore and Ohio rail road, but both being compared at the same time, the two companies were in the incipient stage of advance with their respective works involved in a legal controversy, alluded to in the subjoined extract from the 4th annual report of the Baltimore company.

"The injunction which was obtained at the suit of the Chesapeake and Ohio canal company, prohibiting this company soon after its organization, from proceeding to construct the rail-road along the Potomac r. still remains in force, and has hitherto limited the operations of the Board, to the country estrd. of the point of rocks."

"The Point of Rocks," so called, is the termination of the Cotoctin mtn., on the left bank of Potomac r. 6 ms. above the mouth of the Monocacy. Between Baltimore and the Point of Rocks, the road is to follow a general wstrn. course, with an elliptic curve to the nthrd. pursuing the valley of Patapsco to its forks, thence along the wstrn. branch to the summit of the first mtn. ridge. Thence down Bush creek into Monocacy r., down the latter stream a few ms., and finally in a s. s. w. direction to the Potomac at the Point of Rocks. Entire distance from Baltimore 66 ms. Above the Point of Rocks the right of way along the Potomac, long in litigation, was decided in favor of the Chesapeake and Ohio canal company. Some attempts were made at compromise but hitherto without effect, and has arrested this work at the Point of Rocks.

History.—The first permanent settlement of whites made on the territory now comprised in the state of Md., was made in 1631, under William Claiborne, on Kent Island, now a part of Queen Ann co. The original charter, however, under which the colony was established was granted to Cecilius Calvert (Lord Baltimore) and dated 20th June, 1632. The first emigrants arrived on the N. Bank of Potomac, in 1634. The early settlers were much disturbed by contentions with Clayborne, who resisted the proprietary grant, and finally instigated the Indians to war. These troubles were followed by the inhabitants sharing the political feelings and violence of the revolution in England, about the middle of the 17th century. A civil war distracted the infant colony, which was terminated by the submission of the people to Cromwell's government. The charter of Md. was obtained by a Roman Catholic nobleman, and the settlement made on principles of civil and religious toleration, far in advance of the age; but whilst the republican party, as they were with ineffable absurdity called, prevailed, all the rigor of anti-popery statutes enacted in England were enforced in Md. Such was the intolerance of the times that the restoration of the Stewart family only changed the objects of oppression. Under James II. a quo warranto was sued out against the charter of Md., but before judgement could be had on the writ, the family ceased to reign. Under Cromwell the Calvert family were deprived of the government; were restored by Charles II., and again deprived by William and Mary, and Maryland continued a royal government until 1716. The Calvert who held the claims of his family, finally renounced the Roman Catholic religion, and at the latter epoch was reinstated in his rights.

From 1716 to the revolution, the advance of Md. was slow, affording few incidents for history. In the war of Independence, it is not too much to say that "the Md. Line" was marked with unfading renown. The existing constitution was ratified at Annapolis, 14th Aug. 1776. The federal constitution was adopted by Md. in 1788, and in 1790, that part of the district of Columbia lying to the left of the Potomac was ceded to the general government.

Government.—The legislature is divided into two distinct branches, a senate and house of delegates, styled "The general assembly of Maryland." Senators must be upwards of 25 years of age, 15 in number, 9 for the western, and 6 for the eastern shore: before being elected they must have had three years residence in the state—term of office five years. Delegates or members of the lower house, must, when chosen, be above 21 years of age, and must have resided one year in the co. where chosen, next preceding their election. The governor is chosen by the legislature annually, but eligible only 3 years out of 7; when chosen, he must have resided in the state 5 years, and have attained to the age of 25 years. The council, 5 in number, is elected by the legislature, and must, when elected, have attained 25 years of age, and resided in the state 3 years. Their duties are to advise the governor and assent to or dissent from the executive appointments. The judiciary is formed of a chancellor, superior and district judges. By the 9th section of the amendments to the constitution of Md., passed Nov. 1812, it was divided into 6 judicial districts, 2 E. and 4 W. of Chesapeake bay; over each of these presides one chief, and 2 associate judges, who, during their term of office, must reside in their judicial district, and hold their office during good behavior, removeable by conviction in a court of law, or by address of the general assembly, two thirds of the members voting for the removal. The court of appeals is formed by the chief judges of the districts, of which three form a quorum; but no chief judge can sit as a member of the same court of appeals before whom the original decision was made. The right of suffrage demands only citizenship, 21 years of age, and one year's residence in the co. where the election is held, but is confined to free white males.

Education.—Under the colonial government, as early as 1696, funds were, by legislative enactments, appropriated to education, by means of a college and free schools. As in nearly every other instance in the U. S. either before or since the revolution, the college absorbed the funds, made progress, and left the system of common education neglected. Washington college at Chestertown, Kent co., eastern shore, was established in 1782. On the western shore at Annapolis, St. John's college was established in 1784, and the 2 subsequently formed a university. A Roman Catholic college at Georgetown was also formed in 1784. The medical college was founded in Baltimore, 1807. This latter institution was, in 1812, connected with the faculties of divinity, law, and general sciences, and the whole formed into a body corporate, under the title of "the university of Maryland." Baltimore college, and St. Mary's colleges, are separate institutions. The funds to support these different establishments, are drawn from lands, funded stock, and fees paid by students. Academies with more or less approach to collegiate form exist in most of the principal towns in the state; but no system of common instruction is in operation, though, since 1813, funds for that purpose were provided, amounting to $15,000 per annum, to be equally divided between the cos., whatever might be relative population. This fund is derived from bank stock, and appropriated to free and charity schools. Some few counties have met the provision, and availed themselves of its benefits, whilst others have not received their share of a real benefaction.

Manufactures and commerce.—Though Maryland has not been ranked amongst the manufacturing states, it is doubtful whether in proportion to her population she is not in that respect amongst the first. Numerous woollen and cotton mills, copper and iron rolling mills are in operation near Baltimore, and are also scattered over other parts of the state.

Flour and tobacco have been called the staples of Maryland, but the former so greatly exceeds as to claim pre-eminence. Tobacco is however produced largely, and of excellent quality. Mineral coal, and iron ore abound in some of the western cos. The mineral coal is confined, indeed, to Alleghany co., but there is in inexhaustible abundance the bituminous species. Iron ore is found in most of the cos. W. of the Chesapeake bay, and is extensively wrought into iron and pot metal.

By the annual report, Dec. 30, 1831, of the treasurer of the western shore, it appears that the actual income of the state of Maryland, for the year which ended on the 1st inst., (including $54,106 88, the balance in the treasury of the western shore on the 1st Dec. 1830,) was $294,002 07. The disbursements of the year amounted to $216,824 43, leaving an unexpended balance of $77,177 64. Subject to appropriations uncalled for, $41,810 42, leaving an unappropriated balance in the treasury on the 1st December, 1831, of $35,367 22, which will enable the committee on ways and means, to discharge the entire amount of the public debt, which is payable at the pleasure of the state.

MARYSVILLE, p-v. in the southern part of Campbell co. Va. 20 ms. a little w. of s. Lynchburg, and by p-r. 147 s. w. by w. Richmond.

MARYSVILLE, st. jus. Charlotte co. Va. situated on a branch of Little Roanoke, 30 miles s. e. Lynchburg, and by p-r. 69 ms. s. w. from Richmond. N. lat. 37° 03', long. 1° 52' w. W. C.

MARYSVILLE, p-v. on Licking r. Harrison co. Ky. about 45 ms. northestrd. Frankfort.

MARYSVILLE, p-v. and st. jus. Union co. O. by p-r. 37 ms. N. W. Columbus. Lat. 40° 16', long. 6° 22' w. W. C.

MARYVILLE, p-v. and st. jus. Blount co. Ten. 18 ms. s. w. Knoxville, and by p-r. 161 ms. a little s. of E. Nashville. Lat. 35° 46', and long. 6° 51' w. W. C. It is the seat of the southern and western theological seminary.

MASARD creek and p-o. Crawford co. Ark. by p-r. 226 ms. above, and westward Little Rock.

MASCOMY pond, Grafton co. N. H.; contains 2 or 3,000 acres.

MASCOMY, r. Grafton co. N. H. enters Mascomy pond at Enfield.

MASON, p-t. Hillsboro' co. N. H. 43 ms. s. Concord, 15 s. w. Amherst; is crossed by Souhegan r., and has mills and manufactories. Pop. 1830, 1,403.

MASON, one of the western counties of Va. bounded by Cabell s. w., Kenhawa s. E., Wood N. E., Ohio r. separating it from Meigs co. in O. N., and again by the Ohio r. separating it from Gallia co. O. w. Length 40, mean breadth 22, and area 880 sq. ms. Extending in lat. 38° 32' to 39° 05', and in long. 4° 22' to 5° 12' w. W. C. This co. is washed, if we follow the bends, nearly 60 ms. by the O. r.; and the southern part is traversed in a northwesterly direction by the Great Kenhawa. The surface is very broken, though much of the soil is of good quality. Salt water has been procured by digging wells near Kenhawa r. Chief t. Mount Pleasant. Pop. 1820, 4,868, 1830, 6,534.

MASON, co. Ky. bounded by Bracken w., by Nicholas s. w., Fleming s. and s. E., Lewis E., Ohio r. separating it from Adams co. O. N. E., and again by Ohio r. separating it from Brown co. O. N. Length 20, mean width 13, and area 260 sq. ms. Extending in lat. 38° 28' to 38° 44', and in long. 6° 32' to 7° w. W. C. Though this co. is bounded by Ohio r. on the N., the general slope is wstrd. towards Licking r. Chief ts. Washington and Maysville. Population 1820, 13,588, 1830, 16,199.

MASON, or to preserve the sound more correctly, Masson river of La. is one of the drains of the annually inundated tract between the Miss. and Boeuf branch of Ouachitta, rises near Grand lake and flowing sthrd. about 80 ms. falls into Tensaw 20 or 30 ms. above the junction of the latter with the Ouachitta.

MASON HALL, p-v. in the northwestern part of Orange co. N. C. by p-r. 51 ms. N. W. Raleigh.

MASONTOWN, p-v. Fayette co. Pa. about 20 ms. N. Union Town, and by p-r. 222 ms. N. W. W. C.

MASONVILLE, p-t. Delaware co. N. Y. 24 ms. w. Delhi, furnishes fine grazing; streams run into Del. and Susquehanah rivers. Pop. 1830, 1,145.

MASONVILLE, or Mason's Ferry, p-o. York dist. S. C. situated where the road from Yorkville to Charlotte in N. C. crosses Catawba r. by p-r. 87 ms. a little E. of N. Columbia.

MASONVILLE, p-v. Lauderdale co. Ala. by p-r. 119 ms. northward Tuscaloosa.

MASSABESICK pond, in chester, Rockingham co. N. H. 6 ms. long, by 2 or 300 rods, has an area of 1,500 acres.

MASSACHUSETTS bay, the waters enclosed by the coast of Mass. from cape Ann to cape Cod. It contains many islands, chiefly in Boston harbor, and several ports, the most important of which is Boston. It was visited by the pilgrims before they landed at Plymouth.

MASSACHUSETTS, one of the United States, the oldest and most important state in New England, bounded N. by Vermont and New Hampshire, E. by the Atlantic ocean, s. by the Atlantic, Rhode Island, and Connecticut, and w. by New York. It lies between 40° 23' N. lat., and 3° 38' and 7° 7' E. long. from W. C. It is 60 ms. wide by 130 long, and contains 7,800 square ms. of which, about 4,644,000 acres are land.

The early history of Mass. is that of New England. In this state the first permanent settlements by Englishmen were made. There were tried the first experiments of founding a community on the principles of general virtue and intelligence. A party of emigrants who had fled from England to seek a country in which they might enjoy freedom of conscience, landed at Plymouth, Dec. 22d, 1620. They had sailed for Hudson's river, but were carried by the master of the ship, who had been bribed by the Dutch, to a region far better fitted for the ultimate success of their plans, than that for which they started. In founding their political community, the equal rights and powers of individuals were distinctly recognized. Legislative acts were soon passed, for the instruction of every child in the community. Religion was the first object of care with the colonists, and as early as 1631, the general court decreed that none but church members should enjoy the privilege of voting, &c. From a few of the first colonies on the coast of Massachusetts bay; other settlements were soon formed, so that the same principles, habits, and institutions, extended throughout New England, and have exercised an important influence on it, and on the United States. The first general court or legislative assembly of Mass. composed of 24 representatives from the various settlements, was held in 1634. Before this, all the freemen were accustomed to meet for the transaction of public business, &c. The trial by jury was now adopted. The Pequod war, which threatened the south and west settlements, took place in 1637, and after some unjustifiable cruelties, terminated in the almost entire destruction of the only Indian nation, inimical to the colonies. In 1641 the settlements of New Hampshire were incorporated with Mass. In 1643, the first union took place between the New England colonies, when articles of an offensive and defensive confederacy were agreed to, which enabled them to combine their powers in the subsequent French and Indian wars. In 1652 the province of Maine, placed itself under the protection of Mass., and was called the co. of Yorkshir. In 1664, four royal judges were sent out from England, to determine all causes of every kind, in the colonies, but were not permitted to perform the office assigned, and returned. In 1675 began Philip's war, during which, about 1000 buildings were destroyed, 12 or 13 settlements broken up, and nearly 600 of the colonists were killed. In 1680 New Hampshire was constituted a separate colony by the British cabinet. In 1684 the English high court of chancery, declared the charter of Massachusetts forfeited, but Col. Kirk, who was appointed governor of New England, was prevented by the king's death, from entering on his office. In 1685 Joseph Dudley became president of New England. He was succeeded the next year by Sir Edmund Andross, who was resisted in Mass. and Conn. In 1689 Plymouth was, by royal order, united to Mass., and the old charter of Mass. was confirmed. In 1692 Sir Wm. Phipps, a native of New England, became governor under a new charter, which vested the appointment of governor, lieutenant governor, secretary, and admiralty officers, in the crown, and rendered the governor's assent necessary to every public act, beside giving him the appointment of military and judicial officers, and a negative on all the elections of civil officers, by the general court. In 1720, a controversy commenced between the house of representatives and the governor, in relation to privileges, which continued for some time.

In 1745 the fortress of Louisburg, was captured by New England troops, most of them from Mass. In 1753 was formed the first society for the encouragement of industry, at the celebration of which 300 young women appeared on Boston common, at their spinning wheels, while one working at a loom, was carried on a stage, on men's shoulders. In 1756 began the last French war, in which Mass. and the other colonies, took an active part, and suffered much.

In 1765 measures were first taken by the British government, to raise revenue in the colonies, and at the suggestion of Mass., a congress of delegates assembled at New York, to procure the removal of duties on stamped paper, &c. The stamp act was repealed the following year, but renewed in 1767, with duties on various other articles. Public excitement prevailed against the government, and in 1770, the King's troops being insulted by the people of Boston, killed four of the citizens. In 1773, several ship loads of tea, sent out by the East India company, subject to a duty, were forcibly thrown into the harbor, by the inhabitants of Boston. In 1774, commercial privileges were denied to Boston, and Gen. Gage, who was made commander of the troops, in North America, adopted severe measures, which at length led to a general insurrection, and finally to the establishment of American independence.

In September, 1774, delegates from the colonies, met at Philadelphia; in 1775, Gen. Gage's troops were resisted; April 19th, occurred the battle of Lexington, and June 17th, that of Bunker's hill. In all these contests, the people opposed the encroachments of arbitrary power, and rose in defence of those rights, in strong attachment to which they had been educated. In 1776, Gen. Washington commenced the siege of Boston, and compelled the evacuation of it on the 17th of May. After this time, the soil of Mass., excepting somo islands, remained free from actual invasion; but they contributed powerfully to the success of the American arms, by councils, men, and money.

The constitution went into operation in 1780. In 1786 commenced Shay's rebellion, which greatly agitated the state. It led to no bloodshed except at its close, when 3 of his men were killed, in attempting to take the barracks at Springfield. The federal constitution of the United States, was adopted by the convention of Mass. in 1788.

The present constitution of Mass., is that of 1780, with some amendments, adopted in 1820. The government now consists of three parts. The legislature, called the "general court," is composed of a senate of 40 members, chosen annually, and a house of representatives, of one or more members from each town, consisting in all of 500, or 600 members, when all the towns send the full number to which they are entitled. Each of these branches has a negative on the other. The senate is founded on the representation of property, the house of representatives on the representation of the population in towns; the number of senators, (with a limitation to six) from any district, being proportioned to the amount of its taxes; the number of representatives from any town, depending on the number of its inhabitants. The senate may constitute a court of impeachment; the house of representatives may impeach, originate all money bills, &c. &c. Every bill must be approved and signed by the governor, before it becomes a law, unless after being returned with his objections, it shall have been passed by two thirds of the legislature. The executive is vested in a governor, lieut. governor, and 9 counsellors, who are chosen annually, the two former by the people, the counsellors, by the legislature from the senators. The governor has the power of opposing or rejecting bills passed by the legislature; he is commander-in-chief of the military forces, appoints all judicial officers, and with the council, exercises the pardoning power. The judiciary department consists of a supreme judicial court, and a court of common pleas, each composed of a chief judge, and 3 associates, who hold their offices during good behavior. Beside these, are courts held by justices of the peace, and also probate courts in each county. By an amendment to the constitution, made in 1831, the political year, hereafter, begins on the 1st Wednesday of January. Massachusetts is entitled to 12 representatives in congress.

The state is divided into 14 counties, and 307 towns. In each of the counties, is a registry of deeds, a house of correction, and one or more jails. The soil is various, though generally good, and the face of the state, greatly diversified. Nantucket, Duke's, Barnstable, Plymouth, Suffolk, and Essex counties, on the sea, have much poor soil, but good harbors, valuable fisheries, and much navigation and commerce. Worcester county, the largest in the state, and extending across its breadth, has an irregular surface, with good land, and excellent farms. Franklin, Hampshire, and Hampden counties are divided by Connecticut river, on which are extensive and fertile meadows, and which affords navigation for rafts and boats. Steamboats have recently begun to ply to the upper parts of the river. Berkshire county, which forms the west extremity of the state, is mountainous, being crossed by the Green mountains, and Taughkannic ranges; it is of more recent settlement and has much poor land, though a considerable portion of excellent pasturage; it has good marble quarries, and its mountains abound in iron ore. The soil of the state generally is well adapted to the growth of grass and fruit trees, and produces nearly all the fruits of temperate climates, also indian corn, rye, oats, &c. There is a lead mine in Southampton, the works in which have been for a long time suspended. The middle and E. parts of the state abound in granite of an excellent quality for building. Marble and limestone are found in exhaustless quantities in West Stockbridge, Hinsdale and Lanesborough. Anthracite coal is found in Worcester, and quarries of soap stone in Middlefield.

Near the W. line of the state is the Taughkannic range of mts., which divides the waters of the Hudson and Housatonnic rs. The highest peak is Saddle mtn. in the N. W. angle of the state. Hoosic mts. run nearly parallel, being a continuation of the Green mtn. range of Vt., and dividing the streams of the Housatonnic and Hoosic rs. extend to N. Haven, Conn. Wachusett mtn. in Worcester co., Mt. Tom and Mt. Holyoke, in Hampshire co., and Mt. Toby, in Franklin co., are isolated, but very considerable elevations.

The principal river, is the Conn., the largest and most important in the state, which flows N. and S. through the cos. of Franklin, Hampshire and Hampden. That portion which is included in Mass., affords great advantages for navigation. By its annual floods, though they often injure bridges, crops, &c., it greatly enriches the extensive meadows on its banks, with a deposit of soil; and its fish, particularly the shad, afford a supply of excellent food to the inhabitants. Large sums of money have been expended on dams, locks, canals, &c., by which the navigation is much improved. Merrimack r. touches the N. E. part of the state. The principal tributaries of the Conn. r. in this state, are Deerfield, Westfield, Millers and Chickopee rs., the two last of which rise in Worcester co. Housatonic and Hoosic rs. rise in Berkshire co., the former running S. into Conn., the latter N. into Vt. and N. York. In Worcester co. rise also Quinnebaug r. which runs S. into Conn., Pawtucket r. which runs into R. Island, Charles r. whic hempties at Boston, and Concord and Nashua rs. which join the Merrimack. There are also many smaller streams and ponds which are generally supplied with fish, and most of which afford excellent mill seats for manufacturing, &c.

Population. Mass. in 1800, contained 422,845 inhabitants; in 1810, 472,040; and in 1820, 523,287. In 1830 the pop. was as follows:—to which is prefixed the counties, with their population in 1820.

Counties.	Pop. 1820.	Pop. 1830.
Barnstable,	24,026	28,514
Berkshire,	35,720	37,835
Bristol,	40,908	49,592
Dukes,	3,292	3,517
Essex,	74,655	82,859
Franklin,	29,268	29,501
Hampden,	28,021	31,639
Hampshire,	26,487	30,254
Middlesex,	61,472	77,961
Nantucket,	7,266	7,203
Norfolk,	36,471	41,972
Plymouth,	38,136	43,044
Suffolk,	43,940	62,163
Worcester,	73,625	84,835
Total,	523,287	610,408

Of which were white persons—

	Males.	Females.
Under 5 years of age,	40,644	39,532
From 5 to 15	70,667	67,963
" 15 to 30	91,422	94,934
" 30 to 50	59,116	64,847
" 50 to 70	25,327	31,445
" 70 to 90	7,335	9,701
" 90 and over	174	351
Total	294,685	308,674

Of the above, were deaf and dumb, under 14 years, 56; between 14 and 25, 62; over 25, 138. Blind 218. Aliens 8,787.

Free colored.	Males.	Females.
Under 10 years of age,	794	809
From 10 to 24	889	965
" 24 to 36	725	816
" 36 to 55	626	661
" 55 to 100	316	394
" 100 and over,	10	40
Total,	3,360	3,685

Slaves, males none, females 4. Colored, deaf and dumb, 9. Blind 5.

Recapitulation.

Whites.	Free colored.	Slaves.	Total.
603,359	7,045	4	610,408

Slavery does not exist in this state; a decision made by the supreme court of the state, in 1783, declared that it was abolished by the following clause in the declaration of rights, "all men are born free and equal."

The commerce of Mass. extends to all parts of the globe. In the amount of its shipping it is before any state in the Union, and in the extent of its foreign trade, second only to New York. By the report of the secretary of the treasury, for the year ending September 30, 1830, the amount of Am. and foreign tonnage entered, was 74,741; departed, 97,794. Amount of imports, $572,666; exports, domestic produce, $643,435; foreign, $27,087; total exports, $670,522.

A large amount of shipping is employed in the mackerel, cod, and whale fisheries. The whale fishery was commenced very early. In 1668, James Soper, in petitioning for an exclusive right, stated that he had caught whale for 22 years. In the beginning of the 18th century, whales were constantly taken on the bay shore of Cape Cod. For the last 60 years few have been seen in the bay; but they have been pursued in all parts of the world; and the enterprize, skill and hardihood, fostered by this adventurous business, have contributed to the improvement of American seamen. The ships are chiefly fitted out at Nantucket, and New Bedford. The cod-fishery is carried on on the N. E. coasts of the U. S., and on those of Newfoundland and Labrador. In 1831, in the custom house district including Barnstable, licences were granted to 188 vessels engaged in cod-fishing, each averaging 58 tons, and employing in all, 1,500 men and boys. The proceeds for the year, were about $319,000, or about $120 per share, after deducting owners' portions and incidental expences. The mackerel fishery is chiefly carried on along the coast.

The manufactures of Mass. are extensive and various; those of cotton and woollen are carried on chiefly by large and wealthy companies, and by machinery. In Berkshire co., there is invested for manufacturing purposes, in real estate, buildings and fixtures, $653,625, in machinery and tools, $376,405, and in active capital, $526,650. The value of sheep and wool in the county, is about $591,250, making an aggregate amount invested by the manufacturers and wool growers of Berkshire of $2,087,930. The value of the produce of these establishments from Oct. 1830, to Oct. 1831, was estimated at $2,000,965. Salt is extensively manufactured on the coast from sea water. There are in the state 17,545,760 sq. feet of salt works, of which 12,799,710 sq. feet are in the co. of Barnstable, and cost $1,379,971. The expenses of the state in 1831, were, $381,481 68 cents, receipts $325,055 25 cents, deficit $26,451 45 cents, to be supplied by taxation. The amount of taxable property, May 1, 1832, $208,353,024 45 cents, and the number of polls 159,444. In 1821, property, $153,360,407 54 cts. and polls 122,715.

The taxable property in the cos. of Mass. March 2, 1832, was as follows:—Suffolk, $86,244,261 25; Essex, $24,335,935 57; Middlesex, $21,182,609; Worcester, $21,166,640 68; Hampshire, $5,603,255 87; Hampden, $6,548,342 20; Franklin, $5,452,300; Norfolk, $10,229,111 09; Berkshire, $6,744,648 34; Bristol, $11,346,916 33; Barnstable, $3,500,000; Dukes, $534,166 75; Nantucket, $3,895,288 40; Plymouth, $7,576,932 06. There were in the state in Oct. 1831, seventy chartered banking corporations; capital stock paid in, $21,439,800; bills in circulation, 7,739,317; nett profits on hand, 734,312 33; balances due to other banks, 2,477,615 43; cash deposited, &c. not bearing interest, 4,401,965 62; cash deposited, bearing interest, 4,550,947 68; due from the banks, 41,393,083 33; gold, silver, &c. in banks, 919,959 73; real estate, 683,307 89; bills of banks in this state, 1,104,567 29; bills of banks elsewhere, 270,606 88; balances due from other banks, 2,427,679 37; due to the banks, excepting balances, 36,040,760 76; total resources of the banks, 41,445,700 09; amount of last dividend, 566,715; amount of reserved profits, 409,128 76; debts secured by pledge of stock, 752,312 37; debts due, and considered doubtful, $268,687 81.

Rate of dividend on capital of the banks, 3 per cent. less ⅛ of 1-100th part of 1 per cent.

Eight of the seventy being new banks, made no dividend on the 1st October; one no longer in operation. Four new banks have gone into operation since Oct. 1st, making seventy-two now in existence, of which twenty-two are located in Boston, eighteen in the county of Essex, five in Middlesex, one in Plymouth, seven in Bristol, two in Barnstable, three in Nantucket, two in Norfolk, six in Worcester, three in Hampshire, one in Franklin, and two in Berkshire.

The interests of learning have ever been cherished in Mass. with peculiar care. Many of the learned divines and civilians of England were among its early settlers, and the people have ever been conspicuous for their regard to useful knowledge; to the general diffusion of which they have greatly contributed. Harvard college, at Cambridge, the most liberally endowed institution in the U. S., was founded in 1638, chiefly by a donation of Jno. Harvard; and the first printing press in America, at which all the printing of the colony was done for 30 years, was set up at that place the following year. In 1764, the college buildings, with a library of 5,000 vols. &c. were burnt. The constitution places this institution under a board of overseers, consisting in part of the gov., lieut. gov., council and senate. It now has a choice library of 36,000 vols. There are two other colleges in the state, Williams college, in Williamstown, founded in 1755, and Amherst college, near Northampton, founded in 1821. The constitution makes it the duty of the legislature and magistrates, "to cherish the interests of literature and science, and all seminaries of them, especially the university at Cambridge, public schools, and grammar schools in the towns." There is a flourishing theological institution for Congregationalists at Andover, founded in 1807; and one for Baptists at Newton. There is a medical school at Pittsfield, beside that at Cambridge, and various private literary institutions of highly respectable standing. The number of incorporated academies in the state is 43.

The means of common education are provided for all at the expense of the state. Every town with 50 families is required by law to have a free school for children, in which must be taught the rudiments of learning, at least 6 months in the year. In towns of 100 and 150 families, it must be kept 10 and 12 months, those of 500 families, the history of U. S., book-keeping, geometry, algebra, and surveying must be taught, at least 10 months of the year. In towns of 4,000, in addition to the other branches, must be taught Latin, Greek, history, rhetoric and logic. It is made the duty of all teachers to impress the pupils with the principles of religion and virtue, as the basis of human society and republican institutions. The schools are superintended, and the instructers appointed by committees of the districts.

The internal improvements of the state are numerous. The roads and bridges are many and excellent. The South Hadley canal, round a fall of the Conn. r., was the first work of the kind used in the U. S. That round Miller's falls, near Greenfield, forms a part of the same line of improvements in the navigation of Conn. r. The Middlesex canal connects the Merrimack at Lowell with the Boston harbor. The Blackstone canal, which extends to Providence in R. I., lies partly in this state. The Hampshire and Hampden canal is partly completed, and extends from the river at Northampton to the Farmington canal in Conn.; thus opening a line of boat navigation to New Haven, Conn. Several other canals have been planned and some surveyed, but the modern improvements in rail roads will probably prevent their prosecution. The first rail road constructed in America was that of Quincy, in Norfolk co. which is used to transport granite to the waters of Boston harbor. Others have been projected, but only 2 are now constructing, the Boston and Lowell rail-road, and the Boston and Worcester rail-road.

The state prison at Charlestown is an institution highly creditable to the state. By liberal appropriations from the treasury to the means of experiment in penitentiary regulation and discipline, a mere prison house, for the physical restraint of the body, has been converted into a school of salutary instruction and reform to the minds of the most vicious and abandoned of our fellow men. The demeanor of the convicts has been softened and corrected, and from the admonitions afforded here, and the greater terror inspired abroad, commitments have sensibly diminished. Within the last year, the number of prisoners was reduced from 290, at its commencement, to 256 at its close. Of 256 convicts, 156 were led by intemperance to the commission of offences, 182 had lived in the habitual neglect and violation of the Sabbath; 82 were permitted to grow up without regular employment; 68 had been truants to their parents while in their minority; 61 could not write, and many were wholly unable to read. In 1828, the excess of expenditure was more than $12,000; in 1829, it was between 7 and $8,000, and in 1830, it approached to $7,000, while in 1831, it was only $477 47. A hospital for the insane is now erecting. The state government is doing much by the annual bestowment of a bounty for the education of the destitute deaf and dumb; by liberal encouragement to agricultural societies, and by fulfilling the injunctions of the constitution upon "legislatures and magistrates, in all periods of the commonwealth, to promote by rewards and immunities, agriculture, arts, sciences, trades, manufactures, and a natural history of the country." To promote the culture of silk, the legislature had a concise manuel compiled and circulated, on the growth and culture of the mulberry tree.

The design of obtaining an accurate map of the state from actual surveys and admeasurements upon trigonometrical principles, is in a course of diligent prosecution. The examinations of the country have been mostly made, and the first part of an elaborate scientific report, comprising the economical geology of the state, accompanied with a map, delineating by numbers and colorings, the various minerals and rock formations which prevail, is prepared. The second part is to exhibit the topographical geology; the third, the scientific geology, and the fourth, catalogues of the native mineralogical, botanical, and zoological productions. Arrangements have been made to procure the immediate publication of the first part of the report. There were in Mass. in 1831, 491 Congregational churches, with 423 ordained ministers, of whom 118 are Unitarians; 129 Baptist churches, with 110 ministers, and 12,580 communicants; 71 Methodist preachers, and 8,200 members; 46 Universalist societies; 31 Episcopal ministers; 8 New Jerusalem societies; 9 Presbyterian ministers; 4 Roman Catholic churches, and 4 Shaker societies.

MASSAMETTEE, p-v. marked on the p-o. list as in Shenandoah co. Va. but is probably in the new co. of Page, by p-r. 114 ms. westward W. C.

MASSENA, p-t. St. Lawrence co. N. Y. 43 ms. E. N. E. Ogdensburgh, has a good soil, and Grass and Racket rivers furnish good mill seats. It is opposite Cornwall, Upper Canada. Pop. 1830, 2,068.

MASSERNE, from Mt. Cerne, one of its peaks; a chain of mtns. in the United States and Texas, extending from the state of Mississippi over Arkansas into Texas in a nearly similar direction with the mtn. range of the Appalachian system. The Masserne is traversed by Red and Arkansas rs. and gives source to the Merrimac, Gasconnade, St. Francis, White Ouachitta rs. No scientific survey has ever been made of the Masserne, a remark which might indeed be extended and applied to the Appalachian system. The provincial vulgarism Ozark, the hunters' name for Arkansas, has been given to the Massernes, by some writers and map makers.

MASSILLON, p-v. on the Ohio canal, near the centre of Stark co. Ohio, by p-r. 108 ms. N. E. Columbus. Pop. 1830, 359. The water level in the canal at Massillon, is 942 feet above the mean height of Atlantic tides.

MATAPOISET Harbor, extends from Buzzard's bay, into Rochester, and receives Matapoiset river from Plymouth co. Mass.

MATCHAPUNGO Inlet, on the coast of the Atlantic, between Hog and Prouts islands, Northampton co. Va. It opens into a sheet or small gulf called Broad Water, 28 ms. N. N. E. cape Charles. On Tanner's U. S. it is laid down at lat. 37° 20′.

MATHEWS, co. of Va. bounded by Gloucester co. S. W. and W., by Piankatanck river separating it from Middlesex N., Chesapeake bay E., and Mobjack bay S. Length from Point Comfort to the N. W. angle on Piankatanck bay, 17 miles, mean width 4 ms., and area 68 square ms. Extending in lat. 37° 22′ to 37° 30′ and in long. 0° 33′ to 0° 48′ E. W. C. This co. is commensurate with a small peninsula between Mobjack and Piankatanck bays. Pop. 1830, 7,664.

MATHEWS court house, and p-o. Mathews co. Va. by p-r. 108 ms. though in direct distance about 70 ms. E. Richmond.

MATHEWS' PRAIRIE, and p-o. sthrn. part of Scott co. Mo. by p-r. 256 ms. S. E. Jefferson city, and 150 a little E. of S. St. Louis.

MATHEWSVILLE, p-o. Pocahontas co. Va. by p-r. 205 ms. wstrd. W. C.

MATTAPONY, river of Va. has its extreme source on the eastern border of Orange co. near the Rapid Ann, about 25 ms. westward Fredericksburg, but the most numerous of its creeks are in Spotsylvania. These unite within and traverse Carolina, and thence forming a boundary between King William, and King and Queen, unite with the Pamunky, to form York river, after a comparative southeastern course of 10 ms. The valley of the Mattapony lies between those of the Rappahannoc and Pamunky, and is traversed by N. lat. 38° and the meridian of W. C.

MATTEAWAN, creek, Monmouth co. N. J. runs into Raritan bay, 4 ms. S. E. Amboy, is navigable for vessels of 60 tons to Middletown point.

MATTEAWAN OR FISHKILL MTS., N. Y., called the Highlands of Hudson r., 16 or 18 ms. wide, in cos. of Rockland, Orange, Westchester, Putnam and Dutchess. They are probably connected with the Alleghany, being of primitive rocks; the numerous peaks form the romantic pass of the Highlands in Hudson r.; the range extends from N. J. N. E. to Mass.

MATTOX, or Mattax bridge and p-o. in the nrthwst. part of Westmoreland co. Va., by p-r. 97 ms. N. N. E. Richmond, and 90 S. W. C.

MATTOX'S, p-o. Tatnall co. Geo., by p-r. 131 ms. S. E. Milledgeville.

MATTSVILLE, p-o. Bucks co. Pa., by p-r. 53 ms. nrthrd. Phil.

MAUCH CHUNK, flourishing p-v. on the right bank of Lehigh r., 31 ms. N. W. Bethlehem, and 84 N. N. W. Phil. This very remarkable village has risen amid mtns. and rocks, on ground scarce wide enough to admit a street, from being the depot for the immense strata of anthracite coal found in the mtn. from which the name is taken, and at the foot of which the village is situated. The coal strata, or the most extensive mine yet opened, is about 9 ms. wstrd. from the village, and lies upwards of 1,000 feet above the Lehigh level. Down this descent the coal is brought along a rail-way, and meets a canal, and slack water navigation, at Mauch Chunk. This work, called "The Lehigh navigation," extends along the Lehigh r. 47 ms., with a fall of 364 feet to the Del. at Easton. This navigation every where admits boats of 5 feet draught, through 57 locks of 22 feet wide. (*See arts. Del. and Lehigh rs.*) The village was commenced in 1820 or '21, and in 1830 the number of inhabitants was 1,343.

According to a statement in the Phil. Eve. Post, Jan. 29th, 1830, the quantity of coal shipped from the Lehigh mines, and passing through Mauch Chunk was, in 1825, 28,393 tons; 1826, 31,280; 1827, 30,305; 1828, 30,111; 1829, 25,110; 1830, 42,225; total, 187,424 tons.

MAUHANOY, p-v. Northumberland co. Pa. (*See Mahanoy.*)

MAUKPORT, p-v. on O. r. Harrison co. Ind., by p-r. 152 ms. sthrd. Indianopolis.

MAUMEE, r. of O., Ind., and Mich., the greatest wstrn. confluent of lake Erie. This very remarkable r. is composed of two constituent branches, the St. Mary's and St. Joseph's. St. Mary's rises in Allen, Mercer, and Shelby cos. O., interlocking sources with those of Wabash, Great Miami, and Au Glaize rs.; flowing thence 60 ms. to the N. W., into Allen co. Ind., it unites with the St. Joseph's

r. The latter rising in Hillsdale co. Mich., and assuming a s. w. course, traverses the nrthwstrn. angle of Williams co. O., enters Allen co. Ind., and unites with the St. Mary's as already noticed.

To view those two rivers on a map, their natural course would appear to be down the Wabash, but curving on themselves, the united waters now known as Maumee, assume a N. E. course; flows in that direction 45 ms., to where it receives almost at the same point, Au Glaize r. from the s. and Bean or Tiffen's r. from the N. Continuing N. E. 60 ms. farther, Maumee is lost in the extreme wstrn. angle of lake Erie. This stream, like all others which issue from O. into lake Erie, is obstructed by rapids a few ms. above its mouth; otherwise it is navigable at high water into both its main branches. The Au Glaize, which falls into the Maumee at Defiance, is the most considerable branch, not falling much under the St. Mary's and St. Joseph's united. The valley of Maumee, occupying the whole nrthwstrn. angle of the state of O., is in length from s. w. to N. E. 100 ms., with a mean breadth of at least 50, area 5,000 sq. ms., comprising small fractions in Mich. and Ind. This r. and its branches drain in O. the cos. of Mercer, Allen, Vanwat, Putnam, Hancock, Wood, Henry, Williams, and Paulding. In lat. the valley stretches from 40° 30' to 42°.

MAUMEE, p-v. on Maumee r., where the road crosses from Columbus to Detroit, by p-r. 136 ms. N. N. W. Columbus.

MAUREPAS, lake of, La., between the parishes of St. Helena and St. John Baptiste, receiving the Amite r. from the w., and communicating on the estrd. with lake Ponchartrain, by the pass of Mauchae, a strait of about 6 ms. Lake Maurepas lies in an elliptic form, 12 by 7 ms.; depth generally about 12 feet, though in the pass of Mauchae the water shallows to about 6 feet. Beside the Amite, Maurepas receives New r. from the s. w., and Tickfoha from the N. The country adjacent to this lake is mostly low and marshy.

MAURICE, r. Cumberland co. N. J., rises in Gloucester co. and runs nearly s. about 30 ms., receiving several branches, and enters Del. bay at Maurice cove, through low and swampy banks.

MAURICE RIVER, p-t. Cumberland co. N. J., between Maurice r. and Salem, and Gloucester cos.

MAURY, co. of Ten., bounded by Hickman w.; Dickson N. W.; Williamson N.; Bedford E.; and Giles s. Length 30, mean width 24, and area 720 sq. ms. Extending in lat. 35° 22' to 35° 50' N., and in long. 9° 42' to 10° 18' w. W. C. This co. lies entirely in the valley of Duck r., which winds over it in a N. W. by W. direction, receiving numerous crs. from both sides; soil of first rate quality; staple, cotton. Chief t. Columbia. Pop. 1820, 22,141; 1830, 27,665.

MAXATAWNY, p-o. and tsp. Berks co. Pa., 20 ms. N. N. E. Reading. The tsp. lies on the border of Lehigh co., and on the Sacony branch of Maiden cr. Pop. 1820, 1,847.

MAY (Cape), Cape May co. N. J., the N. point of Del. bay; long. 74° 56' w. (Greenwich,) lat. 39° N. The Del. breakwater is erecting within this cape. Here is the termination of a range of low, sandy, barren coast from Shrewsbury hither. Pop. 1830, 4,936.

MAYBINTON, p-o. Newberry dist. S. C., by p-r. 14 ms. wstrd. Newberry, and 54 N. W. by W. Columbia.

MAYFIELD, r. small stream of wstrn. Ky., rises in Graves co., and first pursuing a northern course, turns abruptly west, separating Graves from MacCracken, and thence MacCracken from Hickman, falling into Miss. a few ms. below the mouth of Ohio.

MAYFIELD, p-t. Montgomery co. N. Y., 8 ms. N. E. Johnstown and 40 N. W. Albany; has good grass and grain soil, and Cranberry, Mayfield, and Fondas crs. with mill seats; 2 post offices. Mayfield mtn. extends to Mohawk r. Pop. 1830, 2,614.

MAYFIELD, p-v. and st. jus. Graves co. Ky., situated on a branch of Mayfield r., about 35 miles s. E. the mouth of Ohio r., and by p-r. 277 miles s. w. by w. Frankfort, and lat. 36° 45', and long. 11° 45' w. W. C.

MAYO, p-v. in the eastern part of Rockingham county N. C., by p-r. 97 ms. N. W. Raleigh.

MAYS LICK, and p-o. in the sthrn. part of Mason co. Ky., about 65 ms. N. E. by E. Frankfort.

MAYSVILLE, p-v. and st. jus. Buckingham co. Va., situated near the centre of the co., on Slate cr., about 35 miles nthestrd. Lynchburg, and by p-r. 287 ms. very nearly due w. of Richmond. Lat. 37° 32', and long. 1° 32', w. W. C.

MAYSVILLE, p-v. on the O. r., nthrn. part of Mason co. Ky., by p-r. 67 ms. N. E. Frankfort. This village was formerly called Limestone, and was amongst the original settlements of the state. The site is on a rather elevated bottom of the Ohio r., 3 ms. from Washington, the co. seat; and by water about 500 ms. below Pittsburg. It is the second t. of Ky. in regard to commercial importance, and contained by the census of 1830, a population of 2,040. It contains a glass manufactory of considerable magnitude, a number of stores and warehouses, and three or four places of public worship. The importance of Maysville has arisen from being the mart of upper Ky., and lying on the direct nrthn. thoroughfare. Lat. 38° 40', long. 6° 40' w. W. C.

MAYTOWN, p-v. near the left bank of the Susquehannah r. Lancaster co. Pa., 22 ms. s. E. Harrisburg, and 15 w. Lancaster.

MAZEVILLE, p-o. Greenbriar co. Va. by p-r. 266 ms. s. w. by. w. W. C.

MEADVILLE, p-v. Halifax co. Va., by p-r. 139 ms. s. w. W. C.

MEANSVILLE, p-o. Union district, S. C., by p-r. 87 ms. N. W. Columbia.

MEARS FARM, and p-o. Hamilton co. Ohio, by p-r. 116 ms. s. w. Columbus.

MECCA, p-v. Trumbull co. O., by p-r. 176 ms. N. E. by E. Columbus.

MECHANICSBURG, p-v. Champaign co. O., by p-r. 39 ms. N. W. by W. Columbus.

MECHANIC'S HALL, p-o. Moore co. N. C., by p-r. 83 ms. s. w. Raleigh.

MECHANIC GROVE, and p-o. Clark co. Ala., by p-r. 132 ms. sthwd. Tuscaloosa.

MECHANICSVILLE, p-v. Bucks co. Pa., by p-r. 39 ms. nthrd. Philadelphia.

MECHANICSVILLE, p-v. Montgomery co. Md., 30 ms. s. w. Baltimore, and 8 N. E. Rockville.

MECHANICSVILLE, p-v. Vanderburg co. Ind. by p-r. 164 ms. s. s. w. Indianopolis.

MECHANICVILLE, p-v. Stillwater, Saratoga co. N. Y.

MECKLENBURG, co. Va., bounded by Halifax W., Charlotte N. W., Lunenburg N., Brunswick E., and by Warren and Granville cos. N. C. S. Length 36, mean width 18, and area 648 sq. ms. Extending in lat. 36° 30′ to 36° 53′, and in long. 1° 08′ to 1° 46′ w. W. C. The junction of Stanton and Dan rivers, to form the Roanoke, is made on the wstrn. side of this co., and the thence fine navigable river winds by a rather sinuous channel over the co. in a sthest. by E. direction, leaving it at the sthestrn. angle. The nthrn. side is drained by Meherin r., flowing nearly parallel to the Roanoke; the slope is of course in the direction of the streams. Staples, grain, flour, cotton, tobacco, &c. Chief town, Boydton. Pop. 1820, 19,786—1830, 20,477.

MECKLENBURG, co. N. C., bounded by Catawba r. separating it from Lincoln N. W., Iredell N., Cabarras N. E., Anson E., Lancaster dist. S. C. S., and York dist. S. C. S. W. Length S. to N. 50, mean width 18, and area 900 sq. ms. Extending in lat. from 34° 48′ to 35° 30′, and in long. 3° 32′ to 4° 06′ w. W. C. The sthestrn. and estrn. boundary of this co. coincides nearly with the dividing ridge between the sources of streams flowing into the Yadkin estrd. and into the Catawba river wstrd. The slope is of course wstrd. or rather a little S. of W. towards the latter r. Chief t. Charlotte. Pop. 1820, 16,895—1830, 20,078.

MECKLENBURG, p-v. Knox co. Ten., 12 ms. sthrd. Knoxville, and by p-r. 177 ms. a little S. of E. Nashville.

MEDFIELD, p-t. Norfolk co. Mass., 9 ms. s. w. Dedham, 17 s. w. Boston, E. side Charles r., was burnt by Indians in 1675. Pop. 1830, 817.

MEDFORD, (formerly Mystic) p-t. Middlesex co. Mass., 4 ms. N. of Boston. Mystic river and Middlesex canal pass through it. Burgoyne's army encamped at Winter Hill after his capture. Pop. 1830, 1,755.

MEDINA, co. of Ohio, bounded by Stark S. E., Wayne S., Lorain W. and N. W., Cuyahoga N., and Portage E. Length 24, mean breadth 24, and area 576 sq. ms. Lat. 41° 10′, long. 4° 48′ w. Slope to the N. towards lake Erie. Chief t. Medina. Pop. 1830, 7,560.

MEDINA, p-v. and st. jus. Medina co. O., by p-r. 111 ms. N. N. E. Columbus. Pop. 1830, 254.

MEDROSTA LAKE, Me. is drained by Spey r. which flows into St. John's r.

MEDWAY, p-t. Norfolk co. Mass., 15 miles s. w. Dedham, 20 from Boston, N. Charles r., contains a number of manufactories. It was incorporated in 1713, before a part of Medfield. Pop. 1830, 1,756.

MEESVILLE, p-v. Roane co. Ten., by p-r. 153 miles estrd. Nashville.

MEETING STREET, p-o. nthrn. part of Edgefield district, S. C., by p-r. 65 miles wstrd. Columbia.

MEHERIN, river of Virginia, and N. C., deriving its most remote sources from Charlotte, but rising principally in Lunenburg and Mecklenburg cos., and uniting on the wstrn. margin of Brunswick. Continuing its original course S. E. by E. over Brunswick and Greensville, and thence separating a part of Greensville from Southampton, it enters N. C. between Northampton and Gates cos., and joins the Nottaway to form the Chowan, between Gates and Hertford cos. The entire comparative course of the Meherin is about 95 miles, but the valley is narrow, not exceeding 20 miles width at any part, (mean width hardly 10) area about 900 sq. ms., lying between the vallies of Roanoke and Nottaway.

MEHERIN GROVE, and p-o. Lunenburg co. Va. by p-r. 92 miles s. w. Richmond.

MEIGS, co. Ohio. bounded S. W. by Gallia, N. W. and N. by Athens, Ohio river separating it from Wood co. Va. N. E., and again by the Ohio river separating it from Mason co. Va. E., S. E. and S. Length from east to west 30, mean breadth 15, and area 450 sq. ms. Lat. 39° and long. 5° w. W. C. intersect in this co. Surface very broken, but some tolerably productive. Chief t. Chester. Pop. 1820, 4,480, and in 1830, 6,158.

MEIGS CREEK, and p-o. Morgan co. O. by p-r. 77 ms. S. E. by E. Columbus.

MEIGSVILLE, p-v. Randolph co. Va. by p-r. 211 ms. wstrd. W. C.

MEIGSVILLE, p-v. Jackson co. Ten. by p-r. 84 ms. N. E. by E. Nashville.

MELMORE, p-v. Seneca co. Ohio by p-r. 80 ms. N. Columbus.

MELTONSVILLE, p-v. in the northeastern part of Anson co. N. C. by p-r. 132 ms. s. w. by w. Raleigh.

MEMPHIS, p-v. and st. jus. Shelby co. Ten. situated on an elevated bluff of the Miss. r. immediately below the mouth of Loosahatche or Wolf r., by p-r. 226 ms. s. w. by w. Nashville. Lat. 35° 06′ and long. 13° 02′ w. W. C.

MEMPHREMAGOG LAKE, partly in Orleans co. Vt. but chiefly in Lower Canada, is 30 or 40 ms. long fron N. to S., and 2 or 3 wide, and communicates by the St. Francis with St. Lawrence river. Only 7 or 8 ms. of the S. end are in Vermont. It lies about half way between Connecticut r. and lake Champlain.

A bay from the s. end extends into Coventry. In Vt. the lake occupies about 15 sq. ms., receiving Clyde, Black and Barton rivers. The "Magog oilstones" are brought from an island 2 ms. N. of the Canada line, and are sold in the seaports for about 50 cents a pound.

MENAN, LITTLE, island Washington co., Me. has a light house, 2 ms. S. S. E. Goldsborough, and S. Steuben.

MENASSAS GAP, and p-o. Frederick co. Va. by p-r. 134 ms. westward W. C.

MENDHAM, p-t. Morris co. N. J. 7 miles W. Morristown, 35 W. N. Y., near the head waters of Passaic r.; has an academy, and a fine hilly surface, with good farms. Pop. 1830, 1,314.

MENDON, p-t. Worcester co. Mass. 39 ms. S. W. Boston. Blackstone river and canal cross the S. W. part of the town, and Mill r. runs through it. There are several manufactories of cotton, and woollen mills; the Blackstone factory is very large. Pop. 1830, 3,152.

MENDON, p-t. Monroe co. N. Y. 15 ms. S. of Rochester. Honeoye outlet and other streams supply mills. Pop. 1830, 3,057.

MENOMONIE, r. of Mich. ter., and confluent of Green Bay, rises in the country of the Menomonie Indians, sthrd. lake Superior, and flowing thence sthestrd. about 100 ms., falls into Green Bay at lat. 45° 28′, and nearly due W., the strait uniting Green bay to lake Mich., and 50 ms. N. E. fort Howard.

MENOMONIE island, in Green Bay, Mich. ter., lying about midway between the mouth of Menomonie r., and the Grand Traverse straits between Green bay and lake Mich.

MENOMONIE, st. of jus. Brown co. Mich. Position uncertain.

MENTOR, p-v. and tsp. Geauga co. O., by p-r. 162 ms. N. E. Columbus. Pop. of the tsp. 1830, 703.

MENTZ, p-t. Cayuga co. N. Y., 8 ms. N. N. W. Auburn. Seneca r. and Owasco outlet furnish mills and navigation. Villages, Montezuma and Bucksville, on Erie canal, and Throopsville. Pop. 1830, 4,143.

MERCER, p-t. Somerset co. Me., S. W. Norridgewock, N. Kennebeck co.; has a large pond in the S. E. Pop. 1830, 1,210.

MERCER co. Pa., bounded N. by Crawford; N. E. by Venango; S. E. by Butler; S. by Beaver; and W. by Trumbull co. O. Length S. to N. 34, mean width 25, and area 850 sq. ms. Extending in lat. 41° 02′ to 41° 28′, and in long. 3° 04′ to 3° 37′ W. W. C. The valley of Shenango r. occupies the far greater part of this co., flowing sthrd. into Big Beaver; consequently the slope of the co. is sthrd.; surface moderately hilly, and soil productive. Chief t. Mercer. Population 1820, 11,681; 1830, 19,731.

MERCER, p-v. borough and st. jus. Mercer co. Pa., situated near the centre of the co., 55 ms. a little W. of N. Pittsburg, and 30 ms. a little W. of S. Meadville; lat. 41° 15′, long. 3° 20′ W. W. C. Pop. 1820, 506.

MERCER, co. of Ky., bounded by Dicks r. separating it from Garrard E.; Lincoln S. E.; Casey S.; Washington W.; Anderson N.; and Ky. r. separating it from Woodford and Jessamine N. E. Length S. to N. 28, mean breadth 13, and area 364 sq. ms. Extending in lat. 37° 32′ to 37° 55′, and in long. 7° 36′, to 7° 56′ W. W. C. The declivity of this co. is nthrd.; the soil highly productive. Chief t. Harrodsburg. Pop. 1820, 15,587; 1830, 17,694.

MERCER, co. O., bounded by Vanwert N.; Allen N. E.; Shelby S. E.; Darke S.; and the state of Ind. W. Length 28, mean breadth 20, and area 560 sq. ms. Lat. 40, 35′, and long. W. C. 7° 38′ W. This co. occupies the table land from which flows St. Mary's branch of Great Maumee, and on which rise the extreme sources of the Wabash. It is remarkable that both rivers assume a parallel N. W. course, which they maintain over Mercer into Indiana, and thence converge into directly opposite courses; the Wabash to the S. W. and Maumee N. E. It is obvious from the foregoing circumstances in the course of its streams, that Mercer is amongst the most elevated tracts between the vallies of O. and St. Lawrence. Chief town, St. Mary's. Pop. 1830, 1,110.

MERCER co. Il. bounded N. by Rock Island co.; N. E. by Henry; S. E. by Knox; S. Warren; and W. Miss. r. Breadth 20, mean length from E. to W. 30, and area 600 sq. ms. Lat. of its centre, 41° 15′, long. 14° 42′ W. W. C. This new co. occupies the space estrd. of the Great bend of the Miss. r., below the rapids of Rock r., and opposite the influx of the Low and Iowa rs., and lies about 200 ms. N. W. Vandalia. By the census of 1830, it contained but 26 inhabitants, and possessed no p-o. Oct. 1831.

MERCER'S BOTTOM, and p-o. on the Ohio r. Mason co. Va., by p-r. 326 ms. wstrd. W. C.

MERCERSBURG, p-v. in the sthwst. part of Franklin co. Pa., 16 ms. S. W. Chambersburg, and by p-r. 89 ms. N. W. W. C.

MERCERSVILLE, p-v. Edgecombe co. N. C. by p-r. 47 ms. estrd. Raleigh.

MEREDITH, p-t. Strafford co. N. H., W. Winnipiseogee lake, and 29 ms. N. Concord. Contains many ponds, a good soil, rich landscapes, and advantages of boat navigation on the lake and streams. The village at the bridge is partly in Guilford, and contains an academy, &c. Pop. 1830, 2,683.

MEREDITH, p-t. Delaware co. N. Y., 8 ms. N. Delhi, and 66 W. Catskill. It sends streams both to Del. and Susquehannah rs., and is half way between both. Hilly, with good soil. Pop. 1830, 1,666.

MERIDEN, p-t. New Haven co. Conn. 17 ms. S. Hartford, 17 N. New Naven, 8 W. Middletown. Has pretty good, but uneven land; in some parts mountainous; and though with few natural advantages, has become an important manufacturing place, by dint of industry. The v. is pleasant, and contains several very fine private houses, and 4 churches, 1 Congregational, 1 Baptist, 1 Episcopal, and 1 Methodist. The streams are small, with little water power; yet about a million of dollars worth of different articles are annually manufactured and sent to other places. One company employs about 230 hands in the manufacture of brittania coffee pots, spoons, coffee mills, waffle irons, signal lanthorns, &c.; value about $200,000 per ann. Other manufactures are wooden clocks, value per ann. about $50,000; ivory, wood, box wood, and horn combs, value per ann. about $40,000; augur bits and rakes, value per ann. about $20,000; tin ware, value per ann. $90,000; and another manufactory of brittania ware, which manufactures $25,000 worth annually. There are others of japanned ware, shoes and boots, &c. &c. Some very useful inventions have originated in this place. The first branch of manufacture extensively engaged in here, was that of tin ware. Pop. 1830, 1,708.

MERIDIAN SPRINGS, and p-v. in the wstrn. part of Hinds co. Miss. about 65 ms. N. E. Natchez, and 31 s. w. Jackson; the seat of government for the state.

MERIDIANVILLE, p-v. Madison co. Ala., 8 ms. N. Huntsville, and by p-r. 136 ms. N. N. E. Tuscaloosa.

MERITT, p-v. Wayne co. N. C. 75 ms. s. E. Raleigh.

MERMENTAU, r. of La., rises within and drains the extensive prairies of Opelousas and wstrn. Attacapas. It is formed by the bayous, Nezpique, Cane, Plaquemine Brulé, and Queue Fortue. General course sthwst. over a country almost a perfect plane, where the smallest fragment of stone is rare, and except narrow lines of woods along the streams, the whole covered with grass. Before reaching its outlet into the gulf of Mexico, timber, with the exception of detached clumps of live oak, entirely ceases. The soil towards the sources of its branches, is but of second rate quality, but contrary to the usual operations of nature, still more deteriorates advancing downwards towards the Gulf. The prevailing timber in the woods is oak, hickory, sweet gum, and pine along the bayous Cane and Plaquemine Brulé; on the Nezpique, pine increases proceeding nrthwstrd., and approaching the waters of Calcasin becomes the common tree. Cypress swamps are frequent near the confluence of Nezpique and Plaquemine Brulé. Below the latter point, live oak appears, and in greater or less quantities is found on this r. thence to the mouth. With the Mermentau, however, live oak terminates in that part of the gulf coast; none is to be seen on either the Calcasin or the Sabine.

In autumn, when the streams are low, the tide ascends this basin into each of the confluent streams; but on the contrary, in winter and spring, when heavy rains have fallen, the flood from the prairies overpowers the low tides of the gulf. Below the union of its branches, this r. expands into a lake, and again contracts into a narrow stream, and finally is lost in the gulf, over a bar affording at common tides about 3 feet of water. The lake is also a shallow sheet of water, not deeper than the outer bar. The greatest length of the Mermentau basin is about 90, mean width 30, and area 2,700 sq. ms. Of this surface, more than four fifths is composed of open grassy plains, water, or sea marsh.

MERMENTAU lake, is an expansion of the r. of the same name, below the union of the different confluent branches. It is a shallow sheet, about 30 by 10 ms., differing in no essential physical characteristic from similar lakes along the coast of La. The shores are low and marshy.

MEROM, p-v. and st. jus. Sullivan co. Ind., by p-r. 115 ms. s. w. Indianopolis, and 30 ms. N. Vincennes. It is situated on the left bank of Wabash r., lat. 39° 04', long. 10° 36' w. W. C.

MERRIMACK, r. N. H., the largest in that state, and one of the principal rivers of New England, is formed of the Pemigewasset, from the White mtns. and Winnipiseogee, which unite near the lower part of the line of Strafford and Grafton cos. It crosses the line of Mass. in Hillsboro' co. near Rockingham co. after a course of 78 ms. nearly s., and soon after runs N. E. 35 ms. to the ocean at Newburyport. The Pemigewasset receives Mad and Baker's rs. and the streams from Squam and Newfound lakes. Winnipiseogee r. comes from Winnipiseogee lake and Great bay. The Merrimack receives Contoocook, Soucook, Suncook, Piscataquog, Souhegan and Nashua rs. There are many falls, the principal of which are dammed, and supply water to important manufactories. Canals have been made round them all, with locks, by which the r. has been navigated in boats, for some years, up to Concord. There are several bridges and many ferries, and the capital and other chief towns stand on its banks. Monomake, its Indian name, means a sturgeon. The Middlesex canal extends from the bend in this river in Mass. to Boston harbor. The following is a list of the chief canals, &c. Bow canal, below Concord, made in 1812, cost $20,000; Hookset, 6 ms. below, $1,500; Amoskeag, 8 ms. below, $50,000; the Union canal embraces 6 falls, and with Cromwell's falls canal cost $50,000. Wicasee, 15 miles below, $14,000.

MERRIMACK, p-t. Hillsboro' co. N. H. 6 ms. from Amherst, 27 s. Concord, on w. side of Merrimack r. Souhegan r. and its branches supply manufactories. Pop. 1830, 1,193.

MERRYMEETING bay, Me. at the junction of Merrimack and Androscoggin rs. 20 ms. from the sea.

MERRITTSTOWN, p-v. on Dunlap's creek, Fayettte co. Pa. 5 ms. a little w. of s. Brownsville, and 10 N. w. Uniontown.

MERRITTSVILLE, p-v. at the foot of Blue Ridge, and in the northern part of Greenville dist. S. C. about 40 ms. s. w. Rutherfordton, N. C. and by p-r. 122 ms. N. w. Columbia.

MERRIWETHER, co. Geo. bounded by Talbot s. E., Harris s. w., Troup w., Coweta N., and Flint r. separating it from Fayette N. E., and

Pike r. It is very nearly a square of 20 ms. each way; area 400 sq. ms. Extending in lat. 32° 53′ to 33° 15′, and in long. 7° 39′ to 8° 2′ w. W. C. The slope of this co. is estwrd. towards Flint r. Chief t. Greenville. Pop. 1830, 4,422.

MERRY HILL and p-o. Bertie co. N. C. by p-r. 145 ms. N. E. by E. Raleigh.

MERRY MEETING BAY, Alton, Strafford co. N. H., is the S. E. arm of Winnipiseogee lake, 1,600 rods long.

MESOPOTAMIA, p-v. Trumbull co. O. by p-r. 168 ms. N. E. Columbus.

METCALF, p-v. Richland co. O. by p-r. 93 ms. N. E. Columbus.

METAWAMKEAG, r. Me. enters Penobscot r. on the E. side.

METETECUNK, r. Monmouth co. N. J. enters the head of Barnegat bay.

METHUEN, p-t. Essex co. Mass. 26 ms. N. Boston, 10 from Lowell, 5 N. Andover, N. Merrimack r., and bordering on N. H., is a large town, with wild scenery. The village, with about 800 inhabitants, is on Spicket r. 2 ms. from its junction with the Merrimack. 150,000 to 200,000 pair shoes are annually manufactured here; and on the Spicket r. are 2 brick cotton factories, owned by the Methuen co., containing 4,400 spindles, 134 looms, and manufacturing annually 1,137,200 yards tickings, drillings, and sheetings. One of these factories is 124 feet long, and 5 stories high. The Spicket has a fall at this place of 40 feet, which turns a wheel 108 feet in circumference, and 14 wide, supposed to be the largest in the country. Pop. 1830, 2,006.

METUCHIN, v. Middlesex co. N. J. 4 ms. N. E. New Brunswick.

MEXICO, great inland sea of North America, having the Mexican states on the N. W., W., S., and S. E., the Cuba channel, island of Cuba and Florida channel E., and the U. S. N. E. and N. Lying between lat. 18° and 30° 31′ N., and in long. from 4° to 20° 30′ w. W. C. The greatest length from Florida point to Tampico bay, about 1,000 ms., with a mean breadth at least 600 ms., and area 660,000 sq. ms. This Mediterranean is remarkable for its great depth. It is an immense reservoir, receiving the current of rotation through the Cuba, and discharging it again by the Flor. channel. The surface of the gulf must consequently be higher than that of the Atlantic ocean. The tides in the Gulf of Mexico, where examined at several distant points along the northern shore, are found about 2 to 3 feet. A steady current sets wstrd. along the coast of La. a phenomenon demonstrated by the debris of the Miss. being entirely borne in that direction, and found scattered along the coast. Of all the inland seas of the earth, the Gulf of Mexico is most compact in its form, and least broken by islands or salient capes, and with an immense periphery of 3,200 ms.; is greatly deficient in good harbors.

MEXICO, p-t. Oxford co. Me. 47 ms. from Augusta; it lies N. of Androscoggin r., and is watered by two of its tributaries. Pop. 1830, 343.

MEXICO, p-t. Oswego co. N. Y. 20 ms. E. Oswego, has many springs and mill streams. Salmon creek, the principal, with Mexico Point and Juliana p-vs. Population 1830, 2,671.

MEXICO, p-v. on the left bank of Juniata r. Mifflin co. Pa. 31 ms. N. W. Harrisburg.

MIAMI, or GREAT MIAMI, r. of O. and Ind., has its extreme sources in Shelby and Darke cos. of the former, and flowing thence S. S. E. 50 ms., over Miami and Montgomery, receives in the latter Mad r. from N. E. Mad r. rising in Logan, traverses Champaign and Clark, unites with Great Miami as already noticed. The two streams unite at Dayton, and assuming a course of S. S. W. flows in that direction by comparative courses 100 ms. to its junction with Ohio r. having received, a few ms. above its mouth, White Water r. from Ind. White Water is a stream of 70 miles comparative course, rising in Darke co. Ohio, but having most of its course in, and deriving its principal tributaries from Ind. The boundary line between Ind. and Ohio, strikes the Ohio r. at the mouth of Great Miami. The valley of Great Miami is in length 120 miles from S. W. to N. E., with a mean breadth of 50 ms., or comprises an area of 6,000 sq. ms. This river has become of increasing consequence since the completion of a navigable canal, extending from Dayton to Cincinnati, opening a water means of transport from the interior table land of O. to the O. r. reaching the latter at the most considerable city in the state, or indeed of all the western states except New Orleans. (*See article Ohio.*)

MIAMI, co. O. bounded by Montgomery S., Darke W., Shelby N., Champaign N. E., and Clark S. E. Length and breadth nearly equal, 20 ms., area 400 sq. ms. Lat. 40° and 7° 15′ w. long. W. C. intersect in this co. It is traversed in a direction from N. N. W. to S. S. E. by the main stream of Great Miami, and also by its S. W. branch. The soil is generally fertile. Chief t. Troy. Pop. 1820, 8,851, 1830, 12,807.

MIAMI, co. Ind. bounded N. E. by Wabash co., S. E. by Grant, Miami's ter. S., Cass co. of Ind. W., and to the N. uncertain. Length from S. to N. 30, mean breadth 10, and area 300 sq. ms. Lat. 40° 50′, and long. 9° w. W. C. intersect in this co. Slope to the W., and in that direction it is traversed by the main Wabash in the centre, by the Mississinewa to the S. and Eel r. to the N. These rivers unite near its western border at Miamisport, the chief t.

MIAMI, p-v. Hamilton co. O. by p-r. 129 ms. S. W. Columbus.

MIAMISBURG, p-v. Montgomery co. O. by p-r. 82 ms. S. W. by W. Columbus.

MIAMISPORT, placed in the p-o. list as in Cass, is really in Miami co., and situated at the junction of Wabash and Mississinewa rs. by p-r. 131 ms. a little E. of N. Indianapolis. As laid down by Tanner it stands at lat. 40° 45', long. 9° 4' w. W. C.

MICCOTOWN, p-v. on Alahapa r. northern part of Hamilton co. Flor. about 90 ms. a little N. of E. Tallahasse.

MICHAELSVILLE, p-o. Hartford co. Md. about 34 ms. N. E. Baltimore.

MICHIGAN, largest lake, which lies entirely in the U. S. Taken in connexion with the general physical geography of the two basins of Mississippi and St. Lawrence, it is evident that lake Michigan fills a part of the great valley, of which Illinois r. is the continuation towards the Gulf of Mexico. On our old maps all the Canadian lakes were delineated too round. They were represented as vast ponds; but more recent observation has increased their length when compared with their breadth, and given them a natural approach to the form of rivers. As now laid down by Tanner, this great sheet of fresh water extends 360 ms. from the mouth of W. Calumick (Calumet) r. to the straits of Michilimakinak. The breadth, opposite the mouth of Kalemazoo r. 65 ms.

The breadth is remarkably uniform, and yields an average of at least 45 miles, and adopting that breadth as a mean, the area will be 16,200 square ms., an area by no means overrated, if Green bay is included. The elevation of its surface, is above the Atlantic ocean, very near 600 feet. In lat. it extends from 41° 40' to 46° 10', and in long. from 7½ to 11° w. W. C. including Green bay. In depth, it is a profound gulf, the bottom far below the level of either the Atlantic ocean or Gulf of Mexico; consequently if a channel existed, similar to the straits of Gibraltar, still the lake would exist as an immense reservoir to the rivers, to which it is a recipient. If we commence on the southern end or what is really the source, Michigan receives from the left the two Calumicks, and Riviere du Chemin, from Indiana; from the peninsula of Michigan, the rivers St. Joseph, Kalemazoo, Grand river, Maskegon, White river, Pent-water, Pere Marquette, Saudy, Monistic, Platte, Carp, Grand Traverse, and some other streams of lesser note. From the opposite or western slope, advancing in the same manner, from the mouth of West Calamick, to the mouth of Green bay, the confluent rivers are mere creeks, of which the Manawakee, 70 ms. long, is the most important. Green bay is, however, the recipient to Fox, Menomonie, and numerous other small rivers. See Fox rivers, for explanation of the peculiar physical geography of the western slope of lake Michigan basin.

The eastern Michigan slope is a triangle of 320 ms. base, perpendicular 110 ms. up the vallies of Kalemazoo, and Grand rivers; area about 17,600 square ms. The western slope, measuring from the southern source of W. Calumick, to the source of Mino Coquien river, has a base of 380 ms. Greatest breadth from the mouth of the western Cheboiegon river, across Fox river of Green bay, and up Wolf river to its source, 170 ms.; mean breadth about 100, and area 19,000 square ms. It may be observed that the whole Michigan basin approaches the form of a parallelogram, as the salient or most acute angles of the two slopes are reversed, the eastern coming to a point on the straits of Michilimakinak and the western, towards the sources of West Calamick.

If the hand of art had cut the channel of Michigan, it could not have been much better placed to constitute part of an immense channel of intercommunication between different sections of the earth. A canal has already been projected, and will no doubt be executed in a few years, to connect lake Michigan with Illinois river, and a rail road has been projected from Indianopolis to the southern extremity of lake Michigan, to be extended in the state of Indiana.

MICHIGAN, territory of the United States. For political purposes, the large territory of Huron, westward of lake Michigan, has been united to the peninsula, properly called Michigan. Mr. Austin E. Wing, the delegate in the present congress, 1831—2, informed the author of this article, that it was probable that as early as 1834, Michigan would become a state, as the population was rapidly augmenting, and, that on the western side the boundary would be a line following the middle of lake Michigan, from the northwestern angle of Indiana to the northern extremity of the lake, and thence due north to lake Superior. If this demarcation is adopted, the state of Michigan will contain the peninsula north of Ohio, and Indiana, together with the co. of Chippeway, or that peninsula, bounded s. by lake Michigan, N. by lake Superior, and E. by the straits or river St. Mary, and lake Huron, and will have outlines, commencing at the point on lake Michigan, separating La Porte co. of Indiana from Berrien co. of Michigan, and thence along western shore of the peninsula, to the mouth of Traverse bay 280 miles; over lake Michigan to its northern coast opposite Beavor islands 45 ms.; across the intermediate land surface between lakes Michigan and Superior to the latter 50 ms.; along the southern shore of lake Superior to the outlet or head of St. Mary's strait 90 ms.; from eastrn. end of Drummond's isl. to the straits and island of Michilimakinak 60 ms.; thence along the s. w. shore of lake Huron to its outlet, or to the head of St. Clair river 250 ms.; thence down St. Clair r., St. Clair lake, Detroit river, and along the western end of lake Erie to the northern boundary of Ohio 136 ms.; thence due west along the northern boundary of Ohio, to the eastern boundary of Indiana 85 ms.; due north along east boundary of Indiana, to the northeast angle of that state 10 ms.; and thence due west along north boundary of Indiana, to place of beginning 110 ms.; having an entire outline of 1,106 ms.; measured either by the rhombs or proportional scale, the area of the peninsular part of Michigan comes out about 34,000 square ms. and of Chippeway co. 3,000, yielding, for what is supposed to be the surface to be included in the state, 37,000 square ms.; or in statute acres, 23,680,000. Extending in lat. from 41° 40′ to 46° 47′, and in long. from 5° 18′ to 10° 35′ w. W. C.

The dividing ridge which separates the sources of Great Miami and Maumee, from those of the Wabash, is continued over Michigan, in a northerly direction, dividing the peninsula into two not very unequal inclined plains. The western or lake Michigan plain is drained by St. Joseph's, Kalamazoo, Grand and numerous other streams. (*See lake Michigan.*) The opposing or eastern plain gives source and course to the rivers Raisin, Huron of Erie, Rouge, Huron of lake St. Clair, Belle river, Black, Saginaw, Thunder, Cheboiegon, and numerous smaller streams.

What might well be called the sea shore of Mich. which, if the inflections of the coasts were included, would far exceed 1,000 ms., are with little exception uniform along lake Mich. only affording harbors in the mouths of the rivers. Lake Superior and Huron coasts are more indented, and Saginaw bay offers a gulf of 60 ms. depth. Taken as a whole, the number of havens are not in a proportion favorable to commerce when compared with the distance of sea line. The phenomena of the rivers, prove the peninsula to be a vast table land, as all the rivers fall over ledges of rock before reaching their recipients. The nthrn. part of the peninsula is sterile when compared with the sthrn. towards Ohio and Ind., and the whole country with very partial exceptions a forest.

Political geography.—By Tanner's improved map, up to the present time, Aug. 1832, it appears that Mich. as delineated in the first part of this article, is subdivided into the counties of:—

Counties.	Pop. 1830,	Counties.	Pop. 1830,
Allegan,		Lapeer,	
Arenac,		Lenawee,	1,491
Barry,		Macomb,	2,413
Berrien,	325	Michilimakinak,	877
Branch,		Midland,	
Calhoun,		Monroe,	3,187
Cass,	919	Montcalm,	
Chippeway,	626	Oakland,	4,911
Clinton,		Oceana,	
Eaton,		Ottawa,	
Gladwin,		Saginaw,	
Gratiot,		St. Clair,	1,114
Hillsdale,		St. Joseph,	1,313
Ingham,		Sanilac,	
Ionia,		Shiawassee,	
Isabella,		Van Buren,	5
Jackson,		Washtenau,	4,042
Kalamazoo,		Wayne,	6,781
Total on peninsular Mich. and Chippeway co.			28,004

In the three Trans-Michigan cos. Brown, Crawford and Iowa, (for the pop. of which, *see article Huron Ter.*) there were in 1830, 3,635 inhabitants, and including these the whole territory contains a pop. of 31,639, of which were white persons:—

	Males.	Females.
Under 5 years of age,	3,023	2,743
From 5 to 10	3,326	2,066
" 10 to 15	1,905	1,686
" 15 to 20	1,543	1,436
" 20 to 30	4,389	2,540
" 30 to 40	2,739	1,399
" 40 to 50	1,232	796
" 50 to 60	658	390
" 60 to 70	264	140
" 70 to 80	64	35
" 80 to 90	20	10
" 90 to 100	4	5
" 100 and upwards,	1	0
Total,	18,168	13,178

Whites who are deaf and dumb, under 14 years, 4; 14 to 25, 7; 25 and upwards, 4; Blind 5.

Colored population as follows:—

	Free colored.		Slaves.	
	Males.	Fem's.	Males.	Fem's.
Under 10 years of age	31	30	2	1
From 10 to 24	43	36	7	3
" 24 to 36	48	26	11	3
" 36 to 55	29	16	1	3
" 55 to 100	8	4	1	0
100 and upwards	0	0	0	0
Total,	159	102	22	10

Of the colored pop. none are either deaf and dumb, or blind.

Recapitulation.

Whites.	Free colored.	Slaves.	Total.
31,346	261	32	31,639

General remarks on Mich.—Volney, and some other writers who knew little of the real geography, and of course still less of the climate, have given a very erroneous idea of the aerial temperature of this peninsula and adjacent countries. The writer of this article, from actual observation, found the winds on lake Erie so excessively prevalent from the wstrd. and n. wstrd., as to bend the whole forest trees in an opposite direction. In making voyages from Detroit to Buffalo, and the reverse, the time demanded differs about as three to one. By reference to the tables under the head of U. S., the excessive severity of winter at Detroit, and Fort Brady, Chippeway co. may be seen.

History.—The first civilized settlements in Mich. were made by the French from Canada; and Detroit (the strait) was founded about 1670, but this region, so remote from the Atlantic coast, was peopled slowly, and at the end of the revolutionary war, when ceded to the U. S. by the treaty of Paris, contained but few inhabitants. Under various pretences the British colonial agents retained Detroit, with all that is now Michigan, until after the treaty of Greenville, and the U. S. did not obtain the country in actual possession until 1796. The territory of Mich. was formed in 1805. The country had to sustain more than a share of the vicissitudes of the last war between the U.S. and Great Britain, and was, in 1812 actually overrun by the troops of the latter, but in the ensuing year was retaken by an army of the U. S. under Gen. Harrison. Relieved from calamities of war, and laid open to Atlantic commerce and emigration by the great watrn. canal of New York, the advance of Michigan has been rapid. By the census of 1820, it contained but 8,896 inhabitants, but as shown by the tables in this article, in 1830, the pop. of the peninsula and Chippeway co. contained a fraction above 28,000, and by information received from Mr. Wing, the inhabitants now, 1832, exceed 50,000.

On the 17th July, 1822, Col. Brady founded Fort Brady on the straits of St. Mary, and commenced the settlement of Chippeway co. The settlements on the peninsula are spreading with great rapidity, as may be seen by the numerous counties formed since the census of 1830 was taken. Those with numbers annexed are those which were organized in 1830, and amount to 13, out of 36 named in the table.

MICHILIMAKINAK, co. of Mich. including the strait and islands of the same name, and the nrthrn. extremity of the peninsula. The limits are undefined on the main land, as the nthrn. settlements are separated by a wilderness from those of the sthrn. and central parts of the territory. Chief town, Fort Mackinac. Pop. 1832, 877.

Through the straits on both sides of Michilimakinak island, a constant and very sensible current flows from lake Mich. into lake Huron. The straits are wider and shorter than any of the other water connexions between the great Canadian lakes, but are in every other respect similar to St. Mary's, St. Clair, Detroit, or Niagara.

MIDDLEBOROUGH, p-t. Plymouth co. Mass. 34 ms. s. from Boston, has a poor sandy soil. Assawampsit and Long ponds are chiefly in this town. Bog iron ore is taken from the former with long tongs, and manufactured here. Cotton is also manufactured. Population 1830, 5,008.

MIDDLEBOURNE, p-v. and st. jus. Tyler co. Va. situated on middle island creek, 45 ms. a little w. of s. of Wheeling, and 258 ms. by p-r. westward W. C. Lat. 39° 32′, long. 3° 55′ w. W. C.

MIDDLEBOURNE, p-v. Guernsey co. Ohio, by p-r. 97 ms. N. E. Columbus.

MIDDLEBROOK, p-v. in the southern part of Augusta co. Va. 10 ms. s. s. w. Staunton, and by p-r. 185 ms. s. w. by w. W. C.

MIDDLEBROOK, p-v. in the western part of Edgefield district S. C. by p-r. 98 ms. watrd. Columbia.

MIDDLEBROOK MILLS, on little Seneca, p-o. Montgomery co. Md. 28 ms. N. W. W. C.

MIDDLEBURGH, p-t. Schoharie co. N. Y. 10 ms. s. of Schoharie, and 35 w. of Albany. Schoharie creek affords mill seats. The Helderburg limestone hills are scattered over the town.

MIDDLEBURG, p-v. Union co. Pa. 15 miles westward Sunbury, and by p-r. 61 ms. N. N. W. Harrisburg, and 6 ms. s. w. New Berlin.

MIDDLEBURG p-v. Frederick co. Md. by p-r. 20 ms. N. E. of Frederick.

MIDDLEBURG, p-v. Hardiman co. Ten. Pop. 1830, 3,278.

MIDDLEBURY, p-t. and st. jus. Addison co. Vt. 33 ms. s. Burlington, and 31 s. w. Montpelier. Generally level. Otter creek and Middlebury river afford mill seats. Much marble is quarried and wrought here, and is fine, white, bluish &c. The village is on Otter creek at the falls. Here is Middlebury college, a jail, academy, court house, state arsenal, &c. The college was incorporated in 1800, and owed its support to private contributions. The medical academy at Castleton is connected with it; a new college building is to be erected. Pop. 1830, 3,468.

MIDDLEBURY, p-t. New Haven co. Conn. 22 ms. N. W. New Haven, 36 ms. from Hartford, 4 ms. by 5; 19 sq. ms; has Hop river, &c. It is hilly, with granite rocks. It produces grass and grain. Pop. 1830, 816.

MIDDLEBURY, p-t. Genesee co. N. Y. 15 ms. s. of Batavia, produces excellent fruits. The village is on Black creek. Pop. 1830, 2,416.

MIDDLEBURY, p-v. on Goose creek in the s. s. w. part of Loudon co. Va. 12 ms. s. w. Leesburg, and by p-r. 44 ms. N. W. by W. W. C.

MIDDLEBURY, p-v. Portage co. O. by p-r. 115 ms. N. E. Columbus.

MIDDLE CREEK, p-o. southern part of Wake co. N. C. by p-r. 12 ms. s. Raleigh. Middle creek is a small branch of Neuse r.

MIDDLEFIELD, p-t. Otsego co. N. Y. 3 ms. E. of Cooperstown, 35 s. E. Utica, E. lake Otsego and Susquehannah river. The great Western turnpike passes through it. The dairy is esteemed. Pop. 1830, 3,323.

MIDDLEFIELD, p-v. Geauga co. O. by p-r. 178 ms. N. E. Columbus.

MIDDLEFORD, p-v. Sussex co. Del. by p-r. 54 ms. sthrd. Dover.

MIDDLE GROVE, and p-o. Ralls co. Mo. about 110 ms. N. W. by W. St. Louis.

MIDDLEPORT, p-v. Schuylkill co. Pa. by p-r. 74 ms. N. E. Harrisburg.

MIDDLESEX, p-t. Washington co. Vt. N. of Onion r., 30 ms. E. Burlington, is rough. A bridge over Onion r. crosses a rocky chasm 30 feet deep. Pop. 1830, 1,156.

MIDDLESEX, co. Mass. in the E. part of the state, bounded by New Hampshire N., Essex co. E., Norfolk s., and Worcester w. Contains 46 towns. Cambridge and Concord are the shire towns, but Charlestown is the most populous. Merrimack, Concord and Nashua rivers are the principal streams. The Middlesex canal, 31 ms. long, 24 feet wide, and 4 feet deep, reaching from Boston harbor to Merrimack river, is wholly within this county. It was begun in 1793, finished in 1804, cost above $700,000, has 13 locks, and 107 feet descent N., and 3 locks and 21 feet descent s. The amount of capital invested in manufacturing is very great. Lowell and Waltham, the two principal manufacturing towns in Massachusetts, are in this county. In Lowell, in April, 1832, $3,129,000 were stated to be invested in real estate and machinery, for manufacturing purposes, without including the Suffolk, Tremont, and Lawrence manufacturing companies. There are numerous paper mills in this county, and an agricultural society which has annual cattle shows, &c.

In June 1832 a census was taken of Lowell, which shows the population to be 10,254; increase in 2 years nearly 2,000. Belvidere village 1,004. It has been recently stated, (July, 1832) that in this county the cotton manufacture employs $3,129,000 capital, vested in real estate and machinery, consumes 6,-913,000 lbs. cotton, produces annually 20,-378,000 yards cloth, employs 3,896 hands, pays in wages $731,750; the woollen manufacture employs $394,000 capital, vested in real estate and machinery, consumes 899,000 lbs. wool, produces annually 849,300 yds. woollen cloth, flannel and carpeting, employs 653 hands, pays in wages $152,000; the manufacture of leather, boots, shoes, hats, paper, glass, sheet lead, lead pipe, iron, starch, gunpowder, soap and candles, drugs, oil of vitriol and other acids, barilla and other chemicals, used in the county by bleachers, dyers, calico printers, soap boilers, and other artists, are more extensive than in any other section of our country of equal extent, employing in these branches, in the aggregate $1,050,255 capital, vested in real estate, machinery, tools, &c. and producing manufactured articles of the annual value of $3,565,613. Pop. 1820, 61,472—1830, 77,961.

MIDDLESEX CANAL. (*See art. Rail Roads and Canals.*)

MIDDLESEX, co. Conn. situated near the middle of the state, is bounded N. by Hartford co., E. by Hartford and New London cos., S. by Long Island sound, and W. by New Haven co. It is 342 square miles in extent, and contains 7 towns. Chatham, Durham, E. Haddam, Haddam, Killingworth, Middletown and Saybrook. It was formed into a county in May, 1785, and then consisted of 6 towns, to which a seventh, taken from New Haven co. was annexed in May, 1799.

The early settlers of Middlesex were almost entirely of English origin and extraction, and its present inhabitants are chiefly their descendants. The first English settlement was commenced in Saybrook in 1635. The several townships of the county were purchased of the Indians, who were formerly numerous here, and in no case were obtained by conquest.

The general surface of Middlesex is uneven. A wide range of hills crosses the county obliquely from S. W. to N. E., and on the W. border of Durham and Middletown are the Wallingford hills. The soil adjacent to the Connecticut river is generally good. The Chatham meadows are of excellent quality—the uplands usually very good. The numerous hills of the co. give rise to a multitude of springs, brooks and streams, which fertilize the land, and many of them are highly valuable for mill seats, &c. The climate is fine, and the region remarkably healthy.

The county is divided by Connecticut river which affords great advantages for navigation, and abounds in valuable fish, particularly shad, which are taken in large numbers. Much ship building is carried on in the co. The foreign trade was formerly extensive, and the coasting trade is still very great. Manufactures are flourishing, particularly on the streams in and near Middletown, where are manufactories of woollen, cotton, fire arms, &c. Valuable quarries have long been wrought at Chatham, which furnish the best free stone to the New York market, and employ many sloops, &c. Building stone is also abundantly supplied from the granite hills bordering the straits of the Connecticut river. Many of the rocks of this county exhibit petrified fish, leaves, &c. partly carbonized, and other indications of bituminous coal, though no bed of that valuable mineral has been discovered. There are also a lead and a cobalt mine, the latter about 5 ms. E. from the head of the strait.

At Middlefield, where is a fall of about 30 feet over a bed of trapp rocks, are found chlorophœite, datholite and iolite.

There are in the county 41 houses of public worship, viz: 19 Congregational, 18 Baptist, 1 Free Will Baptist, 7 Episcopalian, 5 Methodist, and 1 Universalist. For more than a century after the first settlement of the county, the inhabitants were universally Congregationalists.

The courts sit alternately at Middletown and Haddam. Middletown is the principal town. Middlesex comprises 2 senatorial districts. Population of the county 1820, 22,-405—1830, 24,845.

MIDDLESEX, p-t. Yates co. N. Y. 194 miles W. Albany. Pop. 1830, 3,428.

MIDDLESEX, co. N. J. near the centre of the state, is bounded by Essex co. N., the N. Y. line and Monmouth co. E. and S., Burlington and Somerset cos. W. It is divided by the lower part of Raritan r. It contains the upper parts of Millstone and Assanpink rivers, and its principal town is New Brunswick. Pop. 1820, 23,157.

MIDDLESEX, co. of Va. bounded by Piankatank r. which separates it from Gloucester S., and King and Queen W., by Essex N. W., by the bay of Rappahannoc N., separating it from Lancaster, and E. by Chesapeake bay. Length S. E. to N. W. 35 ms., mean width 5, and area 175 sq. ms. Extending in lat. 37° 30′ to 37° 48′, and in long. 0° 13′ to 0° 40′ E. W. C. This county comprises a long and narrow point between the 2 bounding rivers. Chief t. Urbanna. Pop. 1820, 4,057—1830, 4,122.

MIDDLETON, p-t. Strafford co. N. H., has no rivers, ponds nor mountains, except part of Moose mountain, and the soil is rocky; 48 ms. N. E. Concord. Population in 1830, 561.

MIDDLETON, t. Essex co. Mass. 20 ms. N. of Boston, has no village, an uneven surface, and only tolerable soil. Pop. 1830, 607.

MIDDLETOWN, p-t. Rutland co. Vt. 70 ms. S. of Burlington, and 41 N. of Bennington, is crossed by Poultney r., has a soil of gravelly loam, and a pleasant village and several mills, &c. Pop. 1830, 919.

MIDDLETOWN, t. Newport co. R. I. 2 miles N. E. Newport, and 28 S. E. Providence. Pop. 1830, 915.

MIDDLETOWN, city, port of entry, and chief town of Middlesex county, Connecticut. Lat. 41° 35′ N. and long 4° 15′ E. The Indian name was *Mattabeseek*. It is pleasantly situated on the gradually rising ground on the west bank of the Connecticut river, 31 miles above its mouth, 15 miles S. of Hartford, 24 N. E. of New Haven, and 325 from W. C.

The principal street, which runs N. and S. is broad, level and well built, and with those parallel to it, is intersected at right angles by others leading to the river. The wharves are commodious, and two of them are appropriated for steam-boats, by which daily communication is kept up with the cities of New York and Hartford. Population of the city 1820, 2,618, including the town, 6,681, 1830, city 2,965, including the town, 6,892.

The city contains a court house, a custom house, 2 banks, a jail, an almshouse, and 7 places of public worship, 2 of which are Congregational, 1 Episcopalian, 1 Baptist, 1 Methodist, 1 Universalist and 1 African. The Wesleyan university, founded in 1831, is an institution of great promise under the patronage of the Methodist Episcopal church. Its buildings are eligibly situated on a hill adjacent to the city, and command a fine view of the river and the surrounding country. Its officers in 1832, were a president and 4 professors. It possesses a valuable library, cabinet of minerals, chemical and philosophical apparatus, &c. Many of the houses and stores are built with brick, and much taste is displayed in and about the residences of the citizens. Two weekly newspapers are published in the city.

The manufactories of the city and town are numerous. Among them are three for arms for the United States' service, 1 of broadcloth, 1 of cotton, 1 of webbing, 1 of combs, 1 of Gunter's scales, &c., 1 of machinery, 1 of pewter, 1 of axes, 2 of tin ware, 1 paper mill, 1 powder mill, 4 jewelry establishments, &c. &c. One manufactory makes 1,500 rifles annually, milling all the parts; another 2,000 milled muskets; another 1,200 guns which are cast. One company make 45,000 lbs. of cotton yarn, and another 30,000 yards of broad cloth; 200,000 coffee mills are made here every year, and the annual value of manufactures in the place, is about $700,000.

The coasting trade of Middletown is extensive—its foreign trade considerable. In 1816 it owned more shipping than any town in Connecticut. Vessels for Hartford and other towns on the river, are registered here. The river is navigable to Middletown for vessels drawing 10 feet of water. There is a horse boat ferry between this place and Chatham. Two miles above the city is the village of *Middletown Upper Houses*, which contains a post office.

The whole township from N. to S. is about 9 miles long, and it varies in breadth from 4 to 10 miles, and contains about 58 sq. ms. It was settled in 1636, the same year with Hartford and Windsor. The public records of the town commence in 1654. The city was incorporated in 1784.

Middletown rests on secondary red sand stone—the other rocks are pudding stone and bituminous shale, having impressions of leaves, fish, &c. The range of granite hills terminates 2 miles S. of the city, forming the straits of Connecticut river. Valuable minerals are found in various parts of it. During the revolution, a lead mine was wrought some distance E. of the city, on the bank of the Connecticut river, and several shafts were sunk. The ore was found in quartz veins, with some fluor spar. It is now neglected.

MIDDLETOWN, p-t. Delaware co. N. Y. 20 ms. S. E. Delhi, and 68 S. W. Albany, is watered by Papachton river, the E. branch of Delaware river, and tributaries which supply many mill seats. It is very hilly, with vallies of good land, and has a mixed population. Pop. 1830, 2,383.

MIDDLETOWN, p-v. Orange co. N. Y. 23 ms. N. of Newburgh.

MIDDLETOWN, p-v. Saratoga co. N. Y. 3 ms. N. W. Waterford.

MIDDLETOWN, p-t. Monmouth co. N. J. on Raritan bay, and at the mouth of a creek, 50 ms. E. of Trenton, and 30 S. W. N. Y. It has an academy. Pop. 1830, 5,128.

MIDDLETOWN POINT, p-v. Monmouth co. N. J. on Matteawan creek, which falls into Raritan bay, 14 ms. N. W. Shrewsbury, 12 N. N. E. Freehold, and 9 S. E. Amboy.

MIDDLETOWN, p-v. on the left bank of Susquehannah r., above the mouth of Swatara creek, 9 ms. below Harrisburg, and 27 N. W. by W. Lancaster. It is comparatively an ancient village, extending mostly in a single street along the main road. Pop. 1820, 567.

MIDDLETOWN, p-v. upon Appoquinimink creek, in the southwestern part of New Castle co. Del. 25 ms. S. S. W. Wilmington, and 27 by p-r. N. N. W. Dover.

MIDDLETOWN, p-v. Washington co. Pa. (*See West Middletown.*)

MIDDLETOWN, p-v. Frederick co. Md. 8 ms. N. W. by W. Frederick, and 17 S. S. E. Hagerstown.

MIDDLETOWN, p-v. on Cedar creek, near the southwestern border of Frederick co. Va. 16 ms. S. W. Winchester, and by p-r. 83 miles a little N. of W. W. C.

MIDDLETOWN, p-v. in the east part of Hyde co. N. C. 158 ms. eastrd. of Raleigh.

MIDDLETOWN, p-v. in the northeastern part of Jefferson county, Ky. 12 ms. E. of Louisville, and by p-r. 44 ms. a little N. of W. of Frankfort.

MIDDLETOWN, p-v. southern part of Butler co. Ala. by p-r. 165 ms. S. S. E. Tuscaloosa.

MIDDLETOWN, p-v. Butler co. O. by p-r. 90 ms. S. W. by. W. Columbus.

MIDDLETOWN, p-v. Henry co. Ind. by p-r. 49 ms. N. E. by E. Indianopolis.

MIDDLEWAY, p-v. western part of Jefferson co. Va. 85 ms. by p-r. N. W. by W. W. C.

MIDWAY, p-v. Culpepper co. Va. by p-r. 81 ms. S. W. by W. W. C.

MIDWAY, p-v. in the western part of Caldwell co. Ky. by p-r. 216 ms. S. W. by W. Frankfort.

MIDWAY, p-v. Stark co. O. by p-r. 126 ms. N. E. by. E. Columbus.

MIDWAY, p-v. Spencer co. Ind. by p-r. 177 ms. S. S. W. Indianopolis.

MIFFLIN, co. Pa. bounded by Perry S. E. and S., Huntington S. W. and W., Centre N., Union N. E., and the Susquehannah river separating it from Dauphin E. Length 45, mean width 20, and area 900 sq. ms. Extending in lat. 40°, 14′ to 40° 52′, and in long. from the meridian of W. C. to 0° 56′ W. The surface of Mifflin is very much broken by mountain ridges, stretching from S. W. to N. E.; yet much of the soil is excellent. The county is traversed in the greatest part of its length by the Juniata r., flowing first northeast, then east, and finally entering the adjacent county of Perry by a bend to S. S. E. Along this river valley passes the Transverse Division of the Union canal, affording an outlet to the productions of Mifflin co. Chief t. Lewistown. Pop. 1820, 16,818—1830, 21,690.

MIFFLIN, p-v. southwestern part of Henderson co. Ten. by p-r. 143 ms. S. W. by W. Nashville.

MIFFLIN, p-v. Richland co. Ohio, N. E. Columbus.

MIFFLINBURG, p-v. on Buffalo cr. Union co. Pa. 6 ms. N. W. New Berlin, and by p-r. 65 ms. a little W. of N. Harrisburg.

MIFFLINTOWN, p-v. on the left bank of Juniata river, by p-r. 43 ms. N. W. Harrisburg.

MIFFLINSVILLE, in the p-o. list, but Mifflinsburg on Tanner's map, p-v. on the left bank of Susquehannah river, and southeast part of Columbia co. Pa. 18 ms. N. E. by E. Danville, and by p-r. 80. ms. N. N. E. Harrisburg.

MILAN, p-t. Dutchess co. N. Y. 22 ms. N. N. E. Poughkeepsie, a part of Ancrams creek, and streams of Wappingers creek furnish mill sites. Pop. 1830, 1,886.

MILAN, p-v. Huron co. Ohio, by p-r. 117 ms. northward Columbus.

MILBORO' SPRING, and p-o. Bath co. Va. by p-r. 199 ms. S. W. W. C.

MILESBURG, p-v. Centre co. Pa. 2 ms. W. Bellefonte, and by p-r. 87 ms. N. W. Harrisburg.

MILES, cross roads and p-o. Knox co. O. by p-r. 54 ms. N. E. Columbus.

MILESTOWN, p-v. near Phil. city, Phil. co. Pa.

MILFIELD, p-v. Athens co. Ohio, by p-r. 82 ms. S. E. Columbus.

MILFORD, p-t. Hillsboro' co. N. H. on Souhegan river, has mills and factories, 31 ms. S. Concord, and yields good apples, &c. Pop. 1830, 1,302.

MILFORD, p-t. Worcester co. Mass. 18 ms. S. E. Worcester, is supplied with excellent mill seats by Charles and Mill rivers. It is gently swelling in surface, has an academy. Pop. 1830, 1,360.

MILFORD, p-t. New Haven co. Conn. on Long Island sound, 9 ms. S. W. New Haven; has some good land, but is very rocky. It has a harbor for vessels of 200 tons, and a quarry of marble like verde antique, clouded greenish, &c. but not very valuable for working. Pop. 1830, 2,256.

MILFORD, p-t. Otsego co. N. Y. 10 ms. S. Cooperstown, 76 ms. W. Albany, on the north side of Susquehannah river, is hilly, but has excellent soil for grazing. Pop. 1830, 3,025,

MILITARY ACADEMY of the United States. (*See West Point.*)

MILFORD, p-v. and st. jus. Pike co. Pa. situated on the right bank of Del. river, 56 ms. above and a little E. of N. Easton, and 119 ms. northward Philadelphia, lat. 41° 18′, long. 2° 16′ E. W. C.

MILFORD, tsp. of Mifflin co. Pa. opposite Mifflintown, and on Tuscarora creek. Pop. 1820, 1,554.

MILFORD, village and tsp. in the western part of Somerset co. Pa. The village is situated 8 ms. S. W. by W. the borough of Somerset. Pop. of the tsp. 1820, 1,394.

MILFORD, p-v. on Mispillion creek, southeastern part of Kent co. Del. 20 ms. S. S. E. Dover, and by the p-r. 102 ms. a little N. of E. W. C.

MILFORD, p-v. on the left bank of Monongahela river, in Harrison co. Va. 5 ms. S. S. W. Clarksburg.

MILFORD, p-v. in the sthrn. part of Greenville dist. S. C. by p-r. 95 ms. N. W. by W. Columbia.

MILFORD, p-v. Monroe co. Geo. by p-r. 66 ms. W. Milledgeville.

MILFORD, p-v. in the N. W. part of Clermont co. Ohio, by p-r. 98 ms. S. W. Columbus.

MILFORD CENTRE, p-v. sthrn. part of Union co. O. by p-r. 32 ms. N. W. Columbus.

MILITARY GROVE, p-v. Burke co. N. C. by p-r. 220 ms. a little N. of W. Raleigh.

MILLBORO', p-v. in the eastern part of Washington co. Pa. by p-r. 214 ms. N. W. by W. W. C. and 207 W. Harrisburg.

MILBORO', p-v. Sussex co. Va. 4 or 5 ms. S. E. Sussex court house, and by p-r. 55 ms. S. S. E. Richmond.

MILLBORO' SPRING, and p-o. Bath co. Va. by p-r. 214 ms. S. W. by W. W. C.

MILLBROOK, p-v. Wayne co. Ohio, northeastward Columbus.

MILLBURY, p-t. Worcester co. Mass. 40 ms. S. W. Boston, is crossed by Blackstone river and canal. The water taken from the falls feeds the canal to Mendon. Here was formed the first of those lyceums which are now so numerous and useful. The Goodell manufacturing company make woollens, and there are gun and cotton manufactories, &c. &c.

and quarries of granite. Pop. 1830, 1,611.

MILL CREEK, p-o. Berkeley co. Va. by p-r. 93 ms. N. W. W. C.

MILL CREEK, and p-o. Coshocton co. Ohio, by p-r. 93 ms. a little N. of E. Columbus.

MILL CREEK, p-o. Madison co. Indiana, by p-r. 46 ms. N. E. Indianopolis.

MILLEDGEVILLE, p-t., st. jus. for Baldwin co. and of government for the state of Geo. is situated on the right bank of Oconee river, at lat. 33° 05′, long. 6° 17′ w. W. C. This geographic position gives by actual calculation, the bearing from W. C. to Milledgeville s. 41° 09′ w. and a distance of 536 ms. nearly; the post office distance along the post road 662 statute ms. Milledgeville is 92 ms. s. w. by w. along the road through Warrentown and Sparta from Augusta, and 175 ms. N. w. by w. Savannah. Following the windings of the Oconeee and Alatamaha, this place is 312 ms. above the Atlantic ocean. Boats of 25 or 30 tons are navigated to Milledgeville. The site of this town is broken into hills. It contains a state house, a branch of the state bank, several places of public worship, and 2 or 3 printing offices. The state penitentiary is also located in Milledgeville. The latter establishment appears from recent information to support itself. The convicts in 1829, were 92.

MILLER, co. of Ark. bounded N. by Crawford, E. by Clark, s. by Sevier, and w. by the unappropriated western territory, length 50, width 36, and area 1,800 square ms. Extending in lat. 34° to 34° 43′ N., and in long. 16° 50′ to 17° 30′ w. W. C. It is chiefly drained to the southward by the constituent creeks of the Little river of the north. The northern part is mountainous and drained to the northward by the head branches of Potomac river flowing into Arkansas. The co. is therefore in part a table land, between the Red and Arkansas rs. and lies about 100 ms. s. w. by w. Little Rock. Pop. 1830, 356.

MILLER, C. H. and p-o. in the preceding co. is marked on the p-o. list, as distant 1,326 ms. from W. C., and 215 from Little Rock.

MILLER's river, Worcester and Franklin cos. Mass. enters the Connecticut river at Northfield, 35 ms. long, with a fall of 62 feet near its mouth. Just above it a large body of Indians were destroyed in Philip's war, by a small army of volunteers from Northampton, &c.

MILLERSBURG, p-v. in the northeastern part of Bourbon co. Ky. 10 ms. N. E. Paris, and by p-r. 53 ms. a little N. of E. Frankfort.

MILLERSBURG, p-v. and st. jus. Holmes co. Ohio, by p-r. 80 ms. N. E. by E. Columbus, and 46 ms. N. Zanesville, lat. 40° 32′, long. W. C. 4° 57′ w. It is situated on Kilbuck creek, near the centre of the co.

MILLERSBURG, p-v. Ripley co. Indiana, by p-r. 75 ms s. E. Indianopolis.

MILLERSBURG, p-v. in the northern part of Callaway co. Mo. by p-r. 44 ms. N. E. Jefferson city, and about 140 N. w. by w. St. Louis.

MILLERSBURG, p-v. on the left bank of Susquehannah river, on the point above the mouth of Wicomisco creek, 23 ms. N. and above Harrisburg.

MILLERS CREEK, and p-o. Estill co. Ky. by p-r. 81 ms. s. E. by E. Frankfort.

MILLER's Inn, sthwstrn. part of Nelson co. Ky. by p-r. 9 ms. from Bardstown, and 64 ms. s. w. Frankfort.

MILLER's TAVERN, and p-o. Essex co. Va. by p-r. 119 ms. s. W. C.

MILLERSTOWN, p-v. on the left bank of Juniata r. Perry co. Pa. 29 ms. N. N. W. Harrisburg.

MILLERSTOWN, p-v. Grayson co. Ky. 10 ms. from Litchfield, and by p-r. 115 ms. s. w. by w. Frankfort.

MILLERSVILLE, p-v. Lancaster co. Pa. 5 ms. s. w. Lancaster.

MILL FARM, and p-o. Caroline co. Va. by p-r. 92 ms. sthrd. W. C.

MILL GROVE, and p-o. Mecklenburg co. Va. by p-r. 258 ms. s. s. w. W. C., and 135 ms. s. w. Richmond.

MILL GROVE, and p-o. Sumpter dist. S. C. by p-r. 60 ms. E. Columbia.

MILL HALL, p-v. in the nthwstrn. part of Centre co. Pa. 23 ms. N. W. Bellefonte, and 108 ms. N. W. Harrisburg.

MILL HAVEN, p-v. Scriven co. Geo. by p-r. 142 ms. a little s. of E. Milledgeville.

MILLHEIM, p-v. in the sthestrn. part of Centre co. Pa. 20 ms. N. E. by E. Bellefonte, and by p-r. 86 ms. N. N. W. Harrisburg.

MILLINGTON, p-v. Kent co. Md. by p-r. 53 ms. N. E. Annapolis.

MILLINGTON, p-v. Decatur co. Ind. by p-r. 55 ms. s. E. Indianopolis.

MILLPORT, p-v. in the wstrn. part of Mecklenburg co. Ky. by p-r. 190 ms. s. w. by w. Frankfort.

MILL RIVER, p-o. Buncombe co. N. C. by p-r. 250 ms. wstrd. Raleigh.

MILLSBORO', p-v. near the head of Indian r. in the s. E. part of Sussex co. Del. by p-r. 49 ms. s. s. E. Dover.

MILLSFIELD, t. Coos co. N. H. 150 ms. N. Concord. Pop. 1830, 33.

MILLSFORD, p-v. Ashtabula co. O. by p-r. 197 ms. N. E. Columbus.

MILLS POINT, p-v. Hickman co. Ky. by p-r. 338 ms. s. w. by w. Frankfort.

MILL SPRINGS, p-v. in the sthrn. part of Wayne co. Ky. 24 ms. from Monticello, and by p-r. 152 ms. sthrd. Frankfort.

MILLSTONE BROOK, N. J. a branch of Raritan r. rises in Monmouth co., flows N. through Middlesex, receiving Stony brook, and part of Somerset, where it joins the r. 9 ms. above Brunswick. On its banks, where it is crossed by the Princeton and Trenton roads, Washington defeated the British regiment of grenadiers, on his retreat from Lamberton. The Delaware and Raritan canal is now constructing along the course of this stream a part of its length.

MILLSTONE, p-v. Somerset co. N. J. 38 ms. N. E. Trenton.

MILLTOWN, p-v in nthwstrn. part of Bradford co. Pa. by p-r. 146 ms. N. Harrisburg.

MILLTOWN, p-v. Crawford co. Ind. by p-r. 114 ms. sthrd. Indianopolis.

MILLVILLE, p-t. Cumberland co. N. J. 12 ms. E. Bridgetown. It has iron works, which are supplied with water by a short canal from a pond. It is crossed by Maurice r. Pop. 1830, 1,561.

MILLVILLE, p-v. in the nthrn. part of Columbia co. Pa. by p-r. 93 ms. nthrds. Harrisburg.

MILLVILLE, p-v. King George's co. Va. by p-r. 91 ms. s. W. C.

MILLVILLE, p-v. Spartanburg dist. S. C. by p-r. 97 ms. N. N. W. Columbia.

MILLVILLE, p-v. Lincoln co. Ten. about 60 ms. sthrd. Nashville.

MILLVILLE, p-v. Caldwell co. Ky. by p-r. 235 ms. s. w. by w. Frankfort.

MILLVILLE, p-v. Butler co. O. by p-r. 115 ms. s. w. by w. Columbus. Pop. 1830, 196.

MILLWOOD, p-v. Frederick co. Va. 11 ms. s. E. by E. Winchester, and by p-r. 61 ms. N. w. by w. W. C.

MILNERSVILLE, p-v. Guernsey co. O. by p-r. 102 ms. E. Columbus.

MILO, p-t. Yates co. N. Y. 25 ms. s. E. Canandaigua, w. Seneca lake, E. Crooked lake, whose outlet affords mill seats. The soil is rich, argillaceous loam, with some alluvion and warm gravel. Penn Yan p-v. was named from the settlers being Pennsylvanians and N. Englanders. Pop. 1830, 3,610.

MILTON, p-t. Strafford co. N. H. 27 ms. N. Portsmouth, 46 from Concord, is on the w. side of Salmon Falls r., which divides it from Maine. It includes Teneriffe mtn., and is crossed by a stream. Pop. 1830, 1,273.

MILTON, p-t. Chittenden co. Vt. E. side lake Champlain, on Lamoille r. 12 ms. N. Burlington, and 40 N. w. Montpelier, has the advantage of the Great Falls of Lamoille and its branches, plenty of iron ore and limestone, and of a low sand bank, extending to the s. w. corner of S. Hero, by which the lake is fordable most of the year. The Great Falls are curious; an island stands in the channel, where the river descends 150 ft. in 50 rods. Cobble and Rattlesnake hills, 4 or 500 ft. high, are the principal; the surface is gently varied. Pop. 1830, 2,097.

MILTON, p-t. Norfolk co. Mass. 7 ms. s. Boston, s. Neponset r., has various mills, &c. Good tillage in the middle and N. E., but in the s. part, is broken and hilly. Part of the Blue hills are in the t. some of which are 710 ft. above high water. Pop. 1830, 1,576.

MILTON, t. Saratoga co. N. Y. 30 ms. N. Albany, is nearly level, and has a stiff or sandy loam, except a sandy pine tract in the E. Kayderosseras brook crosses it. Slate and limestone lie under the surface, and loose masses of granite, gneiss, limestone, &c. above. It comprehends Ballston springs, and the village of Ballston Spa, celebrated as a fashionable retreat, on account of the value of its waters. There are several chalybeate springs, and one of them is strongly charged with salts. They all rise near the margin of a small valley, probably once a lake; and there is the Spa village, which contains 2 churches, a court house &c., with several boarding houses, the chief of which is the Sans Souci. Milton v. 3 ms. N. w. of the Spa, has 2 churches, limekilns, and several factories. Pop, 1830, 3,079.

MILTON, p-v. Orange co. N. Y. 12 ms. N. Newburgh.

MILTON, v. Middlesex co. N. J. 1 m. w. Rahway.

MILTON, p-v. on the left side of Susquehannah r. at and above the mouth of Limestone run, 12 ms. above Northumberland, and 81 N. Harrisburg.

MILTON, p-v. on Dan r. in the nthesrn. angle of Caswell co. N. C. by p-r. 98 ms. N. w. Raleigh.

MILTON, p-v. Laurens dist. S. C. by p-r. 65 ms. N. w. Columbia.

MILTON, p-v. in the nthwstrn. part of Rutherford co. Ten.

MILTON, p-v. Gallatin co. Ky. by p-r. 83 ms. N. E. Frankfort.

MILTON, p-v. Trumbull co. O. by p-r. 154 ms. N. E. Columbus.

MILTON, p-v. Wayne co. Ind. by p-r. 75 ms. E. Indianopolis.

MINA, p-t. Chatauque co. N. Y. Population 1830, 1,388.

MINDEN, p-t. Montgomery co. N. Y. s. of Mohawk r., 15 ms. w. Johnstown, and 58 w. N. w. Albany. Has gentle hills and rich vallies for wheat, with argil. loam on clay. Otsquaga creek affords mill seats. Fort Plain was on Mohawk r. The inhabitants are German, and speak the German language. Pop. 1830, 2,567.

MINE RIVER, confluent of Missouri, and having its entire course in the state of Mo., rises between the northern sources of Osage river and that part of Missouri river between the influx of Kansas and Grand rs. The valley of Mine r. comprises the northern sections of Lafayette, Saline and Cooper counties, the stream falling into Missouri 4 or 5 ms. above Booneville, the st. jus. of the latter co. The valley of Mine r. does not amount to 70 ms. in its greatest length, but it is a large stream compared to its length, draining a circular valley of 60 ms. diameter, exceeding an area of 2,800 sq. ms.

MINEHEAD, t. Essex co. Vt. N. Conn. r. 60 ms. N. E. Montpelier, 100 from Windsor, and is watered by Nulhegan r. &c. Pop. 1830, 150.

MINERAL POINT, and p-o. Iowa co. Mich., or more correctly in Huron. As laid down on Tanner's improved U. States' map, this place is situated on the head of the w. Fork of Peektano r. 74 ms. s. w. of fort Winnebago, 75 ms. a little s. of E. Prairie du Chien, and 64 N. E. Galena in Il.

MINERSVILLE, p-v. in the northeastern part of Schuylkill co. Pa. by p-r. 71 ms. N. E. Harrisburg, and 179 N. N. E. W. C.

MINERVA, p-t. Essex co. N. Y. 30 ms. s. w. Elizabethtown, is little inhabited, and 14 ms. by 25. The sources of Hudson r. rise in it, and water it well. The surface is irregular, and the soil pretty good. Pop. 1830, 358.

MINERVA, p-v. Mason co. Ky. by p-r. 83 ms. N. E. Frankfort.

MINERVA, p-v. Stark co. O. by p-r. 135 ms. N. E. Columbus.

MINISINK, p-t. Orange co. N. Y. 10 ms. w. Goshen, N. Y. on Wallkill creek, N. E. N. J. and Pennsylvania lines. The Shawangunk mtns. from the Alleganies and Navisink r. cross it. Near the Wallkill are some drowned lands. Soil and surface various. There are 4 p-os. at Minisink, West town, Carpenter's point, and Ridgeburgh. Dolsentown and Brookfield are also villages. The Hudson and Delaware canal crosses the town, meets Delaware r. at Carpenter's point, and proceeds up that stream. Pop. 1830, 4,979.

MINOT, p-t. Cumberland co. Me. w. of Androscoggin r., 33 ms. N. Portland, N. Little Androscoggin r., and S. Oxford co. Population 1830, 2,904.

MINTONSVILLE, p-v. Gates co. N. C. by p-r. 149 ms. N. E. by E. Raleigh.

MIRANDA, p-v. Lincoln co. N. C. by p-r. 163 ms. a little S. of W. Raleigh.

MISSISQUE, (*See Troy.*)

MISSISQUE, bay, an arm of lake Champlain containing 35 sq. ms., reaching 4 or 5 miles into Lower Canada, between Swanton and Highgate, Vt. It is 5 miles wide on the line.

MISSISQUE, r. of Vt. rises in Orleans co., and passing into Lower Canada, traverses Franklin co., enters lake Champlain at Missisque bay. It is wide, slow and shallow, with several falls. It receives Trout river, Black creek, Taylor's branch, &c., draining about 582 sq. ms. in Vermont. It is 75 miles long, and navigable to Swanton falls, 6 ms. in vessels of 50 tons.

MISSISSIPPI, river of the United States. Though the various large constituent streams of the mighty Mississippi will be severally noticed, yet a general view of the great central basin is indispensable in a treatise of the nature of this Gazetteer. Beside many of inferior magnitude, the great constituent rivers which drain the basin, and unite their waters to form the Miss., are the Red, White, Arkansas, Miss., Miss. proper, and Ohio.

A very erroneous opinion of the relative extent of the basin of the Miss. has been fostered by too many geographers of our own country. The true characteristic to determine the comparative importance of rivers, is the area drained, and not mere length of course. To give more correct views of the true rank of the large rivers of the earth, the following table was constructed.

No. 1. Table of the basins of the large rivers of the earth, including the length of course of each great river, exclusive of minute sinuosities.

River Basins.	Length of course.	Mean width of Basin.	Area in sq. ms.
Rio de la Plate	1,600	800	1,280,000
Amazon, inclusive of the Tocantinas,	3,000	980	2,940,000
Orinoco,	1,100	360	396,000
Atlantic slope of N. America, from Flor. point, exclusive of St. Lawrence,	1,800	170	306,000
Miss. including Red, Arkansas, White, Miss. proper, Ohio, Missouri, &c.	2,000	550	1,100,000
St. Lawrence,	1,200	425	510,000
Saskatchawaine,	1,200	200	240,000
Unjiga, or Mackenzies r.	1,400	200	280,000
Euxine Basin,	1,800	550	990,000
White sea Basin,	1,380	700	966,000
Caspian & Arab united basins,	2,500	1,000	2,500,000
Oby,	2,150	600	1,290,000
Yeniseli,	2,100	400	840,000
Lena,	2,070	350	724,000
Amur,	1,820	360	655,000
Yellow river,	1,980	200	396,000
Blue r	2,280	200	456,000
Basin of S. E. Asia,	1,800	150	270,000
Ganges & Buramapootre, united,	1,500	380	589,000
Indus,	1,200	180	216,000
Euphrates & Tigris, united,	1,150	140	161,000
Nile,	1,680	250	420,000
Niger,	2,000	200	400,000

By this table it is shewn, that the surface comprised in the Miss. basin, falls short of that of the Plate, and is only to that of the Amazon as 377 to 1000; yet the enormous extent of the former, though the third in rank amongst the rivers of America, becomes very apparent, when it is seen, that it exceeds in extent all the rivers of the Atlantic slope of North America, including the St. Lawrence, or either the Baltic or Euxine basins; that it far exceeds the united basins of the Indus, Ganges, and Buramapootre, or the great central basin of China.

A line drawn from the Appalachian system, where the sources of Ten. and Great Kenhawa separate in Ashe co. N. C. to the sources of Marias river, the northwestern confluent of Miss. is by calculation N. 55° 40' W. 1,985 statute ms. Another line very nearly at right angles to the preceding, drawn between the sources of Red and Ouisconsin rs., measures 1,100 ms., the mean width of the basin is, however, about 550 ms. The following table exhibits the relative extent and geographic position of the constituent vallies of this great basin.

Nat. Sections.	length.	mean width.	area sq. ms.
Ohio valley,	750	261	196,000
Miss. valley, or Miss. proper,	650	277	180,000
Missouri valley,	1,200	437	523,000
Ohio, including the valleys of White, Arkansas, Red, &c.	1,000	200	200,000

Of these vallies, which drain a territory of 1,099,000 sq. ms., the extent is as follows:

	lat.	long. W. W.C.
Ohio, fm. lat.	34° to 42° 30'	1° to 11° 40'
Miss. proper,	37° " 48°	9° " 20°
Missouri,	37° " 50°	13° " 35°
Ohio, including the vallies of White, Ark. Red, &c.	29° " 42°	11° " 30°

The various sections of this great physical region will be found under their respective heads, but we here notice the general features in order to explain the phenomena of the annual inundations. By reference to table II. it will be seen that the difference of lat. between the extremes is from lat. 29 to 50, or 21 degrees. The relative elevation has never been accurately determined, but may without estimating mountain ridges, be assumed safely at 5,000 feet, or an equivalent to 10 degrees of lat. Combining these elements would give a winter climate to Miss. sources similar to that of Labrador, on the Atlantic coast, of lat. 61°

The basin, if taken as a whole, is composed of two very unequal inclined plains, one, the western and much most extensive, falling from the Chippewayan system, is about 800 miles mean width; the second declines from the Appalachian system westward, and is about 400 miles wide at a mean. The base line, or line of common depression, follows the valley of Illinois and Miss. below the mouth of Miss. The general characters of the Appalachian and Chippawayan systems are communicated to their respective plains. In a state of nature the Appalachian system was a dense forest. This vast body of woods was protruded on one side to the shores of the Atlantic ocean, and westward encroached on the central plains.

The Chippewayan is mostly naked of timbers, as are the immense grassy plains which compose its eastern slope. By reference to the art. Md. it will be seen that, independent of comparative height, the prevalent winds of the continent are from the westward, and that winter cold increases in intensity advancing westward to the summits of the Chippewayan.

Permanent snows cover the earth in winter over the Atlantic slope and Miss. basin as low as lat. 31°, but from the peculiar structure of the vallies, the floods produced by winter snows and spring rains cannot be simultaneously discharged. The gradual discharge is produced by three causes; first, difference of lat.; second, difference of height; and thirdly, contrariety of direction.

The general course of the flood being to the southward, spring advances in a reverse direction, and releases in succession, the waters of the lower valley, then those of O., then those of Miss. In a mean of ten years the swell commences on the Delta, in the end of Feb. and beginning of March, and continues to rise by unequal diurnal accretions to the middle of June, when the waters begin again to depress. But what might excite much surprise to those unacquainted with the cause, the waters of the upper Miss. do not reach the Delta until upwards of a month after the inundation has been abating.

Rising between 42° and 50° and at an elevation of from 1,200 to 5,000 feet, the higher sources of the Miss. are locked in ice and snow long after summer reigns on the Delta. Again the courses of the Yellow Stone river and Miss. are to the northeastward for 5 or 600 ms. from the Chippewayan ridges, giving to their floods a very circuitous route.

To these particular causes of separate discharge one general cause may be added, that is the slow motion of the waters. Amongst the many vulgar errors introduced into our books concerning the Miss. basin, none stands more opposed to fact than the rapid motion of the waters. If in reality the floods moved with half the commonly assigned velocity, the Delta would be annually and totally submerged. The waters of Upper Miss. do not reach the Delta before the beginning of August, about 100 days or 2,400 hours after the breaking up of winter. This supposes a motion of about one mile per hour. Similar to the Russian or northeastern plains of Europe, the Miss. basin is remarkable for the very regular slope of its declivities, and consequently the scarcity of direct falls or even cataracts in its rs. If we allow an elevation of 5,000 feet to the sources of Miss. we find the much greater part of the fall in the vicinity of the Chippewayan, and to estimate the height of the junction of Miss. and Yellow Stone river at 2,000 feet is full more than would be warranted by known elevation at the source of Miss. proper. But allowing 2,000 feet elevation for the mouth of Yellow Stone river, and 2,400 ms. for distance thence to the Delta, we have a fall of only 10 inches to the mile. This estimate, moderate as it appears, is nevertheless too high. Pittsburg is by actual measurement within a small fraction of 700 feet above the surface of the Gulf of Mexico, and distant by the windings of the streams from that recipient, about 1,800 ms. yielding a mean fall of 4 6.10 inches per mile nearly. If in brief, we allow a mean fall of 6 inches to the mile, it is more than sufficient for the mean fall of the waters of the Miss. basin, from their heads to final discharge into the Gulf of Mexico.

The seasons of general inundation are tolerably well known to the inhabitants of the Delta, but so very greatly do the quantity of meteor differ in different years, that no length of experience enables any person to anticipate, with any approach to certainty, the elevation of flood in any given year. Some years, as in 1800–1, the waters do not rise above their channels, of course no inundation takes place.

Connected with the general history of the Miss. Delta, is the mistaken opinion that the main channel is changeable. When the annual inundations occur, the surface of the river is indeed above that of the adjacent country, but the bed or bottom, similar to all other, rivers, is, nevertheless, the deepest valley of the region through which it flows. The author of this article has sounded the Miss. from the efflux of Atchafaluya to the different outlets, and found the stream at the lowest water, from 75 to 80 feet at the head of the Delta, 130 feet near the outlet of La. fourche at Donaldsonville, upwards of 100 feet opposite New Orleans, and from 75 to 80 feet three ms. above the main bars. Lake Pontchartrain is the deepest lake of La. and yet does not average a depth of 18 feet; say its bottom is 25 feet below the general level of the Delta, then would the bottom of the Miss. at New Orleans, be 75 feet below that of the greatest adjacent depression.

The great, and in many cases almost circular bends of the Miss. in and above the Delta, produce a reverse of the current at once on the opposite sides of a neck of land. This neck being composed of alluvion, yields easily to the abrasion of water, and is finally worn away and a new channel opened. Above and below where the isthmus formerly existed, the ancient bed is filled up with sand and earth, whilst the old bed around the point assumes the aspect of a lake, but by its proximity to the parent river and its form proves the origin. Such lakes are Fause Riviere, one near the mouth of Homochitto river, Concordia, St. John's, St. Joseph's, Providence, and Grand lakes, and one forming the mouth of Yazoo. Of these, Fause Riviere, that on the left bank near the mouth of Homochitto, and that also on the left bank at the mouth of Yazoo, have been formed within the period of white settlement. With the exception stated, the volume of the Miss. is as effectually and permanently confined to its channel, as is any other river of the earth. (*See art. Atchafalaya, Lafourche, La. &c.*)

Mississippi, state of the U.S. bounded by the Gulf of Mexico s., La. s. w., Ark. n. w., Tenn. n., and Ala. e. The outlines of this state are, from the southwestern angle of Ala. along that part of the Gulf of Mexico, called lake Borgne, to the mouth of Pearl r. 60 ms.; up Pearl river to lat. 31° 65 ms.; thence due w. along lat. 31° to the bank of the Miss. nearly opposite the outlet of Atchafalaya 105 ms.; thence up the Miss. river to lat. 35° at the southwestern angle of Ten. following the windings 530 ms.; thence due e. along the southern boundary of Ten. to Ten. river, and up that stream to the mouth of Bear cr. 123 ms; thence along the western boundary of Ala. to the place of beginning on the Gulf of Mexico 320 ms. having an entire outline of 1,203 ms. Lying between lat. 30° 08′ and 35°, and between long. 11° 12′ and 14° 42′ w. W. C. Extreme length from s. to n. 337 ms. and the area being 45,760 square ms., the mean breadth is a small fraction above 135 3-4 miles; containing 29,286,400 statute acres.

Natural Geography.—The general declivity of this state is sthrd. but the western side declining by an easy descent s. s. w. towards the Miss. whilst the eastern side declines slightly towards the Tombigbee. Of the rs. of this state the principal is the great stream from which its name is derived. The state of Miss. rises from the river of the same name into a buttress of moderate and undefined general elevation. This interior buttress reaches the stream in a series of crumbling banks, called "The Bluffs." Between the Bluffs and stream, the bottoms are as low and more subject to inundation than are those on the western bank, as in the former case; the hills confine the water which is augmented by the river and creeks flowing from the interior of the state. The bottoms of the Miss. river which exist in the state of Miss. bear a small fractional proportion to the aggregate surface. From the western side of the state, advancing n. to s. flow into the Miss. river in succession, the Yazoo, Big Black, Bayou, Pierre, and Homochitto. Pearl r. rises near the centre of the state, but flows s. s. w. nearly parallel to the Big Black, about 80 ms. and thence curving s. s. e. 150 ms. falls into the pass of rigolets between lakes Pontchartrain and Boyne, after an entire comparative course of 230 ms. draining the central and much of the sthrn. parts of the state. Between Homochitto and Pearl rise in the sthrn. part of the state of Miss. and flow thence into La., the Bogue, Chitto, Tangipoho, Tickfoha, and Amite rivers. Eastward from Pearl, and draining the southeastern angle, and that protruding point between La. and Ala. comprising the cos. of Hancock and Jackson, the various branches of Pascagoula water the space between the vallies of Mobile and Pearl. The northeastern part of the state gives source to the Tombigbee or Great western branch of the Mobile r. whilst the extreme n. is drained by the sources of Wolf, and Big Hatch rs. The northeastern angle is terminated by Ten. r. These rivers are noticed under their proper heads.

It is obvious from the preceding brief notice of its rs. that as far as river navigation extends, this state possesses great advantages. It is true that neither the Pascagoula or Pearl offer navigable facilities in proportion to their comparative magnitude, but the most fertile part and that yet best inhabited and cultivated, have access to the Miss. or streams directly flowing into it as a recipient.

The soil of the state is varient in quality, but the much greater proportion thin if not sterile; the southwestern cos. drained directly into the Miss. river, contain large bodies of excellent land, and tracts of productive soil skirt the streams over the whole state. Cotton, indigo and tobacco, have been, in succession, staples of this state. The soil and climate are favorable to the growth of each of these vegetables, as also to Indian corn, potatoes, and numerous garden vegetables. The peach and fig are the common fruits, though apples are cultivated in some places to advantage. For the last 30 years, the great object of farming operations in this state has been cotton, to the injurious neglect of grain and meadow grasses. Large quantities of Indian corn are indeed annually produced, but too much dependence is placed on supplies from the northward, through the channel of the Miss. In general terms the bluff lands are the best in the state; those next the river alluvion; and the third and least productive, the pine woods. Indigenous trees most common, are the pine, various species and varieties of oak, and hickory, sweet gum, liriodendron, tulipifera, black walnut, persimon, beech, red maple, honey locust, black locust, and numerous other species of trees. Of dwarf trees, the most common are dogwood, chinquipin, papan, spice wood, thorn, &c. Buck eye, a forest tree in the valley of Ohio, is a dwarf in the states of Miss. and La. whilst the chinquipin, a mere bush in the middle states, rises in the southwest to a tree often more than 25 feet high. In the rich bottom lands of the state of Miss. the large reed cane, arundo gigantea abounded, but has in great part disappeared.

Climate.—Compared with the winters of the nrthrn. states, those of Miss. may be regarded as mild, but the seasons of the latter, like those of all the adjacent regions, are variable from each other, and not unfrequently very severe. The temperature near Natchez has afforded a cold of 12° above zero of Fahrenheit. No winter passes without less or more severe frost, and few without snow. The sugar cane and orange tree, can neither be preserved in any part of the state of Miss. above lat. 31°. The summers are, however, very warm, and long droughts frequent, as are, on the contrary, excessive and protracted rains. These are the exceptions to a generally pleasant climate. Along the streams, bilious complaints are frequent in autumn, but taken altogether, the settled cos. of the state of Miss. are healthy. The winters along the Miss. and adjacent places, are from two to three degrees colder than those of corresponding lats. along the Atlantic coasts. This difference is demonstrated by native and exotic vegetation, and by recent thermometical observations. The prevailing winds of the whole sthwstrn. parts of the U. S. are from the wstrn. side of the meridians, and principally from the nrthwst. (*See art. Md.*)

Political Geography.—For political purposes, the state of Miss. is subdivided into the cos. of

	Pop. 1820.	*Pop.* 1830.
Adams,	12,073	14,937
Amite,	6,853	7,934
Claiborne,	5,963	9,787
Copiate,		7,001
Covington,	2,230	2,551
Franklin,	3,821	4,622
Greene,	1,445	1,854
Hancock,	1,594	1,962
Hinds,		8,645
Jackson,	1,682	1,792
Jefferson,	6,822	9,755
Jones,		1,471
Lawrence,	4,916	5,293
Lowndes,		3,173
Madison,		4,973
Marrion,	3,116	3,691
Monroe,	2,721	3,861
Perry,	2,037	2,300
Pike,	4,438	5,402
Rankin,		2,083
Simpson,		2,680
Warren,	2,693	7,861
Washington,		1,976
Wayne,	3,323	2,781
Wilkinson,	9,718	11,686
Yazoo,		6,550

Total population of the state 136,621, of which are white persons,

	Males.	Females.
Under 5 years of age,	7,918	7,319
From 5 to 10,	5,572	5,165
" 10 to 15	4,591	4,169
" 15 to 20	3,683	3,653
" 20 to 30	7,237	6,231
" 30 to 40	4,638	3,090
" 40 to 50	2,419	1,739
" 50 to 60	1,595	983
" 60 to 70	632	436
" 70 to 80	180	149
" 80 to 90	47	34
" 90 to 100	11	7
" 100 and upwards,	60	2
Total,	38,466	31,977

Of which were deaf and dumb, under 14 years, 12; 14 to 25, 10; 25 and upwards, 7. Blind 25.

Colored population as follows:

	Free colored.		Slaves.	
	males.	females.	males.	females.
Under 10 years,	81	72	11,937	10,860
10 to 24	88	51	10,793	10,841
24 to 36	59	45	6,947	6,983
36 to 55	43	49	3,455	3,173
55 to 100	28	14	845	682
100 and over,	1	0	22	21
Total,	288	231	33,099	32,560

Free colored and slaves who are deaf and dumb, 12. Blind, none.

Recapitulation.

Whites.	Free colored.	Slaves.	Total.
70,443	519	65,659	136,621

Comprising the aggregate area of that part of the state of Miss. yet organized into cos., and comparing it with the superficial extent of the state, as given at the head of this article, the reader will perceive how large a portion remains unsettled and uncultivated. Examining the cos. separately, it will again appear, that density of population is in proportion to distance from the original settlements, downwards along the Miss. Bluffs, from Natchez to lat. 31°.

History.—The whole country now included in the states of Ala. and Miss. was held by France, or more correctly, that nation claimed this region as a part of La. from their first settlement on the northern shores of the Gulf of Mexico. In 1716, the French formed a settlement amongst the Natchez Indians, and built a fort where the city of Natchez now stands. In the first instance the Indians were unaware of the consequence, but dissatisfaction soon arose, and ended, in 1723, in open war. Bienville, the governor general, marched a force from New Orleans to Natchez, which the Indians were unable to oppose, and were compelled to submit to terms. In 1729 a man of the name of Chopart was commandant at Natchez, but his injustice and folly so exasperated the natives, and at the same time neglecting the means of defending his colony against their wrath, a massacre was planned, and on the 30th of Nov. 1729, perpetrated; when, with two or three exceptions, the French of both sexes. to the amount of 700, fell victims. The total dispersion of the Natchez nation soon followed, as they were too weak to sustain themselves against the French. The country in the vicinity of Natchez was abandoned by both whites and Indians, and remained long uninhabited. The French still, however, claimed the country until 1763, when it was ceded as part of Florida to Great Britain. Settlers slowly entered the country, and many very respectable British families located themselves in and near Natchez. During the revolutionary war in 1781, governor Galvez of La. invaded and conquered W. Flor., and by the treaty of Paris, 1783, it fell once more to Spain, who held it until 1798, when it was given up to the U. S. By an act of congress passed 7th of April, 1798, the president of the U. S. was authorized to appoint commissioners to adjust the limits between Flor., La., and the acquired territory N. 31st degree of N. lat., and W. of Chattahoochee r. By a subsequent act of the 10th of the same month, provision was made for a territorial government, and what is now comprised in Ala. and Miss. named the Miss. ter. The second grade of government went into operation in the spring of 1801. The 9th of July, 1808, an act of congress was passed to admit a delegate from Miss. ter. into congress. June 17th, the assent of Geo. demanded to the formation of two states from the Miss. ter. Geo. acceded to the demand, but the country remained a territory until December 1817. Previous to the latter date, on the 21st of January, 1815, a petition from the legislature of the Miss. ter. praying admission into the union as a state. This petition was favorably reported on by a committee of congress, December 1816. An act was passed the 1st of March 1817, authorizing the people of the petitioning territory, to a call a convention, which was called and met in July 1817. The convention accepted the act of congress and proceeded to frame a constitution of government. The constitution was adopted on the 15th of August, and in the ensuing December was confirmed by congress, and the new state, with the limits given at the head of this article, took her station as a member of the U. S. (*See Ala. state of.*)

Government.—A governor, with a general assembly, composed of two houses, a senate and house of representatives. To be eligible as a senator, the person must be a citizen of the U. S., shall have been an inhabitant of the state 4 years next preceding his election, and the last year thereof a resident of the district, for which he shall be chosen, and shall have attained to the age of 26 years, and also, he shall hold, in his own right within this state, 300 acres of land, or an interest in real estate of the value of $1,000, at the time of his election, and for 6 months previous thereto. Term 3 years. No person shall be a representative unless he be a citizen of the U. S., and shall have been an inhabitant of this state 2 years next preceding his election, and the last year thereof, a resident of the county, city, or town, for which he shall be chosen, and shall have attained to the age of 21 years, and also unless he shall hold in his own right, within this state, 150 acres of land, or an interest in real estate of the value of $500 at the time of his election, and for six months previous to the term of one year. Every free white male person, of the age of 21 years or upwards, who shall be a citizen of the U. S., and shall have resided in this state 1 year next preceding an election, and the last six months within the county, city, or town, in which he offers to vote, and shall be enrolled in the militia thereof, except exempted by law from military service, or having the aforesaid qualifications of citizenship and residence, shall have paid a state or county tax, shall be deemed a qualified voter. The supreme executive power of this state shall be vested in a governor, who shall be elected by the qualified electors, and shall hold his office for 2 years from the time of his installation, and until his successor be duly qualified. The governor shall be at least 30 years of age, shall have been a citizen of the U. S. 20 years, shall have resided in this state at least 5 years next preceding the day of his election, and shall be seized in his own right of a freehold estate of the value of $2,000 at the time of his election, and 12 mths. previous thereto. The judicial power of this state shall be vested in one supreme court, and such superior and inferior courts of law and equity, as the legislature may, from time to time, direct and establish. There shall be appointed in this state, not less than 4, nor more than 8 judges of the supreme superior courts. The judges of the several courts of this state shall hold their offices during good behavior; removeable by address to the governor of the two thirds of both houses of the legislature, or by impeachment before the senate, brought up by the lower house. No person who shall have arrived at the age of 65 years shall be appointed to or continue in the office of judge in this state. By the 6th article and 7th section, no minister of the gospel or priest of any denomination whatever shall be eligible to the offices of governor, lieutenant governor, or to a seat in either branch of the general assembly. Post masters are the only officers of the general government admitted to office in Miss. Revision provided for when two thirds of the general assembly shall recommend to the qualified voters to vote for or against a convention. Number of the convention equal to that of the general assembly, which convention shall meet within 3 months after the election of its members, for the purpose of revising, amending, or changing the constitution.

MISSOURI, large r. of North America, but in great part included in the U. S. great western territory. So much has been already given of the phenomena of this stream under the general head of Mississippi, as to very much abridge what is necessary to notice under its own head.

The course of discovery has led to the adoption of the name Mississippi, as a generic term for the main stream of the basin, though the Missouri is already a very large river when it approaches and passes the sources of its very inferior rival. In regard to area drained, the Mo. is the largest secondary river of the earth. A direct line drawn along its valley from its junction with the Miss. r. to the head of Marias r. is within a small fraction of 1,400 ms., a length of course, falling but little short of either the Madeira branch of Amazon, or the Paraguay branch of Rio de de la Platte; but by reference to table II. article Miss., it will be seen that the Mo. r. drains 523,000 sq. ms., or a surface more than double that of the whole Atlantic slope of the U. S. between the two St. Johns' rs. inclusive. This fine river derives its sources from the Chippewayan chains between lat. 42½° and 50½°, and about 30° long. w. W. C. From these elevated regions, the general course of the main branches is to the N. E., until they reach nearly the 49th degree of N. lat. Here the Mo. Proper and Yellowstone rs. unite. In either length of course or surface drained there is but little difference between these confluent rivers above their point of union. Though much less extensive than the sthrn. slope, from which fall the numerous branches of Yellowstone and Mo. Proper, there is another northern or counter slope, from which issue the rivers Marias, Brattons, Milk, Porcupine, and several smaller streams, which enter the Mo. above the influx of Yellowstone r. The entire Mo. valley above the mouth of Yellowstone r. is 600 ms. across the sources, and a mean of 300 ms. in the general direction of the streams; area 180,000 sq. ms. This higher valley of Mo. presents a surface on the western side, broken by mountains, and descending the rivers, gradually spreading into plains. The whole country, with partial exceptions along the rivers, is open prairie, exhibiting a great resemblance to the steppes of Asia, in very nearly the same latitude.

After their junction, it is probable that the united waters of the Mo. and Yellowstone form a river as large in volume and as wide and deep as at the reception of the Miss. The Mo. now a powerful volume, rolls on to the N. E. to the mouth of White Earth r. where it has reached its extreme northern bend at lat. 48° 20′. Inflecting to S. E. about 60 ms. by comparative courses it receives Little Mo. from the right. And here it may be remarked that the Moose r. a branch of Assiniboin, rises within 1 m. of the bank of Mo. Continuing S. E. 160 ms. the Mo. reaches the Mandan vs. at lat. 47° 25′. Passing the Mandan towns, this great stream inflects to a southern course, which it maintains upwards of 300 ms. by comparative courses. The structure of the country is such, that in the latter long course through 4½° of lat. the Mo. receives no remarkable tributary from the left, and from the right the comparatively small rivers, Heart, Cannon Ball, Maripa, Wetarhoo, Sarwarcarna, Chayenne and White rs. Sweeping an immense general curve to the northestwrd. and gradually round to sthrd. 300 ms. the Mo. is augmented from the wstrd. by the large river Platte, a stream deriving its sources from the same system of mountains which produced the recipient. Along the great curve above the Platte, the Mo. receives from the nthrd. Jacques, and the Great and Little Sioux rs. Receiving the Platte, the main volume rolls on S. E. 200 ms. to the influx of the Kansas, another very large confluent from the wstrd. The Kansas rises also in the Chippewayan, and flowing eastwardly, joins the Mo. after a general comparative course of upwards of 600 ms. The Platte and Kansas fill the space between the higher valley of Mo. and that of Ark. The length of course not materially different, and the character of country they drain, is mostly open plains, similar to that drained by the Mo. itself.

With the influx of Kansas, the Mo. bends to a general course of a little S. of E. 250 ms. to where its immense volume and name is lost in the inferior stream of Miss. proper. With the entrance of the Kansas, Mo. enters the state of the same name, within which it receives from the right, descending, Mino, Moreau, Osage, and Gasconade rivers, and from the left or N. Grand r., W. Chariton, E. Chariton, and a long series of streams which are merely large creeks.

The entire comparative course of Mo. is 1,870 ms., but following the bends or channel, the length no doubt exceeds 3,000 ms. The real length of this great r. as indeed of all the American rivers, has been overrated. Our knowledge of the valley is general, and except along the main stream, in few places have we exact specific material for these immense regions. As far, however, as explored, the face of the earth is monotonous when compared with extent. From much greater relative elevation, higher lat. and from the peculiar courses of its confluents, the flood of Mo. is the last in order, and occurs after the tide from the Miss. proper, Ohio, Ark., and Red rs. have in great part subsided. (*See arts. Miss. Ark. &c.*)

Missouri, state of the U. S., bounded N. E. and E. by the Miss. r. separating it from the state of Il., S. E. by the Miss. r. separating it from Ky. and Ten., S. by Ark. territory, and W. and N. by the wstrn. unappropriated domain of the U. S.

Having outlines, beginning on the right bank of the Miss., at the mouth of Les Moines r., and thence down the former stream, to where it is intersected by lat. 36°, 550 ms.; due W. and along lat. 36° to the St. Francis r., 50; thence up St. Francis r. to lat. 36½°, 50; thence due W. along the N. boundary of Ark. to a meridian line passing through the junction of Mo. and Kansas rivers intersect lat. 36½°, 200; thence due N. to a point where a line drawn due w. from the Sac village on Lemoine r. will intersect the w. boundary, 273; thence due E. to the Lemoine r., 130; down Lemoine r. to place of beginning, 20; having an entire outline of 1,273 ms. Lying betwen lat. 36° and 40° 36′ and long. w. W. C. 12° 12′, and 17° 28′. Without including the small rhomb between Miss. and St. Francis rivers, the length is 287 ms. The greatest breadth from a little distance below the mouth of Ohio to the wstrn. boundary is 300. The breadth exceeding the length may seem absurd, but the reader will observe that the half degree of lat. extended between St. Francis and Miss. rivers was excluded, and which, if added, would make the entire length 321 ms. The mean breadth 230 ms., would be very nearly represented by a line drawn due w. from Herculaneum in Jefferson co. to the wstrn. boundary.

Measured carefully by the rhomb, the area of Mo. amounts to 64,000 sq. ms. very nearly, or 40,960,000 statute acres. It is the second state of the U. S. in point of superficial extent, only falling short of Va. This state is naturally divided into two unequal slopes. Leaving the Miss. near St. Genevieve, opposite the mouth of the Kaskaskias r., a dividing ridge extends rather w. of s. w. by w. From this ridge issue and flow sthrdly. the sources of St. Francis, Black, White, and Grand r. of Arkansas r. This sthrn. slope has a breadth of about 60 ms. with the entire breadth of the state, 300 ms., or 18,000 sq. ms. As a physical section the sthrn. slope of Mo. belongs to the same inclined plane, down which flows the Miss. below the influx of Ohio r., but in extent amounts to only about the 28-100th of the whole state. The central and nthrn. sections are comprised in the lower slope of the Mo. valley, and incline very nearly due E. By a very circuitous channel, but general course of a little s. of E., the main volume of Mo. r. winds down the central plain, leaving about one third of the state to the nthrd.

The sthestrn. angle of Mo. is a level, and in a great part an annually inundated tract. This submerged section has been too highly estimated in regard to extent; it is about 100 ms. from s. to N. with a width of 40 ms., or 4,000 sq. ms. It is not all, indeed, subject to submersion, and affords parcels of dry arable land over the whole extent. The bottoms along the rivers are subject to casual flood, but taking the whole state into view, it is a hilly, and in many parts a very broken state. The ridge noticed in the first part of this article as dividing the Mo. slope from that of Ark., rises into rocky elevations, which have received the title of mtns. The idea of extended plains is given by the appearance of the landscape from the Miss. r. the usual channel of entrance. It is 28 ms. above the mouth of the O. r. before a rocky eminence shows itself on the Mo. side of the stream. The first rocks are enormous walls of limestone, evidently, if we extend our view estrdly., an extension of the vast limestone formation of Ky., Ind. and Il. They are in Mo. the buttress of the dividing ridge already noticed and extend to an undefined distance wstrd. and nthrd. Under their respective heads will be found noticed, the rivers which water Mo. It is sufficient to observe in this place that the state is washed in all its length by the Miss.; the sthrn. part drained by the heads of St. Francis, Black, and White rivers; central part is deeply cut and channelled by the Mo. and its confluents, the Osage and Gasconade from the right, and the Chariton from the left. Beside the Lemoine and Salt rs., the Miss. above the mouth of the Mo. receives from the state of Mo. a long series of crs. many of which are for a greater or less distance above their mouth navigable streams. From the preceding data it is evident, that in natural commercial facilities Mo. abounds.

Soil.—Climate.—Vegetables.—Minerals.— The soil of Mo. as indeed the face of the country, is a mean between the same objects in the valley of Ohio. Much of the bottom land along the Mo. r. and it confluents are more sandy than that on the Miss. and its tributaries, and this character of soil prevails wherever the alluvion of Mo. r. is deposited. Distant from the streams, the soil is almost invariably gravelly and poor. There are, however, some partial exceptions, and detached spots of upland are found, with a very productive soil, but they are oases. Much of the state is prairie, and the prairie soil, as in the contiguous states, and in La. present the same varieties of soil, with woodland. As the expense of clearing timber was avoided where prairies exist, settlements will be first formed on their margins, and such has been the case in La., Ark., Mo., and Il. "There are scarcely any lands in this state" (Mo.) says Flint, "sufficiently level for cultivation, that have not fertility enough to bring good crops of corn without manure, and in many instances the poorer lands are better for wheat than the richer." The very deep and rich alluvial lands are no where in the central or s. wstrn. states, suitable for wheat, until cultivated several years. Cotton in small quantities can be cultivated on the s.E. section of the state, Indian corn, wheat, rye, oats, &c., are however the staple crops of the state. Apples, peaches, pears, plums, and perhaps some kinds of grape, succeed well. Natural grasses abound, and yet from some cause meadows have not been cultivated to advantage in Mo. The abundance indeed of any natural production is inimical to the artificial culture of analagous species. In fine, it would be safe to say that at least 20,000,000 acres of farming land, sufficiently fertile to produce good crops, exist in Mo. The mineral wealth of the state, particularly lead and iron, is, according to all concurrent testimony, inexhaustible. The tract in an especial manner called "The Mineral Tract," in Madison, Washington, and St. Francis cos., and from which rise the sources of Maramec, and St. Francis rs., is represented as not only abundant in lead, but still more so in iron ore. Mr. Schoolcraft mentions zinc as also amongst the productions of Mo., and in great quantities. Water impregnated with Muriate of soda (common salt) is found in several places. Plaster of Paris is plentiful, and, it is said, produces a more than common effect on the vegetation where used as a manure. Such are, in a rough sketch, the outlines of the resources of this new and extensive state.

The native vegetables, and particularly forest timber, evince an approach towards the prairie region. The peccan hickory is plentiful; wild grapes and plums are plentiful. The crab apple tree, which in La. grows to the height of 30 or 40 feet, is also of large growth in Mo.

The climate is here, as elsewhere in central N. America, the great stumbling block of travellers and geographers. "This state," says Flint, "occupies a medial position and has a temperature intermediate between that of N. Y. and La." Whatever may be the resemblance in the face of Mo. and La., there is but little similarity in their respective seasons. La. is in winter a very cold country, when compared with its lat., but it is tropical when contrasted with Mo. At St. Louis, which may be regarded as a central point between the northern and southern extremes of Mo., the Miss. r. is frozen and passable on the ice by the first of January, in a great majority of years. In the winter of 1831–2, the Miss. was frozen and passable on the ice at Memphis in Ten., nearly a degree of lat. s. of any part of Mo. In article U. S., it may be seen that at the Council Bluffs, lat. 21° 25′, about a degree of lat. N. of Missouri, the mercury has fallen to 21° minus zero; and what is more decisive, by a letter directed to the editor of the Saturday Evening Post, Philadelphia, on the 26th Jan. 1832, the mercury was 18° minus zero at Florence in Ala., lat. 34° 47′, or 1° 43′ s. of any part of Mo. It may be safely stated, that of all sections of the actually inhabited parts of the U. States, no other is so exposed to excessive vicissitudes of atmospheric temperature as is Mo. Open on the westward and northwestern sides to the great plains of grass, with winds prevailing about ⅔ths of the time in all seasons of the year, the cold of the vast central table land of the continent is borne towards the Appalachian system of mountains, and sweeps over Mo., with a severity which, to be known, must be felt. The most accurate observers have acknowledged the dryness of the atmosphere over all the prairie regions of central North America; and Mo. shares the exemption from moisture. "The winter," says Flint, "commences about Christmas, (a month sooner would be nearer the fact,) and is frequently so severe, as to bridge the mighty current of the Mo. so firmly that it may be passed many weeks with loaded teams. In the winter of 1818, this was the case for nine weeks." This author again, after some general observations, comes at last to the rational conclusion and acknowledges that, "on the whole, instead of the climate becoming more mild, as we advance W. on the same parallel, it is believed that the reverse is the case." The reader will find in the article U. S. that the increasing severity of cold, advancing towards the Chippewayan mountains, is not simply believed but demonstrated.

It is worse than idle to speak of the health of such a widely spread and diversified surface as that of Mo. A country containing fens constantly filled with stagnant water, as low as lat. 36°, and high, dry, and airy tracts above lat. 40°, where the human breast is inflated by air coming from regions exempt, as far as the face of earth can any where be exempt, from every source of *miasmata*.

Political Geography.—For political purposes the following counties have been organised out of the territory of Mo., leaving considerable tracts not yet laid out.

Counties.	Pop. 1830.	Counties.	Pop. 1830.
Boon,	8,859	Marion,	4,837
Callaway,	6,159	Montgomery,	3,902
Cape Girardeau,	7,445	New Madrid,	2,350
Chariton,	1,780	Perry,	3,349
Clay,	5,338	Pike,	6,129
Cole,	3,023	Ralls,	4,375
Cooper,	6,904	Randolph,	2,942
Crawford,	1,721	Ray,	2,657
Franklin,	3,484	St. Charles,	4,320
Gasconade,	1,545	St. Francois,	2,366
Howard,	10,854	St. Genevieve,	2,186
Jackson,	2,823	St. Lewis,	14,125
Jefferson,	2,592	Saline,	2,873
La Fayette,	2,912	Scott,	2,136
Lincoln,	4,059	Washington,	6,784
Madison,	2,371	Wayne,	3,264

Total population 140,455, of which were white persons,

	Males.	Females.
Under 5 years of age,	12,531	12,561
From 5 to 10	9,617	9,077
" 10 to 15	7,469	6,794
" 15 to 20	5,639	5,765
" 20 to 30	11,147	8,794
" 30 to 40	7,064	5,121
" 40 to 50	3,642	2,718
" 50 to 60	1,939	1,499
" 60 to 70	927	766
" 70 to 80	334	227
" 80 to 90	60	60
" 90 to 100	14	9
100 & upwards	2	2
Total	61,405	53,390

Of which 12 persons are deaf and dumb under 14 years of age, 5 between 14 and 25 years, and 10 of 25 years and upwards. Blind 27. Of the colored population were

	Free.		Slaves.	
	Males.	Fem's.	Males.	Fem's.
Under 10 years of age	87	77	4,872	4,611
From 10 to 24	76	62	4,364	4,605
" 24 to 36	43	46	2,058	2,199
" 36 to 55	57	63	923	1,014
" 55 to 100	18	34	208	219
100 and over	3	3	14	4
Total,	284	285	12,439	12,652

Of the colored pop. none are either deaf and dumb, or blind.

Recapitulation.

Whites.	Free colored.	Slaves.	Total.
114,795	569	25,091	140,455

Constitution.—Government—Judiciary.—The constitution of Mo. was adopted in convention at St. Louis, the 25th June, 1820; the most important provisions provide, that:

Art. 2. The powers of the government shall be divided into three distinct departments; each of which shall be confided to a separate magistracy; and no person charged with the exercise of powers properly belonging to one of those departments, shall exercise any power properly belonging to either of the others, except in the instances hereinafter expressly directed or permitted.

Art. 3.—Sec. 1. The legislative powers shall be vested in a "general assembly," which shall consist of a "senate," and a "house of representatives." Sec. 2.—The house of representatives shall consist of members to be chosen every 2nd year, by the qualified electors of the several counties. Sec. 3.—No person shall be a member of the house of representatives, who shall not have attained to the age of twenty-four years; who shall not be a free white male citizen of the U. States; who shall not have been an inhabitant of the state two years, and of the county which he represents one year next before his election. Sec. 5.—The senators shall be chosen by the qualified electors, for the term of 4 years. No person shall be a senator who shall not have attained to the age of thirty years; who shall not be a free white male citizen of the U. S.; who shall not have been an inhabitant of this state 4 years, and of the district which he may be chosen to represent one year next before his election. Sec. 13.—No person, while he continues to exercise the functions of a bishop, priest, clergyman, or teacher of any religious persuasion, denomination, society, or sect, whatsoever, shall be eligible to either house of the general assembly; nor shall he be appointed to any office of profit within the state, the office of justice of the peace excepted. Sec. 16.—No senator or representative shall, during the term for which he shall have been elected, be appointed to any civil office under this state, which shall have been created, or the emoluments of which, shall have been increased during his continuance in office, except to such offices as shall be filled by elections of the people.

Art. 4. Sec. 1.—The supreme executive power shall be vested in a chief magistrate, who shall be styled "the governor of the state of Mo." Sec. 2.—The governor shall be at least 35 years of age, and a natural born citizen of the U. S.; or a citizen at the adoption of the constitution of the U. S.; or an inhabitant of that part of La. now included in the state of Mo., at the time of the cession thereof from France to the U. S.; and shall have been a resident of the same at least 4 years next before his election. Sec. 3.—The governor shall hold his office 4 years, and until a successor shall be duly appointed and qualified. He shall be elected in the manner following. At the time and place of voting for members of the house of representatives, the qualified electors shall vote for a governor, and when 2 or more persons shall have an equal number of votes, and a higher number than any (*other*) person, the election shall be decided between them by a joint vote of both houses of the general assembly, at their next session. Sec. 4.—The governor shall be ineligible for the next 4 years after the expiration of his term of service. Sec. 14.—There shall be a lieutenant governor, who shall be elected at the same time, in the same manner, for the same term, and shall possess the same qualifications as the governor. Sec. 15.—The lieutenant governor, shall, by virtue of his office, be president of the senate. In committee of the whole he may debate on all questions; and when there is an equal division, he shall give the casting vote in senate, and also in joint votes of both houses.

Art. 5. Sec. 1.—The judicial powers, as to matters of law and equity, shall be vested in a "supreme court," in a "chancellor," in "circuit courts," and in such inferior tribunals as the general assembly may, from time to time, ordain and establish. Sec. 3.—The supreme court shall have a general superintending power and control over all inferior courts of law. It shall have power to issue writs of *habeas corpus, mandamus, quo warranto, certiorari* and other original remedial writs; and to hear and determine the same.

Except, however, in specified cases provided for in the constitution, the supreme court of Mo. has only appellate jurisdiction.

Right of Suffrage.—This primary right, is, by the10th sec. of the 3rd article, secured to "every free white male citizen of the U. S. who shall have attained to the age of 21 years, and who shall have resided in the state one year before an election, the last 3 months whereof, shall have been in the county or district, in which he offers to vote, shall be deemed a qualified elector, of all elective offices; provided, that no soldier, seaman or marine, in the regular army or navy of the U. S, shall be entitled to vote at any election in this state." The principles set forth in the declaration of rights, general provisions for offices, civil and military, and their duties, powers, and term of office, do not materially differ from other constitutions of, the states of the U. S. Slavery of the blacks is admitted, but the power of the master is placed under control of the legislature, and on trials for capital offences, trial by jury secured to the slave, and no other punishment permitted except what would be inflicted on a free white person in like case; and the courts are required to provide counsel to manage the defence of slaves under a criminal prosecution.

History.—This country was amongst the original discoveries of the French from Canada, who reached the Miss. about 1674. The first civilized settlements made, however, by the French on that great river, were in Il., and St. Louis was not founded until after the treaty of Paris, in 1763. St. Genevieve preceded St. Louis, and was founded by a mining company, styled "Pierre Claude, Maxan and Co." St. Louis was established in 1764, and in 1780 St. Charles, on Mo. The settlements and towns remained feeble and scattered, until after the cession of La. to the U. S. In 1804, the unwieldy La. was divided, and the territory of Mo. created. Emigration, though not very rapid, carried the pop. in 1819 to the constitutional amount to entitle the people to state government. Application was accordingly made to congress at the session of 1819—20, and after a stormy and protracted debate, turning principally on the admission or rejection of slavery, permission was given to the people of Mo. to form a constitution, admitting slavery under certain restrictions. Complying with the conditions, a constitution was formed as already noticed, and on the 10th of Aug. 1821, Mo. became a state of the U. S.

MITCHELL'S Mill, and p-o. Shelby co. Ky., by p-r. 31 ms. wstrd. Frankfort.

MITCHELL'S Store, and p-o. Goochland co. Va. by p-r. 153 ms. s. s. w. W. C., and 50 ms. N. w. by w. Richmond.

MOBILE BAY, in Ala. The estuary of the same name opens from the gulf of Mexico, between Mobile point and Dauphin isl., at lat. 30° 12', long. 11° 10' w. W. C. The following directions to enter Mobile bay, will also serve to aid in giving its geographical features. In running in for the land in the bay of Mobile, should you make it to the westward of the bar, the land will appear broken, as it consists of small islands; if to the eastward, the land is uniform as far as Pensacola E., and covered with timber; the beech is generally sandy, and quite perceptible in clear weather 8 or 10 ms. distant. Dauphin isl. on the west point of the bay, appears high and bluff—Mobile point low, sandy, with a single tree on the extremity, in the form of an umbrella, and thinly wooded for five miles from the point. There are houses on the point, and on Dauphin isl. Before shoaling into 7 fathoms water, bring Mobile point to bear N. ½ w., and the estrd. of Dauphin isl. to bear N. N. w. ½ w., and steer in N. N. w. This course will run you over the bar, on which you will have from 16 to 20 ft. water in good tides. After passing a small *burth* isl. on your larboard, you are over the bar and out of danger, with a shoal on each side of you. Then haul up for the point of Mobile, giving it a *burth* of 3 or 400 yards, and steer up the bay. It is necessary to calculate for the bay currents, as, when the tide is flowing, you will drift to the watrd., and when at ebb to the estrd., until you get near Dog r. bar, which extends across the bay. When in 11 feet water, and 2½ ms. from the watrn. shore, 7 ms. from Mobile, and 15 from Blakeley, come to for a pilot.

The bay is in form of a triangle, of about 32 ms. base from Dauphin isl. to Mobile harbor, the apex formed by the minor bay of Bon Secours, stretching N. E. by E. from Mobile point into the high angle; and between Mobile and Blakeley, the bay is terminated by the different mouths of Mobile r.

On the outer bar, there is, as we have shown, 16 feet water; but on Dog r. bar, 7 ms. below Mobile harbor, 11 feet only can be safely calculated on. Beside the principal entrance between Mobile point and Dauphin isl., there is another inner passage by the pass of Heron. The latter is the strait between Dauphin isl. and the continent, opening from the s. w. angle of Mobile bay into Pascagoula sound. In the pass of Heron there is at mid tide 6 feet water. It is by this passage that steamboats and small sail vessels are navigated between Blakeley and Mobile, to New Orleans, reaching the latter by the rigolets, lake Pontchartrain, and Bayou St. John's. Anchorage in mud, sand, and shells, can be had in any place in this interior chain of lakes, straits, and sounds.

MOBILE BASIN. Under this head is included a very important physical section, comprising 37,120 sq. ms., and drained by the various constituents of Mobile r. Each of those constituents will be formed, noticed and described, under their proper heads, but we here insert a general view of the whole basin. This basin occupies the space between that of Ten. N., Chattahooche E., Cunecut s. E., the Gulf of Mexico s., Pascagoula r. basin s. w., and the sources of the Pearl and Yazoo N. w.

This fine agricultural and navigable basin, lies between lat. 30° 12' and 35° 05', and between long. 7° and 12° w. W. C. It is in a near approach to a triangle, base 400 ms. from the pass of Heron, to the extreme nrthestrn. sources of the Coosa r. in the nrthwstrn. part of Geo., in a direction very nearly from s. w. to N. E. The greatest breadth 230 ms., from the eastern sources of Tallapoosa, to the north watrn. of Tombigbee. The area of the basin measured by the rhombs, 37,120 sq. ms. Though the course of Tombigbee and Mobile declines a few degrees estrd. of s., the general declivity of the basin is about s. s. w. The difference of relative height from the sources of Mobile r. to Mobile bar, cannot fall much, if any, short of 2,000 feet, or an equivalent to 5 degrees of lat. Uniting the actual difference of lat. between the extremes to the allowance for relative elevation, the winter climate must differ about equal to 10° of lat. If the soil suited the growth of that vegetable, sugar might be cultivated near Mobile bay, whilst the nrthrn. part of Geo. has a climate suitable to wheat, rye, &c. Of cultivated vegetables in the U. S., Indian corn, cotton, tobacco, and the peach tree, seem most congenial to this region, though the apple flourishes in the northern, and the fig in the southern extreme.

The soil is extremely variable. Along the streams, are tracts of very productive alluvion, and bordering on the alluvion, extensive bodies of second rate soil; but if taken as a whole, a large proportion of the surface is sterile.

Mobile, co. Ala., bounded N. by Washington, N. E. by Tombigbee r., separating it from Clarke, E. by Mobile r. separating it from Baldwin, S. E. by Mobile bay, S. by the pass of Heron and Pascagoula sound, S. W. by Jackson co. state of Miss., and N. W. by Greene co. state of Miss. The extreme length of Dauphin isl. is 90 ms., mean breadth 25, and area 2,250 sq. ms.; extending in lat. from 30° 12′ to 31° 30′, and in long. from 11° 04′ to 11° 34′. The dividing line of the sources of crs. flowing wstrd. into the basin of Pascagoula, and those flowing estrd. into that of Mobile, divides Mobile co. into two nearly equal portions. Surface towards the Gulf of Mexico waving, but becomes rather hilly in the nrthrn. part; with but partial exception, the whole superfices covered with pine forest, and soil sterile. Chief town, Mobile. Pop. 1820, 2672; 1830, 3,073.

Mobile, city, port of entry, and st. jus. for Mobile co. Ala., is situated on the right bank of Mobile r., near the head of the bay of the same name. Lat. 30° 44′, long. 11° 12′ w. W. C. The harbor admits vessels of 8 ft., but to reach the anchorage, or wharves, with such vessels, it is necessary to pass round a small isl. in front of the town, which compels ships of more than 3 feet draught, to be navigated round the head of the isl. 5 ms. above. By this circuitous entrance, however, all vessels which can pass Dog r. bar, can reach Mobile.

Mobile, r. Locally, this name only applies to the stream, or streams, below tne junction of Alabama and Tombigbee rs., to the head of Mobile bay. The entire water of Alabama does not mingle with the Tombigbee, the former having an outlet above their junction, which outlet, flowing sthrd., joins another and larger, which leaves the united streams about 10 ms. below their confluence. The two outlets, united, form the Tensaw, or eastern Mobile, which flowing sthrd. passes Blakeley, and is lost in the northeastern angle of Mobile bay. The wstrn., the proper Mobile, and main stream, flows along the wstrn. bluffs at a distance of from 3 to 4 or 5 ms. from Tensaw, passes the town of Mobile, and is terminated in the bay about 2 ms. below Mobile harbor. Vessels which can pass Dog r. bar, can reach to either Mobile or Blakeley, and those drawing from 5 to 6 feet can be navigated into either the Tombigbee or Alabama, and up the former to St. Stephen's, and the latter to Claiborne. The junction of Alabama and Tombigbee is at lat. 31° 06′ and long. 11° 05′ w. W. C.

Mockville, p-v. in the northern part of Rowan co. N. C. by p-r. 141 ms. westward Raleigh.

Moffitt's Mills, and p-o. Randolph co. N. C. by p-r. 70 ms. w. Raleigh.

Mohawk, r. Coos co. N. H. rises in Dixville mountains, and enters Conn. r. in Colebrook.

Mohawk, r. N. Y. about 135 ms. long, the principal branch of Hudson r., rises in Oneida co. near the source of Black r., runs 20 ms. s. to Rome, thence E. by s. to Hudson r. at Waterford, between Albany and Saratoga cos. It has many rapids, and falls a little at German Flats, 42 feet at Little Falls, and nearly 70 at the Cahoos. The banks are very level and fertile in some places, particularly at Herkimer, and poor or rocky in others. The navigation for boats was formed some years ago by a canal round the falls, and one from Rome to Wood creek and Oswego river. The Erie canal now passes along its course, (chiefly on the s. bank) to Rome. (*See Erie Canal.*)

Mohegan, Indian village, Conn. w. Thames r., 4 ms. s. of Norwich in the t. of Montville, on a reservation of land for the Mohegan tribe, now reduced to a small number. A church was built here in 1831, on the site of Unca's fort. The government of the U. S. have appropriated $900 for their benefit, and exertions have been recently made, for their instruction, by benevolent individuals. The ancestors of these Indians were faithful friends of the colonists, and assisted them in their wars.

Moira, t. Franklin co. N. Y. Pop. 1830, 791.

Monadnock mountain, (commonly called Grand Monadnock) Cheshire co. N. H., 22 ms. E. Conn. r., 10 ms. N. Mass., is a high ridge, N. E. and S. W., 5 ms. long and 3 wide. The base is said to be 1,452 feet above tide, the top 3,250. The rocks are talc and mica slate, stratified, and sometimes contain schorl, garnets, quartz and feldspar. Plumbago, or black lead, is found on the E. side, and made into crucibles and indifferent pencils. Monadnock mineral spring is near the base, and the top commands a fine view.

Monamet Point, cape, Mass., in Cape Cod bay. Long. 6° 35′ E. W. C., lat. 41° 45′.

Monguago, p-v. in the southeastern part of Wayne co. Mich. 14 ms. s. s. w. Detroit, and by p-r. 512 ms. nrthwstrd. W. C.

Moniteau, p-v. Cole co. Mo. 5 ms. wstrd. Jefferson city, and 139 ms. wstrd. St. Louis.

Monkton, p-t. Addison co. Vt. 18 ms. s. Burlington, and 27 w. Montpelier. Little Otter creek, Pond brook and Lewis creek, tho' small, are the principal streams, and afford few mill sites. It contains a considerable pond, Hogback mtn. and others. Iron ore is very abundant in the s., principally hematite, as well as black oxyde of manganese, and a large bed of porcelain clay. There is also a curious cavern. Pop. 1830, 1,348.

Monmouth, p-t. Kennebec co. Me. 17 ms. w. Augusta, contains an academy. It is N. of Lincoln co. and has small streams flowing into Kennebec r. Pop. 1830, 1,879.

Monmouth, co. N. J. bounded by Middlesex co. and Raritan bay N., Atlantic ocean E., Burlington co. s. and w. Contains much poor pine land, with a scattered population, and only a few villages in the N. The coast is low and sandy, and the scene of frequent shipwrecks. Marl, which is found in different places, is an excellent and lasting manure for the poorest soil. Much pine wood is sent to N. Y. It has 7 large townships, but a great deal of poor sandy soil, with invaluable beds of marl, which makes the richest manure. Sandy Hook is the s. cape of Raritan bay, by which is the communication between the ocean and N. York bay, and the N. E. extremity of this county. Shrewsbury and Navesink rs. (short but broad streams) enter Raritan bay just within the Hook, which they have sometimes isolated by cutting thro' the neck into the sea. At Shrewsbury and Howel, the coast is a sand bank, about 30 ft. with a beautiful white beach, having arable land to the bluff. Below, Barnegat and Little Egg Harbor bays are formed by Squam Isle and Long Beaches, with 2 inlets to the ocean, in this co. Many small streams flow into them, and others rise in the co. which fall into Raritan and Delaware rs. The principal town is Freehold. In the Pines are furnaces for iron, &c. Pop. 1830, 29,233.

Monongahela, r. of the U. S. in Va., Md. and Pa., is formed by Monongahela proper, Tygart's valley r., Cheat r., and the Youghioghany. The Cheat is in fact the main stream, having its remote source in the sthrn. part of Randolph co. Va., at lat. 38° 27′, interlocking sources with those of Green r. and Jackson's branch of James r. The remote sources of Tygart's valley r. are nearly as far s. as those of Cheat, and also in Randolph co. The mountain ridge from which both streams rise is known locally as Green Brier mountain, and the valleys from which the higher sources are derived, must be at least 2,500 feet elevated above tide water in Chesapeake bay.

Monongahela proper is the western branch, rising in Lewis co. Va. with interlocking sources with those of Tygart's valley and little Kenhawa. The three branches near their sources pursue a general northern course, but the two western gradually approach each other, and unite at lat. 39° 28′, where they form a point of separation between Harrison and Monongahela cos. Thence assuming a northern course over the latter county, finally leave Va., and form a junction with Cheat on the boundary between Fayette and Green cos. Pa.

The Cheat in the highest part of its course flows along a mtn. valley in a nrthrn. direction, but gradually inclining to nthwstrd., as already noticed under the head of Cheat r. Below the junction of the main branches, the Monongahela, by a rather circuitous channel, pursues a general nthrn. course over Pa. about 50 ms. comparative length to its junction with Youghioghany, 11 ms. s. E. of Pittsburg.

The Youghioghany is a considerable branch, having its remote sources in the wstrn. part of Alleghany co. Md. Flowing thence nrthrdly. enters Pa., and separating for some few ms. Somerset, from Fayette co., receives a large tributary from the estrd. Casselman's r. and turning to N. N. W. about 50 ms., comparative course, is lost in the Monongahela at MacKeesport. Augmented by the Youghioghany, the Monongahela below the junction assumes the course of the former, 18 ms. by the channel, but only 11 direct distance to Pittsburg, where it unites with the Alleghany to form the Ohio. The general course of the Monongahela is almost exactly N., and almost as exactly along long. 3° w. W. C., 150 ms. by comparative distance. The widest part of its valley lies nearly along the line between Pa. and Va. 80 ms.; the mean width 40, and area 6,000 sq. ms.

If we allow only 1,500 feet elevation to the cultivatable country on the head branches of Cheat, Pittsburg being elevated 678 feet, will give a descent of 822 feet to the valley of Monongahela. The extremes of lat. are thus almost exactly compensated by declivity, and explain why the seasons near Pittsburg and in Randolph co. Va. differ but slightly.

Though the two estrn. branches, Cheat and Youghioghany, rise in mountain vallies, and the whole country drained by all the confluents of Monongahela is very broken, and rocky, direct falls are rare and of no great elevation when they occur. Cheat r. is navigable through Monongahela and Preston, into Randolph co., both branches of Monongahela proper above their junction, and Youghioghany to Ohio pile falls. The whole valley has gained recent increase of importance as being part of the route or routes of proposed lines of canal improvement.

Monongalia, co. Va. bounded E. and S. E. by Preston, S. W. by Tygart's valley river and Buffalo creek, separating it from Harrison, W. by Tyler co., N. W. by Green co. Pa., and N. E. by Fayette co. Pa. Extending in lat. from 39° 17′ to 39° 42′, and in long. from 2° 39′ to 3° 25′ w. W. C. This county declines to the nrthrd. and is traversed by both branches of Monongahela, Cheat to the E., and Monongahela proper to the W. Its length from W. to E. is 38 ms., mean width 15, and area 570 sq. ms. Though very broken, the soil is excellent. Pop. 1820, 11,060, 1830, 14,056.

Monroe, p-t. Waldo co. Me. s. Penobscot co., crossed by Marsh r. a branch of the Penobscot. Pop. 1830, 409.

Monroe, p-t. Fairfield co. Conn. on Housatonic r. 20 ms. W. New Haven.

Monroe, p-t. N. Y. (*See Munroe.*)

Monroe, co. N. Y. 236 ms. W. N. W. Albany, bounded by lake Ontario and Upper Canada N., Ontario co. E., Livingston S., Genesee W.; 20 by 30 ms.; area 600 sq. ms. Contains 16 townships, has a gently varied surface, rich soil and mild climate. It is crossed by Genesee r., the Erie canal, and the Mountain Ridge, which was probably once the shore of lake Ontario, and extends from near York, Upper Canada, to Jefferson co. Bog iron ore, salt springs, and free stone are found in some parts. Rochester is the co. t. Pop. 1830, 49,682.

Monroe, p-o. Bucks co. Pa.

MONROE, co. Va. bounded by Giles s. and w., Greenbrier N., Alleghany N. E., Botetourt E. Length 40, mean width 18, and area 720 sq. ms. Extending in lat. from 37° 22′ to 37° 45′, and in long. from 3° 16′ to 3° 54′ w. W. C. The base of this co. may be regarded as New r. which bounds it on the w., but the general declivity is wstrd. from the Alleghany mtn. The northwestern part is traversed by Greenbrier r., which falls into New r. at the point where meet the angles of Giles, Logan, Greenbrier and Monroe. By actual measurement, the mouth of Greenbrier river is 1,333 feet above the oceanic level, and of course the surface of Monroe co. must be still higher, say from 1,400 to 1,700, or 1,800 feet. Chief t. Union Town. Pop. 1820, 6,620, 1830, 7,798.

MONROE, p-v. Warren co. N. C. by p-r. 70 ms. N. E. Raleigh.

MONROE, co. Geo. bounded by Bibb s. E., Crawford s., Upson w., Butts N., and Oakmulgee r. separating it from Jones N. E. Length diagonally from s. w. to N. E. 30 ms., mean width 12, and area 360 sq. ms. Lat. 33°, and long. 7° w. W. C., intersect very near the centre of this co. Declivity to the s. E., and drained by Chupee and other creeks, flowing into Oakmulgee r. Chief t. Forsyth. Pop. 1830, 16,202.

MONROE, p-v. and st. jus. Walton co. Geo. by p-r. 66 ms. N. N. W. Milledgeville.

MONROE, co. Ala. bounded by Baldwin s. w., Cunecut s. E., Butler N. E., Wilcox N. and the Ala. r. separating it from Clarke w. The greatest length from s. w. to N. 48 ms., mean width 20, and area 960 sq. ms. Extending in lat. from 31° 14′ to 31° 48′, and in long. from 10° 04′ to 10° 50′ w. W. C. Declivity wstrd. towards the Ala. r. Surface generally sterile. Pine wooded land. The banks of the Ala. afford some excellent soil. Staple, cotton. Chief town, Claiborne. Pop. 1820, 8,838, 1830, 8,782.

MONROE, p-v. s. E. part of Perry co. Miss. by p-r. 151 ms. s. E. Jackson, and by the direct road 158 ms. s. E. by E. Natchez.

MONROE, co. Miss. bounded by Lowndes co. same state s., by Tombigbee r. separating it from the Chickasaw ter. w., the Chickasaw ter. again on the N., by Marion co. Ala. N. E., and Lafayette co. Ala. s. E. Length 25 ms., mean breadth 15, and area 375 sq. ms. Central lat. 33° 50′, long. 11° 30′ w. W. C. Slopes s. w., and traversed by Battahatchee and Weaver rs., with some smaller streams. Chief t. Hamilton. Pop. 1830, 3,861.

MONROE, p-v. and st. jus., parish of Washitaw, La., situated on the left bank of Washitaw r., about 80 ms. in a direct line N. N. W. Natchez, and 100 a little E. of N. Alexandria at Rapides. Lat. 32° 32′, long. 15° 10′ w. W. C.

MONROE, co. Ten. bounded by the Cherokee country s. E. and s., McMinn co. w., Roan N., and Ten. r. separating it from Blount N. E. and E. Length 30 ms., mean width 15, and area 450 sq. ms. Extending in lat. from 35° 18′ to 35° 48′, and in long. from 6° 57′ to 7° 33′ w. W. C. The boundary line between McMinn and Monroe cos. follows very nearly the ridge dividing the sources of the creeks flowing s. w. into the Hiwassee, from those flowing in an opposite direction into Tenn.; the declivity therefore of Monroe co. is northestrd. Chief t. Tellico. Pop. 1820, 2,539, 1830, 13,708.

MONROE, p-v. and st. jus. Overton co. Ten. situated on a branch of Obies r. 100 ms. a little N. of E. Nashville, and about 35 a little E. of s. Burkesville in Ky. Lat. 36° 22′, long. 8° 10′ w. W. C.

MONROE, co. Ky. bounded w. by Big Barren r. separating it from Allén, N. by Barren, N. E. by Adair, E. by Cumberland, s. by Jackson co. Ten., and s. w. by Smith co. Ten. Length from E. to w. 30 ms., mean width 20, and area 600 sq. ms. Extending in lat. from 36° 36′ to 36° 53′, and in long. from 8° 19′ to 9° w. W. C. This co. is very nearly commensurate with the higher part of the valley of Big Barren r., but with the exception of the southeastern angle. Into the latter part of the co. the main volume of Cumberland r. enters by one of its sweeping bends, and again abruptly winds back into Cumberland co. The dividing ground between the waters of Cumberland and Green rs. passing from Adair over Monroe into Jackson co. Ten., divides Monroe into two unequal portions. The much larger section, with a N. western declivity, is in the valley of Green r. or sub-valley of Big Barren. The chief t. Tomkinsville, is by p-r. 137 ms. s. s. w. from Frankfort. Pop. 1820, 4,956, 1830, 5,340.

MONROE, p-v. in the southeastern part of Hart co. Ky. by p-r. 96 ms. s. s. w. Frankfort, and 20 N. N. E. Glasgow.

MONROE, co. O. bounded by Washington s., Morgan w., Guernsey N. w., Belmont N., and the O. r. separating it from Ohio co. Va. E. Length from E. to w. 36 ms., mean breadth 16, and area 576 sq. ms. Lat. 39° 40′, long. 4° w. W. C. The central part of this hilly but fertile co. is a real table land, from which Sunfish cr. flows estrd. into Ohio r. Little Muskingum sthrd. also into O. r., but by a s. w. course over Washington co., and the extreme head sources of Will's creek, branch of Muskingum, N. w. into Guernsey co. Chief town, Woodsfield. Pop. 1820, 4,641, 1830, 8,768.

MONROE, p-v. in the eastern part of Butler co. O. by p-r. 25 ms. N. N. E. Cincinnati. Pop. 1830, 119.

MONROE, co. Ind. bounded by Lawrence s., Greene s. w., Owen N. w., Morgan N., and E. uncertain. Length 24 ms., breadth 20, and area 480 sq. ms. Lat. 39° 10′, and long. 9° 38′ w. W. C. The nthrn. part slopes wstrd., and is drained by Bean Blossom creek, a branch of White r., and the southern section slopes to the sthrd., and is drained by Salt creek, a branch of the South fork of White r. Chief t. Bloomington. Pop. 1830, 6,577.

MONROE, co. Il. bounded N. E. and E. by St.

Clair, s. e. by Randolph, s. w. and w. by Miss. r. separating it from Jefferson co. Mo., and n. w. by the Miss. r. separating it from St. Louis co. Mo. Length 30 ms., mean width 12, and area 360 sq. ms. Lat. 38° 15′, long. 13° 12′ w. W. C. This co. stretching estrd. from the Miss. r. to the Kaskaskias r., slopes towards both, the central part being a table land. Chief town, Waterloo. Pop. 1830, 2,000.

Monroe, co. Mo.; situation uncertain.

Monroe, C. H. and p-o. Monroe co. Mo. by p-r. 129 ms. from Jefferson city.

Monroe, co. Mich. bounded by Sandusky co. O. s. e., Wood co. O. s., Lenawee county, Mich. w., Washtenaw co. Mich. n. w., Wayne n. e., and lake Erie e. Length from s. to n. 32 ms., mean width 22, and area 704 sq. ms. Lat. 42°, and long. 6½° w. W. C. intersect in the northeastern part of this co. Slope estrd. and traversed by the river Raisin and Ottawa creek, with other smaller streams. Much of the soil is excellent. The southestrn. part receives Maumee r. from Wood co. O. Chief t. Monroe. Pop. 1820, 1,831, 1830, 3,187.

Monroe, p-v. and st. jus. Monroe co. Mich. by p-r. 36 ms. s. s. w. Detroit, and 490 northwstrd. W. C. It is situated on the right bank of the river Raisin, near its mouth. Vessels of 5 or 6 feet draught can ascend thus far.

Monroeton, p-v. Bradford co. Pa. 126 ms. nthrd. Harrisburg.

Monroeton, p-v. on the right bank of Staunton r., and in the extreme northwestern angle of Pittsylvania co. Va., by direct road about 130 ms., but by p-r. 150 ms. s. w. by w. Richmond.

Monroeville, p-v. Huron co. O. by p-r. 109 ms. nthrd. Columbus.

Monson, p-t. Hampden co. Mass. 17 ms. e. Springfield. 50 s. w. Boston, s. Chickapee r. The soil is good, and in some parts the land is irrigated. It is a pleasant t. and has several manufactories, and an academy with about 100 pupils, a valuable apparatus, and a boarding house connected with the institution. Pop. 1830, 2,263.

Montague, p-t. Franklin co. Mass. e. side Conn. r., 87 ms. n. w. Boston, s. and e. Conn. r., which falls 65 feet, among rude scenery. A little below is an old Indian fort, which was attacked with great slaughter, in Philip's war, by captain Holyoke of Northampton. A rocky island divides the fall, which is dammed 330 yards, and passed by a canal 3 ms. long, 25 feet wide, with 8 locks 75 feet long, 12 deep, and 20 wide. The dam is of timber, and in one place 40 feet high. It was torn down 2 or 3 years ago by a violent flood. 4 ms. above is the dam at Miller's falls, where is a canal cut through a pudding stone of primitive rock. A bridge crosses to Deerfield. Pop. 1830, 1,152.

Montague, p-t. Sussex co. N. J., the most northerly town in the state, with Delaware r. w., the Blue mtns. e., N. Y. state n. e., Pa. n. w., and is connected with it by a bridge over Del. r. Its small streams flow in several directions. Pop. 1830, 990.

Montague, p-v. sthrn. part of Essex co. Va. 72 ms. s. e. by e. Richmond.

Montalban, p-v. in the sthrn. part of Warren co. Miss. by p-r. 81 ms. wstrd. Jackson, and about 60 ms. n. n. e. Natchez.

Montauk Point, Easthampton, Suffolk co. N. Y., the e. end of L. I. The light house is on the extreme point, on an elevation, and is a very important land mark, particularly to vessels bound into L. I. sound. It was erected in 1795. It commands a clear view of Block isl. and the opposite shores of Conn. The road leading to Montauk Point is rough; but no troublesome insects are found there; and there is a convenient tavern on the spot. It is 20 miles from East Hampton. There are a few Indians remaining, but many of them of mixed blood. The soil is rich, and affords pasturage to numerous oxen, horses and sheep; but Napeage beach, 5 ms. in extent, is a sandy tract. The distance from Sandy Hook, in a direct line, is 140 ms.

Montebello, p-v. Hancock co. Il. by p-r. 144 ms. n. w. Vandalia.

Montevallo, p-v. Shelby co. Ala. about 45 ms. nearly due e. Tuscaloosa.

Montezuma, p-v. Mentz, Cayuga co. N. Y., 11 ms. n. w. Auburn, 170 w. Albany; has some salt springs, and is 80 rods from the junction of Erie canal and Seneca r.

Montezuma, p-v. and st. jus. Covington co. Ala. situated on the Connecuh r. by p-r. 176 ms. southestrd. Tuscaloosa. Lat. 31° 22′, long. 9° 40′ w. W. C.

Montezuma Salt Works and p-o. in the estrn. part of Casey co. Ky. by p-r. 81 ms. very nearly due s. Frankfort.

Montezuma, p-v. Parke co. Ind. by p-r. 77 ms. w. Indianopolis.

Montgomery, p-t. Franklin co. Vt. 42 ms. n. Montpelier, 39 n. e. Burlington. Trout r. and its branches afford mill seats. On the principal stream is excellent meadow land; but there is much mountainous country. Pop. 1830, 460.

Montgomery, t. Hampden co. Mass. 10 ms. w. n. w. Springfield, 100 w. Boston, n. e. Westfield r. Pop. 1830, 579.

Montgomery co., N. Y. bounded by Hamilton co. n., Saratoga co. e., Schenectady, Schoharie, and Otsego cos. s., Herkimer co. W. The greatest length 36 miles, breadth 32; about 1,000 sq. ms. The surface is a little varied, being crossed by the Klypse ridge from Sacandaga r. to the Nose, on Mohawk river. It has rich meadows on Mohawk r. and various soils. Crossed by Mohawk r. and E. Canal. Pop., 1830, 43,715.

Montgomery, p-t., Orange co., N. Y., 12 ms. w. Newburg; 12 n. Goshen; 100 from Albany; N. Waalkill r., is irregular in form, varied surface and good soil, yielding much hemp, &c. In the v. sometimes called Wards bridge, and which is on Waalkill cr.

is an academy, &c. The skeleton of the mammoth in the Philadelphia museum was found here. Pop., 1830, 3,885.

MONTGOMERY t., Somerset co., N. J., E. Millstone r., has much handsome swelling land, with a range called Rocky hill. It produces good grass and excellent cider, chiefly from a species of apples called Harrison, Crab, &c. On the s. border is the borough of Princeton, the seat of Nassau Hall, or Princeton college, and a Presbyterian theological seminary. Pop., 1830, 2,834.

MONTGOMERY co., Pa., bounded by Phil. co. s. E.; Del. co. s.; in part an artificial and in part the Schuylkill r. separates it from Chester s. w.; on the N. w. a range of hills or rather a minor chain of mnts. separates it from Bucks, and on the N. E. it is bounded by Bucks. This co. is very nearly a parallelogram of 24 by 16 ms.; area 384 sq. ms. Extending in lat. from 39° 58′ to 40° 27′, and in long. from 1° 16′ to 1° 56′ E. W. C. The declivity of this fine co. is almost due s. and in most part drained by the various confluents of Perkiomen and Wissahiccon crs. The surface is beautifully variegated by hill, dale, and even mtn. scenery. The soil, though naturally not of first rate quality, is no where sterile. The staples, nearly every vegetable production of the U. S. in the same lat. The beautiful marble of White Marsh is also amongst the most valuable staples of the co. Chief t. Norristown. Pop., 1820, 35,793; 1830, 39,406.

MONTGOMERY, p-o. in Montgomery tsp., Montgomery co., Penn. The tsp. lies on the N. boundary of the co., between the source of Perkiomen, Neshaminy, and Wissahiccon crs. The village called Montgomery square stands 20 ms. N. Phil. Pop. of the township, 1820, 751.

MONTGOMERY co., Md., bounded s. E. by Prince George's co. and the Dis. Col., by Potomac r., which separates it from Fairfax s., and Loudon w.; by Frederick N. w., and by Patuxent r., separating it from Ann Arundel N. E. The greatest length of this co. is by a westerly line from the easterly angle on Patuxent to the bend of Potomac r., between the mouths of Seneca and Monocacy, 32 ms.; mean width 18, and area 576 sq. ms. Extending in lat. from 38° 55′ to 39° 21′, and in long. from 0° 09′ E. to 0° 29′ w.W. C. Montgomery comprises two unequal inclined plains; one falling s. s. w. towards the Potomac and the second and least sthestrd. towards the Patuxent. The Potomac plain is drained by the estrn. branch of Potomac, Rock, Watts, and Seneca crs., and contains near two thirds of the co. The surface of Montgomery is moderately hilly. The estrn. part drained by the sources of the estrn. branch of Potomac and Rock crs. rests on primitive gneiss. From this part nrthwetrly. extends the dividing ridge between the waters of Potomac and Patuxent; this ridge is elevated from 400 to 800 ft. above tide water. If taken generally, the soil of Mont. co. is rather sterile, yet much very good land skirts the streams. Staples, grain and tobacco. Chief t. Rockville. Pop., 1820, 16,400; 1830, 19,876.

MONTGOMERY co., Va., bounded by the Blue Ridge, which separates it from Franklin E., and Patrick s. E., Grayson bounds it s. w., Wythe on the w., Walker's mtn. separates it from Giles N. w.; on the N. E. it has Botetourt, diagonally from the Blue Ridge to Walker's mtn.; the length in a northerly direction is 50 ms.; mean width 20, and area 1,000. Extending in lat. fr. 30° 43′ to 37° 24′, and in long. fr. 3° 04′ to 03° 50′ w. W. C. This co. occupies two mtn. vallies, being bounded on one side by the Blue Ridge and on the opposite by Walker's mtn., and traversed at near mid-distance by a minor ridge. The mountains, stretching from s. w. to N. E., present the curious phenomenon of the dividing line of r. source being nearly at right angles to the mtn. chains. This line of river source divides Montgomery into two unequal sections. The larger portion of about two thirds, comprising the sthrn. and sthwstrn. parts, has a nrthwstrn. declivity, and is traversed by the Great Kenhawa, and drained by Little r. and numerous crs. falling into the main stream.

The nrthrn. and smaller section, declines to the N. E. giving source to the extreme head of Roanoke, and to Craigs cr. and other branches of James r. Thus, Montgomery occupies a part of the plateau between the Atlantic slope and Miss. basin.

The mouth of Sinking cr., which enters the Great Kenhawa in Giles co., at the wstrn. foot of Walker's mtn., of course below any part of Montgomery, is found, from actual measurement, elevated 1,585 feet above tide water in James r. The highest spring tributary to Sinking cr. was found 2,509; we may, therefore, very safely assume as the general elevation of Montgomery from 1,800, to 2,500; or a mean exceeding 2,100 feet, or an equivalent to more than five degrees of lat. If then we assume 37° as the mean lat. of Mont. co., Va., the real winter climate will be similar to that on the Atlantic coast in N. lat. 42°. Beside the mountains the whole face of this co. is broken and rocky, yet though so rough and elevated, the streams are bordered with excellent soil.

Chief town, Christiansburg. Pop., 1820, 8,733; 1830, 12,306.

MONTGOMERY co., N. C., bounded w. by Cabarras, N. w. by Davidson, N. E. by Randolph, E. by Moore, s. E. by Richmond, and s. w. by Rocky r. separating it from Anson. The greatest length along the sthrn. border 50; mean width 17, and area 850 sq. ms. Extending in lat. from 35° 10′ to 35° 30′, and in long. from 2° 40′ to 3° 32′ w. W. C. This co. is subdivided into two not very unequal portions by Yadkin r., which traverses it by a sthrdly. course. The general declivity of the co. is also sthrdly. The surface hilly, and in part mountainous. Chief t., Lawrenceville. Pop., 1820, 8,693; 1830, 10,919.

MONTGOMERY co., Geo., bounded by the

Ockmulgee r. separating it from Appling s., Auchenchatchee r. separating it fr. Telfair s. w., Laurens N. w., Great Ohoopee r. separating it from Emanuel N. E. and Tatnall E. and S. E. Length, S. w. to N. E., 38; mean width 22, and area 896 sq. ms. Extending in lat. from 32° to 32° 37', and in long. from 5° 18' to 6° 06' w. W. C. This co. is traversed and subdivided into two unequal sections by Oconee r., which joins the Ockmulgee and forms the Altamaha, on its sthrn. border. The course of the Oconee is here S. S. E., and the other streams of Montgomery flow nearly parallel except the Ockmulgee which flows, in that part of its course immediately above the mouth of Oconee, to the N. E. by E. Chief t., Vernon. Pop., 1820, 1,869; 1830, 1,269.

MONTGOMERY p-v., Green co., Geo., 35 ms. sthrdly. from Milledgeville.

MONTGOMERY co., Ala., bounded S. E. by Pike; S. W. by Butler and Wilcox; W. by Dallas; N. W. and N. by Ala. r., separating it from Autauga, and N. E. by the country of the Creek Indians. Length from E. to W. 50 ms., mean width 30, and area 1,500 sq. ms. Extending in lat. from 31° 57' to 32° 32' and in long. from 9° 07' to 9° 55' w. W. C. The sthrn. boundary extends along the dividing ridge between the sources of Conecuh r., flowing sthwrdly, and numerous crs. flowing nthwstrdly. over this co. into Ala. r. The declivity is of course to the N. W. Staple, cotton. Chief t., Montgomery. Pop., 1820, 6,604; 1830, 12,695.

MONTGOMERY, p-v., and st. just., Montgomery co., Ala., situated on the left bank of Mobile r., 54 ms. by the road E. Cahaba, and by p-r. 104 ms. S. E. Tuscaloosa. Lat. 31° 22', long. 9° 25' w. W. C.

MONTGOMERY co., Tenn., bounded by Robertson E. and S. E.; Dickson S.; Stewart S. W. and W.; Trigg co. of Ky. N. W.; Christian co. of Ky. N.; and Todd co. of Ky. N. E. The greatest length along Tenn. 44 ms.; mean width 15, and area 660 sq. ms. Extending in lat. from 36° 17' to 36° 37. Cumberland r. enters and traverses the southern part of this co., receiving at Clarksville, near the centre, Red r. from the nrthestrd. The general declivity is wstrd. Chief t., Clarksville. Pop., 1820, 12,219; 1830, 14,349.

MONTGOMERY, p-v. and st. just., Morgan co., Tenn., situated on the Sulphur branch of Obies r., about 120 ms. a little N. of E. Nashville, and 68 ms. N. W. Knoxville. Lat. 36° 22', long. 7° 42' w. W. C.

MONTGOMERY, p-v., Sumner co., Tenn., 17 ms. N. E. Gallatin, and 48 in a similar direction from Nashville.

MONTGOMERY co., Ky., bounded by Estill S.; Clark W.; Bourbon and Nicholas N. W.; Bath N. and N. E., and Morgan E. and S. E. Length from S. E. to N. W. 33 ms.; mean width 8, and area 264 sq. ms. Extending in lat. from 37° 46' to 38° 11', and in long. from 6° 38' to 7° 04' w. W. C. This narrow co. extends along the dividing ground between Ky. and Licking rivers, and is drained in a nearly equal proportion by the crs. of those two rivers respectively. Mount Sterling, the st. just., is situated 55 ms. a little N. of E. Frankfort. Pop., 1820, 9,587; 1830, 10,240.

MONTGOMERY co., O., bounded by Warren S. E.; Butler S. W.; Prebble W.; Dark N. W.; Miami N.; Clarke N. E., and Green E. Length 24 ms.; mean breadth 20, and area 480 sq. ms. Lat. 39° 45', long. W. C. 7° 18' w. The main stream of the Great Miami receives Mad. r. in the nrthest. part of this co. at Dayton, the st. just., from whence the Miami canal commences. The general course of the Great Miami, and the slope of the co., is to the S. S. W. The surface finely diversified by hill and dale, and soil fertile. Chief t. Dayton. Pop., 1820, 15,999; 1830, 24,362.

MONTGOMERY, p-v., Hamilton co., O., 11 ms. N. N. E. Cincinnati. Pop., 1830, 219.

MONTGOMERY co., Ind., bounded by Putnam S.; Parke S. W.; Tippecanoe N.; Boon E., and Hendricks S. E. Length 24 ms.; breadth 21, and area 504 sq. ms. Lat. 40°, and long. W. C. 10° w. intersect in this co.; slope S. W., and in that direction it is traversed by Sugar and Raccoon crs., branches of Wabash r. Chief t., Crawfordsville. Pop., 1830, 7,317.

MONTGOMERY co., Il., bounded by Bond S.; Madison S. W.; Macaupin W.; Sangamo N.; Shelby N. E., and Fayette S. E. Length 36 ms.; breadth 24, and area 864 sq. ms. Lat. 38° 15, long. W. C. 12° 30' w. Sthrn. part drained by Shoal cr., branch of Kaskaskias r. flowing S.; wstrn. part by Macaupin cr., branch of Illinois r., flowing W.; and the nrthrn. pt. by the sthrn. confluents of Sangamon r. flowing N. Chief t., Hillsboro'. Pop., 1830, 2,953.

MONTGOMERY co., Mo., bounded by Calloway co. W.; Ralls N. W.; Pike N. E.; Lincoln and St. Charles E., and Mo. r., separating it from Franklin S. E., and Gasconade S. W. Length from E. to W. 38 ms.; mean breadth 30 ms., and area 1,140 sq. ms. Lat. 38° 50', long. W. C. 14° 18' w. The nrtheastrn. section slopes estrd., and is drained in that direction by Cuivre (Copper) r., a small confluent of the Miss. The central, sthrn., and much the most extensive sections decline sthrd. towards the Mo. r. Chief t., Lewistown. Pop., 1830, 3,902.

MONTGOMERY'S Ferry, and p-o. Perry co. Pa., 26 ms. N. W. Harrisburg.

MONTGOMERYVILLE, p-v. Montgomery co. Pa., by p-r. 24 ms. nrthrd. Phil.

MONTICELLO, the seat of the late venerable Thomas Jefferson, 2 ms. estrd. Charlottesville, Albemarle co. Va.

MONTICELLO, p-v. Fairfield dist. S. C., 35 ms. N. Columbia.

MONTICELLO, p-v. and st. jus. Jasper co. Geo., 38 ms. N. W. Milledgeville. Lat. 33° 18' and long. 6° 44' w. W. C.

MONTICELLO, p-v. and st. jus. Jefferson co. Flor., situated in the nrthrn. part of the co., 31 ms. N. E. by E. Tallahassee. Lat. 30° 31', long. 7° 06' w. W. C.

MONTICELLO, p-v. and st. jus. Lawrence co

Miss., situated on the right bank of Pearl r., 80 ms. nearly due E. Natchez, and by p-r. 66 ms. below and sthrd. Jackson. Lat. 31° 27′, long. 13° 12′ w. W. C.

MONTICELLO, p-v. and st. jus. Wayne co. Ky., situated in the nrthwstrn. part of the co., by p-r. 128 ms. nearly due s. Frankfort. Lat. 36° 53′, long. 7° 44′ w. W. C.

MONTICELLO, p-v. Fairfield co. O., by p-r. 35 ms. s. E. Columbus.

MONTMORENCY, p-v. Jefferson co. Pa., by p-r. 242 ms. N. W. W. C., and 171 ms. N. W. by w. Harrisburg.

MONTPELIER, p-t. and st. jus. Washington co. Vt., and capital of the state, is situated 36 ms. s. E. Burlington, 140 N. W. Boston, 524 N. by E. Washington, 120 s. E. Montreal, and at the confluence of the two head branches of Onion r. It is surrounded by rough hills, and on broken ground, and has a wild situation. Here is the state house, court house, jail, bank, academy, churches, and various manufactories. Pop. 1830, 1,792.

MONTPELIER, p-v. Hanover co. Va., 24 ms. nrthrdly. Richmond.

MONTPELIER, p-v. Richmond co. Va., by p-r. 105 ms. s. w. Raleigh.

MONTROSE, p-v. and st. jus. Susquehannah co. Pa. This is a very neat village, occupying a remarkable site. It stands on the elevated table land, encircled on three sides by the Susquehannah r. In the vicinity rise the higher sources of Tunkhannock, Meshoppen, and Wyalusing, flowing sthwstrdly. into Susquehannah, whilst, also from the same vicinity, crs. are discharged nrthestrdly. into the same stream. Montrose is 31 ms. s. E. Oswego, in the state of New-York, 71 N. W. Milford, on Del. r., and by p-r. 163 ms. N. E. by E. Harrisburg. Lat. 41° 51′, long. 1° E. W. C.

MONTVILLE, p-t. Waldo co. Me., 30 ms. N. E. Wiscasset, without considerable streams; has an eminence in the centre. Pop. 1830, 676.

MONTVILLE, p-t. New London co. Conn., 35 ms. s. E. Hartford, next s. Norwich, and N. New London, w. Thames r. and 7 ms. N. its mouth; surface uneven, with good land; contains the Indian reservation of Mohegan, in which are 3,000 acres of very good soil. Pop. 1830, 1,964.

MONTVILLE, p-v. Geauga co. O., by p-r. 178 ms. N. E. Columbus.

MOOERS, p-t. Clinton co. N. Y., 23 ms. N. W. Plattsburgh. Pop. with Ellenburgh, 1830, 1,222.

MOORE, co. N. C., bounded s. w. by Richmond, w. by Montgomery, N. W. by Randolph, N. by Chatham, and E. and s. E. by Cumberland. It lies in form of an isosceles triangle, two sides 44 ms., and base 34, area 748 sq. ms. Extending in lat. from 35° 04′ to 35° 30′, and in long. from 1° 58′ to 2° 44′ w. W. C. Deep r. curves into, and again leaves the nrthrn. boundary of Moore, flowing estrdly. into Haw r., which gives a nrthrn. declivity to this part of the co. The estrn. part is also drained into Haw r., whilst the sthrn. gives source to Lumber r. The st. just., Carthage, is 55 ms. sthwstrdly. from Raleigh. Pop. 1820, 7,128; 1830, 7,745.

MOOREFIELD, p-v. and st. jus. Hardy co. Va., situated on the right bank of the south branch of Potomac, 50 ms. a little s. of w. Winchester, and by p-r. 123 ms. w. W. C. Lat. 39° 02′, long. 2° 02′ w. W. C.

MOOREFIELD, p-v. Nicholas co. Ky., by p-r. 68 ms. N. E. by E. Frankfort.

MOOREFIELD, p-v. Harrison co. O., by p-r. 111 ms. a little N. of E. Columbus.

MOORESBURG, p-v. wstrn. part Columbia co. Pa., 10 ms. N. E. Northumberland, and by p-r. 71 ms. N. Harrisburg.

MOORESFIELD, or Moorestown, p-v. Chester, Burlington co. N. J., 13 ms. E. Philadelphia.

MOORE's Hill, and p-o. Dearborn co. Ind., about 100 ms. s. E. Indianopolis.

MOORE's Ordinary, and p-o. Prince Edward co. Va., by p-r. 137 ms. s. w. by w. Richmond.

MOORE's Prairie, and p-o. Jefferson co. Il., by p-r. 79 ms. a little E. of s. Vandalia.

MOORE's Salt Works, and p-o. Jefferson co. O., by p-r. 147 ms. a little N. of E. Columbus.

MOORESVILLE, p-v. nrthrn. part of Limestone co. Ala., by p-r. 124 ms. nrthrd. Tuscaloosa.

MOORESVILLE, p-v. in the sthrn. part of Maury co. Ten. 16 ms. from Columbia, the co. seat, and by p-r. 61 ms. s. s. w. Nashville.

MOORESVILLE, p-v. Morgan co. Ind., by p-r. 16 ms. s. w. Indianopolis.

MOORLAND, p-v. Wayne co. O., by p-r. 92 ms. N. E. Columbus.

MOORING's Cross Roads, and p-o. wstrn. part of Pitt co. N. C., 10 ms. N. W. Greenville, the co. seat, and by p-r. 95 ms. E. Raleigh.

MOOSE ISL., Me. (*See Eastport.*)

MOOSE R., N. H., rises on the N. side of the White mtns. near Durand, through which it passes, and unites with the Ameriscoggin in Shelburne. Its source is near that of Israel's r., which passes w. into Connecticut.

MOOSE R., N. Y., runs into the E. side of Black r., at the High Falls in Turin.

MOOSEHEAD LAKE, Kennebec co. Me., 60 ms. long; the source of the E. branch of Kennebec r. has an irregular form, and lies in a tract little inhabited.

MOOSEHILLOCK, or Mooshelock, N. H. a noble eminence in the s. E. part of Coventry. The height of the N. peak, as estimated by Capt. Partridge, is 4,636 feet; that of the s. peak, 4,536. Baker's r. has its source on its E. side.

MOOSUP r., joins the Quinnebaug in Plainfield, Ct.

MOREAU, p-t. Saratoga co. N. Y., 21 ms. N. E. Ballston Spa, and 50 N. Albany. Situated in the Great Bend of the Hudson, embracing part of Baker's falls, Glen's falls, and the Great dam at Fort Edward, and possessing extensive water power; soil in general good for farming; timbered with pine, &c. in some parts, and contains likewise a large tract of beech and maple. Here is an extensive manufactory of gunpowder, and a paper mill, beside other manufactories of various kinds. The navigation of the Champlain canal is in the Hudson, along the line of this town; 6 schools, attended 7 months, in 12. Pop. 1830, 1,690.

MOREMAN'S R., Albemarle co. Va. Though called a r., it is only a creek about 10 ms. long, but it is one of the extreme heads of Ravenna r., having its source in the Blue Ridge.

MOREMAN'S r. p-o., on Moreman's r. 10 ms. N. W. Charlotteville, and in the nrthwstrn. angle of Albemarle co. Va.

MORETOWN, p-t. Washington co. Vt., 8 ms. W. Montpelier. Much of this town is mountainous, and incapable of being settled. It is watered by Mad r. which furnishes several mill privileges; 6 school dists. Pop. 1830, 815.

MORGAN, t. Orleans co. Vt., 52 ms. N. E. Montpelier; contains Knowlton's lake, 4 ms. long, and part of Clyde r. Pop. 1830, 331.

MORGAN, co. Va., bounded by Berkeley E. and S. E., Frederick S., Hampshire S. W., Potomac r. separating it from Alleghany co. of Md. N. W., and by Washington co. Md. N. Greatest length along Berkeley 22 ms., mean width 16, and area 352 sq. ms. Extending in lat. from 39° 22′ to 39° 40′, and in long. from 0° 58′ to 1° 25′ w. W. C. The declivity of this mountainous co. is from S. W. to N. E., and drained by Sleepy and Great Cacapon crs. Though very broken and rocky, this co. contains much excellent r. and valley soil. Chief t. Berkeley Springs. Pop. 1820, 2,500; 1830, 2,094.

MORGAN, co. Ala., bounded E. by the Cherokee territory of that state, S. by Blount co., W. by Lawrence, N. W. by Ten. r. separating it from Limestone, and N. E. by Ten. r. separating it from Madison. Length from E. to W. 30 ms., mean breadth 20, and area 600 sq. ms. Extending in lat. from 34° 18′ to 34° 41′, and in long. from 9° 40′ to 10° 13′ w. W. C. The slope of this co. is to the nrthrd., and drained in that direction into Ten. r., by Flint r. and Cotaco cr. Its sthrn. boundary extends along the dividing ridge, between the valley of Ten., and the sources of Mulberry, and branch of Black Warrior. Chief town, Somerville, or Summerville. Pop. 1830, 9,062.

MORGAN, co. of Ten., bounded by Campbell E., Cumberland mtns. separating it from Anderson S. E., Roan S., Bledsoe S. W., Overton W., and Wayne co. Ky. N. Length diagonally S. W. to N. E. 52 ms., mean width 15, and area 760 sq. ms. Extending in lat. from 36° to 36° 35′, and in long. from 7° 14′ to 7° 50′ w. W. C. This co. comprises part of three inclined plains. The nthestrn. is the higher part of the valley of the south fork of Cumberland r. which flowing nrthrd. into Ky. gives that exposure to this section of the co. The opposite extreme S. of Cumberland mtns., declines nthestrdly. and is traversed in that direction by Emery's r. The central section, containing about one half the whole area, has a westerly declivity and gives source to Obies r. Taken as a whole, Morgan co. occupies the plateau between Cumberland and Ten. rs. The surface is broken by mtns. and hills. Cumberland mtn. bounding the co. on the S. E., inclines wstrdly., and traversing the sthrn. part separates the vallies of Emery's and Obies rs., whilst the N. E. and central parts are again separated by the Poplar mtn. Chief t. Montgomery. Pop. 1820, 1,626, 1830, 2,582.

MORGAN, C. H., Morgan co. Ten. (*See Montgomery, Morgan co. Ten.*)

MORGAN, co. O., bounded S. E. by Washington, Athens S. W., Perry W., Muskingum N. W., Guernsey N. E., and Monroe E. Length 32 ms., mean breadth 18, and area 576 sq. ms. Lat. 39° 40′, long. W. C. 4° 50′ w. Slope S. S. E. and traversed in that direction by the Muskingum r. Surface broken and hilly. Chief t. MacConnellsville. Pop. 1820, 5,297, 1830, 11,799.

MORGAN, p-v. Ashtabula co. O. by p-r. 187 ms. N. E. Columbus.

MORGAN, co. Ind. bounded by Monroe S., Owen S. W., Putnam N. W., Hendricks N., Marion N. E., and Johnson E. Length 26 ms., breadth 21, and area 546 sq. ms. Lat. 39° 30′, long. W. C. 9° 30′ w. This co. approaching very nearly to a sq., is entered near the nthestrn. angle by the main stream of White r., which crossing diagonally leaves it at the S. W. angle, after having divided it into two not greatly unequal sections. Chief t. Martinsville. Pop. 1830, 5,593.

MORGAN, co. Il. bounded S. E. by Macaupin, S. W. by Il. r. separating it from Pike, N. W. by Il. r. separating it from Schuyler, N. Sangamon r. separating it from Sangamon co., and again on the E. by Sangamon co. Length from S. to N. 42 ms., mean breadth 30, and area 1,260 sq. ms. Lat. of its centre 39° 50′, long. W. C. 13° 18′ w. Slope almost due W. towards Il. r., which is also the general course of the Sangamon on its nthrn. border. The eastern boundary follows the dividing ridge between the confluents of Il. and Sangamon. Chief town, Jacksonville. Pop. 1830, 12,714.

MORGANFIELD, p-v. and st. jus. Union co. Ky., situated 12 ms. E. Shawneetown, on Ohio r., about an equal distance S. E. from the mouth of Wabash r., and by p-r. 197 ms. a little S. of W. Frankfort. Lat. 37° 41′, long. 11° w. W. C.

MORGAN'S STORE, and p-o. Montgomery co. N. C. by p-r. 121 ms. S. W. by W. Raleigh.

MOGANTOWN, p-v. Berks co. Pa. 10 ms. estrd. Reading, and 58 in a like direction from Harrisburg.

MORGANTOWN, p-v. and st. jus. Monongalia co. Va., situated on a high bottom of the right bank of Monongahela r. 35 ms. below and N. N. E. Clarksburg, about 60 ms. S. Pittsburg, and by p-r. 201 ms. N. W. by W. W. C. Lat. 39° 40′, long. 2° 50′ w. W. C.

MORGANTOWN, p-v. and st. jus. Burke co. N. C., situated near the right bank of Catawba r. 35 ms. N. N. E. Rutherfordton, and 205 ms. almost exactly due W. Raleigh. Lat. 35° 45′, long. 4° 39′ w. W. C.

MORGANTOWN, p-v. wstrn. part of Blount co. Ten., by p-r. 152 ms. E. Nashville.

MORGANTOWN, p-v. and st. jus. Butler co. Ky., situated on the left bank of Green r. 32 ms. N. N. E. Russelville, and by p-r. 144 ms. s. w. by w. Frankfort. Lat. 37° 12', long. 9° 40' w. W. C.

MORGANVILLE, p-o. Nottaway co. Va. by p-r. 56 ms. s. w. Richmond.

MORIAH, p-t. Essex co. N. Y. on the w. shore of lake Champlain. Soil good for grass and well watered. Timbered with maple, beech, ash, basswood, &c. Contains iron ore; has two mill streams; a quarry of white limestone or marble, and some asbestos; 10 ms. s. Elizabethtown and 112 N. Albany. 4 schools, attended 8 months in 12. Pop. 1830, 1,742.

MORNING SUN, p-o. Shelby co. Ten. by p-r. 205 ms. s. w. by w. Nashville.

MORRIS CANAL, N. J. (*See "RailRoads and Canals."*)

MORRIS, co. N. J. is bounded N. E. by Bergen, s. E. by Essex, s. by Somerset, s. w. by Hunterdon, and N. w. by Sussex cos., and contains an area of about 500 sq. ms. It is watered by several streams, Rockaway and other confluents of the Passaic, and some streams flowing s. into the Raritan. The surface of this county is undulating, except in the N. w. part, which is mountainous. The Passaic is formed on its estrn. border by the union of the Pompton and Rockaway rs., about 5 ms. above the falls of the former. The soil is generally very productive in grain, pasturage, and fruits. Chief t. Morristown. Pop. 1820, 21,368, 1830, 23,580.

MORRISANA, Green co. Pa. (*See Ryerson's station, Green co. Pa.*)

MORRIS COVE, p-o. Bedford co. Pa., 5 or 6 ms. N. w. from the borough of Bedford, and by p-r. 132 N. w. W. C.

MORRIS HILL, p-o. Alleghany co. Va. by p-r. 184 ms. w. Richmond.

MORRISON'S BLUFF, and p-o. Pope co. Ark. It is on the Ark. r. 23 ms. above and wstrd. Dwight, and by p-r. 101 ms. above andnthwstrd. Little Rock.

MORRISON'S TAN YARD, and p-o. Mecklenburg co. N. C. by p-r. 148 ms. s. w. by w. Raleigh.

MORRISTOWN, p-t. Orleans co. Vt. 20 ms. N. w. Montpelier, 29 N. E. Burlington. It is very level for an interior t.; diversified, however, with gentle hills and vales. Soil, in general, very good. Timbered with maple, beech, birch, hemlock, &c. Watered on the N. E. part by Lamoille r. 13 school dists. Pop. 1830, 1,315.

MORRISTOWN, p-t. St. Lawrence co. N. Y. on St. Lawrence r. 12 ms. above Ogdensburgh. The shores of the r. hereabout present the most beautiful scenery. 6 schools, attended 4 months in 12. Pop. 1830, 1,600.

MORRISTOWN, p-t. and st. jus. Morris co. N. J. 18 ms. N. w. Newark, 19 w. N. w. Elizabeth t., 28 w. N. w. N. York. The village stands on a fine elevated plain, with steep slopes on two sides of the public square, and picturesque views. The court house is a fine building, and contains the jail. Near it is the bank. It is on the Oswego mail route, and daily stage coaches run on two routes for N. York. The American army wintered here in the revolutionary war, while the British held New Brunswick. The house is standing in which Washington had his quarters. There is a church for Presbyterians, one for Episcopalians, one for Baptists, and one for Methodists. Pop. 1830, 3,636.

MORRISTOWN, p-v. Belmont co. Ohio by p-r. 115 ms. estrd. Columbus. Pop. 1830, 267.

MORRISVILLE, p-v. in the estrn. part of Bucks co. Pa. 25 ms. N. E. Phil.

MORRISVILLE, p-v. in the sthrn. part of Fauquier co. Va. by p-r. 62 ms. s. w. W. C.

MORRISVILLE, p-v. Hickman co. Ky. by p-r. 313 ms. s. w. by w. Frankfort.

MORTONSVILLE, p-v. Woodford co. Ky. about 30 ms. s. s. E. Frankfort.

MORVEN, p-v. Anson co. N. C. by p-r. 132 ms. s. w. Raleigh.

MORVEN, p-v. Shelby co. Ind. by p-r. 42 ms. s. E. Indianopolis.

MOSCOW, t. Somerset co. Me. 28 ms. N. of Norridgewock, E. Kennebec r. and crossed by one of its streams. Pop. 1830, 405.

MOSCOW, p-v. Lafayette co. Tenn. by p-r. 246 ms. s. w. by w. Nashville.

MOSCOW, p-v. Hickman co. Ky. by p-r. 320 ms. s. w. by w. Frankfort.

MOSCOW, p-v. on O. r. in the sthrn. part of Clermont co. O. by p-r. 127 ms. s. w. Columbus. Pop. 1830, 196.

MOSCOW, p-v. Rush co. Ind. by p-r. 52 ms. s. E. by E. Indianopolis.

MOSS CREEK, p-v. Jefferson co. Ten. by p-r. 239 ms. estrd. Nashville.

MOTTE ISLE, Vt. in lake Champlain, 8 miles long and 2 broad.

MOTTS, p-o. Wilcox co. Ala. by p-r. 102 ms. sthrd. Tuscaloosa.

MOTTVILLE, p-v. St. Joseph's co. Mich. by p-r. 151 ms. s. w. by w. ½ w. Detroit.

MOULTON, p-v. and st. jus. Lawrence co. Ala., situated near the head of a creek, flowing nrthrd. into Ten. r., 50 ms. s. w. by w. Huntsville, and by p-r. 116 ms. N. Tuscaloosa. Lat. 34° 33', and long. 10° 28' w. W. C.

MOULTONBOROUGH, p-t. Strafford co. N. H. on lake Winnipiseogee, 50 ms. N. of Concord. Broken by mountains and ponds. Bog ore is found in this town, and there is a mineral chalybeate spring. There is a large spring in the t. which furnishes water sufficient for mills. On the stream thus produced, nearly a mile below its source, is a beautiful waterfall of 70 feet perpendicular. Descending on the left of this fall, a cave is found, containing charcoal, and other evidences of its having been a hiding place for Indians. Soil fruitful, though in some parts rocky. The Ossipee tribe of Indians once resided in this vicinity. Pop. 1830, 1,422.

MOUNDVILLE, p-v. Iowa co. Mich. (*Huron*) as laid down on Tanner's improved U. S. map, situated on the road from Fort Winnebago to both Prairie du Chien, and Galena,

10 ms. E. of the Fork, 52 ms. s. w. by w. from Fort Winnebago, 74 ms. nrthestrd. Galena, and 97 E. of Prairie du Chien, on the ridge between the sources of Peektano, branch of Rock r. and the valley of Ouisconsin r.

MOUNTAIN COVE, p-o. Nicholas co. Va. by p-r. 273 miles wstrd. W. C.

MOUNTAIN CREEK, and p-o. Lincoln co. N. C. by p-r. 175 ms. s. w. by w. Raleigh.

MOUNTAIN CREEK, p-o. Harris co. Geo. by p-r. 145 ms. wstrd. Milledgeville.

MOUNTAIN ISLAND, p-o. Owen co. Ky. 44 ms. nrthrd. Frankfort.

MOUNTAIN SHOALS, and p-o. nrthrn. part of Laurens district S. C. by p-r. 81 ms. N. W. Columbia. The falls or shoals of Ennoree r. from which the place is named, is just below the mouth of Beaver Dam creek, 16 ms. N. of Laurensville.

MOUNT AIRY, p-o. Randolph co. Mo. by p-r. 85 ms. N. N. W. Jefferson city.

MOUNT AIRY, p-v. Pittsylvania co. Va. by p-r. 177 ms. s. w. Richmond.

MOUNT AIRY, p-v. in Surry co. N. C. by p-r. 172 ms. N. W. by W. Raleigh.

MOUNT AIRY, p-v. Tuscaloosa, Ala. 10 ms. from the village of Tuscaloosa.

MOUNT AIRY, p-v. Bledsoe co. Ten. by p-r. 152 ms. s. E. by E. Nashville.

MOUNT ALTO, p-o. in the Blue Ridge, wstrn. part of Albemarle co. Va. by p-r. 104 miles s. w. by w. W. C.

MOUNT ARIEL, p-v. Abbeville district, S. C. by p-r. 128 ms. w. Columbia.

MOUNT BETHEL, p-v. Northampton co. Pa. by p-r. 208 miles N. N. E. W. C.

MOUNT CARBON. (*See Port Carbon.*)

MOUNT CARMEL, p-v. Covington co. Miss. about 110 ms. E. Natchez.

MOUNT CARMEL, p-v. Fleming co. Ky. by p-r. about 85 ms. E. Frankfort.

MOUNT CLEMENS, p-v. and st. jus. Macomb co. Mich. on Clinton r. or Huron of lake St. Clair, about 3 ms. above its mouth, on the road from Detroit to Fort Gratiot, 26 ms. N. N. E. the former, and 45 s. s. w. the latter place. Lat. 42° 35′, long. W. C. 5° 47′ W.

MOUNT CLIO, p-v. on Lynch's creek, estrn. side of Sumpter district, S. C. by p-r. 52 ms. a little N. of E. Columbia.

MOUNT COMFORT, p-v. Hardiman co. Ten. by p-r. 152 ms. s. w. by w. Nashville.

MOUNT CRAWFORD, p-v. in the wstrn. part of Rockingham co. Va. by p-r. 152 ms. s. w. by w. W. C.

MOUNT CROGHAN, p-v. Chesterfield district, S. C. by p-r. 110 ms. N. E. Columbia.

MOUNT DESERT, island and p-t. Hancock co. Me. 15 ms. long and 12 wide. Lat. 44° 12′, and is a peninsula between Union r. and Mt. Desert sound. Pop. 1830, 1,603.

MOUNT EATON, p-v. Wayne co. O. by p-r. 100 ms. N. E. Columbus.

MOUNT EDEN, p-o. in the western part of Spencer co. Ky. 10 ms. w. Taylorsville, and 37 s. w. by w. Frankfort.

MOUNT ELON, p-v. Darlington district, S. C. by p-r. 75 ms. E. Columbia.

MOUNT GALLAGHER, p-v. Laurens district, S. C. by p-r. 90 ms. N. W. Columbia.

MOUNT GILEAD, p-o. in the wstrn. part of Loudon co. Va. by p-r. 43 ms. northwestward from W. C., and 8 in a similar direction from Leesburg.

MOUNT GOULD, p-v. Bertie co. N. C. by p-r. 144 ms. a little N. of E. Raleigh.

MOUNT HENRY, p-o. Montgomery co. Ten. by p-r. 58 ms. N. W. by W. Nashville.

MOUNT HILL, p-o. Abbeville district, S. C. by p-r. 99 ms. w. Raleigh.

MOUNT HOLLY, p-t. Rutland co. Vt. 60 ms. s. Montpelier, and 20 w. Windsor. Mill r. is the only stream of consequence. Better adapted to grass than grain. Here are found *amianthus*, common and ligniform asbestos, and fossil leather. There are 10 school districts. Pop. 1830, 1,318.

MOUNT HOLLY, p-v. and st. jus. Gloucester co. N. J. on Rancocus creek, 7 ms. s. E. Burlington, and 17 E. Philadelphia; it has a handsome court house and jail, a bank, and several churches. The creek is navigable to the village.

MOUNT HOLYOKE, Hadley, Mass., E. Conn. r. 3 ms. s. E. Northampton. It is 830 feet above the level of Conn. r., and affords an extensive and beautiful view of the surrounding country.

MOUNT HOPE BAY, the N. E. arm of Narraganset bay, receives Taunton r.

MOUNT HOPE, on the W. shore of the above bay in Bristol R. I. is a beautiful eminence, and is celebrated as the residence of the famous Wampanoag, Indian king Philip.

MOUNT HOPE, p-v. Lancaster co. Pa. by p-r. 34 ms. estrd. Harrisburg.

MOUNT HOPE, p-v. Williamsburg district, S. C. by p-r. 81 ms. s. E. by E. Columbia.

MOUNT HOPE, p-o. in the southern part of Shenandoah co. Va. by p-r. 98 ms. s. w. by w. W. C.

MOUNT HOPE, p-o. Tuscaloosa co. Ala. (*See Mount Airy, same county and state.*)

MOUNT HOPE, p-o. Lawrence co. Ala. by p-r. 104 ms. N. Tuscaloosa.

MOUNT HOPE, p-v. Lawrence co. Ala. by p-r. 114 ms. nrthrd. Tuscaloosa.

MOUNT HOREB, p-o. Nelson co. Va. by p-r. 111 ms. wstrd. Richmond.

MOUNT HOREB, p-v. Jasper co. Geo. 24 ms. N. W. Milledgeville.

MOUNT INDEPENDENCE, Orwell Vt. about 2 ms. s. E. Ticonderoga fort. It figured as a military position in the early history of our country.

MOUNT ISRAEL, p-v. Albemarle co. Va. by p-r. 145 ms. s. w. W. C.

MOUNT JACKSON, p-o. wstrn. part of Beaver co. Pa. 20 ms. wstrd. Beavertown.

MOUNT JACKSON, p-o. Shenandoah co. Va. 97 ms. wstrd. W. C.

MOUNT JOY, p-o. township, Lancaster co. Pa. between little Chiques and Conewago creeks. The p-o. is 21 ms. s. E. Harrisburg, and about 10 w. Lancaster. Pop. of the township 1820, 1,835.

Mount Laurel, p-o. in the sthwstrn. part of Halifax co. Va. by p-r. 125 ms. s. w. Richmond.

Mount Lebanon, p-v. Augusta co. Va. by p-r. 184 ms. s. w. by w. W. C.

Mount Level, p-v. Dinwiddie co. Va. by p-r. 47 ms. s. s. w. Richmond.

Mount Lewis, p-v. Lycoming co. Pa. 25 ms. wstrly. from Williamsport, and 118 ms. northwardly from Harrisburg.

Mount Lineus, p-v. Monongalia co. Va. by p-r. 240 ms. n. w. by w. W. C.

Mount Meigs, p-o. in the estrn. part of Montgomery co. Ala. by p-r. 110 ms. s. e. of Tuscaloosa.

Mount Meridian, p-v. in the wstrn. part of Augusta co. Va. by p-r. 176 ms. s. w. by w. W. C.

Mount Maria, or Mariah, p-v. on the right bank of Lackawaxen r. and in the extreme nrthrn. part of Pike co. Pa. 24 ms. n. w. by w. Milford, and 144 ms. n. Phil.

Mount Morris, p-t. Livingston co. N. Y. on the Genessee r. 8 ms. s. s. w. Geneseo. Land of a good quality, presenting a pleasing variety of surface, heavily timbered with maple, beach, oak, elm, &c., 6 schools continued 8 months in 12. Pop. 1830, 2,534.

Mount Morris, p-o. Green co. Pa.

Mount Mourne, p-v. Iredell co. N. C. by p-r. 153 ms. w. Raleigh.

Mount Olympus, p-v. Madison co. Miss. by p-r. 140 ms. n. e. Natchez.

Mount Pinson, p-v. Madison co. Ten. by p-r. 166 ms. s. w. by w. Nashville.

Mount Pisgah, p-o. Iredell co. N. C. by p-r. 152 ms. w. Raleigh.

Mount Pisgah, p-v. in the sthrn. part of Wilcox co. Ala. by p-r. 128 ms. s. Tuscaloosa.

Mount Pisgah, p-o. Blount co. Ten. 8 ms. sthrd. Maryville, the co. t. and by p-r. 168 ms. a little s. of e. Nashville.

Mount Pleasant, p-t. Westchester co. N. Y., on the e. side of Hudson r. 33 ms. n. N. Y., 130 s. Albany. The land is of good quality, and the town is abundantly supplied with mill seats. It contains a copper mine, and a marble quarry. The Sing Sing state prison, containing cells for 1,000 prisoners, is here. There are 16 school dists. Pop. 1830, 4,932.

Mount Pleasant, p-v. and tsp. in the s. w. angle of Wayne co. Pa. on the head of Lackawaxen creek. Pop. of the tsp. 1820, 874. (*See Pleasant Mount, Wayne co. Pa.*)

Mount Pleasant, p-v. and tsp. on the waters of Jacob's and Sewickly creeks, in the southern part of Westmoreland co. Pa. The village and p-o. is situated about 11 miles s. Greensburg, the co. t. Pop. of the tsp. in 1820, 874.

Mount Pleasant, p-v. Frederick co. Md. by p-r. 49 ms. n. n. w. W. C.

Mount Pleasant, p-v. Spottsylvania co. Va.

Mount Pleasant, p-v. in the western part of Rockingham co. N. C. 10 ms. w. Wentworth, and 136 n. w. by w. Raleigh.

Mount Pleasant, p-v. in the northern part of Fairfield dist. S. C. 13 ms. northwardly from Winnsborough, and 44 ms. in a similar direction from Columbia.

Mount Pleasant, p-v. Monroe co. Ala.

Mount Pleasant, p-v. Wilkinson co. Miss. 10 ms. northwardly from Woodville, the co. t. and 23 southwardly Natchez.

Mount Pleasant, p-v. in the eastern part of East Baton Rouge, La.

Mount Pleasant, p-v. Williamson co. Ten. about 23 ms. sthrd. Nashville.

Mount Pleasant, p-t. and st. jus. Harlan co. Ky. on the left bank of Cumberland r., about 70 ms. n. n. e. Knoxville in Tenn., and by p-r. 152 ms. s. e. Frankfort. Lat. 36° 47′, long. 6° 21′ w. W. C. This is the most southestrd. co. seat in Ky. The situation is elevated, mountainous and romantic.

Mount Pleasant, p-v. Jefferson co. O. by p-r. 135 ms. a little n. of e. Columbus, 21 s. w. Steubenville, and 273 ms. n. w. by w. W. C. This fine village is situated on a hill, and is chiefly composed of one main street, and contains a printing office, bank, several stores, and schools. The Friend's meeting house is a capacious building, 92 by 62 feet; the Seceders and Methodists have also meeting houses. By the census of 1830, the village contained 554 inhabitants.

Mount Pleasant, tsp. around and comprising the foregoing village, is in the southwestern part of Jefferson co. O., and in 1820, contained 1,468 inhabitants, which had augmented to 2,362 in 1830; in both times including the village. The tsp. is drained by Indian Short creek, and gives by the rapid descent of its branches numerous and excellent sites for mills and manufactories, which are numerous and valuable, consisting of grist and saw mills, paper mills, and cloth factories.

Mount Pleasant, p-v. and st. jus. Martin co. Ind. by p-r. 121 ms. s. s. e. Indianopolis, and 659 ms. wstrd. W. C.

Mount Pleasant, p-v. Union co. Il. by p-r. 167 ms. s. Vandalia.

Mount Pleasant Mills and p-o. on a branch of the Mantango creek, and in the sthestrn. part of Union co. Pa. by p-r. 46 ms. a little w. of n. Harrisburg.

Mount Pocono, p-o. nthrn. part of Northampton co. Pa. by p-r. 221 ms. n. n. e. W. C.

Mount Prairie, p-o. Ralls co. Mo. by p-r. 145 ms., but by direct distance only about 100 n. n. e. Jefferson city, and about a like distance n. w. St. Louis.

Mount Prospect, p-v. Edgecome co. N. C. 15 ms. sthrd. Tarboro', and by p-r. 82 e. Raleigh.

Mount Republic, p-v. in the central part of Wayne co. Pa. by p-r. 164 ms. n. e. Harrisburg, and 127 n. Phil.

Mount Reserve, p-o. Bedford co. Ten. about 35 s. Nashville.

Mount Richardson, p-v. Jackson co. Ten. by p-r. 67 ms. n. e. by e. Nashville.

Mount Salus, p-v. Hinds co. Miss. situated on the main road from Natchez to Florence in Ala. 12 ms. w. Jackson, the seat of government for the state, and 91 ms. N. E. from Natchez.

Mount Seir, p-v. Mecklenburg co. N. C. by p-r. 158 ms. s. w. by w. Raleigh.

Mount Sharon, p-v. Blount co. Ala. 93 ms. N. N. E. Tuscaloosa, and about 40 southwardly from Huntsville.

Mount Sidney, p-v. Augusta co. Va. by p-r. 131 ms. s. w. by w. W. C.

Mount Sterling, p-v. and st. jus. Montgomery co. Ky. on the table land between the sources of creeks flowing northwardly into Licking from those pursuing an opposite direction into Ky. r. 33 ms. E. Lexington, and 57 ms. a little s. of E. Frankfort. Lat. 38° 04', long. 6° 55' w. W. C. Pop. 1830, 561.

Mount Sterling, p-v. Madison co. O. by p-r. wstrd. Columbus.

Mount Sterling, p-v. Switzerland co. Ind. by p-r. 102 ms. s. E. Indianopolis.

Mount Tabor, t. Rutland co. Vt. 26 ms. s. w. Windsor, 36 N. E. Bennington. It is mountainous, and much of it incapable of being settled. Pop. 1830, 210.

Mount Tirza, p-v. Person co. N. C. by p-r. 89 ms. N. N. W. Raleigh.

Mount Tom, Mass. w. Connecticut r., near Northampton, opposite Mt. Holyoke. It gives name to a range of mountains commencing in New Haven, Conn., and extending N. to East Hampton, Mass., where it crosses Conn. r. and unites with the Lyme range at Belchertown.

Mount Vernon, p-t. Kennebec co. Me. 18 ms. N. w. Augusta. Pop. 1830, 1,439.

Mount Vernon, p-t. Hillsborough co. N. H. 28 ms. s. Concord. It occupies a very elevated position. Pop. 1830, 762.

Mount Vernon, p-v. Chester co. Pa. about 45 ms. s. w. by w. Phil., and by p-r. 104 ms. N. E. W. C.

Mount Vernon, p-v. Rowan co. N. C. 11 ms. nthrd. Salisbury, and by p-r. 131 ms. w. Raleigh.

Mount Vernon, p-v. in the western part of Spartanburg dist. S. C. 105 ms. N. W. Columbia, and 9 w. Spartanburg.

Mount Vernon, p-v. and st. jus. Montgomery co. Geo. situated E. from the Oconee r. by p-r. 85 ms. s. s. E. Milledgeville. Lat. 32° 13', long. 5° 39' w. W. C.

Mount Vernon, p-v. on the left bank of Appalachicola r. immediately below the junction of Flint and Chattahooche rs., and is the northwestern angle of Gadsden co. Flor. about 160 ms. a little N. of E. Pensacola, and by p-r. 52 ms. N. W. by w. Tallahassee.

Mount Vernon, p-v. and st. jus. Rock Castle co. Ky. by p-r. 81 ms. s. E. Frankfort. Lat. 37° 22', long. 7° 12' w. W. C.

Mount Vernon, Bullitt co. Ky. (*See Mount Washington, Bullitt co. Ky.*)

Mount Vernon, p-v. Mobile co. Ala. by p-r. 189 ms. s. Tuscaloosa.

Mount Vernon, p-v. Warren co. Miss. about 60 ms. N. N. E. Natchez.

Mount Vernon, p-v. and st. jus. Knox co. O. on the left bank of Owl creek, by p-r. 45 ms. N. E. Columbus. This village contains the usual appendages belonging to a st. jus. of a co., with numerous mills and factories in the vicinity. Pop. 1830, 886. Lat. 40° 24', long. 5° 30' w. W. C.

Mount Vernon, p-v. and st. jus. Posey co. Ind. by p-r. 187 ms. s. w. Indianopolis. It is situated on Ohio r. in the bend above the mouth of Wabash. Lat. 38° 50', long. 11° w. W. C.

Mount Vernon, p-v. and st. jus. Jefferson co. Il. by p-r. 65 ms. s. s. E. Vandalia. Lat. 38° 21', long. 11° 58' w. W. C.

Mount View, p-v. Davidson co. Ten. 16 ms. from Nashville.

Mountville, p-v. Lancaster co. Pa. 6 ms. sthrd. Lancaster, and by p-r. 32 ms. s. E. Harrisburg.

Mountville, p-v. Loudon co. Va. 42 ms. N. w. by w. W. C.

Mount Vintage, p-v. Edgefield dist. S. C. by p-r. 63 ms. s. w. by w. Columbia.

Mount Washington, N. H. (*See White mtns.*)

Mount Washington, t. Berkshire co. Mass. 130 ms. s. w. Boston. This town is situated on the height of land between the Housatonic and Hudson rs., upon the Taghgannuck range, the principal summit of which is in this town, and is about 3,000 feet above the level of the sea. A broken tsp. of scattered habitations. Pop. 1830, 345.

Mount Washington, p-v. eastern part of Bullitt co. Ky. 7 ms. N. E. by E. Shepherdsville, and 62 s. w. by w. Frankfort.

Mount Washington, p-v. Copiah co. Miss. by p-r. about 55 ms. E. Natchez.

Mount Washington, p-v. Catahoola parish, La. by p-r. 263 ms. N. W. New Orleans.

Mount Welcome, p-v. Lincoln co. N. C. by p-r. 159 ms. wstrd. Raleigh.

Mount Willing, p-v. Edgefield dist. S. C. situated on a branch of Little Saluda, 12 ms. N. E. Edgefield, and 40 ms. w. Columbia.

Mount Willing, p-v. East Feliciana, 12 ms. E. St. Francisville.

Mount Wilson, p-v. Fentress co. Ten. about 130 ms. E. Nashville.

Mount Yonah, p-o. Habersham co. Geo. by p-r. 159 ms. N. Milledgeville.

Mount Zion, p-o. nrthrn. part of Hancock co. Geo. 31 ms. N. E. Milledgeville.

Mount Zion, p-v. Monroe co. Miss. by p-r. 163 ms. N. E. Jackson.

Mount Zion, p-v. sthrn. part of Union co. Ky. by p-r. 236 ms. s. w. by w. Frankfort.

Mount Zion, p-v. Lowndes co. Miss. by p-r. 256 ms. N. E. Natchez, and 10 ms. from Columbus, the county seat of Lowndes.

Mouth of Black river, p-o. extreme northern part Lorain co. O. by p-r. 139 ms. N. N. E. Columbus.

Mouth of Paint Rock creek, sthrn. part of Roan co. Ten. by p-r. 10 ms. s. Kingston, the county seat, and 166 ms. E. Nashville.

Mouth of Monocacy, p-o. extreme wstrn.

part of Montgomery co. Md. by p-r. 43 ms. N. W. W. C.

MOUTH OF SANDY creek, and p-o. nrthestrn. part of Henry co. Ten. 94 ms. N. W. byw. ½ W. Nashville.

MOUTH OF TELLICO, p-o. Monroe co. Ten. on Ten. r. where the road from Knoxville to Athens crosses that stream, 42 ms. S. W. of Knoxville, and by p-r. 166 ms. S. E. by E. of Nashville.

MUD CAMP, p-v. Cumberland co. Ky. 152 ms. sthrd. Frankfort.

MUD CREEK, Ontario co. N. York, rises in Bristol, and after a course of about 43 ms. enters the Canandaigua outlet at the village of Lyons. A very valuable stream.

MUDDY RIVER, Ky. rising in Todd and Logan cos. interlocking sources with Red river, branch of Cumberland, and flowing to the nrthrd. leaves Todd and Logan, and for about 12 ms. forms a boundary between Butler and Muhlenburg cos., finally falling into Green r. opposite Ohio co.

MUHLENBURG, co. Ky. bounded S. by Todd, S. W. by Christian, W. by Pond r. separating it from Hopkins, N. by Green r. separating it from Daviess, N. E. by Green r. separating it from Ohio co. and S. E. by Muddy r. separating it from Butler. Length diagonally S. E. to N. W. 38 ms., mean width 13, and area 494 sq. ms. Extending in lat. from 37° 04′ to 37° 32′, and in long. from 9° 47′ to 10° 17′ W. W. C. It will be seen that this co. is bounded on all sides except to the S. by rivers. The declivity is to the N. N. W. Chief t. Greenville. Pop. 1820, 4,979, 1830, 5,340.

MULBERRY, r. of Ala., the nrthwstrn. and main branch of Tuscaloosa or Black Warrior r. having its sources in the table land between the basins of Mobile and Ten. The general course is S., draining the wstrn. half of Blount and all Walker co., and uniting on the wstrn. border of Jefferson, with the Locust fork to form the Black Warrior. The valley of the Mulberry comprises an area of 1,500 sq. ms. lying in form of a triangle, base 60 ms. and altitude 50 ms. The valley is traversed and divided into two not very unequal sections by lat. 34°.

MULBERRY, p-v. in the nrthrn. part of Lincoln co. Ten. about 50 ms. S. Nashville.

MULBERRY, p-v. in the wstrn. part of Autauga co. Ala. by p-r. 81 ms. S. E. Tuscaloosa.

MULBERRY, p-v. Crawford co. Ark. by p-r. 136 ms. wstrd. Little Rock.

MULBERRY GAP, p-v. Claiborne co. Ten. by p-r. 264 ms. estrd. Nashville.

MULBERRY GROVE, and p-o. Harris co. Geo. by p-r. 135 ms. wstrd. Milledgeville.

MULLENSFORD, and p-o. Franklin co. Geo. by p-r. 114 ms. N. Milledgeville.

MULLICUS RIVER, N. J. runs into the Atlantic through New Inlet, 4 ms. E. of Leeds. It is navigable 20 ms. for vessels of 60 tons, and forms the boundary of Burlington and Gloucester cos.

MULLOY'S, p-o. Robertson co. Ten. by p-r. 29 ms. N. W. Nashville.

MUMFORDSVILLE, p-v. and st. jus. Hart co. Ky. situated on the right bank of Green r., 20 ms. N. Glasgow, 32 S. Elizabethtown, and 97 S. W. Frankfort. Lat. 37° 17′, long. 8° 50′ W. W. C. Pop. 1830, 194.

MUNCYTOWN, p-v. and st. jus. Delaware co. Ind. by p-r. 59 ms. N. E. Indianopolis, lat. 40° 13′, and long. W. C. 8° 36′ W.

MUNCY, post township, on both sides of Muncy creek, in the sthestrn. part of Lycoming co. Pa. 80 ms. N. Harrisburg.

MUNROE, p-t. Orange co. N. Y. 19 ms. S. Newburgh, 115 S. Albany, and 50 N. N. Y. Surface broken and hilly, and well watered by numerous streams. The hills or mountains abound with iron ore. Here are extensive iron works; 11 schools, continued 6 mo. in 12. Pop. 1830, 3,671.

MUNSTER, p-v. Cambria co. Pa. eastward Ebensburg, and by p-r. 130 ms. wstrd. Harrisburg.

MURFREESBORO', p-v. Hertford, N. C.

MURFREESBORO', p-v. and st. jus. for Rutherford co. Ten. situated on a branch of Stone r. 30 ms. S. E. Nashville, and 82 ms. a little E. of N. Huntsville in Ala. Lat. 35° 51′, and long. 9° 15′ W. W. C.

MURRAY'S MILLS, and p-o. Dearborn co. Ind. by p-r. 117 ms. S. E. Indianopolis.

MURRAYSVILLE, p-v. Lorain co. O. by p-r, 128 ms. N. N. E. Columbus.

MURRAYSVILLE, p-v. on a branch of Turtle creek and in the wstrn. part of Westmoreland co. Pa., 12 ms. N. W. Greensburg, and 20 ms. a little S. of E. Pittsburg.

MURRILL'S SHOP, and p-o. Nelson co. Va. by p-r. 110 ms. W. Richmond.

MURRINSVILLE, p-v. Butler co. Pa. by p-r. 251 ms. N. W. W. C.

MUSCOGEE, one of the wstrn. cos. of Geo. bounded by Harris N., Talbot N. E., Marion E., Randolph S., and the Chattahoochee r. separating it from the Creek country in Ala. W. Length E. to W. 25 ms., breadth 20, and area 500 sq. ms. Extending in lat. from 31° 17′ to 31° 35′, and in long. from 7° 52′ to 8° 14′ W. W. C. The slope of this co. is wstrd. and drained by the different branches of Upotoi cr. Chief t. Columbus. Pop. 1830, 3,508.

MUSKEGAT, isl. Mass. lying between Nantucket and Martha's Vineyard, in the form of a horse shoe, about 3 ms. in extent.

MUSKINGUM, important river of Ohio, and one of the great branches of the river Ohio, from the right or N. W. side. It is formed by two branches, Tuscarawas from the northeastward, and White Woman's r. to the nrthwestward. White Woman's r. rises near the centre of the state of Ohio, interlocking sources with those of Sciota, Huron of Erie, Vermillion, and Black rs. Composed of two branches, Mohiccon and Killbuck crs., White Woman's r. drains Wayne, Holmes, Richland, Knox, and part of Coshocton counties; general course S. E. joining Tuscarawas in Coshocton co., between the villages of Coshocton and Caldersburg, after a general comparative course of 60 ms.

Tuscarawas has interlocking sources with those of Cuyahoga and Big Beaver. In the higher part of its course for 50 ms. it pursues a sthrn. course, out of Medina and Portage cos. over Stark into Tuscarawas co. Inflecting abruptly to the w. and entering Coshocton, it unites with White Woman's r., as already noticed, after a general comparative course of 60 miles.

It is at the junction of Tuscarawas and White Woman's rs. that the united waters take the name of Muskingum, which flowing s. 10 ms. receives a large estrn. branch, Wills cr., and bending to about s. s. w. 15 ms. receives Licking creek, and falls over a ledge of rocks at Zanesville. Below Zanesville, with large partial bends, the general course is s. E. 50 ms. comparative distance to its influx into O. river at Marietta. The Tuscarawas branch drains all Tuscarawas and Stark, with parts of Harrison, Columbiana, Portage, Medina, Wayne, Holmes, and Coshocton cos. Wills creek drains and its valley is nearly commensurate with Guernsey co. The Muskingum r. properly so called, winds over the southern side of Coshocton, and over Muskingum, Morgan, and Washington cos.

The entire Muskingum valley approaches remarkably near a circle, of 100 ms. diameter; but with allowance for the salient parts, the area is about 8,000 sq. ms. The Ohio and Erie canal enters this valley in Licking co. and is carried N. E. to Coshocton, and thence along the main channel of Tuscarawas to the Portage Summit. (*See art. Rail Roads and Canals.*)

The level of the canal on the Portage summit is 973 feet above the ocean tides, whilst that of Ohio at Marietta, but little if any exceeds 600 feet of similar relative height. The arable soil around the sources of the higher fountains of White Woman's and Tuscarawas rs. must exceed 1,000 feet above the ocean, or the difference of level of the valley amounts to at least an equivalent to a degree of lat. The actual extremes of lat. are 39° 20' and 41° 10'. The soil of the Muskingum valley is of unsurpassed fertility. The surface presenting the usual features of the rs. of Ohio; that is, level at the sources and becoming more and more hilly approaching the main recipient, the Ohio r. The true cause of this inversion of the common character of rs., will be seen by reference to article O. r.

MUSKINGUM, co. Ohio, bounded by Morgan s. E. and s., Perry s. w., Licking w., Coshocton N., and Guernsey E. Length 27 ms., mean breadth 26, and area 700 sq. ms. Lat. 40° and long. W. C. 5° w. intersect almost exactly at the centre of this co. It is traversed from N. to s. and very nearly equally divided by Muskingum r. Surface moderately hilly, and soil fertile. Chief t. Zanesville. Pop. 1820, 17,824, 1830, 29,334.

The northwestern angle is traversed by the Ohio and Erie canal, and in the opposite direction the U. S. road passes over at the greatest breadth.

MUSKONETCUNK, lake, or Hopatcong, 9 ms. long, 14 ms. N. N. W. Morristown N. J., has been dammed at the outlet (South) and supplies Morris canal, through a feeder.

MUSKONETCUNK, r. N. J. rises in Muskonetcunk lake, and flowing s. w. divides Sussex and Warren from Morris and Hunterdon cos., and falls into Delaware r. 5 ms. below Easton. It is a fine mill stream.

MYERS, or Meyers creek, a small stream of Frankfort, Herkimer co. N. Y., which enters the Mohawk near the E. extremity of the long level of the Erie canal.

MYERS, p-o. Venango co. Pa. by p-r. 256 ms. N. W. W. C.

MYERSTOWN, p-v. on a branch of Quitapahilla creek, Lebanon co. Pa., 31 ms. a little N. of E. Harrisburg, and 5 ms. w. from the borough of Lebanon.

MYSTIC, river of Mass., flows into Boston harbor, navigable for sloops to Medford.

N.

NACOUCHY Valley, p-o. (*See Cooperstown, Habersham co. Geo.*)

NAGLESVILLE, formerly Tobyhanna, p-v. southern part of Pike co. Pa. by p-r. N. N. E. W. C.

NAHANT, Essex co. Mass. a peninsula extending from the s. shore of Lynn far into the sea. It is considered a great natural curiosity. It appears once to have been two islands, but is now connected to the main land by two ridges of pebbles and sand thrown up by the water. The surface is broken, and the shores are bold and rocky. It is a place of great resort in the summer. The air is fragrant and cooling; the scenery romantic; the walks round the margin of the cliffs pleasant, and the prospect grand. It is 9 ms. s. of Salem, and 14 N. E. Boston.

NAHUNTA, creek and p-o. northern part of Wayne co. N. C. by p-r. 45 ms. s. E. by E. Raleigh.

NAMASKET, r. Mass. joins Bridgewater r. to form the Taunton.

NANCEVILLE, p-o. Floyd co. Ind. by p-r. 129 ms. a little E. of s. Indianopolis.

NANKIN, p-v. western part Wayne co. Mich. by p-r. 17 ms. w. Detroit.

NANJEMOY, creek, bay, and p-o. in the southwestern part of Charles co. Md. The p-o. is by p-r. 47 ms. nearly due s. W. C. Nanjemoy bay is a small opening from the left bank of Potomac r. at the great bend above Port Tobacco.

NANSEMOND, co. Va. bounded by Black Water r. w. separating it from Southampton, by the Isle of Wight N. w., Hampton Roads

N. E., Norfolk co. E., Pasquotank co. N. C. S. E., and Gates co. S. C. S. Length diagonally N. W. to N. E. 40 ms., mean breadth 16, and area 640 sq. ms. Extending in lat. from 36° 30′ to 36° 54′, and in long. from 0° 6′ to 0° 41′ E. W. C. The northern part has a gentle inclination to the N. N. E., and is drained by the branches of Nansemond r., which stream, or rather bay, extends about 18 ms. towards the centre of the co. The southwestern section has a slight declivity to S. S. W., and is drained into Nottaway r. The southeastern angle is low, marshy, and in part occupied by a small lake called Drummond's pond. From this pond, a small lateral canal has been constructed into the main trunk of the Dismal Swamp canal. Lake Drummond canal answers the double purpose of a feeder, and of a navigable channel; it is 5 ms. in length, 16 feet wide, and 4½ feet in depth. The general surface of Nansemond is level, and contains a good share of productive soil. Chief t. Suffolk. Pop. 1820, 10,494, 1830, 11,784.

NANSEMOND, r. Va. rising in Isle of Wight and Nansemond cos. Va., but chiefly in the latter. It opens by a comparative wide bay from Hampton Roads, and is navigable for vessels of 100 tons draught, something above 20 ms. to Suffolk, the co. t. of Nansemond co.

NANTASKET Road, the entrance into Boston harbor, Mass. It affords safe anchorage in 5 to 7 fathoms water, and was formed in 1831.

NANTICOKE, r. of Del. and Md. is formed from two branches, Nantikoke Proper, and Marshy Hope, both rising in Del. The Nantikoke rises within, and drains the central and western parts of Sussex co. Del., and flowing southwstrd. enters Dorchester co. Md., in which it receives from the N. Marshy Hope. The latter rising in Kent co. Del. traverses the southeastern angle of Caroline co. Md., from which, entering Dorchester, it falls into the Nantikoke. Below the junction of the two branches, the Nantikoke gradually widens into a bay from one to two ms. wide, until finally merged into the still wider Fishing bay. The entire comparative course of Nantikoke, by either branch, is about 50 ms., the valley lying between those of Pocomoke and Choptank.

NANTICOKE, mtn. Luzerne co. Pa. extends along the left bank of Susquehannah r. about 8 ms. downwards from Nanticoke falls.

NANTICOKE Falls, or rather rapids in the Susquehannah r. 6 ms. below Wilkesbarre. The river after having flowed down the Wyoming Valley to the S. W. turns abruptly to the W., and piercing the Nanticoke mtn. again resumes a S. W. course.

NANTICOKE, v. Broome co. N. Y. 155 ms. S. W. Albany.

NANTICOKE, p-o. near Nanticoke falls, 7 ms. S. W. Wilkesbarre, and by p-r. 107 ms. N. E. Harrisburg.

NANTIKOKE, hundred, of Sussex co. Del., and occupies the southwestern part of Sussex co. on Nantikoke r. Pop. 1820, 2,335, 1830, 2,366.

NANTUCKET isl., co. and p-t. situated in the ocean about 20 ms. S. Chatham, Barnstable co. Mass., and about 15 ms. E. Martha's Vineyard, being 100 ms. S. E. Boston on a straight line, and 125 round Cape Cod. It is 15 ms. long, and 11 wide at its greatest breadth. The soil is light and sandy, but in some parts productive. The people are almost all whalemen and seamen, and are considered as among the most skilful and adventurous in the world. The ship masters have, with commendable zeal, established a marine reading room, cabinet, &c.

Nantucket is the name for the island, county and town. The climate is much milder than that of the neighboring continent. There is not a tree of natural growth on the island, though it was formerly well wooded. The exports are spermaceti and right whale oil, whalebone and sperm candles; of these and oil there are 50 manufactories.

There were in 1829, sixty ships employed in whaling from the port. Other ships have since been built. The value of this fleet, as fitted for sea, amounts to about $2,000,000. On the S. E. of the island are Nantucket Shoals, where numerous vessels have been wrecked. They extend 50 ms. in length, and 45 in width. The harbor of Nantucket is safe from all winds, being almost landlocked. There are in Nantucket 7 or 8 houses of religious worship, 2 banks, and 2 insurance offices. There is a bar of sand at its mouth, on which there are 7½ feet of water at low tide. The taxable property of this island in 1832, amounted to $3,895,288 40. Pop. 1830, 7,202.

NANTUCKET Bay, N. J. opposite Bombay Hook.

NAPLES, p-t. Ontario co. N. Y. 20 ms. S. W. Canandaigua. Contains fine groves of pine. 13 school dists.; schools continued 5 months in 12. Pop. 1830, 1,941.

NAPLES, or Henderson bay, extends from Chaumont bay to the S. W. into Henderson. (*See Henderson.*)

NAPLES, p-v. Morgan co. Il. by p-r. 125 ms. N. W. Indianopolis.

NAPOLEON, p-v. Ripley co. Ind. by p-r. 67 ms. S. E. Indianopolis.

NAPOLI, p-t. Cataraugus co. N. Y. Pop. 1830, 852.

NAP's creek, p-o. Pocahontas co. Va. by p-r. 242 ms. a little S. of W. W. C.

NARAGANSET Bay, R. I., sets up from S. to N. between Point Judith on the W., and Point Seaconet on the E. It is about 30 ms. long, and 15 broad. Embracing several very considerable islands, and good harbors, and receiving Providence and Taunton rs. It is accessible from the ocean at all seasons.

NARMARCUNGAWACK, N. H. a branch of the Ameriscoggin, rises in the tsp. of Success, and unites with the main stream in Paulsburgh.

NASH, co. N. C. bounded S. W. by Contentny creek, separating it from Johnson, W. and N. W. by Franklin, N. E. by Fishing creek, separating it from Halifax, and by Edgecombe E. and S. E. Length 36 ms., mean width 18, and area 648 sq. ms. Extending in lat. from 35° 42′ to 36° 13′. The declivity of this co. is to the S. E. by E., and drained by various branches of Tar r. Chief t. Nashville. Pop. 1820, 8,185, 1830, 8,490.

NASH AND SAWYER'S LOCATION, a tract of 2,184 acres, granted May 20th, 1773, to Nash and Sawyer, for exploring a route through the White mnts.

NASHAWN, one of the Elizabeth isls. on the S. E. side of Buzzard's bay, 9 ms. long, and 2 broad.

NASHAWENNA, another of the Elizabeth isls. lying between Cutahunk and Presque Isle.

NASH'S STREAM, N. H. a branch of the Upper Amonoosuck, has its sources in Stratford and the lands E., and unites with the r. in the N. W. part of Piercy.

NASHUA, r. a beautiful stream in the S. part of Hillsborough co., has its source in Worcester co. Mass., and falls into the Merrimack at Dunstable, N. H.

NASHUA, v. on the preceding r. in Dunstastable, Hillsborough co. 11 ms. from Amherst, 36 from Boston, and 32 from Concord. It is a manufacturing village, and a place of considerable business. The r. falls 65 feet in the distance of 2 ms.

NASHVILLE, p-v. and st. jus. Nash co. N. C. situated on Peach Tree creek, by p-r. 44 ms. N. E. by E. Raleigh. Lat. 35° 56′, long. 1° 2′ W. W. C.

NASHVILLE, p-t. and st. jus. Davidson co., and seat of the government of Ten. situated on the left bank of Cumberland r. Lat 36° 05′, long. 9° 43′ W. W. C., and by actual calculation, a small fraction above 565 statute ms. S., 70° W. W. C.; but by p-r. the stated distance between the two places is 709 miles. Nashville is 218 ms. S. W. Frankfort, Ky. 430 N. E. Natchez, and 480 N. N. E. New Orleans. The site is a high bank on the concave side of Cumberland r., the central point of a very fertile and well cultivated country. This flourishing town is accessible to steamboat navigation, and possesses all the features of a commercial depot, having numerous stores, a branch of the bank of the U. S., and two other banks. The university of Ten. is located in its vicinity, as are several manufactories. Pop. 1830, whites, 3,554; colored, 2,012; total, 5,566.

NASSAU, r. of Flor. gaining importance only as giving name to a co. This small stream rises in the angle between St. John's and St. Mary's rs. flows estrd. 30 ms. to its outlet by Nassau inlet to the Atlantic ocean, between Cumberland and Talbot's islands.

NASSAU, p-t. Rensselaer co. N. Y. 18 ms. S. E. Troy, 14 from Albany. Surface uneven. Vallies rich and fertile. 14 schools, continued 9 months in 12. Pop. 1830, 3,255.

NASSAU, northeastern co. of Flor., bounded by Nassau r. separating it from Duval co. S., by Duval co. S. W., St. Mary's r. separating it from Camden co. of Geo. W. and N., and by the Atlantic ocean E. Length from E. to W. 36 ms., mean width 16, and area 576 sq. ms. Extending in lat. from 30° 27′, to 30° 46′, long. from 4° 40′ to 5° 14′ W. W. C. Cumberland isl. constitutes the outer part of this co. towards the Atlantic ocean. The general surface of the co. is level, part marshy. Chief t. Fernandina. Pop. 1830, 1,511.

NATCHAUG, r. Conn. joins the Shetucket in Windham.

NATCHEZ, city, p-t. and st. jus. Adams co. Miss., is situated on the left bank of Miss. r. at lat. 31° 33′, long. 14° 30′ W. W. C. 322 ms. above New Orleans, following the bends of the Miss., but only 157 over lake Pontchartrain, and thence by the road direction nearly N. W., and by p-r. 98 ms. S. W. Jackson, the seat of government.

At Natchez, the bluff reaches the r. and is entirely composed of clay unmixed with the smallest pebble; the whole rising on a substratum of pudding stone rock. The rock, however, lies below the higher level of the r., and is only visible at a very low stage of the water in that stream. It is loose, friable, and much admixed with petrifactions of wood. Above this rock rests the clay superstrata, admixed with sand, and in some places, in digging wells, beds of sand are detected. The surface of the ground on which the city stands, and that of the whole adjacent co. is waving, not unlike a sea in a storm, and curiously contrasted with that of La. on the opposite side of the Miss. The streets of Natchez are extended at right angles; many of the houses are elegant, though generally the style of building is plain. It contains several places of public worship; the prevailing sects are Presbyterian, Roman Catholic, Methodist and Baptist. The public edifices are a court-house, jail, and bank. The Natchez bank, with three branches, is the only one in the state, and by its charter, has a pledge that no other banking institution shall be created by the legislature of the state before 1840.

The pleasantly waving site of Natchez, rising from 100 to 200 feet above high water in Miss. affords an airy, and for 9 months in the year, a healthful, agreeable, and advantageous residence. The author of this article resided many years in Natchez, and from his observation found the city in most seasons healthful to residents. There are, however, casual seasons, when all classes are subject to bilious and remittent fevers. There is perhaps no other city of the U. S. where the amount of manufacturing and commercial business bears so large a proportion to its population. In 1810, the total population was 1,511, in 1820, 2,184, and in 1830, 2,789. In 1820, the exports of cotton exceeded 35,000 bales. The quantity of goods sold here as early as 1800, was very great; and within the last 30 years has been constantly increasing. The city is a corporation, governed by a mayor, aldermen, and city council.

NATCHITOCHES, northwestern parish of La., bounded by the parish of Claiborne N. E., Rapides S. E., Opelousas S., Sabine r. separating it from Texas S. W., and by a meridian line from lat. 32° to 33°, also separating it from Texas N. W., and by Lafayette co. in Ark. N. Length S. to N. 150 ms., mean width 40, and area 6,000 sq. ms. Extending in lat. from 31° to 33°, and in long. from 15° 32′ to 16° 24′ W. W. C. Considerably the largest part of this very extensive parish is barren pine wooded land, or equally sterile oak flats. The alluvion of Red r. is, however, to this character of soil, a complete exception. The lower and southern section of the parish is traversed by Red r. and its numerous outlets, affording some of the finest cotton lands in La. This is the only tolerably well peopled part of the parish, and of the pop. of 7,486, in 1820, the far greater part were resident in the town and vicinity of Natchitoches, the st. jus. Pop. 1830, 7,905.

NATCHITOCHES, p-t. and st. jus. for the parish of Natchitoches, La. is situated on the right bank of Red r. at lat. 31° 44′, long. 16° 10′ W. W. C., 355 ms. by the road through Attacapas and Opelousas, N. W. by W. New Orleans, and as stated on the p-o. list 1,339, S. W. by W. W. C. This v. is built chiefly in one street along the r. at the foot of a bluff. Not quite 1 m. S. of the present town is the spot where the original French settlement was made in 1717.

Natchitoches is the extreme southwestern entrepot of the U. S. towards Texas, and has been consequently a place of importance ever since the acquisition of La. by the U. S. In itself it is a very pleasantly situated village.

NATICK, p-t. Middlesex co. Mass. 17 ms. W. Boston, situated on Charles r.; a pleasant farming town. Here labored the apostolic Elliott among the Natick Indians. By his advice, they adopted the form of government proposed by Jethro to Moses, choosing one ruler of a hundred, two rulers of fifties, and ten rulers of tens. There is an extensive wheel factory at Natick bridge. Pop. 1830, 890.

NATURAL BRIDGE, a fine deviation from the ordinary course of nature in the phenomena of streams. A small water course called Cedar creek in the southern angle of Rockbridge co. Va. before it joins James r., passes under a natural arch of rocks, affording a splendid assemblage of bold and contrasted features in scenery. A visit to the Natural Bridge can be rendered still more interesting from the proximity to the peaks of Otter. This highest part of the Appalachian system S. W. from the Del. rises 10 ms. S. from the Natural Bridge. On Tanner's maps the Natural Bridge is laid down at lat. 37° 35′, long. 2° 34′ W. W. C., 14 ms. S. W. Lexington, and 180 W. Richmond.

NATURAL BRIDGE, p-v. in the southern part of Rockbridge co. Va. 16 or 17 ms. S. W. Lexington, the co. st., 30 ms. N. W. Lynchburg, and by p-r. 224 S. W. by W. W. C., and 176 a very little S. of W. Richmond.

NANDAWAY, r., a confluent of Mo. rises about lat. 42°, interlocking sources with the Racoon fork of Des Moines, Grand, and Nishnebatona rs.; flowing thence by a general course of a little W. of S., falls into Mo. at lat. 39° 55′, about 70 ms. in a direct distance above the influx of Kansas r. The valley of the Nandaway lies between those of Nishnebatona and Grand rs. in long. between 17° and 18° 10′ W. W. C.

NAUGATUCK, r. Conn. rises in the N. W. part of the state, and joins the Housatonic at Derby. Above Waterbury, it is called Mattaluck.

NAYLOR'S Store and p-o. St. Charles co. Mo. by p-r. about 25 ms. wstrd. St. Louis.

NAZARETH, Lower and Upper, two contiguous tsps. of Northampton co. Pa. on Bushkill and Manakissy creeks, about 8 ms. nthwestward Easton. The joint pop. 1820, 1,747, 1830, 2,146.

NAZARETH, p-v. Northampton co. Pa. 7 ms. N. W. Easton, and 10 N. Bethlehem. This v. belongs to the Moravian society, and contains a school of that sect.

NEDDOCK, Cape, York, Me., York co. Long. 6° 20′ E. W. C., lat. 43° 8′. It is a rocky, barren bluff, with a small population of poor fishermen.

NEEDHAM, p-t. Norfolk co. Mass. 12 ms. S. W. Boston, on Charles r. Soil coarse, and surface uneven. Here is a perpendicular fall in the river of 20 feet, at which mills are erected. Pop. 1830, 1,418.

NEFFSVILLE, p-v. Lancaster co. Pa. by p-r. 39 ms. estrd. Harrisburg.

NELSON, p-v. Tioga co. Pa. by p-r. 162 ms. nthrd. Harrisburg.

NELSON, p-t. Cheshire co. N. H. 40 ms. from Concord, on the height of land between the Conn. and Merrimack rs. Surface hilly, but good for grazing. Streams small. Contains mill privileges. Pop. 1830, 875.

NELSON, p-t. Madison co. N. Y. 6 ms. W. Morrisville, 109 W. N. W. Albany. Situation elevated. Soil good and fertile. It is better for grass than grain. 15 schools, continued 7 months in 12. Pop. 1830, 2,445.

NELSON, co. Va. bounded by the Blue Ridge, separating it from Rockbridge W. and Augusta S. W., by Albemarle N. E. and E. James r. separating it from Buckingham S. E., and Amherst S. and S. W. The longest line is a diagonal from the extreme southern to the extreme northern angle, about 40 ms.; the co. is in form of a trapezium; greatest breadth 28 ms., and area 560 sq. ms. Extending in lat. from 37° 32′ to 38° 02′, long. from 1° 50′ to 2° 7′ W. W. C. Declivity S. of S. E., and is drained by the different branches of Rock and Tye rs. The surface hilly, and towards James r. traversed by South mtn. Chief t. Lovington. Pop. 1820, 10,137, and in 1830, 11,251.

NELSON, p-v. Portage co. O. by p-r. 146 ms. N. E. Columbus.

NELSON's p-o. Robeson co. N. C. by p-r. 68 ms. S. S. W. Raleigh.

NELSONVILLE, p-o. Athens co. O. by p-r. 59 ms. S. E. Columbus.

NEMAWHAW, the name of two confluents of Mo. called relatively Great and Little Nemawhaw. Great Nemawhaw rises between the vallies of the Republican fork of Kansaw r., and Platte r., and between lat. 40° and 41°, and about 21° long. W. W. C. Flowing thence by a course of a little S. of E. 170 ms. falls into the Mo. at lat. 40° 05', and by direct distance 70 ms. above, and N. W. from the influx of Nandaway.

Little Nemawhaw, a very inferior stream to the preceding, falls into the right side of Mo. a short distance above the influx of the Nishnebatona, after a general course of about 70 ms. from the northwestward.

NEPONSET, r. Mass. flows into Boston harbor, and is navigable for vessels of 150 tons 4 ms. to Milton.

NEPONSET, v. on both sides of Neponset r. 6 ms. S. Boston. Contains a number of mills and manufacturing establishments.

NESCOPECK mtn. in the southern part of Luzerne co. Pa. between Wapwallopen and Nescopeck creeks. The local name is confined to a ridge of about 12 ms. in length; but it is merely a ridge of the chain which separates the vallies of the Lehigh and Lackawaxen r. from that of the Susquehannah, and which rises into bold peaks to the estrd. of Wilkesbarre.

NESCOPECK, creek, in the southern part of Luzerne co. Pa. interlocking sources with the extreme western creeks of the Lehigh, and flowing wstrd. into the Susquehannah opposite Berwick. The valley of the Nescopeck lies between those of the Catawissa and Wapwallopen creeks.

NESCOPECK, p-v. and tsp. Luzerne co. Pa. The v. stands on the left bank of Susquehannah r., above the mouth of Nescopeck creek, and opposite the borough of Berwick, by p-r. 86 ms. above and N. N. E. Harrisburg.

NESHAMINY, small r. or large creek of Bucks co. Pa., heads partly in Montgomery co., but mostly in the central part of Bucks, interlocking sources with the Tohickon, Perkiomen, and Wissihickon creeks, flows southestrd. into Del. r., which it joins 4 ms. below Bristol, after a comparative course of about 25 ms.

NESHANOCK, creek of Mercer co. Pa., the eastern branch of Shenango. (*See Shenango r.*)

NETHER PROVIDENCE, p-v. Del. co. Pa. by p-r. 124 ms. N. E. W. C.

NETTLE creek and p-o. in the northwestern part of Wayne co. Ind. by p-r. 61 ms. estrd. Indianopolis.

NEUSE, r. N. C. rises in Person and Orange cos. interlocking sources with those of Haw r. branch of Cape Fear r., and Dan r. branch of Roanoke. The different higher constituents unite in the N. W. angle of Wake, and crossing that co. and Johnson in a southeasterly direction, it thence enters Wayne, and assuming an easterly course over the latter, Lenoir and Craven cos. to Newbern. Now gradually opening into a wide bay, curving first S. E. and thence N. E. into Pamlico sound between Beaufort and Carteret cos. The valley of the Neuse lies between those of Cape Fear and Tar rs. The length of the Neuse, by comparative courses, is about 200 ms. The valley, independent of the great bends of the stream, 180, but comparatively narrow, the mean breadth not averaging above 25 ms., and area 4,500 sq. ms., lying between lat. 34° 50' and 36° 22', and between long. 0° 30' E. to 3° 10' W.

NEVERSINK, or NAVISINK, t. Sullivan co. N. Y., 15 ms. N. Monticello, 30 W. Kingston. 9 schools, continued 7 months in 12. Pop. 1830, 1,257.

NEVILLE, p-v. Clermont co. Ohio, by p-r. 123 ms. S. W. Columbus.

NEW ALBANY, p-v. Bradford co. Penn. by p-r. 116 ms. nrthrd. from Harrisburg.

NEW ALBANY, p-v. and st. jus. Floyd co. Ind. by p-r. 121 ms. a little E. of S. Indianopolis. It is situated on the right bank of O. r. at the foot of the rapids, and nearly opposite Shipping port in Ky. Mr. Flint states that the main street is 3-4 of a mile in length. It has a convenient harbor for boats, and is a fine thriving v. Pop. 1830, 1900.

NEW ALBION, t. Cattaraugus co. N. Y. Pop. 1830, 380.

NEW ALEXANDER, p-v. Columbiana co. O. by p-r. 138 ms. N. E. Columbus.

NEW ALEXANDRIA, p-v. Westmoreland co. Penn. 11 ms. N. E. from the borough of Greensburg, & 8 by p-r., 171 ms. wstrd. Harrisburg.

NEW ANTRIM, p-v. Washington co. Va. by p-r. 353 ms. S. W. by W. W. C.

NEWARK, or ARTHUR KULL, bay, N. J. formed by the confluence of the Passaic and Hackinsack rs. and separated from Hudson r. on the E. by Bergen neck. It communicates through the kills, 4 ms. long, with N. Y. bay, and through Staten isl. sound with Amboy bay.

NEWARK, p-t. Tioga co. N. Y. 8 ms. N. N. E. Owego. Pop. 1830, 1027.

NEWARK, p-t. and cap. Essex co. N. J. the most populous t. in the state, is on the W. side of Passaic r. 3 ms. from its mouth, in Newark bay; 9 ms. W. N. Y., 5 N. E. Elizabethtown, and a remarkably beautiful and flourishing place. It is noted for the variety and excellence of its manufactures; particularly carriages, saddlery, leather, shoes and jewelry, which are sold in different parts of the U. S. to a great amount. About 2,000,000 of pairs of shoes are said to be produced annually by one manufactory. There are quarries of excellent free stone in the vicinity, which are extensively worked for N. Y. and other places. The Newark cider, which is made near this place, is produced from two or three sorts of apples, and is of proverbial excellence. The Morris canal, terminating at this place, affords great advantages and has added to its trade, pop. and enterprize. There are a fine C. H., academy, 3 banks, and several churches, for Presbyterians, Episcopalians, Baptists, Methodists and Catholics; some of them are very large and beautiful. The v. is situated on a beautiful level, and principally on a fine street of remarkable breadth and straightness. Pop. 1830, 10,953; 1832, supposed to be more than 12,500.

NEWARK, p-v. in the N. western part of New Castle co. Del. 12 ms. S. W. by W. Wilmington, 52 ms. N. N. W. Dover, and 113 ms. N. N. E. W. C.

NEWARK, p-v. in the S. eastern part of Worcester co. Md. by p-r. 158 ms. S. E. by E. W. C.

NEWARK, p-v. in the S. eastern part of Louisa co. Va. by p-r. 31 ms. N. W. Richmond.

NEWARK, p-v. and st. jus. Licking co. O. by p-r. 34 ms. a little N. of E. Columbus, and 362 ms. a little W. of N. W. by W. W. C. lat. 40° 04', long. W. C. 5° 27' W. It is situated at the main forks of Licking cr. and on the O. and Erie canal, and contains the usual co. buildings, several stores, 2 printing offices, 2 ware houses, market house, 5 or 6 taverns, several schools, and 2 or 3 places of public worship. Pop. 1830, 999. The elevation of the water in the canal at Newark is 834 feet above the mean level of the Atlantic tides, and 360 feet above the mean level of O. r. at the mouth of Sciota.

NEW ASHFORD, t. Berkshire co. Ms. 20 ms. N. Lennox, 121 from Boston. Pop. 1830, 285.

NEW ATHENS, p-v. in the S. eastern part of Harrison co. O. by p-r. 130 ms. a little N. of E. Columbus, and 6 ms. S. Cadiz, the co. seat. Pop. 1830, 198.

NEW BALTIMORE, Greene co. N. Y. 16 ms. N. Catskill, 20 S. Albany. Watered by Coxsackie and Haanekrai crs. which supply mill seats in abundance. Surface, broken; soil, diversified. Has a landing on the Hudson. There is a spring in this t. which is said to rise and fall at certain periods. 10 schools, continued 9 months in 12. Pop. 1830, 2,370.

NEW BALTIMORE, p-v. in the eastern part of Fauquier co. Va. 45 ms. wstrd. W. C.

NEW BARBADOES, t. Bergen co. N. J. W. Hackensack r. Pop. 1830, 1,693. Hackensack, the st. jus. is a v. in this t.

NEW BEDFORD, p-t. and port of entry, Bristol co. Mass. 52 ms. S. Boston, lat. 41° 38', long. 6° 10' E. W. C. It is beautifully situated on the W. side of the Acushnet r., which here empties into Buzzard's bay. It is chiefly built of wood on an inclined plane, and presents a lively and picturesque appearance. This is one of the most flourishing towns in New England, as is indicated by the rapidity of its growth, and the wealth and enterprize of its inhabitants. The citizens are much engaged in commerce, but the whale fishery constitutes the chief business of the place. A steamboat runs to Nantucket, and sometimes is used for towing vessels over the bar. Here are three banks, whose united capital is *nine hundred thousand dollars;* three insurance offices, each with a capital of 250 000 dollars; ten places of public worship, 3 Baptists, 2 Presbyterians, 2 Methodist, 1 Unitarian, 1 Quaker, and 1 Roman. There are seven considerable manufactories of sperm candles, and there are employed fifty thousand tons of shipping in the foreign and whale fishery—forty thousand, probably, engaged in the whale business—about 1,200 tons in the cod and mackerel fishery, and 8,000 tons coastwise. The number of foreign clearances at the port of New Bedford, 1831, was 101, and of foreign entries 83. Of the arrivals 58 were from whaling voyages, importing 41,144 bbls. of spermaceti oil, 53,-145 bbls. whale oil, and 381,000 lbs. whalebone. There remained at sea, on whaling voyages, at the end of the year, 100 ships, 9 barques, and 7 brigs, measuring 35,208 tons, navigated by 2,635 men. Of these vessels, 56 are in the Pacific ocean, and the rest on the Brazil Banks, in the Indian ocean and elsewhere. The whole tonnage of the district is 55,588. Pop. 1820, 3,947; 1830, 7,592.

NEW BEDFORD, p-v. S. W. part of Mercer co. Penn. 15 ms. S. W. from the borough of Mercer, and 55 N. W. Pittsburg.

NEW BEDFORD, p-v. Coshocton co. O. by p-r. 99 ms. a little N. of E. Columbus. Pop. 1830, 51.

NEW BERLIN, p-t. Chenango co. N. Y. on the W. bank of the Unadilla, 7 ms. N. E. Norwich, 93 W. Albany. It is supplied with good mill seats by the Unadilla, and some of its branches. Here are manufactories on a large scale. 14 schools, continued 7 months in 12. Pop. 1830, 2,643.

NEW BERLIN, p-t. and st. jus. Union co. Pa. by p-r. 60 ms. N. N. W. Harrisburg, 11 ms. W. Sunbury, lat. 40° 52', and very nearly on the meridian of W. C.

NEWBERN, p-v. in the western part of Montgomery co. Va. 16 ms. S. W. by W. Christiansburg, and by p-r. 324 ms. S. W. by W. W. C.

NEWBERN, p-t. and st. jus. Craven co. N. C. situated on the point above the union of the Neuse and Trent rs. by p-r. 351 ms. nearly due S. W. C. and 119 S. E. by E. Raleigh. Newbern was long the seat of government of N. C., and is still the largest t. of the state. It is a port of entry, and though large vessels cannot ascend Neuse bay, the trade is considerable in lumber, tar, turpentine, pitch, &c. Pop. 1820, 2,467; 1830, 3,776.

NEWBERRY, p-v. on the point above the junction of Lycoming cr. with Susquehannah r. Lycoming co. Pa. 2 ms. W. Williamsport, and 89 N. N. W. Harrisburg.

NEWBERRY, district of S. C. bounded by Laurens W. and N. W., Union N., Broad river, separating it from Fairfield N. E., Lexington S. E., and the Saluda r. separating it from Edgefield, S. and S. W. Length 26 ms., mean breadth 20, and area, 540 sq. ms. Extending in lat. from 34° 03' to 34° 30', and in long. 4° 20' to 4° 55' W. W. C. The dividing ridge between the sources of waters flowing

s. eastward into Saluda, and N. eastward into Broad and Ennoree rs. traverses this co. and subdivides it into two not very unequal inclined plains. Bush r. and Little r. both rising in Laurens, flow s. eastward over the wstrn. part of Newberry and falling into Saluda. Ennoree r. forming a part of the northern boundary, then enters Newberry, and falls into Broad r. in the N. eastern angle of the district. The N. eastern declivity thus falling towards Ennoree and Broad rs. is drained by Cannon's and Keller's crs. flowing eastward into the latter, and by King's, Indian, and Duncan's crs. flowing N. eastward into the former. There is much excellent soil in Newberry. Staples, cotton, grain &c. Chief town, Newberry. Pop. 1820, 16,104; 1830, 17,441.

NEWBERRY, p.t. and st. jus. Newberry district, S. C. situated near the centre of the district, by p-r. 43 ms. N. W. by W. Columbia, lat. 34° 12', long. 4° 23' W. W. C.

NEWBERRY, p-v. in the s. western part of Geauga co. O. by p-r. 147 ms. N. E. Columbus. Pop. of Newberry t-sp., 1830, 594.

NEWBERRY TOWN, p-v. York co. Pa. 2 ms. N. W. by W. from the borough of York, and 14 ms. S. S. W. of Harrisburg.

NEWBERRY TOWN, (see Newberry, York co. Pa.)

NEWBIGGEN cr., p-o. Pasquotank co. N. C. 9 ms. S. E. Elizabeth city, and by p-r. 190 N. E. by E. Raleigh.

NEW BLOOMFIELD, p-v. and st. jus. Perry co. Pa. by p-r. 36 ms. S. W. Harrisburg.

NEWBORN, p-v. Jasper co. Geo. by p-r. 63 ms. northwestward Milledgeville.

NEW BOSTON, p-t. Hillsborough co. N. H., 9 ms. from Amherst, 22 S. Concord, 57 from Boston. Watered by the S. branch of the Piscataquog, and several other streams. This is a mountainous t. In the S. part is a considerable elevation, on one side of which it is nearly perpendicular. Its height, taken from the road through the notch of the hill, is 572 ft. Pop. 1830, 1,684.

NEW BRAINTREE, p-t. Worcester co. Mass. 18 ms. W. N. W. Worcester, 66 W. Boston. It is excellent grazing land, with fine hills, well watered. Pop. 1830, 825.

NEW BRITAIN, p-v. of Berlin, Hartford co. Conn. 10 ms. S. W. Hartford. Here are various and extensive manufactures of brass, and plated ware, of different kinds; three manufactories of suspenders,—one of silver spoons, and another of machinery for cotton factories, which is operated by steam power. These and similar causes have rendered this one of the most thriving and pleasant villages in the state.

NEW BRITAIN, p-v. Bucks co. Pa. 24 ms. nrthrd. Phil.

NEW BRUNSWICK, city, Middlesex co. N. J., S. W. Raritan r. which is navigable to this place for vessels of 80 tons, 16 ms. N. E. Princeton, 33 S. W. N. Y., 57 N. E. Phil. The situation is low, but it is not unhealthy. There is a bed of peat of great size 2 or 3 ms. E. of this city, and ½ a mile from the Raritan, depth about 11 ft. It is estimated that 5 or 6 millions of chaldrons per annum could be extracted for 25 years. Three chaldrons of this peat are believed to be equal to one of coal. Here is Rutgers college, founded by ministers of the Reformed Dutch church, and likewise a Dutch Reformed theological seminary, partly connected with the college. Pop. 1830, 7,831.

NEW BUFFALO, p-v. Perry co. Pa. by p-r. 20 ms. nrthwstrd. Harrisburg.

NEWBURGH, p-t. Penobscot co. Me. Pop. 1830, 626. N. Waldo co. 54 ms. E. Augusta.

NEWBURGH, p-t. and half shire town, Orange co. N. Y. on the W. bank of the Hudson, 95 ms. S. Albany, and 70 on the stage road N. New York. It is good for farming. Contains mill seats in abundance. The village of Newburgh commands a very extensive trade with the country on the W., and by navigation of the Hudson, with N. Y. It is incorporated and is handsomely laid out in streets and squares. 13 common schools continued 10 months in 12. Here is an academy, and there is an extensive cannon foundry, on Chamber's creek. Pop. 1830, 6,424.

NEWBURG, p-v. sthwstrn. part of Cumberland co. Pa. 19 ms. S. W. by W. Carlisle, and 37 ms. a little S. of W. Harrisburg.

NEWBURGH, p-v. Cuyahoga co. O. 6 ms. S. E. Cleaveland, the co. seat, and by p-r. 144 ms. N. E. Columbus. Pop. of Newburgh township 1830, 869.

NEWBURGH, p-v. nrthwst. part of Warrick co. Ind. by p-r. 181 ms. S. S. W. Indianopolis.

NEW BURLINGTON, p-v. wstrn. part of Hamilton co. Ohio, 12 ms. from Cincinnati, and by p-r. 124 ms. S. W. Columbus.

NEWBURY, p-t. Orange co. Vt. W. Conn. r., 27 ms. E. Montpelier, 47 N. E. Windsor. Well supplied with mill streams. Contains several mineral springs. Two bridges cross the Conn. from different parts of this town. The legislature has holden two sessions in this place; the one in 1787, the other in 1801. Here is the bend in the Conn. denominated the Great Ox Bow. Pop. 1830, 2,252.

NEWBURY, t. Essex co. Mass. S. Merrimack r., opposite Salisbury, with which it is connected by a bridge, 32 ms. N. E. Boston. Land in general of an excellent quality. Parker r. a fine mill stream, falls nearly 50 feet in the course of 1½ ms. in this town. Limestone of a good quality is found here; also marble, serpentine, amianthos, asbestos, and arsenical iron pyrites. Here are two academies. Pop. 1830, 3,603.

NEWBURY, p-v. and tsp. York co. Pa. The village is situated 10 ms. S. S. E. Harrisburg, and 14 N. N. W. from the borough of York.

NEWBURY, district and p-t. S. C. (*See Newberry.*)

NEWBY'S BRIDGE, and p-t. Perquimans co. N. C. by p-r. 209 ms. N. E. by E. Raleigh.

NEWBURYPORT, p-t., port of entry, and one of the shire towns of Essex co. Mass., S. of Merrimack r. 3 ms. from its mouth, 38 N. E.

Boston. It is one of the handsomest towns in the U. S., and the smallest t. for land, containing but 647 acres. A turnpike and bridge connects this t. with Plumb isl. A handsome bridge thrown across the Merrimack and suspended by chains, connects it with Salisbury. It is well situated for ship building, having the advantage of receiving lumber by the Merrimack. The harbor is deep, safe, and spacious, but difficult to enter. The t. suffered severely by the restrictions on commerce, previous to the late war, and by fire in 1811. Here was the only stocking factory in the U. S. in 1831. A small silk factory has likewise been established at this place. Pop. 1830, 6,375.

NEWBY'S CROSS ROADS, and p-o. Culpepper co. Va. by p-r. 70 ms. N. W. by W. W. C.

NEW CANAAN, p-t. Fairfield co. Conn. 8 ms. N. Long Island sound, 77 ms. S. W. Hartford. Surface mountainous. Soil a hard gravelly loam, tolerably well timbered. 9 school districts and 1 academy. Pop. 1830, 1,826.

NEW CANTON, p-v. on the right bank of James r., at the mouth of State creek, and in the nrthestrn. part of Buckingham co. Va., 63 ms. W. and by land from Richmond.

NEW CANTON, p-v. nrthestrn. part of Hawkins co. Ten., by p-r. 244 ms. a little N. of E. Nashville.

NEW CARLISLE, p-v. Clarke co. O. by p-r. 66 ms. wstrd. Columbus, and 23 ms. wstrd. Springfield, the co. seat. Pop. 1830, 343.

NEW CARTHAGE, p-v. Concordia parish, La. by p-r. 284 ms. N. W. New Orleans.

NEW CASTLE, p-t. Lincoln co. Me. W. of Sheepscot r., 7 ms. E. Wiscasset. Pop. 1830, 1,544.

NEW CASTLE, or GREAT ISLAND, isl. and t. Rockingham co. N. H., lat. 43° 5'. It is a rough and rocky isl. in Portsmouth harbor. It is connected with Portsmouth by a handsome bridge. Fishing is here pursued with success. Fort Constitution and the light house stand on this isl. Pop. 1830, 845.

NEW CASTLE, t. West Chester co. N. Y. 37 ms. N. N. Y., 128 S. Albany, 6 W. Bedford; 10 schools continued 7 months in 12. Pop. 1830, 1,336.

NEW CASTLE, p-v. on the peninsula between Shenango and Neshanock creeks, and near the S. border of Mercer co. Pa. 18 ms. S. S. W. from the borough of Mercer, 41 N. N. W. Pittsburg, and 264 ms. N. W. W. C.

NEW CASTLE, nrthrn. co. of the state of Delaware, bounded by Kent co. of the same state S., Kent co. of Md. S. W., Cecil co. of Md. S. W., Chester co. of Pa. N. W., Delaware co. Pa. N., and by Del. r. separating it from Salem co. N. J. E. Length from S. to N. 38 ms., mean breadth 12, and area 456 sq. ms. Extending in lat. from 39° 18' to 39° 50', and long. from 1° 17' to 1° 38' E. W. C. The line of demarcation between Md. and Del. states following, particularly in the northern part, very nearly the dividing ridge or summit, separating the sources of creeks flowing westward into Chesapeake, from these pursuing an eastern course into Delaware bay; the slope of New Castle co. is consequently to the eastward. The northern part is traversed and drained by the different confluents of Brandywine creek, which enters the Del. in the vicinity of Wilmington. Below the Brandywine, flow also into Del. in this co. the Appoquiniminck and Black Bird creeks. Duck creek on the S. separates New Castle from Kent. (*See Chesapeake and Delaware canal.*) Some parts of this co. towards Del. r. are low and marshy, but receding wstrd. and northwestward, the surface rises into waving hills, and though no where much elevated the interior is pleasantly diversified. The soil is mostly productive in grain, grasses and orchard fruit. The falls in the different branches of Brandywine have made the northern part of New Castle a manufacturing county. Chief towns, Wilmington and New Castle. Pop. 1820, 27,899, 1830, 29,710.

NEW CASTLE, p-v. and st. jus. New Castle co. Del. situated on the bank of Del. r. 5 ms. a little W. of S. Wilmington, 32 S. W. Phil., and by p-r. 103 ms. N. E. W. C. Lat. 39° 40', long. 1° 24' E. W. C. The village of New Castle extends lengthwise along the Del., and is tolerably compact and well built. The site is a rising plain, and the Hundred, in 1810, contained a pop. of 2,438, in 1820, 2,671, in 1830, 2,463.

NEW CASTLE, p-v. in the forks of Craig's creek, western part of Botetourt co. Va., 15 ms. a little S. of W. Fincastle, and by p-r. 210 ms. S. W. by W. W. C.

NEW CASTLE, p-v. Wilkes co. N. C. by p-r. 175 ms. a little N. of W. Raleigh.

NEW CASTLE, p-v. and st. jus. Henry co. Ky. 24 ms. N. W. Frankfort, 38 ms. N. E. by E. Louisville, and by p-r. 564 ms. a little S. of W. W. C. Lat. 38° 25', long. 8° 08' W. W. C. Pop. 1830, 538.

NEW CHESTER, p-t. Grafton co. N. H., 16 ms. S. Plymouth, 24 ms. from Concord, 44 from Haverhill, 25 from Hanover, and 86 from Boston. Watered by Pemigewasset and Blackwater rivers, and several small streams. Timbered with white pine, birch, beech, hemlock, maple, &c. Pop. 1830, 1,090.

NEW COLUMBIA, p-v. in the northern part of Union co. Pa. 68 ms. N. N. W. Harrisburg.

NEWCOMB, t. Essex co. N. Y. Pop. 1830, 62.

NEWCOMB, p-v. Preble co. Ohio, by p-r. 8 ms. S. Eaton, the county seat, and 100 ms. a little S. of W. Columbus.

NEWCOMERSTOWN, p-v. in the sthwstrn. angle of Tuscarawas co. Ohio, 96 ms. N. E. by E. Columbus, and 12 ms. E. Coshocton. It is situated on Tuscarawas river, and on the O. and Erie canal. Lat. 40° 16'. Pop. 1830, 100.

NEW COVINGTON, p-v. in the northern part of Luzerne co. Pa. 19 ms. nrthrd. from Wilkesbarre, and by p-r. 144 ms. N. E. Harrisburg.

NEW CUMBERLAND, p-v. on the point above the entrance of Yellow Breeches creek into Susquehannah r., and in the extreme eastern angle of Cumberland co. Pa., 3 ms. s. Harrisburg.

NEW DERRY, p-v. Westmoreland co. Pa. 6 ms. estrd. Greensburg, the co. t., by p-r. 188 ms. N. w. W. C.

NEW DESIGN, p-v. Trigg co. Ky. by p-r. 217 ms. s. w. by w. Frankfort.

NEW DURHAM, p-t. Strafford co. N. H. Surface very uneven, a portion so rocky as to be unfit for cultivation. It is well watered. In this town there is a remarkable cave. Pop. 1830, 1,162.

NEW ENGLAND, a name given to the six states of the Union lying east of New York, viz. Maine, New Hampshire, Vermont, Massachusetts, Rhode Island, and Connecticut. It is bounded N. by Lower Canada, E. by N. Brunswick, s by the Atlantic ocean, and Lond Island sound, and w. by New York. It lies between 41° and 48° 12′ N. lat., and between 2° 45′ and 10° long. E. W. C., and contains 65,475 sq. ms.

The inhabitants are almost exclusively of unmixed English origin, and though never united as a political whole, they have at different periods been connected by their common interests From the earliest settlement of their country they have enjoyed peculiar advantages for literary and religious instruction, and being trained to habits of industry, economy and enterprize, by the circumstances of their peculiar situation, as well as by the dangers of prolonged wars, they present traits of character which are considered as remarkable abroad, as they are common and universal at home.

Some of the first settlements were made in the territory of Maine, which had been visited by Martin Pring, an English navigator, in the years 1603, and 1606; but the most important was that of Massachusetts, which was commenced in 1620, by the Pilgrim forefathers of New England, who had been expelled from England for asserting liberty of conscience, and who found Holland not sufficiently remote from their oppressors, to secure to themselves or their offspring, the civil and religious blessings which they desired. While the French missions, and the English colonies in Maine, have scarcely left any traces of their existence, the principles which were regarded as fundamental by the Plymouth Pilgrims, have produced effects which may be more or less plainly traced in the institutions and condition of all the United States, and have diffused an influence which is felt at the present day in every country of Europe. As early as 1638, Harvard college was founded, and in 1647, the legislature of Mass. passed a law making effectual provision for the instruction of every child in the rudiments of learning. The support of public worship was also legally provided for. In consequence of these and similar enactments, the people are generally well instructed and moral; and from them has been furnished a large portion of the learned and influential men who have figured in other parts of the Union. In Rhode Island, where no provision was made by law for the support of either learning or religion, the experience of many years has induced the people to take measures to secure, as far as possible, the advantages in these respects, enjoyed by the other parts of New England.

A large part of the distinguished men of the U. S., have been educated at Harvard & Yale colleges; and though there are many respectable institutions of learning in other parts of the country, still, many students from the s. and w. are annually taught in the colleges of New England. Teachers of schools, of all descriptions and in different states, are derived from the same quarter of the Union, education being so easily and cheaply obtained, that instructers are to be found in abundance. Many defects have hitherto existed in the systems of popular education in N. England; but notwithstanding all such impediments, she has maintained the superiority in common instruction and general intelligence, not only in the U. S., but probably also in the world. Improvements, however, have been commenced: Mass. with liberal and enlighted views, taking the lead in measures which promise much for the interests of education.

Evidence of the good morals of the New Englanders might be adduced from various facts, did the nature and limits of this work permit. The Pequod war, in 1634, placed in their power the first, and it is believed, the only land ever claimed on the ground of conquest. The laws of the colonies forbade any land to be obtained from the Indians by individuals, and the government frequently paid for the same tracts, several times over, to avoid the imputation of injustice. Crimes have always been comparatively rare, and duelling is almost unknown in their criminal records. Criminals have generally been among those who were least instructed, and the conviction is deep and general in N. England, that the general diffusion of learning and religion is indispensable to the good order of society, and to the existence of a free and popular government.

The intelligence and enterprizing spirit of the people are seen in the expedients to which they resort to obtain a livelihood at home, as well as their judgment and foresight in choosing places to which to emigrate. They have never found the means of accumulating wealth, or even of subsisting, without persevering labor and economy; their soil and climate offered no attractions to adventurers, and their simple habits and strict rules of society, are unpalatable to persons of that class.

The early circumstances of New England obliged its inhabitants to dwell in villages, as the Indians could thus be best resisted; the first settlers were thus confined to a few spots on the coast, long enough to discipline them in the political, intellectual, and religious principles of the pilgrims; so that, tho' the emigrants from England brought over much ignorance, and even vice, they were restrained, if not entirely reformed by the pure and intelligent society to which they were introduced. At every step of their progress in extending their settlements, the colonists carried with them their schools and churches. Had not the population been prevented by circumstances, from spreading too fast, this probably could not have been the case, and it would have degenerated both intellectually and morally. Many of the pilgrim settlers, had been men distinguished for their learning and piety in England, and their influence produced happy and permanent impressions on the community, which they had aided in founding. These influences extended to all the early settlements, and have been still more widely diffused by the amount of emigration which has taken place in later years, from N. England to various parts of the country, especially the western states. The early colonies first spread slowly along the coast, then along Conn. r.; and afterwards, as the strength of the people increased, and their enemies diminished, gradually occupied the remaining territory of Mass., Conn., R. I., and the lower parts of N. H. and Vt. The close of the war of the revolution opened the adjacent states to the colonists of New England, and every opportunity has been improved for extending their settlements. Considerable portions of N Jersey, N. York, and a part of Pennsylvania were settled by New Englanders; and Ohio, which within 30 years has grown up from a wilderness to an important state, derived a large part of its inhabitants, and most of its enterprize and prosperity, from New England emigrants; the same is true to a less extent, of Illinois, Michigan Territory, &c.: and emigrants now proceed every year to those states, to Florida, Texas, and even to the Oregon Territory, with as much readiness and confidence of success, as they once did to N. Y., or in earlier days to the Conn. river.

There are in N. England 12 colleges, 3 in Mass. 3 in Conn, 2 in Me., 2 in Vt., 1 in N. Hampshire, and 1 in R. I.: 6 theological seminaries, 4 of which are in Mass., 1 in Me., 1 in Conn.: 8 medical schools, 2 in Me., 3 in Vt., 2 in Mass., and 1 in Conn.: 3 law schools, 1 in Mass. and 2 in Conn.

The following table will show the number of newspapers and periodicals of the New England states, at different periods:

	1775.	1810.	1828.
Maine,			29
Massachusetts,	7	32	78
New Hampshire,	1	12	17
Vermont,		14	21
Rhode Island,	2	7	14
Connecticut,	4	11	33

The pop. of this portion of the U. S. has been gradually but not rapidly increasing. In 1700 it was about 120,000, and in Martin's London Magazine we find it stated in 1755, at 345,000; the troops in the provinces at that time, not being reckoned. The following is the pop. of the six N. E. states by the censuses of 1820 and 1830:

	1820.	1830.	Increase pr. ct.
Maine,	298,335	399,162	34
New Hampshire,	244,161	269,533	10
Vermont,	235,764	280,679	19
Massachusetts,	523,287	610,014	17
Rhode Island,	83,059	97,210	17
Connecticut,	275,248	297,711	8
Total,	1,659,854	1,954,609	

According to the census of 1830, the increase of the U. S. for the preceding 10 years was about 35 per cent. The average increase in the states of New England, during the same period, was 17½ per cent.

To prevent repetition, the reader is referred to individual states, and to the art. United States, for farther details, in agriculture, manufactures, arts, &c.

New Fairfield, Fairfield co. Conn., 64 ms. s. w. Hartford, 7 n. Danbury. Tsp. broken, soil hard and gravelly. Pop. 1830, 940.

Newfane, p-t. and st. jus. Windham co. Vt. 10 ms. w. Conn. r., 12 n. w. Brattleborough, 110 ms. from Boston, 80 from Albany, 110 from Montpelier, and 50 from Windsor. Well watered and supplied with mill seats. Diversified with high hills and deep vallies. Timbered with rock maple, beech, birch, walnut, oak, &c. and contains a variety of minerals. The centre village, which contains a C. H., jail, and academy, stands on an elevated situation, and affords a very extensive and picturesque prospect. From the meeting house may be seen some part of at least 50 towns, lying in Vt., N. H. and Mass. Here are a county grammar school, and 12 school districts. Pop. 1830, 1,441.

Newfane, p-t. Niagara co. N. Y., 276 ms. w. Albany, 10 n. Lockport. Pop. 1830, 1,448.

Newfield, t. York co. Me., 40 ms. n. w. York, 36 w. n. w. Portland, e. Strafford co. N. H. Pop. 1830, 1,286.

Newfield, p-t. Tompkins co. N. Y., 9 ms. s. w. Ithaca. Well watered; limestone plentiful, and some marle. 15 schools, 5 months in 12. Pop. 1830, 2,664.

Newfound lake, Grafton co. N. H., 6 ms. long from n. to s. and 2 broad. Communicates with the Merrimack at Bridgewater.

Newfound river mills, p-o. Hanover co. Va., 30 ms. northward Richmond, and by p-r. 98 ms. s. s. w. W. C.

New Gaillard, (*see New Gilead, Moore co. N. C.*)

New Garden, p-v. between Red and White Clay crs. Chester co. Pa., 45 ms. s. w. by w. Phil., 12 ms. n. w. by w. Wilmington, Del., and by p-r. 123 ms. n. e. W. C.

New Garden, p-v. Guilford co. N. C. by p-r. 82 ms. n. w. by w. Raleigh.

New Garden, p-v. western part Columbiana co. O., by p-r. 9 ms. w. New Lisbon, the co. st., and 142 ms. n. e. by e. Columbus.

New Garden, p-v. Wayne co. Ind., by p-r. 84 ms. e. Indianopolis.

New Geneva, p-v. on the right bank of Monongahela r. in the s. western part of Fayette co. Pa., 20 ms. by land sthrd. Brownsville, and by p-r. 217 ms. N. W. by W. W. C.

NEW GERMANTOWN, p-v. in the N. western part of Perry co. Pa., by p-r. 46 ms. wstrd. Harrisburg.

NEW GILEAD, formerly New Gaillard, p-v. Moore co. N. C., by p-r. 8 ms. s. westward Carthage, and 63 s. w. Raleigh.

NEW GLASGOW, p-v. N. western part of Amherst co. Va. 20 ms. N. N. E. Lynchburg, and by p-r. 175 ms. s. w. W. C., and 132 ms. nearly due w. Richmond.

NEW GLOUCESTER, p-t. Cumberland co. Me. 23 ms. N. Portland, and is crossed by a small stream flowing to the tide. Pop. 1830, 1,682.

NEW GRANTHAM, t. Cheshire co. N. H., 35 ms. N. W. Concord.

NEW HAMPSHIRE, one of the United States, bounded N. by Lower Canada, E. by Me. and the Atlantic ocean, s. by Mass. and W. by the Conn. r. which separates it from Vt. It lies between 42° 40′ and 45° 20′ N. lat., and between 4° 30′ and 6° 15′ E. long. W. C. Its extreme length is 168 ms., its greatest breadth 90, and its whole area, 9,491 sq. ms.

New Hampshire was first discovered in 1614, by Capt. John Smith, the English navigator, and was afterwards named by John Mason, to whom it was granted in 1622, by a patent in which it is called Laconia. The first settlements were made in the following year at Dover and Portsmouth. In 1629, the territory between the Merrimack and Piscataqua rs. and extending 60 ms. from the sea, which had previously been purchased of the Indians by the Rev. John Wheelwright, was granted to Mason alone, by whom it was then first called New Hampshire. In 1641, all the settlements of the state united themselves to Mass. and formed part of the county of Norfolk. In 1679, they were again constituted a separate province by Charles II., and in 1680, the first assembly convened. From 1689, with the exception of a short period, it was again united with Massachusetts, until 1741, when it was constituted a separate government under the care of Gov. Wentworth. A few settlements were commenced in Coos co. before 1775, but were abandoned until the conclusion of peace. During the war of the revolution, the government of New Hampshire was conducted by a temporary administration; and in 1784, a new constitution was adopted, which, with the amendments of 1792, forms the present constitution of the state. The legislative power of the present government is vested in a senate of 12 members, who are chosen by districts, and a house of 229 representatives from the towns; each branch having a negative on the other. The executive is composed of a governor, and a council of 5 members. The governor is annually elected by the people, and has a negative on both branches of the legislature. The regular time for the annual session of the legislature, is the first Wednesday in June. The judiciary department is composed of a superior court and a court of common pleas, each consisting of three judges, who are removable only by impeachment, except that they are disqualified by attaining 70 years of age.

The surface of the state is nearly level for 20 or 30 ms. from the sea coast, which extends but 18 ms., and is generally a sand beach with salt marshes within; back of this it becomes hilly, and in many parts mountainous. Between the Connecticut and Merrimack rs. lie Monadnock, Sunapee, Kearsarge, Moosehillock, and Carr's mtns. In the lower part of Coos co. is a cluster of mtns., called the White hills, or White mtns., among which are the most elevated peaks in the U. States. This region, which is wild and almost entirely uninhabited, abounds in sublime scenery, and formerly afforded much wild game. There are now many deer, wild cats, and some bears, &c. New Hampshire has been called the granite state, from the quantities of that rock quarried within it; and the Switzerland of America, on account of its wild and picturesque mountain scenery, its lakes, cascades, &c. The largest collection of waters in the state, is Lake Winnipiseogee, which is one of the most varied and beautiful in the U. States, and a favorite resort of travellers. Besides this are Connecticut, Ossipee and Squam lakes, &c. which afford fish and fowl. Lake Umbagog is partly in this state and partly in Maine. The state is remarkably well watered, and five of the principal rivers of New England have their sources within its borders. The air is pure and salubrious, and the climate, though severe, very healthy. The soil of New Hampshire is generally fertile, and mostly capable of cultivation. The best lands are those bordering the rivers, which are enriched by the annual floods. The hills afford excellent pasturage. By far the greatest part of the inhabitants is occupied in agricultural pursuits. The principal productions are maize, wheat, rye, oats, barley, flax, &c. Large quantities of pork, beef, butter, cheese, &c. are annually exported. The state produces excellent timber, much of which is also sent abroad. The white pine attains a very large size. The ginseng, long supposed to grow only in China and Tartary, is found here in abundance and of excellent quality. Apples are abundant, and excellent; pears, plums, cherries, &c. are also produced. Beautiful and fine grained granite is found in various parts of the state, of which large quantities are transported for building stone. Iron and copper ore of excellent quality have been found at Franconia; and very good plumbago or black lead, at Bristol. There are many internal improvements and channels of communication. A large part of the commerce of the lower counties finds its vent by the Merrimack r. into Mass., while most of that from the upper cos. passes E. to Portland, Me. Indeed so important has the road through the White mtns. been considered to that state, that the legislature of Me. have sometimes appropriated money for its improvement. Merrimack r. has been dammed, locked and canalled by the state, at the falls between Concord and Mass., so as to be navigable in boats; and great quantities of lumber, granite, produce, and foreign merchandize, are transported by that channel. Numerous factories are erected at the falls. The Middlesex canal opens a communication between the bend of the Merrimack r. and Boston harbor. Piscataqua r., at the mouth of which is Portsmouth, the port of the state, and a navy yard of the U. S., is rather an arm of the sea, which receives 5 small rs. the principal of which is Salmon Falls r. Androscoggin and Saco rs. which flow into Me., rise in the upper parts of N. Hampshire; the last has its source on Mt. Washington.

By the report of the secretary of the treasury, the amount of American and foreign tonnage entered in N. H. for the year ending Sept. 30th, 1830, was 9,416; departed, 4,632; value of imports, $130,828; exports, domestic, $93,499; foreign, $2,685; total exports, $96,184.

The state is divided into 8 counties and 215 towns; none of which are large. Portsmouth is the chief in size, and Concord is the seat of government. The pop. of New Hampshire has been steadily on the increase. In 1800 the pop. was 183,858, in 1810, 214,460. By the two last censuses the pop. of the counties and state is as follows:

Counties.	Pop. 1820.	Pop. 1830.
Cheshire,	45,376	27,016
Coos,	5,549	8,388
Grafton,	32,989	38,682
Hillsborough,	53,884	37,724
Merrimack,		31,614
Rockingham,	55,246	41,325
Strafford,	51,117	58,910
Sullivan,		19,669
Total,	241,161	269,328

Of the foregoing there were white persons,

	Males.	Females.
Under 5 years of age	19,428	18,538
From 5 to 15	34,258	32,315
" 15 to 30	36,038	39,387
" 30 to 50	25,468	28,586
" 50 to 70	12,277	14,336
" 70 to 90	3,626	4,195
90 and above	89	180
Total,	131,184	137,537

Of these were deaf and dumb, under 14 years of age, 32; between 14 and 25, 55; above 25, 48. Blind 105. Aliens 410. Of the colored population in 1830 there were free, males 279; females, 323. Slaves, males none; females 5. There were 9 colored, deaf and dumb,—blind, none.

The counties of Merrimack and Sullivan have been formed since the census of 1820.

The common schools of New Hampshire are established by law, and are generally well supported; and there are academies and high schools in many of the large towns. Dartmouth college at Hanover is the only one in the state: it was founded in 1770. In the number of its graduates, it is the third in the United States; and the libraries connected with it contain 14,000 vols. There is a state prison at Concord.

There are various religious denominations in the state. The Congregationalists have 146 churches, 116 ministers, and 12,867 communicants; Baptists 75 churches, 61 ministers, and 5,279 com.; Free Will Baptists 67 churches, 51 ministers, and 4,500 com.; Methodists 30 ministers, and 3,180 com.; Presbyterians 11 churches, 9 ministers, and 1,499 com.; Christ-ians have 17 ministers; Friends 13 societies; Universalists 20 congregations; Unitarians 10 ministers; Episcopalians 8 ministers; Catholics 2 churches; Shakers 2 societies, and Sandemanians 1.

New Hampton, p-t., Strafford co., N. H., 30 ms. n. Concord, watered in the w. part by Pemigewasset r. The surface is broken and uneven. The soil remarkably fertile. Here is a flourishing academical institution, with 76 pupils; connected with which is a female department with 124 pupils, about a mile and a half from the other. Pop. 1830, 1,905.

New Hampton, p-v., eastern part Madison co., O., by p-r. 15 ms. w. Columbus.

New Hanover, p-v. Montgomery co. Pa. 24 ms. nthrd. Phil.

New Harmony, p-v. Posey co. Ind. by p-r. 171 ms. s. s.w. Indianopolis and 732 ms. a little s. of w. W. C. lat. 38° 10′, long W. C. 11° west.

This v. has been the scene of some interesting revolutions. It was founded in 1814 by a society of Germans, called "the *Harmonites*," who removed there from their settlement of the same name in Butler co. Pa., on the Conequenessing cr. The principles of their civil polity, as far as developed to the public, was a community of goods, landed and personal. Their civil and religious leader was George Rapp. They were remarkable for industry, quietness, decency, and indeed every moral quality which gives force to a people. With such principles they soon made a garden of New Harmony. But MAN continued to be MAN on the Wabash, as he had done since he came with his partner weeping down from the hill of Eden. Robert Owen of Lanark, who had heard of New Harmony, having discovered, or thought he had discovered, a gold mine in the human heart, came to America and purchased New Harmony for $190,000, and began his experiment on a plan directly the reverse of the Harmonites. With the German reformer all was order and obedience, and of course success in his operations; with the Scotch reformer, all was equality, and the result answered to the means. Robert Owen left New Harmony, covered with the weeds of discord. It is probable all reflecting persons will respond to the humane wish of Mr. Flint. "It is to be hoped that this beautiful village, which has been the theatre of such singular and opposing experiments, will again flourish." The actual population is not given by either the census returns, or by Mr. Flint.

New Harrisburg, p-v. Stark co. O., by p-r. 132 ms. N. E. by E. Columbus.

New Haven co. Conn. bounded N. by Litchfield and Hartford cos., E. by Middlesex co., S. by Long Island sound, and W. by Litchfield co. and the Ousatonic r. which separates it from Fairfield co. Its average length from E. to W. is about 26 ms. and its width from N. to S. 21 ms. Containing 540 square ms. or 345,600 acres. This county, lying on Long Island sound, has a very extensive maratime border, but its foreign trade is chiefly confined to New Haven harbor. Its fisheries of oysters and clams and other fish are valuable. It is intersected by several streams, none of them of very large size, but of some value for their water power and fish. Of these the principal are the Pomperaug and Naugatuck, on the W.; the Quinnepiack, the Menunkatuck, and West and Mill rs. on the E. The Quinnepiack is the largest, and passes through extensive meadows. A part of its course is pursued by the Farmington canal, which passes through this county from N. to S. There is a great variety of soil in this county, as well as of native vegetable and mineral productions. The range of secondary country which extends along Conn. r. as far as Middletown, there leaves that stream, crosses into this county and terminates at New Haven. This intersection of the primitive formation by a secondary ridge, affords a great variety of minerals, and materials for different soils. Considerable tracts on the mountains and sandy plains are of little value. This county contains the largest city in the state, one of its capitals, a seaport with pretty extensive trade, and one of the most beautiful towns in the union. The manufactures are not very numerous. There are however large manufactories of cotton, and buttons, at Humphreysville, in the western part of the co.; an extensive gun manufactory at Whitneyville, near New Haven, and a number of manufactories of various articles at Meriden and other towns. Population of the county in 1820, 39,616, 1830, 43,847.

New Haven, city, seaport, and st. of jus. of New Haven co. Conn., and one of the capitals of the state, is 34 ms. S. W. Hartford, 52 W. New London, 76 N. E. New York, and 301 from W. C. in lat. 41° 17', and long. 3° 58' E. W. C. It is beautifully situated about 4 ms. from Long Island sound, at the head of New Haven bay, on a large and level plain, surrounded, except in the direction of the harbor, by a grand amphitheatre of hills, two of which present bold and perpendicular precipices of rude and naked trap rock. These abrupt eminences, which are called East and West rock, are 350 to 370 feet high, and in connection with the surrounding scenery are said very much to resemble the famous "Salisbury craig" in England. New Haven was first settled by the English in 1638, and was united with the Connecticut colony in 1665. The Indian name was Quinnipiack. The city was incorporated in 1764, is 3 ms. long from E. to W. and 2 wide, and includes the old and new townships, each of which is regularly laid out by right lines which divide it into spacious squares. The central square of the old township, which is 182 yards on each side, is, with its ornaments, one of the finest in the U. S. The city is characterised by an appearance of plainness, neatness and order. Its houses and private edifices, are mostly of wood, not expensive, but neat and convenient. The public square and the principal streets are finely ornamented with large and spreading elms, and other shade trees; and a great part of the houses have gardens attached to them, filled with fruit trees and shrubbery, giving to the city a rural and delightful appearance. The central square is intersected by a beautiful street, overspread by elms. The east section is free from buildings and occupied only by majestic elms. On the west, are situated the new state house, 2 Congregational, 1 Episcopal, and 1 Methodist church. The new state house is a splendid edifice, built after the model of the Parthenon, commanding in its appearance; and for the beauty of its proportions, and the style of its workmanship, it holds a high rank among the best specimens of architecture in the country. It is situated near the centre of the section, and includes a large hall for city and town meetings, the halls of legislature, with committee rooms, court rooms, &c. The Episcopal church is a large Gothic edifice, built of dark stone from East Rock. In the new township, is also a new Episcopal church, in the Gothic style, an elegant Congregational church also lately erected, and a Baptist church of stuccoed stone. The state hospital, erected in 1832, is a fine stuccoed edifice, with a colonnade, standing on an eminence about half a mile S. W. from the centre of the city. This institution is one that must prove highly useful, and honorable to the state. Yale college, one of the oldest and most distinguished literary institutions of the country, is located here. It was founded in 1700, and received donations in books and money, the former from clergymen in Connecticut and others, and the latter chiefly from England. Its name was derived from its principal foreign donor. It was chartered in 1701, was originally located at Killingworth, was removed to Saybrook in 1707, and to New Haven in 1717. The original design of the institution was to afford instruction to young men designed for the ministry. A large proportion of all the youth who have received a classical education in the the U. S. have, however, been instructed here. It has long suffered for want of funds. The whole amount of pecuniary donations received from all sources, since it was founded, is less than $150,000, viz. from the state $75,000, and from individuals about $70,000. It has not an endowed professorship, and its annual income is only about $2,000. The receipts of the students' bills constitute, therefore, the only means of defraying the expenses of instruction, and these have hitherto been insufficient. A subscription has recently been opened, which it is presumed will soon furnish a fund of $100,000, by which the facilities and means of instruction will be greatly increased, and its embarrassements, at least for a time, removed. The general management of the college is committed to the corporation, consisting of its president, the governor and lieutenant governor of the state, the 6 oldest members of the state senate, and the same number of distinguished clergymen of the state, chosen by the corporation. The faculty of the university, to whom is entrusted the government and instruction of the pupils, consists of a president, 14 professors, viz. of law; of the principles and practice of surgery; of chemistry; pharmacy; mineralogy and geology; of the Latin language and literature; of the theory and practice of physic; of materia medica and therapeutics; of didactic theology; of anatomy and physiology; of obstetrics; of sacred literature; of divinity; of rhetoric and oratory; of mathematics and natural philosophy; and of the Greek language and literature; 7 tutors; besides assistants to the professors of law and chemistry, and instructers in elocution, drawing, and perspective, botany, and in the German, French, and Spanish languages. The situation of the college buildings is very fine, healthful, and convenient. They consist of 4 buildings 100 feet by 40, each of 4 stories, and containing 32 rooms for students; a chapel, in which is one story appropriated to the theological school, and one to the college library; with 2 other buildings, called the lyceum and atheneum, appropriated to recitation and lecture rooms, rooms for the professors, and libraries for the literary societies. Those are all built of brick, and are ranged in a line, on a gentle elevation facing the city green, with a broad yard in front, shaded with elms and maples. In the rear of these is another range of buildings, consisting of the chemical laboratory; the commons hall, in the 2d story of which, is an elegant and spacious apartment, fitted up for the mineralogical cabinet; and a third, a neat and tasteful building of stuccoed stone, recently erected, for the reception of a part of Col. Trumbull's paintings, (which have lately become the property of the college) and other pictures. A short distance from these, are the buildings of the law and medical schools. The medical institution is furnished with a library and an anatomical museum. The lectures commence the last week in October and terminate the last week in February. During the course, from 50 to 100 lectures are given by each professor. The library of the college, consists principally of old and valuable books, and contains 9,500 vols. The libraries of the literary societies of the students amount to ten thousand vols. The philosophical and chemical apparatus, is very extensive and valuable. The mineralogical cabinet, contains more than 16,000 specimens, and is the most valuable in the country. Commencement is held on the third Wednesday in August The number of students in 1831, was 469, of whom 331 were in the academical department, and the remainder in preparation for the various learned professions The number of living graduates is 2,506; of alumni 4,609; of degrees conferred 5,138. There are 10 very respectable schools of the higher class for young ladies, in which about 400 pupils are educated.

The Farmington canal, which in connection with the Hampshire and Hampden canal, was designed to afford a communication with Connecticut river at Northampton, terminates here. The harbor of New Haven is well protected from winds, but is shallow and gradually filling up; there being but 7 feet of water on the bar at low tide. To remedy this, a wharf with flood gates has been lately erected, at considerable expense, forming a spacious basin, where the water may be always kept at high tide mark. There is another wharf extending 3,943 feet into the harbor; longer than any other in the U. S. by 2,000 feet. The harbor bridge is half a mile in length, 27 feet wide, and cost $60,000. The foreign commerce of the city was formerly very extensive, but is now principally confined to the West Indies. Its coasting trade is more important. Regular lines of packets run to New York, with which city there is a daily communication, by swift and commodious steamboats. Among other objects of enterprise in the city, are a large carpet manufactory, and a carriage manufactory, one of the most extensive in the U. S. It is a coach making village, comprising the habitations of the workmen, and a large building with a Doric front in which are the workshops, chapel, &c. The ground is laid out for gardens, and the whole occupies a beautiful situation, near the mouth of Quinnipiack river, and the N. side of the harbor. There are in the city, beside what have been mentioned, the county jail, an almshouse, custom house, museum, 3 banks, 2 insurance offices, and 6 printing offices, from which are issued five weekly newspapers, and three other periodicals. The Franklin institution has been lately established for the benefit of the citizens, and at great expense, by an enterprising mechanic of the city. It is an institution for popular lectures, and comprises a spacious lecture room, chemical apparatus, and a mineralogical cabinet. At Whitneyville, a village at the base of East rock and within the limits of the town of New Haven, is a very extensive gun manufactory, established a few years since by an enterprising citizen of New Haven and one of the most ingenious and inventive mechanics of our country; from whom the village derives its name. The burying ground of New Haven is a level spot, regularly laid out in squares and ornamented with rows of poplars. It contains a great number of very handsome monuments, many of them made from ancient models, and is said to be one of the most solemn and impressive spots of the same kind in our country. There are 10 churches in the city, viz. 4 Congregational, 2 Episcopal, 1 Baptist, 1 Methodist, and 2 African. Pop. of the city and town, 1820, 8,326. In 1830, city 10,180; town exclusive of the city, 498; total 10,678.

NEW HAVEN, p-t. Oswego co. N. Y. s. lake Ontario, 10 ms. E. Oswego, has good land, pretty well supplied with mill seats by Catfish creek, &c. Fruit grows very well. Pop. 1830, 1,410.

NEW HAVEN, p-v. Huron co. O. by p-r. 95 ms. nthrd. Columbus. Pop. of the tsp. of New Haven, 1830, 615.

NEW HOLLAND, p-v. southern part of Pickaway co. O. by p-r. 44 ms. sthwrd. Columbus.

NEW HOLLAND, p-v. Lancaster co. Pa. 13 ms. N. E. Lancaster city, and 55 ms. a little N. of W. Phil.

NEW HOPE, p-v. on the right bank of Del. r. Bucks co. Pa. opposite Lambertsville in New Jersey, 11 ms. N. E. Doylestown, and 34 N. N. E. Phil.

NEW HOPE, p-v. Augusta co. Va. by p-r. 114 ms. s. w. by w. W. C.

NEW HOPE, p-v. Spartanburg dist. S. C. by p-r. 112 ms. nrthrd. Columbia.

NEW HOPE, p-v. Hancock co. Geo. by p-r. 26 ms. N. E. Milledgeville.

NEW HOPE, p-v. Lincoln co. Ten. by p-r. 56 ms. sthrd. Nashville.

NEW HOPE, p-v. Brown co. O. by p-r. 97 ms. s. s. w. Columbus.

NEW HOPE, Perquimans co. N. C. (*See Durant's Neck.*)

NEW HOPE FORGE and p-o. Iredell co. N. C. by p-r. 372 ms. s. w. W. C., and 166 wstrd. Raleigh.

NEW IBERIA, p-v. on the right bank of Teche r. parish of St. Martin's, La. about 200 ms. following the p-r., and in a direct course almost due w. New Orleans. It is a small village containing about 200 inhabitants, situated on a bank something higher than those of Teche r. generally. It is 11 ms. s. St. Martinsville, the st. jus. for the parish, and 45 ms. S. S. E. St. Landre in Opelousas.

NEWICHAWANNOCK. (*See Piscataqua r.*)

NEW INLET, N. J. between Brigantine and Tucker's beaches, leads from Great bay into the sea s. of Little Egg harbor. Long. about 3° E. W. C., lat. 39° 50'. It is at the mouth of Mullicus r.

NEW IPSWICH, p-t. Hillsboro' co. N. H. 50 ms. s. w. Concord, 52 N. Boston, 5 ms. by 6; 20,860 acres, crossed by Souhegan r., over which is a stone bridge, on the turnpike road. The first cotton factory in N. H. was built here in 1803. It is fertile. Population 1830, 1,673.

NEW JERSEY, one of the United States, bounded N. by New York, E. by the Atlantic ocean and the Hudson r., which separates it from N. Y., s. by Del. bay, and w. by the Del. r. which separates it from Pa. It is situated between 38° 56' and 41° 21' N. lat., and between 1° 45' and 3° 30' E. long. W. C. Its greatest length is 163 ms., and its width 52, and it contains about 8,320 sq. ms.

The first settlement within the limits of New Jersey, was made by the Danes in 1624, at Bergen, so called from a city of Norway. In 1626, a company was formed in Sweden for the purpose of colonizing some part of America, and in the following year the Swedes and Fins made a settlement on the w. bank of the Del. r. In 1640, the English formed a settlement on the eastern bank of the same river, but they were soon driven out by the Swedes in concert with the Dutch. In 1655, Peter Stuyvesant, governor of the New Netherlands, conquered the country, and transported most of the Swedes to Europe. In 1664, it was taken from the Dutch by King Charles II., and granted by charter to the duke of York. In 1676, after having been reconquered by the Dutch, and restored by treaty, it was divided into East and West Jersey, which were reunited by Queen Anne in 1702. In 1738, they were placed under a separate governor. In the controversies preceding the revolution, New Jersey was early and sincerely attached to the interests of the mother country; but when compelled to seek a separation as the only refuge from arbitrary oppression, she was one of the first to resolve on independence, and the second colony which adopted a constitution for her own government. She was prompt in accepting the present constitution of the U. S., and during the scenes of the revolution was distinguished for her patriotic exertions. The battles of Princeton, Trenton and Monmouth, were fought within her limits. The present constitution was adopted in 1776. The legislature is composed of a legislative council, a house of assembly; the former containing 14 members, one from each county, the latter 50. They are annually chosen, and meet on the 4th Tuesday of October. The governor is also elected annually, by a joint vote of both houses of the legislature. In case of vacancy, the vice president of the legislative council acts as governor. The judiciary consists of a court of chancery, of which the governor is chancellor, a supreme court of three judges, circuit courts, and courts of common pleas in the different counties, beside inferior tribunals. Residence for one year in the co. and payment of taxes, are the qualifications for an elector.

The soil of New Jersey, and face of the country, present every variety. The N. W. and N. portions of the state are hilly and mountainous, but interspersed with rich valleys, and extensive tracts, well adapted for grazing, and the production of all kinds of grain and vegetables. The middle parts are agreeably diversified, and generally of good quality. The more southern counties are of alluvial formation, generally level, with loam or sandy soil, in most parts well improved and highly productive. The lands in the vicinity of New York and Philadelphia produce great quantities of fruit and vegetables for those markets. The apples and cider of N. Jersey are proverbially excellent. Wheat, maize, rye, barley, &c. are staple productions. The great quantities of lime in the northern, and marl in the middle parts of the state, of late years extensively used as manure, have greatly improved the quality and productions of the lands. A part of the southern portion of the state and the sea coast is sandy, and valuable only for fuel and timber; but the quantity of land of this description has been greatly overrated. The principal mineral productions are, iron, copper, copperas, paints and various colors. Iron is abundant, and is extensively manufactured. Ores of gold and silver have been discovered in Warren co. Copper mines were wrought before the revolution. Free stone, limestone, marl, and varieties of fine clay for potters use, large quantities of which are exported, are also found. The state, excepting the N. boundary, is almost surrounded by navigable water. It is intersected by many navigable rivers, and has numerous streams for mills, iron works, and every species of manufactures requiring water power. The principal of these streams are the Raritan, Hackensack, Passaic, Salem, Tom, Cohanzey and Mauriee rs. Raritan bay is an extensive arm of the sea on the E. coast, affording a ready communication at all times between the ocean and Perth Amboy, the principal sea port in the state. The internal communications are generally good. The great thoroughfare between the N. and S. states passes through N. J.; and the advantaages which the state enjoys as it regards distance and facilities in transporting goods to market, are not exceeded by those of any equally extended district of the country. In addition to the natural advantages of water communication, the state enjoys the benefit of many internal improvements. The Morris canal, uniting the Passaic and Del. rs. has been completed. The Delaware and Raritan canal, a splendid work, is in active progress. It will connect those rivers by a channel of 70 feet wide, and 7 deep, adapted for sloop navigation, and completing an internal water communication between Albermarle sound and N. Y. A rail road from Camden, across the state to Amboy, is nearly completed, and others are in progress in several parts of the state.

New Jersey is divided into 14 counties, Bergen, Morris, Sussex, Warren, Essex, Somerset, Hunterdon, Middlesex, Burlington, Monmouth, Gloucester, Salem, Cumberland and Cape May. These are sub-divided into t. ships. Trenton is the capital; Newark is the largest town in the state.

The population of New Jersey, though not rapidly increasing, has been steadily progressive; and its increase has been considerably in advance of some others of the Atlantic states. In 1790 the population was 184,139; in 1800, 211,149; in 1810, 245,562; in 1820, 277,575 and in 1830, 320,823. The latter in detail, as follows:

Cos.	pop.	Cos.	pop.
Bergen,	22,412	Middlesex,	23,157
Burlington,	31,107	Monmouth,	29,233
Cape May,	4,936	Morris,	23,666
Cumberland,	14,093	Salem,	14,155
Esex,	41,911	Somerset,	17,689
Gloucester,	28,431	Sussex,	20,346
Hunterdon,	31,060	Warren,	18,627

Of which there were white persons,

	males.	females.
Under 5 years of age,	25,071	23,927
Between 5 and 15	40,949	38,746
" 15 and 30	44,124	42,601
" 30 and 50	28,274	27,630
" 50 and 70	11,511	12,012
" 70 and 90	2,555	2,746
" 90 and over	45	65
Total	152,529	147,737

Of the above were deaf and dumb, 207; blind, 205; and aliens, 3,365.

There were also in the state 18,303 free persons of color, and 2,254 slaves. Of these were deaf and dumb, 15; blind 22.

Recapitulation.

whites.	free col'd.	slaves.	total.
300,266	18,303	2,254	320,823

The manufactures of the state are extensive and flourishing. They are chiefly of iron, cotton, woollen, paper, leather, carriages, shoes, &c., large quantities of which are sent abroad. There are 13 manufactories where glass is made, of various kinds, chiefly from sand found in the state. Iron is one of the most important articles of manufacture, and the forges, furnaces and mills are very numerous in several of the counties. Chain cables are made at Dover, and cut nails in abundance at Patterson. In 1829 there were in Patterson 487 looms, and 4 machine factories, in one of which, in the preceding year, were made 15,048 spindles. Connected with the last was a foundry, producing annually 35,000 pounds of brass, and 1,020,000 pounds of iron castings. The cotton and flax annually used amount to 2,779,600 pounds, and the quantity of cloth manufactured is 2,604,450 yards. The foreign trade of New Jersey being carried on through the ports of N. York and Philadelphia, its amount cannot be accurately ascertained. The amount of tonnage in 1829 was 32,465 tons, besides about 5,000 tons registered at New York. By the report of the secretary of the treasury, Sept. 30, 1830, the amount of tonnage entered the previous year was 586; departed 627. Value of imports, $13,444; of exports, domestic $8,224; foreign $100. Total exports $8,324.

The system of common school instruction in the state, has hitherto been very defective; but in consequence of the recent efforts of the friends of education, the attention of the public has been called up to the subject, and measures have been commenced which promise important and cheering results. The state possesses a school fund which commenced in 1816. The income from it, which is about $22,000, is annually distributed in small sums to such towns as raise an equal amount for the support of schools. Academies and private schools are numerous and excellent. There are in the state two colleges—Nassau Hall at Princeton, founded in 1746, which enjoys a high reputation; and Rutger's college at New Brunswick, founded in 1770. There is a theological seminary at Princeton, under the superintendence of the general assembly of the Presbyterian church, and a similar institution at New Brunswick under the care of the general synod of the Dutch Reformed church. Twenty-two newspapers are published in this state.

The religious denominations of the state are, Presbyterians, who have 85 churches, 88 ministers and 12,519 communicants; the Methodists 10,730 members; Dutch Reformed 28 ministers and 28 churches; Baptists 34 churches, 21 ministers and 2,324 communicants; Episcopalians 20 ministers, and some Friends and Congregationalists, and Catholics.

New Jerusalem, p-v. Bucks co. Pa. 11 ms. from Reading, and 65 e. Harrisburg.

New Kent, co. Va. bounded by Chickahomina r. separating it from Charles City co. s. and Henrico s. w.; by Hanover n. w.; Matapony r. separating it from King William n.; and by James City co. s.e. Length diagonally from e. to w. 33 ms.; mean width 7, and area 231 sq. ms. Extending in lat. from 37° 19′ to 37° 36′, and in long. from 0° 11′ e. to 0° 24′ w. W. C. It is obvious from the position of New Kent, between two rivers, that it is composed of two narrow inclined plains. The surface hilly. Pop. in 1820, 6,630.

New Kent, C.H. and p-v. near the centre of New Kent co. Va. by p-r. 133 ms. a little w. of s. W. C., and 30 e. Richmond, lat. 37° 26′, long. 0° 06′ w. W. C.

New Lebanon, p-t. Columbia co. N. Y. 30 ms. n. e. Hudson, is of irregular form, with 32 sq. ms., has good land, with Williamstown mtn. on the e., arable hills on the s. and a large rich valley n. e. and n. where flows Lebanon cr. Limestone lies below the soil, with lead and other ores, marl, &c. Pop. in 1830, 2,695. (*See following article.*)

New Lebanon, p-v. Columbia co. N. Y. 24 ms. s. e. Albany and 6 w. Pittsfield, Mass. This is one of the principal watering places of the U. S. The water flows abundantly from the s. side of a fine hill near the n.e. corner of the t. and a hotel 150 feet long, and a small village have been erected near it. The water is always at 72° Fahrenheit, and esteemed for bathing, for which there are accommodations. The water is but slightly impregnated, and very pure. The scenery is far more agreeable than that of Saratoga or Ballston, and the views from near the hotel are very fine and extensive over a variegated and well cultivated country. About 2 ms. distant is a Shaker village, where agricultual and mechanic arts in several branches are conducted with great neatness, economy and success. The property is all held in common by the members of the society.

New Lebanon, p-v. and st. just. Camden co. N. C., situated on one of the head branches of Pasquotank r. at the sthrn. extremity of the Dismal Swamp canal, about 30 ms. s. Norfolk in Va. and by p-r. 201 ms. n. e. by e. Raleigh, lat. 36° 25′, long. 0° 42′ e. W C.

New Lexington, p-v. Perry co. O. by p-r. 54 ms. s. e. Columbus.

New Liberty, p-v. Owen co. Ky. 26 miles northward Frankfort.

New Lisbon, p-t. Otsego co. N. Y. 10 ms. s. w. Cooperstown, 76 w. Albany, has arable and grazing hills and rich valleys, crossed by Butternuts cr. and a branch of Otsego cr. which give mill seats. Pop. 1830, 2,232.

New Lisbon, p-v. and st. jus. Columbiana co. O. situated on Little Beaver, 33 ms. n. Steubenville, 54 ms. n. w. by w. Pittsburg, and by p-r. 152 ms. n. e. by e. Columbus, and 282 ms. nthwstrd. W. C. lat. 40° 47′, long. W. C. 3° 43, w. According to Flint this place contained when he wrote (early in 1832,) the ordinary co. buildings, bank, two places of public worship, 6 taverns, 9 stores. In the tsp. of Centre, in which New Lisbon is situated, there were 4 merchant mills, 4 saw mills, a paper mill, 2 woollen factories, 1 fulling mill and 1 carding machine. Pop. 1830, 1,129.

New London, co. Conn. bounded by Hartford, Tolland and Windham cos. n., by Windham co. and R. I. e., by L. I. sound s., by Middlesex co. w. Length 30 ms.; mean width 20; area about 600 sq. ms. It abounds in harbors, crs. and bays, convenient both for fishing and navigation. The r. Thames formed by the juncture of the Quinebaug and Shetucket, affords steam and sloop navigation to Norwich. Just below Norwich, the Thames receives the waters of the Yantic, and thus furnishes an admirable entrance for vessels. The different branches of the Thames afford excellent water power; and a canal is proposed along the Shetucket valley, into Mass. The Niantic, Pequonock, Mistic, Stonington, Wickatequack, and Pacatuck are among the important and beautiful bays which indent that portion of coast included within the territorial limits of this co. The n. w. part of the co. is mountainous, and much of the surface is hilly and rocky; but much arable and grazing land is found in the co. The soil is generally productive. Fruits, grain, lumber, fish, are the staples. Considerable attention is paid to manufactures. By a return made to the secretary of state at Washington in June, 1832, it appears that there are in the co. 14 cotton mills, which employ 22,688 spindles and 580 looms. Amount of capital invested $746,000; quantity of wool consumed, 1,647,928.lbs.; yarn sold the previous year 2,500 lbs.; yds. cloth sold in same time, 5,048,780. Capital invested in the manufacture of woollens $206,-

000; quantity of wool consumed 271,600 lbs.; annual value of woollen manufactures $187,784. Pop. in 1820, 35,943; in 1830, 42,201.

NEW LONDON, city, port of entry, p-t. and half-shire, N. London co. Conn., on the w. bank of the Thames, and 3 ms. from L. I. sound. It is 42 ms. s. E. Hartford, 53 E. New Haven, and 14 s. Norwich. Lon. 4° 0' 48" E. W. C., lat. 41° 0' 25" N. The town is ¾ of a mile broad, by 4 ms. long, comprising an area of 2,400 acres. The surface is rather hilly, soil good, producing spontaneously the best of oak and walnut timber. Granite is found here in great abundance. The city is situated on the declivity of a hill, and at the head of a harbor bearing its own name. It contains 4 places of public worship; one for Presbyterians, another for Episcopalians, a third for Baptists, and a fourth for Methodists. Beside these and the ordinary co. buildings, there are in the city 2 banks, and an insurance office. The harbor is one of the best in the U. S., being 3 ms. long, and rarely obstructed with ice, and having 5 fathoms water. It is environed by hills, and defended by 2 forts; the one upon its west side is called fort Trumbull, and is delightfully situated about a mile below the city; while the other, upon its east side, is called fort Griswold, and rises from the top of a commanding eminence opposite the city, and in the town of Groton. These advantages, together with its light house, on a point of land projecting from the w. shore, and forming the dividing point between the harbor and sound, render this in every respect a safe and commodious harbor. It has served in a great degree as the port of Conn. r., the impediments in which frequently prevent its being navigated by large vessels fully laden. The commerce is quite considerable, both in the coasting and foreign trade. The whale fishery is also an important branch of commerce. About half a million of dollars is devoted to its prosecution, and not less than 25 ships, which give employment to about 700 seamen, are engaged in this adventurous business. Several vessels are also engaged in sealing. Fort Griswold, to which reference has been made, was, together with the circumjacent country, the seat of a revolutionary struggle, which is commemorated by a noble granite monument, 150 feet high, bearing an embedded marble slab, which contains the names of those who there fell in defending their country, together with the following appropriate and scriptural inscription, "Zebulon and Napthali were a people that jeoparded their lives unto the death in the high places of the field." The pop. of N. London in 1820, was 3,330; in 1830, 4,356.

NEW LONDON, p-v. near the wstrn. border of Campbell co. Va. 11 ms. s. w. Lynchburg, and 191 ms. s. w. by w. Richmond.

NEW LONDON, cross roads, p-v. in the sthrn. part of Chester co. Pa., 40 ms. s. w. by. w. Phil., and by p-r. 93 ms. N. E. W. C. The tsp. of New London in 1820, contained 1,200 inhabitants.

NEW LONDON, p-v. in the sthrn. part of Jefferson co. Ind., situated on Ohio r. 12 miles below Madison, the county seat, and by p-r. 97 ms. s. s. E. Indianopolis.

NEW LONDON, p-v. and st. jus. Ralls co. Mo., by p-r. 167 ms. N. N. E. Jefferson city, and by the common road 105 N. N. W. St. Louis. It is situated on Salt r. of Mo. 39° 32', long. W. C. 14° 21' w.

NEW LYME, p-v. Ashtabula co. Ohio, by p-r. 183 ms. N. E. Columbus.

NEW MADISON, p-v. Dark co. Ohio, by p-r. 110 ms. wstrd. Columbus.

NEW MADRID, s. E. co. Mo., bounded s. by Crittenden co. Ark., s. w. by St. Francis r. separating it from Lawrence co. Ark., N. W. Stoddard co. Mo., N. Scott co. Mo., N. E. Mississippi r. separating it from Hickman co. Ky., Mississippi r. E. separating it from Obion and Dyer counties, Ten. Length from s. to N. 65 ms., mean breadth 25, and area 1,625 sq. ms. Lat. 36½° and long. W. C. 13° w. intersect near the centre of this co. Slope is almost directly s., as in that course flow the Miss. and St. Francis rs. The surface is with very partial exceptions an annually inundated plain. In 1812, this part of the U. States was considerably disturbed and in some places disrupted by an earthquake. Where the soil is of sufficient elevation for cultivation, it is generally very productive, and the climate sufficiently warm in summer to admit the cultivation of cotton. Chief t. New Madrid. Pop. 1830, 2,350.

NEW MADRID, p-v. and st. jus. New Madrid co. Mo., by p-r. 278 ms. s. E. of Jefferson city, and by the most direct road 170 a little E. of s. St. Louis. It is situated on a rather more than usual high alluvial bank, upon the right shore of Miss. r. directly opposite to the extreme s. w. angle of Ky. By the bends of the r. it stands at about 50 miles below the mouth of Ohio. It is an unimportant village, though historically interesting. It was founded by a Mr. Morgan from Pa., with great expectations of future prosperity. But the bank, apparently more stable, because more elevated than the other alluvial Mississippi banks, was really more subject to detorioration by abrasion from superior weight, and most of the surface on which the original village stood has long since been swept away. To this steady cause of destruction, New Madrid seems to have been the centre of mighty convulsions in 1811, and 1812. To examine the place and adjacent country, all seems tranquil, and but little apprehension of danger preceded a commotion which shook with more or less violence perhaps 200,000 sq. ms. Near New Madrid the rivers, lakes, and even the ground heaved like a boiling pot. Water burst in immense jets into the air, lakes were dried in some places and formed in others. Boats were sunk, or hurled with an inconceivable force amongst the foaming surges. The thinness of the population and the log or frame buildings protected human life, though some persons were lost. The trembling of the ground was felt in the city of New Orleans, and what is very remarkable, continues to be occasionally felt in sthrn. Missouri to this time, after a period of 20 years.

NEWMARKET, p-t. Rockingham co. N. H., 12 ms. w. Portsmouth, 38 s. E. Concord, and 9 from Dover, w. from Great Bay; 11,082 acres; is crossed by Piscasset r. and situated on Lamprey r. about one mile from Piscataqua, and vessels of from 80 to 100 tons can come up to the factories. The town contains one place of public worship for Methodists, and one for Congregationalists. The Newmarket manufacturing company have 3 large stone mills, 2 of which are 156 feet long each, and 6 stories high, including basement stories; the other is 100 feet long, and also 6 stories high; and a machine shop; 14,000 spindles, 487 looms, and 660 hands, 500 of whom are females. They consume 2,500 bales of cotton, per annum, which yields three millions five hundred thousand yards. A Wesleyan academy was incorporated here 1818. Pop. 1830, 2,008.

NEW MARKET, p-v. southeastern part of Frederick co. Md., situated on the Baltimore and Frederick turnpike, 36 ms. wstrd. from the latter place, and 11 estrd. from the former, and by p-r. 51 ms. a little w. of N. W. C., direct distance, however, only 35 ms.

NEW MARKET, p-v. sthrn. part of Shenandoah co. Va. 20 ms. s. s. w. Woodstock, and by p-r. 120 ms. s. w. by w. W. C.

NEW MARKET, p-v. nrthwstrn. part of Madison co. Ala. 17 ms. N. w. Huntsville, and by p-r. 172 ms. N. N. E. Tuscaloosa.

NEW MARKET, p-v. nrthwstrn. part of Jefferson co. Ten. 30 ms. N. E. by E. Knoxville, and by p-r. 195 E. Nashville.

NEW MARKET, p-v. Washington co. Ky., situated in the sthrn. part of the co., on the Rolling fork of Salt r., 13 ms. s. s. w. from the co. seat, Springfield, and 62 ms. s. s. w. of Frankfort. Pop. 1830, 43.

NEW MARKET, p-v. Abbeville district, S. C., by p-r. 88 ms. w. Columbia.

NEW MARLBOROUGH, p-t. Berkshire county, Mass., 148 ms. s. w. Boston, 10 s. E. Lenox, and bordering on Conn., is watered by branches of Conkepot and Housatonic rs. It has a pond s. E. Incorporated 1759. Pop. 1830, 1,656.

NEW MILFORD, p-t. Litchfield co. Conn., 48 ms. s. w. Hartford, 6½ ms. by 13; 84 sq. ms.; is one of the largest towns in Conn., crossed by the Housatonic and other rivers, with good mill seats. Mica slate and marble are quarried. Iron and some silver have been found. The land is generally good, and the village is on the Housatonic, in a valley. First settled 1713. Pop. 1830, 3,979.

NEW MILFORD, p-v. nrthwstrn. part of Susquehannah co. Pa., 19 ms. N. w. Montrose, and by p-r. 183 ms. a little E. of N. Harrisburg.

NEWMAN, p-v. and st. jus. Coweta co. Geo., by p-r. 129 ms. N. w. by w. Milledgeville, lat. 33° 26′, long. W. C. 8° w. It is situated on the summit around, between the vallies of Flint and Chattahooche rivers.

NEW ORLEANS, parish of Louisiana, bounded s. and E. by the parish of Plaquemines, w. by Jefferson, N. by lake Ponchartrain, and N. E. by lake Borgne and the pass of Rigolets. Length 32 ms., mean breadth about 5 ms., and area 160 sq. ms. Extending in lat. from 29° 46′ to 30° 12′, and in long. from 12° 30′ to 13° w. W. C. The surface in the greater part a morass, but with highly fertile arable soil, though narrow borders of land rising two or three feet above the high tide level, and stretching along the margins of the Miss. Bayou, Boeuf, and other marsh bayous. In a state of nature it was only the comparatively elevated margins along the streams, which generally produced timber. The morass mostly covered with coarse grass, and flooded by every tide. The soil actually capable of cultivation, produces sugar, cotton, rice, Indian corn, &c. The usually cultivated fruit trees are orange, fig and peach. Of garden vegetables the number of species and abundant quantity may be seen at the vegetable market of New Orleans city. Chief t. city of New Orleans. Population of the parish 1820, 41,351, 1830, 50,103.

NEW ORLEANS, city and port of entry, La., situated on the left bank of the Miss. r., 105 ms. by the channel above the mouth, and 322 by the channel below Natchez. The city stands on lat. 30°, and very nearly 13° w. W. C. By calculation the two cities bear from each other by angle from the reflective meridians, 50° 15′, distant 966 statute ms. within a small fraction, by the p-r. 1,189 miles.

Similar to other parts of the banks of the Miss. in its vicinity, the site of New Orleans is on an inclined plain, the declivity falling very gently from the margin of the river. When the Miss. is in full flood the surface of the water is from 2 to 4 feet above the streets of the city, but at low water the surface of the river is rather below the front street, but still at any stage elevated above the swamps in the rear of the back streets. To prevent constant inundation, a levée or embankment fronts the city. This levée differs only in breadth and solidity from the otherwise similar embankment, extended along the Miss. on both banks above & below N.Orleans. The city is built on the concave side of the river, and including the suburbs extends about three miles along the stream, with a breadth backwards of not quite one third of a mile, lying in form of a crescent, with the city properly so called near the middle of the curve. New Orleans proper is a parallelogram of 4,000 by 2,000 English feet very nearly, streets extending at right angles, and the long side parallel as near as possible to that part of the river opposite. The larger streets proceeding from the river are Levée, Chartres, Bourbon, Dauphin, Burgundy, &c. Above the city are the suburbs (*faubourgs*) of St. Mary, Duplantier, and Annunciation; below are the suburbs, Marigny, and Da Clouet. In the rear of the city is also another but detached suburb on Bayou St. John. The compactness of the buildings, and in a great degree their individual magnitude, is inverse to distance from the harbor. The latter and the bank of the river, particularly opposite the city proper, are commensurate. Any vessels which can pass the bars at the mouths of the Miss., can be laid along side the levée, and at high water are, when loading or unloading, generally attached to the shore with cables and a platform. The materials of architecture in New Orleans are brick and wood generally. The public edifices are a custom house, town house, market house, cathedral, Ursuline convent, a court house, two theatres, two or three Protestant churches, &c. Besides a branch of the bank of the U. S., there are in New Orleans, the bank of La., the parent of the planter's bank of La., the last with branches at Baton Rouge, Donaldson, Opelousas, Alexandria, and St. Francisville; La. state bank, and bank of Orleans. The aggregate bank capital, exclusive of that of the bank of the U. S. branch, $8,500,000.

In 1829 there were imported into N. Orleans from the wstrn. states of the U. States and from Texas, of bacon, assorted, 2,868 hogsheads; bagging, 13,472 pieces; butter 3,995 kegs; beef 5,405 brls.; beeswax, 795 brls.; buffalo robes, 15,210 lbs.; cotton, 269,571 bales; corn meal, 6,849 brls.; corn in ear, 91,882 brls.; flour, 157,323 brls.; lard, 110,206 kegs; pig lead, 146,203 pigs; linseed oil, 2,946 brls.; deer skins, 6,215 packs; bear skins, 159 packs; tobacco, 29,432 hogsheads. The foregoing can be only a part of the imports into New Orleans, as it does not include sugar, lumber, lime, and numerous other articles of great amount and value. In 1830, the amount of cotton alone exported from New Orleans, was 302,852 bales.

The government of the city is under a mayor and city council, elected by the freeholders.

If we turn our attention to the vast regions of the Miss. basin, to the accumulating population, on its innumerable streams, and the navigable facilities afforded by so many channels, we have the means to estimate the resources which must contribute to augment N. Orleans in extent, wealth, and population. Within this century the increase in every respect has been great indeed. In 1800, the inhabitants amounted to between 5,000 and 6,000. It contained in

	1810	1820.	1830.
Free white males	3,586	8,268	11,962
" " females	2,745	5,318	8,082
Free colored persons	4,950	6,237	11,562
Slaves	5,961	7,355	14,476
Total.	17,242	27,178	46,082

New Orleans was laid out in 1717, and named in honor of the then Duke of Orleans, regent of France during the minority of Louis XV. It remained as capital of La. under the French, until in 1769 it was taken possession of by the Spanish general Oreilly, under a treaty of cession made between Spain and France 1762. The Spanish government continued New Orleans as the capital of the colony until receded to France in 1803. The French colonial prefect, Lausalt, by order of his government, gave it up to the U. States, 20th Dec. 1803. Though the seat of legislation has been removed to Donaldsonville, nearly 80 ms. higher up the Miss., still New Orleans remains not only the principal city of La., but also of the great physical section, at the base of which it rises above the great plain of the Delta.

New Oxford, p-v. Adams co. Pa. by p-r. 87 ms. northward W. C.

New Paltz, p-t. Ulster co. N. Y. 15 ms. s. Kingston, 80 s. Albany, 85 n. N. Y., w. Hudson r. Has good land. The people are of Dutch extraction; first settled about 1672. The village is on Wallkill creek, and there are several smaller ones at the landings, &c. Pop. 1830, 4,973.

New Paris, p-v. Preble co. O., by p-r. 104 ms. wstrd. Columbus.

New Petersburg, p-v. Highland co. Ohio, by p-r. 74 ms. s. s. w. Columbus.

New Philadelphia, p-v. and st. jus. Tuscarawas co. Ohio, by p-r. 107 ms. n. e. by e. Columbus. Lat. 40° 30', long. W. C. 4° 31' w. It is situated on Tuscarawas r., and on the Ohio and Erie canal; contains the ordinary county buildings. Pop. 1830, 410.

Newport, t. Penobscot co. Me., 25 ms. w. n. w. Bangor, with Somerset co. n. and w.; contains a large pond, drained by Sebastocook r. Pop. 1830, 897.

Newport, p-t. Sullivan co. N. H., 40 ms. n. w. Concord, 96 from Boston; 25,267 acres. Three branches of Sugar r. unite near the village. There are fine meadows, but much dry gravel, and moist & cold land; it is, however, generally fertile. Two ponds furnish trout, &c. There are also some high hills. Bald, Coit, East & Blueberry mtns. Pop. 1830, 1,913.

Newport, co. R. I., comprises several islands in Narraganset bay, as well as the adjoining land on the continent, in all 136 sq. ms., with 7 towns. It has great advantages for navigation, and contains Newport, one of the best seaports in the northern states, and now fortifying on a large scale for a naval station. The waters are navigated by all the vessels proceeding to and from Providence, and there is daily steamboat communication with N. York.

Rhode Island is the most important, and is a fine agricultural country. In its n. part is a mine of anthracite coal, which has been wrought, but not to good advantage. The rocks are transition, as are those of Prudence, Canonicut and Block islands, which also belong to the co. Fish are taken in great quantities, and from Block isl. are sent to distant places. Pop. 1830, 16,535.

Newport, t. seaport and co. t. Newport co.

R. I., and one of the capitals of the state, 30 ms. s. by E. Providence, 75 s. w. Boston, about 1 m. by 6, with 8 sq. ms., has an uneven but rich soil, especially in the s. and much well cultivated land. The harbor is excellent, near the sea, and accessible in winds which will not permit a ship to enter any other port, in all this part of the coast. The principal population are collected on the west declivity, and at the foot of a hill. The place was formerly a very flourishing sea-port, and, before the revolution, was the fourth t. in size in the colonies. It has for some years resigned its enterprize and prosperity to Providence. Its fine situation, healthful air, excellent fish, &c., render it a favorite resort of visitors from the southern states, in summer. Here is a large lace manufactory. The harbor has Goat isl. in front, nearly closing the entrances, where are fort Wolcott and a military hospital. Forts Green and Adams also defend the harbor. The latter is a new and extensive work, on the s. point, with powerful batteries, and includes 40 acres. The foundation for a monument to the memory of Com. Oliver H. Perry, has been laid. It is to be of grey granite, and 28 feet high. The foundation is to be surrounded by a mound of earth, 160 feet in circumference. Newport was occupied by the British for some time during the revolutionary war, who stripped the island of its fine forest trees and orchards. They were beseiged by the American troops. Pop. 1830, 8,010.

Newport, Herkimer co. N. Y., 13 ms. N. Herkimer, 95 N. W. Albany, E. Oneida cr., has a deep valley in the middle, where is the v., at a good waterfall in W. Canada cr. Pop. 1830, 1,863.

Newport, v. Orleans co. N. Y., on Erie canal, 2½ ms. s. s. E. Gaines.

Newport, p-v. nthrn. part Perry co. Pa., 41 ms. N. W. Harrisburg.

Newport, p-v. on Christiana cr. New-Castle co. Del. 3 ms. wstrdly. Wilmington, and by p-r. 103 N. E. W. C.

Newport, small r. or cr. of Geo., separating Liberty from McIntosh co., and falling into Sapelo sound.

Newport, p-v. and st. jus. Cocke co. Ten., situated on French Broad r., 48 ms. a little s. of E. Knoxville, and by p-r. 210 ms. in a similar direction Nashville; lat. 35° 56′, long. 6° 4′ w. W. C.

Newport, p-v. and st. jus. Campbell co. Ky., situated on O. r. on the point above the mouth of Licking r., and directly opposite the city of Cincinnati. This town contains an academy, a U. S. arsenal, and the ordinary co. buildings. Pop. 1830, 717.

Newport, p-v. estrn. part Washington co. O. It is situated on O. r. by p-r. 16 ms. above and estrd. Marietta. Population of tsp. 1830, 556.

Newport, p-v. and st. jus. Vermillion co. Ind., situated on the point below the junction of Wabash and Little Vermillion rs., by p-r. 86 ms. w. Indianopolis.

Newport, Franklin co. Mo., p-v. on Mo. r., 43 ms. wstrd. St. Louis.

New Portage, p-v. Medina co. O., by p-r. 110 ms. N. E. Columbus.

New Portland, p-t. Somerset co. Me., crossed by Seven-miles brook, a branch of Kennebec r. Pop. 1830, 1,214.

New Providence, p-v. nthrn. part Lancaster co. Pa., 20 ms. N. E. Lancaster city, and by p-r. 129 N. N. E. W. C.

New Providence, p-v. Clarke co. Ind., by p-r. 104 ms. s. s. E. Indianopolis.

New Richmond, p-v. Clermont co. O., by p-r. 132 ms. s. w. Columbus.

New r., the local name of Great Kenhawa r., above the mouth of Gauley r. (*See Great Kenhawa r.*)

New r., p-v. wstrn. part of Monroe co. Va., 26 ms. wstrd. Union, the co. seat, and by p-r. 296 ms. s. w. by w. W. C.

New r. inlet, Onslow co. N. C., about 50 ms. a little s. of w. cape Look-Out, is a strait between two sand islands, and the entrance to the mouth of a small r. called New r., which, rising in Lenoir, and thence traversing Onslow co., opens to the Atlantic ocean by this entrance.

New r., small stream of Beaufort dist. S. C., draining the swamps between Savannah and Coosaw Hatchie rs., flows s. s. E., and near the Atlantic ocean breaks into several branches, one entering the estuary of Savannah r., and another Calibogue sound.

New r., of La., is the drain of the lowlands between the Miss. and Amite rs., and flowing N. E. by E., falls into the s. w. angle of lake Maurepas.

New Rochelle, p-t. Westchester co. N. Y., 20 ms. N. N. Y. city, 5 s. White Plains, 143 from Albany, w. side of East r. and Long Island sound, has level, stony, but pretty good soil, has an academy, and comprises several small islands. It was settled by Hugeanot emigrants from France, whence its name. Pop. 1830, 1,274.

New Rumsey, p-v. Harrison co. O., by p-r. 134 ms. N. E. by E. Columbus.

Newry, t. Oxford co. Me., 24 ms. N. W. Paris, N. Androscoggin r., is mountainous and wild. Population 1830, 345.

Newry, p-v. wstrn. part of Huntingdon co. Pa., 32 ms. N. W. by w. the borough of Huntingdon, and by p-r. 122 ms. a little s. of w. Harrisburg.

New Salem, p-t. Franklin co. Mass., 80 ms. w. Boston, incorporated 1753, is high in the centre, and has Monaduoc N. There is an academy, which was incorporated 1795. Pop. 1830, 1,889.

New Salem, p-v. wstrn. part of Fayette co. Pa., 9 ms. N. W. Union, the st. jus. and by p-r. 207 ms. N. W. by w. W. C.

New Salem, p-v. Harrison co. Va., by p-r. 240 ms. wstrd. W. C.

New Salem, p-v. Randolph co. N. C., by p-r. 77 ms. w. Raleigh.

New Salem, p-v. Sangamon co. Il., 98 ms. N. N. W. Vandalia.

NEW SHARON, p-t. Kennebeck co. Me., s. Somerset co., crossed by Sandy r. a branch of the Kennebec. Pop., 1830, 1,599.

NEW SHOREHAM, t. Newport co. R. I., on Block isl., all which it comprehends, has but little communication with other places, but possesses a valuable fishery. Pop. 1830, 1,185.

NEWSTEAD, t. Erie co. N. Y. Pop. 1830, 1,926.

NEW STOCKBRIDGE, N. Y., a tract of land 6 ms. square, in Augusta, Oneida co., and Smithfield, Madison co. It was granted to the Indians of Stockbridge, Mass., by the Oneida Indians, who were bound to them by ancient treaties and friendship; but a large portion of the tribe have removed to Green Bay within a few years. The Scotch society for promoting Christian knowledge, have for many years supported a missionary among them.

NEW STORE, and p-o. Buckingham co. Va., by p-r. 81 ms. wstrd. W. C.

NEWTON, p-t. Middlesex co. Mass., 9 ms. w. Boston, incorporated 1691, is large, and has Charles r. on 3 sides, along which are broad and rich tracts of meadow. The uplands are hilly and woody. There are 2 falls, and at the lower are manufactories. Nonantum, a hill in this t., was the scene of the apostle Elliot's first exertions to teach the Indians Christianity, in 1746, and witnessed his success. He was ordained as pastor of the church here in 1664. A Baptist theological seminary was founded here a few years since. Pop. 1830, 2,376.

NEWTON, p-t. and st. jus. Sussex co. N. J., 50 ms. N. Trenton, 28 N. W. Morristown; it is mountainous E., and gives some of the head streams to Pequest branch and Pawlin's kill. Much of the soil is excellent, and remarkably well cultivated. There is an academy, and the v. is pleasant and flourishing. Pop. 1830, 3,464.

NEWTON, p-v. Newton, Sussex co. N. J., 28 ms. N. W. Morris, is near the centre of the town, pleasant, with an academy, bank, C. H., a church for Presbyterians, and one for Episcopalians.

NEWTON, p-v. and borough of Bucks co. Pa., 26 ms. N. N. E. Phil., and 11 w. Trenton, N. Jersey.

NEWTON Mills, p-o. Licking co. O., by p-r. 42 ms. estrd. Columbus.

NEWTON'S, p-o. Greene co. Ky., by p-r. 84 ms. s. w. Frankfort.

NEWTOWN, t. Rockingham co. N. H., 40 ms. s. E. Concord, 27 s. w. Portsmouth, N. Mass., 5,250 acres, contains one third of Country pond, and 2 small ones connected with it; with good grass land. First settled 1720. The Baptist church is the oldest of that denomination in N. H. Pop. 1830, 510.

NEWTOWN, p-t. Fairfield co. Conn., s. w. Housatonic r., 48 ms. s. w. Hartford, 26 N. W. N. Haven, is almost triangular, with 50 sq. ms., on high ground, hilly surface, gravelly soil, yields much rye and fruit. It is crossed by Powtatuck r., and has a pond, with several manufactories. A toll bridge crosses to Southbury; 15 school dists.; the v. is near the centre. Pop. 1830, 3,100.

NEWTOWN, p-t. Queen's co. N. Y., on w. part Long Isl., 8 ms. E. N. York, s. East r., opposite Hurl Gate, N. King's co., s. w. Flushing bay; extends near half across the isl., is well cultivated, yields grass, vegetables, &c. for the city, and abounds in fine apples, particularly a fine sort of yellow winter apples called Newtown pippins, which have been sent to various parts of the world. It has a v. near the centre, and some fine country seats. Peat, found near the v. is much used for fuel. It comprises the isls. Two Brothers, and two coves, Hallet's and Riker's. Pop. 1830, 2,610.

NEWTOWN, v. Elmira, Tioga co. N. Y. (*See Elmira.*)

NEWTOWN, p-t. Gloucester co. N. J., 25 ms. s. w. Trenton, s. w. Cooper's cr., s. E. Del. r., opposite Phil., from which it derives the advantage of a market. Camden v. in this t., is on Del. r., and has a ferry to Phil. It is also at the end of the S. Amboy and Camden rail road, which, when completed, will form a most important route across N. J., for travellers and merchandize passing between New York and Phil. Pop. 1830, 3,298.

NEWTOWN, p-v. Worcester co. Md. 8 ms. N. w. Snow Hill, and by p r. 159 ms. s. E. by E. W. C.

NEWTOWN, p-v. northern part King and Queen co. Va., by p-r. 99 ms. a little w. of s. W. C. and 38 ms. N. E. Richmond.

NEWTOWN, p-v. Scott co. Ky. 25 ms. from Frankfort.

NEWTOWN, p-v. Hamilton co. O. by p-r. 120 ms. s. w. Columbus.

NEWTOWN, Hamilton, p-v. Mifflin co. Pa.

NEWTOWN SQUARE, p-v. northern part Del. co. Pa. 15 ms. a little N. of w. Phil.

NEWTOWN STEPHENSBURG, p-v. Frederic co. Va. by p-r. 79 ms. N. w. by w. W. C., and 10 ms. N. w. Winchester.

NEWTOWN TRAP, p-v. in the s. western part Frederic co. Md. about 9 ms. s. w. the city of Frederick, and by p-r. 51 ms. N. w. W. C.

NEW TRENTON, p-v. Franklin co. Ind. by p-r. 81 ms. a little s. of E. Indianopolis.

NEW TRIPOLI, p-v. Lehigh co. Pa. by p-r. 187 ms. N. E. W. C.

NEW TROY, p-v. near the right bank of Susquehannah r. Luzerne co. Pa. 5 ms. above and N. N. E. Wilkesbarre, and 120 ms. N. N. E. Harrisburg.

NEW UTRECHT, t. Kings co. N. Y. w. end of Long isl. opposite the narrows, the entrance into N. Y. harbor, 9 ms. s. city, E. and N. Hudson r. It is hilly and stony w., level interior, with sandy soil, and has suffered from fever and ague, a few years. One of the best shad fisheries is in this t. Bath, on the shore, is resorted to from N. Y. for bathing, fishing, and shooting. The v. about a mile from this spot, is on a plain, in a retired situation. On the N. is the w. end of a long ridge, extending through the isl., N. of the great plains, E. to Southhold. They were important in the revolution. The British army landed at Bath, 1776. Fort Lafayette, triangular with 3 tiers of guns, on a small isl. W. of this t., with the works on the shore, defends the entrance of N. Y. harbor on the E. Nyak point and fort Lewis are on the same shore. Pop. 1830, 1,217.

NEW VERNON, v. Morris co. N. J. 3 ms. S. Morristown.

NEWVILLE, p-v. western part Cumberland co. Pa. and near the right bank of Conedogwinet, 16 ms. W. Carlisle.

NEWVILLE, p-v. Barnwell dist. S. C. by p-r. 14 ms. from Barnwell C. H., and 76 S. E. W. Columbia.

NEWVILLE, p-v. Richland co. O. by p-r. 76 ms. N. E. Columbus.

NEW VINEYARD, t. Somerset co. Me. N. Kennebec co., is rough, and crossed by a branch of Seven-miles brook, branch of Kennebec r. Pop. 1830, 869.

NEW WASHINGTON, p-v. Clarke co. Ind. by p-r. 94 ms. S. S. E. Indianapolis.

NEW YORK, one of the most important of the U. S., bounded N. by Canada, Lake Ontario, and the river St. Lawrence, which separates it from Canada, E. by Vt., Mass. and Conn., S. by the Atlantic, N. Jersey and Pa., and W. by Pa., lake Erie and Niagara river, which separates it from Canada.

It is situated between 40° 30′ and 45° N. lat. and between 3° 45′ E. and 2° 50′ W. W. C., (not including Long Isl., which extends E. to long. 5° 50′,) and contains 46,085 sq. miles, including lakes and islands. Its greatest length from E. to W. (excepting Long Island) is 316 ms., and its greatest breadth from N. to S. 304 ms. The tract now composing N. Y. was originally included in the grant of Virginia, made by Queen Elizabeth; and in the grants of N. and S. Virginia, made in 1606 by James I. This part of the continent, however, was not known to the Europeans until 1609, when it was first discovered by Henry Hudson, an enterprizing English navigator, then in the employ of the Dutch East India company. After sailing 150 ms. up the river that now bears his name, he returned to Europe and communicated to the Company, the result of his voyage. In 1613, a trade with the natives was commenced by the Dutch, and trading establishments were formed at New Amsterdam and fort Orange, (*now* the cities of New York and Albany.) In 1621, the Dutch government, desirous of founding a colony in America, granted to the Dutch West India company, an extensive territory on both sides of the Hudson, called New Netherlands. The boundaries were not accurately defined, but were considered by the company as extending to the Connecticut r. at the north, and the Delaware on the south. In 1623, they erected a fort on the Del., and a few years after, another on the Connecticut. This territory continued in their possession till 1664, when Charles II. denying their right to any portion of it, made a grant to his brother, the duke of York and Albany, of the whole extent of country, from Nova Scotia to Del. bay; and the same year took possession of it by conquest. In 1673, it was recaptured by the Dutch; and the year following was restored by treaty to the English. Thus it remained one of the British colonies till the war of the revolution, and the adoption of a free constitution by the people. In 1790, Vt., previously a part of N. York, became a separate state. In 1821, the amended constitution, was adopted by the state of N. Y. under which it is at present governed. By this constitution the legislative power is vested in a senate and assembly, the former consisting of 32, and the latter of 128 members. The former chosen for 4 years by districts, and the latter elected annually by counties. The executive power is vested in a governor, who holds his office for 2 years. A lieutenant governor is chosen at the same time and for the same term, who is qualified to act as governor whenever that office becomes vacant; and is also president of the senate. The judiciary consists of a court of chancery, a supreme court with 3 judges, 8 circuit courts in as many different circuits, which correspond to the senatorial dists., & the superior court of the city of N. Y. consisting of 3 judges. Beside these are the county and justice courts, &c. The senate, with the chancellor and justices of the supreme court, constitute a court of impeachment and for the correction of errors. The right of suffrage is enjoyed by every male citizen of the age of 21 years, who has resided one year in the state, and six months in the town or county where he offers his vote; having paid a tax within the year, or legally served as a militia man or fireman, or labored upon the public highways, &c.

The surface of the state of N. Y. is greatly diversified, but in general may be considered as an elevated tract, with numerous indentations and depressions, which form the basins of lakes and the valleys of fertilizing streams. There are several ridges of mtns., which in general are considered as continuations of the Alleghany ridges. The Cattskill mtns., the highest in the state, are about 3,800 ft. above the level of the sea. The form of this state is irregular, and it enjoys very great advantages for commerce, agriculture, and manufactures, which have been greatly improved. It has a large extent of sea coast, with all the southern shore of lake Ontario, most of the W. shore of lake Champlain and the E. end of lake Erie, with the whole course of the Hudson, navigable in ships of the line to Hudson, and in sloops to the Mohawk r. The Mohawk, the branches of the Susquehannah, and other streams afford abundance of water power, and at the same time diversify and fertilize the state. In the valley of Genesee r. is some of the best wheat country in the world; and many other parts of the state are remarkable for their fertility. The principal productions are wheat and other grain, flour, provisions, salt, pot and pearl ashes, and lumber. New York also abounds in various natural curiosities and mineral productions: among which, the most remarkable are the Trenton, Cohoes and Glenn's falls on the Mohawk and its branches, and the cataract of Niagara, the largest in the world, by which the waters of lake Erie and the other great lakes, pass into lake Ontario, over a precipice of 160 ft. In the western parts of the state are found large quantities of gypsum, which are used very extensively for agricultural purposes. At Sing Sing, on the Hudson, is an excellent quarry of marble, which is much used in architecture. In the cos. w. of lake Champlain, are vast beds of iron ore, and the iron mines of Columbia co. are also highly valuable and extensively wrought. Traces of other metals, as silver, lead, zinc, and titanium, have also been observed in various parts of the state. And of minerals which may be deemed rather objects of curiosity than of real value, this state affords abundant varieties. There are 56 counties in this state, which, under the apportionment of 1832, are divided into 32 congressional districts, and are entitled to 40 representatives in congress. The number of cities is 7, viz.: Albany (the capital), N. York, the largest and most important city in the U. States, Troy, Hudson, Schenectady, Utica and Buffalo. The number of towns is 764, and of incorporated villages, 102.

The progressive population of the state of New York is almost without a parallel. In 1700 it amounted to about 20,000; and in 1730 to about 50,000; in 1749 there were 100,000 inhabitants; in 1770, 160,000; in 1790, 340,120; in 1800, 586,050; in 1810, 959,049; and in 1820, 1,372,812. In 1830 there were 1,918,608, as follows:

Counties.	Pop. 1820.	Pop. 1830.
Albany,	38,116	53,560
Alleghany,	9,330	26,218
Broome,	11,100	17,582
Cataraugus,	4,090	16,726
Cayuga,	38,897	47,947
Chatauque,	12,568	36,657
Chenango,	31,215	37,404
Clinton,	12,070	19,344
Columbia,	38,330	39,959
Cortland,	16,507	23,693
Delaware,	26,587	32,933
Dutchess,	46,615	50,929
Erie,	15,668	35,710
Essex,	12,811	19,387
Franklin,	4,459	11,312
Genesee,	39,835	51,992
Greene,	22,996	29,525
Hamilton,	1,251	1,325
Herkimer,	31,017	35,869
Jefferson,	32,952	18,515
Kings,	11,187	20,537
Lewis,	9,227	11,958
Livingston,	19,196	27,719
Madison,	32,208	39,027
Monroe,	26,529	49,862
Montgomery,	27,569	43,593
New York,	123,706	203,007
Niagara,	7,322	18,485
Oneida,	71,326	71,326
Onandaga,	41,461	58,974
Ontario,	35,312	40,167
Orange,	41,213	45,372
Orleans,	7,625	18,873
Oswego,	12,374	27,104
Otsego,	44,856	51,372
Putnam,	11,268	12,701
Queens,	21,519	22,278
Renssellaer,	40,153	49,472
Richmond,	6,135	7,084
Rockland,	8,837	9,388
Saratoga,	36,052	38,616
St. Lawrence,	16,037	36,351
Schenectady,	13,081	12,334
Schoharie,	23,154	27,910
Seneca,	17,773	21,081
Steuben,	21,989	33,975
Suffolk,	24,272	26,780
Sullivan,	8,900	12,372
Tioga,	14,716	27,704
Tompkins,	26,178	36,545
Ulster,	30,934	36,550
Warren,	9,453	11,795
Washington,	38,831	42,615
Wayne,	20,319	33,515
West Chester,	32,638	36,459
Yates,	11,025	19,019

Of the above were white males, 951,516; females, 916,670; free colored males, 21,465; females, 23,404; slaves, males 12; females 64. To these is to be added, aliens in the city of New York, not originally returned, 5,477. Included in the foregoing population are deaf and dumb persons, whites 842; colored, 41. Blind, whites 642; colored 82. Aliens 52,488.

This state carries on an extensive foreign commerce with all parts of the world; for the most part through the city and port of N. York. The number of vessels that arrived at New York from foreign ports, during the year ending Dec. 1831, was 1,634; of which 1,264 were American. (*See art. N. Y. city.*) The inland trade is also very thriving, and has been greatly increased by canals, rail roads and other works of public enterprize and improvement. The Erie canal, the longest in the world, being 360 miles in length, connects the navigation of Hudson river with that of lake Erie, the western lakes and the Ohio canal. The Champlain canal connects it with that of lake Champlain; and the Delaware and Hudson canal, with Delaware and Lackawana rs. The canal debt amounted on the first day of January, 1831, to 7,825,035 dollars. 9,653 dollars of the stock has been cancelled, and 240,263 dollars has been borrowed during the past year, for continuing the works upon the Chemung and Crooked Lake canals, so that the debt on the first day of January amounted to 8,055,645 dollars.

The receipts into the treasury of revenue on account of the canal fund during the year ending the 30th day of September, 1831, are: On account of tolls, 722,896 dollars; from other sources, 307,012 dollars. These receipts, however, do not include the collections of tolls and salt duties for the month of September.

Canals are in progress or contemplation to open a communication with the navigable waters of Pennsylvania, and between important streams in this state. A rail road has just been completed between Albany and Schenectady, called the Mohawk and Hudson rail road; and many others are projected. The commerce of lake Erie has rapidly increased; and about 130 vessels of 70 tons each are now employed on the southern shore, besides 16 steamboats, of from 150 to 400 tons. Within the past year 75,000 tons were entered at the Buffalò custom house. 23,467 barrels of flour, 200,802 bushels of wheat, 8,426 bbls. of pork, 1,768 do. ashes, 1,044 do. whiskey, 11,040 barrels of salt, passed through the Welland canal previous to the 30th September. At least 60,000 barrels of salt from the state of New York must have have been transported on lake Erie to various markets, beside large quantities that were detained in different ports, by the early closing of the navigation. Not less than 70,000 barrels of flour, 500,000 bushels of wheat, 5,000 barrels of pork, 3,000 barrels of ashes, 10,000 barrels of lake fish and an incalculable amount of other products of the country, have found their way to market by means of this lake during 1831.

The steamboats in this state in 1831, were estimated at 86; the principal part of which ran from the city of N. York. One of them measured 527 tons. The most rapid boats have run between the 2 cities of N. York and Albany in less than 10 hours. The mail is carried from N. York to Quebec (almost the whole distance in steamboats) in 96 hours.

In this state there are 200 woollen manufactories; 112 cotton mills, with a capital of $4,485,500, manufacturing 21,010,920 yds. valued at $3,530,250, and using 7,961,670 lbs. of cotton annually, and employing 15,970 persons:—above 200 iron works, making $4,000,000 worth:—50 paper mills, making $700,000 worth:—leather made to about the value of $3,458,000; and hats to the value of $3.500,000, &c. &c. per annum. By an act of the state of New York, passed in 1829, every bank thereafter to be chartered or renewed was obliged to contribute one half per cent. per ann. of its capital, in order to establish a safety fund, which should be placed in the hands of commissioners, to provide for the payment of the debts of any banks which should become insolvent. In 1830, there were 29 banks subject to this act, with a capital of $6,294,600. The capital of banks not subject to the safety fund was, $21,323,460.

There are 4 colleges in New York; Columbia, city of N. York; Union, Schenectady; Hamilton, Clinton, Oneida co.; and Geneva, Ontario co.—5 theological seminaries; Episcopal, N. Y. city; Presbyterian, Auburn; Oneida Institute, Whitestown; Hamilton, Madison co.; Hartwick, Otsego co.;—2 medical colleges; one at New York city and one at Fairfield, Herkimer co.—57 academies and seminaries which derive part of their support from the public fund. The regents of the university, 21 in number and appointed by the legislature, may grant degrees in medicine and the arts, and control the income of the literature fund; dividing it among the 8 senate districts, among the incorporated seminaries of learning, except colleges, in proportion to their numbers of students. They are also authorized to report on the colleges and academies to the legislature. 9,333 school districts, furnishing instruction to 497,257 pupils.

The productive capital of the school fund amounts to $1,704,159 40 cents. The revenue actually received on account of this fund, for the 10 months ending September 30, 1831, has been 80,043 86. The receipts are estimated by the comptroller at $96,350, for the coming year.

The unproductive portion of the school fund consists of about 850,000 acres of land, lying principally in the fourth senate district. The value of these lands has been estimated at $400,000; which sum, if added to the present productive capital, would make a total of more than 2,000,000 of dollars.

During the last twelve years, and since the school system has been in fair operation under the act of 1819, the average annual increase of children, between 5 and 16, has been 16,008; and the average increase of scholars instructed has been 16,860 each year.

There are 2 institutions for the instruction of the deaf and dumb; one at the city of New York, and one at Canajoharie. In addition to these institutions, may be mentioned the university of the city of New York; an institution recently chartered by the state legislature; and established on the comprehensive system of the universities of Europe.

This state has led the way in the late important improvements of prison discipline, with such success as to obtain imitators in many other states and receive the approbation of some foreign countries. This system was introduced into the Auburn state prison about 10 years since, and soon after was also adopted in the prison at Sing Sing. In 1830, the number of convicts was 600, the amount of whose labor was $40,341. The whole expenses of the establishment for the same time was $36,-226. The state prison at Sing Sing as yet is less profitable. The expenses of this prison in 1831 were $77,600. The amount of the labor of convicts, $48,000. Number of convicts 990. The marble quarries at which the prisoners are employed require much labor to remove the earth and inferior stone, before the good marble can be obtained, and are therefore less profitable. In 1831, the number of paupers supported by the state was 15,564,—at an expense of $245,433. The portion of this sum saved by the labor of the paupers was $17,-546. The constitution of this state secures to all its citizens, "the free exercise and enjoyment of religious profession and worship, without discrimination or preference"; and accordingly, some of almost every denomination are found within its limits. The following estimate is for 1831. The Presbyterians have 587 churches, 486 ministers, 54,093 communicants. The Dutch Reformed 148 churches, 111 ministers, and 8,672 communicants. Associate Synod of N. Y. 15 congregations, 13 ministers, and 1,668 communicants. Methodists 73,174 members. Baptists 549 churches, 387 ministers, and 43,565 communicants. Episcopalians 129 ministers. Lutherans 27 ministers, and 2,973 communicants. There are some Roman Catholics, Friends, Universalists, Unitarians, Shakers and United Brethren.—(*For further details in internal improvements, see article rail roads and canals.*)

New York City, in the state and county of the same name, is 225 ms. N. E. from Washington City; 90 N. E. Philadelphia; 210 S. W. Boston; 160 S. Albany; and 390 S. Montreal. It is about 16 ms. from the Atlantic ocean, at lat. 40° 42', and long. 2° 54' E. W. C. This is the most populous, wealthy, commercial and important city in the United States. It stands at the junction of the Hudson and East rivers, on the S. end of Manhattan isl., where the population is almost entirely concentrated. The ground is generally high near the middle, from which it slopes eastward to East river, which is the channel of the eastern coasting trade; and westward to Hudson r., which affords the grand route of internal commerce through the Erie canal, and the great northern lakes. The bay of N. York, which is one of the finest harbors in the world, is about 4 ms. by 9; being almost entirely enclosed by land, and much protected by the heights of Staten and Long islands. The shores on both rivers are lined with wharves and slips, where ships of the largest size are laden and discharged. The outer bay affords a convenient retreat and safe anchorage, during the prevalence of certain winds; and is well furnished with light-houses and forts.

By a recent estimate the city contained 30,000 dwelling houses, stores, manufactories, and churches. The first houses were built in 1621, in the southern part of the isl., where there are still some narrow and crooked streets; but the other parts are laid out, with more regard to beauty and convenience. Broadway, the principal street of the city, is 80 feet wide, and perfectly straight, and passes from the S. extremity, through the centre of the city, about 2 ms., and there joins the Fifth avenue, which passes through the island to Harlaem r. Besides this, there are many other streets worthy of notice. Those in the S. part and on the East r., are remarkable as places of business; and those in the upper part are chiefly occupied by private residences. The battery is a fine open public walk, on the S. extremity of the isl.; commanding an extensive view of the bay, and the opposite shore of New Jersey. The City Hall is a marble building, standing in the Park, 216 feet by 105, and 65 high; built at an expense of $500,000. It contains the chambers of the two council-boards, court rooms, offices, &c. The new City Hall in its rear contains the alms-house, court of sessions, and police office. The Merchant's Exchange, in Wall street, contains the Post Office, as well as the Exchange Hall, and various offices. The hospital is an old and very respectable institution, with a library of 3,000 vols. At Bellevue are the alms-house, a hospital, and a penitentiary. The first contained in January, 1832, 1,207 natives of the United States, and 1,049 foreigners. The second, 302 patients, and the third 417 vagrants and other prisoners, of whom 151 were foreigners. The new penitentiary is on Blackwell's island, in East r., about 7 ms. from the city. There are two large stone buildings, with cells for solitary confinement by night. The prisoners are employed in quarrying building stone in the vicinity. The police of the institution is strict. The children receive instruction, and the morals, habits, and minds are improved under the excellent system of the prison. The public school society, have 12 large brick school houses in different parts of the city, each of which contains 2, and some of them 3 large apartments. The public schools received in 1831, from the state, $20,549 38, and from a city tax, $15,661. The number of pupils instructed in 1830, was 24,952. There are also the Protest. Episcopal school, the mechanic's school, besides numerous private schools for both sexes. Columbia college is a venerable institution, founded in 1754; and is now possessed of an estate valued at $400,000. Its officers are a president, 28 professors, and the number of students about 100. This college and the grammar school connected with it enjoy a fine situation, near the centre of the city. The New York university has been but recently chartered, and has not yet commenced its operations; but is an institution founded on the liberal system of the European universities, and is one which promises much benefit to the cause of science and literature in our country. Its funds have been raised by the voluntary subscriptions of individuals. It is governed by a council of 32 members, chosen by the subscribers, together with the mayor and 4 members of the common council of the city. The number of literary, scientific, religious, benevolent, and other societies in the city of N. York, is almost innumerable. The American Bible society buildings are very large and extensive. The receipts of the society, for the year ending May, 1832, were $107,059, of which $40,193 were in payment for books. Number of Bibles and Testaments issued during the year 115,802. The American tract society have also a large building, and published during the same year, 5,471,750 tracts, of 87,622,000 pages :—making the whole number of pages printed since the formation of the society, 288,281,000. Number of pages distributed during the year, 4,927,009. Whole number of societies publications, 614, in 10 different languages. The American home missionary society, received in 1831, $50,299 25; expended $52,808 39; supported 509 missionaries, and assisted 745 congregations. The American education society, in the same year aided 673 young men, and received $41,947. Whole number of young men assisted since its formation in 1826, 1,426. The receipt of the American seaman's friend society, for the same year, amounted to $5,679. Among other benevolent societies, are the New York Sunday school union; general Protestant Episcopal Sunday school union; colonization society; manumission society; numerous temperance societies; institution for the instruction of deaf and dumb; do. for the blind; orphan asylum; Roman Catholic benevolent society; marine society; St. George's society; St. Andrew's society; friendly sons of St. Patrick society; French benevolent society; German society; Humane society; Education society of the Reformed Dutch church; charity school of do.; Sunday school union of do.; societies for the relief of poor widows, of orphan children, of aged indigent females; asylum for the reformation of juvenile delinquents, for the education of Jewish children, for the encouragement of faithful domestics; besides numerous branch societies, &c. &c. Among the literary, scientific, mercantile and other societies, may be enumerated, the American academy of fine arts, (Barclay st.) the National Academy of Design (Clinton hall); Lyceum of Natural History; New York Historical Society; New York Society Library; N. Y. Atheneum; N. Y. Literary and Philosohical Society; Mercantile Library Association; Clinton Hall Association; N. Y. Law Institute; N. Y. Chamber of Commerce; American Institute, (for the encouragement of domestic industry, with annual fairs and exhibitions); N. Y. Chamber of Trade; N. Y. Horticultural Society; N. Y. State Society of Cincinnati: Merchants' Exchange Co.; N. Y. Society of Merchants and Tradesmen; N. Y. University; American Lyceum, &c. &c.

There is no city in the United States, perhaps none in the world, which possesses greater advantages, both for internal and external commerce. From the 1st January to the 31st December, 1831, inclusive, there arrived at New York from foreign ports, 387 ships, 42 barques, 757 brigs, 433 schooners, 1 ketch, 1 galliot, 1 pollucca, 1 felucca, and 11 sloops; in all 1,634, of which 1,264 were American, 273 English, 8 Spanish, 14 Sweedish, 17 German, 25 French, 2 Haytien, 18 Danish, 1 Mexican, 2 Brazillian, 1 Genoese, 1 Russian, and 3 Dutch.

The total number of passengers by these arrivals was 31,739.

According to the report of the secretary of the treasury, during the year ending Sept. 30, 1831, the amount of tonnage entered in the state was 333,778; amount departed, 265,915; value of imports $35,624,070; exports, domestic, $13,618,278; foreign, $6,079,705; total exports, $19,697,983. The amount of duties on imported merchandise, discharged at this port, $20,096,136 60. But the internal commerce with the western states and the interior of the state of N. Y. is a still greater source of wealth and prosperity to the city. The following estimate will show the relative values of real estate in the city of N. Y. during two commercial periods of seven years each; and in some degree perhaps the comparative effects of external and internal commerce. In 1817, the real estate of the city was assessed at $57,799,-435. In 1824, it was assessed at $52,019,730; showing a decrease of $5,779,705 during a period of 7 years, in which foreign commerce was regulated by the tariff of 1816. In 1825, the Erie canal was completed; and the real estate of the city was estimated for this year at $58,425,395; and in 1831, was assessed at $95,716,485; showing an increase of $43,706,-755, during another period of 7 years, after the internal commerce with the Western states had commenced. In 1831, the city inspections of flour, grain and salt, were as follows. Wheat flour 928,281 bbls.; rye flour, 9,222, do.; Indian meal, 31,950 do.; buckwheat flour, 380 do. The amount of wheat inspected was, 466,559 bushels; rye 438,114 do.; corn 1,028,674; oats, 1,067,-693; barley 129,297; malt, 37,018; and of salt, 74,008. The whole amount of grain inspected was 3,267,231 bushels; value $2,-305,687 81. From New York there are lines of regular packet ships to Liverpool, London and Havre; and packet brigs to Hull, Greenock, Belfast, Carthagena and Vera Cruz; besides lines of packet vessels to almost every large port of the United States. There are 19 banks in the city, whose capital amounts $11,311,200, and which in Jan. 1832, had notes in circulation to the amount of $4,396,387 13; with specie on hand to the amount $1,207,363 65. There are also a branch of the United States bank, a savings bank, a seaman's savings bank, and 9 marine and fire insurance companies. The municipal tax for the year 1832 is $550,000; averaging $2,20 cts. to every individual, old and young. The estimated revenue from other sources, is $159,000, and the estimated expenditure $600,475. In 1830, the number of votes for members of congress was 21,000; the proportion of freeholders not known, as property is no longer the basis of representation. There were in July 1832, in the city, 11 daily newspapers, 9 semi-weekly, 29 weekly, of which several are religious; 2 Spanish and 1 French, 3 semi-monthly and 2 monthly. The number of sheets issued annually, is about 10,628,600. The consumption of fuel in 1831, was 297,606 loads of wood, at an aggregate cost of $493,085 86; 26,605 tons of anthracite coal, 11,875 chaldrons Virginia, 12,953 of charcoal, at a cost of $321,-642 34; total $814,728 20. A great amount of Liverpool coal consumed, is not included. The number of deaths in 1805, was 2,252; in 1815, 2,507; in 1820, 3,520; in 1825, 5,018; and in 1830, 5,537. One sixth of these have been from consumption, and, except to those inclined to consumptive disorders, the city generally is very healthy. The number of churches in the city is upwards of 100, embracing some of almost every denomination of christians.

The population of the city of New York increased very moderately during the earlier periods of its history. Among the earlier dates, we find that in 1696 it amounted to 4,302, and in 1786 it had only increased to 23,614, a period of 80 years. Since that time its growth has been exceedingly rapid; in 4 years afterwards, 1790, the population had increased to 33,131, and at subsequent periods it was as follows; in 1800, 60,489; in 1810, 96,373; in 1820, 123,706; in 1825, 166,-086, and in 1830, 202,589.

New York, p-v. western part of Albemarle co. Va. 18 ms. w. Charlotteville, and by p-r. 143 ms. s. w. by w. W. C.

New York, p-v. Switzerland co. Indiana, by p-r. 114 ms. s. e. Indianopolis.

Niagara, r. on the n. w. boundary of New York, and a part of the boundary of the U. S. 36 ms. long, from 1-2 m. to 6 or 7 ms. wide, empties the waters of lake Erie into lake Ontario, has several islands, great rapids, and the cataract of the same name, which is the most remarkable in the world. The shores are low and nearly level from lake Erie to the falls, and but little inhabited, and Grand isl. 12 miles in length, and 7 broad in the widest part, divides its channel a part of its course. Tonawanta and Chippewa creeks empty into the r., the latter from the Canada side; the former supplies Erie canal to Rochester, and serves as the first part of the route. The surface of Niagara r. is smooth to the rapids, where it is broken by ledges of rocks, over a descending bed, for about 3-4 m. and then is precipitated from a perpendicular wall of rock, 160 feet high on the American side, and 174 feet on the Canadian side. The precipice near the middle of the cataract is much higher up the r. than near the shores, and forms an irregular arch, or horse shoe, towards the west side. Goat isl. also occupies a part of the channel and divides the cataract into two unequal parts, but is connected with the American shore by a bridge. The Biddle stair case conducts the visitor in safety from Goat isl. to the rocks below; and there are stair cases on the sides of the cataract, as well as spacious hotels. Several persons have at different times been carried over the precipice, and none have ever survived. Ducks have sometimes been picked up alive after the fall, with legs and wings broken. The waters are precipitated into a gulf, which is constantly kept in a state of commotion, and is covered with white foam, while the rocks overhang it from a great height on three sides, dripping with the moisture which rises in large clouds of mist.

Niagara, co. N. Y. is bounded by lake Ontario or Upper Canada n., Orleans and Genesee co. e., Tonawanta creek or Erie co. s., Niagara river w. 16 ms. by 28, 448 square ms. has 7 towns, is crossed by the mountain ridge, between 7 and 9 ms. from the lake and nearly parallel to the shore. In its highest part this ridge is 330 ft. above the lake. The s. part is agreeably varied. The soil is good. The alluvial way is an inferior elevation, 3 or 4 ms. n. of the mountain ridge, partly in this co. remarkably uniform, and serving for the route of a good road. Tuscarora and Eighteen Miles creeks are the principal streams in the co. but small. The rapids in Niagara river, however, afford good mill sites. Erie canal enters from Orleans co. runs at the foot of a mountain ridge to Lockport, where it surmounts that obstacle by one of the most expensive and splendid works on its whole route. The excavations are great, for the deep cut and the basin; and the vast water power is of great value. Since the formation of the canal, Lockport has been formed from a wilderness to a considerable village. From this place the canal passes to Tonawanta creek, on the s. line of the county. The falls of Niagara are in this co. and the portage of 8 miles round them. Welland canal, (on the Canada side) has been constructed to effect a navigation round the falls. In 1796, there was but one family in the limits of this co. The settlers were from different parts of the country, some from Canada. Pop. 1830, 14,482.

Niagara, p-t. Niagara co. N. Y. 300 ms. w. Albany, 11 s. w. Lockport, 11 n. e. Lewiston, at the falls of Niagara, has a varied and agreeable surface, sloping towards the river, with a light soil, in some parts good. Manchester village is at the rapids, near the falls, 7 ms. from Lewiston. It is small, but has a large hotel for visiters, who annually appear in great numbers, to see this noblest cataract in the world. Several mills are moved by the water at the rapids. A bridge reaches to Goat island, a little above the precipice, and a covered spiral staircase conducts to the foot of the falls on the main land from both; the views of the falls are various and interesting. Gypsum is found in the rocks. Fort Schlosser, a stockade, built after the old French war, stood at the mouth of Gill creek on Niagara river, 1 1-2 miles above the falls. It was surrendered to the United States 1796. Pendleton village is in the s. e. part of this town. Pop. 1830, 1,401.

Niagara, village, Porter, Niagara county, N. Y. 7 miles n. Lewiston, at the mouth of Niagara river, on the shore of lake Ontario. Fort Niagara is on this site.

Nichols, p-t. Tioga co. N. Y. 10 miles w. Owego, s. Susquehannah river. Population 1830, 1,284.

NICHOLAS, co. of Va., bounded by Pocahontas E., Green Brier S. E. and S., New river, separating it from Logan, S. W., Kenhawa W., Lewis N., and Randolph N. E. Length 50 ms., mean width 28, and area, 1,400 sq. ms. Extending in lat. from 38° 4' to 38° 43' and in long. from 3° 18' to 4° 12' W. W. C. The declivity is a little S. of W., drained by Gauley r. on the sthrn., and Elk r. on the nrthrn side. Pop. 1820, 1,853; 1830, 3,349.

NICHOLAS, C. H. and and p-v. Nicholas co. Va. 310 ms. wstrd. W. C., and 268 N. W. by W. Richmond, lat. 38° 18', long. 3° 48' W. W. C.

NICHOLAS, co. of Ky., bounded S. W. by Bourbon, W. and N. W. by Harrison, N. by Bracken, N. E. by Mason, E. by Fleming, and S. E. by Bath. Length 30, mean width 12, and area, 360 sq. ms. Extending in lat. from 38° 12' to 38° 33', and in long. from 6° 47' to 7° 10' W. W. C. The main volume of Licking r. crosses this co. in a nthwstly. direction, following the general declivity. Chief t. Carlisle. Pop. 1820, 7,973; 1830, 8,832.

NICHOLASVILLE, p-v. and st. of jus. Jessamine co. Ky. 30 ms. S. E. Frankfort, and 14 S. S. W. Lexington. Lat. 37° 52', long. 7° 33' W. W. C. Pop. 1830, 408.

NICHOLSON, p-v. nrthwstrn. part of Luzerne co. Pa. 32 ms. above Wilkesbarre, and by p-r. 146 ms. above Harrisburg.

NICHOLSON, p-o. Copiah co. Miss, about 60 ms. a little N. of E. Natchez, and 70 ms. S. S. W. Jackson.

NICKSVILLE, p-v. Lovely co. Ark.

NILES, p-v. sthestrn. part Berrien co. Mich. by p-r. 179 ms. S. W. by W. ½ W. Detroit. Lat. 41° 51', long. 9° 18' W. W. C. As laid down by Tanner on his improved U. S. map it is situated on St. Joseph's r. 5 ms. N. of the boundary between Ind. and Mich.

NIMISILA, p-v. Stark co. O. by p-r. 117 ms. N. E. by E. Columbus.

NINE MILE CREEK, and p-o. Blount co. Ten. by p-r. 155 ms. a little S. of E. Nashville.

NINE MILE PRAIRIE, p-o. Perry co. Ill. by p-r. 127 ms. sthrd. Vandalia.

NINEVEH, p-v. N. part Frederick co. Va. by p-r. 81 ms. N. W. by W. W. C.

NIPPENOSE, p-o. sthrn. part Lycoming co. Pa. by p-r. 104 ms. W. Harrisburg.

NISHNEBATONA, r. confluent of Mo. r. rises at about lat. 42°, flowing thence by a general course of a little W. of S., inclining upon that part of Mo. immediately above and below the mouth of Platte r., and inflecting with the former finally falls into it a short distance below the influx of little Nemawhaw. The valley of Nishnebatona lies between those of Mo. and Naudoway.

NISKAYUNA, t. Schenectady co. N. Y. 12 ms. N. W. Albany, S. W. Mohawk r., N. Albany co., E. Schenectady, is small, with much poor sandy land, but some good on the r. The people are of Dutch origin. Alexander's bridge crosses Mohawk r. in the N. E. corner, where are mills, a dam, and the upper acqueduct, 748 feet long, and 25 high, on which passes the Erie canal; near it are 3 locks of 7 ft. lift. Pop. 1830, 452.

NISKAYUNA, Shaker v. Schenectady co. N. Y. (*See Watervliet.*)

NITTANY, p-v. nrthwstrn. part Centre co. Pa. 16 ms. N. W. Bellefonte, and by p-r. 101 ms. N. W. Harrisburg.

NIXON'S p-o. Randolph co. N. C. by p-r. 94 ms. wstrd. Raleigh.

NOAH'S FORK and p-o. nthwstrn. part Bedford co. Ten. 60 ms. S. E. Nashville.

NOBLEBOROUGH, p-t. Lincoln co. Me. 16 ms. E. N. E. Wiscasset, E. Damariscotta r. Carries on a considerable trade. Pop. 1830, 1,876.

NOBLESBORO' or NOBLESTOWN, p-v. on Robeson's run, in the sthwstrn. part Alleghany co. Pa. 11 ms. S. W. by W. Pittsburg.

NOBLESTOWN, p-v. Alleghany co. Pa. by p-r. 212 ms. W. Harrisburg, and 12 ms. S. W. Pittsburg.

NOBLESVILLE, p-v. and st. jus. Hamilton co. Ind. by p-r. 22 ms. N. N. E. Indianopolis. Lat. 40° 03', long. 9° W. W. C.

NOLACHUCKY, r. of N. C. and Ten., has its remote sources in the western slope of the Blue ridge, and northern part of Buncombe co. opposite the sources of Catawba, and between those of French Broad and Watauga rs.; draining the nrthrn. part of Buncombe, and flowing by a general N. W. course, traverses the mtn. pass between the Bald and Iron mtns., enters Ten., where, passing over Washington and Greene cos. joins French Broad. (*See Ten. r.*)

NOLACHUCKY, p-v. on Nolachucky r. sthrn. part Greene co. Ten. by p-r. 286 ms. E. Nashville.

NOLAND'S FERRY and p-o., Loudon co. Va. by p-r. 43 ms. a little above and N. W. W. C. The p-o. is nearly opposite, though rather above the mouth of Monocacy r.

NOLENSVILLE, p-v. nrthrn. part Williamson co. Ten. 16 ms. a little E. of S. Nashville.

NO-MAN'S-LAND, isl. Dukes co. Mass. S. of the S. W. end of Martha's Vineyard. It is small, and lies in long. 6° 15' E. W. C., lat. 41° 15'.

NORFOLK, co. Mass. bounded N. by Middlesex co., N. E. and E. by Mass. bay and Plymouth co., S. E. and S. by Plymouth and Bristol cos., S. W. by Rhode Island, and W. by Worcester and Middlesex cos. It is principally watered by Nepenset and Charles rs., and their tributaries, and contains 22 towns, of which Dedham is the capital. The two streams mentioned are singularly connected by Mother brook, which thus renders a part of the co. an island. The waters of Stony brook are almost on a level with it. Blue hills on the N. are of considerable elevation, but the surface is not very uneven. The N. part has many fine country seats, belonging chiefly to citizens of Boston; and its vicinity to the city, affords a valuable market to the farmers of the co., in which some of the most beautiful farms and rural scenes in the state are found. That part contiguous to Boston was occupied in the revolutionary war, and during the siege of that place, by the American troops; and remains of military works are still to be found. It is crossed by the great route from Boston to Providence, and a rail road, which is projected between the two places, will, doubtless, greatly increase the already great amount of travel and transportation through the co. The amount of manufactures in this co. is considerable. Its surface is generally diluvial, with rocks of sienite granite, or graywacke. The uplands are rounded gravel from the interior, sand and clay mingled, and the valleys marshy with peat. Large masses of rock lie on the surface, for 10 ms. s. of Blue hills, s. E. of the ledges from which they have been torn; but the sienite and green stone ledges are most remote. Diluvial gravel lies upon the slope from Blue hills to Neponest r. The Quincy rail road, the first constructed in the U. S. is in the E. part of this co. Pop. 1820, 36,471; 1830, 41,972.

NORFOLK, p-t. Litchfield co. Conn. 35 ms. N. W. Hartford, s. Mass.; 4½ ms. by 9; 44 sq. ms.; is on high ground, crossed by granite ridges from N. E. The soil is cold, but the grazing good; and the trees chiefly oak and chestnut, with some maple. 20,000 lbs. of maple sugar were once made here annually. Blackberry, Mad, Sandy and other rs., supply water and mill sites. Much iron is manufactured here. Pop. 1830, 1,485.

NORFOLK, p-t. St. Lawrence co. N. Y. 32 ms. N. E. Ogdensburgh, 224 from Albany. Watered by Racket r. at the falls of which is a v., at the head of boat navigation. Pop. 1830, 1,039.

NORFOLK, co. Va. bounded by Nansemond W., Hampton Roads and Lynhaven bay N., Princess Anne co. E., and Currituck and Camden cos. N. C. s. Length from s. to N. 32, mean width 17 ms., and area 544 sq. ms. Extending in lat. from 36½° to 36° 59′, and in long. from 0° 33′ to 1° 2′ E. W. C. This co. is composed of two inclined planes of very little declivity. The nrthrn. section is drained by the confluents of Elizabeth river. (*See Elizabeth r.*) The southern part includes great part of the Dismal Swamp and lake Drummond, and drained sthwrdly. into Pasquotank r. (*See Dismal Swamp canal.*) Chief t. Norfolk. Pop. 1820, 15,465, including that of the borough of Norfolk, 6,987 exclusive of the borough. In 1830, 24,814.

NORFOLK, borough, p-t. and s-p. of Norfolk co. Va. situated on the right bank of Elizabeth r., 8 ms. above Hampton Roads. Lat. 36° 52′, long. 0° 44′ E. W. C. By p-r. 217 ms. s. s. E. W. C., and 114 s. E. by E. Richmond. Norfolk harbor admits vessels of 18 feet draught, and renders the borough the most commercial depot of Va. It is defended by a fortress on Craney isl. and some other forts. It appears, indeed, from examinations made by the U. S. commissioners, that Hampton Roads, though so extensive, admit of complete defence against foreign attack. The Dismal Swamp opens to Norfolk, the commerce of the great basins of Roanoke and Chowan; in consequence some of the finest sections of Va. and N. C., drained into Albemarle sound. Norfolk contains a theatre, marine hospital, academy, orphan asylum, atheneum, a branch of the U. S. bank, and 2 state banks. It contains also numerous private schools, and several places of public worship.

The site of Norfolk, similar to the surface of the adjacent country, is low, level, and in part marshy: but the streets being well paved, obviates many natural disadvantages. The progressive pop. of this borough presents some curious facts. It contained in

	1810.	1820.	1830.
Whites,	4,776	4,618	5,131
Free colored,	592	599	928
Slaves,	3,825	3,261	3,757
Total,	9,193	8,478	9,816

showing an increase of 1,338, in the latter period. The relative increase of the whites and slaves being very nearly equal, at about 11 per cent.

NORRISTOWN, p-v. borough and st. jus. Montgomery co. Pa. situated on the left bank of Schuylkill r., 16 ms. above and N. W. Phil. and by p-r. 143 ms. N. E. W. C. Lat. 40° 08′, long. 1° 42′ E. W. C. The site of Norristown and the adjacent country are delightful. It contains an academy, some places of public worship, and the county edifices. Pop. 1820, 827.

NORMAN'S KILL, brook, Bethlehem Albany, co. N. Y. enters Hudson r. 2½ ms. s. Albany, 28 ms. long, supplies large mills.

NORRIDGEWOCK, p-t. st. jus. Somerset co. Me. 35 ms. W. by N. Hallowel, 94 N. N. E. Portland, 28 Augusta, has a C. H., jail, &c. and has considerable trade. Pop. 1830, 1,710.

NORRISVILLE, p-o. Wilcox co. Ala. by p-r. 127 ms. s. Tuscaloosa.

NORTH ADAMS, v. Adams, Berkshire co. Mass, 15 ms. N. Pittsfield, 40 E. Albany, 5 E. from Williams college, is a flourishing manufacturing place. There are, in and about the village, 12 cotton and woollen factories, (about 24 in the town,) 2 calico print works, 3 furnaces, and several extensive establishments for making cotton and woollen machinery, &c.

NORTHAMPTON, p-t. Rockingham co. N. H. 7 ms. s. Portsmouth, 50 from Concord, W. Atlantic ocean, has 8,465 acres, 2 small streams, and Little Boar's Head point. Garrison houses were built early, to protect the people against the Indians. Pop. 1830, 766.

NORTHAMPTON, p-t. st. jus. Hampshire co. Mass. 95 ms. W. Boston, 42 N. Hartford, 18 N. Springfield. It is one of the oldest and pleasantest towns in that part of the state. It lies W. Conn. r. with a varied surface, good soil, and between 3,000 and 4,000 acres of excellent meadows. The Hampshire and Hampden canal, partly completed, and connected with the Farmington canal in Conn. is to join the Conn. river here. It contains 4 churches, for Calvinists, Baptists, Episcopalians, and Unitarians. A court house, jail, town house, bank, a fine hotel, &c. in the village, which is large, and was in past days, the residence of president Edwards, David Brainard, Gov. Strong, and other distinguished men. First settled 1664, and for many years the village was surrounded with a palisade. It was attacked and threatened at different times by Indians. A great deal of manufacturing is carried on here, 700 men being engaged in cotton and woollen factories; a woollen factory, 4 1-2 ms. w. of the village, on a fall of 50 feet, works 1,384 spindles, 35 broadcloth and 8 satinet looms, and employs 110 persons, consuming 150,000 lbs. of wool, and making 42,000 yards of broadcloth, and 36,000 of satinet. A button manufactory employs 30 persons.

There are in the town, dwelling houses, 417; stores and shops, 69; barns, 302; mills of various kinds, 26; of tillage, 2,635; bushels of rye raised, 6,257; oats, 5,050; Indian corn, 31,000; acres of mowing, 2,148; tons of hay, 2,394; acres of pasture, 4,060; bbls. of cider, 2,150; acres of woodland, 4,414; horses, 334; oxen, 174; steers and cows, 866; sheep, 4,000; woollen factories, 3; spindles, 1,152; carriages and chaises, 3,525.

Mount Holyoke, a fine eminence on the opposite bank of the river, is a favorite resort and commands one of the most beautiful views in this part of the U. S. The shad fishery is valuable, and steam navigation is to be extended hither by the boats from Hartford. In 1786, during Shay's rebellion, a body of insurgents were dispersed here by the sheriff. Pop. 1830, 3,613.

NORTHAMPTON, p-t. Montgomery co. N. Y. 17 ms. N. E. Johnstown, 42 ms. N. N. W. Albany, 22 N. W. Ballstown Spa, W. Saratoga county, 4 ms. by 8, is crossed by Sacandaga river and has 3 small mill streams, and good level land, but 1,000 acres of the great vlie or swamp. At the fish house where Sir William Johnson sometimes resided is a small village. Pop. 1830, 1,380.

NORTHAMPTON, t., Burlington co., N. J., 7 ms. S. E. Burlington, S. North branch of Rankokus cr., W. Monmouth co. Pop. 1830, 5,516; it includes the v. of Mnt. Holley, the st. jus. of the co.

NORTHAMPTON, co., Pa., bounded by Bucks co. S., Lehigh S. W., Schuylkill W., Luzerne N. W., Pike N., and Del. separating it from Warren co. N. J., E. Length diagonally from the extreme sthrn. angle on Bucks, to the extreme nthn. on Luzerne, 46 ms.; the greatest width wstrd. from Del. r. to the wstrn. angle on Schuylkill and Luzerne, is very near equal to the length. The area being within a trifle of 1,100 sq. ms. The mean breadth is very nearly 24 ms., extending in lat. from 40° 33′ to 41° 10′, and in long. from 0° 50′ to 1° 52′ E. W. C.

Though the Lehigh r. does not rise entirely in Northampton co., having its higher branches in Pike and Luzerne, yet the co. and valley of this branch of Del. are in great part commensurate, and the general slope sthwardly. The surface is, however, greatly diversified, both as to mtn., hill, and dale, and in relative level, independent of mtns. and hills. The Kittatinny mts. ranging something E. of S. W., divide this co. into two unequal sections; about one third lying below, or S. of the Kittatinny, and the residue above, or nthrd. from that chain. The Lehigh r. deriving its numerous sources from the very mountainous region above the Kittatinny, pierces that and numerous other chains at nearly right angles; reaches the nthwstrn. foot of Blue Ridge at Allentown, and turning these to N. E., traverses the mtn. foot to its influx into Del. at Easton.

The lower section of Northampton, though comprising only one third of the whole surface of the co., contained in 1820, 22,030, out of 31,765 inhabitants.

The valley between the Blue Ridge and Kittatinny chains, averages in Northampton a width of about 10 ms.; the sthrn. part toward the Blue Ridge resting on limestone, and the opposite on clay slate. The two extreme southern tsps. below the Blue Ridge, lie also partially on limestone. The whole of this sthrn. and lower section has a mean elevation above tide water, of from 250 to 350 ft. The soil is excellent for grain, pasturage, meadow grasses, and orchard fruits.

Without regard to the mtns., the vallies above Kittatinny rise like terraces, from 600 to upwards of 1,200 ft. above tide level. It is observed that in the seasons of spring, harvest, &c., there are two weeks or more between the extremes of this co., and relative height at once explains the phenomenon. But the greatest difference and most important to the farmer in the respective sections of Northampton, is in quality of soil, which deteriorates gradually, receding to the nrthwstrd. from the Blue Ridge. (*See articles, Lehigh, Mauch Chunk, &c.*)

The Lehigh navigation, and a canal from Easton along Del. r. to Bristol, with the enormous masses of anthracite coal near Mauch Chunk, have given great importance to the sthrn. part of Northampton. The produce of its fields and pastures are also abundant and valuable, and rapidly augmenting in annual amount. Chief ts. Easton, Bethlehem, Mauch Chunk, Hellerstown, and Stroudsburg. Pop. 1820, 31,765; 1830, 39,267, an increase of 24 per ct.

NORTHAMPTON, or Allentown, p-v., borough and st. just., Lehigh co., Pa., situated on the point above the junction of the two main branches of Little Lehigh, and about a mile from the main Lehigh r., 6 ms. S. W. by W. Bethlehem, and 18 in nearly a similar direction from Easton, 50 ms. a little W. of N. Phil., and by p-r. 178 ms. N. N. E. W. C., lat. 40° 36′, long. 1° 30′ E. W. C. It is a very pleasant small town, standing on a swelling hill, surrounded by a fine well cultivated country, and contains a bank, printing office, numerous dry good stores, and in the vicinity several merchant mills.

NORTHAMPTON, co., Va., and the sthrn. of the eastern shore, bounded N. by Accomac co., Va., E. by the Atlantic ocean, and S. & W. by Chesapeake bay. Length from S. to N. 32, mean width, if the Atlantic islands are included, 10 ms.; area, 320 sq. ms., extending in lat. from 37° 05′ to 37 33′, long. from 1° to 1° 28′ E. W. C. The surface of this co. is but little broken by hill and dale, but the margin is excessively indented by small creeks, and covered on the Atlantic side by Paramores, Hog, Prout's, Smith's, and Fisherman's islands, proceeding sthwrdly. to Cape Charles. Chief t., Eastville. Pop. 1820, 7,705; 1830, 8,644.

NORTHAMPTON, co., N. C., bounded N. E. in part by Meherin r., separating it from Gates co., by Hertford E., Bertie S. E., Roanoke r. separating it from Halifax S. W., Brunswick co., Va., N. W., Greenville co. Va., N., and Southampton co. Va., N. E.; length from S. E. to N. W. 42 ms.; mean width 13, and area 546 sq. ms.; extending in lat. from 36° 09′ to 36° 30′, long. from 0° 08′ to 0° 56′ W. W. C. Tho' bounded on one of the longest sides by Roanoke r., the general declivity is eastwrd. toward Meherin and Chowan rs. Much good soil; pop. 1820, 13,242; 1830, 13,103.

NORTHAMPTON, C. H. and p-o., Northampton co., N. C., 70 ms. S. W. Norfolk, Va., and by p-r. 95 N. E. Raleigh. Lat. 36° 24′, long. 0° 27′ W. W. C.

NORTHAMPTON, p-v., Portage co., O., by p-r. 126 ms. N. E. Columbia. Pop. of the tsp. of Northampton, 1830, 293.

NORTH BLOOMFIELD, p-v., Trumbull co., O., by p-r. 173 ms. N. E. Columbus.

NORTHBOROUGH, p-t., Worcester co., Mass., 11 ms. E. Worcester, 36 W. Boston, in a valley; has excellent land. Assabet r. has good meadows and mill seats; cotton, shoes, scythes, leather, &c., are manufactured here. It formerly belonged to Marlboro' and suffered from the Indians. A house in this t. was defended against 24 Indians by a man and a woman in 1704. Pop. 1830, 992.

NORTHBRIDGE, p-t., Worcester co., Mass., 12 ms. S. E. Worcester, 45 S. W. Boston, is crossed by Blackstone r. and canal, and has large meadows, with uneven uplands, yielding good grass, &c. Here are granite quarries, and several cotton and woollen factories. Pop. 1830, 1,053.

NORTH BRIDGEWATER, p-t., Plymouth co., Mass., S. Norfolk co., 20 ms. S. Boston. Pop. 1830, 1,953.

NORTH BROOKFIELD, t., Worcester co., Mass., 68 ms. W. Boston, has good soil, excellent farms, & some factories. Pop. 1830, 1,241.

NORTH CAROLINA, state of the U. S., bounded by S. Carolina S., Georgia S. W., Tenn. W., Va. N., and by the Atlantic ocean E. and S. E., having outlines on

	ms.
S. C., from the Atlantic ocean to Chatuga r., ...	300
Along the nrthrn. boundary of Geo.,	58
Along estrn. border of Tenn.,	185
Along sthrn. boundary of Va.,	330
Along Atlantic ocean, ...	320
Having an entire outline of	1,193

Greatest length from the wstrn. extreme to Cape Hateras, within a very trifling fraction of 500 ms.; area 51,000 sq. ms., as carefully measured by the rhombs, will give a mean breadth of 100. The greatest breadth is, however, 185 ms. from the extreme sthrn. angle at Little r. inlet, to the sthrn. border of Va. In lat. it extends from 33° 50′ to 36° 33′, and in long. from 1° 36′ E. to 7° 12′ W. W. C.

In diversity of surface, soil and climate, N. C. presents very wide extremes, falling in either respect, little if any thing below Geo. Though extending lengthwise from E. to W., the relative height decreases the elevation of temperature advancing from the ocean wstrd. The state similar to S. C. and Geo., is naturally divided into 3 zones; the distinction between the physical sections are far more strongly marked in the former, than in the two latter states. The sea sand alluvial tract of N. C., is from S. W. to N. E., 260 ms. in length, with a mean breadth inland of about 90 ms., but varying in width from 80 to 100 ms. The estrn. part is deeply indented by shallow, though wide sounds, of which the principal are Pamlico, and Albemarle. The sthwstrn. part presents a coast directly the reverse; it is a long inflected line, with a remarkable deficiency of inlets. The entire coast of N. C., indeed, with a distance of 320 ms., is the most defective part of that of the Atlantic border of the U. S. in those valuable commercial entrances. The rivers Chowan and Roanoke rising in Va., and Tar, Neuse, and Cape Fear rivers rising in the state itself, issue from the interior section, and reach the sounds of the sea sand region in a S. E. direction, and their channels are the only furrows which materially break the monotony of the great plain of 23,000 sq. ms. There are slight exceptions, but dead uniformity of surface, is the general character of the ocean section of N. C.

Without any very abrupt marks of distinction between them, the sea sand is followed by the hilly or middle section. Much that might be said on this subject has been anticipated under the art. Geo.; we may here, however, observe that the hilly tract of N. C. comprises 14,000 sq. ms., with a slope to the S. E., and traversed at the extremes by Roanoke, Yadkin and Catawba rs., and in the centre gives source to the numerous tributaries of Tar, Neuse and Cape Fear rivers. The Blue Ridge, on most maps very erroneously is made to represent the outer chain of the Appalachian system in N. C., as in the contiguous states; but so far is this geography from being correct, there are two chains outside or between the Blue Ridge and the ocean, nor is in fact the Blue Ridge correctly drawn over N. C. The counties of Person,

Orange, Chatham, Moore and Richmond, are in general terms the wstrn. sections of the middle region; whilst Caswell, Guilford, Randolph, Davidson, Montgomery and Anson, commence the mtn. tract.

Some slight resemblance marks the two contiguous estrn. sections of N. C., but no contrast can be much greater than exists between the extreme regions of sea sand, and the bold, swelling, and delightful mtn. or wstrn. section. Towards the ocean the eye meets no relief, & fresh water is in many places rare; the rivers and sounds are stagnant, or drag their sluggish streams along their oozy beds; and the surface to large extents marshy and uncultivatable, lie unadorned and useless wastes; but ascend the rivers, traverse the hills, and the outer humble but distinct Appalachian chain, and a country opens, to which the boasted peninsulas of Asia Minor, Greece, Italy or Spain, can offer no spot superior in all that can render the face of the earth a happy residence to man. The streams are the pure productions of living fountains; the soil, if not exuberantly fertile, is sufficiently productive to reward, and with the elastic air over its surface, richly reward human labor.

If we reject the mtn. chains, there still remains a difference of level of at least 1,800 feet between the counties along the ocean border, and that of Ashe, and Buncombe, on the wstrn. extreme between the Blue Ridge and Iron chains. The actual difference of lat. a little exceeds 2½ degrees, and the difference of level is fully equal to 4 degrees, making an actual difference of 6½ degrees of Fahrenheit's thermometer as the mean and extreme temperature. The winters of Upper N. C. are perhaps not as long, perhaps something less severe, but on the mtn. sections of not only N. C., but of S. C. and Geo. also, the inhabitants have with the more nrthrn. states a share, and not a slight share of the rigors of frost, snow, and cold rainy weather in winter.

Agriculture.—The natural vegetables afford often good guides to the estimate of climate suitable to exotics. In the whole three sections of N. C. spread immense forests of terebinthine trees, and there may be said to commence, advancing from the north, those vast collections of pines, amongst which the traveller may pursue his way for days without meeting, except a few scattering stems, any other tree but pine. This forest tree evinces thinness, if not sterility of soil, but is generally attended with good fountain water. Though, however, it is the most common, pine gives place, or is intermixed with nearly every forest tree known in the middle states of the U. S., and the live oak, *quercus sempervirens*, a tree ceasing in La. below N. lat. 30° 30′, is found on Cape Fear r., N. C., as high as 34° 20′, showing a difference of temperature between the Atlantic coast and Mississippi valley, of nearly 4 degrees. A similar relative location is found to distinguish the dwarf palms. From these vegetable criteria, we may decide that on the sea sand alluvion of N. C., cotton may be cultivated with success, as in the valley of the Mississippi cotton flourishes 5 or 6 degrees of lat. above the live oak or the dwarf palm, and sugar cane is cultivated nearly as far nrthwrd. as these trees are found. Ascending to the interior and elevated table land, small grain, meadow grasses, and the apple, follow the cotton. Potatoes succeed well over the state, as do a vast abundance of esculent roots and fruits. Indian corn is the staple grain. The fig tree yields its abundant saccharine fruit on the lower section, and the peach over the state gives its tribute to the hand of the cultivator. In fine, N. C. is not a state of more than medium general fertility, but it is a state of abundant product, where labor is properly applied.

Minerals.—In mineral production, the metals, except iron, are rare. Iron ore, however, abounds beyond any attempt yet made to reduce it to the metallic state. Much of that metal is manufactured, it is true, but immensely more might be produced.

Commerce, rivers.—With a very unfavorable sea-coast, the connexion of N. C. with Europe, or the other states of the U. S. is much below the relative proportion of area or population; but again, a considerable fraction of the commerce of Va. and of S. C., originates in the intermediate state. The whole wstrn. sections of the state discharge their rivers either sthwrd. into S. C., nthwrd. into Va., or wstrd. into Ten. One great branch of Roanoke, Dan r., flows from N. C. into Va., returning its waters, however, by the Roanoke. Catawba and Yadkin rs., rising in the fine vallies of wstrn. or rather central N. C., become navigable streams, and bending their courses sthwrd. carry their volumes and their burthens into S. C. The whole margin of the state beyond Blue Ridge, is drained by the numerous confluents of Ten. r., and has a slope to the N. W.

Many partial canals and side cuts, locks, and drains have been made to meliorate the water means of intercommunication, but as a system, roads and canals are in their incipient condition in that state.

Population.—The pop. of North Carolina at several periods follows, with that of the counties in 1830. In 1790, it contained 393,-950 inhabitants; in 1800, 478,103; in 1810, 555,500; in 1820, 638,829; and in 1830, 738,470.

Counties.	Pop. 1830.	Counties.	Pop. 1830.
Ashe,	6,987	Currituck,	7,655
Anson,	14,095	Caswell,	15,185
Burke,	17,888	Chowan,	6,697
Buncombe,	16,281	Camden,	6,733
Brunswick,	6,516	Chatham,	15,405
Bertie,	12,262	Columbus,	4,141
Beaufort,	10,969	Cumberland,	14,834
Bladen,	7,811	Carteret,	6,597
Craven,	13,734	Duplin,	11,291
Cabarras,	8,810	Davidson,	13,389

Counties.	Pop. 1830.	Counties.	Pop. 1830.
Edgecombe,	14,935	Nash,	8,490
Franklin,	10,665	Onslow,	7,814
Granville,	19,355	Orange,	23,908
Gates,	7,866	Person,	10,027
Guilford,	18,737	Pitt,	12,093
Greene,	6,413	Perquimons,	7,419
Hyde,	6,184	Pasquotank,	8,641
Halifax,	17,739	Richmond,	9,396
Haywood,	4,578	Robeson,	9,433
Hertford,	8,537	Rockingham,	12,935
Iredell,	14,918	Rowan,	20,786
Johnston,	10,938	Rutherford,	17,557
Jones,	5,608	Randolph,	12,406
Lincoln,	22,455	Surry,	14,504
Lenoir,	7,723	Sampson,	11,634
Mecklenburg,	20,073	Stokes,	16,196
Martin,	8,539	Tyrrell,	4,732
Moore,	7,745	Wilkes,	11,968
Macon,	5,333	Wake,	20,398
Montgomery,	10,919	Wayne,	10,331
Northampton,	13,391	Washington,	4,552
New Hanover,	10,959	Warren,	11,877

Of which were white persons,	Males.	Females.
Under 5 years of age,	46,749	43,775
From 5 to 10,	35,959	34,964
" 10 to 15	30,597	28,842
" 15 to 20	25,452	27,398
" 20 to 30	30,428	41,636
" 30 to 40	23,042	24,534
" 40 to 50	14,998	16,428
" 50 to 60	10,536	10,601
" 60 to 70	5,968	5,980
" 70 to 80	2,459	2,496
" 80 to 90	649	747
" 90 to 100	138	158
" 100 and upwards,	28	30
Total,	235,954	236,889

Of the preceding were deaf and dumb, under 14 years of age, 70; from 14 to 25, 81; 25 and over, 79. Blind 223.

The colored population was as follows:

	Free colored.		Slaves.	
	Males.	Females.	Males.	Females.
Under 10 years,	2,438	3,287	45,991	44,847
From 10 to 24	2,955	3,118	38,099	37,508
24 to 36	1,400	1,649	20,212	20,095
36 to 55	1,062	1,179	14,030	13,088
55 to 100	685	720	5,848	5,636
100 and upwards,	21	29	133	114
Total,	9,561	9,982	124,313	121,288

Deaf and dumb colored persons, 93; blind, 161.

Recapitulation.

Whites.	Free colored.	Slaves.	Total.
472,843	19,543	245,601	737,987

Constitution.—Judiciary. The constitution of N. C. was adopted in convention at Halifax, 18th December, 1776. It commences with a bill of rights containing 25 sections, the last of which relates to the boundaries of the state. The constitution is itself divided simply into sections, and provides as follows.

Sec. 1.—That the legislative authority shall be vested in two distinct branches, both dependent on the people, *to wit*, a senate, and house of commons.

Sec. 2.—That the senate shall be composed of representatives, annually chosen by ballot, one for each co. in the state.

Sec. 3.—That the house of commons shall be composed of representatives annually chosen by ballot, two for each co., and one for each of the towns of Edenton, Newbern, Wilmington, Salisbury, Hillsborough and Halifax.

Sec. 4.—That the senate and house of commons, assembled for the purpose of legislation, shall be denominated "The General Assembly."

Sec. 5, provides that the members of the senate must, to be eligible, possess in the co. from which he is elected 300 acres of land in fee; and sec. 6, makes a similar provision in regard to members of the house of commons; except limiting the latter to a property qualification of at least 100 acres of land in fee.

Sec. 7, makes it necessary to possess the right of suffrage, that the elector shall be 21 years of age, have resided 1 year in the co. next preceding the election, and possess therein a freehold of 50 acres of land, to vote for a senator; and by sec. 8, like age and residence, as necessary to vote for a senator, and having paid public taxes, qualifies to vote for a member of the house of commons.

Sec. 13.—That the general assembly shall, by joint ballot of both houses, appoint judges of the supreme courts of law and equity, judges of admiralty, and attornies general, who shall be commissioned by the governor, and hold their offices during good behavior.

Sec. 15, provides for the election of a governor by joint ballot, for 1 year, and eligible only 3 years in 6 successive years, and must be 30 years of age, have resided in the state 5 years, and have in the state a freehold in lands and tenements, above the value of 1,000 pounds.

Sections 31, and 32, read with curious contrast. The former renders ineligible to a seat in either house of the general assembly, or the council, all clergymen or preachers of the gospel of any denomination; whilst the 32d section disqualifies from every office in the state of profit or trust, all persons who deny the being of God, the truth of the Protestant religion, or the divine authority of either the Old or New Testament, &c.

Sec. 39, provides for the release of debtors who give up their estates for the benefit of creditors, and against whom there is not strong presumption of fraud.

History.—The first, but abortive attempt to colonize what is now the two Carolinas, was made by the French in the reign of Charles the IX., from whom the name Carolina was derived. The French colonization was opposed and prevented by the Spaniards. A second, and again disastrous enterprise to form a settlement on the Carolina coast, was made in 1586, under a patent granted by Queen Elizabeth of England, to Sir Walter Raleigh. Under this patent, a small number of adventurers were landed in 1586, who were probably murdered by the natives, as no trace of their existence or fate could ever be procured. The coast, under the name of Carolina, remained again desolate 75 years, when in 1661, a small English colony from Mass. fixed themselves on the banks of Cape Fear r. Granted by the English monarchs to various proprietors, and to their conflicting proceedings was added Locke's scheme of government. Under so many causes of embarrassment, the colony advanced slowly and painfully. In the abandonment of Locke's scheme one impediment was removed, but in 1712 a most sanguinary Indian war broke out, and ravaged the settlements. The proprietary government of Carolina produced so many and so just complaints, that in 1717 it was abolished and the colony became royal, and continued so to the revolution, which separated the Carolinas, with other N. American colonies, from Great Britain. In 1720, the colony of Carolina was found too unwieldy for convenient government, and was separated into two, under the relative names of North Carolina, and South Carolina.

The inaccessible coast of North Carolina, if disadvantageous commercially, has been a real and extended line of fortifications, to protect the state from invasion on the side of the ocean, and consequently no other section of the union has felt the evils of two wars with Great Britain, so little as has N. C. In the revolutionary war, some expeditions made from the side of S. C. reached the interior of N. C., but were of more ultimate injury to the enemy than to the invaded country. But though exposed to little of the danger within, the people of N. C. sought it without, and have borne their full share of the perils, and reaped an ample reward, in sharing with their fellow citizens the glory of independence. They were amongst the first who threw off the British yoke, as may be seen by the date of their constitution, 18th Dec. 1776. Fifty-six years of profound and unambitious tranquillity, in regard to her domestic concerns, has been only broken at long intervals by foreign war, which when ended, the ploughshare was formed from the falchion, and N. C. may be named amongst the most happy communities of the earth.

North Castle, p-t. Westchester co. N. Y., 33 ms. n. N. York, 6 s. Bedford, n. w. Conn., is crossed by Byram r., and has Bronx cr. w. on which are mills. The surface is irregular, but cultivated, and the Heights noted in the history of the revolution. Pop. 1830, 1,653.

North Cove, p-o. Burke co. N. C., by p-r. 179 ms. wstrd. Raleigh.

North Dover, p-v. Cuyahoga co. O., by p-r. 150 ms. n. n. e. Columbus.

North East, p-t. Dutchess co. N. Y., 25 ms. n. e. Poughkeepsie, 95 s. Albany, s. of Columbia co. and Mass., w. of Conn. The town is shaped like a boot, is 10 ms. long, and from 3 to 5 wide, is uneven and stony, and has W. Town mtn. overgrown with trees, but contains much good land. Several brooks supply mills. Wappinger's cr. rises in Hittin's pond. Pop. 1830, 1,689.

North East, p-v. nthrn. part Erie co. Pa., by p-r. 348 ms. n. w. W. C.

North East, p-v. on North East r., Cecil co. Md. 6 ms. a little s. of w. Elkton, and by p-r. 82 ms. n. e. W. C.

North End, p-v. wstrn. part Matthews co. Va., by p-r. 91 ms. e. Richmond.

Northfield, p-t. Washington co. Vt., 10 ms. s. w. Montpelier, 35 s. e. Burlington. It contains 18,515 acres, was first settled 1785, and is crossed by Dog r., which has good mill seats; it bears hemlock, spruce, maple, beach, &c. and has generally a good soil. The surface is uneven, and crossed n. and s. by a range of slate. It contains 2 villages, several manufactories, and 9 school dists. Pop. 1830, 1,411.

Northfield, p-t. Merrimack co. N. H., 16 ms. n. Concord, s. of Winnipiseogee r., and e. of Merrimack r., has 20,000 acres, with some good soil. It contains Chestnut pond e., and Sondogardy s., and is crossed by two ridges of high land. The Winnipiseogee and Pemigewasset rs. join in the n. w., and form Merrimack r. There are several mills, &c. First settled 1760. Pop. 1830, 1,169.

Northfield, p-t. Franklin co. Mass., 94 ms. n. w. Boston, s. of New Hampshire, has much excellent land, and extensive and fertile meadows. It was settled in 1687, was purchased from the Indians for 200 fathoms of wampum, and £57 in merchandize. The settlement was attacked by Indians in 1678, and finally deserted and destroyed. It was resettled, and deserted again, and permanently occupied at last in 1713, after which some of the inhabitants were killed. Fort Dummer was built in Vernon, Vt., just beyond this town, but was intended to be within its limits, and served to protect it in the French wars against the savages. Pop. 1830, 1,757.

Northfield, t. Richmond co. N. Y., 5 ms. n. Richmond. Pop. 1830, 1,262.

Northfield, p-v. Vermillion co. Il., by p-r. 162 ms. n. e. Vandalia.

Northford, p-v. New Haven co. Conn., 10 ms. n. e. New Haven, 26 s. Hartford.

North Fork, p-o. on a branch so called of Licking r. Mason co. Ky., by p-r. 69 ms. n. e. by e. Frankfort.

North Haven, t. New Haven co. Conn., 7 ms. n. New Haven, contains about 17 sq. ms. and is nearly level, with hills e. and w. It is crossed by Quinepiack r., navigable 8 ms. Pop. 1830, 1,282.

North Hampstead, p-t. and st. jus. Queens co. N. Y., 20 ms. e. N. York, on Long Island sound. Pop. 1830, 3,091.

North Hero, p-t. isl. and st. jus. Grand Isle co. Vt., in Lake Champlain, 26 ms. n. Burlington, 6 w. St. Alban's, contains 6,272 acres. First settled 1783. In 1789 a block house was built here by the British, and given up in 1796. There are no important streams or mill sites, but the soil is very good. The v. is small, with a stone C. H. and jail, and 4 school dists.

Northington, p-v. Cumberland co. N. C., by p-r. 10 ms. nthrd. Fayetteville, co. st., and 51 ms. s. w. Raleigh.

North Kingston, p-t. Washington co. R. I., 20 ms. s. w. Providence, w. Narraganset bay, about 7 ms. by 8, 56 sq. ms., is hilly, with level land N., and yields free stone. Several brooks supply mill sites, and afford fish. There is a good harbor at Wickford, rarely shut by ice; and also two others, Cole's and Allen's. A few vessels are employed in fishing on the banks, the shoals, and other parts of the coast, and others are employed in coasting. There are several factories. Pop. 1830, 3,037.

NORTH MIDDLETON, p-v. Bourbon co. Ky., 49 ms. E. Frankfort.

NORTH MORELAND, p-v. of Luzerne co. Pa., 20 ms. N. Wilkes Barre, and 134 ms. N. N. E. Harrisburg.

NORTH NORWICH, p-v. sthrn. part Huron co. O., by p-r. 95 ms. N. N. E. Columbus.

NORTHPORT, p-t. Waldo co. Me., 14 ms. N. W. Castine, 46 E. Augusta, S. of Belfast, W. of Belfast Bay, and opposite Isle Borough. Pop. 1830, 1,083.

NORTH PROVIDENCE, p-t. Providence co. R. I., 4 ms. N. Providence, W. of Seekonk r., which divides it from Mass., has Wanaguatucket r. W., about 2 ms. by 6, 16 sq. ms.; is uneven, with primitive and transition rocks, limestone, &c., and a gravelly soil, bearing oak, walnut, &c., grass, hay, corn, and vegetables, for Providence. There are many mill seats, and some good fisheries. The town is extensively engaged in manufacturing, especially cotton. Pawtucket v. is in the N. E., on the border of Mass., and is a large manufacturing v., with a considerable one on the opposite side of the r. Pop. 1830, 3,503.

NORTH RIDGEVILLE, p-v. Lorain co. O. by p-r. 134 ms. nthrd. Columbus.

NORTH r. N. Y. (*See Hudson r.*)

NORTH r. Plymouth co. Mass., is navigable 18 ms. to Pembroke, in vessels of 300 tons, and in boats to the falls.

NORTH RIVER Meeting House and p-o. Hampshire co. Va. by p-r. 115 ms. N. W. by W. W. C.

NORTH RIVER Mills and p-o. 16 ms. S. E. Romney, and by p-r. 99 ms. N. W. by W. W. C.

NORTH ROYALTON, p-v. Cuyahoga co. O. by p-r. 130 ms. N. E. Columbus.

NORTH SALEM, p-t. Westchester co. N. Y. 53 ms. N. N. Y., 8 N. Bedford, W. of Conn. line, about 4 ms. by 6, E. Croton r. It has mill seats on a branch of it. There is an academy in the town. Pop. 1830, 1,276.

NORTH SEWICKLY, p-v. Beaver co. Pa. by p-r. 12 ms. S. E. borough of Beaver, and 263 ms. N. W. W. C.

NORTH SMITHFIELD, p-v. Bradford co. Pa. by p-r. 142 ms. N. Harrisburg.

NORTH SPRINGFIELD, p-v. Portage co. O. by p-r. 120 ms. N. E. Columbus.

NORTH STONINGTON, p-t. New London co. Conn. 50 ms. S. E. Hartford, N. W. Pawcatuck r. which separates it from R. I., about 6 ms. by 8; 44 sq. ms.; is hilly with granite rocks, but good for grass; and yielding oak, chestnut, &c. There are many mill sites. Pop. 1830, 2,840.

NORTHUMBERLAND, p-t. Coos co. N. H. 130 ms. N. Concord, E. Conn. r., opposite Maidstone, Vt.; has very good soil near the river, without stone or gravel, formerly covered with butternut, with some good upland S. Cape Horn mtn. 1,000 feet high, is near the centre, with Conn. r. on one side, and Upper Amonoosuck r. on another. Below the mouth of the latter is a fall in Conn. r. with a dam, mills, &c., above which the meadows are overflown, in the spring, to a great extent. There is a bridge over Conn. r. There is a small village at the falls. First settled, 1767. There was a fort in the town in the revolution. Pop. 1830, 342.

NORTHUMBERLAND, p-t. Saratoga co. N. Y., 11 ms. N. E. Ballston Spa, 36 N. of Albany, W. of Hudson r. and Washington co., 6 ms. sq., and has a variety of soils. There is a large pine plain W. with much good sandy and argillaceous loam in other parts. Cold creek supplies a few mill seats. Pop. 1830, 1,606.

NORTHUMBERLAND, co. of Pa. bounded by the Mahantango r. separating it from Dauphin S., by the Susquehannah r. separating it from Union W., and part of Lycoming N. W., by another part of Lycoming N., Columbia N. E. and E., and by Schuylkill S. E. Length from S. to N. 40 ms.; and the area being about 440 sq. ms., the mean breadth will be 11; but the width is very unequal. In one place below the E. branch it is 24 ms. wide, and at another about 3 ms.; above the borough of Northumberland, it is confined to a width of less than 4 ms., though at a distance of 16 ms. from the northern extremity. It extends from the meridian of W. C. to 0° 32′ E. The declivity is wstrd. in the direction of the East branch of Susquehannah, where it unites with the western branch between Sunberry and Northumberland, and very nearly at right angles to the course of the main volume along the western margin of the co. Contrary to their general range, the Appalachian ridges where they traverse Northumberland, extend with a very slight declination from E. to W. Though mountainous and rocky in much of the surface, this co. comprises so much river margin as to give it great comparative extent of fertile arable land. Following the inflections of the river, 40 ms. of the eastern branch of the main river, and including both sides, 20 ms. of the eastern branch of Susquehannah, are included in Northumberland; and beside the large rivers, fine vallies extend along Mahantango, Mahanoy and Shamokin creeks below, and Chillisquake, Limestone and Warrior creeks above the mouth of the eastern branch. Chief ts. Sunbury, Northumberland, Milton and Watsonburg. Pop. 1820, 15,424, 1830, 18,170.

NORTHUMBERLAND, p-v. and borough of Northumberland co. Pa. situated on the point above the confluence of the two principal branches of Susquehannah r. 2 ms. above Sunbury, and 54 N. and above Harrisburg. Lat. 40° 55′, long. 0° 8′ E. W. C. Pop. of the tsp. 1820, 1,373.

NORTHUMBERLAND, co. of Va. bounded by Lancaster south and s. w., Richmond w., Westmoreland N. W., the mouth of Potomac r. N. E., and Chesapeake bay E. and S. E. Length 30 ms., mean width 8, and area 240 sq. ms. Extending in lat. from 37° 40′ to 38° 05′, long. from 0° 2′ to 0° 45′ E. W. C. The declivity of this co. is, in the southern part, southestrd. towards the Chesapeake, and in the northern section northestrd. toward Potomac r. Pop. 1820, 8,016, 1830, 7,953.

NORTHUMBERLAND, C. H. and p-o. Northumberland co. Va. by p-r. 151 ms. S. E. W. C., and 92 N. E. Richmond.

NORTH UNION, p-v. Harrison co. O. by p-r. 129 ms. a little N. of E. Columbus.

NORTHVILLE, p-v. northern part of Erie co. Pa. 19 ms. N. E. Erie, and by p-r. 352 ms. N. W. W. C.

NORTH WASHINGTON, p-v. Westmoreland co. Pa. by p-r. 215 ms. N. W. W. C.

NORTH WEST RIVER BRIDGE, p-v. S. E. part of Norfolk co. Va. on a small confluent of Currituck sound, 24 ms. a little E. of S. Norfolk, and about an equal distance N. Elizabeth city in N. C.

NORTH WHITEHALL, p-o. southern part of Lehigh co. Pa. 18 ms. N. W. Allentown.

NORTHWOOD, p-t. Rockingham co. N. H. 20 ms. N. E. Concord, 27 W. N. W. Portsmouth, 20 from Exeter; has 17,075 acres, 6 ponds, and parts of 2 more, some of which supply it with streams. Saddleback mtn. lies between this town and Deerfield, which affords crystals, &c., and give rise to the N. branch of Lamprey r. It was formerly part of Nottingham, and first settled in 1763 from Northampton. Pop. 1830, 1,342.

NORTH YARMOUTH, p-t. Cumberland co. Me. 42 ms. S. S. W. Augusta, 42 N. N. E. Portland, N. Casco bay. It has an academy, and is crossed by a considerable stream. Pop. 1830, 2,606.

NORTON, p-t. Bristol co. Mass. 32 ms. S. Boston, 8 N. W. Taunton. It was formerly part of Taunton, incorporated 1711; has not very good soil, much of which is rented. It is watered by several branches of Taunton r. which afford very good mill seats. Iron is mined here. Winnicunnit pond was much resorted to by Indians in past days, for fish and clams; and they sometimes lived in caves. This town was first settled by a cabin boy, 1670. The Leonard iron works have been long established. Before 1828, 40 young men of this town had been educated at college. Pop. 1830, 1,479.

NORTON, p-v. Del. co. O. by p-r. 36 ms. N. Columbus.

NORWALK, p-t. Fairfield co. Conn. 66 ms. S. W. Hartford, 32 W. New Haven, 48 N. E. N. Y., N. of Long Island sound, W. of Saugatuck r., about 5 ms. by 7, with 34 sq. ms.; is uneven, high N. with much granite rock. The soil is good for grain, grass, and fruit, and bears walnut, chestnut, and other timber, much of which has been taken to the N. Y. market, with which there is constant intercourse by sloops, and a steamboat which plies daily. There are 2 good harbors, one at the mouth of Norwalk r., with water for vessels of 100 tons, and the other at Five Mile r. There are many islands and small streams near the coast. The fisheries of black fish, shell fish, &c. are valuable. The village is considerable, with a Congregational and an Episcopal church, and an academy Settled in 1651. Pop. 1830, 3,792.

NORWALK, p-v. and st. jus. Huron co. O. situated on a branch of Huron r. 20 ms. S. S. E. Sandusky, and by p-r. 113 ms. a little E. of N. Columbus. Lat. 41° 15′, long. 5° 53′ W. W. C. Pop. 1830, 310.

NORWAY, p-t. Oxford co. Me. 47 ms. W. by S. Augusta, 8 S. W. Paris; has a large pond which empties into Little Androscoggin r. It lies N. of Cumberland co. Pop. 1830, 1,713.

NORWAY, p-t. Herkimer co. N. Y. 90 ms. N. W. Albany, 18 N. Herkimer, 6 ms. by 15; has a warm, rich, and dark soil, with a tract of light sand. It has moderate hills, and is rather stony, bearing a variety of trees, and furnished with many mill seats on W. Canada creek. Pop. 1830, 1,152.

NORWICH, p-t. Windsor co. Vt. 21 ms. N. Windsor, 40 S. E. Montpelier, W. Conn. r. opposite Dartmouth college in N. H.; contains about 25,000 acres. Settled, 1762. Conn. r. is here about 120 yards wide, and fordable at low water in 3 places. Ompompanoosuc r., Blood brook, Smalley's creek, &c. water the town, and furnish some mill seats. The surface is uneven, but is generally good for grain and grass, and bears excellent orchards. There are large beds of iron ore. Subterranean sounds were formerly heard near Ompompanoosuc r. The v. is on a pleasant plain, and contains a literary and military academy of captain Partridge, a grammar school, &c. Population 1830, 1,392.

NORWICH, p-t. Hampshire co. Mass. 12 ms. W. Northampton, 105 W. Boston; is crossed by a N. branch of Westfield r. N. and S., and has the main stream S. W. Incorporated, 1772, Pop. 1830, 795.

NORWICH, p-t., city, and half shire, N. London co. Conn. at the head of navigation on Thames r. (formerly called Pequod,) 13 ms. N. New London, 38 S. E. Hartford, 38 S. W. Providence, 50 N. E. New Haven. Lat. 41° 34′ N., long. 4° 55′ E. W. C. The town has Shetucket and Thames rs. on the E. It contains 29 sq. ms., being 3 ms. by 7. The boundary of the town encircles three distinct villages, viz. Norwich falls, Beanhill, and Yanticville. At the falls are 9 establishments for manufacturing purposes, at Bean hill 2, and at Yanticville 1. The aggregate of manufactured goods during the last year, was somewhat over $600,000. The town contains 8 houses for public worship, viz. 3 for Congregationalists, 1 for Episcopalians, 1 for Baptists, 2 for Methodists, and 1 for Universalists. A high school for boys, and a female academy, in which the higher branches of education are taught, have been in operation for a considerable time, and are in flourishing circumstances. A hotel sufficiently spacious to accommodate 200 boarders has been recently built near the court house on the green. A large public building has been erected, which was designed for the use of county and town. In the town are 2 banks, with a capital of $200,000 each; a savings bank, incorporated 1824, whose deposits already exceed $100,000; and two insurance offices. The scenery of the town is in a high degree picturesque and delightful: and its beauty is greatly heightened by a rich and well cultivated soil.

About a mile E. of the city a dam has been recently erected across the Shetucket river, which will, it is calculated, furnish sufficient water power to carry 60,000 spindles. Five large factories, besides 40 or 50 dwelling houses, are being built; and there is little doubt that, in respect to the amount of its water privileges, Norwich is the second town in New England. A rail-road also is contemplated, through the valley of the Quinebaug, to intersect the Boston and Worcester rail-road at Worcester. A charter has been obtained for this object with a capital of one million dollars. A bank has been chartered, with a capital of $500,000, on condition that it shall subscribe for $100,000 of rail-road stock.

That part of Norwich known by the name of The Town, or The Plain, was, in ancient times, the summer residence of the Mohegan Indians, the remnants of whom now reside on the reservation in the adjoining town of Montville. The burying ground of the Uncas family is near the mouth of the Yantic. The township was sold by Uncas in 1659, for about $230. It is reported that Uncas did this out of gratitude to the Narragansett Indians, for provisions which they furnished him during a close seige. Sachem's Plain, near the Shetucket, was the scene of the battle between Uncas and Miantonomoh, and the place of the latter's grave. The settlement of Norwich was begun in 1660, by Rev. Mr. Fitch and a part of his church from Saybrook. Population of the t. in 1820, 3,624; in 1830, 5,161, of which 3,135 resided in the city.

Norwich, p-t. Chenango co. N. Y., 8 ms. N. E. Oxford, 100 w. Albany, 7 ms. by 12; is crossed by Chenango r., which, with several branches, affords mill seats. The soil is good. The post borough, the st. jus. of the co., is in a fine plain at the junction of Canassawacta creek and Chenango river. It contains the co. buildings, a female academy, &c. There is a mineral spring 2 ms. from the borough. Pop. 1830, 3,619.

Norwich, p-v. McKean co. Pa., by p-r. 281 ms. N. N. W. W. C.

Norwich, p-v. estrn. part of Muskingum co. O., by p-r. 71 ms. E. Columbus.

Norwood, p-v. Montgomery co. N. C., by p-r. 159 ms. sthwstrd. Raleigh.

Notch, in the White mountains, N. H.— (*See White Mountains.*)

Nottaway, river of Va. and N. C., has its most remote source in Prince Edward co. Flowing thence S. S. E. between Nottaway and Lunenburg cos., between Dinwiddie and Brunswick, turns to eastward between Greenville and the western part of Sussex. Entering the latter, and first curving nrthrd. winds to S. E., and traversing Sussex and Southampton cos., receives Blackwater r. from the N., and entering Gates co. N. C., bends to S. W. 10 ms. to its junction with Meherrin, to form Chowan r. The entire length of Nottaway by comparative courses is 110 ms. The Nottaway valley is about 100 ms., by 20 mean width, comprising great part of Nottaway, Dinwiddie, Sussex, Surry, and Southampton cos., and a smaller part of Lunenburg, Brunswick, Greenville, Prince George, Isle of Wight, and Nansemond cos. Va., and a minor part of Gates co. N. C.

Nottaway, co. of Va., bounded E. by Dinwiddie, S. by Nottaway r. separating it from Lunenburg, W. by Prince Edward, and N. by Amelia. Length 22 ms., mean width 12, and area 264 sq. ms. Extending in lat. from 36° 54′ to 37° 14′, and in long. from 1° 3′ to 1° 26′ W. W. C. This co. comprises two inclined planes; the nrthrn. falling nrtheastrd., and drained by creeks flowing over Amelia, into Appomatox r., and the sthrn. declining stheastrd. toward Nottaway r. Chief t. Nottaway C. H. or Hendersonville. (*See Hendersonville, Nottaway co. Va.*) Pop. 1820, 9,658, 1830, 10,141.

Nottaway, cr. and p-o., nrthestrn. part of St. Joseph's co. Mich. The p-o. is by p-r. 130 ms. a little S. of W. Detroit. The creek is a nrthrn. branch of St. Joseph of lake Mich.

Nottaway, C. H. (*See Hendersonville, same co. and state.*)

Nottingham, p-t. Rockingham co. N. H., 25 ms. from Concord, 20 from Portsmouth, 55 from Boston; is crossed by North r.; contains several ponds, and gives rise to some small streams. Some of the Blue hills are in the W. part; the surface is generally rough, but the soil is often good. Bog and other iron ores are found. The village called Nottingham square, is pleasant, and stands on a hill. Settled in 1727. Gen. Joseph Cilley was a native of this t. Pop. 1830, 1,157.

Nottingham, West, p-t. Hillsborough co. N. H., 17 ms. S. E. Amherst, 39 S. W. Portsmouth, 55 N. W. Boston, E. Merrimack r.; N. Mass., has good land, with rich meadows on the r., and a broken surface W. The timber is oak, pine, &c. Little Massabesick pond contains 200 acres, and Otternick pond 80. Settled 1710. Pop. 1830, 1,263.

Nottingham, t. Burlington co. N. J., 5 ms. S. Trenton, 9 S. S. W. Princeton; has Del. r. W., Assanpink N., Crosswicks creek S. It is level, lies opposite Duck and Biles islands in the Del., and contains several villages; Lamberton, Sandhills, &c. At Lamberton, Gen. Washington was encamped when threatened by the British at Trenton, and here he commenced the retreat which occasioned the battle of Princeton. Pop. 1830, 3,900.

NOTTINGHAM, p-v. Prince George's co. Va., by p-r. 32 ms. s. E. Richmond.

NULHEGAN, r. Essex co. Vt., rises near Canada, and falls into Conn. r. at Brunswick. It is rapid, and was the channel of navigation for the Indians, between Conn. r. and Memphremagog lake, there being a portage for canoes, of two miles.

NUNDA, p-t. Allegany co. N. Y., 14 ms. N. Angelica, has good grass land, and large and fertile alluvial tracts. It is crossed by Genesee r. which has 2 falls of 50 and 90 feet, 1 mile apart. Pop. 1830, 1,291.

NUTTSVILLE, p-v. in the nrthrn. part of Lancaster co. Va., by p-r. 138 ms. s. s. E. W. C., and 76 N. E. by E. Richmond.

NYACK, village, Rockland co. N. Y., w. Hudson r.

NYESVILLE, p-v. Meigs co. Ohio, by p-r. 102 ms. s. E. Columbus.

O.

OAK FLAT, p-o. wstrn. part Pendleton co. Virginia by post-road 186 miles s. w. by w. W. C.

OAK GROVE, p-o. Lunenburg co. Va. by p-r. 88 ms. s. w. Richmond.

OAK GROVE, and p-o. Edgecombe co. N. C. 72 ms. E. Raleigh.

OAK GROVE, and p-o. Union dist. S. C. by p-r. 91 ms. N. N. W. Columbia.

OAK GROVE, p-o. Jasper co. Geo. 7 miles nthrdly. Monticello, the co. st., and 42 N. W. Milledgeville.

OAK GROVE, and p-o. Jefferson co. Ten. by p-r. 236 ms. E. Nashville.

OAK GROVE, p-o. Christian co. Ky. 14 miles sthrd. Hopkinsville, the co. seat, and by p-r. 220 ms. s. w. by w. Frankfort.

OAK GROVE, furnace and p-o. Perry co. Pa. by p-r. 28 ms. N. W. Harrisburg.

OAKHAM, p-t. Worcester co. Mass. 9 ms. N. W. Worcester, 55 miles s. w. Boston, hilly, with not very good soil, has small streams falling into Chickapee r. Pop. 1830, 1,010.

OAK HILL, p-o. Granville co. N. C. by p-r. 59 ms. nthrd. Raleigh.

OAK HILL, p-o. Fauquier co. Va. by p-r. 58 ms. s. w. by w. W. C.

OAK HILL, p-v. Newton co. Geo. 10 ms. N. W. Covington, the co. st. and 70 ms. in a similar direction from Milledgeville.

OAKINGHAM, p-v. wstrn. part of Laurens district, S. C. by p-r. 74 ms. N. W. Columbia.

OAKLAND, p-o. Morgan co. Va. by p-r. 96 ms. N. W. by W. W. C.

OAKLAND, p-v. Orange co. N. C. by p-r. 49 ms. northwestward Raleigh.

OAKLAND, p-v. parish of St. Tammany La.

OAKLAND, p-o. Christian co. Ky. by p-r. 222 ms. s. w. by w. Frankfort.

OAKLAND, co. Mich. bounded by Macomb co. E., Wayne s., Washtenaw s. w., Shiawassee w. and N. W., and Lapeer co. N. It is a square of 30 ms. each side, area 900 square ms.; lat. 42° 35′, long. W. C. 6° 18′ w. Oakland is a true table land, in the centre flat and full of small lakes, from which issue and flow wstrdly. the sources of Huron of Erie, N. W. Flint river, branch of Saginaw, s. E. the sources of the river Rouge, and E. those of Clinton river, or the sources of Huron of lake St. Clair. Chief town Pontiac. Population 1830, 4,911.

OAKLAND, p-v. Oakland co. Mich. by p-r. 40 ms. N. W. Detroit.

OAKLAND MILLS, and p-o. Ann Arundel co. Md. by p-r. 53 ms. nthrd. W. C. and 45 N. W. Annapolis.

OAKLAND, Mills, and p-o. Juniata co. Pa. by p-r. 41 ms. N. W. Harrisburg.

OAKLEY, p-o. Franklin co. Ky. by p-r. 4 ms. from Frankfort.

OAKMULGEE, river. (*See Ocmulgee river.*)

OAK POINT, and p-o. Randolph co. Mo. by post-road about 100 miles N. W. by W. St. Louis.

OAKTOMIE, p-v. Covington co. Miss. by p-r. 110 ms. E. Natches.

OAK RIDGE, p-v. Guilford co. N. C. by p-r. about 100 ms. N. W. by w. Raleigh.

OAKVILLE, p-v. southwestern part Buckingham co. Va. 49 ms. s. w. by w. New Canton, and 103 wstrd. Richmond.

OAKVILLE, p-v. Mecklenburg co. N. C. by p-r. 125 ms. s. w. by w. Raleigh.

OAKVILLE, p-v. Lawrence co. Ala. by p-r. 111 ms. N. Tuscaloosa.

OAKLAND MILLS, and p-o. western part Loudon co. Va. 37 ms. a little N. of w. W. C.

OAT'S LANDING, and p-o. Marion co. Ten. 121 ms. s. E. by E. Nashville.

OBIES, river of Ten. and Ky. but chiefly of the former, rises in Cumberland mtns. and in Morgan and Overton cos. Ten. deriving some inconsiderable tributaries from Cumberland co. Ky. The course is a little N. of w. 70 ms. to its influx into Cumberland river on the border between Overton and Jackson counties. The valley of Obies r. is nearly commensurate with Morgan and Overton cos.

OCCOQUAN, r. Va. rises in Loudon, Fairfax, and Fauquier cos. traverses and drains the western part of Prince William co. and thence forming the boundary between Prince William and Fairfax cos. falls into the Potomac, about 25 ms. below W. C. and nearly opposite Indian Point.

OCCOQUAN, p-v. N. E. part Prince William co. Va. 23 ms. a little w. of s. W. C.

OCEANA, co. Mich. bounded by ——— N., Montcalm co. E., Kent s., Ottawa s. w., and lake Michigan N. W., lat. 43° 20′, long. 8° 40′ w. W. C. slope s. w. and drained in that direction by White r. and Maskegon r. flowing separate into lake Michigan, and by Rouge r.

a small northern branch of Grand r. This co. has been recently formed and is situated about 150 ms. N. W. by W. Detroit.

OCMULGEE, river, Geo. the watrn. and main constituent branch of Alatamaha, rises in Gwinnett and De Kalb cos. and flowing thence S. S. E. between the Oconee and Flint rs. and nearly parallel to both, by comparative courses 170 ms. curves to S. E. by E. 30 ms. to its junction with Oconee, as will be noticed under the head of the latter, having an entire comparative course of 200 ms. Though the actual length of the streams of the Ocmulgee exceeds that of the Oconee, the vallies of the two streams are remarkably similar in length, width, and direction, and of course in area, each comprising about 4,900 square ms. Taken together, the Oconee and Ocmulgee drain the great central plain of Geo. and water one of the finest sections of the state, and nearly one sixth of the entire surface.

OCOHA, Bridge, and p-o. Covington co. Miss. about 120 ms. E. Natchez.

OCONEE, river, Geo. the estrn. branch of the Alatamaha, having the most remote of its sources in Hall co. within 5 ms. of the main volume of the Chattahoochee, and flowing thence by comparative courses, 175 ms. in a S. S. E. direction, joins the Ocmulgee to form the Alatamaha. The junction is made very nearly on lat. 32° and between Montgomery and Appling cos. The confluents of Oconee are numerous, but relatively small, the valley where widest in Putnam, Jasper, and Greene cos. is only about 40 ms. and the mean width about 26, the area 4,900 square ms. The higher or northern part of Oconee valley, lies between those of Ocmulgee and Savannah, the middle part between those of Ocmulgee and Great Ogechee, and the southern between Ocmulgee and Great Ohoopee.

OCRACOKE Inlet, is the pass from the Atlantic ocean into Pamlico sound, between Cove and Hatteras islands. It admits vessels of 14 feet draught. On Tanner's United States map, lat. 35° and long. 1° E. W. C. intersect about 2 nautical ms. S. W. from the entrance of this inlet.

ODAMSVILLE, p-o. Northampton co. N. C. by p-r. 106 ms. N. E. Raleigh.

OFFICE, Tavern, and p-o. watrn. part of Amelia co. Va. by p-r. 43 ms. S. W. by W. Richmond.

OGDEN, p-t. Monroe co. N. Y. 12 ms. W. Rochester, and containing 32 square ms., is crossed by the mountain ridge, and in the E. by Erie canal. Salmon, Rush, and Little Rush creek, water the town, flowing in several directions. The land is pretty good and uneven. Pop. 1830, 2,401.

OGDEN, p-v. northwestern part of New Madrid co. Mo. by p-r. about 150 ms. S. St. Louis.

OGDENSBURGH, incorporated p-v. port of entry, and st. jus. St. Lawrence co. N. Y. 116 ms. N. Utica, 120 W. Plattsburgh, 209 ms. N. W. Albany, on St. Lawrence r. at the mouth of the Oswegatchie. It is situated on a fine plain, with a good harbor, is regularly laid out. It lies opposite to Prescott, Upper Canada.

OGECHEE, or as commonly called, Great Ogechee, river of Geo. having the remote sources in Greene and Taliaferro cos. about 40 miles N. N. E. Milledgeville ; flowing thence S. S. E. by comparative course 190 ms. falls into Ossabow sound, 20 ms. due S. the city of Savannah. The valley of Great Ogechee lies between those of Alatamaha and Savannah rivers, (*see Cannouchee river.*) The valley of Ogechee, including that of Cannouchee, is about 160 ms. in length, with a mean breadth of 30 ms. and an area of 1,800 square ms.

OGEE's Ferry, and p-o. Joe Daviess co. Illinois, about 320 miles a little west of north Vandalia.

OGLETHORPE, co. Geo. bounded by Taliaferro co. S. S. E., Green S., Clarke W., Madison N., Broad r. separating it from Elbert, N. E., and Wilkes E. and S. E. Length diagonally from southwest to northeast 38 ms., mean width 13, and area four hundred ninety-four square ms. Extending in lat. from 33° 41′ to 34° 02′, and in long. from 5° 44′ to 6° 23′ W. W. C. Though Oglethorpe co. reaches to the Oconee river on the western border, the far greater part of the surface is in the valley of Broad r. and declines estrd. toward the Savannah r. Chief town, Lexington. Population 1820, 14,046, 1830, 13,558.

O'HARRA, p-o. Randolph co. Il. 101 ms. S. S. W. Vandalia.

OHIO river, the great northeastern confluent of the Miss. and in proportion to the extent of land it drains, perhaps the most remarkable river of the earth. The physical section of the earth drained by this fine river lies geographically betwen lat. 34° 12′ and 42° 27′, and long. 1° and 12° W. W. C. The course of the Ohio proper, from the sources of Alleghany to its junction with Miss. is by calculation S. 59° 30′, W. 680 statute ms. This is not, however, the longest, nor in regard to relative space drained, the most central line that can be drawn over the Ohio valley. Another line extended from the sources of Oleans creek, Cataraugus co. New York, to those of Bear Grass creek, Marion co. Al. that is, from the most northern to the most southern sources, amounts by calculation to 750 statute miles, declining from the meridians 40° 37′.

The form of the valley approaches in a very remarkable manner that of a regular ellipse, of which the latter calculated line would be the transverse diameter, and the conjugate diameter, another line extending from the Blue Ridge where the sources of Great Kenhawa and those of Watauga branch of Ten. rise, to the northwestern sources of Wabash, 450 statute ms. Measured by the rhombs following the elements in the following table, the area comes out so very nearly 200,000 square ms. as to admit the adoption of that round number.

Table of the extent in square miles of the valley of Ohio river.

Between lat.	and	Rhombs		sq. ms.
34°	35	2	1-4 Rhombs	8,986
35	36	6	1-8 do	25,655
36	37	7	1-2 do	29,206
37	38	8	1-2 do	32,700
38	39	8	1-2 do	32,250
30	40	8	3-4 do	32,742
40	41	8	do	29,488
41	42	2	1-2 do	9,085
Aggregate extent in square miles				200,111

Allowing the greatest length to be 750 ms. the mean width will be 267 very nearly, or the mean breadth amounts to within a trifling fraction of 1-3 of the greatest length, a compactness seldom equalled in rivers.

If the Alleghany is regarded as the primary and remote constituent of Ohio, this great stream rises by numerous creeks in McKean and Potter cos. Pa., and Alleghany and Cataraugus cos. New York. Becoming navigable near the line of demarcation between the two states, the stream, with partial windings, pursues the general course already stated, to its junction with the Miss. affording a natural navigable channel of between 1,200 and 1,300 ms. The opposing inclined plains of Ohio valley are of unequal extent, nearly in the proportion of 2 to 3, the larger falling from the Appalachian system of mtns. and containing 120,000 square ms.

In their features also the two Ohio plains differ essentially. The southeastern, declining from a mountainous outline, has a comparatively rapid slope. The most elevated table land from which the eastern confluents flow, is that where rise the sources of Clinch, Holston, and Great Kenhawa, about 2,500 feet. The Appalachian table land declines in relative elevation both to N. and S. of this nucleus, but there is no one part from the sources of Alleghany and Genesee to those of Ten. and Coosa through 7° of lat. but which exceeds 1,000 feet.

The elevation of the Ohio at Pittsburg, where the Alleghany and Monongahela unite, is 678 feet, and that of the low water at the confluents of Ohio and Miss. 283 ft.; of course the Ohio below Pittsburg, has a fall of 395 feet in 948 ms., the length of the intermediate channel. The left confluents must have from the preceding data, a descent of from 1,000 to 2,200 feet. Down this rapid declivity, advancing from N. to S. are found the streams of Clarion, Kiskiminitas, Monongahela, Great Kenhawa, Sandy, Ky. Cumberland and Ten. and several of lesser length of course, whose sources do not reach the Appalachian vallies.

It may well excite surprise, that along this steep plain, direct falls are not frequent, and where they do occur, of moderate direct pitch.

The western, or more correctly northwestern plain is directly the reverse of its opposite in respect to apex; the inflected line of river source which separates the valley of Ohio from that of the Great Canadian lakes, is in great part level and marshy. Proceeding from the southern extremity of lake Michigan, and tracing the line from which the Ohio water flows, the face of the country very slowly changes from level to hill and dale, and it is not until reaching the fountains of Alleghany that any protuberance would appear deserving the name of mtn. From this flat, and in winter partially inundated plain, the Big Beaver, Muskingum, Scioto, Miami, and Wabash first slowly descend, gaining more and more rapidity of declivity approaching their recipient, the Ohio.

Ascending the southeastern confluents the scenery becomes rugged and diversified in character, in proportion to proximity to the Appalachian ridges; on the contrary the northwstrn. streams afford the boldest scenery along the immediate margin of Ohio, and the banks become more tame and monotonous until they end in unbroken plains.

To an eye sufficiently elevated, and powers of vision sufficiently enlarged, the whole valley of Ohio would indeed appear one immense declivity, falling very nearly at right angles to the general range of the Appalachian system, and the rivers would appear to have cut deep channels seldom in a direction corresponding to the plain of general descent.

Of these channels that of Ohio would appear as the principal. The author of this article carefully measured the height of the hills, in the vicinity of Pittsburg, and found them about 460 feet above the low water level of the rivers, or 1,138 feet above the level of the Atlantic tides. Above Pittsburg to the hills, which rise like mtns. from lake Erie, the ascent is at least 400 feet, and below Pittsburg the fall to the Miss. has been shown to be 395 feet. Without therefore estimating mtn. ridges, the great inclined plain of Ohio has a descent of upwards of a foot to the statute m. but what is something remarkable, the rivers, and particularly the Ohio itself, do not fall gradually with the plains of their courses. The actual channel from Pittsburg to the mouth is 948 statute miles, and the fall 4,716 inches, or not quite 5 inches per mile.

The waters in effect have abraded their channels, deeper toward their sources than in proportion to length of course. It is this circumstance which has contributed to give to the Ohio proper. the appearance of flowing in a deep and immense ravine. The difference of climate arising from difference of level, frequently exceeding a degree of lat. in less than a mile, and radiated heat, with an exuberant alluvial soil, giving in spring a precocious vegtation along the river bank, have superinduced great misunderstanding respecting the temperature and seasons of this region.

Descending the Ohio, say from Pittsburg, the scenery along the banks and hills, is in an eminent degree picturesque and varied, but these fine features imperceptibly fade away, and long before reaching the Miss. totally disappear, and leave a narrow horizontal ring sweeping round the heavens, formed by the trees along the banks.

As a navigable channel few, if any other rivers of the globe, equal the Ohio. In the higher part of its course the navigation is annually more or less impeded in winter by ice, and in autumn by a want of water. Impediment from ice prevails in all its course, but below the influx of Kenhawa, drought is of less injury, and below the rapids at Louisville, very seldom impedes navigation. The only direct cataract in Ohio was that at Louisville, now in a commercial point of view, removed by a navigable canal. (*See Louisville and Portland canal*, or the latter part of *article Louisville, Ky.*)

The 4 most important of all mineral productions abound in the Ohio valley, limestone, mineral coal, salt, and iron ore. Of all continuous bodies of productive soil on earth, if climate and fertility are combined, the valley of Ohio will, it is probable, sustain the most dense population. I was in it when there did not exist upon its immense surface 20,000 civilized human beings. It now, 1831, sustains about 3,000,000. Can the history of the world afford any parallel to such increase? (*See the different confluents of Ohio under their respective heads.*)

OHIO, state of the United States, bounded by Pa. N. E., E. and S. E. by Ohio r. separating it from Va., S. and S. W. by Ohio r. separating it from Ky., W. by Ind., N. W. by Mich., and N. by lake Erie.

This state bounds on the Ohio r. from the mouth of Little Beaver to that of Great Miami, 440 miles; due N. in common with Ind. 170; due E. along Mich. to lake Erie, 80; thence along the sthrn. shore of Erie, 150; thence due S. in common with Pa. to place of beginning at the mouth of Little Beaver, 93; having an entire outline of 933 miles.

The superficial contents of O. have been generally under-rated; measured carefully by the rhombs on Tanner's and Mitchel's maps, the area comes out within a small fraction of 44,000 sq. ms. or 28,160,000 statute acres.

In lat. this state extends from 37° 25′ at the mouth of Great Sandy r. to 41° 58′, at its extreme north eastern angle, long W. C. 3° 30′ to 7° 48′.

A general idea prevails, that the state of Ohio presents a great uniformity of surface and aerial temperature; but neither its surface or meteorological phenomena sustains such an opinion. The subjoined table will serve to give the reader an idea of its relative and mean height. (*See table, next column.*)

A not very inflected line extended over Ohio S. W. by W. from the W. boundary of Pa., between the sources of Ashtabula r. and those of Shenango branch of Big Beaver, and crossing the summit level of the canal between Massillon and Akron, and thence between the sources of the rivers flowing into Ohio r. and those flowing into lake Erie, would divide the state into two inclined plains of very unequal area, and relative rapidity of descent from their common apex. The northern or Erie plain, not above 25 ms. wide at its N. E extremity, widens to 80 ms. along the S. boundary of Ind. and contains about the 1-4 part of the state.

The Ohio plain, much more extensive in breadth and of course in area, has a much more gentle declivity. By reference to the table below, we find the slope of the nrthn. plain 31 ms. is 405 feet; whilst down the Ohio plain, in a distance of 247 ms. the mean fall is 509 feet. The fall of the nrthn. plain exceeds 13 feet per mile; that of the sthrn. but a small fraction above 2 feet per mile.

The mean elevation of the common apex of those 2 opposing plains may be assumed at 1,000 feet, the positive mean height is, however, perhaps something more. Without a knowledge of the real features of Ohio, it would be natural to place a range of hills along the sources of the stream which flow down the plains of Ohio; but the very reverse is the fact. The central table land is comparatively level, in part marshy, and what is peculiarly remarkable, the hill along the Ohio r. is very nearly of similar elevation with that of the central table land. It is almost demonstrable, that originally the whole sthrn. or Ohio inclined plain, was a vast level with a very slight declination towards what is now the particular valley of Ohio r., and that what appears hills along that great recipient, are the remains of what earth and rock was left as the rivers cut away their beds. A similar

Ascents and descents from the Ohio r. at the mouth of the great Kenhawa; thence down that stream to the mouth of Sciota r. and thence following the Ohio and Erie canal to the latter at Cleveland.

Stations on the route.	dist. in miles.		ascent or descent.	elevation in ft. abv. mn. tide.	
Height of the water level at a mean in the Ohio r. at the mouth of the Great Kenhawa,					535
Mouth of Sciota r.		85	falls	61	474
Thence leaving Ohio r. and up the Sciota to Chillicothe,	52	137	rises	140	614
Circleville,	20	157	do	60	674
Hebron on Licking summit,	52	209	do	219	893
Newark st. just. Licking county,	10	219	falls	69	834
Muskingum valley on the border between Muskingum and Coshocton cos.				90	744
Conhocton village,	42	261	rises	24	768
New Philadelphia,	43	304	do	106	874
Massillon in Stark co. and commencement of Portage summit,	28	332	do	68	942
Akron in Portage co. and nrthrn. extremity of Portage summit,	28	360	do	31	973
Cuyahoga aqueduct,	13	373	falls	269	704
Cleaveland and level of lake Erie,	18	391	do	136	568

remark is applicable to the lake Erie shore of Ohio. Protruding from the extreme wstrn. part of N. Y., the ridge dividing the sources of the confluents of the O. valley from those of lake Erie, rises sloping but abrupt from the latter; and so abrupt that some of the feeders of Chataque lake have their fountains within 5 ms. from the lake shore. Here, at an elevation of 1,300 or 1,400 feet above the ocean tides, flows water on one side which finds a recipient in the Gulf of Mexico; whilst on the opposite side the water in 5 ms. has a fall of near 800 ft., a fall which the O. waters have not reached at the mouth of Muskingum. Sailing on lake Erie from Buffalo, the ridge we have been noticing is seen stretching over the north western angle of Pa. into the state of Ohio, and slowly receding into the interior of the latter state, until from Sandusky bay it ceases to be visible from the lake. This remark is founded on personal observation by the author of this article. The rs. which fall into lake Erie from O. are from their precipitate descent difficult of navigation, and all roll over direct falls or cataracts. Falls or cataracts are on the other side rare, though some of the latter do occur, one in the Muskingum at Zanesville, for instance.

We at once see from the preceding data, that the state of Ohio occupies an immense, and not slightly elevated plateau or table land. Along the sources of its rivers much of the original plain remains unchannelled by rivers, presenting wide spread levels; but receding either to nrth. or sth. the river channels become more and more deep; hills seem to rise as the waters really fall. The dull monotony gradually ceases, and a country is presented which abounds with rich and varied, and in some places even grand scenery. Descending from the central table land, the courses of the rs. with partial inflections, are nrthrd. towards Erie, or sthrd. towards O. The course of that part of O. itself which separates the state of O. from Va. is S. S. W., whilst that part of the same stream separating O. from Ky. is N. W. by W., a bend not far from a right angle taking place at the mouth of Big Sandy r. The whole Ohio r. border of the state of O. approaches astonishingly to a circular curve. Setting one leg of a pair of compasses in Worthington, 9 ms. N. Columbus, and extending the other to the mouth of Great Sandy, 125 ms., the distance will reach the mouth of Great Miami, approach very near the mouth of Little Beaver, and carried from extreme to extreme will sweep along or very near the O r. It is this salient curve of Ohio r. and the little difference between its mean length and breadth, which render this state the most compact, not even excepting Conn., of any state of the U. S. in proportion to outline.

Climate and seasons.—On no other subject connected with the geography of the U. S. has there been so much of palpable mistake as in regard to the climate of, not only O., but of the entire central basin of N. A. Volney, who understood the meteorology and winds of N. A. about as well, and no better than do European travellers in general, the moral and political character of the U. S., propagated the opinion that the central basin was warmer on a given latitude than the Atlantic coast. Consulting neither relative elevation or exposure to prevalent winds; nor awaiting recorded observations with the thermometer, and the inflorescence, foliage, and decay of vegetables, or the freezing of rs. and crr. this foreigner, in a nine month's transit, mostly in summer, thought himself competent to decide a problem which no human sagacity could determine, except by the aid of actual experiment; and experiment long continued and carefully registered.

Due attention being given to the internal structure of the country, especially that of O., will enable the reader to detect the leading cause of an error which sets at nought all the induction drawn from the known laws of nature. The individual channel of Ohio r. at Pittsburg is, by actual measurement, made by myself, 460 feet below the apex of the adjacent hills. This relative height does not materially change on the right side, especially until 50 or 60 miles below the mouth of Wabash. The hills, or more correctly the buttress of the interior table land, reaches the bank, or recedes one or two miles with intervening bottoms, through the entire valley of O. r. from hill to hill; and is above Louisville in few plac- - two ms. wide. Thus that stream flows in a deep chasm, which receives the sun's rays as in a focus, and has besides an actual depression equivalent to a degree of lat. below the surface, one or two miles from its banks. The rapid transition of temperature is [illegible] on vegetation, and is felt and seen in a very striking manner on health. Between the valley of O. r. at Wheeling, which is very nearly opposite the middle lat. of the state of Ohio, and the farms on the hills 8 or 10 ms. distance E. or W. there is a difference of at least 10 days in seed time, harvest, or the inflorescence of fruit trees or ripening of their fruit. Here again I may be permitted to state, that I was bred from a child to a man on the table land of O. r. near Washington in Pa., and resided some years at Wheeling, and give the data from actual observation. But the accuracy of the opinion, that so far from being warmer, the valley of O. is in winter, greatly colder, does not depend alone on my observation or assertion. The following is an extract from the National Intelligencer, Dec. 29th, 1831.—"Wheeling, Dec. 24th, 1831. The mercury in the thermometer (Fah.) on Sunday morning last, (Dec. 18th,) stood at 16° below zero, which was 10 degrees colder than the coldest day last winter." From this extract, the extreme cold of the winter of 1830—31, was 6 minus zero. In the art. U. S. the reader will find this subject more amply discussed.

Soil and Agriculture.—As a general character, the soil of Ohio is eminently productive, and the productive part, perhaps 9-10ths of the whole, very equally distributed over the state. But with extremes of lat. 3½ degrees, and with relative height taken into the account of 5 degrees of Fahrenheit in mean temperature, the effects on vegetation are severely felt. Small grain, Indian corn, salted meat and live-stock, are the staples of the state. Of grains, Indian corn is cultivated in an abundance which might be styled excessive, and is the grain which is indeed most suitable to all parts of the state, as the summers on the highest part of the table land are sufficiently long for its ripening. An intelligent man who removed from the vicinity of Sandy Spring, and who settled in Portage co., informed the author of this article, that the climate was there too cold for the successful culture of winter grain. The apple succeeds well over Ohio, as does the peach in sheltered situations. Flint says, "Fruits of all kinds are raised in the greatest profusion; and apples are as plenty in the cultivated parts of the state, as in any part of the Atlantic country. The markets are amply supplied with peaches, plums, cherries, gooseberries, strawberries, and cultivated grapes. From the fulness and richness of the clusters of cultivated grapes, it is clear, that this ought to be a country of vineyards. The Germans have already made a few establishments of the kind, with entire success. Apricots, nectarines, and quinces, succeed; and the state is the appropriate empire of pomona."*

To the preceding we may add, tobacco, hemp & flax, as these vegetables are in a high degree suitable to the soil and climate of O. Hemp and flax must, from the very nature and analogy of things, become standing staples of Ohio. But we may extend the observations on these two latter vegetables, to the contiguous political sections, Ky., Ind., Mich., Pa. and wstrn. Va.; and perhaps still more suitably to Il. and Mo.

Rivers, Canals, Commerce, Penitentiary.—The principal river of Ohio, is that queen of rivers from which the state derives its name, and which semicircles the state with its channel, unequalled for tranquillity of current, and soft splendid scenery along its banks. Flowing in fact along the base of the enormous platform of Ohio, the r. O. has a very moderate descent, falling only 204 feet from Pittsburg to the mouth of Sciota. In proper seasons when amply supplied with water, the O. is one of the most safely and easily navigated rs. of the earth; but it is impeded in winter by ice, and in summer by drought, and is not, on an average of one year in ten, navigable above the mouth of great Kenhawa, more than half the year. Winter frost impedes the navigation in all its length to a longer or shorter time, a great majority of seasons. It is also liable to excessive and destructive floods. (*See art. Ohio r.*)

Of the rivers flowing from the state into the Ohio, the principal are descending, Muskingum, Hockhocking, Sciota, and Great and Little Miami. These, with numerous creeks, drain the great sthrn. plain of O., and have interlocking sources with the streams flowing in an opposite direction into lake Erie. Advancing from E. to W. the most important of the latter are, Ashtabula, Grand r., Cuyahoga, Huron, Sandusky, and Maumee, with innumerable intermediate creeks.

If the recent settlement (Marietta, the incipient step, was founded 1787,) and its population were to be compared in Europe, the following statistics would demand no ordinary evidence to render them credible; and yet they are true in principle and fact, and are annually fading from memory by other statistics of similar nature, but enlarged magnitude, both as to object and expenditure.

The subjoined extracts are taken from "the Civil Engineer," a weekly paper published at Columbus, O., and dedicated to canals, roads, &c.; and from other authentic documents, and show the financial condition and internal improvements of Ohio, 1831.

Ohio Canals.—"To people out of O., who are not conversant with the localities of this state, the following remarks will not, probably, be unacceptable. The main Erie and Ohio canal, commences at Cleaveland, on the lake shore, follows up the valley of the Cuyahoga r. sthwrdly. above 30 ms., then crosses the Portage Summit, to the Tuscarawas or Muskingum river, whose valley it follows to Dresden, about 14 ms. N. Zanesville; from thence it takes a sthwstrn. direction across the height of land dividing the Muskingum from the Sciota r., into the valley of the latter, about 12 ms. S. from Columbus; thence a sthrn. direction along the valley of the Sciota r., passing Circleville, Chillicothe, and Piketon, to Portsmouth, on the N. bank of the O. r., at the mouth of the Sciota. Total length 306 ms. Nearly two thirds of the whole is under contract; and about 40 miles of the nrthrn. part, from Cleaveland to Akron, is completed and in successful operation.

The Miami canal commences at Cincinnati, and extends nrthwrdly. along the valley of the Great Miami r.; a total distance of 67 ms. It passes the towns of Hamilton, Middletown, Franklin and Miamiesburg to Dayton. This canal is now finished, and in constant use from Cincinnati to Middletown, about 43 ms. The remaining 24 ms. to Dayton, are to be completed in about 3 months.

A route was surveyed, some three years ago, for the future extension of this canal, nrthwrdly. from Dayton, to the Maumee river at fort Defiance; thence nrthestrdly. along the Maumee, to its mouth in the western extremity of lake Erie. But this continuation of the canal from Dayton, (for a number of years hence,) has not entered into the canal polity of the state. But the late grant by Congress, of some 300,000 acres of land, adjacent to this canal line, on condition of its being immediately constructed thro' the Congress Lands, will probably induce the state of Ohio to prosecute its continuance, the ensuing year."

*Flint's Geog. Miss. valley, vol. 1, p. 393--4.

The following extracts from a late message of the governor, which cannot well be abridged, exhibit the condition of the finances and public works of this prosperous member of the confederacy:

The aggregate amount paid into the treasury for state and canal purposes, for the year ending the 15th of November, 1831, is $235,-[illegible] 75, which, added to the balance remaining in the treasury on the 15th of Nov. 1830, viz. $6,280 44, amounts to $242,266 19.

The aggregate amount disbursed at the treasury, for state and canal purposes, for the year ending 15th Nov. 1831, including interest on school funds, is $236,190 81; leaving a balance in the treasury at the last date, of $6,075 38—to which may be added the $2,000, drawn from the treasury, for the repairs of the United States' road.

The tax levied for 1831, for state and canal purposes, is the same as for 1830; and it is estimated that there will be paid into the treasury, from the 15th of Nov. last, to the 1st of March, ensuing, the additional sum of about $220,000, which will be amply sufficient to defray all the expenses of the government for the ensuing year.

The amount of the foreign debt contracted on account of the canals, is $4,400,000. The interest payable annually on that sum, to foreign stockholders, is $260,000.

The amount borrowed from the different school funds, and transferred to the canal fund, up to the 15th Nov. 1831, is $257,128,-08. The annual interest on the last named amount, is $15,427 68, payable to our own citizens for the support of schools. Making the whole canal debt of the state, $4,657,-128 08; and the annual interest payable thereon, $275,427 68.

The amount received into the treasury from the sale of lands granted by Congress to the state of Ohio, for canal purposes, during the year ending the 15th Nov. last, was $55,-090 79. The amount of tolls collected upon the Miami canal, from the first day of Nov. 1830, to the first day of Nov. 1831, is $36,-177 78. The amount of tolls collected upon the Ohio Canal, from the first day of Nov. 1830, to the first day of Nov. 1831, is $63,-934 27 1; making together the sum of $100,-112 05 1; which, after deducting the expense of collection, leaves $94,619 15 1. This net amount of tolls, added to the proceeds of the sales of lands granted for canal purposes, as above stated, is applied towards the payment of the interest of the canal debt.

The navigation of the Erie and Ohio canal has been opened during the past season as far south as Chillicothe, a distance of 259 ms. This, with the Miami canal, and the number of navigable feeders connected with the main line, make an amount of finished canal, now navigable, of about 344 miles.

It is believed by the acting canal commissioners, that that portion of the Ohio canal between Chillicothe and Portsmouth, a distance of about 50 miles, together with the Granville feeder of 6 miles, already in a very advanced state, (but the operations upon which have been considerably retarded by the great quantity of rain during the last summer), will be completed in July next; when Ohio will have of navigable canals, 400 miles.

The influence of these great works is already visible in the increase of commerce and travel. Substantial improvements have been wrought in the country which they traverse, and there has been a regular arrival and departure of packet and freight boats at a season of the year when navigation has been hitherto unknown. Merchants of the South-western and Western states have in many instances had their merchandise transported by this channel. There is a visibly increased demand for the staples of the state itself. The governor expresses a reasonable hope that such a work will tend somewhat to cement the union of the states. The tolls received on the United States' road between Wheeling and Zanesville, (it having been transferred to the state of Ohio,) amount to $2,777.

The Ohio State Journal, in reply to queries addressed to it through the Circleville Herald, has published a long statement embracing various matters of general interest, from which we abstract the following.

CANAL DEBT OF OHIO.

	Amount borrowed.		Interest.
Loan of 1825	$400,000	5	$20,000
" 1826	1,000,000	6	60,000
" 1827	1,200,000	6	72,000
" 1828	1,200,000	6	72,000
" 1830	600,000	6	36,000
Foreign debt	$4,400,000		260,000
School fund	169,460		10,167
Total	$4,569,460		270,167

The school funds borrowed for the use of the canals, amounted on the 15th Nov. last, to $169,460 63, as follows: common school fund, $82,626 31. Virginia military do., $47,014 32. U. S. military do., $27,895 50. Sales of salt reserves, $11,004 20. Ohio university fund, $920 35.

To meet the interest due for 1831, on the canal loans, the following are the sources relied on. Direct tax of 2 mills on a dollar, $121,516. Canal tolls, $80,000. Sales of land granted by congress, $50,000. Donations, interest on deposites, &c., $20,000; amounting to $271,156.

It is believed that the canals will be completed without resort to further foreign loans.

Taxes for 1830.—The gross amount of tax collected in Ohio during the last year is stated as follows: For canal purposes, $129,-551 93. For state purposes, $97,163 95. For county school, and township and road purposes, $350,860 33. Sundry items, $7,-500 00. Total $585,076 21. Averaging about 62 cents to every inhabitant of the state.

Expenses of Government.—The ordinary expenses of the state government are reported by the auditor of the state, to be about $90,000. Extra printing, &c. 1831, $7,500. Penitentiary, $3,000. Balance in the treasury, Nov. 1831, $7,062. Amount of revenue 1831, $107,562.

The message, leaving objects of more con[illegible] import, goes on to observe that, "The penitentiary is, in its present condition, ill calculated to promote its proper objects. It is on the contrary rather a school of crime. More rigid discipline is required, and corresponding alteration of the buildings. In addition to the cost of erection, its average annual expene is $10,000, and this year it is $13,000."

It would seem that the emigration of free colored persons to Ohio, had been found an evil demanding legislative interference.

"I think it my duty to make some suggestions to you, in relation to the colored population within our limits. Much evil has been experienced in various parts of the state from the great influx of this kind of population amongst us. The recent excitement in Va. and other slave holding states, will have a tendency to drive many free people of color from them, and they very naturally seek an asylum in the free states. The feeling of hostility towards them which has been manifesting itself recently, will drive many of them from the slave holding states; and we are in danger, from our proximity to them, of being much annoyed by that kind of population. Our laws relative to these people have not been strictly enforced; and I suggest the propriety of adopting such measures as may guard us against the evils which must inevitably result, unless something be done to secure us against imposition."

Progressive pop. up to 1830 *inclusive.*—It has already been noticed that the settlement of Marietta, 1787, or 45 years ago, commenced Ohio, which had in 1800, 45,365 inhabitants; in 1810, 230,760; and in 1820, 581,434.

The subjoined table exhibits the aggregate by counties, 1830.

Counties.	Pop. 1830.	Counties.	Pop. 1830.
Adams,	12,281	Hamilton,	52,317
Ashtabula,	14,584	Hocking,	4,008
Athens,	9,787	Highland,	16,345
Allen,	578	Harrison,	20,916
Butler,	27,142	Hancock,	813
Belmont,	28,627	Hardin,	210
Brown,	17,867	Henry,	262
Champaign,	12,131	Holmes,	9,135
Clarke,	13,114	Huron,	13,341
Clermont,	20,466	Jefferson,	22,489
Columbiana,	35,592	Jackson,	5,941
Coshocton,	11,161	Knox,	17,085
Cuyahoga,	10,373	Lawrence,	5,367
Crawford,	4,791	Licking,	20,869
Clinton,	11,436	Lorain,	5,696
Dark,	6,204	Logan,	6,440
Delaware,	11,504	Madison,	6,190
Fairfield,	24,786	Marion,	6,551
Fayette,	8,182	Medina,	7,560
Franklin,	14,741	Meigs,	6,158
Gallia,	9,733	Mercer,	1,110
Geauga,	15,813	Miami,	12,807
Green,	14,801	Monroe,	8,768
Guernsey,	18,036	Montgomery,	24,362
Morgan,	11,800	Scioto,	[illegible]
Muskingum,	29,334	Seneca,	[illegible]
Perry,	13,970	Stark,	[illegible]
Pickaway,	16,001	Tuscarawas,	[illegible]
Pike,	6,024	Trumbull,	[illegible]
Portage,	18,826	Union,	[illegible]
Preble,	16,291	Van Wert,	[illegible]
Putnam,	230	Washington,	[illegible]
Paulding,	161	Wayne,	[illegible]
Richland,	24,006	Williams,	[illegible]
Ross,	24,068	Warren,	[illegible]
Sandusky,	2,851	Wood,	[illegible]
Shelby,	3,671		

Of which were white persons:

	Males.	Females.
Under 5 years of age,	96,411	[illegible]
From 5 to 10	74,690	[illegible],851
" 10 to 15	62,151	[illegible]
" 15 to 20	51,138	[illegible],635
" 20 to 30	81,290	75,574
" 30 to 40	49,346	[illegible],894
" 40 to 50	31,112	[illegible],546
" 50 to 60	18,058	15,898
" 60 to 70	10,783	8,293
" 70 to 80	3,632	[illegible],915
" 80 to 90	935	736
" 90 to 100	138	89
" 100 and upwards	29	6
Total,	478,680	[illegible]7,631

Among the preceding who are deaf and dumb, there are under 14 years of age, 148; from 14 to 25, 160; 25 and upwards, 118; blind, 232.

Of free colored persons, there were,

	Males.	Females.
Under 10 years	1,562	1,573
10 to 24	1,440	1,551
24 to 36	808	799
36 to 55	646	611
55 to 100	325	241
100 and upwards	8	4
Total,	4,788	4,779

Slaves.—Males 1, females 5. Deaf and dumb colored, 9.

Recapitulation.

Whites.	Free colored.	Slaves.	Total.
926,311	9,567	6	935,884

Constitution, Government, Judiciary.—The constitution of Ohio was adopted in convention at Chillicothe, 29th Nov. 1802; the most important provisions of which are:

Art. 1.–*Sec.* 1.—The legislative authority of this state shall be vested in a general assembly, which shall consist of a senate and house of representatives, both to be elected by the people.

Sec. 3.—Representatives to be chosen annually.

Sec. 4.—No person shall be a representative, who shall not have attained the age of 25 years, and be a citizen of the U. S., and an inhabitant of this state; shall also have resided within the limits of the co. in which he shall be chosen, one year next preceding his election, unless absent on public business.

Sec. 5.—The senators shall be chosen biennially, &c.

Sec. 7.—No person shall be a senator who has not arrived at the age of 30 years, and who is not a citizen of the U. S., shall have resided 2 years in the co. or district, immediately preceding his election, unless absent on public business, &c.

Sec. 20.—No senator or representative shall, during the time for which he shall have been elected, be appointed to any civil office under this state, which shall have been created, or the emoluments of which shall have been increased, during such time.

Sec. 21.—No money shall be drawn from the treasury, but in consequence of appropriations made by law.

Art. 2d, *Sec.* 1.—The supreme executive power of this state shall be vested in a governor.

Sec. 2.—The governor elected by the qualified electors for the members of the general assembly, and at the same time and place. Holds his office for two years, or until another governor shall be elected and qualified. Eligible only 6 years, in any term of 8 years. He shall be at least 30 years of age, and have been a citizen of the U. S. 12 years, and an inhabitant of this state 4 years next preceding his election.

Sec. 13.—No member of congress, or person holding any office under the U. S. or this state, shall execute the office of governor. The powers of the governor of Ohio are nearly the same as generally vested in governors of states. He can grant pardons and reprieves after conviction, except in cases of impeachment.

Art. 3, *Sec.* 1.—The judicial power of this state, both as to matters of law and equity, shall be vested in a supreme court, in courts of common pleas for each co.; in justices of the peace, and in such other courts as the legislature may, from time to time establish.

Sec. 7.—The judges of the supreme court shall, by virtue of their offices, be conservators of the peace throughout the state. The presidents of the court of common pleas, shall, by virtue of their offices, be conservators of the peace within their respective circuits, and the judges of the court of common pleas shall, by virtue of their offices, be conservators of the peace in their respective cos.

Sec. 8.—The judges of the supreme court, the presidents, and the associate judges of the courts of common pleas, shall be appointed by a joint ballot of both houses of the general assembly, and shall hold their offices for the term of 7 years, if so long they behave well.

This mode of choosing judges of law and equity septennially, is a rather novel, or at least wide deviation from the usual manner of election or term of service.

Art. 4, *Sec.* 1.—In all elections, all white male inhabitants, above the age of 21 years, having resided in the state one year next preceding the election, and who have paid, or are charged with, a state or co. tax, shall enjoy the right of an elector; but no person shall be entitled to vote, except in the county or district in which he shall actually reside, at the time of the election.

Art. 8, contains 28 sections of general principles; amongst which we may notice the following:

Sec. 2.—There shall be neither slavery or involuntary servitude in this state, otherwise than for the punishment of crimes, &c.

Sec. 3.—No preference shall ever be given by law to any religious society, or mode of worship; and no religious test shall be required, as a qualification to any office of trust or profit.

Sec. 6, secures the freedom of the press, and concludes thus: "In prosecutions for any publication respecting the official conduct of men in a public capacity, or where the matter published is proper for public information, the truth thereof may always be given in evidence; and in all indictments for libels, the jury shall have a right to determine the law and the facts, under the direction of the courts, as in other cases."

Sec. 15.—The person of a debtor, where there is not strong presumption of fraud, shall not be continued in prison after delivering up his estate for the benefit of his creditor, or creditors, in such manner as is prescribed by law.

Sec. 17.—That no person shall be liable to be transported out of this state, for any offence committed within the state.

History.—In 1787, what is now the state of Ohio, was included in a territory then created by act of congress, called "The Territory northwest of the r. Ohio," and in the same year preparations were made for the first civilized settlement made within its limits. Gen. Rufus Putnam and the Rev. Menasseh Cutler, led a small colony from Middlesex and Essex counties, Mass., who fixed themselves at Marietta, at the mouth of Muskingum r. Similar to every settlement made on a frontier, exposed to savage war, the first years of Ohio were spent in blood and tears. The treaty of Grenville, in 1795, and the surrender of Mich. in 1796, gave peace to the west, and emigration poured over the mountains, into the Ohio valley. In 1800, O. and Mich. were formed into a separate territory, but having acquired sufficient numbers in O., Mich. was detached April, 1802, and the former authorised to frame a constitution. January, 1802, after every necessary formality was complied with, Ohio was admitted into the Union as a state. Though as a member of the confederacy her history is blended with that of the nation, we cannot omit an expression of admiration at a progress in power, resource and energy that has no parallel in all human history. Forty-six years past all was a wild; now, 1832, with at least a million of inhabitants, her towns, cities, canals, roads, schools, colleges and other improvements mock the pen of the geographer.

Ohio, co. Va. bounded by Washington co.

Pa. N. E., Greene co. Pa. S. E., Tyler co. Va. S., Ohio r. separating it from Monroe co. state of Ohio, S. W., and Belmont co. Ohio, N. W., and by Brooke co. of Va. N. Length from S. to N. 36 ms., mean width 12, and area 432 sq. ms. Extending in lat. from 39° 49' to 40° 14', and in long. from 3° 36' to 3° 55' W. W. C. The declivity is a little N. of W., down which flow into O. r. advancing from N. to S., Short, Wheeling, Grave, Fish, and Fishing creeks, having their sources in Pa. The surface is excessively broken by hills, but with alluvial bottoms of first rate soil. In fact the soil of the highest hills is fertile, and the whole co. a body of excellent land. Chief t. Wheeling. Pop. 1820, 9,182, 1830, 15,590.

Ohio, co. Ky. bounded N. W. by Daviess, N. E. by Hancock, E. by Grayson, S. E. by a part of Butler co., and by Green r. which separates it from a part of Butler S., and from Muhlenburg S. W. This co. lies nearly in form of a square, 24 ms. each side; area 576 sq. ms. Extending in lat. from 37° 12' to 37° 34', and in long. from 9° 37' to 10° 14' W. W. C. The declivity of this co. is wstrd. in the direction of the two main branches of Green r. between which are comprised full ⅔ of all the area. Chief t. Hartford. Pop. 1820, 3,879, 1830, 4,913.

Ohiopyle Falls, in Youghiogany r. is a descent of 7 or 8 feet direct pitch in that stream, where it passes Laurel Hill, 11 or 12 ms. N. E. Uniontown, Fayette co. Pa.

Ohioville, p-v. western part of Beaver co. Pa. by p-r. 11 ms. below, and wstrd. of the borough of Beaver, and 262 ms. N. W. by W. W. C.

Oil Creek, small, but remarkable stream of Pa. rising in the southern part of Crawford, and western part of Warren cos., flows southwardly, enters Venango co. and falls into Alleghany r. about 8 ms. estrd. Franklin. On this creek, and near the border between Venango and Crawford cos. arises a spring of water, on which floats a mineral oil, from which the name of the creek is derived. The oil spring, as laid down on Tanner's Pennsylvania, is 25 ms. a little S. of E. Meadville.

Oil Creek, p-o. and tsp. southeastern angle of Crawford co. Pa. 25 ms. E. Meadville, and 80 a little E. of N. Pittsburg.

Olamon Plantation, Hancock co. Me. Pop. 1830, 222.

Oldbridge, v. Middlesex co. N. J. at the head of navigation on Smith r. 7 ms. S. E. New Brunswick, 2 N. E. Spotswood.

Old Church and p-o. Hanover co. Va. by p-r. 15 ms. N. Richmond.

Oldfield, p-o. Ashe co. N. C. by p-r. 218 ms. N. W. by W. Raleigh.

Old Fort, p-v. Centre co. Pa. by p-r. 75 ms. N. W. Harrisburg.

Old Fort and p-o. Burke co. N. C. by p-r. 235 ms. W. Raleigh.

Oldham, co. Ky., bounded N. by Gallatin, E. by Henry, S. E. by Shelby, S. W. by Jefferson, and W. by Ohio r. separating it from Clark co. Ind. Length 28 ms., mean width 15, and area 420 sq. ms. Extending in lat. from 38° 15' to 38° 40', and in long. from 8° 12' to 8° 37' W. W. C. The surface hilly and rocky. Chief t. Westport. Pop. 1830, 9,563.

Old Mines and p-o. Washington co. Mo. by p-r. 140 ms. S. S. W. St. Louis.

Old Point Comfort, cape and p-o. Elizabeth City co. Va. 12 ms. in a direct line a little W. of N. Norfolk, 3 ms. S. E. Hampton, the co. seat, and by p-r. 202 ms. S. S. E. W. C. The promontory, particularly called Old Point Comfort, is almost exactly on lat. 37°, according to Tanner's U. S., and with the opposing point, Willoughby, on the right shore, forms the real mouth of James r., the intervening strait separating Lynhaven bay from Hampton Roads.

Old Town, p-v. on Potomac r. Alleghany co. Md. 14 ms. below and S. E. Cumberland, and by p-r. 135 ms. N. W. by W. W. C.

Old Town, p-v. nthwstrn. part of Ross co. O. 12 ms. N. W. Chillicothe.

Old Town, p-v. Lowndes co. Ala. by p-r. 131 ms. S. E. Tuscaloosa.

Olean, p-t. Cattaraugus co. N. Y. 20 ms. S. E. Ellicottville, 33 S. W. Angelica, 115 S. W. Geneva, N. of Pa., 8 ms. by 9; is crossed by Olean creek, which runs into Alleghany r., and supplies several mill seats. The timber is chiefly valuable pine, which has occupied the inhabitants in the preparation and transportation of lumber. Olean Point, or Hamilton village, stands just below the confluence of oil creek and Alleghany r., and at the head of navigation. Several ancient mounds were found near this place. Population 1830, 561.

Olean, cr. Cattaraugus co. N. Y. 25 ms. in length.

Oley, tsp. Berks co. Pa. 10 ms. estrd. Reading. Pop. 1820, 1,400, 1830, 1,469.

Oley Furnace and p-o. in Oley tsp. Berks co. Pa. 10 ms. estrd. Reading.

Olive, t. Ulster co. N. Y. 12 ms. W. Kingston, about 10 ms. long, has a rough surface. Pop. 1830, 1,636.

Olive, p-v. Morgan co. O. by p-r. 100 ms. a little S. of E. Columbus.

Olive Green. (*See Ludlow, Morgan co. Ohio.*)

Oliver's p-o. Anderson co. Ten. by p-r. 188 ms. E. Nashville.

Olympian Springs and p-o. Bath co. Ky. 11 ms. S. E. Owingsville, and 49 E. Lexington.

Ompomponoosuc, r. Orange co. Vt., joins Conn. r. at Norwich. It is a good mill stream, about 20 ms. long.

Onancock, creek and p-o. on the Chesapeake shore of Accomac co. Va. 8 ms. S. W. by W. Drummondstown, and by p-r. 210 ms. S. E. W. C.

Oneida Lake, N. Y. in several cos. about the middle of the state, near the shore of lake Ontario, into which it flows. It is about 20 ms. long E. and W., and 4 broad, and receives

Chittenango creek, &c. w., Fish and Wood creeks E., and Oneida creek S. This lake, being the important channel of intercourse between the Canadians and the five nations of Indians, up to the time of the French wars, was fortified by two block houses, fort Brewerton w., and fort Royal at the mouth of Wood creek. The banks are low, with good soil.

ONEIDA, co. N. Y. bounded by Lewis co. and a part of Oswego co. N., Herkimer co. E., Herkimer and Madison cos. S., Madison and Oswego cos. W.; contains 1,136 sq. ms. The head of navigation of Mohawk r. is in this co., a part of Oneida lake. It is 110 W. of Albany, and contains 32 ts.; Sadaguada, Oriskany, and other crs. or streams, tributary to Mohawk r. lie in the middle of this co. Fish and Wood creeks N., Black r. and its branches N. E., W. Canada creek E., with its branches, Steuben and Cincinnati creeks, Oneida creek S. W., and streams of Susquehannah r. S. Mill seats are abundant, and equal to those of any co. in the state. The surface is uneven, and abounds in fine springs. In the N. E. the country is more hilly. Iron ore is found in the S. and N. E., and iron works exist in several places. The Erie canal lies along the S. bank of Mohawk r. to Rome, on the Long Level, which reaches 69½ ms. The old canal from Wood creek to the Mohawk, which connected the navigation of lake Ontario and the Hudson, is in this co. Pop. 1820, 50,997, 1830, 71,326.

ONEIDA, creek, Madison co. N. Y. divides this co. from Oneida co. for 17 ms. and runs N. into the S. E. end of Oneida lake. It formerly supplied the Indians with fine fish, but is now a valuable mill stream, and is crossed by the Erie canal 3½ ms. from its mouth. Its principal tributary, Stanando creek, is also a good mill stream.

ONEIDA, p-v. Vernon, Oneida co. N. Y. 22 ms. W. Utica, 9 S. Oneida lake, is on the borders of the Oneida reservation.

ONEIDA CASTLE, Oneida co. N. Y. was the chief residence of the Oneida Indians, one of the five native nations of the state of N. Y., who have a reservation of about 20,000 acres of valuable land, on which were supported christian missionaries for many years, with some success. The nation gave a portion of their land to the Stockbridge Indians, and another to such of other tribes as chose to settle upon them; and they also enjoyed the benefits of instruction. Within a few years many of these different tribes have emigrated to Green Bay, where they received lands from the Menominee or Rice Indians.

ONE LEG, p-v. Tuscarawas co. O. by p-r. 127 ms. N. E. by E. Columbus.

ONEONTA, t. Oswego co. N. Y. Pop. 1830, 1,759.

ONION, r. Vt. rises in Cabot, Caledonia co. first runs S. and S. W., then turns N. W. and flows through Washington and Chittenden cos. into lake Champlain 5 ms. below Burlington. It is about 70 ms. long, and one of the largest rivers in Vt. Its principal branches are Dog r. and Steven's branch, N. branch at Montpelier, Mad, Waterbury and Huntington rs., and Muddy brook. It passes through a wild and romantic country a considerable part of its course, and has a great descent. On its highest branch is a fall of 500 feet within a distance of 30 feet. In Bolton, Chittenden co., where it crosses the ridge of the Green mtns. the current in the course of ages has worn away the rocks in a remarkable manner. In one place it has cut to a depth of 30 feet, a channel 60 feet wide, and about 270 yards long, through a solid rock. At another, 4 ms. below Waterbury v. it has cut 100 feet down, having one side a perpendicular wall; and there rocks have fallen down so as to form a natural bridge. An artificial bridge has been built over a third place of the kind, three quarters of a mile above the falls, where the channel is 65 feet deep and 70 wide. The water power afforded by this stream is very great. The turnpike road from Royalton to Burlington lies along the course of Onion r., and is one of the best, as well as one of the most romantic roads in this part of the U. S. The great route by which the Indians formerly travelled between Conn. r. and lake Champlain, lay along the courses of White and Onion rs. When the English first settled at Plymouth, an extensive war was carried on in boats of skins by this route; and during the subsequent wars with the French in Canada, many incursions upon the frontiers were made in this way. There is much excellent alluvial land along the banks, especially after its passage of the mountains. Boats go up to the falls about 40 ms., but lake vessels can go only 5 ms.

ONO, p-v. Edgar co. Il. 6 ms. nrthrd. Paris, the co. st. and by p-r. 112 ms. N. E. by E. Vandalia.

ONONDAGA, lake, Salina, Onondaga co. N. Y. 7 ms. from Onondaga, drained by Seneca r. It is about 1½ ms. by 6, with turbid water, and receives many springs of salt water from its banks, besides several streams; Otisco, Onandaga creeks, &c. Its banks have been peculiarly subject to the fever and ague. The surrounding land is low, and often marshy. The great salt manufactories of N. Y. are supplied with water from the shores. It is raised by machinery to an elevation sufficient to convey it in pipes to Syracuse and other places where the manufactories are situated. The branch canal or side cut from the Erie canal, communicates with the lake by locks at Salina, the descent being 38 ft.

ONONDAGA, co. N. Y. 130 ms. W. N. W. Albany, 40 S. S. E. Oswego, 50 W. Utica; bounded by Oswego co. E., Madison and Cortlandt cos. S., Cayuga co. W., Oswego r. on the N. bound., Oneida lake and river E., and Skeneateles lake S. W. Greatest length 32 ms. greatest breadth 28, area, about 334,000 acres. It contains the salt springs of the state, which are of immense value, and quantities of gypsum, limestone and water cement.

with a good soil, and streams highly useful for manufactures and navigation. Besides the waters above mentioned, there are Oswego and Seneca rs., Cross, Onondaga and Otisco lakes, &c. In the s. are several brooks, tributary to Susquehannah r. The manufacture of salt is very extensive and valuable. The Onondaga Indians, though much reduced in numbers, occupy their reservation in this co. The salt springs are owned by the state, and are in the town of Salina. The water is raised by water and steam, conducted into Syracuse, Geddes and Liverpool, and there the salt is made from it, as well as in Salina. It is obtained by solar evaporation, and by artificial heat, in different ways. A branch of the Erie canal extending to Salina, supplies water power. A bushel of salt is obtained from 45 gallons. A tax of 12½ cents a bushel is paid by the manufacturers to the state, for the canal fund. Between 1817 and 1831, including those 2 years, about $1,400,-000 was paid to the treasury from this source. Quantities inspected in different years: 1826, 827,508 bushels; 1827, 983,410; 1828, 1,160,880; 1829, 1,291,280; 1830, 1,-435,446; 1831, 1,514,037 bushels. Of the last, 163,000 bushels were made by solar evaporation, and the residue by solar and artificial heat combined. 189,000 bushels were coarse. Pop. 1820, 41,467; 1830, 58,984.

Onondaga, p-t. and st. jus. Onondaga co. N. Y. 134 ms. w. Albany, 41 s. Oswego, 9 ms. by 10; has an agreeable surface, with very good soil. In the e. is a fine valley, between high hills, through which lies the course of Onondaga creek towards Onondaga lake, and here are many mill seats. Water cement, limestone, marle and gypsum here abound. Onondaga Castle is in this t. 3 ms. s. of the Hollow. Pop. 1830, 5,668.

Onondagas, Indians, Onondaga co. N. Y. These are the remains of the first of the Five Nations of Indians, formerly powerful, but now reduced to a small number. They have a reservation in Onondaga t., 2½ ms. by 5; 3 ms. e. of the Hollow, on which a portion of them reside. They receive an annual payment from the state in money—the interest of the money for which they formerly sold their lands in this and several adjacent cos.

Onslow, an extensive bay of the U. S. on the Atlantic coast, sweeping in a great elliptic curve from cape Fear to cape Look Out. The chord line from cape to cape is, by calculation, 111 statute ms., and deflects from the meridian by an angle of 65° 50′ very nearly. A chain of long, narrow and low sand isls., inflecting, and within from 1 to 2 ms. from the coast, stretch along the curve of Onslow bay. The main shore is also low, and but little broken by either bays or water courses. The inlets between the isls. are numerous, but none admit large vessels. In lat. this bay extends from 33° 54′ to 34° 34′, and in long. from 1° w. to 0° 25′ e. W. C.

Onslow, co. N. C. bounded by New Hanover s. w., Duplin n. w., Jones n. and n. e., Cartaret e., and Onslow bay s. e. and s. Length 40 ms., mean width 18, and area 720 sq. ms. Extending in lat. from 34° 30′ to 35°, long. from 0° 13′ to 0° 40′ w. W. C. The very slight declivity of Onslow is a little e. of s. and drained by New river and Whittock r. Pop. 1820, 7,018; 1830, 7,814.

Ontario, lake, one of the great chain of N. American lakes, lying between N. Y. and Upper Canada. Extends from lat. 43° to 44° and from 0° 40′ e. to 2° 50′ w. long. from W. C. It has the general form of a flat ellipse, with its ends e. and w., and is about 190 ms. long, with an average breadth of 40 ms. The chief supply of its waters is received by Niagara r. which forms part of the w. boundary of N. Y., and after having brought all surplus of the lakes above, down the greatest cataract in the world, pours it into lake Ontario on its s. shore, a little e. of its w. extremity. Its other principal tributaries are the Genesee, Oswego, and Black rs. which flow from N. Y., and there are numerous small streams from both sides. The principal bays on the N. Y. shore are Chaumont e., and Gerondiquot and Braddock's; and the isls., Stony and Grenadier's isls., with Wolfe or Grand isl. at the mouth of the St. Lawrence. This lake is very deep, by some supposed to be 500 ft. It is never closed over with ice. It yields excellent fish. The shores are generally rather low, and in some places marshy, though in others quite elevated, and the land is commonly good. The surface is 334 ft. lower than that of lake Erie. There is a remarkable ridge of land, called the alluvial way, extending in a course generally parallel with the s. shore, from 7 to 10 ms. distant from it in the state of N. Y., which has the appearance of having been thrown up in some manner by the waves, or some current of the lake, when it may have overspread a much greater surface of country. An excellent road is laid out upon the summit of this narrow ridge. The surface of the lake is 231 ft. higher than the tide level of the Hudson at Albany. All this body of water discharges its surplus at its e. extremity into the St. Lawrence r. which is broad and rendered unnavigable, except in boats, by isls. and rapids, for some distance. It is owing to this that most of the commerce of the lake finds its way to N. York. Canals, large enough for schooners of the largest size used on the lakes, have just been completed by the British government, round the falls of Niagara, and those of the St. Lawrence below lake Ontario. The Oswego canal has also been recently constructed in the state of N. Y. from the Erie canal to the mouth of Oswego r.

Ontario, co. N. Y., bounded by Wayne co. n., Seneca co. e., Yates and Steuben cos. s., Livingston co. w., and Monroe co. n. w. The s. part is hilly, and the other parts generally agreeably varied, with a variety of soil, and much good, rich argillaceous loam. Canandaigua lake lies near the middle, and partly on the s. e. boundary; Honeoye and Skeneateles lakes are w. of it, and Hemlock l. is on the w. boundary. The principal streams are Honeoye, Mud, and Flint crs., and Canandaigua outlet, besides West r., flowing into the head of Canandaigua lake, and several other streams—all flowing nrthrdly. There are 13 towns, and manufactories of iron, cotton, woollen, glass, &c. in this co. Pop. 1820, 35,312; 1830, 40,167.

ONTARIO, p-t. Wayne co. N. Y., 208 ms. N. N. W. Albany, 17 N. W. Lyons, s. lake Ontario, is crossed by several small streams running N., and contains a bed of iron ore, and several forges. Pop. 1830, 1,585.

OOSTENALAH r., Ten. and Geo., rises in the former at lat. 35° 05', and is the most nrthrn. fountain, the water of which is conveyed into the gulf of Mexico E. from the Miss. Yet a creek under the name of Connesauga, this stream enters the Cherokee country of Geo., and flowing S. S. W. by comparative courses 60 ms. to New Echota, where it unites with a large confluent, the Rocking Stone or Salequoho, from the nrthestrd. Below New Echota, the Oostenalah continues S. S. W. 25 ms., to its union with the Etowah, to form the Coosa. This stream has interlocking sources with those of Ten., Hiwassee, and Etowah.

OPELOUSAS, parish of La., and the most sthwstrn. section of the U. S., bounded N. W. by the parish of Natchitoches, N. by Rapides, N. E. by Avoyelles, E. by Atchafalaya, a river separating it from point Coupee and West Baton Rouge, S. E. by St. Martin's, the upper parish of Attakapas, S. W. by the gulf of Mexico, and W. by Sabine lake and r., separating it from Texas. The longest line that can be drawn in Opelousas, is a diagonal from the mouth of Sabine, to the nrthestrn. angle of the parish on Atchafalaya r. 16 ms., and measured in that maner, the mean width would be about 50 ms. area 8,000 square ms. Extending in lat. from 29° 26' to 30° 55'. The Opelousas is a part of the great northern plain of the Gulf of Mexico, and the central and western part declines S. S. W. It is drained, advancing from E. to W. in succession by the numerous branches of Mermentou, Calcasin, and Sabine rs. The eastern but much less extensive section has a southeastern declivity, and is drained by the confluents of Atchafalaya, Teche, and Vermillion rs. Though the northern part rises into comparative hills, the whole surface so nearly approaches a level as to admit being designated literally an inclined plain. The declivity is, however, so very slight, as to admit the moderate tides of the Gulf of Mexico above the lakes of Sabine, Calcasin, and Mermentou, and when the Miss. and Atchafalaya are in full flood, the water of Courtableau is rendered stagnant to its head, at the junction of Boeuf and Crocodile creeks.

But though so nearly a curve section of the sphere, Opelousas is far from being uniform in its features. It is divisible into three very distinct physical regions. The sea marsh of the south, the immense prairies or natural meadows in the middle, and the dense forests on the N. and N. E.

The marshy gulf border rises but very little above the ocean level, is clothed with rank and coarse grass with a few clumps of trees, and extends inland from 30 to 40 miles. It is a trembling bog, and irreclaimable for any agricultural purpose, and except along the water channels impassable, perhaps by any human effort.

The prairie and marsh sections along their line of connection blend, the former imperceptibly rising above inundation. Except their monotonous surface, the prairies of Opelousas are seductive to the eye and fancy. There is an oceanic softness impressed on the mind while the vision is swept over those immense plains. The dull uniformity is broken by the lines of woods stretching along the Plaquemines, Brule, Teche, Queue Fortue, Cano, Nezpique, Calcasin and Sabine. The innumerable herds of cattle and droves of horses; the farm houses scattered upon the forest borders, and the bounding horseman give life to this extensive picture. It is here that ideas of relative space are lost in the immensity. From a slight eminence or from the roof of a high house, the surface of many counties of the eastern border of the United States comes on the foreground of the landscape.

Passing the prairies either to the nthrd. or eastward the traveller is plunged into a dense forest; but the northern and eastern forests differ greatly in their features and specific component timber. Towards Rapides and Natchitoches, pine trees so greatly prevail, and grow with so little underwood as truly to deserve their common name "*Pine Woods.*" This vast pine forest, interrupted only by Red r. spreads beyond the bounds of La.

On the eastern border of Opelousas, towards the inundated lands of Courtableau and Atchafalaya pine entirely ceases,but the forest is in an especial manner dense, with an underwood of cane, numerous species of bushes and vines. The prevailing timber, oaks of different species, hickory, different species, linden, and sassafras, sweet gum, and many others. Where actual marsh occurs, cypress and tupeloo prevail.

On the woods along the water courses which traverse the prairies, black oak, white oak, sweet gum and hickory, with an underwood of dogwood and whortleberry prevail.

The arable soil of Opelousas varies in quality to great extremes. In the eastern part of the parish on the waters of Vermillion, Teche, and Courtableau, the land is highly productive, but receding in any direction from this region the soil deteriorates. Eastward the fine arable tract is succeeded by the annually inundated soil in the waters of Mermentou and Calcasin, is thin and followed by sea marsh. To the northwestward the pine forest land is also unproductive when farming is attempted.

Cotton is greatly the prevailing staple of

Opelousas, and is followed in value aggregately by live stock. Peaches, figs, and apples, are the common orchard fruits; the latter do not, however, thrive to much advantage. Indian corn, rice, indigo, tobacco, and many other vegetables are cultivated successfully. Chief town, Opelousas or Saint Landre. Population 1820, 10,085, 1830, 12,591.

OPELOUSAS, p-v. and st. jus. parish of Opelousas, La. (*See Saint Landre.*)

OPPENHEIM, p-t. Montgomery co. N. Y. 56 ms. W. Albany, 15 W. Johnstown, N. Mohawk r., E. of E. Canada creek, is crossed by Zimmermans, Crum and Little Crum creeks, and other small streams. Pop. 1830, 3,660.

OQUAGO village, Windsor, Broome co. N. Y. 16 miles E. Binghampton, on Delaware river.

ORANGE, town, Grafton co. N. H. 16 ms. E. Dartmouth college, 40 ms. N. N. W. Concord, with 22,000 acres, contains lead and iron mines, with a pond in the S. E. on the banks of which are found a kind of yellow paint. Valuable clay is also found, and great quantities of ochre are prepared and sold annually. The surface is uneven, with Cardigan mtn. E. and the soil good. First settled 1773. Pop. 1830, 410.

ORANGE co. Vt. bounded by Caledonia co. N., Conn. river separating it from N. H. E., Windsor county S., Washington and Addison cos. W., is about half way between the N. and S. lines of the state, and contains 650 square ms. Incorporated 1781, and contains 17 towns. It has Wells r. N. E. Ompompanoosuc and Wait's rs. and streams of White and Union rs. In the N. W. is part of the E. range of the Green mountains. Granite rocks abound in the N. and middle parts, which are valuable for building and mill stones; slate is found west, and in the co. are great quantities of iron and lead ores. Pop. 1820, 24,169, 1830, 27,285.

ORANGE, p-t. Orange co. Vt. 13 ms. E. Montpelier, 50 N. Windsor, with 23,040 acres. First settled 1793; has an uneven surface, sometimes rocky, with Knox's mtn. N. E which yields abundance of excellent granite for building. The soil is better for grain than grass. Sail branch is the principal stream, and there are several smaller. Pop. 1830, 1,016.

ORANGE, p-t. Franklin co. Mass. 75 ms. W. Boston, N. Miller's r. which furnishes good mill seats. The surface is rough, and has some manufactories. Pop. 1830, 880.

ORANGE, town, New Haven co. Conn. 5 ms. S. W. New Haven, with Housatonic r. N. W. and Long Island sound S. E. It is crossed by Wopowang r. and is rough and rocky, with some good land. Pop. 1830, 1,341.

ORANGE co. N. Y. bounded by Sullivan and Ulster counties north, Hudson river east, which separates it from Putnam and Duchess counties, Rockingham southeast, New Jersey S. W., Delaware r. W., which separates it from Pa. It contains 609 square ms. and 14 townships. The east part is very mountainous, embracing the principal part of the W. highlands of the Hudson, called the Matteawan mtns. There is also much drowned land, which has a good soil for hemp, &c. when drained. Iron ore, and timber are obtained from the mtns. There is also some marble. Waalkill r. runs N. W. thro' the middle; Shawangunk r. one of its branches, bounds it on the N. W. The Delaware and Navisink, its tributaries, are W. and Murderer's creek flows E. to the Hudson. Ramapo and Sterling rs. flow into the Passaic in N. J. The Shawangunk range of mtns. crosses the N. W. corner. The mtns. in the E. part of this co. were a most important bulwark of the country in the revolutionary war. There were the forts Clinton, Montgomery, and Putnam, (the latter at W. Point) guarding the passage of the river. They were taken by the British, in 1777, who hoped to open a communication with Gen. Burgoyne, but were disappointed. These forts Gen. Arnold covenanted to betray to the enemy. The military academy of the United States, is at West Point, (which see.) The Delaware and Hudson canal enters this co. along the valley of Navisink r. and up the bank of the Del. In the village of Walden are manufactories of cotton, flour, flannel, and broadcloth. Pop. 1820, 41,213, 1830, 45,336.

ORANGE, p-t. Essex co. N. J. N. W. Newark, S. W. Bloomfield, has an academy and a mineral spring. The west part lies on the east acclivity of the Short hills. It is crossed by one or two small streams. Pop. 1830, 3,887.

ORANGE village, Orange, Essex co. N. J. is a flourishing manufacturing and populous village 4 ms. N. W. Newark, and has a bank. The inhabitants manufacture a large amount of hats, shoes, &c.

ORANGE, co. Va. bounded by Spottsylvania, E. and S. E., Louisa S., Albemarle S. W., Blue Ridge separating it from Rockingham N. W., by Conway r. separating it from Madison N., and by Rapidan r. separating it from Culpepper N. Length diagonally from E. to W. 56 ms. mean width 10, and area 560 square ms. Extending in lat. from 38° 07′ to 38° 25′, and in long. from 0° 42′ to 1° 45′ W. W. C. The northern part in its entire length is bounded by and drained into Rapidan, on the southwestern branch of Rappahannoc; declivity eastward. The southeastern angle gives source to N. Anne, and the southwestern to the extreme northern sources of Rivanna r. The surface is hilly and the co. is nearly equally divided by the southwest mtn. Much of the soil is good. Chief towns, Orange court house, Barboursville, and Stannardsville. Pop. 1820, 12,913, 1830, 14,637.

ORANGE, co. N. C. bounded S. E. by Wake, S. by Chatham, W. by Guilford, N. by Caswell and Person, and N. E. by Granville. Length 40 ms. mean width 25, and area 1,000 square ms. Extending in lat. from 35° 53′ to 36° 14′, and in long. from 1° 48, to 2° 34′. Some of the higher sources of Neuse r. rise in the northeastern part of Orange, but the central sthrn. and western sections are drained by Haw r. and its confluents. The general declivity southeastward. Chief town, Hillsboro'. Pop. 1820, 23,492, 1830, 23,908.

ORANGE, co. Indiana, bounded S. by Crawford, S. W. by Dubois, N. W. by Martin, N. by Lawrence, and E. by Washington. It is a square of 21 ms. each way, area 440 square ms. lat. 38° 35', long. W. C. 9° 30' W., slope W. giving source to Patoka r. and to Salt cr. a branch of White river. Chief town, Paoli. Pop. 1830, 7,901.

ORANGE, p-o., Trumbull co., O., by p-r. 165 ms. N. E. Columbus.

ORANGEBURGH, dist. of S. C., bounded S. E. by Charleston and Colleton districts, S. W. by South Edisto r., separating it from Barnwell, W. by Edgefield, N. W. by Lexington, N. by Congaree r., separating it from Richland, and N. E. by Santee r., separating it from Sumpter. Length 76 ms., mean width 24, and area 1824 sq. ms. Extending in lat. from 30° 10' to 30° 53', and in long. from 3° 23' to 4° 41' W. W. C. Declivity to the S. S. E. by E., and mostly drained by the two Edistos; chief t., Orangeburg. Pop. 1820, 15,653; 1830, 18, 453.

ORANGEBURGH, p-v. and st. jus., Orangeburgh dist., S. C., by p-r. 43 ms. S little E. of S. Columbia. Lat. 33° 28', long. 3° 51' W. W. C.

ORANGE, C. H. and p. o., Orange co., Va., at the foot of the S. W. mtn., 92 ms. S. W. by W. W. C.

ORANGE SPRINGS and p-o., S. E. part Orange co., Va., by p-r. 94 ms. S. W. W. C.

ORANGETOWN, t., Rockland co., N. Y., the S. E. corner of the co., 28 ms. N. N. Y., 142 S. Albany; has Tappan bay of Hudson r. E., and N. Jersey S. and W. It contains an academy. Hackensack r. flows S. into N. J., supplying useful water power. It contains the vs. of Tappan, Nyack and Middletown. Dobb's Ferry and Slote Landing are on Hudson r. Pop. 1830, 1,947.

ORANGEVILLE, p-t., Genesee co., N. Y., 22 ms. S. Batavia, gives rise to one of the head streams of Tonewanta cr., and is nearly level, with light loam, bearing beech, maple, linden, &c. Pop. 1830, 1,525.

ORANGEVILLE, p-v., eastrn. part of Columbia co., Pa., 16 ms. nrthestrd. Danville the co., t., and by p-r. 81 ms. N. N. E. Harrisburg.

OREGON, p-v., Franklin co., O.

OREGON, or Columbia, large r. of North America, included in the discoveries of Lewis and Clarke, and usually regarded as part of the domain of the U. S. We introduce the article here in order to give a general view of the country between the Chippewayan or Rocky mtns. and the Pacific ocean, and first notice the great r. from which the name has been derived, and which is slowly but probably securely regaining the Spanish name of Oregon.

Oregano, in Spanish, is the name of wild marjoram, and from that herb, or some other bearing to it a strong resemblance, it is supposed the name arose. The origin of the name Columbia is generally mistaken, as it was not derived from any connexion with the great discoverer of America; but from the Columbia Redivina, an American vessel from Boston in Mass., commanded by Capt. Robert Gray, who entered the Columbia r. on the 11th of May, 1792, and was the real re. discoverer of that important stream.

The Oregon is composed of two great constituent branches, the Clark to the N. and Lewis S. The real remote source of the nrthrn. branch of Clarks r. has not been accurately determined, but it is supposed to rise about lat. 53°, opposite to the sources of Saskatchewaine and Unjiga. Flowing sthrd. 300 ms., it joins a much superior stream from the estrd., Clarks r. proper. The latter heads in the Chippewayan range or system, interlocking sources with the various branches of Mo. lat. 45°. Flowing thence by a general N. W. by W. course, but with a very extended nrthrn. curve, joins the N. Branch as already noticed. Below the junction, the united stream assumes a S. S. W. course 120 ms., to its union with Lewis r., after an entire comparative length of 800 ms. The entire valley of Clark's r., as laid down on Tanner's N. A., embraces an area of 75,000 sq. ms.

If Lewis r. is correctly delineated on either Tanner's N. A., or on the upper margin of his U. S., it is a still superior stream to that of Clarks r.; the former rising amongst the chains of the Chippewayan system at lat. 40°, long. 30 W. W. C. Interlocking sources with those of Rio Grande del Norte of the Gulf of Mexico; the Colerado of the Gulf of Calefornia, Rio Buenaventura of the Pacific, and with those of Arkansas, Platte, and Missouri rivers; Lewis r., the main constituent of Oregon, assumes a N. W. by W. course, receiving large tributary branches from both sides, but particularly from the right. This large stream, after a comparative course of 800 ms., joins the N. branch or Clarks r. The valley of Lewis r. exceeds an area of 100,000 sq. ms. It is a remarkable circumstance that the elevated vallies from which the upper sources of Lewis r. are derived, is, following the general courses of the Rio del Norte and Columbia, almost exactly equidistant from the Gulf of Mexico, and Pacific Ocean; exceeding 1,000 ms. distance in either direction.

Combining the two vallies of Lewis r. and Clark's r., and measuring from the most nrthn. source of the latter to the extreme srthn. source of the former, the breadth of their sources is about 1,000 ms. Below their union the vast volume pursues a general wstrn. course, but with a considerable sthrn. curve 300 ms. to its final efflux into the Pacific ocean, having received the Multnomah below the last chain of mtns. which it traverses, and near the head of tide water.

In a pamphlet published in Boston in 1830, written by Hall J. Kelley, A. M., and entitled a geographical sketch of Oregon Territory, the following notices are given of the main r.: "It is six miles wide at its mouth; at the distance of 175 ms. from the sea, and near the foot of the grand rapids, it meets the tide water, and assumes a new character. The mouth of the r. is spacious and easy of ingress, affording good anchorage, and a number of safe and commodious harbors. There are flats and bars, which extend from Point Adams on the s., nearly across the entrance; but over these flats, there are no less than 20 ft. of water at low tide. The ship channel which lies snug to Cape Disappointment on the N., gives no less than 24 ft. at low water. It has been observed that the tide sets up 175 ms. It rises at Cape Disappointment, about 9 ft., and its reflux at this place, is generally in the spring 5 or 6 knots per hour. The mouth of the river has been particularly surveyed, by Capt. Nash, from whom the following directions were obtained for entering its mouth.

"Bring Chenoke point to bear N. E. by E., at any distance not less than 4 leagues, and steer for it, until Cape Disappointment bears N., then run for the estn. part of Cape D. and pass it at a quarter of a mile distance, and when the sthn. part of it bears w. half s. you may steer nearly E., keeping Chenoke Point a little open on the larboard bow; this will clear the Spit bank, and bring the ship into a fine channel of 6, 9, 12 and 13 fathoms, (should the wind be ahead, you may work up for Cape Disappointment, standing to the wstrd. until the cape bears N. N. E. half E. and to the estrd., until it bears N. half w. in a good channel,) when abreast of Chenoke Point, haul in for Point Ellice, and pass it at half cable's length, when you may bring it to bear s. half s., and steer E. by N. half N., until the Red Cliff bears N. W. half W.; then steer for the low land to the sthwrd. of Tongue Point, until two trees, which stand above the rest of the high woods to the s., are directly over the middle of three trees, that stand near the water, between two red patches; then run for them until you shut a bluff point of sand upon the river, into Tongue Point; then steer for port George, till an old white stump, or withered tree, bears s. E. by s. half s.; then haul in shore, till two trees on the high land, to the N. E. shut just on to Tongue Point, and keep them so, and you may anchor at Fort George, in 7 fathoms mud."

In the Boston pamphlet quoted above, the Multnomah is derived from lake Timpanogos, and from the Rocky mts.; but from maps drawn from actual observation, and communicated by Gen. Ashley to Mr. H. S. Tanner, that r. has a far more brief origin. It appears that sthrd. of the Columbia, extensive open plains spread to a great extent, and in which the Multnomah rises at least 300 ms. wstrd. of the Rocky mts., and pursuing a course of N. N. W. about 350 ms., falls into Columbia or Oregon, near the head of tide water.

So little more than the rough general outlines, are known of these immense regions, which are merely merging into view, that an attempt at specific information would be premature. Under the art. U. S., it will be shown that the climate of that part of N. A. w. of the Chippewayan system, is much milder on a given lat., than on or near the Atlantic coast; and the true cause of the difference will be also shown. The soil of Oregon has been boasted of as in a high degree fertile, which, if assumed as a general character, is not supported by the facts developed by actual discovery.

Oregon. This territory, taken as a physical section, is in great part confined to the basin of Columbia or Oregon r., and is bounded s. by the lat. 42°, or the nrthrn. boundary of Hispano, N. A.; N. it has the Russian territories, E. it is limited by the Chippewayan mtns., and on the w. by the Pacific ocean. Before proceeding to describe the country geographically, it may be well to glance on the international stipulations which have fixed, as far as they are fixed, the N. and s. boundaries. By the treaty of Washington, ratified by the king of Spain, on the 24th Oct., 1820, and the ratifications exchanged at Washington, the 22nd Feb., 1821, the 42°nd of lat. is made the limit between the Mexican provinces and the Oregon territory of the U. S. This closed a tedious and vexatious controversy with Spain, but two far more powerful competitors presented themselves. Great Britain and Russia severally urged their claims.

It was certainly stretching a right too far to claim Oregon as a part of La., nor was such plea at all requisite. The people of the U. S. were the original discoverers after Spain, of the Columbia or Oregon, in both directions, from the sources to the mouth. As early as 1788, some merchants of Boston fitted out two vessels, for the avowed purpose of trading along the N. W. coast of N. A. One of these, the Columbia Redivina, commanded by Capt. Robert Gray, discovered, entered, and gave the prenomen of his ship to the r. Capt. Vancouver was then at Nootka, and the discovery very frankly and fortunately communicated to that great navigator, who sent one of his principal officers to examine the channel, and in his narrative acknowledges the facts; thus placing the right of prior discovery in the U. S. beyond dispute, on British evidence. In addition to the discovery by sea, the expedition of Lewis and Clark from the Mo. was one of those transactions too notorious to admit additional evidence, and the result was a discovery and partial survey of the Oregon regions.

Some attempts have been made by individuals to induce the U. S. government to formally take possession of the Oregon, and in 1810 a private expedition, at the expense and under the direction of John Jacob Astor of N. Y., actually formed an establishment, and named the principal depot Astoria. This colony of 120 men, went out well provided for trade and agriculture. Two years after the first founding of Astoria, they had established themselves at 5 other places beside Astoria. One settlement was on the Multnomah, one at the mouth of Lewis r., and the 3 others in the interior of the country. Had the U. S. government sustained the colony sent out in 1810, it is probable all conflicting claims would long since have been quieted, or perhaps never urged.

In the convention of 1818, between the U. S. and Great Britain, the right of both parties to the country of Oregon, seems to be mutually conceded, and to embarrass the conflicting claims still more, those of Russia were brought forward, though not strenuously asserted, and were, as far as that power and the U. S. were concerned, fixed by the convention of the 5th of April, 1824. By the third article of the above noticed convention, we may consider lat. 54° 40′ as the provisional boundary of the U. S. and Russia on the Pacific ocean; and to the south, since the definitive ratification of the treaty of Washington, lat. 42° has become the determined limit. If we consider the title of the U. S. paramount, then this nation will possess on the Pacific coast 12 degrees and 40 minutes of lat., or 880 statute ms., with a mean breadth inland of 460 ms.; area exceeding 400,000 sq. ms.

The surface of this immense country as far as known, appears to be broken and mountainous. It is bounded on the E. by the vast system of the Chippewayan, and traversed in a similar direction from S. to N., at no great distance from the Pacific ocean by another system, which has hitherto remained without a generic name. The great body of the country spreads between those two systems. As laid down on Tanner's N. A. and U. S., the coast between lat. 42 and the Columbia r. seems rock bound, and very devoid of openings, bays or harbors. On the map of N. A. between lat. 42° and 43°, is placed the mouth of the r. Los Mongos, the outlet of lake Timponogos, and to the N. of its efflux is placed cape Dilligencias of Spanish, the Oxford of English geography. The wide and open bay of Cannaveral lies between lat 44° and 45°, and is terminated at lat 44° 55′ by Cape Fairweather; and that again at 45° 30′ by Cape Xelimak, and r. of the same name. The mouth of Oregon follows at lat 46° 12′. Beyond this important opening the coast is again in great part rock bound to cape Flattery, lat. 48° 30′, or to the mouth of the strait St. Juan de Fuca.

The character of the coast now entirely changes; a deep entrance of near 100 ms. separates the sthrn. extreme of the isl. of Quadra and Van Couver, from the continent. The straits of Fuca extend, a long irregular bay of 80 ms., to the sthrd., reaching to within 50 ms. of the Oregon. On the nrthrn. side the strait again under, first, the name of the Gulf of Georgia, and thence by that of Queen Charlotte's sound, merges into the Pacific ocean at lat. 51°. The isl. of Quadra and Van Couver is near 300 ms. long, in form of an ellipse, broadest part about 80; but mean breadth perhaps under 50 ms. The ocean side is represented as abounding in bays.

From lat. 42° to the straits of St. Juan de Fuca, the coast deviates only by a small angle from due N. and S.; but the isl. of Quadra and Van Couver, and the opposing coast, bears nearly N. W. With the nrthrn. termination of the isl. & entrance of Queen Charlotte's sound, the coast again bends to N. N. W., and preserves that direction to lat. 54° 40′, excessively broken by bays and isls.; the latter not very distinctly grouped, but called generically, Princess Royal, and Pitt's isls. Outside of these groups, with an intervening strait from 20 to 100 ms. wide, extends Queen Charlotte's isl., a body of land 180 ms. in length, with a mean breadth of perhaps 20 ms., and extending nearly N. and S. With the N. cape of Queen Charlotte's isl., opens Dixon's entrance or Vancouver's sound, and also terminates the U. S. territory, as fixed by treaty with Russia, 1824, as already noticed. Imperfect as is our knowledge of this coast, what we possess of the interior country, particularly N. of the Oregon r., is still greatly more defective. On Tanner's and other maps two large rs. are represented as discharging into the gulf of Georgia; the sthrn. called the Caledonia, and the nrthrn. the *Tacoutche Tesse*. The latter name is undoubtedly misapplied. It is derived from the narrative of an Indian to Du Pradt, who, about 1735, published an account of La. The Indian related to the author, that he made the traverse of the continent of N. A. to the great water towards the setting sun. He describes a river down which he accompanied other Indians to the ocean, and named it Tacoutche Tesse. From what we know of the rs. and mtns. between the Miss. and Pacific, two things must be evident to whoever reads Du Pradt; one, that the Indian did really make the journey he related; second, that the r. called Oregon, or Columbia, was his Tacoutche Tesse. The stream now so called, answers in neither course or discharge to the account given by the Indian; whilst on the contrary, in both respects, the Oregon corresponds with his description to convincing exactness.

To name the minor capes and isls. along the Oregon coast, would be to give a confused list, which the several navigators have made utterly unintelligible by imposing different names to the same place, and at the same time leaving the position uncertain.

General remarks on Oregon.—It cannot be deemed presumption to assert that the government of the U. S. has shown a dangerous neglect of this extensive country against European claims, and which of right belongs to her, upon every principle which has guided European nations in regard to discovery, preoccupancy, and of course prior right of soil. The fertility of the land has no doubt been overrated, but the undeniable mildness of the climate perhaps fully compensates for the sterility of soil. In a country comprising a surface more than double that of the whole Atlantic slope of the U. S.; and also more than double that of the valley of Ohio; and a country abounding in rivers, an immense population may be certainly supported, and if estimated at 20 to the sq. m., would amount to 8,000,000. It is probable, however, that 20,-000,000 would fall far short of the capabilities of the country. The commercial advantages of its position give again incalculable interest to Oregon; and we may without danger of successful contradiction, say that, taken in every respect, it is the most important section of the earth on which no civilized settlements have been permanently established. If we turn our eye to that human tide which is flowing wstrd., and augmenting in volume as it flows, where are we to fix its bounds? On the Pacific, will every well informed man answer.

Oriskany, cr. N. Y. a branch of Mohawk r., 23 ms. long, empties between Utica and Rome. Its principal branch is Deane's cr.

Oriskany, p-v. Whitestown, Oneida co. N. Y. 7 ms n. Utica, 3 from Whitesboro', on Oriskany cr. and the Erie canal, has a large woollen factory.

Orland, p-t. Hancock co. Me. 64 ms. e. by n. Augusta, e. Penobscot r. opposite Orphan isl., has a large pond. Pop. 1830, 975.

Orlean, p-v. Fauquier co. Va.

Orleans, co. Vt. bounded by Canada n., Essex co. e., Caledonia and Washington cos. s., Franklin co. w.; lies about half way between the e. and w. bounds of the state. Incorporated 1792. Irasburgh is the st. jus. First settled 1787. The soil is good for grain and grass. It contains much marshy ground, and more ponds than any other co. in Vt. Memphremagog lake lies partly within it. Lamoille r. is in the s., Black, Barton and Clyde rs. in the n. and middle. It is enclosed between the two ranges of the Green mtns. The rocks in the central part are argillite; in the w. mica and chlorite; slate and serpentine are found on Misisque r., with asbestos, magnetic iron, &c. Pop. 1820, 6,976, 1830, 13,980.

Orleans, p-t. Barnstable co. Mass. 85 ms. s. e. Boston, 20 e. Barnstable; is very irregular in form, lying on the ocean, which bounds it on the e. It forms the elbow of the long and crooked point called Cape Cod; and like it is low and sandy with a miserable soil, except on some parts of the neck and isl. There are many creeks and coves along the shore, and Chatham beach shuts in an extent of salt marsh, e. from Chatham harbor, which is large, and lies between this t. and Chatham. The inhabitants are scattered. The fuel, which is partly peat, is bro't from elsewhere. Pleasant Bay, contains several isls., and Stage Harbor is an opening in Chatham beach. The people are almost exclusively devoted to catching fish and clams. Pop. 1830, 1,789.

Orleans, co. N. Y. bounded by lake Ontario n., 23 ms. from Monroe co. e., Genesee co. s., Niagara co. w., is crossed by Oak Orchard cr. which flows through the w. and n. w. part, and enters the lake. Johnson's cr. and several smaller streams flow in the same direction w. of it; and Sandy cr. runs e. into Monroe co. The Ridge road on the alluvial way passes through the middle of the co. from e. to w. nearly parallel to which, at a short distance s., are the Erie canal, and the Mountain Ridge, which is near the s. boundary. The surface of the country is gently undulated, and the variations are remarkably regular, owing to the two singular ridges above mentioned. It is supposed that all the n. part was at some long past age overflown by lake Ontario; and that the alluvial way was formed by a current. The soil is various, generally favorable to grass. It contains 8 towns. Pop. 1820, 7,625; 1830, 17,732.

Orleans, t. Jefferson co. N. Y. 10 ms. n. Watertown, s. St. Lawrence r. opposite Grindstone isl., is crossed s. w. by Chaumont r. which rises near its centre, and runs 12 or 14 ms. into Chaumont bay; with good mill sites. Perch cr. s. runs 13 ms. into Black River bay. The shore of the St Lawrence is indented with many coves. The surface is varied, and the soil is clay and sand. White pine, white oak, beach, hard maple, &c. are abundant. The form is square, with a triangular tract n. on the shore. Pop. 1830, 3,091.

Orleans, parish, La. (*See New Orleans.*)

Orleans, island of. I have introduced this article in order to explain away a misconception. The name of Island of Orleans was given to that part of La. on which the city of New Orleans is situated. It is that part of the state, bounded by the Miss. r., Iberville and Amite rs., lakes Maurepas, and Pontchartrain, the Pass of the Rigolets, lakes Borgne, Chandeleur, and Bayard, and Gulf of Mexico, and stretches from the efflux of Iberville to the mouth of the Miss. 180 ms., differing in width from 3 to 25 ms. Except the narrow arable border along the Miss. and a few other places, the surface is level with the surface of high tide.

Though called an island, this part of the Delta no more deserves the term than do the spaces in other parts, which are enclosed by the numerous bayous, lakes and lagoons, which chequer the whole surface from the mouth of the Pearl to that of Sabine r.

Orleans, p-v. nrthestrn. part of Orange co. Ind. by p-r. 8 ms. n. Paoli, the co. st., and 86 ms. a little w. of s. Vandalia.

Orono, or Indian Old Town, p-t. and isl. Penobscot co. Me. 73 ms. n. e. Augusta, is an isl. in Penobscot r. 12 ms. n. n. e. Bangor, above the Great Falls. It is inhabited by the remains of the Penobscot Indians, who have a Catholic church. Pop. 1830, 1,472.

Orrington, p-t. Penobscot co. Me. 74 ms. n. n. e. Augusta, 5 s. Bangor, 32 n. Castine, e. Penobscot r., indents the n. line of Hancock co. and is separated from the n. e. corner of Waldo co. by Penobscot r. It is crossed by a pond and stream flowing n. w. into that stream. Pop. 1830, 1,234.

Orton, mills and p-o. Brunswick co. N. C. by p-r. 167 ms. s. Raleigh.

Orville, p-v. Manlius, Onondaga co. N. Y. 5 ms. s. w. Manlius, is on a branch of the Erie canal and on Butternut cr.

Orwell, p-t. Rutland co. Vt. 20 ms. n. w. Rutland, 47 s. w. Montpelier, 47 s. Burlington and opposite Ticonderoga, N. Y. It contains 49 sq. ms. including Mt. Independence, where was a picket fort, a battery, &c. in the revolutionary war, as one of the outworks of fort Ticonderoga. The first permanent settlement was made in 1783. There are hills s., but the land is generally almost level. It is fertile and watered by East cr., and Lemonfair r. which afford mill seats. Epsom salts have been made from a spring on the lake shore n. w., and other springs are impregnated with them. Compact lime rocks contain impressions of shells, fish, &c. Zinc has also been found. The lake is here from 1 to 2 ms. wide. Pop. 1830, 1,598.

Orwell, t. Oswego co. N. Y. 139 ms. from Albany, and 12 e. Pulasky, is crossed by Salmon r. in the s., and many branches in all parts, and has pretty good land, with an irregular surface, bearing beach, maple, hemlock, &c. The rocks, which are limestone and slate, abound in petrifactions on hills and in valleys. In Richland there is a fall in Salmon r. 20 ms. from its mouth. The current is first smooth, then it runs over rapids 2 ms., and then falls almost perpendicularly 107 ft. At high water the fall is about 250 feet wide. Pop. 1830, 501.

Orwell, p-v. Bradford co. Pa. by p-r. 16 ms. n. of Towanda, and 141 n. Harrisburg.

Orwell, p-v. Ashtabula co. O. by p-r. 179 ms. n. e. Columbus.

Orwigsburg, p-v. boro' and st. jus. Schuylkill co. Pa. situated on a small branch of Schuylkill r. 7 ms. s. e. Pottsville, 29 ms. n. w. Reading and by p-r. 59 n. e. Harrisburg, lat. 40° 41', long. 0° 48' e. W. C. The site of Orwigsburg is broken, but very pleasant, and though even mountainous in appearance, the adjacent country is well cultivated. The village contains the ordinary co. buildings, with about 120 dwelling houses. It stands on the main road through Reading from Phil. to Sunbury and Northumberland. Pop. about 600.

Osage, r. of the U. S. in the great western territory of Mo., and in the state of Mo. This stream has its remote sources on the great plains between those of Grand r. of Arkansas, and the main channel of Kansas r. composed of two branches, Grand r. on the nrthn. and Osage proper on the sthrn. Both branches assume an eatrn. course, and entering the wstrn. side of the state of Mo. unite after a respective comparative course of 150 ms. Below the junction, the united waters continue eatrd. about 50 ms., first s. e. thence n. e., fall into Mo. r. 12 ms. below the city of Jefferson and 108 ms. w. by the land road from St. Louis. The entire valley of the Osage, is about 250 ms. in length, extending very nearly from e. to w. In its broadest part it is 130 ms., but the mean width not above 60 ms., area 15,000 sq. ms. The lower part of the channel below the junction of the two main branches is excessively crooked, a feature which has contributed to magnify the real size of the stream. Lat. 38 divides this r. valley into two not very unequal sections. The valley lies between Lower Kansas and Mo. r. n., Gasconade s. e., the nrthn. sources of White r. s., and those of Grand r. of Arkansas s. w. It is a fine navigable stream as high as the main fork, and along its shores, and on many of its tributaries spread some of the finest lands of Mo.

Osage, ferry and p-o. nrthwstrn. part of Gasconade co. Mo. The ferry is over Osage r. at its mouth, 12 ms. below Jefferson, and the post office is at the ferry.

Osnaburgh, p-v. Stark co. O. 4 ms. e. Canton, the co. st. and 107 n. e. Columbus.

Ossabau, sound and isl. Bryan co. Geo. The sound is in fact the mouth or estuary of Great Ogeechee r., and opens into the Atlantic ocean 18 ms. s. w. the mouth of Savannah r., lat. 31° 50', long. W. C. 40° 8' w. Ossabau isl. is one of those level small isls. which extend in a chain along the Atlantic coast of Florida, Geo. and S. C. The Ossabau fills the space between Ossabau and St. Catharine's sounds, and is about 10 ms. in length.

Ossian, p-t. Alleghany co. N. Y. 20 ms. n. e. Angelica, is crossed by Canaseraga cr. which affords mill sites. It is in the n. e. corner of the co. Pop. 1830, 812.

Ossipee Lake, Strafford co. N. H., is nearly of an oval form, and covers about 7,000 acres. The water is clear, and it contains no islands. It discharges its surplus water by Ossipee r. through several ponds into Saco r. in Me.

Ossipee mtns., Strafford co. N.H. lie along the e. side of Winnipiseogee lake about 8 ms. The ridge is broken in some parts, but nearly uniform in height, and of considerable elevation. The slope towards the lake is gradual near it, where the land is pretty well cultivated. These mtns. form one of the most conspicuous features in the fine scenery of that region. In easterly storms the wind sometimes comes over the mtns. with destructive violence.

Ossipee, p-t. Strafford co. N. H. 60 ms. n. e. Concord, has part of the range of Ossipee mtns. n. w. and part of Ossipee lake and r. Its form is irregular, and the e. angle almost touches the boundary of Me. Pine r. crosses the e. part, and Bearcamp r. is in the n. w. a tributary of the lake. There are several ponds. A mound of earth 10 feet high, w. of the lake, was found to contain skeletons, tomahawks, &c. Incorporated 1785.

Oswegatchie r. N. Y., lies chiefly in St. Lawrence co. and has 2 branches which, after rising near the s. boundary of the co. and flowing n. unite about 4 ms. from their mouth in St. Lawrence r. The e. branch pursues a remarkably crooked course; running w. over the line of

Jefferson co., then turning back almost into the same channel, flows N. E., and then more W. till it meets the W. branch or Indian r. It has many falls and rapids favorable to manufacturing, and passes through a fertile country. It is about 120 ms. long. At one of its angles is a natural canal communicating with Grassy r. The W. branch flows from Jefferson co. into Black lake, or Oswegatchie lake, on leaving which it joins the E. branch.

OSWEGATCHIE lake, St. Lawrence co. N.Y. receives Indian r. or the W. branch of Oswegatchie r., and a great part of the course of that stream lies through it. It contains several isls., and is chiefly in Morristown. It lies N. E. and S. W., and parallel with St. Lawrence r.

OSWEGATCHIE, p-t. and st. jus. St. Lawrence co. N. Y. 204 ms. N. N. W. Albany, 476 from Washington, lies on the St. Lawrence r. is crossed by Oswegatchie r., and watered by its two branches for a few ms. before their junction. It is opposite Prescott, Upper Canada, and at the end of sloop and steamboat navigation; being a little above the rapids. It is 10 ms. sq. with a varied surface and pretty good soil. Oswegatchie r. is navigable and supplies many valuable factories. An old fort stood at its mouth, built by the French. Ogdensburg, a port of entry, and a large and flourishing village, the st. jus. of the co. is in this t. Fort Van Rensselaer was built here in the last war. Pop. 1830, 3,993.

OSWEGO, r. N. Y., the outlet of the lakes in the middle of the state, flows into lake Ontario near the middle of the S. shore, at Oswego village. The lakes with which it is connected are Fish, Cross, Otisco, Onondaga, Owasco, Skeneateles, Canandaigua, Crooked, Oneida, Cayuga, and Seneca. The communication is by outlets, which unite and form Seneca r., and this soon falls into Oswego r., as do many other small streams. These waters abound in mill sites. It is 24 ms. long, runs N., and serves a very important purpose for the greater part of its course, as the Oswego canal, which connects Erie canal at Syracuse with lake Ontario. 12 ms. from the mouth of the r. is a fall of about 100 feet, round which was formerly a portage of 1 m. for boats. The canal is conducted along the E. bank below the falls, to the village near the mouth of the r. where it descends to the level of the harbor by locks, and affords abundance of water power to many factories. It is crossed by a bridge 700 feet long. It served in early times as the great channel of intercourse between the English and French in Canada, and the Six Nations of Indians, and was a route of invasion in the revolutionary war. The harbor at the river's mouth, has been greatly improved by the government of the U. S., and has much navigation. It is one of the places regularly visited by the lake steamboats.

OSWEGO, co. N.Y. bounded by lake Ontario and Jefferson, Lewis and Oneida cos. E., Madison, Onondaga and Cayuga cos. S. and Cayuga co. W., Oneida lake and r. forming the line on the S. It contains 900 sq. ms. and 18 towns. It is watered by numerous streams, flowing in all directions, the principal of which, after that mentioned, is Oswego r. Salmon r. crosses from the E. line to lake Ontario. Scriba and Bay creeks run S. into Oneida lake, Scott's cr. into Oneida r., Black cr. W. into Oswego r. and Catfish, Little Salmon, Grindstone and Deer crs. N. and W. into lake Ontario. N. E. are some hills, but the co. is nearly level, with good grazing land, some soils favorable to grain, and remarkably adapted to fruit trees. Oswego and Pulaski are the co. towns. The co. contains salt springs and freestone quarries. There are old forts at the mouth of Oswego r. by which was the communication between the Canadians and the Five Nations of Indians before the French wars. Fort Oswego is in this co. as is part of the village of Oswego, the st. of jus. of the co. and a port of entry. The canal from the falls lies along the E. shore, and many manufactories have been erected there. The construction of the Oswego canal affords a navigable connection between lake Ontario and the Erie canal. Population in 1820, 12,374; 1830, 27,119.

OSWEGO, p-t. Oswego co. N. Y. 168 ms. N. W. Albany, 72 N. N. W. Utica, S. lake Ontario, W. Oswego r., is level in some parts, gently varied in others, with soil favorable for grain, grass and fruit, and contains part of the v. of Oswego, the other part being in Scriba. It is at the mouth of Oswego r. and enjoys great advantages as a manufacturing place, from an excellent lake harbor, (being defended from the waves by two long piers built by the government of the U. S.) and from the Oswego canal, which connects the navigation of lake Ontario and the Erie canal. A trading house was built here, 1772, at the mouth of Oswego r., and fort Oswego in 1727, 50 ft. above the level of the water. In 1755 it was enlarged, and fort Oswego was built on the opposite side, in Scriba, on much higher ground. They were captured by Gen. Montcalm, from the English, in 1756, with stores, arms, boats, and ammunition, but abandoned. In 1814, fort Ontario was taken by the British, but evacuated the next day. Pop. 1830, 2,703.

OSWEGO, incorporated v. and port, in the tps. of Oswego and Scriba, st. jus. Oswego co. N.Y. 167 ms. N. W. by W. Albany, is a flourishing place at the mouth of Oswego r. on high ground near the sites of the old forts Oswego and Ontario. It has a good harbor, with 10 feet water on the bar, and is protected from the waves of lake Ontario by two long piers, built by the U. S. A bridge 700 ft. long connects the two banks of the r. Oswego canal here terminates, and communicates with the river, on a level with the lake, by locks on the E. side, affording water for many manufactories. It is carried along the E. bank to the falls, 12 ms. above which it enters the r. Boats pass by this channel to Syracuse, on the Erie canal; and it is the only navigable communication between that canal and the lake. This work has given great importance to the v., which has increased very rapidly within three or four years.

Otego, p-t. Otsego co. N. Y., 86 ms. w. by s. Albany, 20 s. w. Cooperstown, and n. w. Susquehannah r. or Delaware co.; has large and rich meadows on that r., with hills elsewhere, and pretty good soil. It is crossed by Otego creek, which affords fine mill seats, Otsaawa creek and other streams. The timber is valuable, and rafts are sent down the Susquehannah r. Pop. 1830, 1,148.

Otego, r. Otsego co. N. Y., rises near the centre of the co. in 2 branches, near Oak cr., runs s. parallel to it about 28 ms., supplies good mill sites, and falls into Susquehannah r. a little above Huntsville.

Otis, p-t. Berkshire co. Mass., 116 ms. w. Boston, 20 s. e. Lenox; is on high ground, between Farmington and Westfield rs., and has several streams and ponds flowing into them. The surface is uneven. Pop. 1830, 1,012.

Otisco, lake, or pond, Onondaga co. N. Y. near the s. w. corner of the co., 1 mile by 4, and parallel to Skeneateles lake; receives several brooks, and discharges through Otisco creek into Onondaga lake. Its Indian name means "waters much dried away."

Otisco, creek, or Nine-Mile creek, Onondaga co. N. Y., flows from Otisco lake northeastrdly. in a curving course, into the middle of Onondaga lake. It is 15 ms. long, and supplies mill seats.

Otisco, p-t. Onondaga co. N. Y., 134 ms. w. Albany, 8 s. s. w. Onondaga, and 50 from Utica. It is 5½ ms. long n. and s., contains 30 sq. ms.; has high land, sloping w. to Otisco lake, and its inlet and outlet, which bound it s. w. and e., towards Onondaga creek, to which it sends a tributary. The soil is moist and warm, the mill sites are good, and there is some limestone in the n. e. The timber is oak, nutwood, tulip, &c. Pop. 1830, 1,938.

Otisfield, p-t. Cumberland co. Me., 82 ms. s. w. Augusta, 40 n. Portland, s. w. of Oxford co.; has crooked r. e., and part of a large ponds. Pop. 1830, 1,274.

Otsego, lake, Otsego co. N. Y., is a small but pure and picturesque sheet of water, 66 ms. w. Albany, lying n. and s. near the n. line of the co., abounding in salmon trout at all seasons, and giving rise to one of the north streams of Susquehannah r. It is from ½ m. to 3 ms. by 9; and its outlet, on which stands the pleasant village of Cooperstown, affords water power to several large manufactories.

Otsego, co. N. Y., is bounded by Herkimer, Oneida, and Montgomery cos. n., Schoharie co. e., Delaware co. s., Unadilla river, dividing it from Chenango and Madison cos. w. It is of irregular form, with 935 sq. ms. It is 66 ms. w. Albany, with 22 towns, and supplies the principal sources of the n. branch of Susquehannah r. Otsego and Schuyler's lakes, near the n. line, flow sthwstly., as does Charlotte r. in the s. e., and Scheneoas, Cherry Valley, Otego, Otsaawa, Butternut, and Wharton creeks. Branches of Unadilla cr. are on the w. line, which is a large tributary of the Susquehannah. A few brooks in the n. flow into Mohawk r. The co. is elevated and hilly, being crossed by the Susquehannah hills and Kaatsberg range, which passes to the Hudson at Little Falls. The soil is various, and most favorable for grass. Cooperstown, a pleasant village, and the st. jus. of the co., is at the outlet of Otsego lake. Iron ore is found in some places, limestone near Schuyler's lake, marble in Cherry Valley, and sand stone n. There are many cotton factories on the outlet of Otsego lake and Oak cr., the outlet of Schuyler's cr., altogether containing about 8,000 spindles. On Butternut creek are 2 cotton factories with 5,200 spindles, and a woollen factory, besides others of different kinds at Hartwick, &c. &c. Pop. 1820, 44,856, 1830, 51,372.

Otsego, t. Otsego co. N. Y. Pop. 1830, 4,363.

Otselic, p-t. Chenango co. N. Y., 106 ms. w. Albany, 17 n. w. Norwich, and s. of Madison co. It is square, has good soil, pleasantly varied, and is crossed from n. e. to s. w. by Otselic creek. Pop. 1830, 1,236.

Otselic, creek, N. Y., rises in the s. w. part of Madison co., crosses part of Chenango, and enters Tioughnioga creek in Broome co., after a course of 43 ms.

Otsquago, creek, (Osquaga or Otsquaga,) rises in Hamilton co., flows e. into Montgomery co., and enters Mohawk r. at Fort Plain. It has a course of about 23 ms. and affords good mill sites.

Ottawa, co. Michigan, bounded by Allegan co. s., Kent s. e. and e., Oceana n. e. and n., and lake Michigan w. Length from s. to n. 38 ms., mean breadth 18, and area 684 sq. ms. Lat. 43° and long. W. C. 9° w. intersect in this co., about 170 ms. n. w. by w. Detroit.

Otter Bridge, and p-o. Bedford co. Va., 17 ms. s. w. by w. Lynchburg, and 215 ms. s. w. W. C.

Otter Creek, Vt., the largest stream in the state, whose entire course is within its bounds, rises near Bennington co. and flows s. into it, then n. through Rutland co. and empties into lake Champlain near the line of Chittenden co., just above Vergennes. It is about 90 ms. long, and, with its branches, waters 900 sq. ms. The first part of its course, to Middlebury, is smooth; thence to Pittsford 25 ms., it is navigable in boats; and has valuable falls at Middlebury, Weybridge, and Vergennes, where are manufactories. It is navigable for the largest lake vessels for 8 ms. from Vergennes to the lake. In some parts this r. flows thro' fine meadows of great fertility. Its tributaries are numerous; principally Lemonfair, Little West, Mill, New Haven, Leicester, Furnace, Cold rs. &c.

Otter Peaks, the highest part of the Appalachian system, sthwst. of the Delaware. The Peaks of Otter are parts of the Blue Ridge, rising to a height above the Atlantic ocean of 4,260 feet. They are situated 30 ms. by the road from Lynchburg, and between Bedford and Botetourt cos. Va. Direction from Lynchburg a very little N. of due west.

Otto, t. Cattaraugus co. N. Y., 10 ms. N. W. Ellicottville, and S. of Cattaraugus creek. Pop. 1830, 1,224.

Ottsville, p-v. northern part of Bucks co. Pa. by p-r. 38 ms. N. Phil., and 14 in a similar direction from Doylestown.

Ouachita. (*See Washitaw.*)

Oury's p-o. Hamilton co. O. by p-r. 127 ms. S. W. Columbus.

Overall's p-o. Shenandoah co. Va. by p-r. 91 ms. wstrd. W. C.

Overslaugh, a shoal and crooked channel in Hudson r. in Bethlehem, 3 ms. S. Albany; has caused much difficulty to the navigation in sloops and steamboats in times past.

Overton, co. of Ten. bounded by Morgan E., Cumberland mtn. which separates it from Fentress S., Jackson W. and Cumberland co. Ky. N. Length 32, mean width 28, and area 896 sq. ms. Extending in lat. from 36° 10′ to 36° 36′. Cumberland r. in a southwestern direction traverses the northwestern angle of Overton, receiving in that part of its course Obies r., the different branches of which latter stream drain the far greater part of the co.; declivity a little N. of W. Chief town, Monroe. Pop. 1820, 7,128, 1830, 8,242.

Ovid, p-t. and one of the sts. jus. Seneca co. N. Y. 205 ms. W. Albany, 18 S. Waterloo, between Cayuga and Seneca lakes, about 9 ms. E. and W., and nearly 5 N. and S.; has a varied surface, well cultivated, with good soil, and many small mill streams. Ovid v. near the middle of the N. line, is on an eminence descending E. and W. and overlooking both lakes. There is much good wheat land on the shores. Pop. 1830, 2,756.

Owasco lake, Cayuga co. N. Y., near the middle of the co., from 1 to 2 ms. wide, by 11 long, receives Owasco inlet from the S., and discharges Owasco outlet S. It lies nearly equidistant between Cayuga and Skeneateles lakes, and lies nearly N. and S. like almost all the interior lakes of N. Y. It is said to owe its name (which in the native language signifies a bridge,) from a raft formed of timber near the outlet.

Owasco Inlet, N. Y., enters Owasco lake, Cayuga co. in Sempronius.

Owasco Outlet, Cayuga co. N. Y., flows N. from the N. end of Owasco lake, 15 ms. to Seneca r.

Owasco, p-t. Cayuga co. N. Y. 164 ms. W. Albany, 5 S. E. Auburn v., and W. Onondaga co.; lies S. W. on the E. side of Owasco lake, and a short distance on the outlet, has very rich land, and a few mill sites on a small stream. Pop. 1830, 1,350.

Owego, p-t. and half shire, Tioga co. N. Y. 170 ms. W. S. W. Albany, N. Pa. line, E. Owego creek and Susquehannah r., and W. Broome co.; is crossed by Susquehannah r., 7 ms. by 15, has a varied surface and soil, with white pine timber near the river, maple, beech, &c. on the hills. It is favorable to fruit. First settled about 1690. Owego v. is on Susquehannah r. 1½ ms. from Owego creek, 29 S. S. E. Ithaca. Lumber, salt and gypsum are sent to Baltimore by the Susquehannah. Pop. 1830, 3,026.

Owen, co. Ky. bounded by Gallatin N., Grant N. E., Harrison E., Scott S. E., Franklin S., and Ky. r. separating it from Henry W. Length 20 ms., mean breadth 16, and area 320 sq. ms. Extending in lat. from 38° 22′ to 38° 42′, and in long. from 7° 33′ to 8° W. W. C. Though bounding on Ky. r. the body of the co. is drained by Eagle creek, which, rising into Scott and Harrison, flows N. N. W. between Grant and Owen, and thence bending abruptly to wstrd. forms the boundary between Gallatin and Owen, finally falls into Ky. r. Chief t. Owenton. Pop. 1820, 2,031, 1830, 5,786.

Owensboro, p-o., t. and st. jus. Daviess co. Ky. on the left bank of Ohio r., 76 ms. W. Elizabethtown in Hardin co., and 151 ms. a little S. of W. Frankfort. Lat. 37° 48′, long. 10° 09′ W. W. C. Pop. 1830, 229.

Owensville, p-v. Gibson co. Ind. by p-r. 151 ms. S. W. Indianopolis.

Owenton, p-v. and st. jus. Owen co. Ky. about 25 ms. N. N. E. Frankfort. Lat. 38° 30′, long. 7° 42′ W. W. C. Pop. 1830, 143.

Owingsville, p-v. and st. jus. Bath co. Ky. by p-r. 70 ms. E. Frankfort. Lat. 38° 10′, long. 6° 44′ W. W. C. Pop. 1830, 241.

Owl's Head, Thomaston, Lincoln co. Me. the W. cape of the inner part of Penobscot bay, runs S. E. and forms Clam cove. Opposite are Fox isls., and outside of it several others.

Oxbow, a bend in Conn. r. at Newbury, Vt. enclosing 450 acres of fine meadow.

Oxford, co. Me. bounded by the dist. of Three Rivers in Lower Canada N. E., Somerset and Kennebec cos. E., Cumberland and York cos. S., New Hampshire W. It is of an irregular form, elonged N. and S. The surface is rough, and in many places mountainous. Moose and Dead rs. rise here, and flow E. into the Kennebec. Androscoggin and Magolloway, a branch, rise in the N. part. The main stream flows through a chain of large lakes, the last of which is Umbagog, on the line of New Hampshire. The river returns into the co. in the S. part, crosses it, and enters Kennebec co. The small streams are numerous. Saco r. crosses the S. W. corner. The st. jus. is Paris. Pop. 1820, 17,630, 1830, 35,211.

Oxford, t. Oxford co. Me. Pop. 1830, 1,116.

Oxford, p-t. Grafton co. N. H. 17 ms. N. Hanover, 10 S. Haverhill, 60 from Concord, 120 from Boston, with 27,000 acres, E. Conn. r., over which is a bridge. The soil is generally fertile, especially the meadows on Conn. r. Mounts Cuba and Sunday are near the centre. There are 4 or 5 ponds, some of which flow into Conn. r., and others into the Merrimack. Coarse primitive limestone

abounds at the foot of a mountain. Building-granite, soap stone, and lead ore are also found. The village is on a street, in a beautiful valley 1 m. by 5 or 6 ms. long, enclosed by hills, which approach each other very nearly in the middle. A social library was incorporated 1797. First settled 1765. Pop. 1830, 1,829.

OXFORD, p-t., Worcester co., Mass., 55 ms. s. w. Boston, 12 s. Worcester; is divided by Stony or French r. the upper part of Quinebaug r. which flows s. into Connecticut. It affords good mill sites, which are occupied by 7 factories. Here is a large thread factory, and several mills are soon to be erected. In 1686, a French colony settled this t. It consisted of Protestants, who left France on the repeal of the edict of Nantes. Gov. Dudley obtained a grant of land here 8 ms. sq., for this purpose. In the E. part of the t. on a hill, are the remains of their principal fort, which had bastions and a well; and their grapes, currants and asparagus, still grow there. In 1696 an Indian incursion broke up the settlement, and the colonists retired to Boston, where they had a church for some years. A few of them afterwards returned, the place being reoccupied in 1713. Pop. 1830, 2,034.

OXFORD, p-t., New Haven co., Conn., 40 ms. s. w. Hartford, 14 N. w. New Haven, E. Housatonic r., 5 ms. by 8, with 38 sq. ms.; is uneven, with gravelly loam, calcareous w., and generally productive. The trees are chiefly nut. Naugatuck r. and other streams water the tsp., and there are several mills and factories. Pop. 1830, 1,763.

OXFORD, p-t., Chenango co., N. Y., 10 ms. s. w. Norwich, 108 from Albany, 56 from Utica, has good land, crossed by Chenango r., (200 feet wide,) and other streams, which supply mill seats. The soil is good. The remains of an old fort are seen on a high bank of Chenango r., with a ditch 3 feet deep, enclosing about an acre. In 1788 large trees stood on the ground. Pop. 1830 2,943.

OXFORD, an incorporated v., Oxford, Chenango co., N. Y., 108 ms. s. s. w. Albany, 236 N. w. N. Y., 110 w. Catskill, 56 s. by w. Utica; is pleasantly situated on the meadows of Chenango r., (about 1,200 yds. wide,) with handsome swells at a little distance. There is an academy, &c.

OXFORD, p-v., Blooming Grove, Orange co., N. Y., 12 ms. s. w. Hudson, w. West Point, has an academy.

OXFORD, t., Warren co., N. J., E. Delaware r., opposite Northampton co., Pa., is crossed by Pequest cr., and Beaver cr. its branch. It contains Belvidere v., the st. jus. of the co. Pop. 1830, 3,665.

OXFORD FURNACE, v. Warren co., N. J., 5 ms. E. Belvidere.

OXFORD, p-v., sthwstrn. part of Chester co., Pa., 50 ms. s. w. by w. Phil., and by p-r. 92 ms. N. E. W. C.

OXFORD, v. Talbot co., Md., on the estrn. or left side of Tread Haven bay. It is a port of entry and one of the most trading places on the eastern shore of Maryland.

OXFORD, p-v. and st. jus., Granville co., N. C., by p-r. 47 ms. N. Raleigh, lat 36° 20′, long. 1° 40′ w. W. C.

OXFORD, p-v., N. w. part Butler co., O., by p-r. 110 ms s. w. by w. Columbus. Pop. 1830, 737.

OYSTER, r., N. H., flows into Great Bay, through Durham.

OYSTER BAY, p-t., Queen's co., N. Y., Long Isl., 28 ms. E. N. Y., 172 s. Albany, s. Long Isl. sound, N. Atlantic O., has a variety of soil, pretty level surface, and West, Fort & Unkway necks on its s. coast, which extends only 3 ms., and has Jones's inlet opposite, which is a channel through the beaches which here line the coast. At the N. w. corner of the t. is Hempstead harbor, and N. Oyster Bay harbor, a large square sheet of water communicating E. with Cold Spring harbor, which extends along the N. E. boundary. There are several vs. in different parts of this extensive t.; Musqueto, Oyster Bay, Norwich and Wolver Hollow N., Wheatly, and Jerico, and Cold Spring in the middle. Oak and Cove necks extend along Oyster Bay harbor. On Fort Neck were 2 Indian forts at the 1st settlement of the t. The remains of 1 are 30 yards square. Pop. 1830, 5,348.

OYSTER BAY, p-v., Oyster Bay, Suffolk co., N. Y., on the s. w. corner of Oyster Bay, is resorted to in summer for fish, &c.

P.

PACIFIC OCEAN. If we regard the Oregon Territory as appertaining to the U. S., the Pacific ocean ought to be named as one of the great boundaries, and demands a notice with the same propriety as does the Atlantic. Under the art. Oregon, we have already stated that the respective treaties with Spain and Russia, give the U. S. the sovereignty along the Pacific ocean from lat. 42° to 54° 40′, or equal to 880 statute ms. in round numbers. A general view of this coast has already been given under the head of Oregon, and under that of the U. S. The phenomena of the prevailing winds as far as known, will be discussed. We may in this place observe, that in regard to prevailing winds, those of the two bounding oceans of the U. S. present a directly contrary excess. Along the Atlantic from 6 to 7 in 10 of the winds are from the wstrd., of course towards the ocean; on the Pacific coast the prevailing winds are also from the wstrd.; therefore, from the ocean towards the land. This wstrn. current of the winds, which, as will be shown, is an established effect of the laws of nature, must have a most powerful effect on the navigation of the two coasts. The average time of voyages from the U. S. to and from Europe, is about as 21 is to 40, and vice versa. In a numerous series of voyages from the U. S. to Europe, if it demands 21 days, it will demand 40 days to return. From these ascertained comparative elements, it must be evident, that the departure of vessels from the Atlantic coast of the U. S. is as 40 to 21, to the facility of approach; and it must be equally evident, that the very reverse will be the case on the Pacific coast. If indeed we compare the relative width of the two oceans, in the direction of the winds, we might risk the theory, that the proportions will be greater on the Pacific than on the Atlantic coast.

The breadth of the Atlantic ocean along N. lat. 40° between the wstrn. coast of Spain and the estrn. of the U. S., is about equal to 60 degrees of long., whilst the Pacific ocean along the same line of lat. from the wstrn. coast of N. A. to the estrn. coast of Asia, is equal to 105 degrees; or the breadth of the two oceans are, along the line assumed, as 3 to 5 very nearly. In bearing, however, the two oceanic coasts of N. A. are almost at right angles to each other; the Atlantic coast bearing N. E. and S. W., whilst the Pacific coast bears S. E. and N. W. From this structure of the respective coasts, it is clear, that the prevailing winds must leave the estrn. at a very different angle to their impulse on the opposite side of the continent. In art. U. S. it will be seen, that the prevailing winds, being from the land in N. A. and from the ocean on Europe, is the true cause of the difference of climate between the opposing sides of the Atlantic, and that cause once discovered and acknowledged, its application demonstrates also the cause why Oregon, between lat 42° and 54° 40′, has a climate approaching in temperature to that along the Atlantic coast, lat. 35° and 45°. It is remarked in the narative of the passage of Lewis and Clark over the continent of N. A. to the Pacific, that the name was far from appropriate at the mouth of Columbia, and we may at once perceive that from the prevailing wstrn. winds, and the immense body of ocean water, that the wstrn. coast of N. A. between N. lat. 42° and 55°, must be a truly sea-beat shore. (*See Art. U. S.*)

PACOLET, r., N. and S. C., rises from the spurs of Blue ridge and in the sthwstrn. part of Rutherford co., of the former state. Flowing thence estrd. 15 ms., the stream bends to S. E. by E. over Spartanburg and Union dists., falls into Broad r. nearly opposite the S. W. angle of York dist., after a comparative course of 60 ms. This stream heads opposite the sources of French Broad cr., and its valley lies between those of Ennoree and Broad rivers.

PACTOLUS, p-v. on the South Fork of Holston r., and in the wstrn. part of Sullivan co., Tenn., about 80 ms. N. E. by E. Knoxville, and by p-r. 268 ms. a little N. of E. Nashville.

PADDYTOWN, p-v., Hampshire co., Va., on Potomac r., 20 ms. by land road above Cumberland in Md., and by p-r. 135 ms. N. W. by W. W. C.

PADUCAH, p-v., estrn. part of McCracken, Ky., by p-r. 19 ms. estrd. Wilmington, the co. st., and 245 ms. S. W. by W. Frankfort.

PAGE, co., Va., bounded S. by Rockingham, W. Shenandoah, and N. Frederick; Blue Ridge separating it from Culpepper E., and Madison S. E.; length 34 ms., breadth 11 ms., and area 374; lat. 38° 45′, long. W. C. 1° 25′ W. The main and estrn. branch of Shenandoah r. winds to the S. E., traversing this co. in its greatest length; slope of course in the direction of its principal r. The surface is generally hilly, and the co. being bounded on two sides by mtns. gives it the appearance and reality of a rugged valley, though much of the r. soil is fertile and well adapted to farming. The co. of Page corresponds nearly to what is called E. Shenandoah in the census returns, and contained in 1830, a pop. of 8,327; chief t. Luray.

PAGE'S Mill and p-o., Gibson co., Tenn., by p-r. 150 ms. W. Nashville.

PAGESVILLE, p-v., wstrn. part Newberry dist., S. C., by p-r. 75 ms. N. W. by W. Columbia, and 30 ms. sthwstrd. Newberry, the st. just. for the dist.

PAINSVILLE. p-v., wstrn. part Amelia co., Va., 46 ms. S. W. by W. Richmond, and about a similar distance a little N. of W. Petersburg.

PAINESVILLE, p-v., Rockingham co., N. C., by p-r. 106 ms. N. W. by W. Raleigh.

PAINESVILLE, p-v. on Grand r., nrthestrn. part Geauga co., O., 4 ms. S. Fairport on Lake Erie, and by p-r. 161 ms. N. E. Columbus. Pop. of the tsp., 1830, 1,499.

PAINT CR., p-o., Floyd co., Ky., 10 ms. N. Petersburg, the co. st., and by p-r. 161 ms. S. E. by E. Frankfort.

PAINTED POST, p-t., Steuben co., N. Y., 27 ms. S. E. Bath, 234 W. by S. Albany, W. Tioga co., is crossed by Tioga r. from N. W. to S. E., just below the mouth of Conhocton r. and the canal. It is 12 ms. by 20, and had its name from an oaken post, erected, and occasionally painted rod by the Indians, in memory, it is believed, of a great warrior. The land good, and there is much rich alluvian. Locust timber is sent to market. Pop. 1830, 974.

PAINTER'S CROSS ROADS, and p-o., Del. co., Pa., by p-r. 116 ms. N. E. W. C.

PAINTED ROCK, p-v., Jackson co., Ala., by p-r. 152 ms. N. E. Tuscaloosa.

PALATINE, p-t., Montgomery co., N. Y., 10 ms. W. Johnstown, 51 N. N. W. Albany, and N. Mohawk r., is well watered and supplied with mill sites by Garoga cr., &c., and has excellent land. It was settled by Germans, 1724. Stone Arabia is a part 4 ms. from the r., declining S., and remarkably fertile. There was a small paliesaded fort here in the revolution. In 1780 it was garrisoned with 200 men, and here Col. Brown fell. The Indians used to grind their corn in a hole in a rock in the S. E. corner of the t., with a large stone. Hence

Bread cr. derived its name. Palatine bridge is a v. in this t. Pop. 1830, 2,742.

PALATINE HILL and p-o., Monongalia co., Va., 4 ms. sthestrd. Morgantown, and by p-r. 211 ms. N. W. by W. W. C.

PALATKA, v., John's co., Flor., on the left bank of St. John's r., about 85 ms. S. W. St. Augustine.

PALERMO, p-t., Waldo co., Me., 16 ms. E. Augusta, 30 N. E. Wiscasset; has Kennebec co. W. and N. W., and Lincoln co. S., and contains several large ponds, which flow S. W. into Sheepscut r. Pop. 1830, 1,257.

PALESTINE, p-v., Picken's co., Ala., by p-r., 62 ms. wstrd. Tuscaloosa.

PALESTINE, p-v., and st. jus. Crawford co., Il., situated near the right bank of Wabash r., by p-r. E. Vandalia, lat. 39° 02', long. W. C. 10° 40' W.

PALISADO ROCKS, the precipitous W. bank of Hudson r., beginning in Bergen, N. J., and extending into Rockland co., N. Y. The rock is of the trap formation, in some places lying upon red sand stone, which shows itself in horizontal layers at the water level. The height varies, but in some parts is very regular, high and smooth like a wall. The frost gradually splits off fragments, which have accumulated below, and offer an extensive quarry for an inferior kind of stone. As the shore forms many projections, the peculiar form of these precipices gives a very picturesque aspect to this part of the shore of the Hudson, and the steamboats generally pass near the base. Small streams of water sometimes pour down from the neighboring fields, and timber is sometimes slid down to the water, where are numerous little landing places. A few patches of sloping soil, and level arable land, are cultivated and inhabited.

PALMER, p-t., Hampden co., Mass., 14 ms. E. Springfield, 82 S. W. Boston, incorporated 1752, N. and W. Chickapee r.; E. Swift r., crossed by Ware r. These 3 streams unite on the W. line of the t., and afford many facilities for manufacturing. The surface is irregular, and the soil good for farms. This t. was first settled from the North of Ireland. The Three Rivers cotton & woolen manufacturing company was incorporated 1826, with a capital of a million. Population 1830, 1,237.

PALMER'S SPRINGS and p-o., wstrn. part Mecklenburg co., Va., 103 ms. S. W. Richmond.

PALMER'S TAVERN, and p-o., Prince George's co., Md., 30 ms. from W. C.

PALMERSTOWN, mtn., N. Y., rises between lakes Champlain and George, in Washington co., bounds lake George for some distance on the E., crosses Warren co. in the S. E., and enters Saratoga co. It consists of granite and gneiss, is steep, and from 200 to about 1,000 ft. high.

PALMYRA, p-t., Somerset co., Me., 51 ms. N. E Augusta, 28 E. by N. Norridgewock, 215 N. E. Boston, W. Penobscot co., and adjoining Newport, is crossed N. and S. by Sebasticook r., and several of its small streams. Pop. 1830, 902.

PALMYRA, p-v., wstrn. part Lebanon co., Pa., 14 ms. N. E. by E. Harrisburg, and 10 a little S. of W. from the borough of Lebanon.

PALMYRA, p-t., Wayne co., N. Y., 15 ms. N. Canandaigua, 220 N. N. W. Albany, is crossed by Mud cr. which runs E., and furnishes some mill sites, and admits of a little boat navigation. It is 6 ms. by 12, with a good soil, and has also Red cr. Erie canal passes through the tsp., and the v. or borough of Palmyra is on Mud cr., and the canal. Pop. 1830, 3,427.

PALMYRA, incorporated v. Palmyra, Wayne co., N. Y. On Mud cr. and Erie canal, 196 ms. N. N. W. Albany, 13 N. Canandaigua, 15 W. Lyons, and 16 S. Pultneyville, has an academy, several churches, factories, &c., and has considerable trade.

PALMYRA, p-v. and st. just., Fluvanna co., Va., by p-r. 45 ms. N. W. by W. Richmond, and 136 S. W. W. C.; lat. 37° 47', long. 1° 29' W. W. C.

PALMYRA, p-v., sthrn. part Halifax co., N. C., by p-r. 101 ms. N. E. by E. Raleigh.

PALMYRA, p-v., on the left bank of Cumberland r., Montgomery co., Ten., 4 or 5 ms. below, but on the opposite side from Clarksville, and by p-r. 48 ms. S. W. by W. Nashville.

PALMYRA, p-v., Portage co., O., by p-r. 139 ms. N. E. Columbus. Pop. tsp. 1830, 839.

PALMYRA, p-v., and st. jus., Marion co., Mo., 125 ms. N. N. W. St. Louis.; lat. 39° 46', long. W. C. 14° 30' W.

PAMELA, p-t., Jefferson co., N. Y., 166 ms. N. W. Albany, N. Black r., and opposite Watertown, 4 ms. by 8., has an uneven surface, light loamy soil, yielding wheat, corn, &c., remarkably well. It has few springs or brooks. Williamsville, or Williamstown is on Black r. Limestone, which abounds, contains vegetable impressions. Kanady's Grotto is a remarkable cavern, in a rock near the shore of Black r. Pop. 1830, 2,273.

PAMLICO, river, N. C. This name is applied only to the bay of Tar r. below Washington, Beaufort county. It is a sheet of water varying in width from 1 to 8 ms., and about 40 ms. in length, with depth of water admitting any vessel which can be navigated over Pamlico sound. (*See Tar river.*)

PAMLICO POINT, and p-o. Beaufort co. N. C. by p-r. 29 ms. S. E. by E. Washington, and 151 in a similar direction from Raleigh. The point is the cape on the S. side at the entrance of Pamlico r. The name is spelled Pantego in the P. O. list.

PAMLICO SOUND, is an extensive shallow gulf, or more correctly, cape of N. C., the recipient from the W. of Tar, or Pamlico r., and on the S. W. of Neuse r. It is in form of a half moon, stretching 70 ms. from the mouth of Neuse in a N. E. direction, to the strait which unites it with Albemarle sound. The breadth varies from 8 at the northeastern extremity to 30 towards Core sound, and the mouths of Neuse and Pamlico rs. This sound is separated from the Atlantic ocean by Core and Hatteras islands. These islands are, however, mere narrow, low, but very dangerous reefs. Core isl. from Cedar to Occacoke inlet is 22 ms. long. Hatteras isl. is about 66 ms. in length from Occacoke to New Inlet. The land around Pamlico sound is every where low, and in many places marshy.

PAMUNKEY, r. Va., and the principal constituent of York r., is formed by Pamunkey proper and North Anna. The latter rises in Orange, the nrthrn. part of Louisa, and in Spottsylvania cos., and flowing thence southeastward unites with the Pamunkey between Caroline and Hanover cos.

The Pamunkey rises in the south west mtn., on the border between Albemarle and Louisa; drains the sthrn. and central part of Louisa, and traversing Hanover joins the North Anna. Below their junction the united waters, known by the name of Pamunkey, preserves the original course sthestrd. about 45 ms. comparative course, (but perhaps double that distance by the bends,) to its junction with Mattapony to form York r. The entire comparative length of Pamunkey, by either branch, is about 90 ms. The broadest part of the valley but little exceeds 30, and is only about 15 ms. mean width, area 1,300 sq. ms., lying between those of Jas. and Chickahominy on the right, and Mattapony on the left.

PANTHER, cr. and p-o. sthestrn. part of Surry co. N. C., by p-r. 130 ms. s. w. by w. Raleigh.

PANTHER, cr. and p-o. Daviess co. Ky., 10 ms. sthrd. Owensborough, and by p-r. 169 ms. w. of s. s. w. Frankfort.

PANTHER'S GAP, and p-o. Rockbridge co. Va., by p-r. 195 ms. s. w. W. C.

PANTON, p-t. Addison co. Vt., 13 ms. N. w. Middlebury, 25 s. Burlington, w. Otter cr., E. lake Champlain, opposite Elizabethtown, N. Y.; chartered 1764; with 10,530 acres; is very level, and crossed by a sluggish stream of Otter creek running through it N. Pop. 1830, 907.

PAOLI, p-v. Chester co. Pa. 25 ms. wstrd. Philadelphia.

PAOLI, p-v. and st. jus. Orange co. Ind., situated near the centre of the co., by p-r. 94 ms. a little w. of s. Indianopolis; lat. 38° 34'.

PAPACHTON, r. Delaware co. N. Y. The E. branch of the Delaware, rises in many small streams in Stamford and Roxbury, the east towns of the co., flows s. w. through its south towns 48 ms. to the Del. in Hancock co., and receives many tributaries, of which Beaverkill, from N. Jersey, is the principal. It affords mill sites.

PAPERTOWN, p-v. Cumberland co. Pa.

PAPERVILLE, p-v. in the northeastern angle of Sullivan co. Ten., situated on Holstein r. by the road 118 ms. above, and N. E. by E. Knoxville, and by p-r. 274 ms. a little N. of E. Nashville.

PARACLIFTA, p-v. Sevier co. Il., by p-r. 168 ms. s.-w. by w. Little Rock.

PARADISE, p-v. Lancaster co. Pa., by p-r. 44 ms. estrd. Harrisburg.

PARADISE, p-v. Cole co. Il., by p-r. 70 ms. N. E. Vandalia.

PARADOX, lake, Scaroon, Essex co. N. Y., 5 ms. long, and empties into Scaroon r. It is surrounded by high hills, in a wild region, from which the water frequently descends in great quantities. Sometimes the outlet is raised by a shower above the level of the lake, and flows back into it, from which remarkable peculiarity the lake is said to have derived its name.

PARCIPHANY, p-v. Morris co. N. J., 21 ms. N. W. Newark, and 63 from Trenton, on a small branch of Passaic r. A school for the instruction of Africans, was formed here in 1816, under the Presbyterian synods of this state and N. Y., to supply the colony at Liberia and in Hayti, with school teachers and clergymen, but it is not continued.

PARHAM'S STORE, and p-o. Sussex co. Va., by p-r. 50 ms. s. s. E. Richmond.

PARIS, p-t. st. jus. Oxford co. Me., 42 ms. w. Augusta, 46 N. w. Portland, and 160 N. N. E. Boston, is nearly of an oblong form, lying N. W. and s. E. and crossed by Little Androscoggin r., in the upper part of its course, which rises in the adjoining counties. It has a high mtn. s. Pop. 1830, 2,306.

PARIS, p-t. Oneida co. N. Y. 8 ms. w. Utica, 106 ms. N. N. w. Albany, is of regular form, and contains about 100 sq. ms. with a varied surface, and good soil, especially in the vallies, favorable to grain, grass, &c., and bearing maple, beech, birch, elm, &c., with some hemlock and cedar. It is well watered and supplied with mill sites and trout by Oriskany and Sadaguada crs. There is a mill seat on the latter stream, to every 22 yards of its course. Hamilton college is in this t., in the village of Clinton, where is also a seminary, several boarding schools &c. There are 37 school districts in the t. Iron is obtained from ore furnished by the t., and silicious and lime stones are quarried. The Brothertown Indians settled in this t. some years ago, on land given them by the Oneidas. They were from the remnants of the New England tribes, and some from the Delawares. The town is 8 ms. s. of Erie canal, but is crossed by the route of a new canal to be constructed. There are several villages; Clinton, Paris Hill, Paris Furnace, Manchester village, Sanquait village, and Hanover. Moses Foote commenced the settlement, with 10 families, in 1787. Pop. 1830, 1,477.

PARIS, p-v. nrthrn. part Fauquier co. Va., 58 ms. w. W. C.

PARIS, p-v. and st. jus. Henry co. Ten., situated on a small branch of Sandy creek, by p-r. 118 ms. a little N. of w. Nashville, lat. 36° 19', long. 11° 25' w. W. C.

PARIS, p-v. and st. jus. Bourbon co. Ky., situated on the s. fork of Licking r., 40 ms. E. Frankfort, and 20 ms. N. w. Lexington. Lat. 38° 12', long. 7° 13' w. W. C. Pop. 1830, 1,219.

PARIS, p-v. Stark co. O., by p-r. 127 ms. N. E. by E. Columbus.

PARIS, p-v. Jefferson co. Ind. by p-r. 76 ms. S. S. E. Indianopolis.

PARIS, p-v. and st. jus. Edgar co. Il., by p-r. 106 ms. N. E. by E. Vandalia. N. lat. 39° 36', long. W. C. 10° 44' W.

PARISBURG, p-o. and st. jus. Giles co. Va., situated on the left bank of New r., where that stream passes through Peter's mtn., and immediately above the gap, by p-r. 298 ms. S. W. by W. W. C., and 240 ms. a little S. of W. Richmond. Lat. 37° 21', long. 3° 43' W. W. C.

PARISVILLE, p-v. nrthrn. part Baltimore co. Md., by p-r. 26 ms. from Baltimore.

PARISVILLE, p-v. Portage co. O. by p-r. 144 ms. N. E. Columbus.

PARKERSBURGH, p-v. and st. jus. Wood co. Va., situated on the point above the confluence of Ohio and Little Kenhawa rs., 12 ms. below Marietta, Ohio, and by p-r. 299 ms. a little N. of W. W. C. Lat. 39° 15', long. 4° 34' W. W. C.

PARKERSVILLE, p-v. Chester co. Pa.

PARKHEAD, p-v. Washington co. Md., by p-r. 87 ms. N. W. W. C.

PARKINSON'S FERRY, and p-v. on the left bank of Monongahela, directly below the mouth of Pigeon creek, Washington co. Pa., 20 ms. E. from the borough of Washington, and very nearly a similar distance S. Pittsburg.

PARKMAN, p-v. sthestrn. part of Geauga co. Ohio, by p-r. 159 ms. N. E. Columbus. Pop. twp. 1830, 732.

PARKS, p-o. Edgefield district, S. C. by p-r. 111 ms. wstrd. Columbia.

PARMA, p-t. Monroe co. N. Y., 230 ms. N. by W. Albany. Pop. 1830, 2,639.

PARMA, p-v. Cuyahoga co. O., by p-r. 131 ms. N. E. Columbus.

PARSONSFIELD, p-t. York co. Me. 93 ms. S. W. Augusta, 38 N. W. Portland, and 118 N. N. E. Boston; borders W. on Effingham, Strafford co. N. H., and N. on Oxford co. Me., Ossipee r. forming the line. Two ponds empty S. E. into little Ossipee r. Pop. 1830, 2,492.

PARTLOW'S, p-o. Spottsylvania co. Va., by p-r. 79 ms. S. S. W. W. C., and 59 N. Richmond.

PASCAGOULA, river of Miss. and Ala., the much greater share of its valley being in Miss., is formed by 2 branches of the Chickasawhay and Leaf rivers. The Chickasawhay rises in the Choctaw country, lat. 32° 50', and flowing thence by a course of very near S., receives the Leaf r. about 2 ms. below lat. 31°. The valley of Chickasawhay lies between those of Leaf and Ala.

Leaf r. rises about N. lat. 32° 20', and pursuing a southeastern course unites with the Chickasawhay as already noticed. The united water, thence known as the Pascagoula, continues the course of the latter, to the Pascagoula sound, N. lat. 30° 20'. The entire comparative length of the Pascagoula by the main branch, Chickasawhay, is 170 ms.; the mean breadth of the valley is at least 50 ms. and area 8,500 sq. ms. The Pascagoula valley lies between those of Pearl and Ala. and comprises the western part of Mobile and Washington counties, Ala., and all of Jackson, Perry, Greene, Wayne, Jones and Covington, and part of Lawrence, Sampson and Rankin cos. in the state of Miss., with a considerable space in the Choctaw country.

PASCAGOULA SOUND, is a sheet of water spreading along the southwestern border of Alabama, and the southeastern of Mississippi, extending in length 55 ms. from the Pass of Heron W. to the Pass of Christian, with a mean width of about 8 ms. It is separated from the gulf of Mexico, by a chain of low, narrow sand islands, named, advancing from E. to W., Dauphin's, Massacre, Petite Bois, Horn, Dog, Ship and Cat island. The depth of water in the sound is generally about from 10 to 18 feet, but no vessels drawing more than 6 feet can be navigated through the Passes. The depth increases rapidly on the Gulf side of the islands. When the British fleet came on the coast of La., their heaviest ships of the line were anchored close on Cat isl., outside of the Pass of Mariam. This anchorage is in fact the most sheltered on the U. S. coasts of the Gulf of Mexico, where ships of war of the largest class can be safely moored.

PASCAGOULA, p-o. southern part of Jackson co. Miss., about 200 ms. a little E. of S. E. Natchez.

PASCATAQUA, river N. H., empties into the ocean at Portsmouth, on the boundary between that state and Me., and is formed by several small streams, which rise in Rockingham and Strafford cos., and meet a few ms. from the coast. Of these Salmon Fall r., the principal, runs on the boundary of Me. The others are Cocheco, Bellamybank, Oyster, Lamprey, Squamscot and Winnicut rs. The five last fall into a kind of lake, which takes the name of Pascataqua river, and contracting in size below, at the distance of 3 ms. joins the ocean, forming Portsmouth harbor, which is a very good and safe one, and has a navy yard of the U. S. with several islands.

PASQUOTANK, r. N. C. is the drain of the sthrn. part of Dismal Swamp, and after flowing S. S. E. between Camden and Pasquotank cos. opens by a comparative wide bay into Albemarle sound, after a course of 40 ms. including bay and river. The Pasquotank bay admits ordinary coasting vessels to its head at Elizabeth City.

PASQUOTANK, co. N. C. bounded by Pasquotank r. separating it from Camden co. N. E. and E., Albemarle sound S. E., Perquimans co. S. W., Gates N. W., and Nansemond, and Norfolk cos. Va. N. Length from Albemarle sound to the Va. line 40 ms., mean width about 8, and area 320 square ms. Extending in lat. from 36° 03' to 36° 30', and in long. from 0° 23' to 1° E. W. C. The surface is a plain, partly marshy, but with considerable tracts of good soil. The slight declivity is s. s. e. Chief town, Elizabeth City. Pop. 1820, 8,008, 1830, 8,641.

Passadunkeag, town, Penobscot co. Me. Pop. 1830, 269.

Passaic river, N. J. a valuable stream, navigable 10 ms. for sloops, rises in Morris and Somerset cos. and forms the boundary of Essex co. almost on three entire sides, w., n. and e. It receives Pompton r. n. which is formed of Pequannoc and Ramapo rs. which last rises in Rockland co. N. Y. Rockaway r. falls into it on the w. and there are several smaller branches. The Passaic supplies water to the most important manufacturing village in the state, Patterson. It there makes a fall of 72 feet from a precipice, into a deep pool between two rocks, but the current has now been diverted into numerous channels for the supply of the various manufactories, so that the cascade, which was formerly celebrated by its picturesque beauty, is now to be seen only during the wet season. The Passaic is crossed by an aqueduct of Morris canal, 3 ms. above Patterson.

Passamaquoddy bay, partly in Penobscot co. Me. lies principally in New Brunswick. It receives St. Croix r. which forms the e. boundary of the United States for some miles, and the communication with the Atlantic is nearly closed by Campbello island. It is formed by Quoddy Head, in Lubec, Me. and the s. w. corner of New Brunswick, being about 6 ms. by 12, and containing Deer isl. The tide rises from 25 to 33 feet. The water is deep, well stocked with fine fish, and never frozen over. The fish are cod, herring, mackerel, &c.

Passamaquoddy Indians, Me. of whom only a small tribe remain, reside in Perry, on a reservation of 27,000 acres, and have a Roman Catholic church.

Passumpsic river, Vt. rises in a pond in Essex co. and flows s. through a part of Caledonia co. into Connecticut river, at the bend in Barnet. It is rapid till it reaches Lyndon, and then winds slowly through several rich tracts of meadow, with a few falls. It is deep, and has several branches, running a course of 34 ms.

Passyunk, tsp. Phil. co. Pa. adjoining the sthrn. side of the city of Phila. and extending from the Del. to the Schuylkill r.

Patapsco, r. Md. This comparatively small stream has gained great importance from having had the eastern part of Baltimore and Ohio rail road, formed along its valley. The Patapsco rises by numerous creeks from the southeastern foot of the Parr spring or Sugar Loaf ridge of mtns. between Frederick and Baltimore cos. and opposite to Little Pike creek, Linganore, and Bush creek, branches of Monocacy. The main or northern branch rises near Westminster in Baltimore co. and flowing first s. e. about 8 ms. turns to the southward 15 ms. to the forks of Patapsco, where it receives the western branch or Parr's Spring branch. The latter rises near Ridgeville and almost on the Baltimore and Frederick road, and near where that road is crossed by the Baltimore and Ohio rail road. From thence pursuing an eastern course between Baltimore and Ann Arundel cos. 15 ms. to its junction with the northern branch.

The forks of Patapsco on the junction of the two main constituents of that r. is 16 ms. air measure n. w. by w. city of Baltimore. From the forks, with many partial bends and a general southern curve, and a comparative distance of 20 ms. the Patapsco opens to a bay, receiving Gwyns Falls creek on the southwestern side and Jones' Falls creek in the city of Baltimore. Jones' Falls creek bay is in fact the harbor of Baltimore, and the compactly built part of the city does not yet reach the Patapsco bay. The junction of the two latter is made below fort McHenry; from the basin of Baltimore the Patapsco bay stretches fourteen miles southeast, with a width from one to three miles to the Chesapeake between Bodkin and North Points. The basin of Patapsco is in length 40 ms. and mean width 15, area 600 square ms. between lat. 39° 08′ and 39° 38′.

Measuring the plains of descent from Parr's Spring ridge to tide water in Patapsco, we find it within an inconsiderable fraction of 27 ms. The ridge near Westminster is 675, and near Parr's Spring 850 feet elevated above tide water. The water level at the forks is 385 feet, of similar comparative height. The mean height of the ridge 780 feet nearly, consequently the mean fall from the summit to tide water is 28 7-8 feet per mile, or yields a plain of descent or ascent of 3° and 8 minutes. This great and rapid fall renders the Patapsco and its branches highly valuable as mill streams. (*See Baltimore and Ohio rail road.*)

Patoka river, Indiana, rises in Orange and Crawford cos. and entering Dubois, approaches to within 3 ms. of the e. fork of White r. at Portersville, but inflecting thence first s. e. and thence west, crosses Pike, and entering Gibson, falls into Wabash, one or two ms. below the mouth of White r. after an entire comparative course of 80 ms. in a direction from e. to w. It may be remarked, that the corresponding courses of Ohio, Patoka, and White rs. being all from east to west, demonstrate a corresponding uniformity of structure in the country where r. channels have such striking resemblance in their direction. The Patoka valley lies between that of Ohio and White rivers.

Patrick, co. Va. bounded by the Blue Ridge which separates it from Grayson w., and Montgomery n. w., by Franklin n. e., Henry e., Rockingham co. N. C. s. e., Stokes co. North Carolina south, and Surry county, N. C. s. w. Length diagonally from s. w. to n. e. 42, mean width 12, and area 504 square ms. Extending in lat. from 36° 30′ to 30° 47′, and in long. from 2° 56′ to 33° 40′ w. W. C.

The northern part of Patric : declines northeastward, and is drained by Irvine or Smith's r. The sthestrn. angle gives source to Mayo river, the central part gives source to the extreme fountains of Dan r. whilst the western angle, towards Grayson, is drained by the extreme northern sources of the Yadkin. The whole county has a general declivity to the southeastward. Chief town, Taylorsville. Pop. 1820, 5,089, 1830, 7,395.

PATRICK, C. H. (*See Taylorsville, Patrick co. Va.*)

PATRICK'S, p-o. King and Queen co. Va.

PATRICK'S Salt works and p-o. Perry county Ky. by p-r. 125 ms. S. E. Frankfort.

PATRIOT, p-v. Switzerland co. Ind. by p-r. 121 ms. S. E. Indianopolis.

PATTERSON, p-t. Putnam co. N. Y. 6 ms. N. E. Carmel, 22 S. E. Poughkeepsie, 93 S. Albany, S. Duchess co., W. Connecticut, is hilly E. and W. with a broad and fertile valley between, in which is a large swamp, containing an isl. of 12 acres, and extending into Pawlings. This is the source of Croton r. The village is N.

PATTERSON, p-t. Essex co. N. J. 61 ms. N. W. by N. Trenton, 18 N. N. W. New York, at the great falls of Passaic r. is one of the principal manufacturing villages in the U. S. In 1791 the society for establishing useful manufacturers was incorporated, with a capital of $1,000,000, and the right to dig canals, clear rivers, &c. within 6 ms. and authority to form a city and co. with the consent of the inhabitants; the last has never been done. The population 30 years since was only about 300. At this time (1832,) there are about 8,000 inhabitants. There are about 800 dwellings in the place, including 57 stores; nine churches, viz :—Presbyterian, one; Roman Catholic one; Reformed Dutch, three; Episcopal, one; Baptist, one; Reformed Presbyterian, 1; Methodist, 1. A large new Catholic church, of stone, is also building. There is also a bank, with a capital of $125,000 paid in. There are about 20 day schools, 8 for females, instructing together more than 700 children; a free school, supported by the town, in which about 80 poor children are instructed, and an infant school where 150 poor children are gratuitously instructed.

The literary societies are, the mechanics institute, and a philosophical society, which has a respectable library. There are fifteen blacksmiths' shops, besides those immediately connected with the machine shops, twenty-five shoe shops, employing fifty seven hands, ten taverns, two millwright and machine shops, including a blacksmith shop, which employ twenty hands, and four other machine factories, employing two hundred and eighty hands. In one of these, were manufactured last year 15,048 spindles, together with all the necessary frames and fixtures, which, at $12 the spindle, amount to $180,576. Connected with this is an iron and brass foundry, producing annually 1,020,000 pounds of iron and brass castings. Another manufactory for machinery is also erecting.

There is one rolling and slitting mill, and nail factory, employing 23 hands, and producing annually 672,000 pounds of nails, a woollen or satinet factory, 17 cotton factories, with 22,029 spindles. The raw cotton consumed in these factories in 1829, was 2,179,600, producing 1,914,450 pounds of yarn; the raw cotton costing $223,501. There is now manufactured annually in Patterson, upwards of 400,000 yards of cotton duck, and about 200,000 yards of other description of cotton cloth. In 1829 the cotton duck made amounted to 150,000 yards, and of other cotton cloths 1,861,450 yards. The cotton yarn not made into cloth, amounting in eighteen hundred tweny nine to 1,192,400, now to 1,500,000, is sent from Patterson, principally to New York and Philadelphia. There are in operation in the factories 266 power looms, and 26 hand looms. In the town there are employed upwards of 500 hand looms, making at least 800 power and hand looms in operation in the place.

The Phenix duck manufacturing co. employ 1,616 spindles, consume annually 600,000 lbs. of flax, manufacture 450,000 yards of duck & 143,000 yds. of bagging, and employ 395 hands. The total of cotton and flax spindles employed in Patterson are now rising of 40,000; the amount of cotton and flax consumed annually is estimated at 3,200,000; the total of cloth and duck of all kind made annually, in 1829 was 2,604,450 yards, now nearly 3,000,000. The annual amount of manufactured goods in Patterson is about 2,590,000 dollars. The Morris canal passes within sight of the town, and a rail road to Hoboken is in a course of completion.

The beautiful falls of the Passaic r. at this place, attract many visiters. The water power which operates all the machinery we have noticed, is procured from above the falls, by a sluice way cut through the precipice, and is conducted by canal to the several manufactories. The supply is yet more than abundant for the purposes to which it is applied.

PATTERSON'S Mills, and p-o. Washington co. Pa. 249 ms. N. W. W. C.

PATTONSBURG, p-v. on James r. Botetourt co. Va. 12 ms. N. E. by E. Fincastle, the co. seat, and 40 ms. N. W. by W. Lynchburg, by p-r. 223 ms. N. W. W. C. The water level in James r. at low flood, is 806 feet above tide water at Rokett's, below Richmond.

PATTONSVILLE, p-v. Granville co. N. C. by p-r. 38 ms. N. Raleigh.

PATUXENT Forge and p-o. Ann Arundel co. Md. 25 ms. N. E. W. C. and nearly a similar distance a little W. of S. Baltimore.

PATUXENT, r. Md. having its remote source on the southern side of the Sugar Loaf ridge, between Montgomery and Ann Arundel cos. The main stream from its source pursues a S. E. course 40 ms. separating first Montgomery from Ann Arundel, and thence Ann Arundel from Prince George's. Having reached within 6 ms. of Chesapeake bay, the Patuxent inflects to a southern course 30 ms. with Prince George's and Charles' cos. on the right, and Ann Arundel and Calvert on the left, it gradually expands to a wide estuary, and bending again to the S. E. 20 ms. between Calvert and St. Mary's it terminates in Chesapeake bay, after an entire comparative course of 90 ms. It may be noticed as a curious fact, that the Patuxent in the 50 lower miles of its course is in no one place 12 ms. from Chesapeake bay, the mean width of the intervening country being about 8 ms.

The entire valley of the Patuxent is remarkably narrow. The direct distance between the opposing sources is in no part fifteen ms. asunder, and the mean width of the valley is perhaps overrated at 10 ms., area about 900 square ms. The higher part of this confined basin lies between those of Potomac and Patapsco, the lower between Potomac and Chesapeake bay.

PAULINSKILL, r. N. J. rises in Sussex and flows through Warren co. into Delaware r. at Columbiaville, in Knowlton. It has its its principal source in Long pond, in Frankford. Its course is about 25 ms.

PAULUS HOOK, a small peninsula, Bergen, Bergen co. N. J. opposite the city of New York, containing the village of Jersey city. Towards the main land it has a low neck, over which the road is carried on a causeway, being in danger from high tides. There is a steamboat ferry to the city, and several lines of stage coaches proceed hence to Phila. Easton, &c.

PAWCATUCK river, R. I. rises in Washington co. with one of its branches heading just over the boundary of Conn. Its streams are principally supplied from ponds, which still bear the Indian names. The latter part of its course marks the boundary between the two states, near the sea.

PAWLET, p-t. Rutland co. Vt. 21 ms. s. w. Rutland, 33 N. Bennington, 23,040 acres, chartered 1761, first settled 1762, is crossed by Pawlet r. southwesterly, and has Indian r. in the s. w. which has its source in an abundant spring, and abounds in trout; a mountainous range divides it N. and s., in the middle of which is Haystack mtn. The soil is warm, dry, bearing grain and grass, maple, beech, birch, elm, &c. Indian river was once a favorite fishing place of the Indians. There is an academy in the town. Pop. 1830, 1,965.

PAWLET river, rises in Vt. near the line of Bennington and Rutland cos., flows N. w. in Washington co. N. Y. passes through Granville, and falls into Wood creek in Whitehall. It is a valuable mill stream, well supplied with water, and stocked with trout, and above 20 ms. in length.

PAWLING, p-t. Dutchess co. N. Y. 22 ms. s. E. Poughkeepsie, 105 ms. s. Albany, w. Connecticut, N. Putnam co. 8 ms. by about 9. The hills of Dover and Patterson extend through its E. and w. parts, and between them lies a continuation of the same valley, with a part of the large swamp which was mentioned in Patterson. The waters flow partly s. forming the source of Croton r. and partly N. into Ten Mile r. in Dover, a branch of Housatonic river of Conn.; Quaker hill, and West mountain, are considerable eminences. Iron ore is found here, mica, &c. Population, 1830, 1,705.

PAWTUCKET, r. R. I. the principal branch of Seaconk r., rises in Worcester co. Mass. and bears the name of Blackstone r. in that state. It enters R. I. near the N. E. corner, divides Cumberland and Smithfield in Providence co., and supplies water power to many of the principal manufactories in the state, particularly in the village of Pawtucket. The Blackstone canal extends up its valley for most of its length, to Worcester, Mass. and affords great advantages to numerous manufactories in that state as well as in R. I.

PAWTUCKET, p-v. 4 ms. N. Providence, at the falls of Pawtucket r., partly in Providence R. I. and partly in Seekonk, Mass. 4 ms. N. E. Providence, is a large manufacturing village. There are three falls. At the central falls are 3 large cotton factories, with 6,600 spindles, 162 looms, and using 900 bales of cotton annually; there is also a thread factory. At the Upper or Valley falls 4 factories, with about 17,500 spindles, and 140 looms; and at the Lower falls where the greater part of the inhabitants live, are 11 factories, with 18,687 spindles and 430 looms. The population in 1831 was suppos. to be about 4,000. The Blackstone canal passes near the village.

PAWTUCKET FALLS, in Merrimack r. Mass., between Lowell and Dracut. Within the distance of about 300 yards the descent is 30 ft. The water first falls perpendicularly over a ledge of rocks, and then pours foaming down a rough channel. A bridge crosses just at the fall. A canal, 90 ft. broad, is dug on the s. side, and draws off a large volume of water to the great manufactories of Lowell, which is situated at the confluence of Concord and Merrimack rs., about 1 mile below. The country in this vicinity was the seat of the Pawtuckets, in the early history of New England a powerful tribe of Indians, who were governed by Wonnalonset, an old and friendly sachem. After residing here till 1686, they sold their remaining land and retired into the interior.

PAXTON, p-t. Worcester co. Mass. 55 ms. w. Boston, 9 N. w. Worcester; is watered by Nashua r. flowing into the Merrimack, and Chickapee r. flowing into the Connecticut, has good land, with a varied surface, also several fish ponds. Pop. 1830, 597.

PEACHAM, p-t. Caledonia co. Vt. 20 ms. N. E. Montpelier, 18 N. w. Newbury; was chartered 1763, but was much impeded in its growth by the revolutionary war. A grammar school was established in 1795. Pop. 1830, 1,351.

PEACH BOTTOM, p-v. lower part of the south-eastern angle of York co. Pa. by p-r. 36 ms. s. E. by E. from the borough of York, and 80 N. N. E. W. C.

PEAKS OF OTTER. (*See Otter, Peaks of.*)

PEARL, r. Miss. and La. having its remote sources in the Choctaw country and in the former state, about lat. 33° interlocking sources with those of Big Black, Pearl and those of Oaknoxabee branch of Tombigbee. Flowing thence by comparative courses 80 ms., curves gradually to s. s. E. 160 ms. to its mouth into the Rigolets, after an entire compartive course of 24 miles. The valley of the Pearl is narrow, and in all its length the only confluent above the length and volume of an ordinary creek, is the Bouge Chitto. (*See Bouge Chitto.*) The basin of the Pearl on the strict principles of geographical classification contains the valleys of the Tchefoute, Tanchepaha, Tickfah and Amite, as the Rigolets are the common estuary of all those streams, and the Pearl, being lowest in order of discharge and in length of course very greatly the superior volume, is entitled to give name to the basin.

The actual valley of the Pearl, including only with the main stream that of Bogue Chitto, is about 240 miles in length, with a mean width of 30 ms. or 7,200 sq. ms. The Pearl drains a small section in the Choctaw territory, and part of the counties of Madison, Hinds, Rankin, Copiah, Simpson, Lawrence, Pike, Marion and Hancock in the state of Miss., and in La. part of the parishes of Washington and St. Tammany.

As a navigable stream the facilities afforded by the Pearl, bear a very small proportion to the comparative length of its course. The estuary is also impeded by rafts of timber, shallows, and sand bars.

PEARLINGTON, p-v. and st. jus. Hancock co. Miss. situated on the left or estrn. bank of Pearl r. about 150 ms. S. E. Natchez, and 50 ms. N. E. New Orleans, lat. 30° 30′, long. W. C. 12° 38′ w.

PEARMAN'S, ferry and p-o. Dale co., Ala. by p-r. 220 ms. S. E. Tuscaloosa.

PEÇAN, grove and p-o. Washitau parish, La. by p-r. 346 ms. N. W. New Orleans.

PECONERY, p-o. Conway co. Ark. 33 ms. nthwstd. Little Rock.

PEDEE, r. N. and S. C., having its extreme nrthrn. source, however, in the wstrn. part of Patrick co. Va. This river has received the name of Yadkin in N. C. and it is only after entering S. C. that it is known as Pedee.

The Yadkin rises from the sthestrn. valleys of the Blue Ridge, opposite the sources of French Broad, Nolachucky and Great Kenhawa, and on the Atlantic slope, having interlocking sources with those of Dan r. N., and Great Catawba S. Flowing thence nrthestrd. over Wilkes and Surry cos. N.C. and receiving its extreme nrthrn. water from Va., it bends to a course a little E. of S., after having flown about 80 ms. nearly parallel to the Blue Ridge. The last noted inflection is made on the border between Surry and Stokes cos., at lat. 36° 17′, long 3° 30 w. W. C. From this point the Yadkin gains only 30′ of long. in the residue of its course to lat. 34° 48′, where it enters S. C. and looses its name in that of Pedee. In the latter comparative course of about 110 ms. no tributary above the size of a large creek enters from the left; but from the right, Little Yadkin from Iredell and Rowan cos., and Rocky r. from Cabarras, Mecklenburg, Anson and the western part of Montgomery, are considerable streams, which will be described under their proper heads. The valley of the Yadkin is about 135 ms. in length, with a mean width of 55 or a small fraction above 7,400 sq. ms., draining in Va. a small part of Patrick co., and in N.C. all Wilkes, Surry, Rowan, Davidson, Montgomery and Cabarras; with great part of Stokes, Iredell, Randolph, Richmond and Anson cos.

If we regard the Yadkin as one of the constituent streams of the Great Pedee, the latter is formed by the Yadkin, Lynches, Waccamaw, and Little Pedee rs. See the secondary streams under their respective heads.

The main stream enters S. C., between Marlborough and Chesterfield dists., and continuing the general course of Yadkin, over Marion, and thence between Horry and Georgetown dists. opens into Winyaw bay after a comparative S. S. E. course in S. C. of 110 ms. This lower part of the basin is very nearly a square of 100 ms. each side, equal to 10,000 sq. ms., making the whole basin, including the Yadkin valley, equal to 17,400 sq. ms. If the basin is extended from Winyaw Point at Georgetown entrance to the source of Toms cr. in Patrick co. it stretches from lat. 33° 11′, to 36° 35′, and from the estrn. bend of Waccamaw to the extreme wstrn. fountains of Yadkin, from 1° 40′ to 4° 30′ of long. w. W. C.

Without estimating the mtn. ridges, the relative oceanic level of the arable soil along the sthestrn. slope of Blue Ridge in Wilkes, Surry, and Patrick cos. is at least 1,500 feet or an equivalent to 3½° of Fahrenheit, which added to 3° 24′, the difference of lat. yields almost 7° difference in temperature between the higher and lower part of the Pedee basin. This basin lies between those of Santee and Cape Fear rs., and if duly improved would be of immense importance as a navigable and commercial channel.

PEDLAR'S HILL, and p-o. wstrn. part of Chatham co. N. C. 10 ms. wstrd. Pittsboro', and 43 ms. in a similar direction from Raleigh.

PEDLAR'S MILLS, and p-o. sthwstrn. part of Amherst co. Va. by p-r. 198 ms. S. W. W. C. and 135 wstrd. Richmond.

PEEBLES, tavern and p-o. Northampton co. N. C. by p-r. 208 ms. a very little w. of S. W. C., and 94 ms. N. E. by E. Raleigh.

PEEKSKILL, incorporated v. Cortlandt, West Chester co. N. Y. on the east side of Hudson river, near the mouth of Peekskill cr. at the S. entrance of the Highlands, 40 ms. N. New York; has considerable trade, and daily communication with that city, in the warm season, by a steamboat.

PEELING, p-t. Grafton co. N. H. 20 ms. N. Plymouth, 60 N. by w. Concord, with 33,359 acres, is crossed by Pemigewasset r., its three branches uniting here, and has mill seats on several other streams, particularly the sources of Wild Amonoosuc, Baker's rs. &c. It has two ponds and several mtns. of which Cushman's and Blue mtns. are the chief. Settled 1773. Pop. 1830, 292.

PELHAM, p-t. Hillsborough co. N. H. 37 ms. S. by E. Concord, 45 S. W. Portsmouth, 32 N. W.

Boston, with 16,338 acres, contains Gumpas & Isl. ponds, and part of North pond, and is crossed by Beaver r. on which and its branches are fine meadows, bordered by pine lands, good for grain. There is good grazing, orchard and wood land E. and W. Wood, chiefly oak, has been taken down the r. There are several factories, &c. First settled 1772. Pop. 1830, 1,070.

PELHAM, p-t. Hampshire co. Mass. 85 ms. W. Boston, 10 N. E. Northampton, and 5 ms. E. Amherst; is elevated and has a hilly surface, with good grazing land, and is watered in the E. by Swift r. and W. by Fort r. Pop. 1830, 904.

PELHAM, t. Westchester co. N.Y. 18 ms. N.E. N. Y., 9 S. White Plains, N. Long Isl. sound, N. of East r. and E. of Chester cr., is small and terminates in an angle N. The surface is nearly level, with a stony but good soil. Pell's or Rodman's Point is S. It comprehends City, Hart's and High isl. in the sound. Pop. 1830, 334.

PEMBROKE, p-t. Merrimack co. N. H. 60 ms. N. W. Boston, 6 E. Concord; E. Merrimack r., S. E. Soucook r., N. W. Suncook r., with 10,240 acres, has several factories and mills, and a considerable village, with a fine street of 3 ms. parallel to Merrimack r. The roads generally run at right angles. The land near the v. slopes pleasantly to the narrow meadows on the rs. and it contains a town house, and an academy founded by Mr. Blanchard. The Indian name was Suncook, and it was granted, 1727, to Capt. Lovewell and 60 associates, for services against the savages. It was first settled 1728, and much interrupted by their attack. The settlers were of English and Scotch descent. Pop. 1830, 1,312.

PEMBROKE, p-t. Plymouth co. Mass. 23 ms. S. E. Boston. It originally belonged to Duxbury, and had the only saw mill in the Old colony for 40 years. North r. runs between this t. and Hanover, and 2 branches flow from ponds. There are some manufactories. Pop. 1830, 1,325.

PEMBROKE, p-t. Genesee co. N. Y., 10 ms. W. Batavia, 257 W. Albany, E. Erie co.; 8 miles by 14½; is watered by Murder and Tonawanta creeks, with streams of Oak Orchard creek. It contains the Tonawanta Indian village on Tonawanta creek. The land bears maple, elm, beech, hemlock, &c., and is pretty good. Pop. 1830, 3,828.

PEMBROKE, p-v. sthrn. part Todd co. Ky., by p-r. 196 ms. S. W. by W. Frankfort.

PEMBROKE SPRINGS, and p-o. Frederick co. Va. 18 ms. wstrd. Winchester, the co. seat, and by p-r. 89 ms. wstrd. W. C.

PEMIGEWASSET, r. N. H., chiefly in Grafton co., is the W. branch of the Merrimac. Its N. branch rises in Franconia, a few ms S. W. of the white mtns., and meets two others in Peeling.

PENDLETON, co. Va., bounded by a ridge called there locally "The Great North Mountain," separating it from Rockingham E., Augusta S. E., and Bath S. W., by the main spine of the Alleghany separating it from Randolph W., and by Hardy N. E. Length 40 ms., mean width 25, and area 1,000 sq. ms. Extending in lat. from 38° 15′ to 38° 53′, and in long. from 2° to 2° 42′ W. W. C. Pendleton occupies the most elevated part of the table land between its two bounding ridges of mountains, discharging to the S. W. the extreme sources of James r., and in an opposite direction the higher sources of South Branch of Potomac. More than four-fifths of the surface is, however, in the valley of the latter. Comparing the general elevation of Pendleton, with determined height in James r. in Alleghany co., with the whole slope of Bath co. intervening, the level of the arable land from whence flow the sources of James and Potomac rs., must exceed 2,000 feet. Covington in Alleghany, at the junction of Pott's creek with Jackson's river, is 1,222 feet above the mean tide in Chesapeake bay, and at this point the water of Jackson's r. has fallen down a plain of upwards of 50 ms. descent. The surface is generally mountainous, rocky and sterile. Chief t. Franklin. Pop. 1820, 4,836, 1830, 6,271.

PENDLETON, formerly the northwestern district of S. C., has been sub-divided and the name discontinued. The territory formerly comprised in Pendleton, contains the present existing districts of Anderson and Pickens. (*Which see.*)

PENDLETON, p-v. northwestern part of Anderson district, S. C., situated on a branch of Savannah r., by p-r. 143 ms. northwestward Columbia. Lat. 34° 38′, long. 5° 42′ W. W. C.

PENDLETON, co. Ky., bounded by Bracken E., Harrison S. E. and S., Grant W., Campbell N., and Ohio river separating it from Clermont co. in the state of O. N. E. Length 32 ms. by a diagonal from S. W. to N. E., mean width 14, and area 448 sq. ms. Extending in lat. from 38° 30′ to 38° 52′, and in long. from 7° 10′ to 7° 35′ W. W. C. Licking river traverses this co. in a northwestern direction, and very nearly parallel to that of the O. r., where it joins the boundary between this and Clermont co. The declivity is of course in the same direction with the rivers. Chief t. Falmouth. Pop. 1820, 3,086, 1830, 3,863.

PENDLETON, p-v. wstrn. part Madison co. Ind., by p-r. 40 ms. N. E. Indianopolis.

PENFIELD, p-t. Monroe co. N. Y., 10 ms. E. Rochester, 211 W. by N. Albany, S. lake Ontario, W. Ontario in Ontario co.; 67 sq. ms. It has Teoronto bay N. W., into which flows a mill stream, and several other streams run N. into the lake. The surface has an inclination nrthrd., and the soil is poor. The village is S. W. The bay, which is often written Gerundegut, and Irondequot, is 1 mile by 5, and opens by a narrow strait into lake Ontario. Pop. 1830, 4,474.

PENFIELD, p-o. Lorain co. O., by p-r. 116 ms. N. N. E. Columbus.

PENN BRANCH, p-o. Orangeburg district, S. C., by p-r. 65 ms. sthwrd. Columbia.

PENN LINE, p.v. wstrn. part of Crawford co. Pa., 21 ms. northwestward Meadville, and about 100 ms. N. N. W. Pittsburg.

PENNSBORO', p.v. Wood co. Va., by p-r. 268 ms. W. W. C.

PENN'S NECK, Upper, t. Salem co. N. J., 50 ms. S. W. Trenton, and 25 S. W. Philadelphia. It has Old Man's creek N. which separates it from Gloucester co., and Delaware r. W., opposite the mouth of the Brandywine, and Wilmington, Del. It has settlements at Pedrickstown and Sculltown, and the Cove. Population 1830, 1,638.

PENN'S NECK, Lower, t. Salem co. N. J., 58 ms. S. W. Trenton, 30 S. W. Philadelphia, Del. river W., and Salem r. E., which flows into it on the S. line, where there are tracts of marshy land. It lies opposite New Castle and Delaware city, Del., and near the middle of the river is the small island on which was fort Delaware, belonging to the U. S., which was accidentally burnt in the winter of 1831 and 1832. Pop. 1830, 994.

PENN'S STORE, and p-o. estrn. part Patrick co. Va., 17 ms. nrthestrd. Taylorsville, the co. seat, and by p-r. 316 ms. S. W. W. C.

PENNSVILLE, p-v. Bucks co. Pa., 18 ms. N. W. Trenton, N. J., and 26 a little E. of N. Philadelphia.

PENNSVILLE, p-v. Morgan co. O., by p-r. 76 ms. S. E. by E. Columbus.

PENNSYLVANIA, state of the U. S., bounded by N. Y. N. and N. E., N. J. E., Del. S. E., Md. S., Va. S. W., and the state of Ohio N. W.

The name of this state is derived from the surname of William Penn, and sylva, woods; and means, literally, Penn's woods. Though at the epoch when the name was imposed, the real features of the country it was in future to designate, were in great part unknown; to those who were its authors, no term could be more appropriate. Few, if any, regions of equal extent, and in one continuous body, ever bore, in a state of nature, a more dense forest. Pennsylvania was an expanse of woods, in the strictest acceptation of the word.

As now limited, Pennsylvania extends from lat. 39° 43′ to 42° 16′, and from 2° 20′ E. to 3° 36′ W. W. C. It is bounded in common with Delaware, from the Del. r. by a circular line, around New Castle co., to the N. E. limits of Cecil co. Md., 24 ms.; due north to the N. E. angle of Md., 2 ms.; along the northern limit of Md., 203 ms.; in common with Va., from the N. W. angle of Md. to the S. W. angle of Greene co., 59 ms.; due north, in common with Ohio and Brooke cos. of Va. to the Ohio river, 64 ms.; continuing the last noted limit, in common with O. to lake Erie, 91 ms.; along the S. E. shore of lake Erie to the western limit of New York, 39 ms.; due south along Chatauque co. of New York to lat. 42°, 19 ms.; thence due east in common with New York, to the right bank of Del. r., 230 ms.; down the Del. to the N. E. angle of the state of Delaware, 230 ms.; having an *entire* outline of 961 miles.

Its greatest length is due W. from Bristol on Del. r., to the eastern border of Ohio co. Va., through 356 minutes of longitude, along lat. 40° 09′. This distance, on that line of latitude, is equal to 315 American statute ms. The greatest breadth, 176 ms., from the Virginia line to the extreme northern angle on lake Erie; and general breadth, 188 ms.

The area of the state has been variously stated, but propably never very accurately determined. In both Morse's and Worcester's Gazetteers, the superficies is given at 46,000 square miles. Other authorities vary, but comparing the best maps, and from calculating the rhombs, and parts occupied by the state, Pennsylvania includes above 47,000 sq. ms. Rejecting the fractional excess, and using that curve superficies, the state will contain thirty million and eighty thousand statute acres.

Its mountains obtrude themselves at the first glance on a map, as the most prominent of its natural features. No even tolerably good survey having ever been made of the mountains of this region, and many important chains having been entirely omitted, a lucid classification is attended with great difficulty. Some of the collateral chains hitherto overlooked, have been supplied from personal observation; but no doubt much remains to be added or rectified, by future research. The structure and position of its mountains, has given to Pennsylvania an aspect peculiar to itself. The Appalachian system in the United States, generally extends in a direction, deviating not very essentially from S. W. to N. E.; but in Pennsylvania, the whole system is inflected from that course, and passes the state in a serpentine direction. Towards the S. boundary, the mountains lie about N. N. E., gradually inclining more eastwardly as they penetrate northwards; and in the central cos. many of the chains lie nearly east and west; but as they extend towards the northern border of the state, they again imperceptibly incline to the north east, and enter New York and New Jersey, in nearly that direction.

The influence of the mountains in modifying the general features, is very obvious, far beyond where any chains or ridges are sufficiently elevated to be classed as parts of the Appalachian system. It will be, however, shown in the progress of this review, that the mountain system is very much too greatly restricted, not alone in Pennsylvania, but also in Md., Va., N. Y. and N. J. Without attending to minor claims, the mountains of Pa. advancing from the southeast to N. W. are as follows:—though omitted in most maps, a chain enters the south boundary of York co. and cut by the Susquehannah river, rises in and traverses Lancaster county between Pequea and Octorara creeks; and between the sources of the Conestoga and Brandywine, separates for a short distance, Lancaster and Chester cos. Continuing between Berks and Chester, it is interrupted by the Schuylkill above Pottstown Rising again, and stretching N. E. forms, first, the boundary between Montgomery and Berks; thence between Lehigh and Bucks, and separating Northampton from Bucks, reaches the Del. Pursuing a north east course through N. J. separating Sussex from Huntingdon, Morris, and Bergen cos. enters N. Y. between the sources of the Walkill and Passaic rs., and extending in broken ridges, through the s. e. part of Orange co. forms the Highlands near West Point.

The almost uniform neglect of professed geographers respecting this strongly marked feature, attests the infancy of the science in the United States. After having formed the celebrated masses on both sides of the Hudson between Newburg and West Point, the ridge continues n. e. separating Putnam from Duchess co. Inflecting to the n. and forming the separating ridge between the waters of the Hudson and Housatonic rivers, stretches through the eastern part of Dutchess, Columbia, and Rensselaer cos. Along the two latter, however, the ridge forms, in reality, the separating boundary between N. Y. and Mass. and entering the s. w. angle of Vt. continues through that state, by the name of Green mountains, into Lower Canada. Thus prominent and continuous, from the Susquehannah to the n. e., this part of the Appalachian system is equally so through Md., Va. and N. C. Passing over Harford, Baltimore, Ann Arundel, and Montgomery cos. in Md. it forms falls in the Potomac, twelve ms. above Georgetown, and extends into Va. in Fairfax co. Varying in distance from 20 to 30 ms. the Great Kittatinny or Blue Ridge, and the ridge we have been tracing, traverses Va. into N. C. Leaving Va. in Henry, and entering N. C. in Stokes co., there is no doubt but that it is distinctly continued over the Carolinas and Geo. into Ala. Though the structure of the Atlantic slope, decidedly evinces a conformity to the Appalachian system, far below the S. E. mountain, it is the terminating continuous ridge towards the Atlantic. n. w. from, and nearly parallel to, the South mountain, another very remarkable ridge traverses N. J. and Pa. and similar to the former, the latter is unknown in either of these states, by any general name. Its continuation in New York is designated by the Shawangunk. Between the Susquehannah and Potomac, it is termed relatively, the South mountain, and in Virginia and the Carolinas, it forms the Blue ridge, and entering the n. w. part of Geo. is gradually lost amongst the sources of Chattahooche river.

To preserve perspicuity, we have adopted, or rather extended the name, Blue Ridge into Pa. and N. J. This very remarkable chain of the Appalachian system enters Pa. on its southern line, and stretching n. between Adams and Franklin cos. reaches the sthrn. angle of Cumberland, where it turns to northeast, and extending towards the Susquehannah, separates Cumberland from Adams and York cos. About six miles below Harrisburg, the Blue Ridge is pierced or broken by the Susquehannah, and again rising below the mouth of Swatara, crosses the southern angle of Dauphin; thence known as the Conewago hills, it separates Lebanon from Lancaster co. enters Berks, and reaches the Schuylkill at Reading. Continuing through Berks, Lehigh, and Northampton cos., the Blue Ridge passes Allentown, Bethlehem, and Easton, is again interrupted by the Del. below the latter town. Extending through Sussex co. the Blue Ridge enters N. Y. and and is finally terminated in the Shawangunk, on the west side of Hudson r. and amongst the branches of the Walkill. In one respect, the Southeast mtn. and Blue Ridge, in Pa. and N. J. differ from other sections of the Appalachian system. The two chains we have noticed, are formed of links more detached, than are those more remote from the Atlantic; but, otherwise, in respect to component matter, range, and vegetation, are in every place well marked sections of the general system. The very unequal elevation of their various parts, may, perhaps, be also adduced, as a characteristic of the Southeast mountain and Blue Ridge. The former does not, it is probable, in any part of Pa. or N. J. rise to 1,000 feet above the level of the Atlantic, whilst in N. Y. at the Highlands, some of the peaks, particularly Butterhill, exceed 1,500 feet elevation above tide water; and in Mass. and Vermont tower to near 3,000 feet. If taken generally, the Blue Ridge in Pa. and N. J. is more elevated than the Southeast mountain, yet no particular part of the former rises to an equal elevation with the Highlands, on either bank of the Hudson. In Md. the Blue Ridge assumes a very distinctive aspect, and separating Frederick and Washington cos. is broken by the Potomac at Harper's Ferry, below the mouth of Shenandoah. This fine chain crosses, and adorns Va. and N. and S. Carolina. In one remarkable circumstance, the Blue Ridge stands alone amongst the mountain chains of the United States. From the Susquehannah to n. w. angle of S. C., in a distance of upwards of 500 miles, it every where forms a county demarcation.

The third, and in some respects the most remarkable chain of Pa. is the Kittatinny. Known by divers local names, the Kittatinny, in a survey advancing from s. w. to n. e. first rises distinctively in Franklin co., and like other chains in the sthrn. margin of Pa. ranges a little east of north; but inflecting more to the northeast, extends to the Susquehannah, separating Cumberland and Perry cos. Five miles above Harrisburg, the Kittatinny is interrupted by the Susquehannah. Broken also by the Swatara, the Schuylkill, the Lehigh, and Delaware, the Kittatinny enters N. J. through which it passes into N. Y. and forms, by its continuation, the Catsbergs. The general aspect of the Kittaninny is much more continuous than any other mountain chain of Pa. It is, however, very far from being uniform in elevation, varying from

800, to perhaps 1,500 feet above tide water. Northwest from the Kittatinny, though more elevated, the chains are much less distinctly defined. Between the Kittatinny mountain, and the north branch of Susquehannah r. the intermediate country is in a great part composed of high rugged mountains, and narrow, deep, and precipitous valleys. This is the most sterile and least improvable part of Pa. but it is the region producing the most extensive masses of anthracite coal, known on the globe. The confusion in the natural arrangement of the anthracite section of Pa. is more apparent than real. The Kittatinny mountain and Susquehannah r. lie nearly parallel upwards of seventy ms.; distance from each other about 35 ms. The intervening space is filled by lateral chains, rising in many places, far above any part of the Kittatinny. Amongst these chains, two are worthy of particular notice, and serve, preeminently, to elucidate the very peculiar topography of interior Pa.

Bedford and Franklin cos. are separated by a chain, there known as Cove mountain. With a change of name, to Tuscarora mountain, the latter chain separates Franklin from Huntingdon, and Perry from Mifflin, and reaches the Susquehannah nearly opposite the southern extremity of Northumberland co. Rising again below the Mahantango r. and broken into vast links, the chain divides into nearly equal parts, the space between the Kitttatinny mountains and the main branch of Susquehannah r. Broad mountain, passed on the road from Easton and Bethlehem to Berwick, is one of the great links of this central chain.

More accurate surveys would, it is more than probable, identify Sideling hill, of Bedford co., Jack's mountain, of Huntingdon and Mifflin, and the central chains of Union, Columbia, and Luzerne cos. The chain which rises on both banks of the Susquehannah, in Luzerne, is amongst the most interesting features, not only in the U. S. but the world. The very peculiar structure of this valley will be noticed more appropriately, when treating of the rivers of that part of Pa. In the present instance, it is the mountains we have before us, and to which our attention is directed. Below Sunbury, a chain commences, or if my supposition is correct, is continued up the Susquehannah, along its left shore; this chain is crossed by the river above Danville, and again above Catawissa. From the latter place, the chain stretches to the northeast, through Columbia, enters Luzerne by the name of Nescopeck, and mingles ultimately with other chains, and is terminated towards the southern angle of Wayne co. Nearly parallel to the Nescopeck, and with a comparatively narrow intervening valley, another chain leaves the Susquehannah, above the borough of Northumberland, and traversing Northumberland and Columbia cos. enters Luzerne, and is broken by the Susquehanah sixteen ms. below Wilkes-Barre. Skirting the left bank about eight ms. it is again crossed by the r. and continuing its course N. E. passes about two and a half miles from and opposite Wilkes-Barre. Preserving its course N. E. it is for the third and last time, crossed by the Susquehannah, above the mouth of Lackawannock creek ten miles above Wilkes-Barre, and stretching towards the Del. is lost in Wayne co. Beyond the main branch of Susquehannah, to the northwest, the chains lie nearly parallel to those S. E. from that r. The structure of the country on both sides of the Susquehannah nearly the same. The yet discovered mines of anthracite coal, advancing from southeast to northwest, cease, in the chain immediately opposite Wikes-Barre.

To the eye, the region included between the west branch of Susquehannah and the Potomac, bears a strong analogy to that between the west and north branches of Susquehannah, but a minute scrutiny exposes a great change advancing southwest towards the borders of Maryland. Soil and vegetation both differ materially. The beech, hemlock, and sugar-maple forests, are succeeded in the valleys, by oak, hickory, and elm. Thus far the entire drain of Pa. is into the Atlantic ocean. The chain called the Alleghany forms in the southern parts of Pa. the dividing ridge between the Atlantic slope and the valley of Ohio.

Alleghany mountain has, no doubt, from this circumstance, received its preeminence amongst the mountain chains of Pa., Md. and Va. Only about sixty miles of its range in the former state, however, does separate the sources of the streams of the two great natural sections, the Atlantic slope and Ohio valley. The Alleghany chain leaving Alleghany co. in Md., separates Bedford and Somerset cos. and extending in a northerly direction, also separates the N. W. part of Bedford from the S. E. part of Cambria co. At the extreme northern angle of Bedford, the Alleghany turns to northeast, and is thence drained on both sides by the tributary streams of the Susquehannah. Discharging the waters of the west branch to the N. W. and those of the Juniata and Bald Eagle rivers to the S. E., the Alleghany reaches the west branch of Susquehannah at the mouth of Bald Eagle river.

Here, once more, the defect of our maps is strikingly apparent. Lycoming co. is delineated as if no mountain chains traversed its surface. This is not the fact, though too little is known of that part of Pa. to admit a classification of its mountains. If I was to hazard a conjecture, I should make the chain which crosses the Susquehannah in Bradford co. near to, and below Towanda, the continuation of the Alleghany. It may be remarked, that it is only in a few places E. of and those immediately in its spurs, that bituminous coal has been hitherto discovered in Pa. on the Atlantic slope, whilst this mineral abounds N. W. from the Alleghany chain.

This locality of bituminous coal prevails across the whole state, and is found from near Towanda, in Pennsylvania, into Maryland.

How far, and to what extent, the bituminous coal formation spreads into Virginia, we are unable to determine. The Alleghany chain may, in the existing state of our mineralogical knowledge, be viewed as the limit between the two species of coal in Pa.

Whatever may be the elevation of its summit, the base of the Alleghany chain, between Bedford and Somerset, and Cambria cos. constitutes the height of land between the Ohio r. and Atlantic tides, and forms also a similar demarcation in Md. This circumstance is entitled to our serious notice, from this region being the intended route of the Chesapeake and Ohio canal. The summit level, or Cumberland road, as given by Mr. Schriver, is 2,825 feet.

As a mountain chain, the Alleghany yields in grandeur of scenery, and in elevation above its base, to not only the Broad mountain, but to many other chains of the Appalachian system.

Chesnut ridge is the next chain west of the Alleghany, the two chains extending nearly parallel, and about twenty miles asunder. Though comparatively humble in respect to elevation, Chesnut ridge is one of the most extended chains of the system to which it appertains, reaching by various local names over Va., into Ten., and most probably into Ala. As placed on our maps, Chesnut ridge enters Pa. at the N. W. angle of Md., and ranging a little east of north, forms the boundary between Union and Somerset, thence between Westmoreland and Somerset, and finally between the N. E. angle of Westmoreland and the S. W. of Cambria co. At the extreme N. E. angle of Westmoreland, the Chesnut ridge reaches the Kiskiminitas r., and as delineated, its termination. So far from being so in nature, this chain preserves its identity through the state farther N. than any other chain of the Appalachian system.

Laurel hill is the last chain of the system in Pennsylvania. – What has been already observed respecting the comparatively depressed chains nearest the Atlantic, may be repeated respecting the Chesnut ridge, and the Laurel hill; that, though not very elevated, they nevertheless exist as well defined mountain chains. The latter is a very extended branch of the system, reaching from the northern part of Pa. into Ala. This chain traverses Va. by various names; separates Va. from Ky. as Cumberland mountain; traversed Ten., and penetrates Ala. under the latter term, and interrupted by Ten. r. it forms the Muscle Shoals, and is imperceptibly merged into the central hills of Ala. Like many others, this very lengthened chain is delineated defectively in every map of Pa. I have seen. Similar to Chesnut ridge, Laurel hill is terminated on our maps, near the Kiskiminitas, though in reality extending to near the south boundary of New York.

In addition to the great chains we have been surveying, many of minor importance might be noted; but we have deemed a view of the most striking parts sufficient.

If engrouped into one view, the mountains of Pa. exhibit many very interesting points of observation. The Appalachian system is here upwards of one hundred and fifty miles wide. The particular chains do not average more than three miles, if so much, in breadth.

Before proceeding farther in our review, I may be permitted to observe, that mountains are considered as the superlative of hills. In not only Pa. but in the Appalachian system generally, hills and mountains are not only specifically, but generically, distinct features of nature. If this was not the case, the slope would, in most cases, gradually rise from the mouths to the sources of rivers, and no regular ranges of elevated ground could be found crossing the streams obliquely. According to common opinion, the mountains of the U. S. form the dividing ridge between the waters of the Atlantic slope, and those of the Miss. and St. Lawrence basins. So far, however, are the mountains from constituting the separating line of the waters, that the real dividing ridge, if it can be so called, crosses the mountains diagonally.

The Appalachian system is formed, as we have seen, by a number of collateral chains, lying nearly parallel; each chain is again formed by ridges, which interlocking, or interrupted by rivers, extend generally in a similar direction with the chain to which they particularly appertain. The chains differ materially from each other in elevation and in continuity. In some of the chains, at each side of the system, the parts are of very unequal height above their bases, and of tide water. The Southeast mountain and Blue Ridge are prominent examples.

In the correct solution of any question arising out of the advance or distribution of population, the determination of the real surface covered with mountains, would afford extremely satisfactory element. As far as my own personal observation, and the present state of our geographical knowledge afford data, I have estimated the extent of mountain base in Pa.; and on the best maps, carefully measuring every chain, the entire length produced, amounts to a small excess above 2,250 miles. If the latter sum is, however, taken, and three miles allowed for the mean breadth of the chains, the mountain area will be 6,750 square miles, or very nearly one seventh part of the superficies of the state.

The respective r. basins, or rather the sections included in Pa., are of very unequal extent. Delaware, Susquehannah, and Ohio include an immense proportion of the whole state, and subdivide it naturally into the eastern, middle, and western river sections.

The following tables give the respective area of each, and also the smaller sections of Potomac, Genesee, and Erie.

Delaware river drains the counties of

	Square Miles.	Acres.
Berks,	950	608,000
Bucks,	640	409,600
Chester 3-4,	550	352,000
Delaware,	180	115,200
Lebanon 1-8,	40	25,600
Lehigh,	360	230,400
Luzerne,	180	115,200
Montgomery,	450	288,000
Northampton,	1,100	704,000
Philadelphia,	120	76,800
Pike,	850	544,000
Schuylkill 5-8,	500	320,000
Wayne,	790	505,600
	6,710	4,294,400

Susquehannah drains the counties of

	Square Miles.	Acres.
Adams 3-5,	350	224,000
Bedford 3-5,	1,000	640,000
Bradford,	1,260	806,400
Cambria 2-5,	330	211,200
Centre,	1,460	934,400
Chester 1-4,	180	111,200
Clearfield 9-10,	1,450	928,000
Columbia,	630	403,200
Cumberland,	630	403,200
Dauphin,	550	352,000
Franklin 1-3,	280	179,200
Huntingdon,	1,280	819,200
Indiana 1-10,	80	51,200
Lebanon 7-8,	280	179,200
Luzerne 9-10,	1,920	1,228,800
Lycoming,	2510	1,606,400
M'Kean 1-4,	380	243,200
Mifflin,	910	582,400
Northumberland,	500	320,000
Perry,	550	352,000
Potter 5-8,	750	480,000
Schuylkill 3-8,	300	192,000
Susquehannah,	910	582,400
Tioga,	1,180	755,200
Union,	600	384,000
York,	1,120	716,800
	21,390	13,685,600

	Square Miles.	Acres.
Genesee drains 1-8 of Potter	150	96,000

Potomac drains

	Square Miles.	Acres.
Adams 2-5,	220	140,800
Bedford 2-5,	630	403,200
Franklin 2-3,	560	358,400
Somerset 1-5,	180	115,200
	1,590	1,017,600

Lake Erie drains 1-2 of Erie county,

	Square Miles.	Acres.
	380	243,200

Ohio river drains the counties of

	Square Miles.	Acres.
Alleghany,	818	518,400
Armstrong,	1,010	646,400
Beaver,	690	441,600
Butler,	850	544,000
Cambria 3-5,	800	512,000
Clearfield 1-10,	160	102,400
Crawford,	1,040	665,600
Erie 1-2,	380	243,200
Fayette,	900	576,000
Greene,	640	409,600
Indiana 9-10,	680	435,200
Jefferson,	1,280	819,200
M'Kean 3-4,	1,140	729,600
Mercer,	880	563,200
Potter 1-4,	320	204,800
Somerset 5-6,	800	512,000
Venango,	1,200	768,000
Warren,	900	576,000
Washington,	900	576,000
Westmoreland,	1,180	755,200
	16,760	10,598,400

SUMMARY.

	Square Miles.	Acres.
Delaware drains,	6,710	4,294,400
Susquehannah,	21,390	13,685,600
Genesee,	150	96,000
Potomac,	1,590	1,017,600
Ohio,	16,760	10,598,400
Lake Erie,	380	243,200
	46,980	29,935,200

(See articles Delaware, Susquehannah, Potomac, Genesee, Alleghany, Monongahela, Ohio, &c.)

Over a surface of 47,000 sq. ms. traversed by a wide mtn. system, and on which relative level of arable land differs from a surface, barely above tide water, to upwards of 2,000 feet, the varieties of soil must necessarily be very great, and such is the fact. Though as a state, Pa. may be designated fertile, yet, between the river alluvion, on both sides of the system and in the Appalachian vallies, and the rocky slopes of the mountains, the respective quality of soil embraces nearly the extremes of sterility and productiveness. The whole state where at all arable is favorable to grasses, including bread grain. Of fruits, the apple seems best adapted to the climate and soil, though similar to the stone fruits, liable to destruction from untimely frosts.

philosophy of climate, it is only recently that relative height has been duly introduced, and yet without regard to difference of level, no rational deduction respecting climate can be formed on any part of the earth. Under the head of Maryland, to which article the reader is referred, are introduced general tables of mean and extreme temperature, and of prevalent winds. The tables were placed in that article from the central position of Md., amongst the Atlantic states of the U. S.

The following tables, founded on observations made in the city of Phila., and at Ger-

contiguous places.

1. The monthly mean temperature at Phila. is from a series of 20 years observations, made by James Young, from 1807 in-

clusive: that of Germantown is from a series of 10 years observations, by Reuben Haines, from 1819 to 1828 inclusive, Fahrenheit.

	Philadelphia.	Germantown.	Excess in Phil.
Jan.	32° 7	30° 0	2° 7
Feb.	36 32	33 10	3 22
March	46 64	41 22	4 42
April	57 18	49 40	7 78
May	68 01	61 30	6 80
June	78 27	71 20	7 7
July	82 25	75 0	7 25
Aug.	80 06	73 0	7 6
Sep.	73 39	65 0	8 39
Oct.	60 81	53 40	7 41
Nov.	47 34	42 60	4 74
Dec.	37 01	32 60	4 41
mean ann. temp.	58 41	52 37	

2. Table of mean monthly temp. at Phil. and Germantown, for 7 consecutive years, from 1820, 1826 inclusive, being years common to both observers:

	Philadelphia.	Germantown.	Differ.
January,	32.95°	29.68°	3.27°
February,	39.93	31.72	8.21
March,	47.03	40.61	6.42
April,	55.53	50.32	5.21
May,	70.44	61.76	8.68
June,	80.05	70.03	10.02
July,	84.07	75.04	9.03
August,	80.46	72.92	7.54
September,	76.68	64.83	11.85
October,	61.89	53.38	8.51
November,	47.23	41.75	5.48
December,	36.93	31.75	5.18
Mean Annual,	57.08	52	5.08

From the above elements we involuntarily deduce the important fact, that the summer climate of Phil. is from 8° to 10°, and in some instances still higher, above that of the adjacent country; and that in winter the city atmosphere is warmest by from 3½ to 5 degrees. For the prevailing winds of Pa., we may again refer to the article Md., with at the same time observing, that from Mr. Young's observations in Phil., the winds from the true wstrn. points N. W., W. and S. W., amount to 602 thousandths of the whole winds of the year; and in Germantown observations of Mr. Haine's, a similar proportion gives 663 thousandths. Combining therefore, the various observations quoted in this treatise, we are shown that the greatly prevailing winds of the middle states of the U. S., are from the wstrn. sides of the meridians. Again, if we add the intensity of the ærial currents to their respective courses, it would be safe to say, that four fifths of all the atmospheric pressure is estwrd., and in Penn. the almost uniform leaning of forest, and more of orchard trees, demonstrates this physical fact.

The climate of Pa. is relatively influenced by change of level. The surface of the state, with two partial exceptions, is composed of two great plains, declining from the dividing ridge of its waters. The estrn. declivity drained by the Del. and Susquehannah, and their confluents, falls from an elevation of about 2,000 ft. to the level of tide water; but the wstrn. declivity also to the wstrd., and drained by the numerous confluents of Ohio, is upwards of 600 ft. elevated above the ocean tides at the very lowest part. This greater height is one of the most influential causes of the comparative low temperature w. of the mtns., particularly in winter.

Observations made at U. S. military posts and by several other observers, have dissipated the long cherished vulgar error of a superior warmth on like latitudes w. of the mtns., and have established directly the reverse, giving rationality to the theory of our climate, and distributing comparative temperature according to the relative latitude and height.

On both plains of Pa. it is a rare occurrence when the rivers are not frozen and rendered unnavigable in winter, for a longer or shorter period; 40 days would probably approach near a mean of this winter period.

The actual summer or period between frosts, does not exceed, if it amounts to, a mean of 120 days, except in the sthestrn. and lowest part. Receding wstrd., occasional frosts entrench on summer, and on the high mountain vallies, even where farming is conducted to considerable advantage, untimely frosts happen occasionally in every month of the year.

The quantity of rain, or rather more explicitly expressed, of water in rain, hail, snow, &c., which falls at any given place, affords very requisite elements in a theory of the climate. The following table was extracted literally from the Philadelphia Gazette, Jan. 5th, 1831. No. 494.

Statement of the rain fallen from 1810 to 1830, inclusive, the first 14 years by the guage of P. Legarux, of Spring Mill, the following 7 years by that kept at the Pa. hospital.

	Inches.		Inches.
1810,	32.656	1821,	32.182
1811,	34.968	1822,	29.864
1812,	39. 3	1823,	41.815
1813,	35.625	1824,	38.74
1814,	43.135	1825,	29.57
1815,	34.666	1826,	35.14
1816,	27.947	1827,	38.50
1817,	36.005	1828,	37.97
1818,	30.177	1829,	41.85
1819,	23.354	1830,	45.07
1820,	39.609		

"The whole quantity fallen for 21 years is 748.143 inches, which, divided by 21 years, gives 35.626 inches as the annual average for that time."—*Pa. Hospital, 1st mo., 1st*, 1831.

From the whole of these tables we find the discrepancy between the mean and extreme temperature of different years at the same place, to be great, but we also find that moisture falls as unequally. Comparing the tables in my possession, I cannot trace any strong obvious connexion between the prevalence or scarcity of rain and mean temperature. In 1816, a year of unusual low mean temperature, it appears that the quantity of rain fell short of 28 inches; but in 1819, a warm year, the rain amounted only to 23.354 inches; and in 1814, when the rain fallen was so high as 43.135 inches, the mean temperature was high.

Natural productions. Under this head a volume might be written, but a few brief notices can only be admitted into an article necessarily brief. The two great mineral productions of Pa., are iron and fossil coal.

Iron ore is very extensively disseminated, and as greatly diversified in quality and richness. The iron mines in the estrn. part of the state were explored and worked at an early period of colonial settlement in Pa., and had become an interest of great value before the revolution. Since the peace of 1783, with much fluctuation iron has at all times employed much capital and labor.

Next to iron ore, mineral coal is most widely disseminated and is also next in importance. The fossil coal of Pa. is of 2 species: anthracite in the valleys E. of the main spine of the Appalachian system, and bituminous W. of that ridge. This distribution may admit of some exceptions, but if taken generally it is correct. Independent of specific distinction the coal of Pa. has very deeply influenced the improvement of the state at both extremes. The immense canals and locks which have been created on the Lackawannock, Lehigh, Delaware and the Schuylkill, and others in progress or designed in every part of the state, have been more or less the effect of a desire to render accessible these vast mineral deposits. The author of this article has visited the coal regions of Pa. both E. and W. The eastern anthracite is mostly found imbedded in inclining strata; the coal beds themselves of every variety of thickness from less than 6 inches to immmense mountain masses of unknown extent. The largest body yet laid open is that explored and worked on Mauch Chunk mtn. by the Lehigh coal and navigation company. It is probable, however, from numerous indications, that the most extensive deposits of coal E. of the mtns. remain unexplored, perhaps undiscovered.

The strata actually known are vast, and beyond the power of man to exhaust in many succeeding ages. The position of the bituminous coal of the western part of Pa. is level or very nearly so. It is imbedded in horizontal strata, and unlike the anthracite is often so very near a dead level as to admit drainage with difficulty. In using the relative terms *east* and *west* in stating the great deposits of Pa. coal, some modification is necessary. I have myself found bituminous coal on Towanda cr. Bradford co. Pa., and in Alleghany co. Md. near Cumberland. A line from one of these points to the other, runs nearly with the great spine of Alleghany, and divides Pa. into two not very unequal sections. The bituminous deposits prevail from the Alleghany and increase in extent falling down the r. to Pittsburg. In the vicinity of this city the the coal strata are in extent immense; from 3 to 6 feet in thickness, and often from 250 to 350 feet above the high water level of the rs. The number of mines already open, and the ease of reaching the coal from the deep river valleys has contributed to fasten the opinion that coal is in unusual abundance near Pittsburg. That opinion is rendered doubtful by the fact, that in every part of the adjacent country where sufficient pains have been taken, coal has been found, and from analogy we may suspect its existence as underlaying strata far into the state of Ohio. It abounds along the Ohio r. as low down as Cincinnati. Tho' an indefinite number of other minerals have been named as having been discovered in Pa. except iron and coal, limestone is the only one of extensive use and value. Limestone is the prevailing rock in a band spreading N. W. from Blue Ridge, and crossing the whole of Pa. from the Del. into Md. It exists in detached deposits E. of Blue Ridge, and in western Pa. is found in interminable beds, alternating with other rocks. Water impregnated with *muriate of soda*, or common salt, is found on the waters of Conemaugh, and might be, it is probable, discovered in other places in the Ohio valley, by digging to sufficient depth. On the Conemaugh salt works have been many years in operation. Marble, of great variety of shade and tint, and that receives a beautiful polish, is found in the lower cos. of Philadelphia, Chester, Montgomery, &c. These elegant marbles have greatly added to the convenience, cheapness and beauty of domestic architecture.

Internal Improvements.—The foundation of the wealth and improvement of this prosperous state is deeply laid, in her fertility of soil, her iron mines, her coal stratas, and the industry of her population. Under this head we can only give the names of the works, and refer to the respective heads. Belonging to the state is the magnificent line of canals and rail road, entitled the Pa. canal and Columbia rail road. Belonging to joint stock companies, are the Schuylkill navigation; Union canal; Lehigh navigation; consisting of artificial navigation along the Lehigh and the Mauch Chunk rail road; Lackawaxen canal and rail road; Conestoga canal; and the Chesapeake and Del. canal. We include the latter in the works of Pa. from the circumstance of the work having been in great part designed and executed by citizens of that state. Beside numerous others, there is a line of turnpike road extending from the city of Philadelphia through Lancaster, York, Gettysburg, Chambersburg, Bedford, Greensburg and Pittsburg to Washington, where it meets the U. S. road. The latter enters Pa. in the southwestern angle of Somerset, and traversing Fayette and Washington cos. by Union, Brownsville and Washington, passes on to Wheeling in Va.

Without entering into an enumeration of

separate works, the subjoined extract will give some idea of the magnitude of the improvement interest of Pa. at this time, 1831.

The bill making additional appropriations for internal improvements, passed the house of representatives of Pa. on Friday, (Feb. 4, 1831.)—Yeas 56.—Nays 38. The following are the appropriations made by this bill as it finally passed the house. The amount added to previous expenditure, will make an aggregate of about fifteen millions of dollars, which that state has applied to the purpose of improvement by canals and rail roads within the last six years. The present bill appropriates to:—

Phil. and Columbia rail road,	$600,000
Canal from Middletown to Columbia,	116,170
North Branch canal,	100,000
West Branch canal,	200,000
Lewisburg inlet,	25,000
Canal from Huntingdon West, and rail road over the Alleghany,	700,000
French Creek feeder,	60,000
Beaver and Chenango route,	100,000
Southwestern turnpike,	125,000
Amount,	$2,026,170

Vide National Intelligencer, Feb. 8th, 1831, No. 4,582.

According to the report of the auditor general of the treasury, at the beginning of the year 1831, the capital stock paid in of 32 banks, was $12,815,581 83; notes in circulation $7,870,613 90; contingent funds $1,170,068 02; bills discounted $18,454,213 50; specie $3,013,383 84; amount of deposits $7,244,752 95.

By the same report, the commonwealth owns,

In bank stock,	$2,108,700 00
Turnpike stock,	1,911,243 39
Bridge stock,	410,000 00
Canal stock,	200,000 00
	$4,629,943 39

The dividends received by the state on the bank stock, amounted during the last year to $121,716, and on the bridge, canal and turnpike stocks to $29,715, amounting aggregately to $151,431.

Political divisions and population.—Pennsylvania, as has been observed, is naturally divided into three physical sections: first, southeastern section from the Kittatinny; second the central mtns.; and third, the western or that part drained by the constituents of O. r. Along their lines of separation these natural sections blend, but in their physiognomy respectively, they are strongly contrasted. Of these divisions, the first or southeastern contains about 8,028 sq. ms., and a population of 603,864; the 2d mountainous or middle section, 20,850 sq. ms. and pop. 306,214; and the 3d or Ohio section 16,332 sq. ms., and pop. 342,922. The population of the state in 1800 was 602,545; 1810, 810,091; 1820, 1,049,313, and in 1830, 1,348,233. In 1820 and 1830 as follows:

Counties.	Pop. 1820.	Pop. 1830.
Adams,	19,370	21,379
Alleghany,	34,921	50,552
Armstrong,	10,324	17,701
Beaver,	15,340	24,183
Bedford,	20,248	24,502
Berks,	46,275	53,152
Bradford,	11,554	19,746
Bucks,	37,842	45,745
Butler,	10,193	14,581
Cambria,	2,287	7,076
Centre,	13,796	18,879
Chester,	44,451	50,910
Clearfield,	2,342	4,803
Columbia,	17,621	20,059
Crawford,	9,397	16,030
Cumberland,	23,606	29,226
Dauphin,	21,653	25,243
Delaware,	14,810	17,323
Erie,	8,553	17,041
Fayette,	27,285	29,172
Franklin,	31,892	35,037
Greene,	15,554	18,028
Huntingdon,	20,144	27,145
Indiana,	8,882	14,252
Jefferson,	561	2,025
Juniata, included in Mifflin by census 1830.		
Lancaster,	68,336	76,631
Lebanon,	16,988	20,557
Lehigh,	18,895	22,256
Luzerne,	20,027	27,379
Lycoming,	13,517	17,636
MacKean,	728	1,439
Mercer,	11,681	19,729
Mifflin,	16,618	21,690
Montgomery,	35,793	39,406
Northampton,	31,765	39,482
Northumberland,	15,424	18,133
Perry,	11,342	14,261
Phil. city and co.	137,097	188,797
Pike,	2,894	4,843
Potter,	186	1,265
Schuylkill,	11,339	20,744
Somerset,	13,974	17,762
Susquehannah,	9,660	16,787
Tioga,	4,021	8,978
Union,	18,619	20,795
Venango,	4,915	9,470
Warren,	1,976	4,697
Washington,	40,038	42,784
Wayne,	4,127	7,663
Westmoreland,	30,540	38,400
York,	38,759	42,859

Total pop. 1820, 1,049,313, 1830, 1,348,233.

Of the foregoing were white persons,

	Males.	Females.
Under 5 years of age,	117,853	111,947
From 5 to 10	96,199	92,719
" 10 to 15	82,375	80,067
" 15 to 20	73,113	75,976
" 20 to 30	121,359	115,898
" 30 to 40	75,172	69,604
" 40 to 50	46,600	44,485
" 50 to 60	28,032	27,882
" 60 to 70	16,085	16,221
" 70 to 80	6,979	7,084
" 80 to 90	1,775	1,929
" 90 to 100	228	235
" 100 and upwards	42	21
Total,	565,812	614,089

Persons in the foregoing who are deaf and dumb, under 14 years of age, 224; of 14 to 25, 279, and of 25 and upwards 255. Blind, 475.

Colored population as follows:—

	Free.		Slaves.	
	Male.	Female.	Male.	Female.
Under 10 yrs.	5,095	5,054	23	32
From 10 to 24	6,250	6,142	102	106
" 24 to 36	4,069	4,476	25	22
" 36 to 55	2,796	2,742	11	25
" 55 to 100	1,132	1,105	10	42
" 100 and over	35	34	1	4
Total,	18,377	19,553	172	231

Colored deaf and dumb, under 14 years of age, 12; from 14 to 26, 12; 26 and over, 15. Blind, 28.

Recapitulation.

Whites.	Free colored.	Slaves.	Total.
1,309,900	37,930	403	1,348,233

Education.—Literary Institutions.—Pennsylvania has two universities, one in Philadelphia, and another, "the Western university," in Pittsburg; Mount Airy college, Germantown; Dickinson college, Carlisle; Washton, in Washington; Jefferson, in Cannonsburg; Alleghany, in Meadville; Madison, in Union; and numerous academies in the different boroughs. The Phil. library may be, with great propriety, ranked amongst the first literary institutions of the state. The interests of education have received so much legislative attention, as to induce a false opinion of the distributive benefits derived from these enactments. Elementary instruction is in many large sections lamentably neglected. One cause of this evil, may be found, by examining the tables of pop., where the very unequal density of the objects of education is most strikingly apparent. But the inequality of inhabitants on a given space is, however, only one cause why mental culture is neglected.

The Moravian, or United Brethren, have schools at Bethlehem, Nazareth, Litiz, &c. There are theological seminaries at York, at Gettysburg, and in Alleghany town, opposite the city of Pittsburg.

Under a constitutional injunction, legislative provision has been made for gratuitous instruction to the children of indigent parents. The first school dist. of the state, comprising the city and co. of Phil. has received an organization, which, according to the 12th annual report of the comptrollers of public schools in this dist. dated 5th Feb. 1830, has extended instruction to 34,703 children within the 12 preceding years.

Constitution.—Judiciary.—The existing constitution of Pa. was adopted the 23d Sept. 1790. The legislative power is vested in a general assembly, which shall consist of a senate and house of representatives. No person shall be a representative, who shall not have attained the age of 21 years, and have been a citizen and inhabitant of the state three years next preceding his election, and *the last year* thereof an inhabitant of the city or co. in which he shall be chosen; unless he shall have been absent on the public business of the U. S. or of this state. Representatives are chosen annually.

The senators shall be chosen for four years by the citizens of Philadelphia, and of the several cos., at the same time, in the same manner, and at the same place where they shall vote for representatives. No person shall be a senator who shall not have attained the age of 25 years, and have been a citizen and inhabitant of the state four years next before his election, and the last year thereof an inhabitant of the district for which he shall have been chosen; unless he shall have been absent on the public business of the U. S. or of this state.

The senators shall be chosen in dists. to be formed by the legislature; each dist. containing such a number of taxable inhabitants as shall be entitled to elect not more than 4 senators. When a dist. is composed of 2 or more cos., they shall be adjoining. Neither the city of Phil. nor any co. shall be divided, in forming a dist.

In elections by the citizens, every free man of the age of 21 years, having resided in the state two years next before the election, and within that time paid a state or co. tax, which shall have been assessed at least 6 months before the election, shall enjoy the right of an elector.

An enumeration of the taxable inhabitants shall be made separately, in such manner as shall be directed by law. Such enumerations have been made in 1793, 1800, 1807, &c. up to 1828.

The supreme executive power shall be vested in a governor, who shall be chosen on the second Tuesday of October, by the citizens of the commonwealth, at the place where they shall respectively vote for representatives. The person having the highest number of votes shall be governor. But, if two or more shall be equal and highest in votes, one of them shall be chosen governor by the joint vote of the members of both houses. The governor shall hold his office during three years from the third Tuesday of December, next ensuing his election; and shall not be capable of holding it longer than 9 years in any term of twelve years. He shall be at least 30 years of age, and have been a citizen and inhabitant of this state 7 years next before his election; unless he shall have been absent on the public business of the U. S., or of this state. No member of congress, or person holding any office under the U. S. or this state, shall exercise the office of governor.

The governor shall be commander in chief of the army and navy of the commonwealth, and the militia; except when they shall be called into the actual service of the U. S. He shall appoint all officers whose offices are established by this constitution, or shall be established by law, and whose appointments are not herein otherwise provided for. He shall have power to remit fines and forfeitures and grant reprieves and pardons, except in cases of impeachment.

Every bill, which shall have passed both houses of the general assembly, shall be presented to the governor. If he approve, he shall sign it; but if he shall not approve, he shall return it, with his exceptions, to the house in which it shall have originated, and must be first examined in that house, and sent to the other, with the governor's objections, and if approved by two thirds of each house it becomes a law. Any bill sent to the governor and not returned in ten days, Sundays excepted, becomes also a law.

The judicial power is vested in a supreme court; in courts of oyer and terminer and general jail delivery; in a court of common pleas, orphans court, registers court, and a court of quarter sessions of the peace, for each co., in justices of the peace, &c.

The state is divided into sixteen judicial circuits, over which is appointed a president judge. For the supreme court five dists. have been formed. The jurisdiction of the supreme court shall extend over the state, and the judges thereof shall, by virtue of their offices, be justices of oyer and terminer and general jail delivery, in the several cos.

The trial by jury in issues of fact to remain inviolate.

No person who acknowledges the being of a God and a future state of rewards and punishments, shall on account of his religious sentiments, be disqualified to hold any office or place of trust or profit under this commonwealth.

History.—The first settlement made in Pa. by a civilized people, was formed in 1627 or 1628, by a Swedish colony; but remote from a nation, and not qualified by population or wealth to sustain distant settlements, the Swedish colony remained weak, and in 1655, was conquered by the Dutch from N. Y. What is now Del., Pa., and N. J., shared the fate of all New Netherlands, or the Dutch settlements on the Del. and Hudson rs. In Aug. 1764, the whole country was seized by the English in virtue of a previous cession made by the States General. March, 1664, Charles II. granted the New Netherlands, by the name of N. Y. to his brother James, Duke of York. The latter on the 24th June, 1664, granted N. J. to Lord Berkeley, and Sir Geo. Carteret.

In June, 1680, Wm. Penn, son and heir of admiral Sir Wm. Penn, presented a petition to Charles II., stating not only his relationship to the late admiral, but that he was deprived of a debt due from the crown, and praying for a grant of lands lying northward of Md. and wstrd. of Del. His petition was recommended by the Duke of York, and acceded to by Lord Baltimore's agents. It was confirmed Jan. 1681, and in the ensuing May, Markham, an agent and relation of William Penn, was sent over to take possession.

What is now Del. or then called the three lower counties, was conveyed to William Penn by a grant from the Duke of York, Aug 1682, and on the 24th of Oct. of that year he landed in person at New Castle, and found on his arrival, in both his colonies, about 3,000 people Swedes, English, Dutch and Finns. The first deliberative assembly was convened at New Castle, Dec. 4th, and commenced a regular government.

Coaquanock, now Philadelphia, was chosen as the capital and laid out in 1682. Tho' involved in a controversy respecting their common boundaries, Md. and Pa. slowly but solidly advanced. The line between the two colonies was finally fixed in 1762 by actual survey, executed by two eminent English mathematicians, Mason and Dixon.

The second assembly of Pa. was held at Phil. 1683. In 1718 the founder died, and from that period to the revolution in 1775, except fixing the boundary as already noticed, and treaties with the Indians, Pa., happily for its inhabitants, afforded few events for history.

In the events of the revolution and in the subsequent history of the U.S., this great colony has acted a conspicuous part and risen to be, in wealth, pop. and improvement, the second state in the confederacy.

PENN TOWNSHIP, p-o. Pa., township adjoining Phil. to the northward, Phil. co. Pa.

PENN YAN, p-v. Milo, st. jus. Yates co. N. Y., 185 ms. w. Albany, 30 N. Bath, 16 s. Geneva; is in the N. w. corner of the town, on the outlet, ½ mile from Crooked lake, and partly in Benton. The stream affords good mill sites, which are used. The village contains the co. buildings. It derives its name from a combination of parts of the words Pennsylvanians and Yankees, having been settled by New Englanders and Pennsylvanians.

PENOBSCOT, river, Me., the largest which is wholly in that state, rises in the w. part of Somerset co., in numerous small branches, which flow from springs in the height of land on the frontier of the district of Quebec, L. Canada, very near the head waters of Chaudiere river. Some of the streams flow from Bald mountain ridge, which gives rise to some of the head waters of Kennebec r. Others interlock with head streams of St. John's r., so that a portage of 2 miles connects their boat navigation. All the waters of the Penobscot in Somerset co. unite in Chesumcook lake, and leave its s. end on the boundary of Penobscot co. The stream then flows s. E. near the foot of Katahdin mountain, thro' the crooked and irregular lake of Bamedumpkok, and receives the waters of several other lakes and ponds, the principal of which is Millinoket, and afterwards the east branch, which affords a boat navigation to Aroostook river, with only a short portage from the head of the Seboois lakes. Beyond this it bends s. w. receiving several tributaries, the principal of which is the Mattawamkeag, and on crossing the s. line of the co., flows between Lincoln and Waldo cos., forming, with Penobscot bay, their boundary to the ocean. Besides the above mentioned branches it receives in Penobscot co., there are many others, the chief of which is Piscataquis river, on the w. It contains many islands, most of which are small, except Old Town in Penobscot co., and Orphan isl. in Hancock co., and Isleborough in Waldo co. The Fox islands and Little Deer isls. in Hancock co. are in Penobscot bay; and off its mouth are several more, Isle au Haut, Manticus, &c. There are several very good harbors in the bay; and Belfast, nearly at its head, in Lincoln co. is a flourishing place. Vessels of some size go up to Bangor, in Penobscot co., 50 miles from the mouth of the bay. This great stream, with such a multitude of branches spread over a great extent of country, is capable of becoming a most important channel of trade; and, although the districts it waters, are still to a great degree wild and uninhabited, Bangor and Belfast have already become important places. A large part of the business continues to be the timber trade; but the forests annually recede before the axe, and give place to the plough.

PENOBSCOT, co. Maine, bounded by Lower Canada N., Washington co. E., Hancock and Waldo cos. S., and Somerset co. W.; is the largest co. in the state, and contains a large part of the uninhabited land within its boundaries. It was incorporated in 1816. The territory which it embraces, contained, in 1790, only 1,154 inhabitants. In 1820 only 1,143 acres, were improved in the towns, and the inhabitants were 13,870. In 1830 there were 3,582 acres of tillage, 9,476 pasturage, 11,000 mowing, &c. 23,940 acres cultivated, in all; and there were raised 12,957 bushels corn, 25,591 wheat, 1,333 rye, & 2,719 of oats. The streams, lakes, & ponds are too numerous to be all mentioned. St. John's r. crosses the co. E. and W. in the N. part, while its 3 N. branches rise and join it in the co. viz: St. Francis, Madawaska & Green rs. The Allagash is partly in this co., as are the sources of the Aroostic, and most of its course. Temiscouata, Long and Eagle lakes are the largest which are tributary to St. John's r. The S. part of the co. presents a labyrinth of lakes and streams, the sources and principal tributaries of Penobscot r. which, crossing the S. line, divides the cos. of Waldo and Hancock on its way to Penobscot bay and the sea. The largest of these lakes are Millinoket, Benedumpkok, &c. There are numerous mountains in different parts of the county, the principal of which is Katahdin, the highest eminence in the state. The E. and W. lines of the co. run N. and S. about 63 miles apart for nearly 200 ms. The N. and S. boundaries are irregular. The S. part of the co. contains nearly its entire population, and Bangor, which is a large and flourishing village, is on the Penobscot near the S. line. In 1830 there were only 46 towns with names, but many more laid out in ranges, besides plantations. Pop. 1820, 13,870, 1830, 31,530.

PENOBSCOT, p-t. and sea port, Hancock co. Me., 75 ms. E. Augusta, E. Penobscot bay, and opposite Belfast; is penetrated by an arm of the bay in the S. part, and possesses a considerable number of coasting vessels. Pop. 1830, 1,271.

PENOBSCOT, Indians. The remains of this tribe reside at Oldtown, or Orono, Penobscot co. Me., where, though reduced in numbers, they have a Catholic church, having been formerly collected under French missionaries from Canada.

PENSACOLA, fine bay of the U. S. on the nrthrn. shore of the Gulf of Mexico, Escambia co. Florida. The Pensacola bay is united to the Gulf by a narrow entrance, between Barancas Point and the wstrn. end of St. Rose's island. Stretching from the bar N. E. by E. 28 ms., with a mean width of about 3 ms., it receives into the northern part Escambia r. and Yellow Water r. from N. E. The entrance admitting vessels of 21 feet draught, is about 8 ms. S. S. W. from the city of Pensacola. The bar is on lat. 30° 19′, long. 10° 24′ w. W. C. The country around Pensacola bay is in general low, sandy, and barren.

PENSACOLA, city, port of entry, naval station of the U. S., p-o., t. and st. jus. Escambia co. Florida, is situated on the N. W. shore of the bay of the same name, by p-r. 242 ms. a little S. of W. Tuscaloosa; as laid down on Tanner's U. States, at lat. 30° 23′, long. 10° 19′ w. W. C. This city was founded 1699, by Don Andre de la Riola, a Spanish officer. Pop. 1830, about 2,000.

The harbor is safe and commodious, being the deepest haven belonging to the U. S. on the northern shore of the Gulf of Mexico, admitting vessels of 21 feet draught. The anchorage is good, on mud and sand, but towards the shores the water is generally shallow.

PEORIA, lake, between Tazewell and Peoria counties, Il., is an elliptical expansion, of the usual breadth of Illinois r. of about 20 ms. in length, and from half a mile to a mile wide. The lower part of this sheet of water is about 130 ms. a little W. of N. Vandalia.

PEORIA, co. Illinois, bounded by Fulton S. W., Knox N. W., Putnam N., and Illinois river separating it from Tazewell E. and S. E. It is in form of a triangle base, along the western border 40 ms., and perpendicular along the northern border 30 ms.; mean breadth 15 ms., and area 600 sq. ms. The northwestern angle is traversed in a southwestern direction by Spoon r., but the body of the co. slopes S. E. towards Illinois r. This co. was connected with Putnam in taking the census of 1830, and contained an aggregate population of 1,310. For lat. and long. see next art.

PEORIA, p-v. and st. jus. Peoria co. Il., situated on Illinois river, at the lower extremity of Peoria lake, by p-r. 143 ms. a little W. of N. Vandalia, lat. 40° 40′, long. W. C. 12° 35′ w.

PEPPERELL, p-t. Middlesex co. Mass., 40 ms. S. W. Boston, W. Nashua r.; has a soil favourable to grain, grass and fruit. The Nashua and one of its branches offer abundant water power. Maj. Gen. Prescott, who commanded the American troops at Bunker's Hill, was born here. Pop. 1830, 1,440.

PEQUANNOCK, t. Morris co. N. J. 18 ms. N. N. W. Newark, has Pequannock r. N., Rockaway r. S., and Pompton r. E. The Morris canal lies along its S. border, and Copperas brook, and Green meadow mountain, near its W line. Pequannock r. separates it from Bergen co. Pop. 1830, 4,451.

PEQUAWKETT, r. Strafford co. N.H., a small stream flowing N. into Saco r.

PEQUAWKETT, N. H., the Indian name for the proper residence of the Pequawkett Indians, a tract on Saco r. partly in Me. Conoray, N. H., and Fryeburgh, Me., with several other towns, are within its limits. The region is romantic, and was peculiarly adapted to the habits of Indians. Their principal residence was at Fryeburgh, where the Saco pursues a most serpentine course, flowing 36 ms. through rich meadows, in a township six ms. square, and affording a circuitous line of boat navigation of about 100 ms. in connection with the ponds emptying into it. Fish and fowl abounded here, and near where the village now is, was an Indian fort, which overlooked the extensive and fertile plain. This tribe was troublesome to the settlers in Mass. in the early part of the 18th century, and the general court having offered a reward for Indian scalps, a party proceeded through the wilderness, by Winnipiseogee and Ossipee lakes, against the Pequawketts, under the command of capt. Lovell, with Mr. Frye for their chaplain, from whom Fryeburgh had its name. After a desperate fight this party was defeated, and only a few of them found their way back.

PEQUAWKETT, mtn. N. H., a prominent eminence of the second peaks of the White mtns., between Bartlett, Coos co., and Chatham, Strafford co. It was formerly called Kearsearge mtn.

PEQUEST cr., N. J., a good mill stream, rises in Newton, Sussex co., crosses Warren co., and enters Del. r. at Belvidere v., Oxford t., after a course of about 35 ms., watering a narrow valley.

PERCIVAL'S, p-o. nthrn. part Brunswick co. Va., 67 ms. S. W. Richmond.

PERDIDO, r. and bay, forming the boundary between Baldwin co. of Ala., and Escambia of Flor. It rises in Baldwin co., flows sthrdly. about 40 ms., and expands into a narrow and shallow bay. Perdido bay is crooked, and from the intricacy of its entrance, derives its name "*Perdido*" or *Lost Bay*. The country adjacent to the bay, and drained by the Perdido, is mostly barren, and timbered with pine.

PERKIOMEN, r. of Pa., rises in Berks, Lehigh, and Bucks cos., and the constituent crs., uniting in the nrthwstrn. angle of Montgomery, assumes a sthrn. course into the Schuylkill 6 ms. above Norristown.

PERKIOMEN, p-o. on Perkiomen r., Montgomery co. Pa., 25 ms. N. W. Phil.

PERRIN'S mills and p-o., Clermont co. O., about 90 ms. S. W. Columbus.

PERRINGTON, p-t. Monroe co. N. Y., 12 ms. S. E. Rochester, has a good soil, and several streams, the chief of which is Irondequot, or Teorondo cr. The Erie canal crosses it near the middle, and is carried over Irondequot cr. with an embankment, at which is Hartwell's basin. Thomas' creek is a small mill stream. Pop. 1830, 2,183.

PERRY, p-t. Washington co. Me., 184 ms. E. Augusta, 5 N. W. Eastport, opposite Passamaquoddy bay and Deer island, in N. Brunswick, has Cobscook bay S., and is crossed from N. W. to S. E. by the outlet of a pond which empties into the bay. It enjoys facilities for fishing. Pop. 1830, 735.

PERRY, p-t. Genesee co. N. Y., 22 ms. S. E. Batavia, 239 W. Albany, W. Livingston co., about 6 ms. square, has a soil of ordinary quality, and several small streams flowing in different directions, with a part of the outlet of Silver lake in the S. E. Bog iron ore has been found here. Pop. 1830, 2,792.

PERRY, co. Pa., bounded by the Kittatinny mtns., separating it from Cumberland S., by Franklin S. W., Tuscarora mtn. separating it from Mifflin N. W. and N., and by the Susquehannah, separating it from Dauphin E. Length 36 ms., mean width 15, and area 540 sq. ms. Extending in lat. from 40° 12′ to 40° 40′, and in long. from the meridian of W. C. to 0° 42′ W. This co. partly is composed of what was formerly called Sherman's valley, and the sthrn. part is chiefly drained by Sherman's creek. The nrthestrn. section is traversed by Juniata r. The declivity of the whole is estrd. towards the Susquehannah. Though the surface is rocky, and much of it mountainous, the arable soil is excellent for grain, fruit, and pasturage. Chief towns, New Bloomfield, Landisburg, and Millerstown. Pop. 1820, 11,342; 1830, 14,361.

PERRY, p-v. N. W. part Venango co. Pa., 22 ms. from Franklin, the co. st., and about 75 ms. N. Pittsburg.

PERRY, p-v. and st. jus., Houston co. Geo., by p-r. 60 ms. S. W. Milledgeville, lat. 32° 25′, long. W. C. 6° 54′ W.

PERRY, co. Ala., bounded by Dallas S. E. and S., Marengo S. W., Greene N. W., Tuscaloosa and Bibb N., and Autauga E. Length 42 ms., mean width 23, and area 966 sq. ms. Extending in lat. from 32° 17′ to 32° 54′, and in long. from 10° 02′ to 10° 38′ W. W. C. The wstrn. side falls towards the wstrd., and is drained by creeks flowing into Tombigbee and Black Warrior rs., but the central, estrn., and much larger sections of the co., are traversed and drained by the Catawba and its branches. General declivity sthrd. Chief t. Marion, or Perry C. H. Pop. 1830, 11,490.

PERRY, C. H. and p-v., Perry co. Ala., by p-r. 61 ms. S. S. E. Tuscaloosa. Lat. 32° 37′, long. 10° 27′ W. W. C.

PERRY, co. Miss. bounded by Jackson S., Hancock S. W., Marion W., Jones N., Wayne N. E., and Greene E. Length 36, width 30, and area 1,080 sq. ms. Extending in lat. from 30° 55′ to 31° 33′, and in long. from 11° 58′ to 12° 30′ w. W. C. Declivity sthestrd., and drained by Leaf r. and Black cr., branches of Pascagoula r. It is moderately broken, with a soil generally thin, and covered with pine timber. Chief t. Augusta. Pop. 1820, 2,037; 1830, 2,300.

Perry, co. Ten., bounded by Wayne s. e., Hardin s. w., Henderson w., Carroll n. w., Humphreys n., and Hickman e. Length 36 ms., width 30, and area 1,080 sq. ms. Extending in lat. from 35° 27′ to 35° 55′, and the 11th degree w. W. C. passes very nearly over the middle of the co. The main volume of Ten. r. traverses it also in a nthrn. direction, as does the Buffalo branch of Duck r. Ten. flows over the wstrn. and Buffalo over the estrn. side, the latter entering its recipient, Duck r., in the nrthestrn. angle. General declivity nrthrd. Chief t. Barrysville. Pop. 1820, 2,384; 1830, 7,094.

Perry, co. Ky., bounded by Laurel mtn., separating it from Harlan s. e. and s., by Clay w., Estill n. w., Morgan n., Floyd n. e., and Pike e. Length 53 ms., mean width 20, and area 1,060 sq. ms. Extending in lat. from 36° 55′ to 37° 36′, and in long. from 5° 51′ to 6° 30′ w. W. C. Except a few creeks from Pike co., Perry gives source to the higher branches of Ky. These branches leave their mountain vallies in a n. n. w. direction, over Perry, uniting and turning wstrd. in Estill. Chief t. Perry C. H. Population 1830, 3,330.

Perry, C. H. and p-v., Perry co. Ky., by p-r. 114 ms. s. e. by e. Frankfort.

Perry, co. O., bounded by Athens s., Hocking s. w., Fairfield w., Licking n., Muskingum n. e., and Morgan e. and s. e. Length n. to s. 28 ms., mean breadth 18, and area 500 sq. ms. Lat. 39° 45′, long. W. C. 5° 15′ w. It is a table land, between the vallies of Hockhocking and Muskingum rs., and from which creeks of the former flow w. and s. w., and of the latter n. and n. e. The surface is in most parts hilly, and in several places mineral coal has been found. Chief t. Somerset. Pop. 1820, 8,429; 1830, 13,970.

Perry, p-v. Geauga co. O., by p-r. 165 ms. n. e. Columbus.

Perry, co. of Ind., bounded w. by Spencer, n. w. Dubois, n. Crawford, and by the O. r., separating it from Meade co. Ky. e., Breckenridge co. Ky. s., and Hancock co. Ky. s. w. Length from s. to n. 30 ms., mean breadth 15, and area 450 sq. ms. Lat. 38° 08′, long. W. C. 9° 40′ w. Slope sthrd. towards O. r. The surface is very broken, as it comprises a part of the great buttress of O. r., and reaches w. to the dividing ridge between that stream and White r. Chief t. Rome. Pop. 1830, 3,369.

Perry, co. Il., bounded by Jackson s., Randolph w., Washington n., Jefferson n. e., and Franklin s. e. Length from e. to w. 24 ms., mean breadth 18, and area 432 sq. ms. Lat. 38° 05′, long. 12° 24′ w. W. C. Slope sthrd., and drained in that direction by different branches of Muddy creek. Chief t. Pinckneyville. Pop. 1830, 1,215.

Perry, co. Mo., bounded s. by Cape Girondeau co., s. w. Madison, n. w. St. Genevieve, and by the Miss. r., separating it from Randolph co. Il. Lat. 37° 44′, long. W. C. 13° 00′ w. Slope nrthestrd. towards the Miss. r. Chief town, Perryville. Pop. 1830, 3,349.

Perryopolis, p-v. nrthwstrn. part Fayette co. Pa. 16 ms. a little w. of n. Uniontown, 8 n. e. Brownsville, and by p-r. 209 n. w. W. C.

Perry's, Bridge and p-o. on Vermillion r., Lafayette parish, La., by p-r. 217 ms. w. New Orleans.

Perrysburgh, p-t. Cattaraugus co. N. Y., 12 ms. n. w. Ellicottsville, 306 w. Albany; is nearly in the form of a triangle, with its long and irregular side n. e. bounded by Cattaraugus creek and Erie co., and partly by the south and its western branch, on the line of Chatauque co. It has several brooks flowing n. into Cattaraugus creek, and others s., the head streams of Conewango cr. The n. w. corner lies only 5 or 6 ms. from the e. corner of lake Erie, at the mouth of Cattaraugus creek, and is about 500 feet above its level. The soil bears maple, beech, elm, &c., with some evergreens, and is most favorable to grass, though some of it produces grain well. Pop. 1830, 2,440.

Perrysburg, p-v. and st. jus., Wood co. O., situated on the right bank of Maumee r., 15 or 16 ms. above its mouth, and by p-r. 135 ms. a little w. of n. Columbus. Lat. 41° 35′, long. 6° 36′ w. W. C. Pop. 1830, 182.

Perry's Mills, p-o. and st. jus., Tatnall co. Geo., by p-r. 115 ms. s. e. Milledgeville.

Perrysville, p-v. Alleghany co. Pa., 7 ms. a little w. of n. Pittsburg.

Perrysville, p-v. Perry co. Ten.

Perrysville, p-v. Mercer co. Ky. by p-r. 40 ms. sthrd. Frankfort. Pop. 1830, 283.

Perry's, store and p-o. Giles co. Ten., by p-r. 67 ms. sthrd. Nashville.

Perryville, p-v. Sullivan, Madison co. N.

Perryville, p-v. Richland co. O., by p-r. 72 ms. nrthestrd. Columbus.

Perryville, p-v. Vermillion co. Ind., by p-r. 88 ms. n. w. by w. Indianopolis.

Perryville, p-v. and st. jus., Perry co. Mo., by p-r. about 88 ms. s. s. e. St. Louis, and 20 ms. s. s. e. St. Genevieve. Lat. 37° 33′.

Perth Amboy, Middlesex co. N. J. (*See Amboy.*)

Peru, t. Oxford co. Me., s. Androscoggin r., with several small streams. Pop. 1830, 666.

Peru, p-t. Bennington co. Vt., 30 ms. n. e. Bennington, 30 s. w. Windsor, with 23,040 acres; first settled 1773; lies on the range of the Green mtns., in the n. e. corner of the co., and is high, with much broken land. There are 2 ponds of 60 and 40 acres, and some of the streams of W. r. water the e. part; 3 school dists. Pop. 1830, 445.

PERU, p-t. Berkshire co. Mass., 118 ms. W. Boston; is on elevated ground, giving rise to the principal and middle branches of Westfield r. The surface is hilly, being on the declivity of the range of the Green mts. The climate is as cold as that of any part of the state; the soil is hard, but yields grass well. The inhabitants are scattered. Pop. 1830, 729.

PERU, p-t. Clinton co. N. Y. 9 ms. S. W. Plattsburgh, 153 N. Albany, W. lake Champlain, N. Essex co., E. Franklin co. Great Sable r. forms the boundary S. E., and partly S. Little Sable r., whose whole course is in this t. is a good mill stream. From the lake, 10 ms. W., the land is nearly level, and good; the remainder is hilly, woody, and supplied with inexhaustible mines of the best iron ore. There are several iron works, particularly the Etna furnace, 9 ms. from the lake, at Port Kent, which is a landing place. The Russia iron works are on Sable r. Pop. 1830, 4,949.

PERU, p-v. Huron co. O., by p-r. 104 ms. N. N. E. Columbus.

PETERBOROUGH, p-t. Hillsborough co. N. H., 75 ms. W. S. W. Portsmouth, 60 N. N. W. Boston, 40 S. W. Concord, 20 from Amherst, 20 from Keene, with 23,780 acres; has the range of Pack Monadnock hills on the E. line, and is crossed N. by Contoocook r. and contains part of the N. branch, which affords good mill seats. At the falls on the latter are broad meadows, and the soil is generally very good. Pine grows on the S. branch, hard wood in other parts, and large oaks on the hills. A church stands on a hill in the centre of the t. 200 feet above the r. 200 feet higher than this, on the E. hills, is a pond of 9 acres; and lower, one of 33 acres. The Notch in the mtn. is a remarkable pass. Iron ore is found, also ginseng and huck bean. The surface is varied, the t. healthy; and there is a library, several cotton factories, &c. First settled 1739, deserted 1744, reoccupied 1745. Pop. 1830, 1,983.

PETERBOROUGH, p-v. Smithfield, Madison co. N. Y., 29 ms. S. W. Utica, 6 N. Morrisville, 108 W. N. W. Albany, on Oneida creek, 7 ms. S. Erie canal.

PETERSBURGH, p-t. Rensselaer co. N. Y., 25 ms. N. E. Albany, 18 E. Troy, W. Pownal, Vt., and Williamstown Mass.; about 6 ms. by 8; has the range of Bald mtns. E. and the Green woods W. Little Hoosac river, a mill stream, flows N. through the middle, along the course of which is a broad valley. It is included in the great estate of Rensselaerwyck, and the land is leased for about 10 bushels of wheat for 100 acres. The village of Rensselaer's mills, 18 ms. from Troy, is near the centre. Good limestone abounds: pop. 1830 2,011.

PETERSBURG, p-v. Adams co. Pa. 23 ms. a little N. of W. from the borough of York, and 20 S. S. W. Harrisburg.

PETERSBURG, port of entry and p-o. Dinwiddie co. Va., situated in the extreme northeastern angle of the co., on the right or sthrn. bank of Appomattox river, about 12 ms. above the mouth; by p-r. 22 ms. a little E. of S. Richmond, and 144 S, S. W. W. C. Lat. 37° 13′ long. 0° 24′ W. W. C.

If the contiguous villages are regarded as part of the t., and commercially they ought, Petersburg contains a part of 3 cos. Down the Appomattox and adjoining to the estrd., is the village of Blandford in Prince George's, and over the Appomattox & to the nthrd. connected by a bridge, is Powhattan in Chesterfield co.

This depot is well situated to sustain a high commercial rank amongst the ports of Va. The harbor admits vessels of considerable draught, and the adjacent country is well peopled and cultivated. The falls of Appomattox, near which the city stands, affords an illimitable water power, whilst a canal obviates the navigable impediment. The Bank of Va. and the Farmer's Bank of Va., have each a branch at this place. It possesses also an insurance office, and custom house: pop. 1830, 8,322.

PETERSBURGH, p-v. on the point above the junction of Broad and Savannah rs., and in the extreme sthestrn. angle of Elbert co. Geo. 50 ms. by the land road above Augusta, and by p-r. 86 ms. N. E. Milledgeville.

PETERSBURGH, p-v. Boone co. Ky., by p-r. 102 ms. N. Frankfort.

PETERSBURGH, p-v. Lincoln co. Ten., by p-r. 61 ms. sthrd. Nashville.

PETERSBURGH, p-v. Columbiana co. O., by p-r. 173 ms. N. E. by E. Columbus.

PETERSBURGH, p-v. and st. jus. Pike co. Indiana, situated on the left bank of White r., below the main fork, 25 ms. S. E. Vincennes, and by p-r. 119 ms. S. W. Indianopolis. Lat. 38° 32′, long. W. C. 10° 20′ W.

PETER'S CREEK, p-o. Barren co. Ky., by p-r. 104 ms. S. W. Frankfort.

PETERSHAM, p-t. Worcester co. Mass. 66 ms. W. Boston; has a productive soil, and was an early settlement. The Indians had a village here called Nashawang. It was granted 1732, as a reward for services in wars, and suffered hardships in the war of 1755, when they had forts erected for their defence. The village is pleasantly situated on rising ground, and commands a fine view. Population 1830, 1,696.

PETER'S MOUNTAIN, and p-o. Dauphin co. Pa., 20 ms. N. Harrisburg.

PETERSTOWN, p-v. Monroe co. Va., by p-r. 294 ms. S. W. by W. W. C., and 219 W. Richmond.

PETERSVILLE, p-v. northeastern part Frederick co. Md., 25 ms. N. E. Frederick, and 35 N. W. by W. Baltimore, and by p-r. 56 ms. a very little W. of N. W. C.

PETIT GULF, Little gulf, a remarkable bend in the Miss. r. opposite the nrthwstrn. angle of Jefferson co. state of Miss. (*See Rodney, Jefferson co. Miss.*)

PETIT MENAN, isl. Washington co. Me. It lies off a point of Steuben t. with a reef between, and has a light house. Pop. 1830, 11.

Peytonsburgh, p-v. Pittsylvania co. Va. by p-r. 148 ms. s. w. W. C.

Pharsalia, p-t. Chenango co. N. Y., 114 ms. w. Albany, 11 w. Norwich, 45 s. s. w. Utica, 122 w. by n. Cattskill; 6 ms. square, is high, a little uneven, and has several good mill streams and others, the head waters of Canasawacta creek, a branch of Chenango creek, and of tributaries of Tioughnioga cr. First settled 1798. There is a spring charged with sulphuretted hydrogen gas. Pop. 1830, 1,011.

Phelps, p-t. Ontario co. N. Y., 197 ms. w. Albany, 12 e. Canandaigua, 5 n. Geneva, and w. Genesee co. It is about 8 ms. by 10, has a gently varied surface, and excellent soil. A part of Canandaigua creek is in the e. part, which has valuable meadows, and supplies mill seats. It was named after Oliver Phelps, the first purchaser of a large tract of land in this part of the state. Flint creek, a branch of Canandaigua creek, is in the w. The v. of Vienna is at their junction. Orleans is another village in this t. Gypsum is found on the creek. Pop. 1830, 4,876.

Philadelphia, p-t. Jefferson co. N. Y. 170 ms. n. w. Albany; 5 ms. by 8, is nearly level, with a good arable clay and sandy soil, and pretty well watered by Indian river and its branches, which supply mill seats. The timber is oak, beech, bass, &c., with some hemlock and pine. First settled, 1813. Population 1830, 1,167.

Philadelphia, co. Pa., bounded by Del. co. s. w., Montgomery n. w., Bucks n. e., Del. r. separating it from Burlington co. N. J. e., and Gloucester co. N. J. s. Length from s. w. to n. e. 22 ms., mean width 7, and area 154 sq. ms. Extending in lat. from 39° 52′ to 40° 08′, and in long. from 1° 47′ to 2° 08′ e. W. C. Bounded on one side by the Delaware, and traversed by the Schuylkill, the confluence of these two streams is made in the southern part of the co. The general declivity is to the sthestrd. in the direction of the mean course of the Schuylkill, and at right angles to that of Delaware. Though comparatively confined in extent, the features of this co. are very strongly contrasted. The upper and northern part is beautifully broken by hill and dale, whilst the lower and the sthrn. section is composed of recent alluvion, and is an almost dead level. The primitive ledge on the margin of which the city of Philadelphia is built, traverses the co. from s. w. to n. e., arresting the tide in Schuylkill within the precincts of the city. On the Delaware, the scenery is rather tame, but along the Schuylkill and Wissahiccon cr. becomes picturesque, and in many places even wild and bold. The variety of site for country residences in the northern part of the co., adds no little advantage to Philadelphia; and the inexhaustible masses of gneiss, affords more than an ample supply of material for the rougher, more solid, and more durable kinds of architecture.

Besides the city of Philadelphia and places adjoining, the co. contains the boroughs or towns of Frankfort, Germantown, Holmesburg, Bustletown, Smithfield, and some others.

Independent of Philadelphia and places connected with it, the co. in 1820 contained a pop. of 28,288, 1830, 33,373, or at the latter enumeration, upwards of 210 souls to the sq. mile, and including the city, upwards of 1,221 to the square mile.

Philadelphia, city, the second largest in the United States, Philadelphia co. Pa., is situated on the neck between Delaware and Schuylkill rivers, the centre about 5 miles above the junction of these two streams. By reference to the table inserted at page 37 of the Memoir attending Tanner's U. S. map, the state house on Chesnut, between 5th and 6th streets, is on lat. 39° 56′ 51″, long. 75° 10′ 05′ w. of the royal observatory at Greenwich, and 1° 46′ 30″ e. of W. C. From these elements, by a calculation on Mercator's principles, the line between the 2 cities deflects from the meridian 52° 17′ very nearly; distant from each other in statute ms. 120, within a small fraction; and from Harrisburg by the turnpike through Lancaster, 96 ms.

In its natural state, the ground on which Philadelphia stands was an undulating plain, composed of relatively ancient alluvion. The Indian name Coaquanock, was changed to that of Philadelphia, and the city laid out into streets at right angles to each other, extending by a small angle from the true meridians. This regularity does not, however, extend to either of the suburbs or Liberties. In the latter the streets in part correspond to those of the city; in part they are at right angles to each other, but oblique to those of the city; and in part they are irregular, crossing at acute and oblique angles. Dock is the only street of either the city or Liberties which extends in curve lines. Happily the waving surface on which the body of the city is built, has been left untouched by the rage for levelling, and contributes to clear the streets of filth whenever rain falls on them, an advantage madly thrown away in some other places. Philadelphia, like N. York, stands on a superstratum of porous alluvion, based on primitive rock, another cause of cleanliness, or rather dryness.

The environs of Philadelphia, on the Pa. side of Delaware r. have been justly admired for the richness, and along the banks of the Schuylkill for the variety of scenery. Without due attention to the cause it may excite some surprise that the Delaware scenery should present features so much more monotonous than those of Schuylkill, but the geological structure of the country explains the difference. The primitive rock ledge, on the margin of which Philadelphia is situated, is at that city only touched by the Del., whilst on the contrary, it is there that the Schuylkill emerges from the region of hills based on the primitive, and meets the tide. If the two rs. are compared as to their relative connexion with the primitive ledge, the Delaware forms the traverse at Trenton, similar to that of Schuylkill at Philadelphia.

As a commercial port, that of Philadelphia from its great distance from the ocean, and not having counter tides, is more liable to obstruction in winter from ice than is that of N. York. The latter, at some states of water, and with adequate nautical skill and knowledge of the channel, will admit ships of 74 guns, which the former under no circumstances possesses sufficient depth of water. The deep channel of Delaware is, however, at Philadelphia close on the Pa. shore, and vessels of 600 tons can be laid close on the docks, and there laden.

In hopes of receiving a more minute and recent account of this important city in time for insertion in the appendix to this Gazetteer, we refer the reader to that part of our treatise, for the conclusion of the article.

Philadelphia, p-v. northern part Monroe co. Ten., by p-r. 173 ms. s. e. by e. Nashville.

Philadelphus, p-v. Robeson co. N. C., by p-r. 107 ms. s. w. Raleigh.

Philanthropy, p-v. Butler co. O., by p-r. 126 ms. s. w. by w. Columbus.

Philips,, r. Coos co. N. H., a branch of Upper Amonoosuc r.; joins it in Piercy.

Philips, p-t. Somerset co. Me., 53 ms. n. n. w. Augusta, 40 n. Norridgewock, and e. of Berlin. Oxford co. is crossed n. and s. by Sandy r., a branch of Kennebec r. Population 1830, 954.

Philips, p-t. Putnam co. N. Y., 96 ms. s. Albany, opposite West Point, Cornwall, Orange co., about 8 ms. by 12, is very mountainous, and has the principal part of Break Neck hill in the n. one of the chief eminences of the Highlands. It contains also Blue hill, &c. In the south along the branches of Peekskill cr. are handsome meadows. There are other streams, and several ponds. The scenery is bold and varied, but much of the soil is broken, and iron ore abounds, which is mined. The village of Pleasant Valley is nearly opposite to West Point. Above it is Cold Spring, and the principal cannon foundry in the U. States. The Robinson mansion stands on an elevation not far from the shore. The property of the owner was confiscated in the Revolution, and the house was occupied by Arnold when he treacherously deserted the American cause. Pop. 1830, 4,761.

Philips, co. Arkansas, as laid down by Tanner, is bounded by White r. w., Miss. r. e., and St. Francis co. n. It is in form of a triangle, 54 ms. from s. to n.; mean breadth 40 ms., and area 2,160 sq. ms. Lat. 34° 30′, long. W. C. 14° w. Slopes southward. The greatest part of the surface liable to annual submersion by the floods of the two bounding rs., and of the Saint Francis r. which enters the Miss. in the northeastern angle; where the soil admits cultivation, it is exuberantly fertile, and the climate admits the profitable cultivation of cotton. Chief t. Helena. Pop. 1830, 1,152.

Philipsburgh, p-v. Wallkill, Orange co. N. Y., on Wallkill creek, 20 ms. w. Newburgh, and 4 from Goshen; has several manufactories.

Philipsburgh, village, Warren co. N. J., e. side Del. r., opposite Easton, Pa.

Philipsburg, p-v. Centre co. Pa., by p-r. 114 ms. wstrd. Harrisburg.

Philipsburg, p-v. Jefferson co. O., by p-r. 149 ms. n. e. by e. Columbus.

Philips Store, and p-o. Nash co. N. C., by p-r. 61 ms. estrd. Raleigh.

Philipston, p-t. Worcester co. Mass., 65 ms. n. w. Boston, 26 n. w. Worcester; has very good grass land, and many mill sites. Burnshint r. rises here in a fine pond: pop. 1830, 932.

Philipsville, p-v. Erie co. Pa., 345 ms. n. w. W. C.

Philomont, p-v. Loudon co. Va., 41 miles wstrd. W. C.

Phipsburgh, p-t. Lincoln co. Me., 44 ms. s. Augusta, 20 s. w. Wiscasset; forms the s. part of a long and irregular peninsula on the w. side of Kennebec r. at its mouth. It terminates in two points, called Bald Head and Cape Small Point, and has Cape Small Point harbor: pop. 1830, 1,311.

Phoenixville, p-v. wstrn. part Chester co. Pa., about 30 ms. from Phil.

Phyfer's Cross Roads, and p-o. Knox co. O., 60 ms. n. e. Columbus.

Physic Spring, and p-o. Buckingham co. Va., 67 ms. w. Richmond.

Pickaway, co. O., bounded s. by Ross, Fayette s. w., Madison n. w., Franklin n., Fairfield e., and Hocking s. e. Greatest length 28 ms., mean length 24, mean breadth 21, and area 500 sq. ms. Central lat. 39° 37′, long. W. C. 6° w. Slope sthrd., the Sciota river traversing it in that direction, and within its limits receiving Walnut creek from the n. e., and Darby's creek from the n. w. The soil is remarkably diversified, some part being exuberantly fertile, whilst the opposite extreme reaches barrenness. The fertile part has been found so productive as to afford crops of 40 or 45 bushels of wheat per acre, and other grains and fruits in equal proportion. The pop. being in 1820, 13,149, 1830, 16,001, would seem to show that in general fertility, Pickaway must fall short of several other cos. of Ohio. Chief town, Circleville.

Pickens, extreme wstrn. dist. of S. C., bounded by Greenville dist. n. e., Anderson e., Tugalvo r. separating it from Franklin co. Geo. s. w., and Habersham co. Geo. w., Chatuga r. separating it from Rabun co. Geo. n. w., and by Haywood and Buncombe cos. in N. C., n. Length from s. w. to n. e. 40 ms., mean breadth 30, and area 1,200 sq. ms.; lat 34° 50′, long. W. C. 6° w. slope sthrd., and drained chiefly by the various branches of Seneca r. Surface hilly, and in part mountainous; chief t. Pickenville. Pop. 1830, 14,473.

Pickens co. Ala., bounded by Lowndes co. Miss. n. w., Layfayette co. Ala. n., Tuscaloosa co. E., Greene s., and Tombigbee r. separating it from the Choctaw territory, Ala. W. Length from s. to N. 36 ms., mean brdth. 20, and area 720 sq. ms.; lat. 33° 13', long. W. C. 11° 15' W. Slope s. W., and in that direction drained by Sipsey r. and several creeks flowing into Tombigbee r.; chief t. Pickensville. Pop. 1830, 6,622.

PICKENS, C. H. and p-o. Pickens dist. S. C., by p-r. 157 ms. N. W. Columbia.

PICKENSVILLE, p-v. estrn. part Pickens dist. S. C., 43 ms. N. W. by W. Spartanburg, and by p-r. 130 ms. N. W. by W. Columbia.

PICKENSVILLE, p-v. and st. jus. Pickens co. Ala., by p-r. N. W. by W. Tuscaloosa; lat. 32° 20', long. W. C. 11° 16' W.

PICKERING Isl. Hancock co. Me. Pop. 10.

PICKERINGTON, p-v. Franklin co., O.

PICKETT'S valley and p-o., Greenville dist. S. C., by p-r. 123 ms. N. W. Columbia.

PIERCY, t. Coos co. N. H., 5 ms. N. E. Lancaster, 20,000 acres, of irregular form, is crossed by Upper Amonoosuc r., whose N. and s. branches here unite. Piercy's pond is E. The surface is uneven, soil not very good. It contains Mill and Pilot mtns., and Devil's Sliding Place, which has a smooth declivity N., and a precipice of 300 ft. s. First settled 1788. Pop. 1830, 236.

PIERMONT, p-t. Grafton co. N. H., 70 ms. N. N. W. Concord, 132 N. N. W. Boston. It contains 23,000 acres, lies E. Conn. r., has good soil, excellent on the r. with broad meadows. Grain and grass grow well on the adjoining plains; E. of these are hills favorable to grass. White pine grows near the r.; hard maple, birch, elm, &c. E. Eastman's ponds are N. E., flowing into Conn. thro' Eastman's brook, on which are mill seats. Indian brook, also a mill stream, is s.; Barron's Isl. is in Conn. r.; valuable quarries are in the N. First settled 1770. Pop. 1830, 1,042.

PIERPONT, p-t. St. Lawrence co. N. Y., 213 ms. N. N. W. Albany, 28 E. S. E. Ogdensburgh, is crossed in the N. E. corner by Racket r., & in other parts has several small streams of Grassy r. Pop. 1830, 749.

PIERPONT, p-v. Ashtabula co. O., by p-r. 199 ms. N. E. Columbus.

PIG r. Va., rising in the sthestrn. slope of the Blue Ridge, and flowing thence estrd., between Blackwater and Irvine rs., traverses and drains the central part of Franklin co., and entering Pittsylvania, turns to N. E. and falls into Roanoke, after a comparative course of 35 ms.

PIGEON HILL, p-v. York co., Pa., by p-r. 90 ms. N. W. C.

PIGEON ROOST, p-v. Henry co. Ten., by p-r. 106 ms. a little N. of W. Nashville.

PIG POINT, p-v. on the left bank of Patuxent r., opposite the mouth of the W. Branch, and in the sthrn. part of Ann Arundel co. Md., by p-r. 59 ms., but by the common intermediate road, only about 30 ms. S. E. by E. W. C.

PIKE, p-t. Alleghany co., N. Y., 255 ms. W. by S. Albany, 18 N. W. Angelica, and S. *Genesee co., 6 ms.* by 12. It is crossed N. & S. by both branches of Wiscoy cr. Genesee r. flows through the adjoining t. of Portage on the E. The soil is good, bearing much maple, beech, bass, elm, &c. Bog iron ore is found here. Pop. 1830, 2,016.

PIKE co. Pa., bounded by Northampton S.; the nrthrn. branch of Lehigh separating it from Luzerne W., by Wallenpaupack cr. separating it from Wayne N. W., Lackawaxen r. separating it from Wayne N., Del. r. separating it from Sullivan co. N. Y., N. E., and by Del. r. separating it from Sussex co. N. J., E., and Warren co. N. J., S. E. Length crossing diagonally from E. to W. 48 ms., mean breadth 15, and area 720 sq. ms. Lat. 41° 17', long. W. C. 1° 48' E. A mtn. chain traverses this co. from S. W. to N. E., dividing it into two slopes, one estrd. towards that part of Del. below the mouth of Nevesink r., and the other nrthrds. towards that part of Del. between the mouths of Lackawaxen and Nevesink rs.; surface of the co. very broken. Chief t. Milford. Pop. 1830, 4,843.

PIKE, p-v. Bradford co. Pa., by p-r. 149 ms. nrthrd. Harrisburg.

PIKE co. Geo., bounded by Fayette N. W., Henry N., Butts N. E., Monroe S. E., Upson S., and Flint r. separating it from Merriwether W. Length from E. to W. 28 ms., mean width 17, and area 476 sq. ms.; lat. 32° 07', long W. C. 7° 30' W.; slope sthrd. and drained in that direction by Auhau cr. a branch of Flint, and Chupee cr., a branch of Ocmulgee r.; chief t. Zebulon. Pop. 1830, 6,149.

PIKE co. Ala., bounded by Henry S. E., Dale S., Covington S. W., Butler W., Montgomery N. W., and the Cherokee territory N. and N. E.; greatest length along the sthrn. border 68 ms.; mean breadth 25, area 1,700 sq. ms. The extreme nrthrn. angle is drained by Ockfuskee creek, a branch of Tallapoosa r. and slopes to the N. The much greater part of the whole surface, however, slopes S. W., and is drained in that direction by the sources of Conecuh and Choctaw rs. Chief t. Pike court house. Pop. 1830, 7,108.

PIKE co. Miss., bounded S. W. by Amite, N. W. by Franklin, Lawrence N., Marion E., Washington parish, La. S. E. and S., and St. Helena parish, La. S. W.; length 30 ms., mean breadth 28, and area 840. Extending in lat. from 31° to 31° 27', and in long. from 13° 10' to 13° 40' W. This co. is traversed in a S. S. E. direction by the Bogue Chito r., and the sthwstrn. angle gives source to the Tangipao r. The general slope nearly sthrd.; surface in great part open pine woods; chief town Holmesville. Pop. 1830, 5,402.

PIKE co. Ky., bounded S. by Harlan, S. W. and W. by Perry, N. by Floyd, and by Cumberland mtn., which separates it from Tazewell co. Va. E., and from Russell co. Va. S. E. It approaches the form of a triangle, base 55 ms. along Cumberland mtn.; mean breadth 20 ms., 1,100 sq. ms.; lat. 37° 15', long W. C. 5° 40' W. The nrthestrn. angle is traversed nrthwstrly. by the West Fork of Big Sandy r., whilst from the central and sthrn. sections riso the extreme sources of Kentucky r. flowing also to the N. W.; surface very broken; chief t. Piketon. Pop. 1830, 2,677.

PIKE co. O., bounded S. by Sciota, Adams S. W., Highland W., Ross N., and Jackson E.; length from E. to W. 32 ms., mean width 18, and area 576 sq. ms.; lat. 39° and long. W. C. 6° W. intersect in this co. It is traversed in a S. S. W. course by Sciota r.; soil productive in grain, fruits and meadow grasses; chief t. Piketon. Pop. 1820, 4,253; 1830, 6,024.

PIKE co. Ind., bounded by Warrick S., Gibson W., White r. separating it from Knox N. W., the estrn. branch of White r. separating it from Daviess N. E., and by Dubois co. E.; mean length from S. to N. 22 miles, mean breadth 18, and area 396 sq. ms. The slope of this co. is very nearly due W., and in that direction is traversed by Patoka r. The course also of both branches of White r. where they bound the co. is also to the W.; chief t. Petersburgh. Pop. 1830, 2,475.

PIKE co. Il., bounded by Calhoun S., Miss. r. separating it from Pike co. Mo. S. W., Ralls, Mo. W., and Marion Mo. N. W., again by Adams, Il. N., Schuyler N. E., and Il. r. separating it from Morgan E. and Greene S. E.; length from S. to N. 33 ms., mean breadth 30, and area 990 sq. ms. Lat. 39° 35′, long. W. C. 14° W. extending from the Miss. to Il. r.; the principal slope is wstrd. towards the former; chief t. Atlas. Pop. 1830, 2,396.

PIKE co. Mo., bounded by Lincoln S. E., Montgomery S., Ralls S. W., W. and N. W., and the Miss. r. separating it from Pike co. Il. N. E., and from Calhoun co. Il. E.; length from S. to N. 36 ms. mean breadth 20, and area 720 sq. ms.; lat. 39° 20′, long. W. C. 14 W. Salt r. of Mo., enters the Miss. in the nrthrn. angle of this co., which it traverses to the S. E. The nrthrn. branches of Cuivre (Copper) r. rise in the sthrn. section, and also flow S. E.; chief t. Bowling Green. Pop. 1830, 6,129.

PIKE C. H. and p-o. Pike co. Ala., by p-r. 179 ms. S. E. Tuscaloosa.

PIKESVILLE, p-v. Baltimore co. Md. by p-r. 46 ms. N. E. W. C., and 8 ms. from Baltimore.

PIKETON, p-v. and st. jus. Pike co. Ky. situated on the W. Fork of Sandy r., by p-r. 165 ms. S. E. by E. Frankfort.

PIKETON, p-v. and st. jus. Pike co. O. situated on the left bank of Sciota r. 26 ms. above Portsmouth, and by p-r. 65 ms. S. Columbia; lat. 39° 02′, long. W. C. 6° W. Pop. 1830, 271.

PIKEVILLE, p-v. and st. jus. Marion co. Ala. situated on Battahatche r., by p-r. 118 ms. N. N. W. Tuscaloosa; lat. 34° 07′, long. W. C. 11 W.

PIKEVILLE, p-v. and st. jus. Bledsoe co. Ten., situated on Sequatchie r., by p-r. 109 ms. S. E. by E. Nashville; lat. 35° 39′, long. W. C. 8° 12′ W.

PIKEVILLE, p-v. Monroe co. Ky. by p-r. 145 ms. S. S. W. Frankfort.

PILESGROVE, t. Salem co. N. J. 50 ms. S. W. Trenton, 25 S. Phila., has Oldman's cr. N. which separates it from Woolwich, Gloucester co., and is crossed by Salem r., on which are the villages of Sharptown and Woodstown: pop. 1830, 2,150.

PINCKNEY, p-t. Lewis co. N. Y. 153 ms. N. W. Albany, 13 S. E. Watertown, and E. of Rodman in Jefferson co., 6 ms. by 6½, first settled 1805, has nearly a level surface, with much moist, sandy loam, favorable to grain and grass, and bearing a variety of forest trees: pop. 1830, 763.

PINCKNEY, p-v. on the left bank of the Mo. r., Montgomery co. Mo., by p-r. 66 ms. W. St. Louis.

PINCKNEYVILLE, p-v. Union dist. S. C. 92 ms. N. N. W. Columbia.

PINCKNEYVILLE, p-v. Gwinnet co. Geo. by p-r. 106 ms. N. W. Milledgeville.

PINCKNEYVILLE, p-v. Wilkinson co. Miss. 44 ms. S. Natchez.

PINCKNEYVILLE, p-v. and st. jus. Perry co. Il., situated on Boucoup cr. a branch of Muddy cr., by p-r. 129 ms. a little W. of S. Vandalia, lat. 38° 02′, long. W. C. 12° 25′ W.

PINDERTOWN, p-v. and st. jus., Lee co. Geo. situated on Flint r., by p-r. 130 ms. S. S. W. Milledgeville, lat. 31° 40′, long. W. C. 7° 10′ W.

PINE BLUFF, p-o. sthestrn. part Pulaski co. Ark., by p-r. 50 ms. S. E. Little Rock.

PINE cr. or more correctly r. of Pa., rising in Potter and Tioga cos., interlocking sources with those of Tioga, Genesee, and Alleghany rs. The various branches unite in Tioga, from which the united water flows into Lycoming, and falls into the W. Branch of Susquehannah r., after a sthrn. course by comparative distance 60 ms.

PINE cr. p-o. Tioga co. Pa., by p-r. 159 ms. N. Harrisburg.

PINE GROVE, p-v. Schuylkill co. Pa. by p-r. 41 ms. N. E. Harrisburg.

PINE GROVE, p-v. Tyler co. Va. by p-r. 249 ms. wstrd. W. C.

PINE GROVE, mills and p-o., Centre co. Pa. by p-r. 88 ms. nrthwstrd. Harrisburg.

PINE ORCHARD, Catskill, Greene co. N. Y. 8 ms. W. Hudson r., and Catskill v. a small level on the Catskill mtns., a favorite resort of travellers of taste during the hot season. It was originally covered with a grove of pine trees, growing at nearly equal distances, on a surface scattered with broken rocks, and terminating at a projection which overhangs a precipice of some hundreds of ft. At that spot has been erected a splendid hotel, called the Catskill Mountain house, from which the view ranges without interruption over the wide valley of the Hudson, including some of the highlands of Conn., Mass. and Vt. The most distant eminences in sight N. and S. are about 70 ms. apart.

PINE PARK, p-v. Bibb co. Ala. by p-r. 59 ms. estrd. Tuscaloosa.

PINE PLAINS, p-t. Dutchess co. N. Y. 79 ms. S. Albany, 28 N. E. Poughkeepsie, 4 ms. by 10, has a small village; a high hill E. and Stissing mtn. W., on the E. side of which is Stisding pond, with an outlet s. the head stream of Wappinger's creek. Chicome cr. co. crosses the town from s. e. to n. w. flowing into Roeleff Jansen's creek, which crosses the n. w. corner. Pop. 1830, 1,503.

Pine Street, p-v. Clearfield co. Pa. by p-r. 162 ms. n. w. Harrisburg.

Pine Village, p-v. Edgefield dist. S. C. 66 ms. wstrd. Columbia.

Pineville, p-v. nthrn. part of Charleston dist. S. C. 53 ms. n. Charleston, and by p-r. 92 ms. s. e. Columbia.

Pineville, p-v. northern part Clarke co. Ala. by p-r. 107 ms. southward Tuscaloosa.

Piney river, p-v. on a small stream of the same name, sthrn. part Dickson co. Ten. By p-r. the p-o. is 54 ms. wstrd. Nashville.

Pintlalah, or Pintelalah, small river of Montgomery co. Ala. rises on the sthrn. border of the co. and flowing northward falls into the left side of Ala. r.

Pintlalah, p-o. Montgomery co. Ala. 13 ms. from Montgomery, the co. st. and by p-r. 132 ms. s. e. Tuscaloosa.

Piping Tree, p-v. King William co. Va. by p-r. 20 ms. n. e. Richmond.

Piqua, p-v. Washington tsp. northern part of Miami co. Ohio, by p-r. 79 ms. a little n. of w. Columbus, and 8 ms. n. n. w. Troy, the co. st. : pop. 1830, 488.

Piquea, p-v. sthrn. part of Lancaster co. Pa. 27 ms. sthrd. Lancaster, and by p-r. 126 ms. northeastward W. C.

Piscataquay river, N. H. the boundary between N. H. and Me. from the ocean 40 ms. n. n. w. which is the length of its course. Its source is in Wakefield. Its mouth is near Portsmouth, which capital stands on its s. shore. It is a large, deep and important stream the last few miles of its course, spreading out into several bays. Only this part of it is commonly called Piscataqua ; the middle part, from the mouth of Cocheco river to the lower falls in Berwick, bearing the name of Newichawannoc, and the upper part Salmon Falls river. The western branch is formed by several branches, which fall into Great Bay, Swamscot r. from Exeter, Winnicot r. which passes through Greenland, and Lamprey river flowing between Durham and Newmarket. At a smaller bay below, Oyster river comes in from the n. The tide, which flows up to the lower falls in all these streams, affords navigation from them to Portsmouth. The channel, being narrow, though very deep, 7 ms. from the ocean, causes a very rapid rush of water both at the rising and the falling of the tide, so that ice is never formed across. The harbor formed near the mouth of the Piscataquay is very safe and capacious. There is an outer and an inner bay, islands with forts, a light house, &c., and in the inner is a navy yard of the U. S.

Piscataquog river, Hillsborough co. N. H. is formed of 2 branches, and enters Merrimac river on the line of Goffstown and Bedford.

Piscataquog, p-v. Bedford, Hillsborough co. N. H. a pleasant little village on Piscataquog r. near its junction with the Merrimac, with a bridge over the former, 60 feet long. The Union canal here passes the falls, and facilitates the business of the place.

Piscataway, village, Piscataway, Middlesex co. N. J. 3 ms. n. e. New Brunswick, and n. Raritan river.

Piscataway cr. sthrn. part Prince George's county, Md. rises a few ms. w. of Upper Marlborough, and flowing s. w. falls into Potomac at Fort Washington.

Piscataway, p-v. on Piscataway creek, sthrn. part Prince George's co. Md. by p-r. 16 ms. s. W. C.

Pisgah, p-v. Cooper co. Mo. 34 ms. wstrd. Jefferson.

Pitch Landing, and p-o. sthrn. part Hertford co. N. C. 12 ms. s. Winton, the co. st. and 129 ms. n. e. by e. Raleigh.

Pitt, co. N. C. bounded e. by Beaufort, Johnson s., Lenoir s. w., Greene w., Edgecombe n. w., and Martin n. and n. e. Length from e. to w. 44 ms.; greatest breadth 36 ms. but mean breadth 18 ms., and area about 800 square ms. Central lat. 35° 35′. The meridian of W. C. traverses the eastern angle. Contentney or the mtn. branch of Neuse r. traverses the sthrn. angle in a s. e. course, whilst Tar r. in a similar direction winds over the central part. The general slope is to the s. e. Chief town, Greenville : pop. 1830, 12,093.

Pittsboro', p-v. and st. jus. Chatham co. N. C. situated on the road from Raleigh to Ashboro', 33 ms. wstrd. of the former, and 39 estrd. of the latter, lat. 35° 43′, and long. 2° 14′ w. W. C.

Pittsburg, city and p-t. Pa. situated on the point above the junction and between the Alleghany and Monongahela rivers, and where these two streams form the Ohio r. by p-r. 323 ms. southwestward W. C., 201 w. Harrisburg, and 297 ms. a little n. of w. Phila., lat. 40° 28′, long. W. C. 2° 56′ w. That part incorporated and particularly called Pittsburg, lies entirely on the point above noticed, and occupies an alluvial plain and part of the adjacent hill protruded between the two rs. In its form the city of Pittsburg is laid out with a very strong resemblance to N. Y. The streets along the Monongahela are laid out at right angles to each other, and perpendicular or parallel to that stream ; and the same relative arrangement prevails along the Alleghany, and renders the streets of the two sections of the city oblique to each other.

In 1820, if the adjacent villages were included, there were within 1 mile of its centre, about 10,000 persons, of whom 7,248 resided in the corporation. In 1826, the city contained a population of 10,515. By the the census of 1830, the city alone contained 12,568 inhabitants, but in a commercial and social point of view, Alleghany town, Birmingham, Lawrenceville, Bayardstown, and the street along the Monongahela opposite, all belong to Pittsburg as suburbs, and all included in one aggregate, amounted to 18,000, in 1830, and now 1832, no doubt exceeds 20,000. Pittsburg is emphatically the

Birmingham of the Ohio valley, and is in no small proportion made up of manufacturing edifices, and inhabited by manufacturers. Amongst the machinery erected here, may be named as first in utility and efficiency, a high pressure engine of 84 horse power, which raises water from the Alleghany river 116 feet, and can afford a diurnal supply of 1,500,000 gallons. There were according to Flint early in this year, 1832, 11 extensive iron foundries, from which, in 1830, were manufactured from pigs, 5,339 tons. There were 6 rolling mills and nail factories united, which manufactured 7.950 tons of pigs into blooms, and 2,805 tons into nails. There were 4 extensive cotton factories, one of which worked 10,000 spindles. Two glass works, several breweries, and taking altogether upwards of 270 manufacturing establishments. There are 13 churches, for Roman Catholics, Baptists, Covenanters, Seceders, Methodists, German Lutherans, Episcopalians, Presbyterians, Unitarians, &c. In this city is located the Western University of Pa. Pittsburg High school, numerous private schools, Lamdin's museum, a branch of the United States bank, and Pittsburg bank, and also a state prison.

The city is united to the adjacent country beyond the two rivers by a bridge over each. That across the Monongahela is, however, in some measure useless, from being located too high up the stream, and a ferry is still kept up from the point to the great western road, through Washington, Wheeling, Steubenville, and other places. The site is a real amphitheatre formed by the hand of nature. The rivers flow in channels from 450 to 465 feet below the highest peaks of the neighboring hills. The writer of this article measured the height of several hills in the vicinity of Pittsburg, and found them varying between the relative elevations stated. Another geological phenomenon deserves particular notice ; the main coal strata lie something above 300 feet above the level of the streets of that part of the city on the alluvial point, and these strata lie almost exactly on one level. A levelling instrument placed at the mouth of any of the coal beds, if carried round the horizon the circle of vision passes along the openings of all the other mines.

The hills though steep are not, except in a few instances, precipitous, and afford from their slopes and peaks a series of rich and varied landscape. The scenery is in a most interesting manner strengthened in color by the fertility of soil which continues to the very summits. There is nothing of barrenness visible ; vegetation in the forests, meadows, fields, orchards, and gardens, exhibits one theatre of abundance.

The formation is here, as in every other part of the valley of Ohio, *floetz*, or level, so much so, indeed, as to render the draining of the coal mines difficult. Limestone is formed, but the prevailing rock is a porous sand stone as far as the earth has been penetrated.

Few places in the United States combine so great advantages of position as does Pittsburg. The great line of canal and rail road from Philadelphia by the Schuylkill over the Susquehannah valley, and the intervening mountains, terminates for the present in the valley of Ohio at Pittsburg, but its extension down the latter stream is amongst the inevitable effects of its execution to the head of such a navigation. The Ohio is impeded by autumn drought in direct excess with ascent, and is nearly annually rendered unnavigable as low down as Marietta, for one, two or three months before the frosts of winter. It is, however, navigable as high as Wheeling, long after it ceases to be so at the confluence of the Alleghany and Monongahela.

Taken with all its existing business, few if any other places, in either the United States or Europe with an equal population, have transactions to so great amount. The value of its manufactures falls annually but little under $3,000,000, and the objects of manufacture being those of primary necessity, renders the prosperity of the place permanent. The character of the people is stamped by their occupations. Persevering industry and perhaps a rather too overstrained, though natural bent to private interest, may be said of the far greatest part of the population. None are idle, and few are dissipated. I have known this town for the fifty last years, and have perceived its growth less fluctuating, and more solidly based than most of the towns of western United States.

Pittsburgh, p-v. western part of Baldwin co. Geo. 8 ms. from Milledgeville.

Pittsfield, p-t. Merrimack co. N. H., 15 ms. N. W. Concord, with 14,921 acres, is uneven and rocky, with good soil, and is crossed by Suncook r. from N. to S., supplying mill seats. S. E. is Catamount mtn. from which is seen the ocean, and on which is Berry's pond, 300 yards by ½ mile, supplying mill seats with its outlet. There are several other ponds : pop. 1830, 1,276.

Pittsfield, p-t. Rutland co. Vt., 35 ms. S. W. Montpelier, 17 N. E. Rutland, is in the N. E. corner of the co. ; first settted 1786, and organized 1793. Tweed r. a branch of White r., is formed near the centre, by the union of three streams, which afford mill sites. White r. also crosses the E. part. The surface is mountainous, and Wilcox's peak is the highest. 4 school dists. : pop. 1830, 505.

Pittsfield, p-t. Berkshire co. Mass., 125 ms. W. Boston, 38 E. S. E. Albany ; lies between the two mountainous ranges of Taughkannic and the Green mtns., has a varied and beautiful surface, good soil, and, along the 2 main branches of Housatonic r., extensive meadows. The village enjoys an airy situation, on the summit of a hill, with a large public square, in the centre of which is a remarkably fine elm, which was left when the forest was cleared away. It has several streets, with a number of stores and handsome residences, and a bank ; the Berkshire academy, and a seminary for females. In the tsp. are several extensive factories, including

Mr. Pomeroy's, of muskets, where arms are frequently made for the U. S. The settlement began in 1736, and two garrison houses were erected in 1754, but the general occupation of the town has been much more recent. The Indians called it Pontoosuc. During the last war, many British prisoners were cantoned in the village : pop. 1830, 3,515.

PITTSFIELD, p-t. Otsego co. N. Y., 87 ms. w. Albany, and 15 s. w. Cooperstown; has Unadilla w. which separates it from Chenango co., and has an irregular surface, with fertile vallies and arable hills, with fine pastures well watered. Wharton's creek falls into Mead r.: pop. 1830, 1,006.

PITTSFORD, p-t. Rutland co. Vt., 60 miles N. Bennington, 44 s. w. Montpelier, 8 N. Rutland. First settled 1768, from Greenwich, Mass., and had forts Mott and Vengeance, picketed in the revolution; the latter being the most N. frontier point held by Americans in the war, w. of the Green mtns. It contains 25,000 acres. Otter creek flows from s. to N. through the middle, from 40 to 50 yards wide, winding, and slow. Furnace r. is a branch formed of East cr. and Philadelphia r., which have valuable meadows, and good mill sites. A pond s. E. contains twenty acres, and one N. E. 30. A hilly range is on the w. line. The soil is loam, with some sand and clay, bearing oaks, pine, maple, beech, &c. &c., and contains iron ore, yielding 25 per cent. Marble is quarried and sent to Middlebury to be sawn and cut. It is coarse and elastic, so that a thin slab laid horizontally, supported only at the ends, bends in the middle. Oxide of manganese also is found here. There are 14 school districts : pop. 1830, 2,005.

PITTSFORD, p-t. Monroe co. N. Y., 215 ms. N. N. w. Albany, 8 E. Rochester, 22 N. w. Canandaigua; with 22 sq. ms.; has Irondequot cr. s. E., Noyes cr. N. w., with few mill seats, and is crossed by Erie canal from N. w. in a crooked course to s. E., where is the great embankment over Irondequot cr. The village is in the N. w. corner, and almost surrounded by the canal: pop. 1830, 1,831.

PITTSGROVE, p-t. Salem co. N. J., 74 ms. s. s. w. Trenton, 25 s. Philadelphia; has a little of the head of Oldman's cr. N. with Gloucester co., and at its s. angle just touches the N. w. angle of Deerfield, Cumberland county: pop. 1830, 2,216.

PITTSTON, p-t. Kennebec co. Me., 7 ms. s. Augusta, has Lincoln co. E. and s., and Kennebec r. w., into which it sends several small streams. It is a place of some trade : pop. 1830, 1,799.

PITTSTON, p-v. on the bank of the East Branch of Susquehannah r., at the mouth of Lackawannoc r., Luzerne co. Pa., 9 miles above Wilkes-Barre.

PITTSTON FERRY, nearly opposite Pittston, p-v. Luzerne co. Pa., 8 ms. above, but on the opposite side of the East Branch of Susquehannah from Wiles-Barre.

PITTSTOWN, p-t. Rensselaer co. N. Y., 18 ms. N. E. Albany, 15 N. E. Troy, s. Washington co.; with 35,500 acres; first settled 1750; is uneven but arable, with a good soil, bearing oak, maple, beech, ash, &c., and sends wheat, pork, beef, &c. to market. The villages of Pittstown & Tomhanoc, are in pleasant vallies. Hoosac r. is on the N. line. The mill streams are small.

PITTSYLVANIA, co. Va., bounded s. w. by Henry, Franklin w., Roanoke r. separating it from Bedford, N. w., and Campbell N.; it has Halifax on the E., and Caswell and Rockingham cos. N. C. s. Greatest length from s. to N. 40 ms., mean length 36, breadth 28, and area 1,000 sq. ms. Lat. 36° 50', long. 2° 21' w. W. C. This co. is bounded on the N. by Roanoke, in the centre by Banister r., and on the south by Dan r., all of which streams in that part of their respective courses flow estrd., and of course give that slope to the surface. Much of the soil is excellent. Chief town, Competition, usually called Pittsylvania C. H.: pop. 1820, 21,313, 1830, 26,034.

PITTSYLVANIA, C. H. and p-o., or Competition, st. jus. Pittslyvania co. Va., is situated near the centre of the co. on a branch of Banister r., by p-r. 259 ms. s. w. W. C., and 167 ms. s. w. by w. Richmond. Lat. 36° 50', long. W. C. 2° 20' w.

PLACENTIA, island, Hancock co. Me.: pop. 1830, 39.

PLAIN DEALING, p-v. Meade co. Ky., 10 ms. estrd. Brandenburg, the co. seat, and by p-r. 80 ms. wstrd. Frankfort.

PLAINFIELD, t. Washington co. Vt., 55 ms. N. Windsor, 21 N. w. Newbury; first settled about 1794, and has 10,000 acres. Onion r. is in the N. w. part, and is here joined by Great brook, which crosses the town. The village is at the junction, with several mills, &c. There is a trout pond, and a small mineral spring which is resorted to by invalids; the soil is pretty good, the surface hilly, and timber is abundant : pop. 1830, 874.

PLAINFIED, p-t. Sullivan co. N. H., 12 ms. s. w. Dartmouth college, 55 w. N. w. Concord, 111 N. w. Boston, E. Conn. r., and s. of Grafton co. Has pine timber near the river, maple, beech, &c. on the hills. It contains fine meadows, particularly on the river. Harts island, 19 acres, belongs to this town. There is a pleasant village, in which is Union academy, with $40,000, given by David Kimball, the interest of which is to be given partly to a clergyyman, and partly to the education of ministers. First settled 1764. Waterqueechy falls are in this town : pop. 1830, 1,581.

PLAINFIELD, p-t. Hampshire co. Mass., 110 ms. w. Boston, 20 N. w. Northampton; lies on the range of the Green mtns., and supplies the head streams of the N. branch of Westfield r.: pop. 1830, 984.

PLAINFIELD, p-t. Windham co. Conn., lies E. of Quinnebaug r., N. New London co. and is crossed by Moosup r. a branch of the Quinnebaug, with other small streams. The v. is pleasantly situated in the midst of a level.

The town has good soil and many valuable farms. It has also an academy: pop. 1830, 2,290.

PLAINFIELD, p-t. Otsego co. N. Y., 75 ms. w. Albany, 15 N. w. Cooperstown, E. Unadilla r. or Madison co., and s. Oneida county. Several small branches of Unadilla r. are in the town. The surface is level and fertile N. w., and hilly s., with fine pastures. The country is elevated between the head streams of the Mohawk and Susquehannah: pop. 1830, 1,626.

PLAINFIELD, village, Westfield, Essex co. N. J., 16 ms. s. w. Newark, and E. of Greenbrook.

PLAINFIELD, p-v. Coshocton co. O., by p-r. 87 ms. N. E. by E. Columbus.

PLAINFIELD, p-v. St. Clair co. Mich., by p-r. 64 ms. N. E. Detroit.

PLAINSVILLE, p-v. Luzerne co. Pa., by p-r. 119 ms. N. E. Harrisburg.

PLAISTOW, p-t. Rockingham co. N. H., 36 ms. s. E. Concord, 30 s. w. Portsmouth, 35 N. by w. Boston, N. and N. w. Haverhill, Mass.; 6,839 acres; was purchased of the Indians, 1642, as a part of that town, and has a good, black loamy soil, rocky N. w. Some minerals are found here, many springs and a few small streams: pop. 1830, 591.

PLAQUEMINES, (Percimon,) outlet of the Miss. to the right, 96 miles below the mouth of Red r., 8 miles below the outlet of Iberville from the opposite side, and 117 miles above New Orleans. The Plaquemine outlet receives water only when the mississippi is within 8 or 10 feet of its extreme height of flood; but when the main stream has attained its greatest height, large barges and steamboats are safely navigated down the Plaquemine into its recipient the Atchafalaya; and thence by the various interlocking streams to upper Attacapas, and to Opelousas. The channel of Plaquemine, of 15 ms. in length, is very winding, but the banks being steep and composed of alluvial soil, vessels receive but little damage by running on shore. It is a pass of very great importance, as, through it, passes the travelling and commerce of a wealthy and fertile section of La.

PLAQUEMINE, remarkable bend of the Mississippi r. 75 ms. below New Orleans. Fort St. Philip, called in the p-o. list fort Jackson, stands on this bend, and on the left bank of the r., and completely commands the stream, which, opposite the glacis, is only 37 chains, or a fraction less than half a mile wide; of course the opposite shore is within reach of point blank shot. At this place is a post office called Fort Jackson.

PLAQUEMINES, parish of La., bounded w. by Jefferson, St. Bernard N. w., lake Borgne N., Chandeleur bay E., and the gulf of Mexico s. E. and s. Greatest length, following the general comparative course of Miss. r. 85 ms. Greatest breadth from Barataria bay to the pass of Marian 75 ms. The form approaching that of a cross, ends in narrow points at each extreme; the area about 2,500 sq. ms. Extending in lat. from 29° to 30° 10′, and in long. W. C. 12° to 13° w. The surface is the sthestrn. salient part of the great plain of the Mississippi, and is literally a plain, over which no spot rises 10 feet above the level of the gulf of Mexico. This was demonstrated by the hurricane of the 18th and 19th August, 1812, when the water was raised 8 feet above its ordinary level, inundated the whole Plaquemine parish, spreading ruin and death along the cultivated banks of the Miss. r. Houses, fences, horses, cattle, and not a few human beings, were engulfed. The storm was truly terrible over all La., but below the English Turn, 15 ms. below New Orleans, it was a real deluge with all the terrors of such a catastrophe. The writer of this article passed along the scene in April, 1813, when it still appeared as if an enormous weight had been rolled over the whole surface where any timber had stood to meet the fury of the tempest. The wooded, very slightly elevated, and arable margins of the Miss. r. would be fully estimated at 120 sq. ms. in Plaquemine parish; and the residue is one extended grassy marsh. The arable soil is, however, extremely productive. Sugar cane, cotton, Indian corn, rice, the orange and fig tree, with an indefinite list of esculent plants, grow luxuriantly. There is no town in this parish deserving the title: pop. 1820, 2,354, 1830, 4,489.

PLATO, p-v. Lorain co. Ohio, by p-r. 139 ms. N. N. E. Columbus.

PLATTE, large river of the United States, and one of the great wstrn. confluents of Mo. r., rises according to Tanner, in the eastern vallies of the Chippewayan or Rocky mountains, interlocking sources to the southward with those of Arkansas, to the nrthrd. with those of Yellow Stone r., and to the wstrd. with those of Lewis' r. branch of Oregon or Columbia r. The extreme source of Platte, as laid down on Tanner's N. A., is in lat. 40° and a little w. of 30° w. long. W. C., and so nearly due E. is the general course of this large stream, that though traversing 11 degrees of long., its entrance into the Mo. river is at lat. 41° 03′. It is in the higher part of its course composed of two branches; the Padouca or sthrn. and the Platte proper or nrthrn., both deriving their sources along the estrn. slope of the Chippewayan system, and along or near long. W. C. 30° w. The two branches inclining upon each other, unite after a separate course over 5 degrees of long. The name of this r. is derived from the features of its channel, which is disproportionably wide, shallow, and impeded by sand banks and islands. These phenomena, however, the Platte shares in common with all the streams sthrd. from Miss. proper, and which flow from the Chippewayan mtns. or immense plains between that system and the Miss. and gulf of Mexico, none of which maintain throughout the year navigable water, answering in any moderate proportion to their length of volume or surface they respectively drain. This is the case with the Platte, Kansas, Arkansas, Red river, Sabine, Trinity, Brasos, Colerado of the Gulf of Mexico, and in a very striking manner with the Rio Grande del Norte.

The valley of Platte, as laid down on our best maps, lies between those of Mo. and Yellow stone r. to the N., and the Arkansas and Kansas rs. S., and is about 560 ms. from W. to E.; mean breadth 120, and area 67,200 sq. ms. Surface in great part unwooded and in many places desert plains. The series of rs. belonging to the system of which Platte is one, is continued sthrd. to the Rio Grande inclusive, & nrthrd. to the Mo. at the Mandan villages. In the latter direction the streams are rapidly abridged in their length by the peculiar form of the upper valley of Mo. If the volume of the Platte afforded navigable facilities in proportion to the length, and direction of its channel, it would be the most suitable route of intercommunication between the Miss. basin and Pacific ocean, as it heads in the same system of mountains, and at no great distance from the sources of the Timpanogos, Buoneventura, and Colorado of the gulf of California. These latter rs. from the vague knowledge we possess of the region they drain, partake of the navigable defects we have noticed in regard of those streams issuing from the opposite side of the same system of mtns.

PLATTEKILL, p-t. Ulster co. N. Y., 89 ms. S. by W. Albany, 22 S. Kingston, N. Orange co.; with 30 sq. ms.; has an irregular form, few streams, and a village called Pleasant Valley: pop. 1830, 2,044.

PLATTSBURGH, p-t. and st. jus. Clinton co. N. Y., 164 ms. N. Albany, 112 N. Whitehall, 120 E. Ogdensburgh, W. lake Champlain, is crossed by Saranac and Salmon rs., and several smaller streams which furnish good mill seats. The E. part is nearly level, and the W. very hilly and broken. The v. is at the mouth of the Saranac, 13 ms. N. Port Kent, and contains the co. buildings. It was taken by the British twice in the last war. The view upon the lake from the high grounds near, is very fine. Cumberland bay was the scene of McDonough's victory in the American squadron on the lake, on the 11th Sept. 1814, over that of the British general, Sir Geo. Prevost, who was at that time encamped in the N. part of the v. of Plattsburgh, with 14,000 men. The American vessels had a total of 86 guns and 820 men, and the British 95 guns and 1,050 men. The result of the battle was of the highest importance, as it compelled the enemy to retreat, and delivered the country below from the fear of invasion: pop. 1830, 4,913.

PLEASANT, p-v. nrthwst. part Switzerland co. Ind., 93 ms. S. E. Indianopolis.

PLEASANT EXCHANGE, p-v. Henderson co. Tenn., by p-r. 128 ms. S. W. by W. Nashville.

PLEASANT GARDEN, p-v. Burke co., N. C., *by p-r.* 223 ms. W. Raleigh.

PLEASANT GROVE, p-o. Lunenburg co. Va. by p-r. 89 ms. S. W. Richmond.

PLEASANT GROVE, p-o. Orange co. N. C. by p-r. 64 ms. N. W. by W. Raleigh.

PLEASANT GROVE, p-o. Greenville dist. S. C. by p-r. 125 ms. N. W. Columbia.

PLEASANT GROVE, p-o. Henry co. Geo. by p-r. 115 ms. N. W. Milledgeville.

PLEASANT GROVE, p-o., Maury co. Tenn. 10 ms. sthrd. Columbia the co. st., and by p-r. 52 ms. sthrd. Nashville.

PLEASANT GROVE, p-v. Tazewell co. Il. by p-r. 153 ms. N. N. W. Vandalia.

PLEASANT GROVE, p-v. Lafayette co. Mo. by p-r. 286 ms. wstrd. St. Louis.

PLEASANT HILL, p-v. Delaware co. Pa. by p-r. 125 ms. N. E. W. C.

PLEASANT HILL, p-o. Charles co. Md. 26 ms. sthrd. W. C.

PLEASANT HILL, p-v. wstrn. part of Wythe co. Va., by p-r. 344 ms. S. W. by W. W. C.

PLEASANT HILL, p-v. Northampton co. N. C. by p-r. 101 ms. N. E. Raleigh.

PLEASANT HILL, p-v. Lancaster dist. S. C. by p-r. 66 ms. N. N. E. Columbia.

PLEASANT HILL, p-v. Dallas co. Ala. by p-r. 105 ms. S. S. E. Tuscaloosa.

PLEASANT HILL, p-o. Jefferson co. Miss. by p-r. 14 ms. N. Natchez.

PLEASANT HILL, p-o. Crawford co. Ark. by p-r. 139 ms. wstrd. Little Rock.

PLEASANT HILL, p-o. Davidson co. Ten. by p-r. 8 ms. wstrd. Nashville.

PLEASANT MOUNT, p-v. Wayne co. Pa. by p-r. 269 ms. N. N. E. W. C.

PLEASANT PLAINS, p-o. Franklin co. Ten. by p-r. 81 ms. S. E. Nashville.

PLEASANT RIDGE, p-o. Greene co. Ala. by p-r. 67 ms. sthrd. Tuscaloosa.

PLEASANT RIDGE, p-o. Rush co. Ind. 49 ms. S. E. by E. Indianopolis.

PLEASANT SPRING, p-v. Limestone co. Ala. by p-r. 178 ms. N. N. E. Tuscaloosa.

PLEASANT UNITY, p-v. Westmoreland co. Pa., by p-r. 189 ms. N. W. W. C.

PLEASANT VALE, p-v. Pike co. Il., 10 ms. N. Atlas, the co. st., and by p-r. 158 ms. N. W. Vandalia.

PLEASANT VALLEY, p-t. Dutchess co. N. Y. 7 ms. N. E. Poughkeepsie and 82 from Albany, about 6 ms. sq., is nearly level, with good land. The v. is near the centre, on Wappinger's cr. 7 ms. S. E. Poughkeepsie, and contains several factories: pop. 1830, 2,419.

PLEASANT VALLEY, p-o. Bucks co. Pa. about 43 ms. N. Phila.

PLEASANT VALLEY, p-v. Fairfax co. Va. 30 ms. wstrd. W. C.

PLEASANT VALLEY, p-v. Lancaster dist. S. C. by p-r. 96 ms. N. N. E. Columbia.

PLEASANT VALLEY, p-v. Dallas co. Ala. by p-r. 92 ms. S. S. E. Tuscaloosa.

PLEASANT VALLEY, p-v. Washington co. Ind. by p-r. 89 ms. S. Indianopolis.

PLEASANT VIEW, p-v. Henry co. Ten., by p-r. 189 ms. wstrd. Nashville.

PLEASANTVILLE, p-v. Montgomery co. Pa. by p-r. 22 ms. nthrd. Phil.

PLEASANTVILLE, p-v. Rockingham co. N. C. by p-r. 118 ms. N. W. Raleigh.

PLEASANTVILLE, p-v. Fairfield co. O. by p-r. 29 ms. S. E. Columbus: pop. 1830, 34.

PLEASUREVILLE, p-v. Henry co. Ky. by p-r. 34 ms. N. W. Frankfort.

PLUCKAMIN, p-v. Bedminster, Somerset co. N. J., 6 ms. N. Somersville. The range of Pluckamin mtns. begins here, which extends N. E. to the Passaic falls at Patterson.

PLUMB, isl., Mass., between Ipswich and Newburyport, is near the main land, and about 9 ms. in length.

PLUMB isl., Southold, Suffolk co. N. Y., 1 m. by 3, has a few families; it is separated from Oyster Pond point, by a narrow strait. A line drawn nearly N. E. from that point passes through this isl., the Gull isls., the Race and Fishers' isl., where Long Isl. sound appears to have been formerly more nearly closed at its E. extremity than now. The surface is very stony. Some pine wood is found in a swamp.

PLUMB GROVE, p-o. St. Charles' co. Mo., about 40 ms. wstrd. St. Louis.

PLUM ORCHARD, p-o. Fayette co. Ind., by p-r. 60 ms. estrd. Indianopolis.

PLYMOUTH, t. Penobscot co. Me., 44 ms. from Augusta: pop. 1830, 504.

PLYMOUTH, p-t. Grafton co. N. H. 75 ms. N. W. Portsmouth, 40 N. by W. Concord, 31 S. E. Haverhill, W. Pemigewasset r., 16,256 acres, has also Baker's r., 30 ms. long, and several smaller streams, pretty good soil, bearing beech, maple, birch, hemlock and white pine. The uplands, which are mountainous, are seven eighths of the t. The church in the N. E. corner, is on a commanding hill. There is a library. First settled 1764. Baker's r. has its name from a successful attack on the Indians, who dwelt on its meadows, by capt. Baker, from Haverhill, Mass.: pop. 1830, 1,175.

PLYMOUTH, p-t. Windsor co. Vt., 15 ms. W. Windsor, 52 S. Montpelier, 16 S. W. Rutland; settled 1776; gives rise to Black r. which runs S. E., and furnishes mill seats, and has several fish ponds connected with it. Here rise also 2 branches of Queechy r. The surface is broken. Mount Tom, and another mtn. cross the t. parallel to the r.; primitive limestone is quarried and cut here for market; soap stone also abounds. There are several caverns 500 yds. S. W. of the r., one of which, discovered in 1818, is quite extensive. The soil is good for grazing: pop. 1830, 1,667.

PLYMOUTH co. Mass., bounded by Norfolk co. N. W., Massachusetts and Cape Cod bays E., Barnstable and Buzzard's bays S., and Bristol co. W. The surface is uneven, and the soil various; the form is irregular, extending N. to Point Alderton, the S. point of Boston bay, from which to the S. W. extremity is a line of irregular coast, with a short interval of land, where the S. E. boundary crosses the isthmus of Cape Cod. Plymouth bay indents the E. line near the middle, and receives a few brooks. It is the spot first settled by the pilgrim fathers of N. England who landed here on the 22d December, 1620, O. S. The principal stream in the co. is Taunton r., which rises in the N. W. part, and crosses the W. boundary into Bristol co.; there are many ponds and brooks. Plymouth colony remained under a separate colonial government until 1685. It has several harbors, Plymouth, Duxbury, &c., with considerable coasting and some foreign trade; fisheries of value, and some manufactures. It contains 21 tsps.: pop. 1820, 38,136; 1830, 43,044.

PLYMOUTH, sea port, p-t. and st. jus., Plymouth co. Mass., 36 ms. S. E. Boston, 5 ms. by 16, contains the oldest permanent settlement in New England. It stands on Plymouth bay, which is large, but affords but little depth of water. It is almost shut in by two long reaches, formed of sand thrown up by the waves, and is gradually increasing. The government of the U. S. appropriated $2,500 to repair it in 1832. The land is high on the N. & S. sides of the bay, and there are rocky isls. off the harbor. Manumet point, a bold, rocky promontory, lies S. The soil is generally thin and poor, and some portions of it are very good. The v. is near the N. E. part; the principal street runs N. and S. between the head of the harbor, and several sandy hills, which rise at a little distance from the shore. Some foreign trade has been carried on here. One of the principal buildings is Pilgrim's Hall, which was erected by the pilgrim society, for the annual celebration of the landing of the forefathers of New England. This important event occurred here on the 22d of December, 1620, O. S., when the crew of the Mayflower debarked. A large granite, on which they first stepped from the boat, is still preserved. One half of it retains its original position, near the water, which has since been somewhat encroached on by the land, while the other has been removed to the centre of the v. The Indians on this part of the coast had been greatly reduced in numbers before the arrival of the colonists, by the small pox; and Massasoit and his men first presented themselves on Watson's hill. A fort was erected on Burying hill, which also became a grave yard; and several of the stones of the early colonists are still preserved there. The first well dug in N. England is still in existence. The first child born in the colony was Peregrine White. The first mill erected in New England was built here, in 1632. From this spot at different periods, proceeded some of the first settlers of many of the old towns in Massachusetts and Conn. It contains 407 acres of tillage land, 828 mowing, 3,486 of pasturage: pop. 1830, 4,758.

PLYMOUTH, p-t. Litchfield co. Conn., 24 ms. W. Hartford, 30 from New Haven, W. Bristol, Hartford co., and N. of New Haven co., about 5 ms. by 5½, is hilly, with primitive rocks, bearing oak, chestnut, swamp maple, &c. also rye, corn, oats and grass. It has Naugatuck r. W. with mill seats, and other streams: pop. 1830, 2,064.

PLYMOUTH, p-t. Chenango co. N. Y., 107 ms. w. Albany, 7 N. W. Norwich, has an uneven surface, with good land, well watered by Canasawacta cr., whose two branches meet near the centre, at Frankville v. The stream then runs S. E. towards Chenango r., which it meets in the next tsp., Norwich. There are several mill sites. The timber is maple, beech, elm, bass, &c.: pop. 1830, 1,609.

PLYMOUTH, p-o. and tsp. Luzerne co. Pa., opposite Wilkesbarre. The p-o. is 6 ms. from Kingston, and 7 s. w. Wilkes-Barre.

PLYMOUTH, p-v. and st. jus. Washington co., N. C., situated on a small cr. extending sthrd. from the mouth of Roanoke r., by p-r. 128 ms. E. Raleigh, and 35 ms. N. N. E. Washington, in Beaufort co.; lat. 35° 51′, long. W. C. 0° 19′ E.

PLYMOUTH, p-v. Richland co. O., 20 ms. nrthrd. Mansfield, the co. st., and by p-r. 91 ms. N. N. E. Columbus.

PLYMOUTH, p-o. N. W. part Wayne co., Mich., by p-r. about 25 ms. N. W. Detroit.

PLYMPTON, p-t. Plymouth co. Mass., 32 ms. S. E. Boston, has a branch of Taunton r. s., has extensive iron manufactories. The Indian name was Patuxet, or Wanatuxet. It contains 349 acres under tillage, 613 of mowing, and 1,366 of pasturage: pop. 1830, 950.

POCAGON, p-v. in the southwestern angle of Cass co. Mich. It is situated on St. Joseph's r. of lake Michigan, by p-r. 180 ms. a little s. of w. Detroit. Though placed in Cass co. by the p-o. list, it is laid down by Tanner in his improved U. S. map, in the southeastern part of Berrien co. Mich.

POCAHONTAS, co. Va. bounded by Greenbrier s. and s. w., Nicholas w., Randolph N. w. and N., and Alleghany mtn. separating it from Pendleton N. E. and E. Length from s. w. to N. E. 50 ms., mean breadth 20 ms., and area 1,000 sq. ms. Lat. 38° 20′, long. 3° w. W. C. This co. is amongst the most elevated in the U. S. giving source to Cheat r. branch of Monongahela, flowing northwardly, and to Greenbrier r. flowing southwardly. The mean height of the arable land of Greenbrier co. is about 1,700 feet, and of course, being lower down Greenbrier r. than Pocahontas, the lowest part of the latter must exceed that relative oceanic elevation, or rise to a mean exceeding 1,800 feet, or an equivalent to four degrees of lat. or mean winter temperature. Greenbrier mtn. enters and traverses Pocahontas from s. w. to N. E. from the western slopes of which issue the extreme fountains of Gauly and Elk rs. The surface is excessively broken and rocky, and most of the soil sterile. Chief t. Huntersville: pop. 1830, 2,542.

POCKET (The). (*See "The Pocket," p-v. Moore co. N. C.*)

POCOMOKE, r. and bay, Md. The river rises on the border between Sussex co. Del., and Worcester co. Md., from whence by a s. s. w. direction 60 ms. by comparative courses, traversing Worcester co., it opens into a bay of the same name at lat. 38°, and on the line between the eastern shores of Va. and Md. The bay of Pocomoke is a triangular sheet of water, bounded N. W. by Tangier isl., N. by the sthern. shore of Somerset co. Md., and E. by the western shore of Accomac co. Va. To the s. w. it opens into, and is confounded with, Chesapeake bay. Small coasting vessels ascend to Snowhill on Pocomoke r.

POCOTALIGO, r. Va., in Kenhawa co., rises in the N. W. part of the co. interlocking sources with those of the west fork of Little Kenhawa, and flowing thence southwestward 60 ms. by comparative courses, falling into Great Kenhawa r. at the point of separation on that stream, between Kenhawa and Mason cos.

POCOTALIGO, p-o. on Pocotaligo r. Kenhawa co. Va. by p-r. 353 ms. a little s. of w. W. C.

POCOTALIGO, p-v. near the right side of Combahee r. and in the N. E. part of Beaufort dist. S. C. 67 ms. a little s. of w. Charleston, and by p-r. 141 ms. s. Columbia.

POESTEN KILL, Rensselaer co. N. Y. a very good mill stream, which falls into Hudson r. at Troy, after turning much machinery for various manufactures in the vicinity. It rises in Grafton, and has a course of about 20 ms. At the falls, on the side of mount Ida, 1 m. east of Troy, is a small manufacturing village.

POGE, cape, the N. E. end of Chippaquiddick isl., E. of Martha's Vineyard.

POINDEXTER'S Store and p-o. Louisa co. Va. by p-r. 68 ms. N. W. Richmond.

POINT COUPEE, parish of La. bounded s. by West Baton Rouge, w. by Atchafalaya r. separating it from Opelousas or Saint Landry, and from the parish of Avoyelles, N. E. and E. by Miss. r. separating it from West Feliciana, and East Baton Rouge. The outline is triangular, base along the general course of the Atchafalaya 34 ms., perpendicular 30 ms. along the southern border; area 510 sq. ms. Lat. 30° 45′, long. 14° 36′ w. W. C. The very slight inclination sthrd. The whole being a plain, elevated about 4 or 5 feet along the margins of the streams, but depressed from the water courses so as to be annually submerged. It extends from the efflux of Atchafalaya, widening as the two rivers diverge from each other. In its natural state the surface was covered with a very dense forest, and the greatest part remains in that state. The soil, where sufficiently elevated for the plough, is exuberantly fertile. Cotton is the common staple, and in this parish is the highest point in La. where the sugar cane has been cultivated to any advantage. Chief t. Point Coupee: pop. 1820, 4,912, 1830, 5,936.

POINT COUPEE, or Cut Point, p-v. and st. jus. parish of Point Coupee, situated on the left shore of the Miss. r. opposite St. Francisville, and by p-r. 154 ms. above and N. W. New Orleans. Lat. 30° 42′.

POINT HARMER, p-v. Washington co. O. by p-r. 106 ms. S. E. by E. Columbus.

POINT LABADIE, p-v. Franklin co. Mo. by p-r. 43 ms. w. St. Louis.

POINT PLEASANT, p-v. and st. jus. Mason co. Va. situated on the point above the junction of Ohio and Great Kenhawa rs. by p-r. 358 ms. a little s. of w. W. C., and 358 ms. n. w. by w. ½ w. Richmond. Lat. 38° 50', long. 5° 7' w. W. C.

POINT PLEASANT, p-v. southern part of Clermont co. O. by p-r. 19 ms. s. Batavia, the co. st. and 128 s. w. Columbus.

POINT REMOVE, p-v. on Arkansas r. sthrn. part of Conway co. Ark. by p-r. 51 ms. above and n. w. by w. Little Rock. This name is another instance of that propensity so common of accommodating proper names to our own language; it comes from the French *Point Remu*, and that from a counter current in the adjacent r.

POLAND, p-t. Cumberland co. Me. 44 ms. s. s. w. Augusta, 30 n. Portland, s. Little Androscoggin r., borders on Oxford co. on the n. w., and part of a small lake, which, with several ponds and small streams, empties into Little Androscoggin r.: pop. 1830, 1,916.

POLAND, p-v. in the southeastern angle of Trumbull co. O. 20 ms. s. e. Warren, the co. st., and by p-r. 283 ms. n. w. W. C., and 164 n. e. by e. Columbus: pop. of the tsp. of Poland, 1830, 1,186.

POLSLEY'S Mills and p-o. Monongalia co. Va. by p-r. 235 ms. n. w. by w. W. C.

POMFRET, t. Windsor co. Vt. 18 ms. n. Windsor, 40 s. Montpelier, 5½ ms. by 7; first settled, 1770; is uneven, with good soil, and has White r. n. e. and Queechy s. e.; 13 school dists. There is a range of young timber in the forests, 7 or 8 ms. long, and about 500 yards wide, which appears to have grown up after a hurricane, which is supposed to have swept through that region about 120 years ago: pop. 1830, 1,866.

POMFRET, p-t. Windham co. Conn. 40 ms. n. e. Hartford, 30 e. Providence, and w. Quinebaug r., about 6 ms. by 7, with about 42 sq. ms. It is hilly, with primitive rocks, and has a good soil, favorable to grazing. Cotton, woollen, &c. are manufactured to some extent. It is watered by Little r. and several other branches of the Quinebaug. Shad are caught in Quinebaug r. In a wild and solitary part of the town is the famous cavern, in which major general Israel Putnam, who afterwards commanded the American militia at the battle of Bunker's Hill, performed the bold and celebrated feat of killing a wolf. He was an inhabitant of Pomfret from 1739, for many years. He distinguished himself in the French war of 1755 as well as through the revolution: pop. 1830, 1,981.

POMFRET, p-t. Chatauque co. N. Y. 20 ms. n. n. e. Maysville, and s. of lake Erie, has 90 sq. ms., and is crossed n. e. and s. w. by Chatauque ridge, which runs parallel to the lake, 3 or 4 ms. distance, with a smooth alluvial tract of land lying between them, with a good sandy loam; s. is a slaty loam, bearing tulip, maple, beach, hemlock and other trees. Canadawa cr. crosses the t. in a n. w. direction, and there are several smaller streams. Dunkirk v. is on the lake, with a good harbor, 3 ms. n. e. Fredonia, and 45 s. w. Buffalo. There are 7 ft. of water on a reef of rocks at the bar. From this place to Erie the shore is rocky. Fredonia v. stands on Canadawa cr. 22 ms. from Maysville and 45 from Buffalo, is on the Buffalo and Erie road, and is a thriving v. Bear and Cassadaga ponds are on the s. line of the t.: pop. 1830, 3,386.

POMONA, p-v. Wake co. N. C. 14 ms. n. e. Raleigh.

POMPEY, p-t. Onondaga co. N.Y. 11 ms. s. e. Onondaga, 146 w. n. w. Albany, has several excellent mill streams; Butternut cr. w. and 2 branches of Limestone cr. e., all which flow n. to Chitteningo cr. The surface is varied by hills and valleys. Traces of considerable excavations and mounds are perceptible here, of unknown antiquity; and metallic weapons and instruments, and even a church bell, have been dug up from the ground. The first settlement was made in 1788, and no tradition exists which refers to the ancient inhabitants. The mounds are evidently of remote construction. Three of them are traceable near Delphi, the largest of which is a triangle of about 6 acres, with a gateway and picquets. From numerous graves have been dug bones, weapons, utensils, Spanish coins, &c. There are several villages in this t. Pompey v., Pompey w., Hill, and Delphi, s. e. In the town is an academy. On Limestone cr. are two falls, about 100 yards apart, which turn machinery for several manufactories, &c.: pop. 1830, 4,812.

POMPTON, r. N. J. between Morris and Bergen cos. is a branch of Passaic r. formed by the union of Pequannoc, Longpond and Rampo rs. and bears the name of Pompton for only 7 or 8 ms., when it enters the Passaic at the corners of 3 cos. Morris, Bergen and Essex.

POMPTON, t. Bergen co. N. J. 60 ms. n. n. e. Trenton, has N. Y. on the n. e., Sussex co. n. w., and Pequannoc r. s., dividing it from Morris co. It is crossed n. and s. by Kingwood r., and is hilly and mountainous in many parts, being rendered rough by the mountainous range which extends s. w. nearly across the state, and forms a natural line of defence, which was occupied by the American troops during different periods of the revolutionary war. In advance of it, s., is the inferior range of elevated ground called the Short Hills: pop. 1830, 3,085.

POMPTON, p-v. Pequannoc, Morris co. N. J. 5 ms. n. w. Patterson, stands on the s. side of Pompton r. a little n. of Pompton mtns., and n. w. of the plain.

PONTCHARTRAIN, lake of La. between the alluvial Delta, and the comparatively high and hilly interior. Inspection on a map, and still more actual examination of the country adjacent, must convince any person that lakes Borgne, Pontchartrain and Maurepas, are the remains of a deep bay, which in remote ages penetrated upwards of 120 ms. from opposite the mouth of Passagoula r. towards the Miss. and separating the high grounds to the N. from the Delta. This chain of lakes has been formed by alluvial protrusions into the ancient gulf. Pearl r. and the outlets of the Miss. have formed a neck only traversed by the Rigolets and Chef Menteur straits separating lakes Borgne and Pontchartrain; and at the opposite extremity of the latter, similar natural operations have formed a similar neck with the Bayou Manchac, connecting it with lake Maurepas. From the high lands, lake Maurepas receives Amite, and Tickfolah rs. Into Pontchartrain is poured Tangipaha, Tchefuncte, and some smaller creeks; whilst lake Borgne, or rather the Rigolets, receive the different outlets of the large stream of the Pearl. The greatest length of Pontchartrain from the outlet of the Rigolets to the Pass of Manchac is about 45 ms.; greatest breadth 25, but mean breadth 12 ms. The common depth from 16 to 18 feet, but every where shallow along shores, and in no harbor or creek affording a harbor of 9 feet draught. Along the nrthrn. side the banks are low, but in part solid; towards the Delta it is bordered by an uninterrupted marsh. Timber covers the nrthrn. and open grassy plains the sthrn side. Compared with the depth of the Miss. r. at New Orleans, the bottom of lake Pontchartrain is about 50 feet elevated, and the surface about 6 feet depressed below that of the Miss. at mean flood. The tides of the Gulf of Mexico, slight as they are, not exceeding a mean of 2½ feet, flow into Pontchartrain and are diurnally perceptible in the rear of New Orleans.

PONDICHERRY, mtn. between Jefferson and Bretton Woods, Coos co. N. H.

PONTIAC, p-v. and st. jus. Oakland co. Mich. situated on Clinton r. by p-r. 26 ms. N. N. W. Detroit, lat. 42° 37′ long. W. C., 6° 15′ W.

POOLESVILLE, p-v. wstrn. part Montgomery co. Md. 33 ms. N. W. W. C.

POOLESVILLE, p-v. Spartanburg dist. S.C. by p-r. 112 ms. N. W. Columbia.

POOR'S, p-v. Jackson co. O. by p-r. 82 ms. S. S. E. Columbus.

POPE, co. Il. bounded by Johnson W., Gallatin N. and N. E.; Ohio r. separating it from Livingston co. Ky. E., and the Ohio r. again separating it from MacCracken co. Ky. S. Greatest length due N. from the O. r. opposite the mouth of Tennessee r. 40 ms.; mean breadth 20 ms. and area 800 sq. ms., lat. 37° 20′, long. W. C. 11° 36′ W. General slope sthestrd. towards Ohio r. Soil of middling quality; and surface hilly. Chief town, Golconda: pop. 1830, 3,316.

POPE, co. Ark. on Arkansas r. above Pulaskie, and below Crawford, but the outlines of which we have not documents to delineate. Chief t. Scotia: pop. in 1830, 1,483.

POPLAR BRANCH, p-v. Currituck co. N. C., by p-r. 228 ms. N. E. by E. Raleigh.

POPLAR CORNER, p-o. Madison co. Ten. by pr. 10 ms. wstrd. Jackson, the co. st. and 157 ms. S. W. by W. Nashville.

POPLAR GROVE, p-o. Dinwiddie co. Va. by p-r. 39 ms. S. Richmond.

POPLAR GROVE, and p-o. Iredell co. N. C. by p-r. 155 ms. W. Raleigh.

POPLAR GROVE, p-o. Newberry dist. S. C., by p-r. 45 ms. N. W. Columbia.

POPLAR HILL, p-o. Giles co. Va. by p-r. 310 ms. S. W. by W. W. C.

POPLAR MOUNT, p-o. Greenville co. Va. by p-r. 56 ms. S. Richmond.

POPLAR PLAINS, p-v. Fleming co. Ky. by p-r. 84 ms. E. Frankfort.

POPLAR RIDGE, p-v. Scipio, Cayuga co. N.Y. 4 ms. E. Cayuga lake, 14 S. W. Auburn.

POPLAR RIDGE, p-o. Obion co. Ten. 10 ms. from Troy, the co. st. and by p-r. 168 ms. a little N. of W. Nashville.

POPLAR RUN, p-o. Orange co. Va. by p-r. 95 ms. S. W. W. C.

POPLAR SPRINGS, p-v. near the extreme nrthwestern angle of Ann Arundel co. Md. It is situated on the wstrn. turnpike from Baltimore to Frederick, by p-r. 61 ms. N. W. C.

POPLAR SPRING, p-o. Fairfield dist. S. C. 36 ms. nrthrd. Columbia.

POPLARTOWN, p-v. Worcester co. Md. 12 ms. wstrd. Snowhill, the co. st. and by p-r. 152 ms. S. E. by E. W. C.

POPLIN, p-t. Rockingham co. N. H. 24 ms. W. S. W. Portsmouth, E. S. E. Concord, 50 N. N. E. Boston, with 10,320 acres, is watered by Squamscot or Exeter r. and other streams, and has Loon pond N., and Spruce swamp E., has good soil, and no high hills. Incorporated 1764: pop. 1830, 429.

PORPOISE, cape, Kennebunk port, York co. Me., long. 70° 23′ W., lat. 43° 22′, forms Kennebunk harbor, which lies at the mouth of a small stream.

PORTAGE, p-t. Alleghany co. N. Y. 247 ms. W. Albany, and S. of Livingston co., is crossed by Genesee r. which pursues a serpentine course from the S. to the N. line, and passes three falls, of 8, 66, and 110 feet, near which, on the W. side, is situated the village. The shape of the t. is regular, except at the S. W. corner: pop. 1830, 1,839.

PORTAGE, co. O. bounded S.E. by Columbiana; Stark S.; Medina W.; Cuyahoga N. W.; Geauga N.; and Trumbull N. E. The greatest length 30 ms. is from E. to W.; breadth 24, and area 720 sq. ms.; lat. 41° 12′, long. W. C. 4° 20′ W. This co. is a true table land between the valleys of Big Beaver and Cuyahoga rs. The southwestern angle also giving source to Tuscarawas r. or the nrthestrn. constituent of Muskingum r. The peculiar structure of the surface may be more particularly seen by reference to the article Cuyahoga r. The arable surface of Portage co. exceeds a mean of 1,000 feet above tide water in the Atlantic, or rather more than an equivalent to two degrees of lat. The surface is rather level, and in part deficient in good fountain water. The soil moderately fertile. Chief town, Ravenna: pop. 1820, 10,095; 1830, 16,963.

The Ohio and Erie canal traverses this co. in its greatest breadth and near the western border; and within it is the summit level of that work, 973 ft. above the Atlantic tides.

PORTAGE r. O. rising in Hancock co. interlocking sources with those of Blanchard's fork of Maumee r. Formed by numerous creeks which unite in Wood co., and curving to N. E. enters Sandusky co. in which latter it again curves more estrd., finally falling into lake Erie after an entire comparative course of 50 ms. The valley of Portage lies between those of Sandusky and Maumee.

PORTAGE, p-v. sthestrn. part of Wood co. O. by p-r. 136 ms. N. N. W. Columbus.

PORT BAY, p-t. Wayne co. N. Y. 193 ms. from Albany: pop. 1830, 1,082.

PORT BYRON, p-v. Cayuga co. N. Y.

PORT CARBON, flourishing p-v. Schuylkill co. Pa. 10 ms. wstrd. Orwigsburg, the co. st., and by p-r. 177 ms. N. N. E. W. C. and 69 ms. N. E. Harrisburg.

PORT CLINTON. p-v. Schuylkill co. Pa. by p-r. 60 ms. N. E. Harrisburg.

PORT CLINTON, p-v. at the mouth of Portage r., into lake Erie in the nrthrn. part of Sandusky co. O. by p-r. 117 ms. due N. Columbus: pop. 1830, 116.

PORT CONWAY, p-v. and s-p. on Rappahannoc r. sthwstrn. part King George's co. Va. by p-r. 79 ms. sthrd. W. C.

PORT DEPOSIT, p-v. on the left bank of Susquehannah r. at its lowest falls, Cecil co. Md. 37 ms. N. E. Baltimore, and 5 ms. above Havre de Grace, at the mouth of Susquehannah r.

PORTER, t. Oxford co. Me. 91 ms. s. w. Augusta, 34 s. w. Paris, lies E. of N. H., N. Ossipee r. which separates it from York co.: pop. 1830, 841.

PORTER, t. Niagara co. N. Y. 15 ms. N. W. Lockport, s. lake Ontario, and E. Niagara r., contains Youngstown village and fort Niagara. There is a ferry across N. r. at Youngstown. Niagara v. stands on the E. side Niagara r. at its mouth in lake Ontario, opposite Newark, U. Canada. It stands 15 ms. below Niagara falls, and 7 from Lewiston. A palisaded fort was made here in 1679, by the French, which, in 1725, was enlarged into a considerable work. It was surprized by the British, Dec. 19, 1813, and delivered up in March, 1815: pop. 1830, 1,490.

PORTER, p-v. Sciota co. Ohio, by p-r. 100 ms. s. Columbus.

PORTERSVILLE, p-v. northern part of Butler co. Pa. 16 ms. N. of the borough of Butler, and by p-r. 252 ms. N. W. W. C.

PORTERSVILLE, p-v. Franklin county, Miss. about 20 miles s. E. by E. Natchez.

PORTERSVILLE, p-v. and st. jus. Dubois co. Ind. situated on the left bank of the East Fork of White river, by p-r. 124 ms. s. s. w. Indianopolis, lat. 38° 30', long. W. C. 9° 52' w.

PORT GENESEE, or Charlotte p-v. Greece, Monroe co. N. Y. stands at the mouth of Genesee river on the shore of lake Ontario.

PORT GLASGOW, village, Wolcott, Wayne co. N. Y. 22 ms. N. Waterloo, stands on Sodus bay, at the head of navigation; it has a good harbor for lake vessels, and is agreeably situated. A good road leads to Clyde, on the Erie canal, 10 3-4 ms. s.

PORT KENT, village, Chesterfield, Essex co. N. Y. 3 1-2 ms. E. Keeseville, 2 s. Sable river, 13 s. Plattsburgh, and w. Lake Champlain, has a good harbor, with stores and docks, and serves as a landing place for vessels engaged in the transportation of iron from the extensive mines in the neighborhood.

PORTLAND, p-t. and port of entry, Cumberland co. Me. until lately the capital of the state, is beautifully situated on an elevated peninsula in Casco bay. It is 54 ms. N. N. E. Portsmouth, 118 N. N. E. Boston, 542 from Washington, and 258 s. Quebec; lying in lat. 43° 9', and long. 6° 45' E. W. C. Portland is the principal commercial and most populous town in the state, and has an excellent and capacious harbor, and seldom frozen, bounded by cape Elizabeth s. on which is a light house of stone, 70 feet high; the land about the harbor is generally elevated. Numerous islands are in the bay to the E. on two of which are forts which defend the entrance of the harbor. Fort Preble, on Bang's isl. and Fort Scammel, a block-house on House island. Fort Burrows stands under the observatory bluff, on the waters edge. About 45,000 tons of shipping belong to this port, consisting of a large number of ships, brigs, schooners, sloops, and steamboats, and other craft. The town, (formerly Falmouth, called Portland, and incorporated 1786,) is handsomely laid out, and the style of the buildings, generally, is neat and convenient. A fine street, on which are several churches and other buildings of granite, runs along the ridge and extends to the observatory, where formerly was fort Sumner on a commanding eminence. From this point the view is extensive and various, embracing the beautiful island scenery in the vicinity, and in clear weather, the peaks of the White mountains of New Hampshire. The town lies principally on a declivity, and has the appearance of a considerable and flourishing commercial place. Among the public buildings is that formerly the state house, a court house, town hall, theatre, almshouse, 5 banks, beside a branch of the United States bank, a custom house, academy, and an atheneum, to which a large library (of about 3,000 vols.) belongs. Beside these, there are 15 churches, one of which is for mariners. Education is well attended to, and there are numerous schools, including some of a high character. Portland, (then called Falmouth) was burnt by Capt. Mowatt, of the British sloop of war Canceau, Oct. 18, 1775, on the inhabitants refusing to deliver up their arms. The place was first bombarded for about 9 hours, after which torches were applied and about 130 houses, (two thirds the whole number) were consumed. The old church was one of the buildings which remained: pop. 1820, 8,581, 1830, 12,601.

PORTLAND, p-t. Chatauque co. N. Y. 8 ms. N. Maysville, s. lake Erie, 36 sq. ms., is crossed N. E. and s. w. by the Chatauque ridge, a few ms. from the lake shore and parallel to it, with a regular descent towards the water, and unbroken except by the courses of a few streams. On this ridge grow chestnut, beech, maple, hemlock and other forest trees, and on the lower country, with these are found walnut, tulip, cucumber tree, &c. The soil is pretty good, and the mill seats are numerous and valuable. The rocks are often of mica slate. The town is crossed by the portage road from lake Erie to the head of Chatauque lake, 8 ms. passing through Westfield, a p-v. 1 mile from the harbor, and 7 from Maysville. The harbor of Portland is good, and the p-v. stands upon it 8 miles from Maysville. The earth was once bored near this place 600 feet for salt water, without success: pop. 1830, 1,771.

PORTLAND, p-v. Dallas co. Al. by p-r. 112 ms. s. E. Tuscaloosa.

PORTLAND, p-v. in the northern part of Fountain co. Ind. 88 ms. N. w. Indianopolis.

PORT LAWRENCE, p-v. and port, on the left bank of Maumee r. and in the s. E. part of Monroe co. Mich. by p-r. 55 ms. s. s. w. Detroit, and about 3 ms. above the mouth of Maumee into lake Erie.

PORT PENN, p-v. New Castle co. Del. and on the right bank of Delaware r. opposite Ready Island, 15 ms. sthrd. Wilmington, and 121 ms. N. E. W. C.

PORT REPUBLIC, p-v. Rockingham co. Va. by p-r. 143 ms. N. w. by w. W. C.

PORT ROYAL, Caroline co. Va. p-v. on Rappahannoc r. opposite Port Conway, in King George co. about 25 ms. below Fredericsburg, and by p-r. 78 ms. sthrd. W. C.

PORT ROYAL, p-v. in the eastern part of Montgomery co. Ten. situated at the mouth of Sulphur creek into Red r. 20 ms. estrd. Clarksville, the co. st. and by p-r. 42 ms. N. w. Nashville.

PORT ROYAL, p-v. in the northeastern part of Morgan co. Ind. by p-r. 16 ms. s. Indianopolis.

PORTSMOUTH, p-t. and port of entry Rockingham co. N. H.; the most populous town in the state, and the only seaport. It lies on on Piscataqua r. which divides it from Maine, on a fine peninsula about 3 ms. from the ocean, in lat. 43° 5', and long. 6° 23' E. W. C. It is 45 ms. E. Concord, 55 N. by E. Boston, 58 s. w. Portland, and 491 from W. C. The population is chiefly collected near the harbor, on a hill descending towards it N. and E., and from commanding points the view is very fine. It was settled in 1623 under the authority of Sir George and Capt. J. Mason, and incorporated 1633, and never suffered from Indian attacks, the neck on the s. being stockaded. The town originally included all the peninsula formed by the river and the ocean. The harbor of Portsmouth is one of the finest in the world, rarely, or never freezing, owing to the excessive tides, and has 40 feet of water in its channel at low tide. It is well protected from storms, being completely land-locked, admits vessels of the largest class, and is defended by fort Constitution on Great island, fort McClary opposite, fort Sullivan on Trefethen island and fort Washington on Pierce's island. The two latter were garrisoned during the late war. The amount of shipping owned in Portsmouth, which includes nearly all belonging to the state, is quite large, and though not extensive, it has considerable coasting and other trade. (*See article New Hampshire.*)

Portsmouth contains several houses of public worship; a branch of the U. S. bank and 4 others; several markets, insurance offices, and a custom house. Two bridges were built to Kittery, Maine, in 1822, across the Piscataqua, the channel of which is broad, and the current rapid at particular times of tide. The long bridge 1,750 feet in length, extended across water varying from 43 to 45 feet in depth at low tide, a distance of 900 feet, and crossed an island in the river. A water company was formed and commenced operations in 1799, which supplies all the streets with good water, brought a distance of 3 ms. On Great isl. is a light house. On Continental island, which is owned by the U. S. is a navy yard belonging to government, and on Badger's island was constructed the first ship-of-the-line in America. It was built during the revolution, and named the North America. Portsmouth has suffered severely at different periods from fires. Stocking weaving has recently been commenced here: pop. 1820, 7,327, 1830, 8,082.

PORTSMOUTH, p-t. Newport co. R. I. 7 ms. N. w. Newport, about 2 ms. by 8, occupies the N. part of the island of R. I. with water on 3 sides, viz. E. bay E., Mount Hope bay N., and Narraganset bay w., has a moderate elevation, with slopes, a variety of soil, generally good, with slate rocks. Wheat, barley and fruit flourish, and sheep are raised in considerable numbers. Fish abound along the shores. Prudence island, and several others still smaller, belong to Portsmouth. A bed of anthracite coal, at the N. w. corner of the town, has been worked to some extent, on the shore, but has been abandoned; the quality being inferior to that of the Pa. mines: pop. 1830, 1,727.

PORTSMOUTH, p-v. and st. jus. Norfolk co. Va. opposite the borough of Norfolk, 2 1-2 ms. distant, on the left bank of Elizabeth r., and at the mouth of the sthrn. branch, by p-r. 219 ms. s. s. E. W. C.

Portsmouth affords one of the finest harbors in America; ships of the largest class may lay with safety at the wharves. The navy yard is directly on the sthrn. extremity of Portsmouth, and within the boundaries of the town. This part is called Gosport, and resembles the Northern Liberties of Phila. Charlestown, or Newtown, another suburb rapidly improving, stands at the opposite side from Gosport. Pop. 1830, 2,000.

PORTSMOUTH, p-v. and st. jus. Sciota co. O. situated on the point above the junction of O. and Sciota rs. by p-r. 421 ms. a little s. of w. W. C. and 91 ms. s. Columbus, lat. 38° 42', long. W. C. 5° 54' w. Though rather exposed to river floods, from the lowness of its site, this is a flourishing town. Here the Ohio and Erie canal leaves the former, at an elevation of 474 feet above the Atlantic tides. According to Flint it contains a printing office, bookstore, a bank, two churches, 18 stores, 4 commission stores, 1 druggist, 20 mechanical establishments, steam mill, market house, and the ordinary county buildings. In position it has great and enduring advantages, as a commercial depot: pop. 1830. 1,063.

PORT TOBACCO, p-v. and st. jus. Charles co. Md. by p-r. 32 ms. a very little E. of s. W. C. and 69 s. w. Annapolis. It is situated on a small creek or bay, making northward from the Potamac r. at the Great bend opposite King George's co. Va. lat. 38° 30': pop. 1830, 500.

PORT WATSON, v. Cortlandtville, Cortlandt co. N. Y. 3 ms. s. Homer v., w. Tioughnioga r. at the head of boat navigation, just below Cortlandt v., which is on Factory Branch.

PORT WILLIAM, p-v. and st. jus. Gallatin co. Ky., on the point above the junction of Kentucky r. with the Ohio, by the land p-r. 57 ms. below and N. N. w. Frankfort, lat. 38° 40', long W. C. 8° 09' w.: pop. 1830, 323.

PORT WILLIAM, p-v. sthrn. part Lawrence co. Ind. by p-r. 87 ms. s. s. w. Indianopolis.

POSEY, sthwstrn. co. of Ind. bounded N. by Gibson, E. by Vanderburg, Ohio r. s. separating it from Henderson and Union cos. Ky., by Wabash r. separating it from Gallatin co. Il. s. w., and White co. Il. w. Greatest length from s. to N. 32 ms.; mean breadth 16, and area 512 sq. ms.; lat. 38 and long. W. C. 11° w., intersect near Springfield. Though bordered on the sthrd. by Ohio r. the slope of Posey co. is wstrd. towards the Wabash r. The surface is hilly, but soil productive. Chief ts. Springfield, Harmony, and Mount Vernon, the st. jus.: pop. 1820, 4,061; 1830, 6,549.

POTOMAC r. of Va., Md. and Pa. This r. above Blue Ridge, is formed by the north branch, distinctively called Potomac, Patterson's r., South Branch, Cacapon, Back cr., Opequhan, and Shenandoah, from the southwestward, and by a series of bold, tho' comparatively small streams from the nrthrd. The stream to which the name of Potomac is first applied, rises in the Alleghany chain opposite to the sources of Cheat and Youghioghany branches of Monongahela, at lat. 39° 10', long W. C. 2° 30' w. Flowing thence N. E. 30 ms. receives from the N. Savage r., and bending to s. E. 10 ms. traverses one or two minor chains of mtns., and returning to N. E. 18 ms. to the influx of Will's creek from the N. at Cumberland. Now a considerable stream, by a very tortuous channel, but direct distance 15 ms. to s. E. the Potomac below Cumberland, breaks through several chains of mtns. to the influx of South Branch. The latter is in length of course, and area drained, the main branch. The various sources of this mountain r. originate in Pendleton co. Va. lat. 38° 25', between the Alleghany and Kittatinny chains. Assuming a general course of N. E. the branches unite in Hardy co. near Moorfields, below which, in a distance comparative of 40 miles to its union with the North Branch, the South Branch receives no considerable tributary. The volume formed by both branches, breaks through a mtn. chain immediately below their junction and bending to N. E. by comparative distance 25 miles, but by a very winding channel reaches its extreme nrthrn. point at Hancock'stown, lat. 39° 41', and within less than 2 ms. s. of the sthrn. boundary of Pa. Passing Hancock'stown the Potomac again inflects to s. E. and as above winds by a very crooked channel, but by comparative courses 35 ms. to the influx of Shenandoah from the sthrd.

Shenandoah is the longest branch of Potomac, having a comparative length of 130 ms. and brings down a volume of water but little inferior to that of the main stream. Having its most remote sources in Augusta co. Va. interlocking sources with those of Great Calf Pasture branch of James r. and by Blue Ridge separated from those of Rivanna, as far s. as lat. 37° 55', almost exactly due w. of the mouth of Potomac into Chesapeake bay. The elongated valley of Shenandoah is part of the great mtn. valley of Kittatinny, and comprises nearly all the cos. of Augusta, Rockingham, Page, and Shenandoah, with the estrn. sections of Frederick and Jefferson. The upper valley of Potomac including that of Shenandoah is in length from s. w. to N. E. 160 miles, where broadest 75 ms. but having a mean breadth of 50 ms., area 8,000 sq. ms. The water level of Potomac at Harper's Ferry is 288 feet above tide water; therefore we may assume at 350 feet the lowest arable land in the valley above the Blue Ridge. This is equivalent to a degree of lat. on the aerial temperature at the lowest point of depression. So rapid is the rise, however, in crossing the valley to the foot of Alleghany mtn. that an allowance of 1,200 feet is rather too moderate an estimate for the extremes of cultivated soil.

Passing the Blue Ridge, with partial windings, the Potomac continues s. E. by comparative courses 50 ms. to the lower falls and head of ocean tides at Georgetown. Having in the intermediate distance received the Monocacy r. from the N. and some minor creeks from the s. similar to the Delaware, below Trenton, and the higher part of Chesapeake bay below the mouth of Susquehannah, the Potomac meeting the tide bends along the outer margin of the primitive rock. It is indeed very remarkable that the three bends, in the three consecutive rs. follow almost exactly the same geographical line, or flow from head of tide s. w., the Delaware 60, Chesapeake 40, and Potomac 45 ms. The latter, a few miles below where it retires from the primitive, has reached within 6 ms. of Rappahannoc r. below Fredericksburg. Leaving the primitive, the two latter, not far from parallel to each other, assume a comparative course of 75 ms. to the N. E., the intermediate peninsula in no part above 22 ms. wide, and the distance 20 ms. from Smith's Point, on the s. side of the mouth of Potomac to Windmill Point, the N. side of that of the Rappahannoc.

Combining the two sections above and below the Blue Ridge, the whole basin of Potomac embraces an area of 12,950 sq. ms., or in round numbers 13,000, extending from lat. 37° 50′ to 40°, and in long. from W. C. from 0° 45′ E. to 2° 45′ W. The winding of its tide water channel renders the navigation of the Potomac bay (for such it is below Georgetown) tedious though not dangerous. The channel is of adequate depth for ships of the line of 74 guns, to the navy yard at W. C. With its defects and advantages, as a commercial and agricultural section, the basin of the Potomac is a very interesting object in physical and also in political geography. Deriving its sources from the main Appalachian spine the Potomac channel has been worn thro' the intervening chains to their bases, and performed an immense disproportion of the necessary task to effect a water route into the valley of Ohio. Such a route has been commenced under the name of "Chesapeake and Ohio canal."—(*See article rail roads and canals.*)

Potomac, p-v. Montgomery co. Md. Neither position nor distance in p-o. list.

Potosi, p-v. and st. just. Washington co. Mo., situated on the head waters of Big r. branch of Maremac r. 70 ms. s. s. w. St. Louis, and by the road 55 ms. w. St. Genevieve, lat. 37° 56′, long. W. C. 13° 48′ w. This place derives its name from being the central point of the mine dist. When visited by Mr. Schoolcraft in 1818, it contained 80 houses and probably 400 inhabitants, the ordinary co. buildings, 3 stores, 2 distilleries, 2 flour mills, 1 saw-mill, a post office and 9 lead furnaces. This traveller describes the site as a handsome eminence, dry and pleasant.

Potsdam, p-t. St. Lawrence co. N. Y. 25 ms. E. Ogdensburg, 90 w. Plattsburg, and 216 N. N. W. Albany, has a very fertile soil, and the surface agreeably varied. Racket river flows 11 ms. through the t. and on it are situated quarries of stone. The v. stands at the falls of this stream, 3 ms. from the s. boundary. Above it the r. is almost 1 m. across. The manufactories here are various and include some iron works. Water is brought into the village from the bottom of the r. by a forcing pump: pop. 1830, 3,661.

Potter, co. Pa. bounded by Lycoming s., MacKean w., Alleghany co. of N. Y. N., Steuben co. N. Y. N. E., and Tioga co. Pa. E. Length from s. to N. 37 ms., breadth 30, and area 1,110 sq. ms. Lat 41° 43′, and long 1° w. W. C. intersect near the centre of this co. Independent of mtn. chains this is the most elevated co. of Pa. In the northeastern angle rises the Cowanesque r., flowing to the E.; from the nrthrn. side rise the extreme sources of Genesee r., flowing to the N. From the central and nrthwstrn. sections issue the higher fountains of Alleghany r. and the extreme nrthestrn. sources of Ohio valley; and finally from the south side issue the Sinnamahoning and Kettle crs., branches of the West Branch of the Susquehannah. From these elements it is evident that Potter co. is a real table land, giving source to streams flowing into the basin of St. Lawrence N., that of Miss. s. w., and into the Atlantic s. E. Mean elevation at least 1,200 feet. Chief t. Coudersport: pop. 1820, 4,836, including some adjacent cos., and in 1830, Potter co. alone 1,265.

Potter's mills, and p-o. Centre co. Pa. by p-r. 71 ms. N. W. Harrisburg.

Pottsgrove, p-v. Northumberland co. Pa. by p-r. 67 ms. N. Harrisburg.

Pottstown, p-v. on the left bank of Schuylkill r. and in the N. W. angle of Montgomery co. Pa. by p-r. 68 ms. E. Harrisburg, and 16 from Reading.

Pottsville, p-v. Schuylkill co. Pa. on Schuylkill r. 8 ms. N. W. Orwigsburg, the co. st., and by p-r. 67 N. E. Harrisburg: pop. of tsp. 1830, 2,464.

Poughkeepsie, p-t. and st. jus. Dutchess co. N. Y. 75 ms. s. Albany, 74 N. N.Y. and 10 N. Newburgh, about 3½ ms. by 10, lies on the E. side Hudson r. and is nearly level, except w. where the surface is uneven and broken by courses of streams. Gypsum has been useful on the sandy soils. Fall cr. N., Wappinger's cr. E., and a small stream s. furnish excellent mill seats. Barnegat limestone, which is dug and burnt in the s. w. part of the t. is very good. The v. of Poughkeepsie is about 1 m. from the r. and of considerable size. The principal street runs E. and w., the land is level, and there is a bank, an academy, &c. At the landing there is a considerable number of houses, stores, &c. and a number of sloops are engaged in business with New York. The Albany steamboats stop here several times in the day during the season of navigation: pop. 1830, 7,222.

Poultney, r. Rutland co. Vt. is a small stream, rising in Tinmouth, and after running a few ms. w. marks a part of the boundary between this state and N. Y., till it falls into the head of E. bay, an arm of lake Champlain. It is about 25 ms. long, and Castleton and W. Haven rs. are its branches. In 1783, during a high flood, Poultney r. cut through a ridge near E. bay, which had before dammed it up, and made a channel 100 feet deep, destroying for a time all sloop navigation in E. bay. By the force of the current and the works of a company formed for the purpose, the obstructions have been greatly removed.

Poultney, p-t. Rutland co. Vt. 13 ms. s. w. Rutland, 10 ms. E. Whitehall, 46 N. Bennington, 60 s.w. Montpelier, 7 s. Castleton, and E. of Hampton N. Y., is crossed by Poultney r. and its branches, and has 35 sq. ms. It was first settled 1771, is well supplied with mill seats, and has an agreeable surface and a fertile soil, especially in the river meadows. There are 2 vs., a female academy, &c.: pop. 1830, 1,509.

Poundridge, p-t. Westchester co. N. Y., 139 ms. s. Albany, 15 E. Hudson r., 12 N. L. Island sound, and 5 s. E. Bedford; is supplied with mill seats by Mechanus creek on the w. line; some of the streams of Croton r. &c. The surface is uneven, with much stony land: pop. 1830, 1,437.

Powell's Tavern, and p-o. Goochland co. Va., by p-r. 15 ms. w. Richmond.

Powelton, p-v. Richmond co. N. C., 15 ms. sthrd. Rockingham, the co. seat, and by p-r. 128 ms. s. w. Raleigh.

Powelton, p-v. on Great Ogeeche r. in the nthestrn. part Hancock co. Geo. by p-r. 15 ms. N. E. Sparta, the co. seat, and 37 N. E. Milledgeville.

Powerville, village, Morris co. N. J., 8 ms. N. by E. Morristown; on Morris canal and Rockaway r., near the falls.

Powhatan, county Va., bounded by Chesterfield s. E., Appomattox river separating it from Amelia s. w., Cumberland w., & James r. separating it from Goochland N. Length 25 ms., mean breadth 10, and area 250 sq. ms. Lat. 37° 34′, and long. W. C. 1° w. intersect in this co. It contains two opposing slopes; one sthwstrd. towards the Appomattox; but the second, to the nrthestrd. towards James r. includes much the larger section. Chief town, Scottsville: pop. 1820, 8,292, 1830, 8,517.

Powhatan, p-v. Madison co. O.

Powhatan Point, and p-o. Belmont co. O., by p-r. 155 ms. E. Columbus.

Pownal, p-t. Cumberland co. Maine, is of small size and irregular form, 35 ms. s. s. w. Augusta, 18 N. E. Portland: pop. 1830, 1,308.

Pownal, p-t. Bennington co. Vt., 56 ms. s. w. Rutland, 30 w. Brattleboro'; lies N. Williamstown, Mass., E. Hoosac, N. Y.; with 25,000 acres; first settled 1761. It is uneven, with good soil for grass. Hoosac river flows N. w. into N. York, affording good mill sites, and water tracts of meadow land; several brooks in the N. E. which form head water of Wallamsack r.; 13 school dists.: pop. 1830, 1,834.

Powow, r. N. H. a good mill stream, rises in Kingston, and after a devious course, falls into the Merrimac on the line of Amesbury, in which town is its principal fall, where it descends 100 feet, in about 275 yards.

Prairie, from the French language, signifies literally meadows. It is a term occurring so frequently in the geography of the U. States, that we have deemed it requisite to introduce it as an article, in order to explain the true meaning, and describe the features of country intended by the term. Pré in French, means a meadow in the common acceptation of the word, whilst prairie is the superlative, and used for a large and indefinite space covered with grass. The term prairie, therefore, is perfectly applicable to the immense open grassy spaces in N. America, which, with partial interruptions, extend from the Gulf of Mexico to the Artic ocean, and of course traverse the whole territory of the U. States along the great slope falling estrd. from the Chippewayan system towards the Appalachian, though in no place actually reaching the latter. In their external features and relations to the great mtn. systems of the two continents, the steppes of Asia, and prairies of N. America, have a complete specific resemblance. The Asiatic steppes commence in fact in Europe, in the valley of the Wolga, from whence, following the great system, known by the respective names of Altai, Stavonoy, and Yablony mtns., spread across the whole continent of Asia, from the Caspian sea to that of Ochotz. As in N. A., the Asiatic steppes follow the mtn. chains, and are traversed at or near right angles by the rivers.

Called by either name, these grassy spaces partake of all the varieties of soil and surface of regions covered with forest. It is very erroneous to suppose the prairies necessarily plains; the real fact is, that strictly speaking small parts only of the prairies are level plains. In the southwestern part of La., and skirting along the Gulf of Mexico, over the sea border of Texas, the prairies are level plains, but advancing northwards they exhibit every variety of surface and of soil, and when traversed to their termination on the Artic ocean, sink again to level plains.

The La. prairies are perfectly congenial to the growth of every species of forest tree, that the climate will admit, and where fertile, are equally with woodland adapted to every object of agriculture, gardening, or orchard. This statement is made from personal experience; the writer of this article resided 8 years in Opelousas and Attacapas, or in the prairie section of La.

The llanos, or pampas, of South America, spreading along the great eastern slope of the Andes, are specifically prairies, with similar variety of soil and surface.

Prairie, p-o. Perry co. Ala., by p-r. 68 ms. s. E. Tuscaloosa.

Prairie Creek, p-o. sthrn. part Vigo co. Ind. 17 ms. sthrd. Terre Haute, the co. seat, and by p-r. 100 ms. s. w. by. w. Indianopolis.

Prairie de Long, p-o. Monroe co. Il., by p-r. 87 ms. s. w. Vandalia.

Prairie du Chien, p-v. and st. jus. Crawford co. Mich. or more correctly Huron, is situated on the point above their junction, and between the Miss. and Ouisconsin rs., as stated in the post office list, by p-r. 1,060 ms. If we compare the bearing and distance of Galena, which may be seen by reference to that article, we may see that Prairie du Chien bears about N. 70 w. from W. C., and the direct distance within a small fraction of 800 statute miles: pop. 1830, including that of the military station at fort Crawford, 692. Fort Crawford is adjoining to the village of Prairie du Chien.

PRAIRIE RONDE, p-o. Kalamazoo co. Mich., about 140 ms. nearly due w. Detroit.

PRATTSBURGH, p-t. Steuben co. N. Y., 230 ms. w. Albany, 14 N. Bath; has an uneven surface, and is watered by Five Mile creek, and streams of Crooked lake and Conhocton creek, on which are mill seats: pop. 1830, 2,402.

PRATTSBURG, p-v. Warren co. Miss., about 50 ms. above, and by the road N. N. E. Natchez.

PREBLE, p-t. Cordtland co. N. Y., 138 ms. w. Albany, 7 N. Homer, 24 S. Salina, S. Tully, Onondaga co., E. Cayuga co.; 5 ms. square; has brooks of Tioughnioga creek, a hilly surface favorable for grazing, and rich vallies, where the rocks are limestone and slate. First settled 1800, by New Englanders, Germans and Dutch. Maple, beech, bass, elm, nut woods, and some hemlock and pine grow here; and there are a few ponds. Preble Flats, 2 ms. wide, cross the town N. and S.: pop. 1830, 1,435.

PREBLE, co. Ohio, bounded N. by Darke, Montgomery E., Butler S., Union, Ind. S. W., and Wayne, Ind. N. W. Length from S. to N. 24 miles, breadth 18, and area 432 sq. ms. Lat 39° 45′, long. W. C. 7° 40′ w. General slope S. E., and drained in that direction by St. Clair and Franklin creeks, branches of Great Miami. The soil productive. Chief t. Eaton: pop. 1830, 16,291.

PRESCOTT, p-t. Hampshire co. Mass., 76 ms. w. Boston, 15 N. E. Northampton; is watered by several streams of Swift r., and has an uneven surface, with good grass land: pop. 1830, 758.

PRESTON, p-t. New London co. Conn., 44 ms. S. S. W. Hartford, 5 ms. S. W. Norwich, W. and S. W. Thames and Quinebaug rs.; has an irregular form, containing about 30 square miles. It is uneven, rocky, with a pretty good soil. First settled 1686: pop. 1830, 1,934.

PRESTON, p-t. Chenango co. N. Y., 5 miles w. Norwich, 115 w. by S. Albany, has a good soil, bearing maple, beech, bass, elm, &c., and favorable to grain; watered by small streams of Chenango r.: pop. 1830, 1,213.

PRESTON, co. Va., bounded S. by Randolph, Monongalia w., Fayette co. Pa. N., and Alleghany co. Md. E. The greatest length from S. to N. 36 ms., mean breadth 13, and area 468 sq. ms. Lat. 39° 30′, long. W. C. 2° 38′ w. The main Alleghany chain extends nrthrdly. along the eastern border of this county, and the Chesnut ridge separates it from Monongalia on the w. The body of the co. is a mountain valley between the two chains. Cheat r. enters the southern side, and winding to N. N. W., divides it into two not very unequal sections. Though generally broken, rocky, and in part mountainous, Preston contains some excellent soil. Chief town, Kingwood: pop. 1820, 3,428, 1830, 5,144.

PRESTON, p-v. in the southwestern part of Hamilton co. O., by p-r. 127 ms. S. W. Columbus.

PRESTONBURG, p-v. and st. jus. Floyd county Ky., on the W. fork of Sandy r., by p-r. 142 ms. S. E. by E. Frankfort. Lat. 37° 37′, long. W. C. 5° 38′ w.: pop. 1830, 81.

PRESTONVILLE, p-v. Rhea co. Ten., by p-r. 147 ms. S. E. by E. Nashville.

PREWETT'S KNOB, p-o. Barren co. Ky., by p-r. 118 ms. S. W. Frankfort.

PRIESTFORD, p-o. Harford co. Md.

PRINCE EDWARD, co. Va., bounded S. E. by Lunenburg, S. and S. W. Charlotte, W. Campbell, N. W. and N. Buckingham, N. E. Cumberland and Amelia, and E. Nottaway. Length from E. to W. 32 ms., mean breadth 8, and area 256 sq. ms. Lat. 37° 12′, long. W. C. 1° 30′ w. This county is bounded along its whole northern border and greatest length by Appomattox r., and of course slopes in the direction of that stream or eastward. The southern and central parts have a counter slope to N. E., and drained by numerous crs. falling into Appomattox river. This is one of the best peopled and most enlightened cos. of Va. In 1820, the pop. stood at 12,577, and in 1830, at 14,107, or 55 to the sq. mile. Of the latter aggregate 5,039 were whites. There are 12 post offices, and Hampden Sidney college located in this small co. The following information was forwarded to the editor. "This co. derives great advantage from the navigation of the Appomattox. A large part is fertile, well watered, and highly cultivated. Hampden Sidney college, has in this county an elevated, dry, and remarkably healthful situation, 80 ms. S. W. Richmond. The college was founded in 1775. The charter is as liberal and ample as that of any college in the U. States. The following professorships have been established: the president is the professor of mental philosophy, rhetoric, moral philosophy, and natural law; besides which are the chairs of chemistry, natural philosophy, mathematics, and the learned languages. The philosophical apparatus, and libraries of the college, philanthropic, the union, and philosophical societies are ample. The permanent college funds are vested in lands and bank stock. A preparatory academy is annexed to the college, in which those studies only are taught, that are required for admission into the lowest college class.

The amount of annual expense of a student $150, including board, tuition, room rent, washing, and servant's hire.

PRINCE EDWARD, court house, and p-o. Prince Edward co. Va. by p-r. 75 ms. S. W. by W. Richmond.

PRINCE FREDERICKTOWN, p-v. and st. jus. Calvert co. Md. by p-r. 56 ms. S. E. W. C. and 63 ms. S. Annapolis, lat. 38° 32′, long. W. C. 0° 28′ E.

PRINCE GEORGE, co. Va. bounded by Surry S. E., Sussex S., Dinwiddie W., Appomattox r. separating it from Chesterfield N. W., James r. separating it from Charles City co. N. and

N. E. Length from east to west 26 ms., mean breadth 12, and area 312 square ms., lat. 37° 10', and long. W. C. 20' W. Though bordered on two sides by Appomattox and James rs. the far greatest part of the surface slopes S. E. and is drained by the sources of Blackwater river, and some confluents of Nottaway r.; of course its water is tributary to Albemarle sound, by Chowan r. Chief town, City Point: pop. 1820, 8,030, 1830, 8,367.

PRINCE GEORGE'S co. Md. bounded S. by Charles, S. W. Potomac r. separating it from Fairfax county Va., District of Columbia W., Montgomery co. N. W., and the Patuxent river separating it from Ann Arundel co. N. E., and Calvert S. E. Greatest length 40 ms., mean breadth 15, and area 600 square ms. The meridian of Washington city passes along the western border, whilst the northern angle is traversed by latitude 39°. The dividing ridge between the confluents of Potomac and Patuxent traverses this co. from S. to north dividing it into two very nearly equal sections. The surface is generally hilly, though much of the soil is excellent. Chief town, Upper Marlboro': pop. 1820, 20,216, 1830, 20,474.

PRINCESS ANN, p-v. sea port, and st. jus. Somerset co. Md. situated on Manokin river near the head of tide water, by p-r. 144 ms. S. E. by E. W. C. and 107 ms. S. E. Annapolis, lat. 38° 12', long. W. C. 1° 18' E. It is a place of considerable commerce, and contains a bank, the ordinary co. buildings, and several places of public worship. It stands 18 ms. above the mouth of the Manokin into Chesapeake bay.

PRINCESS ANN, sthestrn. co. of Va. bounded by Norfolk co. W., Chesapeake bay N., Atlantic ocean E., and Currituck county, N. C. S. Length from S. to N. 30 ms., mean breadth 12, and area 360 square ms. Lat. 36° 45', and long. W. C. 1° E. intersect near its centre. Though in general level, it is a table land, discharging the sources of the East branch of Elizabeth's river wstrd.; the various branches of Lynhaven r. into Lynhaven bay, northwards, and the confluents of Currituck sound sthrd. Chief town, Princess Ann Court House: pop. 1820, 8,730, 1830, 9,102.

PRINCESS ANN, C. H., p-v. and st. jus. Princess Ann co. Va. by p-r. 23 ms. S. E. by E. Norfolk, lat. 36° 44', long. 0° 57' east W. C.

PRINCESS' BRIDGE, and p-o. eastern part of Chatham county, N. C. 23 ms. a little S. of W. Raleigh.

PRINCETON, p-t. Worcester co. Mass. 52 ms. W. Boston, and 16 N. Worcester, has a soil very favorable to agriculture, but contains Wachusett mtn. The land embraced by the present town bore the same name in the Indian language. This solitary eminence is 3,000 feet above the ocean, and is often ascended on account of the extensive and delightful view enjoyed from its summit. The village is situated a little distance up the side of the mountain: pop. 1830, 1,346.

PRINCETON, p-t. Schenectady co. N. Y. 20 ms. N. W. Albany, 7 W. Schenectady. It is of irregular form and is crossed by Norman's Kill creek E. and has several brooks N. which flow into the Mohawk. The surface and soil are various, and some of the principal eminences afford fine views: pop. 1830, 812.

PRINCETON, borough, N. J. between West Windsor, Middlesex co. and Montgomery, Somerset co., 10 ms. N. E. Trenton, 10 S. W. New Brunswick, 50 S. W. N. Y., 40 ms. N. E. Phila., is a pleasant and populous place, and the seat of Nassau Hall, founded in 1738, (one of the oldest colleges in the U. S.) and also the theological seminary of the Presbyterian church. The buildings of these two institutions are large; the principal edifice of Nassau Hall is old, and venerable in its appearance, shaded by trees, and has 60 apartments for students. The library contains about 8,000 volumes of old books, and there is a cabinet of natural history. The recitation rooms, library, society rooms, &c. are in two adjacent buildings, 30 feet by 60.

The theological seminary of the general assembly of the Presbyterian church, was founded in 1812, and is devoted to the education of young men destined for preachers of the gospel. There are a number of scholarships, endowed with $2,500 each.

PRINCETON, p-v. Washington co. Miss. by p-r. 100 ms. N. N. E. Natchez.

PRINCETON, p-v. Jackson co. Ten. by p-r. 107 ms. N. E. by E. Nashville.

PRINCETON, p-v. and st. jus. Caldwell co. Ky. situated on the summit ground between the vallies of Cumberland and Tradewater rs. by p-r. 229 ms. S. W. by W. Frankfort, and 59 ms. a little N. of W. Russellville, lat. 37° 02', long. W C. 10° 54' west: pop. 1830, 366.

PRINCETON, p-v. Liberty tsp. sthrn. angle of Butler co. O. 20 ms. a little E. of N. Cincinnati, and 98 ms. S. W. by W. Columbus: pop. 1830, 33.

PRINCETON, p-v. and st. jus. Gibson co. Ind. situated on the summit ground between the valley of Patoka r. branch of Wabash r. and the sources of Pigeon creek flowing into O. r. by p-r. 141 ms. S. W. Indianopolis, lat. 38° 22', long. W. C. 10° 38' W.

PRINTER'S RETREAT, and p-o. Switzerland co. Ind. 111 ms. S. E. Indianopolis.

PROSPECT, p-t. Waldo co. Me. 52 ms. E. Augusta, 15 N. W. Castine, and 227 N. N. E. Boston; lies on the west side of Penobscot river, and north of Belfast bay, opposite Orphan island, and includes Brigadier island in the Penobscot. It has a few small streams: pop. 1830, 2,383.

PROSPECT, p-t. New Haven co. Conn. 12 ms. N. by W. New Haven, is crossed by the West Rock range of hills, and has a few small streams which flow into the Quinnipiack and other rivers emptying at New Haven harbor: pop. 1830, 651.

PROSPECT, p-o. Prince Edward co. Va. by p-r. 80 ms. S. W. Richmond.

PROSPECT HILL, and p-o. Fairfax co. Va. 9 ms. wstrd. W. C.

Prospect Hill, p-o. Caswell co. N. C. by p-r. 59 ms. n. w. Raleigh.

Protho's Mills, and p-o. Orangeburgh dist. S. C. 38 ms. s. w. Columbia.

Providence co., R. I. bounded by Norfolk and Worcester cos. Mass. n., Bristol county, Mass. e., Kent co. and Narraganset river s., and the line of Conn. w. It is about 17 ms. broad by 22 long, containing about 380 square miles and is the largest county in the state. It has an irregular surface and is in some parts rough; most of the rocks are primitive, with some transition and limestone. The soil is most favorable to grass, and fruit thrives. The timber is various. Water power is abundant, and there are manufactories of many kinds carried on in different parts of the co. Pawtucket river crosses the co. on the n. e., and Sekonk river, into which it falls, is on the line of Mass. Pawtuxet flows s., and there are other streams, as the Mashasuc and Wanasquatucket. The Blackstone canal, which commences at Providence and soon after strikes along the course of Blackstone or Pawtucket r., pursues it to Worcester, Mass. & affords an important channel of transportation to many manufacturing and agricultural places. Cotton is the principal article of manufacture in this co. The first machinery ever erected in America for this manufacture was first set up in Providence, and afterwards in the village of Pawtucket. The commerce of this co. is extensive, and almost the whole foreign commerce of the state is centered at Providence. The Canton trade of that city is very extensive. Pop. 1820, 35,736, 1830, 47,018.

Providence, city, p-t. and sea port, Providence co. R. I., the most commercial and populous town in the state, and second in pop. in N. England. It is situated in lat. 40° 51', and in long. 5° 37' e. W. C., at the head of Narraganset bay, 30 ms. from Newport and the ocean, 42 s. w. Boston, 58 n. e. New London, 70 e. Hartford, 190 n. e. N. Y., and 394 n. e. Washington. The town contains about 9 sq. ms., is separated from Mass. by Seekonk r., and the two streams which form Providence r. unite within it, and afford an abundance of fish. The city is divided by the latter stream nearly in the centre, which is here navigable for vessels of 900 tons burthen; two fine bridges across it unite the two parts of the city. Providence is well laid out, and viewed from several eminences within the city, or from the bay, its appearance is fine and imposing. The calamities which it has several times suffered by storms, floods, and particularly an extensive fire in 1801, and the great storm of 1815, when 500 buildings were destroyed, have ultimately tended to the improvement of the city, in its streets and buildings. These are mostly of wood, and are uniformly neat; there are many, however, of brick, granite, &c., which are spacious and elegant, and finely situated. The public buildings are numerous, and sev- of them are very handsome. Among these is the arcade, a noble edifice of stone 222 ft. in length, with two fronts of granite, 72 ft. wide, and colonades of 6 columns each, 25 ft. high, the shafts of which are single blocks 22 ft. in length. This building was completed in 1828, and cost $130,000. There are 14 churches, some of which are in fine taste. There are also a state house, the Dexter asylum (for the poor), the building occupied by the Friend's boarding school, and two edifices belonging to Brown University: these are all of brick, and the three latter are finely located on a very commanding eminence. Providence is well situated for commercial enterprise, and internal improvements have recently added much to its prosperity. The foreign and coasting trade are both extensive, as the commerce of the state, which was formerly concentrated at Newport, is now chiefly transferred to this place. Several lines of packets, beside other vessels, run regularly to different parts of the U. S., and the facilities for internal communication are numerous. The trade with Canton has for some years been quite large.—The duties collected in 1831 amounted to $227,000; the imports of the same year to $457,000, and the exports to $329,000. The registered amount of shipping was more than 12,000 tons; there are in the city 4 insurance companies, with a capital of $360,000; a branch of the U. S. bank, with a capital of $800,000, and 16 other banks, with an aggregate capital of $4,602,000, including a bank for savings, the capital of which is $100,000. The Blackstone canal, which extends from Providence to Worcester, passing near numerous manufactories, adds much to the trade of the city, as does the manufacturing village of Pawtucket, one of the most important in the country, and to which leads one of the finest roads in the U. S. The Boston and Providence rail road, which is to be commenced immediately, will probably be productive of still greater benefits. A branch of it is to extend to Taunton, Mass., and another to New London, Conn., so as to connect the land transportation of the N. York and Boston route with Long Island sound at the latter place; the navigation round point Judith being exposed, and sometimes dangerous. Steamboats of the largest and finest kind, keep up a daily communication with N. Y. during the season of navigation, and in connection with them are several lines of stage coaches, which run to Boston in 6 or 7 hours. Providence, as well as the state of which it is the chief t., is distinguished for its numerous manufactories. There are 4 of cotton, with a capital of $327,500, and consuming annually nearly half a million pounds of cotton, from which are woven about 1,500,000 yards of cloth, valued at about $250,000; there are 3 bleacheries, two of which bleach about 3,300,000 pounds of cotton annually, which is equivalent to about 13,200,000 yds.; there are also 4 dye houses, 4 iron founderies, and 7 machine shops, manufacturing annually about $300,000 worth of machinery, chiefly for cotton factories; 3 brass found.

ries, 2 for the manufacture of steam engines; 10 tin, copper, sheet iron, and coal grate establishments; 27 jewellers' establishments, manufacturing jewelry, &c. to the value of about $230,000; and a glass factory, where cut and flint glass, to the amount of $70,000, is annually manufactured; beside these are various other manufactories of combs, oil, soap, candles, hats, boots, shoes, &c. &c. It is estimated that not less than 3,000 persons are regularly employed, or are principally occupied in the larger manufacturing establishments in the city. In addition to all these, a capital of more than $2,000,000, owned in Providence, is invested in various manufactures in other parts of the state. Brown University, transferred from Warren to Providence in 1770, was founded in 1764. Its two buildings are spacious; four stories high each, and one 150, the other 120 ft. long, and contain rooms for the officers and students, library and philosophical rooms, and a chapel. The philosophical apparatus is now very complete; recent efforts have enlarged its means, and the present condition of this institution is comparatively prosperous. Its officers are, a president, and five professors and tutors. Number of students 1831-2, about 100; the several libraries contain about 12,000 vols., of which 6,000 belong to the library of the college. The total number of alumni, 1831-2, was 1,182; commencement is on the 1st. Wednesday in September. This is the principal literary institution in the state; beside it, and the Friends' boarding school (which belongs to the Friends of N. England), there are several grammar, primary, and other schools. There are two daily, and eight other newspapers published here.

Providence was originally founded in 1636, by Roger Williams, who was banished from the Plymouth colony for avowing the doctrine that all denominations of christians are equally entitled to the protection of the civil magistrate. In 1644, this settlement was permitted to establish a government for itself, independent of Mass.; and in 1663, a charter was granted by the king to the Providence plantations, which extended the right of voting to all except Roman Catholics. In 1831, Providence was incorporated as a city, and its municipal government organized: pop. 1810, 10,071; 1820, 11,767; and in 1830, 16,833, of which 8,701 were on the E., and 8,132 on the W. side of the river.

PROVIDENCE, or Narraganset r. Providence co. R. I., is formed by the union of Seekonk r., and another branch which meet at Providence. It is deep enough for large ships, forms the harbor of that town, and a short distance below, falls into Providence bay.

PROVIDENCE, p-t. Saratoga co. N. Y., 42 ms. N. W. Albany, 15 N. W. Ballston Spa, 20 N. Schenectady, E. Montgomery co., 6 ms. by 7, is hilly N. E., crossed by Kayderosseras mtn. and elsewhere uneven, with Sacandaga r. N. W., and several small streams. The soil is favorable for grass: pop. 1830, 1,579.

PROVIDENCE, p-v. Luzerne co. Pa., 10 ms. N. E. Wilkes-Barre, and by p-r. 130 ms. N. E. Harrisburg.

PROVIDENCE, p-v. Mecklenburg co. N. C., 5 ms. S. E. Charlotte, the co. st., and 135 ms. by p-r. S. W. by W. Raleigh.

PROVIDENCE, p-v. in the wstrn. part of Hopkins co. Ky., by p-r. 17 ms. wstrd. Madisonville, the co. st., and 217 ms. S. W. by W. Frankfort.

PROVINCETOWN, p-t. Barnstable co. Mass., 50 ms. S. E. Boston, (116 by land,) is on the N. point of Cape Cod, with the ocean N. and E., and Cape Cod bay W. It is in the form of a hook, being inwards W. and S., and enclosing Provincetown bay, which is almost shut in by land, with an opening S., with water enough for ships of the largest size. The soil is a loose, sterile and shifting sand; and the houses of the v. which stands on the N. W. side of the bay, are very small, and built on piles so that the winds blow under them. The inhabitants are devoted to fishing, and take and cure great quantities of cod annually. There are no wharves in the harbor; the land is not cultivated. This was the place first visited by the N. England pilgrims, who spent a little time here in Dec. 1620, before they proceeded to Plymouth. Lat. 42° 3′ N., long. 70° 9′ W.: pop. 1830, 1,710.

PRUNTYTOWN, p-v. estrn. part of Harrison co. Va., and near the ferry over Tygart's Valley r., 20 ms. N. E. by E. Clarksburg, the co. st., and by p-r. 209 ms. N. W. by W. ½ W. W. C.

PRYOR'S VALE, p-o. Amherst co. Va., by p-r. 191 ms. S. W. Richmond.

PUGHTOWN, p-v. in the nrthrn. part of Chester co. Pa., situated on French cr. 35 ms. N. W. by W. Phila.

PULASKI, p-v. and half shire, Richland, Oswego co. N. Y., 153 ms. from Albany, 27 N. E. Oswego, 30 S. Sacket's Harbor, 36 N. Salina, 60 N. W. Utica, on Salmon creek, 3½ ms. from its mouth in lake Ontario. The harbor at its mouth receives vessels of 60 or 70 tons.

PULASKI, co. Geo. bounded N. by Twiggs, N. E. Lawrens, S. E. Telfair, and by Ockmulgee r. separating it from Dooley S. W. and Houston N. W. Length from S. E. to N. W. 34 ms., breadth 20, and area 680 sq. ms. N. lat. 32° 20′, long. 6° 22′ W. W. C. Slope S. E., and in that direction it is traversed by different branches of Auchenhatchee r. a small confluent of Ockmulgee; and the latter flows in a similar course along the southwestern border. Chief town, Hartford: pop. 1830, 4,906.

PULASKI, p-v. and st. jus. Giles co. Ten. by p-r. 77 ms. a little W. of S. Nashville. N. lat. 35° 08′, long. 10° W. W. C.

PULASKI, co. Ky. bounded W. by Wolf cr., separating it from Russell, Casey N. W., Lincoln N., Rockcastle N. E., Rockcastle creek, separating it from Whitley E., and Cumberland r. separating it from Wayne S. Length from E. to W. 40 ms., mean breadth 16, and area 640 sq. ms. N. lat. 37°, and long. 7° 30′

intersect in the southern part of this co. Slope southward towards Cumberland r. Chief t. Somerset: pop. 1830, 9,580.

PULASKI, a p-v. named in the p-o. list as being situated in Allen co. Ind., but from the distance given from Indianopolis 214 ms. a very extravagant allowance would be requisite to bring it into any part of Ind. There is no point in Allen co. 150 ms. by the road from Indianopolis.

PULASKI, co. of Ark. on both sides of Arkansas r. and around Little Rock, the st. jus. for the co. and capital of the state. So many new cos. have been made in Ark., the limits of which we have had no means to determine, and which have effected the boundaries of Pulaski, that we are compelled to merely state, that for geographical position we must refer the reader to the article Little Rock. By the census of 1830, Pulaski contained a population of 2,395.

PULTENEY, Vt. (*See Poultney.*)

PULTENEY, p-t. Steuben co. N. Y. 230 ms. W. Albany, 16 N. Bath, N. Ontario co., W. Crooked lake, has uneven land E., with beach, maple, &c. for timber, and several small streams, of which Five Mile creek rises N. E. and flows through it S. E. towards Conhocton creek: pop. 1830, 1,724.

PULTENEYVILLE, p-v. Wayne co. N. Y. 16 ms. N. Palmyra, S. lake Ontario.

PUMPKINTOWN, p-v. Pickens dist. S. C. by p-r. 145 ms. N. W. Columbia.

PUMPKINTOWN, p-v. Campbell co. Geo. by p-r. 130 ms. N. W. Milledgeville.

PUNGOTEAGUE, p-v. Accomac co. Va. 10 ms. S. W. Drummondstown, and by p-r. 218 miles S. E. W. C.

PUNXUTAWNY, PUNGATAWNEY, or PUNXETAUNY, (for all these spellings are used,) p-v. on Mahoning creek, southern part of Jefferson co. Penn., by p-r. 216 ms. N. W. W. C.

PURCELL'S STORE and p-o. Loudon co. Va. by p-r. 41 ms. northwstrd. W. C.

PURDY, p-v. and st. jus. McNairy co. Ten. by p-r. 128 ms. S. W. by W. Nashville. N. lat. 35° 13′, long. 11° 36′ W. W. C.

PUTNAM, co. N. Y. bounded by Duchess co. N., Conn. E., West Chester co. S., Hudson r. or Orange co. W., 12 ms. N. and S. by 21, with 252 sq. ms.; is very rough, with mtns. S. W., and but a small proportion of level land. Iron ore is taken from mines here in considerable quantities, and partly smelted in the co. At Cold Spring, in Philipstown, opposite West Point, is the largest cannon foundry in the U. S. Here are 1 blast furnace, making 850 tons of iron annually,; 3 air furnaces, and 3 cupola furnaces, which melt 2,500 tons, making $280,000 worth per annum. Black lead and pyrites are formed in some places. In the co. are 2 paper mills, and 1 woollen manufactory. The 2 branches of Croton r. or creek flow through the co. and unite on the S. line, after receiving the waters of many ponds. Carmel v. the st. jus. of the co. is in N. E. corner of Carmel t. on a small lake: pop. 1820, 11,268, 1830, 12,628.

PUTNAM, p-t. Washington co. N. Y. 30 ms. N. Sandy hill, W. lake Champlain or Vt., E. lake George, about 3½ ms. by 10; is on a narrow tongue of land between these lakes; mountainous, with poor soil. The N. end is 4 ms. from Ticonderoga: pop. 1830, 718.

PUTNAM, co. Geo. bounded S. by Baldwin, Jasper W., Morgan N., and Oconee r. separating it from Greene N. E., and Hancock S. E. Length 24 ms., mean breadth 18, and area 432 sq. ms. N. lat. 33° 20′, long. 6° 27′ W. W. C. Slope sthrd., and drained by Oconee or confluents. Chief t. Eatonton: pop. 1830, 13,261.

PUTNAM, p-v. Muskingum co. O. situated on Muskingum r. opposite Zanesville, and by p-r. 59 ms. E. Columbus: pop. 1830, 758.

PUTNAM, co. O. bounded S. by Allen, Vanwert S. W., Paulding N. W., Henry N., and Hancock E. Length 24, width 24, and area 576 sq. ms. N. lat. 41°, and long. 7° W. intersect in this co. The general slope N. W., the whole surface, with a very small exception, being in the valley of Au Glaize river. Chief t. Sugar Grove. It is a new settlement; the whole pop. in 1830, 230.

PUTNAM, co. Ind. bounded by Owen S., Clay S. W., Parke W. and N. W., Montgomery N., Hendricks N. E. and E., and Morgan S. E. Length 30 ms., breadth 20, and area 600 sq. ms. N. lat. 39° 40′, and long. 10° W. W. C. intersect in this co. Raccoon creek, a branch flowing S. W. over the northwestern angle, flows thence over Parke into Wabash; but the much larger section inclines to S. S. W., and is in that direction drained by the higher constituents of Eel r., branch of the North fork of White r. Chief t. Green Castle: pop. 1830, 8,262.

PUTNAM, co. Il. bounded S. by Peoria, Knox S. W., Henry W., unappropriated territory N., La Salle E., and McLean S. E. As laid down by Tanner on his recently improved map of the U. S., it is about 40 ms. sq.; area 1,600 sq. ms. Central lat. 41° 18′, long. 12° 35′ W. W. C. The northwestern angle is drained into Rock r. Illinois r. enters on the eastern border, and flows S. S. W. over the southestrn. angle, and the much greater part of the surface is drained to the sthrd. direct into Il. or into that stream by Spoon r. Chief ts. Hennipin and Alexandria. Including Putnam and Peoria, the joint population in 1830, was 1,310.

PUTNEY, p-t. Windham co. Vt. 10 ms. N. Brattleboro', 34 N. E. Bennington, 33 S. Windsor, W. Conn. r.; 18,115 acres; settled 1754, from Mass. Great Meadow fort was burnt by Indians in the last French war. Conn. r. bends so as to form part of the S. boundary. Great Meadow N. E., contains about 400 acres of excellent land. A ridge of hills runs N. and S. through the E. part of the town, W. of which is Sacket's brook, with a fall of 75 ft. in 500 yards near the village, where are several dams, and many mills and factories; W. of this the surface is very hilly, and near the W. line is Brooklyne valley. The rocks are mica slate, black limestone, &c., and green fluate of lime has been discovered. Birch, beech, maple, and some hemlock, grow on the hills, and on Conn. r. nut trees and oak: pop. 1830, 1,510.

Q.

QUAKERTOWN, p-v. in the northwestern part of Bucks co. Pa. 38 ms. N. N. W. Phil. It is a neat small village, in a single street along the main road: pop. about 200.

QUANTICO, creek and p-o. in the northwstrn. angle of Somerset co. Md.

QUEECHY, or Waterqueechy r. Windsor co. Vt. rises in Sherburne, Rutland co., crosses Windsor co. and enters Conn. r. 2 ms. above Queechy falls. Two good mill streams enter the Queechy in Bridgewater, and two others in Woodstock. Its course is about 35 ms. over a stony or gravelly bed, and its water is pure. It waters about 212 sq. ms.

QUEEN ANN, co. Md. having on the E. Choptank r. separating it from Caroline, Talbot co. S., Chesapeake bay W., Chester r. separating it from Kent N. W. and N., and Kent co. Del. E. From the southern part of Kent isl. to the northeastern angle of the co. 40 ms.; mean breadth 10 ms., and area 400 sq. miles. N. lat. 39° and 0° 45′ intersect in Queen Ann. Slope S. W. The co. is composed of Kent isl. and a long narrow space between Choptank and Chester rs. Chief t. Centerville: pop. 1820, 14,952, in 1830, 14,397.

QUEEN'S co. N. Y. on Long Island, bounded by East r. and the sound N., Suffolk co. E., the Atlantic S. and King's co. W.; has 6 townships, and about 355½ sq. miles. The greatest breadth of the island in this co. is 22 ms., but the water approaches from opposite sides in one place with 5½. The surface is slightly varied, the greatest elevation, viz. Harbor Hill, in North Hempstead, being 319 feet above high water. A sandy ridge crosses the co. E. and W. The W. has much excellent soil, well cultivated for vegetables, fruit, &c. for N. Y. market. There are many coves, inlets and bays, on which tide mills are erected, principally for flour. On the N. are Cold Spring bay and harbor, Oyster do. do., Hempstead do. do., Cow and Little Neck bays, and Flushing bay and harbor; on the E. r. is Hurl Gate, and several coves and points on N. Y. harbor and bay, and S. part of S. bay, Jamaica bay, Rockaway beach, Hog isl. &c. &c. There is a light house at Sands' point in Long Isl. sound. The co. court house is in N. Hempstead, 21 ms. from N. Y. A large part of the co. was settled by Holland farmers, among whom were many families of French extraction, whose ancestors fled into Holland after the massacre of St. Bartholomews. This co. suffered severely during the war of the revolution. A landing was made here by the British army after it evacuated Boston, on the S. E. shore; and a battle was fought, in consequence of which general Washington was compelled to draw off his army and evacuate the city of N. Y. In this he was favored by a thick mist, and the dilatoriness of the enemy. Queen's co. long remained in the possession of the British: pop. 1820, 21,519, 1830, 22,460.

QUEENSBORO', p-v. estrn. part of Anderson dist. S. C. by p-r. 108 ms. N. W. Columbia.

QUEENSBURY, t. Warren co. N. Y. 58 ms. N. Hudson, 5 N. W. Sandy Hill, 8 S. Caldwell, W. Washington co., N. Hudson r. and Saratoga co., about 6 ms. by 13, has loose, sandy soil W., and loam E. both bearing pine. There are some plains E., and the W. is hilly, with French mtn. and French pond at its foot, 1½ by 2 ms. in length and width. There are also several cranberry marshes.

Halfway brook rises here and flows into Wood cr. Iron ore and lime are found in the t., and an extensive and beautiful quarry of black marble has recently been wrought in considerable quantities. Glens Falls village is pleasantly situated, on a plain near the falls of the same name in the Hudson. This is a favorite spot in the northern tour of travellers, lying on the road from Saratoga springs and lake George, and presenting interesting scenery at the falls. The descent of the Hudson here is 37 feet. A ledge of blackish limestone crosses the channel, over which the water is precipitated perpendicularly into a deep basin. Below, the channel is divided by an isl. of solid rock, in which the floods have worn holes and two singular caverns, in a direction parallel to the strata and across the course of the stream. The neighboring banks are high, rocky and perpendicular; and the whole is seen at great advantage from a bridge which rests upon the isl. A dam crosses the r. at the falls, which supplies several saw mills, and turns part of the water into a branch canal and feeder. This passes along the bank in this t. and through Sandy Hill v. to Kingsbury, where it enters the Champlain canal above fort Edward. A convoy of wagons was attacked on the banks of the r. in this t. during the French war, and a skirmish ensued. At the foot of French mtn. is a defile, in which occurred (1755) the fight between a detachment of the English and colonial troops under Gen. Sir Charles Johnson, on a scout from fort George; the French troops and Indians under Gen. Dieskau. The latter lay in ambush, and the former suffered severely. The famous Mohawk chief, Hendrick, fell among his allies, the English: pop. 1830, 3,080.

QUEENSDALE, p-v. Robeson co. N. C. by p-r. 117 ms. S. W. Raleigh.

QUEENSTOWN, p-v. Queen Ann co. Md. on the S. E. side of Chester bay, 7 ms. S. W. Centerville, the co. st., and by p-r. 62 ms. estrd. W. C.

QUERCUS GROVE, and p-o. Switzerland co. Ind. by p-r. 117 ms. S. E. Indianopolis.

QUIGLE'S MILLS, and p-o. Centre co. Pa. by p-r. 100 ms. wstrd. Harrisburg.

QUINCY, p-t. Norfolk co. Mass. 8 ms. S. E. Boston, S. W. Boston harbor, settled 1625, before Boston or Salem, under the name of Mount Wollaston, as a trading post. On Neponset r. is a salt marsh. Squantum is a peninsula running into Boston harbor. Two ridges of the Blue hills run parallel through the W. part. There are extensive quarries of fine granite wrought here, 3 ms. from tide water, in Neponset r., to which it is conveyed on a rail road, constructed in 1826—the first work of the kind in America. Pine rails, 12 inches deep and 6 wide, were laid, under oaken rails 2 inches by 3, covered by iron plates 3-8 inches thick. These rails are 6 feet apart, on granite blocks 7½ feet long. In this t. are the mansions of John Adams and Josiah Quincy: pop. 1830, 2,201.

QUINCY, p-v. Franklin co. Pa. by p-r. 83 ms. N. N. W. W. C.

QUINCY, p-v. and st. jus. Gadsden co. Flor. situated in the nrthrn. part of the co. 23 ms. N. W. by W. Tallahassee, N. lat. 30° 34', long. W. C. 7° 47' W.

QUINCY, p-v. Munroe co. Miss. by p-r. 168 ms. N. E. Jackson.

QUINCY, p-v. watrn. part Gibson co. Ten. by p-r. 153 ms. wstrd. Nashville.

QUINCY, p-v. and st. jus. Adams co. Il., situated near the left bank of Mississippi r., by p-r. 193 N. W. Vandalia, N. lat. 39° 52' long. W. 14° 18' W.

QUINEBAUG, r. a considerable stream in the E. part of Conn. and a very valuable river for water power. It takes its rise from a pond (Mashapang) in the town of Union, and after making a circuitous course into Massachusetts, unites with French river between Woodstock and Thompson, Conn. After a course of 30 ms. through a rich agricultural district, it forms a junction with the Shetucket, three ms. north of Norwich city; from thence to its junction with the Yantic it bears the latter name. These united streams form the Thames. On the Quinebaug and its tributaries are now (1832) from 85 to 100 cotton and woollen manufactories, containing from one to four thousand spindles each, exclusive of those in the Yantic, Willimantic and Shetucket rivers and their tributaries. The course of the Quinebaug has been proposed as the route for a canal, from tide water in the Thames into Mass.

QUINIPIACK, the ancient Indian name of New Haven, Conn.

QUINIPIACK, or E. r. Con. rises in a pond in the S. part of Farmington, Hartford co. and flows through Southington, between the two ranges of mtns. in that t., then breaks thro' the E. range, and flows at its E. base at the foot of mt. Carmel and of E. Rock, entering New Haven harbor under a long bridge. The Farmington canal lies along the upper part of the course of this stream.

R.

RABUN, nrthestrn. co. Ga. bounded S. W. by Turoree r. separating it from Habersham co., W. by the nrthrn. part of Habersham, N. by Macon co. N. C., N. E. Haywood co. N. C., and E. and S. E. by Chatuga r., separating it from Pickens dist. S. C., length from E. to W. 30 ms., mean breadth 11, and area 330 sq. ms., lat. 34° 53', long. W. C. 6° 24' W. The Blue Ridge passes along the nrthrn. border, and from the northern side of the chain and in Rabun co. rise the extreme sources of Tennessee proper and Hiwassee r. From the sthrn. slope again issue the fountains of Turoree and Chatuga, or the extreme sources of Savannah r. We may add also, that the higher fountains of Chattahooche r. rise in Habersham, within 4 or 5 ms. from the sthwsrn. part of Rabun. From those elements it is evident that Rabun co. occupies a very elevated table land. The farms must be from 1,500 to 2,000 feet above the Atlantic tides, producing an effect on aerial temperature to at least 4 degrees of Fahrenheit's thermometer. Chief town, Claytonsville: pop. 1830, 2,176.

RACCOON, p-o. nrthrn. part of Washington co. Pa., 12 ms. nthrd. the borough of Washington.

RACCOON FORD, and p-o. Culpepper co. Va.

RACE, the E. end of Long Island sound, between Connecticut and Long Island, where a sunken reef renders the surface agitated when the tide is rapidly passing. The reef lies about N. E. and S. W. in a line with Fisher's, Gull and Plumb islands.

RACE POINT, the N. W. extremity of Cape Cod, Provincetown, Barnstable co. Mass., 3 ms. N. W. Provincetown v.

RACKET, r. N. Y. rises in Hamilton co. in a mountainous region, near the head streams of Hudson and Black rs. It is at first deep, slow and crooked, flows through several ponds; but for about 30 ms. during a part of its course, it has a rapid descent and affords many mill sites. At Louisville it again becomes slow and deep, and is navigable in boats of 5 tons to its mouth in the St. Lawrence. The boat navigation continues thence to Montreal. Its whole length is about 120 ms., its descent 200 feet, and its general course N.

RADNOR, p-v. Delaware co. O. by p-r. 30 ms. N. Columbus.

RAGGED MOUNTAINS, N. H. a range about 10 ms. long, running nearly E. and W. on the line between Merrimac and Grafton cos. between Kearsearge mtn. and Pemigewasset r. They are broken and precipitous, and some peaks are nearly 2,000 feet high.

RAGGED ISL., t. Lincoln co. Me.: population 1830, 14.

RAHWAY, p-t. Essex co. N. J. 4 ms. s. w. Elizabethtown, N. Middlesex co., is crossed by Rahway r. N. and s., with a pleasant v., and 10 school houses. The r. enters Staten Island sound 4 ms. below: pop. 1830, 1,983.

RAIL-ROADS AND CANALS.

Before proceeding to arrange the material of this compound article, we must introduce some prefatory remarks. From the nature of our treatise, no regular essay on the subject of either mode of conveyance is in view; nor is it our intention to enter into the contested merits of canals and rail-roads.* We shall, as far as our document will admit, state what has been effected in regard to canals and rail-roads.

In the arrangement, we have generally adopted the basins with the canals, and follow these with the rail-roads, alphabetically. With the map of the United States before us, we advance with the former from south to north.

Louisiana, or Mississippi Delta navigation.—Besides the main volume of Mississippi, this country is traversed by numerous outlets from that stream, and by counter, or interlocking water courses. The whole plain so nearly approaches the curve superficies of the sphere, that the utmost height of any part of the land above the lowest, (beds of rivers and lakes excepted) is very slight. Under the head of Mississippi we have already shown that that river flows in a comparatively deep valley, and cannot by either natural or artificial means be diverted from its bed. Canal works must, therefore, as every where else, be constructed to obviate the defects of river navigation, or to supply an artificial r. where a natural one did not exist. A view of the country itself, or its representation on a map, suggests the idea of prodigious facility of canal construction. This facility is not deceptive, but has been only very partially taken advantage of, for many reasons.

A short cut to admit schooners, sloops, and other small craft into a basin in the rear of N. Orleans, from Bayou St. John, is yet the most important canal in La., or indeed in the United States thus far south. This canal is called *Carondelet*, from governor Carondelet, under whose administration, during the existence of the Spanish government, it was projected. *Lafourche*, a short canal, supplied with water only when the Mississippi is in flood, unites the outlet of Lafourche (the fork) with the chain of lakes and creeks which lead into the lower Teche, and opens the commerce of Attacapas to N. Orleans. This canal leaves the Lafourche, 16 ms. below its own efflux from the Mississippi. Where the *Plaquemine* issues from the Mississippi, a cut has long been made to admit vessels into the former. This is also supplied with water only at high flood. Bayou Iberville, the first outlet of Mississippi river from the left, descending that stream, is exactly of similar nature to the Lafourche, Plaquemine, and Atchafalaya, from the opposite side. Iberville issues from the main stream below the last highlands, which extend s. of Baton Rouge, and 8 ms. above the efflux of Plaquemine. Following the windings, Iberville receives the Amite r. from the N., 20 ms. from the Mississippi. Now a navigable stream admitting vessels of 5 feet draught, the Amite turns to the eastward, and following its very winding channel flows 35 ms. before it opens into lake Maurepas. This lake is again contracted into a creek (bayou) which connects it with the much larger lake Pontchartrain, which is itself joined to lake Borgne, a bay of the Gulf of Mexico, by two channels; the Rigolets and Chef Menteur. Through this chain of lakes, creeks and bays, an immense line of internal navigation has been projected; which is, however, only in project, and mentioned only in this place, as being connected with the design of cutting a canal over the peninsula of Florida. In furtherance of the plan of the Florida canal, an act of congress was passed March 3rd, 1826, authorizing surveys. The surveys were made in virtue of the act of congress, and reported to the board of internal improvement, 8th Feb. 1828.

From the surveyor's report the canal is practicable, but at an expense beyond the reach of accurate estimate. "The elevation of the highest intermediate ridge above the level of the seas, has been found 152 feet at the head of St. Mary's r. near the Geo. line, 158 feet between Kinsley's pond and Little Sta Fe pond, head of Sta Fe river; and 87 feet between the head branches of the Amaxwra and Ocklawaha." The sea shore was found shallow from Tampa bay to Appalachie bay, on a width outwards from the land, varying from 5 to 15 ms. From the latter to cape San Blas, this width diminishes, except at the intervening capes, where extensive shoals project out, but from cape San Blas to lake Pontchartrain, the shore is generally bold, and the coast affords several good harbors. The Atlantic coast "is all along shallow," say the engineers, "and offers no harbors except at the mouth of St. John's r. and St. Augustine."

The shortest distance across the peninsula is about from St. Augustine to a point on the Gulf between the mouths of the Suwannee and Amaxura rivers, 105 ms. The distance in a straight line from the mouth of St. John to that of the Suwannee is 130 ms., and from the mouth of St. John's to that of St. Mark's, 170. The wide and shoaly bank, which obstructs the coast from Espiritu Santo (Tampæ) continues uninterrupted to the bay of Appalachie, where its breadth is reduced to about 3 ms., and a channel formed of 10 feet,

*As one of the compilers of the U. S. Gazetteer, it would be the height of inconsistency in me to prefer rail-roads or canals, having been long convinced that neither is either the cheapest, or best, or indeed will be the ultimate general system. Humble common roads, like common sense, will sustain their value. W. D.

to enter St. Mark's r., and vessels drawing 8 feet can ascend to the town of St. Mark. Along the shore 4 or 5 feet is the general depth, and 10 or 12 ms. out at sea only twelve feet is found. The difference of level between the Gulf of Mexico and the Atlantic ocean, resulting from these surveys, give to the former an elevation of 3 or 4 feet above the latter. Tides in the Gulf about 2 feet at a mean.

The engineers enter into much detail on the various routes, a detail we have not room to insert. It appears from the investigations as far as prosecuted, that 8 feet is the deepest water that can be calculated on as a debouchment to the intended canal on the side of the Gulf, and of course that the canal itself may not necessarily be constructed with a greater draught than can be navigated from sea to sea. This navigation must be so obviously beneficial, and practicable, and the climate offering no winter obstruction to water navigation on its route, that we may regard its actual construction as amongst the improvements which the coming age will carry into effect.

Leaving the Delta of the Mississippi, and passing along the northern coast of the Mexican Gulf, and over the intervening land and rivers, we reach the small basin of Ashley and Cooper, before we behold the natural navigation meliorated by any exertions of man, that deserve particular notice. Setting out from the mouth of St. John's r. of Florida in a distance of 200 ms., and a coast indented by the outlets of St. John's, St. Mary's, Santilla, Alatamaha, Great Ogeechee, Savannah, Coosahatchie, Edisto, and numerous smaller streams, nature has been left to direct, or impede the channels according to her own caprice.

The importance and wealth of Charleston, with the peculiar range of the channel of Santee r. suggested a canal, which was undertaken about the beginning of this century, and in 1802, the harbor of this southern emporium was united to Santee r. by a canal called "*The Santee canal.*" It extends from the head of Cooper r. N. N. W. 22 ms., and is joined to Santee opposite Black Oak island. The Santee canal is 34 feet wide at surface, with 4 feet water, and cost 650,667 dollars. The Santee, Columbia, and Saluda navigation, has been improved above the Santee canal, upwards of 150 ms., combining side cuts and locks, with the r. channels, and about thirty locks overcome 217 feet fall. The foregoing embraced the western branch. Along the Catawba or Wateree, extensive side cuts and locks in Kershaw district, near Camden, at Rocky Mount in Fairfield, and in other places, opened the fine channel of Catawba to the ocean by the Santee canal. But all that has been done in this extensive region, has hardly done more than to demonstrate the utility and necessity of very extended operations. This will be more obvious when we see the extent of country embraced in this navigable physical section, exclusive of that of Flor. and La.

It may be repeated, that along the Atlantic coast at least, the climate opposes no great obstacle to the formation, and permanent use of canals as far as Albemarle sound, in lat. 36°. The almost united mouths of Santee and Pedee, are but little above lat. 33°, therefore all the Atlantic part of the navigation embraced by the foregoing table, is exempt from impediment by ice in winter. It has been projected to connect Charleston with Savannah, by an inshore chain of natural channels and short cuts. This line of improvement is no doubt practicable, and may be effected at an expense of money bearing a small proportion to its immense advantages. The same natural facilities extend in both directions from Savannah r. To the s. w. the inshore navigation may be extended to connect with that of Flor., and advancing westward, reach and join the already vast commercial operations, of the Delta of the Mississippi. On the opposite side, it is true we discover a new character of coast, but the natural channels still seem to invite to canal improvement. Passing the Pedee, the insular coast, so remarkable along Florida, Georgia, and the Carolinas, changes its nature and aspect. Three great elliptic curves sweep from the mouth of Pedee to cape Hatteras, of very nearly equal length, 100 miles each. Defective in deep harbors, as is the coast s. w. of the Pedee, it is still more so to the N. E. of Winyaw bay, or outlet of Pedee. This latter coast of 300 ms. is broken but by one river, that of cape Fear, and in no place admits vessels of 15 feet draught. Beyond cape Hatteras to Chesapeake bay, this latter character of coast continues, but with increased asperity. Between cape Lookout and cape Henry, extend Pamlico, Albemarle, Currituck, and other shallow sounds, not admitting the navigation of vessels drawing 6 feet water. Into this region of shallow sounds, are poured the volumes of Neuse, Pamlico, and Roanoke rivers.

There is not in America, if there is on earth, another range of ocean coast where one canal improvement would more obviously suggest another, until one chain of such works would unite the extremes, than that stretching from the mouth of the Mississippi, to that of Chesapeake bay. We may here remark, that there is no other obstacle opposed to the construction of canals, so formidable, as an extensive shallow sheet of water. Without a correct knowledge of their real character, the North Carolina sounds would be taken as fine expansive bays, and like the Chesapeake, peculiarly fitted for inland navigation; but when actually and carefully examined, the unwelcome fact is disclosed, that the Cape Fear and Chesapeake basins cannot be united by a chain of canals at any expense within human means, unless that chain is carried along the mouths of the rivers, and heads of the sounds. The basins of Albemarle and

Chesapeake are separated by a marshy, and generally dead level peninsula, 60 ms. wide. The name given to its central part, Dismal Swamp, serves as a brief description of this tract. From this dreary region of lakes, marshes, and almost impervious woods, Bennet's creek flows into Chowan river, and the rivers or rather bays of Perquimans, Pasquotank and North river are connected sthrdly. with Albemarle sound. To the nrthrd. the surplus water is carried into James r. by Nansemond and Elizabeth rivers.

The Dismal Swamp canal, is yet the only work of any importance which has improved the navigation of this region of shallows and fens. It commences on the Va. side on Elizabeth r., near the mouth of Deep creek, and stretches over the Dismal Swamp to the mouth of Joyce's cr., a branch of Pasquotank; length 23 ms.; rises only 16½ feet above the Atlantic level; 40 ft. wide at surface, and 6½ feet water. It receives the water of a feeder from lake Drummond of 4½ feet depth, and 5 ms. in length. This work has cost directly or indirectly, about $800,000. Since the construction of Dismal Swamp canal, it has been projected to deepen it to 8 or 10 feet, but that design has been considered useless if effected, unless a similar depth of canal was extended along the heads of the sounds. A line of connected canal and river navigation has, however, been sketched, which, with such modifications as more accurate surveys may point out, will be no doubt effected, at no very distant time. This splendid project is to commence with the deepening and enlargement of the Dismal Swamp canal, and carrying it into the Chowan r., near the mouth of Bennet's cr. Thence using the volume, or following the shores of Chowan into the Roanoke, and up the latter to the port of Williamston. Thence in a direction little w. of s. 22 ms., to Washington, on Pamlico r.; and continuing sthrd. beyond Washington 16 ms. to the navigable water of Neuse r. at Dawson's bridge. Neuse r. offers a natural channel of 35 ms. past Newbern, to the mouth of Adams' cr.; up the latter with 12 feet water for several ms., and by an intermediate canal to North r., and down that stream to Beaufort.

This line could be varied or branched, by going up Neuse r. to its great bend in Lenoir co., and thence into the North branch of Cape Fear r., or leave the Neuse at Newbern, and follow the Trent into Duplin co., and thence into Cape Fear r. Either of the two latter routes would debouch into Cape Fear r. at Wilmington. To carry this navigation forward into Wineyaw bay, several routes have been proposed. The most direct is a canal from opposite Wilmington to the navigable water of Waccamau r. Another plan is to leave the channel of Cape Fear r. at Haywoodsboro', 180 ms. above Wilmington, proceed up the valley of Deep r. as far as requisite, and thence by the most practicable route to the Yadkin near Blakeley in Montgomery co. N. C. This route has received Legislative sanction, and the name of *Cape Fear and Pedee canal.* Combining the whole space from St. John's basin to Roanoke inclusive, the subjoined table will exhibit the great area, included in the physical navigable section of the U. S. s. of Chesapeake bay, and estrd. of the Appalachian system of mtns., with the pop. of 1830.

Sub-basins.	Lgth.	Mean brth.	Area in sq. ms.	Pop. 1830.
Basin of Geo. and S. C.,	380	170	64,600	925,734
Do. Cape Fear r.,	200	40	8,000	740,000
Do. Neuse r.,	180	40	7,900	
Do. Tar r.,	160	25	4,000	
Do. Albemarle,	290	60	17,400	
Aggregate,			101,900	1,665,734

We have thus, exclusive of Florida and Louisiana, a section of the U. S. comprising a fraction above 100,000 sq. ms., and a population exceeding 1,600,000 inhabitants, on which the Santee and Dismal Swamp canals are the only works of that nature of any magnitude, except such as have been executed to meliorate the navigation of rivers. In the latter species of improvement, however, more has been done than is generally supposed.

Obstructions have been removed, though to no very great extent, in the Savannah river. Some expense has been incurred to open the inner channels of Edisto rs. N. & S. Near Columbia, where the Saluda and Broad rs. unite, there are canals or side cuts, called *the Columbia canal*, and *Saluda canal*, made to permit navigation past rapids. These, with other works in connexion, along Saluda and Broad rs., comprised, in 1826, 28 locks, and 150 miles of mixed navigation. The Wateree, (the principal branch of the Santee,) is obstructed, in Kershaw district, S. C., by rapids. A canal has here been extended along its western side, and another constructed for a similar purpose, at Rocky mtn., in Chester district. At an expense of between 2 and 300,000 dolls., the Catawba has been made navigable nearly to its source in N. C. Between Cheraw and Georgetown, a considerable expense on side cuts and other improvements, has shortened the distance, and given a navigation, though a defective one, to the Pedee. The navigation of the Cape Fear r. has been noticed. Much has been done, and much more remains to be done, to render this r. as valuable as a commercial channel as its position relatively demands, and its volume of water will admit. A plan has been suggested for the improvement of the Neuse, Pamlico and Tar rs., to which we have already alluded. To improve the channels of the Roanoke and its confluents, companies have been formed in both N. C. and Va. Sloops ascend the Roanoke to Weldon above Halifax, and the Chowan to Winton. The *Weldon canal*, in a distance of 12 ms., overcomes 100 ft. fall, and as early as Dec. 1828, by a report of the Va. Roanoke company, it appeared that the improvements had been such as to admit steamboat navigation to Salem in Botetourt co., w. of the Blue Ridge. overcoming upwards of 900 ft. fall in 244 ms., following the r. channel. Danville navigation was also (Nov. 1828) so greatly improved, as to admit, by a mixed series of locks, sluices and side cuts, a regular navigation into Rockingham co. N. C., at the village of Leakesville. 152 ms., following the bends of the r. Expenditures of the Roanoke companies to Nov. 1826, $341,283; Nov. 1828, $365,991.

Virginia navigation east of the Appalachian mts. Under this comprehensive head are included the lower part of the deep bay of Chesapeake, and its confluents, James, York, Rappahannoc, and the far greater part of the valley of Potomac. The earth affords no other instance where so great a physical change is effected in so short a distance, as that between the shallow sounds of N. C., and the deep water of the Chesapeake. In the latter, the largest ships of war have adequate depth almost to the very verge of the primitive rock. Ships of the line ascend the main bay to near its head, up the Potomac to Alexandria, some distance into York r., and up James r. to the mouth of Nansemond r. and Hampton Roads. Sloops drawing 6 or 7 feet water penetrate into innumerable creeks on both sides of the Chesapeake. Here, and over the intervening mtns. to the Ohio, Va. possesses the inappreciable advantage of full sovereignty, an advantage, in the prosecuting public works, that nothing beside can equal. In the peculiar direction of their channels, it would appear as if nature intended to lavish her favors on this state, by making her estrn. border a common centre of confluent streams. The rivers of Geo. and the Carolinas from Alatamaha to Cape Fear inclusive, and without much violence we might say to Roanoke inclusive, flow to S. E., or S. S. E. In sthrn. Va., their general course is estrd. to the Susquehannah, which is again almost due S. Much has been done to improve the navigation of the rs. of Va., but what is executed is indeed small, when compared with the extent of the physical section under review.

James r. admits vessels of 125 tons to Rockett's, the port of Richmond. At that city commences the falls or rapids, to pass which by a navigable canal, the old James river company was chartered in 1784, and the works were so far advanced, that tolls were regularly collected in 1794. (*See art. Richmond city.*) The *Richmond canal* enters a basin in the wstrn. side of the city; is 25 ft. wide, and 3 deep, extends 2½ ms. to where it enters the r.; there are 12 locks, and the fall is 80 feet. Three ms. above the first is a second short canal, with 3 locks, overcoming 34 feet fall. These canals and locks, with other slight improvements, opened a navigation at all seasons of 12 inches water to Lynchburg. The James r. company in 1825, Dec. 10th, under an act of assembly, 17th February, 1825, declared a canal navigation complete to the head of the falls, called Maiden's Adventure, Goochland co., 30½ ms. above Richmond. Width of canal 40 feet, depth of water 3½ feet, and expense $623,295; fall overcome, 140½ feet. If the respective dates in Armroyd's treatise are correct, this, in proportion to magnitude, was the most promptly executed work of its kind ever performed in the U. S. Additional expenditure to January, 1828, swelled the amount of expense to $637,607. A section canal to carry a navigation along James r. through the Blue Ridge gap, was commenced in 1824; fall 96 feet; stone locks 10½ feet wide, and 76 feet long; expenditure $365,013.

With these and some other improvements, the navigation of James' r. has been effected into the valley above Blue Ridge. Extensive farther improvements have been proposed, to the amount of $5,750,000, according to the engineer's report, July 1826. These estimates are again swelled by plans of canal construction in the Ohio section of Va.; but in actual peformance the efforts of the state seem to have rather relaxed than augmented.

Below Richmond and the head of tide water, some canal works have been executed. From City Point at its mouth into James r., the Appomatox has been improved 10 ms., to Fisher's bar, and thence by canal round the falls, to 5 or 6 ms. above Petersburg. Thence the channel of the r. has been cleared to Farmville, Prince Edward co. On upper Appomattox, about $100,000 have been expended; below tide water, about $30,000. Vessels of 7 feet draught can ascend to Petersburg.

It would be idle to enumerate the various projects of canals, locks, sluices and other proposed works, involving a certain expense of ten millions of dollars, whilst so much remains to be done to complete what has been commenced; we therefore proceed to an analysis of the fourth annual report of the Chesapeake and Ohio canal company, 4th June, 1832.

From this, it appears that the Chesapeake and Ohio canal company has received from various resources, funds to the amount of $2,065,769 and 80 cts.; and have expended $2,007,875 and 15 cts., leaving a balance on hand of $57,894 and 65 cts. The charter of this company requires, on penalty of forfeiture in case of failure, the completion of 100 ms. of the canal in 5 years from its commencement, which took place 4th July, 1828; of course unless provided for, the 100 ms. of canal must be in operation by the 4th July, 1833, or the company must cease operations. On the subject of this contingency, the report before us tacitly acknowleges the inability of the company to save the charter by a full compliance with its provisions, but observes, that "although the apprehension should not be for a moment indulged, that the charter of the company would be endangered by their failure to construct 100 ms. of canal in 5 years from its commencement, considering the legal obstructions which have impeded its progress for more than three years of that period, yet this provision of the charter, and the interests of the stockholders, impose on the company the obligation of diligently prosecuting their work, to the extent here contemplated. Accordingly, the board have first endeavored to ascertain the competency of the present resources of the company, to complete 100 ms. of canal, by the autumn of 1833, being within 5 years from the time when the first contracts were made, and the work actually begun, in the vicinity of Georgetown.

"From the treasurer's report, 1st May, 1832, it appears that, on the 30th April, 1832, the subscribed stock, payable in money, as contradistinguished from the part payable in the shares of the former Potomac company, amounted to $3,609,200; of this stock there had been then collected $1,959,087, leaving to be collected the farther sum of $1,650,-113. Deduct allowance for bad debts $70,113, affords a balance of $1,580,000. To which add cash on hand, at the date of the treasurer's report, after deducting a sum paid by the corporation of Alexandria, in anticipation of its future instalments, $30,814, and there results a fund of $1,610,814, applicable to the following objects:—1st. Retained for the payment of work done below the Point of Rocks, $40,841. 2nd. Indemnity for lands taken for the construction of the canal, between the Point of Rocks and the mouth of Tiber creek, $30,000. 3rd. Completion of unfinished work, between the Point of Rocks and the mouth of Tiber creek, $170,000. 4th. To the completion of the 12 ms. of canal and their appurtenances, between the Point of Rocks and the Harper's Ferry feeder, including the dam and guard lock at the latter; but exclusive of the sum of $14,629 already expended on this work, according to the estimates of the engineers, modified in some inconsiderable particulars, $310,000. 5th. To the completion of the 24½ ms. of canal, between the Harper's Ferry feeder, and that in the vicinity of Opeccon, according to actual contracts, so far as they extend, and to the estimate of the engineers, modified in some inconsiderable particulars, $788,197. The total amount for these objects being $1,339,038. These being deducted from the available stock, and cash on hand, $1,610,814, leaves the sum of $271,776, which balance is to be applied to the portion of canal between the feeder at Licking creek, and that next to Opeccon. The portion of canal extending from the former, which will be required to make up 100 ms., when added to the part below, need not exceed 14 ms.; for which the above sum affords near 19,500 dollars a mile."

By reference to the subjoined tables, it will be seen in No. 4. that 100 ms. of canal from Georgetown will reach 32 ms. above Harper's Ferry, and 4 ms. above Williamsport. In regard to work actually completed, the report states, that, "the various works on the canal between the Point of Rocks and the basin in Georgetown, which had been permitted to proceed very tardily, for many months, in consequence of their utter inutility without a supply of water, and the remoteness of that supply, in point of time, have, notwithstanding, reached very near their final completion." The account rendered of the tolls of the canal, for the 11 months which expired on the 30th of April, (1832) is $25,108 93, to which may now be added, those for the month of May, $6,400 32, making the tolls for the year amount to $31,509 25, being an excess beyond those of the previous year of $2,367 90. After some estimates, which we have not room to insert, the report adds, "the preceding resources, exclusive of the canal tolls, may, therefore, be safely computed at a sum exceeding $150,000, and if not profitably converted into money, might be pledged, as the basis of a loan, to that amount, in aid of the uncollected stock of the company, if required to construct 100 ms. of canal by the Autumn of 1833; by which period, the contracts last made, require the part of the canal, below Opeccon, to be completed. Those, for the works below the head of Harper's Ferry falls, limit the period of their completion, as has been stated, to the 1st of December next, (1832,) by which time, or at any rate, by the opening of the ensuing spring (1833), it is confidently expected to bring the entire canal into use, from the still water, at the head of the falls, produced by the dam of the U. S. armory, down to the mouth of the Tiber."

The preceding is a general view, and a brief one it is true, of the present state of the canal system in the U. S. along the Atlantic coast, and that of the Gulf of Mexico from the Potomac to the Mississippi, inclusive. It was our intention to have given a summary table of expenditure, but so desultory have been the operations, and so loose have been the registers of expense, that any summary must be extremely defective; but we present the following, which may give some aid in forming comparative estimates of relative expenditure made in the large physical sections of the U. S.

Expenditures on the Santee canal, $650,-667; Roanoke navigation, $365,991; Dismal Swamp canal and feeder, $800,000; James r. navigation, including the Appomattox, &c. say $1,200,000; Chesapeake and Ohio canal, $2,007,875. To which add for all other improvements on the various rs. and inlets, from the Delta of the Mississippi to the Potomac inclusive, $1,000,000. Aggregate amount, $6,024,533. To which amount we may add as already expended on rail-roads in the same natural section, viz. Chesterfield rail-road, $140,000; Petersburg and Roanoke rail-road, say $100,000, and the South Carolina rail-road, which it is supposed will be completed in January, 1833, the whole estimated sum necessary for its complete construction,

$610,000, and machinery, $61,000; in all $911,000, making a total of expenditures in canals and rail-roads of $6,985,533.

I.—Table of the sub-basins and aggregate extent of Chesapeake basin.

Basins.	Length.	Mean Breadth.	Between Latitudes N.		Between Longitudes from W. C.	
James r.	250	40	36°40'	38°20'	1°00' E.	3°40' W.
York,	130	20	37 15	38 16	0 41 E.	1 12 W.
Rappaha'c	140	20	37 34	38 44	0 41 E.	1 25 W.
Potomac, ab'e Blue Ridge,	160	50	37 58	40 05,	0 25 W.	2 45 W.
Potomac, b'w Blue Ridge,	165	30	37 58	39 55,	0 45 E.	1 00 W.
Patuxent, Patapsco, &c.	110	25	38 10	39 42	0 45 E.	0 08 W.
Eastern sh of Chesapeake,	200	25	37 07	40 00	0 40 E.	1 40 E.
Susqueh'h	230	125	39 33	42 53	2 16 E.	1 41 W.
Chesape's bay,	180	20	37 00	39 33	0 26 E.	1 24 E.
Aggregate,	500	138	36 40	42 53	2 16 E.	3 40 W.

II.—Table of the ascents and descents from tide water at Weldon, on Roanoke, by Salem, and thence over the Alleghany chain into the channel of New river, and down that stream and the Great Kenhawa to the Ohio river at Point Pleasant.

Route.	Distances in miles.		Ascent or descent.	Elevation in ft. above mid-tide.	
Tide water to Salem,	224	224	rises.	1002	1002
Salem to forks of Roanoke,	11 3-4	235 3-4	"	176	1178
Mouth of Elliott creek,	11 1-4	247	"	221	1399
Beginning of summit level,	12 3-4	259 3-4	"	650	2049
Over summit lev'l	5 3-4	265 1-2	"		
From wstrn. end of summit level down Meadow cr. and Little r. to New r.	11 1-4	276 3-4	falls.	309	1740
Thence to mouth of Greenbriar r.	83 1-2	360 1-4	"	358	1382
Bowyer's ferry,	45 3-4	406	"	400	982
Foot of Great falls of Kenhawa,	23	429	"	341	641
Mouth of Great Kenhawa at Point Pleasant,	94	522	"	108	533

III.—Ascents and descents from head of tide water in James river at Richmond, along the channel of James, Greenbriar and Kenhawa rivers to the Ohio river at the mouth of Great Kenhawa; and crossing the Appalachian system by way of Covington and Greenbriar rs.

Route.	Distances in miles.		Ascent or descent in feet.	Elevation in feet above mid-tide.	
From tide water to Maiden's Adventure,		29	rises.	140.5	140.5
Columbia,	30	59	"	39.28	179.78
Big Bremo,	11	70	"	29.22	209.
Hardware r.	3 1-2	73 1-2	"	33.27	242.27
Scottsville,	8	81 1-2	rises.	15.11	257.38
Warminster,	19	100 1-2	"	58.37	315.75
Lynchburg,	50	150 1-2	"	185.88	501.63
Blue Ridge,	20	170 1-2	"	103.47	605.1
Through do.	6 1-2	177	"	94.75	699.85
Pattonsburg,	21 1-2	198 1-2	"	106.23	806.08
Covington,	58 1-2	257	"	416.	1222.08
Mouth of Fork Run,	16 1-2	273 1-2	"	432.	1654.08
Beginning of summit level,	2 1-2	276	"	264.	1918.08
Along summit level,	4 1-2	280 1-2	"		
From western end of summit level down Howard cr. to Greenbriar r.	8	288 1-2	falls.	249.	1669.08
Down Greenbriar to its entrance into New river,	49	337 1-2	"	287.	1382.08
Bowyer's ferry	45 1-2	383	"	400.	982.
Foot of falls in Great Kenhawa,	22	405	"	341.	641.
Mouth of Great Kenhawa,	94	499	"	108.	533.

IV.—Table of the ascents and descents from tide water in James river at Richmond, along the channel of James river to the mouth of Catawba creek, thence up the latter and over the intermediate summit into the valley of Roanoke at the forks of the latter above Salem, and thence, as in table II.

Route.	Distances in miles.		Ascent or Descent.	Elevation in ft. above mid-tide.	
Pattonsburg,		198 1-2	rises.		806
Mouth of Catawba,	14 1-4	212 3-4	"	80	886
Forks of Roanoke,	51	263 3-4	"	292	1178
Summit level, table,	24	287 3-4	"	871	2049
Thence to the mouth of Greenbriar,	94 3-4	382 1-2	falls.	667	1382
Thence to the mouth of Great Kenhawa,	161 1-2	544 1-4	"	849	533

V.—Table of the ascents and descents along the channel of Potomac, from tide water at Georgetown to Cumberland, and thence, following the contemplated route of the Chesapeake and Ohio canal, over the Appalachian system by the channels of Youghioghany and Monongahela to the Ohio at Pittsburg.

Route.	Distances in miles.		Ascent or descent.	Elevation in feet above mid-tide.	
Mouth of Monocacy,		44	rises.		224
Harper's ferry, at passage of Potomac thro' the Blue Ri'e, and influx of Shenandoah,	24	68	"	62	286
Williamsport,	28	96	"	69	355
Hancockstown	31	127	"	52	407
Old Town,	42	169	"	82	489
Cumberland,	17	186	"	84	573
Mouth of Little Wills creek,	12 3-4	198 3-4	"	[illegible]	[illegible]

Route.	Distances in miles.		Ascent or descent.	Elevation in ft. above mid-tide.	
Eastern end of summit level,	15 1-4	215	"	1016	1898
Western end of summit level,	53 3-4	268 3-4	"		
Mouth of Middle Fork creek,	16	284 3-4	falls.	216	1682
Mouth of Casselman river,	20	304 3-4	"	490	1268
Connellsville,	27 1-2	332 1-4	"	432	830
Mouth of Youghiogbany,	43 3-4	376	"	158	678
Pittsburg,	14	390	"	35	643

Pennsylvania navigation. Under this head is included the ***Chesapeake and Delaware canal***, since, though not actually in the state, it was with means principally afforded by Pennsylvania, that this work was constructed.

Though only about 14 ms. in length, this canal was built at great expense, owing to its size, the depth of its excavations, and the extent of its embankments. It is of sufficient dimensions for the passage of coasting vessels, and extends across the state of Delaware, from the Delaware r. to the Elk, which falls into Chesapeake bay. In this canal is a deep cut of 3¾ ms., 76½ feet in depth, where the greatest excavation was made. Within the state of Pa. the following are the most important works of this nature which have been executed. The ***Conestoga canal*** passes from Lancaster, about 62 ms. directly w. from Philadelphia, down the Conestoga cr., 18 ms., in nearly a south west direction, to the Susquehannah r. The ***Delaware canal*** commences at its northern extremity at Easton, 55 ms. nearly N. from Phil. on the N. W. bank Delaware r., which, for about 50 ms. S. of this place, is S. E., when it turns nearly S. W. about 30 ms. to Phil. This canal follows the general course of the r., keeping its w. bank to Morrisville, where it bears off from the river to avoid a bend, and proceeds in a nearly direct course to Bristol, on the w. bank of the Delaware, 19½ ms. N. E. from Phil. The Delaware and Hudson canal is described among the canals of N. Y. The ***Lackawaxen canal*** is a continuation of the ***Delaware and Hudson***, up the Lackawaxen r. to the Lackawana coal-mines. The ***Lehigh canal*** commences at the Mauch Chunk coal-mine on the river Lehigh, and runs to Easton on the Delaware. The whole distance of this navigation is 46¾ ms., but a part of it is on the r., the length of the canal being 37 ms. Its eastern termination, at Easton, meets the western termination of the Morris canal in New Jersey. The ***Pennsylvania canal*** commences at Middletown, at the termination of the Union canal, whence it is proposed to proceed up along the Susquehannah, in a westerly direction to the Alleghanies, which are passed by a rail-road, about 50 ms. in length, into the valley of the Ohio, where the canal again commences, and is continued to Pittsburg, a distance, in the whole, of 320 ms. of canal and rail-road. The ***Schuylkill canal*** is constructed on the banks of Schuylkill r., from Phil. about 110 ms. to Mount Carbon, the region of the anthracite coal in Schuylkill co., the general direction being nearly N. W. The ***Schuylkill (Little) canal*** is 27 ms. in length, from the mouth of the Little Schuylkill r. to the coal-mines. The ***Union Canal*** branches off from the Schuylkill canal, a little to the westward of the town of Reading, in Berks co., about 60 ms. from Philadelphia, in a direction generally S. W.; first passing up a branch of the Schuylkill, and then down the valley of the Swatara, somewhat circuitously, about 80 ms., to Middletown, a little above the junction of the Swatara with the Susquehannah.

It has been a question, idly but somewhat warmly mooted, with whom originated the canal system in the United States. Were it practicable to arrive at a satisfactory adjustment of rival claims, the result would be wholly unimportant, since the idea, by whomever conceived, was entirely without originality, having been borrowed from older countries. In Pennsylvania, the first enterprises of any moment in this country, in the way of internal improvement, were undertaken and accomplished. But it was not till, by the completion of the great Erie canal, the immense benefits resulting from such works were fully demonstrated; it was then that the system acquired vigor, and won upon the confidence of the people. New York succeeded, and roused her powerful sister state into action; and that action has produced effects in direct ratio with positive power; giving a lesson to man that all future ages will read, from a book traced on the surface of the earth. Pennsylvania has already expended not much if any less than $40,000,000 on her stupendous internal improvements. Her system of inland navigation has become complex, however, from having adopted rail-roads and canals on the same line; we therefore refer to the head of rail-roads, our further notice of the inland navigation of Pennsylvania.

New Jersey, from the limited extent of its territory, the dry and sandy nature of its soil in the southern part, its mountains on the N., and the general want of commodious harbors on the eastern coast, has not been the scene of very extensive canal operations. The Morris, and the Delaware and Raritan canals, however, are important works, and will prove of very great utility. The ***Delaware and Raritan canal***, authorized by the legislature of New Jersey, by an act passed in Feb., 1830, will connect the navigable waters of the Delaware with those of the Raritan. The canal is 75 feet in width on the water line, and has 7 feet depth of water throughout. The bridges are moveable like those of the Delaware and Chesapeake canal. The locks are 110 feet in length, by 24 in width. Vessels of large burthen may consequently pass through the canal; and its advantages to the coasting trade will be great, as it will complete an internal water communication for masted vessels between N. York and Albemarle sound. The terminating points of the canal, are, on the Raritan, at New Brunswick, and on the Delaware, at Bordentown. It follows the valley of the Raritan, Millstone, and Stony brook; and, crossing the Lawrence Meadows to the valley of the Assanpink, along the valley of that stream to Trenton, and thence down the river, (crossing the Assanpink by an aqueduct,) to the point where Crosswick's cr. comes into the Delaware at Bordentown. The length of the canal is 42½ ms.; the elevation above tide water but 56 feet. It passes the towns of New Brunswick, Boundbrook, Millstone, Griggstown, Kingston, Princeton, Trenton and Lambarton, discharging at Bordentown. The route is through a beautiful and highly cultivated valley, affording great advantages to numerous mills and other water works, on the various streams adjacent. It is supplied by a feeder from the Delaware r., commencing at Bull's island, 26 ms. above Trenton, and passing along the bank of the river to the main canal at Trenton. The feeder is also a canal, 60 feet in width and 5 deep. The works now progressing are under the direction of an able engineer and assistants, and there is no doubt of the completion of the whole work in 1833. The *Morris canal* extends from Philipsburgh, on the Delaware river, to the Passaic at Newark, across the state of New Jersey, through the counties of Warren, Sussex, Morris and Essex, and was constructed chiefly to open a more direct channel of communication by boats, for the transportion of coal from the mines on Lehigh river, Pa., to the city of New York. It is 34 feet wide, 4 deep, and 84 ms. long, including a feeder from Musconetcunk (or –cong) or Hopatcunk (or –cong) lake. The elevation of the summit is nearly 900 feet above tide water, and 700 feet above the Delaware at Easton, Pa., opposite which it joins that river. On account of the scarcity of water, the company were induced to construct inclined planes at some of the principal elevations on the route. There the boats are received in large cars, which are raised or lowered by machinery; the weight of the descending boat being often applied to assist in raising an ascending one. The canal is navigable in boats of 25 tons, many of which are actively engaged in transporting coal, iron ore, produce, lumber and merchandise of different sorts. The country through which it passes has many iron mines, forges and furnaces, numbers of which have been abandoned on account of the scarcity of fuel in their vicinity, or for other causes; but some of them will again be rendered profitable. Large quantities of anthracite coal will find the way to the New York market by this route, and ore from the different mines is transported by this channel to forges in the different places, particularly in the lower parts of New Jersey, to be smelted. The route, after leaving the Delaware, lies near Musconetcong and Pohatcong rs., through Hackotstown and Stanhope, to the summit near Brooklyn, then down to Suckasunny Plains, Dover, Rockaway, along the valley of Rockaway r. across the Raritan on a fine aqueduct 3 ms. above Patterson, thro' Bloomfield, to Newark. The inclined plane at Newark is 1,040 feet long, rising more than 70 feet, and has a double line of tracks, on each of which is a car with eight wheels, large enough to receive a canal boat. This car is connected to a machine turned by a water wheel, 24 feet in diameter, and by a chain strong enough to support 15 tons. A boat may be raised, and another lowered at the same time, in about 8 minutes. Five such operations may be performed in an hour, and 6,000 tons may be passed in a day. It has been estimated, that if locks had been substituted for inclined planes on this canal, the time spent in passing them all would have been 24 hours, while the inclined planes are passed in 2 1-2 hours.

The great basins of the St. Lawrence and the Mississppi are very intimately connected, and no difficulty exists in the way of uniting their navigable waters, by artificial channels. This object is effected, by the two great Ohio canals; to which will soon be added the Wabash and Erie canal, of which a brief notice is given below.

What has been actually completed on the Atlantic slope, and in the cases of N. York and Pennsylvania, the extensions made into the great Canadian basin by the former, and into the Ohio valley by the latter, may well excite astonishment, but if all things are considered and liberally compared, the two great canals of the state of Ohio are the most stupendous undertakings ever achieved on the face of nature by man. Forty years ago the ground now comprising that state was a wilderness, and it is only a few days past forty years since the United States' army was defeated by savages on the very section of this youthful state, where now a canal is navigated. The Ohio state canals were projected about 1823, and may now be regarded as completed, or so nearly so, as to admit a notice admitting their completion. The *Miami canal* commences at Cincinnati, and extends north-north-eastwardly along the valley of the Great Miami, a total distance of 67 ms. It passes the towns of Hamilton, Middletown, Franklin and Miamisburg, to Dayton, the co. seat of Montgomery co. This canal is in full operation, and it is in contemplation to extend it to lake Erie, by the valleys of Miami, Auglaize and Maumee rivers. To secure this latter extension, the congress of the United States made a grant or grants of land to a large amount, conditioned that the Ohio canals be completed within seven years from 1828, or in 1835, and said canals to be and forever remain public high-ways, for the use of the government of the U. S.

The route of the eastern or *Great canal of Ohio*, with its ascents and descents, will be seen by reference to a table in article Ohio, page 371. This canal commences on the O. at Portsmouth, and at the mouth of Sciota r., and thence ascends the Sciota upwards of 70 miles, passing the towns of Piketon, Chillicothe and Circleville. It then, leaving the Sciota, pursues a course a little E. of N. E. to Coshocton, passing the towns of Hebron and Newark, and the summit level between the valleys, of Sciota and Muskingum rs. From Coshocton, the canal follows the valley of Tuscarawas about 100 miles to the summit level between the Ohio valley and basin of Erie. It thence finally falls rapidly 31 miles to the level of lake Erie at Cleaveland. This great canal traverses the counties of Sciota, Pike, Ross, Pickaway, Franklin, Fairfield, Licking, Muskingum, Coshocton, Tuscarawas, Stark, Portage and Cuyahoga, and may, in more than one important circumstance, be regarded as a continuation of the Erie canal. Both the Ohio canals are owned by the state.

This great canal line may be regarded as a continuation of that of the Hudson and Erie canal of N. York. The *Miami canal*, extending 67 ms. following the canal line from Cincinnati to Dayton, is in full operation. The two canals, according to Flint, will cost from 3 to 4 millions of dolls. (*See art. Ohio for further details of its canals.*)

The *Louisville and Portland canal*, for the passage of large vessels round a cataract in the Ohio r. at Louisville in Ky., is the last work of that kind of any considerable importance yet completed in the valley of Ohio, and which remains to be noticed. For its length, the Ohio and Portland canal is perhaps the most important artificial hydraulic work ever executed. It has been in use since the 21st Dec., 1829. The charter was granted Jan. 1825, to "the Louisville and Portland canal company;" stock $600,000, of which, by act of congress, the U. S. took $100,000. The length of this canal is between 2 and 3 ms., overcoming 22½ feet fall, by 5 locks. By a report of the engineer, 3d Jan. 1831, it appeared that the Ohio and Portland canal was then in full operation, and that steamboats had passed since the previous report.

To the foregoing notice of western canals may be added the *Wabash and Erie canal*, of Indiana, a part of which is already under contract. It is to extend over the intermediate table land between the Maumee and Wabash rivers; is undertaken under the authority of the state, and its route will be in Allen co. (*For this co. see Appendix.*)

To the basin of the St. Lawrence belongs all the northern portion of the state of New York, and it has been for the purpose of forming a connection between the waters of this basin and the Atlantic, that the Champlain and Erie canals have been constructed. But beside these splendid works, which were the first to open the eyes of the people of the U. S. to the vast utility of artificial navigation, so many others, of more or less consequence and extent, exist within the limits of the state, either completed or in progress, that we have thought proper to present a succinct and separate account of each, in the following arrangement. *Black river canal;* a canal has been proposed by the canal commissioners, from the High Falls of Black river, to Rome, 36 ms. with the improvement of the navigation of the river from those falls to Carthage, and a navigable feeder of 9 ms. from Boonville, the whole amounting to 76 ms. at an estimated expense of $602,544. The water it is proposed to take from Black river. *Buffalo canals*, at Buffalo in Erie co. are two short canals. One of these is for the passage of lake vessels from Buffalo harbor to the line of the Erie canal. It is about 700 yards in length, 80 feet wide, and 13 feet deep. It commences near the outlet of Buffalo creek. The other is a boat canal, from Big Buffalo creek to Little Buffalo creek 1,606 feet long. *Cayuga and Seneca canal*, extends from Geneva, at the foot of Seneca lake, to Montezuma, on the Erie canal, 20 ms. 44 chains. About half the distance is by slack water navigation, the other by an artificial canal. The descent to the canal is 73 1-2 feet, which is surmounted by 11 wooden locks. The tolls collected on this canal in 1831, amounted to $12,920 39 cents. The tolls reported in July, 1832, amounted to $725 44. *Champlain canal*, extends from the junction with the Erie canal, 8 ms. N. of Albany, to White hall, Washington co. at the S. extremity, or head of lake Champlain, and affords a boat navigation between that lake and Hudson river. It is 72 ms. long, and has 21 locks, with a total rise and fall of 188 feet; 7 of these descend 54 ft. from the summit level N. to the lake, and 14 S. to the level of Hudson river 134 feet. From Albany to West Troy, the Champlain and Erie canals are united; West Troy is 7 ms. and the route lies along the level on the west bank of Hudson river. At the junction it leaves the Erie canal, and crosses the Mohawk river at a ferry; passing through Waterford, Stillwater, Saratoga, Schuylersville, Fort Miller, Fort Edward, and Fort Ann, it terminates at Whitehall. The tolls collected on this canal in 1831, amounted to $102,896 23. This is the route of an extensive and valuable trade, between the shores of lake Champlain, the Hudson and New York. The work was commenced in 1818, and finished in 1823; 46 ms. of the route is dug, 6 1-2 lies in Wood creek from Fort Ann N., the water being raised by a dam. Near Fort Edward a feeder enters from the Hudson, in which is a dam 900 feet long and 27 feet average height. From Fort Edward to Fort Miller, 8 ms. the canal lies in the Hudson, and again 3 ms. above Saratoga Falls. *Chemung canal*. The legislature appropriated $300,000, in 1829, for the construction of this canal, which is now partly completed. It is to extend from Elmira, Tioga county on Tioga or Chemung river, (a tributary of Susquehannah river,) to the head waters of Seneca lake, 18 miles. To this work is to be added a navigable feeder of 13 ms. for the summit level, from the Chemung at Painted Post. The canal is to have 53 locks, all of wood, 70 bridges, 1 dam, 6 culverts and 3 aqueducts. The distance from Elmira to Philadelphia by this route, is 374 ms. and to Baltimore, 334. *Chenango canal.* A canal has been proposed, from a point on the Erie canal in Oneida co. to the Susquehannah at Binghampton, Broome co. through Oriskany and Saquit creeks and Chenango river. The cost is estimated at $944,775. The lockage would be 1,009 feet, the rise from Erie canal to the summit being 706 ft. and from Susquehannah r. 303 feet. It would cross the following towns, New Hartford, Clinton, Madison, Hamilton, Sherburne, Norwich, Oxford, Greene and Chenango forks. *Chitteningo canal,* extends from Chitteningo village, Madison co. 1 1-2 ms. to the Erie canal, and has 4 locks. *Crooked Lake canal,* is to extend from near Penn-Yan, along the outlet of Crooked lake 7 ms. to Seneca lake. There must be 270 feet descent overcome by locks, and the legislature have appropriated $120,000 for the work. *Delaware and Hudson canal.* This canal was commenced in 1825 and completed in 1828, and its entire length is 108 ms. It extends from the Hudson river at a point 90 ms. N. New York, to Port Jervis, on Delaware r. 59 ms. up the E. bank of the latter 24 ms,. and up Lackawaxen r., Pa., to Honesdale, Wayne co. 25 ms. It is here connected with a rail road, 16 ms. long, running to Carbondale, Luzerne co. Pa. where are extensive mines of Lackawana coal. This company has a capital of $1,500,000, one third part of which is in banking capital in New York. The canal is from 32 to 36 feet in breadth at the surface, 4 feet deep, with locks 9 feet by 76, for boats of 25 or 30 tons; 43,200 tons of coal were transported on it in 1830, and 52,000 in 1831. It is re-shipped into vessels at Bolton, on the Hudson, where there is 11 feet water to market. In 1831, 641 vessels were loaded there with coal and other articles, and $19,500 was received in tolls that year at Rondout, exclusive of that paid on coal; 138 boats were devoted to the transportation of coal only. *Erie canal,* or Grand canal of N. Y. extends from Albany to Buffalo, 363 ms. It is 40 feet wide at top, 28 at bottom, depth 4 feet. The tow path is 10 feet wide, and it is fenced, lined and bridged, by the state. It leaves lake Erie at Buffalo, which it has rendered a rich & flourishing place; runs to Black Rock near the lake along Niagara r. 7 ms. in the channel of Tonnewanta creek, 12 ms. through a deep cut in mountain ridge 7½ ms. to Lockport, descends 60 feet by 5 double locks; passes near the ridge road 63 ms. to Rochester, crosses Genesee river, thence to Mohawk river at Rome, passing a little N. of the small lakes, near the course of Seneca r. through the Cayuga marshes & the long level. It then follows the course of the Mohawk to Cohoes bridge, and after uniting with the Champlain canal, terminates at the great basin at Albany. It crosses Genesee r. on a noble aqueduct, the Mohawk three times, and has many aqueducts, dams, feeders, culverts &c. connected with it. The Albany and Schenectady rail-road will carry many of the commodities, and other rail-roads are projected from Schenectady west even as far as Buffalo. Several canals of much importance are branches of the Erie canal; the Oswego canal, which extends from Syracuse to Oswego, on lake Ontario, and the Cayuga and Seneca canals, connecting it with several of the small lakes in the middle of the state. There are others planned, with rail roads in different directions, which will still further increase the vast amount of transportation now carried on through the state. The following is an account of the amounts received in tolls at different places on the canal in 1831.

Albany,	$269,443 73
West Troy,	156,458 19
Schenectady,	35,700 56
Little Falls,	9,685 78
Utica,	41,012 61
Rome,	28,680 79
Syracuse,	66,144 82
Montezuma,	65,570 15
Lyons,	20,539, 46
Palmyra,	55,776 33
Rochester,	174,350 90
Albion,	10,993 94
Brockport,	10,750 82
Lockport,	31,023 19
Buffalo,	66,009 19
Geneva,	27,742 98
Salina,	39,360 30
Total amount,	$1,122,243 74

Besides this, large quantities of flour, ashes, provisions, &c. have been sent to the Canadas, of which no account is here made.

The N. Y. canals were constructed at the expense of the state, and a large debt has been thus contracted, to defray which their income is pledged with the net revenue from the auction and salt duties. In 1837 about 2-5 of the debt is payable. The debt for the Erie and Champlain canals amounted, on the 1st Jan. 1832, to $7,001,035 86; the Oswego, Chemung and Crooked Lake canal debt to $1,054,610 00; so that the whole canal debt of the state then was $8,055,645 86. In 1831 were inspected at Albany, 48,653 bbls. of wheat flour, a large part of which was raised in the fertile counties of the state, and ground at the extensive mills of Rochester. The tolls collected on the Erie and Champlain canals, in July, 1832, amounted to $102,904 98; $3,953 52 less than in 1831. *Haerlem canal,* N. Y. co. extends from East r. to Hudson r. 3 ms. through Manhattanville. The company was incorporated in 1826, and enjoy a perpetual charter, with a capital of $550,000. It is 60 feet wide, and in the middle part of the route 100; 6 or 7 feet deep, and about 3 ms.

long; a street of 50 feet breadth is to be formed on each side; it has guard locks at the ends, and the sides are to be walled with stone. *Hell Gate canal.* A company has been incorporated to construct a sloop canal. 800 yards long, round the dangerous passage in East r. called Hell Gate, on the w. end of Long Island. Delays are frequently caused to the numerous vessels, principally coasters, which navigate that channel, as the passage is unsafe except at particular states of wind and tide. Hallet's Cove, at a short distance, is often crowded with vessels, waiting for an opportunity to pass. It has been estimated that 520 packets pass 22,520 times in a year; 500 trading vessels which pass 11,000 times; and 13 steamboats, most of them large, and employed in transporting numerous passengers and valuable freights, which pass 5,000 times, making 50,000 in all. It is proposed to make the canal 82 feet wide, 18 feet deep, with gates, and to face it all with stone; and the cost is estimated at $70,000, and the purchase of all the lands, about 50 acres, including houses, buildings of all kinds, ferries, rail-ways, quarries, &c. will amount to about fifty thousand dollars more. It is computed that it will pay about twenty-two per cent on the original cost. *Mohawk canals.* The Mohawk river was rendered navigable in boats from Schenectady to Rome, some years ago, by the construction of canals round Little Falls, and Wolf Rift, on the German Flats. A canal of 1 1-2 ms. was also made from the head of the Mohawk to Wood creek, which leads into Oneida lake. *Oswego canal*, in Onondaga and Oswego cos. extends from Syracuse, on the Erie canal, to Oswego, on lake Ontario, 38 ms. For half the distance Oswego r. is used, having been dammed, and supplied with a towing path on the bank. The descent to the lake is 123 feet, which is overcome by 13 locks, all of which are of stone except one. Cost, $525,-000. $16,271 10 was collected on this canal in tolls in 1831. The tolls reported in July, 1832, amounted to $192,62. *Scottsville canal.* A company was incorporated in 1829, with $15,000 to construct a canal from Genesee r. to Scottsville, Munroe co. *Sodus canal*, is to be 24 ms. long, 6 ms. of its route is on Seneca r. and the outlet of Crusoe lake, and 13 ms. to be excavated at the summit level only 10 feet. The descent is 130 feet, of which 114 is near the end of the canal. The locks will afford abundance of water power. Big Sodus harbor, with which this canal is to communicate, is large, and one of the best on lake Ontario. This work is on a route said to be the most direct communication between the waters of lake Champlain and the St. Lawrence.

N.England, possessing in general, the usual rugged character of primitive formations, offers fewer facilities for extensive artificial navigation, than many other portions of the U. S.

In Connecticut, the work of greatest magnitude which has yet been undertaken, is the *Farmington canal.* This extends 58 ms. from New Haven to Southwick ponds, on the boundary of Mass. It passes through Hampden, Cheshire, Southington, Northington parish, Simsbury, Farmington, Granby, to Southwick, where it enters several ponds. From the level of Farmington it rises N. 38 feet by 6 locks; it crosses Farmington r. on an aqueduct of 280 feet, 34 high. It is 36 feet wide at top, 20 at bottom, 4 deep, with 218 feet lockage, all ascending from N. Haven. Farmington r. feeder, 3 ms. long, gives the principal supply, and is a branch. The work began in 1825. The Hampshire and Hampden canal has been constructed in Mass. in continuation of the Farmington canal to Westfield, and was intended to go to Northampton, and even proposed to be extended to Barnet, Vt. It will probably be completed to Northampton soon, where it will communicate with Connecticut r. Beside this in the same state is the *Enfield canal*, extending round Enfield falls, on the w. side of the Connecticut. It is 6 miles long, and is an important improvement in the navigation of that r. The fall is 30 feet, which is overcome by three locks of 10 feet lift each, of hammered stone. Great advantages are afforded by this canal for hydraulic purposes, which yet have been but partially improved.

In Rhode Island, the *Blackstone canal* commences at Providence, and extends about 40 ms. to Worcester in Mass. It follows principally the course of Blackstone, or Pawtucket river, and passes through North Providence and Bristol in R. I., and Mendon, Uxbridge, Northbridge, Sutton, Grafton and Milbury to Worcester where it terminates. It is supplied with water from the Blackstone r. Numerous manufactories lie on and near this route, to which the canal affords great advantages. This canal was built by a company chartered by the states of R. I. and Mass. at an expense of about $700,000, and was completed in 1828. It is 45 ms. long and has a fall of 450 feet, to surmount which there are 48 locks on the route. The canal has a depth of 4 feet, is 34 feet wide at the surface and 18 at the bottom.

In Massachusetts no other canals of magnitude have been constructed, besides the above. The principal are the Middlesex and the Hampshire and Hampden canals; the latter is not yet completed, and has been already noticed under the head of Farmington canal. *Middlesex canal* extends from Charlestown, on the navigable waters of Boston harbor, to Lowell, on the Merrimack, where it communicates with the works on that r. and extends the line of boat navigation from Concord, N. H. It is 27 ms. long, 30 ft. wide and 4 deep, with 20 locks and 7 aqueducts over valleys and streams. There are four levels, each 5 ms. long. At Charlestown the canal terminates in a large mill pond. On the summit level it crosses Concord r. which supplies it with water. 13 locks descend hence to Charlestown, 107 feet, and 3 locks to Merrimack r. above the falls, 21 ft.

This work cost $530,000. The Pawtucket, South Hadley and Wickasee canals are comparatively of inferior importance. *Pawtucket canal* was constructed in 1797. It passes round the Pawtucket falls (in the town of Lowell) in the Merrimack, a distance of 1½ ms. Since its first construction it has been both deepened and widened and affords water power to several manufactories. The falls in the whole distance are about 30 ft. and the canal is now 90 ft. broad, and 4 deep. *South Hadley canal*, is 2 ms. long, and overcomes a descent of 40 ft. in Connecticut r. It was the first canal in the U. S. being commenced in 1792. Near its lower junction with the Conn. is a cut through solid rock, 300 ft. long and 40 deep, through which it passes. *Wickasee canal*, leads boats round the falls at that place in Merrimack r. 3 ms. above Lowell, where the Middlesex canal commences. It cost $14,000.

The remaining canals of New England will occupy but a very brief space.

The *White r. canal*, in Vt. is a small work around a fall in Conn. r., for flat bottomed boats and rafts. The *Bellows Falls canal*, in the same state, is a short but expensive work along the w. shore of Conn. r. round these falls. It is cut through a bed of hard granite; but a part of the excavation was made in ages past, by the current of the stream. Flat bottomed boats, small steamboats and rafts, thus pass a natural obstruction in the navigation.

In N. H. a company was incorporated in 1811, the charter of which has since been renewed, for the purpose of forming a canal with locks from Winnipisseogee lake to Dover, along Cochego r. 27 ms. As the descent is 452 ft., no less than 53 locks would be necessary; and the expense is estimated at $300,000. This work would be of benefit to above 400 sq. ms.; and it has been even proposed to extend a canal to Pemigewasset r. The following works, completed in the same state, constitute with the Pawtucket and Wickasee canals, in Mass. already mentioned, links in a chain of navigation, extending from Boston harbor by the Middlesex canal and the Merrimack r., to the central part of N. H. *Amoskeag canal*, affording a boat navigation round a fall of 45 ft., in the Merrimack, is one mile in length. The fall is 45 feet, and is overcome by 9 locks, which with the canal cost $50,000. *Bow canal*, affords boat navigation of ¼ m. round the falls in Merrimack r. at Bow, of 25 feet descent. The works cost $21,000. The canal commences at the upper landing in Concord, and is the first link in the chain of improved internal transportation, which extends down the Merrimack to Lowell, and thence to Boston. *Hooksett canal*, 50 rods only in length, passes round Hooksett falls, 7 ms. below Amoskeag. These falls are 16 ft., and the canal and locks cost $17,000. They afford a navigation for boats. The *Union canals* pass 7 falls in the Merrimack, and the distance improved by them for the navigation of boats is 9 ms. There are 7 locks on the route, which furnish water power for several manufactories. This and the 3 canals preceding, all in New Hampshire, are a part of a line of navigation long since projected between Boston and the central parts of N. H. The Middlesex canal unites with this improved navigation 27 ms. N. W. E. Boston, at Chelmsford.

In Maine the Cumberland and Oxford canal extends 50 ms., from Portland to Sebago pond. The latter, with Brandy pond and outlets, include 27 ms. of the canal, the balance, 23 ms., being artificial, and having 24 locks. Bridgeton is at the head of the canal.

This completes what we have to say descriptive of canals in the U. S. With regard to the system in general, the expense attendant on the construction of such works, and the amount of profit accruing from them to the proprietors, a few words here may not be out of place. The following observations and statistical detail are extracted from Wood's treatise on rail roads and interior communication in general, edited by Geo. W. Smith.

"The spirit of enterprise has been displayed," says Mr. Smith, "on a scale commensurate with the extensive territory of the U. S. With the exception of Great Britain and Holland, no country on the face of the globe contains so many or as extensive canals as this republic; and the whole of combined Europe has not effected as much during the last 16 years, as the three states of Pennsylvania, New York and Ohio only. The total number of miles of canals in the union is 2,596, including about 264 which are nearly finished, and which will be navigable during the ensuing spring, (1833.) Several extensive canals are in progress, and an immense number of projected or authorized works are not included in the summary just given. Nearly four-fifths of the aggregate amount have been executed in the three states above mentioned."

"The cost of the canals in the U. S., has been about 21,400 on an average, per mile. Although many expensive alterations have been made, a large additional sum will be requisite, for the purpose of completing these works in a permanent and suitable manner. The amount necessary for this purpose cannot be accurately estimated; but, if a judgment may be formed from the brief and limited experience of N. Y. and Pa., (where much expenditure will still be necessary,) the ultimate cost will probably be at least $28,500 per mile. The *navigable* canals of Pa. have already cost $25,185 per mile."

"The cheapest canal (probably in the Union) cost about $5,200 on an average per mile. The Chesapeake and Delaware canal cost nearly $169,000 per mile. The dimensions of this work permit the passage of coasters. It presents one of the cases where canals are decidedly superior to rail-roads—namely, for connecting by a short line an immense extent of navigable waters; although the tolls chargeable on every ton render the cost of transportation ten times greater than on a rail-road of similar extent, and constructed for perhaps one-tenth of the cost of the canal—nevertheless, the expense, delay, and inconvenience of transhipment give a preference to a work which permits of a continuous voyage. A rival rail-road, to connect the same points, has, however, even in this instance, been made, and with great advantage, for the rapid conveyance of light goods, passengers, &c., for which purposes canals are not adapted."

From the above stated cost of $169,000 per mile, the 14 ms. contained in the Chesapeake and Delaware canal, must have cost $2,366,000; the original estimate of the sum this canal would cost, was made in 1824, and stated at $1,129,036 73, or, too low by more than one half. The estimate is followed by the following sentence: "The adopted canal will be 60 feet wide at the water line, 36 at bottom, 8 feet deep, less than 14 ms. long, and lined with stone. $1,129,036 73, divided by 14, gives $80,645 48 cents per mile." The mistake in the estimates most probably saved the enterprise.

"In the U. S., the proprietors of the two thousand five hundred and twenty-five miles of canals, which are in operation or in progress, have not, *in any one solitary instance*, received from the tolls derived from these works the current interest of the country on the capital expended in their construction (including therein, as part of the real cost, *the arrears of unpaid interest on those portions of the capital which were temporarily dormant.*) The Erie and Champlain canals of N. York (now the most productive in the Union,) *have not in any one year, with one exception, paid the expenses of their repairs and management, and the current rate of interest on their actual cost,* although in other respects they have greatly increased the wealth and welfare of that populous state."

The total cost of the N. Y. canals, including the expense attending the repairs and alterations, has been nearly $12,000,000. The following table is an interesting document:

Tolls on the New York canals.

Erie and Champlain canal.

	1830.	1831.	Gain.
April,	$ 75,470	$116,300	$40,820
May,	166,140	213,311	47,171
June,	103,437	142,315	38,878
July,	84,102	106,858	22,057
August,	80,605	114,216	33,611
	$510,101	$693,100	$182,696

Oswego canal.

	1830.	1831.	Gain.
April,	$ 750 13	$1,180 20	$430 19
May,	2,058 95	2,829 06	770 11
June,	1,455 88	2,429 06	973 18
July,	1,238 10	1,790 38	552 28
August,	1,101 09	1,826 64	724 95
	$6,664 15	$10,054 83	$3,450 68

Cayuga and Seneca canal.

	1830.	1831.	Gain.
April,	$956 60	$1,214 19	$257 59
May,	1,905 79	2,663 42	757 63
June,	1,556 43	1,707 37	151 94
July,	1,095 10	1,164 59	69 49
August,	788 06	2,219 36	431 30
	$6,301 98	$7,968 93	$1,667 95

Total gain, ... $187,814 63

"Justice, however," continues Mr. Smith, "requires the remark, that many of the *American* canals have only recently been constructed, and, consequently, that the trade on them is not yet established to the extent which time will create: on a *few* the navigation has not yet commenced."

RAIL-ROADS.

The authorities consulted in the following notices of rail-roads, are chiefly "Smith's Wood," and the "Rail road Journal" of N. York. Those who would see more full accounts of different rail-roads, are referred to these and other and more extended works.

RAIL-ROADS, *completed, commenced, or incorporated.*

ALBION and TONNAWANDA, r-r. N. Y. A company has been incorporated to construct this road.

ALBANY AND SCHENECTADY, r-r. *(See Hudson and Mohawk r-r.)*

ALLEGHANY AND PORTAGE r-r. This is one of the links of the Pennsylvania chain of r-rs. and canals; it extends over the main Alleghany ridge of mtns., from Hollidaysburg on the Juniata r. to Johnstown on the Conemaugh, 36½ ms. It passes over the Alleghany mts. by means of 10 inclined plains, 5 on each side of the mt.; the estrn. slope from Hollidaysburg to the summit being 10 ms., and the wstrn. declivity 26½. It passes a part of the mt. by a tunnel 900 ft. long, 26 high, and 22 wide. There are 4 viaducts, (road ways) of masonry, containing 15,465 perches, estimated cost about $80,000, also a bridge for the passage of a t-pike, cost $1,284; 73 culverts, 11,775 ft., cost $37,000; cost of grading, exclusive of masonry, $499,300; estimate of $89,000 for engines and machinery. This road is not completed, but is in rapid progress, and will be in operation in 1833.

AMBOY r-r. *(See Camden and Amboy r-r.)*

AU SABLE AND LAKE CHAMPLAIN r-r. A company has been incorporated to construct this road from the forks of the great Au Sable r., along the valley of that stream to lake Champlain about 15 ms., with power to make branches to the iron mines of Pa. The object of this plan is to facilitate transportation between navigable water and the mines.

BALTIMORE AND OHIO r-r. The charter for this work was granted by the legislature of Maryland, Feb. 9th, 1827, and the work was commenced the 4th July, 1828. The original design was to unite the city of Baltimore with Ohio r. by a line of double track r-r; and to that effect, permission was obtained from the legislature of Pa. and Va.; but as the extension of the road beyond the point of rocks where it intersected Potomac r., has been prevented by a legal dispute with the Chesapeake and Ohio canal company, we confine our notice to that part either finished or in progress. It commences in the city of Baltimore, and extends to the Point of Rocks 69¼ ms.; with a branch road to Frederick, of 3 406-1000 ms., or 72⅓ ms. very nearly. The road-bed is 26 ft. wide. The line of the road is inflected very considerably along the vallies of the streams; and the road presents several rather abrupt curves. Of the whole distance, about 33 ms. are for the most part straight. Curves varying in radii from 955 ft. to infinity, occupy 3,963 feet; whilst 21 ms. have radii from 395, to 955. A single curve of 1,400 feet long has so small a radius as 318 feet; and another 1,100 feet, extends on a radius of 337 feet. The bridges and viaducts are numerous, and solid, but very expensive structures. The materials on which the rails are laid, are stone blocks and wooden sleepers. Forty ms. of single track, are composed of granite sills 8 inches thick, 15 wide, and of various lengths. These are laid in trenches, filled with broken stone. The estrn. section of 13 ms. was by far most difficult and expensive; costing for only graduation and masonry, above $46,354. The cost of graduation of these 13 ms., amounted to $8,994 more than did the residue of 54⅛ ms.; and on the first 8¾ ms. was expended in masonry, a sum equal to the cost of the remaining 58⅝ ms.; proportion 5⅓ to 1. The average cost of the road when completed, was estimated at $30,000 per mile. This road is in operation, and during the last very severe winter, kept the cost of fuel in Baltimore down to its ordinary price, about 100 per cent below what it was in Phila., New York, &c.

BALTIMORE AND SUSQUEHANNAH r-r. This line of road, designed to connect the city of Baltimore with York Haven on the Susquehannah, will be about 70 ms. in length, but as the whole line has not yet been fixed, the exact length cannot be accurately stated. It has to pass a summit of 1,000 feet. The first division of 6 8-10th ms., commencing at the depot in the city of Baltimore, and terminating on Jones' Falls cr., was completed and opened with one track, on the 4th July, 1831. The road-bed of 22 ft., will admit two tracks.

The second division, 6 3-10th ms., continuing from the first, up the valley of Jones' Falls to the mouth of Rowland's run; thence up the latter, and over the summit between the vallies of Jones' Falls cr. and Gunpowder r., to a point on York t-pike between the 12th and 13th mile stones from Baltimore, is nearly completed.

From the termination of the first division, on the right branch of Jones' Falls cr., the Westminster branch road leaves the main line, and follows the valley of Jones' Falls to its head, 8 ms., and terminates on Reisterstown road, near the 11th mile stone. This work was commenced in the autumn of 1830. It will extend to N. line of the state of Maryland, and thence to York Haven in Pa. A company chartered by the latter state will then continue it to the end; the estimated average expense per mile, is about $11,400; for the whole 21 1-10th ms., $240,000. Another section of the Westminster branch of this road is completed to "Owing's mill," and the cars have already commenced running to that place. The same is true of another division of this road, which extends to the York t-pike road.

BALTIMORE AND WASHINGTON r-r. This r-r. is another, and an important branch of the Baltimore and Ohio r-r., and has been commenced by the same company. The surveys are nearly or quite completed, and thus far are quite favorable. The r. will be 33 ms. long from Washington to Elkridge landing; its stock has been chiefly taken by the state of Maryland, and the Baltimore and Ohio r-r. company.

BLACK RIVER r-r. A company was incorporated by the legislature of N. Y. in 1832, to construct this r-r. from the Erie canal at Rome or Herkimer, to the r. St. Lawrence. Its capital $900,000.

BOSTON r-rs. There are now three r-rs. constructing from Boston in as many different directions; all of which will probably be greatly extended beyond the points at which for the present they will terminate. The road to Providence will undoubtedly be continued to Norwich or New London; that of Worcester to Albany, and that of Lowell to Vt., perhaps to Burlington, or from the opposite shore of the lake to Ogdensburg, N. Y. The Worcester road may possibly be connected with one from Norwich, Ct.; one to Hartford and New Haven, Ct.; and one to the N. W. parts of Mass.

BOSTON AND LAKE ONTARIO r-r. This proposed line embraces the Boston and Lowell r-r. That part of the road to the N. Hampshire line, via Lowell, is now in a vigorous train of execution; and in New Hampshire it is continued 15 ms. by the Port Kent and Au Sable r-r. A company, under the title of "the Boston and lake Ontario r-r. company," has been incorporated by the legislatures of Massachusetts, Vermont and New Hampshire, and it is expected will be by that of N. Y. at the ensuing session.

BOSTON AND LOWELL r-r. This work has already been commenced; it is to be constructed of the most durable materials, stone and iron, with a single track at present, and provision for the addition of another if expedient. It is to commence near Warren bridge, to cross Charles river by a viaduct, thence through Woburn, and terminate at the Merrimack canal at Lowell.

BOSTON AND PROVIDENCE r-r. Regarding this intended line, which is now in progress of location, the only authentic information we possess, is contained in a letter from the engineer engaged in its survey, by which it appears that the route of "the Boston and Providence r-r." developes greater facilities, to execution, than was anticipated. The road will be virtually (for the most part actually) straight; no curve being of necessity greater than of 6,000 feet radius; and under these circumstances, dispensing with the inclined plane which had been projected, the dividing ridge will be passed on an inclination well adapted to the use of locomotive engines. These improvements on the route will probably be effected at a cost considerably within that which was anticipated in the estimate. "The direction of the route is such, that while it will afford the shortest communication between Boston and Providence, it affords great facilities for a connexion also with Taunton, by a branch rail-way (diverging from the main line, say 23 ms. from Boston,) of but 11½ to 12 ms. in length; making the distance, therefore, from Boston to Taunton, but 35 ms.; or exceeding that by the t-pike, only 3 ms. From the public spirit of gentlemen in Stonington, New London and Norwich, the requisite funds have been raised, and surveys are now being prosecuted by officers of the army, who have been detailed to the service, with a view to ascertain the best route for continuing the r-r. (either thro' Providence or Worcester) from Boston to Long Island sound. Whether it should terminate at Stonington, or New London, as the navigation thence would be uninterrupted by the severity of winter, the completion of a r-r. to either place, would render travelling by means of steamboats and locomotives, at all seasons comfortable, cheap, and expeditious.

BOSTON AND WORCESTER r-r. The excavation for this road was commenced, August 1832, at Brighton and at Needham. The whole line from Brighton to Needham, a distance of 8 ms., is divided into 14 sections, including the passing of Charles r., and the high ground in Western, which constitutes the most difficult portion of the road between Boston and Worcester, is under contract on terms below the estimates, and is to be completed by May, 1833. The greatest supposed curve that will be necessary on any part of the road, will have a radius of 1,150 feet, and the greatest degree of inclination from a level will be at the rate of 30 feet in a mile. Few places will occur, where so short a turn, or so great an inclination will be necessary; while a large part of the route will be perfectly straight. The main street in Worcester is found to be 456 feet higher than Charles street in Boston. This elevation must of course be gained by the inclination of the road, making an average of 10½ feet per mile of the whole distance. On the line of road, as it has been located, the whole amount of ascent in proceeding from Boston to Worcester is 554 feet, being only 98 feet greater than the actual elevation of Worcester above Boston. The whole descent, therefore, in passing from Boston to Worcester, is only 98 feet, or an average of 2 3-10ths feet per mile. The length of the road as it is located, is 43¼ ms. This is about 2 ms. longer than a straight line between the points of termination, and about equal in distance to the road which is now most travelled between Boston and Worcester. The iron for the construction of the road will be admitted into the country free of duty; and it is stated that the work will probably be executed at a less expense than the sum estimated, and considerably below the capital of the company. It is thought also that nearly within the time in which a third part of the capital is expended, more than a quarter part, including the most productive part of the r-r., will be opened for use.

BROOKLYN AND JAMAICA r-r. This road is to be constructed on Long Island, to form a r-r. communication between Brooklyn and Jamaica; a company was incorporated for this purpose in 1832, with a capital of $300,000. (*See table.*)

BUFFALO AND ERIE r-r. A company was incorporated in 1832 by the legislature of New York, to construct this r. road, extending from Buffalo, to lake Erie, with a capital of $650,000. (*See table.*)

CAMDEN AND AMBOY rail-road, in New Jersey, commences on the Delaware r. at Camden, opposite Philadelphia, and extending 61 ms. terminates at Amboy, on Amboy bay. So direct is the line of this road that the actual distance between the extremes is not supposed to exceed 60 ms. The curves are few. The first division of 34½ ms. follows the left bank of Delaware r. from Camden to Bordentown; the ground plan nearly level, and few places having an inclination of 20 feet to the mile. From Bordentown to Amboy, the line is generally favorable, but there are some difficulties at Croswicks creek, at South r. and at the hill near Amboy; on the latter section the descent is 45 feet to the mile. The average descent from Bordentown to South Amboy is 27 feet per mile, with one deep cut of 2 ms. long and 60 feet depth in the deepest place. There is a scarcity of good stone, but the culverts and viaducts already constructed, are of that material. A hope is expressed in the official Reports of the company, that both divisions will be in operation in all 1832. This line was located by Major John Wilson, in 1830, and was immediately commenced. During the time embraced by the charter of this company, no other rail-roads will be allowed to be constructed on the route between N. Y. city and Philadelphia. By the terms of the charter, the completion of the road was limited to 9 years; the legislature of the state was permitted to subscribe for 1-4 of the stock, and to take the work after 30 years, on certain conditions.

Total cost of 61 miles double road, estimated at	$1,120,322 14
Real estate, purchase of,	115,792 84
Steamboats,	180,000 00
Locomotives and cars,	41,587 65
Wharves,	8,674 01
Entire estimated cost of the line,	$1,466,376 64

The legislature of New Jersey has authorised extensions of this line to New Brunswick, and to the Hudson r. opposite the city of N. Y. In speaking of this road, a writer remarks, that "in the year 1824, the construction of a r-r. from Boston to New Orleans was proposed. The project was then derided as visionary: nevertheless, in the few years which have elapsed, various unconnected companies have been formed, and a number of their works actually commenced, which, when completed, will constitute 13-17ths of this great line, the largest and most important in the world! The journey which now requires from 2 to 3 weeks, may then be performed in four days."

Cape Fear, and Yadkin r-r. *(See North Carolina Central rail-road.)*

Carbondale and Honesdale rail-road. In 1826 the legislature of Pa. granted a charter for this road, which was commenced in 1826, and completed in 1829. It is in fact a continuation of the canal line extending from Eddyville on the Hudson r. over a part of N. Y., N. J., and Pa., to Honesdale on the Lackawaxen r. It is 16 3-10ths ms. very nearly, and intended as a channel of general trade, but has been hitherto chiefly used in the transportation of coal. When the Lackawanna rail road is completed, the full benefits of this line will be experienced, and the amount of commercial business and travelling along this channel of intercommunication must be immense. The Carbondale and Honesdale rail-road, reaches the summit of Moosic mountain, 920 feet aggregate ascent above the mines, by 7 inclined planes, worked by stationary power, and thence descends to Honesdale 913 feet by 3 self acting machines or planes.

It is calculated that 460 tons of coals or other matter would be conveyed along this line daily, at an expense of $167 45. The average amount carried upon it, however, has been much less, and the total amount from the 20th of March to the 5th Nov. 1831, was 54,328 tons of coal, with a small additional amount of merchandize, say 55,000 tons aggregate amount. Thus in a period of 231 days, the average daily transportation was 238 tons, and a small fraction. Cost of this line, including machinery, wagons, &c. $310,852 21 cents, or a small fraction above $19,070 per mile. *(See Lackawanna rail road.)*

Catskill and Canajoharie r-r. This r-r. which is 75 ms. long, was commenced in 1831, near the Catskill end of it. When finished it will connect Canajoharie on the Mohawk river, with Catskill on the Hudson river.

Central r-r. This r-r. "extends from Pottsville, through the valley of the Shamokin creek to Sunbury, near the junction of the Susquehannah river, with its western branch." *(See Pennsylvania r. roads.)*

Central r-r., N. C. *(See N. C. rail-roads.)*

Charleston and Hamburg r-r. *(See S. C. r. roads, and the table.)*

Chesterfield r-r. takes its name from Chesterfield co. Va. within which it is formed, to connect the bituminous coal strata on James r. with tide water in the same stream, below Manchester and Richmond. It extends 13½ ms. in single track, with several turn outs, and 1½ mile branch roads to the different coal beds. This work was commenced January, 1830, and opened for use on the 1st of July, 1831, and what no canal ever did or perhaps ever will do, afforded a dividend of 10 per cent to the stockholders on the first 6 months. The cost was $8,000 per mile, and including their wagons, horses, &c. the whole disbursements of the company has been about $140,000, or $10,370 per mile.

Dansville and Rochester r-r. A company has been incorporated, a plan been formed, and surveys made preparotory to extending a r-r. from Dansville to Rochester, under the title of the "The Dansville and Rochester rail-road". Seventeen miles of the route have been critically examined, and it is believed that this portion of the road can be graded as cheap or cheaper than any road has been, since this species of improvement came into existence. The surface to be passed over is unusually level and favorable to the work. The first four miles abound in quarries of fine stone, suitable for building culvert walls and covering for the same, and for other purposes requiring the use of this material. Should the remainder of the route prove as favorable as that already passed over, the greatest rise or fall in any mile of the whole distance, will not exceed 8 feet; nor will the road vary far from a direct course.

Danville and Pottsville rail-road. This rather circuitous but highly important line is really a continuation of Mount Carbon rail-roads and of the Schuylkill navigation. The charter was granted to a company by the legislature of Pa. April, 1826. It is made as a public high way, and calculated to open a cheap and expeditious channel of communication between the Schuylkill valley and that of Susquehannah near the junction of the two main branches of the latter. In order to render the description of the whole line more perspicuous, it is necessary to commence with the Mount Carbon road. This latter line was commenced in 1829, and completed in 1831, with a main line and two branches, amounting to an aggregate length of 7 427-000 ms.; at an expense of 118,000 dollars; or the mean expense per m. of $15,888. It begins at the lower landing of Mount Carbon on the Schuylkill canal, about 106 ms. northwestward Philadelphia; and passing through the town of Pottsville, and thence up the Norwegian cr. a small fraction above 1 48-100. A branch of this road extends up the main fork of the creek 1 7-10 ms. and another branch along the west fork within a small fraction of 3 ms. Both branches and the main line are mostly extended in double tracks. From the branch of the Mount Carbon rail-road on the eastern fork of Norwegian cr. extends the central rail-road or the road from Pottsville to Danville on the Susquehannah, by Sunbury. Danville and Pottsville rail-road was chartered by the legislature of Pa. in 1826, but subsequently merged into the Mount Carbon rail-road company. The former leaves the latter road on the eastern, Norwegian at an elevation above Sunbury of 330 feet, and 2 1-2 ms. from Pottsville by a deep cut and tunnel of 1,400 feet, which leads into Mill creek, along the valley of which it is carried to the summit of Broad mtn. 1,040 feet above Sunbury. The height is reached by 4 inclined planes, and the opposite side of the mtn. is descended by a single plane of 400 feet perpendicular elevation. The next stage of 2 1-4 ms. is level. The sixth inclined plane descends to a level of about 4 ms. The line thence ascends to the summit level between the Mahonoy and Shamokin creeks, by the 7th inclined plane, ascending at the rate of from 10 to 30 feet per mile, and descends to Sunbury by two inclined planes. The stock has been subscribed to a sufficient amount to prosecute the work. The entire length of the main line is 47 ms. 174 poles, and the Danville branch 7 miles, the whole 54 54-100 ms. Three ms. comprising the main line and nearly all of the east branch, is finished, and an additional 8 ms. will probably be finished by 1833. The remaining 36 54-100 ms. and the Danville branch of 7 miles, have been located. "The estimated cost of the line from Sunbury to the junction with the Mount Carbon rail-road, is (for the road graded for a double track, and including the present execution of a single track and turn outs,) 675,500 dollars, and $3000 per m. subsequently adding the remainder of the second track. (*See Mount Carbon rail-road.*) *Mill Creek* rail-road is connected with the two preceding, and was the first road of the kind formed in the Upper Schuylkill valley. It is a single track line of 6 turn outs, main line 4 ms. from Mine Hill to Port Carbon, and branches, 9 in number, extend to an aggregate of about 5 miles. Cost 2,500 dollars per mile, or 22,500 dollars. *Mine Hill, and Schuylkill Haven* rail-road is not yet connected with the Central or Danville and Pottsville rail-road; yet as such union is in comtemplation, and as both these roads are in the same vicinity we unite them in one general view. The main line of the Mine Hill and Schulkill Haven rail-road commences at Schuylkill Haven, and stretching along the West branch of Schuylkill r. 10 1-2 ms. passes the Mine Hill gap. At the fork or where the W. W. branch leaves the W. branch, an arm of the rail-road extends along the former 3 1-2 ms. of a double, and 1 m. of single track; making in all 14 ms. of a double, & 1 of a single track road. The com. have disbursed for all expenses on this road a sum of 181,615 dollars, or 12,107 66 per mile. From the preceding accounts we discover that the Schuylkill navigation in its Upper valley is connected with three systems or lines of rail-roads. There are also in the same region several miles of rail-road not included in the above, but which were constructed on private property by individuals. These immense works, in a period comparatively short, have changed regions, once barren, wild, and desolate, into the busy residence of several thousands.

Summary of the Schuylkill rail-roads, noticed under this head.

Danville and Pottsville, 7 427-1000 miles finished; expense	$118,000
Mill Creek, 9 ms. finished; do.	22,500
Mine Hill and Schuylkill Haven, 15 ms. finished; expense	181,615
Private roads, say 5 ms.; expense	25,000
Amount,	$347,115

Detroit and Pontiac r-r. A company has been incorporated and the surveys made for a rail road between Pontiac and Detroit. The length of the road when completed will be 25 ms.

Dutchess County r-r. A company has been incorporated to construct a r-r. from Poughkeepsie Dutchess co. N. Y. to the Connecticut line. The road will be from 20 to 30 ms. in length. Capital of the company, 600,000 dollars.

Elizabeth-town and Somerville r-r. This road which has been surveyed is soon to be commenced (1832), will extend from Somerville to Elizabeth-town. The company was incorporated in 1831, by the New Jersey legislature, with a capital of 200,000 dollars, and liberty to increase it to 400,000.

Elmira and Williamsport r-r. A company was incorporated by the legislature of N. York in 1832, to construct this rail road; its capital 75,000 dollars.

Experiment r-r. (*See North Carolina r. roads.*)

Fayetteville r-r. This road when completed will extend from Campbeltown on the Cape Fear r. to Fayetteville. The company was incorporated in 1830; its capital $20,000.

Germantown r-r. (*See Philadelphia, Germantown and Norristown r-r.*)

Haarlem r-r. This r-r. is entirely within the city of New York, if we regard that city as commensurate with Manhattan island. When finished it will be about 6 ms. in length, one mile of which is now completed, and in operation. The grading of the other parts of the road is rapidly progressing, and will be ready for the rails in 1833. The contemplated New York and Albany r-r. will probably commence at the N. extremity of this road, so that it may be considered as the first link in the grand chain of r-roads, which shall yet connect the city of New York with "the West."

HUDSON AND BERKSHIRE r-r. The legislature of N. Y., in 1832, incorporated a company, with a capital of $350,000, to construct a r-r. to the Massachusetts line, to meet a r-r. authorized by the government of that state.

HUDSON AND MOHAWK r-r. This, which in length is 15 8625-10,000 ms. is a very important r-r., connecting Albany and Schenectady. It was commenced under a charter from the legislature of New York, the 12th of Aug. 1830, near Schenectady. It is calculated for double tracks, one of which is completed and in operation, and the second in progress. The summit is 335 feet above the level of tide water in the Hudson. This is a dead level of 14 ms. in length. At each end of the road there is a stationary engine of 12 horse power, to overcome, by inclined planes, a rise of about 120 feet. Except in one place where there is a cut of 47 feet for a few hundred yards, the road has been easily graded—the road is nearly straight the whole distance—the only deviation from a straight line is 3 or 4 miles from the western end of the road, where the radius of curvature is large—from this place the line is visible the whole way, and the mountains on the east side of the river seen through the vista; the rails are of pine, with a flat bar of iron for the wheels to move on—the work appears well done, and the only objection to it is, the material of which it is constructed. A very heavy locomotive, imported from Europe, was found by its weight, 12,742 lbs., to injure the road; but another locomotive, also, but weighing only 6,758½ lbs., made at West Point, is in use. The mean rate of motion on this road with a load of 8 tons, is 15 ms. hourly.

Expenditure already made on this road, ...	$483,215
Do. necessary to complete the double tracks .	156,693
Amount of expenditure . .	$639,908
Expenditure per m. when finished,	$40,340

Though the expenditures of the Hudson and Mohawk rail-road have been great, still it is probable that the rail-road will not cost one-third as much as the canal which connects the same points. (*See Saratoga and Schenectady rail-road.*) The number of passengers who passed over this road in October, 1831, averaged 387 per day. The company were authorized, in 1832, to construct a branch rail-road from the line of their present rail-road, at or near its intersection with the great western turnpike, to the capitol square in the city of Albany, and from thence, or from some point between the said place of intersection and the Capital square, to the Albany basin; and to transport, take and carry property and persons on the same.

ILLINOIS AND MICHIGAN r-r. This road, which, when completed, will be 96½ ms. in length, is to commence at Chicago on lake Michigan, and after running in a S. W. direction along the valley of the river Des Plaines, to terminate at the Illinois rapids. The summit level will be less than 200 feet above the lowest part of the road.

ITHICA AND GENEVA r-r. The company for the construction of a rail-road between these two towns was incorporated in 1832 by the New York legislature; capital $800,000.

ITHICA AND CATSKILL r-r. The whole length of this road when completed between the two places will be about 167 ms.

ITHACA AND OWEGO r-r. This line of 29½ ms. is intended to connect the village of Ithaca in Tompkins co. with Owego in Tioga co. N. Y. The direction S. S. E. It is the first rail-road line actually commenced which will unite the basins of Chesapeake and St. Lawrence. It was commenced in 1832, but as little advance, and no details have reached us, we can only state, that application has been made by two companies to the New York legislature for permission to extend this road in one direction to Hudson r., and in the other to the head of Seneca lake.

KNOXVILLE AND SOUTHERN r-r. company. (*See North Carolina Central rail-road.*)

LACKAWANNA AND SUSQUEHANNAH r-r. This line is intended to extend from Carbondale coal-mines down the Tunkhannoc valley to the Susquehannah river, and will be a continuation of the Carbondale and Honesdale r-r. The Lackawaxna r-r. was authorised by the legislature of Pennsylvania, the 7th of April, 1826, and by charter required to be a public high way for the conveyance of persons, produce and merchandize. (*See Carbondale and Honesdale rail-road.*)

LAKE CHAMPLAIN AND OGDENSBURG r-r. A company was incorporated in 1832, by the New York legislature, to construct a rail-road between Ogdensburg on the St. Lawrence, St. Lawrence co. and lake Champlain, with a capital of $3,000,000.

LAKE PONTCHARTRAIN r-r. This rail-road, which is about 4½ ms. in length, and consists of a single track, extends from lake Ponchartrain to New Orleans. The company was incorporated in 1830, the road opened in 1831. Whole cost of construction about $70,000.

LEXINGTON AND OHIO r-r. This road is designed to extend from the town of Lexington, in a direction a little N. of W. through Frankfort, Shelbyville, and some other intermediate places, to Louisville. The length will be, when it is completed, somewhere between 75 and 80 ms. "About 7 ms. of the road have been placed under contract, and the grading of them finished. This division of the road is now completed, and an elegant carriage, sufficiently large to accommodate

60 persons, finds constant employment in the conveyance of passengers upon the first two miles of it; and a locomotive steam-engine now constructing, will be placed on the remainder of the first section of the road.

LITTLE SCHUYLKILL r-r. The Little Schuylkill, or the Tamaqua, is the most northern branch of that river, heading with the Nesquehoning, Quakake, and Mauch Chunk crs. of the Lehigh. From its higher fountains it flows southwardly into the main Schuylkill, which it enters at Port Clinton above the Lehigh Water gap. The Little Schuylkill r-r. commences at Port Clinton and mouth of Tamaqua, following the valley of the latter stream 21½ ms., and 1¼ ms. above the town of Tamaqua. A branch leaves the main line of 1 m. from Tamaqua to other mines. The road is graded for double tracks, and a single track has been constructed throughout. The company are authorized to continue this railroad to Reading from Port Clinton in one direction, and to the foot of Broad mtn. in the other. Another company is empowered to extend it to Catawissa on Susquehannah r. 57½ ms. from Port Clinton. Upon these extensions nothing except surveys has yet been executed.

Of this road nearly 23 ms. were ready for use in 14 months from commencement of the work, and cost,

For grading,	$112,572
Bridges and culverts,	21,594
Superstructure,	70,290
Engineering department,	21,099
Amount of actual expenditure,	$225,555
To complete the whole road second track, expenditure supposed necessary, $2,500 per mile,	$57,500
Total amount to complete 23 ms. nearly	$283,055

According to the preceding estimates this line when completed will have cost per mile about $12,306.

LYKIN'S VALLEY r-r., which was commenced in 1831, and expected to be completed in 1832, in a single track extends from a coal basin of Broad mtn, through Bear Creek gap, down the Wiconisco valley, north side of Berry's mtn., to Millersburg on the Susquehannah. Length 16½ ms. This line extends along the N. side of Dauphin co., Pa., whilst the several roads we have been describing as in the Schuylkill valley, are in Schuylkill co.

MAD RIVER AND ERIE r-r. This road when completed will extend from Dayton at the head of Miami cr., in a N. N. E. direction, to Sandusky; and its length will probably be about 140 ms., stretching along the vallies of Mad r., part of Sciota, and thence down that of Sandusky, to its point of nrthrn. termination. The amount of stock desired by the company, has been subscribed, and the first instalment of 10 per cent, paid in. Arrangements have been made for the immediate survey of the route. This is the commencement of a system of r-rs. in the states of Ohio, Indiana, Illinois, Michigan, and we may say, Missouri.

MANCHESTER r-r. This r-r. which is in Chesterfield co., Va., extends from Manchester to the coal mines, about 13 ms. distant. (*See Chesterfield r-r.*)

MAUCH CHUNK r-r. This was one of the first attempts made in the U. S. to introduce the r-r. system. In construction, it shares the imperfections of first efforts, but in point of profit to the company which constructed it, it has been highly successful. The Mauch Chunk r-r. was commenced in the winter of 1826-7, and brought into use in the latter year. Main line 9 ms.; branches 3¾, or near 13 ms. in all. The main line rises from the mine 100 feet in ¾ths of a mile, or 133⅓ feet in a mile. This steep plane is ascended by horse power. Thence in 8 ms. the road descends a plane of 745 feet perpendicular height, which brings the line to the head of a very steep inclined plane, 215 feet perpendicular elevation, on a descent of 745 feet to the Lehigh. *Room Run and Mauch Chunk r-r.* is in fact an arm of the Mauch Chunk r-r., and extends about 5¼ ms. from the coal mines on Room run to the depot at Mauch Chunk. The principle is that of an inclined plane, down which the loaded wagons and mules are to be carried by the power of gravity: the mules drawing the empty wagons back to the mines. The total cost including machinery $76,111.

MILL CREEK r-r. (*See Danville and Pottsville r-r.*)

MINE HILL AND SCHUYLKILL HAVEN r-r. (*See Danville and Pottsville r-r.*)

MORRIS CANAL AND PATTERSON r-r. This work has been authorized by the legistaure of New Jersey; how far the company which was incorporated for its construction have made preparations for the work is not accurately known.

MOUNT CARBON r-r. (*See Danville and Pottsville r-r.*)

NEW CASTLE AND FRENCHTOWN r-r. This line of 16 46-100 ms. reaches from the centre of Front street in New Castle, to a wharf on Elk r. at Frenchtown. A direct line connecting the extremes, measures 15 97-100 ms. the road not being half a mile longer than its chord. In 1827 the charter was obtained, and in August, 1830, the work was commenced. The very successful results are shown below. The road is composed of 6 curves and 6 straight lines; of which the curves occupy 5 16-100 ms.; and the straight lines 11 3-10 ms. The radius of the least curve is 10,560 feet, or 2 ms.; radius of greatest curve 20,000 feet. Road bed, 26 feet, exclusive of side drains. There are 4 bridges or viaducts, and 29 culverts of stone masonry.

	Cost.
Land,	$14,966
Wharves at New Castle and Frenchtown,	10,722
Graduation and drains for double track,	193,215
Culverts and bridges,	22,090
Materials, and laying single track and turnouts,	98,046
Fences and gates,	10,661

Engineering department,	16,784
Sundries,	10,000
Expenditures incurred,	376,484
Estimated cost of second track,	92,046
Do. locomotives, wagons, &c.	40,000
Entire cost when fully completed and supplied with machinery,	$408,530

From the main line of the New Castle and Frenchtown r-r., there is a small branch of about 800 feet, which, added to 16 46-100, gives 16 61-100 as the entire length of the road, which gives about $24,595, as the mean cost per mile. The whole of this road is now in operation, with a single track, and from experiments made with the steam cars by the chief engineer, the most sanguine expectations as to the success of the road are likely to be realized.

New Jersey, Hudson and Delaware river r-r. A company has been incorporated to construct a r-r. under this title, from the Hudson to the Delaware r.

New York and Albany r-r. A company has been incorporated by the legislature of N. Y., to construct a r-r. between these two cities, on the e. side of Hudson r. The state will be at liberty to take the road at any time between 10 and 15 years after its completion, on paying the cost and 14 per ct. interest. Branches may be constructed by the com. to connect with r-rs. made in Mass. or Conn., but no authority is given to communicate with Hudson r. along the route. The length will be about 160 ms. The capital of the company is $2,000,000.

New York and Erie r-r. This line, the most extended ever actually planned in the U. S., was projected as a continuous road from the Hudson r. opposite the city of New York, to some point on lake Erie. A company was incorporated in 1832, with a capital of $6,000,000, to construct the work, and the surveys were to have been made in the same year. They were suspended, however, in consequence of the failure of congress to lend efficient aid, by appropriating to the purpose, an amount considered adequate to the object. It was proposed to commence at Tappan, or at a point above, opposite the mouth of Croton r. From this point it would pass the valley of Ramapo r., to the head waters of Walkill cr., by the Shawangunk mts., &c., wstrd. The length will be about 400 ms. This road will open an uninterrupted communication, throughout the year, between lake Erie and the ocean.

Norristown r-r. (*See Phila., Germantown and Norristown r-r.*)

North Carolina Central r-r. Under this head we shall enter into some detail for the sake of showing the extent that the r-r. interest has gained in the southern states; and also because the r-r. system is admirably adapted to the localities and climate of the Carolinas, Geo., Tennessee, and the adjacent states. Rail-road meetings have been held at several places in the southern states; enquiries made respecting the best routes of land communication between navigable waters, and the southern Atlantic seaports.—Surveys have also been made of the Tenn. and Savannah rs., and information derived from other sources. A competent engineer will probably soon examine several of the routes which have been proposed. The Charleston and Hamburg r-r., the completion of which will essentially aid the projected channel of communication, is nearly or quite finished (1832). The Fayetteville r-r. extending to the western part of the state, is in contemplation, and a company for its construction is chartered. A central r-r. (the title standing at the head of this article) has been proposed, which shall extend from Beaufort, via Raleigh and Salisbury, to the wstrn. part of the state. The company has been incorporated by the state legislature. Tenn. has not been inattentive to her interest in these grand enterprizes; and at the last session, her legislature incorporated the Knoxville and Southern r-r. company. (*See Art.*) This review, though brief, justifies the conclusion that the several communities interested in the undertaking are aware of its great importance and value to all; and if they but observe a proper concert of action, its accomplishment can no longer be deemed problematical. A meeting of delegates from S. Carolina, N. Carolina and Tennessee, has been proposed, to be held at Ashville, to take the subject into more deliberate consideration. The citizens of N. and S. Carolina have been requested to send delegates to the convention, which was fixed for the first Monday in Sept., 1832. The citizens of N. C. seem fully prepared to second the views of the friends of r-r. improvement, and with a view of promoting the success of the Central, by a practical demonstration of the great advantages attendant upon that mode of transportation, a company has been recently organized in Raleigh for the purpose of constructing in the immediate vicinity of that city, an *experimental* r-r. It is to be about 1 mile in length, and in every respect will be a complete model. Nearly the whole amount required for its completion, has been subscribed, and no doubt is entertained of its successful prosecution. Proposals for grading the line of the road, for furnishing materials, &c., have been advertised for in a N. C. paper.

Norwich and Boston r-r. The legislature of Connecticut have chartered the Quinnebaug bank at Norwich, to aid in the construction of a r-r. from that city to Providence or Worcester, to meet the r-rs. which are to be from between those places and Boston, and have also incorporated a company to perform the task. The capital of the bank is $500,000; and that of the r-r. company 1,000,000. The r-r. company may extend their road to steam navigation on L. Island sound, either at New London, Lyme, or N. Haven. (*See Boston r-rs.*)

OTSEGO r.-r. A r.-r. company was incorporated by the N. Y. legislature in 1832, to construct a work of this kind from Cooperstown to Collierville, with a capital of 200,000 dollars.

PATTERSON AND HUDSON RIVER r.-r. This r.-r. is designed to extend from Patterson in New Jersey, to the Hudson r. at Hoboken, opposite the city of New York, a distance of 14 ms. About 7 ms. is partially, and 4½ entirely finished. The part completed extends from Patterson to the village of Aquackanonk, and is now in actual and successful operation between those places. The company have placed upon the road three splendid and commodious cars, each of which will accommodate 20 passengers inside, and from 6 to 12 on the top, and may be drawn by 1 horse, at the rate of a mile in 3 minutes. There is a gradual ascent from Aquackanonk, or the landing, for about 3 ms.; during which the road passes over an embankment, and through a cutting in rocks from 10 to 20 feet deep, for about 150 yds. The summit level extends about ½ of a mile, and thence to Patterson, there is a descent of about 21 ft. per mile.

PENNSYLVANIA r.-r. By a very culpable confusion of names, this term includes a r.-r. of 81 6-10 ms. from Philadelphia to the Susquehannah, and another of 36 69-100 ms. over the Alleghany mtn., separated by r. and canal navigation of 171 ms. The part over the Alleghany mtn. we have already noticed under the head of "Alleghany Portage r.-r." (*which see.*) The estrn. division of the Pennsylvania r.-r., called the ***Philadelphia and Columbia*** r.-r., as indeed the wstrn. division and intermediate canal work, were undertaken in virtue of numerous acts of the legislature of Pa., from 1811 to the 24th of March 1828. It was at the latter date, that the Pa. r.-r., including both sections, was authorized as a state work, so that this *r.-r. is, therefore, the first which was undertaken in any part of the world by a government.* The ***Philadelphia and Columbia*** r.-r. commences in the city of Philadelphia, at the corner of Broad and Vine streets, from whence branches, constructed by the different corporations of the city and continguous places, will diverge, and terminate at the necessary points. The main road leaves the city and vicinity by a line inflected by curves, and straight lines, and thence to a viaduct of 984 feet over the r. Schuylkill below Peter's island. After passing the r., the road in a distance of 2745½ feet ascends an inclined plain of 187 2-10 feet perpendicular height. It thence continues by Downingstown, Coatsville, and Lancaster, to Columbia on the Susquehannah, 81 6-10. On the line there are 31 viaducts, 73 stone culverts, and nearly 500 stone drains. There are 18 common road and farm bridges. The whole road formation is finished, with the exception of 2 viaducts and the deep cut through Mine hill, which is nearly completed. The rails are laid, and travelling commenced on some sections. The country traversed by this road is very uneven, and presented great obstacles to the line being drawn direct, yet the actual length of the r.-r., exceeds but a few ms. that of the common t-pike, between the same points, and is not one half the length of the Schuylkill, Union canal, and Susquehannah water navigation between the same points. If the profile be analyzed, it will be perceived that 71 per cent. of the useful effect will be obtained on this road, which would be attainable on a line perfectly level. The estimated expense of this great line, allowing a mean of 20,000 dolls. per mile, including all expenses to complete double tracks, with their appropriate machinery, wagons, cars, and other contingencies, will be about 1,632,000 dolls. ***Westchester*** r.-r., is a branch of that of Phila. and Columbia, leaving the latter about 2 ms. w. of the Paoli tavern, and follows the general direction of the ridge 9 ms. to the town of Westchester. The road formation is 25 feet wide, and designed ultimately for a double track. The entire road, *single* track, is expected to be in full operation this season. Total cost supposed $81,000, or $9,000 per mile. There are three companies formed to extend branches from the Phila. and Columbia r.-r. One company to construct a branch from the main line near Downingstown to the city of Wilmington; a second to carry a branch via Oxford to Port Deposit on the Susquehannah r.; and a third in Maryland to extend the latter to Baltimore. Neither of those three branches have been commenced. The ***Philadelphia, Germantown and Norristown*** r.-r., as far as executed, now is, and if completed, will be, in fact, a link in the chain of which the Philadelphia and Columbia r.-r. constitutes the main line. The junction of these roads is contemplated; hitherto, however, they are separate. Six ms. of the Phila., Germantown, and Norristown road were located in 1831, and immediately placed under contract. This part commences in the incorporated limits of Spring Garden at the intersection of Ninth street and Spring Garden, and terminates at Welley's factory or Church lane. The total length as originally designed, is 18 7-10 ms.; but little progress has as yet been made, even on the section actually commenced.

PETERSBURG AND ROANOKE r.-r. This very important road commences at Petersburg in Va., and extends 60 ms. a little w. of s. to Weldon in N. C., and to the foot of the falls in Roanoke r. The line is very direct; graduation in no place exceeding 30 feet per mile; and the curves having radii from 2 to 4 ms. The direction is almost at right angles to the ordinary course of the great roads, and in the line of sthrn. travelling must receive great emolument from the transportation of persons. Norfolk has been hitherto regarded as in some measure the depot to the Roanoke valley, an advantage which the road will divert in great part to Petersburg. The work of road formation was begun on this line in 1831.

Estimated cost, when completed with double tracks, 400,000 dollars. It appears "that about 20 ms. commencing at the corporation line, and extending beyond Stony cr., is entirely completed and ready for use. From Stony cr. to Meherrin r. (about 25 ms.), the road has been graded: on the first 10 ms. of which the wooden rails have been laid, and the contractors are engaged in laying down the iron. From the Meherrin to the Roanoke the road is under contract, with the exception of about 2½ ms. The (Petersburg) section, commencing at the depot, at the corner of Union and Washington streets, and connecting with that portion already completed, is also under contract, and about 200 hands actively engaged upon it. From present appearances, it is probable, that before the close of 1833, the entire line will be completed, and the enterprise of the company rewarded, by seeing Petersburg becoming the mart for the rich products of the country bordering on the Roanoke. A locomotive engine, called "The Roanoke," with a tender and wagon, has been imported for this company. Previous to being shipped, the locomotive engine underwent a trial on the Liverpool and Manchester r-r, and gave entire satisfaction, both as to speed and construction. The locomotive and two passenger cars are now, it is believed, on the road. The iron work for about 15 more passenger cars, has also been received. We may safely pronounce the Petersburg and Roanoke r-r. as amongst the great works of our country, the success of which is now placed beyond doubt.

Philadelphia, Germantown and Norristown. (*See Philadelphia and Columbia r-r. under the head of Pennsylvania r-r.*)

Philipsburg and Juniata r-r. A company was incorporated in 1830 by the Pennsylvania legislature, to construct this rail-road from the Pennsylvania canal near the mouth of the Little Juniata r. to the coal-mines near Philipsburg.

Pine Grove r-r. This road, which is about 5 ms. in length, extends from the coal-mines to the Swatara feeder; cost $30,000.

Quincy r-r. This road, extending in a single track, from the granite quarries in the town of Quincy, terminates at Neponset r. which discharges itself into Boston harbor. It is 3 ms. long, the base of the rails is wood, surmounted with plates of wrought iron, on which the cars traverse. It is used principally for the transportation of granite, and was the first experiment of rail-roads in the U. S., having gone into operation in 1827.

Rensselaer and Saratoga r-r. A company was incorporated in 1832, to construct this work, with a capital of $300,000.

Rochester r-r. completed in 1832, extends from the Erie canal at Rochester to the head of navigation in Genesee r. below the falls. It crosses Main street in Rochester, and terminates at the end of the aqueduct, near Ely's mill. (*See Dansville and Rochester rail-road.*)

Room Run and Mauch Chunk. (*See Mauch Chunk rail-road.*)

Saratoga and Fort Edward r-r. A company was incorporated in 1832, for the construction of this rail-road, with a capital of 200,000 dollars.

Saratoga and Schenectady r-r. (*See Hudson and Mohawk rail-road.*) The rail-road from Albany to Schenectady, called the Hudson and Mohawk r-r., is continued by "the Saratoga and Schenectady road, 21 miles in length. The road was opened in July 1832, and though, owing to the cholera, there was a general suspension of travel in the country, and not more than an eighth or tenth the usual number of visitants at the Saratoga springs, still the receipts on the road have much exceeded what was anticipated when it was opened. They have thus far exceeded $75 per day." This r. "will prove one of the most lucrative investments in the state of N.Y." By reference to the article Hudson and Mohawk r-r., and connecting that line with that of the Saratoga and Schenectady r-r., the reader will perceive that rail-road lines extend from Albany to the Saratoga springs; and farther, that when the Saratoga and Fort Edward r-r. is brought into operation, that the Alpine scenery along the upper Hudson will be rendered cheaply and delightfully accessible to the visitants to Ballstown and Saratoga.

Schoharie and Otsego r-r. A company was incorporated in 1832 to construct a rail-road from the Catskill and Canajoharie rail-road, via the Cobleskill and Schanevas crs. to the Susquehannah r., with a capital of 300,000 dollars.

Schuylkill r-rs. Under the heads of Danville and Pottsville and Little Schuylkill r-rs., we have noticed several of the rail-roads of this system, but there still remains the

Schuylkill Valley r-r. Though in the neighborhood of Mount Carbon, and of Danville and Pottsville r-rs. that of Schuylkill valley is unconnected with either. It commences on the Schuylkill r. and head of the Schuylkill canal, at the mouth of Mill creek, about 2 ms. above Pottsville, and extends northeastward up the valley 10 ms. to the town of Tuscarora. It was commenced, 1828, and completed in 1830. The number of branches about 20, extending in the aggregate 12 ms. From the town of Tuscarora, a branch is constructing, with a tunnel, to Cold run, and which is intended to be connected with the Little Schuylkill r-r. On the Schuylkill valley r-r. and branches, including all incidental expences, about 60,000 dollars have been laid out; and when the branch through the intermediate mountain into the valley of Little Schuylkill is completed, the expenditure it is probable will exceed $100,000.

Summary of expenditure on the rail-roads in the valley of Schuylkill r. above Schuylkill Water-gap, and in Schuylkill co. Pa.

Amount of summary under the head of Danville and Pottsville, brought forward,	$347,115

Little Schuylkill r-r. and branches,	283,037 11
Schuylkill valley r-r. and branches,	60,000
Amount expended,	$690,172 11

This great work is going on, and long before all the main lines and branches are complete with double tracks, the expenditures will no doubt far exceed a million of dollars.

SOUTH CAROLINA r-r. All things considered, this is a most important work, both commercially and politically. By its successful execution and beneficial results, it must have a powerful tendency to introduce similar works into a section of the U. S. adapted to their construction, and the inhabitants of which are in the rear of their northern neighbors in road improvement. The South Carolina r-r. extends from the city of Charleston to Hamburg on Savannah r., opposite to Augusta in Georgia. The direction is N. W. by W., main line 135¼ ms. The summit of the ridge, or rather table land, between the Edisto and Savannah rs. 114 ms. from Charleston, is passed by a stationary engine, the only one on the line. The direction is generally straight, and the curves where they occur have large radii. This crosses a great variety of different soils. Over some marshes the road is based on piles. A car has been constructed on the part finished, with a view of transporting horses, cattle, and stock on the rail-road to and from the country. The steam cars travel daily, twice regularly, and an extra trip if passengers offer, to Somerville, 21½ ms.; beyond which the work is progressing. The mile beyond Somerville is nearly completed, and ready for travelling, and the next m. is now capping and railing. Beyond that, the succeeding mile is all capped, and about half the rails on, with all the timber ready; and the next two miles are wholly finished. The distance thence to the Cypress swamp 1½ ms. being mostly on sleepers, has all the ground sills and cross pieces down, and but 5 days' work of piling to join the Cypress contract; which, however formidable it has hitherto appeared, is now piled throughout, and the capping and railing going on briskly. The next 3½ ms. is finished. The Four Hole Swamp is piled through, and the remaining work going on rapidly; 11 miles thence upwards are finished and ironed; and the road as far as to the Edisto is now completed. On all the contracts, the hands as they finish below, are sent up, so as to expedite the work. The whole is under contract to persons belonging to the state, and mostly residing on the line, employing a force of near 600 hands, independent of horses employed by the contractors. The bridge across the Edisto, which is 65 ms. from Charleston, has all the abutments piled. It is to be 60 feet span, and supported by one arch, the carpenter's work of which is now going on. The first 4 ms. beyond the Edisto are now ready for the iron; and the 10 ms. in succession thence are rapidly progressing. The construction of the unfinished part of this road (about 50 ms.), is also rapidly advancing; and from the perusal of several reports of recent date, there is a very great probability of the whole being completed by the first of January, 1833. On the 35 ms. nearest Augusta, a force of more than 500 men were employed in the summer of 1832. On the other 35 ms. 400 men were employed. "The work of the inclined plane will all be so far completed by the 1st of January, 1833, as to be ready for the machinery, which is now in progress for construction. A large number of axles, made of faggotted iron, have been transported from New York, together with wheels; and there are many more in preparation. The receipts from passengers, several weeks in 1832, averaged 200 dollars per week, independent of the conveyance of iron and other materials for the use of the company. The history of this splendid work is short, but interesting. The charter was granted 1828, the work commenced in the autumn of 1830, and in all probability will be completed by January, 1833; and if so soon finished, will be then the longest continuous iron rail-road ever constructed.

	Cost.
Workmanship, materials, Edisto bridge, &c.	$393,377
Iron,	133,800
Spikes,	12,500
Piling machinery,	3,700
Turn outs, the other parts of the road being a single track,	5,000
Inclined plane and double road,	6,000
Engineering department,	45,623
Contingencies, damages, &c.	10,000
Extra, for stationary engine, 6 locomotives, 160 wagons, and water stations,	61,000
Total cost when in full operation with all its machinery, &c.	$671,000

This amount gives an expense per mile of 4,952 dollars, comparatively moderate for a rail-road with even a single track.

TONAWANDA r-r. N. Y. A company has been incorporated to construct a r-r. from Rochester to Utica, under this name. The capital is 500,000 dollars.

TUSCUMBIA, r-r. This r-r. consisting of a single track, was constructed in order to avoid the Muscle shoals, &c. It extends from Decatur to Tuscumbia, at a cost of 3,500 per mile.

UTICA AND SUSQUEHANNAH r-r. The legislature of N. York, in 1832, incorporated a company for the construction of a r-r. from Utica along the valleys of the Susquehannah and Unadilla rs. to the line of the projected New York and Erie r-r. The capital is 1,000,000 dollars.

WARREN COUNTY r-r. A company has been incorporated by the New York legislature to form a r-r. in Warren co. from Glenn's Falls to Caldwell, at the south end of lake George. The N. portion of the line of communication between that lake and New York city, through Albany, to which city the distance will be 64 1-2 ms. Capital 250,000 dollars.

Watertown and Rome r-r. The construction of a r-r. between these two places was authorized in 1832, when the N. York legislature incorporated a co. for the purpose with a capital of 1,000,000 dollars.

West Branch r-r. This r-r. which is 15 ms. long, with 5 ms. of branch roads, extends from Schuylkill Haven to Broad mountain. The main road has a double track. Cost of road and branches about 160,000 dollars.

West Chester r-r. (*See Pennsylvania r. roads.*)

West Feliciana r-r. A company has been incorporated by the legislature of Louisiana, to form a r-r. from the Mississippi r. near St. Francisville, to the boundary line of the state, in the direction of Woodville, Miss.

West Jersey r-r. A company was incorporated by the New Jersey legislature in 1831, to construct a r-r. either from the Delaware r. in Gloucester co. or from the Camden and Amboy r-r. to the Delaware r. in Penn's Neck, Salem co. Capital 500,000 dollars, with liberty to increase to 2,000,000 dollars.

Wilmington and Downington r-r. This road when completed will extend from Wilmington (Del.) to the boundary line of the state, in the direction of Downington (Pa.) The company for its construction was incorporated by the Delaware legislature in 1831, with a capital of 100,000 dollars, with powers to extend it to 150,000 dollars.

The above list of r-rs. finished, commenced or incorporated, is as complete as it has been possible to render it from the published returns which have been made respecting this species of improvement, in various parts of our country. Some r-rs. which may have been inadvertently omitted in the body of the article, will be found in the following table. The details of the manner of construction, or any explanations of the mathematical principles on which r-rs. are calculated to answer the purpose of transportation, &c., have been purposely omitted. Our aim has been to give the reader a brief view of the existing state of r-rs. in the U. States. In general, it may be stated, (without pretending to perfect accuracy in estimates, founded on documents so recently obtained, and of course deficient in connected details) that in 1833 there will be either actually finished or in progress, 2,600 ms. of r-rs. in the U. S., involving an interest exceeding $38,000,000 to the stockholders, and of greater, far greater interest to the public. This estimate is founded on the numbers actually mentioned in our table, without taking any account of those left in blank. This immense amount of property has been entirely invested in this new mode of transportation and intercommunication, within the short period of 6 years; for previous to 1826, rail-roads were regarded both in this country and in Europe, (with very few exceptions) as visionary projects. Those who desire more minute information than we have given on the subject, are referred to such works as "Smith's Wood," and the "N. York r-r. Journal."

Table of rail-roads completed, commenced, or incorporated.

Names.	*Time of incorpora.*	*Miles in l'gth.*	*Present state.*	*Estimated cost.*
Albany & Schenectady	1826	16	Finish'd	$500,006
Albion and Tonawanda				200,000
Alleghany Portage	1830	36 1-2	In prog.	700,000
Amsterdam and Fish-house	1832			250,000
Auburn and Erie	"			150,000
Aurora & Buffalo	"			300,000
Au Sable and L. Champlain	"	15		
Baltimore & Ohio	1827	72	Finish'd	2,000,000
Balt. and Susquehannah	1829	70	In prog.	1,000,000
Balt. & Wash'ton				
Black river	1832			900,000
Boston & lake Ontario				
Boston & Lowell	1830		In prog.	
Bost. & Providence	1831		Began	1,000,000
Bost. & Taunton	"	35	In prog.	1,000,000
Bost. & Worcester	"	43	In prog.	
Brooklyn and Jamaica	1832			300,000
Buffalo and Erie				650,000
Camden & Amboy	1830	61	Finish'd	1,500,000
Cape Fear & Yadkin				
Carbonsdale and Honesdale	1826	16	Finish'd	300,000
Catskill and Canajoharie	1830	75	Began	
Central (N. C.)				
Central (Pa.)				
Charleston & Hamburg		132	Finish'd	
Chesterfield	1829	13 1-2	Finish'd	140,000
Dansville and Rochester		46		300,000
Danville and Pottsville, & branches	1826	54 1-2	Finish'd	840,000
Detroit & Pontiac		25		
Dutchess county	1832			600,000
Elizabethtown and Somerville	1831			400,000
Elmira and Williamsport	1832			75,000
Experiment (N. C.)		1		
Fayetteville	1830			20,000
Harlem	"	6	In prog.	
Hudson and Berkshire	1832			350,000
Hudson & Mohawk	1826	16	Finish'd	500,000
Illinois & Michigan		96 1-2	In prog.	
Ithaca and Catskill		167		
Ithaca and Geneva	1832			800,000
Ithaca & Owego	1830	29 1-2	In prog.	
Knoxville & Southern	1832			
Lackawanna and Susquehannah	1826	16	Finish'd	120,000
L. Champlain and Ogdensburg	1832			3,000,000
L. Ponchartrain	1830	4 1-2	Finish'd	70,000
Lexington & Ohio	"	80	In prog.	1,000,000
Little Schuylkill		23	Finish'd	285,000
Lykins Valley	"	16 1-2	Finish'd	
Mad river and Erie		140		
Manchester		13		
Mauch Chunk and branches	1826	14	Finish'd	100,000
Maysville & Portland	1832			150,000
Mill cr. & branches		9		22,000

Names.	*Time of incorpora.*	*Miles in length.*	*Present state.*	*Estimated cost.*
Mine Hill & Shuylkill Haven		15	Finish'd	$ 181,000
Morris canal and Patterson				
Mount Carbon	1829	7 1-2	Finish'd	110,000
New Castle and Frenchtown	1827	16 1-2	Finish'd	400,000
N. Jersey, Hudson and Delaware r.				
New York & Erie	1832	400		6,000,000
N. York & Albany		160		2,000,000
North Carolina				
Norwich & Boston	"			1,000,000
Otsego	"			200,000
Patterson & Hudson r.		14	In prog.	
Petersburg and Roanoke	1830	60	In prog.	400,000
Phila. & Columbia	1828	82 3-4	Finish'd	1,600,000
Phila. and Del. Co.			In prog.	
Phila., German-t. & Norris-t.	1828	19	In prog.	
Philipsburg & Juniata	1830			
Pine Grove		5	Finish'd	30,000
Quincy	1825	3	Finish'd	
Rensselaer & Saratoga	1832			300,000
Rochester				
Room run & Mauch Chunk		5 1-4		80,000
Saratoga and Fort Edward	"			900,000
Saratoga & Schenectady	1830	22	In prog.	180,000
Schoharie and Otsego	1832			300,000
Schuylkill		13		95,000
Schuylkill valley & branches	1827	22	Finish'd	100,000
South Carolina	1830	135 1-4	In prog.	670,000
Tonnawanda	1832			500,000
Tuscumbia	1830	-	In prog.	
Utica and Susque-Hannah	1832			1,000,000
Warren County		64 1-4		250,000
Watertown and Rome	1832			1,000,000
West Branch and branches		20	In prog.	160,000
Westchester	1828	9	In prog.	81,000
West Feliciana				
West Jersey	1831			2,000,000
Wilmington and Downington	"			150,000

Besides the rail-roads completed, commenced, or merely incorporated, the following are some of the most important which have been projected. Of others we have been unable to obtain information.

From Augusta to Columbus in Geo.—from Augusta to Heshman's lake, of about 50 ms. in length, to avoid the uncertainty of the navigation of the Savannah—from Baltimore to Annapolis—from Bennington to Troy, about 30 ms. and to extend the same to Brattleborough about 42 ms.—from Boston to Brattleborough—from Boston to Ogdensburgh, N. York, the necessary privileges having been granted by the states of N. Y., Vt., and N. H.—from Boston to Salem, which if constructed will probably be extended to the N. boundary of the state—from Buffalo to Cayuga lake—from Buffalo to the line of Pa.—from Cattskill to the Susquehannah (Canajoharrie)—from lake Champlain, near Burlington, thro' the valley of Onion r., and by Montpelier to the Connecticut, opposite Haverhill, N. H., about 80 ms.—from Columbia to some point on the Ten. r.—from Cooperstown to Clairsville—from Geneva to Ithaca—from Lynchburg to New river—from Lynchburg to Knoxville—from Nashville to Franklin—from New Haven to Hartford—from Norristown to Allentown, on the Lehigh—from Richmond to Lynchburg—from Rochester to the Alleghany river—from Rochester to Carthage—from Rutland to Whitehall, as a link in the proposed chain from Boston to Ogdensburgh, or lake Champlain—from Schenectady to Buffalo, through Utica and Salina—from Steubenvile on the Ohio, to the Ohio canal—from Suffolk, Va., to the Roanoke, near Weldon, N. C.—from Troy to Whitehall—from Utica to some point on Cayuga lake—from Utica to Oswego—from West Stockbridge, Mass., to connect with a rail-road from Albany on the N. York line—from Wilmington through Fayetteville and Salisbury, to the iron mine dists. near Statesville—from Wheeling, on the Ohio, to lake Erie, and from the Yadkin to the Catawba.

RAINE's, p-o. Cumberland co. Va., by p-r. 69 ms. wstrd. Richmond.

RAINE's STORE, and p-o. Twiggs co. Geo., by p-r. 31 ms. s. w. Milledgeville.

RAINSBURG, p-v. Bedford co. Pa., by p-r. 113 ms. wstrd. Harrisburg.

RAISIN, river, of Mich., having its extreme sources in Hillsdale and Jackson cos., from whence, flowing 25 ms. N. E. by E., curves to the sthrd. in the s. w. angle of Washtenaw. Continuing sthrd. 25 ms. over Lenawee, inflects in the latter to N. E. by E., enters and traverses Monroe co. to its final discharge into the wstrn. part of lake Erie, after an entire comparative course of 80 ms. Raisin has interlocking sources with Grand, Kalemazoo, and St. Joseph's rivers of lake Michigan; with Huron of Erie, on the N., and Tiffin's and St. Joseph's branches of Great Maumee, s. The mouth affords good entrance and harbor for small vessels of 5 or 6 feet draught.

RAISINVILLE, p-v. Monroe co. Mich., by p-r. 56 ms. s. w. Detroit.

RALEIGH, p-v. and st. jus. Wake co., and of the government of North Carolina, situated near the w. or right bank of Neuse r., by p-r. 286 ms. s. w. W. C. N. lat. 35° 44′, long. W. C. 1° 38′ w. When I wrote the 2nd edition of the Geographical Dictionary, I had occasion to notice the fine state house in Raleigh, and the still finer statue of Washington placed in it, and chiselled by the hand of Canova; but since the hand of an incendiary or one of carelessness, has deprived N. C. of both those monuments of liberality and taste. It contains a bank, theatre, two academies, several schools and places of public worship. The town is built with streets extending at right angles to each other, with a centre sq. of 10 acres. Pop. 1830, 1,700.

RALEIGH, p-v. Shelby co. Ten., by p-r. 217 ms. s. w. by w. Nashville.

RALEIGH, p-v. on the left bank of Ohio r.,

nrthwstrn. part of Union co. Ky., by p-r. 215 ms. a little s. of w. Frankfort.

RALLS, co. Mo., bounded by Montgomery, Callaway, and Boone s., Randolph w., Marion n., Miss. r. separating it from Pike co. Il. n. e., and Pike co. Mo. e. and s. e. On the n. w. boundary uncertain. Length from e. to w. 60 ms., mean breadth 30, and area 1,800 sq. ms. Lat. 39° 25', long. W. C. 14° 35'. Slope a little n. of e., and drained almost entirely by Salt river and its confluents. Chief t. New London. Pop. 1830, 4,375.

RAMAPO, river, rises in the s. e. part of Orange co. N. York, crosses the w. corner of Rockland co., enters New Jersey, and flows across Bergen co. and joining Kingwood and Pequanock rs., forms Pompton r., which falls into the Passaic 6 ms. w. Patterson. It affords valuable water power, and moves much machinery.

RAMAPO, p-t. Rockland co. N. Y., 132 ms. s. Albany. Pop. 1830, 2,837.

RAMAPO WORKS, p-v. Rockland co. N. Y., 30 ms. n. w. New York city, has extensive iron works, a cotton factory, &c., and is a large and flourishing village. It is situated in a secluded valley on Ramapo r.

RAMSAY'S MILL, and p-o. Chatham co. N. C., 40 ms. w. Raleigh.

RAMSBORO', p-v. Guilford co. N. C., by p-r. 96 ms. n. w. by w. Raleigh.

RANDALLSTOWN, p-v. Baltimore co. Md., 10 ms. from Baltimore.

RANDOLPH, t. Coos co. N. H. Pop. 1830, 143.

RANDOLPH, p-t. Orange co. Vt., 23 miles s. Montpelier, 34 n. w. Windsor; 28,596 acres; is crossed by 2 branches of White r., which, with other streams, furnish mill sites. Maple, beech, birch, &c., grow in the forests; the land is high, the soil pretty good, and the town contains 3 villages. The Orange co. grammar school was incorporated here 1806, which affords advantages for education. Pop. 1830, 2,743.

RANDOLPH, p-t. Norfolk co. Mass., 15 ms. s. Boston, gives rise to a good mill stream, which flows into Boston bay between Quincy and Weymouth. Pop. 1830, 2,200.

RANDOLPH, p-t. Cattaraugus co. N. Y., 312 ms. w. by s. Albany, e. Chatauque co., n. Pennsylvania; has Alleghany r. s. e., and several small branches. Pop. 1830, 776.

RANDOLPH, t. Morris co. N. J., 6 miles w. Morristown; has Trowbridge mtn. s. e., and Rockaway river and Morris canal n. Pop. 1830, 1,443.

RANDOLPH, p-v. Crawford co. Pa., 12 miles nrthrd. Meadville, the county seat, and by p-r. 309 ms. n. w. W. C.

RANDOLPH, co. of Va., bounded by Greenbrier s., Nicholas s. w., Lewis w., Harrison n. w., Monongalia and Preston n., Alleghany in Md., and Hardy, Va., n. e., Alleghany mtn. separating it from Pendleton e., and Greenbrier mtn. separating it from Pocahontas s. e. The greatest length from s. w. to n. e. 90 ms., mean breadth 20, and area 1,800 sq. ms. Lat. 39° n., long. W. C. 3° w. The surface is a congeries of mtn. chains, ridges and deep vallies. It gives source to both Tygart's valley, and Cheat branches of Monongahela, both flowing nrthrd. Chief t. Beverly. Pop. 1830, 5,000.

RANDOLPH, county, N. C., bounded s. e. by Moore, Montgomery s., Davidson w., Guilford n., and Chatham e. It is very near a square of 30 ms. each side; 900 sq. ms. in area. Lat. 35° 40', long. 2° 48' w. W. C. Slope sthrd. but drained nearly equally, by Deep r. into Cape Fear r. valley, and by crs. flowing into Yadkin; it is therefore a table land between two river basins. Soil excellent, and surface finely diversified. Chief t. Ashboro'. Pop. 1820, 11,325, and in 1830, 12,406.

RANDOLPH, co. Geo., bounded n. by Muscogee, Marion n. e., Lee e., Baker s. e., Early s., and Chattahooche r. separating it from the Creek country of Ala. w. Length 44 ms. from s. to n., mean breadth 35 ms., and area 1,540 sq. ms. n. lat. 32°, and long. 8° w. W. C., intersect near its centre. The slope is sthrd., the estrn. part drained into Flint, and the wstrn. into Chattahooche river. Population 1830, 2,191.

RANDOLPH, C. H. and p-o. Randolph county, Geo., by p-r. 170 ms. s. w. Milledgeville.

RANDOLPH, p-v. on the Mississippi r., at the mouth of Big Hatchee r., western part Tipton co. Ten., by p-r. 213 ms. s. w. Nashville.

RANDOLPH, p-v. in the sthrn. part of Portage co. O., 10 ms. s. Ravenna, the co. seat, and by p-r. 132 ms. n. e. Columbus.

RANDOLPH, co. Ind., bounded by Wayne s., Henry s. w., Delaware w. and n. w., ——— n., and Darke co. O. e. Length 24, breadth 24, area 576 sq. ms. Lat. 40° 10', long W. C. 8° w. This co. is a real table land, from which flow to the n. w. the higher sources of Mississinniwa, branch of Wabash; the extreme source of White r. rises on the w. border of Darke co. O., and flowing westward traverses Randolph; and finally the whole southern side gives source to, and is drained by the extreme sources of White water, branch of Great Miami. Chief t. Winchester. Pop. 1830, 3,912.

RANDOLPH, co. of Illinois, bounded n. w. by Monroe, St. Clair n., Washington n. e., Perry e., Jackson s. e., and the Mississippi r. separating it from Perry co. Mo. s., St. Genevieve co. Mo. s. w., and Jefferson, Missouri, w. Length from s. to n. 30 ms., mean breadth 20, and area 600 sq. ms. Lat. 38° and long. 13° w. W. C. intersect near the co. seat, Kaskaskias. Slope sthrd., and in that direction traversed by Kaskaskias r. The lower part of this co. near the mouth of Kaskaskias, is one of the most ancient settlements of civilized inhabitants in the basin of the Mississippi, dating as far backwards as 1674. Pop. of the co. 1830, 4,429.

RANDOLPH, co. Mo., bounded by Ralls e.,

Boone S. E., Howard S. W., Chariton W., and unappropriated territory N. Mean length 38 ms., breadth 20, and area 760 sq. ms. N. lat. 38° 30′ and long. W. C. 15° 30′ intersect in this co. It is a table land between Chariton and Salt rivers, the confluents of the former flowing S. S. W. into Missouri r., and those of the latter estrd. over Ralls into the Mississippi. Chief t. Huntsville. Pop. 1830, 2,942.

RANDOM, t. Essex co. Vermont, 48 ms. N. E. Montpelier; is watered by Clyde river and smaller streams, and a part of Knowlton's lake, the sand of which is remarkably white and beautiful, well fitted for glass making. It was chartered in 1781. In 1823 it contained but a single family. Pop. 1830, 105.

RANKIN, co. Miss., bounded S. by Simpson, Pearl r. W. separating it from Hinds co., N. Madison, and E. Choctaw territory in Miss. Length 28 ms., mean breadth 22, and area 616 sq. ms. Lat. 32° 20′, and long. 13° W. W. C. intersect in this co. Slope wstrd. towards Pearl r. Surface generally covered with pine forest. Chief town Brandon. Pop. 1830, 2,083.

RANKIN, p-v. Yazoo co. Miss., by p-r. 85 ms. nrthrd. Jackson.

RANSOM'S BRIDGE, and p-o. eastern part of Nash co. N. C., by p-r. 70 ms. E. Raleigh.

RAPID ANN, river of Va., deriving its remote sources from the Blue Ridge, and flowing thence S. E. 20 ms. across the valley, between Blue Ridge and South East mountain, turns thence N. E. 15 ms. to the influx of Robertson's river from the N. W. Passing South East mountain and inflecting to a general eastern course of 30 ms., joins the Rappahannoc 10 ms. above Fredericsburg, after a comparative course of 65 ms. In nearly the whole of its length Rapid Ann separates Orange co. first 35 ms. from Madison, and thence 25 from Culpepper. At their junction it is superior in volume to Rappahannoc; and exceeding also in length of course the Rapid Ann is the main stream.

RAPID ANN, meeting house, and p-o. wstrn. part of Madison co. Va., by p-r. 104 ms. S. W. W. C.

RAPIDES, parish of La., bounded by Opelousas, or St. Landry S., Natchitoches W. and N. W., Little or Catahoola r. separating it from Catahoola parish N., Black r. or Lower Ouachitta r. separating it from Concordia E., and Red r., and in part an artificial limit separating it from Avoyelles S. E. Length from S. to N. 65 ms., mean breadth 40, and area 2,600 sq. ms. Extending in lat. very nearly from 31° to 32′ N., and in long. between 15° and 16′ W. W. C. Slope S. E., and in that direction drained by the bayous Bœuf and Crocodile, to the S. W.; by the confluents of Catahoola N. E., and nearly centrally traversed by the main volume of Red river. The soil exhibits every variety, from the most fertile r. alluvion to that of sterile pine forest land. The latter, however, greatly prevails, and comprises most of the southwestern and nrthestrn. sections. The eastern and lower part is subject to annual submersion. Along Red river, and bayous Rapide and Bœuf, the soil is of the very first rate.

This parish derives its name from the lower rapids of Red river, which are opposite the town of Alexandria, the st. jus. At high water they are invisible, but at low water very much impede the navigation of the stream. Pop. 1820, 6,065, and in 1830, 7,575.

RAPPAHANNOC, river of Va., formed by two branches, Hedgeman's and Thornton's rivers, both deriving their remote sources from Blue Ridge. Hedgeman's r. after a comparative course of 30 ms. between Fauquier and Culpepper cos., receives Thornton's river from the latter, and the united waters continuing the course of the former S. E. 20 ms., join the Rapid Ann as already noticed under the head of the latter. A navigable river at the junction of its two main branches, the Rappahannoc continues to the S. E. 10 ms. to its lowest falls, where it traverses the primitive ledge, and meets the ocean tides at Fredericsburg. Similar to the Delaware, and all the large western confluents of Chesapeake bay, the Rappahannoc turns along after passing the primitive rock, but after a short curve to the southward, this streams resumes a S. E. course, which with a rather tortuous channel it maintains to Leeds, in Westmoreland co., where it approaches to within 5 miles of Potomac, at the mouth of Mattox cr. Gradually widening, and with the features of a long narrow bay of 55 ms., the Rappahannoc by a S. S. E. course, is lost in Chesapeake bay between Windmill and Stingray points. The tide ascends this channel to the falls at Fredericsburg, something above 100 miles, admitting vessels of considerable tonnage. In all the distance below the union of its two main branches, it does not receive a confluent above the size of a small creek. The entire basin is 140 ms. by a mean width of 20; area 2,800 sq. ms. Extending in lat. from 37° 34′ to 38° 44′, and in long. W. C. from 0° 41′ E. to 1° 22′ W.

RAPPAHANNOC ACADEMY, and p-o. in the nrthestrn. part of Caroline co. Va., by p-r. 72 ms. S. S. W. W. C., and 64 ms. N. N. E. Richmond.

RARITAN BAY, N. J., between Sandy Hook on the E., Monmouth county on the S., and Staten island on the N. W., terminating at Amboy. The channel carries 3½ fathoms to Amboy.

RARITAN, river, New Jersey, is formed by branches which flow through Morris, Hunterdon, Somerset, Middlesex and Monmouth counties, watering a large extent of country. It enters Raritan bay at Amboy, and is navigable for vessels drawing 8 feet water to N. Brunswick, except at low ebb tides, when the water is shallow and the channel narrow in some places. Along the lower part of the stream, the banks are low, flat, and partly marshy. Large steamboats ply daily between New York and New Brunswick, on the principal steamboat and stage route to Philadelphia.

RARITAN LANDING, v. Middlesex co. N. J., at the head of tide water on Raritan river, 2 ms. above New Brunswick. There is a free bridge over the r.

RARITAN, south branch, river, N. J., rises in Budd's pond, Schooley's mountain, Morris co., N. Suckasunny plains, and runs by German valley, Clinton, Flemington, &c., to its junction with the north branch, 4 miles w. Somerville.

RARITAN, north branch, r. N. J., rises 6 ms. N. W. Morristown village, Morris county, and partly in Suckasunny plains, and runs through Somerset co. to its junction with s. branch.

RATTLING GAP, p-o. Lycoming co. Pa., by p-r. 109 ms. nrthrds. Harrisburg.

RAUBSVILLE, p-o. Northampton co. Pa., by p-r. 196 ms. N. N. E. W. C.

RAVENNA, p-v. and st. jus. Portage co. O., by p-r. 127 ms. N. E. Columbus, and 320 ms. northwestward W. C. N. lat. 41° 10′, long. W. C. 4° 12′ w. It is situated on a branch of Cuyahoga river, on a country, the mean height of which is about 1,000 feet above the Atlantic tides. Pop. of Ravenna township, including the village, 1830, 806.

RAWLINGSBURGH, p-v. Rockingham co. N. C., by p-r. 105 ms. N. W. by W. Raleigh.

RAWLINSVILLE, p-o. Lancaster co. Pa., by p-r. 95 ms. N. E. W. C.

RAWSONSVILLE, p-v. Broadalbin, Montgomery co. N. Y., 10 ms. from Johnstown, on Fondas creek.

RAY, p-v. in the nrthrn. part of Macomb co. Mich., by p-r. 58 ms. N. E. Detroit.

RAY, co. Mo., bounded by Missouri r. s., separating it from Lafayette and Jackson; Clay w., and on the other sides boundaries uncertain. Length 24 ms., mean breadth 20, and area 480 sq. ms. Lat. 39° 15′ N., and long. W. C. 17° w. intersect in this county. Slope s. E. towards Missouri r. Chief town, Richmond. Pop. 1830, 2,657.

RAYMOND, p-t. Cumberland co. Me., 75 ms. s. w. Augusta, 24 N. Portland, lies on the N. side of Sebago pond, is crossed by the lower part of Crooked r. s. w., and has several other small ponds and streams. Pop. 1830, 1,756.

RAYMOND, p-t. Rockingham co. N. H., 25 ms. s. s. E. Concord, 25 s. s. w. Portsmouth, 13 w. Exeter, with 16,317 acres; is crossed by Lamprey river, whose two branches here unite. There are also 2 ponds, and part of Patuckaway river. There are fertile meadows on the r. Oak, &c. grow on the uplands. A small cavern in the w. part called the oven. Rattlesnakes formerly abounded. This town furnished 24 soldiers to the continental army in the revolution, besides militiamen. Pop. 1830, 999.

RAYMOND, p-v., and as marked in p-o. list, chief town or st. jus. Hinds co. Miss., 19 ms. from Jackson, but relative position uncertain.

RAYNHAM, p-t. Bristol co. Mass., 32 ms. s. Boston; has Taunton r. s., which forms an arch round that part. First settled 1650, and the first forge erected in North America was built here in 1652, by James and Henry Leonard. King Philip, or Metacom, had a fishing station here. Iron is here manufactured in various forms, nails, bars, hollow ware, &c. Pop. 1830, 1,200.

RAYSVILLE, p-v. Henry co. Ind., by p-r. 36 ms. N. E. by E. Indianopolis.

RAYTOWN, p-v. Wilkes co. Geo., by p-r. 51 ms. N. E. Milledgeville.

READFIELD, p-t. Kennebec co. Me., 7 ms. w. Augusta, is a small town of irregular form, crossed by a long pond, whose outlet forms the principal upper stream of Cobbesseconte r. Pop. 1830, 1,884.

READING, p-t. Windsor co. Vt., 53 ms. south Montpelier, 9 w. Windsor; was chartered in 1781, and contains 23,040 acres. First settled 1772. It is uneven, with a ridge of mountainous land w., from which descend several streams, flowing partly N. E. to Queechy r., partly E. to Connecticut r. at Windsor, and partly s. into Black r., furnishing pretty good mill sites. There are 12 school dists., several mills, &c. The timber is hard wood and spruce. Pop. 1830, 1,409.

READING, p-t. Middlesex co. Mass., 12 ms. N. Boston. Settled 1644; has much good soil, but some uneven and hard. The village is large. Pop. 1830, 1,806.

READING, p-t. Fairfield co. Conn., 60 ms. s. w. Hartford; about 5 ms. by 6½, with 32 sq. ms.; has rocks of granite and primitive limestone, with an irregular surface, and a good soil. Saugatuck river crosses it through the middle N. and s., and Norwalk r. is in the w. part. The forest trees are oak, nut trees, &c. Joel Barlow was born here. Pop. 1830, 1,686.

READING, p-t. Steuben co. N. Y., 223 ms. w. Albany, 25 N. E. Bath, 15 s. E. Penn-Yan, w. Seneca lake, which separates it from Seneca and Tompkins cos. It is a gore of land from 3 ms. to 4½ by 14, with very good, level land, without stones, and watered by Bigstream and Rockstream, which afford valuable mill sites. They flow into the lake, which they enter near each other. Rockstream has a very romantic fall of 140 feet, at the foot of a long rapid. The water is precipitated into a basin, between high banks of clay slate. Stone quarries are situated at different places along the shore of the lake. Pop. 1830, 1,568.

READING, borough, p-t. and st. jus. Berks co. Pa., by p-r. 52 miles a very little N. of E. Harrisburg; 50 ms. N. W. Philadelphia, and 143 ms. a little N. of N. E. W. C. N. lat. 40° 42′, long. W. C. 1° 03′ E. Reading, similar to many of the other borough towns of Pa., was originally laid out after the model of Philadelphia; streets extending at right angles to each other, with two main streets, at the intersection of which the court-house was erected. It is more than commonly compact, and well built, and contains several places of public worship, the ordinary co. buildings, numerous private schools, and one bookstore. Situated on the canal formed along the Schuylkill, and in the midst of a fertile and well cultivated country, Reading is a flourishing commercial depot, as well as place of domestic trade. The original inhabitants were mostly Germans, and eminent for their industrious and economical habits and quiet manners; and such is still the character of the place. Pop. 1820, 4,332, and in 1830, 5,856; having gained upwards of 35 per cent. in 10 years.

Reading, p-v. Sycamore township, Hamilton co. O., by p-r. 11 ms. n. n. e. Cincinnati. Pop. 1830, 200.

Readyville, p-v. in the estrn. part of Rutherford co. Ten., 12 ms. e. Murfreesboro', the co. seat, and by p-r. 45 ms. s. e. by e. Nashville.

Reamstown, p-v. Lancaster co. Pa., 15 ms. n. e. Lancaster, and 15 ms. s. w. Reading. Pop. 1830, 300.

Rebecca Furnace, and p-o. Botetourt co. Va., 220 ms. n. w. W. C.

Rebersburg, p-v. Centre co. Pa., 12 ms. e. Bellefonte, the co. seat, and 93 ms. n. w. Harrisburg.

Rectortown, p-v. Fauquier co. Va., by p-r. 53 ms. s. w. by w. W. C.

Red Bank, p-v. in the nrthrn. part of Armstrong co. Pa., 20 ms. n. Kittanning, the co. st., and by p-r. 235 ms. n. w. W. C.

Red Bird, p-v. in the sthrn. part of Clay co. Ky., 10 ms. sthrd. Manchester, the co. st., and 125 ms. s. e. Frankfort.

Red Bridge, p o. Hawkins co. Ten., 10 ms s. w. Rogersville, the co. seat, and by p-r. 254 ms. a little e. of n. Nashville.

Reddies, or Reddy's river, and p-o. Wilkes co. N. C., by p-r. 188 ms. a little n. of w. Raleigh.

Redfield, p-t. Oswego co. N. Y., 30 ms. n. Rome, 142 n. w. Albany, s. Jefferson co., w. Lewis co.; 6 ms by 14; is crossed in the s. by Salmon r. which flows into lake Ontario, and a branch flows south through the middle. which furnish mill seats. The soil is good, bearing a variety of timber, and the surface nearly level. Pop. 1830, 341.

Red Hill, Moultonborough, Grafton co. N. H., at the n. end of Winnipiseogee lake, commands the finest view of that beautiful sheet of water, and the surrounding country, for a great distance. It has a small stream n., Great Squam lake w., Long Pond and Winnipiseogee lake s. Iron ore is found in the n. Bluff, and bog iron ore in a brook below it.

Red Hill, p-v. sthrn. part of Kershaw dist. S. C. 16 ms. sthrd. Camden, and by p-r. 49 ms. n. e. by e. Columbia.

Red House, p-o. Charlotte co. Va. by p-r. 112 ms. s. w. Richmond.

Red House, p-o. Caswell co. N. C. by p-r. 75 ms. n. w. Raleigh.

Red Hook, p-t. Dutchess co. N. Y. 23 ms. n. Poughkeepsie, 20 s. Hudson, s. Columbia co., e. Hudson r., has a rich loam, more mixed with clay near the river, well cultivated, and varying in surface. It is crossed by Sawkill, which affords good mill seats. There are several landings, at one of which the New York and Albany steamboats touch; and several villages in the interior. Near the river the banks are fine, and ornamented with the residences of several of the Livingston family, and other gentlemen. There are several factories, an academy in the Upper v., &c. Pop. 1830, 2,983.

Red Mountain, p-o. Orange co. N.C. by p-r. 38 ms. n. w. Raleigh.

Red River. There are several streams in the U. S. which bear this name, and following the geographical relative positions we have pursued in this treatise they stand in the following order.

Red r. of Ky. and Ten. has its sources in Christian, Todd, Logan and Simpson counties of the former state, and which flowing south'rd. enters a stream which originates in Summer co. Ten., and which, flowing wstrd. under the name of Red r. traverses Robertson and Montgomery cos. Ten. falls into Cumberland r. at the bend near Clarksville. Red r. has interlocking sources with Big Barren and Muddy r. branches of Green river.

Red r. of Ky. rises in Morgan co. and flowing thence wstrd. over Montgomery, falls into the right bank of Kentucky r. between Clark and Estill counties.

Red r. great sthrn. constituent of Assiniboin r. *(See Red river, article Assiniboin, p. 32.)*

Red r. great sthwstrn. branch of Mississippi r., has its remote sources in the mountainous prairies of N. Mexico, between the sources of Canadian Fork of Arkansas and those of Rio Colorado of the Gulf of Mexico, and between 25° and 33° long. w. W. C. From its source through 11 degrees of long. the general course is very nearly e. From the 23d degree to 17° 30′ w. W. C. the channel forms the boundary between the U. S. and Texas, and with long. 17° 30′ it inflects to s. e. and becomes entirely a stream of the U. S., traverses a small angle of Ark. and thence entering La. over which it winds 300 ms. by comparative course to its final discharge into Mississippi at n. lat. 31° 01′, long. W.C. 14° 40′ w. The higher volume of Red r. is formed by two main branches, Red river proper, and False Ouachitta, both rising in New Mexico, and flowing about 350 ms. before their junction. It is remarkable that in all its course of upwards of 1,100 ms. Red r. receives no tributary of any consequence worthy of notice, but from the north, beside False Ouachitta, and below that stream comes in in succession Blue r., Kimitchie, Vasseux and Little r. of the n. above La., and in the latter state, Dacheet, Black r., Saline, and Ouachitta. *(See Ouachitta.)* Red r. partakes in some measure with Arkansas, the character of a stream of the desert. Along the immense inclined plain between Missouri proper, and the Gulf of Mexico, in the summer and autumn seasons, the moisture and herbage are alike dried up. The beds of the streams, a few months before replenished to overflowing, become in great part dry sandy lines. At no season, however, does Red r. where passing the rapids at the town of Alexandria in La. and where the whole of its volume is confined to one bed, answer to the great comparative length of its course, but this phenomenon is explained by a feature, as far as I know the natural history of rivers, peculiar to Red r. Some distance below where it bends to s. e. and enters La. it divides into numerous channels, spreading their mazes over an elliptical region of low land between the retiring hills. This tract is about 70 ms. in length with a width varying from one to 8 or 10 ms. It is one immense intricacy of interlocking water courses, but without any direct continuous channel. From personal observation the writer of this article is inclined to the theory that this very recent alluvial tract was once a lake, which the abrasion of the river against its banks has at length filled with earth. But what is at the same time in an extraordinary degree remarkable, is the fact that the same cause which filled the river lake with deposit created numerous others. The various crs. or small rivers flowing in between the hills on each side have now become lakes. Their channels and bottoms from hill to hill, for a distance of from 10 to 30 ms. backwards have been supplied with water, which cannot now all escape as their outlets towards the Red river have become so many embankments. It is true, the water in these new lakes rises and falls with the floods of the main stream. In latter summer and autumn much of their valleys become green meadows, supplied with succulent herbage; but as the immense volume of Red river pours down in winter and spring, a reflux takes place and the river water pours rapidly into these great natural reservoirs, and contributes by this flux and reflux to most effectually equalize the discharge of Red river. The Ouachitta and its confluents present similar features. (*See lakes Bistineau, Bodcau, Catahoula, &c.*)

What is called "*The Raft,*" in Red river, has been thus formed, and to call it a raft in the true intent of the term is a very deceptive misnomer. I have personally surveyed both the lakes of Red r. and the Atchafalaya raft, and found that of the latter to be a raft in the literal sense of the word; but between it, and the thicket islands and lake like channels of Red river above Grand Ecor, there is nothing in common. (*See Atchafalaya.*) At Grand Ecor 4 ms. above the town of Natchitoches, the whole volume of Red river is united; but in less than half a mile below, again separates, the Rigolet de Bon Dieu issuing from the left, and does not again enter the main stream for upwards of 30 ms., in which distance it receives Black and Saline rivers from the north. The main stream also which passes the town of Natchitoches is subdivided into numerous channels. Below the rejunction of the Rigolet de Bon Dieu, the river is once more for three or four miles united in one channel, but again dividing, the bayou Rapide issues to the right, and so called from again meeting the main river at the rapids near Alexandria. With the outlet of bayou Rapide, properly speaking, the unity of Red river is destroyed to be restored no more, as in the natural state of the country at high water outlets flowed from bayou Rapide itself, the waters of which flowed down the bayou Bœuf and were conveyed into Atchafalaya by several channels. These issues from bayou Rapide have been embanked, and by the aid of art the whole of Red river is made to pass Alexandria. The solid pine wood land indeed reaches the bank on the left directly opposite Alexandria, but on the right a few ms. below, outlets commence which have their recipient in Atchafalaya.

A recent revolution at the mouth of Red r. ought not to be passed over in silence. Where it entered the Mississippi, the latter by a long curve, first to the west, thence nthrd. and abruptly back to the east formed a peninsula which about 5 ms. a little s. of e. from the mouth of Red r. had not quite a mile in breadth in 1800, and was continually lessening. The actual breach of this isthmus was long foreseen, and actually took place in 1831. With slight cutting the river was made to act upon the yielding soil, and now Red r. has its mouth 5 ms. above where it formerly existed. This may seem contrary to the laws of nature, as water would naturally fall like other bodies when left free, but exactly similar phenomena took place at the mouths of the Yazoo and Homochitta rivers, which, when the bends were cut and the outlets of the rivers changed, the mouths were formed at the upper and not as might have been expected at the lower end of the cut. Red river it is generally supposed would flow down the Atchafalaya, if the communication between it and Mississippi was interrupted, but an irresistible barrier to a permanent stream passing down the Atchafalaya, may be seen stated at the head of that article, and stated from personal observation on the spot. Here is also the place to notice another error which has been sanctioned by official document. It has been stated that changing the bed of Red river has drained 200,000 acres of pine land. Any person acquainted at all with the respective features of La. knows perfectly well that pine land is never overflowed, and they know also, that before they could be overflowed the whole delta would be many feet under water. The pine lands and delta touch, but no two species of soil however distant can differ more specifically, and besides, the pine tracts are every where elevated above any influence from annual floods from the Mississippi or any of its confluents.

Red River, iron works and p-o. Estill co. Ky. by p-r. 75 ms. s e. by e. Frankfort.

Red Shoals, p-o. Stokes co. N. C. by p-r. 143 ms. n. w. by w. Raleigh.

Red Sulphur Springs, and p-o. on Indian cr. in the wstrn. part of Monroe co. Va. by p-r. 240 ms. w. Richmond.

Reedsborough, t. Bennington co. Vt., 12 ms. s. e. Bennington, 18 s.w. Brattleboro', n. Rowe, Mass., is quite mountainous, with large tracts of useless land. Deerfield r. forms the e. boundary, and a branch crosses the t. Both afford mill seats. Pop. 1830, 662.

Reed's Mills, sthrn. part of Jackson co. O. by p-r. 86 ms. s. s. e. Columbus.

Reedsville, p-o. Rutherford co. N. C. by p-r. 213 ms. s. w. by w. Raleigh.

Reedtown, p-v. Seneca co. O. by p-r. 90 ms. nrthrd. Columbus.

Reedy Fork, p-o. on Reedy Fork r. nrthrn. part of Guilford co. N. C. by p r. 92 ms. n. w. by w. Raleigh.

Reedy Fork, r. and p-o. sthrn. part of Greenville district, S. C. by p-r. 119 ms. n.w. Columbia.

Reedy Spring, p-o. Campbell co. Va. by p-r. 196 ms. s. w. W. C.

Rees' Cross Roads, p-v. Woodford co. Ky. 12 ms. s. e. Frankfort.

Reenier's Mills, and p-o. Washington co. O. 111 ms. s. e. Columbus.

Rehrersburg, p-v. in the nrthwstrn. part of Berks co. Pa. 25 ms. n. w. Reading and by p-r. 38 n. e. by e. Harrisburg.

Rehoboth, Bristol co. Mass. 37 ms. s. w. Boston, e. Sekonk r. or R. Island, is nearly level, with a few gentle hills; settled 1643 by Rev. Samuel Newman and part of his church from Weymouth, Mass. In 1646 the Indians burnt 40 dwellings and 30 barns. Anawan's rock was the wild and secluded retreat of king Philip's principal chief. After the death of the latter, and the death or capture of his other captains, Anawan was surprized here by Capt. Church. Pop. 1830, 2,459.

Reidstown, p-v. Union dist. S. C. by p-r. 95 ms. n. w. Columbia.

Reidsville, p-v. Rockingham co. N. C. by p-r. 103 ms. n. w. by w. Raleigh.

Reiley, p-v. Butler co. O. by p-r. 120 ms. s. w. by w. Columbus.

Reisterstown, p-v. Baltimore co. Md. 15 ms. n.w. Baltimore.

Remsen, p-t. Oneida co. N. Y. 90 ms. w. n. w. Albany, 20 n. Utica, s. Lewis co., w. Herkimer co., is crossed by Black r. running w., and has Cincinnati cr. s. The e. line is partly formed by W. Canada cr. and there are other streams which also supply mill seats. The surface is uneven and the soil generally good. First settled 1793. The v. is on Cincinnati cr. 16 ms. n. Utica, 55 Johnstown. Waters rising but ½ m. apart, flow into Black r. and W. Canada cr. Pop. 1830, 1,400.

Rensselaer co. N. Y. bounded by Washington co. n., Vt. and Mass. e., Columbia co. s., Hudson r. w. which separates it from Albany and Saratoga cos., and contains 572 1-2 sq. ms. and 14 tsps. There are high hills e. and the surface is generally broken, with large valleys and some fine meadows. The soil is various, as are the forest trees. Troy, one of the most flourishing cities in the state, is in this co., and is at the head of sloop navigation in Hudson r.; the great dam across that r. affords a communication between Troy and the Erie canal. Fine steamboats owned here, regularly ply to New York; and there is a large amount of business carried on by sloops, as well as by canal boats. Hoosac r. enters the co. from Mass., and receiving Little Hoosac cr., Wallomsac and Tomhanoc crs., after a crooked course, falls into Hudson r. in the n. Poesten kill joins the Hudson at Troy, after supplying valuable mill seats. There are several other streams of less importance. The ancient estate of Rensselaerwyck included all this co. except the 3 n. towns, together with the co. of Albany, and was early settled. It was 24 ms. wide on the Hudson, 42 long, and purchased and granted between 1630 and 1649. The county is transition, except a little secondary. Roofing slate, some iron ores, &c. are found in different places. Under the patronage of Stephen Van Renssalaer, Esq. who bears the ancient title of patroon of Rensselaerwyck, a plan of public instruction has been in operation here, of a practical nature, by which useful knowledge is furnished to young men in agriculture, as well as in other branches: a central instution being established at Troy, with a farm, which operates in different ways in other parts parts of the co. There are cotton factories at Lansingburgh, Scaghticoke, Hoosac, Troy, Nassau, Pittstown; and a few woollen in different places. There are two rolling mills and nail factories 2 ms. s. Troy, on Wynant's kill: at the Albany nail factory 450 tons are made in a year; and at the Troy factory 1,000 tons, partly into spikes. Pop. 1820, 40,153: 1830, 49,424.

Rensselaer, v. Berlin, Rensselaer co. N. Y. 12 ms. e. Albany.

Rensselaerville, p-t. Albany co. N. Y. 24 ms. s. w. Albany, n. Greene co., e. Schoharie co., about 8 ms. by 8 1-2, with 68 sq. ms., is rough with some high hills of the Catsberg range, and large, fertile valleys. It is crossed in the s. w. by Cattskill creek, whose branches supply many mill seats. The land is generally leased. The v. on Ten Mile cr. is in the n. e. 23 ms. w. s. w. Albany, 10 ms. from Cattskill cr. Preston Hollow is a v. s. w. 30 ms. Albany, and 26 Cattskill. Potterville is 2 ms. w. of this. Pop. 1830, 3,685.

Republican Grove, and p-o. Halifax co, Va. by p-r. 149 ms. s. w. Richmond.

Reynoldsburgh, p-v. and st. jus. Humphries co. Ten. by p-r. 78 ms. w. Nashville. It is situated on the right bank of Tennessee r. n. lat. 36° 05', long. W. C. 11° 04' w.

Rhea, co. Tenn. bounded by Hiwassee r. separating it from the Indian country s., Hamilton co. s. w., Walden's ridge separating it from Bledsoe w. and n. w., Roan n. e., and MacMinn s. e. Length from s. w. to n. e. 36 ms.; breadth 26, and area 926 sq. ms. Lat. 36° 25', long. 7° 54' w. W. C. Tennessee r. enters on the nrthestrn. border, and winding over the co. in a sthwstrn. direction divides it into two unequal sections, and opposing slopes. The larger section is to the N. W. falling from Walden's ridge. Chief t. Washington. Population 1820, 4,215, and in 1830, 8,186.

RHEATOWN, p-v. in the estrn. part of Greene co. Ten. 10 ms. E. Greenville, the co. st. and by p-r. 283 ms. E. Nashville.

RHINEBECK, p-t. Dutchess co. N. Y. 67 ms. S. Albany, 17 N. Poughkeepsie, E. Hudson r. on the banks of which are landings, and by which considerable trade is carried on with New York, &c. Mill seats are found on Landtman's and Crom Elbow crs. The name is formed by a combination of those of river Rhine in Europe, and Beckman, one of the earliest purchasers. The first settlements were made by Germans. The land is rather uneven E. and level W., with a fertile plain in the middle, where is the village of Rhinebeck Flats. Wertemburg S. E. has a light soil. Sepascat lake though small yields fish. Pop. in 1830, 2,938.

RHODE ISLAND, one of the U. S., and the smallest state in the union, is bounded N. and E. by Mass., S. by the Atlantic ocean, and W. by Connecticut. It lies between 41° and 42° N. lat., and between 3° 11′ and 4° E. long. W. C.—being about 42 ms. long from N. to S. and 29 ms. wide, and embracing an area of 1,225 sq. ms. of which 130 sq. ms. are included in Narragansett bay. The territory now comprehended in the state of Rhode Island, was found by the first English settlers, chiefly in possession of the Narraganset Indians, from whose language the present names of many places, rivers, &c. have been derived. The W. boundary was the dividing line between this nation and the Pequods of Connecticut. The Wampanoags, a branch of the latter, inhabited the N. E. parts of the state, about Bristol; and their chief, Metacom, or Philip, involved the colonies in a most dangerous and destructive war, between 1675-.77. A rude map of Mass. bay, which embraces the coast of R. I., was published in London in 1634, by a Mr. Wood. The first settlement by white men was made in 1636, by Rev. Roger Williams, who had been banished from Mass. colony for his peculiar religious opinions. He was followed by many others, who with him laid the foundation of the fine city of Providence. In 1638, Mr. Coddington and 17 others being persecuted in Mass. on account of their religious tenets, followed Roger Williams, and settled at Newport. In 1644, a charter was obtained for both the settlements. In 1647 was held the first general assembly, when the executive power was confided to a president and 4 assistants. In 1663 a new charter was granted by Charles II. which with a few changes has formed the basis of the government until the present time. One of the earliest acts of hostility against the British, before the revolutionary war, was committed in this state, whose inhabitants took an active part in that struggle. The island of Rhode Island was for some time in possession of the enemy. The constitution of the U. S. was adopted by this state in 1790, after it had received the assent of all the others. The state government still proceeds under its colonial charter granted in 1663, by Charles II. The legislative and executive departments are mixed. The legislature consists of a senate and house of representatives. The senate is composed of the governor, lieut. governor, and 10 counsellors. There are 72 representatives, elected by the people semi-annually. The legislature convenes 4 times a year. The salary of the governor is $400, of the lieut. gov. $200, of the secretary of state $750 and fees. The judiciary is vested in a supreme court of 3 judges, and a court of common pleas for each of the 5 counties, each court consisting of 5 judges. These judges are annually appointed by the legislature. The right of suffrage is universal.

The surface of the state is varied, but there are no mountains. About one tenth is water, which is a greater proportion than in any other state in the Union. The S. W. part of the state, and the valley of the Narragansett r. have a large proportion of level land. There are many hills, as Mount Hope in Bristol, Hopkins's hill in W. Greenwich, and Woonsocket hill in Smithfield; and much of the land is uneven and rocky. The soil on the continental part of Rhode Island is tolerably fertile, though its cultivation requires much labor. It is well adapted to many kinds of fruit trees. On the islands it is slaty and more productive. Some iron ore, marble, and free stone are found in different places, and there is a mine of anthracite coal on the isl. of Rhode Island, which is not worked. There is much good pasture land, and grain, and orchards are successfully cultivated. Agriculture is, however, generally much less flourishing than in the adjacent states—commerce and manufactures absorbing more the attention of the inhabitants. The island of Rhode Island, has been celebrated for its beautiful, cultivated appearance, abounding in smooth swells, and being divided with great uniformity into well tilled fields. Oak, walnut, chestnut and other trees are abundant in some parts of the state. The climate much resembles that of Mass. and Conn. in its salubrity—the parts of the state adjacent to the sea are favored with refreshing breezes in summer, and in winter are the most mild. The rivers are small, but some of them afford excellent sites for manufactories; particularly the Pawtucket or Blackstone r. (the largest in the state) the lower part of whose course lies along the E. boundary. Among the numerous factories on this r. and its branches are the following; at Mannsville, 11 miles above Providence, 2 for cotton, with 7,000 spindles; on Peter's r. 5 ms. above, 2 built of stone, with 25,000 spindles; on Mill r. ½ m. further, 2 of wood, with 600 spindles; at Woonsocket falls, about 20,000 spindles, &c. At the latter place are also 2 machine shops, a foundry, and about 2,000 inhabitants. Water is abundant in the state, and is extensively applied to use. The chief bays of R. I. are Narragansett bay, which penetrates north into the state more than 30 ms. and is navigable by large ships up to Providence: it connects several good harbors with the ocean. But the most important in a naval point of view is Newport harbor, which is accessible in the most unfavorable winds on the coast, is safe from storms, and strongly protected by forts of the U. S. A variety of fish is obtained from the coves, bays, &c. Several useful turnpike roads lead in different directions; and the principal route of travelling between New York and Boston, during the months when navigation is unimpeded, has been for several years by the Providence steamboat and stage coach line. Regular packets ply between the principal ports of R. I. and of other states; and an important foreign trade is carried on, principally from Providence, (*see Providence*) the merchants of which city have been engaged in an extensive commerce with Canton. The Blackstone canal lies partly in this state. A rail road is to be constructed from Providence to Boston, for the transportation of passengers and merchandize. (*See Boston and Providence rail-road, under the head of Rail-Roads.*) Several islands of some importance belong to this state: the principal of which are Rhode Isl., Conanicut, Prudence, and Block isls.

Rhode Island is divided into 5 counties, Providence, Newport, Washington, Kent, Bristol, and 31 towns, of which Providence is the largest. The population of the state at several periods has been as follows: in 1790, 68,825; 1800, 69,122; 1810, 76,931; 1820, 83,059 and in 1830, 97,199; the two latter as follows:

Counties,	1820.	1830.
Providence,	30,769	47,018
Newport,	16,294	16,535
Washington,	14,962	15,411
Kent,	9,834	12,789
Bristol,	5,072	5,416

Of the population of 1830, were free white persons:

	Males.	Females.
Under 5 years of age	6,733	6,623
From 5 to 15	11,186	10,855
15 to 30	13,779	14,787
30 to 50	8,891	9,780
50 to 70	3,601	4,765
70 to 90	1,115	1,434
90 and upwards	28	44
Total	45,333	48,288

Free colored persons,

Under 10 years of age	334	358
From 10 to 24	500	598
24 to 36	317	415
36 to 55	239	350
55 to 100	151	266
100 and over	3	3
Total	1,544	2,020

Slaves, males 3; females 11.

White persons, deaf and dumb 48; blind 57; aliens 1,103.

Recapitulation.	Whites.	Free col'd.	Slaves	Total.
	93,621	3,564	14	97,199

Rhode Island is the most manufacturing section of the U. S., in proportion to its population. The manufactures are mostly of cotton; though there are many of woollen, cordage, drilling, &c. At Newport is a manufactory of lace. Upon Woonsocket falls alone are more than 20 different factories, producing between two and three millions of yards annually. Warwick is a flourishing manufacturing town, and Pawtucket has by far the largest manufactories in the state. The commercial prosperity of the state has kept pace with its manufactures. The amount of shipping is between 40 and 50,000 tons. The amount of imports for the year ending Sept. 1830, according to the report of the secretary of the treasury, was $488,756; exports, foreign, $71,985; domestic, $206,965; total exports $278,950. Tonnage entered 16,676; departed 14,094. In no part of the U. S. has banking been carried on to such an extent as in R. I. There are in this small state, 51 banks, with an aggregate capital of $6,723,296. Common schools were not early established and aided by legislative support in R. I. as in most of the N. E. states; though the interests of education have not been neglected. There are now but 323 public schools, with 17,034 pupils; toward the support of which the state pays about $10,000 annually. There are flourishing academies in several places in R. I. Brown University was originally founded at Warren, in 1764, whence it was removed in 1770 to Providence. (*See Providence.*) The religious denominations of this state are various. There are 16 Baptist churches, 12 ministers and 2,000 communicants; 10 Methodist preachers, and 1,100 members; 10 Congregational churches, 10 ministers, and 1,000 communicants; 2 Unitarian societies and 2 ministers; about 1,000 Sabbatarian communicants; 8 churches of Six-principle Baptists, and about 800 communicants; the Friends are numerous, and there is 1 Roman Catholic church, and some Universalists, &c. &c.

Rhode Island, isl. Newport co. R. I. about 3 1-2 ms. by 15, has a good soil, excellent in some parts, an agreeably varied surface, well cultivated and presenting a fine agricultural aspect. The fields are generally divided by excellent stone walls. The isl. contains the townships of Newport, Middletown and Portsmouth; and with several adjoining islands, forms the co. of Newport. It has Narragansett bay w., the E. Passage E. and Mount Hope bay N. which abound with fish. A mine of anthracite coal has been wrought to some extent in the N. part of the isl., but is not now used. The S. W. extremity of the isl. projects in such a manner as to protect the harbor of Newport on the E. and S. with its high ground. Towards the ocean it presents a lofty and precipitous bluff, where the waves often dash tumultuously; yet a considerable tract of land there possesses a deep and fertile soil of great value. Easton's bay, with a fine beach, indents the s. shore and nearly isolates the s. w. corner of the isl. While Newport was occupied by British troops, the American lines were for a time drawn across the neck of the peninsula, from Tamony hill to the beach.

RICEBORO', p-v. and st. jus. Liberty co. Geo. 34 ms. s. w. Savannah, and by p-r. 202 ms. s. E. Milledgeville. N. lat. 31° 45', long. W. C. 4° 30' w.

RICE CREEK SPRING, and p-o. nrthestrn. part of Richland dist. S. C. by p-r. 13 ms. N. E. Columbia.

RICHARDSONVILLE, p-o. Edgefield dist. S. C. by p-r. 77 ms. wstrd. Columbia.

RICHARDSVILLE, formerly Smith's tavern, p-o. Culpepper co. Va. by p-r. 71 ms. southwstrd. W. C.

RICHBORO', p-v. Bucks co. Pa. by p-r. 158 ms. N. E. W. C.

RICHFIELD, p-t. Otsego co. N. Y. 72 ms. w. by N. Albany, 13 N. w. Cooperstown, 18 s. w. Utica, 4 ms. by 8, has Herkimer co. N. and w. and Schuyler's, or Caniaderaga lake s. This affords one of the highest sources of Susquehannah r. The t. is rather uneven, with many small streams, of which some in the N. w. flow into Unadilla r.; has pretty good soil. It was first settled in 1791. Pop. 1830, 1,752.

RICHFIELD, p-v. Juniata co. Pa. by p-r. 61 ms. northwstrd. Harrisburg.

RICHFIELD, p-v. northeastern part of Medina co. O. by p-r. 130 ms. N. E. Columbus. Pop. of Richland tsp. 1830, 444.

RICHFORD, p-t. Franklin co. Vt. 50 ms. N. Montpelier, 24 N. E. St. Albans, with 23,040 acres; lies s. of the Canada line, and was chartered 1780. It is hilly E., and is crossed by Missisque r. which enters from Lower Canada, and flows from it into Berkshire, with rich meadows on its banks. It was first settled about 1790. Pop. 1830, 704.

RICHLAND, p-t. Oswego co. N. Y. 60 miles N. w. Utica, 27 N. w. Oswego v., s. Jefferson co., E. lake Ontario; has very good soil, more favorable to grass, bearing oak and chestnut, beech, &c. near the lake, and a large tract of white pine E. with other trees. The surface is uneven. It is crossed by Salmon r., and has also Little Sandy, Deer and Grindstone creeks, all which empty into the lake, and furnish mill seats. The village, sometimes called Pulaski, is 27 ms. N. E. Oswego, 30 s. Sacket's Harbor, 36 N. Salina, and 153 from Albany, is on Salmon creek 3½ ms. from the harbor at its mouth. The falls are just below. Great quantities of salmon and pickerel are caught here. There is a salt spring near the borders of Salina. Pop. 1830, 2,733.

RICHLAND, dist. S. C. bounded s. by Congaree r. separating it from Orangeburg; Congaree below, and Broad r. above Columbia, separating it from Lexington dist., N. Fairfield, and N. E. Kershaw, and E. Wateree, or the lower Catawba. Length diagonally from the junction of the Wateree and Congaree, to the extreme northwestern angle on Broad r. 50 ms., mean breadth 12, and area 600 sq. ms. N. lat. 34°, and long. 4° w. W. C. intersect at Columbia, the co. st. and st. of government of the state. The general slope to the s. s. E. in the direction of its bounding rivers. Much of the soil of this district is amongst the best in S. C. Pop. 1820, 12,321, and in 1830, 14,772.

RICHLAND, co. of O. bounded s. by Knox, Marion s. w., Crawford w., Huron N., Lorain N. E., Wayne E., and Holmes s. E. It is a sq. of 30 ms.; area 900 sq. ms. Lat. 40° 46', long. 5° 33' w. W. C. The northern border extending along N. lat. 40°, and also along the summit ridge between the valley of Ohio and that of Erie, merely gives source to creeks flowing towards the latter. The northwestern angle gives source to the higher fountains of Sandusky, and along the western border rise the extreme sources of Sciota r. Though two sides are thus drained, the body of the co. including at least seven-eights of its surface, is drained by, and gives source to, Mohicon branch of White woman's r., and slopes to the s. E. It is a comparatively elevated and level table land, which when compared with the known height, 768 feet of the water level at Conhocton, at the mouth of White woman's r., the table land of Richland co. must be 1,000 feet above the Atlantic tides. The soil is generally good. Chief t. Mansfield. Pop. 1820, 9,169, and in 1830, 24,006.

RICHLAND HILL, p-v. East Feliciana, parish of La. by p-r. 141 ms. N. W. New Orleans.

RICHLAND'S p-o. Onslow co. N. C. by p-r. 202 ms. s. E. by E. Raleigh.

RICHMOND, p-t. Lincoln co. Me. 15 ms. s. by w. Augusta, s. Kennebec co., w. Kennebec r. Pop. 1830, 1,308.

RICHMOND, p-t. Cheshire co. N. H. 70 ms. s. w. Concord, 72 N. N. w. Boston, 12 s. Keene, with 23,725 acres; is watered by streams of Millers and Ashuelot rs. and has a pretty level surface, with a good soil. Settled from Mass. and R. I. about 1758. Pop. 1830, 1,302.

RICHMOND, p-t. Chittenden co. Vt. 13 ms. s. E. Burlington, 24 N. w. Montpelier; first settled, 1775; deserted during the revolution; is crossed by Onion r. which has meadows on its banks; is joined by Huntington river, flowing in from the s. Several other streams afford mill seats. Pop. 1830, 1,109.

RICHMOND, p-t. Berkshire co. Mass. 130 ms. w. Boston, 6 w. Lenox, E. New York state; is in a pleasant and well cultivated valley, enclosed by the Taughkannuc mtns. The principal street runs through the town N. and s., and a w. branch of Housatonic river crosses the valley. Iron is taken from mines in this town, and wrought in Salisbury. Pop. 1830, 844.

RICHMOND, p-t. Washington co. R. I. 30 ms. s. s. w. Providence, N. Charles r., E. Wood r., about 6 ms. by 7, with 40 sq. ms.; is generally uneven, with some level ground, and is watered by the above-mentioned streams, with several of their branches. They join below, and form Pawcatuck r. The town is well supplied with mill seats, some of which are occupied by manufactories. Pop. 1830, 1,363.

RICHMOND, co. N. Y. which embraces Staten Island, is bounded by Newark bay and the Kills N., Hudson r. or the Narrows, between the outer and inner bays of New York E., Raritan bay S., and Staten Island sound W. The centre of it is about 11 ms. S. W. N. Y. city. It is 14 ms. long, N. E. and S. W., and the greatest breadth 8 ms. It contains 77 sq. ms. The S. end is in N. lat. 40° 29', and the W. 16' W. from N. Y. It contains 4 towns, with an agreeably varied surface, and some good land, which enjoys the advantage of being near a market. The N. Y. quarantine station, with 3 hospitals, and the Sailors' Snug Harbor and the Sailors' Retreat are near the N. E. corner of the island, in a pleasant situation. A little below, at the Narrows, there are fortifications on both sides for the defence of the entrance of N. York harbor. Forts Tomkins, Richmond and Hudson on this side. It is here 1,760 feet from land to land. A steamboat runs from N. Y. to the quarantine; and the steamboats which ply between the city and New Jersey, touch at other points on the N. shore. There is a large dyeing establishment. The st. jus. is the village of Richmond. There are several country houses of citizens on the island; but fevers and agues have prevailed here within a few years. Pop. 1820, 6,135, 1830, 7,082.

RICHMOND, p-v. and st. jus. Southfield, Richmond co. N. Y. 156 ms. S. Albany, 12 S. N. Y. on Staten island, is three quarters of a mile from sloop navigation in the Fresh Kills.

RICHMOND, p-t. Ontario co. N. Y. 232 ms. W. Albany, 16 S. W. Canandaigua, E. Hemlock lake and Livingston co. First settled, 1789, from Mass. It has Honeoye and Canadea lakes, with hilly and broken land. Allen's Hill village is in the N. E. corner. Pop. 1830, 1,876.

RICHMOND, p-v. in the estrn. part of Northampton co. Pa. by p-r. 13 ms. N. N. E. Easton, the co. st., and 203 ms. N. E. W. C.

RICHMOND, port, p-t. and st. jus. Henrico co. and st. of government of Va. situated on the left bank of James r., at the foot of its lowest falls, and head of tide water, by p-r. 122 ms. a little E. of S. S. W. W. C. N. lat. 37° 32', long. 0° 27' W. W. C. Whoever has seen the rounded hilly site of Baltimore, or indeed the northern part of Philadelphia, may have an idea of the rolling ground on which Richmond stands, except that the hills must be supposed higher and bolder in the latter case, than in either of the two others. A deep hollow ground divides Richmond into two unequal sections, the body of the city lying above this depression. The houses are neat rather than splendid. Amongst the public edifices, the capitol has excited the admiration of travellers for its chaste, yet beautiful proportions and commanding position. It has also an advantage in standing alone. Near the capitol are also the other public buildings for legal and political purposes. The public square is 8 acres in extent, and enclosed with a substantial iron railing. In Richmond, the Presbyterians, Episcopalians, Baptists, Methodists, Friends, Roman Catholics and Jews, have their places of public worship. Of the churches, one called the Monumental church, belonging to the Episcopalians, will long attract the spectator, and command a melancholy interest. It is standing on the site of a theatre which was, on the 26th Dec. 1811, consumed by fire, in which perished G. W. Smith, the governor of Va. and 71 other persons. A new theatre has been erected in another part of the city. This city contains also a state penitentiary, Lancasterian school, orphan asylum, poor house, public library, and a museum. As a commercial depot, Richmond is a city of great and increasing importance. It was established in 1742 by an act of assembly, and in 1780, became the seat of state government, and has gradually gained in wealth and population. The seat of government always gains something from being so; but it is only commerce and manufactures that can create to any great extent a city. As early as 1794, a canal was completed along that part of James r. impeded by falls. This opened to tide water the fertile valley above, and 220 miles of navigable channel had its shores improved. In 1794, the canal tolls amounted to $1,764; 1800, to 12,324; 1805, 16,749; 1810, 23,937; 1815, 24,645, and in 1820, 29,245. Vessels drawing 15 feet water can ascend to within 3 ms. below the city, and those of 7 or 8 to Rockets, or the port of the city. The almost uninterrupted health of Richmond has been a subject of true boast. "Richmond is one of the healthiest cities in the U. S., or perhaps in the world," says a writer of that place; the annual amount of deaths on an average, is 1 in 85: it has never been visited by yellow fever, or any violent or desolating disease. The progressive population of this place is, however, the most conclusive proof of its advance in physical and moral consequence. In 1810, it contained 9,735 inhabitants; in 1820, 12,067, and in 1830, 16,060.

RICHMOND, co. Va. bounded N. W. and N. by Westmoreland, Northumberland E., Lancaster S. E., and Rappahannoc r. separating it from Essex S. W. and W. Length 25 miles, mean width 8, and area 200 sq. ms. Lat. 37° 50', long. 0° 18' E. Chief t. Richmond C. House. Pop. 1820, 5,706, and in 1830, 6,055.

RICHMOND, C. H., p-o. and st. jus. Richmond co. Va. by p-r. a little E. of S. W. C. and 56 ms. N. E. by E. Richmond city. N. lat. 37° 55', long. 0° 18' E. W. C.

RICHMOND, co. of N. C. bounded by Marlboro' dist. S. C. S., Yadkin r. separating it from Anson co. N. C. W., Montgomery co. N. C. N., and Lumber river, separating it from

Moore N. E., Cumberland E., and Robeson S. E. Length 30 ms., mean breadth 18, and area 540 sq. ms. Lat. 35° N., and long. 2° 42′ W. W. C. intersect in this co. Slope southward in the direction of its rivers. Little Pedee rises in its southern section, as do several creeks, which flow into S. C. Chief town Rockingham. Pop. 1820, 7,537, and in 1830, 9,396.

RICHMOND, co. of Geo. bounded S. by Mount Beans creek, separating it from Burke co., Brier creek W. separating it from Jefferson, Columbia N. W., and Savannah r., separating it from Edgefield dist. S. C. E. Length from W. to E. 32 ms., mean breadth 12, and area 384 sq. ms. Lat. 33° 25′, and long. 5° W. W. C. intersect in this co. Though Brier creek, which forms its wstrn. boundary, flows S. E., the body of the co. has a slope almost exactly E. towards Savannah r. Chief town, Augusta. Pop. 1820, 8,608, and in 1830, 11,644.

RICHMOND, p-v. Fayette co. Ten. by p-r. 194 ms. S. W. by W. Nashville.

RICHMOND, p-v. and st. jus. Madison co. Ky. by p-r. 50 ms. S. E. Frankfort, and 27 S. S. E. Lexington. N. lat. 37° 43′, long. 7° 13′ W. W. C. Pop. 1830, 947.

RICHMOND, p-v. Jefferson co. O. by p-r. 143 ms. N. E. by E. Columbus.

RICHMOND, p-v. Wayne co. Ind. by p-r. 69 ms. E. Indianopolis.

RICHMOND, p-v. and st. jus. Ray co. Mo. by p-r. 149 ms. above and N. W. by W. Jefferson, and 284 ms. in a similar course from St. Louis.

RICHMOND DALE, p-v. in the southeastern angle of Ross co. O. by p-r. 58 ms. S. S. E. Columbus.

RIDGE, or Alluvial Way, a singular elevation about 30 feet high, in the cos. of Genesee, Monroe, and Niagara, N. Y. It extends about 78 ms. from Niagara r. almost to Genesee r. nearly parallel to the S. shore of lake Ontario, about 139 feet above the level of its waters, which are from 8 to 10 miles distant, and is supposed to have been formed at some long past period, by its waves or currents, when large tracts of country, now dry, were overflown. The ridge varies in breadth, and serves for the route of a good, level road, called the Ridge road, on which are several small villages.

RIDGE (The), p-v. near the extreme sthrn. point of St. Mary's co. Md. by p-r. 32 ms. S. E. Leonardstown, the co. st., and 95 ms. S. E. W. C.

RIDGE (The), p-v. in the eastern part of Edgefield district, S. C., by p-r. 40 ms. W. Columbia.

RIDGEBURY, p-v. Bradford co. Pa., by p-r. 150 ms. N. Harrisburg.

RIDGEFIELD, p-t. Fairfield co. Conn., 70 ms. S. W. Hartford, 10 S. W. Danbury, 55 N. E. N. York city, touches N. York state W., is varied by several ridges, with rocks of granite and limestone, and a good soil for grain and grass. Mill seats are supplied by branches of Saugatuck and Norwalk rs. There are several manufactories in the town, and some lime kilns. The land was purchased from the Indians in 1708. It is elevated, and Long Island sound is visible from different points; 14 ms. distant. The village is pleasantly situated in the 1st society. Pop. 1830, 2,323.

RIDGEVILLE, p-v. nrthrn. part of Warren co. Ohio, by p-r. 78 ms. S. W. by W. Columbus.

RIDGEWAY, p-t. Orleans co. N. Y., 26 miles N. W. Bavaria, E. Niagara co.; is crossed nearly through the centre by the ridge, and well watered by Oak Orchard and Johnson's creeks with branches, and by Erie canal, which lies S. It has a varied surface, and a variety of good soils. Oak Orchard cr. falls 30 feet just below the intersection with the canal. Pop. 1830, 1,972.

RIDGEWAY, p-v. in the nrthestrn. part of Jefferson co. Pa., by p-r. 165 ms. N. W. by W. Harrisburg.

RIGA, p-t. Monroe co. N. Y., 239 ms. W. by N. Albany, 11 W. S. W. Rochester, E. Genesee co.; with very good land; is crossed by Black creek from W. to E., and 2 small branches. Black creek is navigable in boats to West Pulteney village.

RILEY, t. Oxford co. Me., E. Coos co. N. H., 71 ms. W. by N. Augusta, is very rough and mountainous, S. Speckled mts. Pop. 1830, 57.

RINDGE, p-t. Cheshire co. N. H., 56 ms. S. W. Concord, 20 S. E. Keene, 50 N. N. W. Boston; 5 ms. by 7; with 23,838 acres; has a swelling surface, very good soil, formerly covered with beech, maple, birch, hemlock, &c., and contains 13 ponds. Of these Manomonack, Emerson's and Perley's ponds flow into Miller's river of Mass., and Long, Grassy and Bullet, into Contocook r. a branch of the Merrimack; the waters of those two great rivers being separated in one place only by a narrow ridge. Fish are abundant, and rendered these streams favorite resorts by Indians. Iron ore is found in Rindge. First settled 1752. Pop. 1830, 1,269.

RINGOES, p-v. Amwell, Hunterdon co. New Jersey, 17 ms. N. Trenton.

RING'S MILLS, and p-o. Belmont co. O., by p-r. 129 ms. E. Columbus.

RIPLEY, p-t. Somerset co. Me., 60 miles N. E. Augusta, W. Penobscot co.; is crossed by the upper part of Sebasticook r. Pop. 1830, 644.

RIPLEY, p-t. Chatauque co. N. Y., 336 ms. W. Albany, 12 W. Mayville, S. lake Erie, E. Pennsylvania; has a varied surface and soil, bearing oak, nut trees, maple, &c. It is crossed by Chatauque ridge, from 6 to 10 miles distant from the lake, with a gentle declivity of arable land towards the N., with a foundation of mica slate. The lands near the lake are very good, being alluvial, from 1 to 3 ms. wide. It is crossed by Chatauque creek of lake Erie, about 10 miles long. Pop. 1830, 1,647.

RIPLEY, p-v. on the right bank of Ohio river, sthrn. part of Brown co. Ohio, by p-r. 113 ms. s. s. w. Columbus. Pop. 1830, 572.

Ripley, co. Ind., bounded by Jefferson s., Jennings w., Decatur n. w., Franklin n., Dearborn e., and Switzerland s. e. Length 27 ms., mean breadth 16, and area 432 sq. ms. N. lat. 39°, and long. 8° 15′ w. W. C., intersect in this co. The wstrn. part of this co. gives source to the extreme estrn. branches of the South fork of White r., and which flow wstrd.; the residue is drained by crs. flowing sthestrd. into Ohio r. Chief t. Versailles. Pop. 1820, 1,822, and in 1830, 3,989.

Ripleyville, p-v. Huron co. O., by p-r. 101 ms. n, n. e. Columbus.

Rip Point, Nantucket isl., Mass. The n. e. Point of the island at the end of Sandy Point.

Ripton, p-t. Addison co. Vt., 26 ms. s. w. Montpelier; has Middlebury river s., and is mountainous, rough, and with few inhabitants. Pop. 1830, 605.

Ripton, village, Huntington, Fairfield co. Conn.

Rising Sun, p-o. Philadelphia co. Pa., by p-r. 139 ms. n. e. W. C.

Rising Sun, p-o. Cecil co. Md., by p-r. 89 ms. n. e. W. C.

Rising Sun, p-v. on the right bank of Ohio r., and in the sthrn. part of Dearborn county, Ind., by p-r. 112 ms. s. e. Indianopolis.

Ritchieville, p-v. Dinwiddie co. Va., by p-r. 42 ms. s. Richmond.

Rittersville, p-v. Lehigh co. Pa., by p-r. 181 ms. nrthestrd. W. C.

River Bank, p-v. Orange co. Va., by p-r. 104 ms. s. w. W. C.

Riverhead, t., st. jus. Suffolk co. N. Y., 90 ms. e. New York, 234 s. by e. Albany, on the n. side of Long Island, s. Long Island sound. It has Pequanic river and bay s., and Wading creek on a part of the n. w. boundary, where is a small harbor. One mile from the sound is a broken ridge; in other parts the surface is a little varied, bearing pine, with some oak, &c. Coasting vessels take wood and other articles to New York market; and those of 70 tons can go to the mouth of Pequanic creek 2½ ms. from the C. H. There are 6 small villages. Pop. 1830, 2,016.

River Styx, p-v. northern part of Medina co. Ohio, by p-r. 117 ms. n. e. Columbus.

Rives', p-o. in the nrthrn. part of Hall co. Geo., by p-r. 135 ms. nrthrds. Milledgeville.

Rixeyville, p-o. Culpepper co. Va., 67 ms. s. w. by w. W. C.

Roane, co. Ten., bounded s. by Monroe and MacMinn, s. w. Rhea, w. Bledsoe, n. w. Morgan, n. Anderson, e. Knox, and s. e. Holston r., separating it from Blount. Length 50 ms., mean breadth 15, and area 750 sq. ms. Lat. 36° n., and long. 7° 30′ intersect in this county. Holston and Clinch rivers unite to form Tennessee river, very near the centre of this county, and both the branches and the main stream below their junction flow s. w. by w. The nrthrn. section slopes sthrd. and is drained by Emery's r. a branch of Clinch r. Chief t. Kingston. Pop. 1820, 7,895, and in 1830, 11,341.

Roanoke, river of Va. and N. C. Taken in the utmost extent, Roanoke basin is the same as Albemarle, and includes the sub-basins or vallies of Roanoke proper and Chowan r. The latter has been noticed under its appropriate head, and to which the reader is referred. Advancing from s. to n. all the rivers beyond Roanoke, have their most remote fountains on the Atlantic side of Blue Ridge; but with the Roanoke a new feature appears. The Blue Ridge is pierced by that stream, which derives its higher fountains from the main Alleghany chain in Montgomery county, Va., and within 8 miles of the main channel of New river, and at an elevation without estimating the mtn ridges, of at least 2,000 ft. Issuing by numerous creeks from this elevated tract, and uniting into one stream near the border between Montgomery and Botetourt cos. it is here literally "The rapid Roanoke," having at Salem in the latter co. fallen 1,000 feet in little more than 20 ms. At Salem the water level is 1,002 feet by actual admeasurement, above mean Atlantic tide. Below Salem the river inflects 20 ms. in an eastern course, to its passage through Blue Ridge, and thence s. e. 25 ms. to its passage through South East mountain. Passing South East mountain between Bedford and Pittsylvania cos., the now navigable volume sweeps by an elliptical curve to nrthrd. and round to s. e. 50 ms. comparative course to the influx of Dan river, entering its right side from the w. part. (*See Dan river.*) Below the junction of these two rivers, the united waters in a course of a little s. of e. 60 ms. by comparative distance, reach tide water at Weldon, having fallen by a lengthened cataract over the primitive ledge. About midway between the influx of Dan river and Weldon, Roanoke leaves Va. and enters N. C. Mingling with the tide, the Roanoke by a very tortuous channel, but by comparative course flows s. e. 50 ms., and thence estrd. 25 ms. to its junction with Chowan river at the head of Albemarle sound. (*See Albemarle sound.*) The entire valley of Roanoke, if measured along the main stream or Dan r. is 250 ms., but the rs. wind over this space by channels of much greater length. By comparative courses it is 155 miles from Salem to Weldon, whilst from a report made by the Roanoke company, the intermediate channel is 244 ms. Taking these proportions, the length of this river by its meanders is about 400 ms. Including the whole Albemarle basin, it is 290 ms. from its outlet into the Atlantic ocean, to the fountains of Roanoke in Alleghany mtn., but with the Chowan and Dan vallies united to that of the principal river, the basin is comparatively narrow, being only 80 ms. where broadest, and not having a mean breadth above 50 ms., or an area exceeding 14,500 sq. ms. It is not, however, its extent which gives most interest to the Roanoke or Albemarle basin; it is at once a fine physical section and physical limit. The difference of arable level, amounts to at least 2,000 feet, and no two regions of the earth can differ in every feature more than do the truly beautiful hills and vales, on each side of the Appalachian chains, from the stagnant marshes and level plains towards the Atlantic ocean. Along the lower Roanoke commences, advancing from the N. the profitable cultivation of cotton, the fig tree begins to appear, rice can be produced, and in summer the advance towards the tropics is felt, and very distinctly seen on vegetation. Ascending the basin, the aspect of the northern states gradually appears, both on the features of nature and on cultivated vegetables. Wheat, rye, and other small grain, with meadow grasses, and the apple, flourish. The summers are cooler, and the winters have the severity suitable to relative elevation. Though the higher part of Roanoke is annually frozen, and for a shorter or longer period rendered unnavigable in winter, with lower Roanoke commences the region on the Atlantic coast where navigation remains open at all seasons. It is true that even Albemarle sound has been occasionally impeded with ice, but this phenomenon is rare. As a navigable channel following either branch, the importance of this basin is lessened by the shallowness of Albemarle sound—an irremovable impediment. In the progress of improvement, however, there is no doubt, but that by rail-road or canal, a water communication will be opened direct from tide water below Weldon to Chesapeake bay. In its actual state the rivers are navigable for boats to Salem on the Roanoke, and to Danbury in N. C. by Dan r. This was effected by side canals, sluices and other artificial improvements. (*See rail-roads and canals.*)

ROANOKE BRIDGE, and p-o. Charlotte co., Va., by p-r. 89 ms. s. w. Richmond.

ROARING CREEK, p-o. Columbia co. Pa., by p-r. 77 ms. N. Harrisburg.

ROBBINSTON, p-t. Washington co. Me., 192 ms. N. N. E. Augusta, is bounded E. by St. Croix r., and lies opposite St. Andrew's in New Brunswick. It has a few small ponds and streams. Pop. 1830, 616.

ROBBSTOWN, p-v. Westmoreland co. Pa., on the right bank of Youghioghany river, 206 ms. N. W. W. C. It is a small village of a single street along the r. bank.

ROBINS, island, Southold, Suffolk county, N. York.

ROBERTSON, co. Ten., bounded E. by Sumner, Davidson S., Dickson S. W., Montgomery W., and N. by Logan and Simpson counties, Ky. Length 32 ms., mean breadth 20, and area 640 sq. ms. Lat. 36° 25′ N., long. W. C. 9° 32′ W. Slope S. W. and drained by Red r. and other smaller branches of Cumberland r. Chief town, Springfield. Pop. 1820, 9,938, and in 1830, 13,272.

ROBERTSON, p-v. Giles co. Ten., by p-r. 67 ms. S. S. W. Nashville.

ROBERTSON'S STORE, and p-o. Pittsylvania co. Va., by p-r. 252 ms. S. W. W. C.

ROBERT'S STORE, and p-o. Shelby co. Ky., by p-r. 20 ms. W. Frankfort.

ROBERTSVILLE, p-v. Beaufort district, S. C., by p-r. 90 ms. wstrd. Charleston, and 160 ms. S. Columbia.

ROBESON, co. N. C., bounded by Richmond N. W., Cumberland N. and N. E., East Fork of Lumber river separating it from Bladen E., Lumber river separating it from Columbia S. E., Marion district S. C. S. W., and Marlboro' district, S. C., W. Length from S. to N. 50 ms., mean breadth 22, and area 1,100 sq. ms. Lat. 34° 40′, and long. W. C. 2° W. intersect in this co. Slope sthrd. and drained in that direction by Lumber river, or the higher part of Little Pedee. Chief t. Lumberton. Pop. 1820, 8,204, and in 1830, 9,433.

ROCHESTER, p-t. Windsor co. Vt., 30 ms. S. W. Montpelier, 20 S. E. Middlebury; first settled soon after the revolutionary war; is crossed by White r. from N. to S. which receives a branch near the centre, and both supply mill seats. The surface is mountainous, with much good soil. The village is on the east branch of White r.; 13 school districts.

ROCHESTER, p-t. Strafford co. N. H., 40 ms. E. Concord, 22 N. W. Portsmouth, 10 N. N. W. Dover, W. Salmon Falls river; is divided by Cocheco river, and has a part of Isinglass r. S. near its junction with Cocheco r. Norway Plains, near the centre, is a considerable village, and a great thoroughfare on Cocheco r. Both the principal streams afford valuable water power. At the falls of Cocheco river is Squamanagonnic village. The surface of the town is irregular, the soil generally good, with pine plains, some of which are favorable to corn, &c., and a tract of oak land W. A tract calllled Whitehall was burnt in 1761 and '62, when the seasons were very dry, and the soil was ruined. Squamanagonnic hill is the principal elevation. Incorporated 1722; now contains 60,000 acres; first settled 1728; was a frontier town till 1760, and suffered much. 29 soldiers from this town died in the revolution. Pop. 1830, 2,115.

ROCHESTER, p-t. Plymouth co. Mass. 48 ms. S. Boston, N. Buzzard's bay, is crossed by Mattapoiset r. whose branches rise in ponds here and in Middleboro'. This stream empties into Mattapoiset harbor, which puts up from Buzzard's bay. Sipican r. also flows into a small bay, after passing through several ponds. The soil is poor, and the inhabitants scattered. Pop. 1830, 3,556.

ROCHESTER, p-t. and st. jus. Ulster co. N. Y. 16 ms. S. W. Kingston; has Shawangunk mtns. S. E., and Rondout creek W., with several of its branches, which afford mill seats. It has pretty good land. Population, 1830, 2,420.

ROCHESTER, p-v. in Gates and Brighton tps. Monroe co. N. Y. 236 ms. W. N. W. Albany, 63 E. Lockport, 77 E. Lewiston, 7 S. Charlotte, is the most populous and important village in the state. It stands on the W. side of Genesee r. at the falls of that stream, and at the end of the great aqueduct of the

Erie canal. The rail-road which was constructed in 1832, to the foot of the falls, and head of navigation of Genesee r., to which vessels come up from lake Ontario, terminates here. The growth of this place was remarkably rapid, and caused by the opening of the canal, which afforded a channel of transportation, and encouraged the manufacture of flour. The river is now lined on both sides with flour mills, many of them of immense size, and constructed in the most substantial manner, being abundantly supplied with water power from the river. The village is ornamented with many fine buildings, public and private. In 1812 there were but 2 or 3 dwelling houses, of an inferior description, on the place now the seat of a large population, and of an active and lucrative business. The aqueduct of the Erie canal is built of hewn stone, and has 9 arches, each of 50 ft. chord, with an arch of 40 feet chord over the mill canal on each side. A navigable feeder here joins the canal from above the falls. The fall in Genesee r. at Rochester, is 92 feet. The flour mills, and the amount of flour made, increase from 15 to 20 per cent annually. About 1,000 or 1,200 bls. of flour are now made daily; and the mills could make 1,500 or 1,800. The millers employ large capitals, and frequent advances are made by them on the crops before they are gathered. In 12 months, ending in 1832, there were 240,000 barrels of flour manufactured in the village, and during the same period the amount paid for wheat by the millers amounted to $1,160,000. The principal manufactures of the village are

	Capital invested.	Amount manufactured annually.
Flouring mills,	$281,000	$1,331,000
Cotton goods,	50,000	30,000
Woollen do.	70,000	112,000
Leather, &c.	25,000	166,000
Iron work,	24,000	46,000
Rifles, &c.	3,000	5,000
Soap and candles,	6,000	45,000
Groceries, &c.	21,000	32,800
Tobacco,	4,500	18,000
Pail, sash, &c.	2,500	12,000
Boat building,	11,000	40,200
Linseed oil,	3,000	4,000
Globe building factories,	10,000	15,000
	$511,000	$1,857,000

The trade of the village in lumber, beef and pork, pot and pearl ashes, butter, cheese, lard, wool, &c. &c. is very considerable. There are in the place 3 Presbyterian, 2 Episcopal, 2 Methodist, 2 Friends, 1 Baptist, and 1 Roman Catholic churches; 1 daily and 5 weekly newspapers, and about 100 wholesale and retail stores. Population 1830, 9,207.

Rochester, p-v. Warren co. O. by p-r. 81 ms. s. w. by w. Columbus.

Rochester, p-v. northern part of Oakland co. Mich. by p-r. 43 ms. n. w. Detroit.

Rock, r. important stream of Huron and Il. having its remote sources in the former at lat. 44°, long. 10° 40′ w. W. C., and between lakes Huron and Winnebago. Flowing thence s. s. w. by comparative courses 100 ms. between, and very nearly parallel to the two Fox rs. to the influx of Goosekehawn from the n. w. Goosekehawn (*the river on which we live*), has its source a few miles sthrd. of the Portage between Ouisconsin and Fox r. of lake Michigan. Rising on a flat, and in winter and spring, a generally very wet region, the Goosekehawn in most of its course is in reality a congeries of lakes, and is marked on the maps as the Four Lakes; general course s. e., length 50 miles. Below the mouth of Goosekehawn, Rock r. maintains its original course 50 miles to the influx of Sugar creek, or rather Peektano r., also from the n. w. The Peektano is the most considerable branch of Rock r. and rises by numerous branches in Iowa co., Huron, between the Miss. and Ouisconsin. Comparative length, about 100 ms., and general course to the s. e. Sugar creek and Peektano are separate streams in Huron, and do not unite until the latter has flown 50, and the former 25 ms. in the northern part of Il. The main stream also enters Illinois about 25 ms. above the mouth of Peektano. It may be remarked, that Rock r. in all its length receives no tributary from the left above the size of a large creek, and that below the Peektano in a comparative course of s. w. by w. 100 ms. it is augmented by no confluent of consequence. It falls into the Miss. r. in Rock Island co. at lat. 47° 27′ after a comparative course of 250 ms. The utmost breadth of its valley, 110 ms., from the extreme sources of Peektano to those of Kishwaukee creek; but being very narrow at both extremes, the mean width is about 30 ms. and area 7,500 sq. ms. This valley has that of Fox r. of lake Michigan n., Fox r. of Illinois e., Illinois proper s. e., Miss. r. s. w. and w., and Ouisconsin n. w. It is yet but thinly peopled by whites in any place, and far the greatest part is wilderness.

Rock creek, a small stream of Maryland, and of D. C. gains importance only as it separates the city of Washington from Georgetown. This creek has its extreme source about 4 ms. wstrd. Mechanicsville, Montgomery co. Md. heading with the East branch of Potomac r. at an elevation above tide water at Georgetown of 500 feet. The entire length of the creek, following its valley, is about 28 ms. The fall being upwards of 17 feet to the mile, and that fall being in many places far above the mean, renders it an excellent mill-stream.

Rock and Cave, p-v. on Ohio r. extreme southeastern part of Gallatin co. Il. by p-r. 147 ms. s. s. e. Vandalia, and 20 ms. s. Shawneetown. This place takes its name from enormous precipices of limestone rock, which rise from the western bank of Ohio r., and into which extend caves of unknown extent;

one of which yawns an immense and really awful opening, and when seen, as it was by the author of this article, exhibited a most imposing spectacle at the close of day. The walls were then, 1799, sculptured with innumerable names. The adjacent country was then an uncultivated wild on both sides of the river.

ROCKAWAY, Hempstead, N. Y., on Long Island, 29 ms. from New York, a place resorted to for sea bathing. The beach which bears this name is extensive, partly in this town, and partly in Jamaica, and abounds in sea fowl, as the water does in fish. The sea beats up from the s. upon this beach, there being no protection against the waves; and bathing is sometimes attended with considerable risk.

ROCKBRIDGE, co. Va. bounded s. w. by Botetourt, Alleghany co. w., Mill, or more correctly, Kittatinny mtn. separating it from Bath n. w., Augusta n. e., and Blue Ridge, separating it from Nelson n. e., Amherst e., and Bedford s. e. Lat. 37° 45′, and long. 2° 30′ w. W. C. intersect in this co. Slope sthwrd. and drained entirely by North r. branch of James r. and its confluent creeks. The co. occupies a part of the fine valley which flanks Blue Ridge on the n. w., and derives its name from the celebrated natural bridge, which extends over a creek near the sthrn. border. Chief town, Lexington. Pop. 1820, 11,945, 1830, 14,244.

ROCKBRIDGE, p-v. western part of Gwinnett co. Geo. 14 ms. wstrd. Lawrenceville, the co. st., and 107 n. w. Milledgeville.

ROCKCASTLE, co. Ky. bounded by Pulaski s. w., Lincoln w., Garrard n. w., Madison n. and n. e., and Rockcastle creek, separating it from Laurel s. e. It is a square of about 18 ms. each way, area 324 sq. ms. Lat. 37° 20′, and long. 7° 14′ w. W. C. intersect in this co. It is a table land between the vallies of Ky. and Cumberland rs., and gives source on the n. w. side to Dick's r. of the former, and on the opposite section to Bucks and Rockcastle branches of the latter stream. Rockcastle creek, from which the co. takes its name, is the extreme northern fountain of Cumberland r. rising in Laurel and Madison cos., and flowing s. s. w. falls into Cumberland r. between Pulaski and Whitley cos. Chief town of Rockcastle, Mount Vernon. Pop. of the co. 1820, 2,249, 1830, 2,865.

ROCK CREEK, p-o. Orange co. N. C. by p-r. 64 ms. n. w. by w. Raleigh.

ROCK CREEK, p-o. Muscogee co. Geo. by p-r. 130 ms. s. w. by w. Milledgeville.

ROCK CREEK FORD and p-o. Jennings co. Ind. by p-r. 51 ms. s. s. e. Indianopolis.

ROCKDALE, p-v. northwestern part of Crawford co. Pa. 8 ms. n. w. Meadville, the co. st., and by p-r. 305 ms. n. w. W. C.

ROCK FISH, p-v. Duplin co. N. C. by p-r. 136 ms. s. e. Raleigh.

ROCKFORD, p-v. and st. jus. Surry co. N. C. on the Yadkin r. 151 ms. by p-r. n. w. by w. Raleigh. Lat. 36° 18′, long. 3° 40′ w. W. C.

ROCK HALL, p-v. Kent co. Md. by p-r. 68 ms. estrd. W. C.

ROCK HILL, p-o. Bucks co. Pa. by p-r. 56 ms. nthrd. Phil.

ROCKHOLD's Store and p-o. Sullivan co. Ten. 327 ms. a little n. of e. Nashville.

ROCKINGHAM, co. N. H. bounded by Strafford co. n. and n. e., the Atlantic ocean e., Massachusetts s., Hillsboro' co. w., is the only maritime co. in the state. It is of an irregular triangular shape, about 30 ms. by 50, and contains about 1,034 sq. ms. The surface is irregular, but without any more considerable eminences than Saddleback mtn., Fort hill, Bean's hill and Catamount hill. Merrimack r. runs near the bounds of this co. on the w. and s., and several streams flow hence into it, as well as in other directions. In the e. and s. e. are Lamprey, Exeter, Beaver and Spiggot rs. Great bay, in the n. e. is connected with Piscataquay river. There are other sheets of water, as Massabesick pond, Island, Great, Country, Pleasant, Turkey, Long, and Turtle ponds. Agriculture is of an older date, and in a more flourishing condition than in any other part of the state. There is but one sea port, which is also the only one in N. H.; this is Portsmouth. The tonnage owned here in 1831, was 18,243 30. Concord is the capital of the state. The manufactures are numerous and various. Pop. 1820, 55,246, (53.4 to a sq. m.), 1830, 44,325.

ROCKINGHAM, p-t. Windham co. Vt. 85 ms. from Montpelier, 22 Windsor, 25 Brattleboro', w. of Conn. r. and opposite Charlestown, N. H.; contains 24,955 acres; first settled, 1753. The inhabitants for some years neglected agriculture, and attended chiefly to fishing for Salmon at Bellow's falls. The t. is crossed by Williams r., a branch of the Conn., and affords mill seats as well as Saxton's r. The surface is irregular, but the soil good. Bellow's falls are near the s. e. corner of the town. The river flows for some distance with a smooth current through fine meadows, and is about 120 yards wide, till at the falls it is suddenly narrowed into 2 channels, each about 90 feet across. When the water is low, all the stream rushes through a chasm between the granite rocks only 16 ft. wide. There are several sudden descents in the river within a short distance, altogether being about 50 feet; but a canal has been constructed round them on the w. bank, through which pass flat bottomed boats, rafts, and small steamboats. Salmon formerly swam up the river beyond these falls, but shad have never been caught north of this spot. The first bridge ever built over Conn. r. was constructed here in 1785, by col. Enoch Hale; and the second was not erected till about 1792. Some interesting minerals are found near this spot. There are 3 villages, Saxton's r. village, Rockingham, and Bellow's falls. The last is in a picturesque situation, and contains several handsome dwellings. Pop. 1830, 2,272.

Rockingham co., Va. bounded s. w. by Augusta, w. and n. w. by the Great N. mountain, Pendleton w., and Hardy n. w., Shenandoah co. n., Page n. e., and Blue Ridge separating it from Orange e. Breadth 25 ms., mean length 35, and area 875 square ms. Lat. 37° 30′, long. W. C. 1° 45′ w. Slope to the n. e., and entirely drained by the main stream and branches of Shenandoah r. The surface is generally hilly and in part mountainous, but much of the bottom soil excellent. It is a grain district. Chief town, Harrisonburg. Pop. 1820, 14,784, 1830, 20,683.

Rockingham, co. N. C. bounded by Caswell e., Guilford s., Stokes w., Patrick, Va., n. w., Henry, Va., n., and Pittsylvania n. e. Length from e. to w. 30, width 22, and area 660 square ms. Lat. 36° 24′, long. W. C. 2° 48′ w. This co. is a table land, from which issue to the s. e. the extreme sources of Haw r. and of course Cape Fear r. The nthrn. part is traversed in a n. e. by e. direction by Dan r. Chief town, Wentworth. Pop. 1820, 11,474, 1830, 12,935.

Rockingham, p-v. and st. jus. Richmond co. N. C. by p-r. 113 ms. s. w. by w. Raleigh. Lat. 35° 03′, long. W. C. 2° 49′ w.

Rock Island, p-v. Warren co. Ten. by p-r. 87 ms. s. e. by e. Nashville.

Rock Island, co. Il. as laid down by Tanner in his improved map of the U. S. extends along the left bank of Miss. r. above and below the mouth of Rock r., bounded n. and n. e. by S. Ann creek, separating it from Joe Daviess co., s. e. by Rock r. separating it from Henry, s. w. by Mercer, and w. by the Miss. r. Length from s. w. to n. e. 64 ms., mean breadth 10 ms., and area 640 square miles. Extending in lat. from 41° 20′ to 41° 53′ and in long. from W. C. from 13° to 14° w. Slope s. w. in the general direction of both the Miss. and Rock r. Chief town, Fort Armstrong. This co. is not named in either the p-o. list or census table.

Rock Island, p-v. Adams. co. Il. by p-r. about 150 ms. n. w. Vandalia.

Rockland, p-t. Sullivan co. N. Y. s. w. Ulster co., s. e. Delaware co., has not a very good soil, and is generally leased. It is watered by Willimemock, Big and Little Beaver creeks. Pop. 1830, 547.

Rockland co., N. Y. bounded by Hudson r. or Westchester co. e., New Jersey s. w., Orange co. n. w., is in the form of a triangle, and contains 161 square ms. There are 4 towns. The chief is Clarkstown. The surface is mountainous and broken by the Highlands, with large and fertile vallies, and much arable land and pasture on the Uplands. It is crossed by Ramapo r. and has several streams which flow into Hackensack and Passaic rs.; all these furnish valuable mill seats. There are also several fish ponds on the high lands. The Nyak hills furnish good sand stone, of which the state capitol at Albany was chiefly built. This co. comprises a tract which was of considerable importance in the revolutionary war. Ramapo mills form a considerable manufacturing village on Ramapo r. at the w. corner of the co. 30 ms. from N. York, and 14 w. Hudson r. This establishment comprehends 4,000 acres, and about 100 buildings, including dwelling houses, mills, &c. This com. was incorporated in 1824, with a capital of 400,000 dollars. There are a large rolling and slitting mill, a manufactory of cut nails, employing 100 men, a brick cotton mill of 5,000 spindles and 80 power looms, a grist mill, and a saw mill. The v. contains 700 inhabitants. There is a woollen manufactory, and at Haverstraw Messrs. Phelps & Peck's rolling and slitting mill and iron wire mill. Pop. 1825, 8,016, 1830, 9,388.

Rock Mills, p-o. Culpepper co. Va. by p-r. 75 ms. s. w. W. C.

Rock Mills, p-o. Anderson dist. S. C. by p-r. 144 ms. n. w. Columbia.

Rock Mills, p-o. Hancock co. Geo. by p-r. 41 ms. n. e. Milledgeville.

Rock Port, on the sthrn. shore of lake Erie, p-v. in the northwestern angle of Cuyahoga co. O. by p-r. 146 ms. n. n. e. Columbus. Pop. of the tsp. 1830, 361.

Rockport, p-v. and st. jus. Spencer co. Ind. on the right bank of Ohio r. by p-r. 167 ms. s. s. w. Indianopolis. Lat. 37° 57′, long. W. C. 10° 06′ w.

Rock Rest, p-v. Chatham co. N. C. by p-r. 40 ms. wstrd. Raleigh.

Rock Run, p-o. n. e. part Harford co. Md. by p-r. 12 ms. n. e. by e. Belair, the co. st. and 38 n. e. Baltimore.

Rock Shoal, p-o. Estill co. Ky. by p-r. 95 ms. s. e. Frankfort.

Rock Springs, p-v. Cecil co. Md. by p-r. 79 ms. n. e. W. C.

Rock Spring, p-v. Pickens dist. S. C. by p-r. 149 ms. n. w. by w. Columbia.

Rock Spring, p-v. St. Clair co. Il. by p-r. 68 ms. s. w. Vandalia.

Rocktown, p-v. Harrison co. O. by p-r. 143 ms. a little n. of e. Columbus. Pop. of the tsp. 1830, 708.

Rockville, p-v. and st. jus. Montgomery co. Md. 15 ms. n. w. W. C. 37 ms. s. w. by w. Baltimore, and by p-r. 52 ms. a little n. of w. Annapolis. Lat. 39° 05′, long. W. C. 0° 7′ w. It is a neat and rather close built village, but consists chiefly of one street along the main turnpike or what ought to be a turnpike from W. C. to Frederick. Contains the co. buildings, an academy for young men, two printing offices, several stores and taverns, and in 1830, a pop. of 555.

Rockville, p-v. Putnam co. Geo. by p-r. 33 ms. n. w. Milledgeville.

Rockville, p-v. Monroe co. Ten. by p-r. 151 ms. s. e. by e. Nashville.

Rockville, p-v. and st. jus. Parke co. Ind. by p-r. 68 ms. w. Indianopolis. Lat. 39° 40′, long. W. C. 10° 16′ w. Pop. 1830, about 500.

Rocky Comfort, p-v. Gadsden co. Flor. 22 ms. s. w. Tallahassee.

Rocky Hill, p-v. Barren county Ky. by

post-road 138 miles southwest of Frankfort.

Rocky Mount, p-v. and st. jus. Franklin co. Va. on a branch of Pig r. a confluent of Roanoke, by p-r. 263 ms. s. w. W. C. Lat. 36° 57', long. W. C. 2° 50' w.

Rocky Mount, p-v. sthrn. part Nash co. N. C. by p-r. 54 ms. E. Raleigh.

Rocky Mount, p-v. Fairfield dist. S. C. by p-r. 55 ms. a little E. of N. Columbia.

Rocky Spring, p-v. Claiborne co. Miss. about 60 ms. N. E. Natchez.

Rocky Spring, p-v. Granger co. Ten. by p-r. 245 ms. E. Nashville.

Rodman, p-t. Jefferson co. N. Y. 12 ms. E. lake Ontario, 7 s. Black r. is watered by the N. branch of Sandy creek, and small streams flowing into the other branch; adjoins Pinckney, Lewis co. on the E. It was first settled 1801, has a good soil, yielding various crops, and favorable to fruit. There are remains of small ancient mounds, and fragments of utensils, &c. are found, which mark it as once a favorite resort of the Indians. Pop. 1830, 1,901.

Rodney, p-o. Jefferson co. Miss. about 15 ms. nrthd. Natchez.

Roger's, p-o. Sangamon co. Il. by p-r. 94 ms. N. N. W. Vandalia.

Roger's Store, and p-o. Wake co. N. C. 14 ms. from Raleigh.

Rogersville, p-v. Anderson dist. S. C. by p-r. 147 ms. N. W. Columbia.

Rogersville, p-v. Lauderdale co. Ala. by p-r. 150 ms. N. Tuscaloosa.

Rogersville, p-v. and st. jus. Hawkins co. Ten. by p-r. 264 ms. a little N. of E. Nashville. Lat. 36° 24', long. 5° 48' w. W. C.

Rohrsburg, p-o. Columbia co. Pa.

Role's Store, and p-o. Wake co. N. C. by p-r. 15 ms. nthrd. Raleigh.

Rome, town, Kennebec co. Me. 22 ms. N. Augusta, s. Mercer, Somerset co., has several large ponds on its s. and E. borders, which empty by different channels into Kennebec r. Its form is irregular. Pop. 1830, 883.

Rome, p-t. and half capital, Oneida co. N. Y. 110 ms. W. Albany, 16 N. W. Utica, has the head of boat navigation of Mohawk r. and that of Wood creek, connected by a canal, 1 1-2 ms. long, which forms a link between the waters of the Hudson and of lake Ontario. Several brooks flowing into the two streams water different parts of the town, which is of irregular form. The soil is generally uneven and of very good quality; but the borders of Wood creek are low, level and moist. The land is generally held on lease. Fort Stanwix, built here by Great Britain in 1758, and which cost 266,400 dollars, was rebuilt by the Americans in the revolutionary war, and called fort Schuyler. This frontier post was reduced to great straits by an expedition of Canadians and Indians, under Col. Johnson, but defended with great bravery. The ruins of it are now hardly distinguishable. The battle of Oriscany was fought here, in which Gen. Herkimer fell. Here was formerly a carrying place, before the canal was constructed, and on the route passed a considerable amount of Indian merchandize in early times. The village is on the N. side of the canal, and 1-2 m. N. Erie canal. An arsenal of the United States stands three hundred yards north Erie canal, on the height of land between the streams, and was built in 1816, for a subordinate depot, under the ordnance department. There is a building 40 by 96 feet, 3 1-2 stories high, a magazine 19 by 65 feet, with a stone wall, 15 feet high around it, officers' quarters, &c. &c. Pop. 1830, 4,360.

Rome, p-o. Smith co. Ten. about 50 ms. northeastward Nashville.

Rome, p-v. Trumbull co. O. by p-r. 183 ms. N. E. Columbus.

Rome, p-v. and st. jus. Perry co. Ind. on the right bank of the Ohio r. opposite Stephensport, Breckenridge co. Ky. by p-r. 143 ms. a little W. of s. Indianopolis. Lat. 37° 58', long. W. C. 9° 36' w.

Romeo, formerly called Indian village, p-v. Macomb county, Mich., by p-r. 56 ms. N. N. E. Detroit.

Romney, p-v. and st. jus. Hampshire co. Va. on the right bank of the south branch of Potomac, by p-r. 116 ms. a little N. of W. W. C. 39 ms. in a similar direction from Winchester, and 28 s. Cumberland, in Md. Lat. 39° 20', long. W. C. 1° 42' w. Pop. 1830, 346, of whom 100 were colored persons.

Romulus, p-t. Seneca co. N. Y. 6 ms. N. Ovid, 12 s. Waterloo, has Cayuga lake and co. E., Seneca lake and Ontario co. W., has very good land, which is all cultivated, with few mill streams. The surface has a gentle ascent from the lakes, and the rocks beneath are slate and secondary limestone. Appletown, a small village in the N. W. corner, on the lake shore, has its name from the remains of some ancient Indian orchards which still exist. This town includes part of the Cayuga Indian reservation. Population 1830, 2,089.

Root, p-t. Montgomery co. N. Y. 12 ms. s. w. Johnstown, s. Mohawk r., N. Schoharie co., was formed from the E. part of Canajoharie, and W. part of Charlestown, in 1823. The rocky eminence, called the nose, and Mitchell's cave, are in this town. In the N. passes the Erie canal, where it strikes a narrow tract of primitive rocks. There is but one other disclosure of a primitive formation along the whole canal route. Pop. 1830, 2,750.

Rootstown, p-v. Portage co. O. by p-r. 131 ms. N. E. Columbus.

Roscoe, p-v. Jackson tsp. Coshocton co. O. by p-r. 83 ms. N. E. by E. Columbus. Pop. 1830, 81. This place was formerly Caldersburgh.

Rose, p-t. Wayne co. N. Y. Pop. 1830, 1,641.

Rosedale, p-v. Madison co. O. by p-r. 26 ms. W. Columbus.

Rosehill, p-v. Lee co. Va. by p-r. 20 ms. westrd. Jonesville, the co. st., and 412 miles

s. w. by w. W. C. It is the extreme southwestern p-o. in Va.

Rosshill, p-v. Wilkinson co. Miss. by p-r. about 20 ms. sthrd. Natchez.

Rossland, p-v. Cambria co. Pa. by p-r. 182 ms. n. w. W. C.

Rose Mills, and p-o. Amherst co. Va. by p-r. 170 ms. s. w. W. C.

Roseville, p-v. Loudon co. Va. by p-r. 38 ms. wstrd. W. C.

Roseville, p-v. Muskingum co. Va. by p-r. 69 ms. estrd. Columbus.

Rossville, p-v. Parke co. Ind. by p-r. 78 ms. w. Indianopolis.

Ross, co. O. bounded on the s. by Pike, s. w. Highland, n. w. Fayette, n. Pickaway, n. e. Hocking, and s. e. Jackson. Length from e. to w. 34 ms., mean breadth 22, and area 748 square ms. Lat. 39° 20′, and long. W. C. 6° w., intersect near Chilicothe and near the centre of this co., and near the same point the Sciota r. receives from the w. Paint creek. The former traversing the co. in a s. s. e. direction. It is also traversed in all its breadth by the Ohio and Erie canal, which follows the right or wstrn. bank of Sciota. The face of the co. is peculiarly and finely diversified. Soil productive. Chief town, Chilicothe. Population 1820, 20,619, 1830, 24,068.

Ross' p-o. Anderson co. Ten. by p-r. 201 ms. e. Nashville.

Rossie, p-t. St. Lawrence co. N. Y. 29 ms. s. s. w. Ogdensburgh, has Jefferson co. s. w. It is a large triangle, and is crossed through the middle by Indian r. the w. branch of the Oswegatchie, which affords water power to the iron works at the head of Black lake. Oswegatchie r. also runs for some distance in this town. In the w. is Chippeway bay, in St. Lawrence r., which contains numerous little islands belonging to the group called the Thousand islands. Limestone and iron ore abound, with granite, quartz, &c. The surface is various, partly almost mountainous and partly level. The Oswegatchie is a public highway from Streetor's mills in this town to its mouth at Ogdensburgh. At the village the iron works are quite large, and owned by Mr. David Parish. Population 1830, 641.

Rosstraver, tsp. and p-o. watrn. part of Westmoreland co. Pa. by p-r. 212 ms. n. w. W. C. Pop. of the tsp. 1830, 1,721. It is the watrn. tsp. of the co. and lies between the Youghioghany and Monongahela rs.

Rossville, p-v. watrn. part York co. Pa. 15 ms. n. w. by w. of the borough of York, 17 ms. s. Harrisburg, and 100 n. W. C.

Rossville, p-v. Cherokee Nation, Geo., by p-r. 250 ms. n. w. Milledgeville. In the p-o. list it is marked as 56 ms. farther from Milledgeville than is new Echota.

Rossville, p-v. Butler co. O. on Miami r. 25 ms. n. Cincinnati. Pop. 1830, 639.

Rotherwood, p-v. estrn. part of Carroll co. Geo. 143 ms. n. w. by w. Milledgeville.

Rotterdam, p-t. Schenectady co. N. Y., 4 ms. s. w. Schenectady, on the s. side of Mohawk river, n. Albany co., e. Montgomery co. The land is almost all of excellent quality particularly the large meadows on the r's. bank. Nine small islands in the Mohawk are also well cultivated. The inhabitants are of Dutch extraction. The Erie canal passes through the meadows not far from the river, descending from higher ground by 3 locks. There are several manufactories of different kinds. Pop. 1830, 1,481.

Rough Creek Church, and p-o. Charlotte co. Va., by p-r. 105 ms. s. w. Richmond.

Roulette, p-v. Potter co. Pa., by p-r. 292 ms. n. W. C.

Round Prairie, p-o. Callaway co. Mo., 39 ms. n. n. e. Jefferson.

Round Top, the highest eminence of Catskill mtns., Greene co. N. Y.

Rouse's Point, formerly supposed to belong to Champlain, Clinton co. N. Y., was found to lie n. of lat. 45°, and of course it was decided that it lay in Canada.

Rowan, co. N. C., bounded by Montgomery s. e., Cabarras s., Iredell w., Surry n., and Yadkin river separating it from Davidson e. Length from s. to n. 40 miles, mean breadth 20, and area 800 sq. ms. Lat. 35° 45′, long. W. C. 3° 36′ w. The slope almost due east towards the deep valley of the Yadkin. Chief town, Salisbury. Pop. 1830, 20,786. In 1820, Rowan contained the space e. of Yadkin, now Davidson co., and contained then 26,009 inhabitants; the two counties now contain a pop. of 34,175.

Rowanty, p-v. Sussex co. Va., on Rowanty creek, 43 ms. s. Richmond.

Rowe, p-t. Franklin co. Mass., 130 ms. n. w. Boston, s. Vt.; is elevated, and near the head waters of Deerfield river, at the base of Hoosac mountain. Fort Pelham was built here about the year 1744, being one of the line of forts for the protection of the frontier against savage incursions. Pop. 1830, 716.

Rowlandsville, p-o. Cecil co. Md., by p-r. 80 ms. n. e. W. C.

Rowley, p-t. Essex co. Mass., 28 ms. n. e. Boston, 16 n. e. Salem, 6 s. Newburyport, and w. Massachusetts bay; has much sand and salt marsh e., with very good land in other parts, on hills and in vallies; well watered by Rowley and Parker rs., and other streams. The town was settled in 1639, from Yorkshire, Eng. under Rev. E. Rogers. It includes Plumb island, and extends about 4 ms. by 13, including Plumb island sound. The hills w. are the highest land in the co. Population 1830, 2,044.

Roxboro', or Levering's p-o. in Roxboro' township, Philadelphia co. Pa., 8 ms. nrthrd. Phil. Pop. township 1830, 3,334.

Roxboro', p-v. and st. jus. Person co. N. C., by p-r. 60 ms. n. w. Raleigh, and 271 s. s. w. W. C. Lat. 36° 24′, long. W. C. 2° w.

Roxbury, p-t. Cheshire co. N. H., 5 miles e. Keene, 60 w. s. w. Concord, 76 n. w. Boston; is very small, with only 6,000 acres, and separated from Keene by the n. branch of

Ashuelot r. This is joined s. w. by Roaring brook, which is in the s., and has good meadows. Roaring brook pond is E. The surface is uneven, with good grazing. Population 1830, 322.

ROXBURY, t. Washington co. Vt., 15 ms. N. w. Montpelier, 45 N. w. Windsor; first settled 1789; is on the height of land between Onion and White rs., into both which streams several brooks flow. The soil is uneven, but good for grass and grain, bearing hard wood, with some evergreens. Slate, with crystals of pyrites are found E. Pop. 1830, 737.

ROXBURY, p-t. Norfolk co. Mass., 2½ ms. s. w. Boston, lies s. Charles river or bay, on which are 1,000 acres of marsh, and communicates with Boston by a well built street extending along the Neck. In the middle part of the town the soil is fertile and well cultivated; and s. w. is Jamaica Plain, 1 mile by 2, covered with gardens and country seats. About ⅓ part s. E. is rough and rocky land. Settled 1630, by John Pyncheon and others. Rev. John Eliot, called the apostle to the Indians, became pastor of the church here in 1632. He had great success in christianizing and civilizing the savages, and translated and published the scriptures in their language. Gen. Warren. who fell at Bunker's hill, was born here. Jamaica pond, a beautiful sheet of water, supplies the Boston aqueduct. Pop. 1830, 5,247.

ROXBURY, p-t. Litchfield co. Conn., 46 ms. s. s. w. Hartford, 32 N. w. New Haven, N. N. Haven co.; about 4 ms. by 6½; with about 26 square ms.; has a varied surface, bearing nut trees, &c. The rocks are granite, with some variation; and iron ore exists here. Shepaug river, a small branch of the Housatonic, runs nearly s. through the t. Pop. 1830, 1,122.

ROXBURY, p-t. Delaware co. N. Y., 56 ms. s. w. Albany, 22 E. Delhi, 49 w. Cattskill; lies s. of Schoharie co., and w. of Greene co. A pond gives rise to Papachton r., the E. branch of the Delaware, and on it are mill seats. The surface is mountainous, and some of the vallies have good land. It was settled from the eastern states about 1790. Population 1830, 3,234.

ROXBURY, t. Morris co. N. J., 45 miles N. Trenton; is divided on the N. w. and w. from Sussex and Warren cos. by Musconetcong r., which rises N. in Hopatung pond. The surface is elevated and mountainous; Schooley's mountain extending into the w. part. On it is Budd's pond, which, with a smaller one near Hopatung pond, gives rise to the south branch of Raritan r. A very narrow ridge, in this land, therefore, divides the waters of the Hudson and Delaware. Flanders, Draketown, Drakesville and Stanhope, are villages partly in this town. Pop. 1830, 2,262.

ROXBURY, p-v. northern part of Franklin co. Pa., by p-r. 13 ms. N. Chambersburg, the co. seat, and 103 ms. N. N. w. W. C.

ROYAL OAK, p-v. Oakland co. Mich., 14 ms. northward Detroit.

ROYALTON, p-t. Windsor co. Vt., 31 ms. s. Montpelier, 25 N. w. Windsor; first settled 1771, and the buildings were burnt by Indians from Canada in Oct. 1780. Several persons were killed, and 28 of the inhabitants were carried captive to Canada, all of whom except one were ransomed and returned. The soil is good, though the surface is mountainous. The town is crossed by White r. which is here joined by two branches; and along the banks are rich meadows. The v. is in a pleasant situation, on White r. near the centre of the town, and contains an academy, &c., incorporated in 1807. Pop. 1830, 1,893.

ROYALTON, p-t. Niagara co. N. Y., 26 ms. E. Lewiston, 6 E. Lockport, w. Genesee co.; has Tonawanta creek s., which divides it from Erie co. It is crossed by the mountain ridge and Erie canal. Lockport village is situated at the spot where they cross each other. The canal descends by 5 double, combined locks, after passing through a deep rock cutting for a great distance, and affords most valuable mill sites, which are supplied by the waste water, and some of them occupied. The land in this town, though until recently but little occupied, is generally good. Pop. 1830, 3,138.

ROYALTON, p-v. Fairfield co. Ohio, by p-r. 36 ms. s. E. Columbus.

ROYALSTON, p-t. Worcester co. Mass., 70 ms. w. Boston; has a good soil, but an uneven surface, & is watered by several streams, the principal of which is Miller's r. in the s. E., which, as well as Tully's r., affords good mill sites. There are several mills and factories. The Royalston cotton and woollen factory on Miller's river, was incorporated 1813, with a capital of $50,000. Settled 1762. Pop. 1830, 1,493.

ROYSE, mountain, Coos co. N. H., in the ungranted lands N. Chatham, near Me., and the Androscoggin.

RUCKERSVILLE, p-v. Elbert co. Geo., by p-r. 108 ms. N. N. E. Milledgeville.

RUCKMANVILLE, p-v. Bath co. Va., by p-r. 180 ms. s. w. by w. W. C.

RUDDLE'S MILLS, and p-o. Bourbon co. Ky., by p-r. 45 ms. E. Frankfort.

RUGGLES, p-v. sthestrn. part Huron co. O., by p-r. 100 ms. N. N. E. Columbus.

RUMFORD, p-t. Oxford co. Me., 20 ms. N. Paris; has Androscoggin river on its s. boundary, 2 branches E. and w., and several mtns. Pop. 1830, 1,126.

RUMFORD ACADEMY, and p-o. King William co. Va., by p-r. 115 ms. a very little s. of w. W. C., and 32 N. E. Richmond.

RUMNEY, p-t. Grafton co. N. H., 8 ms. N. w. Plymouth, 47 N. by w. Concord, 110 N. N. w. Boston; with 22,475 acres; crossed by Baker's river, and a branch from Stinson's pond N., Stinson's and Webber's mtns. lie E., and part of Rattlesnake or Carr's mtn. N. w. The soil is pretty good, bearing white pine, beech, sugar maple, oak and birch. Settled 1765. Pop. 1830, 993.

RUPERT, p-t. Bennington co. Vt. 26 ms. N. Bennington, 78 ms. s. w. Montpelier, lies E.

N. Y., has Pawlet r. N. E., and gives rise to White cr. The E. part is mountainous, but there are many good farms. Pop. 1830, 1,318.

RUFF'S, p-o. Marion co. O. by p-r. 38 ms. nrthrd. Columbus.

RURAL VALLEY, p-o. Armstrong co. Pa. by p-r. 224 ms. N. W. W. C.

RUSH, p-t. Monroe co. N. Y. Population 1830, 2,101.

RUSH, co. Ind. bounded by Decatur S., Shelby S. W., Hancock N. W., Henry N., Fayette N. E., and Franklin S. E. Length from S. to N. 24 ms., breadth 20, and area 480 sq. ms. Lat. 39° 35′, long. W. C. 8° 30′ W. Slope S. W. and drained by numerous branches of the Driftwood fork of White r. Chief town, Rushville. Pop. 1830, 9,707.

RUSHFORD, p-t. Alleghany co. N. Y. 12 ms. W. Angelica, E. Cattaraugus co. There are few mill streams. Pop. 1830, 1,115.

RUSHVILLE, p-v. sthrn. part Susquehannah co. Pa. by p-r. 265 ms. N. N. E. W. C.

RUSHVILLE, p-v. near the eastern border of Fairfield co. O. by p-r. 38 ms. S. E. by E. Columbus. Pop. 1830, 234.

RUSHVILLE, p-v. and st. jus. Rush co. Ind. by p-r. 46 ms. S. E. by E. Indianopolis. Lat. 39° 36′, long. 8° 27′ W. W. C.

RUSHVILLE, p-v. and st. jus. Schuyler co. Il. by p-r. 172 ms. N. W. Vandalia. Lat. 40° 06′, long. W. C. 13° 33′ W.

RUSSELL, p-t. Hampden co. Mass. 108 ms. S. W. Boston, is crossed by Westfield r. N. W. and S. E., and Little Westfield r. from W. to E. Incorporated 1792. Pop. 1830, 507.

RUSSELL, p-t. St. Lawrence co. N. Y. 28 ms. S. E. Ogdensburgh, N. Herkimer co. The soil is favorable to pasturage; the streams are numerous and small, except Oswegatchie r. S. Mill sites are abundant. First settled 1805. There are quarries of free stone; iron ore, pyrites, &c. are found. Population 1830, 541.

RUSSELL co. Va. bounded S. by Scott, Lee S. W., Cumberland mtn. separating it from Pike co. Ky. N. W., Tazewell co. Va. N. E., and Clinch mtn. separating it from Washington co. Va. S. E. Mean length between Clinch and Cumberland mtns. 40 ms., mean breadth 35, and area 1,400 sq. ms. Lat. 37°, and long W. C. 5° 30′ W., intersect in this co. Though bounded by 2 mtn. chains, Russell co. of Va. is a very remarkable table land, giving source to the west fork of Sandy r. which flows to the N. W. and pierces Cumberland mountain. A minor chain of mountains traverses the co. from N. E. to S. W. parallel to the Clinch and Cumberland chains, dividing it into two not very unequal sections. Clinch r. rising in Tazewell assumes a southwestrn course down the sthestrn. valley of Russell, giving to that section a slope at right angles to that of the section bordered by the Cumberland chain. The whole co. occupies a region which must be elevated, independent of the mtn. ridges, at least from 1,200 to 1,500 feet above the ocean tides. Surface rocky and in great part mountainous. Chief town, Lebanon. Pop. 1830, 6,714.

RUSSELL, co. Ky. bounded S. by Wayne, S. W. Cumberland co., W. and N. W. Adair, N. E. Casey, and E. Wolf cr. separating it from Pulaski. Length from S. W. to N. E. 26 ms.; mean breadth 10, and area 260 sq. ms. Lat. 37° and 8° W. W. C. intersect in this co. Cumberland r. by a very circuitous channel traverses the sthrn. section in a nearly wstrly. direction, the general slope is, however, to the S. towards that stream, though the nrthrn. extreme reaches into the valley of Green r. Chief t., Jamestown. Pop. 1830, 3,879.

RUSSELL, p-v. Geauga co. O. by p-r. 141 ms. N. E. Columbus.

RUSSELL PLACE, p-o. Kershaw dist. S. C. by p-r. 61 ms. N. E. by E. Columbia.

RUSSELLVILLE, p-v. Chester co. Pa. by p-r. 99 ms. N. E. W. C.

RUSSELLVILLE, p-v. and st. jus. Franklin co. Ala. by p-r. 127 ms. N. Tuscaloosa. Lat. 34° 28′, long. W. C. 10° 46′ W. It is situated on a branch of Bear cr. a confluent of Tennessee r.

RUSSELLVILLE, p-v. Claiborne parish, La. by p-r. 441 ms. N. W. by W. New Orleans.

RUSELLVILLE, p-v. and st. jus. Logan co. Ky. situated on the summit ground between the sources of Muddy r. branch of Green r. and those of Red r. a confluent of Cumberland r., by p-r. 171 ms. S. W. by W. Frankfort, and 58 ms. a little W. of N. Nashville in Ten. Lat. 36° 50′, long. W. C. 9° 50′ W. It is a flourishing v., containing besides the co. buildings an academy, some places of public worship, schools, stores, &c. Pop. 1830, 1,358. It is the largest town in sthrn. Ky.

RUSSELLVILLE, p-v. Brown co. O. by p-r. 106 ms. S. S. W. Columbus.

RUSSIA, p-t. Herkimer co. N. Y. 20 ms. N. Herkimer, has streams running in different directions, the principal of which is West Canada cr. Pop. 1830, 2,458.

RUSSIA, iron works, Peru, Clinton co. N. Y. 6 ms. from Keeseville, 23 from Plattsburgh.

RUTHERFORD, co. N. C. bounded by Blue Ridge separating it from Buncombe W.; a spur of the same chain separates it from Burke N.; it has Lincoln E., and Spartanburg dist. S. C. S. Length from E. to W. 42 ms.; mean breadth 28, and area 1,176 sq. ms. Lat. 35° 20′ N., and long. W. C. 5° W. intersect near the centre of this co. The main stream of Broad r. flows along the sthrn. side in a nearly estrn. direction, receiving numerous creeks which enter from the N. the general slope being sthrd. Much of the soil is excellent, but much is also thin and sterile. Chief t. Rutherfordton. Pop. 1820, 15,351; 1830, 17,557.

RUTHERFORD, co. Ten. bounded by Bedford S., Williamson S. W., Davidson N. W., Wilson N., and Warren E. Length from E. to W. 32 ms.; mean breadth 24, and area 768 sq. ms. Lat. 36° and long. 9° 20′ W. W. C. intersect in this co. The outlines are very nearly commensurate with, and the co. contains nearly the whole valley of, Stone's r. a tributary of Cumberland r. The slope N. W. The N. W. angle is about 6 ms. N. E. by E. Nashville. The soil excellent. Chief town, Murfreesboro'. Pop. 1820, 19,552; 1830, 26,134.

RUTHERFORDTON, p-v. and st. jus. Rutherford co. N. C., by p-r. 223 ms. a little s. of w. Raleigh. It is situated on a branch of Broad r., and near the centre of the co.

RUTLAND, co. Vt. bounded by Addison co. N., Windsor co. E., Bennington co. s., Washington co. w., 34 ms. by 42, with 958 sq. ms. Rutland, near the centre, is the chief t. Castleton is another considerable v. Otter cr. crosses the co. from s. to N. Black, White, and Quecchy rs. rise E. and run into Connecticut r. Pawlet r. is s. and Castleton and Hubbardton rs. w. The principal part of the county is hilly and mountainous, but excellent level land is found on Otter cr., and in the s. w. marble is quarried abundantly in a range of granular limestone along Otter cr., and iron ore is found at the base of the Green mtns., whose heights are included in the E. part of the co. Pop. 1820, 29,983; 1830, 31,294.

RUTLAND, p-t. st. jus. Rutland co. Vt. 50 ms. s. w. Montpelier, 60 s. Burlington, 52 N. E. Bennington, is of irregular form, containing above 26,000 acres. It was first settled about 1770. In the revolution 2 picket forts were built here, one of which was near the site of the present court house in the E. v. It is crossed by Otter cr. from s. to N. West r. and East cr. fall into it in this t. and these streams afford mill sites, where several factories are erected. The soil is very various, the rocks being primitive and secondary. Iron, limestone and clay are found. Marble quarries are wrought, both white and blue, in a range extending from Berkshire co. Mass. through a considerable part of Vt. The principal v. is in the E. parish, and contains a court house, and other public buildings. In the w. parish are 2 small villages. Population 1830, 2,753.

RUTLAND, p-t. Worcester co. Mass. 56 ms. w. Boston, is a pleasant agricultural t. with good soil and varied surface, crossed by an E. branch of Ware r. It was purchased of the Indians in 1686. A tract 12 ms. sq. which included this and several adjacent towns, was purchased for £30. Several of the inhabitants were killed by Indians in 1723 and '24. Pop. 1830, 1,276.

RUTLAND, p-t. Jefferson co. N. Y. 170 ms. N. W. Albany, 6 E. Watertown, s. Black r., N. Lewis co., has light soil and favorable to grain and grass. Limestone rocks lie beneath at a considerable depth. The trees are maple, beech, elm, with some white pine, &c. near Black r. The remains of an ancient work like an encampment are seen on a hill, surrounded by a ditch. The place was overgrown with old trees, and human bones are found in the soil. Pop. 1830, 2,339.

RUTLAND, p-v. Tioga co. Pa. by p-r. 148 ms. N. N. W. Harrisburg.

RUTLAND, p-v. Meigs co. O. by p-r. 95 ms. s. E. Columbus.

RUTLEDGE, p-v. and st. jus. Grainger co. Ten. 33 ms. N. E. by E. Knoxville, and by p-r. 232 ms. E. Nashville. Lat. 36° 15′, long. W. C. 6° 16′ w.

RYAL'S, p-o. Montgomery co. Geo. by p-r. 101 ms. s. s. E. Milledgeville.

RYE, t. Rockingham co. N. H. 6 ms. E. Portsmouth, which it separates from the ocean, 51 E. S. E. Concord, has Little Harbor N. E., the Atlantic E., and contains 7,780 acres. It was first settled in 1635, when it belonged to Portsmouth; incorporated 1719. Its name was probably derived from a town in England. The soil is poor and hard, but sea weed is used with great benefit as manure. The sea coast extends 6 ms., nearly one-third of that possessed by the whole state, and embraces Sandy, Jenniss' and Wallis' beaches, which afford bathing places, much resorted to. There is a small harbor for vessels of 70 or 80 tons; and many fish are caught in boats along the coast. A tract of 300 acres has been drained of a fresh pond, which yields salt hay. At Breakfast hill, a party of Indians were surprized, 1696. That t. suffered considerably in the Indian wars. Pop. 1830, 1,172.

RYE, p-t. Westchester co. N. Y. 29 ms. N. E. N. Y., 5 s. E. Whiteplains, 142 s. Albany, w. Conn., N. Long Isl. sound, is small and of irregular form. It has Byram r. for a short distance on the E. boundary. Parsonage Point extends into the sound. There are 2 small villages, Rye and Saw Pits, the latter on the sound, 28 ms. from N. Y. and a place of some trade. Several small isls. in the sound belong to this t. Pop. 1830, 1,602.

RYEGATE, p-t. Caledonia co. Vt. 33 ms. E. Montpelier, 58 N. W. Windsor, 150 N. W. Boston, lies w. Conn. r., N. Orange co. opposite Bath, Grafton co. N. H. and has 32 sq. ms. The original settlers, (except one family,) were from Scotland, a company with £1,000 sterling, being raised in 1772, by farmers of Renfrew and Lanark, and after a selection made by agents, the settlement was commenced in 1774. New colonists were interrupted by the war, but afterwards arrived; and about two-thirds of the population are of Scotch descent. The habits of their ancestors are still in some degree retained: oat meal and barley form important articles of diet; and frugality and industry prevail. The land is uneven, and in the north rough; but there is much pasturage, and very little waste. The western part has rich soil, and on the Conn. are three small meadows. Ticklenaked pond s., contains 64 acres, and its outlet enters Wells r. North pond discharges into Conn. r. over which is a dam, at Canoe falls, and a ferry. A part of Wells r. s. w. affords mill seats. Mill stones are obtained from Blue mtn. the only considerable eminence. Pop. 1830 1,119.

RYERSON'S STATION, p-v. nrthrn. part of Greene co. Pa. by p-r. 16 ms. s. s. w. of the borough of Washington in Washington co. Pa. and 249 ms. N. W by w. W. C.

RYND'S, p-o. Venango co., Pa. by p-r. 288 ms. N. W. W. C.

S.

Sabillisville, p-v. Frederick co. Md. by p-r. 59 ms. n. n. w. W. C.

Sabina, p-v. Clinton co. O. by p-r. 55 ms. s. w. Columbus.

Sabine, r. La. and the Mexican province of Texas, rising in the latter about lat. 33°, and to the sthrd. of the great bend of Red r. and to the n. w. of La. The country round its sources is generally prairies; but before reaching lat. 32°, where it becomes a boundary between La. and Texas, this stream has entered a dense forest. From this point to the mouth it was navigated and surveyed by the author of this article in 1812 and 1813. At lat. 32° it is already a navigable stream for boats of considerable size at high water; the breadth of the stream 60 or 70 yards. Below 32° the Sabine receives no tributary stream above the size of a large creek; of these, however, there are several from both sides. The main stream, with a curve to the estrd., pursues a general southern course over two degrees of lat. to lat. 30°, where it is joined from the n. w. by the Netchez, a branch from the vicinity of Nacogdoches. Before their junction both rivers have merged into prairie, which continues to the Gulf of Mexico. Immediately below the union of the Sabine and Netchez, the united waters expand into a shallow elliptical lake of about 30 ms. long, and from 1 to 7 or 8 ms. wide. At the lower end of the lake the water again contracts into the size of a river of but little more width than above the lake. The whole length of this river, from the source to final outlet, into the Gulf of Mexico, is 70 ms. above and 250 ms. below the point where it is crossed by lat. 32°.

Though when swelled by rains the Sabine is navigable above lat. 32°, it is not of sufficient depth at the mouth or over its lake for vessels of 3 feet draught. Along the wstrn. side a range of high hills stretches with the Sabine, some parts of which are rocky and even precipitous; but along the opposite shore I saw not one high bank, and with the prairies all eminences cease, and one immense plain extends on all sides. Tufts of trees gradually cease, and from the mouth not a shrub is to be seen. The soil, as far as I could judge from the appearance along the banks, is generally sterile. At any considerable distance from the stream pine is the prevailing timber. Taken as a whole it is a river worthy of notice only as having become a political boundary between two great nations.

Sable r. or River au-Sable, N. Y., empties into lake Champlain, after a course of 35 ms. from Essex co. where it has its source. It runs for some distance on the line of Essex and Clinton cos. with mill seats. Little Sable r. empties into the lake 2 ms. n. of it.

Sacandaga, r. N. Y. a branch of the Hudson, 8 ms. long; has its sources in numerous ponds and small streams in Warren and Hamilton cos., and after a crooked course, and receiving a number of good mill streams, joins the Hudson 8 ms. s. w. of lake George.

Sacarappa, p-v. Cumberland co. Me. 59 ms. from Augusta.

Sachem, Grand, mtn. N. Y., the highest of the Highlands on Hudson r. called also the Beacon.

Sacket's Harbor, incorporated p-v. Hounsfield, Jefferson co. N. Y. 161 ms. n. w. Albany, 12 below Watertown, 8 from lake Ontario, is on a large and important bay and harbor, which was made a naval station during the late war, and is very convenient for ship building as well as for anchorage. Forts Tompkins and Pike were built here in the war.

Saco, r. rises in N. H. and enters the Atlantic in Me. Its highest source is near the summit of one of the loftiest peaks of the White mtns.; and during its course to the Notch, it flows in one place within about 200 yards of the Lower Amonoosuc. After winding slowly through a little narrow alluvial level at the foot of the principal peaks, it passes through the Notch, which it appears to have had much agency in reducing to its present form, and instantly changes its character to a furious and foaming little torrent, rushing impetuously down a descent in a continued cascade, with few interruptions for several miles. On the upper part of the Notch it is about 4 feet wide, and yet leaves barely room enough for the road to pass beside it. It pursues a s. course for about 12 ms. through many romantic scenes, and then turning e. in Bartlett receives Ellis's r., and in its s. course of 10 ms. further is swelled by several other small tributaries. At Conway it flows across a level tract, receives Swift r., and then running e. passes into Fryeburgh, Maine, through which it pursues a remarkably tortuous course, running 36 ms. in a town 6 miles square, the ancient favorite habitation of the Pequawket Indians. It then pursues its way to the sea in Me., on the borders of which it makes a sudden descent, at a spot where its channel is divided by an island, on which, and the adjacent banks, large manufactories were erected a few years since, with the prospect of operating with great advantage on account of its convenient communication with navigable tide water, as well as the abundance of water power. The principal buildings were unfortunately destroyed by fire. The r. is subject to sudden floods, especially in its upper parts.

Saco, p-t. and port of entry, York co. Me. 71 ms. s. s. w. Augusta, 15 s. w. Portland, 29 n. e. York, 103 n. n. e. Boston; has Saco r. on the s. w., Cumberland co. n. e., and a bay on the s. e. where the mouth of the river forms a harbor. The falls here afford water power for manufacturing. The village stands on Saco r. at the head of tide water, about 3 ms. from its mouth, and at the falls, the descent of which is nearly 50 feet, and to which vessels of 100 tons come up from sea. The water power is always very abundant, and numerous factories might be erected on the shore. About 20 saw mills are now moved by the water. The York manufacturing com. own a site 34 feet in length, where they have a new factory with 8,000 spindles, and other sites at Calt's island. They have also a rolling mill and nail factory, producing 400 tons of nails annually. There is one Episcopal, one Calvinist, and one Unitarian church, besides congregations of Baptists and Methodists. The number of inhabitants in the village by the last census was 3,219; the number the preceding year was over 3,800. Only a few months before the census was taken, the large cotton mill which had employed 600 persons was burnt. Pop. 1830, 3,219.

Saddle, r. N. J. rises near the boundary in N. Y., and flows s. through Bergen co. into Pompton r. and forms the Passaic.

Saddleback mtn. Oxford co. Me. n. of Androscoggin r. about 4,000 feet above the sea.

Saddle River, t. Bergen co. N. J. 4 miles n. w. Paterson; has Pompton r. on the s. boundary, and much hilly or mountainous land. Paterson v. is at the falls, opposite this town. An aqueduct of the Morris canal crosses the river. Pop. 1830, 3,397.

Sadsburyville, p-v. western part of Chester co. Pa. by p-r. 43 ms. w. Phil.

Saegersville, p-o. Lehigh co. Pa. by p-r. 85 ms. e. Harrisburg.

Sagadahoc, r. Oxford co. Me. falls into the Androscoggin in Rumford, from the n.

Sag Harbor, p-v. and port of entry, Southampton, Suffolk co. N. Y. in the n. e. corner of that town, 100 ms. e. N. Y., 244 from Albany, has a good harbor, and the seat of some trade, as well as of whale fishing, and the manufacture of salt from sea water.

Saginaw, bay of lake Huron, in Mich. As laid down by Tanner in his recently improved map of the U. S. Saginaw bay opens from the lake between Transit point on the s., and Rock point on the n., by a mouth 32 ms. wide, and extending thence s. w. 50 ms., maintaining a general width of about 20 ms., and terminating in a wide base or shore exceeding in fact in width the mean breadth of the bay. This bay receives from the n. w. the rivers Thunder, Sable, Grindstone, and some of lesser note; the inlets are small, and the main confluent of the bay, the Saginaw river, enters the extreme southwestern shore. The bay of Saginaw is chequered with some islands, the principal groups are the Thunder islands off Rock point, and the Shaungum islands between the mouths of Grindstone and Saginaw rs. Vessels drawing 5 or 6 ft. water are navigated into, and some distance up Saginaw r. The distance is about 75 ms. n. n. w. from the outlet of lake Huron into St. Clair r. to Transit point, or southern entrance of the bay of Saginaw.

Saginaw, r. of Mich. is formed by the rs. Cass, Flint, Saginaw Proper, and Tittibawassee. Cass r. rises in Sanilac co., and flowing wstrd. by comparative courses 50 miles, falls into Saginaw nearly opposite to the mouth of Tittibawassee. Flint r. rises in Lapeer and Oakland cos., and flowing thence to the n. w. enters Saginaw co. and joins Saginaw r. 3 or 4 ms. above the mouths of Cass and Tittibawassee rs. Saginaw Proper rises in Oakland, Washtenaw, and Shiawassee cos., and flowing nthrd. into Saginaw, joins Flint r.; as already noticed the comparative length of the two streams above their junction is nearly equal, and each about 55 miles. The Saginaw Proper has interlocking sources with those of Huron of Erie, and Grand r. of Michigan. The course of the Tittibawassee is almost directly opposite to that of Flint r. Rising between the sources of Thunder r. of lake Huron, and Manistic of lake Michigan, it flows s. s. e. by comparative courses about 70 ms., and is the longest of the constituent branches of Saginaw. This stream rises on the unappropriated territory between Saginaw bay and lake Michigan, and in its course to its recipient traverses Gladwin and Midland cos., and is lost in Saginaw r. in Saginaw co. Below the union of its constituent streams, the Saginaw flows n. n. e. by comparative courses to its final discharge into Saginaw bay.

The valley of Saginaw occupies much of the central parts of the Mich. peninsula; the greatest length, 120 ms. from the source of Flint r. to that of Tittibawassee; mean breadth 35 ms., and area 42 sq. ms. Lying between lat. 42° 35′, and 43° 20′. Contrary to ordinary cases the greatest length of this river valley is almost at right angles to the general course from the middle source to point of ultimate discharge.

Saginaw, co. Mich. bounded n. e. and e. by Sanilac co., s. e. Lapeer, s. Shiawassee, w. Gratiot, n. w. Midland, and n. Saginaw bay. Length from s. to n. 38 ms., mean breadth 32, and area 1,216 sq. ms. Extending in lat. from 43° 07′ to 43° 39′, and in long. from 6° 36′ to 7° 21′ w. W. C. Slope a little e. of n. The far greater part of the whole co. is in the valley of Saginaw r., and near its centre the constituent streams of that river converge and unite within 3 or 4 ms. of the same point. Chief t. Saginaw.

Saginaw, st. of jus. Saginaw co. Mich., as laid down by Tanner, is on the left bank of Saginaw r. 20 ms. above its mouth, and 100 ms. n. w. Detroit. Lat. 43° 25′, long. 6° 55′ w. W. C.

Saint Alban's, p-t. Somerset co. Me. 30 ms. e. Norridgewock, 46 n. n. e. Augusta, touches Penobscot co. at the n. e. and s. e. angles, lying w. Corinna, which breaks the line of that co. It is crossed N. E. and S. W. by a stream flowing into Sebasticook r. the main stream forming the W. boundary of this town. Pop. 1830, 920.

SAINT ALBAN'S, p-t. and st. jus. Franklin co. Vt. 27 ms. N. Burlington, 46 N. W. Montpelier, 70 S. Montreal; is situated on St. Alban's bay of lake Champlain on a handsome slope, commencing about 3 ms. from the shore. The streets of the village are regularly laid out, and the public edifices are built about a central square, the co. buildings, academy, churches, &c. The settlement of the town began in the revolutionary war. The streams are insignificant; the soil is good, bearing maple, beech and birch, and near the lake, oak. The trade of the town has been much increased by the opening of Champlain canal. Saint Alban's academy was incorporated in 1799. Pop. 1830, 2,395.

SAINT AUGUSTINE, p-v. Cecil co. Md. by p-r. 99 ms. N. E. W. C.

SAINT AUGUSTINE, East Flor. (*See Augustine, Saint.*)

SAINT BERNARD, parish of La., as laid down by Tanner, extends S. E. from the lower suburbs of New Orleans, and is bounded W. by the parish of Jefferson, N. by the parish of Orleans, N. E. by lake Borgne, and E. and S. E. by the parish of Plaquemines. Length 30 ms. from the vicinity of New Orleans to the junction of the bayous Levy and Terre aux Bœufs; mean breadth 5 ms., area 150 sq. ms. Central lat. 29° 54', long. 12° 46' W. C. This parish contains both banks of the Miss. from the vicinity of New Orleans to the Great Bend above Woodville, and the whole course of Terre aux Bœufs. The margin of these streams comprise the only arable part, the residue being impassable morass. The whole surface is a plain, being a part of the delta. Staples, sugar, rice, and cotton. There is neither co. st. nor p-o. named in this parish on the p-o. list. Pop. 1830, 3,356.

SAINT CHARLES, parish of La. bounded by Jefferson parish E., Lafourche parish S., Saint John Baptist W., and lake Pontchartrain N. Length 34 ms., mean breadth 15, and area 512 sq. ms. Lat. 30°, and long. 13° 18' W. W. C. intersect in this parish. What slight descent exists in the surface is to the S. E. by E. in the direction of the Miss. r., by which it is traversed. The alluvial banks of the Miss. afford most of the arable soil of the parish, which, like other parts of the delta, is a plain liable to annual, and in the present case, even diurnal submersion, except the margin of streams. Pop. 1820, 3,862, 1830, 5,147. Staples of this parish, sugar, rice, and cotton.

SAINT CHARLES, co. Mo. bounded by Mo. r. which separates it from Saint Louis co. S. E. and S., and from Franklin S. W.; it has Montgomery co. Mo. W., and Lincoln N.; above the mouth of Illinois r. it is separated from Calhoun co. Il. by the Miss., and below the mouth of Il. to that of Mo. r. it is separated by the Miss. r. from Greene co. Il. N., and Madison co. Il. E. It occupies the point between the Mo. and Miss. rs., and approaches the form of a triangle; the hypothenuse or greatest length 52 ms. along the general course of the latter stream; perpendicular along Montgomery 21 ms.; area about 500 sq. ms. Lat. 38° 47', long. 13° 35' W. W. C. General slope to the E. Chief town, Saint Charles. Pop. 1820, 3,970, 1830, 4,320.

SAINT CHARLES, p-v. and st. jus. St. Charles co. Mo. situated on the left bank of Mo. r. 20 ms. N. W. St. Louis. It is principally composed of one long street, on a superstratum underlaid by solid limestone. After rising the slope on which the town stands, an immense plain extends, partly covered with woods, but more an open prairie. According to Flint, about one-third of the population is French, and the whole about 1,200. Lat. 38° 45', long. 13° 30' W. C.

SAINT CLAIR, r. of Mich. and Upper Canada, is the discharge of the immense reservoir of lake Huron, or rather it is the drain of the basin of which lake Huron is itself the reservoir. Towards its southern extremity this large sheet of water gradually contracts, and finally terminates in a river almost exactly at lat. 43°. This r. or strait, with a general width of about a half mile, flows by comparative courses 40 ms. a little W. of S. to its entrance into lake Saint Clair. The lower part of St. Clair r. is a real delta, the water separating into numerous channels, with low marshy or sandy intervening islands. The main channel of St. Clair r. admits in all its length the navigation of vessels drawing 7 or 8 feet water.

SAINT CLAIR, lake, is a nearly circular sheet of water; greatest length or breadth 30 ms., and receiving from the N. the r. or strait of the same name. This lake is shallow, and the shores generally low, level, and in part marshy. It receives from Upper Canada the rs. Bear and Thames, and from Mich. Clinton r. It is discharged at the southwestern angle into Detroit r., and is navigable for vessels of 7 or 8 feet draught.

SAINT CLAIR, co. Ala. bounded by Shelby S. W., Jefferson W., Blount N. W., Wills creek or river N. E., and the main Coosa E. and S. E. Length from S. W. to N. E. 42 ms., mean breadth 20, and area 840 sq. ms. Lat. 33° 45', long. 9° 24' W. W. C. Slope eastward towards Coosa r. Chief t. Ashville. Pop. 1830, 5,975.

SAINT CLAIR, co. Mich. bounded by Macomb co. S. W., Lapeer W., Sanilac N., lake Huron N. E., Saint Clair r. E. and S. E., and lake Saint Clair S. Length from S. to N. 55 ms., mean breadth 20, and area 1,100 sq. ms. Lat. 43°, long. 5° 30' W. W. C. Slope S. E., and in that direction drained by the river Dulude and Belle r. Pop. 1830, 1,114.

SAINT CLAIR, p-v. and st. jus. St. Clair co. Mich. by p-r. 59 ms. N. E. Detroit. It is situated on Saint Clair river at the mouth of Pine river. Lat. 42° 47', long. 5° 25' W. W. C.

SAINT CLAIR, co. Il., bounded N. by Madison, N. E. Clinton, Washington S. E., Randolph S., Monroe S. W., and the Mississippi r. separating it from St. Louis co. Missouri, N. W. Length from S. to N. 30 ms., mean breadth 22, and area 660 sq. ms. Lat. 38° 30′, and long. 13° W. W. C. intersect near the centre of this co. The southeastern angle is traversed in a southwestern direction by Kaskaskias river. Silver creek rising in Macaupin, traverses by a southern course Madison and St. Clair, falling into Kaskaskias river in the latter. The general slope of the co. is to the S., though a small section of the northwestern part declines to the westward towards the Mississippi. Chief town, Belleville. Pop. 1820, 5,-253, 1830, 7,078.

SAINT CLAIRSVILLE, p-v. and st. jus. Belmont co. O., by p-r. 11 ms. a little N. of W. Wheeling in Va., 275 ms. N. W. by W. W. C., and 124 ms. E. Columbus. It is situated on a small branch of Indian or West Wheeling creek, in a very hilly but fertile country. This village stands on the U. S. road, contains the common co. buildings, with a printing office, market house, 3 places of public worship, and several private schools. Pop. 1830, 789. Lat. 40° 05′, long. W. C. 3° 51′ W.

SAINT CLEMENT'S BAY, and p-o. S. W. part of Saint Mary's co. Md., by p-r. 57 ms. S. E. W. C.

SAINT CROIX, river Me., rises in a considerable lake on the borders of Washington co. and New Brunswick, and after a devious course of about 80 or 90 miles in a S. E. direction, on the E. boundary of the U. S., falls into Passamaquoddy bay. It also bears the names of Passamaquoddy, Cheputnetecoock, &c. It receives numerous small streams from Washington county, particularly the outlet of the Shordic lakes. It is navigable 12 ms. to the falls at Calais.

SAINT FRANCIS, river, of Mo. and Ark., is composed of two branches, the eastern or White Water, and the western or Saint Francis proper. White Water has its remote sources in Cape Girardeau co. Mo., and derives some of its fountains within 10 ms. from the channel of the Mississippi. Flowing thence by a course a little W. of S. over Cape Girardeau, Stoddard, Scott, and New Madrid cos., Mo., enters Crittenden co. Ark., within which it joins the Saint Francis after a comparative course of 140 ms. Saint Francis rises in the Iron mountains, Saint Francis co. Mo., interlocking sources with those of Black river, branch of White river, on the W.; with those of Big river, branch of Maramec, and with those of Cold Water, Vase, and other small creeks flowing to E. into Miss. From this comparatively elevated and broken region. Saint Francis, in a general southern course of 160 miles, unites with White Water, having traversed the southern part of Saint Francis, the entire breadth of Madison and Wayne, Mo., and part of Lawrence, Monroe and Crittenden counties, Ark. Below the union of its main branches, Saint Francis maintains its southern direction, by comparative courses 80 miles, but with a very sinuous channel, to its entrance into Mississippi at lat. 34° 35′. In its entire course of 240 miles, the Saint Francis flows so nearly parallel to the general course of Black river, and its continuation, White river, that the two streams vary in relative distance from 10 to 40 ms. The mean breadth of Saint Francis valley is about 35 miles; area 8,400 sq. ms., filling the space between the valley of White river, and the opposing part of that of the Miss. The much greater part of Saint Francis valley is a plain, liable to annual submersion. In the lower part of its course it is in appearance, the Mississippi on a smaller scale.

SAINT FRANCIS, co. Missouri, bounded S. by Madison, Washington W., Jefferson N., and Saint Genevieve E. The outline is very irregular, but the greatest length is from S. to N. 30 miles; mean breadth 20 ms., and area 600 sq. ms. Lat. 37° 50′, long. W. C. 13° 30′ W. The irregular eastern and northeastern borders of this co. follow the dividing ridge or table land between the sources of Saint Francis and Maramec rivers, and those of small creeks falling into the Mississippi, after a brief course of 10 or 15 ms. The range of hills or mountains which reach the Mississippi near the village of Saint Genevieve, crosses Saint Francis co. in a western direction, and discharges the sources of Big river, branch of Maramec, N., and those of Saint Francis river S. The surface of the whole co. is hilly, broken, and in part even mountainous. Chief town, Farmington. Pop. 1830, 2,366.

SAINT FRANCIS, co. Arkansas, is situated between Saint Francis and White rivers, to the N. E. by E. of Little Rock, but the boundary is uncertain. Chief town, Franklin, the position of which is also uncertain.

SAINT FRANCIS, p-v. Saint Francis co. Ark., by p-r. 111 ms. N. E. by E. Little Rock.

SAINT FRANCISVILLE, p-v. and st. jus. West Feliciana parish, Louisiana. It is on a hill rising from the Miss. river about 1-4 m. from the mouth of bayou Sara, 64 ms. S. Natchez, and by p-r. 149 ms. N. W. by W. New Orleans. Lat. 30° 42′, long. W. C. 14° 19′ W. It is a neat village, in one street along the road from the mouth of bayou Sara to Fort Adams, Natchez, &c.

SAINT GENEVIEVE, co. of Mo., bounded by Perry S. E., Saint Francis co. S. W. and W., Jefferson N. W., and Mississippi river separating it from Randolph co. Illinois N. E. It approaches to the form of a square of 20 miles each side; area 400 sq. ms. Lat. 37° 50′, long. W. C. 13° 14′ W. The slope is to the N. E. towards the Mississippi river; the western border following the dividing ridge of the sources of Saint Francis and Maramec, and those of creeks flowing over Saint Genevieve into the Miss. Chief t., Saint Genevieve. Pop. 1830, 2,186.

SAINT GENEVIEVE, p-v. and st. jus. Saint Genevieve co. Mo., situated, says Mr. Flint, about one mile from the Mississippi river on Gabourie creek, and at the head of a fine alluvial prairie. According to this author, the population is about 1,500, and yet not more than it was 30 years ago. The French inhabitants are most numerous. The village contains an academy and Catholic church. It is distant 61 ms. below Saint Louis, and 8 westward of Kaskaskias in Il. Lat. 38°, long. W. C. 13° 05 w.

Saint George, p-t. Lincoln co. Me., 38 ms. from Wiscasset, and 57 s. e. Augusta; forms an irregular cape, running s. w. into the sea, with Saint George's river on the n. w. and w. Pop. 1830, 1,643.

Saint George, town, Chittenden co. Vt., 28 ms. e. Montpelier, 8 s. e. Burlington; is of small size; first settled 1784; has an uneven surface, with some high hills; maple, beech, and birch timber, but no considerable streams. Pop. 1830, 135.

Saint Helena, parish of La., bounded by lake Pontchartrain s. e., the Pass of Manchac, lake Maurepas, and the lower part of Amite river separating it from the parishes of Saint John Baptist, Saint James, and Ascension; the Amite river separating it from East Baton Rouge s. w. and w., and New Feliciana n. w.; on the n. it has the county of Amite n Mississippi, and on the e. the Tangipola river separating it from Washington n. e., and Saint Tammany e. Length from s. to n. 50 ms., mean breadth 34, and area 1,700 sq. ms. (*For lat. and long., see Saint Helena, the st. jus.*) The slope is a little e. of s. There is some good soil along the streams, but the much greater part is sterile, and covered with pine timber. Pop. 1820, 3,026, 1830, 4,028.

Saint Helena, p-v. and st. jus. parish of Saint Helena, La., on Tickfah river, by p-r. 98 ms. n. w. New Orleans; and about 45 ms. a little n. of e. Baton Rouge. Lat. 30° 35′, long. W. C. 13° 40′ w.

Saint Inigoes, p-v. on a small river of the same name, southern part of Saint Mary's co. Md., 27 ms. s. e. by e. Leonardstown, the co. seat, and by p-r. 90 ms. s. s. e. W. C.

Saint James, parish of La., bounded e. by Saint John Baptist, s. by the Miss. river separating it from the parish of Assomption, w. by Ascension, and n. by Amite river separating it from Saint Helena. Length 28 ms., mean width 20, and area 560 sq. ms. Lat. 30° 10′, long. W. C. 13° 45′ w. The southern border of this parish rises only from 1 to 4 minutes above lat. 30, & is about the nrthrn. extreme in La. where the orange tree will grow to any advantage, and even here, its existence is precarious. Sugar and cotton are the staples. In surface and soil it resembles other Louisiana parishes along the Mississippi in the delta. Pop. 1820, 5,660, 1830, 7,646.

Saint James, p-o. St. James parish, La.

Saint James' Church, and p-o. Bedford co. Va., by p-r. 217 ms. s. w. W. C.

Saint John Baptist, parish of Louisiana, bounded by St. Charles e., bayou Cabanose separating it from the parish of Lafourche s., Saint James and Assomption w., lake Maurepas n. w., Pass of Manchac n., and lake Pontchartrain n. e. Length from s. to n. 50 ms., mean breadth 12 ms., and area 600 sq. ms. Lat. 40° and long. 13° w. W. C. intersect near the centre of this parish. A remark may be made here which applies to the present article, and all the other parishes of La., in the delta of the Miss., that is, that the area is in great part nominal as regards arable land. The only part sufficiently elevated for the plough is the margin of the streams, but where arable the soil is highly fertile. The Miss. winds over Saint John Baptist from w. to e. Staples, cotton and sugar. Pop. 1820, 3,854, 1830, 5,677.

Saint John's, river, Me., rises in Somerset co. in that state, near the middle of the w. boundary, where its head waters almost interlock with those of the Ghaudiere, which flows into the Saint Lawrence, and with those of the Penobscot, and approach near to some of the sources of the Kennebec. The canoe navigation of the Penobscot is connected with that of the Saint John's, by a portage of only 2 ms. The first course of this great r. is n., then n. e. and e. to near the upper part of the boundary of Penobscot and Washington counties, where it bends southerly, and crossing the latter county, it passes into New Brunswick. The Saint John's presents a bold and noble curve on the map; and is navigable in sloops of 50 tons in N. Brunswick 80 ms. from the Bay of Fundy. Its principal branch, the Aroostic, rises in the w. part of Penobscot county, runs n. e., and enters N. Brunswick.

Saint John's, river of Florida. This very remarkable river has evidently been formed from one of those sounds which exist along the Atlantic coast of the United States. In strictness it cannot be said to have any definite source, as both branches, the Ocklawaha, and Saint John's proper, originate in one immense marsh, rising but very slightly above the level of the Atlantic ocean. As laid down by Tanner, both branches flow northwardly about 60 miles, unite, and the combined water, continuing the original course 70 miles, inflects abruptly e. 20 ms. to its outlet into the Atlantic ocean at lat. 30° 20′. For such vessels as can enter the mouth, 6 or 7 feet draught, it is navigable more than two thirds of its entire course. The region it drains is generally sterile. "The bar at the mouth of this river is shifting; the greatest depth on it is 15 feet at high tide; but, on account of winds, it varies from 12 to 15 feet. At low tide the least depth is 6 feet, the greatest 7½ feet. As there is constantly more or less swell, a vessel drawing more than 11 feet, could not cross the bar with safety, but might ascend easily the river, as far up as the mouth of Black creek. On this distance of 47 ms. the channel is wide, and affords a depth never less than 15 feet." By the same authority, (report of U. S. engineer) before quoted, it appears that the Saint John's river of Florida can be safely navigated by vessels drawing 8 feet water to lake George, 107 ms., following the stream, above the bar at its mouth. The engineers state that freshets (floods) do not exceed a rise of 2 feet, and that "the banks are principally marsh, hammock land, pine barren, and cypress swamp." Soil generally sterile.

SAINT JOHN'S, co. Flor. As laid down by Tanner, this county is very nearly commensurate with the valley of Saint John's r., having extensive marshes s., Seminole Indians s. w., Alachua co. w., Duval co. N. W. and N., and the Atlantic E. Within these limits it is in length from s. to N. 130 ms., with a mean breadth of at least 40 ms., area 5,200 sq. ms. Extending in lat. from 28° 40′ to 30° 20′, and in long. from 4° to 5° 30′ w. W. C. The outlines it must, however, be premised, are arbitrary except to the N., and along the ocean. This wide region is generally open prairie or marsh, with a very sterile soil. It may be remarked, that the alluvion of Florida is as sterile as that of La. is productive. Where sufficiently elevated and fertile, the soil of Saint John's produces sugar cane, cotton, rice, indigo, and an immense number of other valuable vegetables. The climate is sufficiently mild for the orange, olive, and perhaps the date palm. Chief t. Saint Augustine. The pop. 2,538, in 1830, marks the nature of the country; not 1 person to 2 sq. ms.

SAINT JOHN'S BLUFF, and p-o. Duval co., Florida, by p-r. 274 ms., though by direct distance only about 200 miles east Tallahassee.

SAINT JOSEPH'S, river, of Mich., Ohio and Ind., rises in Branch and Hillsdale counties of the former, and flowing s. s. w. traverses Williams co. Ohio, and entering Ind., unites in Allen county with the Saint Mary's river, to form Maumee, after a comparative course of 70 miles. The Saint Joseph's branch of Maumee has interlocking sources with those of the Saint Joseph's of lake Michigan, and Tiffin's river, branch of Maumee.

SAINT JOSEPH'S, river of lake Michigan, has interlocking sources with those of Eel river, branch of Wabash; those of Saint Joseph's of Maumee, Tiffin's of Maumee, the r. Raisin of lake Erie, and with those of Kallamazoo river of lake Michigan. The most remote source is in Hillsdale co. Mich., but the numerous confluents drain nearly all Branch, Saint Joseph, Cass, and Berrien, with part of Calhoun, Kalamazoo, and Van Buren cos. Mich.; and all La Grange and Elkhart, with part of Saint Joseph's and La Porte cos. Ind. The various streams which contribute to form Saint Joseph's river, unite in Mich., and the main stream inclining s. w., enters Indiana in the N. E. part of Elkhart co., and thence sweeping an elliptic curve over the northern sections of Elkhart and Saint Joseph's cos., re-enters Mich. by a northwestern course, which it maintains over Berrien co. to its final entrance into lake Mich. The Saint Joseph's river of lake Michigan is a large stream in proportion to length. The utmost length of the valley it drains is 110 ms., whilst the mean breadth is fully 40 ms.; area 4,400 sq. ms. This valley is also amongst the finest regions of the Saint Lawrence basin. In lat. it extends from 41° 15′ to 42° 20′; and from about 80 to 180 ms. a little s. of w. from Detroit. The vallies of Saint Joseph's, Kalamazoo, and Grand rs. follow each other from s. to N.

SAINT JOSEPH'S, an isl. in the straits of St. Mary, lies between Drummond's and George's island. It is 20 ms. long, with a mean breadth of 8 ms. and in all its length separates the two channels of the straits of St. Mary.

SAINT JOSEPH, co. Mich. bounded w. by Cass, N. by Kalamazoe, E. by Branch, s. by La Grange co. Ind., and s. w. by Elkhart co. Ind. Length from E. to w. 24 ms., breadth 21, and area 504 square ms. Lat. 42°, and long. W. C. 8° 35′ w. intersect in this co. The main volume of St. Joseph's r. enters the estrn. border and winds southwestwardly over the co. leaving it at the southwestern angle. The whole surface is in the valley of St. Joseph's r.

SAINT JOSEPH, co. of Indiana, bounded by Elkhart co. Indiana E., by the Putawatomie country s. E., s. and s. w., La Porte co. Ind., w., Berrien co. Mich. N. W., and Cass county, Mich., N. E. Length from s. to N. 32 ms. width 21, and area 672 square ms. Lat. 41° 35′, long. W. C. 9° 20′ w. Though a level country this co. comprises a table land. The northeastern angle is traversed by and drained into St. Joseph's r. From the northwestern part issue some creeks, which flow into lake Mich. by a wstrn. course. The central and wstrn. sections give source to the Kankakee branch of Il. r. which leaves the co. by a s. w. by w. course, whilst the sthrn. border gives source to the Tippecanoe branch of the Wabash.

SAINT LANDRE′, parish of La. bounded by Sabine r. on the w. separating it from the Mexican province of Texas, Natchitoches, La. N. W., Rapides parish La. N., Avoyelles parish N. E., Atchafalaya r. separating it from the parishes of Point Coupee, and West Baton Rouge east, St. Martin's parish southeast, the Queue Fortue bayou south, separating it from the parish of Lafayette, and by the gulf of Mexico s. w. Length along the western border 100 miles, mean breadth 60 ms. and area 6,000 square miles. Extending in lat. from from 29° 25′ to 31°, and in long. from W. C. 14° 48′ to 17° w. This very extensive parish embraces most part of the region known from the name of an Indian tribe called Opelousas. It is the extreme southwestern angle of the U. S. Few, if any other continuous surfaces of equal extent, differ more in soil, features, and indigenous vegetation. The nthrn. part towards Natchitoches rises into an undulating country, covered generally with pine timber. The central and wstrn. sections assume the aspect of an immense plain, with lines of woods winding with the streams, but the far greater part between the water courses, prairie. The southwestern part between the Mermentau and Sabine is a level sea marsh, with scattered clumps of trees, but mostly flooded prairie. All these three sections have sterile soil. But advancing eastward to the banks of the Teche, Courtableau, and Vermillion, the timber becomes more plentiful, and the soil exuberantly productive. This fine section is again followed by the inundated margin of Atchafalaya, and lower Courtableau. The ecclesiastical name which heads this article was imposed by the first civilized settlers, the French, which nation in numbers still predominate. Staples, cotton, live stock, hides, &c. Chief town, St. Landre'. Pop. of the parish, in 1820, 10,085, 1830, 12,591.

Saint Landre', post village and st. jus. parish of St. Landre', Opelousas, is situated on bayou Bourbee, the extreme head branch of Vermillion r; a branch of the Teche, however, rises immediately in the rear of the v. Though the adjacent country is level, it is in an uncommon degree pleasant. The lines of woods, the farm houses along their margins, with the innumerable flocks of cattle and horses, and the rich products of the soil, yield a very animated picture. This place gains interest from being the most sthwestrn. post village in the U. S. In the p-o. list it is called Opelousas, and in the direction of letters that name ought to be used. (*See Opelousas.*)

St. Lawrence, co. N. Y. bounded by St. Lawrence r. n. w. which separates it from Upper Canada, Franklin co. e., Hamilton and Herkimer cos. s., and Herkimer, Lewis and Jefferson cos. s. w. The line on the St. Lawrence is 65 1-2 ms. without the sinuosities. It contains about 2,000 square ms. and 24 towns; the capital is Oswegatchie. Black lake is long and narrow, and a convenient channel of navigation from Rossie iron works, running nearly n. to the St. Lawrence. Oswegatchie, Grass, St. Regis, and Racket rs. are the principal streams, whose courses are long. The surface is broken and hilly, except s. and s. e. where it is nearly level. The soil below Ogdensburgh is light and productive. Iron ore abounds, and the streams furnish mill sites. Oak, maple, birch, bass, beech, and white and Norway pine form the forests. There are extensive swamps in the s. e. A canal from the St. Lawrence to lake Champlain has been proposed, to cross this co. and Franklin and Clinton cos. There is a state arsenal at Russell. Pop. 1820, 16,037, 1825, 28,000, 1830, 36,354.

Saint Lawrence, p-v. southwestern part Chatham co. N. C.

Saint Leonard's, p-v. on Chesapeake bay, eatsern side of the peninsula between that sheet of water and Patuxent r. and in Calvert co. Md. 12 ms. s. e. Prince Fredericktown, the co. st., and by p-r. 75 ms. s. e. W. C.

Saint Louis, r. of the U. S. in Huron, as laid down by Tanner rises at lat. 48°, and between long. W. C. 15° and 16° w. interlocking sources with water courses flowing into Rainy Lake r. and between the confluents of Miss. r., and those of the northwestern part of lake Superior. Flowing thence by comparative courses s. w. 100 ms. inflects to a s. e. by e. course about an equal distance to its efflux into the extreme western angle of lake Superior. Along the lower course of St. Louis r. and the Savannah r. branch of Miss. is one of the channels of intercommunication between the basin of St. Lawrence and Miss. According to Mr. Schoolcraft, it is, following the stream, 148 ms. from its mouth up St. Louis r. to the Portage into Savannah r., and the Portage plain has an elevation of 652 1-2 feet above the level of lake Superior, or about 1,270 feet above the level of the Atlantic. Such an elevation is fully equivalent to 3 1-2 degrees of lat., therefore the region from which St. Louis r. flows, has a winter climate suitable to lat. 51 to 52° on the Atlantic coast.

Saint Louis, co. Mo. bounded by Jefferson s., Franklin s. w., Missouri r. separating it from St. Charles co. w., northwest and north, Mississippi river separating it from Madison county, Illinois, n. e., St. Clair co. Il. e. and Monroe co. Il. s. e. Greatest length from s. w. to n. e. 40 ms., greatest width 24 ms. but mean width about 12 ms., area 480 square ms. Lat. 38° 36', long. 13° 30', w. W. C. The sthrn. side is in part bounded and in part traversed by Maramec r. in an estrn. direction. The Mo. r. where it bounds St. Louis, flows to the n. e. by e. to its junction with Miss. at the northeastern angle of the co. Below the mouth of Mo. the Miss. flows s. s. w. to the influx of the Maramec. Thus the three rivers render St. Louis co. literally a peninsula, the neck from the Maramec to Mo. being only about 8 ms. wide. Short creeks flow from the centre into the respective rs. The whole resting on a substratum of limestone. Soil productive. Chief town, St. Louis. Population 1820, 10,049, 1830, 14,125.

Saint Louis, p-t. city and st. jus. Saint Louis co. Mo., on the right bank of Miss. river, 20 ms. below the junction of that stream with Mo. river, 68 ms. s. w. by w. from Vandalia, in Il., 116 ms. e. Jefferson, and by p-r. 856 ms. a little s. of w. W. C.; lat. 38° 36', long. W. C. 13° 14' w. This city was founded in 1764, but during the existence of the French and Spanish colonial governments remained a mere village. The site is advantageous, similar to Cincinnati, and rises by two bottoms or plains. The lower on the Miss. is alluvial, from which a limestone bank rises to the level of the adjacent country, which sweeps backwards as far as the eye can reach. The principal street exceeds a mile in length, and is tolerably compact. This place has now gained all the attributes of a commercial depot; and contains a branch of the bank of the U. S., a Catholic cathedral, several other places of public worship, an academy, numerous schools, and a spacious town house. Three or four gazettes are published weekly. In the harbor appears the activity of commerce; the depth of water in the Miss. being always sufficient for the navigation of the largest steamboats, at all seasons, except when the r. is covered with ice, an obstruction which, however, occurs to a longer or shorter period annually. Population 1830, 6,694.

SAINT MARKS, small river of Flor., which has gained importance from its position, rising between the Ocklockonne and Oscilla rs., and to the s. of Tallahassee. From this limestone region it flows 10 or 12 ms. to the S. E., receives the Walkully from the N. E., and takes the name of Appalache at the village of Saint Marks. (*See Appalache.*) If we compare the navigable facilities of Saint Mark's r. by either branch, with the length of the streams, we are struck with the disparity. Neither branch has 35 ms. comparative course from head to entrance into the gulf of Mexico, and yet large boats ascend both branches to near their sources. "In the winter of 1826," says Williams in his Florida, "The Franklin schooner came up to the fort (Saint Marks) drawing 9 feet water; but 7 is as much as can be depended on."

SAINT MARKS, p-v. at the head of Appalache river, and junction of Saint Mark and Walkully rivers, by p-r. 22 ms. S. S. E. Tallahassee. This is the port of Tallahassee for sail vessels, though boats are navigated 10 ms. still higher. (*See art. Saint Mark's r.*)

SAINT MARTIN'S, river, a small stream of the northeastern angle of Worcester co. Md. The extreme source is in Sussex co. Del., but flowing S. E. it enters Worcester co. Md., and falls into the northern arm of Sinepuxent bay, and opposite Fenwick's isl.

SAINT MARTINS, p-v., nrthestrn. part Worcester co. Md., on Saint Martin's r., 20 miles N. N. E. Snowhill, the co. seat, and by p-r. 144 ms. S. E. by E. W. C.

SAINT MARTINS, upper parish of Attacapas, La., bounded by Lafayette parish S. W., Saint Landré or Opelousas W. N. W. and N., Atchafalaya r., separating it from West Baton Rouge N. E., parish of Iberville E., Ascension S. E., and Saint Mary's or lower Attacapas S. Length from E. to W. 60 ms, mean breadth 30, and area 1,800 sq. ms. Lat. 30° and long. W. C. 15° W., intersect near New Iberia, the port of the parish. The eastern part of this parish between Atchafalaya r. and the prairies E. of Teche r., is liable to annual submersion, and covered with a dense forest of such trees as are natural to inundated land, such as water white oak, willow, bitter nut hickory, sweet gum, tupeeloo gum, cypress, &c. This section is uninhabitable, but with the prairie land a different soil and aspect is presented. The central section is traversed by the Teche, and the western by Vermillion r., both streams flowing sthrdly., and upon both, extend along their banks narrow lines of wood land; the intermediate surface prairie. Near the r. banks the soil is exuberantly fertile, and sufficiently elevated for cultivation. Sugar cane has been cultivated on the Teche in this parish, but is evidently too far N., and too much exposed to the sweeping nrthwst. winds to be made a profitable staple. Cotton succeeds well, as also Indian corn, rice, indigo, &c. The peach and fig are the principal fruits. Live stock is, however, the most valuable staple. Chief t. Saint Martinsville. Pop. 1830, 6,442, of whom 4,301 were slaves.

SAINT MARTINSVILLE, p-v. and st. jus. Saint Martin's parish, La., is on the right bank of Teche r. 35 ms. a little E. of S. Saint Landré, 11 ms. S. New Iberia, and by p-r. 176 ms. a very little N. of W. New Orleans. This village rose, after the establishment of the U. S. government, around Saint Martin's church. It extends chiefly in one street along the high bank of Teche. Pop. about 300. Lat. 30° 09′, long. 14° 56′ W. W. C.

SAINT MARY'S, an important river of the Atlantic slope of the U. S. in Geo. and Flor. As delineated on our maps, this river has its extreme northwestern source in Ware county, Geo., at lat. 31°, between the Santilla r., and the Alapapaha, a branch of Suwannee. Flowing thence by a course a little E. of S., and receiving large accessions of water from the westward, the Saint Mary's reaches its great bend at lat. 30° 21′, having become a boundary at lat. 30° 36′ (nearly) between Geo. and Flor. Inflecting abruptly to the E. about 5 ms., turns again equally abruptly to the northward, and flows about 35 miles comparative course almost reverse to its original direction, to the influx of Spanish r. from the nrthrd. and from Geo. Once more this singular r. bends at very nearly right angles, and by a comparative course of a little S. of E., reaches the Atlantic between Amelia and Cumberland isl. The basin of Saint Mary's river, extending from lat. 30° to 31°, is about 80 ms. in length from E. to W., with a mean breadth of 30 ms.; area 2,400 ms.; having the basin of Saint John's S. E., Suwannee W., and Santilla N. The following description is given by the U. S. engineers. "Saint Mary's river takes its rise out of the extensive swamps which are on the Geo. line, and stretch between the head branches of Saint Mary's and Suwannee rs. These swamps, called emphatically dismal swamps, are generally covered with a thick growth of bay trees, vines and undergrowth. At some places, short bay bushes, at others sedge grass, are the only growth. No lake or natural reservoir of importance is to be found; but on account of the great extent of the swamps, draining both ways, into the Suwannee and Saint Mary's rs., these streams are subject to high freshets during the rainy season, or after a sudden heavy rain. At the upper fork of Saint Mary's r., the rise of freshets is about six feet. In following the windings of the r., the distance from the Atlantic to the very head of the stream, 13 ms. above the upper fork, is about 105 ms. The summit point of the ridge between the Atlantic ocean and the gulf of Mexico, has been found 152 feet. Saint Mary's r. itself flows generally through narrow strips of wet hammocks. The banks immediately adjoining are high, their soil sandy, their growth pine. The greatest depth of water on the bar, at the entrance of Saint Mary's harbor, is as much as 22 feet during spring tides, when easterly winds have blown for a considerable time, and 13½ feet only at low tide. The tide is felt as far up as Barbour's plantations, 50 ms. from the mouth of the r. The commodious harbor at the mouth of Saint Mary's r., presenting on the bar a depth of 13½ feet at low water, and 19½ feet at common high tide, is susceptible of defence, and derives a great importance from the circumstance of being the only good harbor from the boundaries of Geo. to Flor. Point."

SAINT MARY'S, river of Ohio and Ind., has its remote source in Shelby co. Ohio, and assuming a northwestern course traverses Mercer and Vanwert counties of Ohio, enters Indiana, and mingles at Fort Wayne, in Allen co., with the Saint Joseph's river to form the Maumee, after a comparative course of 60 ms. The valley of Saint Mary's lies between those of upper Wabash, and au Glaize, branch of Maumee.

SAINT MARY'S strait, between Chippeway co. Mich., and Upper Canada, unites lakes Huron and Superior. Taken in its utmost extent, the straits of Saint Mary extend N. W. and S. E. 75 ms. from the passage between Drummond's and Saint Joseph's isls., to Maple isls. in lake Superior. From Maple isls. to the cataract of Saint Mary, is about 30 ms. In this higher section the strait gradually narrows, & is but little interspersed with isls. At the *Sault* or Chute, between fort Brady, and the British Hudson's Bay company's factory, the water is contracted to about ½ mile, and rushes over a ledge of rocks. This part of the strait is navigable, with some more difficulty than real danger, by vessels of 6 feet water. Below the cataract the strait becomes and continues to be divided into two channels by George and Saint Joseph's isls. These channels are similar to the Belts in Denmark which unite the Baltic and Scaggerac seas. The southwestern channel on the side of Michigan, is again subdivided by Sugar isl. 20 ms. long, and some other smaller islands, but after winding 50 ms. opens into lake Huron, between Drummond's isl. & the promontory of The True Detour. The eastern channel or strait along the Canada shore, stretches from the cataract of Saint Mary 40 miles, and terminates in Manitou bay of lake Huron. The entire fall from the level of lake Superior to that of Huron is about 23 feet. From recent and accurate observation, it has been clearly established that the rocks in Saint Mary's strait are slowly yielding to the impression of floods and ice, and that the surface of lake Superior is lowering. (*See art. Sault de Saint Mary.*)

SAINT MARY'S, p-o. Chester co. Pa., by p-r. 139 ms. N. E. W. C.

SAINT MARY'S, co. Md., bounded by the Potomac S. which separates it from Northumberland and Westmoreland counties, Va., Charles co. Md. W. and N. W., Patuxent river separating it from Calvert co. Md. S. E., and Chesapeake bay E. Length from Point Lookout at the mouth of Potomac to the northwestern angle 38 ms., mean breadth 10, and area 380 sq. ms. Lat. 38° 03' to 38° 30', long. W. C. 0° 12' to 0° 41' E. Tho' bounded by the Patuxent on the N. E., the slope is southward towards the Potomac, and in that direction flow the Saint Mary's, Britton's and Wicomico rivers. These brief but important water courses are navigable bays for some distance from the Potomac, and the Wicomico by its relative course with the Patuxent, nearly insulates the co. This point or peninsula, now Saint Mary's co., was the cradle of Md.; it was there, that in 1632, Calvert's colony was founded, and where the seat of the government of Md. continued 67 years, until in 1699 it was permanently fixed at Annapolis. Chief town, Leonardtown. Pop. 1820, 12,974, 1830, 13,459.

SAINT MARY'S, sea-port and p-v. at the mouth of Saint Mary's river, and in the sthestrn. angle of Camden co. Geo. It is situated directly W. of the entrance between Cumberland and Amelia isls., 80 ms. by land, a little W. of S. Darien, and 235 ms. S. S. E. Milledgeville. Lat. 30° 42', long. W. C. 4° 48' W. This place from its position must become one of great importance, but hitherto the want of pop. on the basin of Saint Mary's r. has retarded its increase.

SAINT MARY'S, or Lower Attacapas, parish of La., bounded N. W. and N. by Saint Martin's parish, Atchafalaya r. N. E. separating it from Ascension N. E., and Assomption E.; parish of Terre Bonne S. E., the Gulf of Mexico S., and Vermillion bay separating it from the parish of Lafayette W. Length parallel to the Teche river 50 ms., and independent of the deep indentings of Vermillion, Cote Blanch, & Atchafalaya bays, and the surface of lake Chetimaches, the mean breadth of the land surface is about 20 ms.; and area 1,000 sq. ms. Central lat. 29° 45', long. W. C. 30° 40' W. Lying entirely below lat. 30°, Saint Mary's parish is in all its extent within the climate suitable to the growth of sugar cane, and the soil being without exception highly fertile, where of sufficient elevation to admit culture; sugar is a standing staple of the parish. The Atchafalaya river and Teche r. afford a navigable channel of 8 feet to the centre of Saint Mary's parish, and of 5 or 6 feet to New Iberia in Saint Martin's. The general course of the Teche from New Iberia to its mouth into Atchafalaya is S. E. by E., with a channel sweeping very large bends. The lines of woodland along this stream narrow until near the mouth; soil on both banks first rate. As in Saint Martin's, the annually inundated part of Saint Mary's towards Atchafalaya, is covered with a dense forest. On the contrary side of Teche towards the Gulf of Mexico, the general surface is prairie; near the Teche, and some other streams comparatively high and arable, but sinking into immense grassy morasses near the Gulf bays; timber, where found, mostly stands in detached clumps. Along the Gulf shore of this parish, occur those remarkable hills, called Petite Anse, Grand Cote, Cote Blanche, and Belle Isle. These hills rise on the shores of the bays, and though surrounded by marsh, rise far above any other land s. of Upper Opelousas. They are composed of a very productive soil, and in their natural state were covered with dense forests. Neither of these hills exceed 1½ ms. in length; their timber distinct (except live oak which abounds on both), from that along the Teche, and other streams of Saint Mary's. Sugar, cotton, rice, indigo, tobacco and live stock, are the staples; fruits, fig, peach, and some apples. Chief t. Franklin. Pop. 1830, 6,442.

Saint Mary's, p-v. on Saint Mary's r., and in the eastern part of Mercer co. Ohio, by p-r. 111 ms. n. w. by w. Columbus. Population 1830, 92.

Saint Michael's, small river, or creek, of Talbot co. Md., extends first southward 10 ms. past the town of Saint Michael's, until within little more than 1 mile from Tread Haven bay, when it bends 5 or 6 miles towards the n. e. To the w. of Saint Michael's river extends a peninsula, to which that name is often applied.

Saint Michael's, sea-port, and p-v. on the western side of Saint Michael's bay, and in Talbot co. Md., by p-r 12 ms. n. w by w. Easton, the co. seat, and 72 a very little s. of e. W. C.

Saint Regis, village, Saint Lawrence co. N. Y., 45 ms. e. n. e. Ogdensburgh, on Saint Regis r. at the mouth in the Saint Lawrence. Lat. 45°.

Saint Stephens, p-v. and st. jus. Washington co. Ala., on the right bank of Tombigbee river, 70 ms. n. Mobile, and by p-r. 162 a little w. of s. Tuscaloosa. Lat. 31° 33′, long. W. C. 11° 10′ w. It stands at the head of schooner navigation in Tombigbee, is the seat of an academy, contains a printing office, and a pop. of 1,000, or 1,200.

Saint Tammany, parish of La., bounded n. w. by Tangipola r. separating it from Saint Helena, n. by the parish of Washington, e. Pearl river, separating it from Hancock co. Miss., and s. by lake Pontchartrain. Length 70 miles, from the mouth of Pearl river to the extreme northwestern angle on Tangipola river; mean breadth 15 ms.; and area 1,050 sq. ms. Lat. 30° and long. 13° w. W. C. intersect very near the centre of this parish. Slope s. e., and traversed by Chifuncte river. The surface rises from lake Pontchartrain into hills covered with pine and other timber. Soil generally sterile. Chief t. Covington. Pop. 1820, 1,723, 1830, 2,864.

Saint Thomas, township, and p-v. Franklin co. Pa., 9 ms. w. of Chambersburg, and by p-r. 97 ms. n. w. W. C. Pop. township 1830, 1,771.

Salem, p t. Rockingham co. N. H., 30 ms. s. w. Portsmouth, 30 s. e. Concord; with 15,600 acres; has an uneven surface, a soil generally good, and is crossed by Spiggot river, which, with its numerous tributaries, affords many mill seats. Policy, World's End, and Captain's ponds are the principal sheets of water. Pop. 1830, 1,302.

Salem, town, Orleans co. Vt., 49 miles n. Montpelier; first settled 1798; contains 17,330 acres, and is crossed by Clyde r., which falls into Salem or Derby pond, on the boundary line. South bay of lake Memphremagog enters the town on the w. There are no mill sites. The soil is generally good, and the surface level. The trees are various. Pop. 1830, 230.

Salem, p-t. seaport, and capital of Essex co. Mass. 14 ms. n. n. e. Boston, 24 s. Newburyport, 4 ms. n. w. Marblehead, and 450 n. e. W. C. is the second town in the state for population, wealth, and commercial importance, being inferior only to Boston. Its lat. is 42° 30′ n., and its long. 6° e. W. C.

Salem is the oldest settlement in New England except Plymouth, having been settled in 1628. Its Indian name was Naumkeag, or Naumkeek, by which title it was long designated. Its settlement was commenced by John Endicott, for a company in England, which had purchased the place of the Plymouth company. He erected dwellings, &c. and in 1629, ships to the number of 11 came out, bringing 1,500 persons, by whom were commenced the settlements at Boston, Charlestown, Dorchester, &c. John Winthrop was appointed governor, and Thomas Dudley, deputy governor, by charter, and as they resided at Boston, that place became the seat of government for the colony. Two hundred of the settlers died at Salem in the first winter. The territory then included the present townships of Danvers, Beverly and Marblehead. The first cases of witchcraft, which excited public attention so much in the early periods of New England, occured here. The persons first tried on this singular accusation lived in what now is Danvers. Many trials took place in Salem, and many executions on the neighboring eminence called Witch-hill. Roger Williams, who colonized Rhode Island, was once a pastor here, and Bowditch, the mathematician, and Timothy Pickering, secretary of state of the U. S. were both natives of Salem. The British authority was resisted in Salem before the battle of Lexington, for Col. Leslie, who had come from Boston to remove some cannon thither, was prevented from entering the town by the removal of a draw bridge, &c. so that he was unable to accomplish his object.

The town, though low, is pleasantly situated at the head of the bay formed by two inlets from the sea, and including a peninsula running e. The soil is generally poor. About the neck of the peninsula are collected the principal part of the inhabitants, on a surface of about 1-2 a mile by 1 1-2 ms. The streets are generally beautiful and well built, mostly with wood, though with many brick, and some elegant buildings. They cross each other at right angles, with large open squares, bordering on which are the public buildings, some of which make a fine appearance. The common is beautiful, and planted with fine shade trees. The commercial prosperity of the place during the successful prosecution of an active trade with the East Indies and China, some years ago, adorned Salem with many splendid edifices. This trade is still extensive though not so much so as formerly. N. and S. rivers are two arms of the bay between which the peninsula extends. They are crossed by bridges which unite Salem to two considerable villages, or suburbs. The bridge over North r. connecting Salem with Beverly is 1,500 feet long. The harbor has too little water at ebb tide to allow all desirable facilities for commerce; vessels of large burden not being able to lie at the wharves, and those drawing more than 12 feet water being commonly lightened before coming up to the t. An important work has been commenced, by which the town will be supplied with considerable water power. By a dam across the N. river, and a canal across the neck, the tide may be made to be put in motion a large amount of machinery. By this enterprising improvement, the manufactures of Salem will doubtless be much increased. Among the public buildings, are a court-house, the market-house, the atheneum, the orphan asylum, churches, &c. The atheneum, alms-house, hospital, and 2 forts are on the neck. The atheneum has a library of 5,000 volumes. The marine museum is a valuable collection of rare curiosities from all parts of the world, contributed by the members of the East India marine society, who are all nautical, or commercial men, and who established the society with a view to promote a knowledge of East Indian navigation and trade, and to aid indigent members and their families. There are in Salem 9 banks, which in 1831, made half yearly dividends of from 2 1-2 to 3 1-2 per cent. There are also 6 insurance offices, 15 churches, school-houses, &c. The schools, of which there are between 20 and 30, are flourishing and well supported, and common education is placed on a footing highly creditable to the people, and proportioned to its real importance. The town also contains 16 tanneries, 11 twine and cordage factories, and 2 white lead manufactories, to be moved by water power; one of them makes 600 tons annually, beside a large quantity of the sugar of lead, and the other 1,000,000 lbs., half of which is called German white lead, manufactured on a secret plan, for the knowledge of which $10,000 were paid. Ten thousand gallons of oil are consumed in the preparation of the German white lead alone. The iron company makes 500 tons into hoops annually. There are 15 or 20 vessels employed in the coasting trade, and many others in lumber, wood, &c. In 1781, 52 vessels, mounting 746 guns, which were engaged in privateering, were owned in this place. The churches of Salem are 4 Unitarian, 3 Congregational, 2 Baptists, 1 Episcopal, 1 Roman Catholic, 1 Methodist, 1 Quaker, 1 Christ-ians, and 1 Universalist. Population 1810, 12,613, 1820, 12,731, and in 1830, 13,895.

SALEM, p-t. New London co. Conn. 29 ms. S. E. Hartford, lies E. of East Haddam, Middlesex co., has Gadner's lake on the N. E. border, and a few small streams flowing S. W. into Conn. r. The surface is uneven. Pop. 1830, 958.

SALEM, p-t. and half capital, Washington co. N. Y. 46 ms. N. E. Albany, 21 S. E. Sandy hill, W. Vermont, and has Battenkill creek S. Several streams flow into this, and the town is well supplied with mill sites. The land near the streams in some places presents fine meadows. Pop. 1830, 2,972.

SALEM, p-v. and incorporated village, Salem, Washington co. N. Y. 46 ms. from Albany.

SALEM, co. N. J. bounded by Gloucester co. N. and N. E., Cumberland co. S. E. and S., Delaware bay W., which separates it from Pa. contains 9 tsps.; the capital is Salem. Oldman's creek forms the north boundary, Maurice r. part of the E. and Stow creek the S. E. Salem r. rises in the E. part, flows W. by N. then S. and empties into Salem cove, on the Delaware, a little below fort Delaware. South of this, and opposite Reedy island, is the mouth of Alloway's creek which, as well as Salem r. has a tract of marshy land along the lower part of its course. This co. enjoys the advantage of navigation on the W. where it is washed by the Delaware. The streams supply mill sites, which are used. A small canal extends from the bend of Salem river to the Delaware. Pop. 1820, 12,791, 1830, 14,155.

SALEM, p-t. and st. jus. Salem co. N. J. 65 ms. S. W. Trenton, 20 ms. N. W. Bridgetown, and 37 S. W. Phila., is a small town at the head of navigation for vessels of 50 tons, on Salem river, 3 1-2 ms. from its mouth in Delaware bay. It has Fenwick's cr. N. and Salem r. on a part of its W. boundary. Pop. 1830, 1,570.

SALEM, p-v. sthrn. part of Botetourt co. Va., 60 ms. a little S. of W. Lynchburg, and by p-r. 256 ms. S. W. W. C. It is situated in the great valley between the Blue Ridge and Kittatinny or North mountain, at an elevation of about 1,020 feet above the level of the Atlantic. The Roanoke is thus far navigable for boats. (*See article roads and canals, head of Roanoke.*)

SALEM, p-v. sthrn. part Stokes co. N. C. by p-r. 113 ms. a little N. of W. Raleigh.

SALEM, p-v. sthrn. part of Sumpter district, S. C. by p-r. 72 ms. S. E. Columbia.

SALEM, p-v. sthrn. part of Clarke co. Geo. by p-r. 58 ms. N. Milledgeville.

SALEM, p-v. sthrn. part of Franklin co. Ten. by p-r. 10 ms. s. w. Winchester, the co. st. and 92 ms. s. e. by e. Nashville.

SALEM, p-v. and st. jus. Livingston co. Ky. by p-r. 245 miles s. w. by w. 1-2 w. Frankfort, and 35 ms. s. Shawneetown in Il. lat. 37° 15', long. W. C. 11° 20' west. Pop. 1830, 281.

SALEM, p-v. Columbiana county Ohio, by p-r. 10 ms. n. w. New Lisbon, the co. st. and 157 miles n. e. by e. Columbus. Pop. 1830, 56.

SALEM, p-v. and st. jus. Washington co. Indiana, by p-r. 91 miles s. Indianopolis, and 33 miles n. w. Louisville, in Ky. Lat. 38° 37', long. W. C. 9° 06' w.

SALEM, p-v. and st. jus. Marion co. Il. by p-r. 26 ms. s. e. Vandalia, and 74 miles e. St. Louis in Mo. Lat. 38° 40', long. W. C. 12° w.

SALEM CROSS ROADS, and p-o. western part of Westmoreland co. Pa. 8 miles westward Greensburg, the co. st. and by p-r. 200 ms. n. w. W. C.

SALEM FAUQUIER, p-v. northern part Fauquier co. Va. by p-r. 63 miles w. W. C.

SALINA, p-t. and st. jus. Onondaga county N. Y. 130 ms. w. Albany, and 5 miles n. Onondaga, is of irregular form, though bounded by right lines, and includes the lower part of Onondaga river, and all Onondaga or Salina lake, with many of the salt springs on its shore, and extends to Oswego river. The manufacture of salt by artificial heat is carried on to a great extent in this town, and creates an active business on the branch canal which extends from Syracuse to the village. The navigation has been opened in the opposite direction, by the Oswego canal, to the mouth of Oswego river and lake Ontario. The pumps by which water is supplied to the salt works here, are at Syracuse and Geddes, which, as well as Liverpool, are salt making villages, in this town. Syracuse is large and very flourishing; indeed the general increase of inhabitants and wealth in this town since this branch began to be extensively carried on, has been remarkably great. The number of manufactories of salt by artificial heat in Salina, is 135, containing 3,076 kettles. The manufactories making salt by solar heat or evaporation, are, the Onondaga salt company, the Syracuse salt company and Henry Gifford's works; in all, consisting of 1,303,-024 superficial feet of lots. In the 4 villages above mentioned there are (1832,) 125 manufactories of salt, besides two companies whose vats for solar evaporation would extend in a continuous line about 15 miles each. In 1831, there were nearly a million and a half of bushels of salt manufactured. The great salt spring is situated on the edge of the Oswego canal, at a short distance from the shore of the lake, in the village of Salina. It is in a soft alluvial soil, and was formerly a marsh, till the surface of the lake was lowered a few years ago. On the bank of the canal there is a large building, containing the immense reservoir which supplies the manufacturers for several miles around. Two immense iron pipes, on an inclined plane, throw up the water by two forcing pumps, which are worked by a large water-wheel, driven by water taken from the canal. The spring supplies three pumps with water. The reservoir, the house, and the buildings attached, have a singular bronzed appearance, interspersed with salt incrustations. Close to this building there is another reservoir and set of forcing pumps making, which will be ready for use whenever they are required by the manufactures. The old spring in use a few years ago, is now superseded by a new one, recently discovered, which is much stronger and better than the old one. There is at Syracuse, a court house, several churches, large hotels, and handsome private edifices, with an active country trade. Pop. Salina, 1830, 6,929.

SALINE, r. of La. rises in Claiborne parish, between Dugdomen and Black rs., and flowing southward 50 miles, falls into the Rigolet de Bordien, 8 or 9 miles e. of the village of Natchitoches.

SALINE, river of Arkansas, draws its most remote sources from the Masserne mountains, about 20 miles w. little Rock and about lat. 34° 45'. Flowing thence by a course of a little e. of s. 120 miles falls into the left bank of Ouachitta, lat. 33° 10'. The valley of the Saline lies between those of the main Ouachitta, main Arkansas, and Barthelemy.

SALINE, river of Illinois, rising by numerous branches in Johnson, Franklin, and Hamilton counties, which unite in Gallatin and fall into Ohio river about 5 miles below Shawneetown, after a general estrn. course of 55 miles. The country drained by it is very broken. The U. S. possess extensive salt works on it and from which the name is derived.

SALINE, county, Mo. bounded s. e. by Cooper, s. uncertain, Lafayette w., and the Mo. river on all other sides; independent of an uncertain southern extension that part on Mo. is about equal to a square of 30 miles each side, 900 square miles. Latitude 39°, and long. W. C. 16° w. intersect in this co. The Mo. river semicircling the co. gives it a border of near 70 miles on that stream, the lower part of which flows to the west of south. It is towards this bend of Mo. that the slope of the co. falls; it is consequently to the eastward, and in that direction drained by Mine river and its branches. Chief town, as given in the p-o. list, Walnut Farm. Pop. 1830, 2,873.

SALINE, p-v. sthrn. part Washtenau county, Mich., on the head of a creek of the same name, by p-r. 52 miles a little south of west Detroit.

SALISBURY, p-t. Merrimack county, N. H. 15 miles n. Concord, 78 n. n. w. Boston, w

Pemigewasset and Merrimack rivers. The latter is formed here by the junction of the Pemigewasset and Winnipiseogee rivers, near which point is the head of boat navigation. Black river is in the west. Black and yellow oak, white, pitch, and Norway pine abound, and formerly the hills, which are now chiefly devoted to pasturage, were covered with maple, beech, birch, &c. There are valuable meadows on Blackwater river, and 300 acres in a bend of the Merrimack. Kearsearge mountain in the N. W. corner, has its summit a little beyond the line. It is a mass of granite, which rock prevails through the town. First settled, 1750. Several of the inhabitants were carried captive to Canada in the last French war. Pop. 1830, 1,379.

SALISBURY, p-t. Addison co. Vt. 34 miles S. W. Montpelier, 40 S. Burlington, is small and in the centre of the co. First settled, 1775, by a single family. Otter creek bounds it W., Middlebury river is N., and Leicester river S. Lake Dunmore, about 2 ms. by 4, lies partly in this town. Its outlet, Leicester river, supplies water power to a manufacturing village. The soil is good, the surface uneven except W. where are meadows. The Green mountains are E. There are several swamps, and a large cavern. Pop. 1830, 907.

SALISBURY, p-t. Essex co. Mass. 35 miles N. E. Boston, S. New Hampshire, W. Atlantic, N. of Merrimack river, and E. of Powow river, enjoys great advantages for trade, agriculture and manufactures. It has 2 long and expensive bridges, one leading to Newbury and the other to Newburyport. One of the villages at the Point, has been a place of much ship building. The village of Amesbury mills stands on both sides of the river and part of it is in this town. These manufactories here make excellent flannels, and a large amount of cotton goods, &c. The descent of the river at the falls is nearly 40 feet in about 220 yards, and the sites for machinery are very valuable. The soil of the town is generally good; in the E. is an extensive salt marsh, and a beach on the shore, which is frequented for bathing, &c. First settled, 1638, the first spot on Merrimack river inhabited by whites. The general court sat here in 1737, in relation to the boundary. Pop. 1830, 2,519.

SALISBURY, p-t. Litchfield co. Conn. 47 ms. N. W. Hartford, 60 N. N. W. New Haven, is rough and mountainous, and contains valuable iron mines and many forges, furnaces, and iron manufactories. Housatonic river which bounds it east, affords abundant water power, descending in one place 30, and another 60 feet perpendicularly, and flows over rapids below about 550 yards. Scythes, anchors, screws, gun barrels, &c. are made here, while a large amount of ore is smelted annually, and considerable quantities transported. It is about 6 miles by 9, with 58 square miles. The vallies are generally limestone, and the hills granite. Much of the soil is good, and bears a great deal of wheat, while it is also favorable to other crops. There are 4 fish ponds, and Salmon river which crosses the town S. E., affords valuable mill seats. First settled 1720 by three Dutch families, from the state of New York. Pop. 1830, 2,580.

SALISBURY, p-t. Herkimer county, N. Y., 21 miles northeast Utica, northeast Hamilton co. The West branch of East Canada creek, and the East branch of West Canada creek rise here, and East Canada creek forms part of the southeast boundary. Spruce creek is a valuable mill stream. Pop. 1830, 1,999.

SALISBURY MILLS, p-v. Blooming Grove, Orange co. N. Y. on Murderer's creek 6 ms. west New Windsor.

SALISBURY, p-v. Lancaster county, in a tsp. of the same name, 12 miles east Lancaster, and by p-r. 123 miles northeast W. C.

SALISBURY, p-v. on the extreme northeastern margin of Somerset co. Md. 17 ms. N. N. E. Princess Ann, the co. st. and by p-r. 128 miles S. E. by E. W. C.

SALISBURY, p-v. and st. jus. Rowan county, North Carolina, by p-r. 118 miles west Raleigh, and 51 ms. a little N. of E. Lincolnton; lat. 36° 40′, long. W. C. 3° 24′ W. Pop. 1830, 1,613.

SALISBURY, p-v. Meigs co. Ohio, by p-r. 106 miles southeast Columbus.

SALMON, r. Conn. rises in Tolland co. and with many small tributaries from that co., N. London, Hartford and Middlesex cos., flows southerly through a rough and romantic country to Connecticut r. which it enters in East Haddam. It has a fall of 70 feet in that town, where it moves the machinery of several factories. The banks of one of its branches were in ancient times the residence of a tribe called the Moodus Indians, who were famed as magicians.

SALMON r. Oswego co. N. Y. crosses this co. and falls into lake Ontario, where it forms a good harbor. Its course is about 45 ms. It is navigable 1 mile from the mouth, and in boats, at high water, 14 ms. to the falls in Orwell, at which place it is about 180 yards wide.

SALMON, cr., N. Y. There are several streams bearing this name in the state, one in Cayuga co. 19 ms. long; one in Oswego co. 28 ms.; one in Franklin co. entering the St. Lawrence; another in Monroe co. entering Braddock's bay, &c. &c.

SALMON FALLS r. N. H. a part of the Piscataquay, from its source to Berwick falls.

SALMON, p-v. Franklin co. Ind. by p-r. 89 ms. S. E. by E. Indianopolis.

SALT, r. of Ky. This stream is composed of two main and numerous minor branches. The main branches are Salt r. proper, and the Rolling Fork. Both branches have their extreme sources in Casey co., but thence diverge. Salt r. flows a little N. of W. parallel to and within 5 or 6 ms. of Kentucky r. about 36 ms. over Mercer into Nelson co., and thence assuming a wstrn. direction by comparative courses 60 ms. receiving from the nrthrd. Broshear's and Floyd's forks, finally receives or rather unites with the Rolling fork between Meade and Bullitt cos. Rolling fork is composed of two nearly equal branches, Rolling fork proper and Chaplin's fork; both having a s. w. by w. course of about 60 ms. above their junction, and 20 ms. below to the union of their waters with that of Salt r. Below the union of its 2 constituent branches Salt r. flows about 15 ms. comparative course to the N. W., and to its influx into Ohio r. at Shepherdsville, and almost on lat. 38° and long. W. C. 9° w. Salt r. is a large stream in proportion to its length, and drains a triangle of 80 ms. base, with a perpendicular of 50 ms., area 2,000 sq. ms.: comprising all the cos. of Washington, Nelson, Bullitt, Spencer, and Shelby; with part of Meade, Hardin, Casey, Mercer, Oldham, and Jefferson; of course one of the finest regions of Ky.

SALT, r. of Mo., has indeed its most remote source in the N. W. territory to the N. of Mo. between the valleys of the Des Moines and Chariton rs. as high as lat. 40° 50'. Flowing sthrd. about 20 or 30 ms. it enters Mo., and inclining to a direction a little E. of s. by comparative courses 100 ms. into Ralls co., where it receives numerous large creeks from the w. and s. and bends to an estrn. course, which it maintains about 50 ms. to its entrance into the Miss. in the northern part of Pike co. after an entire comparative course of 180 ms. The valley of the Mo. Salt r. has that of Des Moines N. E.; Miss. E.; Missouri s.; and Chariton w.

SALT CREEK, p-v. Muskingum co. O. by p-r. 10 ms. s. E. Zanesville the co. st. and 69 ms. E. Columbus.

SALT SULPHUR, springs and p-o. Munroe co. Va. by p-r. 270 ms. s. w. W. C.

SALTZBURG, p-v. wstrn. part Indiana co. Pa. situated on Kiskiminitas r. and on the Pennsylvania canal, by p-r 197 ms. N. W. W. C.

SALUBRITY, p-v. Gadsden co. Flor. by p-r. 14 ms. s. w. Tallahassee.

SALVAGES, a reef of rocks off Sandy Bay, on the north shore of Cape Ann, Mass.

SALVISA, p-v. Mercer co. Ky. by p-r. 21 ms. s. Frankfort. Pop. 1830, 39.

SAMPSON, co. of N. C. bounded by New Hanover s., the estrn. branch of Cape Fear r. separating it from Bladen s. w., and Cumberland w. and N. W., by Johnson and Wayne N., and Duplin E. Length 40 ms., mean breadth 22, and area 880 sq. ms. Lat. 35° and long. W. C. 1° 20' w. intersect in this co. Slope nearly due s. and drained by Black r. branch of Little Cape Fear r. Chief t., Clinton. Pop. 1820, 8,903; 1830, 11,634.

SAMPTOWN, v. Piscataway, Middlesex co. N. J. on a small branch of the Raritan, near the N. W. corner of the co.

SAM'S CREEK, p-v. cntrl. part Frederick co. Md. about 20 ms. N. E. Frederick, and by p-r. 63 ms. a little w. of N. W. C.

SANBORNTON, p-t. Strafford co. N. H. 20 ms. from Concord, 9 from Guilford, and 60 from Portsmouth, occupies a peninsula, formed by Great and Little bays, and Winnipiseogee r. E. and s., and Pemigewasset r. w. The union of these 2 streams in the s. w. angle of the t. forms the Merrimack. Salmon brook N. W. is the principal stream in the t. and affords a few mill seats, as does Winnipiseogee r. The land is rough, but good, and almost all fit for cultivation. Sanbornton mtns. lie in the N. There is a remarkable chasm 38 feet deep, and a mile long, in a rocky ridge; and on the banks of the Winnipiseogee are remains of an Indian fort. This work was formed of six stone walls, enclosing a piece of ground, within which implements of war, &c. have been found. The tsp. was first settled 1765. There is an academy, one or two social libraries, and a fund for the support of the preaching of the gospel. Pop. 1830, 2,866.

SANCOTY HEAD, the east point of Nantucket isl. Mass. in lat 41° 16', long. 7° 5' E. W. C.

SANDERS, p-v. Limestone co. Ala. by p-r. 149 ms. N. Tuscaloosa.

SANDERS, p-v. Grant co. Ky. by p-r. 54 ms. s. s. w. Frankfort.

SANDERSON'S, p-o. Goochland co. Va. by p-r. 161 ms. s. s. w. W. C.

SANDERSVILLE, p-v. Chester dist. S. C. by p-r. 67 ms. N. Columbia.

SANDERSVILLE, p-v. and st. jus. Washington co. Geo. by p-r. 27 ms. s. E by E. Milledgeville. Lat. 32° 52', long. W. C. 5° 55' w.

SANDERSVILLE, p-v. Vanderburgh co. Ind. by p-r. 158 ms. s. w. Indianopolis.

SANDFORD, p-t. Broome co. N. Y. 24 ms. E. Chenango point, or Binghampton, and s. Chenango co., has a hilly and stony surface, favorable to pasturage, with some good vales. A small stream in this t. flows into Cookquago cr., a branch of Del. r. on the s. E. line. Pop. 1830, 931.

SANDGATE, p-t. Bennington co. Vt. 20 ms. N. Bennington, 31 s. w. Rutland, E. N.Y. state, has a broken surface, with Sheltarack and Bald mtns. N. W., Swearing hill s. w., Red mtn. s. E., and part of Equinox mtn. N. E. It is watered by tributaries of White cr. and Battenkill, but ill supplied with mill seats. Pop. 1830, 933.

SANDIGE'S, p-o. Amherst co. Va. by p-r. 142 ms. s. w. W. C.

SANDISFIELD, p-t. Berkshire co. Mass. 112 ms. s. w. Boston, 22 s. E. Lenox, N. Conn., is crossed by Farmington r. s. E., on both sides of which rise steep and romantic banks. The soil is favorable to agriculture, and scattered with farm houses. Maple sugar is manufactured here in considerable quantities. Pop. 1830, 1,655.

SANDISTON, p-t. Sussex co. N. J. 63 ms. N. Trenton, has Del. r. on the N. W. line, and the Blue mtn. ridge along the E. boundary. Big and Little Flat crs. cross the t. Pop. 1830, 1,097.

SAND LAKE, p-t. Rensselaer co. N. Y., 11 ms. E. Troy, has a rough surface, with many hills and much waste land. Crooked and Glass lakes are the principal of 6 ponds. There are few mill seats. Marle is found in considerable quantities. There are 3 vs. Sand Lake, Rensselaer and Poestenkill. Pop. 1830, 3,650.

SANDOVER, p-v. Abbeville dist. S. C. by p-r. 90 ms. wstrd. Columbus.

SANDOWN, p-t. Rockingham co. N. H. 31 ms. S. E. Concord, is small, with only 8,532 acres. It is uneven, favorable to grain and grass, with several ponds, one of which, Phillip's, gives rise to Squamscot r. This stream sometimes flows backwards towards its source. First settled 1736, then a part of Kingston. Pop. 1830, 557.

SANDS' POINT, North Hempstead, Queen's co. N. Y., the extremity of Cow Neck, a cape running into Long Isl. sound, has a light house, a little E. from Cow Bay.

SANDTON, p-v. Kershaw dist. S. C. by p-r. 55 ms. N. E. Columbus.

SANDUSKY, r. O. having its remote sources in Marion, Crawford, and Richland cos. interlocking sources on the W. with those of Blanchard's fork of au Glaize r., on the S. with those of Sciota; and on the E. with those of White Woman's r. or the nrthwstrn. sources of Muskingum. Issuing from this table land the Sandusky assumes a nrthrn. course, and after traversing Crawford and Seneca cos., enters Sandusky co. where inflecting to the E. it opens into an oblong sheet of water from 1 to 3 miles wide, and about 20 in length. This small gulf is called Sandusky bay, but closed by two projecting points; on the estrn. extreme the water is confined to a narrow channel, admitting vessels of 6 or 7 feet draught. (*See Sandusky vil.*)

SANDUSKY, co. O. bounded by Huron co. E., Seneca S., Wood W., Monroe co. of Mich. N. W., and lake Erie N. It is very nearly a square of 28 miles each side, area 784 sq. ms. Lat. 41° 25′, and long. 6° 06′ W. intersect in this co. The sthrn. section slopes to the N., but towards the centre all the streams which traverse its surface curve in common to N. E. This is the case with Sandusky, Muddy, Portage and Toussaint rs. Between Sandusky bay and lake Erie extends a peninsula which is nearly equally divided between Sandusky and Huron cos. There are tracts of good land in Sandusky, but the general features of its surface are low, and it is consequently wet. The asperity of soil is shown by the progressive population, which was in 1820, 852; 1830, 2,851. Chief t., Lower Sandusky.

SANDUSKY, seaport and p-v. on the S. side of Sandusky bay, Huron co. O. by p-r. 115 ms. a little E. of N. Columbus, and 415 ms. N. W. by W. W. C. Lat. 41° 28′, long. W. C. 5° 40′ W. The rise and progress of this flourishing place is one of those fine creations made by the extension of commerce and agriculture in the interior of "the great west." The writer of this article was on the spot 1818, a few months after the first establishment, and according to Mr. Flint, early in 1832, it contained 9 wharves, 10 stores, a ship yard and rope walk, also a printing office, several private schools, numerous mechanics' shops, hotels, taverns, and in brief all the substance of a great entrepot. The amount of merchandize which was landed there was, $1,319,-823. In 1830, upwards of 500 arrivals, in the port, and the arrival and departure of 2,000 wagons, evinced the importance of its mercantile transactions. The site is high, dry and pleasant. A turnpike is constructing to connect it with Columbus. Pop. 1830, 593.

SANDUSKY Cross Roads and p-o. Knox co. O. by p-r. 48 ms. N. N. E. Columbus.

SANDWICH, p-t. Strafford co. N. H. 70 miles from Portsmouth, and 50 from Concord; has part of the Sandwich mtn., a high ridge which terminates at Chocorua Peak in Burton. There are several other mountains, particularly Squam mtn. Bearcamp pond, part of Squam lake and r., and Red Hill r. are the chief waters in the town, and there are several mills. Pop. 1830, 2,744.

SANDWICH, p-t. Barnstable co. Mass. 54 ms. S. E. Boston, occupies the isthmus of the long and crooked peninsula of Cape Cod, and lies between Barnstable and Buzzard's bays. The soil is generally light and sandy, with extensive meadows, and the town is the most devoted to agriculture of any in the co. It has been heretofore proposed to open a canal navigable for ships through this town, to save the long and dangerous circuit of the Cape to the coasting trade. Between Manumet and Scusset rs. the distance is short and the land low, so that the work might be accomplished at a moderate expense. There is a large pond near the centre, and mills are supplied by a fall. Here is a large manufactory of glass, owned by the Boston and Sandwich glass co., by which more than one quarter of the population are supported; 96 of the workmen are heads of families, and nearly 200 men and boys are constantly employed. The first settlement was made from Lynn in 1637, under a grant from Plymouth colony. Much salt is made here from sea water. The town is much resorted to for trout fishing. There are 4 churches, Methodist, Calvinist, Unitarian, and Roman Catholic. Pop. 1830, 3,361.

SANDY, r. Me. rises near the bounds of Somerset and Penobscot cos. flows S. into Kennebec co., turns N. E. and enters Kennebec r. at Starks, Somerset co. 6 ms. N. Norridgewock.

SANDY, river, stream of Virginia and Kentucky, composed of two branches, called relatively East fork and West fork. East fork, the main constituent of Sandy, rises in the Appalachian valleys, interlocking sources with those of Great Kenhawa to the E., and with those of Holston and Clinch branches of Ten. r. to the S. E. Issuing from this elevated region, and draining part of Tazewell and Logan cos, Va., the Sandy r. pursues a N. W. direction by comparative courses 50 ms. to its passage through Cumberland mtn. Becoming a boundary between Va. and Ky. below the Cumberland chain, Sandy assumes a direction of N. N. W. 70 ms. separating Logan and Cabell cos. of Va. from Floyd, Lawrence, and Greenup cos. of Ky. to its final influx into Ohio r. opposite Burlington, O. West Sandy rises in Russell and Tazewell cos. Va., and assuming a N. W. direction pierces the Cumberland chain, enters Ky., and after traversing Pike and Floyd counties bends to the nthrd. and joins East Sandy in Lawrence co. The valley of Sandy r. has that of Ten. r. S., Ky. S. W., Licking W., that of Ohio N., Guyandot E., and Great Kenhawa S. E. It is about 100 ms. long, mean width 35, and area 3,500 sq. ms.

SANDY, p-o. Columbiana co. O. about 140 ms. N. E. Columbus.

SANDY BAY, v. Gloucester, Essex co. Mass. is near the E. extremity of Cape Ann, and has a convenient harbor, exposed on the N. E., but improved by a breakwater constructed by the U. S. The inhabitants are devoted to fishing, in which about 100 vessels are employed.

SANDY BLUFF and p-o. on Il. r. western part of Morgan co. Il. by p-r. 131 ms. N. W. Vandalia.

SANDY Bridge and p-o. eastern part of Carroll co. Ten. by p-r. 94 ms. W. Nashville.

SANDY Creek, Genesee co. N. Y. enters lake Ontario at Murray.

SANDY Furnace and p-o. Venango co. Pa. by p-r. 283 ms. N. W. W. C.

SANDY GROVE, p-o. Chatham co. N. C. by p-r. 64 ms. W. Raleigh.

SANDY HILL, p-v. Kingsbury, Washington co. N. Y. near Baker's falls, just E. of Hudson r. on a pleasant level, 52 ms. N. Albany, near the junction of Champlain canal and Hudson r. It is a half capital of the co.

SANDY HILL, p-v. southeastern part of Worcester co. Md. 10 ms. sthrd. Snow Hill, the co. st. and by p-r. 174 ms. S. E. by E. W. C.

SANDY HOOK, Shrewsbury, Monmouth co. N. J., the S. cape of Raritan bay, through which is the entrance to N. Y. bay. It is about 3 ms. long, with Shrewsbury r. and Sandy Hook bay W.; and has a light-house erected by the U. S. The sand fast extends the cape N. so that two light-houses have been rendered useless by being left by the water. There are only 2 dwellings on the hook, with a few trees; but the bay is often useful in E. storms. Shrewsbury r. sometimes flows across and isolates the hook.

SANDY HOOK, p-v. Culpepper co. Va. by p-r. 85 ms. S. W. W. C.

SANDY MOUNT, p-v. southern part of Greenville co. Va. by p-r. 75 ms. S. Richmond.

SANDY POINT, the N. extremity of Nantucket isl., Mass., 70° W. long., 41° 23′ N. lat.

SANDY POINT, N. E. extremity of Barnstable co. Mass. 69° 35′ W. long., 41° 24′ N. lat.

SANDY River Church and p-o. sthrn. part of Prince Edward co. Va. by p-r. 79 ms. S. W. by W. Richmond. This place takes its name from a small confluent of Appomattox r.

SANDY SPRING, or Stabler's p-o. and Friends meeting house, northwestern part of Montgomery co. Md. The meeting house stands near the main road from Baltimore to Rockville, 28 ms. S. W. of the former, 9 ms. N. E. of the latter place, and 19 ms. nearly due N. W.C. The adjacent country is peculiarly healthful and pleasant, and is elevated above tide water about from 450 to 500 feet. It derives its name from a spring.

SANDY SPRING, p-v. S. E. part of Adams co. O. by p-r. 111 ms. sthrd. Columbus.

SANDYVILLE, p-v. northeastern part of Tuscarawas co. O. by p-r. 119 ms. N. E. by E. Columbus.

SANFORD, p-t. York co. Me. 94 ms. from Boston, 20 ms. north from York, is of irregular form, with Kennebunk r. N., and a pond S., which flows into a tributary of the Piscataquay. Pop. 1830, 3,485.

SANFORD'S Store and p-o. Hancock co. Geo. 14 ms. N. E. Milledgeville.

SANGAMON, r. of Il. and branch of Il. r. This stream, the name of which is pronounced as if written Sangamo, has its remote fountains on the plains, from which flow to the N. W. the Vermillion branch of Illinois, to the N. the Pickmink branch of Illinois, to the S. E. the Vermillion branch of Wabash, and sthrd. the extreme sources of Kaskaskias r. Flowing from this plain the Sangamon flows sthrd. about 30 ms., and thence sweeping an elliptic curve to the S. W., S. and W. about 100 ms., attains its greatest sthrn. bend a few ms. above the influx from the S. of the Mowawequa r. deflecting to the N. W. 30 ms. to the influx from the N. E. of Sugar creek. Below the mouth of Sugar creek the Sangamon, turning to wstrd. 30 ms., falls into Illinois in Morgan co. The entire comparative length of Sangamon may be stated at 200 ms. The valley is in form of a triangle, base 110 ms., by a line from head to mouth of the main stream; shortest side 50 ms. from the mouth of Sangamon to source of Mowawequa r.; perpendicular 65 ms., area about 3,570 sq. ms. The Mowawequa or southern branch of Sangamon rises in Shelby and Montgomery cos., and flowing N. W. joins the main stream in Sangamon co.; its valley is the southern salient angle of the Sangamon valley. Embosomed in the long curve of the main stream of Sangamon, and to the nrthrd. of that channel, the country is drained by Sugar creek or the northern confluent of Sangamon. Sugar creek has a general western course of 70 ms., and joins the main stream in the northwestern part of Sangamon co. Much of the soil of Sangamon valley has been represented as first rate; but taken as a whole, too much of the surface is composed of low and wet prairie.

SANGAMON OR SANGAMO co. bounded S. E. by Shelby, Montgomery S., Macaupin S. W., Morgan W., Tazewell N., MacLean N. E., and Macon E. Length from S. to N. 50 ms., mean breadth 40, and area 2,000 sq. ms. Extending in lat. from 39° 30′ to 40° 13′, and in long. from 2° 10′ to 13° W. W. C. This co. embraces the central part of the valley of the river from which the name is derived. The main Sangamon traverses it by a curve, first wstrd. and thence round to northwstrd.; and as Sugar creek traverses the northern part to the wstrd. that course may be regarded as that of the general slope, though that of the western side is to the N. of N. W. Much of the surface is flat, and of course wet, except after long drought; but the soil is generally very highly productive. Chief t. Springfield. Pop. 1830, 12,960.

SANGERSFIELD, p-t. Oneida co. N. Y. 15 ms. S. by W. Utica, 94 W. N. W. Albany, N. and E. Madison co.; contains head streams of Chenango and Oriskany creeks, is on high land, with hills S. and E., and good soil. Limestone rocks abound, with impressions of organized substances. The land is generally owned in fee. There is a large pine and cedar swamp. There are 2 villages, Sangerfield and Waterville. Pop. 1830, 2,272.

SANGERVILLE, t. Penobscot co. Me. 70 ms. N. N. E. Augusta, and 35 N. W. Bangor; has Somerset co. W. and Piscataquis r. N. There are several ponds, one of which sends a head stream to Sebasticook r. Population 1830, 776.

SANILAC, co. Mich. as laid down on Tanner's improved map of the U. S. is bounded S. E. by Saint Clair co., S. W. Lapeer, N. W. Saginaw bay, and N. E. and E. lake Huron. Length along the southern boundary 62 ms., mean breadth 32. The area may be assumed in round numbers at 2,000 sq. ms. It is a table land, from the centre of which the waters flow like radii from a common centre. From the southwestern angle flows the higher branches of Cass r. a tributary of Saginaw r. From the southern side issue the nthrn. sources of Flint r. another branch of Saginaw. The river Delude has its higher fountains in the sthestrn. angles, whilst Elm and Black rs. flow N. E. into lake Huron, and Sugar r. and other streams N. W. into Saginaw bay. The northern extremity of the co. is Transit point, or the southern entrance into Saginaw bay.

SAPPONY, creek, Cross Roads and p-o. in the southeastern part of Dinwiddie co. Va. The creek is a branch of Stony creek, and the latter a tributary of Nottaway r. Sappony Cross Roads p-o. is by p-r. 22 ms. S. S. W. Petersburg.

SARACTA, p-v. Duplin co. N. C. by p-r. 115 ms. S. E. Raleigh.

SARANAC, r. N. Y., rises in several ponds in the S. part of Franklin co., flows N. E. through Franklin and Clinton cos. and falls into lake Champlain at Plattsburgh village, S. Cumberland head. Its head streams are near those of Racket, Saint Regis and Grass rs. and the N. branch of the Hudson.

SARANAC, p-v. Lenawa co. Mich. by p-r. 70 ms. S. W. Detroit.

SARATOGA, co. N. Y. bounded by Warren co. N., Hudson r. E. dividing it from Washington and Rensselaer cos., Mohawk r. S. separating it from Albany and Schenectady cos., and Montgomery co. W., with about 772 sq. ms. The Hudson borders this co. N. E. and E. for nearly 70 ms., and Sacandaga r. flows through the N. part. In the middle part is Kayderosseras r. or creek and Fish creek. Anthony's and Snook's kills also flow into the Hudson. Mill sites on its numerous streams are abundant. Two primitive mountainous ranges are in the N. W., Kayderosseras and Palmerstown, while there are valuable meadows on the Hudson, &c. and sandy plains in the S. E. The secondary country is most extensive, though there are also transition tracts. The river hills and meadows bear oak, walnut, chestnut, &c.; the loamy plains, beech, maple, ash, &c., and white and yellow pine grow on the sandy plains. Good sandstone for building is found at Greenfield, &c. Large beds of marle lie under the transition and secondary formations. Saratoga lake, Ballston lake, Round and Owl ponds are the principal sheets of water. This W. side of Hudson r. was an important military route in the early as well as the late French war; and in the revolution it was the scene of important operations. Gen. Burgoyne, in 1777, after two battles on the heights of Saratoga, retreated to Fish cr., and there surrendered. At Milton, Moreau, Mechanicsville and Schuylersville are manufactories of woollen and cotton; and there are 5 oil mills in different parts of the co. The Champlain canal enters this co. at Miller's falls, and passes along the bank of the Hudson to Mohawk r. The Schenectady and Saratoga rail-road runs chiefly in this co. Pop. 1820, 33,147, 1830, 38,679.

SARATOGA, p-t. Saratoga co. N. Y., 32 ms. N. Albany, 15 E. Ballstown Spa; has the t. of Saratoga Springs and Fish creek N., Hudson river E., and Saratoga lake W. Fish creek affords valuable mill seats. White and yellow pine grow on the light soil near Saratoga lake; and oak, walnut, &c., in the neighborhood of Hudson river. The surface is pleasantly diversified with fine ranges of hills. The Quaker Springs are in this town, but the other sources of mineral waters are in the adjoining t. of Saratoga Springs. Champlain canal passes along the bank of the Hudson. The remains of fort Hardy are to be seen near the mouth of Fish creek, where Gen. Burgoyne surrendered in 1777. The p-v. of Schuylersville, situated at that spot, is a place of some importance. Pop. 1830, 2,461.

SARATOGA SPRINGS, p-t. Saratoga co. N. Y., 32 ms. N. Albany, 5 N. E. Ballston Spa; contains the famous sources of mineral waters, and is the annual resort of many visitors from all parts of the country. The village built at that spot, in the N. part of the town, contains many lodging houses, several of which are very extensive. The surface of the town is nearly level, with a poor sandy soil, bearing pines, but capable of being much improved by gypsum or marle, which latter is found in different parts. Limestone prevails near the springs. Part of Palmerstown mts. is also in this t. Kayderosseras, Fish, and Ellis' creeks, with some smaller streams, water different parts. In this township are the famous Saratoga springs, situated 7 ms. N. E. Ballston Spa. The village is built on a low, sandy plain, beneath which is a limestone rock. The street runs on the west side of a narrow marshy tract, in which the springs are found; there are numerous houses for the accommodation of visitors, who resort here annually in great numbers, particularly in July and August. Congress hall, U. S. hall, Union hall, and the Pavilion, are the principal. The most important springs are the Congress, Hamilton, Round Rock, and Flat Rock. Pop. Saratoga Springs township 1830, 2,204.

SARATOGA LAKE, Saratoga co. N. Y., 6 ms. S. E. village of Saratoga Springs, 6 N. E. Ballston Spa; about 3 ms. by 9; has handsome, swelling and cultivated banks; receives Kayderosseras cr. W., and discharges into Hudson r. by Fish creek, which affords valuable mill seats. This lake lies partly in 4 townships. Fish and fowl are abundant. The scenery is very agreeable, and it is a favorite resort during the summer months.

SARDINIA, p-t. Erie co. N. Y., 30 ms. S. E. Buffalo, 273 from Albany, W. of Genesee co., and N. of Cattaraugus co., from which it is divided by Cattaraugus cr. Small streams of Cazenove and Seneca creeks flow in different parts of the t. Pop. 1830, 1,453.

SAUGERTIES, p-t. Ulster co. N. Y. 52 ms. S. Albany, 13 N. Kingston, 113 N. New York, S. Greene co., W. Hudson river, and E. Greene co., and is crossed by Esopus creek. One mile W. of it is the v., and at its mouth is a manufacturing village, supplied with water power by a canal cut deep through a rock round the head of the falls, and which leads into an artificial basin. The water is drawn thence to supply a large foundry, a paper mill, saw mill, &c. The mouth of the creek is navigable in sloops to these mills. There is a horse boat ferry across the Hudson from this spot, to Upper Red Hook landing. The land is high and level, and the soil light and good, along much of the Hudson's bank in this town. The inhabitants were generally of Dutch origin. Pop. 1830, 3,747.

SAULT DE SAINT MARIE, p-v. and st. justice, Chippeway co. Mich., on the right bank of Saint Mary's strait, at the lower extremity of the cataract or falls of Saint Mary, and as stated in the p-o. list, 326 ms. N. W. Detroit. This place was founded on the 17th July, 1822, by a detachment of U. S. troops from Detroit, under command of Col. Brady. Lat. 46° 31′, long. W. C. 7° 20′ W. (*See article Saint Mary's river, Mich. and Upper Canada.*) Vessels of 6 feet draught can be navigated to this village, and it has been stated on good authority, that at an inconsiderable expense, (when compared with the advantages) vessels of 10 feet might be enabled to ascend to the foot of the falls.

SAVANNAH, town, Wayne co. N. Y. Pop. 1830, 886.

SAVANNAH, seaport, p-t. and st. jus. Chatham co. Geo., on the right bank of Savannah river, about 15 miles above the mouth of Savannah river into the Atlantic, 100 ms. S. W. Charleston, and by p-r. 167 miles S. E. by E. Milledgeville. Lat. 32° 05′, long. W. C. 4° 10′ W. Vessels drawing 12 feet water are navigated to Savannah. The site formerly unhealthy, was very much meliorated by the effect of an act of assembly in 1817. The legislature voted $70,000, to induce the planters in the vicinity to abandon the wet, and adopt the dry mode of cultivating rice. Cotton, rice, sugar and tobacco, are the most valuable staples exported from this port. The number of wooden buildings exposed this place to the ravages of fire, and in 1820, a most desructive conflagration consumed an amount of property valued at $4,000,000. It contains a number of fine public buildings, the most conspicuous of which are the Exchange, Academy, and Presbyterian church. In all there are 8 or 9 places of public worship, and 10 public squares. In 1820, it contained 7,523, and in 1830, 7,423 inhabitants. It has not yet entirely recovered from the disaster of 1820.

SAVANNAH, two small, but from their position, important rivers of the territory of Huron. One is a branch of Saint Louis river of lake Superior, and the other a branch of Mississippi river; both are links in the chain of navigable streams by which lake Superior is united to the upper Miss., by the Saint Louis river route.

SAVANNAH, p-v. and st. jus. Hardin co. Ten. by p-r. 112 ms. S. W. by W. Nashville.

SAVANNAHVILLE, p-v. Macon co. N. C., by p-r. 319 ms. a little S. of W. Raleigh.

SAUGUS, p-t. Essex co. Mass., 7 ms. N. E. Boston; has much rocky and irregular land, with a large salt marsh S., and fine fresh water meadows along the banks of Saugus river, which flows through them, as well as the salt meadows, with a very crooked and picturesque course. This t. formerly belonged to Lynn, which was one of the earliest settlements, and bore the name of Saugus. Pop. 1830, 960.

SAVOY, p-t. Berkshire co. Mass., 120 ms. N. W. Boston, and 20 N. E. Lenox. It is on the S. base of Hoosic mtn., and gives rise to Hoosic and Deerfield rs. Pop. 1830, 927.

SAW PITS, p-v. Rye, West Chester co. New York, 28 ms. N. E. New York, 5 S. E. White plains, 142 S. Albany, and near Connecticut, on Long Island sound.

SAYBROOK, p-t. Middlesex co. Conn., 40 ms. S. E. Hartford, 18 W. New London, 34 E. N. Haven, and N. Long Island sound, on the W. side of Conn. river, at its mouth. It extends 6 ms. E. and W., and 11 N. and S., with 70 sq. ms.; is uneven and stony, but has some extensive levels, and tracts of rich soil, particularly about Saybrook v. The soil is generally good for grass. Some of the hills near the Conn. have good granite quarries, convenient to navigable water. Pettipaug and Chester are the principal streams, tributaries of Conn. r. which cross this town. There are several small harbors on the sound, and on Connecticut r., at Saybrook Point and Pettipaug, at the last of which much ship building has been carried on. The bar at the mouth of this great stream offers an unfortunate impediment to navigation, for even vessels of a moderate draught of water are often obliged to pass it with but a part of their cargoes. An important and lucrative trade was formerly carried on from this river to the West Indies, and New London often served in some degree as the port. Saybrook harbor is at the mouth of a handsome cove, making up from Conn. river w., almost to Saybrook village, and is often resorted to by coasting vessels in bad weather. Great quantities of fish are caught in this town. Sea fish are taken to other markets, and the shad fisheries are numerous and lucrative. The first settlement in the bounds of this state by Europeans, was made at Saybrook fort in 1635. A small fort was erected on the Point, on a spot supposed to have been a little s. e. of the present fort, now encroached upon by the water, and in advance of the monument of Lady Arabella Fenwick. It was supposed that the Point would have become an important commercial place; and the ground on that sandy peninsula was early laid out for a city, in right lines, as is still to be seen. The garrison of the fort were several times closely beset by the Indians until after the Pequod war; and a palisade fence was kept up across the isthmus many years after. Yale college was seated in this town for several years after its removal from Killingworth, and a house on the Point was appropriated to its use. The present fort, which is a mere redoubt of earth, is no longer used. During the last war, the borough of Pettipaug was occupied a few hours by a detachment of British, who proceeded up in boats from the squadron in Long Island sound. Pop. 1830, 5,018.

Saybrook, p-v. Ashtabula co. Ohio, by p-r. 183 ms. n. e. Columbus.

Saysville, p-v. estrn. part Morgan co. O., by p-r. 106 ms. s. e. by e. Columbus.

Scaghticoke, p-t. Rensselaer co. N. Y., 16 ms. n. Albany, 10 n. Troy, e. Hudson river, s. Washington co.; has a gently varied surface, with soil good for grain and grass, particularly on the Flats. The form is irregular. Hoosac river n. affords many mill sites, as well as its branch, Tomhanoc cr. An early settlement was made on the Flats, by several Dutch and German families. Pop. 1830, 3,002.

Scarborough, p-t. Cumberland co. Me., 65 ms. s. s. w. Augusta, 10 s. w. Portland; lies n. w. Atlantic ocean, n. e. Saco, York county, with one or two small streams, and Prout's neck running into the sea. Pop. 1830, 2,106.

Scarsdale, town, West Chester co. N. Y. 25 ms. n. e. New York, 3 s. White Plains, has Bronx r. on the w. line, and is small, containing only 8 sq. ms., with pretty good soil. Pop. 1830, 317.

Scaroon, p-t. Essex co. N. Y., 25 ms. s. s. w. Elizabethtown, and n. of Warren co.; contains about half of Scaroon lake, with Paradox lake, &c. The rocks are limestone, often with vegetable impressions. Beech, maple, pine, hemlock, &c., formed the forests. The surface is rough and mountainous. Pop. 1830, 1,614.

Scaroon, lake, Essex and Warren cos. N. Y., 12 ms. w. from the n. end of lake George; is about 1 mile by 8; forms part of the n. e. branch of Hudson r.; abounds with fish, and discharges by Scaroon r.

Scaroon, river, Warren co. N. Y.; is the outlet of Scaroon lake, and forms the n. e. branch of Hudson river, falling into the main branch, after a short course, in the same co.

Schall's Store, and p-o. Berks co. Pa., by p-r. 157 ms. nrthestrd. W. C.

Schellsburg, p-v. Bedford co. Pa., on the main road from Bedford to Pittsburg, 9 ms. wstrd. of the former, and by p-r. 135 ms. n. w. W. C. It is a small v. in a single street along the road. Pop. 1830, 200.

Schenectady, city, and st. jus. Schenectady co. N. Y., 15½ ms. n. w. Albany; has the Mohawk r. and Albany co. n., and contains extensive alluvial meadows, with handsome uplands, and a sandy loam upon clay slate. Sand kill, flowing into the Mohawk, affords mill sites, some of which are occupied by mills and manufactories. The Erie canal crosses the n. part, near the Hudson, but on account of the circuitous route, and the numerous locks between this place and Albany, much of the navigation stops here. Packet boats run hence in numerous lines to Utica, and on as far as Buffalo, and many still extend to Albany. Thus a vast amount of merchandize annually passes through this city. The Albany and Schenectady rail-road greatly facilitates the communication with the Hudson; and the Saratoga and Schenectady rail-road will tend to increase the travelling, especially during the warmer seasons. Numerous lines of stage coaches also pass thro' this city. Union college, which stands a short distance from the centre of the city, is a respectable and flourishing institution. It was founded by the Regents of the University in 1794. The principal college buildings are each 200 feet long, and 4 stories high, built of brick and covered with white stucco. The institution possesses a library, cabinet, philosophical and chemical apparatus, &c. The spot was the site of a Mohawk village. The streets of the city are regular, and paved, but rather narrow; 8 of them are crossed diagonally by the Erie canal. Schenectady was early settled by a few Dutch, but on the night of Feb. 8th, 1690, the village then containing 63 houses and a church, was suddenly attacked and burnt, by French and Indians from Canada. 60 of the people were killed, 27 carried captive, and 27 of the remainder lost limbs by exposure to the cold, in attempting to reach Albany. In 1748, 70 of the inhabitants were massacred by savage invaders from the same quarter, and in 1819, 170 buildings were burned to the ground. Pop. 1830, 4,268.

SCHENECTADY, co. N. Y. bounded by Montgomery and Saratoga counties north and east, Albany county s., and Schoharie county west, is of a very irregular form, and is crossed by Mohawk river and the Erie canal, while Albany and Schenectady rail-road meet here at the ycity of Schenectady. Sand kill and Eel Place kill are the principal mill streams. Along the Mohawk the soil is a rich alluvion, and on the uplands a light sandy loam, on clay state, with an undulating surface, well watered by springs. In Duanesburgh, &c. the soil is argillaceous, and the surface more hilly; streams flow N. to Mohawk river, E. to Hudson river, and W. to Schoharie creek, on the W. line, which affords good mill seats. The Schenectady manufacturing company at Rotterdam, is the only incorporated company for manufacturing purposes in the county. They make about 400,000 yards of cotton goods annually, and 20 or 30,000 lbs. of yarn. It has 2,000 spindles, and 50 looms. There are also satinet, paper, and carpet manufactories, and oil mill and iron foundries, all large; and in other places 11 tanneries, 1 foundry. Pop. 1820, 13,081, 1830, 12,347.

SCHLOSSER, fort Niagara, Niagara county, N. Y. ancient work, long disused.

SCHODAC, p-t. Rensselaer county, N. Y. 9 miles S. Albany, 15 S. Troy, N. Columbia co. W. Hudson r. separating it from Albany county, has a variety of soils, generally good, with some pine plains. The inhabitants are of Dutch extraction. Moordenars' kill, &c. supply mill seats on their course to the Hudson. There are 2 landings, with post villages. Hogeberg or High Hill is on the bank of the Hudson, 9 miles below Albany. Pop. 1830, 3,794.

SCHOHARIE, county, New York, bounded by Montgomery county N., Schenectady and Albany counties E., Greene and Delaware counties S., and Otsego county W., is partly broken by a range of the Catskill and Helderberg hills, and crossed centrally by Schoharie creek. It has also Cobuskill creek and in the E. rises Catskill creek. The rocks are of limestone, and on Schoharie creek, are extensive and very fertile meadows, 26 miles long, where settlements were begun by Germans and Dutch about 100 years since, while the 3 townships are inhabited by people from the E. states. Schoharie village was destroyed by the English and Indians in the revolutionary war. There are 10 townships. Schoharie, the st. jus. of the county, stands on the meadows. In the county are 1 furnace, 1 paper mill, 1 woollen, and 2 leather manufactories. Pop. 1820, 23,154, 1830, 27,902.

SCHOHARIE, p-t. and st. jus. Schoharie co. N. Y. 32 miles W. Albany, 22 miles S. W. Schenectady, 24 S. Johnston, S. Montgomery county, and W. of Schenectady and Albany counties, is crossed by Helderberg hills, and Schoharie cr. which here receives Cobuskill and Fox creek. On the Schoharie are very rich meadows, which have been under constant culture for 100 years. The inhabitants are of Dutch and German extraction. Here are 3 villages, Schoharie, Esperance, and Sloansville. Pop. 1830, 5,157.

SCHOHARIE, creek, or kill, N. Y. rises on the W. side of the Catskill mtns. Greene co. and after winding 23 miles, enters Schoharie co. flows N. 40 miles and empties into the Mohawk in Montgomery county, opposite Tribe's hill. It is rapid, has several branches, and waters some fine alluvial meadows, as well as some hilly regions.

SCHOODIC, or St. Croix river Maine.

SCHOOLEY'S, mountain, N. J. a high range in Washington and Roxbury, Morris county, forming a part of the mountainous region in that part of the state.

SCHOOLEY'S, mountain, p-v. and mineral springs, Washington, Morris county, N. J., 56 miles N. Trenton, 50 N. W. W. New York, 20 S. Newton, and 70 N. Philadelphia, is on an elevation on Schooley's mountain, where the air is pure and the scenery bold and varied. The place is a favorite resort for health and pleasure during the summer months, and there are two large hotels for visitors, besides more private accommodations. The roads are rough, but a line of stage coaches runs daily to the place from Elizabethtown Point, connected with the New York steamboat, and passing through Morristown. The water of the spring, holds in solution muriate of soda, magnesia and lime, sulphate of lime, and oxide of iron.

SCHROON, river, New York. (*See Scaroon river.*)

SCHROON, lake, N. Y. (*See Scaroon lake.*)

SCHROON, p-t. Essex county, New York. (*See Scaroon.*)

SCHULTZ'S, range, and p-o. Wood county, Virginia, by p-r. 324 miles wstrd. W. C.

SCHUYLER, p-t. Herkimer county, N. Y. 86 ms. W. Albany, 8 N. W. Herkimer, 6 ms. E. from Utica, N. of Mohawk river, and E. of Oneida county, has several small mill streams, a good soil, and is somewhat hilly. Pop. 1830, 2,074.

SCHUYLER, county of Illinois, bounded by Pike S., Adams W., Hancock N. W., Macdonough N., Fulton N. E., and Illinois river separating it from Morgan E., and S. E. Length from S. to N. 30 ms., mean breadth 22, and area 660 square ms. Lat. 40° and long. W. C. 13° 40′ W. intersect in this county. It is traversed and drained by Crooked creek, a confluent of Illinois river. Slope to the S. E. It is represented by recent travellers, as amongst the finest counties of Illinois. Chief town, Rushville. In the census returns for 1830, Schuyler and Macdonough counties are comprised under one head, and contained together a pop. of 2,959.

SCHUYLERSVILLE, p-v. Saratoga county, N. Y. 6 ms. W. Union village, N. Fish creek, on the W. bank of Hudson river, and upon the

Champlain canal. On the meadows adjoining the village, the army of Gen. Burgoyne surrendered to the Americans in 1777, after their defeat on the heights of Saratoga, 7 miles below.

SCHUYLKILL, river, Pa. great southwestern branch of Delaware river. The valley of Schuylkill has that of Susquehannah S., S. W., W., and N. W., that of Lehigh N., and that of Delaware above tide N. E. The range of the valley is from N. W. to S. E. 90 ms. in length. The breadth above Blue Ridge about 35 ms., but below that chain the utmost breadth is 25 ms. and mean width about 12 ms. The mean breadth of the entire valley about 22, and area 1,980 square ms. The tide ascends this river about 5 ms. to the primitive ledge in the city of Phila., from whence a chain of canals, locks and rail-roads have been constructed along this stream to near the utmost sources, opening the fine country along and near its banks to the Atlantic tide water, and providing a means to bring to market the immense masses of mineral coal drawn from the bowels of the earth along its higher tributaries. (*See articles Pa., Delaware river, and roads and canals.*)

SCHUYLKILL, county, Pa., bounded S. W. by Dauphin, W. by Northumberland, Columbia N. W., Luzerne N., Northampton N. E., and the Kittatinny mtn. separating it from Lehigh co. E. and Berks S. E. Length from S. W. to N. E. 37 ms., mean breadth 18, and area 660 square ms. Lat. 40° 40′, and long. W. C. 0° 47′ E. Though along the border of this county contiguous to Dauphin, Northumberland, Columbia, and Luzerne counties, creeks rise which have the Susquehannah as their recipient, the body of the county is drained into Schuylkill river, and slopes southeastward. The face of the county is perhaps more diversified by valley, hill, and mountain, than any other in Pa. The mean elevation of the arable soil is about 800 feet above tide water, and with all its mountainous appearance much of the soil is excellent. But what renders this co. an object of peculiar interest, is the vast deposits of mineral coal it contains. Since 1806, upwards of a million of dollars have been expended to facilitate the transportation of this fuel to the Atlantic markets. (*See article roads and canals.*) Chief town, Orwigsburg. Population 1820, 11,339, 1830, 20,744.

SCHUYLKILL, p-o. northern part of Chester county, Pa., by p-r. 134 ms. N. E. W. C.

SCHUYLKILL HAVEN, p-v. Schuylkill county, Pa. 55 ms. N. E. Harrisburg, and 171 N. N. E. W. C.

SCIO, p-t. Alleghany county, N. Y. 14 ms. S. Angelica, N. Pa. Pop. 1830, 602.

SCIOTA, r., O., having its remote sources in Richland, Marion, Crawford, and Hardin coe. It is composed of two branches, Whetstone on the E., and Sciota proper W. Both branches issuing from Marion county, assume a nearly parallel course to S. S. E., traversing Delaware and uniting in Franklin county, between the towns of Columbus and Franklin, after each branch having flowed by comparative courses 70 miles. Below Columbus the general course is almost exactly S., and comparative length 100 ms. to its influx into Ohio river between the villages of Alexandria and Portsmouth. The Sciota valley, lying between lat. 38° 42′ and 40° 50′, and cut into two very nearly equal sections by long. W. C. 6° W., is about 150 ms. long, and 60 miles wide, area 9,000 square ms. Below Columbus the main stream traverses the counties of Franklin, Pickaway, Ross, Pike and Sciota. Though without direct falls, the Sciota is a very rapid stream. (*See article rail-roads and canals.*) The Sciota valley lies between those of Great Miami and Muskingum, and has that of Sandusky N., and Maumee N. W.

SCIOTA, co. Ohio, bounded by Adams W., Pike N., Jackson N. E., Lawrence E., and O. river separating it from Greenup county, Ky., S., and Lewis county, Ky., S. W. Length from E. to W. 34 ms., mean breadth 15, and area 512 square ms. Lat 38° 50′, and long. W. C. 6° W. intersect near its centre. It is divided into two not very unequal sections by Sciota river which traverses it from N. to S. The general slope is southward; surface hilly, and soil tolerably fertile. Chief town, Portsmouth. Pop. 1820, 5,749, 1830, 8,740.

SCIOTA, p-v. Sciota co. Ohio, by p-r. 92 ms. S. Columbus.

SCIPIO, p-t. Cayuga co. N. Y. 180 miles W. Albany, and 11 S. of Auburn, is bounded W. by Cayuga Lake which separates it from Seneca co., has Owaco lake E., and includes a part of the Cayuga Indian reserved lands. The inhabitants are generally farmers. Salmon creek and other brooks supply many mill seats, but the springs are affected by drought. Slate rock lies under the soil. It has Aurora and other small villages. Pop. 1830, 2,691.

SCIPIO, p-v. Seneca co. Ohio, by p-r. 88 ms. N. Columbus.

SCITUATE, p-t. Plymouth co. Mass. 17 ms. S. Boston on the Atlantic coast. Its harbor is protected against the storms by small islands, and it has some coasting trade. It is crossed by Satuit brook, whence it derives its name. It was an early settlement, and in 1676, during Philip's war, had 19 houses and barns burnt by the savages. Thomas Clapp, President of Yale College, Conn. was born here, 1703. Pop. 1830, 3,468.

SCITUATE, p-t. Providence co. R. I. 12 ms. W. Providence, about 6 ms. by 8, has a rocky and varied surface, with good building stone in the W.; soil generally favorable to grass. It has 2 small streams of Pawtuxet r. The mackerel fishery here is important; 21 vessels were engaged in 1832. It also contains several cotton factories, a bank, a foundry of bells and cannon, and an academy. Pop. 1830, 3,394.

SCONONDOA, p-v. Oneida co. N. Y. 23 ms. W. Utica, 11 S. Rome, and 1 S. Erie canal, stands on Scononda creek.

SCOTCH PLAINS, p-v. Westfield, Essex co. N. J. on Green Brook, 14 ms. s. w. Newark, and near the borders of Somerset co.

SCOTCHTOWN, p-v. Wallkill, Orange county, N. Y. 6 ms. N. w. Goshen.

SCOTIA, p-v., and as named in the p-o. list of 1831, st. jus. Pope county, Arkansas, by p-r. 81 ms. northwestward Little Rock. Exact position uncertain.

SCOTT, p-t. Cortlandt co. N. Y. 18 ms. s. Skeneateles, 9 from Cortlandt, s. Onondaga co., and E. Cayuga co., has small streams of Tioughnioga creek, and an inlet of Skeneateles lake, with ridges of land extending N. and s. and a productive soil, bearing grass best on the hills. The soil is held in fee simple. Pop. 1830, 1,452.

SCOTT, p-v. Wayne co. Pa. by p-r. 283 ms. N. N. E. W. C.

SCOTT, co. Va., bounded by Russell county, Va., N. and N. N. E., Washington co. Va. E., Sullivan and Hawkins cos. Tenn. s., and Lee co. Va. w. and N. w. Length along Ten. 40 ms., mean width 15, and area 600 square ms. Lat. 36° 47', long. W. C. 5° 40' w. Slope s. w. and traversed in that direction by the main volume of Clynch, and N. fork of Holston rivers, and between those streams by Clinch mountain. The surface is broken and soil of middling quality. Chief town, Estillville. Population 1820, 4,263, and in 1830, 5,724.

SCOTT, co. Ky. bounded by Lafayette s. E., Woodford s. w.. Franklin w., Owen N. w., and Harrison N. and N. E. Length 18 ms., mean breadth 14, and area 252 square ms. Lat. 38° 15', long. W. C. 7° 40' w. Slope N. w. and drained by Elkhorn and Eagle rs. confluents of Kentucky river; soil excellent. Chief town, Georgetown. Pop. 1820, 12,219, 1830, 14,677.

SCOTT, co. of Indiana, bounded by Clark s., Washington w., Jackson N. w., Jennings N., and Jefferson N. E., and E. Length 20 ms., mean width 10, and area 200 square miles. Lat. 40° 40', long. W. C. 8° 45' w. Slope N. w. by w., and in that direction drained by creeks falling into Graham's Fork of White river. Chief town, New Lexington. Pop. 1820, 2,334, 1830, 3,092.

SCOTT, co. Mo. bounded by New Madrid s. w., Stoddard w., Cape Girardeau N. w., Mississippi r. above the mouth of Ohio, separating it from Alexander co. Il. N., and the Miss. r. below the mouth of Ohio separating it from MacCracken and Hickman cos. Ky. E. Length from the Miss. r. on the s. E. to the border of Cape Girardeau co. 50 ms.; mean breath 18, and area 900 sq. ms. Lat. 37° and long. 12° 30' w. intersect in this co. Slope sthrd. The nrthwstrn. angle traversed by White water branch of St. Francis, and the residue by crs. flowing into Miss. r. Chief t., Benton. Pop. 1830, 2,136.

SCOTT, p-v. Adams co. O. by p-r. 94 ms. a little w. of s. Columbus.

SCOTTSBURGH p-v. Halifax co. Va. by p-r. 235 ms. s. s. w. W. C.

SCOTT'S FERRY, p-o. Albermarle co. Va. by p-r. 150 ms. s. w. W. C.

SCOTTSVILLE, p-v. Wheatland, Monroe co. N.Y. 12 ms. s. Rochester, and 1 from Genesee r., stands on Allan's cr.

SCOTTSVILLE, p-v. nrthwstrn. part of Luzerne co. Pa. 40 ms. N. N. w. Wilkes-Barre.

SCOTTSVILE, p-v. and st. jus. Powhatan co. Va. 32 ms. w. Richmond, and by p-r. 138 ms. s. s. w. W. C. Lat. 37° 32', long. W. C. 0° 56' w.

SCOTTSVILLE, p-o. Orange co. N. C. by p-r. 56 ms. N. N. w. Raleigh.

SCOTTSVILLE, p-v. and st. jus. Allen co. Ky. situated on a branch of Green r. by p-r. 151 ms. s. & w. Frankfort, and 67 ms. N. E. Nashville, Ten. Lat. 36° 45', long. W. C. 9° 06' w. Pop. 1830, 180.

SCRIBA, p-t. Oswego co. N. Y. 173 ms. N. w. Albany, 60 w. N. w. Rome, s. of lake Ontario, and N. E. Oswego r., has a nearly level surface and good soil, with few mill sites. Oswego fort is in this town. It has a triangular form, enclosing 3 or 4 acres, 50 feet above the lake, and was the first military work erected at the mouth of Oswego r. in the old French wars, in 1727. Fort Oswego was afterward erected near it: and both were captured by the French in 1756. It was surrendered to the Americans by the British under Jay's treaty, in 1796. The British, during the late war, once landed here, and occupied the v. at the mouth of the r. for a few hours. Pop. 1830, 2,073.

SCRIVEN, co. of Geo. bounded by Effingham s. E., Great Ogeechee r. separating it from Bullock s. w., and Emanuel w., Burke N. w., and Savannah r. separating it from Barnwell dist. S. C. N. E., and Beaufort dist. S. C. E. Lat. 32° 40', long. W. C. 4° 30' w. The nrthrn. section of this co. is traversed in a sthestrn. direction by Brier cr. branch of Savannah r.; but the sthrn. and central sections slope sthrd. and are drained into Great Ogeechee r. Length in the direction of its bounding rivers, that is, from s. E. to N. w. 34 ms., mean breadth 22, and area 748 sq. ms. Pop. 1820, 3,941; 1830, 4,776.

SCROGGSFIELD, p-v. Columbiana co. O. by p-r. 146 ms. N. E. Columbus.

SCUFFLETOWN, p-v. nrthrn. part of Laurens dist. S. C. about 10 ms. N. N. E. Laurensville, and by p-r. 85 ms. N. w. Columbia.

SCULL CAMP, p-v. nrthwstrn. part Surry co. N. C. by p-r. 182 ms. N. w. by w. W. C.

SCULL SHOALS, and p-o. Greene co. Geo. by p-r. 58 ms. N. Milledgeville.

SEABROOK, t. Rockingham co. N. H. 17 ms. s. s. w. Portsmouth, 7 N. Newburyport, forms the s. E. corner of the state, having the Atlantic ocean E. and Mass. s. First settled 1638. It is watered by Black, Brown's, and Walton's rs., and on many of the brooks is found bog iron ore. The building of whale boats has been extensively carried on here, and the inhabitants are chiefly sailors and mechanics. Pop. 1830, 1,093.

SEACONNET, point and rocks, Newport,

Newport co. R. I. the s. end of the E. shore of Narragansett bay, 6 ms. E s. E. Newport.

SEAFORD, p-v. on Nanticoke r. nrthwstrn. part Sussex co. Del. by p-r. 107 ms. a little s. E. W. C.

SEARCY'S, p-o. Montgomery co. Ten. by p-r. 58 ms. N. W. by W. Nashville.

SEARIGHT, p-o. Fayette co. Pa. by p-r. 199 ms. N. W. W. C.

SEARSBURGH, t. Bennington co. Vt. 12 ms. E. Bennington, has a rough surface and much poor soil, so that it sustains but few families. Pop. 1830, 40.

SEARSMONT, p-t. Waldo co. Me. 25 ms. W. Castine, 30 E. Augusta, has a large pond in the centre, which discharges s. by an outlet into St. George r. The form of the town is irregular. Pop. 1830, 1,151.

SEAY'S, p-o. Merriwether co. Geo. by p-r. 119 ms. W. Milledgeville.

SEBAGO, lake, Cumberland co. Me. is 13 ms. long, and about 20 wide in the broadest parts, but nearly divided by a long and narrow cape, extending s. W. from the E. shore in Raymond. It forms a part of the boundary of 5 tsps. clustered around it, Standish, Baldwin, Sebago, Raymond and Windham. Crooked r. falls into the lake on the N., into the lower part of whose course, (which bears the name of Sungo,) empties Long lake, in the N. part of the co. Presumpscut r. flows from the s. E., part of the lake s. E. into Casco bay. Boat navigation extends by this route to Portland.

SEBAGO, p-t. Cumberland co. Me. 65 ms. s. W. Augusta, lies on the N. W. side of Sebago lake, with Oxford co. W., has an irregular form, and is watered by small streams flowing into the lake. Pop. 1830, 586.

SEBASTICOOK r. Me. rises in Penobscot and Somerset cos. flows across the s. E. corner of the latter, and passing into the N. E. corner of Kennebec co. falls into Kennebec r. in Winslow opposite Waterville.

SEBEC, p-t. Penobscot co. Me. 87 ms. N. E. Augusta, embraces the end of Sebec pond and the head of Sebec r. which rises in it, and is well watered by these and Piscataquis r. on the s. line. Pop. 1830, 906.

SECOND FORK, p-o. Clearfield co. Pa. by p-r. 154 ms. N. W. Harrisburg.

SECTION CREEK, and p-o. Clay co. Ky. by p-r. 106 ms. s. E. Frankford. On Tanner's map of the U. S. this cr. is named *Sexton's*, which is probably the real name, but in directing letters, the p-o. list perhaps ought to be followed.

SEDGWICK, p-t. Hancock co. Me. 6 ms. E. Castine, 87 E. by s. Augusta, has Blue Hill bay E. and a strait s. which separates it from Deer isl., being situated principally on a peninsula, with a coast made irregular by points, coves, &c. Pop. 1830, 1,604.

SEECATCHEE, v. Mass. on the E. shore of Nantucket, and on the verge of the ocean.

SEEKONK, p-t. Bristol co. Mass. 38 ms. s. E. Boston, N. Barrington, R. I., and E. Providence r., there the line of the same state. It is an important maufacturing town. Pop. 1830, 2,133.

SELBY'S, store and p-o. Wake co. N. C. by p-r. 20 ms. sthwstrd. Raleigh.

SELIN'S GROVE, and p-o. Union co. Pa. situated on the right bank of the Susquehannah r., between Penn's and Middle creek, by p-r. 50 ms. above and nthrd. Harrisburg and 4 ms. below Sunbury.

SELLER'S, tavern and p-o. Bucks co. Pa. about 30 ms. N. Phil.

SELMA, p-v. on the right bank of Alabama r. nrthrn. part Dallas co. Ala. by p-r. 86 ms. s. s. E. Tuscaloosa.

SELMA, p-v. Jefferson co. Mo. by p-r. 30 ms. sthrd. St. Louis.

SEMINOLE, Agency and p-o. on Ocklawaha r. Alachua co. Flor. about 80 ms. s. W. by W. St. Augustine, and by p-r. 238 ms. s. E. by E. Tallahassee.

SEMPRONIUS, p-t. Cayuga co. N. Y. 15 ms. s. E. Auburn, 160 W. Albany, has Onondaga co. N., Skeneateles lake E., Onondaga and Cortlandt cos. E., with many hills, some extensive valleys, and a soil generally rich and arable. Owasco lake in the W. has an inlet in the s. part, whose streams afford mill seats, as do other brooks running in different directions. There are several marshes, the largest of which is along the lower part of Owasco inlet. Owasco flats s.w. are fertile, and contain Moravia v. One mile distant is Montville where are mills. Pop. 1830, 5,705.

SENECA LAKE, N. Y. lies between 4 counties, W. of Cayuga lake, and in one part only 6 ms. distant. It is about 35 ms. long. N. and s., from 2 to 4 wide and of great depth. Its outlet, Seneca r., runs from the N. end E. to Cayuga lake. There is a great marsh s. chiefly in Tioga co. through which run several small streams. On the W. side, the outlet of Crooked lake falls into Seneca lake. The surface of this sheet of water is 431 feet above the level of tide water at Albany. Geneva, one of the prettiest vs. in the state, is situated at the N. W. corner of the lake, partly on the low ground, and partly on the elevated bank. The water has a gradual periodical rise and fall, once in several years, the cause of which has never been ascertained. The water never freezes, which is probably owing to its depth. The land gradually rises for several miles, by those broad, natural terraces or successive parallel ridges, running N. and s. over a considerable tract of country. The view from the height of land between Seneca and the adjacent lakes is extensive and agreeable. The region has the appearance of having been swept by a powerful current of water from the N.

SENECA r. N. Y. rises at the N. end of Seneca lake, and crosses Seneca, Cayuga and Onondaga cos. 60 ms. to Oswego r. in Cicero. Its branches are Cayuga, Canandagua, Owasco, Skeneateles and Onondaga outlets. It is rendered navigable by a canal and locks by Waterloo to the Erie canal. At Montezuma it is 371 ft. higher than the Hudson is at Albany.

SENECA r. of N. C. and Geo., has its remote sources in Blue Ridge, Haywood co. N. C. but it is a mere creek where it leaves that state and enters Pickens dist. S. C. Thence augmented by numerous crs. from both sides, the Seneca flows by comparative courses 45 ms. in a direction a little E. of S. to its junction with Tugaloo to form Savannah r. This r. and its confluents drain the greater part of Pickens dist. It is a mtn. stream, and compared with length of course contains a large volume of water.

SENECA, co. N. Y. bounded by Wayne co. N., Cayuga co. E., Tompkins co. S., Ontario and part of Stuben cos. W. It lies chiefly between Cayuga and Seneca lakes, and is crossed in the N. by a part of Seneca r., which here runs from the foot of Seneca lake to the foot of Cayuga lake, and then N. The other streams are small. It contains 10 towns, of which Ovid and Waterloo are the chief. It is about 187 ms. W. Albany, has an agreeably varied surface, with a calcareous loam and vegetable mould. There are some salt springs, iron ore and limestone. The village of Seneca falls has rapidly increased in business and population. In 1825 there were 265 inhabitants, and in 1830, 1,610. The fall is 46 feet, and affords abundant power, part of which is employed in 4 flour mills, 1 grist mill, 1 cotton factory with 4,000 spindles, 1 paper mill, 1 tannery, 1 sash factory, 2 furnaces and 1 oil mill. At Waterloo are 5 flour mills, 2 saw mills, 1 clover seed mill, 1 hemp factory, 1 patent pail factory, 1 tub factory, 1 paper mill, 1 oil mill, 3 carding mills, a lath factory and 1 forge. At Ovid, 1 steam flour mill and 1 carding mill. Population 1820, 23,619; 1830, 21,041.

SENECA, p-t. Ontario co. N. Y. 176 ms. W. Albany, 12 E. Canandaigua, W. Seneca lake and co., and is crossed by the road from Albany to Buffalo. The land is arable and favorable to grass, and the surface s. hilly. The v. of Geneva, one of the pleasantest in the state, is at the N. E. corner of the lake. Pop. 1830, 6,161.

SENECA, co. O. bounded by Crawford S., Hancock S. W., Wood N. W., Sandusky N., and Huron E. Length from E. to W. 32 ms., breadth 20, and area 640 sq. ms. Lat. 41° 10', long. W. C. 6° 06' W. Sandusky r. traverses this co. flowing to the nrthrd. The general slope is of course in that direction, but from the S. W. angle issues the extreme fountains of Blanchard's branch of au Glaize r. Chief town, Tiffin. Pop. 1830, 5,159.

SENECA Falls, p-t. Seneca co. N. Y. 167 ms. W. Albany; contains a flourishing manufacturing village, which has increased in population from 265 to 1,610 between 1825 and 1830. The water falls here 42 feet, and affords abundance of power for several mills and factories. (*See Seneca co.*) A canal here passes round the falls. Here are 18 dry goods stores in the village, besides 2 hardware do., 2 druggist's do., 5 flouring mills, 1 large cotton factory, 1 woollen do., 1 paper mill, 1 distillery, 1 large tannery, 1 sash factory, 1 carriage factory, besides numerous other smaller manufacturing establishments. The prosperity of the village is owing to its valuable hydraulic privileges. Pop. 1830, 2,603.

SENECAS, Indians, N. Y. hold several reservations in the state, but their principal settlement is near Buffalo, on a tract 7 ms. by 18, on Buffalo creek.

SENECA Mills and p-o. Montgomery co. Md. by p-r. 23 ms. N. W. W. C.

SENECAVILLE, p-v. Guernsey co. O. by p-r. 99 ms. E. Columbus.

SENNET, p-t. Cayuga co. N. Y. Pop. 1830, 2,297.

SETAUKET, p-v. Brookhaven, Suffolk co. N. Y. 58 ms. E. N. Y.

SETZLER'S Store and p-o. Chester co. Pa. by p-r. 138 ms. N. E. W. C.

SEVEN MILE FORD and p-o. eastern part of Washington co. Va. 362 ms. S. W. by W. W. C.

SEVENTY SIX, p-v. Beaver co. Pa. by p-r. 256 ms. N. W. W. C.

SEVERN, creek and p-o. Owen co. Ky. by p-r. 21 ms. nrthrd. Frankfort.

SEVIER, co. Ten. bounded S. W. and W. by Blount, Knox N. W., Jefferson N. E., Cocke E., and the Iron mtn. separating it from Haywood co. N. C. S. E. Length from S. E. to N. W. 28 ms., mean breadth 18, and area 500 sq. ms. Lat. 35° 45', long. 6° 25' W. W. C. The northern part of this co. is traversed in a westerly direction by the Nolechucky r.; but the southern and much the most extensive section of the co. is drained to the N. W. by Little Pigeon r. and its confluents, flowing from the Iron mtns. into Nolechucky r. Chief t. Sevierville, or Sevier C. H. Pop. 1820, 4,772, 1830, 5,717.

SEVIERVILLE or SEVIER C. H., p-v. and st. jus. Sevier co. Ten. situated on Little Pigeon r. 25 ms. S. E. by E. Knoxville, and by p-r. 225 ms. a little S. of E. Nashville. Lat. 35° 50', long. 6° 21' W. W. C.

SEWELL creek and mtn. western part of Greenbrier co. Va. Sewell cr. is one of the extreme sthrn. sources of Gauly r.

SEWELL mtns., p-o. western part of Greenbrier co. Va. by p-r. 294 ms. S. W. by W. W. C.

SEWELL Valley, p-o. western part of Greenbrier co. Va. by p-r. 288 ms. S. W. by W. W. C.

SEWICKLEY, the name of three creeks of western Pa. The most considerable is a stream of Westmoreland co. rising opposite to the Loyalhanna r., and flowing wstrd. into Youghioghany r. The second a small creek, though relatively called Big Sewickley, and for a few ms. constituting part of the boundary between Alleghany and Beaver cos. The third or Little Sewickley, is a mere brook of Alleghany co. The two latter Sewickleys fall into the right side of Ohio r.

SEWICKLEY BOTTOM, p-o. wstrn. part of Alleghany co. Pa. 14 ms. N. W. Pittsburg.

Sexton's p-v. western part of Boone co. Mo. by p-r. 64 ms. N. w. Jefferson.

Shade creek, one of the higher branches of Conemaugh r. flowing from the Alleghany mtn. in the northwestern part of Somerset co. Pa.

Shade mtn., a ridge extending from the great bend of Juniata r., below Lewiston, and separating Juniata from Mifflin co. Pa.

Shade, p-o. on Shade creek, N. E. part of Somerset co. Pa., about 20 ms. N. E. the borough of Somerset, and by p-r. 160 ms. N. w. W. C.

Shade Gap and p-o. eastern part of Huntingdon co. Pa. 117 ms. N. w. W. C.

Shady Dale, p-o. Jasper co. Geo. by p-r. 43 ms. N. w. Milledgeville.

Shady Grove, p-o. Franklin co. Va. by p-r. 305 ms. s. w. W. C.

Shady Grove, p-o. Buncombe co. N. C. by p-r. 277 ms. w. Raleigh.

Shady Grove, p-o. Union dist. S. C. by p-r. 86 ms. N. w. Columbia.

Shafer's p-o. Northampton co. Pa. by p-r. 210 ms. N. E. W. C.

Shaferstown, p-v. eastern part of Lebanon co. Pa. 9 ms. E. Lebanon, and by p-r. 129 ms. N. N. E. W. C.

Shaftsbury, p-t. Bennington co. Vt. 97 ms. s. w. Montpelier, 46 from Rutland, 31 from Brattleboro'; first settled about 1763; lies E. N. Y. between Walloomsac and Battenkill creeks, and has no large streams. W. mtn. extends into this town about 3 ms. The soil is generally good, and excellent in the s. w. Iron ore and marble are found in the town. There is a fund of $10,000 for the support of schools. Pop. 1830, 2,142.

Shakleford's, p-o. King and Queen co. Va. by p-r. 160 ms. s. W. C.

Shalersville, p-v. northern part of Portage co. O. 5 ms. N. Ravenna, the co. st., and by p-r. 132 ms. N. E. Columbus. Pop. of the tsp. 1830, 757.

Shallow Ford and p-o. Anderson district, S. C. by p-r. 145 ms. N. w. Columbia.

Shamokin, creek and p-o. central part of Northumberland co. Pa. by p-r. 64 ms. N. Harrisburg. The Shamokin creek falls into the left side of Susquehannah r. immediately below the borough of Sunbury.

Shandakan, p-t. Ulster co. N. Y. 20 ms. w. Kingston, 83 s. by w. Albany, lies s. Greene co., N. Sullivan co., and E. Delaware co. It is mountainous, and several streams flow hence to Del. r. and Esopus creek. Pine hill mtn. lies on the borders of Del. co. Pop. 1830, 966.

Shane's Crossings, over St. Mary's r., or as marked on Tanner's map, *Shanesville*, p-v. on St. Mary's r. northern part of Mercer co. O. 18 ms. N. w. St. Mary's, the co. st., and by p-r. 129 ms. N. w. Columbus. Pop. 1830, 46.

Shanesville, p-v. Tuscarawas co. O. by p-r. 96 ms. N. E. by E. Columbus. Pop. 1830, 160.

Shannon, p-v. Mason co. Ky. by p-r. 55 ms. N. E. by E. Frankfort.

Shannon Hill, p-o. Goochland co. Va. by p-r. 147 ms. s. s. w. W. C.

Shannon's Store and p-o. Randolph co. Il. by p-r. 74 ms. s. w. Vandalia.

Shannonville, p-v., and named in p-o. list as st. jus. Perry co. Ten., by p-r. 114 ms. s. w. by w. Nashville.

Shapleigh, p-t. York co. Me. 163 ms. s. w. Augusta, 35 N. w. York, E. N. H.; contains several ponds, one of which, partly in the adjoining state, gives rise to Salmon Falls r. Pop. 1830, 1,479.

Sharon, t. Hillsborough co. N. H. 18 ms. from Amherst, 48 from Concord, and E. of Cheshire co., gives rise in the s. E. to branches of Contoocook r., but is almost destitute of mill seats. Boundary mtn. 200 feet high, is on the E. boundary. Pop. 1830, 371.

Sharon, p-t. Windsor co. Vt. 22 ms. N. Windsor, lies N. White r. Population 1830, 1,459.

Sharon, p-t. Norfolk co. Mass. 18 ms. s. Boston, is at the head of Neponset r. which furnishes good mill seats, occupied by several manufactories. The Sharon cotton manufacturing company was incorporated 1811, with $100,000; and the Mass. file manufacturing company have a factory here. Mashapoag pond gives rise to one of the chief branches of Neponset r. and gave the Indian name to the town. Pop. 1830, 1,023.

Sharon, p-t. Litchfield co. Conn. 47 ms. w. Hartford, lies w. Housatonic r., and E. N. Y. It is hilly E. with granite rocks. The soil is various, generally stony, with fine calcareous levels w. Grain succeeds better than in most other parts of the state. Pop. 1830, 2,615.

Sharon, p-t. Schoharie co. N. Y. 45 miles from Albany, 16 N. w. Schoharie, s. Montgomery co., and E. Otsego co.; has some low ridges of the Helderbergs. The soil is favorable to wheat. Cobuskill creek rises here and supplies mill seats. The inhabitants are of German descent. Pop. 1830, 4,247.

Sharon, tsp. and p-v. wstrn. part of Mercer co. Pa. The p-v. is very near the border between Mercer co. Pa. and Trumbull of O., and stands on Shenango cr. about 16 ms. w. of the borough of Mercer.

Sharon, p-v. Morgan co. O. by p-r. 99 ms. s. E. by E. Columbus.

Sharonville, p-v. Hamilton co. O. by p-r. 14 ms. N. E. Cincinnati.

Sharpe's Store and p-o. Lowndes co. Geo. by p-r. 203 ms. s. Milledgeville.

Sharpsburg, p-v. sthrn. part Washington co. Md. on the left bank of Potomac r., 18 ms. s. Hagerstown, the co. st., and by p-r. 66 ms. N. w. W. C.

Sharpsburg, p-v. wstrn. part Bath co. Ky. 11 ms. wstrd. Owingsville, the co. st., and 62 ms. E. Frankfort.

Sharp's Mills, and p-o. Indiana co. Pa. by p-r. 197 ms. N. w. W. C.

Shartlesville, p-v. Berks co. Pa. by p-r. 156 ms. N. N. E. W. C.

SHAUCK'S, p-o. Richland co. O. by p-r. 57 ms. N. N. E. Columbus.

SHAVER'S cr. and p-o. nrthrn. part Huntingdon co. Pa. 10 ms. N. the borough of Huntingdon, and by p-r. 152 ms. a little N. of N. W. W. C.

SHAWANGUNK, p-t. Ulster co. N. Y. 91. ms. from Albany, 26 S. W. Kingston, 17 W. N. W. Newburgh, has Montgomery co. S., and reaches the base of Shawangunk mtn. W. Shaw cr. W. and Wallkill cr. E. meet near the N. boundary. The soil is strong loam, with some clay, and the surface nearly level. Oak prevails in the woods. Mill stones are obtained here. The skeleton of the mammoth in Peal's museum, Philadelphia, was taken from a swamp here, and 9 others have been found in this and an adjoining t. The inhabitants are of Dutch origin. Population 1830, 3,681.

SHAWANGUNK mtns. N. Y. cross Ulster and Orange cos. being a spur of the small range of the Catsbergs.

SHAWNEETOWN, p-v. on Ohio r. estrn. part Gallatin co. Il. 9 ms. below the mouth of Wabash r. and by p-r. 127 ms. S. E. Vandalia. Lat. 37° 42', long. W. C. 11° 14' W. It is a flourishing v. containing a bank, printing office, land office, and a number of taverns, stores, &c. It is the depot for the U. S. Saline near the v. of Equality, 12 ms. wstrd.

SHAW'S MEADOWS, and p-o. nrthrn. part Northampton co. Pa. about 36 ms. N. of Easton, the co. st., and 226 ms. N. N. E. W. C.

SHEBOYGON r. of Huron Ter. rises to the estrd. and near the sthrn. end of Winnebago lake, interlocking sources with Rock r. and flowing thence estrdly. into lake Michigan.

SHEEPSCOT, r. Lincoln co. Me. runs a short distance in Kennebec co. and empties into the Atlantic, at Wiscasset, meeting some of those arms of the sea which form so many isls. on that part of the coast.

SHEETZ'S MILL and p-o. Hampshire co. Va. by p-r. 126 ms. N. W. by W. W. C.

SHEFFIELD, p-t. Caledonia co. Vt. 35 ms. N. Montpelier, 40 N. Newbury, with 22,607 acres. First settled 1792. It is on the height of lands dividing the waters of Conn. r. and lake Champlain, containing head streams of Barton and Passumpsic rs. on which are mill seats. Pop. 1830, 720.

SHEFFIELD, p-t. Berkshire co. Mass. 125 ms. W. Boston, N. Conn., was incorporated 1733, 6 years before any other t. in this co. The surface is agreeably varied, and there is much good land, watered by Housatonic r. and several of its branches. The v. is situated in a valley surrounded by several eminences, of which Taughkannic mtn. W. is the loftiest, being about 3,000 feet high. Along the course of the Housatonic, here slow and crooked, are extensive and valuable meadows, on the W. side of which runs the principal street 4½ ms. A grant was made by the general court of Mass. in 1720, which included part of two neighboring towns, and left a reserved tract for the Indians. The settlement was soon commenced from Westfield. Pop. 1830, 2,382.

SHEFFIELD, p-v. on lake Erie, nrthrn. part Lorain co. O. by p-r. 14 ms. N. Elyria, the co. st., and 144 ms. N. N.E. Columbus. Pop. twp. 1830, 215.

SHEGAG'S, store and p-o. about 60 ms. wstrd. Nashville.

SHELBURNE, p-t. Coos co. N. H. 111 ms. from Concord, W. of Maine, is crossed by Androscoggin r. which receives Rattle r. &c. and has good soil on its banks; but the land is generally rough, and often useless for cultivation. Mt. Moriah, of the White mtn. range, is in the S. Moses' rock is a singular block of stone, 90 feet long and 60 high. First settled 1775. Pop. 1830, 312.

SHELBURN, p-t. Chittenden co. Vt. 33 ms. W. Montpelier, and 26 N. W. Middlebury, was first settled before the revolution by Logan and Pottier, on points in the lake which still bear their names. They and ten other families soon after abandoned the place, but after the war it was occupied by settlers from Connecticut. There is a bay of the lake, named after the t. into the head of which falls Laplatte r. Shelburn pond in the N. E. covers about 600 acres. The soil is very good, timber hard wood. Pop. 1830, 1,122.

SHELBURNE, p-t. Franklin co. Mass. 100 ms. N. W. Boston, N. E. Deerfield r., comprizing a valuable fall of 20 ft., has a pleasant situation, and was formerly a part of Deerfield. The schools, library, &c. have proved particularly useful. Mr. Fisk, missionary to Palestine, was born here, 1792. Pop. 1830, 995.

SHELBY, p-t. Orleans co. N. Y. 263 ms. from Albany, 14 N. N. W. Batavia, E. Niagara co., and N. Tonawanta reservation, is watered by Oak Orchard cr. and its branches, crossed by the Mtn. Ridge N. and touched N. W. by Erie canal. There are several mills, &c. Pop. 1830, 2,043.

SHELBY, co. Ky. bounded S. E. by Anderson, Spencer S., Jefferson W., Oldham N. W., Henry N., and Franklin E. Length from E. to W. 26 ms., breadth 17, area 442 sq. ms. Lat. 38° 15', long. W. C. 8° 10' W. Though the estrn. border approaches very near Kentucky r. the slope of this co. is S. W., and in that direction is drained by different confluents of Salt r. Pop. 1830, 19,030. Chief t., Shelbyville.

SHELBY, the extreme sthwstrn. co. of Ten. bounded by Tipton co. Ten. N., and Lafayette co. Ten. E., on the S. it has the Chickasaw territory in the state of Miss., and on the W. the Miss. r. separating it from Crittenden co. Ark. Lat. 35° 15' and long. W. C. 13° W. intersect in the wstrn. part of this co. The slope is wstrd. and in that direction is traversed and drained by the various confluents of Wolf r. and Nanconnah cr. The high land of the interior reaches the Miss. r. at the N.W. angle and at the mouth of Wolf r. in this co. These hills are called Chickasaw Bluffs from the Indian nation who formerly owned and inhabited the country. The soil is good. Staple, cotton. Chief t., Memphis. Pop. 1820, 354; 1830, 5,648.

SHELBY, co. Ohio, bounded by Miami co. S., Dark co. S. W., Mercer N. W., Allen N., Logan N. E., and Champaign S. E. It is about 20 ms. each side; area 400 sq. ms. Lat. 40° 20′, long. W. C. 7° 12′ w. The extreme sources of Saint Mary's and au Glaize rivers rise on the northwestern and northern borders, but the much greater part slopes southward, and gives source to great Miami r. The extreme higher sources of Wabash rise also in Mercer co., very near the northwestern angle of Shelby. The latter comprises, therefore, a part of the high and flat table land of Ohio. Chief town, Sidney. Pop. 1820, 2,106, 1830, 3,671.

SHELBY, co. Ind., bounded S. E. by Decatur, Bartholomew S., Johnson W., Marion N. W., Hancock N., and Rush E. Length from S. to N. 24 ms., breadth 18, and area 432 sq. ms. Lat. 39° 30′, long. W. C. 8° 45′ w. Slope S. S. W., and in that direction drained by different branches of Driftwood fork of White r. Chief t. Shelbyville. Pop. 1830, 6,295.

SHELBY, co. Il., bounded S. E. by Effingham, Fayette S., Montgomery W., Sangamo N. W., Macon N., and Coles E. Length from E. to W. 40 ms., width 32, and area 1,280 sq. ms. Lat. 39° 22′, long. W. C. 11° 45′ w. The northwestern angle gives source to the Mowawequa branch of Sangamon river, and slopes to the N. W.; and the opposite or southeastern angle gives source to Little Wabash, and slopes to the sthrd. Full 9-10ths of the co. is, however, drained by the Kaskaskias, and branches; the main stream traversing it diagonally from N. N. E. to S. S. W. Chief t. Shelbyville. Pop. 1830, 2,972.

SHELBY, co. Ala., bounded by Autauga co. S., Bibb S. W., Jefferson N. W. and N., St. Clair N. E., and Coosa river separating it from the Creek country E. Length from S. to N. 50 ms., mean breadth 22, and area 1,100 sq. ms. Lat. 33° and long. W. C. 10° w. intersect in the southwestern angle of this co. The East fork of Cahaba r., rising in Jefferson and Saint Clair counties, enters and traverses the northwestern side of Shelby, flowing in a S. S. W. direction into Bibb co. Between the Cahaba and Coosa vallies extends a ridge, from which creeks flow eastward towards the Coosa. The Coosa slope comprises full two thirds of the whole surface of the co. Chief town, Shelbyville. Pop. 1830, 5,704.

SHELBY, p-v. northeastern part Macomb co. Mich., 11 ms. northward Mount Clemens, the co. seat, and 37 ms. N. N. E. Detroit.

SHELBYVILLE, p-v. and st. jus. Shelby co. Ala., by p-r. 73 ms. a little N. of E. Tuscaloosa. Lat. 33° 16′, long. W. C. 9° 52′ w.

SHELBYVILLE, p-v. and st. jus. Bedford co. Ten., situated on Duck r., by p-r. 52 ms. S. S. E. Nashville. Lat. 35° 28′, long. W. C. 9° 24′ w.

SHELBYVILLE, p-v. and st. jus. Shelby co. Ky., on a branch of Salt r., 21 ms. W. Frankfort. Lat. 38° 11′, long. W. C. 8° 12′ w. It is a flourishing village. Pop. 1830, 1,201.

SHELBYVILLE, p-v. and st. jus. Shelby co. Ind., by p-r. 30 ms. S. E. Indianopolis. Lat. 39° 32′, long. W. C. 8° 46′ w.

SHELBYVILLE, p-v. and st. jus. Shelby co. Il., on Kaskaskias r., 40 ms. above and N. N. E. Vandalia. Lat. 39° 22′, long. W. C. 11° 52′ w.

SHELDON, p-t. Franklin co. Vt., 46 ms. N. W. Montpelier, 32 N. E. Burlington; was settled 1790, and is watered by Missisque r., and Black r. its branch, on the latter of which are mill seats. The surface is varied, and the soil generally good. Pop. 1830, 1,427.

SHELDON, p-t. Genesee co. N. Y., 270 ms. W. Albany. 24 S. W. Batavia, and E. of Erie co.; is watered by Tonawanta cr. and two branches of Buffalo cr. The land is high, but moist, and more favorable to grass than grain. Pop. 1830, 1,731.

SHELTER ISLAND, t. Suffolk co. N. Y., lies off the E. end of Long Island, 100 miles E. New York, and 250 from Albany, by the common route. This town is formed of two isls., which lie in the bay between Southold and Southampton. Shelter island contains 8,000 acres, of varied surface, with a soil generally light and sandy, but in some parts rich, level and well cultivated. Hog Neck isl. ½ a mile distant, and connected by a ferry, has a ship channel all round it. In the revolutionary war, the British deprived this isl. of its timber. Pop. 1830, 330.

SHENANDOAH, river of Va., and one of the great southern branches of Potomac river, is composed of two branches, called with no great relative correctness, North Branch and South Branch. The southern and main branch rises in Augusta co., as far south as lat. 38°, and long. 2° w. W. C. Flowing thence northeastward along the northwestern slope of Blue Ridge, over Augusta, Rockingham, and Page counties, receives the North Branch in the southern angle of Frederick co., after a comparative course of 90 ms.

The North Branch of Shenandoah river has its source in Rockingham co., from which it flows by comparative courses N. N. E. 50 ms. over Rockingham and Shenandoah counties, enters Frederick, bends to the eastward, and joins the South Branch as already noticed. Below the junction of its two branches, the Shenandoah flows N. E. along the northwest slope of Blue Ridge 40 ms. to its junction with the Potomac at Harper's Ferry. (*See art. Potomac.*)

SHENANDOAH, county, Va., bounded S. W. by Rockingham, Hardy W. and N. W., Frederick N. and N. E., and Page E. and S. E. Length from S. W. to N. E. 32 ms., mean breadth 12, and area 384 sq. ms. Lat. 38° 50′, long. W. C. 1° 30′ w. The whole co. is a part of the valley of the North fork of Shenandoah r. Since the census of 1830, Page co. was detached from Shenandoah, which latter formerly comprised upwards of 1,000 sq. ms. In the census tables of 1830, what is now Shenandoah co., is called West Shenandoah, and contained a pop. of 11,423. Both cos., or the original Shenandoah, contained in 1820, an aggregate pop. of 18,926.

SHEPHERDSTOWN, p-v. eastern part Cumberland co. Pa., by p-r. 8 miles from Harrisburg, and 102 ms. N. W. C.

SHEPHERDSTOWN, p-v. on the Potomac r., northeastern part Jefferson co. Va., 10 miles above Harper's Ferry, and by p-r. 62 ms. N. W. W. C.

SHEPHERDSVILLE, p-v. and st. jus. Bullitt co. Ky., on the North fork of Salt r., 23 ms. S. Louisville, and by p-r. a little S. of W. Frankfort. Lat. 37° 58′, long. W. C. 8° 42′ W. Pop. 1830, 278.

SHERBURNE, town, Rutland co. Vt., 22 miles N. W. Windsor, and 9 N. E. Rutland. First settled 1785. It gives rise to Queechy river N. W., and has several small mill streams, particularly Thundering brook, which rises in one of the ponds. There is some meadow land on Queechy r., but the surface is generally mountainous. Killington peak, of the Green mtns., is south and 3,924 feet high. Pop. 1830, 432.

SHERBURNE, p-t. Middlesex co. Mass., 21 ms. S. W. Boston, W. Charles river, E. and N. E. Nashua river. The town is agricultural, and possesses a good soil. Pop. 1830, 899.

SHERBURNE, p-t. Chenango co. N. Y., 98 ms. W. Albany, 11 N. Norwich, lies S. Madison co., and is crossed by Chenango river, on which are rich meadows. The soil generally is good for both grain and grass. The v. is on the E. bank of the r. Pop. 1830, 2,601.

SHERBURNE, p-v. Beaufort district, S. C., by p-r. 165 ms. S. Columbia.

SHERBURNE MILLS, and p-o. Fleming co. Ky., by p-r. 84 ms. estrd. Frankfort.

SHERIDAN, p-t. Chatauque co. N. Y., 319 miles from Albany. Pop. 1830, 1,666.

SHERMAN, p-t. Fairfield co. Conn., 60 ms. S. W. Hartford, is in the S. W. corner of the co., with Litchfield co. N. and N. Y. W. The surface is hilly, the soil various, and some iron ore is found. Several streams flow into Housatonic r. Pop. 1830, 947.

SHERMAN, p-v. sthrn. part Huron co. O., by p-r. 96 ms. a little E. of N. Columbus. Pop. 1830, 153.

SHERMAN, p-v. Saint Joseph's co. Mich., by p-r. 145 ms. S. W. by W. Detroit.

SHERRARD'S STORE, and p-o. Hampshire co. Va., by p-r. 95 ms. N. W. W. C.

SHERRILL'S FORD, and p-o. Lincoln co. N. C., by p-r. 148 ms. wstrd. Raleigh.

SHESHEQUIN, p-o. Bradford co. Pa., by p-r. 136 ms. nrthrd. Harrisburg.

SHETUCKET, river, Conn., is formed by the junction of Willimantic and Mount Hope rs., and after flowing S. E. joins the Quinebaug, and at Norwich takes the name of the Thames.

SHICKSHINNY, mountain, rises above and stretches along the right bank of Susquehannah r., sthrn. part Luzerne co. Pa.

SHICKSHINNY, p-o. near the right bank of Susquehannah r., southern part Luzerne co. Pa., by p-r. 101 ms. N. E. Harrisburg.

SHILOAH, p-v. Camden co. N. C., by p-r. 200 ms. N. E. by E. Raleigh.

SHILOH, p-v. sthrn. part Marengo co. Ala., by p-r. 97 ms. S. Tuscaloosa.

SHINERSVILLE, p-v. Lycoming co. Pa., by p-r. 224 ms. nrthrd. W. C.

SHINNSTON, p-v. Harrison co. Va., by p-r. 236 ms. wstrd. W. C.

SHIPPEN, p-v. MacKean co. Pa., by p-r. 293 ms. N. W. W. C.

SHIPPENSBURG, borough and p-v. Cumberland co. Pa., by p-r. 39 ms. S. W. by W. Harrisburg, and 100 ms. N. N. W. W. C. Pop. 1830, 1,621. It is a close built v., principally of one street along the main road.

SHIPPENSVILLE, p-v. Venango co. Pa., by p-r. 256 ms. N. W. W. C.

SHIPPINGPORT, p-v. on the Ohio r., 2 miles below the centre of Louisville, and at the lower end of the Rapids, Jefferson co. Ky. Though a separate p-v. and under a different corporate establishment, it is commercially a suburb of Louisville. Pop. 1830, 606.

SHIPPINGPORT, p-v. Tazewell co. Il., about 150 ms. N. N. W. Vandalia.

SHIREMANTOWN, p-v. Cumberland co. Pa. by p-r. 4 ms. from Harrisburg.

SHIRLY, p-t. Middlesex co. Mass. 38 ms. N. W. Boston, S. W. Nashua river, with Squanicook, a branch of it, on the N. on both of which streams are rich meadows. Chairs have been made here to a great amount. Pop. 1830, 991.

SHIRLEYSBURG, p-v. Huntingdon co. Pa. 20 ms. S. S. E. the borough of Huntingdon.

SHIVER'S MILLS, and p-o. Warren co. Geo., by p-r. 45 ms. N. E. by E. Milledgeville.

SHOALS OF OGEECHEE, p-v. Hancock county, Geo., by p-r. 47 ms. N. E. Milledgeville.

SHOBER'S MILLS, and p-o. Jefferson county, Ohio, by p-r. 142 ms. N. E. by E. Columbus.

SHOREHAM, p-t. Addison co. Vt. 12 ms. S. W. Middlebury, 49 ms. S. Burlington, & on the E. side of lake Champlain, has a surface nearly level, with good soil, and is one of the best farming towns in the state. It lies opposite fort Ticonderoga, and commands a view of the ruins of that fortress, and the interesting scenery in its vicinity. The lake is generally about a 1-2 mile wide here, and there is a ferry across it. The shore is generally a little elevated, and the rocks of dark calcareous stone, containing impressions of shells, &c. A variety of fish are taken from the lake. Pop. 1830, 2,137.

SHORT MOUNTAIN, p-o. 111 ms. westward Little Rock, Arkansas, given in the p-o. list as in Crawford co.

SHORT PUMP, p-v. Henrico county, Va. by p-r. 12 ms. from Richmond.

SHREWSBURY, river, Monmouth co. N. J. divided into the North or Navesink and South rivers, is navigable for vessels of 50 tons, and navigated by a steamboat from N. Y. twice a day. This river formerly discharged into the sea 9 ms. S. Sandy Hook, but the outlet was closed by a storm in 1810, and the river discharged into Raritan or Sandy Hook bay. The outlet is again open and the waters discharged by both channels, (1832.)

SHREWSBURY, p-t. Rutland co. Vt. 22 ms. w. Windsor, 9 ms. s. E. Rutland, lies chiefly on the Green mnts. and is very high E. Shrewsbury peak N. 4,100 feet high, is one of the most lofty summits of the range. Mill river s. w. and Cold river N., are mill streams. Pearl's and Ashley's ponds lie s. The soil is good for grass. Pop. 1830, 1,289.

SHREWSBURY, p-t. Worcester co. Mass. 30 ms. w. Boston, 5 E. Worcester, is varied by hills and vallies, and divided near the middle, N. and s. by a high ridge. The soil is fertile, and the inhabitants farmers. Quinsigamond or Long pond, nearly 4 ms. long, and from 50 to 70 feet deep, lies between this town and Worcester, and is crossed by a floating bridge, principally of hewn timber, 525 feet long. Artemas Ward, the first major general of the United States, died here, in 1800. Pop. 1830, 1,386.

SHREWSBURY, p-t. Monmouth co. N. J. 25 ms. from N. Y. city and 14 s. E. Middletown Point. Here are several large Peach orchards, two of which are said to be the largest in the United States, covering together 150 acres, and containing 22,000 trees, the first of which were planted about 1822. The fruit is principally carried to New York market. The soil is even inferior to that of the adjacent pine plains, yet the fruit is remarkably fine. Beds of marle are found in some parts of this town, containing bones, shark's teeth, &c. Pop. 1830, 4,700.

SHREWSBURY, tsp. and p-v. southern side of York county, Pa. The p-o. is about 14 miles southwardly from the borough of York, and by p-r. 72 miles N. W. C. Pop. of the township 1820, 1,983, 1830, 2,571.

SHUTESBURY, p-t. Franklin co., Mass., 82 ms. w. Boston, is very rocky, stony, and hilly, and unfavorable to agriculture. It is crossed by the w. branch of Swift r. N. and s. which furnishes valuable mill seats. Settled 1754, from Sudbury. Pop. 1830, 986.

SIASCONSET, village, Mass., on the E. side of Nantucket, a little s. of Sicacache.

SIDNEY, p-t. Kennebec county, Me., 8 ms. N. Augusta, has Kennebec river E. and a small lake on the w. line, which is connected with several others, and flows by an outlet into the Kennebec. Pop. 1830, 2,191.

SIDNEY, p-t. Delaware county, N. Y. 95 ms. s. w. Albany, 24 ms. w. Delhi; has Oswego river N., which separates it from Otsego co.; Chenango county is w. The Susquehannah is N. w. and Ouleout creek N. E. and both have fine meadows on their banks, though the surface of the town is generally hilly. Pop. 1830, 1,110.

SIDNEY, p-v. and st. jus. Shelby county, O., by p-r. 86 miles N. W. by W. Columbia, on Great Miami river at lat. 40° 17', long. 7° 8' w. Pop. 1830, 240.

SILVAN GROVE, p-v. Morgan co. Illinois, by post-road 138 miles northwest Vandalia.

SILVER CREEK, p-v. on Silver creek, wstrn. side Madison county, Ky., by p-r. 8 miles westward Richmond, the co. st. and 58 miles s. E. Frankfort.

SILVER GLADE, p-v. Anderson district, S. C., by p-r. 148 miles N. W. Columbia.

SILVER LAKE, p-o. Susquehannah county, Pa. by p-r. 280 miles N. N. E. W. C. 6 miles N. W. Montrose, the co. st. Pop. of the tsp. 1820, 456, 1830, 516.

SILVER SPRING, p-v. western part of Wilson county, Ten., 22 miles eastward Nashville.

SIMPSON, county, Miss., bounded by Covington S. E., Lawrence s. w., Copiah w., Hinds N. W., and Rankin N.; length from E. to w. 36 miles, breadth 28, and area 1,008 square miles. Lat. 32° and long. W. C. 13° w. intersect in this county near Westville, the co. st. The slope is to the southward, and in that direction it is traversed by Pearl river and drained by several of its branches. Pop. 1830, 2,680.

SIMPSON, county, Ky., bounded w. and N. w. by Logan, Warren N., Allen E., Sumner co. Tennessee, southeast, and Robertson county, Ten., s. w. The length along Ten. 35 ms., mean breath 12, and area 420 square miles. Lat. 36° 45', long. W. C. 9° 35' w. The estrn. part of this county slopes northward, and in that direction is drained by some of the southern branches of Big Barren river; whilst from the western section issue the extreme northeastern source of Red river, branch of the Cumberland. The county is therefore a table land between the vallies of Green and Cumberland rivers. Chief town, Franklin. Pop. 1820, 4,852, 1830, 5,815.

SIMPSONVILLE, p-v. Montgomery county, Md., by p-r. 7 miles from W. C.

SIMPSONVILLE, p-v. Shelby county, Ky., by p-r. 8 miles westward Shelbyville, the co. st., and 29 ms. westrd. Frankfort. Pop. 1830, 77.

SIMSBURY, p-t. Hartford county, Conn., 12 miles N. w. Hartford, first settled 1670, from Windsor, deserted in 1676, and then burnt by Indians, but was soon after reoccupied. It is crossed by Farmington river, and has a rocky range E., on which is much useless land, and which is broken through by the river on a part of whose course lie some rich meadows. Salmon and shad formerly abounded in this stream but have now deserted it. Pop. 1830, 2,221.

SINEPUXENT INLET, on the Atlantic coast of Maryland, Worcester county. It is the entrance between Assateague and Fenwick's islands, into a long narrow sound which bears the same name with the entrance, and admits small coasting vessels.

SINGSING, p-v. Mount Pleasant, Westchester county, N. Y., is situated at the foot and on the acclivity of the steep bank of Hudson r. just below the mouth of Croton river. It has 4 churches, a male and female academy, and 2 landings, with one of which a fine steamboat communicates twice daily on the way between Peekskill and New York. There are several handsome country seats, and many fine points of view. One of the state prisons is situated a little s. from the village, on the bank of the river, is built of white marble, forming 3 sides of a square, with 1,000 cells for convicts, ranged in 5 stories, a chapel, 2 hospitals, superintendants' and keepers', dwellings, work sheds, and a wharf. The system of discipline is essentially that of the Auburn prison. The convicts are employed in quarrying and working white marble from a quarry in the rear, on the grounds belonging to the prison. They are guarded by 24 centinels, and work without chains or bonds of any kind. There is not even a wall or fence about the quarry. From the high grounds in the upper parts of the village, the eye embraces a view of Hudson river for about 30 ms., including Haverstraw and Tappan bays, with a large part of the Highland range, &c.

Sinking Cane, p-o. Overton county, Ten., by p-r. 113 miles eastward Nashville.

Sinking Spring, p-v. Highland county, O., by p-r. 78 miles s. s. w. Columbus.

Sinking Valley Mills, and p-o. Huntingdon county Pa., by p-r. 170 ms. n. n. w. W. C.

Sinnamahoning, river, Pa., the extreme northwestern branch of the West Branch of Susquehannah. This river, formed by numerous branches flowing from Potter, Mac Kean, Jefferson, and Clearfield counties, is a true mountain stream joining the main w. branch in the western part of Lycoming co. It interlocks sources to the n. with those of Genesee river; to the n. w. with those of Alleghany river and W. Clarion river.

Sinnamahoning, p-o. western part of Lycoming county, Pa., about 120 miles n. w. Harrisburg.

Sipican, p-v. Rochester, Plymouth county, Mass. on Sipican river, which empties into Buzzard's bay.

Sistersville, p-v. Tyler county, Va., by p-r. 274 miles a little n. of w. W. C. on Ohio river about 50 miles n. w. by w. Clarksburg.

Six Nations, of Indians, N. Y. This general name includes the principal tribes or rather nations of Indians, who formerly occupied the principal part of the present state of New York, and exercised authority far into New England. They have gradually become scattered and reduced, having sold most of their land. But numbers of them all, except the Mohawks, still dwell upon small tracts of reserved land in different cos. of the state. The names of these were Onondagas, Senecas, Cayugas, Oneidas, Mohawks, and Tuscaroras. The Onondagas were highest in authority. The Tuscaroras came from the south, and were in modern times admitted into the confederacy. The language of this tribe is said to bear no resemblance to the others beyond the general principles of construction, in which all the tongues and dialects of the American savages agree (with two exceptions.) The other 5 nations, though differing in this respect, speak languages which have a greater affinity.

Skaneateles Lake, N. Y., lies chiefly in Onondaga county, and is 15 miles long, by 1-2 to 1 1-2 wide. It yields trout and other small fish, and its outlet flows from the n. end into Seneca river 10 miles, supplying many mill sites.

Skaneateles, p-t. Onondaga county, N. Y., 149 miles from Albany. Population 1830, 3,812.

Skaneateles, p-v. Marsellus, Onandaga county, N. Y., at the outlet of Skaneateles lake, 145 west Albany, contains several mills, &c.

Skinner's Eddy, and p-o. Luzerne county, Pa., on Susquehannah r. 44 ms. above Wilkes Barre, and 267 miles n. n. e. W. C.

Skippack, creek and p-o. The creek is the eastern branch of Perkiomen river, Montgomery county, Pa., by p-r. the p-o. is about 25 miles n. w. Phila.

Slab Point, and p-o. western part of Montgomery county, Illinois, by pr. 46 ms. n. w. Vandalia.

Slabtown, p-v. Anderson district, S. C., by p-r. 139 miles n. w. Columbia.

Slate, p-o. Bath county, Ky., by p-r. 80 ms. e. Frankfort.

Slate Mills, and p-o. Culpepper county, Va., by p-r. 91 miles s. w. W. C.

Slaterville, village, s. Oxford and Dudley, 6 miles from Uxbridge, and 6 miles from Douglass, contains seven mills, two of stone, three of brick, and two of wood. Five of these derive their power from French river, the other two are in the centre of the village, and obtain their power from Slater's lake; the Indian of which is *Chargoggagoggmanchoggo*. It is four miles long, and never failing. They use 6,000 spindles, 90 looms, and employ 190 hands, and work up 1,000 bales of cotton, which produces 15,000 yards a week, beside large quantities of satinet warps, and sewing thread. They manufacture, also, broadcloths, cassimeres, and satinets. In this branch of their business, they use 600 lbs. of wool a day, or 180,000 lbs. a year.

These factories are owned by Messrs. Slater & Sons. Mr. Samuel Slater, who resides here, has been said to have a larger amount of property vested in manufactures, than any other man in the United States. He invented cotton thread, in 1794.

Slinkard's Mills, and p-o. Greene county, Indiana, by p-r. 88 miles s. w. Indianopolis.

Slippery Rock, creek, the northern branch of Conequenessing river. It rises in Mercer and Butler counties, and flowing s. s. w. about 35 miles joins the Conequenessing about 3 or 4 miles above the influx of the latter into Big Beaver. (*See Conequenessing.*)

Slippery Rock, township and p-o. northwestern part of Butler county, Pa. The p-o. is 18 miles n. w. the borough of Butler, and by p-r. 254 miles n. w. W. C.

Sloanesville, p-v. Schoharie, Schoharie co. New York, 4 miles w. Esperance.

Sloanesville, p-o. Mecklenburg co. N. C., 146 miles s. w. by w. W. C.

Smelsor's Mills, and p-o. Rush co. Ind., by p-r. 46 miles s. e. by e. Indianopolis.

Smicksburg, p-v. Indiana county, Pa., by p-r. 212 miles n. w. W. C.

Smith, county, Ten., bounded by Jackson e., White s. e., Warren south, Wilson southwest, Sumner west, Allen county, Kentucky, northwest, and Monroe county, Ky., n. e. Length from s. to n. 40 miles, mean breadth 16, and area 640 square miles. Lat. 36° 25', and long. W. C. 9° w. intersect in this county. The extreme border on Ky. slopes northward, and is drained by small confluents of Big Beaver river. The much greater part is in the valley of Cumberland river, and the main volume of that stream traverses it from e. to w. Chieftown, Carthage. Population 1820, 17,580, 1830, 19,906.

Smithborough, p-v. Tioga county, N. Y., 10 miles w. Owego.

Smithdale, p-o. Amite county, Miss., about 30 miles s. e. by e. Natchez.

Smithfield, p-t. Providence co. R. I., 9 ms. n. w. Providence, has Blackstone r. n. e. and Mass. n.; has a varied surface, and lime-stone rocks below, which are quarried for the extensive manufacture of lime. Whet stones are found in large quantities. The soil is generally good. Blackstone river, one of its branches, and several smaller streams, supply mill sites. The manufactures of this t. are very important. At Woonsocket falls, on the Blackstone, is also a large manufacturing place. Pop. 1830, 6,857.

Smithfield, p-t. Madison co. N. Y., 108 ms. w. n. w. Albany, 6 n. Morrisville, and has Oneida co. e. Has excellent soil, and is well watered by the head streams of Oneida creek, with other streams of Oneida lake, and Chenango river, and was settled from New England. New Stockbridge, a tract of land given to the remains of eastern tribes by the Oneidas, is in this t. Pop. 1830, 2,636.

Smithfield, p-v. s. w. part Fayette co. Pa., by p-r. 202 ms. n. w. by w. W. C.

Smithfield, p-v. nrthrn. part Isle of Wight co. Va., by p-r. 80 ms. s. e. by e. Richmond. It is situated on a small creek or bay of James river, 15 ms. above Hampton Roads.

Smithfield, p-v. and st. jus. Johnson co. N. C., by p-r. 29 ms. s. e. Raleigh, on the left bank of Neuse r. Lat. 35° 31', long. W. C. 1° 20' w.

Smithfield, p-v. Hamilton co. Ten., by p-r. 135 ms. s. e. by e. Nashville.

Smithfield, p-v. Jefferson co. O., by p-r. 136 ms. n. e. by e. Columbus.

Smithfield, p-v. Delaware co. Ind., by p-r. 66 ms. n. e. Indianopolis.

Smithfield, p-v. on the Ohio r. immediately below the mouth of Cumberland r., Livingston co. Ky., by p-r. 260 ms. s. w. by w. Frankfort. Pop. 1830, 388.

Smithport, as in p-o. list, though usually Smethport, p-v. and st. jus. MacKean co. Pa., by p-r. 200 ms. n. w. Harrisburg. It is situated on one of the highest branches of Alleghany r., 25 ms. s. Hamilton, Cattaraugus co. N. Y. Lat. 41° 50', long. W. C. 1° 32' w.

Smith's, r. Grafton co. N. H., rises in several ponds, flows about 15 ms., and enters the Pemigewasset between New Chester and Bristol.

Smithsburgh, p-v. western part Washington co. Md., by p-r. 76 ms. n. w. W. C., and 12 ms. westward Hagerstown.

Smith's Creek, and p-o. nrthrn. part Rockingham co. Va. The p-o. is by p-r. 130 ms. a little s. of w. W. C. The creek is the extreme southern branch of the West fork of Shenandoah.

Smith's Cross Roads, and p-o. Rhea county, Ten., by p-r. 126 ms. s. e. by e. Nashville.

Smith's Farm, and p-o. Alleghany co. Md., by p-r. 160 ms. n. w. W. C.

Smith's Ford, and p-o. York dist., S. C., by p-r. 92 ms. n. Columbia.

Smith's Grove, and p-v. Warren co. Ky., by p-r. 133 ms. s. w. Frankfort.

Smith's Island, mouth of Cape Fear river, Brunswick co. N. C., is a long narrow sandy slip which divides the r. into two channels. The light house stands s. e. from Smithville, on the western side of the isl., and on the main channel. Cape Fear, the extreme salient point of the isl., is at lat. 33° 54½', long. W. C. 1° 01' w.

Smith's Mills, and p-o. Clearfield co. Pa., by p-r. 178 ms. n. w. W. C.

Smith's Mills, and p-o. Henderson county, Ky., by p-r. 191 ms. a little s. of w. Frankfort.

Smith's Store, and p-o. Pittsylvania county, Va., by p-r. 244 ms. s. w. W. C.

Smith's Store, and p-o. Montgomery co. N. C., by p-r. 159 ms. s. w. by w. Raleigh.

Smith's Store, and p-o. Spartenburgh dist. S. C., by p-r. 104 ms. n. w. Columbia.

Smith's Store, and p-o. Jackson co. Ala., by p-r. 141 ms. n. e Tuscaloosa.

Smithsville, p-v. Powhattan co. Va., by p-r. 38 ms. westward Richmond.

Smithsville, p-v. Dickson co. Ten., by p-r. 57 ms. wstrd. Nashville.

Smithtown, p-t. Suffolk co. N. Y., 53 ms. e. N. Y., and s. of Long Island sound; contains several small vs., with a pond s. which flows into the sound. It has some coasting trade. Pop. 1830, 1,686.

Smithville, p-t. Chenango co. N. Y., 13 ms. s. w. Norwich, and e. of Broome co. It is supplied with mill seats by Chenango river and branches. This town includes some rich meadows, the principal of which is called the Big Flats. Pop. 1830, 1,686.

Smithville, p-v. and st. jus. Brunswick co. N. C., by p-r 178 ms. s. s. e. Raleigh. It is a seaport of some importance, situated on the right side of the western channel of Cape Fear r., one or two miles above its mouth into the Atlantic, and almost exactly on lat. 34°.

Smockville, p-v. Jefferson co. Ind., by p-r. 94 ms. s. s. e. Indianopolis.

Smyrna, p-t. Chenango co. N. Y., 13 miles n. n. w. Norwich, 105 w. Albany, and s. Madison county. It is hilly n. e., where waters of Chenango r. afford many mill seats. The surface is generally uneven, but the vallies are large and fertile, and favorable to hemp.

First settled 1792. Population in 1830, 1,839.

SMYRNA, formerly Duck creek cross roads, p-v. on Duck creek, Kent county, Delaware, 12 miles a little w. of N. Dover, and by p-r. 102 miles N. E. by E. W. C.

SMYRNA, p-v. Harrison county, Ohio, by p-r. 106 miles a little N. of E. Columbus.

SNEEDSBORO', p-v. on Yadkin river, and in the S. E. angle of Anson county, N. C., by p-r. 14 miles S. S. E. Wadesboro', the co. st. and 134 miles S. W. by W. Raleigh.

SNICKERSVILLE, p-v. western part Loudon county, Virginia, by p-r. 49 miles westward W. C. and 21 miles eastward Winchester.

SNODDYVILLE, p-o. Jefferson county, Ten., about 240 miles E. Nashville.

SNOW CAMP, p-v. Orange county, N. C., by p-r. 53 miles N. W. Raleigh.

SNOW HILL, p-v. and st. jus. Worcester co., Maryland, on Pocomoke river, 164 miles S. E. W. C. Latitude 38° 12', longitude W. C. 1° 36' E.

SNOW HILL, p-v. eastern part of Clinton co., Ohio, by p-r. 71 miles S. W. Columbus, and about an equal distance N. E. by E. Cincinnati.

SNYDERSVILLE, p-v. Northampton county, Pa., about 20 miles northward Easton.

SOCIAL CIRCLE, p-o. Walton county, Geo., by p-r. 71 miles N. W. Milledgeville.

SOCIETY HILL, p-o. on Great Pedee river, and in the extreme N. E. angle of Darlington district, South Carolina, 14 ms. N. Darlington, the st. jus. and 101 ms. N. E. by E. Columbia.

SOCIETY LAND, town, Hillsboro' county, N. H., 17 miles from Amherst, and 33 miles from Concord, has Contoocock river W., and a surface generally uneven. Crotched mtn. is an eminence S. It is destitute of mill sites. Pop. 1830, 164.

SODDY, p-v. Hamilton county, Ten., by p-r. 144 miles S. E. by E. Nashville.

SODUS BAY, Wayne county, New York, a bay of lake Ontario, and the best harbor on this shore, is about 6 miles long, and from 2 to 4 miles wide, with good depth of water.

SODUS, p-t. Wayne county, New-York, 208 miles W. by N. Albany, 30 miles N. Geneva, lies S. lake Ontario, and has a surface varied by N. and S. ridges, with good soil, well watered and timbered. Great Sodus bay is N. E., and Sodus village stands on a point projecting into it; 2 miles from this is the mouth of Sodus creek, which affords mill seats. Iron ore is found in the town. Pop. 1830, 3,528.

SOLON, p-t. Somerset county, Maine, 44 miles N. Augusta, 18 miles N. Norridgewock, 44 W. by N. Hallowell, E. Kennebec river, and has a pond E. with small streams empting into that river. Pop. 1830, 768.

SOLON, p-t. Cortlandt county, New York, 132 miles W. Albany, 10 E. Cortlandt, 31 S. Salina, has Tioughnioga creek N. W., Otselic creek S. E., and other smaller streams with mill sites. The soil is good for farms, and the timber, maple, beech, elm, ash, &c. Pop. 1830, 2,033.

SOMERFIELD, p-v. Somerset county, Pa., by p-r. 170 miles N. W. W. C.

SOMERS, p-t. Tolland county, Conn., 22 ms. N. E. Hartford, 12 miles S. E. Springfield, 56 miles W. Providence, S. Mass. and W. Hartford county. It is nearly level W. with few stones, but hilly E. with some eminences which command a view of Hartford and the valley of Conn. river. Orchard grass flourishes well, and it is crossed by Scantic river and other streams. Pop. 1830, 1,429.

SOMERS, p-t., Westchester county, N. Y., 50 miles N. E. New York, 120 miles S. Albany, lies S. Putnam county, with Croton r. for the N. boundary, whose branches afford mill seats. The soil is good, and the village, 1 1-2 miles from the N. line, is a great market for lean cattle and sheep to supply the places of the fat taken for N. Y. market. Pop. 1830, 1,997.

SOMERSET, county, Me., bounded by Lower Canada N. and N. W., Penobscot county, E., a corner of Waldo county S. E., Kennebec county S., and Oxford county W., is the second county in respect to size in the state. Its N. boundary is the range of highland dividing the waters of the St. Lawrence from those of the Atlantic, and it gives rise to the principal rivers of the state, the St. John's, the Penobscot and the Kennebec almost interlocking their head streams near the middle of the W. boundary line.

The 2 first mentioned afford a connected boat navigation, interrupted only by a portage of 2 ms., and between a branch of the Penobscot and the Allagash, a branch of St. John's, there is a portage of similar length. Moosehead lake, the source of Kennebec river, and which receives Moose river from the W., is long, irregular, and contains several considerable islands. Sebasticook lake, through which flows the Penobscot, is nearly of equal length, and extending nearly N. W. and S. E. ends on the line of Penobscot county. There are several lakes also on the course of the Allagash, chiefly in this county; a large part of the surface is uneven and even mountainous; several considerable clusters and ridges of mountains breaking the surface, especially S. and W. There are Mount Abraham, Mount Bigelow, Bald Mountain ridge, &c. N. and N. W. the surface is more uniform, but the land very high. The S. half of the county is indeed much varied by eminences, lakes, ponds, and streams, and there are collected almost all the inhabitants; the upper half being little known. The new road to Quebec crosses this county, following the course of the Kennebec a considerable distance and passing through a wilderness across the height of land to the sources of the Chaudiere in Lower Canada, then pursuing the course of that stream towards the St. Lawrence. The Kennebec and its branches watering a large part of this county, and that part which embraces nearly the whole population, is an important channel of trade, and has offered a route for the transportation of vast quantities of lumber. It contains 52 townships (some of which are only laid out and not named,) besides a number of plantations, purchases, &c. The county town, is Norridgewock. Pop. 1820, 21,787, 1830, 35,787.

SOMERSET, town, Windham county, Vt., 14 miles N. E. Bennington, 16 N. W. Brattleborough, is very mountainous, and crossed by Deerfield river N. and S. and the Moose branch in the W. Mount Pisgah, the highest range in the town, is in the E. Pop. 1830, 245.

SOMERSET, p-t. Bristol county, Mass., 13 miles S. Taunton, 42 S. Boston, on Taunton r. Pop. 1830, 1,023.

SOMERSET, p-t. Niagara county, N. Y., 15 miles N. E. Lockport, S. lake Ontario, W. Genesee county, is watered by Golden Hill creek and Keg harbor and Fish creeks. Pop. 1830, 871.

SOMERSET, county, N. J. bounded by Morris county N., Essex and Middlesex cos. E., Middlesex and Hunterdon counties S. and W. Raritan river runs through it from W. to E., part of it and one of its branches also forming much of the E. bound. The N. and S. branches and Millstone river, another branch, are partly in this county. There is a handsome variety of surface, with much good land, particularly on the level borders of Raritan river. Agriculture is more flourishing than in many other parts of N. J. Among the eminences in different parts are Rock mountain, Rocky hill, Stone mountain, Basking ridge, &c. In the revolutionary war the battle of Princeton was fought near the S. W. boundary of this county, and Gen. Charles Lee was captured by a small party of British dragoons at Basking Ridge. Chief town, Somerville. Pop. 1820, 16,506, 1830, 17,689.

SOMERSET, county, Pa., bounded W. by Fayette, Westmoreland N. W., N. by Cambria, E. by Bedford, and S. by Alleghany county, Md. Length 38 miles, mean breadth 28, and area 1,064 square miles. N. lat. 40° and long. W. C. 2° W., intersect near its centre. Somerset is a real mountain valley between Alleghany mountain and Laurel hill. The southern part slopes westward, and is traversed by Cassellman's river, a branch of Youghiogheny river. The northern section slopes to the northward, and in that direction is drained by the higher sources of the Kiskiminitas river. The surface of Somerset is much less hilly and the soil better than could be generally expected amid mountain chains. The mean elevation above the Atlantic tides must exceed 1,500 feet, as Smithfield on Youghiogheny river near its southwestern angle exceeds 1,400 feet above the ocean tides. It is a grain and pasture region. Chief town, Somerset. Pop. 1820, 13,374, and in 1830, 17,762.

SOMERSET, borough, p-v. and st. jus. Somerset county, Pa., is situated on a small branch of Cassellman's river, by p-r. 58 miles S. E. by E. Pittsburg, 143 miles a little S. of W. Harrisburg, and 165 miles northwestward W. C. Pop. 1820, 442, and in 1830, 649.

SOMERSET, county, Md., bounded S. by Pocomoke bay, S. W. Chesapeake bay, N. W. Fishing bay and Nanticoke river, the latter separating it from Dorchester county, N. Sussex county, Del., Worcester county, Md. E., and the mouth of Pocomoke river separating it from Accomac county, Va., S. E. Greatest length 40 miles, mean breadth 13 miles, and area 540 square miles. Lat. 38° 16', long. W. C. 1° 20' E. Slope southwestward. Beside the mouth of Pocomoke and Nanticoke rivers which form part of its boundaries, this county is comparatively deeply penetrated by Manokin, and Wicomico rivers. Chief town, Princess Ann. Pop. 1820, 19,579, and in 1830, 20,168.

SOMERSET, p-v. and st. jus. Pulaski county, Ky., by p-r. 85 miles S. S. E. Frankfort. It is situated about 5 miles N. Cumberland river, between Fighting and Pitman's creeks, N. lat. 37° 03', long. W. C. 7° 30' W. Pop. 1830, 231.

SOMERSET, p-v. and st. jus. Perry county, Ky., by p-r. 46 ms. E. Columbus, and 18 S. W. Zanesville, N. lat. 39° 52', long. W. C. 5° 20' W. Pop. 1830, 576.

SOMERSET, p-v. in the southwestern part of Franklin county, Indiana, by p-r. 15 miles N. W. by W. Brookville, the co. st., and 55 S. E. by E. Indianopolis.

SOMERSWORTH, p-t. Strafford county, N. H., 11 miles from Portsmouth, 45 Concord, has Salmon Falls river N. E. which separates it from Berwick, Me., Fresh creek and Cocheco rivers. Otis' hill commands a view of the White mountains, and Portsmouth. The soil bears oak, pine, walnut, &c. and is favorable to grain and grass. The river has water for vessels of 250 tons to within 1 mile of Quamphegan falls, where are several large factories. Iron ore and ochre are found. First settled 1750, and suffered in the French war. The village of Great Falls has been formed out of a wilderness within 8 years. It had then one house and a saw mill. It now contains five large factory mills, two large hotels, ten blocks (three stories high) of brick, and about one hundred frame dwelling houses, three churches, and eight or ten stores, and about two thousand inhabitants. There are four cotton and one woollen mills. The cotton mills contain thirty-one thousand spindles, with preparations sufficient to supply nine hundred looms, which produce six millions of yards of cotton cloth per annum. These mills consume annually, above 3,000 bales of cotton, weighing 1,250,000 lbs. The largest mill is 400 feet long and 6 stories high, and contains 22,000 spindles and 650 looms. The cotton mills alone give employment to 90 men, over 100 boys, and 600 females. They use from 7 to 8,000 gallons of oil, 200 tons of anthracite coal, 500 bbls. of flour for sizing, and 300 sides of leather. The mills, which are of brick, are arranged along a fine canal, 30 feet wide and from 6 to 7 feet deep, extending from the dam at the north of the village to the southern extremity of it.

The woollen mill is a fine 6 story brick building, 220 feet in length, containing machinery for the manufacture of from 120 to 130,000 yards of fine broadcloth yearly. This is said to be the largest woollen manufactory in America. The consumption of the raw material, and various articles of commerce, is immense. Upwards of 200,000 pounds of wool, 5,000 gallons of oil, 150 tons of anthracite coal, annually giving employment within the establishment to 300 individuals. Connected with the woollen, is a carpet manufactory, where the best description of ingrain carpeting is made. This factory is capable of producing 150,000 yards annually. This company, "The Great Falls Manufactory," have a capital one million of dollars, and own most of the property in and around the village. The churches are on rising ground south of the village, one each for Congregationalists, Methodists, and Baptists. The cotton mills give employ to 90 men, 100 boys, and 600 females. The capital of the company owning these establishments is one million of dollars. Pop. 1830, 3,090.

SOMERTON, p-v. northern part of Philadelphia county, Pa., 15 miles northward Philadelphia.

SOMERTON, or Somertown, p-v. near the southern side of Nansemond county, Va., following the road about 40 miles S. W. Norfolk, and by p-r. 120 miles S. E. Richmond.

SOMERTON, p-v. Belmont county, Ohio, by p-r. 139 miles E. Columbus.

SOMERVILLE, p-v. and st jus. Somerset co., N. J. in the town of Bridgewater, 38 miles N. Trenton, 1 N. Raritan river, 11 N. N. W. New Brunswick, contains a church, court house, and academy, and is a thriving and populous village.

SOMERVILLE, p-v. Fauquier county, Va., by p-r. 73 miles westward W. C.

SOMERVILLE, p-o. Orange county, N. C. by p-r. 66 miles N. W. Raleigh.

SOMERVILLE, p-v. and st. jus. Lafayette co., Ten., by p-r. 184 miles S. W. by W. Nashville. N. lat. 35° 12', long. W. C. 12° 25' W. It is situated on one of the head branches of Loosahatchie river.

SOUHEGAN, river, Hillsboro' county, N H., which, after receiving numerous tributaries, falls into the Merrimack river in Merrimack.

SOUTH AMBOY, p-t. Middlesex county, N. J., is of an irregular oblong shape, with the lower part of Raritan river and Raritan bay N. and N. E., South river N. W., and Monmouth co., S. W. It has the advantage of navigation, and is crossed by several streams of Raritan river. The New York and New Brunswick steamboats touch at the landing daily. The Amboy and Camden rail-road commences there, at the landing, and will render the town a great thoroughfare. Pop. 1830, 3,782.

SOUTHAMPTON, p-t. Rockingham county, N. H., 50 miles from Concord, 18 miles from Portsmouth, 45 from Boston, has Amesbury Mass. S., with a surface nearly level, good soil, and excellent mill seats on Powow river, which crosses it. Population 1830, 487

SOUTHAMPTON, p-t. Hampshire co., Mass., 110 miles W. Boston, 9 S. W. Northampton, is twice crossed by Manhan river, which affords good mill sites. There is a lead mine N. where a variety of interesting minerals are found. The E. part of this town is crossed by the Hampshire and Hampden canal. There are living 21 ministers of the gospel who are natives of this town. Pop. 1830, 1,244.

SOUTHAMPTON, p-t. Suffolk county, N. Y., on Long Island, 98 miles E. New York, lies N. of the Atlantic, with Peconet river and bay N. Sag Harbor is in the N. E. corner, the capital of Suffolk county. It is about 4 1-2 miles by 23. The soil is light, but improved by manure, for which fish are used. First settled about 1639, chiefly from Lynn, Mass., as a separate colony. In 1644 it came under the jurisdiction of Connecticut. Deer are found in the unsettled tracts of this extensive township. There are 4 principal settlements. Pop. 1830, 4,850.

SOUTHAMPTON, p-v. Somerset co. Pa.

SOUTHAMPTON, co. Va. bounded S. W. by Sussex, Surry N., Blackwater r. separating it from Isle of Wight co. E., and Nansemond co. S. E., Hertford and Northampton cos. N. C. S., and Meherin r. separating it from Greenville S. W. Length 40, mean breadth 15, and area 600 sq. ms. N. lat. 36° 40', and the meridian of W. C. intersect in this co. Slope sthestrd. and in that direction it is traversed by the Nottaway r. Chief t., Jerusalem. Pop. 1820, 14,170; and in 1830, 16,074.

SOUTH BAY, Brookhaven, Suffolk co. N. Y., on Long Island.

SOUTH BAY, Dresden, Washington co. N. Y., sets up from lake Champlain, between the S. part of the latter and lake George.

SOUTH BEND, p-v. Allen co. Ind., by p-r. about 200 ms. N. E. Indianopolis.

SOUTH BERWICK, p-t. York co. Me., 91 ms. S. W. Augusta, is of an irregular triangular form, bordered S. W. by Salmon Falls r., and is crossed by a small stream flowing into it. Pop. 1830, 1,577.

SOUTH BLOMFIELD, p-v. Pickaway co. O., by p-r. 17 ms. S. Columbus.

SOUTHBOROUGH, p-t. Worcester co. Mass. 30 ms. W. Boston, 15 E. Worcester, formerly a part of Marlboro', contains 8,350 acres of good soil, with a small stream of Concord r. Pop. 1830, 1,080.

SOUTHBRIDGE, p-t. Worcester co. Mass. 65 ms. S. W. Boston, N. Woodstock, Conn., has several manufactories on Quinebaug r., as the Southbr. woollen manufac., the Woolcott do. Pop. 1830, 1,444.

SOUTHBRIDGE, p-v. Southbridge, Worcester co. Mass., 61 ms. S. W. Boston, is a flourishing manufacturing place, deriving water power from Quinebaug r. There are 5 cotton and 3 wollen factories, and 2 more are nearly completed.

SOUTH BRUNSWICK, t. Middlesex co. N. J., has Somerset co. N. W., Sandhills N., with small streams of the Raritan, and is crossed in the N. by the South Amboy and Camden rail-road. Pop. 1830, 2,557.

SOUTHBURY, p-t. New Haven co. Ct., 40 ms. S. W. Hartford, on the N. side Housatonic r. and S. of Litchfield co. H. r. separates it from Fairfield co. The surface is gently varied, and the soil pretty good. Shad are caught in the Housatonic, and Shepaug and Pomperaug rs. its branches, afford mill sites. Pop. 1830, 1,557.

SOUTH CANAAN, p-v. Wayne co. Pa., by p-r. 248 ms. N. N. E. W. C.

SOUTH CAROLINA, state of the U. S., bounded by the Savannah r. separating it from Geo. S. W., it has N. C. on the N. W. N. and N. E., and the Atlantic ocean S. E. The ocean border reaches 185 ms. from Little Inlet on the N. E. to the mouth of Savannah r. S. W. Along the Savannah, Tugaloo, and Chatuga rs. in common with Geo. 270 ms.; and in common with N. C. 300 ms. Entire outline 755 ms. The longest line that can be drawn over S.C. is from Little r. inlet, to the watrn. angle of Pickens dist. 275 ms. The area of S. C., even by the author of this article, has been hitherto underrated. Measured carefully on the recent state map of that state, it comes out from the rhombs to so near 33,000 sq. ms. as to justify the adoption of that superficies. The mean width is 120 ms. The state extends in lat. from 32° 01′ to 35° 10′ N., and in long. from W. C. 1° 44′ to 6° 20′ W. To the S. W. of the Susquehannah r. and Chesapeake bay, the Atlantic slope of the U. S. is divided into three zones, which at their margins mingle their respective features, but at or near their individual central lines are very distinct in soil and natural vegetable production. These zones merely perceptible in Va. are bold and prominent in the Carolinas and Geo. The first next to the Atlantic, is that of sea sand alluvion, below the lower falls of the rs. about 60 ms. wide in S. C., and in most part penetrated by the tide. The second commences along or near the lower falls and primitive ledge. The sea-sand zone is very nearly a dead plain, but at its inner margin hills begin to appear, springs of water become plentiful, the soil meliorates, and the whole face of nature assumes an agreeable diversity of surface. The third, or what may be called the mountainous zone, though but little of it is really mountainous, comprises the nrthwstrn. part of the state, and lies based on the Blue Ridge chain. The Atlantic zone comprising the districts of Beaufort, Colleton, Charleston, Georgetown, Marlborough, Horry, and Marion, is near the ocean, cut by innumerable interlocking water courses; in considerable part it is marshy. The entrances are numerous and in no one, however, admit large vessels. Beside many of lesser note, this coast is accessible, advancing from S. W. to N. E. into the Savannah, by Port Royal entrance, St. Helena sound, South and North Edistos, Stono, Charleston, Santee r. by two mouths, and the Georgetown entrance, or estuary of Pedee and Waccamaw rs. The insular character of the coast and interlocking of the streams cease before reaching the middle zone. In a state of nature the sea-sand alluvial coast of South Carolina was covered with a dense forest, amongst which rose the gigantic palm or cabbage tree. Great part of the middle zone is composed of what is called "the sand hills." Here the arable land or at least the best and most extensive part of it skirts the streams; pine timber abounds. It contains the dists. of Barnwell, Orangeburg, Lexington, Sumner, Darlington, Marlborough, with part of Richland, Kershaw, and Chesterfield. The great primitive ledge, so remarkable in the states to the nrthestrd. crosses the middle zone of S. C., passing the Wateree near Camden, the Congaree at Columbia, and the Savannah near Hamburg, and Augusta. Above this ledge and the river falls, the face of the country changes to that hill and dale character, which so very finely distinguishes the whole zone of the U. S. to the sthestrd. of the Appalachian chains. The eye now every where meets the hills bold, swelling, and varied in form. The rivers wind their way amid smiling valleys, and by their rapid and rippling currents show the descent of the plain down which they flow. Here we discover in rapid succession the meadow, orchard, and field of small grain.

There is no straining to suit a theory in stating that S. C. has its temperate and torrid zone. The extremes of lat. exceed 3°, and 2 more may be added for difference of level, giving to the whole state extremes of 5° of Fahrenheit in temperature. Objects of agriculture, are controlled in quantity and position; cotton and rice are staples near the ocean; cotton admixed with small grain in the middle zone; and the latter and the apple in the mtns. region. Indian corn succeeds well over all sections of the state. In Beaufort, or the extreme S., sugar cane has been cultivated with success.

Taken under one sweep of view S. C. is a fine physical and political section, and a prosperous state. The indigenous vegetation combines the oaks and palms; the pines and hickorys; and in exotic plants, nearly every species cultivated in the U. S. The sea coast offer no deep harbor, of course excludes heavy ships of war; but it is open at numerous pours to an active coasting commerce. Similar to every section of the Atlantic slope S. W. of the Susquehannah and Chesapeake, the rs. of S.C. are more navigable at the centre of the state than near the sea or ocean coast. Under the article roads and canals, the reader will be able to see what has been effected in water and rail-road improvement in S. C. *History and progressive pop.*—The name of both Carolinas is derived from that of Charles IX. king of France, and imposed by a colony of Frenchmen who made an abortive attempt to form a settlement on the coast. As early as 1670, a century, however, after the reign of Charles IX, settlements of English began to be formed, and about 1680, a few settlers fixed themselves between Ashley and Cooper rs. and founded Charleston. In 1662, Charles II. granted the whole of what is now both Carolinas to Lord Clarendon and others, which with Locke's imperfect plan of government, retarded the settlement and distracted the country, until 1719, when the two Carolinas were definitively separated. Amid political contests a most salutary revolution was effected by the introduction of rice in 1695. Indigo, and cotton, were introduced subsequently, and laid the foundation of wealth and independence. Though her frontier felt the frequent and and severe wounds inflicted by savage war, her advance was steady to the revolutionary war. In that contest S. C. was an illustrious actor and sufferer. Perhaps no other section of the U. S. felt the evils of that struggle so long and bitterly. Many of her most distinguished sons fell martyrs to the cause, and to its consummation their survivors met the storm with unbending courage. A halo of glory was indeed thrown round the state by the actions of such men as Hayne, Marion, Lee and Sumpter. It is only this year (1832) that Sumpter went to rest with 97 years of honor pressing on his head. The character of Marion is that of history and romance commingled; he was a hero worthy of the richest pages of either.

In 1790, S. C. contained 240,073 inhabitants; in 1800 they amounted to 345,591; in 1810, 415,115; in 1820, 501,154, and in 1830, they had augmented to 581,185, or at the ratio of 242 per cent. in 40 years. The state is subdivided into the following districts, the population of which for 1820 and 1830 is annexed.

	Pop. 1820.	Pop. 1830
Anderson,	18,000	17,169
Abbeville,	23,189	28,149
Barnwell,	14,750	19,236
Beaufort,	32,199	37,032
Charleston,	80,212	86,338
Chester,	14,379	17,182
Chesterfield,	6,645	8,472
Colleton,	26,373	27,256
Darlington,	10,949	13,728
Edgefield,	21,309	30,509
Fairfield,	17,174	21,546
Georgetown,	17,603	19,943
Greenville,	14,530	16,476
Horry,	5,025	5,245
Kershaw,	12,142	13,545
Lancaster,	8,716	10,361
Laurens,	17,682	20,263
Lexington,	8,083	9,065
Marion,	10,201	11,008
Marlborough,	6,425	8,582
Newbury,	16,104	17,441
Orangeburg,	15,653	18,453
Pickens,	9,022	14,473
Richland,	12,321	14,772
Spartanburg,	16,989	21,150
Sumpter,	25,369	28,277
Union,	14,126	17,906
Williamsburg,	8,716	9,018
York,	14,936	17,790
Total,	501,154	581,185

Of the population of 1830, were white persons—

	Males.	Females.
Under 5 years of age	25,132	23,691
5 to 10	20,259	19,043
10 to 15	16,497	15,632
15 to 20	13,961	15,122
20 to 30	22,164	21,866
30 to 40	13,969	13,438
40 to 50	8,334	8,468
50 to 60	5,644	5,455
60 to 70	3,042	2,929
70 to 80	1,210	1,181
80 to 90	298	361
90 to 100	66	80
100 and upwards	14	17
Total	130,590	127,273

Of which were deaf and dumb under 14 years of age, 60; 14 to 25, 52; 25 and upwards 62. Blind 102. Of the colored population were—

	Free.		Slaves.	
	Males.	Females.	Males.	Fem.
Under 10 years of age	1,314	1,378	51,820	51,524
From 10 to 24	958	1,175	44,600	45,517
24 to 36	622	746	29,710	32,689
36 to 55	424	545	21,671	22,006
55 to 100	335	399	7,567	8,112
100 and upwards	19	6	98	84
Total	3,672	4,249	155,469	159,932

Free colored persons deaf and dumb under 14 years of age, 9; 14 to 25, 27; 25 and upwards 23. Blind, 136.

Recapitulation.

Whites.	Free col'd.	Slaves.	Total.
257,863	7,921	315,401	581,185

It may be noticed, that S. C. is the only state in the Union, in which the slave population exceeds in number the free.

Constitution.—Government.—Education.— The constitution of S. C. was adopted the 3d of June, 1790, and under the provisions of the 11th art., was amended the 17th Dec. 1808, and on the 19th Dec. 1816.

The legislative body, under the name of general assembly, is composed of two houses. The senators are chosen for four years; and to be eligible to a seat in the senate, demands the candidate to be a white man of 30 years of age, resident in the state 5 years previous to election; he may be elected whether resident in or out of the district for which he is elected; but if a resident, he must be possessed in the district of a settled freehold estate of £300 sterling, clear of debt; and if non-resident, he must with similar other requisites, possess an estate of £1,000 sterling.

Members of the house of representatives, must be a white man of 21 years of age, resident in the state 3 years immediately before the election; must, if resident in the district from which elected, possess a freehold of 500 acres of land, or 10 negroes, or a real estate of £150 sterling, clear of debt; or, if non-resident, all other requisites, and a clear freehold estate of £500 sterling, clear of debt.

The governor is chosen for 2 years, by

joint ballot of both houses of the legislature; and is ineligible for the next 4 years succeeding his term. He must be a citizen white man of 30 years of age, and a resident in the state 10 years next preceding his election. When elected, he must possess a settled estate within the state, in his own right, of £1,500 sterling, clear of debt. The lieutenant governor is chosen at the same time, for a like term of office, and must be rendered eligible by similar qualifications as the governor.

To exercise the right of suffrage, demands the person to be a white man of 21 years of age; paupers, and non-commissioned officers, and privates of the U. S. army excepted; must have resided in the state 2 years, immediately before the day of election, have a freehold of 50 acres of land, or a town lot, of which property he must be seised and possessed 6 months before the day of election; or, not having such freehold property, he must have been a resident of the election district at least 6 months immediately before the day of election, at which he gives his vote.

The constitution provides that no convention of the people shall be called, unless by the concurrence of two thirds of both branches of the whole representation; that no part of this constitution shall be altered, unless a bill to alter the same shall have been read three times in the house of representatives, and three times in the senate, and agreed to by two thirds of both branches of the whole representation; neither shall any alteration take place until the bill so agreed to, be published three months previous to a new election for members to the house of representatives; and if the alteration proposed by the legislature shall be agreed to in their first session, by two thirds of the whole representation in both branches of the legislature, after the same shall have been read three times, on three several days in each house; then, and not otherwise, the same shall become a part of the constitution. The interests of education have not been neglected in S. C. "The college of South Carolina," located at Columbia, was established in 1801, by the legislature of the state, and has been supported in great part by legislative bounty The edifices, libraries, philosophical apparatus, with some other contingencies, have subjected the state to an expenditure of $200,000 at least, and an annual appropriation of $15,000. The Charleston college, in Charleston, was established in 1785, and of consequence is more ancient than the state seminary. This institution (1832,) has 111 students, and a library of 3,000 vols.; and the Charleston seminary 61 students, and a library of 3,000 vols. Free schools have been established, and are supported at the expense of the state, or more correctly by the people. In 1828, by a report of the commissioners of free schools, there were then established 840 schools, in which 9,036 pupils were taught, at an annual expense of $39,716. In 1829, the appropriation for free schools was $37,200. By reference to the tables in this article, the reader may see that in 1830, there were in S. C. 100,614 white persons, from 5 to 20 yrs. of age inclusive; in the previous year, about 37 2-10 cents had been appropriated for their instruction, admit all to have claimed a share. In this unequal and inadequate provision for the greatest of all human interests, S. C. is far from being alone, and very far from deserving censure not applicable to a great majority of the states of the U. S.

South Charleston, p-v. in the sthestrn. part of Clarke co. O., by p-r. 40 ms. s. w. by w. Columbus.

South East, p-t. Putnam co. N. Y., 18 ms. e. West Point, has Connecticut e., Westchester co. s., with a hilly surface and pretty good soil for grain, &c., well watered, abounding in iron ore, and crossed by Croton r., which furnishes mill sites. Joe's hill, a mountainous ridge, extends from the centre into Connecticut. There are 5 ponds in this t. Pop. 1830, 2,036.

South Farms, p-v. and parish, Litchfield, Litchfield co. Conn., 36 ms. w. Hartford. Part of Great lake, or Litchfield pond, lies in the parish, which gives rise to Bantam r., a branch of the Housatonic.

Southfield, parish of Sandisfield, Berkshire co. Mass., has the boundary of Conn. s.

Southfield, t. Richmond co. N. Y., 9 ms. s. New York. on Staten island, has the narrows e., Raritan bay s., with a coast on those sides of about 10 ms. It is level s., with good soil. Clams are caught at the Great Kills, and shad and other fish at the narrows. At Old Town was formerly a defensive work, erected for protection against Indians; Richmond v. w. contains the co. buildings, and sloops come up the Fresh Kills to within ½ m. of the v. On the e. side, opposite Long isl. are erected forts Richmond, Tompkins, and Hudson. On the heights, near the latter forts, are also the telegraphs erected to communicate with New York and vessels in the offing. Pop. 1830, 971.

South Florence, p-v. on the left bank of Tennessee r. Franklin co. Ala., nearly opposite Florence in Lauderdale co., and by p-r. 145 ms. n. Tuscaloosa.

South Hadley, p-t. Hampshire co. Mass., 90 ms. w. Boston, on the e. bank of Conn. r., where is a fall of 40 ft. in about 80 yards. There are several manufactories at the falls, and it is a place of considerable business. Here also great quantities of shad are caught, the falls generally causing those fish to stop here in great numbers in the spring. There is a canal round the falls here, two miles long, 300 feet distance of which is cut to the depth of 40 feet through solid rock. It was the first constructed on this river. Pop. 1830, 1,185.

South Hanover, p-v. Jefferson co. Ind. by p-r. 90 ms. s. e. Indianopolis.

South Hero, p-t. Grand Isle co. Vt., 12 ms. n. w. Burlington, 16 s. w. St. Albans, has lake Champlain on all sides except the N., and contains 9,065 acres. First settled 1784. A sand bar extending to Chittenden, renders the lake in that place fordable a part of the year. The rocks are limestone, as well as those of the neighboring isls. The soil is very good, and the surface nearly level. Marl, with shells, is found even on the higher parts; and the lime stone is sometimes burnt, and sometimes used for building. These islands were formerly a favorite resort of Indians; and they made implements of quartz, &c. on the shore, bringing the stone from a distance. Pop. 1830, 717.

SOUTHINGTON, p-t. Hartford co. Conn. 18 ms. S. W. Hartford, 21 N. New Haven, lies in the S. W. corner of the co., with N. Haven co. S. and W., about 6 ms. sq., with an uneven surface, agreeably varied, except the Greenstone range in the E. and some considerable eminences W. The soil is various, but generally good for rye, maize, &c. It is crossed by 2 branches of Quinipiack r., which supply mill seats. The Farmington canal crosses the t. N. and S., and affords a convenient channel of transportation. Pop. 1830, 1,844.

SOUTHINGTON, p-v. Trumbull co. O. by p-r. 166 ms. N. E. Columbus.

SOUTH KILLINGLY, v. Killingly, Windham co. Conn., 44 ms. E. by N. Hartford, and 9 N. Plainfield; is a flourishing manufacturing place, at the confluence of Five Mile and Quinaboug rs. Here, at the falls at the mouth of Five Mile r., the Danielson manufacturing company have a mill of 1,840 spindles, 44 looms, employ 65 hands, consume 90,000 lbs. of cotton, and make 350,000 yards of 4-4 sheetings.

Cundall and Woodruff have a small factory of broadcloths, connected with their dressing and fulling mill, and make 4,500 yards of broadcloths. On the Quineboug, about 100 yards below, Comfort Tiffany has a cotton mill of 1,000 spindles, and 24 looms, and makes 150,000 yards of 7-8 shirtings.

At Chesnut Hill, five ms. N. E. of South Killingly, on Whitestone brook, Ebenezer Young owns a stone mill, in which he runs 2,100 spindles, and 36 looms, and consumes 100,000 lbs. of cotton.

SOUTH KINGSTON, p-t. and st. jus. Washington county, R. I., 30 miles S. W. Providence, has the Atlantic and Narraganset bay E., and the Atlantic S. It has an uneven surface, with primitive rocks, and a soil generally rich and strong, favorable to grass and grain. There are several small streams, many fresh water ponds, (one of 3,000 or 4,000 acres,) and one salt water pond, called Point Judith pond. Northeast on Narraganset bay is a good harbor; and considerable coasting trade is carried on, as well as much fresh and salt water fishing. The first settlement was made about 1670, when the town was connected with N. Kingston. Pop. 1830, 3,663.

SOUTHOLD, p-t. Suffolk county, N. Y., 103 miles E. New York, embraces the N. E. corner of Long Island, which extends in a long neck bounding the sound on the S. E. On the S. W. side of the point are several bays, channels, and islands, and Plumb island lies off the extremity of the cape, in a line with the Gull islands, the Race and Fisher's isl. These islands belong to the town with Ram and Robin's island. The soil is various, with few stones and much sand. The coast is generally a sand bank. Fish are taken in great numbers. There are several villages, Mattatuc, Cutchogue, Southold town, &c. The two Gull islands are small, 3 miles E. by N. of Plumb island. Great Gull contains 14 acres, and Little Gull 1, chiefly rocks, on which are erected a light house of the United States, with a house for the keeper. These, with walls for protection against the sea, were built at the expense of $24,000, of stone brought from the Connecticut shore. The dashing of the waves in an E. storm shakes the very foundation of this fabric. The light being in the entrance of Long Island sound, is a very important one. Pop. 1830, 2,900.

SOUTH LANDING, p-v. Cabell county, Va., and by p-r. 349 miles westward W. C.

SOUTH PLYMOUTH, p-v. northwestern part of Wayne county, Mich., by p-r. 22 miles N. W. Detroit.

SOUTHPORT, town, Tioga county, N. Y., 5 miles S. W. Elmira, is in the S. W. corner of the county, with Pennsylvania S. and Steuben co. W. Chemung river which flows through it, has a large quantity of excellent meadow land on its banks. Pop. 1830, 1,454.

SOUTH QUAY, p-v. Nansemond county, Va., 95 miles S. S. E. Richmond.

SOUTH READING, p-t. Middlesex county, Mass., 10 miles N. E. Boston, contains a pleasant village and a pond near it. Pop. 1830, 1,311.

SOUTH RIVER, river, Middlesex county, N. J., formed by the Manalapan and Matchepo-nix which rise in Monmouth county, and unite at Spotswood. It enters the Raritan 5 miles below New Brunswick, and is navigable for sloops 6 miles.

SOUTH SALEM, p-t. Westchester county, N. Y., 50 miles N. New York, 6. N. Bedford, and has Conn. river on the E. It is of irregular form, with several ponds, and is bordered W. by Croton river. Population 1830, 1,537.

SOUTH UNION, p-v. Jasper county, Georgia, by p-r. 24 miles N. W. Milledgeville.

SOUTH UNION, p-v. Logan county, Ky., by p-r. 157 miles S. W. by W. Frankfort.

SOUTH WARREN, p-v. Bradford county, Pa., by p-r. 270 miles northward W. C.

SOUTH WHITE HALL, p-v. Lehigh county, Pa., by p-r. 179 miles N. N. E. W. C.

SOUTHWICK, p-t. Hampden county, Mass., 110 miles S. by W. Boston, and N. of Connecticut line, and contains several large ponds, that serve as a part of the route of the Hampshire and Hampden canal, which passes through the town. It is crossed by a considerable stream. Pop. 1830, 1,355.

SPAFFORD, p-t. Onondaga county, N. Y., 18 miles s. s. w. Onondaga, 14 N. Homer, lies N. of Cortlandt county, and E. Cayuga county or Skaneateles lake. The surface is varied, with a rich soil, bearing maple, beech, bass, &c. Slopes rapidly E. to Otisco r. the valley of which lies partly in this town. The inlet of Otisco lake forms part of the E. line, and there are several smaller streams. First settled about 1806 from the E. states. Pop. 1830, 2,647.

SPANISH GROVE, p-v. Mecklenburg county, Va., by p-r. 116 miles s. w. Richmond.

SPARTA, p-t. Livingston county, N. Y., 25 miles s. w. Canandaigua, 13 s. Geneseo, has Steuben county E., Steuben and Alleghany counties s., and Alleghany county w., is watered by a branch of Hemlock lake and Canaseraga creek; has generally a poor soil. Pop. 1830, 3,777.

SPARTA, p-v. Washington county, Pa., by p-r. 10 miles westward the borough of Washington, and 229 miles N. w. by w. W. C.

SPARTA, p-v. Caroline county, Va., by p-r. 89 miles a little w. of s. W. C.

SPARTA, p-v. Edgecombe county, N. C., by p-r. 81 miles N. E. by E. Raleigh.

SPARTA, p-v. and st. jus. Conecuh county, Alabama, on Murder creek, another branch of Conecuh river, about 85 miles N. E. Mobile, and by p-r. 205 miles a little E. of s. Tuscaloosa. Lat. 31° 20', long. W. C. 10° 10' w.

SPARTA, p-v. and st. jus. White county, Ten., by p-r. 92 miles s. E. by E. Nashville.

SPARTANBURG, district, S. C., bounded N. E. by Broad river, separating it from York district, Union E. and s. E., Ennoree river separating it from Laurens s., Greenville s. w. and w., and Rutherford county, N. C., N. Length from s. to N. 40 miles, mean breadth 28, and area 1,120 square miles. Lat. 35° and long. 5° w. intersect near the centre of this district. Slope s. E. and in that direction advancing from s. to N. it is drained by the rivers Ennoree, Tyger, Hair Forest, and Pacolet. Chief town, Spartanburg. Pop. 1820, 16,989, 1830, 21,150.

SPARTANBURG, p-v. and st. jus. Spartanburg district, S. C., by p-r. 104 miles N. w. Columbia. Lat. 34° 56', long. W. C. 5° w.

SPECKLED MOUNTAIN, Oxford county, Me., N. of Androscoggin river, on the N. line of Riley, is one of the highest eminences in the state, and supposed to be about 4,000 feet above the sea.

SPEEDWELL, p-v. Barnwell district, S. C., by p-r. 111 miles southwestward Columbia.

SPEEDWELL, p-v. western part of Claiborne county, Ten., by p-r. 238 miles a little N. of E. Nashville.

SPEIGHT'S BRIDGE, and p-o. Greene county, N. C., by p-r. 74 ms. s. E. by E. Raleigh.

SPENCER, p-t. Worcester county, Mass., 51 miles w. Boston, 11 s. w. Worcester, has a pleasant variety of surface, and a good soil. It is watered by branches of Chicopee river which furnish mill sites. The land is elevated 880 feet above Connecticut r. at Springfield, and 950 above Boston harbor. On a route surveyed for a canal between those two points, this was the summit level. Pop. 1830, 1,618.

SPENCER, p-t. and st. jus. Tioga county, N. Y., 190 ms. w. by s. Albany, and 18 N. w. Owego, has Cayuta on the w. line, and a pond N. which gives rise to Catetant creek, with Cayuga inlet. The surface and soil are various. Pop. 1830, 1,278.

SPENCER, p-v. Davidson county, N. C., by p-r. 92 ms. westward Raleigh.

SPENCER, co., Ky., bounded s. by Nelson, Bullitt w., Jefferson N. w., Shelby N., and Anderson E. Length 22 ms., mean breadth 12, and area 264 square ms. Lat. 38°, and long. W. C. 8° 14' w., intersect in this co. Slope westward, and in that direction drained by Salt river. Chief town, Taylorsville. Pop. 1830, 6,812.

SPENCER, county, Indiana, bounded w. by Little Pigeon river separating it from Warrick, Dubois N., Anderson's creek E., separating it from Perry, and the Ohio river separating it from Hancock county, Ky., s. E., Daviess county, Ky., s., and Henderson co., Ky., s. w. Length from s. to N. 32 miles, mean width 14, and area 448 square ms. Lat. 38° and long. 10° w. intersect in the southeastern part of this county, general slope is southward towards Ohio river. Chief town, Rockport, no unapt name, as the whole county is hilly, and in part rocky. Pop. 1820, 1,882, 1830, 3,196.

SPENCER, p-v. and st. jus. Owen county, Indiana, by p-r. 52 ms. s. w. Indianopolis. It is situated on Kaskaskias river. Lat. 39° 17', long. W. C. 9° 48' w.

SPENCERTOWN, p-v. Austerlitz, Columbia county, New York, 30 miles s. w. Albany.

SPERMACETI COVE, Monmouth county, New Jersey, a safe and convenient harbor, for vessels of light draught of water, at the s. w. part of Sandy Hook, E. of the Highlands.

SPESUTIA, p-v. Harford county, Md., by p-r. 65 ms. N. E. W. C.

SPINNERSTOWN, p-v. Bucks county, Pa., by p-r. 171 ms. N. E. W. C.

SPLIT ROCK, p-v. Essex, Essex county, N. Y., 2 ms. s. Essex village, on the w. side of lake Champlain.

SPOON ISLAND, Hancock county, Maine.

SPOTSWOOD, p-v. Middlesex county, New-Jersey, 9 ms. s. E. New Brunswick, 10 w. by s. Middletown Point, on the N. side of South r. a branch of the Raritan. Snuff and powder are manufactured here.

SPOTTEDVILLE, p-v. Stafford county, Va., by p-r. 88 ms. s. s. w. W. C.

SPOTTSYLVANIA, county, Va., bounded by Caroline s. E., North Anna river separating it from Hanover s., and Louisa s. w., Orange N. w., Rapid Ann river separating it from Culpepper N., and Rappahannoc river separating it from Stafford N. E. Length from s. w. to N. E. 24 miles, mean breadth 17 miles, and area 408 square ms. Lat. 38° 12', long. W. C. 0° 40' w. The extreme sources of Matapony river rise mostly in this county, and flow to the southeastward as do the two bounding streams North Anna and Rappahannoc rivers; the slope of the county is of course in the direction of its waters. Surface though hilly, is pleasantly and in many places finely diversified. Chief town, Fredericksburg. Pop. 1820, 14,254, 1830, 15,134.

SPREAD EAGLE, p-v. northwestern part of Del. co. Pa. by p-r. 136 ms. N. E. W. C.

SPRING BANK, p-v. Wayne co. N. C. by p-r. 51 ms. S. E. Raleigh.

SPRINGBORO', p-v. Warren co. O. by p-r. 88 ms. N. W. by W. Columbus.

SPRING COTTAGE, p-v. near the southwstrn. angle of Hancock co. Miss. about 120 ms. S. E. by E. Natchez.

SPRING CREEK, p-v. Warren co. Pa. by p-r. 335 ms. N. W. W. C.

SPRING CREEK, p-o. Greenbrier co. Va. by p-r. 254 ms. S. W. by W. W. C.

SPRING CREEK, p-o. Madison co. Ten. by p-r. 132 ms. S. W. by W. Nashville.

SPRING DALE, p-v. Alleghany co. Pa. by p-r. 235 ms. N. W. W. C.

SPRING DALE, p-v. Hamilton co. O. by p-r. 111 ms. S. W. by W. Columbus.

SPRING FARM and p-o. Augusta co. Va. by p-r. 151 ms. S. W. by W. W. C.

SPRINGFIELD, p-t. Sullivan co. N. H. 35 ms. from Concord, 90 from Boston, lies in the N. E. corner of the co. with Grafton co. N., and Merrimack co. E., and contains 28,330 acres. It gives rise to a branch of Sugar r., and one of the streams of Blackwater r., thus dividing the waters of Connecticut and Merrimack rs. It contains several small ponds, and has a rough surface, but a pretty good soil. There is a quarry of valuable stone E. First settled, 1772. Pop. 1830, 1,192.

SPRINGFIELD, p-t. Windsor co. Vt. 13 ms. S. Windsor, 68 from Montpelier, 30 N. Brattleboro', W. Conn. r., and is crossed S. E. by Black r. There are fine meadows on Conn. r.; a village at the falls of Black r., and another in the N. W. part. Pop. 1830, 1,498.

SPRINGFIELD, p-t. and st. jus. Hampden co. Mass., lies on the E. side of Conn. r. 87 ms. W. Boston, 26 N. Hartford, 47 W. S. W. Worcester, and 20 S. Northampton. It is one of the most thriving towns in the state, containing a court-house, jail, bank, 2 insurance offices, besides other public buildings; among these are 4 churches, 1 each for Congregationalists, Unitarians, Baptists and Methodists. It also contains the largest armory of the U. S., and many elegant private edifices, highly creditable to the taste of its inhabitants. In this town is Chickapee, an important manufacturing village, which contains about 1,300 inhabitants, a Congregational and Methodist church, post office, and about 100 houses of brick, belonging to the manufacturers. In the town in 1830, there were 6,784 inhabitants, and in 1831 there were 1,453 polls, 722 dwelling houses, 118 stores, warehouses and mechanic shops, 580 barns and other buildings, 3 cotton factories, 370 looms, and 13,824 spindles, 1 bleachery, 3 paper mills, 5 printing offices, 5 grist mills, 7 saw mills, 2 card factories, 1 carding machine, 1 fulling mill, 2 breweries, 2 distilleries, 3 tan-houses, 5,301 acres of tillage land, 1,807 acres of mowing land, 389 horses, 321 oxen, 474 cows, 237 steers and heifers, and 954 sheep. A bridge, 1,234 ft. long, crosses the Conn. to W. Springfield. Stage coaches run daily to Boston, Albany, Northampton and Hartford; and small steamboats also to Hartford, carrying passengers, &c. &c. The armory of the U. S. was established in 1795, at an expense, with additions from time to time since, of $251,857. From that time up to 1821, there were expended in work and materials $2,553,352. The annual expense is $180,000. The present production is 16,500 muskets a year, and there are on hand, prepared for distribution, more than one hundred thousand stand. The number of workmen is nearly 300. According to the statement of the superintendent, about two-thirds of the amount appropriated to this establishment is paid for labor, and one-third for stock and materials. Of the latter the following comprise the most important items, viz:—165 tons of iron, $23,100; 49,500 lbs. of steel, $7,820; 16,500 files, $3,300; 140,000 bushels charcoal, $9,100; 10,000 bushels pit coal, $3,500; 100 tons Lehigh coal, $1,000. Total, $47,820. There are employed in the estimate 275 men, whose pay amounts for the year to $120,000, leaving for stock and materials $60,000. Total $180,000. The number of arms manufactured per year, with all appendages, is 16,500. Amount of permanent improvements, miscellaneous expenses, &c. say $12,000; leaving for the manufacture of arms, gun boxes, screwdrivers, wipers, ball screws, spring vices, and all the appendages, say $168,000.

The greater part of the buildings belonging to the armory are situated on the hill half a mile E. of the river; the water-shops connected with the arsenal lying one mile S. on Mill r. Chickapee v., which has been mentioned, is situated on a river, from which it derives its name, and which affords abundant water power for its manufactures. It is about 4 ms. N. of the village of Springfield, and contains 4 large cotton factories and a bleaching establishment. Three of the manufactories employ 600 persons, using 900,000 lbs. of cotton annually, and making 3,300, 00 yds. of printing cottons, fine sheetings and shirtings. The other factory is 254 ft. long, and 4 stories high. There are also iron works in the village.

SPRINGFIELD, p-t. Otsego co. N. Y. 58 ms. W. Albany, 12 N. Cooperstown, in the N. W. corner of the co., has Otsego lake and co. W., and is of a varied surface, with a soil generally rich. There are several small mill streams. The town was settled and deserted before the revolution, and has since been settled by English, Scotch and Irish. Pop. 1830, 2,816.

SPRINGFIELD, t. Burlington co. N. J. 18 ms. s. Trenton and 18 from Burlington; has Assiscunk creek N., and is crossed by a small branch. The soil is good and very well cultivated. Pop. 1830, 1,534.

SPRINGFIELD, p-t. Essex co. N. J. 7 ms. w. Newark, 15 w. N. Y., 6 N. w. Elizabethtown; has a pleasant village on Rahway r. which flows through it and affords several mill sites. The Short hills cross the N. part S. E. and N. W. The surface is varied, the soil good, and the inhabitants farmers. A large British foraging party was resisted and stopped at the river during the war; but on another occasion the village was occupied by the enemy one night, and burnt after the battle of Springfield, which was severely contested, and was continued, as the Americans retreated, E. of the village to the Short hills, where the British received a check. This place the enemy were unable to retain possession of, and they retreated the next day. Population 1830, 1,656.

SPRINGFIELD, p-v. Bradford co. Pa. by p-r. 255 ms. N. W. C.

SPRINGFIELD, p-v. Hampshire co. Va. by p-r. 118 ms. N. W. by W. W. C.

SPRINGFIELD, p-v. Greene co. Ala. by p-r. 61 ms. sthrd. Tuscaloosa.

SPRINGFIELD, p-v. southeastern part of St. Helena parish, La., by p-r. 11 ms. S. E. St. Helena, the st. jus. of the parish, and via Madisonville about 80 ms. N. W. New Orleans.

SPRINGFIELD, p-v. and st. jus. Robertson co. Ten. by p-r. 25 ms. a little W. of N. Nashville. Lat. 36° 30', long. 9° 54' W. W. C.

SPRINGFIELD, p-v. and st. jus. Washington co. Ky. by p-r. 50 ms. S. W. Frankfort. Lat. 37° 42', long. 8° 16' W. W. C. Pop. 1830, 618.

SPRINGFIELD, p-v. and st. jus. Clarke co. O. 43 ms. almost due W. Columbus, and 25 N. E. by E. Dayton. Lat. 39° 51', long. 6° 48' W. W. C. It is situated on a small branch of Mad r., contains the ordinary co. buildings, with several manufactures. Pop. 1830, 1,080.

SPRINGFIELD, p-v. Franklin co. Ind. by p-r. 77 ms. S. E. by E. Indianopolis.

SPRINGFIELD, p-v. and st. jus. Sangamon co. Il. by p-r. 79 ms. N. W. Vandalia, and as laid down by Tanner, about 5 ms. wstrd. of the junction of Sangamon proper with the Mowawequa. Lat. 39° 48', long. 12° 40' W. W. C. It is one of the new towns rising as if by miracle from the wilds of the west.

SPRINGFIELD Cross Roads and p-o. Erie co. Pa. by p-r. 330 ms. N. W. W. C.

SPRINGFIELD FURNACE, and p-o. Huntingdon co. Pa., by p-r. 150 ms. N. N. W. W. C.

SPRING FOUR CORNERS, p v. southern part Susquehannah co. Pa., about 6 miles sthrd. Montrose, the co. seat. In directing letters it ought to be observed that though in the same township, Springville and Springville Four Corners are different offices.

SPRING GARDEN, p-v. Pittsylvania co. Va., by p-r. 250 ms. S. W. W. C.

SPRING GARDEN, p-v. Rockingham co. N. C., by p-r. 124 ms. N. W. by W. Raleigh.

SPRING GROVE, p-o. Lancaster co. Pa., by p-r. nrthestrd. W. C.

SPRING GROVE, p-o. Iredell co. N. C., by p-r. 137 ms. wstrd. Raleigh.

SPRING GROVE, p-o. Laurens district, S. C., by p-r. 69 ms. N. W. Columbia.

SPRING GROVE, p-o. Alachua co. Florida, by p-r. 193 ms. S. E. Tallahassee.

SPRING HILL, p-o. Fayette co. Pa., by p-r. 221 ms. N. W. W. C.

SPRING HILL, p-o. Lewis co. N. C., by p-r. 67 ms. S. E. by E. Raleigh.

SPRING HILL, p-o. York district, S. C., by p-r. 97 ms. N. Columbia.

SPRING HILL, p-v. Monroe co. Geo., by p-r. 7 ms. W. Milledgeville.

SPRING HILL, p-v. Maury co. Ten., 30 ms. S. S. W. Nashville.

SPRING HILL, p-o. Decatur co. Ten., 53 ms. S. E. Indianopolis.

SPRING HOUSE, p-o. Montgomery co. Pa., 19 ms. N. Philadelphia.

SPRING MILL, p-v. Lawrence co. Ind., by p-r. 82 ms. S. S. W. Indianopolis. This is the same place formerly called Arcole. (*See the latter art. first column, page 88.*)

SPRING MILLS, p-v. Centre co. Pa., by p-r. 187 ms N. N. W. W. C.

SPRING MOUNT, p-o. eastern part Dyer co. Ten. by p-r. 16 ms. W. Nashville, and 8 ms. E. Dyersburg, the co. seat.

SPRING PLACE, p-o. Cherokee Nation, by p-r. 212 ms. N. W. Milledgeville, and 623 ms. S. W. by W. W. C.

SPRINGPORT, town, Cayuga co. N. Y., 10 ms. S. W. Auburn; has Cayuga lake and Seneca co. W.; contains the v. of Union Springs. Pop. 1830, 1,528.

SPRING ROCK, p-o. York district, S. C., by p-r. 80 ms. N. Columbia.

SPRING'S MILLS, p-o. Lincoln co. N. C., by p-r. 172 ms. westward Raleigh.

SPRINGTOWN, village, Morris co. N. J., 18 ms. W. Morristown, on the acclivity of Schooley's mountain.

SPRINGTOWN, p-v. near the northwestern border of Bucks co. Pa., 7 ms. S. E. Bethlehem, and 43 ms. N. Philadelphia.

SPRINGVILLE, p-v. Susquehannah co. Pa., by p-r. 261 ms. N. N. E. W. C.

SPRINGVILLE, p-v. Darlington dist. S. C., by p-r. 91 ms. eastward Columbia.

SPRINGVILLE, p-v. Lawrence co. Ind., by p-r. 62 ms. S. S. W. Indianopolis.

SPRINGWATER, p-t. Livingston co. N. Y., 18 ms. S. E. Geneseo; has Ontario co. N. and E., and Steuben co. W.; a pretty good soil, tho' a rough surface, and is watered by several small streams. Pop. 1830, 2,253.

SPRING WELLS, p-v. Wayne co. Mich., by p-r. 10 ms. northward Detroit.

SQUAM, lake, Grafton and Coos counties, N. H., 3 ms. by 6; is a beautiful sheet of water, almost surrounded by high hills, and diversified with coves, capes and islands, and affording fine trout. It extends over about 6,000 acres. A canal from this to Winnipiseogee lake 2 ms. distant, has been proposed.

SQUAM, r. Grafton co. N. H., the outlet of Squam lake, crosses part of Holderness, and falls into Pemigewasset r.

SQUAM, village, Gloucester, Essex county, Mass., on the N. shore of Cape Ann; has an excellent and convenient harbor, and a population devoted to fishing and commerce.

SQUAMANAGONICK, v. Rochester, Strafford co. N. H., at the falls of Cocheco r.

SQUAM, bay, Essex co. Mass., on the north shore of Cape Ann, opposite Gloucester harbor, with which it is connected by a short canal, which crosses the isthmus of that cape or peninsula. This bay is also called Squam harbor.

SQUAM BEACH, Morris co. N. J., on the sea coast, S. of Manasquam r.

SQUAMSCOT, or Exeter r., N. H.

STAFFORD, p-t. Tolland co. Conn., 26 ms. N. E. Hartford, and 74 W. S. W. Boston; is an elevated tract, with Mass. line N. The surface is rough, with some pleasant vallies, and much wild scenery. The rocks are primitive, and the soil generally favorable to grass. Iron ore, (chiefly bog ore,) is found in many places, and there are several mines and forges. Willimantic r. and Roaring brook afford mill seats; and there are several manufactories. Straw braiding is carried on to some extent by females. The v. is on a pleasant elevated plain, with a large open square in the centre, affording an extensive view over a varied country. The springs are situated in a narrow valley, 1 mile W. of the v., on the bank of the Willimantic, and in the midst of picturesque scenery, about 100 yards S. of the turnpike road from Boston to Hartford. Near at hand is a large house for the accommodation of visitors. One of the springs is a feeble chalybeate, and the other is impregnated with sulphuretted hydrogen. First settled about 1718. The Indians were acquainted with the valuable properties of the springs, which they made known to the settlers. Pop. 1830, 2,515.

STAFFORD, p-t. Genesee co. N. Y., 6 ms. E. Batavia, has a slightly varied surface, with pretty good soil, watered by Black cr. Pop. 1830, 2,308.

STAFFORD, t. Monmouth co. N. J., forms the S. angle of the co., and is nearly in the form of a triangle. The W. part comprehends a great part of Little Egg Harbor, with several isls. and much of Long Beach, which shuts it in from the ocean. There is also a considerable extent of swamps on the borders of the harbor, through which Manahocking r. and other small streams discharge. Pop. 1830, 2,059.

STAFFORD, co. Va. bounded by King George S. E., Rappahannoc r. separating it from Caroline S., Spottsylvania co. S. W., and Culpepper W., on the N. W. it has Fauquier, King William N., and the Potomac r. separating it from Charles co. Md. E. Lat. 38° 25′, long. W.C. 0° 22′ W. Length 20 ms., mean breadth 12, and area 240 sq. ms., surface hilly. Chief town, Falmouth. Pop. 1820, 9,517; 1830, 9,362.

STAFFORD, C. H. p-o. and st. jus. Stafford co. Va., by p-r. 76 ms. a little E. of N. Richmond, and 46 S. W. W. C.

STAFFORD, springs and p-o. wstrn. part Stafford co. Va.

STAGVILLE, p-v. Orange co. N. C.

STAHLER'S, p-o. Lehigh co. Pa., 10 ms. nrthrd. Allentown, the co. st.

STALLING'S, store and p-o. Monroe co. Geo., 49 ms. W. Milledgeville.

STAMFORD, t. Bennington co. Vt., 9 ms. S. E. Bennington, 21 S. Brattleborough, N. Mass., is uneven, and has much waste land. Some of the head streams of Hoosac r. rise in the S., and in the N. part are Moose, Fish, and other ponds, from which waters run into Walloomsac r. Pop. 1830, 563.

STAMFORD, p-t. Fairfield co. Conn., 76 ms. S.W. Hartford, 42 S. S. W. New Haven, 43 N.E. New York, has N. Y. N. W. and Long Island sound S. W., is crossed by two or three mill streams, and is penetrated by a bay from the sound. The surface is agreeably varied, the soil is fertile and favorable to cultivation. At the mouth of Mill r. is a harbor, with 8½ feet of water at common tides, and a place of some coasting trade. There are two smaller harbors in the town, and here are two large flour mills. Pop. 1830, 3,712.

STAMFORD, p-t. Delaware co. N. Y., 12 ms. E. Delhi, 50 W. Catskill, 60 W. S. W. Albany, has the head stream of Delaware r. N., and Schoharie co. E. The surface is broken, the mill sites good, and it contains two or three small villages. Pop. 1830, 1,597.

STANDING PEACH TREE, and p-o. Dekalb co. Geo., by p-r. 127 ms. N. W. Milledgeville.

STANDING STONE, p-o. Bradford co. Pa., on the Susqnehannah r., 6 ms. N. Towanda, the co. st. and by p-r. 245 ms. N. W. C.

STANDISH, p-t. Cumberland co. Me., 6 ms. S. W. Augusta, 21 N. W. Portland, has Saco r. and York co. S. W., and Sebago pond N. E., and contains several ponds, connected by a stream flowing into it. Pop. 1830, 2,023.

STANFORD, p-t. Dutchess co. N. Y., 18 ms. N. E. Poughkeepsie, has a surface a little varied, with pretty good soil, and is well watered by a branch of Wappinger's cr. Pop. 1830, 2,521.

STANFORD, p-v. and st. jus. Lincoln co. Ky. situated on a small branch of Dick's r. by p-r. 51 ms. a little E. of S. Frankfort. Lat. 37° 32′, long. W. C. 7° 32′ W. Pop. 1830, 363.

STANFORD'S Cross Roads, and p-o. Putnam co. Geo. 21 ms. N. Milledgeville.

STANHOPE, p-v. Sussex co. N. J., on the Muskonetcong cr. and Morris canal, 16 ms. N. W. Morristown and 12 S. Newton, is the seat of extensive iron works.

STANHOPE, p-v. nrthrn. part Northampton co. Pa., by p-r. 32 ms. nthrd. Easton, the co. st., and 222 ms. N. N. E. W. C.

STANNARDSVILLE, p-v. in the extreme western part Orange co. Va., by p-r. 114 ms. S. W.

by w. W. C., and 92 ms. N. w. by w. Richmond.

STANTONSBURGH, p-v. sthrn. part Edgecomb co. N. C., by p-r. 66 ms. a little s. of E. Raleigh.

STANTONVILLE, p-v. in the nrthrn. part of Anderson dist. S. C., by p-r. 113 ms. N. w. Columbia.

STAR, p-v. Hocking co. O., by p-r. 57 ms. s. E. Columbus.

STARK, co. O., bounded by Jefferson s. E., Tuscarawas s., Holmes s. w., Wayne w., Medina N. w., Portage N., and Columbiana E. Length along eastern border 33 ms., mean breadth 25, and area 825 sq. ms. Lat. 40° 30', long. W. C. 4° 26' w. The extreme source of Big Beaver rises in the nrthest. angle of this co., but the far greater part of the surface is drained by the Tuscarawas and its branches; general slope sthrd. The main volume of Tuscarawas crosses the co. from N. to s. nearly, along the valley of which the Ohio and Erie canal has been constructed. The level of the canal at Massillon, near the middle of the co. is 942 feet above tide water; the whole arable surface no doubt exceeding a mean of 1,000 feet of similar comparative height. The soil is fertile in grain, pasturage and fruit. Chief town, Canton. Pop. 1820, 14,506; 1830, 26,588.

STARKEY, p-t. Yates co. N. Y., 10 ms. s. E. Penn Yan. Pop. 1830, 2,285.

STARKS, p-t. Somerset co. Me., 7 ms. w. Norridgewock, and 37 N. N. w. Augusta, forms nearly a complete square, the s. w. corner of which almost touches the co. of Kennebec, and the N. E. boundary is formed by Kennebec r. Pop. 1830, 1,471.

STARKSBOROUGH, p-t. Addison co. Vt., 22 ms. s. w. Montpelier, and 20 s. E. Burlington, first settled 1788, from Conn. and N. Y. Lewis cr. and Huntington r. are the principal streams, and mill seats are abundant, especially on the former where are several factories. Hogback mtn. lies on the west boundary, and East mtn. crosses the middle. The soil is loam, and the timber chiefly hard wood. Pop. 1830, 1,342.

STARUCCA, p-v. Wayne co. Pa., by p-r. 19 ms. nrthrd. Bethany, the co. st., and 284 ms. N. N. E. W. C.

STATE-LINE, p-v. sthrn. part Franklin co. Pa., by p-r. 64 ms. N. w. W. C.

STATEN ISLAND, N. Y., forms the county of Richmond, 9 ms. s. w. N. Y., is 14 ms. long, and 8 wide, bounded by New York bay N., the Narrows E., which separate it from Long Island south, by Raritan bay w., and by the Kills lying opposite New Jersey on the 2 last mentioned sides. (*See Richmond co., and Richmond.*)

STATESBURGH, p-v. nrthwstrn. part of Sumpter dist. S. C., 10 ms. N. w. by w. Sumpterville, the st. jus., and by p-r. 32 ms. a little s. of E. Columbia.

STATESVILLE, p-v. and st. jus. Iredell co. N. C., by p-r. 40 ms. s. s. E. Wilkesville, and 146 ms. w. Raleigh. Lat. 35° 13', long. W. C. 3° 54' w.

STATESVILLE, p-v. estrn. part Wilson co. Ten., by p-r. 48 ms. estrd. Nashville.

STATION CAMP, p-v. Estill co. Ky., by p-r. 6 ms. sthrd. Irvine, the st. jus., and by p-r. 75 ms. s. E. by E. Frankfort.

STAUNTON, p-v. New Castle co. Del., situated at the junction of Red Clay and White Clay crs., 6 ms. s. w. by w. Wilmington, and by p-r. 51 ms. N. E. W. C.

STAUNTON, p-v. and st. jus. Augusta co. Va., on one of the extreme head branches of the E. fork of Shenandoah r. 36 ms. a little N. of w. of Charlotteville, and by p-r. 163 ms. s. w. by w. W. C., and 121 ms. N. w. by w. Richmond. Lat. 38° 09', long W. C. 2° 03' w. It stands on the fine valley between the Blue Ridge and Kittatinny, or as there expressed between Blue Ridge and North mtn. chains, a little north Madison's Cave. It contains 3 or 4 places of public worship, numerous stores, taverns and mechanics' shops. It is a corporate town, and contains houses for the chancery, circuit and corporation courts. The population is not given in the census tables, but is probably about 1,000. It is on the whole one of the most flourishing interior towns of Va.

STEELE CREEK, p-o. Mecklenburg co. N. C., 8 ms. sthrd. Charlotte, the co. st., and by p-r. 158 ms. s. w. by w. Raleigh.

STEEL'S Mills and p-o. sthrn. part Richmond co. N. C., 12 ms. sthrd. Rockingham, the co. st., and by p-r. 125 s. w. Raleigh.

STEELE'S Mills and p-o. sthrn. part Randolph co. Il., by p-r. 111 ms. s. w. Vandalia.

STEELE'S Tavern and p-o. Augusta co. Va., by p-r., 180 ms. s. w. by w. W C.

STEEN'S cr. and p-o. Rankin co. Miss., by p-r. 126 ms. N. E. Natchez.

STEPHENSPORT, p-v. on Ohio r., just below the mouth of Sinking cr. and in the extreme nrthrn. angle of Breckenridge co. Ky., 16 ms. N. N. w. Hardinsburg, the co. st., and by p-r. 118 ms. a little s. of w. Frankfort. Pop. 1830, 64.

STEPHENTOWN, p-t. Rensselaer co. N. Y., 20 ms. s. E. Albany, has the boundary of Mass. on the E. line, and Columbia co. s., with a very hilly surface E., and various soils. There is a broad valley in the middle, in which rises a branch of Lebanon cr. w.; the soil is poor, and occupied by the green woods. Limestone abounds and is wrought. Pop. 1830, 2,716.

STERLING, t. Franklin co. Vt., 24 ms. N. E. Burlington, and 24 N. w. Montpelier, first settled 1799, has no large streams. Sterling peak, s. E., is one of the highest eminences of the Green mtns., and the surface is generally very rough. Pop. 1830, 183.

STERLING, p-t. Worcester co. Mass., 46 ms. w. Boston and 12 N. Worcester, was chiefly purchased of the Indians in 1701, and is crossed by Still r. a branch of Nashua r. It has a surface generally hilly, with a pine plain s. w. Chairs and hats have been made here in great numbers for some years. A battle was fought here in 1707, between some troops from Lancaster and Marlborough, and a party of Indians, in which the latter were defeated. A variety of minerals are found in the rocks. First settled, 1720. Pop. 1830, 1,794.

STERLING, p-t. Windham co. Conn., 44 ms. E. Hartford, has the boundary of R. I. E., an uneven surface, with some pine plains, and a light soil, best appropriate to grain. Quanduck r. is a small stream, but there are several manufactories in the town. Near the centre is the Devil's Den, a remarkable cavern in a ledge of rocks. Pop. 1830, 1,240.

STERLING, p-t. Cayuga co. N. Y., 28 ms. N. Auburn, has lake Ontario N., and Oswego co. E., with Little Sodus bay N. W., into which flows Nine-mile creek. Pop. 1830, 1,436.

STERLING, p-v. sthrn. part Wayne co. Pa., by p-r. 237 ms. N. N. E. W. C.

STERRETT's Gap, and p-o. nrthrn. part Cumberland co. Pa., by p-r. 25 ms., but by direct distance 18 ms. almost due W. Harrisburg, and 8 ms. a little E. of N. Carlisle. This is one of those remarkable depressions in the Appalachian chains called "Gaps." The particular gap here described is highly worthy a visit from the traveller. From it the whole of Cumberland co. seems to spread an immense map, and in a clear day the cupola of the state house in Harrisburg is distinctly to be seen. On the contrary side, or to the N. and W. the congeries of mtn. chains seem to extend in endless variety. It is amongst the finest positions in the U. S. in regard to perspective, having 2 immense landscapes connected in the eye of the spectator.

STEUBEN, p-t. Washington co. Me., 35 ms. W. Machias, has Narragaugus r. on the N.E. line, and Hancock W., with the Atlantic S., from which 3 long bays extend far N. into this t. Dyer's bay in the middle, and Goldsboro' harbor W. Pop. 1830, 695.

STEUBEN, co. N. Y., bounded by Livingston and Ontario cos. N., Seneca lake E. which separates it from Seneca and Tompkins cos., Tioga co Pa. S., and Alleghany co. W. Several streams of the Tioga or Chemung cr. (which is a branch of the Susquehannah) spread over this co. and afford a navigation for boats. Conhocton, Canisteo and Tioga are the principal of these, and afford boat navigation to Bath, Hornellsville and Tyrone. The route of the Chemung canal extends 18 miles from Elmira, on the Chemung r. to the head waters of Seneca lake, and a navigable feeder is brought from the Chemung at Painted Post, 13 ms. to the summit level. The land in this co. is very uneven, and often mountainous, with some large and fertile meadows. The banks of the streams are generally steep and covered with evergreen forests. It is a remarkable fact that at some of the fords the ice forms in winter on the bottom of swift streams, in such a degree as to render the the passage difficult and even dangerous. This co. contains 24 towns. Pop. in 1820, 21,989; 1830, 33,851.

STEUBEN, p-t. Oneida co. N. Y., 20 ms. N. Utica, and 110 W. Albany, was nearly all granted to Frederick William, Baron de Steuben, an officer of the revolutionary army, and named after him, and was for some years his residence. The surface is varied, the land high, and the soil moist and favorable for grass. Steuben and Cincinnati crs. are the principal streams, but there are few mill seats. Baron Steuben died here in 1796. Pop. 1830, 2,094.

STEUBEN, p-v. estrn. part Huron co. O., by p-r. 100 ms. N. N. E. Columbus.

STEUBENVILLE, p-v. corporate town, and st. jus. Jefferson co. O., by p-r. 149 ms. a little S. of N. E. by E. Columbus, 39 ms. by land and 70 by water from Pittsburg, 260 ms. S. W. by W. W. C. Lat. 40° 21′, long. W. C. 3° 45′ W.

The site of this fine town has something peculiar amongst those along the Ohio river. Generally, the first rise from that stream has a depression backwards towards the hills of considerable depth; at Steubenville the acclivity from the river ascends with very little depression. Along this slope the town was laid out in 1798, in streets running at right angles. The opposite side of the river rises into abrupt and even precipitous banks, of from 400 to 460 feet elevation. Though laboring under the disadvantage of not lying in the great western thoroughfare, Steubenville has become a flourishing place with an enlightened society. According to Flint, it contained early in 1832, two printing offices, an academy, market house, woollen factory, cotton factory, steam paper and flour mill; 27 mercantile stores, air foundry and other mechanical establishments. It contains three churches and several private schools. Pop. 1820, 2,539; 1830, 2,937.

STEVEN's, r. Caledonia co. Vt., a mill stream which falls into Conn. r. at Barnet.

STEVENSBURG, p-v. sthrn. part Culpepper co. Va., 30 ms. nrthwstrd. Fredericksburg, and by p-r. 83 ms. S. W. W. C.

STEVENSBURG, p-v. Hardin co. Ky. by p-r. 90 ms. wstrd. Frankfort.

STEVENSBURG, p-v. Hamilton co. Ind., by p-r. 31 ms. nrthrd. Indianopolis.

STEVENSVILLE, p-v. King and Queen co. Va., by p-r. 30 ms. a little N. of E. Richmond, and 130 ms. S. W. C.

STEWART, co. Ten., bounded by Montgomery N. E., Dickson S. E., Humphries S., Ten. r. separating it from Henry S. W. and W., and from Calloway co. Ky. N. W., on the N. it has Trigg co. Ky. Length from S. to N. 28 ms., mean breadth 20, and area 560 sq. ms. Lat. 36° 25′, and long. W. C. 11° W. intersect near its centre. Cumberland r. enters the extreme estrn. angle, and flowing to the N. W. traverses the co. in nearly its greatest length. Tennessee r., where forming part of its boundary, also flows to the N. W.; the general slope is of course in the direction of its two large rs. Chief town, Dover. Pop. 1830, 6,968.

STEWART's Mills, and p-o. Guilford co. N. C., by p-r. 99 ms. N. W. by W. Raleigh.

Stewartstown, p-t. Coos co. N. H., 150 ms. from Concord, 170 from Portsmouth, and 150 from Portland, with 27,000 acres; has Conn. r. w., here about 80 yards wide. Little and Great Diamond ponds discharge into a branch of Androscoggin r., while several brooks flow into Conn. r. There are some hills, the soil is rich on the meadows, and pretty good on the uplands. Pop. 1830, 529.

Stewartsville, p-v. near the western border of Westmoreland co. Pa., 13 ms. n. w. by w. Greensburg, 19 s. e. Pittsburg, and by p-r. 204 ms. n. w. by w. W. C.

Stewartsville, p-v. Richmond co. N. C., by p-r. 112 ms. s. w. Raleigh.

Stillwater, p-t. Saratoga co. N. Y. 22 ms. n. Albany, and 10 s. e. Ballston Spa, on the w. side of Hudson r., has Round lake and its outlet on the s. line, is generally almost level, and traversed by Champlain canal parallel to the r., Bemis' Heights, where the battle of Saratoga was fought in 1777. Anthony's kill, the outlet of Round lake, is led off to the Hudson by a short canal, which affords mill seats. At Mechanicville, in the s. e. corner, are a manufactory and several mills. The scenery in the n. is very pleasant, beautifully swelling in fine ridges, the highest of which, and those nearest the Hudson, are Bemis' Heights. Pop. 1830, 2,601.

Stillwater, p-t. Sussex co. N. J., 78 ms. n. Trenton, has Warren co. s. w., the Blue Hills w., and is crossed in the e. by Pawling's kill, into which flows Swartwout's pond, which lies in this t. Pop. 1830, 1,381.

Stillwater, p-v. Stillwater, Sussex co. N. J., 76 ms. n. Trenton, and 6 s. w. Newton, on Pawling's kill.

Still Water, r. or creek of O., is the s. w. branch of Great Miami, rising in Randolph co. Ind., and Dark co. O., and flowing estrd. over the latter, enters Miami co. O.; inflects to s. s. e. over Miami and Montgomery, and falls into Great Miami a short distance above Dayton, after a comparative course of 50 ms.

Still Water, p-v. on the preceding cr., and in the nrthwstrn. angle of Miami co. O., about 15 ms. n. w. Troy, the co. st., and by p-r. 86 ms. a little n. of w. Columbus.

Stillwell, p-v. Perry co. Ten., by p-r. 112 ms. s. w. by w. Nashville.

Stockbridge, p-t. Windsor co. Vt., 26 ms. n. w. Windsor, and 36 s. w. Montpelier; first settled 1784, has White r. n. w., which here flows in a very narrow channel, and there are but few mill seats. Pop. 1830, 1,333.

Stockbridge, p-t. Berkshire co. Mass. 130 ms. w. Boston, 5 s. Lenox, is divided by Housatonic r., on the banks of which are fine meadows; and the scenery is various and beautiful, while the soil is generally very productive. Marble and limestone for burning, are quarried here; and there are several manufactories on the Housatonic, which affords much water power. The v. is beautifully situated, on the n. side of the r. A tract of land 6 ms. square was formerly reserved by the state for a tribe of Indians, amongst whom a mission was established in 1734. They afterwards removed to New Stockbridge, Oneida co. N. Y., where land was given them by the Oneida Indians; and some of the few survivors have since emigrated to Green Bay and other places. In the last French war, the settlement here was twice attacked by Indians. Pop. 1830, 1,580.

Stockbridge Indians, originally a tribe of the Moheekanuk, or Indians of the race generally spread over New England, and residing at Stockbridge, Mass., afterwards removed to New Stockbridge, Oneida co., and since have nearly all emigrated to Green Bay. They have had missionaries among them many years, and have been in a considerable degree civilized.

Stockertown, p-v. Northampton co. Pa., 7 ms. above Easton.

Stockholm, p-t. St. Lawrence co. N. Y., 30 ms. e. Ogdensburg, is watered by several branches of St. Regis r., has a rich soil, and was settled in 1803, from Mass. Pop. 1830, 1,944.

Stockholm, p-v. Jefferson, Morris co. N. J., 83 ms. n. by e. Trenton, on Pequannock r., at the foot of the Wallkill mtns., and on the borders of Bergen co.

Stockport, p-v. situated on the right bank of Del. r., about 3 ms. below the junction of the Coquago and Popachton branches, 20 ms. n. Bethany, the co. st., and by p-r. 291 ms. n. n. e. W. C.

Stockton, p-t. Chatauque co. N. Y., 6 ms. e. Mayville, has Cosdaga lake e., and part of the outlet, and an undulated surface, with rich vallies. Pop. 1830, 1,605.

Stock Township, p-v. Harrison co. O., by p-r. 134 ms. n. e. by e. Columbus. In the direction of letters, care must be taken to write this name as at the head of this article; such is the title in the p-o. list.

Stoddard, p-t. Cheshire co. N. H., 14 ms. from Keene, 42 s. s. w. Concord, 20 from Charlestown, and w. Hillsboro' co., with 35,925 acres, is elevated, rocky, and mountainous, with a deep and cold clayey soil, unfavorable to Indian corn, but good for grazing. Streams rise here, flowing into Conn. and Merrimack rs. Here are 14 ponds, some of them large. First settled 1769. Pop. 1830, 1,159.

Stoddartsville, p-v. on a branch of Lehigh r., on the extreme sthestrn. border of Luzerne co. Pa., on the direct road from Easton on Del. r., to Wilkes-Barre on Susquehannah r., 32 ms. n. w. the former, and 20 s. e. by e. the latter borough. This v. is situated in a region comparatively alpine, being elevated 1,384 feet above the mean level of the Atlantic tides, at lat. 41° 8', long. 1° 14' e. W. C. In regard to relative climate, the elevation of Stoddartsville is equivalent to at least 3½ degrees of lat., placing it above 44° 30' when compared with places on the sea coast.

Stokeley, p-v. sthwstrn. part Rutherford co. Ten. by p-r. 45 ms. s. e. Nashville.

Stokes, co. N. C. bounded by Rockingham

N. E., Guilford S. E., Davidson S., Yadkin r., separating it from Rowan S. W., and from the southern part of Surry W., by the northern part of Surry N. W., and by Patrick co. Va. N. Length from S. to N. 38 ms., mean width 22, and area 836 sq. ms. Extending in lat. from 36° 02′ to 36° 33′, and in long. from 3° 2′ to 3° 32′ W. W. C. Stokes co. comprises part of a table land, from which the streams are discharged like radii from the centre of a circle. Bounded on the S. W. by Yadkin, that river receives in that direction a number of creeks from the southeastern angle; the extreme western sources of Haw r. flow to the southestrd. Dan r. rising in Patrick co. Va. sweeps an elliptic curve over the northern part of Stokes and Rockingham, and again into Va. in Pittsylvania co. Surface rather hilly, but soil fertile in grain, pasturage and fruit. Chief town, Germantown. Pop. 1820, 14,033, 1830, 16,196.

STONE CHURCH and p-o. Northampton co. Pa. by p-r. 205 ms. N. E. W. C.

STONE FORT, p-v. on one of the extreme higher branches of Duck r. in the northern part of Franklin co. Ten. 10 ms. N. Winchester, the co. seat, and by p-r. 65 ms. S. E. Nashville.

STONEHAM, p-t. Middlesex co. Mass. 10 ms. N. Boston, is uneven, rocky, with some good soil. Pop. 1830, 732.

STONERSTOWN, p-v. on Raystown branch of Juniata r. northeastern part of Bedford co. Pa. by p-r. 124 ms. N. W. W. C.

STONESVILLE, p-v. Greenville dist. S. C. by p-r. 128 ms. N. W. Columbia.

STONES RIVER of Ten. falls into the left side of Cumberland r. a short distance above Nashville. Rutherford co. is nearly commensurate with the valley of this stream.

STONEY CREEK or SHRYOCK, p-v. on Stoney creek, the North fork of Shenandoah r., and in Shenandoah co. Va. about 8 ms. S. S. W. Woodstock, the co. st., 35 ms. in a similar direction from Winchester, and 105 a little S. of W. W. C.

STONEY CREEK, p-o. on a creek of that name, a branch of Shenandoah r., S. W. angle of Shenandoah co. Va. by p-r. 105 ms. S. W. by W. W. C.

STONEY CREEK, a considerable northern branch of Nottaway r. rising in Dinwiddie co. and falling into the Nottaway Sussex.

STONEY CREEK, p-o. on a creek of the same name in the northwestern part of Orange co. N. C. by p-r. 81 ms. N. W. by W. Raleigh.

STONEY CREEK, p-o. northwestern part of Oakland co. Mich. by p-r. 44 ms. N. W. Detroit.

STONE WALL MILLS and p-o. southwestern part Buckingham co. Va. by p-r. 108 ms. a little S. of W. Richmond.

STONEY BATTERY, p-v. sthestrn. part Newberry dist. S. C. 10 ms. S. E. Newberry C. H., and by p-r. 36 ms. N. W. by W. Columbia.

STONEY FORK, p-o. sthrn. part Montgomery co. Va. 25 ms. sthrd. Christiansburg, by p-r. 307 ms. S. W. by W. W. C., and 229 a little S. of W. Richmond.

STONEY POINT, Haverstraw, Orange co. N. Y., is a high and rocky peninsula, stretching into Hudson r., the channel of which it commands, near the head of Haverstraw bay, on which a fort was erected in the revolutionary war, which was taken by storm from the British in 1779, in the night, by Gen. Wayne, at the head of American troops. There is now a light-house on its summit, on the site of the old fort.

STONEY POINT, p-v. northestrn. part Albemarle co. Va. 71 ms. N. W. by W. Richmond.

STONEY POINT, p-v. wstrn. part Iredell co. N. C. 14 ms. westerly Statesville, the co. st., and by p-r. 160 ms. in a similar direction from Raleigh.

STONEY POINT, p-v. Abbeville dist. S. C. by p-r. 100 ms. westerly Columbia.

STONEY POINT, Mills and p-o. sthwstrn. part Cumberland co. Va. by p-r. 61 ms. S. W. by W. Richmond.

STONINGTON, p-t. and borough, New London co. Conn. 12 ms. E. New London, 55 S. E. Hartford, and 62 E. New Haven, has Pawtucket r. E., the boundary of Rhode Island, Fisher's island sound, and Pawtucket bay S. and Mystic r. W. The surface is uneven, the soil, though rough and stony, favorable to grazing, &c. It has a harbor, whence a considerable amount of business is carried on, particularly sealing in the Pacific ocean, and fishing on the coast. Ten sealing vessels, wholly or partly owned here, brought in, in 1831, skins worth $100,000. First settled, 1658, from Rehoboth, Mass. On the 9th Aug. 1814, the borough was bravely defended by the inhabitants against an attack from a British squadron of one 74, a frigate, an 18 gun ship, and a bomb vessel. Pop. 1830, 3,397.

STOREY'S Mills and p-o. wstrn. part Jackson co. Geo. by p-r. 114 ms. N. N. W. Milledgeville.

STOUGHSTOWN, p-v. Cumberland co. Pa. 13 ms. S. W. by W. Carlisle, and 31 in a similar direction from Harrisburg.

STOUGHTON, p-t. Norfolk co. Mass. 17 ms. S. Boston, gives rise to the head waters of Neponset r., and was formerly the residence of some Christian Indians, who removed from Dorchester. There is some manufacturing carried on here. Pop. 1830, 1,591.

STOW, p-t. Washington co. Vt. 15 ms. N. W. Montpelier. Pop. 1830, 1,570.

STOW, p-t. Middlesex co. Mass. 30 ms. N. W. Boston, has a surface but little elevated, with sandy plains, and very ordinary soil on the uplands. Pop. 1830, 1,220.

STOW, p-v. Portage co. O. by p-r. 120 ms. N. E. Columbus.

STOW CREEK, t. Cumberland co. N. J. 55 ms. S. S. W. Trenton, is a small town with Stow creek N. W., which divides it from Salem co. and Newport creek, its tributary, S. There is a swampy tract along the streams. Pop. 1830, 791.

STOWESVILLE, p-v. Lincoln co. N. C. by p-r. 165 ms. a little S. of W. Raleigh.

STOYSTOWN, p-v. on the great watrn. road from Phil. to Pittsburg, 28 ms. westerly Bedford, 11 N. E. the borough of Somerset, and by p-r. 155 ms. N. W. by W. W. C., and 133 W. Harrisburg.

STRABANA, p-v. estrn. part Lenoir co. N. C. by p-r. 92 ms. S. E. by E. Raleigh.

STRAFFORD co. N. H. bounded by Coos co. N., the state of Maine E., Rockingham co. S. and S. W., and Pemigewasset r. W. which separates it from Grafton and Hillsboro' cos., and Grafton co. N. W., 33 ms. by 63 greatest dimensions, with 1,345½ sq. ms. The mtns. are Chocorna, Sandwich, Osipee, Effingham, Gunstock, Moose, &c. Red hill, between Winnipiseogee and Squam lakes, commands one of the finest views in the country. There are several smaller lakes, and Merrymeeting, Long and Great bays connected with Winnipiseogee lake. The principal rs. are Piscataqua, Salmon Falls, Saco, Cocheco and Swift rs. The soil, as well as the surface, is very various, but generally good, and productive when well cultivated. There are many manufactories of cotton and woollen. Dover is a considerable manufacturing town. There are several incorporated academies in different towns. The first settlement was made in the co. in 1623 at Dover, and the second at Portsmouth, in the same year. The co. was formed in 1771. Pop. 1820, 51,117, 1830, 58,910.

STRAFFORD, p-t. Strafford co. N. H. 25 ms. from Concord, 15 from Dover, and 56 from Boston, has Bow pond S. which flows into Isinglass r. and several other small ponds. The Blue hills cross the N. W. part. The soil is generally good. Pop. 1830, 2,201.

STRAFFORD, p-t. Orange co. Vt. 30 ms. S. E. Montpelier, 30 N. Windsor, stands on a branch of Ompompanoosuc r., affords abundance of disintegrated pyrites, from which large quantities of copperas are manufactured. Pop. 1830, 1,935.

STRASBURG, p-v. Lancaster co. Pa. on a branch of Pecquea creek, 8 ms. S. E. by E. the city of Lancaster, and 55 ms. W. Phil.

STRASBURG, p-v. nthrn. part Shenandoah co. Va. on the road from Woodstock to Winchester, 15 ms. N. E. the former, 22 S. W. the latter place, and by p-r. 89 W. W. C.

STRASBURG, p-v. sthrn. part Fairfield co. O. by p-r. 37 ms. S. S. E. Columbus.

STRATFORD, r. or Housatonic r. rises in Berkshire co. Mass., crosses Conn. and empties into Long Island sound between Milford and Stratford.

STRATFORD, p-t. Coos co. N. H. 133 ms. N. Concord, has a broad and valuable tract of meadows on the E. bank of Conn. r. The E. and N. parts are mountainous, with cold and rocky or gravelly soil. The peaks in this town are conspicuous from a distance. Nash's stream, Bog brook, &c. water the town. Pop. 1830, 443.

STRATFORD, p-t. Fairfield co. Conn. 13 ms. S. W. New Haven, has Housatonic r. E., and Long Island sound S., with a level surface, few stones, and a very rich alluvial tract of meadows on the river and harbor. Shad and shell fish are taken in abundance, and there is coasting trade. The borough of Bridgeport is on the W. side of a small arm of the sound, forming a harbor. A draw bridge crosses it, through which sloops can pass. A natural canal connects this harbor with Housatonic r. which might be made more useful to navigation. Pop. 1830, 1,814.

STRATFORD, t. Montgomery co. N. Y. 15 ms. N. W. Johnstown, occupies the N. W. corner of the co. having Hamilton co. N., and Herkimer co. W. It has much marshy ground and many hills, with a poor soil, bearing evergreens. Pop. 1830, 552.

STRATHAM, p-t. Rockingham co. N. H. 39 ms. from Concord, 3 from Exeter, 51 from Boston, on the E. side of the W. branch of Piscataqua r., N. of Piscataqua bay, and is about 8 ms. from the ocean. The soil is good. There is a large peat swamp E. Pop. 1830, 939.

STRATHER'S Mills and p-o. nthrn. part Fayette co. Geo. by p-r. 119 ms. N. W. by W. Milledgeville.

STRATTON, t. Windham co. Vt. 18 ms. N. E. Bennington, 22 N. W. Brattleboro'; was settled from Mass. It gives rise to Bald mtn., a branch of West r. E., which furnishes mill sites, and to Deerfield r. W. Holman's and Jones's ponds, each of about 100 acres, discharge in different directions. Pop. 1830, 312.

STRATTONSVILLE, p-v. Armstrong co. Pa. by p-r. 249 ms. N. W. W. C.

STRAWNTOWN, p-v. on Tohiccon creek, nthwstrn. part Bucks co. Pa. 40 ms. a little W. of N. Phil. and 20 S. Easton.

STREETSBORO', p-v. Portage co. O. by p-r. 134 ms. N. E. Columbus.

STRICKERSVILLE, p-v. Chester co. Pa. by p-r. 99 ms. N. E. W. C.

STRONG, t. Somerset co. Me. 24 ms. N. W. Norridgewock, has Kennebec co. on the S. line, is crossed by Sandy r., and has a pond on the E. boundary, which flows into Seven-Mile brook. This t. is quite hilly on the S. Pop. 1830, 985.

STRONGSVILLE, p-v. Cuyahoga co. O. by p-r. 123 ms. N. E. Columbus.

STROUDSBURG, p-v. in the forks of Broadhead's creek, and in the northeastrn. part Northampton co. Pa. 3 ms. W. Del. Water gap, and 23 ms. N. Easton. It is built on one long street, and is the third village in size in the co.

STRYKERSVILLE. (*See Strickersville, Chester co. Pa.*)

STUMPSTOWN, p-v. nthrn. part Lebanon co. Pa. 8 ms. N. Lebanon borough, and by p-r. 29 ms. N. E. by E. Harrisburg.

STURBRIDGE, p-t. Worcester co. Mass. 70 ms. S. W. Boston, 22 S. W. Worcester, in the S. W. corner of the co., having Hampden co. W., and Conn. S., is crossed by Quinebaug r. and other streams, and has several manufactories, with a large supply of water power.

The surface is rough, the soil rocky and hard to cultivate. Pop. 1830, 1,625.

STURGEONVILLE, p-v. nthestrn. part Brunswick co. Va. by p-r. 60 ms. s. s. w. Richmond.

STUYVESANT, t. Columbia co. N. Y. 12 ms. N. Hudson, has Hudson r. w.; contains Kinderhook landing, and is crossed s. E. by Kinderhook creek, which forms the boundary in the s. w. Pop. 1830, 2,331.

SUBLETT'S Tavern and p-o. eastern part Powhatan co. Va. 23 ms. s. w. Richmond.

SUCCESS, t. Coos co. N. H. 143 ms. from Concord, w. Maine; contains 2 or 3 ponds, and several mountains, and gives rise to Narmarcungawack and Live rs. Pop. 1830, 14.

SUCKASUNNY, p-v. Morris co. N. J. 63 ms. N. by E. Trenton, 10 N. w. Morristown, on Suckasunny plain, has some large iron mines in the vicinity, particularly Dickerson's, which yields excellent ore in great quantities. The opening of the Morris canal offers great advantages for transportation.

SUDBURY, p-t. Rutland co. Vt., 47 miles s. Burlington, 65 N. Bennington, and 43 s. w. Montpelier; was settled from Connecticut. Otter creek touches it E. There are several ponds and small streams; the surface is uneven, with a rich soil. It is crossed by a ridge of high land. In the w. is a small v. Pop. 1830, 812.

SUDBURY, p-t. Middlesex co. Mass., 20 ms. w. Boston, has Concord r. on the E. boundary, and is crossed by one of its branches. First settled 1635. In 1676 a party of 70 men under captain Wadsworth, were ambushed here by 500 Indians, who killed 26 of them, and took most of the others. A monument of this event is 1 mile s. of the church. Pop. 1830, 1,423.

SUDLER'S CROSS ROADS, and p-o. northern part Queen Ann co. Md., on the road from Centreville to Elkton, 45 ms. a little s. of E. Baltimore, and by p-r. 47 N. E. by E. Annapolis.

SUFFIELD, p-t. Hartford co. Conn., 17 ms. N. Hartford, and 10 s. Springfield; has the boundary of Mass. on the N. line, and Conn. river E.; about 5 ms. by 8; has a variety of soil and surface, and a beautiful village, the principal street of which is long, broad and strait, running N. and s. on the ridge of a fine hill, which slopes gradually E. towards the r., 2 ms. distant, and w. commands a view over a diversified country. In the N. w. are part of the Greenstone mountainous range, and part of two Southwick ponds. There is a spring, called Suffield pool, near the s. line, impregnated with sulphuretted hydrogen gas, where a house of entertainment has been erected. There are several manufactories in this t. Pop. 1830, 2,690.

SUFFOLK, co. Mass., bounded by Middlesex co. N. and w., Massachusetts and Boston bays E., and Boston bay and a small part of Norfolk co. s. It is the smallest county in the state, but the most important, embracing Boston and Chelsea. It was incorporated in 1643. (*See Boston and Chelsea, Mass.*) Pop. 1820, 43,940, 1830, 62,163.

SUFFOLK, co. N. Y., comprises about 2-3ds of Long Island, and is bounded by Long Island sound N., the Atlantic E. and s., and w. by Queens co.; is 83 ms. by 20½, greatest dimensions, contains about 798 sq. ms., and includes several islands, the most remote of which is Fisher's isl. on the Conn. coast. It contains 9 townships, and Great, South, and and Drowned Meadow bays, and several smaller ones on the south side. The points, coves, &c., are numerous. The principal islands are Long, Gardiner's, Shelter, Plumb, Great-Hog-Neck, Robins', &c. The surface is broken N., and more level in the middle and s., where are extensive, and almost barren plains, and much salt marsh, with abundance of pine, which is sent in great quantities to New York. Salt is made by evaporation on the Atlantic shore, and there is a considerable number of coasting vessels employed. At Sag Harbor is a port of entry, a considerable village, and foreign trade and whaling. A light house was erected on Montauk point, the E. extremity of Long Island, in 1796. There are light houses also on Eaton's neck, Old Field point, and Little Gull isl. The first settlement was made is 1640 at Southold. Most of the first inhabitants came from New England. Pop. 1820, 24,756, 1830, 26,780.

SUFFOLK, p-v. and st. jus. Nansemond co. Va., on the right bank of Nansemond r., 28 ms. N. w. by w. Norfolk, and by p-r. 102 miles s. E. by E. Richmond, and 224 a little E. of s. W. C. Lat. 36° 43', long. 0° 27' E. W. C.

SUGAR CREEK, p-v. Crawford co. Pa., 12 ms. s. E. by E. Meadville, and by p-r. 291 ms. N. w. W. C.

SUGAR CREEK, stream of the state of Illinois and of Huron territory, the main nrthrn. confluent of Peektano, branch of Rock river. Sugar creek, or more correctly river, rises in Huron near the southern side of Ouisconsin river, flows in 2 branches by a general sthrn. course about 45 ms. to their junction, 2 or 3 ms. above the northern boundary of Il., bending thence s. E. enters Il., and unites with the Peektano, after an entire comparative course of 60 ms. It drains the space between Gooskehawn and Peektano rs.

SUGAR CREEK, p-v. on a creek of the same name, Hancock co. Ind., 15 ms. s. Indianopolis.

SUGAR CREEK, p-v. northern part Sangamon co. Il., 23 ms. N. of Springfield, the co. seat, and by p-r. 65 ms. N. N. w. Vandalia.

SUGAR GROVE, p-v. northern part Warren co. Pa., 14 ms. N. w. Warren, the co. seat, and by p-r. 327 ms. N. w. W. C.

SUGAR GROVE, p-v. Putnam co. Ohio, by p-r. 148 ms. N. w. Columbus.

SUGAR LAKE, and p-o. Crawford co. Pa., by p-r. 307 ms. N. w. W. C.

SUGAR LOAF, p-v. northern part Columbia co. Pa., 91 ms. N. Harrisburg.

SUGAR TREE, p-v. Pittsylvania co. Va., 20 ms. southwestward Competition or Pittsylvania C. H., and by p-r. 280 ms. s. s. w. W. C., and 187 s. w. by w. Richmond.

SUGAR VALLEY, p-v. Centre co. Pa., by p-r. 210 ms. N. W. W. C.

SUGGSVILLE, p-v. Clark co. Ala., by p-r. 159 ms. s. Tuscaloosa.

SULLIVAN, p-t. Hancock co. Me., 30 ms. E. Castine, 93 E. Augusta; has Hog and Taunton bays, and an arm of Frenchman's bay s. w. and w. A bridge, 1,400 feet long, crosses Hog bay to Hancock. Pop. 1830, 538.

SULLIVAN, co. N. H., bounded by Grafton co. N., Merrimack co. E., Cheshire co. s., and Conn. r. w. which separates it from Vermont. Sugar r. which rises partly in Sunapee lake, on the borders of Merrimack co., flows w. into Conn. r., and there are several other streams. It has been newly formed. Pop. 1830, 19,-669.

SULLIVAN, t. Cheshire co. N. H., 42 miles from Concord, and 6 from Keene; has Ashuelot r. s., but no very striking natural features. Pop. 1830, 557.

SULLIVAN, co. N. Y., bounded by Delaware co. N., Ulster co. E., Orange co. s., and Delaware river w., which separates it from Pennsylvania; has a broken surface, with fertile vallies, several ponds or small lakes, and Navisink, Mongaup, Collakoon, Beaver, Willivemock, and Ten Mile creeks. In the town of Thompson are 3 large tanneries. Pop. 1820, 8,900, 1830, 12,364.

SULLIVAN, p-t. Madison co. N. Y., 129 ms. N. W. Albany, has Oneida lake N., and Onondaga co. w.; first settled about 1798. It is hilly s. and level N.; watered by Canasaraga and Chitteningo crs., which furnish good mill seats. Much gypsum is found here, as well as iron ore, limestone and water lime. Pop. 1830, 4,077.

SULLIVAN, p-v. Tioga co. Pa., by p-r. 142 ms. a little w. of N. Harrisburg.

SULLIVAN, one of the nrthestrn. cos. of Ten., bounded by Carter E. and S. E., Washington s., Hawkins w., Scott co. of Va. N. w., and Washington co. of Va. N. E. Length 43 ms., mean width 12, and area 516 sq. ms. Extending in lat. from 36° 22′ to 36° 35′, and in long. from 4° 48′ to 5° 30′ w. W. C. The declivity is westward, and traversed by the main or middle branch of Holston. This stream enters the northeastern angle of the co., flows s. w. 25 ms., receives the Watauga from the E., and inflecting to the N. w., unites with the North fork of Holston on the border between Sullivan and Hawkins cos. It is a mountainous tract. Chief town, Blountville. Population 1820, 7,015.

SULLIVAN, p-v. Iredell co. N. C., by p-r. 160 ms. a little s. of w. Raleigh.

SULLIVAN, p-v. Lorain co. O., by p-r. 101 ms. N. N. E. Columbus.

SULPHUR SPRINGS, p-o. Union co. Ky., by p-r. 210 ms. a little s. of w. Frankfort.

SUMMERFIELD, p-v. Monroe co. O., by p-r. 112 ms. eastward Columbus.

SUMMERFIELD, p-v. Guilford co. N. C.

SUMMERVILLE. *(See Somerville, seat jus. Fayette co. Ten.)*

SUMMERSVILLE, p-v. & st. jus. Nicholas co. Va., on a branch of Gauley r., by p-r. 310 ms. s. w. by w. W. C., and 268 ms. s. w. by w. Richmond. Lat. 38° 19′, long. 3° 47′ w. W. C.

SUMMIT, p-t. Schoharie co. N. Y., 16 ms. w. Schoharie, has Otsego and Delaware cos. w., and is elevated, with a few streams which flow into the Susquehannah. Pop. 1830, 1,-733.

SUMMIT BRIDGE, and p-o., 17 ms. s. w. Wilmington, 33 a little w. of N. Dover, and by p-r. 112 ms. N. E. W. C. The bridge which heads this article extends over the Chesapeake and Delaware canal, at the Deep Cut through the summit level, between the waters of Chesapeake and Delaware bays.

SUMNER, p-t. Oxford co. Me., 6 ms. N. E. Paris, 44 w. Augusta; has several ponds, and is crossed by a small tributary of Androscoggin r. Pop. 1830, 1,098.

SUMNER, co. Ten., bounded by Smith E., Cumberland r. separating it from Wilson s., Manscoes creek, separating it from Davidson s. w., by Robertson w., Simpson co. of Ky. N. w., and by Allen co. of Ky. N. E. Length diagonally from s. w. to N. E. 40 ms., mean width 16, and area 640 sq. ms. Extending in lat. from 36° 12′ to 36° 37′, and in long. from 9° 08′ to 9° 42′ w. W. C. Sumner occupies a part of the table land between Cumberland and Big Barren, branch of Green river, the two declivities falling from each other in a northern and southern direction. Surface waving rather than hilly. Soil excellent. Chief town, Gallatin. Pop. 1820, 19,211, 1830, 20,569.

SUMNERSVILLE, p-v. Gates co. N. C., by p-r. 152 ms. N. E. by E. Raleigh.

SUMPTER, district, S. C., bounded E. and S. E. by Williamsburg, s. by Santee river, separating it from Charleston, Santee river s. w., separating it from Orangeburg, Wateree river w. separating it from Richland, Kershaw district N. w., and Lynches creek separating it from Darlington N. E. Length northwardly from Santee river to the northern angle 62 ms., mean width 20, and area 1,240 sq. ms. Extending in lat. from 33° 23′ to 34° 17′, and in long. from 2° 51′ to 3° 38′ w. W. C. The central part is drained by Black river, flowing similar to Santee and Lynches rs., in a southeastwardly direction. Chief t. Sumpterville. Pop. 1820, 25,369, and in 1830, 28,277.

SUMPTERVILLE, p-v. and st. jus. Sumpter district, S. C., situated between the branches of Black river, 44 ms. a little s. of E. Columbia, and by p-r. 481 ms. s. s. w. W. C. N. lat. 33° 53°, long. 3° 22′ w. W. C.

SUMNEYTOWN, p-v. Montgomery co. Pa., by p-r. 30 ms. northwestward Phil.

SUMRALL'S CHURCH, and p-o. Perry co. Mississippi, about 140 ms. s. E. by. E. Natchez.

SUNAPEE, lake, Hillsborough and Sullivan cos. N. H.; 1½ ms. by 9; discharges w. by Sugar r. The centre is in lat. 43° 22′. The level is more than 820 feet above Connecticut and Merrimack rs.

SUNBURY, p-v., borough, and st. jus. Northumberland co. Pa., on the left bank of Susquehannah river, on the point above the mouth of Shamokin creek, 2 ms. below the borough of Northumberland, and the junction of the two main branches of Susquehannah river. Lat. 40° 53′, long. 0° 10′ E. W. C. It is distant 52 ms. N. Harrisburg, and by p-r. 162 ms. a very little E. of N. W. C.

SUNBURY, p-v. and seaport, on Medway river, Liberty co. Geo., 10 ms. E. Riceboro', the co. seat, and by p-r. 212 ms. S. E. by E. Milledgeville. Lat. 31° 45′, long. 4° 22′ W. W. C. It stands about 8 miles above the open ocean. The harbor is wide, but is defended on the sea side by the northern point of Saint Catharine's isl. It is the seat of an academy.

SUNBURY, p-v. Gates co. N. C., by p-r. 160 ms. N. E. by E. Raleigh.

SUNBURY, p-v. southeastern part Delaware co. Ohio, by p-r. 22 ms. N. N. E. Columbus.

SUNCOOK, river, N. H., rises in a pond near the top of one of the Suncook mountains, 900 feet high. After receiving several branches, it enters the Merrimack between Allenstown and Pembroke.

SUNDERLAND, p-t. Bennington co. Vt., 15 ms. N. E. Bennington, 87 S. W. Montpelier; first settled 1765; has Battenkill river N. W., on which are fine meadows, and Roaring brook E. Lead ore is found here. Population 1830, 463.

SUNDERLAND, p-t. Franklin co. Mass., 90 ms. W. Boston, has Connecticut river on the W. boundary, parallel to which lies the principal street of the v. A large tract of meadows borders that stream. Mount Toby is near the line of this town, and Leverett. The minerals are various. Pop. 1830, 666.

SUNFISH, p-o. on a creek of the same name, falling into Ohio river, northeastern part Monroe co. Ohio.

SURGOINSVILLE, p-v. on Holston r., Hawkins co. Ten., 76 ms. above and N. E. by E. Knoxville, 11 ms. N. E. Rogersville, the co. st., and by p-r. 274 a little N. of E. Nashville.

SURRY, p-t. Hancock co. Me., 18 ms. N. E. Castine, 87 W. by N. Augusta; has Union r. E., Newbury neck S. E. stretching into it, and 2 or 3 large ponds which are connected, and divide the town near the middle. Pop. 1830, 561.

SURRY, town, Cheshire co. N. H., 54 miles from Concord; is crossed by Ashuelot river, which has valuable meadows on its banks. East of this stream is a pond of three acres, 25 feet deep, on the summit of a mountain. First settled 1764. Pop. 1830, 539.

SURRY, co. Va., bounded by Isle of Wight co. E. and S. E., Southampton S., Blackwater river, separating it from Sussex S. W., Prince George W. and N. W., and James river separating it from Charles City N. W., and James City N. and N. E. Length and breadth nearly equal, or 18 ms., area 324 sq. ms. Extending in lat. from 36° 50′ to 37° 11′, and in long. from 0° 19′ E. to 0° 08′ W. W. C. The sthrn. and western part of Surry slopes to the southeastward, and is drained into Blackwater r.; the northeastern part declines in that direction towards James r. Chief town, Surry C. H. Pop. 1820, 6,594, 1830, 7,109.

SURRY, co. N. C., bounded by Stokes N. E., Yadkin separating it from the southern part of Stokes S. E., Rowan S., Iredell S. W., Wilkes W., the Blue Ridge separating it from Ashe N. W., and Grayson and Patrick cos. of Va. N. Length from S. to N. 33 ms., mean width 22, and area 726 sq. ms. Extending in lat. from 36° 04′ to 36° 33′, and in long. from 3° 26′ to 3° 58′ W. W. C. This county is divided into two very nearly equal sections by Yadkin river, which traverses it in a direction a little N. of E. Both sections are drained by creeks falling into Yadkin. The general declivity eastward; surface broken, and in part mountainous, with much excellent soil. Chief towns, Rockford and Huntsville. Pop. 1820, 12,320, 1830, 14,501.

SURRY, C. H., p-v. and st. jus. Surry county, Va., by p-r. 60 ms. S. E. by E. Richmond, and 183 a very little E. of S. W. C.

SURVEYORSVILLE, p-v. Mecklenburg co. N. C., by p-r. 136 ms. S. W. by W. Raleigh.

SUSQUEHANNH, river of N. Y., Pa., and Md. Obeying the correct principles of physical geography, Chesapeake bay ought to be regarded as the continuation of Susquehannah river, but custom has restricted the name to that part of the river above tide water. Under the articles Chesapeake, James river, Potomac, &c. the lower part of the basin will be found noticed in this treatise; the present article will be restricted to a survey of Susquehannah proper. Measured by the rhombs on Tanner's United States, the valley of Susquehannah above the head of Chesapeake bay comes out 28,600 square ms. Extending in lat. from 39° 33′ to 42° 55′, and in long. from 2° 25′ E. to 1° 50′ W. W. C. A small fraction of about 350 square ms. comprising the lower part of this valley is in Md. Above lat. 42° and in the state of N. Y. spreads 7,600 square ms. drained by the two northern branches and their numerous confluents. But the main part of the valley, comprising 20,650 square ms., lies within and forms the central and upwards of four tenths of the whole state of Pa. The Susquehannah is formed by two main branches called, with some inconsistency, the northern and western branches. The northern and principal branch rises in Otsego county, N. Y., in two confluents, the Unadilla and Chenango. The extreme northern sources of the Unadilla rise within less than 5 ms. of the Mohawk river at the Little Falls, but other sources rise from the Catsberg mountains opposite those of the Schoharie, flow generally to the southwestward, unite between Delaware and Chenango counties, and turning southward approach to within 12 ms. of the Coquago branch of Del., enters Pa., and curving to the W. and thence N. W. over Susquehannah county, enters N. Y., receiving the Chenango at Binghamton in Broome county, and winding over Broome and Tioga by an elliptic curve, gradually assumes a southwestern course and again returns into Pa., about 3 ms. within which it receives the Tioga branch from the N. W. The Tioga or Chemung is composed of 3 branches, the Tioga proper, Canisteo, and Conhocton. The Tioga river rises in and drains the northern part of Tioga county, Pa., and flowing northward enters Steuben co., N. Y., within which it first receives the Canisteo from the westward, and next the Conhocton from the N. W. The two latter drain the larger part of Steuben county, on the eastern side of which, as has been stated, they unite with the Tioga. The river thus formed, assuming the name of Tioga, flows a little E. of S. E., enters Pa., and joining the Susquehannah at Athens, or Tioga Point, the now large stream turns to nearly due S. In the latter direction it flows about 5 ms., gradually inclining eastward to the mouth of Towanda creek 10 ms. farther to the northwestern limit of the Appalachian system. This higher section of Susquehannah valley presents some very remarkable features. The sources interlock on the W. with those of Alleghany branch of O., on the northwest with those of Genesee, on the N. with those of Seneca, N. E. with those of Mohawk and Schoharie, and E. with those of the Delaware. Spreading like the head of a tree along a line of 170 ms. on the secondary formation, the declivity of the plain not from, but directly towards, an extensive system of mountains, affording a decisive proof that the Appalachian system does not form the dividing ridge of the water courses of the United States. When the Susquehannah has reached the mountain base, it has drained upwards of 8,000 square miles and is a large navigable river. The country above the mountains is in general composed of high but rounded hills, and deep fertile vallies. The rivers are rapid in their courses, but without direct falls. There is another circumstance in the natural features and relative connection of the upper Susquehannah valley, which deserves particular notice. The mean water level at Tioga Point is 723 feet above the Atlantic ocean, and from the latter point to Newton, or Elmira on Tioga river, the rise is 103, giving to the water level at Newton a comparative elevation of 826 feet. Though the hills are very high in the vicinity of Newtown, there is a natural valley stretching from the Tioga northwards to the head of Seneca lake. The middle ground, or summit level of this valley is only 59 feet above the Tioga river, but falls so rapidly toward Seneca as to have a descent into that lake of 445 feet in 10 or 11 ms. The summit level is 885 feet above the ocean, but is the lowest gap in the Appalachian system, admitting a canal to be formed southward from the valley of the Mohawk to lower Georgia. A single glance at a map of this physical region will serve to exhibit the singular natural navigable facilities afforded by the depression of the summit level of the vallies between them, and the approximation of the lakes of the St. Lawrence basin, to the northern streams of that of Susquehannah. After its entrance into the mtns. the Susquehannah flows about 50 ms. to the S. E. by a direct comparative course, but with a very sinuous and obstructed channel, to its entrance into Wyoming valley, at the mouth of Lackawannock river. Here this stream bends nearly at right angles, and again by a channel of about 70 ms. comparative course S. W., winds its way down the mountain vallies to the entrance of the West branch at the borough of Northumberland. The W. branch is entirely a river of Pa., having its most remote western fountain in Indiana, but deriving sources in a line of 80 miles from Cambria, Clearfield, and McKean counties. The general course of the confluents is eastward by comparative courses 150 ms. on the western secondary formation, to where it passes the main Appalachian chain between Williamsport and Pennsboro', thence bends to nearly due S. 25 ms. to its junction with the northern branch, as already stated. Canals have been designed along both branches, and their routes partially designated. The main trunk is to leave the traverse division of the Pa. canal at Duncan's island near the mouth of the Juniata, and follow the Susquehannah valley to the N. Y. line, distance 204 ms. with a rise of 423 feet. The West branch trunk commences at Northumberland, and follows the valley of the latter stream 70 miles to Dunnstown, at the mouth of Eagle creek, rise 109 feet. Entire elevation of water level at Dunnstown 540 feet. The two principal branches having united between the boroughs of Northumberland and Sunbury, assumes a course of a little W. of S. 40 ms. to the influx of Juniata, from the wstrd. (*See Juniata.*)

Augmented by the last of its large tributaries, the Susquehannah inflects to S. E. 80 miles, receiving from the right Sherman's, Conedogwinet, Yellow Breeches, Conewago, Codorus, and Deer creeks, and from the left Swatara, Conestoga, Pequea, and Octoraro, with numerous smaller streams, finally is lost in Chesapeake bay, after falling over the lower primitive ledge of the Appalachian system. Viewing the entire valley of Susquehannah, we have before us some very remarkable features of the physical geography of the U. S. This great stream, deriving its most remote sources from the western secondary, both in N. Y. and Pa., and in the course of its great confluents and main volume traversing obliquely the whole Appalachian system at the widest part, presents no one direct fall of sufficient pitch to prevent navigation. In reality the mountain chains stretch along the declivity of the Susquehannah valley. With innumerable partial windings, the large and even many of the smaller streams, flow in channels which pursue the mountain vallies in the general direction of the chains, or traverse the latter at right angles. This gives a striking physiognomy to the courses of the rivers which can only be understood by a view of a good map. In its course the Susquehannah traverses also all the great formations of the earth. Rising on the horizontal, or as technically denominated, the secondary or floetz, and breaking immense gaps through the mnts. of transition and primitive rocks, makes its final exit on the inner margin of sea sand alluvion. The relative height of the extremes of this valley deserve particular notice, as element in a theory of its climate. Rejecting the mtn. ridges, the arable soil beyond the principal spine of the Appalachian system, is from 600 to perhaps 1,200 feet, the mean height rather less than a mean term of the extremes of elevation, affording as has been however shown, a valley from the Atlantic to the St. Lawrence lakes, in its highest part falling below 900 feet. The mineral productions of the Susquehannah valley yet explored, have amongst numerous other specimens, presented immense masses of iron ore and fossil coal. The former even more widely disseminated than the latter. (*See articles Juniata and Pennsylvania.*)

SUSQUEHANNAH, co. Pa., bounded by Wayne co. E., Luzerne S., Bradford W., and Broome co. of N. Y. N. Length 35 ms. from E. to W., width 25, and area 875 square ms. Extending in lat. from 41° 40′ to 42°, and in long. from 0° 50′ to 1° 32′ E. W. C. The northern branch of Susquehannah r. enters and again retires from the northern border of this co., and hence by a curve of 80 ms. again approaches the S. W. angle to within one mile. Thus encircling the co. on three sides and receiving its numerous creeks like radii from a common centre. The surface is hilly and broken, but soil excellent. Chief town, Montrose. Pop. 1820, 996, 1830, 16,677.

SUSSEX, co. N. J., the N. co. of the state, bounded by N. Y. state N. E., Bergen and Morris cos. S. E., Warren co. S. W., and Delaware r. N. W., is hilly and mountainous, with many good dairy farms, and well tilled land near Del. r., abounds in valuable iron mines. It is the highest land in the state, and gives rise to Wallkill creek of Hudson r., Pequannock r., Pequest creek, and Paulins kill of the Del. and has the whole course of Flat kill. Hopatung pond, which supplies the summit level of Morris canal, is on the S. E. bound. The Blue mtns. cross the N. W. part parallel to Del. r., between it and which flows Flat kill. Chief town, Newton. Pop. 1820, 32,752, 1830, 20,346.

SUSSEX, southernmost co. of the state of Del., bounded N. by Kent co. Del., N. E. by Del. bay, E. by the Atlantic, S. by Worcester co. Md., S. W. by Somerset, Md., W. by Dorchester, Md., and N. W. by Caroline, Md. Length from W. to E. 35 ms., mean width 25, and area 875 square ms. Extending in lat. from 38° 27′ to 38° 58′, and in long. from 1° 14′ to 1° 58′ E. W. C. Though the surface of this co. is level and in part marshy, it is nevertheless a table land, from which flow southwestwardly the sources of Nantikoke r., sthrdly. those of Pocomoke, estrdly. the various confluents of Rehoboth bay, and northeastward creeks falling into Del. bay. Chief towns, Georgetown and Lewis. Pop. 1820, 24,057, and in 1830, 27,115.

SUSSEX, co. of Va. bounded by Southampton S. E. and S., by Greensville S. W., Dinwiddie W., Prince George N. W., Blackwater r. separating it from a part of Surry N., and by the southern angle of Surry N. E. Length from S. W. to N. E. 37 ms., mean width 16, and area 592 square ms. Extending in lat. from 36° 42′ to 37° 07′, and in long. from 0° 02′ E. to 0° 46′ W. W. C. The southern and central parts are drained by the Nottaway, and the northern by Blackwater river. Chief town, Sussex C. H. Pop. 1820, 11,884, 1830, 12,720.

SUSSEX, C. H. p-v. and st. jus. Sussex co. Va., by p-r. 50 ms. S. S. E. Richmond, and 172 a little W. of S. W. C.

SUTHERLAND, p-v. Trumbull co. Ohio, by p-r. 157 ms. N. E. Columbus.

SUTHERLAND'S, p-o. Edgar co. Il., by p-r. 97 ms. N. E. Vandalia.

SUTTON, p-t. Merrimack co. N. H. 25 ms. from Concord, 17 from Hopkinton, 65 from Portsmouth, and 85 from Boston, has the S. branch of Warner r. S., and is crossed by the N. branch nearly in the centre. On these streams are good mill sites, and valuable meadows; there are several other streams, and a few ponds. Kearsearge is a lofty mtn. in the E. part, which gives rise to several streams. King's hill W., also affords an extensive view. Valuable stone quarries and clay beds exist in this town. The surface is rough, and the soil various. The forest trees were of many different kinds. First settled 1769. Pop. 1830, 1,424.

SUTTON, p-t. Caledonia co. Vt., 54 ms. from Montpelier. Pop. 1830, 1,005.

SUTTON, p-t. Worcester co. Mass. 46 ms. S. W. Boston, was purchased of the Indians 1704, and included Millburg. There are many good mill sites, and many manufactories. The town is crossed by Blackstone river and canal. Wilkinsonville, N. W., contains manufactories which derive water power from the r. Granite is quarried in the town in great quantities; S. E. is a large and curious chasm in the rocks, sometimes called purgatory. Pop. 1830, 2,186.

SUTTONSVILLE, p-v. southern part Nicholas co. Va. by p-r. 312 ms. S. W. by W. W. C. and 300 ms. N. W. by W. Richmond.

SWAINSBORO', p-v. and st. jus. Emanuel co. Geo., by p-r. 79 ms. S. E. by E. Milledgeville. Lat. 32° 40′, long. 5° 28′ W. W. C.

SWANANO, p-v. estrn. part Buncombe co. N. C. 22 ms. N. E. Asheville, the co. st., and 247 W. Raleigh.

SWANKESVILLE, p-v. Putnam co. Il., by p-r. 57 ms. W. Indianopolis.

SWANSBORO', p-v. and sea port of Onslow co. N. C., situated at the mouth of Whittock r. opposite Boyne inlet, by p-r. 160 ms. S. E. Raleigh, and 377 a little W. of S. W. C.

SWANSEY, p-t. Cheshire co. N. H., 60 ms. S. W. Concord, 6 from Keene, and 68 from

Boston, is crossed by Ashuelot r. and its s. branch. Nearly one third of the town is level, and free from stones. There is some iron ore, and a mineral spring, several manufactories and mills. Between 1741 and 1747, this town suffered much from Indian attacks. The settlement was on this account abandoned for 3 years, and the dwellings burnt by the savages. Pop. 1830, 1,816.

SWANSEY, p-t. Bristol co. Mass. 47 ms. s. Boston, has Rhode Island s. and w., and enjoys a pleasant situation on Cole's r. which flows into Taunton r. and is navigable for small vessels. It was early settled by a number of Baptists from Rehoboth, under a grant from Plymouth colony, and was the first town attacked by the Indians in Philip's war, 1675. Here are several manufactories. Pop. 1830, 1,678.

SWANTON, p-t. Franklin co. Vt., 28 ms. N. Burlington, 50 ms. N. W. Montpelier, E. lake Champlain, opposite North Hero, was first settled 1787, when it was occupied by St. Francis Indians. Missisque creek crosses this town and has meadows on its banks, while a fall of 20 feet supplies mill sites. From this fall to the lake the r. is navigable for vessels of 50 tons. Mc Quam creek and several smaller streams also water this town. There are marshes N. W. much resorted to by wild fowl. Iron and marble are found here. The marble is cut at the falls, and transported to N. Y. &c. The v. of Missisque stands on both sides of the r. 6 ms. from its mouth, and 1 mile in a strait line from the lake. Boats which navigate the lake, Champlain canal and Hudson r., come up to the v. Pop. 1830, 2,158.

SWANVILLE, t. Waldo co. Me. 15 ms. N. W. Castine, and N. Belfast, is of irregular form, bounded by straight lines, and crossed by a small stream flowing into Belfast bay. Pop. 1830, 633.

SWATARA, r. Pa. rises by numerous branches from the mtn. vallies in the sthrn. part of Schuylkill co. It thence traverses the wstrn. part of Lebanon and the sthestrn. of Dauphin, falling into Susquehannah 8 ms. below Harrisburg, after a sthwstrn. comparative course of 40 ms. For nearly one-half of the course of this stream the Union canal follows the channel. This artificial navigation is in full operation.

SWEDEN, t. Oxford co. Me. 20 ms. s. w. Paris, has the boundary of Cumberland co. on the s. w. line, and contains several ponds which discharge by an outlet into Loud pond. A tributary of Sunapee lake crosses the w. part. Pop. 1830, 487.

SWEDEN, p-t. Monroe co. N. Y. 16 ms. w. Rochester, lies N. and w. of Genesee co., and is on elevated land, crossed by the Mountain ridge and Erie canal. It gives rise to Salmon creek. Brockport v. on the canal, is in the N. Pop. 1830, 2,938.

SWEDEN, p-v. N. W. part Potter co. Pa. by p-r. 290 ms. N. N. W. W. C.

SWEEDSBURGH, p-v. Woolwich, Gloucester co. N. J. 20 ms. s. Phil. on Raccoon creek.

SWEETZER'S Bridge and p-o. Ann Arundel co. Md. by p-r. 42 ms. from W. C. and 26 from Annapolis.

SWEET SPRINGS, p-v. and watering place, northestrn. part Monroe co. Va. These springs are situated in one of the mountain vallies, from which flow the western sources of James r. at an elevation of about 2,400 feet above the Atlantic tides, 84 ms. N. W. by W. Lynchburg, 263 s. w. by w. W. C. and 204 w. Richmond.

SWIFT, r. N. H. a branch of Saco r. falls into that stream in Conway, after a rapid course.

SWIFT Creek Bridge and p-o. nthrn. part Craven co. N. C. 17 ms. N. Newbern, and by p-r. 137 ms. N. E. by E. Raleigh.

SWINDELL, p-o. Hyde co. N. C. by p-r. 195 ms. E. Raleigh.

SYCAMORE Alley and p-o. sthrn. part Halifax co. N. C. 22 ms. s. Halifax, the co. st., and 84 N. E. by E. Raleigh.

SYCAMORE, creek and p-o. nthrn. part Crawford co. O. by p-r. 74 ms. N. Columbus.

SYLVAN Hill and p-o. sthrn. part Hancock co. Geo. by p-r. 16 ms. E. Milledgeville.

SYLVANIA, p-v. nthrn. part Bradford co. Pa. by p-r. 147 ms. N. Harrisburg.

SYLVANUS, p-v. Hillsdale co. Mich. by p-r. 108 ms. s. w. by w. Detroit.

SYRACUSE, p-v. Salina, st. jus. Onondaga co. N. Y. 4 ms. N. Onondaga, 133 w. Albany, is situated on the Erie canal, adjoining a vast collection of salt pans, and at the junction of the canal with the branch to Salina v. and the Oswego canal. It is a large, handsome, and flourishing village, and has attained a most rapid growth, having been of insignificant size before the opening of Erie canal.

T.

TABERG, p-v. Annsville, Oneida co. N. Y. 112 ms. w. Albany, 7 N. Erie canal, 11 w. Rome, 27 N. W. Utica, is the seat of extensive iron works.

TABOR Church and p-o. in the wstrn. part Iredell co. N. C. by p-r. 159 ms. wstrd. Raleigh.

TAFTON, p-v. in the nthrn. part Pike co. Pa. by p-r. 271 ms. N. N. E. W. C.

TAGHKANIC, p-t. Columbia co. N. Y. w. Mass., is watered by Claverack, Ancram, Rocleff and Jansen's crs., which supply many mill seats. The Taghkanic mtns. rise here, but their greatest elevations are in Mass. The land is held on lease. Iron ore is found in plenty. Pop. 1830, 1,654.

TALBOT, one of the Eastern Shore cos. of Md., bounded s. and s. E. by Choptank r. separating it from Dorchester, E. by Choptank and Tuckahoe rs. separating it from Caroline, N. by St. Michael's bay, separating it from Queen Anne, and w. and s. w. by Chesapeake bay. Length from s. to N. 25 ms., mean width 10, and area 250 sq. ms. Extending in lat. from 38° 34′ to 38° 56′ N., and in long. from 0° 42′ to 1° 10′ E. W. C. This co. is a real peninsula between Choptank r. and Chesapeake bay; and is again cut into three minor peninsulas by Treadhaven and St. Michael's bays. What little declivity exists is to the sthrd. Chief t. Easton. Pop. 1820, 14,389, and in 1830, 12,947.

TALBOT, co. of Geo. bounded s. by Marion, s. w. by Muscogee, w. by Harris, N. w. by Merriwether, and by Flint r. separating it from Upson N. E., and Crawford E. Length along the sthrn. boundary 40 ms., mean width 16, and area 600 sq. ms. Extending in lat. from 32° 35′ to 32° 54′ N., and in long. from 7° 10′ to 7° 54′ w. W. C. The wstrn. part gives source to some creeks which flow southwstrd. towards the Chattahoochee; but the greatest part of the co. slopes eastwardly toward Flint r. Chief t. Talbotton. Pop. 1830, 5,940.

TALBOTTON, p-v. and st. jus. Talbot co. Geo. situated on a small creek of Flint r. by p-r. 112 ms. s. w. by w. Milledgeville. N. lat. 32° 43′, long. 7° 36′ w. W. C.

TALCOT, mtn. Hartford co. Conn., a part of the ridge which extends many miles on the w. of Conn. r., dividing its waters from those of Farmington r.

TALIAFERRO, co. of Geo. bounded N. by Oglethorpe co., N. E. and E. by Wilkes, s. E. and s. by Hancock and w. by Greene. Length from s. to N. 17 ms., mean width 8, and area 136 sq. ms. Extending in lat. from 33° 28′ to 33° 43′ N. In long. it is traversed by 6° w. W. C. Declivity southestrd., and traversed in that direction by the higher branches of Little r. and those of Great Ogechee Chief town, Crawfordsville. Pop. 1830, 4,934.

TALLAHASSEE, p-t. and st. jus. Leon co., and of government Flor., situated about 30 ms. inland and northwards from Ocklockonne bay, about 200 ms. N. w. St. Augustine, a similar distance a little N. of E. Pensacola, and by a calculation on Mercator's principles, s. 36° 10′, w. 725 statute miles, but by the post list 896 from W. C. N. lat. 30° 27′, long. 7° 30′ w. W. C. The city is recent; the buildings were commenced in the summer of 1824. The site is comparatively elevated, affording a good view of the vicinity. The adjacent country is rolling rather than hilly; the soil excellent. A pleasant mill stream formed by fine springs winds along the eastern border of the town, from whence it is precipitated over a fall of 15 feet, and disappears in the calcareous strata. Springs of good water abound, and well water is obtained by digging from 6 or 10 to 30 feet. The first legislature sat in this new-born city the first winter after its erection, or in 1824–5. It was incorporated as a city in 1825. When Mr. John Lee Williams published his View of West Florida, in 1827, he estimated the population at 800. He observes, "few towns in America have increased more rapidly; and population and improvement continue without any abatement. It must in a few years become a charming place of residence, though it will probably never be a place of great commercial importance."

TALLAPOOSA, r. of Geo. and Ala. rises in the Cherokee territory, and in the northwestern part of the former, lat. 34°, between the Etowah and Chattahoochee rs. Flowing s. s. w. it enters Ala., and continuing that course 130 ms., turns abruptly to the w. 25 ms., and falls into the Coosa, or rather, from the great difference of volume, joins the Coosa to form Ala. The junction is made between Montgomery and Autauga cos.

The valley of the Tallapoosa lies entirely between those of Coosa and Chattahoochee. It is about 150 miles in length, with a mean width of 25, area 3,750 sq. ms. Lying between latitudes 32° and 34°, and long. 8° and 9° 20′ w. W. C.

TALLMANSVILLE, p-o. Wayne co. Pa., by p-r. 278 ms. N. E. W. C.

TALLYHO, p-v. northern part of Granville co. N. C., by p-r. 57 ms. N. N. E. Raleigh.

TALMADGE, p-v. Portage co. O., by p-r. 115 ms. N. E. Columbus.

TAMAQUA, p-o. northern part of Schuylkill co. Pa., by p-r. 191 ms. N. N. E. W. C., and 83 ms. N. E. Harrisburg.

TAMAQUA, the Indian name of Little Schuylkill, and on which the p-o. of the same name is situated.

TAMWORTH, p-t. Strafford co. N. H., 58 ms. from Concord, 58 from Portland, 30 from Gilford, 120 from Boston; has part of the Burton mtns., and south part of the Ossipee. Bearcamp r. runs through it E. into Ossipee lake, after receiving 2 branches which rise here, and afford many mill sites. First settled 1771. Pop. 1830, 1,554.

TANEYTOWN, p-v. in the northeastern part of Frederick co. Md., 22 ms. N. N. E. the city of Frederick, and 68 ms. a little w. of N. W. C.

TANGEPAO, river of La. and Miss., has its most remote sources in Amite and Pike cos. of the latter, and flowing s. s. E. enters La., separating the parish of Saint Helena from Washington and Saint Tammany, and falls into the northwestern part of lake Pontchartrain, after a comparative course of between 70 and 80 ms. The valley of Tangipao lies between those of Amite and Bogue Chito, in the higher part of its course, but in La. between the Tchefonte and Tickfah.

TANGIER, islands and sound. The Tangier islands is a group of small islands in Chesapeake bay, evidently an extension of the peninsula between Choptank and Nantikoke rs. They follow each other from N. to s., and are partly in Somerset co. Md., and Accomac co. Va. The sound spreads between the isls.

and main shore. This group lies opposite the mouth of Potomac r.

TANNER'S STORE, and p-o. Mecklenburg co. Va., by p-r. 215 ms. s. s. w. W. C.

TAN YARD, and p-o. Northumberland co. Va., by p-r. s. s. E. W. C.

TAPPAHANNOC, p-v. and st. of jus. Essex co. Va., situated on the right bank of Rappahannock river, by p-r. 109 ms. a little E. of s. W. C., and 50 N. E. Richmond. N. lat. 37° 58', long. 0° 10' E. W. C. The site is low and flat, and in summer the inhabitants are liable to fevers and agues; it is, however, a place of considerable trade, as even large merchant vessels can ascend far above, and here find a safe harbor, which is about 50 ms. from the open Chesapeake bay.

TAPPAN, p-v. Orangetown, Rockland co. N. Y., 28 ms. N. New York, is on the w. side of Hudson r., which is there 4 ms. across.

TAR, or, in the lower part of its course, Pamlico, river of N. C., having the extreme higher fountain in Person co., interlocking sources with Neuse and the lower creeks of Dan r. Flowing thence by a general course of s. E. by E. over Granville, Franklin, Nash, Edgecombe, and Pitt cos., and receiving large accessions from Warren and Halifax, opens into a wide bay, below the harbor of Washington, in Beaufort co. (*See Pamlico bay.*)

The valley of Tar river, including Pamlico bay, is 160 miles in length, with a mean width of 30 ms., area 4,800 sq. ms.; and lying between those of Neuse and Roanoke. Extending in lat. from 35° 15' to 36° 25' N., and in long. from 0° 25' E. to 2° 15' w. W. C. It is navigable for vessels of nine feet draught to Washington, and for river boats to Tarboro', at the confluence of the two main branches.

TARBORO', p-v. and st. of jus. Edgecombe co. N. C., situated on the right bank of Tar r., below the influx of Fishing creek, by p-r. 72 ms. a little N. of E. Raleigh, and 252 a little w. of s. W. C. N. Lat. 35° 53', long. 0° 36' w. W. C.

TARENTUM, p-v. Alleghany co. Pa., by p-r. 231 ms. N. W. W. C.

TARIFF, p-v. Butler co. Ohio, by p-r. 122 ms. s. w. by w. Columbus.

TARIFFVILLE, p-v. Simsbury, Hartford co. Conn., is a manufacturing village, pleasantly situated at the falls of Farmington r., at the w. base of the hilly range which crosses that part of the state, at the spot where the river bursts through it, between two precipitous banks. The carpet manufactory here employs 95 male weavers, and 367 were immediately dependant on it in 1831. The capital invested is $123,000; 237,000 pounds of wool, and 24,000 pounds of yarn are manufactured, producing about 114,000 yards of Ingrain or Kidderminster carpeting. Above $30,000 is paid for labor annually.

TARLTON, p-v. sthestrn. part of Pickaway co. O., by p-r. 36 ms. s. s. E. Columbus.

TARPAULIN COVE, Martha's Vineyard, Ms., is a convenient little harbor for vessels bounded w. in contrary winds. It is 9 ms. N. N. W. Holmes' Hole.

TARRYTOWN, p-v. Greensburgh, Westchester co. N. Y., 30 ms. N. New York, on the E. side Hudson r., has a landing in a cove between two points, where a steamboat touches daily from and for New York.

TARVER'S Store and p-o., in the sthrn. part of Twiggs co. Geo., 10 ms. from Marion, the co. st., and 47 ms. s. w. Milledgville.

TATNALL, co. of Geo., bounded by Montgomery w., Emanuel N., Cannouchee r. separating it from Bullock, N. E. and E., Liberty S. E., and Altamaha river, separating it from Appling s. and s. w. Length 52 ms., mean width 24, and area 1,248 sq. ms. Extending in lat. from 31° 48' to 32° 26' N., and in long. from 4° 44' to 5° 38' w. W. C. The western part is drained into the Altamaha by the Great Ohoopee and other streams, whilst the estrn. section is in the valley of Cannouchee. The Altamaha is formed by the union of the Oconee and Ocmulgee rs., at the extreme western angle of Tatnall. General declivity S. E. C. H. at Percy's mills. Pop. 1820, 2,644; and in 1830, 2,039.

TAUNTON r. Mass., navigable 20 ms. from Narragansett bay to Taunton, in sloops. It has its rise in Plymouth co., and its course is about s. w.

TAUNTON, p-t. and one of the sts. jus. Bristol co. Mass., is pleasantly situated on Taunton r., which is navigable to this place for sloops. It was first settled in 1637, and was called Cohannet by the Indians: within the present limits of the town was the Indian v. Teticut. Taunton is 32 ms. s. Boston, and 20 N. of E. Providence. It contains a bank, several churches, an academy, and one or two county buildings. Canoe, Rumford, and Taunton rs. unite here, and furnish excellent water privileges. The first extensive iron works in America were erected in this town, in 1652, and at present it is famous for its manufactures. The nail factories make from 8 to 10 tons daily. It has 7 cotton factories—1 rolling and slitting mill—1 forge—1 shovel factory—1 copper and lead rolling mill—1 paper mill—1 carding and fulling mill—1 calico printing establishment, which furnishes from 4 to 6,000 pieces a week—2 breweries—1 large factory of britannia ware, and many other establishments of different kinds; besides 8 or 9,000,000 of brick are manufactured annually. Pop. 1830, 6,042.

TAXAHAW, p-o. Lancaster dist. S. C., 19 ms. N. W. Lancaster, and by p-r. 91 ms. N. N. E. Columbia.

TAYLOR'S store and p-o. Franklin co. Va., 12 ms. estrd. Rocky Mount, the co. st., and by p-r. 173 ms. s. w. by w. Richmond.

TAYLOR'S store and p-o., Anson co. N. C., by p-r. 160 ms. s. w. by w. Raleigh.

TAYLORSVILLE, p-o. Bucks co. Pa., by p-r. 36 ms. nrthrd. Philadelphia.

TAYLORSVILLE, p-o. Hanover co. Va., 28 ms. nrthrd. Richmond.

TAYLORSVILLE, or Patrick C. H., p-o. and st. jus. Patrick co. Va., situated on Mays r., 90 ms. s. w. Lynchburg, 35 a little E. of s.

Christiansburg, and by p-r. 241 ms. s. w. by w. Richmond, and 333 s. w. W. C., N. lat. 36° 38′, long. 3° 14′ w. W. C.

TAYLORSVILLE, p-v. and st. jus. Spencer co. Ky., situated on Salt r., 35 ms. s. E. Louisville, by p-r. 35 ms. s. w. by w. Frankfort, and 586 a little s. of w. W. C.; N. lat. 38°, long. 8° 20′ w. W. C.

TAZEWELL, co. of Va., bounded N. by Tug Fork of Sandy r., separating it from Logan, N. E. by Giles, E. and S. E. by Walker's mountains, separating it from Wythe, s. by Clinch mtn., separating it from Washington, s. w. by Russel, and w. by Floyd co. Ky. Length from w. to E. 80 ms., mean width 20, and area 1,600 sq. ms. Extending in lat. from 36° 54′ to 37° 32′ N., and in long. from 4° to 5° 12′ w. W. C. The central part of this co. is a very elevated mtn. table land. The estrn. part declining nrthestrd., and drained by the confluents of Great Kenhawa; the southern gives source to Clinch and Holston rs., the extreme nrthrn. constituents of Tennessee r.; whilst the western and most extensive section has a nrthwstrn. declivity, and gives source to the highest branches of Sandy r. Compared with the ascertained elevation of the water in Great Kenhawa at the influx of Greenbrier, 1,333 feet, the lowest elevation that can be given to the central mountain vallies of Tazewell, must be 1,500 feet; and the mean relative height of the arable soil of the co., must be, at the lowest estimate, 1,200 ft. Chief town, Jeffersonville. Pop. 1820, including a part of what now constitutes Logan, 2,916; that of Tazewell proper in 1830, 5,749.

TAZEWELL, p-v. and st. jus. Claiborne co. Ten., situated between the rs. Clinch and Powell's Valley r., by p-r. 248 ms. a little N. of E. Nashville, and 63 ms. N. E. Knoxville. Lat. 36° 31′, long. W. C. 6° 20′ w.

TAZEWELL, co. Il., bounded by MacLean E., Sangamo s., the Illinois r., separating it from Fulton w., and Peoria N. w.; on the N. it has the sthestrn. angle of Putnam. As laid down by Tanner, in his improved map of the U. S., it lies nearly in form of a right angled triangle, hypothenuse parallel to the general course of Il. r., 66 ms., base along Sangamo, and perpendicular along MacLean equal, or 50 ms. each; area 1,250 sq. ms.; N. lat. 40° 40′, long. W. C. 12° 30′ w. The general course of Il. r. along this co. is about s. w. dilating into Peoria and Mackinaw lakes. The slope of the co. is nearly to the w. (*See Mackinaw r.*) The general surface is level, and part liable to annual submersion, though it contains much good soil. Chief t. Mackinaw. Pop. 1830, 4,716.

TAZEWELL, C. H. (*See Jeffersonville, Tazewell co. Va.*)

TEAZE'S VALLEY, p-o. in the western part of Kenhawa co. Va., 20 ms. westward Charleston, the co. st., and by p-r. 376 ms. a little s. of w. W. C.

TECHE, r. of La., rises from the northern prairies of Opelousas, N. lat. 30° 40′. The drains of those savannahs, after flowing 7 or 8 ms., divide into 2 channels; one flows northwardly into Courtableau, and the other pursues a sthestrn. course. This separation of currents is the head of the stream called Teche, or the sthestrn. branch. Flowing between the waters of the Courtableau and Vermillion 10 ms., it receives an inlet from the latter, and enters Attakapas. The residue of the course of Teche, presents a stream with great specific resemblance to the Miss. in the delta. Though on a very reduced scale, the Teche, similar to its immense prototype, flows in long sweeping bends, with banks above any other part of the adjacent country. From this feature the streams flow from the very margin, and in a channel of upwards of 180 ms., no water course is discharged into the Teche. With slight selvedges of wood, prairies extend along the entire right, and, for more than half the higher part of its course, along the left bank of this interesting river. The channel is comparatively very deep, and the tide rises to New Iberia, N. lat. 30° 02′, upwards of 100 ms. above the mouth, affording one very remarkable contrast to the Mississippi. New Iberia, at the head of tide water in Teche, is a port of entry, and vessels of 7 feet draught can ascend there in safety. The Teche falls into Atchafalaya, after a comparative course of 120, but falling little, if any, short of 200 ms. by the bends. The banks present two continuous zones of the very first rate soil, between latitudes 29° 44′ and 30° 40′ N.

TEKATOKO, p-o. Crawford co. Ark., situated near the Dardanelles mountains, by p-r. 76 ms. N. w. by w. Little Rock.

TELFAIR, co. of Geo., bounded N. E. by Montgomery, E. S. E. and s. by Appling, s. w. by Ocmulgee r., separating it from Irwin, w. by Dooley, and N. w. by Pulaski. Length from s. to N. 28 ms, mean width 22, and area 836 sq. ms. Extending in lat. from 31° 39′ to 32° 12′ N., and in long. from 5° 46′ to 6° 20′ w. W. C. The southern part of this co. slopes to the sthestrd. giving source to many of the higher branches of Santilla, which rise almost on the margin of Ocmulgee. The latter stream, forming the sthwstrn. border, thence traverses the co. in a nrthestrn. direction, serving as a common recipient for the confluents which drain the nrthrn. section towards Pulaski. Chief town, Jacksonville. Pop. 1820, 2,104, and in 1830, 2,146.

TELLICO, p-v. Monroe co. Ten. (*See Madisonville, Monroe co. Ten.*)

TELLICO PLAINS, p-o. Monroe co. Ten., 15 ms. southward Madisonville, the st. of just. of the co., and by p-r. 183 ms. s. E. by E. Nashville.

TELLICO, (Mouth of,) p-o. at the mouth of Tellico cr., 12 ms. N. E. Madisonville, the co. seat, and by p-r. 180 ms. s. E. by E. Nashville.

TEMPERANCE, p-v. Greene co. Geo. by p-r. 53 ms. nrthrd. Milledgeville.

TEMPERANCE RIDGE, p-o. Yazoo co. Miss., by p-r. about 120 ms. N. N. E. Natchez.

TEMPLE, p-t. Kennebec co. Me., 40 ms. N.

w. Augusta, in the N. W. corner of the co., has Oxford co. w., and Somerset co. N., and is mountainous, having part of Blue mtn. w. Pop. 1830, 795.

TEMPLE, p-t. Hillsborough co. N. H., 40 ms. Concord, 12 Amherst, gives rise to several branches of Souhegan r. The situation is high, with a fine and extensive view E. and S., a rocky surface, and pretty good soil. Here is a social library. Pop. 1830, 648.

TEMPLETON, p-t. Worcester co. Mass., 60 ms. w. Boston, has an uneven surface, with rich vallies, and a soil generally good, watered by several streams, which flow partly into the Chickapee, and partly into Miller's r., and afford mill seats The v. is neat and pleasant. It was granted to soldiers who had served in Philip's war, under the name of Narragansett, No. 6. Pop. 1830, 1,552.

TEMPLE OF HEALTH, p-o. in the wstrn. part of Abbeville dist. S. C., by p-r. 114 ms. w. Columbia.

TEMPLETON, p-v. Prince George's co. Va., 36 ms. southeastward Richmond.

TEN MILE STAND, and p-o. Rhea co. Tenn., by p-r. 171 ms. S. E. by E. Nashville.

TENNESSEE, r. of the states of Tenn., N. C., and Geo., though a very minor branch, is the stream from which the general name has been, by custom, arising from the route of original discovery, communicated to the great recipient. Tenn. proper rises in Raban co. Geo., by its extreme sthestrn. source, quickly entering Macon, and receiving numerous creeks from Haywood co. N. C., and flowing N. W. passes the Unika mtn. into Tenn. Within the latter state it continues N. W. 40 ms., between Blount and Monroe, joins the Holston on the southeastern border of Roan co., after a comparative course of about 85 ms., of which 5 are in Geo., and 40 in each of the other two states. Though so much inferior in volume and length of course to the Holston, the name of Tennessee is perpetuated below their union.

TENNESSEE, r. of the state of the same name, and of the states of Ky., Miss., Ala., Geo., N. C., and Va., is the great sthestrn. constituent of the Ohio. Under the respective heads of Clinch, Holston, French Broad, Tenn. proper, and Duck rivers, the constituents of Tenn. will be found described. The very peculiar features of the valley of Tenn., demand a general and particular notice. This valley is naturally divided into two physical sections; the higher or mountainous, and the lower or hilly. The most remote sources of Tenn. are found in those of Clinch in Tazewell, and of Holston in Wythe, cos. of Va., interlocking sources with those of Sandy and Great Kenhawa. From this elevated origin, the main confluents pursue a sthwstrn. course between the two parallel chains of the Appalachian system, Cumberland, and the main spine, both stretching in a similar direction with the rivers, at a mean distance of about 70 ms. asunder. Besides this principal valley, another of less width between the main chain and Blue Ridge, is also drained by the constituents of Tenn.; but this more eastern and more elevated valley slopes to the N. W., at right angles to the mtn. chains. The latter mtn. valley comprises the N. C. and Geo. part of the valley of Tenn., and will be found noticed under the heads of Macon, Haywood, and Buncombe counties, of the former state, and under the heads of Tenn. proper, and French Broad rivers. Including both minor vallies, upper Tenn. drains an elongated ellipse of 350 ms. longer axis; shorter axis 120 ms. from the Blue Ridge at the sources of French Broad, to Cumberland mtn., where it separates the sources of Powell's river from those of Cumberland: mean breadth 80 ms., and area 24,000 sq. ms. Descending from the extreme fountains in Va., the valley widens as the mountain chains recede from each other, and again contracts as the same chains gradually re-approach each other at the northwestern angle of Geo., and nrthestrn. of Ala. At the latter point, well known by the name of Nickajack, all the large confluents have united, and the Blue Ridge and Cumberland chains have inclined to within less than 40 ms. of each other. Below Nickajack, the now large volume of Tenn. continues S. W. 60 ms., without receiving a single creek of 20 ms. course, the two bounding mountain chains still inclining upon each other, till their approaching bases force the river through the Cumberland chain. To one whose eye first glanced on the volume of Tenn., below its passage through Cumberland mtn., without previous knowledge of the valley above, no adequate idea would occur, that before it, flowed the accumulated waters of a mountainous region of 24,000 sq. ms. extent. In fact, to an observer, thus placed, the main volume of Tenn. would appear as one of the constituents of a river valley below the Cumberland chain. About 20 ms. below the passage of Tenn r. through it, the Cumberland mountain receives the Blue Ridge, if such a term can be correctly applied to the merging of two mtn. chains. Here, along the nrthrn. sources of Mobile basin, the Appalachian system changes its distinctive character, and the confused masses of hills follow each other wstrdly. toward the Miss. The Tenn. river deflects rather more than does the mtn. system, and flows N. W. by W. by comparative courses 120 ms., to the nrthwstrn. angle of Ala., and the nrthestrn. of Miss., where this large stream again bends at nearly right angles, and pursues a course of a very little W. of N. 150 ms., to its entrance into the Ohio, after an entire comparative course of 680 ms.

The second great section of Tenn., and the lower part of the first, below Nickajack, are comprised in the fine northern valley of Ala. The main volume flowing along the base of a physical, extending from the Ohio valley in the vicinity of Pittsburg, to the nrthrn. part of the basin of Mobile. The very striking coincidence of the river inflections between the extremes of this region, must appear to the most inattentive observer of a good map of that part of the U. S. This regularity of structure is evinced by the great inflections of Ohio, Kenhawa, Kentucky, Green, Cumberland, and Tennessee rivers. The Tenn. itself literally occupies the base of the physical region indicated, as in all its comparative course below Nickajack, or its entrance into Ala., of 330 ms., it does not receive a single confluent above the size of a large creek, nor does the outer selvedge of its valley on the left, in Ala., Miss., Tenn., and Ky., exceed a mean breadth of 20 ms. On the right, embosomed between Tenn. and Cumberland rivers, and comprising central Tenn., and northern Ala., spreads a physical region, extending from Cumberland mtn. to the lower reach of Tenn. r., 130 ms., with a mean breadth of 80 ms., and an area of 10,400 sq. ms. This beautiful tract is semicircled by the main volume of Tenn., and drained by Elk r., Duck r., and innumerable creeks. Below Duck r., however, Tenn. receives no confluent from either side of any magnitude worthy notice in a general view. Including all its sections, the lower valley of Tenn. comprises an area of 17,600 sq. ms.; and the whole valley embraces a superficies of 41,600 sq. ms. This extent of Tenn. valley, if compared with the whole valley of Ohio, spreads over very nearly 1-5 part, and gives to Tenn. the first rank among the confluents of Ohio. Amongst the peculiar features of the course of Tenn., the most remarkable is, that rising as far N. as lat. 37° 10′, and curving thence southward to lat. 34° 23′, it again recurves back to its original lat., and falls into the Ohio r. almost exactly due W. from its primitive springs in Tazewell co.; thus embosoming nearly the whole large valley of Cumberland, and part of that of Green river. Geographically, Ten. valley lies between N. lat. 34° 10′ and 37° 10′, and in long. between 4° 15′ and 11° 40′ W. W. C. It is the first and largest, advancing from the S., of those streams gushing from the elevated slopes of the Appalachian ridges, and which flow wstrd. into the great basin of the Miss. In relative height, there is above 1,700 feet difference between the highest and lowest extremes of Tenn. valley. The arable surface of Tazewell and Wythe cos., from where the fountains of Kenhawa and Holston have their origin, must be at least 2,000 feet above the Atlantic tides; whilst that of Ohio r., at the influx of Tenn., but little exceeds 300 feet. The difference is fully an equivalent for 4° of lat., and accounts for the rapid changes of climate experienced on lines of lat. in Tenn. The current of every branch of Tenn. is very rapid, though direct falls are rare, and even dangerous shoals are not common. Of the latter, those particularly called Muscle Shoals, between Lauderdale and Lawrence cos. Al., are most remarkable and difficult to navigate. The whole river, however, having a mean fall exceeding 2 feet to the mile, is only favorable to down stream navigation, which it admits in most of its branches *to near their sources.*

Tennessee, state of the U. S., bounded by N. C. E., Geo. S. E., Ala. S., state of Miss. S. W., river Miss. separating it from Ark. W., and state of Mo. N. W., state of Kentucky N., and Va. N. E. If we commence the outline of this state on the southern boundary of Va. it will thence have a boundary, in common with N. C., along the main spine of the Appalachian mtns. to the northwestrn angle of Macon co. 168 ms.; due S. along the western boundary of Macon co. to the northern boundary of Georgia, 20 ms.; due W. along the northern boundary of Geo. and N. lat. 35°, to the northwestern angle of Alabama, 90 ms.; continuing the last noted line along the northern boundary of Ala. to Ten. river, and to the north eastern angle of the state of Miss. 145 ms.; still continuing due W. along the northern boundary of the state of Miss. to the Miss. river, 110 ms.; thence up the latter stream by comparative courses, opposite the Territory Ark. and sthestrn. angle of the state of Mo. 100 ms.; continuing up the Miss. river to the northwestern angle of Ten. and to the southwestern of Kentucky, 70 ms.; thence due E. along the southern boundary of Kentucky to Tennessee river, 80 ms.; thence up Tennessee r., 12 ms.; thence by a line a little S. of E. along the sthrn. boundary of Ky. to Cumberland mtns. and to the S. W. angle of Virginia, 268 miles; thence along the southern boundary of Va. and to place of beginning, 108 ms.; having an entire outline of 1,171 ms. Lying between lat. 35° and 36° 37′ N., and long. 4° 39′ and 13° 14′ W. W. C. The longest line that can be drawn on any state of the U. S. is a diagonal over Ten., from the nrthestrn. to the sthwstrn. angle, by calculation, S. 77°, W. or N. 77°, E. within a fraction of 500 ms. The mean length is about 400 ms., and the mean width being 114, the area of the state comes out 45,600 sq. ms., equal to 29,184,000 statute acres. This area exceeds what is commonly assigned to Ten., but following the most recent and accurate delineations on Tanner's map, is very near the real superficies of that state. By reference to our notice of Ten. river and valley, it will be seen how much the physiognomy of the state of the same name is influenced by the peculiar course of its rivers. Dividing this state into physical sections, and taking the mtns. as lines of demarcation, it presents two unequal sections; one the smaller above, and the second and larger below, the Cumberland chain. The higher and inferior section is entirely in the valley of Ten., and in length diagonally from S. W. to N. E. 280 ms., with a mean width of 57, and area 15,960, or very nearly one third of the state. This comparatively elevated and diversified region, is, in air, water, and surface, amongst the most delightful portions of the U. S. The soil is also much of it excellent, but the relative elevation gives to vegetable life a more northern effect than that found on similar lat. S., either on the Atlantic coast, or on the wstrn. section of Ten. near the Miss. On lower Ten., cotton is a staple production, whilst the climate of the upper section is more congenial to grasses, including the bread grain, or cerealia. The declivity of upper Ten. is to the s. w., and as already shown, by a rather rapid descent. Lower or western Ten. is subdivided by its rivers into two sections. That part comprised in the valley of Ten. river, has been noticed under the head of that stream, but to the nrthrd. of Ten. valley, the state embraces a large and very important section of that of Cumberland river. The latter tract is 250 ms. in length, along the line of demarcation between the states of Ky. and Ten., with a mean width of 40 ms., or 10,000 sq. ms. The area comprised in the valley of Ten. is about 170 ms. long., with a mean breadth of 70, or embracing an area of 11,900 sq. ms. Including the part of Ten. comprised in both the valleys of Ten. and Cumberland, below Cumberland mtn., we have an area of 21,900 sq. ms., which added to 15,960 comprised in upper Ten. yield 37,860 sq. ms. in the eastern and middle sections of the state. The general declivity of central or middle Ten. is wstrd., though the course of Ten. r. is here almost to the due N. Advancing still wstrd. of the valley of Ten. we arrive on a slope drained by numerous small streams direct into the Miss. This wstrn. inclined plane, comprising 7,740 sq. ms. may be both politically and naturally denominated wstrn. Ten. It is drained by Obion, Forked Deer, Big Hatchee, and Wolf rivers. These streams have corresponding curves, first flowing northwestardly, thence w. and s. w., giving a general western declivity to the plain of descent, which commencing about 25 ms. from the main channel of Ten. falls gently toward the Mississippi. In its natural state Ten. was covered with a dense forest. The great features along its very elongated declivity of 500 ms., are varied and strongly contrasted. E. Ten. mountainous or very hilly, with excellent river soil, presents a most seductive region to the eye: middle, or central Ten. less bold in its physiognomy, but with a much larger proportion of productive soil, is followed by the western section; the features of nature from the Cumberland chain, imperceptibly softening, until finally sunk into the annually inundated banks of the Miss. The whole state has sufficient soil to admit a dense population.. Agreeable to the returns of the recent congress of 1830, Middle and Western Ten., containing, as stated in this article, 29,640 sq. ms., has a pop. of 488,448, having had in 1820 only 287,-501, exhibiting a gain in the 10 years, from 1820 to 1830, of almost 70 per cent. The prodigious capacity for future increase may be estimated by the fact, that the existing pop. of the two lower sections of Ten. is distributively only 16 to the sq. mile; and this on a region, over which 10 fold more on an equal surface, would be far from too great density for the soil.

Political subdivisions.—Tennessee is divided into the counties of:

Counties.	Pop. 1820.	Pop. 1830.
Amoi,		
Anderson,	4,668	5,312
Bedford,	16,012	30,444
Bledsoe,	4,005	6,448
Blount,	11,258	11,027
Campbell,	4,214	5,110
Carroll,		9,378
Carter,	4,835	6,418
Cherokee Nation,		
Claiborne,	5,508	8,470
Cocke,	4,892	6,048
Davidson,	20,154	28,122
Dickson,	5,190	7,261
Dyer,		1,904
Fayette,		8,654
Fentress,		2,760
Franklin,	16,571	15,644
Gibson,		5,801
Giles,	12,558	18,920
Grainger,	7,651	10,066
Greene,	11,221	14,410
Hardiman,		11,628
Hamilton,	821	2,274
Hardin,	1,462	4,867
Hawkins,	10,949	13,683
Haywood,		5,366
Henderson,		8,741
Henry,		12,239
Hickman	6,080	8,132
Humphries,	4,067	6,189
Jackson,	7,593	9,902
Jefferson,	8,953	11,799
Knox,	13,034	14,498
Lawrence,	3,271	5,412
Lincoln,	14,761	22,086
McMinn,	6,623	14,497
McNairy,		5,697
Madison,		11,750
Marion,	3,888	5,516
Maury,	22,141	28,153
Monroe,	2,529	13,709
Montgomery,	12,219	14,365
Morgan,	1,676	2,582
Obion,		2,099
Overton,	7,188	8,246
Perry,	2,384	7,038
Rhea,	4,215	8,182
Rhoan,	7,895	11,340
Robertson,	7,270	13,802
Rutherford,	19,552	26,133
Sevier,	4,772	5,117
Shelby,	354	5,652
Smith,	17,580	21,492
Stewart,	8,397	6,988
Sullivan,	7,015	10,073
Sumner,	19,211	20,606
Tipton,		5,317
Warren,	10,348	15,351
Washington,	9,557	10,995
Wayne,	2,459	6,013
Weakly,		4,796
White,	8,701	9,967
Williamson,	20,640	26,608
Wilson,	18,730	25,477

Of whom in 1830, there were white persons—

	Males.	Females.
Under 5 years of age	59,576	55,399
From 5 to 10	45,336	42,975
10 to 15	36,044	33,556

From 15 to 20	29,247	30,616
20 to 30	44,982	42,970
30 to 40	25,111	23,545
40 to 50	15,110	15,261
50 to 60	11,158	9,279
60 to 70	5,513	4,511
70 to 80	2,102	1,855
80 to 90	657	542
90 to 100	105	114
100 and upwards	32	28
Total	275,068	260,680

Of which were deaf and dumb under 14 years of age, 129; 14 to 25, 59; 25 and upwards 54. Blind 176. Of the colored population were—

	Free.		Slaves.	
	Male.	Female.	Male.	Fem.
Under 10 years of age	842	272	27,713	26,568
From 10 to 24	583	626	23,431	24,145
24 to 36	361	359	11,260	12,223
36 to 55	321	285	6,020	6,519
55 to 100	216	187	1,729	1,891
100 and upwards	7	6	63	41
Total	2,330	2,225	70,216	71,387

Of the colored pop. were deaf and dumb under 14 years of age, 13; from 14 to 25, 9; 25 and upwards 6. Blind, 37.

Recapitulation.

Whites.	Free col'd.	Slaves.	Total.
535,748	4,555	141,603	681,906

History.—The territory now comprised in Ten. was included in the 2d charter of N.C., granted by Charles II. in 1664, but no settlement of whites was made so far westward until 1754, when a few families fixed themselves on Cumberland river, but were driven away by the savages. The first permanent settlement in Ten. was made by the founding of fort London in 1757. According to Flint, fort London stood on Little Ten., a mile above the mouth of Tellico. This place is now included in Blount co. Before me lies Pownall's map, founded on Evan's; the latter published in 1755. On this sheet it is noted that the farthest settlements of Va. westward in 1755, were on the heads of Blue Stone branch of Great Kenhawa, and those of Clinch and Holston. Ten. was then one wide wilderness. As noticed in the article Ten. Proper, the course of original settlement was from N. C. into the valley of that stream, and fort London was the cradle. This fort was attacked, however, and taken by the Indians in 1760, when upwards of 200 men, women, and children were massacred. In 1761, the important campaign under Col. Grant broke the power of the savages. A treaty was made which encouraged emigrants. About 1765, settlements began on Holston and gradually increased. Though harrassed by Indian warfare, the hardy frontier men penetrated deeper and deeper into the forest, and at the opening of the revolutionary war, were sufficiently strong to meet their savage enemies. Col. John Sevier was the Tennessean hero of that period. In June, 1776, the inhabitants, aided by a few Virginia soldiers, defeated the Indians. Hostilities continued nevertheless between the parties through the revolutionary war. As early as 1776, when the first repulican constitution of N. C. was framed and went into operation, deputies from Ten. appeared in the first state assembly. Though many previous, but abortive attempts had been made to settle w. Ten., the country around where Nashville now stands, was found a wilderness in 1779. The militia of Ten. gave themselves consequence in the eyes of their countrymen by the share they had on Oct. 7th, 1780, in defeating the British and tories at King's Mountain. In 1783, a land office was opened; courts of justice had been established and opened the previous year. In 1784, by a law of N. C. a provisional cession of what now constitutes Ten. was made to the U. S. This act was repealed, but had permanent effect, as under its influence the people formed an incipient independent state government, under the name of Frankland. These steps led to anarchy. N. C. claimed jurisdiction, as did also the constituted authorities of the state of Frankland. In the contest power prevailed, and the state of Frankland disappeared. The struggle led to many acts of civil commotion, which were not terminated until after 1790, when Ten. was finally ceded to the U. S. In May 1790, by a law of congress, the country was made a territory by the name of "the Territory s. of the river Ohio." In Nov. 1791, the first printing press was established at Rogersville, and on the 5th of the same month was issued the first newspaper, the Knoxville Gazette. On June 1st, 1796, Ten. was formally admitted into the Union as a state of the confederacy. Since her introduction into the family of republics, the advance of Ten. in population and wealth has been constant and peaceable. In the late war her troops acted a most honorable part, as they have in reality since the original settlement in the middle of the last century. *Government.*—Vested in a biennally chosen general assembly, composed of senators and members of assembly; who to be eligible must have resided in the state three years, and in the co. whence selected one year next before their election; and must have in possession, in full right, 200 acres of land. The number of representatives never to exceed 40, and the senators never to be more than one half, or less than one third of the representatives. The executive power is vested in a governor biennially elected, and eligible 6 years in 8; and to be eligible must possess, in full right, a free hold of 500 acres of land, have arrived at the age of 35 years, and have been a resident in the state 4 years next preceding his election. The judiciary is vested in such superior and inferior courts as the legislature may, from time to time appoint. Judges appointed by joint ballot of both houses of the general assembly, hold their offices during good behavior, and removable by impeachment. The right of suffrage secured to every free white male citizen of 21 years of age and upwards, who either possesses a free hold in the county where he offers to vote, or who has resided in the county six months previous to the election day. *Staple productions.*—To enumerate the staples of this state would be to give a list of nearly every vegetable and metallic substance produced in the U. S. The higher part of the state is most favorable to grain; the lower to cotton. Iron is made in several places. The Cumberland river is navigated by steamboats to Nashville, and all the large rivers of the state, for down boats to near their sources. *Education.*—For the advancement of the higher branches of education, the principal seminaries in Ten. are the Nashville university, at Nashville; East Ten. college at Knoxville; Greenville college, at Greenville, Greene co.; and at Maryville, the st. jus. Blount co., the sthrn. and wstrn. Theological seminary.

TENNESSEE RIVER, p-o. Haywood co. N. C., situated in the nrthrn. part of the co., by p-r. 343 ms. w. Raleigh.

TENNESSEE iron works, and p-o. Dickson co. Ten., 50 ms. wstrd. Nashville.

TENSAW, r. of Ala. The Tensaw is an outlet from Mobile river, about 8 ms. below the junction of Ala. and Tombigbee rivers. It is about 35 ms. comparative length, winds along, or near the eastern margin of the innundated tract above Mobile bay; is navigable, passes Blakely, and is lost in Mobile bay 4 or 5 ms. E. the town of Mobile.

TENSAW, r. of La., has its extreme source from Grand lake, and in the southeastern angle of Chicot co. and of the territory of Ark., but immediately enters Ouachita parish, La., and flowing a little S. of S. W. and nearly parallel to the general course of the Miss., by comparative courses about 110 ms. to its junction with Ouachita to form Black r. The Tensaw is the drain of the inundated tract w. the Miss. in the parishes of Ouachita and Concordia.

TENSAW, p-o. on the last noted river, Baldwin co. Ala., by p-r. 32 ms. N. Blakely, and 196 a little w. of s. Tuscaloosa.

TEORONTO BAY, Penfield, Monroe co. N.Y., makes up from lake Ontario, 1 m. by 5.

TERRE BONNE (*good or fertile land*,) parish of La., bounded by Atchafalaya bay, and parish of St. Mary's w., La Fourche Interior (Interior La Fourche) N. N. E. and E., and by the Gulf of Mexico S. E. S. and S. W. Greatest length from the mouth of La Fourche r. to Point au Fer, at the sthrn. entrance of Atchafalaya bay 90 ms., mean breadth 20, and area 1,800 sq.ms. Extending in lat. from N. lat. 29° to 29° 42′, and in long. W. C. 13° 08′ to 14° 35′. The surface very near that of a dead plain, the slight elevation of the alluvial banks of some of the streams excepted; and with the same partial exception devoid of timber. What very slight slope exists is sthrd., and in that direction it is traversed by Terre Bonne, Grande and Petite Cailloux, and Bayou Bœuf. Where the soil is arable, it is of exuberant fertility, and the climate completely within the range of sugar cane. Except the small islets at the S. W. Pass of the Mississippi, the cape of Terre Bonne is the most sthrn. part of La. Chief t., Williamsburgh. Pop. 1830, 2,121.

TERRE COUPEE, p-v. St. Joseph's co. Ind. In the p-o. list it is stated at 245 ms. from Indianopolis by the p-r., though the actual distance between the two places falls short of 150 ms., direction very nearly N. and S.

TERRE HAUTE, p-v. and st. jus. Vigo co. Ind., by p-r. 83 ms. S. W. by W. Indianopolis, and 60 ms. by the land road above and N. Vincennes. N. lat. 39° 30′, long. W. C. 10° 27′ W.

TERRYSVILLE, p-v. Abbeville dist. S. C., by p-r. 116 ms. wstrd. Columbia.

TERRYTOWN, p-v. Bradford co. Pa., by p-r. 142 ms. nrthrd. Harrisburg.

TEWKSBURY, p-t. Middlesex co. Mass., 20 ms. N. W. Boston, has Merrimack r. N., and Concord r. W., which flows into it, and separates the town from Chelmsford and Lowell. N. it is hilly, and stony, with pretty good soil; in other parts nearly level and poor. The Merrimack, at Hunt's falls, descends 40 feet in one-fourth mile, and this is the head of navigation on that stream. When the water is high, rafts easily descend. Above this fall the r. is joined by the Middlesex canal. Belvidere v. is in the N. W. corner of the t. at the junction of Concord and Merrimack rs. It is connected with the great and flourishing manufacturing v. of Lowell by a bridge across the former stream. Pop. 1830, 1,527.

TEWKSBURY, p-t. Hunterdon co. N. J., 45 ms. N. Trenton, has Morris co. N., Somerset co. E., with a range of hills crossing it, and its eastern boundary line formed by Allamatong river, a branch of the Raritan. It contains the village of New Germantown. Pop. 1830, 1,659.

THAMES, river, Conn., is formed by the confluence of the Quinebaug and Shetucket, & at Norwich takes the name of the Thames. It flows thence to New London harbor 14 ms., and affords sloop navigation from Norwich Landing to Long Isl. sound. A steamboat plies between New York and Norwich. The banks of this stream are pleasant and variegated. It flows through the old Mohegan country, and the reserved lands of that tribe lie on its W. banks, a little below Norwich. At the W. point of New London harbor is a light house, and within it are 2 forts, one of which, in Groton, was the scene of British cruelty in the revolutionary war, and is now marked by an obelisk, erected to the memory of the defenders.

THE POCKET, p-v. Moore co. N. C., by p-r. 82 ms. S. W. by W. Raleigh. Uncouth as this name may sound, it is that given in the p-o. list, thus "(The) Pocket."

THETFORD, p-t. Orange co. Vt., 34 ms. S. E. Montpelier, 28 N. E. Windsor; first settled about 1764; is crossed by Ompompanoosuc river, which here receives a branch, and both these streams afford mill sites. North is one half of Fairlee lake. There are other ponds, one of which, containing nine acres, is only about 25 feet from the bank of Conn. river, which is 100 feet below. It has neither inlet nor outlet, falls two or three feet in summer, and abounds in fish. Between the pond and the river passes a road. Galena is found here, which yields 75 per cent of lead. The surface is uneven, and somewhat rocky. An academy was established here 1819. There are several villages. Pop. 1830, 2,113.

THIBADEAUXVILLE, p-v. and st. of jus. parish of La Fourche, interior La.; situated on the left bank of La Fourche r., about 35 ms. s. E. and below Donaldsonville; N. lat. 29° 46', long. 13° 48' w. W. C.

THICKETY FORK, and p-o. northeast part of Spartanburg district, S. C., by p-r. 115 ms. N. N. W. Columbia.

THOMAS, co. of Geo., bounded w. by Decatur, N. W. by Baker, N. by Irwin, E. by Lowndes, s. by Jefferson co. Flor., and s. w. by Leon co. Flor. Length from s. to N. 50 miles, mean breadth 30, and area 1,500 sq. ms. N. lat. 31° and long. 7° w. W. C., intersect very near the centre of this co. The declivity southward; the western part drained by Ocklockonnee, and the eastern by Suwanee river. Chief town, Thomasville. Pop. 1830, 3,299.

THOMASTON, p-t. Lincoln co. Me., 49 ms. s. E. Augusta, 85 E. N. E. Portland, 36 E. Wiscasset; is of irregular form, with Waldo co. N., Penobscot bay E., and Saint George's river on part of the w. line; is of irregular form. It contains abundant quarries of lime stone, which is burnt in great quantities, and known in the ports of the United States for its good quality. About 150,000 casks have been sent out of Thomaston for 20 years past. The Saint George is navigable to this t. in large ships, 12 ms. from the ocean. The state prison is situated on the bank of this stream, in a tract of 10 acres, including a marble quarry. Excellent bluish granite is brought up from quarries below, on the river, which the convicts are employed in cutting. The plan of the building, and the system of discipline, are conformed to those of Auburn, Sing Sing, &c. The keeper's house is 30 feet by 40, the hospital 23 by 48, and there are 50 cells in the prison, all of stone, surrounded by a stone wall. A considerable number of vessels are owned here, and there is an active coasting trade to different parts of the country, chiefly for the transportation of lime; a bank, &c. The seat of the late Gen. Knox, is one of the finest in this part of the country. Pop. 1830, 4,214.

THOMASTON, late Upson C. H., p-v. and st. jus. Upson co. Geo., by p-r. 87 ms. a little s. of w. Milledgeville; N. lat. 32° 52', long. W. C. 7° 27' w.

THOMASVILLE, p-v. and st. of jus. Thomas co. Geo., situated in the forks of Ocklockonnee river, about 160 miles in a direct line, but by p-r. 235 miles s. s. w. Milledgeville; N. lat. 30° 58', long. 7° 04' w. W. C.

THOMPSON, p-t. Windham co. Conn., 46 ms. E. N. E. Hartford, 26 N. w. Providence; has Massachusetts N., and Rhode Island E., and is crossed by Quinebaug river, and French and Five Mile rivers, its branches; on which streams are excellent mill seats, and which afford fish. The surface is hilly. On French river is, first, Mr. Wilson's sattinet factory, with 14 looms, making 65,000 yards annually; then, 3 miles below, Messrs. Andrews and Fisher's, a stone factory, 100 feet long, with 2,200 spindles, and 52 looms, making 350,000 yards of printing cloths for calico. Maconville, 1 mile below, contains Masons and Thatcher's factory, with 2,436 spindles, and 60 looms, making 250,000 yards of shirtings. Near the junction of French river and the Quinebaug is Randall & Co's. factory, with 900 spindles, and 18 looms. Pop. 1830, 3,383.

THOMPSON, p-t. and st. jus. Sullivan co. N. Y., 113 ms. s. s. w. Albany, 34 from Newburgh; has Orange co. s.; watered by Navisink, Mongaup, Sheldrake and other creeks, with a variety of surface, soil, and timber. Monticello village contains the county buildings, and is 110 ms. from Albany. Pop. 1830, 2,457.

THOMPSON, p-v. Geauga co. Ohio, by p-r. 184 ms. N. E. Columbus.

THOMPSON'S, p-o. western part of Fairfield district, S. C., by p-r. 24 ms. N. N. w. Columbia.

THOMPSON'S CROSS ROADS, and p-o. Louisa co. Va., by p-r. 45 ms. N. w. Richmond.

THOMPSON'S STORE, and p-o. in the northern part of Hanover co. Va., by p-r. 46 ms. northward Richmond.

THOMPSON'S STORE, and p-o. southern part of Anderson district, S. C., by p-r. 121 ms. N. w. by w. Columbia.

THOMPSONTOWN, and p-o. Mifflin co. Pa., by p-r. 34 ms. N. N. w. Harrisburg.

THOMPSONSVILLE, p-o. Culpepper co. Va., by p-r. 75 miles s. w. by w. W. C.

THORNBURGH, p-v. Spotsylvania co. Va., by p-r. 70 ms. s. w. W. C.

THORNDIKE, p-t. Waldo co. Me., 40 ms. N. E. Augusta; is bounded by right lines, crossed by the branches of a stream flowing N. w. into Sebasticook river, and approaches nearly to Penobscot co. N. E. Pop. 1830, 652.

THORNBURY, p-v. eastern part of Chester co. Pa., by p-r. 119 miles N. E. W. C., and 18 ms. southwestward Phil.

THORN HILL, p-v. Orange co. N. C., by p-r. 92 ms. N. w. by w. Raleigh.

THORNTON, p-t. Grafton co. N. H., 58 ms. from Concord, 12 from Plymouth, 120 from Boston; is crossed by Pemigewasset river N. and s., and has Mad river and several other small streams. There are valuable meadows, and no high hills; first settled 1770. Pop. 1830, 1,049.

THORNTON, p-v. Delaware co. Pa., by p-r. 119 ms. N. E. W. C.

THORNTON'S GAP, & p-o. in the Blue Ridge, western part of Culpepper co. Va., by p-r. 102 miles N. w. by w. W. C., and 120 N. w. Richmond.

THORNTOWN, p-v. Boone co. Ind., by p-r. 62 ms. N. W. Indianopolis.

THORNVILLE, p-v. in the northwest angle of Perry co. Ohio, by p-r. 37 ms. E. Columbus.

THOROUGHFARE, p-o. Prince William county, Va., by p-r. 47 miles S. W. W. C.

THREE FORGES, and p-o. Bedford co. Pa., by p-r. 140 ms. N. W. W. C.

THREE FORKS, p-o. Barren county, Ky., by p-r. 122 ms. S. S. W. Frankfort.

THREE SPRINGS, and p-o. southeastern part of Huntingdon county, Pa., by p-r. 73 miles westward Harrisburg.

THROG'S NECK, or Point, Westchester, Westchester co. N. Y., the N. point of East river, and marks the W. termination of Long Island sound. It presents a low, broken, sandy bank E., on which the U. S. government have erected a light house. It forms a small peninsula.

THROOPSVILLE, p-v. Mentz, Cayuga co. N. Y., 3 ms. N. Auburn, on Owasco inlet.

THOUSAND ISLES, in the r. St. Lawrence, extend from the E. end of lake Ontario 30 ms. down that stream, and are of various size and form. The principal are Carleton and Welles isls. in N. York and Grand isl. U. Canada.

TICK CREEK, and p-o. southern part Chatham co. N. C., 44 ms. southwestward Raleigh.

TICONDEROGA, p-t. Essex co. N. Y., 96 ms. N. Albany, 3 S. Elizabethtown, has lake Champlain and Vt. E., Warren co. S., and includes the lower part of lake George. Fine levels extend near the lake, with fine swells rising behind, and several high mtns. rise in different parts of the town, some of which, as well as several spots on the lower ground, have been rendered interesting by historical events of importance. The ruins of fort Ticonderoga occupy the S. point of a promontory, below which bends lake Champlain, just before it spreads N. to a greater breadth than before, and opposite are mounts Defiance and Independence, with the narrow part of the lake between them, the former an abrupt elevation, 720 feet high, covered with forests, in this town, and the latter of inferior height and in Vt. The outlet of lake Geo. 3 ms long, flows in the S. part of Ticonderoga, and enters the lake between mount Defiance, and fort Ticonderoga at Swords point, between a tract of beautiful meadows. It has 3 falls, in all 157 feet, and turns some mills. The promontory is now overgrown by young timber, which has grown since the desertion of the fortress. About 500 acres were inclosed by a breastwork across the isthmus, by the French, who defended it against Gen. Abercrombie in 1748. Gen. Amherst took the fortress, the following year. It was taken by surprise in 1775, by a small party of men from Vt., commanded by Ethan Allen, but evacuated in 1777, by the American troops, on the approach of Gen. Burgoyne, who took his cannon to the top of mount Defiance, and thus secured the command of the place, from a position before considered inaccessible. A ferry crosses the lake to Shoreham, Vermont. Pop. 1830, 1 thousand 9 hundred ninety-six.

TIFFIN, p-v. and st. Seneca co. O., by p-r. 85 ms. N. Columbus. It is situated on the right bank of Sandusky r. Lat. 41° 08′, long. W. C. 4° 10′ W. Pop. 1830, 248.

TIFFINS, r. of Mich. and O. rising in the former, interlocking sources with those of the r. Raisin of lake Erie, and St. Joseph of Maumee. Formed by crs. issuing from Lenawee and Hillsdale cos. Mich., Tiffins r. flows S., enters O. traversing Henry and Williams cos., and falling into Maumee r. at fort Defiance just above, but on the contrary side from the influx of au Glaize r. Comparative length 50 ms.

TIMBALLIER, bay of La. extends westward from the mouth of La Fourche r. about 30 ms. with a width of from 3 to 6 ms. It has the same mouth with La Fourche, and is separated from the gulf of Mexico by a long low peninsula or island. It is shallow, with a sandy or muddy bottom.

TIMBERLAKE'S p-o. Campbell co. Ky., by p-r. 71 ms. N. N. E. Frankfort.

TIMPSON'S CREEK, and p-o. sthrn. part Rabun co. Geo. by p-r. 9 ms. S. Clayton, the co. st., and 165 ms. N. Milledgeville.

TINICUM, island and creek, and also tsp. Bucks co. Pa. The creek falls into Del. r. opposite the island, and the tsp. lies along both sides of the creek and on the Del. river between the tsps. of Noxamixon and Plumpstead, about 38 ms. northward Phila.

TINICUM, island and tsp. Del. co. Pa. The island lies in Del. r. below the mouth of Darley creek. Both the island and the adjacent shores are flat, and employed principally as grazing farms.

TINMOUTH, p-t. Rutland co. Vt., 41 ms. N. Bennington, 8 S. Rutland, 81 from Montpelier, first settled 1770, is crossed by Furnace brook; N. Little West r., a branch of Otter creek, which flows between two ranges of mountains. There are several quarries of marble, and plenty of iron ore, which supplies several furnaces and forges in this town. Pop. 1830, 1,049.

TIOGA, r. or Chemung, a W. branch of Susquehannah r. rises in Pa., runs N. into Steuben co. N. Y., which it crosses to Painted Post, where it meets Conhocton r., and then turning back into Pa., meets the E. branch at Tioga point. It flows about 50 ms. in N. Y., and with its branches, is navigable in boats. The Chemung canal connects this stream with Seneca lake. (*See Susquehannah r.*)

TIOGA, co. N. Y., bounded by Tompkins co. and parts of Steuben and Cortlandt cos. N., Broome co. E., Pa. S., the boundary being the 42d degree of lat., and Steuben co. W., about 180 miles W. Albany, contains 18 townships, is crossed in the S. E. by Susquehannah river, and Chemung or Tioga river, S. W., which meet 3 ms. S. of the Penn. line. Owego cr. on the E. line, and Catetant and Cayuta crs. which cross this co. afford boat and raft navigation. There are few manufactories, 2 woollen, and 1 furnace. The surface is hilly. Pop. 1820, 14,716, 1830, 27,690.

TIOGA, p-t. Tioga co. N. Y. 10 ms. W. Owego, 180 from Albany, has Pa. S., is crossed by the E. branch of Susquehannah, and has Cayuta creek W. The surface is generally broken, and the soil poor, with some rich but narrow vallies. Pop. 1830, 1,411.

TIOGA, co. Pa., bounded E. by Bradford, S. E. and S. by Lycoming, W. by Potter, and N. by Steuben co. N. Y. Length 36 miles, mean width 32, and area 1,152 square miles. Extending in lat. from 41° 32′ to 42° and in long. from 0° 04′ E., to 0° 40′ W. W. C. Surface composed of two declivities; that comprising the central and northern sections, and drained by the Tioga r., falls to the N. E. The opposite declivity slopes to the S. W., and is drained by the sources of Pine creek. The mean arable surface of Tioga, exceeds 1,000 feet above the Atlantic level. Chief town, Wellsboro. Pop. 1820, 4,021, 1830, 9,071.

TIONESTA, p-v. Armstrong co. Pa. New name, without relative distances on the general post office list.

TIOUGHNIOGA, creek, N. Y., rises in Onondaga co., and flows through parts of Cortlandt and Broome cos. to Chenango river after a course of about 55 ms.

TIPTON, co. Ten. bounded by Dyer N., Haywood E., Lafayette S. E., Shelby S. and S. W., and the Mississippi river, separating it from Crittenden co. Ark. W. Length 30 ms., mean width 23, and area 600 square ms. Extending in lat. from 35° 23′ to 35° 48′, and in long. from 12° 32′ to 13° W. W. C. The declivity is westward toward the the Miss. r. the northern part traversed by Forked Deer, and the southern by Big Hatchee rs. Chief town, Covington. Pop. 1830, 5,317.

TIPTONSPORT, p-v. and st. jus. Carroll co. Indiana, on Wabash r. by p-r. 94 ms. a little W. of N. Indianopolis. Lat. 40° 31,′ long. 9° 40′ W. W. C.

TICO, p-v. Richland co. O., by p-r. 20 ms. N. N. W. Mansfield, the co. st., and 83 ms. N. Columbus.

TISBURY, p-t. Duke's co. Mass. 85 ms. S. E. Boston, on Martha's Vineyard, has the Vineyard sound on the N. W. with the harbor of Holme's Hole N., where vessels often enter which are prevented from proceeding round Cape Cod shoal by contrary winds. This harbor is safe, and frequently affords protection to foreign ships as well as coasters. Near the harbor is a small village. Pop. 1830, 1,317.

TITUS' STORE, and p-o. Harrison co. O., by p-r. 116 ms. a little N. of E. Columbus.

TIVERTON, p-t. Newport co. R. I., 24 ms. S. E. Providence, 13 N. E. Newport, has the E. passage and Mount Hope bay W., and Mass. N. and E. It is generally pleasantly varied, with some rocky parts, enjoys considerable advantages in fisheries and navigation, and was connected with Rhode Island, some years since, by a stone bridge about 1,000 feet long. Pop. 1830, 2,905.

TOBY, p-v. northwestern angle of Armstrong co. Pa., about 55 ms. a little E. of N. Pittsburg, and by p-r. 236 ms. N. W. W. C.

TOBY'S CREEK. (*See Clarion river.*)

TODD, co. Ky., bounded W. by Christiana, N. by Muhlenburg, E. by Logan, and S. by Montgomery co. Ten. Length from S. to N. 36 ms., mean width 17, and area 612 square ms. Extending in lat. from 36° 37′ to 37° 06′, and in long. from 10° 04′ to 10° 22′ W. W. C. Todd occupies a part of the summit ground from which the waters flow northwardly into Green river and southwardly into Cumberland. Chief town, Elkton. Pop. 1820, 5,089, 1830, 8,683.

TOLLAND, p-t. Hampden co. Mass. 125 ms. S. W. Boston, has the line of Connecticut S. The surface is varied by large swells, and is crossed by Farmington river. Pop. 1830, 723.

TOLLAND, co. Conn., bounded by Mass. N., Windham co. E., New London co. S., Hartford co. W., about 15 ms. by 22, with 337 square ms. and 12 townships. The W. part is nearly level, with a light but good soil, and few stones; in the E. is the granite range, which is mountainous, and but partially cultivated. Scantic, Salmon, and Hockanum rivers, and their branches water the W. and S., and Willimantic and Hop rivers the E. Pop. 1820, 14,330, 1830, 18,702.

TOLLAND, p-t. and st. jus. Tolland co. Conn., 17 ms. N. E. Hartford, 52 N. E. New Haven, 42 N. W. New London, has Willimantic river on the E. line, and is generally rough and stony, with good grazing land. Oak and chestnut prevail in the forests. The earth affords granite, and iron ore. Snipsic pond is 2 ms. but narrow. The village is in the centre, on a plain, with the county buildings, bank, &c. Pop. 1830, 1,698.

TOMBIGBEE, river of Miss. and Ala., is the great western constituent of Mobile, and is formed by 2 branches, Tombigbee proper, and Black Warrior. (*See article Black Warrior.*) Tombigbee has its most remote source in the territory of the Chickasaw Indians, northern part of the state of Miss., interlocking sources with those of Bear creek, branch of Ten., Big Hatchee, and Yazoo. Augmented by numerous creeks from both sides, this river pursues a course of S. S. E. 110 ms., leaves the state of Miss., and enters that of Alabama. Preserving the original course 60 ms. farther, it receives the Black Warrior from the northeastward. This higher and particular valley of Tombigbee is in length 160 ms., with a mean width of 60, and comprising an area of 9,600 square ms., or if added to that of Black Warrior, will give 14,850 as the entire surface drained by the two confluents above their junction. Below the union of the two great branches, the now considerable stream, retaining the name of Tombigbee, assumes a southern and very tortuous course, but comparatively only about 100 ms. to its junction with Alabama, to form the Mobile. The entire length of Tombigbee, by the main stream is 270, and by the Black Warrior 240 miles. Below the mouth of Black Warrior the volume is but slightly augmented, as it receives only creeks of moderate size, and the valley does not exceed 40 ms. width, or 4,000 square ms., giving an entire area to the whole valley of 18,850 square ms. Geographically, the Tombigbee valley extends from lat. 31° 06′ to 34° 45′, and in long. from 9° 24′ to 12° 24′ w. W. C. It has the vallies of Yazoo, and Big Hatchee N. W., Ten. N., Coosa, Cahawba, and Alabama, N. E. E. and S. E., and those of Pearl river and Pascagoula S. W.

TOMLINSON'S, p-o. on the U. S. road, Allеghany co. Md., 20 ms. wstrd. Cumberland, and by p-r. 152 ms. N. W. by W. W. C.

TOMOKA, st. of jus. Mosquito co. Florida. Situation uncertain.

TOMPKINS, co. N. Y., bounded by Seneca and Cayuga cos. N., Cortlandt co. E., Tioga co. S., Steuben co. W., on the line of which lies Seneca lake, 170 ms. W. Albany. It contains 10 townships, 2 ms. of the head of Cayuga lake, which also forms the W. line for 6 miles. Fall creek, Cayuga inlet, Six Miles creek, and Cascadilla, Halsey's and Salmon creeks, &c. afford water power to a great amount. The land rises gradually near Cayuga lake 400 or 500 feet, and the soil is generally favorable to cultivation. There are 2 woollen factories at Ithaca, and 1 cotton factory, besides large flour mills, &c. Besides these there are in other parts of the co. 1 cotton and 1 woollen factory, 2 oil mills, 1 powder mill, 2 rifle manufactories, 3 furnaces, &c. Pop. 1820, 32,747, 1830, 36,545.

TOMPKINS, p-t. Delaware co. N. Y., 100 ms. from Albany, 30 S. W., Delhi, has Broome co. and a corner of Pa. W., and is crossed by Delaware r. which afterwards forms part of the W. line, and receives several branches. The surface is hilly and broken. The village of Deposit, at the bend in the r. and on the W. borders of the co. is a spot of importance in the lumber trade of the r. Pop. 1830, 1,774.

TOMPKINSVILLE, incorporated v., Richmond, Richmond co. N. Y., near the N. E. extremity of Staten island, opposite the quarantine ground of New York harbor, at the N. part of the narrows, occupies the shore and the acclivity of a hill, and has rapidly increased within a few years. It contains the quarantine buildings of the state, and hospitals belonging to the United States. The hospitals are 3 in number, each about 100 feet in length, and all surrounded by a wall, enclosing 30 acres. A Presbyterian church has recently been erected. The pavilion, on an eminence half a mile in the rear of the v. commands an extensive, varied and beautiful panoramic view over the New York bays, the city, the narrows, forts and part of Long and Staten islands, and out upon the Atlantic. The Seamen's Retreat has been recently erected about 1 mile below this village in Southampton near the shore, and affords accommodations for 100 or 150 invalid seamen. The sailor's snug harbor, on the N. shore of Staten island in the town of Richmond, and not far from this village, is a charitable institution long established in New York. Steamboats ply between Tompkinsville and New York city almost every hour. The shore is lined with good wharves, and the ground rises in bold and handsome swells from the water.

TOMPKINSVILLE, p-v. and st. jus. Monroe co. Ky., situated on the extreme head of Big Barren river, 87 miles northeastward Nashville, and by p-r. 144 ms. a little W. of S. Frankfort. Lat. 36° 43′, long. 8° 36′ W. W. C.

TONAWANTA, creek, N. Y., rises in Genesee county, and flowing between Niagara and Erie cos., falls into Niagara river opposite the middle of Grand island, 12 ms. N. Buffalo. It is about 90 ms. long, and has a sluggish current in the lower part of its course, so that it is made to serve as part of the Erie canal, which enters it 11 miles from the lake, and leaves it at Green Haven, near the shore, passing thence along the bank to Buffalo.

TONAWANTA, island N. Y., in Niagara river lies between the mouth of the river of that name, and Grand island. It is 3-4 m. long.

TONAWANTA, reservation, N. Y., is chiefly in Genesee co. and partly in Erie co. on Tonawanta creek, and is a rich tract of low land belonging to the Seneca Indians, who have a village on the creek.

TOPSFIELD, p-t. Essex co. Mass., 21 ms. N. E. Boston, has a varied surface, and is crossed by Ipswich river, on whose banks are fine meadows. First settled 1638. Pop. 1830, 1,010.

TOPSHAM, p-t. and st. jus. Lincoln co. Me., 31 ms. from Augusta, 27 W. Wiscasset, lies in the bend of Androscoggin river, which bounds it E. S. and W., separating it from Bath E., and Brunswick S. It is the seat of several mills and factories, contains the court house, &c. Pop. 1830, 1,567.

TOPSHAM, p-t. Orange co. Vt., 19 ms. S. E. Montpelier, 47 N. Windsor, first settled about 1781, is supplied with mill streams by head waters of Wait's river. The surface is uneven, with granite rocks. Pop. 1830, 1,384.

TORBERTVILLE, p-o. western part Upson co. Geo., by p-r. 92 ms. westward Milledgeville, and 5 ms. westward Thomaston, the co. st.

TORRINGTON, p-t. Litchfield county, Conn., 23 ms. N. W. Hartford, 7 N. E. Litchfield; is uneven, generally with good soil, favorable to grazing; crossed by the E. and W. branches of Waterbury river. Woolcotville is a manufacturing v. Pop. 1830, 1,654.

TOTTEN'S WELLS, and p-o. eastern part Obion co. Ten., by p-r. 11 ms. E. Troy, the county seat, and 161 ms. westward Nashville.

TOWAMENSING, p-v. western part Northampton county, Pa., by p-r. 194 ms. N. N. E. W. C.

TOWANDA, p-v. and st. jus. Bradford co. Pa., situated on the right bank of Susquehannah river, above the entrance of Towanda creek, 65 ms. above and N. W. Wilkes-Barre, and 15 below Tioga-point, and by p-r. 239 ms. a little E. of N. W. C.; N. lat. 41° 47, long. 0° 30′ E. W. C. The village is small, and composed of one street along the main road. In 1820, the township contained a population of 1,024. The adjacent country is in a high degree varied and romantic.

TOWN CREEK MILLS, Lawrence co. Ala. (*See Brickville, same co. and state.*)

TOWNSEND, p-t. Middlesex co. Mass., 45 ms. N. W. Boston; has the line of New Hampshire N., has much of its surface varied by gentle hills, with some pine plains. The soil is of secondary quality; fruit flourishes.—Townsend Harbor is a pleasant village. Pop. 1830, 1,506.

TOWNSEND, p-v. Sandusky county, Ohio, by p-r. 119 ms. northwards Columbus.

TOWNSHEND, p-t. Windham co. Vt., 28 ms. N. E. Bennington, 12 N. W. Brattleboro'; first settled 1761; is uneven, with many steep hills, and is crossed by West river, which has good meadows on its course. Other streams afford mill sites. Pop. 1830, 1,386.

TRACY'S LANDING, and p-o. Ann Arundel county, Md.

TRANSYLVANIA, University, Lexington, Ky., was founded in 1798, and according to the sub-article, Education, in the general article U. S. American edition Brewster's Encyclopœdia, contained in the present year, 1832, 143 students, with 2,350 vols. in the college library, and 1,500 in that of the student's. Mr. Flint in his western geography, says of this institution, "It has 12 professors and tutors, and in the academical, medical, and law classes, 376 students. Its library contains 4,500 volumes of standard works on medicine. All the libraries connected with the University, number 14,100 vols. The law school has 25 pupils, and the medical class 211."

This institution stands at the head of the various seminaries of education in the western states of the U. S.

TRANSYLVANIA, village, on the Ohio, in the extreme northern angle of Jefferson county, Ky.

TRANSYLVANIA, p-v. western part Greene co. Ohio, by p-r. 7 ms. westward Xenia, the co. seat, and 64 ms. S. W. by W. Columbus.

TRAP, p-v. southern part Talbot co. Md., 9 ms. S. Easton, the co. seat, and 93 ms. a little S. of E. W. C.

TRAP, p-v. Montgomery co. Pa., 9 ms. N. W. by W. Norristown, the st. jus., and by p-r. 152 ms. N. E. W. C.

TRAVELLER'S REPOSE, p-o. eastern part Pocahontas co. Va., by p-r. 221 ms. S. W. by W. W. C.

TRAVELLER'S REST, and p-o. Shelby co. Ky., 20 ms. westward Frankfort.

TRAYLORSVILLE, p-o. Henry co. Va., 6 ms. from Martinsville, the co. seat, and by p-r. 305 ms. S. W. W. C.

TRENT, small r. of N. C., rises in Lenoir, and traversing Jones into Craven co., falls into Neuse r., at and below Newbern.

TRENT BRIDGE, and p-o. on Trent r., Jones county, N. C., 7 ms. westward Trenton, the co. seat, and 133 S. E. by E. Raleigh.

TRENTON, p-t. Hancock co. Me., 87 ms. E. Augusta, 30 N. E. Castine; has Skilling's r. E., Union r. W., with the strait S., which separates Mount Desert island from the main land. Pop. 1830, 794.

TRENTON, p-t. Oneida co. N. Y., 13 ms. N. Utica; has West Canada creek E., which divides it from Herkimer co.; has pretty good soil, and many small streams. West Canada creek presents many romantic scenes along its course, having its bed deep into the rocks of dark colored lime-rock, which underlays the soil. In one place the banks are about 140 feet perpendicular, and for a distance of 2 or 3 ms. there is a succession of gulfs, rapids, and cascades, frequently narrowed by rocky precipices, & overhung by forest trees, which render this part of the course of the stream one of the most picturesque regions in the country, and one of the favorite objects among travellers in this state. The rocks abound in curious vegetable and animal remains, and the stream with fine trout. Steuben and Cincinnati creeks unite at the v. Pop. 1830, 3,221.

TRENTON, city and p-t. Hunterdon co. New Jersey, the capital of the state, is situated on the E. bank of the Delaware river, at the head of steamboat and sloop navigation. It is in lat. 40° 14′ N., and in long. 2° 16′ E. W. C., 11 ms. S. W. of Princeton, 27 S. W. of New Brunswick, 60 S. W. of New York, 30 N. of Philadelphia, and 166 N. E. of W. C. The navigation of the r. beyond this place by sloops, &c., is limited by the rapids, and by a fine wooden bridge of 5 arches, and more than 1,000 feet in length, which is the first above its mouth. The town is of considerable size and importance; it is in the S. W. corner of the county, near the Assanpink creek. The principal streets are regularly laid out, and contain many good dwelling houses and numerous stores. Among the public buildings are the state house, 2 banks, and 6 churches in the city and town. The ground on which the city is situated, as well as the surface of the town in general, is considerably varied. The Delaware and Raritan canal, extending from Trenton to New Brunswick, crosses the city, and is here joined by the feeder which enters the river above the falls. These falls afford water power for extensive manufacturing privileges; and in the city and its neighborhood are 10 manufactories and mills, several of which are of cotton goods. These are all supplied with water, either from the Delaware river, or the Assanpink creek. In 1831, a company was incorporated with a capital of $60,000 to construct a dam on the Delaware river near Wells' falls, and a race way on the E. side to any point not more than 1½ ms. below Trenton falls, with authority to sell the water, &c., for manufacturing purposes. The expense is estimated at $100,000, the water power to be obtained equal to that of 529 horses, and another dam, costing but $5,000, would increase the power to that of 1,170 horses. The Delaware river is navigable by boats far up into the state of N. Y., and affords communication with an extensive and fertile country, yielding grain, lumber, anthracite coal, &c. Large steamboats come up from Philadelphia as far as this place, except when the water is very low. The bed of the river here is covered with round stones, which are taken in great quantities to Philadelphia for paving, and supplied by the spring floods yearly. The feeder of the Delaware and Raritan canal enters the city from 20 ms. above, and is 40 feet wide, and 6 feet deep. The canal itself, which will be completed in 1833, will afford sloop navigation across the state from New York to Philadelphia. This place is memorable from its being captured from the British and Hessians by surprise, on the evening of the 25th Dec., 1776, by Gen. Washington. It was occupied by the British in Jan., 1777, when the American army under Washington, effected their celebrated retreat from Lamberton, on the opposite side of the Assanpink creek, and gained the N. part of the state. Pop. in 1820, 3,925.

Trenton, p-v. and st. jus. Jones co. N. C., situated on the small river Trent, 21 miles a little s. of w. Newbern, and by p-r. 140 ms. s. e. by e. Raleigh; lat. 35° 2′, long. 0° 26′ w.

Trenton, p-v. and st. of jus. Gibson county, Ten., situated on the North fork of Forked Deer river, by p-r. 139 ms. w. Nashville. Lat. 35° 57′, long. 12° w. W. C.

Trenton, p-v. extreme southern part Todd co. Ky., by p-r. 200 ms. s. w. by w. Frankfort.

Trenton, p-v. northeastern part Butler co. Ohio, by p-r. 39 ms. n. n. e. Cincinnati, and 93 s. w. by w. Columbus.

Trescot, t. Washington co. Me., e. by n. Augusta; is the last town on the coast, in the state, except Lubec, which is on the e. boundary of the U. S. It has the Atlantic, or rather the strait between Grand Menan isl. and the main, s.; contains Haycock's harbor, and Moose cove, and has a cove called Bailey's Mistake, on the e. line. There are no considerable streams. Pop. 1830, 480.

Trexlertown, p-v. Lehigh co. Pa., on Little Lehigh, 8 ms. s. w. Allentown or Northampton, the co. seat, and by p-r. 170 ms. n. e. W. C.

Triadelphia, p-v. and Cotton Factory, on Patuxent r., northeastern part Montgomery co. Md., 26 ms. n. W. C., and about an equal distance s. w. by w. Baltimore. This place is worthy of a visit for the richness and variety of its scenery.

Triana, p-v. situated on the right bank of Ten. river, and in the southern part of Ala., 15 ms. s. w. Huntsville, and by p-r. 145 ms. n. n. e. Tuscaloosa.

Triangle, p-t. Broome co. N. Y., 132 ms. from Albany, 7 from Lisle village, between Onondaga and Chenango rivers; was formed in 1831.

Trigg, co. of Ky., bounded by Ten. river separating it from Calloway s. w., Caldwell n. w. and n., Christian e., and Montgomery and Stewart cos. Ten. s. It is very nearly a right angled triangle, perpendicular n. from the Ten. line, 32 ms., mean width 16, and area 512 sq. ms. Extending in lat. from 35° 37′ to 36° 04′, and in long. from 10° 42′ to 11° 13′ w. W. C. The western part is traversed in a northerly direction by Cumberland r., and the eastern part drained into the latter stream by Little river. Chief t. Cadiz. Pop. 1820, 3,874, 1830, 5,916.

Trimble's Iron Works, and p-o. Greenup co. Ky., by p-r. 142 ms. a little n. of e. Frankfort.

Triplett, p-v. Fleming co. Ky., by p-r. 97 ms. eastward Frankfort.

Troublesome Iron Works, and p-o., on Troublesome creek, southern part of Rockingham co. N. C., by p-r. 100 ms. sthwstrd. Raleigh.

Trough Creek, and p-o. near the centre of Huntingdon co. Pa., about 9 ms. s. s. e. the borough of Huntingdon, and by p-r. 133 ms. n. n. w. W. C.

Troup, co. of Geo., bounded n. by Carroll, e. by Merriwether, s. by Harris, and w. by the Creek territory in Ala. Length 24 ms., mean width 18, and area 432 sq. ms. Extending in lat. from 32° 55′ to 33° 15′, and in long. from 8° 02′ to 8° 24′ w. W. C. The declivity is to the s. s. w., traversed in that direction by the main volume of Chattahoochee river, which, entering on the northern border, leaves the co. near the s. w. angle. Chief t. La Grange. Pop. 1830, 5,799.

Troup, C. H. (*See La Grange.*)

Troupsburgh, p-t. Steuben co. N. Y., 30 ms. s. w. Bath, has Pennsylvania on the south line, and Allegheny co. west. First settled, 1805; is supplied with mill sites by Tuscarora and Troup's cr. The soil is favorable to grazing, and iron ore is found. Pop. 1830, 666.

Trousdale, p-v. northwestern part Stewart co. Ten., by p-r. 97 ms. n. w. by w. Nashville.

Trout Run, p-o. near the northern border of Lycoming co. Pa., 14 ms. n. w. Williamsport, and by p-r. 101 ms. n. n. w. Harrisburg. Trout run is a branch of Lycoming cr.

Trout Run, p-o. eastern part Hardy county, Va. by p-r. 101 ms. w. W. C.

Troy, p-t. Waldo co. Me., 39 ms. n. e. Augusta; borders on Penobscot co. e. and n. e., and its streams, which are small, flow from the town in different directions, but empty into the Sebasticook. Pop. 1830, 803.

Troy, p-t. Cheshire co. N. H., 60 ms. from Concord; has but few mill sites. Population, 1830, 676.

Troy, p-t. Orleans co. Vt., 47 miles n. e. Montpelier, 51 from Burlington; first settled 1800, from Conn.; was almost deserted in the late war with Great Britain. Missisque crosses w. and falls 70 feet in a rocky and romantic pass. The soil is good for both grain and grass, the surface generally level, particularly on the river meadows. Population, 1830, 608.

TROY, p-t. Bristol co. Mass., lies on the w. side Taunton r., and is divided by Wahupper pond. Fall River village in this town, and near Taunton r., at the head of Mount Hope bay, is a place of extensive manufactures. The river falls here about 800 feet in a distance of about 27 rods, and 9 dams, with each a fall of about 14 feet, supply 13 cotton factories. These manufacture about 9,160,000 yards annually. The largest (Massasoit) runs 10,000 spindles, 350 looms, employs 400 hands, and consumes annually 810,000 lbs. cotton. The whole run upwards of 31,500 spindles, and 1,050 looms, employ 1,276 hands, and manufacture 2,290,000 lbs. of cotton annually. Here is also a satinet factory, employing 150 persons, and a print factory employing 260; iron works manufacturing 1,000 tons annually, and also two machine shops employing about 60 hands. Nearly all the investments have been made within seven years: the village now (1832) contains about 5,000 inhabitants, and 7 places of public worship. Pop. of the town, exclusive of the village of Fall River, in 1830, was 4,159, that of the latter 3,431.

TROY, city and st. jus. Rensselaer co. N. Y., on the E. bank of the Hudson r., 6 ms. N. of Albany, 156 N. of N. York, and 383 N. E. of W. C., is in N. lat. 42° 43', and in 3° 15' E. long. W. C. It is built on a handsome and somewhat elevated plain, extending from the shore of the r. to the foot of a range of hills, about 1 m. w., down which flow several mill streams. The city is regularly laid out, the principal streets being parallel with the river; and these as well as many of the cross streets are compactly and handsomely built, chiefly with brick. Most of the business is transacted near the river, where the stores are mostly located,—some of the private dwelling houses are commodious and elegant. Many of the streets are adorned with fine shade trees; and strangers generally are struck with the neatness and elegance of the city. Among the public buildings are, the court house, which is of stone and in the Grecian style of architecture; the jail, the house of industry, the Episcopal church, which is an elegant Gothic edifice, and 6 other handsome churches, the market house, 3 banks, the lyceum of natural history, connected with which is a mineralogical cabinet, &c. The Rensselaer school, a literary institution for the practical instruction of young men, established by the Hon. S. Van Rensselaer, has been for some years in this city, but is to be removed to some other part of the county. The library apparatus &c. of this institution cost $5,000 or $6,000. A classical department is to be added to it, and the "manual labor system" to be introduced. Another institution is the Troy female seminary, which has acquired a high reputation. The building is large and well situated in an eligible part of the city. The number of its pupils is usually about 200. There are also other good schools and academies. Troy enjoys a very fine situation for trade and manufactures. Its communications with the interior are numerous and good. The river is navigable to this place by steamboats and large sloops; and a water communication is opened with the Erie and Champlain canals by a dam across the Hudson, a branch canal, locks, a basin, &c. Daily lines of steamboats run to New York; and trade with Boston, and other eastern towns is kept up by sloops, and regular packets, as well as across the country. There is a macadamized road commencing opposite to the city and extending to Albany, upon which hourly stages run to that city. The water power afforded by the Poestenkill, and Wynautskill rs.,—small streams which take their rise on the eminences near the city—is profitably employed, and numerous manufactories of iron, cotton &c. are carried on in the vicinity. The scenery in the neighborhood of Troy is interesting; and the eminence in the rear of the city, called mount Ida, is a beautiful and romantic spot. The view of the neighboring cities, of the Hudson r. stretching to the south, and generally of the country for miles around, is very fine. Pop. in 1830, 11,405.

TROY, p-v. Bradford co. Pa., 20 ms. northwardly from Towanda, and by p-r. 148 ms. above and northward Harrisburg.

TROY, p-v. and st. jus. Obion co. Ten., situated near the centre of the co., by p-r. 161 ms. a little N. of w. Nashville. Lat. 36° 16', long. 12° 17' w. W. C. Troy in Obion is the most northwesterly st. jus. in the state of Ten.

TROY, p-v. and st. jus. Miami co. O., on Stillwater branch of Great Miami, 21 ms. N. Dayton, and by p-r. 78 ms. w. Columbus. Lat. 40° 03', long. W. C. 7° 14' w. Pop. 1830, 504.

TROY, p-v. Oakland co. Mich., by p-r. 36 ms. nrthwstrd. Detroit.

TROY, p-v. on Ohio r., at the mouth of Anderson's cr., sthwstrn. angle of Perry co. Ind., by p-r. 148 ms. a little w. of s. Indianopolis.

TROY, p-v. and st. jus. Lincoln co. Mo., situated towards the sthrn. side of the co., 53 ms. N. w. St. Louis. Lat. 38° 53', long. W. C. 13° 56' w.

TROY'S Store, and p-o. wstrn. part Randolph co. N. C., 65 ms. w. Raleigh.

TRUCKSVILLE, p-v. Luzerne co. Pa., by p-r. 6 ms. Wilkes-Barre and 120 N. E. Harrisburg.

TRUMANSBURG, p-v. Tompkins co. N. Y., 11 ms. N. w. Utica.

TRUMBAURSVILLE, p-v. wstrn. part Bucks co. Pa., 24 ms. nrthwrdly. Phil.

TRUMBULL, p-t. Fairfield co. Conn., 17 ms. w. New Haven, 4½ from Bridgeport, 55 from Hartford, about 4 ms. by 5½, is uneven, with good soil, and primitive rocks, watered by Pequannock r. Pop. 1830, 1,242.

TRUMBULL, co. of O., bounded s. by Columbiana, s. w. and w. Portage, N. w. Geauga, N. Ashtabula, N. E. Crawford, Pa., and E. and S. E. Mercer co. Pa. Length 36 ms., breadth

25, and area 930 sq. ms. Lat. 41° 15′, long. W. C. 3° 45′ w. Slope sthestrd. and drained by the Mahoning or western constituent of Big Beaver, and its branches. The soil is generally good. Chief t., Warren. Pop. 1820, 15,546; 1830, 26,153.

TRUMBULL, p-v. nrthrn. part Ashtabula co. O., 182 ms. N. E. Columbus.

TRUXVILLE, p-v. nrthrn. part Richland co. O., by p-r. 83 ms. N. Columbus.

TUCKASAGA, p-v. sthrn. part Mecklenburg co. N. C., by p-r. 160 ms. s. w. by w. Raleigh.

TRURO, p-t. Barnstable co. Mass., 65 ms. s. E. Boston by water, 107 by land. It is surrounded by water except on the N. w., being on a peninsula connected on that side with Truro, by a narrow isthmus. Cape Cod bay and Provincetown harbor are w. of this town, and the Atlantic E. The surface is uneven and the soil sandy. Pamet r. is an inlet 3 ms. long and from 1-4 to 3-4 m. wide. This inlet or bay almost insulates the t. There are 2 small vs. The inhabitants depend principally on fishing. The Indian name was Peeshawn; and it was visited by some of the Plymouth pilgrims before they went up Cape Cod bay. They had landed at Provincetown harbor, and here obtained some corn, which they planted the next season. Settled in 1700. Pop. 1830, 1,547.

TRUXTON, p-t. Cortlandt co. N. Y., 142 ms. w. Albany, 14 N. E. Homer, has good soil, well watered and supplied with mill seats, with a pleasant village. Pop. 1830, 3,885.

TUCKER'S HOLE, or Robinson's hole, Barnstable co. Mass., the passage between Nashawn and Presque isls. into Buzzard's bay.

TUCKERSVILLE, v. Wayne co. Geo.; on Tanner's map this place is marked as the st. jus. of Wayne co. Geo., but in the post list of 1831 there is only one p-o. named in that co., and that Waynesville, marked as the C. H. also. (*See Waynesville, Wayne co. Geo.*)

TUCKERSVILLE, p-v. Crawford co. Ind., by p-r. 108 ms. s. Indianopolis.

TUFTONBOROUGH, p-t. Strafford co. N. H., 50 ms. Concord, N. E. Winnipiseogee lake, has several ponds and brooks, with a varying surface and soil, and scenery enriched by several bays and coves of the lake. First settled about 1780. Pop. 1830, 1,375.

TULL'S cr. and p-o. Currituck co. N. C., by p-r. 221 ms. N. E. by E. Raleigh.

TULLY, p-t. Onondaga co. N. Y., 14 ms. s. Onondaga, 50 Utica, N. Cortlandt co., contains some of the head streams of Onondaga, Tioughnioga and Chenango crs. and Susquehannah r. It is diversified with hilly ridges and broad and fertile valleys. Tioughnioga creek has its source in two ponds of 100 and 400 acres. Pop. 1830, 1,640.

TULLYTON, p-v. Greenville dist. S. C., by p-r. 113 ms. N. w. Columbia.

TULLYTOWN, p-v. Buck co. Pa., by p-r. about 25 ms. nrthrd. Phila.

TUMBLING Shoals, and p-o. Laurens dist. S. C., by p-r. 92 ms. N. w. Columbia.

TUNBRIDGE, t. Orange co. Vt., 30 ms. N. Windsor, 26 s. E. Montpelier, first settled about 1776, is crossed by a branch of White r. N. and s., on which are mill sites, and has a good soil especially on the r., but the surface is uneven. There is a mineral spring west. Pop. 1830, 1,920.

TUNKHANNOCK, mtn. of Pa. and N. Y. The mtn. chains of Pa. are delineated on our maps in masses of confusion. On many maps, the very distinctive chain of Tunkhannock is omitted. It is traversed by the estrn. branch of Susquehannah, below the mouths of Bowman's and Tunkhannock creeks; and is known in Luzerne co. as Bowman's mtn. to the right, and Tunkhannock to the left of the river. It leaves the w. border of Luzerne, and in a s. w. by w. direction separates Lycoming from Columbia and Northumberland counties, and is traversed by the west branch of Susquehannah below Pennsboro'. It thence inflects to the w. s. w. and s. s. w. with the other Appalachian chains and is known locally in Pa. as the White Deer mtn., Nittany mtn., Tussey's mtn. and Evil's mtn., and again traverses Md. Va. and Ten. as a distictinctive chain. Towards the state of New York, though bearing no distinctive name, the continuation of Tunkhannock, passes between the two upper branches of Delaware r., turns to the nrthard. is traversed by the Mohawk at Little Falls, and bears there the local name of Sacandaga mtn.

TUNKHANNOCK, r. of Pa., rising in Susquehannah co., and flowing s. w. along the northwestern base of Tunkhannock mtn., enters Luzerne co. and falls into Susquehannah r. at the village of Tunkhannock, after an entire comparative course of 30 ms.

TUNKHANNOCK, p-v. situated on a beautiful site above the mouth of Tunkhannock cr. and on the bank of Susquehannah r., 28 ms. by the p-r. above Wilkes-Barre, and 142 N. N. E. Harrisburg.

TUPPER'S PLAINS, p-o. Meigs co. O., by p-r. 102 ms. s. E. Columbus.

TURBOTVILLE, p-v. Northumberland co. Pa.

TURMAN'S cr., p-o. Sullivan co. Ind., by p-r. 103 ms. s. w. by w. Indianopolis.

TURIN, p-t. Lewis co. N. Y., 145 ms. N. w. Albany, 15 N. Rome, 46 N. w. Utica, has Black r. E., Oneida co. s. and Oswego co. w. Boat navigation from the high falls, near the south line to Wilna, 45 ms. Fish cr. has its source in this town. At High falls, Black r. descends 63 ft. The inhabitants came principally from the E. states. Pop. 1830, 1,561.

TURKEY, cr. and p-o. wstrn. part of Buncombe co. N. C., 14 ms. wstrd. Ashville, the co. st., and by p-r. 273 ms. a little s. of w. Raleigh.

TURKEY FOOT, p-v. between Laurel Hill cr. and Castleman's r. in the sthwstrn. angle of Somerset co. Pa., 22 ms. s. w. from the borough of Somerset, and by p-r. 185 ms., but by the common travelled direct road 160 ms. N. w. by w. W. C.

TURKEY FOOT, p-v. Scott co. Ky., 16 ms. from Georgetown, the st. jus., and 27 ms. N. E. Frankfort.

TURMEL VIEW, p-o. Ind. co. Pa., 10 ms. wstrd. Blairsville and by p-r. 199 ms. N. W. by W. W. C.

TURNER, p-t. Oxford co. Me., 28 ms. W. Augusta, 18 E. Paris, 155 N. N. E. Boston, has Androscoggin r. on the E. border, which separates it from Kennebec co. and Cumberland co. S. It is crossed by a small tributary of Androscoggin. Population 1830, 2,220.

TURNER'S Cross Roads, and p-o. Bertie co. N. C., 16 ms. S. E. Windsor, the co. st., and by p-r. 114 ms. estrd. Raleigh.

TURNER'S Store and p-o. Caroline co. Va., 37 ms. nrthrd. Richmond.

TURNERSVILLE, p-v. nrthwstrn. part Robertson co. Ten., 35 ms. N. W. Nashville.

TUSCALOOSA, r. of Ala. (*See articles Black Warrior and Tombigbee.*)

TUSCALOOSA, co. Ala., bounded W. by Pickens, N. by Lafayette, N. E. by Jefferson, E. by Bibb, S. E. by Perry and S. W. by Greene. Greatest length diagonally from S. W. to N. E. 58 ms., mean width 24, area 1,392 sq. ms. Extending in lat. from 32° 53′ to 33° 28′, and in long. from 10° 10′ to 11° 03′ W. W.C. This very large co. is divided into two not very unequal sections by the Black Warrior r. which entering on the northern border, winds over it by a very circuitous channel in a general S. S. W. direction. The wstrn. part is drained by the Sipsey, which traverses the co. in a direction nearly parallel to the Black Warrior. The estrn. border is the dividing ridge between the valleys of Cahawba and Black Warrior; two thirds of the whole surface being in the latter valley and general slope S. S. W. This co. contains large tracts of excellent river soil. Chief t., Tuscaloosa. Pop. 1820, 8,229; 1830, 13,646.

TUSCALOOSA, p-t. st. jus. Tuscaloosa co. Ala. and seat of government for that state, is situated on the left bank of Black Warrior r. near the centre of Tuscaloosa co. Lat. 33° 12′, long. 10° 43′ W. W. C., by p-r. 155 ms. S. S. W. Huntsville, 226 a little N. of E. Mobile; and by the p-o. list 858 ms. S., 77° W. W. C. By calculation the course deflects 56° 46′ from the meridians, and the distance comes out 720 3-4 statute ms.

TUSCARAWAS, r. of O. (*See Muskingum r.*)

TUSCARAWAS, co. O., bounded by Harrison E. and S. E., Guernsey S., Coshocton S. W., Holmes N. W., and Stark N. Length from S. to N. 30 ms., mean breadth 23 and area 690 sq. ms. N. lat. 40° 30′, and long. W. C. 4° 30′ W. intersect in this co. The slope of the nrthrn. part is to the S., but inflects with the course of Tuscarawas r., which in the sthrn. part of the co. bends to the S. W. by W. Tuscarawas r. enters at the extreme nrthrn. angle, and flowing S. and thence inflecting gradually to S. S. W. divides it into two nearly equal sections, and has along its entire course the Ohio and Erie canal. The level of the canal near the centre of this co. is 874 feet above the ocean level; the arable soil averages from about 850 to above 1,000 feet of similar relative height. Chief t., New Philadelphia. Pop. 1820, 8,328; 1830, 14,298.

TUSCARORA, cr. Niagara co. N. Y., rises in the Tuscarora reservation, flows N. and N. E. 15 ms. to Lake Ontario.

TUSCARORA, Indian v. Lewiston, Niagara co. N. Y., is the residence of the Tuscarora tribe, which formed the 6th of the Six Nations of Indians in N. York. They came from the S., and speak a language very unlike those of the other nations. There is a church in the v., and a successful mission among them. The lands reserved for them by the state are 1 m. by 3.

TUSCARORA, mtns. of Pa. Similar remarks made on Tunkhannock mtn., might be repeated on the Tuscarora chain. It is known distinctively as the Tuscarora mtn., between Huntingdon and Franklin, and between Perry and Mifflin, on both sides of Juniata. East of the Susquehannah r. it is the Mahantango, between the counties of Dauphin and Northumberland; and towards the Potomac, it is the Cone mtn., between Franklin and Bedford counties. In the latter region it touches almost, but does not merge in the Kittatinny, and after being traversed by the Potomac, is evidently perpetuated in the Sideling hill of Morgan, Hampshire, and Hardy cos., Va.; and if carefully and scientifically examined, would, in all rational probability, fully sustain in both directions, that identity which constitutes the most remarkable characteristic of the Appalachian chains.

TUSCARORA, cr. and valley. This valley, watered by a cr. of the same name, lies between Tuscarora and Shade mtns., and constitutes the southwestern part of Mifflin co., Pa. The Tuscarora cr. however, rises in the sthestrn. part of Huntingdon, but quickly entering Mifflin, flows down the beautiful vale to which it gives name, and falls into Juniata r. below Mifflintown.

TUSCARORA Valley, p-o. is situated in the S. W. part of Mifflin co. 53 ms. wstrd. Harrisburg.

TUSCAMBIA, p-v. nrthrn. part Franklin co. Ala., 3 ms. a little E. of S. Florence, on Ten. r., and 122 ms. N. Tuscaloosa.

TUSCAWILLA, p-v. Leon co. Florida, 10 ms. sthrd. Tallahassee.

TUSCUMBIA, p-v. near the left bank of Ten. r., nrthrn. part Franklin co. Ala., by p-r. 3 ms. S. Florence, in Lauderdale co., and 141 ms. N. Tuscaloosa.

TUTHILLTOWN, p-v. Ulster co. N. Y., 22 ms. S. Kingston, on Sawangunk cr.

TWENTY MILE STAND, p-v. Warren co. O., by p-r. 91 ms. S. W. by W. Columbus.

TWIGGS, co. of Geo. bounded by Jones N., Wilkinson N. E. and E., Pulaski S. E. and S., and Ockmulgee r. separating it from Houston S. W., and Bibb W. Length 26, mean width 16, and area 416 sq. ms. Extending in lat. from 32° 30′ to 32° 56′, and in long. from 6° 18′ to 6° 41′ W. W. C. Narrow as is this co., it is a table land, as from the estrn. border the water courses flow sthestrd., towards the

Oconee, while the body of the co. has a s. w. declivity, towards Ockmulgee r. Chief t. Marion. Pop. 1820, 10,447; 1830, 8,031.

TWIN BLUFFS, p-v. on the left bank of the Miss. r., Warren co. Miss., about 80 ms. N. N. E. Natchez.

TWINSBURG, p-v. Portage co. O., by p-r. 142 ms. N. E. Columbus.

TWINTOWN, p-v. Ross co. O., by p-r. 56 ms. sthrd. Columbus.

TWITCHELL'S, Mills and p-o. Pope co. Il., by p-r. 149 ms. s. s. E. Vandalia.

TWYMAN'S, Store and p-o. Spottsylvania co. Va., by p-r. 89 ms. s. w. W. C.

TYE r., small r. of Va., rising in the Blue Ridge, and flowing southeastward into James r., after draining part of Nelson and Amherst counties, and by one of its constituents, Piney r., forming for some few miles the boundary between those cos.

TYE r. mills and p-o., nrthwstrn. part of Nelson co. Va., by p-r. 131 ms. a little N. of W. Richmond.

TYE r. warehouse and p-o. sthrn. part Nelson co. Va., by p-r. 108 ms. w. Richmond.

TYLER, co. of Va., bounded by Ohio co. Va. N., Greene co. Pa., and Monongalia co. Va. N. E., Harrison E. and S. E., Wood s. w., and Ohio r. separating it from Washington co. O. w., and Munroe co. O. N. w. Length 45 ms. diagonally from s. w. to N. E., mean width 18, and area 810 sq. ms. Extending in lat. from 39° 13′ to 39° 42′, and in long. from 3° 25′ to 4° 12′ w. W. C. This co. has a wstrn. declivity, drained into O. r. by Middle Island and Fishing creeks. The surface is excessively hilly, but soil excellent. Chief town, Middlebourne. Pop. 1820, 2,314; 1830, 4,104.

TYMOCHTEE, cr. and p-o. Crawford co. O. The Tymochtee cr. is the sthwstrn. branch of Sandusky r., rises in Marion co., and flowing northwards, enters and traverses Crawford to near its nrthrn. border, where it falls into the main Sandusky at the village of Tymochtee, which latter is by p-r. 73 ms. a little w. of N. Columbus.

TYNGSBOROUGH, p-t. Middlesex co. Mass., 30 ms. N. w. Boston, has the New Hampshire line N., is divided by Merrimack r. N. and s., on the w. side of which is a v. The r. is navigated with boats and rafts, and is here a broad stream. Pop. 1830, 822.

TYRE, p-t. Seneca co. N. Y., 171 ms. w. Albany. Pop. 1830, 1,482.

TYREE, Springs and p-o., wstrn. part Sumner co. Ten., 19 ms. N. Nashville.

TYRINGHAM, p-t. Berkshire co. Mass., 116 ms. w Boston, contains 2 ponds, which give rise to Conkepot r., a branch of the Housatonic. Pop. 1830, 1,350.

TYRONE, p-t. Steuben co. N. Y., 16 ms. N. E. Bath, 194 Albany, is hilly, but favorable both to grass and grain. Little lake, ½ m. by 3, has beautiful shores, cultivated to the water's edge. Its outlet runs ½ m. to Mud lake, from which flows Mud cr., navigable in boats from the falls to Conhocton cr., and the Susquehannah. The cr. affords mill sites, and the lakes are stocked with fish. Pop. 1830, 1,880.

TYRREL, co. of N. C., bounded by Hyde s., Washington w., Albemarle sound N., and the Atlantic ocean E. Length from E. to w. including the islands along the Atlantic coast, 52 ms., mean width 20, and area of land surface about 750 sq. ms. Extending in lat. from 35° 34′ to 35° 57′ N., and in long. from 0° 36′ to 1° 30′. Surface a dead, and in part inundated, plain, deeply indented from Albemarle sound by Alligator r., and by the strait between Pamlico and Albemarle sounds. Chief t. Columbia. Pop. 1830, 4,732.

TYSON'S, Store and p-o. Moore co. N. C., by p-r. 55 ms. s. w. Raleigh.

U.

ULSTER, co. N. Y., bounded by Delaware and Greene cos. N., Duchess co. E., from which it is separated by Hudson r., Orange co. s., and Sullivan co. w., contains about 966 sq. ms., and 14 tsps. It is broken by the Kaatsbergs, called also the Blue and Shawangunk mtns. The rocks are transition, and the soil various. Wallkill cr. crosses the co., receiving the Shawangunk, Rondout, Esopus, Plattekill, Sawkill, &c. There are extensive tracts of meadows, and other rich levels, and the uplands are often good. Marble, of remarkable hardness, is found in the co.; mill stones, limestone, &c. are also obtained in considerable quantities. Mammoth bones have been found in this co. First settled 1616. The early inhabitants were Dutch and Germans. Pop. 1820, 30,934; 1830, 36,550.

ULSTER, p-v. Bradford co. Pa., 7 ms. above Towanda, and by p-r. 246 ms. nrthrd. W. C.

ULYSSES, t. Tompkins co. N. Y., 174 ms. w. Albany, 6 N. w. Ithaca, has Cayuga lake E., and Seneca co. N., has a fertile soil, and plenty of mill seats on Halsey's cr., which has a fall in one place of 210 feet. This t. contains a woollen factory, numerous mills, &c. Trumansburgh is a v. on the Ithaca and Geneva turnpike road. Jacksonville, a v. on the Newberg and Geneva turnpike. Pop. 1830, 3,130.

UMBAGOG, lake in Maine and N. H., about 10 ms. by 18, flows w. into Androscoggin r., through Errol. It lies partly in Oxford co. Me., and Coos co. N. H.

UNADILLA, p-t. Otsego co. N. Y., 100 ms. w. Albany, 36 s. w. Cooperstown, has Susquehannah r. s. E., which separates it from Delaware co. The surface is hilly, but the soil of the uplands good, as well as along some of the streams. Grindstones are quarried in the t. The v. is on the Susquehannah, in a pleasant situation. Pop. 1830, 2,313.

UNDERHILL. t. Chittenden co. Vt., 15 ms. N. E. Burlington, 26 N. W. Montpelier, first settled about 1786, has several small streams, and generally an uneven surface. Pop. 1830, 1,051.

UNDERWOOD, Store and p-o., Chatham co. N. C., 54 ms. wstrd. Raleigh.

UNIKA mtn., local name given to that section of the central Appalachian chain, which separates N. C. from Ten., which lies s. w. from Ten. r., and between Haywood co. of N. C., and Monroe co. of Ten.

UNION, p-t. Lincoln co. Me., 40 ms. E. S. E. Augusta, has Waldo co. N. E., Muscongus r. N. W., crossed by St. George r., which connects several ponds in this t., and has its surface varied by hills. Pop. 1830, 1,612.

UNION, p-t. Tolland co. Conn. 33 ms. N. E. Hartford, 67 from N. Haven, occupies elevated ground, with Mass. N., has an uneven surface, with granite rocks, and affords iron ore. Breakneck and Mashapaug ponds, are the chief sources of Quinebaug r., and abound in fish. Pop. 1830, 711.

UNION, p-t. Broome co. N. Y., 140 ms. w. Cattskill, 6 w. Binghampton, 150 from Albany, is crossed by Susquehannah r., which here receives Nanticoke cr. Other streams afford mill seats. The soil is favorable to grain, bearing white pine, some oak, &c. Maple and beech grow at some distance from the r. Pop. 1830, 2,121.

UNION, p-t. Essex co. N. J., 47 ms. N. E. Trenton, has Rahway r. w., and a small stream on the E. boundary. It reaches N. to the Short hills, and lies N. W. and N. of Elizabethtown, and s. w. Newark. Pop. 1830, 1,405.

UNION, (College,) Schenectady, Schenectady co. N. Y., was founded in 1795. The faculty consist of a president, professors of Greek and Latin, moral philosophy and rhetoric, natural philosophy and mathematics, oriental literature, several assistant professors and a tutor. Annual expense, $112,50. Number of graduates up to 1831, 1,370.

UNION, tsp. and p-o. Luzerne co. Pa. The tsp. extends from Susquehannah r., to the estrn. boundary of Lycoming. The p-o. is 14 ms. s. w. Wilkes-Barre, and 88 N. N. E. Harrisburg.

UNION, co. Pa., bounded s. and s. w. by Mifflin, N. W. by Centre, N. by Lycoming, N. E. by the w. branch of Susquehannah r., separating it from the nrthrn. part of Northumberland, and E. and s. E. by the main volume of Susquehannah, separating it from Dauphin. The greatest length is from s. to N. 30 ms. parallel to the general course of Susquehannah r., and nearly on the meridian of W. C., mean width 20 ms. and area 520 sq. ms. Extending in lat. from 40° 40′ to 41° 06′, and in long. from 0° 10′ E. to 0° 22′ w. W. C. Declivity estrd., and drained by Buffalo, Penn's, Middle, and western Mahantango crs. Surface hilly, or rather mountainous, but soil excellent. Chief town, New Berlin. Pop. 1820, 18,619, 1830, 20,749.

UNION, p-v. and st. jus. Monroe co. Va., situated to the nrthestrd. from Peter's mtn., in Green Brier valley, about 40 ms. a little w. of N. from Christianburg, and by p-r. 208 ms. w. Richmond, and 267 ms. s. w. by w. W. C. Lat. 37° 34′, and long. 3° 32′ w. W. C.

UNION, dist. S. C., bounded by Spartanburg dist. w. and N. W., Broad r. separating it from York, on the N. E., Chester E., and Fairfield s. E., by Newberry s., and by Ennoree r. separating it from Laurens s. w. The greatest length, parallel to the general course of Broad r., 42 ms., mean width 15, and area 630 sq. ms. Extending in lat. from 34° 28′ to 35° 03′, and in long. from 4° 27′ to 4° 52′ w. W. C. Declivity to the sthestrd. and traversed by Pacolet, and Tyger rs. The outline indeed approaches a triangle, longest side on Broad r., and base on Ennoree r. Chief t. Unionville. Pop. 1820, 14,126; 1830, 17,906.

UNION, co. Ark., not laid down on Tanner's U. States; situation and boundaries uncertain. Chief t. Ecora Fabra.

UNION, p-v. Humphries co. Ten., by p-r. 91 ms. w. Nashville.

UNION, p-v. Boone co. Ky.

UNION, co. Ky., bounded by Hopkins s. E., Trade-water r., separating it from Livingston s., Ohio r., separating it from Gallatin co. Il. w., the Ohio r., separating it from Posey co. Ind. N., and on the N. E. and E. it has Henderson co. Ky. Length from s. to N. 30 ms., mean breadth 18, and area 540 sq. ms. Lat. 37° 35′, and long. W. C. 11° w. intersect near the centre of this co. Slope wstrd. towards Ohio r. The body of the co. lies opposite to, and sthrd. from the mouth of the Wabash r. Chief town, Morganfield. Pop. 1820, 3,470; 1830, 4,764.

UNION, p-v. northern part Montgomery co. O. by p-r. 78 ms. w. Columbus.

UNION co. O. bounded by Franklin s. E., Madison s., Champaign s. w., Logan w., Hardin N. W., Marion N. E., and Delaware E. Length 27 ms., breadth 17, and area 460 sq. ms. Lat. 40° 20′, long. 6° 30′ w. W. C. Slope southestrd., and drained by numerous confluent creeks of the Sciota r. Chief t. Marysville. Population 1820, 1,996, 1830, 3,192.

UNION, one of the estrn. cos. of Indiana, bounded by Franklin s., Fayette w., Wayne N., Prebble co. O. N. E. and E., and Butler co. O. s. Length 14 ms., breadth 11, and area 154 sq. ms. Slope southward, and traversed and drained by White Water r. Chief town, Liberty. Pop. 1830, 7,944.

UNION, p-v. Hendricks co. Ind. watrd. from Indianopolis.

UNION, co. Il. bounded by Jackson N., Franklin N. E., Johnson E., Alexander s., and Miss. r. separating it from Cape Girardeau co. Mo. w. Length 22 ms., breadth 20, and area 440 sq. ms. Lat. 37° 30′, long. 12° 20′ W. C. General slope southwstrd. towards the Miss. r., but the creeks flow from its nthrn. side like radii from a common centre. Chief t. Jonesboro'. Pop. 1820, 2,362, 1830, 3,239.

UNION, p-v. Vermillion co. Il. by p-r. 170 ms. N. E. Vandalia.

Union, p-v. and st. jus. Franklin co. Mo. by p-r. 54 ms. w. St. Louis.

Union Bridge and p-o. northeastern part Frederick co. Md., by p-r. 70 ms. a little w. of n. W. C.

Union Furnace and p-o. Huntingdon co. Pa. by p-r. 160 ms. n. w. W. C.

Union Hall, p-v. Franklin co. Va. by p-r. 276 ms. s. w. W. C.

Union Hill, p-o. Upson co. Geo. by p-r. 79 ms. wstrd. Milledgeville.

Union Iron Works and p-o. Berks co. Pa. by p-r. 60 ms. estrd. Harrisburg.

Union Meeting House and p-o. northwstrn. part Baltimore co. Md. 30 ms. n. w. Baltimore.

Union Mills and p-o. southestrn. part Erie co. Pa. 6 ms. s. e. by e. Waterford, and by p-r. 319 ms. n. w. W. C.

Union Mills and p-o. on Little Pipe creek, northestrn. part Frederick co. Md. 18 miles n. e. from the city of Frederick, and by p-r. 73 ms. a little w. of n. W. C.

Union Mills and p-o. Fluvanna co. Va. by p-r. 68 ms. n. w. by w. Richmond.

Union Square and p-o. Montgomery co. Pa. by p-r. 25 ms. n. Phil.

Uniontown, p-v., borough, and st. jus. Fayette co. Pa. situated on Red Stone creek 4 ms. wstrd. Laurel Hill chain of mountains, 186 ms. a little s. of w. Harrisburg, and 193 ms. n. w. by w. W. C. Lat. 39° 54', long. 2° 45' w. W. C. This borough was founded in 1775, by Jacob and Henry Beeson. It is chiefly composed of one street extending along the U. S. road. There are, however, houses on other streets, particularly on that towards Morgantown in Va.

Uniontown, p-v. nrthestrn. part Frederick co. Md., 35 ms. n. w. Baltimore, and by p-r. 73 ms. n. W. C.

Uniontown, p-t. Belmont co. O. by p-r. 126 ms. e. Columbus.

Unionville, t. Dutchess co. N. Y. 105 ms. s. Albany, has the n. end of the Matteawan mtns. on the e. boundary, and gives rise to Fishkill creek. Pop. 1830, 1,833.

Unionville, p-v. Chester co. Pa. 8 or 9 ms. s. w. from West Chester, and by p-r. 107 ms. n. e. W. C.

Unionville, p-v. Frederick co. Md. by p-r. 58 ms. n. n. w. W. C.

Unionville, p-v. and st. jus. Union district, S. C., on a small branch of Tyger r., by p-r. 27 ms. s. e. Spartanburgh, and 77 n. w. Columbia. Lat. 34° 42', long. 4° 39' w. W. C.

Unionville, p-v. Geauga co. O. by p-r. 176 ms. n. e. Columbus.

Unison, p-v. sthrn. part Luzerne co. Pa. by p-r. 98 ms. n. e. Harrisburg.

Unison, p-v. nthrn. angle Loudon co. Va. by p-r. 51 ms. above and n. w. W. C.

Unison, p-v. nthrn. part Delaware co. O. by p-r. 33 ms. nthrd. Columbus.

Unitia, p-o. Blount co. Ten. by p-r. 194 ms. s. e. by e. Nashville.

United States, of North America. The United States are bounded n. by the British and Russian dominions, e. by the Atlantic ocean, s. by the Gulf of Mexico, and w. by the Mexican territory and the Pacific ocean. The country extends from 25° to 54° n. lat., and from 66° 50' to 125° long. w. from Greenwich, and from 9° 35' e. to 48° 20' w. W. C., containing within its territory an area of more than 2,000,000 sq. ms. This extensive country has outlines in common with Cabotia, or British North America, from the mouth of the Saint Croix r., to the Rocky, or Chippewayan mtns., 3,000 miles; with Russian N. America from the Rocky mtns. to the Pacific ocean, 1,100 miles; along the Pacific ocean, from Dixon's entrance, or Vancouvre's sound, about 880 miles; in common with the republic of Mexico, from lat. 42° n. on the Pacific ocean, along that curve of lat. to the Rocky mtns., and thence to the mouth of the Sabine r. into the Gulf of Mexico, 2,300 ms.; along the Gulf of Mexico to Florida point, 1,000 ms.; along the Atlantic ocean to the mouth of the Saint Croix r., 1,850 ms.; so that the entire outline is not far from 10,130 ms. This territory extends in one immense zone from ocean to ocean. The longest line which can be drawn, entirely over land, without traversing the sea, in this region, stretches from Cape Canaveral, in Florida, to the northern end of Queen Charlotte's island, a distance of 3,214 statute miles. This line being assumed as a base, the mean breadth will be about 700 ms.; so that the whole area would equal a square, each side of which should exceed 1,490 ms. This is nearly one twentieth part of the land surface of the earth, and is capable of subsisting at least one fifteenth of its population. If the whole earth therefore sustain 1,000 millions, the United States would sustain 66,666,666; a number which at the present rate of increase of pop., it will contain within the current century.

The original number of states was 13; the present number is 24, with 3 organized territories, each of which are represented in congress by a delegate. The extensive tract between the Mississippi r. and lake Michigan, will probably soon be organized, and called Ouisconsin, or Huron territory. The subjoined table exhibits the sectional and aggregate extent, of the portion already distributed into states and territories, and their aggregate pop., according to the census of 1830.

State, &c.	Area in sq. ms.	Free pop.	Aggregate pop.	Federal or polit. pop.
New England, or Eastern States.				
Maine,	33,223	399,431	399,437	399,434
N. H.	9,491	269,323	269,328	269,326
Vt.	8,000	280,657	280,657	280,657
Mass.	7,800	610,404	610,408	610,406
R. I.	1,200	97,185	97,199	97,193
Conn.	4,764	297,650	297,675	297,665
Middle States.				
N. Y.	46,085	1,918,532	1,918,608	1,918,577
N. J.	8,320	318,569	320,823	319,922
Penn.	47,000	1,347,830	1,348,230	1,348,072
Del.	2,100	73,456	76,148	75,431
Md.	9,356	344,046	447,040	405,842
Southern States.				
Va.	68,600	741,648	1,211,405	1,023,502
N. C.	51,000	492,386	737,987	639,747
S. C.	33,000	265,784	581,185	455,025
Geo.	62,983	299,292	516,823	429,810

State, &c.	Area in sq. ms.	Free pop.	Aggregate pop.	Federal or polit. pop.
		Western States.		
Ky.	40,500	522,704	687,917	621,832
Oh.	44,000	935,878	935,884	935,882
Ind.	36,670	343,028	343,031	343,030
Il.	53,480	156,698	157,445	157,147
Mo.	64,000	115,364	140,455	130,419
		South Western States.		
Ten.	45,600	540,300	681,903	625,263
Ala.	51,770	191,978	309,527	262,508
Miss.	45,760	70,962	136,621	110,358
La.	48,320	106,151	215,739	171,904
		Territories, &c.		
Mich.	34,000	31,607	31,639	31,625
Ark.	50,000	25,812	30,388	28,557
Flor.	55,000	19,229	34,730	28,529
D. C.	100	33,715	39,834	37,389
Huron	100,000			
Total,	1,061,222	10,849,620	12,858,670	12,055,050

From this table we find that the people of the United States, have, with more or less of compactness, extended their settlements over more than 1,000,000 square miles, or over a surface exceeding that of all Europe w. of the rivers Vistula and Bog, and s. of the Baltic; a surface on which now exist upwards of 150,000,000 of inhabitants. With all this density, the people of that part of Europe more than double in 50 years. If the people of the U. S. double in 30 years, their number will exceed 100,000,000, during the current century. The various classes of the population, by the census of 1830, are as follows:

	Whites.	
	Males.	Females.
Under 5 years of age,	972,980	921,934
From 5 to 10	728,075	750,741
" 10 to 15	669,734	638,856
" 15 to 20	573,196	596,254
" 20 to 30	956,487	918,411
" 30 to 40	592,535	555,531
" 40 to 50	367,840	356,046
" 50 to 60	229,284	223,504
" 60 to 70	135,082	131,307
" 70 to 80	57,772	58,336
" 80 to 90	15,806	17,434
" 90 to 100	2,011	2,523
100 and upwards	301	238

Of the foregoing, were deaf and dumb, under 14 years of age, 1,652; of 14 and under 25, 1,905; of 25 and upwards, 1,806. Blind, 3,974. Aliens, or foreigners not naturalized, 107,832.

Of the colored pop. of the United States, there were:—

	Of Free persons.		Slaves.	
	Males.	Females.	Males.	Females.
Under 10 yrs. of age	48,675	47,329	353,498	347,665
Fm. 10 to 24	43,079	48,138	312,567	308,770
" 24 to 36	27,650	32,541	185,585	185,786
" 36 to 55	22,271	24,327	118,880	111,887
" 55 to 100	11,509	13,425	41,545	41,436
100 and over	269	386	748	676

Recapitulation.

	Whites.	Free colored.	Slaves.	Total.
Males	5,357,102	153,443	1,012,822	6,523,367
Females	5,172,942	166,133	996,228	6,335,303
Total,	10,530,044	319,576	1,009,050	12,858,670

Progressive population from 1790, to 1830, inclusive.

		Increase per cent.
1790,	3,929,827	
1800,	5,305,941	" " " 35
1810,	7,239,814	" " " 36
1820,	9,638,191	" " " 33
1830,	12,866,020	" " " 33

From the best data, we may regard the regular increase as nearly one third, decennially. The greater increment per cent. which appears in the two first periods, is satisfactorily accounted for, from each succeeding enumeration, being more correctly made, and of course the real pop. being more fully represented in the returns. We find from the above table, that white males under 5 years of age, exceeded females of like age in 1830, 51,046, an excess of about 5 per cent. Of white persons above the age of 70, we find 75,920 males, and 78,531 females, or of persons who have passed the ordinary limits of human life, there are 2,611 more females than males. In the class of free colored persons we again discover very nearly similar results; whilst the number of aged male slaves exceeds by a small fraction the number of aged female slaves. Uniting all classes, we discover that the *common law* of birth, is rigidly obeyed by nature, in the U. S., there being an excess of males, but the chances of protracted life being decidedly in favor of females.

Mountains. The face of the country is of course very much varied. For details in this particular, we refer to articles on the different states. Beside the minor chains of mnts., however, the country is traversed by two great chains which are nearly parallel with the coasts of the oceans which they respectively approach. The Appalachian or Atlantic chain extends from s. w. to n. e., whilst the Chippewayan or Pacific range runs from s. s. e. to n. n. w. By these mountains, the United States territory is divided into two great ocean slopes, and an immense interior valley. By another physical division it may be regarded as separated into 4 great inclined planes; the eastern falling from the Appalachian chain, to the Atlantic ocean; the western from the Chippewayan, to the Pacific ocean, and the two central planes having a common line of deepest depression along the lower part of the channel of the Mississippi river, the Illinois river and lakes of Canada. The southwestern Appalachian chains rise abruptly from the Hudson valley, near lat. 41° 30′ n. where the Hudson passes between enormous walls of primitive rock. These precipices rise almost perpendicularly from 1,200 to 1,500 feet, their bases being washed by the tides. From hence, with a breadth from 80 to 100 ms., the range passes southwestward through the United States until gradually lost among the hills between the southern sources of Tennessee river and Appalachicola and Mobile. The intervening vallies rise from 500 to 2,000 feet, discharging on one side the numerous streams which traverse and adorn the Atlantic slope, and from the other supply innumerable fountains to the great central river; a remarkable feature of the Appalachian chain is the fact, that very few falls are found in its rs. In a state of nature, the Atlantic system rose as the central and most elevated section, of perhaps the most extended continuous forest that ever existed on the earth. From the Atlantic border far beyond the Ohio, (with few exceptions) spread countless millions of trees, amongst which the oak, pine and hickory, predominated; but intermixed with numerous other varieties, winding amongst these primeval woods, rose and flowed those streams now decked with farms, studded with cities, or rivalled by canals fed from their own bosoms. The St. Lawrence basin, a part of the same physical section, also sustained its share of this vast forest, stretching almost uninterruptedly from the sources of Ottawa and Saguenai, to the gulf of Mexico. Passing westward from the Appalachian towards the Chippewayan range, timber gradually ceases, and many hundred miles before reaching the sources of the mighty Missouri, its banks are entirely destitute of trees. Here are extensive level prairies, where, far as the eye beholds, nothing like forest trees are visible, and as on the ocean, earth and heaven seem to meet in the distance. In southwestern Louisiana, the extremes meet, and in a few hours the traveller may pass from the deep gloom of forests untouched by the axe, into plains unbroken by the plough. Contrasts like these, though on a larger scale, strongly mark the two great mountain systems of the United States. Towards the Atlantic, spots of grassy glades are interspersed amongst the mtn. chains. Amid the Chippewayan ridges, forest trees are seen in strips or clumps.

Bays, Gulfs, Capes, and Rivers.—All these subjects are mentioned particularly in the articles on the states where they are found, or under their respective heads. It will be sufficient here to say, that no country in the world is intersected by as many navigable rivers as the United States; that the longest river is the Missouri, which, including the Mississippi, is the longest in the world, being 4,490 miles in length; that the gulf of Mexico, on the s. boundary, is the largest in the United States territory, and that the bays of the coast are numerous, and several of them navigable by vessels of the largest size. In the northern part, the Atlantic coast, which is bold and rocky, is indented by numerous inlets, and broken into headlands. Towards the s. the shore is more level, and generally alluvial.

Lakes.—North America, beyond every other country, is distinguished for the immense extent and number of its fresh water lakes. Several of these are on the n. boundary of the United States. Lake Superior is the largest body of fresh water on the globe, its surface containing 35,000 square miles. Lake Huron contains 20,000 square ms., lake Erie 10,350, and lake Ontario, 7,200 square miles. Some of these have been the scenes of important naval engagements.

Soil.—The soil of a district so extensive as the United States, is of course marked by almost every variety. Under the great Appalachian forest on both sides of the main chain, and also in its most elevated vallies, or table lands, the soil offers a full reward for its cultivation, and allows the choice of objects of culture. The western slope of the Appalachian chain exceeds the eastern in fertility, though this disparity is compensated by the numerous rivers on the e. slope opening channels of direct intercommunication with the Atlantic, and by the more diligent cultivation of the soil. In point of extent, the two mtn. systems of the United States are as two to one, very nearly; the Appalachian having about 700,000 and the Chippewayan upwards of 1,400,000 square miles. With but partial exceptions the inhabited parts are as yet on the Appalachian section.

Climate.—The climate of the United States is remarkable for its variety and its sudden changes from extreme heat to cold, and the contrary. In the n. part is the cold and dreary winter of Canada, and in the extreme s. parts, the summer is almost uninterrupted from one end of the year to the other. The climate differs from that of Europe, in the same latitudes. The level portions of the southern states have more moisture, and a less salubrious atmosphere; their noxious effluvia are more constantly formed, and their marshes more numerous. Those parts however which are elevated, more commonly enjoy a temperate and delightful climate. The mean annual temperature of the middle states is the same as in the corresponding European region, though it is differently distributed. In the Atlantic states the climate is marked by extremes, the summers being usually very hot, and the winters though often short, much colder than European winters in the same latitude. Deep and abiding snows are usual in winter on the Atlantic coast, and on the Mississippi, and if we advance to lat. 38° n., or attain an elevation where the temperature is the same as there, there are very few winters, in the course of which, from December to February inclusive, the earth is not covered with ice or snow, and the mean temperature of the nights being below the freezing point. With the n. w. winds the snows are most abundant, and often much drifted. In the winter of 1831, and 1832, the Mississippi r. was frozen and passable on the ice as low as lat. 35° n., and the spring floods of 1832, were very great. The summers of the United States, though often excessively warm, are as agreeable, if not more so, than those of southern Europe, and in autumn no part of the globe possesses a season more congenial to human life, or more charming to the senses. In general the weather is variable, and subject to sudden changes. The climate throughout the country is greatly modified by the mountains, lakes, &c. Facts which our limits forbid us to insert, sustain us in the following general conclusions respecting the climate. For the more minute local peculiarities of climate, the reader is referred to the articles on the respective states. These general conclusions are, that all places of similar latitude and elevation, have like climates, that the United States territory, comprising a zone of N. A., generally has along its opposing coasts similar climates to those prevailing on the opposing zone of the eastern continent. Along the Pacific coast, even beyond Bhering's strait, a mild and moist climate prevails, whilst along the Atlantic coast, the winters are intensely cold & summers as intensely warm; that advancing from the Atlantic coast inland, the thermometer indicates a depression of temperature, according to relative height and exposure; falling occasionally, even in N. lat. 35° and E. of the Mississippi river to 18° below zero of Fahrenheit,—that about 400 ft. elevation is fully equivalent to a degree of lat. on Fahrenheit's thermometer,—that as low as N. lat. 35°, and with no allowance for any difference of elevation, the winters present a season of from 60 to 120 days, say 90 days in which the rivers are frozen. This excess of course increasing with elevation, and progressing northward,—that the prevailing winds which have been mentioned are in frequency about as 7 in 10, and in intensity, at least as 8 in 10, of all aerial currents, over the United States and western Europe, and consequently that this great current, which carries the moist and uniform air of the Atlantic on Europe, at the same time bears the frozen air of an immense continent over the eastern part of the United States,—that observation of prevailing rains shows, that the mean annual amount of rain in the United States is about 37 1-2 inches, whilst the mean annual rains of N. W. Europe amount to only 31 1-3 inches. In the United States rain falls from 140 to 150 days, including snow, sleet, &c. leaving about 220 fair days, annually, while in N. W. Europe, the days of rain, or rather of heavy mist, are on an average 220 days annually; and lastly, that the climate of the United States appears to be nearly stationary, or if subject to any changes, they are small. Clearing of land, if it produces any effect, will probably, as in Scotland, lower the temperature.

Winds.—The winds which prevail about 7-10 of the time throughout the whole N. temperate zone are from the N. W., W. and S. W. This prevalence is said to be so great as to bend the forests on both sides of the Atlantic toward the E. or S. E. Not only are these winds most frequent, but also most violent; and the effects of their constancy and violence may be generally traced by this inclination of the trees of the country, from N. England to the mouth of the Oregon, and even into the Arctic ocean, to Melville island. These winds are however very much modified, by the elevated lands in the mountainous parts, and by the sea breezes, &c. on the coast.

Minerals.—Most kinds of minerals have been discovered more or less extensively, in the United States. Gold has been found in North Carolina, Georgia, and other states. Silver in small quantities in several places; iron in numerous and inexhaustible beds, especially along the Appalachian range of mtns., from New Hampshire to Georgia; copper on lake Superior, in pure metallic masses, and in the ore in several places; lead in several places, and the mines of Missouri, the annual produce of which is estimated at more than 3,000,000 lbs. are among the richest in the world; mercury or quicksilver, which though a rare metal, had been found in small quantities on the borders of 4 of the great lakes, and other metals, as cobalt, bismuth, and antimony. Limestone in all its varieties, slate, sandstone, and building stone of various kinds, are abundant. Coal has been discovered, the anthracite in inexhaustless quantities, and bituminous in abundance. Salt springs are found in several of the states, some of which are profitably worked, and salt is also manufactured from sea water. Within a few years, gypsum, or plaster of Paris, has been extensively quarried in the state of New York. Mineral springs are found in most of the states; some of them highly valuable.

Productions and Internal Improvements.—On these subjects, see articles on the respective states, and on rail-roads and canals.

Agriculture.—Nearly one fifth of all the inhabitants of the United States are engaged in agricultural pursuits. The annual cotton crop is estimated from 300 to 350 millions of pounds. The flour and meal actually inspected at 11 different places in 1830, amounted to 2,851,876 barrels of wheat flour, 41,351 of rye flour, 18,372 hhds. and 35,070 barrels of corn meal. The eastern states are mostly devoted to grazing and the dairy; the middle and western, to the production of various kinds of grain; the southern to raising rice, sugar, tobacco, cotton, &c.

Manufactures.—The manufactures of the United States are considerable, and gradually increasing, to a great extent; they have been noticed in the articles on the different states, though some statements still remain to be made. The eastern and middle states, which are most abundantly supplied with water power, are most extensively engaged in manufactures, especially of cotton, woollen, iron, glass, paper, wood, &c. In 1810, the value of annual manufactures in the United States, was estimated at $172,762,676; the present annual value is computed at $500,000,000; and the capital invested in all the manufactories of the Union is estimated at more than $1,000,000,000. Most of the American manufactures are designed for home consumption, yet in 1831, domestic manufactures were exported to the amount of $7,861,740. More than two thirds of the clothing used by those engaged in agricultural pursuits, are of domestic production.

Commerce.—The United States are among the most commercial countries in the world. In the year ending September 1831, the imports amounted to $103,191,124, of which $93,962,110, were imported in American, and $9,229,014 in foreign vessels. The exports of the same year, amounted to $81,310,-583, of which $61,277,057 were domestic, and $20,033,526 foreign articles. Of the domestic exports, $1,889,472 were the product of the sea; $4,263,477 of the forest; $48,261,233 of agriculture; and $7,862,675 of manufactures. Of domestic articles, $49,671,239 were exported in American, and $11,605,818, in foreign vessels. Of the foreign articles $15,874,942 were exported in American, and $4,158,584 in foreign vessels. In the same year 922,952 tons of Amer., and 281,948 tons of foreign shipping were entered; and 972,504 tons of American, and 271,-994 tons of foreign shipping cleared from the ports of the United States. The whole amount of the registered, enrolled, and licensed tonnage, including fishing vessels, in the United States in 1830, was 1,191,776 tons; of which 38,911 were engaged in the whale fishery. The amount of tonnage built in 1830, was more than 58,000 tons. The most important article of export, was cotton, which amounted to $25,289,492; the exports of tobacco, were $5,269,960; of rice, $2,620,696; of flour, biscuit, &c. $4,464,774; of swine and their products, $1,495,830; of corn and rye meal, $881,894; of cattle and their products, including butter and cheese, $896,316; of the imports, $13,456,625 were free of duty; $61,534,965 were subject to duties "ad valorem"; and $28,199,533 were subject to to specific duties. The number of seamen in the United States is about 50,000, exclusive of the navy, and of those engaged in internal navigation. The greatest export trade is from New Orleans; the greatest import to New York. A great proportion of the shipping of the United States is owned in New England and New York. For additional details, see tables in the apendix.

Fisheries.— Most of the fisheries are carried on from the New England states and by N. E. ships. The cod fishery is the most important, that of the whale next. The annual value of fish exported is $1,889,472. The whole amount of tonnage engaged in the fisheries in 1831, was 98,322 tons.

Public Lands.—These lands consist of the territory belonging to the United States at the time of their independence, of tracts ceded to the general government by individual states, and of tracts acquired by treaty or purchase. They are mostly within the limits of the Western states, and are to a great extent occupied by Indians, who are regarded as the owners, until their title shall have been extinguished by purchase. The aggregate amount of all these lands, is 1,090,871,753 acres, the value of which, at the fixed minimum price of sale, a dollar and a quarter per acre, amounts to the enormous sum of $1,363,589,691. For the title to these lands the United States have paid on the Louisiana purchase, principal and interest, $23,514,225; on the Florida purchase $6,251,016; on the Georgia, Yazoo, and other contracts, $18,312,219 :—total, $48,077,551. The amount of all their sales up to September 1831, has been $37,272,713. The amount of sales is gradually on the increase; in 1831, it was $3,000,000. All sales are for cash. Salt springs and lead mines are reserved by government; and one thirty-sixth part of all public lands, are reserved and applied for the perpetual support of common schools. Three fifths of the value of all sales is applied by congress, for internal improvements in the states where the lands are located, and the remaining two fifths is applied by the states for the promotion of learning. Up to the present time, rather more than 150,000,-000 of acres have been surveyed, about 20,-000,000 of acres have been sold, and the same quantity granted by congress for the purposes of education, internal improvement, &c. and there are now about 110,000,000 of acres surveyed and unsold, of which 80,000,000 are now in market. Of the unsold lands, 340,-871,753 acres are within the limits of the new states and territories, and 750,000,000 acres beyond these limits.

Revenue, expenditure, and national debt.—As there is at present no direct taxation by the general government, the revenue is chiefly derived, 1, from duties on imports; 2, from the public lands; 3, from its bank stock; 4, from post offices, lead mines, &c. Of these the duties on imports are by far the largest. The estimated revenue for 1832. is, from customs, $26,500,000; public lands, $3,000,000; bank dividends, $490,000; other sources, $110,000; total, $30,100,000. The expenditures for the same year, exclusive of payments on the public debt, are estimated at $13,365,202, which, being deducted from the estimated receipts, will leave a balance of $16,734,797. Between March, 1829, and the 2nd of January, 1832 more than $40,000,-000 have been applied to the extinguishment of the public debt, which at the last date amounted to $24,322,235. It is intended to reduce it to $2,302,686 by January, 1833; to pay it off entirely by March of the same year. The whole amount of the disbursements of government made in all the states between 1789 and 1831, for fortifications, light houses, public debt, internal improvements, and revolutionary pensions, is $222,976,821.

Banks.—The present bank of the U. S. was chartered by congress in 1816, for 20 years, with a capital stock of $35,000,000, of which government owns one fifth. The debts of the bank may in no case exceed its deposits by more than $35,000,000. The actual circulation is about $42,000,000; and the average dividends 6 or 7 per cent. The bank is located at Philadelphia, and it has 25 branches in the principal cities of the Union. Besides the U. S. bank, there are in the different states, nearly 400 banks, with capitals of from $3,000,000 downwards, amounting in all to about $200,000,000, including the U. S. bank.

Mint.—The mint was established at Philadelphia in 1792, and the amount of its coinage has been constantly on the increase. During the first 10 years of its establishment, ending in 1801, the amount of silver coinage alone was $1,574,000; from 1801 to 1811, it was $4,858,000; from 1811 to 1821, $6,180,000; and from 1821 to 1831, $18,325,000. The whole coinage of 1831, amounted to $3,923,473, of which $714,270 were of gold, $3,175,600 of silver, and $35,603 of copper. The expense of the mint for the same year was $28,000. Of the gold coined, $518,000 were from the gold regions of the U. S. Gold and silver are coined without expense to the owners. The probable supply of bullion for the next period of 10 years, is estimated at $6,000,000 annually. The metallic currency of the U. S. is estimated at $30,000,000.

Post-office.—The first-post in America was established in New York in 1710, under the old colonial government. In 1789, the exclusive direction of posts, &c. was conferred by the constitution on congress. At that time there were but 75 post-offices in the country. In 1831, the number of post-offices was 8,686; the extent of post-roads 115,176 miles; and the yearly transportation of mails equal to 15,468,692 miles. The expenses of the post-office department, for the year 1830, were $1,959,109; the receipts, $1,919,300; balance against the department, $39,809.

Army and navy.—The standing army of the U. S. is limited by law to 6,442 men; it consists of 7 regiments of infantry, and 4 of cavalry, commanded by one major general and two brigadier generals, beside inferior officers. The estimated expense of the army for 1832, including fortifications, armories, arsenals, &c. is $6,648,099. Beside the standing army, the militia of the country in 1830, amounted to 1,262,315. The navy consists of 12 ships of the line, 17 frigates, 16 sloops, and 7 schooners; total 52, including those which are building, of which 20 are in commission. The total number of officers and men is 6,345. The estimated expenses of the navy for 1832, including the sum for gradual improvement, are $3,907,618.

Salaries, pensions, &c.—The largest salary is that of the president, which is $25,000 annually; ministers plenipotentiary, 9,000 a year, with the same sum for an outfit; the secretaries of state, the navy, treasury and war, and the post-master general, $6,000; the vice president and the chief justice, $5,000; associate judges of the supreme court, and charge d' affairs, $4,500; and members of congress $8 per day. The annual revolutionary and other pensions amount to $1,363,296. There are no sinecures in the U. S.

Newspapers.—No country in the world equals the U. S. in the number of its newspapers. The first newspaper in America was printed at Boston in 1704, by the name of the *Boston News Letter.* In 1720, there were but 7 newspapers in all the North American colonies; in 1810 there were 359 in the U. S.; in 1826, 640; in 1828, 802; and at the present time there are more than 1,000, of which between 50 and 60 are issued daily. Of the 802 newspapers issued in 1828, 192 were in New England, 409 in the middle, 88 in the southern, 115 in the western, and 33 in the southwstrn. states; 5 were in the territories, 9 in the district of Columbia, and 1 in the Cherokee nation and language. The whole number of periodical sheets annually issued is estimated at 64,000,000.

Slavery.—Slavery exists in 12 states, Delaware, Maryland, Virginia, N. and S. Carolina, Georgia, Alabama, Louisiana, Tennessee, Kentucky, Mississippi, and Missouri; also in the territories of Arkansas, Florida and Michigan. Maine, New Hampshire, Vermont, Massachusetts, Ohio, and Indiana, have no slaves. In Rhode Island, Connecticut, New York, New Jersey, Pennsylvania, and Illinois, there are a few; but as slavery is abolished in all of them, it will cease with the death or exportation of the slaves now in them. Whole number of slaves in the U. S. 2,010,436.

Indians.—The whole number of Indians in the U. S. and their territories, is 129,266. The whole number of tribes is 58. In New England the number of Indians is 2,526; in New York, 5,143; in Virginia and S. Carolina, 497; in Ohio, 2,350; in Indiana, Illinois, Georgia, Alabama, Tennessee and Mississippi, 66,004; in Michigan territory, 28,316; in Louisiana and Missouri, 7,113; and in the Florida and Arkansas territories, 17,107.

Education.—As a general government, the U. S. have done little for the interests of public instruction, except that they reserve for this purpose one section in every township of their new lands, besides other reservations for colleges. This highly important subject has, however, probably been much better attended to by being left to the individual states and to private citizens. The chief details of what has been done will be found in the articles on the respective states. In general it may be remarked that the colonists of New England adopted a most admirable system of common school instruction. As early as 1628, a law was passed for the education of every child in the colonies; and in 1647, a school was established by law in every town or neighborhood of 50 families, and a school for the higher branches, for every 100 families. But for more particular accounts, see articles on New England, Massachusetts, and the other states. There are in the U. S. (*as will be seen from the tables in the appendix, which see,*) 66 colleges, the whole number of whose alumni, previous to 1831, was 22,653, of which about one quarter were graduates of Harvard, and nearly the same number of Yale college. The whole number of instructors at that date was about 450; volumes in college libraries, 190,056, and in the students' society libraries, 87,190. Yale has a greater number of students than any other college: Harvard is most richly endowed. Thirty-nine of the sixty-six colleges have risen during the present century; though many of the foundations, now entitled colleges, were respectable academies before the change of their names, with which change in some cases, there has been no corresponding change of studies. From the table of students, (*see tables in appendix*,) it appears that (exclusive of the West Point military academy), there were in 1831, nearly 6000 young men of the U. S. receiving a liberal classical education. Beside the colleges, there are in the U. S. 27 theological seminaries, (*see appendix*,) the number of whose graduates amount to nearly 1,900. Beside those included in the table, there are in the U. S. 5 Roman Catholic seminaries. There are also 18 medical schools, and 9 law schools, for the names, locations, &c. &c., of which *see appendix*. Most of the states of the union have made some legislative provision for common school instruction; and in some states large funds are set apart for this purpose. Private schools and academies of the higher order are quite numerous, especially in New England, so that few grow up without enjoying the means of elementary instruction, or if they desire it, of a more extended liberal education. In the Sabbath schools of the U. S., which are doing much for the intellectual as well as moral improvement of the young, about 600,000 children are weekly instructed, by more than 80,000 teachers.

Religion.—There is no established church in the U. S., but all sects are alike allowed free toleration; nor is any legislative provision made for the support of religion in any of the states, except that in Massachusetts every citizen is obliged to be connected with, or pay taxes to some religious denomination. In the articles on the several states, may be found the details of the different denominations within their limits. For the different denominations, their churches, ministers, communicants, &c. as they were in 1830 in the U. S. *see table in appendix*. Their numbers since that date are much increased, though their relative numbers are not materially changed. The number of churches in the U. S., at the present time, is not far from 12,000.

Constitution and government.—The present constitution of the U. S. was adopted in 1787, though it has since been amended. The form of government which it establishes is a confederated republic, composed of all the states. The legislative branch consists of a senate and house of representatives. The senate is composed of two senators from each state, chosen every two years, for a period of six years, so that one-third of the senate is renewed biennially. Every senator must have been 9 years a citizen, and 30 years of age. The vice-president is president of the senate; and all trials for impeachment are conducted before that body. The number of senators is at present 42. The members of the house of representatives are chosen every two years; they are proportioned to the population of the states from which they come, 5 slaves being counted as 3 freemen in the slave states. The rate of apportionment after March 1833, is one to every 47,700 inhabitants. The qualifications are, that a representative shall have been 7 years a citizen, and be at least 25 years of age. All bills for raising revenue must originate in the house of representatives; and any bill *vetoed* by the president, will, notwithstanding, become a law, if afterward approved by two-thirds of both houses. The number of representatives in 1833, will be 240. Congress has the power to impose taxes, both direct and indirect, regulate commerce, and the coining of money, make bankrupt laws, provide for common defence, borrow money, establish post-offices and post-roads, punish felonies, piracies and counterfeiters, secure copy and patent rights, declare war, borrow money on public credit, raise and provide for an army and navy, call out the militia, execute the laws of the U. S., &c. &c. The judiciary is composed of a supreme court of 1 chief and 6 associate judges; of 31 district courts of a single judge each, except that 6 of the states are each divided into 2 districts; and of 7 circuit courts, composed of the judge of the district, and one of the judges of the supreme court. The judges are appointed by the president with the consent of the senate, and are removeable only by impeachment. The executive power is vested in a president, chosen for 4 years, and eligible for a second term. He must be a native citizen, or have been a citizen at the adoption of the constitution, 35 years of age, and have resided in the U. States 14 years. He exercises a qualified negative; by consent of the senate makes treaties, appoints ambassadors and public officers, and exercises the pardoning power. In case of his death he is succeeded by the vice president; both these officers are removeable only on *conviction* of bribery, treason, or other high crimes, &c. The cabinet of the president consists of the secretaries of state, treasury, war, navy, post master general, and attorney general. The electors of the president and vice president, are, in each state, equal to the number of both its senators and representatives in congress. The whole number of electors is 288, of which New York has the largest number, 42, and Delaware the smallest, 3. (*See appendix*.) If no choice is made by a *majority* of the votes of the electors, the house of representatives, *voting by states*, choose a president from the three candidates having the greatest number of votes. If no president is chosen, the vice president performs the duties of the office. If a vice president be not chosen, the senate choose one from the two highest candidates.

History.—For the settlement and early history of the different states, *see articles on them.* From the first English settlement in 1607, until 1775, the present U. S. were under the colonial government of Great Britain. On July 4th, 1776, independence was declared by a congress of delegates from 13 states, met at Philadelphia. A confederation of the states took place Nov. 15, 1777. In 1783, a treaty of peace was signed, and Great Britain acknowledged the independence of the U. S. The present constitution, (excepting some slight amendments) was formed in 1787, and adopted in 1789. The 13 states which adopted it were, New Hampshire, Massachusetts, Rhode Island, Connecticut, New York, New Jersey, Pennsylvania, Delaware, Maryland, Virginia, N. and S. Carolina and Georgia. Vermont was admitted to the union in 1791, Kentucky in 1792, Tennessee in 1796, Ohio in 1802, Louisiana in 1812, Indiana in 1816, Mississippi in 1821, Illinois in 1818, Alabama and Maine in 1820, and Missouri in 1821. The presidents of the U. S. have been as follows: George Washington, from 1789 to 1797; John Adams, 1797 to 1801; Thomas Jefferson, 1801 to 1809; James Madison, 1809 to 1817; James Munroe, 1817 to 1825; John Quincy Adams, 1825 to 1829; Andrew Jackson, 1829.

UNITY, p-t. Waldo co. Me. 30 ms. N. E. Augusta, has Kennebec co. W., and is crossed by a small branch of Sebasticook r. which is partly supplied by a large pond which encroaches on the N. boundary. Pop. 1830, 1,199.

UNITY, p-t. Sullivan co. N. H. 43 ms. from Concord, 88 from Portsmouth, and 90 from Boston, has Whortleberry pond N., which gives rise to Little Sugar r., a small branch of Connecticut, which crosses this town and Charlestown. The surface is uneven and rocky, the soil good for grazing and flax. A bed of copper ore has been discovered in a ledge of rocks. First settled, 1769. Pop. 1830, 1,258.

UNITY, p-v. northeastern part Montgomery co. Md. 27 ms. N. W. C.

UNIVERSITY OF VIRGINIA and p-o. Albemarle co. Va. situated 1 m. wstrd. Charlotteville, and by p-r. 124 ms. a little W. of S. W. W. C., and 82 N. W. by W. Richmond. This institution has been briefly noticed under the head of Charlotteville.

UPATOIE, as in p-o. list, Upotoie on Tanner's U. S. map, creek and p-v. Muscogee co. Geo. The Upatoie creek rises in Talbot and Marion, and flowing wstrd. enters and traverses Muscogee co., the far greater part of which it drains, and is finally lost in Chattahooche r. The p-v. of Upatoie is on the creek of same name, by p-r. 140 miles wstrd. Milledgeville.

UPPER BLACK EDDY, p-v. on Delaware r. Bucks co. Pa. by p-r. 191 ms. N. E. W. C.

UPPER BLUE LICK and p-o. sthrn. part Fleming co. Ky.

UPPER DUBLIN, tsp. and p-o. Montgomery co. Pa. between White Marsh and Horsham, 17 ms. N. N. W. Phil.

UPPER FLAT LICK and p-o. Knox co. Ky. by p-r. 129 ms. S. E. Frankfort.

UPPER HANOVER, tsp. and p-v. Montgomery co. Pa. situated on the Perkiomen creek, in the N. W. angle of the co. 37 ms. N. W. Phil.

UPPER HUNTING, creek and p-o. southern part Caroline co. Md. by p-r. 95 ms. a little S. of E. W. C.

UPPER MARLBORO', p-v. and st. jus. Prince George's co. Md. situated on a cr. called the Western Branch of Patuxent, 18 ms. S. E. by E. W. C., 23 S. W. Annapolis, and 36 a little W. of S. Baltimore. Lat. 38° 49', and long. 0° 15' E. W. C.

UPPER MERION, tsp. and p-o. Montgomery co. Pa. 11 ms. N. W. by W. Phil. The tsp. is the higher of the two tsps. of Montgomery co. on the right bank of Schuylkill r.

UPPER MIDDLETOWN, tsp. and p-o. Fayette co. Pa. on Red Stone cr., 10 ms. E. Brownsville, and by p-r. 13 ms. northwards Uniontown.

UPPER PEACH TREE, p-v. southwstrn. part Wilcox co. Ala. by p-r. 117 ms. S. Tuscaloosa.

UPPER SANDUSKY, p-v. wstrn. part Crawford co. O. by p-r. 64 ms. a little W. of N. Columbus. The tsp. of Sandusky contains two p-vs., called relatively Upper and Lower Sandusky. Pop. of the tsp. 1830, 579.

UPPER STRASBURG, p-v. Franklin co. Pa. 9 ms. northwards Chambersburg, and by p-r. 99 ms. N. N. W. W. C. This place was from its relative situation in the co. formerly called Upperville.

UPPERVILLE, p-v. in the extreme nrthwstrn. angle of Fauquier co. Va. by p-r. 54 miles W. W. C.

UPSON, co. of Geo. bounded by Pike N., Monroe E., Crawford S. E., and Flint r., separating it from Talbot S. and S. W. Length 25 ms., mean width 12, and area 300 sq. ms. Extending in lat. from 32° 45' to 33°, and in long. from 7° 14' to 7° 39' W. W. C. Declivity S. S. W towards Flint r. Chief t. Thomaston. Pop. 1830, 7,013.

UPSON C. H. (*See Thomaston.*)

UPTON, p-t. Worcester co. Mass. 38 ms. S. S. W. Boston, 10 S. E. Worcester, is partly hilly and partly level, with soil favorable to grass, and gives rise to West r., a branch of the Blackstone. Pop. 1830, 1,167.

URBANA, p-t. Steuben co. N.Y. 207 ms. from Albany, 7 N. N. E. Bath; contains 5 or 6 ms. of Crooked lake, has a rough surface, with the principal part of its soil inferior. Pop. 1830, 1,288.

URBANNA, p-v. and st. jus. Middlesex co. Va. situated on the right bank of Rappahannoc r., by p-r. 83 ms. a little N. of E. Richmond, and 142 a little E. of S. W. C. It is a seaport 18 ms. above the mouth of the r.

URBANNA, p-v. and st. jus. Champaign co. O. by p-r. 50 ms. a little N. of W. Columbus, and 42 ms. N. N. E. Dayton. Lat. 40° 05', long. 6° 44' W. W. C. It is situated on a small branch of Mad r., and contains a printing office, a Methodist and Presbyterian church, market house, 9 or 10 stores, and the common co. buildings. Population 1830, 1,102.

Urquhart's Store and p-o. Southampton co. Va. by p-r. 79 ms. s. s. e. Richmond.

Utica, p-t. and city, Oneida co. N. Y., is situated in 43° 10′ n. lat., and 2° 42′ e. long. W. C.; 96 ms. n. n. w. Albany, 15 s. e. Rome, 246 n. w. New York, and 383 n. e. W. C. It is pleasantly situated on the s. side of the Mohawk r., where formerly stood Old Fort Schuyler, at the point where the Erie canal, the great western road, and the river meet. This situation gives it unusual facilities for intercourse with the large cities, and with the interior, in consequence of which its increase has been remarkably rapid. The town and village are of the same extent, the township being small. The soil is alluvial, of good quality, with a gradual ascent from the river, and formerly was covered with maple, beech, elm, and helmlock forests. These within less than 40 years have given way to the flourishing town which now occupies their place. The city is large, regularly and well built, wealthy and active. The streets are straight, some of them broad, neatly and elegantly built, and adorned with shade trees. In 1794, there were on this spot only a log tavern, and two or three other buildings. Now, among its public buildings are 3 banks, several handsome churches, a college, the court house, an academy, &c. There is also a museum, and several hotels. The principal street is crossed at right angles by the Erie canal, over which are several very good bridges. The bridge over the Mohawk r. is also worthy of notice. The central situation of Utica gives it superior advantages for business, and its already flourishing trade is gradually increasing. Several packet boats pass to and from the city daily; and stage coaches and freight boats constantly arriving and departing, give to the city the air of great enterprize and activity. Numerous manufactories are in operation in the neighborhood of Utica, as of cotton, wool, glass, iron, &c. In Oneida co. are 21 manufactories of cotton goods, which are chiefly owned in this city. The country about Utica is fertile, and the scenery delightful. Trenton falls, within 14 miles, are yearly visited by numbers of travellers, attracted thither by its scenery, which in romantic beauty and sublimity are almost unrivalled. The West Canada creek, on which these falls are situated, here passes through a deep channel of limestone, where the chasm is 150 feet deep. There are 4 principal cataracts, the highest of which is 48 feet high. In another part of the stream is a successive series of beautiful cascades. Other curiosities in the vicinity often engage the notice of travellers. Utica was incorporated as a village in 1798, and a city in 1832. Pop. 1830, 8,323.

Utica, p-v. northern part Licking co. Ohio, by p-r. 47 ms. n. e. by e. Columbus.

Utica, p-v. Clark co. Ind., by p-r. 113 ms. s. s. e. Indianopolis.

Uwchland, p-o. Chester co. Pa., 13 ms. n. w. West Chester, and 35 n. w. by w. Philadelphia.

Uxbridge, p-t. Worcester co. Mass., 38 ms. w. Boston; contains many manufactories. It is crossed near the middle by Blackstone river and canal, and has the line of R. Island s. Blackstone river here receives West and Mumford rs., both which streams afford numerous mill sites. Iron and granite are found in the town. The surface in the centre is nearly level, but hilly in other parts. There was formerly an Indian village here called Wacuntug. Pop. 1830, 2,086.

V.

Vacasausa, bay of Florida, spreading in a circular form about 20 ms. in diameter, to the s. e. of the mouth of Suwannee r. To the s. it opens into the Gulf of Mexico, having Cedar Keys w., and Saint Martin's or Pagoi Keys, s. e. Oyster banks obstruct the bay of Vacasausa, and the Suwannee river cannot be entered with any vessel above 5½ feet water, and with so much only at high tide. The tides are more dependent on the winds than on the moon, and vary along this coast from 18 to 36 inches, and reach 4 feet only after long and high s. w. winds.

The distance from the bay of Espiritu Santo to the mouth of Suwannee river, is 135 ms.; and the wide and shoaly bank which obstructs the coast from Espiritu Santo to Vacasausa bay, continues uninterrupted to the bay of Appalachie, where its breadth is reduced to about 3 ms., and a channel found to enter the river Saint Mark. This channel is accessible to vessels drawing 10 feet, and affords good anchorage 8 ms. from the town of Saint Mark, and vessels drawing 8 feet can reach the t. itself. The distance along the coast from Vacasausa bay, or mouth Suwannee river, is about 95 ms. to the mouth of Saint Mark river, and the channel to the latter is the only good entrance to be found from the bay of Espiritu Santo, or an extent of 230 ms.

Vallie's Mines, and p-o. Jefferson co. Mo., by p-r. 145 ms. s. e. by e. Jefferson City.

Vadensburg, p-o. Chesterfield co. Va., by p-r. 20 ms. southward Richmond.

Valley, p-o. northern part of Mifflin co. Pa., by p-r. 64 ms. n. n. w. Harrisburg.

Valley Forge, p-o. northeastern part Chester co. Pa., about 20 ms. n. w. Phil. It is situated on the Schuylkill, near the mouth of Valley cr.

Valley Hill, p-v. Chester co. Pa.

VALLEYTOWN, and p-o. in Amoi dist., or the Cherokee territory in Ten., by p-r. 621 ms. s. w. by w. W. C., and 228 s. e. by e. Nashville.

VALLONA, p-v. Jackson co. Ind., 4 ms. s. of Brownstown, the co. seat, and by p-r. 73 ms. s. Indianopolis.

VAN BUREN, co. Mich., bounded N. by Allegan co., Kalamazoo co. E., Cass co. Mich. S., Berrien s. w., and lake Michigan N. w. Except an elongation containing about 40 square miles, on the northwestern part, which reaches lake Michigan, the body of the co. is a square of 24 ms. each way; area 616 sq. ms. Lat. 42° 15′ and long. W. C. 9° w. intersect in this co. The slope is westward, and chiefly drained by the Papau, branch of Saint Joseph's river of lake Michigan. From the northern border, however, streams flow northwardly into Kalamazoo r. It is named in the census returns of 1830, but then contained only 5 inhabitants. The central part is about 160 ms. nearly due w. from Detroit.

VAN BUREN, p-v. Vermillion co. Il., by p-r. 185 ms. N. E. Vandalia.

VAN BUREN, p-v. Crawford co. Ark..

VANCEBURG, p-v. on the left bank of O. r., in the northern part of Lewis co. Ky., by p-r. 99 ms. N. E. by. E. Frankfort.

VANCE'S FERRY, and p-o. Orangeburg dist., S. C., 68 ms. by p-r. from Columbia.

VANDALIA, p-v. Wayne co. Ind., by p-r. 53 ms. E. Indianopolis.

VANDALIA, p-v. st. jus. Fayette co., and seat of government, state of Illinois, is situated on the right bank of Kaskaskias river, 80 ms. N. E. by E. Saint Louis, in Mo., about 200 ms. s. w. by w. Indianopolis, and by p-r. 781 ms. w. W. C., and 127 ms. N. N. W. Shawneetown on Ohio r. Lat. 38° 56′, long. W. C. 12° 08′ w. It is of recent foundation, but contains a pop. of about 500. The buildings, public and private, are respectable, if we regard the few years which have elapsed since the site was a wilderness.

VANDERBURG, co. Ind., bounded w. by Posey, Gibson N., Warrick E., and the O. river S. separating it from Henderson co. Ky. N. lat. 38° and long. W. C. 10° 40′ w. intersect in the southern part of this co. Slope southwestward, and drained into the Ohio and Wabash rs. The surface very hilly, and pretty rocky, but soil fertile. Chief t. Evansville. Pop. 1820, 1,798, 1830, 2,611.

VAN HOOK'S STORE, and p-o. Person co. N. C., by p-r. 56 ms. N. N. W. Raleigh.

VANSVILLE, p-o. nrthrn. part Prince George's co. Md., 14 ms. N. E. W. C.

VARRENNES, p-v. western part Anderson dist., S. C., about 20 ms. s. Pendleton, and by p-r. 123 ms. N. W. by W. Columbia.

VARIETY MILLS, and p-o. eastern part Nelson co. Va., by p-r. 112 ms. a little N. of W. Richmond.

VASSALBOROUGH, p-t. Kennebec co. Me., 8 ms. N. Augusta; has Kennebec river on its w. line, and contains part of a large pond, and several small ones, whose waters are discharged into that r. Pop. 1830, 2,761.

VENICE, p-t. Cayuga co. N. Y., 20 ms. s. Auburn. Pop. 1830, 2,445.

VASSAUSA BAY. (*See Vacasausa bay.*)

VENANGO, co. Pa., bounded N. W. by Crawford, N. and N. E. by Warren, E. by Jefferson, S. E. by Clarion river, separating it from Armstrong, S. W. by Butler, and W. by Mercer. Length from E. to W. 40 ms., mean width 28, and area 1,120 sq. ms. Extending in lat. from 40° 10′ to 41° 37′, and in long. from 2° 16′ to 3° 04′ w. W. C. Alleghany r. enters this co. from the N., and winding in a northwestwardly course receives French cr. at Franklin. The united stream thence flows S. S. E., by a very tortuous channel, to its egress from the co., where it receives Clarion river from the eastward. The general declivity of the eastern and central parts is to the S. W. by W., having the channel of the Alleghany and French creek as base. A triangle of about 150 sq. ms. lies to the right of the Alleghany, and slopes eastward towards that stream. Chief t. Franklin. Pop. 1820, 4,915, 1830, 9,469.

VENANGO FURNACE, and p-o. Venango co. Pa., by p-r. 275 ms. northwestward W. C.

VENUS, p-v. on the left bank of Miss. river, northwestern part Hancock co. Il. by p-r. 133 ms. N. W. by W. Vandalia.

VERDIERVILLE, p-o. Orange co. Va., by p-r. 81 ms. S. W. W. C.

VERDON, p-v. Hanover co. Va., 33 miles northwards Richmond.

VERGENNES, city, Addison co. Vt., lies on Otter creek, at the head of navigation, and embraces an area of 400 rods by 480. It was incorporated 1788. The first settler within the limits came in 1766; the others were from Mass. and Conn. The creek falls 37 feet, and affords many good mill sites, some of which are occupied. Above the falls the stream is about 500 feet wide, and at the descent is divided into 3 parts by 2 isls. The largest vessels on lake Champlain come up 7 miles, and the shores are very bold, but the channel is very crooked. Commodore Mac Donough's flotilla was fitted out here in 1814; and the large lake steamboats have wintered here. Considerable trade is carried on, the surrounding country being fertile, and the place advantageous for ship building. The city has 2 school dists. Pop. 1830, 999.

VERMILLION, bay, or more correctly lake, as it differs in no essential respect from similar sheets of water on the La. coast, at the mouths of Sabine, Mermentou, Calcasin, Atchafalaya, and La Fourche, spreads from the Vermillion sthestrd., enclosed on the gulf side by a chain of low, long and narrow marshy islands, terminated towards, and separated from Atchafalaya bay, by Point Chevreuil. The eastern part of Vermillion bay is called locally Cote Blanche bay, but it is only the same sheet of water, with its northern shore indented by Point Cypriere Mort. This bay is in depth about from 10 to 12 feet, but as in respect to the river, the bars admit no vessels with a draught above 5 feet.

VERMILLION, river of La., has its source in the vicinity of the village of Saint Landre, in Opeloussas. Known there as bayou Bourbee, it flows S. S. E. about 12 ms. to where it is connected with the Teche by bayou Fusilier, and thence assuming the name of Vermillion, is gradually augmented by the drain of the prairies on each side, pursues a general southern course of 60 ms. to its final efflux into the Gulf of Mexico. In the superior part of its course, the banks of the Vermillion are clothed with forest timber, which gradually becomes scarcer advancing towards the Gulf, and before reaching the lake or bay, ceases, if we except small detached clumps of live oak, and some other trees. The land along the Vermillion, where of adequate elevation, is every where highly productive, and towards the mouth, the climate below lat. 30° admits the growth of sugar. The tide rises in the Vermillion upwards of 50 ms., but the bars and lake admit only vessels of 5 feet draught.

VERMILLION, small river of Ohio, rising in Lorain and Huron cos., and flowing nrthrdly. nearly along the dividing line of these two cos., falls into lake Erie, after a comparative course of about 30 ms.

VERMILLION, river of Il. and Ind., rising in the former, interlocking sources with those of Kaskaskias, Sangamon, and Pickmink rivers, and flowing thence S. E. by comparative courses 60 ms. over Vermillion co. of Il., and Vermillion of Ind., falls into Wabash river at lat. 40°.

VERMILLION, p-v. Huron co. Ohio, by p-r. 130 ms. N. N. E. Columbus.

VERMILLION, co. Ind., bounded by Warren co. N., Wabash river separating it from Fountain co. N. E., and Parke S. E.; it has Vigo co. S., Edgar co. Il. S. W., and Vermillion co. Il. N. W. N. lat. 40°, long. 10° 30′ w. W. C. Slope eastward towards Wabash river, and in that direction it is drained by Vermillion r. and numerous other streams. Length from S. to N. 38 ms., mean breadth 8, and area 304 sq. ms. Chief town, Newport. Pop. 1830, 5,692.

VERMILLION, co. of Il., bounded by Warren N. E., Vermillion co. Ind. S. E., Edgar Il. S., and Cole S. W. On the other sides it is bounded by unappropriated territory. Length from S. to N. 38 ms., breadth 32, and area 1,216 sq. ms. Lat. 40° and long. W. C. 11° W., intersect in this county. It is very nearly commensurate with the higher part of the valley of Vermillon r. Slope S. E. by E. Chief t. Danville. Pop. 1830, 5,836.

VERMILLIONVILLE, p-v. on the right bank of Vermillion r., Lafayette parish, La., about 30 ms. S. W. by W. New Iberia, and 48 ms. S. St. Landre.

VERMONT, one of the U. S. of America, bounded N. by Lower Canada, E. by the Conn. r. which separates it from New Hampshire, S. by Massachusetts, and W. by New York, and lake Champlain. It lies between 42° 44′ and 45° lat., and 3° 31′ and 5° E. long. from W. C. Its greatest length N. and S. is 157½ ms., and greatest breadth 90 ms.; medial breadth 57 ms., and area 10,200 sq. ms. It is divided into 13 counties, and 245 towns, generally about 6 ms. square, and 2,000 school districts.

Population.—In 1790, Vermont contained 85,539 inhabitants; in 1800, 154,465; in 1810, 217,865; in 1820, 235,764, and in 1830, 280,657, the latter in detail as follows:

Counties.	Pop. 1830	Counties.	Pop. 1830.
Addison,	24,940	Orleans,	13,985
Bennington,	17,168	Orange,	27,285
Caledonia,	20,967	Rutland,	31,294
Chittenden,	21,765	Washington,	21,378
Essex,	3,981	Windham,	28,748
Franklin,	21,525	Windsor,	40,625
Grand Isle,	21,765		

Of the foregoing were white persons—

	Males.	Females.
Under 5 years of age	21,700	21,338
From 5 to 15	37,003	35,513
15 to 30	39,989	40,933
30 to 50	26,168	27,298
50 to 70	12,251	11,879
70 to 90	2,821	2,728
90 and upwards	51	91
Total	139,986	139,790

Of which were deaf and dumb, 153; blind, 51; aliens, 3,364; colored population (there are no slaves) as follows—

	Males.	Females.
Under 10 years of age	122	121
From 10 to 24	113	131
24 to 36	80	74
36 to 55	61	71
55 to 100	47	56
100 and upwards	3	2
	426	455

Recapitulation.

Whites.	Colored persons.	Total.
279,776	881	280,657

The Green mtns., from which the state derives its name, on account of the evergreens with which they are covered, occupy a large part of the state, and most of the surface is very uneven. The range passes through the whole length of the state, about half way between the Connecticut and lake Champlain. It divides the cos. of Windham, Windsor and Orange, from Bennington, Rutland and Addison, with one lofty ridge, through which there is no opening, and no channel of a stream, so that 5 turnpike roads cross at considerable elevations. The range is divided in the S. part of Washington co. The loftier ridge runs along the east line of Chittenden and Franklin cos., and the other, the " the height of lands " runs N. E. into Caledonia co. This ridge is of nearly uniform elevation, and divides the streams of lakes Champlain and Memphremagog, from the tributaries of the Connecticut. The W. ridge presents a more broken outline, and is cut through by Onion and Lamoille rs. In this range are the loftiest peaks in the state. 12 rivers flow from

Vt. into Connecticut r. and 9 into lakes Champlain and Memphremagog, all of which are small. The springs and brooks are so numerous, that every spot in the state appears to be supplied with pure running water. There are mineral springs in different parts of the state, impregnated either with sulphuric acid gas, or iron, some of which are resorted to by invalids. A small part of Memphremagog lake lies in Vt. and the remainder in L. Canada. North Hero, South Hero and Lamotte are three large islands in lake Camplain, belonging in this state; and there are several others of smaller size. The rocks are generally primitive, but there is a transition range, 10 or 15 ms. wide along lake Champlain. The soil on the borders of the streams is chiefly alluvial and is the richest in the state, but some of the uplands are almost equal to it in fertility. A large part of the useful soil is uneven or stony, and better fitted for grazing than tillage. The climate of Vt. is variable and cold, but healthful. The extremes are between 27° below 0, and 100° above, of Fahrenheit. From Dec. 1st till April, the ground is usually covered with snow. The principal indigenous forest trees, are the hemlock, spruce and fir, which are found upon the mtns.; the oak, elm, pine, nut, sugar-maple, beach and birch which occupy the meadows and more cultivated tracts; and the cedar which abounds in the swamps. Moose of very large size, deer, bears, catamounts and wolves were formerly very common in Vt., but have now almost entirely disappeared. Agriculture and grazing form the chief employment of the people. Wheat is most cultivated w. of the mtns., but fruit trees, especially apples, are raised and flourish in all parts. Great numbers of cattle, horses and sheep are annually sent out of the state. Pot and pearl ashes, bar and cast iron and maple sugar are important articles of export. Water power is abundant in most parts of the state, and is applied to some extent to the manufacture of woollen, cotton and iron. There are several quarries of durable and handsome marble. Great quantities of timber were formerly floated down the Connecticut r. in the spring floods, until the legislature required that it should be sawn in the state. Lake Champlain on the w. affords great advantages for navigation, especially since the construction of the Champlain canal, which opens a navigable route to New York city. A company has been recently incorporated for constructing a railroad from Bennington to Troy, N. Y. On the E. boundary Connecticut r. is navigable with rafts, and small steamboats have lately ascended as far as Windsor.

History.—Lake Champlain was discovered by a Frenchman in 1609, but no settlement was made in the state until 1724, when fort Dummer in Windham co. was built by the colony of Mass. In 1731, the French built a fort at Crown Point, and made a settlement on the Vt. shore, at Chimney Point. The Indian and revolutionary wars retarded the population very much. The territory of Vt. was claimed by New Hampshire and New York; and the disputes which this occasioned also impeded the progress of improvement. These contests respecting the territory of Vt., continued for years, between many of the first grantees, and others who purchased the land of New York. The sheriffs both of N. H. and N. Y. were resisted, and at length a system of opposition was commenced under Col. Ethan Allen, Seth Warner, &c. In 1774, New York passed very severe laws on the subject; but the commencement of the revolution suspended the contest, and on the 16th of Jan. 1777, a convention from many of its towns declared the tract of country usually called "the New Hampshire grant" a separate state, by the name of Vermont. In July 1777, a constitution was adopted by another convention at Windsor, and the government was organized, March 13th, 1778. A party of Vermonters, under the command of Ethan Allen, surprized fort Ticonderoga, in 1775, at the same time Crown Point was taken by Seth Warner, and Sheensboro', (now Whitehall,) by another body of them. The people of the state rendered important services to the country during the revolutionary war.

In 1786 the constitution of the state was revised. In 1790 the controversy with New York was terminated, by paying her $30,000, and in Feb. 1791, Vt. was admitted into the Union. In 1793 the constitution of Vt. was again revised.

Government.—The legislative power is vested in a house of representatives, called the general assembly. With the council they appoint the judges of the courts annually, and the higher military of officers when required. The executive power is vested in a governor, deputy gov. and council of 12, annually chosen by the people. The gov. and council have no negative on bills passed by the house, but may postpone them one session. 13 censors are chosen annually by the people, to see that the constitution is not infringed, &c. The supreme court has three judges; each county also has a court of three judges, sitting twice a year; and each probate district has a court composed of 1 judge and justices of the peace. The supreme court sits annually in in each county except Grand Isle; and the judges of it form the court of chancery. The state prison is at Windsor.

Education.—The higher institutions of learning and science are, the Vt. university, Middlebury college, and the Vt. academy of medicine. Most of the cos. have academies, and every town is divided into school districts, in which schools for all classes are kept, usually but part of the year. The university was founded at Burlington, 1791, by the legislature. The gov., speaker of the house of representatives, and president of the university, are ex-officio members of the corporation; and 28 others are appointed by the legislature. Middlebury college was founded in 1800. The academy of medicine in 1818, at Castleton.

Religious denominations.—The Congregationalists have 13 associations, 203 churches, 155 ministers and 17,236 communicants; Baptists 105 churches, 64 ministers, and 8,478 communicants; Methodists 44 ministers and 8,577 communicants; Episcopalians 15 ministers; Unitarians 3 societies and 1 minister; there are some Freewill Baptists, Christians, and Universalists.

VERNON, p-t. Windham co. Vt., 35 ms. E. Bennington, 50 S. Windsor, is in the S. E. corner of the state, with Connecticut r. on its E. boundary. It was one of the first settlments in Vt., the settlers coming from Northfield and Northampton, Mass., and suffering much from Indian attacks and murders. Startwell's fort was built in 1740. White Lilly pond covers 100 acres. The streams are small, the soil is thin and stony, much of the surface mountainous, with small meadows on Conn. river. There are pitch-pine plains E. and slate is quarried W. Pop. 1830, 681.

VERNON, p-t. Tolland co. Conn., 12 ms. N. E. Hartford, about 3½ ms. by 5, is generally uneven, lying on the high lands between the waters of Connecticut and Thames rs. Quarries of micaceous schistus are wrought in the S. W. part, which forms an excellent pavement for side walks, and is extensively used. Hockanum and Tankerooson are good mill streams, and supply water to several mills and factories. Pop. 1830, 1,164.

VERNON, p-t. Oneida co. N. Y., 17 ms. W. Utica, 112 ms. W. by N. Albany, has Oneida cr. W. which separates it from Madison co. It is of irregular form, and comprehends the principal Oneida settlement in the state. About 1-3 of the town belongs to the Oneida and Tuscarora reservations. First settled, 1797. The surface is pleasantly varied, the soil good and watered by Skanando cr. and its branches. It contains the villages of Vernon, Oneida and Castleton. Population 1830, 3,045.

VERNON, incorporated v. Vernon, Oneida co. N. Y., 13 ms. W. by N. Albany, 13 Utica, on Skanando cr., has several manufactories.

VERNON, p-t. Sussex co. N. J., 21 ms. N. E. Newton, 88 N. by E. Trenton, has the state of N. Y. on the N. E., Wawayanda mtn. E., Pochuck mtn. W., and other eminences of the Wallkill range; is crossed by Wallkill cr. W. in the upper part of its course. Pop. 1830, 2,377.

VERNON, p-v. in the sthwstrn part of Kent co. Del., 22 ms. S. S. W. Dover, and by p-r. 96 E. W. C.

VERNON, p-v. on the right bank of Ala. r., in the sthrn. part of Autauga co. Ala., 10 ms. W. Washington, the co. st., and by p-r. 124 ms. S. E. Tuscaloosa.

VERNON, p-v. and st. jus., Hickman co. Ten., situated in the nrthrn. part of the co., in direct distance 40 ms., but by p-r. 66 ms. S. W. by W. Nashville, N. lat. 35° 48′, long. 10° 31′ W. W. C.

VERNON, p-v. Madison co. Miss., by p-r. 38 ms. nrthrd. Jackson.

VERNON, p-v. nrthestrn. part Trumbull co. O., by p-r. 180 ms. N. E. Columbus.

VERNON, p-v. and st. jus., Jennings co. Ind., by p-r. 64 ms. S. S. E. Indianopolis, N. lat. 39°, long. W. C. 8° 36′ W.

VERONA, p-t. Oneida co. N. Y., 113 ms. W. by N. Albany, 12 W. Rome, has Wood cr. N., and Oneida cr., separating it from Madison co. and Oneida lake. It was purchased from the Oneida Indians in 1796. The surface is nearly level, with much swampy land and good soil. Erie canal crosses it N., near the cr. Forts Bull and Rickey were on that stream. Pop. 1830, 3,739.

VERSAILLES, p-v. and st. jus., Woodford co. Ky., 12 ms. W. Lexington, and 13 S. E. Frankfort, N. lat. 38° 02′, long. W. C. 7° 40′ W. Pop. 1830, 904.

VERSAILLES, p-v. and st. jus., Ripley co. Ind., by p-r. 79 ms. S. E. Indianopolis, N. lat. 39° 05′, long. W. C 8° 36′ W.

VERSHIRE, p-t. Orange co. Vt., 25 ms. S. E. Montpelier, 35 N. Windsor, first settled 1780, is uneven and often stony, and watered by the head streams of the Ompompanoosuc. Pop. 1830, 1,260.

VEVAY, p-v. and st. jus., Switzerland co. Ind., by p-r. 105 ms. S. E. Indianopolis, and 45 ms. below Cincinnati. It is situated on the Ohio r., and contains about 1,500 inhabitants. It was founded in 1804, by a small Swiss colony, and now contains the common co. buildings, a printing office, a branch of the bank of Ind., some other public buildings, and in the vicinity, the most extensive vineyard in the U. S. Mr. Flint speaks in high terms of the inhabitants of Vevay, and says, "They are every year improving on the vintage of the past. They are the simple and interesting inhabitants that we might expect, from the prepossessions of early reading, to find from the vine clad hills of Switzerland." There are in Vevay a literary society, and public library. The situation is fine locally, and also commercially with the r. Ohio and interior country.

VESTAL, t. Broome co. N. Y., 150 ms. from Albany, 8 ms. S. Binghampton, has Pennsylvania on the S., and Tioga co. W., has Susquehannah r. N., and includes several islands in that stream. Chocunut cr. flows N., nearly across the whole t. Pop. 1830, 946.

VETERAN, p-t. Tioga co. N. Y., 12 ms. N. Elmira. Pop. 1830, 1,616.

VICKSBURG, p-v. and st. jus., Warren co. Miss., by p-r. 50 ms. N. N. E. Natchez.

VICTOR, p-t. Ontario co. N. Y., 10 ms. N. W. Canandaigua, 203 W. Albany, has Monroe co. N. and W., supplied with mill sites by Mud and Teronto crs.; N. W. is a large cedar swamp, and plains bearing oak timber. Pop. 1830, 2,270.

VICTORY, t. Essex co. Vt., in the S. W. part of the co., is crossed by Moose r. from N. W. to S. E. Pop. 1830, 53.

VICTORY, p-t. Cayuga co. N. Y., 167 ms. W. Albany, 24 N. Auburn, 10 N. Erie canal, has Seneca co. W., has small streams, but good mill seats. Pop. 1830, 1,819.

VIELLEBORO', p-v. in the nrthrn. part of Caroline co. Va., 8 ms. N. Bowling Green, the st. jus. of the co., and 70 ms. S. S. W. W. C.

VIENNA, p-t. Kennebec co. Me., 26 ms. N. W. Augusta, has on its W. boundary a small stream running N. into Sandy r. Pop. 1830, 722.

VIENNA, p-t. Oneida co. N. Y., 125 ms. W. by N. Albany, 12 W. Rome, has Oneida lake S., and Oswego co. W., has good land on the courses of Fish and Wood creeks, but inferior in other parts. First settled 1802. Erie canal is from 5 to 10 ms. distant S. Pop. 1830, 1,766.

VIENNA, v. Phelps, Ontario co. N. Y., 12 ms. E. Canandaigua, stands at the junction of Flint cr. and the Canandaigua outlet, and contains several mills, with large gypsum beds.

VIENNA, p-v. and sea port, on the right bank of Nantikoke r., and in the estrn. part of Dorchester co. Md., about 17 ms. S. E. by E. Cambridge, the co. seat, and by p-r. 118 ms. in the same direction from W. C.

VIENNA, p-v. in the nrthrwstrn. part of Pickens co. Ala., by p-r. 18 ms. N. W. Pickensville, the co. seat, and 66 in the same direction from Tuscaloosa.

VIENNA, p-v. Trumbull co. O., by p-r. 165 ms. N. E. Columbus.

VIENNA, p-v. and st. jus., Johnson co. Il., by p-r. 167 ms. S. Vandalia. N. lat. 37° 27', long. W. C. 12° W.

VIGO, co. Ind., bounded by Vermillion co. Ind. N., Parke N. E., Clay E., Sullivan S., Wabash r., separating it in part from Clarke co. Il. S. W. Length from S. to N. 26, breadth 18, and area 468 sq. ms. N. lat. 39° 30', long. W. C. 10° 30' W. The nrthwstrn. part is traversed by the Wabash r. The general slope S. S. W. Chieftown, Terre Haute. Pop. 1820, 3,390, and in 1830, 5,766.

VILLAGE GREEN, p-v. Delaware co. Pen., 4 ms. wstrd. Chester, the co. seat, 16 ms. S. W. Philadelphia, and by p-r. 126 N. E. W. C.

VILLAGE SPRINGS, and p-o. in the sthrn. part of Blount co. Ala., by p-r. 81 ms. N. E. Tuscaloosa, and about 70 S. Huntsville, in Madison county.

VILLANOVA, p-t. Chatauque co. N. Y., 318 ms. W. Albany, 20 N. E. Maysville, has a few streams running into Walnut and Canandaway crs. Pop. 1830, 1,126.

VILLA RICCA, p-v. Carroll co. Geo., by p-r. 178 ms. N. W. by W. Milledgeville.

VILLEMONT, p-v. and st. jus. Chicot co. Ark., situated on the right bank of Miss. r., about 30 ms. in a direct line below the mouth of Ark. r., and by p-r. 184 ms. S. E. Little Rock. N. lat. 33° 23', and long. 14° 07' W. W. C.

VINALHAVEN, p-t. Hancock co. Me., 73 ms. S. E. Augusta, 13 S. Castine; embraces the Fox isls. in Penobscot bay. Population 1830, 1,794.

VINCENNES, p-v. and st. jus. Knox co. Ind., situated on Wabash r., by p-r. 126 ms. S. W. Indianapolis, and about 110 ms. a little S. of E. from Vandalia. Lat. 38° 42' N., long. W. C. 10° 35' W. This is amongst the early settlements of the French from Canada. It is rapidly improving, and contains a bank, academy, a Roman Catholic, and Presbyterian church, two printing offices, land office, and some other public buildings. Pop. 1830, 1,500.

VINCENT, p-v., tsp. of Chester co. Pa., on the Schuylkill, between East Nantmill and Pikeland, 26 ms. N. W. Phil.

VINEYARD, p-t. Grand Isle co. Vt., 28 ms. N. W. Burlington, 13 W. Saint Albans, 85 from Montpelier; is an island in lake Champlain, containing 4,620 acres. It was first settled about 1785. The rocks are good building limestone; a marsh which crosses it is overgrown with cedar. There are 2 school districts. Pop. 1830, 459.

VINEYARD, p-v. Washington co. Ark., by p-r. 187 ms. N. W. Little Rock.

VIRGIL, p-t. Cortlandt co. N. Y., 148 ms. W. by S. Albany, 10 S. Homer; has Broome and Tioga cos. S., and Cayuga co. W.; has Tioughnioga creek N. E., and some streams of Fall creek N. W., which flow into Cayuga lake. These, with a branch of Owego cr. S., supply mill seats; and there is boat navigation on the Tioughnioga to the Susquehannah. This t. has a good soil. Pop. 1830, 3,912.

VIRGINIA, state of the U. S., bounded S. E. by the Atlantic ocean, S. by North Carolina, S. W. by Tennessee, W. by Kentucky, N. W. by the O. r. separating it from the state of Ohio, N. by the southwestern part of Pennsylvania, and the Potomac separating it from the western part of Maryland, and N. E. also by the Potomac, and a part of Chesapeake bay, separating it from central and eastern Maryland.

Having an outline along the Atlantic ocean from the southeastern angle of Md., to the northeastern of N. C., 112 ms.; westward in common with N. C., 340 ms.; along the Iron mountains from the extreme northwestern angle of N. C., to the extreme northeastern angle of Ten., 4 ms.; westward in common with Ten. to Cumberland mountains, and the extreme southwestern angle, 110 ms.; along Cumberland mountains in common with Ky., to Tug Fork of Sandy river, 110 ms.; down Sandy river in common with Ky., to Ohio r., 70 ms.; up Ohio river opposite the state of Ohio, to the western boundary of Pa., 355 ms.; south along west boundary of Pa., and to the southwestern angle of that state, 64 ms.; east along Pa. to the northwestern angle of Md., 58 ms.; south to the head of the North Branch of Potomac, and southwestern angle of Md., 36 ms.; down Potomac r. opposite Md. to the outlet of that stream into Chesapeake bay, 320 ms.; thence over Chesapeake bay, and along the southeastern boundary of Md. to the Atlantic ocean, and place of beginning, 60 ms.; having an entire outline of 1,639 ms. Extending in lat. from 36° 32' to 40° 38' N., and in long. from 1° 46' E. to 6° 33' W. W. C.

The southern boundary of Va. is nearly commensurate with its greatest length, 450 ms. The area of this state is usually underzated; as by a careful measurement by the rhombs, the superficies are within a fraction of 70,000 sq. ms., which, divided by 450, yields 155¼ very nearly, as the mean width. A geographical error exists as respects the southern boundary of Va. That boundary on most maps is laid down as a line along lat. 36½; but it leaves the Atlantic ocean on 36° 32′ nearly, and gradually inclining to the northwards, when it strikes the Iron mountains it is above lat. 36° 33′. The northern boundary of Ten., between the two states, leaves the Iron mountains lat. 36° 05′ nearly, and maintains that curve to the Cumberland mountains. From these elements we see, that the southern boundary of Va. lies at a mean of about 36° 33½′ N., and that it is not even a continued line between the extremes.

Natural Sections.—Virginia is the most extensive of the states of the U. S., and perhaps the most strongly contrasted in its physical features. Similar to Md. and N. C., Va. is sub-divided into three distinct sections. Sea and alluvial section below the head of tide water; the middle and hilly section, and the central or mountainous; but in the case of Va., a fourth and very important natural section may be superadded. This latter section may be very properly called the western or Ohio section, as it is drained into that stream, as a common recipient. These four natural sections are in their respective features and outlines, so distinct as to be recognized in the legislation of the state, and indeed must ever have political and moral effects. The following is a summary of 4 tables, formed by a member of the convention, recently held to form a constitution for the state, and deserves some attention, as upon them in some degree depended the apportionment of representation as it now stands, under the present constitution of the state.

Summery.

	Supposed pop. 1829. Whites.	Slaves.	Total.	sq. ms. in each division.
1. Below head of tide water.	165,227	175,847	341,074	11,805
2. Between that, & the Blue Ridge	201,219	226,991	427,210	15,386
3. Between the Blue Ridge & the Alleghany	166,994	37,857	203,871	13,072
4. Westward of the Alleghany to Ohio river	153,522	12,831	166,353	26,337
Total,	685,962	452,596	1,138,506	68,600

The population of Virginia after several periods has been as follows:—1790, 747,610; 1800, 880,200; 1810, 974,622; in 1820, 1,065,366; and in 1830, 1,211,375; at the latter period the pop. by cos. was as follows.

Eastern District.

Counties.	Population	Counties.	Population.
Accomac	16,656	Buckingham	18,351
Albemarle	22,618	Campbell	20,350
Amelia	11,036	Caroline	17,760
Amherst	12,071	Charles City	5,500
Bedford	20,246	Charlotte	15,252
Brunswick	15,767	Chesterfield	18,637

Counties.	Population.	Counties	Population.
Culpepper	24,027	Mecklenburg	20,477
Cumberland	11,690	Middlesex	4,122
Dinwiddie	21,901	Nansemond	11,784
Elizabeth City	5,053	Nelson	11,254
Essex	10,521	New Kent	6,458
Fairfax	9,204	Norfolk	24,806
Fauquier	26,086	Northampton	8,641
Fluvanna	8,221	Northumberland	7,953
Franklin	14,911	Nottaway	10,130
Gloucester	10,608	Orange	14,637
Goochland	10,369	Patrick	7,395
Greensville	7,117	Pittsylvania	26,034
Halifax	28,034	Powhatan	8,517
Hanover	16,253	Prince Edward	14,107
Henrico	28,797	Prince George	8,367
Henry	7,100	Prince William	9,330
Isle of Wight	10,517	Princess Anne	9,102
James City	3,838	Richmond	6,056
King and Queen	11,644	Southampton	16,074
King George	6,397	Spottsylvania	15,134
King William	9,812	Stafford	9,362
Lancaster	4,801	Surry	7,109
Loudon	21,939	Sussex	12,720
Louisa	16,151	Warwick	1,570
Lunenburg	11,957	Westmoreland	8,396
Madison	9,236	York	5,354
Matthews	7,664		

Western District.

Counties.	Population	Counties.	Population.
Alleghany	2,816	Monroe	7,798
Augusta	19,926	Montgomery	12,306
Bath	4,002	Morgan	2,694
Berkeley	10,518	Nicholas	3,346
Bottetourt	16,354	Ohio	15,584
Brooke	7,041	Page (formerly E. Shenandoah)	8,327
Cabell	5,884	Pendleton	6,271
Frederick	25,046	Pocahontas	2,542
Giles	5,274	Preston	5,144
Grayson	7,675	Randolph	5,000
Greenbrier	9,006	Rockbridge	14,244
Harrison	14,722	Rockingham	20,683
Hampshire	11,279	Russell	6,714
Hardy	6,798	Scott	5,724
Jefferson	12,927	Shenandoah	11,423
Kenhawa	9,326	Tazewell	5,749
Lee	6,461	Tyler	4,104
Lewis	6,241	Washington	15,614
Logan	3,680	Wood	6,429
Monongalia	14,056	Wythe	12,163
Mason	6,534		

Total population of Eastern Va. 832,980; Western, 378,425.

Of the preceding were white persons,

	Males.	Females.
Under 5 years of age	65,793	62,411
From 5 to 10	51,805	49,964
" 10 to 15	43,287	41,936
" 15 to 20	36,947	40,479
" 20 to 30	60,911	62,044
" 30 to 40	36,539	36,456
" 40 to 50	23,381	23,750
" 50 to 60	15,261	15,447
" 60 to 70	8,971	8,765
" 70 to 80	3,674	3,857
" 80 to 90	1,108	1,098
" 90 to 100	184	158
" 100 and upwards	26	98
Total.	347,887	346,383

Of the colored population, were

	Free.		Slaves.	
	Male.	Female.	Male.	Fem.
Under 10 years of age	8,236	8,002	84,000	83,270
From 10 to 24	6,126	7,031	68,917	66,021
24 to 36	3,546	4,501	43,189	40,927
36 to 55	2,721	3,379	30,683	27,206
55 to 100	1,731	2,024	12,155	12,275
100 and upwards	27	24	133	144
Total,	22,387	24,961	239,077	230,680

Recapitulation.

Whites.	Free col'd.	Slaves.	Total.
691,270	47,348	469,757	1,211,375

Features.—Comparatively there is little of Virginia actually level. Such character is only found in the two counties of Accomac and Northampton, E. of Chesapeake bay, and to Princess Anne, Norfolk and Nansemond, with an aggregate area of 2,200 square ms. or less than the thirty first part of the state. West of the Chesapeake bay the country gradually rises into hill and dale, though much marshy and flat land skirts the wide mouths of the rivers. Virginia and Maryland occupy the central part of that physical section of the Atlantic coast so remarkable for deep and wide rivers. Except in extent and position Chesapeake bay differs in nothing essential, besides its greater depth, from Pamlico and Albemarle sounds, on the S. and Del. to the N. In Virginia and Maryland the confluents of the Chesapeake seem to imitate that great reservoir, and Pocomoke, Nantikoke, Choptank, and Chester rivers on the E., and James, York, Rappahannoc, Potomac, Patuxent, and Patapsco on the W., widen into expansive bays before their final discharge. These minor bays become gradually more shallow and more confined in width approaching the head of tide water, but they all retain the distinctive character of bays as far as the ocean tides penetrate inland. The 1st section in the foregoing summary exhibits the counties which may be strictly designated alluvial. Though where approaching the primitive ledge which terminates the tides, the face of the country is diversified by waving hills, still in their structure they are alluvial, of that species called ancient. The far greater part of the substrata are composed of sand and pebbles. Large masses of rock in its original position is rare except at great depths. The Blue Ridge traverses Va. 260 ms. in a direction from S. W. to N. E. and except where traversed by Roanoke and James rs. is a continuous ridge, and a county limit in all its range in that state. Falling from this finely delineated chain, is an inclined plain containing 15,386 square miles, terminated by the head of the Atlantic tides. This truly beautiful section, if we merely regard the fall of water, has a declivity of from about 300 to 500 feet, but the descent of the water gives but a defective idea of the slope in the arable soil, which latter towards the Blue Ridge rises in many places, to at least 1,000 feet in the intermediate spaces between the rivers. The face of nature, though exhibiting little of grandeur, is rich and pleasing in the endless outline of hill, valley, and river scenery. In the higher part, beside the elegant back ground of the Blue Ridge, other detached mountain chains rise and give intimation that the solid structure of the country is Appalachian, and that the outer ridges of that system influence the great bends of the rivers. This mountain influence is seen in the courses of Roanoké, James, Rappahannoc, and Potomac rivers.

Section 2d, contains the 29 counties embraced by what might be with propriety called the Blue Ridge section of Virginia. The 3d and Great Valley section, is in some respects the most remarkable of the natural sections of Virginia. Extending from the Iron mountain at the northeastern angle of Ten. to the northern bend of the Potomac at Hancock's town, the mean length is within a trifle of 300 ms., the mean distance between the Blue Ridge and Alleghany mountain is about 43 ms. This valley is the continuation of the Kittatinny of Pa., and is a true table land, or mountain plateau. The rise is abrupt, as there is a difference of from 200 to 300 feet in the mean level, on the two sides of the Blue Ridge. The elevation of Lynchburg is about 500 feet, whilst that of Staunton at the sources of Shenandoah is 1,152 feet; Lexington in Rockbridge county 902 feet; Salem on the Roanoke, in Botetourt co. 1,002, the Warm Springs in Bath county 1,782 feet, and the mean elevation of the farms on the whole extent no doubt exceed 1,000 feet. The Blue Ridge is in Virginia as in New Jersey, Pa., and Maryland, bounded on the northwestern side by a calcareous band, of more or less breadth. The surface of the Great Virginia valley is in an especial manner broken and diversified, but every where containing zones of highly productive soil, abounding, with some exceptions, in good water, and so rich in scenery, it affords an endless variety of delightful landscape. In regard to declivity, the Great valley presents some curious phenomena. The northern and nearly one half of the whole surface declines to the N. E. towards the Potomac, and is drained by the Shenandoah, Cacapon, and South branch of Potomac. Southward from the sources of Potomac and Shenandoah is a middle valley, drained eastwardly through the Blue Ridge by James and Roanoke rivers. The extreme southern part falls to the N. W. and gives descent to New river or Great Kenhawa. We thus perceive that this table land is partly on the Atlantic slope and partly in the Ohio valley, and that the inflected line that separates the sources of James and Roanoke of the former from those of Great Kenhawa of the latter river system passes the mountain valley obliquely. Passing the table land between Blue Ridge and Alleghany mountains on the third natural section, brings us on the fourth or Ohio section of Virginia. The extreme length of this western slope is within a small fraction of 300 ms. from the northern boundary of Ten. to the the extreme northern angle of Brooke co.

The greatest breadth is nearly along the general course of Great Kenhawa, 135 ms., but both extremes are narrow, and the mean width is about 94, and the area 28,337 sq. ms. This great space is politically subdivided into 23 counties. The surface is in the far greater part mountainous, and in all parts very broken. The ridges or chains of the Apalachian system stretch over it very nearly parallel to the Ohio, in that part of the course of that stream which bounds Virginia. The soil is as various as the surface, or even more so, as every grade of fertility and of sterility may be found. The elevation of the water at the junction of Ohio and Great Kenhawa, being 533 feet, and that point being only about 40 direct ms. from the extreme lowest point of Western Virginia at the mouth of Great Sandy river, we may regard all land surface of the Ohio section as rising above 500 feet. The oceanic elevation of Wheeling is 634 feet, and the Ohio as a base to the great inclined plain and a recipient for the waters of the Western section of Virginia, rises upwards of five hundred and sixty feet, or very nearly on a level with lake Erie. The dividing ridge of the waters of Ohio and the Atlantic, is the apex of the plain before us, and has its highest elevation in the mountain vallies, from which rise on one side the sources of Roanoke and James rivers, and on the other those of Great Kenhawa. Under the heads of Giles, Pocahontas, and Monroe counties of Virginia, which occupy the highest part of the plain we are surveying, it may be seen that the mean elevation of the arable soil exceeds one thousand six hundred feet. A similar if not a higher mean height might in fact be assigned to the sources of Great Kenhawa, from those of Greenbrier to those of New river. From these elevated vallies the Ohio sources flow like radii from a common centre. The different branches of the Monongahela rise in Lewis and Randolph counties, and flowing northwardly over Harrison, Monongalia and Preston counties, enter Pa. and uniting the mingled waters continue northward to meet those of the Alleghany, to form the Ohio at Pittsburg. The Ohio from Pittsburg first sweeps a curve to the northwestward, thence westward and finally southward upwards of one hundred miles, in a remarkable manner parallel to the general course of the Monongahela, the two streams flowing in opposite directions. From the large curve of Ohio below Pittsburg to the influx of Little Kenhawa, there is only a narrow inclined plain of about 30 ms. width between the Ohio river and the sources of creeks flowing estrd. into Monongahela. Down this confined slope flow, Harman's Cross, Buffalo, Wheeling, Fish, Fishing, Middle Island, and some other creeks of lesser note. With Little Kenhawa the plain widens; and the declivity inclines from W. to N. W. This declination is maintained beside in Little Kenhawa, in the vallies of Great Kenhawa, Great and Sandy rivers. The extreme southern part of the Ohio section of Virginia, though also drained into that recipient, the tributary waters are borne from the elevated plateau between the sources of the Great Kenhawa and Ten., and before their discharge make the immense semicircular curve of the latter. From this rapid outline of the Ohio section of Virginia it must be evident that the climate of the whole must vary materially from that of similar latitudes on the Atlantic coast, and from difference of relative level the climate along the high vallies of the Appalachian chains must be very different from that on the greatly lower and locally deep ravine of the Ohio.

This is so obviously the case that early vegetation is often far advanced at Wheeling, at the same time when little or no appearance of spring is perceptible along the dividing ridge of the waters of Ohio and Monongahela. The difference of level between the high water mark in Ohio river and the ridge we have noticed is about a mean of 850 feet; but this ridge is only the first in a series of plains which rise one above another until a mean height of between 1,800 and 2,000 feet is attained in central Virginia. If we assume lat. 38° 10′ as the central lat. it will, on long. 3° w. W. C., correspond nearly with the greatest elevation, and allowing 400 feet as an equivalent to a degree of lat. will give to the counties along the mountainous section of Virginia a winter climate similar in temperature to that of N. lat. 43° on the Atlantic coast. If from the foregoing elements we embrace the whole of Virginia, we have before us a large section of the United States, extending over a small fraction more than 4° of lat., and 8° 34 of long. differing in relative level upward of 2,000 feet, without estimating mountain peaks or ridges. If we suppose the actually settled parts of the United States to be 630,000 sq. miles, Virginia will embrace the one ninth part. It is as we have seen traversed from S. W. to N. E. by the Appalachian system of mountains, in lateral chains. Of these the Blue Ridge is only the most distinctively defined, as it is one of six or seven chains which can be traced and identified over the state. One of these chains, though omitted on some maps and broken into fragments on others, is really in nature very little less obvious than the Blue Ridge, and is distinct over Virginia. This neglected Appalachian chain stretches at a distance of from 15 to 30 miles southeastward from the Blue Ridge. It is known in New Jersey as Schooley's mtn., and though perfectly prominent over that state has received no distinctive name in Pa. In Md. it is called the Parr Spring Ridge and rendered very conspicuous where it is traversed by the Potomac from the fine conical peak, the Sugar Loaf. In Virginia it traverses Loudon, Fauquier, Culpepper, Orange, Albemarle, Nelson, Amherst, Bedford, Franklin and Henry counties. West of the Blue Ridge the mountain chains are also very confusedly delineated on our maps, though they are far from being so in reality. Even on Tanner's

United States, the continuous chains are terminated by the Alleghany, whilst Western Virginia is traversed by three distinctive chains w. from its main spine. In point of fact, the whole state from the head of tide water to Ohio river, is formed of a series of mountain chains and intervening vallies. This structure is obvious to any person who examines its map with a due previous study of the influence of the mountain system on the inflections of the streams. Amongst the mountain chains, however, the Blue Ridge must always remain the most important, physically and politically. This chain stands in a remarkable manner detached; in the peaks of Otter, Botetourt county, it presents the highest land in the Appalachian system s. w. Delaware river, and it is in all parts of its length a county line. When discovered and colonized by Europeans, the region now comprised in Virginia, was one continued dense and very partially broken forest. A few savage tribes were found along the tide waters, but the interior was scarcely inhabited even by savages. It may be remarked, that though the soil increases in fertility advancing from the sea board, still density of population is in a near ratio to proximity to the place of original settlement on James river. If we make every just allowance for the space actually occupied by mountains, and other unproductive tracts, still there would remain 50,000 square miles at least, capable of sustaining a mean distributive population, equal to any one of the best inhabited of its existing counties, say Henrico, including the city of Richmond. Such a ratio would give Virginia upwards of *five millions of inhabitants*, a number far below the number which it could support.

History.—The first charter of Virginia was granted by Queen Elizabeth, in 1583, to Sir Humphrey Gilbert, who perished at sea in an attempt to avail himself of his patent. By this original grant the name of Virginia was imposed on the whole Atlantic coast of North America, claimed by the English. In future time the general name became restricted to what is now Virginia. Sir Walter Raleigh, maternal brother of Sir Humphrey Gilbert, obtained a renewal of the first charter, May 25th, 1584. Under Raleigh's patent a colony was planted on Roanoke, but after repeated attempts and disasters, the enterprise entirely failed. In August, 1587, the governor sailed to England, in quest of supplies; the few persons left, were never again heard of, and the patent of Raleigh was vacated by his attainder. In the early part of the reign of James I, Mr. Hackluyt revived the intention of settlement on the coast of North America, and himself and others obtained by petition a patent dated April 10th, 1606, for that part of the coast extending from N. lat. 34° to 45°. This great zone of 14° of lat. was granted to 2 companies, under the relative name of North Virginia and South Virginia. South Virginia was given to the London company, North Virginia to the Plymouth company. The London company effected a settlement April, 1607, at Jamestown, on Powhatan, or as since called, James river, and commenced not only Virginia but the U. S. So vague were the views of the company, or the adventurers under the patent, that discord and wretchedness compelled the colonists to break up their establishment in 1610, and an entire abandonment was only prevented by the timely arrival of Lord Delaware. In 1612, the 2d charter was granted, and in 1619, the 1st legislative assembly met. The following year negroes were introduced as slaves. They were brought in by a Dutch ship. A state of natural distrust had existed, from their first arrival against the colonists on the part of the natives, which in 1722, eventuated in a massacre, in which 347 whites perished. To the horrors of Indian warfare was added the arbitrary and vexatious regulations of the London company. Royal power interposed, and in 1624 the government of the colony was seized by the king, and administered by commission. The royal governors were as unsteady in their measures as those of the former company, but to the number of these ignorant tyrants Sir William Berkeley was an exception. This nobleman ruled with moderation, and in 1729 restored the legislature by assembling the Burgesses. As an English colony, Virginia was royal in its political features and feelings. In the long revolutionary struggle in England, from 1642, to 1660, the Virginians sided with the royal party. Compelled by force to submit to the parliament, they seized the first moment to exhibit their real sentiments, and Charles II was ackowledged in Virginia before he was in England. The restoration was alike a failure in both countries, but weak and distant, the colony of Virginia suffered most, and the monopolising spirit of the government of the mother country was felt for upwards of a century, to the revolution in 1775. The church of England was established by law in 1662, which added to exactions in trade, large grants of land to royal favorites, and the caprice of royal governors, kept up a spirit of resistance and state of irritation which prepared the public mind to throw off the yoke, and meet the minions of power in arms. As early as 1732, the future hero of the U. S. was born in Va., and had become mature in years when his services were demanded to teach the kings and people of the earth the most salutary lesson either ever received. In the colonial war, commenced in 1755, George Washington and the Virginians were truly distinguished. In the war of the revolution it was, in many respects, the leading state, and the illustrious Washington was only one of many of her sons who shone in that day of events. Since the revolution no great event particular to Virginia occurred, until October, 1829, when a convention met to revise the constitution.

Government.—The first constitution of Virginia was adopted July 5th, 1776, but as settlements extended westward, the provisions were regarded as partial and opppressive, and after many abortive attempts eventuated in a convention, which, on the 14th of January, 1830, reported the existing constitution, which was ratified by a majority of 10,492. The right of suffrage under this constitution is secured under very complex provisions. The right of voting is extended to every white male citizen of the commonwealth, and resident therein, who has attained the age of 21 years and upwards, and who would have been entitled to vote under the former constitution; or if owner of a freehold of $25 value; or if the holder of a joint interest in a freehold to the amount of $25, or who has a life estate in, or title in reversion to, land of $50 value, and had been in full possession of such an estate or reversionary title six months before the election at which he offers to vote; or who shall own, and be in the actual occupation of a leasehold estate, have put such title on record two months before he shall offer to vote—original term at least 5 years, and rent value $200; or who has been a housekeeper and head of a family, 12 months before offering to vote, and shall have paid a tax within the preceding year. The legislative power is vested in a senate and house of delegates, which together are styled the general assembly of Virginia. The house of delegates consists of 134 members, chosen annually, and apportioned in the 4 districts as follows:—36 from the Tide water district, 42 from that above Tide water and below Blue Ridge, 25 from the Third, or mountain district, and 31 from that of Ohio, or the Western district. Senate 32 members—19 from the east, and 13 west from Blue Ridge. Senators elected for 4 years, one fourth going out of office annually. Reappointment of the relative members from the districts of the members of both houses to take place in 1841, and decennially afterward, but the entire number of senators never to exceed 36, nor delegates 150. The executive power is vested in a governor, elected for 4 years, by a joint vote of both houses of the general assembly, and ineligible for the next three years, after the expiration of his term of office. A council of state elected by joint ballot of the 2 houses, consisting of 3 members, term 3 years, 1 member vacating his seat annually. The senior counsellor is lieutenant governor. Judiciary vested in supreme court of appeals, and superior and inferior courts, judges of the court of appeals and superior courts, elected by joint ballot of both houses, term during good behavior, or until removed by a concurrent vote of both houses, two thirds of the members present voting for removal.

Staple Productions.—From what has been stated under the section of natural features and extent, the great variety of soil and climate over Va., will at once suggest a corresponding variety in the staple productions. This is so far correct, that every vegetable from cotton to wheat, and from the fig to the apple, can be produced in abundance. The lower tide water counties, from depression of surface, and from proximity to large masses of water, enjoy comparatively a tropical temperature. This high temperature abates, rising towards and on the central table lands. On the latter the grasses, including bread grains, flourish. Falling from the mountain vallies to that of Ohio river, the temperature again rises, but I have already shown, that on the two extremes of Va., though on similar latitudes, a greater degree of cold prevails on the western side. Of minerals, the state produces limestone, gypsum, iron ore, and muriate of soda, or common salt. The limestone exists in immense masses or zones, in different parts of the state. Iron ore is also found widely disseminated. Water, holding in solution common salt, is found by digging, in the lower part of the valley of Great Kenhawa, and in lesser quantity in some other places.

Internal Improvement.—In works of internal improvement, Va. has fallen behind either Pa. or N. Y., even when relative population is made the basis of calculation. A Board of public works, consisting of 13 members, has the management of funds devoted to internal improvement, to an amount exceeding 2,000,000 of dollars. As stated in the American Almanac for 1831, this fund contributes 3-5ths of the stock, and the board meets annually on the first Monday of January. The following incorporated companies have received aid from the fund.

	Capital.
Upper Appomatox company,	$61,100
Lower Appomattox com.	40,000
Ashby's Gap com.	130,050
Fairfax com.	13,750
Lynchburg and Salem com	103,900
Leesburg com.	84,000
Little River com.	———
Manchester and Petersburg com.	———
Rappahannoc Navigation com.	50,000
Richmond Dock com.	250,000
Roanoke Navigation com.	412,000
Staunton and James river com.	50,000
Sheppardstown and Smithfield com.	46,000
Snicker's Gap com.	85,000
Swift Run com.	119,800
Tye River com.	6,000
Wellsburg and Washington com.	16,650

The most important chain of internal improvement yet attempted in Va., is that undertaken by the James River navigation company, on the line of James and Kenhawa rivers, and the intervening space. On this route has been expended 1,274,583 dollars; of which were laid out on lower James river canal, $638,883 86; mountain section of the canal, $365,207 02; Kenhawa river, $87,389 81; on turnpike roads and bridges, from Covington to the Kenhawa, $171,982 49.

By a recent act of the legislature, extending the Kenhawa road, loans were authorized for $50,000. Balancing the interest on the sums borrowed and expended, it appears that the disbursements exceed the receipts by $37,727 26; and that the general income of the fund for internal improvement, is taxed with the deficiency. The whole capital invested by Va., exclusive of that belonging to the Manchester & Petersburg turnpike com., and Little River Turnpike com., amounts to $3,263,811. In aid of improvements in Va., the Dismal Swamp canal company received from the United States $200,000; and the Roanoke navigation company received from N. C. $50,000.

Education.—In order of date, the venerable college of William and Mary was founded at Williamsburg, 1693, and next to Harvard, is the most ancient literary institution in the United States; Hampden Sidney college, in Prince Edward co., 1774; Washington, at Lexington, Rockbridge co., 1812; and the University of Virginia, at Charlotteville, Albemarle co., 1819. By the statutes of Va., all property arising from escheats, confiscations, lands forfeited for non-payment of taxes, and sums refunded by the national government for services rendered by Va. in the war of 1812, revert to the literary fund. This fund was created in 1809, and possesses an available capital exceeding *one million, two hundred and thirty thousand dollars.* Of the interest on this fund, the University of Va. receives an annual appropriation of $15,000. To the education of the poor of each county, an appropriation of $45,000 is annually applied, and divided amongst the counties in a ratio of white population, under the management of commissioners appointed by the court of each county.

VISALIA, p-v. Campbell co. Ky., by p-r. 97 ms. northward Frankfort.

VOLNEY, p-t. Oswego co. N. Y., 159 ms. N. N. W. Albany, 15 S. E. Oswego, 50 W. Rome; has Oneida and Oswego rivers S., which separate it from Onondaga co., and Oswego co. W.; has a surface nearly level; good soil. Scotts, Catfish and Black creeks, as well as the streams aforesaid, afford mill sites. The falls of the Oswego, particularly, offer abundant water power. At the p-v. of Oswego Falls are several mills and factories, as well as a quarry of freestone. Pop. 1830, 3,629.

VOLUNTOWN, p-t. Windham co. Conn., 54 ms. S. S. E. Hartford, has Rhode Island on the E. line; about 4 ms. by 9; has some pine plains, but is generally hilly, with a light and poor soil. Paucamack pond, partly in R. I., gives rise to Pochaug river, which crosses this town, generally with a slow current, yet affording some mill sites, and empties into Quinebaug r. First settled 1696. It has its name from having been granted to volunteers in the Narraganset war. Pop. 1830, 1,304.

VULCAN, p-v. Randolph co. Il., by p-r. 81 ms. S. W. Vandalia.

W.

WABASH, r. of the U. S. in O., Ind. and Il., and the great northwestern constituent of the O. r. Beside many minor streams, the Wabash is composed of three main branches, Little Wabash on the S. W., Wabash proper in the centre, and White r. on the eastern side of the valley. Little Wabash rises in Shelby co. Il., interlocking sources with those of Kaskaskia r., and flowing thence in a S. S. E. direction over Fayette, Clay, Wayne, and White cos. Il., falls into the main channel of the Wabash, between White and Gallatin cos., about 10 ms. direct course above the influx of Wabash into Ohio r. The entire comparative course of Little Wabash is about 110 ms., mean breadth of its valley 25, and area 2,750 sq. ms.; between lat. 37° 50′, and 40° 30′. Entire valley in the state of Illinois. Embarras r. is another branch of Wabash, the whole valley of which lies in the state of Il. The Embarras has its extreme source in Vermillion co. Il., interlocking sources with those of Vermillion, Kaskaskia, and Little Wabash. Flowing nearly parallel to the latter, over Edgar, Clark, and Crawford cos., falls into the Wabash 10 ms. below Vincennes, after a comparative course of about 100 ms. Mean width of the valley 20 ms., and area 2,000 sq. ms. The valley of the Embarras occupies the space between the higher sources of Kaskaskia, and the main Wabash. In the distance of 100 ms., air measure, from the influx of Embarras to that of Vermillion, the Wabash does not receive a stream from the right, or from the state of Il., above the size of a large cr. Vermillion r. rises in the state of Il., to the nrthrds. of Vermillion co., interlocking sources with the Embarras and Kaskaskia, and with those of the Sangamon and Pickmink branches of Il. r. Flowing by comparative courses 60 ms. to the sthestrd., it falls into the main channel of the Wabash, after having traversed Vermillion co. of Il., and Vermillion co. of Ind. Tippecanoe, as laid down by Tanner, is the extreme northern source of Wabash, rising at lat. 41° 30′, and long. 9° W. W. C., interlocking sources with those of Kankakee branch of Il. r., and with the Elkhart, or southern branch of the St. Joseph's r., of lake Michigan. Flowing by comparative courses 70 ms., first to the S. W., and thence curving S., it traverses Carroll co., and falls into the Wabash in the nrthrn. margin of La Fayette co. The whole left inclined plane of the Wabash valley, is in length about 330 ms., the breadth in no place extends to 60, and is about a mean of 35 ms. from the main channel; area 11,550 sq. ms. Wabash proper rises on the great plateau, or table land between the Ohio r., and lakes Erie and Michigan, and within 5 ms. of the junction of St. Joseph's and St. Mary's rs. The country from which the Wabash rises, is amongst the most remarkable on the earth. The two main constituents of the Maumee, the St. Joseph rising in Michigan, and St. Mary's rising in the state of Ohio, flow each for a comparative distance of 70 ms., in complete accordance with the confluents of Wabash; but uniting at Fort Wayne, Allen co. Ind., the united waters, ,in place of continuing what would be apparently the natural course, down the channel of the Wabash, turn in a directly opposite direction, and form the Maumee, which flowing nrthestrdly. 110 ms., is lost in the sthwstrn. bay of lake Erie. The central plain is indeed so nearly a dead level, as to admit but little current in the streams. That which is laid down by Tanner as the main source of Wabash, rises in Mercer and Darke cos. O., flowing thence N. W. by W., enters Ind., and after a comparative course of 60 ms., receives Little r., from the central table land in Allen co. It is the sources of the latter stream, which so nearly approach the junction of St. Mary's and St. Joseph's rs., and it is along its channel, that a canal has been proposed to unite the Wabash and Maumee rs. The Wabash, already a navigable r., at the influx of Little r., inflects to a course of a little s. of w. 50 ms., receiving the Salamanic and Mississinewa from the S. E., and Eel r. from the N. W. Below the influx of Eel r., the main channel inflects to s. w. 70 ms., receiving in that distance, 30 ms. below the mouth of Eel r., Tippecanoe, as already noticed. At the lower end of the last mentioned course, according to Tanner's U. S., the channel of Wabash is only about 10 ms. from the estrn. boundary of Il., but deflecting to a course of a very little w. of s., continues 60 ms. entirely in Ind., to a point between Vigo co. of the latter, and Clarke co. of Il. From hence the main channel continues a general comparative course 120 ms., forming a boundary between the two states, and receiving the Little Wabash from the N. W., and the White r. from the nrthestrd. White r. is the most considerable branch of Wabash, draining the large space between the main stream above their confluence, and that part of Ohio r. between the mouths of Miami and Wabash. The valley of White r., comprising an area of 11,000 sq. ms., is drained by innumerable smaller streams, which first unite in two branches, which again by their union form White r. White r. proper, or the nrthrn. branch, has its extreme source in Randolph co. Ind., but almost on the wstrn. border of Darke co. O. Flowing thence wstrd. by comparative courses 70 ms., over Randolph, Delaware, and Madison, into Hamilton co., inflecting to s. w., and traversing Hamilton, Marion, Morgan, Owen, and Greene cos., and thence separating Daviess from Knox co., receives the East Fork, after an entire comparative course of upwards of 200 ms. The East Fork, though not having an equal length of course, drains, however, very little, if any, less surface than the main branch. The former rises in Henry and Hancock cos., and flowing by a general sthwstrn. course, drains the cos. of Henry, Hancock, Rush, Shelby, Decatur, Bartholomew, Jennings, Scott, Jackson, Monroe, Lawrence, Martin, Orange, and part of Jefferson, Dubois, and Daviess. It may be noticed as a remarkable peculiarity of the valley of White r., that the extreme sthestrn. source in Jefferson co., rises within less than 1 m. of the bank of O. r., and flows directly from that great stream, into which the waters, thus singularly turned by the features of the country, are poured, 160 ms. air measure, lower down both rs. A ridge of hills extends entirely over Ind., from the mouth of Great Miami, to that of Wabash, across the cos. of Dearborn, Switzerland, Jefferson, Scott, Washington, Orange, Crawford, Dubois, Spencer, Warrick, Gibson, Vanderburg, and Posey. From this ridge creeks are discharged on each side, into the Ohio and White rs. respectively, the two streams flowing very nearly parallel in a direction s. w. by w. The entire valley of Wabash approaches the form of an ellipsis the longer axis 300 ms. from the extreme sthwstrn. sources of Little Wabash, to the nrthrn. fountains of Eel river. The greatest breadth 200 ms., from the sources of Graham's Fork near Madison in Jefferson co. Ind., to the nrthwstrn. fountains of Vermillion r., in the state of Il. The whole area of the valley about 40,000 sq. ms., exceeding by a small fraction, the one fifth part of the superficies of the whole Ohio valley. In fixing the relative extent of the confluents of Ohio, Wabash is the third in length of course, and second in regard to area drained; being in the former case exceeded by Ten. and Cumberland, but in the latter by Ten. only. As a navigable channel, Wabash is a very important stream. It is but slightly impeded by falls and rapids, and its course seems to be almost artificially drawn to form a part of the line of commercial connexion between the Miss. r. and lake Erie, by the most direct route. As an agricultural section, it may be doubted whether any other of equal continuous extent on earth exceeds the Wabash valley. The surface is in part hilly, in no part mountainous, nor in any part, to a considerable extent, a dead level. The northern extreme approaches, and mingles with the prairie physical section of N. America, but the prairies of Ind. are of moderate extent, when compared with those more wstrd., even those of the contiguous state, Il. If peopled only equal to some of the eastern cos. of the U. S., of far inferior soil, and without any town of note, the valley of Wabash would sustain a population of 4,000,000. Geographically, this fine portion of the U. S. extends from lat. 37° 47′ to 41° 30′, and in long. from 7° 35′ to 11° 55′ w. The difference of level between the arable extremes, is not far above or below 1,000 feet, or an equivalent to about 2½ degrees of lat., or adding the result of the difference of height to that of the lat., the real difference of temperature will be about 6 degrees of Fahrenheit.

Wabash, co. of Ind., bounded by Carroll n. w., Miami nation n. e., Hamilton e., Hendricks s., Montgomery s. w., and Tippecanoe w. Length 38 ms., mean width 25, and area 950 sq. ms. Extending in lat. from 39° 57′ to 40° 28′ n., and in long. from 9° to 9° 40′ w. W. C. The eastern border of this co. is on the table land between the main or western branch of White r. and the Wabash, but slopes wstrd., and is drained by creeks flowing in that direction, towards the latter r. On the p-o. list of 1831, the chief town is called Elk Heart Plain; on Tanner's U. S. is a village named Thorntown, 40 ms. n. w. Indianopolis. Pop. uncertain.

Wabash, one of the southeastern cos. of Il., bounded w. by Edwards, n. by Lawrence, n. e. by Wabash r, separating it from Knox co. in Ind., and by the Wabash r. s. e. separating it from Gibson co. of Ind. Extending in lat. from 38° 17′ to 38° 36′, and in long. from 10° 44′ to 11° 04′ w. W. C. Length 24 ms., mean width 12, and area 288 sq. ms. This co. lies along the Wabash, opposite the mouth of White r., general slope to the sthrd. Mount Carmel, the co. st., is situated on the Wabash, directly opposite the mouth of White r., 30 ms. below Vincennes in Ind., and by p-r. 109 s. e. Vandalia. Lat. 38° 28′, long. 10° 48′ w. W. C. Pop. 1830, 2,710.

Wachovia, name formerly given to a tract of country in N. C., now included in Surry and Stokes cos. It was purchased in 1751, by the Moravians, settled by that society, and named from an estate of Count Zinzendorf in Austria. In 1755, by an act of the assembly of N. C., it was named Dobb's parish. The names are now obsolete, and the tract only known from the villages of Salem, Bethabara, &c.

Wachusett, mtn. Princeton, Worcester co. Mass., more than 2,000 feet higher than the ocean, affords a fine and extensive view, and is a favorite resort, not being difficult of ascent.

Waddington, p-v. Madrid, St. Lawrence co. N. Y., 222 ms. n. w. Albany, 18 from Ogdensburgh, on the shore of the St. Lawrence, opposite Ogden's isl., is situated just above the Long Falls in that stream. A dam extends to the isl., which supplies several mills with water.

Waddle's Ferry, and p-o. Moore co. N. C., by p-r. 75 ms. sthwstrd. Raleigh.

Wadesborough, p-v. and st. jus., Anson co. N. C., near the centre of the co., by p-r. 134 ms. s. w. by w. Raleigh, and by the common road 80 ms. w. Fayetteville, n. lat. 35° 03′, long. 3° 12′ w. W. C.

Wadesboro', p-v. and st. jus., Calloway co. Ky., situated on Clark's r., about 120 ms. n. w. by w. Nashville in Ten., and by p-r. 262 ms. s. w. by w. Frankfort, about 35 ms. s. e. the junction of Ten. and Ohio rs. Lat. 36° 43′, long. 11° 28′ w. W. C.

Wadsworth, p-v. Medina co. O., by p-r. 108 ms. n. e. Columbus.

Wait's r., Vt., rises in several heads, affords many good mill seats, and enters the Connecticut in Bradford.

Waitsfield, p-t. Washington co. Vt., 11 ms. s. w. Montpelier, 30 s. e. Burlington, first settled 1789, has generally an excellent soil, yielding a variety of crops, particularly grass. Mad r. pursues a serpentine course through the s. part, between extensive and fertile meadows. Iron ore and clay are found in this t. Pop. 1830, 957.

Wake, co. N. C., bounded n. by Granville co., n. e. by Franklin, s. e. by Johnson, s. by Cumberland, s. w. and w. by Chatham, and n. w. by Orange. Length 38 ms., mean width 30, and area 1,140 sq. ms. Extending in lat. from 35° 30′ to 36° 07′, and in long. from 1° 14′ to 2° 02′ w. W. C. A small angle along the wstrn. part, is drained sthrdly. into cape Fear r., but the body of the co. is contained in the valley of the Neuse, with a declivity to the s. e. The Neuse, deriving its sources from Person, Granville, and Orange cos., is formed into a r. on the nrthwstrn. angle of Wake, and winding thence sthestrdly., crosses the co. into Johnson. Chief t. Raleigh, the capital of the state. Pop. 1820, 20,102; 1830, 20,398.

Wakefield, p-t. Strafford co. N. H., 50 ms. from Concord, 30 from Dover, and 100 from Boston, lies w. Maine, and contains several ponds. Lovewell's pond s., is famous for a bloody engagement which took place on its banks in 1724, between an expedition from Mass. of nearly 100 men, under the command of capt. Lovewell, and the Pickwaket Indians, whose residence was at a short distance, in which the latter suffered greatly, and the former were almost entirely cut off. This sheet of water is about 2 ms. long; Wakefield pond about 1 m.; East pond is the source of Piscataqua r. The soil of this t. is generally good, but most favorable to grass. There are a few mills and factories. Pop. 1830, 1,470.

Wake Forest, p-v. Wake co. N. C., 14 ms. from Raleigh.

Walden, p-t. Caledonia co. Vt., 22 ms. n. e. Montpelier. First settled, 1789, lies between the head waters of Onion and Lamoille rs. The surface is agreeably varied n. and good soil, other parts are little cultivated. Cole's pond is n. e. and Liffords's s. e. Pop. 1830, 827.

Walden, p-v. Orange co. N. Y. 90 ms. s. by w. Albany, 11 ms. w. Newburgh, is a flourishing manufacturing village, commenced in 1823. It stands on the Wallkill, where the stream makes a descent of 32 feet. The Franklin company have here the largest manufactory of flannel in New York, consuming 65,000 or 70,000 lbs. of wool in a year, and producing about 240,000 yards of flannel, white and colored. Capital, $100,000. The Wallkill cotton company, consume about 120,000 lbs., and make 360,000 yards of sheeting. The Orange company make 30,000 yards of low priced broadcloth. There are also 1 flour and 1 saw mill. There is a wire bridge of 150 feet across the creek, just below the fall. The village contains an Episcopal church, a library, and schools. The surrounding scenery is varied and picturesque. Pop. 1830, about 800.

WALDO, co. Me., bounded by Somerset and Penobscot cos. N., Hancock E., separated from it by Penobscot river and bay, Lincoln co. S. and S. W., and Kennebec co. W., contains 26 towns, and enjoys great commercial advantages. Belfast, is the chief town. It has a number of ponds, one considerable island in the Penobscot, and a number of small streams, some falling into that river, and others into Sebasticook river. It has been, recently formed. Population 1830, 29,788.

WALDO, p-t. Waldo co. Me., 44 ms. E. Augusta, adjoins Belfast on the N. W. side, and is crossed by a small stream flowing into Belfast bay. It is of small size. Pop. 1830, 534.

WALDOBOROUGH, p-t. and port of entry, Lincoln co. Me., 37 ms. S. E. Augusta, 22 N. E. Wiscasset, 180 N. E. Boston, at the head of Muscongus bay, is crossed N. and S. by Muscongus river, has a considerable amount of shipping, and enjoys an active coasting trade. Pop. 1830, 3,113.

WALES, p-t. Lincoln, co. Me., 20 ms. S. W. Augusta, 26 N. W. Wiscasset, has Kennebec co. N. and W., and part of a large pond S. W., which discharges S. through an outlet extending to the bend of Penobscot river. Pop. 1830, 612.

WALES, p-t. Erie co. N. Y., 268 ms. W. Albany, 22 E. S. E. Buffalo, has Genesee co. E.; it contains a part of the Seneca reservation; has a gently varied surface, and is crossed N. E. by Buffalo creek. Pop. 1830, 1,470.

WALKER, p-o. Nittany valley, Centre co. Pa., 5 ms. N. E. Bellefonte, and by p-r. 93 ms. N. W. Harrisburg, and 200 N. W. C.

WALKER, co. Ala., bounded by Blount E., Jefferson S. E., Lafayette S. W., Marion W., Franklin N. W., and Lawrence N. Length from S. to N. 50 ms., mean width 30, and area 1,500 square ms. Extending in lat. from 33° 35′ to 34° 17′, and in long. from 9° 56′ to 10° 43′ W. W. C. Declivity southeastward and drained by the numerous western branches of Mulberry r. The latter stream, formed in this co. by two main branches, Sipsey and Blackwater, flows a little W. of S. along the southeastern border, receiving the drain of the western part of Mulberrry river valley. Chief town, Walker C. H.

WALKER, court house, p-o. and st. jus. Walker co. Ala., by p-r. 47 ms. northward Tuscaloosa.

WALKER'S, p-o. Colleton district, S. C., by p-r. 92 ms. a little E. of S. Columbia, and 4 miles from Walterboro', the st. jus. for the district.

WALKER CHURCH, and p-o. southern part of Prince Edward co. Va., by p-r. 88 ms. S. W by W. Richmond.

WALKERSVILLE, p-v. Frederick co. Md., by p-r. 49 ms. N. N. W. W. C.

WALKERSVILLE, p-o. Mecklenburg co. N. C., by p-r. 109 ms. S. W. by W. Raleigh.

WALKERTON, p-v. on the left bank of Mattapony river, King and Queen co. Va., 19 ms. below and S. E. Dunkirk, the co. seat, and by p-r. 30 ms. N. E. Richmond, 123 S. W. C.

WALLINGFORD, p-t. Rutland co. Vt., 42 ms. N. E. Bennington, 10 S. Rutland. First settled 1773, from Conn., is crossed from S. to N. by Otter creek and has several small streams, and many mill sites. Lake Hiram or Spectacle pond, 350 acres, is on a mountain in the S. E., and there are 2 others of 50 and 100 acres. The Green mountains are E., and the highest summit is called the White Rocks. There is a limestone range W. with marble quarries, and Green hill, in the centre, consists chiefly of quartz. There are several natural and perennial ice houses in caves, at the foot of the White Rocks. Near the N. line is a handsome village near Otter creek. Pop. 1830, 1,741.

WALLINGFORD, p-t. New Haven co. Conn., 13 ms. N. E. New Haven, 23 S. W. Hartford, has Middlesex co. E., 6 ms. by 7, is crossed by a branch of Quinipiack river, on which are extensive meadows, and has the main stream on the N. W. boundary. The E. part is mountainous, but the soil is generally rich, except the plain, which is about 4 ms. long, and so sandy as often to fill the air with clouds of dust in dry seasons. There are several mills and manufactories on the Quinipiack, and shad are taken in this stream. The village is a pleasant situation on a hill, and is of considerable size; contains an academy, &c. Pop. 1830, 2,418.

WALLKILL, river or creek, N. Y., rises in Sussex co. N. J., flows N. E. through Orange into Ulster co. N. Y., to Rondout creek, which it enters near Esopus. It is about 80 miles long, 65 of which it runs in N. Y. It affords very good mill seats.

WALLKILL, p-t. Orange co. N. Y., 101 ms. S. Albany, 20 W. Newburgh, 6 N. Goshen, is of a triangular form, with Sullivan co. W., and is crossed in the E. part by the Wallkill, whose streams supply mill seats. The soil is various, and favorable to agriculture. It contains 2 woollen factories, and several villages; Scotchtown, Middletown, Mount Hope, and Mechanictown. Pop. 1830, 4,056.

WALLSVILLE, p-o. in the northern part of Luzerne co. Pa., by p-r. 114 ms. N. E. Harrisburg.

WALNUT, large creek of Ohio, rises in and drains the northern part of Fairfield co. from which, flowing westward, traverses the S. E. angle of Franklin, and inflecting to S. E. falls into Sciota in Pickaway co.

WALNUT, p-v. Fairfield co. Ohio, by p-r. 38 ms. S. E. by E. Columbus

WALNUT BRANCH, and p-o. Fauquier co. Va., by p-r. 55 ms. a little S. of W. W. C.

WALNUT FARM, and p-o. also st. jus. Saline co. Mo., by p-r. 85 ms. W. Jefferson.

WALNUT FLAT, and p-v. Lincoln co. Ky., by p-r. 5 ms. southward Stanford, the county st., and 56 a little E. of S. Frankfort.

WALNUT GROVE, p-o. in the westward part of Kenhawa co. Va., 23 ms. westrd. Charleston, the co. st., and by p-r. 379 ms. S. W. by W. W. C.

WALNUT GROVE, p-o. in the western part of Cabarras, co. N. C., 11 ms. from Concord, the co. st., and by p-r. 152 ms. a little S. of W. Raleigh.

WALNUT GROVE, p-o. in the southern part of Spartanburg district, S. C., by p-r. 92 ms. N. W. Columbia.

WALNUT GROVE, village, in the N. E. angle of Mercer co. Ky., 20 ms. S. Frankfort.

WALNUT HILL, and p-o. Marion co. Il., by p-r. 55 ms. S. S. E. Vandalia.

WALOOMSCOIC, or Waloomsac, river Vt., a branch of Hoosic r., formed in Bennington by several branches, was rendered famous in the revolution, by the victory gained on its banks, by the militia of Vermont and Mass., in a battle with the Hessian troops, sent by Gen. Burgoyne to seize the public stores at Bennington, 1777.

WALPACK, p-t. Sussex co. N. J., 82 ms. N. Trenton, 50 N. W. New Brunswick, is of a long and narrow shape, with Delaware river on the N. W. boundary, the Blue mountains on the S. E., and Flatkill running through the town, between them. Pop. 1830, 660.

WALPOLE, p-t. Cheshire co. N. H., 60 ms. W. by S. Concord, 48 S. by W. Dartmouth college, 90 ms. N. W. Boston, with Connecticut r. and Vt. on the W. line; it is much varied by hills and vales, with a good soil, especially on the meadows, and devoted to agriculture. Cold r. in the N. part flows into Conn. r. Fall mtn. belonging to the Mt. Toby range, is here 7 or 800 feet above the r. Near its foot is the village, on land sufficiently elevated to command an extensive view upon the neighboring country, especially the meadows of Westminster, opposite. There are 2 toll bridges across the river in the town, which is a great thoroughfare for travelling N. and S., and E. and W. Bellows falls are within the limits of this town, as New Hampshire extends to the W. shore of Conn. r. The town first settled 1749 by Col. Bellows, who built a frontier fort, which was taken by Indians from St. Francis, in 1755, and retaken by him in a few hours. Pop. 1830, 1,979.

WALPOLE, p-t. Norfolk co. Mass., 18 ms. S. by W. Boston. Three branches of Neponset r. unite in this town, through which also passes the road from Boston to Providence, one of the principal thoroughfares in the state. Pop. 1830, 1,442.

WALTERBORO', p-v. and st. jus. Colleton district, S. C., 47 ms. W. Charleston, and by p-r. 93 ms. a little E. of S. Columbia. Lat. 32° 53′, long. 3° 43′ W. W. C.

WALTHAM, town, Addison co. Vt., 24 ms. S. Burlington, 9 N. W. Middlebury, 3 ms. square. It was first settled just before the revolution, abandoned, and settled again about the close of the war. It has Otter creek on the W. line, and Buck mountain in the centre, which commands a fine view. The soil is generally good, with excellent meadows on the creek. Pop. 1830, 330.

WALTHAM, p-t. Middlesex co. Mass., 11 ms. W. N. W. Boston, is bounded S. by Charles r. The Plain, 1 mile by 2 1-2 in the S. E. has a good well cultivated soil, and is thickly peopled. The Waltham factories are on the S. part of the Plain, and form one of the principal manufacturing villages in the United States. The surface in the W. part is hilly, and most of the soil in the town, poor. Prospect Hill is 470 feet in height, and commands a view of Boston. Pop. 1830, 1,857.

WALTON, p-t. Delaware co. N. Y., 85 ms. S. W. Albany, is hilly or mountainous, with rich vallies on the streams, and good grazing on the uplands. It is crossed by the W. branch of Delaware r. or the Cooquago, which affords a channel for transporting great quantities of lumber. Pop. 1830, 1,663.

WALTON, co. Geo., bounded by Morgan S. E., Newton S. W., Gwinnett N. W., and Appalachee branch of Oconee, separating it from Jackson N. E., and Clark E. Length from S. E. to N. W. 20 ms., mean width 16, and area 320 square ms. Central lat. 33° 50′, and long. W. C. 6° 50′ W. From it issue some of the higher branches of both Oconee and Ocmulgee rivers. General slope southward. Chief town, Monroe. Pop. 1830, 10,929.

WALTON, co. Florida, bounded by Choctawhatchee bay, or Gulf of Mexico S., Escambia co. Florida W., Covinton and Dale cos. Ala. N., and Choctawhatchee r. separating it from Jackson E. Length along the Ala. line 52 ms., mean width 30, and area 1,560 sq. ms. Extending in lat. from 30° 22′ to 31°. Long. from 9° to 9° 52′ W. W. C. The general declivity is to the southward, and drained by Yellow Water and Choctawhatchee rivers. Chief town, Allaqua. Pop. 1830, uncertain. "The Yellow Water settlement is in the N. W. part of the co. on the banks of the river of that name. Here is a small body of excellent land, very well improved for a new country. Cotton and corn are their principal crops, the pine lands for 6 ms. from the river, produce equally well with the river bottoms. 12 ms. S. there is another settlement, commencing on Shoal r. There they have a similar tract of land, founded on the same kind of soap stone as on the Allaqua. (*See Allaqua river.*) Nearly one third of Walton co. is good tillable upland; the rest is pine barren."—*Williamson's Florida.*

WALTONHAM, p-o. St. Louis co. Mo.

WANBORO', p-v. Edwards co. Il., by p-r. 91 ms. S. E. by E. Vandalia.

WANTAGE, p-t. Sussex co. N. J., 83 ms. N. by E. Trenton, 15 N. Newton, has N. Y. state line on the N. boundary, the Blue mtns. on the W., the upper part of Wallkill cr. E., and is traversed by Pappakating cr., one of its branches. Deckertown is a v. near its centre, on that cr. Pop. 1830, 4,034.

WANTON, p-v. Alachua co. Flor., by p-r. 212 ms. sthestrd. Tallahassee.

WAPAHKONETTA, p-v. sthrn. part Allen co. O., on au Glaize r., and signifies in the Shawnee language, "*Kingstown.*" By p-r. it stands 110 ms. N. W. Columbus, and 65 ms. above and s. fort Defiance.

WAPPINGER'S, cr. Duchess co. N. Y., has a course of 33 ms. through several rich agricultural tsps., to which it supplies mill seats, and enters the Hudson 8 ms. s. Poughkeepsie village.

WAQUOIT bay, Barnstable co. Mass., on the N. side of Cape Cod, between Marshpee and and Falmouth, is several miles long. It is connected with a large pond, W. of which are several other ponds adjacent to each other, but communicating only with the ocean.

WARD, p-t. Worcester co. Mass., 50 ms. s. W. Boston, 7 s. Worcester, has an uneven surface, rich soil, and is crossed by French r. which has a serpentine course, and here receives several branches, furnishing mill seats. Pop. 1830, 690.

WARD'S, p-o. Holmes co. O., by p-r. 95 ms. N. E. Columbus.

WARDSBOROUGH, p-t. Windham co. Vt., 20 ms. N. E. Bennington, 15 N. W. Brattleboro', settled 1780, has a range of hills on the south line, with a good soil for grass, and some mill sites on a branch of West r. Some rare minerals are found here. Pop. 1830, 1,148.

WARE, r. Mass., a branch of Chicapee r., rises in Worcester co. and meets Swift and Chicapee near the line of Hampshire and Hampden cos. It is a good mill stream, and affords water power to the manufacturing v. of Ware. At the junction of these 3 streams is a small p-v. called Three Rivers, where is a large stone manufactory.

WARE, p-t. Hampshire co. Mass., 70 ms. W. Boston, has a hilly surface, and soil of secondary quality. Swift r. forms its W. boundary. On the E. is Ware river, on which is a large manufacturing village, contains about 50 tenements, the value of which, with that of the manufacturing buildings, is about $300,000. Other capital employed in manufacturing amounts to about $140,000. About 330 persons are employed in the factories. 120,000 lbs. of wool, and 1,100 bales of cotton are annually manufactured; the woollens estimated at $120,000, and the cottons at $180,000 per ann. The machinery used in these factories is estimated to be worth about $20,000. Much attention is paid to the instruction of children in the village; there are several day schools, an infant school, and one for the older children under the care of a male teacher. The Sabbath schools are well organized; intemperate persons are not employed in the factories, and the temperance society consists of 416 members. There are 5 stores, a large and well kept hotel and stage house. Pop. 1830, 2,045.

WARE, co. of Geo., as laid down on Tanner's U. S., is bounded by Lowndes W., Appling N., Wayne N. E., Camden E. and S. E., and Hamilton co. in Flor. S. Greatest length from S. to N. 80 ms., mean breadth 43, and area 3,440 sq. ms. Extending in lat. from 30° 20′ to 31° 30′ N., and in long. from 5° 09′ to 6° 06′ W. Declivity sthestrd. The N. side is drained by the numerous sources of the Santilla river. The central and southern sections are drained by the tributaries of St. Mary's river. The latter stream has its higher sources in that tract vaguely called the Okefinoke swamp, which occupies the southern part of Ware co. The various branches oozing from this extended flat unite on, or very near, the line between Geo. and Flor., flows thence southward about 20 ms., curves rapidly E. and thence N. E. and still winding, assumes nearly a northern course of 40 ms. The point on the Florida boundary which separates Ware from Camden co. is at the head of this great bend of St. Mary's river. The surface of Ware co. is flat in the valley of St. Mary's, and level in that of St. Illa. Much of the soil is productive, but exposed to submersion in spring and early summer. Chief town, Waresboro'. Pop. 1830, 1,205.

WAREHAM, p-t. Plymouth co. Mass., 39 ms. S. E. Boston, has Plymouth and Buttermilk bays E., and Buzzard's bay S. It is crossed by two small streams from Plymouth, Agawam and Wankinquog rs., which flow into Buzzard's bay, and afford some mill sites. The soil is generally thin on sand, but is better near the streams and the coast. Pop. 1830, 1,885.

WARESBORO', p-v. and st. jus. Ware co. Geo., is situated on the left bank of Santilla river; very nearly mid-distance between Savannah and Tallahassee, and about 140 ms. from each, 75 ms. N. W. St. Mary's, and by p-r. 161 ms. S. S. E. Milledgeville. N. lat. 31° 18′ long. 5° 41′ W. W. C.

WARMINSTER, p-v. on the left bank of James r. estrn. part of Nelson co. Va., by p-r. 160 ms. S. W. W. C., and 100 W. Richmond.

WARM SPRINGS and p-o. near the French Broad river, nrthestrn. part of Buncombe co. N. C., 34 ms. N. N. W. Asheville, the co. st., and by p-r. 220 ms. W. Raleigh.

WARM SPRINGS, new co. of Ark., embracing the country round the Warm Springs on the higher part of the Washitau valley, and lies about 60 ms. a little S. of W. Little Rock. Boundaries uncertain. Chief town, Warm Springs. Pop. unknown.

WARM SPRINGS, p-v. and st. jus. Warm Springs co. Ark., 60 ms. S. W. by W. Little Rock. Lat. 34° 32′, long. 16° W. W. C. This village has risen from the celebrity of its springs, and is now a place of much resort.

WARNER, p-t. Merrimack co. N. H., 15 ms. from Concord, 4 from Hopkinton, and 72 from Boston, is supplied with mill sites by Warner r. which runs through the middle of it, and a branch of Contoocook r. The surface is broken, the soil good for grass and grain. Mink hills are W. and abound in orchards and pasturage. There are 4 ponds, one of which, Pleasant pond, has no known inlet or outlet, yet overflows in the driest seasons. First settled 1762. Pop. 1830, 2,222.

WARNERSVILLE, p-v. Hardiman co. Ten., 12 ms. nthrdly. Bolivar, the co. st.

WARREN, p-t. Lincoln co. Me., 44 ms. S. E. Augusta, 30 E. Wiscasset, has Waldo co. N. E. and is crossed by St. George r. with several ponds flowing into it. Sloops navigate the r. to this t. There is an academy in the v. Pop. 1830, 2,030.

WARREN, p-t. Grafton co. N. H., 63 ms. from Concord, and 10 from Haverhill; a corner is crossed by Baker's r. which furnishes mill seats S., where are Clement's mills. It is mountainous S. E., a great part of Carr's mtn. being on that boundary. Pop. 1830, 702.

WARREN, p-t. Washington co. Vt., 31 ms. S. E. Burlington, and 16 S. W. Montpelier, settled 1797, lies between the 2 ranges of the Green mtns., and is crossed by Mad r. Pop. 1830, 765.

WARREN, p-t. Bristol co. R. I., 11 ms. S. E. Providence, 19 from Newport, has Palmer's r. N. and W., Mass. E., is a small t. containing only about 4 sq. ms., but has a rich soil, a handsomely varied surface, and an active commerce, coasting and foreign. The v. is pleasantly situated on the S. E. side of Warren r. with a harbor for vessels of 300 tons; contains a bank, academy, insurance office, &c. Pop. 1830, 1,800.

WARREN, p-t. Litchfield co. Conn., 38 ms. W. Hartford, and 45 from N. Haven, is mountainous and hilly, with granite rocks, and quarries of micaceous schistus. It is crossed in the E. by Shepaug r., a branch of Housatonic, and has other small streams. Pop. 1830, 986.

WARREN, co. N. Y., bounded by Essex co. N., Washington co. E, Saratoga co. S., and Hamilton co. W., occupies elevated land, near the head springs of Hudson r., and has a surface generally very irregular, and much covered with forests. It contains about half of lake George, and part of Scaroon lake. Several other lakes and ponds are wholly within this co. On the shores of lake George the eminences are from 500 to 1,200 ft. high; and the co. crossed N. and S. by the Kayderosseras range. The co. contains 9 towns. Pop. 1820, 9,453; 1830, 11,796.

WARREN, p-t. Herkimer co. N. Y., 68 ms. W. Albany, 15 N. Otsego, and 10 S. Herkimer, has Otsego co. S., and is an elevated tract of ground, at the head of the lakes which form the sources of Susquehannah. The surface is agreeably varied, with fertile vallies, small cedar swamps, and lime rocks. Pop. 1830, 2,084.

WARREN, co. N. J., bounded by Sussex co. N. E., Morris co. E., Hunterdon co. S. E., and Delaware r. W. which separates it from Pa. The Blue mountains rise N. W. near the Delaware, and parallel to its course; while several ridges, following the same direction, cross the S. E. part from Pa. to Sussex co. Musconetcong creek forms the whole W. and S. W. boundary, just W. of which is the line of Morris canal, which enters the valley of Pohatcong creek, and pursues it to the Delaware, at Philipsburgh. There are 7 towns. Pop. 1830, 18,627.

WARREN, p-t. Somerset co. N. J., 41 miles from Trenton; has Dead or Passaic river N. separating it from Morris co., &c., Green brook S. on the line of Middlesex county, and Essex co. E. It is rendered hilly by Rocky Hill ridge; has copper mines, and the village of Bound Brook S. Pop. 1830, 1,561.

WARREN, co. Pa., bounded E. by MacKean, S. E. by Jefferson, S. by Venango, W. by Crawford, N. W. by Erie, N. by Chatauque co. N. Y., and N. E. by Cattaraugus co. N. Y. It lies in form of a parallelogram, 32 ms. from E. to W., and 28 from S. to N.; area 896 sq. ms. Extending in lat. 41° 37′ to 42°, and in long. from 2° 03′ to 2° 43′ W. W. C. Declivity to the sthwrd. Alleghany r. forms for a few miles the northeastern boundary, and thence entering, winds over this county into Venango, in a southwestern direction, receiving at the borough of Warren, the Conewango, a considerable tributary from the northwards, and at the Great Bend 5 miles below Warren, Brokenstraw, a large cr., comes in from the westward. Surface broken, and though yet thinly populated, much of its soil is excellent. Chief town, Warren. Population 1820, 1,976, 1830, 4,766.

WARREN, p-v. and st. jus. Warren co. Pa., is situated on the right bank of Alleghany r., on the point below the mouth of Conewango creek, about 120 ms. N. N. E. Pittsburg, and by p-r. 240 ms. northwestward Harrisburg, and 213 N. W. W. C. Lat. 41° 50′, long. 2° 17′ W.

WARREN, p-v. on the left bank of James r., at the mouth of Battinger's cr., and in the southern angle of Albemarle co. Va., 25 ms. S. S. W. Charlotteville, and 89 N. W. by W. Richmond.

WARREN, co. N. C., bounded N. E. by Roanoke r. separating it from Northampton, E. by Halifax, S. and S. W. by Franklin, W. by Granville, and N. by Mecklenburg co. Virginia. Length 23 ms., mean width 17, and area 391 sq. ms. Extending in lat. from 36° 07′ to 36° 32′, and in long. from 0° 56′ to 1° 21′ W. W. C. This co. is a table land. A little more than one third declines to the nrthrd., and is drained into the Roanoke; the sthrn. slope of the valley of which, is here only about 8 ms. wide. Beyond this narrow inclined plane rise the extreme sources of Fishing cr., and a branch of Tar r. draining the central and southern part of Warren, flowing southeastward, and uniting between Halifax and Nash cos. The soil is generally good. Chief t. Warrenton. Pop. 1820, 11,158, 1830, 11,877.

WARREN, co. of Geo., bounded N. E. and E. by Columbia, S. E. & S. by Jefferson, by Great Ogechee r. separating it from Washington S. W., Hancock W., and by Wilkes N. Extending in lat. from 33° 07′ to 33° 34′, and in long. from 5° 26′ to 5° 52′ W. W. C. Length 28 ms., mean width 20, and area 560 sq. miles.

Declivity of the southern and western parts to the sthestrd., and drained by Great Ogechee and Brier cr.; the northern part slopes toward the N. E., and is drained by some branches of Little r. into Savannah r. Chief t. Warrenton. Pop. 1820, 10,630, 1830, 10,-946.

WARREN, co. Miss., bounded N. by Washington, N. E. by Yazoo co., E. by Big Black river separating it from Hinds, S. E. and S. by Big Black river separating it from Claiborne, and W. by the Miss. river separating it from Concordia parish in La. Length 40 miles, mean width 15, and area 600 sq. ms. Extending in lat. from 32° 03′ to 32° 35′, and in long. from 13° 42′ to 14° 13′ W. W. C. The Miss. river bounding this co. on the westward, receives the Yazoo and Big Black rivers from the N. E.; the general declivity is therefore to the S. W. The eastern part is broken into hills, which in one or two places reach the Miss., forming clay bluffs. Along that great river, however, the bottoms are liable to submersion; but over the whole co. where the soil is sufficiently elevated to admit cultivation, it is highly fertile. Principal staple, cotton. Chief towns, Vicksburg and Warrenton. Pop. 1820, 2,693, 1830, 7,861.

WARREN, co. Ten., bounded by Franklin S., Bedford S. W., Rutherford W., Wilson N. W., Smith N., White N. E. and E., and Cumberland mountain separating it from Bledsoe S. E. Greatest length from the southern to the northern angle 48 ms., mean breadth 20, and area 960 sq. ms. Extending in lat. from 35° 28′ to 36° 06′, and in long. from 8° 19′ to 9° 04′ W. W. C. Declivity N. E. and commensurate with the western and larger section of the valley of Caney Fork river. Chief town, McMinnville. Pop. 1820, 10,348, 1830, 15,-210.

WARREN, co. Ky., bounded by Edmondson N., Barren E., Allen S. E., Simpson S., Logan W., and Butler N. W. Length from E. to W. 36 ms., mean width 17, and area 612 sq. ms. Extending in lat. from 36° 50′ to 37° 11′, and in long. from 9° 02′ to 9° 38′ W. W. C. Declivity N. N. W., and traversed in that direction by Big Barren river, which unites with Green river at the extreme N. W. angle of the county. Chief t. Bowling Green. Pop. 1820, 11,776, 1830, 10,949.

WARREN, co. Ohio, bounded S. by Clermont, S. W. by Hamilton, Butler W., Montgomery N. W., Greene N. E., and Clinton E. Length 24 ms. mean width 20, and area 480 sq. ms. Extending in lat. from 39° 14′ to 39° 37′, and in long. from 6° 55′ to 7° 22′ W. W. C. The Miami river and canal cross the N. W. angle of this co.; whilst the central parts are traversed by Little Miami. The course of both rs., and the slope of the co. to the S. E. Surface rolling and soil excellent. Besides at Lebanon, the co. seat, there were in 1830, post offices at Deerfieldville, Edwardsville, Franklin, Hopkinsville, Kirkwood, Red Lion, Ridgeville, Rochester, Springboro', Twenty Mile Stand, and Waynesville. Lebanon p-v. and st. jus. for this co., is situated near the centre of the co. 31 ms. N. E. Cincinnati, and by p-r. 83 ms. S. W. by W. Columbus. Lat. 39° 25′, long. 7° 12′ W. W. C. Pop. 1830, 21,-468.

WARREN, p-v. and st. jus. Trumbull co. O., situated on the Mahoning branch of Big Beaver river, 70 miles north west Pittsburg, 70 miles north Steubenville, and by post road 157 ms. N. E. by E. Columbus, and 297 N. W. by W. W. C. It is a thriving village, with a population of about 500. Lat. 41° 17′, long. 3° 50′ W. W. C.

WARREN, co. Indiana, bounded by the Indian country N. W., Tippecanoe co. N. E., Wabash river separating it from Fountain S. E., Vermillion co. S. W., and Vermillion co. of Il. W. Greatest length as laid-down by Tanner, 26 ms., mean breadth 18, and area 468 sq. ms. Extending in lat. from 40° 10′ to 40° 30′, and in long. from 10° 06′ to 10° 40′ W. W. C. The slope of this co. is to the S. E. toward the Wabash. Williamsport, the co. seat, lies about 80 ms. N. W. by W. Columbus. Pop. 1830, 2,861.

WARREN, co. Il., bounded by Mercer N., Knox E., Fulton S. E., Macdonough S., Hancock S. W., and Miss. r. W. This county was formed out of a part of the Bounty Lands between the Il. and Miss. rivers, and similar to the adjacent cos., is laid out agreeably to the cardinal points. Breadth from S. to N. 32 ms., mean breadth from E. to W. 30, and area 960 sq. ms. Extending in lat. from 40° 37′ to 41° 04′, and in long. from 13° 26′ to 14° 06′ W. This county comprises a part of the table land between the vallies of Il. and Miss. rs. From the southeastern angle issues Swan creek, branch of Spoon river, a confluent of Il. river. The central, and much the larger part of the surface, is drained by the confluents of Henderson's river, and other streams flowing westward into the Miss. By a note inserted after the name of Warren, Il., in the post office list, it contained no office on Oct. 1st, 1830. The centre of this new county is about 160 ms. northwestward Vandalia. Pop. 1830, 308.

WARREN, C. H., p-v. and st. jus. Warren co. Il., about 160 miles N. W. Vandalia.

WARREN FERRY, and p-o. wstrn. part Buckingham co. Va., 10 ms. westward Buckingham C. H., and by p-r. 87 ms. wstrd. Richmond.

WARRENSBURG, town, Warren co. N. Y., 7 ms. N. W. Caldwell; is watered by the north branch of Hudson river and Scaroon creek; has some good soil, and iron ore. Pop. 1830, 1,191.

WARRENSBURG, p-v. western part Greene co. Ten., 12 ms. S. W. Greensville, the county seat, and by p-r. 256 ms. E. Nashville.

WARREN'S STORE, and p-o. northwestern part Halifax co. Va., by p-r. 115 ms. S. W. by W. Richmond.

WARREN TAVERN, and p-o. nrtheastrn. part Chester co. Pa., 20 ms. N. W. by W. Phil., and by p-r. 131 ms. N. E. W. C.

WARRENTON, p-v. and st. jus. Fauquier co. Va., by p-r. 51 ms. s. w. by w. W. C. Lat. 38° 41', long. 0° 46' w. W. C.

WARRENTON, p-v. and st. jus. Warren co. N. C., situated near the centre of the co., 57 ms. N. E. Raleigh, and about 115 ms. s. w. Richmond, Va. Lat. 36° 21', long. 1° 10' w. W. C.

WARRENTON, p-v. Abbeville district, S. C., 6 ms. from Abbeville, the co. seat, and by p-r. 106 ms. westward Columbia.

WARRENTON, p-v. and st. jus. Warren county, Geo., 50 ms. N. E. by E. Milledgeville, & 42 ms. w. Augusta. Lat. 33° 23', long. 5° 40' w. W. C.

WARRENTON, p-v. and formerly st. justice Warren co. Miss., situated on the left bank of the Miss. r., 60 miles by land above Natchez, and by p-r. 54 ms. w. Jackson. Lat. 32° 17', long. 14° 2' w. W. C.

WARRICK, co. Ind., bounded by Vanderburg w., Gibson N. w., Pike N., Dubois N. E., Spencer E. and S. E., and Ohio r. separating it from Henderson co. Ky. s. Length 25 ms., mean width 13, and area 325 sq. ms. Extending in lat. from 37° 54' to 38° 15', and in long. from 10° 04' to 10° 33' w. The northern boundary of this co. is on the table land between the vallies of O. r. and the Patoka, a branch of the Wabash; but nearly the whole of its surface slopes sthrdly. toward the former river. Surface hilly, but soil productive. Chief t. Boonsville, by p-r. 187 ms. s. s. w. Indianopolis. Pop. 1830, 2,877.

WARRIOR'S MARK, p-o. western part Huntingdon co. Pa., by p-r. 20 ms. from Huntingdon borough, and 168 ms. N. N. W. W. C.

WARSAW, p-t. Genesee co. N. Y., 20 miles N. Batavia; is supplied with mill seats by Allen's cr., on which is the v. The surface is varied, and the soil generally free from stone. Pop. 1830, 2,474.

WARSON'S, p-v. Morgan co. Il., by p-r. 123 ms. N. W. Vandalia.

WARTHEN'S STORE, and p-o. northern part Washington co. Geo., by p-r. 27 ms. E. Milledgeville.

WARWASING, p-t. Ulster co. N. Y., 25 ms. s. w. Kingston, in the s. w. corner of the co.; lies w. Shawangunk mountain, at the base of which flows Rondout creek, receiving several branches. Pop. 1830, 2,738.

WARWICK, p-t. Franklin co. Mass., 80 ms. w. by N. Boston; has the line of N. H. N. It has an uneven surface, good soil, and no large streams. Pop. 1830, 1,150.

WARWICK, p-t. and st. jus. Kent co. R. I., 10 ms. s. s. w. Providence; has Narragansett bay E.; is hilly s., and level E. The branches of Pawtucket river unite in this t., which enters the bay at Pawtucket village. Apponang harbor, 1 mile distant, is the principal one in this town, and vessels of 50 tons come up to the v. This is one of the principal manufacturing towns in the country; the fisheries are also important. Pop. 1830, 5,-529.

WARWICK, p-t. Orange co. N. Y., 116 ms. from Albany, 10 s. Goshen, 54 N. New York; has N. J. s. w., and is of triangular form. On the s. boundary are many mtns., from which several large ponds pour their waters s. into the Passaic. The Wallkill flows N. E. The soil is favorable to fruit, and generally good. The Sterling iron works, and others, manufacture a large amount of iron annually. Pop. 1830, 5,009.

WARWICK, p-v. near the southeastern angle of Cecil co. Md., 15 ms. s. Elkton, the county seat, and by p-r. 82 miles N. E. Washington City.

WARWICK, co. Va., bounded by James City co. N. W., York N. and N. E., Elizabeth City co. E., and James r. separating it from Isle of Wight co. s. Length diagonally from s. E. to N. W. 18 ms., mean width 5, and area 90 sq. ms. Extending in lat. from 37° 03' to 37° 13', and in long. from 0° 22' to 0° 38' E. W. C. It occupies a part of the narrow peninsula between James and York rivers, and slopes southward toward the former. Chief town, Warwick C. H. Pop. 1820, 1,608, 1830, 1,570.

WARWICK C. H., and p-o. Warwick county, Va., by p-r. 184 ms. a little E. of s. W. C., and 81 miles south east by east Richmond.

WASHINGTON, co. Me., bounded by Lower Canada N., New Brunswick E., the Atlantic ocean s., and Hancock and Penobscot counties w. It presents a most singular figure on the map, as drawn according to the E. boundary as claimed by the United States, bearing a resemblance to a rudely hewn gun stock. It is crossed in its upper part, by the Saint John's and Aroostic rivers; has the St. Croix on the E. boundary, with the lakes and bays connected with it; and has several streams running into bays which make up from the ocean. There are 32 named townships, many others numbered, besides plantations, islands, &c. Pop. 1830, 21,294.

WASHINGTON, p-t. Lincoln co. Me., 35 ms. from Augusta; has Waldo co. E. Pop. 1830, 1,135.

WASHINGTON, p-t. Sullivan co. N. H., 35 ms. from Concord, 22 from Keene, 20 from Charlestown, and 80 from Boston; is hilly, abounding in springs, brooks and ponds, and contains Lovewell's mountain, which is small, and of a conical form. Island pond, $1\frac{1}{2}$ ms. by 2, is filled with islands. Ashuelot pond, 1 mile by $1\frac{1}{2}$, gives rise to one of the chief branches of Ashuelot r. Long pond is five ms. in length, and like the others, contains fish. Several ponds E. give rise to Contoocook r. The soil is deep and moist, favorable to grass, and bearing white maple, black ash, birch, beech, elm, &c., &c. Clay and peat abound, and here is some iron ore. The v. is in a pleasant situation, and there are good mill sites in the town. First settled 1768. Pop. 1830, 1,135.

WASHINGTON, co. Vt., bounded by Orleans co. N., Caledonia co. E., Orange co. s. E., Addison co. s. w., and Chittenden co. w.; lies chiefly between the 2 ranges of Green mtns., and nearly in the centre of the state. The surface is very uneven; there is much good granite E., but W. the rocks are argillaceous, mica and chlorite slate. Onion river and its branches spread over the irregular surface. There are 18 towns. Pop. 1820, 14,725, 1830, 21,378.

WASHINGTON, p-t. Orange co. Vt., 15 miles S. E. Montpelier, 43 N. Windsor; is watered by a small branch of Onion, one of Wait's, and one of White river, and the trees are principally maple. Pop. 1830, 1,374.

WASHINGTON, p-t. Berkshire co. Mass., 120 ms. W. Boston; has Housatonic river on the W. line, and 2 small branches rise in the town. Pop. 1830, 701.

WASHINGTON, co. R. I., bounded by Kent co. N., Narragansett bay E., the Atlantic S., Conn. W.; about 18 by 20 ms.; has a slightly varied surface, with primitive rocks. Some hills N. and plains S., with much good grazing land, as well as soil in many parts favorable to different branches of agriculture. Wickford and Pawtucket are the principal harbors. The coast on the ocean and the bay extends about 50 ms. There are valuable fisheries. The county contains several ponds, fresh and salt, with many small streams, forming Charles and Wood rivers, branches of the Pawcatuck, which forms a part of the west boundary of the state. The Narragansett country, formerly the seat of a powerful Indian nation, & since celebrated for a small race of pacing horses, said to have been derived from France, is included within the bounds of this co., as also the Shannock country, which had a superior kind of horned cattle. Pop. 1820, 15,687, 1830, 15,411.

WASHINGTON, p-t. Litchfield co. Conn., 40 ms. S. W. Hartford; is crossed by several ranges of primitive mountains, or lofty hills, and limestone is found in the vallies, from which marble is obtained. Aspetuck and Bantam rs. water different parts of the town. There are several iron forges, &c. Population 1830, 986.

WASHINGTON, co. N. Y., bounded by Essex co. N., Vermont E. from which it is separated N. E. by lake Champlain, Rensselaer co. S., Hudson r. and lake George W. which separate it from Saratoga and Warren counties. Wood cr. rises and terminates in this co., and in the E. Hoosac and Battenkill rs. which rise in Vt. flow into this co. and into lake Champlain. Poultney r. from Vt. forms a part of the N. boundary. Numerous mill sites are supplied by these streams. The surface and soil are very various. The lower parts of the co. are pretty well cultivated, but the N. parts are very mountainous and abound with timber, which affords much lumber. Iron, marble and slate are found in different places. The Champlain canal extends from Hudson r. to Wood cr., and along its bed to lake Champlain. This line was formerly an important military route, and here are found remains of forts erected and garrisoned at different periods from the early French wars to the revolution; and important military events have occurred here, on the land carriage between the navigable waters of the N. and the S. which approach so nearly. There are several cotton, woollen, and iron factories; and at Sandy Hill, a cotton bagging factory. The co. contains 17 towns. Pop. 1820, 38,831; 1830, 42,635.

WASHINGTON, p-t. Duchess co. N. Y., 80 ms. S. Albany, 15 E. N. E. Poughkeepsie, has Wappinger's cr. &c. N. W., with mill seats, has an irregular surface, with Tower hill E. and Chesnut ridge S. E., and good soil. Mechanic village is near the centre, with a Friends' boarding school, &c. Pop. 1830, 3,036.

WASHINGTON, v. Watervliet, Albany co. N. Y., 5 ms. N. Albany, on the W. side of Hudson r.

WASHINGTON, p-t. Morris co. N. Y., has the S. branch of Raritan r. on the E. line, Musconetcong r. W., and Schooley's mtn. between them. It borders S. on Hunterdon co., and W. on Warren co. It has the villages of Pleasant Grove and Spring t. with Schooley's mtn. springs. Pop. 1830, 2,188.

WASHINGTON, t. Burlington co. N. J., 30 ms. S. by E. Trenton, is of an irregular form, with Gloucester co. S. W., from which it is divided by the main branch of Little Egg Harbor cr., and is watered in different parts by several of its branches, the principal of which is Wading r. Pop. 1830, 1,315.

WASHINGTON, co. of Pa., bounded N. W. by Beaver co., N. and N. E. by Alleghany, E. by Monongahela r. separating it from Westmoreland and Fayette, S. by Greene co., S. W. by Ohio co. Va., and W. by Brooke co. Va. The longest line that can be drawn in this co. is diagonal in a N. W. direction from the mouth of Ten Mile cr. to the N. W. angle on Va. 45 ms., mean breadth in a similar direction 22 ms., and area 1,000 sq. ms. Extending in lat. from 39° 58′ to 40° 36′ N., and in long. from 2° 52′ to 3° 35′ W. W. C. The central part of this co. 3 or 4 ms. sthrdly. from the borough of Washington, is an elevated, and might be called a mountainous region, from which the waters flow like radii from a common centre. From hence issue the sources of Ten Mile, Pigeon, Chartier's, Buffalo and Wheeling creeks. The borough of Washington, situated in a valley, is by actual measurement 1,406 feet above tide water; and the mean elevation of the farms might be safely assumed at 1,400 feet, though no doubt many are more elevated by a difference of 300 or 400 feet. Hillsborough on the U. S. road, stands on a height of 1,750 feet. Indeed the lowest point in the co. on the Monongahela between Williamsport and Elizabethtown, is elevated at least 900 feet above tide water. The face of the co. is very broken, but soil almost uniformly productive. Limestone and sandstone are the prevailing rocks *in situ*. Mineral coal of the bituminous species abounds, and of very fine quality. It is found near Washington and Cannonsburg, at an elevation of at least 1,200 feet. The difference of climate between the central parts of this co. and that on either the Monongahela or Ohio r. is very perceptible, particularly in spring at opening leaf, and in the season of reaping small grain. The excellence of its soil is seen on inspection of the progressive pop. The first civilized settlement was made about 1770, and in 1800, with the existing limits, it contained 28,298; in 1810, 36,289; in 1820, 40,038; and in 1830, 42,909 inhabitants.

WASHINGTON borough, p-t. and st. jus. Washington co. Pa., situated near the centre of the co., on 1 of the head branches of Chartier's cr., and on the U. S. road, 26 ms. s. w. the city of Pittsburg, 22 N. w. by w. Brownsville, and by p-r. 212 ms. a little s. of w. Harrisburg, and 229 N. w. by w. W. C., N. lat. 40° 11', long. 3° 19' w. W. C. Though elevated as stated in the previous article, 1,406 feet above the Atlantic ocean, the site of this fine village is in a comparative valley; but the ground beautifully rolling. The town extends up a gentle acclivity, the main street rather closely built. It contains the co. buildings, 2 or 3 places of public worship; and numerous stores, taverns, and mechanics' shops. To the E. of the body of the place stands Washington college, sufficiently spacious to accommodate 150 students. This institution is under the direction of a president and two professors. The U. S. road enters Washington from the E., turns up the main street, and passes out of the upper end of the village. Pop. 1830, 1,816.

WASHINGTON, co. Md., bounded by Alleghany co. in the same state w., Bedford co. Pa. N. w., Franklin co. Pa. N., Frederick co. Md. E. and S. E., and by the Potomac r. separating it from Jefferson, Berkley, and Morgan cos. Va. s. w. Length along the southern boundary of Pa. 44 ms.; the breadth differs greatly, as along the South mtn. in common with Frederick, it is upwards of 30 ms. in width, whilst near Hancock'stown, the width falls short of 3 ms. The winding course of the Potomac renders the outline very irregular along that stream, but the mean breadth is very nearly 10 ms., and area 410 sq. ms. Extending in lat. from 39° 19' to 39° 42' N., and in long. from 0° 26' to 1° 18' w. W. C. The declivity is very nearly due s., and traversed by the Antietam, Conecocheague, and numerous lesser streams. The surface is broken, and in part mountainous, with much excellent river and valley soil. Chief t., Hagerstown. Pop. 1820, 23,075; 1830, 25,263.

WASHINGTON, city, the st. of the general government of the U. S. of America, and cap. of the Dist. of Columbia, is situated on the left, or Maryland side of the Potomac, near the head of tide water, and by the river and Chesapeake bay, 290 ms. from the Atlantic. It is 38 ms. s. w. from Baltimore, 136 from Phila., 225 from New York, 432 from Boston, 595 from Augusta, Me., 546 from Detroit, Mich., 1,068 from Little Rock, Ark., 856 from St. Louis, 1,203 from New Orleans, 662 from Savannah, Geo., and 544 from Charleston, S. C. The capitol stands in lat. 38° 52' 45", long. w. from the observatory at Greenwich 76° 55' 30". The site of Washington is a basin, environed by gently swelling hills; the soil is generally sterile, mixed with pebles and sand. Length of the city from S. E. to N. W. 4½ ms., mean width 2½, containing a fraction less than 8½ sq. ms. The city was laid out under the supervision of Washington (then president of the U. S.) in 1791. The principal streets are 10 in number, called avenues, and are named after different states of the Union. These diverge, 5 of them from the capitol, and 5 from the President's house, and a direct line of communication between these two edifices is formed by Pennsylvania avenue, the principal and finest street of the city. The avenues are crossed by streets running N. and S. and others running E. and W. Many of these are shaded and all of them are very broad, the former being from 120 to 160 feet in width, and the latter from 70 to 110. The buildings are much scattered, and but a small part of the city is yet compactly built. The greater part of these are on, or contiguous to, Pennsylvania avenue, including Capitol hill. The number of buildings erected in the city in 1830, was 178, 86 of which were of brick and 92 of wood. The total number of buildings in 1831 was 3,560; of these there were, public, 65; dwelling, 3,233, and 262 shops and warehouses. The value of the real and personal estate in the city Dec. 31, 1830, was, buildings, $3,125,038; lots, $3,488,032; personal property, $[illegible]00,200; total, $7,213,350. The population of Washington has increased rapidly, and from its being the seat of the government of the country, and its salubrious and healthy location, it must continue to augment in numbers. In 1800 its population was 3,210; in 1803, 4,352; in 1807, 5,652; in 1810, 8,208; in 1817, 11,299; in 1820, 13,247; and in 1830, 18,827. Of the latter there were in 1830—

	White persons.	Free col'd.	Slaves.	Total.
Males	6,581	1,342	1,010	8,933
Females	6,798	1,787	1,309	9,894
Total.	13,379	3,129	2,319	18,827

The public buildings in Washington are numerous and many of them elegant; among these the first in rank is the capitol, the most elegant edifice in the U. S. It is built of free stone, after the Corinthian order, cost rising of $2,000,000, and is altogether imposing in appearance. It stands on a commanding eminence, and has a front of 350 feet, including the wings. The rotunda, in the centre, has a diameter of 90 feet; its heighth, to the top of the dome, is the same. In this are the splendid historical paintings, executed by Col. Trumbull. The senate chamber and representatives hall are semi-circular in form, the former 74 feet in length; the latter 95 ft. and 60 in height. The dome and galleries of the hall are supported by pillars of variegated marble, from the banks of the Potomac: this apartment is truly magnificent. The library of congress occupies one apartment in this building, and contains 16,000 volumes. The president's house, which is built of white free stone, is 2 stories high, 186 ft. long, and 85 in width. It is an elegant edifice, and its location commands a fine view, particularly to the s.: it stands about 1½ ms. from the capitol. Other government buildings, are the general post office, on Pa. avenue, in which is the patent office; 4 buildings, on quadrangular bases, 2 stories high, of brick, 2 to the eastward, and 2 to the westward of the president's house, in which are kept the principal departments of the government, with their subordinate offices; a magazine, arsenal and work shops, marine barracks, navy yard, navy hospital and a penitentiary. Other public buildings are the city hall, a fine building 250 ft. by 50; 19 places of public worship, 4 well supplied market houses, an infirmary, female orphan asylum, jail, theatre, &c. There are also 4 banks, 4 extensive hotels, a foundry, breweries, museum, a city library, &c. &c. Columbian college, incorporated by congress, is about 2 ms. N. of the city. Regular lines of steamboats ply from Washington to Alexandria, Baltimore, Norfolk, &c., and numerous stages run to other places, among which are 8 daily coaches to Baltimore alone. The territory now Washington was formerly a part of Prince George co. Md., and was ceded to the U. S. in 1790. In 1800 it became the seat of government, and in 1802 was incorporated as a city. In 1812 it was remoddled, and finally chartered 1815. The government is composed of a mayor, 12 aldermen, and a common council of 18 members; these are elected by the citizens, the latter for one, and the mayor and aldermen for 2 years. During the last war with Great Britain, the city was taken by an army under General Ross, Aug. 24th, 1814, and the capitol, president's house and other public buildings were burnt. A very valuable library belonging to congress was at that time destroyed. These buildings were rebuilt soon after.

Washington, co. dist. Columbia, bounded N. W. and N. by Montgomery co. Md., N. E. by Prince George's co. Md., S. E. by the estrn. branch of Potomac, and S. W. by the main stream of Potomac. In form it approaches near a parallelogram, 8 ms. in length from S. E. to N. W., mean breadth 5¼ ms., area 42 sq. ms., or 42-100ths of the whole dist. Extending in lat. from 38° 51′ to 38° 58′ nearly, and in long. from 0° 6′ 6″ E. to 0° 03′ W. the capitol. The surface of this co. is very finely diversified by hill and dale. Rock cr. enters near the northern angle, and meandering in a general direction from N. to S. enters Potomac between the city of Washington and Georgetown. The slope of the whole co. is indeed from N. to S., and the descent very rapid. The soil generally thin, tho' some very favorable exceptions exist. Exclusive of W. C. and Georgetown, it contained in 1830, a population of 2,994. For distributive pop. see article Washington city, table 2. The entire population of the two cities and the county was in 1830, 30,262.

Washington, p-v. at the sthestrn. foot of Blue Ridge, and at the head of Thornton's r., wstrn. part of Culpepper co. Va., by p-r. 81 ms. a little S. of W. W. C.

Washington, co. Va., bounded W. by Scott, by Clinch mtn. separating it from Russell N. W. and Tazewell N., by Wythe co. E., by Blue Ridge separating it from Grayson S. E., by Carter co. Ten. S., and Sullivan, Ten. S. W. Length between Wythe and Scott cos. 50 ms. mean breadth 17, and area 850 sq. ms. Extending in lat. from 36° 35′ to 36° 55′, and in long. from 4° 30′ to 5° 19′ W. W. C. This co. occupies part of the valley between the Blue Ridge and Clinch mtns. Those chains extend in this region from S. W. by W. to N. E. by E., with minor lateral ridges. The slope of the co. is to the S. W. by W., and traversed by the S. E., middle, and north branches of Holston. All these streams have their source in Wythe, and sub-divide Washington into as many fine fertile valleys. It may, however, excite some reflection when told that in this large and well populated co. there were in 1831, but two post offices, at Abingdon, the capital, and Seven Mile Ford. Pop. 1820, 12,444; 1830, 15,614.

Washington, co. N.C., bounded by Tyrrell E., Hyde S., Martin W., and Albemarle sound N. It lies in form of a parallelogram, 20 ms. by 18, area 360 sq.ms. Extending in lat. from 35° 40′ to 35° 56′, and in long. from 0° 12′ to 0° 38′ E. W. C. What very little declivity this co. presents is from S. to N. toward Albemarle sound, but the surface is nearly a dead, and in good part, a swampy level. Chief t., Plymouth. Pop. 1820, 3,986; 1830, 4,552.

Washington, seaport, p-v. and st. jus. Beaufort co. N. C., situated on the left bank of Tar r., at or near the point where that stream assumes the name of Pamlico sound, by p-r. 122 ms. a little S. of E. Raleigh, and 312 ms. almost directly S. W. C. Lat. 35° 32′, long. 0° 13′ W. W. C. Washington is at the head of such ship navigation as Pamlico sound will admit, and having the fine valley of Tar r. in the rear, is a place of considerable note. (*See Beaufort.*)

Washington, co. Geo., bounded N. W. by Baldwin co., N. by Hancock, N. E. and E. by Jefferson, S. E. by Emanuel, S. W. by Lawrens, and W. by Oconee r. separating it from Wilkinson. Extending in lat. from 32° 42′ to 33° 13′, and in long. from 5° 36′ to 6° 11′ W. W. C. Though bounded on the W. by Oconee, this co. is a table land. It is bounded on the N. E. by the main stream and gives source to several confluents of Great Ogechee; this section falling to the sthestrd. The general declivity is nevertheless to the sthrd. discharging creeks into Oconee. Much of the soil is good, some excellent, but in general thin. The greatest length is from the sthrn. angle on Oconee to the nthrn. on Great Ogechee 38 ms., mean breadth 20, and area 760 sq. ms. Pop. 1820, 10,627; 1830, 9,820.

WASHINGTON, p-v. and st. jus. Wilkes co. Geo., 51 ms. w. N. w. Augusta, and by p-r. 64 ms. N. E. Milledgeville. Lat. 33° 42', long. 5° 45' w. W. C. This place contains an academy, and about 800 inhabitants.

WASHINGTON, co. Flor., as laid down on Tanner's U. S., is bounded on the N. w. by Choctawhatchee bay and r. separating it from Walton co., on the N. by Jackson co., E. by Appalachicola r., separating it from Gadsden co., and s. E., s. and s. w. by the Gulf of Mexico. Length from the entrance of Choctawhatchee bay to the mouth of Appalachicola r. 110 ms., mean breadth 22, and area 2,420 sq. ms. Extending in lat. from cape St. George 29° 20' to 30° 40', and in long. from 8° to 9° 36' w. W. C. Williams in his View of West Florida, says—"It is a mis-shapen tract of worthless land in general; a few hammocks on St. Andrew's bay, the south edges of Oak and Hickory hills, a part of Holmes' valley, and the borders of Econfina r. are valuable exceptions." St. Andrew's bay opens into and occupies the central parts of this co., and is a fine sheet of water, which according to Williams' map has 18 ft. water on its shallowest bar. Chief t., Holmes Valley. Pop. uncertain.

WASHINGTON, co. Ala., bounded by Chickasawhay r. separating it from Wayne co. Miss. w., by the Choctaw territory Ala. N., by Tombigbee r. separatung it from Clark co., Ala. E., and by Mobile co. s. The greatest length is along the estrn. border, 42 ms. by the general course of Tombigbee r., mean breadth about 20 ms., area 840 sq. ms. Extending in lat. from 31° 23' to 32°, and in long. from 11° 03' to 11° 37' w. W. C. The mere wstrn. border of this co. is in the valley of Chickasawhay r., but the far greater part slopes to the estrd. toward Tombigbee r. Chief ts., Washington and St. Stephens. Pop. 1830, 3,474.

WASHINGTON, p-v. and st. jus. Washington co. Ala., on the small r. or cr. Sinta Bogue, 16 ms. N. w. St. Stephens, and by p-r. 146 ms. s. s. w. Tuscaloosa. Lat. 31° 39', long. 11° 18' w. W. C.

WASHINGTON, p-v. and st. jus. Autauga co. Ala., on the right bank of Ala. r., by p-r. 129 ms. s. E. by E. Tuscaloosa. Lat. 32° 23', long. 9° 35' w. W. C.

WASHINGTON, p-v. Adams co. Miss. situated on St Catherine cr. 6 ms. E. Natchez. This place was many years the seat of government for the Miss. Territory, and afterwards for the state of Miss. Jefferson college was located here in 1802, but has not flourished as a literary institution beyond the ordinary routine of a common academy. The site of the town is high, dry and pleasant.

WASHINGTON, parish of La., bounded by Pike co., Miss. N. w., Marion co. Miss. N., Pearl r. separating it from Hancock co. Miss. E., St. Tammany parish La. s., and Tangipao r. separating it from St. Helena, parish of La. w. Greatest length a diagonal from the s. E. to the N. w. angle 66 ms., mean breadth 15, and area within a small fraction of 1,000 sq. ms. Extending in lat. from 30° 34' to 31°, and in long. from 12° 36' to 13° 34' w. W. C. The declivity of this co. is to the s. s. E., and in that direction it is bounded by the Pearl R. and Tangipao w. The Bogue Chito rising in Lawrence and Pike cos. Miss., traverses Washington parish, which, also giving source to the Chifuncte r., discharges the former into Pearl r., and the latter, over St. Tammany, into the N. side of lake Pontchartrain. The far greater part of the surface of Washington parish is composed of open and sterile pine woods. Where the land admits cultivation the staple is cotton. Chief town, Franklinton. Pop. 1820, 2,517; 1830, 2,286.

WASHINGTON, p-v. and st. jus. Rhea co. Ten. on the right bank of Ten. r., about 70 ms. below and s. w. by w. Knoxville, and by p-r. 129 ms. s. E. by E. Nashville. Lat. 35° 38', long. 7° 48' w. W. C.

WASHINGTON. co. Ten., bounded by Green w, Sullivan N., Carter E., and by Blue Ridge separating it from Buncombe co. N. C. s. Length from s. to N. 30 ms., mean width 20, and area 600 sq. ms. Extending in lat. from 35° 57' to 36° 24', and in long. from 5° 10' to 5° 35' w. W. C. A small section of this co. slopes to the nrthrd., and is drained by small creeks flowing into Watanga r.; but the sthrn. central and much most extensive sections are commensurate with the higher valley of Nolechucky r. and sthwstrdly. Chief town, Jonesborough. Pop. 1820, 9,557; 1830, 10,995.

WASHINGTON, co. of Ky., bounded w. by Hardin, N. w. by Chaplin's fork of Salt river, separating it from Nelson, N. E. and E. by Mercer, s. E. by Casey, and s. by the s. fork of Salt r. Length 28 ms., mean breadth 18, and area about 500 sq. ms. Extending in lat. from 36° 30' to 36° 52' N., and in long. from 7° 58' to 8° 32' w. W. C. Comprised within the two main branches and drained by numerous crs. of Salt r., the declivity of this co. is to the wstrd. Chief t., Springfield. Pop. 1820, 15,947; 1830, 19,130.

WASHINGTON, p-v. and st. jus. Mason co. Ky., 4 ms. from O. r. at Maysville, and by p-r. 63 ms. N. E. by E. Frankfort. Lat. 38° 37', and long. 6° 43' w. W. C. It contains beside the co. buildings, an academy, and three or four places of public worship.

WASHINGTON, co. Ohio, bounded by Athens s. w., and w., Morgan N. w., Monroe N. E., O. r. separating it from Tyler co. Va. E., and from Wood co. Va. s. Extending in lat. from 39° 15' to 39° 40', and in long. from 4° to 4° 54' w. W. C. Length on the northern border along Monroe and Morgan cos., 50 ms. It lies in a wide resemblance to a triangle, mean breadth 15, and area 750 square ms. The slope is southward towards the Ohio r. The Muskingum enters the northwestern angle, and winding thence estrd. to near the centre of the co., inflects to the s. and falls into the Ohio river at Marietta. The surface of this county is generally very hilly, but soil productive. By the post list of 1831, beside at Marietta, the co. seat, there were offices at Belpre, Bent's, Brown's Mills, Carroll, Fearing, Little Hockhocking, Lower Salem, Newport, Point Harmar, Waterford, Watertown, and Wesley. Pop. 1820, 10,425, 1830, 11,731.

WASHINGTON, p-v. Guernsey co. Ohio, 10 ms. E. Cambridge, the co. st., and 91 miles estrd. Columbus. Pop. 1830, 372.

WASHINGTON, p-v. and st. jus., Fayette co. Ohio, by p-r. 45 ms. S. W. Columbus. Lat. 39° 30', long. W. C. 6° 24' W. Pop. 1830, 299.

WASHINGTON, co. Indiana, bounded S. by Harrison, S. W. by Crawford, W. by Orange, N. W. by Lawrence, N. by the E. Fork of White r. separating it from Jackson, N. E. by Scott, E. by Clark, and S. E. by Floyd. Length from E. to W. 26 ms., mean breadth 20, and area 520 square ms. Extending in lat. from 38° 27' to 38° 47', and in long. from 8° 54' to 9° 20' W. W. C. This co. extends sthrdly. from the bank of the E. branch of White river, over the table land between that stream and Ohio river. Blue river, a small confluent of the Ohio, rises in and drains the sthrn. part, whilst from the western flow small creeks into the E. Fork of White river. Surface broken, hilly, and soil fertile. Chief town, Salem. By the post-office list of 1831, beside at Salem there were offices in this co. at Claysville, Livonia, Martinsburg, and Pleasant Valley. Pop. 1820, 9,039, 1830, 13,064. Salem, the co. seat, is situated near the centre of the co. 91 ms. S. Indianopolis. Lat. 38° 36', long. 9° 06' W. W. C.

WASHINGTON, p-v. and st. jus. Daviess co. Indiana, by p-r. 106 ms. S. W. Indianopolis, and 20 ms. a little S. of E. Vincennes. It is situated 5 or 6 ms. N. N. E. from the junction of the two main branches of White r. Lat. 38° 40', long. W. C. 10° 12' W.

WASHINGTON, co. Illinois, bounded S. by Perry, S. W. by Randolph, W. by St. Clair, N. by Clinton, and E. by Jefferson. Length from E. to W. 30 ms., mean breadth 18, and area 540 square ms. Extending in lat. from 38° 13' to 38° 30' and in long. from 12° 10' to 12° 44' W. W. C. Kaskaskia r. forms a part of the N. W. boundary of this co. separating it from Clinton, and the slope of the contiguous part is nrthwstrd. towards that stream. The southern side declines to the southward, and gives source to the northwestern branches of Muddy creek. In 1831, by the post list there was no office at Nashville, the co. st., but there were at Beaucoup, Covington, and Elkhorn. Covington is a p-v. on Kaskaskia river 47 ms. a little S. of E. St. Louis, in Mo., and by p-r. 40 ms. S. S. W. Vandalia. Pop. 1830, 1,675.

WASHINGTON, co. Mo., bounded by Madison S. E., St. Francis co. E., Jefferson N. E., Franklin N., and as laid down by Tanner by territory not laid out into counties on the S. and W. Length from S. to N. 40 ms., mean width 25, and area 1,000 square ms. Extending in lat. from 37° 35' to 38° 10', and in long. from 13° 36' to 14° 08' W. W. C. This co. as delineated by Tanner, has a natural boundary on the S. in the Iron mountains, from the southern slopes of which, and the border of the co. rise the extreme sources of St. Francis and Black rs. The body of the co. however, declines almost due N. and is drained by the fountains of both main branches of Maramec r. The tract of country embraced by Washington co. is an important section of the state of Mo. Schoolcraft, who visited the lead mines at, and contiguous to Potosi, speaks thus of the country. "Washington co. although the seat of the principal lead mines is at the same time not deficient in farming land. Big r. (the eastern branch of Maramec) in its whole course, which is long and devious, and most completely subtends the N. E. and S. boundaries of Washington co., affords the finest of farming lands. The principal farming tracts of this co. although detached, with ridges of poor land intervening, taken in the aggregate, bear a respectable proportion to its whole number of square ms. and exalt its agricultural character above that of the other mining cos. of Mo., St. Genevieve excepted. Bellevue abounds in granite and iron ore. The iron of Bellevue is a subject of universal notoriety. In the richness of the ore, and extent of the beds or mines, it is no where paralleled. The most noted place is called the Iron mountain, where the ore is piled in such enormous masses as to constitute the entire sthrn. extremity of a lofty ridge, which is elevated 5 or 600 feet above the plain." Mr. Schoolcraft denominates the species of iron ore to be that called *micaceous oxyd of iron*, and very rich in quality. This author enumerates zinc also as amongst the minerals of Washington co., but lead was then, February 1819, the only ore worked to any great amount, if we except 3 salt petre caves. In 1831, by the post office list, beside at Potosi, the co. st., there were offices at Caledonia, Harmony, and Old Mines. Potosi, the st. jus., is situated on a branch of Big r. about 60 miles S. W. St. Louis, 40 W. St. Genevieve, and by p-r. 127 ms. S. E. by E. Jefferson, the seat of government of the state. Lat. 37° 56', long. 13° 48' W. W. C. Pop. 1830, 6,784.

WASHINGTON, co. Ark., as laid down on Tanner's United States, is bounded N. E. by the county of Izard, S. by Crawford, W. by the Osage territories, and N. by the state of Mo. The extent on the map exceeds 3,000 square ms. but the country is too imperfectly known to admit a detailed description. By the list of 1831, there were three post offices, namely, Cane Hill, Fayetteville, and Vineyard, neither of which are, however, marked on the map. The extreme sources of White r. are delineated as rising near the centre and flowing northeastwardly, out of this co. into Mo., and from the latter, curving back into Ark. Pop. 1830, 2,182.

WASHINGTON, or Hempstead court house, p-v. and st. jus. Hempstead co. Ark., by p-r. 117 ms. S. W. Little Rock. Lat. 33° 45', long. 16° 36' W. W. C.

WASHINGTON, p-v. Macomb co. Mich., by p-r. 50 ms. N. N. E. Detroit.

WASHINGTONVILLE, p-v. Columbia co. Pa., 7 ms. N. Danville, the co. st., and 72 ms. N. Harrisburg.

WASHITAU, or according to French orthography, Ouachitta r. of Ark. and La. The most remote sources of this stream are in the Masserne mountain and in Pulaski co. Ark., and within 4 or 5 ms. from the Ark. r. This northern confluent is the Saline Fork. The middle branches rise also from the Masserne, but more to the southwestward in Clark co. and advancing still farther to S. W. and from the same chain issues the Little Mo. The Ouachitta proper and Little Mo. flow each by comparative courses 70 ms. to their junction between Hempstead and Clark cos. Below their union, the united waters continue to the S. E. 50 ms. to the influx of Saline, or the northeastern branch. The general course of Saline is a little E. of S. 120 ms. Now a fine navigable r. the Washitau assumes a course of very little E. of N. which it maintains all the residue of its channel of 140 ms. to its union with Red r. The entire comparative course of Washitau, is about 260 ms., but the channel being very tortuous the navigable length is usually estimated at upwards of 400 miles. About 15 miles below the influx of Saline, Ouachitta enters La. within which it receives from the westward Saluta, Terre Bonne, and Little r. and from the eastward Barthelenny, Boeuf, and Tensaw rs. The greatest length of the Washitau valley from the sources of Saline to Red r. is 260 ms., the mean breadth at least 80 ms., and area 20,800 square ms. In lat. it extends from 31° 20′ to 34° 45′, and in long. from 14° 18′ to 16° 50′ W. W. C.

WASHITAU, parish, La., W. Miss. r., bounded S. E. by the parish of Concordia, S. by the parish of Ocatahoola, S. W. by the parishes of Rapides and Natchitoches, W. by the parish of Claiborne, and N. by the cos. of Lafayette, and Chicot in Ark. Length from E. to W. 90 ms., mean width 60, and area 5,400 square ms. Extending in lat. from 31° 48′ to 33° and in long. from 14° 15′ to 15° 46′ W. W. C. The surface of this extensive region presents very striking varieties of soil. The general declivity is to the southward, but the western part declines southeastward towards Washitau r. and is generally a pine forest, drained by the branches of Terre Bonne, Saluta, and Little r. A similar character is again prevalent E. of Washitau on that part drained by Barthelenny. Advancing however to the eastward on the vallies of Boeuf and Tensaw the inundated lands of the Miss. are reached. Scattered over every part, comparatively small, but very productive zones of soil are found. E. of Washitau, lie scattered some small detached prairies, with a soil varying in quality similar to other parts of the parish. Where the soil will admit of cultivation, cotton is the common staple. Chief town, Monroe. Pop. 1820, 2,896, 1830, 5,140.

WASHTENAW, co. Mich., bounded N. W. by Ingham co., N. by Shiawassee, N. E. by Oakland, E. by Wayne, S. E. by Monroe, S. W. by Lenawee, and W. by Jackson. Length from S. to N. 36 miles, mean breadth 30, and area 1,080 square ms. Extending in lat. from 42° 06′ to 42° 39′. From the northwestern border issue the extreme sources of Shiawassee, one of the branches of Saginaw r. flowing to the nrthrd. Huron of lake Erie rising in Oakland county flows S. W. into Washtenaw, forms a sweeping curve towards the centre and thence bends to S. E. into Wayne county. The southern section is drained eastward by the N. branch of the river Raisin. The surface of Washtenaw is level and rather flat, general slope southeastward towards lake Erie. The western border is, however, on the table land between lakes Erie and Mich., and gives source to the extreme fountains of Grand r. of the latter lake. Chief town, Ann Arbour. Pop. 1830, 4,042.

WASQUE POINT, Dukes co. Mass., the N. E. extremity of Martha's Vineyard, is formed by the meeting of 2 beaches, nearly at right angles.

WATAUGA, r. of N. C. and Ten. rises in Ashe co. of the former state, from the northwestern vallies of Blue Ridge opposite the sources of Catawba and Yadkin, and interlocking sources with those of New r. branch of Great Kenhawa. These higher creeks of Watauga flowing to the W. of N. W. traverse the Iron mountain and unite in Carter county, Ten. Carter county is indeed very nearly commensurate with the lower valley of Watauga, in the western angle of which the various branches unite, and entering Sullivan fall into the S. E. branch of Holston.

WATERBOROUGH, p-t. York co. Me., 36 ms. N. York, 81 ms. S. W. Augusta, has Little Ossippee r. on the N. line, and S. a head stream of the Kennebunk. Pop. 1830, 1,814.

WATERBURY, river, Washington county, Vt., a branch of Onion river, 16 ms. long.

WATERBURY, p-t. Washington co. Vt., 12 ms. N. W. Montpelier, 24 S. E. Burlington, has Onion river on the S. line. It was first settled 1784, and has a surface generally level or gently varied, with good soil, especially on the rich meadows of Onion river, bearing hard wood with some spruce and hemlock. Waterbury r. and Thatcher's branch afford mill seats. In the S. W. is a deep cut, about 100 feet wide through rocks, where Onion river has forced its passage, and the land above appears to have been the bed of the lake. The rocks in the bed of the stream form in one place a natural bridge, and in another a cavern. Pop. 1830, 1,650.

WATERBURY, p-t. New Haven co. Conn. 20 ms. N. New Haven, has a varied surface, is crossed N. and S. by Naugatuck river which, with other streams, affords mill sites. Pop. 1830, 3,071.

WATEREE, local name given to the Catawba r. in the lower part of its course in Kershaw, Richland, and Sumpter districts, S. C., (*See Catawba and Congaree rivers.*)

WATERFORD, p-t. Oxford co. Me., 57 ms. from Augusta, 12 s. w. Paris, is square, with Cumberland co. on the s. E., and has several ponds, some of which flow into Crooked lake, and others into Crooked r. a tributary of Sebago pond. Pop. 1830, 1,123.

WATERFORD, p-t. Caledonia co. Vt. 32 ms. E. Montpelier, 21 N. Newbury, with Connecticut river s. E., first settled 1787, has Passumpsic r. N. w., and is touched by Moose r. in one part. Fifteen miles fall in the Conn. is partly opposite this town. The meadows on the r. are not overflown by the floods. The land is rough and stony. Pop. 1830, 1,538.

WATERFORD, p-t. New London, Conn., 4 ms. N. London, 37 s. E. Hartford, lies on the w. side of Thames r. and N. Long Island sound, is uneven, watered by Niantic and Jordan rs. &c. Many fish of different sorts are caught here. Pop. 1830, 2,463.

WATERFORD, p-t. Saratoga co. N. Y., 10 ms. N. Albany, lies on the w. side of the Hudson, and is separated from Albany county by the Mohawk, the banks of which from the falls are high, rocky, and nearly perpendicular to its entrance into the Hudson. A view of these falls (the Cohoes) from the bridge which crosses the river below, is very fine. Near its mouth is a dam, above which the Champlain canal crosses the stream. This canal crosses the town also, descending to the level of the river by locks, and here forms a junction with the Erie canal.—The soil of the low grounds is clay, and the more elevated parts are sandy. The village, which is 20 ms. s. s. E. Balston Spa, and 26 s. by E. Saratoga Springs, is one of the neatest in the state, and a great thoroughfare, being on the canal, and on one of the great routes from Albany to Whitehall, and also on the route from the former place to the Springs. It stands on a fine alluvial level, has a good soil, is partly surrounded by fine hills, and has some commerce, being situated at the head of sloop navigation. A bridge across the Hudson connects the town with Lansingburg. Pop. 1830, 1,473.

WATERFORD, town, Gloucester co. N. J., 30 ms. s. s. w. Trenton, is of irregular elongated form, with Burlington co. N. E., Delaware r. N. w., and Cooper's creek on part of the s. w. line. Pety's island lies opposite, in the Delaware, just below which is the city of Philadelphia. Pop. 1830, 3,088.

WATERFORD, formerly Le Boeuf, situated on Le Boeuf cr. Erie co. Pa., 15 ms. a little E. of s. from the borough of Erie, 100 ms. very nearly due N. Pittsburg, and by p-r. 333 ms. N. w. W. C. At seasons of high water, a down stream navigation is practicable from this place. Pop. 1830, ——. In 1820, the tsp. contained 570 inhabitants.

WATERFORD, p-v. in the nrthrn. part of Loudon co. Va., 10 ms. N. Leesburg, and by p-r. 37 ms. N. w. W. C. A fine flourishing village.

WATERFORD, p-v. on the right bank of Muskingum r., nrthwstrn. part of Washington co. O., by p-r. 18 ms. N. w. Marietta, and 88 ms. s. E. by E. Columbus. Population tsp. 1830, 906.

WATERLOO, p-v. Junius, Seneca co. N. Y., capital of the co., stands at a fall of Seneca r., 7 ms. E. Geneva, 4 w. Seneca falls, contains the co. buildings, and large mills.

WATERLOO, p-v. in the extreme sthwstrn. angle of Mifflin co. Pa., by p-r. 70 ms. N. w. Harrisburg.

WATERLOO, tavern and p-o. Anne Arundel co. Md., on the main road from W. C. to Baltimore, 25 ms. from the former, and 13 from the latter city.

WATERLOO, formerly Bullock's, p-o. Granville co. N. C., by p-r. 63 ms. N. Raleigh.

WATERLOO, p-o. Laurens dist. S. C., by p-r. 11 ms. sthrd. Laurensville, and 75 N. w. by w. Columbia.

WATERLOO, p-v. Lauderdale co. Ala., by p-r. 176 ms. N. Tuscaloosa.

WATERLOO, p-v. Fayette co. Ind., by p-r. 73 ms. s. E. by E. Indianopolis.

WATERLOO, p-v. and st. jus., Monroe co. Il., by p-r. 99 ms. s. w. Vandalia.

WATER STREET, p-v. Huntingdon co. Pa., 9 ms. wstrd. from the borough of Huntingdon, and by p-r. 157 ms. N. w. W. C.

WATERTOWN, p-t. Middlesex co. Mass., 7 ms. w. N. w. Boston, is situated on Charles r., on the N. side of which is the v., with a large cotton factory, &c. Sloops come up to this place, to which there is 7 feet of water at spring tides. A bridge crosses the r. just above the v., 1½ m. below; on the N. bank is an arsenal of the U. S. The surface is agreeably varied, the soil is dry and good, and almost entirely under cultivation. Fresh pond, partly in this t., is much resorted to. First settled 1630, by the sons of Sir Richard Saltonstall and others. The provincial congress sat here in 1775, and were in session during the battle of Bunker's hill. Pop. 1830, 1,641.

WATERTOWN, p-t. Litchfield co. Conn., 30 ms. w. Hartford, 26 N. w. New Haven, has Naugatuck r. and West Branch E., about 4 ms. by 6, is hilly, with some level tracts, generally granite rocks, some limestone, soil favorable to grass, well watered. Pop. 1830, 1,500.

WATERTOWN, p-t. and st. jus. Jefferson co. N. Y., lies 160 ms. N. of w. Albany, and 8 ms. from the mouth of Black r., which forms its N. boundary. It has an uneven surface, and a soil of brown loam mixed with pebbles, and limestone beneath. The village is at the falls of Black r., whence it derives an immense water power. There are 2 cotton factories, one, the Jefferson, is the largest in the state, being 250 feet long, 50 wide, and 4 stories high. 120,000 to 130,000 pounds of cotton are manufactured here, about 40,000 lbs. at the Black r. factory, and about 20,000 lbs. of wool at the Watertown woollen factory, annually. For several miles extent there are favorable sites for factories, along the banks of Black r. There are in the t. 2 Presbyterian churches, 1 Baptist, 1 Methodist, and

1 Universalist, a court house, a bank with a capital of 400,000 dollars, 2 machine shops, a tannery, morocco manufactory, paper mill, &c. &c. The village is very pleasantly located, and contains many handsome private edifices, some of them of stone, and very good taste. The prosperity of the place will be very greatly promoted by the construction of a proposed rail road to the Erie canal. Pop. 1830, 4,768.

WATERTOWN, p-v. Washington co. Ohio, by p-r. 94 ms. S. E. by E. Columbus.

WATERVILLE, p-t. Kennebec co. Me., 18 ms. N. by E. Augusta, 20 N. Hallowell, 15 S. E. Norridgewock; has Kennebec r. E., Somerset co. N., and contains part of two large ponds, which flow into Kennebec r. Here is established the Wesleyan seminary, the students of which contribute to their support by manual labor. It possesses a philosophical and chemical apparatus, with two buildings, each containing 32 rooms for students. Pop. 1830, 2,216.

WATERVILLE, p-v. Delaware co. N. Y., 17 ms. N. E. Delhi, 56 ms. S. W. Albany, and 51 from Catskill, on Delaware r.

WATERVILLE, p-v. Wood co. Ohio, by p-r. 142 ms. N. N. W. Columbus.

WATERVLIET, p-t. Albany co. N. Y., 6 ms. N. Albany; has Hudson river E. and N. In the N. E. corner of the county, on the Hudson, are extensive meadows. Some of the hills have good soil, but there are sandy tracts W. The land is principally leased, belonging to the Manor of Rensselaerwick. The Erie canal crosses the Mohawk in this t., descends by double locks to the level of the Champlain canal, which crosses the Mohawk near its mouth, joins the Erie canal, 2½ ms. above Gibbonsville. From Gibbonsville a branch canal crosses above the dam, to Troy. There is also the U. S. arsenal. The main building faces Hudson river, and the grounds enclosed by the wall extend back to the canal, which affords a convenient channel of transportation. In the W. part is Niskayuna, a settlement of Shakers, on a handsome level. Pop. 1830, 4,962.

WATKIN'S STORE, and p-o. in the southwestern angle of Pittsylvania co. Va., by p-r. 192 ms. S. W. by W. Richmond.

WATKINSVILLE, p-v. southwestern part of Goochland co. Va., 36 ms. westward Richmond.

WATKINSVILLE, p-v. and st. jus. Clark county, Geo., situated between Oconee and Apalache rivers, by p-r. 69 ms. a little W. of N. Milledgeville. Lat. 33° 50', long. 6° 28' W. W. C.

WATSON, t. Lewis co. N. Y., 128 ms. N. W. Albany, in the N. E. part of the co.; has Black river E., and the falls are 63 feet high. Pop. 1830, 909.

WATSON'S STORE, and p-o. Columbia county, Geo., 5 ms. westward Applingville, and by p-r. 88 ms. N. E. by E. Milledgeville.

WATSONTOWN, p-v. on the left bank of Susquehannah river, northern part Northumberland co. Pa., 71 ms. northward Harrisburg.

WATTSBORO', p-o. western part of Lunenburg co. Va., by p-r. 97 ms. S. W. Richmond.

WATTSBURG, or Wattsville, p-v. Erie co. Pa., 18 ms. S. E. from the borough of Erie, 10 N. E. Waterford, and about 120 ms. a very little N. of E. Pittsburg.

WAUGH'S FERRY, and p-o. Amherst co. Va., by p-r. 205 ms. S. W. W. C.

WAUKENAH, p-v. Jefferson co. Florida, 22 ms. E. Tuscaloosa.

WAVERLY, p-v. Pike co. Ohio, by p-r. 61 ms. S. Columbus.

WAVERLY HALL, and p-o. Harris co. Geo., by p-r. 119 ms. W. Milledgeville.

WAXHAW, large creek, rising in the southern part of Mecklenburg co. N. C., flows thence southwestward into Lancaster dist., S. C., falling into the right side of Catawba r., opposite Patton's isl.

WAXHAW, p-o. on Waxhaw cr., Lancaster district, S. C., by p-r. 84 ms. a little E. of N. Columbia, and 11 ms. N. N. W. Lancaster C. H.

WAYLANDSBURG, p-v. Culpepper co. Va., by p-r. 84 ms. a little S. of W. W. C.

WAYNE, p-t. Kennebec co. Me., 20 ms. W. Augusta; has Androscoggin co. N. W., and contains part of a large pond which flows into Androscoggin r. Pop. 1830, 1,153.

WAYNE, co. N. Y., bounded by lake Ontario N., Cayuga co. E., Seneca and Ontario cos. S., and Monroe co. W.; is bounded by nearly strait lines on the land sides; has Sodus bay near the middle of the lake shore, and Little Sodus, East Bays, E. of it. Mud creek enters the co. from the S., near the S. W. angle, crosses five of the south line of towns, and leaves it near the south east angle, bearing the name of Clyde river, from where it receives the Canadagua outlet. Salmon cr., and others, flow N. into the lake. The soil is generally very good. Erie canal passes thro' the S. part of this co., and the Clyde is navigable in boats. In Ontario, in this co., are 2 forges, and two blast furnaces. It contains 15 towns. Pop. 1820, 20,310, 1830, 33,643.

WAYNE, p-t. Steuben co. N. Y. 14 ms. N. E. Bath; has Crooked lake W., with a nearly level surface, and pretty good soil. Population, 1830, 1,172.

WAYNE, co. Pa., bounded S. E. and S. by Pike co., S. W. by Luzerne, W. by Susquehannah, N. by Broome co. N. Y., by Delaware r. separating it from Delaware co. N. Y. N. E., and from Sullivan county N. Y. E. Greatest length is along its western border, 54 ms. in common with Luzerne and Susquehannah counties, mean breadth 12, and area 648 sq. ms. Extending in lat. from 41° 13' to 42°, and in long. from 1° 30' to 1° 58' E. W. C. The general declivity is eastward, towards Delaware r. The southern and central sections, embracing full two thirds of the whole area, are drained by the various creeks of Lackawaxen r. The western border along the Lackawaxen mountain, gives source to the higher sources of Lackawannoc r., flowing southwestwardly into Susquehannah river, in Wyoming valley, and to the Starucca creek, flowing northwestwardly into the Susquehannah r. at the head of the Great Bend. The northeastern part is drained into the Delaware, by numerous short creeks above the Lackawaxen r. The surface is either mountainous or hilly, with, however, much excellent soil. Though bordering on the Delaware river, Wayne co. Pa. is comparatively a new settlement. In 1820 it contained but 4,127, but in 1830 the population had risen to 7,674, having gained 86 per cent. in 10 years. For the causes of such prosperity, see Lackawaxen, Honesdale, and Carbondale. Chief towns, Bethany st. jus., Honesdale, Damascus, and Stockport.

Wayne, co. N. C., bounded E. by Pitt, S. E. by Lenoir, S. by Duplin, S. W. by Sampson, W. by Johnson, N. by Nash, and N. E. by Contentny creek separating it from Edgecombe. Length 36 ms., mean breadth 20, and area 720 sq. ms. Extending in lat. from 35° 12′ to 35° 41′, and in long. from 0° 51′ to 1° 21′ W. W. C. It is entirely in the valley of Neuse river. Declivity S. E. by E. The main stream of Neuse enters from Johnson co., and passing Waynesboro', divides Wayne co. into 2 unequal sections. Pop. 1820, 9,040, 1830, 10,331. Chief t. Waynesboro'.

Wayne, co. Geo., bounded by Glynn E., Camden S., Ware S. W., Appling N. W., and Altamahah river separating it from MacIntosh N. Length 45 ms., mean breadth 16, and area 720 sq. ms. Extending in lat. from 31° 07′ to 31° 44′, and in long. from 4° 44′ to 5° 18′ W. W. C. Declivity southeastward, in the direction of the courses of Altamahah and Santilla rivers. The latter stream traverses the southwestern parts of the co. Surface generally low, flat, and in part marshy. Chief town, Waynesville. Pop. 1820, 1,010, 1830, 963.

Wayne, co. Miss., bounded by Greene S., Perry S. W., Jones W., the Choctaw country N., and Washington co. Ala. E. Length 32 ms., mean breadth 28, and area 896 sq. ms. Extending in lat. from 31° 26′ to 31° 53′, and in long. from 11° 37′ to 12° 05′ W. W. C. This co. declines to the sthrd., and is drained in that direction by the Chickasawhay r. The general surface, pine forest, moderately hilly, with sterile soil. Chief t. Winchester. Pop. 1820, 3,323, 1830, 2,781.

Wayne, co. Ten., bounded W. by Hardin, N. by Perry, N. E. by Hickman, E. by Lawrence, and S. by Lauderdale co. of Ala. Length 24 ms., breadth 21, and area 504 sq. ms. Extending in lat. from 35° to 35° 20′, and traversed by the 11th degree of long. W. W. C. Tenn. river sweeps in a semicircle round Wayne, and touching it on the northwestern angle, receives from it creeks like radii from a common centre. The surface hilly, and soil good. Pop. 1820, 2,459, 1830, 6,013. Chief t. Waynesboro'.

Wayne, co. Ky., bounded by Cumberland co. same state W., Cumberland river separating it from Russell N. W., Pulaski N., Whitby E., and Morgan co. in Tenn. S. Length 40 ms., mean breadth 22, and area 880 sq. ms. Extending in lat. from 36° 36′ to 37°, and in long. from 7° 16′ to 8° W. W. C. Declivity a little W. of N. towards Cumberland r., and drained in that direction by the South Fork of Cumberland and some other streams. Chief t. Monticello. Pop. 1820, 7,951, 1830, 8,731.

Wayne, p-v. Ashtabula co. O., by p-r. 187 ms. N. E. Columbus.

Wayne, co. Ohio, bounded by Stark E., Holmes S., Richland W., Lorain N. W., and Medina N. Length from E. to W. 30 miles, mean breadth 24, and area 720 sq. ms. Lat. 40° 50′ and long. W. C. 5° W. intersect in this co. Slope sthrd., and drained by numerous branches of Tuscarawas, Kilbuck, and Mohiccon rivers. The northern border approaches very near the dividing summit level, between the waters of Ohio river and lake Erie. It is comparatively an elevated tract, being upwards of 500 feet above lake Erie at a mean, and the arable surface exceeding 1,000 feet above the Atlantic tides. Chief t. Wooster. Pop. 1830, 23,333.

Wayne, co. Mich., bounded by Monroe S., Washtenaw W., Oakland N., Macomb N. E., and lake St. Clair and Detroit r. E. Breadth 24 ms. in the western part, greatest length along the northern border 38 miles, mean length 28, and area 672 sq. ms. Central lat. 42° 20′, long. 6° 15′ W. W. C. Slope S. E., and in that direction drained by the Huron river, Riviere Rouge, and several lesser streams. The surface is generally level. Chief town, Detroit. Pop. 1820, 3,574, 1830, 6,781.

Wayne, co. Ind., bounded S. by Union, Fayette S. W., Henry W., Randolph N., Darke co. Ohio N. E., and Preble co. O. S. E. It is very near a square of 20 ms., area 400 sq. miles. Lat. 39° 50′, long. W. C. 8° W. Slope sthrd., and drained by the higher branches of White Water r., a confluent of Great Miami river. Chief t. Centreville. Pop. 1830, 18,571.

Wayne, co. Il., bounded E. by Edwards, S. E. White, S. Hamilton, S. W. Jefferson, N. W. Marion, and N. Clay. It is a square of 24 ms., 576 sq. ms. in area. Central lat. 38° 25′, long. W. C. 11° 36′ W. Slope S. E., and drained by different branches of Little Wabash. Chief t. Fairfield. Pop. 1830, 2,553.

Wayne, co. Mo., embraces a rather extensive and mostly undefined region, on the head waters of White, Gasconade, and Maramec rivers, extending to the wstrd. of Stoddard, Madison, and Washington cos. Chief town, Greenville. Pop. 1830, 3,264.

Waynesboro', p-v. southeastern angle of Franklin co. Pa., 14 ms. S. E. Chambersburg, and 79 N. N. W. W. C.

Waynesboro', p-v. western foot of Blue Ridge, and on South river Augusta co. Va., 12 ms. E. S. E. Staunton, and 30 ms. a little N. of W. Charlotteville.

Waynesboro', p-v. and seat jus. Wayne co. N. C., on the left bank of Neuse r., immediately below the mouth of Little river, by p-r.

51 ms. s. e. Raleigh. Lat. 35° 21', long. 1° w. W. C.

Waynesboro', p-v. and st. jus. Burke county, Geo., situated on a branch of Brier cr., 30 ms. s. Augusta, and by p-r. 87 ms. e. Milledgeville. It is the seat of an academy. Lat. 33° 05', long. 5° w. W. C.

Waynesboro', p-v. and st. jus. Wayne co. Ten., situated on Ryan's creek, by p-r. 92 ms. s. w. Nashville; lat. 35°10', long. 11° w. W. C.

Waynesburg, p-v. and st. jus. Greene co. Pa., by p-r. 229 ms. n. w. by w. W. C. It is situated on a branch of Ten Mile creek. Lat. 39° 54', long. W. C. 3° 16' w. Pop. 1830, of the twp. of Wayne including the borough, 1,-130.

Waynesburg, p-v. southern part Lincoln co. Ky., by p-r. 16 ms. southward Stanford, the co. seat, and 67 a little e. of s. Frankfort.

Waynesburg, p-v. Stark co. Ohio, by p-r. 125 ms. n. e. by e. Columbus. Pop. 1830, 98.

Waynesville, p-v. and st. jus. Haywood co. N. C., on a branch of French Broad r., about 70 ms. a little n. of w. Rutherfordton, and by p-r. 295 ms. a little s. of w. Raleigh. Lat. 35° 28', long. W. C. 5° 54' w.

Waynesville, usually called Tuckersville, or Wayne C. H., Wayne co. Geo., about 70 ms. s. w. Savannah, and by p-r. 190 ms. s. e. Milledgeville.

Waynesvivle, p-v. northeastern part Warren co. Ohio, 9 ms. n. e. Lebanon, the st. justice, and by p-r. 71 ms. s. w. by w. Columbus. Pop. 1830, 439.

Weakly, co. Ten., bounded e. by Henry, s. e. by Carroll, s. w. by Gibson, w. by Obion, by Hickman co. Ky. n. w., and Graves co. Ky. n. e. Length 30 ms., breadth 28, and area 840 sq. ms. Extending in lat. from 36° 06' to 36° 30', and from long. 11° 38' to 11° 04' w. W. C. This co. is entirely in the valley of Obion r., and the declivity wstrd. towards the Miss. Chief t. Dresden. Pop. 1830, 4,797.

Weare, p-t. Hillsborough co. N. H.; is crossed in a winding course by the north west branch of Piscataquog, which affords mill seats. There are several manufactories in the town, and several small ponds. Mountains Misery and William are of no great elevation. Rattlesnake hill is near the centre. The soil is various; that of the uplands good and well watered. The surface is broken, and there are small swamps and some meadows. Pop. 1830, 2,432.

Weathersfield, p-t. Windsor co. Vt., 61 ms. s. Montpelier, 50 n. e. Bennington; has Conn. river e., on which is the Bow, an extensive and beautiful tract of fine meadows, where is situated the large and valuable farm of Mr. William Jarvis. These meadows were overgrown with a heavy forest when the settlements began. The people came from New Haven, Conn., and the town was organized 1778. Black river affords mill sites, and there are other streams. Ascutney mtn. lies partly in the n. of this t. There are several small villages. Pop. 1830, 2,213.

Weathersfield, Conn. (*See Wethersfield.*)

Weaver's Mill, and p-o. Fauquier co. Va., by p-r. 59 ms. s. w. by w. W. C.

Webb's, p-o. northern part of Stokes co. N. C., by p-r. 148 ms. n. w. by w. Raleigh.

Webbville, p-v. Jackson co. Flor., 9 miles n. n. w. Mariana, the co. seat, and by p-r. 86 ms. n. w. by w. Tallahassee.

Webster's, p-o. Richland co. Ohio. by p-r. 11 ms. n. Mansfield, the co. seat, and 62 ms. n. n. e. Columbus.

Webster's Store, and p-o. Lancaster co. Pa., by p-r. 54 ms. eastward Harrisburg.

Weedsport, p-v. Cayuga co. N. Y., 7 ms. n. Auburn, 87 w. Utica; on Erie canal; has a basin, and is a place of considerable business.

Weld, p-t. Oxford co. Me., 53 ms. n. w. Augusta, 25 n. Paris; has Somerset & Kennebec cos. e., with several high eminences, particularly the Blue mtns. e., and a large pond which empties into Androscoggin river. Pop. 1830, 765.

Weldon, p-v. on the right bank of Roanoke river, at the lower end of its falls, Halifax co. N. C., by p-r. 65 ms. n. e. Raleigh. The Roanoke navigation by canals, sluices, and river channel, is completed from Weldon to Salem, in Botetourt co. Va. Distance from Weldon to Salem, 244 miles, following Roanoke and Staunton rs. Below Weldon the navigation is again open by Roanoke r. and Albemarle sound, to the Atlantic and Dismal Swamp canal.

Welfleet, p-t. Barnstable co. Mass., 97 ms. s. by e. Boston; has the Atlantic e., and Cape Cod bay w., being situated on Cape Cod. Welfleet bay s. w., makes a good harbor, being separated from Cape Cod bay, by Beach-hill, Griffin's and Poundbrook islands. The harbor is almost encircled by rounded sand hills. The inhabitants live by fishing, the soil being almost entirely waste. A cotton and woollen manufactory was established here in 1815. Pop. 1830, 2,046.

Wellington, p-v. Bristol co. Mass., 37 ms. s. Boston, on Taunton r. A cotton manufactory was established here in 1814.

Wellington, p-v. sthrn. part Lorain co. O. by p-r. 111 ms. n. n. e. Columbus.

Wells, r. Vt. a small tributary of Conn. r. rises in Kettle pond, in the s. part of Caledonia co., flows through Long pond, &c.; and after receiving 2 branches, terminates a little s. of the line, in Newbury, Orange co. It has several falls, affords very good mill sites, and moves the machinery of several factories.

Wells, p-t. York co. Me. 32 ms. s. w. Portland, 13 n. n. e. York, 85 s. w. Augusta; has the Atlantic on the e., Kennebunk r. n., Bald Head s., with a harbor. The village was a very early settlement, and designed for a large city. The ground was laid out with regularity, and is a fine level on the borders of the sea. Pop. 1830, 2,978.

Wells, p-t. Rutland co. Vt. 40 ms. n. Bennington, 65 s. w. Montpelier, 13 s. w. Rutland, is small, rocky E. and level w., with one-third of Wells pond, or St. Augustin, within its limits; a sheet of water about 5 ms. long, and in some parts 1½ ms. wide. The soil is good, but the surface is often too rough for cultivation. First settled about 1768. Pop. 1830, 880.

WELLS, t. Hamilton co. N. Y. 72 ms. N. N. W. Albany, is mountainous, with swamps, and gives rise to head streams of Saranac, Sacandaga and Hudson rs., and contains Pezeeco lake. Pop. 1830, 340.

WELLSBORO', p-v. and st. jus. Tioga co. Pa. situated near the centre of the co. on a small creek flowing into Crooked creek, branch of Tioga r., about 45 ms. N. N. W. Williamsport on the West branch of Susquehannah r., and by p-r. 147 ms. in a similar direction from Harrisburg. Lat. 41° 45', long. 0° 22' w. W. C.

WELLSBURGH, p-v. and st. jus. Brooke co. Va. situated on the left bank of O. r. immediately above the mouth of Buffalo creek, 16 ms. above Wheeling, and by p-r. 280 ms. N. W. by W. W. C. Lat. 40° 18', long. 3° 36' w. W. C. It is a small village, mostly in one street along the river.

WELLSVILLE, p-v. on Ohio r. southeastern part Columbiana co. O. 16 ms. s. s. E. New Lisbon, the co. st., and 186 ms. N. E. by E. Columbus. Pop. 1830, 169.

WELSH RUN, p-o. sthrn. part Franklin co. Pa. by p-r. 20 ms. s w. Chambersburg, the co. st., and 82 N. W. W. C.

WENDALL, t. Sullivan co. N. H. 35 miles from Concord, 80 from Portsmouth, and w. Merrimack co.; contains 15,666 acres, of which 3,000 are water. Sunapee lake lies partly in this town and partly in Merrimack co. Sugar r. rises from its w. part, and flows across this town. First settled, 1772, from Rhode Island. Pop. 1830, 637.

WENDALL, p-t. Franklin co. Mass. 85 ms. w. Boston, has Miller's r. on the N. line, a good soil and uneven surface, with no considerable village. Pop. 1830, 874.

WENHAM, p-t. Essex co. Mass. 21 ms. N. E. Boston, is nearly level, and has a good soil, and no considerable village. There is a pond on the s. line, and a large swamp N. W. It was an early settlement. Population 1830, 611.

WENLOCK, t. Essex co. Vt. 53 ms. N. E. Montpelier, gives rise to the principal branch of Nulhegan r. Pop. 1830, 24.

WENTWORTH, p-t. Grafton co. N. H. 52 ms. from Concord, and 15 from Plymouth. Baker's r. has a fall of about 20 feet, where is a village, a bridge, and various mills and factories. The South branch of Baker's r. is s. There are several ponds, which contain all sorts of fish found in the state. Carr's mtn. E. affords valuable granite. Limestone is obtained from mount Cuba w., and iron ore is found in the town. The soil near the streams is excellent, and elsewhere generally good. This town was named after general Benning Wentworth. First settled after the revolution. Pop. 1830, 924.

WENTWORTH, p-v. and st. jus. Rockingham co. N. C., situated on the summit level between the vallies of Haw and Dan r., by p-r. 292 ms. s. w. W. C., and 108 ms. N. W. by w. Raleigh. Lat. 36° 24', and long. 2° 46' w. W. C.

WESLEY, p-v. Haywood co. Ten. by p-r. 186 ms. s. w. by w. Nashville.

WESLEY, p-v. Washington co. O. by p-r. 99 ms. s. E. by E. Columbus.

WEST ALEXANDRIA, p-v. on the wstrn. border of Washington co. Pa. 16 ms. s. w. by w. from the borough of Washington, and 14 E. Wheeling, Va.

WEST ALEXANDRIA, p-v. Preble co. O., by p-r. 87 ms. a little s. of w. Columbus.

WEST BEDFORD, p-v. Coshocton co. O. by p-r. 71 ms. N. E. by E. Columbus.

WEST BERLIN, p-v. Frederick co. Md. by p-r. 58 ms. nthwstrd. W. C.

WESTBOROUGH, p-t. Worcester co. Mass. 34 ms. w. s. w. Boston, 10 E. Worcester, is on high ground, giving rise to Concord and Blackstone rs. There is a large pond N. The soil is good, and there are several small streams. It formerly belonged to Marlborough. In 1704 several persons were captured by Indians. Pop. 1830, 1,438.

WEST BOYLSTON, p-t. Worcester co. Mass. 42 ms. w. Boston, N. Worcester. First settled, 1720, from Marlboro'; is crossed by Nashua r., which is formed in the N. W. part by the union of Quinepoxet and Still Water rs. Almost the entire town lies in the valley of the Nashua, and has a rich and well cultivated soil. The mill seats are very good, and some of them occupied by large manufactories. There is some iron ore, and a mineral spring. Pop. 1830, 1,045.

WEST BRIDGEWATER, p-t. Plymouth co. Mass. 24 ms. s. Boston, has a soil of inferior quality, and is watered by a few small streams. Pop. 1830, 1,042.

WESTBROOK, t. Cumberland co. Me. 3 ms. w. Portland, is crossed N. W. by Presumscot r. Pop. 1830, 2,238.

WEST BROOK, p-v. sthrn. part Bladen co. N. C. 18 ms. from Elizabethtown, the co. st., and by p-r. 117 ms. s. Raleigh.

WEST BRUNSWICK, p-t. Herkimer co. N. Y. 22 ms. N. Herkimer, has much rough and inferior land. Pop. 1830, 713.

WEST BUCKINGHAM, p-o. Washington co. Pa.

WEST CAMBRIDGE, t. Middlesex co. Mass., 5 ms. N. W. Boston, w. Charles r. bay, was formerly a part of Cambridge. It has rocky and broken land N., low and some swampy land s., with good pasturage and tillage in the middle part. Craigie's bridge connects this t. with Boston. Pop. 1830, 1,230.

WEST CANAAN, p-v. Madison co. O., by p-r. 26 ms. wstrd. Columbus.

WEST CARLISLE, p-v. Coshocton co. O., by p-r. 68 ms. N. E. by E. Columbus.

WEST CHARLESTON, p-v. Miami co. O. by p-r. about 80 ms. w. Columbus.

West Castle, p-o. Caswell co. N. C., by p-r. 96 ms. n. w. Raleigh.

Westchester, co. N. Y., bounded by Putnam co. n., Conn. e., Long Island sound and East r. s., Harlaem and Hudson rs w., is of irregular form, something triangular, contains about 480 sq. ms., and is crossed sthwstrly. by Croton and Peekskill crs.; Saw mill, Bronx, and Byram crs., are smaller streams. The rocks are generally primitive; there are mountains in the n. w., and a high ridge extends through the co. n. and s. Marble is quarried at Mount Pleasant, by the convicts in the state prison, where is also an old silver mine and a copper mine. There are 3 cotton factories in this co., 4 paper mills, 5 woollen factories, and 2 iron foundries. There are 21 ts. Pop. 1820, 32,638; 1830, 43,594.

Westchester, p-t. Westchester co. N. Y., 12 ms. n. by e. New York, 140 s. Albany, has Long Isl. s. e., with clay soil, generally stony, watered by Bronx and W. Chester crs. The manufactures are various, but not very extensive. The villages are Westchester, and West Farms. Pop. 1830, 2,362.

West Chester, borough, p-v. and st. jus., Chester co. Pa., situated 28 ms. almost exactly due w. from the city of Philadelphia, 75 s. e. by e. Harrisburg, and by p-r. 115 ms. n. e. W. C. Lat. 39° 58', long. 1° 28' e. W. C. This is a very flourishing borough, in a well cultivated country. Pop. 1830, 1,258. Besides the usual co. buildings, West Chester contains several places of public worship, a bank, printing office, well filled schools, and numerous stores and public houses.

West Chester, p-v. Butler co. O., by p-r. 87 ms. s. w. by w. Columbus.

Westerlo, p-t. Albany co. N. Y., 21 ms. s. w. Albany, has Greene co. s., is crossed by low ridges, with pretty good soil in the vallies. The w. part belongs to the manor of Rensselaerwyck. First settled 1759. It is watered by small streams. Pop. 1830, 3,321.

Westerly, p-t. Washington co. R. I., 36 ms. s. w. Providence, 35 w. by s. Newport, has Pawcatuck r. n. and w., the Atlantic s., and adjoins Conn. The surface is broken, the soil various, generally favorable to grazing. Vessels of 80 tons go 4 ms. up Pawcatuck r., and the v. of Pawcatuck is 6 ms. from the ocean. Pop. 1830, 1,904.

Western, p-t. Worcester co. Mass., 22 ms. s. w. Worcester. Pop. 1830, 1,119.

Western, p-t. Oneida co. N. Y., 20 ms. n. Utica, 8 n. e. Rome, is well supplied with mill seats by head streams of Mohawk r. Pop. 1830, 2,419.

Western Star, p-o. Medina co. O., by p-r. 112 ms. n. e. Columbus.

West Fairfield, p-o. Westmoreland co. Pa., by p-r. 184 ms. n. w. W. C.

Western Ford, and p-o. watrn. part Randolph co. Va., by p-r. 240 ms. w. W. C.

Western Port, p-v. on Potomac r. Alleghany co. Md., 24 ms. above, and s. w. Cumberland, and by p-r. 141 ms. n. w. by w. W. C.

West Farmington, p-v. Oakland co. Mich., by p-r. 42 ms. n. w. Detroit.

West Farms, p-v. Westchester, Westchester co. N. Y., 12 ms. n. by e. New York, 140 from Albany, on Bronx cr., at the head of navigation, 3 ms. from Long Island sound, has several small manufactories.

West Fairlee, t. Orange co. Vt., 28 ms. s. e. Montpelier, 35 n. e. Windsor, is crossed in the s. w. by Ompompanoosuc r., and part of West Fairlee lake. The surface is uneven. Pop. 1830, 841.

Westfield, t. Orleans co. Vt., 42 ms. n. Montpelier, 44 n. e. Burlington, has a small part of Missisque r. s. e., into which flow 3 mill streams from this t. The land is good e., but mountainous w., having the Green mtns. in that part, through which Hazen's Notch affords a passage. Few settlers came here before 1800. Pop. 1830, 353.

Westfield, p-t. Hampden co. Mass., 105 ms. w. by s. Boston, 6 ms. w. Springfield, is crossed by Westfield r. which affords mill sites; it is a furious stream when raised by floods. The Hampshire and Hampden canal crosses the t., and is carried across this r. on an aqueduct, being brought down to the level of the meadows by locks, and up again to Hungry plain. In the hollow is the v., which is a pleasant and busy place, with a large and flourishing academy, incorporated in 1793. In 1832 it contained 375 pupils. Tuition $3 00 per quarter in summer, and $3 25 in autumn. Lectures are given weekly upon natural philosophy every term—upon chemistry during the fall term—and upon natural history and to school teachers when required. All lectures *gratis*, except chemistry. The town was first settled 1667, being called Warranoake by the Indians, and was attacked by savages in Philip's war. It was long a frontier settlement. Pop. 1830, 2,940.

Westfield, t. Richmond co. N. Y., 3½ ms. s. w. Richmond v., has Raritan bay s., and Staten isl. sound w., which separates it from N. Jersey. It occupies the s. w. part of the island, and has Prince's bay s., whence great quantities of oysters are derived for the New York market. They are brought from Virginia, and placed here to grow. Many of the inhabitants are employed in this business. The land is good, and there are many good farms. There are extensive meadows at the head of the Fresh Kills. Pop. 1830, 1,733.

Westfield, p-t. Chatauque co. N. Y., 7 ms. n. Mayville, 1 from Portland harbor, 28 from Erie, Pa. Pop. 1830, 2,477.

Westfield, p-t. Essex co. N. J., 7 ms. w. Elizabethtown, has Rahway r. e., Morris co. n. w., Somerset co. w., and Middlesex co. s. It is crossed in the n. part by the Short hills. Pop. 1830, 2,492.

Westfield, p-v. western part Tioga co. Pa., by p-r. 180 ms. n. n. w. Harrisburg.

Westfield, p-v. Delaware co. Ohio, by p-r. 32 ms. northward Columbus.

West Finley, or Findlay, p-v. northwestern part of Washington co. Pa., by p-r. 248 ms. n. w. W. C.

Westford, post town, Chittenden county,

Vermont, 13 miles northeast Burlington, 32 N. W. Montpelier, first settled soon after the revolution, is crossed by Brown's r. from S. to N. The surface is irregular. Pop. 1830, 1,291.

WESTFORD, p-t. Middlesex county, Mass., 28 ms. N. W. Boston, and 8 from Concord, is on high ground, with a fertile soil, favorable to grass, grain, and fruit. The v. occupies a fine elevation, near the centre of the town, commanding a view of Monadnock, Kearsearge and Watchusett mountains, and contains an academy, incorporated in 1793. Pop. 1830, 1,329.

WESTFORD, p-t. Otsego co. N. Y., 9 ms. E. S. E. Cooperstown, 11 S. Cherry Valley, and 56 W. Albany, is hilly, with rich vallies, and good grazing on the uplands, and watered by Elk r. and other streams. Pop. 1830, 1,645.

WEST FRIENDSHIP, p-v. on the Frederick turnpike, and in the northwestern part of Ann Arundel co. Md., by p-r. 57 ms. N. W. C., and 49 N. W. Annapolis.

WEST GREENVILLE, p-v. northwestern part Mercer co. Pa., 14 ms. N. W. from the borough of Mercer.

WEST GREENWICH, town, Kent co. R. I., 18 ms. S. W. Providence, has Washington co. S. and Conn. W., gives rise to the S branch of Pawtuxet r. and has Wood r. W. Hopkins' hill, is a conspicuous eminence. Pop. 1830, 1,818.

WEST GROVE, p-v. Chester co. Pa., by p-r. 71 ms. estrd. Harrisburg.

WEST HAMPTON, p-t. Hampshire co. Mass., 100 ms. W. Boston; formerly a part of Northampton; gives rise to several streams of Manhan r., a branch of the Conn., and is devoted to agriculture. Pop. 1830, 918.

WEST HANOVER, p-o. Dauphin co. Pa., 16 ms. N. E. by E. Harrisburg.

WEST HAVEN, p-t. Rutland co. Vt., 86 ms. from Montpelier, has Poultney r. and N. Y. state S., and lake Champlain W. Hubbardton r. and Cogman's creek afford mill seats. Pop. 1830, 722.

WEST LIBERTY, p-v. O. co. Va., by p-r. 276 ms. N. W. by W W C.

WEST LIBERTY, p-v. and st. jus. Morgan co. Ky., by p-r. 107 ms. a little S. of E. Frankfort.

WEST MIDDLETOWN, p-v. western part Washington co. Pa., 13 ms. N. W. the borough of Washington.

WEST MILTON, p-v. Miami co. O., by p-r. 84 ms. W. Columbus.

WESTMINSTER, post-town Windham county, Vermont, 37 miles N. E. Bennington, 82 ms. S. Montpelier, 27 ms. from Windsor, has Conn. river E. on the borders of which is a tract of fine meadows. The first permanent settlements here, were made from Northfield, Mass., and from Conn. about 1741. The v. is on a large level considerably above the adjacent meadows, shut in at some distance, by hills which touch the r. both above and below. There are no mill streams. The legislature of Vt. was held here several times soon after the formation of the state. Pop. 1830, 1,737.

WESTMINSTER, p-t. Worcester co. Mass., 54 ms. W. by N. Boston, lies in the fork of Nashua river and a western branch, and is a pleasant agricultural town, yielding grass and fruit. The land divides some of the waters of the Connecticut and Merrimack rs. It was granted to soldiers of Philip's war as "Naragansett No. 2." Pop. 1830, 1,696.

WESTMINSTER, p-v. on the very eastern border of Frederick co. Md., 29 ms. N. W. Baltimore, 566 ms. N. W. C.

WESTMORE, town, Orleans co. Vt., 43 ms. N. E. Montpelier, is uneven, with mtns. Hor, Pisgah and Pico for the principal summits, and Willoughby's lake, about 1 1-2 ms. by 6, within its boundaries. Willoughby's L. is the outlet of this lake, and the head streams of Clyde and Passumpsic river rise in this town. Pop. 1830, 353.

WESTMORELAND, p-t. Cheshire co. N. H., 65 ms. from Concord, and 100 from Boston, lies on the E. side of Connecticut r. which separates it from Vermont, and has several small streams running into that river. One flows from Spafford's lake, in Chesterfield, and affords mill sites. Pop. 1830, 1,647.

WESTMORELAND, p-t. Oneida co. N. Y., 9 ms. W. Utica, 105 from Albany, and 8 S. Rome, has Oriskany creek a few ms. E., and a few small mill streams. The surface is nearly level, the soil very good. Erie canal is about 6 ms. N. of the centre. Pop. 1830, 3,303.

WESTMORELAND, co. Pa., bounded by Laurel Hill separating it from Somerset S. E., by Lafayette S., by Monongahela r. separating it from Washington co. S. W., by Youghioghany r. separating it from the extreme sthrn. part of Alleghany co. W., by the central part of Alleghany N. W., by the Alleghany separating it from the northern part of Alleghany co. N., and by Conemaugh r. separating it from Armstrong and Indiana N. E. The longest line is a diagonal, 50 ms. from the sthrn. angle on Laurel Hill to the northern at the mouth of Conemaugh, mean breadth 21 ms., and area 1,050 square ms. Extending in lat. from 40° 03′ to 40° 43′, and in long. from 2° to 2° 56′ W. W. C. This co. is composed of two inclined plains E. of Monongahela. The northeastern declines to the northward, and is drained by Loyalhannah and other creeks into Conemaugh r. The second or central plain has a western declivity towards the Monongahela and Alleghany rivers. Beside these two sections, Westmoreland contains two turnpikes on the peninsula between the Youghiogany and Monongahela rivers. The surface is every where broken by hills or mountains, yet few counties in the U. S. even in proportion to surface, have more good land. Fruits, grasses, and indeed vegetables of every kind suitable to the climate, grow abundantly. Chief town, Greensburg. Pop. 1820, 30,540, 1830, 38,400, having gained upwards of 25 per cent in 10 years.

WESTMORELAND, co. Va., bounded S. E. by Northumberland, S. by Richmond, S. W. by Rappahannoc r. separating it from Essex, N. W. by King George, and by Potomac r. separating it from Charles co. in Md. N., and St. Mary's of Md. N. E. Length along Potomac

r. 30 ms., mean breadth 5, and area 150 square ms. Extending in lat. from 38° to 38° 16′ N., and in long. from the meridian of W. C. to 0° 30′ E. Chief town, Westmoreland court house. Pop. 1820, 6,900, 1830, 8,411.

WESTMORELAND, court house, and p-o. Westmoreland co. Va., by p-r. 116 ms. S. E. W. C. and 70 N. E. Richmond.

WEST NEWBURY, p-t. Essex co. Mass., 34 ms. N. E. Boston, 6 w. Neburyport, has Merrimac r. on the N. line, with many hills and vallies, and excellent soil. Although the inhabitants are generally farmers, carriages, shoes, combs, &c. are manufactured in great numbers. A bridge crosses the Merrimack, to Rock's v. in Haverhill. Pop. 1830, 1,586.

WESTON, town, Windsor co. Vt., 66 ms. S. Montpelier, 22 S. W. Windsor, is crossed by West r. which affords mill sites, and passes 2 villages. Pop. 1830, 972.

WESTON, p-t. Middlesex co. Mass., 15 ms. w. Boston. Pop. 1830, 1,091.

WESTON, p-t. Fairfield co., Conn., 61 ms. S. W. Hartford, 8 ms. from Long Island sound, 6 ms. by 9, is hilly, with primitive rocks, and is supplied with mill sites by Saugatuck river, Mill r. &c. There is an academy well endowed. Pop. 1830, 2,997.

WESTON, p-v. and st. jus. Lewis co. Va., situated on the West fork of Monongahela r. about 70 ms. a little S. of E. Marietta, in the state of Ohio, and by p-r. 249 ms. w. W. C.

WEST PENN, p-v. Schuylkill co. Pa., by p-r. 76 ms. N. E. Harrisburg.

WEST PHILADELPHIA, p-o. west side Schuylkill, Phila. co. Pa. 2 ms. from the centre of the city.

WEST POINT, Cornwall, Orange co. N. Y., the site of the military academy of the U. States, is a high and rocky projection which turns the course of Hudson r. estrd. a little below the north entrance of the Highlands, and occupies a commanding point on that stream, on which account it became an important position in the revolutionary war. A fort was built on the brow of the hill, and another on the opposite shore, and Mount Independence, afterwards called fort Putnam, now in ruins, was erected on the top of a steep eminence in the rear, its guns bearing upon the river above and below, and upon the mouth of a defile here opening through the mountains on the w. A chain was stretched across the r. to prevent the passage of ships, but was taken up by the British in 1777, when they forced the pass of the highlands. The military academy of the U. S. was founded here in 1802, and buildings have been constructed under an appropriation of $12,000 made in 1812, on 250 acres of land ceded by the state for the use of the institution. The course of instruction occupies 4 years; the 1st and 2d years are devoted to mathematics, French and drawing; the 3d to philosophy, natural and experimental, chemistry, drawing and artillery; and the 4th to engineering, ethics, civil and military, belles lettres, and national law, artillery and tactics. There are professors and teachers in different departments, with many assistants, some of whom are taken from among the pupils.

WEST POINT, p-v. Hardin co. Ky., by p-r. 72 ms. S. W. by W. Frankfort.

WESTPORT, p-t. Lincoln co. Me., 29 ms. S. Augusta, occupies an isl. on Sheepscot river, separated from the ocean by one or two other isls. Pop. 1830, 554.

WESTPORT, p-t. Bristol co. Mass. 60 ms. S. Boston, 8 S. W. New Bedford, borders S. upon the Atlantic, and W. on R. I., and is crossed by 2 branches of Acouscet r. which are navigable for some distance. There are 3 villages. The soil is good for grazing, and yields many articles for the New Bedford market. Pop. 1830, 2,779.

WESTPORT, p-t. Essex co. N. Y. 123 ms. N. Albany, 5 E. Elizabethtown, has lake Champlain E., contains N. W. Bay, with good land, and abounds in iron ore, of which much is obtained for manufacture. Pop. 1830, 1,513.

WEST PORT, p-v. and st. jus. Oldham co. Ky., situated on the left bank of O. r. about 25 ms. by the land road above Louisville, and by p-r. 44 ms. N. W. by W. Frankfort. Lat. 38° 27′, long. 29° 30′ W. W. C.

WEST QUODDY HEAD, Lubec, Me., is the S. W. point of Passamaquoddy Bay, and has a light house.

WEST RIVER, Windham county Vt., rises in the N. W. corner and flowing S. E. empties into Connecticut river in Brattleborough, receiving the waters of about 440 square miles, through Bald mountain, Meadow, South and Smith's branches. It affords few mill sites, but its branches are many.

WEST RIVER, Worcester co. Mass., is a mill stream of the Blackstone, and has 2 cotton factories, each with 500 spindles.

WEST RIVER, a small bay of the main Chesapeake bay, making into the southeastern shore of Ann Arundel co. Md.

WEST RIVER, p-v. on the western bank of West Bay, Ann Arundel co. Md., 12 ms. S. S. W. Annapolis, and 49 ms. E. W. C.

WEST SPINGFIELD, p-t. Hampden co. Mass., 100 ms. W. by S. Boston, is on the W. bank of Connecticut r. and crossed by Westfield river. The surface is irregular, but there is a fine and extensive meadow E., with a beautiful level, on which the village is situated. The streets are bordered by noble elms, and some of the houses are very fine. There are high hills or mountains N., and sandy plains S. This town formerly belonged to Springfield, which was one of the earliest settlements on the r. Some lead ore has been found here. Pop. 1830, 3,270.

WEST STOCKBRIDGE, town, Berkshire co. Mass., 130 ms. W. Boston, has the New York line for its W. boundary, and is crossed by Williams river (a branch of Housatonic,) on whose banks are fine meadows. The soil is good for grazing, but the Taughkannuck mtns. encroach upon the W. part of the town. Iron mines are wrought here, as well as quarries of white and clouded marble. A railroad is authorized to be extended to the Hudson. Pop. 1830, 1,209.

WEST TAMIAQUA, p-v. on a branch of Little Schuylkill r. Schuylkill co. Pa., by p-r. 81 ms. N. E. Harrisburg.

WEST TURIN, town, Lewis co. N. Y., 120 ms. from Albany. Pop. 1830, 1,534.

WEST UNION, p-v. and st. jus. Adams co. O., by p-r. 101 ms. s. s. w. Columbus, and 39 ms. a little N. of w. Portsmouth, at the mouth of the Sciota. Lat. 38° 48′, long. W. C. 6° 27′ w. Pop. 1830, 429.

WEST UNION, p-v. on Wheeling creek, Ohio co. Va., 5 ms. s. E. by E. Wheeling, and by p-r. 266 ms. s. w. by w. W. C.

WESTVILLE, p-t. Franklin co. N. Y., 8 ms. N. N. E. Owego. Pop. 1830, 619.

WESTVILLE, p-v. and st. jus., Simpson co. Miss., situated on Strong river, about 100 ms. N. E. by E. Natchez, and by p-r. 56 miles s. E. Jackson. Lat. 31° 58′, long. 13° 02′ w. W. C.

WEST WHITELAND, township and p-o. Chester co. Pa., 25 ms. w. Phila., and 5 E. Downingstown.

WETHERED'S, p-o. southwestern part Shelby co. Ten., by p-r. 223 ms. s. w. by w. Nashville.

WETHERSFIELD, p-t. Hartford co. Conn., 4 ms. s. Hartford, 34 N. New Haven; has Connecticut r. E. which makes a beautiful cove in this town, affords navigation, and valuable shad fisheries. Extensive and fertile meadows border the stream, and a broad and higher level tract, with light but rich soil, lies w., on which the principal village is built, about 1 mile from the r. On this level great quantities of onions are raised, which are sent to different parts of the United States, and to some foreign countries. The v. is remarkably pleasant, having broad streets planted with elms. The township has an agreeably varied surface. Rocky Hill, or Stepney, the s. parish, has a pleasant village, on an elevated situation, with a landing at some distance, where considerable commerce and ship building were formerly carried on. The river at that place, leaves the fine meadows through which it has flowed for many miles, and enters a narrower channel between hills & high sandy banks. The first attempt to form a settlement on Conn. river, was made here by white men, in 1634, when a few persons spent the winter. The next year this town, Hartford, and Windsor, were occupied by colonists, but their sufferings during the cold weather were so great, that numbers of them travelled by land to Saybrook fort, then the nearest place of refuge. The capture of 2 white women here by Indians soon afterwards, was one of the arguments for the Pequod war. The first settlements were made at the Point. Pop. 1830, 3,853.

The state prison of Connecticut has been erected in Wethersfield within a few years. It stands on the margin of the cove, and is surrounded by a stone wall. The building is on the plan of the Auburn prison, and the discipline of the same general description. The number of convicts being comparatively small, the discipline is more perfect and effectual; the inmates are instructed, and evident improvement is produced in their manners and character.

WETHERSFIELD, p-t. Genesee co. N. Y., 258 ms. w. Albany, and 28 s. Batavia; has Alleghany co. on the s. line. Pop. 1830, 1,179.

WETHERSFIELD, p-v. Trumbull co. Ohio, by p-r. 169 ms. N. E. Columbus.

WEXFORD, p-v. Alleghany co. Pa., 14 miles westward Pittsburg.

WEYBRIDGE, t. Addison co. Vt., 80 miles N. Bennington, 30 s. Burlington; has Otter cr. N. and E., with several mill seats; was first settled about the beginning of the revolution, deserted, and afterwards settled again, from Mass. Lemonfair river w. is a slow stream. Snake mtn. is also in the w. Pop. 1830, 850.

WEYMOUTH, p-t. Norfolk co. Mass., 10 ms. E. by s. Boston; has Boston harbor on the N.; has a good soil, favorable to grass, a mill stream, and a point extending into the harbor. The Indian name of the place was Wessaguscus, and it was one of the oldest settlements in New England, but was at first occupied by a set of dissolute Englishmen, who were saved from massacre by the Indians, only through timely aid sent from Plymouth. Pop. 1830, 2,837.

WEYMOUTH, t. Gloucester co. N. J.; has Little Egg Harbor river on the N. E. line, and Tuckahoe creek w. and s., which divides it from Cumberland county. The s. E. corner, which is formed by the confluence of the two streams, is a large swamp, connected with those in the neighboring townships, round the head of Little Egg Harbor. Pop. 1830, 1,270.

WEYMOUTH, p-v. Medina co. Ohio, by p-r. 117 ms. N. E. Columbus.

WHARTON'S, p-o. Morgan co. Ohio, by p-r. 83 ms. s. E. by E. Columbus.

WHARTON'S MILLS, and p-o. Bedford county, Va., by p-r. 154 ms. s. w. W. C.

WHATELEY, p-t. Franklin co. Mass., 100 ms. w. Boston, lies w. of Connecticut r., N. Hampshire co.. and has some meadow land, but more mountainous, and fine soil, though some of that in the interior of the t. is good. It is crossed by two small streams, and contains Sugar Loaf hill, at the foot of which a bloody battle was fought in Phillip's war, between Capt. Lothrop's company and several hundred Indians. This town first belonged to Hadley, and afterwards to Hatfield. Pop. 1830, 1,111.

WHEATLAND, t. Monroe co. N. Y., 15 ms. s. s. w. Rochester; has Genesee r. E., Livingston co. s., and Genesee co. w.; has good land, an undulated surface, and is crossed by Allan's creek, a stream of the Genesee. On it stands the village, 1 mile from the mouth. Pop. 1830, 2,239.

WHEATLY, p-v. Fauquier co. Va., by p-r. 64 ms. s. w. by w. W. C.

WHEELER, p-t. Steuben co. N. Y., 10 miles N. Bath; crossed by Five Mile creek, and other streams of Conhocton creek; has an uneven surface. Population, 1830, 1,389.

Wheeler's Springs, and p-o. 79 ms. s. w. Richmond.

Wheeling, large creek of Pa. and Va., rises in Washington and Green cos. of the former state, and flowing northwestward, enters Ohio co. of the latter state, falls into Ohio r. immediately below the city of Wheeling.

Wheeling, p-o. city, and st. jus. for Ohio co. Va., situated on the left bank of Ohio r., 56 ms. s. w. Pittsburg, 31 s. w. by w. Washington in Pa., and by p-r. 264 ms. n. w. by w. W. C. Lat. 40° 07', long. 4° 36' w. W. C. The narrowness of the bottoms between the river and a high and steep hill, confines the town to a single street. This street ranges along on high land in the upper part, and a second, 15 or 20 feet lower, towards the mouth of Wheeling creek. The origin of the place was Wheeling fort, built early in the revolutionary war, which stood on the breast of a high bank, and where the U. S. road reaches the Ohio r. Wheeling advanced at first but slowly. It was laid out as a village early after the peace of 1783, and in 1820 contained only 1,567 inhabitants; in the last ten years the advance has been rapid, and in 1830, the pop. was, white males, 2,667; females do., 2,349; free blacks, 94; slaves, 101; total 5,211.

Wheelock, p-t. Caledonia co. Vt., 30 miles n. e. Montpelier; first settled 1780; was granted to Dartmouth college, and named after its president. Several small streams afford mill seats. Wheelock mtn. w. is a part of the e. range of the Green mountains. The land is generally stony. Pop. 1830, 834.

Wheelwright's Pond, Strafford co. N. H., in Lee, gives rise to Oyster r. A battle was fought on its banks, 1690, between some Indians and Americans.

Whetstone, river, Ohio, rising in Richland and Crawford cos., and flowing thence s. w. about 20 ms., inflects to the southward, and maintains the latter course over Marion and Delaware counties, and to near the centre of Franklin county, where it unites with the Sciota r., after a comparative course of 70 ms. It has interlocking sources with those of Mohiccon, branch of Muskingum, those of Sandusky, and the Sciota.

Whetstone, p-v. eastern part Marion co. Ohio, by p-r. 50 ms. n. Columbus.

White, r. Vt., rises in Rutland co. near the division in the 2 branches of the Green mtns., flows s. e., crosses Windsor co., and joins the Connecticut, 5 ms. above the mouth of Queechy river. From the n. it receives 1st, 2nd and 3rd branches, each about 20 miles long, and supply mill sites. White river is about 55 miles long, waters about 680 sq. ms. In one place it just crosses a corner of Addison county.

White, river, a large stream of the state of Mo., and territory of Arkansas, formed by the confluence of two streams, White r. proper, and Black r. The following description is founded on the delineations on Tanner's U. S. White river proper, rises in Washington co. Ark., about 30 miles northeastward from the junction of Arkansas and Canadian rs. Flowing thence about 60 ms. to the northeastward, it enters the country of the Delaware Indians in the southwestern angle of Missouri. Curving to e. and s. e. 40 ms., it enters the territory of Arkansas, within which it pursues a course of s. e. by e., by comparative distance 120 miles, and joins Black river in Independence county. In its entire comparative course of 220 ms., White r. receives few tributary streams of any considerable length of volume. The northeastern confluent, Black river, rises in Wayne county, Mo., interlocking sources with those of Maramec and Saint Francis. Flowing thence by a general southern course, but an elliptical curve to the e. about 100 ms., unites with an equal, if not superior confluent, Current r. The latter rises also in Wayne co. Mo., to the westward of the sources of Black river, and interlocking sources with the Maramec and Gasconade rivers. The general course, curves, and length of Current river, is remarkably similar to similar phenomena in Black river. The now navigable Black river, assuming a southwestern course, 15 ms., receives at Davidsonville, st. jus. for Lawrence co., from the n. w. a large accumulation, by the united streams of Eleven Points and Spring rs. Below Davidsonville, Black river flows by comparative courses, 45 ms., entering in that distance Independence co., and joining White river, as already noted. The general course of White river, below the influx of Black r., is a little e. of s., by comparative courses 126 miles, to its influx into the Mississippi, receiving in the latter distance Red river from the w., and Caché river from the e. Taken as a whole, the valley of the White river lies between those of Ark. to the s. w., St. Francis to the e., the southern sources of Osage r. to the n. w., and those of Maramec to the n. The form of this fine valley approaches that of a triangle, 270 ms. base, from the mouth of White r. to the sources of Black r., 170 ms. perpendicular; area 22,950 sq. ms. Extending in lat. from 33° 56' to 37° 40', and in long. from 13° 20' to 17° 20' w. W. C. Rising in a mountainous region, the valley of White river exhibits every variety of soil, from the barren rock, and almost equally sterile prairie, to the rich, but annually submerged alluvion towards the Arkansas; have their respective points of discharge within 10 or 12 miles of each other, and are also connected by an interlocking, and in seasons of high water, navigable stream, many miles above their mouths. White river is navigable by both its great branches far above their junction. It is, however, an example of a stream greatly overrated, by estimating its length from the partial windings, in place of by the general comparative distances along the vallies. By that of White r. proper, the valley is 340, and by Black r. 28 miles long, pursuing the great curves of the rs.

WHITE, r. of Ind., the great estrn. branch of Wabash r. (*Which see.*)

WHITE MOUNTAINS, of N. H., Coos co. N. H., form a group of the loftiest summits in North America, N. and E. of the Rocky mtns. and Mexico, and are famous for the wildness and sublimity of their scenery. They are about 70 ms. N. Concord, 25 S. E. Lancaster, and extend 8 or 10 ms. in breadth, and 20 or 25 ms. from S. W. to N. E. The highest summits are covered with snow during the whole year, except about 2 months, and are visible from a great distance when the air is clear. They are often seen from sea, and sometimes before any intermediate land, although they are about 60 miles from the coast. The Indians, who called them Agiocochook, regarded them with superstitious reverence, and believed that no person could ever ascend them and return in safety. It is supposed that Neal, Jocelin and Field, who visited them in 1632, were the first white men who entered those wild and romantic regions. They called them the Crystal hills. It was long before any settlement was attempted, though Indians and white hunters often ranged thro' the forests in pursuit of the moose, fallow deer, bears, &c., which abounded. Until within about 40 years, moose were killed there, in great numbers, for their hides and tallow only. Bears, wild cats, fallow deer, &c. are still common; and within a few years, the cariboo, or rein deer, has made its appearance here, from the north. The soil is almost all incapable of improvement. There are a few small meadows, which have been cleared and subjected to culture; but the shortness of the summer, and the variations of the climate, render most crops very uncertain. Vegetation, as in the boreal zones, is very rapid and luxuriant in the spring; and the little patches here and there cultivated generally, present a cheering contrast to the surrounding wildness, during the months of July and August. Apple trees flourish well as high up as the elder Crawford's, 12 miles S. of the Notch, and grain sometimes ripens well. At the Notch House, however, 8 ms. N. of that place, the soil, though good, is not worth cultivating. The timber N. of the Notch is generally very different from that S. of it, in species; the former being such as is usually found in much colder climates.

There are seven principal eminences in the White mtns., collected in one majestic group, generally of a steep but uniform ascent, but most of them presenting, towards the N. E., a frightful precipice. They are formed of granite, and scattered with loose fragments of that stone to their peaks, as if they had once been of greater height. The 7 eminences are of the following heights. Mount Adams, 5,385 feet above the level of Connecticut, at Lancaster; Jefferson, 5,281; Madison, 5,039; Monroe, 4,932; Franklin, 4,470; Pleasant, 4,339. Proceeding from the remarkable pass called the Notch, in the mountains, the traveller may enjoy a fine display of natural scenery, by passing in succession over the summits, or along the sides of the 3 first eminences, and thence ascending Mount Washington. The following is the order in which the peaks rise, generally 1 mile apart; the first S. is Mount Pleasant, then Franklin, Monroe, Washington, Jefferson and Adams; Mount Madison being more E. Mt. Washington is the highest in the group; being 5,-850 feet above Conn. river at Lancaster, and 6,428 feet above the level of the sea.

The summit of Mount Washington being the highest land in the United States, and commanding an extensive and most sublime view, is annually sought by travellers, though the ascent is attended with much fatigue. It has been surmounted in a few instances, by ladies. The sides of the mountains, except where they are too precipitous, are overgrown with a thick forest of different trees, to which succeeds a belt of firs, which forms a well defined line round these eminences, at the same elevation. The size of these trees diminishes as the visitor advances, until they are very much stunted, and finally disappear. The surface is then covered with short shrubs, which gradually give place to mosses and lichens; and finally the loose rocks are quite bare. Numerous mtns., hills and vallies, are seen on all sides, with many lakes, and the vales, through which wind numerous streams of water. The works and the habitations of man are generally so far removed, as to form but an insignificant feature in this scene, which abounds with the magnificent features of nature. North are Mounts Adams and Jefferson, E. Mount Madison, S. and S. E. a plain of more than 40 acres extends on the side of the mountain, from which a ridge of eminences reaches along the course of Saco r., whose head spring is on the E. side of Mt. Washington. The highest source of Amonoosuc r. is at no great distance, but the streams approach each other much nearer, 3 or 4 miles below. Among the White mountains, rise also the Androscoggin and the Pemigewasset; so that within a small circuit, rise four considerable streams, 1 of which flows into Connecticut river, 2 into the Atlantic ocean, and 1 into the Merrimac. The Notch in the White mountains, is sometimes understood as applying to a narrow pass about 2 ms. long beginning at the Notch meadow, and following the course of the Saco S., down a rapid descent; and sometimes a remarkable narrow chasm in the rocks, apparently cut thro' by a powerful current of water. It barely affords room for the diminutive channel of the Saco, about 4 feet wide, and a road 12 feet. This is the only practicable passage through this mountainous region, nearer than Adams N., and Franconia S. W. The pass has been considerably improved by the artificial removal of rocks. In 1826, there was a sudden fall of rain at night, which produced tremendous effects, the evidence of which must remain for ages. Immense masses of earth and rocks, with acres of forests which covered them, were torn from the mountain sides, and heaped in confusion in the valleys, while the streams rose to rivers; ponds, and lakes were formed, and the banks of the Saco were strewn with heaps of timber as far down as Conway, while its channel was ploughed out in some places to a great depth by the force of the current. The Amonoosuc presented similar effects. A family of 11 persons inhabiting the Notch House, 4 ms. s. of the Notch, having fled in the night to seek safety out of doors, all lost their lives. The Flume is a little stream which flows from an eminence 250 or 300 feet high, and crosses the road about ¾ m. s. of the Notch; and the Silver Cascade, about 1½ ms. below, comes down a precipitous descent about 800 feet high. The turnpike road from Portland, Me. to Lancaster, N. H., passes through this long and romantic valley, through which the Saco flows, from the Notch to Conway. It has often been injured by the falling of rocks and the washing of torrents; but was long rendered useless by the devastating flood just mentioned.

WHITE, co. Ten., bounded by Smith w., Jackson N.W. and N., Fentress E., Cumberland mtn. separating it from Bledsoe S. E., and Caney Fork r. separating it from Warren S. and S. W. Length 42 ms., mean breadth 16, and area 672 sq. ms. Extending in lat. from 35° 40′ to 36° 17′, and in long. from 8° 10′ to 8° 50′ w. W. C. The slope of this co. is wstrd. and entirely drained by different branches of Caney Fork r. Chief t., Sparta. Pop. 8,701, in 1820, but in that number were included a part of what is now contained in Fentress co. In 1830, the pop. of White co. as now limited amounted to 9,967.

WHITE, co. Il., bounded by Gallatin S., Hamilton W., Wayne N. W., Edwards N., and Wabash r. separating it from Gibson co. Ind. N. E., and Posey co. Ind. E. Lat. 38° 08′, long. W. C. 11° 15′ w. Slope sthrd., and in that direction traversed by Little Wabash. Greatest length from the junction of Wabash and Little Wabash 28 ms., mean breadth 18, and area 504 sq. ms. Chief t., Carmi. As this co. st. was omitted under its proper head we insert a notice of it under that of the co to which it belongs. Carmi, is situated on a branch of Little Wabash, 29 ms. N. Shawanoetown on Ohio r., and by p-r. 94 ms. S. E. Vandalia. Lat. 38° 06′, long. W. C. 11° 18′ w. W. C. Pop. co. 1830, 6,091.

WHITE CHIMNEYS, p-v. Caroline co. V., by p-r. 30 ms. nrthrd. Richmond.

WHITE CREEK, p-t. Washington co. N. Y., 42 ms. from Albany, on a small branch of Walloomscoic cr., has a diversified surface, with Hoosac r. on the S. W. line, and Little White cr. and Walloomscoic cr. S. It touches Vt. E. where it is hilly. The v. contains an academy, and there are several mills and factories. Pop. 1830, 2,446.

WHITEFIELD, p-t. Lincoln co. Me., 16 miles from Augusta, 15 N. Wiscasset, has Sheepscot r. E., and small streams W. flowing S. W. into the Kennebec. It has Kennebec co. N. and W. Pop. 1830, 2,020.

WHITE DAY, p-o. estrn. part Monongalia co. Va., 10 ms. estrd. Morgantown, and by p-r. 205 ms. S. W. by W. W. C.

WHITE DEER mtn., one of the Appalachian chains, extending from the right bank of Susquehannah r. between Union and Lycoming cos. about 10 ms., and thence along the sthrn. side of the latter, and finally merging into other chains in Centre co.

WHITE DEER. There are two tsps. of that name, one the extreme sthestrn. of Lycoming, W. Susquehannah r., and another the nthestrn. of Union co., with White Deer mtn. intervening.

WHITE DEER, p-o. in White Deer tsp. Lycoming co. Pa., by p-r. 74 ms. a little W. of N. Harrisburg.

WHITE EYES PLAINS, and p-o. estrn. part Coshocton co. O., by p-r. 92 ms. N. E. by E. Columbus.

WHITEFIELD, p-t. Coos co. N. H., 120 miles from Concord, has a light soil, of pretty good quality, with spruce swamps N. It is crossed by John's r., down which the pine timber which here abounds, has been sent in great quantities to the Connecticut. It contains several ponds, and was settled about 1774. Pop. 1830, 684.

WHITEHALL, p-t. Washington co. N. Y., 71 ms. N. Albany, 21 N. Sandy Hill, has N. and S. bays and Vermont N., and lake George on part of the W. line. Poultney r. is on the line between this t. and Vt. The rocks N. W. are primitive, and in the E. transition; here are marble and limestone for burning. The soil is generally clayey. Wood cr. and Pawlet r. join in this t. and afford mill sites. Skeene's mtn. on the E. side of the lake, is high and rough. The former name was Skeenesboro'; and it was important as embracing the landing place at the mouth of Wood cr. at the S. end of lake Champlain, on the route between Canada and Hudson r. It was the scene of military movements in the French and revolutionary wars. In 1777, Gen. Burgoyne pursued the retreating American army this way, after dislodging it from Ticonderoga, but remained here so long to form a log road, &c. that the Americans found time to rally and prepare to oppose him, as they effectually did at Bemis' heights, in Stillwater. There is an extensive level tract along Wood creek. The v. is now an important place, as the Champlain canal commences there, and the steamboats which navigate the lake to St. John's, in Lower Canada, come up to the wharf. Pop. 1830, 2,889.

WHITEHALL, p-v. Whitehall, Washington co. N. Y., at the S. end of lake Champlain, at the mouth of Wood cr., and the N. end of the Champlain canal, is a place of considerable importance, and active business. The surrounding country presents rough eminences, with an extensive, open and marly tract along the course of Wood cr. whose bed serves for some miles as the route of the canal. There is a fall over the rocks at its mouths, where mills and factories are erected, and 3 locks are constructed which raise boats 31 ft. to the level of the canal. There is also an academy, church, bridge, hotels, &c., and the place is a great thoroughfare during the warm seasons.

WHITE HALL, p-v. Columbia co. Pa., by p-r. 81 ms. N. Harrisburg, and 12 in a similar direction from Danville, the co. st.

WHITE HALL, p-o. Frederick co. Va., by p-r. 79 ms. a little N. of W. W. C.

WHITE HALL, p-v. Mecklenburg co. N. C., by p-r. 160 ms. N. W. by W. Raleigh.

WHITE HALL, p-o. Abbeville dist. S.C., about 5 ms. sthrd. Abbeville, the st. jus. for the dist. and by p-r. 95 ms. from Columbia.

WHITE HALL, p-v. Marengo co. Ala., by p-r. 13 ms. sthrd. Linden, the co. st., and 91 ms. S. S. W. Tuscaloosa.

WHITE HALL, p-v. nrthwstrn. part Green co. Il., 10 ms. nrthrd. Carrollton, the co. st., and 116 ms. N. W. by W. Vandalia.

WHITE HAVEN, p-v. on Wicomico r., wstrn. part Somerset co. Md., 10 ms. N. N. W. Princes Ann, the co. st., and 106 ms. S. E. by E. W. C.

WHITE HORSE, tavern and p-o. estrn. part Somerset co. Pa., by p-r. 149 ms. nrthwstrd. W. C.

WHITE HOUSE, p-o. sthwstrn. part Mecklenburg co. Va., by p-r. 137 ms. S. W. Richmond.

WHITELEY, co. Ky. (*See Whitly.*)

WHITELEY, cr. and p-o. S.E. part Greene co. Pa., by p-r. 225 ms. S. W. by W. W. C.

WHITELEY, C. H. and p-o. Whitely or Whitly co. Ky., by p-r. 130 ms. S. S. E. Frankfort.

WHITELEYSBURG, p-v. wstrn. part Kent co. Del., about 20 ms. S. W Dover.

WHITE MARSH, p-v. Montgomery co. Pa., 12 ms. nrthrd. Phila. It has been long famed for the beauty and abundance of its variegated marbles.

WHITE OAK, p-v. wstrn. part Rutherford co. N.C., 10 ms. wstrd. from Rutherfordton, the co. st., and by p-r. 233 ms. a little S. of W. Raleigh.

WHITE OAK, p-v. estrn. part Humphries co. Ten., 63 ms. W. Nashville.

WHITE OAK, p-o. Columbia co. Geo., by p-r. 88 ms. N. E. by E. Milledgeville.

WHITE OAK GROVE, and p-o. Bedford co. Va., by p-r. 227 ms. S. W. W. C.

WHITE PIGEON PRAIRIE, p-v. and st. jus. St. Joseph co. Mich., about 150 ms. a little S. of W. Detroit. Position in the co. uncertain.

WHITE PLAINS, p-t. and half cap. Westchester co. N. Y., 30 ms. from N. York, 140 S. Albany, 6 E. Hudson r., and 14 S. Bedford, with 8½ sq. ms., has Bronx cr. on the W. line, Mamaroneck cr. E., and mill sites on both. The soil is good, and the surface varied by a range of hills, running N. and S. some of which were rendered interesting in the revolution by the battle fought here, Oct. 28th, 1776. Pop. 1830, 759.

WHITE PLAINS, v. White Plains, Westchester co. N. Y., 28 ms. N. New York, 7 from Sawpits, 7 E. Hudson r., ¾ E. Bronx cr.

WHITE PLAINS, p-v. Brunswick co. Va., by p-r. 94 ms. S. S. W. Richmond.

WHITE PLAINS, p-v. Jackson co. Ten., by p-r. 110 ms. N. E. by E. Nashville.

WHITE PLAINS, and p-o. Greene co. Geo., by p-r. 48 ms. nrthrd. Milledgeville.

WHITE PLAINS, and p-o. Lawrence co. Miss., by p-r. about 75 ms. E. Natchez.

WHITE POST, p-v. in Frederick co. Va., 8 ms. S. E. Winchester, and by p-r. 71 ms. a little N. of W. W. C.

WHITE RIVER, p-v. estrn. part Arkansas co. Ark., by p-r. 136 ms. below and S. E. by E. Little Rock.

WHITE'S, p-o. Elbert co. Geo., by p-r. 108 ms. a little E. of N. Milledgeville.

WHITE SAND, p-v. Lawrence co. Miss., by p-r. 100 ms. S. Jackson.

WHITESBOROUGH, p-v. half cap. Whitestown, Oneida co. N. Y., 4 ms. N. W. Utica, 100 W. N. W. Albany, is pleasant and well built, on Erie canal. It has an academy.

WHITESBURG, p-v. wstrn. part Madison co. Ala., 10 ms. wstrd. Huntsville, and by p-r. 115 ms. N. N. E. Tuscaloosa.

WHITE'S MILLS, and p-o. nrthwstrn. part Chester dist. S. C., by p-r. 70 ms. N.N. W. Columbia.

WHITE'S STORE, and p-o. nrthrn. part York dist. S. C., by p-r. 85 ms. N. N. W. Columbia.

WHITESTOWN, p-t. Oneida co. N. Y., 98 ms. W. N. W. Albany, 5 W. Utica, 16 S. E. Rome, has Mohawk r. N. E., and Herkimer co. E., with very good soil, swelling surface, and fertile meadows along Mohawk r., Oriskany and Sadaquada crs. The Erie canal passes along the Mohawk N E. The town contains the vs. of Whitestown and Oriskany, and its settlement was commenced in 1784, by Canvass White, the first settler in the W. part of the state of N. Y. Pop. 1830, 4,410. On Lanaquoit, a small stream uniting with the Mohawk in this town, are found the principal manufactories of Oneida co. The Oriskany woollen factory is on a cr. of the same name, 4 ms. from Whitestown—spindles, 1,510—looms, 40—hands, 136—using annually 120,000 lbs. wool, and manufacturing goods to the amount of $155,000. On the Lanaquoit, the Oneida cotton factory, running 2,500 spindles and 84 looms, work 300 bales cotton and make 780,000 yds. annually;—at the York mills, two stone edifices, 150 and 130 ft. long, are run 8,328 spindles and 260 looms, employing 350 hands, and making 900,000 yards sheeting annually;—at the Whitestown cotton factory are run 2,900 spindles, 70 looms, 85 hands are employed and 300 bales cotton per annum used;—at the Utica cotton factory 2,600 spindles and 60 looms are run, (thirty of the latter on ticking,) 112 hands are employed and 450 bales cotton consumed; a woollen factory makes 3,500 yards broadcloth, and there is a callico printing establishment;—the New Hartford manuf. co. run 2,500 spindles, 64 looms, employ 80 hands, and work ann. 300 bales cotton;—the Eagle cotton factory run 1,600 spindles, 40 looms, employ 75 hands, and use 200 bales;—the Franklin com. run 3,000 spindles, 76 looms, employ 120 hands, and use 300 bales cotton;—and the Paris cotton man. com. run 1,500 spindles, 60 looms and employ 70 hands. These with several machine shops are all on the Lanaquoit, and within 8 ms. of each other.

WHITESTOWN, p-v. Butler co. Pa., 10 ms. from the borough of Butler, and about 60 ms. N. Pittsburg.

WHITESVILLE, p-v. and st. jus. Columbus co. N. C., situated near the extreme head of Waccamaw r., about 40 ms. N. E. Conwaysborough, and by p-r. 138 ms. S. Raleigh. Lat. 34° 13', long. 1° 48' w. W. C.

WHITE SULPHUR Springs, watering place and p-o. estrn. part Greenbrier co. Va., 9 ms. S. E. by E. Lewisburg, the co. st., and by p-r. 254 ms. s. w. W. C., and 212 w. Richmond.

WHITING, p-t. Washington co. Me., contains several ponds, some of which flow w. into Machias bay, and others E. into Cobscook bay. The town is separated from the sea coast by the town of Cutler. Pop. 1830, 309.

WHITING, p-t. Addison co. Vt., 40 ms. s. w. Montpelier, 42 s. Burlington, 70 N. Bennington, has Otter cr. E., and was first settled 1772, but deserted in the revolutionary war. In 1819, pickerel were placed in Otter cr. above Middlebury falls, which rapidly increased, so that 500 lbs. of this kind of fish, (before unknown in this stream,) were caught in this town in 1823. The soil is good both for grass and grain. Pop. 1830, 653.

WHITINGHAM, p-t. Windham co. Vt., 18 ms. S. E. Bennington, 20 N. w. Greenfield, has the Mass. line on the s. boundary, and was first settled 1770. Deerfield r. flows through the w. part, with rich meadows on its course. Sawdawda pond has land gradually forming on the surface, and 70 or 80 acres now rise and fall with the water. The soil is good. Good limestone abounds w. which is burnt. Pop. 1830, 1,477.

WHITLEY, co. Ky., bounded w. by Wayne co. of the same state, N. w. by Rock Castle cr. separating it from Pulaski, N. by Laurel, E. by Knox, and s. by Campbell co. in Ten. Length from N. to S. 30 ms., mean breadth 20, area 600 sq. ms. Extending in lat. from 36° 35' to 37° 01', and in long. from 6° 48' to 7° 14' w. W. C. The main volume of Cumberland r. winds, by a very circuitous channel over Whitley from S. E. to N. w. Chief t., Whitley C. H. Pop. 1830, 3,806.

WHITTLES', Mills and p-o. Mecklenburg co. Va., by p-r. 105 ms. s. w. Richmond.

WICKFORD, p-v. North Kingston, Washington co. R. I., 22 ms. s. w. Providence, 15 N. w. Newport, is on a peninsula on w. side of Narraganset bay. Here is Washington academy, 30 ft. by 60, with a library; and there is considerable commerce.

WICOMICO, small r. which rises in the south-western angle of Sussex co. Del., flows thence sthwstrdly. between Manakin and Nantikoke rs., and in Somerset co. Md., falling into Fishing bay, 10 ms. w. Princess Anne.

WICOMICO, or Great Wicomico, small r. of Northumberland co. Va., falling into Chesapeake bay 12 ms. S. E. Bridgetown, the county seat.

WICOMICO, church and p-o. Northumberland co. Va., about 6 ms. wstrd. Smith's point and 9 ms. estrd. Bridgetown, by p-r. 160 ms. S. S. E. W. C.

WIESESBURG, p-v. Baltimore co. Md., 24 ms. N. w. Baltimore, and 8 ms. a little w. of N. Reisterstown.

WILBRAHAM, p-t. Hampden co. Mass., 89 ms. w. Boston, has the Conn. line for its s. boundary, and Chickapee r. for its N. The surface is agreeably varied, and the soil is generally very good. It contains a Wesleyan academy, on the manual labor system, for the education of indigent children of the Methodists. The Chickapee is a very rapid stream, and makes considerable descent at the rapids in this place. Chickapee was a word used by Indian mothers to crying children, meaning "be still." The v. has a pleasant situation near the E. borders of the elevated plain which extends w. to the armory at Springfield. It has a hilly and picturesque country E. and N. of it. Pop. 1830, 2,034.

WILCOX, co. Ala., bounded s. by Monroe, s. w. by Clarke, N. w. by Marengo, N. and N. E. by Dallas, E. by Montgomery and S. E. by Butler. Length from E. to w. 60 ms., mean breadth 20, and area 1,200 sq. ms. Extending in lat. from 31° 49' to 32° 15', and in long. 9° 56' to 10° 56' w. W. C. Declivity s. w., and traversed in that direction by Alabama r. Chief t., Canton. Pop. 1820, 2,917; 1830, 9,548.

WILDERNESS, p-o. on the nrthwstrn. border of Spottsylvania co. Va. 15 ms. w. Fredericksburg, and 71 S. S. w. W. C.

WILSONG'S, mill and p-o. Lincoln co. N. C., by p-r. 182 ms. s. w. by w. Raleigh.

WILKES-BARRE, boro', p-t. and st. jus. Luzerne co. Pa., stands on a high bank right side of Susquehannah r., about 120 ms. N. N. w. Phila., and by p-r. 222 N. N. E. W. C. and 114 N. E. Harrisburg. Lat. 41° 13', long 1° 07' E. W. C. Wilkes-Barre was laid out about the year 1775, by Col. John Durkee, who imposed the compound name as a grateful tribute to two eminent members of the British parliament, for their exertions in favor of the North American colonies. The plan is perhaps entirely singular. The streets form a parallelogram, extending along or at right angle to the r. In the centre is a public square containing the co. buildings, but this square stands at an angle of 45° to the streets, form of the latter extending from each corner of the former. The wstrn. angle of the square is opposite a bridge over the Susquehannah, with a portion of the main street intervening. The bridge connects

Wilkes-Barre with the village of Kingston. Pop. of Wilkes-Barre in 1830, 2,233.

WILKES-BARRE, valley of, usually called the valley of Wyoming, is amongst the natural scenes in the U. S. that richly deserve a visit. The Susquehannah r. may be said to rush into, and break through the Appalachian system of mountains. Passing the first great chain at Towanda, the large volume of water in its rocky bed rolls through several other chains in quick succession, at length reaches Wyoming valley at the mouth of Lackawannoc r. by a very striking mtn. gorge. Inflecting at right angles, and turning from S. E. to S. W. the stream with very gentle partial windings flows down the Wyoming valley 9 ms., passes Wilkes-Barre and Kingston, and 6 ms. farther leaves the valley by another mountain pass. The bed of Susquehannah merely touches the wstrn. verge of this fine vale, which is indeed extended up the Lackawannoc, and to the sthwstrd. some miles below where it is abandoned by the r. The valley is distinct therefore 25 ms. above and 7 or 8 below the borough of Wilkes-Barre, exceeding 30 ms. in length, but with a width that does not at the utmost exceed a mean of 2½ ms. Enclosed between mtns. every where steep and rugged, in many places precipitous and in some rising into naked summits, spread alluvial flats of exuberant fertility. Here as along the Susquehannah generally, there are two stages of bottoms. The lower, and of course most recent, are much the most productive, and least admixed with rounded pebbles, but are still subject to casual submersion. The higher stages, on one of which stands Wilkes-Barre, are in the existing order of things above all floods, but both have been evidently once actually under water. This conclusion is almost irresistible to any observer in the vicinity of Wilkes-Barre. In brief, it may be asserted, that many of our citizens who admire natural scenery, know the wealth of the Alps in objects of taste infinitely better than they do regions at their door. The Wyoming is only one of innumerable pictures, along the Appalachian system, where are combined every feature from the most stern to the most soft and seducing. Again in the vicinity of Wilkes-Barre and Kingston the mineral curiosities are not the least attractive. The formation is transition or leaning; the inclination S. E. Embeded in strata from one to twenty or more feet in thickness lie masses of anthracite coal, which appear more and more vast as they are better explored.

WILKES, co. N. C., bounded N. E. and E. by Surry, S. E. by Iredell, S. W. by Burke, and W. N. W. and N. by Blue Ridge separating it from Ashe. Length from S. W. to N. E. 48 ms., mean breadth 18, and area 864 sq. ms. Extending in lat. from 35° 56′ to 36° 24′, and in long. from 3° 51′ to 4° 35′ W. W. C. This co. is a real mtn. valley, environed on every side but the N. E. by the Blue Ridge and adjacent chains. It is commensurate with the extreme higher valley of Yadkin r. by the confluents of which it is entirely drained. Declivity nrthestrd. Chief town, Wilkesville. Pop. 1820, 9,967; 1830, 11,968.

WILKESBORO', in the p-o. list, but Wilkesville on Tanner's U. S. map, p-v. and st. jus. Wilkes co. N. C., situated on the right bank of Yadkin r. 51 ms. N. E. Morgantown in Burke co., and by p-r. 175 ms. a little N. of W. Raleigh. Lat. 36° 10′, long. 4° 08′ W. W. C.

WILKINSON, co. Geo., bounded S. E. by Laurens, S. W. by Twiggs, N. W. by Jones, N. by Baldwin, and E. by Oconee r. separating it from Washington. Length from N. W. to S. E. 24 ms., mean breadth 18, and area 432 sq. ms. Extending in lat. from 32° 37′ to 33° 02′, and in long. from 6° 02′ to 6° 30′ W. W. C. Declivity sthestrd. towards the Oconee. Chief t., Irwington. Pop. 1820, 6,992; 1830, 14,237.

WILKINSON, co. Miss., bounded by the Homochitto r. separating it from Adams co. on the N., and Franklin N. E., by Amite co. E., by the parish of East Feliciana in La. S. E., by West Feliciana in Louisiana S., and by the Miss. r. separating it from the parish of Avoyelles in La. W., and the parish of Concordia La. N. W. Greatest length from E. to W. 30 ms., mean breadth 20, and area 600 sq. ms. Extending in lat. from 31° to 31° 14′, and in long. from 14° 12′ to 14° 46′ W. W. C. The general declivity of this co. is wstrd., but the extreme sthrn. border declines in a sthrn. direction giving source to Thompson's cr. and Bayou Sarah. Buffaloe cr. rises on the estrn. side and flowing wstrd. divides the co. into two nearly equal sections. The surface is very much broken by hills, however, of no great elevation. Soil excellent. Principal staple, cotton. Chief t., Woodville. Pop. 1820, 9,718; 1830, 11,686.

WILLET, p-t. Cortlandt co. N. Y., 139 ms. from Albany, 19 S. E. Cortlandt v., has Chenango co. E., Broome co. S., is crossed by Otselic cr. with few mill seats. Pop. 1830, 840.

WILLIAMS' r. Windham co. Vt., flows S. E. 15 ms. into Connecticut in Rockingham. At its mouth the Rev. Mr. Williams, in 1704, on his way from Deerfield, Mass., to Canada, while a captive in the power of the Indians, preached to his fellow prisoners.

WILLIAMSBOROUGH, p-v. northeastern part Granville co. N. C., 12 ms. N. E. by E. Oxford, the co. st., and 59 N. E. Raleigh.

WILLIAMSBURG, p-t. Penobscot co. Me. 94 ms. from Augusta, 40 N. W. Bangor, has several small streams flowing into Pleasant and Lubec rs., branches of the Penobscot, and has a quarry of marble. Pop. 1830, 227.

WILLIAMSBURG, v. Bushwick, King's co. N. Y. on the E. end of Long Island, opposite N. Y., has a steam ferry to that city.

WILLIAMSBURG, p-v. situated on Franktown branch of Juniata r. 9 ms. W. from the borough of Huntingdon co. Pa.

WILLIAMSBURG, p-t. and st. jus. James City co. Va. situated on the summit level between York and James rs., by p-r. 60 ms. S. E. by E.

Richmond. Lat. 37° 16′, and long. 0° 20′ E. W. C. The p-r. distance from W. C. 163 ms. This little city, though it has not advanced much in wealth or population, has many very interesting claims on the student of U. S. geography. It was the cradle of our political existence, and for a long period, the seat of government of "Infant Virginia." The college of William and Mary, in Williamsburg, was founded in 1693, and with various fortunes of advance and recession, has continued to exist as a respectable literary institution. By the original charter this college was endowed with a clear and certain revenue of £3000 per annum. Recent attempts have been made to revive the former prosperous condition of this seminary.

WILLIAMSBURG, dist. S. C., bounded N. W. by Sumpter, N. E. by Lynches cr. separating it from Marion, E. and S. E. by Georgetown dist., and S. W. by Santee r., separating it from Charleston dist. Length between Santee r. and Lynches creek 40 ms., mean breadth 30, and area 1,200 sq. ms. Extending in lat. from 33° 15′ to 34° 02′, and in long. from 2° 24′ to 3° 12′. The declivity is southeastward, in the direction of Lynches creek, Santee r. and Black r. The latter stream rising in Sumpter, traverses Williamsburg at a mean distance of 16 or 17 ms. from Santee r. Chief t. Kingtree. Pop. 1820, 8,716, 1830, 9,018.

WILLIAMSBURG, p-v. and st. jus. Covington co. Miss. situated on a branch of Leaf river, about 120 ms. E. Natches, and by p-r. 83 ms. S. E. Jackson. Lat. 31° 40′, long. 12° 38′ W. W. C.

WILLIAMSBURG, p-v. Mason co. Ky. by p-r. 75 ms. N. E. Frankfort.

WILLIAMSBURGH, p-t. Hampshire co. Mass. 100 ms. W. Boston, 9 N. W. Northampton, is crossed N. W. and S. E. by a tributary of the Connecticut, which affords mill sites. Pop. 1830, 1,236.

WILLIAMSBURGH, p-v. Groveland, Livingston co. N. Y. 3½ ms. from Geneseo v. on Genesee r.

WILLIAMSON, p-t. Wayne co. N. Y. 206 ms. from Albany, 20 N. by E. Canandaigua, crossed by the Ridge road, has a descent N., whither flow some small streams to lake Ontario. Pop. 1830, 1,806.

WILLIAMSTON, p-v. and st. jus. Martin co. N. C. situated on the right bank of Roanoke r. 23 ms. N. Washington on Pamlico r., and by p-r. 106 ms. E. Raleigh. Lat. 35° 49′, and long. 0° 06′ W. W. C.

WILLIAMSTOWN, p-t. Orange co. Vt. 11 ms. S. W. Montpelier, 45 N. W. Windsor; first settled, 1784, occupies the high ground between Onion and White rs., to each of which it sends a brook. Along the courses of these small branches passes the Gulf road from Royalton to Montpelier, through remarkably wild scenes, and often a very narrow passage. In some places the road is supported by artificial walls of stone, where there is barely room enough for it and the stream. Pop. 1830, 1,487.

WILLIAMSTOWN, p-t. Berkshire co. Mass. 130 ms. N. by W. Boston, is in the N. W. corner of the state, with Vermont N. and N. Y. W., and occupies a fine valley, through which flows Hoosic r. into Vt. The land bordering on the valley is mountainous, and some of the peaks are very high. The town was named after colonel Ephraim Williams, who was commander of the line of forts W. of Conn. r. during the old French war, from 1740 till 1748, and resided for some time at Hoosic fort. He was killed at the battle of fort George, 1755, and left by will, his property for the foundation of a grammar school in Williamstown. Williams college, in this town, was commenced in 1791, and became a college in 1793. Two townships of land in Maine were afterwards granted it by Mass. one of which was sold for $10,000. There are two buildings of brick, 100 feet long, 40 wide, and 4 stories high, containing a chapel, library, philosophical chamber, and 60 students' rooms. The necessary expenses are from $60 to $105 per annum; 18 young men received the degree of A. B. in 1832. An academy was incorporated here in 1828. Pop. 1830, 2,134.

WILLIAMSTOWN, p-t. Oswego co. N. Y. 137 ms. from Albany, 31 E. Oswego, has Oneida co. E., has a soil good for grass, and pretty good for grain, watered by streams of Fish creek, which furnish mill sites. The surface is nearly level, with much moist land. Pop. 1830, 606.

WILLIAMSTOWN, p-v. northestrn. part Lancaster co. Pa. 13 ms. N. E. from the city of Lancaster.

WILLIAMSTOWN, on the p-o. list, but Williamsville on Tanner's U. S., p-v. and st. jus. Grant co. Ky., situated on the right bank of Eagle creek, about 40 ms S. W. Cincinnati, and 44 a little E. of N. Frankfort. Lat. 38° 41′, long. 7° 42′ W. W. C.

WILLIAMSVILLE, p-v. Erie co. N. Y. 11 ms. N. E. Buffalo, is on Ellicott's creek at the falls.

WILLIAMSVILLE, p-v. Kent co. Del. 25 ms. S. Dover.

WILLIAMSVILLE, p-v. nthrn. part Person co. N. C., by p-r. 68 ms. N. W. by N. Raleigh.

WILLIMANTIC, r. Tolland co. Connecticut, a stream of small size, flows through Stafford, and falls into Natchaug r. and forms the Shetucket.

WILLIMANTIC, p-v. Windham co. Conn. 3 ms. from Windham v., 24 E. Hartford, on Willimantic r.; contains 5 large manufactories, 3 churches, several schools, &c.

WILLINBOROUGH, t. Burlington co. N. J. 14 ms. N. E. Phil., has the Del. r. N. W., Rancocus creek S. W., and Burlington N. A small branch of Rancocus cr. crosses the t. Pop. 1830, 782.

WILLINGTON, p-t. Tolland co. Conn. 26 ms. N. E. Hartford, 26 N. Norwich; about 4 ms. by 8, is hilly, with primitive rocks, and contains some iron ore. It has Willimantic r. E., and some manufactories. Population 1830, 1,305.

WILLINGTON, p-v. sthrn. part Abbeville dist. S. C. 17 ms. s. s. w. Abbeville, and by p-r. 97 ms. a little N. of w. Columbia.

WILLISTON, p-t. Chittenden co. Vt. 27 ms. N. w. Montpelier, has Onion r. N. and Muddy brook w., and was first settled 1774, but soon deserted until the close of the war. The surface is uneven, but generally favorable to agriculture, with a good soil. Pop. 1830, 1,606.

WILLOUGBY, p-v. and st. jus. Effingham co. Geo., by p-r. 181 ms. s. E. by E. Milledgeville.

WILLOW GROVE, p-v. Montgomery co. Pa. 14 ms. N. Phil.

WILLOW GROVE, p-o. Lincoln co. N. C. by p-r. 164 ms. s. w. by w. Raleigh.

WILLOW GROVE, p-o. nthestrn. part Sumpter dist. S. C., about 20 ms. N. E. Sumpterville, and by p-r. 64 ms. a little N. of E. Columbia.

WILLSBOROUGH, p-t. Essex co. N. Y. 157 ms. N. Albany, 13 N. E. Elizabethtown, has lake Champlain or Vt. E., towards which the land is level. Peru bay extends s. into this town from the lake 5 or 6 ms.; and there are in it Bouquet's and Gilliland's creeks, on whose falls are very good mill sites. Pop. 1830, 1,316.

WILLSTOWN, Indian village, on a creek of the same name, a branch of Coosa r. This place, as located by Tanner, is situated in the Cherokee territory 50 ms. s. E. by E. Huntsville, and about 150 ms. N. E. Tuscaloosa. Lat. 34° 26', long. 8° 53' w. W. C.

WILMINGTON, p-t. Windham co. Vt. 17 ms. E. Bennington, 46 s. w. Windsor, 138 from Montpelier; was first settled just before the revolution. The E. and w. branches of Greenfield r. unite here. Pop. 1830, 1,034.

WILMINGTON, t. Middlesex co. Mass. 16 ms. N. Boston, has a light soil, but has produced great quantities of hops. The Middlesex canal crosses the town through the middle; and Ipswich r. rises here. Pop. 1830, 731.

WILMINGTON, p-t. Essex co. N. Y. 17 miles N. w. Elizabethtown, has Franklin and Clinton cos. E., and Franklin co. w., and contains Sable and White Face mtns., and parts of Palmer and Hamlin mtns.; White Face mtn. is about 2,600 feet high, and commands a view of Montreal, 80 ms. distant. The town is crossed by the w. branch of Sable r., &c. which furnish mill seats. Pop. 1830, 695.

WILMINGTON, city, p-o. and port of entry, New Castle co. Del. situated on the point above the junction of Brandywine and Christiana creeks, 28 ms. s. w. Phil., 47 N. Dover, and by p-r. 108 ms. N. E. W. C. The site is similar to those of Phil., Baltimore, Georgetown and Richmond, on the outer edge of the primitive rock, and on the inner of the sea sand alluvion. The site of Wilmington is less variegated than either of the above named cities. In the vicinity of Wilmington the falls of Brandywine afford a water power which has been rendered available to a great extent by the erection of machinery applied to grist mills, saw mills, powder and paper mills, cloth factories, both of cotton and woollen, and other manufactories of different kinds. The town is incorporated and governed by two burgesses and six assistants, annually elected. The trade of the place is extensive; the buildings generally good, and many elegant. It contains an hospital and poor house. The hospital is a large edifice built on a healthy eminence. Wilmington is much the largest town of the state in which it is situated, and after the city of Phil. the most extensive mart in the basin of Del. r. In 1820, the pop. amounted to 5,268, and in 1830, 6,628.

WILMINGTON, p-v. situated on Fluvanna r. Fluvanna co. Va. 14 ms. above the mouth of the stream on which it stands, 55 ms. N. w. by w. Richmond.

WILMINGTON, p-t. seaport and st. jus. New Hanover co. N. C., situated on the left bank of Cape Fear r. 88 ms. below and along the land route below Fayetteville, and by p-r. 149 ms. s. s. E. Raleigh. Lat. 34° 20', and long. almost on the meridian of W. C. Pop. 1830, 3,000.

WILMINGTON, p-v. and st. jus. MacCracken co. Ky. situated 25 ms. a little N. of E. from the mouth of Ohio r., and by p-r. 289 ms. s. w. by w. Frankfort. Lat. 37° 02', and long. 11° 52' w. W. C.

WILMOT, t. Merrimack co. N. H. 30 miles from Concord, 87 from Boston. Some of the streams of Blackwater r. afford mill seats. The surface is rough, with Kearseargo mtn. on the s. boundary. Pop. 1830, 835.

WILNA, p-t. Jefferson co. N. Y. 151 ms. N. w. Albany, 57 s. s. w. Ogdensburgh, has a light soil, well watered by Indian r. a few miles in the N. E., and by Black r. for a short distance s. w. including the rapids or Long falls. Above this spot the river is navigable 45 ms. Here is the village of Carthage, where are several iron works, ore being found in the town. Pop. 1830, 1,602.

WILSON, p-t. Niagara co. N. Y. 294 miles w. Albany, is nearly square, with lake Ontario N. and watered by Howel's and Tuscarora creek, the former joining the latter in this town, and flowing into the lake. Pop. 1830, 913.

WILSON, co. Ten. bounded by Smith E., Warren s. E., Rutherford s., Davidson w., and Cumberland r. N., separating it from Sumner. Length 38 ms., mean breadth 14, and area 432 sq. ms. Extending in lat. from 35° 58' to 36° 20', and in long. from 9° to 9° 44' w. W. C. Declivity a little w. of N. towards Cumberland r. Chief t. Lebanon. Population 1820, 18,730, 1830, 25,472.

WILSON'S, p-v. Anderson co. Ten., by p-r. 201 ms. E. Nashville.

WILSON'S creek and p-o. Graves co. Ky. 14 ms. southwestwardly from Mayfield.

WILSONVILLE, p-v. southwestrn. part Bath co. Va., by p-r. 178 ms. N. w. by w. Richmond, and 220 ms. s. w. by w. W. C.

WILSONVILLE, p-v. Lincoln co. N. C. by p-r. 182 ms. s. w. by w. Raleigh.

WILTON, p-t. Hillsborough co. N. H. 37 ms. from Concord, 9 from Amherst, and 58 from Boston, is crossed by Souhegan river, whose branches afford mill sites. The soil is rocky, but fertile, bearing oak, pine, beech, birch, hemlock and chestnut. Clay and building stone are found in different places. First settled, 1738. Pop. 1830, 1,039.

WILTON, p-t. Fairfield co. Conn., 34 ms. s. w. New Haven, 6 N. Norwalk, has the boundary of N. York on the w. line, it is 4 ms. by 6, is crossed by 2 ridges N. and s., with soil favorable to grain, and has 2 small streams from Norwalk r. near the centre. Pop. 1830, 2,095.

WILTON, p-t. Saratoga co. N. Y., 42 ms. N. Albany, lies N. of Saratoga, and near the v. of Saratoga Springs. There is an extensive sandy plain in the t., and the streams of a small tributary of the Hudson. Pop. 1830, 1,373.

WILTON, p-v. in the sthrn. part of Granville co. N. C., 14 ms. sthrd. Oxford.

WINCHENDON, p-t. Worcester co. Mass., 60 ms. w. N. w. Boston, 30 N. by w. Worcester, has an uneven surface, a rough and stony, but strong soil, and affords some building granite. There is a chalybeate spring in the N. part. Miller's r. affords valuable mill seats. There are 2 villages. Pop. 1830, 1,463.

WINCHESTER, p-t. Cheshire co. N. H., 70 ms. from Concord, 15 from Keene, is crossed by Ashuelot r., which receives Muddy, Broad, and several other brooks. The s. E. part is very level, elsewhere uneven, with a good soil, bearing pine, chestnut, oak, hard maple, &c. Great quantities of shingles, staves, &c. have been made here. There are 2 villages on Ashuelot r., that in the w. contains several manufactories. First settled 1732; it was burnt in the French war, by the Indians. Pop. 1830, 2,052.

WINCHESTER, p-t. Litchfield co. Conn., 27 ms. N. w. Hartford, about 5 ms. by 6½, mountainous in some parts, with primitive rocks; it has a soil favorable to grazing, bearing maple, beech, oak and birch; it is supplied with mill sites by Mad and Still rs., which unite in the E. part. A lake ¾ m. by 3½, is situated on the top of one of the mtns., and sends an outlet about ¼ m. down a steep descent into Mad r. The scenery in this t. is in many places quite wild and romantic. There are many factories, forges, mills, &c. Iron from Salisbury is smelted and wrought here. Pop. 1830, 1,766.

WINCHESTER, a flourishing p-t. and st. jus., Frederick co. Va., situated on a branch of Opequan cr., 34 ms. s. w. Harper's Ferry, 71 ms. N. w. by w. W. C., and 150 ms. N. N. w. Richmond. Lat. 39° 10′, and long. 1° 10′ w. W. C. It is a very flourishing inland town, and contains many public buildings, some of them very fine. In 1826, Winchester contained a white population of 2,575, free colored 270, slaves 644, total 3,489; 23 attorneys, 8 physicians, 35 mercantile stores, 3 iron stores, 2 book stores, 2 printing offices, 12 taverns, 4 tanneries, 1 distillery, 1 pottery, 1 bookbindery, 3 silver smiths, watch repairer's, and jeweller's shops, 1 clock and mathematical instrument maker, 1 rope maker, 1 tinner, 1 confectioner, 1 tobacconist, 3 brick makers, 1 saddle tree maker, 1 upholsterer, 3 shoe stores, 1 wheel maker, 2 gun smiths, 1 white smith, 2 cabinet makers' shops, with numerous weavers, saddlers, shoemakers, &c.; 7 houses of public worship completed, and one more in progress of erection, for Episcopalians, Presbyterians, Lutherans, Methodists, Baptists, Roman Catholics, and Quakers. A law school of eminence, under chancellor Tucker; an incorporated medical school, and an incorporated academy; the latter flourishing, and averaging 50 pupils. Two female seminaries, with an average of 30 or 40 pupils each, with numerous private schools, and two banks. Pop. 1830, not in the census.

WINCHESTER, p-v. and st. jus., Wayne co. Miss., situated on the Chickasawhay r., about 90 ms. N. N. w. Mobile, 180 E. Natchez, and by p-r. s. E. by E. Jackson. Lat. 31° 40′, long. 11° 48′ w. W. C.

WINCHESTER, p-v. and st. jus., Franklin co. Ten., situated on or near the left bank of Elk r., 50 ms. N. E. Huntsville in Ala., and by p-r. 82 ms. s. E. Nashville. Lat. 35° 14′, long. 9° 02′ w. W. C.

WINCHESTER, p-v. and st. jus., Clark co. Ky., 45 ms. N. E. by E. Frankfort, and 20 in a similar direction from Lexington. Lat. 37° 58′, and long. 7° 07′ w. W. C.

WIND GAP, p-o. Northampton co. Pa., by p-r. 19 ms. a little E. of N. Bethlehem. This place is designated Williamsburg on Tanner's map. The Gap is one of the passes in the Kittatinny mtns.

WINDHAM, p-t. Cumberland co. Me., 67 ms. s. w. Augusta, 16 N. w. Portland, borders on the s. E. side of Sebago pond, and has its outlet. Presumscot r. on its w. line. It has 2 or 3 ponds on its boundaries, and is crossed by a branch of the stream above mentioned. Pop. 1830, 2,182.

WINDHAM, p-t. Rockingham co. N. H., 34 ms. from Concord, and 45 from Portsmouth, contains part of Policy pond, with Golden pond s., and Mitchell's N. E. On Beaver cr., the w. boundary, is excellent land; and the t. is generally well watered. Pop. 1830, 998.

WINDHAM, co. Vt., bounded by Windsor co. N., Conn. r. E., which separates it from N. H., Mass. s., Bennington co. w., is 28 ms. by 36, with 780 sq. ms. It has an irregular surface. Williams's and Saxton's rs. are in the N. E. part, West r. in the middle, and Deerfield r. s. w. There are mountains w. The rocks are all primitive. Manicnung is the highest eminence. Limestone is found and quarried in several towns. On the Conn. are some rich meadows. Pop. 1820, 28,457; 1830, 28,748.

WINDHAM, p-t. Windham co. Vt., 31 ms. N.

E. Bennington, 25 S. W. Windsor, has small streams, a large pond, and various minerals. Pop. 1830, 847.

WINDHAM, co. Conn., bounded by Worcester co. Mass. N., Rhode Island E., New London co. S. and S. W., and Tolland co. W. It is about 21 ms. by 29, with about 620 sq. ms., and contains 13 towns. The surface is varied, hilly W. and S. with much stony land, good for grazing, generally a gravelly soil formed from primitive rocks. It is crossed by Quinebaug r. E., and Shetucket r. W., with several of their branches, which run southerly, and unite in N. London co. to form the Thames. These streams afford many good mill seats, and supply shad and other fish. On the alluvial meadows on their shores, is much excellent land, favorable to grain, &c. Surveys have been made for a canal, to extend from tide water at Norwich, N. London co., into Mass., across this co. Extensive manufactories have been erected, and Windham co. is now far more extensively engaged in manufactures, than any other co in the state. By a recent estimate, it was computed that there were in the co. 47 cotton factories, running 62,550 spindles, and 1,462 looms, manufacturing per ann. 37,500 lbs. of yarn, (sold in that state,) and 11,000,000 yards of cotton goods, consuming 1,537,500 lbs. of cotton, and employing a capital of $1,537,500. At the same time other factories were erecting, which were to run more than 12,500 spindles. The woollen factories, by the same estimate, manufactured goods valued at $133,600; the sum invested in them stated at $127,550. A carpet factory, at Moosup, was also manufacturing that article at the rate of 25,000 yards per ann. Other very considerable manufactures are carried on in the co. Pop. 1820, 25,331; 1830, 27,082.

WINDHAM, p-t. Windham co. Conn., 14 ms. N. Norwich, 30 E. Hartford, 44 W. Providence, N. Shetucket r., has an irregular form, with about 46 sq. ms., and contains much good land, generally hilly, particularly E., with primitive rocks; a sandy soil prevails in the W. part. The timber is oak, walnut, chestnut, &c. Willimantic and Nachaug rs., after flowing some distance in the t., unite and form the Shetucket. Each of these streams affords mill sites, and several kinds of fish, particularly shad. The v. composing the 1st society, is large, and contains some public buildings, besides a number of stores, &c. The land was given by a son of the Mohegan sachem, Uncas, to John Mason and 13 others, in 1676, when it was surveyed, but it was not settled till 1686. Incorporated 1692. The town contains several considerable villages, beside the above; Willimantic, Scotland, &c. There are 6 churches, 3 of which are Congregational, 1 Methodist, and 1 Baptist. The surface is undulated; stone walls are generally used for the division of fields. Willimantic is a very flourishing village, 3 ms. W. of the 1st society, in which are several valuable mill sites on the river, which has a considerable descent for about a mile. Here are 7 cotton factories, 5 of stone, and some of them quite extensive. In them all, 13,150 spindles are run, and 288 looms. There is also a satinet factory, with 200 spindles and 8 looms, and an extensive paper mill. Pop. of town 1830, 2,812.

WINDHAM, p-t. Greene co. N. Y., 44 ms. S. W. Albany, 26 W. Catskill, has the Catskill mtns. N., the S. ridge of which is on the line, and Del. co. W. The surface is generally mountainous, with beech, maple, hemlock, bass, and other timber, and is watered by Schoharie cr. Pop. 1830, 3,471.

WINDHAM, p-v. in the nrthrn. part of Bradford co. Pa., by p-r. 153 ms. nrthrd. Harrisburg.

WINDSOR, t. Kennebec co. Me., 6 ms. E. Augusta, which adjoins it on the W., and has Lincoln co. E. and S. It is crossed by a branch of Sheepscot r., which touches it on the S. E. Pop. 1830, 1,485.

WINDSOR, t. Hillsborough co. N. H., is of a triangular form, and has a good soil, favorable to pasturage and grain, with an agreeably varied surface. Pop. 1830, 226.

WINDSOR, co. Vt., bounded by Orange co. N., Conn. r. E., which separates it from N. Hampshire, Windham co. S., Rutland co. W. It is crossed in the N. by White r., and has Queechy r. in the middle, and Black r. S., with some of the sources of West and Williams's rs. S. W. It lies on the E. declivity of the Green mtns., and has a rough surface, but a soil very favorable to grass. In the W. part are several quarries of soap stone, in Plymouth, Bridgewater, and Bethel; good granite is found in the S. E. part, and much lime is made from primitive limestone in the S. W. It contains 24 townships. Pop. 1820, 38,233; 1830, 40,625.

WINDSOR, p-t. and st. jus., Windsor co. Vt., 55 ms. S. Montpelier, 55 N. E. Bennington, 95 N. W. Boston, 420 N. Washington, first settled 1764. It is hilly, fertile, and well watered, having Conn. r. on the E. line, and Mill r. S., which supplies mill sites. Ascutney mtn., partly in this t., is one of the most conspicuous eminences in this part of the state. It is about 3,320 feet above tide water, and its summit, (which divides this t. from Weathersfield,) has two peaks, from which it is said to have derived its name: a word of this sound, in the Indian language, meaning the Two Brothers. This mtn. is almost clothed in evergreens, except on its S. side, which is bare. It is composed of granite. The v. is handsome, and is very pleasantly situated on the W. bank of Conn. r., surrounded by rich and picturesque scenery, in which mt. Ascutney forms a striking feature. Pulk Hole brook empties into Conn r. N., and Mill brook S. of the village. There are several handsome streets, the principal of which runs N. and S., in an irregular line, with many good dwellings, stores, churches, bank, court house, &c. The opposite shore of the Conn. is high and almost mountainous; there is a beautiful meadow near the v. At the mouth of Mill brook is a fall, which is dammed, and made to supply water to several factories; and near the same spot is a bridge which crosses the Conn. The state prison is in the s. w. part of the village. Pop. 1830, 3,134.

WINDSOR, p-t. Berkshire co. Mass., 120 ms. w. Boston, is on high ground, separating the waters of the Westfield, Deerfield, Hoosic and Housatonic rs. A swamp of 500 acres gives rise to a stream flowing into Westfield r., and another near by sends a stream to Deerfield r. Pop. 1830, 1,042.

WINDSOR, p-t. Hartford co. Ct., 6 ms. N. Hartford, 41 N. N. E. New Haven, about 6½ ms. by 8, with 50 sq. ms., was one of the first settlements made by white men in Ct. (1636.) For many years it embraced East Windsor. It lies on the w. bank of the Connecticut, and is crossed by Farmington or Tunxis r. There is a very extensive tract of fertile and beautiful meadows in this town, which afford a wide and delightful view from the numerous hills which rise in other parts. There is a higher plain, of light sand, much of which is waste. Farmington r. is navigable in sloops to the v. bridge, during the spring floods, and for flat bottomed boats at all seasons. Fish are caught in considerable quantities. Agriculture is here very flourishing. The principal v. extends for 2 or 3 miles along a broad and level street, much shaded by elms, and contains the mansion of the late chief justice Oliver Ellsworth. Pop. 1830, 3,220.

WINDSOR, p-t. Broome co. N. Y., 128 ms. s. s. w. Albany, 15 E. Chenango point, has Pennsylvania s., and is watered by Susqehannah r. and other streams, flowing through meadows. The uplands afford very good sheep pasture. Valuable locust timber is cut near Oquago, and sent down the r. to Philadelphia and Baltimore, for ship building. Pop. 1830, 2,180.

WINDSOR, p-v. York co. Pa., by p-r. 10 ms. E. York, and 36 s. E. Harrisburg.

WINFIELD, p-t. Herkimer co. N. Y., 75 ms. w. N. w. Albany, 15 s. w. Herkimer, 10 s. Utica, has Otsego co. E. and s., and Oneida co. w., and is supplied with mill seats by the Unadilla &c. Pop. 1830, 1,778.

WINHALL, p-t. Bennington co. Vt., 25 ms. N. E. Bennington, 33 s. w. Windsor, 102 Montpelier, was first settled during the revolution, and is supplied with mill sites by Winhall r. Pop. 1830, 571.

WINNICUT, or Winnicónett r. Merrimack co. N. H., is a small stream running N. into Great Bay of the Piscataqua r.

WINNIPISEOGEE, lake, Coos co. N. H., is one of the most picturesque sheets of water in the eastern states, and forms an interesting feature in the fine natural scenery usually embraced in a tour to the White mtns. It is about 22 ms. long from N. w. to s. E., and varies in breadth from 1 to 10 ms. Several long capes stretch far into its bosom from different sides almost dividing it into several parts. Three beautiful bays are thus formed on the w. side, 3 E., and 1 N. Merry-meeting bay which forms the s. E. extremity, may be almost regarded as a separate lake. The fine shores of the Winnipiseogee present a charming variety of surface, rendered still more attractive by a ride over the undulating country through which the roads pass on both sides, and the innumerable islands scattered over the surface. A company has been incorporated, who design to place a steamboat on the lake, to ply between Alton, at the s. E. extremity, and Centre harbor, in the N. w. in connection with lines of stage coaches, to form a regular channel of travelling between Boston and Lancaster. The summit of Mt. Washington is visible on that route until intercepted by the land when within 5 ms. of Centre harbor. Some of the isls. are large, and contain several farms. One of them has about 500 acres. They are, however, of almost every size and form, down to mere rocks. The water of this lake is remarkably pure, and abounds in fish, which are often caught through the ice in the winter, and sometimes sent to Boston market. Red mtn. near Centre harbor, commands a delightful view upon the lake; Squam l. and many of the mtns. N. The lake is navigated by a few sail boats. It receives a few small streams, and discharges s. w. by Winnipiseogee r.

WINNIPISEOGEE r. Coos co. N. H., the outlet of Winnipiseogee lake, after a short course enters Great bay, and afterwards flows thro' 2 smaller lakes, making a part of the boundary between Merrimack and Coos cos. and falls into Pemigewasset r. below Webster's falls. Its whole descent is 232 feet, and there are many good mill sites on its banks.

WINNS, p-o. Hall co. Geo., by p-r. 10 ms. sthrd. Gainesville, the st. jus., and 133 a little w. of N. Milledgeville.

WINNSBOROUGH, p-v. and st. jus. Fairfield dist. S. C., situated 29 ms. almost due N. Columbia, and 25 ms. a little E. of s. Chesterfield. Lat. 34° 24′, and long. 4° 07′ w. W. C.

WINN'S Tavern, and p-o. in the watrn. part of Fluvanna co. Va., 68 ms. N. w. by w. Richmond.

WINSLOW, p-t. Kennebec co. Me., 16 ms. N. E. Augusta, has Sebasticook and Kennebec rs. on the w. boundary, with 2 or 3 ponds and streams flowing into them. Pop. 1830, 1,263.

WINTERSVILLE, p-o. Lincoln co. N. C.

WINTHROP, p-t. Kennebec co. Me., 12 ms. s.w. Augusta. It is crossed N. and s. by 2 large ponds, connected and discharging into a third on the s. E. boundary, which forms the Cobbosseconte, a tributary of Kennebec r. Pop. 1830, 1,888.

WINTON, p-v. and st. jus. Hertford co. N.C., is situated on the right bank of Chowan r., about 2 ms. below the junction of Meherrin and Nottaway rs., about 60 ms. s. w. by w. Norfolk in Va., and by p-r. 129 ms N. E. by E. Raleigh. Lat. 36° 24′, and almost on the meridian of W. C.

WINYAW, bay, estuary of Black r., Great Pedee, and Waccamaw rs., Georgetown dist.

S. C. This sheet of water opens into the Atlantic ocean from Georgetown entrance. Lat. 33° 10′, long. 2° 14′ w. W. C. If we consider the head to be the junction of Black r. and Great Pedee at or near Georgetown, the length of Winyaw bay thence to Georgetown entrance, will be 14 ms., the mean breadth about 2 ms., and depth of water sufficient to admit large merchant vessels to Georgetown.

Wiscasset, p-t. port of entry, and st. jus. Lincoln co. Me., 24 ms. s. Augusta, 14 n. e. Bath, 49 n. e. Portland, 167 n. n. e. Boston, on the w. side of Sheepscot r., has a large and safe harbor, always open at some distance from the sea, with a considerable amount of shipping. Pop. 1830, 2,255.

Wisenburg, p-v. Lehigh co. Pa., by p-r. 180 ms. n. e. W. C.

Woburn, p-t. Middlesex co. Mass., 10 ms. n. w. Boston, has generally a good soil, with some hills, and is crossed by Middlesex canal. Horn pond affords a natural route for the canal for some distance, and is a favorite resort on account of the beauty of its scenery. First settled 1641. Pop. 1830, 1,977.

Wolcott, p-t. Orleans co. Vt., 22 ms. n. Montpelier, 37 n. e. Burlington. It is crossed by Lamoille r., which receives Green r. and Wildbranch. Fish pond is in the n. e. Pop. 1830, 492.

Wolcott, p-t. Wayne co. N. Y., 184 ms. w. Albany, 22 n. Waterloo, 9 n. Erie canal. has lake Ontario n., Cayuga co. e., and contains the greater part of Great Sodus bay and its isls. with East and Port bays. The surface is varied, the streams supplying mill seats. Fish and water foul abound in Sodus bay, which forms a good harbor at Port Glasgow. Pop. 1830, 1,085.

Wolcottville, p-v. Torrington, Litchfield co. Conn., 24 ms. w. by n. Hartford, has a cotton factory, &c.

Wolf r. and p-o. Hardin co. Ten., about 120 ms. s. w. by w. Nashville.

Wolfborough, p-t. Strafford co. N. H., 45 ms. from Concord, 45 from Portsmouth, and 105 from Boston, and n. e. Winnipiseogee lake, has a level surface, a rocky but valuable soil, bearing oak, &c. Smith's r. a small stream, rises in a pond, and empties into the lake near the v., which contains an academy, with a fund of $5,000. First settled 1770. Gov. Wentworth had once a splendid summer residence 5 ms. e. of the v. There is a mineral spring in the t. Pop. 1830, 1,928.

Wolfsville, p-v. nrthrn. part Frederick co. Md., by p-r. 54 ms. n. n. w. W. C.

Womack's, p-v. estrn. part Wilcox co. Ala., by p-r. s. s. e. Tuscaloosa.

Womelsdorf, p-v. and flourishing borough, Berks co. Pa., on the Union canal, 38 ms. a little n. of e. Harrisburg, 15 ms. n. w. by w. Reading.

Wonasquatocket, r. Providence co. R. I., a small stream which supplies a remarkable number of manufactories with water power. There are 25 factories and mills of different kinds and dimensions, which employ about 1,300 persons. The 1st power looms ever used in R. I. were placed on the banks of this stream. There are 2 reservoirs of water owned by the mill companies, 8 feet deep, and together extending over 200 acres, which supply manufactories at all seasons. Eight cotton factories it is estimated run 17,900 spindles, 590 looms, employ 650 hands, and manufacture of sheetings, shirtings, (some of very fine quality) and of goods for printing, more than 3,000,000 of yds. annually. A wollen factory, with buildings for dyeing &c., runs 600 spindles and 21 broadcloth looms, and manufactures 2,225 yds. per ann. There is also an oil, and a brown paper manufactory, and a manufactory of hat bodies, where 200 lbs. of wool are used per day, and 300,000 hat bodies are made annually.

Wood cr., Washington co. N. Y., runs n. 23 ms. by the v. of Fort Ann, to the s. end of lake Champlain, terminating at the village of Whitehall. It receives Pawlet r. from Vt., and now serves as the channel of the Champlain canal for some miles, the water being set back by damming. It was formerly used for batteaux and canoe navigation on the route between the lake and Hudson r. It is locked at its mouth by three large locks, by which boats are raised from the lake.

Wood cr. Oneida co. N. Y, receives Fish creek, and flows into Oneida lake. It has long served as part of the route for boat navigation between Mohawk r. and lake Ontario, there being a carrying place from its banks to that stream. A canal has since been dug across.

Wood, co. Va., bounded n. e. by Tyler and Harrison cos., e. by Lewis, s. e. by Kenhawa, s. w. by Mason, and by the O. r. which separates it from Meigs and Athens cos. O. on the w., and from Washington co. O. n. Length from s. w. to n. e. 36 ms., mean breadth 30, and area 1,080 sq. ms. Extending lat. from 38° 50′ to 39° 22′, and in long. from 3° 52′ to 4° 10′ w. W. C. The far greater part of Wood is included in the valley of Little Kenhawa, with a nrthwstrn. declivity. Surface excessively broken, with much good soil. Chief t., Parkersburgh. Pop. 1820, 5,860, in 1830, 6,429.

Woodbridge, t. New Haven co. Conn., 7 ms. n. w. New Haven, 40 s. w. Hartford, about 4 ms. by 10, is hilly and rough, with much good timber, and watered by West and Wapawaug rs. Pop. 1830, 844.

Woodbridge, p-t. Middlesex co. N. J., 42 ms. n. e. Trenton, 3 w. n. w. Amboy, has Sussex co. n., Rahway r. and the Kills e., and Raritan r. s. Pop. 1830, 3,969.

Woodbury, p-t. Caledonia co. Vt., 15 ms. n. e. Montpelier, first settled about 1800, contains more ponds than any other t. in the state, and is watered by streams of Lamoille and Onion rs. Pop. 1830, 824.

Woodbury, p-t. Litchfield co. Conn., 36 ms. s. w. Hartford, 25 n. w. New Haven, 15 from Litchfield. It has about about 41 sq.

ms., an irregular surface, with a rich soil, favorable to grain, fruit &c., with a variety of timber. The branches of Pomperaug r. afford mill seats and unite in this t. Pop. 1830, 2,049.

WOODBURY, st. jus. and p-v. Deptfort, Gloucester co. N. J., 39 ms. s. by w. Trenton, 9 s. Philadelphia, is on Wondury r., near the Delaware.

WOODBURY, p-v. in the nrthrn. part of Bedford co. Pa., 17 ms. N. N. E. Bedford, and by p-r. 136 ms. N. W. W. C.

WOODCOCK, p-v. in the sthestrn. part of Crawford co. Pa., by p-r. 305 ms. N. W. W. C.

WOODCOCK valley and p-o. Huntingdon co. Pa.

WOODFORD, t. Bennington co. Vt., 6 ms. E. Bennington, 24 w. Brattleborough, 50 s. Rutland, first settled after the revolution, has a pond of 100 acres near the centre, which gives rise to a branch of Walloomscoic r. Other streams water different parts. The surface is mountainous, and much of it is useless. Pop. 1830, 395.

WOODFORD, co. Ky., bounded by Franklin N.W., Scott N. E., Lafayette E., Jessamine S. E., Ky. r. separating it from Mercer S. W., and Anderson W. Length from s. to N. 22, mean breadth 7, and area 154 sq. ms. Extending in lat. from 38° 53′ to 39° 11′, and in long. from 7° 36′ to 7° 50′ w. W. C. The declivity of this narrow co. is wstrd. towards Ky. r. The soil is generally excellent. Chief town, Versailles. Pop. 1820, 12,207; and in 1830, 12,294.

WOOD GROVE, and p-o. in the nrthrn. part of Loudon co. Va., by p-r. 44 ms. N. W. W. City.

WOOD GROVE, and p-o. N. C., 13 ms. Salisbury, and by p-r. 131 ms. a little s. of w. Raleigh.

WOODHULL, t, Steuben co. N. Y., 236 ms. s. s. w. Albany. Pop. 1830, 501.

WOODLANDS, p-o. in the nrthrn. part of O. co. Va., 271 ms. N. w. by w. W. C.

WOODLAWN, p-o. in the nrthrn. part of Hanover co. Va., 30 ms. nrthrd. Richmond, and by p-r. 105 ms. w. of N. W. C.

WOODLAWN, p-o. in the wstrn. part of Edgefield dist. S. C., by p-r. 123 ms. wstrd. Columbia, and 45 ms. wstrd. Edgefield court house.

WOODPECKER'S LEVEL, and p-o. in the western part of Franklin co. Va., 23 ms. wstrdly. from Rocky Mount, the co. st., and by p-r. 286 ms. s. w. W. C.

WOODRUFF'S, p-o. near Bethel meeting house in the sthrn. part of Spartanburg dist. S. C., 18 ms. a little w. of s. Spartanburg C. H., and by p-r. 92 ms. N. W. Columbia.

WOOD'S, p-o. Knox co. Ten., by p-r. 213 ms. E. Nashville.

WOODSBOROUGH, p-v. Frederick co. Md., 10 ms. N. N. E. Frederick, and by p-r. 54 ms. N. N. W. W. C.

WOOD'S FERRY and p-o. in the sthrn. part of Green co. Ten., 10 ms. from Greenville, and by p-r. 262 ms. E. Nashville.

WOOD'S HILL, p-o. Roane co. Ten.

WOODSTOCK, p-t. Oxford co. Me., 40 ms. w. Augusta, 6 N. Paris, contains mountainous hills, with several ponds emptying s. into little Androscoggin r. Pop. 1830, 573.

WOODSTOCK, p-t. and st. jus., Windsor co. Vt., 11 ms. N. W. Windsor, 46 s. Montpelier, first settled 1768, was exposed to Indian incursions in the revolutionary war, and often to wild beasts. The surface is varied, the soil good, and agriculture flourishing. Here is a bank. Apples thrive remarkably well. Queechy r. and two of its branches afford mill sites. There are 2 villages. Pop. 1830, 3,044.

WOODSTOCK, p-t. Windham co. Conn., 45 ms. a little N. of E. Hartford, 33 N. W. Providence, 66 s. w. Boston, and has the Mass. line on the N. boundary. It is about 7 by 8 ms., has an irregular surface, a soil good for grazing, &c., is watered by Muddy brook and other streams, and has several factories. Pop. 1830, 2,915.

WOODSTOCK, p-t. Ulster co. N. Y., 57 ms. s. Albany, 14 N. W. Kingston, has Greene co. N., is mountainous. There is a good fish pond. The N. Y. crown and cylinder glass co. here manufacture 1,500 boxes of window-glass monthly, employing 50 persons. Pop. 1830, 1,375.

WOODSTOCK, p-v. and st. jus. Shenandoah co. Va., situated wstrd. of the north fork of Shenandoah r., 32 ms. s. s. w. Winchester, and by p-r. 100 ms. a little s. of w. W. C. Lat. 38° 51′, and long. 1° 34′ w. W. C.

WOOD'S Store and p-o. Coweta co. Geo., by p-r. 145 ms. N. W. by w. Milledgeville.

WOOD'S Store and p-o. Carroll co. Ten., by p-r. 136 ms. w. Nashville.

WOODSTOWN, p-v. Pilesgrove, Salem co. N. J., 55 ms. s. w. Trenton, 12 N. E. Salem, 26 s. s. w. Philadelphia.

WOODVILLE, p-v. in the wstrn. angle of Culpepper co. Va., by p-r. 97 ms. s. w. by w. W. C., and 115 N. W. Richmond.

WOODVILLE, p-v. in the nrthrn. part of Perquimans co. N.C., by p-r. 271 ms. a little E. of s. W. C., and 177 N. E. by E. Raleigh.

WOODVILLE. A place of this name is given in the p-o. list as a p-v. and st. jus., Jackson co. Ten., whilst in the same table, Bellefonte is marked also as a st. just. in the same co.; the former 185 ms. N. E. Tuscaloosa.

WOODVILLE, p-v. and st. jus., Wilkinson co. Miss., 38 ms. a little E. of s. Natchez, and 180 ms. N. W. New Orleans. Lat 31° 07′, long. 14° 27′ w. W. C. This village is situated in a fine rolling country, which is one of the most productive cotton districts in the U. S. Pop. about 500.

WOODWARD'S, Store and p-o. in the estrn. part of the parish of East Feliciana, La., by p-r. 118 ms. N. W. by w. New Orleans.

WOOLWICH, p-t. Lincoln co. Me., 32 ms. s. Augusta, 7 w. Wiscasset village, has Kennebec r. w., and a strait connected with it on the s., with one or two small streams. Pop. 1830, 1,495.

WOOLWICH, t. Gloucester co. N. J., 55 ms.

s. w. Trenton, has Oldman's cr. s., Raccoon cr. N., and Del. r. w. in which are several isls. It is opposite Marcus Hook, Pa. Pop. 1830, 3,033.

WOONSOCKET FALLS, village in the towns of Smithfield and Cumberland, Providence co. R. I., 17 ms. N. Providence, is a flourishing manufacturing village, on Blackstone r., near the line of Mass. There are 2 cotton factories, one of them of stone, with 4,000 spindles; another of 2,500 spindles; another of stone with 4,000; another with 2,500 spindles, &c. &c. The whole number of spindles in this place, is about 20,000. 2,617,000 yards of cotton goods are annually manufactured, besides 30,000 yards of satinet in a single establishment. There is also a furnace, where castings are annually made amounting to $35,000; a whetstone manufactory and 4 machine shops. There is also a Bank in the village. Pop. 1830, about 2,000.

WORCESTER, town, Washington co. Vt., 10 ms. N. Montpelier, 31 E. Burlington; first settled 1797, watered by the N. branch of Onion river which affords mill sites, and along its course is a road through the mtns. It is mountainous w. and rough in other parts. Pop. 1830, 432.

WORCESTER, county, Mass., bounded by N. Hampshire N.; Middlesex and Norfolk cos. E., Connecticut s., Hampden, Hampshire and Franklin cos. w. It is the largest in the state, with a varied surface, a soil generally good, and many flourishing agricultural townships and manufacturing villages. Its limits comprehend some of the places first settled after the early colonists of N. England began to leave the sea coast, and several of these were scenes of blood in Philip's war. It is watered by numerous streams, some of which form Nashua river N. E., others Pawtucket r. s. E., Quinebaug s., Chickopee r. s. w., and Miller's r. N. w. The mill sites are numerous, many of which are occupied by large manufactories, particularly at Ware, along the Blackstone, &c. Blackstone canal affords boat navigation from Worcester, to Providence, R. I. A rail-road has been commenced from Worcester to Boston, and other rail-roads are proposed from the same point. The trade of this co. is chiefly carried on with Boston, some with Providence, and is somewhat diverted in other directions. It contains 55 towns. Pop. 1820, 73,625, 1830, 84,365.

WORCESTER, p-t. and st. jus., Worcester co. Mass., is situated 40 ms. w. by s. Boston, 40 N. N. w. Providence, and 60 E. N. E. Hartford. The soil of the town is generally fertile, is well cultivated, and its surface is pleasant and finely varied. It was first settled in 1685, and suffered much in its earlier history from the attacks of the natives. Tatmuck and Bogachoak hills were once the sites of Indian villages. The head waters of Blackstone r. unite in this t. The village of Worcester is one of the most flourishing and beautiful in New England, and is a great thoroughfare for travellers, some important roads passing through it. It lies principally on one street, about a mile in length, broad, lined with trees, and nearly on a level. Other streets diverge, some of them at right angles, from it. The village contains several public buildings, a court house, which cost $20,000, a jail, 32 feet by 64, and 3 stories high, 3 handsome churches, a bank, the library and cabinet of the American historical society, &c. Many of the dwellings are elegant, and display much taste in their exterior, and there are some fine hotels. There are also several fine country seats in the immediate vicinity of the village. The Mass. lunatic hospital is located a little out, on a commanding eminence, and is a spacious structure of brick, 256 feet in length, consisting of a centre and wings. Its interior arrangement is admirable, and it is calculated for the accommodation of 110 to 120 patients. The printing of books was extensively carried on here after the revolution, by Isaiah Thomas, who published in 1791, the first folio Bible printed in the U. S. The Blackstone canal, extending to Providence, terminates here, and affords boat navigation to that place; this canal with a rail-road to Boston, already commenced, must prove of great advantage to Worcester. It is also designed to connect with the Boston rail-road, one to New London, Conn., and another to Springfield, on the Conn. In the village and town are many and various manufactories. Pop. 1830, 4,172.

WORCESTER, p-t. Otsego co. N. Y., 59 ms. s. w. Albany, 16 s. E. Cooperstown. It has Schoharie co. N E., Delaware co. s., and is crossed by Shenevas creek, which flows through rich meadows. It has good soil and mill sites. Pop. 1830, 2,093.

WORCESTER, p-o. Montgomery co. Pa., about 21 miles N. w. Philadelphia. The tsp. of Worcester lies on the E. side of Skippack creek, between Gwynned and Norriston.

WORCESTER, co. Md., the extreme southeastern co. of that state, bounded by Accomac county, Va., s., Somerset co. Md., w., Sussex co. Del. N., and by the Atlantic E. Length from s. to N. 30 ms., mean width including its islands 25, excluding these 20, the area with that of the islands about 700 square ms. Most maps extend this co., and of course the state of Md. to lat. 58°; this is a mistake; the sthrn. boundary is not along a curve of lat. but is on the Atlantic ocean about 3 minutes, and on Pocomoke bay upwards of 1 minute N. of lat. 38. The northern boundary is on lat. 38° 28', and the co. lies between long. 1° 24' and 1° 55' E. W. C. The declivity is to the s. w., and it is drained in that direction by Pocomoke r. which is discharged into Chesapeake bay. The surface is level, and soil sandy. Chief town, Snow Hill. Pop. 1820, 17,421, and in 1830, 18,271.

WORTHINGTON, p-t. Hampshire co. Mass., 110 ms. w. Boston. It is situated on the E. declivity of the Green mtn. range, has Westfield river on the s. w. line, into which smaller streams flow. The surface is agreeably diversified, and the soil good. Several kinds of minerals are found in this town, titanium, &c. Pop. 1830, 1,179.

WORTHINGTON, p-o. Muhlenburg co. Ky., by post-road 177 miles southwest by w. Frankfort.

WRENTHAM, p-t. Norfolk co. Mass., 24 ms. w. s. w. Boston, has the R. I. line s. with several ponds and streams, some of which flow into Charles, and others into Taunton and Neponset rivers. There are several manufactories in this town, an academy, &c. Pop. 1830, 2,698.

WRIGHTSBORO', p-v. in the western part of Columbia co. Geo., by p-r. 78 ms. N. E. by E. Milledgeville.

WRIGHTS MILLS, and p-o. by p-r. 315 ms. s. w. by w. W. C. and 315 ms. N. w. by w. Richmond.

WRIGHTSVILLE, p-v. on Susquehannah river, York co. Pa., 11 ms. N. E. by E. from York, and 11 a little s. of w. from the city of Lancaster. It stands directly opposite Columbia, in Lancaster co.

WYALUSING, large cr. of Pa., rising in the western part of Susquehannah co., from which flowing southwestward falls into Susquehannah r. in Bradford.

WYALUSING, township, lies along and contiguous to the Susquehannah r. above and below the mouth of Wyalusing cr. In this township, there are two post offices, Wyalusing, and Wyalusing centre, the former near the mouth of the creek, and the latter 6 ms. above, on the cr. Wyalusing is marked on the p-o. list as distant from W. C. 254 ms. and from Harrisburg 143 ms.

WYATT'S FERRY, and p-o. Randolph co. Va., by p-r. 229 ms. w. W. C.

WYE RIVER, or rather creek, separating Talbot from Queen Anne county, Maryland.

WYE RIVER, mills and p-o. in the northwestern part of Talbot co. Md., 12 miles N. Easton, the co. st., and 30 ms. a little s. of E. Annapolis.

WYNANT'S KILL, creek, Rensselaer co. N. Y., enters Hudson r. at Troy, after supplying valuable mill seats, 2 ms. s. Poesten kill. It is about 15 ms. long.

WYOMING. (*See Wilkes-Barre.*)

WYOMING, p-v. in the sthrn. part of Dinwiddie co. Va., by p-r. 54 ms. sthrd. Richmond.

WYSOX, small creek of Bradford co. Pa.

WYSOX, p-o. is on the Wysox creek E. of the Susquehannah r., 5 ms. N. E. Towanda, and by p-r. 130 ms. N. Harrisburg.

WYTHE, co. of Va., bounded s. w. by Washington, N. w. by Tazewell, N. by Giles, N. E. by Montgomery, and by the Iron mountains separating it from Grayson s. E. and s. It contains a part of a mtn. valley. Greatest length along the valley from s. w. to N. E. 48 ms., breadth about 22, and area 1,056 square ms. Extending in lat. from 36° 40′ to 37° 08′, and in long. from 3° 36′ to 4° 32′ w. W. C. Wythe valley is an elevated table land. From the southwestern part issue the extreme fountains of the Middle Fork of Holston r., interlocking sources with those of various creeks flowing into New r. Comparing the elevation of Wythe, with that of Giles co., gives to the former an elevation exceeding a mean of 1,600 feet. The cultivated land of Wythe must indeed rise between 1,600 and 2,000 feet above the ocean level.

WYTHE, C. H. (*See Evansham.*)

Y.

YADKIN, river. (*See Great Pedee river.*)

YADLEYVILLE, p-v. Bucks co. Pa., by p-r. 29 ms. northwardly Phila.

YANCEY'S MILLS, and p-o. Albemarle co. Va., by p-r. 97 ms. s. w. by w. W. C.

YANTIC, river, New London co. is a small branch of the Thames, falling into the head of the cove, in Norwich, which communicates with the main stream at the landing. At its mouth it descends an abrupt ledge of solid granite, into which it has worn deep holes. The water is received into a deep and still pool at the foot of the falls, which is overshadowed by a rocky bank 60 or 80 feet high, from which a body of Mohegan Indians once precipitated themselves, when pursued by their enemies, the Narragansetts. This fall has been dammed, and a canal on the N. bank leads the water to several large manufactories, around which has recently been built a large and flourishing village, chiefly situated under the steep bank at the head of the cove. The Thames manufacturing co. on Yantic r. have a brick cotton factory 47 feet by 120, 5 stories, with 3,200 spindles, 120 looms, and 150 hands, using 750 bales of cotton annually, and making a million of yards of shirting and sheeting. They have an iron foundry, a rolling and slitting mill, and a nail factory, and use 750 tons of iron annually. Near them the Williams manufacturing company run 1,800 spindles, and a paper mill makes 60 reams a day, on 4 Foudieneir machines. Another factory makes 50,000 yards of flannel per annum. At an upper fall the Norwich manufacturing company make 18,000 yards of carpet per annum. Another cotton factory is to be built at the falls of Shetucket r. for 2,000 spindles. There is also a comb, button, and oil mill.

YARMOUTH, p-t. Barnstable co. Mass., 70 ms. s. E. Boston, occupies the breadth of Cape Cod, at a narrow part, with Cape Cod bay N., and the Atlantic s. It has a very poor soil, consisting of loose sand, and the town is very poor; a considerable quantity of salt, and glauber salts are made here annually from sea water. At the s. w. corner is Lewis' bay, on which is Hyannis harbor, a place of some coasting trade. Pop. 1830, 2,251.

YATES, co. N. Y., bounded by Ontario co. N. and W., Seneca lake or Seneca co. E., and Steuben co. S. The N. ends of Crooked lake penetrate the co. from the S., and Canandaigua lies on the N. W. corner. The surface and soil are various. Bluff point is a tongue of land extending S. between the N. arms of Crooked lake. The outlet of that lake, Flint creek, &c. water different parts of this co. It contains 7 towns, and Penn-Yan, 191 ms. W. Albany, is the st. jus. It was formed in 1823, out of Ontario co. Pop. 1830, 19,009.

YATES, p-t. Orleans co. N. Y., 30 ms. N. W. Batavia, has lake Ontario N., and Niagara co. W. The surface is slightly varied, is crossed by the Ridge road, and is watered by Johnson's creek, &c. Pop. 1830, 1,538.

YAZOO, river of the state of Miss., having its remote sources in the northern part of the state, lat. 34° 45', interlocking sources with those of Tombigbee branch of Ala., Silver cr. flowing into Ten. r., and Wolf r. flowing into Miss. r. in the state of Ten. Winding from this elevated tract the Yazoo r. winds to the S. W. entering the Miss. at lat. 32° 22', after a course of about 200 ms. As laid down on Tanner's U. S., there is an outlet represented as leaving the left bank of the Miss. 25 or 30 ms. above the mouth of St. Francis r. This outlet, after flowing to S. E. 10 or 12 ms., separates into two channels. The left or estrn. is continued to S. S. E. 70 ms. to its union with the Yazoo, about 125 ms. above the mouth of the latter. The right or western channel after a comparative course of 150 ms. is also united to the Yazoo, at the north-eastern angle of Warren co. 25 ms. above the mouth. If these delineations are correct, there is included in the state of Miss., an elliptical annually inundated tract of 170 ms. long, between the Yazoo and Miss. rs., with a breadth where widest, of 70 miles, and a mean breadth of at least 40, with an area of 6,800 square ms.

YELLOW BRANCH, p-o. in the western part of Campbell co. Va., by p-r. 135 ms. a little S. of W. Richmond.

YELLOW CREEK, Furnace, and p-o., Montgomery co. Ten., 16 ms. wstrd. Clarksville, the co. st., and by p-r. 64 miles N. W. by W. Nashville.

YELLOW STONE, r. (*See Missouri.*)

YELLOW SPRINGS, and p-o. in the northern part of Huntingdon co. Pa. 15 ms. N. W. Huntingdon, and by p-r. 163 ms. N. N. W. W. C.

YELLOW SPRINGS, p-o. Claiborne co. Ten., by p-r. 255 ms. a little N. of E. Nashville.

YELLOW WATER, river of Ala. and Flor., rises in Covington co. of the former, between Choctaw, and Cunecuh rivers, and flowing thence to the S. W., enters Walton co. of Flor., which it traverses, and inclining more to the westrd. enters Escambia co., there receiving Shoal river from the E. it is finally lost in the estrn. arm of Pensacola bay, after a course of 80 ms. (*See Escambia, Cunecuh, &c.*)

YOCOM'S, p-o. in the northern part of Washington co. Ky., 38 ms. S. W. Frankfort.

YONGUESVILLE, p-o. northern part of Fairfield dist., S. C., 12 ms. northward Winnsboro', the st. jus., and by p-r. 41 ms. in a similar direction from Columbia.

YONKERS, p-t. Westchester co. N. Y., 18 ms. N. N. York, 10 S. W. Whiteplains, 131 S. Albany, has Hudson r. W., Bronx cr. E., and N. York co. S. It extends about 8 ms. on the Hudson, and 3 ms. is its average width; it is crossed by Saw mill creek with a branch of the Bronx N. E., on which are mill seats. The ground is uneven, and there are several eminences in the town celebrated for events in the revolution; part of Valentine's hill, Boar hill, Tetard's hill, and the heights of Fordham, with the site of fort Independence. Philipsburgh, now called Yonker's, is a small village, with a landing at the mouth of Bronx cr. where a steamboat touches daily. Pop. 1830, 1,761.

YORK r., York co. Me., is a small stream emptying into the Atlantic, with a broad mouth, and depth of water for vessels of 200 tons, forming a good harbor.

YORK, co. Me., bounded by Oxford co. W., Cumberland co. N. E., the Atlantic E., and N. Hampshire W. and S. W. It has Ossipee r. on the N. line, Saco r. on part of the N. E. line, and Salmon falls r. and Piscataqua r. W. and S. W. Saco r. crosses the N. E. part and enters the sea at Saco. Below this are Kennebunk and York rs. besides several smaller streams flowing into the ocean. The coast is generally rocky and waste, with cape Porpoise, Fletcher's Neck, Bald Head, cape Neddock, and Kittery Point, and several harbors at Saco, Kennebunk, Wells and York. Portsmouth harbor is near the south boundary of this co. Some of the early settlements in New England were made at York, Wells, &c., and suffered greatly from the fear and violence of savages. Considerable foreign trade has long been carried on, chiefly with the West Indies; and the coasting trade and fisheries are valuable. There is a great diversity of soil, and a considerable variety of surface; but there is no very elevated land in this co. It contains 24 townships, and its capitals are York and Alfred. Pop. 1820, 46,283; 1830, 51,722.

YORK, p-t. port of entry and one of the sts. jus. York co. Me., 99 ms. S. W. Augusta, 42 S. W. Portland, 9 N. N. E. Portsmouth, 67 N. E. Boston, is a place of considerable trade. York r. crosses it and empties into the ocean affording a good harbor for vessels of 200 tons. The Agamenticus hills are in the N. W. part, and much of the sea coast is rocky, barren, and incapable of cultivation. Cape Neddock, about on the E. line, is an iron bound promontory. Near the v. the soil is very good and the land level. This place was laid out for a large city by its early inhabitants, and the lines run at right angles. Pop. 1830, 3,485.

YORK, p-t. Livingston co. N. Y., 237 ms.

Albany, 7 N. N. W. Geneseo, has Genesee co. N. and W., Genesee r. E. on which is the Conewago Reservation. The streams are small and mill sites few, but the soil is generally good, especially on the Genesee. Pop. 1830, 2,636.

YORK, one of the southern counties of Pa., bounded by Adams co. W., Cumberland N. W., Susquehannah r. separating it from Dauphin N. and Lancaster N. E. and E., by Hartford co. in Md. S. E., Baltimore co. in Md. S., and Frederick co. Md. S. W. York co. bounds on Md. 42 ms., but the longest part is a line parallel to the general course of Susquehannah r. 48 ms., mean width 18, and area 864 sq. ms. Extending in lat. from 39° 42′ to 40° 43′, and in long. from 0° 04′ W. to 0° 46′ E. W. C. The nrthrn. and central sections comprising the much larger part of this co., decline to the nrthestrd. and are drained into the Susquehannah, in that direction by Cadorus, Conewago, and Yellow Breeches crs. The extreme sthestrn. angle has an estrn. declivity. The face of the co. though broken by hills and decorated by some minor mtn. ridges, has much sameness of character. The soil is generally good, and much of it excellent. Staples, grain, livestock, &c. Chief t., the borough of York. Pop. 1820, 38,759, and in 1830, 42,858.

YORK, p-t. borough and st. jus. York co. Pa., situated on Cadorus cr., 22 ms. a little S. of W. Lancaster, and by p-r. 87 ms. a little E. of N. W. C. and 24 S. S. E. Harrisburg. Lat. 39° 57′, and long. 0° 17′ E. from the meridian of W. C. The site of York is a plain, in part liable to occasional submersion. The streets are extended at right angles to each other, and beside the ordinary co. buildings the borough contains an almshouse, academy, several schools, and places of public worship. Pop. 1830, 4,216.

YORK, r. of Va., formed by 2 main branches, Pamunkey and Mattapony. *(See Pamunky and Mattapony.)* Below the union of its constituent streams, York r. is rather a bay, varying from 2 to 3 ms. in width, extending to the S. E. 27 ms., and thence E. 12 ms. into Chesapeake, between York and Gloucester cos. Below the junction of Pamunkey and Mattapony rivers, York bay does not receive a tributary above the size of a small cr. It admits ships of any size to or near the Great Bend at Yorktown, but above admits only coasting vessels. Incluing all its confluents the valley of York r. lies between those of James and Rappahannoc. The greatest length 120 ms. from the mouth of York r. to the extreme source of North Anna r. in South West Mtn.; but, if taken with this extent the mean width would not exceed 20 ms., and at the utmost breadth, only about 45 ms. The area 2,600 sq. ms. Extending in lat. from 37° 15′ to 38° 16′, and in long. from 0° 41′ E. to 1° 22′ W. W. C.

YORK, co. Va., bounded by Elizabeth City co. S. E., Warwick S., James City W. and N. W., York r. separating it from Gloucester N. and Chesapeake bay E. Greatest length along York bay 34 ms., mean width 6 and area 204 sq. ms. Extending in lat. from 37° 08′ to 37° 23′, and in long. from 0° 12′ to 0° 46′ E. W. C. Narrow as is this comparatively lengthened co., it occupies nearly one half of the width of the peninsula between James and York rs. Declivity nrthrd. towards the latter. The soil is generally good. Chief town, Yorktown. Pop. 1820, 14,384; and in 1830, 5,334.

YORK, dist. S.C., bounded by Lancaster dist. E., Chester dist. S., on the W. by Broad r. separating it from Union and Spartanburg dist., on the N. by Lincoln co. N.C., and N. E. by Catawba r. separating it from Mecklenburg co. N. C. Length from E. to W. 35, breadth 23, and area about 800 sq. ms. Extending in lat. from 34° 48′ to 35° 07′, and in long. from 3° 55′ to 4° 40′ W. W. C. This district occupying the whole space between Catawba and Broad rs. contains part of two inclined plains. The Broad r. plain, or that on the wstrd., declines a little W. of S., and is drained by Turkey, Bullock's, Kings, and Buffaloe creeks. The opposing or estrn. plain declines sthestrd. and discharges in that direction Fishing, Alisons, and Crowder's creeks. Surface hilly. Chief town, Yorkville. Pop. 1820, 14,936; 1830, 17,790.

YORK C. H. (See Yorkville, York dist. S. C.)

YORK HAVEN, p-v. on the Susquehannah r. opposite East Conewago, and in the nrthestrn. part of York co. Pa., by p-r. 10 ms. N. from the borough of York, and 14 ms. below and S. E. Harrisburg.

YORKSHIRE, t. Cataraugus co. N. Y., 274 ms. from Albany, 14 N. N. E. Ellicottville, it has Cataraugus cr. N. which separates it from Erie co., with a slightly uneven surface. Limestone lake, 1 m. by 2, in the E. gives rise to a tributary of Cataraugus cr. Pop. 1830, 823.

YORK, Sulphur springs and p-o. in the nrthrn. part of Adams co. Pa., by p-r. 20 ms. S. S. W. Harrisburg, and 14 N. N. E. Gettysburg.

YORKTOWN, p-t. Westchester co. N. Y., 116 ms. S. Albany, 45 N. New York, 8 N. W. Bedford, it has Putnam co. N. and is generally hilly, with a pretty good soil, often stony, and much improved by the use of gypsum. Pop. 1830, 2,141.

YORKTOWN, p-v. port of entry and st. jus. for York co. Va., situated on the right bank of York r., 11 ms. above the mouth, 33 N. W. Norfolk, by p-r. 175 ms. S. S. E. W. C., and 72 S. E. by E. Richmond. Lat. 37° 14′, long. 0° 30′ E. W. C.

YORKVILLE, p-v. and st. jus. York dist. S. C., 22 ms. N. Chesterville, and by p-r. 78 ms. a little W. of N. Columbia. Lat. 34° 58′, long. 4° 18′ W. W. C.

YOUGH GLADES, p-o. in the wstrn. part of Alleghany co. Md., 33 ms. wstrd. Cumberland and by p-r. 165 ms. N. W. by W. W. C.

YOUGHIOGHANY, r. of Pa., Md. and Va., having its most remote source in Preston co. of the latter state, but deriving its most numerous sthrn. tributaries from the valley between the Back Bone and Laurel mtns. Alleghany co. Md. From this elevated tract the main stream flows nearly due N. 35 ms., enters Pa. between Fayette and Somerset cos., within which it thence flows about 8 ms. direct course to where it is joined by Castleman's r., an equal or probably a superior stream, entering from the N. E. Some of the southern fountains of Castleman's r. rise in Alleghany co. Md., but the greater part of its tributaries flow from Somerset co. Pa., and rise in the same valley with the confluents of Youghioghany. Below the union of the 2 main branches the Youghioghany assuming a nrthwstrn. course, continues in that direction 60 ms. to its junction with the Monongahela at MacKees port, in Alleghany co. Where Youghioghany is traversed by the U. S. road at Smithfield, the water level is 1,405 feet above that of the Atlantic. The extreme heads of this stream have an elevation exceeding 2,500 feet; the mouth being elevated about 700 feet, the entire fall must be 1,800 feet. The whole valley of Youghioghany is either mountainous or very hilly and broken.

YOUNG'S Store and p-o. in the wstrn. part of Laurens dist S. C., by p-r. 88 ms. northwestwardly Columbia.

YOUNG'S, cross roads and p-o. in the wstrn. part of Lauderdale co. Ala., 14 ms. wstrdly. Florence, the co. seat, and 160 a little W. of N. Tuscaloosa.

YOUNGSTOWN, p-v. Niagara co. N. Y., 304 ms. W. Albany, 6 N. Lewistown, is situated on Niagara r.

YOUNGSTOWN, p-v. (on the great western road from Philadelphia to Pittsburg) Westmoreland co. Pa., 10 ms. E. Greensburg, the co. st., and by p-r. 182 ms. N. W. W. C. It is a small v., of a single street along the road.

YOUNGSVILLE, p-v. in the nrthwstrn. part of Warren co. Pa., 17 ms. N. W. Warren, and by p-r. 330 ms. N. W. W. C.

YOUNG WOMANSTOWN, p-v. on Young Womans cr., and in the nrthwstrn. part of Lycoming co. Pa., by p-r. 50 ms. N. W. Williamsport, the co. st., and 138 N. N. W. Harrisburg.

Z.

ZANESVILLE, p-t. and st. jus. Muskingum co. O., on the E. side Muskingum r., 58 ms. E. Columbus and 84 W. Wheeling. It is a flourishing t., containing a court house and other co. buildings, several churches, stores, and fine dwellings. Two wire bridges connect the town with Putnam and W. Zanesville, and the Cumberland road passes through it. At the falls, where is a superior water power, are several mills—saw, oil and rolling mills, a woollen and nail factory, &c. The river connects with Ohio canal. Pop. 1830, 3,094.

ZEBULON, p-v. and st. jus. Pike co. Geo., by p-r. 86 ms. almost due W. Milledgeville. N. lat. 33° 04', long. 7° 26' W. W. C.

ZIDON, p-v. Spartanburg dist. S. C., by p-r. 116 ms. N. W. Columbia.

ZION, p-v. Iredell co. N. C., by p-r. 16 ms. W. Raleigh.

ZOAR, t. Berkshire co. Mass., is one of the smallest tows in the state, and was incorporated in 1822. It is crossed by Deerfield r. Pop. 1830, 129.

APPENDIX,

CONTAINING SEVERAL NEW COUNTIES, OMISSIONS EITHER OF TOWNS OR THEIR POPULATION, &C. &C.

A.

ALLEGAN, a new co. of Mich. bounded by Ottaway co N, Kent N. E, Barry E., Kalamazoo co S. E., Van Buren S. and S W., and lake Huron W Breadth 21 ms, mean length from W. to E. 38 ms., and area about 900 sq ms. Lat. 42° 35', and long 9° W W. C. intersect near the centre of this co, which lies about 150 ms in a direct line a little N. of W. Detroit Slope wstrd., and in that direction traversed by Kalamazoo r., which stream enters lake Mich. on the wstrn. border of the co. Pop. uncertain.

ALLEN, co. of O. bounded S. E. by Logan, S. by Shelby, S. W. and W. by Mercer, N. W. by Vanwert, N. by Putnam, and E. by Hardin. Length from E. to W. 26 ms, breadth 24, and area 624 sq. ms. Lat. 40° 35', and long. 7° 10' W. W. C. Slope to the N W It is drained into the sthrn. branch of Au Glaize r., and contains the large Wapahkonetta Reserve. Chief t. Wapahkonetta. Pop. 1830, 578.

ALLEN, co. Ind. bounded E. by Paulding co. O.; it lies to the N. E. from Huntingdon co. Ind To the W., N. and S. the bounding cos. uncertain. It is a square of 26 ms. each side; area 676 sq. ms. For central lat. *see Fort Wayne.* Under the respective articles Maumee and Wabash rs. the peculiar structure of the country now included in Allen co. may be seen. A canal to cross this co., and to unite the navigable waters of Maumee and Wabash is in actual progress. (*See article rail-roads and canals, col. first of page* 447.) Chief t. Fort Wayne. Pop. 1830, 996.

ARENA, new co. Mich. bounded s. w. by Midland, by Gladwin w., on the N. by the unappropriated part of the territory, and on the E. and S. E. by Saginaw bay. Breadth from S. to N. 24 ms., mean length 30, and area 720 sq. ms. Lat. 44°, and long. 7° W. W. C. The centre of the co lying a little E. of N. N. W. 120 ms. from Detroit. Slope estrd. Population uncertain.

ARKANSAS. When the article Arkansas was printed, the returns of the population had not been published. We insert here that of 1830, by sexes, ages, and classes.

	White Persons.	
	Males.	Females.
Under 5 years of age,	3,020	2,782
From 5 to 10,	2,021	1,897
10 to 15,	1,626	1,494
15 to 20,	1,272	1,225
20 to 30,	2,835	2,012
30 to 40,	1,820	1,067
40 to 50,	876	528
50 to 60,	434	301
60 to 70,	209	107
70 to 80,	69	31
80 to 90,	12	9
90 to 100,	1	3
100 and upwards,	0	0
Total,	14,195	11,476

Of these 10 were deaf and dumb, and 8 blind.

	Colored Population.			
	Free.		Slaves.	
	Males.	Fems.	Males.	Fems
Under 10 years of age,	27	17	846	803
From 10 to 24,	17	13	814	836
24 to 36,	23	10	395	399
36 to 55,	17	7	192	193
55 to 100,	3	6	47	51
100 and upwards,	1	0	0	1
Total,	88	53	2,293	2,283

Of colored persons, 4 were deaf and dumb, and 2 blind.

Recapitulation,

Whites.	Free col.	Slaves.	Total.
25,671	141	4,576	30,388

ASSANPINK, r. or cr. N. J., the boundary of Burlington and Hunterdon co. rises in Monmouth, and falls into the Del. at Trenton. This stream separated the British and American armies in 1777, the day previous to the battle of Princeton.

ACCORD, p-v. Ulster co., for N. J read N. Y.—ACRA, p-v. Greene co., for N. J. read N. Y.—ACTON, p-t. York co. Me. Pop. 1,398.—"ACWORTH," stated as in *Cheshire* co. is in *Sullivan* co. N. H.—ALBION, p-t. Kennebec co. Me. Pop. 1,393.—ATHENS, Greene co. N. Y. Pop. 2,425.—AUGUSTA, Geo., for 'seat of government,' so printed in a few copies, read '*formerly*' seat of government.—AVON, stated as in Erie, is in Livingston co N. Y.

B.

BALTIMORE, p-v. Fairfield co Ohio, by p-r. 30 ms. S. E. from Columbus. It is in the sthrn. part of the co., and on the Ohio and Erie canal, has risen rapidly from the advantages of position, and contains upwards of 200 houses, and 500 inhabitants.

BAPTIST TOWN, p-v. Hunterdon co. N. J. 10 ms. w. Flemington.

BARNEGAT, v. Monmouth co. N. J., on Barnegat bay opposite the inlet.

BARRY, new co. Mich. bounded by Calhoun S. E., Kalamazoo co. S. W., Allegan W., Kent N. W., Ionia N. E., and Eaton E. Length from S. to N. 24, breadth 21, and area 504 sq. ms. Lat 42° 35′, long. 8° 20′ W. W C. The sthrn. part slopes to the S., and is drained by creeks flowing into Kalamazoo r., but the body of the co. slopes to S. W. by W., and in that direction is drained by Apple r. branch of Grand r. The central part is about 120 ms. a little N. of W. from Detroit. Pop. uncertain.

BEATTYSTOWN, v. Warren co. N. J. on the Muskonetcunk, 2 and a half ms. below Hackettstown.

BERRIEN, co. of Mich. bounded N. E. by Van Buren co., by Cass co. E., St. Joseph's co. Ind. S. E., La Porte co. Indiana S. W., and lake Michigan W. Length from S. to N. 33 ms., mean breadth 18, and area 594 sq. ms. Lat. 42°, and long. 9° 25′ W. W. C. Slope to the northwestrd., and in that direction is traversed by St. Joseph's r. which enters at the south eastern angle, and winding to the N. W., receiving tributaries from the northestrd., and falling into lake Michigan at the st. jns., Saranac. Pop. 1830, 325.

BLACK RIVER, v. Morris co. N. J. on a stream of that name, which runs into the N. branch of Raritan r. 14 ms. W. Morristown.

BLAZING STAR, a noted ferry on the sound between N J. and Staten Island, 5 ms S. of Elizabeth, 7 ms. N. of Amboy. Previous to the revolution on the main route between Phil. and N. Y.

BLOOMFIELD, v. Essex co. N. J. in the town of that name, distinguished by E. and W. Bloomfield, two flourishing villages on a small stream which affords power for many manufactories of woollen, paper, &c. &c. There is a large and well conducted academy, 4 ms. N. of Newark. The Morris canal passes through the village.

BLOOMSBURY, v. Burlington co. N. J., lies adjoining the city of Trenton, and includes all the buildings S. of the Assanpink creek. The bridge over the Delaware is in this town.

BLOOMSBURY, v. Hunterdon co N J. on the Muskonetcunk 6 ms. E. Easton, 40 W. N. W. New Brunswick.

BOONETON, v Morris co. N J., on the Rockaway r., where there are a succession of falls affording great water power, and on which several extensive manufactories of iron are erected, 10 ms N. N. E. of Morristown. The Morris canal passes by this place.

BRANCH, one of the southern cos. of Michigan, bounded W. by St Joseph co., N. by Calhoun, Hillsdale E., and La Grange co. of Ind. S. W. It is a sq. of 21 ms. each way, 441 sq ms. Lat. 42°, and long. 8° W. W. C. intersect in this co. Slope wstrd, and in that direction drained by various branches of St. Joseph's r. of lake Mich. The central part is 110 ms. a little S. of W. Detroit. Pop. uncertain

BROOKLYN or BROOKLAND iron works, Morris co. N. J. at the outlet of the Muskonetcunk lake, 14 ms. N. N. W. of Morris Town. A dam is erected here to raise the lake for the supply of the Morris canal.

BARNARDSTOWN, p-t. Franklin co. Mass. Pop. 945.—BARRE, p-t. Worcester co. Mass. Pop. 2,503.—BARRE, Orleans co. N. Y. Pop. stated 2,503, is 4,801.—BARRINGTON, stated in Steuben co is in Yates co. N. Y.—BARTON, p-t. Orleans co. Vt. Pop. 729.—BARTON, Tioga co. N. Y. Pop.

972.—BEEKMAN, Duchess co. N Y. Pop. 1,564. —BELFAST, Waldo co. Me. Pop stated, 1,743, is 3,077 —BETHEL, p-t. Oxford co. Me. Pop. 1,620. —BOLIVAR, t. Alleghany co. N. Y. Pop. 449.— BORDENTOWN, N. J., for p-t. read p-v.— "BOW," stated as in *Rockingham* co. is in Merrimack co. N. H.—BREWCER, p-t. Lincoln co. Me. Pop. 770.—BREWSTER, Barnstable co. Mass. Pop. stated 118, is 1,418.—BROOKS, t. Waldo co. Me. Pop. 601.

C.

CALHOUN, co. of Il. comprising the point between the Mississippi proper and Illinois rs. bounded by Il. r., separating it from Greene co. E., and the Miss. r., separating it from St. Charles co. Mo. s., Lincoln co. Mo. w., and Pike co. Mo. N. Length nearly parallel to both the bounding rivers 40 ms., mean breadth 6 ms., and area 240 sq. ms. Extending in lat. from 38° 52′ to 39° 38′, long. from 13° 26′ to 13° 52′ w. W. C. Chief t. Gilead. Pop. 1830, 1,090.

CALHOUN, co. Mich. bounded by Jackson E., Hillsdale s. E., Branch s. and s. w., Kalamazoo w., Barry N. w., and Eaton N. E. Length from E. to w. 30 ms., breadth 24, and area 720 sq. ms. N. lat. 42° 15′, and long. 8° w. W. C. intersect in this co Slope wstrd., sthrn. part drained by branches of St. Joseph's r. of lake Mich., and the nthrn. by the Kalamazoo r. The central part is about 110 ms. w. Detroit. Pop. uncertain.

CALLAWAY, co. Mo. bounded by Boone co. w., Ralls N., Montgomery E., and the Mo. r., separating it from Gasconade co. s. E., and from Cole co. s. w. Length from s. to N. 50 ms., mean breadth 23, and area 1,150 sq. ms. Lat. 39°, and long. 15° w. W. C. intersect in this co. Though bordering on the Mo. r. it contains a table land, from which flow northwardly the southern sources of Salt r., eastwardly the extreme wstrn. sources of Cuivre (Copper) r., and numerous crs. to the s. E., s., and s.w. into Mo. r. Chief town, Fulton. Pop. 1830, 6,159.

CARMI, p-v. and st. jus. White co. Il. (*See White co. Il.*)

CASS co. Ind. bounded E. by Miami co., s. w. by Carroll co., on other sides boundaries uncertain. As laid down in Tanner's improved map of the U. S. it is in length 24 ms. from E. to w., mean breadth 18 ms., and area 432 sq. ms. Lat. 40° 45′, long. 9° 12′ w. W. C. Slope very nearly due w. The main stream of Wabash receives Eel r. from the right at Loganport, the st. jus. of this co., the village standing on the point between the two rivers. Pop 1830, 1,162.

CASS, co. Mich. bounded by Berrien w., Van Buren N., St. Joseph E., Elkhart co. Ind. s. E., and St. Joseph, Ind. s. w. It is very nearly a sq of 24 ms. each way, and area 576 sq. ms. N. lat. 42°, and long. 9° w. W. C. intersect very near the centre of this co. It is entirely drained by tributary streams of St. Joseph's r. of lake Mich , but from the great general sthrn. curve of that stream, the creeks of Cass co. flow like radii from the central part of the co. Chief t. Edwardsburg. Pop. 1830, 919.

CASSVILLE, p-v. on the left bank of Miss. r. opposite the mouth of Upper Ioway r. and the western part of Ioway co., Huron, or western Mich. It is situated 27 ms. s. of Prairie du Chien, by p-r. 1,028 ms N. w. by w. W. C., and 36 ms. above and N. w. Galena in Il.

CEDAR Creek, v. Monmouth co. N. J. near the sea, on a stream of the same name which runs into Barnegat bay 6 ms. s. Toms r.

CHEBOIGONG r. (*See Sheboigen r. in this Addenda.*)

CLINTON, new co. Mich. bounded E. by Shiawassee, Ingham s. E., Eaton s. w., Ionia w., and Gratiot N. It is a square of 24 ms. each way, area 576 sq. ms. Lat. 43°, and long. 7° 42′ w. W. C. intersect in this co. 100 ms a little w. of N. w. Detroit. Slope westward, and traversed by different branches of Grand r. Pop. uncertain.

COOK, co. Il. bounded N. E. by lake Mich. s. E. by the northwestern angle of the state of Ind., s. by Vermillion co. Il., and w. by Lasalle co. On the N. boundary uncertain. The outlines of this co. are too vague to admit any statement of its area. Though bounded by lake Michigan, nearly the entire slope is to the southwstrd., in which direction flows Plain r. branch of Illinois. A canal has been designed to unite the navigable water of the Illinois to lake Michigan at Chicago, mouth of Chicago r. Cook co. Chicago is the st. jus. This co. formed since the census of 1830.

CORDELERA. In article Chippewayan, p. 102, first col., line 7 from the top, for Corelebra read Cordelera.

CALHOUN, p-t. Orange co. N. Y. Pop. 1,535. —CAMDEN, Waldo co. Me. Pop. stated 674, is 2,200.—CANAAN, p-t. Columbia co. N. Y. Pop. 2,064.—CANANDAIGUA, Ontario co. N. Y. Pop. stated 1830, is 5,162.—CARLISLE, p-t. Schoharie co. N. Y. Pop. 1,748.—CARLTON, p-t. Orleans co. N. Y. Pop. 1,168.—CARMEL, p-t. Putnam co. N. Y. Pop. 2,379.—CARVER, p-t. Plymouth co. Mass. Pop. 970.—CECILIUS, t. Cataraugus co. N. Y. Pop. 378.—CHENANGO, p-t. Broome co. N. Y. Pop. 3,716.—CHERRY CREEK, t. Chatauque co. N. Y. Pop. 574.—CHESTERVILLE, Kennebec co. Me. Pop. 923.—CHICHESTER, stated as in *Rockingham* co. N. H., is in *Merrimack* co. Pop. 1,084.—CHILMACK, Dukes co. Mass. Pop. stated 2,010, is 691.—CLINTON, Duchess co. N. Y. Pop. stated 19,344, is 2,130.—COCHECTON, t. Sullivan co. N. Y. Pop. 438.—COHOCTON, p-t. Steuben co. N. Y. Pop. 2,711.—COLUMBUS, p-t. Chenango co. N. Y. Pop. 1,744.—"CORNISH," stated as in *Cheshire* co., is in *Sullivan* co. N. H.— CORINNA, Somerset co. Me., erroneously spelt "Cornina," and thus arranged in the alphabet.—COVERT, Seneca co. N. Y. Pop. 1,791.—"CROYDON," stated as in Cheshire co., is in Sullivan co. N. H.

D.

DANSVILLE, p-t. Steuben co. N. Y. 46 ms. s. Rochester, 11 N. Arkport; the village is on the highest navigable point of the Chemung, a branch of Susquehannah r. A rail-road from this place to Rochester. (*See rail-roads and canals, article Dansville and Rochester.*) Pop. 1830, 1,728.

DAVIESS, co. Ind. bounded N. by Greene co., E. by Martin, East fork of White r., separating it from Dubois s. E., and Pike s. w., and by the w. or Main White r. separating it from Knox co. w. Length from s. to N. 30 ms., mean breadth 18, and area 540 sq. ms. Lat. 38° 45′, long. 10° 12′ W. C. This co. occupies the peninsula between the two branches of White r., but the slope is s. w. by w. towards the western or main fork. Chief t. Washington. Pop. 1830, 4,543.

DES MOINES, r. of the northwestrn. territory, and state of Mo. rises in the former near lat. 43° 40', long. 18° w. W. C., interlocking sources with those of Little Sioux branch of Mo. r., Blue Earth or the southern branch of St. Peter's r., and those of Upper Ioway r., and flowing thence S. E. 300 ms., falls into the Miss. r. opposite Hancock co. Il. Beside numerous smaller confluents, this stream receives at about 150 ms. below its source, the North fork from the left, and 50 ms. lower, a much more considerable addition by the Raccoon fork from the right. The latter has a comparative length of about 150 ms. The valley of Des Moines is 300 ms. in length, but is comparatively narrow, being rather liberally estimated at a mean breadth of 50 ms., area 15,000 sq. ms. About 20 ms. comparative course of the lower part of this r. it forms the extreme northeastrn. boundary of Mo. It is the longest, and probably the most abundant tributary of Miss. proper from the right.

DANBY, Tompkins co. N. Y. Pop. 2,481.—DEER ISLE, Hancock co. Maine. Pop. 2,207.—DELHI, Delaware co. N. Y. Pop. stated 435, is 2,114.—DEPAU, p-t. St. Lawrence co. N. Y. Pop. 668.—DEPEYSTER, p-t. St. Lawrence co. N. Y. Pop. 814.—DOVER, p-t. Windham co. Vt. Pop. 831.—DUANESBURGH, Schenectady co. N. Y. Pop. 2,837.—DUKES co. Mass. Pop. stated 1,768, is 3,518.—DUTTON, t. Penobscot co. Me. Pop. 443.

E.

EATON, co. Mich., bounded by Ingham E., Jackson S. E., Calhoun S. W., Barry W., Ionia N. W., and Clinton N. E. It is a square of 24 ms. each way; area 576 sq. ms. Lat. 42° 35', and long. W. C. 8° W. intersect in this co. The main stream of Grand r. winds along the estrn. border, in a nrthrly. direction, but the body of the co. slopes watrd., giving source to Apple r., branch of Grand r., and to Battle r., branch of Kalamazoo r. The centre is about 100 ms. a little N. of W. Detroit. Pop. uncertain.

EDGAR, co. Il., bounded by Clarke S., Coles W., Vermillion co. of Il. N., Vermillion co. of Ind. N. E., and Vigo co. Ind. S. E. Length from S. to N. 30, breadth 24, and area 720 sq. ms. Lat. 39° 40', long. W. C. 10° 45' W. General slope sthestrd., and drained by crs. flowing into the main Wabash. Chief t. Paris. Pop. 1830, 4,071.

ESPIRITU SANTO, often called Tampa bay, fine haven on the W. coast of the peninsula of Florida, opens from the gulf of Mexico, at lat. 27° 45', long. W. C. 6° 50' W.

EAST HARTFORD, Hartford co. Conn. Pop. stated 3,537, is 2,237.—EASTHAMPTON, Suffolk co. N. Y. Pop. 1,668.—EASTON, Washington co. N. Y. Pop 3,753.—EATON, p-t. Madison co. N. Y. Pop. 3,558.—EAST WINDSOR, Hartford co. Conn. Population stated 2,129, is 3,537.—EDGARTOWN, Dukes co. Mass. Pop. 1,509.—EDINBURGH, Saratoga co. N. Y. Pop. 1,571.—EDWARDS, p-t. St Lawrence co. N. Y. Pop. 683.—"EPSOM," stated as in Rockingham co., is in Merrimack co. N. H.—ERIN, p-t. Tioga co. N. Y. Pop. 976.—ERVING'S GRANT, Franklin co. Mass. Pop. 429.—ERWIN, p-t. Steuben co. N. Y. Pop. 795.—ESOPUS, p-t. Ulster co. N. Y. Pop. 1,770.

F.

FENWICK'S ISLAND, a long narrow sand bar, extending along the shore of the Atlantic ocean, from Sinepuxent inlet, Worcester co. Md., to Rehoboth bay, or mouth of Indian r., Sussex co. Del., length about 28 miles.

FLORENCE, formerly Briceland's cross roads, p-o. Washington co. Pa. The name has been changed recently.

FORT BRADY. (*See St. Mary's strait.*)—FAIRFIELD, Franklin co. Vt. Pop. 2,270.—FALL RIVER, v. Bristol co. Mass. Pop. 3,431.—FALLSBURGH, p-t. Sullivan co. N. Y. Pop. 1,173.—FALMOUTH, p-t. Cumberland co. Me. Pop. 1,966.—FARMINGTON, Hartford co. Conn. Pop. should be 3,500.—FAYETTE, Seneca co. N. Y. Pop. 3,216.—FENNER, Madison co. N. Y. Pop. 2,017.—FRANKLIN, t. Hancock co. Me. Pop. 382.—FRANKLIN, p-t. Merrimack co. N. H. Pop. 1,370.—FRANKLINVILLE, p-t. Cataraugus co. N. Y. Pop. 903.—FRENCH CREEK, t. Chatauque co. N. Y. Pop. 420.—FULTON, p-t. Schoharie co. N. Y. Pop. 1,592.

G.

GOOSEKENHAWN r. (*See Rock r.*)

GOSPORT, navy yard near Norfolk, Norfolk co. Va.

GRAND, r. of O. This stream has its extreme source in Trumbull co. about 5 or 6 ms. N. W. of Warren, and flowing thence nearly due N. 30 ms., to Ashtabula co., and within 10 ms. of lake Erie, bends at right angles to the W., and continues that course 30 ms. to the lake, which it enters in Geauga co., between the villages of Fairport and Newmarket, leaving a peninsula of 22 miles in length, and from 10 to 2 miles wide, between the lower course of the river and the lake.

GRANT, co Ind., bounded N. W. by Miami co., N. by Wabash and Huntingdon cos., S. E. by Del. co., and S. by Madison. Breadth 18 ms., mean length 20, and area 360 sq. ms. Lat. 40° 30', long. W. C. 8° 38' W. The Mississinewa r. enters near the sthestrn. and leaves it at the nrthwstrn. angle, flowing to the N. W., the general slope of the co. being in that direction. The extreme nrthrn. sources of White r. are, however, in the sthestrn. angle of this co., and flow to the S. W., and from the opposite angle issues a branch of Salamanic r., flowing to the N. W. This co. is named neither in the p-o. list or census tables.

GULF OF MEXICO. (*See Mexico inland sea.*)

GALEN, stated in Seneca co., is in Wayne co. N. Y. Pop. 3,631.—GALLATIN, p-t. Columbia co. N. Y. Pop. 1,568.—GENESEE, t. Alleghany co. N. Y. Pop. 219.—GILL, Franklin co. Mass. Pop. stated 1,407, is 864.—GILMAN POND, t. Somerset co. Me. Pop. 335.—GOULDSBOROUGH, p-t. Hancock co. Me. Pop 880.—GOUVERNEUR, t. St. Lawrence co. N. Y. Pop. 552.—GREAT BARRINGTON, p-t. Berkshire co. Mass. Pop. 2,276.—GREENE, p-t. Chenango co. N. Y. Pop. 2,962.—GREENWOOD, p-t. Steuben co. N. Y. Pop. 796.—GROVE, p-t. Alleghany co. N. Y. Pop. 1,388.

H.

HOT SPRINGS, co. of Ark. This co., which takes its name from the remarkable fountain so called, on the head branches of Washitau r., lies s. w. by w. from Little Rock, but the outlines are uncertain. As laid down by Tanner, the Hot Springs are situated 47 ms. a little s. of w. of Little Rock. Lat. 31° 32', long. W. C. 15° 58' w. The Washitau springs considerably exceed blood heat, are much resorted to, and have been found very efficacious in many disorders, particularly chronic. The dryness and elevation of the adjacent country, render the place a most delightful asylum from the low and warmer plains in its vicinity. Pop. 1830, 458.

HUNTINGTON, new co. of Ind., bounded by Grant s. w., Wabash co. w., and on the other sides, boundaries uncertain. Length from s. to N. 21 ms., breadth 15, and area 360 sq. ms. Lat. 40° 50', long. W. C. 8° 30' w. The general slope of this co. is wstrd., though the Salamanic, and Wabash proper, traverse it in a N. w. direction, but the latter receives near the centre of the co., Little r. from Allen co., flowing from the N. w., and some smaller confluents from the N. Below the junction of Wabash and Little r., the united streams fall over rapids, at the foot of which the r. becomes navigable. The centre of this co. is about 90 ms. N. N. E. Indianopolis. Pop. uncertain.

HAIGHT, t. Alleghany co. N. Y. Pop. 655.—HAMPDEN, p-t. Delaware co. N. Y. Pop. 1,210.—HAMPTONBURGH, p-t. Orange co. N. Y. Pop. 1,365.—HAMMOND, p-t. St. Lawrence co. N. Y. Pop. 767.—HANCOCK, p-t. Hancock co. Me. Pop. 653.—HANCOCK, p-t. Addison co. Vt. Pop. 472.—HARRINGTON, p-t. Washington co. Me. Pop. 1,118.—HARTFORD, p-t. Windsor co. Vt. Pop. 2,044.—HASTINGS, p-t. Oswego co. N. Y. Pop. 1,494.—HORNBY, p-t. Steuben co. N. Y. Pop. 1,463.—HOULTON, Washington co. Me. Pop. 579.—HOWLAND, t. Penobscot co. Me. Pop. 329.—HUDSON, p-t. Hillsborough co. N. H. Pop. 1,282.

I.

INDIAN STREAM, t. Coos co. N. H. Pop. 301.

INGHAM, co. Mich., bounded by Jackson s., Eaton w., Clinton N. w., Shiawassee N. E., and Washtenaw s. E. It is a square of 24 ms. each way, area 576 sq. ms. Lat. 42° 35', long. W. C. 7° 24' w Slope N. w., and drained by the two main and higher branches of Grand r. The central part 75 ms. N. w. by w. of Detroit. Pop. uncertain.

IONIA, co of Mich., bounded by Kent w., Barry s. w., Eaton s. E., Clinton E., and Montcalm N. Similar to Clinton and Kent cos., it is a square of 24 ms. each way, area 576 sq. ms. Lat. 43°, and long. W. C. 8° w., intersect in this co. about 110 ms. N. w. by w. of Detroit. The slope is wstrd., and in that direction it is traversed by Grand r., the principal constituents of which unite near its estrn. border. Chief t. Generau. It is recently formed, and does not appear on the p-o list, or census tables.

IOWAY, the name of two rs., called relatively Upper Ioway and Lower Ioway. Upper Ioway rises about lat. 44°, long. W. C. 17 1-2 w., interlocking sources with those of Des Moines r., and with those of Blue Earth branch of St. Peter's r., and flowing thence to the estrd. 160 ms., falls into the right side of Miss. r., at lat. 43° 30', and about 40 ms. above Prairie du Chien, at the mouth of Ouisconsin r. Lower Ioway rises about lat. 43°, long. W. C. 16° w., interlocking sources with those of the North Fork of Des Moines r., and flowing thence s. E. by comparative courses 160 ms., falls into the right side of Miss. r., about 30 ms. below Rock Island rapids, and nearly opposite the s. w. angle of Mercer co. Il. The Des Moines, and Lower Ioway rs., flow nearly parallel, and about 10 ms. asunder.

ISABELLA, new co. of Mich., bounded by Midland E., Gratiot s. E., Montcalm s. w., and by unappropriated territory on the other sides. It is a square of 24 ms. each way, area 576 sq. ms. It probably occupies a share of the middle table land of the Mich. peninsula. The central part is about 130 ms. in direct line N. w. Detroit. Pop. uncertain.

ISLE AU HAUT, t. Hancock co. Me. Pop. 315.

ITALY, stated in Ontario co. N. Y., is in Yates co. Pop. 1,092.

J.

"JACKSON, or HITCHCOCK," stated to be in Hancock co., is in Waldo co. Pop. 493.

JACKSON, p-t. Coos co. N. H. Pop. 515.

JASPER, p-t. Steuben co. N. Y. Pop. 557.

K.

KENNEBUNK PORT, p-t. York co. Me. Pop. 2,763.

KENT, new co. of Mich., bounded E. by Ionia, s. E. by Barry, s. w. Allegan, w. Ottawa, and N. Oceana. It is a square of 24 ms. each side, area 576 sq. ms. Lat. 43°, and long. 8° 36' w. The main volume of Grand r. enters on the estrn. border, and winds wstrd. over the co., but by a very circuitous channel, receiving within it Rouge and Flat rs. from the N. E., and Apple r. from the s. The body of the co. is about 150 ms. N. w. by w. of Detroit. Pop. uncertain.

KENTUCKY, state of. In the body of the Gazetteer, the tables of population of this state for 1830, by sexes, ages, and classes, were omitted, and are as follows:—

	White persons.	
	Males.	Females.
Under 5 years of age,	54,116	50,835
From 5 to 10,	41,073	39,439
" 10 to 15	34,222	32,197
" 15 to 20	29.017	29,623
" 20 to 30	45,913	41,936
" 30 to 40	26,289	23,463
" 40 to 50	15,966	15,476
" 50 to 60	10,843	9,499
" 60 to 70	6,253	5,315
" 70 to 80	2,585	2,195
" 80 to 90	699	575
" 90 to 100	119	97
" 100 and upwards,	28	14
Total,	267,123	250,664

Of these were deaf and dumb under 14 years of age, 100; of 14 and under 25, 113; and of 25 and upwards, 90; total deaf and dumb, 303. Blind 169.

Of colored persons there were,

	Free colored.		Slaves.	
	Males.	Females.	Males.	Fem.
Under 10 years of age,	717	639	31,513	30,990
From 10 to 24	570	497	27,488	27,224
" 24 to 36	391	357	13,386	14,177
" 36 to 55	478	389	7,513	8,119
" 55 to 100	386	358	2,286	2,560
" 100 and upwards,	17	17	45	49
Total,	2,559	2,257	82,231	83,119

Deaf and dumb, colored, 42; blind, 78.
Recapitulation.

Whites.	Free col'd.	Slaves.	Total.
517,787	4,816	165,350	687,953

KIRKLAND, p-t. Oneida co. N. Y. Population 2,505.

L.

LA GRANGE, new co. of Ind., bounded by Elkhart co. W., St. Joseph co. Mich. N. W. and N., Branch co Mich. N. E., other boundaries uncertain. Length from W. to E. 30 ms., breadth 18, and area 540 sq. ms. Lat. 41° 40′ N., long. W. C. 8° 30′ W. Slope N. W. by W. and in that direction drained by Pigeon r. and other confluents of St. Joseph's r. of lake Mich. The central part is about 140 ms. a little E. of N. Indianopolis. Pop. uncertain.

LA PORTE, new co. of Ind., bounded by St. Joseph's co. Ind. E., Berrien co. Mich N., lake Michigan N. W., and the Indian country to the S. of lake Mich. on the other sides. It is, with the exception of the N. W. angle, a square of 24 ms. each side, area about 560 sq. ms. Lat. 41° 35′, long. W. C. 9° 42′ W. The extreme source of Kankakee r branch of Illinois r. rises in St. Joseph's co. Ind., and traversing the sthrn. part of La Porte co. gives it a wstrn. slope. The northwestern section declines N. W. towards lake Michigan. Surface generally a plain. The port on lake Michigan, from which this co. is named, is by direct line about 140 ms. N. N. W. from Indianopolis. Over this space a rail-road has been projected. Pop. uncertain.

LES MOINES r. (*See art. Des Moines r. in this Addenda.*)—LAFAYETTE, p-t. Onondaga co. N. Y. Pop. 2,560.—LA GRANGE, p-t Duchess co. N. Y. Pop. 2,044.—LAWRENCE, p-t. St. Lawrence co. N. Y. Pop. 1,097.—LENOX, t. Waldo co. Me. Pop. 666.—LE RAY, p-t Jefferson co. N. Y. Pop. 3,430.—LIBERTY, p-t. Waldo co. Me. Pop. 676.—LINCKLAEN, p-t. Chenango co. N. Y. Pop. 1,425.—LINCOLN, t., stated as in Hancock co., is in Penobscot co. Me. Pop. 404.—"LINCOLNVILLE," stated to be in Hancock co, should be spelt "*Lincolnsville*," and is in Waldo co. Me. Pop. 1,702.—LISBON, p-t Grafton co. N. H. Pop. 1,485.—LITTLE FALLS, Herkimer co. N. Y. Pop. stated 1,500, is 2,539.—LOCKPORT, p-t. Niagara co. N. Y. Pop. stated 1,801, is 3,823.—LODI, p-t. Seneca co. N. Y. Pop. 1,786.—LYNDON, t. Cataraugus co. N. Y. Pop. 271.

M.

MACDONOUGH, co. of Il., bounded by Fulton E., Schuyler S., Hancock W., and Warren N. It is a square of 24 ms. each way, area 576 sq. ms. N. lat. 40° 35′, long. W. C. 13° 38′ W. Slope S. W., and drained by numerous branches of Crooked cr. Chief town, Macomb, at which, says the p-o. list, there was no office Oct. 1st, 1830. It is situated about 150 ms. N. W. Vandalia. Pop. uncertain.

MIDLAND, new co. of Mich., bounded S. E. and S. by Saginaw, Gratiot S. W., Isabella W., Gladwin N. W., Arena N. E., and Saginaw bay N. E. Length from E. to W. 40 ms., mean breadth 20, and area 800 sq. ms. Lat. 43° 40′, long. W. C. 7° 18′ W. Slope S. E., and traversed from Gladwin co. by the Tittabewassee r. or nrthrn. branch of Saginaw r. The central part is about 120 ms. N. W. Detroit. Saginaw r. enters Saginaw bay at the point where the boundary between Midland and Saginaw cos. reaches that bay. Pop. uncertain.

MONROE, co. of Ark., situated between St. Francis and Black rs., about 100 ms. N. E. by E. from Little Rock, exact boundaries uncertain. Pop. 1830, 461.

MONTCALM, new co. of Mich., bounded by Gratiot E., Ionia S., Oceana W., and Isabella N. It is a square of 24 ms. each side, area 576 sq. ms. Lat. 43° 20′, and long. 8° W. W. C. Slope S. W. and drained by numerous confluents of Grand r. The central part about 120 ms. N. W. by W. Detroit.

MACHIAS, t. Cataraugus co. N. Y. Pop. 737.—MADAWASCA, p-t. Penobscot co. Me. Pop. 2,487.—MADISON, p-t. New Haven co. Conn. Pop 1,809.—MARION, p-t. Wayne co. N. Y. Pop. 1,981.—MIDDLEBURGH, p-t. Schoharie co. N. Y. Pop. 3,266.—MIDDLEFIELD, p-t. Hampshire co. Mass. Pop 721.—MILAN, t. Coos co. N. H. Pop. 243.—MILBURN, p-t. Somerset co. Me. Pop. 1,006.—MILO, t. Penobscot co. Me. Pop. 381.—MENDON, p-t. Rutland co. Vt. Pop. 432.—MONROE, pop. stated 409, is 1,081.—MONROE, t. Franklin co. Mass. Pop. 265.—MONROE, Fairfield co. Conn. Pop. 1,522.—MONTPELIER, v. Washington co. Vt. Pop. 1,193.—MONTVILLE, Waldo co. Me. Pop. stated 676, is 1,743.—MUNSON, t. Somerset co. Me. Pop. 411.—MURRAY, p-t. Orleans co. N. Y. Pop. 3,138.

N.

NEW EGYPT, v. Monmouth co. N. J., on Crosswick's cr. 18 ms. S. W. Freehold.

NEW GUILFORD, p-v. Coshocton co. O., by p-r. 63 ms. N. E by E. Columbus.

NEW HAGERSTOWN p-v. in the estrn. part of Tuscarawas co. O., by p-r. 124 ms. N. E. by E. Columbus, and 18 ms S. E. by E. New Philadelphia, the co st. Pop. 1830, 102.

NEW HARTFORD, p-t Litchfield co. Conn., 20 ms. N. W. Hartford. Watered by the Farmington or Tunxis r. which affords numerous sites for mills, &c. It is likewise watered by numerous small streams, is hilly and mountainous, and timbered with deciduous trees, except in the N. part of the town, where the perennial or evergreen region of Connecticut commences. Pop. 1830, 1,766.

NEW HARTFORD, p-t. Oneida co. N. Y., 100 ms. N. N. W. Albany. Pop. 3,549.

NEW HAVEN, p-t. Addison co. Vt., 26 ms. S. Burlington, 31 W. Montpelier. It is watered by Otter cr., Little Otter cr. and New Haven r., and contains good mill privileges. Quarries of excellent marble are found in almost every part; timber, maple, beech, birch, elm, basswood, &c. Pop. 1830, 1,834.

NEW LONDON, p-t. Merrimack co. N. H., E. Sunapee lake, containing 17,000 acres, it is 33 ms. N. W. Concord, 75 from Portsmouth, and 9 from Boston, and has 3 ponds, good deep soil, maple, birch, &c. Incorporated 1779. Pop. 1830, 913.

New Wilmington, p-v. in the southwestern part of Mercer co. Pa., by p-r. 274 ms. n. w. W. C.

New Windsor, p-t. Orange co. N. Y., 100 ms. s Albany, 65 n. N. York, and 5 s. w. Newburgh on the w. side Hudson r. It is 4 ms. by 8, containing about 30 sq. ms., has some small streams, and good soil. In the w. part are some of the highlands. The v. is on Hudson r. 2 1-2 ms. s. Newburgh. A part of the t. is called Little Britain. Pop. 1830, 2,310.

New Windsor, p-v. Frederick co. Md., by p-r. 67 ms. a little w. of n. W. C.

NEWBURY, p-t. Orange co. Vt. Pop. 2,252.—NEW HEMSTEAD, p-t. Queens co. N. Y. Pop. 3,062.—NEWINGTON, p-t. Rockingham co. N. H. Pop. 549.—NEW PORTLAND, p-t. Somerset co. Me. Pop. 1,215.—NEW SALEM, stated in Ontario co., is in Yates co. N. Y. Pop. 2,783.—NEW SHOREHAM, p-t. Newport co. R. I. Pop. stated 1,185, is 1,885.—NIAGARA, co. N. Y. Pop. stated 14,482, is 18,482.—NORTH SALEM, t. Somerset co. Me. Pop. 389.

O.

Oxford, p-t. Grafton co. N. H. Pop. 1,829.
Ossipee, p-t. Strafford co. N. H. Pop. 1,935.
Otis, t. Hancock co. Me. Pop. 350.

P.

Peektano, branch of Rock r. (*See Rock r.*) In article Iowa co. 2d line from the bottom, this river is erroneously spelled Peektans.

Perkinsville, a manufacturing v. Weathersfield, Windsor co. Vt. on Black r. at the falls It contains a woollen factory, a machine shop, saw mill, a manufactory of ivory black, and a cassimere factory. The v. has grown up within 6 years, and contains a church and a number of neat dwellings.

Philadelphia, a brief notice of its geographical location, &c. was inserted in the body of this work. The city was founded in 1682, by Wm. Penn; its original form was a parallelogram, extending 2 ms. w. from Delaware r., crossing the Schuylkill, and over a mile, n. and s. Its present limits embrace several suburbs, which will be named hereafter. It is accessible by the Delaware for the largest merchant vessels, and the Schuylkill is also navigable for smaller ones from its junction with the former, to Permanent bridge. That part of the city in the vicinity of the Delaware, was formerly the principal seat of business; but since the coal mines in the interior have been opened, and become a source of such immense importance, these, with the facilities offered in its vicinity for internal communication, (*see articles rail-roads and canals,*) have attracted much attention towards the Schuylkill, and numerous stores and other buildings have been erected near it, and the banks lined with wharves. Philadelphia is situated 120 ms. from the Atlantic, by the course of the Delaware, and is distant from Harrisburg 98 ms.; from New York 89; from Baltimore 98; and 136 from Washington. The city is laid out with remarkable regularity, the streets crossing each other at right angles. There are nearly 600 of these, generally paved, with good walks of brick. Some of them are broad, and are fine promenades. There also several public squares, one of them containing 10 acres. Among the most prominent of the public buildings are, the U. S. bank, a magnificient white marble edifice, copied after the Parthenon, at Athens; it is 161 by 87 feet, including porticos; the bank of Pennsylvania, also of white marble, moddled after the temple of Minerva, is 125 feet by 51; Girard's, and the Philadelphia bank, are also handsome edifices. Of banks, there are 13 in the city and suburbs. The Philadelphia library, and hall of the philosophical society, are plain but spacious structures. Besides these, are the university of Pennsylvania, the arcade, (in which is Peal's museum,) the state house, (from whence the declaration of independence was promulgated,) the academy of fine arts, the U. S. mint, (a splendid edifice,) masonic hall, hospital, alms-house, arsenal, exchange, orphans' asylum, widows' do, carpenter's hall, custom house, old and new penitentiary, 3 prisons, 4 theatres, and several markets, one of which is very extensive; in 1830, there were 90 churches in the city and suburbs, of which Christ's church is the oldest and most venerable, having been commenced in 1727. Of these, were Presbyterians, 20; Episcopal, 9; Friends, 7; Methodists, 10; Baptists 6; Lutherans, 5; Roman Catholic, 4; and African 10; other denominations, 19. There are many elegant private edifices in the city, and the general uniformity and neatness of those on many of the principal streets, is often remarked. Philadelphia is noted for the benevolent disposition of its citizens, and for the number, variety and extent of its charitable and literary institutions. Among these may be mentioned, the Pennsylvania hospital, founded in 1750, the alms-house, the dispensaries, Friend's asylum for the insane, humane society, orphans', and indigent widows and single women's asylum, institution for the deaf and dumb, and the abolition, savings fund, and fuel saving societies, besides many other similar moral and religious establishments. Among the literary, learned, and other institutions, are the university of Pa., the American philosophical society, the academy of natural sciences, the medical society, marine asylum, the college of physicians, college of pharmacy, association of druggists and apothecaries, law academy, academy of fine arts, atheneum, several fine libraries, Franklin institute, &c. The city is well supplied with public schools, and academies, and has lately received a most munificent bequest by the will of the late Stephen Girard, for the establishment of a college for orphans. The magnificent water works of Philadelphia, by means of which the city is supplied from the Schuylkill with pure and wholesome water, are without a parallel on this side the Atlantic. The total amount expended on this object is $1,443,583, and the annual receipts are $60,000. The water is raised from the river and conveyed into reservoirs, elevated 56 feet above the highest ground in the city; 60 miles of pipe conveys it through the city and districts. 3,000,000 of gallons is about the average daily supply. From this fountain the fire companies are plenteously supplied in times of fire; of these, there are about 30, and 16 or 18 hose companies. There are 2 bridges across the Schuylkill. Permanent bridge, thrown across from the w. end of Market-street, 1,300 feet long, and the Fairmount bridge, of a single arch, and 340 feet span. These are beautiful and substantial structures. Philadelphia and the vicinity, abounds in manufactures of different kinds, and vast quantities of useful and fancy articles are annually sent abroad. The internal trade of the city is very great, particularly with the western states. In 1830, 473,876 bbls. of wheat flour were inspected. The commerce of Philadelphia is also extensive; the number of arrivals in 1831, were 3,602, of which 396 were foreign, and 3,206 coastwise; the tonnage of vessels built during the same year was 3,525. There are several extensive ship yards for building merchant vessels, and the U. S. have also a navy yard here. The population of the city for several periods, has been as follows:—in 1731, 12,000, 1753, 18,000, 1790, 42,500, 1800, 70,287, 1810, 96,664, and in 1820, 119,325. In 1830, the population of the city and suburbs was as follows:—

Suburbs N.,	Nthn. Liberties,	31,376	
	Spring Garden,	11,141	
	Penn Township,	2,507	
	Kingston,	13,326	58,350
Suburbs, S.	Southwark, E.	10,361	
	Do. W.	10,379	
	Moyamensing,	6,822	
	Passyunk,	1,441	29,003
	City,		80,458
	Total,		167,811

PACKMAN, p-t. Somerset co. Me. Pop. 803.—PANTON, p-t. Addison co. Vt. Pop. 605.—PARIS, p-t. Oneida county, N. Y. Pop. stated 1,477, is 2,765.—PARISH, p-t. Oswego co. N. Y. Pop. 968.—PARISHVILLE, p-t. St. Lawrence co. N. Y. Pop. 1,479.—PATTERSON, Putnam county, N. Y. Pop. 1,536.—PATRICKTOWN, town, Lincoln county, Me. Pop 382. PAWTUCKET, Bristol co. Mass. Pop. 1,458.—PENDLETON, town, Niagara co. N. Y. Pop. 577.—"PHIPSBURGH," in the Gaz. should be *Phillipsburgh*, Maine, Oxford co.—PITCHER, p-t. Chenango co. N. Y. Pop. 1,214.—PITTSFIELD, town, Somerset co. Me. Pop. 609.—PITTSTOWN, p-t. Rensselaer co. N. Y. Pop. 3,702.

R.

RIGA, p-t. Munroe county, N. Y. Population 1830, 1,908.

ROCHESTER, p-t. Windsor county, Vt. Pop. 1,392

ROCKSTREAM, p-v. Reading, Steuben co. N. Y., 17 miles from Penn-Yan, is at the falls of Rockstream, where a descent of 140 feet affords water power. Above the falls is a rapid, a mile long, and below it the stream falls into Seneca lake.

RUM RIVER, considerable stream of Huron territory, rising at lat. 47°, and long. W. C. 17° w.; its sources are between the Miss. and St. Louis r of lake Superior, flowing sthrd. 30 ms. it expands into Spirit lake, about 30 ms. by 8 wide, and continuing sthrd. 100 ms. falls into the left side of Miss. r. about 90 ms. direct course N. W of the mouth of St. Peter's r. The valley of Rum river lies between those of Upper Miss. and St. Croix rs.

S.

SAINT ALBAN'S, p-t. and st. jus., Franklin co. Vt., 23 ms. N. Burlington, and 46 N. W. Montpelier. Pop. 1830, 2,395.

SAINT CROIX, river of Huron, rises about lat. 46, long. W. C. from 14° to 16° w., interlocking sources on the w. with the confluents of Rum r., on the E. with those of Chippeway r., and N. with numerous small streams flowing into lake Superior. This is a large r. compared with its length of course, which is about 110 ms. to the S. S. W., and the breadth across its sources, is very little less than the length of its valley. St. Croix enters the Mississippi about half way between the mouths of St. Peter's and Chippeway rs. at lat. 44° 45', its valley occupying the space between those of Rum r. and Chippeway rs. This stream has numerous branches, and much of the land is spoken of by travellers, as excellent. The mouth following the land route E. of Miss. r. is about 200 ms. above Prairie du Chien

SAINT GEORGE, town, Chittenden co. Vt., 8 miles S. E. Burlington, 28 w. Montpelier, was first settled 1784, it is uneven, with only small streams. Pop. 1830, 135.

SAINT JOHNSBURY, p-t. Caledonia co. Vt., 31 ms N. E. Montpelier, 26 N. Newbury, first settled 1788; it is hilly, and is crossed by Passumpsic r. N. and S. on which are fine meadows. The plain is a pleasant village, near the centre. Pop. 1830, 1,592.

SAINT PETER'S, a large and important confluent of the Miss. r. rises at lat. 45° 40', long. W. C. 20° w. interlocking sources with the highest fountains of Red r. branch of the Assiniboin; it flows thence 160 ms. to the S. E by E. and at a mean distance of about 55 ms. in a very remarkable manner parallel to the course of the Miss. r. Having reached its most sthrn. bend at the influx of Blue Earth r. St. Peter's inflects very nearly at right angles, and assuming a northeastern direction 70 ms. falls into the right side of Miss. r. immediately below the falls of St. Anthony, after an entire course of 230 ms. There is a very short portage between Bigstone lake of St. Peter's, and lake Traverse of Red r., and the two, in the advance of civilized settlement, may afford a highly important channel of commercial intercommunication between the two great slopes of central N. A. On examination it is evident, that the Red r. branch of Assiniboin, St. Peter's, and Des Moines rs., flow along sections of one great valley. The importance of these streams is just becoming visible.

SARANAC, the 2d article of that name, at the bottom of the 1st column of page 501, is given as in Lenawee co. Mich., but on Tanner's map, it is laid down on the left side of the mouth of St. Joseph's r. western part of Berrien co., about 200 ms. a little s. of w. Detroit.

SHEBOIGON, river of the northern part of Mich. peninsula, rises with sources interlocking those of Ottawa r., and flowing thence N. falls into lake Huron, after a comparative course of 70 ms.

SAUGATUCK, p-v. Fairfield, Fairfield county, Conn. It is a place of considerable coasting trade, near L. I. sound, on a harbor formed by the river of its name.

SAUGATUCK, r. Fairfield co. Conn., flows between Fairfield and Norwalk, and forms a good sloop harbor.

SANDY CREEK, p-t. Oswego county, N. Y. Pop 1,839.—SARANAC, town, Clinton county, N. Y. Pop. 1830, 316.—SAUGUS, p-t. Essex county, Mass. Pop. 1830, 960.—SCITUATE, Providence co. R. I. Pop. stated 3,394, is 6,853. —SMITHFIELD, Providence co. R. I. Pop. stated, 6,857, is 3,994.—SOUTHAMPTON, p-t. Rockingham co. N. H. Pop. 487.—SOUTHAMPTON, p-t. Suffolk co. N. Y. Pop. 4,850. —SOUTHOLD, p-t. Suffolk co. N. Y. Pop. 2,900.—SOUTHWICK, Hampden co. Mass. Pop. 1,855.—SPRINGFIELD, Hampden co. Mass. Pop. stated 2,816, is 6,784.—STARKS, p-t. Herkimer co. N. Y. Pop. 1,781.

T.

TAMPA. This name is frequently applied to the whole bay of Espiritu Santo, and it probably was the original name of all that sheet of water; it is now confined to the northern cove or minor bay of Espiritu Santo, w. of the mouth of Hillsboro r.

TAUGHKANNUC, a branch of the Green mts. which commences near Middlebury, Vt., and extends across the lower part of that state, and the w. parts of Mass, and Conn., to the shore of L. I. sound, gradually diminishing in elevation towards the south. The highest eminences are Taughkannuc and Saddle mountains in Mass., the former of which is about 3,000 feet.

U.

UNCASVILLE, p-v. Montville, New London co. Conn., 40 ms. s. w. Hartford, 7 s. Norwich, 6 n. New London, is a manufacturing village on a small tributary of the Thames; it has a cotton factory of stone, 4 stories high, with 2,200 spindles and 56 looms; it employs 60 persons, making 450,000 yards of sheetings annually. 7 ms. above another is erecting, to contain 1,000 spindles.

W.

WASHINGTON, v. Middlesex co. N. J., on South r., 5 ms. s. e. New Brunswick.

WATERFORD, v. Uxbridge, Worcester co. Mass., contains a cotton and a wollen factory; the latter with 1,600 spindles, and 68 looms, the former 2,000 spindles and 32 looms. Above, the same company have a stone factory, 312 ft. long, with 10,000 spindles and 300 looms, making 2,500,000 yards of printing goods annually; these besides other factories.

WATER GAP, Warren co. N. J., the passage of Delaware r. through the Blue mtns. 20 ms. above Easton. The scenery is quite romantic.

WATERLOO, p t. and half shire, Seneca co. N. Y., 173 ms. w. Albany. Pop. 1830, 1,847.

WHIPPANY, v. Morris co. N. J., 4 ms. e. Morristown, on a stream of the same name, which falls into Passaic r.

WILKES, co. Geo, bounded by Warren s., Taliaferro s. w., Oglethorpe w. and s. w., Broad r. separating it from Elbert n., Lincoln co. e., and Little r. separating it from Columbia s. e. Length 32, mean breadth 19, and area about 600 sq. ms. Lat. 34° 40′, long. W. C. 5° 46′ w. General slope estrd. towards Savannah r., from which it is separated only by the narrow co. of Lincoln. Chief t., Washington. Pop. 1820, 16,912; 1830, 14,237; at the former epoch Wilkes contained what is now Taliaferro co, and in 1830, the aggregate of both cos. amounted to 19,171.

WILLIAMS, nrthwstrn. co. of Ohio, bounded by Henry e., Paukling s., the state of Ind. w., Hillsdale co. Mich. n., and Lenawee co. Mich. n. e. Length from s. to n. 27 ms., breadth 24, and area 648. Lat. 41° 28′, long. W. C. 7° 36′ w. Slope sthrd., and in that direction traversed by St. Joseph's and Tiffin's rs., branches of Maumee. Chief t, Defiance. Pop. 1830, 387.

WILLIAMSON, co. Ten., bounded by Bedford s. e., Maury s., Hickman s w., Dickson n. w., Davidson n., and Rutherford e. Length along the sthrn. border 40 ms., mean breadth 12, and area 480 sq. ms Lat. 36° and long. 10° w. W. C., intersect in the sthwstrn. part of this co. The sthrn. border follows the dividing ridge between the valleys of Duck and Harpeth rs., and the co. sloping to the n. w., is almost entirely drained by the confluent creeks of the latter r. Chief t., Franklin. Pop. 1820, 20,640; and in 1830, 26,638.

WILLIAMSPORT, borough and st. jus. Lycoming co., Pa, situated on the left bank of the west branch of Susquehannah r., by the p-r. 87 ms. above Harrisburg, and 196 ms. a very little w. of n. W. C. N. lat. 41° 15′, long. W. C. 0° 07′ w. Pop. 800.

WILLIAMSPORT, flourishing p-v. of Washington co. Md., situated on the left bank of Potomac r. on the point, below the mouth of Conecocheague r., 8 ms. s. w. Hagerstown, and by p-r. 74 ms. n. w. W. C. Pop. 500.

WILLIAMSPORT, p-v. and st. jus. Warren co. Ind., situated on the right bank of Wabash r., at the mouth of Pine cr., by p-r. about 80 ms. n. w. by w. Indianopolis. Lat. 40° 20′, long. W. C. 10° 16′ w.

WILMINGTON, p-v. and st. jus. Clinton co. O., by p-r. 67 ms. s. w. Columbus. Lat. 39° 24′, long. W. C. 6° 46′ w. Pop. 1830, 616.

WINCHESTER, p-v. and st. jus. Randolph co. Ind, situated on Wabash r., by p-r. 97 ms. n. e. by e. Indianopolis. Lat. 40° 11′, long. W. C. 7° 04′ w.

WOODSFIELD, p-v. and st. jus. Monroe co. O. It is situated on Sunfish cr., 26 ms. s. s. w. St. Clairsville, and by p-r. 140 ms. estrd. Columbus. Lat. 39° 48′, long. W. C. 4° 04′ w. Pop. 1820, 157.

WOOD, co. O., bounded by Sandusky co. e., Seneca s. e., Hancock s., Henry w., Lenawee co. Mich. n. w., and Monroe co. Mich. n. It is a parallelogram of the same length and breadth as the adjacent co. of Henry, 32 ms. from s. to n., with a breadth of 27 ms., area 864 sq. ms. Lat. 41° 26′, long. W. C. 6° 38′ w. Slope southeastward, and traversed in that direction by Portage r. to the s. e., and Maumee n. w. Chief t., Perrysburgh. Pop. 1830, 1,102.

WOOSTER, p-v. and st. jus. Wayne co. O., on Killbuck branch of White Woman's r., 86 ms. n. e. Columbus. Lat. 40° 48′, long. W. C. 5° w. Pop. 1830, 1,000.

X.

XENIA, p-v. and st. jus. Greene co. O., by p-r. 57 ms. s. w. by w. Columbus, and 55 ms. n. n. e. Cincinnati. It is situated on a branch of Little Miami. Lat. 39° 40′, long. W. C. 6° 53′ w. It contains the co. buildings, 2 printing offices, 3 churches, 10 or 12 mercantile stores, and in 1830 had 917 inhabitants.

UNIVERSITIES AND COLLEGES IN THE UNITED STATES.

State.	Name.	Place.	Founded.	Instructors.	Academ. Students.	Med. Students.	Total.	Vols. in College Library.	Vols. in Students' Library.
Maine,	Waterville,	Waterville,	1820	5	59	28	87	2500	600
	Bowdoin,	Brunswick,	1794	6	156	99	255	8500	5200
N. H.	Dartmouth,	Hanover,	1770	10	178	98	276	6000	8000
Vermont,	University of Vt.	Burlington,	1791	4	36	40	76	1600	300
	Middlebury,	Middlebury,	1800	5	99	62	161	1846	2322
Mass.	Harvard University,	Cambridge,	1638	94	236	95	331	35000	4000
	Williams,	Williamstown,	1793	7	115	85	200	2550	2000
	Amherst,	Amherst,	1821	10	197		197	2389	4515
R. Island,	Brown University,	Providence,	1764	6	114		114	6100	6000
Conn.	Yale,	New Haven,	1700	15	354	60	415	8500	9000
	Washington,	Hartford,	1826	9	70		70	5000	1900
	Wesleyan University,	Middletown,	1831	5					
N. York,	Union,	Schenectady,	1795	9	205		205	5150	8450
	Geneva,	Geneva,	1823	6	31		31	500	900
	Columbia,	New York,	1754	9	100		100	8000	6000
	New York University,	do. do.	1831				93		
	Hamilton,	Clinton,	1812	6	93			2900	3000
N. Jersey,	College of New Jersey,	Princeton,	1746	10	105		105	8000	4000
	Rutgers,	New Brunswick,	1770	5	70		70		
Pa.	University of Penn.	Philadelphia,	1755	9	125	410	535		
	Jefferson,	Canonsburg,	1802	7	120	121	241	700	1800
	Western University,	Pittsburg,	1820	4	53		53		50
	Madison,	Uniontown,	1829	5	70		70		
	Alleghany,	Meadville,	1815	3	6		6	8000	
	Dickinson,	Carlisle,	1783	4	21		21	2000	5000
	Franklin,	Lancaster,	1787						
	Lafayette,	Easton,	1826						
	Girard,	Penn Township,	1831						
	Washington,	Washington,	1806	4	47		47	400	525
Maryland,	St. Mary's,	Baltimore,	1799	18	147		147	10000	
	University of Md.	do.	1812	11					
	St. John's,	Annapolis,	1784	5	76		76	2100	
	Mount St Mary's,	Near Emmitsburg,	1830	25	130		136	7000	
D. C.	Columbian	Washington,	1821	4	50		50	4000	
	Georgetown,	Georgetown,	1799	19	140		140	7000	
Virginia,	William and Mary,	Williamsburg,	1693	7	60		60	3600	600
	Hampden Sidney,	Prince Edward Co.	1774	6	54		54		
	Washington,	Lexington,	1812		23		23	700	1500
	University of Va.	Charlottesville,	1819	9	130		130	8000	
	Randolph Macon,	Boydton,		4					
N. C.	University of N. C.	Chapel Hill,	1791	9	69		69	1800	3000
S. C.	Charleston,	Charleston,	1785	7	61		61	3000	1000
	College of S. C.	Columbia,	1801	9	111		111	8000	
Georgia,	Univ. of Georgia,	Athens,	1785	7	114		114	2000	2250
Alabama,	Univ. of Alabama,	Tuscaloosa,	1820	6	100		100	1000	
	La Grange,	Franklin co.	1830						
Miss.	Jefferson,	Washington,	1802	10	160		160		
La.	Louisiana,	Jackson,							
	New Orleans,	New Orleans,							
Ten.	Greenville,	Greenville,	1794		32		32	3500	
	Univers. of Nashville,	Nashville,	1806	4	95		95	2500	750
	East Tennessee,	Knoxville,		2	21		21	340	200
Kentucky,	Transylvania,	Lexington,	1798	6	141	211	352	2350	1500
	Centre,	Danville,	1822	4	66		66	1258	108
	Augusta,	Augusta,	1823	7	98		98	1500	550
	Cumberland,	Princeton,	1825	3	57		57	1000	600
	St. Joseph's,	Bardstown,	1819	15	150		150	1300	
	Georgetown,	Georgetown,	1830	7	75		75	500	
Ohio,	University of Ohio,	Athens,	1802	4	57		57	1000	1000
	Miami University,	Oxford,	1824	11	82		82	1000	1200
	Western Reserve,	Hudson,	1826	4	25		25	1000	100
	Kenyon,	Gambier,	1828	4	80		80		
	Franklin,	New Athens,	1824	3	40		40		
Indiana,	Indiana,	Bloomington,	1827	3	51		51	182	50
Illinois,	Illinois,	Jacksonville	1830	3	35		35	600	
Missouri,	St. Louis,	St. Louis,	1829	6	125		125	1200	

Law Schools.—There are in the United States nine Law Schools, one at Cambridge, Massachusetts, with two professors and forty-one students; one at New Haven, Conn. with two professors and 33 students; one at Litchfield, Conn.; one at Philadelphia, Pa.; one at Baltimore, Md. with twenty-two students; one at Williamsburg, and one at Staunton, Va.; one at Charleston, S. C., and one at Lexington, Ken.

THEOLOGICAL SEMINARIES IN THE UNITED STATES.

Name.	Place.	Denomination.	Founded.	Students, 1831.	Vols. in Libraries.	No. of Professors.
Bangor Theol. Seminary,	Bangor, Me.	Cong.	1816	14	1200	
Theol. Seminary,	Andover, Mass.	Cong.	1808	139	10000	4
Theol. School,	Cambridge, do.	Cong. Unit.	1824	33		4
Mass. Episc. Theol. School,	do. do.	Episcopal.	1831			4
Theol. Institution,	Newton, do.	Baptist,	1825	22	1000	2
Theol. School, Yale Col.	New Haven, Conn.	Cong.	1822	48		3
Theol. Inst. Epis. Church,	New York, N. Y.	Prot. Episco.	1819	26	3000	4
Theol. Sem. of Auburn,	Auburn, do.	Presbyt.	1821	51	4000	3
Hamilton Theol. Instit.	Hamilton, do.	Baptist.	1820	80	1600	4
Hartwick Seminary,	Hartwick, do.	Lutheran,	1816			
Theol. Sem. Dutch Ref. Church,	New Brunswick, N. J.	Dutch Ref.		24		
Theol. Sem. Pres. Church, U. S.	Princeton, do.	Presbyt.	1812	92	6000	3
Sem. Lutheran Church, U. S.	Gettysburg, Pa.	Evang. L.	1826	43	6900	2
German Reformed Sem.	York, do.	G. Ref. Ch.	1825	14		2
Western Theol. Sem.	Alleghany town, do.	Presbyt.	1828	22	3964	2
Epis. Theol. School, Va.	Fairfax co. Va.	Prot. Episc.		19	1500	3
Union Theol. Sem.	Prince Edward co. do.	Presbyt.	1824	42	3000	3
Southern Theol. Sem.	Columbia, S. C.	do.	1829	9		2
South Western Theol. Sem.	Maryville, Ten.	do.	1821	22	5500	3
Lane Sem.	Cincinnati, Ohio.	do.	1829			
Rock Spring Sem.	Rock Spring, Il.	Baptist,	1827	5	1200	1

Beside those included in the table, there are in the U. S. 5 Roman Catholic Theological Seminaries, 2 in Maryland, 1 in South Carolina, 2 in Kentucky, and 1 in Missouri.

RELIGIOUS DENOMINATIONS IN THE UNITED STATES.

Denominations.	Population.	Communicants.	Churches or Congregations.	Ministers.
Calvinistic Baptist,	2,743,453	304,827	4384	2914
Methodist Episcopal Church,	2,600,000	476,000		1777
Presbyterian, *General Assembly*,	1,800,000	182,017	2253	1801
Congregationalist, *Orthodox*,	1,260,000	140,000	1270	1000
Protestant Episcopal Church,	600,000		700	558
Universalists,	500,000		300	150
Roman Catholics,	500,000			
Lutherans,	400,000	44,000	1200	205
Christ-ians,	275,000	25,000	800	200
German Reformed,	200,000	17,000	400	84
Friends or Quakers,	200,000		400	
Unitarians, *Congregational*,	176,000		193	160
Associate and other Methodists,	175,000	35,000		350
Free-will Baptists,	150,000	16,000	400	300
Dutch Reformed,	125,000	17,888	194	159
Mennonites,	120,000	30,000		200
Associate Presbyterians,	100,000	15,000.	144	74
Cumberland do.	100,000	8,000	75	50
Tunkers,	30,000	3,000	40	40
Free Communion Baptists,	30,000	3,500		30
Seventh Day Baptists,	20,000	2,000	40	30
Six Principle Baptists,	20,000	1,800	30	25
United Brethren, or Moravians,	7,000	2,000	23	23
Millennial Church or Shakers,	6,000		15	45
New Jerusalem Church,	5,000		28	30
Emancipators, *Baptists*,	4,500	600		15
Jews and others,	50,000		150	

MEDICAL AND LAW SCHOOLS IN THE UNITED STATES.

Name.	Place.	Professors.	Students.
Maine Med. School,	Brunswick,	4	99
Waterville Med. School,	Waterville,	4	28
New Hampshire Med. School,	Hanover,	3	98
Vt. Med. School, Univ. Vt.	Burlington,	3	40
Vt. Acad. of Med. Mid. Col.	Castleton,		
Mass. Med. School, Harv. Univ.	Boston,	5	95
Berkshire Med. Inst. Wms. Col.	Pittsfield,	6	85
Medical School, Yale Col.	New Haven,	5	69
Col. Phys. and Surg. N. Y.	New York,	7	180
New York School of Med.	do. do.		
Col. Phys. and Surg. Western Dist.	Fairfield,	5	170
Med. Dep. Univ. Pen.	Philadelphia,	9	410
Med. Dep. Jef. Col.	do.	5	121
Med. Dep. Univ. Md.	Baltimore,	7	
Med. Dep. Univ. Va.	Charlottesville.	3	
Med. Col. Charleston, S. C.	Charleston,	7	150
Med. Col. Transylvania Univ.	Lexington,	6	200
Med. Col. of Ohio,	Cincinnati,	8	112

STATISTICAL VIEW of the Commerce of the United States, exhibiting the value of every description of Imports from, and the value of articles of every description of Exports to, each Foreign country; also, the tonnage of American and Foreign vessels arriving from, and departing to, each Foreign country during the year ending on the 30th day of September, 1831.

COUNTRIES.	COMMERCE.				NAVIGATION.			
	Value of imports.	Value of exports.			Amern. tonnage.		Foreign tonn.	
		Domestic produce.	Foreign produce.	Total.	Entered into U. S.	Departed from U. S.	Entered into U. S.	Departed from U. S.
	Dollars.				Tons.			
Russia,	1,668,398	114,852	347,914	462,766	8,931	4,310	577	
Prussia,	50,970	27,043	-	27,043	709	387		
Sweden and Norway,	901,812	190,511	86,519	277,030	11,346	3,232	2,999	472
Swedish West Indies,	218,918	251,937	11,111	263,048	4,793	7,199	262	532
Denmark,	575	178,333	176,883	355,216	-	3,060		
Danish West Indies,	1,651,641	1,421,075	224,502	1,645,577	27,501	41,739	2,887	2,708
Netherlands,	969,837	1,707,292	212,860	1,920,152	24,076	23,168	349	1,994
Dutch West Indies,	343,799	370,857	45,274	416,131	11,296	11,430	312	194
Dutch East Indies,	319,395	128,884	631,442	760,326	2,533	6,498		
England,	41,854,323	28,841,430	2,367,439	31,208,869	223,345	235,345	84,394	83,461
Scotland,	1,977,839	1,185,142	5,567	1,190,709	5,674	6,312	11,098	9,102
Ireland,	261,564	589,941	-	589,941	4,386	7,838	7,029	2,206
Gibraltar,	150,517	429,087	165,786	594,873	3,509	11,703	-	256
British African ports,	-	6,064	-	6,064	-	121		
British East Indies,	1,544,273	132,442	675,390	807,832	5,342	6,481		
British West Indies,	1,303,301	1,417,291	23,962	1,441,253	38,046	40,922	23,769	17,903
Newfoundland, &c.	-	-	-	-	275	277	736	
British Am. Colonies,	864,909	4,026,392	35,446	4,061,838	92,672	79,364	82,557	94,776
Other British Colonies,	-	-	-	-	248	434		
Hanse towns,	3,493,301	1,812,241	779,931	2,592,172	15,934	17,147	12,175	17,487
France on the Atlantic,	12,876,977	4,963,557	3,228,452	8,192,009	40,849	48,022	8,686	3,722
France on the Mediter.	1,188,766	671,867	300,926	972,793	13,774	15,459	493	1,477
French West Indies,	671,842	704,833	13,044	717,877	26,704	35,334	2,793	2,254
Spain on the Atlantic,	566,072	235,584	63,428	299,012	6,760	4,596	-	1,068
Spain on the Mediter.	709,022	75,121	7,198	82,319	9,563	1,906	-	536
Teneriffe & other Canaries,	125,159	34,931	3,446	38,377	1,963	1,418		
Manilla and Philippine Isls.	348,995	15,994	16,830	32,824	2,938	249		
Cuba,	8,371,797	3,634,144	1,259,698	4,893,842	139,830	132,222	19,639	17,816
Other Spanish W. I.	1,580,156	261,801	53,245	315,046	24,060	8,272	3,117	1,051
Portugal,	124,446	39,149	2,356	41,505	5,043	1,396	1,451	
Madeira,	177,369	171,563	5,728	177,291	2,514	5,163	-	131
Fayal and other Azores,	32,092	10,549	6,049	16,598	660	475	397	251
Cape De Verd Isls.	63,643	45,432	13,557	58,989	875	1,200	-	236
Italy,	1,704,264	371,515	323,010	694,525	10,683	9,190	159	
Sicily,	144,047	2,369	-	2,369	2,080	378		
Trieste, &c.	161,062	276,561	262,808	539,369	1,920	4,215		
Turkey,	521,598	38,503	298,304	336,807	3,918	2,935		
Hayti,	1,580,578	1,126,698	191,677	1,318,375	26,446	27,807	699	1,006
Mexico,	5,166,745	1,091,489	5,086,729	6,178,218	22,377	22,303	11,498	10,619
Central Repub. of Am.	198,504	141,179	165,318	306,497	2,891	3,315		
Colombia,	1,207,154	375,319	282,830	658,149	9,174	7,188	58	
Honduras,	44,463	46,233	13,732	59,965	1,456	1,449	600	223
Brazil,	2,375,829	1,652,193	423,902	2,076,095	29,855	36,892	1,360	203
Argentine Republic,	928,103	415,489	244,290	659,779	9,632	8,169		
Cisplatine Republic,	-	-	-	-	274	356		
Peru,	917,788	8,560	7,616	16,176	2,577	523		
Chili,	413,758	849,493	518,662	1,368,155	3,729	11,145		
South Am. generally,	4,924	19,922	15,731	35,653	703	1,018	94	948
Cape of Good Hope,	-	-	-	-	929	891		
China,	3,083,205	244,790	1,046,045	1,290,835	4,316	5,061		
Asia, generally,	77,861	48,268	251,126	299,394	1,171	2,447		
East Indies, generally,	-	-	-	-	-	669		
West Ind. generally,	10,691	628,153	7,474	635,627	2,903	17,839	-	400
Europe, generally,	-	25,702	15	25,717	4,169	560	2,020	
Africa, generally,	146,932	175,166	69,891	245,057	2,511	5,098	-	148
South Seas,	51,186	16,910	8,963	25,873	29,581	39,470		
N. W. Coast of Am.	67,635	27,206	51,420	78,626	375	783		
Uncertain,	11,168	-	-	-	80	-		
Total,	103,191,124	61,277,057	20,033,526	81,310,583	922,952	972,504	281,948	271,994

IMPORTS AND EXPORTS, FROM 1822 TO 1831 INCLUSIVE.

Year.	Imports.	Exports.	Year.	Imports.	Exports.
1822	$83,241,541	$72,160,281	1827	$79,484,068	$82,324,827
1823	77,579,267	74,699,030	1828	88,509,824	72,264,686
1824	80,549,007	75,986,657	1829	74,492,527	72,358,671
1825	96,340,075	99,535,388	1830	70,876,920	73,849,508
1826	84,974,477	77,595,322	1831	103,191,124	81,310,583

Statement of the Commerce of each State and Territory, commencing on the 1st day of October, 1830, and ending on the 30th day of September, 1831.

	Value of imports.	Value of Exports.		
		Domestic prod.	Foreign prod.	Total.
Maine,	941,407	799,748	5,825	805,573
N. H.	146,205	109,456	1,766	111,222
Vt.	166,206	925,127	-	925,127
Mass.	14,269,056	4,027,201	3,706,562	7,733,763
R. I.	562,161	348,250	19,215	367,465
Conn.	405,066	482,073	810	482,883
N. Y.	57,077,417	15,726,118	9,809,026	25,535,144
N. J.	-	11,430	-	11,430
Penn.	12,124,083	3,594,302	1,919,411	5,513,713
Del.	21,656	34,514	-	34,514
Md.	4,826,577	3,730,506	578,141	4,308,647
Dist.Col.	193,565	1,207,517	13,458	1,220,975
Vir.	488,522	4,149,986	489	4,150,475
N. C.	196,356	340,973	167	341,140
S. C.	1,238,163	6,528,605	46,596	6,575,201
Geo.	399,940	3,957,245	2,568	3,959,813
Ala.	224,435	2,412,862	1,032	2,413,894
Miss.				
La.	9,766,693	12,835,531	3,926,458	16,761,989
Ohio,	617	14,728		14,728
Flor.	115,710	28,493	2,002	30,495
Mich.	27,299	12,392	-	12,392
Total,	103,191,124	61,277,057	20,033,526	81,310,583

NAVIGATION.

	Amt. Amn. tonnage.		Amt. Foreign tonnage.		Total Am. and For. tonnage.	
	Entered.	Departed.	Entered.	Departed.	Entered.	Departed.
Maine,	51635	61582	49819	49872	101454	111454
N. H.	7198	4362	-	-	7198	4362
Vt.	20201	20201	-	-	20201	20201
Mass.	182459	157530	9760	7483	192219	165013
R. I.	23845	22787	100	-	23945	22787
Conn.	17750	20139	-	-	17750	20139
N. Y.	315972	254331	77719	72444	393691	326775
N. J.	369	703	-	-	369	703
Penn.	71232	65149	8826	7596	80058	72745
Del.	1550	799	2186	965	3736	1764
Md.	55371	65370	10455	10276	65826	75646
Dist.Col.	4796	19362	872	878	5668	20240
Vir.	22933	48719	9985	11879	32918	60598
N. C.	16773	30450	1729	1990	18502	32530
S. C.	24379	48426	29011	29045	53390	77471
Geo.	15543	35747	13491	14307	29034	50054
Ala.	10126	14707	11840	10953	22166	25660
Miss.						
La.	76231	96752	55541	53558	131772	150311
Ohio,	91	91	138	138	229	229
Flor.	4455	5163	476	610	4931	5773
Mich.	43	43	-	-	43	43
Total,	922952	972504	281948	271994	1204900	1244498

A condensed view of the Tonnage of the several districts of the United States, on the last day of December, 1830.

DISTRICTS.	Registered.	Enr. & lic.	Total.
Passamaquoddy, Me.	7,636	2,850	10,486
Machias,	195	3,904	4,099
Frenchman's Bay,	2,612	3,478	6,090
Penobscot,	3,575	15,601	19,177
Belfast,	2,053	11,192	13,245
Waldoborough,	2,802	18,986	21,789
Wiscasset,	2,232	5,716	7,949
Bath,	16,313	10,355	26,668
Portland,	29,317	13,400	42,717
Saco,	953	2,387	3,340
Kennebunk,	2,789	1,999	4,789
York,	103	853	957
Portsmouth, N. H.	9,753	8,490	18,243
Newburyport, Mass.	9,714	6,862	16,577
Ipswich,	140	2,191	2,331
Gloucester,	2,098	9,642	11,741
Salem,	21,510	6,684	28,195
Marblehead,	1,196	5,742	6,949
Boston,	100,214	34,794	135,009
Plymouth,	11,090	8,386	19,476
Dighton,	301	3,360	3,661
New Bedford,	46,086	9,169	55,256
Barnstable,	2,409	22,775	25,184
Edgartown,	2,012	780	2,792
Nantucket,	18,854	3,473	22,327
Providence, R. I.	9,876	4,523	14,400
Bristol,	6,654	1,431	8,086
Newport,	4,879	3,543	8,423
Middletown, Conn.	1,604	7,429	9,033
New London,	10,004	6,208	16,213
New Haven,	2,954	4,174	7,128
Fairfield,	425	8,462	8,887
Vermont, Vt.	877	-	877
Champlain, N. York,	2,417	-	2,417
Sacket's Harbor,	-	942	942
Oswego,	505	612	1,118
Niagara,	-	-	-
Genesee,	585	1,082	1,668
Oswegatchie,	128	17	145
Buffalo Creek,	28	2,272	2,300
Sag Harbor,	4,465	2,808	7,274
New York,	101,946	154,710	256,557
Cape Vincent,	85	187	273
Perth Amboy, N. J.	458	7,746	8,205
Bridgetown,	115	10,169	10,284
Burlington,	-	2,393	2,393
Little Egg Harbor,	-	2,619	2,619
Great Egg Harbor,	-	9,481	9,481
Philadelphia, Pa.	47,935	23,754	71,689
Presque Isle,	44	481	525
Wilmington, Delaware,	143	12,326	12,469
Baltimore, Md.	23,941	11,678	35,621
Oxford,	-	9,135	9,135
Vienna,	345	10,340	10,685
Snow Hill,	143	3,996	4,140
Annapolis,	20	3,091	3,111
St. Mary's,	-	1,672	1,672
Georgetown, D. C.	1,760	3,564	5,324
Alexandria,	4,462	3,937	8,400
Norfolk, Virginia,	3,937	6,364	10,301
Petersburg,	1,600	1,604	3,205
Richmond,	1,904	1,105	3,009
Yorktown,	-	4,407	4,407
Tappahannock,	1,898	3,700	5,599
Folly Landing,	79	2,558	2,637
Cherry Stone,	154	1,946	2,100
East River,	487	2,119	2,606
Wilmington, N. C.	8,309	414	8,724
Newbern,	1,357	1,986	3,343
Washington,	1,067	1,618	2,685
Edenton,	993	2,730	3,724
Camden,	2,261	2,575	4,837
Beaufort,	530	847	1,377
Plymouth,	240	263	503
Ocracoke,	516	1,172	1,688
Charleston, S. C.	6,659	6,695	13,354
Georgetown,	383	1,447	1,831
Beaufort,	-		
Savannah, Georgia,	3,849	2,280	6,130
Sunbury,	-		
Hardwick,	-		
Brunswick,	509	280	789
St. Mary's,	-	450	450
Miami, Ohio,	-		
Cuyahoga,	-	1,029	1,029
Sandusky,	94	868	963
Detroit, Mich.	98	1,233	1,331
Michilimackinac,	-	114	114
Mobile, Ala.	1,585	3,778	5,364
Blakely,	-	-	
Pearl River, Miss.	-	870	870
New Orleans, La.	13,894	31,793	45,687

Continued from preceding page.

DISTRICTS.	Registered.	Enr. & lic.	Total.
Teche,	-		
Pensacola, Flor.	243	1,037	1,281
St. Augustine,	450	155	606
St. Mark's,	151	70	222
Key West,	1,094	-	1,094
Total,	576,475	615,301	1,191,776

Statemement of the number of vessels, with the amount of tonnage, and the number of seamen employed in navigating the same, (including their repeated voyages) which entered into, and departed from, each State and Territory, in the year ending on the 30th September, 1830.

State or territory.	No. of vessels.	Amount of tonnage.		No. of seamen employed.
		Entered.	Departed.	
Maine,	536	69363	91629	2949
New Hampshire,	38	9116	4632	284
Vermont,	121	29741	19290	871
Massachusetts,	912	168243	148124	9118
Rhode Island,	87	16676	14094	836
Connecticut,	93	16171	18285	1103
New York,	1382	298434	229341	14298
New Jersey,	3	586	627	23
Pennsylvania,	365	72009	63022	3907
Delaware,	9	1691	962	81
Maryland,	90	55317	55020	908
Dist. of Columbia,	54	10458	13803	446
Virginia,	93	25997	43715	843
North Carolina,	235	27757	36592	1482
South Carolina,	115	50859	52464	927
Georgia,	79	19249	50394	772
Alabama,	66	10490	22277	484
Louisiana,	451	83270	106017	4323
Florida,	15	1444	1366	93
Ohio,	1	56	56	3
Michigan,	1	-	50	3
	4745	967227	971760	43756

A comparative view of the registered, enrolled, and licensed tonnage of the U. S. from 1820 to 1830 inclusive.

Yrs.	Registered.	Enr. and lic.	Total.
1820	619,047	661,118	1,280,166
1821	619,096	679,062	1,298,958
1822	628,150	696,548	1,324,699
1823	639,920	696,644	1,336,565
1824	669,972	719,190	1,389,163
1825	700,787	722,323	1,423,111
1826	737,978	796,212	1,534,190
1827	747,170	873,437	1,620,607
1828	812,619	928,772	1,741,391
1829	650,142	610,654	1,260,977
1830*	576,475	615,301	1,191,776

* 89,307 tons cancelled, sold or lost, 1830; making an actual increase that year of 20,286 tons.

	Tons.
Registered vessels employed in the foreign trade at the close of the year 1830,	576,475
Enrolled vessels in the coasting trade,	496,639
Licensed vessels under 20 tons,	20,339
Enrolled vessels employed in the cod fishery,	59,042
do. mackeral fishery,	35,973
do. whale fishery,	793
Licensed vessels under 20 tons, employed in the cod fishery,	3,515
Total,	1,191,776

Registered tonnage employed other than in the whale fishery, 1830,	537,563
Employed in the whale fishery,	38,912
Total,	576,475

COMMERCE OF EACH STATE AND TERRITORY IN THE U. S.

Statement of the Commerce of each state and territory, commencing on the 1st day of October, 1830, and ending on the 30th day of September, 1831.

States and Territories.	Value of Imports.	Value of Exports.	Tonnage entered.
Maine,	941,407	805,573	101,454
New Hampshire,	146,205	111,222	7,198
Vermont,	166,206	925,127	20,201
Massachusetts,	14,269,056	7,733,763	192,219
Rhode Island,	562,161	367,465	23,945
Connecticut,	405,066	482,883	17,750
New York,	57,077,417	25,535,144	393,691
New Jersey,		11,430	369
Pennsylvania,	12,124,083	5,513,713	80,038
Delaware,	21,656	34,514	3,736
Maryland,	4,826,577	4,308,647	65,826
Dist. of Columbia,	193.555	1,220,975	5,668
Virginia,	488,522	4,150,475	32,918
North Carolina,	196,356	341,140	18,502
South Carolina,	1,238,163	6,575,201	53,390
Georgia,	399,940	3,959,813	29,034
Alabama,	224,435	2,413,894	29,166
Mississippi,			
Louisiana,	9,766,693	16,761,989	131,772
Ohio,	617	14,728	229
Florida Territory,	115,710	30,495	4,931
Mich. Territory,	27,299	12,392	43
Total,	103,191,124	81,310,583	1,204,900

Number of representatives of each state, with their number of electors.

	Reps.	Elec.		Reps.	Elec.
Maine,	8	10	N. C.	13	15
N. H.	5	7	S. C.	9	11
Vt.	5	7	Geo.	9	11
Mass.	12	14	Ala.	5	7
R. I.	2	4	Miss.	2	4
Conn.	6	8	La.	3	5
N. Y.	40	42	Ten.	13	15
N. J.	6	8	Ken.	13	15
Penn.	28	30	Ohio,	19	21
Del.	1	3	Ind.	7	9
Md.	8	10	Il.	3	5
Vir.	21	23	Mo.	2	4

Total representatives 240; electors 288. Each state sends two senators to congress.